The Elements

Name	Symbol	Atomic Number	Atomic Mass*
Actinium	Ac	89	(227)
Aluminum	Al	13	26.98
Americium	Am	95	(243)
Antimony	Sb	51	121.8
Argon	Ar	18	39.95
Arsenic	As	33	74.92
Astatine	At	85	(210)
Barium	Ba	56	137.3
Berkelium	Bk	97	(247)
Beryllium	Be	4	9.012
Bismuth	Bi	83	209.0
Bohrium	Bh	107	(267)
Boron	B	5	10.81
Bromine	Br	35	79.90
Cadmium	Cd	48	112.4
Calcium	Ca	20	40.08
Californium	Cf	98	(249)
Carbon	C	6	12.01
Cerium	Ce	58	140.1
Cesium	Cs	55	132.9
Chlorine	Cl	17	35.45
Chromium	Cr	24	52.00
Cobalt	Co	27	58.93
Copernicium	Cn	112	(285)
Copper	Cu	29	63.55
Curium	Cm	96	(247)
Darmstadtium	Ds	110	(281)
Dubnium	Db	105	(262)
Dysprosium	Dy	66	162.5
Einsteinium	Es	99	(254)
Erbium	Er	68	167.3
Europium	Eu	63	152.0
Fermium	Fm	100	(253)
Flevorium	Fl	114	(289)
Fluorine	F	9	19.00
Francium	Fr	87	(223)
Gadolinium	Gd	64	157.3
Gallium	Ga	31	69.72
Germanium	Ge	32	72.61
Gold	Au	79	197.0
Hafnium	Hf	72	178.5
Hassium	Hs	108	(277)
Helium	He	2	4.003
Holmium	Ho	67	164.9
Hydrogen	H	1	1.008
Indium	In	49	114.8
Iodine	I	53	126.9
Iridium	Ir	77	192.2
Iron	Fe	26	55.85
Krypton	Kr	36	83.80
Lanthanum	La	57	138.9
Lawrencium	Lr	103	(257)
Lead	Pb	82	207.2
Lithium	Li	3	6.941
Livermorium	Lv	116	(293)
Lutetium	Lu	71	175.0
Magnesium	Mg	12	24.31
Manganese	Mn	25	54.94
Meitnerium	Mt	109	(268)
Mendelevium	M	[illegible]	[illegible]
Mercury	Hg	80	200.6
Molybdenum	Mo	42	95.94
Moscovium	Mc	115	(288)
Neodymium	Nd	60	144.2
Neon	Ne	10	20.18
Neptunium	Np	93	(244)
Nickel	Ni	28	58.70
Nihonium	Nh	113	(284)
Niobium	Nb	41	92.91
Nitrogen	N	7	14.01
Nobelium	No	102	(253)
Oganesson	Og	118	(294)
Osmium	Os	76	190.2
Oxygen	O	8	16.00
Palladium	Pd	46	106.4
Phosphorus	P	15	30.97
Platinum	Pt	78	195.1
Plutonium	Pu	94	(242)
Polonium	Po	84	(209)
Potassium	K	19	39.10
Praseodymium	Pr	59	140.9
Promethium	Pm	61	(145)
Protactinium	Pa	91	(231)
Radium	Ra	88	(226)
Radon	Rn	86	(222)
Rhenium	Re	75	186.2
Rhodium	Rh	45	102.9
Roentgenium	Rg	111	(272)
Rubidium	Rb	37	85.47
Ruthenium	Ru	44	101.1
Rutherfordium	Rf	104	(263)
Samarium	Sm	62	150.4
Scandium	Sc	21	44.96
Seaborgium	Sg	106	(266)
Selenium	Se	34	78.96
Silicon	Si	14	28.09
Silver	Ag	47	107.9
Sodium	Na	11	22.99
Strontium	Sr	38	87.62
Sulfur	S	16	32.07
Tantalum	Ta	73	180.9
Technetium	Tc	43	(98)
Tellurium	Te	52	127.6
Tennessine	Ts	117	(294)
Terbium	Tb	65	158.9
Thallium	Tl	81	204.4
Thorium	Th	90	232.0
Thulium	Tm	69	168.9
Tin	Sn	50	118.7
Titanium	Ti	22	47.88
Tungsten	W	74	183.9
Uranium	U	92	238.0
Vanadium	V	23	50.94
Xenon	Xe	54	131.3
Ytterbium	Yb	70	173.0
Yttrium	Y	39	88.91
Zinc	Zn	30	65.41
Zirconium	Zr	40	91.22

*All atomic masses are given to four significant figures. Values in parentheses represent the mass number of the most stable isotope.

Silberberg ~ Amateis

CHEMISTRY

The Molecular Nature of Matter and Change

Advanced Topics

8e

McGraw Hill Education

CHEMISTRY: THE MOLECULAR NATURE OF MATTER AND CHANGE WITH ADVANCED TOPICS, EIGHTH EDITION

Published by McGraw-Hill Education, 2 Penn Plaza, New York, NY 10121.

This book is printed on acid-free paper.

1 2 3 4 5 6 7 8 9 LWI 21 20 19 18 17

ISBN 978-1-259-74109-8
MHID 1-259-74109-5

Chief Product Officer, SVP Products & Markets: *G. Scott Virkler*
Vice President, General Manager, Products & Markets: *Marty Lange*
Vice President, Content Design & Delivery: *Betsy Whalen*
Managing Director: *Thomas Timp*
Director: *David Spurgeon, Ph.D.*
Director, Product Development: *Rose Koos*
Associate Director of Digital Content: *Robin Reed*
Marketing Manager: *Matthew Garcia*
Market Development Manager: *Shannon O'Donnell*
Director of Digital Content: *Shirley Hino, Ph.D.*
Digital Product Developer: *Joan Weber*
Director, Content Design & Delivery: *Linda Avenarius*
Program Manager: *Lora Neyens*
Content Project Managers: *Laura Bies, Tammy Juran & Sandy Schnee*
Buyer: *Sandy Ludovissy*
Design: *David W. Hash*
Content Licensing Specialists: *Ann Marie Jannette & Lorraine Buczek*
Cover Image: © *Don Farrall/Photographer's Choice RF/Getty Images*
Compositor: *Aptara®, Inc.*
Printer: *LSC Communications*

All credits appearing on page or at the end of the book are considered to be an extension of the copyright page.

Library of Congress Cataloging-in-Publication Data

Names: Silberberg, Martin S. (Martin Stuart), 1945- | Amateis, Patricia.
Title: Chemistry : the molecular nature of matter and change : with advanced topics / Silberberg, Amateis.
Description: 8e [8th edition, revised]. | New York, NY : McGraw-Hill Education, [2018] | Includes index.
Identifiers: LCCN 2017009580| ISBN 9781259741098 (alk. paper) | ISBN 1259741095 (alk. paper)
Subjects: LCSH: Chemistry—Textbooks.
Classification: LCC QD33.2 .S55 2018b | DDC 540—dc23 LC record available at https://lccn.loc.gov/2017009580

mheducation.com/highered

To Ruth and Daniel, with all my love and gratitude.
MSS

To Ralph, Eric, Samantha, and Lindsay:
you bring me much joy.
PGA

BRIEF CONTENTS

Preface xx
Acknowledgments xxxii

1 *Keys to Studying Chemistry: Definitions, Units, and Problem Solving* 2
2 *The Components of Matter* 42
3 *Stoichiometry of Formulas and Equations* 94
4 *Three Major Classes of Chemical Reactions* 144
5 *Gases and the Kinetic-Molecular Theory* 204
6 *Thermochemistry: Energy Flow and Chemical Change* 256
7 *Quantum Theory and Atomic Structure* 294
8 *Electron Configuration and Chemical Periodicity* 330
9 *Models of Chemical Bonding* 368
10 *The Shapes of Molecules* 404
11 *Theories of Covalent Bonding* 442
12 *Intermolecular Forces: Liquids, Solids, and Phase Changes* 470
13 *The Properties of Mixtures: Solutions and Colloids* 532
14 *Periodic Patterns in the Main-Group Elements* 584
15 *Organic Compounds and the Atomic Properties of Carbon* 632
16 *Kinetics: Rates and Mechanisms of Chemical Reactions* 690
17 *Equilibrium: The Extent of Chemical Reactions* 746
18 *Acid-Base Equilibria* 792
19 *Ionic Equilibria in Aqueous Systems* 842
20 *Thermodynamics: Entropy, Free Energy, and Reaction Direction* 894
21 *Electrochemistry: Chemical Change and Electrical Work* 938
22 *The Elements in Nature and Industry* 996
23 *Transition Elements and Their Coordination Compounds* 1036
24 *Nuclear Reactions and Their Applications* 1072

Appendix A Common Mathematical Operations in Chemistry A-1
Appendix B Standard Thermodynamic Values for Selected Substances A-5
Appendix C Equilibrium Constants for Selected Substances A-8
Appendix D Standard Electrode (Half-Cell) Potentials A-14
Appendix E Answers to Selected Problems A-15
Glossary G-1
Index I-1

DETAILED CONTENTS

© Fancy Collection/SuperStock RF

CHAPTER 1 Keys to Studying Chemistry: Definitions, Units, and Problem Solving 2

1.1 Some Fundamental Definitions 4
The States of Matter 4
The Properties of Matter and Its Changes 5
The Central Theme in Chemistry 8
The Importance of Energy in the Study of Matter 8

1.2 Chemical Arts and the Origins of Modern Chemistry 10
Prechemical Traditions 10
The Phlogiston Fiasco and the Impact of Lavoisier 11

1.3 The Scientific Approach: Developing a Model 12

1.4 Measurement and Chemical Problem Solving 13
General Features of SI Units 13
Some Important SI Units in Chemistry 14
Units and Conversion Factors in Calculations 18
A Systematic Approach to Solving Chemistry Problems 19
Temperature Scales 25
Extensive and Intensive Properties 27

1.5 Uncertainty in Measurement: Significant Figures 28
Determining Which Digits Are Significant 29
Significant Figures: Calculations and Rounding Off 30
Precision, Accuracy, and Instrument Calibration 32

CHAPTER REVIEW GUIDE 33

PROBLEMS 37

CHAPTER 2 The Components of Matter 42

2.1 Elements, Compounds, and Mixtures: An Atomic Overview 44

2.2 The Observations That Led to an Atomic View of Matter 46
Mass Conservation 46
Definite Composition 47
Multiple Proportions 49

2.3 Dalton's Atomic Theory 50
Postulates of the Atomic Theory 50
How the Theory Explains the Mass Laws 50

2.4 The Observations That Led to the Nuclear Atom Model 52
Discovery of the Electron and Its Properties 52
Discovery of the Atomic Nucleus 54

2.5 The Atomic Theory Today 55
Structure of the Atom 55
Atomic Number, Mass Number, and Atomic Symbol 56
Isotopes 57
Atomic Masses of the Elements 57

TOOLS OF THE LABORATORY:
MASS SPECTROMETRY 60

2.6 Elements: A First Look at the Periodic Table 61

2.7 Compounds: Introduction to Bonding 64
The Formation of Ionic Compounds 64
The Formation of Covalent Substances 66

2.8 Compounds: Formulas, Names, and Masses 68
Binary Ionic Compounds 68
Compounds That Contain Polyatomic Ions 71
Acid Names from Anion Names 74
Binary Covalent Compounds 74
The Simplest Organic Compounds: Straight-Chain Alkanes 76
Molecular Masses from Chemical Formulas 76
Representing Molecules with Formulas and Models 78

2.9 Mixtures: Classification and Separation 81
An Overview of the Components of Matter 81

TOOLS OF THE LABORATORY:
BASIC SEPARATION TECHNIQUES 83

CHAPTER REVIEW GUIDE 84

PROBLEMS 86

Source: NASA

CHAPTER 3 Stoichiometry of Formulas and Equations 94

3.1 The Mole 95
Defining the Mole 95
Determining Molar Mass 96
Converting Between Amount, Mass, and Number of Chemical Entities 97
The Importance of Mass Percent 102

3.2 Determining the Formula of an Unknown Compound 104
Empirical Formulas 105
Molecular Formulas 106
Chemical Formulas and Molecular Structures; Isomers 110

3.3 Writing and Balancing Chemical Equations 111

3.4 Calculating Quantities of Reactant and Product 116
Stoichiometrically Equivalent Molar Ratios from the Balanced Equation 116
Reactions That Occur in a Sequence 120
Reactions That Involve a Limiting Reactant 122
Theoretical, Actual, and Percent Reaction Yields 127

CHAPTER REVIEW GUIDE 130
PROBLEMS 135

CHAPTER 4 Three Major Classes of Chemical Reactions 144

4.1 Solution Concentration and the Role of Water as a Solvent 145
The Polar Nature of Water 146
Ionic Compounds in Water 146
Covalent Compounds in Water 150
Expressing Concentration in Terms of Molarity 150
Amount-Mass-Number Conversions Involving Solutions 151
Preparing and Diluting Molar Solutions 152

4.2 Writing Equations for Aqueous Ionic Reactions 155

4.3 Precipitation Reactions 157
The Key Event: Formation of a Solid from Dissolved Ions 157
Predicting Whether a Precipitate Will Form 157
Stoichiometry of Precipitation Reactions 162

4.4 Acid-Base Reactions 165
The Key Event: Formation of H_2O from H^+ and OH^- 167
Proton Transfer in Acid-Base Reactions 168
Stoichiometry of Acid-Base Reactions: Acid-Base Titrations 172

4.5 Oxidation-Reduction (Redox) Reactions 174
The Key Event: Movement of Electrons Between Reactants 174
Some Essential Redox Terminology 175
Using Oxidation Numbers to Monitor Electron Charge 176
Stoichiometry of Redox Reactions: Redox Titrations 179

4.6 Elements in Redox Reactions 181
Combination Redox Reactions 181
Decomposition Redox Reactions 182
Displacement Redox Reactions and Activity Series 184
Combustion Reactions 186

4.7 The Reversibility of Reactions and the Equilibrium State 188

CHAPTER REVIEW GUIDE 190
PROBLEMS 196

CHAPTER 5 Gases and the Kinetic-Molecular Theory 204

5.1 An Overview of the Physical States of Matter 205

5.2 Gas Pressure and Its Measurement 207
Measuring Gas Pressure: Barometers and Manometers 208
Units of Pressure 209

5.3 The Gas Laws and Their Experimental Foundations 210
The Relationship Between Volume and Pressure: Boyle's Law 211
The Relationship Between Volume and Temperature: Charles's Law 212
The Relationship Between Volume and Amount: Avogadro's Law 214
Gas Behavior at Standard Conditions 215
The Ideal Gas Law 216
Solving Gas Law Problems 217

5.4 Rearrangements of the Ideal Gas Law 222
The Density of a Gas 222
The Molar Mass of a Gas 224
The Partial Pressure of Each Gas in a Mixture of Gases 225
The Ideal Gas Law and Reaction Stoichiometry 228

5.5 The Kinetic-Molecular Theory: A Model for Gas Behavior 231
How the Kinetic-Molecular Theory Explains the Gas Laws 231
Effusion and Diffusion 236
The Chaotic World of Gases: Mean Free Path and Collision Frequency 238

CHEMICAL CONNECTIONS TO ATMOSPHERIC SCIENCE:
HOW THE GAS LAWS APPLY TO EARTH'S ATMOSPHERE 239

5.6 Real Gases: Deviations from Ideal Behavior 241
Effects of Extreme Conditions on Gas Behavior 241
The van der Waals Equation: Adjusting the Ideal Gas Law 243

CHAPTER REVIEW GUIDE 244
PROBLEMS 247

CHAPTER 6 Thermochemistry: Energy Flow and Chemical Change 256

6.1 Forms of Energy and Their Interconversion 257
Defining the System and Its Surroundings 258
Energy Change (ΔE): Energy Transfer to or from a System 258
Heat and Work: Two Forms of Energy Transfer 258
The Law of Energy Conservation 261
Units of Energy 261
State Functions and the Path Independence of the Energy Change 262
Calculating Pressure-Volume Work (*PV* Work) 263

6.2 Enthalpy: Changes at Constant Pressure 265
The Meaning of Enthalpy 265
Comparing ΔE and ΔH 265
Exothermic and Endothermic Processes 266

6.3 Calorimetry: Measuring the Heat of a Chemical or Physical Change 268
Specific Heat Capacity 268
The Two Major Types of Calorimetry 269

6.4 Stoichiometry of Thermochemical Equations 273

6.5 Hess's Law: Finding ΔH of Any Reaction 275

6.6 Standard Enthalpies of Reaction (ΔH°_{rxn}) 277
Formation Equations and Their Standard Enthalpy Changes 277
Determining ΔH°_{rxn} from ΔH°_f Values for Reactants and Products 279

CHEMICAL CONNECTIONS TO ENVIRONMENTAL SCIENCE: **THE FUTURE OF ENERGY USE 281**

CHAPTER REVIEW GUIDE 285

PROBLEMS 288

CHAPTER 7 Quantum Theory and Atomic Structure 294

7.1 The Nature of Light 295
The Wave Nature of Light 296
The Particle Nature of Light 299

7.2 Atomic Spectra 302
Line Spectra and the Rydberg Equation 302
The Bohr Model of the Hydrogen Atom 303
The Energy Levels of the Hydrogen Atom 305

TOOLS OF THE LABORATORY: **SPECTROMETRY IN CHEMICAL ANALYSIS 308**

7.3 The Wave-Particle Duality of Matter and Energy 310
The Wave Nature of Electrons and the Particle Nature of Photons 310
Heisenberg's Uncertainty Principle 313

7.4 The Quantum-Mechanical Model of the Atom 314
The Schrödinger Equation, the Atomic Orbital, and the Probable Location of the Electron 314
Quantum Numbers of an Atomic Orbital 319
Quantum Numbers and Energy Levels 321
Shapes of Atomic Orbitals 323
The Special Case of Energy Levels in the Hydrogen Atom 325

CHAPTER REVIEW GUIDE 326

PROBLEMS 329

CHAPTER 8 Electron Configuration and Chemical Periodicity 330

8.1 Characteristics of Many-Electron Atoms 332
The Electron-Spin Quantum Number 332
The Exclusion Principle 333
Electrostatic Effects and Energy-Level Splitting 333

8.2 The Quantum-Mechanical Model and the Periodic Table 335
Building Up Period 1 335
Building Up Period 2 336
Building Up Period 3 338
Building Up Period 4: The First Transition Series 339
General Principles of Electron Configurations 340
Intervening Series: Transition and Inner Transition Elements 342
Similar Electron Configurations Within Groups 342

8.3 Trends in Three Atomic Properties 345
Trends in Atomic Size 345
Trends in Ionization Energy 348
Trends in Electron Affinity 351

8.4 Atomic Properties and Chemical Reactivity 353
Trends in Metallic Behavior 353
Properties of Monatomic Ions 355

CHAPTER REVIEW GUIDE 361

PROBLEMS 363

© Chip Clark/Fundamental Photographs, NYC

CHAPTER 9 Models of Chemical Bonding 368

9.1 **Atomic Properties and Chemical Bonds 369**
The Three Ways Elements Combine 369
Lewis Symbols and the Octet Rule 371

9.2 **The Ionic Bonding Model 372**
Why Ionic Compounds Form: The Importance of Lattice Energy 373
Periodic Trends in Lattice Energy 375
How the Model Explains the Properties of Ionic Compounds 377

9.3 **The Covalent Bonding Model 379**
The Formation of a Covalent Bond 379
Bonding Pairs and Lone Pairs 380
Properties of a Covalent Bond: Order, Energy, and Length 380
How the Model Explains the Properties of Covalent Substances 383

TOOLS OF THE LABORATORY: **INFRARED SPECTROSCOPY 384**

9.4 **Bond Energy and Chemical Change 386**
Changes in Bond Energy: Where Does ΔH°_{rxn} Come From? 386
Using Bond Energies to Calculate ΔH°_{rxn} 386
Bond Strengths and the Heat Released from Fuels and Foods 389

9.5 **Between the Extremes: Electronegativity and Bond Polarity 390**
Electronegativity 390
Bond Polarity and Partial Ionic Character 392
The Gradation in Bonding Across a Period 394

9.6 **An Introduction to Metallic Bonding 395**
The Electron-Sea Model 395
How the Model Explains the Properties of Metals 396

CHAPTER REVIEW GUIDE 397
PROBLEMS 399

CHAPTER 10 The Shapes of Molecules 404

10.1 **Depicting Molecules and Ions with Lewis Structures 405**
Applying the Octet Rule to Write Lewis Structures 405
Resonance: Delocalized Electron-Pair Bonding 409
Formal Charge: Selecting the More Important Resonance Structure 411
Lewis Structures for Exceptions to the Octet Rule 413

10.2 **Valence-Shell Electron-Pair Repulsion (VSEPR) Theory 417**
Electron-Group Arrangements and Molecular Shapes 418
The Molecular Shape with Two Electron Groups (Linear Arrangement) 419
Molecular Shapes with Three Electron Groups (Trigonal Planar Arrangement) 419
Molecular Shapes with Four Electron Groups (Tetrahedral Arrangement) 420
Molecular Shapes with Five Electron Groups (Trigonal Bipyramidal Arrangement) 421
Molecular Shapes with Six Electron Groups (Octahedral Arrangement) 422
Using VSEPR Theory to Determine Molecular Shape 423
Molecular Shapes with More Than One Central Atom 426

10.3 **Molecular Shape and Molecular Polarity 428**
Bond Polarity, Bond Angle, and Dipole Moment 428
The Effect of Molecular Polarity on Behavior 430

CHEMICAL CONNECTIONS TO SENSORY PHYSIOLOGY: **MOLECULAR SHAPE, BIOLOGICAL RECEPTORS, AND THE SENSE OF SMELL 431**

CHAPTER REVIEW GUIDE 432
PROBLEMS 437

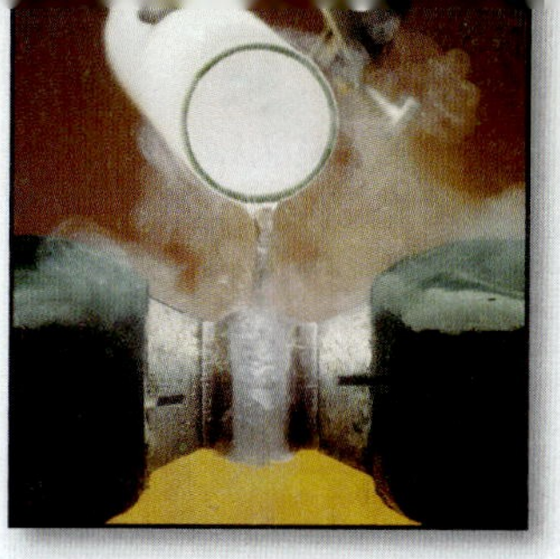

CHAPTER 11 Theories of Covalent Bonding 442

11.1 Valence Bond (VB) Theory and Orbital Hybridization 443
The Central Themes of VB Theory 443
Types of Hybrid Orbitals 444

11.2 Modes of Orbital Overlap and the Types of Covalent Bonds 451
Orbital Overlap in Single and Multiple Bonds 451
Orbital Overlap and Rotation Within a Molecule 455

11.3 Molecular Orbital (MO) Theory and Electron Delocalization 455
The Central Themes of MO Theory 455
Homonuclear Diatomic Molecules of Period 2 Elements 458
Two Heteronuclear Diatomic Molecules: HF and NO 462
Two Polyatomic Molecules: Benzene and Ozone 463

CHAPTER REVIEW GUIDE 464

PROBLEMS 466

CHAPTER 12 Intermolecular Forces: Liquids, Solids, and Phase Changes 470

12.1 An Overview of Physical States and Phase Changes 471

12.2 Quantitative Aspects of Phase Changes 474
Heat Involved in Phase Changes 475
The Equilibrium Nature of Phase Changes 478
Phase Diagrams: Effect of Pressure and Temperature on Physical State 482

12.3 Types of Intermolecular Forces 484
How Close Can Molecules Approach Each Other? 484
Ion-Dipole Forces 485
Dipole-Dipole Forces 485
The Hydrogen Bond 486
Polarizability and Induced Dipole Forces 487
Dispersion (London) Forces 488

12.4 Properties of the Liquid State 490
Surface Tension 491
Capillarity 491
Viscosity 492

12.5 The Uniqueness of Water 493
Solvent Properties of Water 493
Thermal Properties of Water 493
Surface Properties of Water 494
The Unusual Density of Solid Water 494

12.6 The Solid State: Structure, Properties, and Bonding 495
Structural Features of Solids 495
TOOLS OF THE LABORATORY: **X-RAY DIFFRACTION ANALYSIS AND SCANNING TUNNELING MICROSCOPY 502**
Types and Properties of Crystalline Solids 503
Amorphous Solids 506
Bonding in Solids: Molecular Orbital Band Theory 506

12.7 Advanced Materials 509
Electronic Materials 509
Liquid Crystals 511
Ceramic Materials 514
Polymeric Materials 516
Nanotechnology: Designing Materials Atom by Atom 521

CHAPTER REVIEW GUIDE 523

PROBLEMS 525

© amnat11/Shutterstock.com

CHAPTER 13 The Properties of Mixtures: Solutions and Colloids 532

13.1 Types of Solutions: Intermolecular Forces and Solubility 534
Intermolecular Forces in Solution 534
Liquid Solutions and the Role of Molecular Polarity 535
Gas Solutions and Solid Solutions 537

13.2 Intermolecular Forces and Biological Macromolecules 539
The Structures of Proteins 539
Dual Polarity in Soaps, Membranes, and Antibiotics 541
The Structure of DNA 542

13.3 Why Substances Dissolve: Breaking Down the Solution Process 544
The Heat of Solution and Its Components 544
The Heat of Hydration: Dissolving Ionic Solids in Water 545
The Solution Process and the Change in Entropy 547

13.4 Solubility as an Equilibrium Process 549
Effect of Temperature on Solubility 549
Effect of Pressure on Solubility 551

13.5 Concentration Terms 552
Molarity and Molality 552
Parts of Solute by Parts of Solution 554
Interconverting Concentration Terms 556

13.6 Colligative Properties of Solutions 557
Nonvolatile Nonelectrolyte Solutions 558
Using Colligative Properties to Find Solute Molar Mass 563
Volatile Nonelectrolyte Solutions 564
Strong Electrolyte Solutions 564
Applications of Colligative Properties 566

13.7 The Structure and Properties of Colloids 568

CHEMICAL CONNECTIONS TO ENVIRONMENTAL ENGINEERING:
SOLUTIONS AND COLLOIDS IN WATER PURIFICATION 570

CHAPTER REVIEW GUIDE 572

PROBLEMS 576

CHAPTER 14 Periodic Patterns in the Main-Group Elements 584

14.1 Hydrogen, the Simplest Atom 585
Where Hydrogen Fits in the Periodic Table 585
Highlights of Hydrogen Chemistry 586

14.2 Trends Across the Periodic Table: The Period 2 Elements 587

14.3 Group 1A(1): The Alkali Metals 590
Why the Alkali Metals Are Unusual Physically 590
Why the Alkali Metals Are So Reactive 592

14.4 Group 2A(2) 592
How the Alkaline Earth and Alkali Metals Compare Physically 593
How the Alkaline Earth and Alkali Metals Compare Chemically 593
Diagonal Relationships: Lithium and Magnesium 595

14.5 Group 3A(13): The Boron Family 595
How the Transition Elements Influence This Group's Properties 595
Features That First Appear in This Group's Chemical Properties 595
Highlights of Boron Chemistry 597
Diagonal Relationships: Beryllium and Aluminum 598

14.6 Group 4A(14): The Carbon Family 598
How Type of Bonding Affects Physical Properties 598
How Bonding Changes in This Group's Compounds 601
Highlights of Carbon Chemistry 601
Highlights of Silicon Chemistry 603
Diagonal Relationships: Boron and Silicon 604

14.7 Group 5A(15): The Nitrogen Family 604
The Wide Range of Physical Behavior 606
Patterns in Chemical Behavior 606
Highlights of Nitrogen Chemistry 607
Highlights of Phosphorus Chemistry 610

14.8 Group 6A(16): The Oxygen Family 612
How the Oxygen and Nitrogen Families Compare Physically 612
How the Oxygen and Nitrogen Families Compare Chemically 614
Highlights of Oxygen Chemistry: Range of Oxide Properties 615
Highlights of Sulfur Chemistry 615

14.9 Group 7A(17): The Halogens 617
Physical Behavior of the Halogens 617
Why the Halogens Are So Reactive 617
Highlights of Halogen Chemistry 619

14.10 Group 8A(18): The Noble Gases 622
How the Noble Gases and Alkali Metals Contrast Physically 622
How Noble Gases Can Form Compounds 624

CHAPTER REVIEW GUIDE 624

PROBLEMS 625

© Miroslav Hlavko/Shutterstock.com

CHAPTER 15 Organic Compounds and the Atomic Properties of Carbon 632

15.1 The Special Nature of Carbon and the Characteristics of Organic Molecules 633
The Structural Complexity of Organic Molecules 634
The Chemical Diversity of Organic Molecules 634

15.2 The Structures and Classes of Hydrocarbons 636
Carbon Skeletons and Hydrogen Skins 636
Alkanes: Hydrocarbons with Only Single Bonds 639
Dispersion Forces and the Physical Properties of Alkanes 641
Constitutional Isomerism 641
Chiral Molecules and Optical Isomerism 642
Alkenes: Hydrocarbons with Double Bonds 644
Restricted Rotation and Geometric (*cis-trans*) Isomerism 645
Alkynes: Hydrocarbons with Triple Bonds 646
Aromatic Hydrocarbons: Cyclic Molecules with Delocalized π Electrons 647
Variations on a Theme: Catenated Inorganic Hydrides 648
TOOLS OF THE LABORATORY: **NUCLEAR MAGNETIC RESONANCE (NMR) SPECTROSCOPY 649**

15.3 Some Important Classes of Organic Reactions 651
Types of Organic Reactions 651
The Redox Process in Organic Reactions 653

15.4 Properties and Reactivities of Common Functional Groups 654
Functional Groups with Only Single Bonds 654
Functional Groups with Double Bonds 659
Functional Groups with Both Single and Double Bonds 662
Functional Groups with Triple Bonds 666

15.5 The Monomer-Polymer Theme I: Synthetic Macromolecules 668
Addition Polymers 668
Condensation Polymers 669

15.6 The Monomer-Polymer Theme II: Biological Macromolecules 670
Sugars and Polysaccharides 670
Amino Acids and Proteins 672
Nucleotides and Nucleic Acids 674
CHEMICAL CONNECTIONS TO GENETICS AND FORENSICS: **DNA SEQUENCING AND FINGERPRINTING 679**

CHAPTER REVIEW GUIDE 681

PROBLEMS 683

CHAPTER 16 Kinetics: Rates and Mechanisms of Chemical Reactions 690

16.1 Focusing on Reaction Rate 691

16.2 Expressing the Reaction Rate 694
Average, Instantaneous, and Initial Reaction Rates 694
Expressing Rate in Terms of Reactant and Product Concentrations 696

16.3 The Rate Law and Its Components 698
Some Laboratory Methods for Determining the Initial Rate 699
Determining Reaction Orders 699
Determining the Rate Constant 704

16.4 Integrated Rate Laws: Concentration Changes over Time 708
Integrated Rate Laws for First-, Second-, and Zero-Order Reactions 708
Determining Reaction Orders from an Integrated Rate Law 710
Reaction Half-Life 712

16.5 Theories of Chemical Kinetics 716
Collision Theory: Basis of the Rate Law 716
Transition State Theory: What the Activation Energy Is Used For 719

16.6 Reaction Mechanisms: The Steps from Reactant to Product 722
Elementary Reactions and Molecularity 722
The Rate-Determining Step of a Reaction Mechanism 724
Correlating the Mechanism with the Rate Law 725

16.7 Catalysis: Speeding Up a Reaction 729
The Basis of Catalytic Action 730
Homogeneous Catalysis 730
Heterogeneous Catalysis 731
Kinetics and Function of Biological Catalysts 732
CHEMICAL CONNECTIONS TO ATMOSPHERIC SCIENCE: **DEPLETION OF EARTH'S OZONE LAYER 735**

CHAPTER REVIEW GUIDE 736

PROBLEMS 740

© hxdbzxy/Shutterstock.com

CHAPTER 17 Equilibrium: The Extent of Chemical Reactions 746

17.1 The Equilibrium State and the Equilibrium Constant 747

17.2 The Reaction Quotient and the Equilibrium Constant 750
- The Changing Value of the Reaction Quotient 750
- Writing the Reaction Quotient in Its Various Forms 751

17.3 Expressing Equilibria with Pressure Terms: Relation Between K_c and K_p 756

17.4 Comparing Q and K to Determine Reaction Direction 757

17.5 How to Solve Equilibrium Problems 760
- Using Quantities to Find the Equilibrium Constant 760
- Using the Equilibrium Constant to Find Quantities 763
- Problems Involving Mixtures of Reactants and Products 768

17.6 Reaction Conditions and Equilibrium: Le Châtelier's Principle 770
- The Effect of a Change in Concentration 770
- The Effect of a Change in Pressure (Volume) 773
- The Effect of a Change in Temperature 775
- The Lack of Effect of a Catalyst 777
- Applying Le Châtelier's Principle to the Synthesis of Ammonia 779

CHEMICAL CONNECTIONS TO CELLULAR METABOLISM: **DESIGN AND CONTROL OF A METABOLIC PATHWAY 781**

CHAPTER REVIEW GUIDE 782

PROBLEMS 785

CHAPTER 18 Acid-Base Equilibria 792

18.1 Acids and Bases in Water 794
- Release of H^+ or OH^- and the Arrhenius Acid-Base Definition 794
- Variation in Acid Strength: The Acid-Dissociation Constant (K_a) 795
- Classifying the Relative Strengths of Acids and Bases 797

18.2 Autoionization of Water and the pH Scale 798
- The Equilibrium Nature of Autoionization: The Ion-Product Constant for Water (K_w) 799
- Expressing the Hydronium Ion Concentration: The pH Scale 800

18.3 Proton Transfer and the Brønsted-Lowry Acid-Base Definition 803
- Conjugate Acid-Base Pairs 804
- Relative Acid-Base Strength and the Net Direction of Reaction 805

18.4 Solving Problems Involving Weak-Acid Equilibria 808
- Finding K_a Given Concentrations 809
- Finding Concentrations Given K_a 810
- The Effect of Concentration on the Extent of Acid Dissociation 811
- The Behavior of Polyprotic Acids 813

18.5 Molecular Properties and Acid Strength 816
- Acid Strength of Nonmetal Hydrides 816
- Acid Strength of Oxoacids 816
- Acidity of Hydrated Metal Ions 817

18.6 Weak Bases and Their Relation to Weak Acids 818
- Molecules as Weak Bases: Ammonia and the Amines 818
- Anions of Weak Acids as Weak Bases 820
- The Relation Between K_a and K_b of a Conjugate Acid-Base Pair 821

18.7 Acid-Base Properties of Salt Solutions 823
- Salts That Yield Neutral Solutions 823
- Salts That Yield Acidic Solutions 823
- Salts That Yield Basic Solutions 824
- Salts of Weakly Acidic Cations and Weakly Basic Anions 824
- Salts of Amphiprotic Anions 825

18.8 Generalizing the Brønsted-Lowry Concept: The Leveling Effect 827

18.9 Electron-Pair Donation and the Lewis Acid-Base Definition 827
- Molecules as Lewis Acids 828
- Metal Cations as Lewis Acids 829
- An Overview of Acid-Base Definitions 830

CHAPTER REVIEW GUIDE 831

PROBLEMS 834

© Joe Scherschel/Getty Images

CHAPTER 19 Ionic Equilibria in Aqueous Systems 842

19.1 Equilibria of Acid-Base Buffers 843
What a Buffer Is and How It Works: The Common-Ion Effect 843
The Henderson-Hasselbalch Equation 848
Buffer Capacity and Buffer Range 849
Preparing a Buffer 851

19.2 Acid-Base Titration Curves 853
Strong Acid–Strong Base Titration Curves 853
Weak Acid–Strong Base Titration Curves 855
Weak Base–Strong Acid Titration Curves 859
Monitoring pH with Acid-Base Indicators 860
Titration Curves for Polyprotic Acids 862
Amino Acids as Biological Polyprotic Acids 863

19.3 Equilibria of Slightly Soluble Ionic Compounds 864
The Ion-Product Expression (Q_{sp}) and the Solubility-Product Constant (K_{sp}) 864
Calculations Involving the Solubility-Product Constant 865
Effect of a Common Ion on Solubility 868
Effect of pH on Solubility 869
Applying Ionic Equilibria to the Formation of a Limestone Cave 870
Predicting the Formation of a Precipitate: Q_{sp} vs. K_{sp} 871
Separating Ions by Selective Precipitation and Simultaneous Equilibria 874

CHEMICAL CONNECTIONS TO ENVIRONMENTAL SCIENCE:
THE ACID-RAIN PROBLEM 875

19.4 Equilibria Involving Complex Ions 877
Formation of Complex Ions 877
Complex Ions and the Solubility of Precipitates 879
Complex Ions of Amphoteric Hydroxides 881

CHAPTER REVIEW GUIDE 883
PROBLEMS 887

CHAPTER 20 Thermodynamics: Entropy, Free Energy, and Reaction Direction 894

20.1 The Second Law of Thermodynamics: Predicting Spontaneous Change 895
The First Law of Thermodynamics Does Not Predict Spontaneous Change 896
The Sign of ΔH Does Not Predict Spontaneous Change 896
Freedom of Particle Motion and Dispersal of Kinetic Energy 897
Entropy and the Number of Microstates 898
Quantitative Meaning of an Entropy Change–Measuring Thermodynamic Variables 900
Entropy and the Second Law of Thermodynamics 905
Standard Molar Entropies and the Third Law 905
Predicting Relative $S°$ of a System 906

20.2 Calculating the Change in Entropy of a Reaction 910
Entropy Changes in the System: Standard Entropy of Reaction ($\Delta S°_{rxn}$) 910
Entropy Changes in the Surroundings: The Other Part of the Total 912
The Entropy Change and the Equilibrium State 914
Spontaneous Exothermic and Endothermic Changes 915

20.3 Entropy, Free Energy, and Work 916
Free Energy Change and Reaction Spontaneity 916
Calculating Standard Free Energy Changes 917
The Free Energy Change and the Work a System Can Do 919
The Effect of Temperature on Reaction Spontaneity 920
Coupling of Reactions to Drive a Nonspontaneous Change 924

CHEMICAL CONNECTIONS TO BIOLOGICAL ENERGETICS:
THE UNIVERSAL ROLE OF ATP 925

20.4 Free Energy, Equilibrium, and Reaction Direction 926

CHAPTER REVIEW GUIDE 932
PROBLEMS 936

© Griffin Technology

CHAPTER 21 Electrochemistry: Chemical Change and Electrical Work 938

21.1 Redox Reactions and Electrochemical Cells 939
A Quick Review of Oxidation-Reduction Concepts 939
Half-Reaction Method for Balancing Redox Reactions 940
An Overview of Electrochemical Cells 944

21.2 Voltaic Cells: Using Spontaneous Reactions to Generate Electrical Energy 945
Construction and Operation of a Voltaic Cell 946
Notation for a Voltaic Cell 948
Why Does a Voltaic Cell Work? 949

21.3 Cell Potential: Output of a Voltaic Cell 950
Standard Cell Potential (E°_{cell}) 950
Relative Strengths of Oxidizing and Reducing Agents 953
Using $E^\circ_{half\text{-}cell}$ Values to Write Spontaneous Redox Reactions 954
Explaining the Activity Series of the Metals 958

21.4 Free Energy and Electrical Work 959
Standard Cell Potential and the Equilibrium Constant 959
The Effect of Concentration on Cell Potential 961
Following Changes in Potential During Cell Operation 963
Concentration Cells 964

21.5 Electrochemical Processes in Batteries 968
Primary (Nonrechargeable) Batteries 968
Secondary (Rechargeable) Batteries 969
Fuel Cells 970

21.6 Corrosion: An Environmental Voltaic Cell 972
The Corrosion of Iron 972
Protecting Against the Corrosion of Iron 973

21.7 Electrolytic Cells: Using Electrical Energy to Drive Nonspontaneous Reactions 974
Construction and Operation of an Electrolytic Cell 974
Predicting the Products of Electrolysis 976
Stoichiometry of Electrolysis: The Relation Between Amounts of Charge and Products 980

***CHEMICAL CONNECTIONS TO BIOLOGICAL ENERGETICS:* CELLULAR ELECTROCHEMISTRY AND THE PRODUCTION OF ATP 982**

CHAPTER REVIEW GUIDE 984

PROBLEMS 987

CHAPTER 22 The Elements in Nature and Industry 996

22.1 How the Elements Occur in Nature 997
Earth's Structure and the Abundance of the Elements 997
Sources of the Elements 1000

22.2 The Cycling of Elements Through the Environment 1002
The Carbon Cycle 1002
The Nitrogen Cycle 1004
The Phosphorus Cycle 1005

22.3 Metallurgy: Extracting a Metal from Its Ore 1008
Pretreating the Ore 1009
Converting Mineral to Element 1010
Refining and Alloying the Element 1012

22.4 Tapping the Crust: Isolation and Uses of Selected Elements 1014
Producing the Alkali Metals: Sodium and Potassium 1014
The Indispensable Three: Iron, Copper, and Aluminum 1015
Mining the Sea for Magnesium 1021
The Sources and Uses of Hydrogen 1022

22.5 Chemical Manufacturing: Two Case Studies 1025
Sulfuric Acid, the Most Important Chemical 1025
The Chlor-Alkali Process 1028

CHAPTER REVIEW GUIDE 1029

PROBLEMS 1030

© PjrStudio/Alamy

CHAPTER 23 Transition Elements and Their Coordination Compounds 1036

23.1 Properties of the Transition Elements 1037
Electron Configurations of the Transition Metals and Their Ions 1038
Atomic and Physical Properties of the Transition Elements 1040
Chemical Properties of the Transition Elements 1042

23.2 The Inner Transition Elements 1044
The Lanthanides 1044
The Actinides 1045

23.3 Coordination Compounds 1046
Complex Ions: Coordination Numbers, Geometries, and Ligands 1046
Formulas and Names of Coordination Compounds 1048
Isomerism in Coordination Compounds 1051

23.4 Theoretical Basis for the Bonding and Properties of Complex Ions 1055
Applying Valence Bond Theory to Complex Ions 1055
Crystal Field Theory 1056

***CHEMICAL CONNECTIONS TO NUTRITIONAL SCIENCE:* TRANSITION METALS AS ESSENTIAL DIETARY TRACE ELEMENTS 1063**

CHAPTER REVIEW GUIDE 1065

PROBLEMS 1067

CHAPTER 24 Nuclear Reactions and Their Applications 1072

24.1 Radioactive Decay and Nuclear Stability 1073
Comparing Chemical and Nuclear Change 1074
The Components of the Nucleus: Terms and Notation 1074
The Discovery of Radioactivity and the Types of Emissions 1075
Modes of Radioactive Decay; Balancing Nuclear Equations 1075
Nuclear Stability and the Mode of Decay 1079

24.2 The Kinetics of Radioactive Decay 1083
Detection and Measurement of Radioactivity 1083
The Rate of Radioactive Decay 1084
Radioisotopic Dating 1087

24.3 Nuclear Transmutation: Induced Changes in Nuclei 1090
Early Transmutation Experiments; Nuclear Shorthand Notation 1090
Particle Accelerators and the Transuranium Elements 1091

24.4 Ionization: Effects of Nuclear Radiation on Matter 1093
Effects of Ionizing Radiation on Living Tissue 1093
Background Sources of Ionizing Radiation 1095
Assessing the Risk from Ionizing Radiation 1096

24.5 Applications of Radioisotopes 1098
Radioactive Tracers 1098
Additional Applications of Ionizing Radiation 1100

24.6 The Interconversion of Mass and Energy 1101
The Mass Difference Between a Nucleus and Its Nucleons 1101
Nuclear Binding Energy and Binding Energy per Nucleon 1102

24.7 Applications of Fission and Fusion 1104
The Process of Nuclear Fission 1105
The Promise of Nuclear Fusion 1109

***CHEMICAL CONNECTIONS TO COSMOLOGY:* ORIGIN OF THE ELEMENTS IN THE STARS 1110**

CHAPTER REVIEW GUIDE 1112

PROBLEMS 1114

Appendix A Common Mathematical Operations in Chemistry A-1
Appendix B Standard Thermodynamic Values for Selected Substances A-5
Appendix C Equilibrium Constants for Selected Substances A-8
Appendix D Standard Electrode (Half-Cell) Potentials A-14
Appendix E Answers to Selected Problems A-15
Glossary G-1
Index I-1

LIST OF SAMPLE PROBLEMS *(Molecular-scene problems are shown in color.)*

Chapter 1
1.1 Visualizing Change on the Atomic Scale 6
1.2 Distinguishing Between Physical and Chemical Change 7
1.3 Converting Units of Length 20
1.4 Converting Units of Volume 21
1.5 Converting Units of Mass 22
1.6 Converting Units Raised to a Power 23
1.7 Calculating Density from Mass and Volume 24
1.8 Converting Units of Temperature 27
1.9 Determining the Number of Significant Figures 29
1.10 Significant Figures and Rounding 32

Chapter 2
2.1 Distinguishing Elements, Compounds, and Mixtures at the Atomic Scale 45
2.2 Calculating the Mass of an Element in a Compound 48
2.3 Visualizing the Mass Laws 51
2.4 Determining the Numbers of Subatomic Particles in the Isotopes of an Element 57
2.5 Calculating the Atomic Mass of an Element 58
2.6 Identifying an Element from Its *Z* Value 62
2.7 Predicting the Ion an Element Forms 66
2.8 Naming Binary Ionic Compounds 69
2.9 Determining Formulas of Binary Ionic Compounds 70
2.10 Determining Names and Formulas of Ionic Compounds of Metals That Form More Than One Ion 71
2.11 Determining Names and Formulas of Ionic Compounds Containing Polyatomic Ions (Including Hydrates) 73
2.12 Recognizing Incorrect Names and Formulas of Ionic Compounds 73
2.13 Determining Names and Formulas of Anions and Acids 74
2.14 Determining Names and Formulas of Binary Covalent Compounds 75
2.15 Recognizing Incorrect Names and Formulas of Binary Covalent Compounds 75
2.16 Calculating the Molecular Mass of a Compound 77
2.17 Using Molecular Depictions to Determine Formula, Name, and Mass 77

Chapter 3
3.1 Converting Between Mass and Amount of an Element 98
3.2 Converting Between Number of Entities and Amount of an Element 99
3.3 Converting Between Number of Entities and Mass of an Element 99
3.4 Converting Between Number of Entities and Mass of a Compound I 100
3.5 Converting Between Number of Entities and Mass of a Compound II 101
3.6 Calculating the Mass Percent of Each Element in a Compound from the Formula 102
3.7 Calculating the Mass of an Element in a Compound 104
3.8 Determining an Empirical Formula from Amounts of Elements 105
3.9 Determining an Empirical Formula from Masses of Elements 106
3.10 Determining a Molecular Formula from Elemental Analysis and Molar Mass 107
3.11 Determining a Molecular Formula from Combustion Analysis 108
3.12 Balancing a Chemical Equation 114
3.13 Writing a Balanced Equation from a Molecular Scene 115
3.14 Calculating Quantities of Reactants and Products: Amount (mol) to Amount (mol) 118
3.15 Calculating Quantities of Reactants and Products: Amount (mol) to Mass (g) 119
3.16 Calculating Quantities of Reactants and Products: Mass to Mass 120
3.17 Writing an Overall Equation for a Reaction Sequence 121
3.18 Using Molecular Depictions in a Limiting-Reactant Problem 123
3.19 Calculating Quantities in a Limiting-Reactant Problem: Amount to Amount 125
3.20 Calculating Quantities in a Limiting-Reactant Problem: Mass to Mass 125
3.21 Calculating Percent Yield 128

Chapter 4
4.1 Using Molecular Scenes to Depict an Ionic Compound in Aqueous Solution 148
4.2 Determining Amount (mol) of Ions in Solution 149
4.3 Calculating the Molarity of a Solution 150
4.4 Calculating Mass of Solute in a Given Volume of Solution 151
4.5 Determining Amount (mol) of Ions in a Solution 151
4.6 Preparing a Dilute Solution from a Concentrated Solution 153
4.7 Visualizing Changes in Concentration 154
4.8 Predicting Whether a Precipitation Reaction Occurs; Writing Ionic Equations 159
4.9 Using Molecular Depictions in Precipitation Reactions 160
4.10 Calculating Amounts of Reactants and Products in a Precipitation Reaction 162
4.11 Solving a Limiting-Reactant Problem for a Precipitation Reaction 163
4.12 Determining the Number of H^+ (or OH^-) Ions in Solution 166
4.13 Writing Ionic Equations for Acid-Base Reactions 167
4.14 Writing Proton-Transfer Equations for Acid-Base Reactions 171
4.15 Calculating the Amounts of Reactants and Products in an Acid-Base Reaction 172
4.16 Finding the Concentration of an Acid from a Titration 173
4.17 Determining the Oxidation Number of Each Element in a Compound (or Ion) 177
4.18 Identifying Redox Reactions and Oxidizing and Reducing Agents 178
4.19 Finding the Amount of Reducing Agent by Titration 180
4.20 Identifying the Type of Redox Reaction 187

Chapter 5
5.1 Converting Units of Pressure 210
5.2 Applying the Volume-Pressure Relationship 217
5.3 Applying the Volume-Temperature and Pressure-Temperature Relationships 218
5.4 Applying the Volume-Amount and Pressure-Amount Relationships 218
5.5 Applying the Volume-Pressure-Temperature Relationship 219
5.6 Solving for an Unknown Gas Variable at Fixed Conditions 220
5.7 Using Gas Laws to Determine a Balanced Equation 221
5.8 Calculating Gas Density 223
5.9 Finding the Molar Mass of a Volatile Liquid 225
5.10 Applying Dalton's Law of Partial Pressures 226
5.11 Calculating the Amount of Gas Collected over Water 228

5.12 Using Gas Variables to Find Amounts of Reactants or Products I 229
5.13 Using Gas Variables to Find Amounts of Reactants or Products II 230
5.14 Applying Graham's Law of Effusion 236

Chapter 6
6.1 Determining the Change in Internal Energy of a System 262
6.2 Calculating Pressure-Volume Work Done by or on a System 264
6.3 Drawing Enthalpy Diagrams and Determining the Sign of ΔH 267
6.4 Relating Quantity of Heat and Temperature Change 269
6.5 Determining the Specific Heat Capacity of a Solid 270
6.6 Determining the Enthalpy Change of an Aqueous Reaction 270
6.7 Calculating the Heat of a Combustion Reaction 272
6.8 Using the Enthalpy Change of a Reaction (ΔH) to Find the Amount of a Substance 274
6.9 Using Hess's Law to Calculate an Unknown ΔH 276
6.10 Writing Formation Equations 278
6.11 Calculating ΔH°_{rxn} from ΔH°_f Values 280

Chapter 7
7.1 Interconverting Wavelength and Frequency 297
7.2 Interconverting Energy, Wavelength, and Frequency 301
7.3 Determining ΔE and λ of an Electron Transition 307
7.4 Calculating the de Broglie Wavelength of an Electron 311
7.5 Applying the Uncertainty Principle 313
7.6 Determining ΔE and λ of an Electron Transition Using the Particle-in-a-Box Model 316
7.7 Determining Quantum Numbers for an Energy Level 320
7.8 Determining Sublevel Names and Orbital Quantum Numbers 321
7.9 Identifying Incorrect Quantum Numbers 322

Chapter 8
8.1 Correlating Quantum Numbers and Orbital Diagrams 337
8.2 Determining Electron Configurations 344
8.3 Ranking Elements by Atomic Size 347
8.4 Ranking Elements by First Ionization Energy 350
8.5 Identifying an Element from Its Ionization Energies 351
8.6 Writing Electron Configurations of Main-Group Ions 356
8.7 Writing Electron Configurations and Predicting Magnetic Behavior of Transition Metal Ions 358
8.8 Ranking Ions by Size 360

Chapter 9
9.1 Depicting Ion Formation 372
9.2 Predicting Relative Lattice Energy from Ionic Properties 376
9.3 Comparing Bond Length and Bond Strength 382
9.4 Using Bond Energies to Calculate ΔH°_{rxn} 389
9.5 Determining Bond Polarity from EN Values 393

Chapter 10
10.1 Writing Lewis Structures for Species with Single Bonds and One Central Atom 407
10.2 Writing Lewis Structures for Molecules with Single Bonds and More Than One Central Atom 408
10.3 Writing Lewis Structures for Molecules with Multiple Bonds 409
10.4 Writing Resonance Structures and Assigning Formal Charges 412
10.5 Writing Lewis Structures for Octet-Rule Exceptions 416
10.6 Examining Shapes with Two, Three, or Four Electron Groups 425
10.7 Examining Shapes with Five or Six Electron Groups 426
10.8 Predicting Molecular Shapes with More Than One Central Atom 427
10.9 Predicting the Polarity of Molecules 429

Chapter 11
11.1 Postulating Hybrid Orbitals in a Molecule 449
11.2 Describing the Types of Orbitals and Bonds in Molecules 454
11.3 Predicting Stability of Species Using MO Diagrams 457
11.4 Using MO Theory to Explain Bond Properties 461

Chapter 12
12.1 Finding the Heat of a Phase Change Depicted by Molecular Scenes 477
12.2 Applying the Clausius-Clapeyron Equation 480
12.3 Using a Phase Diagram to Predict Phase Changes 483
12.4 Drawing Hydrogen Bonds Between Molecules of a Substance 487
12.5 Identifying the Types of Intermolecular Forces 489
12.6 Determining the Number of Particles per Unit Cell and the Coordination Number 497
12.7 Determining Atomic Radius 500
12.8 Determining Atomic Radius from the Unit Cell 501

Chapter 13
13.1 Predicting Relative Solubilities 537
13.2 Calculating an Aqueous Ionic Heat of Solution 546
13.3 Using Henry's Law to Calculate Gas Solubility 552
13.4 Calculating Molality 553
13.5 Expressing Concentrations in Parts by Mass, Parts by Volume, and Mole Fraction 555
13.6 Interconverting Concentration Terms 556
13.7 Using Raoult's Law to Find ΔP 559
13.8 Determining Boiling and Freezing Points of a Solution 561
13.9 Determining Molar Mass from Colligative Properties 563
13.10 Depicting Strong Electrolyte Solutions 565

Chapter 15
15.1 Drawing Hydrocarbons 637
15.2 Naming Hydrocarbons and Understanding Chirality and Geometric Isomerism 646
15.3 Recognizing the Type of Organic Reaction 652
15.4 Predicting the Reactions of Alcohols, Alkyl Halides, and Amines 658
15.5 Predicting the Steps in a Reaction Sequence 661
15.6 Predicting Reactions of the Carboxylic Acid Family 665
15.7 Recognizing Functional Groups 667

Chapter 16
16.1 Expressing Rate in Terms of Changes in Concentration with Time 697
16.2 Determining Reaction Orders from Rate Laws 701
16.3 Determining Reaction Orders and Rate Constants from Rate Data 705
16.4 Determining Reaction Orders from Molecular Scenes 706
16.5 Determining the Reactant Concentration After a Given Time 709
16.6 Using Molecular Scenes to Find Quantities at Various Times 713
16.7 Determining the Half-Life of a First-Order Reaction 714
16.8 Determining the Energy of Activation 718
16.9 Drawing Reaction Energy Diagrams and Transition States 721
16.10 Determining Molecularities and Rate Laws for Elementary Steps 723
16.11 Identifying Intermediates and Correlating Rate Laws and Reaction Mechanisms 726

Chapter 17

17.1 Writing the Reaction Quotient from the Balanced Equation 752
17.2 Finding K for Reactions Multiplied by a Common Factor or Reversed and for an Overall Reaction 754
17.3 Converting Between K_c and K_p 757
17.4 Using Molecular Scenes to Determine Reaction Direction 758
17.5 Using Concentrations to Determine Reaction Direction 759
17.6 Calculating K_c from Concentration Data 762
17.7 Determining Equilibrium Concentrations from K_c 763
17.8 Determining Equilibrium Concentrations from Initial Concentrations and K_c 763
17.9 Making a Simplifying Assumption to Calculate Equilibrium Concentrations 766
17.10 Predicting Reaction Direction and Calculating Equilibrium Concentrations 768
17.11 Predicting the Effect of a Change in Concentration on the Equilibrium Position 772
17.12 Predicting the Effect of a Change in Volume (Pressure) on the Equilibrium Position 774
17.13 Predicting the Effect of a Change in Temperature on the Equilibrium Position 776
17.14 Determining Equilibrium Parameters from Molecular Scenes 778

Chapter 18

18.1 Classifying Acid and Base Strength from the Chemical Formula 798
18.2 Calculating $[H_3O^+]$ or $[OH^-]$ in Aqueous Solution 800
18.3 Calculating $[H_3O^+]$, pH, $[OH^-]$, and pOH for Strong Acids and Bases 802
18.4 Identifying Conjugate Acid-Base Pairs 805
18.5 Predicting the Net Direction of an Acid-Base Reaction 807
18.6 Using Molecular Scenes to Predict the Net Direction of an Acid-Base Reaction 807
18.7 Finding K_a of a Weak Acid from the Solution pH 809
18.8 Determining Concentration and pH from K_a and Initial [HA] 810
18.9 Finding the Percent Dissociation of a Weak Acid 812
18.10 Calculating Equilibrium Concentrations for a Polyprotic Acid 814
18.11 Determining pH from K_b and Initial [B] 819
18.12 Determining the pH of a Solution of A^- 822
18.13 Predicting Relative Acidity of Salt Solutions from Reactions of the Ions with Water 824
18.14 Predicting the Relative Acidity of a Salt Solution from K_a and K_b of the Ions 826
18.15 Identifying Lewis Acids and Bases 830

Chapter 19

19.1 Calculating the Effect of Added H_3O^+ or OH^- on Buffer pH 846
19.2 Using Molecular Scenes to Examine Buffers 850
19.3 Preparing a Buffer 851
19.4 Finding the pH During a Weak Acid–Strong Base Titration 857
19.5 Writing Ion-Product Expressions 865
19.6 Determining K_{sp} from Solubility 866
19.7 Determining Solubility from K_{sp} 867
19.8 Calculating the Effect of a Common Ion on Solubility 869
19.9 Predicting the Effect on Solubility of Adding Strong Acid 870
19.10 Predicting Whether a Precipitate Will Form 871
19.11 Using Molecular Scenes to Predict Whether a Precipitate Will Form 872
19.12 Separating Ions by Selective Precipitation 874
19.13 Calculating the Concentration of a Complex Ion 878
19.14 Calculating the Effect of Complex-Ion Formation on Solubility 880

Chapter 20

20.1 Calculating the Change in Entropy During an Isothermal Volume Change of an Ideal Gas 902
20.2 Calculating the Change in Entropy During a Phase Change 903
20.3 Calculating the Entropy Change Resulting from a Change in Temperature 904
20.4 Predicting Relative Entropy Values 909
20.5 Calculating the Standard Entropy of Reaction, ΔS°_{rxn} 911
20.6 Determining Reaction Spontaneity 913
20.7 Calculating ΔG°_{rxn} from Enthalpy and Entropy Values 917
20.8 Calculating ΔG°_{rxn} from ΔG°_f Values 919
20.9 Using Molecular Scenes to Determine the Signs of ΔH, ΔS, and ΔG 921
20.10 Determining the Effect of Temperature on ΔG 922
20.11 Finding the Temperature at Which a Reaction Becomes Spontaneous 923
20.12 Exploring the Relationship Between ΔG° and K 927
20.13 Using Molecular Scenes to Find ΔG for a Reaction at Nonstandard Conditions 928
20.14 Calculating ΔG at Nonstandard Conditions 930

Chapter 21

21.1 Balancing a Redox Reaction in Basic Solution 942
21.2 Describing a Voltaic Cell with a Diagram and Notation 948
21.3 Using $E^\circ_{half\text{-}cell}$ Values to Find E°_{cell} 951
21.4 Calculating an Unknown $E^\circ_{half\text{-}cell}$ from E°_{cell} 953
21.5 Writing Spontaneous Redox Reactions and Ranking Oxidizing and Reducing Agents by Strength 956
21.6 Calculating K and ΔG° from E°_{cell} 961
21.7 Using the Nernst Equation to Calculate E_{cell} 962
21.8 Calculating the Potential of a Concentration Cell 966
21.9 Predicting the Electrolysis Products of a Molten Salt Mixture 977
21.10 Predicting the Electrolysis Products of Aqueous Salt Solutions 979
21.11 Applying the Relationship Among Current, Time, and Amount of Substance 981

Chapter 23

23.1 Writing Electron Configurations of Transition Metal Atoms and Ions 1040
23.2 Finding the Number of Unpaired Electrons 1045
23.3 Finding the Coordination Number and Charge of the Central Metal Ion in a Coordination Compound 1049
23.4 Writing Names and Formulas of Coordination Compounds 1050
23.5 Determining the Type of Stereoisomerism 1054
23.6 Ranking Crystal Field Splitting Energies (Δ) for Complex Ions of a Metal 1059
23.7 Identifying High-Spin and Low-Spin Complex Ions 1061

Chapter 24

24.1 Writing Equations for Nuclear Reactions 1078
24.2 Predicting Nuclear Stability 1080
24.3 Predicting the Mode of Nuclear Decay 1082
24.4 Calculating the Specific Activity and the Decay Constant of a Radioactive Nuclide 1085
24.5 Finding the Number of Radioactive Nuclei 1086
24.6 Applying Radiocarbon Dating 1089
24.7 Calculating the Binding Energy per Nucleon 1103

ABOUT THE AUTHORS

Courtesy of Martin S. Silberberg

Martin S. Silberberg received a B.S. in Chemistry from the City University of New York and a Ph.D. in Chemistry from the University of Oklahoma. He then accepted a position as research associate in analytical biochemistry at the Albert Einstein College of Medicine in New York City, where he developed methods to study neurotransmitter metabolism in Parkinson's disease and other neurological disorders. Following six years in neurochemical research, Dr. Silberberg joined the faculty of Bard College at Simon's Rock, a liberal arts college known for its excellence in teaching small classes of highly motivated students. As head of the Natural Sciences Major and Director of Premedical Studies, he taught courses in general chemistry, organic chemistry, biochemistry, and liberal-arts chemistry. The small class size and close student contact afforded him insights into how students learn chemistry, where they have difficulties, and what strategies can help them succeed. Dr. Silberberg decided to apply these insights in a broader context and established a textbook writing, editing, and consulting company. Before writing his own texts, he worked as a consulting and development editor on chemistry, biochemistry, and physics texts for several major college publishers. He resides with his wife Ruth in the Pioneer Valley near Amherst, Massachusetts, where he enjoys the rich cultural and academic life of the area and relaxes by traveling, gardening, and singing.

Courtesy of Patricia G. Amateis

Patricia G. Amateis graduated with a B.S. in Chemistry Education from Concord University in West Virginia and a Ph.D. in Analytical Chemistry from Virginia Tech. She has been on the faculty of the Chemistry Department at Virginia Tech for 31 years, teaching General Chemistry and Analytical Chemistry. For the past 16 years, she has served as Director of General Chemistry, responsible for the oversight of both the lecture and lab portions of the large General Chemistry program. She has taught thousands of students during her career and has been awarded the University Sporn Award for Introductory Teaching, the Alumni Teaching Award, and the William E. Wine Award for a history of university teaching excellence. She and her husband live in Blacksburg, Virginia and are the parents of three adult children. In her free time, she enjoys biking, hiking, competing in the occasional sprint triathlon, and playing the double second in Panjammers, Blacksburg's steel drum band.

PREFACE

Chemistry is so crucial to an understanding of medicine and biology, environmental science, and many areas of engineering and industrial processing that it has become a requirement for an increasing number of academic majors. Furthermore, chemical principles lie at the core of some of the key societal issues we face in the 21st century—dealing with climate change, finding new energy options, and supplying nutrition and curing disease on an ever more populated planet.

SETTING THE STANDARD FOR A CHEMISTRY TEXT

The eighth edition of *Chemistry: The Molecular Nature of Matter and Change* maintains its standard-setting position among general chemistry textbooks by evolving further to meet the needs of professor and student. The text still contains the most accurate molecular illustrations, consistent step-by-step worked problems, and an extensive collection of end-of-chapter problems. And changes throughout this edition make the text more readable and succinct, the artwork more teachable and modern, and the design more focused and inviting. The three hallmarks that have made this text a market leader are now demonstrated in its pages more clearly than ever.

Visualizing Chemical Models—Macroscopic to Molecular

Chemistry deals with observable changes caused by unobservable atomic-scale events, requiring an appreciation of a size gap of mind-boggling proportions. One of the text's goals coincides with that of so many instructors: to help students visualize chemical events on the molecular scale. Thus, concepts are explained first at the macroscopic level and then from a molecular point of view, with pedagogic illustrations always placed next to the discussions to bring the point home for today's visually oriented students.

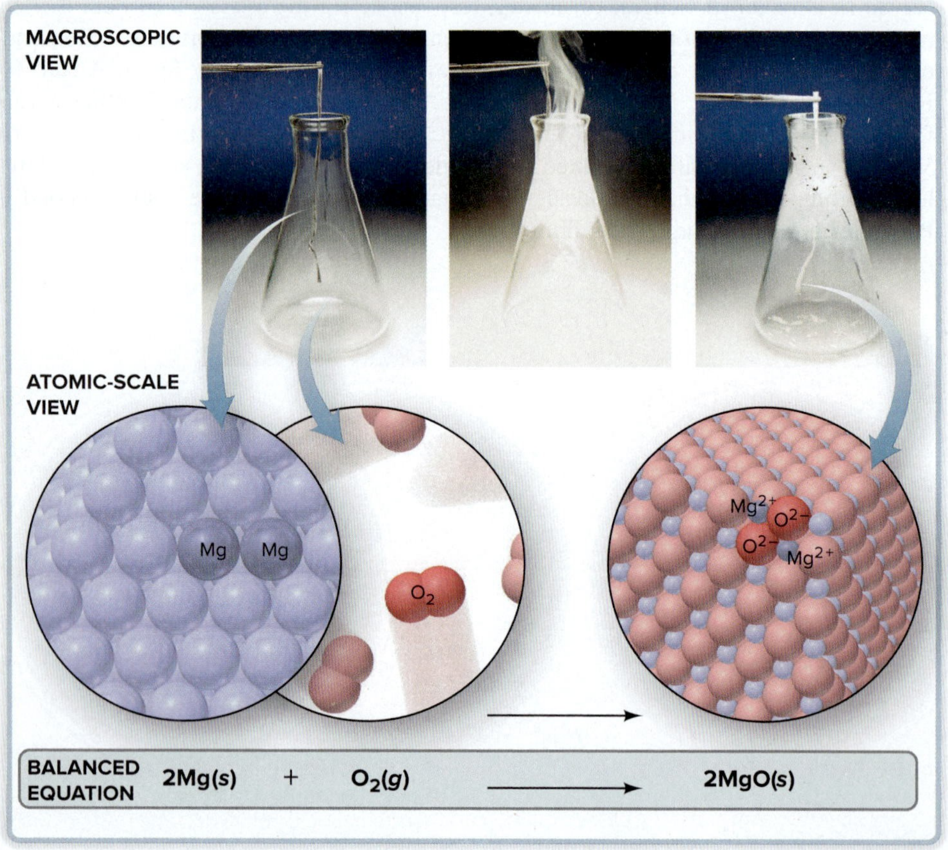

(three photos): © McGraw-Hill Education/Charles Winters/Timeframe Photography, Inc.

Thinking Logically to Solve Problems

The problem-solving approach, based on the four-step method widely accepted by experts in chemical education, is introduced in Chapter 1 and employed *consistently* throughout the text. It encourages students to *plan* a logical approach to a problem, and only then proceed to *solve* it. Each sample problem includes a *check,* which fosters the habit of "thinking through" both the chemical and the quantitative reasonableness of the answer. Finally, for *practice* and reinforcement, each sample problem is followed immediately by *two* similar follow-up problems. And, *Chemistry* marries problem solving to visualizing models with molecular-scene problems, which appear not only in homework sets, as in other texts, but also in the running text, where they are worked out stepwise.

SAMPLE PROBLEM 3.9 Determining an Empirical Formula from Masses of Elements

Problem Analysis of a sample of an ionic compound yields 2.82 g of Na, 4.35 g of Cl, and 7.83 g of O. What are the empirical formula and the name of the compound?

Plan This problem is similar to Sample Problem 3.8, except that we are given element *masses* that we must convert into integer subscripts. We first divide each mass by the element's molar mass to find the amount (mol). Then we construct a preliminary formula and convert the amounts (mol) to integers.

Solution Finding amount (mol) of each element:

$$\text{Amount (mol) of Na} = 2.82\ \cancel{\text{g Na}} \times \frac{1\ \text{mol Na}}{22.99\ \cancel{\text{g Na}}} = 0.123\ \text{mol Na}$$

$$\text{Amount (mol) of Cl} = 4.35\ \cancel{\text{g Cl}} \times \frac{1\ \text{mol Cl}}{35.45\ \cancel{\text{g Cl}}} = 0.123\ \text{mol Cl}$$

$$\text{Amount (mol) of O} = 7.83\ \cancel{\text{g O}} \times \frac{1\ \text{mol O}}{16.00\ \cancel{\text{g O}}} = 0.489\ \text{mol O}$$

Constructing a preliminary formula: $Na_{0.123}Cl_{0.123}O_{0.489}$

Converting to integer subscripts (dividing all by the smallest subscript):

$$Na_{\frac{0.123}{0.123}}Cl_{\frac{0.123}{0.123}}O_{\frac{0.489}{0.123}} \longrightarrow Na_{1.00}Cl_{1.00}O_{3.98} \approx Na_1Cl_1O_4, \quad \text{or} \quad NaClO_4$$

The empirical formula is $NaClO_4$; the name is sodium perchlorate.

Check The numbers of moles seem correct because the masses of Na and Cl are slightly more than 0.1 of their molar masses. The mass of O is greatest and its molar mass is smallest, so it should have the greatest number of moles. The ratio of subscripts, 1/1/4, is the same as the ratio of moles, 0.123/0.123/0.489 (within rounding).

FOLLOW-UP PROBLEMS

of an unknown compound is found to contain 1.23 g of H, 12.64 g of
of O. What is the empirical formula and the name of the compound?

wn metal M reacts with sulfur to form a compound with the formula
of M reacts with 2.88 g of S, what are the names of M and M_2S_3? [*Hint:*
mount (mol) of S, and use the formula to find the amount (mol) of M.]

PROBLEMS 3.42(b), 3.43(b), 3.46, and 3.47

SAMPLE PROBLEM 2.3 Visualizing the Mass Laws

Problem The scenes below represent an atomic-scale view of a chemical reaction:

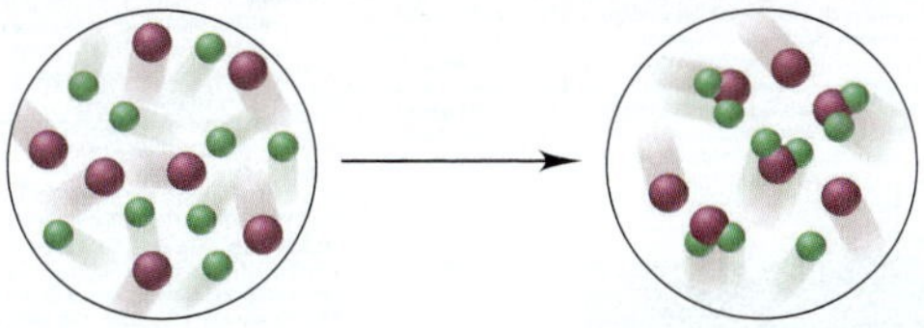

Which of the mass laws—mass conservation, definite composition, and/or multiple proportions—is (are) illustrated?

Plan From the depictions, we note the numbers, colors, and combinations of atoms (spheres) to see which mass laws pertain. If the numbers of each atom are the same before and after the reaction, the total mass did not change (mass conservation). If a compound forms that always has the same atom ratio, the elements are present in fixed parts by mass (definite composition). If the same elements form different compounds and the ratio of the atoms of one element that combine with one atom of the other element is a small whole number, the ratio of their masses is a small whole number as well (multiple proportions).

Solution There are seven purple and nine green atoms in each circle, so mass is conserved. The compound formed has one purple and two green atoms, so it has definite composition. Only one compound forms, so the law of multiple proportions does not pertain.

FOLLOW-UP PROBLEMS

2.3A The following scenes represent a chemical change. Which of the mass laws is (are) illustrated?

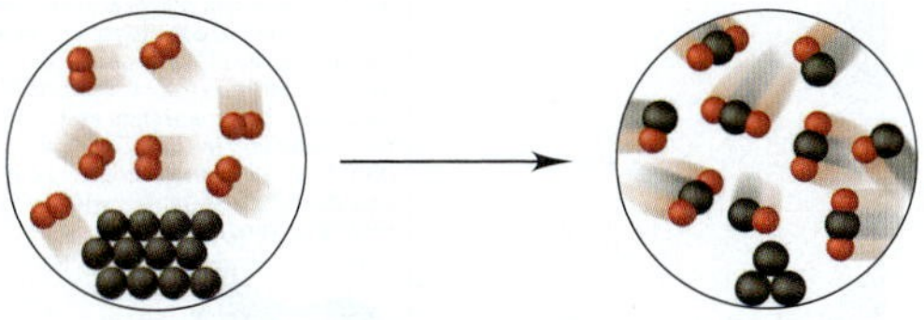

2.3B Which sample(s) best display(s) the fact that compounds of bromine *(orange)* and fluorine *(yellow)* exhibit the law of multiple proportions? Explain.

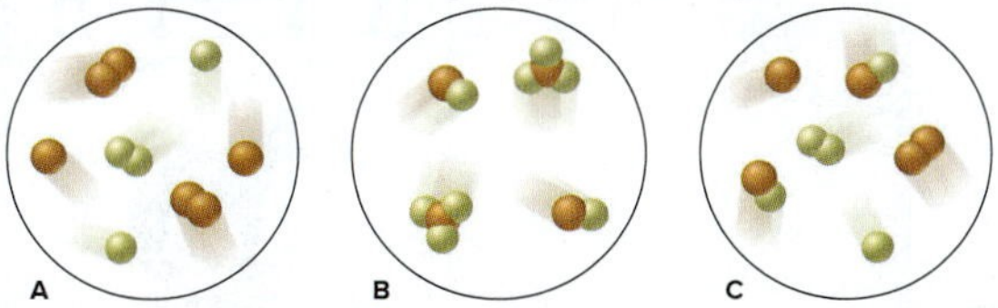

A B C

SOME SIMILAR PROBLEMS 2.14 and 2.15

Applying Ideas to the Real World

As the most practical science, chemistry should have a textbook that highlights its countless applications. Moreover, today's students may enter emerging chemistry-related hybrid fields, like biomaterials science or planetary geochemistry, and the text they use should point out the relevance of chemical concepts to such related sciences. The *Chemical Connections* and *Tools of the Laboratory* boxed essays (which include problems for added relevance), the more pedagogic margin notes, and the many applications woven into the chapter content are up-to-date, student-friendly features that are directly related to the neighboring content.

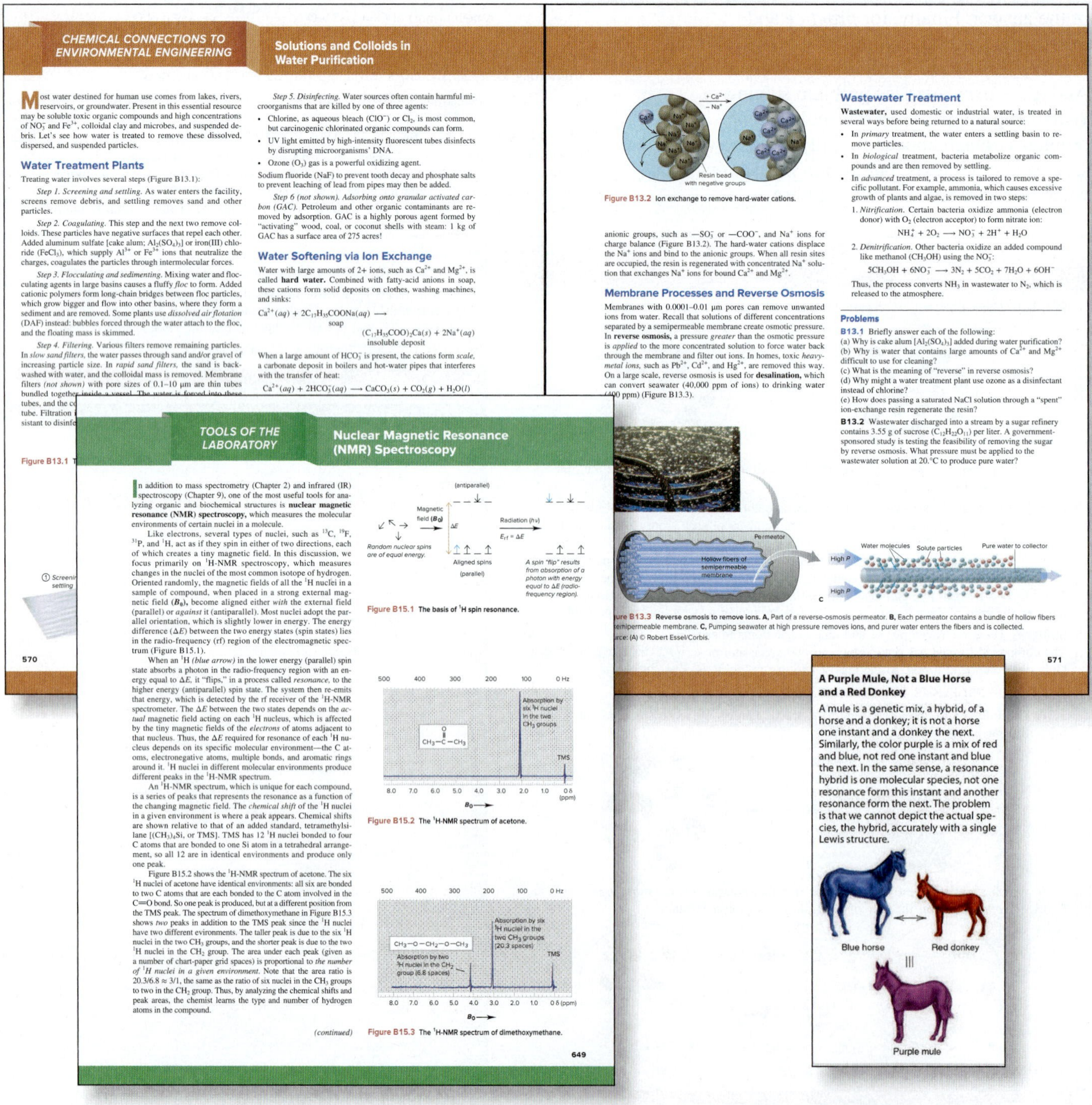

Reinforcing through Review and Practice

A favorite feature, the section summaries that conclude every section restate the major ideas concisely and immediately (rather than postponing such review until the end of the chapter).

A rich catalog of study aids ends each chapter to help students review the content:

- **Learning Objectives,** with section and/or sample problem numbers, focus on the concepts to understand and the skills to master.
- **Key Terms,** boldfaced and defined within the chapter, are listed here by section (with page numbers), as well as being defined in the *Glossary.*
- **Key Equations and Relationships** are highlighted and numbered within the chapter and listed here with page numbers.
- **Brief Solutions to Follow-up Problems** triple the number of worked problems by providing multistep calculations at the end of the chapter, rather than just numerical answers at the back of the book.

› Summary of Section 9.1

- › Nearly all naturally occurring substances consist of atoms or ions bonded to others. Chemical bonding allows atoms to lower their energy.
- › Ionic bonding occurs when metal atoms transfer electrons to nonmetal atoms, and the resulting ions attract each other and form an ionic solid.
- › Covalent bonding is most common between nonmetal atoms and usually results in individual molecules. Bonded atoms share one or more pairs of electrons that are localized between them.
- › Metallic bonding occurs when many metal atoms pool their valence electrons into a delocalized electron "sea" that holds all the atoms in the sample together.
- › The Lewis electron-dot symbol of a main-group atom shows valence electrons as dots surrounding the element symbol.
- › The octet rule says that, when bonding, many atoms lose, gain, or share electrons to attain a filled outer level of eight (or two) electrons.

CHAPTER REVIEW GUIDE

Learning Objectives

Relevant section (§) and/or sample problem (SP) numbers appear in parentheses.

Understand These Concepts

1. The quantitative meaning of solubility (§13.1)
2. The major types of intermolecular forces in solution and their relative strengths (§13.1)
3. How the like-dissolves-like rule depends on intermolecular forces (§13.1)
4. Why gases have relatively low solubilities in water (§13.1)
5. General characteristics of solutions formed by various combinations of gases, liquids, and solids (§13.1)
6. How intermolecular forces stabilize the structures of proteins, the cell membrane, and DNA (§13.2)
7. The enthalpy components of a solution cycle and their effect on ΔH_{soln} (§13.3)
8. The dependence of ΔH_{hydr} on ionic charge density and the factors that determine whether ionic solution processes are exothermic or endothermic (§13.3)
9. The meaning of entropy and how the balance between the change in enthalpy and the change in entropy governs the solution process (§13.3)
10. The distinctions among saturated, unsaturated, and supersaturated solutions, and the equilibrium nature of a saturated solution (§13.4)
11. The relation between temperature and the solubility of solids (§13.4)
12. Why the solubility of gases in water decreases with a rise in temperature (§13.4)
13. The effect of gas pressure on solubility and its quantitative expression as Henry's law (§13.4)
14. The meaning of molarity, molality, mole fraction, and parts by mass or by volume of a solution, and how to convert among them (§13.5)
15. The distinction between electrolytes and nonelectrolytes in solution (§13.6)
16. The four colligative properties and their dependence on number of dissolved particles (§13.6)
17. Ideal solutions and the importance of Raoult's law (§13.6)
18. How the phase diagram of a solution differs from that of the pure solvent (§13.6)
19. Why the vapor over a solution of a volatile nonelectrolyte is richer in the more volatile component (§13.6)
20. Why strong electrolyte solutions are not ideal and the meanings of the van't Hoff factor and ionic atmosphere (§13.6)
21. How particle size distinguishes suspensions, colloids, and solutions (§13.7)
22. How colloidal behavior is demonstrated by the Tyndall effect and Brownian motion (§13.7)

Master These Skills

1. Predicting relative solubilities from intermolecular forces (SP 13.1)
2. Calculating the heat of solution for an ionic compound (SP 13.2)
3. Using Henry's law to calculate the solubility of a gas (SP 13.3)
4. Expressing concentration in terms of molality, parts by mass, parts by volume, and mole fraction (SPs 13.4, 13.5)
5. Interconverting among the various terms for expressing concentration (SP 13.6)
6. Using Raoult's law to calculate the vapor pressure lowering of a solution (SP 13.7)
7. Determining boiling and freezing points of a solution (SP 13.8)
8. Using a colligative property to calculate the molar mass of a solute (SP 13.9)
9. Calculating the composition of vapor over a solution of volatile nonelectrolyte (§13.6)
10. Calculating the van't Hoff factor (i) from the magnitude of a colligative property (§13.6)
11. Using a depiction to determine colligative properties (SP 13.10)

Key Terms

Page numbers appear in parentheses.

alloy (538)	freezing point depression (ΔT_f) (561)	mass percent [% (w/w)] (554)	soap (541)
amino acid (539)	hard water (570)	miscible (534)	solubility (S) (534)
boiling point elevation (ΔT_b) (559)	heat of hydration (ΔH_{hydr}) (545)	molality (m) (553)	solute (534)
charge density (545)	heat of solution (ΔH_{soln}) (544)	mole fraction (X) (554)	solvation (545)
colligative property (557)	Henry's law (551)	mononucleotide (543)	solvent (534)
colloid (568)	hydration (545)	nonelectrolyte (557)	supersaturated solution (549)
desalination (571)	hydration shell (534)	nucleic acid (542)	suspension (568)
dipole–induced dipole force (535)	ideal solution (558)	osmosis (562)	Tyndall effect (569)
double helix (543)	ion exchange (570)	osmotic pressure (Π) (562)	unsaturated solution (549)
electrolyte (557)	ionic atmosphere (565)	protein (539)	vapor pressure lowering (ΔP) (558)
entropy (S) (547)	ion–induced dipole force (534)	Raoult's law (558)	volume percent [% (v/v)] (554)
fractional distillation (564)	like-dissolves-like rule (534)	reverse osmosis (571)	wastewater (571)
	lipid bilayer (542)	saturated solution (549)	water softening (570)
		semipermeable membrane (562)	

Key Equations and Relationships

Page numbers appear in parentheses.

13.1 Dividing the general heat of solution into component enthalpies (544):

$$\Delta H_{soln} = \Delta H_{solute} + \Delta H_{solvent} + \Delta H_{mix}$$

13.2 Dividing the heat of solution of an ionic compound in water into component enthalpies (545):

$$\Delta H_{soln} = \Delta H_{lattice} + \Delta H_{hydr\ of\ the\ ions}$$

13.3 Relating gas solubility to its partial pressure (Henry's law) (551):

$$S_{gas} = k_H \times P_{gas}$$

13.4 Defining concentration in terms of molarity (552):

$$\text{Molarity } (M) = \frac{\text{amount (mol) of solute}}{\text{volume (L) of solution}}$$

13.5 Defining concentration in terms of molality (553):

$$\text{Molality } (m) = \frac{\text{amount (mol) of solute}}{\text{mass (kg) of solvent}}$$

13.6 Defining concentration in terms of mass percent (554):

$$\text{Mass percent } [\%\ (w/w)] = \frac{\text{mass of solute}}{\text{mass of solution}} \times 100$$

13.7 Defining concentration in terms of volume percent (554):

$$\text{Volume percent } [\%\ (v/v)] = \frac{\text{volume of solute}}{\text{volume of solution}} \times 100$$

13.8 Defining concentration in terms of mole fraction (554):

Mole fraction (X)

$$= \frac{\text{amount (mol) of solute}}{\text{amount (mol) of solute} + \text{amount (mol) of solvent}}$$

13.9 Expressing the relationship between the vapor pressure of solvent above a solution and its mole fraction in the solution (Raoult's law) (558):

$$P_{solvent} = X_{solvent} \times P^\circ_{solvent}$$

13.10 Calculating the vapor pressure lowering due to solute (558):

$$\Delta P = X_{solute} \times P^\circ_{solvent}$$

13.11 Calculating the boiling point elevation of a solution (560):

$$\Delta T_b = K_b m$$

13.12 Calculating the freezing point depression of a solution (561):

$$\Delta T_f = K_f m$$

13.13 Calculating the osmotic pressure of a solution (562):

$$\Pi = \frac{n_{solute}}{V_{soln}} RT = MRT$$

BRIEF SOLUTIONS TO FOLLOW-UP PROBLEMS

13.1A (a) 1-Butanol has one —OH group/molecule, while 1,4-butanediol has two —OH groups/molecule. 1,4-Butanediol is more soluble in water because it can form more H bonds.
(b) Chloroform is more soluble in water because of dipole-dipole forces between the polar $CHCl_3$ molecules and water. The forces between nonpolar CCl_4 molecules and water are weaker dipole–induced dipole forces, which do not effectively replace H bonds between water molecules.

13.1B (a) Chloroform dissolves more chloromethane due to similar dipole-dipole forces between the polar molecules of these two substances. CH_3Cl molecules do not exhibit H bonding and so do not effectively replace H bonds between methanol molecules.
(b) Hexane dissolves more pentanol due to dispersion forces between the hydrocarbon chains in each molecule.

13.2A From Equation 13.2, we have

$$\Delta H_{soln} \text{ of } KNO_3 = \Delta H_{lattice} \text{ of } KNO_3 + (\Delta H_{hydr} \text{ of } K^+ + \Delta H_{hydr} \text{ of } NO_3^-)$$
$$34.89 \text{ kJ/mol} = 685 \text{ kJ/mol} + (\Delta H_{hydr} \text{ of } K^+ + \Delta H_{hydr} \text{ of } NO_3^-)$$
$$\Delta H_{hydr} \text{ of } K^+ + \Delta H_{hydr} \text{ of } NO_3^- = 34.89 \text{ kJ/mol} - 685 \text{ kJ/mol} = -650. \text{ kJ/mol}$$

13.2B From Equation 13.2, we have

$$\Delta H_{soln} \text{ of NaCN} = \Delta H_{lattice} \text{ of NaCN} + (\Delta H_{hydr} \text{ of } Na^+ + \Delta H_{hydr} \text{ of } CN^-)$$
$$1.21 \text{ kJ/mol} = 766 \text{ kJ/mol} + (-410. \text{ kJ/mol} + \Delta H_{hydr} \text{ of } CN^-)$$
$$\Delta H_{hydr} \text{ of } CN^- = 1.21 \text{ kJ/mol} - 766 \text{ kJ/mol} + 410. \text{ kJ/mol} = -355 \text{ kJ/mol}$$

13.3A The partial pressure of N_2 in air is the volume percent divided by 100 times the total pressure (Dalton's law, Section 5.4): $P_{N_2} = 0.78 \times 1 \text{ atm} = 0.78 \text{ atm}$.

$$S_{gas} = k_H \times P_{gas}$$
$$S_{N_2} = (7\times10^{-4} \text{ mol/L}\cdot\text{atm})(0.78 \text{ atm}) = 5\times10^{-4} \text{ mol/L}$$

13.3B In a mixture of gases, the volume percent of a gas divided by 100 times the total pressure equals the gas's partial pressure (Dalton's law, Section 5.4):

$P_{gas} = 0.40 \times 1.2 \text{ atm} = 0.48 \text{ atm}$.

$$k_H = \frac{S_{gas}}{P_{gas}} = \frac{1.2\times10^{-2} \text{ mol/L}}{0.48 \text{ atm}} = 2.5\times10^{-2} \text{ mol/L}\cdot\text{atm}$$

13.4A Convert mass (g) of ethanol to kg, multiply by the molality to obtain amount (mol) of glucose, and then multiply amount (mol) of glucose by the molar mass to obtain mass of glucose.

Amount (mol) of glucose

$$= 563 \text{ g ethanol} \times \frac{1 \text{ kg}}{10^3 \text{ g}} \times \frac{2.40\times10^{-2} \text{ mol glucose}}{1 \text{ kg ethanol}} = 1.35\times10^{-2} \text{ mol glucose}$$

$$\text{Mass (g) glucose} = 1.35\times10^{-2} \text{ mol } C_6H_{12}O_6 \times \frac{180.16 \text{ g } C_6H_{12}O_6}{1 \text{ mol } C_6H_{12}O_6} = 2.43 \text{ g glucose}$$

13.4B Convert mass (g) of I_2 to amount (mol) and amount (mol) of $(CH_3CH_2)_2O$ to mass (kg). Then divide moles of I_2 by kg of $(CH_3CH_2)_2O$.

$$\text{Amount (mol) of } I_2 = 15.20 \text{ g } I_2 \times \frac{1 \text{ mol } I_2}{253.8 \text{ g } I_2} = 5.989\times10^{-2} \text{ mol } I_2$$

Mass (kg) of $(CH_3CH_2)_2O$

$$= 1.33 \text{ mol } (CH_3CH_2)_2O \times \frac{74.12 \text{ g } (CH_3CH_2)_2O}{1 \text{ mol } (CH_3CH_2)_2O} \times \frac{1 \text{ kg}}{10^3 \text{ g}} = 9.86\times10^{-2} \text{ kg } (CH_3CH_2)_2O$$

$$\text{Molality } (m) = \frac{5.989\times10^{-2} \text{ mol}}{9.86\times10^{-2} \text{ kg}} = 0.607\ m$$

Finally, an exceptionally large number of qualitative, quantitative, and molecular-scene problems end each chapter. Four types of problems are presented—three by chapter section, with comprehensive problems following:

- **Concept Review Questions** test qualitative understanding of key ideas.
- **Skill-Building Exercises** are grouped in similar pairs, with one of each pair answered in the back of the book. A group of similar exercises may begin with explicit steps and increase in difficulty, gradually weaning the student from the need for multistep directions.
- **Problems in Context** apply the skills learned in the skill-building exercises to interesting scenarios, including realistic examples dealing with industry, medicine, and the environment.
- **Comprehensive Problems,** mostly based on realistic applications, are more challenging and rely on material from any section of the current chapter or any previous chapter.

PROBLEMS

Problems with **colored** numbers are answered in Appendix E and worked in detail in the Student Solutions Manual. Problem sections match those in the text and give the numbers of relevant sample problems. Most offer Concept Review Questions, Skill-Building Exercises (grouped in pairs covering the same concept), and Problems in Context. The Comprehensive Problems are based on material from any section or previous chapter.

Depicting Molecules and Ions with Lewis Structures

(Sample Problems 10.1 to 10.5)

Concept Review Questions

10.1 Which of these atoms *cannot* serve as a central atom in a Lewis structure: (a) O; (b) He; (c) F; (d) H; (e) P? Explain.

10.2 When is a resonance hybrid needed to adequately depict the bonding in a molecule? Using NO_2 as an example, explain how a resonance hybrid is consistent with the actual bond length, bond strength, and bond order.

10.3 In which of these structures does X obey the octet rule?

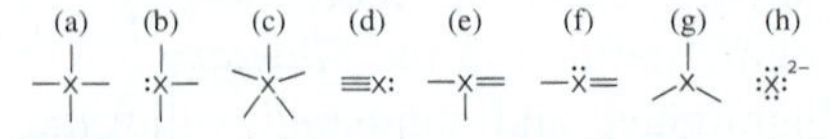

10.4 What is required for an atom to expand its valence shell? Which of the following atoms can expand its valence shell: F, S, H, Al, Se, Cl?

Skill-Building Exercises (grouped in similar pairs)

10.5 Draw a Lewis structure for (a) SiF_4; (b) $SeCl_2$; (c) COF_2 (C is the central atom).

10.6 Draw a Lewis structure for (a) PH_4^+; (b) C_2F_4; (c) SbH_3.

10.7 Draw a Lewis structure for (a) PF_3; (b) H_2CO_3 (both H atoms are attached to O atoms); (c) CS_2.

10.8 Draw a Lewis structure for (a) CH_4S; (b) S_2Cl_2; (c) $CHCl_3$.

10.9 Draw Lewis structures of all the important resonance forms of (a) NO_2^+; (b) NO_2F (N is central).

10.10 Draw Lewis structures of all the important resonance forms of (a) HNO_3 ($HONO_2$); (b) $HAsO_4^{2-}$ ($HOAsO_3^{2-}$).

10.11 Draw Lewis structures of all the important resonance forms of (a) N_3^-; (b) NO_2^-.

10.12 Draw Lewis structures of all the important resonance forms of (a) HCO_2^- (H is attached to C); (b) $HBrO_4$ ($HOBrO_3$).

10.13 Draw the Lewis structure with lowest formal charges, and determine the charge of each atom in (a) IF_5; (b) AlH_4^-.

10.14 Draw the Lewis structure with lowest formal charges, and determine the charge of each atom in (a) OCS; (b) NO.

10.15 Draw the Lewis structure with lowest formal charges, and determine the charge of each atom in (a) CN^-; (b) ClO^-.

10.16 Draw the Lewis structure with lowest formal charges, and ... ; (b) ClNO.

... e form of each ion ... w the charges, and ... ; (b) SO_3^{2-}.

10.18 Draw a Lewis structure for a resonance form of each ion with the lowest possible formal charges, show the charges, and give oxidation numbers of the atoms: (a) AsO_4^{3-}; (b) ClO_2^-.

10.19 These species do not obey the octet rule. Draw a Lewis structure for each, and state the type of octet-rule exception:
(a) BH_3 (b) AsF_4^- (c) $SeCl_4$

10.20 These species do not obey the octet rule. Draw a Lewis structure for each, and state the type of octet-rule exception:
(a) PF_6^- (b) ClO_3 (c) H_3PO_3 (one P—H bond)

10.21 These species do not obey the octet rule. Draw a Lewis structure for each, and state the type of octet-rule exception:
(a) BrF_3 (b) ICl_2^- (c) BeF_2

10.22 These species do not obey the octet rule. Draw a Lewis structure for each, and state the type of octet-rule exception:
(a) O_3^- (b) XeF_2 (c) SbF_4^-

Problems in Context

10.23 Molten beryllium chloride reacts with chloride ion from molten NaCl to form the $BeCl_4^{2-}$ ion, in which the Be atom attains an octet. Show the net ionic reaction with Lewis structures.

10.24 Despite many attempts, the perbromate ion (BrO_4^-) was not prepared in the laboratory until about 1970. (In fact, articles were published explaining theoretically why it could never be prepared!) Draw a Lewis structure for BrO_4^- in which all atoms have lowest formal charges.

10.25 Cryolite (Na_3AlF_6) is an indispensable component in the electrochemical production of aluminum. Draw a Lewis structure for the AlF_6^{3-} ion.

10.26 Phosgene is a colorless, highly toxic gas that was employed against troops in World War I and is used today as a key reactant in organic syntheses. From the following resonance structures, select the one with the lowest formal charges:

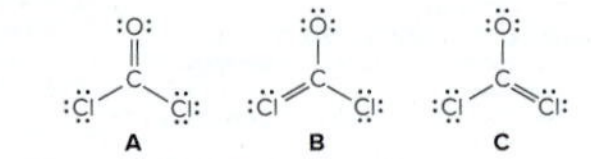

Valence-Shell Electron-Pair Repulsion (VSEPR) Theory

(Sample Problems 10.6 to 10.8)

Concept Review Questions

10.27 If you know the formula of a molecule or ion, what is the first step in predicting its shape?

10.28 In what situation is the name of the molecular shape the same as the name of the electron-group arrangement?

10.29 Which of the following numbers of electron groups can give rise to a bent (V shaped) molecule: two, three, four, five, six? Draw an example for each case, showing the shape classification (AX_mE_n) and the ideal bond angle.

10.30 Name all the molecular shapes that have a tetrahedral electron-group arrangement.

Comprehensive Problems

2.119 Helium is the lightest noble gas and the second most abundant element (after hydrogen) in the universe.
(a) The radius of a helium atom is 3.1×10^{-11} m; the radius of its nucleus is 2.5×10^{-15} m. What fraction of the spherical atomic volume is occupied by the nucleus (V of a sphere $= \frac{4}{3}\pi r^3$)?
(b) The mass of a helium-4 atom is 6.64648×10^{-24} g, and each of its two electrons has a mass of 9.10939×10^{-28} g. What fraction of this atom's mass is contributed by its nucleus?

2.120 From the following ions (with their radii in pm), choose the pair that forms the strongest ionic bond and the pair that forms the weakest:

Ion:	Mg^{2+}	K^+	Rb^+	Ba^{2+}	Cl^-	O^{2-}	I^-
Radius:	72	138	152	135	181	140	220

2.121 Give the molecular mass of each compound depicted below, and provide a correct name for any that are named incorrectly.

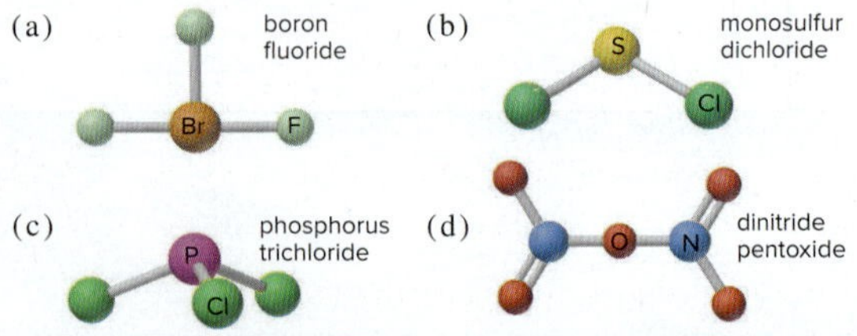

OPTIMIZING THE TEXT

The modern chemistry student's learning experience is changing dramatically. To address the changes that students face, a modern text partnered with a suite of robust digital tools must continue to evolve. With each edition, students and instructors alike have been involved in refining this text. From one-on-one interviews, focus groups, and symposia, as well as extensive chapter reviews and class tests, we learned that everyone praises the pioneering molecular art, the stepwise problem-solving approach, the abundant mix of qualitative, quantitative, and applied end-of-chapter problems, and the rigorous *and* student-friendly coverage of mainstream topics.

Global Changes to Every Chapter

Our revision for the eighth edition focused on continued optimization of the text. To aid us in this process, we were able to use data from literally thousands of student responses to questions in LearnSmart, the adaptive learning system that assesses student knowledge of course content. The data, such as average time spent answering each question and the percentage of students who correctly answered the question on the first attempt, revealed the learning objectives that students found particularly difficult. We utilized several approaches to present these difficult concepts in a clearer, more straightforward way in the eighth edition of *Chemistry: The Molecular Nature of Matter and Change.*

Making the concepts clearer through digital learning resources. Students will be able to access over 2,000 digital learning resources throughout this text's SmartBook. These learning resources present summaries of concepts and worked examples, including over 400 videos of chemistry instructors solving problems or modeling concepts that students can view over and over again. Thus, students can have an "office hour" moment at any time.

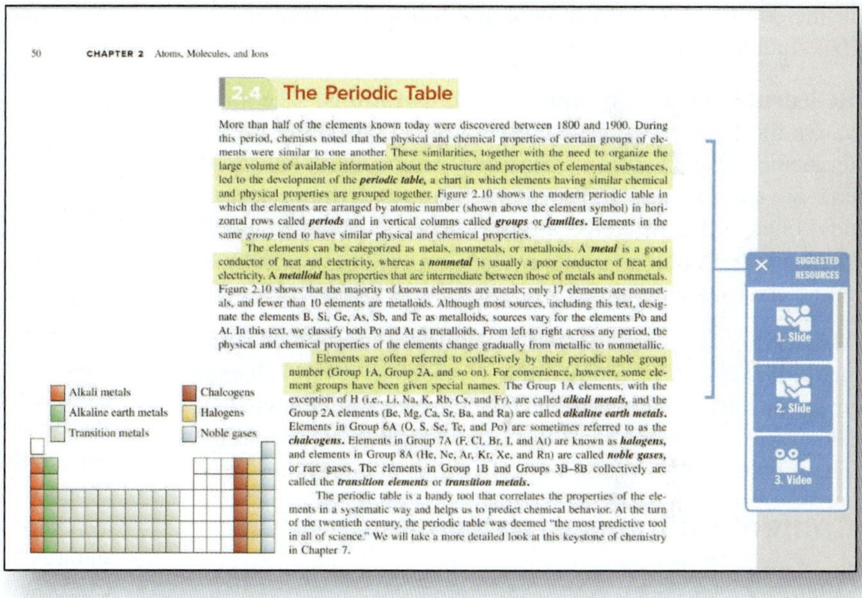

50 **CHAPTER 2** Atoms, Molecules, and Ions

2.4 The Periodic Table

More than half of the elements known today were discovered between 1800 and 1900. During this period, chemists noted that the physical and chemical properties of certain groups of elements were similar to one another. These similarities, together with the need to organize the large volume of available information about the structure and properties of elemental substances, led to the development of the ***periodic table,*** a chart in which elements having similar chemical and physical properties are grouped together. Figure 2.10 shows the modern periodic table in which the elements are arranged by atomic number (shown above the element symbol) in horizontal rows called ***periods*** and in vertical columns called ***groups*** or ***families.*** Elements in the same *group* tend to have similar physical and chemical properties.

The elements can be categorized as metals, nonmetals, or metalloids. A ***metal*** is a good conductor of heat and electricity, whereas a ***nonmetal*** is usually a poor conductor of heat and electricity. A ***metalloid*** has properties that are intermediate between those of metals and nonmetals. Figure 2.10 shows that the majority of known elements are metals; only 17 elements are nonmetals, and fewer than 10 elements are metalloids. Although most sources, including this text, designate the elements B, Si, Ge, As, Sb, and Te as metalloids, sources vary for the elements Po and At. In this text, we classify both Po and At as metalloids. From left to right across any period, the physical and chemical properties of the elements change gradually from metallic to nonmetallic.

Elements are often referred to collectively by their periodic table group number (Group 1A, Group 2A, and so on). For convenience, however, some element groups have been given special names. The Group 1A elements, with the exception of H (i.e., Li, Na, K, Rb, Cs, and Fr), are called ***alkali metals,*** and the Group 2A elements (Be, Mg, Ca, Sr, Ba, and Ra) are called ***alkaline earth metals.*** Elements in Group 6A (O, S, Se, Te, and Po) are sometimes referred to as the ***chalcogens.*** Elements in Group 7A (F, Cl, Br, I, and At) are known as ***halogens,*** and elements in Group 8A (He, Ne, Ar, Kr, Xe, and Rn) are called ***noble gases,*** or rare gases. The elements in Group 1B and Groups 3B–8B collectively are called the ***transition elements*** or ***transition metals.***

The periodic table is a handy tool that correlates the properties of the elements in a systematic way and helps us to predict chemical behavior. At the turn of the twentieth century, the periodic table was deemed "the most predictive tool in all of science." We will take a more detailed look at this keystone of chemistry in Chapter 7.

NEW! Student Hot Spot

We are very pleased to incorporate real student data points and input, derived from thousands of our LearnSmart users, to help guide our revision. LearnSmart Heat Maps provided a quick visual snapshot of usage of portions of the text and the relative difficulty students experienced in mastering the content. With these data, we were able to both hone our text content when needed and, for particularly challenging concepts, point students to the learning resources that can elucidate and reinforce those concepts. You'll see these marginal features throughout the text. Students should log into Connect and view the resources through our SmartBook.

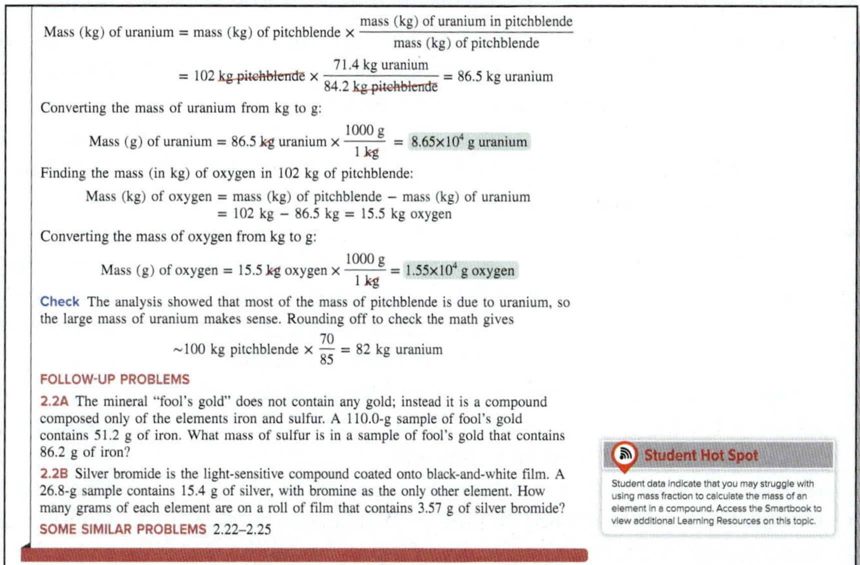

$$\text{Mass (kg) of uranium} = \text{mass (kg) of pitchblende} \times \frac{\text{mass (kg) of uranium in pitchblende}}{\text{mass (kg) of pitchblende}}$$

$$= 102\ \cancel{\text{kg pitchblende}} \times \frac{71.4\ \text{kg uranium}}{84.2\ \cancel{\text{kg pitchblende}}} = 86.5\ \text{kg uranium}$$

Converting the mass of uranium from kg to g:

$$\text{Mass (g) of uranium} = 86.5\ \cancel{\text{kg}}\ \text{uranium} \times \frac{1000\ \text{g}}{1\ \cancel{\text{kg}}} = 8.65\times10^4\ \text{g uranium}$$

Finding the mass (in kg) of oxygen in 102 kg of pitchblende:

$$\text{Mass (kg) of oxygen} = \text{mass (kg) of pitchblende} - \text{mass (kg) of uranium}$$
$$= 102\ \text{kg} - 86.5\ \text{kg} = 15.5\ \text{kg oxygen}$$

Converting the mass of oxygen from kg to g:

$$\text{Mass (g) of oxygen} = 15.5\ \cancel{\text{kg}}\ \text{oxygen} \times \frac{1000\ \text{g}}{1\ \cancel{\text{kg}}} = 1.55\times10^4\ \text{g oxygen}$$

Check The analysis showed that most of the mass of pitchblende is due to uranium, so the large mass of uranium makes sense. Rounding off to check the math gives

$$\sim 100\ \text{kg pitchblende} \times \frac{70}{85} = 82\ \text{kg uranium}$$

FOLLOW-UP PROBLEMS

2.2A The mineral "fool's gold" does not contain any gold; instead it is a compound composed only of the elements iron and sulfur. A 110.0-g sample of fool's gold contains 51.2 g of iron. What mass of sulfur is in a sample of fool's gold that contains 86.2 g of iron?

2.2B Silver bromide is the light-sensitive compound coated onto black-and-white film. A 26.8-g sample contains 15.4 g of silver, with bromine as the only other element. How many grams of each element are on a roll of film that contains 3.57 g of silver bromide?

SOME SIMILAR PROBLEMS 2.22–2.25

Applying ideas with enhanced problems throughout the chapters. The much admired four-part problem-solving format (plan, solution, check, follow-up) is retained in the eighth edition, in both data-based and molecular-scene *Sample Problems*. Two *Follow-up Problems* are included with each sample problem, as well as a list of *Similar Problems* within the end-of-chapter problem set. *Brief Solutions* for all of the follow-up problems appear at the end of each chapter (rather than providing just a numerical answer in a distant end-of-book appendix, as is typical). The eighth edition has over 250 sample problems and over 500 follow-up problems. In almost every chapter, several sample and follow-up problems (and their brief solutions) were revised in this edition with two goals in mind. We sought to provide students with a variety of problems that would clearly elucidate concepts and demonstrate problem solving techniques, while giving students the opportunity to be challenged and gain competence. We also included more intermediate steps in the solutions to both sample and follow-up problems so that students could more easily follow the solutions.

Re-learning ideas with annotated illustrations. The innovative three-level figures and other art that raised the bar for molecular visualization in chemistry textbooks is still present. Several existing figures have been revised and several new ones added to create an even better teaching tool. We continue to streamline figure legends by placing their content into clarifying annotations with the figures themselves.

Mastering the content with abundant end-of-chapter problem sets. New problems were added to several chapter problem sets, providing students and teachers with abundant choices in a wide range of difficulty and real-life scenarios. The problem sets are more extensive than in most other texts.

Content Changes to Individual Chapters

In addition to the general optimization of concept explanations and problem solutions throughout the text, specific improvements were made to most chapters:

- **Chapter 1** has a revised table of decimal prefixes and SI units to make conversion among SI units clearer, a revised discussion on intensive and extensive properties, and a revised sample problem on density.
- **Chapter 2** includes revised sample problems on mass percent and naming of compounds.
- **Chapter 3** has several new end-of-chapter problems: one new problem on the determination of a molecular formula, two new problems on writing a balanced reaction and determining the limiting reactant from molecular scenes, and two new stoichiometric problems involving limiting reactants.
- **Chapter 4** includes a new figure illustrating the activity series of the halogens. Sample problems on stoichiometry in precipitation and acid-base reactions were revised to include reactions that do not have 1:1 mole ratios.
- **Chapter 5** has two revised sample problems that provide students with additional opportunities for pressure unit conversions and stoichiometry calculations for gas reactions.
- **Chapter 6** has a clearer and more detailed discussion on pressure-volume work and a revised sample problem on the calorimetric determination of heat of combustion. Also included are new end-of-chapter problems on the calculation of enthalpy change for an aqueous reaction and determination of heat of combustion with bomb calorimetry.
- **Chapter 7** contains a new table summarizing the relationships between the quantum numbers and orbitals for the first four main energy levels.
- **Chapter 8** contains a new figure on electron spin; orbital diagrams have been added to the solutions of several sample problems.
- **Chapter 9** has improvements to several figures, a more detailed discussion of relationship between difference in electronegativity and ionic character, and some new follow-up problems.
- **Chapter 10** includes more detailed examples of depicting molecules with double bonds and ions with Lewis structures. Sample and follow-up problems have been revised to provide more opportunities to calculate formal charges and use those to evaluate resonance structures.
- **Chapter 11** has new art to illustrate formation of sigma and pi bonds and a new figure to show the placement of lone pairs in hybrid orbitals.
- **Chapter 12** includes additional information about viscosity and intermolecular forces.
- **Chapter 13** includes a more challenging sample problem on Henry's law, as well as revisions to several follow-up problems. There are new problems on the calculation of molar mass from freezing point depression.
- **Chapter 15** incorporates new art to make nomenclature clearer and a revised figure to show the key stages in protein synthesis.
- **Chapter 16** has a revised sample problem using the first-order integrated rate law, a revised figure on reaction mechanisms, and a new molecular scene problem on first-order reactions.
- **Chapter 17** contains a revised table on concentration ratios in an equilibrium system and two new sample problems, one on finding the equilibrium constant for an overall reaction, and the other on converting between K_p and K_c.
- **Chapter 18** has a new table on magnitude of K_a and percent dissociation and two revised sample problems.
- **Chapter 19** has a revised sample problem on buffer pH that reflects a more realistic lab procedure, a new molecular scene problem involving buffer solutions, a clearer presentation of pH calculations during acid-base titrations, and revised figures of pH titration curves. The section on acid-base indicators has been expanded, including the addition of a new figure about choosing an indicator for each type of acid-base titration. The discussion of aqueous solutions of metal sulfides was simplified.
- **Chapter 20** incorporates a new table that summarizes Q, K, ΔG, and reaction spontaneity.
- **Chapter 21** has several revised follow-up problems.
- **Chapter 23** has a new figure illustrating chelate complex ions and several revised figures. A new equation for calculating the charge of the metal ion in a complex ion has been added.

- **Chapter 24** has a new table summarizing changes in mass and atomic numbers during radioactive decay; a table on stability of even vs. odd numbers of nucleons has been revised. The discussion about mode of decay and neutron/proton ratio has been expanded.

Addition of Advanced Topics

In this special version of the 8th edition, advanced topics have been added to three chapters for use in classes in which a deeper and more rigorous level of discussion is appropriate. Problems on these advanced topics have been added to the end-of-chapter problem sets and to the online homework question bank.

- **Chapter 7** includes an expanded discussion on the development of the Schrödinger equation and the particle-in-a-box model. A new sample problem gives students an opportunity to apply the particle-in-a-box model to electron transitions.
- **Chapter 16** incorporates the calculus involved in the derivation of the integrated rate laws for zero-, first-, and second-order reactions. Also now included are discussions of pseudo-first-order reactions, steady-state approximation, and the Michaelis-Menten equation for enzyme kinetics.
- **Chapter 20** has a significantly expanded section on entropy. The calculations of entropy changes during isothermal gas expansion or contraction, phase changes, and changes in temperature have been added to enhance the current content; three new sample problems demonstrating these entropy change calculations are included.

Innovative Topic and Chapter Presentation

While the topic sequence coincides with that used in most mainstream courses, built-in flexibility allows a wide range of differing course structures:

For courses that follow their own topic sequence, the general presentation, with its many section and subsection breaks and bulleted lists, allows topics to be rearranged, or even deleted, with minimal loss of continuity.

For courses that present several chapters, or topics within chapters, in different orders:

- Redox balancing by the oxidation-number method (formerly covered in Chapter 4) has been removed from the text, and the half-reaction method is covered with electrochemistry in Chapter 21, but it can easily be taught with Chapter 4.
- Gases (Chapter 5) can be covered in sequence to explore the mathematical modeling of physical behavior or, with no loss of continuity, just before liquids and solids (Chapter 12) to show the effects of intermolecular forces on the three states of matter.

For courses that use an atoms-first approach for some of the material, Chapters 7 through 13 move smoothly from quantum theory (7) through electron configuration (8), bonding models (9), molecular shape (10), VB and MO bonding theories (11), intermolecular forces in liquids and solids (12), and solutions (13). Immediate applications of these concepts appear in the discussions of periodic patterns in main-group chemistry (Chapter 14) and in the survey of organic chemistry (Chapter 15). Some instructors have also brought forward the coverage of transition elements and coordination compounds (23) as further applications of bonding concepts. (Of course, Chapters 14, 15, and 23 can just as easily remain in their more traditional placement later in the course.)

For courses that emphasize biological/medical applications, many chapters highlight these topics, including the role of intermolecular forces in biomolecular structure (12), the chemistry of polysaccharides, proteins, and nucleic acids (including protein synthesis, DNA replication, and DNA sequencing) (15), as well as introductions to enzyme catalysis (16), biochemical pathways (17), and trace elements in protein function (23).

For courses that stress engineering applications of physical chemistry topics, Chapters 16 through 21 cover kinetics (16), equilibrium in gases (17), acids and bases (18), and aqueous ionic systems (19) and entropy and free energy (20) as they apply to electrochemical systems (21), all in preparation for coverage of the elements in geochemical cycles, metallurgy, and industry in Chapter 22.

McGraw-Hill Create™ is another way to implement innovative chapter presentation. With Create, you can easily rearrange chapters, combine material from other content sources, and quickly upload content you have written, such as your course syllabus or teaching notes. Find the content you need in Create by searching through thousands of leading McGraw-Hill textbooks. Create even allows you to personalize your book's appearance by selecting the cover and adding your name, school, and course information. Order a Create book, and you'll receive a complimentary print review copy in 3–5 business days or a complimentary electronic review copy (eComp) via e-mail in minutes. Go to www.mcgrawhillcreate.com today and register to experience how McGraw-Hill Create empowers you to teach *your* students *your* way. **www.mcgrawhillcreate.com**

McGraw-Hill Tegrity® records and distributes your class lecture with just a click of a button. Students can view it anytime and anywhere via computer, iPod, or mobile device. Tegrity indexes as it records your PowerPoint® presentations and anything shown on your computer, so students can use key words to find exactly what they want to study. Tegrity is available as an integrated feature of McGraw-Hill Connect® Chemistry and as a stand-alone product.

connect®

Required=Results

© Hero Images/Getty Images

McGraw-Hill Connect® Learn Without Limits

Connect is a teaching and learning platform that is proven to deliver better results for students and instructors.

Connect empowers students by continually adapting to deliver precisely what they need, when they need it, and how they need it, so your class time is more engaging and effective.

> 73% of instructors who use **Connect** require it; instructor satisfaction **increases** by 28% when **Connect** is required.

Connect's Impact on Retention Rates, Pass Rates, and Average Exam Scores

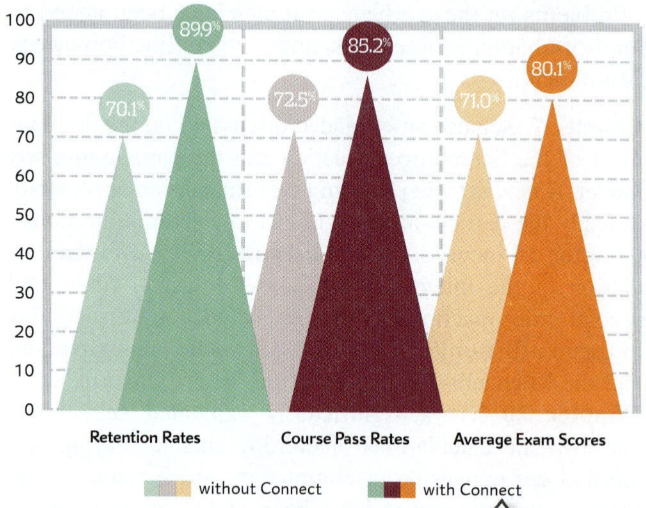

Using **Connect** improves retention rates by **19.8%**, passing rates by **12.7%**, and exam scores by **9.1%**.

Analytics

Connect Insight®

Connect Insight is Connect's new one-of-a-kind visual analytics dashboard that provides at-a-glance information regarding student performance, which is immediately actionable. By presenting assignment, assessment, and topical performance results together with a time metric that is easily visible for aggregate or individual results, Connect Insight gives the user the ability to take a just-in-time approach to teaching and learning, which was never before available. Connect Insight presents data that helps instructors improve class performance in a way that is efficient and effective.

Impact on Final Course Grade Distribution

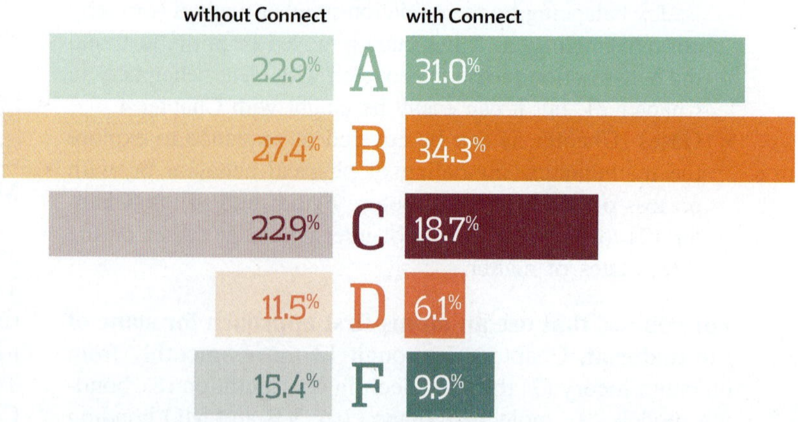

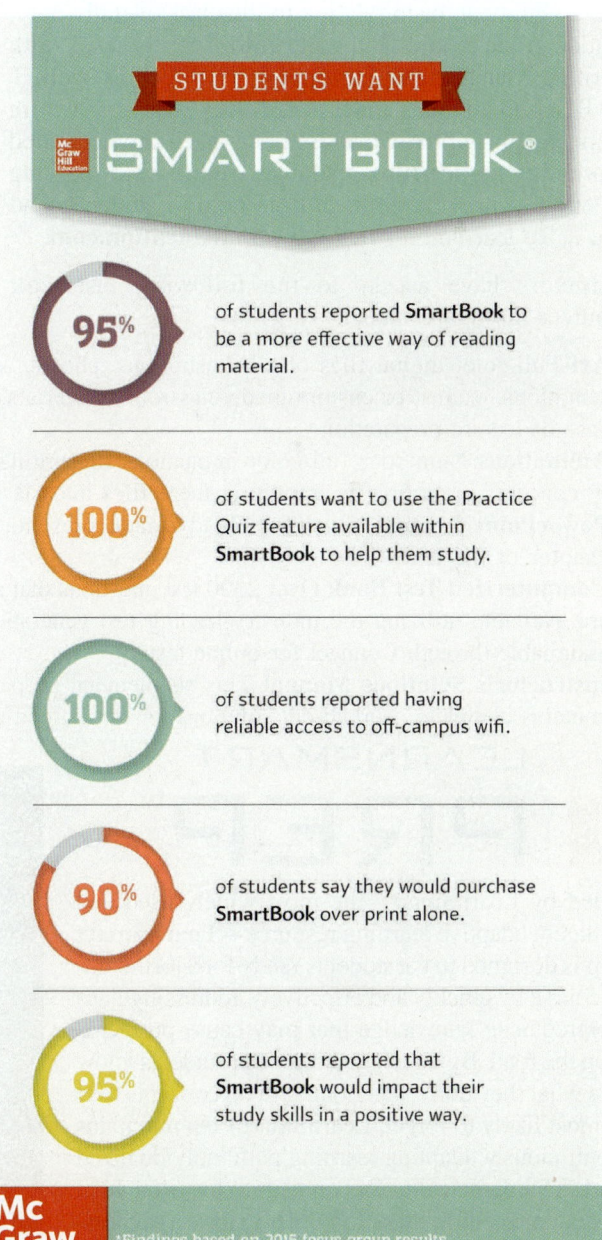

ADDITIONAL INSTRUCTOR AND STUDENT RESOURCES FOR YOUR COURSE!

McGraw Hill Education connect® CHEMISTRY

MCGRAW-HILL CONNECT CHEMISTRY

A robust set of questions, problems, and interactive figures are presented and aligned with the textbook's learning goals. The integration of **ChemDraw by PerkinElmer,** the industry standard in chemical drawing software, allows students to create accurate chemical structures in their online homework assignments. As an instructor, you can edit existing questions and write entirely new problems. Track individual student performance—by question, assignment, or in relation to the class overall—with detailed grade reports. Integrate grade reports easily with Learning Management Systems (LMS), such as WebCT and Blackboard—and much more. Also available within Connect, our adaptive SmartBook has been supplemented with additional learning resources tied to each learning objective to provide point-in-time help to students who need it. To learn more, visit **www.mheducation.com.**

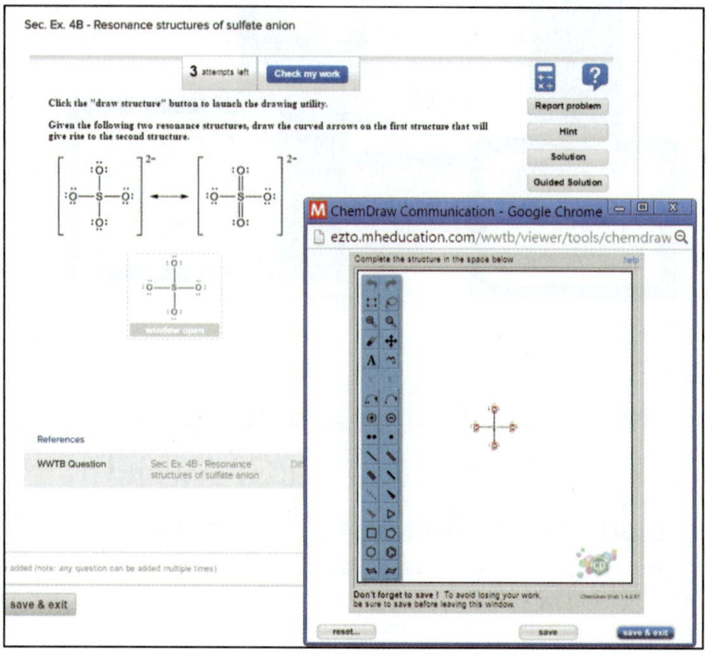

Instructors have access to the following instructor resources through Connect.

- **Art** Full-color digital files of all illustrations, photos, and tables in the book can be readily incorporated into lecture presentations, exams, or custom-made classroom materials. In addition, all files have been inserted into PowerPoint slides for ease of lecture preparation.
- **Animations** Numerous full-color animations illustrating important processes are also provided. Harness the visual impact of concepts in motion by importing these files into classroom presentations or online course materials.
- **PowerPoint Lecture Outlines** Ready-made presentations that combine art and lecture notes are provided for each chapter of the text.
- **Computerized Test Bank** Over 2300 test questions that accompany *Chemistry: The Molecular Nature of Matter and Change* are available utilizing the industry-leading test generation software TestGen. These same questions are also available and assignable through Connect for online tests.
- **Instructor's Solutions Manual** This supplement, prepared by Mara Vorachek-Warren of St. Charles Community College, contains complete, worked-out solutions for *all* the end-of-chapter problems in the text.

McGraw Hill Education LEARNSMART PREP™

Fueled by LearnSmart—the most widely used and intelligent adaptive learning resource—**LearnSmart Prep** is designed to get students ready for a forthcoming course by quickly and effectively addressing gaps in prerequisite knowledge that may cause problems down the road. By distinguishing what students know from what they don't, and honing in on concepts they are most likely to forget, LearnSmart Prep maintains a continuously adapting learning path individualized for each student, and tailors content to focus on what the student needs to master in order to have a successful start in the new class.

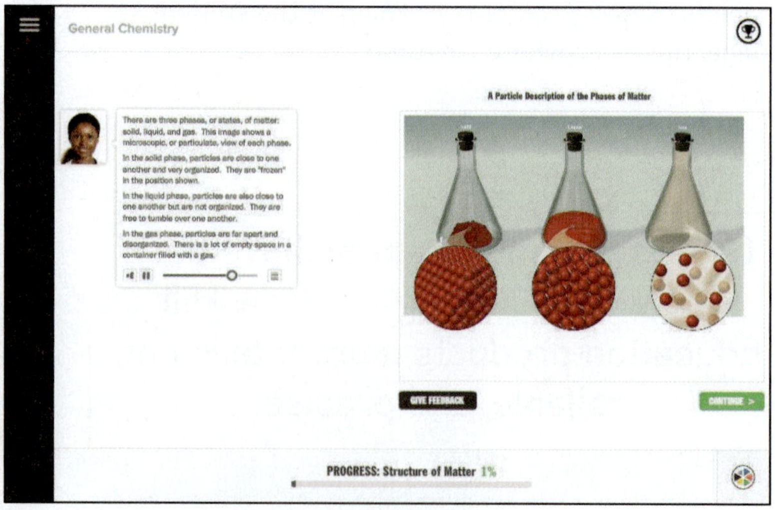

THE VIRTUAL LAB EXPERIENCE

Based on the same world-class, superbly adaptive technology as LearnSmart, **McGraw-Hill LearnSmart Labs** is a must-see, outcomes-based lab simulation. It assesses a student's knowledge and adaptively corrects deficiencies, allowing the student to learn faster and retain more knowledge with greater success. First, a student's knowledge is adaptively leveled on core learning outcomes: questioning reveals knowledge deficiencies that are corrected by the delivery of content that is conditional on a student's response. Then, a simulated lab experience requires the student to think and act like a scientist: recording, interpreting, and analyzing data using simulated equipment found in labs and clinics. The student is allowed to make mistakes—a powerful part of the learning experience! A virtual coach provides subtle hints when needed, asks questions about the student's choices, and allows the student to reflect on and correct those mistakes. Whether your need is to overcome the logistical challenges of a traditional lab, provide better lab prep, improve student performance, or make students' online experience one that rivals the real world, LearnSmart Labs accomplishes it all.

Mc Graw Hill Education LEARNSMART LABS™

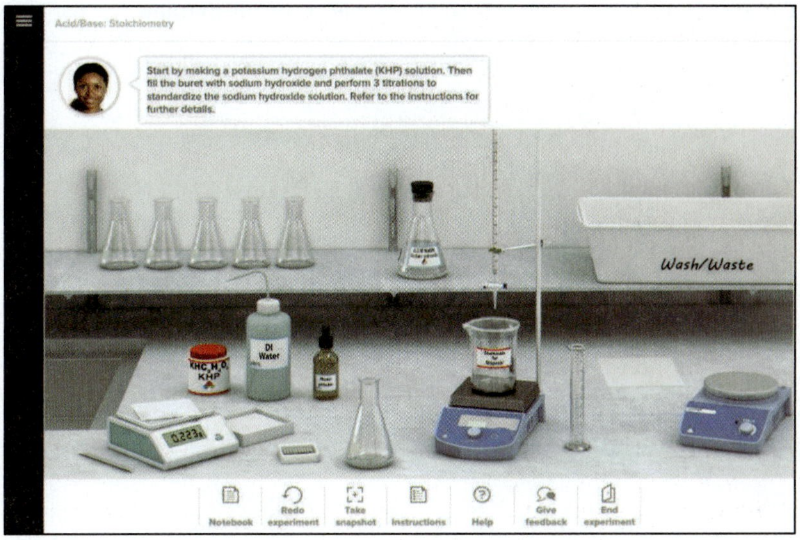

COOPERATIVE CHEMISTRY LABORATORY MANUAL

Prepared by Melanie Cooper of Clemson University, this innovative manual features open-ended problems designed to simulate experience in a research lab. Working in groups, students investigate one problem over a period of several weeks, so they might complete three or four projects during the semester, rather than one preprogrammed experiment per class. The emphasis is on experimental design, analytic problem solving, and communication.

STUDENT SOLUTIONS MANUAL

This supplement, prepared by Mara Vorachek-Warren of St. Charles Community College, contains detailed solutions and explanations for all problems in the main text that have colored numbers.

ACKNOWLEDGMENTS

It would be nearly impossible to put together a more professional, talented, and supportive publishing team than our colleagues at McGraw-Hill Education: Managing Director Thomas Timp, Director of Chemistry David Spurgeon, Ph.D., Associate Director of Digital Content Robin Reed, Program Manager Lora Neyens, Content Project Manager Laura Bies, Designer David Hash, Marketing Manager Matthew Garcia, and Director of Digital Content Shirley Hino. It is a pleasure to work with them; their leadership, knowledge, and encouragement have helped to make this latest edition a reality.

Mara Vorachek-Warren of St. Charles Community College provided a thorough accuracy check of all the new sample problems, follow-up problems, and end-of-chapter problems as part of her superb preparation of both the Student and Instructor's Solutions Manuals.

The following individuals helped write and review learning goal-oriented content for **LearnSmart for General Chemistry:** Margaret Ruth Leslie, Kent State University and Adam I. Keller, Columbus State Community College.

Several expert freelancers contributed as well. Jane Hoover did her usual excellent job in copyediting the text, and Lauren Timmer and Louis Poncz followed with meticulous proofreading. And many thanks to Jerry Marshall, who patiently researched new stock and studio photos.

CHEMISTRY

Advanced Topics

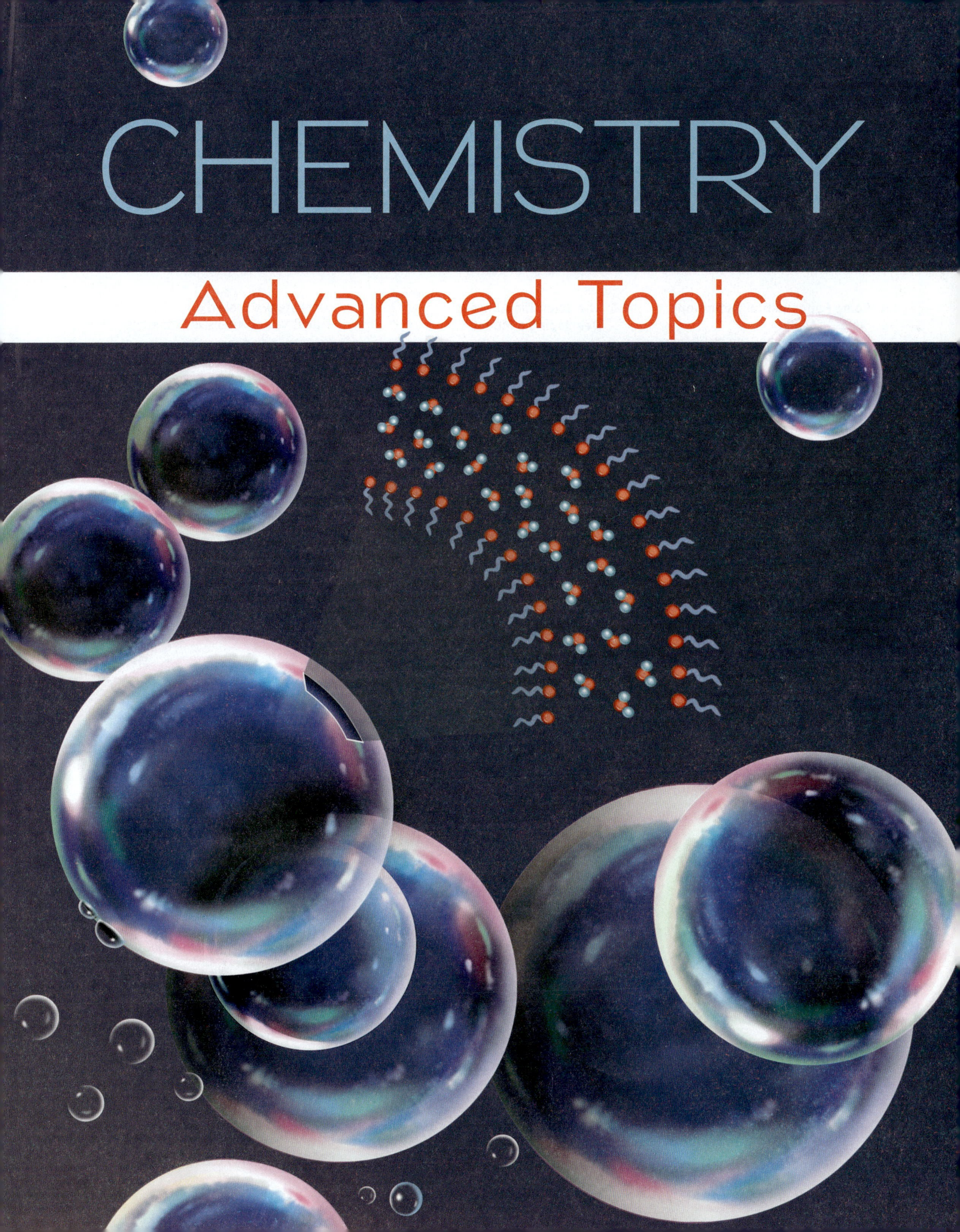

1 Keys to Studying Chemistry: Definitions, Units, and Problem Solving

1.1 Some Fundamental Definitions
States of Matter
Properties of Matter and Its Changes
Central Theme in Chemistry
Importance of Energy

1.2 Chemical Arts and the Origins of Modern Chemistry
Prechemical Traditions
Impact of Lavoisier

1.3 The Scientific Approach: Developing a Model

1.4 Measurement and Chemical Problem Solving
Features of SI Units
SI Units in Chemistry
Units and Conversion Factors
Systematic Problem-Solving Approach
Temperature Scales
Extensive and Intensive Properties

1.5 Uncertainty in Measurement: Significant Figures
Determining Significant Digits
Calculations and Rounding Off
Precision, Accuracy, and Instrument Calibration

Source: © Fancy Collection/SuperStock

Concepts and Skills to Review Before You Study This Chapter

- exponential (scientific) notation (Appendix A)

Maybe you're taking this course because chemistry is fundamental to understanding other natural sciences. Maybe it's required for your medical or engineering major. Or maybe you just want to learn more about the impact of chemistry on society or even on your everyday life. For example, does the following morning routine (described in chemical terms) sound familiar? You are awakened by the buzzing of your alarm clock, a sound created when molecules align in the liquid-crystal display of your clock and electrons flow to create a noise. You throw off a thermal insulator of manufactured polymer (blanket) and jump in the shower to emulsify fatty substances on your skin and hair with purified water and formulated detergents. Next you adorn yourself in an array of pleasant-smelling pigmented gels, dyed polymeric fibers, synthetic footwear, and metal-alloy jewelry. After a breakfast of nutrient-enriched, spoilage-retarded carbohydrates (cereal) in a white emulsion of fats, proteins, and monosaccharides (milk) and a cup of hot aqueous extract containing a stimulating alkaloid (coffee), you abrade your teeth with a colloidal dispersion of artificially flavored, dental-hardening agents (toothpaste), grab your portable electronic device containing ultrathin, microetched semiconductor layers powered by a series of voltaic cells (laptop), collect some objects made from processed cellulose and plastic, electronically printed with light- and oxygen-resistant inks (books), hop in your hydrocarbon-fueled, metal-vinyl-ceramic vehicle, electrically ignite a synchronized series of controlled gaseous explosions (start your car), and take off for class!

But the true impact of chemistry extends much farther than the commercial products of daily life. The truth is that the most profound biological and environmental questions ultimately have chemical answers: How does an organism reproduce, grow, and age? What are the underlying explanations for health and disease? How can we sustain a planetary ecosystem in which plant, animal, and human populations thrive? Is there life on other worlds?

So, no matter what your reason for studying chemistry, you're going to learn some amazing things. And, this course comes with a bonus for developing two mental skills. The first, common to all science courses, is the ability to solve problems systematically. The second is specific to chemistry, for as you comprehend its ideas, you begin to view a hidden reality, one filled with incredibly minute particles moving at fantastic speeds, colliding billions of times a second, and interacting in ways that allow your brain to translate fluxes of electric charge into thoughts and that determine how all the matter inside and outside of you behaves. This chapter holds the keys to unlock and enter this new world.

IN THIS CHAPTER . . . *We discuss some central ideas about matter and energy, the process of science, units of measurement, and how scientists handle data.*

- We begin with fundamental concepts about matter and energy and their changes.
- A brief discussion of chemistry's origins, including some major missteps, leads to an overview of how scientists build models to study nature.
- We examine modern units for mass, length, volume, density, and temperature and apply systematic chemical problem solving to unit conversions.
- We see that data collection always includes some uncertainty and examine the distinction between accuracy and precision.

1.1 SOME FUNDAMENTAL DEFINITIONS

A good place to begin our exploration of chemistry is by defining it and a few central concepts. **Chemistry** is *the scientific study of matter and its properties, the changes that matter undergoes, and the energy associated with those changes.* **Matter** is the "stuff" of the universe: air, glass, planets, students—*anything that has mass and volume.* (In Section 1.4, we discuss the meanings of mass and volume in terms of how they are measured.) Chemists want to know the **composition** of matter, *the types and amounts of simpler substances that make it up.* A *substance* is a type of matter that has a defined, fixed composition.

The States of Matter

Matter occurs commonly in *three physical forms* called **states:** solid, liquid, and gas. On the macroscopic scale, each state of matter is defined by the way the sample fills a container (Figure 1.1, *flasks at top*):

- A **solid** has a fixed shape that does not conform to the container shape. Solids are *not* defined by rigidity or hardness: solid iron is rigid and hard, but solid lead is flexible, and solid wax is soft.
- A **liquid** has a varying shape that conforms to the container shape, but only to the extent of the liquid's volume; that is, a liquid has *an upper surface.*
- A **gas** also has a varying shape that conforms to the container shape, but it fills the entire container and, thus, does *not* have a surface.

On the atomic scale, each state is defined by the relative positions of its particles (Figure 1.1, *circles at bottom*):

- In a *solid,* the particles lie next to each other in a regular, three-dimensional pattern, or *array.*
- In a *liquid,* the particles also lie close together but move randomly around each other.
- In a *gas,* the particles have large distances between them and move randomly throughout the container.

Solid
Particles are close together and organized.
Liquid
Particles are close together but disorganized.
Gas
Particles are far apart and disorganized.

Figure 1.1 The physical states of matter.

The Properties of Matter and Its Changes

We learn about matter by observing its **properties,** *the characteristics that give each substance its unique identity.* To identify a person, we might observe height, weight, hair and eye color, fingerprints, and, now, even DNA pattern, until we arrive at a unique identification. To identify a substance, we observe two types of properties, physical and chemical, which are closely related to two types of change that matter undergoes.

Physical Change: No Change in Composition **Physical properties** are characteristics a substance shows *by itself, without changing into or interacting with another substance.* These properties include color, melting point, electrical conductivity, and density. A **physical change** occurs when a substance *alters its physical properties,* ***not*** *its composition.* For example, when ice melts, several physical properties change, such as hardness, density, and ability to flow. But the composition of the sample does *not* change: it is still water. The photograph in Figure 1.2A shows what this change looks like in everyday life. The "blow-up" circles depict a magnified view of the particles making up the sample. In the icicle, the particles lie in the repeating pattern characteristic of a solid, whereas they are jumbled in the liquid droplet; however, *the particles are the same* in both states of water.

Physical change (same substance before and after):

$$\text{Water (solid state)} \longrightarrow \text{water (liquid state)}$$

All changes of state of matter are physical changes.

Chemical Change: A Change in Composition **Chemical properties** are characteristics a substance shows *as it changes into or interacts with another substance (or substances).* Chemical properties include flammability, corrosiveness, and reactivity with acids. A **chemical change,** also called a **chemical reaction,** occurs when *one or more substances are converted into one or more substances with different composition and properties.* Figure 1.2B shows the chemical change (reaction) that occurs when you pass an electric current through water: the water decomposes (breaks down) into two other substances, hydrogen and oxygen, that bubble into the tubes. The composition *has* changed: the final sample is no longer water.

Chemical change (different substances before and after):

$$\text{Water} \xrightarrow{\text{electric current}} \text{hydrogen} + \text{oxygen}$$

Let's work through a sample problem that uses atomic-scale scenes to distinguish between physical and chemical change.

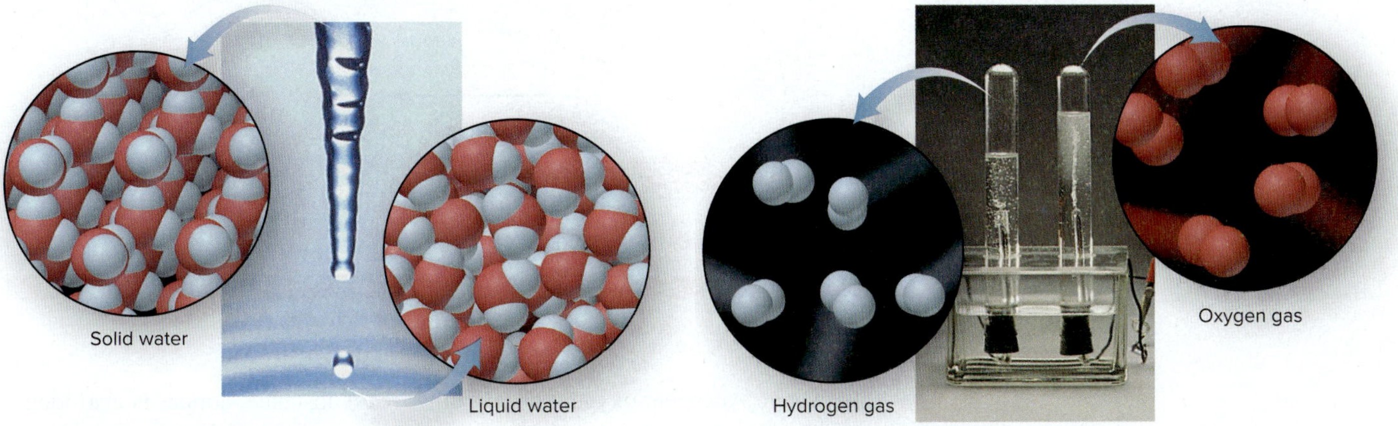

A Physical change:
Solid state of water becomes liquid state. Particles before and after remain the same, which means composition did ***not*** *change.*

B Chemical change:
Electric current decomposes water into different substances (hydrogen and oxygen). Particles before and after are different, which means composition ***did*** *change.*

Figure 1.2 The distinction between physical and chemical change.

Source: (A) © Paul Morrell/Stone/Getty Images; (B) © McGraw-Hill Education/Stephen Frisch, photographer

SAMPLE PROBLEM 1.1 **Visualizing Change on the Atomic Scale**

Problem The scenes below represent an atomic-scale view of a sample of matter, A, undergoing two different changes, left to B and right to C:

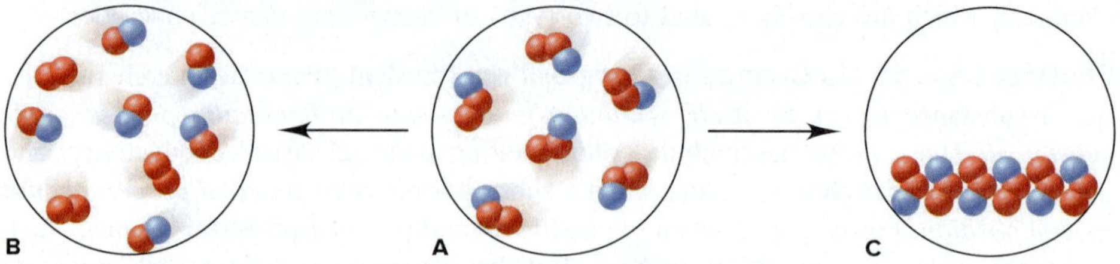

Decide whether each depiction shows a physical or a chemical change.

Plan Given depictions of two changes, we have to determine whether each represents a physical or a chemical change. The number and colors of the little spheres that make up each particle tell its "composition." Samples with particles of the *same* composition but in a different arrangement depict a *physical* change, whereas samples with particles of a *different* composition depict a *chemical* change.

Solution In A, each particle consists of one blue and two red spheres. The particles in A change into two types in B, one made of red and blue spheres and the other made of two red spheres; therefore, they have undergone a chemical change to form different particles. The particles in C are the same as those in A, but they are closer together and arranged in a regular pattern; therefore, they have undergone a physical change.

FOLLOW-UP PROBLEMS

*Brief Solutions for **all** Follow-up Problems appear at the end of the chapter.*

1.1A Is the following change chemical or physical?

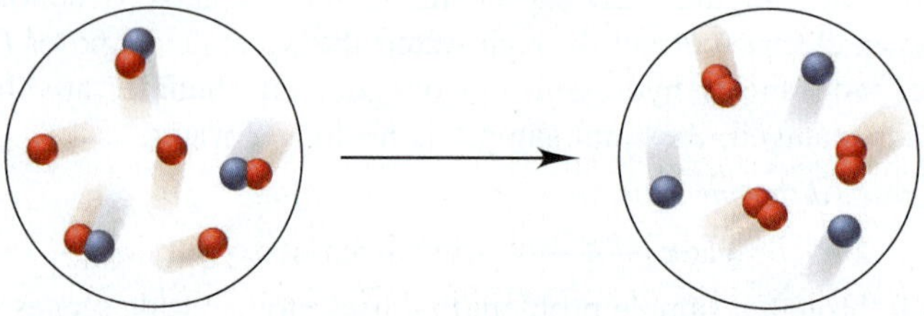

1.1B Is the following change chemical or physical?

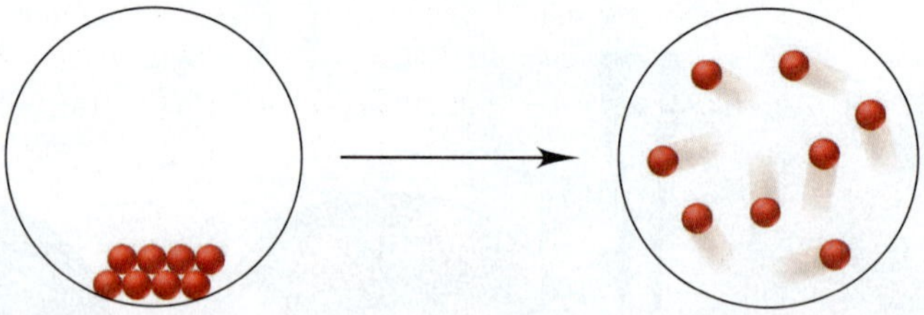

SOME SIMILAR PROBLEMS 1.1 and 1.84

Like water, hydrogen, oxygen, or any other real substance, copper is also identified by *its own set* of physical and chemical properties (Table 1.1).

Temperature and Changes in Matter Depending on the temperature and pressure of the surroundings, many substances can exist in each of the three physical states and undergo changes in state as well. For example, as the temperature increases, solid water melts to liquid water, which boils to gaseous water (also called *water vapor*).

Table 1.1 Some Characteristic Properties of Copper

Physical Properties	Chemical Properties
Easily hammered into sheets (malleable) and drawn into wires (ductile) 	Slowly forms a blue-green carbonate in moist air 
Can be melted and mixed with zinc to form brass 	Reacts with nitric or sulfuric acid 
Density = 8.95 g/cm^3 Melting point = 1083°C Boiling point = 2570°C	Slowly forms deep-blue solution in aqueous ammonia 

Source: (copper) © McGraw-Hill Education/Mark Dierker, photographer; (candlestick) © Ruth Melnick; (copper carbonate, copper reacting with acid, copper and ammonia) © McGraw-Hill Education/Stephen Frisch, photographer

Similarly, as the temperature drops, water vapor condenses to liquid water, and with further cooling, the liquid freezes to ice:

$$\text{Ice} \xrightarrow{\textit{heating}} \text{Liquid water} \xrightarrow{\textit{heating}} \text{Water vapor}$$

$$\text{Ice} \xleftarrow{\textit{cooling}} \text{Liquid water} \xleftarrow{\textit{cooling}} \text{Water vapor}$$

In a steel plant, solid iron melts to liquid (molten) iron and then cools to the solid again. And, far beyond the confines of a laboratory or steel plant, lakes of molten sulfur (a solid on Earth at room temperature) lie on Jupiter's moon Io *(see photo),* which is capped by poles of frozen hydrogen sulfide, a gas on Earth.

The main point is that *a physical change caused by heating can generally be reversed by cooling*. This is *not* generally true for a chemical change. For example, heating iron in moist air causes a chemical reaction that yields the brown, crumbly substance known as rust. Cooling does not reverse this change; rather, another chemical change (or series of them) is required.

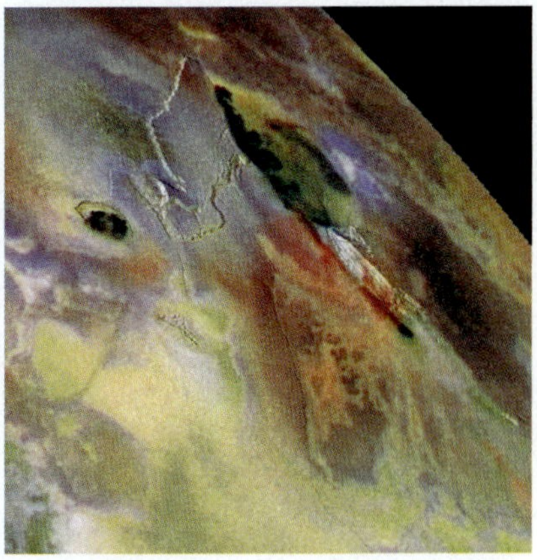

Many substances that are common on Earth occur in unusual states on other worlds.

Source: JPL-NASA

The following sample problem provides practice in distinguishing some familiar examples of physical and chemical change.

SAMPLE PROBLEM 1.2 Distinguishing Between Physical and Chemical Change

Problem Decide whether each of the following processes is primarily a physical or a chemical change, and explain briefly:

(a) Frost forms as the temperature drops on a humid winter night.
(b) A cornstalk grows from a seed that is watered and fertilized.
(c) A match ignites to form ash and a mixture of gases.
(d) Perspiration evaporates when you relax after jogging.
(e) A silver fork tarnishes slowly in air.

Plan The basic question we ask to decide whether a change is chemical or physical is, "Does the substance change composition or just change form?"

Solution **(a)** Frost forming is a physical change: the drop in temperature changes water vapor (gaseous water) in humid air to ice crystals (solid water).
(b) A seed growing involves chemical change: the seed uses water, substances from air, fertilizer, and soil, and energy from sunlight to make complex changes in composition.
(c) The match burning is a chemical change: the combustible substances in the match head are converted into other substances.
(d) Perspiration evaporating is a physical change: the water in sweat changes its state, from liquid to gas, but not its composition.
(e) Tarnishing is a chemical change: silver changes to silver sulfide by reacting with sulfur-containing substances in the air.

FOLLOW-UP PROBLEMS

1.2A Decide whether each of the following processes is primarily a physical or a chemical change, and explain briefly:
(a) Purple iodine vapor appears when solid iodine is warmed.
(b) Gasoline fumes are ignited by a spark in an automobile engine's cylinder.
(c) A scab forms over an open cut.

1.2B Decide whether each of the following processes is primarily a physical or a chemical change, and explain briefly:
(a) Clouds form in the sky.
(b) Old milk turns sour.
(c) Butter is melted to use on popcorn.

SOME SIMILAR PROBLEMS 1.6 and 1.7

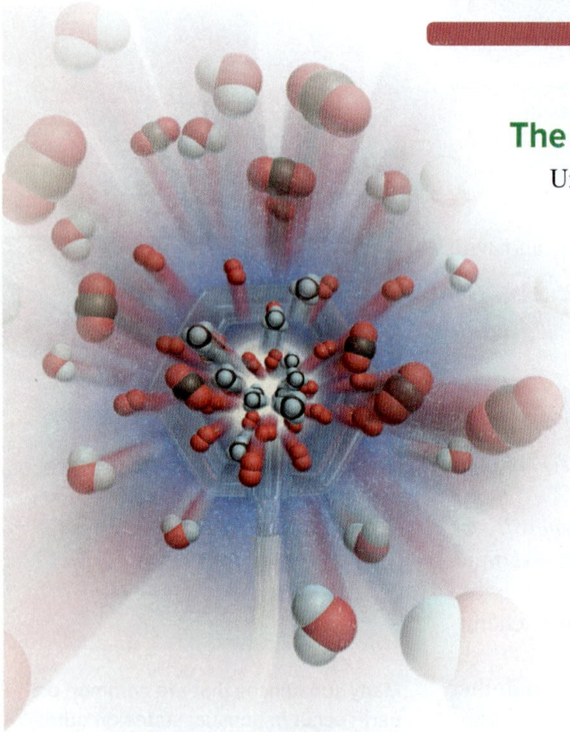

Methane and oxygen form carbon dioxide and water in the flame of a lab burner. (Carbon is black, oxygen red, and hydrogen blue.)

The Central Theme in Chemistry

Understanding the properties of a substance and the changes it undergoes leads to the central theme in chemistry: *macroscopic-scale* properties and behavior, those we can see, are the results of *atomic-scale* properties and behavior that we cannot see. The distinction between chemical and physical change is defined by composition, which we study macroscopically. But composition ultimately depends on the makeup of substances at the atomic scale. Similarly, macroscopic properties of substances in any of the three states arise from atomic-scale behavior of their particles. Picturing a chemical event on the molecular scale, even one as common as the flame of a laboratory burner *(see margin)*, helps clarify what is taking place. What is happening when water boils or copper melts? What events occur in the invisible world of minute particles that cause a seed to grow, a neon light to glow, or a nail to rust? Throughout the text, we return to this central idea:

*We study **observable** changes in matter to understand their **unobservable** causes.*

The Importance of Energy in the Study of Matter

Physical and chemical changes are accompanied by energy changes. **Energy** is often defined as *the ability to do work.* Essentially, all work involves moving something. Work is done when your arm lifts a book, when a car's engine moves the wheels, or when a falling rock moves the ground as it lands. The object doing the work (arm, engine, rock) transfers some of the energy it possesses to the object on which the work is done (book, wheels, ground).

The total energy an object possesses is the sum of its potential energy and its kinetic energy.

- **Potential energy** is *the energy due to the **position** of the object relative to other objects.*
- **Kinetic energy** is *the energy due to the **motion** of the object.*

Let's examine four systems that illustrate the relationship between these two forms of energy: a weight raised above the ground, two balls attached by a spring, two electrically charged particles, and a fuel and its waste products. Two concepts central to all these cases are

1. *When energy is converted from one form to the other, it is conserved, not destroyed.*
2. *Situations of lower energy are more stable and are favored over situations of higher energy, which are less stable.*

The four cases are

- *A weight raised above the ground* (Figure 1.3A). The energy you exert to lift a weight against gravity increases the weight's potential energy (energy due to its position). When you drop the weight, that additional potential energy is converted to kinetic energy (energy due to motion). The situation with the weight elevated and higher in potential energy is *less stable,* so the weight will fall when released, resulting in a situation that is lower in potential energy and *more stable.*
- *Two balls attached by a spring* (Figure 1.3B). When you pull the balls apart, the energy you exert to stretch the relaxed spring increases the system's potential energy. This change in potential energy is converted to kinetic energy when you release the balls. The system of balls and spring is less stable (has more potential energy) when the spring is stretched than when it is relaxed.
- *Two electrically charged particles* (Figure 1.3C). Due to interactions known as *electrostatic forces, opposite charges attract each other, and like charges repel each other.* When energy is exerted to move a positive particle away from a negative one, the potential energy of the system increases, and that increase is converted to

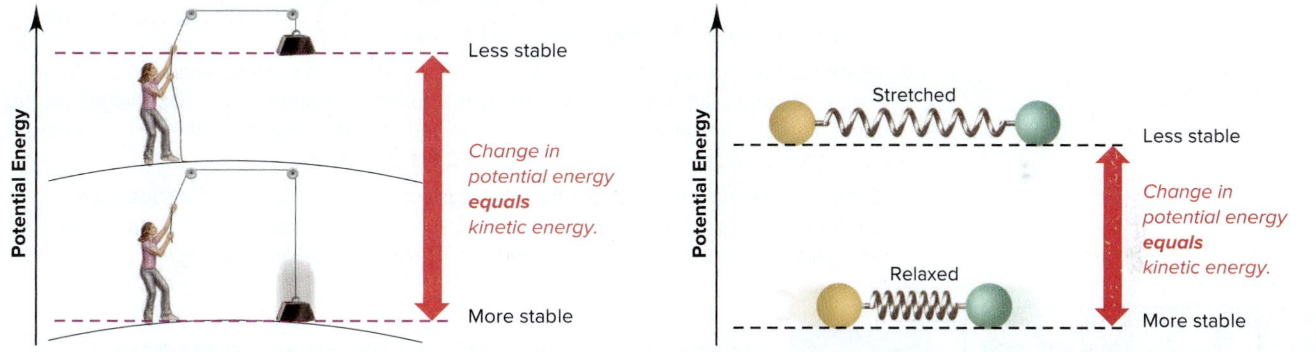

A A gravitational system. *Potential energy is gained when a weight is lifted. It is converted to kinetic energy as the weight falls.*

B A system of two balls attached by a spring. *Potential energy is gained when the spring is stretched. It is converted to the kinetic energy of the moving balls as the spring relaxes.*

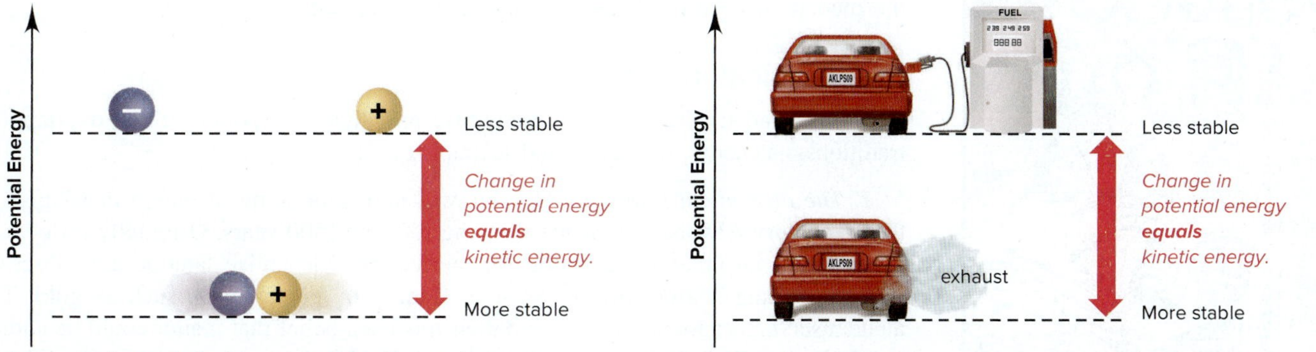

C A system of oppositely charged particles. *Potential energy is gained when the charges are separated. It is converted to kinetic energy as the attraction pulls the charges together.*

D A system of fuel and exhaust. *A fuel is higher in chemical potential energy than the exhaust. As the fuel burns, some of its potential energy is converted to the kinetic energy of the moving car.*

Figure 1.3 Potential energy is converted to kinetic energy. The dashed horizontal lines indicate the potential energy of each system before and after the change.

kinetic energy when the particles are pulled together by the electrostatic attraction. Similarly, when energy is used to move two positive (or two negative) particles together, their potential energy increases and changes to kinetic energy when they are pushed apart by the electrostatic repulsion. Charged particles move naturally to a more stable situation (lower energy).

- *A fuel and its waste products* (Figure 1.3D). Matter is composed of positively and negatively charged particles. *The chemical potential energy of a substance results from the relative positions of its particles and the attractions and repulsions among them.* Some substances are higher in potential energy than others. For example, gasoline and oxygen have more chemical potential energy than the exhaust gases they form. This difference is converted into kinetic energy, which moves the car, heats the interior, makes the lights shine, and so on. Similarly, the difference in potential energy between the food and air we take in and the wastes we excrete enables us to move, grow, keep warm, study chemistry, and so on.

› Summary of Section 1.1

- › Chemists study the composition and properties of matter and how they change.
- › Matter exists in three physical states—solid, liquid, and gas. The behavior of each state is due to the arrangement of the particles.
- › Each substance has a unique set of *physical* properties (attributes of the substance itself) and *chemical* properties (attributes of the substance as it interacts with or changes to other substances). Changes in matter can be *physical* (different form of the same substance) or *chemical* (different substance).
- › A physical change caused by heating may be reversed by cooling. But a chemical change caused by heating can be reversed only by other chemical changes.
- › Macroscopic changes result from submicroscopic changes.
- › Changes in matter are accompanied by changes in energy.
- › An object's potential energy is due to its position; an object's kinetic energy is due to its motion. Energy used to lift a weight, stretch a spring, or separate opposite charges increases the system's potential energy, which is converted to kinetic energy as the system returns to its original condition. Energy changes form but is conserved.
- › Chemical potential energy arises from the positions and interactions of a substance's particles. When a higher energy (less stable) substance is converted into a more stable (lower energy) substance, some potential energy is converted into kinetic energy.

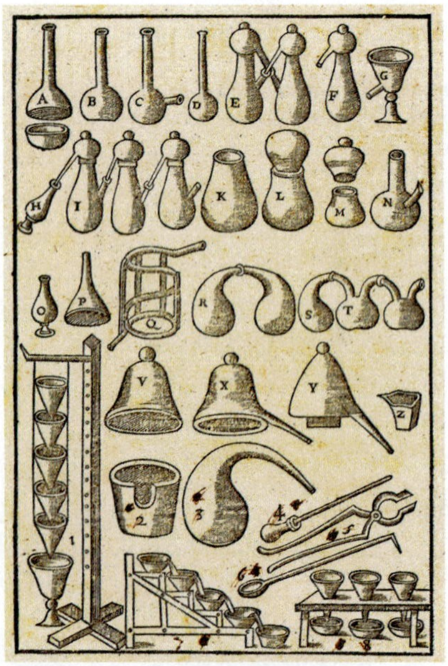

Figure 1.4 Alchemical apparatus.
Source: © SSPL/The Image Works

1.2 CHEMICAL ARTS AND THE ORIGINS OF MODERN CHEMISTRY

This brief overview of early breakthroughs, and a few false directions, describes how the modern science of chemistry arose and progressed.

Prechemical Traditions

Chemistry had its origin in a prescientific past that incorporated three overlapping traditions—alchemy, medicine, and technology:

1. *The alchemical tradition. Alchemy* was an occult study of nature that began in the 1st century AD and dominated thinking for over 1500 years. Originally influenced by the Greek idea that matter strives for "perfection," alchemists later became obsessed with converting "baser" metals, such as lead, into "purer" ones, such as gold. The alchemists' names for substances and their mistaken belief that matter could be altered magically persisted for centuries. Their legacy to chemistry was in technical methods. They invented distillation, percolation, and extraction and devised apparatus still used routinely today (Figure 1.4). But perhaps even more important was that alchemists encouraged observation and experimentation, which replaced the Greek approach of explaining nature solely through reason.

2. *The medical tradition.* Alchemists also influenced medical practice in medieval Europe. And ever since the 13th century, distillates and extracts of roots, herbs, and other plant matter have been used as sources of medicines. The alchemist and physician Paracelsus (1493–1541) considered the body to be a chemical system and illness an imbalance that could be restored by treatment with drugs. Although many early prescriptions were useless, later ones had increasing success. Thus began the alliance between medicine and chemistry that thrives today.

3. *The technological tradition.* For thousands of years, pottery making, dyeing, and especially metallurgy contributed greatly to people's experience with materials. During the Middle Ages and the Renaissance, books were published that described how to purify, assay, and coin silver and gold, how to use balances, furnaces, and crucibles, and how to make glass and gunpowder. Some of the books introduced quantitative measurement, which was lacking in alchemical writings. Many creations from those times are still marveled at throughout the world. Nevertheless, the skilled artisans showed little interest in *why* a substance changes or *how to predict* its behavior.

The Phlogiston Fiasco and the Impact of Lavoisier

Chemical investigation in the modern sense—inquiry into the causes of changes in matter—began in the late 17th century. At that time, most scientists explained **combustion,** the process of burning a material in air, with the *phlogiston theory*. It proposed that combustible materials contain *phlogiston,* an undetectable substance released when the material burns. Highly combustible materials like charcoal were thought to contain a lot of phlogiston, and slightly combustible materials like metals only a little. But inconsistencies continuously arose.

Phlogiston critics: Why is air needed for combustion, and why does charcoal stop burning in a closed vessel?
Phlogiston supporters: Air "attracts" phlogiston out of the charcoal, and burning stops when the air in the vessel is "saturated" with phlogiston.

Critics also noted that when a metal burns, it forms a substance known as its *calx,* which weighs more than the metal, leading them to ask,

Critics: How can the *loss* of phlogiston cause a *gain* in mass?
Supporters: Phlogiston has negative mass.

As ridiculous as these responses seem now, it's important to remember that, even today, scientists may dismiss conflicting evidence rather than abandon an accepted idea.

The conflict over phlogiston was resolved when the young French chemist Antoine Lavoisier (1743–1794) performed several experiments:

1. Heating mercury calx decomposed it into two products—mercury and a gas—whose *total mass equaled the starting mass of the calx.*
2. Heating mercury with the gas reformed the calx, and, again, *the total mass remained constant.*
3. Heating mercury in a measured volume of air yielded mercury calx and left four-fifths of the air remaining.
4. A burning candle placed in the remaining air was extinguished.

Lavoisier named the gas *oxygen* and gave metal calxes the name *metal oxides.* His explanation of his results made the phlogiston theory irrelevant:

- Oxygen, a normal component of air, combines with a substance when it burns.
- In a closed container, a combustible substance stops burning when it has combined with all the available oxygen.
- A metal calx (metal oxide) weighs more than the metal because its mass includes the mass of the oxygen.

This new theory triumphed because it relied on *quantitative, reproducible measurements,* not on strange properties of undetectable substances. Because this approach is at the heart of science, many propose that the *science* of chemistry began with Lavoisier.

› Summary of Section 1.2

› Alchemy, medicine, and technology placed little emphasis on objective experimentation, focusing instead on mystical explanations or practical experience, but these traditions contributed some apparatus and methods that are still important.

› Lavoisier overthrew the phlogiston theory by showing, through quantitative, reproducible measurements, that oxygen, a component of air, is required for combustion and combines with a burning substance.

1.3 THE SCIENTIFIC APPROACH: DEVELOPING A MODEL

Our prehistoric ancestors survived through *trial and error,* gradually learning which types of stone were hard enough to use for shaping other types, which plants were edible and which poisonous, and so forth. Unlike them, we employ the *quantitative theories* of chemistry to understand materials, make better use of them, and create new ones: specialized drugs, advanced composites, synthetic polymers, and countless others (Figure 1.5).

To understand nature, scientists use an approach called the **scientific method.** It is not a stepwise checklist, but rather a process involving creative proposals and tests aimed at objective, verifiable discoveries. There is no single procedure, and luck often plays a key role in discovery. In general terms, the scientific approach includes the following parts (Figure 1.6):

- ***Observations.*** These are the facts our ideas must explain. The most useful observations are quantitative because they can be analyzed to reveal *trends*. Pieces of quantitative information are **data.** When the same observation is made by many investigators in many situations with no clear exceptions, it is summarized, often in mathematical terms, as a **natural law.** The observation that mass remains constant during chemical change—made by Lavoisier and numerous experimenters since—is known as the law of mass conservation (Chapter 2).
- ***Hypothesis.*** Whether derived from observation or from a "spark of intuition," a hypothesis is a proposal made to explain an observation. A sound hypothesis need not be correct, but it must be *testable by experiment.* Indeed, a hypothesis is often the reason for performing an experiment: if the results do not support it, the hypothesis must be revised or discarded. *Hypotheses can be altered, but experimental results cannot.*
- ***Experiment.*** A set of procedural steps that tests a hypothesis, an experiment often leads to a revised hypothesis and new experiments to test it. An experiment typically contains at least two **variables,** quantities that can have more than one value. A well-designed experiment is **controlled** in that it measures the effect of one variable on another while keeping all other variables constant. Experimental results must be *reproducible* by others. Both skill and creativity play a part in experimental design.

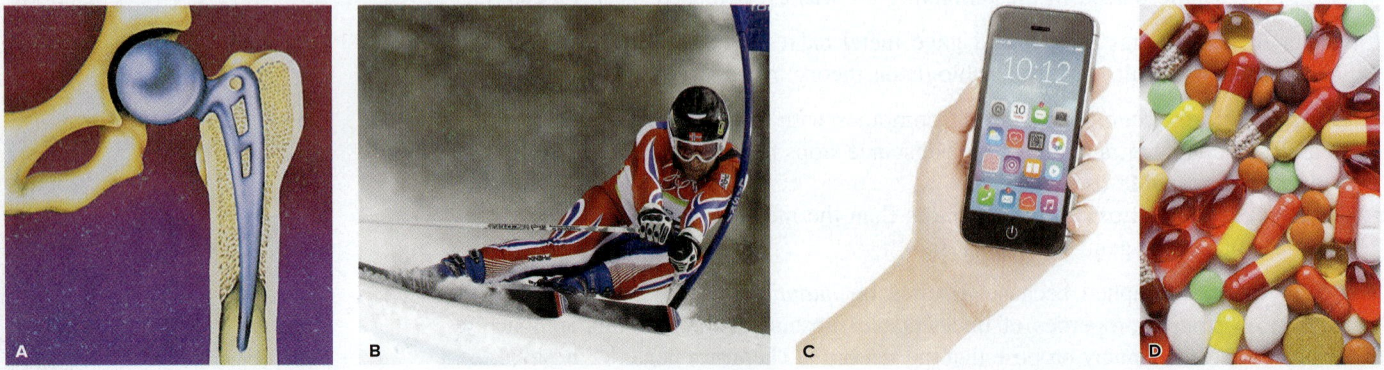

Figure 1.5 Modern materials in a variety of applications. A, High-tension polymers in synthetic hip joints. **B,** Specialized polymers in clothing and sports gear. **C,** Liquid crystals and semiconductors in electronic devices. **D,** Medicinal agents in pills.

Source: (A) © George Haling/Science Source; (B) © Javier Soriano/AFP/Getty Images; (C) © Alexey Boldin/Shutterstock.com; (D) © Didecs/Shutterstock.com

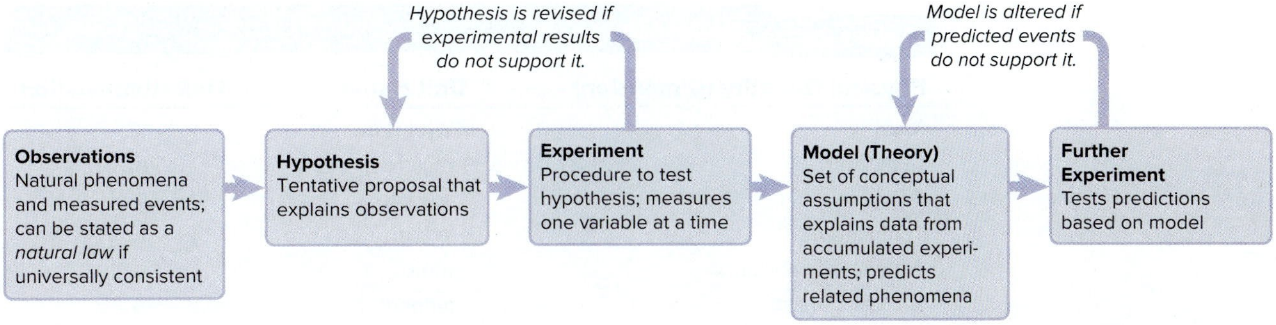

Figure 1.6 The scientific approach to understanding nature. Hypotheses and models are mental pictures that are revised to match observations and experimental results, *not* the other way around.

- ***Model.*** Formulating conceptual models, or **theories,** based on *experiments* that test *hypotheses* about *observations* distinguishes scientific thinking from speculation. As hypotheses are revised according to experimental results, a model emerges to explain how a phenomenon occurs. A model is a *simplified,* not an exact, representation of some aspect of nature that we use to *predict* related phenomena. Ongoing experimentation refines the model to account for new facts.

Lavoisier's overthrow of the phlogiston theory demonstrates the scientific method of thinking. *Observations* of burning and smelting led to the *hypothesis* that combustion involved the loss of phlogiston. *Experiments* showing that air is required for burning and that a metal gains mass during combustion led Lavoisier to propose a new *hypothesis,* which he tested repeatedly with quantitative *experiments.* Accumulating evidence supported his developing *model (theory)* that combustion involves combination with a component of air (oxygen). Innumerable *predictions* based on this theory have supported its validity, and Lavoisier himself extended the theory to account for animal respiration and metabolism.

› Summary of Section 1.3

- › The scientific method is a process designed to explain and predict phenomena.
- › Observations lead to hypotheses about how a phenomenon occurs. When repeated with no exceptions, observations may be expressed as a natural law.
- › Hypotheses are tested by controlled experiments and revised when necessary.
- › If reproducible data support a hypothesis, a model (theory) can be developed to explain the observed phenomenon. A good model predicts related phenomena but must be refined whenever conflicting data appear.

1.4 MEASUREMENT AND CHEMICAL PROBLEM SOLVING

Measurement has a rich history characterized by the search for *exact, invariable standards.* Measuring for purposes of trade, building, and surveying began thousands of years ago, but for most of that time, it was based on standards that could vary: a yard was the distance from the king's nose to the tip of his outstretched arm, and an acre was the area tilled in one day by a man with a pair of oxen. Our current system of measurement began in 1790 in France, when a committee, of which Lavoisier was a member, developed the *metric system.* Then, in 1960, another committee in France revised the metric system and established the universally accepted **SI units** (from the French **S**ystème **I**nternational d'Unités).

General Features of SI Units

The SI system is based on seven **fundamental units,** or **base units,** each identified with a physical quantity (Table 1.2, *next page*). All other units are **derived units,** combinations

Table 1.2 SI Base Units

Physical Quantity (Dimension)	Unit Name	Unit Abbreviation
Mass	kilogram	kg
Length	meter	m
Time	second	s
Temperature	kelvin	K
Amount of substance	mole	mol
Electric current	ampere	A
Luminous intensity	candela	cd

of the seven base units. For example, the derived unit for speed, meters per second (m/s), is the base unit for length (m) divided by the base unit for time (s).

For quantities much smaller or larger than the base unit, we use decimal prefixes and exponential (scientific) notation (Table 1.3). For example, the prefix *kilo-* (symbolized by k) indicates that the unit is one thousand times larger than a base unit, and the prefix *milli-* (m) indicates that the unit is one-thousandth the size of a base unit:

$$1 \text{ kilosecond } (1 \text{ ks}) = 1000 \text{ seconds} = 1\times10^{3} \text{ seconds}$$
$$1 \text{ millisecond } (1 \text{ ms}) = 0.001 \text{ second} = 1\times10^{-3} \text{ second}$$

Note that a mathematically equivalent statement of the second relationship is

$$1000 \text{ milliseconds (ms)} = 1 \text{ second}$$

Because the prefixes are based on powers of 10, SI units are easier to use in calculations than English units. (If you need a review of exponential notation, see Appendix A.)

Some Important SI Units in Chemistry

In this chapter, we discuss units for length, volume, mass, time, density, and temperature; other units are presented in later chapters. Table 1.4 shows some SI quantities for length, volume, and mass, along with their English-system equivalents.

Table 1.3 Common Decimal Prefixes Used with SI Units*

Prefix*	Symbol	Conventional Notation	Exponential Notation	Example [using gram (g)† or meter (m)††]
tera	(T)	1,000,000,000,000	1×10^{12}	1 teragram (Tg) = 1×10^{12} g
giga	(G)	1,000,000,000	1×10^{9}	1 gigagram (Gg) = 1×10^{9} g
mega	(M)	1,000,000	1×10^{6}	1 megagram (Mg) = 1×10^{6} g
kilo	(k)	1000	1×10^{3}	1 kilogram (kg) = 1×10^{3} g
hecto	(h)	100	1×10^{2}	1 hectogram (hg) = 1×10^{2} g
deka	(da)	10	1×10^{1}	1 dekagram (dag) = 1×10^{1} g
—	—	1	1×10^{0}	
deci	(d)	0.1	1×10^{-1}	1 decimeter (dm) = 1×10^{-1} m
centi	(c)	0.01	1×10^{-2}	1 centimeter (cm) = 1×10^{-2} m
milli	(m)	0.001	1×10^{-3}	1 millimeter (mm) = 1×10^{-3} m
micro	(μ)	0.000001	1×10^{-6}	1 micrometer (μm) = 1×10^{-6} m
nano	(n)	0.000000001	1×10^{-9}	1 nanometer (nm) = 1×10^{-9} m
pico	(p)	0.000000000001	1×10^{-12}	1 picometer (pm) = 1×10^{-12} m
femto	(f)	0.000000000000001	1×10^{-15}	1 femtometer (fm) = 1×10^{-15} m

*The prefixes most frequently used by chemists appear in bold type.
†The gram is a unit of mass.
††The meter is a unit of length.

Table 1.4 Common SI-English Equivalent Quantities

Quantity	SI Units	SI Equivalents	English Equivalents	English to SI Equivalent
Length	1 kilometer (km)	1000 (10^3) meters	0.6214 mile (mi)	1 mile = 1.609 km
	1 meter (m)	100 (10^2) centimeters	1.094 yards (yd)	1 yard = 0.9144 m
	1 meter (m)	1000 millimeters (mm)	39.37 inches (in)	1 foot (ft) = 0.3048 m
	1 centimeter (cm)	0.01 (10^{-2}) meter	0.3937 inch	1 inch = 2.54 cm (exactly)
Volume	1 cubic meter (m^3)	1,000,000 (10^6) cubic centimeters	35.31 cubic feet (ft^3)	1 cubic foot = 0.02832 m^3
	1 cubic decimeter (dm^3)	1000 cubic centimeters	0.2642 gallon (gal)	1 gallon = 3.785 dm^3
	1 cubic decimeter (dm^3)	1000 cubic centimeters	1.057 quarts (qt)	1 quart = 0.9464 dm^3
				1 quart = 946.4 cm^3
	1 cubic centimeter (cm^3)	0.001 dm^3	0.03381 fluid ounce	1 fluid ounce = 29.57 cm^3
Mass	1 kilogram (kg)	1000 grams (g)	2.205 pounds (lb)	1 pound = 0.4536 kg

Length The SI base unit of length is the **meter (m).** In the metric system, it was originally defined as 1/10,000,000 of the distance from the equator to the North Pole, and later as the distance between two fine lines engraved on a corrosion-resistant metal bar. More recently, the first exact, unchanging standard was adopted: 1,650,763.73 wavelengths of orange-red light from electrically excited krypton atoms. The current standard is exact and invariant: 1 meter is the distance light travels in a vacuum in 1/299,792,458 of a second.

A meter is a little longer than a yard (1 m = 1.094 yd); a centimeter (10^{-2} m) is about two-fifths of an inch (1 cm = 0.3937 in; 1 in = 2.54 cm). Biological cells are often measured in micrometers (1 μm = 10^{-6} m). On the atomic scale, nanometers (10^{-9} m) and picometers (10^{-12} m) are used. Many proteins have diameters of about 2 nm; atomic diameters are about 200 pm (0.2 nm). An older unit still in use is the angstrom (1Å = 10^{-10} m = 0.1 nm = 100 pm).

Volume In chemistry, the significance of length appears when a sample of matter is measured in three dimensions, which gives its **volume (*V*),** the amount of space it occupies. The derived SI unit of volume is the **cubic meter (m^3).** Since the volume of some objects can be calculated using the relationship length × width × height, a length unit cubed (meters × meters × meters = m^3) is a unit of volume. In chemistry, we often use non-SI units, the **liter (L)** and the **milliliter (mL)** (note the uppercase L), to measure volume. Physicians and other medical practitioners measure body fluids in cubic decimeters (dm^3), which are equivalent to liters:

$$1 \text{ L} = 1 \text{ dm}^3 = 10^{-3} \text{ m}^3$$

And 1 mL, or $\frac{1}{1000}$ of a liter, is equivalent to 1 cubic centimeter (cm^3):

$$1 \text{ mL} = 1 \text{ cm}^3 = 10^{-3} \text{ dm}^3 = 10^{-3} \text{ L} = 10^{-6} \text{ m}^3$$

A liter is slightly larger than a quart (qt) (1 L = 1.057 qt; 1 qt = 946.4 mL); 1 fluid ounce ($\frac{1}{32}$ of a quart) equals 29.57 mL (29.57 cm^3).

Figure 1.7 *(next page)* is a depiction of the two 1000-fold decreases in volume from 1 dm^3 to 1 cm^3 and then to 1 mm^3. The edge of a 1-m^3 cube would be just a little longer than a yardstick.

Figure 1.8A *(next page)* shows some laboratory glassware for working with volumes. Erlenmeyer flasks and beakers are used to contain liquids. Graduated cylinders, pipets, and burets are used to measure and transfer them. Volumetric flasks and pipets have a fixed volume indicated by a mark on the neck. Solutions are prepared quantitatively in volumetric flasks, and specific amounts are put into cylinders, pipets, or burets to transfer to beakers or flasks for further steps. In Figure 1.8B, an automatic pipet transfers liquid accurately *and* quickly.

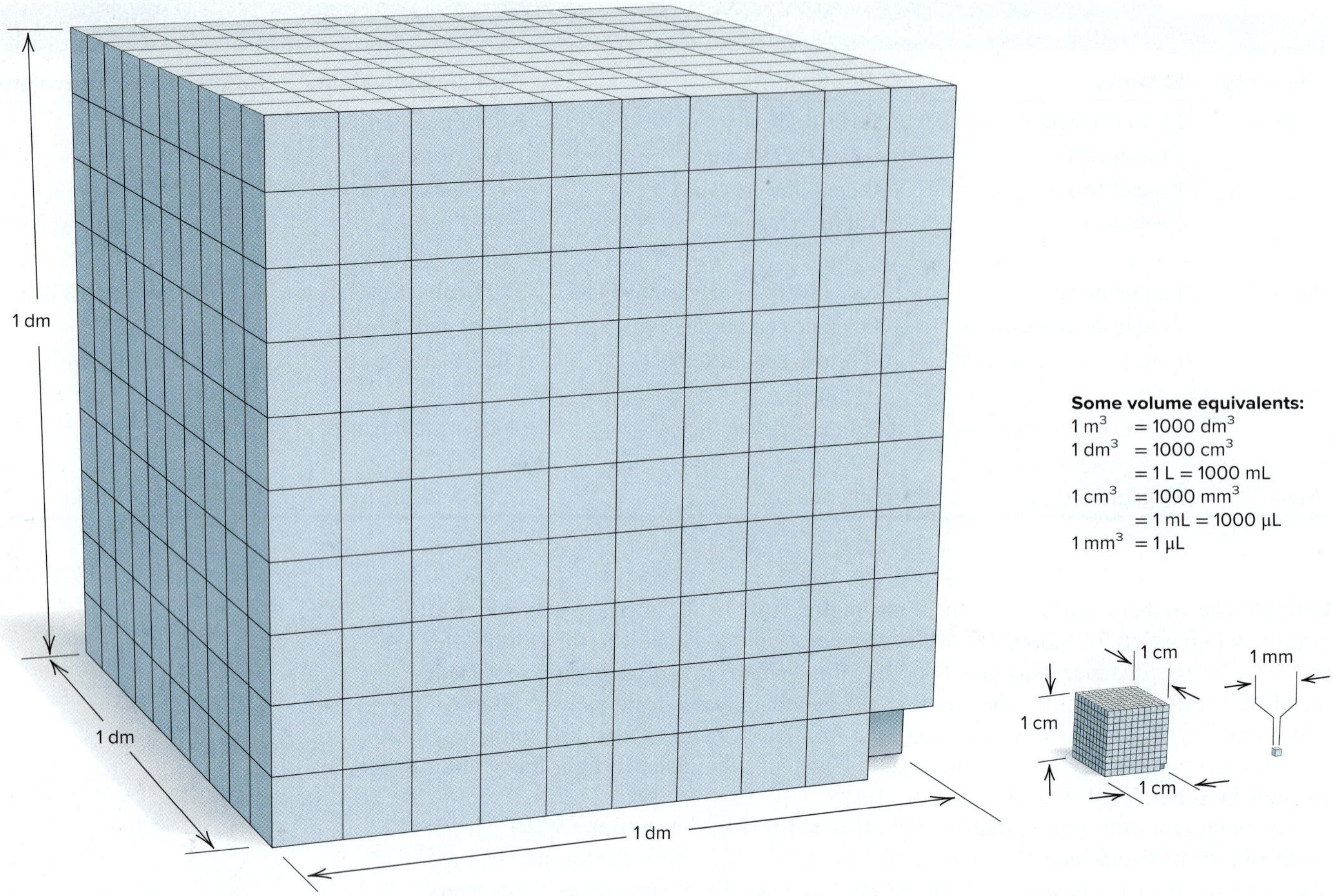

Figure 1.7 Some volume relationships in SI: From cubic decimeter (dm^3) to cubic centimeter (cm^3) to cubic millimeter (mm^3).

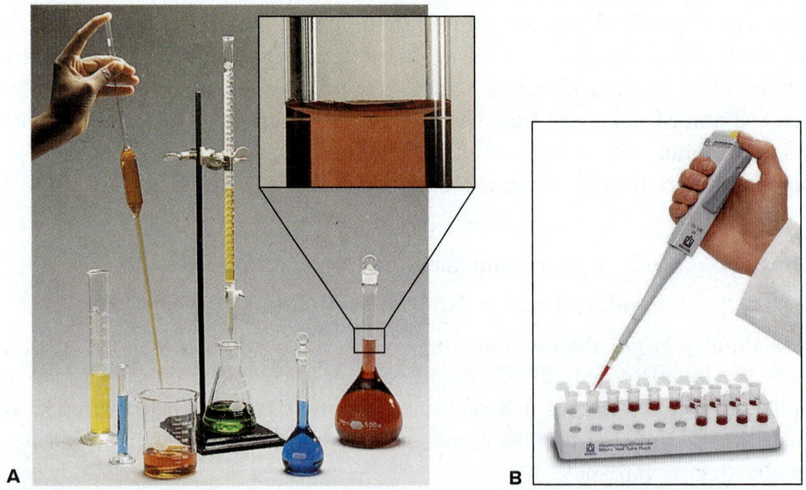

Figure 1.8 Common laboratory volumetric glassware. A, From left to right are two graduated cylinders, a pipet being emptied into a beaker, a buret delivering liquid to an Erlenmeyer flask, and two volumetric flasks. **Inset,** In contact with the glass neck of the flask, the liquid forms a concave meniscus (curved surface). **B,** An automatic pipet delivers a given volume of liquid to each test tube.

Source: (A) © McGraw-Hill Education/Stephen Frisch, photographer; (B) © BrandTech Scientific, Inc.

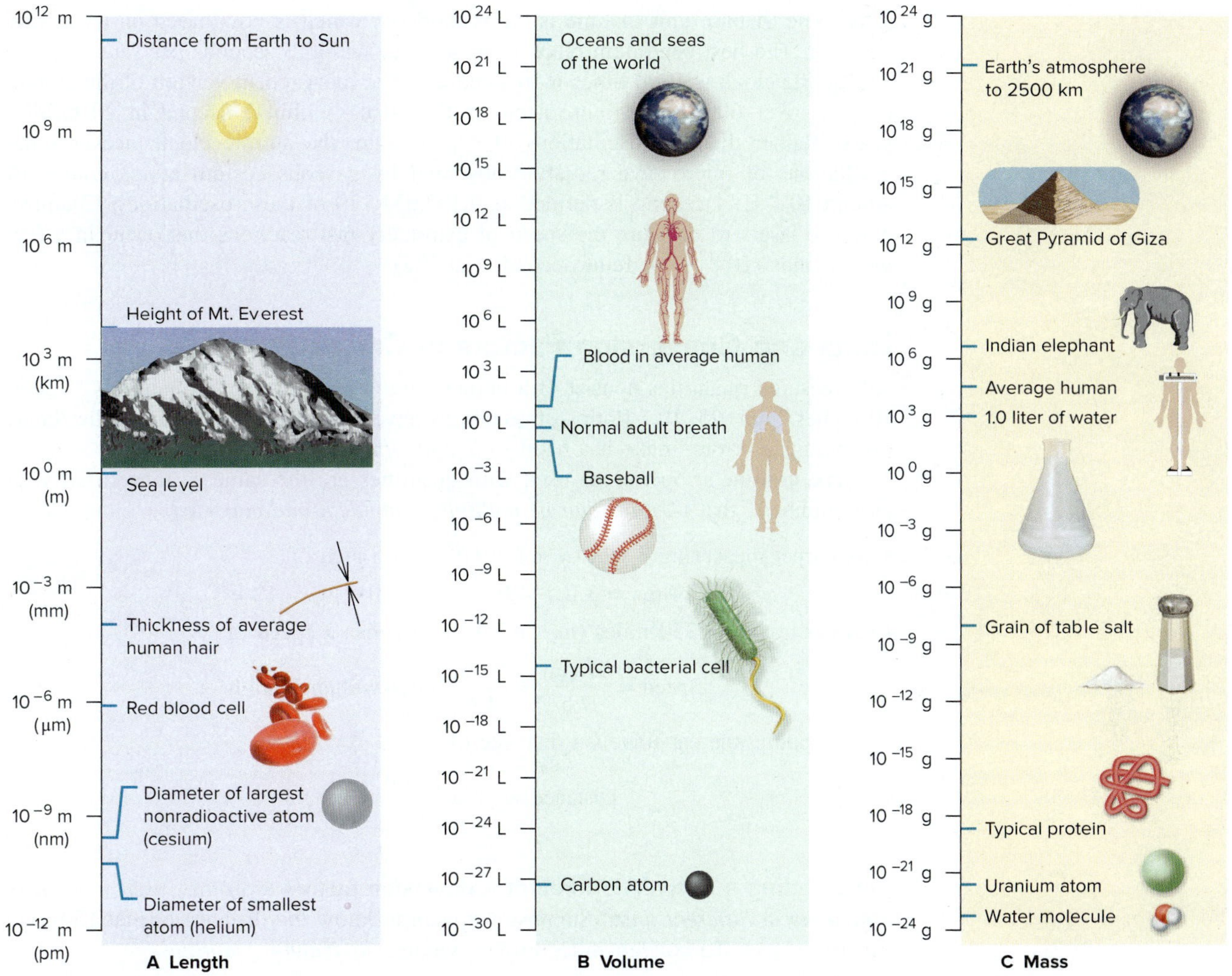

Figure 1.9 **Some interesting quantities of length (A), volume (B), and mass (C).** The vertical scales are exponential.

Mass The quantity of matter an object contains is its **mass.** The SI unit of mass is the **kilogram (kg),** the only base unit whose standard is an object—a platinum-iridium cylinder kept in France—and the only one whose name has a prefix.*

The terms *mass* and *weight* have distinct meanings:

- *Mass is constant* because an object's quantity of matter cannot change.
- **Weight** *is variable* because it depends on the local gravitational field acting on the object.

Because the strength of the gravitational field varies with altitude, you (and other objects) weigh slightly less on a high mountain than at sea level.

Does this mean that a sample weighed on laboratory balances in Miami (sea level) and in Denver (about 1.7 km above sea level) give different results? No, because these balances measure mass, not weight. (We actually "mass" an object when we weigh it on a balance, but we don't use that term.) Mechanical balances compare the object's mass with masses built into the balance, so the local gravitational field pulls on them equally. Electronic (analytical) balances generate an electric field that counteracts the local field, and the current needed to restore the pan to zero is converted to the equivalent mass and displayed.

Figure 1.9 shows the ranges of some common lengths, volumes, and masses.

*The names of the other base units are used as the root words, but for units of mass we attach prefixes to the word "gram," as in "microgram" and "kilogram"; thus, we say "milligram," never "microkilogram."

Time The SI base unit of time is the **second (s)**, which is now based on an atomic standard. The best pendulum clock is accurate to within 3 seconds per year, and the best quartz clock is 1000 times more accurate. The most recent version of the atomic clock is over 6000 times more accurate than that—within 1 second in 20 million years! Rather than the oscillations of a pendulum, the atomic clock measures the oscillations of microwave radiation absorbed by gaseous cesium atoms cooled to around 10^{-6} K: 1 second is defined as 9,192,631,770 of these oscillations. Chemists now use lasers to measure the speed of extremely fast reactions that occur in a few picoseconds (10^{-12} s) or femtoseconds (10^{-15} s).

Units and Conversion Factors in Calculations

All measured quantities consist of a number *and* a unit: a person's height is "5 feet, 10 inches," not "5, 10." Ratios of quantities have ratios of units, such as miles/hour. To minimize errors, make it a habit to *include units in all calculations.*

The arithmetic operations used with quantities are the same as those used with pure numbers; that is, *units can be multiplied, divided, and canceled:*

- A carpet measuring 3 feet by 4 feet (ft) has an area of

$$\text{Area} = 3\text{ ft} \times 4\text{ ft} = (3 \times 4)(\text{ft} \times \text{ft}) = 12\text{ ft}^2$$

- A car traveling 350 miles (mi) in 7 hours (h) has a speed of

$$\text{Speed} = \frac{350\text{ mi}}{7\text{ h}} = \frac{50\text{ mi}}{1\text{ h}} \text{ (often written 50 mi}\cdot\text{h}^{-1}\text{)}$$

- In 3 hours, the car travels a distance of

$$\text{Distance} = 3\cancel{\text{h}} \times \frac{50\text{ mi}}{1\cancel{\text{h}}} = 150\text{ mi}$$

Constructing a Conversion Factor **Conversion factors** are *ratios used to express a quantity in different units.* Suppose we want to know the distance of that 150-mile car trip in feet. To convert miles to feet, we use *equivalent quantities,*

$$1\text{ mi} = 5280\text{ ft}$$

from which we can construct two conversion factors. Dividing both sides by 5280 ft gives one conversion factor (shown in blue):

$$\frac{1\text{ mi}}{5280\text{ ft}} = \frac{5280\cancel{\text{ft}}}{5280\cancel{\text{ft}}} = 1$$

And, dividing both sides by 1 mi gives the other conversion factor (the inverse):

$$\frac{1\cancel{\text{mi}}}{1\cancel{\text{mi}}} = \frac{5280\text{ ft}}{1\text{ mi}} = 1$$

Since the numerator and denominator of a conversion factor are equal, multiplying a quantity by a conversion factor is the same as multiplying by 1. Thus, *even though the number and unit change, the size of the quantity remains the same.*

To convert the distance from miles to feet, we choose the conversion factor with miles in the denominator, because it cancels miles and gives the answer in feet:

$$\text{Distance (ft)} = 150\cancel{\text{mi}} \times \frac{5280\text{ ft}}{1\cancel{\text{mi}}} = 792{,}000\text{ ft}$$

$$\text{mi} \Rightarrow \text{ft}$$

Choosing the Correct Conversion Factor It is easier to convert if you first decide whether the answer expressed in the new units should have a larger or smaller number. In the previous case, we know that a foot is *smaller* than a mile, so the distance in feet should have a *larger* number (792,000) than the distance in miles (150). The conversion factor has the larger number (5280) in the numerator, so it gave a larger number in the answer.

Most importantly, the *conversion factor you choose must cancel all units except those you want in the answer.* Therefore, set the unit you are converting *from* (beginning unit) in the *opposite position in the conversion factor* (numerator or denominator) so that it cancels and you are left with the unit you are converting *to* (final unit):

$$\cancel{\text{beginning unit}} \times \frac{\text{final unit}}{\cancel{\text{beginning unit}}} = \text{final unit} \qquad \text{as in} \qquad \cancel{\text{mi}} \times \frac{\text{ft}}{\cancel{\text{mi}}} = \text{ft}$$

Or, in cases that involve units raised to a power:

$$(\cancel{\text{beginning unit}} \times \cancel{\text{beginning unit}}) \times \frac{\text{final unit}^2}{\cancel{\text{beginning unit}^2}} = \text{final unit}^2$$

$$\text{as in} \qquad (\cancel{\text{ft}} \times \cancel{\text{ft}}) \times \frac{\text{mi}^2}{\cancel{\text{ft}^2}} = \text{mi}^2$$

Or, in cases that involve a ratio of units:

$$\frac{\cancel{\text{beginning unit}}}{\text{final unit}_1} \times \frac{\text{final unit}_2}{\cancel{\text{beginning unit}}} = \frac{\text{final unit}_2}{\text{final unit}_1} \qquad \text{as in} \qquad \frac{\cancel{\text{mi}}}{\text{h}} \times \frac{\text{ft}}{\cancel{\text{mi}}} = \frac{\text{ft}}{\text{h}}$$

Converting Between Unit Systems We use the same procedure to convert between one system of units and another, for example, between the English (or American) unit system and the International System. Suppose we know that the height of Angel Falls in Venezuela (the world's highest) is 3212 ft; we can find its height in miles as

$$\text{Height (mi)} = 3212\ \cancel{\text{ft}} \times \frac{1\ \text{mi}}{5280\ \cancel{\text{ft}}} = 0.6083\ \text{mi}$$

$$\text{ft} \Rightarrow \text{mi}$$

Now, we want its height in kilometers (km). The equivalent quantities (see Table 1.4) are

$$1.609\ \text{km} = 1\ \text{mi}$$

Because we are converting *from* miles *to* kilometers, we use the conversion factor with miles in the denominator in order to cancel miles:

$$\text{Height (km)} = 0.6083\ \cancel{\text{mi}} \times \frac{1.609\ \text{km}}{1\ \cancel{\text{mi}}} = 0.9788\ \text{km}$$

$$\text{mi} \Rightarrow \text{km}$$

Notice that kilometers are *smaller* than miles, so this conversion factor gave us an answer with a *larger* number (0.9788 is larger than 0.6083).

If we want the height of Angel Falls in meters (m), we use the equivalent quantities 1 km = 1000 m to construct the conversion factor:

$$\text{Height (m)} = 0.9788\ \cancel{\text{km}} \times \frac{1000\ \text{m}}{1\ \cancel{\text{km}}} = 978.8\ \text{m}$$

$$\text{km} \Rightarrow \text{m}$$

In longer calculations, we often string together several conversion steps:

$$\text{Height (m)} = 3212\ \cancel{\text{ft}} \times \frac{1\ \cancel{\text{mi}}}{5280\ \cancel{\text{ft}}} \times \frac{1.609\ \cancel{\text{km}}}{1\ \cancel{\text{mi}}} \times \frac{1000\ \text{m}}{1\ \cancel{\text{km}}} = 978.8\ \text{m}$$

$$\text{ft} \Rightarrow \text{mi} \Rightarrow \text{km} \Rightarrow \text{m}$$

The use of conversion factors in calculations is also referred to as the factor-label method, or **dimensional analysis** (because units represent physical dimensions). We use this approach throughout the text.

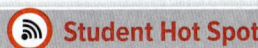

Student Hot Spot

Student data indicate that you may struggle with conversion factors. Access the Smartbook to view additional Learning Resources on this topic.

A Systematic Approach to Solving Chemistry Problems

The approach used in this book to solve problems emphasizes reasoning, not memorizing, and is based on a simple idea: plan how to solve the problem *before* you try to solve it, then check your answer, and practice with similar follow-up problems. In general, the sample problems consist of several parts:

1. **Problem.** This part states all the information you need to solve the problem, usually framed in some interesting context.

2. **Plan.** This part helps you *think* about the solution *before* juggling numbers and pressing calculator buttons. There is often more than one way to solve a problem, and the given plan is one possibility. The plan will
   - Clarify the known and unknown: what information do you have, and what are you trying to find?
   - Suggest the steps from known to unknown: what ideas, conversions, or equations are needed?
   - Present a road map (especially in early chapters), a flow diagram of the plan. The road map has a box for each intermediate result and an arrow showing the step (conversion factor or operation) used to get to the next box.
3. **Solution.** This part shows the calculation steps in the same order as in the plan (and the road map).
4. **Check.** This part helps you check that your final answer makes sense: Are the units correct? Did the change occur in the expected direction? Is it reasonable chemically? To avoid a large math error, we also often do a rough calculation and see if we get an answer "in the same ballpark" as the actual result. Here's a typical "ballpark" calculation from everyday life. You are in a clothing store and buy three shirts at \$14.97 each. With a 5% sales tax, the bill comes to \$47.16. In your mind, you know that \$14.97 is about \$15, and 3 times \$15 is \$45; with the sales tax, the cost should be a bit more. So, your quick mental calculation *is* in the same ballpark as the actual cost.
5. **Comment.** This part appears occasionally to provide an application, an alternative approach, a common mistake to avoid, or an overview.
6. **Follow-up Problems.** This part presents similar problems that require you to apply concepts and/or methods used in solving the sample problem.
7. **Some Similar Problems.** This part lists a few more problems (found at the end of each chapter) for practice.

Of course, you can't learn to solve chemistry problems, any more than you can learn to swim, by reading about it, so here are a few suggestions:

- Follow along in the sample problem with pencil, paper, and calculator.
- Try the follow-up problems as soon as you finish the sample problem. A very useful feature called Brief Solutions to Follow-up Problems appears at the end of each chapter, allowing you to compare your solution steps and answers.
- Read the sample problem and text again if you have trouble.
- The end-of-chapter problems review and extend the concepts and skills in the chapter, so work as many as you can. (Answers are given in Appendix E in the back of the book for problems with a colored number.)

Let's apply this systematic approach in some unit-conversion problems. Refer to Tables 1.3 and 1.4 as you work through these problems.

SAMPLE PROBLEM 1.3 Converting Units of Length

Road Map

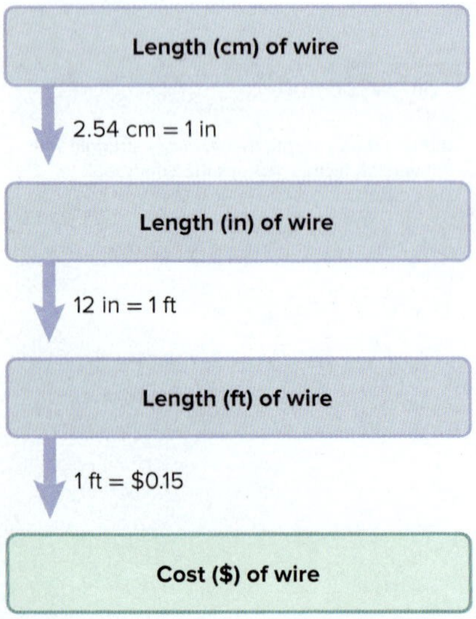

Problem To wire your stereo equipment, you need 325 centimeters (cm) of speaker wire that sells for \$0.15/ft. How much does the wire cost?

Plan We know the length of wire in centimeters (325 cm) and the price in dollars per foot (\$0.15/ft). We can find the unknown cost of the wire by converting the length from centimeters to inches (in) and from inches to feet. The price gives us the equivalent quantities (1 ft = \$0.15) that allow us to convert from feet of wire to cost in dollars. The road map starts with the known quantity of 325 cm and moves through the calculation steps to the unknown.

Solution Converting the known length from centimeters to inches: The equivalent quantities beside the road map arrow are needed to construct the conversion factor.

We choose 1 in/2.54 cm, rather than the inverse, because it gives an answer in inches:

$$\text{Length (in)} = \text{length (cm)} \times \text{conversion factor}$$
$$= 325\ \text{cm} \times \frac{1\ \text{in}}{2.54\ \text{cm}} = 128\ \text{in}$$

Converting the length from inches to feet:

$$\text{Length (ft)} = \text{length (in)} \times \text{conversion factor}$$
$$= 128\ \text{in} \times \frac{1\ \text{ft}}{12\ \text{in}} = 10.7\ \text{ft}$$

Converting the length in feet to cost in dollars:

$$\text{Cost (\$)} = \text{length (ft)} \times \text{conversion factor}$$
$$= 10.7\ \text{ft} \times \frac{\$0.15}{1\ \text{ft}} = \$1.60$$

Check The units are correct for each step. The conversion factors make sense in terms of the relative unit sizes: the number of inches is *smaller* than the number of centimeters (an inch is *larger* than a centimeter), and the number of feet is *smaller* than the number of inches. The total cost seems reasonable: a little more than 10 ft of wire at \$0.15/ft should cost a little more than \$1.50.

Comment **1.** We could also have strung the three steps together:

$$\text{Cost (\$)} = 325\ \text{cm} \times \frac{1\ \text{in}}{2.54\ \text{cm}} \times \frac{1\ \text{ft}}{12\ \text{in}} \times \frac{\$0.15}{1\ \text{ft}} = \$1.60$$

2. There are usually alternative sequences in unit-conversion problems. Here, for example, we would get the same answer if we first converted the cost of wire from \$/ft to \$/cm and kept the wire length in cm. Try it yourself.

FOLLOW-UP PROBLEMS

1.3A A chemistry professor can walk a mile in 15 minutes. How many minutes will it take the professor to walk 10,500 meters (m)? Draw a road map to show how you planned the solution.

1.3B The rhinovirus, one cause of the common cold, has a diameter of 30 nm. How many of these virus particles could line up side by side on a line that is 1.0 in long? Draw a road map to show how you planned the solution.

SOME SIMILAR PROBLEMS 1.26–1.29

SAMPLE PROBLEM 1.4 Converting Units of Volume

Problem The volume of an irregularly shaped solid can be determined from the volume of water it displaces. A graduated cylinder contains 19.9 mL of water. When a small piece of galena, an ore of lead, is added to the cylinder, it sinks and the volume increases to 24.5 mL. What is the volume of the piece of galena in cm^3 and in L?

Plan We have to find the volume of the galena from the change in volume of the cylinder contents. The volume of galena in mL is the difference before (19.9 mL) and after (24.5 mL) adding it. Since mL and cm^3 represent identical volumes, the volume in mL equals the volume in cm^3. We then use equivalent quantities (1 mL = 10^{-3} L) to convert mL to L. The road map shows these steps.

Road Map

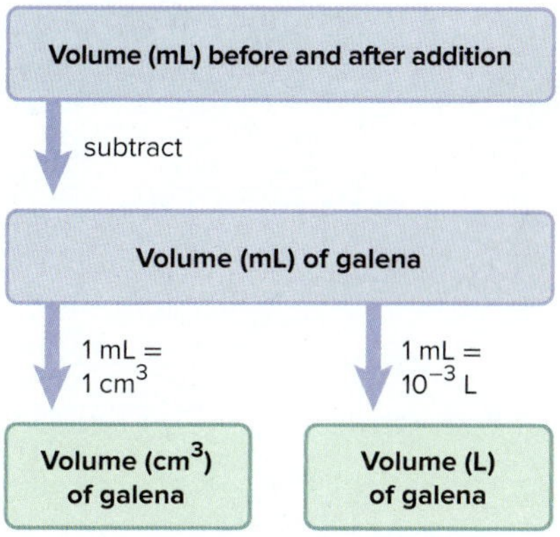

Solution Finding the volume of galena:

$$\text{Volume (mL)} = \text{volume after} - \text{volume before} = 24.5\ \text{mL} - 19.9\ \text{mL} = 4.6\ \text{mL}$$

Converting the volume from mL to cm^3:

$$\text{Volume (cm}^3) = \text{volume (mL)} \times \text{conversion factor} = 4.6\ \text{mL} \times \frac{1\ \text{cm}^3}{1\ \text{mL}} = 4.6\ \text{cm}^3$$

Converting the volume from mL to L:

$$\text{Volume (L)} = \text{volume (mL)} \times \text{conversion factor} = 4.6\ \text{mL} \times \frac{10^{-3}\ \text{L}}{1\ \text{mL}} = 4.6\times10^{-3}\ \text{L}$$

Check The units and magnitudes of the answers seem correct, and it makes sense that the volume in mL would have a number 1000 times larger than the same volume in L.

FOLLOW-UP PROBLEMS

1.4A Within a cell, proteins are synthesized on particles called ribosomes. Assuming ribosomes are spherical, what is the volume (in dm^3 and μL) of a ribosome whose average diameter is 21.4 nm (V of a sphere $= \frac{4}{3}\pi r^3$)? Draw a road map to show how you planned the solution.

1.4B During an eruption, the famous geyser Old Faithful in Yellowstone National Park can expel as much as 8400 gal of water. What is this volume in liters (L)? Draw a road map to show how you planned the solution.

SOME SIMILAR PROBLEMS 1.36 and 1.37

SAMPLE PROBLEM 1.5 Converting Units of Mass

Road Map

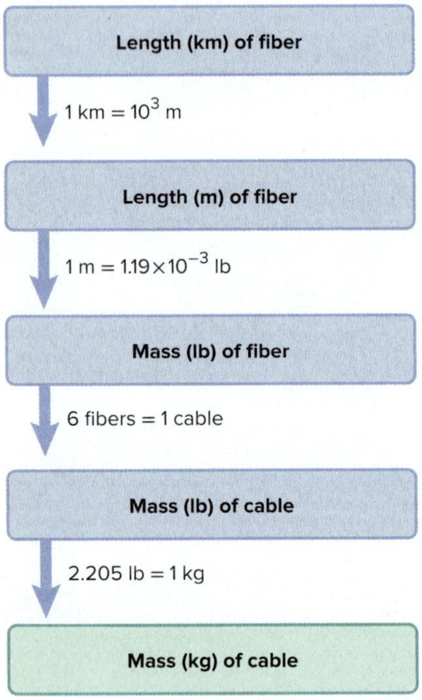

Problem Many international computer communications are carried by optical fibers in cables laid along the ocean floor. If one strand of optical fiber weighs 1.19×10^{-3} lb/m, what is the mass (in kg) of a cable made of six strands of optical fiber, each long enough to link New York and Paris (8.84×10^3 km)?

Plan We have to find the mass (in kg) of a known length of cable (8.84×10^3 km); we are given equivalent quantities for mass and length of a fiber (1.19×10^{-3} lb = 1 m) and for number of fibers and a cable (6 fibers = 1 cable), which we can use to construct conversion factors. Let's first find the mass of one fiber and then the mass of cable. As shown in the road map, we convert the length of one fiber from km to m and then find its mass (in lb) by converting m to lb. Then we multiply the fiber mass by 6 to get the cable mass, and finally convert lb to kg.

Solution Converting the fiber length from km to m:

$$\text{Length (m) of fiber} = 8.84\times10^3\ \cancel{\text{km}} \times \frac{10^3\ \text{m}}{1\ \cancel{\text{km}}} = 8.84\times10^6\ \text{m}$$

Converting the length of one fiber to mass (lb):

$$\text{Mass (lb) of fiber} = 8.84\times10^6\ \cancel{\text{m}} \times \frac{1.19\times10^{-3}\ \text{lb}}{1\ \cancel{\text{m}}} = 1.05\times10^4\ \text{lb}$$

Finding the mass of the cable (lb):

$$\text{Mass (lb) of cable} = \frac{1.05\times10^4\ \text{lb}}{1\ \cancel{\text{fiber}}} \times \frac{6\ \cancel{\text{fibers}}}{1\ \text{cable}} = 6.30\times10^4\ \text{lb/cable}$$

Converting the mass of cable from lb to kg:

$$\text{Mass (kg) of cable} = \frac{6.30\times10^4\ \cancel{\text{lb}}}{1\ \text{cable}} \times \frac{1\ \text{kg}}{2.205\ \cancel{\text{lb}}} = 2.86\times10^4\ \text{kg/cable}$$

Check The units are correct. Let's think through the relative sizes of the answers to see if they make sense: The number of m should be 10^3 larger than the number of km. If 1 m of fiber weighs about 10^{-3} lb, then about 10^7 m should weigh about 10^4 lb. The cable mass should be six times as much as that, or about 6×10^4 lb. Since 1 lb is about $\frac{1}{2}$ kg, the number of kg should be about half the number of lb.

FOLLOW-UP PROBLEMS

1.5A An intravenous nutrient solution is delivered to a hospital patient at a rate of 1.5 drops per second. If a drop of solution weighs 65 mg on average, how many kilograms are delivered in 8.0 h? Draw a road map to show how you planned the solution.

1.5B Nutritional tables give the potassium content of a standard apple (about 3 apples per pound) as 159 mg. How many grams of potassium are in 3.25 kg of apples? Draw a road map to show how you planned the solution.

SOME SIMILAR PROBLEMS 1.32 and 1.33

SAMPLE PROBLEM 1.6 Converting Units Raised to a Power

Problem A furniture factory needs 31.5 ft^2 of fabric to upholster one chair. Its Dutch supplier sends the fabric in bolts that hold exactly 200 m^2. How many chairs can be upholstered with 3 bolts of fabric?

Plan We are given a known number of fabric bolts (3 bolts) which can be converted to amount of fabric in m^2 using the given equivalent quantities (1 bolt = 200 m^2 fabric). We convert the amount of fabric from m^2 to ft^2 and use the equivalent quantities (31.5 ft^2 of fabric = 1 chair) to find the number of chairs (see the road map).

Solution Converting from number of bolts to amount of fabric in m^2:

$$\text{Amount (m}^2\text{) of fabric} = 3\ \cancel{\text{bolts}} \times \frac{200\ \text{m}^2}{1\ \cancel{\text{bolt}}} = 600\ \text{m}^2$$

Converting the amount of fabric from m^2 to ft^2: Since 0.3048 m = 1 ft, we have $(0.3048)^2\ m^2 = (1)^2\ ft^2$, so

$$\text{Amount (ft}^2\text{) of fabric} = 600\ \cancel{\text{m}^2} \times \frac{1\ \text{ft}^2}{(0.3048)^2\ \cancel{\text{m}^2}} = 6460\ \text{ft}^2$$

Finding the number of chairs:

$$\text{Number of chairs} = 6460\ \cancel{\text{ft}^2} \times \frac{1\ \text{chair}}{31.5\ \cancel{\text{ft}^2}} = 205\ \text{chairs}$$

Check Since 1 ft = 0.3048 m, 1 ft is a little less than $\frac{1}{3}$ of a meter, which means that 3.3 ft is about the same as 1 m. We are using squared length units, and $(3.3\ ft)^2$, or 11 ft^2, is about the same as 1 m^2. Multiplying the amount of fabric in m^2 by 11 gives 6600 ft^2 of fabric. Each chair requires about 30 ft^2 of fabric, so 6600 ft^2 will upholster 220 chairs, a number close to our answer.

Comment When using conversion factors raised to a power, be certain to raise *both* the number *and* unit to that power. In this problem,

Correct: 0.3048 m = 1 ft

so $(0.3048)^2\ m^2 = (1)^2\ ft^2$ or $0.09290\ m^2 = 1\ ft^2$

Incorrect: $0.3048\ m^2 = 1\ ft^2$ (squaring only the units and not the numbers)

Road Map

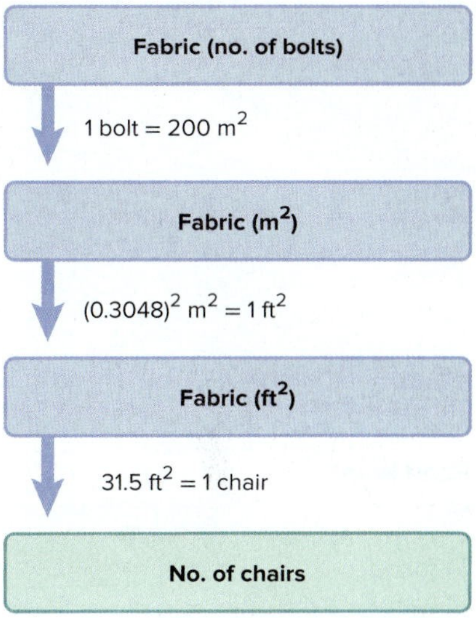

FOLLOW-UP PROBLEMS

1.6A A landowner wants to spray herbicide on a field that has an area of 2050 m^2. The herbicide comes in bottles that hold 16 fluid ounces (fl oz), and 1.5 fl oz mixed with 1 gal of water will treat 300 ft^2. How many bottles of herbicide will the landowner need? Draw a road map to show how you planned the solution.

1.6B From 1946 to 1970, 75,000 kg of mercury was discharged into a lake in New York. The lake has a surface area of 4.5 mi^2 and an average depth of 35 ft. What mass (in g) of mercury is contained in each mL of lake water in 1970? Draw a road map to show how you planned the solution.

SOME SIMILAR PROBLEMS 1.30, 1.31, and 1.34

Student Hot Spot

Student data indicate that you may struggle with the conversion of units raised to a power. Access the Smartbook to view additional Learning Resources on this topic.

Density: A Combination of Units as a Conversion Factor Derived units, combinations of two or more units, can serve as conversion factors (previously we mentioned that speed is length/time). In chemistry, a very important example is **density (*d*)**, the mass of a sample of a substance divided by its volume (mass/volume ratio):

$$\text{Density} = \frac{\text{mass}}{\text{volume}} \qquad \textbf{(1.1)}$$

We isolate each of these variables by treating density as a conversion factor:

$$\text{Mass} = \text{volume} \times \text{density} = \cancel{\text{volume}} \times \frac{\text{mass}}{\cancel{\text{volume}}}$$

or,

$$\text{Volume} = \text{mass} \times \frac{1}{\text{density}} = \cancel{\text{mass}} \times \frac{\text{volume}}{\cancel{\text{mass}}}$$

Table 1.5 Densities of Some Common Substances*

Substance	Physical State	Density (g/cm³)
Hydrogen	gas	0.0000899
Oxygen	gas	0.00133
Grain alcohol	liquid	0.789
Water	liquid	0.998
Table salt	solid	2.16
Aluminum	solid	2.70
Lead	solid	11.3
Gold	solid	19.3

*At room temperature (20°C) and normal atmospheric pressure (1 atm).

Because volume can change with temperature, so can density. But, at a given temperature and pressure, the *density of a substance is a characteristic physical property and, thus, has a specific value.*

The SI unit of density is kilograms per cubic meter (kg/m^3), but in chemistry, density has units of g/L (g/dm^3) or g/mL (g/cm^3) (Table 1.5). The densities of gases are much lower than the densities of liquids or solids (see Figure 1.1).

Road Map

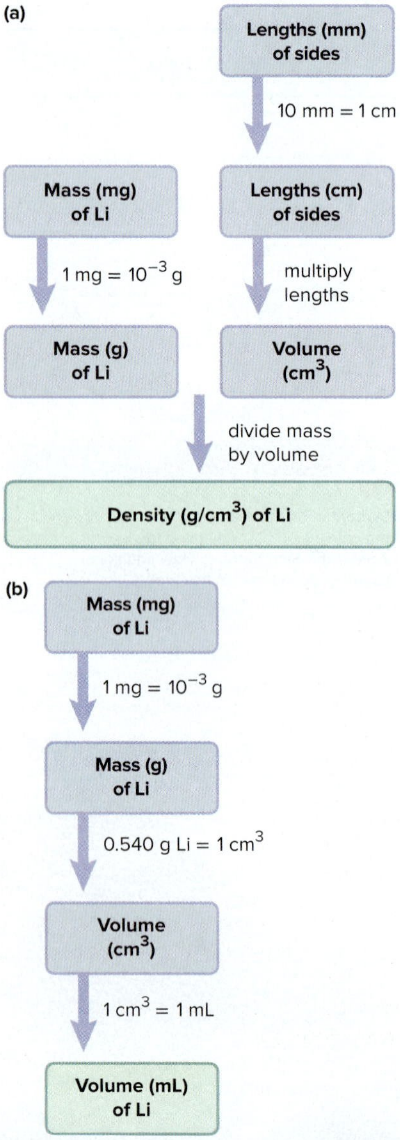

SAMPLE PROBLEM 1.7 Calculating Density from Mass and Volume

Problem **(a)** Lithium, a soft, gray solid with the lowest density of any metal, is a key component of advanced batteries, such as the one in your laptop. A slab of lithium weighs 1.49×10^3 mg and has sides that are 20.9 mm by 11.1 mm by 11.9 mm. Find the density of lithium in g/cm^3. **(b)** What is the volume (in mL) of a piece of lithium with a mass of 895 mg?

Plan **(a)** To find the density in g/cm^3, we need the mass of lithium in g and the volume in cm^3. The mass is 1.49×10^3 mg, so we convert mg to g. We convert the lengths of the three sides from mm to cm, and then multiply them to find the volume in cm^3. Dividing the mass by the volume gives the density (see the road map). **(b)** We have a known mass of lithium (895 mg); we convert mg to g and then use the density from part (a) as a conversion factor between the mass of lithium in g and its volume in cm^3. Finally, we convert cm^3 to mL (see the road map).

Solution **(a)** Converting the mass from mg to g:

$$\text{Mass (g) of lithium} = 1.49\times10^3\ \cancel{\text{mg}}\left(\frac{10^{-3}\ \text{g}}{1\ \cancel{\text{mg}}}\right) = 1.49\ \text{g}$$

Converting side lengths from mm to cm:

$$\text{Length (cm) of one side} = 20.9\ \cancel{\text{mm}} \times \frac{1\ \text{cm}}{10\ \cancel{\text{mm}}} = 2.09\ \text{cm}$$

Similarly, the other side lengths are 1.11 cm and 1.19 cm.

Multiplying the sides to get the volume:

$$\text{Volume (cm}^3\text{)} = 2.09\ \text{cm} \times 1.11\ \text{cm} \times 1.19\ \text{cm} = 2.76\ \text{cm}^3$$

Calculating the density:

$$\text{Density of lithium} = \frac{\text{mass}}{\text{volume}} = \frac{1.49\ \text{g}}{2.76\ \text{cm}^3} = \boxed{0.540\ \text{g/cm}^3}$$

(b) Converting the mass from mg to g:

$$\text{Mass (g) of lithium} = 895\ \cancel{\text{mg}} \times \frac{10^{-3}\ \text{g}}{1\ \cancel{\text{mg}}} = 0.895\ \text{g}$$

Converting the mass of lithium to volume (mL):

$$\text{Volume (mL)} = 0.895\ \cancel{\text{g}} \times \frac{1\ \cancel{\text{cm}^3}}{0.540\ \cancel{\text{g}}} \times \frac{1\ \text{mL}}{1\cancel{\text{cm}^3}} = 1.66\ \text{mL}$$

Check **(a)** Since 1 cm = 10 mm, the number of cm in each length should be $\frac{1}{10}$ the number of mm. The units for density are correct, and the size of the answer (~0.5 g/cm^3) seems correct since the number of g (1.49) is about half the number of cm^3 (2.76). Also, the problem states that lithium has a very low density, so this answer makes sense. **(b)** Since the density of lithium is 0.540 g/cm^3 or 0.540 g/mL, about 0.5 g of lithium has a volume of 1 mL. Therefore, a mass of about 0.9 g would have a volume of about 1.8 mL.

FOLLOW-UP PROBLEMS

1.7A The density of Venus is similar to that of Earth. The mass of Venus is 4.9×10^{24} kg, and its diameter is 12,100 km. Find the density of Venus in g/cm^3. The volume of a sphere is given by $\frac{4}{3}\pi r^3$. Draw a road map to show how you planned the solution.

1.7B The piece of galena in Sample Problem 1.4 has a volume of 4.6 cm^3. If the density of galena is 7.5 g/cm^3, what is the mass (in kg) of that piece of galena? Draw a road map to show how you planned the solution.

SOME SIMILAR PROBLEMS 1.38–1.41

Student Hot Spot

Student data indicate that you may struggle with using density as a conversion factor. Access the Smartbook to view additional Learning Resources on this topic.

Temperature Scales

Temperature is a frequently measured quantity in chemistry; there is a noteworthy distinction between temperature and heat:

- **Temperature (*T*)** is *a measure* of how hot or cold one object is relative to another.
- **Heat** is *the energy* that flows from an object with a higher temperature to an object with a lower temperature. When you hold an ice cube, it feels like the "cold" flows into your hand, but actually, heat flows from your hand to the ice.

In the laboratory, we measure temperature with a **thermometer,** a graduated tube containing a fluid that expands when heated. When the thermometer is immersed in a substance hotter than itself, heat flows from the substance through the glass into the fluid, which expands and rises in the thermometer tube. If a substance is colder than the thermometer, heat flows to the substance from the fluid, which contracts and falls within the tube.

We'll consider three temperature scales: the Celsius (°C, formerly called centigrade), the Kelvin (K), and the Fahrenheit (°F) scales. The SI base unit of temperature is the **kelvin (K)** (note that no degree sign is used with the symbol for this unit). Figure 1.10 shows some interesting temperatures in the Kelvin scale, which is preferred in scientific work (although the Celsius scale is still used frequently). In the United States, the Fahrenheit scale is used for weather reporting, body temperature, and so forth.

The three scales differ in the size of the unit and/or the temperature of the zero point. Figure 1.11 on the next page shows the freezing and boiling points of water on the three scales.

- The **Celsius scale** sets water's freezing point at 0°C and its boiling point (at normal atmospheric pressure) at 100°C. Thus, the size of a Celsius degree is $\frac{1}{100}$ of the difference between the freezing and boiling points of water.

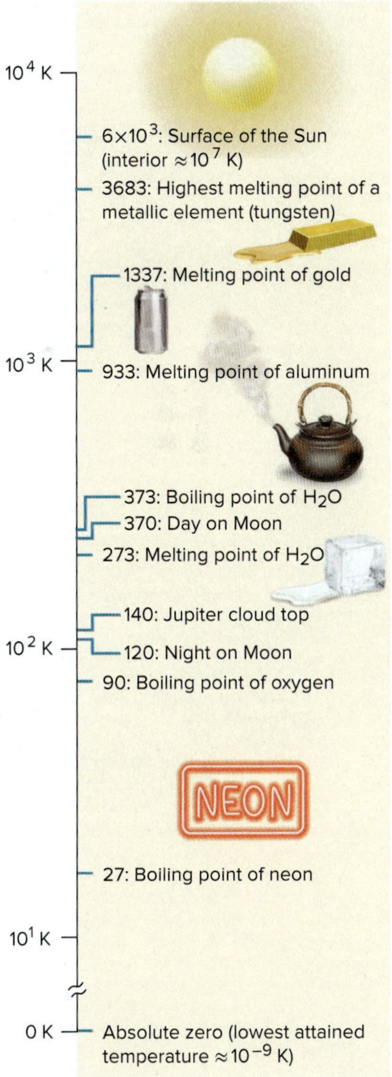

Figure 1.10 Some interesting temperatures.

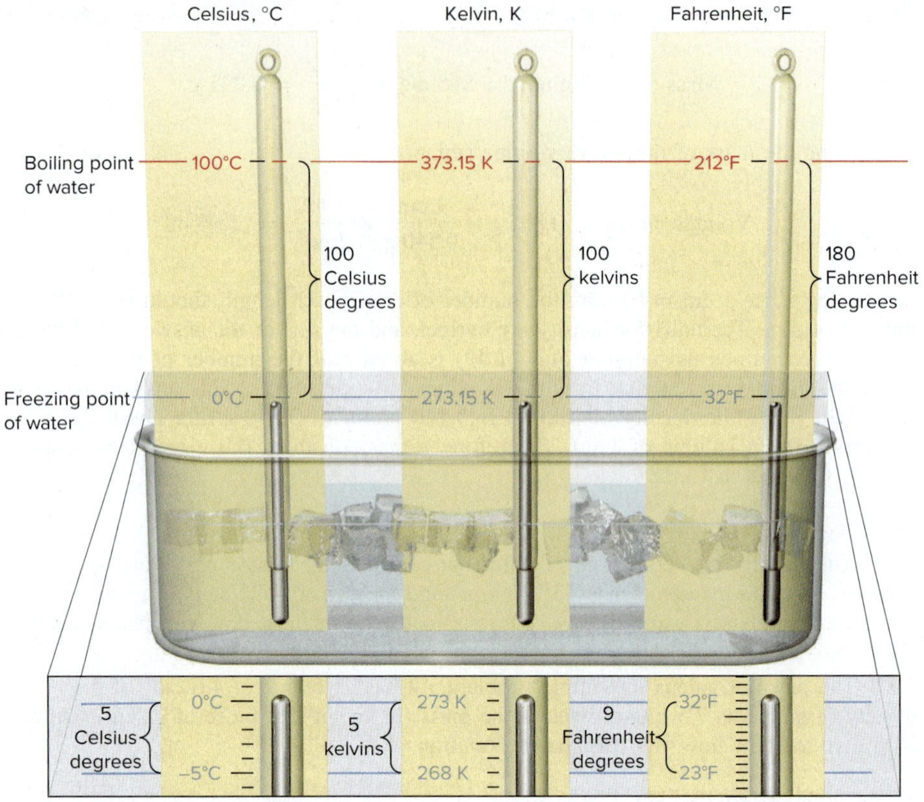

Figure 1.11 **Freezing and boiling points of water in the Celsius, Kelvin (absolute), and Fahrenheit scales.** At the bottom of the figure, a portion of each of the three thermometer scales is expanded to show the sizes of the units.

- The **Kelvin (absolute) scale** uses the *same size degree* as the Celsius scale; the difference between the freezing point (+273.15 K) and the boiling point (+373.15 K) of water is again 100 degrees, but these temperatures are 273.15 degrees higher on the Kelvin scale because it has a *different zero point. Absolute zero,* 0 K, equals −273.15°C. Thus, on the Kelvin scale, *all temperatures are positive.*

  We convert between the Celsius and Kelvin scales by remembering the different zero points: 0°C = 273.15 K, so

$$T \text{ (in K)} = T \text{ (in °C)} + 273.15 \qquad \textbf{(1.2)}$$

  And, therefore,

$$T \text{ (in °C)} = T \text{ (in K)} - 273.15 \qquad \textbf{(1.3)}$$

- The Fahrenheit scale differs from the other scales in its zero point *and* in the size of its degree. Water freezes at 32°F and boils at 212°F. Therefore, 180 Fahrenheit degrees (212°F − 32°F) represents the same temperature change as 100 Celsius degrees (or 100 kelvins). Because 100 Celsius degrees equal 180 Fahrenheit degrees,

$$1 \text{ Celsius degree} = \frac{180}{100} \text{ Fahrenheit degrees} = \frac{9}{5} \text{ Fahrenheit degrees}$$

  To convert a temperature from °C to °F, first change the degree size and then adjust the zero point (0°C = 32°F):

$$T \text{ (in °F)} = \frac{9}{5} T \text{ (in °C)} + 32 \qquad \textbf{(1.4)}$$

To convert a temperature from °F to °C, do the two steps in the opposite order: adjust the zero point and then change the degree size. In other words, solve Equation 1.4 for T (in °C):

$$T \text{ (in °C)} = [T \text{ (in °F)} - 32]\frac{5}{9} \quad (1.5)$$

Table 1.6 The Three Temperature Scales

Scale	Unit	Size of Degree (Relative to K)	Freezing Point of H_2O	Boiling Point of H_2O	T at Absolute Zero	Conversion
Kelvin (absolute)	kelvin (K)	—	273.15 K	373.15 K	0 K	to °C (Equation 1.2)
Celsius	Celsius degree (°C)	1	0°C	100°C	−273.15°C	to K (Equation 1.3) to °F (Equation 1.4)
Fahrenheit	Fahrenheit degree (°F)	$\frac{5}{9}$	32°F	212°F	−459.67°F	to °C (Equation 1.5)

Table 1.6 compares the three temperature scales.

SAMPLE PROBLEM 1.8 Converting Units of Temperature

Problem A child has a body temperature of 38.7°C, and normal body temperature is 98.6°F. Does the child have a fever? What is the child's temperature in kelvins?

Plan To see if the child has a fever, we convert from °C to °F (Equation 1.4) and compare it with 98.6°F. Then, to convert the child's temperature in °C to K, we use Equation 1.2.

Solution Converting the temperature from °C to °F:

$$T \text{ (in °F)} = \tfrac{9}{5}T \text{ (in °C)} + 32 = \tfrac{9}{5}(38.7\text{ °C}) + 32 = 101.7\text{°F}$$ Yes, the child has a fever.

Converting the temperature from °C to K:

$$T \text{ (in K)} = T \text{ (in °C)} + 273.15 = 38.7\text{°C} + 273.15 = 311.8\text{ K}$$

Check From everyday experience, you know that 101.7°F is a reasonable temperature for someone with a fever. In the second step, we can check for a large error as follows: 38.7°C is almost 40°C, and 40 + 273 = 313, which is close to our answer.

FOLLOW-UP PROBLEMS

1.8A Mercury melts at 234 K, lower than any other pure metal. What is its melting point in °C and °F?

1.8B The temperature in a blast furnace used for iron production is 2325°F. What is this temperature in °C and K?

SOME SIMILAR PROBLEMS 1.42 and 1.43

Extensive and Intensive Properties

The variables we measure to study matter fall into two broad categories of properties:

- **Extensive properties** are *dependent* on the amount of substance present; mass and volume, for example, are extensive properties.
- **Intensive properties** are *independent* of the amount of substance; density is an intensive property.

Thus, a gallon of water has four times the mass of a quart of water, but it also has four times the volume, so the density, the *ratio* of mass to volume, is the same for both samples; this concept is illustrated for copper in Figure 1.12. Another important example concerns heat, an extensive property, and temperature, an intensive property: a vat of

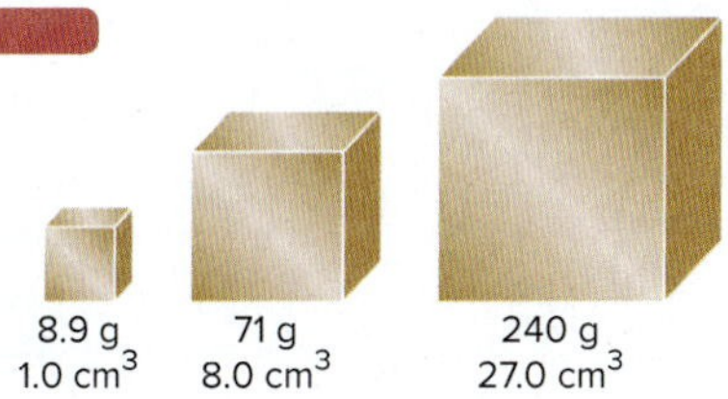

The mass and volume of the three cubes of copper are different; mass and volume are extensive properties.

For these three cubes of copper,

$$\text{density} = \frac{8.9\text{ g}}{1.0\text{ cm}^3} = \frac{71\text{ g}}{8.0\text{ cm}^3} = \frac{240\text{ g}}{27.0\text{ cm}^3} \approx 8.9\text{ g/cm}^3$$

The density remains the same regardless of sample size; density is an intensive property.

Figure 1.12 Extensive and intensive properties of matter.

boiling water has more heat, that is, more energy, than a cup of boiling water, but both samples have the same temperature.

Some intensive properties, like color, melting point, and density are characteristic of a substance, and thus, are used to identify it.

› Summary of Section 1.4

- The SI unit system consists of seven base units and numerous derived units.
- Exponential notation and prefixes based on powers of 10 are used to express very small and very large numbers.
- The SI base unit of length is the meter (m); on the atomic scale, the nanometer (nm) and picometer (pm) are used commonly.
- Volume (V) units are derived from length units, and the most important volume units are the cubic meter (m^3) and the liter (L).
- The *mass* of an object—the quantity of matter in it—is constant. The SI unit of mass is the kilogram (kg). The *weight* of an object varies with the gravitational field.
- A measured quantity consists of a number and a unit.
- A conversion factor is a ratio of equivalent quantities (and, thus, equal to 1) that is used to express a quantity in different units.
- The problem-solving approach used in this book has four parts: (1) plan the steps to the solution, which often includes a flow diagram (road map) of the steps, (2) perform the calculations according to the plan, (3) check to see if the answer makes sense, and (4) practice with similar, follow-up problems and compare your solutions with the ones at the end of the chapter.
- Density (d), a characteristic physical property of a substance, is the ratio of the mass of a sample to its volume.
- Temperature (T) is a measure of the relative hotness of an object. Heat is energy that flows from an object at higher T to one at lower T.
- Temperature scales differ in the size of the degree unit and/or the zero point. For scientific uses, temperature is measured in kelvins (K) or degrees Celsius (°C).
- Extensive properties, such as mass, volume, and energy, depend on the amount of a substance. Intensive properties, such as density and temperature, do not.

1.5 UNCERTAINTY IN MEASUREMENT: SIGNIFICANT FIGURES

All measuring devices—balances, pipets, thermometers, and so forth—are made to limited specifications, and we use our imperfect senses and skills to read them. Therefore, we can *never* measure a quantity exactly; put another way, every measurement includes some **uncertainty.** The device we choose depends on how much uncertainty is acceptable. When you buy potatoes, a supermarket scale that measures in 0.1-kg increments is acceptable:

$$\text{Measured mass: } 2.0 \pm 0.1 \text{ kg} \longrightarrow \text{actual mass: between } 1.9 \text{ and } 2.1 \text{ kg}$$

The "± 0.1 kg" term expresses the uncertainty in the mass. Needing more certainty than that to weigh a substance, a chemist uses a balance that measures in 0.001-kg increments:

$$\text{Measured mass: } 2.036 \pm 0.001 \text{ kg} \longrightarrow \text{actual mass: between } 2.035 \text{ and } 2.037 \text{ kg}$$

The greater number of digits in this measurement means we know the mass of the substance with *more certainty* than we know the mass of the potatoes.

We *always estimate the rightmost digit* of a measurement. The uncertainty can be expressed with the ± sign, but generally we drop the sign and *assume an uncertainty of one unit in the rightmost digit.* The digits we record, both the certain and the uncertain ones, are called **significant figures.** There are four significant figures

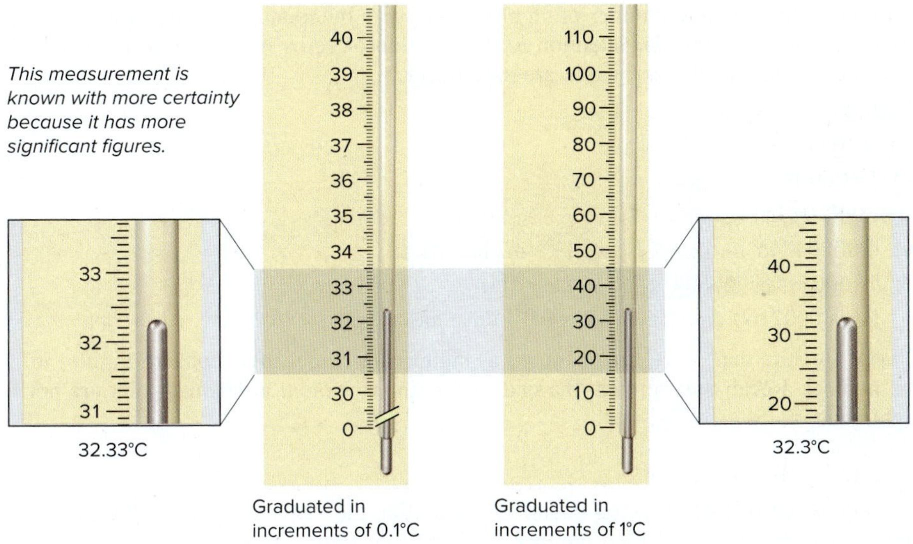

Figure 1.13 The number of significant figures in a measurement.

in 2.036 kg and two in 2.0 kg. *The greater the number of significant figures, the greater is the certainty of a measurement.* Figure 1.13 shows this point for two thermometers.

Determining Which Digits Are Significant

When you take a measurement or use one in a calculation, you must know the number of digits that are significant: *all digits are significant, except zeros used only to position the decimal point.* The following procedure applies this point:

1. Make sure the measurement has a decimal point.
2. Start at the left, and move right until you reach the first nonzero digit.
3. Count that digit and every digit to its right as significant, including zeros between nonzero digits.

Thus, 2.033 has four significant figures, and 0.000562 has only three because a leading zero is never significant and the other three zeros are used only to position the decimal point.

A complication can arise when zeros end a number:

- If there *is* a decimal point and the zeros lie either after or before it, they *are* significant: 1.1300 g has five significant figures and 6500. has four.
- If there is *no* decimal point, we assume that the zeros are *not* significant, unless exponential notation clarifies the quantity: 5300 L is *assumed* to have two significant figures, but 5.300×10^3 L has four, 5.30×10^3 L has three, and 5.3×10^3 L has two.
- A terminal decimal point indicates that zeros are significant: 500 mL has one significant figure, but 500. mL has three (as do 5.00×10^2 mL and 0.500 L).

SAMPLE PROBLEM 1.9 Determining the Number of Significant Figures

Problem For each of the following quantities, underline the zeros that are significant figures (sf) and determine the total number of significant figures. For (e) to (g), express each quantity in exponential notation first.

(a) 0.0030 L **(b)** 0.1044 g **(c)** 53,069 mL
(d) 3040 kg **(e)** 0.00004715 m **(f)** 57,600. s
(g) 0.0000007160 cm^3

Plan We determine the number of significant figures by counting digits, as just discussed, paying particular attention to the position of zeros in relation to the decimal point, and underline the zeros that are significant.

Solution **(a)** 0.0030 L has 2 sf

(b) 0.1044 g has 4 sf

(c) 53,069 mL has 5 sf

(d) 3040 kg has 3 sf

(e) 0.00004715 m, or 4.715×10^{-5} m, has 4 sf

(f) 57,600. s, or 5.7600×10^{4} s, has 5 sf

(g) 0.0000007160 cm^3, or 7.160×10^{-7} cm^3, has 4 sf

Check Be sure that every zero counted as significant comes after nonzero digit(s) in the number. Recall that zeros at the end of a number *without* a decimal point are not significant.

FOLLOW-UP PROBLEMS

1.9A For each of the following quantities, underline the zeros that are significant figures and determine the total number of significant figures (sf).
(a) 31.070 mg **(b)** 0.06060 g **(c)** 850.°C

1.9B For each of the following quantities, underline the zeros that are significant figures and determine the total number of significant figures (sf). Express each quantity in exponential notation first.
(a) 200.0 mL **(b)** 0.0000039 m **(c)** 0.000401 L

SOME SIMILAR PROBLEMS 1.52 and 1.53

Significant Figures: Calculations and Rounding Off

Measuring several quantities typically results in data with differing numbers of significant figures. In a calculation, we keep track of the number in each quantity so that we don't have more significant figures (more certainty) in the answer than in the data. If we do have too many significant figures, we must **round off** the answer.

The general rule for rounding is that *the least certain measurement sets the limit on certainty for the entire calculation and determines the number of significant figures in the final answer.* Suppose you want to find the density of a new ceramic. You measure the mass of a piece of it on a precise laboratory balance and obtain 3.8056 g; you measure the volume as 2.5 mL by displacement of water in a graduated cylinder. The mass has five significant figures, but the volume has only two. Should you report the density as 3.8056 g/2.5 mL = 1.5222 g/mL or as 1.5 g/mL? The answer with five significant figures implies more certainty than the answer with two. But you didn't measure the volume to five significant figures, so you can't possibly know the density with that much certainty. Therefore, you report 1.5 g/mL, the answer with two significant figures.

Rules for Arithmetic Operations

The two rules in arithmetic calculations are

1. *For multiplication and division.* The answer contains the same number of significant figures as there are in the measurement with the *fewest significant figures.* Suppose you want to find the volume of a sheet of a new graphite composite. The length (9.2 cm) and width (6.8 cm) are obtained with a ruler, and the thickness (0.3744 cm) with a set of calipers. The calculation is

$$\text{Volume (cm}^3) = \underset{2}{9.2\text{ cm}} \times \underset{2}{6.8\text{ cm}} \times \underset{4}{0.3744\text{ cm}} = 23.422464\text{ cm}^3 = \underset{2}{23\text{ cm}^3}$$

No. of significant figures: 2 2 4 2

Even though your calculator may show 23.422464 cm^3, you report 23 cm^3, the answer with two significant figures, the same as in the measurements with the lower number of significant figures. After all, if the length and width have two significant figures, you can't possibly know the volume with more certainty.

2. *For addition and subtraction.* The answer has the same number of *decimal places* as there are in the measurement with the *fewest decimal places*. Suppose you want the total volume after adding water to a protein solution: you have 83.5 mL of solution in a graduated cylinder and add 23.28 mL of water from a buret. The calculation is shown in the margin. Here the calculator shows 106.78 mL, but you report the volume as 106.8 mL, because the measurement with fewer decimal places (83.5 mL) has one decimal place *(see margin)*.

$$\begin{array}{r} 83.5\ \text{mL} \\ +\ 23.28\ \text{mL} \\ \hline 106.78\ \text{mL} \end{array}$$

Answer: Volume = 106.8 mL

Note that the answer, 106.8 mL, has four significant figures, while the volume of the protein solution, 83.5 mL, has only three significant figures. In addition and subtraction, the number of significant figures is determined by the number of decimal places, not the total number of significant figures, in the measurements.

Rules for Rounding Off You usually need to round off the final answer to the proper number of significant figures or decimal places. Notice that in calculating the volume of the graphite composite, we removed the extra digits, but in calculating the total volume of the protein solution, we removed the extra digit and increased the last digit by one. The general rule for rounding is that *the least certain measurement sets the limit on the certainty of the final answer.* Here are detailed rules for rounding off:

1. If the digit removed is *more than 5,* the preceding number increases by 1: 5.379 rounds to 5.38 if you need three significant figures and to 5.4 if you need two.
2. If the digit removed is *less than 5,* the preceding number remains the same: 0.2413 rounds to 0.241 if you need three significant figures and to 0.24 if you need two.
3. If the digit removed *is 5,* the preceding number increases by 1 if it is odd and remains the same if it is even: 17.75 rounds to 17.8, but 17.65 rounds to 17.6. If the 5 is followed only by zeros, rule 3 is followed; if the 5 is followed by nonzeros, rule 1 is followed: 17.6500 rounds to 17.6, but 17.6513 rounds to 17.7.
4. *Always carry one or two additional significant figures through a multistep calculation and round off the final answer* ***only.*** Don't be concerned if you string together a calculation to check a sample or follow-up problem and find that your answer differs in the last decimal place from the one in the book. To show you the correct number of significant figures in text calculations, *we round off intermediate steps,* and that process may sometimes change the last digit.

Note that, unless you set a limit on your calculator, it gives answers with too many figures and you must round the displayed result.

Significant Figures in the Lab The measuring device you choose determines the number of significant figures you can obtain. Suppose an experiment requires a solution made by dissolving a solid in a liquid. You weigh the solid on an analytical balance and obtain a mass with five significant figures. It would make sense to measure the liquid with a buret or a pipet, which measures volumes to more significant figures than a graduated cylinder. If you do choose the cylinder, you would have to round off more digits, and some certainty you attained for the mass value would be wasted (Figure 1.14). With experience, you'll choose a measuring device based on the number of significant figures you need in the final answer.

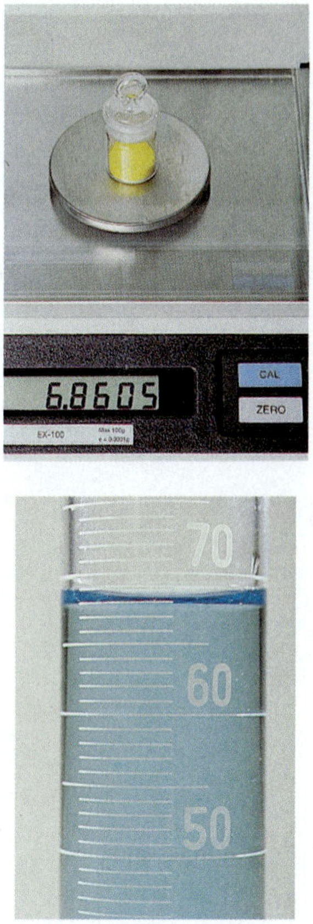

Figure 1.14 Significant figures and measuring devices. The mass measurement (6.8605 g) has more significant figures than the volume measurement (68.2 mL).

Source: (both) © McGraw-Hill Education/Stephen Frisch, photographer

Exact Numbers **Exact numbers** have no uncertainty associated with them. Some are part of a unit conversion: by definition, there are exactly 60 minutes in 1 hour, 1000 micrograms in 1 milligram, and 2.54 centimeters in 1 inch. Other exact numbers result from actually counting items: there are exactly 3 coins in my hand, 26 letters in the English alphabet, and so forth. Therefore, unlike measured quantities, *exact numbers do not limit the number of significant figures in a calculation.*

SAMPLE PROBLEM 1.10 Significant Figures and Rounding

Problem Perform the following calculations and round the answers to the correct number of significant figures:

(a) $\dfrac{16.3521\ \text{cm}^2 - 1.448\ \text{cm}^2}{7.085\ \text{cm}}$ (b) $\dfrac{(4.80\times10^4\ \text{mg})\left(\dfrac{1\ \text{g}}{1000\ \text{mg}}\right)}{11.55\ \text{cm}^3}$

Plan We use the rules just presented in the text: **(a)** We subtract before we divide. **(b)** We note that the unit conversion involves an exact number.

Solution (a) $\dfrac{16.3521\ \text{cm}^2 - 1.448\ \text{cm}^2}{7.085\ \text{cm}} = \dfrac{14.904\ \text{cm}^{\cancel{2}}}{7.085\ \cancel{\text{cm}}} = 2.104\ \text{cm}$

(b) $\dfrac{(4.80\times10^4\ \cancel{\text{mg}})\left(\dfrac{1\ \text{g}}{1000\ \cancel{\text{mg}}}\right)}{11.55\ \text{cm}^3} = \dfrac{48.0\ \text{g}}{11.55\ \text{cm}^3} = 4.16\ \text{g/cm}^3$

Check Note that in (a) we lose a decimal place in the numerator, and in (b) we retain 3 sf in the answer because there are 3 sf in 4.80. Rounding to the nearest whole number is always a good way to check: **(a)** $(16 - 1)/7 \approx 2$; **(b)** $(5\times10^4/1\times10^3)/12 \approx 4$.

FOLLOW-UP PROBLEMS

1.10A Perform the following calculation and round the answer to the correct number of significant figures:

$$\frac{25.65\ \text{mL} + 37.4\ \text{mL}}{73.55\ \text{s}\left(\dfrac{1\ \text{min}}{60\ \text{s}}\right)}$$

1.10B Perform the following calculation and round the answer to the correct number of significant figures:

$$\frac{154.64\ \text{g} - 35.26\ \text{g}}{4.20\ \text{cm} \times 5.12\ \text{cm} \times 6.752\ \text{cm}}$$

SOME SIMILAR PROBLEMS 1.56–1.59, 1.66, and 1.67

Precision, Accuracy, and Instrument Calibration

We may use the words "precision" and "accuracy" interchangeably in everyday speech, but for scientific measurements they have distinct meanings. **Precision,** or *reproducibility,* refers to how close the measurements in a series are to each other, and **accuracy** refers to how close each measurement is to the actual value. These terms are related to two widespread types of error:

1. **Systematic error** produces values that are *either* all higher or all lower than the actual value. This type of error is part of the experimental system, often caused by a faulty device or by a consistent mistake in taking a reading.
2. **Random error,** in the absence of systematic error, produces values that are higher *and* lower than the actual value. Random error *always* occurs, but its size depends on the measurer's skill and the instrument's precision.

Precise measurements have low random error, that is, small deviations from the average. *Accurate measurements have low systematic error and, generally, low random error.* In some cases, when many measurements have a high random error, the *average* may still be accurate.

Suppose each of four students measures 25.0 mL of water in a preweighed graduated cylinder and then weighs the water *plus* cylinder on a balance. If the density of water is 1.00 g/mL at the temperature of the experiment, the *actual* mass of 25.0 mL of water is 25.0 g. Each student performs the operation four

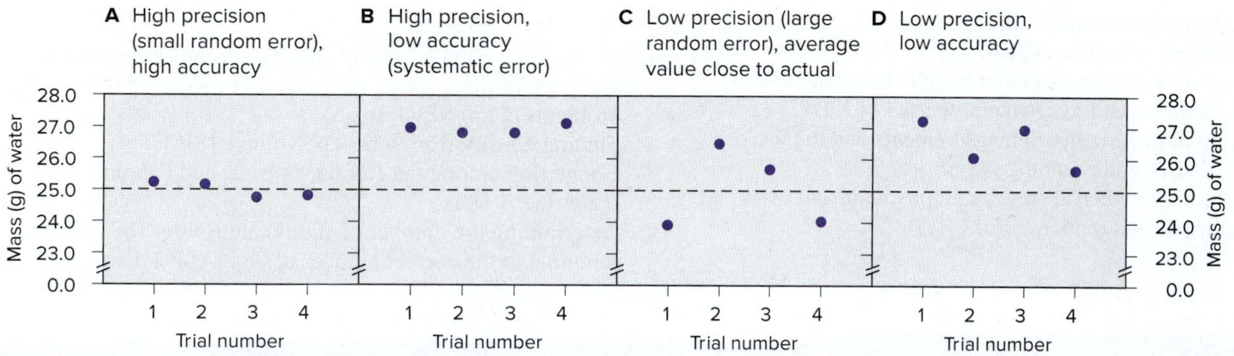

Figure 1.15 Precision and accuracy in a laboratory calibration.

times, subtracts the mass of the empty cylinder, and obtains one of four graphs (Figure 1.15):

- In graph A, random error is small; that is, precision is high (the weighings are reproducible). Accuracy is high as well, as all of the values are close to 25.0 g.
- Random error is also small and precision high in graph B, but accuracy is low; there is systematic error, with all of the weighings above 25.0 g.
- In graph C, random error is large and precision is low. But since the average of the scattered values is close to the actual value, systematic error is low.
- Graph D also exhibits large random error, but note that there is also significant systematic error in this case, resulting in low accuracy (all the values are high).

Systematic error can be taken into account through **calibration,** comparing the measuring device with a known standard. The systematic error in graph B, for example, might be caused by a poorly manufactured cylinder that reads "25.0" when it actually contains about 27 mL. If that cylinder had been calibrated, the students could have adjusted all volumes measured with it. The students also should calibrate the balance with standardized masses.

› Summary of Section 1.5

- › The final digit of a measurement is always estimated. Thus, all measurements have some uncertainty, which is expressed by the number of significant figures.
- › The certainty of a calculated result depends on the certainty of the data, so the answer has as many significant figures as in the least certain measurement.
- › Excess digits are rounded off in the final answer according to a set of rules.
- › The choice of laboratory device depends on the certainty needed.
- › Exact numbers have as many significant figures as the calculation requires.
- › Precision refers to how close values are to each other, and accuracy refers to how close values are to the actual value.
- › Systematic errors give values that are either all higher or all lower than the actual value. Random errors give some values that are higher and some that are lower than the actual value.
- › Precise measurements have low random error; accurate measurements have low systematic error and low random error.
- › A systematic error is often caused by faulty equipment and can be compensated for by calibration.

CHAPTER REVIEW GUIDE

Learning Objectives

Relevant section (§) and/or sample problem (SP) numbers appear in parentheses.

Understand These Concepts

1. The defining features of the states of matter (§1.1)
2. The distinction between physical and chemical properties and changes (§1.1; SPs 1.1, 1.2)
3. The nature of potential and kinetic energy and their interconversion (§1.1)
4. The process of approaching a phenomenon scientifically and the distinctions between observation, hypothesis, experiment, and model (§1.3)

5. The common units of length, volume, mass, and temperature and their numerical prefixes (§1.4)
6. The distinctions between mass and weight, heat and temperature, and intensive and extensive properties (§1.4)
7. The meaning of uncertainty in measurements and the use of significant figures and rounding (§1.5)
8. The distinctions between accuracy and precision and between systematic and random error (§1.5)

Master These Skills

1. Using conversion factors in calculations and a systematic approach of plan, solution, check, and follow-up for solving problems (§1.4; SPs 1.3–1.6)
2. Finding density from mass and volume (SP 1.7)
3. Converting among the Kelvin, Celsius, and Fahrenheit scales (SP 1.8)
4. Determining the number of significant figures (SP 1.9) and rounding to the correct number of digits (SP 1.10)

Key Terms

Page numbers appear in parentheses.

accuracy (32)
base (fundamental) unit (13)
calibration (33)
Celsius scale (25)
chemical change (chemical reaction) (5)
chemical property (5)
chemistry (4)
combustion (11)
composition (4)
controlled experiment (12)
conversion factor (18)
cubic meter (m^3) (15)
data (12)
density (*d*) (23)
derived unit (13)
dimensional analysis (19)
energy (8)
exact number (31)
experiment (12)
extensive property (27)
gas (4)
heat (25)
hypothesis (12)
intensive property (27)
Kelvin (absolute) scale (26)
kelvin (K) (25)
kilogram (kg) (17)
kinetic energy (8)
liquid (4)
liter (L) (15)
mass (17)
matter (4)
meter (m) (15)
milliliter (mL) (15)
model (theory) (13)
natural law (12)
observation (12)
physical change (5)
physical property (5)
potential energy (8)
precision (32)
property (5)
random error (32)
round off (30)
scientific method (12)
second (s) (18)
SI unit (13)
significant figures (28)
solid (4)
state of matter (4)
systematic error (32)
temperature (*T*) (25)
thermometer (25)
uncertainty (28)
variable (12)
volume (*V*) (15)
weight (17)

Key Equations and Relationships

Page numbers appear in parentheses.

1.1 Calculating density from mass and volume (23):

$$\text{Density} = \frac{\text{mass}}{\text{volume}}$$

1.2 Converting temperature from °C to K (26):

$$T\,(\text{in K}) = T\,(\text{in °C}) + 273.15$$

1.3 Converting temperature from K to °C (26):

$$T\,(\text{in °C}) = T\,(\text{in K}) - 273.15$$

1.4 Converting temperature from °C to °F (26):

$$T\,(\text{in °F}) = \tfrac{9}{5}\,T\,(\text{in °C}) + 32$$

1.5 Converting temperature from °F to °C (27):

$$T\,(\text{in °C}) = [T\,(\text{in °F}) - 32]\,\tfrac{5}{9}$$

BRIEF SOLUTIONS TO FOLLOW-UP PROBLEMS

1.1A Chemical. The red-and-blue and separate red particles on the left become paired red and separate blue particles on the right.

1.1B Physical. The red particles are the same on the right and on the left, but they have changed from being close together in the solid state to being far apart in the gaseous state.

1.2A (a) Physical. Solid iodine changes to gaseous iodine. (b) Chemical. Gasoline burns in air to form different substances. (c) Chemical. In contact with air, substances in torn skin and blood react to form different substances.

1.2B (a) Physical. Gaseous water (water vapor) changes to droplets of liquid water.
(b) Chemical. Different substances are formed in the milk that give it a sour taste. (c) Physical. Solid butter changes to a liquid.

1.3A The known quantity is 10,500 m; start the problem with this value. Use the conversion factor 1 mi/15 min to convert distance in miles to time in minutes.

Time (min)

$$= 10{,}500\ \cancel{\text{m}} \times \frac{1\ \cancel{\text{km}}}{1000\ \cancel{\text{m}}} \times \frac{1\ \cancel{\text{mi}}}{1.609\ \cancel{\text{km}}} \times \frac{15\ \text{min}}{1\ \cancel{\text{mi}}} = 98\ \text{min}$$

See Road Map 1.3A.

1.3B Start the problem with the known quantity of 1.0 in; use the conversion factor 1 virus particle/30 nm to convert from length in nm to number of virus particles.

No. of virus particles

$$= 1.0\ \cancel{\text{in}} \times \frac{2.54\ \cancel{\text{cm}}}{1\ \cancel{\text{in}}} \times \frac{1{\times}10^7\ \cancel{\text{nm}}}{1\ \cancel{\text{cm}}} \times \frac{1\ \text{virus particle}}{30\ \cancel{\text{nm}}}$$

$$= 8.5{\times}10^5\ \text{virus particles}$$

See Road Map 1.3B.

Road Map 1.3A

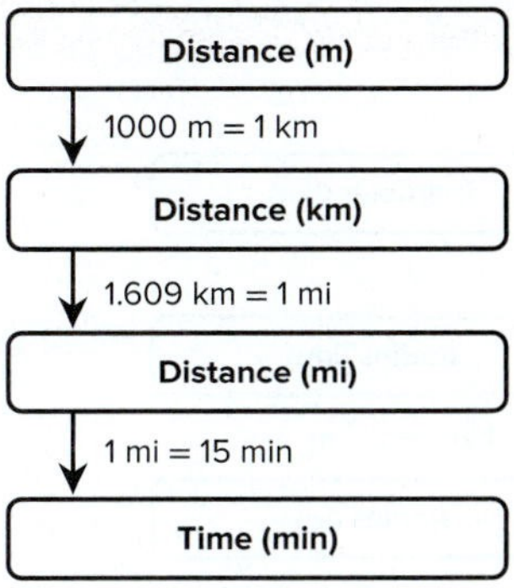

Road Map 1.3B

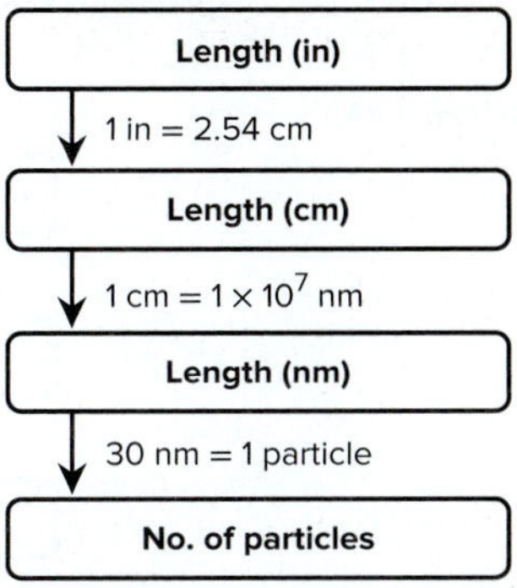

1.4A Radius of ribosome (dm) $= \dfrac{21.4\ \cancel{\text{nm}}}{2} \times \dfrac{1\ \text{dm}}{10^8\ \cancel{\text{nm}}}$

$= 1.07\times10^{-7}$ dm

Volume of ribosome (dm^3) $= \frac{4}{3}\pi r^3 = \frac{4}{3}(3.14)(1.07\times10^{-7}\ \text{dm})^3$

$= 5.13\times10^{-21}\ \text{dm}^3$

Volume of ribosome (μL)

$$= (5.13\times10^{-21}\ \cancel{\text{dm}^3})\left(\frac{1\ \cancel{\text{L}}}{1\ \cancel{\text{dm}^3}}\right)\left(\frac{10^6\ \mu\text{L}}{1\ \cancel{\text{L}}}\right)$$

$= 5.13\times10^{-15}\ \mu\text{L}$

See Road Map 1.4A.

1.4B Volume (L) $= 8400\ \cancel{\text{gal}} \times \dfrac{3.785\ \cancel{\text{dm}^3}}{1\ \cancel{\text{gal}}} \times \dfrac{1\ \text{L}}{1\ \cancel{\text{dm}^3}}$

$= 32{,}000$ L

See Road Map 1.4B.

Road Map 1.4A

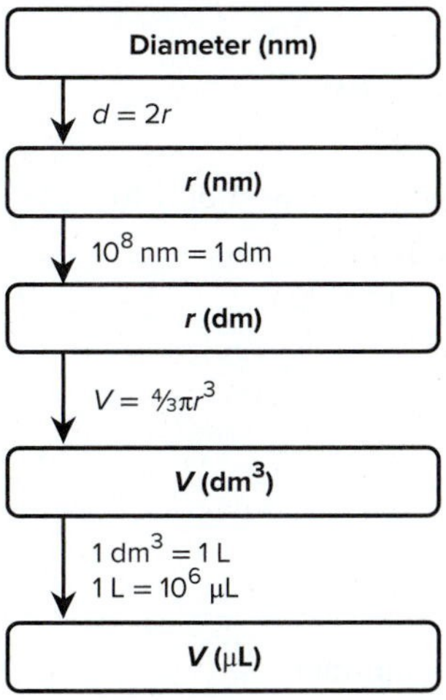

Road Map 1.4B

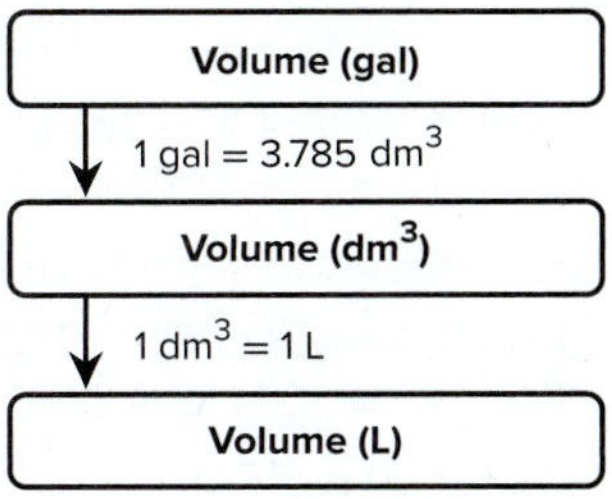

1.5A Start the problem with the known value of 8.0 h. The conversion factors are constructed from the equivalent quantities given in the problem: 1 s = 1.5 drops; 1 drop = 65 mg.

Mass (kg) of solution

$$= 8.0\ \cancel{\text{h}} \times \frac{60\ \cancel{\text{min}}}{1\ \cancel{\text{h}}} \times \frac{60\ \cancel{\text{s}}}{1\ \cancel{\text{min}}} \times \frac{1.5\ \cancel{\text{drops}}}{1\ \cancel{\text{s}}}$$

$$\times \frac{65\ \cancel{\text{mg}}}{1\ \cancel{\text{drop}}} \times \frac{1\ \cancel{\text{g}}}{10^3\ \cancel{\text{mg}}} \times \frac{1\ \text{kg}}{10^3\ \cancel{\text{g}}}$$

$= 2.8$ kg

See Road Map 1.5A.

1.5B Start the problem with the known quantity of 3.25 kg of apples. The conversion factors are constructed from the equivalent quantities given in the problem: 1 lb = 3 apples; 1 apple = 159 mg potassium.

$$\text{Mass (g)} = 3.25\ \cancel{\text{kg}} \times \frac{1\ \cancel{\text{lb}}}{0.4536\ \cancel{\text{kg}}}$$

$$\times \frac{3\ \cancel{\text{apples}}}{1\ \cancel{\text{lb}}} \times \frac{159\ \cancel{\text{mg}}\ \text{potassium}}{1\ \cancel{\text{apple}}} \times \frac{1\ \text{g}}{10^3\ \cancel{\text{mg}}}$$

$= 3.42$ g potassium

See Road Map 1.5B.

Road Map 1.5A

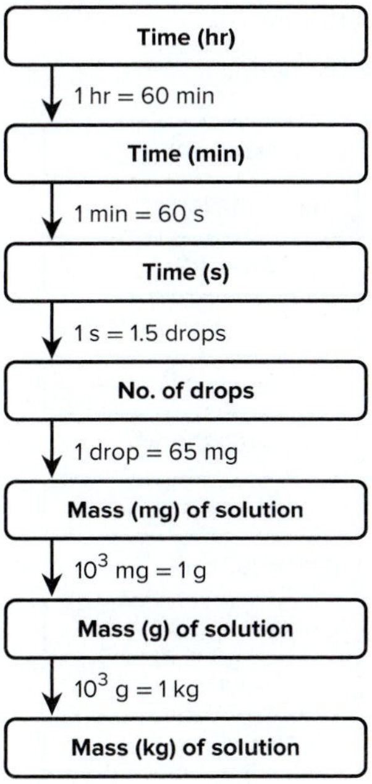

Road Map 1.5B

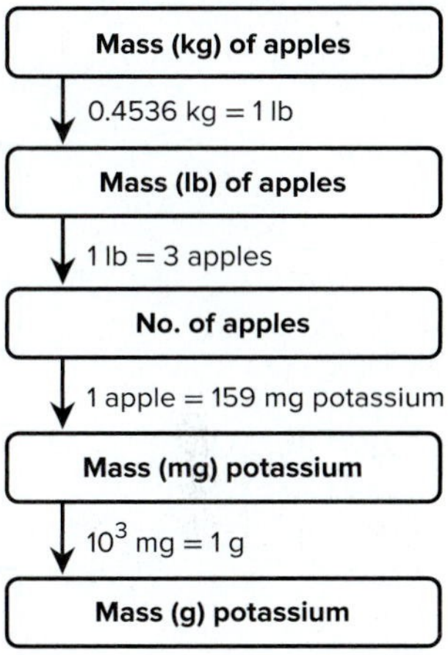

1.6A Start the problem with the known value of 2050 m^2. The conversion factors are constructed from 300 ft^2 = 1.5 fl oz and 16 fl oz = 1 bottle.

$$\text{No. of bottles} = 2050\ \cancel{\text{m}^2} \times \frac{1\ \cancel{\text{ft}^2}}{(0.3048)^2\ \cancel{\text{m}^2}} \times \frac{1.5\ \cancel{\text{fl oz}}}{300\ \cancel{\text{ft}^2}} \times \frac{1\ \text{bottle}}{16\ \cancel{\text{fl oz}}}$$

$= 7$ bottles

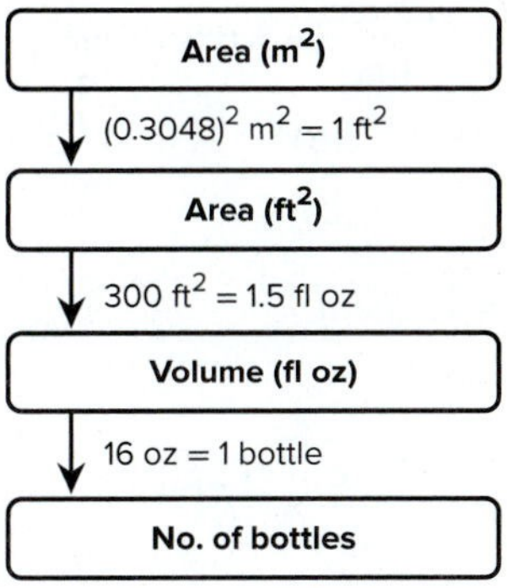

BRIEF SOLUTIONS TO FOLLOW-UP PROBLEMS (continued)

1.6B $\text{Mass (g)} = 75{,}000\ \cancel{\text{kg}} \times \dfrac{1000\ \text{g}}{1\ \cancel{\text{kg}}} = 7.5\times10^{7}\ \text{g}$

$$\text{Volume (mL)} = 4.5\ \cancel{\text{mi}^2} \times \frac{(5280)^2\ \cancel{\text{ft}^2}}{1\ \cancel{\text{mi}^2}} \times 35\ \cancel{\text{ft}} \times \frac{0.02832\ \cancel{\text{m}^3}}{1\ \cancel{\text{ft}^3}} \times \frac{1\times10^{6}\ \cancel{\text{cm}^3}}{1\ \cancel{\text{m}^3}} \times \frac{1\ \text{mL}}{1\ \cancel{\text{cm}^3}}$$

$$= 1.2\times10^{14}\ \text{mL}$$

$$\text{Mass (g) of mercury per mL} = \frac{7.5\times10^{7}\ \text{g}}{1.2\times10^{14}\ \text{mL}} = 6.2\times10^{-7}\ \text{g/mL}$$

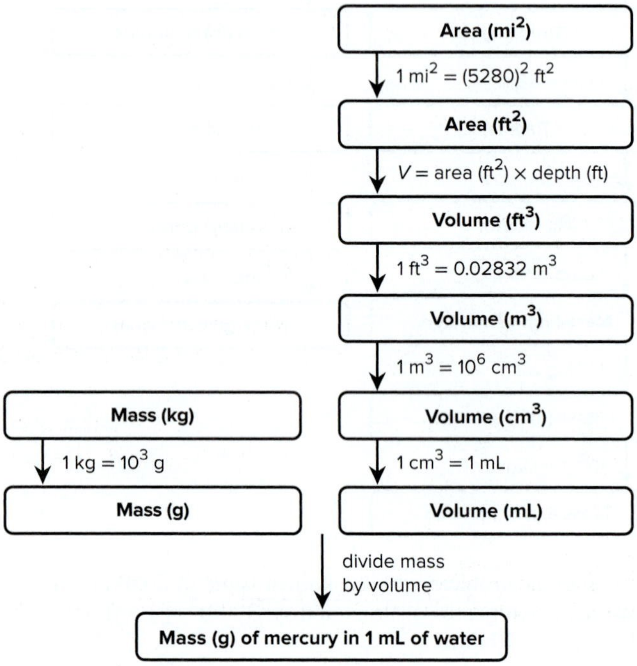

1.7A $\text{Mass (g)} = 4.9\times10^{24}\ \cancel{\text{kg}} \times \dfrac{10^{3}\ \text{g}}{1\ \cancel{\text{kg}}} = 4.9\times10^{27}\ \text{g}$

$$\text{Radius (cm)} = \frac{12{,}100\ \cancel{\text{km}}}{2} \times \frac{10^{3}\ \cancel{\text{m}}}{1\ \cancel{\text{km}}} \times \frac{10^{2}\ \text{cm}}{1\ \cancel{\text{m}}} = 6.05\times10^{8}\ \text{cm}$$

$$\text{Volume (cm}^3) = \tfrac{4}{3}\pi r^3 = \tfrac{4}{3}(3.14)(6.05\times10^{8}\ \text{cm})^3 = 9.27\times10^{26}\ \text{cm}^3$$

$$\text{Density (g/cm}^3) = \frac{4.9\times10^{27}\ \text{g}}{9.27\times10^{26}\ \text{cm}^3} = 5.3\ \text{g/cm}^3$$

See Road Map 1.7A.

1.7B $\text{Mass (kg) of sample} = 4.6\ \cancel{\text{cm}^3} \times \dfrac{7.5\ \cancel{\text{g}}}{1\ \cancel{\text{cm}^3}} \times \dfrac{1\ \text{kg}}{10^{3}\ \cancel{\text{g}}}$

$$= 0.034\ \text{kg}$$

See Road Map 1.7B.

Road Map 1.7A

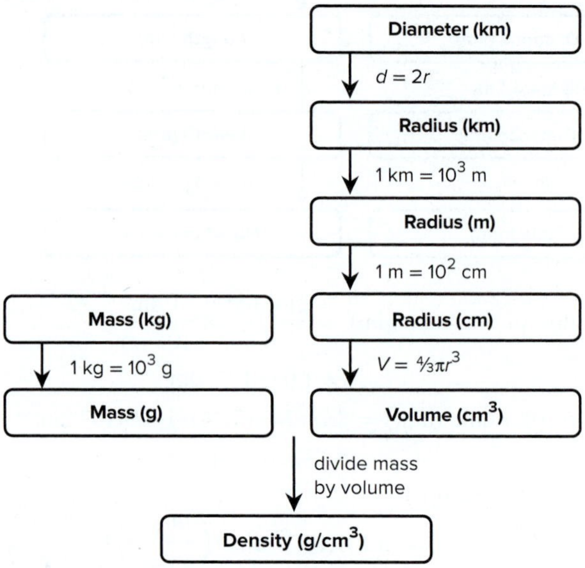

Road Map 1.7B

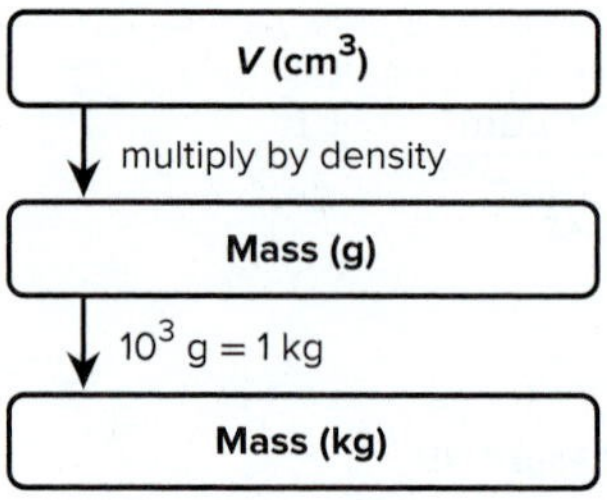

1.8A $T\ (\text{in }°\text{C}) = 234\ \text{K} - 273.15 = -39°\text{C}$
$T\ (\text{in }°\text{F}) = \tfrac{9}{5}(-39°\text{C}) + 32 = -38°\text{F}$

Answers contains two significant figures (see Section 1.5).

1.8B $T\ (\text{in }°\text{C}) = (2325°\text{F} - 32)\tfrac{5}{9} = 1274°\text{C}$
$T\ (\text{in }°\text{K}) = 1274°\text{C} + 273.15 = 1547\ \text{K}$

1.9A (a) $31.\underline{0}7\underline{0}$ mg, 5 sf
(b) $0.06\underline{0}6\underline{0}$ g, 4 sf
(c) $85\underline{0}.$°C, 3 sf

1.9B (a) $2.\underline{000}\times10^{2}$ mL, 4 sf
(b) 3.9×10^{-6} m, 2 sf
(c) $4.\underline{0}1\times10^{-4}$ L, 3 sf

1.10A $$\frac{25.65\ \text{mL} + 37.4\ \text{mL}}{73.55\ \cancel{\text{s}}\left(\dfrac{1\ \text{min}}{60\ \cancel{\text{s}}}\right)} = 51.4\ \text{mL/min}$$

1.10B $$\frac{154.64\ \text{g} - 35.26\ \text{g}}{4.20\ \text{cm} \times 5.12\ \text{cm} \times 6.752\ \text{cm}} = 0.823\ \text{g/cm}$$

PROBLEMS

Problems with **colored** numbers are answered in Appendix E and worked in detail in the Student Solutions Manual. Problem sections match those in the text and give the numbers of relevant sample problems. Most offer Concept Review Questions, Skill-Building Exercises (grouped in pairs covering the same concept), and Problems in Context. The Comprehensive Problems are based on material from any section.

Some Fundamental Definitions

(Sample Problems 1.1 and 1.2)

Concept Review Question

1.1 Scenes A–D represent atomic-scale views of different samples of substances:

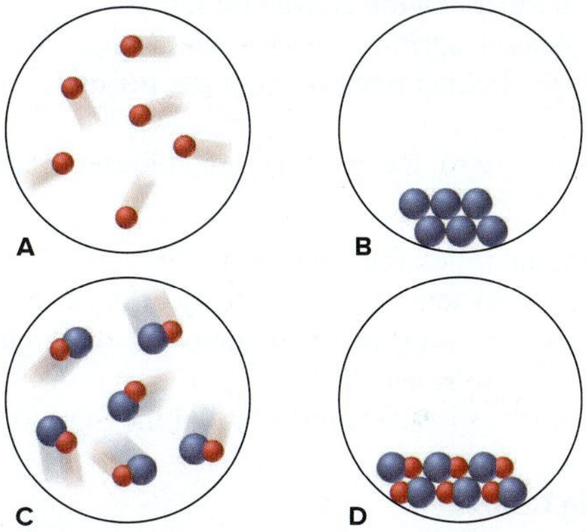

(a) Under one set of conditions, the substances in A and B mix, and the result is depicted in C. Does this represent a chemical or a physical change?
(b) Under a second set of conditions, the same substances mix, and the result is depicted in D. Does this represent a chemical or a physical change?
(c) Under a third set of conditions, the sample depicted in C changes to that in D. Does this represent a chemical or a physical change?
(d) After the change in part (c) has occurred, does the sample have different chemical properties? Physical properties?

Skill-Building Exercises (grouped in similar pairs)

1.2 Describe solids, liquids, and gases in terms of how they fill a container. Use your descriptions to identify the physical state (at room temperature) of the following: (a) helium in a toy balloon; (b) mercury in a thermometer; (c) soup in a bowl.

1.3 Use your descriptions from Problem 1.2 to identify the physical state (at room temperature) of the following: (a) the air in your room; (b) tablets in a bottle of vitamins; (c) sugar in a packet.

1.4 Define *physical property* and *chemical property*. Identify each type of property in the following statements:
(a) Yellow-green chlorine gas attacks silvery sodium metal to form white crystals of sodium chloride (table salt).
(b) A magnet separates a mixture of black iron shavings and white sand.

1.5 Define *physical change* and *chemical change*. State which type of change occurs in each of the following statements:
(a) Passing an electric current through molten magnesium chloride yields molten magnesium and gaseous chlorine.
(b) The iron in discarded automobiles slowly forms reddish brown, crumbly rust.

1.6 Which of the following is a chemical change? Explain your reasoning: (a) boiling canned soup; (b) toasting a slice of bread; (c) chopping a log; (d) burning a log.

1.7 Which of the following changes can be reversed by changing the temperature: (a) dew condensing on a leaf; (b) an egg turning hard when it is boiled; (c) ice cream melting; (d) a spoonful of batter cooking on a hot griddle?

1.8 For each pair, which has higher potential energy?
(a) The fuel in your car or the gaseous products in its exhaust
(b) Wood in a fire or the ashes after the wood burns

1.9 For each pair, which has higher kinetic energy?
(a) A sled resting at the top of a hill or a sled sliding down the hill
(b) Water above a dam or water falling over the dam

Chemical Arts and the Origins of Modern Chemistry

Concept Review Questions

1.10 The alchemical, medical, and technological traditions were precursors to chemistry. State a contribution that each made to the development of the science of chemistry.

1.11 How did the phlogiston theory explain combustion?

1.12 One important observation that supporters of the phlogiston theory had trouble explaining was that the calx of a metal weighs more than the metal itself. Why was that observation important? How did the phlogistonists respond?

1.13 Lavoisier developed a new theory of combustion that overturned the phlogiston theory. What measurements were central to his theory, and what key discovery did he make?

The Scientific Approach: Developing a Model

Concept Review Questions

1.14 How are the key elements of scientific thinking used in the following scenario? While making toast, you notice it fails to pop out of the toaster. Thinking the spring mechanism is stuck, you notice that the bread is unchanged. Assuming you forgot to plug in the toaster, you check and find it is plugged in. When you take the toaster into the dining room and plug it into a different outlet, you find the toaster works. Returning to the kitchen, you turn on the switch for the overhead light and nothing happens.

1.15 Why is a quantitative observation more useful than a nonquantitative one? Which of the following is (are) quantitative? (a) The Sun rises in the east. (b) A person weighs one-sixth as much on the Moon as on Earth. (c) Ice floats on water. (d) A hand pump cannot draw water from a well more than 34 ft deep.

1.16 Describe the essential features of a well-designed experiment.

1.17 Describe the essential features of a scientific model.

Measurement and Chemical Problem Solving

(Sample Problems 1.3 to 1.8)

Concept Review Questions

1.18 Explain the difference between mass and weight. Why is your weight on the Moon one-sixth that on Earth?

1.19 When you convert feet to inches, how do you decide which part of the conversion factor should be in the numerator and which in the denominator?

1.20 For each of the following cases, state whether the density of the object increases, decreases, or remains the same:
(a) A sample of chlorine gas is compressed.
(b) A lead weight is carried up a high mountain.
(c) A sample of water is frozen.
(d) An iron bar is cooled.
(e) A diamond is submerged in water.

1.21 Explain the difference between heat and temperature. Does 1 L of water at 65°F have more, less, or the same quantity of energy as 1 L of water at 65°C?

1.22 A one-step conversion is sufficient to convert a temperature in the Celsius scale to the Kelvin scale, but not to the Fahrenheit scale. Explain.

1.23 Describe the difference between intensive and extensive properties. Which of the following properties are intensive: (a) mass; (b) density; (c) volume; (d) melting point?

Skill-Building Exercises (grouped in similar pairs)

1.24 Write the conversion factor(s) for
(a) in^2 to m^2 (b) km^2 to cm^2
(c) mi/h to m/s (d) lb/ft^3 to g/cm^3

1.25 Write the conversion factor(s) for
(a) cm/min to in/s (b) m^3 to in^3
(c) m/s^2 to km/h^2 (d) gal/h to L/min

1.26 The average radius of a molecule of lysozyme, an enzyme in tears, is 1430 pm. What is its radius in nanometers (nm)?

1.27 The radius of a barium atom is 2.22×10^{-10} m. What is its radius in angstroms (Å)?

1.28 What is the length in inches (in) of a 100.-m soccer field?

1.29 The center on your school's basketball team is 6 ft 10 in tall. How tall is the player in millimeters (mm)?

1.30 A small hole in the wing of a space shuttle requires a 20.7-cm^2 patch. (a) What is the patch's area in square kilometers (km^2)? (b) If the patching material costs NASA $3.25/$in^2$, what is the cost of the patch?

1.31 The area of a telescope lens is 7903 mm^2. (a) What is the area in square feet (ft^2)? (b) If it takes a technician 45 s to polish 135 mm^2, how long does it take her to polish the entire lens?

1.32 Express your body weight in kilograms (kg).

1.33 There are 2.60×10^{15} short tons of oxygen in the atmosphere (1 short ton = 2000 lb). How many metric tons of oxygen are present in the atmosphere (1 metric ton = 1000 kg)?

1.34 The average density of Earth is 5.52 g/cm^3. What is its density in (a) kg/m^3; (b) lb/ft^3?

1.35 The speed of light in a vacuum is 2.998×10^8 m/s. What is its speed in (a) km/h; (b) mi/min?

1.36 The volume of a certain bacterial cell is 2.56 μm^3. (a) What is its volume in cubic millimeters (mm^3)? (b) What is the volume of 10^5 cells in liters (L)?

1.37 (a) How many cubic meters of milk are in 1 qt (946.4 mL)? (b) How many liters of milk are in 835 gal (1 gal = 4 qt)?

1.38 An empty vial weighs 55.32 g. (a) If the vial weighs 185.56 g when filled with liquid mercury (d = 13.53 g/cm^3), what volume of mercury is in the vial? (b) How much would the vial weigh if it were filled with the same volume of water (d = 0.997 g/cm^3 at 25°C)?

1.39 An empty Erlenmeyer flask weighs 241.3 g. When filled with water (d = 1.00 g/cm^3), the flask and its contents weigh 489.1 g. (a) What is the volume of water in the flask? (b) How much does the flask weigh when filled with the same volume of chloroform (d = 1.48 g/cm^3)?

1.40 A small cube of aluminum measures 15.6 mm on a side and weighs 10.25 g. What is the density of aluminum in g/cm^3?

1.41 A steel ball-bearing with a circumference of 32.5 mm weighs 4.20 g. What is the density of the steel in g/cm^3 (V of a sphere = $\frac{4}{3}\pi r^3$; circumference of a circle = $2\pi r$)?

1.42 Perform the following conversions:
(a) 68°F (a pleasant spring day) to °C and K
(b) −164°C (the boiling point of methane, the main component of natural gas) to K and °F
(c) 0 K (absolute zero, theoretically the coldest possible temperature) to °C and °F

1.43 Perform the following conversions:
(a) 106°F (the body temperature of many birds) to K and °C
(b) 3410°C (the melting point of tungsten, the highest for any metallic element) to K and °F
(c) 6.1×10^3 K (the surface temperature of the Sun) to °F and °C

Problems in Context

1.44 A 25.0-g sample of each of three unknown metals is added to 25.0 mL of water in graduated cylinders A, B, and C, and the final volumes are depicted in the circles below. Given their densities, identify the metal in each cylinder: zinc (7.14 g/mL), iron (7.87 g/mL), or nickel (8.91 g/mL).

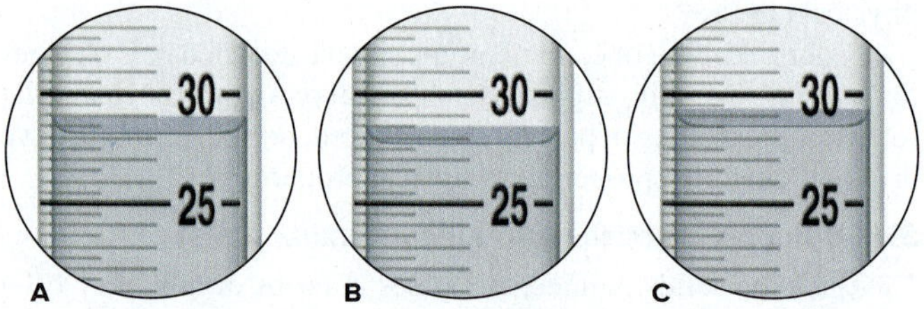

1.45 The distance between two adjacent peaks on a wave is called the *wavelength.*
(a) The wavelength of a beam of ultraviolet light is 247 nanometers (nm). What is its wavelength in meters?
(b) The wavelength of a beam of red light is 6760 pm. What is its wavelength in angstroms (Å)?

1.46 Each of the beakers depicted below contains two liquids that do not dissolve in each other. Three of the liquids are designated A, B, and C, and water is designated W.

(a) Which of the liquids is (are) more dense than water and less dense than water?
(b) If the densities of W, C, and A are 1.0 g/mL, 0.88 g/mL, and 1.4 g/mL, respectively, which of the following densities is possible for liquid B: 0.79 g/mL, 0.86 g/mL, 0.94 g/mL, or 1.2 g/mL?

1.47 A cylindrical tube 9.5 cm high and 0.85 cm in diameter is used to collect blood samples. How many cubic decimeters (dm^3) of blood can it hold (V of a cylinder $= \pi r^2 h$)?

1.48 Copper can be drawn into thin wires. How many meters of 34-gauge wire (diameter $= 6.304\times10^{-3}$ in) can be produced from the copper in 5.01 lb of covellite, an ore of copper that is 66% copper by mass? (*Hint:* Treat the wire as a cylinder: V of cylinder $= \pi r^2 h$; d of copper $= 8.95\ g/cm^3$.)

Uncertainty in Measurement: Significant Figures

(Sample Problems 1.9 and 1.10)

Concept Review Questions

1.49 What is an exact number? How are exact numbers treated differently from other numbers in a calculation?

1.50 Which procedure(s) decrease(s) the random error of a measurement: (1) taking the average of more measurements; (2) calibrating the instrument; (3) taking fewer measurements? Explain.

1.51 A newspaper reported that the attendance at Slippery Rock's home football game was 16,532. (a) How many significant figures does this number contain? (b) Was the actual number of people counted? (c) After Slippery Rock's next home game, the newspaper reported an attendance of 15,000. If you assume that this number contains two significant figures, how many people could actually have been at the game?

Skill-Building Exercises (grouped in similar pairs)

1.52 Underline the significant zeros in the following numbers: (a) 0.41; (b) 0.041; (c) 0.0410; (d) 4.0100×10^4.

1.53 Underline the significant zeros in the following numbers: (a) 5.08; (b) 508; (c) 5.080×10^3; (d) 0.05080.

1.54 Round off each number to the indicated number of significant figures (sf): (a) 0.0003554 (to 2 sf); (b) 35.8348 (to 4 sf); (c) 22.4555 (to 3 sf).

1.55 Round off each number to the indicated number of significant figures (sf): (a) 231.554 (to 4 sf); (b) 0.00845 (to 2 sf); (c) 144,000 (to 2 sf).

1.56 Round off each number in the following calculation to one fewer significant figure, and find the answer:

$$\frac{19 \times 155 \times 8.3}{3.2 \times 2.9 \times 4.7}$$

1.57 Round off each number in the following calculation to one fewer significant figure, and find the answer:

$$\frac{10.8 \times 6.18 \times 2.381}{24.3 \times 1.8 \times 19.5}$$

1.58 Carry out the following calculations, making sure that your answer has the correct number of significant figures:

(a) $\dfrac{2.795\ m \times 3.10\ m}{6.48\ m}$

(b) $V = \frac{4}{3}\pi r^3$, where $r = 17.282$ mm

(c) 1.110 cm + 17.3 cm + 108.2 cm + 316 cm

1.59 Carry out the following calculations, making sure that your answer has the correct number of significant figures:

(a) $\dfrac{2.420\ g + 15.6\ g}{4.8\ g}$ (b) $\dfrac{7.87\ mL}{16.1\ mL - 8.44\ mL}$

(c) $V = \pi r^2 h$, where $r = 6.23$ cm and $h = 4.630$ cm

1.60 Write the following numbers in scientific notation: (a) 131,000.0; (b) 0.00047; (c) 210,006; (d) 2160.5.

1.61 Write the following numbers in scientific notation: (a) 282.0; (b) 0.0380; (c) 4270.8; (d) 58,200.9.

1.62 Write the following numbers in standard notation. Use a terminal decimal point when needed.
(a) 5.55×10^3; (b) 1.0070×10^4; (c) 8.85×10^{-7}; (d) 3.004×10^{-3}.

1.63 Write the following numbers in standard notation. Use a terminal decimal point when needed.
(a) 6.500×10^3; (b) 3.46×10^{-5}; (c) 7.5×10^2; (d) 1.8856×10^2.

1.64 Convert the following into correct scientific notation: (a) 802.5×10^2; (b) 1009.8×10^{-6}; (c) 0.077×10^{-9}.

1.65 Convert the following into correct scientific notation: (a) 14.3×10^1; (b) 851×10^{-2}; (c) 7500×10^{-3}.

1.66 Carry out each calculation, paying special attention to significant figures, rounding, and units (J = joule, the SI unit of energy; mol = mole, the SI unit for amount of substance):

(a) $\dfrac{(6.626\times10^{-34}\ J\cdot s)(2.9979\times10^8\ m/s)}{489\times10^{-9}\ m}$

(b) $\dfrac{(6.022\times10^{23}\ molecules/mol)(1.23\times10^2\ g)}{46.07\ g/mol}$

(c) $(6.022\times10^{23}\ atoms/mol)(1.28\times10^{-18}\ J/atom)\left(\dfrac{1}{2^2} - \dfrac{1}{3^2}\right)$,

where the numbers 2 and 3 in the last term are exact

1.67 Carry out each calculation, paying special attention to significant figures, rounding, and units:

(a) $\dfrac{4.32\times10^7\ g}{\frac{4}{3}(3.1416)(1.95\times10^2\ cm)^3}$ (The term $\frac{4}{3}$ is exact.)

(b) $\dfrac{(1.84\times10^2\ g)(44.7\ m/s)^2}{2}$ (The term 2 is exact.)

(c) $\dfrac{(1.07\times10^{-4}\ mol/L)^2(3.8\times10^{-3}\ mol/L)}{(8.35\times10^{-5}\ mol/L)(1.48\times10^{-2}\ mol/L)^3}$

1.68 Which statements include exact numbers?
(a) Angel Falls is 3212 ft high.
(b) There are 8 known planets in the Solar System.
(c) There are 453.59 g in 1 lb.
(d) There are 1000 mm in 1 m.

1.69 Which of the following include exact numbers?
(a) The speed of light in a vacuum is a physical constant; to six significant figures, it is 2.99792×10^8 m/s.
(b) The density of mercury at 25°C is 13.53 g/mL.
(c) There are 3600 s in 1 h.
(a) In 2016, the United States had 50 states.

Problems in Context

1.70 How long is the metal strip shown below? Be sure to answer with the correct number of significant figures.

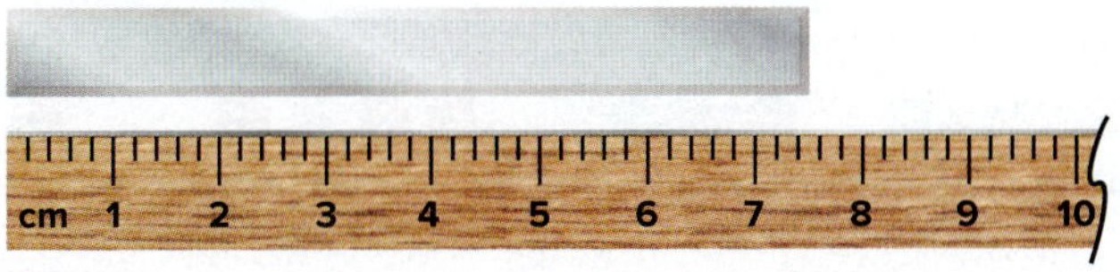

1.71 These organic solvents are used to clean compact discs:

Solvent	Density (g/mL) at 20°C
Chloroform	1.492
Diethyl ether	0.714
Ethanol	0.789
Isopropanol	0.785
Toluene	0.867

(a) If a 15.00-mL sample of CD cleaner weighs 11.775 g at 20°C, which solvent does the sample most likely contain?
(b) The chemist analyzing the cleaner calibrates her equipment and finds that the pipet is accurate to ±0.02 mL, and the balance is accurate to ±0.003 g. Is this equipment precise enough to distinguish between ethanol and isopropanol?

1.72 A laboratory instructor gives a sample of amino-acid powder to each of four students, I, II, III, and IV, and they weigh the samples. The true value is 8.72 g. Their results for three trials are
I: 8.72 g, 8.74 g, 8.70 g   II: 8.56 g, 8.77 g, 8.83 g
III: 8.50 g, 8.48 g, 8.51 g   IV: 8.41 g, 8.72 g, 8.55 g
(a) Calculate the average mass from each set of data, and tell which set is the most accurate.
(b) Precision is a measure of the average of the deviations of each piece of data from the average value. Which set of data is the most precise? Is this set also the most accurate?
(c) Which set of data is both the most accurate and the most precise?
(d) Which set of data is both the least accurate and the least precise?

1.73 The following dartboards illustrate the types of errors often seen in measurements. The bull's-eye represents the actual value, and the darts represent the data.

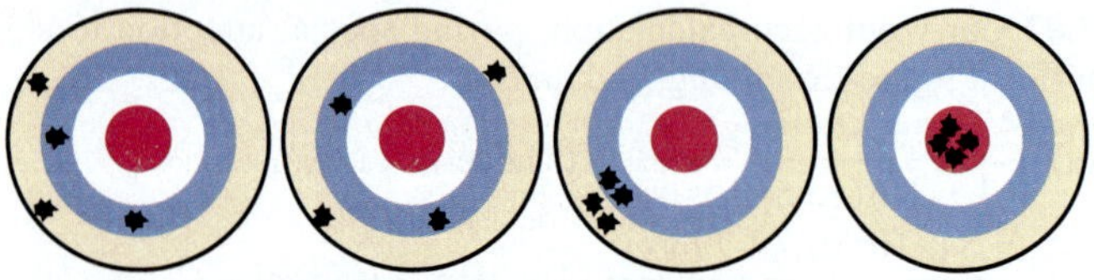

Exp. I   Exp. II   Exp. III   Exp. IV

(a) Which experiments yield the same average result?
(b) Which experiment(s) display(s) high precision?
(c) Which experiment(s) display(s) high accuracy?
(d) Which experiment(s) show(s) a systematic error?

Comprehensive Problems

1.74 Two blank potential energy diagrams appear below. Beneath each diagram are objects to place in the diagram. Draw the objects on the dashed lines to indicate higher or lower potential energy and label each case as more or less stable:

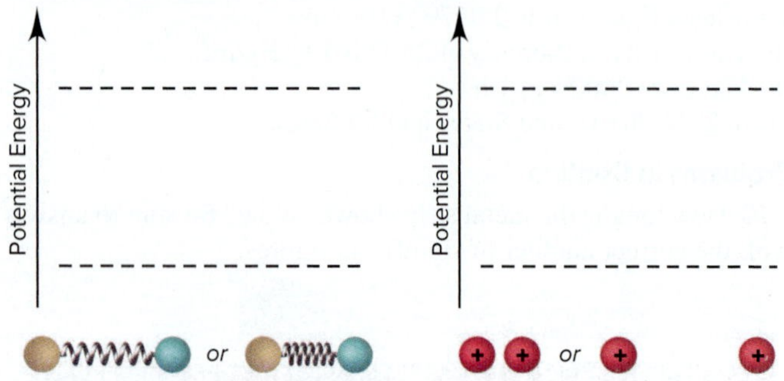

(a) Two balls attached to a relaxed *or* a compressed spring   **(b)** Two positive charges near *or* apart from each other

1.75 The scenes below illustrate two different mixtures. When mixture A at 273 K is heated to 473 K, mixture B results.

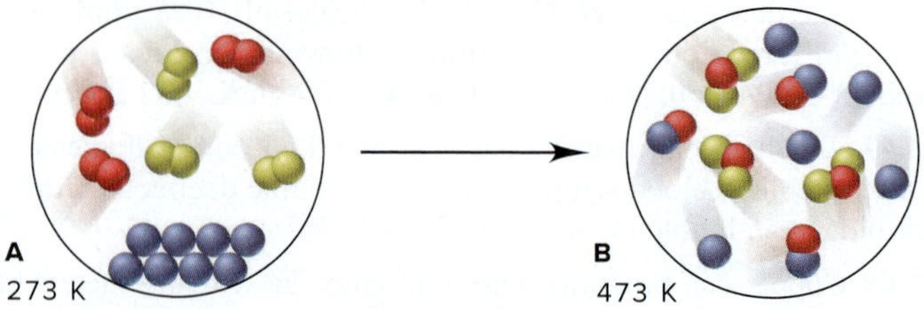

(a) How many different chemical changes occur?
(b) How many different physical changes occur?

1.76 Bromine is used to prepare the pesticide methyl bromide and flame retardants for plastic electronic housings. It is recovered from seawater, underground brines, and the Dead Sea. The average concentrations of bromine in seawater (d = 1.024 g/mL) and the Dead Sea (d = 1.22 g/mL) are 0.065 g/L and 0.50 g/L, respectively. What is the mass ratio of bromine in the Dead Sea to that in seawater?

1.77 An Olympic-size pool is 50.0 m long and 25.0 m wide. (a) How many gallons of water (d = 1.0 g/mL) are needed to fill the pool to an average depth of 4.8 ft? (b) What is the mass (in kg) of water in the pool?

1.78 At room temperature (20°C) and pressure, the density of air is 1.189 g/L. An object will float in air if its density is less than that of air. In a buoyancy experiment with a new plastic, a chemist creates a rigid, thin-walled ball that weighs 0.12 g and has a volume of 560 cm^3.
(a) Will the ball float if it is evacuated?
(b) Will it float if filled with carbon dioxide (d = 1.830 g/L)?
(c) Will it float if filled with hydrogen (d = 0.0899 g/L)?
(d) Will it float if filled with oxygen (d = 1.330 g/L)?
(e) Will it float if filled with nitrogen (d = 1.165 g/L)?
(f) For any case in which the ball will float, how much weight must be added to make it sink?

1.79 Asbestos is a fibrous silicate mineral with remarkably high tensile strength. But it is no longer used because airborne asbestos particles can cause lung cancer. Grunerite, a type of asbestos, has a tensile strength of 3.5×10^2 kg/mm^2 (thus, a strand of grunerite with a 1-mm^2 cross-sectional area can hold up to 3.5×10^2 kg). The tensile strengths of aluminum and Steel No. 5137 are 2.5×10^4 lb/in^2 and 5.0×10^4 lb/in^2, respectively. Calculate the cross-sectional areas (in mm^2) of wires of aluminum and of Steel No. 5137 that have the same tensile strength as a fiber of grunerite with a cross-sectional area of 1.0 μm^2.

1.80 Earth's oceans have an average depth of 3800 m, a total surface area of 3.63×10^8 km^2, and an average concentration of dissolved gold of 5.8×10^{-9} g/L. (a) How many grams of gold are in the oceans? (b) How many cubic meters of gold are in the oceans? (c) Assuming the price of gold is \$1595/troy oz, what is the value of gold in the oceans (1 troy oz = 31.1 g; d of gold = 19.3 g/cm^3)?

1.81 Brass is an alloy of copper and zinc. Varying the mass percentages of the two metals produces brasses with different properties. A brass called *yellow zinc* has high ductility and strength and is 34–37% zinc by mass. (a) Find the mass range (in g) of copper in 185 g of yellow zinc. (b) What is the mass range (in g) of zinc in a sample of yellow zinc that contains 46.5 g of copper?

1.82 Liquid nitrogen is obtained from liquefied air and is used industrially to prepare frozen foods. It boils at 77.36 K. (a) What is this temperature in °C? (b) What is this temperature in °F? (c) At the boiling point, the density of the liquid is 809 g/L and that of the gas is 4.566 g/L. How many liters of liquid nitrogen are produced when 895.0 L of nitrogen gas is liquefied at 77.36 K?

1.83 A jogger runs at an average speed of 5.9 mi/h. (a) How fast is she running in m/s? (b) How many kilometers does she run in 98 min? (c) If she starts a run at 11:15 am, what time is it after she covers 4.75×10^4 ft?

1.84 Scenes A and B depict changes in matter at the atomic scale:

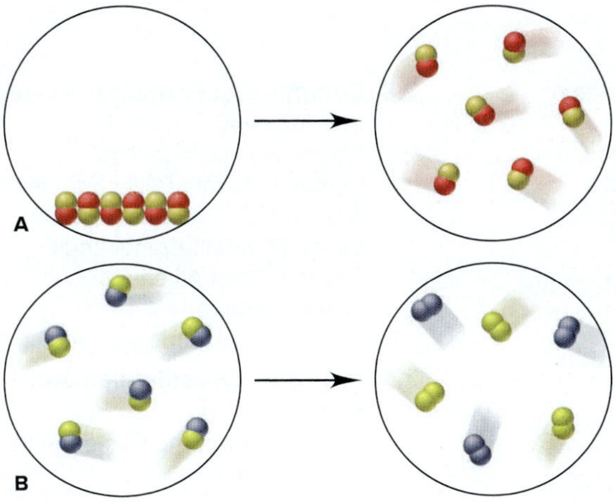

(a) Which show(s) a physical change?
(b) Which show(s) a chemical change?
(c) Which result(s) in different physical properties?
(d) Which result(s) in different chemical properties?
(e) Which result(s) in a change in state?

1.85 If a temperature scale were based on the freezing point (5.5°C) and boiling point (80.1°C) of benzene and the temperature difference between these points was divided into 50 units (called °X), what would be the freezing and boiling points of water in °X? (See Figure 1.11.)

1.86 Earth's surface area is 5.10×10^8 km^2; its crust has a mean thickness of 35 km and a mean density of 2.8 g/cm^3. The two most abundant elements in the crust are oxygen (4.55×10^5 g/t, where t stands for "metric ton";1 t = 1000 kg) and silicon (2.72×10^5 g/t), and the two rarest nonradioactive elements are ruthenium and rhodium, each with an abundance of 1×10^{-4} g/t. What is the total mass of each of these elements in Earth's crust?

2 The Components of Matter

2.1 Elements, Compounds, and Mixtures: An Atomic Overview

2.2 The Observations That Led to an Atomic View of Matter
Mass Conservation
Definite Composition
Multiple Proportions

2.3 Dalton's Atomic Theory
Postulates of the Theory
Explanation of the Mass Laws

2.4 The Observations That Led to the Nuclear Atom Model
Discovery of the Electron
Discovery of the Nucleus

2.5 The Atomic Theory Today
Structure of the Atom
Atomic Number, Mass Number, and Atomic Symbol
Isotopes
Atomic Masses

2.6 Elements: A First Look at the Periodic Table

2.7 Compounds: Introduction to Bonding
Formation of Ionic Compounds
Formation of Covalent Substances

2.8 Compounds: Formulas, Names, and Masses
Binary Ionic Compounds
Compounds with Polyatomic Ions
Acid Names from Anion Names
Binary Covalent Compounds
Straight-Chain Alkanes
Molecular Masses
Formulas and Models

2.9 Mixtures: Classification and Separation
An Overview of the Components of Matter

Source: © M. Shcherbyna/Shutterstock.com; © Raul Gonzalez/Science Source

Concepts and Skills to Review Before You Study This Chapter

- physical and chemical change (Section 1.1)
- states of matter (Section 1.1)
- attraction and repulsion between charged particles (Section 1.1)
- meaning of a scientific model (Section 1.3)
- SI units and conversion factors (Section 1.4)
- significant figures in calculations (Section 1.5)

Look closely at almost any sample of matter—a rock, a piece of wood, a butterfly wing—and you'll see that it's made of smaller parts. With a microscope, you'll see still smaller parts. And, if you could zoom in a billion times closer, you'd find, on the atomic scale, the ultimate particles that make up all things.

Modern scientists are certainly not the first to wonder what things are made of. The philosophers of ancient Greece did too, and most believed that everything was made of one or, at most, a few elemental substances (elements), whose hotness, wetness, hardness, and other properties gave rise to the properties of everything else. Democritus (c. 460–370 BC), the father of atomism, took a different approach, and his reasoning went something like this: if you cut a piece of, say, aluminum foil smaller and smaller, you must eventually reach a particle of aluminum so small that it can no longer be cut. Therefore, matter is ultimately composed of indivisible particles with nothing but empty space between them. He called the particles *atoms* (Greek *atomos*, "uncuttable") and proclaimed: "According to convention, there is a sweet and a bitter, a hot and a cold, and ... there is order. In truth, there are atoms and a void." But, Aristotle, one of the greatest and most influential philosophers of Western culture, said it was impossible for "nothing" to exist, and the concept of atoms was suppressed for 2000 years.

Finally, in the 17th century, the English scientist Robert Boyle argued that, by definition, an element is composed of "simple Bodies, not made of any other Bodies, of which all mixed Bodies are compounded, and into which they are ultimately resolved," a description remarkably close to our idea of an element, with atoms being the "simple Bodies." The next two centuries saw rapid progress in chemistry and the development of a "billiard-ball" image of the atom. Then, an early 20th-century burst of creativity led to our current model of an atom with a complex internal structure.

IN THIS CHAPTER . . . *We examine the properties and composition of matter on the macroscopic and atomic scales.*

- We relate the three types of observable matter—elements, compounds, and mixtures—to the atoms, ions, and molecules they consist of.
- We see how the defining properties of the types of matter relate to 18th-century laws concerning the masses of substances that react with each other.
- We examine the 19th-century atomic model proposed to explain these laws.
- We describe some 20th- and 21st-century experiments that have led to our current understanding of atomic structure and atomic mass.
- We focus on the general organization of the elements in the periodic table and introduce the two major ways that elements combine.
- We derive names and formulas of compounds and calculate their masses on the atomic scale.
- We examine some of the ways chemists depict molecules.
- We classify mixtures and see how to separate them in the lab.
- We present a final overview of the components of matter.

2.1 ELEMENTS, COMPOUNDS, AND MIXTURES: AN ATOMIC OVERVIEW

Based on its composition, matter can be classified into three types—elements, compounds, and mixtures. Elements and compounds are called **substances,** matter with a *fixed* composition; mixtures are not substances because they have a variable composition.

1. *Elements.* An **element** is *the simplest type of matter with unique physical and chemical properties. It consists of only one kind of atom* and, therefore, cannot be broken down into a simpler type of matter by any physical or chemical methods. Each element has a name, such as silicon, oxygen, or copper. A sample of silicon contains only silicon atoms. The *macroscopic* properties of a piece of silicon, such as color, density, and combustibility, are different from those of a piece of copper because the *submicroscopic* properties of silicon atoms are different from those of copper atoms; that is, *each element is unique because the properties of its atoms are unique.*

In nature, most elements exist as populations of atoms, either separated or in contact with each other, depending on the physical state. Figure 2.1A shows atoms of an element in its gaseous state. Several elements occur in molecular form: a **molecule** is an independent structure of two or more atoms bound together (Figure 2.1B). Oxygen, for example, occurs in air as *diatomic* (two-atom) molecules.

2. *Compounds.* A **compound** *consists of two or more different elements that are bonded chemically* (Figure 2.1C). That is, the elements in a compound are not just mixed together: their atoms have joined in a chemical reaction. Many compounds, such as ammonia, water, and carbon dioxide, consist of molecules. But many others, like sodium chloride (which we'll discuss shortly) and silicon dioxide, do not.

All compounds have three defining features:

- *The elements are present in fixed parts by mass* (have a fixed mass ratio). This is so because *each unit of the compound consists of a fixed number of atoms of each element.* For example, consider a sample of ammonia. It is 14 parts nitrogen by mass and 3 parts hydrogen by mass *because* 1 nitrogen atom has 14 times the mass of 1 hydrogen atom, and each ammonia molecule consists of 1 nitrogen atom and 3 hydrogen atoms:

  Ammonia gas is 14 parts N by mass and 3 parts H by mass.
  1 N atom has 14 times the mass of 1 H atom.
  Each ammonia molecule consists of 1 N atom and 3 H atoms.

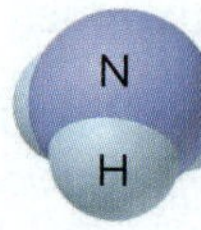

- *A compound's properties are different from the properties of its elements.* Table 2.1 shows a striking example: soft, silvery sodium metal and yellow-green, poisonous chlorine gas are very different from the compound they form—white, crystalline sodium chloride, or common table salt!
- *A compound, unlike an element, can be broken down into simpler substances—its component elements—by a chemical change.* For example, an electric current breaks down molten sodium chloride into metallic sodium and chlorine gas.

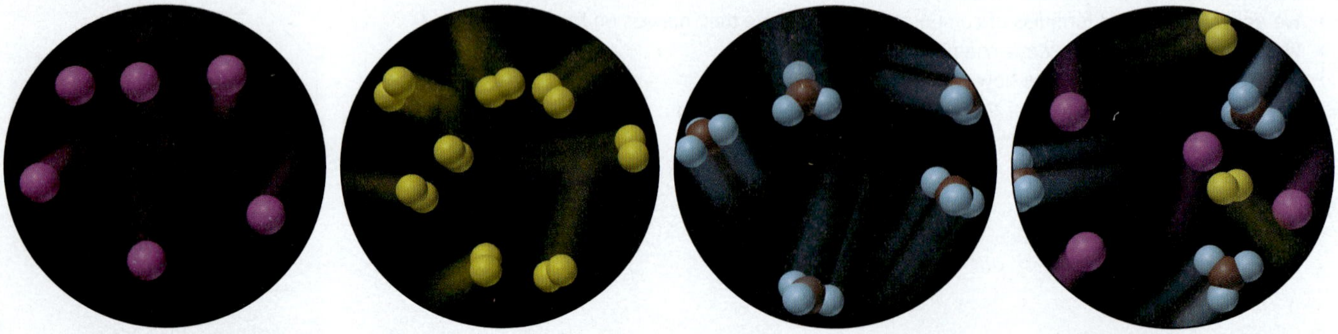

A Atoms of an element **B** Molecules of an element **C** Molecules of a compound **D** Mixture of two elements and a compound

Figure 2.1 Elements, compounds, and mixtures on the atomic scale. The samples depicted here are gases, but the three types of matter also occur as liquids and solids.

Table 2.1 Some of the Very Different Properties of Sodium, Chlorine, and Sodium Chloride

Property	Sodium	+	Chlorine	→	Sodium Chloride	
Melting point	97.8°C	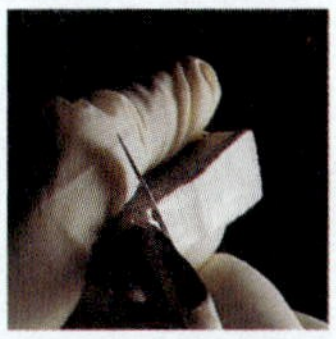	−101°C		801°C	
Boiling point	881.4°C		−34°C		1413°C	
Color	Silvery		Yellow-green		Colorless (white)	
Density	0.97 g/cm^3		0.0032 g/cm^3		2.16 g/cm^3	
Behavior in water	Reacts		Dissolves slightly		Dissolves freely	

Source: (Sodium, Chlorine, Sodium chloride) © McGraw-Hill Education/Stephen Frisch, photographer

3. *Mixtures.* A **mixture** *consists of two or more substances (elements and/or compounds) that are physically intermingled,* ***not*** *chemically combined.* Mixtures differ from compounds as follows:

- *The components of a mixture* ***can*** *vary in their parts by mass.* A mixture of the compounds sodium chloride and water, for example, can have many different parts by mass of salt to water.
- *A mixture retains many of the properties of its components.* Saltwater, for instance, is colorless like water and tastes salty like sodium chloride. The component properties are maintained because on the atomic scale, a mixture consists of the individual units of its component elements and/or compounds (Figure 2.1D).
- *Mixtures, unlike compounds, can be separated into their components by physical changes; chemical changes are not needed.* For example, the water in saltwater can be boiled off, a physical process that leaves behind solid sodium chloride.

The following sample problem will help you to differentiate these types of matter.

SAMPLE PROBLEM 2.1 Distinguishing Elements, Compounds, and Mixtures at the Atomic Scale

Problem The scenes below represent atomic-scale views of three samples of matter:

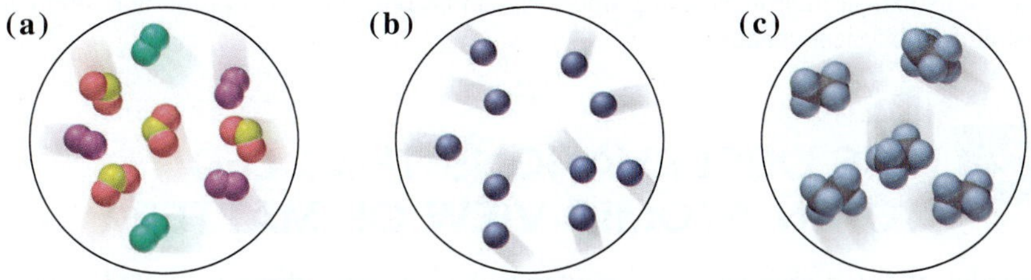

Describe each sample as an element, compound, or mixture.

Plan We have to determine the type of matter by examining the component particles. If a sample contains only one type of particle, it is either an element or a compound; if it contains more than one type, it is a mixture. Particles of an element have only one kind of atom (one color of sphere in an atomic-scale view), and particles of a compound have two or more kinds of atoms.

Solution **(a)** Mixture: there are three different types of particles. Two types contain only one kind of atom, either green or purple, so they are elements, and the third type contains two red atoms for every one yellow, so it is a compound.

(b) Element: the sample consists of only blue atoms.

(c) Compound: the sample consists of molecules that each have two black and six blue atoms.

FOLLOW-UP PROBLEMS

*Brief Solutions for **all** Follow-up Problems appear at the end of the chapter.*

2.1A Does each of the following scenes best represent an element, a compound, or a mixture?

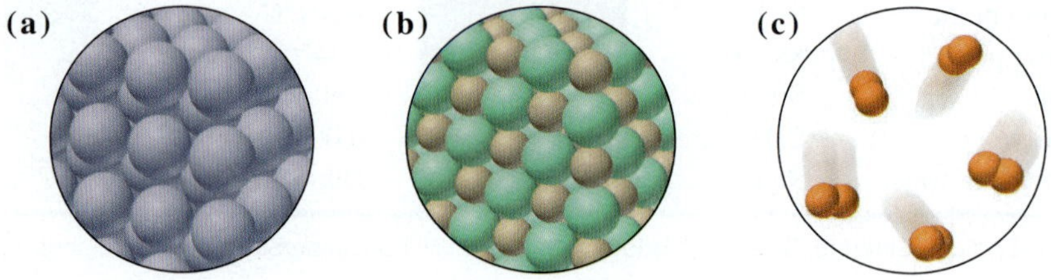

2.1B Describe the following representation of a reaction in terms of elements, compounds, and mixtures.

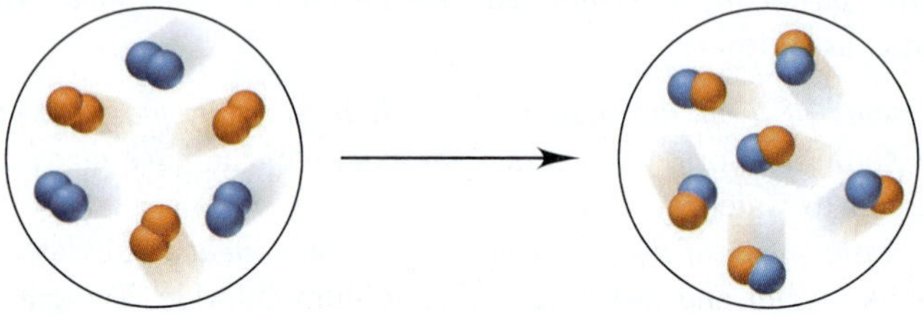

SOME SIMILAR PROBLEMS 2.3, 2.4, and 2.9

› Summary of Section 2.1

- All matter exists as either elements, compounds, or mixtures.
- Every element or compound is a substance, matter with a fixed composition.
- An element consists of only one type of atom and occurs as a collection of individual atoms or molecules; a molecule consists of two or more atoms chemically bonded together.
- A compound contains two or more elements chemically combined and exhibits different properties from its component elements. The elements occur in fixed parts by mass because each unit of the compound has a fixed number of each type of atom. Only a chemical change can break down a compound into its elements.
- A mixture consists of two or more substances mixed together, *not* chemically combined. The components exhibit their individual properties, can be present in any proportion, and can be separated by physical changes.

2.2 THE OBSERVATIONS THAT LED TO AN ATOMIC VIEW OF MATTER

Any model of the composition of matter had to explain three so-called mass laws: the *law of mass conservation,* the *law of definite (or constant) composition,* and the *law of multiple proportions.*

Mass Conservation

The first mass law, stated by Lavoisier on the basis of his combustion experiments, was the most fundamental chemical observation of the 18th century:

- **Law of mass conservation:** *the total mass of substances does not change during a chemical reaction.*

The *number* of substances may change and, by definition, their properties must, but the *total amount* of matter remains constant. Figure 2.2 illustrates mass conservation because the lead nitrate and sodium chromate solutions *(left)* have the same mass as the solid lead chromate in sodium nitrate solution *(right)* that forms after their reaction.

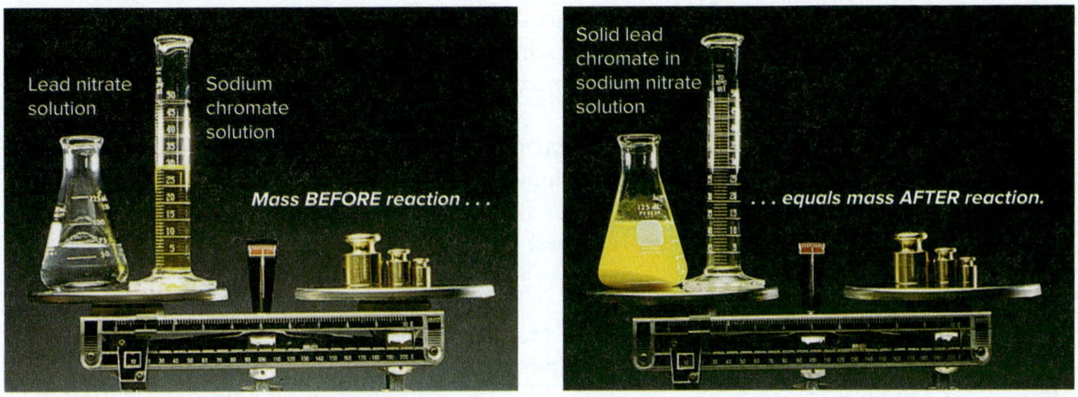

Figure 2.2 The law of mass conservation.
Source: © McGraw-Hill Education/Stephen Frisch, photographer

Even in a complex biochemical change, such as the metabolism of the sugar glucose, which involves many reactions, mass is conserved. For example, in the reaction of, say, 180 g of glucose, with oxygen, we have

$$\begin{aligned} 180\text{ g glucose} + 192\text{ g oxygen gas} &\longrightarrow 264\text{ g carbon dioxide} + 108\text{ g water} \\ 372\text{ g material before} &\longrightarrow 372\text{ g material after} \end{aligned}$$

Mass conservation means that, based on all chemical experience, *matter cannot be created or destroyed.*

To be precise about it, however, we now know, based on the work of Albert Einstein (1879–1955), that the mass before and after a reaction is not *exactly* the same. Some mass is converted to energy, or vice versa, but the difference is too small to measure, even with the best balance. For example, when 100 g of carbon burns, the carbon dioxide formed weighs 0.000000036 g (3.6×10^{-8} g) less than the sum of the carbon and oxygen that reacted. Because the energy changes of *chemical* reactions are so small, for all practical purposes, mass *is* conserved. Later in the text, you'll see that energy changes in *nuclear* reactions are so large that mass changes are easy to measure.

Definite Composition

The sodium chloride in your salt shaker is the same substance whether it comes from a salt mine in Pakistan, a salt flat in Argentina, or any other source. This fact is expressed as the second mass law:

- **Law of definite (or constant) composition:** *no matter what its source, a particular compound is composed of the same elements in the same parts (fractions) by mass.*

The **fraction by mass (mass fraction)** is the part of the compound's mass that each element contributes. It is obtained by dividing the mass of each element in the compound by the mass of the compound. The **percent by mass (mass percent, mass %)** is the fraction by mass expressed as a percentage (multiplied by 100):

$$\text{Mass fraction} = \frac{\text{mass of element X in compound A}}{\text{mass of compound A}}$$

$$\text{Mass percent} = \text{mass fraction} \times 100$$

For an everyday example, consider a box that contains three types of marbles: yellow ones weigh 1.0 g each, purple 2.0 g each, and red 3.0 g each. Each type makes up a fraction of the total mass of marbles, 16.0 g. The *mass fraction* of yellow marbles is their number times their mass divided by the total mass:

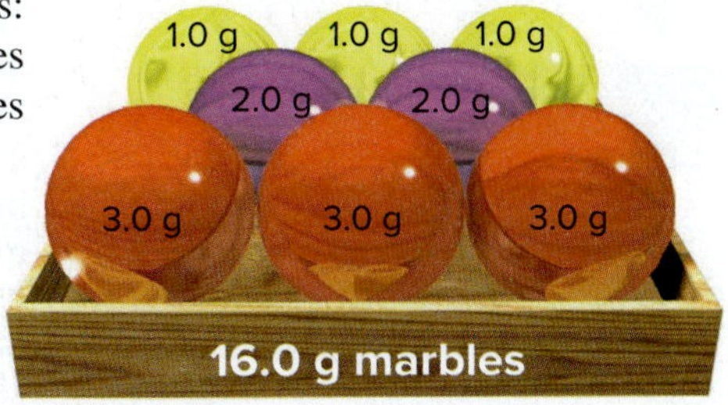

$$\text{Mass fraction of yellow marbles} = \frac{\text{no. of yellow marbles} \times \text{mass of yellow marbles}}{\text{total mass of marbles}}$$

$$= \frac{3\ \cancel{\text{marbles}} \times 1.0\text{ g}/\cancel{\text{marble}}}{16.0\ \cancel{\text{g}}} = 0.19$$

Figure 2.3 The law of definite composition. Calcium carbonate occurs in many forms (such as marble, *top*, and coral, *bottom*).

Source: (top): © Punchstock RF; (bottom): © Alexander Cherednichenko/Shutterstock.com

The *mass percent* (parts per 100 parts) of yellow marbles is 0.19 × 100 = 19% by mass. Similarly, the purple marbles have a mass fraction of 0.25 and represent 25% by mass of the total, and the red marbles have a mass fraction of 0.56 and represent 56% by mass of the total.

In the same way, each element in a compound has a *fixed* mass fraction (and mass percent). For example, calcium carbonate, the major compound in seashells, marble, and coral, is composed of three elements—calcium, carbon, and oxygen. The following results are obtained from a mass analysis of 20.0 g of calcium carbonate:

Analysis by Mass (grams/20.0 g)	Mass Fraction (parts/1.00 part)	Percent by Mass (parts/100 parts)
8.0 g calcium	0.40 calcium	40% calcium
2.4 g carbon	0.12 carbon	12% carbon
9.6 g oxygen	0.48 oxygen	48% oxygen
20.0 g	1.00 part by mass	100% by mass

The mass of each element depends on the mass of the sample—that is, more than 20.0 g of compound would contain more than 8.0 g of calcium—but *the mass fraction is fixed no matter what the size of the sample.* The sum of the mass fractions (or mass percents) equals 1.00 part (or 100%) by mass. The law of definite composition tells us that pure samples of calcium carbonate, no matter where they come from, always contain 40% calcium, 12% carbon, and 48% oxygen by mass (Figure 2.3).

Because a given element always constitutes the same mass fraction of a given compound, we can use that mass fraction to find the actual mass of the element in any sample of the compound:

$$\text{Mass of element} = \text{mass of compound} \times \text{mass fraction}$$

Or, more simply, we can skip the need to find the mass fraction first and use the results of mass analysis directly:

$$\text{Mass of element in sample} = \text{mass of compound in sample} \times \frac{\text{mass of element in compound}}{\text{mass of compound}} \quad \textbf{(2.1)}$$

For example, knowing that a 20.0-g sample of calcium carbonate contains 8.0 g of calcium, the mass of calcium in a 75.0-g sample of calcium carbonate can be calculated:

$$\text{Mass (g) of calcium} = 75.0\ \cancel{\text{g calcium carbonate}} \times \frac{8.0\ \text{g calcium}}{20.0\ \cancel{\text{g calcium carbonate}}} = 30.0\ \text{g calcium}$$

SAMPLE PROBLEM 2.2 Calculating the Mass of an Element in a Compound

Problem Pitchblende is the most important compound of uranium. Mass analysis of an 84.2-g sample of pitchblende shows that it contains 71.4 g of uranium, with oxygen the only other element. How many grams of uranium and of oxygen are in 102 kg of pitchblende?

Plan We have to find the masses (in g) of uranium and of oxygen in a known mass (102 kg) of pitchblende, given the mass of uranium (71.4 g) in a different mass of pitchblende (84.2 g) and knowing that oxygen is the only other element present. The mass *ratio* of uranium to pitchblende in grams is the same as it is in kilograms. Therefore, using Equation 2.1, we multiply the mass (in kg) of the pitchblende sample by the ratio of uranium to pitchblende (in kg) from the mass analysis. This gives the mass (in kg) of uranium, and we convert kilograms to grams. To find the mass of oxygen, the only other element in the pitchblende, we subtract the calculated mass of uranium (in kg) from the given mass of pitchblende (102 kg) and convert kilograms to grams.

Road Map

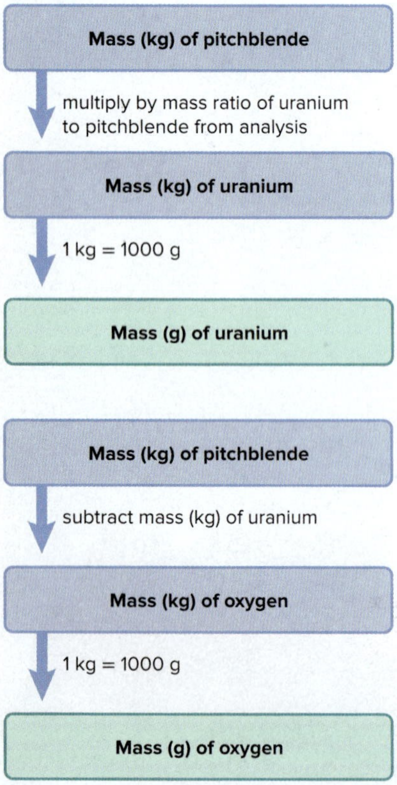

Solution Finding the mass (kg) of uranium in 102 kg of pitchblende:

$$\text{Mass (kg) of uranium} = \text{mass (kg) of pitchblende} \times \frac{\text{mass (kg) of uranium in pitchblende}}{\text{mass (kg) of pitchblende}}$$

$$= 102 \cancel{\text{kg pitchblende}} \times \frac{71.4 \text{ kg uranium}}{84.2 \cancel{\text{kg pitchblende}}} = 86.5 \text{ kg uranium}$$

Converting the mass of uranium from kg to g:

$$\text{Mass (g) of uranium} = 86.5 \cancel{\text{kg}} \text{ uranium} \times \frac{1000 \text{ g}}{1 \cancel{\text{kg}}} = 8.65 \times 10^4 \text{ g uranium}$$

Finding the mass (in kg) of oxygen in 102 kg of pitchblende:

$$\text{Mass (kg) of oxygen} = \text{mass (kg) of pitchblende} - \text{mass (kg) of uranium}$$
$$= 102 \text{ kg} - 86.5 \text{ kg} = 15.5 \text{ kg oxygen}$$

Converting the mass of oxygen from kg to g:

$$\text{Mass (g) of oxygen} = 15.5 \cancel{\text{kg}} \text{ oxygen} \times \frac{1000 \text{ g}}{1 \cancel{\text{kg}}} = 1.55 \times 10^4 \text{ g oxygen}$$

Check The analysis showed that most of the mass of pitchblende is due to uranium, so the large mass of uranium makes sense. Rounding off to check the math gives

$$\sim 100 \text{ kg pitchblende} \times \frac{70}{85} = 82 \text{ kg uranium}$$

FOLLOW-UP PROBLEMS

2.2A The mineral "fool's gold" does not contain any gold; instead it is a compound composed only of the elements iron and sulfur. A 110.0-g sample of fool's gold contains 51.2 g of iron. What mass of sulfur is in a sample of fool's gold that contains 86.2 g of iron?

2.2B Silver bromide is the light-sensitive compound coated onto black-and-white film. A 26.8-g sample contains 15.4 g of silver, with bromine as the only other element. How many grams of each element are on a roll of film that contains 3.57 g of silver bromide?

SOME SIMILAR PROBLEMS 2.22–2.25

Student Hot Spot

Student data indicate that you may struggle with using mass fraction to calculate the mass of an element in a compound. Access the Smartbook to view additional Learning Resources on this topic.

Multiple Proportions

It's quite common for the same two elements to form more than one compound—sulfur and fluorine do this, as do phosphorus and chlorine and many other pairs of elements. The third mass law we consider applies in these cases:

- **Law of multiple proportions:** *if elements A and B react to form two compounds, the different masses of B that combine with a fixed mass of A can be expressed as a ratio of small whole numbers.*

Consider two compounds of carbon and oxygen; let's call them carbon oxides I and II. These compounds have very different properties: the density of carbon oxide I is 1.25 g/L, whereas that of II is 1.98 g/L; I is poisonous and flammable, but II is not. Mass analysis shows that

Carbon oxide I is 57.1 mass % oxygen and 42.9 mass % carbon
Carbon oxide II is 72.7 mass % oxygen and 27.3 mass % carbon

To demonstrate the phenomenon of multiple proportions, we use the mass percents of oxygen and of carbon to find their masses in a given mass, say 100 g, of each compound. Then we divide the mass of oxygen by the mass of carbon in each compound to obtain the mass of oxygen that combines with a fixed mass of carbon:

	Carbon Oxide I	**Carbon Oxide II**
g oxygen/100 g compound	57.1	72.7
g carbon/100 g compound	42.9	27.3
g oxygen/g carbon	$\frac{57.1}{42.9} = 1.33$	$\frac{72.7}{27.3} = 2.66$

If we then divide the grams of oxygen per gram of carbon in II by that in I, we obtain a ratio of small whole numbers:

$$\frac{2.66 \text{ g oxygen/g carbon in II}}{1.33 \text{ g oxygen/g carbon in I}} = \frac{2}{1}$$

The law of multiple proportions tells us that in two compounds of the same elements, the mass fraction of one element relative to that of the other element changes in *increments based on ratios of small whole numbers.* In this case, the ratio is 2/1—for a given mass of carbon, compound II contains *2 times* as much oxygen as I, not 1.583 times, 1.716 times, or any other intermediate amount. In the next section, we'll discuss the model that explained the mass laws on the atomic scale and learn the identities of the two carbon oxides.

› Summary of Section 2.2

- The law of mass conservation states that the total mass remains constant during a chemical reaction.
- The law of definite composition states that any sample of a given compound has the same elements present in the same parts by mass.
- The law of multiple proportions states that, in different compounds of the same elements, the masses of one element that combine with a fixed mass of the other can be expressed as a ratio of small whole numbers.

2.3 DALTON'S ATOMIC THEORY

With over 200 years of hindsight, it may be easy to see how the mass laws could be explained by the idea that matter exists in indestructible units, each with a particular mass and set of properties, but it was a major breakthrough in 1808 when John Dalton (1766–1844) presented his atomic theory of matter.

Postulates of the Atomic Theory

Dalton expressed his theory in a series of postulates, presented here in modern terms:

1. All matter consists of **atoms,** tiny indivisible units of an element that cannot be created or destroyed. (This derives from the "eternal, indestructible atoms" proposed by Democritus more than 2000 years earlier and reflects mass conservation as stated by Lavoisier.)
2. Atoms of one element *cannot* be converted into atoms of another element. In chemical reactions, the atoms of the original substances recombine to form different substances. (This rejects the alchemical belief in the magical transmutation of elements.)
3. Atoms of an element are identical in mass and other properties and are different from atoms of any other element. (This contains Dalton's major new ideas: *unique mass and properties* for the atoms of a given element.)
4. Compounds result from the chemical combination of a specific ratio of atoms of different elements. (This follows directly from the law of definite composition.)

How the Theory Explains the Mass Laws

Let's see how Dalton's postulates explain the mass laws:

- *Mass conservation.* Atoms cannot be created or destroyed (postulate 1) or converted into other types of atoms (postulate 2). Therefore, a chemical reaction cannot possibly result in a mass change because atoms are just combined differently.
- *Definite composition.* A compound is a combination of a *specific* ratio of different atoms (postulate 4), each of which has a particular mass (postulate 3). Thus, each element in a compound must constitute a fixed fraction of the total mass.
- *Multiple proportions.* Atoms of an element have the same mass (postulate 3) and are indivisible (postulate 1). The masses of element B that combine with a fixed mass of element A must give a small, whole-number ratio because different numbers of B atoms combine with each A atom in different compounds.

The *simplest* arrangement consistent with the mass data for carbon oxides I and II in our earlier example is that one atom of oxygen combines with one atom of carbon in compound I (carbon monoxide) and that two atoms of oxygen combine with one atom of carbon in compound II (carbon dioxide):

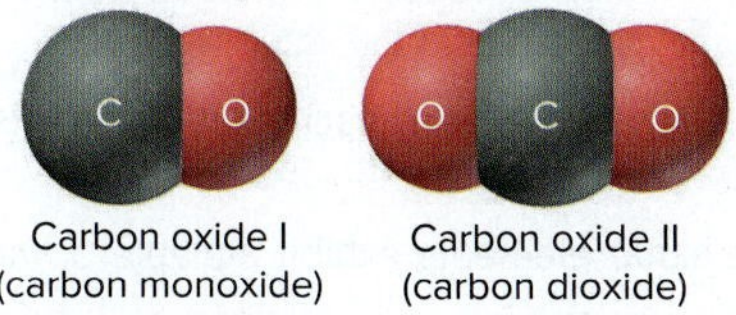

Let's work through a sample problem that reviews the mass laws.

SAMPLE PROBLEM 2.3 Visualizing the Mass Laws

Problem The scenes below represent an atomic-scale view of a chemical reaction:

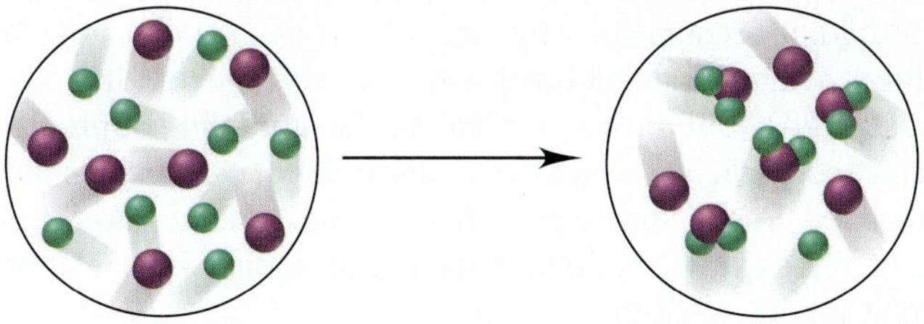

Which of the mass laws—mass conservation, definite composition, and/or multiple proportions—is (are) illustrated?

Plan From the depictions, we note the numbers, colors, and combinations of atoms (spheres) to see which mass laws pertain. If the numbers of each atom are the same before and after the reaction, the total mass did not change (mass conservation). If a compound forms that always has the same atom ratio, the elements are present in fixed parts by mass (definite composition). If the same elements form different compounds and the ratio of the atoms of one element that combine with one atom of the other element is a small whole number, the ratio of their masses is a small whole number as well (multiple proportions).

Solution There are seven purple and nine green atoms in each circle, so mass is conserved. The compound formed has one purple and two green atoms, so it has definite composition. Only one compound forms, so the law of multiple proportions does not pertain.

FOLLOW-UP PROBLEMS

2.3A The following scenes represent a chemical change. Which of the mass laws is (are) illustrated?

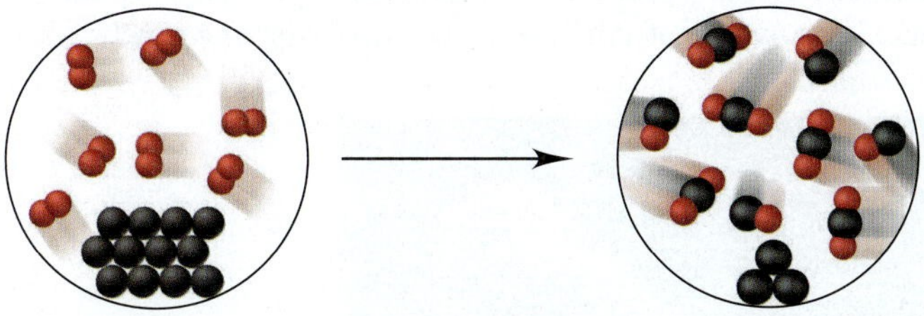

2.3B Which sample(s) best display(s) the fact that compounds of bromine *(orange)* and fluorine *(yellow)* exhibit the law of multiple proportions? Explain.

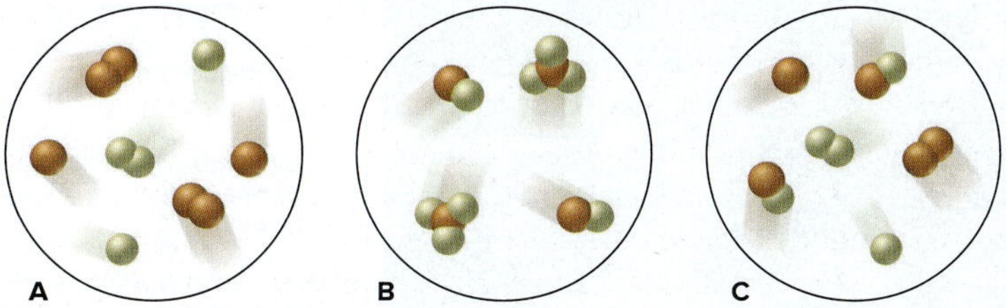

SOME SIMILAR PROBLEMS 2.14 and 2.15

› Summary of Section 2.3

- › Dalton's atomic theory explained the mass laws by proposing that all matter consists of indivisible, unchangeable atoms of fixed, unique mass.
- › Mass is conserved during a reaction because the atoms retain their identities but are combined differently.
- › Each compound has a fixed mass fraction of each of its elements because it is composed of a fixed number of each type of atom.
- › Different compounds of the same elements exhibit multiple proportions because they each consist of whole atoms.

2.4 THE OBSERVATIONS THAT LED TO THE NUCLEAR ATOM MODEL

Dalton's model established that masses of elements reacting could be explained in terms of atoms, but it did not account for why atoms bond as they do: why, for example, do two, and not three, hydrogen atoms bond with one oxygen atom in a water molecule?

Moreover, Dalton's "billiard-ball" model of the atom did not predict the existence of subatomic charged particles, which were observed in experiments at the turn of the 20th century that led to the discovery of *electrons* and the atomic *nucleus*. Let's examine some of the experiments that resolved questions about Dalton's model and led to our current model of atomic structure.

Effect of electric charges.

Source: © McGraw-Hill Education/Bob Coyle, photographer

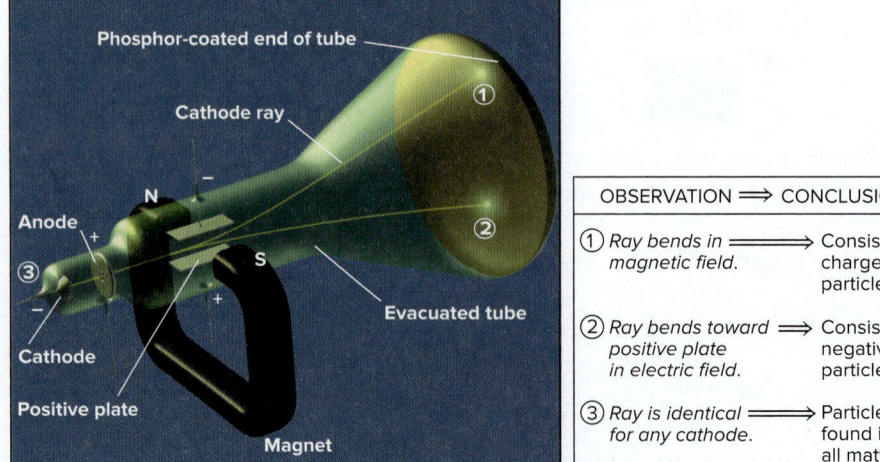

Discovery of the Electron and Its Properties

For many years, scientists had known that matter and electric charge were related. When amber is rubbed with fur, or glass with silk, positive and negative charges form—the same charges that make your hair crackle and cling to your comb on a dry day *(see photo)*. Scientists also knew that an electric current could decompose certain compounds into their elements. But they did not know what a current was made of.

Cathode Rays To discover the nature of an electric current, some investigators tried passing it through nearly evacuated glass tubes fitted with metal electrodes. When the electric power source was turned on, a "ray" could be seen striking the phosphor-coated end of the tube, which emitted a glowing spot of light. The rays were called **cathode rays** because they originated at the negative electrode (cathode) and moved to the positive electrode (anode).

Figure 2.4 shows some properties of cathode rays based on these observations. The main conclusion was that *cathode rays consist of negatively charged particles*

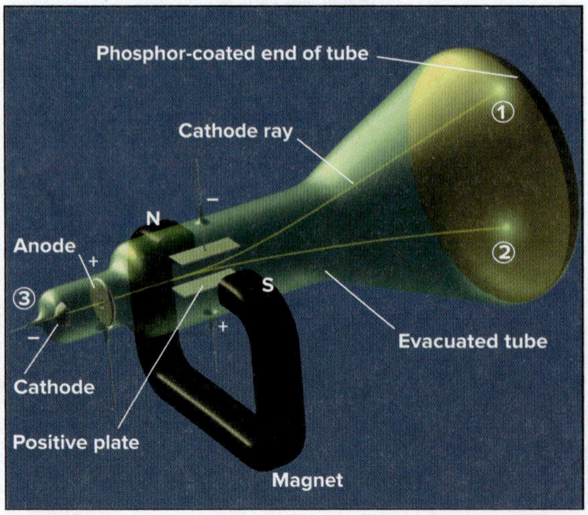

OBSERVATION ⟹ CONCLUSION	
① *Ray bends in magnetic field.* ⟹	Consists of charged particles
② *Ray bends toward positive plate in electric field.* ⟹	Consists of negative particles
③ *Ray is identical for any cathode.* ⟹	Particles found in all matter

Figure 2.4 Observations that established the properties of cathode rays.

found in all matter. The rays become visible as their particles collide with the few remaining gas molecules in the evacuated tube. Cathode ray particles were later named *electrons*. There are many familiar cases of the effects of charged particles colliding with gas particles or hitting a phosphor-coated screen:

- In a "neon" sign, electrons collide with the gas particles in the tube, causing them to give off light.
- The aurora borealis, or northern lights *(see photo)*, occurs when Earth's magnetic field bends streams of charged particles coming from the Sun, which then collide with gas particles in the atmosphere to give off light.
- In older televisions and computer monitors, the cathode ray passes back and forth over the phosphor-coated screen, creating a pattern that we see as a picture.

Aurora borealis display.

Source: © Roman Crochuk/Shutterstock.com

Mass and Charge of the Electron Two classic experiments and their conclusion revealed the mass and charge of the electron:

1. *Mass/charge ratio*. In 1897, the British physicist J. J. Thomson (1856–1940) measured the ratio of the mass of a cathode ray particle to its charge. By comparing this value with the mass/charge ratio for the lightest charged particle in solution, Thomson estimated that the cathode ray particle weighed less than $\frac{1}{1000}$ as much as hydrogen, the lightest atom! He was shocked because this implied that, contrary to Dalton's atomic theory, *atoms contain even smaller particles*. Fellow scientists reacted with disbelief to Thomson's conclusion; some even thought he was joking.

2. *Charge*. In 1909, the American physicist Robert Millikan (1868–1953) measured the *charge* of the electron. He did so by observing the movement of tiny oil droplets in an apparatus that contained electrically charged plates and an x-ray source (Figure 2.5). X-rays knocked electrons from gas molecules in the air within the apparatus, and the electrons stuck to an oil droplet falling through a hole in a positively charged plate. With the electric field off, Millikan measured the mass of the now negatively charged droplet from its rate of fall. Then, by adjusting the field's strength, he made the droplet slow and hang suspended, which allowed him to measure its total charge.

After many tries, Millikan found that the total charge of the various droplets was always some *whole-number multiple of a minimum charge*. If different oil droplets picked up different numbers of electrons, he reasoned that this minimum charge

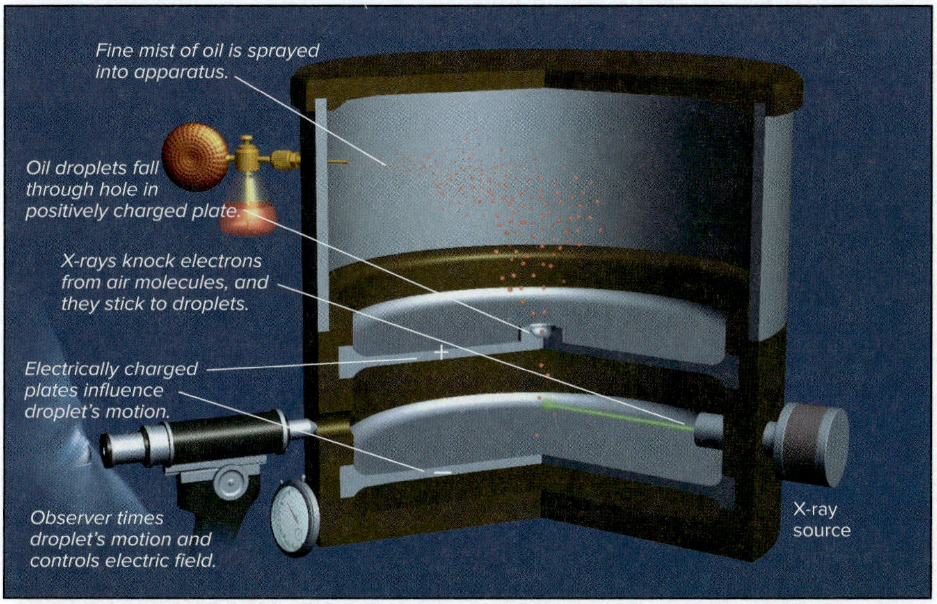
Fine mist of oil is sprayed into apparatus.
Oil droplets fall through hole in positively charged plate.
X-rays knock electrons from air molecules, and they stick to droplets.
Electrically charged plates influence droplet's motion.
+
−
Observer times droplet's motion and controls electric field.
X-ray source

Figure 2.5 Millikan's oil-drop experiment for measuring an electron's charge. The total charge on an oil droplet is some whole-number multiple of the charge of the electron.

must be the charge of the electron itself. Remarkably, the value that he calculated over a century ago was within 1% of the modern value of the electron's charge, -1.602×10^{-19} C (C stands for *coulomb,* the SI unit of charge).

3. *Conclusion: calculating the electron's mass.* The electron's mass/charge ratio, from work by Thomson and others, multiplied by the value for the electron's charge gives the electron's mass, which is *extremely* small:

$$\text{Mass of electron} = \frac{\text{mass}}{\cancel{\text{charge}}} \times \cancel{\text{charge}} = \left(-5.686\times10^{-12}\,\frac{\text{kg}}{\cancel{\text{C}}}\right)(-1.602\times10^{-19}\,\cancel{\text{C}})$$
$$= 9.109\times10^{-31}\text{ kg} = 9.109\times10^{-28}\text{ g}$$

Discovery of the Atomic Nucleus

The presence of electrons in all matter brought up two major questions about the structure of atoms. Matter is electrically neutral, so atoms must be also. But if atoms contain negatively charged electrons, what positive charges balance them? And if an electron has such a tiny mass, what accounts for an atom's much larger mass? To address these issues, Thomson proposed his "plum-pudding" model—a spherical atom composed of diffuse, positively charged matter with electrons embedded in it like "raisins in a plum pudding."

In 1910, New Zealand–born physicist Ernest Rutherford (1871–1937) tested this model and obtained a very unexpected result (Figure 2.6):

1. *Experimental design.* Figure 2.6A shows the experimental setup, in which tiny, dense, positively charged alpha (α) particles emitted from radium are aimed at gold foil. A circular, zinc-sulfide screen registers the deflection (scattering angle) of the α particles emerging from the foil by emitting light flashes when the particles strike it.
2. *Hypothesis and expected results.* With Thomson's model in mind (Figure 2.6B), Rutherford expected only minor, if any, deflections of the α particles because they

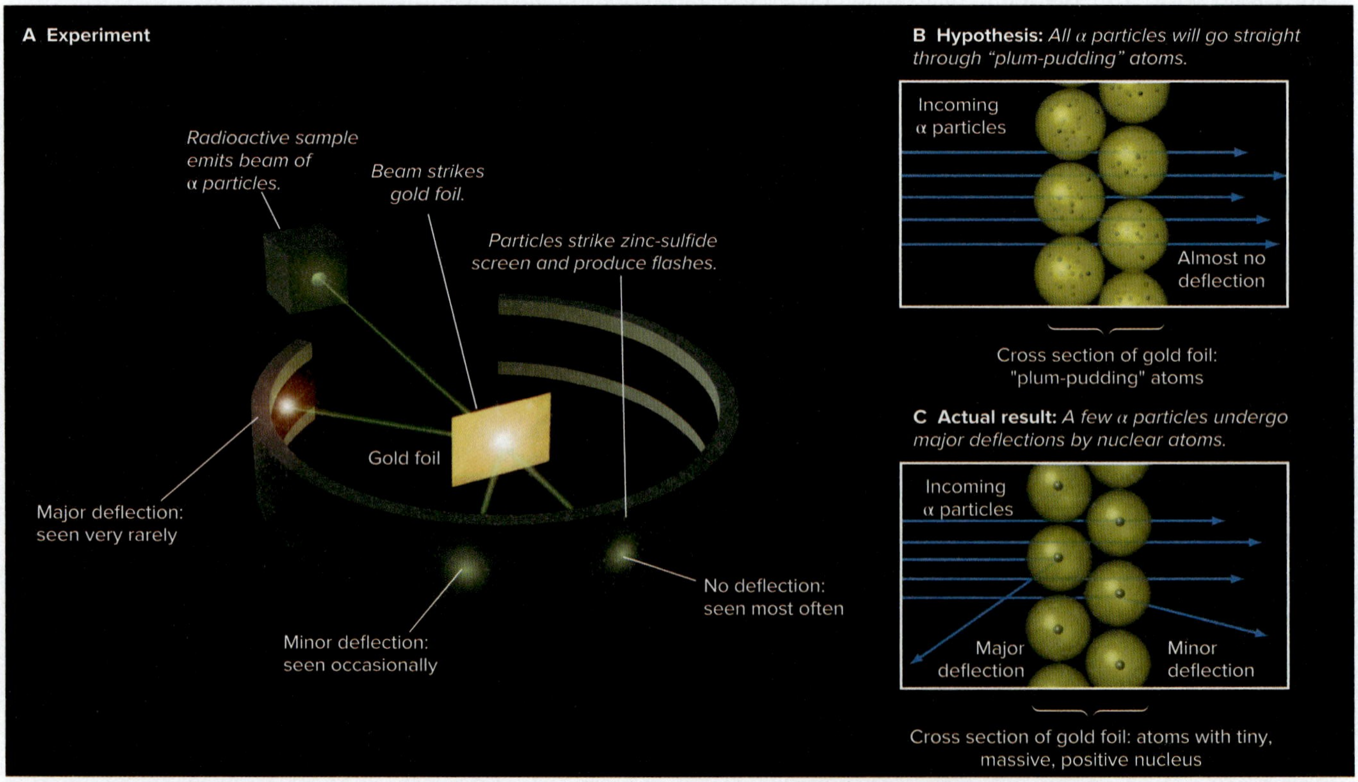

Figure 2.6 Rutherford's α-scattering experiment and discovery of the atomic nucleus.

should act as bullets and go right through the gold atoms. After all, he reasoned, an electron should not be able to deflect an α particle any more than a Ping-Pong ball could deflect a baseball.

3. *Actual results.* Initial results were consistent with this hypothesis, but then the unexpected happened (Figure 2.6C). As Rutherford recalled: "I remember two or three days later Geiger [one of his coworkers] coming to me in great excitement and saying, 'We have been able to get some of the α particles coming backwards . . . ' It was quite the most incredible event that has ever happened to me in my life. It was almost as incredible as if you fired a 15-inch shell at a piece of tissue paper and it came back and hit you." The data showed very few α particles deflected at all and only 1 in 20,000 deflected by more than 90° ("coming backwards").
4. *Rutherford's conclusion.* Rutherford concluded that these few α particles were being repelled by something small, dense, and positive within the gold atoms. Calculations based on the mass, charge, and velocity of the α particles and the fraction of these large-angle deflections showed that
   - An atom is mostly space occupied by electrons.
   - In the center is a tiny region, which Rutherford called the **nucleus,** that contains all the positive charge and essentially all the mass of the atom.

   He proposed that positive particles lay within the nucleus and called them *protons.*

Rutherford's model explained the charged nature of matter, but it could not account for all the atom's mass. After more than 20 years, in 1932, James Chadwick (1891–1974) discovered the *neutron,* an uncharged dense particle that also resides in the nucleus.

› Summary of Section 2.4

- › Several major discoveries at the turn of the 20th century resolved questions about Dalton's model and led to our current model of atomic structure.
- › Cathode rays were shown to consist of negative particles (electrons) that exist in all matter. J. J. Thomson measured their mass/charge ratio and concluded that they are much smaller and lighter than atoms.
- › Robert Millikan determined the charge of the electron, which he combined with other data to calculate its mass.
- › Ernest Rutherford proposed that atoms consist of a tiny, massive, positively charged nucleus surrounded by electrons.

2.5 THE ATOMIC THEORY TODAY

Dalton's model of an indivisible particle has given way to our current model of an atom with an elaborate internal architecture of subatomic particles.

Structure of the Atom

An *atom* is an electrically neutral, spherical entity composed of a positively charged central nucleus surrounded by one or more negatively charged electrons (Figure 2.7, *next page*). The electrons move rapidly within the available volume, held there by the attraction of the nucleus. (To indicate their motion, they are often depicted as a cloudy blur of color, darkest around a central dot—the nucleus—and becoming paler toward the outer edge.) An atom's diameter ($\sim 1 \times 10^{-10}$ m) is about 20,000 times the diameter of its nucleus ($\sim 5 \times 10^{-15}$ m). The nucleus contributes 99.97% of the atom's mass, but occupies only about 1 quadrillionth of its volume, so it is incredibly dense: about 10^{14} g/mL! ›

An atomic nucleus consists of protons and neutrons (the only exception is the simplest nucleus, that of hydrogen, which is just a proton). The **proton (p^+)** has a positive charge, and the **neutron (n^0)** has no charge; thus, the positive charge of the nucleus results from its protons. The *magnitudes* of the charges possessed by a proton and by an **electron (e^-)** are equal, but the *signs* of the charges are opposite. *An atom is neutral because the number of protons in the nucleus equals the number*

The Tiny, Massive Nucleus

A few analogies can help you grasp the incredible properties of the atomic nucleus. A nucleus the size of the period at the end of a sentence would weigh about 100 tons, as much as 50 cars! An atom the size of the Houston Astrodome would have a nucleus the size of a green pea that would contain virtually all the stadium's mass. If the nucleus were about 1 cm in diameter, the atom would be about 200 m across, or more than the length of two football fields!

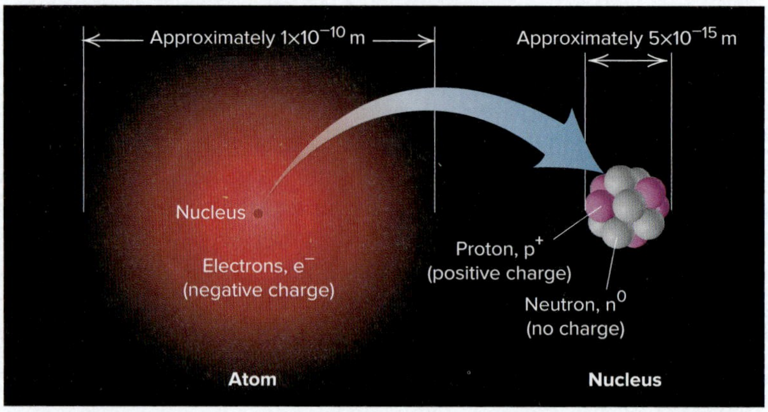

Figure 2.7 General features of the atom.

of electrons surrounding the nucleus. Some properties of these three subatomic particles are listed in Table 2.2.

Atomic Number, Mass Number, and Atomic Symbol

The **atomic number (*Z*)** of an element equals the number of protons in the nucleus of each of its atoms. *All atoms of an element have the same atomic number, and the atomic number of each element is different from that of any other element.* All carbon atoms ($Z = 6$) have 6 protons, all oxygen atoms ($Z = 8$) have 8 protons, and all uranium atoms ($Z = 92$) have 92 protons. There are currently 118 known elements, of which 90 occur in nature and 28 have been synthesized by nuclear scientists.

The **mass number (*A*)** is the total number of protons and neutrons in the nucleus of an atom. Each proton and each neutron contributes one unit to the mass number. Thus, a carbon atom with 6 protons and 6 neutrons in its nucleus has a mass number of 12, and a uranium atom with 92 protons and 146 neutrons in its nucleus has a mass number of 238.

The **atomic symbol** (or *element symbol*) of an element is based on its English, Latin, or Greek name, such as C for carbon, S for sulfur, and Na for sodium (Latin *natrium*). Often written with the symbol are the atomic number (*Z*) as a left *sub*script and the mass number (*A*) as a left *super*script, so element X would be ${}^{A}_{Z}\text{X}$ (Figure 2.8A). Since the mass number is the sum of protons and neutrons, the number of neutrons (*N*) equals the mass number minus the atomic number:

$$\text{Number of neutrons} = \text{mass number} - \text{atomic number}, \quad \text{or} \quad N = A - Z \qquad \textbf{(2.2)}$$

Thus, a chlorine atom, symbolized ${}^{35}_{17}\text{Cl}$, has $A = 35$, $Z = 17$, and $N = 35 - 17 = 18$. Because each element has its own atomic number, we also know the atomic number from the symbol. For example, instead of writing ${}^{12}_{6}\text{C}$ for carbon with mass number 12, we can write ${}^{12}\text{C}$ (spoken "carbon twelve"), with $Z = 6$ understood. Another way to name this atom is carbon-12.

Table 2.2 Properties of the Three Key Subatomic Particles

	Charge		Mass		
Name (Symbol)	**Relative**	**Absolute (C)***	**Relative (amu)†**	**Absolute (g)**	**Location in Atom**
Proton (p^+)	1+	$+1.60218\times10^{-19}$	1.00727	1.67262×10^{-24}	Nucleus
Neutron (n^0)	0	0	1.00866	1.67493×10^{-24}	Nucleus
Electron (e^-)	1−	-1.60218×10^{-19}	0.00054858	9.10939×10^{-28}	Outside nucleus

*The coulomb (C) is the SI unit of charge.
†The atomic mass unit (amu) equals 1.66054×10^{-24} g; discussed later in this section.

Isotopes

All atoms of an element have the same atomic number but not the same mass number. **Isotopes** of an element are atoms that have *different numbers of neutrons* and therefore different mass numbers. Most elements occur in nature in a particular *isotopic composition,* which specifies the proportional abundance of each of its isotopes. For example, all carbon atoms ($Z = 6$) have 6 protons and 6 electrons, but only 98.89% of naturally occurring carbon atoms have 6 neutrons ($A = 12$). A small percentage (1.11%) have 7 neutrons ($A = 13$), and even fewer (less than 0.01%) have 8 ($A = 14$). These are carbon's three naturally occurring isotopes—^{12}C, ^{13}C, and ^{14}C. *A natural sample of carbon has these three isotopes in these relative proportions.* Five other carbon isotopes—^{9}C, ^{10}C, ^{11}C, ^{15}C, and ^{16}C—have been created in the laboratory. Figure 2.8B depicts the atomic number, mass number, and symbol for four atoms, two of which are isotopes of the element uranium.

A key point is that the chemical properties of an element are primarily determined by the number of electrons, so *all isotopes of an element have nearly identical chemical behavior,* even though they have different masses.

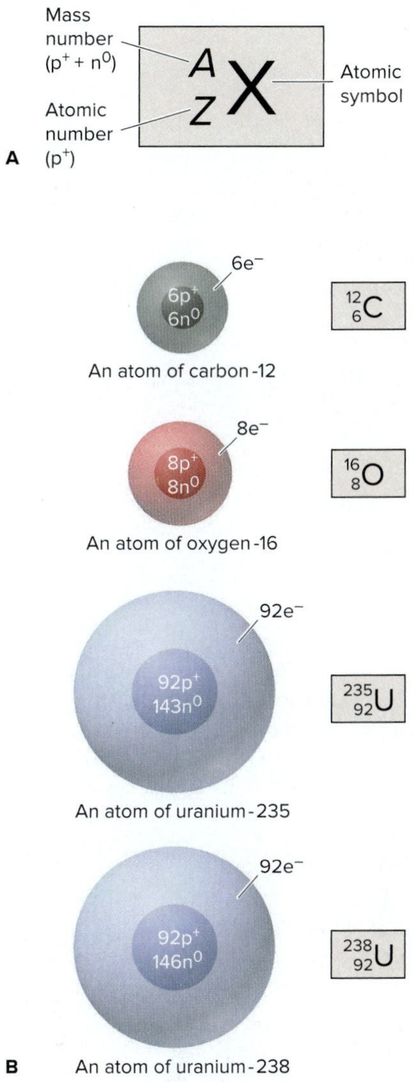

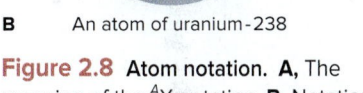

Figure 2.8 Atom notation. A, The meaning of the $^A_Z X$ notation. **B,** Notations and spherical representations for four atoms. (The nuclei are not drawn to scale.)

SAMPLE PROBLEM 2.4 **Determining the Numbers of Subatomic Particles in the Isotopes of an Element**

Problem Silicon (Si) is a major component of semiconductor chips. It has three naturally occurring isotopes: ^{28}Si, ^{29}Si, and ^{30}Si. Determine the numbers of protons, electrons, and neutrons in an atom of each silicon isotope.

Plan The mass number (*A;* left superscript) of each of the three isotopes is given, which is the sum of protons and neutrons. From the List of Elements on the text's inside front cover, we find the atomic number (*Z,* number of protons), which equals the number of electrons. We obtain the number of neutrons by subtracting *Z* from *A* (Equation 2.2).

Solution From the elements list, the atomic number of silicon is 14. Therefore,

$$^{28}\text{Si has } 14p^+,\ 14e^-, \text{ and } 14n^0(28 - 14)$$
$$^{29}\text{Si has } 14p^+,\ 14e^-, \text{ and } 15n^0(29 - 14)$$
$$^{30}\text{Si has } 14p^+,\ 14e^-, \text{ and } 16n^0(30 - 14)$$

FOLLOW-UP PROBLEMS

2.4A Titanium, the ninth most abundant element, is used structurally in many objects, such as electric turbines, aircraft bodies, and bicycle frames. It has five naturally occurring isotopes: ^{46}Ti, ^{47}Ti, ^{48}Ti, ^{49}Ti, and ^{50}Ti. How many protons, electrons, and neutrons are in an atom of each isotope?

2.4B How many protons, electrons, and neutrons are in an atom of each of the following, and which elements are represented by Q, R, and X? **(a)** $^{11}_{5}Q$ **(b)** $^{41}_{20}R$ **(c)** $^{131}_{53}X$

SOME SIMILAR PROBLEMS 2.39–2.42

Atomic Masses of the Elements

The mass of an atom is measured *relative* to the mass of an atomic standard. The modern standard is the carbon-12 atom, whose mass is defined as *exactly* 12 atomic mass units. Thus, the **atomic mass unit (amu)** is $\frac{1}{12}$ the mass of a carbon-12 atom. Based on this standard, the ^{1}H atom has a mass of 1.008 amu; in other words, a ^{12}C atom has almost 12 times the mass of an ^{1}H atom. We will continue to use the term *atomic mass unit* in the text, although the name of the unit has been changed to the **dalton (Da);** thus, one ^{12}C atom has a mass of 12 daltons (12 Da, or 12 amu). The atomic mass unit is a relative unit of mass, but it is equivalent to an absolute mass of 1.66054×10^{-24} g.

Finding Atomic Mass from Isotopic Composition The isotopic composition of an element is determined by **mass spectrometry,** a method for measuring the relative masses and abundances of atomic-scale particles very precisely (see the Tools of the Laboratory essay at the end of this section). For example, using a mass spectrometer, we measure the mass ratio of ^{28}Si to ^{12}C as

$$\frac{\text{Mass of } ^{28}\text{Si atom}}{\text{Mass of } ^{12}\text{C standard}} = 2.331411$$

From this mass ratio, we find the **isotopic mass** of the ^{28}Si atom, the mass of this silicon isotope relative to that of ^{12}C:

$$\begin{aligned}\text{Isotopic mass of } ^{28}\text{Si} &= \text{measured mass ratio} \times \text{mass of } ^{12}\text{C} \\ &= 2.331411 \times 12 \text{ amu} = 27.97693 \text{ amu}\end{aligned}$$

Along with the isotopic mass, the mass spectrometer gives the relative abundance as a percentage (or fraction) of each isotope in a sample of the element. For example, the relative abundance of ^{28}Si is 92.23% (or 0.9223), which means that 92.23% of the silicon atoms in a naturally occurring sample of silicon have a mass number of 28.

From such data, we can obtain the **atomic mass** (also called *atomic weight*) of an element, the *average* of the masses of its naturally occurring isotopes weighted according to their abundances. Each naturally occurring isotope of an element contributes a certain portion to the atomic mass. For instance, multiplying the isotopic mass of ^{28}Si by its fractional abundance gives the portion of the atomic mass of Si contributed by ^{28}Si:

$$\text{Portion of Si atomic mass from } ^{28}\text{Si} = 27.97693 \text{ amu} \times 0.9223 = 25.8031 \text{ amu}$$

(retaining two additional significant figures)

Similar calculations give the portions contributed by ^{29}Si with a relative abundance of 4.67% (28.976495 amu × 0.0467 = 1.3532 amu) and by ^{30}Si with a relative abundance of 3.10% (29.973770 amu × 0.0310 = 0.9292 amu). Adding the three portions together (rounding to two decimal places at the end) gives the atomic mass of silicon:

$$\begin{aligned}\text{Atomic mass of Si} &= 25.8031 \text{ amu} + 1.3532 \text{ amu} + 0.9292 \text{ amu} \\ &= 28.0855 \text{ amu} = 28.09 \text{ amu}\end{aligned}$$

Thus, the average atomic mass of an element can be calculated by multiplying the mass of each naturally occurring isotope by its fractional abundance and summing those values:

$$\text{Atomic mass} = \Sigma(\text{isotopic mass})(\text{fractional abundance of isotope}) \quad \textbf{(2.3)}$$

where the fractional abundance of an isotope is the percent natural abundance of the isotope divided by 100.

The atomic mass is an average value; that is, while no individual silicon atom has a mass of 28.09 amu, in the laboratory, we consider a sample of silicon to consist of atoms with this average mass.

SAMPLE PROBLEM 2.5 Calculating the Atomic Mass of an Element

Problem Silver (Ag; $Z = 47$) has 46 known isotopes, but only two occur naturally, ^{107}Ag and ^{109}Ag. Given the following data, calculate the atomic mass of Ag:

Isotope	Mass (amu)	Abundance (%)
^{107}Ag	106.90509	51.84
^{109}Ag	108.90476	48.16

Plan From the mass and abundance of the two Ag isotopes, we have to find the atomic mass of Ag (weighted average of the isotopic masses). We divide each percent abundance by 100 to get the fractional abundance and then multiply that by each isotopic mass to find the portion of the atomic mass contributed by each isotope. The sum of the isotopic portions is the atomic mass (Equation 2.3).

Road Map

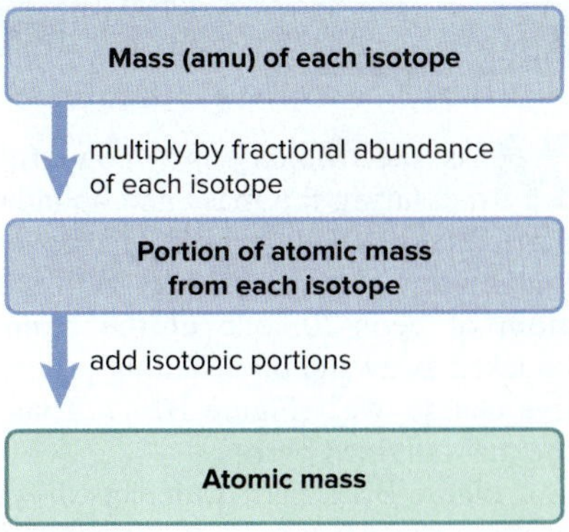

Solution Finding the fractional abundances:

$$\text{Fractional abundance of } {}^{107}\text{Ag} = 51.84/100 = 0.5184;\ \text{similarly, } {}^{109}\text{Ag} = 0.4816$$

Finding the portion of the atomic mass from each isotope:

$$\begin{aligned}\text{Portion of atomic mass from } {}^{107}\text{Ag} &= \text{isotopic mass} \times \text{fractional abundance}\\ &= 106.90509 \text{ amu} \times 0.5184 = 55.42 \text{ amu}\\ \text{Portion of atomic mass from } {}^{109}\text{Ag} &= 108.90476 \text{ amu} \times 0.4816 = 52.45 \text{ amu}\end{aligned}$$

Finding the atomic mass of silver:

$$\text{Atomic mass of Ag} = 55.42 \text{ amu} + 52.45 \text{ amu} = \boxed{107.87 \text{ amu}}$$

Or, in one step using Equation 2.3:

$$\begin{aligned}\text{Atomic mass of Ag} &= (\text{isotopic mass of } {}^{107}\text{Ag})(\text{fractional abundance of } {}^{107}\text{Ag})\\ &\quad + (\text{isotopic mass of } {}^{109}\text{Ag})(\text{fractional abundance of } {}^{109}\text{Ag})\\ &= (106.90509 \text{ amu})(0.5184) + (108.90476 \text{ amu})(0.4816)\\ &= 107.87 \text{ amu}\end{aligned}$$

Check The individual portions seem right: $\sim$100 amu × 0.50 = 50 amu. The portions should be almost the same because the two isotopic abundances are almost the same. We rounded each portion to four significant figures because that is the number of significant figures in the abundance values. This is the correct atomic mass (to two decimal places); in the List of Elements (inside front cover), it is rounded to 107.9 amu.

FOLLOW-UP PROBLEMS

2.5A Neon, a gas found in trace amounts in air, emits light in the visible range when electricity is passed through it. Use the isotopic abundance data from the Tools of the Laboratory essay at the end of this section and the following isotopic masses to determine the atomic mass of neon: ^{20}Ne has a mass of 19.99244 amu, ^{21}Ne has a mass of 20.99385 amu, and ^{22}Ne has a mass of 21.99139 amu.

2.5B Boron (B; Z = 5) has two naturally occurring isotopes. Find the percent abundances of ^{10}B and ^{11}B given these data: atomic mass of B = 10.81 amu, isotopic mass of ^{10}B = 10.0129 amu, and isotopic mass of ^{11}B = 11.0093 amu. (*Hint:* The sum of the fractional abundances is 1. If x = abundance of ^{10}B, then $1 - x$ = abundance of ^{11}B.)

SOME SIMILAR PROBLEMS 2.47–2.50

The Atomic-Mass Interval From the time Dalton proposed his atomic theory through much of the 20th century, atomic masses were considered constants of nature, like the speed of light. However, in 1969, the International Union of Pure and Applied Chemistry (IUPAC) rejected this idea because results from more advanced mass spectrometers showed consistent variations in isotopic composition from source to source.

In 2009, IUPAC proposed that an *atomic-mass interval* be used for 10 elements with exceptionally large variations in isotopic composition: hydrogen (H), lithium (Li), boron (B), carbon (C), nitrogen (N), oxygen (O), silicon (Si), sulfur (S), chlorine (Cl), and thallium (Th). For example, because the isotopic composition of hydrogen from oceans, rivers, and lakes, from various minerals, and from organic sediments varies so much, its atomic mass is now given as the interval [1.00784; 1.00811], which means that the mass is greater than or equal to 1.00784 and less than or equal to 1.00811. It's important to realize that *the mass of any given isotope of an element is constant, but the* ***proportion*** *of isotopes varies from source to source.*

TOOLS OF THE LABORATORY

Mass Spectrometry

Mass spectrometry is a powerful technique for measuring the mass and abundance of charged particles from their mass/charge ratio *(m/e)*. When a high-energy electron collides with a particle, say, an atom of neon-20, one of the atom's electrons is knocked away and the resulting particle has one positive charge, Ne^+ (Figure B2.1). Thus, its *m/e* equals the mass divided by 1+.

Figure B2.2, part A, depicts the core of one type of mass spectrometer. The sample is introduced (and vaporized if liquid or solid), then bombarded by high-energy electrons to form positively charged particles, in this case, of the different neon isotopes. These are attracted toward a series of negatively charged plates with slits in them. Some particles pass through the slits into an evacuated tube exposed to a magnetic field. As the particles enter this region, their paths are bent, the lightest particles (lowest *m/e*) are deflected most and the heaviest particles (highest *m/e*) least. At the end of the magnetic region, the particles strike a detector, which records their relative positions and abundances (Figure B2.2, parts B and C).

Mass spectrometry is now employed to measure the mass of virtually any atom, molecule, or molecular fragment. The technique is being used to study catalyst surfaces, forensic materials, fuel mixtures, medicinal agents, and many other samples, especially proteins (Figure B2.2, part D); in fact, John B. Fenn and Koichi Tanaka shared part of the 2002 Nobel Prize in chemistry for developing methods to study proteins by mass spectrometry.

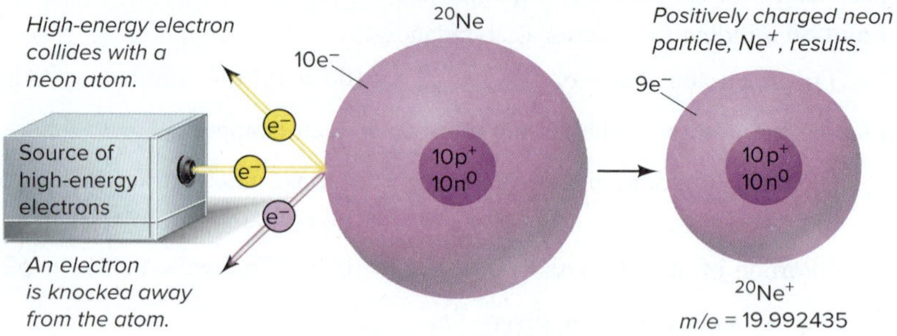

Figure B2.1 Formation of a positively charged neon particle (Ne^+).

PROBLEMS

B2.1 Chlorine has two naturally occurring isotopes, ^{35}Cl (abundance 76%) and ^{37}Cl (abundance 24%), and it occurs as diatomic (two-atom) molecules. In a mass spectrum, peaks are seen for the molecule and for the separated atoms.

a. How many peaks are in the mass spectrum?

b. What is the *m/e* value for the heaviest particle and for the lightest particle?

B2.2 When a sample of pure carbon is analyzed by mass spectrometry, peaks X, Y, and Z are obtained. Peak Y is taller than X and Z, and Z is taller than X. What is the *m/e* value of the isotope responsible for peak Z?

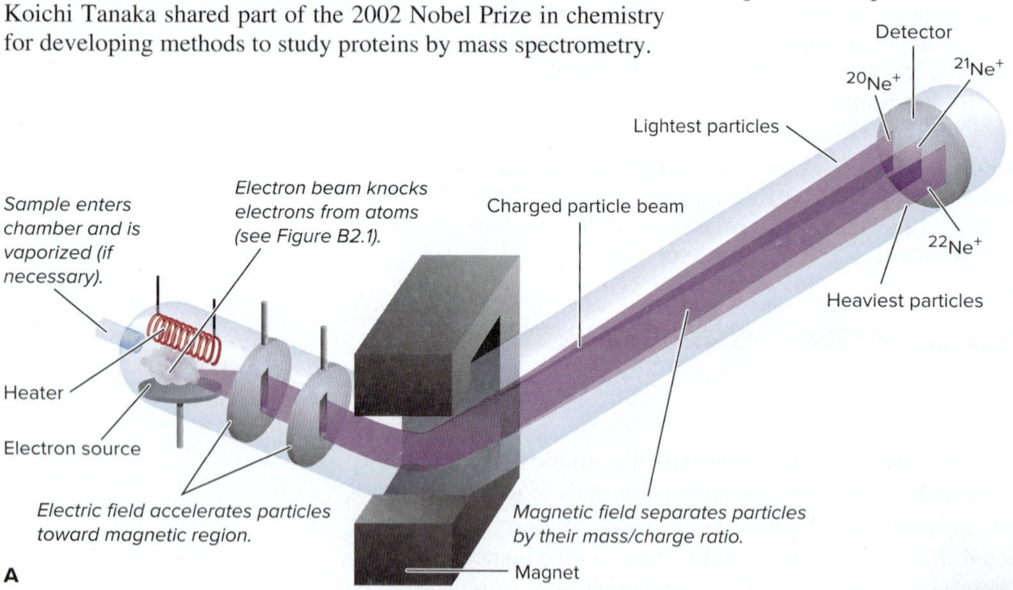

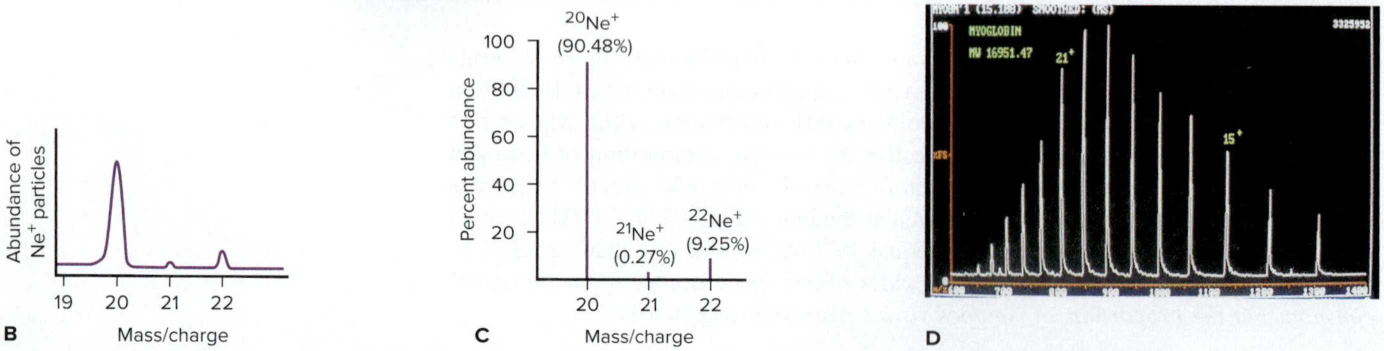

Figure B2.2 The mass spectrometer and its data. **A,** Particles are separated by their *m/e* values. **B,** The relative abundances of three Ne isotopes. **C,** The percent abundance of each isotope. **D,** The mass spectrum of a protein molecule. Each peak represents a molecular fragment. Source: © James King-Holmes/Oxford Centre for Molecular Sciences/Science Source.

The atomic-mass interval is important for very precise work, but because this text uses four significant figures for atomic masses in calculations—for example, 1.008 amu for the atomic mass of H—the change does not affect our discussions (see the List of Elements inside the front cover). Also, such small differences in composition do not overturn the basic idea of the law of definite composition: elements occur in a fixed proportion by mass in a compound, no matter what the source (Section 2.2).

› Summary of Section 2.5

- An atom has a central nucleus, which contains positively charged protons and uncharged neutrons and is surrounded by negatively charged electrons. An atom is neutral because the number of electrons equals the number of protons.
- An atom is represented by the notation $^{A}_{Z}X$, in which Z is the atomic number (number of protons), A the mass number (sum of protons and neutrons), and X the atomic symbol.
- An element occurs naturally as a mixture of isotopes, atoms with the same number of protons but different numbers of neutrons. Each isotope has a mass relative to the ^{12}C mass standard.
- The atomic mass of an element is the average of its isotopic masses weighted according to their natural abundances. It is determined using a mass spectrometer.

2.6 ELEMENTS: A FIRST LOOK AT THE PERIODIC TABLE

At the end of the 18th century, Lavoisier compiled a list of the 23 elements known at that time; by 1870, 65 were known; by 1925, 88; today, there are 118! By the mid-19th century, enormous amounts of information concerning reactions, properties, and atomic masses of the elements had been accumulated. Several researchers noted recurring, or *periodic,* patterns of behavior and proposed schemes to organize the elements according to some fundamental property.

In 1871, the Russian chemist Dmitri Mendeleev (1836–1907) published the most successful of these organizing schemes as a table of the elements listed by increasing atomic mass and arranged so that elements with similar chemical properties fell in the same column. The modern **periodic table of the elements,** based on Mendeleev's version (but arranged by *atomic number,* not mass), is one of the great classifying schemes in science and an indispensable tool to chemists—and chemistry students.

Organization of the Periodic Table One common version of the modern periodic table appears inside the front cover and in Figure 2.9 *(next page).* It is formatted as follows:

1. Each element has a box that contains its atomic number, atomic symbol, and atomic mass. (A mass in parentheses is the mass number of the most stable isotope of that element.) The boxes lie, from left to right, in order of *increasing atomic number* (number of protons in the nucleus).
2. The boxes are arranged into a grid of **periods** (horizontal rows) and **groups** (vertical columns). Each period has a number from 1 to 7. Each group has a number from 1 to 8 *and* either the letter A or B. A newer system, with group numbers from 1 to 18 but no letters, appears in parentheses under the number-letter designations. (The text uses the number-letter system and shows the newer group number in parentheses.)
3. The eight A groups (two on the left and six on the right) contain the *main-group elements.* The ten B groups, located between Groups 2A(2) and 3A(13), contain the *transition elements.* Two horizontal series of *inner transition elements,* the lanthanides and the actinides, fit *between* the elements in Group 3B(3) and Group 4B(4) and are placed below the main body of the table.

MAIN–GROUP ELEMENTS (1A–2A; 3A–8A) · TRANSITION ELEMENTS (3B–2B) · INNER TRANSITION ELEMENTS

Key: 4 — Atomic number; Be — Atomic symbol; 9.012 — Atomic mass (amu)

Legend: Metals (main-group); Metals (transition); Metals (inner transition); Metalloids; Nonmetals

Period	1A (1)	2A (2)	3B (3)	4B (4)	5B (5)	6B (6)	7B (7)	8B (8)	8B (9)	8B (10)	1B (11)	2B (12)	3A (13)	4A (14)	5A (15)	6A (16)	7A (17)	8A (18)
1	1 H 1.008																	2 He 4.003
2	3 Li 6.941	4 Be 9.012											5 B 10.81	6 C 12.01	7 N 14.01	8 O 16.00	9 F 19.00	10 Ne 20.18
3	11 Na 22.99	12 Mg 24.31											13 Al 26.98	14 Si 28.09	15 P 30.97	16 S 32.06	17 Cl 35.45	18 Ar 39.95
4	19 K 39.10	20 Ca 40.08	21 Sc 44.96	22 Ti 47.87	23 V 50.94	24 Cr 52.00	25 Mn 54.94	26 Fe 55.85	27 Co 58.93	28 Ni 58.69	29 Cu 63.55	30 Zn 65.38	31 Ga 69.72	32 Ge 72.63	33 As 74.92	34 Se 78.96	35 Br 79.90	36 Kr 83.80
5	37 Rb 85.47	38 Sr 87.62	39 Y 88.91	40 Zr 91.22	41 Nb 92.91	42 Mo 95.96	43 Tc (98)	44 Ru 101.1	45 Rh 102.9	46 Pd 106.4	47 Ag 107.9	48 Cd 112.4	49 In 114.8	50 Sn 118.7	51 Sb 121.8	52 Te 127.6	53 I 126.9	54 Xe 131.3
6	55 Cs 132.9	56 Ba 137.3	57 La 138.9	72 Hf 178.5	73 Ta 180.9	74 W 183.8	75 Re 186.2	76 Os 190.2	77 Ir 192.2	78 Pt 195.1	79 Au 197.0	80 Hg 200.6	81 Tl 204.4	82 Pb 207.2	83 Bi 209.0	84 Po (209)	85 At (210)	86 Rn (222)
7	87 Fr (223)	88 Ra (226)	89 Ac (227)	104 Rf (265)	105 Db (268)	106 Sg (271)	107 Bh (270)	108 Hs (277)	109 Mt (276)	110 Ds (281)	111 Rg (280)	112 Cn (285)	113 Nh (284)	114 Fl (289)	115 Mc (288)	116 Lv (293)	117 Ts (294)	118 Og (294)

INNER TRANSITION ELEMENTS

6	Lanthanides	58 Ce 140.1	59 Pr 140.9	60 Nd 144.2	61 Pm (145)	62 Sm 150.4	63 Eu 152.0	64 Gd 157.3	65 Tb 158.9	66 Dy 162.5	67 Ho 164.9	68 Er 167.3	69 Tm 168.9	70 Yb 173.1	71 Lu 175.0
7	Actinides	90 Th 232.0	91 Pa (231)	92 U 238.0	93 Np (237)	94 Pu (244)	95 Am (243)	96 Cm (247)	97 Bk (247)	98 Cf (251)	99 Es (252)	100 Fm (257)	101 Md (258)	102 No (259)	103 Lr (262)

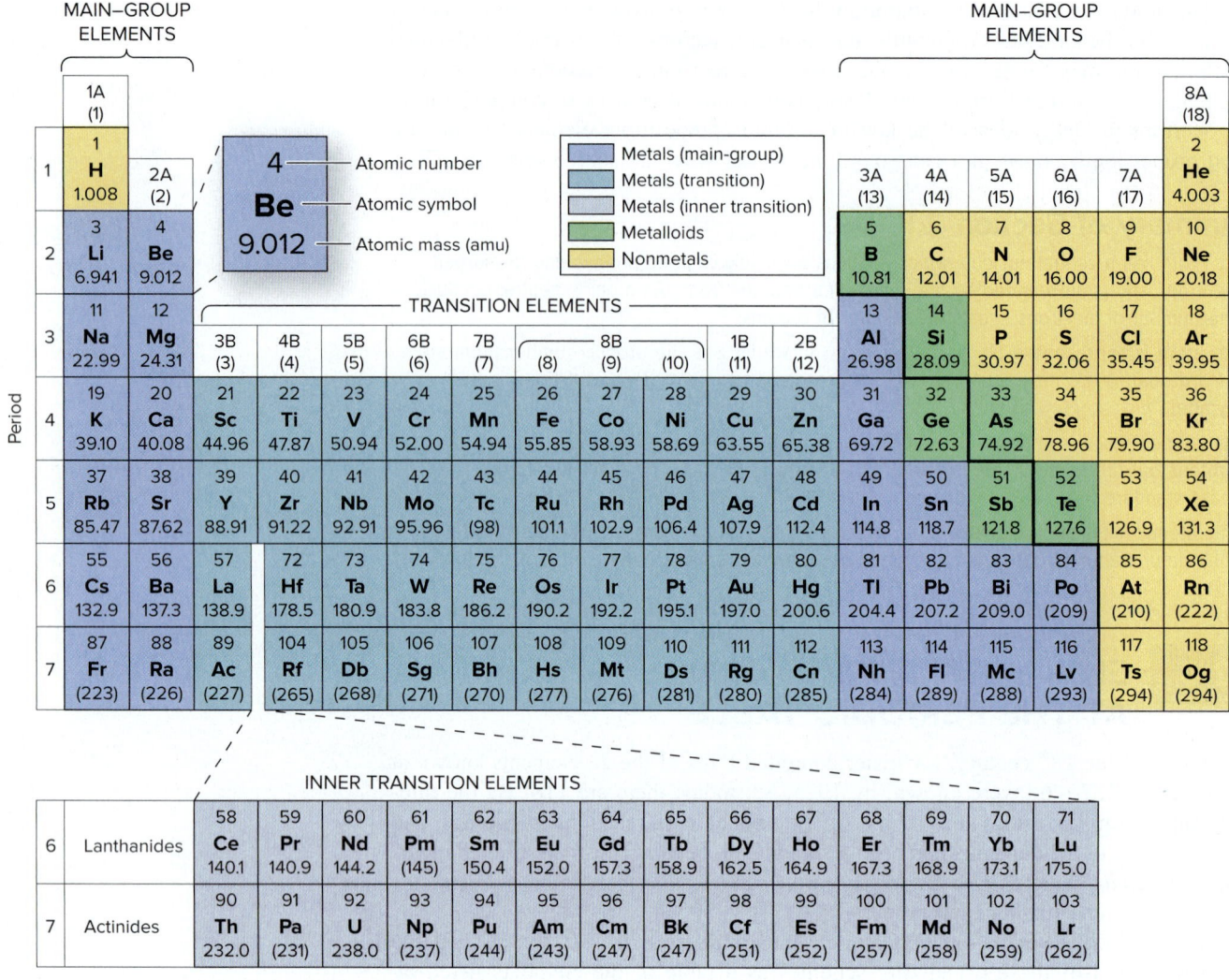

Figure 2.9 The modern periodic table.

Classifying the Elements One of the clearest ways to classify the elements is as metals, nonmetals, and metalloids. The "staircase" line that runs from the top of Group 3A(13) to the bottom of Group 6A(16) is a dividing line:

- The **metals** (three shades of blue in Figure 2.9) lie in the large lower-left portion of the table. About three-quarters of the elements are metals, including many main-group elements and all the transition and inner transition elements. They are generally shiny solids at room temperature (mercury is the only liquid) that conduct heat and electricity well. They can be tooled into sheets (are malleable) and wires (are ductile).
- The **nonmetals** (yellow) lie in the small upper-right portion of the table (with the exception of the nonmetal hydrogen in the upper-left corner). They are generally gases or dull, brittle solids at room temperature (bromine is the only liquid) and conduct heat and electricity poorly.
- The **metalloids** (green; also called **semimetals**), which lie along the staircase line, have properties between those of metals and nonmetals.

Figure 2.10 shows examples of these three classes of elements.

Two major points to keep in mind:

1. In general, elements in a group have *similar* chemical properties, and elements in a period have *different* chemical properties.
2. Despite the classification of elements into three types, in reality, there is a gradation in properties from left to right and top to bottom.

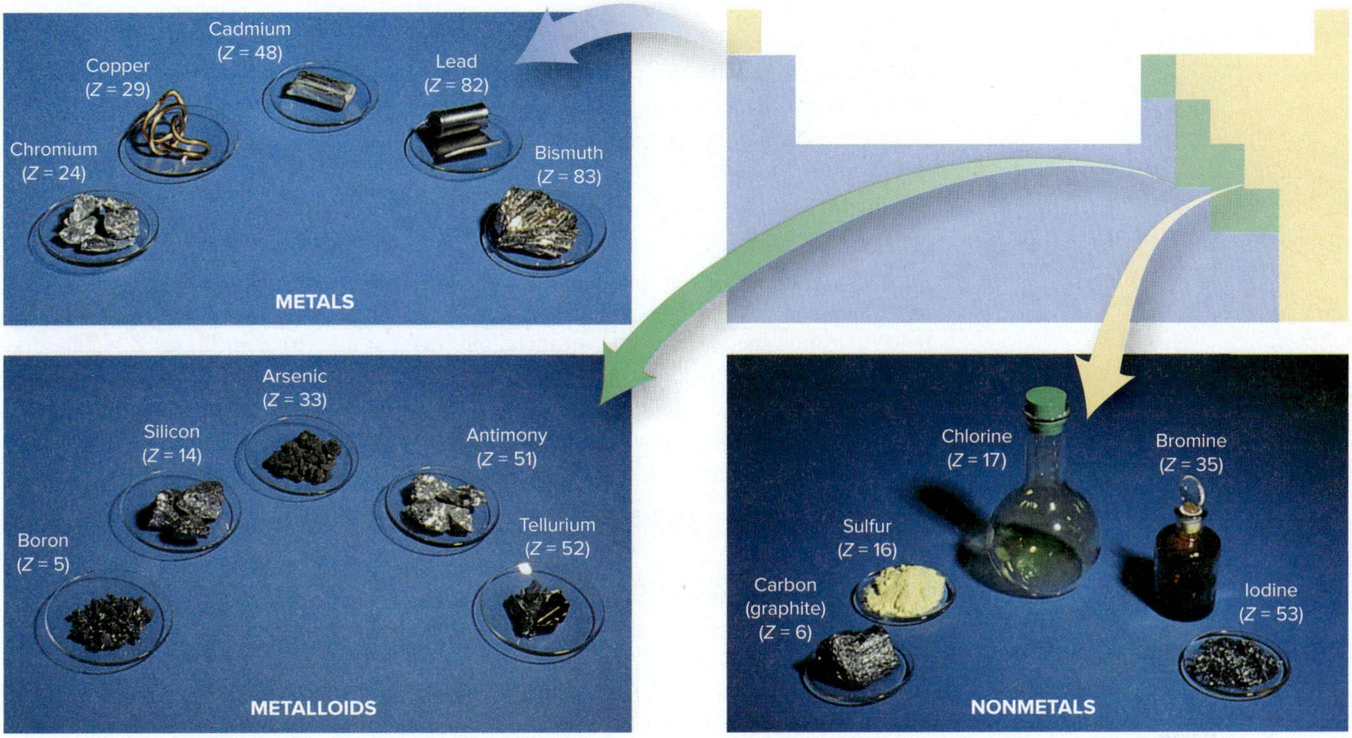

Figure 2.10 **Some metals, metalloids, and nonmetals.** Source: © McGraw-Hill Education/Stephen Frisch, photographer

It is important to learn some of the group (family) names:

- Group 1A(1) (except for hydrogen)—*alkali metals* (reactive metals)
- Group 2A(2)—*alkaline earth metals* (reactive metals)
- Group 7A(17)—*halogens* (reactive nonmetals)
- Group 8A(18)—*noble gases* (relatively nonreactive nonmetals)

Other main groups [3A(13) to 6A(16)] are often named for the first element in the group; for example, Group 6A(16) is the *oxygen family*.

Two of the major branches of chemistry have traditionally been defined by the elements that each studies. *Organic chemistry* studies the compounds of carbon, specifically those that contain hydrogen and often oxygen, nitrogen, and a few other elements. This branch is concerned with fuels, drugs, dyes, and the like. *Inorganic chemistry,* on the other hand, focuses on the compounds of all the other elements and is concerned with catalysts, electronic materials, metal alloys, mineral salts, and the like. With the explosive growth in biomedical and materials sciences, the line between these branches has, in practice, virtually disappeared.

SAMPLE PROBLEM 2.6 Identifying an Element from Its *Z* Value

Problem From each of the following *Z* values, give the name, symbol, and group and period numbers of the element, and classify it as a main-group metal, transition metal, inner transition metal, nonmetal, or metalloid: **(a)** $Z = 38$; **(b)** $Z = 17$; **(c)** $Z = 27$.

Plan The *Z* value is the atomic number of an element. The List of Elements inside the front cover is alphabetical, so we look up the name and symbol of the element. Then we use the periodic table to find the group number (top of the column) and the period number (left end of the row) in which the element is located. We classify the element from the color coding in the periodic table.

Solution

(a) Strontium, Sr, is in Group 2A(2) and Period 5, and it is a main-group metal.
(b) Chlorine, Cl, is in Group 7A(17) and Period 3, and it is a nonmetal.
(c) Cobalt, Co, is in Group 8B(9) and Period 4, and it is a transition metal.

Check You can work backwards by starting at the top of the group and moving down to the period to check that you get the correct Z value.

Comment Strontium is one of the alkaline earth metals, and chlorine is a halogen.

FOLLOW-UP PROBLEMS

2.6A Identify each of the following elements from its Z value; give the name, symbol, and group and period numbers, and classify it as a main-group metal, transition metal, inner transition metal, nonmetal, or metalloid: **(a)** $Z = 14$; **(b)** $Z = 55$; **(c)** $Z = 54$.

2.6B Identify each of the following elements from its Z value; give the name, symbol, and group and period numbers, and classify it as a main-group metal, transition metal, inner transition metal, nonmetal, or metalloid: **(a)** $Z = 12$; **(b)** $Z = 7$; **(c)** $Z = 30$.

SOME SIMILAR PROBLEMS 2.56 and 2.57

› Summary of Section 2.6

- In the periodic table, the elements are arranged by atomic number into horizontal periods and vertical groups.
- Nonmetals appear in the upper-right portion of the table, metalloids lie along a staircase line, and metals fill the rest of the table.
- Elements within a group have similar behavior, whereas elements within a period have dissimilar behavior.

2.7 COMPOUNDS: INTRODUCTION TO BONDING

Aside from a few exceptions, *the overwhelming majority of elements occur in nature in compounds combined with other elements.* Only a few elements occur free in nature. The noble gases—helium (He), neon (Ne), argon (Ar), krypton (Kr), xenon (Xe), and radon (Rn)—occur in air as separate atoms. In addition to occurring in compounds, oxygen (O), nitrogen (N), and sulfur (S) occur in their most common elemental form as the molecules O_2, N_2, and S_8, and carbon (C) occurs in vast, nearly pure deposits of coal. And some metals—copper (Cu), silver (Ag), gold (Au), and platinum (Pt)—are also sometimes found uncombined.

Elements combine in two general ways, and both involve *the electrons of the atoms of interacting elements:*

1. *Transferring electrons* from atoms of one element to atoms of another to form **ionic compounds**
2. *Sharing electrons* between atoms of different elements to form **covalent compounds**

These processes generate **chemical bonds,** the forces that hold the atoms together in a compound. This section introduces compound formation, which we'll discuss in much more detail in later chapters.

The Formation of Ionic Compounds

Ionic compounds are composed of **ions,** charged particles that form when an atom (or small group of atoms) gains or loses one or more electrons. The simplest type of ionic compound is a **binary ionic compound,** one composed of ions of two elements. It typically forms *when a metal reacts with a nonmetal:*

- Each metal atom *loses* one or more electrons and becomes a **cation,** a positively charged ion.
- Each nonmetal atom *gains* one or more of the electrons lost by the metal atom and becomes an **anion,** a negatively charged ion.

In effect, the metal atoms *transfer electrons* to the nonmetal atoms. The resulting large numbers of oppositely charged cations and anions attract each other by *electrostatic forces* and form the ionic compound. A cation or anion derived from a single atom is called a **monatomic ion;** we'll discuss *polyatomic ions,* those derived from a small group of atoms, later.

The Case of Sodium Chloride *All binary ionic compounds are solid arrays of oppositely charged ions.* The formation of the binary ionic compound sodium chloride, common table salt, from its elements is depicted in Figure 2.11. In the electron transfer, a sodium atom loses one electron and forms a sodium cation, Na^+. (The charge on the ion is written as a *right superscript.* For a charge of 1+ or 1−, the 1 is not written; for charges of larger magnitude, the sign is written *after* the number: 2+.) A chlorine atom *gains* the electron and becomes a chloride anion, Cl^-. (The name change when the nonmetal atom becomes an anion is discussed in Section 2.8.) The oppositely charged ions (Na^+ and Cl^-) attract each other, and the similarly charged ions (Na^+ and Na^+, or Cl^- and Cl^-) repel each other. The resulting solid aggregation is a regular array of alternating Na^+ and Cl^- ions that extends in all three dimensions. Even the tiniest visible grain of table salt contains an enormous number of sodium and chloride ions.

Coulomb's Law The strength of the ionic bonding depends to a great extent on the net strength of these attractions and repulsions and is described by *Coulomb's law,* which can be expressed as follows: *the energy of attraction (or repulsion) between two particles is directly proportional to the product of the charges and inversely proportional to the distance between them.*

$$\text{Energy} \propto \frac{\text{charge 1} \times \text{charge 2}}{\text{distance}}$$

In other words, as is summarized in Figure 2.12 *(next page),*

- Ions with higher charges attract (or repel) each other more strongly than do ions with lower charges.
- Smaller ions attract (or repel) each other more strongly than do larger ions, because the charges are closer to each other.

Predicting the Number of Electrons Lost or Gained *Ionic compounds are neutral* because they contain equal numbers of positive and negative *charges.* Thus, there are equal numbers of Na^+ and Cl^- ions in sodium chloride, because both ions are singly charged. But there are two Na^+ ions for each oxide ion, O^{2-}, in sodium oxide because it takes two 1+ ions to balance one 2− ion.

Can we predict the number of electrons a given atom will lose or gain when it forms an ion? For A-group elements, we usually find that metal atoms lose electrons and nonmetal atoms gain electrons to *form ions with the same number of electrons as in an atom of the nearest noble gas* [Group 8A(18)]. Noble gases have a stability that is related to their number (and arrangement) of electrons. Thus, a sodium atom ($11e^-$) can attain the stability of a neon atom ($10e^-$), the nearest noble gas, by losing one electron. Similarly, a chlorine atom ($17e^-$) attains the stability of an argon atom ($18e^-$), its nearest noble gas, by gaining one electron. Thus, in general, when an element located near a noble gas forms a monatomic ion,

- *Metals lose electrons:* elements in Group 1A(1) lose one electron, elements in Group 2A(2) lose two, and aluminum in Group 3A(13) loses three.
- *Nonmetals gain electrons:* elements in Group 7A(17) gain one electron, oxygen and sulfur in Group 6A(16) gain two, and nitrogen in Group 5A(15) gains three.

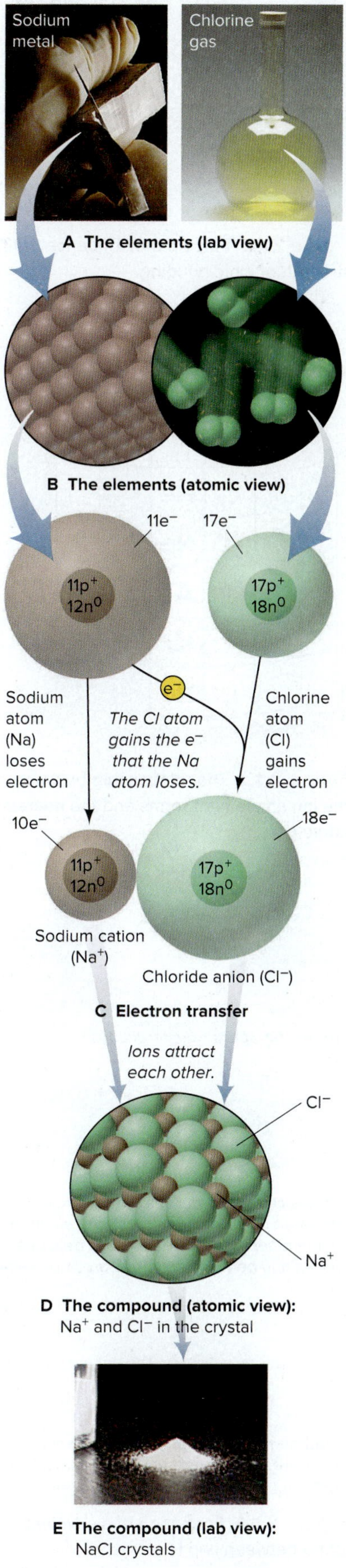

Figure 2.11 **The formation of an ionic compound. A,** The two elements as seen in the laboratory. **B,** The elements on the atomic scale. **C,** The electron transfer from Na atom to Cl atom to form Na^+ and Cl^- ions, shown schematically. **D,** Countless Na^+ and Cl^- ions attract each other and form a regular three-dimensional array. **E,** Crystalline NaCl occurs naturally as the mineral halite.

Source: A(1–2), E: © McGraw-Hill Education/Stephen Frisch, photographer

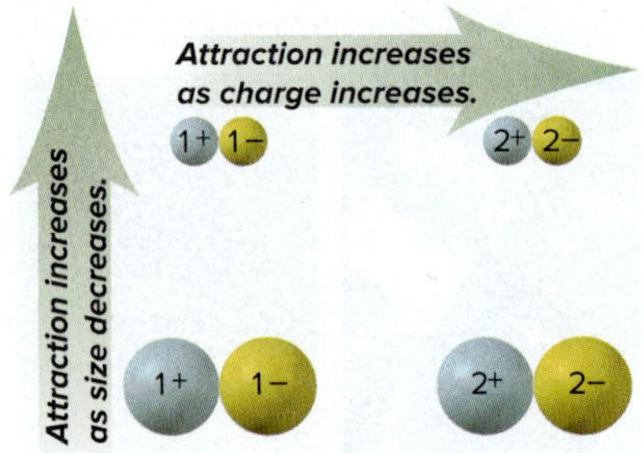
Attraction increases as charge increases.
Attraction increases as size decreases.
1+ 1−
2+ 2−
1+ 1−
2+ 2−

Figure 2.12 Factors that influence the strength of ionic bonding.

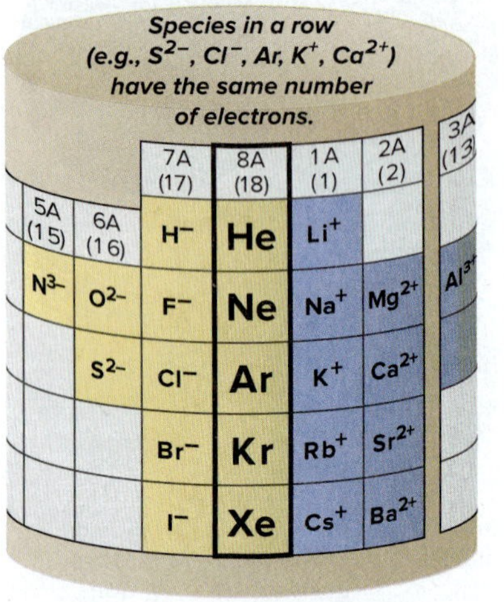
Species in a row (e.g., S^{2-}, Cl^-, Ar, K^+, Ca^{2+}) have the same number of electrons.
5A (15)
6A (16)
7A (17)
8A (18)
1A (1)
2A (2)
3A (13)
H^- He Li^+
N^{3-} O^{2-} F^- Ne Na^+ Mg^{2+} Al^{3+}
S^{2-} Cl^- Ar K^+ Ca^{2+}
Br^- Kr Rb^+ Sr^{2+}
I^- Xe Cs^+ Ba^{2+}

Figure 2.13 The relationship between the ion an element forms and the nearest noble gas.

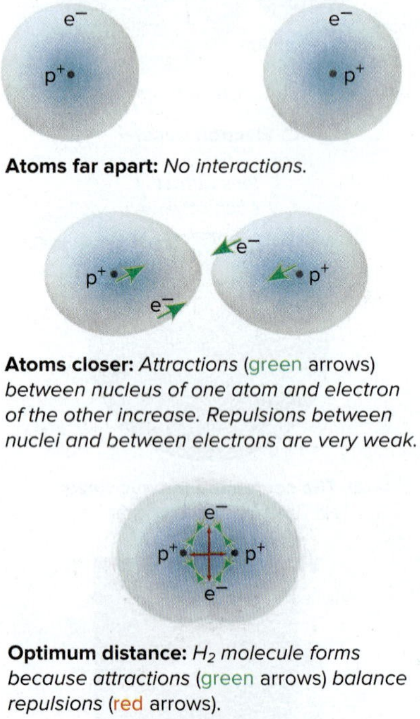
e^- p^+
e^- p^+
Atoms far apart: *No interactions.*
p^+ e^- e^- p^+
Atoms closer: *Attractions (green arrows) between nucleus of one atom and electron of the other increase. Repulsions between nuclei and between electrons are very weak.*
e^- p^+ p^+ e^-
Optimum distance: *H_2 molecule forms because attractions (green arrows) balance repulsions (red arrows).*

Figure 2.14 Formation of a covalent bond between two H atoms.

In the periodic table in Figure 2.9, it looks like the elements in Group 7A(17) are "closer" to the noble gases than the elements in Group 1A(1). In truth, both groups are only one electron away from having the number of electrons in the nearest noble gas. Figure 2.13 shows a periodic table of monatomic ions that has the left and right sides joined so that it forms a cylinder. Note that fluorine (F; $Z = 9$) has one electron *fewer* than the noble gas neon (Ne; $Z = 10$) and sodium (Na; $Z = 11$) has one electron *more;* thus, they form the F^- and Na^+ ions. Similarly, oxygen (O; $Z = 8$) gains two electrons and magnesium (Mg; $Z = 12$) loses two to form the O^{2-} and Mg^{2+} ions and attain the same number of electrons as neon. In Figure 2.13, species in a row have the same number of electrons.

SAMPLE PROBLEM 2.7 **Predicting the Ion an Element Forms**

Problem What monatomic ions would you expect the following elements to form?
(a) Iodine ($Z = 53$) **(b)** Calcium ($Z = 20$) **(c)** Aluminum ($Z = 13$)

Plan We use the given Z value to find the element in the periodic table and see where its group lies relative to the noble gases. Elements in Groups 1A, 2A, and 3A *lose* electrons to attain the same number as the nearest noble gas and become positive ions; those in Groups 5A, 6A, and 7A *gain* electrons and become negative ions.

Solution **(a)** I^- Iodine ($_{53}I$) is in Group 7A(17), the halogens. Like any member of this group, it gains 1 electron to attain the same number as the nearest Group 8A(18) member, in this case, $_{54}Xe$.
(b) Ca^{2+} Calcium ($_{20}Ca$) is in Group 2A(2), the alkaline earth metals. Like any Group 2A member, it loses 2 electrons to attain the same number as the nearest noble gas, $_{18}Ar$.
(c) Al^{3+} Aluminum ($_{13}Al$) is a metal in the boron family [Group 3A(13)] and thus loses 3 electrons to attain the same number as its nearest noble gas, $_{10}Ne$.

FOLLOW-UP PROBLEMS

2.7A What monatomic ion would you expect each of the following elements to form: **(a)** $_{16}S$; **(b)** $_{37}Rb$; **(c)** $_{56}Ba$?

2.7B What monatomic ion would you expect each of the following elements to form: **(a)** $_{38}Sr$; **(b)** $_{8}O$; **(c)** $_{55}Cs$?

SOME SIMILAR PROBLEMS 2.70 and 2.71

The Formation of Covalent Substances

Covalent substances form when atoms of elements share electrons, which usually occurs between nonmetals.

Covalent Bonding in Elements and Some Simple Compounds The simplest case of electron sharing occurs not in a compound but in elemental hydrogen, between two hydrogen atoms (H; $Z = 1$). Imagine two separated H atoms approaching each other (Figure 2.14). As they get closer, the nucleus of each atom attracts the electron of the other atom more and more strongly. As the separated atoms begin to interpenetrate each other, repulsions between the nuclei and between the electrons begin to increase. At some optimum distance between the nuclei, the two atoms form a **covalent bond,** a pair of electrons mutually attracted by the two nuclei. The result is a hydrogen molecule, in which each electron no longer "belongs" to a particular H atom: the two electrons are *shared* by the two nuclei. A sample of hydrogen gas consists of these diatomic molecules (H_2)—pairs of atoms that are chemically bound, each pair behaving as a unit—*not* separate H atoms. Figure 2.15 shows other nonmetals that exist as molecules at room temperature.

Atoms of ***different*** *elements share electrons to form the molecules of a covalent compound.* A sample of hydrogen fluoride, for example, consists of molecules in

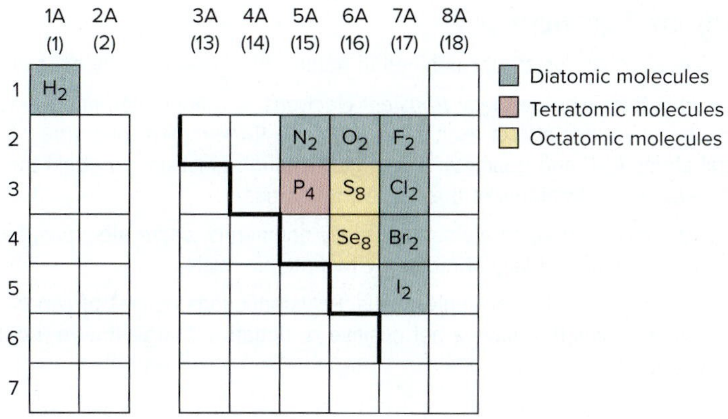

Figure 2.15 Elements that occur as molecules.

which one H atom forms a covalent bond with one F atom; water consists of molecules in which one O atom forms covalent bonds with two H atoms:

Distinguishing the Entities in Covalent and Ionic Substances There are two key distinctions between the chemical entities in covalent substances and those in ionic substances.

1. *Most covalent substances consist of molecules.* A cup of water, for example, consists of individual water molecules lying near each other. In contrast, under ordinary conditions, *there are no molecules in an ionic compound.* A piece of sodium chloride, for example, is a continuous array in three dimensions of oppositely charged sodium and chloride ions, *not* a collection of individual sodium chloride "molecules."
2. *The nature of the particles attracting each other in covalent and in ionic substances is fundamentally different.* Covalent bonding involves the mutual attraction between two (positively charged) nuclei and the two (negatively charged) electrons that reside between them. Ionic bonding involves the mutual attraction between positive and negative ions.

Polyatomic Ions: Covalent Bonds Within Ions Many ionic compounds contain **polyatomic ions,** which consist of two or more atoms bonded *covalently* and have a net positive or negative charge. For example, Figure 2.16 shows that a crystalline form of calcium carbonate *(left)* occurs on the atomic scale *(center)* as an array of polyatomic carbonate anions and monatomic calcium cations. The carbonate ion *(right)* consists of a carbon atom *covalently* bonded to three oxygen atoms, and two additional electrons give the ion its 2− charge. In many reactions, the polyatomic ion stays together as a unit.

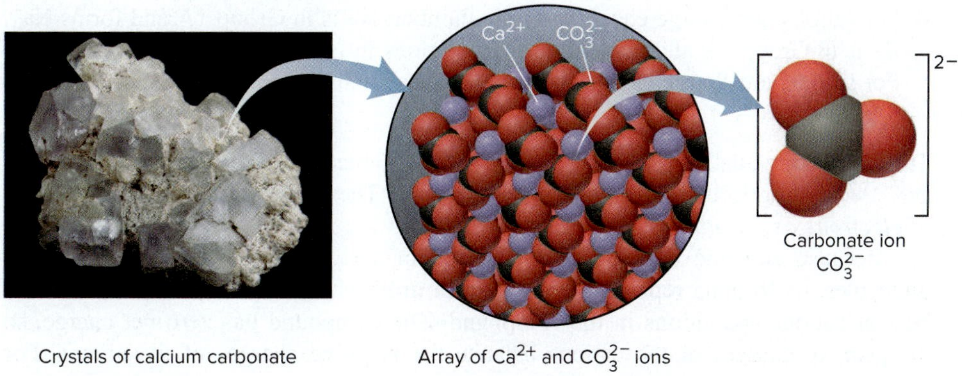

Figure 2.16 The carbonate ion in calcium carbonate.
Source: (calcium carbonate crystals) © papa1266/Shutterstock.com

› Summary of Section 2.7

› Although a few elements occur uncombined in nature, the great majority exist in compounds.

› Ionic compounds form when a metal *transfers* electrons to a nonmetal, and the resulting positive and negative ions attract each other to form a three-dimensional array. In many cases, metal atoms lose and nonmetal atoms gain enough electrons to attain the same number of electrons as in atoms of the nearest noble gas.

› Covalent compounds form when elements, usually nonmetals, *share* electrons. Each covalent bond is an electron pair mutually attracted by two atomic nuclei.

› Monatomic ions are derived from single atoms. Polyatomic ions consist of two or more covalently bonded atoms that have a net positive or negative charge due to a deficit or excess of electrons.

2.8 COMPOUNDS: FORMULAS, NAMES, AND MASSES

In a **chemical formula,** element symbols and, often, numerical subscripts show the type and number of each atom in the smallest unit of the substance. In this section, you'll learn how to write the names and formulas of ionic and simple covalent compounds, how to calculate the mass of a compound from its formula, and how to visualize molecules with three-dimensional models. To make learning the names and formula of compounds easier, we'll rely on various rules, so be prepared for a bit of memorization and a lot of practice.

Binary Ionic Compounds

Let's begin with two general rules:

- For *all* ionic compounds, *names and formulas give the positive ion (cation) first and the negative ion (anion) second.*
- For all *binary* ionic compounds, *the name of the cation is the name of the metal, and the name of the anion has the suffix -ide added to the root of the name of the nonmetal.*

For example, the anion formed from brom*ine* is named brom*ide* (brom+ide). Therefore, the compound formed from the metal calcium and the nonmetal bromine is named *calcium bromide.*

In general, if the metal of a binary ionic compound is a main-group element (A groups), it usually forms a single type of ion; if it is a transition element (B groups), it often forms more than one. We discuss each case in turn.

Compounds of Elements That Form One Ion The periodic table presents some key points about the formulas of main-group monatomic ions (Figure 2.17):

- Monatomic ions of elements in the same main group have the same ionic charge; the alkali metals—Li, Na, K, Rb, Cs, and Fr—form ions with a 1+ charge; the halogens—F, Cl, Br, and I—form ions with a 1− charge; and so forth.
- For cations, ion charge equals A-group number: Na is in Group 1A and forms Na^+, Ba is in Group 2A and forms Ba^{2+}. (Exceptions in Figure 2.17 are Sn^{2+} and Pb^{2+}.)
- For anions, ion charge equals A-group number minus 8; for example, S is in Group 6A ($6 - 8 = -2$) and thus forms S^{2-}.

Try to memorize the A-group monatomic ions in Table 2.3 (all except Ag^+, Zn^{2+}, and Cd^{2+}) according to their positions in Figure 2.17. These ions have *the same number of electrons as an atom of the nearest noble gas.*

Because an ionic compound consists of an array of ions rather than separate molecules, its formula represents the **formula unit,** which contains the *relative* numbers of cations and anions in the compound. The compound has zero net charge, so the positive charges of the cations balance the negative charges of the anions. For

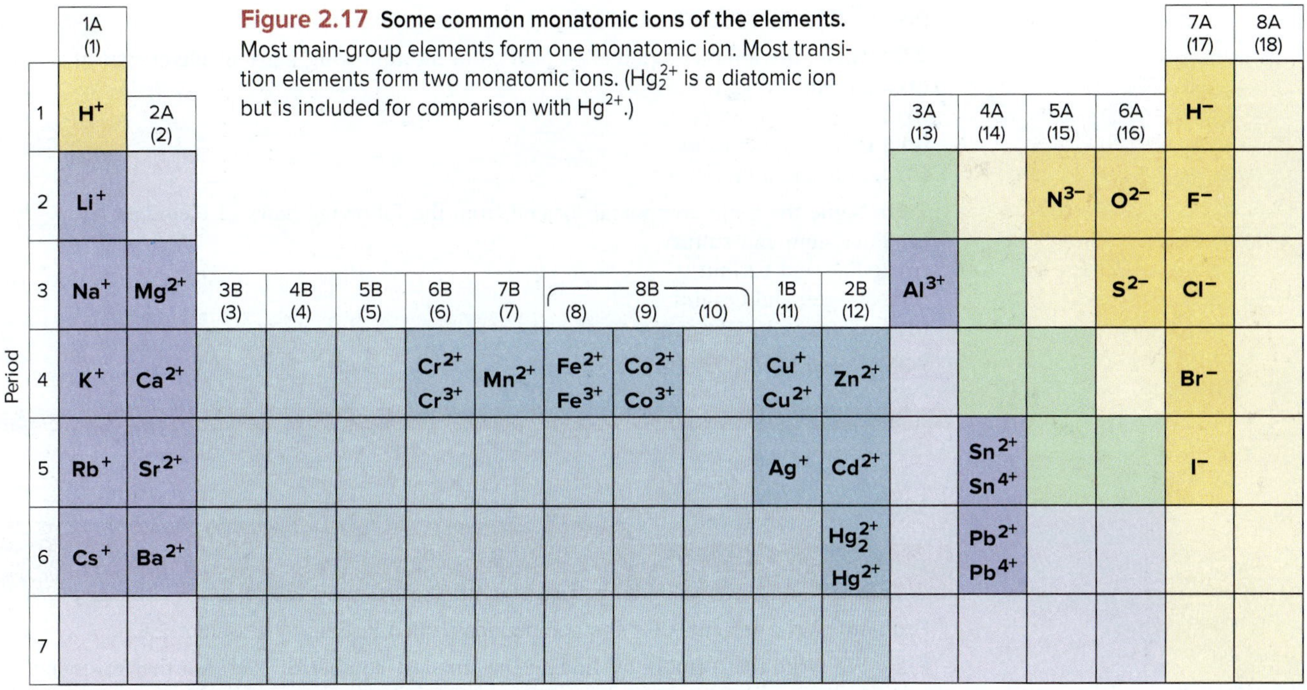

Figure 2.17 **Some common monatomic ions of the elements.** Most main-group elements form one monatomic ion. Most transition elements form two monatomic ions. (Hg_2^{2+} is a diatomic ion but is included for comparison with Hg^{2+}.)

example, calcium bromide is composed of Ca^{2+} ions and Br^- ions, so two Br^- balance each Ca^{2+}. The formula is $CaBr_2$, not Ca_2Br. In this and all other formulas,

- The subscript refers to the element symbol *preceding* it.
- The *subscript 1 is understood* from the presence of the element symbol alone (that is, we do not write Ca_1Br_2).
- The charge (without the sign) of one ion becomes the subscript of the other:

$$Ca^{2+} \quad Br^{1-} \qquad \text{gives} \qquad Ca_1Br_2 \qquad \text{or} \qquad CaBr_2$$

- The subscripts are reduced to the smallest whole numbers that retain the ratio of ions. Thus, for example, for the Ca^{2+} and O^{2-} ions in calcium oxide, we get Ca_2O_2, which we reduce to the formula CaO.*

The following two sample problems apply these rules. In Sample Problem 2.8, we name the compound from its elements, and in Sample Problem 2.9, we find the formula.

SAMPLE PROBLEM 2.8 **Naming Binary Ionic Compounds**

Problem Name the ionic compound formed from the following pairs of elements:
(a) Magnesium and nitrogen
(b) Iodine and cadmium
(c) Strontium and fluorine
(d) Sulfur and cesium

Plan The key to naming a binary ionic compound is to recognize which element is the metal and which is the nonmetal. When in doubt, check the periodic table. We place the cation name first, add the suffix *-ide* to the nonmetal root, and place the anion name second.

Solution **(a)** Magnesium is the metal; *nitr-* is the nonmetal root: magnesium nitride.
(b) Cadmium is the metal; *iod-* is the nonmetal root: cadmium iodide.
(c) Strontium is the metal; *fluor-* is the nonmetal root: strontium fluoride. (Note the spelling is fl*uo*ride, not fl*ou*ride.)
(d) Cesium is the metal; *sulf-* is the nonmetal root: cesium sulfide.

*Compounds of the mercury(I) ion, such as Hg_2Cl_2, and peroxides of the alkali metals, such as Na_2O_2, are the only two common exceptions to this step: reducing the subscripts for these compounds would give the incorrect formulas HgCl and NaO.

Table 2.3 **Common Monatomic Ions***

Charge	Formula	Name
Cations		
1+	H^+	hydrogen
	Li^+	**lithium**
	Na^+	**sodium**
	K^+	**potassium**
	Cs^+	cesium
	Ag^+	**silver**
2+	Mg^{2+}	**magnesium**
	Ca^{2+}	**calcium**
	Sr^{2+}	strontium
	Ba^{2+}	**barium**
	Zn^{2+}	**zinc**
	Cd^{2+}	cadmium
3+	Al^{3+}	aluminum
Anions		
1−	H^-	hydride
	F^-	**fluoride**
	Cl^-	**chloride**
	Br^-	**bromide**
	I^-	**iodide**
2−	O^{2-}	**oxide**
	S^{2-}	**sulfide**
3−	N^{3-}	nitride

*Listed by charge; those in **boldface** are most common.

FOLLOW-UP PROBLEMS

2.8A Name the ionic compound formed from the following pairs of elements:
(a) Zinc and oxygen
(b) Bromine and silver
(c) Lithium and chlorine
(d) Sulfur and aluminum

2.8B Name the ionic compound formed from the following pairs of elements:
(a) Potassium and sulfur
(b) Iodine and barium
(c) Nitrogen and cesium
(d) Sodium and hydrogen

SOME SIMILAR PROBLEMS 2.84–2.87

SAMPLE PROBLEM 2.9 Determining Formulas of Binary Ionic Compounds

Problem Write formulas for the compounds named in Sample Problem 2.8.

Plan We write the formula by finding the smallest number of each ion that gives the neutral compound. These numbers appear as *right subscripts* to the element symbol.

Solution **(a)** Mg^{2+} and N^{3-}; three Mg^{2+} ions (6+) balance two N^{3-} ions (6−): Mg_3N_2
(b) Cd^{2+} and I^-; one Cd^{2+} ion (2+) balances two I^- ions (2−): CdI_2
(c) Sr^{2+} and F^-; one Sr^{2+} ion (2+) balances two F^- ions (2−): SrF_2
(d) Cs^+ and S^{2-}; two Cs^+ ions (2+) balance one S^{2-} ion (2+): Cs_2S

Comment **1.** The subscript 1 is understood and so not written; thus, in (b), we do *not* write Cd_1I_2.
2. Ion charges do *not* appear in the compound formula; thus, in (c), we do *not* write $Sr^{2+}F_2^-$.

FOLLOW-UP PROBLEMS

2.9A Write the formula of each compound named in Follow-up Problem 2.8A.

2.9B Write the formula of each compound named in Follow-up Problem 2.8B.

SOME SIMILAR PROBLEMS 2.84–2.87

Compounds of Metals That Form More Than One Ion As noted earlier, many metals, particularly the transition elements (B groups), can form more than one ion. Table 2.4 lists some examples; see Figure 2.17 for their placement in the periodic table. Names of compounds containing these elements include a *roman numeral within parentheses* immediately after the metal ion's name to indicate its ionic charge. For example, iron can form Fe^{2+} and Fe^{3+} ions. Iron forms two compounds with chlorine: $FeCl_2$, named iron(II) chloride (spoken "iron two chloride"), which contains Fe^{2+}; and $FeCl_3$, named iron(III) chloride, which contains Fe^{3+}.

We are focusing here on systematic names, but some common (trivial) names are still used. In common names for certain metal ions, the Latin root of the metal is followed by either of two suffixes (see Table 2.4):

- The suffix *-ous* for the ion with the lower charge
- The suffix *-ic* for the ion with the higher charge

Thus, iron(II) chloride is also called ferr*ous* chloride and iron(III) chloride is ferr*ic* chloride. (Memory aid: there is an *o* in *-ous* and *lower,* and an *i* in *-ic* and *higher.*)

Table 2.4 Some Metals That Form More Than One Monatomic Ion*

Element	Ion Formula	Systematic Name	Common (Trivial) Name
Chromium	Cr^{2+}	chromium(II)	chromous
	Cr^{3+}	**chromium(III)**	chromic
Cobalt	Co^{2+}	cobalt(II)	
	Co^{3+}	cobalt(III)	
Copper	**Cu^{+}**	**copper(I)**	cuprous
	Cu^{2+}	**copper(II)**	cupric
Iron	**Fe^{2+}**	**iron(II)**	ferrous
	Fe^{3+}	**iron(III)**	ferric
Lead	**Pb^{2+}**	**lead(II)**	
	Pb^{4+}	lead(IV)	
Mercury	Hg_2^{2+}	mercury(I)	mercurous
	Hg^{2+}	**mercury(II)**	mercuric
Tin	**Sn^{2+}**	**tin(II)**	stannous
	Sn^{4+}	tin(IV)	stannic

*Listed alphabetically by metal name; the ions in **boldface** are most common.

SAMPLE PROBLEM 2.10 Determining Names and Formulas of Ionic Compounds of Metals That Form More Than One Ion

Problem Give the systematic name for the formula or the formula for the name:
(a) Tin(II) fluoride **(b)** CrI_3 **(c)** Ferric oxide **(d)** CoS

Solution **(a)** Tin(II) ion is Sn^{2+}; fluoride is F^-. Two F^- ions balance one Sn^{2+} ion: tin(II) fluoride is SnF_2. (The common name is stannous fluoride.)
(b) The anion is I^-, iodide, and the formula shows three I^-. Therefore, the cation must be Cr^{3+}, chromium(III) ion: CrI_3 is chromium(III) iodide. (The common name is chromic iodide.)
(c) *Ferric* is the common name for iron(III) ion, Fe^{3+}; oxide ion is O^{2-}. To balance the charges, the formula is Fe_2O_3. [The systematic name is iron(III) oxide.]
(d) The anion is sulfide, S^{2-}, which requires that the cation be Co^{2+}. The name is cobalt(II) sulfide.

FOLLOW-UP PROBLEMS

2.10A Give the systematic name for the formula or the formula for the name:
(a) lead(IV) oxide (a component of car batteries); **(b)** Cu_2S; **(c)** $FeBr_2$; **(d)** mercuric chloride

2.10B Give the systematic name for the formula or the formula for the name:
(a) copper(II) nitride; **(b)** PbI_2; **(c)** chromic sulfide; **(d)** FeO

SOME SIMILAR PROBLEMS 2.88 and 2.89

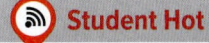

Student Hot Spot

Student data indicate you may struggle with formulas of ionic compounds that include Roman numerals. Access the Smartbook to view additional Learning Resources on this topic.

Compounds That Contain Polyatomic Ions

Many ionic compounds contain polyatomic ions. Table 2.5 on the next page lists some common polyatomic ions. Remember that *the polyatomic ion stays together as a charged unit.* The formula for potassium nitrate is KNO_3: each K^+ balances one NO_3^-. The formula for sodium carbonate is Na_2CO_3: two Na^+ balance one CO_3^{2-}. *When two or more of the same polyatomic ion are present in the formula unit, that ion appears in parentheses with the subscript written outside.* For example, calcium nitrate contains

Table 2.5 Common Polyatomic Ions*

Formula	Name
Cations	
NH_4^+	ammonium
H_3O^+	hydronium
Anions	
CH_3COO^- (or **$C_2H_3O_2^-$**)	**acetate**
CN^-	cyanide
OH^-	**hydroxide**
ClO^-	hypochlorite
ClO_2^-	chlorite
ClO_3^-	**chlorate**
ClO_4^-	**perchlorate**
NO_2^-	nitrite
NO_3^-	**nitrate**
MnO_4^-	**permanganate**
CO_3^{2-}	**carbonate**
HCO_3^-	**hydrogen carbonate** (or **bicarbonate**)
CrO_4^{2-}	chromate
$Cr_2O_7^{2-}$	**dichromate**
O_2^{2-}	peroxide
PO_4^{3-}	**phosphate**
HPO_4^{2-}	hydrogen phosphate
$H_2PO_4^-$	dihydrogen phosphate
SO_3^{2-}	sulfite
SO_4^{2-}	**sulfate**
HSO_4^-	hydrogen sulfate (or bisulfate)

*__Boldface__ ions are the most common.

one Ca^{2+} and two NO_3^- ions and has the formula $Ca(NO_3)_2$. Parentheses and a subscript are *only* used if *more than one* of a given polyatomic ion is present; thus, sodium nitrate is $NaNO_3$, *not* $Na(NO_3)$.

Families of Oxoanions As Table 2.5 shows, most polyatomic ions are **oxoanions** (or *oxyanions*), those in which an element, usually a nonmetal, is bonded to one or more oxygen atoms. There are several families of two or four oxoanions that differ only in the number of oxygen atoms. The following naming conventions are used with these ions.

With two oxoanions in the family:

- The ion with *more* O atoms takes the nonmetal root and the suffix *-ate*.
- The ion with *fewer* O atoms takes the nonmetal root and the suffix *-ite*.

For example, SO_4^{2-} is the sulf*ate* ion, and SO_3^{2-} is the sulf*ite* ion; similarly, NO_3^- is nitr*ate*, and NO_2^- is nitr*ite*.

With four oxoanions in the family (a halogen bonded to O) (Figure 2.18):

- The ion with the *most* O atoms has the prefix *per-*, the nonmetal root, and the suffix *-ate*.
- The ion with *one fewer* O atom has just the root and the suffix *-ate*.
- The ion with *two fewer* O atoms has just the root and the suffix *-ite*.
- The ion with the *least (three fewer)* O atoms has the prefix *hypo-*, the root, and the suffix *-ite*.

No. of O atoms (↑)	Prefix	Root	Suffix
	per	*root*	ate
		root	ate
		root	ite
	hypo	*root*	ite

Figure 2.18 Naming oxoanions. Prefixes and suffixes indicate the number of O atoms in the anion.

For example, for the four chlorine oxoanions,

ClO_4^- is *per*chlor*ate*, ClO_3^- is chlor*ate*, ClO_2^- is chlor*ite*, ClO^- is *hypo*chlor*ite*

Hydrated Ionic Compounds Ionic compounds called **hydrates** have a specific number of water molecules in each formula unit. The water molecules are shown after a centered dot in the formula and named with a Greek numerical prefix before the word *hydrate* (Table 2.6). To give just two examples,

Epsom salt: $MgSO_4 \cdot 7H_2O$ magnesium sulfate *hepta*hydrate
(seven water molecules in each formula unit)

Washing soda: $Na_2CO_3 \cdot 10H_2O$ sodium carbonate *deca*hydrate
(ten water molecules in each formula unit)

The water molecules, referred to as "waters of hydration," are part of the hydrate's structure. Heating can remove some or all of them, leading to a different substance. For example, when heated strongly, blue copper(II) sulfate pentahydrate ($CuSO_4 \cdot 5H_2O$) is converted to white copper(II) sulfate ($CuSO_4$).

Copper(II) sulfate pentahydrate

Copper(II) sulfate

Source: © McGraw-Hill Education/Charles Winters/Timeframe Photography, Inc.

SAMPLE PROBLEM 2.11 Determining Names and Formulas of Ionic Compounds Containing Polyatomic Ions (Including Hydrates)

Problem Give the systematic name for the formula or the formula for the name:
(a) $Fe(ClO_4)_2$ **(b)** Sodium sulfite **(c)** $Ba(OH)_2{\cdot}8H_2O$

Solution (a) ClO_4^- is perchlorate, which has a 1− charge, so the cation must be Fe^{2+}. The name is iron(II) perchlorate. (The common name is ferrous perchlorate.)
(b) Sodium is Na^+; sulfite is SO_3^{2-}, and two Na^+ ions balance one SO_3^{2-} ion. The formula is Na_2SO_3.
(c) Ba^{2+} is barium; OH^- is hydroxide. There are eight (*octa-*) water molecules in each formula unit. The name is barium hydroxide octahydrate.

FOLLOW-UP PROBLEMS

2.11A Give the systematic name for the formula or the formula for the name:
(a) cupric nitrate trihydrate; **(b)** zinc hydroxide; **(c)** LiCN.

2.11B Give the systematic name for the formula or the formula for the name:
(a) ammonium sulfate; **(b)** $Ni(NO_3)_2{\cdot}6H_2O$; **(c)** potassium bicarbonate.

SOME SIMILAR PROBLEMS 2.90 and 2.91

Table 2.6 Numerical Prefixes for Hydrates and Binary Covalent Compounds*

Number	Prefix
1	mono-
2	di-
3	tri-
4	tetra-
5	penta-
6	hexa-
7	hepta-
8	octa-
9	nona-
10	deca-

*It is common practice to drop the final *a* from the prefix when naming binary covalent oxides (see the Comment in Sample Problem 2.14).

Student Hot Spot

Student data indicate that you may struggle with writing formulas of ionic compounds containing polyatomic ions. Access the Smartbook to view additional Learning Resources on this topic.

SAMPLE PROBLEM 2.12 Recognizing Incorrect Names and Formulas of Ionic Compounds

Problem Explain what is wrong with the name or formula at the end of each statement, and correct it:
(a) $Ba(C_2H_3O_2)_2$ is called barium diacetate.
(b) Sodium sulfide has the formula $(Na)_2SO_3$.
(c) Iron(II) sulfate has the formula $Fe_2(SO_4)_3$.
(d) Cesium carbonate has the formula $Cs_2(CO_3)$.

Solution (a) The charge of the Ba^{2+} ion *must* be balanced by *two* $C_2H_3O_2^-$ ions, so the prefix *di-* is unnecessary. For ionic compounds, we do not indicate the number of ions with numerical prefixes. The correct name is barium acetate.
(b) Two mistakes occur here. The sodium ion is monatomic, so it does *not* require parentheses. The sulfide ion is S^{2-}, *not* SO_3^{2-} (which is sulfite). The correct formula is Na_2S.
(c) The roman numeral refers to the charge of the ion, *not* the number of ions in the formula. Fe^{2+} is the cation, so it requires one SO_4^{2-} to balance its charge. The correct formula is $FeSO_4$. [$Fe_2(SO_4)_3$ is the formula for iron(III) sulfate.]
(d) Parentheses are *not* required when only one polyatomic ion of a particular kind is present. The correct formula is Cs_2CO_3.

FOLLOW-UP PROBLEMS

2.12A State why the formula or name at the end of each statement is incorrect, and correct it:
(a) Ammonium phosphate is $(NH_3)_4PO_4$.
(b) Aluminum hydroxide is $AlOH_3$.
(c) $Mg(HCO_3)_2$ is manganese(II) carbonate.

2.12B State why the formula or name at the end of each statement is incorrect, and correct it:
(a) $Cr(NO_3)_3$ is chromic(III) nitride.
(b) $Ca(NO_2)_2$ is cadmium nitrate.
(c) Potassium chlorate is $P(ClO_4)$.

SOME SIMILAR PROBLEMS 2.92 and 2.93

Acid Names from Anion Names

Acids are an important group of hydrogen-containing compounds that have been used in chemical reactions since before alchemical times. In the laboratory, acids are typically used in water solution. When naming them and writing their formulas, we consider acids as anions that have one or more hydrogen ions (H^+) added to them to give a neutral compound. The two common types of acids are binary acids and oxoacids:

1. *Binary acid* solutions form when certain gaseous compounds dissolve in water. For example, when gaseous hydrogen chloride (HCl) dissolves in water, it forms hydrochloric acid, which is named as follows:

   Prefix *hydro-* + nonmetal *root* + suffix *-ic* + separate word *acid*
   hydro + chlor + ic + acid

   or *hydrochloric acid.* This naming pattern holds for many compounds in which hydrogen combines with an anion that has an *-ide* suffix.
2. *Oxoacid* names are similar to those of the oxoanions, except for two suffix changes:
   - *-ate* in the anion becomes *-ic* in the acid
   - *-ite* in the anion becomes *-ous* in the acid

   Thus, following this pattern:

   Oxyanion *root* + suffix -ic or -ous + separate word *acid*

   IO_2^- is iod*ite*, and HIO_2 is iod*ous* acid.

   NO_3^- is nitr*ite*, and HNO_3 is nitr*ic* acid.

The oxoanion prefixes *hypo-* and *per-* are retained:

BrO_4^- is *per*brom*ate*, and $HBrO_4$ is *per*brom*ic* acid.

Note that the prefix hydro- is *not* used when naming oxoacids.

SAMPLE PROBLEM 2.13 Determining Names and Formulas of Anions and Acids

Problem Name each of the following anions, and give the name and formula of the acid derived from it: **(a)** Br^-; **(b)** IO_3^-; **(c)** CN^-; **(d)** HSO_4^-.

Solution **(a)** The anion is bromide; the acid is hydrobromic acid, HBr.
(b) The anion is iodate; the acid is iodic acid, HIO_3.
(c) The anion is cyanide; the acid is hydrocyanic acid, HCN.
(d) The anion is hydrogen sulfate; the acid is sulfuric acid, H_2SO_4. (In this case, the suffix is added to the element name *sulfur,* not to the root, *sulf-*.)

FOLLOW-UP PROBLEMS

2.13A Write the formula and name for the anion of each acid: **(a)** chloric acid; **(b)** hydrofluoric acid; **(c)** acetic acid; **(d)** nitrous acid.

2.13B Name each of the following acids, and give the name and formula of its anion: **(a)** H_2SO_3 (two anions); **(b)** HBrO; **(c)** $HClO_2$; **(d)** HI.

SOME SIMILAR PROBLEMS 2.94 and 2.95

Binary Covalent Compounds

Binary covalent compounds are typically formed by the combination of two nonmetals. Some are so familiar that we use their common names, such as ammonia (NH_3), methane (CH_4), and water (H_2O), but most are named systematically:

- The element with the lower group number in the periodic table comes first in the name. The element with the higher group number comes second and is named with its root and the suffix *-ide*. For example, nitrogen [Group 5A(15)] and fluorine

[Group 7A(17)] form a compound that has three fluorine atoms for every nitrogen atom. The name and formula are nitrogen trifluoride, NF_3. (*Exception:* When the compound contains oxygen and any of the halogens chlorine, bromine, or iodine, the halogen is named first.)

- If both elements are in the same group, the one with the higher period number is named first. Thus, one compound that the Group 6A(16) elements sulfur (Period 3) and oxygen (Period 2) form is sulfur dioxide, SO_2.
- Covalent compounds use Greek numerical prefixes (see Table 2.6) to indicate the number of atoms of each element. The first element in the name has a prefix *only* when more than one atom of it is present; the second element *usually* has a prefix.

SAMPLE PROBLEM 2.14 Determining Names and Formulas of Binary Covalent Compounds

Problem **(a)** What is the formula of carbon disulfide?
(b) What is the name of PCl_5?
(c) Give the name and formula of the compound whose molecules each consist of two N atoms and four O atoms.

Solution **(a)** The prefix *di-* means "two." The formula is CS_2.
(b) P is the symbol for phosphorus; there are five chlorine atoms, which is indicated by the prefix *penta-*. The name is phosphorus pentachloride.
(c) Nitrogen (N) comes first in the name (lower group number). The compound is dinitrogen tetroxide, N_2O_4.

Comment Note that, as the first elements in the name, carbon in (a) and phosphorus in (b) do not have the prefix *mono-*. In (c), the *a* at the end of the Greek prefix is dropped when the prefix is added to *-oxide.* Similarly, "hexa-oxide" becomes *hexoxide,* "deca-oxide" becomes *decoxide,* and so forth.

FOLLOW-UP PROBLEMS

2.14A Give the name or formula for **(a)** SO_3; **(b)** SiO_2; **(c)** dinitrogen monoxide; **(d)** selenium hexafluoride.

2.14B Give the name or formula for **(a)** SCl_2; **(b)** N_2O_5; **(c)** boron trifluoride; **(d)** iodine tribromide.

SOME SIMILAR PROBLEMS 2.98 and 2.99

SAMPLE PROBLEM 2.15 Recognizing Incorrect Names and Formulas of Binary Covalent Compounds

Problem Explain what is wrong with the name or formula at the end of each statement, and correct it:
(a) SF_4 is monosulfur pentafluoride.
(b) Dichlorine heptoxide is Cl_2O_6.
(c) N_2O_3 is dinitrotrioxide.

Solution **(a)** There are two mistakes. *Mono-* is not needed if there is only one atom of the first element, and the prefix for four is *tetra-,* not *penta-*. The correct name is sulfur tetrafluoride.
(b) The prefix *hepta-* indicates seven, not six. The correct formula is Cl_2O_7.
(c) The full name of the first element is needed, and a space separates the two element names. The correct name is dinitrogen trioxide.

FOLLOW-UP PROBLEMS

2.15A Explain what is wrong with the name or formula at the end of each statement, and correct it:
(a) S_2Cl_2 is disulfurous dichloride.
(b) Nitrogen monoxide is N_2O.
(c) $BrCl_3$ is trichlorine bromide.

2.15B Explain what is wrong with the name or formula at the end of each statement, and correct it:

(a) P_4O_6 is tetraphosphorous hexaoxide.

(b) SF_6 is hexafluorosulfide.

(c) Nitrogen tribromide is $NiBR_3$.

SOME SIMILAR PROBLEMS 2.100 and 2.101

The Simplest Organic Compounds: Straight-Chain Alkanes

Organic compounds typically have complex structures that consist of chains, branches, and/or rings of carbon atoms that are also bonded to hydrogen atoms and, often, to atoms of oxygen, nitrogen, and a few other elements. At this point, we'll name just the simplest organic compounds. Rules for naming other organic compounds are provided in Chapter 15.

Hydrocarbons, the simplest type of organic compound, contain *only* carbon and hydrogen. *Alkanes* are the simplest type of hydrocarbon; many function as important fuels, such as methane, propane, butane, and the mixture that makes up gasoline. The simplest alkanes to name are the *straight-chain alkanes* because the carbon chains have no branches. Alkanes are named with a *root,* based on the number of C atoms in the chain, followed by the suffix *-ane.*

Table 2.7 gives names, molecular formulas, and space-filling models (discussed shortly) for the first 10 straight-chain alkanes. Note that the roots of the four smallest alkanes are new, but those for the larger ones are the same as the Greek prefixes shown in Table 2.6 (with the final *-a* dropped).

Table 2.7 The First 10 Straight-Chain Alkanes

Name (Formula)	Model
Methane (CH_4)	
Ethane (C_2H_6)	
Propane (C_3H_8)	
Butane (C_4H_{10})	
Pentane (C_5H_{12})	
Hexane (C_6H_{14})	
Heptane (C_7H_{16})	
Octane (C_8H_{18})	
Nonane (C_9H_{20})	
Decane ($C_{10}H_{22}$)	

Molecular Masses from Chemical Formulas

In Section 2.5, we calculated the atomic mass of an element. Using the periodic table and the formula of a compound, we calculate the **molecular mass** (also called the *molecular weight*) of a molecule or formula unit of the compound as the sum of the atomic masses:

$$\text{Molecular mass} = \text{sum of atomic masses} \quad \textbf{(2.4)}$$

The molecular mass of a water molecule (using atomic masses to four significant figures from the periodic table) is

$$\begin{aligned}\text{Molecular mass of } H_2O &= (2 \times \text{atomic mass of H}) + (1 \times \text{atomic mass of O})\\ &= (2 \times 1.008 \text{ amu}) + 16.00 \text{ amu} = 18.02 \text{ amu}\end{aligned}$$

Ionic compounds don't consist of molecules, so the mass of a formula unit is sometimes called the **formula mass** instead of *molecular mass.* For example, for calcium chloride, we have

$$\begin{aligned}\text{Formula mass of } CaCl_2 &= (1 \times \text{atomic mass of Ca}) + (2 \times \text{atomic mass of Cl})\\ &= 40.08 \text{ amu} + (2 \times 35.45 \text{ amu}) = 110.98 \text{ amu}\end{aligned}$$

We can use atomic masses, not ionic masses, because electron loss equals electron gain, so electron mass is balanced.

To calculate the formula mass of a compound with a polyatomic ion, *the number of atoms of each element inside the parentheses is multiplied by the subscript outside the parentheses.* For barium nitrate, $Ba(NO_3)_2$,

Formula mass of $Ba(NO_3)_2$

$$\begin{aligned}&= (1 \times \text{atomic mass of Ba}) + (2 \times \text{atomic mass of N}) + (6 \times \text{atomic mass of O})\\ &= 137.3 \text{ amu} + (2 \times 14.01 \text{ amu}) + (6 \times 16.00 \text{ amu}) = 261.3 \text{ amu}\end{aligned}$$

In the next two sample problems, the name or molecular depiction is used to find a compound's molecular or formula mass.

SAMPLE PROBLEM 2.16 Calculating the Molecular Mass of a Compound

Problem Using the periodic table, calculate the molecular (or formula) mass of
(a) Tetraphosphorus trisulfide
(b) Ammonium nitrate

Plan We first write the formula, then multiply the number of atoms (or ions) of each element by its atomic mass (from the periodic table), and find the sum.

Solution
(a) The formula is P_4S_3.

$$\begin{aligned}\text{Molecular mass} &= (4 \times \text{atomic mass of P}) + (3 \times \text{atomic mass of S})\\ &= (4 \times 30.97\ \text{amu}) + (3 \times 32.06\ \text{amu})\\ &= 220.06\ \text{amu}\end{aligned}$$

(b) The formula is NH_4NO_3. We count the total number of N atoms even though they belong to different ions:

Formula mass

$$\begin{aligned}&= (2 \times \text{atomic mass of N}) + (4 \times \text{atomic mass of H}) + (3 \times \text{atomic mass of O})\\ &= (2 \times 14.01\ \text{amu}) + (4 \times 1.008\ \text{amu}) + (3 \times 16.00\ \text{amu})\\ &= 80.05\ \text{amu}\end{aligned}$$

Check You can often find large errors by rounding atomic masses to the nearest 5 and adding: **(a)** $(4 \times 30) + (3 \times 30) = 210 \approx 220.06$. The sum has two decimal places because the atomic masses have two. **(b)** $(2 \times 15) + 4 + (3 \times 15) = 79 \approx 80.05$.

FOLLOW-UP PROBLEMS

2.16A What is the molecular (or formula) mass of each of the following?
(a) Hydrogen peroxide
(b) Cesium carbonate

2.16B What is the molecular (or formula) mass of each of the following?
(a) Sulfuric acid
(b) Potassium sulfate

SOME SIMILAR PROBLEMS 2.104 and 2.105

SAMPLE PROBLEM 2.17 Using Molecular Depictions to Determine Formula, Name, and Mass

Problem Each scene represents a binary compound. Determine its formula, name, and molecular (or formula) mass.

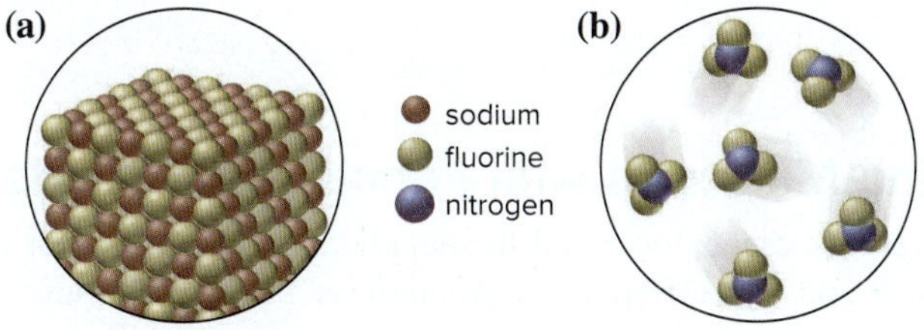

Plan Each of the compounds contains only two elements, so to find the formula, we find the simplest whole-number ratio of one atom to the other. From the formula, we determine the name and the molecular (or formula) mass.

Solution (a) There is one brown sphere (sodium) for each yellow sphere (fluorine), so the formula is NaF. A metal and nonmetal form an ionic compound, in which the metal is named first: sodium fluoride.

$$\text{Formula mass} = (1 \times \text{atomic mass of Na}) + (1 \times \text{atomic mass of F})$$
$$= 22.99\ \text{amu} + 19.00\ \text{amu} = 41.99\ \text{amu}$$

(b) There are three yellow spheres (fluorine) for each blue sphere (nitrogen), so the formula is NF_3. Two nonmetals form a covalent compound. Nitrogen has a lower group number, so it is named first: nitrogen trifluoride.

$$\text{Molecular mass} = (1 \times \text{atomic mass of N}) + (3 \times \text{atomic mass of F})$$
$$= 14.01\ \text{amu} + (3 \times 19.00\ \text{amu}) = 71.01\ \text{amu}$$

Check (a) For binary ionic compounds, we predict ionic charges from the periodic table (see Figure 2.13). Na forms a 1+ ion, and F forms a 1− ion, so the charges balance with one Na^+ per F^-. Also, ionic compounds are solids, consistent with the picture. **(b)** Covalent compounds often occur as individual molecules, as in the picture. Rounding in (a) gives $25 + 20 = 45$; in (b), we get $15 + (3 \times 20) = 75$, so there are no large errors.

FOLLOW-UP PROBLEMS

2.17A Determine the name, formula, and molecular (or formula) mass of the compound in each scene below:

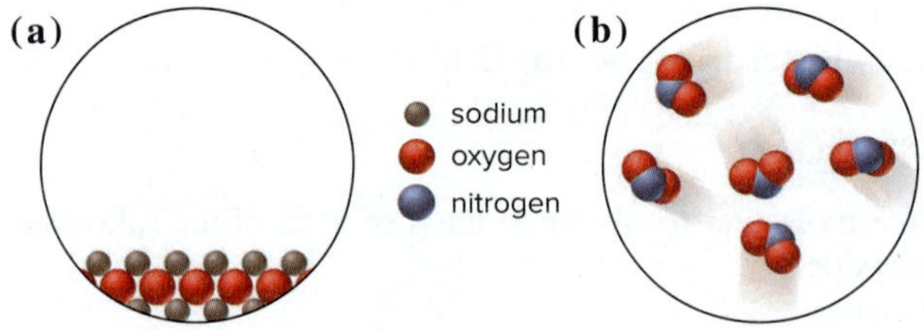

2.17B Determine the name, formula, and molecular (or formula) mass of the compound in each scene below:

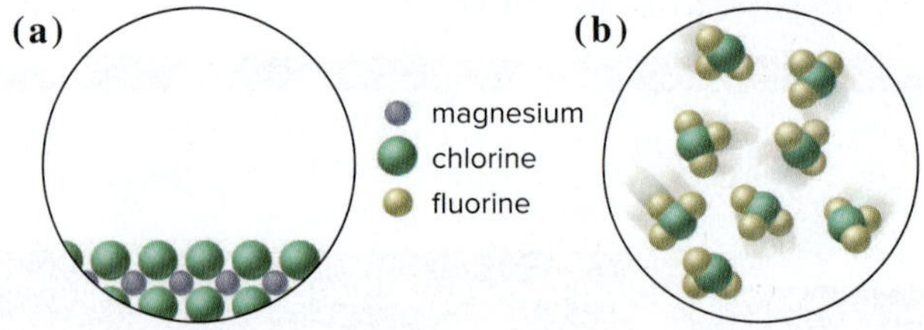

SOME SIMILAR PROBLEMS 2.108 and 2.109

Representing Molecules with Formulas and Models

In order to represent objects too small to see, chemists employ a variety of formulas and models. Each conveys different information, as shown for water:

- A *molecular formula* uses element symbols and, often, numerical subscripts to give the *actual* number of atoms of each element in a molecule of the compound. (Recall that, for ionic compounds, the *formula unit* gives the *relative* number of each type of ion.) The molecular formula of water is H_2O: there are two H atoms and one O atom in each molecule.

$$H_2O$$

- A *structural formula* shows the relative placement and connections of atoms in the molecule. It uses symbols for the atoms *and* either a pair of dots *(electron-dot formula, left)* or a line *(bond-line formula, right)* to show the bonds between the atoms. In water, each H atom is bonded to the O atom, but not to the other H atom.

H:O:H   H—O—H

In models shown throughout this text, colored balls represent atoms.

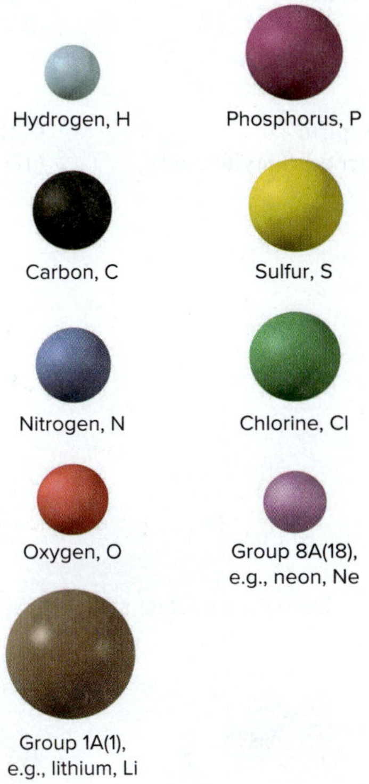

- A *ball-and-stick model* shows atoms as balls and bonds as sticks, and the angles between the bonds are accurate. Note that water is a bent molecule (with a bond angle of 104.5°). This type of model doesn't show the bonded atoms overlapping (see Figure 2.14) or their accurate sizes, so it exaggerates the distance between them.

- A *space-filling model* is an accurately scaled-up image of the molecule, so it shows the relative sizes of the atoms, the relative distances between the nuclei (centers of the spheres), and the angles between the bonds. However, bonds are not shown, and it can be difficult to see each atom in a complex molecule.

Every molecule is minute, but the range of molecular sizes, and thus molecular masses, is enormous. Table 2.8 on the next page shows the two types of formulas and models for some diatomic and small polyatomic molecules, as well as space-filling models of portions of two extremely large molecules, called *macromolecules,* deoxyribonucleic acid (DNA) and nylon.

› Summary of Section 2.8

- An ionic compound is named with cation first and anion second: metal name nonmetal root + *-ide*. For metals that can form more than one ion, the charge is shown with a roman numeral: metal name(charge) nonmetal root + *-ide*.
- Oxoanions have suffixes, and sometimes prefixes, attached to the root of the element name to indicate the number of oxygen atoms.
- Names of hydrates have a numerical prefix indicating the number of associated water molecules.
- Acid names are based on anion names.
- For binary covalent compounds, the first word of the name is the element farther left or lower down in the periodic table, and prefixes show the numbers of each atom.
- The molecular (or formula) mass of a compound is the sum of the atomic masses.
- Chemical formulas give the number of atoms (molecular) or the arrangement of atoms (structural) of one unit of a compound.
- Molecular models convey information about bond angles (ball-and-stick) and relative atomic sizes and distances between atoms (space-filling).

Table 2.8 Representing Molecules

Name	Molecular Formula (Molecular Mass, in amu)	Bond-Line Formula	Ball-and-Stick Model	Space-Filling Model
Carbon monoxide	CO (28.01)	C≡O		
Nitrogen dioxide	NO_2 (46.01)	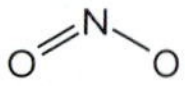	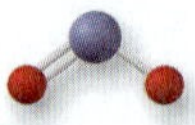	
Butane	C_4H_{10} (58.12)	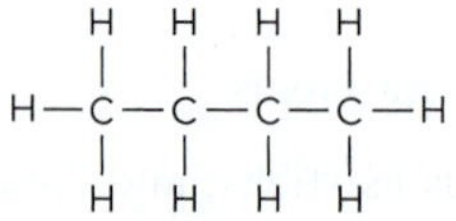	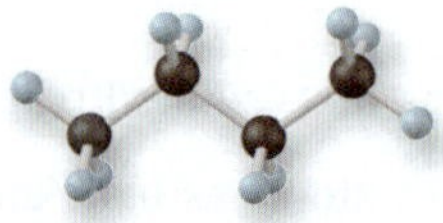	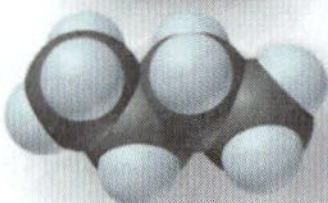
Aspirin (acetylsalicylic acid)	$C_9H_8O_4$ (180.15)	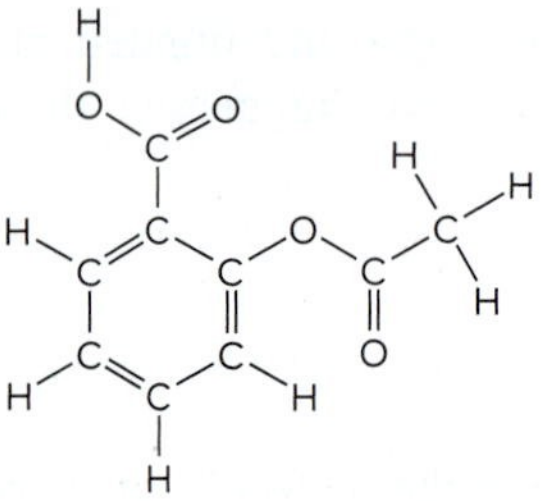	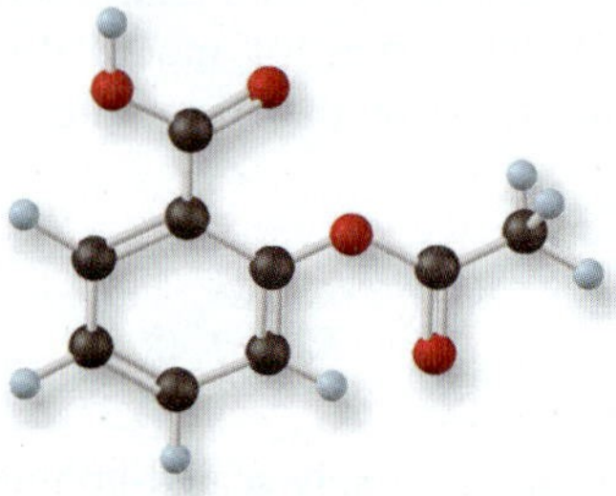	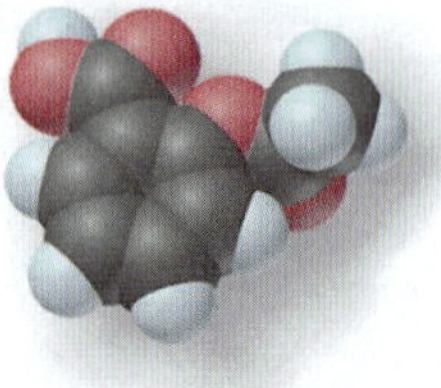

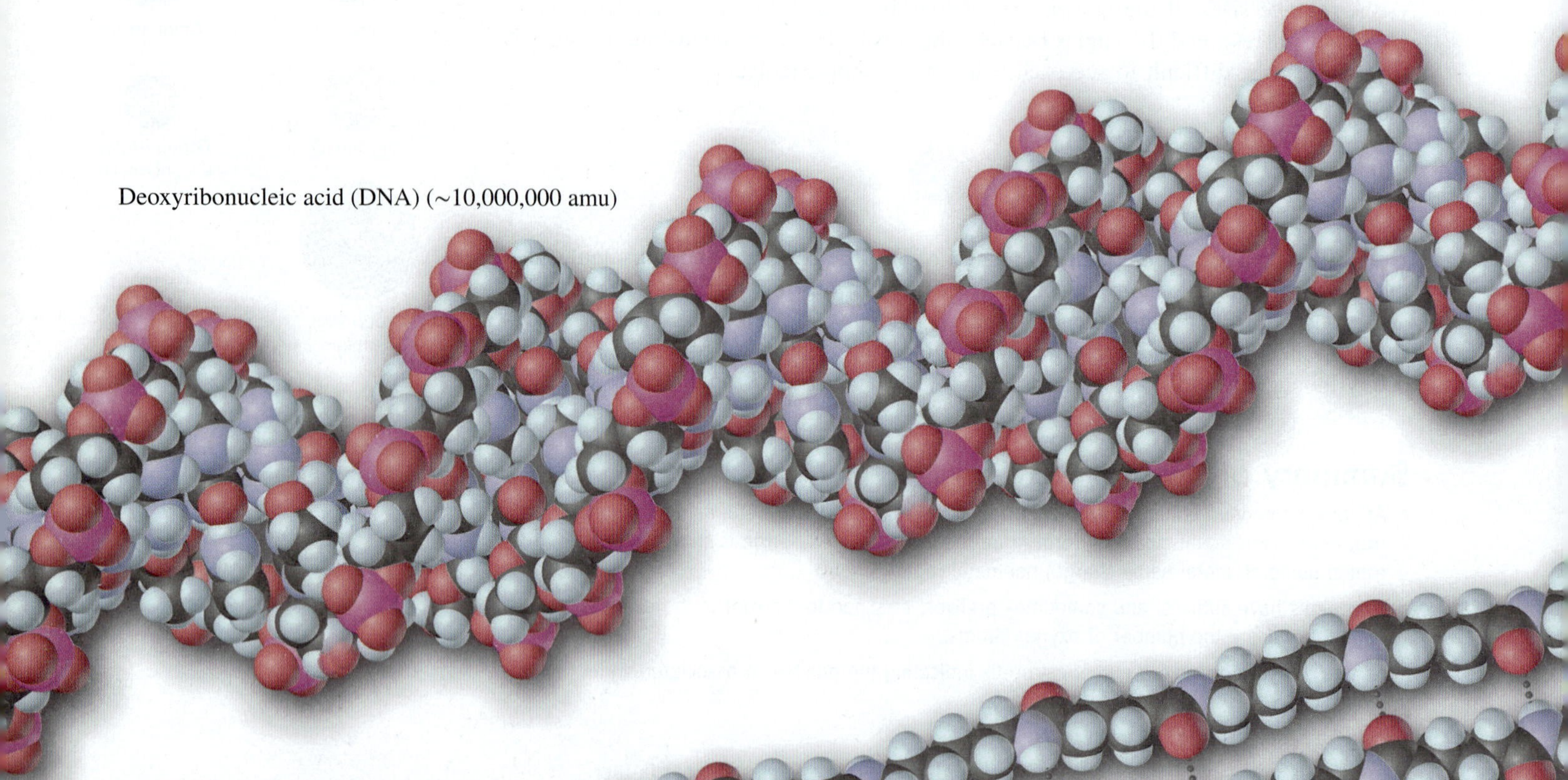

Deoxyribonucleic acid (DNA) (~10,000,000 amu)

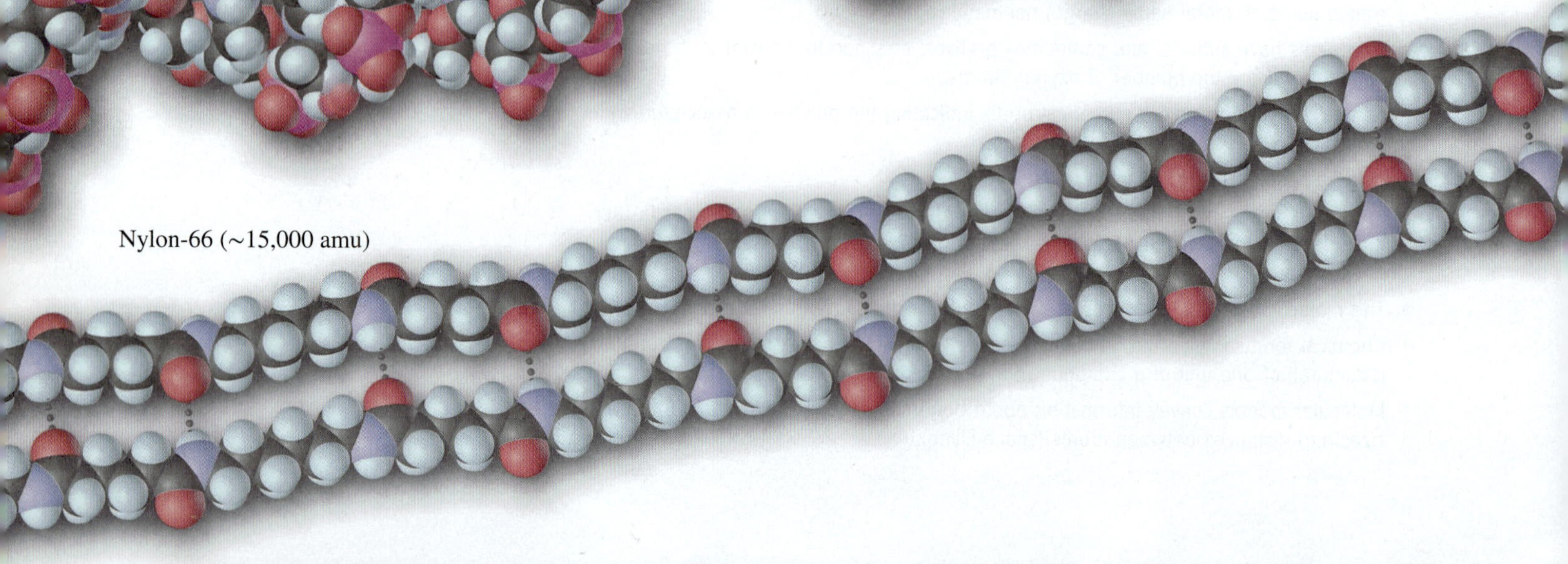

Nylon-66 (~15,000 amu)

2.9 MIXTURES: CLASSIFICATION AND SEPARATION

In the natural world, *matter usually occurs as mixtures.* A sample of clean air, for example, consists of many elements and compounds physically mixed together, including O_2, N_2, CO_2, the noble gases [Group 8A(18)], and water vapor (H_2O). The oceans are complex mixtures of dissolved ions and covalent substances, including Na^+, Mg^{2+}, Cl^-, SO_4^{2-}, O_2, CO_2, and of course H_2O. Rocks and soils are mixtures of numerous compounds, including calcium carbonate ($CaCO_3$), silicon dioxide (SiO_2), aluminum oxide (Al_2O_3), and iron(III) oxide (Fe_2O_3). And living things contain thousands of substances—carbohydrates, lipids, proteins, nucleic acids, and many simpler ionic and covalent compounds.

There are two broad classes of mixtures:

- A **heterogeneous mixture** has one or more visible boundaries among the components. Thus, its composition is *not* uniform, but rather varies from one region of the mixture to another. Many rocks are heterogeneous, having individual grains of different minerals. In some heterogeneous mixtures, such as milk and blood, the boundaries can be seen only with a microscope.
- A **homogeneous mixture** (or **solution**) has no visible boundaries because the components are individual atoms, ions, or molecules. Thus, its composition *is* uniform. A mixture of sugar dissolved in water is homogeneous, for example, because the sugar molecules and water molecules are uniformly intermingled on the molecular level. We have no way to tell visually whether a sample of matter is a substance (element or compound) or a homogeneous mixture.

Although we usually think of solutions as liquid, they exist in all three physical states. For example, air is a gaseous solution of mostly oxygen and nitrogen molecules, and wax is a solid solution of several fatty substances. Solutions in water, called **aqueous solutions,** are especially important in chemistry and comprise a major portion of the environment and of all organisms.

Recall that mixtures differ from compounds in three major ways:

1. The proportions of the components of a mixture can vary.
2. The individual properties of the components in a mixture are observable.
3. The components of a mixture can be separated by physical means.

The difference between a mixture and a compound is well illustrated using iron and sulfur as components (Figure 2.19). Any proportions of iron metal filings and powdered sulfur form a mixture. The components can be separated with a magnet because iron metal is magnetic. But if we heat the container strongly, the components will form the compound iron(II) sulfide (FeS). The magnet can then no longer remove the iron because it exists as Fe^{2+} ions chemically bound to S^{2-} ions.

Chemists have devised many procedures for separating a mixture into its components, and the Tools of the Laboratory essay at the end of this section describes some common ones.

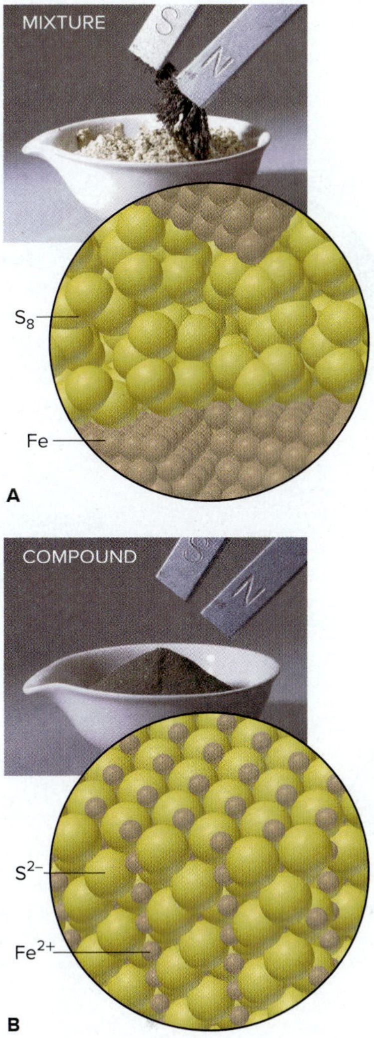

Figure 2.19 The distinction between mixtures and compounds. A, A mixture of iron and sulfur consists of the two elements. **B,** The compound iron(II) sulfide consists of an array of Fe^{2+} and S^{2-} ions.

Source: © McGraw-Hill Education/Stephen Frisch, photographer

An Overview of the Components of Matter

Understanding matter at the observable and atomic scales is the essence of chemistry. Figure 2.20 on the next page is a visual overview of many key terms and ideas in this chapter.

› Summary of Section 2.9

- › Heterogeneous mixtures have visible boundaries among the components.
- › Homogeneous mixtures (solutions) have no visible boundaries because mixing occurs at the molecular level. They can occur in any physical state.
- › Components of mixtures (unlike those of compounds) can have variable proportions, can be separated physically, and retain their properties.
- › Separation methods are based on differences in physical properties and include filtration (particle size), crystallization (solubility), distillation (volatility), and chromatography (solubility).

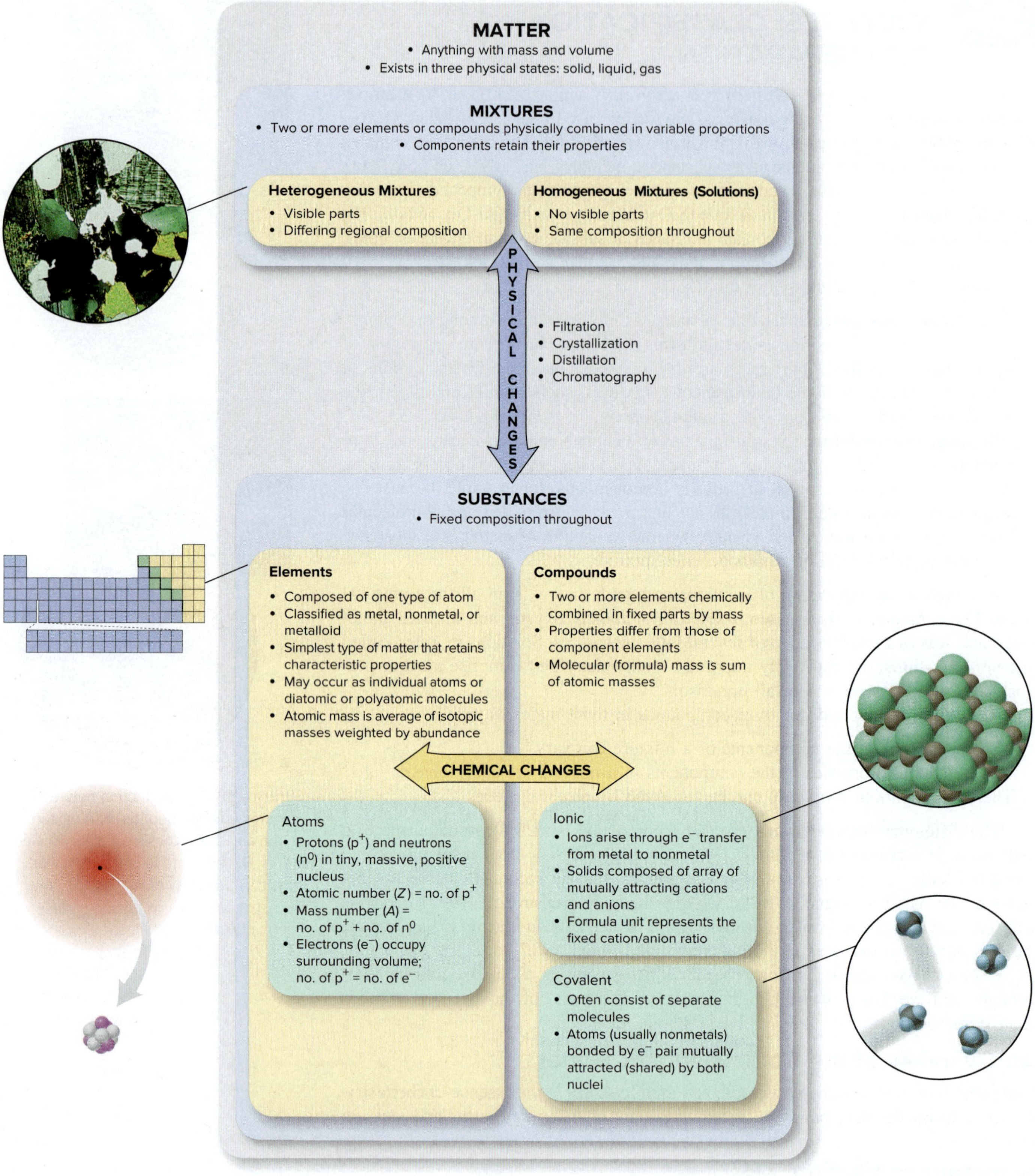

Figure 2.20 The classification of matter from a chemical point of view.

Basic Separation Techniques

Some of the most challenging laboratory procedures involve separating mixtures and purifying the components. All of the techniques described here depend on the *physical properties* of the substances in the mixture; no chemical changes occur.

Filtration is based on *differences in particle size* and is often used to separate a solid from a liquid, which flows through the tiny holes in filter paper as the solid is retained. In vacuum filtration, reduced pressure below the filter speeds the flow of the liquid through it. Filtration is used in the purification of tap water.

Crystallization is based on *differences in solubility.* The *solubility* of a substance is the amount that dissolves in a fixed volume of solvent at a given temperature. Since solubility often increases with temperature, the impure solid is dissolved in hot solvent and when the solution cools, the purified compound solidifies (crystallizes). A key component of computer chips is purified by a type of crystallization.

Distillation separates components through *differences in* **volatility,** the tendency of a substance to become a gas. *Simple* distillation separates components with *large* differences in volatility, such as water from dissolved ionic compounds (Figure B2.3). As the mixture boils, the vapor is richer in the more volatile component, in this case, water, which is condensed and collected separately. *Fractional distillation* (discussed in Chapter 13) uses many vaporization-condensation steps to separate components with small volatility differences, such as those in petroleum.

Chromatography is also based on *differences in solubility.* The mixture is dissolved in a gas or liquid, and the components are separated as this solution, called the *mobile phase,* flows through a solid (or viscous liquid) called the *stationary phase.* A component with lower solubility in the stationary phase moves through it faster than a component with a higher solubility. Figure B2.4 depicts the separation of inks by *column chromatography.*

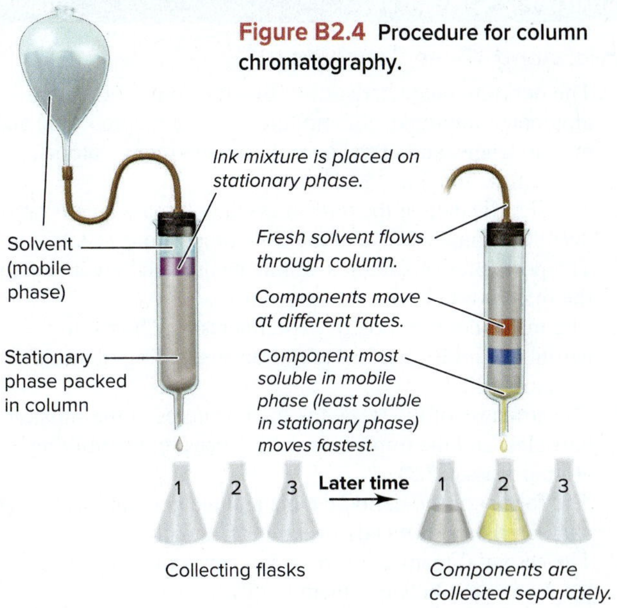
Solvent (mobile phase)
Ink mixture is placed on stationary phase.
Stationary phase packed in column
Fresh solvent flows through column.
Components move at different rates.
Component most soluble in mobile phase (least soluble in stationary phase) moves fastest.
1 2 3
Later time
1 2 3
Collecting flasks
Components are collected separately.

Figure B2.4 Procedure for column chromatography.

In a related technique called *gas-liquid chromatography (GLC),* the mobile phase is an inert gas, such as helium, that carries a gaseous mixture of components into a long tube packed with the stationary phase (Figure B2.5, part A). The components emerge separately and reach a detector. A chromatogram has numerous peaks, each representing the amount of a specific component (Figure B2.5, part B). *High-performance (high-pressure) liquid chromatography (HPLC)* is similar to GLC, but the mixture need not be vaporized, so more heat-sensitive components can be separated.

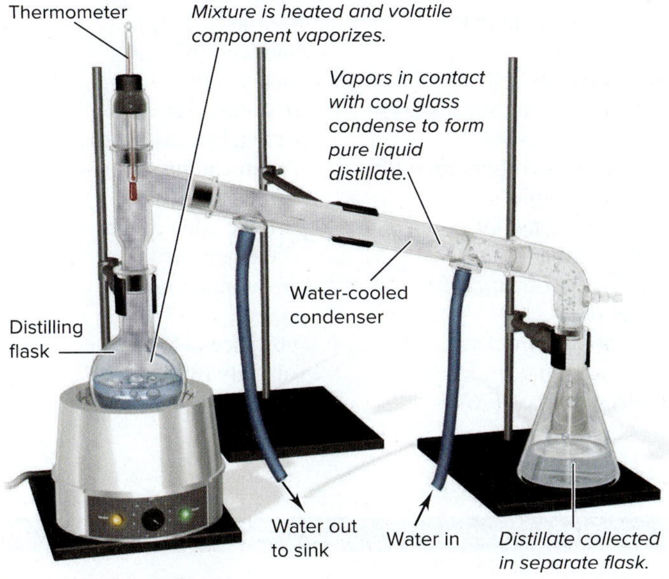
Thermometer
Mixture is heated and volatile component vaporizes.
Vapors in contact with cool glass condense to form pure liquid distillate.
Water-cooled condenser
Distilling flask
Water out to sink
Water in
Distillate collected in separate flask.

Figure B2.3 Distillation.

Problem

B2.3 Name the technique(s) and briefly describe the procedure for separating each of the following mixtures into pure components: (a) table salt and pepper; (b) drinking water contaminated with soot; (c) crushed ice and crushed glass; (d) table sugar dissolved in ethanol; (e) two pigments (chlorophyll *a* and chlorophyll *b*) from spinach leaves.

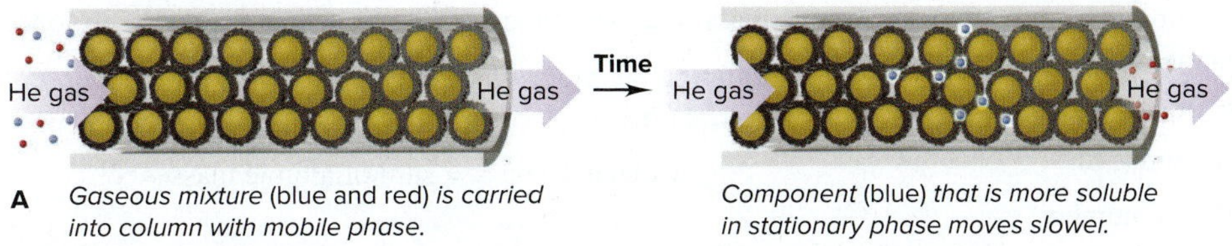
He gas
He gas
Time
He gas
He gas
A
Gaseous mixture (blue and red) *is carried into column with mobile phase.*
Component (blue) *that is more soluble in stationary phase moves slower.*

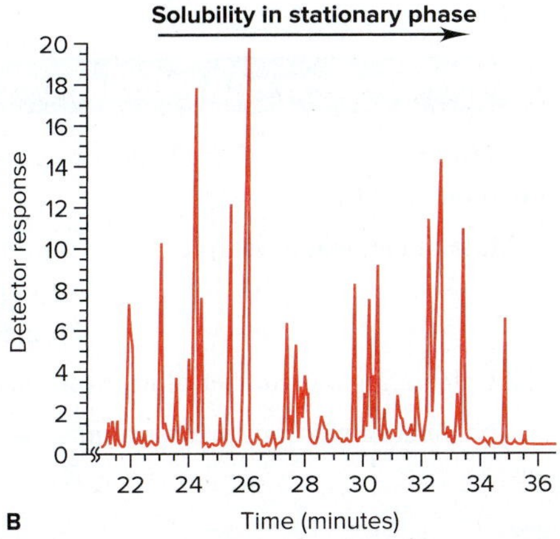
Solubility in stationary phase
Detector response
0
2
4
6
8
10
12
14
16
18
20
22
24
26
28
30
32
34
36
Time (minutes)
B

Figure B2.5 Principle of gas-liquid chromatography (GLC). The stationary phase is shown as a viscous liquid (*gray circles*) coating the solid beads (*yellow*) of an inert packing.

CHAPTER REVIEW GUIDE

Learning Objectives

Relevant section (§) and/or sample problem (SP) numbers appear in parentheses.

Understand These Concepts

1. The defining characteristics of the three types of matter—element, compound, and mixture—on the macroscopic and atomic levels, and of the particles within them—atoms, molecules, and ions (§2.1)
2. The significance of the three mass laws—mass conservation, definite composition, and multiple proportions (§2.2)
3. The postulates of Dalton's atomic theory and how it explains the mass laws (§2.3)
4. The major contribution of experiments by Thomson, Millikan, and Rutherford to our understanding of atomic structure (§2.4)
5. The structure of the atom, the main features of the subatomic particles, and the importance of isotopes in determining atomic mass (§2.5)
6. The format of the periodic table and the general location and characteristics of metals, metalloids, and nonmetals (§2.6)
7. The essential features of ionic and covalent compounds and the distinction between them (§2.7)
8. The types of mixtures and their properties (§2.9)

Master These Skills

1. Distinguishing elements, compounds, and mixtures at the atomic scale (SP 2.1)
2. Using the mass ratio of element to compound to find the mass of an element in a compound (SP 2.2)
3. Visualizing the mass laws (SP 2.3)
4. Determining the numbers of subatomic particles in the isotopes of an element (SP 2.4)
5. Calculating an atomic mass from isotopic composition (SP 2.5)
6. Identifying an element from its Z value (SP 2.6)
7. Predicting the ion an element forms (SP 2.7)
8. Naming and writing the formula of an ionic compound formed from the ions in Tables 2.3–2.5 (SPs 2.8–2.12, 2.17)
9. Naming and writing the formula of an acid and its anion (SP 2.13)
10. Naming and writing the formula of a binary covalent compound (SPs 2.14, 2.15, 2.17)
11. Calculating the molecular or formula mass of a compound (SP 2.16, 2.17)

Key Terms

Page numbers appear in parentheses.

anion (64)
aqueous solution (81)
atom (50)
atomic mass (58)
atomic mass unit (amu) (57)
atomic number (Z) (56)
atomic symbol (56)
binary covalent compound (74)
binary ionic compound (64)
cathode ray (52)
cation (64)
chemical bond (64)
chemical formula (68)
chromatography (83)
compound (44)
covalent bond (66)
covalent compound (64)
crystallization (83)
dalton (Da) (57)
distillation (83)
electron (e^-) (55)
element (44)
filtration (83)
formula mass (76)
formula unit (68)
fraction by mass (mass fraction) (47)
group (61)
heterogeneous mixture (81)
homogeneous mixture (solution) (81)
hydrate (72)
ion (64)
ionic compound (64)
isotope (57)
isotopic mass (58)
law of definite (or constant) composition (47)
law of mass conservation (46)
law of multiple proportions (49)
mass number (A) (56)
mass spectrometry (58)
metal (62)
metalloid (semimetal) (62)
mixture (45)
molecular mass (76)
molecule (44)
monatomic ion (65)
neutron (n^0) (55)
nonmetal (62)
nucleus (55)
oxoanion (72)
percent by mass (mass percent, mass %) (47)
period (61)
periodic table of the elements (61)
polyatomic ion (67)
proton (p^+) (55)
substance (44)
volatility (83)

Key Equations and Relationships

Page numbers appear in parentheses.

2.1 Finding the mass of an element in a given mass of compound (48):

$$\text{Mass of element in sample} = \text{mass of compound in sample} \times \frac{\text{mass of element in compound}}{\text{mass of compound}}$$

2.2 Calculating the number of neutrons in an atom (56):

$$\text{Number of neutrons} = \text{mass number} - \text{atomic number}$$

or

$$N = A - Z$$

2.3 Calculating the average atomic mass of the isotopes of an element (58):

$$\text{Atomic mass} = \Sigma(\text{isotopic mass})(\text{fractional abundance of isotope})$$

2.4 Determining the molecular mass of a formula unit of a compound (76):

$$\text{Molecular mass} = \text{sum of atomic masses}$$

BRIEF SOLUTIONS TO FOLLOW-UP PROBLEMS

2.1A (a) There is only one type of atom (blue) present, so this is an element. (b) Two different atoms (brown and green) appear in a fixed ratio of 1/1, so this is a compound. (c) These molecules consist of one type of atom (orange), so this is an element.

2.1B There are two types of particles reacting (left circle), one with two blue atoms and the other with two orange; the depiction shows a mixture of two elements. In the product (right circle), all the particles have one blue atom and one orange; this is a compound.

2.2A Mass (g) of fool's gold

$$= 86.2\ \cancel{\text{g iron}} \times \frac{110.0\text{ g fool's gold}}{51.2\ \cancel{\text{g iron}}}$$

$$= 185\text{ g fool's gold}$$

$$\text{Mass (g) of sulfur} = 185\ \cancel{\text{g fool's gold}} \times \frac{(110.0 - 51.2\text{ g sulfur})}{110.0\ \cancel{\text{g fool's gold}}}$$

$$= 98.9\text{ g sulfur}$$

2.2B Mass (g) of silver

$$= 3.57\ \cancel{\text{g silver bromide}} \times \frac{15.4\text{ g silver}}{26.8\ \cancel{\text{g silver bromide}}}$$

$$= 2.05\text{ g silver}$$

Mass (g) of bromine

$$= 3.57\ \cancel{\text{g silver bromide}} \times \frac{(26.8 - 15.4\text{ g bromine})}{26.8\ \cancel{\text{g silver bromide}}}$$

$$= 1.52\text{ g bromine}$$

2.3A There are 12 black atoms and 14 red atoms in each circle (mass conservation). In the right circle, there are molecules of two compounds—one compound has one black and one red atom, and the other has one black and two red atoms (multiple proportions). Each compound has a fixed ratio of black-to-red atoms (definite composition).

2.3B Sample B. Two bromine-fluorine compounds appear. In one, there are three fluorine atoms for each bromine; in the other, there is one fluorine for each bromine. Therefore, the masses of fluorine that combine with a given mass of bromine are in a 3/1 ratio.

2.4A Titanium has an atomic number of 22. Mass number − 22 = number of neutrons.

^{46}Ti has $22p^+$, $22e^-$, and $(46 - 22) = 24n^0$.

^{47}Ti has $22p^+$, $22e^-$, and $(47 - 22) = 25n^0$.

^{48}Ti has $22p^+$, $22e^-$, and $(48 - 22) = 26n^0$.

^{49}Ti has $22p^+$, $22e^-$, and $(49 - 22) = 27n^0$.

^{50}Ti has $22p^+$, $22e^-$, and $(50 - 22) = 28n^0$.

2.4B (a) $5p^+$, $5e^-$, $(11 - 5) = 6n^0$; Q = B

(b) $20p^+$, $20e^-$, $(41 - 20) = 21n^0$; R = Ca

(c) $53p^+$, $53e^-$, $(131 - 53) = 78n^0$; X = I

2.5A First, divide the percent abundance value (Figure B2.2C, Tools of the Laboratory) by 100 to obtain the fractional value for each isotope. Multiply each isotopic mass by the fractional value, and add the resulting masses to obtain neon's atomic mass:

Atomic mass of Ne

$$\begin{aligned} &= (\text{isotopic mass of }^{20}\text{Ne})(\text{fractional abundance of }^{20}\text{Ne}) \\ &\quad + (\text{isotopic mass of }^{21}\text{Ne})(\text{fractional abundance of }^{21}\text{Ne}) \\ &\quad + (\text{isotopic mass of }^{22}\text{Ne})(\text{fractional abundance of }^{22}\text{Ne}) \\ &= (19.99244\text{ amu})(0.9048) + (20.99385\text{ amu})(0.0027) \\ &\quad + (21.99139\text{ amu})(0.0925) \\ &= 20.18\text{ amu} \end{aligned}$$

2.5B $10.0129x + [11.0093(1 - x)] = 10.81$; $0.9964x = 0.1993$; $x = 0.2000$ and $1 - x = 0.8000$; percent abundance of ^{10}B = 20.00%; percent abundance of ^{11}B = 80.00%

2.6A (a) Silicon, Si; Group 4A(14) and Period 3; metalloid

(b) Cesium, Cs; Group 1A(1) and Period 6; main-group metal

(c) Xenon, Xe; Group 8A(18) and Period 5; nonmetal

2.6B (a) Magnesium, Mg; Group 2A(2) and Period 3; main-group metal

(b) Nitrogen, N; Group 5A(15) and Period 2; nonmetal

(c) Zinc, Zn; Group 2B(12) and Period 4; transition metal

2.7A (a) S^{2-}; (b) Rb^+; (c) Ba^{2+}

2.7B (a) Sr^{2+}; (b) O^{2-}; (c) Cs^+

2.8A (a) Zinc oxide; (b) silver bromide; (c) lithium chloride; (d) aluminum sulfide

2.8B (a) Potassium sulfide; (b) barium iodide; (c) cesium nitride; (d) sodium hydride

2.9A (a) ZnO; (b) AgBr; (c) LiCl; (d) Al_2S_3

2.9B (a) K_2S; (b) BaI_2; (c) Cs_3N; (d) NaH

2.10A (a) PbO_2; (b) copper(I) sulfide (cuprous sulfide); (c) iron(II) bromide (ferrous bromide); (d) $HgCl_2$

2.10B (a) Cu_3N_2; (b) lead(II) iodide; (c) Cr_2S_3; (d) iron(II) oxide

2.11A (a) $Cu(NO_3)_2{\cdot}3H_2O$; (b) $Zn(OH)_2$; (c) lithium cyanide

2.11B (a) $(NH_4)_2SO_4$; (b) nickel(II) nitrate hexahydrate; (c) $KHCO_3$

2.12A (a) $(NH_4)_3PO_4$; ammonium is NH_4^+ and phosphate is PO_4^{3-}.

(b) $Al(OH)_3$; parentheses are needed around the polyatomic ion OH^-.

(c) Magnesium hydrogen carbonate; Mg^{2+} is magnesium and can have only a 2+ charge, so (II) is not needed in the name; HCO_3^- is hydrogen carbonate (or bicarbonate).

2.12B (a) Chromium(III) nitrate; the *-ic* ending is not used with roman numerals; NO_3^- is nitr*ate*.

(b) Calcium nitrite; Ca^{2+} is calcium and NO_2^- is nitr*ite*.

(c) $KClO_3$; potassium is K^+ and chlorate is ClO_3^-; parentheses are not needed when only one polyatomic ion is present.

2.13A (a) ClO_3^-, chlorate; (b) F^-, fluoride; (c) CH_3COO^- (or $C_2H_3O_2^-$), acetate; (d) NO_2^-, nitrite

2.13B (a) Sulfurous acid; add one H^+ ion to hydrogen sulfite, HSO_3^-, or two H^+ to sulfite, SO_3^{2-}; (b) hypobromous acid, hypobromite, BrO^-; (c) chlorous acid, chlorite, ClO_2^-; (d) hydriodic acid, iodide, I^-

2.14A (a) Sulfur trioxide; (b) silicon dioxide; (c) N_2O; (d) SeF_6

2.14B (a) Sulfur dichloride; (b) dinitrogen pentoxide; (c) BF_3; (d) IBr_3

2.15A (a) Disulfur dichloride; the *-ous* suffix is not used.

(b) NO; the name indicates one nitrogen.

(c) Bromine trichloride; Br is in a higher period in Group 7A(17), so it is named first.

2.15B (a) Tetraphosphorus hexoxide; the name of the element phosphorus ends in *-us* not *-ous*, and the prefix *hexa-* is shortened to *hex-* before *oxide*. (b) Sulfur hexafluoride; sulfur has a lower group number so it comes first, and fluorine gets the *-ide* ending. (c) NBr_3; nitrogen's symbol is N, and the second letter of a symbol is lowercase: bromine is Br.

2.16A (a) Hydrogen peroxide is H_2O_2.

Molecular mass

$$= (2 \times \text{atomic mass of H}) + (2 \times \text{atomic mass of O})$$
$$= (2 \times 1.008 \text{ amu}) + (2 \times 16.00 \text{ amu}) = 34.02 \text{ amu}$$

(b) Cesium carbonate is Cs_2CO_3.

Formula mass

$$= (2 \times \text{atomic mass of Cs}) + (1 \times \text{atomic mass of C}) + (3 \times \text{atomic mass of O})$$
$$= (2 \times 132.9 \text{ amu}) + 12.01 \text{ amu} + (3 \times 16.00 \text{ amu})$$
$$= 325.8 \text{ amu}$$

2.16B (a) Sulfuric acid is H_2SO_4.

Molecular mass

$$= (2 \times \text{atomic mass of H}) + (1 \times \text{atomic mass of S}) + (4 \times \text{atomic mass of O})$$
$$= (2 \times 1.008 \text{ amu}) + 32.06 \text{ amu} + (4 \times 16.00 \text{ amu})$$
$$= 98.08 \text{ amu}$$

(b) Potassium sulfate is K_2SO_4.

Formula mass

$$= (2 \times \text{atomic mass of K}) + (1 \times \text{atomic mass of S}) + (4 \times \text{atomic mass of O})$$
$$= (2 \times 39.10 \text{ amu}) + 32.06 \text{ amu} + (4 \times 16.00 \text{ amu})$$
$$= 174.26 \text{ amu}$$

2.17A (a) Na_2O. This is an ionic compound, so the name is sodium oxide.

Formula mass

$$= (2 \times \text{atomic mass of Na}) + (1 \times \text{atomic mass of O})$$
$$= (2 \times 22.99 \text{ amu}) + 16.00 \text{ amu} = 61.98 \text{ amu}$$

(b) NO_2. This is a covalent compound, and N has the lower group number, so the name is nitrogen dioxide.

Molecular mass

$$= (1 \times \text{atomic mass of N}) + (2 \times \text{atomic mass of O})$$
$$= 14.01 \text{ amu} + (2 \times 16.00 \text{ amu}) = 46.01 \text{ amu}$$

2.17B (a) Magnesium chloride, $MgCl_2$

Formula mass

$$= (1 \times \text{atomic mass of Mg}) + (2 \times \text{atomic mass of Cl})$$
$$= (1 \times 24.31 \text{ amu}) + (2 \times 35.45 \text{ amu}) = 95.21 \text{ amu}$$

(b) Chlorine trifluoride, ClF_3

Molecular mass

$$= (1 \times \text{atomic mass of Cl}) + (3 \times \text{atomic mass of F})$$
$$= 35.45 \text{ amu} + (3 \times 19.00 \text{ amu}) = 92.45 \text{ amu}$$

PROBLEMS

Problems with **colored** numbers are answered in Appendix E and worked in detail in the Student Solutions Manual. Problem sections match those in the text and give the numbers of relevant sample problems. Most offer Concept Review Questions, Skill-Building Exercises (grouped in pairs covering the same concept), and Problems in Context. The Comprehensive Problems are based on material from any section or previous chapter.

Elements, Compounds, and Mixtures: An Atomic Overview

(Sample Problem 2.1)

Concept Review Questions

2.1 What is the key difference between an element and a compound?

2.2 List two differences between a compound and a mixture.

2.3 Which of the following are pure substances? Explain.
(a) Calcium chloride, used to melt ice on roads, consists of two elements, calcium and chlorine, in a fixed mass ratio.
(b) Sulfur consists of sulfur atoms combined into octatomic molecules.
(c) Baking powder, a leavening agent, contains 26–30% sodium hydrogen carbonate and 30–35% calcium dihydrogen phosphate by mass.
(d) Cytosine, a component of DNA, consists of H, C, N, and O atoms bonded in a specific arrangement.

2.4 Classify each substance in Problem 2.3 as an element, compound, or mixture, and explain your answers.

2.5 Explain the following statement: The smallest particles unique to an element may be atoms or molecules.

2.6 Explain the following statement: The smallest particles unique to a compound cannot be atoms.

2.7 Can the relative amounts of the components of a mixture vary? Can the relative amounts of the components of a compound vary? Explain.

Problems in Context

2.8 The tap water found in many areas of the United States leaves white deposits when it evaporates. Is this tap water a mixture or a compound? Explain.

2.9 Each scene below represents a mixture. Describe each one in terms of the number(s) of elements and/or compounds present.

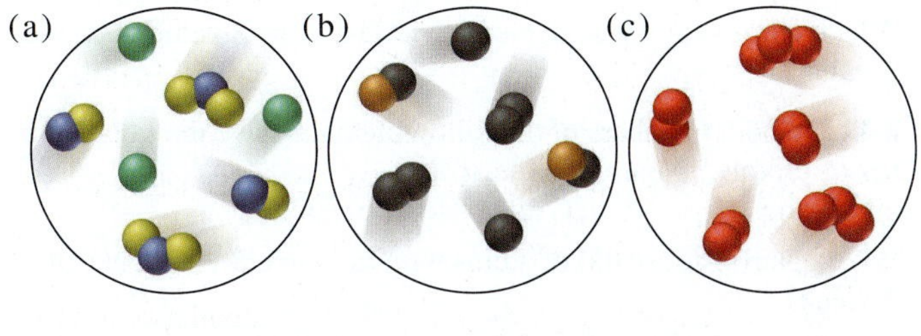

2.10 Samples of illicit "street" drugs often contain an inactive component, such as ascorbic acid (vitamin C). After obtaining a sample of cocaine, government chemists calculate the mass of vitamin C per gram of drug sample and use it to track the drug's distribution. For example, if different samples of cocaine obtained on the streets of New York, Los Angeles, and Paris all contain 0.6384 g of vitamin C per gram of sample, they very likely come from a common source. Do these street samples consist of a compound, element, or mixture? Explain.

The Observations That Led to an Atomic View of Matter

(Sample Problem 2.2)

Concept Review Questions

2.11 Why was it necessary for separation techniques and methods of chemical analysis to be developed before the laws of definite composition and multiple proportions could be formulated?

2.12 To which class(es) of matter—elements, compounds, and/or mixtures—do the following apply: (a) law of mass conservation; (b) law of definite composition; (c) law of multiple proportions?

2.13 In our modern view of matter and energy, is the law of mass conservation still relevant to chemical reactions? Explain.

2.14 Identify the mass law that each of the following observations demonstrates, and explain your reasoning:
(a) A sample of potassium chloride from Chile contains the same percent by mass of potassium as one from Poland.
(b) A flashbulb contains magnesium and oxygen before use and magnesium oxide afterward, but its mass does not change.
(c) Arsenic and oxygen form one compound that is 65.2 mass % arsenic and another that is 75.8 mass % arsenic.

2.15 Which of the following scenes illustrate(s) the fact that compounds of chlorine (green) and oxygen (red) exhibit the law of multiple proportions? Name the compounds.

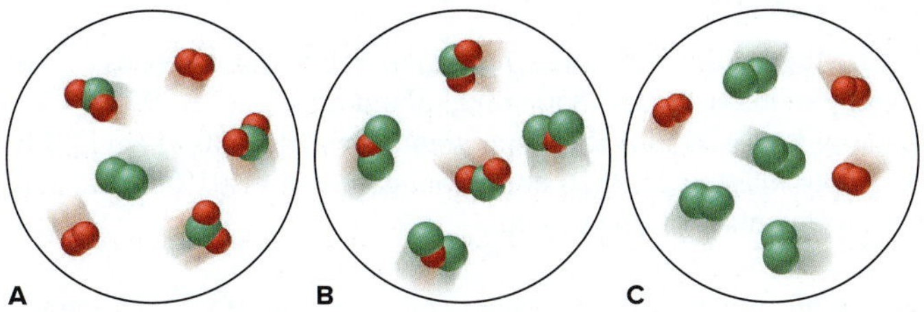

2.16 (a) Does the percent by mass of each element in a compound depend on the amount of compound? Explain.
(b) Does the mass of each element in a compound depend on the amount of compound? Explain.

2.17 Does the percent by mass of each element in a compound depend on the amount of that element used to make the compound? Explain.

Skill-Building Exercises (grouped in similar pairs)

2.18 State the mass law(s) demonstrated by the following experimental results, and explain your reasoning:

Experiment 1: A student heats 1.00 g of a blue compound and obtains 0.64 g of a white compound and 0.36 g of a colorless gas.

Experiment 2: A second student heats 3.25 g of the same blue compound and obtains 2.08 g of a white compound and 1.17 g of a colorless gas.

2.19 State the mass law(s) demonstrated by the following experimental results, and explain your reasoning:

Experiment 1: A student heats 1.27 g of copper and 3.50 g of iodine to produce 3.81 g of a white compound; 0.96 g of iodine remains.

Experiment 2: A second student heats 2.55 g of copper and 3.50 g of iodine to form 5.25 g of a white compound; 0.80 g of copper remains.

2.20 Fluorite, a mineral of calcium, is a compound of the metal with fluorine. Analysis shows that a 2.76-g sample of fluorite contains 1.42 g of calcium. Calculate the (a) mass of fluorine in the sample; (b) mass fractions of calcium and fluorine in fluorite; (c) mass percents of calcium and fluorine in fluorite.

2.21 Galena, a mineral of lead, is a compound of the metal with sulfur. Analysis shows that a 2.34-g sample of galena contains 2.03 g of lead. Calculate the (a) mass of sulfur in the sample; (b) mass fractions of lead and sulfur in galena; (c) mass percents of lead and sulfur in galena.

2.22 Magnesium oxide (MgO) forms when the metal burns in air.
(a) If 1.25 g of MgO contains 0.754 g of Mg, what is the mass ratio of magnesium to magnesium oxide?
(b) How many grams of Mg are in 534 g of MgO?

2.23 Zinc sulfide (ZnS) occurs in the zincblende crystal structure.
(a) If 2.54 g of ZnS contains 1.70 g of Zn, what is the mass ratio of zinc to zinc sulfide?
(b) How many kilograms of Zn are in 3.82 kg of ZnS?

2.24 A compound of copper and sulfur contains 88.39 g of metal and 44.61 g of nonmetal. How many grams of copper are in 5264 kg of compound? How many grams of sulfur?

2.25 A compound of iodine and cesium contains 63.94 g of metal and 61.06 g of nonmetal. How many grams of cesium are in 38.77 g of compound? How many grams of iodine?

2.26 Show, with calculations, how the following data illustrate the law of multiple proportions:

Compound 1: 47.5 mass % sulfur and 52.5 mass % chlorine

Compound 2: 31.1 mass % sulfur and 68.9 mass % chlorine

2.27 Show, with calculations, how the following data illustrate the law of multiple proportions:

Compound 1: 77.6 mass % xenon and 22.4 mass % fluorine

Compound 2: 63.3 mass % xenon and 36.7 mass % fluorine

Problems in Context

2.28 Dolomite is a carbonate of magnesium and calcium. Analysis shows that 7.81 g of dolomite contains 1.70 g of Ca. Calculate the mass percent of Ca in dolomite. On the basis of the mass percent of Ca, and neglecting all other factors, which is the richer source of Ca, dolomite or fluorite (see Problem 2.20)?

2.29 The mass percent of sulfur in a sample of coal is a key factor in the environmental impact of the coal because the sulfur combines with oxygen when the coal is burned and the oxide can then be incorporated into acid rain. Which of the following coals would have the smallest environmental impact?

	Mass (g) of Sample	Mass (g) of Sulfur in Sample
Coal A	378	11.3
Coal B	495	19.0
Coal C	675	20.6

Dalton's Atomic Theory

(Sample Problem 2.3)

Concept Review Questions

2.30 Which of Dalton's postulates about atoms are inconsistent with later observations? Do these inconsistencies mean that Dalton was wrong? Is Dalton's model still useful? Explain.

2.31 Use Dalton's theory to explain why potassium nitrate from India or Italy has the same mass percents of K, N, and O.

The Observations That Led to the Nuclear Atom Model

Concept Review Questions

2.32 Thomson was able to determine the mass/charge ratio of the electron but not its mass. How did Millikan's experiment allow determination of the electron's mass?

2.33 The following charges on individual oil droplets were obtained during an experiment similar to Millikan's: -3.204×10^{-19} C; -4.806×10^{-19} C; -8.010×10^{-19} C; -1.442×10^{-18} C. Determine a charge for the electron (in C, coulombs), and explain your answer.

2.34 Describe Thomson's model of the atom. How might it account for the production of cathode rays?

2.35 When Rutherford's coworkers bombarded gold foil with α particles, they obtained results that overturned the existing (Thomson) model of the atom. Explain.

The Atomic Theory Today

(Sample Problems 2.4 and 2.5)

Concept Review Questions

2.36 Define *atomic number* and *mass number*. Which can vary without changing the identity of the element?

2.37 Choose the correct answer. The difference between the mass number of an isotope and its atomic number is (a) directly related to the identity of the element; (b) the number of electrons; (c) the number of neutrons; (d) the number of isotopes.

2.38 Even though several elements have only one naturally occurring isotope and all atomic nuclei have whole numbers of protons and neutrons, no atomic mass is a whole number. Use the data from Table 2.2 to explain this fact.

Skill-Building Exercises (grouped in similar pairs)

2.39 Argon has three naturally occurring isotopes, ^{36}Ar, ^{38}Ar, and ^{40}Ar. What is the mass number of each isotope? How many protons, neutrons, and electrons are present in each?

2.40 Chlorine has two naturally occurring isotopes, ^{35}Cl and ^{37}Cl. What is the mass number of each isotope? How many protons, neutrons, and electrons are present in each?

2.41 Do both members of the following pairs have the same number of protons? Neutrons? Electrons?
(a) $^{16}_{8}$O and $^{17}_{8}$O (b) $^{40}_{18}$Ar and $^{41}_{19}$K (c) $^{60}_{27}$Co and $^{60}_{28}$Ni
Which pair(s) consist(s) of atoms with the same Z value? N value? A value?

2.42 Do both members of the following pairs have the same number of protons? Neutrons? Electrons?
(a) $^{3}_{1}$H and $^{3}_{2}$He (b) $^{14}_{6}$C and $^{15}_{7}$N (c) $^{19}_{9}$F and $^{18}_{9}$F
Which pair(s) consist(s) of atoms with the same Z value? N value? A value?

2.43 Write the $^{A}_{Z}$X notation for each atomic depiction:

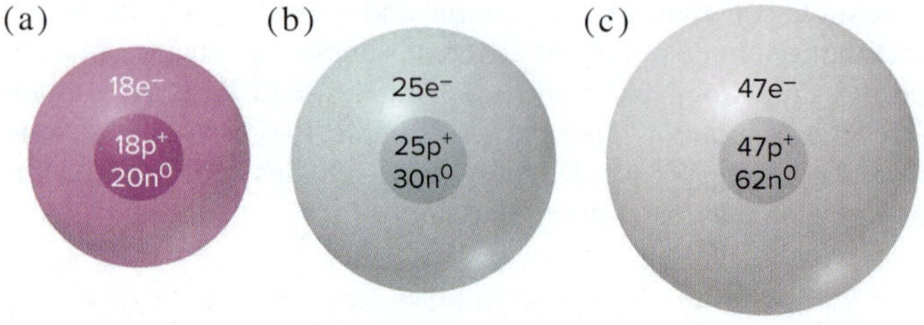

2.44 Write the $^{A}_{Z}$X notation for each atomic depiction:

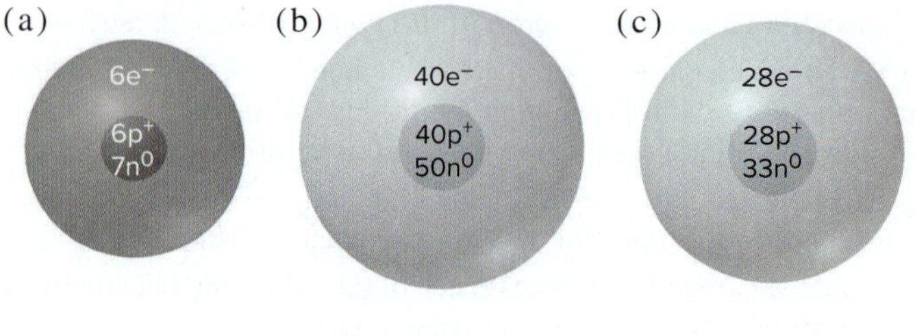

2.45 Draw atomic depictions similar to those in Problem 2.43 for (a) $^{48}_{22}$Ti; (b) $^{79}_{34}$Se; (c) $^{11}_{5}$B.

2.46 Draw atomic depictions similar to those in Problem 2.43 for (a) $^{207}_{82}$Pb; (b) $^{9}_{4}$Be; (c) $^{75}_{33}$As.

2.47 Gallium has two naturally occurring isotopes, ^{69}Ga (isotopic mass = 68.9256 amu, abundance = 60.11%) and ^{71}Ga (isotopic mass = 70.9247 amu, abundance = 39.89%). Calculate the atomic mass of gallium.

2.48 Magnesium has three naturally occurring isotopes, ^{24}Mg (isotopic mass = 23.9850 amu, abundance = 78.99%), ^{25}Mg (isotopic mass = 24.9858 amu, abundance = 10.00%), and ^{26}Mg (isotopic mass = 25.9826 amu, abundance = 11.01%). Calculate the atomic mass of magnesium.

2.49 Chlorine has two naturally occurring isotopes, ^{35}Cl (isotopic mass = 34.9689 amu) and ^{37}Cl (isotopic mass = 36.9659 amu). If chlorine has an atomic mass of 35.4527 amu, what is the percent abundance of each isotope?

2.50 Copper has two naturally occurring isotopes, ^{63}Cu (isotopic mass = 62.9296 amu) and ^{65}Cu (isotopic mass = 64.9278 amu). If copper has an atomic mass of 63.546 amu, what is the percent abundance of each isotope?

Elements: A First Look at the Periodic Table

(Sample Problem 2.6)

Concept Review Questions

2.51 How can iodine ($Z = 53$) have a higher atomic number yet a lower atomic mass than tellurium ($Z = 52$)?

2.52 Correct each of the following statements:
(a) In the modern periodic table, the elements are arranged in order of increasing atomic mass.
(b) Elements in a period have similar chemical properties.
(c) Elements can be classified as either metalloids or nonmetals.

2.53 What class of elements lies along the "staircase" line in the periodic table? How do the properties of these elements compare with those of metals and nonmetals?

2.54 What are some characteristic properties of elements to the left of the elements along the "staircase"? To the right?

2.55 The elements in Groups 1A(1) and 7A(17) are all quite reactive. What is a major difference between them?

Skill-Building Exercises (grouped in similar pairs)

2.56 Give the name, atomic symbol, and group number of the element with each Z value, and classify it as a metal, metalloid, or nonmetal:
(a) $Z = 32$ (b) $Z = 15$ (c) $Z = 2$ (d) $Z = 3$ (e) $Z = 42$

2.57 Give the name, atomic symbol, and group number of the element with each Z value, and classify it as a metal, metalloid, or nonmetal:
(a) $Z = 33$ (b) $Z = 20$ (c) $Z = 35$ (d) $Z = 19$ (e) $Z = 13$

2.58 Fill in the blanks:
(a) The symbol and atomic number of the heaviest alkaline earth metal are ________ and ________.
(b) The symbol and atomic number of the lightest metalloid in Group 4A(14) are ________ and ________.
(c) Group 1B(11) consists of the *coinage metals*. The symbol and atomic mass of the coinage metal whose atoms have the fewest electrons are ________ and ________.
(d) The symbol and atomic mass of the halogen in Period 4 are ________ and ________.

2.59 Fill in the blanks:
(a) The symbol and atomic number of the heaviest nonradioactive noble gas are ________ and ________.
(b) The symbol and group number of the Period 5 transition element whose atoms have the fewest protons are ________ and ________.
(c) The elements in Group 6A(16) are sometimes called the *chalcogens*. The symbol and atomic number of the first metallic chalcogen are ________ and ________.
(d) The symbol and number of protons of the Period 4 alkali metal atom are ________ and ________.

Compounds: Introduction to Bonding

(Sample Problem 2.7)

Concept Review Questions

2.60 Describe the type and nature of the bonding that occurs between reactive metals and nonmetals.

2.61 Describe the type and nature of the bonding that often occurs between two nonmetals.

2.62 How can ionic compounds be neutral if they consist of positive and negative ions?

2.63 Given that the ions in LiF and in MgO are of similar size, which compound has stronger ionic bonding? Use Coulomb's law in your explanation.

2.64 Are molecules present in a sample of BaF_2? Explain.

2.65 Are ions present in a sample of P_4O_6? Explain.

2.66 The monatomic ions of Groups 1A(1) and 7A(17) are all singly charged. In what major way do they differ? Why?

2.67 Describe the formation of solid magnesium chloride ($MgCl_2$) from large numbers of magnesium and chlorine atoms.

2.68 Describe the formation of solid potassium sulfide (K_2S) from large numbers of potassium and sulfur atoms.

2.69 Does potassium nitrate (KNO_3) incorporate ionic bonding, covalent bonding, or both? Explain.

Skill-Building Exercises (grouped in similar pairs)

2.70 What monatomic ions would you expect potassium ($Z = 19$) and bromine ($Z = 35$) to form?

2.71 What monatomic ions would you expect radium ($Z = 88$) and selenium ($Z = 34$) to form?

2.72 For each ionic depiction, give the name of the parent atom, its mass number, and its group and period numbers:

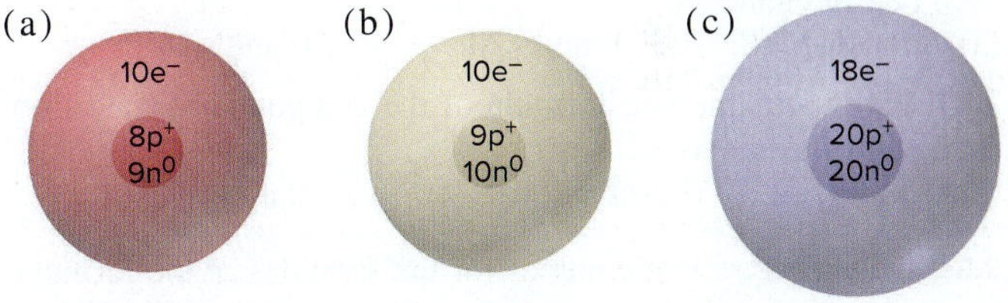

2.73 For each ionic depiction, give the name of the parent atom, its mass number, and its group and period numbers:

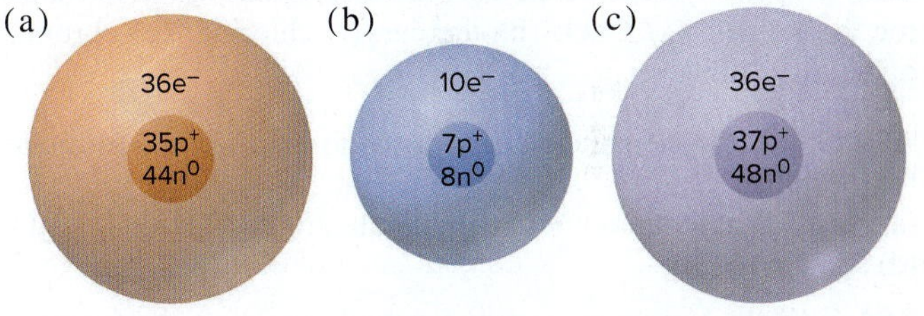

2.74 An ionic compound forms when lithium ($Z = 3$) reacts with oxygen ($Z = 8$). If a sample of the compound contains 8.4×10^{21} lithium ions, how many oxide ions does it contain?

2.75 An ionic compound forms when calcium ($Z = 20$) reacts with iodine ($Z = 53$). If a sample of the compound contains 7.4×10^{21} calcium ions, how many iodide ions does it contain?

2.76 The radii of the sodium and potassium ions are 102 pm and 138 pm, respectively. Which compound has stronger ionic attractions, sodium chloride or potassium chloride?

2.77 The radii of the lithium and magnesium ions are 76 pm and 72 pm, respectively. Which compound has stronger ionic attractions, lithium oxide or magnesium oxide?

Compounds: Formulas, Names, and Masses

(Sample Problems 2.8 to 2.17)

Concept Review Questions

2.78 What information about the relative numbers of ions and the percent masses of elements is contained in the formula MgF_2?

2.79 How is a structural formula similar to a molecular formula? How is it different?

2.80 Consider a mixture of 10 billion O_2 molecules and 10 billion H_2 molecules. In what way is this mixture similar to a sample containing 10 billion hydrogen peroxide (H_2O_2) molecules? In what way is it different?

2.81 For what type(s) of compound do we use roman numerals in the names?

2.82 For what type(s) of compound do we use Greek numerical prefixes in the names?

2.83 For what type of compound are we unable to write a molecular formula?

Skill-Building Exercises (grouped in similar pairs)

2.84 Give the name and formula of the compound formed from each pair of elements: (a) sodium and nitrogen; (b) oxygen and strontium; (c) aluminum and chlorine.

2.85 Give the name and formula of the compound formed from each pair of elements: (a) cesium and bromine; (b) sulfur and barium; (c) calcium and fluorine.

2.86 Give the name and formula of the compound formed from each pair of elements:
(a) ${}_{12}L$ and ${}_{9}M$ (b) ${}_{30}L$ and ${}_{16}M$ (c) ${}_{17}L$ and ${}_{38}M$

2.87 Give the name and formula of the compound formed from each pair of elements:
(a) ${}_{37}Q$ and ${}_{35}R$ (b) ${}_{8}Q$ and ${}_{13}R$ (c) ${}_{20}Q$ and ${}_{53}R$

2.88 Give the systematic names for the formulas or the formulas for the names:
(a) tin(IV) chloride; (b) $FeBr_3$; (c) cuprous bromide; (d) Mn_2O_3.

2.89 Give the systematic names for the formulas or the formulas for the names: (a) CoO; (b) mercury(I) chloride; (c) chromic oxide; (d) $CuBr_2$.

2.90 Give the systematic names for the formulas or the formulas for the names:
(a) Na_2HPO_4; (b) ammonium perchlorate; (c) $Pb(C_2H_3O_2)_2 \cdot 3H_2O$; (d) sodium nitrite.

2.91 Give the systematic names for the formulas or the formulas for the names: (a) $Sn(SO_3)_2$; (b) potassium dichromate; (c) $FeCO_3$; (d) potassium carbonate dihydrate.

2.92 Correct each of the following formulas:
(a) Barium oxide is BaO_2.
(b) Iron(II) nitrate is $Fe(NO_3)_3$.
(c) Magnesium sulfide is $MnSO_3$.

2.93 Correct each of the following names:
(a) CuI is cobalt(II) iodide.
(b) $Fe(HSO_4)_3$ is iron(II) sulfate.
(c) $MgCr_2O_7$ is magnesium dichromium heptaoxide.

2.94 Give the name and formula for the acid derived from each of the following anions:
(a) hydrogen carbonate (b) IO_4^- (c) cyanide (d) HS^-

2.95 Give the name and formula for the acid derived from each of the following anions:
(a) perchlorate (b) NO_3^- (c) bromite (d) $H_2PO_4^-$

2.96 Many chemical names are similar at first glance. Give the formulas of the species in each set:
(a) Ammonium ion and ammonia
(b) Magnesium sulfide, magnesium sulfite, and magnesium sulfate
(c) Hydrochloric acid, chloric acid, and chlorous acid
(d) Cuprous bromide and cupric bromide

2.97 Give the formulas of the compounds in each set:
(a) Lead(II) oxide and lead(IV) oxide
(b) Lithium nitride, lithium nitrite, and lithium nitrate
(c) Strontium hydride and strontium hydroxide
(d) Magnesium oxide and manganese(II) oxide

2.98 Give the name and formula of the compound whose molecules consist of two sulfur atoms and four fluorine atoms.

2.99 Give the name and formula of the compound whose molecules consist of two chlorine atoms and one oxygen atom.

2.100 Correct the name to match the formula of the following compounds: (a) calcium(II) dichloride, $CaCl_2$; (b) copper(II) oxide, Cu_2O; (c) stannous tetrafluoride, SnF_4; (d) hydrogen chloride acid, HCl.

2.101 Correct the formula to match the name of the following compounds: (a) iron(III) oxide, Fe_3O_4; (b) chloric acid, HCl; (c) mercuric oxide, Hg_2O; (d) potassium iodide, P_2I_3.

2.102 Write the formula of each compound, and determine its molecular (formula) mass: (a) ammonium sulfate; (b) sodium dihydrogen phosphate; (c) potassium bicarbonate.

2.103 Write the formula of each compound, and determine its molecular (formula) mass: (a) sodium dichromate; (b) ammonium perchlorate; (c) magnesium nitrite trihydrate.

2.104 Calculate the molecular (formula) mass of each compound: (a) dinitrogen pentoxide; (b) lead(II) nitrate; (c) calcium peroxide.

2.105 Calculate the molecular (formula) mass of each compound: (a) iron(II) acetate tetrahydrate; (b) sulfur tetrachloride; (c) potassium permanganate.

2.106 Give the number of atoms of the specified element in a formula unit of each of the following compounds, and calculate the molecular (formula) mass:
(a) Oxygen in aluminum sulfate, $Al_2(SO_4)_3$
(b) Hydrogen in ammonium hydrogen phosphate, $(NH_4)_2HPO_4$
(c) Oxygen in the mineral azurite, $Cu_3(OH)_2(CO_3)_2$

2.107 Give the number of atoms of the specified element in a formula unit of each of the following compounds, and calculate the molecular (formula) mass:
(a) Hydrogen in ammonium benzoate, $C_6H_5COONH_4$
(b) Nitrogen in hydrazinium sulfate, $N_2H_6SO_4$
(c) Oxygen in the mineral leadhillite, $Pb_4SO_4(CO_3)_2(OH)_2$

2.108 Give the formula, name, and molecular mass of the following molecules:

(a)

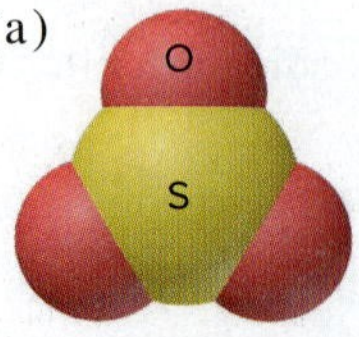

(b)

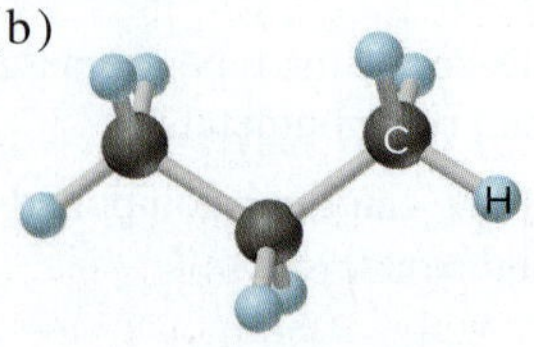

2.109 Give the formula, name, and molecular mass of the following molecules:

(a) 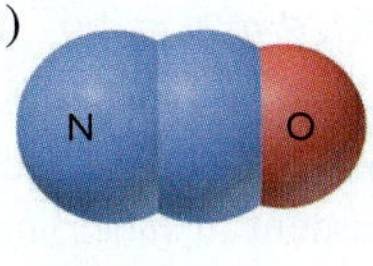

(b) 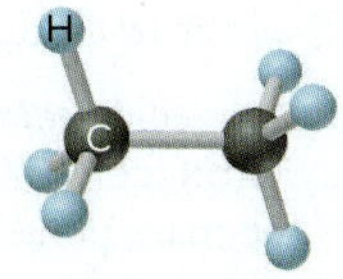

Problems in Context

2.110 Before the use of systematic names, many compounds had common names. Give the systematic name for each of the following:
(a) Blue vitriol, $CuSO_4{\cdot}5H_2O$
(b) Slaked lime, $Ca(OH)_2$
(c) Oil of vitriol, H_2SO_4
(d) Washing soda, Na_2CO_3
(e) Muriatic acid, HCl
(f) Epsom salt, $MgSO_4{\cdot}7H_2O$
(g) Chalk, $CaCO_3$
(h) Dry ice, CO_2
(i) Baking soda, $NaHCO_3$
(j) Lye, NaOH

2.111 Each circle contains a representation of a binary compound. Determine its name, formula, and molecular (formula) mass.

(a) (b)
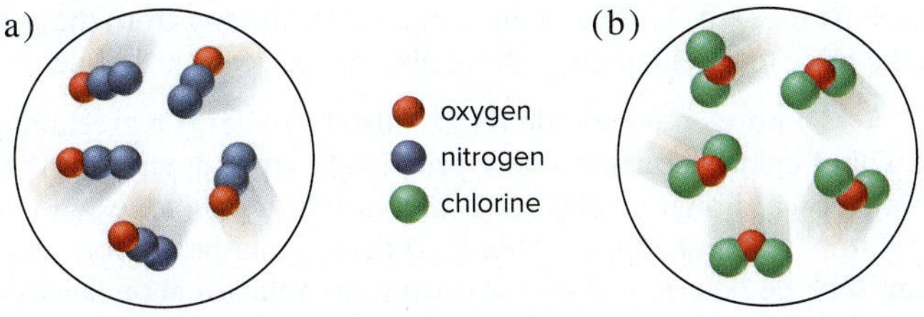

Mixtures: Classification and Separation

Concept Review Questions

2.112 In what main way is separating the components of a mixture different from separating the components of a compound?

2.113 What is the difference between a homogeneous and a heterogeneous mixture?

2.114 Is a solution a homogeneous or a heterogeneous mixture? Give an example of an aqueous solution.

Skill-Building Exercises (grouped in similar pairs)

2.115 Classify each of the following as a compound, a homogeneous mixture, or a heterogeneous mixture: (a) distilled water; (b) gasoline; (c) beach sand; (d) wine; (e) air.

2.116 Classify each of the following as a compound, a homogeneous mixture, or a heterogeneous mixture: (a) orange juice; (b) vegetable soup; (c) cement; (d) calcium sulfate; (e) tea.

Problems in Context

2.117 Which separation method is operating in each of the following procedures?
(a) Pouring a mixture of cooked pasta and boiling water into a colander
(b) Removing colored impurities from raw sugar to make refined sugar

2.118 A quality-control laboratory analyzes a product mixture using gas-liquid chromatography. The separation of components is more than adequate, but the process takes too long. Suggest two ways, other than changing the stationary phase, to shorten the analysis time.

Comprehensive Problems

2.119 Helium is the lightest noble gas and the second most abundant element (after hydrogen) in the universe.
(a) The radius of a helium atom is $3.1{\times}10^{-11}$ m; the radius of its nucleus is $2.5{\times}10^{-15}$ m. What fraction of the spherical atomic volume is occupied by the nucleus (V of a sphere $= \frac{4}{3}\pi r^3$)?
(b) The mass of a helium-4 atom is $6.64648{\times}10^{-24}$ g, and each of its two electrons has a mass of $9.10939{\times}10^{-28}$ g. What fraction of this atom's mass is contributed by its nucleus?

2.120 From the following ions (with their radii in pm), choose the pair that forms the strongest ionic bond and the pair that forms the weakest:

Ion:	Mg^{2+}	K^+	Rb^+	Ba^{2+}	Cl^-	O^{2-}	I^-
Radius:	72	138	152	135	181	140	220

2.121 Give the molecular mass of each compound depicted below, and provide a correct name for any that are named incorrectly.

(a) (b) (c) (d)
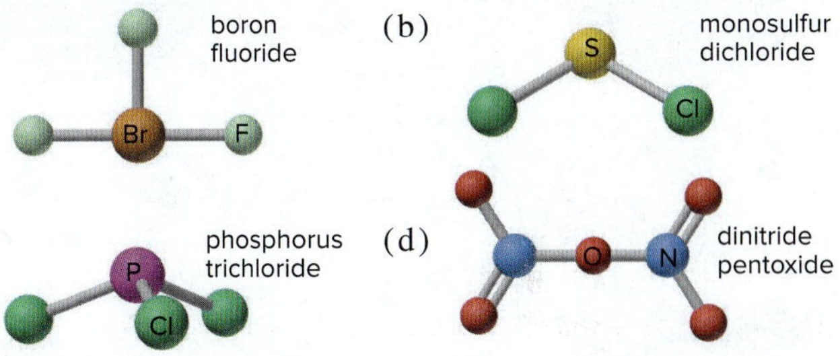

2.122 Polyatomic ions are named by patterns that apply to elements in a given group. Using the periodic table and Table 2.5, give the name of each of the following: (a) SeO_4^{2-}; (b) AsO_4^{3-}; (c) BrO_2^-; (d) $HSeO_4^-$; (e) TeO_3^{2-}.

2.123 Ammonium dihydrogen phosphate, formed from the reaction of phosphoric acid with ammonia, is used as a crop fertilizer as well as a component of some fire extinguishers. (a) What are the mass percentages of N and P in the compound? (b) How much ammonia is incorporated into 100. g of the compound?

2.124 Nitrogen forms more oxides than any other element. The percents by mass of N in three different nitrogen oxides are (I) 46.69%, (II) 36.85%, and (III) 25.94%. For each compound, determine (a) the simplest whole-number ratio of N to O and (b) the number of grams of oxygen per 1.00 g of nitrogen.

2.125 The number of atoms in 1 dm^3 of aluminum is nearly the same as the number of atoms in 1 dm^3 of lead, but the densities of these metals are very different (see Table 1.5). Explain.

2.126 You are working in the laboratory preparing sodium chloride. Consider the following results for three preparations of the compound:

Case 1: 39.34 g Na + 60.66 g Cl_2 $\longrightarrow$ 100.00 g NaCl

Case 2: 39.34 g Na + 70.00 g Cl_2 $\longrightarrow$
100.00 g NaCl + 9.34 g Cl_2

Case 3: 50.00 g Na + 50.00 g Cl_2 $\longrightarrow$
82.43 g NaCl + 17.57 g Na

Explain these results in terms of the laws of conservation of mass and definite composition.

2.127 Scenes A–I on the next page depict various types of matter on the atomic scale. Choose the correct scene(s) for each of the following:
(a) A mixture that fills its container

(b) A substance that cannot be broken down into simpler ones
(c) An element with a very high resistance to flow
(d) A homogeneous mixture
(e) An element that conforms to the walls of its container and displays an upper surface
(f) A gas consisting of diatomic particles
(g) A gas that can be broken down into simpler substances
(h) A substance with a 2/1 ratio of its component atoms
(i) Matter that can be separated into its component substances by physical means
(j) A heterogeneous mixture
(k) Matter that obeys the law of definite composition

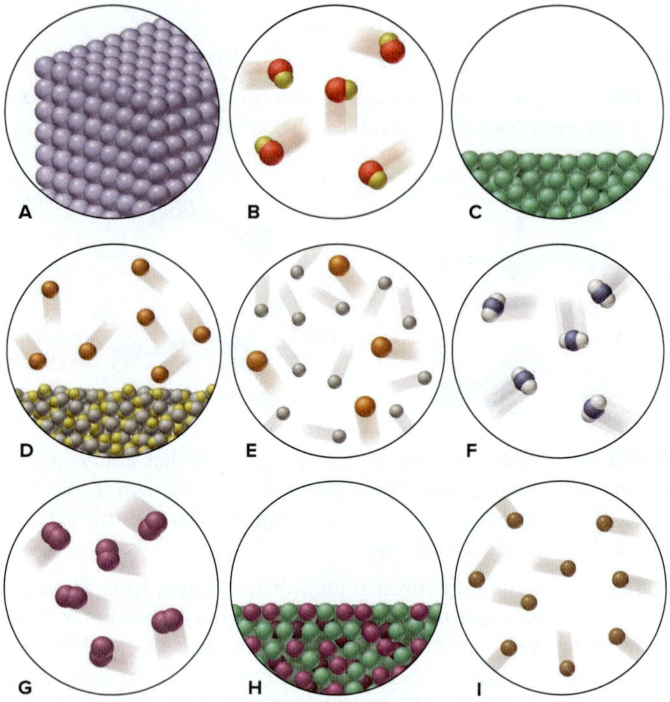

2.128 The seven most abundant ions in seawater make up more than 99% by mass of the dissolved compounds. Here are their abundances in units of mg ion/kg seawater: chloride 18,980; sodium 10,560; sulfate 2650; magnesium 1270; calcium 400; potassium 380; hydrogen carbonate 140.
(a) What is the mass % of each ion in seawater?
(b) What percent of the total mass of ions is represented by sodium ions?
(c) How does the total mass % of alkaline earth metal ions compare with the total mass % of alkali metal ions?
(d) Which make up the larger mass fraction of dissolved components, anions or cations?

2.129 The following scenes represent a mixture of two monatomic gases undergoing a reaction when heated. Which mass law(s) is (are) illustrated by this change?

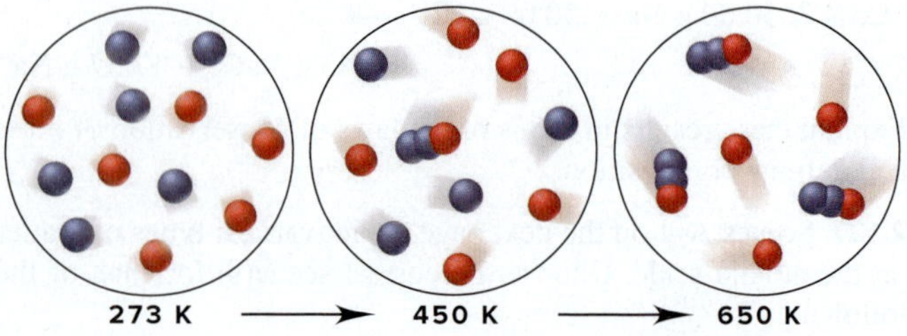

2.130 When barium (Ba) reacts with sulfur (S) to form barium sulfide (BaS), each Ba atom reacts with an S atom. If 2.50 cm^3 of Ba reacts with 1.75 cm^3 of S, are there enough Ba atoms to react with the S atoms (d of Ba = 3.51 g/cm^3; d of S = 2.07 g/cm^3)?

2.131 Succinic acid (*below*) is an important metabolite in biological energy production. Give the molecular formula, molecular mass, and the mass percent of each element in succinic acid.

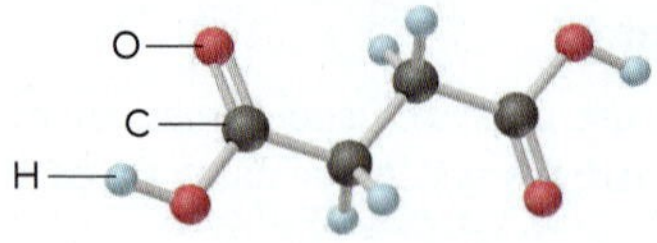

2.132 Fluoride ion is poisonous in relatively low amounts: 0.2 g of F^- per 70 kg of body weight can cause death. Nevertheless, in order to prevent tooth decay, F^- ions are added to drinking water at a concentration of 1 mg of F^- ion per L of water. How many liters of fluoridated drinking water would a 70-kg person have to consume in one day to reach this toxic level? How many kilograms of sodium fluoride would be needed to treat an 8.50×10^7-gal reservoir?

2.133 Antimony has many uses, for example, in infrared devices and as part of an alloy in lead storage batteries. The element has two naturally occurring isotopes, one with mass 120.904 amu and the other with mass 122.904 amu. (a) Write the ${}^{A}_{Z}X$ notation for each isotope. (b) Use the atomic mass of antimony from the periodic table to calculate the natural abundance of each isotope.

2.134 Dinitrogen monoxide (N_2O; nitrous oxide) is a greenhouse gas that enters the atmosphere principally from natural fertilizer breakdown. Some studies have shown that the isotope ratios of ^{15}N to ^{14}N and of ^{18}O to ^{16}O in N_2O depend on the source, which can thus be determined by measuring the relative abundances of molecular masses in a sample of N_2O.
(a) What different molecular masses are possible for N_2O?
(b) The percent abundance of ^{14}N is 99.6%, and that of ^{16}O is 99.8%. Which molecular mass of N_2O is least common, and which is most common?

2.135 Use the box color(s) in the periodic table below to identify the element(s) described by each of the following:

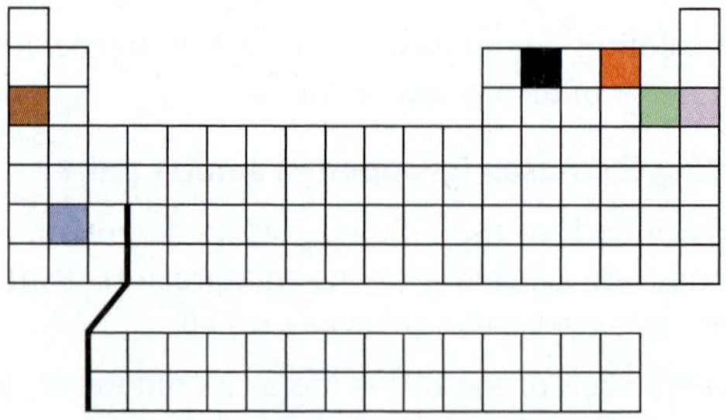

(a) Four elements that are nonmetals
(b) Two elements that are metals
(c) Three elements that are gases at room temperature
(d) Three elements that are solid at room temperature
(e) One pair of elements likely to form a covalent compound
(f) Another pair of elements likely to form a covalent compound
(g) One pair of elements likely to form an ionic compound with formula MX
(h) Another pair of elements likely to form an ionic compound with formula MX
(i) Two elements likely to form an ionic compound with formula M_2X

(j) Two elements likely to form an ionic compound with formula MX_2
(k) An element that forms no compounds
(l) A pair of elements whose compounds exhibit the law of multiple proportions

2.136 The two isotopes of potassium with significant abundances in nature are ^{39}K (isotopic mass = 38.9637 amu, 93.258%) and ^{41}K (isotopic mass = 40.9618 amu, 6.730%). Fluorine has only one naturally occurring isotope, ^{19}F (isotopic mass = 18.9984 amu). Calculate the formula mass of potassium fluoride.

2.137 Boron trifluoride is used as a catalyst in the synthesis of organic compounds. When this compound is analyzed by mass spectrometry (see Tools of the Laboratory: Mass Spectrometry, following Section 2.5), several different 1+ ions form, including ions representing the whole molecule as well as molecular fragments formed by the loss of one, two, and three F atoms. Given that boron has two naturally occurring isotopes, ^{10}B and ^{11}B, and fluorine has one, ^{19}F, calculate the masses of all possible 1+ ions.

2.138 Nitrogen monoxide (NO) is a bioactive molecule in blood. Low NO concentrations cause respiratory distress and the formation of blood clots. Doctors prescribe nitroglycerin, $C_3H_5N_3O_9$, and isoamyl nitrate, $(CH_3)_2CHCH_2CH_2ONO_2$, to increase the blood level of NO. If each compound releases one molecule of NO per atom of N it contains, calculate the mass percent of NO in each.

2.139 TNT (trinitrotoluene; *below*) is used as an explosive in construction. Calculate the mass of each element in 1.00 lb of TNT.

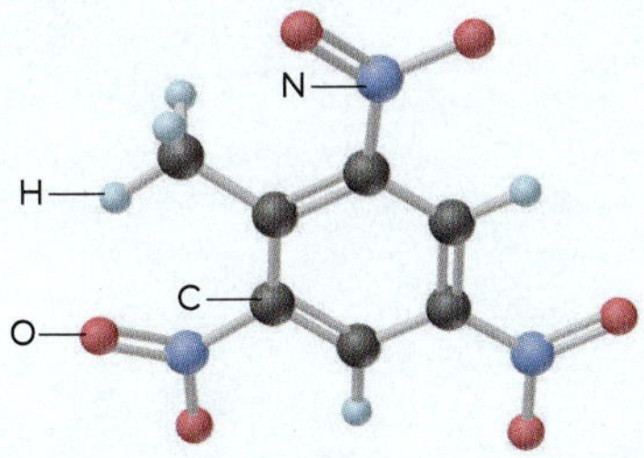

2.140 Nuclei differ in their stability, and some are so unstable that they undergo radioactive decay. The ratio of the number of neutrons to number of protons (*N/Z*) in a nucleus correlates with its stability. Calculate the *N/Z* ratio for (a) ^{144}Sm; (b) ^{56}Fe; (c) ^{20}Ne; (d) ^{107}Ag. (e) The radioactive isotope ^{238}U decays in a series of nuclear reactions that includes another uranium isotope, ^{234}U, and three lead isotopes, ^{214}Pb, ^{210}Pb, and ^{206}Pb. How many neutrons, protons, and electrons are in each of these five isotopes?

2.141 The anticancer drug Platinol (cisplatin), $Pt(NH_3)_2Cl_2$, reacts with a cancer cell's DNA and interferes with its growth. (a) What is the mass % of platinum (Pt) in Platinol? (b) If Pt costs $51/g, how many grams of Platinol can be made for $1.00 million (assume that the cost of Pt determines the cost of the drug)?

2.142 From the periodic table below, give the name, symbol, atomic number, atomic mass, period number, and group number of (a) the *building-block elements (red),* which occur in nearly every biological molecule, and (b) the *macronutrients (green),* which are either essential ions in cell fluids or are part of many biomolecules.

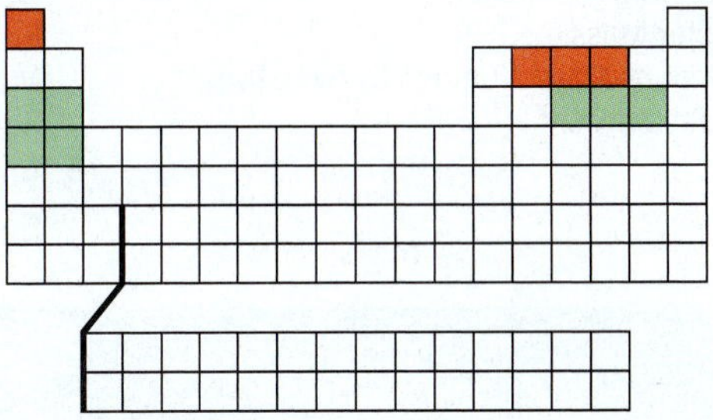

2.143 The block diagram below classifies the components of matter on the macroscopic scale. Identify blocks (a)–(d).

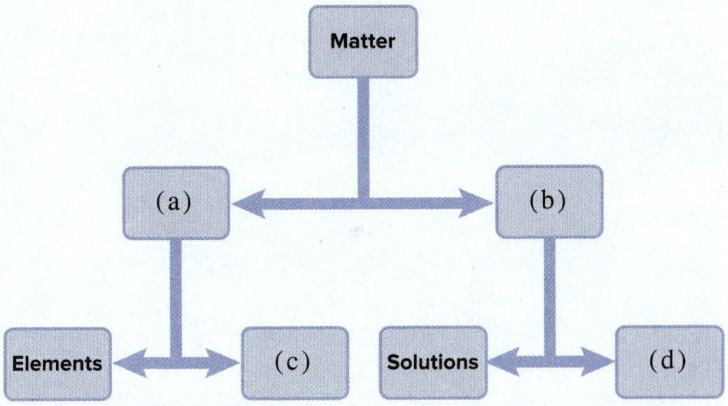

2.144 Which of the steps in the following process involve(s) a physical change and which involve(s) a chemical change?

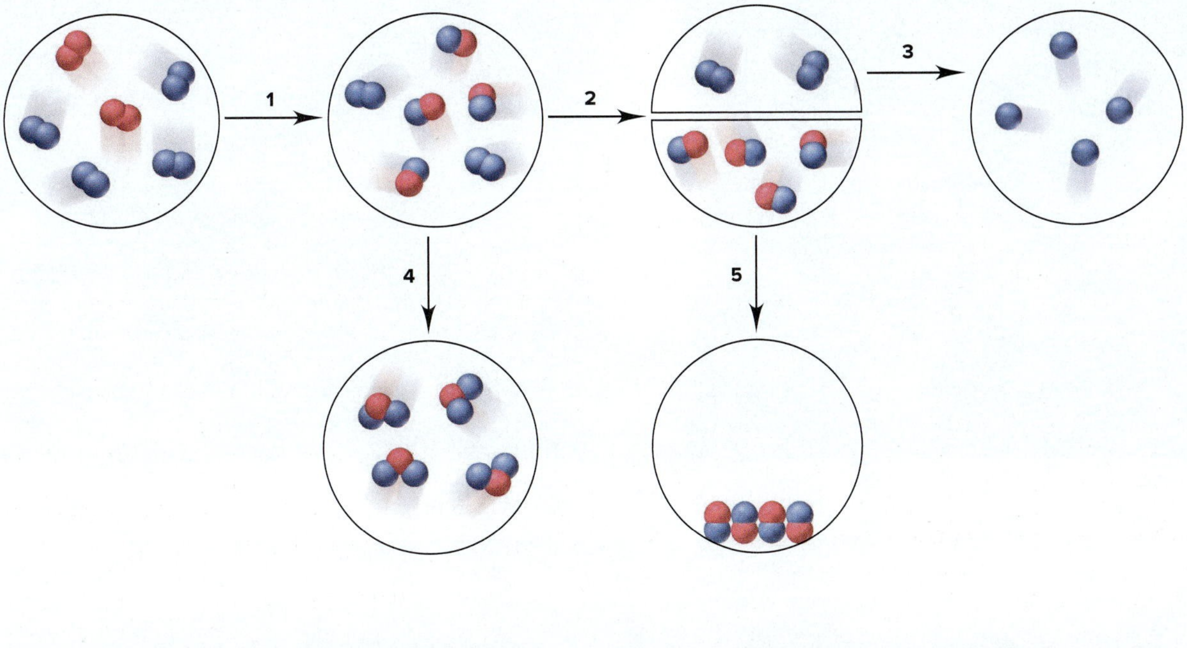

3 Stoichiometry of Formulas and Equations

3.1 The Mole
Defining the Mole
Molar Mass
Amount-Mass-Number Conversions
Mass Percent

3.2 Determining the Formula of an Unknown Compound
Empirical Formulas
Molecular Formulas
Formulas and Structures

3.3 Writing and Balancing Chemical Equations

3.4 Calculating Quantities of Reactant and Product
Molar Ratios from Balanced Equations
Reaction Sequences
Limiting Reactants
Reaction Yields

Source: NASA

Concepts and Skills to Review Before You Study This Chapter

- atomic mass (Section 2.5)
- names and formulas of compounds (Section 2.8)
- molecular (or formula) mass (Section 2.8)
- molecular and structural formulas and ball-and-stick and space-filling models (Section 2.8)

Chemistry is, above all, a practical science. Imagine that you're a biochemist who has extracted a substance with medicinal activity from a tropical plant: what is its formula, and what quantity of metabolic products will establish a safe dosage level? Or, suppose you're a chemical engineer studying rocket-fuel thrust *(see photo)*: what amount of propulsive gases will a fuel produce? Perhaps you're on a team of environmental chemists examining coal samples: what quantity of air pollutants will a sample produce when burned? Or, maybe you're a polymer chemist preparing a plastic with unusual properties: how much of this new material will the polymerization reaction yield? You can answer countless questions like these with a knowledge of **stoichiometry** (pronounced "stoy-key-AHM-uh-tree"; from the Greek *stoicheion,* "element or part," and *metron,* "measure"), the study of the quantitative aspects of formulas and reactions.

IN THIS CHAPTER . . . *We relate the mass of a substance to the number of chemical entities comprising it (atoms, ions, molecules, or formula units) and apply this relationship to formulas and equations.*

- We discuss the mole, the chemist's unit for amount of a substance, and use it to convert between mass and number of entities.
- We also use the mole concept to derive a chemical formula from the results of mass analysis.
- We compare three types of chemical formulas.
- We learn how to write chemical equations and how to balance them in terms of the amounts of substances reacting and produced.
- We calculate the amounts of reactants and products in a reaction and see why one of the reactants limits the amount of product that can form and, thus, the reaction yield.

3.1 THE MOLE

In daily life, we often measure things by weighing or by counting: we weigh coffee beans or bananas, but we count eggs or pencils. And we use mass units (a kilogram of coffee beans) or counting units (a dozen pencils) to express the amount. Similarly, the daily routine in the laboratory involves measuring substances. We want to know the numbers of chemical entities—atoms, ions, molecules, or formula units—that react with each other, but how can we possibly count or weigh such minute objects? As you'll see, chemists have devised a unit, called the *mole,* to *count chemical entities by weighing a very large number of them.*

Defining the Mole

The **mole** (abbreviated **mol**) is the SI unit for *amount of substance*. It is defined as *the amount of a substance that contains the same number of entities as the number* *of atoms in 12 g of carbon-12.* This number, called **Avogadro's number** (in honor of the 19th-century Italian physicist Amedeo Avogadro), is enormous:

$$\text{One mole (1 mol) contains } 6.022{\times}10^{23} \text{ entities (to four significant figures)} \quad (3.1)$$

Imagine a Mole of . . .

A mole of any ordinary object is a staggering amount: a mole of periods (.) lined up side by side would equal the radius of our galaxy; a mole of marbles stacked tightly together would cover the continental United States 70 miles deep. However, atoms and molecules are not ordinary objects: you can swallow a mole of water molecules (about 18 mL) in one gulp!

A counting unit, like *dozen,* tells you the number of objects but not their mass; a mass unit, like *kilogram,* tells you the mass of the objects but not their number. The mole tells you both—the *number* of objects in a given *mass* of substance:

1 mol of carbon-12 contains $6.022{\times}10^{23}$ carbon-12 atoms *and* has a mass of 12 g

What does it mean that the mole unit allows you to count entities by weighing the sample? Suppose you have a sample of carbon-12 and want to know the number of atoms present. You find that the sample weighs 6 g, so it is 0.5 mol of carbon-12 and contains $0.5(6.022{\times}10^{23})$, or $3.011{\times}10^{23}$ atoms:

6 g of carbon-12 is 0.5 mol of carbon-12 and contains $3.011{\times}10^{23}$ atoms

Knowing the amount (in moles), the mass (in grams), and the number of entities becomes very important when we mix different substances to run a reaction. The central relationship between masses on the atomic scale and on the macroscopic scale is the same for elements and compounds:

- *Elements.* The mass in *atomic mass units (amu)* of one atom of an element is the *same numerically* as the mass in *grams (g)* of 1 mole of atoms of the element. Recall from Chapter 2 that each atom of an element is considered to have the *atomic mass* given in the periodic table. Thus,

16
S
32.06

1 atom of S	has a mass of	32.06 amu	and	1 mol ($6.022{\times}10^{23}$ atoms) of S	has a mass of	32.06 g
1 atom of Fe	has a mass of	55.85 amu	and	1 mol ($6.022{\times}10^{23}$ atoms) of Fe	has a mass of	55.85 g

Note, also, that since atomic masses are relative, 1 Fe atom weighs 55.85/32.06 as much as 1 S atom, and 1 mol of Fe weighs 55.85/32.06 as much as 1 mol of S.

- *Compounds.* The mass in *atomic mass units (amu)* of one molecule (or formula unit) of a compound is the *same numerically* as the mass in *grams (g)* of 1 mole of the compound. Thus, for example,

1 molecule of H_2O	has a mass of	18.02 amu	and	1 mol ($6.022{\times}10^{23}$ molecules) of H_2O	has a mass of	18.02 g
1 formula unit of NaCl	has a mass of	58.44 amu	and	1 mol ($6.022{\times}10^{23}$ formula units) of NaCl	has a mass of	58.44 g

Here, too, because masses are relative, 1 H_2O molecule weighs 18.02/58.44 as much as 1 NaCl formula unit, and 1 mol of H_2O weighs 18.02/58.44 as much as 1 mol of NaCl.

Figure 3.1 One mole ($6.022{\times}10^{23}$ entities) of some familiar substances. From left to right: 1 mol of copper (63.55 g), of liquid H_2O (18.02 g), of sodium chloride (table salt, 58.44 g), of sucrose (table sugar, 342.3 g), and of aluminum (26.98 g).

The two key points to remember about the importance of the mole unit are

- The *mole* lets us relate the *number* of entities to the *mass* of a sample of those entities.
- The mole maintains the *same numerical relationship* between mass on the atomic scale (atomic mass units, amu) and mass on the macroscopic scale (grams, g).

In everyday terms, a grocer *does not* know that there are 1 dozen eggs from their weight or that there is 1 kilogram of coffee beans from their count, because eggs and coffee beans do not have fixed masses. But, by weighing out 63.55 g (1 mol) of copper, a chemist *does* know that there are $6.022{\times}10^{23}$ copper atoms, because all copper atoms have an atomic mass of 63.55 amu. Figure 3.1 shows 1 mole of some familiar elements and compounds.

Determining Molar Mass

The **molar mass ($\mathcal{M}$)** of a substance is the mass of a mole of its entities (atoms, molecules, or formula units) and has units of grams per mole (g/mol). The periodic table is indispensable for calculating molar mass:

1. *Elements.* To find the molar mass, look up the atomic mass and note whether the element is monatomic or molecular.

- *Monatomic elements.* The molar mass of a monatomic element is the periodic table value in grams per mole.* For example, the molar mass of neon is 20.18 g/mol, and the molar mass of gold is 197.0 g/mol.
- *Molecular elements.* You must know the formula of a molecular element to determine the molar mass (see Figure 2.15). For example, in air, oxygen exists most commonly as diatomic molecules, so the molar mass of O_2 is twice that of O:

$$\text{Molar mass } (\mathcal{M}) \text{ of } O_2 = 2 \times \mathcal{M} \text{ of O} = 2 \times 16.00 \text{ g/mol} = 32.00 \text{ g/mol}$$

The most common form of sulfur exists as octatomic molecules, S_8:

$$\mathcal{M} \text{ of } S_8 = 8 \times \mathcal{M} \text{ of S} = 8 \times 32.06 \text{ g/mol} = 256.5 \text{ g/mol}$$

2. *Compounds. The molar mass of a compound is the sum of the molar masses of the atoms in the formula.* Thus, from the formula of sulfur dioxide, SO_2, we know that 1 mol of SO_2 molecules contains 1 mol of S atoms and 2 mol of O atoms:

$$\mathcal{M} \text{ of } SO_2 = \mathcal{M} \text{ of S} + (2 \times \mathcal{M} \text{ of O}) = 32.06 \text{ g/mol} + (2 \times 16.00 \text{ g/mol}) = 64.06 \text{ g/mol}$$

Similarly, for ionic compounds, such as potassium sulfide (K_2S), we have

$$\mathcal{M} \text{ of } K_2S = (2 \times \mathcal{M} \text{ of K}) + \mathcal{M} \text{ of S} = (2 \times 39.10 \text{ g/mol}) + 32.06 \text{ g/mol} = 110.26 \text{ g/mol}$$

Thus, *subscripts in a formula refer to individual atoms (or ions) as well as to moles of atoms (or ions).* Table 3.1 summarizes these ideas for glucose, $C_6H_{12}O_6$, the essential sugar in energy metabolism.

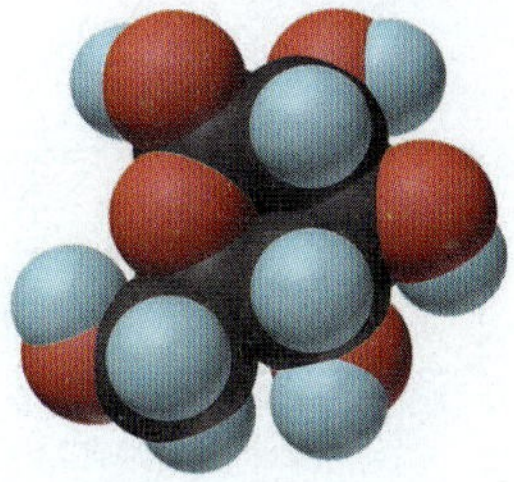

Glucose

Table 3.1 Information Contained in the Chemical Formula of Glucose, $C_6H_{12}O_6$ ($\mathcal{M}$ = 180.16 g/mol)

	Carbon (C)	Hydrogen (H)	Oxygen (O)
Atoms/molecule of compound	6 atoms	12 atoms	6 atoms
Moles of atoms/mole of compound	6 mol of atoms	12 mol of atoms	6 mol of atoms
Atoms/mole of compound	$6(6.022\times10^{23})$ atoms	$12(6.022\times10^{23})$ atoms	$6(6.022\times10^{23})$ atoms
Mass/molecule of compound	6(12.01 amu) = 72.06 amu	12(1.008 amu) = 12.10 amu	6(16.00 amu) = 96.00 amu
Mass/mole of compound	72.06 g	12.10 g	96.00 g

Converting Between Amount, Mass, and Number of Chemical Entities

One of the most common skills in the lab—and on exams—is converting between amount (mol), mass (g), and number of entities of a substance.

1. *Converting between amount and mass.* If you know the amount of a substance, you can find its mass, and vice versa. The molar mass ($\mathcal{M}$), which expresses the equivalence between 1 mole of a substance and its mass in grams, is the conversion factor between amount and mass:

$$\frac{\text{no. of grams}}{1 \text{ mol}} \quad \text{or} \quad \frac{1 \text{ mol}}{\text{no. of grams}}$$

- *From amount (mol) to mass (g),* multiply by the molar mass to cancel the mole unit:

$$\text{Mass (g)} = \text{amount } (\cancel{\text{mol}}) \times \frac{\text{no. of grams}}{1 \cancel{\text{mol}}} \qquad \textbf{(3.2)}$$

*The mass value in the periodic table has no units because it is a *relative* atomic mass, given by the atomic mass (in amu) divided by 1 amu ($\frac{1}{12}$ mass of one ^{12}C atom in amu):

$$\text{Relative atomic mass} = \frac{\text{atomic mass } (\cancel{\text{amu}})}{\frac{1}{12} \text{ mass of } ^{12}\text{C } (\cancel{\text{amu}})}$$

Therefore, you use the same number (with different units) for the atomic mass and for the molar mass.

- *From mass (g) to amount (mol),* divide by the molar mass (multiply by $1/\mathcal{M}$) to cancel the mass unit:

$$\text{Amount (mol)} = \text{mass } (\cancel{\text{g}}) \times \frac{1 \text{ mol}}{\cancel{\text{no. of grams}}} \qquad \textbf{(3.3)}$$

2. *Converting between amount and number.* Similarly, if you know the amount (mol), you can find the number of entities, and vice versa. Avogadro's number, which expresses the equivalence between 1 mole of a substance and the number of entities it contains, is the conversion factor between amount and number of entities:

$$\frac{6.022{\times}10^{23} \text{ entities}}{1 \text{ mol}} \quad \text{or} \quad \frac{1 \text{ mol}}{6.022{\times}10^{23} \text{ entities}}$$

- *From amount (mol) to number of entities,* multiply by Avogadro's number to cancel the mole unit:

$$\text{No. of entities} = \text{amount } (\cancel{\text{mol}}) \times \frac{6.022{\times}10^{23} \text{ entities}}{1 \cancel{\text{ mol}}} \qquad \textbf{(3.4)}$$

- *From number of entities to amount (mol),* divide by Avogadro's number to cancel the number of entities:

$$\text{Amount (mol)} = \cancel{\text{no. of entities}} \times \frac{1 \text{ mol}}{6.022{\times}10^{23} \cancel{\text{ entities}}} \qquad \textbf{(3.5)}$$

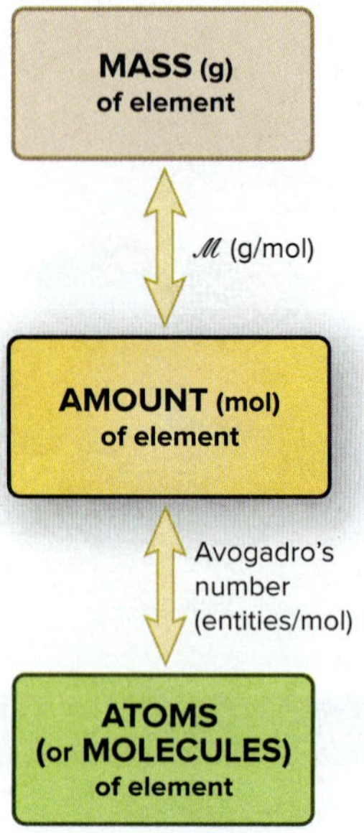

Figure 3.2 Mass-mole-number relationships for elements.

Amount-Mass-Number Conversions Involving Elements We begin with amount-mass-number relationships of elements. As Figure 3.2 shows, *convert mass or number of entities (atoms or molecules) to amount (mol) first.* For molecular elements, Avogadro's number gives *molecules* per mole.

Let's work through several sample problems that show these conversions for some elements.

SAMPLE PROBLEM 3.1 Converting Between Mass and Amount of an Element

Problem Silver (Ag) is used in jewelry and tableware but no longer in U.S. coins. How many grams of Ag are in 0.0342 mol of Ag?

Plan We know the amount of Ag (0.0342 mol) and have to find the mass (g). To convert units of *moles* of Ag to *grams* of Ag, we multiply by the *molar mass* of Ag, which we find in the periodic table (see the road map).

Road Map

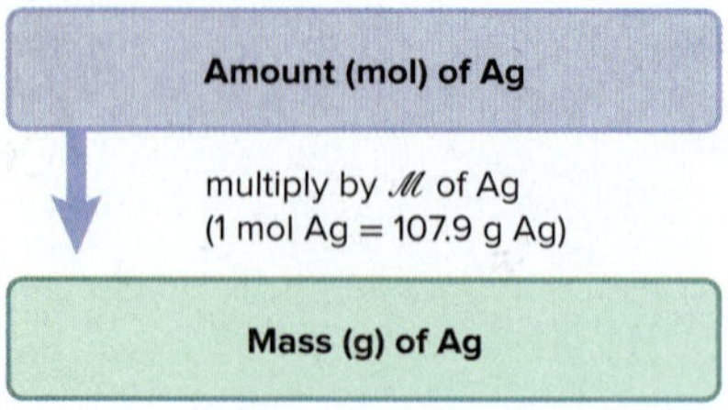

Solution Converting from amount (mol) of Ag to mass (g):

$$\text{Mass (g) of Ag} = 0.0342 \cancel{\text{ mol Ag}} \times \frac{107.9 \text{ g Ag}}{1 \cancel{\text{ mol Ag}}} = 3.69 \text{ g Ag}$$

Check We rounded the mass to three significant figures because the amount (in mol) has three. The units are correct. About 0.03 mol × 100 g/mol gives 3 g; the small mass makes sense because 0.0342 is a small fraction of a mole.

FOLLOW-UP PROBLEMS

Brief Solutions for ***all*** *Follow-up Problems appear at the end of the chapter.*

3.1A Graphite is the crystalline form of carbon used in "lead" pencils. How many moles of carbon are in 315 mg of graphite? Include a road map that shows how you planned the solution.

3.1B A soda can contains about 14 g of aluminum (Al), the most abundant element in Earth's crust. How many soda cans can be made from 52 mol of Al? Include a road map that shows how you planned the solution.

SOME SIMILAR PROBLEMS 3.12(a) and 3.13(a)

SAMPLE PROBLEM 3.2 Converting Between Number of Entities and Amount of an Element

Problem Gallium (Ga) is a key element in solar panels, calculators, and other light-sensitive electronic devices. How many Ga atoms are in 2.85×10^{-3} mol of gallium?

Plan We know the amount of gallium (2.85×10^{-3} mol) and need the number of Ga atoms. We multiply amount (mol) by Avogadro's number to find number of atoms (see the road map).

Road Map

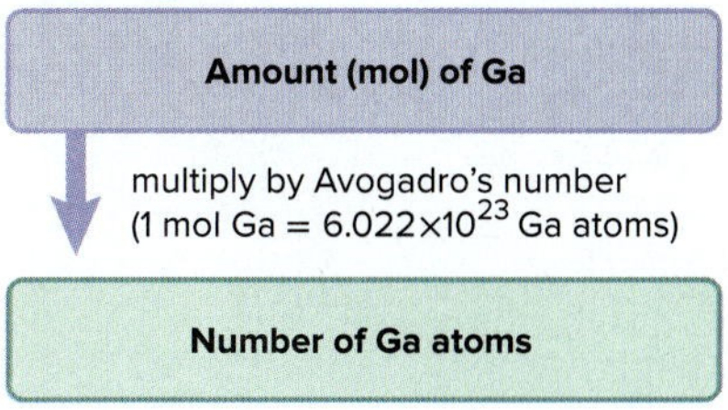

Solution Converting from amount (mol) of Ga to number of atoms:

$$\text{No. of Ga atoms} = 2.85\times10^{-3}\ \cancel{\text{mol Ga}} \times \frac{6.022\times10^{23}\ \text{Ga atoms}}{1\ \cancel{\text{mol Ga}}}$$
$$= 1.72\times10^{21}\ \text{Ga atoms}$$

Check The number of atoms has three significant figures because the number of moles does. When we round amount (mol) of Ga and Avogadro's number, we have $\sim(3\times10^{-3}\ \text{mol})(6\times10^{23}\ \text{atoms/mol}) = 18\times10^{20}$, or 1.8×10^{21} atoms, so our answer seems correct.

FOLLOW-UP PROBLEMS

3.2A At rest, a person inhales 9.72×10^{21} nitrogen molecules in an average breath of air. How many moles of nitrogen *atoms* are inhaled? (*Hint:* In air, nitrogen occurs as a diatomic molecule.) Include a road map that shows how you planned the solution.

3.2B A tank contains 325 mol of compressed helium (He) gas. How many He atoms are in the tank? Include a road map that shows how you planned the solution.

SOME SIMILAR PROBLEMS 3.12(b) and 3.13(b)

For the next sample problem, note that mass and number of entities relate directly to amount (mol), but *not* to each other. Therefore, *to convert between mass and number, first convert to amount.*

SAMPLE PROBLEM 3.3 Converting Between Number of Entities and Mass of an Element

Problem Iron (Fe) is the main component of steel and, thus, the most important metal in industrial society; it is also essential in the human body. How many Fe atoms are in 95.8 g of Fe?

Plan We know the mass of Fe (95.8 g) and need the number of Fe atoms. We cannot convert directly from mass to number of atoms, so we first convert to amount (mol) by dividing the mass of Fe by its molar mass. Then, we multiply amount (mol) by Avogadro's number to find number of atoms (see the road map).

Road Map

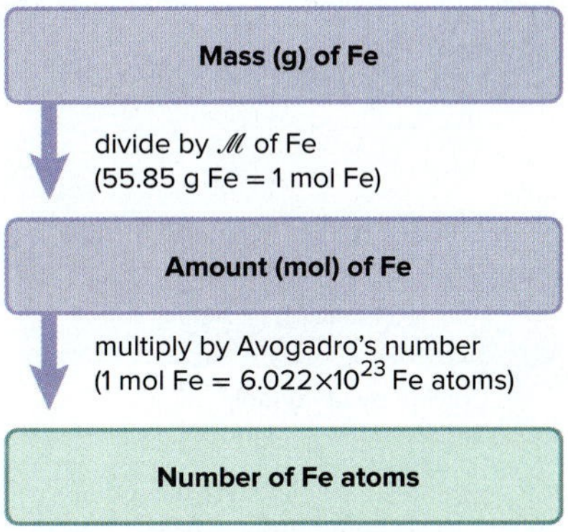

Solution Converting from mass (g) of Fe to amount (mol):

$$\text{Amount (mol) of Fe} = 95.8\ \cancel{\text{g Fe}} \times \frac{1\ \text{mol Fe}}{55.85\ \cancel{\text{g Fe}}} = 1.72\ \text{mol Fe}$$

Converting from amount (mol) of Fe to number of Fe atoms:

$$\text{No. of Fe atoms} = 1.72\ \cancel{\text{mol Fe}} \times \frac{6.022\times10^{23}\ \text{atoms Fe}}{1\ \cancel{\text{mol Fe}}}$$
$$= 1.04\times10^{24}\ \text{atoms Fe}$$

Check Rounding the mass and the molar mass of Fe, we have $\sim100\ \text{g}/(\sim60\ \text{g/mol}) = 1.7$ mol. Therefore, the number of atoms should be a bit less than twice Avogadro's number: $<2(6\times10^{23}) = <1.2\times10^{24}$, so the answer seems correct.

Comment The two steps can be combined into one:

$$\text{Amount (mol) of Fe} = 95.8\ \cancel{\text{g Fe}} \times \frac{1\ \cancel{\text{mol Fe}}}{55.85\ \cancel{\text{g Fe}}} \times \frac{6.022\times10^{23}\ \text{atoms Fe}}{1\ \cancel{\text{mol Fe}}} = 1.04\times10^{24}\ \text{atoms Fe}$$

FOLLOW-UP PROBLEMS

3.3A Manganese (Mn) is a transition element essential for the growth of bones. What is the mass in grams of 3.22×10^{20} Mn atoms, the number found in 1 kg of bone? Include a road map that shows how you planned the solution.

3.3B Pennies minted after 1982 are made of zinc plated with a thin coating of copper (Cu); the copper layer on each penny has a mass of 0.0625 g. How many Cu atoms are in a penny? Include a road map that shows how you planned the solution.

SOME SIMILAR PROBLEMS 3.12(c) and 3.13(c)

Amount-Mass-Number Conversions Involving Compounds Only one new step is needed to solve amount-mass-number problems involving compounds: we need the chemical formula to find the molar mass and the amount of each element in the compound. The relationships are shown in Figure 3.3, and Sample Problems 3.4 and 3.5 apply them to compounds with simple and more complicated formulas, respectively.

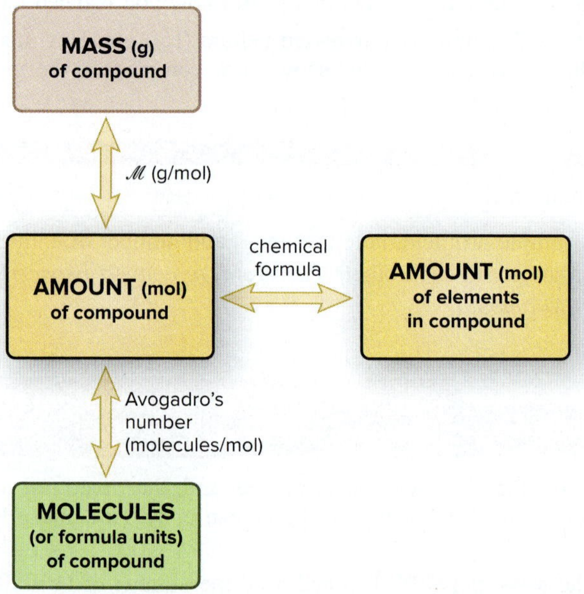

Figure 3.3 **Amount-mass-number relationships for compounds.** Use the chemical formula to find the amount (mol) of each element in a compound.

SAMPLE PROBLEM 3.4 Converting Between Number of Entities and Mass of a Compound I

Problem Nitrogen dioxide is a component of urban smog that forms from gases in car exhaust. How many molecules are in 8.92 g of nitrogen dioxide?

Plan We know the mass of compound (8.92 g) and need to find the number of molecules. As you just saw in Sample Problem 3.3, to convert mass to number of entities, we have to find the amount (mol). To do so, we divide the mass by the molar mass ($\mathcal{M}$), which we calculate from the molecular formula (see Sample Problem 2.16). Once we have the amount (mol), we multiply by Avogadro's number to find the number of molecules (see the road map).

Road Map

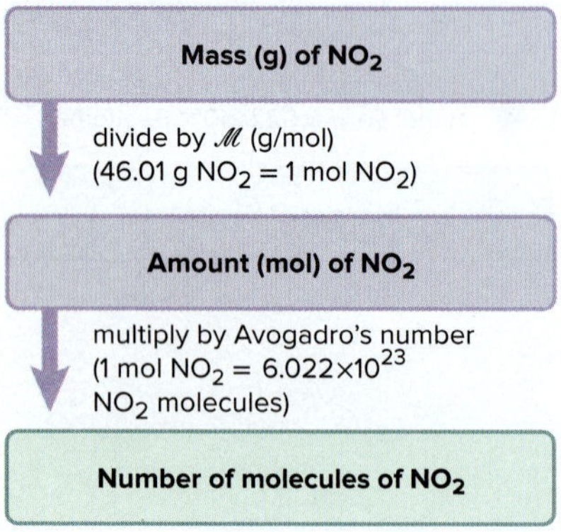

Solution The formula is NO_2. Calculating the molar mass:

$$\begin{aligned}\mathcal{M} &= (1 \times \mathcal{M} \text{ of N}) + (2 \times \mathcal{M} \text{ of O})\\ &= 14.01 \text{ g/mol} + (2 \times 16.00 \text{ g/mol})\\ &= 46.01 \text{ g/mol}\end{aligned}$$

Converting from mass (g) of NO_2 to amount (mol):

$$\text{Amount (mol) of } NO_2 = 8.92 \text{ g } NO_2 \times \frac{1 \text{ mol } NO_2}{46.01 \text{ g } NO_2}$$
$$= 0.194 \text{ mol } NO_2$$

Converting from amount (mol) of NO_2 to number of molecules:

$$\text{No. of molecules} = 0.194 \text{ mol } NO_2 \times \frac{6.022\times10^{23} \text{ } NO_2 \text{ molecules}}{1 \text{ mol } NO_2}$$
$$= 1.17\times10^{23} \text{ } NO_2 \text{ molecules}$$

Check Rounding, we get ($\sim$0.2 mol)(6×10^{23}) = 1.2×10^{23}, so the answer seems correct.

FOLLOW-UP PROBLEMS

3.4A Fluoride ion is added to drinking water to prevent tooth decay. What is the mass (g) of sodium fluoride in a liter of water that contains 1.19×10^{19} formula units of the compound? Include a road map that shows how you planned the solution.

3.4B Calcium chloride is applied to highways in winter to melt accumulated ice. A snow-plow truck applies 400 lb of $CaCl_2$ per mile of highway. How many formula units of the compound are applied per mile? Include a road map that shows how you planned the solution.

SOME SIMILAR PROBLEMS 3.14–3.19

SAMPLE PROBLEM 3.5 Converting Between Number of Entities and Mass of a Compound II

Problem Ammonium carbonate is a white solid that decomposes with warming. It has many uses, for example, as a component in baking powder, fire extinguishers, and smelling salts.
(a) How many formula units are in 41.6 g of ammonium carbonate?
(b) How many O atoms are in this sample?

Plan **(a)** We know the mass of compound (41.6 g) and need to find the number of formula units. As in Sample Problem 3.4, we find the amount (mol) and then multiply by Avogadro's number to find the number of formula units. (A road map for this step would be the same as the one in Sample Problem 3.4.) **(b)** To find the number of O atoms, we multiply the number of formula units by the number of O atoms in one formula unit (see the road map).

Road Map

Number of formula units of $(NH_4)_2CO_3$

↓ multiply by number of O atoms in one formula unit [1 formula unit of $(NH_4)_2CO_3$ = 3 O atoms]

Number of O atoms

Solution **(a)** The formula is $(NH_4)_2CO_3$ (see Table 2.5). Calculating the molar mass:

$$\mathcal{M} = (2 \times \mathcal{M} \text{ of N}) + (8 \times \mathcal{M} \text{ of H}) + (1 \times \mathcal{M} \text{ of C}) + (3 \times \mathcal{M} \text{ of O})$$
$$= (2 \times 14.01 \text{ g/mol N}) + (8 \times 1.008 \text{ g/mol H}) + 12.01 \text{ g/mol C} + (3 \times 16.00 \text{ g/mol O})$$
$$= 96.09 \text{ g/mol } (NH_4)_2CO_3$$

Converting from mass (g) to amount (mol):

$$\text{Amount (mol) of } (NH_4)_2CO_3 = 41.6 \text{ g } (NH_4)_2CO_3 \times \frac{1 \text{ mol } (NH_4)_2CO_3}{96.09 \text{ g } (NH_4)_2CO_3}$$
$$= 0.433 \text{ mol } (NH_4)_2CO_3$$

Converting from amount (mol) to formula units:

$$\text{Formula units of } (NH_4)_2CO_3 = 0.433 \text{ mol } (NH_4)_2CO_3$$
$$\times \frac{6.022\times10^{23} \text{ formula units } (NH_4)_2CO_3}{1 \text{ mol } (NH_4)_2CO_3}$$
$$= 2.61\times10^{23} \text{ formula units } (NH_4)_2CO_3$$

(b) Finding the number of O atoms:

$$\text{No. of O atoms} = 2.61\times10^{23} \text{ formula units } (NH_4)_2CO_3 \times \frac{3 \text{ O atoms}}{1 \text{ formula unit } (NH_4)_2CO_3}$$
$$= 7.83\times10^{23} \text{ O atoms}$$

Check In (a), the units are correct. Since the mass is less than half the molar mass (~42/96 < 0.5), the number of formula units should be less than half Avogadro's number ($\sim 2.6\times10^{23}/6.0\times10^{23} < 0.5$).

Comment A *common mistake* is to forget the subscript 2 outside the parentheses in $(NH_4)_2CO_3$, which would give a much lower molar mass.

FOLLOW-UP PROBLEMS

3.5A Tetraphosphorus decoxide reacts with water to form phosphoric acid, a major industrial acid. In the laboratory, the oxide is a drying agent.
(a) What is the mass (g) of 4.65×10^{22} molecules of tetraphosphorus decoxide?
(b) How many P atoms are present in this sample?

3.5B Calcium phosphate is added to some foods, such as yogurt, to boost the calcium content and is also used as an anticaking agent.
(a) How many formula units are in 75.5 g of calcium phosphate?
(b) How many phosphate ions are present in this sample?

SOME SIMILAR PROBLEMS 3.14–3.19

Student Hot Spot

Student data indicate that you may struggle with converting between mass, moles, and number of entities. Access the Smartbook to view additional Learning Resources on this topic.

The Importance of Mass Percent

For many purposes, it is important to know how much of an element is present in a given amount of compound. A biochemist may want the ionic composition of a mineral nutrient; an atmospheric chemist may be studying the carbon content of a fuel; a materials scientist may want the metalloid composition of a semiconductor. In this section, we find the composition of a compound in terms of mass percent and use it to find the mass of each element in the compound.

Determining Mass Percent from a Chemical Formula Each element contributes a fraction of a compound's mass, and that fraction multiplied by 100 gives the element's mass percent. Finding the mass percent is similar on the molecular and molar scales:

- *For a molecule (or formula unit) of compound,* use the molecular (or formula) mass and chemical formula to find the mass percent of any element X in the compound:

$$\text{Mass \% of element X} = \frac{\text{atoms of X in formula} \times \text{atomic mass of X (amu)}}{\text{molecular (or formula) mass of compound (amu)}} \times 100$$

- *For a mole of compound,* use the molar mass and formula to find the mass percent of each element on a mole basis:

$$\text{Mass \% of element X} = \frac{\text{moles of X in formula} \times \text{molar mass of X (g/mol)}}{\text{mass (g) of 1 mol of compound}} \times 100 \quad \textbf{(3.6)}$$

As always, the individual mass percents add up to 100% (within rounding). In Sample Problem 3.6, we determine the mass percent of each element in a compound.

SAMPLE PROBLEM 3.6 Calculating the Mass Percent of Each Element in a Compound from the Formula

Problem The effectiveness of fertilizers depends on their nitrogen content. Ammonium nitrate is a common fertilizer. What is the mass percent of each element in ammonium nitrate?

Plan We know the relative amounts (mol) of the elements from the formula, and we have to find the mass % of each element. We multiply the amount of each element by its molar mass to find its mass. Dividing each element's mass by the mass of 1 mol of ammonium nitrate gives the mass fraction of that element, and multiplying the mass fraction by 100 gives the mass %. The calculation steps for any element (X) are shown in the road map.

Road Map

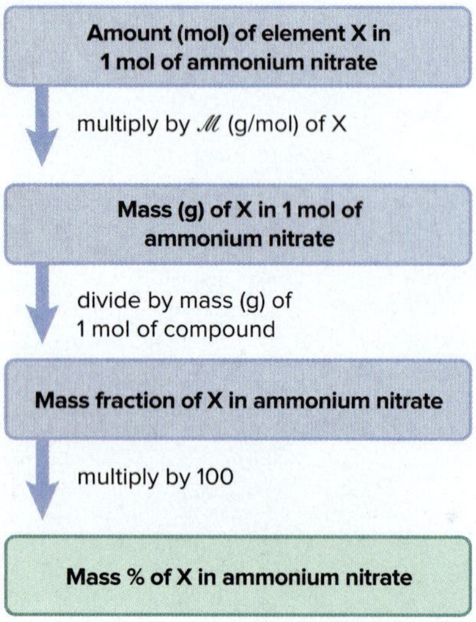

Solution The formula is NH_4NO_3 (see Table 2.5). In 1 mol of NH_4NO_3, there are 2 mol of N, 4 mol of H, and 3 mol of O.

Converting amount (mol) of N to mass (g): We have 2 mol of N in 1 mol of NH_4NO_3, so

$$\text{Mass (g) of N} = 2 \text{ mol N} \times \frac{14.01 \text{ g N}}{1 \text{ mol N}} = 28.02 \text{ g N}$$

Calculating the mass of 1 mol of NH_4NO_3:

$$\begin{aligned}\mathcal{M} &= (2 \times \mathcal{M} \text{ of N}) + (4 \times \mathcal{M} \text{ of H}) + (3 \times \mathcal{M} \text{ of O}) \\ &= (2 \times 14.01 \text{ g/mol N}) + (4 \times 1.008 \text{ g/mol H}) + (3 \times 16.00 \text{ g/mol O}) \\ &= 80.05 \text{ g/mol } NH_4NO_3\end{aligned}$$

Finding the mass fraction of N in NH_4NO_3:

$$\text{Mass fraction of N} = \frac{\text{total mass of N}}{\text{mass of 1 mol } NH_4NO_3} = \frac{28.02 \text{ g N}}{80.05 \text{ g } NH_4NO_3} = 0.3500$$

Changing to mass %:

$$\begin{aligned}\text{Mass \% of N} &= \text{mass fraction of N} \times 100 = 0.3500 \times 100 \\ &= 35.00 \text{ mass \% N}\end{aligned}$$

Combining the steps for each of the other elements in NH_4NO_3:

$$\text{Mass \% of H} = \frac{\text{mol H} \times \mathcal{M} \text{ of H}}{\text{mass of 1 mol } NH_4NO_3} \times 100 = \frac{4 \text{ mol H} \times \frac{1.008 \text{ g H}}{1 \text{ mol H}}}{80.05 \text{ g } NH_4NO_3} \times 100$$

$$= 5.037 \text{ mass \% H}$$

$$\text{Mass \% of O} = \frac{\text{mol O} \times \mathcal{M} \text{ of O}}{\text{mass of 1 mol } NH_4NO_3} \times 100 = \frac{3 \text{ mol O} \times \frac{16.00 \text{ g O}}{1 \text{ mol O}}}{80.05 \text{ g } NH_4NO_3} \times 100$$

$$= 59.96 \text{ mass \% O}$$

Check The answers make sense. The mass % of O is greater than that of N because there are more moles of O in the compound and the molar mass of O is greater. The mass % of H is small because its molar mass is small. The sum of the mass percents is 100.00%.

Comment From here on, you should be able to determine the molar mass of a compound, so that calculation will no longer be shown.

FOLLOW-UP PROBLEMS

3.6A In mammals, lactose (milk sugar) is metabolized to glucose ($C_6H_{12}O_6$), the key nutrient for generating chemical potential energy. Calculate the mass percent of C in glucose.

3.6B For many years, compounds known as *chlorofluorocarbons* were used as refrigerants, until it was discovered that the chlorine atoms in these compounds destroy ozone molecules in the atmosphere. The compound CCl_3F is a chlorofluorocarbon with a high chlorine content. Calculate the mass percent of Cl in CCl_3F.

SOME SIMILAR PROBLEMS 3.20–3.23

Determining the Mass of an Element from Its Mass Fraction Sample Problem 3.6 shows that *an element always constitutes the same fraction of the mass of a given compound* (see Equation 3.6). We can use that fraction to find the mass of element in any mass of a compound:

$$\text{Mass of element} = \text{mass of compound} \times \frac{\text{mass of element in 1 mol of compound}}{\text{mass of 1 mol of compound}} \quad \textbf{(3.7)}$$

For example, to find the mass of oxygen in 15.5 g of nitrogen dioxide, we have

$$\text{Mass (g) of O} = 15.5 \text{ g } NO_2 \times \frac{2 \text{ mol} \times \mathcal{M} \text{ of O (g/mol)}}{\text{mass (g) of 1 mol } NO_2}$$

$$= 15.5 \text{ g } \cancel{NO_2} \times \frac{32.00 \text{ g O}}{46.01 \text{ g } \cancel{NO_2}} = 10.8 \text{ g O}$$

SAMPLE PROBLEM 3.7 Calculating the Mass of an Element in a Compound

Problem Use the information in Sample Problem 3.6 to determine the mass (g) of nitrogen in 650. g of ammonium nitrate.

Plan To find the mass of N in the sample of ammonium nitrate, we multiply the mass of the sample by the mass of 2 mol of N divided by the mass of 1 mol of ammonium nitrate.

Solution Finding the mass of N in a given mass of ammonium nitrate:

$$\text{Mass (g) of N} = \text{mass (g) of } NH_4NO_3 \times \frac{2 \text{ mol N} \times \mathcal{M} \text{ of N (g/mol)}}{\text{mass (g) of 1 mol } NH_4NO_3}$$

$$= 650. \text{ g } \cancel{NH_4NO_3} \times \frac{28.02 \text{ g N}}{80.05 \text{ g } \cancel{NH_4NO_3}} = 228 \text{ g N}$$

Check Rounding shows that the answer is "in the right ballpark": N accounts for about one-third of the mass of NH_4NO_3 and $\frac{1}{3}$ of 700 g is 233 g.

FOLLOW-UP PROBLEMS

3.7A Use the information in Follow-up Problem 3.6A to find the mass (g) of C in 16.55 g of glucose.

3.7B Use the information in Follow-up Problem 3.6B to find the mass (g) of Cl in 112 g of CCl_3F.

SOME SIMILAR PROBLEMS 3.27 and 3.28

› Summary of Section 3.1

- › A mole of substance is the amount that contains Avogadro's number (6.022×10^{23}) of chemical entities (atoms, ions, molecules, or formula units).
- › The mass (in grams) of a mole of a given entity (atom, ion, molecule, or formula unit) has the same numerical value as the mass (in amu) of the entity. Thus, the mole allows us to count entities by weighing them.
- › Using the molar mass ($\mathcal{M}$, g/mol) of an element (or compound) and Avogadro's number as conversion factors, we can convert among amount (mol), mass (g), and number of entities.
- › The mass fraction of element X in a compound is used to find the mass of X in a given amount of the compound.

3.2 DETERMINING THE FORMULA OF AN UNKNOWN COMPOUND

In Sample Problems 3.6 and 3.7, we used a compound's formula to find the mass percent (or mass fraction) of each element in it *and* the mass of each element in any size sample of it. In this section, we do the reverse: we use the masses of elements in a compound to find the formula. Then, we look briefly at the relationship between molecular formula and molecular structure.

Let's compare three common types of formula, using hydrogen peroxide as an example:

- The **empirical formula** is derived from mass analysis. It shows the *lowest* whole number of moles, and thus the *relative* number of atoms, of each element in the compound. For example, in hydrogen peroxide, there is 1 part by mass of hydrogen for every 16 parts by mass of oxygen. Because the atomic mass of hydrogen is 1.008 amu and that of oxygen is 16.00 amu, there is one H atom for every O atom (a 1/1 H/O atom ratio). Thus, the empirical formula is HO.

Recall from Section 2.8 that

- The **molecular formula** shows the *actual* number of atoms of each element in a molecule: the molecular formula of hydrogen peroxide is H_2O_2, twice the empirical formula. Notice that the molecular formula exhibits the same 1/1 H/O atom ratio as in the empirical formula.
- The **structural formula** also shows the relative *placement and connections of atoms* in the molecule: the structural formula of hydrogen peroxide is H—O—O—H.

Let's focus on how to determine empirical and molecular formulas.

Empirical Formulas

A chemist studying an unknown compound goes through a three-step process to find the empirical formula:

1. Determine the mass (g) of each component element.
2. Convert each mass (g) to amount (mol), and write a preliminary formula.
3. Convert the amounts (mol) mathematically to whole-number (integer) subscripts. To accomplish this math conversion,
   - Divide each subscript by the smallest subscript, and
   - If necessary, multiply through by the *smallest integer* that turns all subscripts into integers.

Sample Problem 3.8 demonstrates these steps.

SAMPLE PROBLEM 3.8 Determining an Empirical Formula from Amounts of Elements

Problem A sample of an unknown compound contains 0.21 mol of zinc, 0.14 mol of phosphorus, and 0.56 mol of oxygen. What is the empirical formula?

Plan We are given the amount (mol) of each element as a fraction. We use these fractional amounts directly in a preliminary formula as subscripts of the element symbols. Then, we convert the fractions to whole numbers.

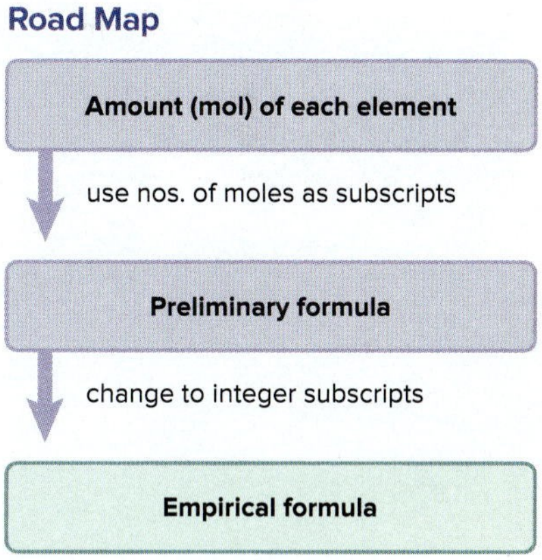

Solution Using the fractions to write a preliminary formula, with the symbols Zn for zinc, P for phosphorus, and O for oxygen:

$$Zn_{0.21}P_{0.14}O_{0.56}$$

Converting the fractions to whole numbers:

1. Divide each subscript by the smallest one, which in this case is 0.14:

$$Zn_{\frac{0.21}{0.14}}P_{\frac{0.14}{0.14}}O_{\frac{0.56}{0.14}} \longrightarrow Zn_{1.5}P_{1.0}O_{4.0}$$

2. Multiply through by the *smallest integer* that turns all subscripts into integers. We multiply by 2 because that makes 1.5 (the subscript for Zn) into an integer:

$$Zn_{(1.5\times2)}P_{(1.0\times2)}O_{(4.0\times2)} \longrightarrow Zn_{3.0}P_{2.0}O_{8.0}, \quad \text{or} \quad Zn_3P_2O_8$$

Check The integer subscripts must be the smallest integers with the same ratio as the original fractional numbers of moles: 3/2/8 is *the same ratio* as 0.21/0.14/0.56.

Comment A more conventional way to write this formula is $Zn_3(PO_4)_2$; this compound is zinc phosphate, formerly used widely as a dental cement.

FOLLOW-UP PROBLEMS

3.8A A sample of a white solid contains 0.170 mol of boron and 0.255 mol of oxygen. What is the empirical formula?

3.8B A sample of an unknown compound contains 6.80 mol of carbon and 18.1 mol of hydrogen. What is the empirical formula?

SOME SIMILAR PROBLEMS 3.42(a), 3.43(a), and 3.50

Sample Problems 3.9–3.11 show how other types of compositional data are used to determine chemical formulas.

SAMPLE PROBLEM 3.9 Determining an Empirical Formula from Masses of Elements

Problem Analysis of a sample of an ionic compound yields 2.82 g of Na, 4.35 g of Cl, and 7.83 g of O. What are the empirical formula and the name of the compound?

Plan This problem is similar to Sample Problem 3.8, except that we are given element *masses* that we must convert into integer subscripts. We first divide each mass by the element's molar mass to find the amount (mol). Then we construct a preliminary formula and convert the amounts (mol) to integers.

Solution Finding amount (mol) of each element:

$$\text{Amount (mol) of Na} = 2.82\ \cancel{\text{g Na}} \times \frac{1\ \text{mol Na}}{22.99\ \cancel{\text{g Na}}} = 0.123\ \text{mol Na}$$

$$\text{Amount (mol) of Cl} = 4.35\ \cancel{\text{g Cl}} \times \frac{1\ \text{mol Cl}}{35.45\ \cancel{\text{g Cl}}} = 0.123\ \text{mol Cl}$$

$$\text{Amount (mol) of O} = 7.83\ \cancel{\text{g O}} \times \frac{1\ \text{mol O}}{16.00\ \cancel{\text{g O}}} = 0.489\ \text{mol O}$$

Constructing a preliminary formula: $Na_{0.123}Cl_{0.123}O_{0.489}$

Converting to integer subscripts (dividing all by the smallest subscript):

$$Na_{\frac{0.123}{0.123}}Cl_{\frac{0.123}{0.123}}O_{\frac{0.489}{0.123}} \longrightarrow Na_{1.00}Cl_{1.00}O_{3.98} \approx Na_1Cl_1O_4, \quad \text{or} \quad NaClO_4$$

The empirical formula is $NaClO_4$; the name is sodium perchlorate.

Check The numbers of moles seem correct because the masses of Na and Cl are slightly more than 0.1 of their molar masses. The mass of O is greatest and its molar mass is smallest, so it should have the greatest number of moles. The ratio of subscripts, 1/1/4, is the same as the ratio of moles, 0.123/0.123/0.489 (within rounding).

FOLLOW-UP PROBLEMS

3.9A A sample of an unknown compound is found to contain 1.23 g of H, 12.64 g of P, and 26.12 g of O. What is the empirical formula and the name of the compound?

3.9B An unknown metal M reacts with sulfur to form a compound with the formula M_2S_3. If 3.12 g of M reacts with 2.88 g of S, what are the names of M and M_2S_3? [*Hint:* Determine the amount (mol) of S, and use the formula to find the amount (mol) of M.]

SOME SIMILAR PROBLEMS 3.42(b), 3.43(b), 3.46, and 3.47

Molecular Formulas

If we know the molar mass of a compound, we can use the empirical formula to obtain the molecular formula, which uses as subscripts the *actual* numbers of moles of each element in 1 mol of compound. For some compounds, such as water (H_2O), ammonia (NH_3), and methane (CH_4), the empirical and molecular formulas are identical, but for many others, the molecular formula is a *whole-number multiple* of the empirical formula. As you saw, hydrogen peroxide has the empirical formula HO. Dividing the molar mass of hydrogen peroxide (34.02 g/mol) by the empirical formula mass of HO (17.01 g/mol) gives the whole-number multiple:

$$\text{Whole-number multiple} = \frac{\text{molar mass (g/mol)}}{\text{empirical formula mass (g/mol)}} = \frac{34.02\ \cancel{\text{g/mol}}}{17.01\ \cancel{\text{g/mol}}} = 2.000 = 2$$

Multiplying the empirical formula subscripts by 2 gives the molecular formula:

$$H_{(1\times2)}O_{(1\times2)} \text{ gives } H_2O_2$$

Since the molar mass of hydrogen peroxide is twice as large as the empirical formula mass, the molecular formula has twice the number of atoms as the empirical formula.

Instead of giving compositional data as masses of each element, analytical laboratories provide mass percents. We use this kind of data as follows:

1. Assume 100.0 g of compound to express each mass percent directly as a mass (g).
2. Convert each mass (g) to amount (mol).
3. Derive the empirical formula.
4. Divide the molar mass of the compound by the empirical formula mass to find the whole-number multiple and the molecular formula.

SAMPLE PROBLEM 3.10 Determining a Molecular Formula from Elemental Analysis and Molar Mass

Problem During excessive physical activity, lactic acid ($\mathcal{M}$ = 90.08 g/mol) forms in muscle tissue and is responsible for muscle soreness. Elemental analysis shows that this compound has 40.0 mass % C, 6.71 mass % H, and 53.3 mass % O.
(a) Determine the empirical formula of lactic acid.
(b) Determine the molecular formula.

(a) Determining the empirical formula

Plan We know the mass % of each element and must convert each to an integer subscript. The mass of the sample of lactic acid is not given, but the mass percents are the same for any sample of it. Therefore, we assume there is 100.0 g of lactic acid and express each mass % as a number of grams. Then, we construct the empirical formula as in Sample Problem 3.9.

Solution Expressing mass % as mass (g) by assuming 100.0 g of lactic acid:

$$\text{Mass (g) of C} = \frac{40.0 \text{ parts C by mass}}{100 \text{ parts by mass}} \times 100.0 \text{ g} = 40.0 \text{ g C}$$

Similarly, we have 6.71 g of H and 53.3 g of O.

Converting from mass (g) of each element to amount (mol):

$$\text{Amount (mol) of C} = \text{mass of C} \times \frac{1}{\mathcal{M} \text{ of C}} = 40.0 \cancel{\text{g C}} \times \frac{1 \text{ mol C}}{12.01 \cancel{\text{g C}}} = 3.33 \text{ mol C}$$

Similarly, we have 6.66 mol of H and 3.33 mol of O.

Constructing the preliminary formula: $C_{3.33}H_{6.66}O_{3.33}$

Converting to integer subscripts by dividing each subscript by the smallest subscript:

$$C_{\frac{3.33}{3.33}}H_{\frac{6.66}{6.66}}O_{\frac{3.33}{3.33}} \longrightarrow C_{1.00}H_{2.00}O_{1.00} = C_1H_2O_1\text{, the empirical formula is } \boxed{CH_2O}$$

Check The numbers of moles seem correct: the masses of C and O are each slightly more than 3 times their molar masses (e.g., for C, 40 g/(12 g/mol) > 3 mol), and the mass of H is over 6 times its molar mass of 1.

(b) Determining the molecular formula

Plan The molecular formula subscripts are whole-number multiples of the empirical formula subscripts. To find this multiple, we divide the molar mass given in the problem (90.08 g/mol) by the empirical formula mass, which we find from the sum of the elements' molar masses. Then we multiply each subscript in the empirical formula by the multiple.

Solution The empirical formula mass is 30.03 g/mol. Finding the whole-number multiple:

$$\text{Whole-number multiple} = \frac{\mathcal{M} \text{ of lactic acid}}{\mathcal{M} \text{ of empirical formula}} = \frac{90.08 \cancel{\text{g/mol}}}{30.03 \cancel{\text{g/mol}}} = 3.000 = 3$$

Determining the molecular formula:

$$C_{(1\times3)}H_{(2\times3)}O_{(1\times3)} = \boxed{C_3H_6O_3}$$

Check The calculated molecular formula has the same ratio of moles of elements (3/6/3) as the empirical formula (1/2/1) and corresponds to the given molar mass:

$$\begin{aligned}\mathcal{M} \text{ of lactic acid} &= (3 \times \mathcal{M} \text{ of C}) + (6 \times \mathcal{M} \text{ of H}) + (3 \times \mathcal{M} \text{ of O}) \\ &= (3 \times 12.01 \text{ g/mol}) + (6 \times 1.008 \text{ g/mol}) + (3 \times 16.00 \text{ g/mol}) \\ &= 90.08 \text{ g/mol}\end{aligned}$$

FOLLOW-UP PROBLEMS

3.10A One of the most widespread environmental carcinogens (cancer-causing agents) is benzo[*a*]pyrene ($\mathcal{M}$ = 252.30 g/mol). It is found in coal dust, cigarette smoke, and even charcoal-grilled meat. Analysis of this hydrocarbon shows 95.21 mass % C and 4.79 mass % H. What is the molecular formula of benzo[*a*]pyrene?

3.10B Caffeine ($\mathcal{M}$ = 194.2 g/mol) is a stimulant found in coffee, tea, many soft drinks, and chocolate. Elemental analysis of caffeine shows 49.47 mass % C, 5.19 mass % H, 28.86 mass % N, and 16.48 mass % O. What is the molecular formula of caffeine?

SOME SIMILAR PROBLEMS 3.44, 3.45, and 3.51

Combustion Analysis of Organic Compounds Still another type of compositional data is obtained through **combustion analysis,** a method commonly used to measure the amounts of carbon and hydrogen in a combustible organic compound. The unknown compound is burned in an excess of pure O_2; during the combustion, the compound's carbon and hydrogen react with the oxygen to form CO_2 and H_2O, respectively, which are absorbed in separate containers (Figure 3.4). By weighing the absorbers before and after combustion, we find the masses of CO_2 and H_2O produced and use them to find the masses of C and H in the compound; from these results, we find the empirical formula. Many organic compounds also contain oxygen, nitrogen, or a halogen. As long as the third element doesn't interfere with the absorption of H_2O and CO_2, we calculate its mass by subtracting the masses of C and H from the original mass of the compound.

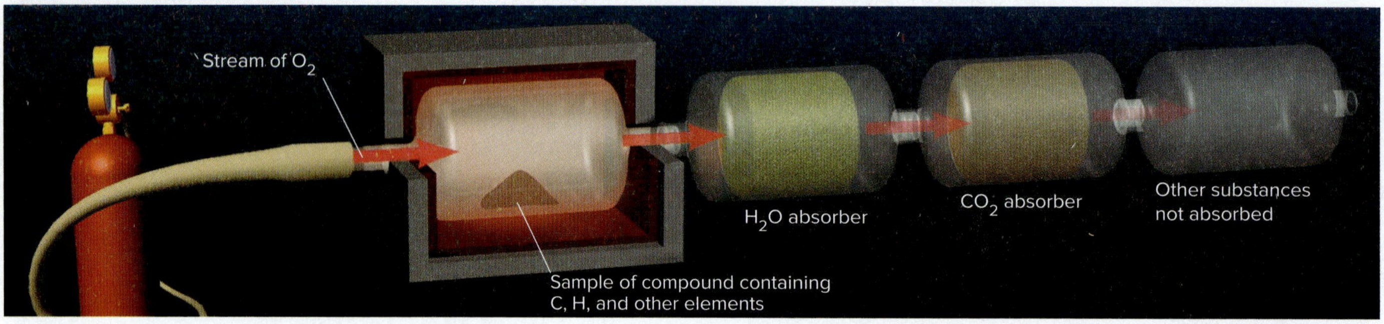

Figure 3.4 Combustion apparatus for determining formulas of organic compounds. A sample of an organic compound is burned in a stream of O_2. The resulting H_2O is absorbed by $Mg(ClO_4)_2$, and the CO_2 is absorbed by NaOH on asbestos.

SAMPLE PROBLEM 3.11 Determining a Molecular Formula from Combustion Analysis

Problem Vitamin C ($\mathcal{M}$ = 176.12 g/mol) is a compound of C, H, and O found in many natural sources, especially citrus fruits. When a 1.000-g sample of vitamin C is burned in a combustion apparatus, the following data are obtained:

$$\text{Mass of } CO_2 \text{ absorber after combustion} = 85.35 \text{ g}$$
$$\text{Mass of } CO_2 \text{ absorber before combustion} = 83.85 \text{ g}$$
$$\text{Mass of } H_2O \text{ absorber after combustion} = 37.96 \text{ g}$$
$$\text{Mass of } H_2O \text{ absorber before combustion} = 37.55 \text{ g}$$

What is the molecular formula of vitamin C?

Plan We find the masses of CO_2 and H_2O by subtracting the mass of each absorber before the combustion from its mass after combustion. From the mass of CO_2, we use Equation 3.7 to find the mass of C. Similarly, we find the mass of H from the mass of H_2O. The mass of vitamin C (1.000 g) minus the sum of the masses of C and H gives the mass of O, the third element present. Then, we proceed as in Sample Problem 3.10: calculate the amount (mol) of each element using its molar mass, construct the empirical formula, determine the whole-number multiple from the given molar mass, and construct the molecular formula.

Solution Finding the masses of combustion products:

$$\text{Mass (g) of } CO_2 = \text{mass of } CO_2 \text{ absorber after} - \text{mass before}$$
$$= 85.35\text{ g} - 83.85\text{ g} = 1.50\text{ g } CO_2$$
$$\text{Mass (g) of } H_2O = \text{mass of } H_2O \text{ absorber after} - \text{mass before}$$
$$= 37.96\text{ g} - 37.55\text{ g} = 0.41\text{ g } H_2O$$

Calculating masses (g) of C and H using Equation 3.7:

$$\text{Mass of element} = \text{mass of compound} \times \frac{\text{mass of element in 1 mol of compound}}{\text{mass of 1 mol of compound}}$$

$$\text{Mass (g) of C} = \text{mass of } CO_2 \times \frac{1\text{ mol C} \times \mathcal{M}\text{ of C}}{\text{mass of 1 mol } CO_2} = 1.50\text{ }\cancel{\text{g } CO_2} \times \frac{12.01\text{ g C}}{44.01\text{ }\cancel{\text{g } CO_2}}$$
$$= 0.409\text{ g C}$$

$$\text{Mass (g) of H} = \text{mass of } H_2O \times \frac{2\text{ mol H} \times \mathcal{M}\text{ of H}}{\text{mass of 1 mol } H_2O} = 0.41\text{ }\cancel{\text{g } H_2O} \times \frac{2.016\text{ g H}}{18.02\text{ }\cancel{\text{g } H_2O}}$$
$$= 0.046\text{ g H}$$

Calculating mass (g) of O:

$$\text{Mass (g) of O} = \text{mass of vitamin C sample} - (\text{mass of C} + \text{mass of H})$$
$$= 1.000\text{ g} - (0.409\text{ g} + 0.046\text{ g}) = 0.545\text{ g O}$$

Finding the amounts (mol) of elements: Dividing the mass (g) of each element by its molar mass gives 0.0341 mol of C, 0.046 mol of H, and 0.0341 mol of O.

Constructing the preliminary formula: $C_{0.0341}H_{0.046}O_{0.0341}$

Determining the empirical formula: Dividing through by the smallest subscript gives

$$C_{\frac{0.0341}{0.0341}}H_{\frac{0.046}{0.0341}}O_{\frac{0.0341}{0.0341}} = C_{1.00}H_{1.3}O_{1.00}$$

We find that 3 is the smallest integer that makes all subscripts into integers:

$$C_{(1.00\times3)}H_{(1.3\times3)}O_{(1.00\times3)} = C_{3.00}H_{3.9}O_{3.00} \approx C_3H_4O_3$$

Determining the molecular formula:

$$\text{Whole-number multiple} = \frac{\mathcal{M}\text{ of vitamin C}}{\mathcal{M}\text{ of empirical formula}} = \frac{176.12\text{ }\cancel{\text{g/mol}}}{88.06\text{ }\cancel{\text{g/mol}}} = 2.000 = 2$$

$$C_{(3\times2)}H_{(4\times2)}O_{(3\times2)} = C_6H_8O_6$$

Check The element masses seem correct: carbon makes up slightly more than 0.25 of the mass of CO_2 (12 g/44 g > 0.25), as do the masses in the problem (0.409 g/1.50 g > 0.25). Hydrogen makes up slightly more than 0.10 of the mass of H_2O (2 g/18 g > 0.10), as do the masses in the problem (0.046 g/0.41 g > 0.10). The molecular formula has the same ratio of subscripts (6/8/6) as the empirical formula (3/4/3) and the preliminary formula (0.0341/0.046/0.0341), and it gives the known molar mass:

$$(6 \times \mathcal{M}\text{ of C}) + (8 \times \mathcal{M}\text{ of H}) + (6 \times \mathcal{M}\text{ of O}) = \mathcal{M}\text{ of vitamin C}$$
$$(6 \times 12.01\text{ g/mol}) + (8 \times 1.008\text{ g/mol}) + (6 \times 16.00\text{ g/mol}) = 176.12\text{ g/mol}$$

Comment The subscript we calculated for H was 3.9, which we rounded to 4. But, if we had strung the calculation steps together, we would have obtained 4.0:

$$\text{Subscript of H} = 0.41\text{ g } H_2O \times \frac{2.016\text{ g H}}{18.02\text{ g } H_2O} \times \frac{1\text{ mol H}}{1.008\text{ g H}} \times \frac{1}{0.0341\text{ mol}} \times 3 = 4.0$$

FOLLOW-UP PROBLEMS

3.11A A dry-cleaning solvent ($\mathcal{M}$ = 146.99 g/mol) that contains C, H, and Cl is suspected to be a cancer-causing agent. When a 0.250-g sample was studied by combustion analysis, 0.451 g of CO_2 and 0.0617 g of H_2O were formed. Find the molecular formula.

3.11B Anabolic steroids are sometimes used illegally by athletes to increase muscle strength. A forensic chemist analyzes some tablets suspected of being a popular steroid. He determines that the substance in the tablets contains only C, H, and O and has a molar mass of 300.42 g/mol. When a 1.200-g sample is studied by combustion analysis, 3.516 g of CO_2 and 1.007 g of H_2O are collected. What is the molecular formula of the substance in the tablets?

SOME SIMILAR PROBLEMS 3.48 and 3.49

Table 3.2 Some Compounds with Empirical Formula CH_2O (Composition by Mass: 40.0% C, 6.71% H, 53.3% O)

Name	Molecular Formula	Whole-Number Multiple	$\mathcal{M}$ (g/mol)	Use or Function
Formaldehyde	CH_2O	1	30.03	Disinfectant; biological preservative
Acetic acid	$C_2H_4O_2$	2	60.05	Acetate polymers; vinegar (5% solution)
Lactic acid	$C_3H_6O_3$	3	90.08	Causes milk to sour; forms in muscles during exercise
Erythrose	$C_4H_8O_4$	4	120.10	Forms during sugar metabolism
Ribose	$C_5H_{10}O_5$	5	150.13	Component of many nucleic acids and vitamin B_2
Glucose	$C_6H_{12}O_6$	6	180.16	Major nutrient for energy in cells

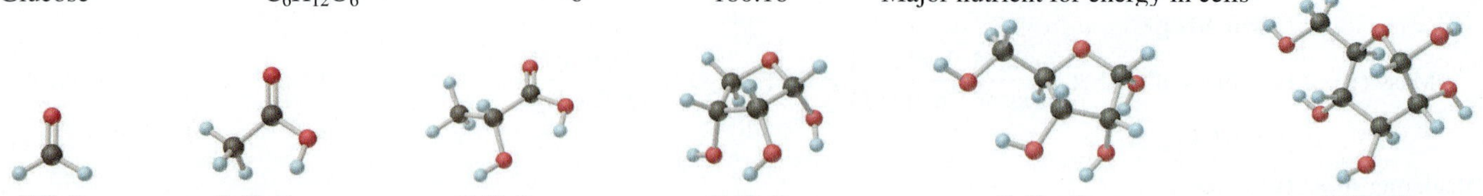

CH_2O $C_2H_4O_2$ $C_3H_6O_3$ $C_4H_8O_4$ $C_5H_{10}O_5$ $C_6H_{12}O_6$

Chemical Formulas and Molecular Structures; Isomers

A formula represents a real, three-dimensional object. The structural formula makes this point, with its relative placement of atoms, but do empirical and molecular formulas contain structural information?

1. *Different compounds with the same empirical formula.* The empirical formula tells nothing about molecular structure because it is based solely on mass analysis. In fact, different compounds can have the *same* empirical formula. NO_2 and N_2O_4 are inorganic cases, and there are numerous organic ones. For example, many organic compounds have the empirical formula CH_2 (the general formula is C_nH_{2n}, with n an integer greater than or equal to 2), such as ethylene (C_2H_4) and propylene (C_3H_6), starting materials for two common plastics. Table 3.2 shows some biological compounds with the same empirical formula, CH_2O.

2. *Isomers: Different compounds with the same molecular formula.* A molecular formula also tells nothing about structure. Different compounds can have the *same* molecular formula because their atoms can bond in different arrangements to give more than one *structural formula.* **Isomers** are compounds with the same molecular formula, and thus molar mass, but different properties. *Constitutional,* or *structural, isomers* occur when the atoms link together in different arrangements. Table 3.3 shows

Table 3.3 Two Pairs of Constitutional Isomers

	C_4H_{10}		C_2H_6O	
Property	**Butane**	**2-Methylpropane**	**Ethanol**	**Dimethyl Ether**
$\mathcal{M}$ (g/mol)	58.12	58.12	46.07	46.07
Boiling point	−0.5°C	−11.6°C	78.5°C	−25°C
Density (at 20°C)	0.00244 g/mL (gas)	0.00247 g/mL (gas)	0.789 g/mL (liquid)	0.00195 g/mL (gas)
Structural formula	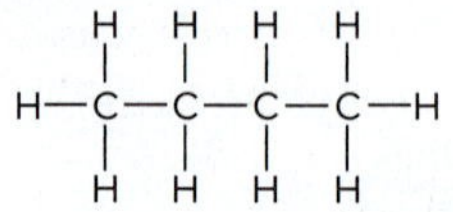	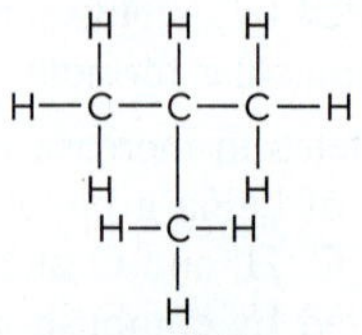	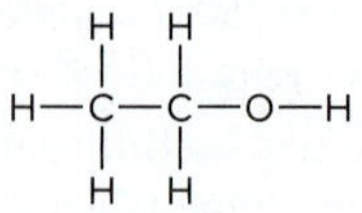	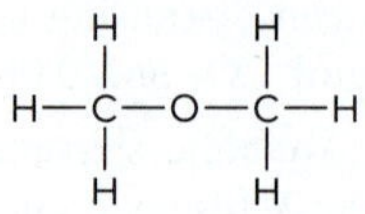
Space-filling model	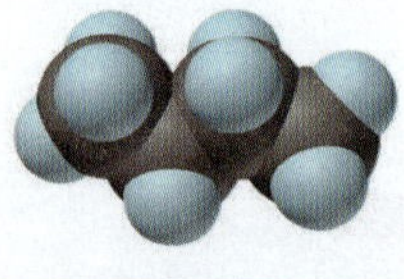	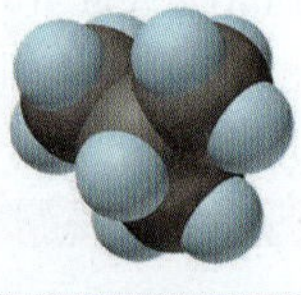	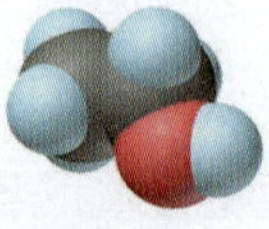	

two pairs of examples. The left pair, butane and 2-methylpropane, share the molecular formula C_4H_{10}. One has a four-C chain and the other a one-C branch off a three-C chain. Both are small alkanes, so their properties are similar, but not identical. The two compounds with the molecular formula C_2H_6O have very different properties; indeed, they are different classes of organic compound—one is an alcohol and the other an ether.

As the numbers of the different kinds of atoms increase, the number of constitutional isomers—that is, the number of structural formulas that can be written for a given molecular formula—also increases: C_2H_6O has two structural formulas (Table 3.3), C_3H_8O has three, and $C_4H_{10}O$ seven. Imagine how many there are for $C_{16}H_{19}N_3O_4S$! Of all the possible isomers with this molecular formula, only one is the antibiotic ampicillin (Figure 3.5). We'll discuss constitutional and other types of isomerism fully later in the text.

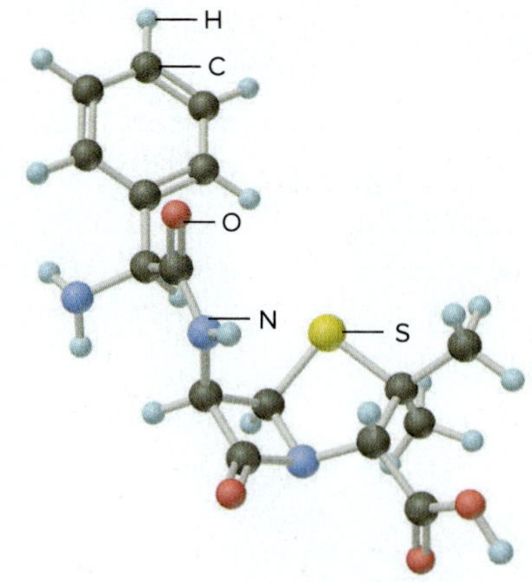

Figure 3.5 The antibiotic ampicillin.

› Summary of Section 3.2

- › From the masses of elements in a compound, their relative numbers of moles are found, which gives the empirical formula.
- › If the molar mass of the compound is known, the molecular formula, the actual numbers of moles of each element, can also be determined, because the molecular formula is a whole-number multiple of the empirical formula.
- › Combustion analysis provides data on the masses of carbon and hydrogen in an organic compound, which are used to obtain the formula.
- › Atoms can bond in different arrangements (structural formulas). Two or more compounds with the same molecular formula are constitutional isomers.

3.3 WRITING AND BALANCING CHEMICAL EQUATIONS

Thinking in terms of amounts, rather than masses, allows us to view reactions as interactions among large populations of particles rather than as involving grams of material. For example, for the formation of HF from H_2 and F_2, if we weigh the substances, we find that

Macroscopic level (grams): 2.016 g of H_2 and 38.00 g of F_2 react to form 40.02 g of HF

This information tells us little except that mass is conserved. However, if we convert these masses (g) to amounts (mol), we find that

Macroscopic level (moles): 1 mol of H_2 and 1 mol of F_2 react to form 2 mol of HF

This information reveals that an enormous number of H_2 molecules react with just as many F_2 molecules to form twice as many HF molecules. Dividing by Avogadro's number gives the reaction between individual molecules:

Molecular level: 1 molecule of H_2 and 1 molecule of F_2 react to form 2 molecules of HF

Thus, *the macroscopic (molar) change corresponds to the submicroscopic (molecular) change* (Figure 3.6). This information forms the essence of a **chemical equation,** a

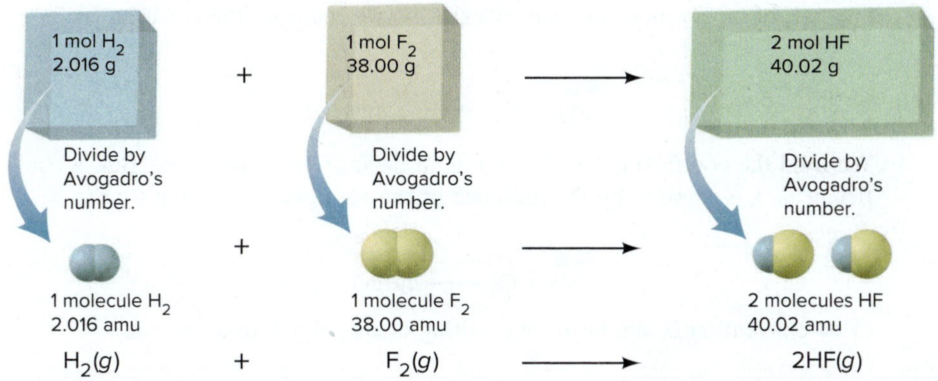

Figure 3.6 The formation of HF on the macroscopic and molecular levels and written as a balanced chemical equation.

statement that uses formulas to express the identities and quantities of substances involved in a chemical or physical change.

Steps for Balancing an Equation To present a chemical change quantitatively, an equation must be *balanced: the same number of each type of atom must appear on both sides.* As an example, here is a description of a chemical change that occurs in many fireworks and in a common lecture demonstration: a magnesium strip burns in oxygen gas to yield powdery magnesium oxide. (Light and heat are also produced, but we are concerned here only with substances.) Converting this description into a balanced equation involves the following steps:

1. *Translating the statement.* We first translate the chemical statement into a "skeleton" equation: the substances present *before* the change, called **reactants,** are placed to the left of a yield arrow, which points to the substances produced *during* the change, called **products:**

$$\overbrace{\_\_\mathrm{Mg} \quad + \quad \_\_\mathrm{O_2}}^{\text{reactants}} \xrightarrow{\text{yield}} \overset{\text{product}}{\_\_\mathrm{MgO}}$$

magnesium and oxygen yield magnesium oxide

At the beginning of the balancing process, we put a blank *in front of* each formula to remind us that we have to account for its atoms.

2. *Balancing the atoms.* By shifting our attention back and forth, we *match the numbers of each type of atom on the left and the right of the yield arrow.* In each blank, we place a **balancing (stoichiometric) coefficient,** a numerical multiplier of *all the atoms* in the formula that follows it. In general, balancing is easiest when we
   - Start with the most complex substance, the one with the largest number of different types of atoms.
   - End with the least complex substance, such as an element by itself.

   In this case, MgO is the most complex, so we place a coefficient 1 in that blank:

$$\_\_\mathrm{Mg} + \_\_\mathrm{O_2} \longrightarrow \underline{1}\mathrm{MgO}$$

   To balance the Mg in MgO, we place a 1 in front of Mg on the left:

$$\underline{1}\,\mathrm{Mg} + \_\_\mathrm{O_2} \longrightarrow \underline{1}\mathrm{MgO}$$

   The O atom in MgO must be balanced by one O atom on the left. One-half an O_2 molecule provides one O atom:

$$\underline{1}\,\mathrm{Mg} + \underline{\tfrac{1}{2}}\mathrm{O_2} \longrightarrow \underline{1}\mathrm{MgO}$$

   In terms of numbers of each type of atom, the equation is balanced.

3. *Adjusting the coefficients.* There are several conventions about the final coefficients:
   - In most cases, *the smallest whole-number coefficients are preferred.* In this case, one-half of an O_2 molecule cannot exist, so we multiply the equation by 2:

$$2\mathrm{Mg} + 1\mathrm{O_2} \longrightarrow 2\mathrm{MgO}$$

   - We used the coefficient 1 to remind us to balance each substance. But, a coefficient of 1 is implied by the presence of the formula, so we don't write it:

$$2\mathrm{Mg} + \mathrm{O_2} \longrightarrow 2\mathrm{MgO}$$

   (This convention is similar to not writing a subscript 1 in a formula.)

g	for gas
l	for liquid
s	for solid
aq	for aqueous solution

4. *Checking.* After balancing and adjusting the coefficients, we always check that the equation is balanced:

$$\text{Reactants (2 Mg, 2 O)} \longrightarrow \text{products (2 Mg, 2 O)}$$

5. *Specifying the states of matter.* The final equation also indicates the physical state of each substance or whether it is dissolved in water. The abbreviations used for these states are shown in the margin. From the original statement, we know that the Mg "strip" is solid, O_2 is a gas, and "powdery" MgO is also solid. The balanced equation, therefore, is

$$2\text{Mg}(s) + \text{O}_2(g) \longrightarrow 2\text{MgO}(s)$$

As you saw in Figure 3.6, *balancing coefficients refer to both individual chemical entities and moles of entities.* Thus,

2 atoms of Mg and 1 molecule of O_2 yield 2 formula units of MgO
2 moles of Mg and 1 mole of O_2 yield 2 moles of MgO

Figure 3.7 depicts this reaction on three levels:

- *Macroscopic level (photos),* as it appears in the laboratory
- *Atomic level (blow-up circles),* as chemists imagine it (with darker colored atoms representing the stoichiometry)
- *Symbolic level,* in the form of the balanced chemical equation

Keep in mind several key points about the balancing process:

- A coefficient operates on *all* the atoms in the formula that follows it:

2MgO	*means* 2 × (MgO),	or 2 Mg atoms + 2 O atoms
$2Ca(NO_3)_2$	*means* 2 × $[Ca(NO_3)_2]$,	or 2 Ca atoms + 4 N atoms + 12 O atoms

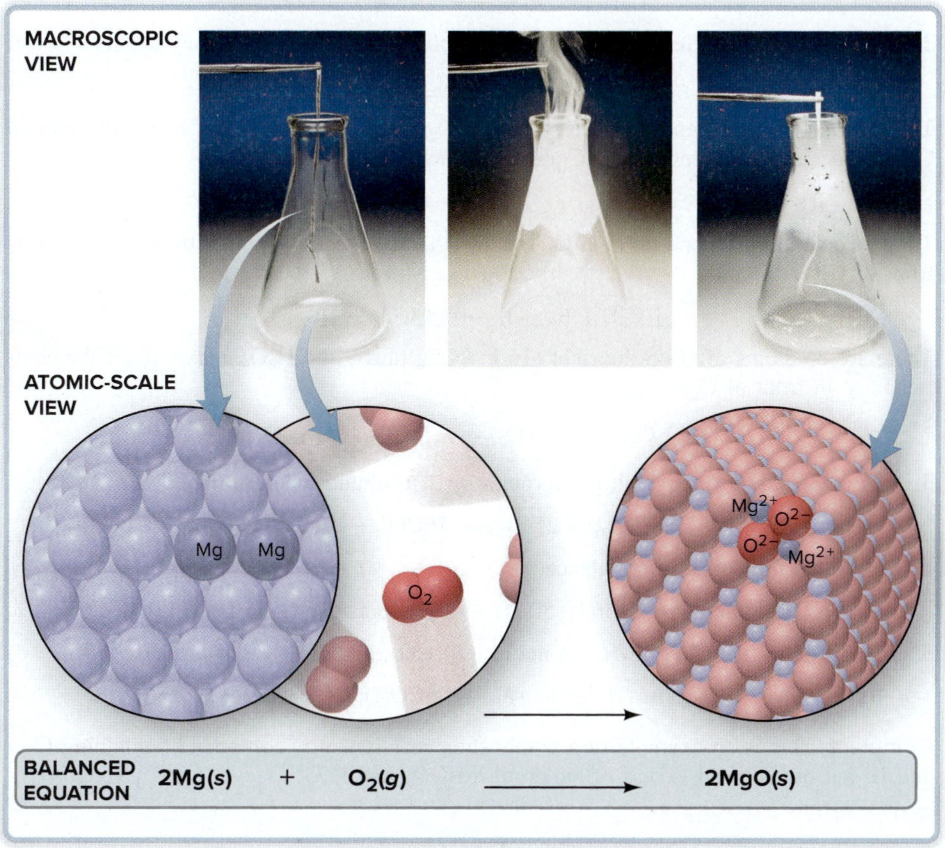

Figure 3.7 A three-level view of the reaction between magnesium and oxygen.
Source: (all): © McGraw-Hill Education/Charles Winters/Timeframe Photography, Inc.

- Chemical formulas *cannot* be altered. Thus, in step 2 of the example, we *cannot* balance the O atoms by changing MgO to MgO_2 because MgO_2 is a different compound.
- Other reactants or products *cannot* be added. Thus, we *cannot* balance the O atoms by changing the reactant from O_2 molecules to O atoms or by adding an O atom to the products. The description of the reaction mentions oxygen gas, which consists of O_2 molecules, *not* separate O atoms.
- A balanced equation remains balanced if you multiply all the coefficients by the same number. For example,

$$4Mg(s) + 2O_2(g) \longrightarrow 4MgO(s)$$

is also balanced because the coefficients have all been multiplied by 2. However, *by convention,* we balance an equation with the *smallest* whole-number coefficients.
- While atoms must be conserved in a chemical equation, each side of the equation may *not* have the same number of molecules or moles. In the reaction between magnesium and oxygen, three moles of reactants (Mg and O_2) produce only two moles of product (MgO).

SAMPLE PROBLEM 3.12 Balancing a Chemical Equation

Problem Within the cylinders of a car's engine, the hydrocarbon octane (C_8H_{18}), one of many components of gasoline, mixes with oxygen from the air and burns to form carbon dioxide and water vapor. Write a balanced equation for this reaction.

Solution

1. *Translate* the statement into a skeleton equation (with coefficient blanks). Octane and oxygen are reactants; "oxygen from the air" implies molecular oxygen, O_2. Carbon dioxide and water vapor are products:

$$\_\_C_8H_{18} + \_\_O_2 \longrightarrow \_\_CO_2 + \_\_H_2O$$

2. *Balance the atoms.* Start with the most complex substance, C_8H_{18}, and balance O_2 last:

$$\underline{1}\,C_8H_{18} + \_\_O_2 \longrightarrow \_\_CO_2 + \_\_H_2O$$

The C atoms in C_8H_{18} end up in CO_2. Each CO_2 contains one C atom, so 8 molecules of CO_2 are needed to balance the 8 C atoms in each C_8H_{18}:

$$\underline{1}\,C_8H_{18} + \_\_O_2 \longrightarrow \underline{8}\,CO_2 + \_\_H_2O$$

The H atoms in C_8H_{18} end up in H_2O. The 18 H atoms in C_8H_{18} require the coefficient 9 in front of H_2O:

$$\underline{1}\,C_8H_{18} + \_\_O_2 \longrightarrow \underline{8}\,CO_2 + \underline{9}\,H_2O$$

There are 25 atoms of O on the right (16 in $8CO_2$ plus 9 in $9H_2O$), so we place the coefficient $\frac{25}{2}$ in front of O_2:

$$\underline{1}\,C_8H_{18} + \underline{\tfrac{25}{2}}\,O_2 \longrightarrow \underline{8}\,CO_2 + \underline{9}\,H_2O$$

3. *Adjust the coefficients.* Multiply through by 2 to obtain whole numbers:

$$2C_8H_{18} + 25O_2 \longrightarrow 16CO_2 + 18H_2O$$

4. *Check* that the equation is balanced:

$$\text{Reactants (16 C, 36 H, 50 O)} \longrightarrow \text{products (16 C, 36 H, 50 O)}$$

5. *Specify* states of matter. C_8H_{18} is liquid; O_2, CO_2, and H_2O vapor are gases:

$$2C_8H_{18}(l) + 25O_2(g) \longrightarrow 16CO_2(g) + 18H_2O(g)$$

Comment This is an example of a combustion reaction. *Any* compound containing C and H that burns in an excess of air produces CO_2 and H_2O.

FOLLOW-UP PROBLEMS

3.12A Write a balanced equation for each of the following:

(a) A characteristic reaction of Group 1A(1) elements: chunks of sodium react violently with water to form hydrogen gas and sodium hydroxide solution.

(b) The destruction of marble statuary by acid rain: aqueous nitric acid reacts with calcium carbonate to form carbon dioxide, water, and aqueous calcium nitrate.
(c) Halogen compounds exchanging bonding partners: phosphorus trifluoride is prepared by the reaction of phosphorus trichloride and hydrogen fluoride; hydrogen chloride is the other product. The reaction involves gases only.

3.12B Write a balanced equation for each of the following:
(a) Explosive decomposition of dynamite: liquid nitroglycerine ($C_3H_5N_3O_9$) explodes to produce a mixture of gases—carbon dioxide, water vapor, nitrogen, and oxygen.
(b) A reaction that takes place in a self-contained breathing apparatus: solid potassium superoxide (KO_2) reacts with carbon dioxide gas to produce oxygen gas and solid potassium carbonate.
(c) The production of iron from its ore in a blast furnace: solid iron(III) oxide reacts with carbon monoxide gas to produce solid iron metal and carbon dioxide gas.

SOME SIMILAR PROBLEMS 3.58–3.63

Visualizing a Reaction with a Molecular Scene A great way to focus on the rearrangement of atoms from reactants to products is by visualizing an equation as a molecular scene. Here's a representation of the combustion of octane we just balanced in Sample Problem 3.12:

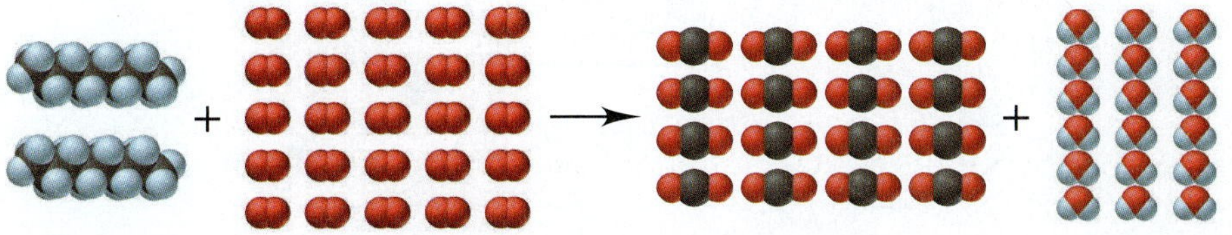

$$2C_8H_{18}(l) + 25O_2(g) \longrightarrow 16CO_2(g) + 18H_2O(g)$$

Now let's work through a sample problem to do the reverse—derive a balanced equation from a molecular scene.

SAMPLE PROBLEM 3.13 Writing a Balanced Equation from a Molecular Scene

Problem The following molecular scenes depict an important reaction in nitrogen chemistry (nitrogen is blue; oxygen is red):

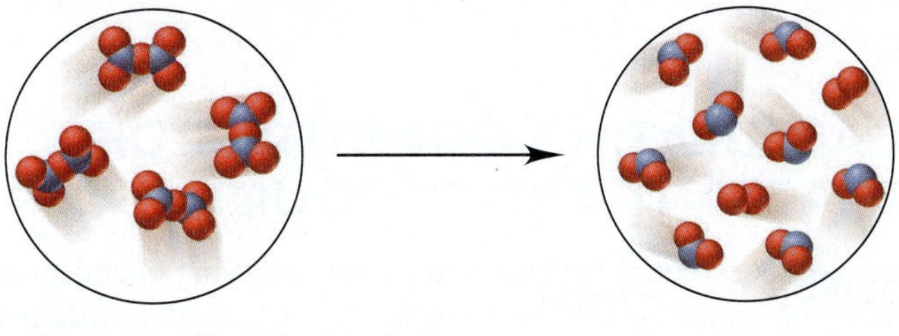

Write a balanced equation for this reaction.

Plan To write a balanced equation, we first have to determine the formulas of the molecules and obtain coefficients by counting the number of each type of molecule. Then, we arrange this information in the correct equation format, using the smallest whole-number coefficients and including states of matter.

Solution The reactant circle shows only one type of molecule. It has two N and five O atoms, so the formula is N_2O_5; there are four of these molecules. The product circle shows two different types of molecules, one type with one N and two O atoms, and the other with two O atoms; there are eight NO_2 and two O_2. Thus, we have

$$4N_2O_5 \longrightarrow 8NO_2 + 2O_2$$

Writing the balanced equation with the smallest whole-number coefficients and all substances as gases:

$$2N_2O_5(g) \longrightarrow 4NO_2(g) + O_2(g)$$

Check Reactant (4 N, 10 O) $\longrightarrow$ products (4 N, 8 + 2 = 10 O)

FOLLOW-UP PROBLEMS

3.13A Write a balanced equation for the important atmospheric reaction depicted below (carbon is black; oxygen is red):

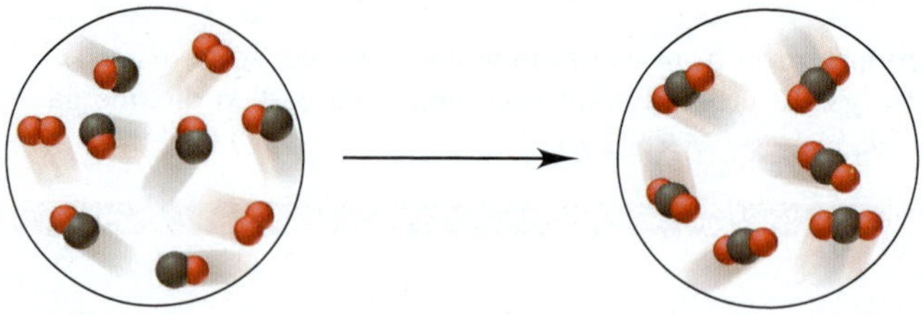

3.13B Write a balanced equation for the important industrial reaction depicted below (nitrogen is dark blue; hydrogen is light blue):

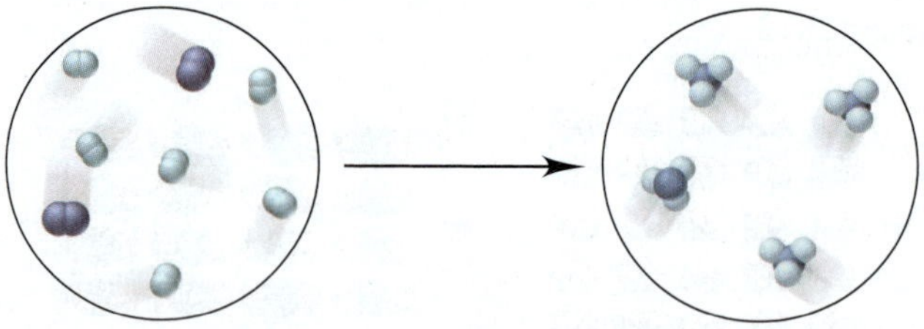

SOME SIMILAR PROBLEMS 3.56 and 3.57

› Summary of Section 3.3

- A chemical equation has reactant formulas on the left of a yield arrow and product formulas on the right.
- A balanced equation has the same number of each type of atom on both sides.
- Balancing coefficients are integer multipliers for *all* the atoms in a formula and apply to the individual entities or to moles of entities.

3.4 CALCULATING QUANTITIES OF REACTANT AND PRODUCT

A balanced equation is essential for all calculations involving chemical change: *if you know the number of moles of one substance, the balanced equation tells you the numbers of moles of the others.*

Stoichiometrically Equivalent Molar Ratios from the Balanced Equation

In a balanced equation, *the amounts (mol) of substances are stoichiometrically equivalent to each other,* which means that a specific amount of one substance is formed from, produces, or reacts with a specific amount of the other. The quantitative relationships are expressed as *stoichiometrically equivalent molar ratios* that we use as conversion factors to calculate the amounts. For example, consider the equation for the combustion of propane, a hydrocarbon fuel used in cooking and water heating:

$$C_3H_8(g) + 5O_2(g) \longrightarrow 3CO_2(g) + 4H_2O(g)$$

If we view the reaction quantitatively in terms of C_3H_8, we see that

1 mol of C_3H_8 reacts with 5 mol of O_2
1 mol of C_3H_8 produces 3 mol of CO_2
1 mol of C_3H_8 produces 4 mol of H_2O

Therefore, in this reaction,

1 mol of C_3H_8 is stoichiometrically equivalent to 5 mol of O_2
1 mol of C_3H_8 is stoichiometrically equivalent to 3 mol of CO_2
1 mol of C_3H_8 is stoichiometrically equivalent to 4 mol of H_2O

We chose to look at C_3H_8, but any two of the substances are stoichiometrically equivalent to each other. Thus,

3 mol of CO_2 is stoichiometrically equivalent to 4 mol of H_2O
5 mol of O_2 is stoichiometrically equivalent to 3 mol of CO_2

and so on. A balanced equation contains a wealth of quantitative information relating individual chemical entities, amounts (mol) of substances, and masses of substances, and Table 3.4 presents the quantitative information contained in the equation for the combustion of propane.

Here's a typical problem that shows how stoichiometric equivalence is used to create conversion factors: in the combustion of propane, how many moles of O_2 are consumed when 10.0 mol of H_2O are produced? To solve this problem, we have to find the molar ratio between O_2 and H_2O. From the balanced equation, we see that for every 5 mol of O_2 consumed, 4 mol of H_2O is formed:

5 mol of O_2 is stoichiometrically equivalent to 4 mol of H_2O

As with any equivalent quantities, we can construct two conversion factors, depending on the quantity we want to find:

$$\frac{5\ \text{mol}\ O_2}{4\ \text{mol}\ H_2O} \quad \text{or} \quad \frac{4\ \text{mol}\ H_2O}{5\ \text{mol}\ O_2}$$

Table 3.4 Information Contained in a Balanced Equation

Viewed in Terms of	**Reactants** $C_3H_8(g)$ + $5O_2(g)$	⟶	**Products** $3CO_2(g)$ + $4H_2O(g)$
Molecules	1 molecule C_3H_8 + 5 molecules O_2	⟶	3 molecules CO_2 + 4 molecules H_2O
	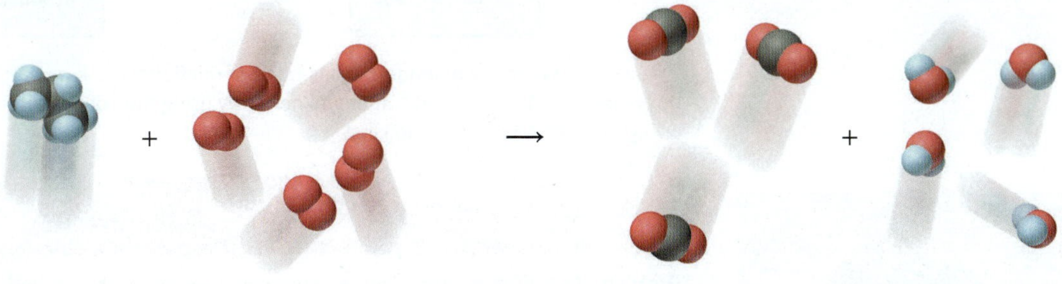		
Amount (mol)	1 mol C_3H_8 + 5 mol O_2	⟶	3 mol CO_2 + 4 mol H_2O
Mass (amu)	44.09 amu C_3H_8 + 160.00 amu O_2	⟶	132.03 amu CO_2 + 72.06 amu H_2O
Mass (g)	44.09 g C_3H_8 + 160.00 g O_2	⟶	132.03 g CO_2 + 72.06 g H_2O
Total mass (g)	204.09 g	⟶	204.09 g

Since we want to find the amount (mol) of O_2 and we know the amount (mol) of H_2O, we choose "5 mol O_2/4 mol H_2O" to cancel "mol H_2O":

$$\text{Amount (mol) of } O_2 \text{ consumed} = 10.0\ \cancel{\text{mol } H_2O} \times \frac{5 \text{ mol } O_2}{4\ \cancel{\text{mol } H_2O}} = 12.5 \text{ mol } O_2$$

$$\text{mol } H_2O \xrightarrow[\text{conversion factor}]{\text{molar ratio as}} \text{mol } O_2$$

Be sure to note that the 5/4 ratio between O_2 and H_2O is a *molar* ratio and not a *mass* ratio:

$$5 \text{ mol } O_2 = 4 \text{ mol } H_2O \quad \text{but} \quad 5 \text{ g } O_2 \neq 4 \text{ g } H_2O$$

*You **cannot** solve this type of problem without the balanced equation.* Here is an approach for solving *any* stoichiometry problem that involves a reaction:

1. Write the balanced equation.
2. When necessary, convert the known mass (or number of entities) of one substance to amount (mol) using its molar mass (or Avogadro's number).
3. Use the molar ratio in the balanced equation to calculate the unknown amount (mol) of the other substance.
4. When necessary, convert the amount of that other substance to the desired mass (or number of entities) using its molar mass (or Avogadro's number).

Figure 3.8 summarizes the possible relationships among quantities of substances in a reaction, and Sample Problems 3.14–3.16 apply three of them in the first chemical step of converting copper ore to copper metal.

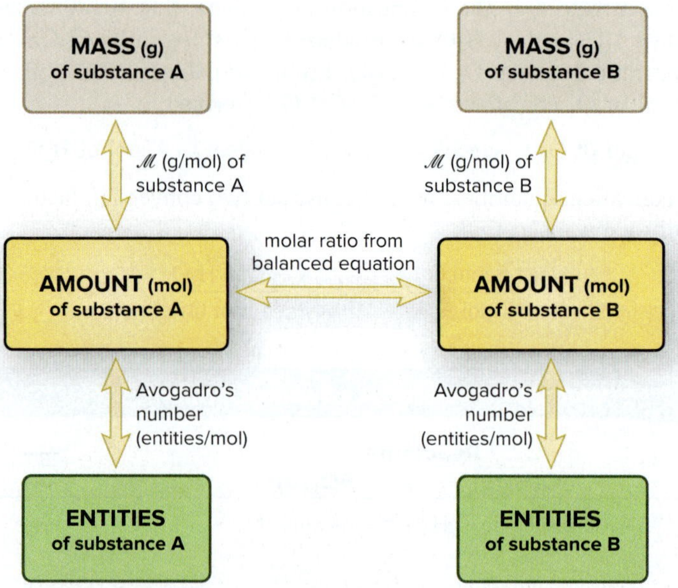

Figure 3.8 Summary of amount-mass-number relationships in a chemical equation. Start at any box (known) and move to any other (unknown) by using the conversion factor on the arrow. As always, convert to amount (mol) first.

Source: © McGraw-Hill Education/Stephen Frisch, photographer

SAMPLE PROBLEM 3.14 Calculating Quantities of Reactants and Products: Amount (mol) to Amount (mol)

Problem In a lifetime, the average American uses more than a half ton (>500 kg) of copper in coins, plumbing, and wiring. Copper is obtained from sulfide ores, such as chalcocite [copper(I) sulfide] by a multistep process. After initial grinding, the ore is "roasted" (heated strongly with oxygen gas) to form powdered copper(I) oxide and gaseous sulfur dioxide. How many moles of oxygen are required to roast 10.0 mol of copper(I) sulfide?

Plan We *always* write the balanced equation first. The formulas of the reactants are Cu_2S and O_2, and the formulas of the products are Cu_2O and SO_2, so we have

$$2Cu_2S(s) + 3O_2(g) \longrightarrow 2Cu_2O(s) + 2SO_2(g)$$

We know the amount of Cu_2S (10.0 mol) and must find the amount (mol) of O_2 that is needed to roast it. The balanced equation shows that 3 mol of O_2 is needed to roast 2 mol of Cu_2S, so the conversion factor for finding amount (mol) of O_2 is 3 mol O_2/2 mol Cu_2S (see the road map).

Road Map

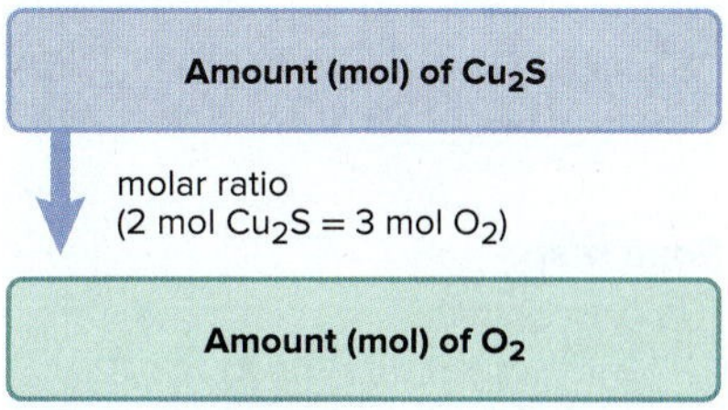

Solution Calculating the amount of O_2:

$$\text{Amount (mol) of } O_2 = 10.0\ \cancel{\text{mol } Cu_2S} \times \frac{3\text{ mol } O_2}{2\ \cancel{\text{mol } Cu_2S}} = 15.0\text{ mol } O_2$$

Check The units are correct, and the answer is reasonable because this molar ratio of O_2 to Cu_2S (15/10) is identical to the ratio in the balanced equation (3/2).

Comment A *common mistake* is to invert the conversion factor; that calculation would be

$$\text{Amount (mol) of } O_2 = 10.0\text{ mol } Cu_2S \times \frac{2\text{ mol } Cu_2S}{3\text{ mol } O_2} = \frac{6.67\text{ mol}^2\ Cu_2S}{1\text{ mol } O_2}$$

The strange units should alert you that an error was made in setting up the conversion factor. Also note that this answer, 6.67, is *less* than 10.0, whereas the equation shows that there should be *more* moles of O_2 (3 mol) than moles of Cu_2S (2 mol). Be sure to think through the calculation when setting up the conversion factor and canceling units.

FOLLOW-UP PROBLEMS

3.14A Thermite is a mixture of iron(III) oxide and aluminum powders that was once used to weld railroad tracks. It undergoes a spectacular reaction to yield solid aluminum oxide and molten iron. How many moles of iron(III) oxide are needed to form 3.60×10^3 mol of iron? Include a road map that shows how you planned the solution.

3.14B The tarnish that forms on objects made of silver is solid silver sulfide; it can be removed by reacting it with aluminum metal to produce silver metal and solid aluminum sulfide. How many moles of aluminum are required to remove 0.253 mol of silver sulfide from a silver bowl? Include a road map that shows how you planned the solution.

SOME SIMILAR PROBLEMS 3.69(a) and 3.70(a)

SAMPLE PROBLEM 3.15 Calculating Quantities of Reactants and Products: Amount (mol) to Mass (g)

Problem During the roasting process, how many grams of sulfur dioxide form when 10.0 mol of copper(I) sulfide reacts?

Plan We use the balanced equation in Sample Problem 3.14, but here we are given amount of reactant (10.0 mol of Cu_2S) and need the mass (g) of product (SO_2) that forms. We find the amount (mol) of SO_2 using the molar ratio (2 mol SO_2/2 mol Cu_2S) and then multiply by its molar mass (64.07 g/mol) to find the mass (g) of SO_2 (see the road map).

Road Map

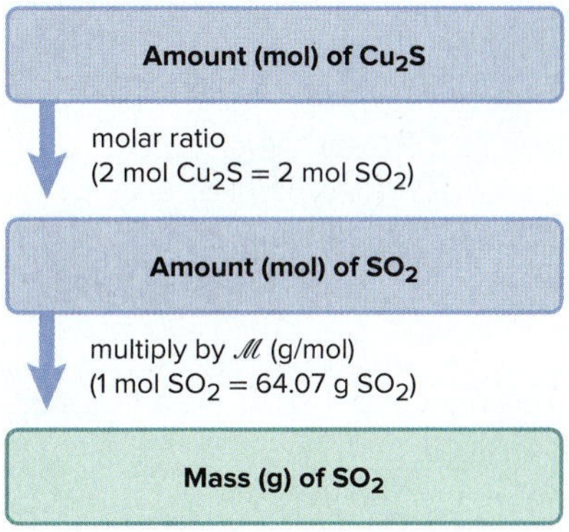

Solution Combining the two conversion steps into one calculation, we have

$$\text{Mass (g) of } SO_2 = 10.0\ \cancel{\text{mol } Cu_2S} \times \frac{2\ \cancel{\text{mol } SO_2}}{2\ \cancel{\text{mol } Cu_2S}} \times \frac{64.07\text{ g } SO_2}{1\ \cancel{\text{mol } SO_2}} = 641\text{ g } SO_2$$

Check The answer makes sense, since the molar ratio shows that 10.0 mol of SO_2 is formed and each mole weighs about 64 g. We rounded to three significant figures.

FOLLOW-UP PROBLEMS

3.15A In the thermite reaction (see Follow-up Problem 3.14A), what amount (mol) of iron forms when 1.85×10^{25} formula units of iron(III) oxide react? Include a road map that shows how you planned the solution.

3.15B In the reaction that removes silver tarnish (see Follow-up Problem 3.14B), how many moles of silver are produced when 32.6 g of silver sulfide reacts? Include a road map that shows how you planned the solution.

SOME SIMILAR PROBLEMS 3.69(b), 3.70(b), 3.71(a), and 3.72(a)

SAMPLE PROBLEM 3.16 Calculating Quantities of Reactants and Products: Mass to Mass

Problem During the roasting of chalcocite (see Sample Problem 3.14), how many kilograms of oxygen are required to form 2.86 kg of copper(I) oxide?

Plan In this problem, we know the mass of the product, Cu_2O (2.86 kg), and we need the mass (kg) of O_2 that reacts to form it. Therefore, we must convert from mass of product to amount of product to amount of reactant to mass of reactant. We convert the mass of Cu_2O from kg to g and then to amount (mol). Then, we use the molar ratio (3 mol O_2/2 mol Cu_2O) to find the amount (mol) of O_2 required. Finally, we convert the amount of O_2 to g and then kg (see the road map).

Road Map

Mass (kg) of Cu_2O

↓ 1 kg = 10^3 g

Mass (g) of Cu_2O

↓ divide by $\mathcal{M}$ (g/mol) (143.10 g Cu_2O = 1 mol Cu_2O)

Amount (mol) of Cu_2O

↓ molar ratio (2 mol Cu_2O = 3 mol O_2)

Amount (mol) of O_2

↓ multiply by $\mathcal{M}$ (g/mol) (1 mol O_2 = 32.00 g O_2)

Mass (g) of O_2

↓ 10^3 g = 1 kg

Mass (kg) of O_2

Solution Converting from kilograms of Cu_2O to moles of Cu_2O: Combining the mass unit conversion with the mass-to-amount conversion gives

$$\text{Amount (mol) of } Cu_2O = 2.86\ \cancel{\text{kg } Cu_2O} \times \frac{10^3\ \cancel{g}}{1\ \cancel{kg}} \times \frac{1\text{ mol } Cu_2O}{143.10\ \cancel{g\ Cu_2O}} = 20.0\text{ mol } Cu_2O$$

Converting from moles of Cu_2O to moles of O_2:

$$\text{Amount (mol) of } O_2 = 20.0\ \cancel{\text{mol } Cu_2O} \times \frac{3\text{ mol } O_2}{2\ \cancel{\text{mol } Cu_2O}} = 30.0\text{ mol } O_2$$

Converting from moles of O_2 to kilograms of O_2: Combining the amount-to-mass conversion with the mass unit conversion gives

$$\text{Mass (kg) of } O_2 = 30.0\ \cancel{\text{mol } O_2} \times \frac{32.00\ \cancel{g}\ O_2}{1\ \cancel{\text{mol } O_2}} \times \frac{1\text{ kg}}{10^3\ \cancel{g}} = 0.960\text{ kg } O_2$$

Check The units are correct. Rounding to check the math, for example, in the final step, ~30 mol × 30 g/mol × 1 kg/10^3 g = 0.90 kg. The answer seems reasonable: even though the amount (mol) of O_2 is greater than the amount (mol) of Cu_2O, the mass of O_2 is less than the mass of Cu_2O because $\mathcal{M}$ of O_2 is less than $\mathcal{M}$ of Cu_2O.

Comment The three related sample problems (3.14–3.16) highlight the main point for solving stoichiometry problems: *convert the information given into amount (mol).* Then, use the appropriate molar ratio and any other conversion factors to complete the solution.

FOLLOW-UP PROBLEMS

3.16A During the thermite reaction (see Follow-up Problems 3.14A and 3.15A), how many atoms of aluminum react for every 1.00 g of aluminum oxide that forms? Include a road map that shows how you planned the solution.

3.16B During the reaction that removes silver tarnish (see Follow-up Problems 3.14B and 3.15B), how many grams of aluminum react to form 12.1 g of aluminum sulfide? Include a road map that shows how you planned the solution.

SOME SIMILAR PROBLEMS 3.71(b), 3.72(b), and 3.73–3.76

Reactions That Occur in a Sequence

In many situations, a product of one reaction becomes a reactant for the next in a sequence of reactions. For stoichiometric purposes, when the same substance forms in one reaction and reacts in the next (it is common to both reactions), we eliminate that substance in an **overall (net) equation.** The steps in writing the overall equation are

1. Write the sequence of balanced equations.
2. Adjust the equations arithmetically to cancel the common substance(s).
3. Add the adjusted equations together to obtain the overall balanced equation.

Sample Problem 3.17 shows the approach by continuing with the copper recovery process that started in Sample Problem 3.14.

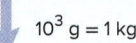

SAMPLE PROBLEM 3.17 Writing an Overall Equation for a Reaction Sequence

Problem Roasting is the first step in extracting copper from chalcocite. In the next step, copper(I) oxide reacts with powdered carbon to yield copper metal and carbon monoxide gas. Write a balanced overall equation for the two-step sequence.

Plan To obtain the overall equation, we write the individual equations in sequence, adjust coefficients to cancel the common substance (or substances), and add the equations together. In this case, only Cu_2O appears as a product in one equation and a reactant in the other, so it is the common substance.

Solution Writing the individual balanced equations:

$$2Cu_2S(s) + 3O_2(g) \longrightarrow 2Cu_2O(s) + 2SO_2(g) \quad \text{[equation 1; see Sample Problem 3.14]}$$

$$Cu_2O(s) + C(s) \longrightarrow 2Cu(s) + 2CO(g) \quad \text{[equation 2]}$$

Adjusting the coefficients: Since 2 mol of Cu_2O form in equation 1 but 1 mol of Cu_2O reacts in equation 2, we double *all* the coefficients in equation 2 to use up the Cu_2O:

$$2Cu_2S(s) + 3O_2(g) \longrightarrow 2Cu_2O(s) + 2SO_2(g) \quad \text{[equation 1]}$$

$$2Cu_2O(s) + 2C(s) \longrightarrow 4Cu(s) + 2CO(g) \quad \text{[equation 2, doubled]}$$

Adding the two equations and canceling the common substance: We keep the reactants of both equations on the left and the products of both equations on the right:

$$2Cu_2S(s) + 3O_2(g) + \cancel{2Cu_2O(s)} + 2C(s) \longrightarrow \cancel{2Cu_2O(s)} + 2SO_2(g) + 4Cu(s) + 2CO(g)$$

or,
$$2Cu_2S(s) + 3O_2(g) + 2C(s) \longrightarrow 2SO_2(g) + 4Cu(s) + 2CO(g)$$

Check Reactants (4 Cu, 2 S, 6 O, 2 C) $\longrightarrow$ products (4 Cu, 2 S, 6 O, 2 C)

Comment **1.** Even though Cu_2O *does* participate in the chemical change, it is not involved in the reaction stoichiometry. An overall equation *may not* show which substances actually react; for example, C(*s*) and $Cu_2S(s)$ do not interact directly in this reaction sequence, even though both are shown as reactants.
2. The SO_2 formed in copper recovery contributes to acid rain, so chemists have devised microbial and electrochemical methods to extract metals without roasting sulfide ores. Such methods are examples of *green chemistry,* a topic discussed further at the end of this section.
3. These reactions were shown to explain how to obtain an overall equation. The actual extraction of copper is more complex, as you'll see in Chapter 22.

FOLLOW-UP PROBLEMS

3.17A The SO_2 formed in copper recovery reacts in air with oxygen and forms sulfur trioxide. This gas, in turn, reacts with water to form a sulfuric acid solution that falls in rain. Write a balanced overall equation for this process.

3.17B During a lightning strike, nitrogen gas can react with oxygen gas to produce nitrogen monoxide. This gas then reacts with the gas ozone, O_3, to produce nitrogen dioxide gas and oxygen gas. The nitrogen dioxide that is produced is a pollutant in smog. Write a balanced overall equation for this process.

SOME SIMILAR PROBLEMS 3.77 and 3.78

Reaction Sequences in Organisms Multistep reaction sequences called *metabolic pathways* occur throughout biological systems. (We discuss them again in Chapter 17.) For example, in most cells, the chemical energy in glucose is released through a sequence of about 30 individual reactions. The product of each reaction step is the reactant of the next, so once all the common substances are canceled out, the overall equation is

$$C_6H_{12}O_6(aq) + 6O_2(g) \longrightarrow 6CO_2(g) + 6H_2O(l)$$

We eat food that contains glucose, inhale O_2, and excrete CO_2 and H_2O. In our cells, these reactants and products are many steps apart: O_2 never reacts *directly* with glucose, and CO_2 and H_2O are formed at various, often distant, steps along the sequence of reactions. Even so, the molar ratios in the overall equation are the same as if the glucose burned in a combustion chamber filled with O_2 and formed CO_2 and H_2O directly (Figure 3.9).

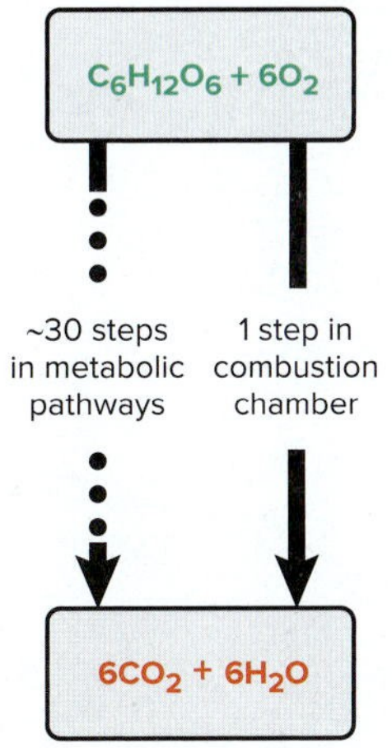

Figure 3.9 An overall equation equals the sum of the individual steps.

Reactions That Involve a Limiting Reactant

In problems up to now, the amount of *one* reactant was given, and we assumed there was enough of the other reactants to react with it completely. For example, using the chemical reaction for the roasting of copper ore (see Sample Problem 3.14), suppose we want the amount (mol) of SO_2 that forms when 5.2 mol of Cu_2S reacts with O_2:

$$2Cu_2S(s) + 3O_2(g) \longrightarrow 2Cu_2O(s) + 2SO_2(g)$$

We assume the 5.2 mol of Cu_2S reacts with as much O_2 as needed. Because all the Cu_2S reacts, its initial amount of 5.2 mol determines, or *limits,* the amount of SO_2 that can form, no matter how much more O_2 is present. In this situation, we call Cu_2S the **limiting reactant** (or *limiting reagent*).

Suppose, however, you know the amounts of both Cu_2S *and* O_2 and need to find out how much SO_2 forms. You first have to determine whether Cu_2S *or* O_2 is the limiting reactant—that is, which one is completely used up—because that reactant limits how much SO_2 can form. The reactant that is *not* limiting is present *in excess,* which means the amount that doesn't react is left over. ‹

Limiting "Reactants" in Everyday Life

Limiting-"reactant" situations arise in business all the time. A car assembly-plant manager must order more tires if there are 1500 car bodies and only 4000 tires, and a clothes manufacturer must cut more sleeves if there are 320 sleeves for 170 shirt bodies. You've probably faced such situations in daily life as well. A muffin recipe calls for 2 cups of flour and 1 cup of sugar, but you have 3 cups of flour and only 3/4 cup of sugar. Clearly, the flour is in excess and the sugar limits the number of muffins you can make. Or, you're in charge of making cheeseburgers for a picnic, and you have 10 buns, 12 meat patties, and 15 slices of cheese. Here, the number of buns limits how many cheeseburgers you can make. Or, there are 26 students and only 23 microscopes in a cell biology lab. You'll find that limiting-"reactant" situations are almost limitless.

To determine which is the limiting reactant, we use the molar ratios in the balanced equation to perform a series of calculations to see *which reactant forms* ***less*** *product.*

Determining the Limiting Reactant Let's clarify these ideas in a much more appetizing situation. Suppose you have a job making ice cream sundaes. Each sundae requires two scoops (12 oz) of ice cream, one cherry, and 50 mL of syrup:

$$2 \text{ scoops (12 oz)} + 1 \text{ cherry} + 50 \text{ mL syrup} \longrightarrow 1 \text{ sundae}$$

A mob of 25 ravenous school kids enters, and each one wants a sundae with vanilla ice cream and chocolate syrup. You have 300 oz of vanilla ice cream (at 6 oz per scoop), 30 cherries, and 1 L of syrup: can you feed them all? A series of calculations based on the balanced equation shows the number of sundaes you can make from each ingredient:

$$\text{Ice cream: No. of sundaes} = 300 \text{ oz} \times \frac{1 \text{ scoop}}{6 \text{ oz}} \times \frac{1 \text{ sundae}}{2 \text{ scoops}} = 25 \text{ sundaes}$$

$$\text{Cherries: No. of sundaes} = 30 \text{ cherries} \times \frac{1 \text{ sundae}}{1 \text{ cherry}} = 30 \text{ sundaes}$$

$$\text{Syrup: No. of sundaes} = 1000 \text{ mL syrup} \times \frac{1 \text{ sundae}}{50 \text{ mL syrup}} = 20 \text{ sundaes}$$

Of the reactants (ice cream, cherry, syrup), the syrup forms the *least* amount of product (sundaes), so it is the limiting "reactant." When all the syrup has been used up, some ice cream and cherries are "unreacted" so they are in excess:

$$300 \text{ oz (50 scoops)} + 30 \text{ cherries} + 1 \text{ L syrup} \longrightarrow \mathbf{20\ sundaes} + 60 \text{ oz (10 scoops)} + 10 \text{ cherries}$$

Figure 3.10 shows a similar example with different initial (starting) quantities.

Using Reaction Tables in Limiting-Reactant Problems A good way to keep track of the quantities in a limiting-reactant problem is with a *reaction table*. The balanced equation appears at the top for the column heads. The table shows the

- *Initial* quantities of reactants and products *before* the reaction
- *Change* in the quantities of reactants and products *during* the reaction
- *Final* quantities of reactants and products remaining *after* the reaction

For example, for the ice cream sundae "reaction," the reaction table would be

Quantity	12 oz (2 scoops)	+	1 cherry	+	50 mL syrup	⟶	1 sundae
Initial	300 oz (50 scoops)		30 cherries		1000 mL syrup		0 sundaes
Change	−240 oz (40 scoops)		−20 cherries		−1000 mL syrup		+20 sundaes
Final	60 oz (10 scoops)		10 cherries		0 mL syrup		20 sundaes

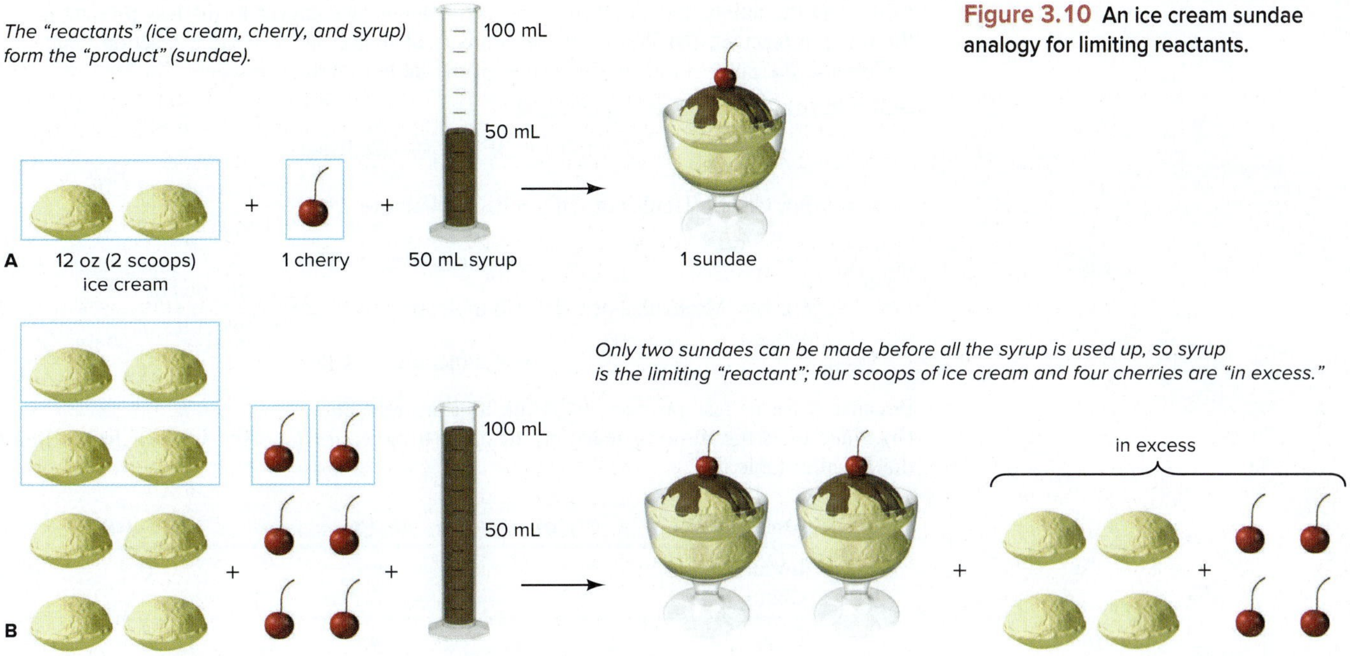
The "reactants" (ice cream, cherry, and syrup) form the "product" (sundae).
100 mL
50 mL
A
12 oz (2 scoops) ice cream
1 cherry
50 mL syrup
1 sundae
Only two sundaes can be made before all the syrup is used up, so syrup is the limiting "reactant"; four scoops of ice cream and four cherries are "in excess."
100 mL
50 mL
in excess
B

Figure 3.10 An ice cream sundae analogy for limiting reactants.

The body of the table shows the following important points:

- In the **Initial** line, "product" has not yet formed, so the entry is "0 sundaes."
- In the **Change** line, since the reactants (ice cream, cherries, and syrup) are used during the reaction, their quantities decrease, so the changes in their quantities have a *negative* sign. At the same time, the quantity of product (sundaes) increases, so the change in its quantity has a *positive* sign.
- For the **Final** line, we *add* the Change and Initial lines. Notice that some reactants (ice cream and cherries) are in excess, while the limiting reactant (syrup) is used up.

Solving Limiting-Reactant Problems In limiting-reactant problems, *the amounts of two (or more) reactants are given, and we first determine which is limiting*. To do this, just as we did with the ice cream sundaes, we use the balanced equation to solve a series of calculations to see how much product forms from the given amount of each reactant: the limiting reactant is the one that yields the *least* amount of product.

The following problems examine these ideas from several aspects. In Sample Problem 3.18, we solve the problem by looking at a molecular scene; in Sample Problem 3.19, we start with the amounts (mol) of two reactants; and in Sample Problem 3.20, we start with masses of two reactants.

SAMPLE PROBLEM 3.18 Using Molecular Depictions in a Limiting-Reactant Problem

Problem Nuclear engineers use chlorine trifluoride to prepare uranium fuel for power plants. The compound is formed as a gas by the reaction of elemental chlorine and fluorine. The circle shows a representative portion of the reaction mixture before the reaction starts (chlorine is *green;* fluorine is *yellow*).

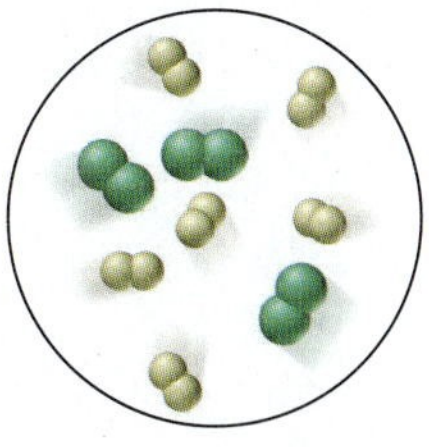

(a) Find the limiting reactant.
(b) Write a reaction table for the process.
(c) Draw a representative portion of the mixture after the reaction is complete. (*Hint:* The ClF_3 molecule has Cl bonded to three individual F atoms.)

Plan (a) We have to find the limiting reactant. The first step is to write the balanced equation, so we need the formulas and states of matter. From the name, chlorine trifluoride, we know the product consists of one Cl atom bonded to three F atoms, or ClF_3. Elemental chlorine and fluorine are the diatomic molecules Cl_2 and F_2, and all three substances are gases. To find the limiting reactant, we find the number of molecules of product that would form from the numbers of molecules of each reactant: whichever forms less product is the limiting reactant. **(b)** We use these numbers of molecules to write a reaction table. **(c)** We use the numbers in the Final line of the table to draw the scene.

Solution **(a)** The balanced equation is

$$Cl_2(g) + 3F_2(g) \longrightarrow 2ClF_3(g)$$

$$\text{For } Cl_2\text{: Molecules of } ClF_3 = 3 \cancel{\text{molecules of } Cl_2} \times \frac{2 \text{ molecules of } ClF_3}{1 \cancel{\text{molecule of } Cl_2}}$$

$$= 6 \text{ molecules of } ClF_3$$

$$\text{For } F_2\text{: Molecules of } ClF_3 = 6 \cancel{\text{molecules of } F_2} \times \frac{2 \text{ molecules of } ClF_3}{3 \cancel{\text{molecules of } F_2}}$$

$$= 4 \text{ molecules of } ClF_3$$

Because it forms less product, F_2 is the limiting reactant.

(b) Since F_2 is the limiting reactant, all of it (6 molecules) is used in the Change line of the reaction table:

Molecules	$Cl_2(g)$	+	$3F_2(g)$	$\longrightarrow$	$2ClF_3(g)$
Initial	3		6		0
Change	−2		−6		+4
Final	1		0		4

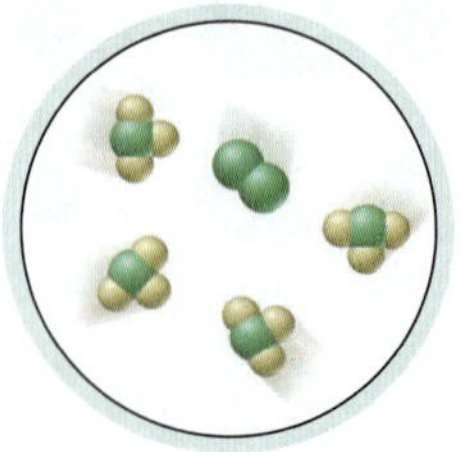

(c) The representative portion of the final reaction mixture includes 1 molecule of Cl_2 (the reactant in excess) and 4 molecules of product ClF_3.

Check The equation is balanced: reactants (2 Cl, 6 F) → products (2 Cl, 6 F). And, as shown in the circles, the numbers of each type of atom before and after the reaction are equal. Let's think through our choice of limiting reactant. From the equation, one Cl_2 needs three F_2 to form two ClF_3. Therefore, the three Cl_2 molecules in the circle depicting reactants need nine (3 × 3) F_2. But there are only six F_2, so there is not enough F_2 to react with the available Cl_2; or put the other way, there is too much Cl_2 to react with the available F_2. From either point of view, F_2 is the limiting reactant.

Comment Notice that there are fewer Cl_2 molecules than F_2 molecules initially. The limiting reactant is *not* the reactant present in the smaller amount, but the reactant that forms *less product.*

FOLLOW-UP PROBLEMS

3.18A B_2 (B is *red*) reacts with AB as shown below:

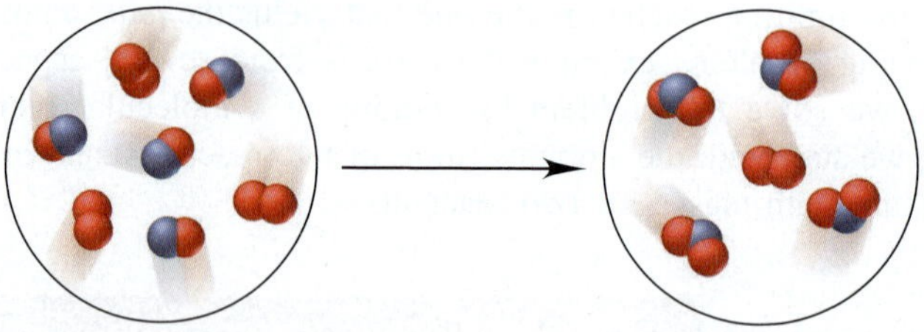

Write a balanced equation for the reaction, and determine the limiting reactant.

3.18B Sulfur dioxide gas reacts with oxygen gas to produce sulfur trioxide, as shown below (sulfur is *yellow*; oxygen is *red*):

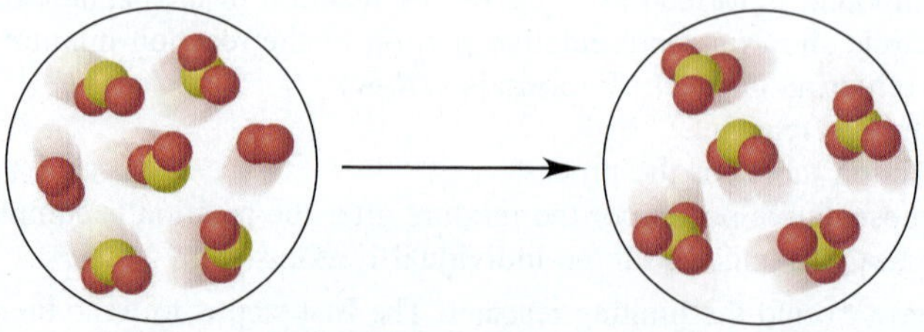

Write a balanced equation for the reaction, and determine the limiting reactant.

SOME SIMILAR PROBLEMS 3.79 and 3.80

SAMPLE PROBLEM 3.19 Calculating Quantities in a Limiting-Reactant Problem: Amount to Amount

Problem In another preparation of ClF_3 (see Sample Problem 3.18), 0.750 mol of Cl_2 reacts with 3.00 mol of F_2. **(a)** Find the limiting reactant. **(b)** Write a reaction table.

Plan (a) We find the limiting reactant by calculating the amount (mol) of ClF_3 formed from the amount (mol) of each reactant: the reactant that forms fewer moles of ClF_3 is limiting.

(b) We enter those values into the reaction table.

Solution (a) Determining the limiting reactant:

Finding amount (mol) of ClF_3 from amount (mol) of Cl_2:

$$\text{Amount (mol) of ClF}_3 = 0.750\ \cancel{\text{mol Cl}_2} \times \frac{2\ \text{mol ClF}_3}{1\ \cancel{\text{mol Cl}_2}} = 1.50\ \text{mol ClF}_3$$

Finding amount (mol) of ClF_3 from amount (mol) of F_2:

$$\text{Amount (mol) of ClF}_3 = 3.00\ \cancel{\text{mol F}_2} \times \frac{2\ \text{mol ClF}_3}{3\ \cancel{\text{mol F}_2}} = 2.00\ \text{mol ClF}_3$$

In this case, Cl_2 is limiting because it forms fewer moles of ClF_3.

(b) Writing the reaction table, with Cl_2 limiting:

Amount (mol)	$Cl_2(g)$	+	$3F_2(g)$	$\longrightarrow$	$2ClF_3(g)$
Initial	0.750		3.00		0
Change	−0.750		−2.25		+1.50
Final	0		0.75		1.50

Check Let's check that Cl_2 is the limiting reactant by assuming, for the moment, that F_2 is limiting. If that were true, all 3.00 mol of F_2 would react to form 2.00 mol of ClF_3. However, based on the balanced equation, obtaining 2.00 mol of ClF_3 would require 1.00 mol of Cl_2, and only 0.750 mol of Cl_2 is present. Thus, Cl_2 must be the limiting reactant.

Comment A major point to note from Sample Problems 3.18 and 3.19 is that the relative amounts of reactants *do not* determine which is limiting, but rather the amount of product formed, which is based on the *molar ratio in the balanced equation*. In both problems, there is more F_2 than Cl_2. However,

- Sample Problem 3.18 has an F_2/Cl_2 ratio of 6/3, or 2/1, which is less than the required molar ratio of 3/1, so F_2 is limiting and Cl_2 is in excess.
- Sample Problem 3.19 has an F_2/Cl_2 ratio of 3.00/0.750, which is greater than the required molar ratio of 3/1, so Cl_2 is limiting and F_2 is in excess.

FOLLOW-UP PROBLEMS

3.19A In the reaction in Follow-up Problem 3.18A, how many moles of product form from 1.5 mol of each reactant?

3.19B In the reaction in Follow-up Problem 3.18B, 4.2 mol of SO_2 reacts with 3.6 mol of O_2. How many moles of SO_3 are produced?

SOME SIMILAR PROBLEMS 3.81 and 3.82

SAMPLE PROBLEM 3.20 Calculating Quantities in a Limiting-Reactant Problem: Mass to Mass

Problem A fuel mixture used in the early days of rocketry consisted of two liquids, hydrazine (N_2H_4) and dinitrogen tetroxide (N_2O_4), which ignite on contact to form nitrogen gas and water vapor.
(a) How many grams of nitrogen gas form when 1.00×10^2 g of N_2H_4 and 2.00×10^2 g of N_2O_4 are mixed?
(b) How many grams of the excess reactant remain unreacted when the reaction is over?
(c) Write a reaction table for this process.

Road Map

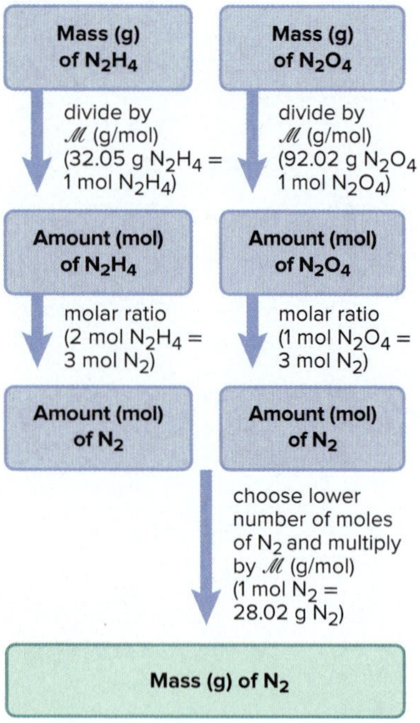

Plan The amounts of two reactants are given, which means this is a limiting-reactant problem. **(a)** To determine the mass of product formed, we must find the limiting reactant by calculating which of the given masses of reactant forms *less* nitrogen gas. As always, we first write the balanced equation. We convert the grams of each reactant to moles using that reactant's molar mass and then use the molar ratio from the balanced equation to find the number of moles of N_2 each reactant forms. Next, we convert the lower amount of N_2 to mass (see the road map). **(b)** To determine the mass of the excess reactant, we use the molar ratio to calculate the mass of excess reactant that is required to react with the given amount of the limiting reactant. We subtract that mass from the given amount of excess reactant; this difference is the mass of unreacted excess reactant. **(c)** We use the values based on the limiting reactant for the reaction table.

Solution **(a)** Writing the balanced equation:

$$2N_2H_4(l) + N_2O_4(l) \longrightarrow 3N_2(g) + 4H_2O(g)$$

Finding the amount (mol) of N_2 from the amount (mol) of each reactant:

$$\text{For } N_2H_4\text{: Amount (mol) of } N_2H_4 = 1.00{\times}10^2\ \cancel{\text{g } N_2H_4} \times \frac{1\ \text{mol } N_2H_4}{32.05\ \cancel{\text{g } N_2H_4}} = 3.12\ \text{mol } N_2H_4$$

$$\text{Amount (mol) of } N_2 = 3.12\ \cancel{\text{mol } N_2H_4} \times \frac{3\ \text{mol } N_2}{2\ \cancel{\text{mol } N_2H_4}} = 4.68\ \text{mol } N_2$$

$$\text{For } N_2O_4\text{: Amount (mol) of } N_2O_4 = 2.00{\times}10^2\ \cancel{\text{g } N_2O_4} \times \frac{1\ \text{mol } N_2O_4}{92.02\ \cancel{\text{g } N_2O_4}} = 2.17\ \text{mol } N_2O_4$$

$$\text{Amount (mol) of } N_2 = 2.17\ \cancel{\text{mol } N_2O_4} \times \frac{3\ \text{mol } N_2}{1\ \cancel{\text{mol } N_2O_4}} = 6.51\ \text{mol } N_2$$

Thus, N_2H_4 is the limiting reactant because it yields less N_2.

Converting from amount (mol) of N_2 to mass (g):

$$\text{Mass (g) of } N_2 = 4.68\ \cancel{\text{mol } N_2} \times \frac{28.02\ \text{g } N_2}{1\ \cancel{\text{mol } N_2}} = \boxed{131\ \text{g } N_2}$$

(b) Finding the mass (g) of N_2O_4 that reacts with $1.00{\times}10^2$ g of N_2H_4:

$$\text{Mass (g) of } N_2O_4 = 1.00{\times}10^2\ \cancel{\text{g } N_2H_4} \times \frac{1\ \cancel{\text{mol } N_2H_4}}{32.05\ \cancel{\text{g } N_2H_4}} \times \frac{1\ \cancel{\text{mol } N_2O_4}}{2\ \cancel{\text{mol } N_2H_4}} \times \frac{92.02\ \text{g } N_2O_4}{1\ \cancel{\text{mol } N_2O_4}}$$
$$= 144\ \text{g } N_2O_4$$

$$\text{Mass (g) of } N_2O_4 \text{ in excess} = \text{initial mass of } N_2O_4 - \text{mass of } N_2O_4 \text{ reacted}$$
$$= 2.00{\times}10^2\ \text{g } N_2O_4 - 144\ \text{g } N_2O_4 = \boxed{56\ \text{g } N_2O_4}$$

(c) With N_2H_4 as the limiting reactant, the reaction table is

Amount (mol)	$2N_2H_4(l)$	+	$N_2O_4(l)$	$\longrightarrow$	$3N_2(g)$	+	$4H_2O(g)$
Initial	3.12		2.17		0		0
Change	−3.12		−1.56		+4.68		+6.24
Final	0		0.61		4.68		6.24

Check There are more grams of N_2O_4 than N_2H_4, but there are fewer moles of N_2O_4 because its $\mathcal{M}$ is much higher. Rounding for N_2H_4: 100 g N_2H_4 × 1 mol/32 g ≈ 3 mol; ~3 mol × $\frac{3}{2}$ ≈ 4.5 mol N_2; ~4.5 mol × 30 g/mol ≈ 135 g N_2.

Comment **1.** Recall this *common mistake* in solving limiting-reactant problems: The limiting reactant is not the *reactant* present in fewer moles (or grams). Rather, it is the reactant that forms fewer moles (or grams) of *product.*

2. An *alternative approach* to finding the limiting reactant compares "How much is needed?" with "How much is given?" That is, based on the balanced equation,

- Find the amount (mol) of each reactant needed to react with the other reactant.
- Compare that *needed* amount with the *given* amount in the problem statement. There will be *more* than enough of one reactant (excess) and *less* than enough of the other (limiting).

For example, the balanced equation for this problem shows that 2 mol of N_2H_4 reacts with 1 mol of N_2O_4. The amount (mol) of N_2O_4 needed to react with the given 3.12 mol of N_2H_4 is

$$\text{Amount (mol) of } N_2O_4 \text{ needed} = 3.12 \cancel{\text{mol } N_2H_4} \times \frac{1 \text{ mol } N_2O_4}{2 \cancel{\text{mol } N_2H_4}} = 1.56 \text{ mol } N_2O_4$$

The amount of N_2H_4 needed to react with the given 2.17 mol of N_2O_4 is

$$\text{Amount (mol) of } N_2H_4 \text{ needed} = 2.17 \cancel{\text{mol } N_2O_4} \times \frac{2 \text{ mol } N_2H_4}{1 \cancel{\text{mol } N_2O_4}} = 4.34 \text{ mol } N_2H_4$$

We are given 2.17 mol of N_2O_4, which is *more* than the 1.56 mol of N_2O_4 needed, and we are given 3.12 mol of N_2H_4, which is *less* than the 4.34 mol of N_2H_4 needed. Therefore, N_2H_4 is limiting, and N_2O_4 is in excess.

FOLLOW-UP PROBLEMS

3.20A How many grams of solid aluminum sulfide can be prepared by the reaction of 10.0 g of aluminum and 15.0 g of sulfur? How many grams of the nonlimiting reactant are in excess?

3.20B Butane gas (C_4H_{10}) is used as the fuel in disposable lighters. It burns in oxygen to form carbon dioxide gas and water vapor. What mass of carbon dioxide is produced when 4.65 g of butane is burned in 10.0 g of oxygen? How many grams of the excess reactant remain unreacted when the reaction is over?

SOME SIMILAR PROBLEMS 3.83–3.90

Student Hot Spot

Student data indicate that you may struggle with limiting reactant calculations. Access the Smartbook to view additional Learning Resources on this topic.

Figure 3.11 provides an overview of all of the stoichiometric relationships we've discussed in this chapter.

Theoretical, Actual, and Percent Reaction Yields

Up until now, we've assumed that 100% of the limiting reactant becomes product, that ideal methods exist for isolating the product, and that we have perfect lab technique and collect all the product. In theory, this may happen, but in reality, it doesn't, and chemists recognize three types of reaction yield:

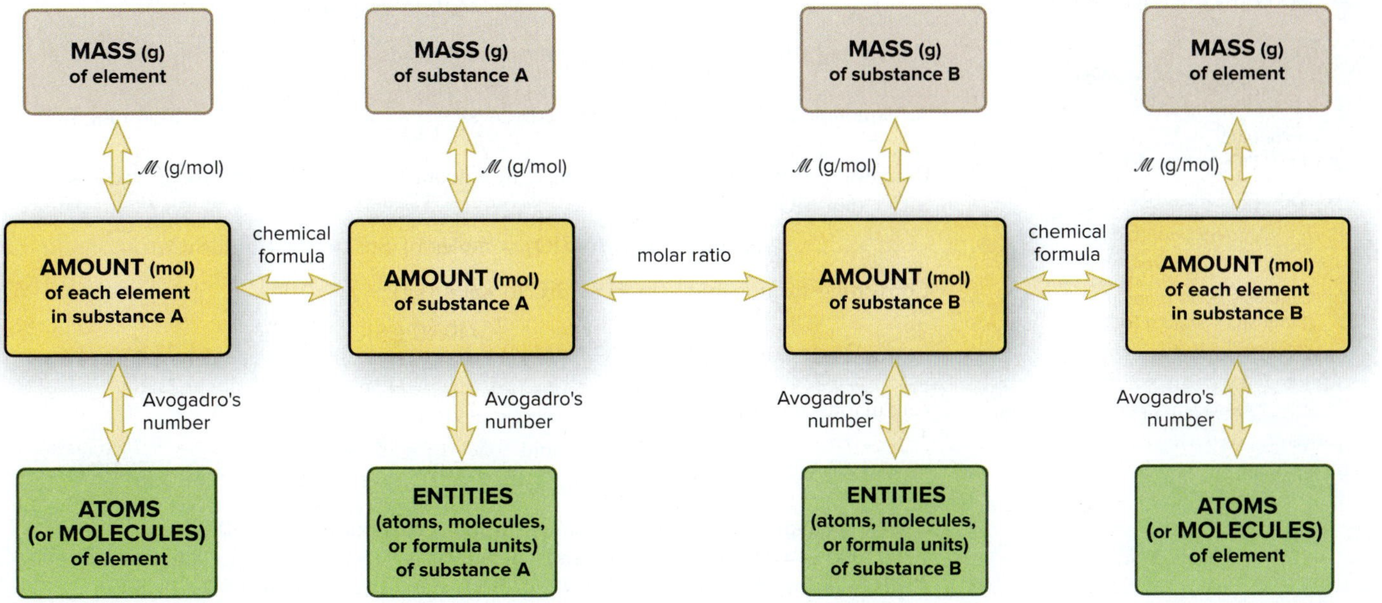

Figure 3.11 An overview of amount-mass-number stoichiometric relationships.

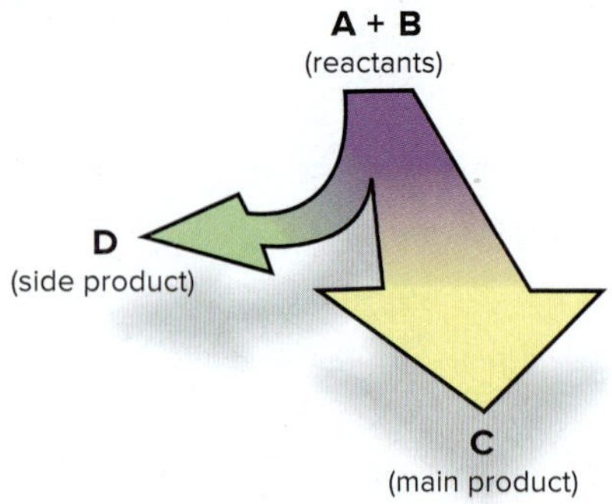

Figure 3.12 The effect of a side reaction on the yield of the main product.

1. *Theoretical yield.* The amount of product calculated from the molar ratio in the balanced equation is the **theoretical yield.** But, there are several reasons why the theoretical yield is *never* obtained:
   - Reactant mixtures often proceed through **side reactions** that form different products (Figure 3.12). In the rocket fuel reaction in Sample Problem 3.20, for example, the reactants might form some NO in the following side reaction:
   $$N_2H_4(l) + 2N_2O_4(l) \longrightarrow 6NO(g) + 2H_2O(g)$$
   This reaction decreases the amounts of reactants available for N_2 production.
   - Even more important, many reactions seem to stop before they are complete, so some limiting reactant is unused. (We'll see why in Chapter 4.)
   - Physical losses occur in every step of a separation (see Tools of the Laboratory, Section 2.9): some solid clings to filter paper, some distillate evaporates, and so forth. With careful technique, you can minimize, but never eliminate, such losses.
2. *Actual yield.* Given these reasons for obtaining less than the theoretical yield, the amount of product *actually* obtained is the **actual yield.** Theoretical and actual yields are expressed in units of amount (moles) or mass (grams).
3. *Percent yield.* The **percent yield (% yield)** is the actual yield expressed as a percentage of the theoretical yield:

$$\%\ \text{yield} = \frac{\text{actual yield}}{\text{theoretical yield}} \times 100 \qquad \textbf{(3.8)}$$

By definition, the actual yield is less than the theoretical yield, so the percent yield is *always* less than 100%.

SAMPLE PROBLEM 3.21 Calculating Percent Yield

Problem Silicon carbide (SiC) is an important ceramic material made by reacting sand (silicon dioxide, SiO_2) with powdered carbon at a high temperature. Carbon monoxide is also formed. When 100.0 kg of sand is processed, 51.4 kg of SiC is recovered. What is the percent yield of SiC from this process?

Plan We are given the actual yield of SiC (51.4 kg), so we need the theoretical yield to calculate the percent yield. After writing the balanced equation, we convert the given mass of SiO_2 (100.0 kg) to amount (mol). We use the molar ratio to find the amount of SiC formed and convert it to mass (kg) to obtain the theoretical yield. Then, we use Equation 3.8 to find the percent yield (see the road map).

Road Map

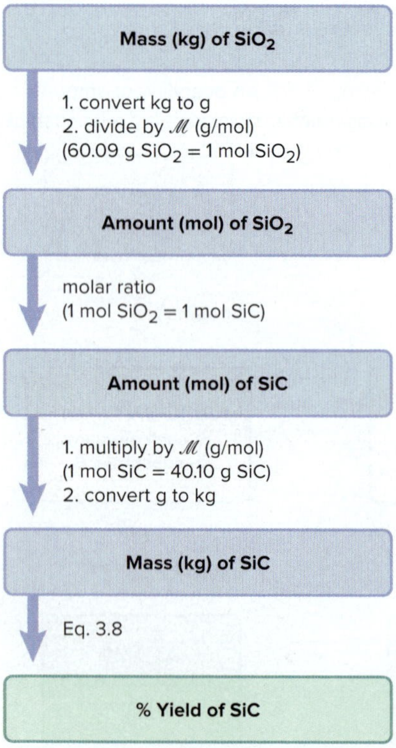

Solution Writing the balanced equation:

$$SiO_2(s) + 3C(s) \longrightarrow SiC(s) + 2CO(g)$$

Converting from mass (kg) of SiO_2 to amount (mol):

$$\text{Amount (mol) of } SiO_2 = 100.0\ \cancel{\text{kg } SiO_2} \times \frac{1000\ \cancel{\text{g}}}{1\ \cancel{\text{kg}}} \times \frac{1\ \text{mol } SiO_2}{60.09\ \cancel{\text{g } SiO_2}} = 1664\ \text{mol } SiO_2$$

Converting from amount (mol) of SiO_2 to amount (mol) of SiC: The molar ratio is 1 mol SiC/1 mol SiO_2, so

$$\text{Amount (mol) of } SiO_2 = \text{moles of SiC} = 1664\ \text{mol SiC}$$

Converting from amount (mol) of SiC to mass (kg):

$$\text{Mass (kg) of SiC} = 1664\ \cancel{\text{mol SiC}} \times \frac{40.10\ \cancel{\text{g}}\ \text{SiC}}{1\ \cancel{\text{mol SiC}}} \times \frac{1\ \text{kg}}{1000\ \cancel{\text{g}}} = 66.73\ \text{kg SiC}$$

Calculating the percent yield:

$$\%\ \text{yield of SiC} = \frac{\text{actual yield}}{\text{theoretical yield}} \times 100 = \frac{51.4\ \cancel{\text{kg SiC}}}{66.73\ \cancel{\text{kg SiC}}} \times 100 = 77.0\%$$

Check Rounding shows that the mass of SiC seems correct: ~1500 mol × 40 g/mol × 1 kg/1000 g = 60 kg. The molar ratio of SiC/SiO_2 is 1/1, and ℳ of SiC is about two-thirds ($\sim\frac{40}{60}$) of ℳ of SiO_2, so 100 kg of SiO_2 should form about 66 kg of SiC.

FOLLOW-UP PROBLEMS

3.21A Marble (calcium carbonate) reacts with hydrochloric acid solution to form calcium chloride solution, water, and carbon dioxide. Find the percent yield of carbon dioxide if 3.65 g is collected when 10.0 g of marble reacts.

3.21B Sodium carbonate, also known as *soda ash,* is used in glassmaking. It is obtained from a reaction between sodium chloride and calcium carbonate; calcium chloride is the other product. Calculate the percent yield of sodium carbonate if 92.6 g is collected when 112 g of sodium chloride reacts with excess calcium carbonate.

SOME SIMILAR PROBLEMS 3.93–3.96

Yields in Multistep Syntheses In the multistep synthesis of a complex compound, the overall yield can be surprisingly low, even if the yield of each step is high. For example, suppose a six-step synthesis has a 90.0% yield for each step. To find the overall percent yield, *express the yield of each step as a decimal, multiply all the decimal amounts together, and then convert back to a percentage.* The overall yield in this case is only slightly more than 50%:

$$\text{Overall \% yield} = (0.900 \times 0.900 \times 0.900 \times 0.900 \times 0.900 \times 0.900) \times 100 = 53.1\%$$

Such multistep sequences are common in laboratory syntheses of medicines, dyes, pesticides, and many other organic compounds. For example, the antidepressant Sertraline is prepared from a simple starting compound in six steps, with yields of 80%, 80%, 50%, 100%, 48%, and 30%, respectively, and an overall percent yield of only 4.6% (Figure 3.13). Because a typical synthesis begins with large amounts of inexpensive, simple reactants and ends with small amounts of expensive, complex products, the overall yield greatly influences the commercial potential of a product.

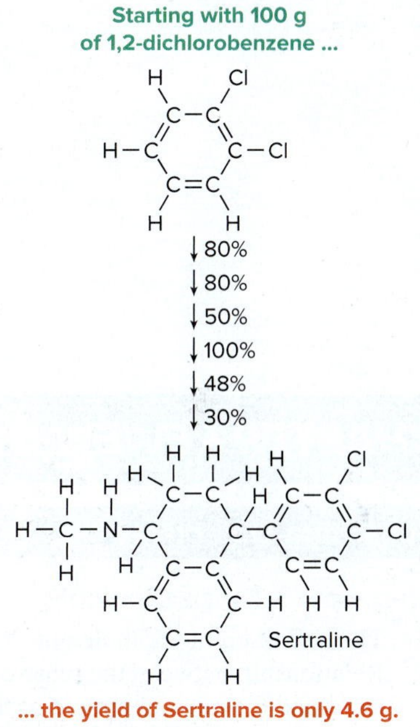
Starting with 100 g of 1,2-dichlorobenzene ...
80%
80%
50%
100%
48%
30%
Sertraline
... the yield of Sertraline is only 4.6 g.

Figure 3.13 Low overall yield in a multistep synthesis.

Atom Economy: A Green Chemistry Perspective on Yield In the relatively new field of **green chemistry**, academic, industrial, and government chemists develop methods that reduce or prevent the release of harmful substances into the environment and the wasting of energy resources.

One way that green chemists evaluate a synthetic route is to focus on its *atom economy,* the proportion of reactant atoms that end up in the desired product. The efficiency of a synthesis is quantified in terms of the *percent atom economy:*

$$\text{\% atom economy} = \frac{\text{no. of moles} \times \text{molar mass of desired product}}{\text{sum of (no. of moles} \times \text{molar mass) for all products}} \times 100$$

Consider two synthetic routes—one starting with benzene (C_6H_6), the other with butane (C_4H_{10})—for the production of maleic anhydride ($C_4H_2O_3$), a key substance in the manufacture of polymers, dyes, medicines, pesticides, and other products:

$$\text{Route 1.}\quad 2C_6H_6(l) + 9O_2(g) \longrightarrow \cdots \longrightarrow 2C_4H_2O_3(l) + 4H_2O(l) + 4CO_2(g)$$

$$\text{Route 2.}\quad 2C_4H_{10}(g) + 7O_2(g) \longrightarrow \cdots \longrightarrow 2C_4H_2O_3(l) + 8H_2O(l)$$

Let's compare the efficiency of these routes in terms of percent atom economy:

Route 1:

$$\text{\% atom economy} = \frac{2 \times \mathcal{M} \text{ of } C_4H_2O_3}{(2 \times \mathcal{M} \text{ of } C_4H_2O_3) + (4 \times \mathcal{M} \text{ of } H_2O) + (4 \times \mathcal{M} \text{ of } CO_2)} \times 100$$

$$= \frac{2 \times 98.06\ \text{g}}{(2 \times 98.06\ \text{g}) + (4 \times 18.02\ \text{g}) + (4 \times 44.01\ \text{g})} \times 100$$

$$= 44.15\%$$

Route 2:

$$\text{\% atom economy} = \frac{2 \times \mathcal{M} \text{ of } C_4H_2O_3}{(2 \times \mathcal{M} \text{ of } C_4H_2O_3) + (8 \times \mathcal{M} \text{ of } H_2O)} \times 100$$

$$= \frac{2 \times 98.06\ \text{g}}{(2 \times 98.06\ \text{g}) + (8 \times 18.02\ \text{g})} \times 100$$

$$= 57.63\%$$

From the perspective of atom economy, route 2 is preferable because a larger percentage of reactant atoms end up in the desired product. It is also a "greener" approach than route 1 because it avoids the use of the toxic reactant benzene and does not produce CO_2, a gas that contributes to global warming.

› Summary of Section 3.4

- The substances in a balanced equation are related to each other by stoichiometrically equivalent molar ratios, which are used as conversion factors to find the amount (mol) of one substance given the amount of another.
- In limiting-reactant problems, the quantities of two (or more) reactants are given, and the limiting reactant is the one that forms the lower quantity of product. Reaction tables show the initial and final quantities of all reactants and products, as well as the changes in those quantities.
- In practice, side reactions, incomplete reactions, and physical losses result in an actual yield of product that is less than the theoretical yield (the quantity based on the molar ratio from the balanced equation), giving a percent yield less than 100%. In multistep reaction sequences, the overall yield is found by multiplying the yields for each step.
- Atom economy, or the proportion of reactant atoms found in the product, is one criterion for choosing a "greener" reaction process.

CHAPTER REVIEW GUIDE

Learning Objectives

Relevant section (§) and/or sample problem (SP) numbers appear in parentheses.

Understand These Concepts

1. The definition of the mole unit (§3.1)
2. Relationship between the mass of a chemical entity (in amu) and the mass of a mole of that entity (in g) (§3.1)
3. The relationships among amount of substance (in mol), mass (in g), and number of chemical entities (§3.1)
4. Mole-mass-number information in a chemical formula (§3.1)
5. The difference between empirical and molecular formulas of a compound (§3.2)
6. How more than one substance can have the same empirical formula or the same molecular formula (isomers) (§3.2)
7. The importance of balancing equations for the quantitative study of chemical reactions (§3.3)
8. Mole-mass-number information in a balanced equation (§3.4)
9. The relationship between amounts of reactants and amounts of products (§3.4)
10. Why one reactant limits the amount of product (§3.4)
11. The causes of lower-than-expected yields and the distinction between theoretical and actual yields (§3.4)

Master These Skills

1. Calculating the molar mass of any substance (§3.1; SPs 3.4–3.6)
2. Converting between amount of substance (in moles), mass (in grams), and number of chemical entities (SPs 3.1–3.5)
3. Using mass percent to find the mass of an element in a given mass of compound (SPs 3.6, 3.7)
4. Determining empirical and molecular formulas of a compound from mass percents and molar masses of elements (SPs 3.8–3.10)
5. Determining a molecular formula from combustion analysis (SP 3.11)
6. Converting a chemical statement or a molecular depiction into a balanced equation (SPs 3.12, 3.13)
7. Using stoichiometrically equivalent molar ratios to convert between amounts of reactants and products in reactions (SPs 3.14–3.16)
8. Writing an overall equation for a reaction sequence (SP 3.17)
9. Solving limiting-reactant problems for reactions (SPs 3.18–3.20)
10. Calculating percent yield (SP 3.21)

Key Terms

Page numbers appear in parentheses.

actual yield (128)
Avogadro's number (96)
balancing (stoichiometric) coefficient (112)
chemical equation (111)
combustion analysis (108)
empirical formula (104)
green chemistry (129)
isomer (110)
limiting reactant (122)
mole (mol) (95)
molar mass ($\mathcal{M}$) (96)
molecular formula (105)
overall (net) equation (120)
percent yield (% yield) (128)
product (112)
reactant (111)
side reaction (128)
stoichiometry (95)
structural formula (105)
theoretical yield (128)

Key Equations and Relationships

Page numbers appear in parentheses.

3.1 Number of entities in one mole (96):

1 mol contains 6.022×10^{23} entities (to 4 sf)

3.2 Converting amount (mol) to mass (g) using $\mathcal{M}$ (97):

$$\text{Mass (g)} = \text{amount (mol)} \times \frac{\text{no. of grams}}{1\text{ mol}}$$

3.3 Converting mass (g) to amount (mol) using $1/\mathcal{M}$ (98):

$$\text{Amount (mol)} = \text{mass (g)} \times \frac{1\text{ mol}}{\text{no. of grams}}$$

3.4 Converting amount (mol) to number of entities (98):

$$\text{No. of entities} = \text{amount (mol)} \times \frac{6.022\times10^{23}\text{ entities}}{1\text{ mol}}$$

3.5 Converting number of entities to amount (mol) (98):

$$\text{Amount (mol)} = \text{no. of entities} \times \frac{1\text{ mol}}{6.022\times10^{23}\text{ entities}}$$

3.6 Calculating mass % (102):

Mass % of element X

$$= \frac{\text{moles of X in formula} \times \text{molar mass of X (g/mol)}}{\text{mass (g) of 1 mol of compound}} \times 100$$

3.7 Finding the mass of an element in any mass of compound (103):

$$\text{Mass of element} = \text{mass of compound} \times \frac{\text{mass of element in 1 mol of compound}}{\text{mass of 1 mol of compound}}$$

3.8 Calculating percent yield (128):

$$\%\text{ yield} = \frac{\text{actual yield}}{\text{theoretical yield}} \times 100$$

BRIEF SOLUTIONS TO FOLLOW-UP PROBLEMS

3.1A $\text{Amount (mol) of C} = 315\ \cancel{\text{mg C}} \times \frac{1\text{ g}}{10^3\ \cancel{\text{mg}}} \times \frac{1\text{ mol C}}{12.01\ \cancel{\text{g C}}} = 2.62\times10^{-2}\text{ mol C}$

See Road Map 3.1A.

3.1B 14 g Al = 1 soda can provides a conversion factor between mass of Al and number of soda cans:

$$\text{Number of cans} = 52\ \cancel{\text{mol Al}} \times \frac{26.98\ \cancel{\text{g Al}}}{1\ \cancel{\text{mol Al}}} \times \frac{1\text{ can}}{14\ \cancel{\text{g Al}}} = 100\text{ cans}$$

See Road Map 3.1B.

Road Map 3.1A

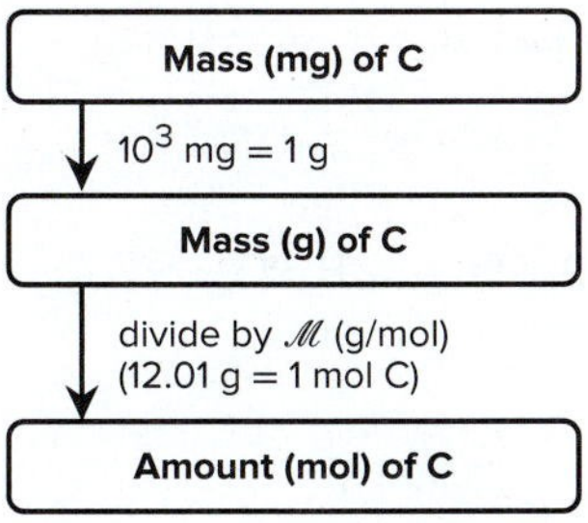

Road Map 3.1B

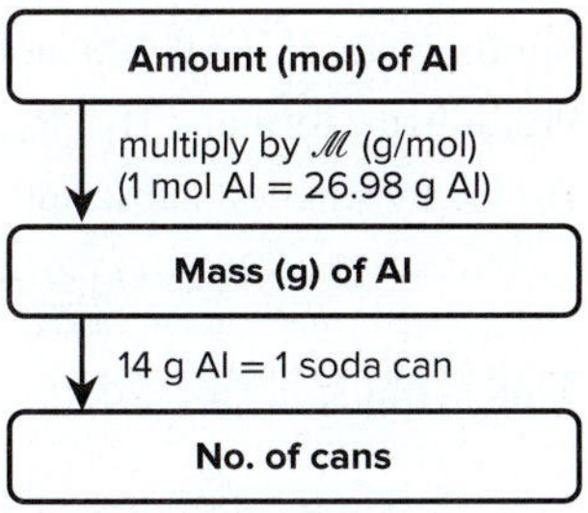

3.2A

$$\text{Amount (mol) of N} = 9.72\times10^{21}\ \cancel{\text{N}_2\text{ molecules}} \times \frac{1\ \cancel{\text{mol N}_2\text{ molecules}}}{6.022\times10^{23}\ \cancel{\text{N}_2\text{ molecules}}} \times \frac{2\text{ mol N atoms}}{1\ \cancel{\text{mol N}_2\text{ molecules}}} = 3.23\times10^{-2}\text{ mol N atoms}$$

See Road Map 3.2A.

3.2B $\text{No. of He atoms} = 325\ \cancel{\text{mol He}} \times \frac{6.022\times10^{23}\text{ He atoms}}{1\ \cancel{\text{mol He}}} = 1.96\times10^{26}\text{ He atoms}$

See Road Map 3.2B.

Road Map 3.2A

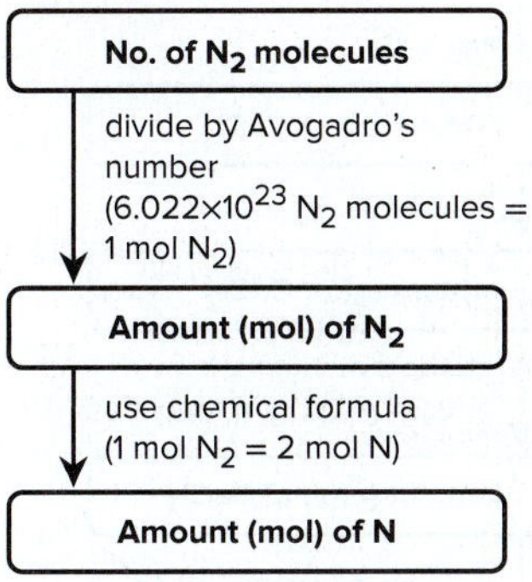

Road Map 3.2B

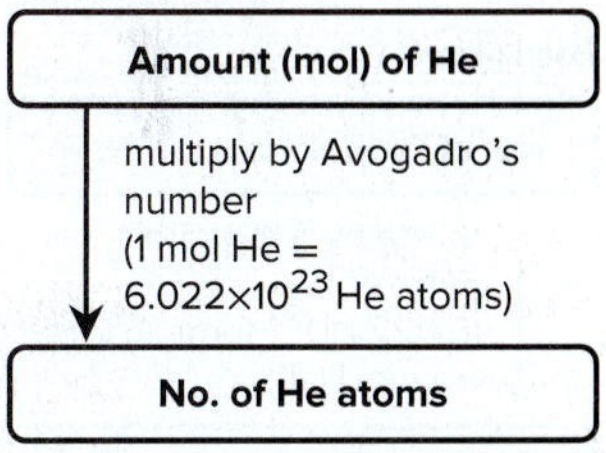

3.3A

$$\text{Mass (g) of Mn} = 3.22\times10^{20}\ \cancel{\text{Mn atoms}} \times \frac{1\ \cancel{\text{mol Mn}}}{6.022\times10^{23}\ \cancel{\text{Mn atoms}}} \times \frac{54.94\text{ g Mn}}{1\ \cancel{\text{mol Mn}}} = 2.94\times10^{-2}\text{ g Mn}$$

See Road Map 3.3A.

3.3B

$$\text{No. of Cu atoms} = 0.0625\ \cancel{\text{g Cu}} \times \frac{1\ \cancel{\text{mol Cu}}}{63.55\ \cancel{\text{g Cu}}} \times \frac{6.022\times10^{23}\text{ Cu atoms}}{1\ \cancel{\text{mol Cu}}} = 5.92\times10^{20}\text{ Cu atoms}$$

See Road Map 3.3B.

Road Map 3.3A

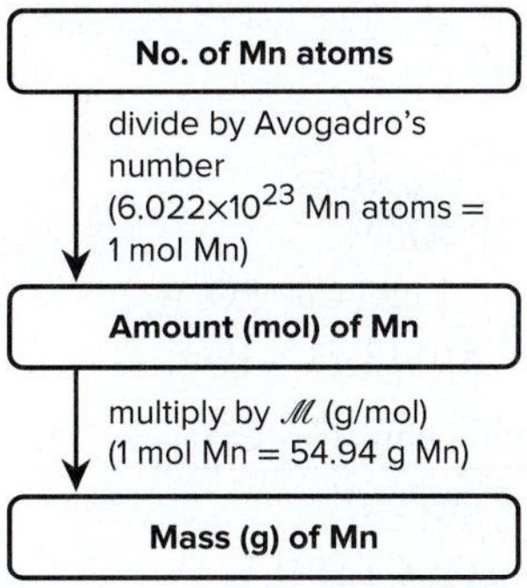

Road Map 3.3B

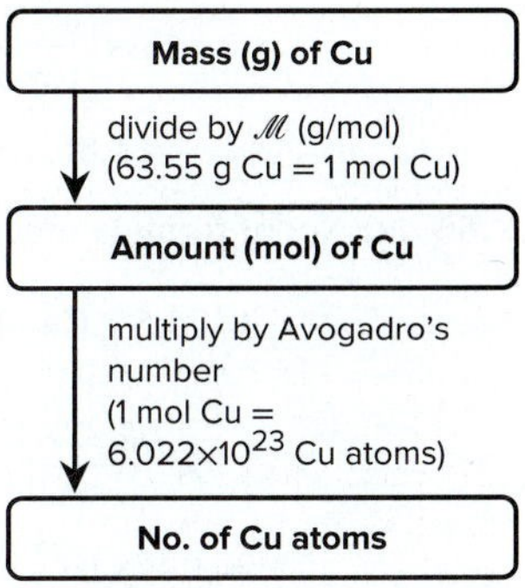

3.4A $\mathcal{M} = (1 \times \mathcal{M} \text{ of Na}) + (1 \times \mathcal{M} \text{ of F})$
$= 22.99 \text{ g/mol} + 19.00 \text{ g/mol} = 41.99 \text{ g/mol}$

Mass (g) of NaF

$$= 1.19\times10^{19}\ \cancel{\text{NaF formula units}} \times \frac{1\ \cancel{\text{mol NaF}}}{6.022\times10^{23}\ \cancel{\text{NaF formula units}}} \times \frac{41.99\text{ g NaF}}{1\ \cancel{\text{mol NaF}}}$$
$$= 8.30\times10^{-4}\text{ g NaF}$$

See Road Map 3.4A.

3.4B $\mathcal{M} = (1 \times \mathcal{M} \text{ of Ca}) + (2 \times \mathcal{M} \text{ of Cl})$
$= 40.08 \text{ g/mol} + (2 \times 35.45 \text{ g/mol}) = 110.98 \text{ g/mol}$

No. of $CaCl_2$ formula units $= 400\ \cancel{\text{lb } CaCl_2}$

$$\times \frac{453.6\ \cancel{\text{g}}}{1\ \cancel{\text{lb}}} \times \frac{1 \text{ mol } CaCl_2}{110.98\ \cancel{\text{g } CaCl_2}} \times \frac{6.022\times10^{23}\ CaCl_2 \text{ formula units}}{1 \text{ mol } CaCl_2}$$
$$= 1\times10^{27}\ CaCl_2 \text{ formula units}$$

See Road Map 3.4B.

Road Map 3.4A

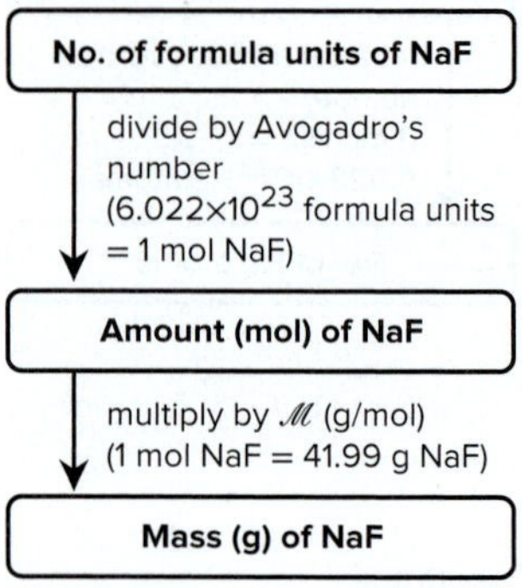

Road Map 3.4B

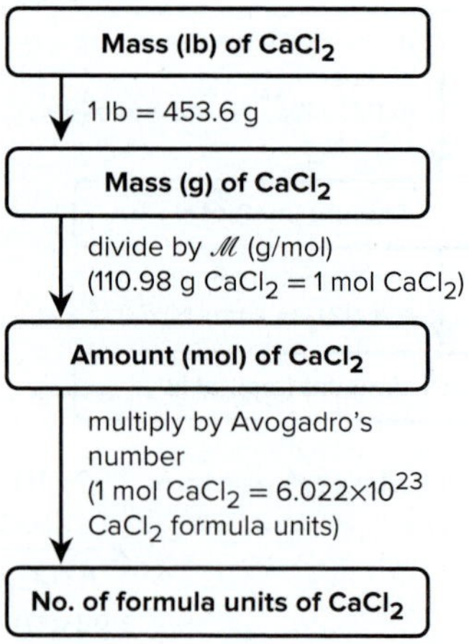

3.5A (a) Mass (g) of P_4O_{10}

$$= 4.65\times10^{22}\ \cancel{\text{molecules } P_4O_{10}} \times \frac{1\ \cancel{\text{mol } P_4O_{10}}}{6.022\times10^{23}\ \cancel{\text{molecules } P_4O_{10}}} \times \frac{283.88 \text{ g } P_4O_{10}}{1\ \cancel{\text{mol } P_4O_{10}}}$$
$$= 21.9 \text{ g } P_4O_{10}$$

(b) No. of P atoms $= 4.65\times10^{22}\ \cancel{\text{molecules } P_4O_{10}}$

$$\times \frac{4 \text{ atoms P}}{1\ \cancel{\text{molecule } P_4O_{10}}}$$
$$= 1.86\times10^{23} \text{ P atoms}$$

3.5B (a) No. of formula units of $Ca_3(PO_4)_2$

$$= 75.5\ \cancel{\text{g } Ca_3(PO_4)_2} \times \frac{1\ \cancel{\text{mol } Ca_3(PO_4)_2}}{310.18\ \cancel{\text{g } Ca_3(PO_4)_2}} \times \frac{6.022\times10^{23} \text{ formula units } Ca_3(PO_4)_2}{1\ \cancel{\text{mol } Ca_3(PO_4)_2}}$$
$$= 1.47\times10^{23} \text{ formula units } Ca_3(PO_4)_2$$

(b) No. of PO_4^{3-} ions $= 1.47\times10^{23}\ \cancel{\text{formula units } Ca_3(PO_4)_2}$

$$\times \frac{2\ PO_4^{3-} \text{ ions}}{1\ \cancel{\text{formula unit } Ca_3(PO_4)_2}}$$
$$= 2.94\times10^{23}\ PO_4^{3-} \text{ ions}$$

3.6A $$\text{Mass \% of C} = \frac{6\ \cancel{\text{mol C}} \times \dfrac{12.01 \text{ g C}}{1\ \cancel{\text{mol C}}}}{180.16 \text{ g } C_6H_{12}O_6} \times 100$$
$$= 40.00 \text{ mass \% C}$$

3.6B $$\text{Mass \% of Cl} = \frac{3\ \cancel{\text{mol Cl}} \times \dfrac{35.45 \text{ g Cl}}{1\ \cancel{\text{mol Cl}}}}{137.36 \text{ g } CCl_3F} \times 100$$
$$= 77.42 \text{ mass \% Cl}$$

3.7A $$\text{Mass (g) of C} = 16.55\ \cancel{\text{g } C_6H_{12}O_6} \times \frac{72.06 \text{ g C}}{180.16\ \cancel{\text{g } C_6H_{12}O_6}}$$
$$= 6.620 \text{ g C}$$

3.7B $$\text{Mass (g) of Cl} = 112\ \cancel{\text{g } CCl_3F} \times \frac{106.35 \text{ g Cl}}{137.36\ \cancel{\text{g } CCl_3F}}$$
$$= 86.7 \text{ g Cl}$$

3.8A Preliminary formula: $B_{0.170}O_{0.255}$

Divide by smaller subscript: $B_{\frac{0.170}{0.170}}O_{\frac{0.255}{0.170}} = B_{1.00}O_{1.50}$

Multiply by 2: $B_{(2\times1.00)}O_{(2\times1.50)} = B_{2.00}O_{3.00} = B_2O_3$

3.8B Preliminary formula: $C_{6.80}H_{18.1}$

Divide by smallest subscript: $C_{\frac{6.80}{6.80}}O_{\frac{18.1}{6.80}} = C_{1.00}H_{2.66}$

Multiply by 3: $C_{(3\times1.00)}H_{(3\times2.66)} = C_{3.00}H_{7.98} = C_3H_8$

3.9A $$\text{Amount (mol) of H} = 1.23\ \cancel{\text{g H}} \times \frac{1 \text{ mol H}}{1.008\ \cancel{\text{g H}}}$$
$$= 1.22 \text{ mol H}$$

Similarly, there are 0.408 mol P and 1.63 mol O.

Preliminary formula: $H_{1.22}P_{0.408}O_{1.63}$

Divide by smallest subscript:

$$H_{\frac{1.22}{0.408}}P_{\frac{0.408}{0.408}}O_{\frac{1.63}{0.408}} = H_{2.99}P_{1.00}O_{4.00} = H_3PO_4$$

This is phosphoric acid.

3.9B $$\text{Amount (mol) of S} = 2.88\ \cancel{\text{g S}} \times \frac{1 \text{ mol S}}{32.06\ \cancel{\text{g S}}}$$
$$= 0.0898 \text{ mol S}$$

$$\text{Amount (mol) of M} = 0.0898\ \cancel{\text{mol S}} \times \frac{2 \text{ mol M}}{3\ \cancel{\text{mol S}}} = 0.0599 \text{ mol M}$$

$$\text{Molar mass of M} = \frac{3.12 \text{ g M}}{0.0599 \text{ mol M}} = 52.1 \text{ g/mol}$$

M is chromium, and M_2S_3 is chromium(III) sulfide.

3.10A Assuming 100.00 g of compound, we have 95.21 g of C and 4.79 g of H:

$$\text{Amount (mol) of C} = 95.21\ \cancel{\text{g C}} \times \frac{1 \text{ mol C}}{12.01\ \cancel{\text{g C}}} = 7.928 \text{ mol C}$$

Similarly, there is 4.75 mol H.

Preliminary formula: $C_{7.928}H_{4.75} = C_{\frac{7.928}{4.75}}H_{\frac{4.75}{4.75}} = C_{1.67}H_{1.00}$

Empirical formula: $C_{(3\times1.67)}H_{(3\times1.00)} = C_5H_3$

$$\text{Whole-number multiple} = \frac{\mathcal{M}\text{ of benzo[a]pyrene}}{\mathcal{M}\text{ of empirical formula}} = \frac{252.30\ \cancel{\text{g/mol}}}{63.07\ \cancel{\text{g/mol}}} = 4$$

Molecular formula: $C_{(5\times4)}H_{(3\times4)} = C_{20}H_{12}$

3.10B Assuming 100.00 g of compound, we have 49.47 g C, 5.19 g H, 28.86 g N, and 16.48 g O:

$$\text{Amount (mol) of C} = 49.47\ \cancel{\text{g C}} \times \frac{1\text{ mol C}}{12.01\ \cancel{\text{g C}}} = 4.119\text{ mol C}$$

Similarly, there are 5.15 mol H, 2.060 mol N, and 1.030 mol O.

Preliminary formula: $C_{4.119}H_{5.15}N_{2.060}O_{1.030}$

$$= C_{\frac{4.119}{1.030}}H_{\frac{5.15}{1.030}}N_{\frac{2.060}{1.030}}N_{\frac{1.030}{1.030}} = C_{4.00}H_{5.00}N_{2.00}O_{1.00}$$

Empirical formula: $C_4H_5N_2O$

$$\text{Whole-number multiple} = \frac{\mathcal{M}\text{ of caffeine}}{\mathcal{M}\text{ of empirical formula}} = \frac{194.2\ \cancel{\text{g/mol}}}{97.10\ \cancel{\text{g/mol}}} = 2$$

Molecular formula: $C_{(4\times2)}H_{(5\times2)}N_{(2\times2)}O_{(1\times2)} = C_8H_{10}N_4O_2$

3.11A $\text{Mass (g) of C} = 0.451\ \cancel{\text{g } CO_2} \times \frac{12.01\text{ g C}}{44.01\ \cancel{\text{g } CO_2}} = 0.123\text{ g C}$

Similarly, there is 0.00690 g H.

Mass (g) of Cl = 0.250 g − (0.123 g + 0.00690 g) = 0.120 g Cl

Amount (mol) of elements: 0.0102 mol C; 0.00685 mol H; 0.00339 mol Cl

Empirical formula: $C_{0.0102}H_{0.00685}Cl_{0.00339}$

$$= C_{\frac{0.0102}{0.00339}}H_{\frac{0.00685}{0.00339}}Cl_{\frac{0.00339}{0.00339}} = C_3H_2Cl$$

$$\text{Whole-number multiple} = \frac{\mathcal{M}\text{ of compound}}{\mathcal{M}\text{ of empirical formula}} = \frac{146.99\ \cancel{\text{g/mol}}}{73.50\ \cancel{\text{g/mol}}} = 2$$

Molecular formula: $C_{(3\times2)}H_{(2\times2)}Cl_{(1\times2)} = C_6H_4Cl_2$

3.11B $\text{Mass (g) of C} = 3.516\ \cancel{\text{g } CO_2} \times \frac{12.01\text{ g C}}{44.01\ \cancel{\text{g } CO_2}} = 0.9595\text{ g C}$

Similarly, there is 0.1127 g H.

Mass (g) of O = 1.200 g − (0.9595 g + 0.1127 g) = 0.128 g O

Amount (mol) of elements: 0.07989 mol C; 0.1118 mol H; 0.00800 mol O

Empirical formula: $C_{0.07989}H_{0.1118}O_{0.00800}$

$$= C_{\frac{0.07989}{0.00800}}H_{\frac{0.1118}{0.00800}}O_{\frac{0.00800}{0.00800}} = C_{10}H_{14}O$$

$$\text{Whole-number multiple} = \frac{\mathcal{M}\text{ of compound}}{\mathcal{M}\text{ of empirical formula}} = \frac{300.42\ \cancel{\text{g/mol}}}{150.21\ \cancel{\text{g/mol}}} = 2$$

Molecular formula: $C_{(10\times2)}H_{(14\times2)}O_{(1\times2)} = C_{20}H_{28}O_2$

3.12A (a) $2Na(s) + 2H_2O(l) \longrightarrow H_2(g) + 2NaOH(aq)$

(b) $2HNO_3(aq) + CaCO_3(s) \longrightarrow CO_2(g) + H_2O(l) + Ca(NO_3)_2(aq)$

(c) $PCl_3(g) + 3HF(g) \longrightarrow PF_3(g) + 3HCl(g)$

3.12B (a) $4C_3H_5N_3O_9(l) \longrightarrow 12CO_2(g) + 10H_2O(g) + 6N_2(g) + O_2(g)$

(b) $4KO_2(s) + 2CO_2(g) \longrightarrow 3O_2(g) + 2K_2CO_3(s)$

(c) $Fe_2O_3(s) + 3CO(g) \longrightarrow 2Fe(s) + 3CO_2(g)$

3.13A From the depiction, we have

$$6CO + 3O_2 \longrightarrow 6CO_2$$

Or,

$$2CO(g) + O_2(g) \longrightarrow 2CO_2(g)$$

3.13B From the depiction, we have

$$2N_2 + 6H_2 \longrightarrow 4NH_3$$

Or,

$$N_2(g) + 3H_2(g) \longrightarrow 2NH_3(g)$$

3.14A $Fe_2O_3(s) + 2Al(s) \longrightarrow Al_2O_3(s) + 2Fe(l)$

$$\text{Amount (mol) of } Fe_2O_3 = 3.60\times10^3\ \cancel{\text{mol Fe}} \times \frac{1\text{ mol } Fe_2O_3}{2\ \cancel{\text{mol Fe}}} = 1.80\times10^3\text{ mol } Fe_2O_3$$

See Road Map 3.14A.

3.14B $3Ag_2S(s) + 2Al(s) \longrightarrow 6Ag(s) + Al_2S_3(s)$

$$\text{Amount (mol) of Al} = 0.253\ \cancel{\text{mol } Ag_2S} \times \frac{2\text{ mol Al}}{3\ \cancel{\text{mol } Ag_2S}} = 0.169\text{ mol Al}$$

See Road Map 3.14B.

Road Map 3.14A

Amount (mol) of Fe

↓ molar ratio ($2\text{ mol Fe} = 1\text{ mol } Fe_2O_3$)

Amount (mol) of Fe_2O_3

Road Map 3.14B

Amount (mol) of Ag_2S

↓ molar ratio ($3\text{ mol } Ag_2S = 2\text{ mol Al}$)

Amount (mol) of Al

3.15A Amount (mol) of Fe

$$= 1.85\times10^{25}\ \cancel{\text{formula units } Fe_2O_3} \times \frac{1\ \cancel{\text{mol } Fe_2O_3}}{6.022\times10^{23}\ \cancel{\text{formula units } Fe_2O_3}} \times \frac{2\text{ mol Fe}}{1\ \cancel{\text{mol } Fe_2O_3}} = 61.4\text{ mol Fe}$$

See Road Map 3.15A.

3.15B Amount (mol) of Ag

$$= 32.6\ \cancel{\text{g } Ag_2S} \times \frac{1\ \cancel{\text{mol } Ag_2S}}{247.9\ \cancel{\text{g } Ag_2S}} \times \frac{6\text{ mol Ag}}{3\ \cancel{\text{mol } Ag_2S}} = 0.263\text{ mol Ag}$$

See Road Map 3.15B.

Road Map 3.15A

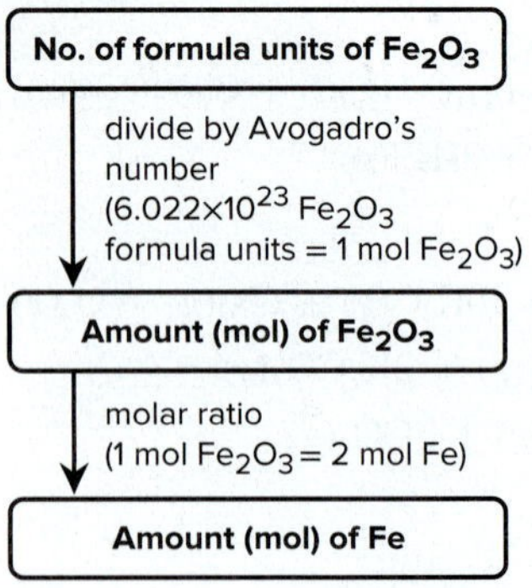

Road Map 3.15B

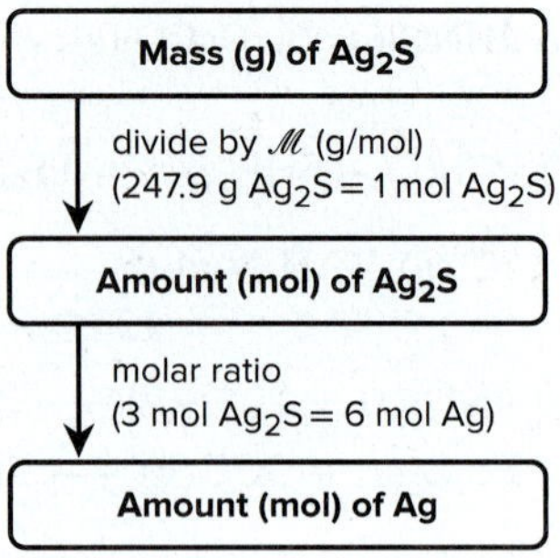

3.16A

$$\text{No. of Al atoms} = 1.00\ \cancel{\text{g Al}_2\text{O}_3} \times \frac{1\ \cancel{\text{mol Al}_2\text{O}_3}}{101.96\ \cancel{\text{g Al}_2\text{O}_3}} \times \frac{2\ \cancel{\text{mol Al}}}{1\ \cancel{\text{mol Al}_2\text{O}_3}} \times \frac{6.022\times10^{23}\ \text{Al atoms}}{1\ \cancel{\text{mol Al}}}$$
$$= 1.18\times10^{22}\ \text{Al atoms}$$

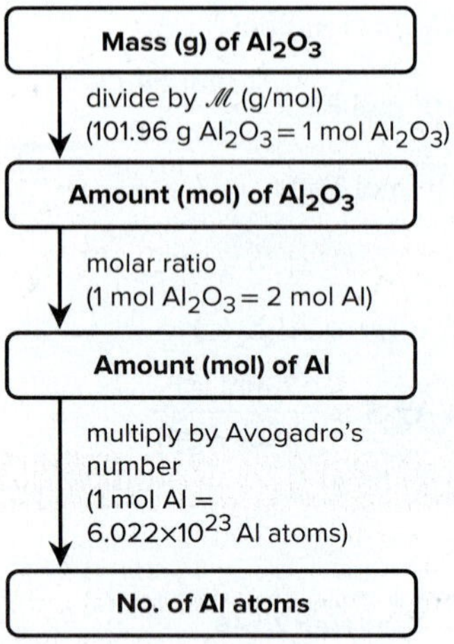

3.16B

$$\text{Mass (g) of Al} = 12.1\ \cancel{\text{g Al}_2\text{S}_3} \times \frac{1\ \cancel{\text{mol Al}_2\text{S}_3}}{150.14\ \cancel{\text{g Al}_2\text{S}_3}} \times \frac{2\ \cancel{\text{mol Al}}}{1\ \cancel{\text{mol Al}_2\text{S}_3}} \times \frac{26.98\ \text{g Al}}{1\ \cancel{\text{mol Al}}}$$
$$= 4.35\ \text{g Al}$$

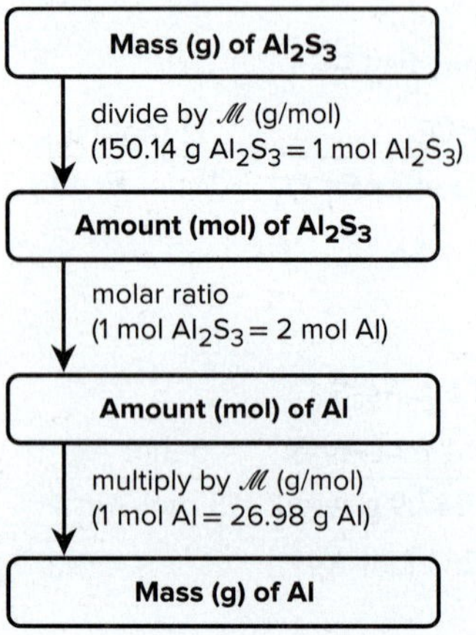

3.17A

$$2SO_2(g) + O_2(g) \longrightarrow \cancel{2SO_3(g)}$$
$$\cancel{2SO_3(g)} + 2H_2O(l) \longrightarrow 2H_2SO_4(aq)$$
$$\overline{2SO_2(g) + O_2(g) + 2H_2O(l) \longrightarrow 2H_2SO_4(aq)}$$

3.17B

$$N_2(g) + \cancel{O_2(g)} \longrightarrow \cancel{2NO(g)}$$
$$\cancel{2NO(g)} + 2O_3(g) \longrightarrow 2NO_2(g) + \cancel{2}O_2(g)$$
$$\overline{N_2(g) + 2O_3(g) \longrightarrow 2NO_2(g) + O_2(g)}$$

3.18A $4AB(g) + 2B_2(g) \longrightarrow 4AB_2(g)$

Or $2AB(g) + B_2(g) \longrightarrow 2AB_2(g)$

For AB: Molecules of AB_2

$$= 4\ \cancel{\text{molecules of AB}} \times \frac{2\ \text{molecules of AB}_2}{2\ \cancel{\text{molecules of AB}}}$$
$$= 4\ \text{molecules of AB}_2$$

For B_2: Molecules of AB_2

$$= 3\ \cancel{\text{molecules of B}_2} \times \frac{2\ \text{molecules of AB}_2}{1\ \cancel{\text{molecule of B}_2}}$$
$$= 6\ \text{molecules of AB}_2$$

Thus, AB is the limiting reactant; one B_2 molecule is in excess.

3.18B $4SO_2(g) + 2O_2(g) \longrightarrow 4SO_3(g)$

Or $2SO_2(g) + O_2(g) \longrightarrow 2SO_3(g)$

For SO_2: Molecules of SO_3

$$= 5\ \cancel{\text{molecules of SO}_2} \times \frac{2\ \text{molecules of SO}_3}{2\ \cancel{\text{molecules of SO}_2}}$$
$$= 5\ \text{molecules of SO}_3$$

For O_2: Molecules of SO_3

$$= 2\ \cancel{\text{molecules of O}_2} \times \frac{2\ \text{molecules of SO}_3}{1\ \cancel{\text{molecule of O}_2}}$$
$$= 4\ \text{molecules of SO}_3$$

Thus, O_2 is the limiting reactant; one SO_2 molecule is in excess.

3.19A Amount (mol) of AB_2

$$= 1.5\ \cancel{\text{mol AB}} \times \frac{2\ \text{mol AB}_2}{2\ \cancel{\text{mol AB}}} = 1.5\ \text{mol AB}_2$$

Amount (mol) of AB_2

$$= 1.5\ \cancel{\text{mol B}_2} \times \frac{2\ \text{mol AB}_2}{1\ \cancel{\text{mol B}_2}} = 3.0\ \text{mol AB}_2$$

Therefore, 1.5 mol of AB_2 can form.

3.19B Amount (mol) of SO_3 $= 4.2\ \cancel{\text{mol SO}_2} \times \frac{2\ \text{mol SO}_3}{2\ \cancel{\text{mol SO}_2}}$

$$= 4.2\ \text{mol SO}_3$$

Amount (mol) of SO_3 $= 3.6\ \cancel{\text{mol O}_2} \times \frac{2\ \text{mol SO}_3}{1\ \cancel{\text{mol O}_2}}$

$$= 7.2\ \text{mol SO}_3$$

Therefore, 4.2 mol of SO_3 is produced.

3.20A $2Al(s) + 3S(s) \longrightarrow Al_2S_3(s)$

Mass (g) of Al_2S_3 formed from 10.0 g of Al

$$= 10.0\ \cancel{\text{g Al}} \times \frac{1\ \cancel{\text{mol Al}}}{26.98\ \cancel{\text{g Al}}} \times \frac{1\ \cancel{\text{mol Al}_2\text{S}_3}}{2\ \cancel{\text{mol Al}}} \times \frac{150.14\ \text{g Al}_2\text{S}_3}{1\ \cancel{\text{mol Al}_2\text{S}_3}}$$
$$= 27.8\ \text{g Al}_2\text{S}_3$$

Mass (g) of Al_2S_3 formed from 15.0 g of S

$$= 15.0\ \cancel{g\ S} \times \frac{1\ \cancel{mol\ S}}{32.06\ \cancel{g\ S}} \times \frac{1\ \cancel{mol\ Al_2S_3}}{3\ \cancel{mol\ S}} \times \frac{150.14\ g\ Al_2S_3}{1\ \cancel{mol\ Al_2S_3}}$$

$$= 23.4\ g\ Al_2S_3$$

Thus, S is the limiting reactant, and 23.4 g of Al_2S_3 forms.

Mass (g) of Al in excess

= initial mass of Al − mass of Al reacted

$$= 10.0\ g\ Al - \left(15.0\ \cancel{g\ S} \times \frac{1\ \cancel{mol\ S}}{32.06\ \cancel{g\ S}} \times \frac{2\ \cancel{mol\ Al}}{3\ \cancel{mol\ S}} \times \frac{26.98\ g\ Al}{1\ \cancel{mol\ Al}}\right)$$

$$= 1.6\ g\ Al$$

(We would obtain the same answer if sulfur were shown more correctly as S_8.)

3.20B $2C_4H_{10}(g) + 13O_2(g) \longrightarrow 8CO_2(g) + 10H_2O(g)$

Mass (g) of CO_2 formed from 4.65 g of C_4H_{10}

$$= 4.65\ \cancel{g\ C_4H_{10}} \times \frac{1\ \cancel{mol\ C_4H_{10}}}{58.12\ \cancel{g\ C_4H_{10}}} \times \frac{8\ \cancel{mol\ CO_2}}{2\ \cancel{mol\ C_4H_{10}}} \times \frac{44.01\ g\ CO_2}{1\ \cancel{mol\ CO_2}}$$

$$= 14.1\ g\ CO_2$$

Mass (g) of CO_2 formed from 10.0 g of O_2

$$= 10.0\ \cancel{g\ O_2} \times \frac{1\ \cancel{mol\ O_2}}{32.00\ \cancel{g\ O_2}} \times \frac{8\ \cancel{mol\ CO_2}}{13\ \cancel{mol\ O_2}} \times \frac{44.01\ g\ CO_2}{1\ \cancel{mol\ CO_2}}$$

$$= 8.46\ g\ CO_2$$

Thus, O_2 is the limiting reactant, and 8.46 g of CO_2 forms.

Mass (g) of C_4H_{10} in excess

= initial mass of C_4H_{10} − mass of C_4H_{10} reacted

$$= 4.65\ g - \left(10.0\ \cancel{g\ O_2} \times \frac{1\ \cancel{mol\ O_2}}{32.00\ \cancel{g\ O_2}} \times \frac{2\ \cancel{mol\ C_4H_{10}}}{13\ \cancel{mol\ O_2}} \times \frac{58.12\ g\ C_4H_{10}}{1\ \cancel{mol\ C_4H_{10}}}\right)$$

$$= 1.86\ g\ C_4H_{10}$$

3.21A $CaCO_3(s) + 2HCl(aq) \longrightarrow CaCl_2(aq) + H_2O(l) + CO_2(g)$

$$\text{Theoretical yield (g) of } CO_2 = 10.0\ \cancel{g\ CaCO_3} \times \frac{1\ \cancel{mol\ CaCO_3}}{100.09\ \cancel{g\ CaCO_3}} \times \frac{1\ \cancel{mol\ CO_2}}{1\ \cancel{mol\ CaCO_3}} \times \frac{44.01\ g\ CO_2}{1\ \cancel{mol\ CO_2}}$$

$$= 4.40\ g\ CO_2$$

$$\%\ \text{yield} = \frac{\text{actual yield}}{\text{theoretical yield}} \times 100 = \frac{3.65\ \cancel{g\ CO_2}}{4.40\ \cancel{g\ CO_2}} \times 100 = 83.0\%$$

3.21B $2NaCl(aq) + CaCO_3(s) \longrightarrow Na_2CO_3(s) + CaCl_2(aq)$

Theoretical yield of Na_2CO_3

$$= 112\ \cancel{g\ NaCl} \times \frac{1\ \cancel{mol\ NaCl}}{58.44\ \cancel{g\ NaCl}} \times \frac{1\ \cancel{mol\ Na_2CO_3}}{2\ \cancel{mol\ NaCl}} \times \frac{105.99\ g\ Na_2CO_3}{1\ \cancel{mol\ Na_2CO_3}}$$

$$= 102\ g\ Na_2CO_3$$

$$\%\ \text{yield} = \frac{\text{actual yield}}{\text{theoretical yield}} \times 100 = \frac{92.6\ \cancel{g\ Na_2CO_3}}{102\ \cancel{g\ Na_2CO_3}} \times 100 = 90.8\%$$

PROBLEMS

Problems with **colored** numbers are answered in Appendix E and worked in detail in the Student Solutions Manual. Problem sections match those in the text and give the numbers of relevant sample problems. Most offer Concept Review Questions, Skill-Building Exercises (grouped in pairs covering the same concept), and Problems in Context. The Comprehensive Problems are based on material from any section or previous chapter.

The Mole

(Sample Problems 3.1 to 3.7)

Concept Review Questions

3.1 The atomic mass of Cl is 35.45 amu, and the atomic mass of Al is 26.98 amu. What are the masses in grams of 3 mol of Al atoms and of 2 mol of Cl atoms?

3.2 (a) How many moles of C atoms are in 1 mol of sucrose ($C_{12}H_{22}O_{11}$)?
(b) How many C atoms are in 2 mol of sucrose?

3.3 Why might the expression "1 mol of chlorine" be confusing? What change would remove any uncertainty? For what other elements might a similar confusion exist? Why?

3.4 How is the molecular mass of a compound the same as the molar mass, and how is it different?

3.5 What advantage is there to using a counting unit (the mole) for amount of substance rather than a mass unit?

3.6 You need to calculate the number of P_4 molecules that can form from 2.5 g of $Ca_3(PO_4)_2$. Draw a road map for solving this and write a Plan, without doing any calculations.

3.7 Each of the following balances weighs the indicated numbers of atoms of two elements:

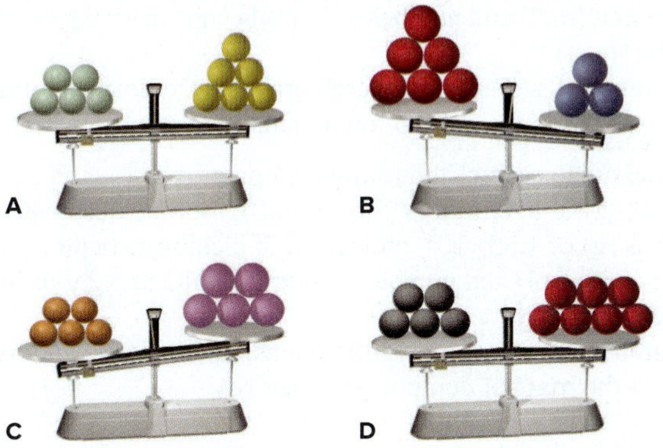

For each balance, which element—left, right, or neither,
(a) Has the higher molar mass?
(b) Has more atoms per gram?
(c) Has fewer atoms per gram?
(d) Has more atoms per mole?

Skill-Building Exercises (grouped in similar pairs)

3.8 Calculate the molar mass of each of the following:
(a) $Sr(OH)_2$ (b) N_2O_3 (c) $NaClO_3$ (d) Cr_2O_3

3.9 Calculate the molar mass of each of the following:
(a) $(NH_4)_3PO_4$ (b) CH_2Cl_2 (c) $CuSO_4{\cdot}5H_2O$ (d) BrF_3

3.10 Calculate the molar mass of each of the following:
(a) SnO (b) BaF_2 (c) $Al_2(SO_4)_3$ (d) $MnCl_2$

3.11 Calculate the molar mass of each of the following:
(a) N_2O_4 (b) C_4H_9OH (c) $MgSO_4{\cdot}7H_2O$ (d) $Ca(C_2H_3O_2)_2$

3.12 Calculate each of the following quantities:
(a) Mass (g) of 0.346 mol of Zn
(b) Number of F atoms in 2.62 mol of F_2
(c) Number of Ca atoms in 28.5 g of Ca

3.13 Calculate each of the following quantities:
(a) Amount (mol) of Mn atoms in 62.0 mg of Mn
(b) Amount (mol) for $1.36{\times}10^{22}$ atoms of Cu
(c) Mass (g) of $8.05{\times}10^{24}$ Li atoms

3.14 Calculate each of the following quantities:
(a) Mass (g) of 0.68 mol of $KMnO_4$
(b) Amount (mol) of O atoms in 8.18 g of $Ba(NO_3)_2$
(c) Number of O atoms in $7.3{\times}10^{-3}$ g of $CaSO_4{\cdot}2H_2O$

3.15 Calculate each of the following quantities:
(a) Mass (kg) of $4.6{\times}10^{21}$ molecules of NO_2
(b) Amount (mol) of Cl atoms in 0.0615 g of $C_2H_4Cl_2$
(c) Number of H^- ions in 5.82 g of SrH_2

3.16 Calculate each of the following quantities:
(a) Mass (g) of $6.44{\times}10^{-2}$ mol of $MnSO_4$
(b) Amount (mol) of compound in 15.8 kg of $Fe(ClO_4)_3$
(c) Number of N atoms in 92.6 mg of NH_4NO_2

3.17 Calculate each of the following quantities:
(a) Total number of ions in 38.1 g of SrF_2
(b) Mass (kg) of 3.58 mol of $CuCl_2{\cdot}2H_2O$
(c) Mass (mg) of $2.88{\times}10^{22}$ formula units of $Bi(NO_3)_3{\cdot}5H_2O$

3.18 Calculate each of the following quantities:
(a) Mass (g) of 8.35 mol of copper(I) carbonate
(b) Mass (g) of $4.04{\times}10^{20}$ molecules of dinitrogen pentoxide
(c) Amount (mol) and number of formula units in 78.9 g of sodium perchlorate
(d) Number of sodium ions, perchlorate ions, chlorine atoms, and oxygen atoms in the mass of compound in part (c)

3.19 Calculate each of the following quantities:
(a) Mass (g) of 8.42 mol of chromium(III) sulfate decahydrate
(b) Mass (g) of $1.83{\times}10^{24}$ molecules of dichlorine heptoxide
(c) Amount (mol) and number of formula units in 6.2 g of lithium sulfate
(d) Number of lithium ions, sulfate ions, sulfur atoms, and oxygen atoms in the mass of compound in part (c)

3.20 Calculate each of the following:
(a) Mass % of H in ammonium bicarbonate
(b) Mass % of O in sodium dihydrogen phosphate heptahydrate

3.21 Calculate each of the following:
(a) Mass % of I in strontium periodate
(b) Mass % of Mn in potassium permanganate

3.22 Calculate each of the following:
(a) Mass fraction of C in cesium acetate
(b) Mass fraction of O in uranyl sulfate trihydrate (the uranyl ion is UO_2^{2+})

3.23 Calculate each of the following:
(a) Mass fraction of Cl in calcium chlorate
(b) Mass fraction of N in dinitrogen trioxide

Problems in Context

3.24 Oxygen is required for the metabolic combustion of foods. Calculate the number of atoms in 38.0 g of oxygen gas, the amount absorbed from the lungs in about 15 min when a person is at rest.

3.25 Cisplatin *(right)*, or Platinol, is used in the treatment of certain cancers. Calculate (a) the amount (mol) of compound in 285.3 g of cisplatin; (b) the number of hydrogen atoms in 0.98 mol of cisplatin.

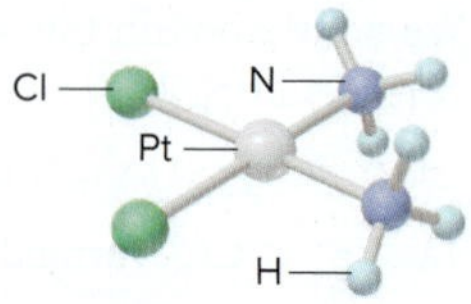

3.26 Allyl sulfide *(below)* gives garlic its characteristic odor. Calculate (a) the mass (g) of 2.63 mol of allyl sulfide; (b) the number of carbon atoms in 35.7 g of allyl sulfide.

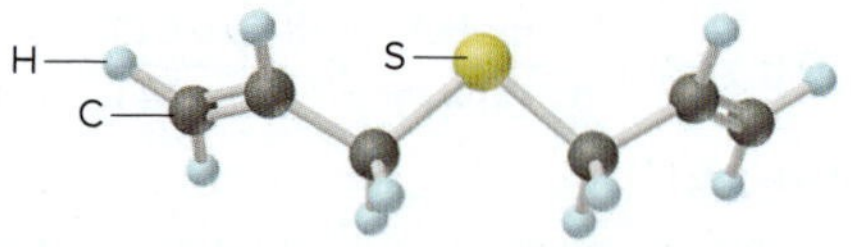

3.27 Iron reacts slowly with oxygen and water to form a compound commonly called rust ($Fe_2O_3{\cdot}4H_2O$). For 45.2 kg of rust, calculate (a) moles of compound; (b) moles of Fe_2O_3; (c) grams of Fe.

3.28 Propane is widely used in liquid form as a fuel for barbecue grills and camp stoves. For 85.5 g of propane, calculate (a) moles of compound; (b) grams of carbon.

3.29 The effectiveness of a nitrogen fertilizer is determined mainly by its mass % N. Rank the following fertilizers, most effective first: potassium nitrate; ammonium nitrate; ammonium sulfate; urea, $CO(NH_2)_2$.

3.30 The mineral galena is composed of lead(II) sulfide and has an average density of 7.46 g/cm^3. (a) How many moles of lead(II) sulfide are in 1.00 ft^3 of galena? (b) How many lead atoms are in 1.00 dm^3 of galena?

3.31 Hemoglobin, a protein in red blood cells, carries O_2 from the lungs to the body's cells. Iron (as ferrous ion, Fe^{2+}) makes up 0.33 mass % of hemoglobin. If the molar mass of hemoglobin is $6.8{\times}10^4$ g/mol, how many Fe^{2+} ions are in one molecule?

Determining the Formula of an Unknown Compound

(Sample Problems 3.8 to 3.11)

Concept Review Questions

3.32 What is the difference between an empirical formula and a molecular formula? Can they ever be the same?

3.33 List three ways compositional data may be given in a problem that involves finding an empirical formula.

3.34 Which of the following sets of information allows you to obtain the molecular formula of a covalent compound? In each case that allows it, explain how you would proceed (draw a road map and write a plan for a solution).

(a) Number of moles of each type of atom in a given sample of the compound
(b) Mass % of each element and the total number of atoms in a molecule of the compound
(c) Mass % of each element and the number of atoms of one element in a molecule of the compound
(d) Empirical formula and mass % of each element
(e) Structural formula

3.35 Is $MgCl_2$ an empirical or a molecular formula for magnesium chloride? Explain.

Skill-Building Exercises (grouped in similar pairs)

3.36 What is the empirical formula and empirical formula mass for each of the following compounds?
(a) C_2H_4 (b) $C_2H_6O_2$ (c) N_2O_5 (d) $Ba_3(PO_4)_2$ (e) Te_4I_{16}

3.37 What is the empirical formula and empirical formula mass for each of the following compounds?
(a) C_4H_8 (b) $C_3H_6O_3$ (c) P_4O_{10} (d) $Ga_2(SO_4)_3$ (e) Al_2Br_6

3.38 Give the name, empirical formula, and molar mass of the compound depicted in Figure P3.38.

3.39 Give the name, empirical formula, and molar mass of the compound depicted in Figure P3.39.

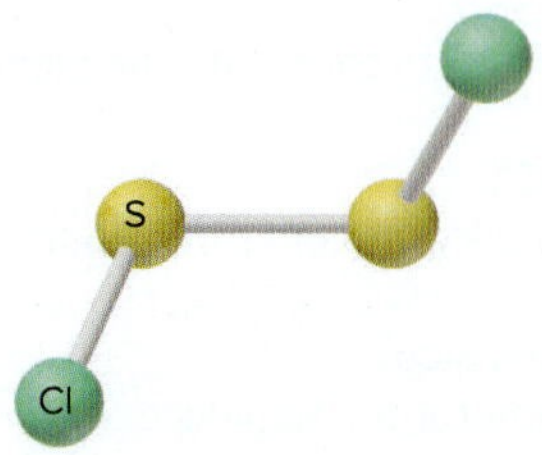

Figure P3.38

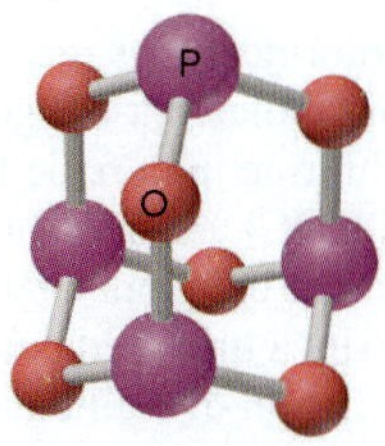

Figure P3.39

3.40 What is the molecular formula of each compound?
(a) Empirical formula CH_2 ($\mathcal{M}$ = 42.08 g/mol)
(b) Empirical formula NH_2 ($\mathcal{M}$ = 32.05 g/mol)
(c) Empirical formula NO_2 ($\mathcal{M}$ = 92.02 g/mol)
(d) Empirical formula CHN ($\mathcal{M}$ = 135.14 g/mol)

3.41 What is the molecular formula of each compound?
(a) Empirical formula CH ($\mathcal{M}$ = 78.11 g/mol)
(b) Empirical formula $C_3H_6O_2$ ($\mathcal{M}$ = 74.08 g/mol)
(c) Empirical formula HgCl ($\mathcal{M}$ = 472.1 g/mol)
(d) Empirical formula $C_7H_4O_2$ ($\mathcal{M}$ = 240.20 g/mol)

3.42 Find the empirical formula of each of the following compounds: (a) 0.063 mol of chlorine atoms combined with 0.22 mol of oxygen atoms; (b) 2.45 g of silicon combined with 12.4 g of chlorine; (c) 27.3 mass % carbon and 72.7 mass % oxygen

3.43 Find the empirical formula of each of the following compounds: (a) 0.039 mol of iron atoms combined with 0.052 mol of oxygen atoms; (b) 0.903 g of phosphorus combined with 6.99 g of bromine; (c) a hydrocarbon with 79.9 mass % carbon

3.44 An oxide of nitrogen contains 30.45 mass % N. (a) What is the empirical formula of the oxide? (b) If the molar mass is 90 ± 5 g/mol, what is the molecular formula?

3.45 A chloride of silicon contains 79.1 mass % Cl. (a) What is the empirical formula of the chloride? (b) If the molar mass is 269 g/mol, what is the molecular formula?

3.46 A sample of 0.600 mol of a metal M reacts completely with excess fluorine to form 46.8 g of MF_2.
(a) How many moles of F are in the sample of MF_2 that forms?
(b) How many grams of M are in this sample of MF_2?
(c) What element is represented by the symbol M?

3.47 A 0.370-mol sample of a metal oxide (M_2O_3) weighs 55.4 g.
(a) How many moles of O are in the sample?
(b) How many grams of M are in the sample?
(c) What element is represented by the symbol M?

3.48 Menthol ($\mathcal{M}$ = 156.3 g/mol), the strong-smelling substance in many cough drops, is a compound of carbon, hydrogen, and oxygen. When 0.1595 g of menthol was burned in a combustion apparatus, 0.449 g of CO_2 and 0.184 g of H_2O formed. What is menthol's molecular formula?

3.49 The compound dimethyl phthalate, used in insect repellants, is composed of carbon, hydrogen, and oxygen and has a molar mass of 194.2 g/mol. A 0.2572-g sample of the compound was burned in a combustion apparatus and 0.583 g of CO_2 and 0.119 g of H_2O were collected. What is the molecular formula of dimethyl phthalate?

Problems in Context

3.50 Nicotine is a poisonous, addictive compound found in tobacco. A sample of nicotine contains 6.16 mmol of C, 8.56 mmol of H, and 1.23 mmol of N [1 mmol (1 millimole) = 10^{-3} mol]. What is the empirical formula of nicotine?

3.51 Cortisol ($\mathcal{M}$ = 362.47 g/mol) is a steroid hormone involved in protein synthesis. Medically, it has a major use in reducing inflammation from rheumatoid arthritis. Cortisol is 69.6% C, 8.34% H, and 22.1% O by mass. What is its molecular formula?

3.52 Acetaminophen *(below)* is a popular nonaspirin pain reliever. What is the mass % of each element in acetaminophen?

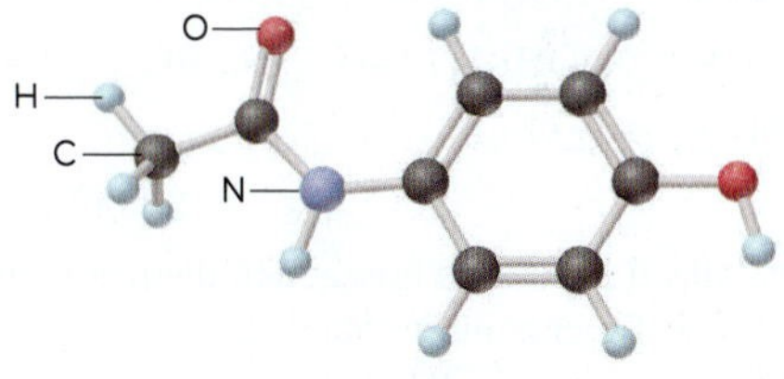

Writing and Balancing Chemical Equations

(Sample Problems 3.12 and 3.13)

Concept Review Questions

3.53 What three types of information does a balanced chemical equation provide? How?

3.54 How does a balanced chemical equation apply the law of conservation of mass?

3.55 In the process of balancing the equation

$$Al + Cl_2 \longrightarrow AlCl_3$$

Student I writes: $Al + Cl_2 \longrightarrow AlCl_2$
Student II writes: $Al + Cl_2 + Cl \longrightarrow AlCl_3$
Student III writes: $2Al + 3Cl_2 \longrightarrow 2AlCl_3$

Is the approach of Student I valid? Student II? Student III? Explain.

Skill-Building Exercises (grouped in similar pairs)

3.56 The scenes below represent a chemical reaction between elements A *(red)* and B *(green):*

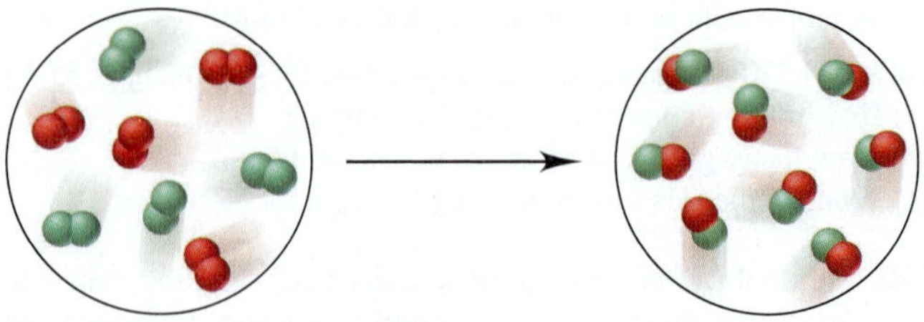

Which best represents the balanced equation for the reaction?

(a) $2A + 2B \longrightarrow A_2 + B_2$

(b) $A_2 + B_2 \longrightarrow 2AB$

(c) $B_2 + 2AB \longrightarrow 2B_2 + A_2$

(d) $4A_2 + 4B_2 \longrightarrow 8AB$

3.57 Write a balanced equation for the following gas phase reaction (carbon is black, hydrogen is light blue, and fluorine is green):

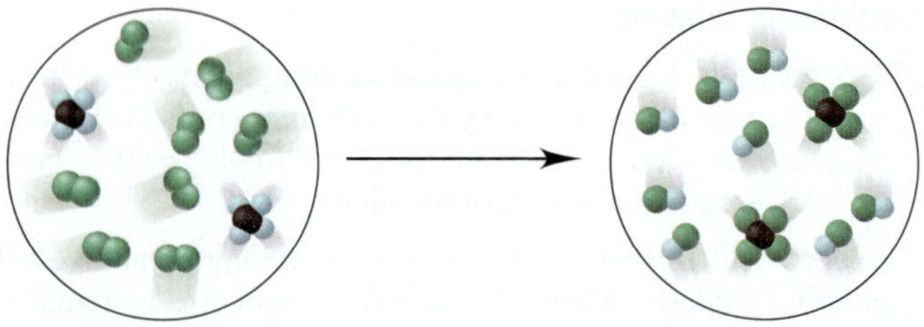

3.58 Write balanced equations for each of the following by inserting the correct coefficients in the blanks:

(a) $\_Cu(s) + \_S_8(s) \longrightarrow \_Cu_2S(s)$

(b) $\_P_4O_{10}(s) + \_H_2O(l) \longrightarrow \_H_3PO_4(l)$

(c) $\_B_2O_3(s) + \_NaOH(aq) \longrightarrow \_Na_3BO_3(aq) + \_H_2O(l)$

(d) $\_CH_3NH_2(g) + \_O_2(g) \longrightarrow \_CO_2(g) + \_H_2O(g) + \_N_2(g)$

3.59 Write balanced equations for each of the following by inserting the correct coefficients in the blanks:

(a) $\_Cu(NO_3)_2(aq) + \_KOH(aq) \longrightarrow \_Cu(OH)_2(s) + \_KNO_3(aq)$

(b) $\_BC1_3(g) + \_H_2O(l) \longrightarrow \_H_3(BO_3(s) + \_HC1(g)$

(c) $\_CaSiO_3(s) + \_HF(g) \longrightarrow \_SiF_4(g) + \_CaF_2(s) + \_H_2O(l)$

(d) $\_(CN)_2(g) + \_H_2O(l) \longrightarrow \_H_2C_2O_4(aq) + \_NH_3(g)$

3.60 Write balanced equations for each of the following by inserting the correct coefficients in the blanks:

(a) $\_SO_2(g) + \_O_2(g) \longrightarrow \_SO_3(g)$

(b) $\_Sc_2O_3(s) + \_H_2O(l) \longrightarrow \_Sc(OH)_3(s)$

(c) $\_H_3PO_4(aq) + \_NaOH(aq) \longrightarrow \_Na_2HPO_4(aq) + \_H_2O(l)$

(d) $\_C_6H_{10}O_5(s) + \_O_2(g) \longrightarrow \_CO_2(g) + \_H_2O(g)$

3.61 Write balanced equations for each of the following by inserting the correct coefficients in the blanks:

(a) $\_As_4S_6(s) + \_O_2(g) \longrightarrow \_As_4O_6(s) + \_SO_2(g)$

(b) $\_Ca_3(PO_4)_2(s) + \_SiO_2(s) + \_C(s) \longrightarrow \_P_4(g) + \_CaSiO_3(l) + \_CO(g)$

(c) $\_Fe(s) + \_H_2O(g) \longrightarrow \_Fe_3O_4(s) + \_H_2(g)$

(d) $\_S_2Cl_2(l) + \_NH_3(g) \longrightarrow \_S_4N_4(s) + \_S_8(s) + \_NH_4Cl(s)$

3.62 Convert the following into balanced equations:

(a) When gallium metal is heated in oxygen gas, it melts and forms solid gallium(III) oxide.

(b) Liquid hexane burns in oxygen gas to form carbon dioxide gas and water vapor.

(c) When solutions of calcium chloride and sodium phosphate are mixed, solid calcium phosphate forms and sodium chloride remains in solution.

3.63 Convert the following into balanced equations:

(a) When lead(II) nitrate solution is added to potassium iodide solution, solid lead(II) iodide forms and potassium nitrate solution remains.

(b) Liquid disilicon hexachloride reacts with water to form solid silicon dioxide, hydrogen chloride gas, and hydrogen gas.

(c) When nitrogen dioxide is bubbled into water, a solution of nitric acid forms and gaseous nitrogen monoxide is released.

Problem in Context

3.64 Loss of atmospheric ozone has led to an ozone "hole" over Antarctica. The loss occurs in part through three consecutive steps:

(1) Chlorine atoms react with ozone (O_3) to form chlorine monoxide and molecular oxygen.

(2) Chlorine monoxide forms ClOOCl.

(3) ClOOCl absorbs sunlight and breaks into chlorine atoms and molecular oxygen.

(a) Write a balanced equation for each step.

(b) Write an overall balanced equation for the sequence.

Calculating Quantities of Reactant and Product

(Sample Problems 3.14 to 3.21)

Concept Review Questions

3.65 What does the term *stoichiometrically equivalent molar ratio* mean, and how is it applied in solving problems?

3.66 Percent yields are generally calculated from masses. Would the result be the same if amounts (mol) were used instead? Why?

Skill-Building Exercises (grouped in similar pairs)

3.67 Reactants A and B form product C. Draw a road map and write a Plan to find the mass (g) of C when 25 g of A reacts with excess B.

3.68 Reactants D and E form product F. Draw a road map and write a Plan to find the mass (g) of F when 27 g of D reacts with 31 g of E.

3.69 Chlorine gas can be made in the laboratory by the reaction of hydrochloric acid and manganese(IV) oxide:

$$4HCl(aq) + MnO_2(s) \longrightarrow MnCl_2(aq) + 2H_2O(g) + Cl_2(g)$$

When 1.82 mol of HCl reacts with excess MnO_2, how many (a) moles of Cl_2 and (b) grams of Cl_2 form?

3.70 Bismuth oxide reacts with carbon to form bismuth metal:

$$Bi_2O_3(s) + 3C(s) \longrightarrow 2Bi(s) + 3CO(g)$$

When 0.607 mol of Bi_2O_3 reacts with excess carbon, how many (a) moles of Bi and (b) grams of CO form?

3.71 Potassium nitrate decomposes on heating, producing potassium oxide and gaseous nitrogen and oxygen:

$$4KNO_3(s) \longrightarrow 2K_2O(s) + 2N_2(g) + 5O_2(g)$$

To produce 56.6 kg of oxygen, how many (a) moles of KNO_3 and (b) grams of KNO_3 must be heated?

3.72 Chromium(III) oxide reacts with hydrogen sulfide (H_2S) gas to form chromium(III) sulfide and water:

$$Cr_2O_3(s) + 3H_2S(g) \longrightarrow Cr_2S_3(s) + 3H_2O(l)$$

To produce 421 g of Cr_2S_3, how many (a) moles of Cr_2O_3 and (b) grams of Cr_2O_3 are required?

3.73 Calculate the mass (g) of each product formed when 43.82 g of diborane (B_2H_6) reacts with excess water:

$$B_2H_6(g) + H_2O(l) \longrightarrow H_3BO_3(s) + H_2(g) \quad \text{[unbalanced]}$$

3.74 Calculate the mass (g) of each product formed when 174 g of silver sulfide reacts with excess hydrochloric acid:

$$Ag_2S(s) + HCl(aq) \longrightarrow AgCl(s) + H_2S(g) \text{ [unbalanced]}$$

3.75 Elemental phosphorus occurs as tetratomic molecules, P_4. What mass (g) of chlorine gas is needed to react completely with 455 g of phosphorus to form phosphorus pentachloride?

3.76 Elemental sulfur occurs as octatomic molecules, S_8. What mass (g) of fluorine gas is needed to react completely with 17.8 g of sulfur to form sulfur hexafluoride?

3.77 Solid iodine trichloride is prepared in two steps: first, a reaction between solid iodine and gaseous chlorine to form solid iodine monochloride; second, treatment of the solid with more chlorine gas.
(a) Write a balanced equation for each step.
(b) Write a balanced equation for the overall reaction.
(c) How many grams of iodine are needed to prepare 2.45 kg of final product?

3.78 Lead can be prepared from galena [lead(II) sulfide] by first roasting the galena in oxygen gas to form lead(II) oxide and sulfur dioxide. Heating the metal oxide with more galena forms the molten metal and more sulfur dioxide.
(a) Write a balanced equation for each step.
(b) Write an overall balanced equation for the process.
(c) How many metric tons of sulfur dioxide form for every metric ton of lead obtained?

3.79 The scene below represents a mixture of A_2 (blue) and B_2 (green) before they react to form AB_3.

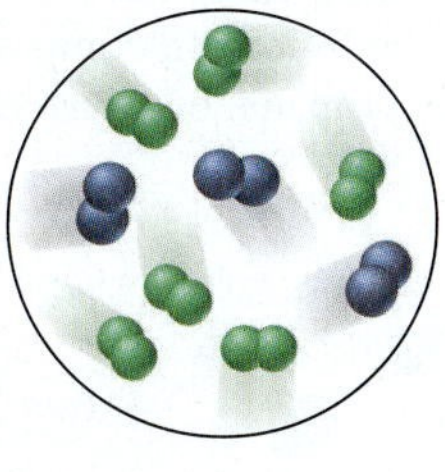

(a) What is the limiting reactant?
(b) How many molecules of product can form?

3.80 The following scene shows a mixture of nitrogen (dark blue) and hydrogen (light blue) before they react to produce ammonia, NH_3. What is the limiting reactant? Draw a scene that shows the number of reactant and product molecules present after reaction.

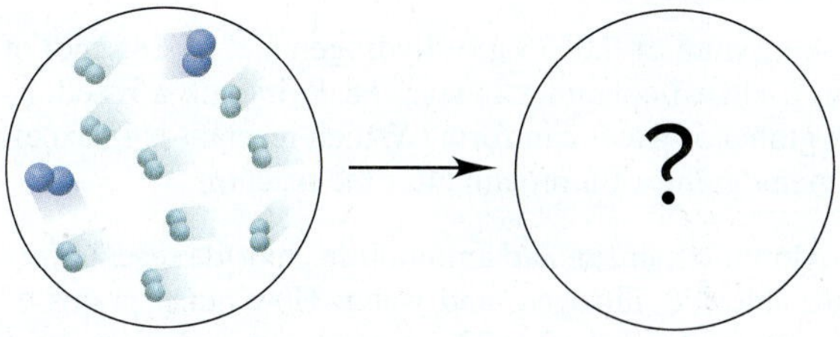

3.81 Methanol, CH_3OH, is produced from the reaction of carbon monoxide and hydrogen:

$$CO(g) + 2H_2(g) \longrightarrow CH_3OH(l)$$

How many moles of methanol can be produced when 4.5 mol of CO reacts with 7.2 mol of H_2?

3.82 In the following reaction, 2.35 mol of NH_3 reacts with 2.75 mol of O_2; how many moles of water form?

$$4NH_3(g) + 5O_2(g) \longrightarrow 4NO(g) + 6H_2O(l)$$

3.83 Many metals react with oxygen gas to form the metal oxide. For example, calcium reacts as follows:

$$2Ca(s) + O_2(g) \longrightarrow 2CaO(s)$$

You wish to calculate the mass (g) of calcium oxide that can be prepared from 4.20 g of Ca and 2.80 g of O_2.
(a) What amount (mol) of CaO can be produced from the given mass of Ca?
(b) What amount (mol) of CaO can be produced from the given mass of O_2?
(c) Which is the limiting reactant?
(d) How many grams of CaO can be produced?

3.84 Metal hydrides react with water to form hydrogen gas and the metal hydroxide. For example,

$$SrH_2(s) + 2H_2O(l) \longrightarrow Sr(OH)_2(s) + 2H_2(g)$$

You wish to calculate the mass (g) of hydrogen gas that can be prepared from 5.70 g of SrH_2 and 4.75 g of H_2O.
(a) What amount (mol) of H_2 can be produced from the given mass of SrH_2?
(b) What amount (mol) of H_2 can be produced from the given mass of H_2O?
(c) Which is the limiting reactant?
(d) How many grams of H_2 can be produced?

3.85 Calculate the maximum numbers of moles and grams of iodic acid (HIO_3) that can form when 635 g of iodine trichloride reacts with 118.5 g of water:

$$ICl_3 + H_2O \longrightarrow ICl + HIO_3 + HCl \text{ [unbalanced]}$$

How many grams of the excess reactant remain?

3.86 Calculate the maximum numbers of moles and grams of H_2S that can form when 158 g of aluminum sulfide reacts with 131 g of water:

$$Al_2S_3 + H_2O \longrightarrow Al(OH)_3 + H_2S \text{ [unbalanced]}$$

How many grams of the excess reactant remain?

3.87 When 0.100 mol of carbon is burned in a closed vessel with 8.00 g of oxygen, how many grams of carbon dioxide can form? Which reactant is in excess, and how many grams of it remain after the reaction?

3.88 A mixture of 0.0375 g of hydrogen and 0.0185 mol of oxygen in a closed container is sparked to initiate a reaction. How many grams of water can form? Which reactant is in excess, and how many grams of it remain after the reaction?

3.89 Aluminum nitrite and ammonium chloride react to form aluminum chloride, nitrogen, and water. How many grams of each substance are present after 72.5 g of aluminum nitrite and 58.6 g of ammonium chloride react completely?

3.90 Calcium nitrate and ammonium fluoride react to form calcium fluoride, dinitrogen monoxide, and water vapor. How many grams of each substance are present after 16.8 g of calcium nitrate and 17.50 g of ammonium fluoride react completely?

3.91 Two successive reactions, $A \longrightarrow B$ and $B \longrightarrow C$, have yields of 73% and 68%, respectively. What is the overall percent yield for conversion of A to C?

3.92 Two successive reactions, $D \longrightarrow E$ and $E \longrightarrow F$, have yields of 48% and 73%, respectively. What is the overall percent yield for conversion of D to F?

3.93 What is the percent yield of a reaction in which 45.5 g of tungsten(VI) oxide (WO_3) reacts with excess hydrogen gas to produce metallic tungsten and 9.60 mL of water ($d = 1.00$ g/mL)?

3.94 What is the percent yield of a reaction in which 200. g of phosphorus trichloride reacts with excess water to form 128 g of HCl and aqueous phosphorous acid (H_3PO_3)?

3.95 When 20.5 g of methane and 45.0 g of chlorine gas undergo a reaction that has a 75.0% yield, what mass (g) of chloromethane (CH_3Cl) forms? Hydrogen chloride also forms.

3.96 When 56.6 g of calcium and 30.5 g of nitrogen gas undergo a reaction that has a 93.0% yield, what mass (g) of calcium nitride forms?

Problems in Context

3.97 Cyanogen, $(CN)_2$, has been observed in the atmosphere of Titan, Saturn's largest moon, and in the gases of interstellar nebulas. On Earth, it is used as a welding gas and a fumigant. In its reaction with fluorine gas, carbon tetrafluoride and nitrogen trifluoride gases are produced. What mass (g) of carbon tetrafluoride forms when 60.0 g of each reactant is used?

3.98 Gaseous dichlorine monoxide decomposes readily to chlorine *(green)* and oxygen *(red)* gases.

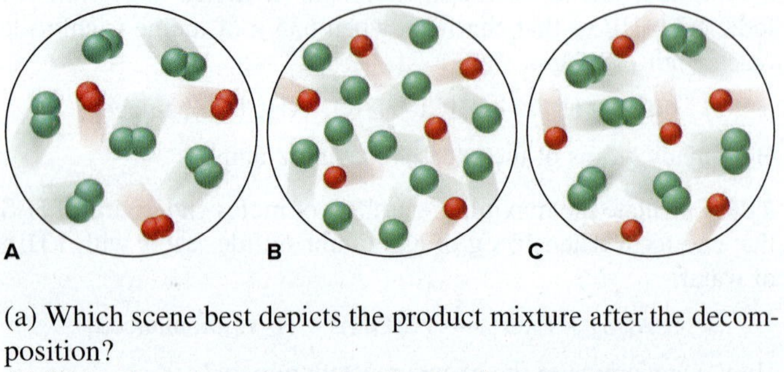

A B C

(a) Which scene best depicts the product mixture after the decomposition?
(b) Write the balanced equation for the decomposition.
(c) If each oxygen atom represents 0.050 mol, how many molecules of dichlorine monoxide were present before the decomposition?

3.99 An intermediate step in the production of nitric acid involves the reaction of ammonia with oxygen gas to form nitrogen monoxide and water. How many grams of nitrogen monoxide can form in the reaction of 485 g of ammonia with 792 g of oxygen?

3.100 Butane gas is compressed and used as a liquid fuel in disposable cigarette lighters and lightweight camping stoves. Suppose a lighter contains 5.50 mL of butane ($d = 0.579$ g/mL).
(a) How many grams of oxygen are needed to burn the butane completely?
(b) How many moles of H_2O form when all the butane burns?
(c) How many total molecules of gas form when the butane burns completely?

3.101 Sodium borohydride ($NaBH_4$) is used industrially in many organic syntheses. One way to prepare it is by reacting sodium hydride with gaseous diborane (B_2H_6). Assuming an 88.5% yield, how many grams of $NaBH_4$ can be prepared by reacting 7.98 g of sodium hydride and 8.16 g of diborane?

Comprehensive Problems

3.102 The mole is defined in terms of the carbon-12 atom. Use the definition to find (a) the mass in grams equal to 1 atomic mass unit; (b) the ratio of the gram to the atomic mass unit.

3.103 The first sulfur-nitrogen compound was prepared in 1835 and has been used to synthesize many others. In the early 1980s, researchers made another such compound that conducts electricity like a metal. Mass spectrometry of the compound shows a molar mass of 184.27 g/mol, and analysis shows it to contain 2.288 g of S for every 1.000 g of N. What is its molecular formula?

3.104 Hydroxyapatite, $Ca_5(PO_4)_3(OH)$, is the main mineral component of dental enamel, dentin, and bone. Coating the compound on metallic implants (such as titanium alloys and stainless steels) helps the body accept the implant. When placed in bone voids, the powder encourages natural bone to grow into the void. Hydroxyapatite is prepared by adding aqueous phosphoric acid to a dilute slurry of calcium hydroxide. (a) Write a balanced equation for this preparation. (b) What mass (g) of hydroxyapatite could form from 100. g of 85% phosphoric acid and 100. g of calcium hydroxide?

3.105 Narceine is a narcotic in opium that crystallizes from solution as a hydrate that contains 10.8 mass % water and has a molar mass of 499.52 g/mol. Determine x in narceine·xH_2O.

3.106 Hydrogen-containing fuels have a "fuel value" based on their mass % H. Rank the following compounds from highest fuel value to lowest: ethane, propane, benzene, ethanol, cetyl palmitate (whale oil, $C_{32}H_{64}O_2$).

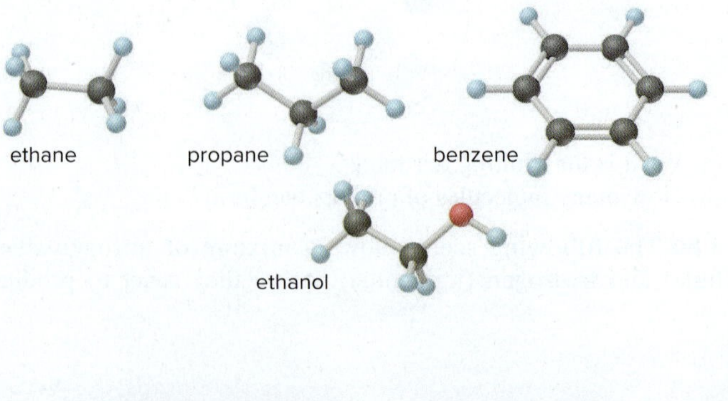

3.107 Serotonin ($\mathscr{M}$ = 176 g/mol) transmits nerve impulses between neurons. It contains 68.2% C, 6.86% H, 15.9% N, and 9.08% O by mass. What is its molecular formula?

3.108 In 1961, scientists agreed that the atomic mass unit (amu) would be defined as $\frac{1}{12}$ the mass of an atom of ^{12}C. Before then, it was defined as $\frac{1}{16}$ the *average* mass of an atom of naturally occurring oxygen (a mixture of ^{16}O, ^{17}O, and ^{18}O). The current atomic mass of oxygen is 15.9994 amu. (a) Did Avogadro's number change after the definition of an amu changed and, if so, in what direction? (b) Did the definition of the mole change? (c) Did the mass of a mole of a substance change? (d) Before 1961, was Avogadro's number 6.02×10^{23} (to three significant figures), as it is today?

3.109 Convert the following descriptions into balanced equations:
(a) In a gaseous reaction, hydrogen sulfide burns in oxygen to form sulfur dioxide and water vapor.
(b) When crystalline potassium chlorate is heated to just above its melting point, it reacts to form two different crystalline compounds, potassium chloride and potassium perchlorate.
(c) When hydrogen gas is passed over powdered iron(III) oxide, iron metal and water vapor form.
(d) The combustion of gaseous ethane in air forms carbon dioxide and water vapor.
(e) Iron(II) chloride is converted to iron(III) fluoride by treatment with chlorine trifluoride gas. Chlorine gas is also formed.

3.110 Isobutylene is a hydrocarbon used in the manufacture of synthetic rubber. When 0.847 g of isobutylene was subjected to combustion analysis, the gain in mass of the CO_2 absorber was 2.657 g and that of the H_2O absorber was 1.089 g. What is the empirical formula of isobutylene?

3.111 The multistep smelting of ferric oxide to form elemental iron occurs at high temperatures in a blast furnace. In the first step, ferric oxide reacts with carbon monoxide to form Fe_3O_4. This substance reacts with more carbon monoxide to form iron(II) oxide, which reacts with still more carbon monoxide to form molten iron. Carbon dioxide is also produced in each step. (a) Write an overall balanced equation for the iron-smelting process. (b) How many grams of carbon monoxide are required to form 45.0 metric tons of iron from ferric oxide?

3.112 One of the compounds used to increase the octane rating of gasoline is toluene (ball-and-stick model is shown).

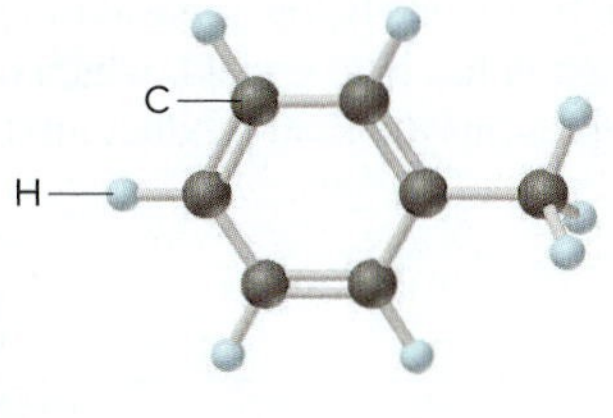

Suppose 20.0 mL of toluene (d = 0.867 g/mL) is consumed when a sample of gasoline burns in air.
(a) How many grams of oxygen are needed for complete combustion of the toluene? (b) How many total moles of gaseous products form? (c) How many molecules of water vapor form?

3.113 During studies of the reaction in Sample Problem 3.20,

$$2N_2H_4(l) + N_2O_4(l) \longrightarrow 3N_2(g) + 4H_2O(g)$$

a chemical engineer measured a less-than-expected yield of N_2 and discovered that the following side reaction occurs:

$$N_2H_4(l) + 2N_2O_4(l) \longrightarrow 6NO(g) + 2H_2O(g)$$

In one experiment, 10.0 g of NO formed when 100.0 g of each reactant was used. What is the highest percent yield of N_2 that can be expected?

3.114 A 0.652-g sample of a pure strontium halide reacts with excess sulfuric acid. The solid strontium sulfate formed is separated, dried, and found to weigh 0.755 g. What is the formula of the original halide?

3.115 The following scenes represent a chemical reaction between AB_2 and B_2:

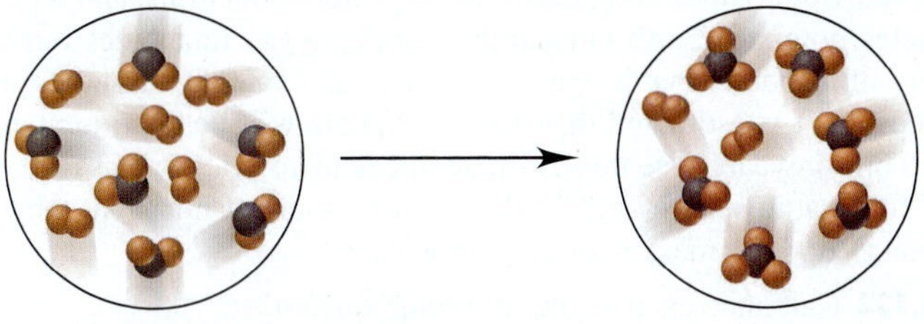

(a) Write a balanced equation for the reaction. (b) What is the limiting reactant? (c) How many moles of product can be made from 3.0 mol of B_2 and 5.0 mol of AB_2? (d) How many moles of excess reactant remain after the reaction in part (c)?

3.116 Which of the following models represent compounds having the same empirical formula? What is the empirical formula mass of this common formula?

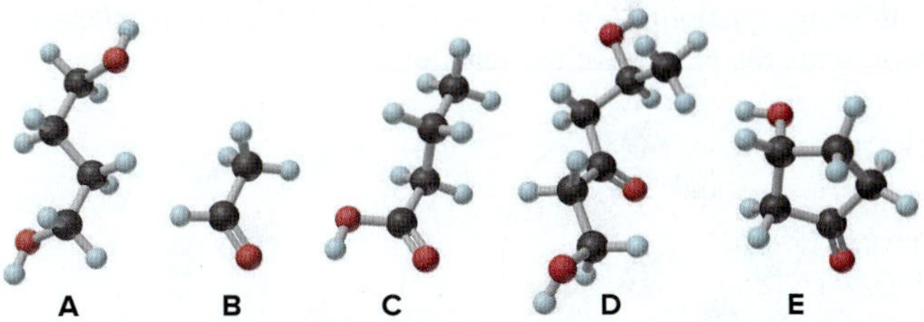

A B C D E

3.117 The zirconium oxalate $K_2Zr(C_2O_4)_3(H_2C_2O_4){\cdot}H_2O$ was synthesized by mixing 1.68 g of $ZrOCl_2{\cdot}8H_2O$ with 5.20 g of $H_2C_2O_4{\cdot}2H_2O$ and an excess of aqueous KOH. After 2 months, 1.25 g of crystalline product was obtained, along with aqueous KCl and water. Calculate the percent yield.

3.118 Seawater is approximately 4.0% by mass dissolved ions, 85% of which are from NaCl. (a) Find the mass % of NaCl in seawater. (b) Find the mass % of Na^+ ions and of Cl^- ions in seawater.

3.119 Is each of the following statements true or false? Correct any that are false.
(a) A mole of one substance has the same number of atoms as a mole of any other substance.
(b) The theoretical yield for a reaction is based on the balanced chemical equation.
(c) A limiting-reactant problem is being stated when the available quantity of one of the reactants is given in moles.
(d) The empirical and molecular formulas of a compound are always different.

3.120 In each pair, choose the larger of the indicated quantities or state that the samples are equal:
(a) Entities: 0.4 mol of O_3 molecules or 0.4 mol of O atoms
(b) Grams: 0.4 mol of O_3 molecules or 0.4 mol of O atoms
(c) Moles: 4.0 g of N_2O_4 or 3.3 g of SO_2
(d) Grams: 0.6 mol of C_2H_4 or 0.6 mol of F_2
(e) Total ions: 2.3 mol of sodium chlorate or 2.2 mol of magnesium chloride
(f) Molecules: 1.0 g of H_2O or 1.0 g of H_2O_2
(g) Grams: 6.02×10^{23} atoms of ^{235}U or 6.02×10^{23} atoms of ^{238}U

3.121 For the reaction between solid tetraphosphorus trisulfide and oxygen gas to form solid tetraphosphorus decoxide and sulfur dioxide gas, write a balanced equation. Show the equation (see Table 3.4) in terms of (a) molecules, (b) moles, and (c) grams.

3.122 Hydrogen gas is considered a clean fuel because it produces only water vapor when it burns. If the reaction has a 98.8% yield, what mass (g) of hydrogen forms 105 kg of water?

3.123 Solar winds composed of free protons, electrons, and α particles bombard Earth constantly, knocking gas molecules out of the atmosphere. In this way, Earth loses about 3.0 kg of matter per second. It is estimated that the atmosphere will be gone in about 50 billion years. Use this estimate to calculate (a) the mass (kg) of Earth's atmosphere and (b) the amount (mol) of nitrogen, which makes up 75.5 mass % of the atmosphere.

3.124 Calculate each of the following quantities:
(a) Amount (mol) of 0.588 g of ammonium bromide
(b) Number of potassium ions in 88.5 g of potassium nitrate
(c) Mass (g) of 5.85 mol of glycerol ($C_3H_8O_3$)
(d) Volume (L) of 2.85 mol of chloroform ($CHCl_3$; $d = 1.48$ g/mL)
(e) Number of sodium ions in 2.11 mol of sodium carbonate
(f) Number of atoms in 25.0 μg of cadmium
(g) Number of atoms in 0.0015 mol of fluorine gas

3.125 Elements X *(green)* and Y *(purple)* react according to the following equation: $X_2 + 3Y_2 \longrightarrow 2XY_3$. Which molecular scene represents the product of the reaction?

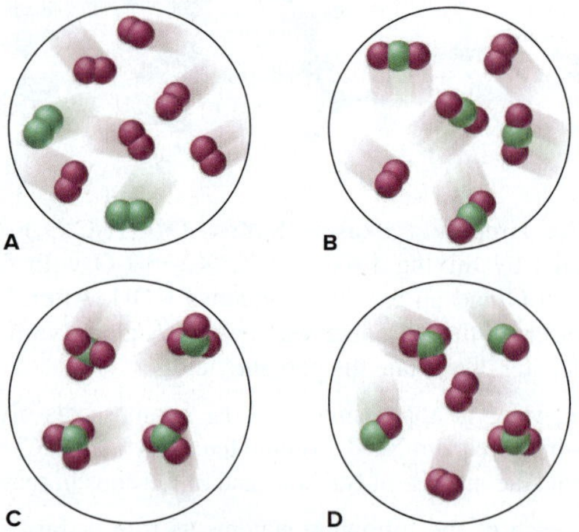

3.126 Hydrocarbon mixtures are used as fuels. (a) How many grams of $CO_2(g)$ are produced by the combustion of 200. g of a mixture that is 25.0% CH_4 and 75.0% C_3H_8 by mass? (b) A 252-g gaseous mixture of CH_4 and C_3H_8 burns in excess O_2, and 748 g of CO_2 gas is collected. What is the mass % of CH_4 in the mixture?

3.127 Nitrogen (N), phosphorus (P), and potassium (K) are the main nutrients in plant fertilizers. By industry convention, the numbers on a label refer to the mass percents of N, P_2O_5, and K_2O, in that order. Calculate the N/P/K ratio of a 30/10/10 fertilizer in terms of moles of each element, and express it as *x*/*y*/1.0.

3.128 What mass percents of ammonium sulfate, ammonium hydrogen phosphate, and potassium chloride would you use to prepare 10/10/10 plant fertilizer (see Problem 3.127)?

3.129 Ferrocene, synthesized in 1951, was the first organic iron compound with Fe—C bonds. An understanding of the structure of ferrocene gave rise to new ideas about chemical bonding and led to the preparation of many useful compounds. In the combustion analysis of ferrocene, which contains only Fe, C, and H, a 0.9437-g sample produced 2.233 g of CO_2 and 0.457 g of H_2O. What is the empirical formula of ferrocene?

3.130 When carbon-containing compounds are burned in a limited amount of air, some $CO(g)$ as well as $CO_2(g)$ is produced. A gaseous product mixture is 35.0 mass % CO and 65.0 mass % CO_2. What is the mass % of C in the mixture?

3.131 Write a balanced equation for the reaction depicted below:

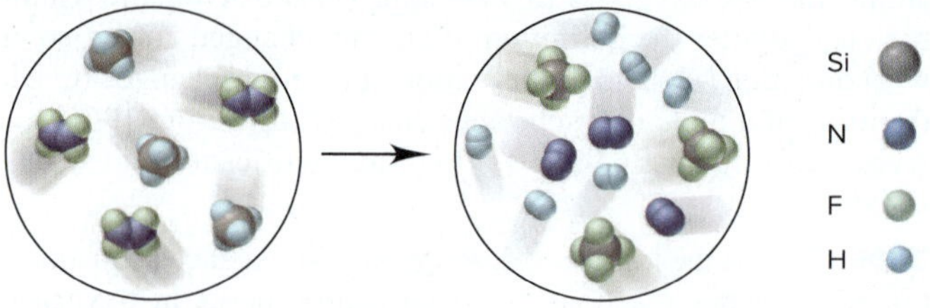

If each reactant molecule represents 1.25×10^{-2} mol and the reaction yield is 87%, how many grams of Si-containing product form?

3.132 Citric acid *(below)* is concentrated in citrus fruits and plays a central metabolic role in nearly every animal and plant cell. (a) What are the molar mass and formula of citric acid? (b) How many moles of citric acid are in 1.50 qt of lemon juice ($d = 1.09$ g/mL) that is 6.82% citric acid by mass?

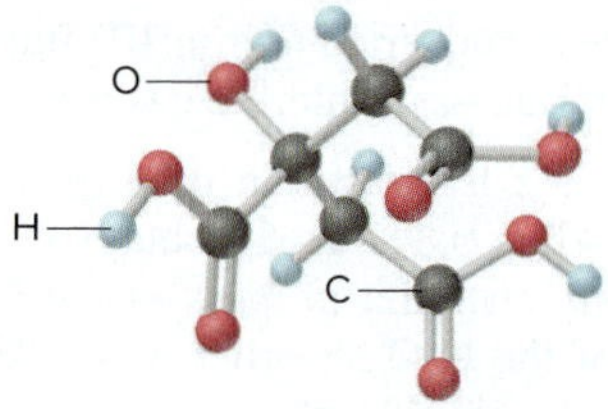

3.133 Various nitrogen oxides, as well as sulfur oxides, contribute to acidic rainfall through complex reaction sequences. Nitrogen and oxygen combine during the high-temperature combustion of fuels in air to form nitrogen monoxide gas, which reacts with more oxygen to form nitrogen dioxide gas. In contact with water vapor, nitrogen dioxide forms aqueous nitric acid and more nitrogen monoxide. (a) Write balanced equations for these reactions. (b) Use the equations to write one overall balanced equation that does *not* include nitrogen monoxide and nitrogen dioxide. (c) How many metric tons (t) of nitric acid form when 1350 t of atmospheric nitrogen is consumed (1 t = 1000 kg)?

3.134 Nitrogen monoxide reacts with elemental oxygen to form nitrogen dioxide. The first scene below represents an initial mixture of reactants. If the reaction has a 66% yield, which of the lower scenes (A, B, or C) best represents the final product mixture?

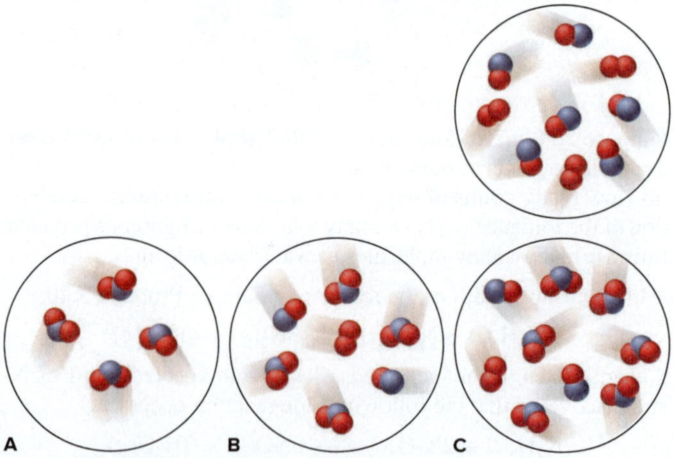

3.135 Fluorine is so reactive that it forms compounds with several of the noble gases.
(a) When 0.327 g of platinum is heated in fluorine, 0.519 g of a dark red, volatile solid forms. What is its empirical formula?
(b) When 0.265 g of this red solid reacts with excess xenon gas, 0.378 g of an orange-yellow solid forms. What is the empirical formula of this compound, the first to contain a noble gas?
(c) Fluorides of xenon can be formed by direct reaction of the elements at high pressure and temperature. Under conditions that produce only the tetra- and hexafluorides, 1.85×10^{-4} mol of xenon reacted with 5.00×10^{-4} mol of fluorine, and 9.00×10^{-6} mol of xenon was found in excess. What are the mass percents of each xenon fluoride in the product mixture?

3.136 Hemoglobin is 6.0% heme ($C_{34}H_{32}FeN_4O_4$) by mass. To remove the heme, hemoglobin is treated with acetic acid and NaCl, which forms hemin ($C_{34}H_{32}N_4O_4FeCl$). A blood sample from a crime scene contains 0.65 g of hemoglobin. (a) How many grams of heme are in the sample? (b) How many moles of heme? (c) How many grams of Fe? (d) How many grams of hemin could be formed for a forensic chemist to measure?

3.137 Manganese is a key component of extremely hard steel. The element occurs naturally in many oxides. A 542.3-g sample of a manganese oxide has an Mn/O ratio of 1.00/1.42 and consists of braunite (Mn_2O_3) and manganosite (MnO). (a) How many grams of braunite and of manganosite are in the ore? (b) What is the Mn^{3+}/Mn^{2+} ratio in the ore?

3.138 The human body excretes nitrogen in the form of urea, NH_2CONH_2. The key step in its biochemical formation is the reaction of water with arginine to produce urea and ornithine:

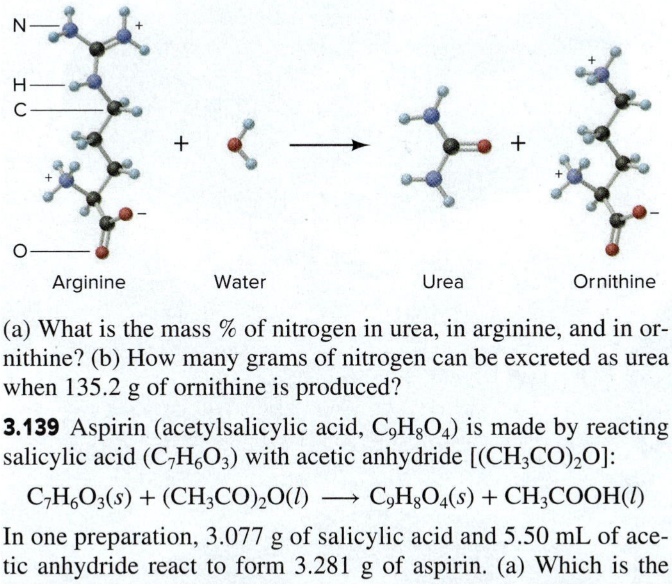

(a) What is the mass % of nitrogen in urea, in arginine, and in ornithine? (b) How many grams of nitrogen can be excreted as urea when 135.2 g of ornithine is produced?

3.139 Aspirin (acetylsalicylic acid, $C_9H_8O_4$) is made by reacting salicylic acid ($C_7H_6O_3$) with acetic anhydride [$(CH_3CO)_2O$]:

$$C_7H_6O_3(s) + (CH_3CO)_2O(l) \longrightarrow C_9H_8O_4(s) + CH_3COOH(l)$$

In one preparation, 3.077 g of salicylic acid and 5.50 mL of acetic anhydride react to form 3.281 g of aspirin. (a) Which is the limiting reactant (the density of acetic anhydride is 1.080 g/mL)? (b) What is the percent yield of this reaction? (c) What is the percent atom economy of this reaction?

3.140 The rocket fuel hydrazine (N_2H_4) is made by the three-step Raschig process, which has the following overall equation:

$$NaOCl(aq) + 2NH_3(aq) \longrightarrow N_2H_4(aq) + NaCl(aq) + H_2O(l)$$

What is the percent atom economy of this process?

3.141 Lead(II) chromate ($PbCrO_4$) is used as the yellow pigment for marking traffic lanes but is banned from house paint because of the risk of lead poisoning. It is produced from chromite ($FeCr_2O_4$), an ore of chromium:

$$4FeCr_2O_4(s) + 8K_2CO_3(aq) + 7O_2(g) \longrightarrow 2Fe_2O_3(s) + 8K_2CrO_4(aq) + 8CO_2(g)$$

Lead(II) ion then replaces the K^+ ion. If a yellow paint is to have 0.511% $PbCrO_4$ by mass, how many grams of chromite are needed per kilogram of paint?

3.142 Ethanol (CH_3CH_2OH), the intoxicant in alcoholic beverages, is also used to make other organic compounds. In concentrated sulfuric acid, ethanol forms diethyl ether and water:

$$2CH_3CH_2OH(l) \longrightarrow CH_3CH_2OCH_2CH_3(l) + H_2O(g)$$

In a side reaction, some ethanol forms ethylene and water:

$$CH_3CH_2OH(l) \longrightarrow CH_2CH_2(g) + H_2O(g)$$

(a) If 50.0 g of ethanol yields 35.9 g of diethyl ether, what is the percent yield of diethyl ether? (b) If 45.0% of the ethanol that did not produce the ether reacts by the side reaction, what mass (g) of ethylene is produced?

3.143 When powdered zinc is heated with sulfur, a violent reaction occurs, and zinc sulfide forms:

$$Zn(s) + S_8(s) \longrightarrow ZnS(s) \text{ [unbalanced]}$$

Some of the reactants also combine with oxygen in air to form zinc oxide and sulfur dioxide. When 83.2 g of Zn reacts with 52.4 g of S_8, 104.4 g of ZnS forms.
(a) What is the percent yield of ZnS?
(b) If all the remaining reactants combine with oxygen, how many grams of each of the two oxides form?

3.144 Cocaine ($C_{17}H_{21}O_4N$) is a natural substance found in coca leaves, which have been used for centuries as a local anesthetic and stimulant. Illegal cocaine arrives in the United States either as the pure compound or as the hydrochloride salt ($C_{17}H_{21}O_4NHCl$). At 25°C, the salt is very soluble in water (2.50 kg/L), but cocaine is much less so (1.70 g/L).
(a) What is the maximum mass (g) of the hydrochloride salt that can dissolve in 50.0 mL of water?
(b) If the solution from part (a) is treated with NaOH, the salt is converted to cocaine. How much more water (L) is needed to dissolve it?

3.145 High-temperature superconducting oxides hold great promise in the utility, transportation, and computer industries.
(a) One superconductor is $La_{2-x}Sr_xCuO_4$. Calculate the molar masses of this oxide when $x = 0$, $x = 1$, and $x = 0.163$.
(b) Another common superconducting oxide is made by heating a mixture of barium carbonate, copper(II) oxide, and yttrium(III) oxide, followed by further heating in O_2:

$$4BaCO_3(s) + 6CuO(s) + Y_2O_3(s) \longrightarrow 2YBa_2Cu_3O_{6.5}(s) + 4CO_2(g)$$

$$2YBa_2Cu_3O_{6.5}(s) + \tfrac{1}{2}O_2(g) \longrightarrow 2YBa_2Cu_3O_7(s)$$

When equal masses of the three reactants are heated, which reactant is limiting?
(c) After the product in part (b) is removed, what is the mass % of each reactant in the remaining solid mixture?

4 Three Major Classes of Chemical Reactions

4.1 Solution Concentration and the Role of Water as a Solvent
Polar Nature of Water
Ionic Compounds in Water
Covalent Compounds in Water
Expressing Concentration in Terms of Molarity
Amount-Mass-Number Conversions Involving Solutions
Preparing and Diluting Molar Solutions

4.2 Writing Equations for Aqueous Ionic Reactions

4.3 Precipitation Reactions
The Key Event: Formation of a Solid
Predicting Whether a Precipitate Will Form
Stoichiometry of Precipitation Reactions

4.4 Acid-Base Reactions
The Key Event: Formation of Water
Proton Transfer in Acid-Base Reactions
Stoichiometry of Acid-Base Reactions: Acid-Base Titrations

4.5 Oxidation-Reduction (Redox) Reactions
The Key Event: Movement of Electrons
Redox Terminology
Oxidation Numbers
Stoichiometry of Redox Reactions: Redox Titrations

4.6 Elements in Redox Reactions
Combination Reactions
Decomposition Reactions
Displacement Reactions and Activity Series
Combustion Reactions

4.7 The Reversibility of Reactions and the Equilibrium State

Source: © Nicolas Rung/age fotostock

Concepts and Skills to Review Before You Study This Chapter

- names and formulas of compounds (Section 2.8)
- nature of ionic and covalent bonding (Section 2.7)
- amount-mass-number conversions (Section 3.1)
- balancing chemical equations (Section 3.3)
- calculating quantities of reactants and products (Section 3.4)

Water is a precious commodity; the United States uses about 408 billion gallons of water every day, and you probably use about 80–100 of those gallons! Water treatment plants use chemical reactions to provide safe, uncontaminated drinking water and to remove dissolved heavy metals and phosphates from polluted groundwater and industrial wastewater. These reactions convert dissolved substances into solid particles, which can then be removed from the water by filtration. The reactions involved in water purification are just a few examples of the myriad of important reactions that take place in aqueous solution. In nature, aqueous reactions occur unceasingly in the gigantic containers we know as oceans. And, in every cell of your body, thousands of reactions taking place right now enable you to function. With millions of reactions occurring in and around you, it would be impossible to describe them all. Fortunately, it isn't necessary because when we survey even a small percentage of reactions, especially those in aqueous solution, a few major classes emerge.

IN THIS CHAPTER . . . *We examine the underlying nature of three classes of reactions and the complex part that water plays in aqueous chemistry.*

- We visualize the molecular structure of water to learn its role as solvent, how it interacts with ionic and covalent compounds, and the nature of electrolytes.
- We discuss molarity as a way to express the concentration of an ionic or covalent compound in solution.
- We write ionic equations that represent the species in solution and the actual change that occurs when ionic compounds dissolve and react in water.
- We describe precipitation reactions and learn how to predict when they will occur.
- We see how acids and bases act as electrolytes and how they neutralize each other. We view acid-base reactions as proton-transfer processes and discuss how to quantify these reactions through titration.
- We examine oxidation-reduction (redox) reactions, which can form ionic or covalent substances. We learn how to determine oxidation numbers and use them to determine oxidizing and reducing agents and how to quantify these reactions through titration. We describe several important types of redox reactions, one of which is used to predict the relative reactivity of metals.
- An introduction to the reversible nature of all chemical change and the nature of equilibrium previews coverage of this central topic in later chapters.

4.1 SOLUTION CONCENTRATION AND THE ROLE OF WATER AS A SOLVENT

In Chapter 2, we defined a solution as a homogeneous mixture with a uniform composition. A solution consists of a smaller quantity of one substance, the **solute,** dissolved in a larger quantity of another, the **solvent;** in an aqueous solution, water serves as the solvent. For any reaction in solution, the solvent plays a key role that depends on its chemical nature. Some solvents passively disperse the substances into individual molecules. But water is much more active, interacting strongly with the substances and even reacting with them in some cases. Let's focus on how the water molecule interacts with both ionic and covalent solutes.

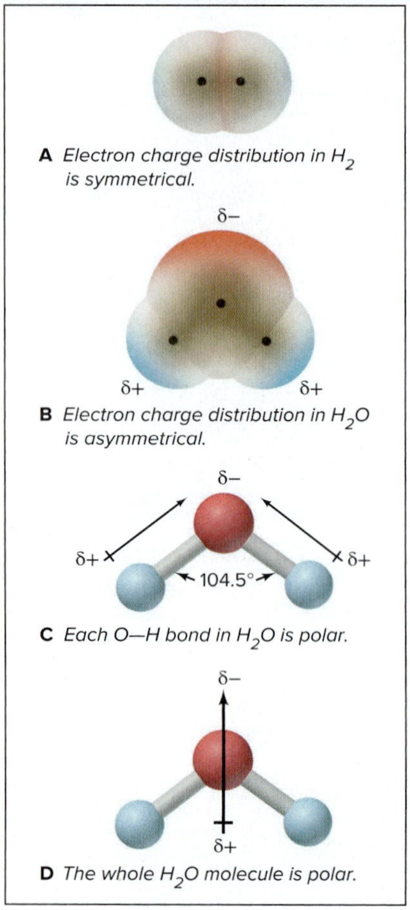

Figure 4.1 Electron distribution in molecules of H_2 and H_2O.

The Polar Nature of Water

On the atomic scale, water's great solvent power arises from the *uneven distribution of electron charge* and its *bent molecular shape,* which create a *polar molecule:*

1. *Uneven charge distribution.* Recall from Section 2.7 that the electrons in a covalent bond are shared between the atoms. In a bond between identical atoms—as in H_2, Cl_2, O_2—the sharing is equal and electron charge is distributed evenly between the two nuclei (symmetrical shading in the space-filling model in Figure 4.1A). In covalent bonds between different atoms, the sharing is uneven because one atom attracts the electron pair more strongly than the other atom does.

For example, in each O—H bond of water, the shared electrons are closer to the O atom because an O atom attracts electrons more strongly than an H atom does. (We'll discuss the reasons in Chapter 9.) This uneven charge distribution creates a *polar bond,* one with partially charged "poles." In Figure 4.1B, the asymmetrical shading shows this distribution, and the δ symbol is used to indicate a partial charge. The O end is partially negative, represented by red shading and δ−, and the H end is partially positive, represented by blue shading and δ+. In the ball-and-stick model in Figure 4.1C, the *polar arrow* points to the negative pole, and the arrow's tail, shaped like a plus sign, marks the positive pole. There are two polar O—H bonds in each water molecule.

2. *Bent molecular shape.* The sequence of the H—O—H atoms in water is not linear: the water molecule is bent with a bond angle of 104.5° (Figure 4.1C).

3. *Molecular polarity.* The combination of polar bonds and bent shape makes water a **polar molecule:** the region near the O atom is partially negative (there is a higher electron density), and the region between the H atoms is partially positive (there is a lower electron density) (Figure 4.1D). The H_2O molecule is still electrically neutral overall, although its electrons are distributed asymmetrically.

Ionic Compounds in Water

In this subsection, we consider two closely related aspects of aqueous solutions of ionic compounds—how they occur and how they behave. We also use a compound's formula to calculate the amount (mol) of each ion in solution.

How Ionic Compounds Dissolve: Replacement of Charge Attractions In an ionic solid, oppositely charged ions are held together by electrostatic attractions (see Figure 1.3C and Section 2.7). Water separates the ions in an ionic compound by *replacing these attractions with others between several water molecules and each ion.* Picture a granule of a soluble ionic compound in water: the negative ends of some water molecules are attracted to the cations, and the positive ends of other water molecules are attracted to the anions (Figure 4.2). Dissolution occurs because the

Figure 4.2 An ionic compound dissolving in water. The inset shows the polar arrow and partial charges (not shown in the rest of the scene) for a water molecule.

attractions between each type of ion and several water molecules outweigh the attractions between the ions themselves. Gradually, all the ions separate (dissociate), become **solvated**—surrounded closely by solvent molecules—and then move randomly in the solution. The solvent molecules prevent the cations and anions from recombining.

For an ionic compound that doesn't dissolve in water, the attraction between its cations and anions is *greater* than the attraction between the ions and water. Actually, these so-called insoluble substances *do* dissolve to a very small extent, usually several orders of magnitude less than so-called soluble substances. For example, NaCl (a "soluble" compound) is over 4×10^4 times more soluble in water than AgCl (an "insoluble" compound):

$$\text{Solubility of NaCl in } H_2O \text{ at } 20°C = 365 \text{ g/L}$$
$$\text{Solubility of AgCl in } H_2O \text{ at } 20°C = 0.009 \text{ g/L}$$

In Chapter 13, you'll see that dissolving involves more than a contest between the relative energies of attraction of ions for each other or for water. It is also favored by the greater freedom of motion the ions have when they leave the solid and disperse randomly through the solution.

How Ionic Solutions Behave: Electrolytes and Electrical Conductivity When an ionic compound dissolves, the solution's *electrical conductivity,* the flow of electric current, increases dramatically. When electrodes are immersed in distilled water (Figure 4.3A) or pushed into an ionic solid (Figure 4.3B), no current flows, as shown by the unlit bulb. But when the electrodes are submerged in an aqueous solution of the compound, a large current flows, as shown by the lit bulb (Figure 4.3C). Current flow implies the *movement of charged particles:* when the ionic compound dissolves,

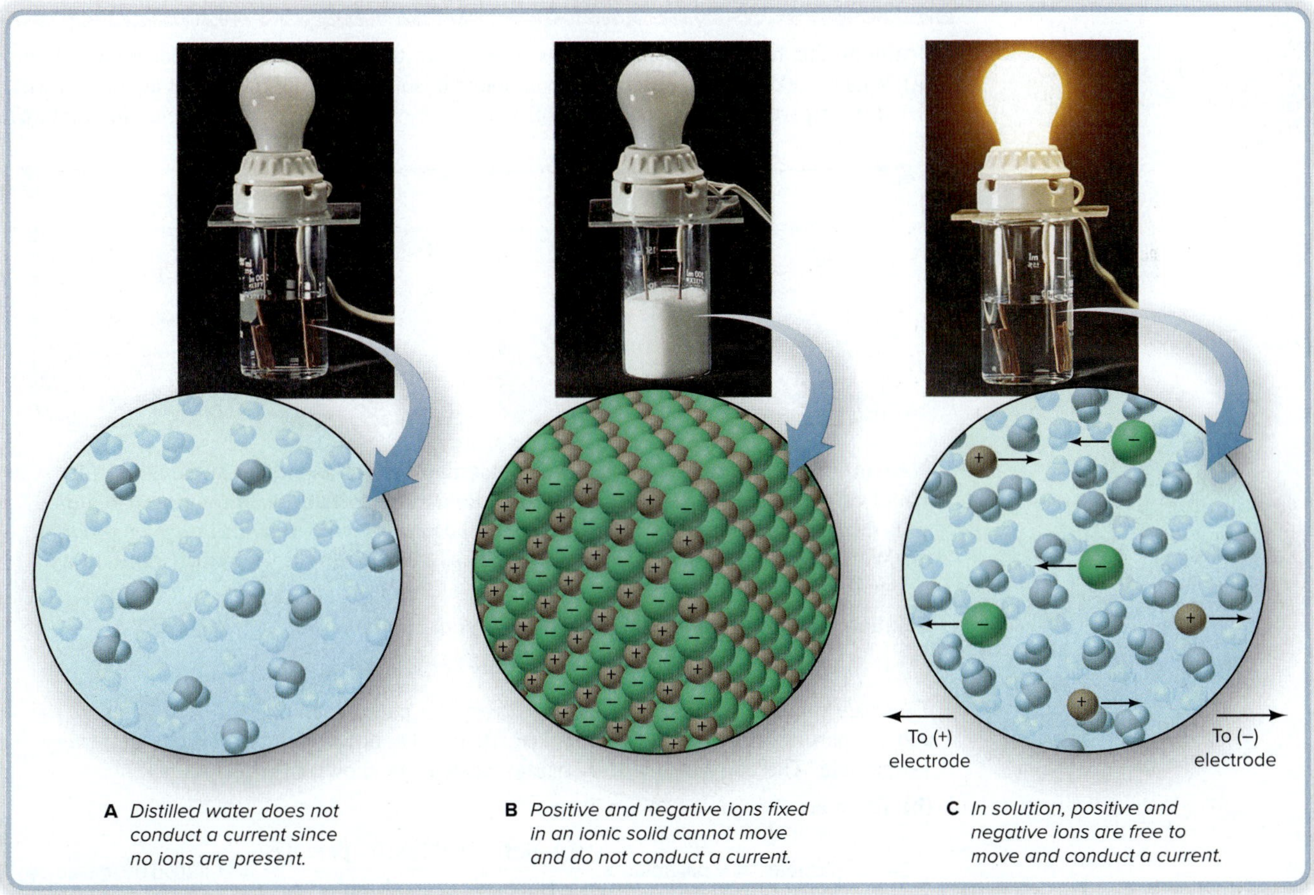
To (+)
electrode
To (–)
electrode
A Distilled water does not conduct a current since no ions are present.
B Positive and negative ions fixed in an ionic solid cannot move and do not conduct a current.
C In solution, positive and negative ions are free to move and conduct a current.

Figure 4.3 The electrical conductivity of ionic solutions.
Source: © McGraw-Hill Education/Stephen Frisch, photographer

the separate solvated ions move toward the electrode of opposite charge. A substance that conducts a current when dissolved in water is an **electrolyte.** Soluble ionic compounds are *strong* electrolytes because they dissociate completely and conduct a large current.

Calculating the Number of Moles of Ions in Solution *From the formula of the soluble ionic compound, we know the number of moles of each ion in solution.* For example, the equations for dissolving KBr, $CaBr_2$, and $Fe(NO_3)_3$ in water to form solvated ions are

$$KBr(s) \xrightarrow{H_2O} K^+(aq) + Br^-(aq)$$
$$CaBr_2(s) \xrightarrow{H_2O} Ca^{2+}(aq) + 2Br^-(aq)$$
$$Fe(NO_3)_3(s) \xrightarrow{H_2O} Fe^{3+}(aq) + 3NO_3^-(aq)$$

("H_2O" above the arrows means that water is the solvent, not a reactant.) The subscripts of the ions in the formula of an ionic compound indicate the moles of each ion produced when the compound dissociates. Thus, 1 mol of KBr dissociates into 2 mol of ions—1 mol of K^+ and 1 mol of Br^-—and 1 mol of $CaBr_2$ dissociates into 3 mol of ions—1 mol of Ca^{2+} and 2 mol of Br^-. (*not* 1 mol of Ca^{2+} and 1 mol of Br_2^-). Note that the NO_3^- polyatomic ion remains intact so that 1 mol of $Fe(NO_3)_3$ dissociates into 4 mol of ions—1 mol of Fe^{3+} and 3 mol of NO_3^-. Polyatomic ions, whose atoms are covalently bonded, stay together as a charged unit. Sample Problems 4.1 and 4.2 apply these ideas, first with molecular scenes and then in calculations.

SAMPLE PROBLEM 4.1 Using Molecular Scenes to Depict an Ionic Compound in Aqueous Solution

Problem The beakers contain aqueous solutions of the strong electrolyte potassium sulfate. **(a)** Which beaker best represents the compound in solution (water molecules are not shown)? **(b)** If each particle represents 0.1 mol, what is the total number of particles in solution?

Plan (a) We determine the formula of potassium sulfate and write an equation for 1 mol of compound dissociating into ions. Potassium sulfate is a strong electrolyte, so it dissociates completely, but, in general, *polyatomic ions remain intact in solution.* **(b)** We count the number of separate particles, and then multiply by 0.1 mol and by Avogadro's number.

Solution (a) The formula is K_2SO_4, so the equation is

$$K_2SO_4(s) \xrightarrow{H_2O} 2K^+(aq) + SO_4^{2-}(aq)$$

There are two separate 1+ particles for every 2− particle; Beakers A and B show 1+ and 2− particles that are not dissociated. Beaker D shows two 2− particles for every 1+ particle. The correct ratio of separate particles is shown in beaker C.

(b) There are 9 particles so we have

$$\text{No. of particles} = 9 \text{ particles} \times \frac{0.1 \text{ mol}}{1 \text{ particle}} \times \frac{6.022\times10^{23} \text{ particles}}{1 \text{ mol}} = 5.420\times10^{23} \text{ particles}$$

Check Rounding to check the math in (b) gives 9 × 0.1 × 6 = 5.4, so the answer seems correct. The number of particles is an exact number since we actually counted them; thus, the answer can have as many significant figures as in Avogadro's number.

FOLLOW-UP PROBLEMS

*Brief Solutions for **all** Follow-up Problems appear at the end of the chapter.*

4.1A **(a)** Which strong electrolyte is dissolved in water (water molecules not shown) in the beaker shown: LiBr, Cs_2CO_3, or $BaCl_2$? **(b)** If each particle represents 0.05 mol, what mass (g) of compound was dissolved?

4.1B **(a)** Make a drawing similar to that in Follow-up Problem 4.1A that shows two formula units of the strong electrolyte sodium phosphate dissolved in water (without showing any water molecules). **(b)** When 0.40 mol of sodium phosphate is dissolved in water, how many moles of ions result?

SOME SIMILAR PROBLEMS 4.5 and 4.6

SAMPLE PROBLEM 4.2 Determining Amount (mol) of Ions in Solution

Problem What amount (mol) of each ion is in each solution?
(a) 5.0 mol of ammonium sulfate dissolved in water
(b) 78.5 g of cesium bromide dissolved in water
(c) 7.42×10^{22} formula units of copper(II) nitrate dissolved in water

Plan We write an equation that shows 1 mol of compound dissociating into ions. **(a)** We multiply the number of moles of ions by 5.0. **(b)** We first convert grams to moles. **(c)** We first convert formula units to moles.

Solution **(a)** $(NH_4)_2SO_4(s) \xrightarrow{H_2O} 2NH_4^+(aq) + SO_4^{2-}(aq)$

Calculating amount (mol) of NH_4^+ ions:

$$\text{Amount (mol) of } NH_4^+ = 5.0\ \cancel{\text{mol } (NH_4)_2SO_4} \times \frac{2\text{ mol } NH_4^+}{1\ \cancel{\text{mol } (NH_4)_2SO_4}} = 10.\text{ mol } NH_4^+$$

The formula shows 1 mol of SO_4^{2-} per mole of $(NH_4)_2SO_4$, so 5.0 mol SO_4^{2-} is also present.

(b) $CsBr(s) \xrightarrow{H_2O} Cs^+(aq) + Br^-(aq)$

Converting from mass (g) to amount (mol):

$$\text{Amount (mol) of CsBr} = 78.5\ \cancel{\text{g CsBr}} \times \frac{1\text{ mol CsBr}}{212.8\ \cancel{\text{g CsBr}}} = 0.369\text{ mol CsBr}$$

Thus, 0.369 mol of Cs^+ and 0.369 mol of Br^- are present.

(c) $Cu(NO_3)_2(s) \xrightarrow{H_2O} Cu^{2+}(aq) + 2NO_3^-(aq)$

Converting from formula units to amount (mol):

$$\text{Amount (mol) of } Cu(NO_3)_2 = 7.42\times10^{22}\ \cancel{\text{formula units } Cu(NO_3)_2} \times \frac{1\text{ mol } Cu(NO_3)_2}{6.022\times10^{23}\ \cancel{\text{formula units } Cu(NO_3)_2}} = 0.123\text{ mol } Cu(NO_3)_2$$

$$\text{Amount (mol) of } NO_3^- = 0.123\ \cancel{\text{mol } Cu(NO_3)_2} \times \frac{2\text{ mol } NO_3^-}{1\ \cancel{\text{mol } Cu(NO_3)_2}} = 0.246\text{ mol } NO_3^-$$

0.123 mol of Cu^{2+} is also present.

Check Round off to check the math and see if the relative numbers of moles of ions are consistent with the formula. For instance, in (a), 10. mol NH_4^+/5.0 mol SO_4^{2-} = 2 NH_4^+/1 SO_4^{2-}, or $(NH_4)_2SO_4$.

FOLLOW-UP PROBLEMS

4.2A What amount (mol) of each ion is in each solution?
(a) 2 mol of potassium perchlorate dissolved in water
(b) 354 g of magnesium acetate dissolved in water
(c) 1.88×10^{24} formula units of ammonium chromate dissolved in water

4.2B What amount (mol) of each ion is in each solution?
(a) 4 mol of lithium carbonate dissolved in water
(b) 112 g of iron(III) sulfate dissolved in water
(c) 8.09×10^{22} formula units of aluminum nitrate dissolved in water

SOME SIMILAR PROBLEMS 4.18–4.21

Covalent Compounds in Water

Water dissolves many covalent (molecular) compounds also. Table sugar (sucrose, $C_{12}H_{22}O_{11}$), beverage (grain) alcohol (ethanol, CH_3CH_2OH), and automobile antifreeze (ethylene glycol, $HOCH_2CH_2OH$) are some familiar examples. All contain their own polar bonds, which are attracted to the polar bonds of water. However, most soluble covalent substances *do not* separate into ions, but remain intact molecules. For example,

$$HOCH_2CH_2OH(l) \xrightarrow{H_2O} HOCH_2CH_2OH(aq)$$

As a result, their aqueous solutions do not conduct an electric current, and these substances are **nonelectrolytes.** (As you'll see shortly, however, a small group of H-containing molecules that act as acids in aqueous solution *do* dissociate into ions.) Many other covalent substances, such as benzene (C_6H_6) and octane (C_8H_{18}), do not contain polar bonds, and these substances do not dissolve appreciably in water.

Expressing Concentration in Terms of Molarity

When working quantitatively with any solution, it is essential to know the **concentration**—*the quantity of solute dissolved in a given quantity of solution (or of solvent).*

Concentration is an *intensive* property (like density or temperature; Section 1.4) and thus is independent of the solution volume: a 50-L tank of a solution has the *same concentration* (solute quantity/solution quantity) as a 50-mL beaker of the solution. **Molarity (*M*)** is the most common unit of concentration (Chapter 13 covers others). It expresses the concentration in units of *moles of solute per liter of solution:*

$$\text{Molarity} = \frac{\text{moles of solute}}{\text{liters of solution}} \quad \text{or} \quad M = \frac{\text{mol solute}}{\text{L soln}} \qquad \textbf{(4.1)}$$

SAMPLE PROBLEM 4.3 Calculating the Molarity of a Solution

Problem Glycine ($C_2H_5NO_2$) has the simplest structure of the 20 amino acids that make up the proteins in the human body. What is the molarity of a solution that contains 53.7 g of glycine dissolved in 495 mL of solution?

Plan The molarity is the number of moles of solute in each liter of solution. We convert the mass (g) of glycine (53.7 g) to amount (mol) by dividing by the molar mass. We divide that number of moles by the volume (495 mL) and convert the volume to liters to find the molarity (see the road map).

Road Map

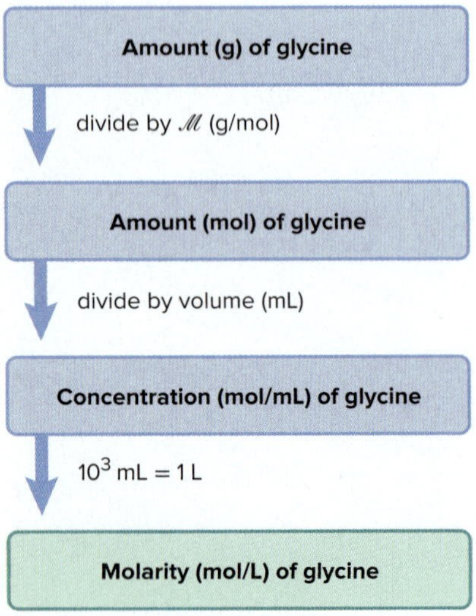

Solution Finding the amount (mol) of glycine:

$$\text{Amount (mol) of glycine} = 53.7\ \cancel{\text{g glycine}} \times \frac{1\ \text{mol glycine}}{75.07\ \cancel{\text{g glycine}}} = 0.715\ \text{mol glycine}$$

Dividing by the volume and converting the volume units to obtain molarity:

$$\text{Molarity} = \frac{0.715\ \text{mol glycine}}{495\ \cancel{\text{mL soln}}} \times \frac{1000\ \cancel{\text{mL}}}{1\ \text{L}} = 1.44\ M\ \text{glycine}$$

Check A quick look at the math shows about 0.7 mol of glycine in about 0.5 L of solution, so the concentration should be about 1.4 mol/L, or 1.4 *M*.

FOLLOW-UP PROBLEMS

4.3A Calculate the molarity of the solution made when 6.97 g of KI is dissolved in enough water to give a total volume of 100. mL.

4.3B Calculate the molarity of a solution that contains 175 mg of sodium nitrate in a total volume of 15.0 mL.

SOME SIMILAR PROBLEMS 4.22(b), 4.23(c), 4.24(b), and 4.25(a)

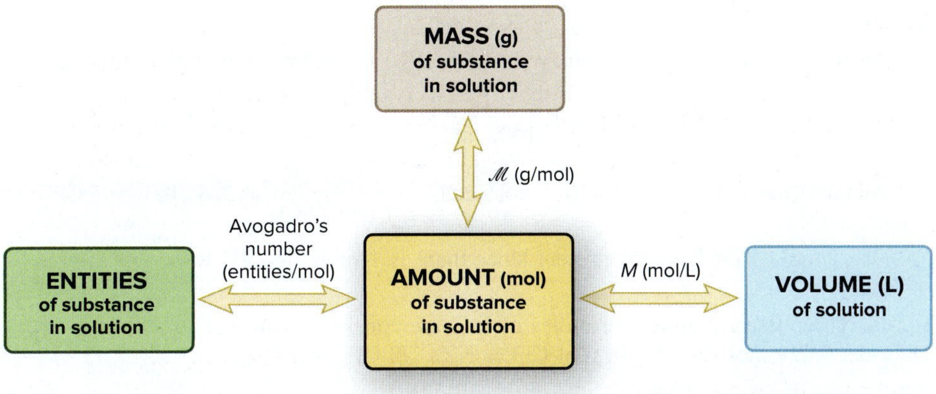

Figure 4.4 Summary of amount-mass-number relationships in solution. The amount (mol) of a substance in solution is related to the volume (L) of solution through the molarity (*M*; mol/L). As always, convert the given quantity to amount (mol) first.

Amount-Mass-Number Conversions Involving Solutions

Like many intensive properties, molarity can be used as a *conversion factor* between volume (L) of solution and amount (mol) of solute:

$$\frac{\text{mol solute}}{\text{L solution}} \quad \text{or} \quad \frac{\text{L solution}}{\text{mol solute}}$$

From moles of solute, we can find the mass or the number of entities of solute (Figure 4.4), as applied in Sample Problems 4.4 and 4.5.

SAMPLE PROBLEM 4.4 Calculating Mass of Solute in a Given Volume of Solution

Problem Biochemists often study reactions in solutions containing phosphate ion, which is commonly found in cells. How many grams of solute are in 1.75 L of 0.460 *M* sodium hydrogen phosphate?

Plan We need the mass (g) of solute, so we multiply the known solution volume (1.75 L) by the known molarity (0.460 *M*) to find the amount (mol) of solute and convert it to mass (g) using the solute's molar mass (see the road map).

Road Map

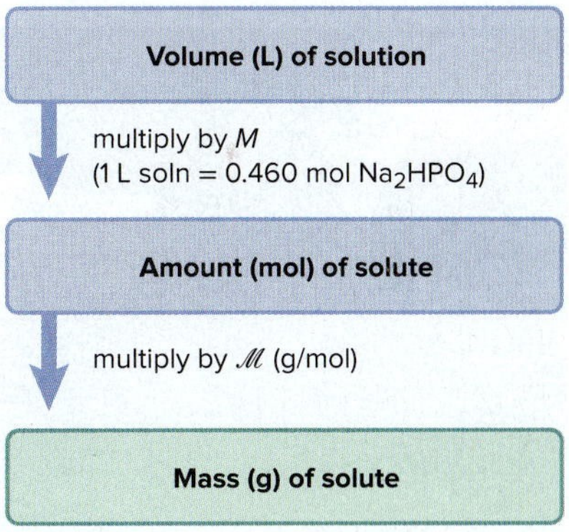

Solution Calculating amount (mol) of solute in solution:

$$\text{Amount (mol) of } Na_2HPO_4 = 1.75\ \cancel{\text{L soln}} \times \frac{0.460 \text{ mol } Na_2HPO_4}{1\ \cancel{\text{L soln}}} = 0.805 \text{ mol } Na_2HPO_4$$

Converting from amount (mol) of solute to mass (g):

$$\text{Mass (g) } Na_2HPO_4 = 0.805\ \cancel{\text{mol } Na_2HPO_4} \times \frac{141.96 \text{ g } Na_2HPO_4}{1\ \cancel{\text{mol } Na_2HPO_4}} = 114 \text{ g } Na_2HPO_4$$

Check The answer seems to be correct: ~1.8 L of 0.5 mol/L solution contains 0.9 mol, and 150 g/mol × 0.9 mol = 135 g, which is close to 114 g of solute.

FOLLOW-UP PROBLEMS

4.4A In biochemistry laboratories, solutions of sucrose (table sugar, $C_{12}H_{22}O_{11}$) are used in high-speed centrifuges to separate the parts of a biological cell. How many liters of 3.30 *M* sucrose contain 135 g of solute? Include a road map that shows how you planned the solution.

4.4B A chemist adds 40.5 mL of a 0.128 *M* solution of sulfuric acid to a reaction mixture. How many moles of sulfuric acid are being added? Include a road map that shows how you planned the solution.

SOME SIMILAR PROBLEMS 4.22(a), 4.22(c), 4.23(a), 4.24(a), 4.25(b), and 4.25(c)

SAMPLE PROBLEM 4.5 Determining Amount (mol) of Ions in a Solution

Problem What amount (mol) of each ion is in 35 mL of 0.84 *M* zinc chloride?

Plan We write an equation that shows 1 mol of compound dissociating into ions. We convert molarity and volume to moles of zinc chloride and use the dissociation equation to convert moles of compound to moles of ions.

Solution $ZnCl_2(aq) \longrightarrow Zn^{2+}(aq) + 2Cl^-(aq)$

Converting from volume (mL) and molarity (mol/L) to amount (mol) of compound:

$$\text{Amount (mol) of } ZnCl_2 = 35\ \cancel{mL} \times \frac{1\ \cancel{L}}{1000\ \cancel{mL}} \times \frac{0.84\ \text{mol } ZnCl_2}{1\ \cancel{L}} = 2.9\times10^{-2}\ \text{mol } ZnCl_2$$

$$\text{Amount (mol) of } Cl^- = 2.9\times10^{-2}\ \cancel{\text{mol } ZnCl_2} \times \frac{2\ \text{mol } Cl^-}{1\ \cancel{\text{mol } ZnCl_2}} = 5.8\times10^{-2}\ \text{mol } Cl^-$$

2.9×10^{-2} mol of Zn^{2+} is also present since there is 1 mole of Zn^{2+} for every 1 mole of $ZnCl_2$.

Check The relative numbers of moles of ions are consistent with the formula for zinc chloride: 0.029 mol Zn^{2+}/0.058 mol Cl^- = 1 Zn^{2+}/2 Cl^-, or $ZnCl_2$.

FOLLOW-UP PROBLEMS

4.5A What amount (mol) of each ion is in 1.32 L of 0.55 *M* sodium phosphate?

4.5B What is the molarity of aluminum ion in a solution that contains 1.25 mol of aluminum sulfate in 875 mL?

SOME SIMILAR PROBLEMS 4.23(b), 4.24(c), 4.26, 4.27

Preparing and Diluting Molar Solutions

Notice that the volume term in the denominator of the molarity expression in Equation 4.1 is the *solution* volume, **not** the *solvent* volume. This means that you *cannot* dissolve 1 mol of solute in 1 L of solvent to make a 1 *M* solution. Because the solute volume adds to the solvent volume, the total volume (solute + solvent) would be *more* than 1 L, so the concentration would be *less* than 1 *M*.

Preparing a Solution Correctly preparing a solution of a solid solute requires four steps. Let's prepare 0.500 L of 0.350 *M* nickel(II) nitrate hexahydrate [$Ni(NO_3)_2{\cdot}6H_2O$]:

1. *Weigh the solid.* Calculate the mass of solid needed by converting from volume (L) to amount (mol) and then to mass (g):

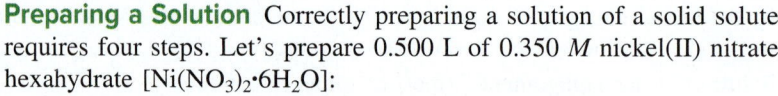

$$\text{Mass (g) of solute} = 0.500\ \cancel{L\ soln} \times \frac{0.350\ \cancel{\text{mol } Ni(NO_3)_2{\cdot}6H_2O}}{1\ \cancel{L\ soln}} \times \frac{290.82\ \text{g } Ni(NO_3)_2{\cdot}6H_2O}{1\ \cancel{\text{mol } Ni(NO_3)_2{\cdot}6H_2O}}$$
$$= 50.9\ \text{g } Ni(NO_3)_2{\cdot}6H_2O$$

2. *Transfer the solid.* We need 0.500 L of solution, so we choose a 500-mL volumetric flask (a flask with a fixed volume indicated by a mark on the neck), add enough distilled water to fully dissolve the solute (usually about half the final volume, or 250 mL of distilled water in this case), and transfer the solute. Wash down any solid clinging to the neck with some solvent.
3. *Dissolve the solid.* Swirl the flask until all the solute is dissolved. If necessary, wait until the solution is at room temperature. (As we discuss in Chapter 13, the solution process may be accompanied by heating or cooling.)
4. *Add solvent to the final volume.* Add distilled water to bring the solution volume to the line on the flask neck; cover and mix thoroughly again.

Figure 4.5 shows the last three steps.

Figure 4.5 Laboratory preparation of molar solutions.
Source: © McGraw-Hill Education/Stephen Frisch, photographer

Diluting a Solution A concentrated solution (higher molarity) is converted to a dilute solution (lower molarity) by adding solvent, which means the solution volume increases but the amount (mol) of solute

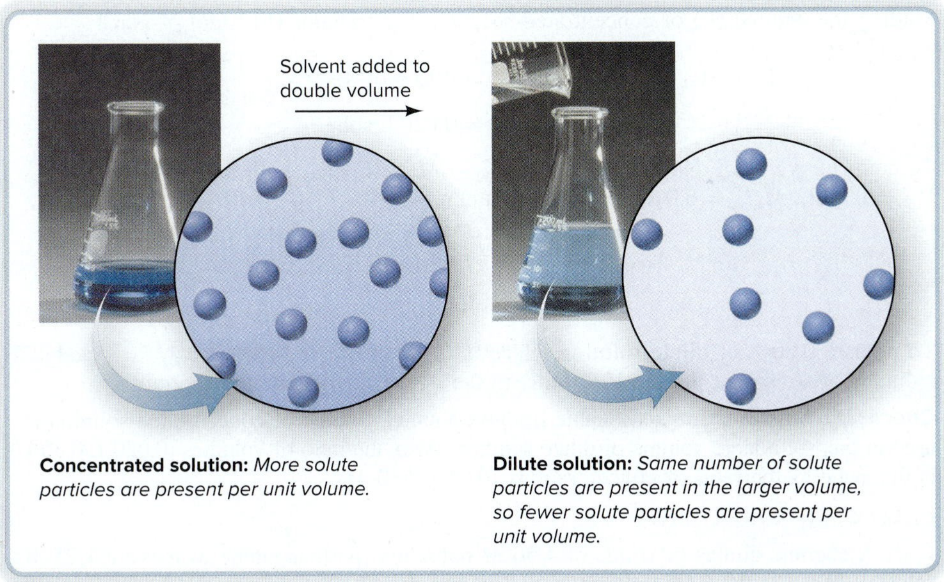
Solvent added to double volume
Concentrated solution: More solute particles are present per unit volume.
Dilute solution: Same number of solute particles are present in the larger volume, so fewer solute particles are present per unit volume.

Figure 4.6 **Converting a concentrated solution to a dilute solution.**
Source: © McGraw-Hill Education/Stephen Frisch, photographer

stays the same. As a result, the dilute solution contains *fewer solute particles per unit volume* and, thus, has a lower concentration than the concentrated solution (Figure 4.6). Chemists often prepare and store a concentrated solution *(stock solution)* and then dilute it as needed.

Solving Dilution Problems To solve dilution problems, we use the fact that the amount (mol) of solute does not change during the dilution process. Therefore, once we calculate the amount of solute needed to make a particular volume of a dilute solution, we can then calculate the volume of concentrated solution that contains that amount of solute. This two-part calculation can be combined into one step using the following relationship:

$$M_{\text{dil}} \times V_{\text{dil}} = \text{amount (mol) of solute} = M_{\text{conc}} \times V_{\text{conc}} \quad \textbf{(4.2)}$$

where M and V are the molarity and volume of the *dil*ute (subscript "dil") and *con*centrated (subscript "conc") solutions. In Sample Problem 4.6, we first use a two-part calculation, followed by the one-step relationship.

SAMPLE PROBLEM 4.6 Preparing a Dilute Solution from a Concentrated Solution

Problem Isotonic saline is 0.15 *M* aqueous NaCl. It simulates the total concentration of ions in many cellular fluids, and its uses range from cleaning contact lenses to washing red blood cells. How would you prepare 0.80 L of isotonic saline from a 6.0 *M* stock solution?

Plan To dilute a concentrated solution, we add only solvent, so the *moles of solute are the same in both solutions.* First, we will use a two-part calculation. We know the volume (0.80 L) and molarity (0.15 *M*) of the dilute (dil) NaCl solution needed, so we find the amount (mol) of NaCl it contains. Then, we find the volume (L) of concentrated (conc; 6.0 *M*) NaCl solution that contains the same amount (mol) (see the road map). We will confirm this answer by using Equation 4.2. Once the final volume is calculated, we add solvent *up to* the final volume.

Road Map

Volume (L) of dilute solution

↓ multiply by *M* of dilute solution (1 L soln = 0.15 mol NaCl)

Amount (mol) of NaCl in dilute solution = amount (mol) of NaCl in concentrated solution

↓ divide by *M* of concentrated solution (6 mol NaCl = 1 L soln)

Volume (L) of concentrated solution

Solution Two-part calculation: Finding amount (mol) of solute in dilute solution:

$$\text{Amount (mol) of NaCl in dil soln} = 0.80\ \cancel{\text{L soln}} \times \frac{0.15\ \text{mol NaCl}}{1\ \cancel{\text{L soln}}}$$
$$= 0.12\ \text{mol NaCl}$$

Finding amount (mol) of solute in concentrated solution: Because we add only solvent to dilute the solution,

$$\text{Amount (mol) of NaCl in dil soln} = \text{amount (mol) of NaCl in conc soln}$$
$$= 0.12\ \text{mol NaCl}$$

Finding the volume (L) of concentrated solution that contains 0.12 mol of NaCl:

$$\text{Volume (L) of conc NaCl soln} = 0.12\ \cancel{\text{mol NaCl}} \times \frac{1\ \text{L soln}}{6.0\ \cancel{\text{mol NaCl}}}$$
$$= 0.020\ \text{L soln}$$

To confirm, we use Equation 4.2 $M_{dil} \times V_{dil} = M_{conc} \times V_{conc}$, with $M_{dil} = 0.15\ M$, $V_{dil} = 0.80$ L, $M_{conc} = 6.0\ M$, and V_{conc} = unknown:

$$V_{conc} = \frac{M_{dil} \times V_{dil}}{M_{conc}} = \frac{0.15\ M \times 0.80\ \text{L}}{6.0\ M}$$
$$= 0.020\ \text{L soln}$$

To prepare 0.80 L of dilute solution, place 0.020 L of 6.0 *M* NaCl in a 1.0-L graduated cylinder, add distilled water (~780 mL) to the 0.80-L mark, and stir thoroughly.

Check The answer seems reasonable because a small volume of concentrated solution is used to prepare a large volume of dilute solution. Also, the ratio of volumes (0.020 L/0.80 L) is the same as the ratio of concentrations (0.15 *M*/6.0 *M*).

FOLLOW-UP PROBLEMS

4.6A A chemist dilutes 60.0 mL of 4.50 *M* potassium permanganate to make a 1.25 *M* solution. What is the final volume of the diluted solution?

4.6B A chemical engineer dilutes a stock solution of sulfuric acid by adding 25.0 m^3 of 7.50 *M* acid to enough water to make 500. m^3. What is the concentration of sulfuric acid in the diluted solution, in g/mL?

SOME SIMILAR PROBLEMS 4.28 and 4.29

In the next sample problem, we use a variation of Equation 4.2, with molecular scenes showing numbers of particles, to visualize changes in concentration.

SAMPLE PROBLEM 4.7 Visualizing Changes in Concentration

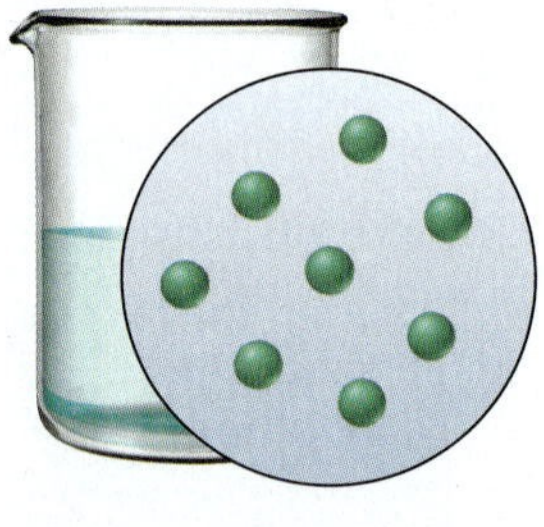

(a)

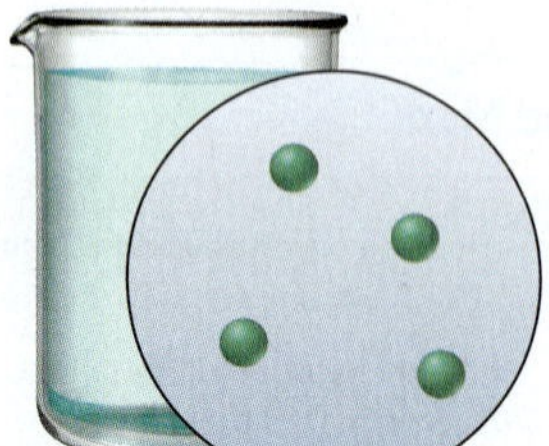

(b)

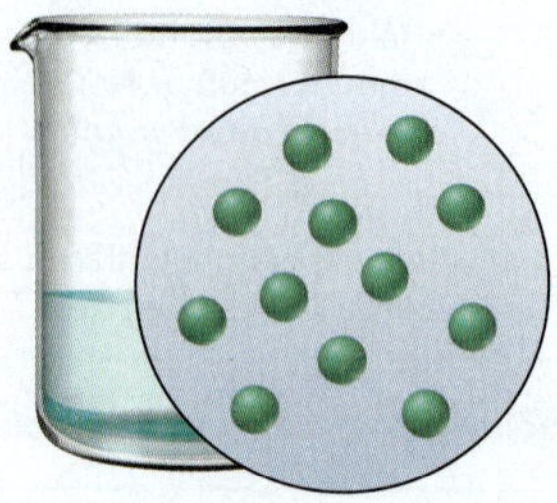

Problem The top circle at left represents a unit volume of a solution. Draw a circle representing a unit volume of the solution after each of these changes:
(a) For every 1 mL of solution, 1 mL of solvent is added.
(b) One-third of the solvent is boiled off.

Plan Given the starting solution, we have to find the number of solute particles in a unit volume after each change. The number of particles per unit volume, *N*, is directly related to the number of moles per unit volume, *M*, so we can use a relationship similar to Equation 4.2 to find the number of particles.

(a) The volume increases, so the final solution is more dilute—fewer particles per unit volume. **(b)** One-third of the solvent is lost, so the final solution is more concentrated—more particles per unit volume.

Solution (a) Finding the number of particles in the dilute solution, N_{dil}:

$$N_{dil} \times V_{dil} = N_{conc} \times V_{conc}$$

where N_{conc} = 8 particles, V_{conc} = 1 mL, and V_{dil} = 2 mL; thus,

$$N_{dil} = N_{conc} \times \frac{V_{conc}}{V_{dil}} = 8\ \text{particles} \times \frac{1\ \text{mL}}{2\ \text{mL}} = 4\ \text{particles}$$

(b) Finding the number of particles in the concentrated solution, N_{conc}:

$$N_{dil} \times V_{dil} = N_{conc} \times V_{conc}$$

where N_{dil} = 8 particles, V_{dil} = 1 mL, and $V_{conc} = \frac{2}{3}$ mL; thus,

$$N_{conc} = N_{dil} \times \frac{V_{dil}}{V_{conc}} = 8\ \text{particles} \times \frac{1\ \text{mL}}{\frac{2}{3}\ \text{mL}} = 12\ \text{particles}$$

Check In (a), the volume is doubled (from 1 mL to 2 mL), so the number of particles should be halved; $\frac{1}{2}$ of 8 is 4. In (b), the volume is $\frac{2}{3}$ of the original, so the number of particles should be $\frac{3}{2}$ of the original; $\frac{3}{2}$ of 8 is 12.

Comment In (b), we assumed that only solvent boils off. This is true with nonvolatile solutes, such as ionic compounds, but in Chapter 13, we'll encounter solutions in which both solvent *and* solute are volatile.

FOLLOW-UP PROBLEMS

4.7A The circle labeled A represents a unit volume of a solution. Explain the changes that must be made to A to obtain the unit volumes in B and C.

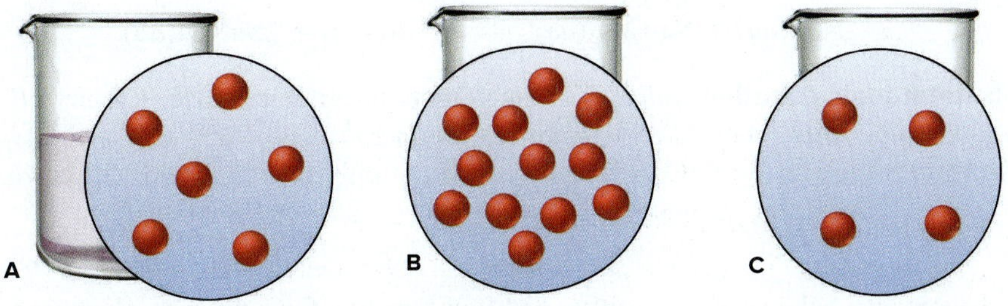

4.7B Circle A represents a unit volume of 100. mL of a solution. Which circle (B, C, or D) best represents the unit volume after 300. mL of solvent has been added?

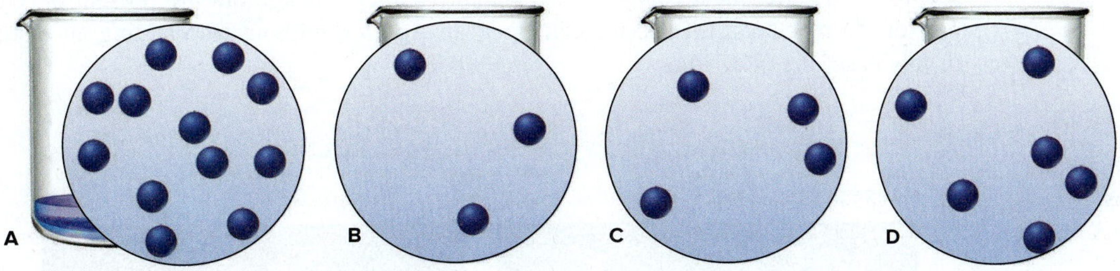

SOME SIMILAR PROBLEMS 4.10 and 4.12

› Summary of Section 4.1

- › Because of polar bonds and a bent shape, the water molecule is polar, and water dissolves many ionic and covalent compounds.
- › When an ionic compound dissolves, the attraction between each ion and water molecules replaces the attraction between ions. Soluble ionic compounds are electrolytes because the ions are free to move and, thus, the solution conducts electricity.
- › The formula of a soluble ionic compound shows the number of moles of each ion in solution per mole of compound dissolved.
- › Water dissolves many covalent substances that contain polar bonds. These compounds are nonelectrolytes because the molecules remain intact, and, thus, the solution does not conduct electricity.
- › The concentration of a solute in a solution can be expressed by molarity which is the number of moles of solute dissolved in 1 liter of solution. A more concentrated solution (higher molarity) is converted to a more dilute solution (lower molarity) by adding solvent.
- › Molarity can be used as a conversion factor between volume of solution and amount (mol) of solute, from which you can find mass of solute.

4.2 WRITING EQUATIONS FOR AQUEOUS IONIC REACTIONS

Chemists use three types of equations to represent aqueous ionic reactions. Let's examine a reaction to see what each type shows. When solutions of silver nitrate and sodium chromate are mixed, brick-red, solid silver chromate (Ag_2CrO_4) forms.

Figure 4.7 depicts the reaction at the macroscopic level *(photos),* the atomic level *(blow-up circles),* and the symbolic level with the three types of equations (reacting ions are in red type):

- The **molecular equation** *(top)* reveals the least about the species that are actually in solution because *it shows all the reactants and products as if they were intact, undissociated compounds.* Only the designation for solid, (*s*), tells us that a change has occurred:

$$2AgNO_3(aq) + Na_2CrO_4(aq) \longrightarrow Ag_2CrO_4(s) + 2NaNO_3(aq)$$

- The **total ionic equation** *(middle)* is much more accurate because *it shows all the soluble ionic substances as they actually exist in solution, where they are dissociated into ions.* The $Ag_2CrO_4(s)$ stands out as the only undissociated substance:

$$2Ag^+(aq) + 2NO_3^-(aq) + 2Na^+(aq) + CrO_4^{2-}(aq) \longrightarrow Ag_2CrO_4(s) + 2Na^+(aq) + 2NO_3^-(aq)$$

The charges also balance: four positive and four negative for a net zero charge on the left side, and two positive and two negative for a net zero charge on the right.

Notice that $Na^+(aq)$ and $NO_3^-(aq)$ appear unchanged on both sides of the equation. These are called **spectator ions** (shown with pale colors in the atomic-level scenes). They are not involved in the actual chemical change but are present only as part of the reactants; that is, we can't add an Ag^+ ion without also adding an anion, in this case, the NO_3^- ion.

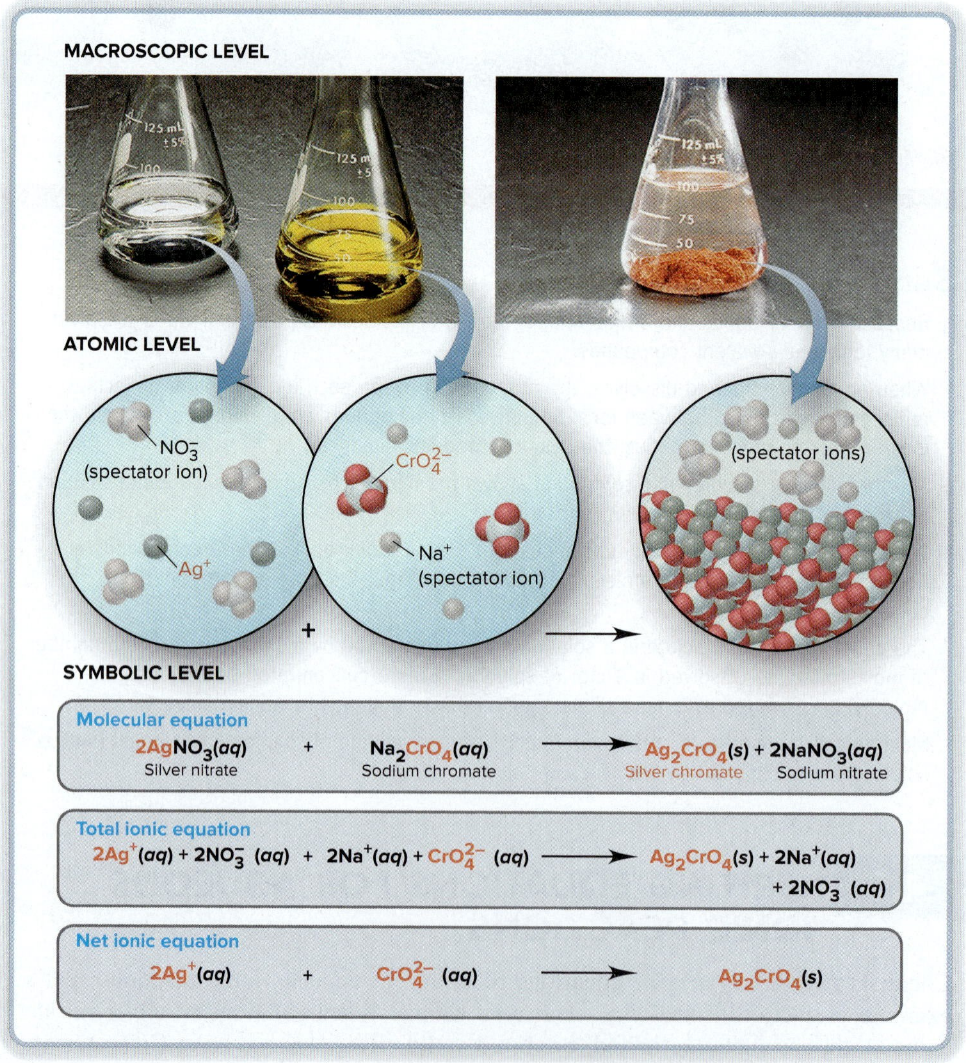

Figure 4.7 An aqueous ionic reaction and the three types of equations.
Source: © McGraw-Hill Education/Stephen Frisch, photographer

- The **net ionic equation** *(bottom)* is very useful because *it eliminates the spectator ions and shows only the actual chemical change:*

$$2Ag^+(aq) + CrO_4^{2-}(aq) \longrightarrow Ag_2CrO_4(s)$$

The formation of solid silver chromate from silver ions and chromate ions *is* the only change. To make that point clearly, suppose we had mixed solutions of silver acetate $AgC_2H_3O_2(aq)$ and potassium chromate, $K_2CrO_4(aq)$, instead of silver nitrate and sodium chromate. The three equations for the reaction would then be

Molecular: $2AgC_2H_3O_2(aq) + K_2CrO_4(aq) \longrightarrow Ag_2CrO_4(s) + 2KC_2H_3O_2(aq)$

Total ionic: $2Ag^+(aq) + 2C_2H_3O_2^-(aq) + 2K^+(aq) + CrO_4^{2-}(aq) \longrightarrow Ag_2CrO_4(s) + 2K^+(aq) + 2C_2H_3O_2^-(aq)$

Net ionic: $2Ag^+(aq) + CrO_4^{2-}(aq) \longrightarrow Ag_2CrO_4(s)$

Thus, the same change would have occurred, and only the spectator ions would differ—$K^+(aq)$ and $C_2H_3O_2^-(aq)$ instead of $Na^+(aq)$ and $NO_3^-(aq)$.

Next, we'll apply these types of equations to three important classes of chemical reactions—precipitation, acid-base, and oxidation-reduction.

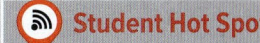

Student Hot Spot

Student data indicate that you may struggle with writing total and net ionic equations. Access the Smartbook to view additional Learning Resources on this topic.

› Summary of Section 4.2

- › A molecular equation shows all substances as intact and undissociated into ions.
- › A total ionic equation shows all soluble ionic compounds as separate, solvated ions. Spectator ions appear unchanged on both sides of the equation.
- › A net ionic equation eliminates the spectator ions and, thus, shows only the actual chemical change.

4.3 PRECIPITATION REACTIONS

Precipitation reactions occur commonly in both nature and commerce. Coral reefs and some gems and minerals form, in part, through this process. And the chemical industry employs precipitation methods to make several important inorganic compounds.

The Key Event: Formation of a Solid from Dissolved Ions

In a **precipitation reaction,** two soluble ionic compounds react to form an insoluble product, a **precipitate.** The reaction between silver nitrate and sodium chromate you saw in Section 4.2 is one example. Precipitates form for the same reason that some ionic compounds don't dissolve: the electrostatic attraction between the ions outweighs the tendency of the ions to remain solvated and move throughout the solution. When the two solutions are mixed, the ions collide and stay together, and a solid product "comes out of solution." Thus, the key event in a precipitation reaction is *the formation of an insoluble product through the net removal of ions from solution.* Figure 4.8 *(on the next page)* shows the process for calcium fluoride.

In an aqueous solution of $CaCl_2$, Ca^{2+} and Cl^- ions are dissociated and solvated (surrounded by water molecules); the attraction between each ion and water molecules is greater than the attraction between the ions. The same is true for an aqueous solution of NaF, which consists of solvated Na^+ and F^- ions. When the four ions are mixed, however, the attraction between Ca^{2+} and F^- ions is greater than the attraction between those ions and water molecules; CaF_2 is insoluble in water.

Predicting Whether a Precipitate Will Form

To predict whether a precipitate will form when we mix two aqueous ionic solutions, we refer to the short list of solubility rules in Table 4.1 *(on page 159)*; the solubility rules tell us which ionic compounds are soluble, and thus dissociate in water, and which ionic compounds are insoluble in water.

Let's see how to apply these rules. According to Table 4.1, all compounds containing Group 1A(1) ions and all compounds containing the nitrate ion are soluble,

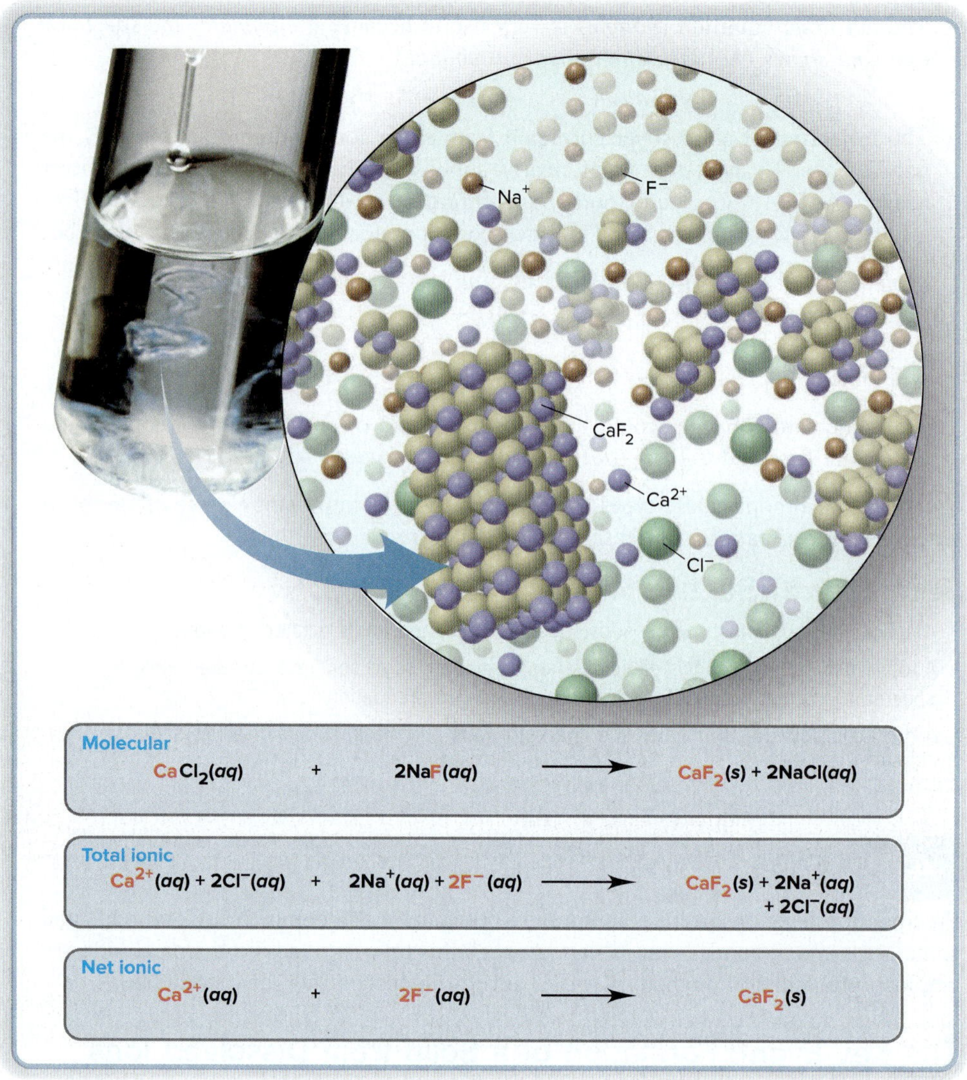

Figure 4.8 The precipitation of calcium fluoride. When aqueous solutions of NaF (from pipet) and $CaCl_2$ (in test tube) react, solid CaF_2 forms (water molecules are omitted for clarity).
Source: © McGraw-Hill Education/Richard Megna, photographer

so sodium iodide and potassium nitrate each dissolve in water to form solutions of solvated, dispersed ions:

$$NaI(s) \xrightarrow{H_2O} Na^+(aq) + I^-(aq)$$

$$KNO_3(s) \xrightarrow{H_2O} K^+(aq) + NO_3^-(aq)$$

Three steps help us predict if a precipitate forms when these two solutions are combined:

1. *Note the ions in the reactants.* The reactant ions are

$$Na^+(aq) + I^-(aq) + K^+(aq) + NO_3^-(aq) \longrightarrow ?$$

2. *Consider all possible cation-anion combinations.* In addition to NaI and KNO_3, which we know are soluble, the other cation-anion combinations are $NaNO_3$ and KI.
3. *Decide whether any combination is insoluble.* Remember that Table 4.1 tells us that all compounds of Group 1A(1) ions and all nitrate compounds are soluble. Therefore, all possible cation-anion combinations—NaI, KNO_3, $NaNO_3$, and KI—are soluble and no reaction occurs:

$$Na^+(aq) + I^-(aq) + K^+(aq) + NO_3^-(aq) \longrightarrow Na^+(aq) + NO_3^-(aq) + K^+(aq) + I^-(aq)$$

All the ions are spectator ions, so when we eliminate them, we have no net ionic equation.

Now, what happens if we substitute a solution of lead(II) nitrate, $Pb(NO_3)_2$, for the KNO_3 solution? The reactant ions are Na^+, I^-, Pb^{2+}, and NO_3^-. In addition to the two soluble reactants, NaI and $Pb(NO_3)_2$, the other two possible cation-anion combinations are

Table 4.1 Solubility Rules for Ionic Compounds in Water

Soluble Ionic Compounds	Insoluble Exceptions
All common compounds of Group 1A(1) ions (Li^+, Na^+, K^+, etc.)	None
All common compounds of ammonium ion (NH_4^+)	None
All common nitrates (NO_3^-), acetates (CH_3COO^- or $C_2H_3O_2^-$), and perchlorates (ClO_4^-)	None
All common chlorides (Cl^-), bromides (Br^-), and iodides (I^-)	Chlorides, bromides, and iodides of Ag^+, Pb^{2+}, Cu^+, and Hg_2^{2+}
All common fluorides (F^-)	PbF_2 and fluorides of Group 2A(2)
All common sulfates (SO_4^{2-})	$CaSO_4$, $SrSO_4$, $BaSO_4$, Ag_2SO_4, $PbSO_4$
Insoluble Ionic Compounds	**Soluble Exceptions**
All common metal hydroxides	Group 1A(1) hydroxides and $Ca(OH)_2$, $Sr(OH)_2$, and $Ba(OH)_2$
All common carbonates (CO_3^{2-}) and phosphates (PO_4^{3-})	Carbonates and phosphates of Group 1A(1) and NH_4^+
All common sulfides	Sulfides of Group 1A(1), Group 2A(2), and NH_4^+

$NaNO_3$ and PbI_2. According to Table 4.1, $NaNO_3$ is soluble, but PbI_2 is *not*. The total ionic equation shows the reaction that occurs as Pb^{2+} and I^- ions collide and form a precipitate:

$$2Na^+(aq) + 2I^-(aq) + Pb^{2+}(aq) + 2NO_3^-(aq) \longrightarrow 2Na^+(aq) + 2NO_3^-(aq) + PbI_2(s)$$

And the net ionic equation confirms it:

$$Pb^{2+}(aq) + 2I^-(aq) \longrightarrow PbI_2(s)$$

A Type of Metathesis Reaction The molecular equation for the reaction between $Pb(NO_3)_2$ and NaI shows *the ions exchanging partners* (Figure 4.9):

$$2NaI(aq) + Pb(NO_3)_2(aq) \longrightarrow PbI_2(s) + 2NaNO_3(aq)$$

Such reactions are called *double-displacement reactions,* or **metathesis** (pronounced *meh-TA-thuh-sis*) **reactions.** Some that involve precipitation are important in industry, including the preparation of silver bromide used in black-and-white film:

$$AgNO_3(aq) + KBr(aq) \longrightarrow AgBr(s) + KNO_3(aq)$$

The reactions that form Ag_2CrO_4 (Figure 4.7) and CaF_2 (Figure 4.8) are also examples of metathesis reactions, as are acid-base reactions (Section 4.4). Sample Problems 4.8 and 4.9 provide practice in predicting if a precipitate forms.

$2NaI(aq) + Pb(NO_3)_2(aq) \longrightarrow PbI_2(s) + 2NaNO_3(aq)$

Figure 4.9 The precipitation of PbI_2, a metathesis reaction.

Source: © McGraw-Hill Education/Stephen Frisch, photographer

SAMPLE PROBLEM 4.8 Predicting Whether a Precipitation Reaction Occurs; Writing Ionic Equations

Problem Does a reaction occur when each of these pairs of solutions is mixed? If so, write balanced molecular, total ionic, and net ionic equations, and identify the spectator ions.

(a) Potassium fluoride(*aq*) + strontium nitrate(*aq*) ⟶

(b) Ammonium perchlorate(*aq*) + sodium bromide(*aq*) ⟶

Plan We note the reactant ions, write the cation-anion combinations, and refer to Table 4.1 to see if any are insoluble. For the molecular equation, we predict the products and write them all as intact compounds. For the total ionic equation, we write the soluble compounds as separate ions. For the net ionic equation, we eliminate the spectator ions.

Solution **(a)** In addition to the reactants, the two other ion combinations are strontium fluoride and potassium nitrate. Table 4.1 shows that strontium fluoride is insoluble, so a reaction *does* occur. Writing the molecular equation:

$$2KF(aq) + Sr(NO_3)_2(aq) \longrightarrow SrF_2(s) + 2KNO_3(aq)$$

Writing the total ionic equation:

$$2K^+(aq) + 2F^-(aq) + Sr^{2+}(aq) + 2NO_3^-(aq) \longrightarrow SrF_2(s) + 2K^+(aq) + 2NO_3^-(aq)$$

Writing the net ionic equation:

$$Sr^{2+}(aq) + 2F^-(aq) \longrightarrow SrF_2(s)$$

The spectator ions are K^+ and NO_3^-.

(b) The other ion combinations are ammonium bromide and sodium perchlorate. Table 4.1 shows that ammonium, sodium, and perchlorate compounds are soluble, and all bromides are soluble except those of Ag^+, Pb^{2+}, Cu^+, and Hg_2^{2+}. Therefore, ***no*** **reaction occurs.** The compounds remain as solvated ions:

$$NH_4^+(aq) + ClO_4^-(aq) + Na^+(aq) + Br^-(aq) \longrightarrow NH_4^+(aq) + Br^-(aq) + Na^+(aq) + ClO_4^-(aq)$$

FOLLOW-UP PROBLEMS

4.8A Predict whether a reaction occurs, and if so, write balanced total and net ionic equations:
(a) Iron(III) chloride(aq) + cesium phosphate(aq) ⟶
(b) Sodium hydroxide(aq) + cadmium nitrate(aq) ⟶
(c) Magnesium bromide(aq) + potassium acetate(aq) ⟶

4.8B Predict whether a reaction occurs, and if so, write balanced total and net ionic equations:
(a) Silver nitrate(aq) + barium chloride(aq) ⟶
(b) Ammonium carbonate(aq) + potassium sulfide(aq) ⟶
(c) Nickel(II) sulfate(aq) + lead(II) nitrate(aq) ⟶

SOME SIMILAR PROBLEMS 4.41–4.46

SAMPLE PROBLEM 4.9 Using Molecular Depictions in Precipitation Reactions

Problem The molecular views below depict reactant solutions for a precipitation reaction (with ions shown as colored spheres and water molecules omitted for clarity):

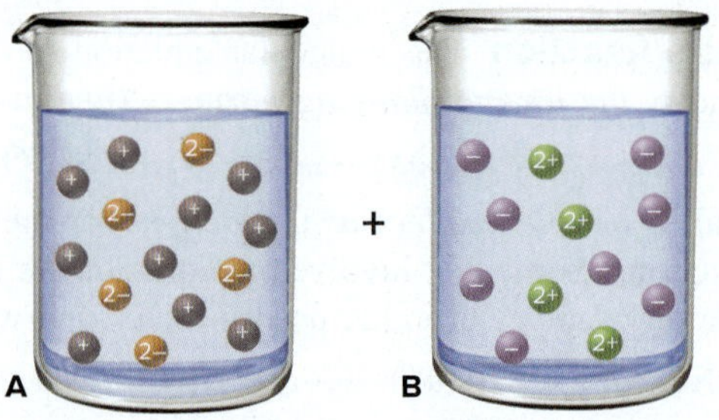

(a) Which compound is dissolved in beaker A: KCl, Na_2SO_4, $MgBr_2$, or Ag_2SO_4?
(b) Which compound is dissolved in beaker B: NH_4NO_3, $MgSO_4$, $Ba(NO_3)_2$, or CaF_2?
(c) Name the precipitate and the spectator ions when solutions A and B are mixed, and write balanced molecular, total ionic, and net ionic equations for any reaction.
(d) If each particle represents 0.010 mol of ions, what is the maximum mass (g) of precipitate that can form (assuming complete reaction)?

Plan **(a)** and **(b)** From the depictions, we note the charge and number of each kind of ion and use Table 4.1 to determine the ion combinations that are soluble. **(c)** Once we know the combinations, Table 4.1 tells which two ions form the solid, so the other two are spectator ions. **(d)** This part is a limiting-reactant problem because the amounts of two species are involved. We count the number of each kind of ion that forms the solid. We multiply the number of each reactant ion by 0.010 mol and calculate the amount (mol) of product that forms from each. Whichever ion forms less is limiting, so we use the molar mass of the precipitate to find mass (g).

Solution **(a)** In solution A, there are two 1+ particles for each 2− particle. Therefore, the dissolved compound cannot be KCl or $MgBr_2$. Of the remaining two choices, Ag_2SO_4 is insoluble, so the dissolved compound must be Na_2SO_4.
(b) In solution B, there are two 1− particles for each 2+ particle. Therefore, the dissolved compound cannot be NH_4NO_3 or $MgSO_4$. Of the remaining two choices, CaF_2 is insoluble, so the dissolved compound must be $Ba(NO_3)_2$.
(c) Of the two possible ion combinations, $BaSO_4$ and $NaNO_3$, $BaSO_4$ is insoluble, so Na^+ and NO_3^- are spectator ions.

Molecular: $Ba(NO_3)_2(aq) + Na_2SO_4(aq) \longrightarrow BaSO_4(s) + 2NaNO_3(aq)$

Total ionic: $Ba^{2+}(aq) + 2NO_3^-(aq) + 2Na^+(aq) + SO_4^{2-}(aq) \longrightarrow BaSO_4(s) + 2NO_3^-(aq) + 2Na^+(aq)$

Net ionic: $Ba^{2+}(aq) + SO_4^{2-}(aq) \longrightarrow BaSO_4(s)$

(d) Finding the ion that is the limiting reactant:
For Ba^{2+}:

$$\text{Amount (mol) of } BaSO_4 = 4\ \cancel{Ba^{2+}\text{ particles}} \times \frac{0.010\ \cancel{\text{mol }Ba^{2+}\text{ ions}}}{1\ \cancel{Ba^{2+}\text{ particle}}} \times \frac{1\text{ mol }BaSO_4}{1\ \cancel{\text{mol }Ba^{2+}\text{ ions}}}$$
$$= 0.040\text{ mol }BaSO_4$$

For SO_4^{2-}:

$$\text{Amount (mol) of } BaSO_4 = 5\ \cancel{SO_4^{2-}\text{ particles}} \times \frac{0.010\ \cancel{\text{mol }SO_4^{2-}\text{ ions}}}{1\ \cancel{SO_4^{2-}\text{ particle}}} \times \frac{1\text{ mol }BaSO_4}{1\ \cancel{\text{mol }SO_4^{2-}\text{ ions}}}$$
$$= 0.050\text{ mol }BaSO_4$$

Therefore, Ba^{2+} ion is the limiting reactant.

Calculating the mass (g) of product ($\mathcal{M}$ of $BaSO_4$ = 233.4 g/mol):

$$\text{Mass (g) of } BaSO_4 = 0.040\ \cancel{\text{mol }BaSO_4} \times \frac{233.4\text{ g }BaSO_4}{1\ \cancel{\text{mol }BaSO_4}} = 9.3\text{ g }BaSO_4$$

Check Counting the number of Ba^{2+} particles allows a more direct calculation for a check: four Ba^{2+} particles means the maximum mass of $BaSO_4$ that can form is

$$\text{Mass(g) of } BaSO_4 = 4\ \cancel{Ba^{2+}\text{ particles}} \times \frac{0.010\ \cancel{\text{mol }Ba^{2+}\text{ ions}}}{1\ \cancel{Ba^{2+}\text{ particle}}} \times \frac{1\ \cancel{\text{mol }BaSO_4}}{1\ \cancel{\text{mol }Ba^{2+}\text{ ions}}} \times \frac{233.4\text{ g }BaSO_4}{1\ \cancel{\text{mol}}}$$
$$= 9.3\text{ g }BaSO_4$$

FOLLOW-UP PROBLEMS

4.9A Molecular views of the reactant solutions for a precipitation reaction are shown below (with ions represented as spheres and water molecules omitted):

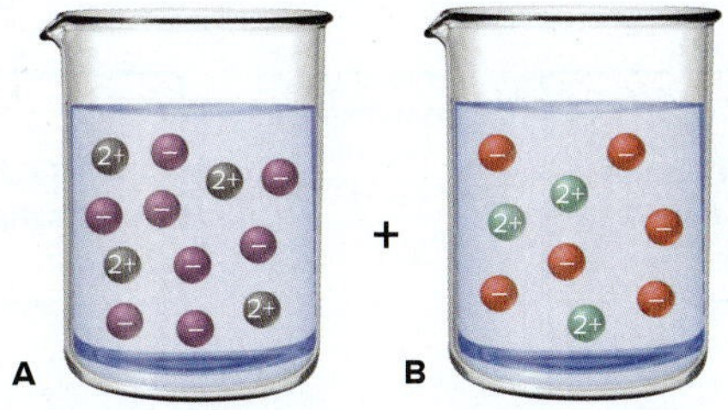

(a) Which compound is dissolved in beaker A: $Zn(NO_3)_2$, KCl, Na_2SO_4, or $PbCl_2$?
(b) Which compound is dissolved in beaker B: $(NH_4)_2SO_4$, $Cd(OH)_2$, $Ba(OH)_2$, or KNO_3?
(c) Name the precipitate and the spectator ions when solutions A and B are mixed, and write balanced molecular, total ionic, and net ionic equations for the reaction.
(d) If each particle represents 0.050 mol of ions, what is the maximum mass (g) of precipitate that can form (assuming complete reaction)?

4.9B Molecular views of the reactant solutions for a precipitation reaction are shown below (with ions represented as spheres and water molecules omitted):

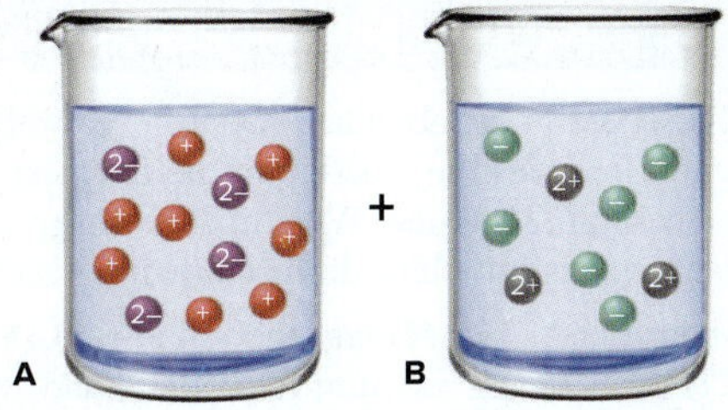

(a) Which compound is dissolved in beaker A: Li_2CO_3, NH_4Cl, Ag_2SO_4, or FeS?
(b) Which compound is dissolved in beaker B: $Ni(OH)_2$, $AgNO_3$, $CaCl_2$, or $BaSO_4$?
(c) Name the precipitate and the spectator ions when solutions A and B are mixed, and write balanced molecular, total ionic, and net ionic equations for the reaction.
(d) If each particle represents 0.20 mol of ions, what is the maximum mass (g) of precipitate that can form (assuming complete reaction)?

SOME SIMILAR PROBLEMS 4.51 and 4.52

Stoichiometry of Precipitation Reactions

In Chapter 3, we saw that the amounts (mol) of reactants and products in a reaction are stoichiometrically equivalent to each other and that the molar ratios between them can be used as conversion factors to calculate the amount of one substance that reacts with, produces, or is formed from a specific amount of another substance. Solving stoichiometry problems for any reaction that takes place in solution, such as a precipitation reaction, requires the additional step of using the solution molarity (mol/L) to convert the volume of reactant or product in solution to amount (mol):

1. Write a balanced equation.
2. Find the amount (mol) of one substance using the volume and molarity (for a substance in solution) or using its molar mass (for a pure substance).
3. Use the molar ratio to relate that amount to the stoichiometrically equivalent amount of another substance.
4. Convert to the desired units.

Figure 4.10 summarizes the possible relationships among quantities of substances in a reaction occurring in solution. Sample Problems 4.10 and 4.11 apply these relationships to precipitation reactions, and we'll also use them for the other classes of reactions.

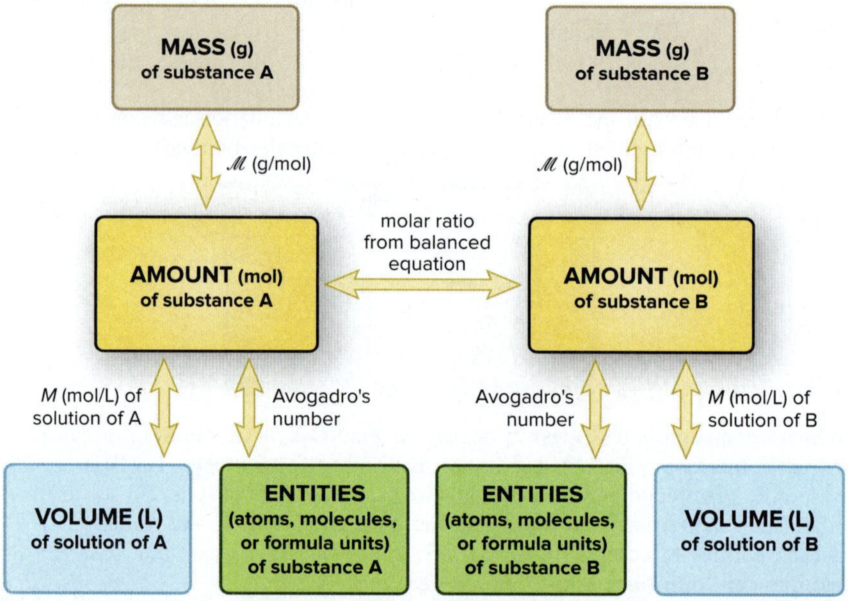

Figure 4.10 Summary of amount-mass-number relationships for a chemical reaction in solution.

SAMPLE PROBLEM 4.10 **Calculating Amounts of Reactants and Products in a Precipitation Reaction**

Road Map

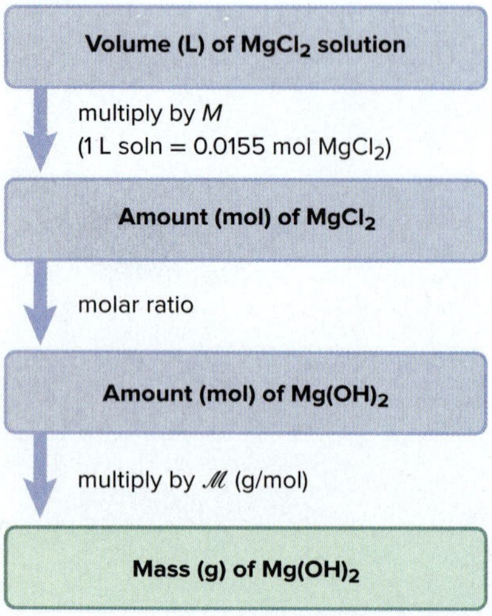

Problem Magnesium is the second most abundant metal in seawater, after sodium. The first step in its industrial extraction involves the reaction of the magnesium ion with calcium hydroxide to precipitate magnesium hydroxide. What mass of magnesium hydroxide is formed when 0.180 L of 0.0155 *M* magnesium chloride reacts with excess calcium hydroxide?

Plan We are given the molarity (0.0155 *M*) and the volume (0.180 L) of magnesium chloride solution that reacts with excess calcium hydroxide, and we must find the mass of precipitate. After writing the balanced equation, we find the amount (mol) of magnesium chloride from its molarity and volume and use the molar ratio to find the amount (mol) of magnesium hydroxide that is produced. Finally, we use the molar mass to convert from amount (mol) to mass (g) of magnesium hydroxide (see the road map).

Solution Writing the balanced equation:

$$MgCl_2(aq) + Ca(OH)_2(aq) \longrightarrow Mg(OH)_2(s) + CaCl_2(aq)$$

Finding the amount (mol) of $MgCl_2$:

$$\text{Amount (mol) of } MgCl_2 = 0.180 \cancel{\text{L } MgCl_2} \times \frac{0.0155 \text{ mol } MgCl_2}{1 \cancel{\text{L } MgCl_2}} = 0.00279 \text{ mol } MgCl_2$$

Using the molar ratio to convert amount (mol) of $MgCl_2$ to amount (mol) of $Mg(OH)_2$:

$$\text{Amount (mol) of } Mg(OH)_2 = 0.00279 \cancel{\text{mol } MgCl_2} \times \frac{1 \text{ mol } Mg(OH)_2}{1 \cancel{\text{mol } MgCl_2}}$$
$$= 0.00279 \text{ mol } Mg(OH)_2$$

Converting from amount (mol) of $Mg(OH)_2$ to mass (g):

$$\text{Mass (g) of } Mg(OH)_2 = 0.00279 \cancel{\text{mol } Mg(OH)_2} \times \frac{58.33\ Mg(OH)_2}{1 \cancel{\text{mol } Mg(OH)_2}} = 0.163 \text{ g } Mg(OH)_2$$

Check The answer seems reasonable; rounding to check the math shows that we have (0.2 L) × (0.02 *M*) = 0.004 mol of $MgCl_2$; due to the 1/1 molar ratio, 0.004 mol of $Mg(OH)_2$ is produced.

FOLLOW-UP PROBLEMS

4.10A It is desirable to remove calcium ion from hard water to prevent the formation of precipitates known as *boiler scale* that reduce heating efficiency. The calcium ion is reacted with sodium phosphate to form solid calcium phosphate, which is easier to remove than boiler scale. What volume of 0.260 *M* sodium phosphate is needed to react completely with 0.300 L of 0.175 *M* calcium chloride?

4.10B To lift fingerprints from a crime scene, a solution of silver nitrate is sprayed on a surface to react with the sodium chloride left behind by perspiration. What is the molarity of a silver nitrate solution if 45.0 mL of it reacts with excess sodium chloride to produce 0.148 g of precipitate?

SOME SIMILAR PROBLEMS 4.47, 4.48, 4.53, and 4.54

Except for the additional step of finding amounts (mol) in solution, limiting-reactant problems for precipitation reactions in solution are handled just like other such problems.

SAMPLE PROBLEM 4.11 Solving a Limiting-Reactant Problem for a Precipitation Reaction

Problem Iron(III) hydroxide, used to adsorb arsenic and heavy metals from contaminated soil and water, is produced by reacting aqueous solutions of iron(III) chloride and sodium hydroxide.
(a) What mass of iron(III) hydroxide is formed when 0.155 L of 0.250 *M* iron(III) chloride reacts with 0.215 L of 0.300 *M* sodium hydroxide?
(b) Write a reaction table for this process.

Plan This is a limiting-reactant problem because *the quantities of two reactants are given.* After balancing the equation, we determine the limiting reactant. From the molarity and volume of each solution, we calculate the amount (mol) of each reactant. Then, we use the molar ratio to find the amount of product [$Fe(OH)_3$] that each reactant forms. The limiting reactant forms fewer moles of $Fe(OH)_3$, which we convert to mass (g) of $Fe(OH)_3$ using its molar mass (see the road map). We use the amount of $Fe(OH)_3$ formed from the limiting reactant in a reaction table (see Section 3.4).

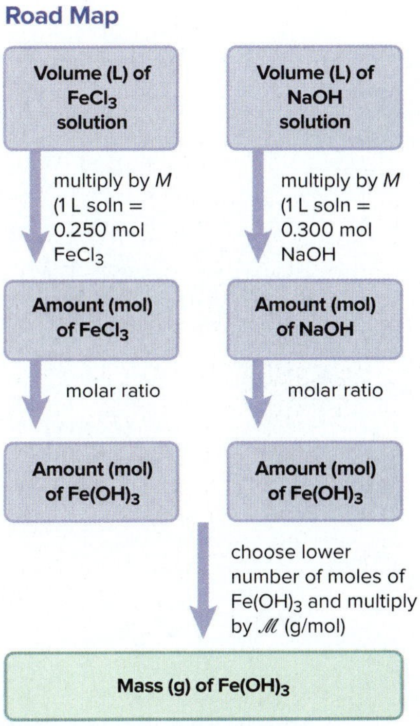

Solution **(a)** Writing the balanced equation:

$$FeCl_3(aq) + 3NaOH(aq) \longrightarrow Fe(OH)_3(s) + 3NaCl(aq)$$

Finding the amount (mol) of $Fe(OH)_3$ formed from $FeCl_3$: Combining the steps gives

$$\text{Amount (mol) of } Fe(OH)_3 = 0.155 \cancel{\text{L soln}} \times \frac{0.250 \cancel{\text{mol } FeCl_3}}{1 \cancel{\text{L soln}}} \times \frac{1 \text{ mol } Fe(OH)_3}{1 \cancel{\text{mol } FeCl_3}}$$
$$= \mathbf{0.0388 \text{ mol } Fe(OH)_3}$$

Finding the amount (mol) of $Fe(OH)_3$ from NaOH: Combining the steps gives

$$\text{Amount (mol) of } Fe(OH)_3 = 0.215 \cancel{\text{L soln}} \times \frac{0.300 \cancel{\text{mol NaOH}}}{1 \cancel{\text{L soln}}} \times \frac{1 \text{ mol } Fe(OH)_3}{3 \cancel{\text{mol NaOH}}}$$
$$= \mathbf{0.0215 \text{ mol } Fe(OH)_3}$$

NaOH is the limiting reactant because it forms fewer moles of $Fe(OH)_3$.

Converting the amount (mol) of $Fe(OH)_3$ formed from NaOH to mass (g):

$$\text{Mass (g) of } Fe(OH)_3 = 0.0215 \text{ mol } Fe(OH)_3 \times \frac{106.87 \text{ g } Fe(OH)_3}{1 \text{ mol } Fe(OH)_3} = 2.30 \text{ g } Fe(OH)_3$$

(b) With NaOH as the limiting reactant, the reaction table is

Amount (mol)	$FeCl_3(aq)$	+	$3NaOH(aq)$	⟶	$Fe(OH)_3(s)$	+	$3NaCl(aq)$
Initial	0.0388		0.0645		0		0
Change	−0.0215		−0.0645		+0.0215		+0.0645
Final	0.0173		0		0.0215		0.0645

A large excess of $FeCl_3$ remains after the reaction. Note that the amount of NaCl formed is three times the amount of $FeCl_3$ consumed, as the balanced equation shows.

Check As a check on our choice of the limiting reactant, let's use the alternative method outlined in the Comment in Sample Problem 3.20.

Finding amounts (mol) of reactants given:

$$\text{Amount (mol) of } FeCl_3 = 0.155 \text{ L soln} \times \frac{0.250 \text{ mol } FeCl_3}{1 \text{ L soln}} = 0.0388 \text{ mol } FeCl_3$$

$$\text{Amount (mol) of NaOH} = 0.215 \text{ L soln} \times \frac{0.300 \text{ mol NaOH}}{1 \text{ L soln}} = 0.0645 \text{ mol NaOH}$$

The molar ratio of the reactants is 1 $FeCl_3$/3 NaOH. Therefore, NaOH is limiting because there is less of it than the $3 \times 0.0388 = 0.116$ mol we would need to react with all of the available $FeCl_3$.

FOLLOW-UP PROBLEMS

4.11A Despite the toxicity of lead, many of its compounds are still used to make pigments. **(a)** When 268 mL of 1.50 *M* lead(II) acetate reacts with 130. mL of 3.40 *M* sodium chloride, how many grams of solid lead(II) chloride can form? **(b)** Using the abbreviation "Ac" for the acetate ion, write a reaction table for the process.

4.11B Mercury and its compounds have uses from fillings for teeth (as a mixture with silver, copper, and tin) to the production of chlorine. Because of their toxicity, however, soluble mercury compounds, such as mercury(II) nitrate, must be removed from industrial wastewater. One removal method reacts the wastewater with sodium sulfide solution to produce solid mercury(II) sulfide and sodium nitrate solution. In a laboratory simulation, 0.050 L of 0.010 *M* mercury(II) nitrate reacts with 0.020 L of 0.10 *M* sodium sulfide. **(a)** What mass of mercury(II) sulfide is formed? **(b)** Write a reaction table for this process.

SOME SIMILAR PROBLEMS 4.49, 4.50, and 4.56

› Summary of Section 4.3

- In a precipitation reaction, an insoluble ionic compound forms when solutions of two soluble ones are mixed. The electrostatic attraction between certain pairs of solvated ions is strong enough to overcome the attraction of each ion for water molecules.
- Based on a set of solubility rules, we can predict the formation of a precipitate by noting which of all possible cation-anion combinations is insoluble.
- When chemical changes such as precipitation reactions occur in solution, amounts of reactants and products are given in terms of concentration and volume.
- By using molarity as a conversion factor, we can apply the principles of stoichiometry to precipitation reactions in solution.

4.4 ACID-BASE REACTIONS

Aqueous acid-base reactions occur in processes as diverse as the metabolic action of proteins and carbohydrates, the industrial production of fertilizer, and the revitalization of lakes damaged by acid rain.

These reactions involve water as reactant or product, in addition to its common role as solvent. Of course, an **acid-base reaction** (also called a **neutralization reaction**) occurs when an acid reacts with a base, but the definitions of these terms and the scope of this reaction class have changed over the years. For our purposes at this point, we'll use definitions that apply to substances found commonly in the lab:

- *An* ***acid*** *is a substance that produces H^+ ions when dissolved in water.*

$$HX \xrightarrow{H_2O} H^+(aq) + X^-(aq)$$

- *A* ***base*** *is a substance that produces OH^- ions when dissolved in water.*

$$MOH \xrightarrow{H_2O} M^+(aq) + OH^-(aq)$$

(Other definitions are presented later in this section and in Chapter 18.)

Acids and the Solvated Proton *Acidic solutions arise when certain covalent H-containing molecules dissociate into ions in water.* In every case, these molecules contain a *polar bond to H* in which the other atom pulls much more strongly on the electron pair. A good example is HBr. The Br end of the H—Br bond is partially negative, and the H end is partially positive. When hydrogen bromide gas dissolves in water, the poles of H_2O molecules are attracted to the oppositely charged poles of HBr molecules. The bond breaks, with H becoming the solvated cation $H^+(aq)$ and Br becoming the solvated anion $Br^-(aq)$:

$$HBr(g) \xrightarrow{H_2O} H^+(aq) + Br^-(aq)$$

The solvated H^+ ion is a very unusual species. The H atom is a proton surrounded by an electron, so H^+ is just a proton. With a full positive charge concentrated in such a tiny volume, H^+ attracts the negative pole of water molecules so strongly that it forms a covalent bond to one of them. We can show this interaction by writing the solvated H^+ ion as an H_3O^+ ion (**hydronium ion**) that also is solvated:

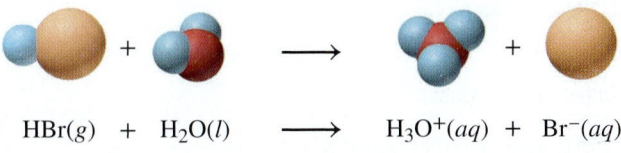

$$HBr(g) + H_2O(l) \longrightarrow H_3O^+(aq) + Br^-(aq)$$

The hydronium ion, which we write as H_3O^+ [or $(H_2O)H^+$], associates with other water molecules to give species such as $H_5O_2^+$ [or $(H_2O)_2H^+$], $H_7O_3^+$ [or $(H_2O)_3H^+$], $H_9O_4^+$ [or $(H_2O)_4H^+$], and so forth. Figure 4.11 shows $H_7O_3^+$, an H_3O^+ ion associated *(dotted lines)* with two H_2O molecules. As a general notation for these various species, we write $H^+(aq)$, but later in this chapter and in much of the text, we'll use $H_3O^+(aq)$.

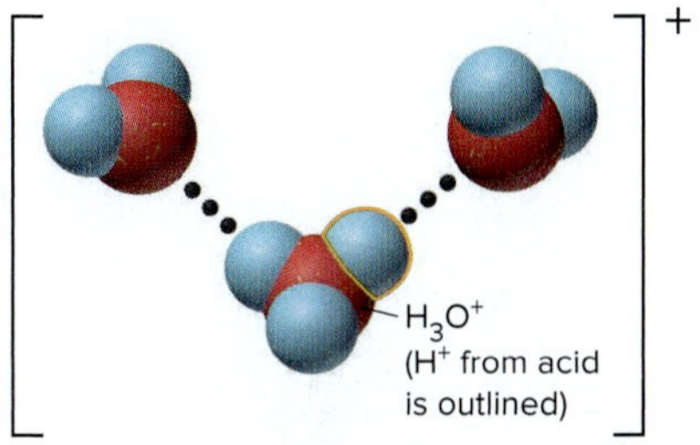

Figure 4.11 The H^+ ion as a solvated hydronium ion.

Acids and Bases as Electrolytes Acids and bases are categorized in terms of their "strength," the degree to which they dissociate into ions in water:

- *Strong acids and strong bases dissociate completely into ions.* Therefore, like soluble ionic compounds, they are *strong* electrolytes and conduct a large current, as shown by the brightly lit bulb (Figure 4.12A, on the next page).

- *Weak acids and weak bases dissociate very little into ions.* Most of their molecules remain intact. Therefore, they are weak electrolytes, which means they conduct a small current (Figure 4.12B).

Table 4.2 lists the strong acids and bases and a few examples of weak acids and bases. Because a strong acid (or strong base) dissociates completely, we can find the molarity

Table 4.2 Strong and Weak Acids and Bases

Acids

Strong
Hydrochloric acid, HCl
Hydrobromic acid, HBr
Hydriodic acid, HI
Nitric acid, HNO_3
Sulfuric acid, H_2SO_4
Perchloric acid, $HClO_4$

Weak (a few of many examples)
Hydrofluoric acid, HF
Phosphoric acid, H_3PO_4
Acetic acid, CH_3COOH (or $HC_2H_3O_2$)

Bases

Strong
Group 1A(1) hydroxides:
Lithium hydroxide, LiOH
Sodium hydroxide, NaOH
Potassium hydroxide, KOH
Rubidium hydroxide, RbOH
Cesium hydroxide, CsOH

Heavy Group 2A(2) hydroxides:
Calcium hydroxide, $Ca(OH)_2$
Strontium hydroxide, $Sr(OH)_2$
Barium hydroxide, $Ba(OH)_2$

Weak (one of many examples)
Ammonia, NH_3

A Strong acid (or base) = strong electrolyte

B Weak acid (or base) = weak electrolyte

Figure 4.12 Acids and bases as electrolytes.

Source: © McGraw-Hill Education/Stephen Frisch, photographer

of H^+ (or OH^-) and the amount (mol) or number of each ion in solution. (You'll see how to determine these quantities for weak acids in Chapter 18.)

SAMPLE PROBLEM 4.12 Determining the Number of H^+ (or OH^-) Ions in Solution

Problem Nitric acid is a major chemical in the fertilizer and explosives industries. How many $H^+(aq)$ ions are in 25.3 mL of 1.4 *M* nitric acid?

Plan We know the volume (25.3 mL) and molarity (1.4 *M*) of the nitric acid, and we need the number of $H^+(aq)$. We convert from mL to L and multiply by the molarity to find the amount (mol) of acid. Table 4.2 shows that nitric acid is a strong acid, so it dissociates completely. With the formula, we write a dissociation equation, which shows the amount (mol) of H^+ ions per mole of acid. We multiply that amount by Avogadro's number to find the number of $H^+(aq)$ ions (see the road map).

Road Map

Volume (mL) of HNO_3

↓ 10^3 mL = 1 L

Volume (L) of HNO_3

↓ multiply by *M* (1 L soln = 1.4 mol HNO_3)

Amount (mol) of HNO_3

↓ mol of H^+/mol of HNO_3

Amount (mol) of H^+ ions

↓ multiply by Avogadro's number

No. of H^+ ions

Solution Finding the amount (mol) of nitric acid:

$$\text{Amount (mol) of HNO}_3 = 25.3\ \text{mL} \times \frac{1\ \text{L}}{1000\ \text{mL}} \times \frac{1.4\ \text{mol}}{1\ \text{L}} = 0.035\ \text{mol}$$

Nitric acid is HNO_3, so we have:

$$HNO_3(l) \xrightarrow{H_2O} H^+(aq) + NO_3^-(aq)$$

Finding the number of H^+ ions:

$$\text{No. of H}^+\ \text{ions} = 0.035\ \text{mol HNO}_3 \times \frac{1\ \text{mol H}^+}{1\ \text{mol HNO}_3} \times \frac{6.022\times10^{23}\ \text{H}^+\ \text{ions}}{1\ \text{mol H}^+}$$

$$= 2.1\times10^{22}\ \text{H}^+\ \text{ions}$$

Check The number of moles seems correct: 0.025 L × 1.4 mol/L = 0.035 mol, and multiplying by 6×10^{23} ions/mol gives 2×10^{22} ions.

FOLLOW-UP PROBLEMS

4.12A How many $OH^-(aq)$ ions are present in 451 mL of 0.0120 *M* calcium hydroxide?

4.12B How many $H^+(aq)$ ions are present in 65.5 mL of 0.722 *M* hydrochloric acid?

SOME SIMILAR PROBLEMS 4.63 and 4.64

Structural Features of Acids and Bases A key structural feature appears in common laboratory acids and bases:

- *Acids.* Strong acids, such as HNO_3 and H_2SO_4, and weak acids, such as HF and H_3PO_4, have one or more H atoms as part of their structure, which are either completely released (strong) or partially released (weak) as protons in water.

- *Bases.* Strong bases have either OH^- (e.g., NaOH) or O^{2-} (e.g., K_2O) as part of their structure. The oxide ion is not stable in water and reacts to form OH^- ion:

$$O^{2-}(s) + H_2O(l) \longrightarrow 2OH^-(aq) \quad \text{so} \quad K_2O(s) + H_2O(l) \longrightarrow 2K^+(aq) + 2OH^-(aq)$$

Weak bases, such as ammonia, do not contain OH^- ions, but, as you'll see in later chapters, they all have an electron pair on a nitrogen atom. That nitrogen atom attracts one of the H atoms in a water molecule. The loss of an H^+ ion from the H_2O molecule produces OH^- ions:

$$NH_3(g) + H_2O(l) \rightleftharpoons NH_4^+(aq) + OH^-(aq)$$

(The reaction arrow in the preceding equation indicates that the reaction proceeds in both directions; we'll discuss this important idea further in Section 4.7.)

The Key Event: Formation of H_2O from H^+ and OH^-

To see the key event in acid-base reactions, we'll focus on the reaction between the strong acid HCl and the strong base $Ba(OH)_2$ and write the three types of aqueous ionic equations (with color):

- The molecular equation is

$$2HCl(aq) + Ba(OH)_2(aq) \longrightarrow BaCl_2(aq) + 2H_2O(l)$$

- HCl and $Ba(OH)_2$ dissociate completely, so the total ionic equation is

$$2H^+(aq) + 2Cl^-(aq) + Ba^{2+}(aq) + 2OH^-(aq) \longrightarrow Ba^{2+}(aq) + 2Cl^-(aq) + 2H_2O(l)$$

- In the net ionic equation, we eliminate the spectator ions, $Ba^{2+}(aq)$ and $Cl^-(aq)$:

$$2H^+(aq) + 2OH^-(aq) \longrightarrow 2H_2O(l) \quad \text{or} \quad H^+(aq) + OH^-(aq) \longrightarrow H_2O(l)$$

Thus, *the key event in aqueous reactions between a strong acid and a strong base is that an H^+ ion from the acid and an OH^- ion from the base form a water molecule.* Only the spectator ions differ from one strong acid–strong base reaction to another.

Like precipitation reactions, acid-base reactions occur through *the electrostatic attraction of ions and their removal from solution as the product.* In this case, rather than an insoluble ionic solid, the product is H_2O, which consists almost entirely of undissociated molecules. Actually, water molecules dissociate *very* slightly (which, as you'll see in Chapter 18, is very important), but the formation of water in an acid-base reaction results in an enormous net removal of H^+ and OH^- ions.

The molecular and total ionic equations above show that if you evaporate the water, the spectator ions remain: the ionic compound that results from the reaction of an acid and a base is called a **salt,** which in this case is barium chloride. Thus, in an aqueous neutralization reaction, *an acid and a base form a salt solution and water:*

$$\underset{\text{acid}}{HX(aq)} + \underset{\text{base}}{MOH(aq)} \longrightarrow \underset{\text{salt}}{MX(aq)} + \underset{\text{water}}{H_2O(l)}$$

Note that *the cation of the salt comes from the base and the anion from the acid.*

Like precipitation reactions, *acid-base reactions are metathesis (double-displacement) reactions.* The reaction of aluminum hydroxide, the active ingredient in some antacids, with HCl, the major component of stomach acid, is another example:

$$3HCl(aq) + Al(OH)_3(s) \longrightarrow AlCl_3(aq) + 3H_2O(l)$$

SAMPLE PROBLEM 4.13 Writing Ionic Equations for Acid-Base Reactions

Problem Write balanced molecular, total ionic, and net ionic equations for each of the following acid-base reactions and identify the spectator ions:

(a) Hydrochloric acid(*aq*) + potassium hydroxide(*aq*) $\longrightarrow$

(b) Strontium hydroxide(*aq*) + perchloric acid(*aq*) $\longrightarrow$

(c) Barium hydroxide(*aq*) + sulfuric acid(*aq*) $\longrightarrow$

Plan All are strong acids and bases (see Table 4.2), so the actual reaction is between H^+ and OH^-. The products are H_2O and a salt solution of spectator ions. In (c), we note that the salt ($BaSO_4$) is insoluble (see Table 4.1), so there are no spectator ions.

Solution

(a) Writing the molecular equation:

$$HCl(aq) + KOH(aq) \longrightarrow KCl(aq) + H_2O(l)$$

Writing the total ionic equation:

$$H^+(aq) + Cl^-(aq) + K^+(aq) + OH^-(aq) \longrightarrow K^+(aq) + Cl^-(aq) + H_2O(l)$$

Writing the net ionic equation:

$$H^+(aq) + OH^-(aq) \longrightarrow H_2O(l)$$

$K^+(aq)$ and $Cl^-(aq)$ are the spectator ions.

(b) Writing the molecular equation:

$$Sr(OH)_2(aq) + 2HClO_4(aq) \longrightarrow Sr(ClO_4)_2(aq) + 2H_2O(l)$$

Writing the total ionic equation:

$$Sr^{2+}(aq) + 2OH^-(aq) + 2H^+(aq) + 2ClO_4^-(aq) \longrightarrow Sr^{2+}(aq) + 2ClO_4^-(aq) + 2H_2O(l)$$

Writing the net ionic equation:

$$2OH^-(aq) + 2H^+(aq) \longrightarrow 2H_2O(l) \quad \text{or} \quad OH^-(aq) + H^+(aq) \longrightarrow H_2O(l)$$

$Sr^{2+}(aq)$ and $ClO_4^-(aq)$ are the spectator ions.

(c) Writing the molecular equation:

$$Ba(OH)_2(aq) + H_2SO_4(aq) \longrightarrow BaSO_4(s) + 2H_2O(l)$$

Writing the total ionic equation:

$$Ba^{2+}(aq) + 2OH^-(aq) + 2H^+(aq) + SO_4^{2-}(aq) \longrightarrow BaSO_4(s) + 2H_2O(l)$$

This is a neutralization *and* a precipitation reaction, so the net ionic equation is the same as the total ionic. There are no spectator ions.

FOLLOW-UP PROBLEMS

4.13A Write balanced molecular, total ionic, and net ionic equations for the reaction between aqueous solutions of calcium hydroxide and nitric acid.

4.13B Write balanced molecular, total ionic, and net ionic equations for the reaction between aqueous solutions of hydriodic acid and lithium hydroxide.

SOME SIMILAR PROBLEMS 4.65(a) and 4.66(a)

Proton Transfer in Acid-Base Reactions

When we take a closer look (with color) at the reaction between a strong acid and strong base, as well as several related reactions, a unifying pattern appears. Let's examine three types of reaction to gain insight into this pattern.

Reaction Between a Strong Acid and a Strong Base When HCl gas dissolves in water, the H^+ ion ends up bonded to a water molecule. Thus, hydrochloric acid actually consists of solvated H_3O^+ and Cl^- ions:

$$HCl(g) + H_2O(l) \longrightarrow H_3O^+(aq) + Cl^-(aq)$$

If we add NaOH solution, the total ionic equation shows that H_3O^+ *transfers a proton* to OH^- (leaving a water molecule written as H_2O, and forming a water molecule written as HOH):

H^+ transfer

$$[H_3O^+(aq) + Cl^-(aq)] + [Na^+(aq) + OH^-(aq)] \longrightarrow H_2O(l) + Cl^-(aq) + Na^+(aq) + HOH(l)$$

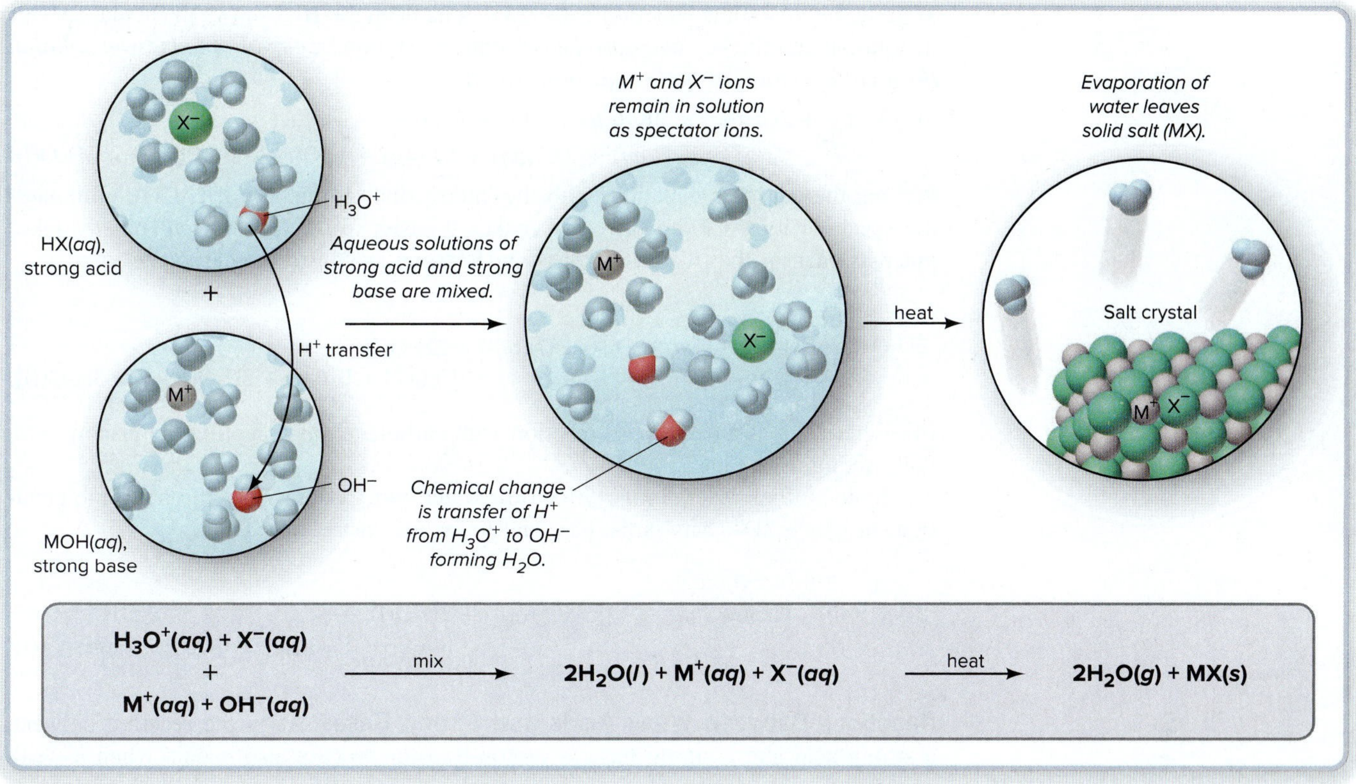

Figure 4.13 An aqueous strong acid–strong base reaction as a proton-transfer process.

Without the spectator ions, the net ionic equation shows more clearly the *transfer of a proton* from H_3O^+ to OH^-:

$$\overset{H^+ \text{ transfer}}{H_3O^+(aq) + OH^-(aq)} \longrightarrow H_2O(l) + HOH(l) \qquad [\text{or } 2H_2O(l)]$$

This equation is identical to the one we saw earlier in this section for the reaction of HCl and $Ba(OH)_2$,

$$H^+(aq) + OH^-(aq) \longrightarrow H_2O(l)$$

but with an additional H_2O molecule left over from the H_3O^+. Figure 4.13 shows this process on the atomic level and also shows that, if the water is evaporated, the spectator ions, Cl^- and Na^+, crystallize as the salt NaCl.

Thus, *an acid-base reaction is a proton-transfer process.* In the early 20th century, the chemists Johannes Brønsted and Thomas Lowry stated:

- *An acid is a molecule (or ion) that donates a proton.*
- *A base is a molecule (or ion) that accepts a proton.*

Therefore, in an aqueous reaction between strong acid and strong base, *H_3O^+ ion acts as the acid and donates a proton to OH^- ion, which acts as the base and accepts it.* (We discuss the Brønsted-Lowry concept thoroughly in Chapter 18.)

Gas-Forming Reactions: Acids with Carbonates (or Sulfites) When an ionic carbonate, such as K_2CO_3, is treated with an acid, such as HCl, one of the products is carbon dioxide, as the molecular equations show:

$$2HCl(aq) + K_2CO_3(aq) \longrightarrow 2KCl(aq) + [H_2CO_3(aq)]$$

$$[H_2CO_3(aq)] \longrightarrow H_2O(l) + CO_2(g)$$

Square brackets around a species, in this case H_2CO_3, mean it is very unstable: H_2CO_3 decomposes immediately into water and carbon dioxide. Combining these two equations (see Sample Problem 3.17) cancels $[H_2CO_3]$ and gives the overall equation:

$$2HCl(aq) + K_2CO_3(aq) \longrightarrow 2KCl(aq) + H_2O(l) + CO_2(g)$$

Writing the total ionic equation with H_3O^+ ions from the HCl shows the actual species in solution and the key event in a gas-forming reaction, *removal of ions from solution through formation of both a gas **and** water:*

$$2H_3O^+(aq) + 2Cl^-(aq) + 2K^+(aq) + CO_3^{2-}(aq) \longrightarrow$$
$$2Cl^-(aq) + 2K^+(aq) + CO_2(g) + H_2O(l) + 2H_2O(l) \quad [\text{or } 3H_2O(l)]$$

Writing the net ionic equation, with the intermediate formation of H_2CO_3, eliminates the spectator ions, Cl^- and K^+, and makes it easier to see that *proton transfer* takes place as each of the two H_3O^+ ions transfers one proton to the carbonate ion:

2H⁺ transfer

$$2H_3O^+(aq) + CO_3^{2-}(aq) \longrightarrow [H_2CO_3(aq)] + 2H_2O(l) \longrightarrow$$
$$CO_2(g) + H_2O(l) + 2H_2O(l) \quad [\text{or } 3H_2O(l)]$$

In essence, this is an acid-base reaction with carbonate ion accepting the protons and, thus, acting as the base.

Ionic sulfites react similarly to form water and gaseous SO_2; the net ionic equation, in which SO_3^{2-} acts as the base and forms the unstable H_2SO_3, is

2H⁺ transfer

$$2H_3O^+(aq) + SO_3^{2-}(aq) \longrightarrow [H_2SO_3(aq)] + 2H_2O(l) \longrightarrow$$
$$SO_2(g) + H_2O(l) + 2H_2O(l) \quad [\text{or } 3H_2O(l)]$$

Reactions Between Weak Acids and Strong Bases As in the reaction between a strong acid and a strong base, a proton-transfer process also occurs when a weak acid reacts with a strong base, but the total ionic equation is written differently.

For example, when solutions of the weak acid acetic acid (CH_3COOH, Table 4.2) and the strong base sodium hydroxide are mixed, the molecular equation is

$$CH_3COOH(aq) + NaOH(aq) \longrightarrow CH_3COONa(aq) + H_2O(l)$$

Since acetic acid is weak and, thus, dissociates very little, it appears on the left side of the total ionic equation as an *undissociated, intact molecule* instead of as ions:

$$CH_3COOH(aq) + Na^+(aq) + OH^-(aq) \longrightarrow CH_3COO^-(aq) + Na^+(aq) + H_2O(l)$$

The net ionic equation reveals that proton transfer occurs directly from the weak acid rather than from H_3O^+ (as is the case with a strong acid):

H⁺ transfer

$$CH_3COOH(aq) + OH^-(aq) \longrightarrow CH_3COO^-(aq) + H_2O(l)$$

There is only one spectator ion, $Na^+(aq)$, the cation of the strong base. For the reaction of *any weak acid with any strong base,* the net ionic equation has the form

$$HX(aq) + OH^-(aq) \longrightarrow X^-(aq) + H_2O(l)$$

Only the spectator ion from the strong base differs from one weak acid–strong base reaction to another. Table 4.3 compares the reactions of strong and weak acids with a strong base.

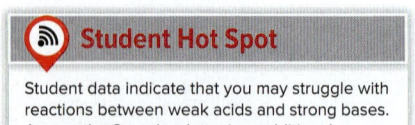

Student Hot Spot

Student data indicate that you may struggle with reactions between weak acids and strong bases. Access the Smartbook to view additional Learning Resources on this topic.

Table 4.3 Reactions of Strong and Weak Acids with a Strong Base

	Strong Acid and Strong Base	Weak Acid and Strong Base
Molecular equation	$HCl(aq) + NaOH(aq) \longrightarrow NaCl(aq) + H_2O(l)$	$CH_3COOH(aq) + NaOH(aq) \longrightarrow CH_3COONa(aq) + H_2O(l)$
Total ionic equation	$H_3O^+(aq) + Cl^-(aq) + Na^+(aq) + OH^-(aq) \longrightarrow H_2O(l) + Na^+(aq) + Cl^-(aq) + H_2O(l)$	$CH_3COOH(aq) + Na^+(aq) + OH^-(aq) \longrightarrow CH_3COO^-(aq) + Na^+(aq) + H_2O(l)$
Net ionic equation	$H^+(aq) + OH^-(aq) \longrightarrow H_2O(l)$	$CH_3COOH(aq) + OH^-(aq) \longrightarrow CH_3COO^-(aq) + H_2O(l)$

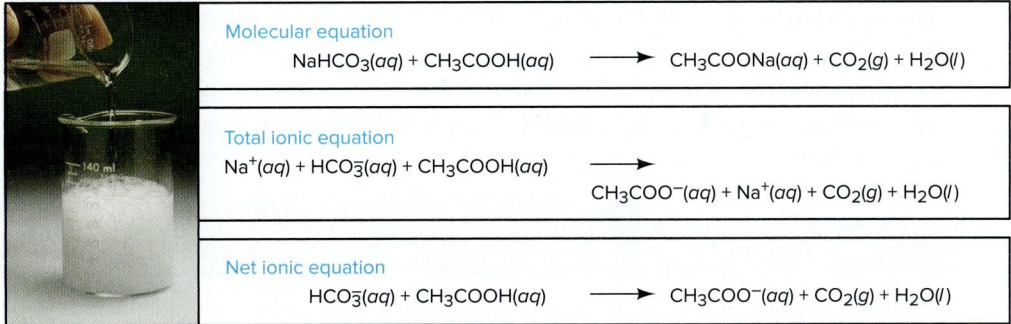

Molecular equation

$$NaHCO_3(aq) + CH_3COOH(aq) \longrightarrow CH_3COONa(aq) + CO_2(g) + H_2O(l)$$

Total ionic equation

$$Na^+(aq) + HCO_3^-(aq) + CH_3COOH(aq) \longrightarrow CH_3COO^-(aq) + Na^+(aq) + CO_2(g) + H_2O(l)$$

Net ionic equation

$$HCO_3^-(aq) + CH_3COOH(aq) \longrightarrow CH_3COO^-(aq) + CO_2(g) + H_2O(l)$$

Figure 4.14 A gas-forming reaction with a weak acid. Note that the two ionic equations include acetic acid because it does *not* dissociate appreciably into ions.

Source: © McGraw-Hill Education/Charles Winters/Timeframe Photography, Inc.

Weak acids also react with carbonates to form CO_2, but as we have seen, the weak acid is written as an intact molecule, instead of as ions, because it dissociates very little in solution. Figure 4.14 shows the reaction between vinegar (a 5% solution of acetic acid) and baking soda (sodium hydrogen carbonate) solution. In fact, when stomach acid (mostly HCl) builds up, your duodenum releases the hormone secretin, which stimulates your pancreas to release HCO_3^- ions that react with the excess acid in an analogous reaction.

SAMPLE PROBLEM 4.14 Writing Proton-Transfer Equations for Acid-Base Reactions

Problem Write balanced total and net ionic equations for the following reactions and use curved arrows to show how the proton transfer occurs. For (a), give the name and formula of the salt present when the water evaporates. For (b), note that propanoic acid (CH_3CH_2COOH) is a weak acid and identify the spectator ion(s).

(a) Hydriodic acid(*aq*) + calcium hydroxide(*aq*) ⟶

(b) Potassium hydroxide(*aq*) + propanoic acid(*aq*) ⟶

Plan **(a)** The reactants are a strong acid and a strong base (Table 4.2), so the acidic species is H_3O^+, which transfers a proton to the OH^- from the base. The products are H_2O and a solution of spectator ions that becomes a solid salt when the water evaporates. **(b)** Since the acid is weak, it occurs as intact molecules in solution and transfers its proton to the OH^- from the base. The only spectator ion is the cation of the base.

Solution **(a)** Writing the total ionic equation:

$$2H_3O^+(aq) + 2I^-(aq) + Ca^{2+}(aq) + 2OH^-(aq) \longrightarrow 2I^-(aq) + Ca^{2+}(aq) + 4H_2O(l)$$

Writing the net ionic equation:

$$2H_3O^+(aq) + 2OH^-(aq) \longrightarrow 4H_2O(l) \quad \text{or} \quad H_3O^+(aq) + OH^-(aq) \longrightarrow 2H_2O(l)$$

$I^-(aq)$ and $Ca^{2+}(aq)$ are spectator ions, so the salt is calcium iodide, CaI_2.

(b) Writing the total ionic equation:

$$K^+(aq) + OH^-(aq) + CH_3CH_2COOH(aq) \longrightarrow K^+(aq) + H_2O(l) + CH_3CH_2COO^-(aq)$$

Writing the net ionic equation:

$$OH(aq) + CH_3CH_2COOH(aq) \longrightarrow H_2O(l) + CH_3CH_2COO^-(aq)$$

$K^+(aq)$ is the only spectator ion.

FOLLOW-UP PROBLEMS

4.14A Write balanced total and net ionic equations for the reaction between aqueous solutions of nitrous acid (a weak acid) and strontium hydroxide. Use a curved arrow in the net ionic equation to show the proton transfer. Give the name and formula of the salt that is present when the water evaporates, and identify the spectator ion(s).

4.14B Write balanced total and net ionic equations for the reaction between aqueous solutions of calcium hydrogen carbonate and hydrobromic acid. Use a curved arrow in the net ionic equation to show the proton transfer. Give the name and formula of the salt that is present when the water evaporates.

SOME SIMILAR PROBLEMS 4.65(b), 4.66(b), 4.67, and 4.68

Stoichiometry of Acid-Base Reactions: Acid-Base Titrations

As we saw for precipitation reactions, the molar ratios in an acid-base reaction can also be combined with concentration and volume information to quantify the reactants and products involved in the reaction. Sample Problem 4.15 demonstrates this type of calculation.

SAMPLE PROBLEM 4.15 Calculating the Amounts of Reactants and Products in an Acid-Base Reaction

Problem Specialized cells in the stomach release HCl to aid digestion. If they release too much, the excess can be neutralized with a base in the form of an antacid. Magnesium hydroxide is a common active ingredient in antacids. As a government chemist testing commercial antacids, you use 0.10 *M* HCl to simulate the acid concentration in the stomach. How many liters of this "stomach acid" will react with a tablet containing 0.10 g of magnesium hydroxide?

Road Map

Mass (g) of $Mg(OH)_2$

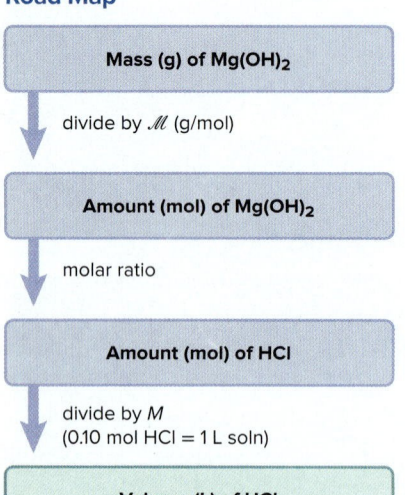

↓ divide by $\mathcal{M}$ (g/mol)

Amount (mol) of $Mg(OH)_2$

↓ molar ratio

Amount (mol) of HCl

↓ divide by *M* (0.10 mol HCl = 1 L soln)

Volume (L) of HCl

Plan We are given the mass (0.10 g) of magnesium hydroxide, $Mg(OH)_2$, that reacts with the acid. We also know the acid concentration (0.10 *M*) and must find the acid volume. After writing the balanced equation, we convert the mass (g) of $Mg(OH)_2$ to amount (mol) and use the molar ratio to find the amount (mol) of HCl that reacts with it. Then, we use the molarity of HCl to find the volume (L) that contains this amount (see the road map).

Solution Writing the balanced equation:

$$\mathrm{Mg(OH)_2}(s) + 2\mathrm{HCl}(aq) \longrightarrow \mathrm{MgCl_2}(aq) + 2\mathrm{H_2O}(l)$$

Converting from mass (g) of $Mg(OH)_2$ to amount (mol):

$$\text{Amount (mol) of Mg(OH)}_2 = 0.10\ \cancel{\text{g Mg(OH)}_2} \times \frac{1\ \text{mol Mg(OH)}_2}{58.33\ \cancel{\text{g Mg(OH)}_2}}$$
$$= 1.7\times10^{-3}\ \text{mol Mg(OH)}_2$$

Converting from amount (mol) of $Mg(OH)_2$ to amount (mol) of HCl:

$$\text{Amount (mol) of HCl} = 1.7\times10^{-3}\ \cancel{\text{mol Mg(OH)}_2} \times \frac{2\ \text{mol HCl}}{1\ \cancel{\text{mol Mg(OH)}_2}}$$
$$= 3.4\times10^{-3}\ \text{mol HCl}$$

Converting from amount (mol) of HCl to volume (L):

$$\text{Volume (L) of HCl} = 3.4\times10^{-3}\ \cancel{\text{mol HCl}} \times \frac{1\ \text{L}}{0.10\ \cancel{\text{mol HCl}}}$$
$$= 3.4\times10^{-2}\ \text{L}$$

Check The size of the answer seems reasonable: a small volume of dilute acid (0.034 L of 0.10 *M*) reacts with a small amount of antacid (0.0017 mol).

FOLLOW-UP PROBLEMS

4.15A Another active ingredient in some antacids is aluminum hydroxide. What mass of aluminum hydroxide is needed to react with the volume of 0.10 *M* HCl calculated in Sample Problem 4.15?

4.15B The active ingredient in an aspirin tablet is acetylsalicylic acid, $HC_9H_7O_4$, which is mixed with inert binding agents and fillers. An aspirin tablet dissolved in water requires 14.10 mL of 0.128 *M* NaOH for complete reaction of the acetylsalicylic acid. What mass of acetylsalicylic acid is in the tablet? (*Hint:* Only one proton in acetylsalicylic acid reacts with the base.)

SOME SIMILAR PROBLEMS 4.69, 4.70, 4.76, and 4.77

Quantifying Acid-Base Reactions by Titration Acid-base reactions are studied quantitatively in a laboratory procedure called a *titration*. In any **titration,** *the known concentration of one solution is used to determine the unknown concentration of another.* In a typical acid-base titration, a *standardized* solution of base, one whose concentration is *known,* is added to a solution of acid whose concentration is *unknown* (or vice versa).

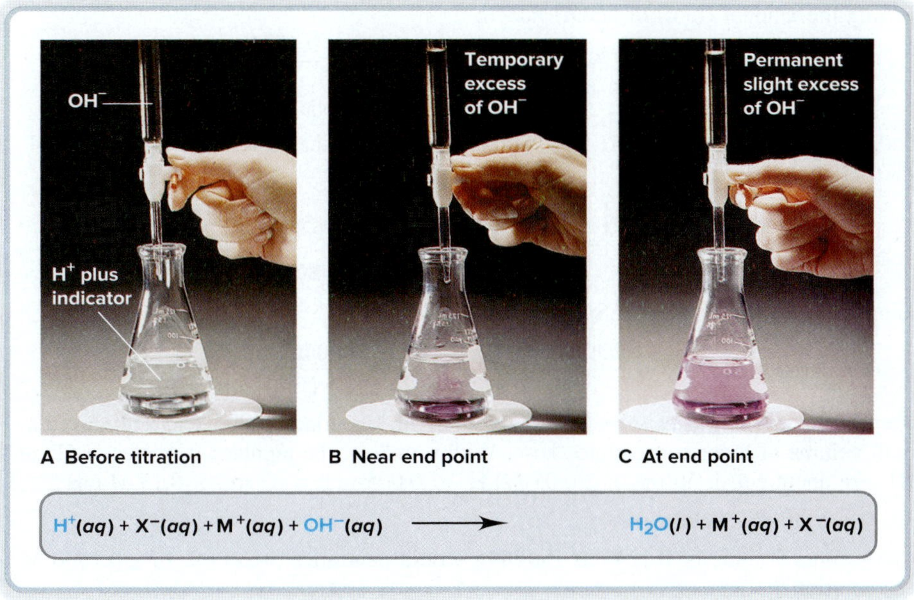

Figure 4.15 An acid-base titration.
Source: © McGraw-Hill Education/Stephen Frisch, photographer

Figure 4.15A shows the laboratory setup for an acid-base titration with a known volume of acid and a few drops of indicator in a flask. An *acid-base indicator* is a substance whose color is different in acid than in base; the indicator used in the figure is phenolphthalein, which is pink in base and colorless in acid. (We examine indicators in Chapters 18 and 19.) Base is added from a buret, and the OH^- ions react with the H^+ ions. As the titration nears its end (Figure 4.15B), the drop of added base creates a temporary excess of OH^-, causing some indicator molecules to change to the basic color; they return to the acidic color when the flask is swirled. There are two key stages in the titration:

- The **equivalence point** occurs when *the amount (mol) of H^+ ions in the original volume of acid has reacted with the same amount (mol) of OH^- ions from the buret:*

  Amount (mol) of H^+ (originally in flask) = amount (mol) of OH^- (added from buret)

- The **end point** occurs when a tiny excess of OH^- ions changes the indicator permanently to its basic color (Figure 4.15C).

In calculations, such as in Sample Problem 4.16, we assume that this tiny excess of OH^- ions is insignificant and that *the amount of base needed to reach the end point is the same as the amount needed to reach the equivalence point.*

SAMPLE PROBLEM 4.16 Finding the Concentration of an Acid from a Titration

Problem To standardize an H_2SO_4 solution, you put 50.00 mL of it in a flask with a few drops of indicator and put 0.1524 *M* NaOH in a buret. The buret reads 0.55 mL at the start and 33.87 mL at the end point. Find the molarity of the H_2SO_4 solution.

Plan We have to find the molarity of the acid from the volume of acid (50.00 mL), the initial (0.55 mL) and final (33.87 mL) volumes of base, and the molarity of the base (0.1524 *M*). First, we balance the equation. The volume of added base is the difference in buret readings, and we use the base's molarity to calculate the amount (mol) of base. Then, we use the molar ratio from the balanced equation to find the amount (mol) of acid originally present and divide by the acid's original volume to find the molarity (see the road map).

Road Map

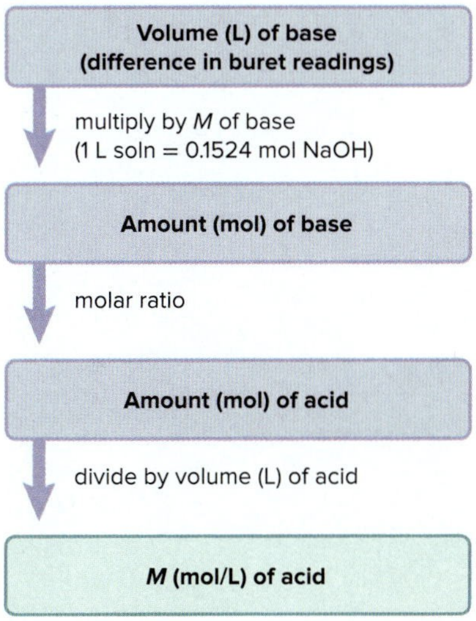

Solution Writing the balanced equation:

$$2NaOH(aq) + H_2SO_4(aq) \longrightarrow Na_2SO_4(aq) + 2H_2O(l)$$

Finding the volume (L) of NaOH solution added:

$$\text{Volume (L) of solution} = (33.87\ \cancel{\text{mL}}\ \text{soln} - 0.55\ \cancel{\text{mL}}\ \text{soln}) \times \frac{1\ \text{L}}{1000\ \cancel{\text{mL}}}$$

$$= 0.03332\ \text{L soln}$$

Finding the amount (mol) of NaOH added:

$$\text{Amount (mol) of NaOH} = 0.03332 \cancel{\text{L soln}} \times \frac{0.1524 \text{ mol NaOH}}{1 \cancel{\text{L soln}}}$$
$$= 5.078\times10^{-3} \text{ mol NaOH}$$

Finding the amount (mol) of H_2SO_4 originally present: Since the molar ratio is 2 mol NaOH/1 mol H_2SO_4,

$$\text{Amount (mol) of } H_2SO_4 = 5.078\times10^{-3} \cancel{\text{mol NaOH}} \times \frac{1 \text{ mol } H_2SO_4}{2 \cancel{\text{mol NaOH}}}$$
$$= 2.539\times10^{-3} \text{ mol } H_2SO_4$$

Calculating the molarity of H_2SO_4:

$$\text{Molarity of } H_2SO_4 = \frac{2.539\times10^{-3} \text{ mol } H_2SO_4}{50.00 \cancel{\text{mL}}} \times \frac{1000 \cancel{\text{mL}}}{1 \text{ L}} = 0.05078 \, M \, H_2SO_4$$

Check The answer makes sense: a large volume of less concentrated acid neutralized a small volume of more concentrated base. With rounding, the numbers of moles of H^+ and OH^- are about equal: 50 mL × 2(0.05 M) H^+ = 0.005 mol = 33 mL × 0.15 M OH^-.

FOLLOW-UP PROBLEMS

4.16A What volume of 0.1292 M $Ba(OH)_2$ would neutralize 50.00 mL of a 0.1000 M HCl solution?

4.16B Calculate the molarity of a solution of KOH if 18.15 mL of it is required for the titration of a 20.00-mL sample of a 0.2452 M HNO_3 solution.

SOME SIMILAR PROBLEMS 4.71 and 4.72

› Summary of Section 4.4

- In an acid-base (neutralization) reaction between an acid (an H^+-yielding substance) and a base (an OH^--yielding substance), H^+ and OH^- ions form H_2O.
- Strong acids and bases dissociate completely in water (strong electrolytes); weak acids and bases dissociate slightly (weak electrolytes).
- An acid-base reaction involves the transfer of a proton from an acid (a species that donates H^+) to a base (a species that accepts H^+).
- A gas-forming acid-base reaction occurs when an acid transfers a proton to a carbonate (or sulfite), forming water and a gas that leaves the reaction mixture.
- Since weak acids dissociate very little, an ionic equation shows a weak acid as an intact molecule transferring its proton to the base.
- Molarity can be used as a conversion factor to solve stoichiometric problems involving acid-base reactions.
- In a titration, the known concentration of one solution is used to determine the concentration of the other.

4.5 OXIDATION-REDUCTION (REDOX) REACTIONS

Oxidation-reduction (redox) reactions include the formation of a compound from its elements (and the reverse process), all combustion processes, the generation of electricity in batteries, the production of cellular energy, and many others. In fact, redox reactions are so widespread that many do not occur in solution at all. In this section, we examine the key event in the redox process, discuss important terminology, see how to determine whether a reaction involves a redox process, and learn how to quantify redox reactions. (The redox reactions in this section can be balanced by methods you learned in Chapter 3; you'll learn how to balance more complex ones when we discuss electrochemistry in Chapter 21.)

The Key Event: Movement of Electrons Between Reactants

The key chemical event in an **oxidation-reduction** (or **redox**) **reaction** is the *net movement of electrons from one reactant to another*. The movement occurs from the

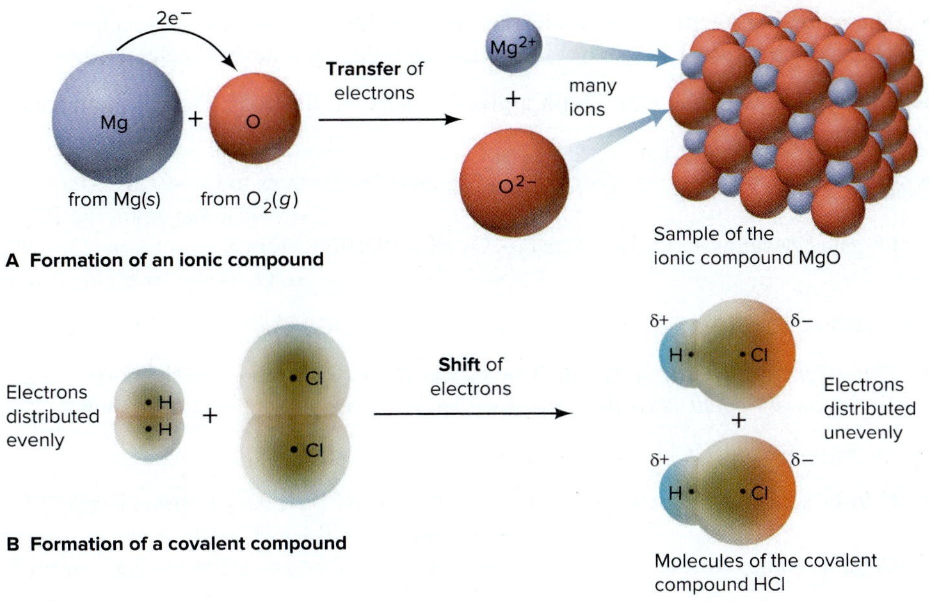

Figure 4.16 The redox process in the formation of (A) ionic and (B) covalent compounds from their elements.

reactant (or atom in the reactant) with *less* attraction for electrons to the reactant (or atom) with *more* attraction for electrons.

This process occurs in the formation of both ionic and covalent compounds:

- *Ionic compounds: transfer of electrons.* In the reaction that forms MgO from its elements (see Figure 3.7), the balanced equation is

$$2Mg(s) + O_2(g) \longrightarrow 2MgO(s)$$

Figure 4.16A shows that each Mg atom loses two electrons and each O atom gains them. This loss and gain is a ***transfer** of electrons* away from each Mg atom to each O atom. The resulting Mg^{2+} and O^{2-} ions aggregate into an ionic solid.

- *Covalent compounds: shift of electrons.* During the formation of a covalent compound from its elements, there is more a ***shift** of electrons* rather than a full transfer. Thus, *ions do not form.* Consider the formation of HCl gas:

$$H_2(g) + Cl_2(g) \longrightarrow 2HCl(g)$$

To see the electron movement, we'll compare the electron distribution in reactant and product. As Figure 4.16B shows, in H_2 and Cl_2, the electrons are shared equally between the atoms *(symmetrical shading).* Because the Cl atom attracts electrons more than the H atom, in HCl, the electrons are shared unequally *(asymmetrical shading).* Electrons shift away from H and toward Cl, so the Cl atom has more negative charge (*red* and δ−) in HCl than it had in Cl_2, and the H atom has less negative charge (*blue* and δ+) than it had in H_2. Thus, H_2 loses some electron charge and Cl_2 gains some electron charge.

Some Essential Redox Terminology

Certain key terms describe the process and explain the name of this reaction class:

- **Oxidation** is the *loss* of electrons.
- **Reduction** is the *gain* of electrons.

During the formation of MgO, Mg undergoes oxidation (loss of electrons), and O_2 undergoes reduction (gain of electrons). The loss and gain are simultaneous, but we can imagine them occurring separately:

Oxidation (electron loss by Mg): $Mg \longrightarrow Mg^{2+} + 2e^-$

Reduction (electron gain by O_2): $\frac{1}{2}O_2 + 2e^- \longrightarrow O^{2-}$

(Throughout this discussion, blue type indicates oxidation, and red type reduction.)

- The **oxidizing agent** is the species doing the oxidizing (causing the electron loss).
- The **reducing agent** is the species doing the reducing (causing the electron gain).

One reactant acts on the other, with a give-and-take of electrons. During the reaction that forms MgO,

- Mg loses electrons → Mg is *oxidized* → Mg provides electrons to O_2 → Mg *reduces* O_2 → Mg is the *reducing agent*
- O_2 gains electrons → O_2 is *reduced* → O_2 takes electrons from Mg → O_2 *oxidizes* Mg → O_2 is the *oxidizing agent*

This means that

- *The oxidizing agent is reduced:* it takes electrons (and, thus, gains them).
- *The reducing agent is oxidized:* it gives up electrons (and, thus, loses them).

In the formation of HCl,

- H loses some electron charge → H_2 is oxidized → H provides electron charge to Cl → H_2 *reduces* Cl_2 → H_2 is the *reducing agent*
- Cl gains some electron charge → Cl_2 is reduced → Cl takes electron charge from H → Cl_2 *oxidizes* H_2 → Cl_2 is the *oxidizing agent*

Using Oxidation Numbers to Monitor Electron Charge

Chemists have devised a "bookkeeping" system to monitor which atom loses electron charge and which atom gains it: each atom in a molecule (or formula unit) is assigned an **oxidation number (O.N.),** or *oxidation state,* which is the charge the atom would have *if* its electrons were transferred completely, not shared.

Each element in a binary *ionic* compound has a full charge because the atom transferred its electron(s), and so the atom's oxidation number equals the ionic charge. But, each element in a *covalent* compound (or in a polyatomic ion) has a partial charge because the electrons shifted away from one atom and toward the other. For these cases, we determine oxidation number by a set of rules (presented in Table 4.4; you'll learn the atomic basis of the rules in Chapters 8 and 9).

An O.N. has the sign *before* the number (e.g., +2), whereas an ionic charge has the sign *after* the number (e.g., 2+). Also, unlike a unitary ionic charge, as in Na^+ or Cl^-, an O.N. of +1 or −1 retains the numeral. For example, we don't write the sodium ion as Na^{1+}, but the O.N. of the Na^+ ion is +1, not +.

Table 4.4 Rules for Assigning an Oxidation Number (O.N.)

General Rules

1. For an atom in its elemental form (Na, C, O_2, Cl_2, P_4, etc.): O.N. = 0
2. For a monatomic ion: O.N. = ion charge (with the sign *before* the numeral)
3. The sum of O.N. values for the atoms in a molecule or formula unit of a compound equals zero. The sum of O.N. values for the atoms in a polyatomic ion equals the ion's charge.

Rules for Specific Atoms or Periodic Table Groups

1. For Group 1A(1):	O.N. = +1 in all compounds
2. For Group 2A(2):	O.N. = +2 in all compounds
3. For hydrogen:	O.N. = +1 in combination with nonmetals
	O.N. = −1 in combination with metals and boron (e.g. NaH)
4. For fluorine:	O.N. = −1 in all compounds
5. For oxygen:	O.N. = −1 in peroxides (e.g. H_2O_2)
	O.N. = −2 in all other compounds (except with F)
6. For Group 7A(17):	O.N. = −1 in combination with metals, nonmetals (except O), and other halogens lower in the group

SAMPLE PROBLEM 4.17 **Determining the Oxidation Number of Each Element in a Compound (or Ion)**

Problem Determine the oxidation number (O.N.) of each element in these species:
(a) Zinc chloride **(b)** Sulfur trioxide **(c)** Nitric acid **(d)** Dichromate ion

Plan We determine the formulas and consult Table 4.4, noting the general rules that the O.N. values for a compound add up to zero and those for a polyatomic ion add up to the ion's charge.

Solution **(a)** $ZnCl_2$. The O.N. of each chloride ion is −1 (rule 6 in Table 4.4), for a total of −2, so the O.N. of Zn must be +2 since the sum of O.N.s must equal zero for a compound.

$$\overset{+2\ -1}{ZnCl_2}$$
$$+2 + 2(-1) = 0$$

(b) SO_3. The O.N. of each oxygen is −2 (rule 5 in Table 4.4), for a total of −6. The O.N.s must add up to zero, so the O.N. of S is +6.

$$\overset{+6\ -2}{SO_3}$$
$$+6 + 3(-2) = 0$$

(c) HNO_3. The O.N. of H is +1 (rule 3 in Table 4.4) and the O.N. of each O is −2 (rule 5 in Table 4.4), for a total of −6. Therefore, the O.N. of N is +5.

$$\overset{+1\ +5\ -2}{HNO_3}$$
$$+1 + (+5) + 3(-2) = 0$$

(d) $Cr_2O_7^{2-}$. The sum of the O.N. values in a polyatomic ion equals the ion's charge. The O.N. of each O is −2 (rule 5 in Table 4.4), so the total for seven O atoms is −14. Therefore, each Cr must have an O.N. of +6 in order for the sum of the O.N.s to equal the charge of the ion:

$$\overset{+6\ \ -2}{Cr_2O_7^{2-}}$$
$$2(+6) + 7(-2) = -2$$

FOLLOW-UP PROBLEMS

4.17A Determine the O.N. of each element in the following:
(a) Scandium oxide (Sc_2O_3) **(b)** Gallium chloride ($GaCl_3$)
(c) Hydrogen phosphate ion **(d)** Iodine trifluoride

4.17B Determine the O.N. of each element in the following:
(a) Potassium carbonate **(b)** Ammonium ion
(c) Calcium phosphide (Ca_3P_2) **(d)** Sulfur tetrachloride

SOME SIMILAR PROBLEMS 4.84–4.91

Using O.N.s to Identify Redox Reactions and Oxidizing and Reducing Agents A redox reaction is defined as one in which the oxidation numbers of the species change. By assigning an oxidation number to each atom, we can see which species was oxidized and which reduced and, thus, which is the oxidizing agent and which the reducing agent:

- If an atom has a higher (more positive or less negative) O.N. in the product than it had in the reactant, the reactant that contains that atom was oxidized (lost electrons) and is the reducing agent. Thus, *oxidation is shown by an increase in O.N.*
- If an atom has a lower (more negative or less positive) O.N. in the product than it had in the reactant, the reactant that contains that atom was reduced (gained electrons) and is the oxidizing agent. Thus, *reduction is shown by a decrease in O.N.*

Sample Problem 4.18 introduces the *tie-line,* a useful device for keeping track of changes in oxidation numbers.

Student Hot Spot

Student data indicate that you may struggle with identifying the reducing and oxidizing agents in a redox reaction. Access the Smartbook to view additional Learning Resources on this topic.

SAMPLE PROBLEM 4.18 **Identifying Redox Reactions and Oxidizing and Reducing Agents**

Problem Use oxidation numbers to decide which of the following are redox reactions. For each redox reaction, identify the oxidizing agent and the reducing agent:

(a) $2Al(s) + 3H_2SO_4(aq) \longrightarrow Al_2(SO_4)_3(aq) + 3H_2(g)$

(b) $H_2SO_4(aq) + 2NaOH(aq) \longrightarrow Na_2SO_4(aq) + 2H_2O(l)$

(c) $PbO(s) + CO(g) \longrightarrow Pb(s) + CO_2(g)$

Plan To determine whether a reaction is an oxidation-reduction process, we use Table 4.4 to assign each atom an O.N. and then note whether it changes as the reactants become products. If the O.N. of an atom changes during the reaction, we draw a tie-line from the atom on the left side to the same atom on the right side and label the change. For those reactions that are redox reactions, a reactant is the reducing agent if it contains an atom that is oxidized (O.N. *increases* from left to right in the equation). A reactant is the oxidizing agent if it contains an atom that is reduced (O.N. *decreases*). Note that products are never oxidizing or reducing agents.

Solution **(a)** Assigning oxidation numbers and marking the changes:

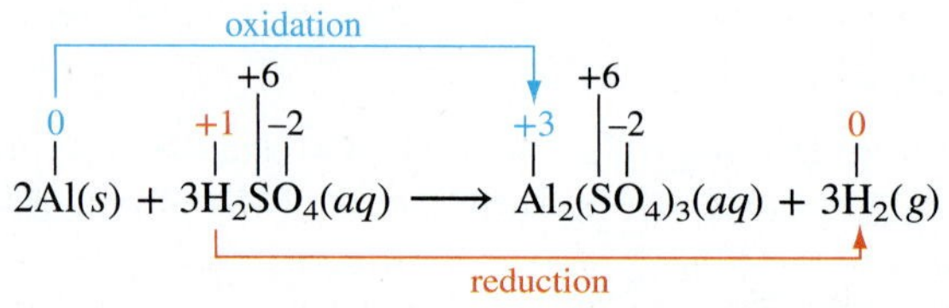

In this case, the O.N.s of Al and H change, so this is a redox reaction. The O.N. of Al changes from 0 to +3 (Al loses electrons), so Al is oxidized; Al is the reducing agent. The O.N. of H decreases from +1 to 0 (H gains electrons), so H^+ is reduced; H_2SO_4 is the oxidizing agent.

(b) Assigning oxidation numbers:

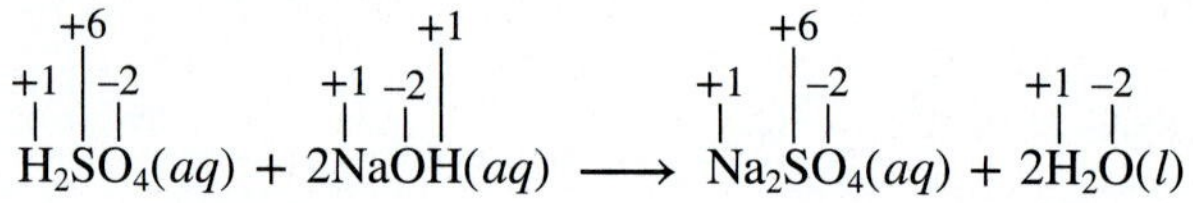

Because each atom in the products has the same O.N. that it had in the reactants, we conclude that this is *not* a redox reaction.

(c) Assigning oxidation numbers and marking the changes:

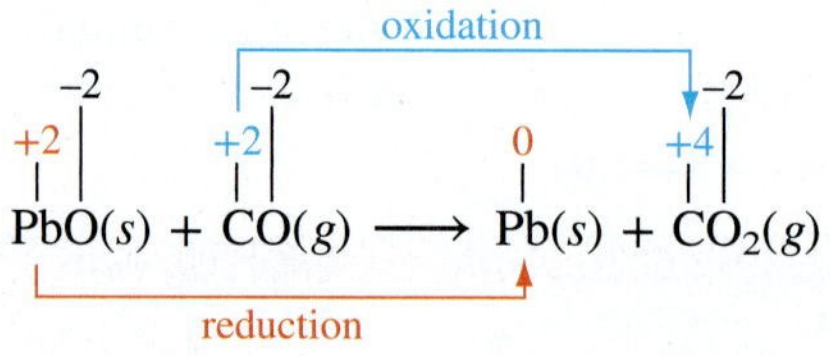

In this case, the O.N.s of Pb and C change, so this is a redox reaction. The O.N. of Pb decreases from +2 to 0, so Pb is reduced; PbO is the oxidizing agent. The O.N. of C increases from +2 to +4, so CO is oxidized; CO is the reducing agent.

In general, when a substance (such as CO) changes to one with more O atoms (such as CO_2), it is oxidized; and when a substance (such as PbO) changes to one with fewer O atoms (such as Pb), it is reduced.

Comment The reaction in part (b) is an acid-base reaction in which H_2SO_4 transfers two H^+ ions to two OH^- ions to form two H_2O molecules. In the net ionic equation,

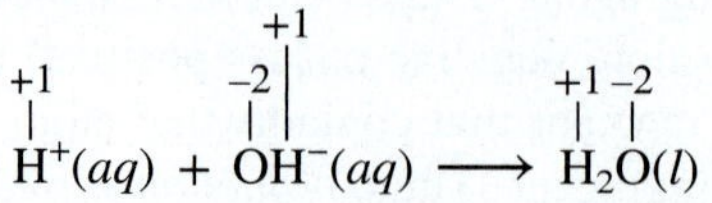

we see that the O.N.s remain the same on both sides. Therefore, *an acid-base reaction is **not** a redox reaction.*

FOLLOW-UP PROBLEMS

4.18A Use oxidation numbers to decide which of the following are redox reactions. For all redox reactions, identify the oxidizing agent and the reducing agent:
(a) $NCl_3(l) + 3H_2O(l) \longrightarrow NH_3(aq) + 3HOCl(aq)$
(b) $AgNO_3(aq) + NH_4I(aq) \longrightarrow AgI(s) + NH_4NO_3(aq)$
(c) $2H_2S(g) + 3O_2(g) \longrightarrow 2SO_2(g) + 2H_2O(g)$

4.18B Use oxidation numbers to decide which of the following are redox reactions. For all redox reactions, identify the oxidizing agent and the reducing agent:
(a) $SiO_2(s) + 4HF(g) \longrightarrow SiF_4(g) + 2H_2O(l)$
(b) $2C_2H_6(g) + 7O_2(g) \longrightarrow 4CO_2(g) + 6H_2O(g)$
(c) $5CO(g) + I_2O_5(s) \longrightarrow 5CO_2(g) + I_2(s)$

SOME SIMILAR PROBLEMS 4.92–4.95

Be sure to remember that *transferred electrons are never free because the reducing agent loses electrons and the oxidizing agent gains them* ***simultaneously.*** In other words, a complete reaction *cannot* be "an oxidation" *or* "a reduction"; it must be an oxidation-reduction. Figure 4.17 summarizes redox terminology.

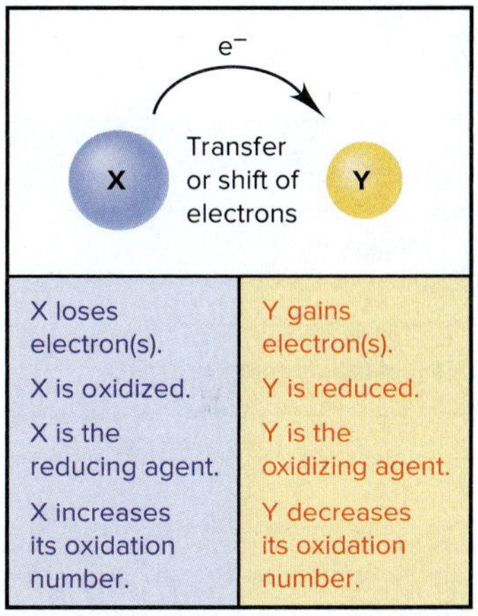

Figure 4.17 A summary of terminology for redox reactions.

Stoichiometry of Redox Reactions: Redox Titrations

Just as for precipitation and acid-base reactions, stoichiometry is important for redox reactions; a special application of stoichiometry that we focus on here is redox titration. In an acid-base titration, a known concentration of a base is used to find an unknown concentration of an acid (or vice versa). Similarly, in a redox titration, a known concentration of oxidizing agent is used to find an unknown concentration of reducing agent (or vice versa). This kind of titration has many applications, from measuring the iron content in drinking water to quantifying vitamin C in fruits and vegetables.

The permanganate ion, MnO_4^-, is a common oxidizing agent because it is deep purple, and so also serves as an indicator. In Figure 4.18, MnO_4^- is used to determine oxalate ion ($C_2O_4^{2-}$) concentration. As long as $C_2O_4^{2-}$ is present, it reduces the added MnO_4^- to faint pink (nearly colorless) Mn^{2+} (Figure 4.18, *left*). When all the $C_2O_4^{2-}$ has been oxidized, the next drop of MnO_4^- turns the solution light purple (Figure 4.18, *right*). This color change indicates the *end point,* which we assume is the same as the *equivalence point,* the point at which the electrons lost by the oxidized species ($C_2O_4^{2-}$) equal the electrons gained by the reduced species (MnO_4^-). We can calculate the $C_2O_4^{2-}$ concentration from the known volume of $Na_2C_2O_4$ solution and the known volume and concentration of $KMnO_4$ solution.

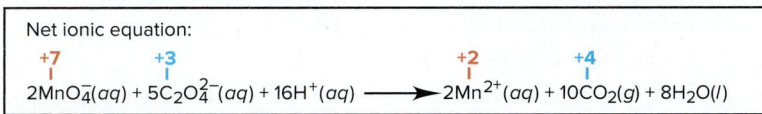

Net ionic equation:

$$\overset{+7}{2MnO_4^-(aq)} + \overset{+3}{5C_2O_4^{2-}(aq)} + 16H^+(aq) \longrightarrow \overset{+2}{2Mn^{2+}(aq)} + \overset{+4}{10CO_2(g)} + 8H_2O(l)$$

Figure 4.18 The redox titration of $C_2O_4^{2-}$ with MnO_4^-.
Source: © McGraw-Hill Education/Stephen Frisch, photographer

Road Map

Volume (mL) of $KMnO_4$ solution

↓ 10^3 mL = 1 L

Volume (L) of $KMnO_4$ solution

↓ multiply by *M* (1 L soln = 4.88×10^{-4} mol $KMnO_4$)

Amount (mol) of $KMnO_4$

↓ molar ratio

Amount (mol) of CaC_2O_4

↓ ratio of elements in chemical formula

Amount (mol) of Ca^{2+}

SAMPLE PROBLEM 4.19 Finding the Amount of Reducing Agent by Titration

Problem Calcium ion (Ca^{2+}) is necessary for the clotting of blood and many other physiological processes. To measure the Ca^{2+} concentration in 1.00 mL of human blood, $Na_2C_2O_4$ solution is added, which precipitates the Ca^{2+} as CaC_2O_4. This solid is dissolved in dilute H_2SO_4 to release $C_2O_4^{2-}$, and 2.05 mL of 4.88×10^{-4} *M* $KMnO_4$ is required to reach the end point. The balanced equation is

$$2KMnO_4(aq) + 5CaC_2O_4(s) + 8H_2SO_4(aq) \longrightarrow 2MnSO_4(aq) + K_2SO_4(aq) + 5CaSO_4(s) + 10CO_2(g) + 8H_2O(l)$$

Calculate the amount (mol) of Ca^{2+} in 1.00 mL of blood.

Plan We have to find the amount (mol) of Ca^{2+} from the volume (2.05 mL) and concentration (4.88×10^{-4} *M*) of $KMnO_4$ used to oxidize the CaC_2O_4 precipitated from 1.00 mL of blood. We find the amount (mol) of $KMnO_4$ needed to reach the end point and use the molar ratio to find the amount (mol) of CaC_2O_4. Then we use the chemical formula, which shows 1 mol of Ca^{2+} for every mol of CaC_2O_4 (see the road map).

Solution Converting volume (mL) and concentration (*M*) to amount (mol) of $KMnO_4$:

$$\text{Amount (mol) of } KMnO_4 = 2.05 \cancel{\text{mL soln}} \times \frac{1 \cancel{\text{L}}}{1000 \cancel{\text{mL}}} \times \frac{4.88\times10^{-4} \text{ mol } KMnO_4}{1 \cancel{\text{L soln}}} = 1.00\times10^{-6} \text{ mol } KMnO_4$$

Using the molar ratio to convert amount (mol) of $KMnO_4$ to amount (mol) of CaC_2O_4:

$$\text{Amount (mol) of } CaC_2O_4 = 1.00\times10^{-6} \cancel{\text{mol } KMnO_4} \times \frac{5 \text{ mol } CaC_2O_4}{2 \cancel{\text{mol } KMnO_4}} = 2.50\times10^{-6} \text{ mol } CaC_2O_4$$

Finding the amount (mol) of Ca^{2+}:

$$\text{Amount (mol) of } Ca^{2+} = 2.50\times10^{-6} \cancel{\text{mol } CaC_2O_4} \times \frac{1 \text{ mol } Ca^{2+}}{1 \cancel{\text{mol } CaC_2O_4}} = 2.50\times10^{-6} \text{ mol } Ca^{2+}$$

Check A very small volume of dilute $KMnO_4$ is needed, so 10^{-6} mol of $KMnO_4$ seems reasonable. The molar ratio of CaC_2O_4 to $KMnO_4$ is 5/2, which gives 2.5×10^{-6} mol of CaC_2O_4 and thus 2.5×10^{-6} mol of Ca^{2+}.

Comment 1. When blood is donated, the receiving bag contains $Na_2C_2O_4$ solution, which precipitates the Ca^{2+} ion to prevent clotting.

2. Normal Ca^{2+} concentration in human adult blood is 9.0–11.5 mg Ca^{2+}/100 mL. If we multiply the amount of Ca^{2+} in 1 mL by 100 and then multiply by the molar mass of Ca (40.08 g/mol), we get 1.0×10^{-2} g, or 10.0 mg Ca^{2+} in 100 mL:

$$\text{Mass (mg) of } Ca^{2+} = \frac{2.50\times10^{-6} \cancel{\text{mol } Ca^{2+}}}{1 \cancel{\text{mL}}} \times 100 \cancel{\text{mL}} \times \frac{40.08 \cancel{\text{g}} \text{ } Ca^{2+}}{1 \cancel{\text{mol } Ca^{2+}}} \times \frac{1000 \text{ mg}}{1 \cancel{\text{g}}} = 10.0 \text{ mg } Ca^{2+}$$

FOLLOW-UP PROBLEMS

4.19A When 2.50 mL of low-fat milk is treated with sodium oxalate and the precipitate dissolved in H_2SO_4, 6.53 mL of 4.56×10^{-3} *M* $KMnO_4$ is required to reach the end point. **(a)** Calculate the molarity of Ca^{2+} in the milk sample. **(b)** What is the concentration of Ca^{2+} in g/L? Is this value consistent with the typical value for milk of about 1.2 g Ca^{2+}/L?

4.19B The amount of iron in an iron ore is determined by converting the Fe to $FeSO_4$, followed by titration with a solution of $K_2Cr_2O_7$. The balanced equation is

$$6FeSO_4(aq) + K_2Cr_2O_7(aq) + 7H_2SO_4(aq) \longrightarrow 3Fe_2(SO_4)_3(aq) + Cr_2(SO_4)_3(aq) + 7H_2O(l) + K_2SO_4(aq)$$

(a) What is the molarity of Fe^{2+} if 21.85 mL of 0.250 *M* $K_2Cr_2O_7$ is required for the titration of 30.0 mL of $FeSO_4$ solution? **(b)** If the original sample of iron ore had a mass of 2.58 g, what is the mass percent of iron in the ore?

SOME SIMILAR PROBLEMS 4.96 and 4.97

› Summary of Section 4.5

› When one reactant has a greater attraction for electrons than another, there is a net movement of electrons, and a redox reaction takes place. Electron gain (reduction) and electron loss (oxidation) occur simultaneously.

› Assigning oxidation numbers to all atoms in a reaction is a method for identifying a redox reaction. The species that is oxidized (contains an atom that increases in oxidation number) is the reducing agent; the species that is reduced (contains an atom that decreases in oxidation number) is the oxidizing agent.

› A redox titration determines the concentration of an oxidizing agent from the known concentration of the reducing agent (or vice versa).

4.6 ELEMENTS IN REDOX REACTIONS

In many redox reactions, such as those in parts (a) and (c) of Sample Problem 4.18, *atoms occur as an element on one side of an equation and as part of a compound on the other.** One way to classify these reactions is by comparing the *numbers* of reactants and products. With that approach, we have three types—*combination, decomposition,* and *displacement;* one other type involving elements is *combustion.* In this section, we survey each type and consider several examples.

Combination Redox Reactions

In a combination redox reaction, two or more reactants, at least one of which is an element, form a compound:

$$X + Y \longrightarrow Z$$

Combining Two Elements Two elements may react to form binary ionic or covalent compounds. Here are some important examples:

1. *Metal and nonmetal form an ionic compound.* Figure 4.19 *(on the next page)* shows a typical example of a combination redox reaction, between the alkali metal potassium and the halogen chlorine, on the macroscopic, atomic, and symbolic levels. Note the change in oxidation numbers in the balanced equation: K is oxidized, so it is the reducing agent; Cl_2 is reduced, so it is the oxidizing agent. In fact, whenever a binary ionic compound forms from its elements, *the metal is the reducing agent and the nonmetal is the oxidizing agent.*

In another example of metal reacting with nonmetal, aluminum reacts with O_2, as does nearly every metal, to form an ionic oxide:

$$\overset{0}{4\text{Al}}(s) + 3\overset{0}{\text{O}_2}(g) \longrightarrow 2\overset{+3}{\text{Al}_2}\overset{-2}{\text{O}_3}(s)$$

The Al_2O_3 layer on the surface of an aluminum pot prevents oxygen from reacting with the next layer of Al atoms, thus protecting the metal from further oxidation.

2. *Two nonmetals form a covalent compound.* In one of thousands of examples, ammonia forms from nitrogen and hydrogen in a reaction that occurs in industry on an enormous scale:

$$\overset{0}{\text{N}_2}(g) + 3\overset{0}{\text{H}_2}(g) \longrightarrow 2\overset{-3}{\text{N}}\overset{+1}{\text{H}_3}(g)$$

Halogens react with many other nonmetals, as in the formation of phosphorus trichloride, a reactant in the production of pesticides and other organic compounds:

$$\overset{0}{\text{P}_4}(s) + 6\overset{0}{\text{Cl}_2}(g) \longrightarrow 4\overset{+3}{\text{P}}\overset{-1}{\text{Cl}_3}(l)$$

*Many other redox reactions, however, do not involve elements (see Sample Problem 4.19).

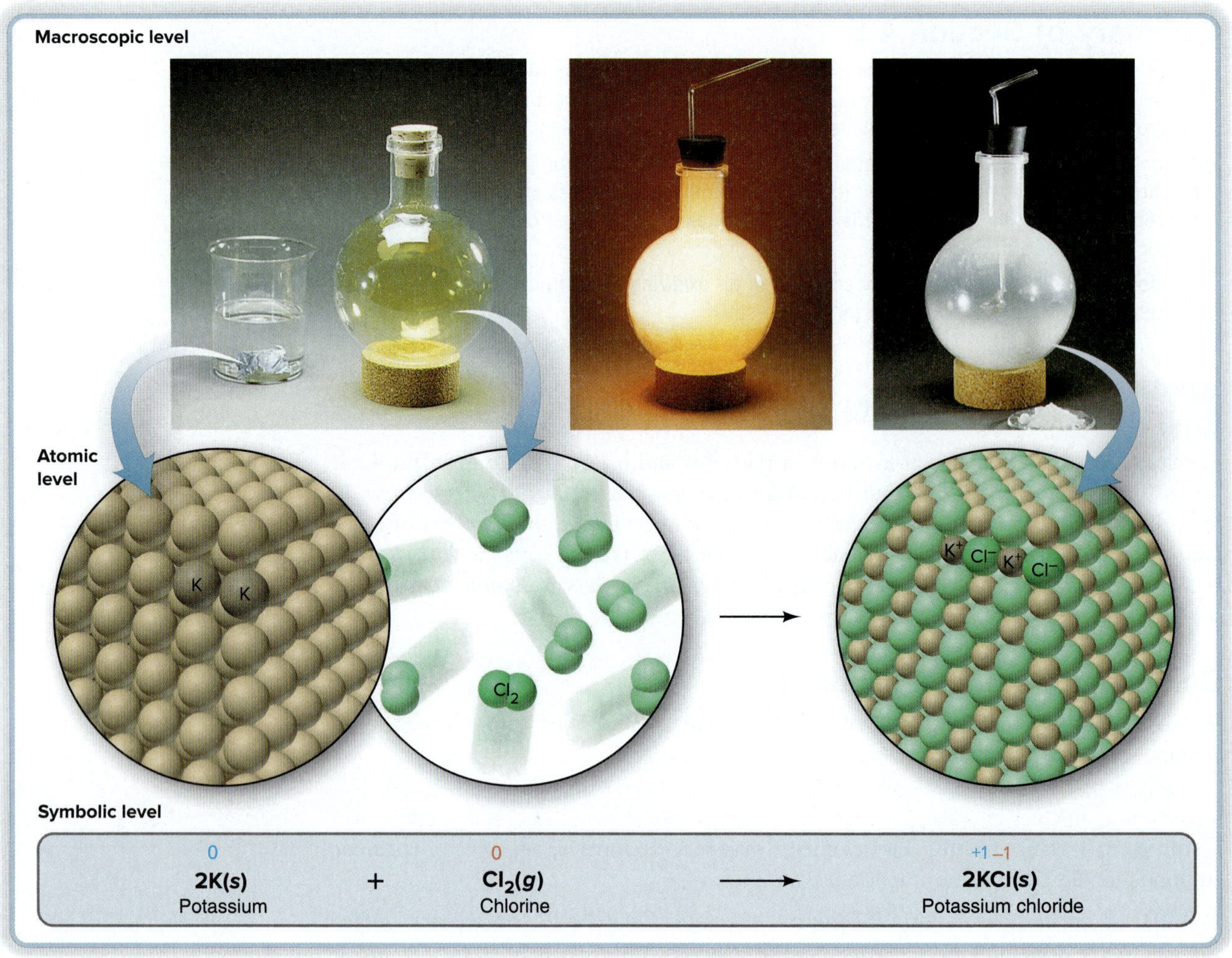

Figure 4.19 Combining elements to form an ionic compound: potassium metal and chlorine gas form solid potassium chloride. Darker atoms in the atomic view indicate the stoichiometry.

Source: © McGraw-Hill Education/Stephen Frisch, photographer

Nearly every nonmetal reacts with O_2 to form a covalent oxide, as when nitrogen monoxide forms from the nitrogen and oxygen in air at the very high temperatures created by lightning and in a car's engine:

$$\overset{0}{N_2}(g) + \overset{0}{O_2}(g) \longrightarrow 2\overset{+2}{N}\overset{-2}{O}(g)$$

Combining a Compound and an Element Many binary covalent compounds react with nonmetals to form larger compounds. Many nonmetal oxides react with additional O_2 to form "higher" oxides (those with more O atoms in each molecule). For example,

$$2\overset{+2}{N}\overset{-2}{O}(g) + \overset{0}{O_2}(g) \longrightarrow 2\overset{+4}{N}\overset{-2}{O_2}(g)$$

Similarly, many nonmetal halides combine with additional halogen to form "higher" halides:

$$\overset{+3}{P}\overset{-1}{Cl_3}(l) + \overset{0}{Cl_2}(g) \longrightarrow \overset{+5}{P}\overset{-1}{Cl_5}(s)$$

Decomposition Redox Reactions

In a *decomposition redox reaction,* a compound forms two or more products, at least one of which is an element:

$$Z \longrightarrow X + Y$$

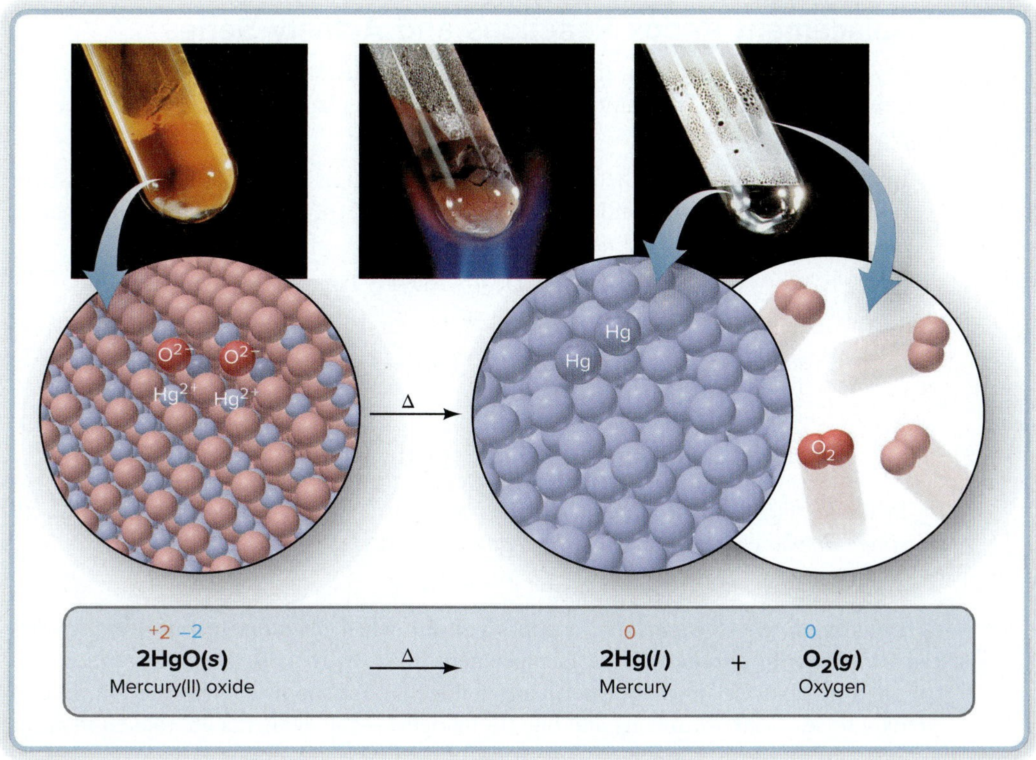

Figure 4.20 Decomposition of the compound mercury(II) oxide to its elements.

Source: © McGraw-Hill Education/ Stephen Frisch, photographer

In any decomposition reaction, the reactant absorbs enough energy for one or more bonds to break. The energy can take several forms, but the most important are decomposition by heat (thermal) and by electricity (electrolytic). The products are either elements or elements and smaller compounds.

Thermal Decomposition When the energy absorbed is heat, the reaction is called a *thermal decomposition*. (A Greek delta, Δ, above the yield arrow indicates strong heating is required for the reaction to occur.) Many metal oxides, chlorates, and perchlorates release oxygen when strongly heated. Heating potassium chlorate is a method for forming small amounts of oxygen in the laboratory; the same reaction occurs in some explosives and fireworks. In this reaction, Cl in potassium chlorate is reduced, while oxygen in the reactant is oxidized to elemental O_2:

$$\overset{+1}{2K}\overset{+5}{Cl}\overset{-2}{O_3}(s) \xrightarrow{\Delta} 2\overset{+1}{K}\overset{-1}{Cl}(s) + 3\overset{0}{O_2}(g)$$

The decomposition of mercury(II) oxide, used by Lavoisier in his classic experiments, is shown in Figure 4.20. Notice that, in these thermal decomposition reactions, the lone reactant is the oxidizing *and* the reducing agent. For example, in the reactant HgO, the O^{2-} ion reduces the Hg^{2+} ion, and at the same time, Hg^{2+} oxidizes O^{2-}.

Electrolytic Decomposition In the process of electrolysis, a compound absorbs electrical energy and decomposes into its elements. In the early 19th century, the observation of the electrolysis of water was crucial for establishing atomic masses:

$$2\overset{+1}{H_2}\overset{-2}{O}(l) \xrightarrow{\text{electricity}} 2\overset{0}{H_2}(g) + \overset{0}{O_2}(g)$$

Many active metals, such as sodium, magnesium, and calcium, are produced industrially by electrolysis of their molten halides:

$$\overset{+2}{Mg}\overset{-1}{Cl_2}(l) \xrightarrow{\text{electricity}} \overset{0}{Mg}(l) + \overset{0}{Cl_2}(g)$$

(We examine the details of electrolysis in Chapter 21 and its role in the industrial recovery of several elements in Chapter 22.)

Displacement Redox Reactions and Activity Series

In any *displacement reaction,* the number of substances on the two sides of the equation remains the same, but atoms (or ions) exchange places. There are two types of displacement reactions:

1. In *double*-displacement (metathesis) reactions, such as precipitation and acid-base reactions (Sections 4.3 and 4.4), atoms (or ions) of two *compounds* exchange places; these reactions are *not* redox processes:

$$AB + CD \longrightarrow AD + CB$$

$$\text{Precipitation:}\quad \overset{+1\ +4\ -2}{Na_2CO_3(aq)} + \overset{+2\ -1}{CaCl_2(aq)} \longrightarrow \overset{+2\ +4\ -2}{CaCO_3(s)} + \overset{+1\ -1}{2NaCl(aq)}$$

$$\text{Acid-Base:}\quad \overset{+1\ -1}{HCl(aq)} + \overset{+1\ -2\ +1}{KOH(aq)} \longrightarrow \overset{+1\ -2}{H_2O(l)} + \overset{+1\ -1}{KCl(aq)}$$

2. In *single*-displacement reactions, one of the substances is an *element;* therefore, *all single-displacement reactions **are** redox processes:*

$$X + YZ \longrightarrow XZ + Y$$

In solution, single-displacement reactions occur when an atom of one element displaces the ion of another: if the displacement involves metals, the atom *reduces* the ion; if it involves nonmetals (specifically halogens), the atom *oxidizes* the ion. In two *activity series*—one for metals and one for halogens—the elements are ranked in order of their ability to displace hydrogen (for metals) and one another.

The Activity Series of the Metals Metals are ranked by their ability to displace H_2 from various sources and another metal from solution. In all displacements of H_2, the metal is the reducing agent (O.N. increases), and water or acid is the oxidizing agent (O.N. of H decreases). The activity series of the metals is based on these facts:

- *The most reactive metals displace H_2 from liquid water.* Group 1A(1) metals and Ca, Sr, and Ba from Group 2A(2) displace H_2 from water (Figure 4.21).

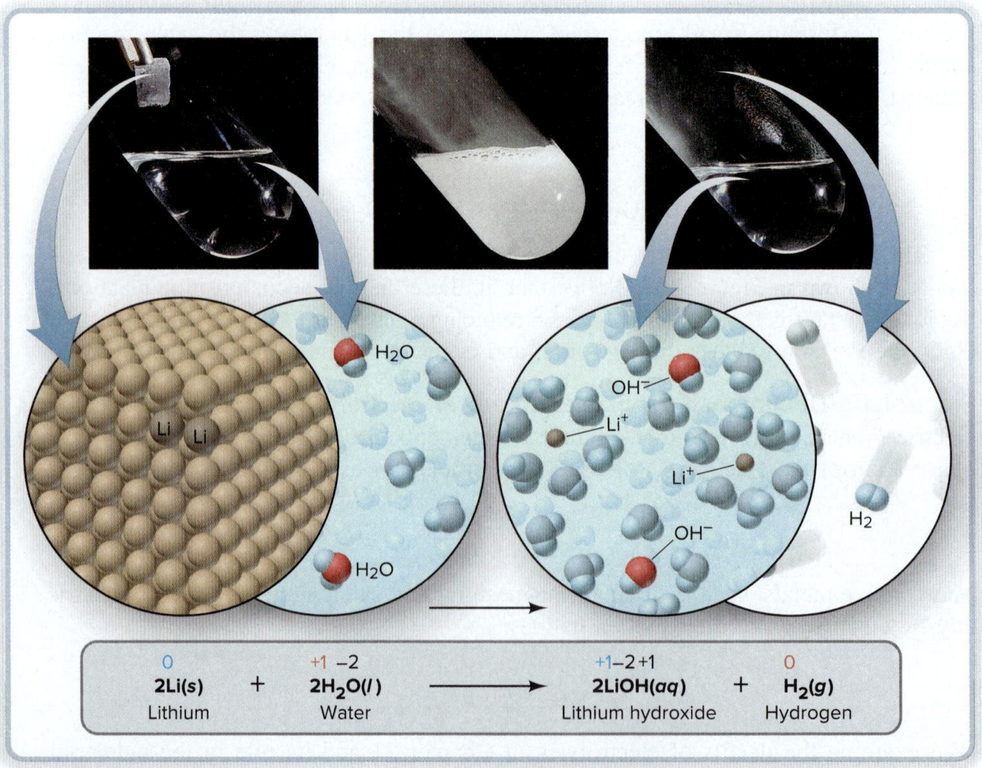

Figure 4.21 The active metal lithium displaces hydrogen from water. Only water molecules involved in the reaction are red and blue.

Source: © McGraw-Hill Education/Stephen Frisch, photographer

- *Slightly less reactive metals displace H_2 from steam.* Heat supplied by steam is needed for less reactive metals such as Al and Zn to displace H_2:

$$\overset{0}{2\text{Al}}(s) + 6\overset{+1}{\text{H}}_2\overset{-2}{\text{O}}(g) \xrightarrow{\Delta} 2\overset{+3}{\text{Al}}(\overset{-2}{\text{O}}\overset{+1}{\text{H}})_3(s) + 3\overset{0}{\text{H}}_2(g)$$

- *Still less reactive metals displace H_2 from acids.* The higher concentration of H^+ in acid solutions is needed for even less reactive metals such as Ni and Sn to displace H_2 (Figure 4.22). For nickel, the net ionic equation is

$$\overset{0}{\text{Ni}}(s) + 2\overset{+1}{\text{H}}{}^+(aq) \longrightarrow \overset{+2}{\text{Ni}}{}^{2+}(aq) + \overset{0}{\text{H}}_2(g)$$

- *The least reactive metals cannot displace H_2 from any source.* Fortunately, precious metals, such as silver, gold, and platinum, do *not* react with water or acid.
- *An atom of one metal displaces the ion of another.* Comparisons of metal reactivity show, for example, that Zn metal displaces Cu^{2+} ion from aqueous $CuSO_4$:

$$\overset{+2}{\text{Cu}}{}^{2+}(aq) + \overset{+6}{\text{S}}\overset{-2}{\text{O}}{}_4^{2-}(aq) + \overset{0}{\text{Zn}}(s) \longrightarrow \overset{0}{\text{Cu}}(s) + \overset{+2}{\text{Zn}}{}^{2+}(aq) + \overset{+6}{\text{S}}\overset{-2}{\text{O}}{}_4^{2-}(aq)$$

  And Figure 4.23 shows that Cu metal displaces silver (Ag^+) ion from solution; therefore, zinc is more reactive than copper, which is more reactive than silver.

Figure 4.22 The displacement of H_2 from acid by nickel.

Source: © McGraw-Hill Education/Stephen Frisch, photographer

Many such reactions form the basis of the **activity series of the metals.** Note these two points from Figure 4.24 *(on the next page)*:

- Elements higher on the list are stronger reducing agents than elements lower down; in other words, any metal can reduce, or displace, the ions of metals below it. For example, since Mn is higher in the list than Ni, Mn(*s*) can reduce Ni^{2+} ions, displacing them from solution:

$$\text{Mn}(s) + \text{NiCl}_2(aq) \longrightarrow \text{MnCl}_2(aq) + \text{Ni}(s)$$

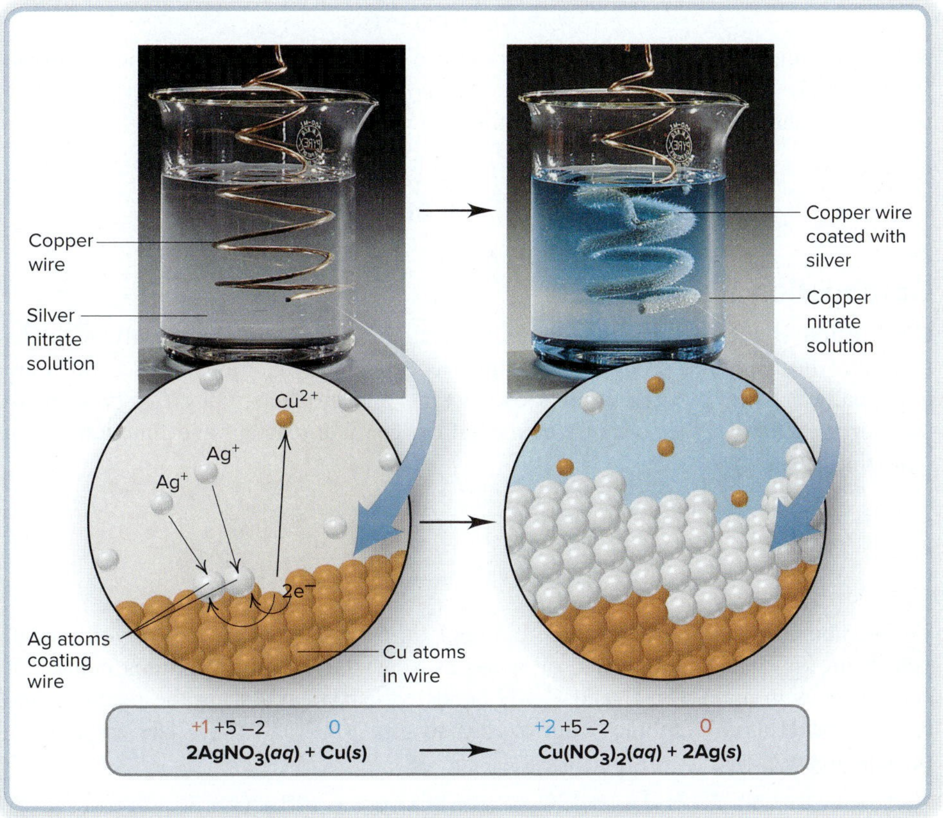

Figure 4.23 A more reactive metal (Cu) displacing the ion of a less reactive metal (Ag^+) from solution.

Source: © McGraw-Hill Education/Stephen Frisch, photographer

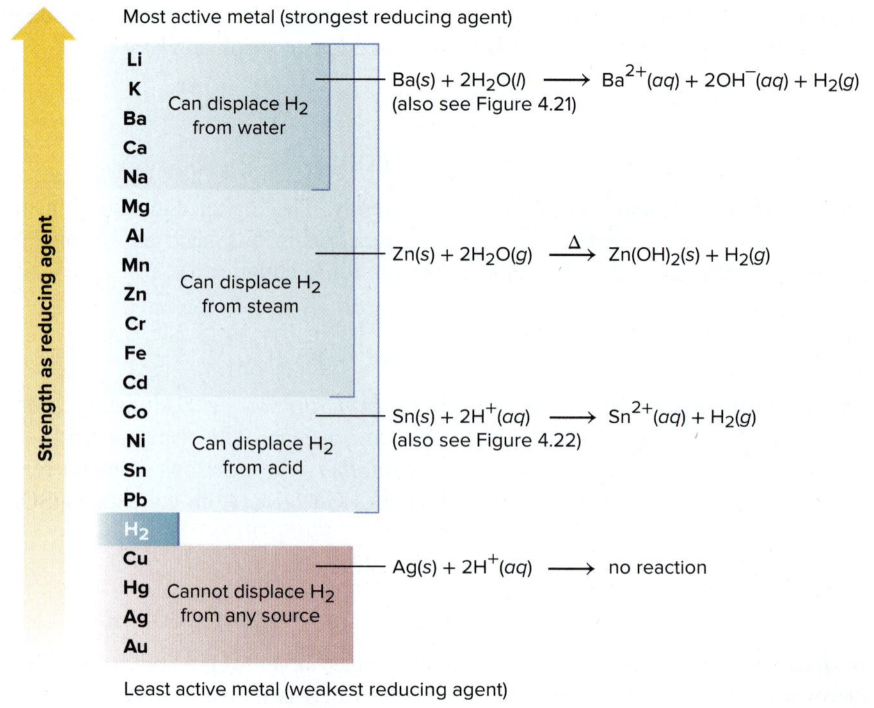

Figure 4.24 The activity series of the metals.

but $Ni(s)$ cannot reduce Mn^{2+} ions:

$$Ni(s) + MnCl_2(aq) \longrightarrow \text{no reaction}$$

- From higher to lower, a metal can displace H_2 (reduce H^+) from water, steam, acid, or not at all.

The Activity Series of the Halogens Reactivity of the elements increases up Group 7A(17), so we have

$$F_2 > Cl_2 > Br_2 > I_2$$

A halogen higher in the group is a stronger oxidizing agent than one lower down (Figure 4.25). Thus, elemental chlorine can oxidize bromide ions *(below)* or iodide ions from solution, and elemental bromine can oxidize iodide ions:

$$\overset{-1}{2Br^-(aq)} + \overset{0}{Cl_2(aq)} \longrightarrow \overset{0}{Br_2(aq)} + \overset{-1}{2Cl^-(aq)}$$

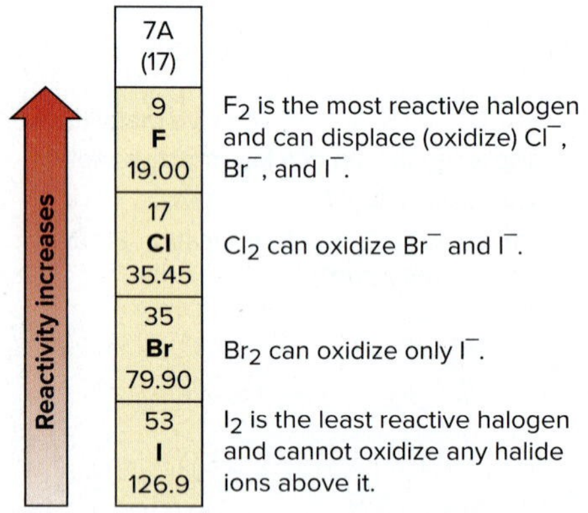

Figure 4.25 Reactivity of the halogens.

Combustion Reactions

Combustion is the process of combining with oxygen, most commonly with the release of heat and the production of light, as in a flame. Combustion reactions are not usually classified by the number of reactants and products, but *all of these reactions are redox processes* because elemental oxygen is a reactant and functions as an oxidizing agent:

$$2\overset{+2}{C}\overset{-2}{O}(g) + \overset{0}{O_2}(g) \longrightarrow 2\overset{+4}{C}\overset{-2}{O_2}(g)$$

The combustion reactions that we commonly use to produce energy involve coal, petroleum, gasoline, natural gas, or wood as a reactant. These mixtures consist of substances with many C—C and C—H bonds, which break during the reaction, and each C and H atom combines with oxygen to form CO_2 and H_2O. The combustion of butane is typical:

$$2C_4H_{10}(g) + 13O_2(g) \longrightarrow 8CO_2(g) + 10H_2O(g)$$

Biological *respiration* is a multistep combustion process that occurs within our cells when we "burn" foodstuffs, such as glucose, for energy:

$$C_6H_{12}O_6(s) + 6O_2(g) \longrightarrow 6CO_2(g) + 6H_2O(g) + \text{energy}$$

SAMPLE PROBLEM 4.20 Identifying the Type of Redox Reaction

Problem Classify each of these redox reactions as a combination, decomposition, or displacement reaction, write a balanced molecular equation for each, as well as total and net ionic equations for part (c), and identify the oxidizing and reducing agents:

(a) Magnesium(s) + nitrogen(g) $\longrightarrow$ magnesium nitride(s)
(b) Hydrogen peroxide(l) $\longrightarrow$ water + oxygen gas
(c) Aluminum(s) + lead(II) nitrate(aq) $\longrightarrow$ aluminum nitrate(aq) + lead(s)

Plan To decide on reaction type, recall that combination reactions have fewer products than reactants, decomposition reactions have more products than reactants, and displacement reactions have the same number of reactants and products. The oxidation number (O.N.) becomes more positive for the reducing agent and less positive for the oxidizing agent.

Solution (a) Combination: two substances form one. This reaction occurs, along with formation of magnesium oxide, when magnesium burns in air, which is mostly N_2:

$$3\overset{0}{Mg}(s) + \overset{0}{N_2}(g) \longrightarrow \overset{+2}{Mg_3}\overset{-3}{N_2}(s)$$

Mg is the reducing agent; N_2 is the oxidizing agent.

(b) Decomposition: one substance forms two. Because hydrogen peroxide is very unstable and breaks down from heat, light, or just shaking, this reaction occurs within every bottle of this common household antiseptic:

$$2\overset{+1}{H_2}\overset{-1,\,-1}{O_2}(l) \longrightarrow 2\overset{+1}{H_2}\overset{-2}{O}(l) + \overset{0}{O_2}(g)$$

H_2O_2 is both the oxidizing *and* the reducing agent. The O.N. of O in peroxides is −1. It is shown in blue *and* red because it both increases to 0 in O_2 and decreases to −2 in H_2O.

(c) Displacement: two substances form two others. As Figure 4.24 shows, Al is more active than Pb and, thus, displaces it from aqueous solution:

$$2\overset{0}{Al}(s) + 3\overset{+2}{Pb}(\overset{+5}{N}\overset{-2}{O_3})_2(aq) \longrightarrow 2\overset{+3}{Al}(\overset{+5}{N}\overset{-2}{O_3})_3(aq) + 3\overset{0}{Pb}(s)$$

Al is the reducing agent; $Pb(NO_3)_2$ is the oxidizing agent.

The total ionic equation is

$$2Al(s) + 3Pb^{2+}(aq) + 6NO_3^-(aq) \longrightarrow 2Al^{3+}(aq) + 6NO_3^-(aq) + 3Pb(s)$$

The net ionic equation is

$$2Al(s) + 3Pb^{2+}(aq) \longrightarrow 2Al^{3+}(aq) + 3Pb(s)$$

FOLLOW-UP PROBLEMS

4.20A Classify each of the following redox reactions as a combination, decomposition, or displacement reaction, write a balanced molecular equation for each, as well as total and net ionic equations for parts (b) and (c), and identify the oxidizing and reducing agents:

(a) $S_8(s) + F_2(g) \longrightarrow SF_4(g)$
(b) $CsI(aq) + Cl_2(aq) \longrightarrow CsCl(aq) + I_2(aq)$
(c) $Ni(NO_3)_2(aq) + Cr(s) \longrightarrow Ni(s) + Cr(NO_3)_3(aq)$

4.20B Classify each of the following redox reactions as a combination, decomposition, or displacement reaction, write a balanced molecular equation for each, as well as total and net ionic equations for part (a), and identify the oxidizing and reducing agents:

(a) $Co(s) + HCl(aq) \longrightarrow CoCl_2(aq) + H_2(g)$

(b) $CO(g) + O_2(g) \longrightarrow CO_2(g)$

(c) $N_2O_5(s) \longrightarrow NO_2(g) + O_2(g)$

SOME SIMILAR PROBLEMS 4.103–4.106

› Summary of Section 4.6

- A reaction that has an element as reactant or product is a redox reaction.
- In combination redox reactions, elements combine to form a compound, or a compound and an element combine.
- In decomposition redox reactions, a compound breaks down by absorption of heat or electricity into elements or into a different compound and an element.
- In displacement redox reactions, one element displaces the ion of another from solution.
- Activity series rank elements in order of ability to displace each other. A more reactive metal can displace (reduce) hydrogen ion or the ion of a less reactive metal from solution. A more reactive halogen can displace (oxidize) the ion of a less reactive halogen from solution.
- Combustion releases heat through a redox reaction of a substance with O_2.

4.7 THE REVERSIBILITY OF REACTIONS AND THE EQUILIBRIUM STATE

So far, we have viewed reactions as transformations of reactants into products that proceed until completion, that is, until the limiting reactant is used up. However, many reactions stop before this happens because two opposing reactions are taking place simultaneously: the forward (left-to-right) reaction continues, but the reverse (right-to-left) reaction occurs just as fast (at the same rate). At this point, *no further changes appear in the amounts of reactants or products,* and the reaction mixture has reached **dynamic equilibrium.** On the macroscopic level, the reaction is *static,* but it is *dynamic* on the molecular level.

Reversibility of Calcium Carbonate Decomposition In principle, *every reaction is reversible and will reach dynamic equilibrium if all the reactants and products are present.* Consider the reversible breakdown and formation of calcium carbonate. When heated, this compound breaks down to calcium oxide and carbon dioxide:

$$CaCO_3(s) \xrightarrow{\Delta} CaO(s) + CO_2(g) \text{ [breakdown]}$$

And it forms when calcium oxide and carbon dioxide react:

$$CaO(s) + CO_2(g) \longrightarrow CaCO_3(s) \text{ [formation]}$$

Let's study this process with two experiments:

1. We place $CaCO_3$ in an *open* steel container and heat it to approximately 900°C (Figure 4.26A). The $CaCO_3$ starts breaking down to CaO and CO_2, and the CO_2 escapes from the open container. The reaction goes to completion because the reverse reaction (formation) cannot occur in the absence of CO_2.
2. We perform the same experiment in a *closed* container from which the CO_2 produced cannot escape (Figure 4.26B). The breakdown (forward reaction) starts as before, but soon the CaO and CO_2 start reacting with each other (formation). Gradually, with time, as the amounts of CaO and CO_2 increase, the formation reaction speeds up. Eventually, the reverse reaction (formation) occurs as fast as the forward reaction (breakdown), and the amounts of $CaCO_3$, CaO, and CO_2 no

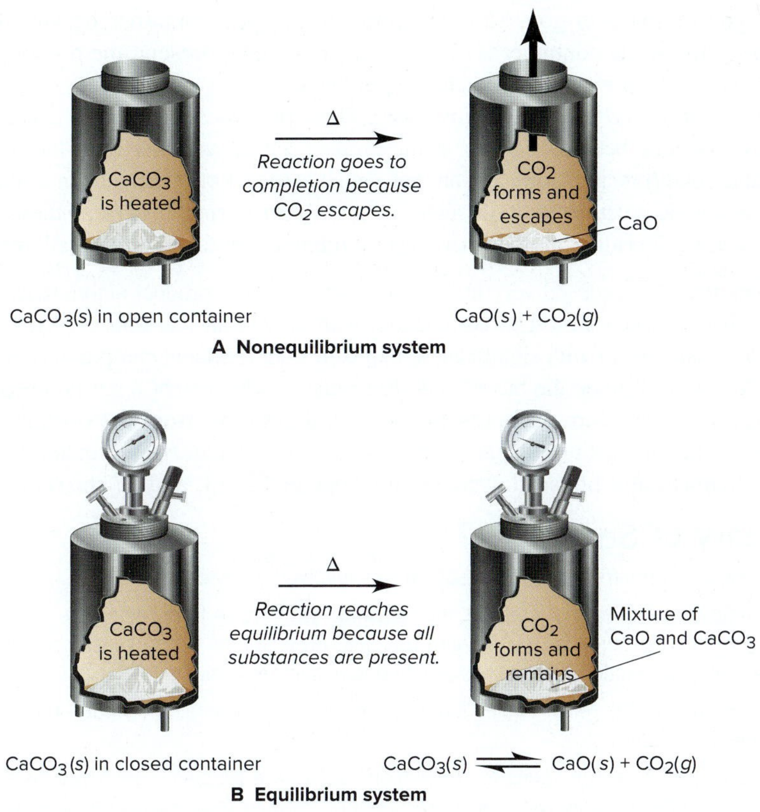

Figure 4.26 **The equilibrium state.**

longer change: the system has reached a state of equilibrium. We indicate this condition with a pair of arrows pointing in opposite directions:

$$CaCO_3(s) \rightleftharpoons CaO(s) + CO_2(g)$$

In contrast to the reaction in the open container, equilibrium *is* reached because *all the substances remain in contact with each other.*

Some Other Types of Reversible Reactions Reaction reversibility applies to other systems as well:

1. *Weak acids in water.* Weak acids dissociate very little into ions in water because the dissociation becomes balanced by a reassociation of H_3O^+ with the anion to re-form the weak acid and water. For example, when acetic acid dissolves, some CH_3COOH molecules transfer a proton to H_2O and form H_3O^+ and CH_3COO^- ions. As more ions form, the ions react with each other more often to re-form acetic acid and water:

$$CH_3COOH(aq) + H_2O(l) \rightleftharpoons H_3O^+(aq) + CH_3COO^-(aq)$$

In fact, in 0.1 *M* CH_3COOH at 25°C, only about 1.3% of acid molecules are dissociated! Each weak acid dissociates to its own extent. For example, under the same conditions, propionic acid (CH_3CH_2COOH) is 1.1% dissociated, and hydrofluoric acid (HF) is 8.6% dissociated.

2. *Weak bases in water.* Similarly, the weak base ammonia reacts with water to form NH_4^+ and OH^- ions. As more ions form, they interact more often to re-form ammonia and water, and the rates of the forward and reverse reactions soon balance:

$$NH_3(aq) + H_2O(l) \rightleftharpoons NH_4^+(aq) + OH^-(aq)$$

The weak base methylamine (CH_3NH_2) reacts with water to a greater extent than ammonia does before reaching equilibrium, and the weak base aniline ($C_6H_5NH_2$) reacts to a lesser extent.

3. *Gas-forming reactions.* Acids react with carbonates (see the subsection on Gas-Forming Reactions in Section 4.4):

$$2HCl(aq) + K_2CO_3(aq) \longrightarrow 2KCl(aq) + H_2O(l) + CO_2(g)$$

This type of reaction goes to completion in an open container because the CO_2 escapes. But, if the container is closed to keep the CO_2 present, the reverse reaction takes place and the mixture reaches equilibrium.

4. *Other acid-base and precipitation reactions.* These processes seem to "go to completion" because the ions become tied up in the product, either as water (acid-base) or as an insoluble solid (precipitation), and are not available to re-form reactants. In truth, though, water and ionic precipitates do dissociate to an extremely small extent, so these reactions also reach equilibrium, but the final reaction mixture consists of *almost* all product.

Thus, some reactions proceed very little, producing very little product before reaching equilibrium, others proceed almost to completion, with almost all reactants depleted, and still others reach equilibrium with significant amounts of both reactants *and* products present. In Chapter 20, we'll examine the factors that determine to what extent a reaction proceeds.

Dynamic equilibrium occurs in many natural systems, from the cycling of water in the environment to the nuclear processes occurring in stars. We examine equilibrium in chemical and physical systems in Chapters 12, 13, and 17 through 21.

› Summary of Section 4.7

- Every reaction is reversible if all the substances involved are present.
- As the amounts of products increase, the reactants begin to re-form. When the reverse reaction occurs as rapidly as the forward reaction, the amounts of the substances no longer change, and the reaction mixture has reached dynamic equilibrium.
- Weak acids and bases reach equilibrium in water with a very small proportion of their molecules dissociated.
- An aqueous ionic reaction "goes to completion" because a product is removed (as a gas) or because a product (usually water or a precipitate) is formed that reacts very slightly in a reverse reaction.

CHAPTER REVIEW GUIDE

Learning Objectives

Relevant section (§) and/or sample problem (SP) numbers appear in parentheses.

Understand These Concepts

1. Why water is a polar molecule and how it dissociates ionic compounds into ions (§4.1)
2. The differing nature of the species present when ionic and covalent compounds dissolve in water (§4.1)
3. The distinctions among strong electrolytes, weak electrolytes, and nonelectrolytes (§4.1, §4.4)
4. The meanings of *concentration* and *molarity* (§4.1)
5. The effect of dilution on the concentration of a solute (§4.1)
6. The three types of equations that specify the species and the chemical change in an aqueous ionic reaction (§4.2)
7. The key event in three classes of aqueous ionic reactions (§4.3–§4.5)
8. How to decide whether a precipitation reaction occurs (§4.3)
9. The distinction between strong and weak aqueous acids and bases (§4.4)
10. How acid-base reactions are proton transfers; how carbonates (and sulfites) act as bases in gas-forming reactions (§4.4)
11. The distinction between a transfer and a shift of electrons in the formation of ionic and covalent compounds (§4.5)
12. The relation between changes in oxidation number and identities of oxidizing and reducing agents (§4.5)
13. The presence of elements in combination, decomposition, and displacement redox reactions; the meaning of an activity series (§4.6)
14. The balance between forward and reverse reactions in dynamic equilibrium; why some acids and bases are weak (§4.7)

Master These Skills

1. Using molecular scenes to depict an ionic compound in aqueous solution (SP 4.1)
2. Using the formula of a compound to find the number of moles of ions in solution (SP 4.2)
3. Calculating molarity or mass of a solute in solution (SPs 4.3–4.5)
4. Preparing a dilute solution from a concentrated one (SP 4.6)
5. Using molecular scenes to visualize changes in concentration (SP 4.7)
6. Predicting whether a precipitation reaction occurs (SP 4.8)
7. Using molecular scenes to understand precipitation reactions (SP 4.9)
8. Using molar ratios to convert between amounts of reactants and products for reactions in solution (SPs 4.10, 4.11, 4.15)
9. Determining the number of H^+ (or OH^-) ions in an aqueous acid (or base) solution (SP 4.12)
10. Writing ionic equations to describe precipitation and acid-base reactions (SPs 4.8, 4.9, 4.13)
11. Writing proton-transfer equations for acid-base reactions (SP 4.14)
12. Calculating an unknown concentration from an acid-base or redox titration (SPs 4.16, 4.19)
13. Determining the oxidation number of any element in a compound or ion (SP 4.17)
14. Identifying redox reactions and oxidizing and reducing agents from changes in O.N.s (SP 4.18)
15. Distinguishing among combination, decomposition, and displacement redox reactions (SP 4.20)

Key Terms

Page numbers appear in parentheses.

acid (165)
acid-base (neutralization) reaction (165)
activity series of the metals (185)
base (165)
concentration (150)
dynamic equilibrium (188)
electrolyte (148)
end point (173)
equivalence point (173)
hydronium ion (165)
metathesis reaction (159)
molarity (M) (150)
molecular equation (156)
net ionic equation (157)
nonelectrolyte (150)
oxidation (175)
oxidation number (O.N.) (or oxidation state) (176)
oxidation-reduction (redox) reaction (174)
oxidizing agent (176)
polar molecule (146)
precipitate (157)
precipitation reaction (157)
reducing agent (176)
reduction (175)
salt (167)
solute (145)
solvated (147)
solvent (145)
spectator ion (156)
titration (172)
total ionic equation (156)

Key Equations and Relationships

Page numbers appear in parentheses.

4.1 Defining molarity (150):

$$\text{Molarity} = \frac{\text{moles of solute}}{\text{liters of solution}} \quad \text{or} \quad M = \frac{\text{mol solute}}{\text{L soln}}$$

4.2 Diluting a concentrated solution (153):

$$M_{\text{dil}} \times V_{\text{dil}} = \text{amount (mol) of solute} = M_{\text{conc}} \times V_{\text{conc}}$$

BRIEF SOLUTIONS TO FOLLOW-UP PROBLEMS

4.1A (a) There are three 2+ particles and twice as many 1− particles. The compound is $BaCl_2$.

(b) Mass (g) of $BaCl_2$ = 9 particles × $\frac{0.05 \text{ mol particles}}{1 \text{ particle}}$ × $\frac{1 \text{ mol } BaCl_2}{3 \text{ mol particles}}$ × $\frac{208.2 \text{ g } BaCl_2}{1 \text{ mol } BaCl_2}$ = 31.2 g $BaCl_2$

4.1B (a) $2Na_3PO_4(s) \xrightarrow{H_2O} 6Na^+(aq) + 2PO_4^{3-}(aq)$

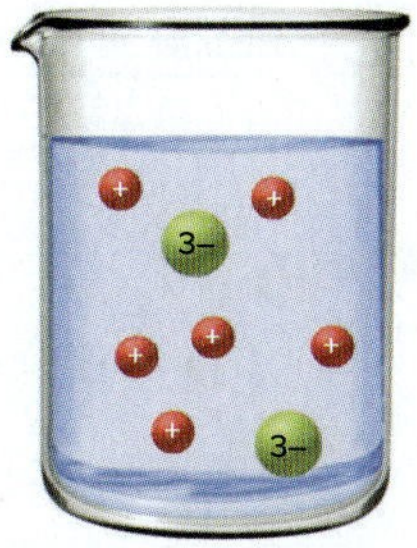

(b) Amount (mol) of ions = 0.40 mol Na_3PO_4 × $\frac{4 \text{ mol ions}}{1 \text{ mol } Na_3PO_4}$ = 1.6 mol ions

4.2A (a) $KClO_4(s) \xrightarrow{H_2O} K^+(aq) + ClO_4^-(aq)$;
2 mol of K^+ and 2 mol of ClO_4^-

(b) $Mg(C_2H_3O_2)_2(s) \xrightarrow{H_2O} Mg^{2+}(aq) + 2C_2H_3O_2^-(aq)$;

Amount (mol) of Mg^{2+}

$$= 354 \text{ g } Mg(C_2H_3O_2)_2 \times \frac{1 \text{ mol } Mg(C_2H_3O_2)_2}{142.40 \text{ g } Mg(C_2H_3O_2)_2} \times \frac{1 \text{ mol } Mg^{2+}}{1 \text{ mol } Mg(C_2H_3O_2)_2} = 2.49 \text{ mol } Mg^{2+}$$

Amount (mol) of $C_2H_3O_2^-$

$$= 354 \text{ g } Mg(C_2H_3O_2)_2 \times \frac{1 \text{ mol } Mg(C_2H_3O_2)_2}{142.40 \text{ g } Mg(C_2H_3O_2)_2} \times \frac{2 \text{ mol } C_2H_3O_2^-}{1 \text{ mol } Mg(C_2H_3O_2)_2} = 4.97 \text{ mol } C_2H_3O_2^-$$

(c) $(NH_4)_2CrO_4(s) \xrightarrow{H_2O} 2NH_4^+(aq) + CrO_4^{2-}(aq)$;

Amount (mol) of NH_4^+ = 1.88×10^{24} formula units $(NH_4)_2CrO_4$

$$\times \frac{1 \text{ mol } (NH_4)_2CrO_4}{6.022\times10^{23} \text{ formula units } (NH_4)_2CrO_4} \times \frac{2 \text{ mol } NH_4^+}{1 \text{ mol } (NH_4)_2CrO_4} = 6.24 \text{ mol } NH_4^+$$

Amount (mol) of CrO_4^{2-} = 1.88×10^{24} formula units $(NH_4)_2CrO_4$

$$\times \frac{1 \text{ mol } (NH_4)_2CrO_4}{6.022\times10^{23} \text{ formula units } (NH_4)_2CrO_4} \times \frac{1 \text{ mol } CrO_4^{2-}}{1 \text{ mol } (NH_4)_2CrO_4} = 3.12 \text{ mol } CrO_4^{2-}$$

4.2B (a) $Li_2CO_3(s) \xrightarrow{H_2O} 2Li^+(aq) + CO_3^{2-}(aq)$;
8 mol of Li^+ and 4 mol of CO_3^{2-}

(b) $Fe_2(SO_4)_3(s) \xrightarrow{H_2O} 2Fe^{3+}(aq) + 3SO_4^{2-}(aq)$;
Amount (mol) of Fe^{3+}

$$= 112 \text{ g } Fe_2(SO_4)_3 \times \frac{1 \text{ mol } Fe_2(SO_4)_3}{399.88 \text{ g } Fe_2(SO_4)_3} \times \frac{2 \text{ mol } Fe^{3+}}{1 \text{ mol } Fe_2(SO_4)_3} = 0.560 \text{ mol } Fe^{3+}$$

BRIEF SOLUTIONS TO FOLLOW-UP PROBLEMS (continued)

Amount (mol) of SO_4^{2-}

$$= 112 \text{ g } Fe_2(SO_4)_3 \times \frac{1 \text{ mol } Fe_2(SO_4)_3}{399.88 \text{ g } Fe_2(SO_4)_3} \times \frac{3 \text{ mol } SO_4^{2-}}{1 \text{ mol } Fe_2(SO_4)_3}$$

$$= 0.840 \text{ mol } SO_4^{2-}$$

(c) $Al(NO_3)_3(s) \xrightarrow{H_2O} Al^{3+}(aq) + 3NO_3^-(aq)$;

Amount (mol) of Al^{3+} $= 8.09\times10^{22}$ formula units $Al(NO_3)_3$

$$\times \frac{1 \text{ mol } Al(NO_3)_3}{6.022\times10^{23} \text{ formula units } Al(NO_3)_3} \times \frac{1 \text{ mol } Al^{3+}}{1 \text{ mol } Al(NO_3)_3}$$

$$= 0.134 \text{ mol } Al^{3+}$$

Amount (mol) of NO_3^- $= 8.09\times10^{22}$ formula units $Al(NO_3)_3$

$$\times \frac{1 \text{ mol } Al(NO_3)_3}{6.022\times10^{23} \text{ formula units } Al(NO_3)_3} \times \frac{3 \text{ mol } NO_3^-}{1 \text{ mol } Al(NO_3)_3}$$

$$= 0.403 \text{ mol } NO_3^-$$

4.3A Amount (mol) of KI $= 6.97 \text{ g KI} \times \frac{1 \text{ mol KI}}{166.0 \text{ g KI}}$

$$= 0.0420 \text{ mol KI}$$

$$M = \frac{0.0420 \text{ mol KI}}{100. \text{ mL soln}} \times \frac{1000 \text{ mL}}{1 \text{ L}} = 0.420\ M$$

4.3B Amount (mol) of $NaNO_3$ $= 175 \text{ mg } NaNO_3 \times \frac{1 \text{ g}}{1000 \text{ mg}}$

$$\times \frac{1 \text{ mol } NaNO_3}{85.00 \text{ g } NaNO_3}$$

$$= 2.06\times10^{-3} \text{ mol } NaNO_3$$

$$M = \frac{2.06\times10^{-3} \text{ mol } NaNO_3}{15.0 \text{ mL soln}} \times \frac{1000 \text{ mL}}{1 \text{ L}} = 0.137\ M$$

4.4A

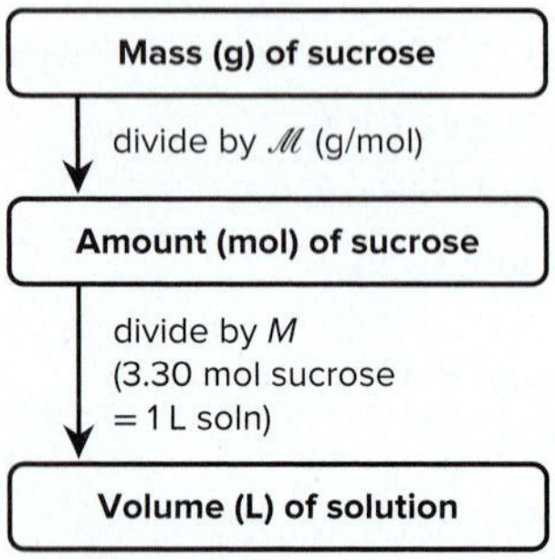

Volume (L) of sucrose soln

$$= 135 \text{ g sucrose} \times \frac{1 \text{ mol sucrose}}{342.30 \text{ g sucrose}} \times \frac{1 \text{ L soln}}{3.30 \text{ mol sucrose}}$$

$$= 0.120 \text{ L soln}$$

4.4B

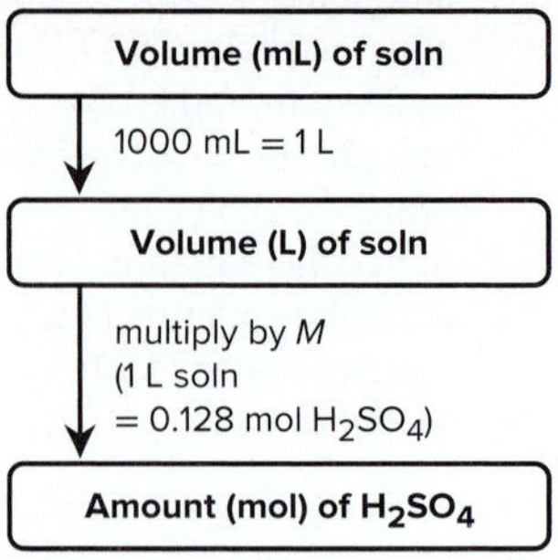

Amount (mol) of H_2SO_4

$$= 40.5 \text{ mL soln} \times \frac{1 \text{ L}}{1000 \text{ mL}} \times \frac{0.128 \text{ mol } H_2SO_4}{1 \text{ L soln}}$$

$$= 5.18\times10^{-3} \text{ mol } H_2SO_4$$

4.5A $Na_3PO_4(aq) \longrightarrow 3Na^+(aq) + PO_4^{3-}(aq)$

Amount (mol) of Na_3PO_4

$$= 1.32 \text{ L} \times \frac{0.55 \text{ mol } Na_3PO_4}{1 \text{ L}} \times \frac{3 \text{ mol } Na^+}{1 \text{ mol } Na_3PO_4}$$

$$= 2.2 \text{ mol } Na^+$$

Similarly, there are 0.73 mol PO_4^{3-}.

4.5B $Al_2(SO_4)_3(aq) \longrightarrow 2Al^{3+}(aq) + 3SO_4^{2-}(aq)$

Amount (mol) of Al^{3+}

$$= 1.25 \text{ mol } Al_2(SO_4)_3 \times \frac{2 \text{ mol } Al^{3+}}{1 \text{ mol } Al_2(SO_4)_3} = 2.50 \text{ mol } Al^{3+}$$

$$M = \frac{2.50 \text{ mol } Al^{3+}}{0.875 \text{ L}} = 2.86\ M$$

4.6A $V_{dil} = \frac{M_{conc} \times V_{conc}}{M_{dil}} = \frac{4.50\ M \times 60.0 \text{ mL}}{1.25\ M} = 216 \text{ mL}$

4.6B M of diluted H_2SO_4 $= \frac{M_{conc} \times V_{conc}}{V_{dil}} = \frac{7.50\ M \times 25.0 \text{ m}^3}{500. \text{ m}^3}$

$$= 0.375\ M\ H_2SO_4$$

$$\text{Mass (g) of } H_2SO_4\text{/mL soln} = \frac{0.375 \text{ mol } H_2SO_4}{1 \text{ L soln}} \times \frac{1 \text{ L}}{1000 \text{ mL}} \times \frac{98.08 \text{ g } H_2SO_4}{1 \text{ mol } H_2SO_4}$$

$$= 3.68\times10^{-2} \text{ g/mL soln}$$

4.7A To obtain B, the total volume of solution A was reduced by half:

$$V_{conc} = V_{dil} \times \frac{N_{dil}}{N_{conc}} = 1.0 \text{ mL} \times \frac{6 \text{ particles}}{12 \text{ particles}} = 0.50 \text{ mL}$$

To obtain C, $\frac{1}{2}$ of a unit volume of solvent was added to the unit volume of A:

$$V_{dil} = V_{conc} \times \frac{N_{conc}}{N_{dil}} = 1.0 \text{ mL} \times \frac{6 \text{ particles}}{4 \text{ particles}} = 1.5 \text{ mL}$$

4.7B $N_{dil} = N_{conc} \times \frac{V_{conc}}{V_{dil}} = 12 \text{ particles} \times \frac{100. \text{ mL}}{400. \text{ mL}}$

$= 3 \text{ particles}$

Circle B is the best representation of the unit volume.

4.8A (a) $Fe^{3+}(aq) + 3Cl^-(aq) + 3Cs^+(aq) + PO_4^{3-}(aq) \longrightarrow FePO_4(s) + 3Cl^-(aq) + 3Cs^+(aq)$

$Fe^{3+}(aq) + PO_4^{3-}(aq) \longrightarrow FePO_4(s)$

(b) $2Na^+(aq) + 2OH^-(aq) + Cd^{2+}(aq) + 2NO_3^-(aq) \longrightarrow 2Na^+(aq) + 2NO_3^-(aq) + Cd(OH)_2(s)$

$2OH^-(aq) + Cd^{2+}(aq) \longrightarrow Cd(OH)_2(s)$

(c) No reaction occurs.

4.8B (a) $2Ag^+(aq) + 2NO_3^-(aq) + Ba^{2+}(aq) + 2Cl^-(aq) \longrightarrow 2AgCl(s) + 2NO_3^-(aq) + Ba^{2+}(aq)$

$Ag^+(aq) + Cl^-(aq) \longrightarrow AgCl(s)$

(b) No reaction occurs.

(c) $Ni^{2+}(aq) + SO_4^{2-}(aq) + Pb^{2+}(aq) + 2NO_3^-(aq) \longrightarrow Ni^{2+}(aq) + 2NO_3^-(aq) + PbSO_4(s)$

$SO_4^{2-}(aq) + Pb^{2+}(aq) \longrightarrow PbSO_4(s)$

4.9A (a) There are 2+ and 1− ions. $PbCl_2$ is insoluble, so beaker A contains a solution of $Zn(NO_3)_2$.
(b) There are 2+ and 1− ions. $Cd(OH)_2$ is insoluble, so beaker B contains a solution of $Ba(OH)_2$.
(c) The precipitate is zinc hydroxide, and the spectator ions are Ba^{2+} and NO_3^-.

Molecular: $Zn(NO_3)_2(aq) + Ba(OH)_2(aq) \longrightarrow Zn(OH)_2(s) + Ba(NO_3)_2(aq)$

Total ionic:

$Zn^{2+}(aq) + 2NO_3^-(aq) + Ba^{2+}(aq) + 2OH^-(aq) \longrightarrow Zn(OH)_2(s) + Ba^{2+}(aq) + 2NO_3^-(aq)$

Net ionic: $Zn^{2+}(aq) + 2OH^-(aq) \longrightarrow Zn(OH)_2(s)$

(d) The OH^- ion is limiting.

$\text{Mass (g) of } Zn(OH)_2 = 6 \text{ OH}^- \text{ particles} \times \frac{0.050 \text{ mol OH}^- \text{ ions}}{1 \text{ OH}^- \text{ particle}} \times \frac{1 \text{ mol Zn(OH)}_2}{2 \text{ mol OH}^- \text{ ions}} \times \frac{99.40 \text{ g Zn(OH)}_2}{1 \text{ mol Zn(OH)}_2}$

$= 15 \text{ g Zn(OH)}_2$

4.9B (a) There are 1+ and 2− ions. Ag_2SO_4 is insoluble, so beaker A contains a solution of Li_2CO_3.
(b) There are 2+ and 1− ions. $Ni(OH)_2$ is insoluble, so beaker B contains a solution of $CaCl_2$.
(c) The precipitate is $CaCO_3$, and the spectator ions are Li^+ and Cl^-.

Molecular:

$Li_2CO_3(aq) + CaCl_2(aq) \longrightarrow 2LiCl(aq) + CaCO_3(s)$

Total ionic: $2Li^+(aq) + CO_3^{2-}(aq) + Ca^{2+}(aq) + 2Cl^-(aq) \longrightarrow 2Li^+(aq) + 2Cl^-(aq) + CaCO_3(s)$

Net ionic: $Ca^{2+}(aq) + CO_3^{2-}(aq) \longrightarrow CaCO_3(s)$

(d) The Ca^{2+} ion is limiting.

$\text{Mass (g) of } CaCO_3 = 3 \text{ Ca}^{2+} \text{ particles} \times \frac{0.20 \text{ mol Ca}^{2+} \text{ ions}}{1 \text{ Ca}^{2+} \text{ particle}} \times \frac{1 \text{ mol CaCO}_3}{1 \text{ mol Ca}^{2+} \text{ ions}} \times \frac{100.09 \text{ g CaCO}_3}{1 \text{ mol CaCO}_3}$

$= 60. \text{ g CaCO}_3$

4.10A $3CaCl_2(aq) + 2Na_3PO_4(aq) \longrightarrow Ca_3(PO_4)_2(s) + 6NaCl(aq)$

$\text{Volume (L) of } Na_3PO_4 = 0.300 \text{ L CaCl}_2 \text{ soln} \times \frac{0.175 \text{ mol CaCl}_2}{1 \text{ L CaCl}_2 \text{ soln}} \times \frac{2 \text{ mol Na}_3\text{PO}_4}{3 \text{ mol CaCl}_2} \times \frac{1 \text{ L Na}_3\text{PO}_4 \text{ solution}}{0.260 \text{ mol Na}_3\text{PO}_4}$

$= 0.135 \text{ L soln}$

4.10B $AgNO_3(aq) + NaCl(aq) \longrightarrow AgCl(s) + NaNO_3(aq)$

$\text{Amount (mol) of } AgNO_3 = 0.148 \text{ g AgCl} \times \frac{1 \text{ mol AgCl}}{143.4 \text{ g AgCl}} \times \frac{1 \text{ mol AgNO}_3}{1 \text{ mol AgCl}}$

$= 0.00103 \text{ mol AgNO}_3$

$\text{Molarity of } AgNO_3 = \frac{0.00103 \text{ mol AgNO}_3}{45.0 \text{ mL}} \times \frac{10^3 \text{ mL}}{1 \text{ L}}$

$= 0.0229 \, M$

4.11A (a) $Pb(C_2H_3O_2)_2(aq) + 2NaCl(aq) \longrightarrow PbCl_2(s) + 2NaC_2H_3O_2(aq)$

For $Pb(C_2H_3O_2)_2$: Amount (mol) of $Pb(C_2H_3O_2)_2$

$= 0.268 \text{ L} \times \frac{1.50 \text{ mol Pb(C}_2\text{H}_3\text{O}_2)_2}{1 \text{ L}}$

$= 0.402 \text{ mol Pb(C}_2\text{H}_3\text{O}_2)_2$

$\text{Amount (mol) of } PbCl_2 = 0.402 \text{ mol Pb(C}_2\text{H}_3\text{O}_2)_2 \times \frac{1 \text{ mol PbCl}_2}{1 \text{ mol Pb(C}_2\text{H}_3\text{O}_2)_2}$

$= 0.402 \text{ mol PbCl}_2$

$\text{For NaCl: Amount (mol) of NaCl} = 0.130 \text{ L} \times \frac{3.40 \text{ mol NaCl}}{1 \text{ L}}$

$= 0.442 \text{ mol NaCl}$

$\text{Amount (mol) of } PbCl_2 = 0.442 \text{ mol NaCl} \times \frac{1 \text{ mol PbCl}_2}{2 \text{ mol NaCl}}$

$= 0.221 \text{ mol PbCl}_2$

Thus, NaCl is the limiting reactant.

$\text{Mass (g) of } PbCl_2 = 0.221 \text{ mol PbCl}_2 \times \frac{278.1 \text{ g PbCl}_2}{1 \text{ mol PbCl}_2}$

$= 61.5 \text{ g PbCl}_2$

(b) Using "Ac" for acetate ion and with NaCl limiting:

Amount (mol)	$Pb(Ac)_2$ +	2NaCl	⟶	$PbCl_2$ +	2NaAc
Initial	0.402	0.442		0	0
Change	−0.221	−0.442		+0.221	+0.442
Final	0.181	0		0.221	0.442

4.11B (a) $Hg(NO_3)_2(aq) + Na_2S(aq) \longrightarrow HgS(s) + 2NaNO_3(aq)$

For $Hg(NO_3)_2$:
Amount (mol) of $Hg(NO_3)_2$

$= 0.050 \text{ L Hg(NO}_3)_2 \times \frac{0.010 \text{ mol Hg(NO}_3)_2}{1 \text{ L Hg(NO}_3)_2}$

$= 5.0 \times 10^{-4} \text{ mol Hg(NO}_3)_2$

4.95 Identify the oxidizing and reducing agents in the following reactions:

(a) $8H^+(aq) + Cr_2O_7^{2-}(aq) + 3SO_3^{2-}(aq) \longrightarrow 2Cr^{3+}(aq) + 3SO_4^{2-}(aq) + 4H_2O(l)$

(b) $NO_3^-(aq) + 4Zn(s) + 7OH^-(aq) + 6H_2O(l) \longrightarrow 4Zn(OH)_4^{2-}(aq) + NH_3(aq)$

Problems in Context

4.96 The active agent in many hair bleaches is hydrogen peroxide. The amount of H_2O_2 in 14.8 g of hair bleach was determined by titration with a standard potassium permanganate solution:

$$2MnO_4^-(aq) + 5H_2O_2(aq) + 6H^+(aq) \longrightarrow 5O_2(g) + 2Mn^{2+}(aq) + 8H_2O(l)$$

(a) How many moles of MnO_4^- were required for the titration if 43.2 mL of 0.105 M $KMnO_4$ was needed to reach the end point?
(b) How many moles of H_2O_2 were present in the 14.8-g sample of bleach?
(c) How many grams of H_2O_2 were in the sample?
(d) What is the mass percent of H_2O_2 in the sample?
(e) What is the reducing agent in the redox reaction?

4.97 A person's blood alcohol (C_2H_5OH) level can be determined by titrating a sample of blood plasma with a potassium dichromate solution. The balanced equation is

$$16H^+(aq) + 2Cr_2O_7^{2-}(aq) + C_2H_5OH(aq) \longrightarrow 4Cr^{3+}(aq) + 2CO_2(g) + 11H_2O(l)$$

If 35.46 mL of 0.05961 M $Cr_2O_7^{2-}$ is required to titrate 28.00 g of plasma, what is the mass percent of alcohol in the blood?

Elements in Redox Reactions

(Sample Problem 4.20)

Concept Review Questions

4.98 Which type of redox reaction leads to each of the following?
(a) An increase in the number of substances
(b) A decrease in the number of substances
(c) No change in the number of substances

4.99 Why do decomposition redox reactions typically have compounds as reactants, whereas combination redox and displacement redox reactions have one or more elements as reactants?

4.100 Which of the types of reactions discussed in Section 4.6 commonly produce more than one compound?

4.101 Are all combustion reactions redox reactions? Explain.

4.102 Give one example of a combination reaction that is a redox reaction and another that is not a redox reaction.

Skill-Building Exercises (grouped in similar pairs)

4.103 Balance each of the following redox reactions and classify it as a combination, decomposition, or displacement reaction:
(a) $Ca(s) + H_2O(l) \longrightarrow Ca(OH)_2(aq) + H_2(g)$
(b) $NaNO_3(s) \longrightarrow NaNO_2(s) + O_2(g)$
(c) $C_2H_2(g) + H_2(g) \longrightarrow C_2H_6(g)$

4.104 Balance each of the following redox reactions and classify it as a combination, decomposition, or displacement reaction:
(a) $HI(g) \longrightarrow H_2(g) + I_2(g)$
(b) $Zn(s) + AgNO_3(aq) \longrightarrow Zn(NO_3)_2(aq) + Ag(s)$
(c) $NO(g) + O_2(g) \longrightarrow N_2O_4(l)$

4.105 Balance each of the following redox reactions and classify it as a combination, decomposition, or displacement reaction:
(a) $Sb(s) + Cl_2(g) \longrightarrow SbCl_3(s)$ (b) $AsH_3(g) \longrightarrow As(s) + H_2(g)$
(c) $Zn(s) + Fe(NO_3)_2(aq) \longrightarrow Zn(NO_3)_2(aq) + Fe(s)$

4.106 Balance each of the following redox reactions and classify it as a combination, decomposition, or displacement reaction:
(a) $Mg(s) + H_2O(g) \longrightarrow Mg(OH)_2(s) + H_2(g)$
(b) $Cr(NO_3)_3(aq) + Al(s) \longrightarrow Al(NO_3)_3(aq) + Cr(s)$
(c) $PF_3(g) + F_2(g) \longrightarrow PF_5(g)$

4.107 Predict the product(s) and write a balanced equation for each of the following redox reactions:
(a) $Sr(s) + Br_2(l) \longrightarrow$ (b) $Ag_2O(s) \xrightarrow{\Delta}$
(c) $Mn(s) + Cu(NO_3)_2(aq) \longrightarrow$

4.108 Predict the product(s) and write a balanced equation for each of the following redox reactions:
(a) $Mg(s) + HCl(aq) \longrightarrow$ (b) $LiCl(l) \xrightarrow{\text{electricity}}$
(c) $SnCl_2(aq) + Co(s) \longrightarrow$

4.109 Predict the product(s) and write a balanced equation for each of the following redox reactions:
(a) $N_2(g) + H_2(g) \longrightarrow$ (b) $NaClO_3(s) \xrightarrow{\Delta}$
(c) $Ba(s) + H_2O(l) \longrightarrow$

4.110 Predict the product(s) and write a balanced equation for each of the following redox reactions:
(a) $Fe(s) + HClO_4(aq) \longrightarrow$ (b) $S_8(s) + O_2(g) \longrightarrow$
(c) $BaCl_2(l) \xrightarrow{\text{electricity}}$

4.111 Predict the product(s) and write a balanced equation for each of the following redox reactions:
(a) Cesium + iodine $\longrightarrow$
(b) Aluminum + aqueous manganese(II) sulfate $\longrightarrow$
(c) Sulfur dioxide + oxygen $\longrightarrow$
(d) Butane + oxygen $\longrightarrow$
(e) Write a balanced net ionic equation for (b).

4.112 Predict the product(s) and write a balanced equation for each of the following redox reactions:
(a) Pentane (C_5H_{12}) + oxygen $\longrightarrow$
(b) Phosphorus trichloride + chlorine $\longrightarrow$
(c) Zinc + hydrobromic acid $\longrightarrow$
(d) Aqueous potassium iodide + bromine $\longrightarrow$
(e) Write a balanced net ionic equation for (d).

4.113 How many grams of O_2 can be prepared from the thermal decomposition of 4.27 kg of HgO? Name and calculate the mass (in kg) of the other product.

4.114 How many grams of chlorine gas can be produced from the electrolytic decomposition of 874 g of calcium chloride? Name and calculate the mass (in g) of the other product.

4.115 In a combination reaction, 1.62 g of lithium is mixed with 6.50 g of oxygen.
(a) Which reactant is present in excess?
(b) How many moles of product are formed?
(c) After reaction, how many grams of each reactant and product are present?

4.116 In a combination reaction, 2.22 g of magnesium is heated with 3.75 g of nitrogen.
(a) Which reactant is present in excess?
(b) How many moles of product are formed?
(c) After reaction, how many grams of each reactant and product are present?

4.117 A mixture of $KClO_3$ and KCl with a mass of 0.950 g was heated to produce O_2. After heating, the mass of residue was 0.700 g. Assuming all the $KClO_3$ decomposed to KCl and O_2, calculate the mass percent of $KClO_3$ in the original mixture.

4.118 A mixture of $CaCO_3$ and CaO weighing 0.693 g was heated to produce gaseous CO_2. After heating, the remaining solid weighed 0.508 g. Assuming all the $CaCO_3$ broke down to CaO and CO_2, calculate the mass percent of $CaCO_3$ in the original mixture.

Problems in Context

4.119 Before arc welding was developed, a displacement reaction involving aluminum and iron(III) oxide was commonly used to produce molten iron. Called the thermite process, this reaction was used, for example, to connect sections of iron rails for train tracks. Calculate the mass of molten iron produced when 1.50 kg of aluminum reacts with 25.0 mol of iron(III) oxide.

4.120 Iron reacts rapidly with chlorine gas to form a reddish-brown, ionic compound (A), which contains iron in the higher of its two common oxidation states. Strong heating decomposes compound A to compound B, another ionic compound, which contains iron in the lower of its two oxidation states. When compound A is formed by the reaction of 50.6 g of Fe and 83.8 g of Cl_2 and then heated, how much compound B forms?

4.121 A sample of impure magnesium was analyzed by allowing it to react with excess HCl solution. After 1.32 g of the impure metal was treated with 0.100 L of 0.750 *M* HCl, 0.0125 mol of HCl remained. Assuming the impurities do not react, what is the mass % of Mg in the sample?

The Reversibility of Reactions and the Equilibrium State

Concept Review Questions

4.122 Why is the equilibrium state said to be "dynamic"?

4.123 In a decomposition reaction involving a gaseous product, what must be done for the reaction to reach equilibrium?

4.124 Describe what happens on the molecular level when acetic acid dissolves in water.

4.125 When either a mixture of NO and Br_2 or pure nitrosyl bromide (NOBr) is placed in a reaction vessel, the product mixture contains NO, Br_2, and NOBr. Explain.

Problem in Context

4.126 Ammonia is produced by the millions of tons annually for use as a fertilizer. It is commonly made from N_2 and H_2 by the Haber process. Because the reaction reaches equilibrium before going completely to product, the stoichiometric amount of ammonia is not obtained. At a particular temperature and pressure, 10.0 g of H_2 reacts with 20.0 g of N_2 to form ammonia. When equilibrium is reached, 15.0 g of NH_3 has formed. (a) Calculate the percent yield. (b) How many moles of N_2 and H_2 are present at equilibrium?

Comprehensive Problems

4.127 Nutritional biochemists have known for decades that acidic foods cooked in cast-iron cookware can supply significant amounts of dietary iron (ferrous ion). (a) Write a balanced net ionic equation, with oxidation numbers, that supports this fact. (b) Measurements show an increase from 3.3 mg of iron to 49 mg of iron per $\frac{1}{2}$-cup (125-g) serving during the slow preparation of tomato sauce in a cast-iron pot. How many ferrous ions are present in a 26-oz (737-g) jar of the tomato sauce?

4.128 Limestone ($CaCO_3$) is used to remove acidic pollutants from smokestack flue gases. It is heated to form lime (CaO), which reacts with sulfur dioxide to form calcium sulfite. Assuming a 70.% yield in the overall reaction, what mass of limestone is required to remove all the sulfur dioxide formed by the combustion of 8.5×10^4 kg of coal that is 0.33 mass % sulfur?

4.129 The brewing industry uses yeast to convert glucose to ethanol. The baking industry uses the carbon dioxide produced in the same reaction to make bread rise:

$$C_6H_{12}O_6(s) \xrightarrow{\text{yeast}} 2C_2H_5OH(l) + 2CO_2(g)$$

How many grams of ethanol can be produced from 100. g of glucose? What volume of CO_2 is produced? (Assume 1 mol of gas occupies 22.4 L at the conditions used.)

4.130 A chemical engineer determines the mass percent of iron in an ore sample by converting the Fe to Fe^{2+} in acid and then titrating the Fe^{2+} with MnO_4^-. A 1.1081-g ore sample was dissolved in acid and then titrated with 39.32 mL of 0.03190 *M* $KMnO_4$. The balanced equation is

$$8H^+(aq) + 5Fe^{2+}(aq) + MnO_4^-(aq) \longrightarrow 5Fe^{3+}(aq) + Mn^{2+}(aq) + 4H_2O(l)$$

Calculate the mass percent of iron in the ore.

4.131 Mixtures of $CaCl_2$ and NaCl are used to melt ice on roads. A dissolved 1.9348-g sample of such a mixture was analyzed by using excess $Na_2C_2O_4$ to precipitate the Ca^{2+} as CaC_2O_4. The CaC_2O_4 was dissolved in sulfuric acid, and the resulting $H_2C_2O_4$ was titrated with 37.68 mL of 0.1019 *M* $KMnO_4$ solution.
(a) Write the balanced net ionic equation for the precipitation reaction.
(b) Write the balanced net ionic equation for the titration reaction. (See Sample Problem 4.19.)
(c) What is the oxidizing agent?
(d) What is the reducing agent?
(e) Calculate the mass percent of $CaCl_2$ in the original sample.

4.132 You are given solutions of HCl and NaOH and must determine their concentrations. You use 27.5 mL of NaOH to titrate 100. mL of HCl and 18.4 mL of NaOH to titrate 50.0 mL of 0.0782 *M* H_2SO_4. Find the unknown concentrations.

4.133 The flask represents the products of the titration of 25 mL of sulfuric acid with 25 mL of sodium hydroxide.

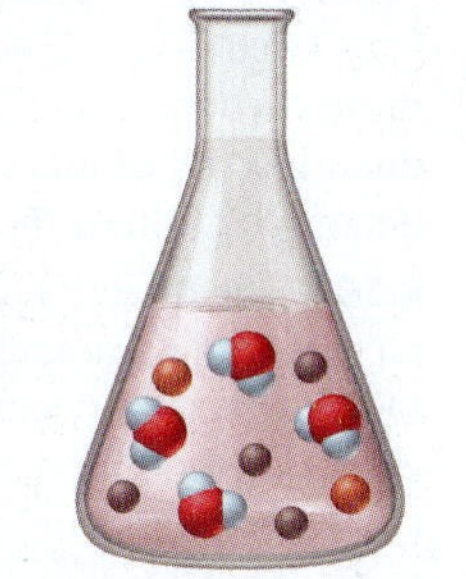

(a) Write balanced molecular, total ionic, and net ionic equations for the reaction.
(b) If each orange sphere represents 0.010 mol of sulfate ion, how many moles of acid and of base reacted?
(c) What are the molarities of the acid and the base?

4.134 To find the mass percent of dolomite [$CaMg(CO_3)_2$] in a soil sample, a geochemist titrates 13.86 g of soil with 33.56 mL of 0.2516 *M* HCl. What is the mass percent of dolomite in the soil?

4.135 On a lab exam, you have to find the concentrations of the monoprotic (one proton per molecule) acids HA and HB. You are given 43.5 mL of HA solution in one flask. A second flask contains 37.2 mL of HA, and you add enough HB solution to it to reach a final volume of 50.0 mL. You titrate the first HA solution with 87.3 mL of 0.0906 *M* NaOH and the mixture of HA and HB in the second flask with 96.4 mL of the NaOH solution. Calculate the molarity of the HA and HB solutions.

4.136 Nitric acid, a major industrial and laboratory acid, is produced commercially by the multistep Ostwald process, which begins with the oxidation of ammonia:

Step 1. $4NH_3(g) + 5O_2(g) \longrightarrow 4NO(g) + 6H_2O(l)$
Step 2. $2NO(g) + O_2(g) \longrightarrow 2NO_2(g)$
Step 3. $3NO_2(g) + H_2O(l) \longrightarrow 2HNO_3(l) + NO(g)$

(a) What are the oxidizing and reducing agents in each step?
(b) Assuming 100% yield in each step, what mass (in kg) of ammonia must be used to produce 3.0×10^4 kg of HNO_3?

4.137 Various data can be used to find the composition of an alloy (a metallic mixture). Show that calculating the mass % of Mg in a magnesium-aluminum alloy (d = 2.40 g/cm^3) using each of the following pieces of data gives the same answer (within rounding): (a) a sample of the alloy has a mass of 0.263 g (d of Mg = 1.74 g/cm^3; d of Al = 2.70 g/cm^3); (b) an identical sample reacting with excess aqueous HCl forms 1.38×10^{-2} mol of H_2; (c) an identical sample reacting with excess O_2 forms 0.483 g of oxide.

4.138 In 1995, Mario Molina, Paul Crutzen, and F. Sherwood Rowland shared the Nobel Prize in chemistry for their work on atmospheric chemistry. One reaction sequence they proposed for the role of chlorine in the decomposition of stratospheric ozone (we'll see another sequence in Chapter 16) is

(1) $Cl(g) + O_3(g) \longrightarrow ClO(g) + O_2(g)$
(2) $ClO(g) + ClO(g) \longrightarrow Cl_2O_2(g)$
(3) $Cl_2O_2(g) \xrightarrow{\text{light}} 2Cl(g) + O_2(g)$

Over the tropics, O atoms are more common in the stratosphere:

(4) $ClO(g) + O(g) \longrightarrow Cl(g) + O_2(g)$

(a) Which, if any, of these are oxidation-reduction reactions?
(b) Write an overall equation combining reactions 1–3.

4.139 Sodium peroxide (Na_2O_2) is often used in self-contained breathing devices, such as those used in fire emergencies, because it reacts with exhaled CO_2 to form Na_2CO_3 and O_2. How many liters of respired air can react with 80.0 g of Na_2O_2 if each liter of respired air contains 0.0720 g of CO_2?

4.140 A student forgets to weigh a mixture of sodium bromide dihydrate and magnesium bromide hexahydrate. Upon strong heating, the sample loses 252.1 mg of water. The mixture of anhydrous salts reacts with excess $AgNO_3$ solution to form 6.00×10^{-3} mol of solid AgBr. Find the mass % of each compound in the original mixture.

4.141 A typical formulation for window glass is 75% SiO_2, 15% Na_2O, and 10.% CaO by mass. What masses of sand (SiO_2), sodium carbonate, and calcium carbonate must be combined to produce 1.00 kg of glass after carbon dioxide is driven off by thermal decomposition of the carbonates?

4.142 The quantity of dissolved oxygen (DO) in natural waters is an essential parameter for monitoring survival of most aquatic life. DO is affected by temperature and the amount of organic waste. An earlier method for determining DO involved a two-step process:

(1) The water sample is treated with KI,

$$O_2(aq) + 4KI(aq) + 2H_2SO_4(aq) \longrightarrow 2I_2(aq) + 2H_2O(l) + 2K_2SO_4(aq)$$

(2) The I_2 is titrated with sodium thiosulfate,

$$I_2(aq) + 2Na_2S_2O_3(aq) \longrightarrow Na_2S_4O_6(aq) + 2NaI(aq)$$

A 50.0-mL water sample is treated with KI, and then 15.75 mL of 0.0105 M $Na_2S_2O_3$ is required to reach the end point.
(a) Which substance is oxidized in step 1, and which is reduced in step 2?
(b) What mass (g) of O_2 is dissolved in the water sample?

4.143 Physicians who specialize in sports medicine routinely treat athletes and dancers. Ethyl chloride, a local anesthetic commonly used for simple injuries, is the product of the combination of ethylene with hydrogen chloride:

$$C_2H_4(g) + HCl(g) \longrightarrow C_2H_5Cl(g)$$

Assume that 0.100 kg of C_2H_4 and 0.100 kg of HCl react. (a) How many molecules of gas (reactants plus products) are present when the reaction is complete? (b) How many moles of gas are present when half the product forms?

4.144 Thyroxine ($C_{15}H_{11}I_4NO_4$) is a hormone synthesized by the thyroid gland and used to control many metabolic functions in the body. A physiologist determines the mass percent of thyroxine in a thyroid extract by igniting 0.4332 g of extract with sodium carbonate, which converts the iodine to iodide. The iodide is dissolved in water, and bromine and hydrochloric acid are added, which convert the iodide to iodate.
(a) How many moles of iodate form per mole of thyroxine?
(b) Excess bromine is boiled off and more iodide is added, which reacts as shown in the following equation:

$$IO_3^-(aq) + 6H^+(aq) + 5I^-(aq) \longrightarrow 3I_2(aq) + 3H_2O(l)$$

How many moles of iodine are produced per mole of thyroxine? (*Hint:* Be sure to balance the charges as well as the atoms.) What are the oxidizing and reducing agents in the reaction?
(c) The iodine reacts completely with 17.23 mL of 0.1000 M thiosulfate as shown in the following *unbalanced* equation:

$$I_2(aq) + S_2O_3^{2-}(aq) \longrightarrow I^-(aq) + S_4O_6^{2-}(aq)$$

What is the mass percent of thyroxine in the thyroid extract?

4.145 Over time, as their free fatty acid (FFA) content increases, edible fats and oils become rancid. To measure rancidity, the fat or oil is dissolved in ethanol, and any FFA present is titrated with KOH dissolved in ethanol. In a series of tests on olive oil, a stock solution of 0.050 M ethanolic KOH was prepared at 25°C, stored at 0°C, and then placed in a 100-mL buret to titrate oleic acid [an FFA with formula $CH_3(CH_2)_7CH{=}CH(CH_2)_7COOH$] in the oil. Each of four 10.00-g samples of oil took several minutes to titrate: the first required 19.60 mL, the second 19.80 mL, and the third and fourth 20.00 mL of the ethanolic KOH.
(a) What is the apparent acidity of each sample, in terms of mass % of oleic acid? (*Note:* As the ethanolic KOH warms in the buret, its volume increases by a factor of 0.00104/°C.)
(b) Is the variation in acidity a random or systematic error? Explain.
(c) What is the actual acidity? How would you demonstrate this?

4.146 A chemist mixes solid AgCl, $CuCl_2$, and $MgCl_2$ in enough water to give a final volume of 50.0 mL.
(a) With ions shown as spheres and solvent molecules omitted for clarity, which scene best represents the resulting mixture?

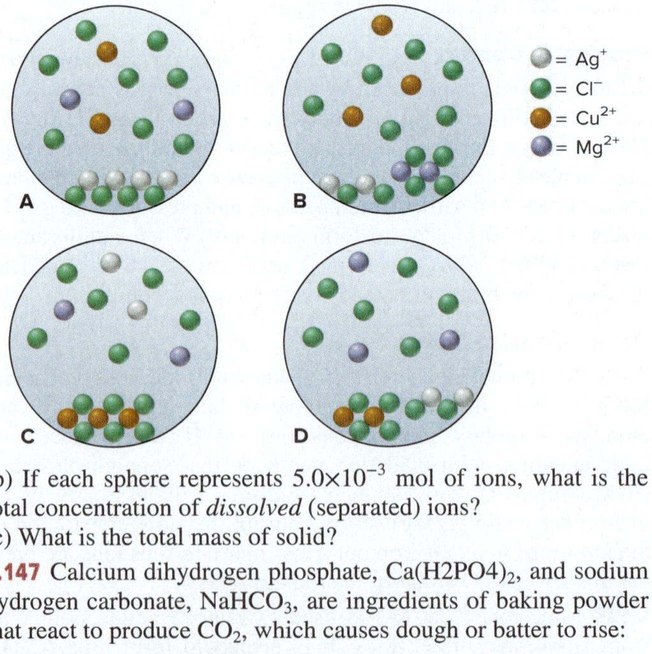

(b) If each sphere represents 5.0×10^{-3} mol of ions, what is the total concentration of *dissolved* (separated) ions?
(c) What is the total mass of solid?

4.147 Calcium dihydrogen phosphate, Ca(H2PO4)$_2$, and sodium hydrogen carbonate, $NaHCO_3$, are ingredients of baking powder that react to produce CO_2, which causes dough or batter to rise:

$$Ca(H_2PO_4)_2(s) + NaHCO_3(s) \longrightarrow CO_2(g) + H_2O(g) + CaHPO_4(s) + Na_2HPO_4(s)$$

[unbalanced]

If the baking powder contains 31% $NaHCO_3$ and 35% $Ca(H_2PO_4)_2$ by mass:

(a) How many moles of CO_2 are produced from 1.00 g of baking powder?
(b) If 1 mol of CO_2 occupies 37.0 L at 350°F (a typical baking temperature), what volume of CO_2 is produced from 1.00 g of baking powder?

4.148 In a titration of HNO_3, you add a few drops of phenolphthalein indicator to 50.00 mL of acid in a flask. You quickly add 20.00 mL of 0.0502 *M* NaOH but overshoot the end point, and the solution turns deep pink. Instead of starting over, you add 30.00 mL of the acid, and the solution turns colorless. Then, it takes 3.22 mL of the NaOH to reach the end point.
(a) What is the concentration of the HNO_3 solution?
(b) How many moles of NaOH were in excess after the first addition?

4.149 The active compound in Pepto-Bismol contains C, H, O, and Bi.
(a) When 0.22105 g of the compound was burned in excess O_2, 0.1422 g of bismuth(III) oxide, 0.1880 g of carbon dioxide, and 0.02750 g of water were formed. What is the empirical formula of the compound?
(b) Given a molar mass of 1086 g/mol, determine the molecular formula.
(c) Complete and balance the acid-base reaction between bismuth(III) hydroxide and salicylic acid ($HC_7H_5O_3$), which is used to form this compound.
(d) A dose of Pepto-Bismol contains 0.600 mg of active ingredient. If the yield of the reaction in part (c) is 88.0%, what mass (in mg) of bismuth(III) hydroxide is required to prepare one dose?

4.150 Two aqueous solutions contain the ions indicated below.

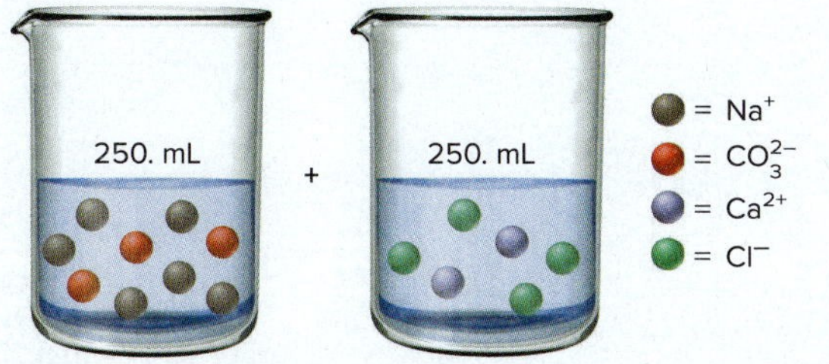

(a) Write balanced molecular, total ionic, and net ionic equations for the reaction that occurs when the solutions are mixed.
(b) If each sphere represents 0.050 mol of ion, what mass (in g) of precipitate forms, assuming 100% yield?
(c) What is the concentration of each ion in solution after reaction?

4.151 In 1997 and 2009, at United Nations conferences on climate change, many nations agreed to expand their research efforts to develop renewable sources of carbon-based fuels. For more than a quarter century, Brazil has been engaged in a program to replace gasoline with ethanol derived from the root crop manioc (cassava).
(a) Write separate balanced equations for the complete combustion of ethanol (C_2H_5OH) and of gasoline (represented by the formula C_8H_{18}).
(b) What mass (g) of oxygen is required to burn completely 1.00 L of a mixture that is 90.0% gasoline (d = 0.742 g/mL) and 10.0% ethanol (d = 0.789 g/mL) by volume?
(c) If 1.00 mol of O_2 occupies 22.4 L, what volume of O_2 is needed to burn 1.00 L of the mixture?
(d) Air is 20.9% O_2 by volume. What volume of air is needed to burn 1.00 L of the mixture?

4.152 In a car engine, gasoline (represented by C_8H_{18}) does not burn completely, and some CO, a toxic pollutant, forms along with CO_2 and H_2O. If 5.0% of the gasoline forms CO:
(a) What is the ratio of CO_2 to CO molecules in the exhaust?
(b) What is the mass ratio of CO_2 to CO?
(c) What percentage of the gasoline must form CO for the mass ratio of CO_2 to CO to be exactly 1/1?

4.153 The amount of ascorbic acid (vitamin C; $C_6H_8O_6$) in tablets is determined by reaction with bromine and then titration of the hydrobromic acid with standard base:

$$C_6H_8O_6(aq) + Br_2(aq) \longrightarrow C_6H_6O_6(aq) + 2HBr(aq)$$
$$HBr(aq) + NaOH(aq) \longrightarrow NaBr(aq) + H_2O(l)$$

A certain tablet is advertised as containing 500 mg of vitamin C. One tablet was dissolved in water and reacted with Br_2. The solution was then titrated with 43.20 mL of 0.1350 *M* NaOH. Did the tablet contain the advertised quantity of vitamin C?

4.154 In the process of *salting-in,* protein solubility in a dilute salt solution is increased by adding more salt. Because the protein solubility depends on the total ion concentration as well as the ion charge, salts containing doubly charged ions are often more effective at increasing the protein solubility than those containing singly charged ions.
(a) How many grams of $MgCl_2$ must dissolve to equal the ion concentration of 12.4 g of NaCl?
(b) How many grams of CaS must dissolve?
(c) Which of the three salt solutions would dissolve the most protein?

4.155 At liftoff, a space shuttle uses a solid mixture of ammonium perchlorate and aluminum powder to obtain great thrust from the volume change of solid to gas. In the presence of a catalyst, the mixture forms solid aluminum oxide and aluminum trichloride and gaseous water and nitrogen monoxide.
(a) Write a balanced equation for the reaction, and identify the reducing and oxidizing agents.
(b) How many total moles of gas (water vapor and nitrogen monoxide) are produced when 50.0 kg of ammonium perchlorate reacts with a stoichiometric amount of Al?
(c) What is the change in volume from this reaction? (d of NH_4ClO_4 = 1.95 g/cc, d of Al = 2.70 g/cc, d of Al_2O_3 = 3.97 g/cc, and d of $AlCl_3$ = 2.44 g/cc; assume 1 mol of gas occupies 22.4 L.)

4.156 A *reaction cycle* for an element is a series of reactions beginning and ending with that element. In the following copper reaction cycle, copper has either a 0 or a +2 oxidation state. Write balanced molecular and net ionic equations for each step.
(1) Copper metal reacts with aqueous bromine to produce a green-blue solution.
(2) Adding aqueous sodium hydroxide forms a blue precipitate.
(3) The precipitate is heated and turns black (water is released).
(4) The black solid dissolves in nitric acid to give a blue solution.
(5) Adding aqueous sodium phosphate forms a green precipitate.
(6) The precipitate forms a blue solution in sulfuric acid.
(7) Copper metal is recovered from the blue solution when zinc metal is added.

4.157 Two 25.0-mL aqueous solutions, labeled A and B, contain the ions indicated:

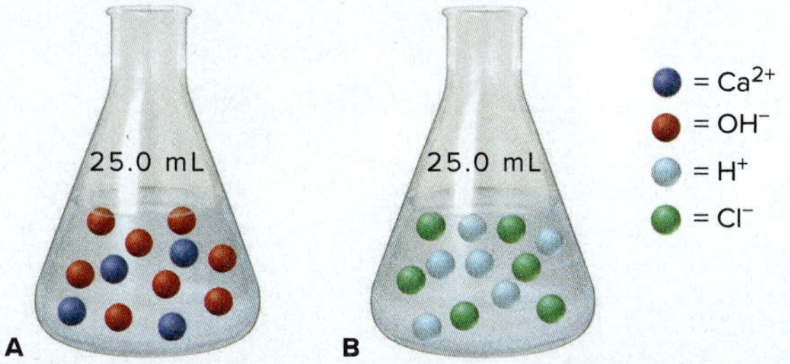

(a) If each sphere represents 1.0×10^{-3} mol of ion, will the equivalence point have been reached when all of B has been added to A?
(b) What is the molarity of B?
(c) What additional volume (mL) of B must be added to reach the equivalence point?

5 Gases and the Kinetic-Molecular Theory

5.1 An Overview of the Physical States of Matter

5.2 Gas Pressure and Its Measurement
Measuring Pressure
Units of Pressure

5.3 The Gas Laws and Their Experimental Foundations
Relationship Between Volume and Pressure: Boyle's Law
Relationship Between Volume and Temperature: Charles's Law
Relationship Between Volume and Amount: Avogadro's Law
Gas Behavior at Standard Conditions
The Ideal Gas Law
Solving Gas Law Problems

5.4 Rearrangements of the Ideal Gas Law
Density of a Gas
Molar Mass of a Gas
Partial Pressure of a Gas
Reaction Stoichiometry

5.5 The Kinetic-Molecular Theory: A Model for Gas Behavior
How the Theory Explains the Gas Laws
Effusion and Diffusion
Mean Free Path and Collision Frequency

5.6 Real Gases: Deviations from Ideal Behavior
Effects of Extreme Conditions
The van der Waals Equation: Adjusting the Ideal Gas Law

Source: © Rich Carey/Shutterstock.com

Concepts and Skills to Review Before You Study This Chapter

- physical states of matter (Section 1.1)
- SI unit conversions (Section 1.4)
- amount-mass-number conversions (Section 3.1)

Trained scuba divers know not to hold their breath underwater, especially while ascending. These divers are familiar with one of the properties of gases—as the water pressure decreases during an ascent, air in the lungs expands in volume, causing lung damage. Divers also know that the gas pressure in a warm scuba tank will drop when the tank is submerged in cold water; that drop in pressure is the reason scuba tanks are sometimes overfilled. Other properties of gases are at work in the inflation of a car's air bag, the rising of a loaf of bread, and the operation of a car engine. People have been studying the behavior of gases and the other states of matter throughout history; in fact, three of the four "elements" of the ancient Greeks were air (gas), water (liquid), and earth (solid). Yet, despite millennia of observations, many questions remain. In this chapter and its companion, Chapter 12, we examine the physical states and their interrelations. Here, we highlight the gaseous state, the one we understand best.

Gases are everywhere. Our atmosphere is a colorless, odorless mixture of 18 gases, some of which—O_2, N_2, H_2O vapor, and CO_2—take part in life-sustaining cycles of redox reactions throughout the environment. And several other gases, such as chlorine and ammonia, have essential roles in industry. Yet, in this chapter, we put aside the *chemical* behavior unique to any particular gas and focus instead *on the **physical** behavior common to all gases.*

IN THIS CHAPTER . . . *We explore the physical behavior of gases and the theory that explains it. In the process, we see how scientists use mathematics to model nature.*

- We compare the behaviors of gases, liquids, and solids.
- We discuss laboratory methods for measuring gas pressure.
- We consider laws that describe the behavior of a gas in terms of how its volume changes with a change in (1) pressure, (2) temperature, or (3) amount. We focus on the ideal gas law, which encompasses these three laws, and apply it to solve gas law problems.
- We rearrange the ideal gas law to determine the density and molar mass of an unknown gas, the partial pressure of any gas in a mixture, and the amounts of gaseous reactants and products in a chemical change.
- We see how the kinetic-molecular theory explains the gas laws and accounts for other important behaviors of gas particles.
- We apply key ideas about gas behavior to Earth's atmosphere.
- We find that the behavior of real, not ideal, gases, especially under extreme conditions, requires refinements of the ideal gas law and the kinetic-molecular theory.

5.1 AN OVERVIEW OF THE PHYSICAL STATES OF MATTER

Most substances can exist as a solid, a liquid, or a gas under appropriate conditions of pressure and temperature. In Chapter 1, we used the relative position and motion of the particles of a substance to distinguish how each state fills a container (see Figure 1.1):

- A gas adopts the container shape and fills it, because its particles are far apart and move randomly.
- A liquid adopts the container shape to the extent of its volume, because its particles stay close together but are free to move around each other.
- A solid has a fixed shape regardless of the container shape, because its particles are close together and held rigidly in place.

Gas: Particles are far apart, move freely, and fill the available space.
Liquid: Particles are close together but move around one another.
Solid: Particles are close together in a regular array and do not move around one another.

Figure 5.1 The three states of matter. Many pure substances, such as bromine (Br_2), can exist under appropriate conditions of pressure and temperature as a gas, liquid, or solid.

Source: A–C: © McGraw-Hill Education/ Stephen Frisch, photographer

Figure 5.1 focuses on the three states of bromine.

Several other aspects of their behaviors distinguish gases from liquids and solids:

1. *Gas volume changes significantly with pressure.* When a sample of gas is confined to a container of variable volume, such as a cylinder with a piston, *increasing* the force on the piston *decreases* the gas volume; thus, gases are compressible—the gas particles can be forced closer together into a smaller volume. Removing the external force allows the volume to increase again. Gases under pressure can do a lot of work: rapidly expanding compressed air in a jackhammer breaks rock and cement; compressed air in tires lifts the weight of a car. In contrast, liquids and solids exhibit very little or no compressibility; the volume of a liquid or a solid does not change significantly under pressure.
2. *Gas volume changes significantly with temperature.* When a sample of gas is heated, it expands; when it is cooled, it shrinks. This volume change is 50 to 100 times greater for gases than for liquids or solids. The expansion that occurs when gases are rapidly heated can have dramatic effects, like lifting a rocket into space, and everyday ones, like popping corn.
3. *Gases flow very freely.* Gases flow much more freely than liquids and solids. This behavior allows gases to be transported more easily through pipes, but it also means they leak more rapidly out of small holes and cracks.
4. *Gases have relatively low densities.* Gas density is usually measured in units of grams per liter (g/L), whereas liquid and solid densities are in grams per milliliter (g/mL), about 1000 times as dense (see Table 1.5). For example, compare the density values of a gas, liquid, and solid at 20°C and normal atmospheric pressure:

	Density
$O_2(g)$	1.3 g/**L** or 0.0013 g/**mL**
$H_2O(l)$	1.0 g/**mL**
$NaCl(s)$	2.2 g/**mL**

When a gas cools, its density *increases* because its volume *decreases:* on cooling from 20°C to 0°C, the density of $O_2(g)$ increases from 1.3 to 1.4 g/L.

5. *Gases form a solution in any proportions.* Air is a solution of 18 gases. Two liquids, however, may or may not form a solution: water and ethanol do, but water and gasoline do not. Two solids generally do not form a solution unless they are melted and mixed while liquids, then allowed to solidify (as is done to make the alloy bronze from copper and tin).

Like the way a gas completely fills a container, these macroscopic properties—changing volume with pressure or temperature, great ability to flow, low density, and ability to form solutions—arise because the particles in a gas are much farther apart than those in either a liquid or a solid at ordinary pressures.

› Summary of Section 5.1

- › The volume of a gas can be altered significantly by changing the applied force or the temperature. Corresponding changes for liquids and solids are *much* smaller.
- › Gases flow more freely and have much lower densities than liquids and solids.
- › Gases mix in any proportions to form solutions; liquids and solids generally do not.
- › Differences in the physical states are due to the greater average distance between particles in a gas than in a liquid or a solid.

5.2 GAS PRESSURE AND ITS MEASUREMENT

Gas particles move randomly within a container with relatively high velocities, colliding frequently with the container's walls. The force of these collisions with the walls, called the *pressure* of the gas, is the reason you can blow up a balloon or pump up a tire. **Pressure (*P*)** is defined as the force exerted per unit of surface area:

$$\text{Pressure} = \frac{\text{force}}{\text{area}}$$

The gases in the atmosphere exert a force (or *weight*) uniformly on *all* surfaces; the resulting pressure is called *atmospheric pressure* and is typically about 14.7 pounds per square inch (lb/in^2; psi) of surface. Thus, a pressure of 14.7 lb/in^2 exists on the outside of your room (or your body), and it equals the pressure on the inside.

What would happen if the inside and outside pressures on a container were *not* equal? Consider the empty can attached to a vacuum pump in Figure 5.2. With the pump off *(left),* the can maintains its shape because the pressure caused by gas particles in the room colliding with the outside walls of the can is equal to the pressure caused by gas particles in the can colliding with the inside walls. When the pump is turned on (*right*), it removes much of the air inside the can; fewer gas particles inside the can mean fewer collisions with its inside walls, decreasing the internal pressure greatly. The external pressure of the atmosphere then easily crushes the can. Vacuum-filtration flasks and tubing used in chemistry labs have thick walls that withstand the relatively higher external pressure.

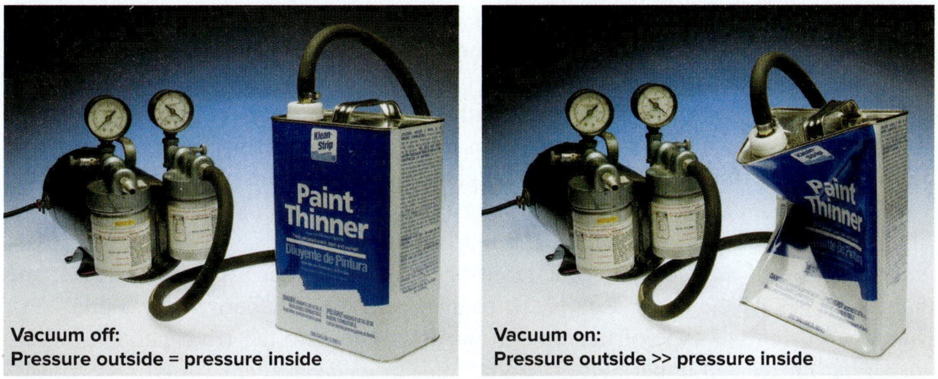

Figure 5.2 **Effect of atmospheric pressure on a familiar object.**

Source: A–B: © McGraw-Hill Education/Charles Winters/Timeframe Photography, Inc.

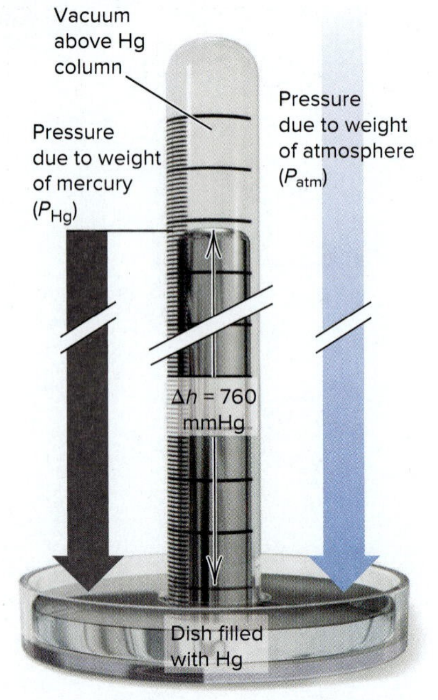

Figure 5.3 A mercury barometer. The pressure of the atmosphere, P_{atm}, balances the pressure of the mercury column, P_{Hg}.

Measuring Gas Pressure: Barometers and Manometers

The **barometer** is used to measure atmospheric pressure. The device is still essentially the same as it was when invented in 1643 by the Italian physicist Evangelista Torricelli: a tube about 1 m long, closed at one end, filled with mercury (atomic symbol, Hg), and inverted into a dish containing more mercury. When the tube is inverted, some of the mercury flows out into the dish, and a vacuum forms above the mercury remaining in the tube (Figure 5.3). At sea level, under ordinary atmospheric conditions, the mercury stops flowing out when the surface of the mercury in the tube is about 760 mm above the surface of the mercury in the dish. At that height, *the column of mercury exerts the same pressure (weight/area) on the mercury surface in the dish as the atmosphere does:* $P_{Hg} = P_{atm}$. Likewise, if you evacuate a closed tube and invert it into a dish of mercury, the atmosphere pushes the mercury up to a height of about 760 mm.

Notice that we did not specify the diameter of the barometer tube. If the mercury in a 1-cm diameter tube rises to a height of 760 mm, the mercury in a 2-cm diameter tube will rise to that height also. The *weight* of mercury is greater in the wider tube, but so is the area; thus, the *pressure,* the *ratio* of weight to area, is the same.

Because the pressure of the mercury column is directly proportional to its height, a unit commonly used for pressure is millimeters of mercury (mmHg). We discuss other units of pressure shortly. *At sea level and 0°C, normal atmospheric pressure is 760 mmHg;* at the top of Mt. Everest (elevation 29,028 ft, or 8848 m), the atmospheric pressure is only about 270 mmHg. Thus, *pressure decreases with altitude:* the column of air above the sea is taller, so it weighs more than the column of air above Mt. Everest.

Laboratory barometers contain mercury because its high density (13.6 g/mL) allows a barometer to be a convenient size. If a barometer contained water (d = 1.0 g/mL) instead, it would have to be more than 34 ft high, because the pressure of the atmosphere equals the pressure of a column of water whose height is about 13.6 × 760 mm = 10,300 mm (almost 34 ft). For a given pressure, the ratio of heights (h) of the liquid columns is inversely related to the ratio of the densities (d) of the liquids:

$$\frac{h_{H_2O}}{h_{Hg}} = \frac{d_{Hg}}{d_{H_2O}}$$

Interestingly, several centuries ago, people thought a vacuum had mysterious "suction" powers, and they didn't understand why a suction pump could remove water from a well only to a depth of 34 feet. We know now, as the great 17th-century scientist Galieo explained, that a vacuum does not suck mercury up into a barometer tube, a suction pump does not suck water up from a well, the vacuum you create in a straw does not suck the drink into your mouth, and the vacuum pump in Figure 5.2 does not suck in the walls of the crushed can. Instead, the atmospheric gases exert a force, pushing the mercury up into the tube, the water up from the well, and the drink up into the straw, and crushing the can.

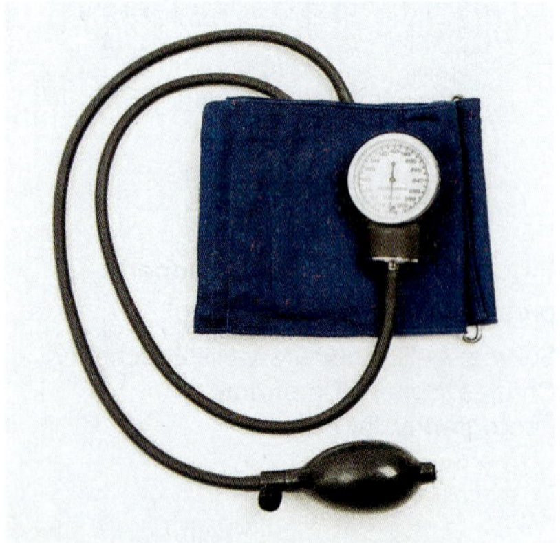

The device used to measure blood pressure is based on a manometer apparatus.

Source: © Fotosearch Premium/Getty Images RF

Manometers are devices used to measure the pressure of a gas in an experiment. Figure 5.4 shows two types of manometers. In the *closed-end manometer (left side),* a mercury-filled, curved tube is *closed* at one end and attached to a flask at the other. When the flask is evacuated, the mercury levels in the two arms of the tube are the same because no gas exerts pressure on either mercury surface. When a gas is in the flask, it pushes down the mercury level in the near arm, causing the level to rise in the far arm. The *difference* in column heights (Δh) equals the gas pressure.

The *open-end manometer* (*right side* of Figure 5.4) also consists of a curved tube filled with mercury, but one end of the tube is *open* to the atmosphere and the other is connected to the gas sample. The atmosphere pushes on one mercury surface, and the gas pushes on the other. Again, Δh equals the difference between two pressures. But, when using this type of manometer, we must measure the atmospheric pressure with a barometer and either add or subtract Δh from that value.

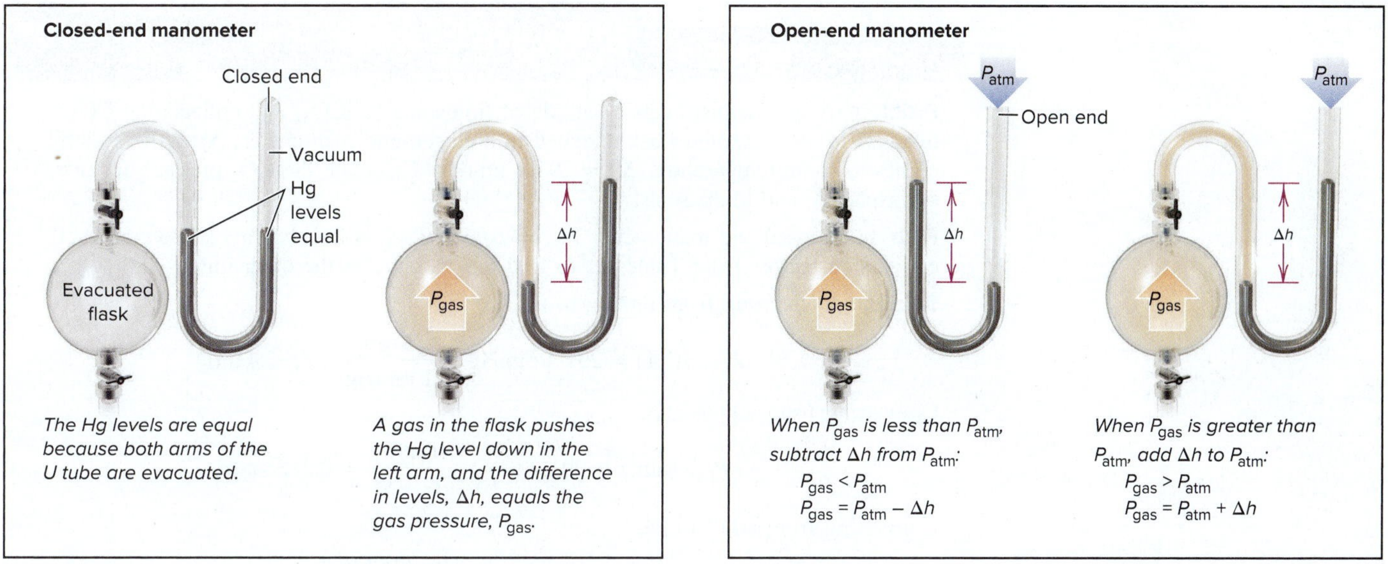

Figure 5.4 **Two types of manometer.**

Units of Pressure

Pressure results from a force exerted on an area. The SI unit of force is the newton (N): 1 N = 1 kg·m/s^2 (about the weight of an apple). The SI unit of pressure is the **pascal (Pa),** which equals a force of one newton exerted on an area of one square meter:

$$1 \text{ Pa} = 1 \text{ N/m}^2$$

A much larger unit is the **standard atmosphere (atm),** the average atmospheric pressure measured at sea level and 0°C. It is defined in terms of the pascal:

$$1 \text{ atm} = 101.325 \text{ kilopascals (kPa)} = 1.01325\times10^5 \text{ Pa}$$

Another common unit is the **millimeter of mercury (mmHg),** mentioned earlier; in honor of Torricelli, this unit has been renamed the **torr:**

$$1 \text{ torr} = 1 \text{ mmHg} = \frac{1}{760} \text{ atm} = \frac{101.325}{760} \text{ kPa} = 133.322 \text{ Pa}$$

The *bar* is coming into more common use in chemistry:

$$1 \text{ bar} = 1\times10^2 \text{ kPa} = 1\times10^5 \text{ Pa}$$

Despite a gradual change to SI units, many chemists still express pressure in torrs and atmospheres, so those units are used in this book, with reference to pascals and bars. Table 5.1 lists some important pressure units with the corresponding values for normal atmospheric pressure.

Table 5.1 **Common Units of Pressure**

Unit	Normal Atmospheric Pressure at Sea Level and 0°C
pascal (Pa); kilopascal (kPa)	1.01325×10^5 Pa; 101.325 kPa
atmosphere (atm)	1 atm*
millimeters of mercury (mmHg)	760 mmHg*
torr	760 torr*
pounds per square inch (lb/in^2 or psi)	14.7 lb/in^2
bar	1.01325 bar

*These are exact quantities; in calculations, we use as many significant figures as necessary.

SAMPLE PROBLEM 5.1 Converting Units of Pressure

Problem A geochemist heats a sample of limestone ($CaCO_3$) and collects the CO_2 released in an evacuated flask attached to a closed-end manometer. After the system comes to room temperature, Δh = 291.4 mmHg. Calculate the CO_2 pressure in torrs, atmospheres, and kilopascals.

Plan In a closed-end manometer, $P_{gas} = \Delta h$, so P_{CO_2} = 291.4 mmHg. We construct conversion factors from Table 5.1 to find the pressure in the other units.

Solution Converting from mmHg to torr:

$$P_{CO_2}\text{ (torr)} = 291.4\ \cancel{\text{mmHg}} \times \frac{1\text{ torr}}{1\ \cancel{\text{mmHg}}} = 291.4\text{ torr}$$

Converting from torr to atm:

$$P_{CO_2}\text{ (atm)} = 291.4\ \cancel{\text{torr}} \times \frac{1\text{ atm}}{760\ \cancel{\text{torr}}} = 0.3834\text{ atm}$$

Converting from atm to kPa:

$$P_{CO_2}\text{ (kPa)} = 0.3834\ \cancel{\text{atm}} \times \frac{101.325\text{ kPa}}{1\ \cancel{\text{atm}}} = 38.85\text{ kPa}$$

Check There are 760 torr in 1 atm, so ~300 torr should be <0.5 atm. There are ~100 kPa in 1 atm, so <0.5 atm should be <50 kPa.

Comment **1.** In the conversion from torr to atm, we retained four significant figures because this unit conversion factor involves *exact* numbers; that is, 760 torr has as many significant figures as the calculation requires (see the footnote to Table 5.1). **2.** From here on, except in particularly complex situations, *unit canceling will no longer be shown.*

FOLLOW-UP PROBLEMS

*Brief Solutions to **all** Follow-up Problems appear at the end of the chapter.*

5.1A The CO_2 released from another limestone sample is collected in an evacuated flask connected to an open-end manometer. If the barometer reading is 753.6 mmHg and P_{gas} is less than P_{atm}, giving Δh = 174.0 mmHg, calculate P_{CO_2} in torrs, pascals, and lb/in^2.

5.1B A third sample of limestone is heated, and the CO_2 released is collected in an evacuated flask connected to an open-end manometer. If the atmospheric pressure is 0.9475 atm and P_{gas} is greater than P_{atm}, giving Δh = 25.8 torr, calculate P_{CO_2} in mmHg, pascals, and lb/in^2.

SOME SIMILAR PROBLEMS 5.10–5.13

› Summary of Section 5.2

- Gases exert pressure (force/area) on all surfaces they contact.
- A barometer measures atmospheric pressure based on the height of a mercury column that the atmosphere can support (760 mmHg at sea level and 0°C).
- Closed-end and open-end manometers are used to measure the pressure of a gas sample.
- Pressure units include the atmosphere (atm), torr (identical to mmHg), and pascal (Pa, the SI unit).

5.3 THE GAS LAWS AND THEIR EXPERIMENTAL FOUNDATIONS

The physical behavior of a sample of gas can be described completely by four variables: pressure (P), volume (V), temperature (T), and amount (number of moles, n). The variables are interdependent, which means that *any one of them can be determined by measuring the other three.* Three key relationships exist among the four gas variables—Boyle's, Charles's, and Avogadro's laws. Each of these *gas laws expresses the effect of one variable on another, with the remaining two variables held constant.* Because gas volume is so easy to measure, the laws are expressed as the effect on gas volume of a change in the pressure, temperature, or amount of the gas.

The individual gas laws are special cases of a unifying relationship called the *ideal gas law,* which quantitatively describes the behavior of an **ideal gas,** one that exhibits linear relationships among volume, pressure, temperature, and amount. Although *no ideal gas actually exists,* most simple gases, such as N_2, O_2, H_2, and the noble gases, behave nearly ideally at ordinary temperatures and pressures. We discuss the ideal gas law after the three individual laws.

The Relationship Between Volume and Pressure: Boyle's Law

Following Torricelli's invention of the barometer, the great 17th-century English chemist Robert Boyle studied the effect of pressure on the volume of a sample of gas.

1. *The experiment.* Figure 5.5 illustrates the setup Boyle might have used in his experiments (parts A and B), the data he might have collected (part C), and graphs of the data (parts D and E). Boyle sealed the shorter leg of a J-shaped glass tube and poured mercury into the longer open leg, thereby trapping some air (the gas in the experiment) in the shorter leg. He calculated the gas volume (V_{gas}) from the height of the trapped air and the diameter of the tube. The total pressure, P_{total}, applied to the trapped gas is the pressure of the atmosphere, P_{atm} (760 mm, measured with a barometer), plus the difference in the heights of the mercury columns (Δh) in the two legs of the J tube, 20 mm (Figure 5.5A); thus, P_{total} is 780 torr. By adding mercury, Boyle increased P_{total}, and the gas volume decreased. In Figure 5.5B, more mercury has been added to the tube, increasing Δh from its original value of 20 mm to 800 mm, so P_{total} *doubles* to 1560 torr; note that V_{gas} is *halved* from 20 mL to 10 mL. In this way, by keeping the temperature and amount of gas constant, Boyle was able to measure the effect of the applied pressure on gas volume.

Note the following results in Figure 5.5:

- The product of corresponding P and V values is a constant (part C, rightmost column).
- V is *inversely* proportional to P—the greater the pressure, the smaller the gas volume (part D).
- V is *directly* proportional to $1/P$ (part E), and a plot of V versus $1/P$ is linear. This *linear relationship between two gas variables* is a hallmark of ideal gas behavior.

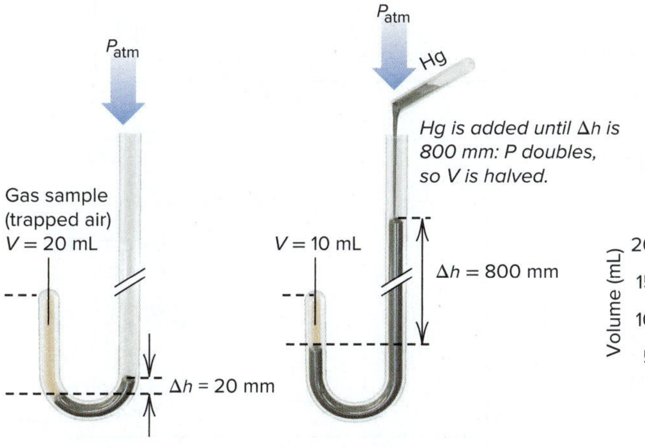

A $P_{total} = P_{atm} + \Delta h$
= 760 torr + 20 torr
= 780 torr

B $P_{total} = P_{atm} + \Delta h$
= 760 torr + 800 torr
= 1560 torr

C

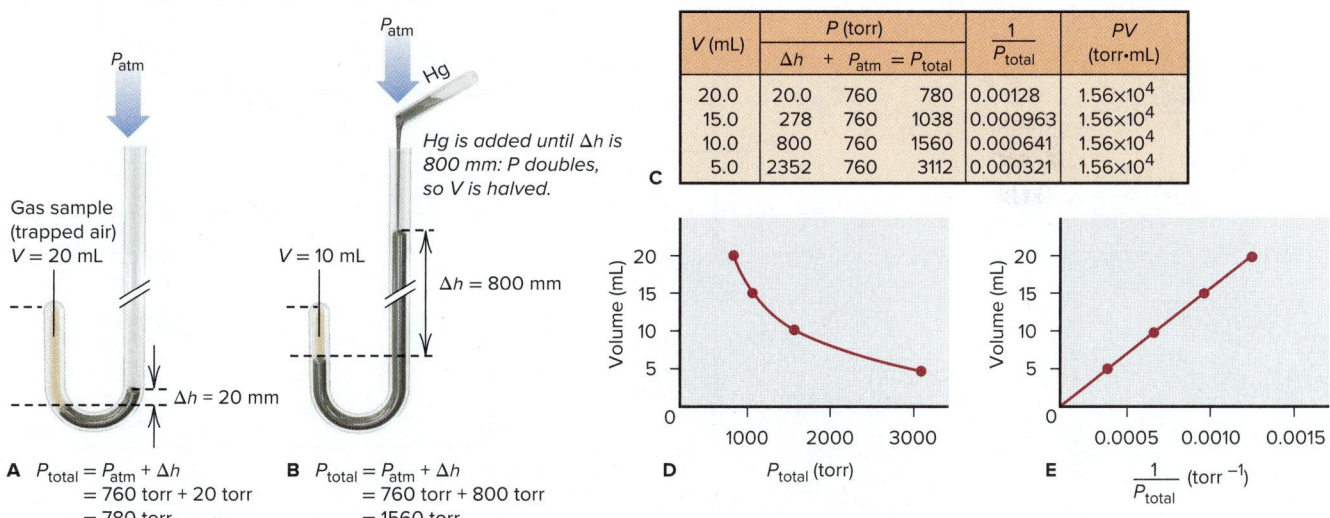

V (mL)	P (torr) Δh	+ P_{atm}	= P_{total}	$\frac{1}{P_{total}}$	PV (torr•mL)
20.0	20.0	760	780	0.00128	1.56×10^4
15.0	278	760	1038	0.000963	1.56×10^4
10.0	800	760	1560	0.000641	1.56×10^4
5.0	2352	760	3112	0.000321	1.56×10^4

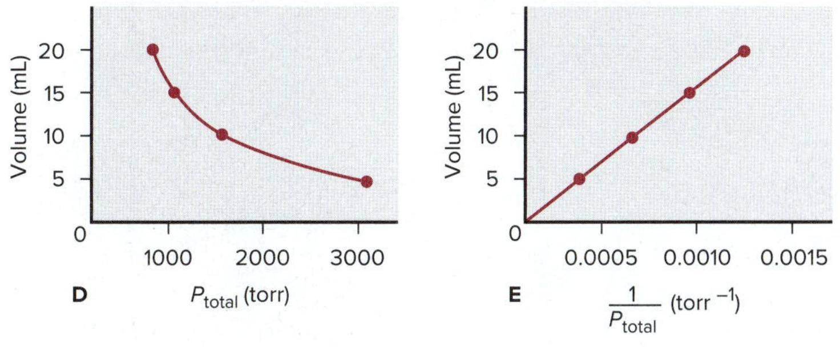

Figure 5.5 Boyle's law, the relationship between the volume and pressure of a gas.

2. *Conclusion and statement of the law.* The generalization of Boyle's observations is known as **Boyle's law:** *at constant temperature, the volume occupied by a fixed amount of gas is **inversely** proportional to the applied (external) pressure,* or

$$V \propto \frac{1}{P} \qquad [T \text{ and } n \text{ fixed}] \tag{5.1}$$

This relationship can also be expressed as

$$V = \frac{\text{constant}}{P} \qquad \text{or} \qquad PV = \text{constant} \qquad [T \text{ and } n \text{ fixed}]$$

That is, at fixed T and n, $P\uparrow, V\downarrow$ and $P\downarrow, V\uparrow$

The constant is the same for most simple gases under ordinary conditions. Thus, tripling the external pressure reduces the volume of a gas to a third of its initial value; halving the pressure doubles the volume; and so forth.

The wording of Boyle's law focuses on *external* pressure. But, notice that the mercury level rises as mercury is added, until the pressure of the trapped gas *on* the mercury increases enough to stop its rise. At that point, the pressure exerted *on* the gas equals the pressure exerted *by* the gas (P_{gas}). Thus, in general, at constant temperature, if V_{gas} increases, P_{gas} decreases, and vice versa.

The Relationship Between Volume and Temperature: Charles's Law

Boyle's work showed that the pressure-volume relationship holds only at constant temperature, but why should that be so? It would take more than a century, until the work of French scientists J. A. C. Charles and J. L. Gay-Lussac, for the relationship between gas volume and temperature to be understood.

1. *The experiment.* Let's examine this relationship by measuring the volume at different temperatures of a fixed amount of a gas under constant pressure. A straight tube, closed at one end, traps a fixed amount of gas (air) under a small mercury plug. The tube is immersed in a water bath that is warmed with a heater or cooled with ice. After each change of temperature, we calculate the gas volume from the length of the gas column and the diameter of the tube. The total pressure exerted on the gas is constant because the mercury plug and the atmospheric pressure do not change (Figure 5.6, parts A and B).

Figure 5.6 Charles's law, the relationship between the volume and temperature of a gas.

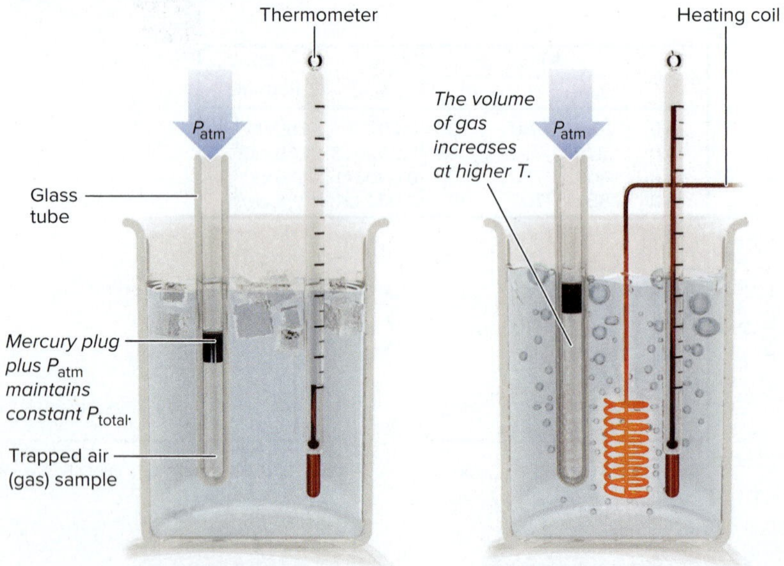

A Ice water bath (0°C, or 273 K)

B Boiling water bath (100°C, or 373 K)

V (L)	T (°C)	T (K)	$\frac{V}{T}$ (L/K)
0.8965	0	273	3.28×10^{-3}
0.9786	25	298	3.28×10^{-3}
1.061	50	323	3.28×10^{-3}
1.143	75	348	3.28×10^{-3}
1.225	100	373	3.28×10^{-3}

C

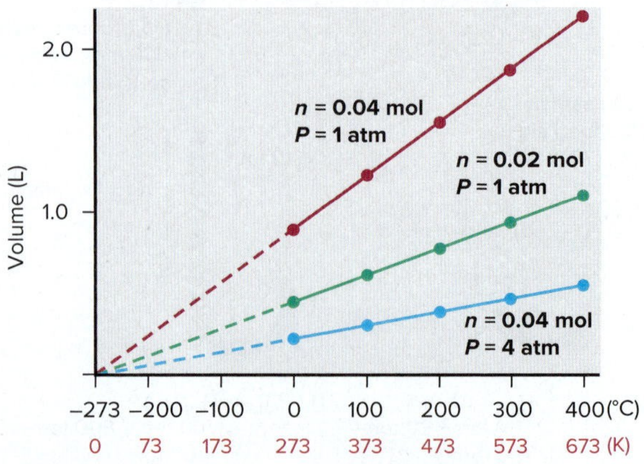

D

Note the following results in Figure 5.6:

- Typical data (for a 0.04-mol sample of gas at 1 atm) show that dividing the volume of gas by the corresponding Kelvin temperature results in a constant (part C, right-most column).
- V is *directly* proportional to T in kelvins (red line in part D).
- There is a *linear relationship between V and T.*
- Extrapolating the red line in the graph in part D to lower temperatures (dashed portion) shows that, in theory, the gas occupies zero volume at −273.15°C (the intercept on the temperature axis).
- Plots for a different amount of gas (green line in part D) or for a different gas pressure (blue line) have different slopes, but they converge at −273.15°C.

2. *Conclusion and statement of the law.* Above all, note that *the volume-temperature relationship is linear,* but, unlike volume and pressure, volume and temperature are *directly* proportional. This behavior is incorporated into the modern statement of the volume-temperature relationship, which is known as **Charles's law:** *at constant pressure, the volume occupied by a fixed amount of gas is* ***directly*** *proportional to its absolute (Kelvin) temperature,* or

$$V \propto T \qquad [P \text{ and } n \text{ fixed}] \tag{5.2}$$

This relationship can also be expressed as

$$V = \text{constant} \times T \qquad \text{or} \qquad \frac{V}{T} = \text{constant} \qquad [P \text{ and } n \text{ fixed}]$$

That is, at fixed P and n, $T\uparrow, V\uparrow$ and $T\downarrow, V\downarrow$

If T increases, V increases, and vice versa. As with Boyle's law, the constant is the same for most simple gases under ordinary conditions and for any given P and n.

William Thomson (Lord Kelvin) used the linear relationship between gas volume and temperature to devise the absolute temperature scale (Section 1.4). Absolute zero (0 K or −273.15°C) is the temperature at which an ideal gas would have zero volume. (Absolute zero has never been reached, but physicists have attained 10^{-9} K.) In reality, no sample of matter can have zero volume, and every real gas condenses to a liquid at some temperature higher than 0 K. Nevertheless, the linear dependence of volume on absolute temperature holds for most common gases over a wide temperature range. This dependence of gas volume on the *absolute* temperature adds a practical requirement for chemists (and chemistry students): *the Kelvin scale must be used in gas law calculations.* For instance, if the temperature changes from 200 K to 400 K, the volume of gas doubles because the absolute temperature doubles. But, if the temperature doubles from 200°C to 400°C, the volume increases by a factor of 1.42, not be a factor of 2; that is,

$$\left(\frac{400°\text{C} + 273.15}{200°\text{C} + 273.15}\right) = \frac{673}{473} = 1.42$$

Other Relationships Based on Boyle's and Charles's Laws Two other important relationships arise from Boyle's and Charles's laws:

1. *The pressure-temperature relationship.* Charles's law is expressed as the effect of temperature on gas *volume* at constant pressure. But volume and pressure are interdependent, so a similar relationship can be expressed for the effect of temperature on pressure (sometimes referred to as *Amontons's law*). You can examine this relationship for yourself: measure the pressure in your car or bike tires before and after a long ride and you will find that the air pressure in the tires increases. The air temperature inside the tires also increases from heating due to friction between the tires and the road; a tire's volume can't increase very much, but the increased temperature results in increased pressure. So we observe that, *at constant volume, the pressure exerted by a fixed amount of gas is directly proportional to the absolute temperature:*

$$P \propto T \qquad [V \text{ and } n \text{ fixed}] \tag{5.3}$$

or

$$P = \text{constant} \times T \qquad \text{or} \qquad \frac{P}{T} = \text{constant} \qquad [V \text{ and } n \text{ fixed}]$$

That is, at fixed V and n, $T\uparrow, P\uparrow$ and $T\downarrow, P\downarrow$

If T increases, V increases, and vice versa.

2. *The combined gas law.* Combining Boyle's and Charles's laws gives the *combined gas law,* which applies to cases when changes in *two* of the three variables (V, P, T) affect the third:

$$\text{Boyle's law: } V \propto \frac{1}{P}; \text{ Charles's law: } V \propto T$$

$$\text{Combined gas law: } V \propto \frac{T}{P} \qquad \text{or} \qquad V = \text{constant} \times \frac{T}{P} \qquad \text{or} \qquad \frac{PV}{T} = \text{constant}$$

The Relationship Between Volume and Amount: Avogadro's Law

Let's see why both Boyle's and Charles's laws specify a fixed amount of gas.

1. *The experiment.* Figure 5.7 shows an experiment that involves two small test tubes, each fitted to a much larger piston-cylinder assembly. We add 0.10 mol (4.4 g) of dry ice (solid CO_2) to the first tube (A) and 0.20 mol (8.8 g) to the second tube (B). As the solid CO_2 warms to room temperature, it changes to gaseous CO_2, and the volume increases until $P_{gas} = P_{atm}$. At constant temperature, when all the solid has changed to gas, cylinder B has twice the volume of cylinder A.

2. *Conclusion and statement of the law.* Thus, *at fixed temperature and pressure, the volume occupied by a gas is directly proportional to the amount (mol) of gas:*

$$V \propto n \qquad [P \text{ and } T \text{ fixed}] \qquad \textbf{(5.4)}$$

That is, as n increases, V increases, and vice versa. This relationship is also expressed as

$$V = \text{constant} \times n \qquad \text{or} \qquad \frac{V}{n} = \text{constant} \qquad [P \text{ and } T \text{ fixed}]$$

That is, at fixed P and T, $n\uparrow, V\uparrow$ and $n\downarrow, V\downarrow$

Student Hot Spot

Student data indicate that you may struggle with applying the relationships between the pressure, volume, temperature, and amount of a gas. Access the Smartbook to view additional Learning Resources on this topic.

The constant is the same for all simple gases at ordinary temperature and pressure. This relationship is another way of expressing **Avogadro's law,** which states that *at fixed temperature and pressure, equal volumes of* ***any*** *ideal gas contain equal numbers (or moles) of particles.*

Familiar Applications of the Gas Laws The gas laws apply to countless familiar phenomena. In a car engine, a reaction occurs in which fewer moles of gasoline and O_2 form more moles of CO_2 and H_2O vapor, and heat is released. The increase in n

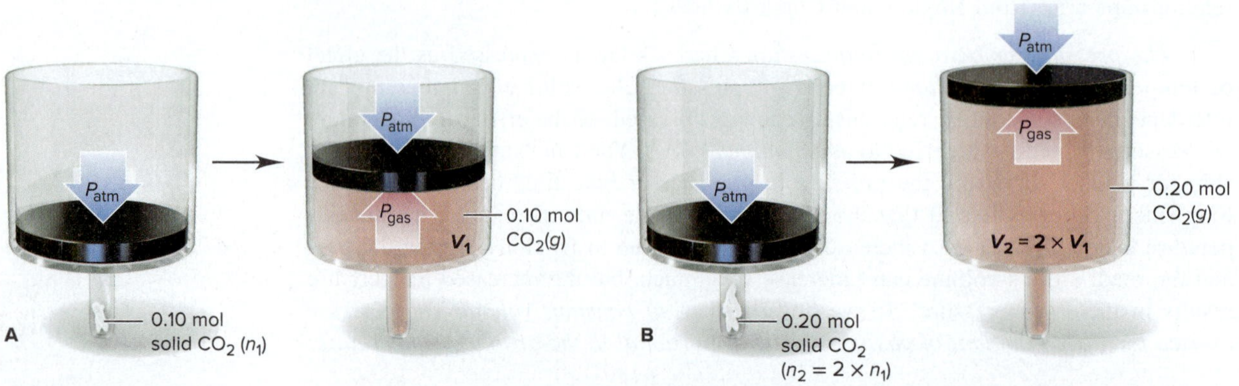

Figure 5.7 Avogadro's law, the relationship between the volume and amount of a gas.

(Avogadro's law) and T (Charles's law) increases V, and the pistons in the engine are pushed back. Dynamite is a solid that forms more moles of hot gases very rapidly when it reacts (Avogadro's and Charles's laws). Dough rises because yeast digests sugar, which creates bubbles of CO_2 (Avogadro's law); the dough expands more as the bread bakes in a hot oven (Charles's law).

No application of the gas laws can be more vital or familiar than breathing (Figure 5.8). When you inhale, muscles move your diaphragm down and your rib cage out *(blue)*. This coordinated movement increases the volume of your lungs, which decreases the air pressure inside them (Boyle's law). The inside pressure is 1–3 torr *less* than atmospheric pressure, so air rushes in. The greater amount of air stretches the elastic tissue of the lungs and expands the volume further (Avogadro's law). The air also expands as it warms from the external temperature to your body temperature (Charles's law). When you exhale, the diaphragm moves up and the rib cage moves in, so your lung volume decreases *(red)*. The inside pressure becomes 1–3 torr *more* than the outside pressure (Boyle's law), so air rushes out.

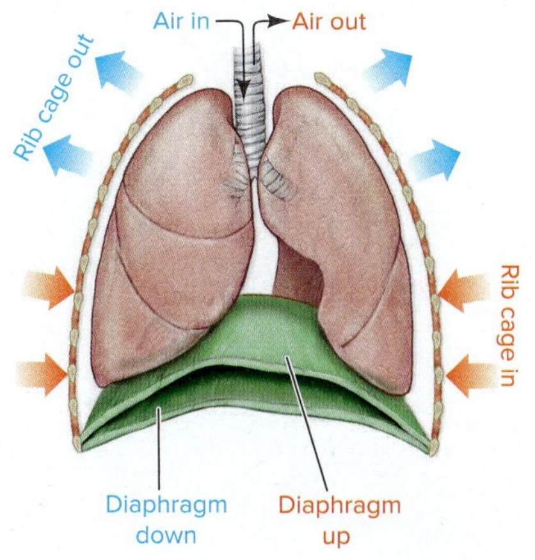

Figure 5.8 The process of breathing applies the gas laws.

Gas Behavior at Standard Conditions

To better understand the factors that influence gas behavior, chemists have assigned a baseline set of *standard conditions* called **standard temperature and pressure (STP):**

$$\text{STP:} \quad 0°\text{C } (273.15 \text{ K}) \text{ and } 1 \text{ atm } (760 \text{ torr}) \qquad \textbf{(5.5)}$$

Under these conditions, the volume of 1 mol of an ideal gas is called the **standard molar volume:**

$$\text{Standard molar volume} = 22.4141 \text{ L or } 22.4 \text{ L [to 3 sf]} \qquad \textbf{(5.6)}$$

At STP, helium, nitrogen, oxygen, and other simple gases behave nearly ideally (Figure 5.9). Note that the mass, and thus the density (d), depends on the specific gas, but 1 mol of any of them occupies 22.4 L at STP.

Figure 5.10 compares the volumes of some familiar objects with the standard molar volume of an ideal gas.

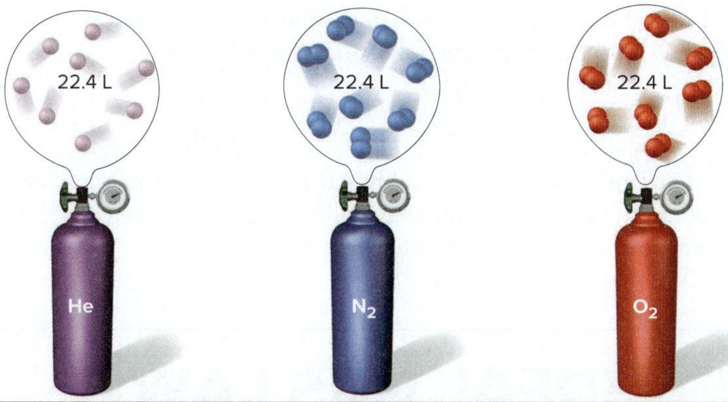

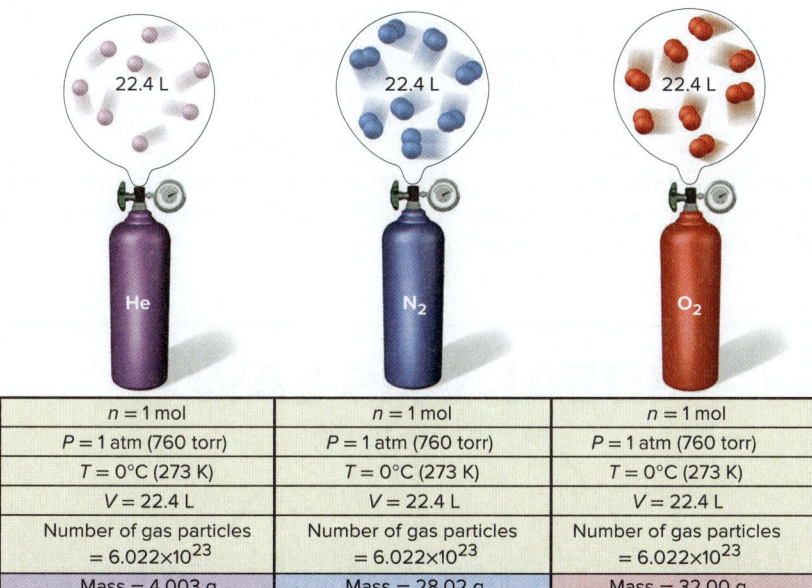

$n = 1$ mol	$n = 1$ mol	$n = 1$ mol
$P = 1$ atm (760 torr)	$P = 1$ atm (760 torr)	$P = 1$ atm (760 torr)
$T = 0°C$ (273 K)	$T = 0°C$ (273 K)	$T = 0°C$ (273 K)
$V = 22.4$ L	$V = 22.4$ L	$V = 22.4$ L
Number of gas particles $= 6.022 \times 10^{23}$	Number of gas particles $= 6.022 \times 10^{23}$	Number of gas particles $= 6.022 \times 10^{23}$
Mass = 4.003 g	Mass = 28.02 g	Mass = 32.00 g
$d = 0.179$ g/L	$d = 1.25$ g/L	$d = 1.43$ g/L

Figure 5.9 Standard molar volume. One mole of an ideal gas occupies 22.4 L at STP (0°C and 1 atm).

Figure 5.10 The volumes of 1 mol (22.4 L) of an ideal gas and of some familiar objects: 1 gal of milk (3.79 L), a basketball (7.50 L), and 2.00 L of a carbonated drink.

Source: © McGraw-Hill Education/Charles Winters/Timeframe Photography, Inc.

The Ideal Gas Law

Each of the three gas laws shows how one of the other three gas variables affects gas volume:

- Boyle's law focuses on pressure ($V \propto 1/P$).
- Charles's law focuses on temperature ($V \propto T$).
- Avogadro's law focuses on amount (mol) of gas ($V \propto n$).

By combining these individual effects, we obtain the **ideal gas law** (or *ideal gas equation*):

$$V \propto \frac{nT}{P} \quad \text{or} \quad PV \propto nT \quad \text{or} \quad PV = \text{constant} \times nT \quad \text{or} \quad \frac{PV}{nT} = R$$

where R is a proportionality constant known as the **universal gas constant.** Rearranging gives the most common form of the ideal gas law:

$$PV = nRT \qquad \textbf{(5.7)}$$

We obtain a value of R by measuring the volume, temperature, and pressure of a given amount of gas and substituting the values into the ideal gas law. For example, using standard conditions for the gas variables and 1 mol of gas, we have

$$R = \frac{PV}{nT} = \frac{1\ \text{atm} \times 22.4141\ \text{L}}{1\ \text{mol} \times 273.15\ \text{K}} = 0.082058\ \frac{\text{atm·L}}{\text{mol·K}} = 0.0821\ \frac{\text{atm·L}}{\text{mol·K}}\ [3\ \text{sf}] \qquad \textbf{(5.8)}$$

This numerical value of R corresponds to P, V, and T expressed *in these units; R has a different numerical value when different units are used.* For example, in the equation for root-mean-square speed in Section 5.5, R has the value 8.314 J/mol·K (J stands for joule, the SI unit of energy).

Figure 5.11 makes a central point: the ideal gas law *becomes* one of the individual gas laws when two of the four variables are kept constant. When initial conditions (subscript 1) change to final conditions (subscript 2), we have

$$P_1V_1 = n_1RT_1 \quad \text{and} \quad P_2V_2 = n_2RT_2$$

Thus,

$$\frac{P_1V_1}{n_1T_1} = R \quad \text{and} \quad \frac{P_2V_2}{n_2T_2} = R, \quad \text{so} \quad \frac{P_1V_1}{n_1T_1} = \frac{P_2V_2}{n_2T_2}$$

Notice that if, for example, the two variables P and T remain constant, then $P_1 = P_2$ and $T_1 = T_2$, and we obtain an expression for Avogadro's law:

$$\frac{\cancel{P_1}V_1}{n_1\cancel{T_1}} = \frac{\cancel{P_2}V_2}{n_2\cancel{T_2}} \quad \text{or} \quad \frac{V_1}{n_1} = \frac{V_2}{n_2}$$

As you'll see next, you can use a similar approach to solve gas law problems. Thus, by keeping track of the initial and final values of the gas variables, you avoid the need to memorize the three individual gas laws.

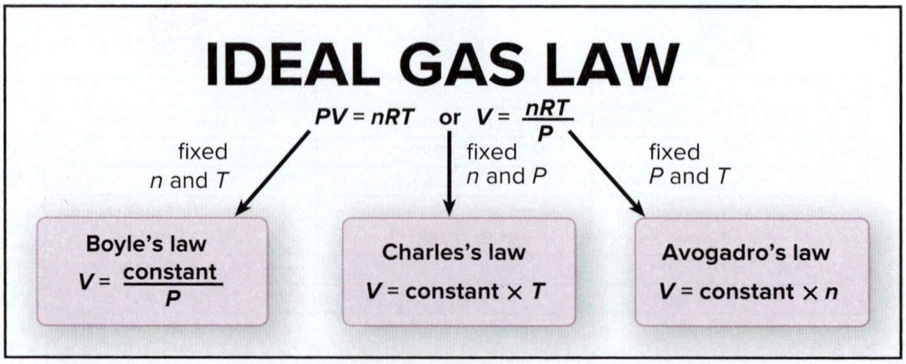

Figure 5.11 The individual gas laws as special cases of the ideal gas law.

Solving Gas Law Problems

Gas law problems are phrased in many ways but can usually be grouped into two types:

1. *There is a change in one of the four variables which causes a change in another, while the two other variables remain constant.* In this type, the ideal gas law reduces to one of the individual gas laws when you omit the variables that remain constant; you then solve for the new value of the affected variable. Units must be consistent and T must always be in kelvins, but R is not involved. Sample Problems 5.2 to 5.4 are of this type. A variation on this type, shown in Sample Problem 5.5, involves the combined gas law when simultaneous changes in two of the variables cause a change in a third.

2. *One variable is unknown, but the other three are known*, as is R, *and no change occurs.* In this type, exemplified by Sample Problem 5.6, you apply the ideal gas law directly to find the unknown, and the units must conform to those in R.

Solving these problems requires a systematic approach:

- Summarize the changing gas variables—knowns and unknown—and those held constant.
- Convert units, if necessary.
- Rearrange the ideal gas law to obtain the needed relationship of variables, and solve for the unknown.

SAMPLE PROBLEM 5.2 Applying the Volume-Pressure Relationship

Problem Boyle's apprentice finds that the air trapped in a J tube occupies 24.8 cm^3 at 851 torr. By adding mercury to the tube, he increases the pressure on the trapped air to 2.64 atm. Assuming constant temperature, what is the new volume of air (in L)?

Plan We must find the final volume (V_2) in liters, given the initial volume (V_1), initial pressure (P_1), and final pressure (P_2). The temperature and amount of gas are fixed. We must use consistent units of pressure, so we convert the unit of P_1 from torr to atm. We then convert the unit of V_1 from cm^3 to mL and then to L, rearrange the ideal gas law to the appropriate form, and solve for V_2. (Note that the road map has two parts, unit conversion and the gas law calculation.)

Road Map

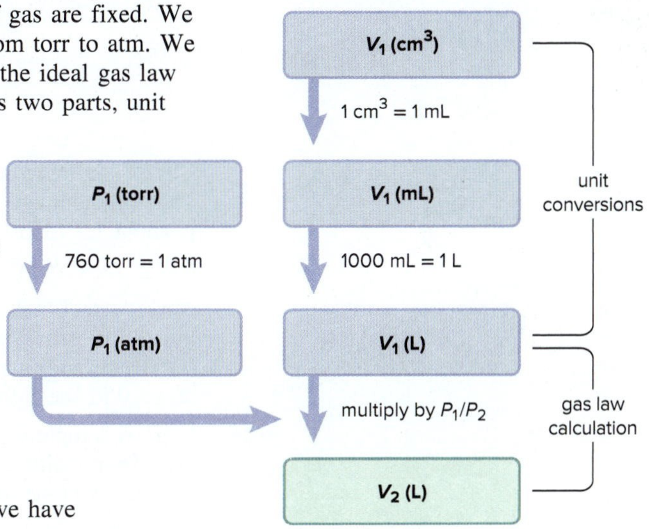

Solution Summarizing the gas variables:

P_1 = 851 torr (convert to atm) P_2 = 2.64 atm

V_1 = 24.8 cm^3 (convert to L) V_2 = unknown T and n remain constant

Converting P_1 from torr to atm:

$$P_1\ (\text{atm}) = 851\ \text{torr} \times \frac{1\ \text{atm}}{760\ \text{torr}} = 1.12\ \text{atm}$$

Converting V_1 from cm^3 to L:

$$V_1\ (\text{L}) = 24.8\ \text{cm}^3 \times \frac{1\ \text{mL}}{1\ \text{cm}^3} \times \frac{1\ \text{L}}{1000\ \text{mL}} = 0.0248\ \text{L}$$

Rearranging the ideal gas law and solving for V_2: At fixed n and T, we have

$$\frac{P_1V_1}{n_1T_1} = \frac{P_2V_2}{n_2T_2} \quad \text{or} \quad P_1V_1 = P_2V_2$$

$$V_2 = V_1 \times \frac{P_1}{P_2} = 0.0248\ \text{L} \times \frac{1.12\ \text{atm}}{2.64\ \text{atm}} = 0.0105\ \text{L}$$

Check The relative values of P and V can help us check the math: P more than doubled, so V_2 should be less than $\frac{1}{2}V_1$ ($0.0105/0.0248 < \frac{1}{2}$).

Comment **1.** Predicting the direction of the change provides another check on the problem setup: since P increases, V will decrease; thus, V_2 should be less than V_1. To make $V_2 < V_1$, we must multiply V_1 by a number *less than* 1. This means the ratio of pressures must be *less than* 1, so the larger pressure (P_2) must be in the denominator, or P_1/P_2. **2.** Since R was not involved in this problem, any unit of pressure can be used, as long as the same unit is used for both P_1 and P_2. We used atm, but we could have used torr instead; however, both torr and atm cannot be used.

FOLLOW-UP PROBLEMS

5.2A A sample of argon gas occupies 105 mL at 0.871 atm. If the volume of the gas is increased to 352 mL at constant temperature, what is the final pressure of the gas (in kPa)?

5.2B A tank contains 651 L of compressed oxygen gas at a pressure of 122 atm. Assuming the temperature remains constant, what is the volume of the oxygen (in L) at 745 mmHg?

SOME SIMILAR PROBLEMS 5.24 and 5.25

SAMPLE PROBLEM 5.3 Applying the Volume-Temperature and Pressure-Temperature Relationships

Problem A balloon is filled with 1.95 L of air at 25°C and then placed in a car sitting in the sun. What is the volume of the balloon when the temperature in the car reaches 90°C?

Road Map

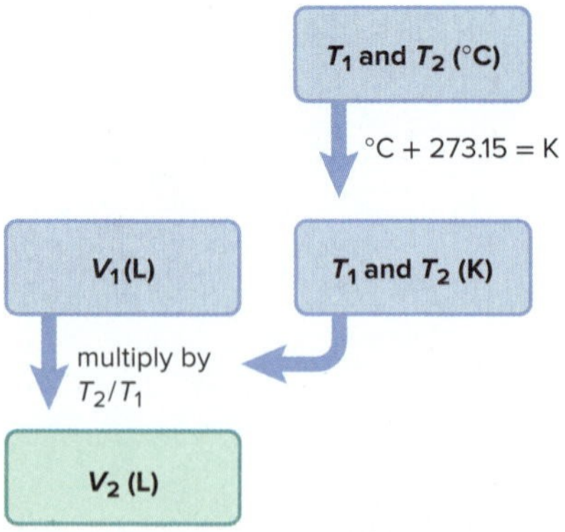

Plan We know the initial volume (V_1) and the initial (T_1) and final temperatures (T_2) of the gas; we must find the final volume (V_2). The pressure of the gas is fixed since the balloon is subjected to atmospheric pressure, and n is fixed since air cannot escape or enter the balloon. We convert both T values to kelvins, rearrange the ideal gas law, and solve for V_2 (see the road map).

Solution Summarizing the gas variables:

$V_1 = 1.95$ L
$T_1 = 25°C$ (convert to K)
P and n remain constant

$V_2 =$ unknown
$T_2 = 90°C$ (convert to K)

Converting T from °C to K:

$$T_1 \text{ (K)} = 25°\text{C} + 273.15 = 298 \text{ K} \qquad T_2 \text{ (K)} = 90°\text{C} + 273.15 = 363 \text{ K}$$

Rearranging the ideal gas law and solving for V_2: At fixed n and P, we have

$$\frac{\cancel{P_1}V_1}{\cancel{n_1}T_1} = \frac{\cancel{P_2}V_2}{\cancel{n_2}T_2} \quad \text{or} \quad \frac{V_1}{T_1} = \frac{V_2}{T_2}$$

$$V_2 = V_1 \times \frac{T_2}{T_1} = 1.95 \text{ L} \times \frac{363 \text{ K}}{298 \text{ K}} = 2.38 \text{ L}$$

Check Let's predict the change to check the math: because $T_2 > T_1$, we expect $V_2 > V_1$. Thus, the temperature ratio should be greater than 1 (T_2 in the numerator). The T ratio is about 1.2 (363/298), so the V ratio should also be about 1.2 (2.4/2.0 ≈ 1.2).

FOLLOW-UP PROBLEMS

5.3A A steel tank used for fuel delivery is fitted with a safety valve that opens if the internal pressure exceeds 1.00×10^3 torr. The tank is filled with methane at 23°C and 0.991 atm and placed in boiling water at 100.°C. What is the pressure in the heated tank? Will the safety valve open?

5.3B A sample of nitrogen occupies a volume of 32.5 L at 40°C. Assuming that the pressure remains constant, what temperature (in °C) will result in a decrease in the sample's volume to 28.6 L?

SOME SIMILAR PROBLEMS 5.26–5.29

SAMPLE PROBLEM 5.4 Applying the Volume-Amount and Pressure-Amount Relationships

Problem A scale model of a blimp rises when it is filled with helium to a volume of 55.0 dm^3. When 1.10 mol of He is added to the blimp, the volume is 26.2 dm^3. How many more grams of He must be added to make it rise? Assume constant T and P.

Plan We are given the initial amount of helium (n_1), the initial volume of the blimp (V_1), and the volume needed for it to rise (V_2), and we need the additional mass of helium to make it rise. So we first need to find n_2. We rearrange the ideal gas law to the appropriate form, solve for n_2, subtract n_1 to find the additional amount ($n_{\text{add'l}}$), and then convert moles to grams (see the road map).

Solution Summarizing the gas variables:

$n_1 = 1.10$ mol $\quad\quad n_2 =$ unknown (find, and then subtract n_1)
$V_1 = 26.2$ dm^3 $\quad\quad V_2 = 55.0$ dm^3
P and T remain constant

Rearranging the ideal gas law and solving for n_2: At fixed P and T, we have

$$\frac{\cancel{P_1}V_1}{n_1\cancel{T_1}} = \frac{\cancel{P_2}V_2}{n_2\cancel{T_2}} \quad \text{or} \quad \frac{V_1}{n_1} = \frac{V_2}{n_2}$$

$$n_2 = n_1 \times \frac{V_2}{V_1} = 1.10 \text{ mol He} \times \frac{55.0 \text{ dm}^3}{26.2 \text{ dm}^3}$$
$$= 2.31 \text{ mol He}$$

Finding the additional amount of He:

$$n_{\text{add'l}} = n_2 - n_1 = 2.31 \text{ mol He} - 1.10 \text{ mol He}$$
$$= 1.21 \text{ mol He}$$

Converting amount (mol) of He to mass (g):

$$\text{Mass (g) of He} = 1.21 \text{ mol He} \times \frac{4.003 \text{ g He}}{1 \text{ mol He}}$$
$$= 4.84 \text{ g He}$$

Check We predict that $n_2 > n_1$ because $V_2 > V_1$: since V_2 is about twice V_1 ($55/26 \approx 2$), n_2 should be about twice n_1 ($2.3/1.1 \approx 2$). Since $n_2 > n_1$, we were right to multiply n_1 by a number greater than 1 (that is, V_2/V_1). About 1.2 mol × 4 g/mol ≈ 4.8 g.

Comment **1.** Because we did not use R, notice that any unit of volume, in this case dm^3, is acceptable as long as V_1 and V_2 are expressed in the same unit.

2. A different sequence of steps should give you the same answer: first find the additional volume ($V_{\text{add'l}} = V_2 - V_1$), and then solve directly for $n_{\text{add'l}}$. Try it for yourself.

3. You saw that Charles's law ($V \propto T$ at fixed P and n) becomes a similar relationship between P and T at fixed V and n. Follow-up Problem 5.4A demonstrates that Avogadro's law ($V \propto n$ at fixed P and T) becomes a similar relationship at fixed V and T.

FOLLOW-UP PROBLEMS

5.4A A rigid plastic container holds 35.0 g of ethylene gas (C_2H_4) at a pressure of 793 torr. What is the pressure if 5.0 g of ethylene is removed at constant temperature?

5.4B A balloon filled with 1.26 g of nitrogen gas has a volume of 1.12 L. Calculate the volume of the balloon after 1.26 g of helium gas is added while T and P remain constant.

SOME SIMILAR PROBLEMS 5.30 and 5.31

Road Map

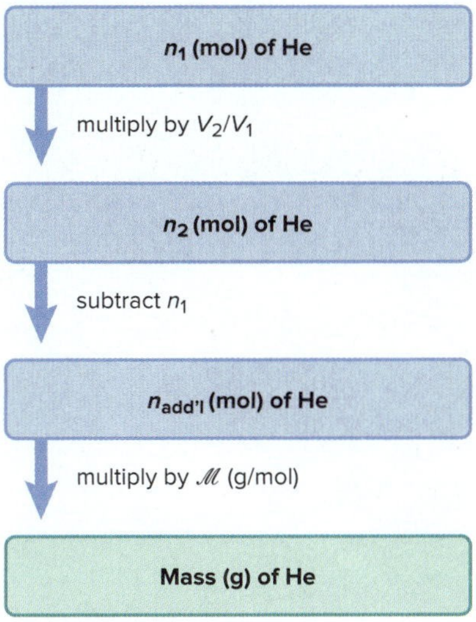

SAMPLE PROBLEM 5.5 Applying the Volume-Pressure-Temperature Relationship

Problem A helium-filled balloon has a volume of 15.8 L at a pressure of 0.980 atm and 22°C. What is its volume at the summit of Mt. Hood, Oregon's highest mountain, where the atmospheric pressure is 532 mmHg and the temperature is 0°C?

Plan We know the initial volume (V_1), pressure (P_1), and temperature (T_1) of the gas; we also know the final pressure (P_2) and temperature (T_2), and we must find the final volume (V_2). Since the amount of helium in the balloon does not change, n is fixed. We convert both T values to kelvins, the final pressure to atm, rearrange the ideal gas law, and solve for V_2. This is a combined gas law problem.

Solution Summarizing the gas variables:

$V_1 = 15.8$ L $\quad\quad V_2 =$ unknown
$T_1 = 22°C$ (convert to K) $\quad\quad T_2 = 0°C$ (convert to K)
$P_1 = 0.980$ atm $\quad\quad P_2 = 532$ mmHg (convert to atm)
n remains constant

Converting T from °C to K:

$$T_1 \text{ (K)} = 22°\text{C} + 273.15 = 295 \text{ K} \qquad T_2 \text{ (K)} = 0°\text{C} + 273.15 = 273 \text{ K}$$

Converting P_2 to atm:

$$P_2 \text{ (atm)} = 532 \text{ mmHg} \times \frac{1 \text{ atm}}{760 \text{ mmHg}} = 0.700 \text{ atm}$$

Rearranging the ideal gas law and solving for V_2: At fixed n, we have

$$\frac{P_1V_1}{\cancel{n_1}T_1} = \frac{P_2V_2}{\cancel{n_2}T_2} \quad \text{or} \quad \frac{P_1V_1}{T_1} = \frac{P_2V_2}{T_2}$$

$$V_2 = V_1 \times \frac{P_1T_2}{P_2T_1} = 15.8 \text{ L} \times \frac{(0.980 \text{ atm})(273 \text{ K})}{(0.700 \text{ atm})(295 \text{ K})} = 20.5 \text{ L}$$

Check Let's predict the change to check the math: because $T_2 < T_1$, we expect $V_2 < V_1$; but because $P_2 < P_1$, we expect $V_2 > V_1$. The temperature ratio (T_2 in the numerator) is about 0.93 (273/295), and the pressure ratio (P_1 in the numerator) is about 1.4 (0.980/0.700), so the V ratio should be about 0.93 × 1.4, or 1.3 (20.5/15.8 ≈ 1.3). The pressure decreases by a larger factor than the temperature increases, so there is an overall increase in volume.

FOLLOW-UP PROBLEMS

5.5A A sample of nitrogen gas occupies a volume of 2.55 L at 755 mmHg and 23°C. Its volume increases to 4.10 L, and its temperature decreases to 18°C. What is the final pressure?

5.5B A diver has 2.2 L of air in his lungs at an air temperature of 28°C and a pressure of 0.980 atm. What is the volume of air in his lungs after he dives, while holding his breath, to a depth of 10 m, where the temperature is 21°C and the pressure is 1.40 atm?

SOME SIMILAR PROBLEMS 5.32 and 5.33

SAMPLE PROBLEM 5.6 Solving for an Unknown Gas Variable at Fixed Conditions

Problem A steel tank has a volume of 438 L and is filled with 0.885 kg of O_2. Calculate the pressure of O_2 at 21°C.

Plan We are given V, T, and the mass of O_2, and we must find P. Since the conditions are not changing, we apply the ideal gas law without rearranging it. We use the given V in liters, convert T to kelvins and mass (kg) of O_2 to amount (mol), and solve for P.

Solution Summarizing the gas variables:

V = 438 L T = 21°C (convert to K)
n = 0.885 kg O_2 (convert to mol) P = unknown

Converting T from °C to K:

$$T(\text{K}) = 21°\text{C} + 273.15 = 294 \text{ K}$$

Converting from mass (g) of O_2 to amount (mol):

$$n = \text{mol of } O_2 = 0.885 \text{ kg } O_2 \times \frac{1000 \text{ g}}{1 \text{ kg}} \times \frac{1 \text{ mol } O_2}{32.00 \text{ g } O_2} = 27.7 \text{ mol } O_2$$

Solving for P (note the unit canceling here):

$$P = \frac{nRT}{V} = \frac{27.7 \cancel{\text{mol}} \times 0.0821 \dfrac{\text{atm}\cdot\cancel{\text{L}}}{\cancel{\text{mol}}\cdot\cancel{\text{K}}} \times 294 \cancel{\text{K}}}{438 \cancel{\text{L}}}$$
$$= 1.53 \text{ atm}$$

Check The amount of O_2 seems correct: ~900 g/(30 g/mol) = 30 mol. To check the approximate size of the final calculation, round off the values, including that for R:

$$P = \frac{30 \text{ mol } O_2 \times 0.1 \dfrac{\text{atm}\cdot\text{L}}{\text{mol}\cdot\text{K}} \times 300 \text{ K}}{450 \text{ L}} = 2 \text{ atm}$$

which is reasonably close to 1.53 atm.

FOLLOW-UP PROBLEMS

5.6A The tank in Sample Problem 5.6 develops a slow leak that is discovered and sealed. The new pressure is 1.37 atm. How many grams of O_2 remain?

5.6B A blimp is filled with 3950 kg of helium at 731 mmHg and 20°C. What is the volume of the blimp under these conditions?

SOME SIMILAR PROBLEMS 5.34–5.37

Finally, in a picture problem, we apply the gas laws to determine the balanced equation for a gaseous reaction.

SAMPLE PROBLEM 5.7 Using Gas Laws to Determine a Balanced Equation

Problem The piston-cylinder below is depicted before and after a gaseous reaction that is carried out in it at constant pressure: the temperature is 150 K before and 300 K after the reaction. (Assume the cylinder is insulated.)

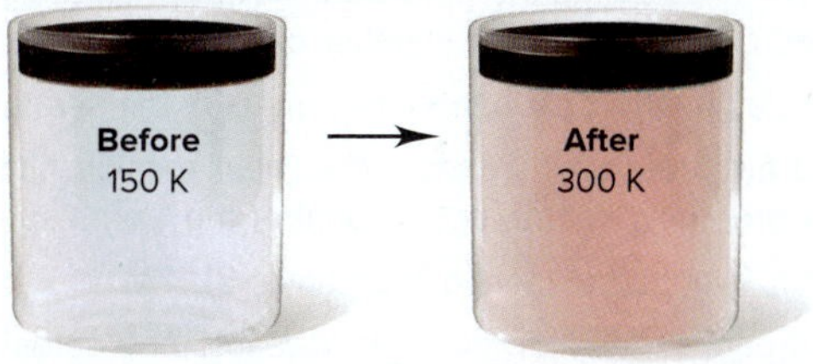

Which of the following balanced equations describes the reaction?

(1) $A_2(g) + B_2(g) \longrightarrow 2AB(g)$ **(2)** $2AB(g) + B_2(g) \longrightarrow 2AB_2(g)$

(3) $A(g) + B_2(g) \longrightarrow AB_2(g)$ **(4)** $2AB_2(g) \longrightarrow A_2(g) + 2B_2(g)$

Plan We are shown a depiction of the volume and temperature of a gas mixture before and after a reaction and must deduce the balanced equation. The problem says that P is constant, and the picture shows that, when T doubles, V stays the same. If n were also constant, Charles's law tells us that V should double when T doubles. But, since V does not change, n cannot be constant. From Avogadro's law, the only way to maintain V constant, with P constant and T doubling, is for n to be halved. So we examine the four balanced equations and count the number of moles on each side to see in which equation n is halved.

Solution In equation (1), n does not change, so doubling T would double V.
In equation (2), n decreases from 3 mol to 2 mol, so doubling T would increase V by one-third.
In equation (3), n decreases from 2 mol to 1 mol. Doubling T would exactly balance the decrease from halving n, so V would stay the same.
In equation (4), n increases, so doubling T would more than double V.
Therefore, equation (3) is correct:

$$A(g) + B_2(g) \longrightarrow AB_2(g)$$

FOLLOW-UP PROBLEMS

5.7A The piston-cylinder below shows the volumes of a gaseous reaction mixture before and after a reaction that takes place at constant pressure and an initial temperature of −73°C.

If the *unbalanced* equation is $CD(g) \longrightarrow C_2(g) + D_2(g)$, what is the final temperature (in °C)?

5.7B The piston-cylinder below shows a gaseous reaction mixture before and after a reaction that takes place at constant pressure.

Which of the following balanced equations describes the reaction?

(1) $2X_2(g) + Y_2(g) \longrightarrow 2X_2Y(g)$ **(2)** $2XY_3(g) \longrightarrow X_2(g) + 3Y_2(g)$

(3) $X_2Y_2(g) \longrightarrow 2X(g) + Y_2(g)$ **(4)** $2XY(g) \longrightarrow X_2(g) + Y_2(g)$

SOME SIMILAR PROBLEMS 5.19 and 5.67

› Summary of Section 5.3

- › Four interdependent variables define the physical behavior of an ideal gas: volume (V), pressure (P), temperature (T), and amount (number of moles, n).
- › Most simple gases display nearly ideal behavior at ordinary temperatures and pressures.
- › Boyle's, Charles's, and Avogadro's laws refer to the linear relationships between the volume of a gas and the pressure, temperature, and amount of the gas, respectively.
- › At STP (0°C and 1 atm), 1 mol of an ideal gas occupies 22.4 L.
- › The ideal gas law incorporates the individual gas laws into one equation: $PV = nRT$, where R is the universal gas constant.

5.4 REARRANGEMENTS OF THE IDEAL GAS LAW

The ideal gas law can be mathematically rearranged to find gas density, molar mass, the partial pressure of each gas in a mixture, and the amount of gaseous reactant or product in a reaction.

The Density of a Gas

One mole of any gas behaving ideally occupies the same volume at a given temperature and pressure, so differences in gas density ($d = m/V$) depend on differences in molar mass (see Figure 5.9). For example, at STP, 1 mol (32.00 g) of O_2 occupies 22.4 L, the same volume occupied by 1 mol (28.02 g) of N_2 so

$$d \text{ of } O_2 = \frac{32.00 \text{ g}}{22.4 \text{ L}} = 1.43 \text{ g/L} \qquad \text{and} \qquad d \text{ of } N_2 = \frac{28.02 \text{ g}}{22.4 \text{ L}} = 1.25 \text{ g/L}$$

O_2 is denser because each O_2 molecule has a greater mass (32.00 amu) than each N_2 molecule (28.02 amu). Thus,

$$d \text{ of } O_2 = \frac{32.00}{28.02} \times d \text{ of } N_2 = \frac{32.00}{28.02} \times 1.25 \text{ g/L} = 1.43 \text{ g/L}$$

We can rearrange the ideal gas law to calculate the density of a gas from its molar mass. Recall that the number of moles (n) is the mass of substance (m) divided by its molar mass ($\mathcal{M}$), $n = m/\mathcal{M}$. Substituting for n in the ideal gas law gives

$$PV = \frac{m}{\mathcal{M}}RT$$

Rearranging to isolate m/V gives

$$\frac{m}{V} = d = \frac{P \times \mathcal{M}}{RT} \qquad \textbf{(5.9)}$$

Three important ideas are expressed by Equation 5.9:

- *The density of a gas is directly proportional to its molar mass.* The volume of a given amount of a heavier gas equals the volume of the same amount of a lighter gas at the same conditions of temperature and pressure (Avogadro's law), so the density of the heavier gas is higher (as you just saw for O_2 and N_2).
- *The density of a gas is inversely proportional to the temperature.* As the volume of a gas increases with temperature (Charles's law), the same mass occupies more space, so the density of the gas is lower.
- *The density of a gas is directly proportional to the pressure.* As the volume of a gas decreases with increasing pressure (Boyle's law), the same mass occupies less space, so the density of the gas is higher.

We use Equation 5.9 to find the density of a gas at any temperature and pressure near standard conditions.

SAMPLE PROBLEM 5.8 Calculating Gas Density

Problem To apply a green chemistry approach, a chemical engineer uses waste CO_2 from a manufacturing process, instead of chlorofluorocarbons, as a "blowing agent" in the production of polystyrene. Find the density (in g/L) of CO_2 and the number of molecules per liter **(a)** at STP (0°C and 1 atm) and **(b)** at room conditions (20.°C and 1.00 atm).

Plan We must find the density (d) and the number of molecules of CO_2, given two sets of P and T data. We find $\mathcal{M}$, convert T to kelvins, and calculate d with Equation 5.9. Then we convert the mass per liter to molecules per liter with Avogadro's number.

Solution (a) Density and molecules per liter of CO_2 at STP. Summary of gas properties:

$$T = 0°\text{C} + 273.15 = 273\text{ K} \qquad P = 1\text{ atm} \qquad \mathcal{M}\text{ of } CO_2 = 44.01\text{ g/mol}$$

Calculating density (note the unit canceling here):

$$d = \frac{P \times \mathcal{M}}{RT} = \frac{1.00\ \cancel{\text{atm}} \times 44.01\text{ g}/\cancel{\text{mol}}}{0.0821\ \frac{\cancel{\text{atm}}\cdot\text{L}}{\cancel{\text{mol}}\cdot\cancel{\text{K}}} \times 273\ \cancel{\text{K}}} = \boxed{1.96\text{ g/L}}$$

Converting from mass/L to molecules/L:

$$\text{Molecules } CO_2\text{/L} = \frac{1.96\text{ g } CO_2}{1\text{ L}} \times \frac{1\text{ mol } CO_2}{44.01\text{ g } CO_2} \times \frac{6.022\times10^{23}\text{ molecules } CO_2}{1\text{ mol } CO_2}$$
$$= \boxed{2.68\times10^{22}\text{ molecules } CO_2\text{/L}}$$

(b) Density and molecules of CO_2 per liter at room conditions. Summary of gas properties:

$$T = 20.°\text{C} + 273.15 = 293\text{ K} \qquad P = 1.00\text{ atm} \qquad \mathcal{M}\text{ of } CO_2 = 44.01\text{ g/mol}$$

Calculating density:

$$d = \frac{P \times \text{M}}{RT} = \frac{1.00\text{ atm} \times 44.01\text{ g/mol}}{0.0821\ \frac{\text{atm}\cdot\text{L}}{\text{mol}\cdot\text{K}} \times 293\text{ K}} = \boxed{1.83\text{ g/L}}$$

Converting from mass/L to molecules/L:

$$\text{Molecules } CO_2\text{/L} = \frac{1.83\text{ g } CO_2}{1\text{ L}} \times \frac{1\text{ mol } CO_2}{44.01\text{ g } CO_2} \times \frac{6.022\times10^{23}\text{ molecules } CO_2}{1\text{ mol } CO_2}$$
$$= \boxed{2.50\times10^{22}\text{ molecules } CO_2\text{/L}}$$

Check Round off to check the density values; for example, in (a), at STP:

$$\frac{50\text{ g/mol} \times 1\text{ atm}}{0.1\ \frac{\text{atm}\cdot\text{L}}{\text{mol}\cdot\text{K}} \times 250\text{ K}} = 2\text{ g/L} \approx 1.96\text{ g/L}$$

At the higher temperature in (b), the density should decrease, which can happen only if there are fewer molecules per liter, so the answer is reasonable.

Comment 1. An *alternative approach* for finding the density of most simple gases, but *at STP only,* is to divide the molar mass by the standard molar volume, 22.4 L:

$$d = \frac{\mathcal{M}}{V} = \frac{44.01\ \text{g/mol}}{22.4\ \text{L/mol}} = 1.96\ \text{g/L}$$

Once you know the density at one temperature (0°C), you can find it at any other temperature with the following relationship: $d_1/d_2 = T_2/T_1$ (if the pressure is constant).
2. Note that we have different numbers of significant figures for the pressure values. In (a), "1 atm" is part of the definition of STP, so it is an exact number. In (b), we specified "1.00 atm" to allow three significant figures in the answer.

FOLLOW-UP PROBLEMS

5.8A Compare the density of CO_2 at 0°C and 380. torr with its density at STP.

5.8B Nitrogen oxide (NO_2) is a reddish-brown gas that is a component of smog. Calculate its density at 0.950 atm and 24°C. How does the density of NO_2 compare to that of dry air (d = 1.13 g/L at the same conditions)?

SOME SIMILAR PROBLEMS 5.45–5.48

Gas Density and the Human Condition A few applications demonstrate the wide-ranging relevance of gas density:

- *Engineering.* Architectural designers and heating engineers place heating ducts near the floor so that the warmer, and thus less dense, air coming from the ducts will rise and mix with the cooler room air.
- *Safety and air pollution.* In the absence of mixing, a less dense gas will lie above a more dense one. Fire extinguishers that release CO_2 are effective because this gas is heavier than air: it sinks onto the fire and keeps more O_2 from reaching the fuel. The dense gases in smog that blankets urban centers, such as Mexico City, Los Angeles, and Beijing, contribute to respiratory illnesses (see Follow-up Problem 5.8B).
- *Toxic releases.* During World War I, poisonous phosgene gas ($COCl_2$) was used against ground troops because it was dense enough to sink into their trenches. In 1984, the accidental release of poisonous methylisocyanate gas from a Union Carbide plant in India killed thousands of people as it blanketed nearby neighborhoods. In 1986, CO_2 released naturally from Lake Nyos in Cameroon suffocated thousands of people as it flowed down valleys into villages. Some paleontologists suggest that the release of CO_2 from volcanic lakes may have contributed to widespread dying off of dinosaurs.
- *Ballooning.* When the gas in a hot-air balloon is heated, its volume increases and the balloon inflates. Further heating causes some of the gas to escape. Thus, the gas density decreases and the balloon rises. In 1783, Jacques Charles (of Charles's law) made one of the first balloon flights, and 20 years later, Joseph Gay-Lussac (who studied the pressure-temperature relationship) set a solo altitude record that held for 50 years.

Source: © Charles F. McCarthy/Shutterstock.com

The Molar Mass of a Gas

Through another rearrangement of the ideal gas law, we can determine the molar mass of an unknown gas or a volatile liquid (one that is easily vaporized):

$$n = \frac{m}{\mathcal{M}} = \frac{PV}{RT} \quad \text{so} \quad \mathcal{M} = \frac{mRT}{PV} \tag{5.10}$$

Notice that this equation is just a rearrangement of Equation 5.9.

SAMPLE PROBLEM 5.9 Finding the Molar Mass of a Volatile Liquid

Problem An organic chemist isolates a colorless liquid from a petroleum sample. She places the liquid in a preweighed flask and puts the flask in boiling water, which vaporizes the liquid and fills the flask with gas. She closes the flask and reweighs it. She obtains the following data:

Volume (V) of flask = 213 mL T = 100.0°C P = 754 torr
Mass of flask + gas = 78.416 g Mass of flask = 77.834 g

Calculate the molar mass of the liquid.

Plan We are given V, T, P, and mass data and must find the molar mass ($\mathcal{M}$) of the liquid. We convert V to liters, T to kelvins, and P to atmospheres, find the mass of gas by subtracting the mass of the flask from the mass of the flask plus gas, and use Equation 5.10 to calculate $\mathcal{M}$.

Solution Summarizing and converting the gas variables:

$$V\text{ (L)} = 213\text{ mL} \times \frac{1\text{ L}}{1000\text{ mL}} = 0.213\text{ L} \qquad T\text{ (K)} = 100.0°\text{C} + 273.15 = 373.2\text{ K}$$

$$P\text{ (atm)} = 754\text{ torr} \times \frac{1\text{ atm}}{760\text{ torr}} = 0.992\text{ atm} \qquad m = 78.416\text{ g} - 77.834\text{ g} = 0.582\text{ g}$$

Calculating $\mathcal{M}$:

$$\mathcal{M} = \frac{mRT}{PV} = \frac{0.582\text{ g} \times 0.0821\ \frac{\text{atm·L}}{\text{mol·K}} \times 373.2\text{ K}}{0.992\text{ atm} \times 0.213\text{ L}} = 84.4\text{ g/mol}$$

Check Rounding to check the arithmetic, we have

$$\frac{0.6\text{ g} \times 0.08\ \frac{\text{atm·L}}{\text{mol·K}} \times 375\text{ K}}{1\text{ atm} \times 0.2\text{ L}} = 90\text{ g/mol} \qquad \text{(which is close to 84.4 g/mol)}$$

FOLLOW-UP PROBLEMS

5.9A An empty 149-mL flask weighs 68.322 g before a sample of volatile liquid is added. The flask is then placed in a hot (95.0°C) water bath; the barometric pressure is 740. torr. The liquid vaporizes and the gas fills the flask. After cooling, flask and condensed liquid together weigh 68.697 g. What is the molar mass of the liquid?

5.9B An empty glass bulb has a volume of 350. mL and a mass of 82.561 g. When filled with an unknown gas at 733 mmHg and 22°C, the bulb has a mass of 82.786 g. Is the gas argon (Ar), methane (CH_4), or nitrogen monoxide (NO)?

SOME SIMILAR PROBLEMS 5.49 and 5.50

The Partial Pressure of Each Gas in a Mixture of Gases

The ideal gas law holds for virtually any gas at ordinary conditions, whether it is a pure gas or a mixture of gases such as air, because

- Gases mix homogeneously (form a solution) in any proportions.
- Each gas in a mixture behaves as if it were the only gas present (assuming no chemical interactions).

Dalton's Law of Partial Pressures The second point above was discovered by John Dalton during his lifelong study of humidity. He observed that when water vapor is added to dry air, the total air pressure increases by the pressure of the water vapor:

$$P_{\text{humid air}} = P_{\text{dry air}} + P_{\text{added water vapor}}$$

He concluded that each gas in the mixture exerts a **partial pressure** equal to the pressure it would exert *by itself.* Stated as **Dalton's law of partial pressures,** his discovery was that *in a mixture of unreacting gases, the total pressure is the sum of the partial pressures of the individual gases:*

$$P_{\text{total}} = P_1 + P_2 + P_3 + \cdots \quad \textbf{(5.11)}$$

As an example, suppose we have a tank of fixed volume that contains nitrogen gas at a certain pressure, and we introduce a sample of hydrogen gas into the tank. Each gas behaves independently, so we can write an ideal gas law expression for each:

$$P_{N_2} = \frac{n_{N_2}RT}{V} \qquad \text{and} \qquad P_{H_2} = \frac{n_{H_2}RT}{V}$$

Because each gas occupies the same total volume and is at the same temperature, the pressure of each gas depends only on its amount, n. Thus, the total pressure is

$$P_{\text{total}} = P_{N_2} + P_{H_2} = \frac{n_{N_2}RT}{V} + \frac{n_{H_2}RT}{V} = \frac{(n_{N_2} + n_{H_2})RT}{V} = \frac{n_{\text{total}}RT}{V}$$

where $n_{\text{total}} = n_{N_2} + n_{H_2}$.

Each component in a mixture contributes a fraction of the total number of moles in the mixture; this portion is the **mole fraction (X)** of that component. Multiplying X by 100 gives the mole percent. The sum of the mole fractions of all components must be 1, and the sum of the mole percents must be 100%. For N_2 in our mixture, the mole fraction is

$$X_{N_2} = \frac{\text{mol of } N_2}{\text{total amount (mol)}} = \frac{n_{N_2}}{n_{N_2} + n_{H_2}}$$

If the total pressure is due to the total number of moles, the partial pressure of gas A is the total pressure multiplied by the mole fraction of A, X_A:

$$P_A = X_A \times P_{\text{total}} \quad \textbf{(5.12)}$$

For example, if $\frac{1}{4}$ (0.25) of a gas mixture is gas A, then gas A contributes $\frac{1}{4}$ of the total pressure.

Equation 5.12 is a very useful result. To see that it is valid for the mixture of N_2 and H_2, we recall that $X_{N_2} + X_{H_2} = 1$; then we obtain

$$P_{\text{total}} = P_{N_2} + P_{H_2} = (X_{N_2} \times P_{\text{total}}) + (X_{H_2} \times P_{\text{total}}) = (X_{N_2} + X_{H_2})P_{\text{total}} = 1 \times P_{\text{total}}$$

SAMPLE PROBLEM 5.10 Applying Dalton's Law of Partial Pressures

Problem In a study of O_2 uptake by muscle tissue at high altitude, a physiologist prepares an atmosphere consisting of 79 mole % N_2, 17 mole % $^{16}O_2$, and 4.0 mole % $^{18}O_2$. (The isotope ^{18}O will be measured to determine O_2 uptake.) The total pressure is 0.75 atm to simulate high altitude. Calculate the mole fraction and partial pressure of $^{18}O_2$ in the mixture.

Plan We must find $X_{^{18}O_2}$ and $P_{^{18}O_2}$ from P_{total} (0.75 atm) and the mole % of $^{18}O_2$ (4.0). Dividing the mole % by 100 gives the mole fraction, $X_{^{18}O_2}$. Then, using Equation 5.12, we multiply $X_{^{18}O_2}$ by P_{total} to find $P_{^{18}O_2}$ (see the road map).

Road Map

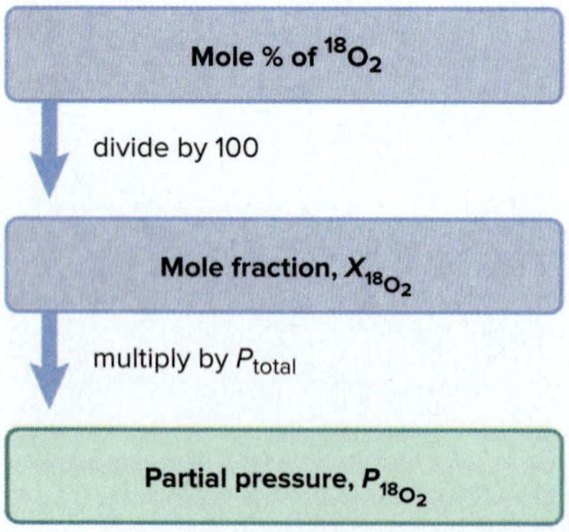

Solution Calculating the mole fraction of $^{18}O_2$:

$$X_{^{18}O_2} = \frac{4.0 \text{ mol \% } ^{18}O_2}{100} = 0.040$$

Solving for the partial pressure of $^{18}O_2$:

$$P_{^{18}O_2} = X_{^{18}O_2} \times P_{\text{total}} = 0.040 \times 0.75 \text{ atm} = 0.030 \text{ atm}$$

Check $X_{^{18}O_2}$ is small because the mole % is small, so $P_{^{18}O_2}$ should be small also.

Comment At high altitudes, specialized brain cells that are sensitive to O_2 and CO_2 levels in the blood trigger an increase in rate and depth of breathing for several days, until a person becomes acclimated.

FOLLOW-UP PROBLEMS

5.10A To prevent air from interacting with highly reactive chemicals, noble gases are placed over the chemicals to act as inert "blanketing" gases. A chemical engineer places a mixture of noble gases consisting of 5.50 g of He, 15.0 g of Ne, and 35.0 g of Kr in a piston-cylinder assembly at STP. Calculate the partial pressure of each gas.

5.10B Trimix is a breathing mixture of He, O_2, and N_2 used by deep-sea scuba divers; He reduces the chances of N_2 narcosis and O_2 toxicity. A tank of Trimix has a total pressure of 204 atm and a partial pressure of He of 143 atm. What is the mole percent of He in the mixture?

SOME SIMILAR PROBLEMS 5.51 and 5.52

Collecting a Gas over Water Whenever a gas is in contact with water, some of the water vaporizes into the gas. The water vapor that mixes with the gas contributes a pressure known as the *vapor pressure,* a portion of the total pressure that depends only on the water temperature (Table 5.2). A common use of the law of partial pressures is to determine the amount of a water-insoluble gas formed in a reaction: the gaseous product bubbles through water, some water vaporizes into the bubbles, and the mixture of product gas and water vapor is collected into an inverted container (Figure 5.12).

To determine the amount of gas formed, we look up the vapor pressure (P_{H_2O}) at the temperature of the experiment in Table 5.2 and subtract it from the total gas pressure (P_{total}, corrected for barometric pressure) to get the partial pressure of the gaseous product (P_{gas}). With V and T known, we can calculate the amount of product.

Table 5.2 Vapor Pressure of Water (P_{H_2O}) at Different T

T (°C)	P_{H_2O} (torr)	T (°C)	P_{H_2O} (torr)	T (°C)	P_{H_2O} (torr)	T (°C)	P_{H_2O} (torr)
0	4.6	20	17.5	40	55.3	75	289.1
5	6.5	22	19.8	45	71.9	80	355.1
10	9.2	24	22.4	50	92.5	85	433.6
12	10.5	26	25.2	55	118.0	90	525.8
14	12.0	28	28.3	60	149.4	95	633.9
16	13.6	30	31.8	65	187.5	100	760.0
18	15.5	35	42.2	70	233.7		

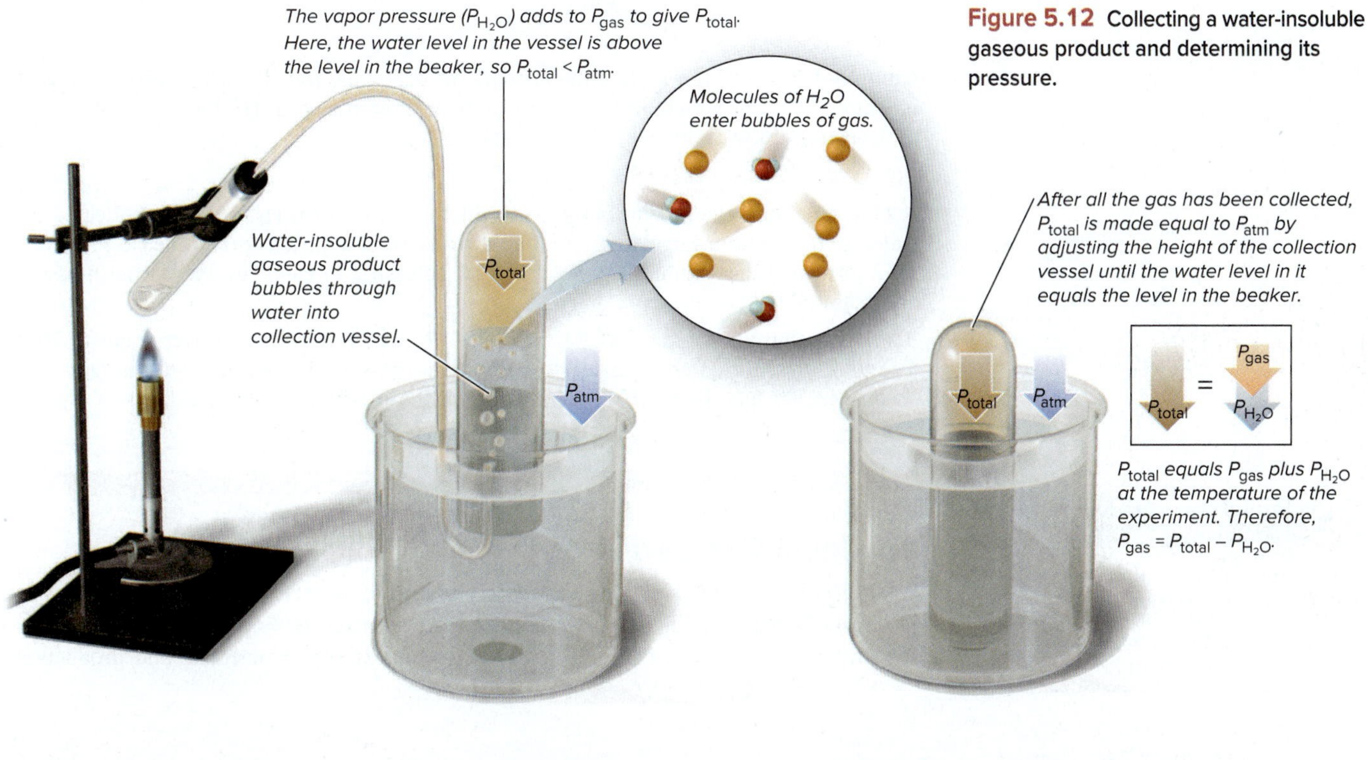
The vapor pressure (P_{H_2O}) adds to P_{gas} to give P_{total}. Here, the water level in the vessel is above the level in the beaker, so $P_{total} < P_{atm}$.
Molecules of H_2O enter bubbles of gas.
Water-insoluble gaseous product bubbles through water into collection vessel.
P_{total}
P_{atm}
After all the gas has been collected, P_{total} is made equal to P_{atm} by adjusting the height of the collection vessel until the water level in it equals the level in the beaker.
P_{total}
P_{atm}
P_{total} = P_{gas} P_{H_2O}
P_{total} equals P_{gas} plus P_{H_2O} at the temperature of the experiment. Therefore, $P_{gas} = P_{total} - P_{H_2O}$.

Figure 5.12 Collecting a water-insoluble gaseous product and determining its pressure.

Road Map

P_{total}

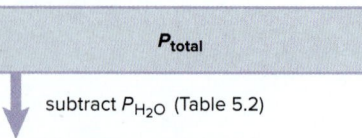

↓ subtract P_{H_2O} (Table 5.2)

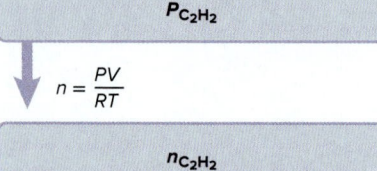

$P_{C_2H_2}$

↓ $n = \frac{PV}{RT}$

$n_{C_2H_2}$

↓ multiply by $\mathcal{M}$ (g/mol)

Mass (g) of C_2H_2

SAMPLE PROBLEM 5.11 Calculating the Amount of Gas Collected over Water

Problem Acetylene (C_2H_2), an important fuel in welding, is produced in the laboratory when calcium carbide (CaC_2) reacts with water:

$$CaC_2(s) + 2H_2O(l) \longrightarrow C_2H_2(g) + Ca(OH)_2(aq)$$

For a sample of acetylene collected over water, total gas pressure (adjusted to barometric pressure) is 738 torr and the volume is 523 mL. At the temperature of the gas (23°C), the vapor pressure of water is 21 torr. How many grams of acetylene are collected?

Plan In order to find the mass of C_2H_2, we first need to find the number of moles of C_2H_2, $n_{C_2H_2}$, which we can obtain from the ideal gas law by calculating $P_{C_2H_2}$. The barometer reading gives us P_{total}, which is the sum of $P_{C_2H_2}$ and P_{H_2O}, and we are given P_{H_2O}, so we subtract to find $P_{C_2H_2}$. We are also given V and T, so we convert to consistent units, and find $n_{C_2H_2}$ from the ideal gas law. Then we convert moles to grams using the molar mass from the formula for acetylene, as shown in the road map.

Solution Summarizing and converting the gas variables:

$$P_{C_2H_2}\ (\text{torr}) = P_{total} - P_{H_2O} = 738\ \text{torr} - 21\ \text{torr} = 717\ \text{torr}$$

$$P_{C_2H_2}\ (\text{atm}) = 717\ \text{torr} \times \frac{1\ \text{atm}}{760\ \text{torr}} = 0.943\ \text{atm}$$

$$V\ (\text{L}) = 523\ \text{mL} \times \frac{1\ \text{L}}{1000\ \text{mL}} = 0.523\ \text{L}$$

$$T\ (\text{K}) = 23°\text{C} + 273.15 = 296\ \text{K}$$

$$n_{C_2H_2} = \text{unknown}$$

Solving for $n_{C_2H_2}$:

$$n_{C_2H_2} = \frac{PV}{RT} = \frac{0.943\ \text{atm} \times 0.523\ \text{L}}{0.0821\ \frac{\text{atm·L}}{\text{mol·K}} \times 296\ \text{K}} = 0.0203\ \text{mol}$$

Converting $n_{C_2H_2}$ to mass (g):

$$\text{Mass (g) of } C_2H_2 = 0.0203\ \text{mol } C_2H_2 \times \frac{26.04\ \text{g } C_2H_2}{1\ \text{mol } C_2H_2} = 0.529\ \text{g } C_2H_2$$

Check Rounding to one significant figure, a quick arithmetic check for n gives

$$n \approx \frac{1\ \text{atm} \times 0.5\ \text{L}}{0.08\ \frac{\text{atm·L}}{\text{mol·K}} \times 300\ \text{K}} = 0.02\ \text{mol} \approx 0.0203\ \text{mol}$$

Comment The C_2^{2-} ion (called the *carbide,* or *acetylide, ion*) is an interesting anion. It is simply $^-C{\equiv}C^-$, which acts as a base in water, removing an H^+ ion from two H_2O molecules to form acetylene, $H{-}C{\equiv}C{-}H$.

FOLLOW-UP PROBLEMS

5.11A A small piece of zinc reacts with dilute HCl to form H_2, which is collected over water at 16°C into a large flask. The total pressure is adjusted to barometric pressure (752 torr), and the volume is 1495 mL. Use Table 5.2 to calculate the partial pressure and mass of H_2.

5.11B Heating a sample of $KClO_3$ produces O_2, which is collected over water at 20.°C. The total pressure, after adjusting to barometric pressure, is 748 torr, and the gas volume is 307 mL. Use Table 5.2 to calculate the partial pressure and mass of O_2.

SOME SIMILAR PROBLEMS 5.57 and 5.58

Student Hot Spot

Student data indicate that you may struggle with problems that combine stoichiometry and the ideal gas law. Access the Smartbook to view additional Learning Resources on this topic.

The Ideal Gas Law and Reaction Stoichiometry

As you saw in Chapters 3 and 4, and in the preceding discussion of collecting a gas over water, many reactions involve gases as reactants or products. From the balanced equation for such a reaction, you can calculate the amounts (mol) of reactants and products and convert these quantities into masses or numbers of molecules. Figure 5.13 shows how you use the ideal gas law to convert between gas variables (*P*, *T*, and *V*) and amounts (mol) of gaseous reactants and products. In effect, you combine a gas law problem with a stoichiometry problem, as you'll see in Sample Problems 5.12 and 5.13.

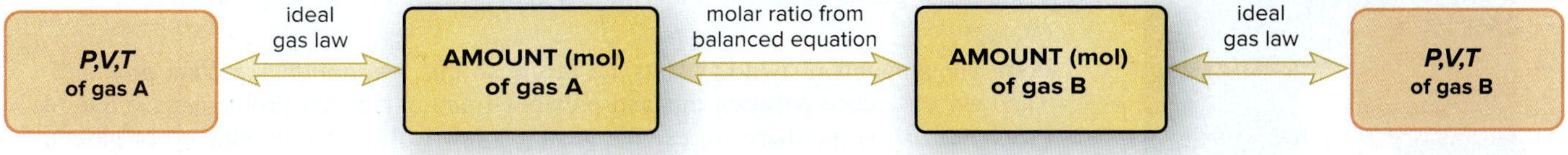

Figure 5.13 The relationships among the amount (mol, *n*) of gaseous reactant (or product) and the gas pressure (*P*), volume (*V*), and temperature (*T*).

SAMPLE PROBLEM 5.12 Using Gas Variables to Find Amounts of Reactants or Products I

Problem Solid lithium hydroxide is used to "scrub" CO_2 from the air in spacecraft and submarines; it reacts with the CO_2 to produce lithium carbonate and water. What volume of CO_2 at 23°C and 716 torr can be removed by reaction with 395 g of lithium hydroxide?

Plan This is a stoichiometry *and* a gas law problem. To find V_{CO_2}, we first need n_{CO_2}. We write and balance the equation. Next, we convert the given mass (395 g) of lithium hydroxide, LiOH, to amount (mol) and use the molar ratio to find amount (mol) of CO_2 that reacts (stoichiometry portion). Then, we use the ideal gas law to convert moles of CO_2 to liters (gas law portion). A road map is shown, but you are familiar with all the steps.

Road Map

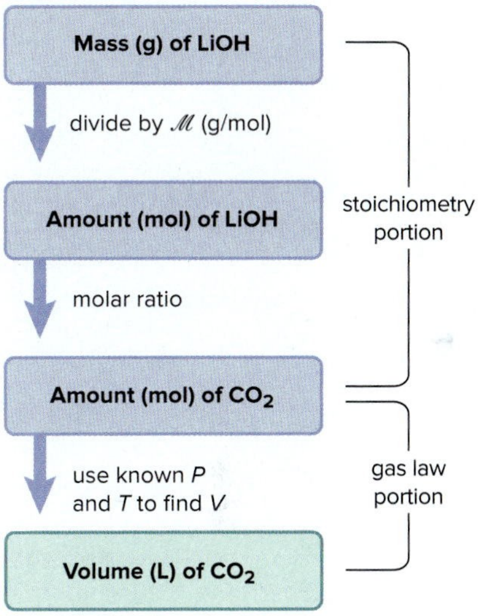

Solution Writing the balanced equation:

$$2\text{LiOH}(s) + \text{CO}_2(g) \longrightarrow \text{Li}_2\text{CO}_3(s) + \text{H}_2\text{O}(l)$$

Calculating n_{CO_2}:

$$n_{CO_2} = 395 \text{ g LiOH} \times \frac{1 \text{ mol LiOH}}{23.95 \text{ g LiOH}} \times \frac{1 \text{ mol CO}_2}{2 \text{ mol LiOH}} = 8.25 \text{ mol CO}_2$$

Summarizing and converting other gas variables:

$$V = \text{unknown} \qquad P\text{ (atm)} = 716 \text{ torr} \times \frac{1 \text{ atm}}{760 \text{ torr}} = 0.942 \text{ atm}$$

$$T\text{ (K)} = 23°\text{C} + 273.15 = 296 \text{ K}$$

Solving for V_{CO_2}:

$$V = \frac{nRT}{P} = \frac{8.25 \text{ mol} \times 0.0821 \dfrac{\text{atm·L}}{\text{mol·K}} \times 296 \text{ K}}{0.942 \text{ atm}} = 213 \text{ L}$$

Check One way to check the answer is to compare it with the molar volume of an ideal gas at STP (22.4 L at 273.15 K and 1 atm). One mole of CO_2 at STP occupies about 22 L, so about 8 mol occupies 8 × 22 = 176 L. *T* is a little higher than 273 K and *P* is slightly lower than 1 atm, so *V* should be somewhat larger than 176 L.

Comment The main point here is that the stoichiometry provides one gas variable (*n*), two more are given (*P* and *T*), and the ideal gas law is used to find the fourth (*V*).

FOLLOW-UP PROBLEMS

5.12A Sulfuric acid reacts with sodium chloride to form aqueous sodium sulfate and hydrogen chloride gas. How many milliliters of gas form at STP when 0.117 kg of sodium chloride reacts with excess sulfuric acid?

5.12B Engineers use copper in absorbent beds to react with and remove oxygen impurities from ethylene that is used to make polyethylene. The beds are regenerated when hot H_2 reduces the copper(II) oxide, forming the pure metal and H_2O. On a laboratory scale, what volume of H_2 at 765 torr and 225°C is needed to reduce 35.5 g of copper(II) oxide?

SOME SIMILAR PROBLEMS 5.53 and 5.54

Chlorine gas reacting with potassium.
Source: © McGraw-Hill Education/Stephen Frisch, photographer

SAMPLE PROBLEM 5.13 Using Gas Variables to Find Amounts of Reactants or Products II

Problem The alkali metals [Group 1A(1)] react with the halogens [Group 7A(17)] to form ionic metal halides. What mass of potassium chloride forms when 5.25 L of chlorine gas at 0.950 atm and 293 K reacts with 17.0 g of potassium *(see photo)*?

Plan The amounts of two reactants are given, so this is a limiting-reactant problem. The only difference between this and previous limiting-reactant problems (see Sample Problem 3.20) is that here we use the ideal gas law to find the amount (n) of gaseous reactant from the known V, P, and T. We first write the balanced equation and then use it to find the limiting reactant and the amount and mass of product.

Solution Writing the balanced equation:

$$2K(s) + Cl_2(g) \longrightarrow 2KCl(s)$$

Summarizing the gas variables for $Cl_2(g)$:

$$P = 0.950 \text{ atm} \qquad V = 5.25 \text{ L}$$
$$T = 293 \text{ K} \qquad n = \text{unknown}$$

Solving for n_{Cl_2}:

$$n_{Cl_2} = \frac{PV}{RT} = \frac{0.950 \text{ atm} \times 5.25 \text{ L}}{0.0821 \dfrac{\text{atm·L}}{\text{mol·K}} \times 293 \text{ K}} = 0.207 \text{ mol}$$

Converting from mass (g) of potassium (K) to amount (mol):

$$\text{Amount (mol) of K} = 17.0 \text{ g K} \times \frac{1 \text{ mol K}}{39.10 \text{ g K}} = 0.435 \text{ mol K}$$

Determining the limiting reactant: If Cl_2 is limiting,

$$\text{Amount (mol) of KCl} = 0.207 \text{ mol } Cl_2 \times \frac{2 \text{ mol KCl}}{1 \text{ mol } Cl_2} = 0.414 \text{ mol KCl}$$

If K is limiting,

$$\text{Amount (mol) of KCl} = 0.435 \text{ mol K} \times \frac{2 \text{ mol KCl}}{2 \text{ mol K}} = 0.435 \text{ mol KCl}$$

Cl_2 is the limiting reactant because it forms less KCl. Converting from amount (mol) of KCl to mass (g):

$$\text{Mass (g) of KCl} = 0.414 \text{ mol KCl} \times \frac{74.55 \text{ g KCl}}{1 \text{ mol KCl}} = 30.9 \text{ g KCl}$$

Check The gas law calculation seems correct. At STP, 22 L of Cl_2 gas contains about 1 mol, so a 5-L volume will contain a bit less than 0.25 mol of Cl_2. Moreover, since P (in numerator) is slightly lower than STP, and T (in denominator) is slightly higher than STP, these should lower the calculated n further below the ideal value. The mass of KCl seems correct: less than 0.5 mol of KCl gives <0.5 mol × $\mathcal{M}$ (~75 g/mol), and 30.9 g < 0.5 mol × 75 g/mol.

FOLLOW-UP PROBLEMS

5.13A Ammonia and hydrogen chloride gases react to form solid ammonium chloride. A 10.0-L reaction flask contains ammonia at 0.452 atm and 22°C, and 155 mL of hydrogen chloride gas at 7.50 atm and 271 K is introduced. After the reaction occurs and the temperature returns to 22°C, what is the pressure inside the flask? (Neglect the volume of the solid product.)

5.13B What volume of gaseous iodine pentafluoride, measured at 105°C and 0.935 atm, can be prepared by the reaction of 4.16 g of solid iodine with 2.48 L of gaseous fluorine at 18°C and 0.974 atm?

SOME SIMILAR PROBLEMS 5.55 and 5.56

› Summary of Section 5.4

- Gas density is inversely related to temperature and directly related to pressure: higher T and lower P causes lower d, and vice versa. At the same P and T, gases with larger values of $\mathcal{M}$ have higher values of d.
- In a mixture of gases, each component contributes its partial pressure to the total pressure (Dalton's law of partial pressures). The mole fraction of each component is the ratio of its partial pressure to the total pressure.
- When a gaseous reaction product is collected by bubbling it through water, the total pressure is the sum of the gas pressure and the vapor pressure of water at the given temperature.
- By converting the variables P, V, and T for a gaseous reactant (or product) to amount (n, mol), we can solve stoichiometry problems for gaseous reactions.

5.5 THE KINETIC-MOLECULAR THEORY: A MODEL FOR GAS BEHAVIOR

The **kinetic-molecular theory** is the model that accounts for macroscopic gas behavior at the level of individual particles (atoms or molecules). Developed by some of the great scientists of the 19th century, most notably James Clerk Maxwell and Ludwig Boltzmann, the theory was able to explain the gas laws that some of the great scientists of the 18th century had arrived at empirically. The theory draws quantitative conclusions based on a few postulates (assumptions), but our discussion will be largely qualitative.

How the Kinetic-Molecular Theory Explains the Gas Laws

Let's address some questions the theory must answer, then state the postulates, and draw conclusions that explain the gas laws and related phenomena.

Questions Concerning Gas Behavior Observing gas behavior at the macroscopic level, we must derive a molecular model that explains it:

1. *Origin of pressure.* Pressure is a measure of the force a gas exerts on a surface. How do individual gas particles create this force?
2. *Boyle's law* ($V \propto 1/P$). A change in gas pressure in one direction causes a change in gas volume in the other. What happens to the particles when external pressure compresses the gas volume? And why aren't liquids and solids compressible?
3. *Dalton's law* ($P_{total} = P_1 + P_2 + P_3 + \cdots$). The pressure of a gas mixture is the sum of the pressures of the individual gases. Why does each gas contribute to the total pressure in proportion to its number of particles?
4. *Charles's law* ($V \propto T$). A change in temperature causes a corresponding change in volume. What effect does higher temperature have on gas particles that increases gas volume? This question raises a more fundamental one: what does temperature measure on the molecular scale?
5. *Avogadro's law* ($V \propto n$). Gas volume depends on the number of moles present, not on the chemical nature of the gas. But shouldn't 1 mol of heavier particles exert more pressure, and thus take up more space, than 1 mol of lighter ones?

Postulates of the Kinetic-Molecular Theory The theory is based on three postulates:

Postulate 1. *Particle volume.* A gas consists of a large collection of individual particles with empty space between them. The volume of each particle is so small compared with the volume of the whole sample that it is assumed to be zero; each particle is essentially a point of mass.

Postulate 2. *Particle motion.* The particles are in constant, random, straight-line motion, except when they collide with the container walls or with each other.

Postulate 3. *Particle collisions.* The collisions are *elastic,* which means that, like minute billiard balls on a frictionless billiards table, the colliding molecules exchange energy but do not lose any energy through friction. Thus, *their total kinetic energy (E_k) is constant.* Between collisions, the molecules do not influence each other by attractive or repulsive forces.

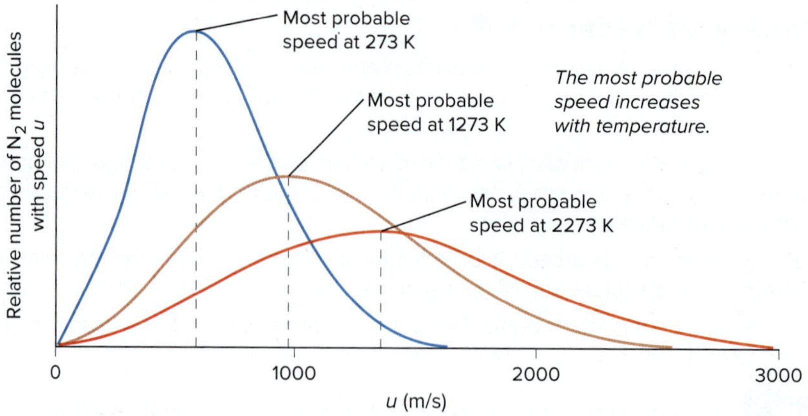

Figure 5.14 Distribution of molecular speeds for N_2 at three temperatures.

Imagine what a sample of gas in a container looks like. Countless minute particles move in every direction, smashing into the container walls and each other. Any given particle changes its speed often—at one moment standing still from a head-on collision and the next moment zooming away from a smash on the side.

In the sample as a whole, *each particle has a molecular speed (u); most are moving near the most probable speed, but some are much faster and others much slower.* Figure 5.14 depicts this distribution of molecular speeds for N_2 gas at three temperatures. Two observations can be made from the graph:

- the curves flatten and spread at higher temperatures
- the *most probable speed (the peak of each curve) increases as the temperature increases.*

The increase in most probable speed occurs because the average kinetic energy of the molecules, which is related to the most probable speed, is proportional to the absolute temperature:

$$\overline{E_k} \propto T \quad \text{or} \quad \overline{E_k} = c \times T$$

where $\overline{E_k}$ is the average kinetic energy of the molecules (an overbar indicates the average value of a quantity) and c is a constant that is the same for any gas. (We'll return to this equation shortly.) Thus, a major conclusion based on the distribution of speeds, which arises directly from postulate 3, is that *at a given temperature, all gases have the same average kinetic energy.*

A Molecular View of the Gas Laws Let's keep visualizing gas particles in a container to see how the theory explains the macroscopic behavior of gases and answers the questions we just posed:

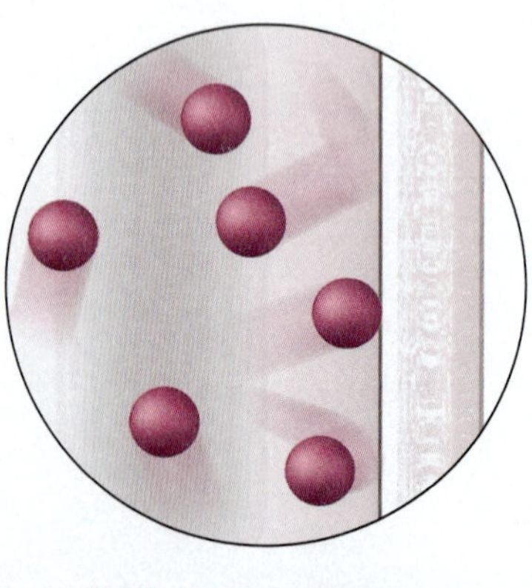

Figure 5.15 Pressure arises from countless collisions between gas particles and walls.

1. *Origin of pressure* (Figure 5.15). From postulates 1 and 2, each gas particle (point of mass) colliding with the container walls (and bottom of piston) exerts a force. Countless collisions over the inner surface of the container result in a pressure. The greater the number of particles, the more frequently they collide with the container, and so the greater the pressure.

2. *Boyle's law ($V \propto 1/P$)* (Figure 5.16). The particles in a gas are points of mass with empty space between them (postulate 1). Before any change in pressure, the pressure exerted *by* the gas (P_{gas}) equals the pressure exerted *on* the gas (P_{ext}), and there is some average distance (d_1) between the particles and the container walls. As P_{ext} increases at constant temperature, the average distance (d_2) between the particles and the walls decreases (that is, $d_2 < d_1$), and so the sample volume decreases. Collisions of the particles with the walls become more frequent over the shorter average distance, which causes P_{gas} to increase until it again equals P_{ext}. The fact that liquids and solids cannot be compressed implies there is little, if any, free space between their particles.

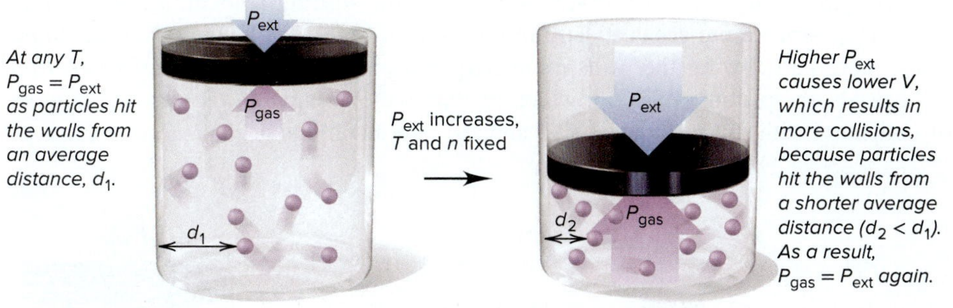
At any T, $P_{gas} = P_{ext}$ as particles hit the walls from an average distance, d_1.
P_{ext}
P_{gas}
d_1
P_{ext} increases, T and n fixed
P_{ext}
d_2
P_{gas}
Higher P_{ext} causes lower V, which results in more collisions, because particles hit the walls from a shorter average distance ($d_2 < d_1$). As a result, $P_{gas} = P_{ext}$ again.

Figure 5.16 A molecular view of Boyle's law.

3. *Dalton's law of partial pressures* ($P_{total} = P_A + P_B$) (Figure 5.17). Adding a given amount (mol) of gas A to a given amount of gas B causes an increase in the total number of particles, in proportion to the particles of A added. This increase causes a corresponding increase in the total number of collisions with the walls per second (postulate 2), which causes a corresponding increase in the total pressure of the gas mixture (P_{total}). Each gas exerts a fraction of P_{total} in proportion to its fraction of the total number of particles (or equivalently, its fraction of the total number of moles, that is, the mole fraction).

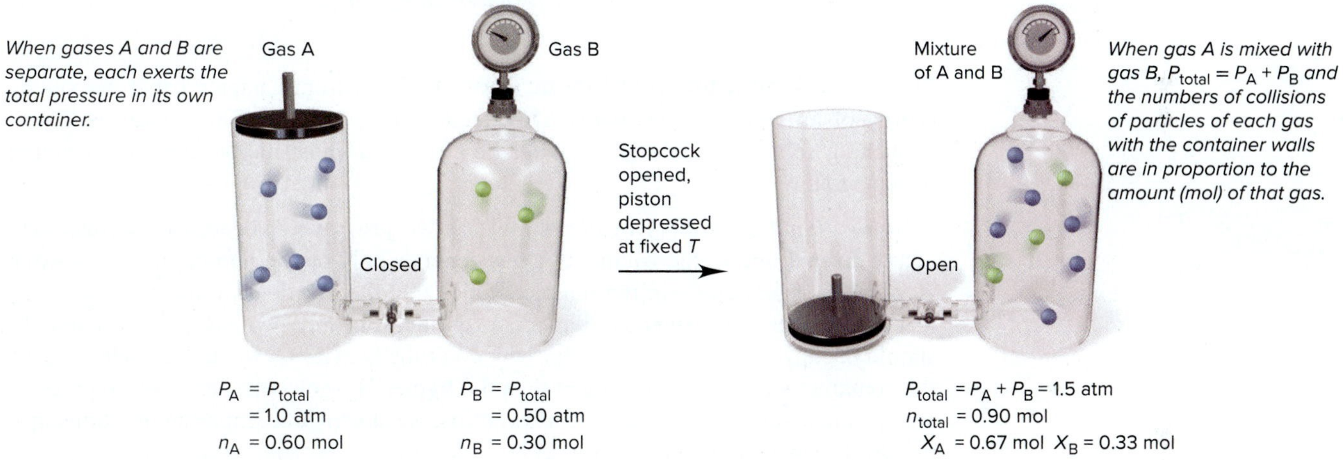
When gases A and B are separate, each exerts the total pressure in its own container.
Gas A
Gas B
Closed
Stopcock opened, piston depressed at fixed T
Mixture of A and B
Open
When gas A is mixed with gas B, $P_{total} = P_A + P_B$ and the numbers of collisions of particles of each gas with the container walls are in proportion to the amount (mol) of that gas.
$P_A = P_{total} = 1.0$ atm
$n_A = 0.60$ mol
$P_B = P_{total} = 0.50$ atm
$n_B = 0.30$ mol
$P_{total} = P_A + P_B = 1.5$ atm
$n_{total} = 0.90$ mol
$X_A = 0.67$ mol $X_B = 0.33$ mol

Figure 5.17 A molecular view of Dalton's law.

4. *Charles's law* ($V \propto T$) (Figure 5.18). At some starting temperature, T_1, the external (atmospheric) pressure (P_{atm}) equals the pressure of the gas (P_{gas}). When the gas is heated and the temperature increases to T_2, the most probable molecular speed and the average kinetic energy increase (postulate 3). Thus, the particles hit the walls more frequently *and* more energetically. This change temporarily increases P_{gas}. As a result, the piston moves up, which increases the volume and lowers the number of collisions with the walls until P_{atm} and P_{gas} are again equal.

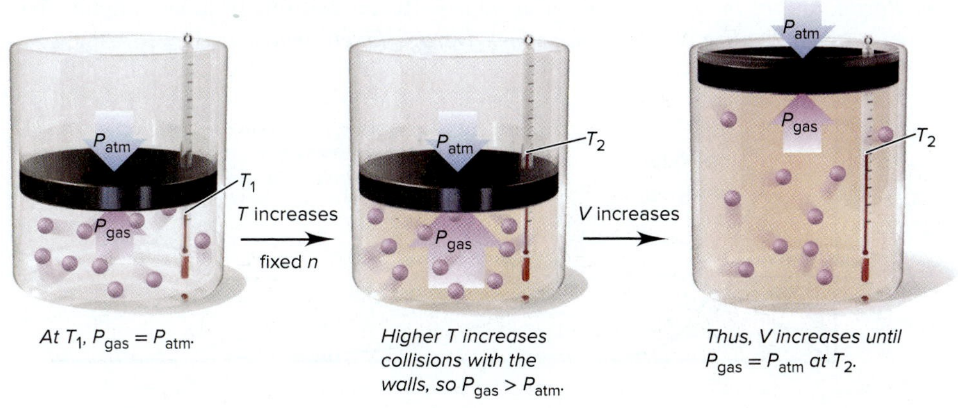
P_{atm}
T_1
P_{gas}
T increases
fixed n
P_{atm}
T_2
P_{gas}
V increases
P_{atm}
P_{gas}
T_2
At T_1, $P_{gas} = P_{atm}$.
Higher T increases collisions with the walls, so $P_{gas} > P_{atm}$.
Thus, V increases until $P_{gas} = P_{atm}$ at T_2.

Figure 5.18 A molecular view of Charles's law.

5. *Avogadro's law ($V \propto n$)* (Figure 5.19). At some starting amount, n_1, of gas, P_{atm} equals P_{gas}. When more gas is added from the attached tank, the amount increases to n_2. Thus, more particles hit the walls more frequently, which temporarily increases P_{gas}. As a result, the piston moves up, which increases the volume and lowers the number of collisions with the walls until P_{atm} and P_{gas} are again equal.

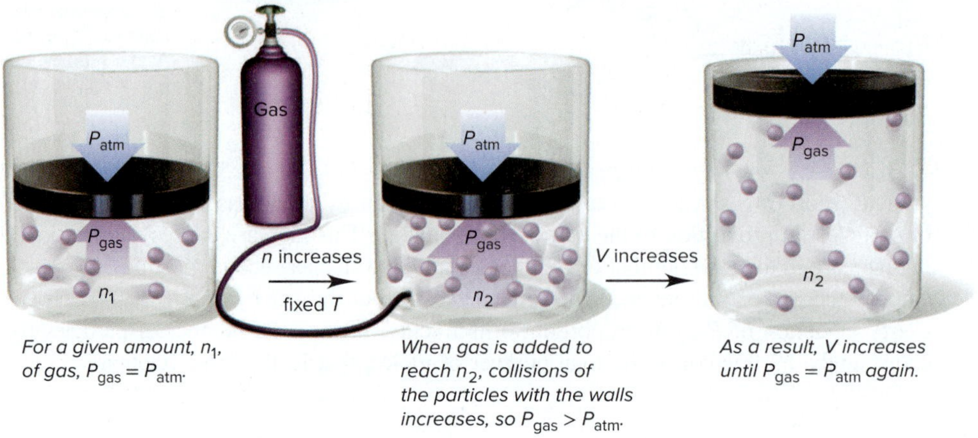
Gas
P_{atm}
P_{gas}
n_1
n increases
fixed T
n_2
V increases
For a given amount, n_1, of gas, $P_{gas} = P_{atm}$.
When gas is added to reach n_2, collisions of the particles with the walls increases, so $P_{gas} > P_{atm}$.
As a result, V increases until $P_{gas} = P_{atm}$ again.

Figure 5.19 A molecular view of Avogadro's law.

The Central Importance of Kinetic Energy Recall from Chapter 1 that the kinetic energy of an object is the energy associated with its motion. This energy is key to explaining some implications of Avogadro's law and, most importantly, the meaning of temperature.

1. *Key implication of Avogadro's law.* As we just saw, Avogadro's law says that, at any given *T* and *P*, the volume of a gas depends only on the number of moles—that is, number of particles—in the sample. Notice that the law doesn't mention the chemical nature of the gas, so equal numbers of particles of any two gases, say O_2 and H_2, should occupy the same volume. But why don't the heavier O_2 molecules, which strike the container walls with more force than the lighter H_2 molecules, exert more pressure and thus take up more volume? To answer this, we'll use three important relationships, the third of which you've already seen:

- *Molecular speed.* Figure 5.20 shows that, for several gases at a particular temperature, the most probable molecular speed (top of the curve) increases as the molar mass (number in parentheses) decreases: *lighter molecules move faster, on average, than heavier molecules.*
- *Kinetic energy.* One way to express kinetic energy mathematically is
$$E_k = \frac{1}{2}\text{ mass} \times \text{speed}^2$$
This equation means that, for a given E_k, an object's mass and speed are inversely related: *a heavy object moving slower can have the same kinetic energy as a light object moving faster.*
- *Postulate 3 of the kinetic-molecular theory.* Recall that this postulate implies that, *at a given T, all gases have the same average kinetic energy.*

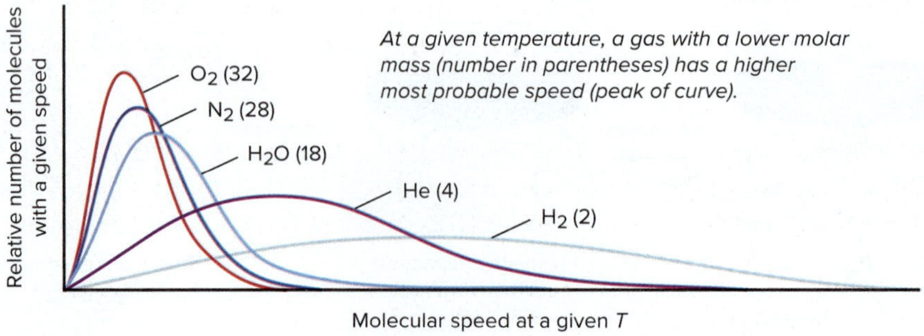
Relative number of molecules with a given speed
Molecular speed at a given T
At a given temperature, a gas with a lower molar mass (number in parentheses) has a higher most probable speed (peak of curve).
O_2 (32)
N_2 (28)
H_2O (18)
He (4)
H_2 (2)

Figure 5.20 The relationship between molar mass and molecular speed.

Their higher most probable speed means that H_2 molecules collide with the walls of a container more often than O_2 molecules do, but their lower mass means that each collision has less force. Therefore, because the molecules of any gas hit the walls with the same kinetic energy, *at a given T, equimolar samples of any gases exert the same pressure (force per unit area) and, thus, occupy the same volume.*

2. *The meaning of temperature.* Closely related to these ideas is the central relation between kinetic energy and temperature. Earlier we saw that the average kinetic energy of the particles ($\overline{E_k}$) equals the absolute temperature times a constant; that is, $\overline{E_k} = c \times T$. The definitions of velocity, momentum, force, and pressure let us express this relationship as the following equation (although we won't derive it):

$$\overline{E_k} = \tfrac{3}{2}\left(\frac{R}{N_A}\right)T$$

where R is the gas constant and N_A is the symbol for Avogadro's number. This equation makes the essential point that *temperature is a* ***measure*** *of the average kinetic energy of the particles:* as T goes up, $\overline{E_k}$ increases, and as T goes down, $\overline{E_k}$ decreases. Temperature is an intensive property (Section 1.4), so it is not related to the *total* energy of motion of the particles, which depends on the size of the sample, but to the *average* energy.

Thus, for example, in the macroscopic world, we heat a beaker of water over a flame and see the mercury rise inside a thermometer in the beaker. We see this because, in the molecular world, kinetic energy transfers, in turn, from higher energy gas particles in the flame to lower energy particles in the beaker glass, the water molecules, the particles in the thermometer glass, and the atoms of mercury.

Root-Mean-Square Speed Finally, let's derive an expression for the speed of a gas particle that has the average kinetic energy of the particles in a sample. From the general expression for kinetic energy of an object,

$$E_k = \tfrac{1}{2}\text{mass} \times \text{speed}^2$$

the average kinetic energy of each particle in a large population is

$$\overline{E_k} = \tfrac{1}{2}m\overline{u^2}$$

where m is the particle's mass (atomic or molecular) and $\overline{u^2}$ is the average of the squares of the molecular speeds. Setting this expression for average kinetic energy equal to the earlier one gives

$$\tfrac{1}{2}m\overline{u^2} = \tfrac{3}{2}\left(\frac{R}{N_A}\right)T$$

Multiplying through by Avogadro's number, N_A, gives the average kinetic energy for a mole of gas particles:

$$\tfrac{1}{2}N_A\, m\overline{u^2} = \tfrac{3}{2}RT$$

Avogadro's number times the molecular mass, $N_A \times m$, is the molar mass, $\mathcal{M}$, and solving for $\overline{u^2}$, we have

$$\overline{u^2} = \frac{3RT}{\mathcal{M}}$$

The square root of $\overline{u^2}$ is the root-mean-square speed, or **rms speed (u_{rms}):** *a particle moving at this speed has the average kinetic energy.** That is, taking the square root of both sides of the previous equation gives

$$u_{rms} = \sqrt{\frac{3RT}{\mathcal{M}}} \qquad \textbf{(5.13)}$$

where R is the gas constant, T is the absolute temperature, and $\mathcal{M}$ is the molar mass. (Because we want u in m/s and R includes the joule, which has units of kg·m^2/s^2, we use the value 8.314 J/mol·K for R and express $\mathcal{M}$ in kg/mol.)

*The rms speed, u_{rms}, is proportional to, but slightly higher than, the most probable speed; for an ideal gas, $u_{rms} = 1.09 \times$ average speed.

Thus, as an example, the root-mean-square speed of an O_2 molecule ($\mathcal{M}$ = 3.200×10^{-2} kg/mol) at room temperature (20°C, or 293 K) in the air you're breathing right now is

$$u_{\text{rms}} = \sqrt{\frac{3RT}{\mathcal{M}}} = \sqrt{\frac{3(8.314\ \text{J/mol·K})(293\ \text{K})}{3.200\times10^{-2}\ \text{kg/mol}}}$$

$$= \sqrt{\frac{3(8.314\ \cancel{\text{kg}}\cdot\text{m}^2/\text{s}^2/\cancel{\text{mol}}\cdot\cancel{\text{K}})(293\ \cancel{\text{K}})}{3.200\times10^{-2}\ \cancel{\text{kg}}/\cancel{\text{mol}}}}$$

$$= 478\ \text{m/s (about 1070 mi/hr)}$$

Effusion and Diffusion

The movement of a gas into a vacuum and the movement of gases through one another are phenomena with some vital applications.

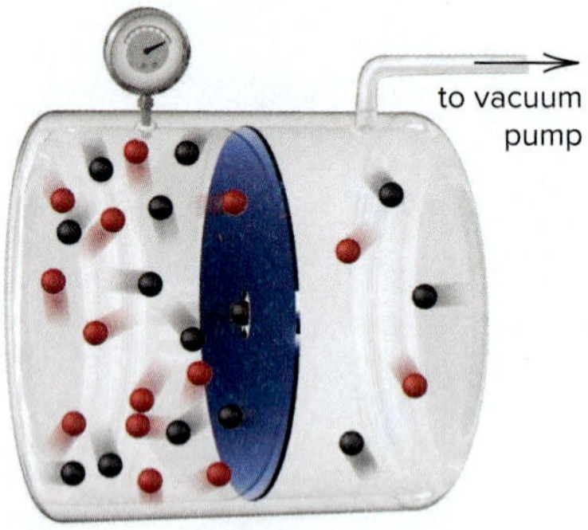

Figure 5.21 Effusion. Lighter (*black*) particles effuse faster than heavier (*red*) particles.

The Process of Effusion One of the early triumphs of the kinetic-molecular theory was an explanation of **effusion,** the process by which a gas escapes through a tiny hole in its container into an evacuated space. In 1846, Thomas Graham studied the effusion rate of a gas, the number of molecules escaping per unit time, and found that it was inversely proportional to the square root of the gas density. But, density is directly related to molar mass, so **Graham's law of effusion** is stated as follows: *the rate of effusion of a gas is inversely proportional to the square root of its molar mass,* or

$$\text{Rate of effusion} \propto \frac{1}{\sqrt{\mathcal{M}}}$$

From Equation 5.13, we see that, at a given temperature and pressure, a gas with a lower molar mass has a greater rms speed than a gas with a higher molar mass; *since the rms speed of its atoms is higher, more atoms of the lighter gas reach the hole and escape per unit time* (Figure 5.21). At constant T, the ratio of the effusion rates of two gases, A and B, is

$$\frac{\text{Rate}_\text{A}}{\text{Rate}_\text{B}} = \frac{u_{\text{rms of A}}}{u_{\text{rms of B}}} = \frac{\sqrt{3RT/\mathcal{M}_\text{A}}}{\sqrt{3RT/\mathcal{M}_\text{B}}} = \frac{\sqrt{\mathcal{M}_\text{B}}}{\sqrt{\mathcal{M}_\text{A}}} = \sqrt{\frac{\mathcal{M}_\text{B}}{\mathcal{M}_\text{A}}} \qquad \textbf{(5.14)}$$

Graham's law can be used to determine the molar mass of an unknown gas. By comparing the effusion rate of gas X with that of a known gas, such as He, we can solve for the molar mass of X:

$$\frac{\text{Rate}_\text{X}}{\text{Rate}_\text{He}} = \sqrt{\frac{\mathcal{M}_\text{He}}{\mathcal{M}_\text{X}}}$$

Squaring both sides and solving for the molar mass of X gives

$$\mathcal{M}_\text{X} = \mathcal{M}_\text{He} \times \left(\frac{\text{rate}_\text{He}}{\text{rate}_\text{X}}\right)^2$$

SAMPLE PROBLEM 5.14 Applying Graham's Law of Effusion

Problem A mixture of helium (He) and methane (CH_4) is placed in an effusion apparatus. **(a)** Calculate the ratio of the effusion rates of the two gases. **(b)** If it takes 7.55 min for a given volume of CH_4 to effuse from the apparatus, how long will it take for the same volume of He to effuse?

Plan **(a)** The effusion rate is inversely proportional to $\sqrt{\mathcal{M}}$, so we find the molar mass of each substance from the formula and take its square root. The inverse of the ratio of the square roots is the ratio of the effusion rates. **(b)** Once we know the ratio of effusion rates, we can apply that ratio to the time.

Solution (a) $\mathcal{M}$ of CH_4 = 16.04 g/mol $\mathcal{M}$ of He = 4.003 g/mol

Calculating the ratio of the effusion rates:

$$\frac{\text{Rate}_{He}}{\text{Rate}_{CH_4}} = \sqrt{\frac{\mathcal{M}_{CH_4}}{\mathcal{M}_{He}}} = \sqrt{\frac{16.04 \text{ g/mol}}{4.003 \text{ g/mol}}} = \sqrt{4.007} = 2.002$$

A ratio of 2.002 indicates that He atoms effuse a bit more than twice as fast as CH_4 molecules under the same conditions.

(b) Since He atoms effuse twice as fast as CH_4 molecules, a volume of He will effuse in half the time it takes for an equal volume of CH_4:

$$\frac{7.55 \text{ min}}{2.002} = 3.77 \text{ min}$$

Check A ratio greater than 1 makes sense because the lighter He should effuse faster than the heavier CH_4. Because the molar mass of CH_4 is about four times the molar mass of He, He should effuse about twice as fast as CH_4 ($\sqrt{4}$) and therefore take half as long for a given volume to effuse from the apparatus.

FOLLOW-UP PROBLEMS

5.14A If it takes 1.25 min for 0.010 mol of He to effuse, how long will it take for the same amount of ethane (C_2H_6) to effuse?

5.14B If 7.23 mL of an unknown gas effuses in the same amount of time as 13.8 mL of argon under the same conditions, what is the molar mass of the unknown gas?

SOME SIMILAR PROBLEMS 5.75, 5.76, 5.79, and 5.80

A Key Application of Effusion: Preparation of Nuclear Fuel By far the most important application of Graham's law is in the preparation of fuel for nuclear energy reactors. The process of *isotope enrichment* increases the proportion of fissionable, but rarer, ^{235}U (only 0.7% by mass of naturally occurring uranium) relative to the nonfissionable, more abundant ^{238}U (99.3% by mass). Because the two isotopes have identical chemical properties, they are extremely difficult to separate chemically. But, one way to separate them takes advantage of a difference in a physical property—the effusion rate of gaseous compounds. Uranium ore is treated with fluorine to yield a gaseous mixture of $^{238}UF_6$ and $^{235}UF_6$ that is pumped through a series of chambers separated by porous barriers. Molecules of $^{235}UF_6$ are slightly lighter ($\mathcal{M}$ = 349.03) than molecules of $^{238}UF_6$ ($\mathcal{M}$ = 352.04), so they move slightly faster and effuse through each barrier 1.0043 times faster. Many passes must be made, each one increasing the fraction of $^{235}UF_6$, until the mixture obtained is 3–5% by mass $^{235}UF_6$. This process was developed during the latter years of World War II and produced enough ^{235}U for two of the world's first atomic bombs. Today, a less expensive centrifuge process is used more often. The ability to enrich uranium has become a key international concern, as more countries aspire to develop nuclear energy and nuclear arms.

The Process of Diffusion Closely related to effusion is the process of gaseous **diffusion,** the movement of one gas (or fluid) through another (for example, the movement of gaseous molecules from perfume through air). Diffusion rates are also described generally by Graham's law:

$$\text{Rate of diffusion} \propto \frac{1}{\sqrt{\mathcal{M}}}$$

For two gases at equal pressures, such as NH_3 and HCl, moving through another gas or a mixture of gases, such as air, we find

$$\frac{\text{Rate}_{NH_3}}{\text{Rate}_{HCl}} = \sqrt{\frac{\mathcal{M}_{HCl}}{\mathcal{M}_{NH_3}}}$$

The reason for this dependence on molar mass is the same as for effusion rates: *lighter molecules have higher average speeds than heavier molecules, so they move farther in a given time.*

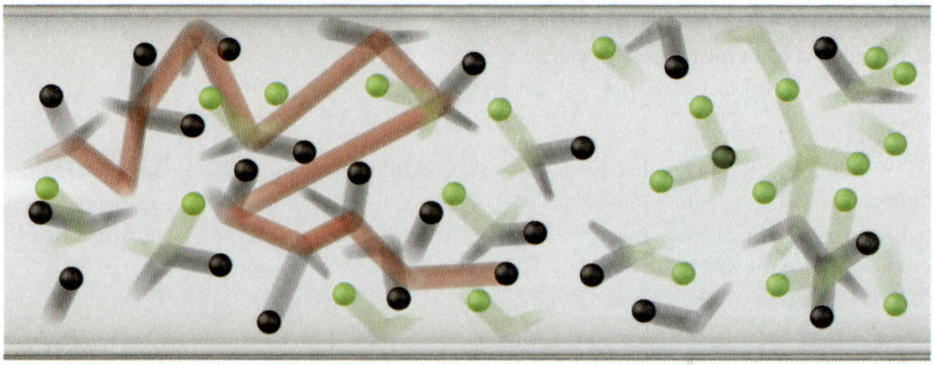

Figure 5.22 Diffusion of gases. Different gases (*black*, from the left, and *green*, from the right) move through each other in a tube and mix. For simplicity, the complex path of only one black particle is shown (in *red*). In reality, all the particles have similar paths.

If gas molecules move at hundreds of meters per second (see Figure 5.14), why does it take a second or two after you open a bottle of perfume to smell it? Although convection plays an important role in this process, another reason for the time lag is that a gas particle does not travel very far before it collides with another particle (Figure 5.22). Thus, a perfume molecule travels slowly because it collides with countless molecules in the air. The presence of so many other particles means that *diffusion rates are much lower than effusion rates.* Think about how quickly you could walk through an empty room compared to a room crowded with other moving people.

Diffusion also occurs when a gas enters a liquid (and even to a small extent a solid). However, the average distances between molecules are so much shorter in a liquid that collisions are much more frequent; thus, diffusion of a gas through a liquid is *much* slower than through a gas. Nevertheless, this type of diffusion is a vital process in biological systems, for example, in the movement of O_2 from lungs to blood.

The Chaotic World of Gases: Mean Free Path and Collision Frequency

Refinements of the basic kinetic-molecular theory provide a view into the chaotic molecular world of gases. Try to visualize an "average" N_2 molecule in the room you are in now. It is continually changing speed as it collides with other molecules—at one instant, going 2500 mi/h, and at another, standing still. But these extreme speeds are *much* less likely than the most probable one and those near it (see Figure 5.14). At 20°C and 1 atm pressure, the N_2 molecule is hurtling at an average speed of 470 m/s (rms speed = 510 m/s), or nearly 1100 mi/h!

Mean Free Path From a particle's diameter, we can obtain the **mean free path,** the average distance it travels between collisions at a given temperature and pressure. The N_2 molecule (3.7×10^{-10} m in diameter) has a mean free path of 6.6×10^{-8} m, which means it travels an average of 180 molecular diameters before smashing into a fellow traveler. (An N_2 molecule the size of a billiard ball would travel an average of about 30 ft before hitting another molecule.) Therefore, even though gas molecules are *not* points of mass, it is still valid to assume that a gas sample *is* nearly all empty space. Mean free path is a key factor in the rate of diffusion and the rate of heat flow through a gas.

Danger in a Molecular Amusement Park

To really appreciate the astounding events in the molecular world, let's use a two-dimensional analogy to compare the N_2 molecule moving in air at 1 atm with a bumper car you are driving in an enormous amusement park ride. To match the collision frequency of the N_2 molecule, you would need to be traveling 2.8 billion mi/s (4.5 billion km/s, much faster than the speed of light!) and would smash into another bumper car every 700 yd (640 m).

Collision Frequency Divide the most probable speed (meters per second) by the mean free path (meters per collision) and you obtain the **collision frequency,** the average number of collisions per second that each particle undergoes, whether with another particle or with the container. As you can see, the N_2 molecule experiences, on average, an enormous number of collisions every second:

$$\text{Collision frequency} = \frac{4.7\times10^{2}\ \text{m/s}}{6.6\times10^{-8}\ \text{m/collision}} = 7.0\times10^{9}\ \text{collision/s}$$

Distribution of speed (and kinetic energy) and collision frequency are essential ideas for understanding the speed of a reaction, as you'll see in Chapter 16. And, as the Chemical Connections essay below shows, many of the concepts we've discussed so far apply directly to our planet's atmosphere.

How the Gas Laws Apply to Earth's Atmosphere

An **atmosphere** is the envelope of gases that extends continuously from a planet's surface outward, thinning gradually until it is identical with outer space. A sample of clean, dry air at sea level on Earth contains 18 gases (Table B5.1). Under standard conditions, they behave nearly ideally, so volume percent equals mole percent (Avogadro's law), and the mole fraction of a component relates directly to its partial pressure (Dalton's law). Let's see how the gas laws and kinetic-molecular theory apply to our atmosphere, first with regard to variations in pressure and temperature, and then as explanations of some very familiar phenomena.

The Smooth Variation in Pressure with Altitude

Because gases are compressible (Boyle's law), the pressure of the atmosphere *increases* smoothly as we approach Earth's surface, with a more rapid increase at lower altitudes (Figure B5.1, *left*). No boundary delineates the beginning of the atmosphere from the end of outer space, but the densities and compositions are identical at an altitude of about 10,000 km (6000 mi). Yet, about 99% of the atmosphere's mass lies within 30 km (almost 19 mi) of the surface, and 75% lies within the lowest 11 km (almost 7 mi).

Table B5.1 Composition of Clean, Dry Air at Sea Level

Component	Mole Fraction
Nitrogen (N_2)	0.78084
Oxygen (O_2)	0.20946
Argon (Ar)	0.00934
Carbon dioxide (CO_2)	0.00040
Neon (Ne)	1.818×10^{-5}
Helium (He)	5.24×10^{-6}
Methane (CH_4)	2×10^{-6}
Krypton (Kr)	1.14×10^{-6}
Hydrogen (H_2)	5×10^{-7}
Dinitrogen monoxide (N_2O)	5×10^{-7}
Carbon monoxide (CO)	1×10^{-7}
Xenon (Xe)	8×10^{-8}
Ozone (O_3)	2×10^{-8}
Ammonia (NH_3)	6×10^{-9}
Nitrogen dioxide (NO_2)	6×10^{-9}
Nitrogen monoxide (NO)	6×10^{-10}
Sulfur dioxide (SO_2)	2×10^{-10}
Hydrogen sulfide (H_2S)	2×10^{-10}

The Zig-Zag Variation in Temperature with Altitude

Unlike pressure, temperature does *not* change smoothly with altitude above Earth's surface. The atmosphere is classified into regions based on the direction of temperature change, and we'll start at the surface (Figure B5.1, *right*).

1. *The troposphere.* In the troposphere, which extends from the surface to between 7 km (23,000 ft) at the poles and 17 km (60,000 ft) at the equator, the temperature *drops* 7°C per kilometer to −55°C (218 K). This region contains about 80% of the total mass of the atmosphere, with 50% of the mass in the lower 5.6 km (18,500 ft). All weather occurs in this region, and nearly all, except supersonic, aircraft fly within it.

(*continued*)

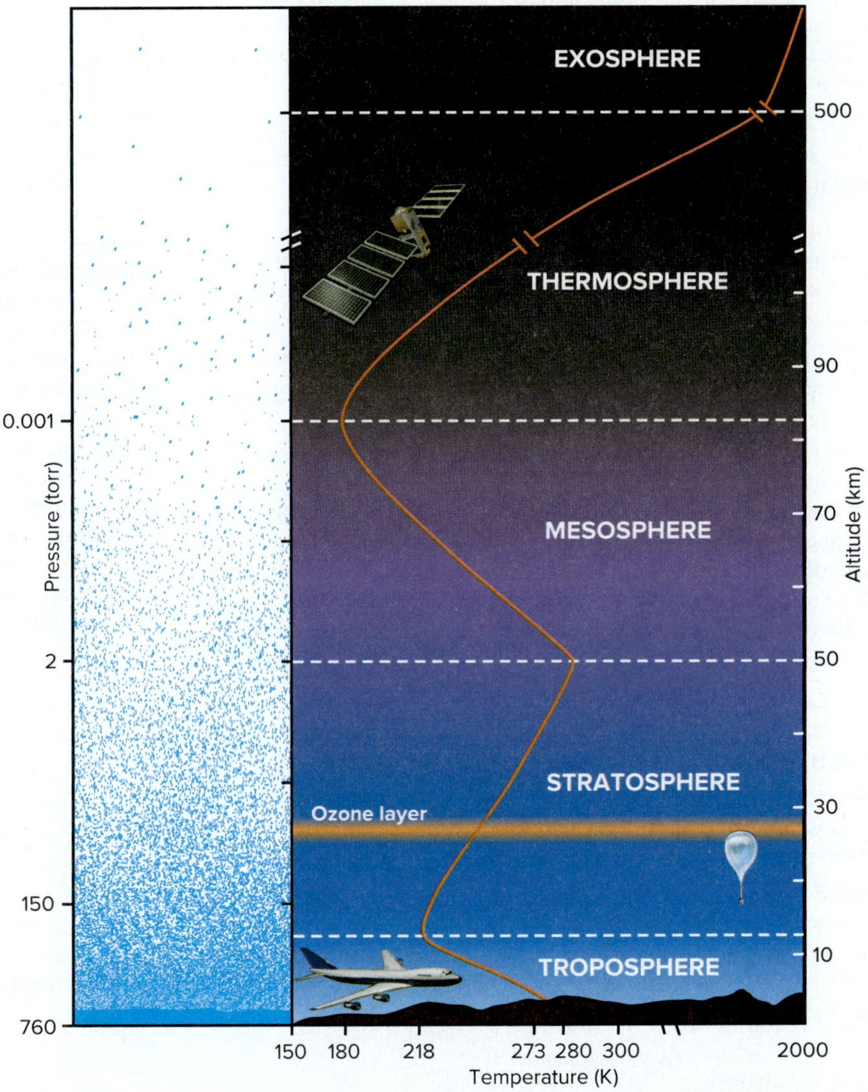

Figure B5.1 Variations in pressure and temperature with altitude in Earth's atmosphere.

2. *The stratosphere and the ozone layer.* In the stratosphere, the temperature *rises* from −55°C to about 7°C (280 K) at 50 km. This rise is due to a variety of complex reactions, mostly involving ozone (O_3), another molecular form of oxygen, which are initiated by the absorption of solar radiation. Most high-energy radiation is absorbed by upper levels of the atmosphere, but some that reaches the stratosphere breaks O_2 into O atoms. The energetic O atoms collide with more O_2 to form O_3:

$$O_2(g) \xrightarrow{\text{high-energy radiation}} 2O(g)$$
$$M + O(g) + O_2(g) \longrightarrow O_3(g) + M + \textit{heat}$$

where M is any particle that can carry away excess energy. This reaction releases heat, which is why stratospheric temperatures increase with altitude. Over 90% of atmospheric ozone remains in a thin layer within the stratosphere, its thickness varying both geographically (greatest at the poles) and seasonally (greatest in spring in the northern hemisphere and in fall in the southern). Stratospheric ozone is vital to life because it absorbs over 95% of the harmful ultraviolet (UV) solar radiation that would otherwise reach the surface. In Chapter 16, we'll discuss the destructive effect of certain industrial chemicals on the ozone layer.

3. *The mesosphere.* In the mesosphere, the temperature *drops* again to −93°C (180 K) at around 80 km.

4. *The outer atmosphere.* Within the *thermosphere,* which extends to around 500 km, the temperature *rises* again, but varies between 700 and 2000 K, depending on the intensity of solar radiation and sunspot activity. The *exosphere,* the outermost region, maintains these temperatures and merges with outer space.

What does it actually mean to have a temperature of 2000 K at 500 km (300 mi) above Earth's surface? Would a piece of iron (melting point = 1808 K) glow red-hot within a minute or so and melt in the thermosphere, as it does if heated to 2000 K in the troposphere? The answer involves the relation between temperature and the time it takes to transfer kinetic energy. Our use of the words "hot" and "cold" refers to measurements near the surface. There, the collision frequency of gas particles with a thermometer is enormous, and so the transfer of their kinetic energy is very fast. But, at an altitude of 500 km, where the density of gas particles is a million times *less,* collision frequency is extremely low, and a thermometer, or any object, experiences *very slow transfer of kinetic energy.* Thus, the object would not become "hot" in the usual sense in any reasonable time. But the high-energy solar radiation that *is* transferred to the few particles present in these regions makes their average kinetic energy extremely high, as indicated by the high absolute temperature.

Convection in the Lower Atmosphere

Why must you take more breaths per minute on a high mountaintop than at sea level? Because at the higher elevation, there is a smaller *amount* of O_2 in each breath. However, the *proportion* (mole percent) of O_2 throughout the lower atmosphere remains about 21%. This uniform composition arises from *vertical (convective) mixing,* and the gas laws explain how it occurs.

Let's follow an air mass from ground level, as solar heating of Earth's surface warms it. The warmer air mass expands (Charles's law), which makes it less dense, so it rises. As it does so, the pressure on it decreases, making it expand further (Boyle's law). Pushing against the surrounding air requires energy, so the temperature of the air mass decreases and thus the air mass shrinks slightly (Charles's law). But, as its temperature drops, its water vapor condenses (or solidifies), and these changes of state release heat; therefore, the air mass becomes warmer and rises further. Meanwhile, the cooler, and thus denser, air that was above it sinks, becomes warmer through contact with the surface, and goes through the same process as the first air mass. As a result of this vertical mixing, the composition of the lower atmosphere remains uniform.

Warm air rising from the ground, called a *thermal,* is used by soaring birds and glider pilots to stay aloft. Convection helps clean the air in urban areas, because the rising air carries up pollutants, which are dispersed by winds. Under certain conditions, however, a warm air mass remains stationary over a cool one. The resulting *temperature inversion* blocks normal convection, and harmful pollutants build up, causing severe health problems.

Problems

B5.1 Suggest a reason why supersonic aircraft are kept well below their maximum speeds until they reach their highest altitudes.

B5.2 Gases behave nearly ideally under Earth's conditions. Elsewhere in the Solar System, however, conditions are very different. On which planet would you expect atmospheric gases to deviate most from ideal behavior, Saturn (4×10^6 atm and 130 K) or Venus (90 atm and 730 K)? Explain.

B5.3 What is the volume percent and partial pressure (in torr) of argon in a sample of dry air at sea level?

B5.4 Earth's atmosphere is estimated to have a mass of 5.14×10^{15} t (1 t = 1000 kg).
(a) If the average molar mass of air is 28.8 g/mol, how many moles of gas are in the atmosphere?
(b) How many liters would the atmosphere occupy at 25°C and 1 atm?

› Summary of Section 5.5

- › The kinetic-molecular theory postulates that gas particles have no volume, move in random, straight-line paths between elastic (energy-conserving) collisions, and have average kinetic energies proportional to the absolute temperature of the gas.
- › This theory explains the gas laws in terms of changes in distances between particles and the container walls, changes in molecular speed, and the energy of collisions.
- › Temperature is a measure of the average kinetic energy of the particles.
- › Effusion and diffusion rates are *inversely* proportional to the square root of the molar mass (Graham's law) because they are *directly* proportional to molecular speed.
- › Molecular motion is characterized by a temperature-dependent, most probable speed (within a range of speeds), which affects mean free path and collision frequency.
- › The atmosphere is a complex mixture of gases that exhibits variations in pressure and temperature with altitude. High temperatures in the upper atmosphere result from absorption of high-energy solar radiation. The lower atmosphere has a uniform composition as a result of convective mixing.

5.6 REAL GASES: DEVIATIONS FROM IDEAL BEHAVIOR

A fundamental principle of science is that simpler models are more useful than complex ones, as long as they explain the data. With only a few postulates, the kinetic-molecular theory explains the behavior of most gases under ordinary conditions of temperature and pressure. But, two of the postulates are useful approximations that do not reflect reality:

1. *Gas particles are* **not** *points of mass with zero volume* but have volumes determined by the sizes of their atoms and the lengths and directions of their bonds.
2. *Attractive and repulsive forces* **do** *exist among gas particles* because atoms contain charged subatomic particles and many substances are polar. (As you'll see in Chapter 12, such forces lead to changes of physical state.)

These two features cause deviations from ideal behavior under *extreme conditions of low temperature and high pressure.* These deviations mean that we must alter the simple model and the ideal gas law to predict the behavior of real gases.

Effects of Extreme Conditions on Gas Behavior

At ordinary conditions—relatively high temperatures and low pressures—most real gases exhibit nearly ideal behavior. Yet, even at STP (0°C and 1 atm), gases deviate *slightly* from ideal behavior. Table 5.3 shows that the standard molar volumes of several gases, when measured to five significant figures, do not equal the ideal value of 22.414 L/mol. Note that gases with higher boiling points exhibit greater deviations.

The phenomena that cause slight deviations under standard conditions exert more influence as temperature decreases and pressure increases. Figure 5.23 on the next page shows a plot of PV/RT versus external pressure (P_{ext}) for 1 mol of several real gases and an ideal gas. The PV/RT values range from normal (at $P_{ext} = 1$ atm, $PV/RT = 1$) to very high (at $P_{ext} \approx 1000$ atm, $PV/RT \approx 1.6$ to 2.3). For the *ideal* gas, PV/RT is 1 at any P_{ext}.

The PV/RT curve for methane (CH_4) is typical of most gases: it decreases *below* the ideal value at moderately high P_{ext} and then rises *above* the ideal value as P_{ext} increases to very high values. This shape arises from two overlapping effects:

- At moderately high P_{ext}, PV/RT values are lower than ideal values (less than 1) because of *interparticle attractions.*
- At very high P_{ext}, PV/RT values are greater than ideal values (more than 1) because of *particle volume.*

Table 5.3 Molar Volume of Some Common Gases at STP (0°C and 1 atm)

Gas	Molar Volume (L/mol)	Boiling Point (°C)
He	22.435	−268.9
H_2	22.432	−252.8
Ne	22.422	−246.1
Ideal gas	**22.414**	—
Ar	22.397	−185.9
N_2	22.396	−195.8
O_2	22.390	−183.0
CO	22.388	−191.5
Cl_2	22.184	−34.0
NH_3	22.079	−33.4

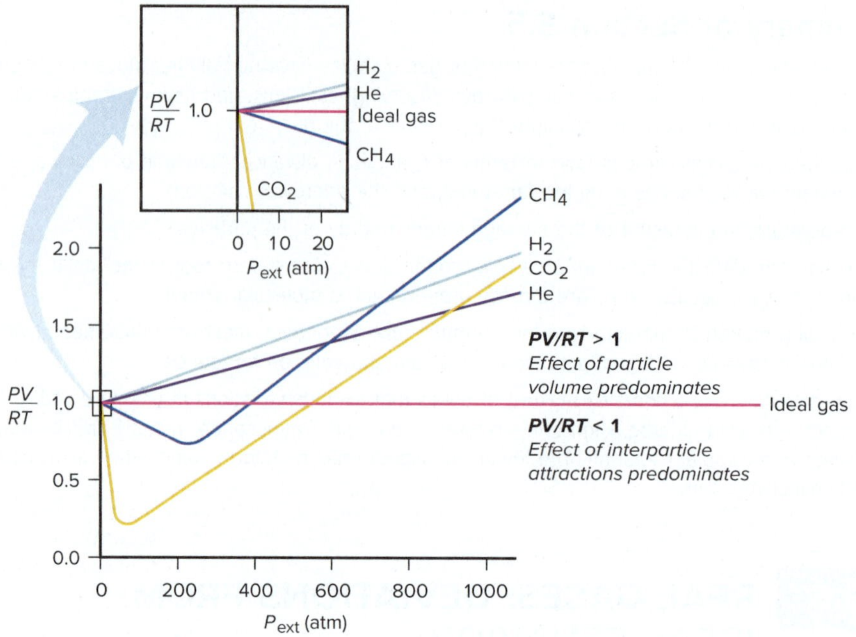

Figure 5.23 Deviations from ideal behavior with increasing external pressure. The horizontal line shows that, for 1 mol of ideal gas, $PV/RT = 1$ at all P_{ext}. At very high P_{ext}, real gases deviate significantly from ideal behavior, but small deviations appear even at ordinary pressures (expanded portion).

Let's examine these effects on the molecular level:

1. *Effect of interparticle attractions.* Interparticle attractions occur between separate atoms or molecules and are caused by imbalances in electron distributions. They are important only over *very* short distances and are *much* weaker than the covalent bonding forces that hold a molecule together. At normal P_{ext}, the spaces between the particles are so large that attractions between the gas particles are negligible and the gas behaves nearly ideally. As conditions of pressure and temperature become more extreme, however, the interparticle attractions become more important:

- *High pressure.* As P_{ext} rises, the volume of the sample decreases and the particles get closer together, so interparticle attractions have a greater effect. As a particle approaches the container wall under these higher pressures, nearby particles attract it, which lessens the force of its impact (Figure 5.24). *Repeated throughout the sample, this effect results in decreased gas pressure and, thus, a smaller numerator in PV/RT.*
- *Low temperature.* Lowering the temperature slows the particles, so they attract each other for a longer time. (As you'll see in Chapter 12, given a low enough temperature,

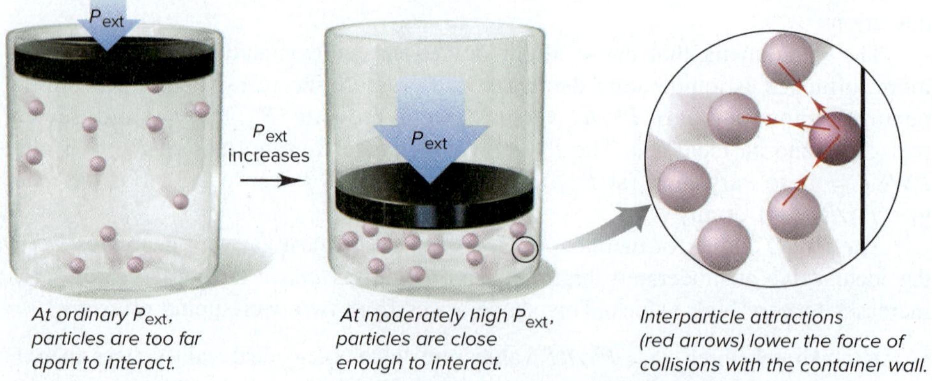

Figure 5.24 The effect of interparticle attractions on measured gas pressure.

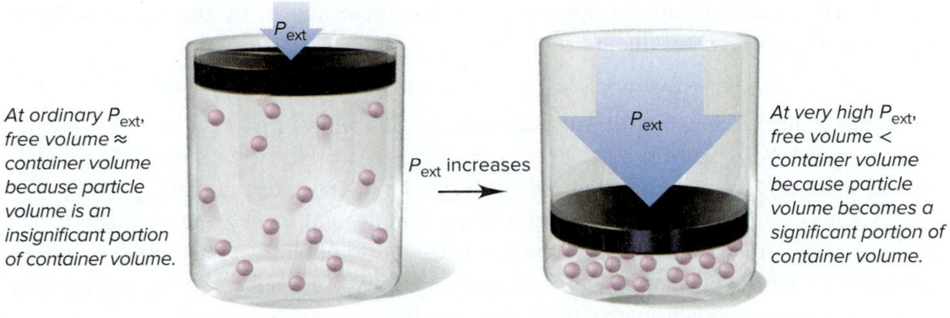
At ordinary P_{ext}, free volume ≈ container volume because particle volume is an insignificant portion of container volume.
P_{ext}
P_{ext} increases
P_{ext}
At very high P_{ext}, free volume < container volume because particle volume becomes a significant portion of container volume.

Figure 5.25 The effect of particle volume on measured gas volume.

the interparticle attractions cause the gas particles to lose enough kinetic energy that they condense to a liquid.)

2. *Effect of particle volume.* At normal P_{ext}, the space between particles (free volume) is enormous compared with the volume of the particles *themselves* (particle volume); thus, the free volume is essentially equal to V, the container volume in PV/RT. At *moderately* high P_{ext} and as free volume decreases, the particle volume makes up an increasing proportion of the container volume (Figure 5.25). At *extremely* high pressures, the space taken up by the particles themselves makes the free volume significantly *less* than the container volume. Nevertheless, we continue to use the container volume for V in PV/RT, which causes the numerator, and thus the ratio, to become artificially high. This particle volume effect increases as P_{ext} increases, eventually outweighing the effect of interparticle attractions and causing PV/RT to rise above the ideal value.

In Figure 5.23, note that the H_2 and He curves do not show the typical dip at moderate pressures. These gases consist of particles with such weak interparticle attractions that the particle volume effect predominates at all pressures.

The van der Waals Equation: Adjusting the Ideal Gas Law

To describe real gas behavior more accurately, we need to adjust the ideal gas equation in two ways:

1. Adjust P *up* by adding a factor that accounts for interparticle attractions.
2. Adjust V *down* by subtracting a factor that accounts for particle volume.

In 1873, Johannes van der Waals revised the ideal gas equation to account for the behavior of real gases. The **van der Waals equation** for n moles of a real gas is

$$\underbrace{\left(P + \frac{n^2a}{V^2}\right)}_{\text{adjusts } P \text{ up}}\underbrace{(V - nb)}_{\text{adjusts } V \text{ down}} = nRT \qquad (5.15)$$

where P is the measured pressure, V is the known container volume, n and T have their usual meanings, and a and b are **van der Waals constants,** experimentally determined and specific for a given gas (Table 5.4). The constant a depends on the number and distribution of electrons, which relates to the complexity of a particle and the strength of its interparticle attractions. The constant b relates to particle volume. For instance, CO_2 is both more complex and larger than H_2, and the values of their constants reflect this.

Here is a typical application of the van der Waals equation. A 1.98-L vessel contains 215 g (4.89 mol) of dry ice. After standing at 26°C (299 K), the $CO_2(s)$ changes to $CO_2(g)$. The pressure is measured (P_{real}) and then calculated by the ideal

Table 5.4 Values of the van der Waals Constants for Some Common Gases

Gas	$a\left(\frac{\text{atm·L}^2}{\text{mol}^2}\right)$	$b\left(\frac{\text{L}}{\text{mol}}\right)$
He	0.034	0.0237
Ne	0.211	0.0171
Ar	1.35	0.0322
Kr	2.32	0.0398
Xe	4.19	0.0511
H_2	0.244	0.0266
N_2	1.39	0.0391
O_2	1.36	0.0318
Cl_2	6.49	0.0562
CH_4	2.25	0.0428
CO	1.45	0.0395
CO_2	3.59	0.0427
NH_3	4.17	0.0371
H_2O	5.46	0.0305

gas law (P_{IGL}) and, using the appropriate values of a and b, by the van der Waals equation (P_{VDW}). The results are revealing:

$$P_{real} = 44.8 \text{ atm} \qquad P_{IGL} = 60.6 \text{ atm} \qquad P_{VDW} = 45.9 \text{ atm}$$

Comparing the real value with each calculated value shows that P_{IGL} is 35.3% greater than P_{real}, but P_{VDW} is only 2.5% greater than P_{real}. At these conditions, CO_2 deviates so much from ideal behavior that the ideal gas law is not very useful.

A key point to realize: According to kinetic-molecular theory, the constants a and b are zero for an ideal gas because the gas particles do not attract each other and have no volume. Yet, even in a real gas at ordinary pressures, the particles are very far apart. This large average interparticle distance has two consequences:

- Attractive forces are miniscule, so $P + \frac{n^2a}{V^2} \approx P$.
- The particle volume is a minute fraction of the container volume, so $V - nb \approx V$

Therefore, *at ordinary conditions, the van der Waals equation* ***becomes*** *the ideal gas equation.*

Student Hot Spot

Student data indicate that you may struggle with the van der Waals equation. Access the Smartbook to view additional Learning Resources on this topic.

Summary of Section 5.6

- At very high P or low T, all gases deviate significantly from ideal behavior.
- As external pressure increases, most real gases exhibit first a lower and then a higher PV/RT; for 1 mol of an ideal gas, this ratio remains constant at 1.
- The deviations from ideal behavior are due to (1) attractions between particles, which lower the pressure (and decrease PV/RT), and (2) the volume of the particles themselves, which takes up an increasingly larger fraction of the container volume (and increases PV/RT).
- The van der Waals equation includes constants specific for a given gas to correct for deviations from ideal behavior. At ordinary P and T, the van der Waals equation reduces to the ideal gas equation.

CHAPTER REVIEW GUIDE

Learning Objectives

Relevant section (§) and/or sample problem (SP) numbers appear in parentheses.

Understand These Concepts

1. How gases differ in their macroscopic properties from liquids and solids (§5.1)
2. The meaning of pressure and the operation of a barometer and a manometer (§5.2)
3. The relations among gas variables expressed by Boyle's, Charles's, and Avogadro's laws (§5.3)
4. How the individual gas laws are incorporated into the ideal gas law (§5.3)
5. How the ideal gas law can be used to study gas density, molar mass, and amounts of gases in reactions (§5.4)
6. The relation between the density and the temperature and pressure of a gas (§5.4)
7. The meaning of Dalton's law and the relation between partial pressure and mole fraction of a gas; how Dalton's law applies to collecting a gas over water (§5.4)
8. How the postulates of the kinetic-molecular theory are applied to explain the origin of pressure and the gas laws (§5.5)
9. The relations among molecular speed, average kinetic energy, and temperature (§5.5)
10. The meanings of *effusion* and *diffusion* and how their rates are related to molar mass (§5.5)
11. The relations among mean free path, molecular speed, and collision frequency (§5.5)
12. Why interparticle attractions and particle volume cause gases to deviate from ideal behavior at low temperatures and high pressures (§5.6)
13. How the van der Waals equation corrects the ideal gas law for extreme conditions (§5.6)

Master These Skills

1. Interconverting among the units of pressure (atm, mmHg, torr, kPa) (SP 5.1)
2. Reducing the ideal gas law to the individual gas laws (SPs 5.2–5.6)
3. Applying gas laws to choose the correct balanced chemical equation (SP 5.7)
4. Rearranging the ideal gas law to calculate gas density (SP 5.8) and molar mass of a volatile liquid (SP 5.9)
5. Calculating the mole fraction and the partial pressure of a gas in a mixture (SP 5.10)
6. Using the vapor pressure of water to correct for the amount of a gas collected over water (SP 5.11)
7. Applying stoichiometry and gas laws to calculate amounts of reactants and products (SPs 5.12, 5.13)
8. Using Graham's law to solve problems involving gaseous effusion (SP 5.14)

Key Terms

Page numbers appear in parentheses.

atmosphere (239)
Avogadro's law (214)
barometer (208)
Boyle's law (212)
Charles's law (213)
collision frequency (238)
Dalton's law of partial pressures (225)
diffusion (237)
effusion (236)
Graham's law of effusion (236)
ideal gas (211)
ideal gas law (216)
kinetic-molecular theory (231)
manometer (208)
mean free path (238)
millimeter of mercury (mmHg) (209)
mole fraction (X) (226)
partial pressure (225)
pascal (Pa) (209)
pressure (P) (207)
rms speed (u_{rms}) (235)
standard atmosphere (atm) (209)
standard molar volume (215)
standard temperature and pressure (STP) (215)
torr (209)
universal gas constant (R) (216)
van der Waals constants (243)
van der Waals equation (243)

Key Equations and Relationships

Page numbers appear in parentheses.

5.1 Expressing the volume-pressure relationship (Boyle's law) (212):

$$V \propto \frac{1}{P} \quad \text{or} \quad PV = \text{constant} \quad [T \text{ and } n \text{ fixed}]$$

5.2 Expressing the volume-temperature relationship (Charles's law) (213):

$$V \propto T \quad \text{or} \quad \frac{V}{T} = \text{constant} \quad [P \text{ and } n \text{ fixed}]$$

5.3 Expressing the pressure-temperature relationship (Amontons's law) (213):

$$P \propto T \quad \text{or} \quad \frac{P}{T} = \text{constant} \quad [V \text{ and } n \text{ fixed}]$$

5.4 Expressing the volume-amount relationship (Avogadro's law) (214):

$$V \propto n \quad \text{or} \quad \frac{V}{n} = \text{constant} \quad [P \text{ and } T \text{ fixed}]$$

5.5 Defining standard temperature and pressure (215):

STP: 0°C (273.15 K) and 1 atm (760 torr)

5.6 Defining the volume of 1 mol of an ideal gas at STP (215):

Standard molar volume = 22.4141 L or 22.4 L [to 3 sf]

5.7 Relating volume to pressure, temperature, and amount (ideal gas law) (216):

$$PV = nRT \quad \text{and} \quad \frac{P_1V_1}{n_1T_1} = \frac{P_2V_2}{n_2T_2}$$

5.8 Calculating the value of R (216):

$$R = \frac{PV}{nT} = \frac{1 \text{ atm} \times 22.4141 \text{ L}}{1 \text{ mol} \times 273.15 \text{ K}}$$

$$= 0.082058 \frac{\text{atm·L}}{\text{mol·K}} = 0.0821 \frac{\text{atm·L}}{\text{mol·K}} \quad [3 \text{ sf}]$$

5.9 Rearranging the ideal gas law to find gas density (222):

$$PV = \frac{m}{\mathcal{M}} RT \quad \text{so} \quad \frac{m}{V} = d = \frac{P \times \mathcal{M}}{RT}$$

5.10 Rearranging the ideal gas law to find molar mass (224):

$$n = \frac{m}{\mathcal{M}} = \frac{PV}{RT} \quad \text{so} \quad \mathcal{M} = \frac{mRT}{PV}$$

5.11 Relating the total pressure of a gas mixture to the partial pressures of the components (Dalton's law of partial pressures) (226):

$$P_{total} = P_1 + P_2 + P_3 + \cdots$$

5.12 Relating partial pressure to mole fraction (226):

$$P_A = X_A \times P_{total}$$

5.13 Defining rms speed as a function of molar mass and temperature (235):

$$u_{rms} = \sqrt{\frac{3RT}{\mathcal{M}}}$$

5.14 Applying Graham's law of effusion (236):

$$\frac{\text{Rate}_A}{\text{Rate}_B} = \frac{\sqrt{\mathcal{M}_B}}{\sqrt{\mathcal{M}_A}} = \sqrt{\frac{\mathcal{M}_B}{\mathcal{M}_A}}$$

5.15 Applying the van der Waals equation to find the pressure or volume of a gas under extreme conditions (243):

$$\left(P + \frac{n^2a}{V^2}\right)(V - nb) = nRT$$

BRIEF SOLUTIONS TO FOLLOW-UP PROBLEMS

5.1A P_{CO_2} (torr) $= P_{atm} - \Delta h$

$$= (753.6 \text{ mmHg} - 174.0 \text{ mmHg}) \times \frac{1 \text{ torr}}{1 \text{ mmHg}}$$

$$= 579.6 \text{ torr}$$

$$P_{CO_2} (\text{Pa}) = 579.6 \text{ torr} \times \frac{1 \text{ atm}}{760 \text{ torr}} \times \frac{1.01325 \times 10^5 \text{ Pa}}{1 \text{ atm}}$$

$$= 7.727 \times 10^4 \text{ Pa}$$

$$P_{CO_2} (\text{lb/in}^2) = 579.6 \text{ torr} \times \frac{1 \text{ atm}}{760 \text{ torr}} \times \frac{14.7 \text{ lb/in}^2}{1 \text{ atm}}$$

$$= 11.2 \text{ lb/in}^2$$

5.1B P_{atm} (torr) $= 0.9475 \text{ atm} \times \frac{760 \text{ torr}}{1 \text{ atm}} = 720.1$ torr

$$P_{CO_2} (\text{mmHg}) = P_{atm} + \Delta h$$

$$= (720.1 \text{ torr} + 25.8 \text{ torr}) \times \frac{1 \text{ mmHg}}{1 \text{ torr}}$$

$$= 745.9 \text{ mmHg}$$

$$P_{CO_2} (\text{Pa}) = 745.9 \text{ mmHg} \times \frac{1 \text{ atm}}{760 \text{ mmHg}} \times \frac{1.01325 \times 10^5 \text{ Pa}}{1 \text{ atm}}$$

$$= 9.945 \times 10^4 \text{ Pa}$$

$$P_{CO_2} (\text{lb/in}^2) = 745.9 \text{ mmHg} \times \frac{1 \text{ atm}}{760 \text{ mmHg}} \times \frac{14.7 \text{ lb/in}^2}{1 \text{ atm}}$$

$$= 14.4 \text{ lb/in}^2$$

BRIEF SOLUTIONS TO FOLLOW-UP PROBLEMS (continued)

5.2A $P_2 = P_1 \times \frac{V_1}{V_2}$

$$P_2\ (\text{atm}) = 0.871\ \text{atm} \times \frac{105\ \text{mL}}{352\ \text{mL}} = 0.260\ \text{atm}$$

$$P_2\ (\text{kPa}) = 0.260\ \text{atm} \times \frac{101.325\ \text{kPa}}{1\ \text{atm}} = 26.3\ \text{kPa}$$

5.2B $P_2\ (\text{atm}) = 745\ \text{mmHg} \times \frac{1\ \text{atm}}{760\ \text{mmHg}} = 0.980\ \text{atm}$

$$V_2 = V_1 \times \frac{P_1}{P_2}$$

$$V_2\ (\text{L}) = 651\ \text{L} \times \frac{122\ \text{atm}}{0.980\ \text{atm}} = 8.10\times10^4\ \text{L}$$

5.3A $P_1\ (\text{torr}) = 0.991\ \text{atm} \times \frac{760\ \text{torr}}{1\ \text{atm}} = 753\ \text{torr}$

$$P_2 = P_1 \times \frac{T_2}{T_1} = 753\ \text{torr} \times \frac{373\ \text{K}}{296\ \text{K}} = 949\ \text{torr}$$

$P_2 < 1.00\times10^3$ torr, so the safety valve will not open.

5.3B $T_2\ (\text{K}) = T_1 \times \frac{V_2}{V_1} = 313\ \text{K} \times \frac{28.6\ \text{L}}{32.5\ \text{L}} = 275\ \text{K}$

$$T_2\ (°\text{C}) = 275\ \text{K} - 273.15 = 2°\text{C}$$

5.4A $P_2\ (\text{torr}) = P_1 \times \frac{n_2}{n_1} = 793\ \text{torr} \times \frac{35.0\ \text{g} - 5.0\ \text{g}}{35.0\ \text{g}} = 680.\ \text{torr}$

(There is no need to convert mass to moles because the ratio of masses equals the ratio of moles.)

5.4B Amount (mol) of $N_2 = 1.26\ \text{g}\ N_2 \times \frac{1\ \text{mol}\ N_2}{28.02\ \text{g}\ N_2}$

$$= 0.0450\ \text{mol}\ N_2$$

$$\text{Amount (mol) of He} = 1.26\ \text{g He} \times \frac{1\ \text{mol He}}{4.003\ \text{g He}} = 0.315\ \text{mol He}$$

$$V_2 = V_1 \times \frac{n_2}{n_1} = 1.12\ \text{L} \times \frac{0.0450\ \text{mol} + 0.315\ \text{mol}}{0.0450\ \text{mol}}$$

$$= 8.96\ \text{L}$$

5.5A

$$P_2\ (\text{mmHg}) = P_1 \times \frac{V_1T_2}{V_2T_1} = 755\ \text{mmHg} \times \frac{(2.55\ \text{L})(291\ \text{K})}{(4.10\ \text{L})(296\ \text{K})}$$

$$= 462\ \text{mmHg}$$

5.5B $V_2\ (\text{L}) = V_1 \times \frac{P_1T_2}{P_2T_1} = 2.2\ \text{L} \times \frac{(0.980\ \text{atm})(294\ \text{K})}{(1.40\ \text{atm})(301\ \text{K})}$

$$= 1.5\ \text{L}$$

5.6A $n = \frac{PV}{RT} = \frac{1.37\ \text{atm} \times 438\ \text{L}}{0.0821\ \frac{\text{atm·L}}{\text{mol·K}} \times 294\ \text{K}} = 24.9\ \text{mol}\ O_2$

$$\text{Mass (g) of}\ O_2 = 24.9\ \text{mol}\ O_2 \times \frac{32.00\ \text{g}\ O_2}{1\ \text{mol}\ O_2} = 7.97\times10^2\ \text{g}\ O_2$$

5.6B

$$\text{Amount (mol) of He} = 3950\ \text{kg He} \times \frac{1000\ \text{g}}{1\ \text{kg}} \times \frac{1\ \text{mol He}}{4.003\ \text{g He}}$$

$$= 9.87\times10^5\ \text{mol He}$$

$$V = \frac{nRT}{P} = \frac{9.87\times10^5\ \text{mol} \times 0.0821\ \frac{\text{atm·L}}{\text{mol·K}} \times 293\ \text{K}}{731\ \text{mmHg} \times \frac{1\ \text{atm}}{760\ \text{mmHg}}}$$

$$= 2.47\times10^7\ \text{L}$$

5.7A The balanced equation is $2CD(g) \longrightarrow C_2(g) + D_2(g)$, so n does not change. The volume has doubled. Therefore, given constant P, the temperature, T, must also double: $T_1 = -73°\text{C} + 273.15 = 200\ \text{K}$; so $T_2 = 400\ \text{K}$, or $400\ \text{K} - 273.15 = 127°\text{C}$.

5.7B $T_1 = 199°\text{C} + 273.15 = 472\ \text{K}$
$T_2 = -155°\text{C} + 273.15 = 118\ \text{K}$

Temperature has been reduced by $\frac{1}{4}$; if n were constant, V would also be reduced by $\frac{1}{4}$. But, since V has been halved, n must be doubled. The number of moles of gas doubles (from 2 mol to 4 mol) in equation (2).

5.8A $d\ (\text{at}\ 0°\text{C and 380 torr}) = \frac{\frac{380\ \text{torr}}{760\ \text{torr/atm}} \times 44.01\ \text{g/mol}}{0.0821\ \frac{\text{atm·L}}{\text{mol·K}} \times 273\ \text{K}}$

$$= 0.982\ \text{g/L}$$

The density is lower at the smaller P because V is larger. In this case, d is lowered by one-half because P is one-half as much.

5.8B $d\ (\text{at}\ 24°\text{C and 0.950 atm}) = \frac{0.950\ \text{atm} \times 46.01\ \text{g/mol}}{0.0821\ \frac{\text{atm·L}}{\text{mol·K}} \times 297\ \text{K}}$

$$= 1.79\ \text{g/L}$$

NO_2 is more dense than dry air.

5.9A Mass of gas $= 68.697\ \text{g} - 68.322\ \text{g} = 0.375\ \text{g}$

$$\mathcal{M} = \frac{mRT}{PV} = \frac{0.375\ \text{g} \times 0.0821\ \frac{\text{atm·L}}{\text{mol·K}} \times (273.15 + 95.0)\ \text{K}}{\frac{740.\ \text{torr}}{760\ \text{torr/atm}} \times \frac{149\ \text{mL}}{1000\ \text{mL/L}}}$$

$$= 78.1\ \text{g/mol}$$

5.9B Mass of gas $= 82.786\ \text{g} - 82.561\ \text{g} = 0.225\ \text{g}$

$$\mathcal{M} = \frac{mRT}{PV} = \frac{0.225\ \text{g} \times 0.0821\ \frac{\text{atm·L}}{\text{mol·K}} \times (273.15 + 22)\ \text{K}}{\frac{733\ \text{mmHg}}{760\ \text{mmHg/atm}} \times \frac{350.\ \text{mL}}{1000\ \text{mL/L}}}$$

$$= 16.1\ \text{g/mol}$$

This is the molar mass of CH_4.

5.10A $n_{\text{He}} = 5.50\ \text{g He} \times \frac{1\ \text{mol He}}{4.003\ \text{g He}} = 1.37\ \text{mol He}$

Similarly, there are 0.743 mol of Ne and 0.418 mol Kr.

$$n_{total} = 1.37 \text{ mol He} + 0.743 \text{ mol Ne} + 0.418 \text{ mol Kr} = 2.53 \text{ mol}$$

$$P_{He} = X_{He} \times P_{total} = \frac{1.37 \text{ mol He}}{2.53 \text{ mol gas}} \times 1 \text{ atm} = 0.542 \text{ atm}$$

$P_{Ne} = 0.294$ atm $\quad P_{Kr} = 0.165$ atm

5.10B $X_{He} = \frac{P_{He}}{P_{total}} = \frac{143 \text{ atm}}{204 \text{ atm}} = 0.701$

Mole % He = 0.701 × 100 = 70.1%

5.11A P_{H_2} (torr) $= P_{total} - P_{H_2O} = 752$ torr $- 13.6$ torr $= 738$ torr

$$P_{H_2} \text{ (atm)} = 738 \text{ torr} \times \frac{1 \text{ atm}}{760 \text{ torr}} = 0.971 \text{ atm}$$

$$n_{H_2} = \frac{PV}{RT} = \frac{0.971 \text{ atm} \times 1.495 \text{ L}}{0.0821 \frac{\text{atm·L}}{\text{mol·K}} \times 289 \text{ K}} = 0.0612 \text{ mol } H_2$$

$$\text{Mass (g) of } H_2 = 0.0612 \text{ mol } H_2 \times \frac{2.016 \text{ g } H_2}{1 \text{ mol } H_2} = 0.123 \text{ g } H_2$$

5.11B $P_{O_2} = P_{total} - P_{H_2O} = 748$ torr $- 17.5$ torr $= 730.$ torr

$$P_{O_2} \text{ (atm)} = 730. \text{ torr} \times \frac{1 \text{ atm}}{760 \text{ torr}} = 0.961 \text{ atm}$$

$$n_{O_2} = \frac{PV}{RT} = \frac{0.961 \text{ atm} \times 0.307 \text{ L}}{0.0821 \frac{\text{atm·L}}{\text{mol·K}} \times 293 \text{ K}} = 0.0123 \text{ mol } O_2$$

$$\text{Mass (g) of } O_2 = 0.0123 \text{ mol } O_2 \times \frac{32.00 \text{ g } O_2}{1 \text{ mol } O_2} = 0.394 \text{ g } O_2$$

5.12A $H_2SO_4(aq) + 2NaCl(s) \longrightarrow Na_2SO_4(aq) + 2HCl(g)$

$$n_{HCl} = 0.117 \text{ kg NaCl} \times \frac{10^3 \text{ g}}{1 \text{ kg}} \times \frac{1 \text{ mol NaCl}}{58.44 \text{ g NaCl}} \times \frac{2 \text{ mol HCl}}{2 \text{ mol NaCl}} = 2.00 \text{ mol HCl}$$

$$\text{At STP, } V \text{ (mL)} = 2.00 \text{ mol} \times \frac{22.4 \text{ L}}{1 \text{ mol}} \times \frac{10^3 \text{ mL}}{1 \text{ L}} = 4.48{\times}10^4 \text{ mL}$$

5.12B $CuO(s) + H_2(g) \longrightarrow Cu(s) + H_2O(g)$

$$n_{H_2} = 35.5 \text{ g CuO} \times \frac{1 \text{ mol CuO}}{79.55 \text{ g CuO}} \times \frac{1 \text{ mol } H_2}{1 \text{ mol CuO}} = 0.446 \text{ mol } H_2$$

$$V = \frac{nRT}{P} = \frac{0.446 \text{ mol} \times 0.0821 \frac{\text{atm·L}}{\text{mol·K}} \times 498 \text{ K}}{765 \text{ torr} \times \frac{1 \text{ atm}}{760 \text{ torr}}} = 18.1 \text{ L}$$

5.13A $NH_3(g) + HCl(g) \longrightarrow NH_4Cl(s)$

$$n_{NH_3} = \frac{PV}{RT} = \frac{0.452 \text{ atm} \times 10.0 \text{ L}}{0.0821 \frac{\text{atm·L}}{\text{mol·K}} \times 295 \text{ K}} = 0.187 \text{ mol } NH_3$$

$$n_{HCl} = \frac{PV}{RT} = \frac{7.50 \text{ atm} \times 0.155 \text{ L}}{0.0821 \frac{\text{atm·L}}{\text{mol·K}} \times 271 \text{ K}} = 0.0522 \text{ mol HCl}$$

$n_{NH_3} = 0.187$ mol and $n_{HCl} = 0.0522$ mol; thus, HCl is the limiting reactant.

n_{NH_3} after reaction

$$= 0.187 \text{ mol } NH_3 - \left(0.0522 \text{ mol HCl} \times \frac{1 \text{ mol } NH_3}{1 \text{ mol HCl}}\right) = 0.135 \text{ mol } NH_3$$

$$P = \frac{nRT}{V} = \frac{0.135 \text{ mol} \times 0.0821 \frac{\text{atm·L}}{\text{mol·K}} \times 295 \text{ K}}{10.0 \text{ L}} = 0.327 \text{ atm}$$

5.13B $I_2(s) + 5F_2(g) \longrightarrow 2IF_5(g)$

For I_2:

$$\text{Amount (mol) of } IF_5 = 4.16 \text{ g } I_2 \times \frac{1 \text{ mol } I_2}{253.8 \text{ g } I_2} \times \frac{2 \text{ mol } IF_5}{1 \text{ mol } I_2} = 0.0328 \text{ mol } IF_5$$

For F_2:

$$n_{F_2} = \frac{PV}{RT} = \frac{0.974 \text{ atm} \times 2.48 \text{ L}}{0.0821 \frac{\text{atm·L}}{\text{mol·K}} \times 291 \text{ K}} = 0.101 \text{ mol } F_2$$

$$\text{Amount (mol) of } IF_5 = 0.101 \text{ mol } F_2 \times \frac{2 \text{ mol } IF_5}{5 \text{ mol } F_2} = 0.0404 \text{ mol } IF_5$$

Thus, I_2 is the limiting reactant.

$$\text{Volume (L) of } IF_5 = \frac{nRT}{P} = \frac{0.0328 \text{ mol} \times 0.0821 \frac{\text{atm·L}}{\text{mol·K}} \times 378 \text{ K}}{0.935 \text{ atm}} = 1.09 \text{ L}$$

5.14A $\frac{\text{Rate of He}}{\text{Rate of } C_2H_6} = \sqrt{\frac{30.07 \text{ g/mol}}{4.003 \text{ g/mol}}} = 2.741$

Time for C_2H_6 to effuse = 1.25 min × 2.741 = 3.43 min

5.14B $\mathcal{M}$ of unknown gas $= \mathcal{M}_{Ar} \times \left(\frac{\text{rate of Ar}}{\text{rate of unknown gas}}\right)^2$

$$= \left(\frac{13.8 \text{ mL/time}}{7.23 \text{ mL/time}}\right)^2 \times 39.95 \text{ g/mol} = 146 \text{ g/mol}$$

PROBLEMS

Problems with **colored** numbers are answered in Appendix E and worked in detail in the Student Solutions Manual. Problem sections match those in the text and give the numbers of relevant sample problems. Most offer Concept Review Questions, Skill-Building Exercises (grouped in pairs covering the same concept), and Problems in Context. The Comprehensive Problems are based on material from any section or previous chapter.

An Overview of the Physical States of Matter

Concept Review Questions

5.1 How does a sample of gas differ in its behavior from a sample of liquid in each of the following situations?
(a) The sample is transferred from one container to a larger one.
(b) The sample is heated in an expandable container, but no change of state occurs.
(c) The sample is placed in a cylinder with a piston, and an external force is applied.

5.2 Are the particles in a gas farther apart or closer together than the particles in a liquid? Use your answer to explain each of the following general observations:
(a) Gases are more compressible than liquids.
(b) Gases have lower viscosities than liquids.
(c) After thorough stirring, all gas mixtures are solutions.
(d) The density of a substance in the gas state is lower than in the liquid state.

Gas Pressure and Its Measurement

(Sample Problem 5.1)

Concept Review Questions

5.3 How does a barometer work? Is the column of mercury in a barometer shorter when it is on a mountaintop or at sea level? Explain.

5.4 How can a unit of length such as millimeter of mercury (mmHg) be used as a unit of pressure, which has the dimensions of force per unit area?

5.5 In a closed-end manometer, the mercury level in the arm attached to the flask can never be higher than the mercury level in the other arm, whereas in an open-end manometer, it *can* be higher. Explain.

Skill-Building Exercises (grouped in similar pairs)

5.6 On a cool, rainy day, the barometric pressure is 730 mmHg. Calculate the barometric pressure in centimeters of water (cmH_2O) (d of Hg = 13.5 g/mL; d of H_2O = 1.00 g/mL).

5.7 A long glass tube, sealed at one end, has an inner diameter of 10.0 mm. The tube is filled with water and inverted into a pail of water. If the atmospheric pressure is 755 mmHg, how high (in mmH_2O) is the column of water in the tube (d of Hg = 13.5 g/mL; d of H_2O = 1.00 g/mL)?

5.8 Convert the following:
(a) 0.745 atm to mmHg
(b) 992 torr to bar
(c) 365 kPa to atm
(d) 804 mmHg to kPa

5.9 Convert the following:
(a) 76.8 cmHg to atm
(b) 27.5 atm to kPa
(c) 6.50 atm to bar
(d) 0.937 kPa to torr

5.10 In Figure P5.10, what is the pressure of the gas in the flask (in atm) if the barometer reads 738.5 torr?

5.11 In Figure P5.11, what is the pressure of the gas in the flask (in kPa) if the barometer reads 765.2 mmHg?

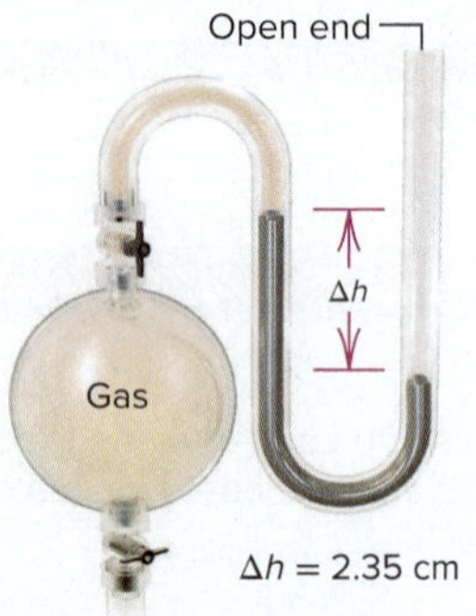

Figure P5.10

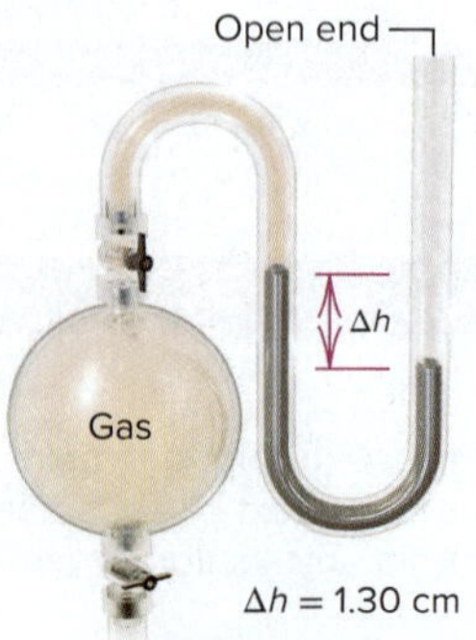

Figure P5.11

5.12 If the sample flask in Figure P5.12 is open to the air, what is the atmospheric pressure (in atm)?

5.13 In Figure P5.13, what is the pressure (in Pa) of the gas in the flask?

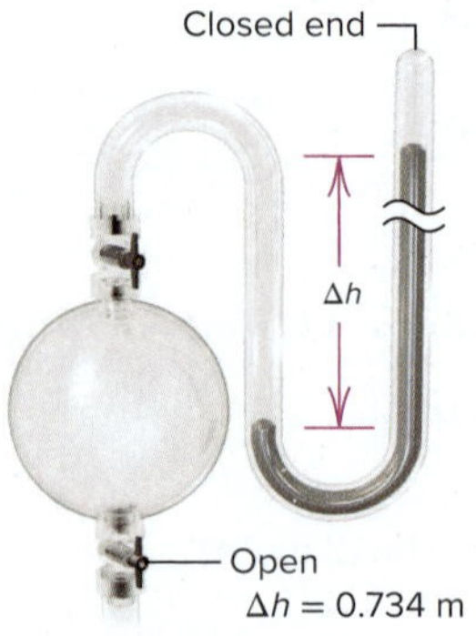

Figure P5.12

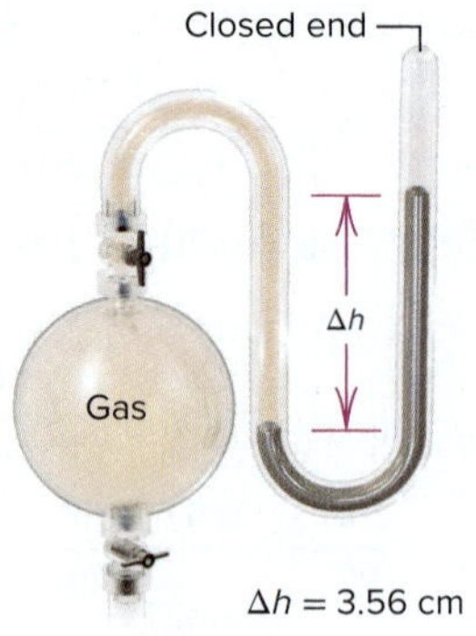

Figure P5.13

Problems in Context

5.14 Convert each of the pressures described below to atm:
(a) At the peak of Mt. Everest, atmospheric pressure is only 2.75×10^2 mmHg.
(b) A cyclist fills her bike tires to 86 psi.
(c) The surface of Venus has an atmospheric pressure of 9.15×10^6 Pa.
(d) At 100 ft below sea level, a scuba diver experiences a pressure of 2.54×10^4 torr.

5.15 The gravitational force exerted by an object is given by $F = mg$, where F is the force in newtons, m is the mass in kilograms, and g is the acceleration due to gravity (9.81 m/s^2).
(a) Use the definition of the pascal to calculate the mass (in kg) of the atmosphere above 1 m^2 of ocean.
(b) Osmium ($Z = 76$) is a transition metal in Group 8B(8) and has the highest density of any element (22.6 g/mL). If an osmium column is 1 m^2 in area, how high must it be for its pressure to equal atmospheric pressure? [Use the answer from part (a) in your calculation.]

The Gas Laws and Their Experimental Foundations

(Sample Problems 5.2 to 5.7)

Concept Review Questions

5.16 A student states Boyle's law as follows: "The volume of a gas is inversely proportional to its pressure." How is this statement incomplete? Give a correct statement of Boyle's law.

5.17 In the following relationships, which quantities are variables and which are fixed: (a) Charles's law; (b) Avogadro's law; (c) Amontons's law?

5.18 Boyle's law relates gas volume to pressure, and Avogadro's law relates gas volume to amount (mol). State a relationship between gas pressure and amount (mol).

5.19 Each of the following processes caused the gas volume to double, as shown. For each process, tell how the remaining gas variable changed or state that it remained fixed:
(a) T doubles at fixed P.
(b) T and n are fixed.

(c) At fixed T, the reaction is $CD_2(g) \longrightarrow C(g) + D_2(g)$.
(d) At fixed P, the reaction is $A_2(g) + B_2(g) \longrightarrow 2AB(g)$.

Skill-Building Exercises (grouped in similar pairs)

5.20 What is the effect of the following on the volume of 1 mol of an ideal gas?
(a) The pressure is tripled (at constant T).
(b) The absolute temperature is increased by a factor of 3.0 (at constant P).
(c) Three more moles of the gas are added (at constant P and T).

5.21 What is the effect of the following on the volume of 1 mol of an ideal gas?
(a) The pressure changes from 760 torr to 202 kPa, and the temperature changes from 37°C to 155 K.
(b) The temperature changes from 305 K to 32°C, and the pressure changes from 2 atm to 101 kPa.
(c) The pressure is reduced by a factor of 4 (at constant T).

5.22 What is the effect of the following on the volume of 1 mol of an ideal gas?
(a) Temperature decreases from 800 K to 400 K (at constant P).
(b) Temperature increases from 250°C to 500°C (at constant P).
(c) Pressure increases from 2 atm to 6 atm (at constant T).

5.23 What is the effect of the following on the volume of 1 mol of an ideal gas?
(a) The initial pressure is 722 torr, and the final pressure is 0.950 atm; the initial temperature is 32°F, and the final temperature is 273 K.
(b) Half the gas escapes (at constant P and T).
(c) Both the pressure and temperature decrease to one-fourth of their initial values.

5.24 A weather balloon is filled with helium to a volume of 1.61 L at 734 torr. What is the volume of the balloon after it has been released and its pressure has dropped to 0.844 atm? Assume that the temperature remains constant.

5.25 A sample of methane is placed in a 10.0-L container at 25°C and 725 mmHg. The gas sample is then moved to a 7.50-L container at 25°C. What is the gas pressure in the second container?

5.26 A sample of sulfur hexafluoride gas occupies 9.10 L at 198°C. Assuming that the pressure remains constant, what temperature (in °C) is needed to reduce the volume to 2.50 L?

5.27 A 93-L sample of dry air cools from 145°C to −22°C while the pressure is maintained at 2.85 atm. What is the final volume?

5.28 A gas cylinder is filled with argon at a pressure of 177 atm and 25°C. What is the gas pressure when the temperature of the cylinder and its contents are heated to 195°C by exposure to fire?

5.29 A bicycle tire is filled to a pressure of 110. psi at a temperature of 30.0°C. At what temperature will the air pressure in the tire decrease to 105 psi? Assume that the volume of the tire remains constant.

5.30 A balloon filled with 1.92 g of helium has a volume of 12.5 L. What is the balloon's volume after 0.850 g of helium has leaked out through a small hole (assume constant pressure and temperature)?

5.31 The average person takes 500 mL of air into the lungs with each normal inhalation, which corresponds to approximately 1×10^{22} molecules of air. Calculate the number of molecules of air inhaled by a person with a respiratory problem who takes in only 350 mL of air with each breath. Assume constant pressure and temperature.

5.32 A sample of Freon-12 (CF_2Cl_2) occupies 25.5 L at 298 K and 153.3 kPa. Find its volume at STP.

5.33 A sample of carbon monoxide occupies 3.65 L at 298 K and 745 torr. Find its volume at −14°C and 367 torr.

5.34 A sample of chlorine gas is confined in a 5.0-L container at 328 torr and 37°C. How many moles of gas are in the sample?

5.35 If 1.47×10^{-3} mol of argon occupies a 75.0-mL container at 26°C, what is the pressure (in torr)?

5.36 You have 357 mL of chlorine trifluoride gas at 699 mmHg and 45°C. What is the mass (in g) of the sample?

5.37 A 75.0-g sample of dinitrogen monoxide is confined in a 3.1-L vessel. What is the pressure (in atm) at 115°C?

Problems in Context

5.38 In preparation for a demonstration, your professor brings a 1.5-L bottle of sulfur dioxide into the lecture hall before class to allow the gas to reach room temperature. If the pressure gauge reads 85 psi and the temperature in the classroom is 23°C, how many moles of sulfur dioxide are in the bottle? (*Hint:* The gauge reads zero when the gas pressure in the bottle is 14.7 psi.)

5.39 A gas-filled weather balloon with a volume of 65.0 L is released at sea-level conditions of 745 torr and 25°C. The balloon can expand to a maximum volume of 835 L. When the balloon rises to an altitude at which the temperature is −5°C and the pressure is 0.066 atm, will it have expanded to its maximum volume?

Rearrangements of the Ideal Gas Law
(Sample Problems 5.8 to 5.13)

Concept Review Questions

5.40 Why is moist air less dense than dry air?

5.41 To collect a beaker of H_2 gas by displacing the air already in the beaker, would you hold the beaker upright or inverted? Why? How would you hold the beaker to collect CO_2?

5.42 Why can we use a gas mixture, such as air, to study the general behavior of an ideal gas under ordinary conditions?

5.43 How does the partial pressure of gas A in a mixture compare to its mole fraction in the mixture? Explain.

5.44 The scene at right represents a portion of a mixture of four gases A (*purple*), B (*black*), C (*green*), and D_2 (*orange*). (a) Which gas has the highest partial pressure? (b) Which has the lowest partial pressure? (c) If the total pressure is 0.75 atm, what is the partial pressure of D_2?

Skill-Building Exercises (grouped in similar pairs)

5.45 What is the density of Xe gas at STP?

5.46 Find the density of Freon-11 ($CFCl_3$) at 120°C and 1.5 atm.

5.47 How many moles of gaseous arsine (AsH_3) occupy 0.0400 L at STP? What is the density of gaseous arsine?

5.48 The density of a noble gas is 2.71 g/L at 3.00 atm and 0°C. Identify the gas.

5.49 Calculate the molar mass of a gas at 388 torr and 45°C if 206 ng occupies 0.206 μL.

5.50 When an evacuated 63.8-mL glass bulb is filled with a gas at 22°C and 747 mmHg, the bulb gains 0.103 g in mass. Is the gas N_2, Ne, or Ar?

5.51 After 0.600 L of Ar at 1.20 atm and 227°C is mixed with 0.200 L of O_2 at 501 torr and 127°C in a 400-mL flask at 27°C, what is the pressure in the flask?

5.52 A 355-mL container holds 0.146 g of Ne and an unknown amount of Ar at 35°C and a total pressure of 626 mmHg. Calculate the number of moles of Ar present.

5.53 How many grams of phosphorus react with 35.5 L of O_2 at STP to form tetraphosphorus decoxide?

$$P_4(s) + 5O_2(g) \longrightarrow P_4O_{10}(s)$$

5.54 How many grams of potassium chlorate decompose to potassium chloride and 638 mL of O_2 at 128°C and 752 torr?

$$2KClO_3(s) \longrightarrow 2KCl(s) + 3O_2(g)$$

5.55 How many grams of phosphine (PH_3) can form when 37.5 g of phosphorus and 83.0 L of hydrogen gas react at STP?

$$P_4(s) + H_2(g) \longrightarrow PH_3(g) \quad \text{[unbalanced]}$$

5.56 When 35.6 L of ammonia and 40.5 L of oxygen gas at STP burn, nitrogen monoxide and water form. After the products return to STP, how many grams of nitrogen monoxide are present?

$$NH_3(g) + O_2(g) \longrightarrow NO(g) + H_2O(l) \quad \text{[unbalanced]}$$

5.57 Aluminum reacts with excess hydrochloric acid to form aqueous aluminum chloride and 35.8 mL of hydrogen gas over water at 27°C and 751 mmHg. How many grams of aluminum reacted?

5.58 How many liters of hydrogen gas are collected over water at 18°C and 725 mmHg when 0.84 g of lithium reacts with water? Aqueous lithium hydroxide also forms.

Problems in Context

5.59 The air in a hot-air balloon at 744 torr is heated from 17°C to 60.0°C. Assuming that the amount (mol) of air and the pressure remain constant, what is the density of the air at each temperature? (The average molar mass of air is 28.8 g/mol.)

5.60 On a certain winter day in Utah, the average atmospheric pressure is 650. torr. What is the molar density (in mol/L) of the air if the temperature is −25°C?

5.61 A sample of a liquid hydrocarbon known to consist of molecules with five carbon atoms is vaporized in a 0.204-L flask by immersion in a water bath at 101°C. The barometric pressure is 767 torr, and the remaining gas weighs 0.482 g. What is the molecular formula of the hydrocarbon?

5.62 A sample of air contains 78.08% nitrogen, 20.94% oxygen, 0.05% carbon dioxide, and 0.93% argon, by volume. How many molecules of each gas are present in 1.00 L of the sample at 25°C and 1.00 atm?

5.63 An environmental chemist sampling industrial exhaust gases from a coal-burning plant collects a CO_2-SO_2-H_2O mixture in a 21-L steel tank until the pressure reaches 850. torr at 45°C.
(a) How many moles of gas are collected?
(b) If the SO_2 concentration in the mixture is 7.95×10^3 parts per million by volume (ppmv), what is its partial pressure? [*Hint:* ppmv = (volume of component/volume of mixture) × 10^6.]

5.64 "Strike anywhere" matches contain the compound tetraphosphorus trisulfide, which burns to form tetraphosphorus decoxide and sulfur dioxide gas. How many milliliters of sulfur dioxide, measured at 725 torr and 32°C, can be produced from burning 0.800 g of tetraphosphorus trisulfide?

5.65 Freon-12 (CF_2Cl_2), widely used as a refrigerant and aerosol propellant, is a dangerous air pollutant. In the troposphere, it traps heat 25 times as effectively as CO_2, and in the stratosphere, it participates in the breakdown of ozone. Freon-12 is prepared industrially by reaction of gaseous carbon tetrachloride with hydrogen fluoride. Hydrogen chloride gas also forms. How many grams of carbon tetrachloride are required for the production of 16.0 dm^3 of Freon-12 at 27°C and 1.20 atm?

5.66 Xenon hexafluoride was one of the first noble gas compounds synthesized. The solid reacts rapidly with the silicon dioxide in glass or quartz containers to form liquid $XeOF_4$ and gaseous silicon tetrafluoride. What is the pressure in a 1.00-L container at 25°C after 2.00 g of xenon hexafluoride reacts? (Assume that silicon tetrafluoride is the only gas present and that it occupies the entire volume.)

5.67 In the four piston-cylinder assemblies below, the reactant in the left cylinder is about to undergo a reaction at constant T and P:

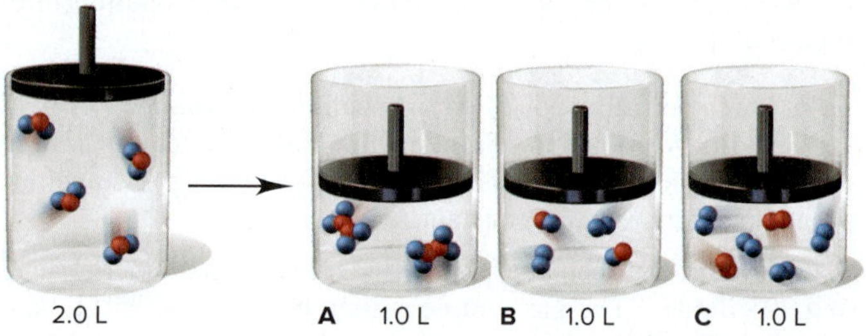

Which of the other three depictions best represents the products of the reaction?

5.68 Roasting galena [lead(II) sulfide] is a step in the industrial isolation of lead. How many liters of sulfur dioxide, measured at STP, are produced by the reaction of 3.75 kg of galena with 228 L of oxygen gas at 220°C and 2.0 atm? Lead(II) oxide also forms.

5.69 In one of his most critical studies on the nature of combustion, Lavoisier heated mercury(II) oxide and isolated elemental mercury and oxygen gas. If 40.0 g of mercury(II) oxide is heated in a 502-mL vessel and 20.0% (by mass) decomposes, what is the pressure (in atm) of the oxygen that forms at 25.0°C? (Assume that the gas occupies the entire volume.)

The Kinetic-Molecular Theory: A Model for Gas Behavior

(Sample Problem 5.14)

Concept Review Questions

5.70 Use the kinetic-molecular theory to explain the change in gas pressure that results from warming a sample of gas.

5.71 How does the kinetic-molecular theory explain why 1 mol of krypton and 1 mol of helium have the same volume at STP?

5.72 Is the rate of effusion of a gas higher than, lower than, or equal to its rate of diffusion? Explain. For two gases with molecules of approximately the same size, is the ratio of their effusion rates higher than, lower than, or equal to the ratio of their diffusion rates? Explain.

5.73 Consider two 1-L samples of gas: one is H_2 and the other is O_2. Both are at 1 atm and 25°C. How do the samples compare in terms of (a) mass, (b) density, (c) mean free path, (d) average molecular kinetic energy, (e) average molecular speed, and (f) time for a given fraction of molecules to effuse?

5.74 Three 5-L flasks, fixed with pressure gauges and small valves, each contain 4 g of gas at 273 K. Flask A contains H_2, flask B contains He, and flask C contains CH_4. Rank the flask contents in terms of (a) pressure, (b) average kinetic energy of the particles, (c) diffusion rate after the valve is opened, (d) total kinetic energy of the particles, (e) density, and (f) collision frequency.

Skill-Building Exercises (grouped in similar pairs)

5.75 What is the ratio of effusion rates for the lightest gas, H_2, and the heaviest known gas, UF_6?

5.76 What is the ratio of effusion rates for O_2 and Kr?

5.77 The graph below shows the distribution of molecular speeds for argon and helium at the same temperature.

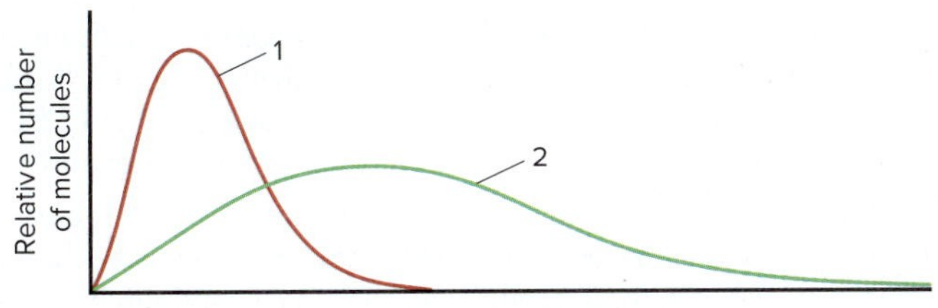

(a) Does curve 1 or 2 better represent the behavior of argon?
(b) Which curve represents the gas that effuses more slowly?
(c) Which curve more closely represents the behavior of fluorine gas? Explain.

5.78 The graph below shows the distribution of molecular speeds for a gas at two different temperatures.

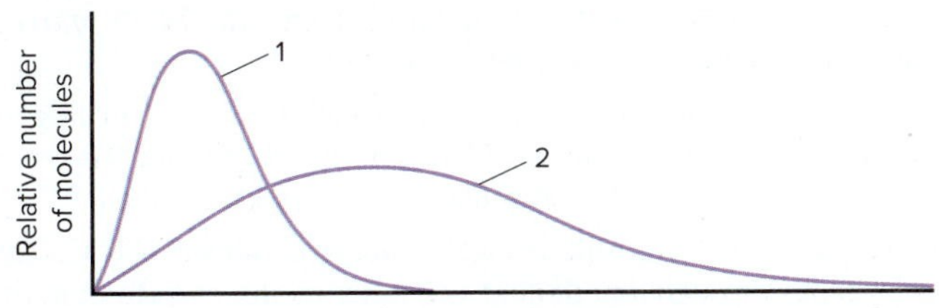

(a) Does curve 1 or 2 better represent the behavior of the gas at the lower temperature?
(b) Which curve represents the gas when it has a higher $\overline{E_k}$?
(c) Which curve is consistent with a higher diffusion rate?

5.79 At a given pressure and temperature, it takes 4.85 min for a 1.5-L sample of He to effuse through a membrane. How long does it take for 1.5 L of F_2 to effuse under the same conditions?

5.80 A sample of an unknown gas effuses in 11.1 min. An equal volume of H_2 in the same apparatus under the same conditions effuses in 2.42 min. What is the molar mass of the unknown gas?

Problems in Context

5.81 White phosphorus melts and then vaporizes at high temperatures. The gas effuses at a rate that is 0.404 times that of neon in the same apparatus under the same conditions. How many atoms are in a molecule of gaseous white phosphorus?

5.82 Helium (He) is the lightest noble gas component of air, and xenon (Xe) is the heaviest. [For this problem, use $R = 8.314$ J/(mol·K) and express $\mathcal{M}$ in kg/mol.]
(a) Find the rms speed of He in winter (0.°C) and in summer (30.°C).
(b) Compare the rms speed of He with that of Xe at 30.°C.
(c) Find the average kinetic energy per mole of He and of Xe at 30.°C.
(d) Find the average kinetic energy per molecule of He at 30.°C.

5.83 A mixture of gaseous disulfur difluoride, dinitrogen tetrafluoride, and sulfur tetrafluoride is placed in an effusion apparatus.
(a) Rank the gases in order of increasing effusion rate.
(b) Find the ratio of effusion rates of disulfur difluoride and dinitrogen tetrafluoride.
(c) If gas X is added, and it effuses at 0.935 times the rate of sulfur tetrafluoride, find the molar mass of X.

Real Gases: Deviations from Ideal Behavior

Skill-Building Exercises (grouped in similar pairs)

5.84 Do interparticle attractions cause negative or positive deviations from the *PV/RT* ratio of an ideal gas? Use Table 5.3 to rank Kr, CO_2, and N_2 in order of increasing magnitude of these deviations.

5.85 Does particle volume cause negative or positive deviations from the *PV/RT* ratio of an ideal gas? Use Table 5.3 to rank Cl_2, H_2, and O_2 in order of increasing magnitude of these deviations.

5.86 Does N_2 behave more ideally at 1 atm or at 500 atm? Explain.

5.87 Does SF_6 (boiling point = 16°C at 1 atm) behave more ideally at 150°C or at 20°C? Explain.

Comprehensive Problems

5.88 An "empty" gasoline can with dimensions 15.0 cm by 40.0 cm by 12.5 cm is attached to a vacuum pump and evacuated. If the atmospheric pressure is 14.7 lb/in^2, what is the total force (in pounds) on the outside of the can?

5.89 Hemoglobin is the protein that transports O_2 through the blood from the lungs to the rest of the body. To do so, each molecule of hemoglobin combines with four molecules of O_2. If 1.00 g of hemoglobin combines with 1.53 mL of O_2 at 37°C and 743 torr, what is the molar mass of hemoglobin?

5.90 A baker uses sodium hydrogen carbonate (baking soda) as the leavening agent in a banana-nut quickbread. The baking soda decomposes in either of two possible reactions:

(1) $2NaHCO_3(s) \longrightarrow Na_2CO_3(s) + H_2O(l) + CO_2(g)$
(2) $NaHCO_3(s) + H^+(aq) \longrightarrow H_2O(l) + CO_2(g) + Na^+(aq)$

Calculate the volume (in mL) of CO_2 that forms at 200.°C and 0.975 atm per gram of $NaHCO_3$ by each of the reaction processes.

5.91 A weather balloon containing 600. L of He is released near the equator at 1.01 atm and 305 K. It rises to a point where conditions are 0.489 atm and 218 K and eventually lands in the northern hemisphere under conditions of 1.01 atm and 250 K. If one-fourth of the helium leaked out during this journey, what is the volume (in L) of the balloon at landing?

5.92 Chlorine is produced from sodium chloride by the electrochemical chlor-alkali process. During the process, the chlorine is collected in a container that is isolated from the other products to prevent unwanted (and explosive) reactions. If a 15.50-L container holds 0.5950 kg of Cl_2 gas at 225°C, calculate:
(a) P_{IGL}
(b) P_{VDW} $\left(\text{use } R = 0.08206\ \frac{\text{atm·L}}{\text{mol·K}}\right)$

5.93 In a certain experiment, magnesium boride (Mg_3B_2) reacted with acid to form a mixture of four boron hydrides (B_xH_y), three as liquids (labeled I, II, and III) and one as a gas (IV).
(a) When a 0.1000-g sample of each liquid was transferred to an evacuated 750.0-mL container and volatilized at 70.00°C, sample I had a pressure of 0.05951 atm; sample II, 0.07045 atm; and sample III, 0.05767 atm. What is the molar mass of each liquid?
(b) Boron was determined to be 85.63% by mass in sample I, 81.10% in II, and 82.98% in III. What is the molecular formula of each sample?
(c) Sample IV was found to be 78.14% boron. The rate of effusion for this gas was compared to that of sulfur dioxide; under identical conditions, 350.0 mL of sample IV effused in 12.00 min and 250.0 mL of sulfur dioxide effused in 13.04 min. What is the molecular formula of sample IV?

5.94 Three equal volumes of gas mixtures, all at the same T, are depicted below (with gas A *red*, gas B *green*, and gas C *blue*):

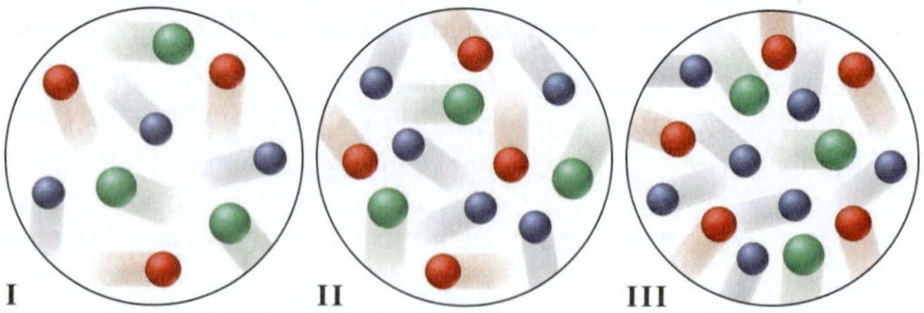

I II III

(a) Which sample, if any, has the highest partial pressure of A?
(b) Which sample, if any, has the lowest partial pressure of B?
(c) In which sample, if any, do the gas particles have the highest average kinetic energy?

5.95 Will the volume of a gas increase, decrease, or remain unchanged with each of the following sets of changes?
(a) The pressure is decreased from 2 atm to 1 atm, while the temperature is decreased from 200°C to 100°C.
(b) The pressure is increased from 1 atm to 3 atm, while the temperature is increased from 100°C to 300°C.
(c) The pressure is increased from 3 atm to 6 atm, while the temperature is increased from −73°C to 127°C.
(d) The pressure is increased from 0.2 atm to 0.4 atm, while the temperature is decreased from 300°C to 150°C.

5.96 When air is inhaled, it enters the alveoli of the lungs, and varying amounts of the component gases exchange with dissolved gases in the blood. The resulting alveolar gas mixture is quite different from the atmospheric mixture. The following table presents selected data on the composition and partial pressure of four gases in the atmosphere and in the alveoli:

	Atmosphere (sea level)		Alveoli	
Gas	**Mole %**	**Partial Pressure (torr)**	**Mole %**	**Partial Pressure (torr)**
N_2	78.6	—	—	569
O_2	20.9	—	—	104
CO_2	00.04	—	—	40
H_2O	00.46	—	—	47

If the total pressure of each gas mixture is 1.00 atm, calculate:
(a) The partial pressure (in torr) of each gas in the atmosphere
(b) The mole % of each gas in the alveoli
(c) The number of O_2 molecules in 0.50 L of alveolar air (volume of an average breath of a person at rest) at 37°C

5.97 Radon (Rn) is the heaviest, and only radioactive, member of Group 8A(18) (noble gases). It is a product of the disintegration of heavier radioactive nuclei found in minute concentrations in many common rocks used for building and construction. In recent years, there has been growing concern about the cancers caused from inhaled residential radon. If 1.0×10^{15} atoms of radium (Ra) produce an average of 1.373×10^{4} atoms of Rn per second, how many liters of Rn, measured at STP, are produced per day by 1.0 g of Ra?

5.98 At 1450. mmHg and 286 K, a skin diver exhales a 208-mL bubble of air that is 77% N_2, 17% O_2, and 6.0% CO_2 by volume.
(a) What would the volume (in mL) of the bubble be if it were exhaled at the surface at 1 atm and 298 K?
(b) How many moles of N_2 are in the bubble?

5.99 Nitrogen dioxide is used industrially to produce nitric acid, but it contributes to acid rain and photochemical smog. What volume (in L) of nitrogen dioxide is formed at 735 torr and 28.2°C by reacting 4.95 cm^3 of copper (d = 8.95 g/cm^3) with 230.0 mL of nitric acid (d = 1.42 g/cm^3, 68.0% HNO_3 by mass)?

$$Cu(s) + 4HNO_3(aq) \longrightarrow Cu(NO_3)_2(aq) + 2NO_2(g) + 2H_2O(l)$$

5.100 In the average adult male, the *residual volume* (RV) of the lungs, the volume of air remaining after a forced exhalation, is 1200 mL.
(a) How many moles of air are present in the RV at 1.0 atm and 37°C?
(b) How many molecules of gas are present under these conditions?

5.101 In a bromine-producing plant, how many liters of gaseous elemental bromine at 300°C and 0.855 atm are formed by the reaction of 275 g of sodium bromide and 175.6 g of sodium bromate in aqueous acid solution? (Assume that no Br_2 dissolves.)

$$5NaBr(aq) + NaBrO_3(aq) + 3H_2SO_4(aq) \longrightarrow 3Br_2(g) + 3Na_2SO_4(aq) + 3H_2O(g)$$

5.102 In a collision of sufficient force, automobile air bags respond by electrically triggering the explosive decomposition of sodium azide (NaN_3) to its elements. A 50.0-g sample of sodium azide was decomposed, and the nitrogen gas generated was collected over water at 26°C. The total pressure was 745.5 mmHg. How many liters of dry N_2 were generated?

5.103 An anesthetic gas contains 64.81% carbon, 13.60% hydrogen, and 21.59% oxygen, by mass. If 2.00 L of the gas at 25°C and 0.420 atm weighs 2.57 g, what is the molecular formula of the anesthetic?

5.104 Aluminum chloride is easily vaporized above 180°C. The gas escapes through a pinhole 0.122 times as fast as helium at the same conditions of temperature and pressure in the same apparatus. What is the molecular formula of aluminum chloride gas?

5.105 (a) What is the total volume (in L) of gaseous *products*, measured at 350°C and 735 torr, when an automobile engine burns 100. g of C_8H_{18} (a typical component of gasoline)?

(b) For part (a), the source of O_2 is air, which is 78% N_2, 21% O_2, and 1.0% Ar by volume. Assuming all the O_2 reacts, but no N_2 or Ar does, what is the total volume (in L) of the engine's gaseous *exhaust?*

5.106 An atmospheric chemist studying the pollutant SO_2 places a mixture of SO_2 and O_2 in a 2.00-L container at 800. K and 1.90 atm. When the reaction occurs, gaseous SO_3 forms, and the pressure falls to 1.65 atm. How many moles of SO_3 form?

5.107 The thermal decomposition of ethylene occurs during the compound's transit in pipelines and during the formation of polyethylene. The decomposition reaction is

$$CH_2{=}CH_2(g) \longrightarrow CH_4(g) + C(\text{graphite})$$

If the decomposition begins at 10°C and 50.0 atm with a gas density of 0.215 g/mL and the temperature increases by 950 K,
(a) What is the final pressure of the confined gas (ignore the volume of graphite and use the van der Waals equation)?
(b) How does the *PV/RT* value of CH_4 compare to that in Figure 5.23? Explain.

5.108 Ammonium nitrate, a common fertilizer, was used by terrorists in the tragic explosion in Oklahoma City in 1995. How many liters of gas at 307°C and 1.00 atm are formed by the explosive decomposition of 15.0 kg of ammonium nitrate to nitrogen, oxygen, and water vapor?

5.109 An environmental engineer analyzes a sample of air contaminated with sulfur dioxide. To a 500.-mL sample at 700. torr and 38°C, she adds 20.00 mL of 0.01017 *M* aqueous iodine, which reacts as follows:

$$SO_2(g) + I_2(aq) + H_2O(l) \longrightarrow HSO_4^-(aq) + I^-(aq) + H^+(aq) \quad \text{[unbalanced]}$$

Excess I_2 reacts with 11.37 mL of 0.0105 *M* sodium thiosulfate:

$$I_2(aq) + S_2O_3^{2-}(aq) \longrightarrow I^-(aq) + S_4O_6^{2-}(aq) \quad \text{[unbalanced]}$$

What is the volume % of SO_2 in the air sample?

5.110 Canadian chemists have developed a modern variation of the 1899 Mond process for preparing extremely pure metallic nickel. A sample of impure nickel reacts with carbon monoxide at 50°C to form gaseous nickel carbonyl, $Ni(CO)_4$.
(a) How many grams of nickel can be converted to the carbonyl with 3.55 m^3 of CO at 100.7 kPa?
(b) The carbonyl is then decomposed at 21 atm and 155°C to pure (>99.95%) nickel. How many grams of nickel are obtained per cubic meter of the carbonyl?
(c) The released carbon monoxide is cooled and collected for reuse by passing it through water at 35°C. If the barometric pressure is 769 torr, what volume (in m^3) of CO is formed per cubic meter of carbonyl?

5.111 Analysis of a newly discovered gaseous silicon-fluorine compound shows that it contains 33.01 mass % silicon. At 27°C, 2.60 g of the compound exerts a pressure of 1.50 atm in a 0.250-L vessel. What is the molecular formula of the compound?

5.112 A gaseous organic compound containing only carbon, hydrogen, and nitrogen is burned in oxygen gas, and the volume of each reactant and product is measured under the same conditions of temperature and pressure. Reaction of four volumes of the compound produces four volumes of CO_2, two volumes of N_2, and ten volumes of water vapor. (a) How many volumes of O_2 were required? (b) What is the empirical formula of the compound?

5.113 Containers A, B, and C are attached by closed stopcocks of negligible volume.

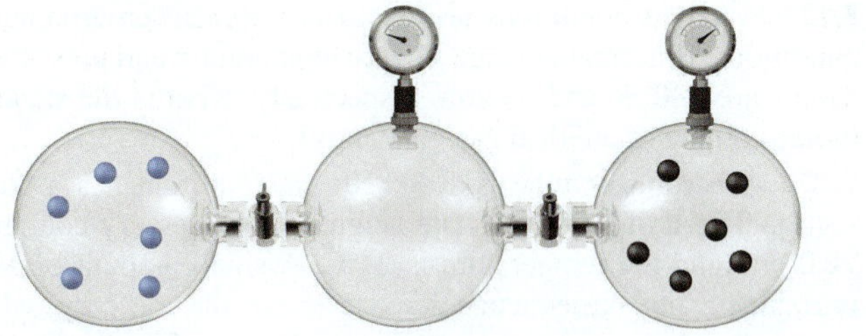

If each particle shown in the picture represents 10^6 particles,
(a) How many blue particles and black particles are in B after the stopcocks are opened and the system reaches equilibrium?
(b) How many blue particles and black particles are in A after the stopcocks are opened and the system reaches equilibrium?
(c) If the pressure in C, P_C, is 750 torr before the stopcocks are opened, what is P_C afterward?
(d) What is P_B afterward?

5.114 Combustible vapor-air mixtures are flammable over a limited range of concentrations. The minimum volume % of vapor that gives a combustible mixture is called the *lower flammable limit* (LFL). Generally, the LFL is about half the stoichiometric mixture, which is the concentration required for complete combustion of the vapor in air. (a) If oxygen is 20.9 vol % of air, estimate the LFL for *n*-hexane, C_6H_{14}. (b) What volume (in mL) of *n*-hexane (d = 0.660 g/cm^3) is required to produce a flammable mixture of hexane in 1.000 m^3 of air at STP?

5.115 By what factor would a scuba diver's lungs expand if she ascended rapidly to the surface from a depth of 125 ft without inhaling or exhaling? If an expansion factor greater than 1.5 causes lung rupture, how far could she safely ascend from 125 ft without breathing? Assume constant temperature (d of seawater = 1.04 g/mL; d of Hg = 13.5 g/mL).

5.116 When 15.0 g of fluorite (CaF_2) reacts with excess sulfuric acid, hydrogen fluoride gas is collected at 744 torr and 25.5°C. Solid calcium sulfate is the other product. What gas temperature is required to store the gas in an 8.63-L container at 875 torr?

5.117 Dilute aqueous hydrogen peroxide is used as a bleaching agent and for disinfecting surfaces and small cuts. Its concentration is sometimes given as a certain number of "volumes hydrogen peroxide," which refers to the number of volumes of O_2 gas, measured at STP, that a given volume of hydrogen peroxide solution will release when it decomposes to O_2 and liquid H_2O. How many grams of hydrogen peroxide are in 0.100 L of "20 volumes hydrogen peroxide" solution?

5.118 At a height of 300 km above Earth's surface, an astronaut finds that the atmospheric pressure is about 10^{-8} mmHg and the temperature is 500 K. How many molecules of gas are there per milliliter at this altitude?

5.119 (a) What is the rms speed of O_2 at STP? (b) If the mean free path of O_2 molecules at STP is 6.33×10^{-8} m, what is their collision frequency? [Use R = 8.314 J/(mol·K) and express $\mathcal{M}$ in kg/mol.]

5.120 Acrylic acid ($CH_2{=}CHCOOH$) is used to prepare polymers, adhesives, and paints. The first step in making acrylic acid involves the vapor-phase oxidation of propylene ($CH_2{=}CHCH_3$) to acrolein ($CH_2{=}CHCHO$). This step is carried out at 330°C and 2.5 atm in a large bundle of tubes around which a heat-transfer agent circulates. The reactants spend an average of 1.8 s in the tubes, which have a void space of 100 ft^3. How many pounds of propylene must be added per hour in a mixture whose mole fractions are 0.07 propylene, 0.35 steam, and 0.58 air?

5.121 Standard conditions are based on relevant environmental conditions. If normal average surface temperature and pressure on Venus are 730. K and 90 atm, respectively, what is the standard molar volume of an ideal gas on Venus?

5.122 A barometer tube is 1.00×10^2 cm long and has a cross-sectional area of 1.20 cm^2. The height of the mercury column is 74.0 cm, and the temperature is 24°C. A small amount of N_2 is introduced into the evacuated space above the mercury, which causes the mercury level to drop to a height of 64.0 cm. How many grams of N_2 were introduced?

5.123 What is the molarity of the cleaning solution formed when 10.0 L of ammonia gas at 33°C and 735 torr dissolves in enough water to give a final volume of 0.750 L?

5.124 The Hawaiian volcano Kilauea emits an average of 1.5×10^3 m^3 of gas each day, when corrected to 298 K and 1.00 atm. The mixture contains gases that contribute to global warming and acid rain, and some are toxic. An atmospheric chemist analyzes a sample and finds the following mole fractions: 0.4896 CO_2, 0.0146 CO, 0.3710 H_2O, 0.1185 SO_2, 0.0003 S_2, 0.0047 H_2, 0.0008 HCl, and 0.0003 H_2S. How many metric tons (t) of each gas are emitted per year (1 t = 1000 kg)?

5.125 To study a key fuel-cell reaction, a chemical engineer has 20.0-L tanks of H_2 and of O_2 and wants to use up both tanks to form 28.0 mol of water at 23.8°C. (a) Use the ideal gas law to find the pressure needed in each tank. (b) Use the van der Waals equation to find the pressure needed in each tank. (c) Compare the results from the two equations.

5.126 For each of the following, which shows the greater deviation from ideal behavior at the same set of conditions? Explain.
(a) Argon or xenon (b) Water vapor or neon
(c) Mercury vapor or radon (d) Water vapor or methane

5.127 How many liters of gaseous hydrogen bromide at 29°C and 0.965 atm will a chemist need if she wishes to prepare 3.50 L of 1.20 *M* hydrobromic acid?

5.128 A mixture consisting of 7.0 g of CO and 10.0 g of SO_2, two atmospheric pollutants, has a pressure of 0.33 atm when placed in a sealed container. What is the partial pressure of CO?

5.129 Sulfur dioxide is used to make sulfuric acid. One method of producing it is by roasting mineral sulfides, for example,

$$FeS_2(s) + O_2(g) \xrightarrow{\Delta} SO_2(g) + Fe_2O_3(s) \quad \text{[unbalanced]}$$

A production error leads to the sulfide being placed in a 950-L vessel with insufficient oxygen. Initially, the partial pressure of O_2 is 0.64 atm, and the total pressure is 1.05 atm, with the balance due to N_2. The reaction is run until 85% of the O_2 is consumed, and the vessel is then cooled to its initial temperature. What is the total pressure in the vessel and the partial pressure of each gas in it?

5.130 A mixture of CO_2 and Kr weighs 35.0 g and exerts a pressure of 0.708 atm in its container. Since Kr is expensive, you wish to recover it from the mixture. After the CO_2 is completely removed by absorption with NaOH(*s*), the pressure in the container is 0.250 atm. How many grams of CO_2 were originally present? How many grams of Kr can you recover?

5.131 When a car accelerates quickly, the passengers feel a force that presses them back into their seats, but a balloon filled with helium floats forward. Why?

5.132 Gases such as CO are gradually oxidized in the atmosphere, not by O_2 but by the hydroxyl radical, •OH, a species with one fewer electron than a hydroxide ion. At night, the •OH concentration is nearly zero, but it increases to 2.5×10^{12} molecules/m^3 in polluted air during the day. At daytime conditions of 1.00 atm and 22°C, what is the partial pressure and mole percent of •OH in air?

5.133 Aqueous sulfurous acid (H_2SO_3) was made by dissolving 0.200 L of sulfur dioxide gas at 19°C and 745 mmHg in water to yield 500.0 mL of solution. The acid solution required 10.0 mL of sodium hydroxide solution to reach the titration end point. What was the molarity of the sodium hydroxide solution?

5.134 In the 19th century, J. B. A. Dumas devised a method for finding the molar mass of a volatile liquid from the volume, temperature, pressure, and mass of its vapor. He placed a sample of such a liquid in a flask that was closed with a stopper fitted with a narrow tube, immersed the flask in a hot water bath to vaporize the liquid, and then cooled the flask. Find the molar mass of a volatile liquid from the following:

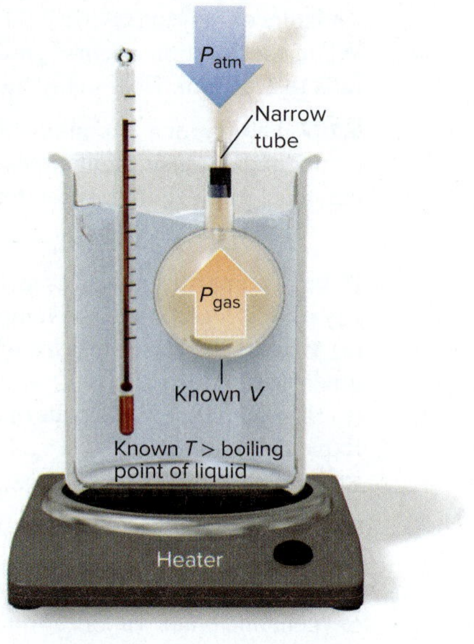

Mass of empty flask = 65.347 g
Mass of flask filled with water at 25°C = 327.4 g
Density of water at 25°C = 0.997 g/mL
Mass of flask plus condensed unknown liquid = 65.739 g
Barometric pressure = 101.2 kPa
Temperature of water bath = 99.8°C

5.135 During World War II, a portable source of hydrogen gas was needed for weather balloons, and solid metal hydrides were the most convenient form. Many metal hydrides react with water to generate the metal hydroxide and hydrogen. Two candidates were lithium hydride and magnesium hydride. What volume (in L) of gas is formed from 1.00 lb of each hydride reacting with excess water at 750. torr and 27°C?

5.136 The lunar surface reaches 370 K at midday. The atmosphere consists of neon, argon, and helium at a total pressure of only 2×10^{-14} atm. Calculate the rms speed of each component in the lunar atmosphere. [Use $R = 8.314$ J/(mol·K) and express $\mathcal{M}$ in kg/mol.]

5.137 A person inhales air richer in O_2 and exhales air richer in CO_2 and water vapor. During each hour of sleep, a person exhales a total of about 300 L of this CO_2-enriched and H_2O-enriched air. (a) If the partial pressures of CO_2 and H_2O in exhaled air are each 30.0 torr at 37.0°C, calculate the mass (g) of CO_2 and of H_2O exhaled in 1 h of sleep. (b) How many grams of body mass does the person lose in 8 h of sleep if all the CO_2 and H_2O exhaled come from the metabolism of glucose?

$$C_6H_{12}O_6(s) + 6O_2(g) \longrightarrow 6CO_2(g) + 6H_2O(g)$$

5.138 Popcorn pops because the horny endosperm, a tough, elastic material, resists gas pressure within the heated kernel until the pressure reaches explosive force. A 0.25-mL kernel has a water content of 1.6% by mass, and the water vapor reaches 170°C and 9.0 atm before the kernel ruptures. Assume that water vapor can occupy 75% of the kernel's volume. (a) What is the mass (in g) of the kernel? (b) How many milliliters would this amount of water vapor occupy at 25°C and 1.00 atm?

5.139 Sulfur dioxide emissions from coal-burning power plants are removed by *flue-gas desulfurization.* The flue gas passes through a scrubber, and a slurry of wet calcium carbonate reacts with it to form carbon dioxide and calcium sulfite. The calcium sulfite then reacts with oxygen to form calcium sulfate, which is sold as gypsum. (a) If the sulfur dioxide concentration is 1000 times higher than its mole fraction in clean dry air (2×10^{-10}), how much calcium sulfate (kg) can be made from scrubbing 4 GL of flue gas (1 GL = 1×10^{9} L)? A state-of-the-art scrubber removes at least 95% of the sulfur dioxide. (b) If the mole fraction of oxygen in air is 0.209, what volume (L) of air at 1.00 atm and 25°C is needed to react with all the calcium sulfite?

5.140 Many water treatment plants use chlorine gas to kill microorganisms before the water is released for residential use. A plant engineer has to maintain the chlorine pressure in a tank below the 85.0-atm rating and, to be safe, decides to fill the tank to 80.0% of this maximum pressure. (a) How many moles of Cl_2 gas can be kept in an 850.-L tank at 298 K if she uses the ideal gas law in the calculation? (b) What is the tank pressure if she uses the van der Waals equation for this amount of gas? (c) Did the engineer fill the tank to the desired pressure?

5.141 At 10.0°C and 102.5 kPa, the density of dry air is 1.26 g/L. What is the average "molar mass" of dry air at these conditions?

5.142 In A, the picture shows a cylinder with 0.1 mol of a gas that behaves ideally. Choose the cylinder (B, C, or D) that correctly represents the volume of the gas after each of the following changes. If none of the cylinders is correct, specify "none."
(a) P is doubled at fixed n and T.
(b) T is reduced from 400 K to 200 K at fixed n and P.
(c) T is increased from 100°C to 200°C at fixed n and P.
(d) 0.1 mol of gas is added at fixed P and T.
(e) 0.1 mol of gas is added and P is doubled at fixed T.

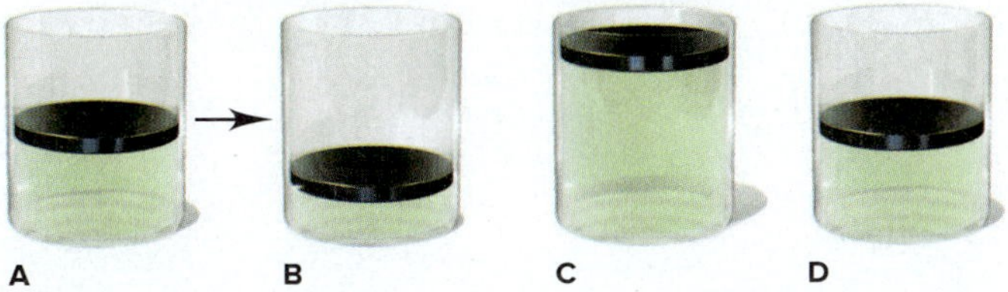

5.143 Ammonia is essential to so many industries that, on a molar basis, it is the most heavily produced substance in the world. Calculate P_{IGL} and P_{VDW} (in atm) of 51.1 g of ammonia in a 3.000-L container at 0°C and 400.°C, the industrial temperature. (See Table 5.4 for the values of the van der Waals constants.)

5.144 A 6.0-L flask contains a mixture of methane (CH_4), argon, and helium at 45°C and 1.75 atm. If the mole fractions of helium and argon are 0.25 and 0.35, respectively, how many molecules of methane are present?

5.145 A large portion of metabolic energy arises from the biological combustion of glucose:

$$C_6H_{12}O_6(s) + 6O_2(g) \longrightarrow 6CO_2(g) + 6H_2O(g)$$

(a) If this reaction is carried out in an expandable container at 37°C and 780. torr, what volume of CO_2 is produced from 20.0 g of glucose and excess O_2?
(b) If the reaction is carried out at the same conditions with the stoichiometric amount of O_2, what is the partial pressure of each gas when the reaction is 50% complete (10.0 g of glucose remains)?

5.146 What is the average kinetic energy and rms speed of N_2 molecules at STP? Compare these values with those of H_2 molecules at STP. [Use $R = 8.314$ J/(mol·K) and express $\mathcal{M}$ in kg/mol.]

5.147 According to government standards, the *8-h threshold limit value* is 5000 ppmv for CO_2 and 0.1 ppmv for Br_2 (1 ppmv is 1 part by volume in 10^6 parts by volume). Exposure to either gas for 8 h above these limits is unsafe. At STP, which of the following would be unsafe for 8 h of exposure?
(a) Air with a partial pressure of 0.2 torr of Br_2
(b) Air with a partial pressure of 0.2 torr of CO_2
(c) 1000 L of air containing 0.0004 g of Br_2 gas
(d) 1000 L of air containing 2.8×10^{22} molecules of CO_2

5.148 One way to prevent emission of the pollutant NO from industrial plants is by a catalyzed reaction with NH_3:

$$4NH_3(g) + 4NO(g) + O_2(g) \xrightarrow{\text{catalyst}} 4N_2(g) + 6H_2O(g)$$

(a) If the NO has a partial pressure of 4.5×10^{-5} atm in the flue gas, how many liters of NH_3 are needed per liter of flue gas at 1.00 atm? (b) If the reaction takes place at 1.00 atm and 365°C, how many grams of NH_3 are needed per kiloliter (kL) of flue gas?

5.149 An equimolar mixture of Ne and Xe is accidentally placed in a container that has a tiny leak. After a short while, a very small proportion of the mixture has escaped. What is the mole fraction of Ne in the effusing gas?

5.150 From the relative rates of effusion of $^{235}UF_6$ and $^{238}UF_6$, find the number of steps needed to produce a sample of the enriched fuel used in many nuclear reactors, which is 3.0 mole % ^{235}U. The natural abundance of ^{235}U is 0.72%.

5.151 A slight deviation from ideal behavior exists even at normal conditions. If it behaved ideally, 1 mol of CO would occupy 22.414 L and exert 1 atm pressure at 273.15 K. Calculate P_{VDW} for 1.000 mol of CO at 273.15 K. $\left(\text{Use } R = 0.08206 \dfrac{\text{atm·L}}{\text{mol·K}}.\right)$

5.152 In preparation for a combustion demonstration, a professor fills a balloon with equal molar amounts of H_2 and O_2, but the demonstration has to be postponed until the next day. During the night, both gases leak through pores in the balloon. If 35% of the H_2 leaks, what is the O_2/H_2 ratio in the balloon the next day?

5.153 Phosphorus trichloride is important in the manufacture of insecticides, fuel additives, and flame retardants. Phosphorus has only one naturally occurring isotope, ^{31}P, whereas chlorine has two, ^{35}Cl (75%) and ^{37}Cl (25%). (a) What different molecular masses (in amu) can be found for PCl_3? (b) Which is the most abundant? (c) What is the ratio of the effusion rates of the heaviest and the lightest PCl_3 molecules?

5.154 A truck tire has a volume of 218 L and is filled with air to 35.0 psi at 295 K. After a drive, the air heats up to 318 K. (a) If the tire volume is constant, what is the pressure (in psi)? (b) If the tire volume increases 2.0%, what is the pressure (in psi)? (c) If the tire leaks 1.5 g of air per minute and the temperature is constant, how many minutes will it take for the tire to reach the original pressure of 35.0 psi ($\mathcal{M}$ of air = 28.8 g/mol)?

5.155 *Allotropes* are different molecular forms of an element, such as dioxygen (O_2) and ozone (O_3). (a) What is the density of each oxygen allotrope at 0°C and 760 torr? (b) Calculate the ratio of densities, d_{O_3}/d_{O_2}, and explain the significance of this number.

5.156 When gaseous F_2 and solid I_2 are heated to high temperatures, the I_2 sublimes and gaseous iodine heptafluoride forms. If 350. torr of F_2 and 2.50 g of solid I_2 are put into a 2.50-L container at 250. K and the container is heated to 550. K, what is the final pressure (in torr)? What is the partial pressure of I_2 gas?

6 Thermochemistry: Energy Flow and Chemical Change

6.1 Forms of Energy and Their Interconversion
Defining System and Surroundings
Energy Change (ΔE)
Heat and Work
Law of Energy Conservation
Units of Energy
State Functions
Calculating Pressure-Volume Work

6.2 Enthalpy: Changes at Constant Pressure
The Meaning of Enthalpy
Comparing ΔE and ΔH
Exothermic and Endothermic Processes

6.3 Calorimetry: Measuring the Heat of a Chemical or Physical Change
Specific Heat Capacity
The Two Major Types of Calorimetry

6.4 Stoichiometry of Thermochemical Equations

6.5 Hess's Law: Finding ΔH of Any Reaction

6.6 Standard Enthalpies of Reaction (ΔH°_{rxn})
Formation Equations
Determining ΔH°_{rxn} from ΔH°_f Values

Source: © Maya Kruchankova/Shutterstock.com

Concepts and Skills to Review Before You Study This Chapter

- interconverting potential and kinetic energy (Section 1.1)
- distinction between heat and temperature (Section 1.4)
- nature of chemical bonding (Section 2.7)
- calculations of reaction stoichiometry (Section 3.4)
- properties of the gaseous state (Section 5.1)
- relation between kinetic energy and temperature (Section 5.5)

All matter contains energy, so *whenever matter undergoes a change, the quantity of energy that the matter contains also* changes. Both chemical and physical changes are involved when a gas stove is used to boil the water for your morning cup of tea. In the chemical change, energy is *released* as heat when higher energy natural gas and O_2 react to form lower energy products. The heat energy released by this chemical change is *absorbed* by lower energy liquid water in the tea kettle as it vaporizes to higher energy gaseous water.

This interplay of matter and energy is profoundly important and has an enormous impact on society. On an everyday level, many familiar materials release, absorb, or alter the flow of energy. Fuels such as natural gas and oil release energy to cook our food, warm our homes, or power our vehicles. Fertilizers help crops convert solar energy into food. Metal wires increase the flow of electrical energy, and polymer fibers in winter clothing slow the flow of thermal energy away from our bodies.

Thermodynamics is the study of energy and its transformations, and three chapters in this text address this central topic. Our focus in this chapter is on **thermochemistry,** the branch of thermodynamics that deals with heat in chemical and physical change. In Chapter 20, we consider *why* reactions occur, and in Chapter 21, we apply those insights to electrical energy.

IN THIS CHAPTER . . . *We see how heat, or thermal energy, flows when matter changes, how to measure the quantity of heat for a given change, and how to determine the direction and magnitude of heat flow for any reaction.*

- We see that energy always flows between a system and its surroundings in the form of heat or work, so the total energy is conserved.
- We discuss the units of energy and show that the total size of an energy change does not depend on how the change occurs.
- We identify the heat of a reaction in an open container as a change in enthalpy, which can be negative (exothermic reaction) or positive (endothermic reaction).
- We describe how a calorimeter measures heat and how the quantity of heat in a reaction is proportional to the amounts of substances.
- We define standard conditions in order to compare enthalpies of reactions and see how to obtain the change in enthalpy for any reaction.
- We discuss some current and future energy sources and describe the critical relationship between energy demand and climate change.

6.1 FORMS OF ENERGY AND THEIR INTERCONVERSION

In Chapter 1, we saw that all energy is either potential or kinetic, and that these forms are interconvertible. An object has potential energy by virtue of its position and kinetic energy by virtue of its motion. Let's re-examine these ideas by considering a weight raised above the ground. As the weight is lifted by your muscles or a motor, its potential energy increases; this energy is converted to kinetic energy as the weight falls (see Figure 1.3). When it hits the ground, some of that kinetic energy appears as *work* done to move the soil and pebbles slightly, and some appears as *heat* when it warms them slightly. Thus, in this situation, *potential energy is converted to kinetic energy, which appears as work and heat.*

Several other forms of energy—solar, electrical, nuclear, and chemical—are examples of potential and kinetic energy on the atomic scale. No matter what the form of energy or the situation, *when energy is transferred from one object to another, it appears as work and/or heat.* In this section, we examine this idea in terms of the release or absorption of energy during a chemical or physical change.

Figure 6.1 A chemical system and its surroundings.
Source: © McGraw-Hill Education. Stephen Frisch, photographer

Defining the System and Its Surroundings

In any thermodynamic study, including measuring a change in energy, the first step is to define the **system**—the part of the universe we are focusing on. And the moment we define the system, everything else is defined as the **surroundings.**

Figure 6.1 shows a typical chemical system—the contents of a flask, usually substances undergoing physical or chemical change. The flask itself, the air surrounding the flask, other equipment, and perhaps the rest of the laboratory are the surroundings. In principle, the rest of the universe is the surroundings, but in practice, we consider only the parts that are relevant to the system: it's not likely that a thunderstorm in central Asia or a methane blizzard on Neptune will affect the contents of the flask, but the temperature and pressure of the lab might. Thus, the experimenter defines the system and the relevant surroundings. An astronomer defines a galaxy as the system and nearby galaxies as the surroundings; a microbiologist defines a given cell as the system and neighboring cells and the extracellular fluid as the surroundings; and so forth.

Energy Change (ΔE): Energy Transfer to or from a System

Each particle in a system has potential energy and kinetic energy, and the sum of all these energies is the **internal energy, *E*,** of the system (some texts use the symbol *U*). When the reactants in a chemical system change to products, the system's internal energy has changed. This change, ΔE, is the difference between the internal energy *after* the change (E_{final}) and the internal energy *before* the change (E_{initial}), where Δ (Greek *delta*) means "change (or difference) in" calculated as the *final state* ***minus*** *the initial state.* Thus, ΔE is the final quantity of energy of the system **minus** the initial quantity:

$$\Delta E = E_{\text{final}} - E_{\text{initial}} = E_{\text{products}} - E_{\text{reactants}} \quad \textbf{(6.1)}$$

Because the universe consists of only system and surroundings, *a change in the energy of the system must be accompanied by an* ***equal*** *and* ***opposite*** *change in the energy of the surroundings.* In an *energy diagram,* the final and initial states are horizontal lines along a vertical energy axis, with ΔE being the difference in the heights of the lines. A system can change its internal energy in one of two ways:

- By releasing some energy in a transfer *to* the surroundings (Figure 6.2A):

$$E_{\text{final}} < E_{\text{initial}} \quad \text{so} \quad \Delta E < 0$$

- By absorbing some energy in a transfer *from* the surroundings (Figure 6.2B):

$$E_{\text{final}} > E_{\text{initial}} \quad \text{so} \quad \Delta E > 0$$

Thus, ΔE is a *transfer* of energy from system to surroundings, or vice versa.

Heat and Work: Two Forms of Energy Transfer

Energy transferred from system to surroundings, or vice versa, appears in two forms:

1. *Heat.* **Heat,** or *thermal energy* (symbolized by q), is the energy transferred as a result of a difference in temperature between the system and the surroundings. For example, energy in the form of heat is transferred *from* hot coffee (system) *to* the mug, your hand, and air (surroundings) because they are at a lower temperature, while heat is transferred *from* a hot stove (surroundings) *to* an ice cube (system) because the ice is at a lower temperature.

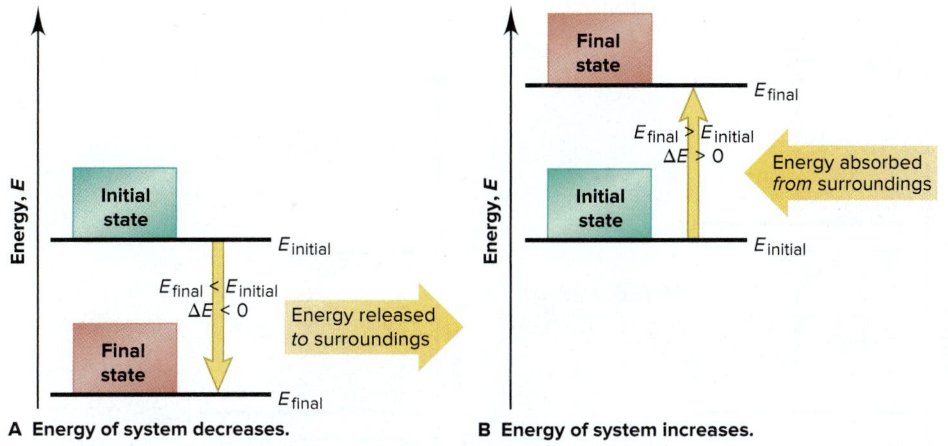

Figure 6.2 **Energy diagrams for the transfer of internal energy (E) between two systems and their surroundings.** **A,** When the system releases energy, ΔE ($E_{final} - E_{initial}$) is negative. **B,** When the system absorbs energy, ΔE ($E_{final} - E_{initial}$) is positive. (The vertical yellow arrow always has its tail at the initial state.)

2. *Work.* All other forms of energy transfer involve some type of **work (w),** the energy transferred when an object is moved by a force. When you (system) kick a football, energy is transferred as work because the force of the kick moves the ball and air (surroundings). When you pump up a ball, energy is transferred as work because the added air (system) exerts a force on the inner wall of the ball (surroundings) and moves it outward.

The total change in a system's internal energy is the sum of the energy transferred as heat and/or as work:

$$\Delta E = q + w \quad (6.2)$$

The values of q and w (and, therefore, of ΔE) can have either a positive or negative sign. *We define the sign of the energy change from the* ***system's*** *perspective:*

- Energy transferred *into* the system is *positive* because the *system ends up with more* energy.
- Energy transferred *out from* the system is *negative,* because the *system ends up with less* energy.

Innumerable combinations of heat and/or work can change a system's internal energy. In the rest of this subsection, we'll examine the four simplest cases—two that involve only heat and two that involve only work.

Energy Transferred as Heat Only For a system that transfers energy only as heat (q) and does no work ($w = 0$), we have, from Equation 6.2, $\Delta E = q + 0 = q$. There are two ways this transfer can happen:

1. *Heat flowing* ***out*** *from a system* (Figure 6.3A, *next page*). Suppose hot water is the system, and the beaker holding it and the rest of the lab are the surroundings. The water transfers energy as heat outward until the temperature of the water and the surroundings are equal. Since heat flows *out* from the system, the final energy of the system is less than the initial energy.

System *releases* energy as heat ⟶ q is *negative* ⟶ $E_{final} < E_{initial}$ ⟶ ΔE is *negative*

2. *Heat flowing* ***into*** *a system* (Figure 6.3B, *next page*). If the system consists of ice water, the surroundings transfer energy as heat *into* the system, once again until the ice melts and the temperature of the water and the surroundings become equal. In this case, heat flows *in,* so the final energy of the system is higher than its initial energy. ›

System *absorbs* energy as heat ⟶ q is *positive* ⟶ $E_{final} > E_{initial}$ ⟶ ΔE is *positive*

Energy Transferred as Work Only For a system that transfers energy only as work, $q = 0$; therefore, $\Delta E = 0 + w = w$. There are two ways this transfer can happen:

1. Work done ***by*** *a system* (Figure 6.4A, *next page*). Consider the aqueous reaction between zinc and hydrochloric acid:

$$Zn(s) + 2HCl(aq) \longrightarrow ZnCl_2(aq) + H_2(g)$$

Thermodynamics in the Kitchen

The functioning of two familiar kitchen appliances can clarify the sign of q. The air in a refrigerator (surroundings) has a lower temperature than a newly added piece of food (system), so the food releases energy as heat to the refrigerator air, $q < 0$. The air in a hot oven (surroundings) has a higher temperature than a newly added piece of food (system), so the food absorbs energy as heat from the oven air, $q > 0$.

Figure 6.3 **The two cases where energy is transferred as heat only. A,** The system releases heat. **B,** The system absorbs heat.

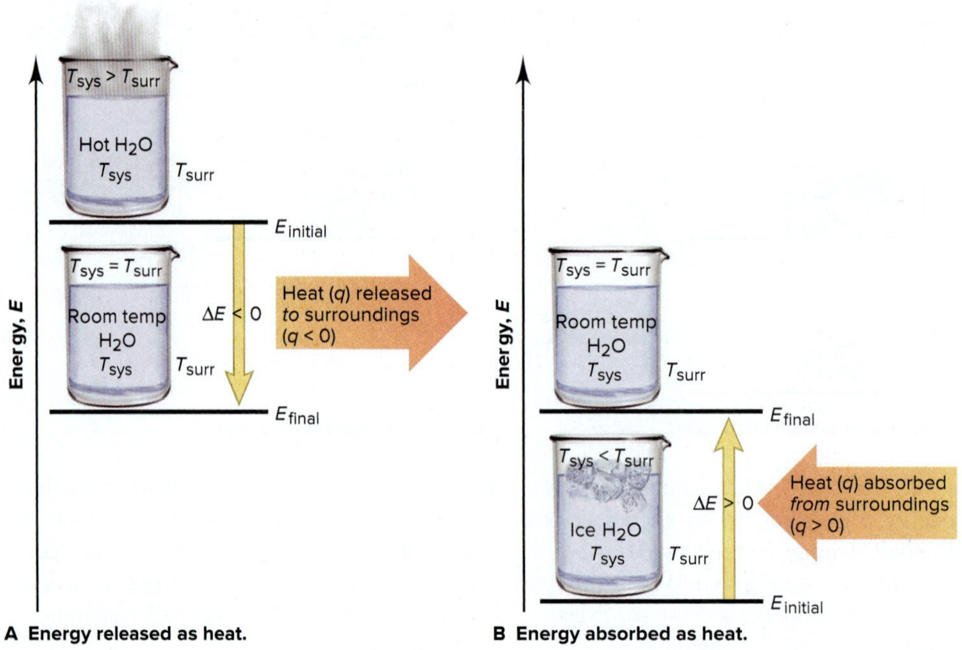

The reaction takes place in a nearly evacuated (narrow P_{sys} arrow), insulated container attached to a piston-cylinder assembly. (The container is insulated so that heat does not flow.) We define the system as the reaction mixture, and the container, piston-cylinder, outside air, and so forth as the surroundings. In the initial state, the internal energy is the energy of the reactants, and in the final state, it is the energy of the products. Hydrogen gas (shown in purple) forms, pushing back the piston as it expands. Thus, energy is transferred as work done *by* the system *on* the surroundings.

$$\text{System } \textit{releases} \text{ energy as work done by it} \longrightarrow w \text{ is } \textit{negative} \longrightarrow$$
$$E_{\text{final}} < E_{\text{initial}} \longrightarrow \Delta E \text{ is } \textit{negative}$$

(The work done here is not very useful because it just pushes back the piston and outside air. But, if the reaction mixture is gasoline and oxygen and the surroundings are an automobile engine, much of the internal energy is transferred as the work done to move the car.)

2. *Work done* ***on*** *a system* (Figure 6.4B). Suppose that, after the reaction is over, we increase the pressure of the surroundings (wider P_{surr} arrow) so that the piston moves in, compressing the hydrogen gas into a smaller volume. Energy is transferred as work done *by* the surroundings *on* the system.

$$\text{System } \textit{absorbs} \text{ energy as work done on it} \longrightarrow w \text{ is } \textit{positive} \longrightarrow$$
$$E_{\text{final}} > E_{\text{initial}} \longrightarrow \Delta E \text{ is } \textit{positive}$$

Table 6.1 summarizes the sign conventions for q and w and the effect on the sign of ΔE.

Figure 6.4 **The two cases where energy is transferred as work only. A,** The system does work *on* the surroundings. **B,** The system has work done on it *by* the surroundings.

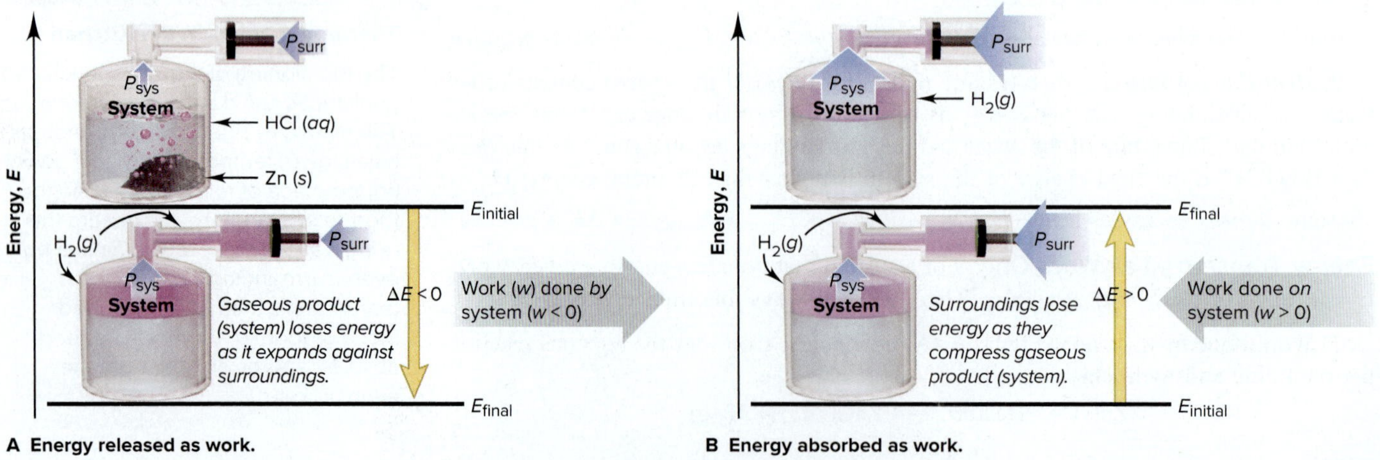

Table 6.1 The Sign Conventions* for q, w, and ΔE

q	+	w	=	ΔE
+ (heat *absorbed*)		+ (work done *on*)		+ (energy *absorbed*)
+ (heat *absorbed*)		− (work done *by*)		Depends on the *sizes* of q and w
− (heat *released*)		+ (work done *on*)		Depends on the *sizes* of q and w
− (heat *released*)		− (work done *by*)		− (energy *released*)

*From the perspective of the system.

The Law of Energy Conservation

As you've seen, when a system absorbs energy, the surroundings release it, and when a system releases energy, the surroundings absorb it. Energy transferred between system and surroundings can be in the form of heat and/or various types of work—mechanical, electrical, radiant, chemical, and so forth.

Indeed, energy is often converted from one form to another during transfers. For example, when gasoline burns in a car engine, the reaction releases energy that is transferred as heat and work. The heat warms the car parts, passenger compartment, and surrounding air. The work is done when mechanical energy turns the car's wheels and belts. That energy is converted into the electrical energy of the sound system, the radiant energy of the headlights, the chemical energy of the battery, and so forth. The sum of all these forms equals the change in energy between reactants and products as the gasoline burns and the exhaust forms.

Complex biological processes follow the same general pattern. During photosynthesis, green plants use solar energy to convert the chemical energy of the bonds in CO_2 and H_2O into the energy of the bonds in starch and O_2; when you digest starch, the energy in its bonds is converted into the muscular (mechanical) energy you need to run a marathon or engage in any activity.

The most important point to realize in these, and all other situations, is that the form of energy may change, but energy does not appear or disappear—energy cannot be created or destroyed. Put another way, *energy is conserved: the total energy of the system plus the surroundings remains constant.* The **law of conservation of energy,** also known as the **first law of thermodynamics,** restates this basic observation: *the total energy of the universe is constant.* It is expressed mathematically as

$$\Delta E_{\text{universe}} = \Delta E_{\text{system}} + \Delta E_{\text{surroundings}} = 0 \qquad \textbf{(6.3)}$$

This law applies, as far as we know, to all systems, from a burning match to continental drift, from the pumping of your heart to the formation of the Solar System.

Units of Energy

The SI unit of energy is the **joule (J),** a derived unit composed of three base units:

$$1\ \text{J} = 1\ \text{kg}{\cdot}\text{m}^2/\text{s}^2$$

Both heat and work are expressed in joules. Let's see why the joule is the unit for work. The work (w) done on a mass is the force (F) times the distance (d) that the mass moves: $w = F \times d$. A *force* changes the velocity of a mass over time; that is, a force *accelerates* a mass. Velocity has units of meters per second (m/s), so acceleration (a) has units of meters per second per second (m/s^2). Force, therefore, has units of mass (m, in kilograms) times acceleration:

$$F = m \times a \quad \text{has units of} \quad \text{kg}{\cdot}\text{m/s}^2$$

Therefore, $$w = F \times d \quad \text{has units of} \quad (\text{kg}{\cdot}\text{m/s}^2) \times \text{m} = \text{kg}{\cdot}\text{m}^2/\text{s}^2 = \text{J}$$

The **calorie (cal)** is an older unit defined originally as the quantity of energy needed to raise the temperature of 1 g of water by 1°C (specifically, from 14.5°C to 15.5°C). The calorie is now defined in terms of the joule:

$$1\ \text{cal} \equiv 4.184\ \text{J} \qquad \text{or} \qquad 1\ \text{J} = \frac{1}{4.184}\ \text{cal} = 0.2390\ \text{cal}$$

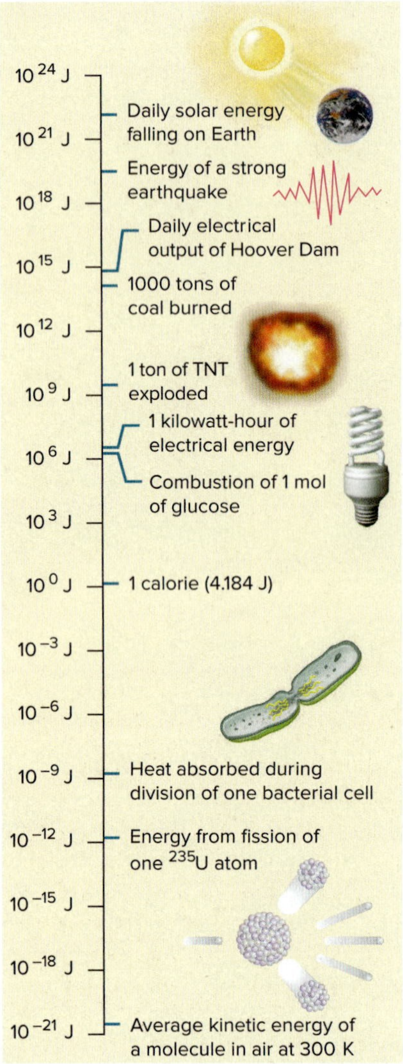

Figure 6.5 Some quantities of energy. The vertical scale is exponential.

Since the quantities of energy involved in chemical reactions are usually quite large, chemists use the kilojoule (kJ) or, in earlier sources, the kilocalorie (kcal):

$$1 \text{ kJ} = 1000 \text{ J} = 0.2390 \text{ kcal} = 239.0 \text{ cal}$$

The nutritional Calorie (note the capital C), the unit that measures the energy available from food, is actually a kilocalorie:

$$1 \text{ Cal} = 1000 \text{ cal} = 1 \text{ kcal} = 4184 \text{ J}$$

The *British thermal unit (Btu),* a unit that you may have seen used for the energy output of appliances, is the quantity of energy required to raise the temperature of 1 lb of water by 1°F; *1 Btu is equivalent to 1055 J.* In general, the SI unit (J or kJ) is used in this text. Some interesting quantities of energy are featured in Figure 6.5.

SAMPLE PROBLEM 6.1 Determining the Change in Internal Energy of a System

Problem When gasoline burns in a car engine, the heat released causes the gaseous products CO_2 and H_2O to expand, which pushes the pistons outward. Excess heat is removed by the radiator. If the expanding gases do 451 J of work on the pistons and the system releases 325 J to the surroundings as heat, calculate the change in energy (ΔE) in J, kJ, and kcal.

Plan We define system and surroundings to choose signs for q and w, and then we calculate ΔE with Equation 6.2. The system is the reactants and products, and the surroundings are the pistons, the radiator, and the rest of the car. Heat is released by the system, so q is negative. Work is done by the system to push the pistons outward, so w is also negative. We obtain the answer in J and then convert it to kJ and kcal.

Solution Calculating ΔE (from Equation 6.2) in J:

$$q = -325 \text{ J}$$
$$w = -451 \text{ J}$$
$$\Delta E = q + w = -325 \text{ J} + (-451 \text{ J}) = -776 \text{ J}$$

Converting from J to kJ:

$$\Delta E = -776 \text{ J} \times \frac{1 \text{ kJ}}{1000 \text{ J}} = -0.776 \text{ kJ}$$

Converting from J to kcal:

$$\Delta E = -776 \text{ J} \times \frac{1 \text{ cal}}{4.184 \text{ J}} \times \frac{1 \text{ kcal}}{1000 \text{ cal}} = -0.185 \text{ kcal}$$

Check The answer is reasonable: combustion of gasoline releases energy from the system, so $E_{\text{final}} < E_{\text{initial}}$ and ΔE should be negative.

FOLLOW-UP PROBLEMS

*Brief Solutions for **all** Follow-up Problems appear at the end of the chapter.*

6.1A A sample of liquid absorbs 13.5 kJ of heat and does 1.8 kJ of work when it vaporizes. Calculate ΔE (in J).

6.1B In a reaction, gaseous reactants form a liquid product. The heat absorbed by the surroundings is 26.0 kcal, and the work done on the system is 15.0 Btu. Calculate ΔE (in kJ).

SOME SIMILAR PROBLEMS 6.9–6.12

State Functions and the Path Independence of the Energy Change

The internal energy (E) of a system is called a **state function,** a property dependent only on the *current* state of the system (its composition, volume, pressure, and temperature), *not* on the path the system takes to reach that state. The energy change of a system can occur through countless combinations of heat (q) and/or work (w).

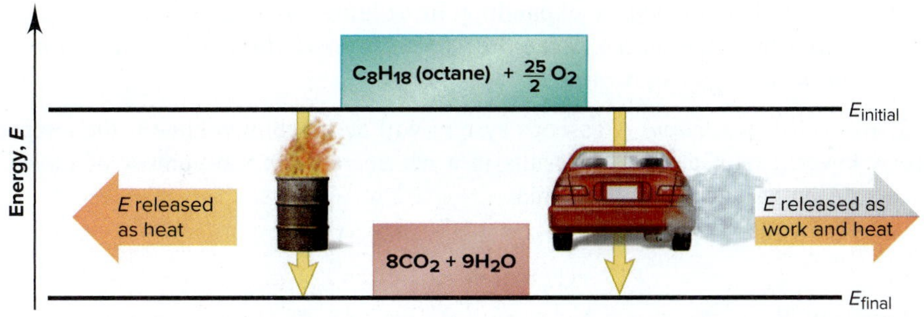

Figure 6.6 Two different paths for the energy change of a system. Even though q and w for the two paths are different, ΔE is the same.

However, because E is a state function, the overall ΔE is the same no matter what the specific combination may be. That is, *ΔE does **not** depend on how the change takes place, but only on the **difference** between the final and initial states.*

As an example, let's define a system in its initial state as 1 mol of octane (a component of gasoline) together with enough O_2 to burn it completely. In its final state, the system is the CO_2 and H_2O that form:

$$C_8H_{18}(l) + \tfrac{25}{2}O_2(g) \longrightarrow 8CO_2(g) + 9H_2O(g)$$

initial state (E_{initial}) final state (E_{final})

Energy is transferred *out* from the system as heat and/or work, so ΔE is negative. Figure 6.6 shows just two of the many ways the change can occur. If we burn the octane in an open container *(left),* ΔE is transferred almost completely as heat (with a small quantity of work done to push back the atmosphere). If we burn it in a car engine *(right),* a much larger portion (~30%) of ΔE is transferred as work that moves the car, with the rest transferred as heat that warms the car, exhaust gases, and surrounding air. If we burn the octane in a lawn mower or a motorcycle, ΔE appears as other combinations of work and heat. Thus, q and w are *not* state functions because their values *do* depend on the path the system takes, but ΔE (the *sum* of q and w) *does not.* ›

Your Personal Financial State Function

The *balance* in your checking account is a state function of your personal financial system. You can open a new checking account with a birthday gift of \$50 or with a deposit of a \$100 paycheck and then write two \$25 checks. The two paths to the balance are different, but the balance (current state) is the same.

As a final example, consider temperature, another state function. If you add enough heat to raise the temperature of a sample of water from an initial temperature of 10°C to a final temperature of 60°C, $\Delta T = \Delta T_{\text{final}} - T_{\text{initial}} = 60°C - 10°C = 50°C$. If instead, you add enough heat to raise the temperature of the water from 10°C to 80°C and then you cool the water to 60°C, ΔT is still 60°C – 10°C = 50°C. Pressure (P) and volume (V) are some other state functions. Path independence means that *changes in state functions—ΔE, ΔT, ΔP, and ΔV—depend only on the initial and final states.*

Calculating Pressure-Volume Work (*PV* Work)

The two most important types of chemical work are electrical work, done by moving charged particles (Chapter 21), and a type of mechanical work called **pressure-volume work (PV work),** done when the volume of the system changes in the presence of an external pressure (P). The quantity of PV work equals P times the change in volume (ΔV, or $V_{\text{final}} - V_{\text{initial}}$):

$$w = -P\Delta V \qquad (6.4)$$

The volume of the system may increase (expand) or decrease (contract).

1. *Work of Expansion.* The volume of a gaseous system will increase if the temperature is raised (Charles's law) or if a chemical reaction results in a net increase in the number of moles of gas (Avogadro's law), as in the combustion of propane:

$$C_3H_8(g) + 5O_2(g) \longrightarrow 3CO_2(g) + 4H_2O(g)$$

6 (1 + 5) mol of gas 7 (3 + 4) mol of gas

In an open flask (or in a cylinder with a weightless, frictionless piston, Figure 6.7, *next page*), a system of an expanding gas does PV work *on* the surroundings as it pushes against the atmosphere, so the work has a negative sign. Note that ΔV is

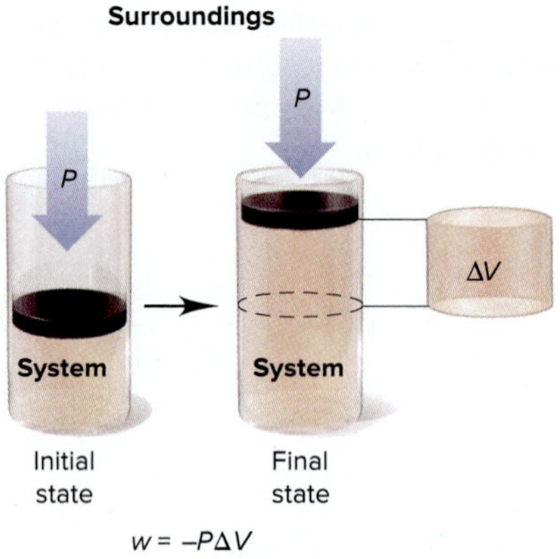

Figure 6.7 Pressure-volume work. An expanding gas pushing back the atmosphere does *PV* work ($w = -P\Delta V$).

positive for a system that is expanding in volume because $V_{final} > V_{initial}$. The negative sign in Equation 6.4 is required because *work done by the system means the system lost energy as work.*

2. *Work of Contraction.* A gaseous system will decrease in volume if the temperature is lowered or if a reaction results in a net decrease in the number of moles of gas, as in the production of ammonia:

$$3H_2(g) + N_2(g) \longrightarrow 2NH_3(g)$$

4 (3 + 1) mol of gas   2 mol of gas

In these situations, *the system has work done on it* ***by*** *the surroundings* as the gas is contracted into a smaller volume by the external pressure. Since $V_{final} < V_{initial}$, ΔV is negative, and the work has a positive sign, which means *the system has gained energy as work.*

When using Equation 6.4, *P* is expressed in atmospheres and ΔV in liters, resulting in a unit of atm·L for work; an important relationship is that 101.3 J of energy is equivalent to 1 atm·L.

SAMPLE PROBLEM 6.2 Calculating Pressure-Volume Work Done by or on a System

Problem A reaction taking place in a container with a piston-cylinder assembly at constant temperature produces a gas, and the volume increases from 125 mL to 652 mL against an external pressure of 988 torr. Calculate the work done (in J; 1 atm·L = 101.3 J).

Plan We are given the external pressure (988 torr) and initial (125 mL) and final volumes (652 mL) and have to find the work done by the gas. We subtract the initial *V* from the final *V* to find ΔV and convert it from mL to L. We convert the given *P* from torr to atm and use Equation 6.4 to calculate *w*. Then we convert the answer from atm·L to J (see the road map).

Road Map

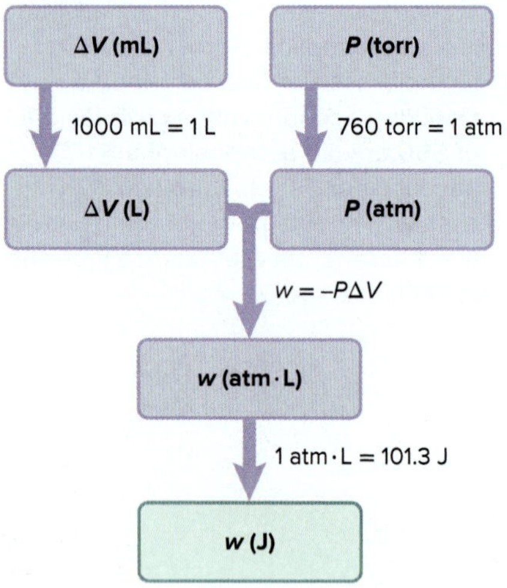

Solution Calculating ΔV:

$$\Delta V\ (\text{mL}) = V_{final} - V_{initial} = 652\ \text{mL} - 125\ \text{mL} = 527\ \text{mL}$$

Converting ΔV from mL to L:

$$\Delta V\ (\text{L}) = 527\ \text{mL} \times \frac{1\ \text{L}}{1000\ \text{mL}} = 0.527\ \text{L}$$

Converting *P* from torr to atm:

$$P\ (\text{atm}) = 988\ \text{torr} \times \frac{1\ \text{atm}}{760\ \text{torr}} = 1.30\ \text{atm}$$

Calculating *w* (using Equation 6.4):

$$w\ (\text{atm·L}) = -P\Delta V = -(1.30\ \text{atm} \times 0.527\ \text{L}) = -0.685\ \text{atm·L}$$

Converting from atm·L to J:

$$w\ (\text{J}) = -0.685\ \text{atm·L} \times \frac{101.3\ \text{J}}{1\ \text{atm·L}} = -69.4\ \text{J}$$

Check Since gas is produced, the system expands and does work on the surroundings (energy is released as work), so the negative sign is correct. Rounding shows that the size of the answer is reasonable: $w \approx -1\ \text{atm·L} \times 0.5\ \text{L} = -0.5\ \text{atm·L}$.

FOLLOW-UP PROBLEMS

6.2A A gas is compressed at constant temperature from a volume of 5.68 L to a volume of 2.35 L by an external pressure of 732 torr. Calculate the work done (in J).

6.2B A gas-producing reaction occurs in a container with a piston-cylinder assembly under an external pressure of 5.5 atm. At constant temperature, the volume increases from 10.5 L to 16.3 L. Calculate the work done (in J).

SOME SIMILAR PROBLEMS 6.15 and 6.16

› Summary of Section 6.1

- › Internal energy (E) is transferred as heat (q) when system and surroundings are at different temperatures or as work (w) when an object is moved by a force.
- › Heat absorbed by a system ($q > 0$) or work done on a system ($w > 0$) increases the system's E; heat released by a system ($q < 0$) or work done by a system ($w < 0$) decreases its E. The change in the internal energy is the sum of the heat and work: $\Delta E = q + w$. Heat and work are measured in joules (J).
- › Energy is always conserved: it can change from one form to another and be transferred into or out of a system, but the total quantity of energy in the universe (system *plus* surroundings) is constant.
- › The internal energy of a system is a state function and, thus, is independent of how the system attained that energy; therefore, the same overall ΔE can occur through any combination of q and w.
- › A system can lose energy by doing work of expansion (increase in volume) or gain energy by having work of contraction done on it (decrease in volume), both of which are forms of pressure-volume (PV) work.

6.2 ENTHALPY: CHANGES AT CONSTANT PRESSURE

Most physical and chemical changes occur at nearly constant atmospheric pressure—a reaction in an open flask, the freezing of a lake, a biochemical process in an organism. In this section, we discuss *enthalpy,* a thermodynamic variable that relates directly to energy changes at constant pressure.

The Meaning of Enthalpy

To determine ΔE, we must measure both heat and work. But, for reactions *at constant pressure,* a thermodynamic quantity called **enthalpy (*H*)** eliminates the need to measure PV work. The enthalpy of a system is defined as the internal energy *plus* the product of the pressure and volume:

$$H = E + PV$$

The **change in enthalpy (Δ*H*)** is the change in the system's internal energy *plus* the product of the pressure, which is constant, and the change in volume (ΔV):

$$\Delta H = \Delta E + P\Delta V \quad \textbf{(6.5)}$$

Combining Equations 6.2 ($\Delta E = q + w$) and 6.5 gives

$$\Delta H = \Delta E + P\Delta V = q + w + P\Delta V$$

At constant pressure, we denote q as q_P, giving $\Delta H = q_P + w + P\Delta V$. According to Equation 6.4, $w = -P\Delta V$, so $-w$ can be substituted for $P\Delta V$ to give

$$\Delta H = q_P + w + P\Delta V = q_P + w - w$$

which simplifies to:

$$\Delta H = \Delta E + P\Delta V = q_P \quad \textbf{(6.6)}$$

Thus, *the change in enthalpy equals the heat absorbed or released at* **constant** *pressure*. For most changes occurring at constant pressure, ΔH is more relevant than ΔE *and* easier to obtain: *to find* ΔH, *measure* q_P only; work need not be measured. We discuss the laboratory method for doing this in Section 6.3.

Comparing ΔE and ΔH

Knowing the *enthalpy* change of a system tells us a lot about its *energy* change as well. In fact, because many reactions involve little (if any) PV work, most (or all) of the energy change is due to a transfer of heat. Here are three cases:

1. *Reactions that do not involve gases.* Gases do not appear in precipitation, many acid-base, many redox reactions, and so forth. For example,

$$2KOH(aq) + H_2SO_4(aq) \longrightarrow K_2SO_4(aq) + 2H_2O(l)$$

   Because liquids and solids undergo *very* small volume changes, $\Delta V \approx 0$; thus $P\Delta V \approx 0$; and $\Delta H = \Delta E + P\Delta V \approx \Delta E + 0 \approx \Delta E$.

2. *Reactions in which the amount (mol) of gas does* ***not*** *change.* When the total amount of gaseous reactants equals the total amount of gaseous products, the volume is constant, $\Delta V = 0$, so $P\Delta V = 0$ and $\Delta H = \Delta E$. For example,

$$N_2(g) + O_2(g) \longrightarrow 2NO(g)$$

3. *Reactions in which the amount (mol) of gas* ***does*** *change.* In these cases, $P\Delta V \neq 0$. However, q_P is usually *much* larger than $P\Delta V$, so ΔH is still very close to ΔE. For instance, in the combustion of H_2, 3 mol of gas yields 2 mol of gas:

$$2H_2(g) + O_2(g) \longrightarrow 2H_2O(g)$$

   In this reaction, $\Delta H = -483.6$ kJ and $P\Delta V = -2.5$ kJ, so (using Equation 6.5),

$$\Delta H = \Delta E + P\Delta V$$

   and

$$\Delta E = \Delta H - P\Delta V = -483.6 \text{ kJ} - (-2.5 \text{ kJ}) = -481.1 \text{ kJ}$$

   Since most of ΔE occurs as energy transferred as heat, $\Delta H \approx \Delta E$.

Thus, *for most reactions,* ΔH *equals, or is very close to,* ΔE.

Exothermic and Endothermic Processes

Because H is a combination of the three state functions E, P, and V, it is also a state function. Therefore, ΔH equals H_{final} ***minus*** H_{initial}. For a reaction, H_{final} is H_{products} and H_{initial} is $H_{\text{reactants}}$, so the enthalpy change of a reaction is

$$\Delta H = H_{\text{final}} - H_{\text{initial}} = H_{\text{products}} - H_{\text{reactants}}$$

Since H_{products} can be either more or less than $H_{\text{reactants}}$, the *sign* of ΔH indicates whether heat is absorbed or released during the reaction. We determine the sign of ΔH *by imagining the heat as a "reactant" or a "product."* The two possible types of processes are

1. *Exothermic process: heat as a "product."* An **exothermic process** *releases* heat (*exothermic* means "heat out") and results in a *decrease* in the enthalpy of the system:

$$\text{Exothermic:} \qquad H_{\text{products}} < H_{\text{reactants}} \qquad \text{so} \qquad \Delta H < 0$$

For example, when methane burns in air, heat flows *out of* the system into the surroundings, so we show it as a "product":

$$CH_4(g) + 2O_2(g) \longrightarrow CO_2(g) + 2H_2O(g) + \textit{heat}$$

The reactants (1 mol of CH_4 and 2 mol of O_2) release heat during the reaction, so they originally had more enthalpy than the products (1 mol of CO_2 and 2 mol of H_2O):

$$H_{\text{reactants}} > H_{\text{products}} \qquad \text{so} \qquad \Delta H\ (H_{\text{products}} - H_{\text{reactants}}) < 0$$

This exothermic change is shown in the **enthalpy diagram** in Figure 6.8A. The ΔH arrow *always* points from reactants to products in this kind of diagram.

2. *Endothermic process: heat as a "reactant."* An **endothermic process** *absorbs* heat (*endothermic* means "heat in") and results in an *increase* in the enthalpy of the system:

$$\text{Endothermic:} \qquad H_{\text{products}} > H_{\text{reactants}} \qquad \text{so} \qquad \Delta H > 0$$

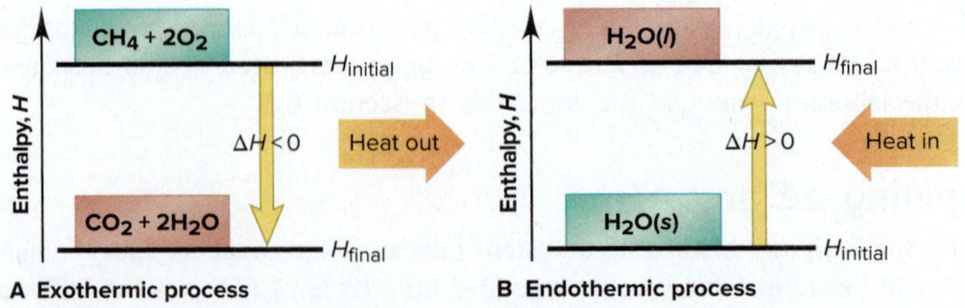

Figure 6.8 Enthalpy diagrams for exothermic and endothermic processes. **A,** The combustion of methane is exothermic: $\Delta H < 0$. **B,** The melting of ice is endothermic: $\Delta H > 0$.

When ice melts, for instance, heat flows *into* the ice from the surroundings, so we show the heat as a "reactant":

$$heat + H_2O(s) \longrightarrow H_2O(l)$$

Because heat is absorbed, the enthalpy of the product (water) must be higher than that of the reactant (ice):

$$H_{water} > H_{ice} \quad \text{so} \quad \Delta H\ (H_{water} - H_{ice}) > 0$$

This endothermic change is shown in Figure 6.8B. (In general, the value of an enthalpy change is determined with reactants and products at the same temperature.)

SAMPLE PROBLEM 6.3 **Drawing Enthalpy Diagrams and Determining the Sign of ΔH**

Problem In each of the following cases, determine the sign of ΔH, state whether the reaction is exothermic or endothermic, and draw an enthalpy diagram:

(a) $H_2(g) + \frac{1}{2}O_2(g) \longrightarrow H_2O(l) + 285.8\ \text{kJ}$

(b) $40.7\ \text{kJ} + H_2O(l) \longrightarrow H_2O(g)$

Plan In each equation, we note whether heat is a "product" (exothermic; $\Delta H < 0$) or a "reactant" (endothermic; $\Delta H > 0$). For exothermic reactions, reactants are above products on the enthalpy diagram; for endothermic reactions, reactants are below products. The ΔH arrow *always* points from reactants to products.

Solution **(a)** Heat is a product (on the right), so $\Delta H < 0$ and the reaction is exothermic. Heat is released during the reaction. See the enthalpy diagram *(top)*.

(b) Heat is a reactant (on the left), so $\Delta H > 0$ and the reaction is endothermic. Heat is absorbed during the reaction. See the enthalpy diagram *(bottom)*.

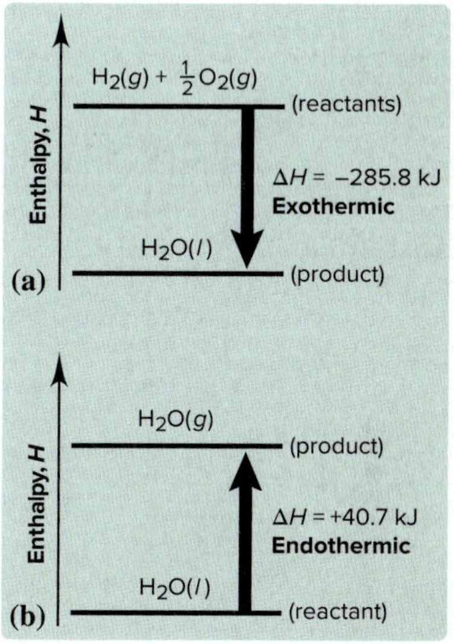

Check Substances that are on the same side of the equation as the heat have less enthalpy than substances on the other side, so make sure those substances are placed on the lower line of the diagram.

FOLLOW-UP PROBLEMS

6.3A Nitroglycerine decomposes through a violent explosion that releases 5.72×10^3 kJ of heat per mole:

$$C_3H_5(NO_3)_3(l) \longrightarrow 3CO_2(g) + \tfrac{5}{2}H_2O(g) + \tfrac{1}{4}O_2(g) + \tfrac{3}{2}N_2(g)$$

Is the decomposition of nitroglycerine exothermic or endothermic? Draw an enthalpy diagram for the process.

6.3B Ammonium nitrate is used in some "instant" cold packs because the salt absorbs 25.7 kJ per mole as it dissolves:

$$NH_4NO_3(s) \xrightarrow{H_2O} NH_4^+(aq) + NO_3^-(aq)$$

Is dissolving ammonium nitrate an exothermic or endothermic process? Draw an enthalpy diagram for the process.

SOME SIMILAR PROBLEMS 6.24–6.29

› Summary of Section 6.2

- Enthalpy (H) is a state function, so any change in enthalpy (ΔH) is independent of how the change occurred. At constant P, the value of ΔH equals ΔE plus the PV work, which occurs when the volume of the system changes in the presence of an external pressure.
- ΔH equals q_P, the heat released or absorbed during a chemical or physical change that takes place at constant pressure.
- In most cases, ΔH is equal, or very close, to ΔE.
- A change that releases heat is exothermic ($\Delta H < 0$); a change that absorbs heat is endothermic ($\Delta H > 0$).

6.3 CALORIMETRY: MEASURING THE HEAT OF A CHEMICAL OR PHYSICAL CHANGE

Data about energy content and usage are everywhere, from the caloric value of a slice of pizza to the gas mileage of a car. In this section, we consider the basic ideas used to determine such values.

Specific Heat Capacity

To find the energy change during a process, we measure the quantity of heat released or absorbed by relating it to the change in temperature. You know from everyday experience that the more you heat an object, the higher its temperature, and the more you cool it, the lower its temperature; in other words, the quantity of heat (q) absorbed or released by an object is proportional to its temperature change:

$$q \propto \Delta T \qquad \text{or} \qquad q = \text{constant} \times \Delta T \qquad \text{or} \qquad \frac{q}{\Delta T} = \text{constant}$$

What you might not know is that the temperature change depends on the object. Every object has its own **heat capacity (*C*),** the quantity of heat required to change its temperature by 1 K. Heat capacity is the proportionality constant in the preceding equation:

$$\text{Heat capacity } (C) = \frac{q}{\Delta T} \quad \text{[in units of J/K]}$$

A related property is **specific heat capacity (*c*)** (note the lowercase letter), the quantity of heat required to change the temperature of 1 *gram* of a substance or material by 1 K:

$$\text{Specific heat capacity } (c) = \frac{q}{\text{mass} \times \Delta T} \quad \text{[in units of J/g·K]}$$

If we know c for the object being heated (or cooled), we can measure the mass and the temperature change and calculate the heat absorbed (or released):

$$q = c \times \text{mass} \times \Delta T \qquad \textbf{(6.7)}$$

Equation 6.7 says that when an object gets hotter, that is, when ΔT ($T_{\text{final}} - T_{\text{initial}}$) is positive, $q > 0$ (the object absorbs heat). And when an object gets cooler, that is, when ΔT is negative, $q < 0$ (the object releases heat). Table 6.2 lists the specific heat capacities of some representative substances and materials. Notice that metals have relatively low values of c and water has a very high value: for instance, it takes over 30 times as much energy to increase the temperature of a gram of water by 1 K as it does a gram of gold!

Closely related to the specific heat capacity (but reserved for substances) is the **molar heat capacity (C_m),** the quantity of heat required to change the temperature of 1 *mole* of a substance by 1 K:

$$\text{Molar heat capacity } (C_m) = \frac{q}{\text{amount (mol)} \times \Delta T} \quad \text{[in units of J/mol·K]}$$

To find C_m of liquid H_2O, we multiply c of liquid H_2O (4.184 J/g·K) by the molar mass of H_2O (18.02 g/mol):

$$C_m \text{ of } H_2O(l) = 4.184 \frac{\text{J}}{\text{g·K}} \times \frac{18.02 \text{ g}}{1 \text{ mol}} = 75.40 \frac{\text{J}}{\text{mol·K}}$$

Liquid water has a *very* high specific heat capacity (~4.2 J/g·K), about six times that of rock (~0.7 J/g·K). If Earth had no oceans *(see photo),* it would take one-sixth as much energy for the Sun to heat the rocky surface, and the surface would give off six times as much heat after sundown. Days would be scorching and nights frigid. Water's high specific heat capacity means that large bodies of water require a lot more heat for a given increase in temperature than land masses do. Earth's oceans help regulate its weather and climate.

Table 6.2 Specific Heat Capacities (c) of Some Elements, Compounds, and Materials

Substance	c (J/g·K)*
Elements	
Aluminum, Al	0.900
Graphite, C	0.711
Iron, Fe	0.450
Copper, Cu	0.387
Gold, Au	0.129
Compounds	
Water, $H_2O(l)$	4.184
Ethyl alcohol, $C_2H_5OH(l)$	2.46
Ethylene glycol, $(CH_2OH)_2(l)$	2.42
Carbon tetrachloride, $CCl_4(l)$	0.862
Materials	
Wood	1.76
Cement	0.88
Glass	0.84
Granite	0.79
Steel	0.45

*At 298 K (25°C)

Earth without oceans.

SAMPLE PROBLEM 6.4 **Relating Quantity of Heat and Temperature Change**

Problem A layer of copper welded to the bottom of a skillet weighs 125 g. How much heat is needed to raise the temperature of the copper layer from 25°C to 300.°C? The specific heat capacity (c) of Cu is given in Table 6.2.

Plan We know the mass (125 g) and c (0.387 J/g·K) of Cu and can find ΔT in °C, which equals ΔT in K. We then use Equation 6.7 to calculate the heat required for the change.

Solution Calculating ΔT and then q:

$$\Delta T = T_{\text{final}} - T_{\text{initial}} = 300.°\text{C} - 25°\text{C} = 275°\text{C} = 275\ \text{K}$$

$$q = c \times \text{mass (g)} \times \Delta T = 0.387\ \text{J/g·K} \times 125\ \text{g} \times 275\ \text{K} = 1.33\times10^4\ \text{J}$$

Check Heat is absorbed by the copper bottom (system), so q is positive. Rounding shows the arithmetic is reasonable:

$$q \approx 0.4\ \text{J/g·K} \times 100\ \text{g} \times 300\ \text{K} = 1.2\times10^4\ \text{J}.$$

Comment Remember that the Kelvin temperature scale uses the same size degree as the Celsius scale; for this reason, ΔT has the same value regardless of which of the two temperature scales is used to express T_{inital} and T_{final}. In this problem, 300.°C = 573 K, 25°C = 298 K, and ΔT = 573 K − 298 K = 275 K, as shown above.

FOLLOW-UP PROBLEMS

6.4A Before baking a pie, you line the bottom of an oven with a 7.65-g piece of aluminum foil and then raise the oven temperature from 18°C to 375°C. Use Table 6.2 to find the heat (in J) absorbed by the foil.

6.4B The ethylene glycol (d = 1.11 g/mL; also see Table 6.2) in a car radiator cools from 37.0°C to 25.0°C by releasing 177 kJ of heat. What volume of ethylene glycol is in the radiator?

SOME SIMILAR PROBLEMS 6.37–6.40

The Two Major Types of Calorimetry

How do we know the heat of an acid-base reaction or the Calories in a teaspoon of sugar? In the lab, we construct "surroundings" that retain the heat as reactants change to products, and then note the temperature change. These "surroundings" take the form of a **calorimeter,** a device used to measure the heat released (or absorbed) by a physical or chemical process. Let's look at two types of calorimeter—one designed to measure the heat at constant pressure and the other at constant volume.

Constant-Pressure Calorimetry For processes that take place at constant pressure, the heat transferred (q_P) is often measured in a *coffee-cup calorimeter* (Figure 6.9). This simple apparatus can be used to find the heat of an aqueous reaction, the heat accompanying the dissolving of a salt, or even the specific heat capacity of a solid that does not react with or dissolve in water.

Let's examine the third application first. The solid (the system) is weighed, heated to some known temperature, and added to a known mass and temperature of water (surroundings) in the calorimeter. After stirring, the final water temperature is measured, which is also the final temperature of the solid. Assuming no heat escapes the calorimeter, the heat released by the system ($-q_{\text{sys}}$ or $-q_{\text{solid}}$) as it cools is equal in magnitude but opposite in sign to the heat absorbed by the surroundings ($+q_{\text{surr}}$, or $+q_{H_2O}$) as they warm:

$$-q_{\text{solid}} = q_{H_2O}$$

Substituting from Equation 6.7 on each side of this equation gives

$$-(c_{\text{solid}} \times \text{mass}_{\text{solid}} \times \Delta T_{\text{solid}}) = c_{H_2O} \times \text{mass}_{H_2O} \times \Delta T_{H_2O}$$

All the quantities are known or measured except c_{solid}:

$$c_{\text{solid}} = -\frac{c_{H_2O} \times \text{mass}_{H_2O} \times \Delta T_{H_2O}}{\text{mass}_{\text{solid}} \times \Delta T_{\text{solid}}}$$

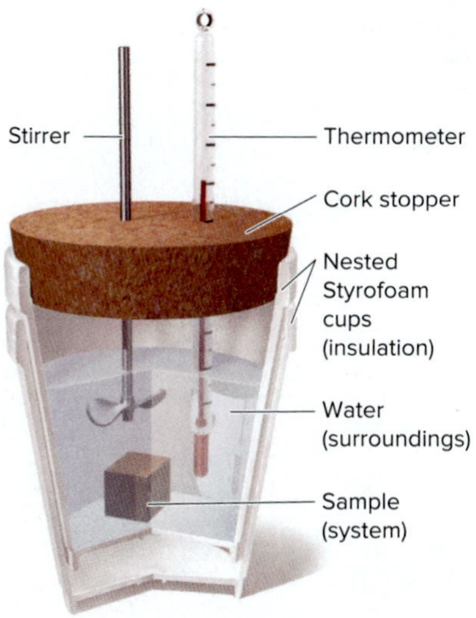

Figure 6.9 Coffee-cup calorimeter. This device measures the heat transferred at constant pressure (q_P).

SAMPLE PROBLEM 6.5 Determining the Specific Heat Capacity of a Solid

Problem You heat 22.05 g of a solid in a test tube to 100.00°C and then add the solid to 50.00 g of water in a coffee-cup calorimeter. The water temperature changes from 25.10°C to 28.49°C. Find the specific heat capacity of the solid.

Plan We are given the masses of the solid (22.05 g) and of H_2O (50.00 g), and we can find the temperature changes of the water and of the solid by subtracting the given values, always using $T_{final} - T_{initial}$. Using Equation 6.7, we set the heat released by the solid ($-q_{solid}$) equal to the heat absorbed by the water (q_{H_2O}). The specific heat of water is known, and we solve for c_{solid}.

Solution Finding ΔT_{solid} and ΔT_{H_2O}:

$T_{initial}$ (solid) = 100.00°C; $T_{initial}$ (H_2O) = 25.10°C; T_{final} (solid and H_2O) = 28.49°C

$$\Delta T_{H_2O} = T_{final} - T_{initial} = 28.49°C - 25.10°C = 3.39°C = 3.39 \text{ K}$$

$$\Delta T_{solid} = T_{final} - T_{initial} = 28.49°C - 100.00°C = -71.51°C = -71.51 \text{ K}$$

Solving for c_{solid}:

$$c_{solid} = -\frac{c_{H_2O} \times \text{mass}_{H_2O} \times \Delta T_{H_2O}}{\text{mass}_{solid} \times \Delta T_{solid}} = -\frac{4.184 \text{ J/g·K} \times 50.00 \text{ g} \times 3.39 \text{ K}}{22.05 \text{ g} \times (-71.51 \text{ K})} = 0.450 \text{ J/g·K}$$

Check Rounding gives −4 J/g·K × 50 g × 3 K/[20 g × (−70°C)] = 0.4 J/g·K, so the answer seems correct.

FOLLOW-UP PROBLEMS

6.5A A 12.18-g sample of a shiny, orange-brown metal is heated to 65.00°C and then added to 25.00 g of water in a coffee-cup calorimeter. The water temperature changes from 25.55°C to 27.25°C. What is the unknown metal (see Table 6.2)?

6.5B A 33.2-g titanium bicycle part is added to 75.0 g of water in a coffee-cup calorimeter at 50.00°C, and the temperature drops to 49.30°C. What was the initial temperature (in °C) of the metal part (*c* of titanium is 0.228 J/g·K)?

SOME SIMILAR PROBLEMS 6.41–6.46

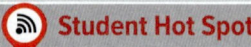

Student Hot Spot

Student data indicate you may struggle with constant-pressure calorimetry calculations. Access the Smartbook to view additional Learning Resources on this topic.

In the next sample problem, the calorimeter is used to study the transfer of heat during an aqueous acid-base reaction. Recall that, if a reaction takes place at constant pressure, the heat of the reaction (q_{rxn}) is equal to its enthalpy change (ΔH).

SAMPLE PROBLEM 6.6 Determining the Enthalpy Change of an Aqueous Reaction

Problem You place 50.0 mL of 0.500 *M* NaOH in a coffee-cup calorimeter at 25.00°C and add 25.0 mL of 0.500 *M* HCl, also at 25.00°C. After stirring, the final temperature is 27.21°C. [Assume that the total volume is the sum of the individual volumes and that the final solution has the same density (1.00 g/mL) and specific heat capacity (Table 6.2) as water.] **(a)** Calculate q_{soln} (in J). **(b)** Calculate the change in enthalpy, ΔH, of the reaction (in kJ/mol of H_2O formed).

(a) Calculate q_{soln}

Plan The solution is the surroundings, and as the reaction takes place, heat flows into the solution. To find q_{soln}, we use Equation 6.7, so we need the mass of solution, the change in temperature, and the specific heat capacity. We know the solutions' volumes (25.0 mL and 50.0 mL), so we find their masses with the given density (1.00 g/mL). Then, to find q_{soln}, we multiply the total mass by the given *c* (4.184 J/g·K) and the change in *T*, which we find from $T_{final} - T_{initial}$.

Solution Finding mass_{soln} and ΔT_{soln}:

$$\text{Total mass (g) of solution} = (25.0 \text{ mL} + 50.0 \text{ mL}) \times 1.00 \text{ g/mL} = 75.0 \text{ g}$$
$$\Delta T = 27.21°C - 25.00°C = 2.21°C = 2.21 \text{ K}$$

Finding q_{soln}:

$$q_{soln} = c_{soln} \times \text{mass}_{soln} \times \Delta T_{soln} = (4.184\ \text{J/g·K})(75.0\ \text{g})(2.21\ \text{K}) = 693\ \text{J}$$

Check The value of q_{soln} is positive because the reaction releases heat to the solution. Rounding to check the size of q_{soln} gives 4 J/g·K × 75 g × 2 K = 600 J, which is close to the answer.

(b) Calculate the change in enthalpy (ΔH)

Plan The heat of the surroundings is q_{soln}, and it is the negative of the heat of the reaction (q_{rxn}), which equals ΔH. And dividing q_{rxn} by the amount (mol) of water formed in the reaction gives ΔH in kJ per mole of water formed. To calculate the amount (mol) of water formed, we write the balanced equation for the acid-base reaction and use the volumes and the concentrations (0.500 *M*) to find amount (mol) of each reactant (H^+ and OH^-). Since the amounts of two reactants are given, we determine which is limiting, that is, which gives less product (H_2O).

Solution Writing the balanced equation:

$$HCl(aq) + NaOH(aq) \longrightarrow H_2O(l) + NaCl(aq)$$

Finding amounts (mol) of reactants:

$$\text{Amount (mol) of HCl} = 0.500\ \text{mol HCl/L} \times 0.0250\ \text{L} = 0.0125\ \text{mol HCl}$$

$$\text{Amount (mol) of NaOH} = 0.500\ \text{mol NaOH/L} \times 0.0500\ \text{L} = 0.0250\ \text{mol NaOH}$$

Finding the amount (mol) of product: All the coefficients in the equation are 1, which means that the amount (mol) of reactant yields that amount of product. Therefore, HCl is limiting because it yields less product: 0.0125 mol of H_2O.

Finding ΔH: Heat absorbed by the solution was released by the reaction; that is,

$$q_{soln} = -q_{rxn} = 693\ \text{J} \quad \text{so} \quad q_{rxn} = -693\ \text{J}$$

$$\Delta H\ (\text{kJ/mol}) = \frac{q_{rxn}}{\text{mol } H_2O} \times \frac{1\ \text{kJ}}{1000\ \text{J}} = \frac{-693\ \text{J}}{0.0125\ \text{mol } H_2O} \times \frac{1\ \text{kJ}}{1000\ \text{J}} = -55.4\ \text{kJ/mol } H_2O$$

Check We check for the limiting reactant: The volume of H^+ is half the volume of OH^-, but they have the same concentration and a 1/1 stoichiometric ratio. Therefore, the amount (mol) of H^+ determines the amount of product. Rounding and taking the negative of q_{soln} to find ΔH gives −600 J/0.012 mol = -5×10^4 J/mol, or −50 kJ/mol, so the answer seems correct.

FOLLOW-UP PROBLEMS

6.6A When 25.0 mL of 2.00 *M* HNO_3 and 50.0 mL of 1.00 *M* KOH, both at 21.90°C, are mixed, a reaction causes the temperature of the mixture to rise to 27.05°C.
(a) Write balanced molecular and net ionic equations for the reaction.
(b) Calculate q_{rxn} (in kJ). (Assume that the final volume is the sum of the volumes being mixed and that the final solution has d = 1.00 g/mL and c = 4.184 J/g·K.)

6.6B After 50.0 mL of 0.500 *M* $Ba(OH)_2$ and the same volume of HCl of the same concentration react in a coffee-cup calorimeter, you find q_{soln} to be 1.386 kJ. Calculate ΔH of the reaction in kJ/mol of H_2O formed.

SOME SIMILAR PROBLEMS 6.47 and 6.48

Constant-Volume Calorimetry Constant-volume calorimetry is carried out in a *bomb calorimeter,* a device commonly used to measure the heat of combustion reactions, usually involving fuels or foods. With a coffee-cup calorimeter, we simply assume that all the heat is absorbed by the water, but in reality, some is absorbed by the stirrer, thermometer, cups, and so forth. With the much more precise bomb calorimeter, the *heat capacity of the entire calorimeter* is known (or can be determined).

Figure 6.10 (*next page*) depicts the preweighed combustible sample in a metal-walled chamber (the bomb), which is filled with oxygen gas and immersed in an insulated water bath fitted with motorized stirrer and thermometer. A heating coil connected to an

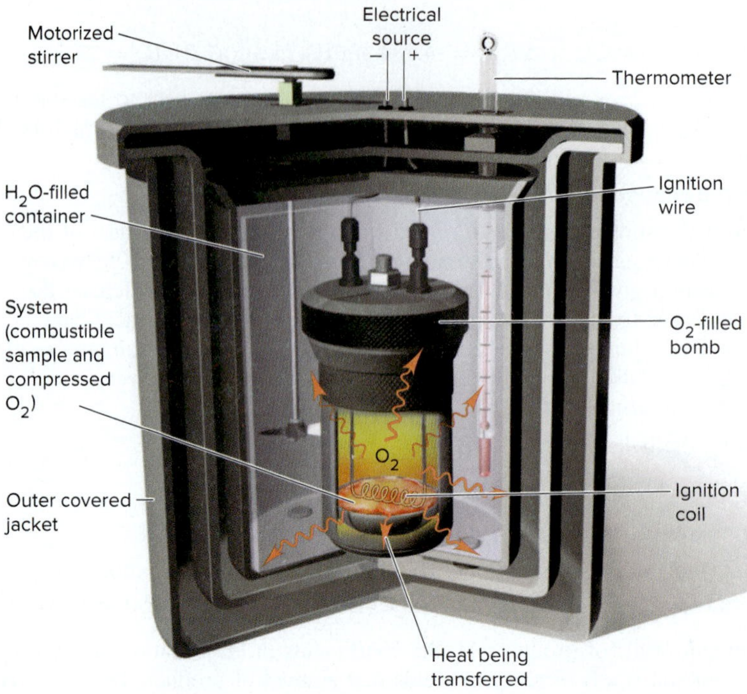

Figure 6.10 **A bomb calorimeter.** This device measures the heat released at constant volume (q_V).

electrical source ignites the sample, and the heat released raises the temperature of the bomb, water, and other calorimeter parts.

Recall from Equation 6.7 that the heat absorbed by the calorimeter can be calculated using the relationship

$$q_{\text{calorimeter}} = c_{\text{calorimeter}} \times \text{mass}_{\text{calorimeter}} \times \Delta T_{\text{calorimeter}}$$

The heat capacity of the entire calorimeter, $C_{\text{calorimeter}}$, in units of J/K or kJ/K, combines $c_{\text{calorimeter}}$ and $\text{mass}_{\text{calorimeter}}$. Thus, Equation 6.7 can be simplified to

$$q_{\text{calorimeter}} = C_{\text{calorimeter}} \times \Delta T_{\text{calorimeter}} \qquad \textbf{(6.8)}$$

Because we know the heat capacity of the entire calorimeter, we can use the measured ΔT to calculate the heat released during the combustion of the sample:

$$-q_{\text{rxn}} = q_{\text{calorimeter}}$$

Note that the steel bomb is tightly sealed, not open to the atmosphere as is the coffee cup, so the pressure is *not* constant. And the volume of the bomb is fixed, so $\Delta V = 0$, and thus $P\Delta V = 0$. Thus, the energy change measured is the *heat released at constant volume* (q_V), which equals ΔE, not ΔH:

$$\Delta E = q + w = q_V + 0 = q_V$$

Recall from Section 6.2, however, that even though the number of moles of gas may change, $P\Delta V$ is usually *much* less than ΔE, so ΔH is very close to ΔE. For example, ΔH is only 0.5% larger than ΔE for the combustion of H_2 and only 0.2% smaller than ΔE for the combustion of octane.

SAMPLE PROBLEM 6.7 Calculating the Heat of a Combustion Reaction

Problem Oxyacetylene torches produce such high temperatures that they are often used to weld and cut metal. When 2.50 g of acetylene (C_2H_2) is burned in a bomb calorimeter with a heat capacity of 10.85 kJ/K, the temperature increases from 23.488°C to 34.988°C. What is ΔE (in kJ/mol) for this combustion reaction?

Plan To find the heat released during the combustion of 2.50 g of acetylene, we find ΔT, multiply it by the given heat capacity of the calorimeter (10.85 kJ/K), and change the sign since $-q_{rxn} = q_{calorimeter}$. We use the molar mass of acetylene to calculate the heat released by 1 mole of acetylene (ΔE).

Solution Finding ΔT:

$$\Delta T = T_{final} - T_{initial} = 34.988°C - 23.488°C = 11.500°C = 11.500\ K$$

Calculating the heat absorbed by the calorimeter:

$$q_{calorimeter} = C_{calorimeter} \times \Delta T = 10.85\ kJ/K \times 11.500\ K = 124.8\ kJ$$

Therefore, $q_{rxn} = -124.8$ kJ for 2.50 g of acetylene.

$$\Delta E = \frac{-124.8\ kJ}{2.50\ g\ C_2H_2} \times \frac{26.04\ g\ C_2H_2}{1\ mol\ C_2H_2} = -1.30\times10^3\ kJ/mol\ C_2H_2$$

Check A quick math check shows that the answer is reasonable: 11 kJ/K × 12 K = 132 kJ for combustion of a 2.50-g sample. Since 1 mol of C_2H_2 has about ten times that mass, ΔE should be 10 × −132 kJ = −1320 kJ.

FOLLOW-UP PROBLEMS

6.7A A chemist burns 0.8650 g of graphite (a form of carbon) in a bomb calorimeter, and CO_2 forms. If 393.5 kJ of heat is released per mole of graphite and ΔT is 3.116 K, what is the heat capacity of the bomb calorimeter?

6.7B A manufacturer claims that its new dietetic dessert has "fewer than 10 Calories per serving." To test the claim, a chemist at the Department of Consumer Affairs places one serving in a bomb calorimeter and burns it in O_2. The initial temperature is 21.862°C, and the temperature rises to 26.799°C. If the heat capacity of the calorimeter is 8.151 kJ/K, is the manufacturer's claim correct?

SOME SIMILAR PROBLEMS 6.49, 6.50, 6.51, and 6.53

Student Hot Spot

Student data indicate you may struggle with constant-volume calorimetry calculations. Access the Smartbook to view additional Learning Resources on this topic.

› Summary of Section 6.3

- We calculate ΔH of a process by measuring the energy transferred as heat at constant pressure (q_P). To do this, we determine ΔT and multiply it by the substance's mass and by its specific heat capacity (c), which is the quantity of energy needed to raise the temperature of 1 g of the substance by 1 K.
- Calorimeters measure the heat released (or absorbed) during a process either at constant pressure (coffee cup; $q_P = \Delta H$) or at constant volume (bomb; $q_V = \Delta E$).

6.4 STOICHIOMETRY OF THERMOCHEMICAL EQUATIONS

A **thermochemical equation** is a balanced equation that includes the enthalpy change of the reaction (ΔH). Keep in mind that a given ΔH refers only to the *amounts (mol) of substances **and** their states of matter in that equation.* The enthalpy change of any process has two aspects:

- *Sign.* The sign of ΔH depends on whether the reaction is exothermic (−) or endothermic (+). A forward reaction has the *opposite* sign of the reverse reaction. Decomposition of 2 mol of water to its elements (endothermic):

$$2H_2O(l) \longrightarrow 2H_2(g) + O_2(g) \qquad \Delta H = 572\ kJ$$

  Formation of 2 mol of water from its elements (exothermic):

$$2H_2(g) + O_2(g) \longrightarrow 2H_2O(l) \qquad \Delta H = -572\ kJ$$

- *Magnitude.* The magnitude of ΔH is *proportional to the amount of substance.* Formation of 1 mol of water from its elements (half the preceding amount):

$$H_2(g) + \tfrac{1}{2}O_2(g) \longrightarrow H_2O(l) \qquad \Delta H = -286\ kJ$$

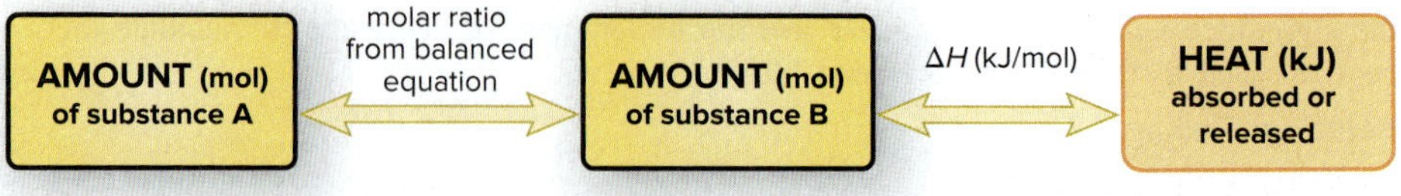

Figure 6.11 The relationship between amount (mol) of substance and the energy (kJ) transferred as heat during a reaction.

Two key points to understand about thermochemical equations are

1. *Balancing coefficients.* When necessary, we use fractional coefficients to balance an equation, because we are specifying the magnitude of ΔH for a *particular amount,* often 1 mol, *of a substance,* for example, gaseous SO_2:

$$\tfrac{1}{8}S_8(s) + O_2(g) \longrightarrow SO_2(g) \qquad \Delta H = -296.8 \text{ kJ}$$

2. *Thermochemical equivalence.* For a *particular reaction,* a certain amount of substance is thermochemically equivalent to a certain quantity of energy. Just as we use stoichiometrically equivalent molar ratios to find amounts of substances, we use thermochemically equivalent quantities to find the ΔH of a reaction for a given amount of substance. Some examples from the previous two reactions are

296.8 kJ is thermochemically equivalent to $\tfrac{1}{8}$ mol of $S_8(s)$
286 kJ is thermochemically equivalent to $\tfrac{1}{2}$ mol of $O_2(g)$
286 kJ is thermochemically equivalent to 1 mol of $H_2O(l)$

Also, just as we use molar mass (in g/mol) to convert an amount (mol) of a substance to mass (g), we use the ΔH (in kJ/mol) to convert an amount (mol) of a substance to an equivalent quantity of heat (in kJ). Figure 6.11 shows this new relationship, and Sample Problem 6.8 applies it.

SAMPLE PROBLEM 6.8 Using the Enthalpy Change of a Reaction (ΔH) to Find the Amount of a Substance

Problem The major source of aluminum in the world is bauxite (mostly aluminum oxide). Its thermal decomposition can be written as

$$Al_2O_3(s) \xrightarrow{\Delta} 2Al(s) + \tfrac{3}{2}O_2(g) \qquad \Delta H = 1676 \text{ kJ}$$

If aluminum is produced this way (see Comment), how many grams of aluminum can form when 1.000×10^3 kJ of heat is transferred?

Plan From the balanced equation and the enthalpy change, we see that 1676 kJ of heat is thermochemically equivalent to 2 mol of Al. With this equivalent quantity, we convert the given number of kJ to amount (mol) formed and then convert amount to mass in g (see the road map).

Road Map

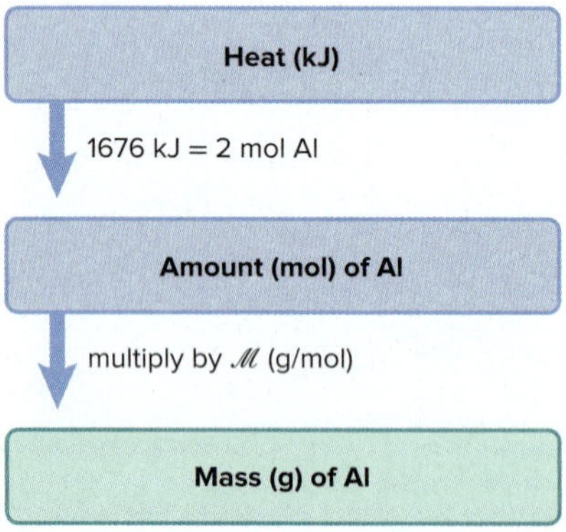

Solution Combining steps to convert from heat transferred to mass of Al:

$$\text{Mass (g) of Al} = (1.000\times10^3 \text{ kJ}) \times \frac{2 \text{ mol Al}}{1676 \text{ kJ}} \times \frac{26.98 \text{ g Al}}{1 \text{ mol Al}} = 32.20 \text{ g Al}$$

Check The mass of aluminum seems correct: ~1700 kJ forms about 2 mol of Al (54 g), so 1000 kJ should form a bit more than half that amount (27 g).

Comment In practice, aluminum is not obtained by heating bauxite but by supplying electrical energy (as you'll see in Chapter 22). Because H is a state function, however, the total energy required for the change is the same no matter how it occurs.

FOLLOW-UP PROBLEMS

6.8A Hydrogenation reactions, in which H_2 and an "unsaturated" organic compound combine, are used in the food, fuel, and polymer industries. In the simplest case, ethene (C_2H_4) and H_2 form ethane (C_2H_6). If 137 kJ is given off per mole of C_2H_4 reacting, how much heat is released when 15.0 kg of C_2H_6 forms?

6.8B In the searing temperatures reached in a lightning bolt, nitrogen and oxygen form nitrogen monoxide, NO. If 180.58 kJ of heat is absorbed per mole of N_2 reacting, how much heat is absorbed when 3.50 metric tons (t; 1 t = 10^3 kg) of NO forms?

SOME SIMILAR PROBLEMS 6.57–6.60

› Summary of Section 6.4

- › A thermochemical equation shows a balanced reaction *and* its ΔH value. The sign of ΔH for a forward reaction is opposite that for the reverse reaction. The magnitude of ΔH is specific for the given equation.
- › The amount of a substance and the quantity of heat specified in a thermochemical equation are thermochemically equivalent and form a conversion factor for finding the quantity of heat transferred when any amount of the substance reacts.

6.5 HESS'S LAW: FINDING ΔH OF ANY REACTION

In some cases, a reaction is difficult, even impossible, to carry out individually: it may be part of a complex biochemical process, or take place under extreme conditions, or require a change in conditions to occur. Even if we can't run a reaction in the lab, we can still find its enthalpy change. In fact, the state-function property of enthalpy (H) allows us to find ΔH of *any* reaction for which we can write an equation.

This application is based on **Hess's law,** which states that *the enthalpy change of an overall process is the sum of the enthalpy changes of its individual steps:*

$$\Delta H_{\text{overall}} = \Delta H_1 + \Delta H_2 + \cdots + \Delta H_n \qquad \textbf{(6.9)}$$

This law follows from the fact that ΔH for a process depends only on the difference between the final and initial states. To apply Hess's law, we

- Imagine that an overall (target) reaction occurs through a series of individual reaction steps, whether or not it actually does. (Adding the steps must give the overall reaction.)
- Choose individual reaction steps, with each one having a known ΔH.
- Adjust, as needed, the equations for the individual steps, as well as their ΔH values, to obtain the overall equation.
- Add the ΔH values for the steps to get the unknown ΔH of the overall reaction. We can also find an unknown ΔH of any step by subtraction, if we know the ΔH values for the overall reaction and all the other steps.

Let's apply Hess's law to the oxidation of sulfur to sulfur trioxide, the key change in the industrial production of sulfuric acid and in the formation of acid rain. (To introduce this approach, we'll use S as the formula for sulfur, rather than the more correct S_8.) When we burn S in an excess of O_2, sulfur dioxide (SO_2) forms (equation 1), *not* sulfur trioxide (SO_3). After a change in conditions, we add more O_2 and oxidize SO_2 to SO_3 (equation 2). Thus, we cannot put S and O_2 in a calorimeter and find ΔH for the overall reaction of S to SO_3 (equation 3). But, we *can* find it with Hess's law.

$$\begin{array}{llll}
\text{Equation 1:} & S(s) + O_2(g) & \longrightarrow SO_2(g) & \Delta H_1 = -296.8\ \text{kJ} \\
\text{Equation 2:} & 2SO_2(g) + O_2(g) & \longrightarrow 2SO_3(g) & \Delta H_2 = -198.4\ \text{kJ} \\
\text{Equation 3:} & S(s) + \tfrac{3}{2}O_2(g) & \longrightarrow SO_3(g) & \Delta H_3 = ?
\end{array}$$

If we can adjust equation 1 and/or equation 2, *along with their ΔH value(s),* so that these equations add up to equation 3, then their ΔH values will add up to the unknown ΔH_3.

First, we identify our "target" equation, the one whose ΔH we want to find, and note the amount (mol) of each reactant and product; our "target" is equation 3. Then, we adjust equations 1 and/or 2 to make them add up to equation 3:

- Equations 3 and 1 have the same amount of S, so we don't change equation 1.
- Equation 3 has half as much SO_3 as equation 2, so we multiply equation 2 ***and*** $\boldsymbol{\Delta H_2}$ by $\frac{1}{2}$; that is, we always treat the equation and its ΔH value the same.

- With the targeted amounts of reactants and products present, we add equation 1 to the halved equation 2 and *cancel terms that appear on both sides:*

$$\begin{array}{lrcll}
\text{Equation 1:} & S(s) + O_2(g) & \longrightarrow & SO_2(g) & \Delta H_1 = -296.8\ \text{kJ} \\
\frac{1}{2}(\text{Equation 2}): & SO_2(g) + \frac{1}{2}O_2(g) & \longrightarrow & SO_3(g) & \frac{1}{2}(\Delta H_2) = \ -99.2\ \text{kJ} \\
\hline
\text{Equation 3:} & S(s) + O_2(g) + \cancel{SO_2(g)} + \frac{1}{2}O_2(g) & \longrightarrow & \cancel{SO_2(g)} + SO_3(g) & \\
\text{or,} & S(s) + \frac{3}{2}O_2(g) & \longrightarrow & SO_3(g) & \Delta H_3 = -396.0\ \text{kJ}
\end{array}$$

Because ΔH depends only on the difference between H_{final} and H_{initial}, Hess's law tells us that the difference between the enthalpies of the reactants (1 mol of S and $\frac{3}{2}$ mol of O_2) and that of the product (1 mol of SO_3) is the same, whether S is oxidized directly to SO_3 (impossible) or through the intermediate formation of SO_2 (actual).

To summarize, calculating an unknown ΔH involves three steps:

1. Identify the target equation, the step whose ΔH is unknown, and note the amount (mol) of each reactant and product.
2. Adjust each equation with known ΔH values so that the target amount (mol) of each substance is on the correct side of the equation. Remember to:
   - Change the sign of ΔH when you reverse an equation.
   - Multiply amount (mol) and ΔH by the same factor.
3. Add the adjusted equations and their resulting ΔH values to get the target equation and its ΔH. All substances except those in the target equation must cancel.

Student Hot Spot

Student data indicate that you may struggle with the application of Hess's law. Access the Smartbook to view additional Learning Resources on this topic.

SAMPLE PROBLEM 6.9 Using Hess's Law to Calculate an Unknown ΔH

Problem Two pollutants that form in automobile exhaust are CO and NO. An environmental chemist must convert these pollutants to less harmful gases through the following:

$$CO(g) + NO(g) \longrightarrow CO_2(g) + \tfrac{1}{2}N_2(g) \qquad \Delta H = ?$$

Given the following information, calculate the unknown ΔH:

$$\begin{array}{lrcll}
\text{Equation A:} & CO(g) + \frac{1}{2}O_2(g) & \longrightarrow & CO_2(g) & \Delta H = -283.0\ \text{kJ} \\
\text{Equation B:} & N_2(g) + O_2(g) & \longrightarrow & 2NO(g) & \Delta H = \ 180.6\ \text{kJ}
\end{array}$$

Plan We note the amount (mol) of each substance and on which side each appears in the target equation. We adjust equations A and/or B *and* their ΔH values as needed and add them together to obtain the target equation and the unknown ΔH.

Solution Noting substances in the target equation: For reactants, there are 1 mol of CO and 1 mol of NO; for products, there are 1 mol of CO_2 and $\frac{1}{2}$ mol of N_2.

Adjusting the given equations:

- Equation A has the same amounts of CO and CO_2 on the same sides of the arrow as in the target, so we leave that equation as written.
- Equation B has twice as much N_2 and NO as the target, and they are on the opposite sides in the target. Thus, we reverse equation B, change the sign of its ΔH, and multiply both by $\frac{1}{2}$:

$$\tfrac{1}{2}[2NO(g) \longrightarrow N_2(g) + O_2(g)] \qquad \Delta H = -\tfrac{1}{2}(\Delta H) = -\tfrac{1}{2}(180.6\ \text{kJ})$$

or

$$NO(g) \longrightarrow \tfrac{1}{2}N_2(g) + \tfrac{1}{2}O_2(g) \qquad \Delta H = -90.3\ \text{kJ}$$

Adding the adjusted equations to obtain the target equation:

$$\begin{array}{lrcll}
\text{Equation A:} & CO(g) + \cancel{\frac{1}{2}O_2(g)} & \longrightarrow & CO_2(g) & \Delta H = -283.0\ \text{kJ} \\
\frac{1}{2}(\text{Equation B reversed}) & NO(g) & \longrightarrow & \frac{1}{2}N_2(g) + \cancel{\frac{1}{2}O_2(g)} & \Delta H = \ -90.3\ \text{kJ} \\
\hline
\text{Target equation:} & CO(g) + NO(g) & \longrightarrow & CO_2(g) + \frac{1}{2}N_2(g) & \Delta H = -373.3\ \text{kJ}
\end{array}$$

Check Obtaining the desired target equation is a sufficient check. Be sure to remember to change the *sign* of ΔH for any equation you reverse.

FOLLOW-UP PROBLEMS

6.9A Nitrogen oxides undergo many reactions in the environment and in industry. Given the following information, calculate ΔH for the overall equation, $2NO_2(g) + \frac{1}{2}O_2(g) \longrightarrow N_2O_5(s)$:

$$N_2O_5(s) \longrightarrow 2NO(g) + \tfrac{3}{2}O_2(g) \qquad \Delta H = 223.7\ \text{kJ}$$
$$NO(g) + \tfrac{1}{2}O_2(g) \longrightarrow NO_2(g) \qquad \Delta H = -57.1\ \text{kJ}$$

6.9B Use the following reactions to find ΔH when 1 mol of HCl gas forms from its elements:

$$N_2(g) + 3H_2(g) \longrightarrow 2NH_3(g) \qquad \Delta H = -91.8\ \text{kJ}$$
$$N_2(g) + 4H_2(g) + Cl_2(g) \longrightarrow 2NH_4Cl(s) \qquad \Delta H = -628.8\ \text{kJ}$$
$$NH_3(g) + HCl(g) \longrightarrow NH_4Cl(s) \qquad \Delta H = -176.2\ \text{kJ}$$

SOME SIMILAR PROBLEMS 6.70 and 6.71

› Summary of Section 6.5

- Because H is a state function, we can use Hess's law to determine ΔH of any reaction by assuming that it is the sum of other reactions.
- After adjusting the equations of the other reactions and their ΔH values to match the substances in the target equation, we add the adjusted ΔH values to find the unknown ΔH.

6.6 STANDARD ENTHALPIES OF REACTION (ΔH°_{rxn})

In this section, we see how Hess's law is used to determine the ΔH values of an enormous number of reactions. Thermodynamic variables, such as ΔH, vary somewhat with conditions. Therefore, in order to study and compare reactions, chemists have established a set of specific conditions called **standard states:**

- For a *gas,* the standard state is 1 atm* and ideal behavior.
- For a substance in *aqueous solution,* the standard state is 1 *M* concentration.
- For a *pure substance* (element or compound), the standard state is usually the most stable form of the substance at 1 atm and the temperature of interest. In this text (and in most thermodynamic tables), that temperature is usually 25°C (298 K).

The standard-state symbol (shown as a superscript degree sign) indicates that the variable has been measured with *all the substances in their standard states.* For example, when the enthalpy change of a reaction is measured at the standard state, it is the **standard enthalpy of reaction,** ΔH°_{rxn} (also called the *standard heat of reaction*).

Formation Equations and Their Standard Enthalpy Changes

In a **formation equation,** 1 mol of a compound forms from its elements. The **standard enthalpy of formation (ΔH°_f)** (or *standard heat of formation*) is the enthalpy change for the formation equation when all the substances are in their standard states. For instance, the formation equation for methane (CH_4) is

$$C(\text{graphite}) + 2H_2(g) \longrightarrow CH_4(g) \qquad \Delta H^\circ_f = -74.9\ \text{kJ}$$

*The definition of the standard state for gases has been changed to 1 bar, a slightly lower pressure than the 1 atm standard on which the data in this book are based (1 atm = 101.3 kPa = 1.013 bar). For most purposes, this makes very little difference in the standard enthalpy values.

Table 6.3 Selected Standard Enthalpies of Formation at 298 K

Formula	ΔH°_f (kJ/mol)*
Bromine	
$Br_2(l)$	0
$Br_2(g)$	30.9
Calcium	
$Ca(s)$	0
$CaO(s)$	−635.1
$CaCO_3(s)$	−1206.9
Carbon	
C(graphite)	0
C(diamond)	1.9
$CO(g)$	−110.5
$CO_2(g)$	−393.5
$CH_4(g)$	−74.9
$CH_3OH(l)$	−238.6
$HCN(g)$	135
$CS_2(l)$	87.9
Chlorine	
$Cl_2(g)$	0
$Cl(g)$	121.0
$HCl(g)$	−92.3
Hydrogen	
$H_2(g)$	0
$H(g)$	218.0
Mercury	
$Hg(l)$	0
$Hg(g)$	61.3
Nitrogen	
$N_2(g)$	0
$NH_3(g)$	−45.9
$NO(g)$	90.3
Oxygen	
$O_2(g)$	0
$O_3(g)$	143
$H_2O(g)$	−241.8
$H_2O(l)$	−285.8
Silver	
$Ag(s)$	0
$AgCl(s)$	−127.0
Sodium	
$Na(s)$	0
$Na(g)$	107.8
$NaCl(s)$	−411.1
Sulfur	
S_8(rhombic)	0
S_8(monoclinic)	0.3
$SO_2(g)$	−296.8
$SO_3(g)$	−396.0

*Values given to one decimal place.

Fractional coefficients are often used with reactants in a formation equation to obtain 1 mol of the product:

$$Na(s) + \tfrac{1}{2}Cl_2(g) \longrightarrow NaCl(s) \qquad \Delta H^\circ_f = -411.1 \text{ kJ}$$
$$2C(\text{graphite}) + 3H_2(g) + \tfrac{1}{2}O_2(g) \longrightarrow C_2H_5OH(l) \qquad \Delta H^\circ_f = -277.6 \text{ kJ}$$

Standard enthalpies of formation have been tabulated for many substances. The values in Table 6.3 make two points (see Appendix B for a more extensive table):

1. *For an element in its standard state,* $\Delta H^\circ_f = 0$.
   - The standard state for an element is the state of matter (solid, liquid, or gas) that exists at standard conditions. For most metals, like sodium, the standard state is the solid:

     $$Na(s)\text{: } \Delta H^\circ_f = 0 \qquad Na(g)\text{: } \Delta H^\circ_f = 107.8 \text{ kJ/mol}$$

     An additional 107.8 kJ of heat is needed to vaporize 1 mol of solid Na to gaseous Na.

     For the metal mercury and the nonmetal bromine, however, the standard state is the liquid phase since these elements exist as liquids at standard conditions:

     $$Hg(l)\text{: } \Delta H^\circ_f = 0 \qquad Hg(g)\text{: } \Delta H^\circ_f = 61.3 \text{ kJ/mol}$$
     $$Br_2(l)\text{: } \Delta H^\circ_f = 0 \qquad Br_2(g)\text{: } \Delta H^\circ_f = 30.9 \text{ kJ/mol}$$

   - The standard state for molecular elements, such as the halogens, is the molecular form, not separate atoms:

     $$Cl_2(g)\text{: } \Delta H^\circ_f = 0 \qquad Cl(g)\text{: } \Delta H^\circ_f = 121.0 \text{ kJ/mol}$$
     $$H_2(g)\text{: } \Delta H^\circ_f = 0 \qquad H(g)\text{: } \Delta H^\circ_f = 218.0 \text{ kJ/mol}$$

     An additional 121.0 kJ of heat is required to break the bonds in 1 mol of Cl_2 to produce separate Cl atoms and 218.0 kJ is required to break the bonds in 1 mol of H_2.
   - Some elements exist in different forms (called *allotropes;* Chapter 14), but only one is the standard state:

     $$C(\text{graphite})\text{: } \Delta H^\circ_f = 0 \qquad C(\text{diamond})\text{: } \Delta H^\circ_f = 1.9 \text{ kJ/mol}$$
     $$O_2(g)\text{: } \Delta H^\circ_f = 0 \qquad O_3(g)(\text{ozone})\text{: } \Delta H^\circ_f = 143 \text{ kJ/mol}$$
     $$S_8(\text{rhombic})\text{: } \Delta H^\circ_f = 0 \qquad S_8(\text{monoclinic})\text{: } \Delta H^\circ_f = 0.3 \text{ kJ/mol)}$$

2. *Most compounds have a negative* ΔH°_f. That is, most compounds have exothermic formation reactions: *under standard conditions, heat is released when most compounds form from their elements.*

SAMPLE PROBLEM 6.10 Writing Formation Equations

Problem Write a balanced formation equation for each of the following and include the value of ΔH°_f: **(a)** $AgCl(s)$; **(b)** $CaCO_3(s)$; **(c)** $HCN(g)$.

Plan We write the elements as the reactants and 1 mol of the compound as the product, being sure all substances are in their standard states. Then, we balance the equations, keeping a coefficient of 1 for the product, and find the ΔH°_f values in Appendix B.

Solution **(a)** $Ag(s) + \tfrac{1}{2}Cl_2(g) \longrightarrow AgCl(s) \qquad \Delta H^\circ_f = -127.03 \text{ kJ}$

(b) $Ca(s) + C(\text{graphite}) + \tfrac{3}{2}O_2(g) \longrightarrow CaCO_3(s) \qquad \Delta H^\circ_f = -1206.9 \text{ kJ}$

(c) $\tfrac{1}{2}H_2(g) + C(\text{graphite}) + \tfrac{1}{2}N_2(g) \longrightarrow HCN(g) \qquad \Delta H^\circ_f = 135 \text{ kJ}$

FOLLOW-UP PROBLEMS

6.10A Write a balanced formation equation for each of the following and include the value of ΔH°_f: **(a)** $CH_3OH(l)$; **(b)** $CaO(s)$; **(c)** $CS_2(l)$.

6.10B Write a balanced formation equation for each of the following and include the value of ΔH°_f: **(a)** $CHCl_3(l)$; **(b)** $NH_4Cl(s)$; **(c)** $PbSO_4(s)$.

SOME SIMILAR PROBLEMS 6.80 and 6.81

Determining ΔH°_{rxn} from ΔH°_f Values for Reactants and Products

We can use ΔH°_f values to determine ΔH°_{rxn} for any reaction. By applying Hess's law, we can imagine the reaction occurring in two steps (Figure 6.12):

Step 1. Each reactant decomposes to its elements. This is the *reverse* of the formation reaction for the *reactant,* so the standard enthalpy change is $-\Delta H^\circ_f$.
Step 2. Each product forms from its elements. This step *is* the formation reaction for the *product,* so the standard enthalpy change is ΔH°_f.

According to Hess's law, we add the enthalpy changes for these steps to obtain the overall enthalpy change for the reaction (ΔH°_{rxn}). Suppose we want ΔH°_{rxn} for

$$TiCl_4(l) + 2H_2O(g) \longrightarrow TiO_2(s) + 4HCl(g)$$

We write this equation as though it were the sum of four individual equations, one for each compound. The first two show step 1, the decomposition of the reactants to their elements (*reverse* of their formation); the second two show step 2, the formation of the products from their elements:

$$\begin{aligned} TiCl_4(l) &\longrightarrow Ti(s) + 2Cl_2(g) & -\Delta H^\circ_f[TiCl_4(l)] \\ 2H_2O(g) &\longrightarrow 2H_2(g) + O_2(g) & -2\Delta H^\circ_f[H_2O(g)] \\ Ti(s) + O_2(g) &\longrightarrow TiO_2(s) & \Delta H^\circ_f[TiO_2(s)] \\ 2H_2(g) + 2Cl_2(g) &\longrightarrow 4HCl(g) & 4\Delta H^\circ_f[HCl(g)] \end{aligned}$$

$$TiCl_4(l) + 2H_2O(g) + \cancel{Ti(s)} + \cancel{O_2(g)} + \cancel{2H_2(g)} + \cancel{2Cl_2(g)} \longrightarrow \cancel{Ti(s)} + \cancel{2Cl_2(g)} + \cancel{2H_2(g)} + \cancel{O_2(g)} + TiO_2(s) + 4HCl(g)$$

or $$TiCl_4(l) + 2H_2O(g) \longrightarrow TiO_2(s) + 4HCl(g)$$

Note that when titanium(IV) chloride and water react, the reactants don't *actually* decompose to their elements, which then recombine to form the products. But the great usefulness of Hess's law and the state-function concept is that ΔH°_{rxn} is the difference between two state functions, $H^\circ_{products}$ minus $H^\circ_{reactants}$, so it doesn't matter how the reaction *actually* occurs. We add the individual enthalpy changes to find ΔH°_{rxn} (Hess's law):

$$\begin{aligned} \Delta H^\circ_{rxn} &= \Delta H^\circ_f[TiO_2(s)] + 4\Delta H^\circ_f[HCl(g)] + \{-\Delta H^\circ_f[TiCl_4(l)]\} + \{-2\Delta H^\circ_f[H_2O(g)]\} \\ &= \underbrace{\{\Delta H^\circ_f[TiO_2(s)] + 4\Delta H^\circ_f[HCl(g)]\}}_{\text{Products}} - \underbrace{\{\Delta H^\circ_f[TiCl_4(l)] + 2\Delta H^\circ_f[H_2O(g)]\}}_{\text{Reactants}} \end{aligned}$$

For this example, the arithmetic gives $\Delta H^\circ_{rxn} = -25.39$ kJ. When we generalize the pattern, we see that *the standard enthalpy of reaction is the sum of the standard enthalpies of formation of the* ***products*** *minus the sum of the standard enthalpies of formation of the* ***reactants*** (see Figure 6.12):

$$\Delta H^\circ_{rxn} = \Sigma m\Delta H^\circ_{f(products)} - \Sigma n\Delta H^\circ_{f(reactants)} \quad \textbf{(6.10)}$$

where Σ means "sum of," and m and n are the amounts (mol) of the products and reactants given by the coefficients in the balanced equation.

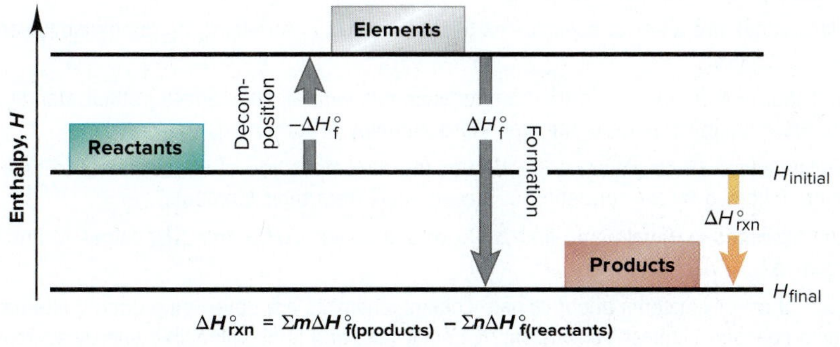

Figure 6.12 The two-step process for determining ΔH°_{rxn} from ΔH°_f values.

SAMPLE PROBLEM 6.11 Calculating ΔH°_{rxn} from ΔH°_f Values

Problem Nitric acid is used to make many products, including fertilizers, dyes, and explosives. The first step in its production is the oxidation of ammonia:

$$4NH_3(g) + 5O_2(g) \longrightarrow 4NO(g) + 6H_2O(g)$$

Calculate ΔH°_{rxn} from ΔH°_f values.

Plan We use values from Appendix B and apply Equation 6.10.

Solution Calculating ΔH°_{rxn}:

$$\begin{aligned}\Delta H^\circ_{rxn} &= \Sigma m\Delta H^\circ_{f\,(products)} - \Sigma n\Delta H^\circ_{f\,(reactants)}\\ &= \{4\Delta H^\circ_f[NO(g)] + 6\Delta H^\circ_f[H_2O(g)]\} - \{4\Delta H^\circ_f[NH_3(g)] + 5\Delta H^\circ_f[O_2(g)]\}\\ &= [(4\text{ mol})(90.29\text{ kJ/mol}) + (6\text{ mol})(-241.840\text{ kJ/mol})]\\ &\quad - [(4\text{ mol})(-45.9\text{ kJ/mol}) + (5\text{ mol})(0\text{ kJ/mol})]\\ &= [361.2\text{ kJ} + (-1451.04\text{ kJ})] - [-184\text{ kJ} + 0\text{ kJ}] = -906\text{ kJ}\end{aligned}$$

Check We write formation equations, with ΔH°_f values for the amounts of each compound, in the correct direction (forward for products and reverse for reactants) and find the sum:

$4NH_3(g) \longrightarrow 2N_2(g) + 6H_2(g)$	$-4\Delta H^\circ_f =$	$-4(-45.9\text{ kJ}) =$	184 kJ
$2N_2(g) + 2O_2(g) \longrightarrow 4NO(g)$	$4\Delta H^\circ_f =$	$4(90.29\text{ kJ}) =$	361.2 kJ
$6H_2(g) + 3O_2(g) \longrightarrow 6H_2O(g)$	$6\Delta H^\circ_f =$	$6(-241.840\text{ kJ}) =$	−1451.04 kJ
$4NH_3(g) + 5O_2(g) \longrightarrow 4NO(g) + 6H_2O(g)$		$\Delta H^\circ_{rxn} =$	−906 kJ

Comment In this problem, we know the individual ΔH°_f values and find the sum, ΔH°_{rxn}. In Follow-up Problem 6.11B, we know the sum and want to find one of the ΔH°_f values.

FOLLOW-UP PROBLEMS

6.11A Tetraphosphorus decoxide reacts vigorously with water to form phosphoric acid, one of the top-10 industrial chemicals:

$$P_4O_{10}(s) + 6H_2O(l) \longrightarrow 4H_3PO_4(l)$$

Calculate ΔH°_{rxn} from ΔH°_f values in Appendix B.

6.11B Use the following to find ΔH°_f of methanol $CH_3OH(l)$:

$$CH_3OH(l) + \tfrac{3}{2}O_2(g) \longrightarrow CO_2(g) + 2H_2O(g) \quad \Delta H^\circ_{rxn} = -638.6\text{ kJ}$$

$$\Delta H^\circ_f \text{ of } CO_2(g) = -393.5\text{ kJ/mol}$$

$$\Delta H^\circ_f \text{ of } H_2O(g) = -241.826\text{ kJ/mol}$$

SOME SIMILAR PROBLEMS 6.82–6.85

The upcoming Chemical Connections essay applies ideas from this chapter to new approaches for energy utilization.

› Summary of Section 6.6

- Standard states are a set of specific conditions used for determining thermodynamic variables for all substances.
- When 1 mol of a compound forms from its elements with all substances in their standard states, the enthalpy change is the standard enthalpy of formation, ΔH°_f.
- Hess's law allows us to picture a reaction as the decomposition of the reactants to their elements followed by the formation of the products from their elements.
- We use tabulated ΔH°_f values to find ΔH°_{rxn} or use known ΔH°_{rxn} and ΔH°_f values to find an unknown ΔH°_f.
- Because of major concerns about climate change, chemists are developing energy alternatives, including coal and biomass conversion, hydrogen fuel, and noncombustible energy sources.

The Future of Energy Use

Out of necessity, we are totally rethinking our global use of energy. The dwindling supplies of our most common fuels and the environmental impact of their combustion products threaten human well-being and the survival of many other species. Energy production presents scientists, engineers, and political leaders with some of the greatest challenges of our time.

A changeover in energy sources from wood to coal and then petroleum took place over the 20th century. The **fossil fuels**—coal, petroleum, and natural gas—remain our major sources, but they are *nonrenewable* because natural processes form them many orders of magnitude slower than we consume them. In the United States, ongoing research seeks new approaches to using coal and nuclear energy, and utilizing *renewable* sources—biomass, hydrogen, sunlight, wind, geothermal heat, and tides—is also the focus of intensive efforts.

This discussion considers converting coal and biomass to cleaner fuels, understanding the effect of carbon-based fuels on climate, developing hydrogen as a fuel, utilizing solar energy, and conserving energy.

Converting Coal to Cleaner Fuels

U.S. reserves of coal are enormous, but the burning of coal, in addition to forming CO_2, produces SO_2 and particulates, and releases mercury. Exposure to SO_2 and particulates causes respiratory disease, and SO_2 can be oxidized to H_2SO_4, the key component of acid rain (Chapter 19). The trace amounts of Hg, a neurotoxin, spread as Hg vapor and bioaccumulate in fish.

Two processes are used to reduce the amounts of SO_2:

Desulfurization. The removal of sulfur dioxide from flue gases is done with devices called *scrubbers* that heat powdered limestone ($CaCO_3$) or spray lime-water slurries [$Ca(OH)_2$]:

$$CaCO_3(s) + SO_2(g) \xrightarrow{\Delta} CaSO_3(s) + CO_2(g)$$
$$2CaSO_3(s) + O_2(g) + 4H_2O(l) \longrightarrow 2CaSO_4{\cdot}2H_2O(s;\text{gypsum})$$
$$2Ca(OH)_2(aq) + 2SO_2(g) + O_2(g) + 2H_2O(l) \longrightarrow 2CaSO_4{\cdot}2H_2O(s)$$

A wallboard (drywall) plant built next to the power plant uses the gypsum (1 ton per customer each year); about 20% of the drywall produced each year is made with this synthetic gypsum.

Gasification. In **coal gasification,** solid coal is converted to sulfur-free gaseous fuels. In this process, the sulfur in the coal is reduced to H_2S, which is removed through an acid-base reaction with a base, such as ethanolamine ($HOCH_2CH_2NH_2$). The resulting salt is heated to release H_2S, which is converted to elemental sulfur by the Claus process (Chapter 22) and sold. Several reactions yield mixtures with increasing fuel value, that is, with a lower C/H ratio:

- Pulverized coal reacts with limited O_2 and water at 800–1500°C to form approximately a 2/1 mixture of CO and H_2.
- Alternatively, in the *water-gas reaction* (or *steam-carbon reaction*), an exothermic oxidation of C to CO is followed by the endothermic reaction of C with steam to form a nearly 1/1 mixture of CO and H_2 called *water gas:*

$$C(s) + \tfrac{1}{2}O_2(g) \longrightarrow CO(g) \qquad \Delta H^\circ_{rxn} = -110\text{ kJ}$$
$$C(s) + H_2O(g) \longrightarrow CO(g) + H_2(g) \qquad \Delta H^\circ_{rxn} = 131\text{ kJ}$$

However, water gas has a much lower fuel value than methane (CH_4). For example, a mixture of 0.5 mol of CO and 0.5 mol of H_2 releases about one-third as much energy as 1.0 mol of methane ($\Delta H = -802$ kJ/mol):

$$\tfrac{1}{2}H_2(g) + \tfrac{1}{4}O_2(g) \longrightarrow \tfrac{1}{2}H_2O(g) \qquad \Delta H^\circ_{rxn} = -121\text{ kJ}$$
$$\tfrac{1}{2}CO(g) + \tfrac{1}{4}O_2(g) \longrightarrow \tfrac{1}{2}CO_2(g) \qquad \Delta H^\circ_{rxn} = -142\text{ kJ}$$
$$\tfrac{1}{2}H_2(g) + \tfrac{1}{2}CO(g) + \tfrac{1}{2}O_2(g) \longrightarrow \tfrac{1}{2}H_2O(g) + \tfrac{1}{2}CO_2(g) \qquad \Delta H^\circ_{rxn} = -263\text{ kJ}$$

- In the *CO-shift* (or *water-gas shift*) *reaction*, the H_2 content of water gas is increased to produce synthesis gas *(syngas):*

$$CO(g) + H_2O(g) \longrightarrow CO_2(g) + H_2(g) \qquad \Delta H^\circ_{rxn} = -41\text{ kJ}$$

- To produce CH_4, syngas that is a 3/1 mixture of H_2 and CO from which the CO_2 has been removed is employed:

$$CO(g) + 3H_2(g) \longrightarrow CH_4(g) + H_2O(g) \qquad \Delta H^\circ_{rxn} = -206\text{ kJ}$$

Drying the product gives **synthetic natural gas (SNG).**

Syngas is used in some newer methods as well. In *integrated gasification combined cycle (IGCC),* electricity is produced in two coordinated steps—by burning syngas in a combustion turbine and then using the hot product gases to generate steam to power a steam turbine. In the *Fischer-Tropsch process,* syngas is used to form liquid hydrocarbon fuels that have higher molar masses than methane:

$$nCO(g) + (2n + 1)H_2(g) \longrightarrow C_nH_{2n+2}(l) + nH_2O(g)$$

Converting Biomass to Fuels

Nearly half the world's people rely on wood for energy. In principle, wood is renewable, but widespread deforestation has resulted from its use for fuel, lumber, and paper. Therefore, great emphasis has been placed on **biomass conversion** of vegetable and animal waste. In one process, chemical and/or microbial methods convert vegetable (sugarcane, corn, switchgrass) and tree waste into fuel, mainly ethanol (C_2H_5OH). In the United States, ethanol is mixed with gasoline to form *gasohol;* in Brazil, it substitutes directly for gasoline.

Methanogenesis is the production of methane by anaerobic (oxygen-free) biodegradation of organic materials such as manure and vegetable waste or wastewater. China's biogas-generating facilities apply this technique, and similar ones in the United States use garbage and sewage. In addition to producing methane, the process yields residues that improve soil; it also prevents the wastes from polluting natural waters.

Another biomass conversion process changes vegetable oils (soybean, cottonseed, sunflower, canola, and even used cooking oil) into a mixture called *biodiesel,* whose combustion produces less CO, SO_2, and particulate matter than fossil-fuel–based diesel.

(continued)

The CO_2 produced in combustion of biodiesel has minimal impact on the net atmospheric amount because the carbon is taken up by the plants grown as sources of the vegetable oils.

The Greenhouse Effect and Global Warming

Combustion of all carbon-based fuels releases CO_2, and over the past two decades, it has become clear that the atmospheric buildup of this gas from our increased use of these fuels is changing the climate. Let's examine the basis of climate change, its current effects, and the long-term outlook.

The natural greenhouse effect. Throughout Earth's history, CO_2 has played a key temperature-regulating role in the atmosphere. Much of the sunlight that shines on Earth is absorbed by the land and oceans and converted to heat (IR radiation; Figure B6.1). Like the glass of a greenhouse, CO_2 does not absorb visible light from the Sun, but it absorbs and re-emits some of the heat radiating from Earth's surface and, thus, helps warm the atmosphere. This process is called the *natural greenhouse effect* (Figure B6.1, *left*). Over several billion years, due to the spread of plant life, which uses CO_2 in photosynthesis, the amount of CO_2 originally present in the atmosphere (from volcanic activity) decreased to 0.028% by volume. With this amount, Earth's average temperature was about 14°C (57°F); without CO_2, it would be −17°C (2°F)!

The enhanced greenhouse effect. As a result of intensive use of fossil fuels for the past 150 years, the amount of CO_2 in the atmosphere increased to 0.040%. Thus, although the same quantity of solar energy passes through the atmosphere, more is being trapped as heat, which has created an *enhanced greenhouse effect* that is changing the climate through *global warming* (Figure B6.1, *right*). Based on current trends, CO_2 amounts will increase to between 0.049% and 0.126% by 2100 (Figure B6.2).

Modeling global warming and its effects. If this projected increase in CO_2 occurs, there are two closely related questions that must be addressed by models: (1) How much will the temperature rise? (2) How will this rise affect life on Earth? Despite constantly improving models, answers are difficult to obtain. Natural fluctuations in temperature must be taken into account, as well as cyclic changes in solar activity. Moreover, as the amount of CO_2 increases from fossil-fuel burning, so do the amounts of particulate matter and SO_2-based aerosols, which may block sunlight and have a cooling effect. Water vapor also traps heat, and as temperatures rise, more water evaporates. The increased amounts of water vapor may thicken the cloud cover and lead to cooling as well.

Despite such opposing factors, all of the best models predict net warming, and we are already observing the effects. The average temperature has increased by 0.6 ± 0.2°C since the late 19th century and 0.2–0.3°C over the past 25 years. Globally, the decade from 2001 to 2010 was the warmest on record and the current decade is continuing this trend. December 2014 was the 358th month in a row with an average temperature above the 20th-century norm, and 2014 was the hottest year on record, with severe droughts and damaging storms in many areas. Snow cover and glacial size in the northern hemisphere and floating ice in the Arctic Ocean have decreased dramatically. Antarctica is also experiencing widespread breakup of icebound regions. Globally, sea

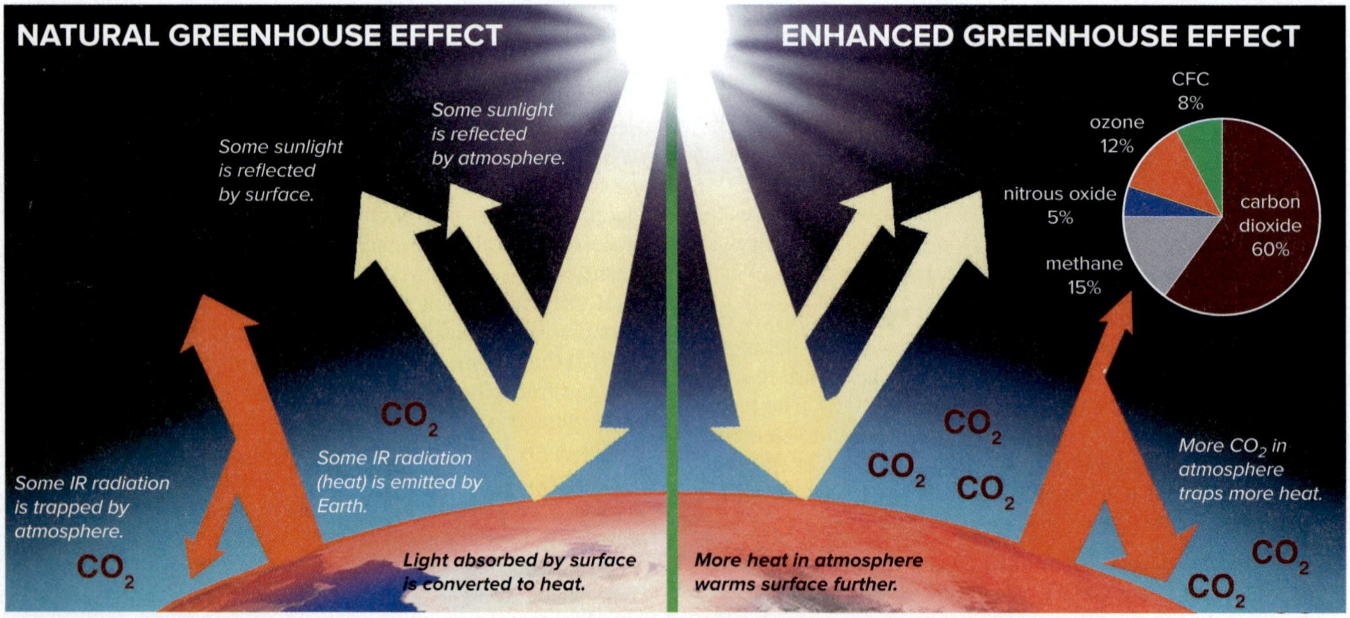

Figure B6.1 The trapping of heat by the atmosphere. Of the total sunlight reaching Earth, some is reflected and some is absorbed and converted to IR radiation (heat). Some heat emitted by the surface is trapped by atmospheric CO_2, creating a *natural* greenhouse effect (*left*) that has been essential to life. But, largely as a result of human activity in the past 150 years, and especially the past several decades, the buildup of CO_2 and several other greenhouse gases (*pie chart*) has created an *enhanced* greenhouse effect (*right*).

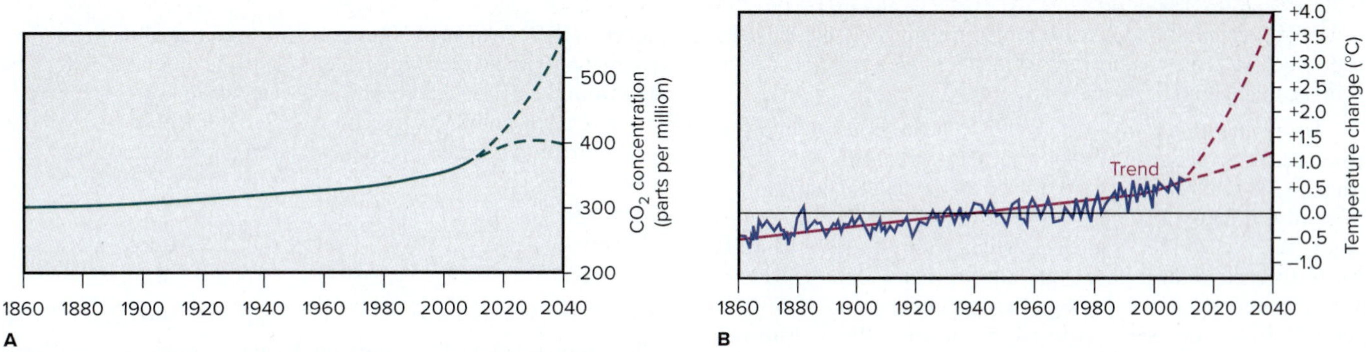

Figure B6.2 Evidence for the enhanced greenhouse effect. **A,** Since the mid-19th century, atmospheric CO_2 has increased. **B,** Since the mid-19th century, average global temperature has risen 0.6°C. (Zero is the average from 1957 to 1970.) The projections in the graphs *(dashed lines)* assume that current fossil-fuel consumption and deforestation continue *(upper line)* or slow *(lower line).*

level has risen an average of 17 cm (6.7 in) over the past century, and flooding and other extreme weather events have increased.

Fifteen years ago, the best models predicted a temperature rise of 1.0–3.5°C; today, their predictions are more than 50% higher. Such increases would alter rainfall patterns and crop yields throughout the world and increase sea level as much as 1 m, thus flooding regions such as the Netherlands, half of Florida, much of southern Asia, coastal New England and New York, and many Pacific island nations. To make matters worse, as we burn fossil fuels that *release* CO_2, we cut down forests that *absorb* it.

Addressing the challenge of climate change. In addition to ways to reduce fossil-fuel consumption, some of which we discuss below, researchers are studying methods of *carbon capture and storage (CCS),* including *CO_2 sequestration,* both naturally by maintaining forests and jungles and industrially by liquefying CO_2 that is produced and either burying it underground or using it to produce mineral carbonates.

The 1997 United Nations Conference on Climate Change in Kyoto, Japan, created a treaty with legally binding targets to limit greenhouse gases. It was ratified by 189 countries, but not by the United States, then the largest emitter. The 2009 conference in Copenhagen, Denmark, was attended by representatives from 192 countries (including the United States), but did not result in legally binding emission targets or financial commitments. In late 2010, however, a conference in Cancun, Mexico, resulted in financial commitments to help developing nations rely on alternative energy sources. And most recently, at the United Nations Framework Convention on Climate Change in December 2015, 195 countries adopted a legally binding global climate action plan aimed at limiting global warming to 1.5°C by reducing emission of greenhouse gases. Rising standards of living in China and India, the two most populous nations, are leading to a need for more energy, which makes large and sustained global reductions in CO_2 even more urgent.

Hydrogen

The use of H_2 as an energy source holds great promise but presents several problems. The simplest element and most plentiful gas in the universe always occurs on Earth in compounds, most importantly with oxygen in water and carbon in hydrocarbons. Once freed, however, H_2 is an excellent energy source because of several properties:

- It has the highest energy content per *mass* unit of any fuel—nearly three times as much as gasoline.
- The energy content per *volume* unit of H_2 is low at STP but increases greatly when the gas is liquefied under extremely high *P* and low *T*.
- The combustion of H_2 does not produce CO_2, particulates, or sulfur oxides. It is such a clean energy source that H_2-fuel cells operate electrical systems aboard the space shuttles, and the crews drink the pure water product.

A hydrogen-based economy is the dream of many because, even though combustion of H_2 ($\Delta H^\circ_{rxn} = -242$ kJ/mol) produces less than one-third as much energy per mole as combustion of CH_4 ($\Delta H^\circ_{rxn} = -802$ kJ/mol), it yields nonpolluting water vapor. The keys to realizing the dream, however, involve major improvements in both the production of H_2 and its transportation and storage.

Production of H_2. Hydrogen is produced by a number of methods, some more widely used than others.

- *Fossil-fuel byproduct.* In the United States today, almost all H_2 is produced by the *steam-reforming process,* in which methane or natural gas is treated with steam in high-temperature reactions (Chapter 22).
- *Electrolysis of liquid water.* Formation of H_2 from the decomposition of liquid water is endothermic ($\Delta H^\circ_{rxn} = 286$ kJ/mol), and most direct methods that utilize electricity are costly. However, energy from flowing water, wind, and geothermal sources can provide the needed electricity. An exciting new *photoelectrochemical approach* uses a photovoltaic panel coupled to an electrolysis unit: the energy of sunlight powers the decomposition of water.
- *Biological methods.* Green algae and cyanobacteria store the energy of sunlight in the form of starch or glycogen for their own metabolism. Under certain conditions, however, these microbes can be made to produce H_2 gas instead of these large organic molecules. For example, eliminating sulfur from the

(continued)

diet of green algae activates a long dormant gene and the cells produce H_2. But, major improvements in yields are needed to make this method cost-effective.

Transportation and storage of H_2. The absence of an infrastructure for transporting and storing large quantities of hydrogen is still a roadblock to realizing a hydrogen-based economy. Currently, the cost of energy to cool and compress the gas to liquefy it is very high. Hydrogen is carried in pipelines, but its ability to escape through metals and to make them brittle requires more expensive piping. Storage in the form of solid metal hydrides has been explored, but heating the hydrides to release the stored hydrogen requires energy. Use of H_2 in fuel cells is a major area of electrochemical research (Chapter 21). Tests are underway to commercialize fuel-cell technology using the hydrogen that is a byproduct of the petrochemical and coal industries.

Solar Energy

The Sun's energy drives global winds and ocean currents, the cycle of evaporation and condensation, and many biological processes, especially photosynthesis. More energy falls on Earth's surface as sunlight in 1 h [1.2×10^5 TW (terawatts); 1 TW = 1×10^{12} W] than is used in all human activity in 1 year. Solar energy is, however, a dilute source (only 1 kW/m^2 at noon); in 1 h, about 600–1000 TW strikes land at sites suitable for collecting. Nevertheless, covering 0.16% of that land with today's photovoltaic systems would provide 20 TW/h, equivalent to the output of 20,000 nuclear power plants producing 1 GW (gigawatt) each. Clearly, solar energy is the largest carbon-free option among renewable sources.

Three important ways to utilize solar energy are with electronic materials, biomass, and thermal systems:

Electronic materials. A **photovoltaic cell** converts sunlight directly into electricity by employing certain combinations of elements (usually metalloids; we discuss the behavior of elements with *semiconductor* properties in Chapter 12). Early cells relied primarily on crystalline silicon, but their *yield,* the proportion of electrical energy produced relative to the radiant energy supplied, was only about 10%. More recent devices have achieved much higher yields by using thin films of polycrystalline materials that incorporate combinations of other elements, such as cadmium telluride (CdTe cells, 16%), gallium arsenide (GaAs cells, 18%), and copper indium gallium selenide (CIGS cells, 20%). And ongoing research on nanometer-sized semiconductor devices, called *quantum dots,* suggests the possibility of yields as high as 70%.

Biomass. Solar energy is used to convert CO_2 and water via photosynthesis to plant material (carbohydrate) and oxygen. Plant biomass can be burned directly as a fuel or converted to other fuels, such as ethanol or hydrogen, as discussed in an earlier section in this Chemical Connections (see "Converting Biomass to Fuels").

Thermal systems. The operation of a photovoltaic cell creates a buildup of heat, which lowers its efficiency. A *photovoltaic thermal (PVT) hybrid cell* thus combines such a cell with conductive metal piping (filled with water or antifreeze) that transfers the heat from the cell to a domestic hot-water system.

Because sunlight is typically available for only 6–8 h a day and is greatly reduced in cloudy weather, another thermal system focuses on storing solar energy during the day and releasing it at night. For example, when the ionic hydrate $Na_2SO_4{\cdot}10H_2O$ is warmed by sunlight to over 32°C, the 3 mol of ions dissolve in the 10 mol of water in an endothermic process:

$$Na_2SO_4(aq){\cdot}10H_2O(s) \xrightarrow{>32^\circ C} Na_2SO_4(aq) \qquad \Delta H^\circ_{rxn} = 354\ kJ$$

When cooled below 32°C after sunset, the solution recrystallizes, releasing the absorbed energy for heating:

$$Na_2SO_4(aq) \xrightarrow{<32^\circ C} Na_2SO_4(aq){\cdot}10H_2O(s) \qquad \Delta H^\circ_{rxn} = -354\ kJ$$

Unfortunately, as of the middle of the second decade of the 21st century, solar energy is still not competitive with highly subsidized fossil fuels. Nevertheless, changes in economic policy as well as improvements in photovoltaic technology and biomass processing should greatly increase solar energy's contribution to a carbon-free energy future.

Nuclear Energy

Despite ongoing problems with disposal of radioactive waste, energy from nuclear *fission*—the splitting of large, unstable atomic nuclei—is used extensively, especially in France and other parts of northern Europe. Nuclear *fusion*—the combining of small atomic nuclei—avoids the problems of fission but has been achieved so far only with a net consumption of power. (We'll discuss these processes in detail in Chapter 24.)

Energy Conservation: More from Less

All systems, whether organisms or factories, waste energy, which is, in effect, a waste of fuel. For example, the production of one aluminum beverage can requires the amount of energy that comes from burning 0.25 L of gasoline. Energy conservation lowers costs, extends our fuel supply, and reduces the effects of climate change. Following are three examples of conservation.

Passive solar design. Heating and cooling costs can be reduced by as much as 50% when buildings are constructed with large windows and materials that absorb the Sun's heat in the day and release it slowly at night.

Residential heating systems. A high-efficiency gas-burning furnace channels hot waste gases through a system of baffles to transfer more heat to the room and then to an attached domestic hot-water system. While moving through the system, the gases cool to 100°C, so the water vapor condenses, thus releasing about 10% more heat:

$$\begin{array}{lll} CH_4(g) + 2O_2(g) \longrightarrow CO_2(g) + \cancel{2H_2O(g)} & \Delta H^\circ_{rxn} = -802\ kJ \\ \cancel{2H_2O(g)} \longrightarrow 2H_2O(l) & \Delta H^\circ_{rxn} = -88\ kJ \\ \hline CH_4(g) + 2O_2(g) \longrightarrow CO_2(g) + 2H_2O(l) & \Delta H^\circ_{rxn} = -890\ kJ \end{array}$$

Modern lighting. Incandescent bulbs give off light when the material in their filament reaches a high temperature. They are very inefficient because less than 7% of the electrical energy is converted to light, the rest being wasted as heat. *Compact fluorescent lamps (CFLs)* have an efficiency of about 18% and are having a major economic impact: in many parts of Europe, incandescent bulbs are no longer sold. *Light-emitting diodes (LEDs)* are even more efficient (about 30%) and are already being used in car indicator lamps and tail lights. Light-emitting devices based on organic semiconductors (OLEDs) are used in flat-panel displays, and recent use of Al-In-Ga-P materials and Ga and In nitrides in LEDs has improved their brightness and color richness. Over 20% of electricity generated in the United States is used for lighting, and burning of fossil fuels in power plants creates smog, particulates, and CO_2. By 2020, use of LEDs for lighting can cut electricity usage by 50%, eliminating the emission of 2.8×10^7 metric tons of carbon per year.

Engineers and chemists will continue to explore alternatives for energy production and use, but a more hopeful energy future ultimately depends on our dedication to conserving planetary resources.

Problems

B6.1 To make use of an ionic hydrate for storing solar energy, you place 500.0 kg of sodium sulfate decahydrate on your house roof. Assuming complete reaction and 100% efficiency of heat transfer, how much heat (in kJ) is released to your house at night?

B6.2 In one step of coal gasification, coal reacts with superheated steam:

$$C(coal) + H_2O(g) \longrightarrow CO(g) + H_2(g) \quad \Delta H^\circ_{rxn} = 129.7\ kJ$$

(a) Use this reaction and the following two to write an overall reaction for the production of methane:

$$CO(g) + H_2O(g) \longrightarrow CO_2(g) + H_2(g) \quad \Delta H^\circ_{rxn} = -41\ kJ$$
$$CO(g) + 3H_2(g) \longrightarrow CH_4(g) + H_2O(g) \quad \Delta H^\circ_{rxn} = -206\ kJ$$

(b) Calculate ΔH°_{rxn} for this overall change.

(c) Using the value from part (b) and a calculated value of ΔH°_{rxn} for the combustion of methane, find the total ΔH for gasifying 1.00 kg of coal and burning the methane formed (assume that water forms as a gas and $\mathcal{M}$ of coal is 12.00 g/mol).

CHAPTER REVIEW GUIDE

Learning Objectives

Relevant section (§) and/or sample problem (SP) numbers appear in parentheses.

Understand These Concepts

1. The distinction between a system and its surroundings (§6.1)
2. The transfer of energy to or from a system as heat and/or work (§6.1)
3. The relation of internal energy change, heat, and work (§6.1)
4. The meaning of energy conservation (§6.1)
5. The meaning of a state function and why ΔE is constant even though q and w vary (§6.1)
6. The meaning of enthalpy and the relation between ΔE and ΔH (§6.2)
7. The meaning of ΔH and the distinction between exothermic and endothermic processes (§6.2)
8. The relation between specific heat capacity and heat (§6.3)
9. How constant-pressure (coffee-cup) and constant-volume (bomb) calorimeters work (§6.3)
10. The relation between ΔH and amount of substance (§6.4)
11. The calculation of ΔH values with Hess's law (§6.5)
12. The meaning of a formation equation and the standard enthalpy of formation (§6.6)
13. How a reaction can be viewed as the decomposition of reactants followed by the formation of products (§6.6)

Master These Skills

1. Determining the change in a system's internal energy in different units (SP 6.1)
2. Calculating PV work done by or on a system (SP 6.2)
3. Drawing enthalpy diagrams for chemical and physical changes (SP 6.3)
4. Solving problems involving specific heat capacity and heat transferred in a reaction (SPs 6.4–6.7)
5. Relating the heat transferred in a reaction to the amounts of substances changing (SP 6.8)
6. Using Hess's law to find an unknown ΔH (SP 6.9)
7. Writing formation equations and using ΔH°_f values to find ΔH°_{rxn} (SPs 6.10, 6.11)

Key Terms

Page numbers appear in parentheses.

biomass conversion (281)
calorie (cal) (261)
calorimeter (269)
change in enthalpy (ΔH) (265)
coal gasification (281)
endothermic process (266)
enthalpy (H) (265)
enthalpy diagram (266)
exothermic process (266)
formation equation (277)
fossil fuel (281)
heat (q) (258)
heat capacity (268)
Hess's law (275)
internal energy (E) (258)
joule (J) (261)
law of conservation of energy (first law of thermodynamics) (261)
methanogenesis (281)
molar heat capacity (C_m) (268)
photovoltaic cell (284)
pressure-volume work (PV work) (263)
specific heat capacity (c) (268)
standard enthalpy of formation (ΔH_f°) (277)
standard enthalpy of reaction (ΔH_{rxn}°) (277)
standard state (277)
state function (262)
surroundings (258)
synthetic natural gas (SNG) (281)
system (258)
thermochemical equation (273)
thermochemistry (257)
thermodynamics (257)
work (w) (259)

Key Equations and Relationships

Page numbers appear in parentheses.

6.1 Defining the change in internal energy (258):

$$\Delta E = E_{final} - E_{initial} = E_{products} - E_{reactants}$$

6.2 Expressing the change in internal energy in terms of heat and work (259):

$$\Delta E = q + w$$

6.3 Stating the first law of thermodynamics (law of conservation of energy) (261):

$$\Delta E_{universe} = \Delta E_{system} + \Delta E_{surroundings} = 0$$

6.4 Determining the work due to a change in volume at constant pressure (PV work) (263):

$$w = -P\Delta V$$

6.5 Relating the enthalpy change to the internal energy change at constant pressure (265):

$$\Delta H = \Delta E + P\Delta V$$

6.6 Identifying the enthalpy change with the heat absorbed or released at constant pressure (265):

$$\Delta H = \Delta E + P\Delta V = q_P$$

6.7 Calculating the heat absorbed or released when a substance undergoes a temperature change or a reaction occurs (268):

$$q = c \times \text{mass} \times \Delta T$$

6.8 Calculating heat absorbed by a bomb calorimeter during a combustion reaction (272):

$$q_{calorimeter} = C_{calorimeter} \times \Delta T_{calorimeter}$$

6.9 Calculating the overall enthalpy change of a reaction (Hess's law) (275):

$$\Delta H_{overall} = \Delta H_1 + \Delta H_2 + \cdots + \Delta H_n$$

6.10 Calculating the standard enthalpy of reaction (279):

$$\Delta H_{rxn}^\circ = \Sigma m \Delta H_{f(products)}^\circ - \Sigma n \Delta H_{f(reactants)}^\circ$$

BRIEF SOLUTIONS TO FOLLOW-UP PROBLEMS

6.1A The liquid (system) *absorbs* heat, so q is +13.5 kJ; the system *does* work, so w is –1.8 kJ.

$$\Delta E = q + w$$
$$= [13.5 \text{ kJ} + (-1.8 \text{ kJ})] \times \frac{1000 \text{ J}}{1 \text{ kJ}}$$
$$= 1.17\times10^4 \text{ J}$$

6.1B The system *releases* heat (the surroundings absorb heat), so $q = -26.0$ kcal; the system has work *done on* it, so $w = +15.0$ Btu.

$$\Delta E = q + w$$
$$= \left(-26.0 \text{ kcal} \times \frac{4.184 \text{ kJ}}{1 \text{ kcal}}\right) + \left(15.0 \text{ Btu} \times \frac{1.055 \text{ kJ}}{1 \text{ Btu}}\right)$$
$$= -93 \text{ kJ}$$

6.2A $P \text{ (atm)} = 732 \text{ torr} \times \dfrac{1 \text{ atm}}{760 \text{ torr}} = 0.963 \text{ atm}$

$$w \text{ (atm·L)} = -P\Delta V = -0.963 \text{ atm} \times (2.35 \text{ L} - 5.68 \text{ L})$$
$$= 3.21 \text{ atm·L}$$
$$w \text{ (J)} = 3.21 \text{ atm·L} \times \frac{101.3 \text{ J}}{1 \text{ atm·L}} = 325 \text{ J}$$

6.2B $w \text{ (atm·L)} = -P\Delta V = -5.5 \text{ atm} \times (16.3 \text{ L} - 10.5 \text{ L})$

$$= -32 \text{ atm·L}$$
$$w \text{ (J)} = -32 \text{ atm·L} \times \frac{101.3 \text{ J}}{1 \text{ atm·L}} = -3.2\times10^3 \text{ J}$$

6.3A The reaction is exothermic (heat energy is released); $H_{reactants} > H_{products}$.

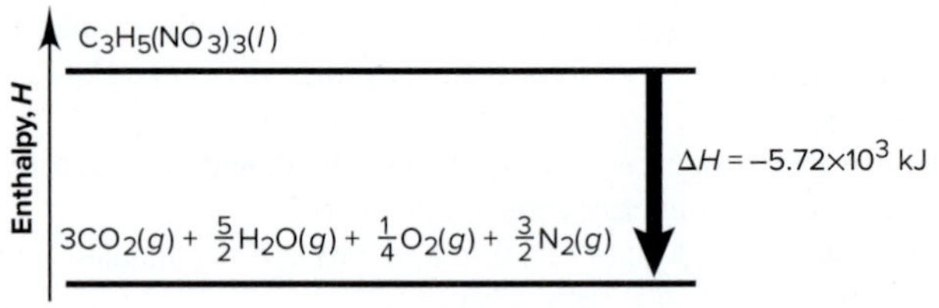

6.3B The process is endothermic (heat energy is absorbed); $H_{products} > H_{reactants}$.

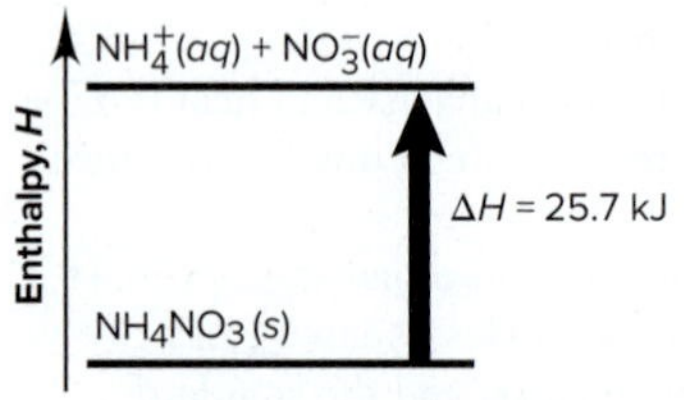

6.4A $\Delta T = 375°\text{C} - 18°\text{C} = 357°\text{C} = 357 \text{ K}$

$$q = c \times \text{mass} \times \Delta T = (0.900 \text{ J/g·K})(7.65 \text{ g})(357 \text{ K})$$
$$= 2.46\times10^3 \text{ J}$$

6.4B $\Delta T = 25.0°C - 37.0°C = -12.0°C = -12.0$ K

$q = -177$ kJ since heat is released.

$q = c \times \text{mass} \times \Delta T$

$$\text{Mass} = \frac{q}{c \times \Delta T} = \frac{-177 \text{ kJ}}{(2.42 \text{ J/g·K})(-12.0 \text{ K})} \times \frac{1000 \text{ J}}{1 \text{ kJ}} = 6.10\times10^3 \text{ g}$$

$$\text{Volume (L)} = 6.10\times10^3 \text{ g} \times \frac{1 \text{ mL}}{1.11 \text{ g}} \times \frac{1 \text{ L}}{1000 \text{ mL}} = 5.50 \text{ L}$$

6.5A ΔT (solid) $= 27.25°C - 65.00°C = -37.75°C = -37.75$ K

ΔT (H_2O) $= 27.25°C - 25.55°C = 1.70°C = 1.70$ K

$$c_{\text{solid}} = -\frac{c_{H_2O} \times \text{mass}_{H_2O} \times \Delta T_{H_2O}}{\text{mass}_{\text{solid}} \times \Delta T_{\text{solid}}}$$
$$= -\frac{4.184 \text{ J/g·K} \times 25.00 \text{ g} \times 1.70 \text{ K}}{12.18 \text{ g} \times (-37.75 \text{ K})}$$
$$= 0.387 \text{ J/g·K}$$

From Table 6.2, the metal is copper.

6.5B ΔT (H_2O) $= 49.30°C - 50.00°C = -0.70°C = -0.70$ K

$-c_{\text{solid}} \times \text{mass}_{\text{solid}} \times \Delta T_{\text{solid}} = c_{H_2O} \times \text{mass}_{H_2O} \times \Delta T_{H_2O}$

$$\Delta T_{\text{solid}} = \frac{c_{H_2O} \times \text{mass}_{H_2O} \times \Delta T_{H_2O}}{-c_{\text{solid}} \times \text{mass}_{\text{solid}}}$$
$$= \frac{(4.184 \text{ J/g·K})(75.0 \text{ g})(-0.70 \text{ K})}{-(0.228 \text{ J/g·K})(33.2 \text{ g})}$$
$$= 29 \text{ K} = 29°C$$

$\Delta T_{\text{solid}} = T_{\text{final}} - T_{\text{initial}}$

$T_{\text{initial}} = T_{\text{final}} - \Delta T_{\text{solid}} = 49.30°C - 29°C = 20.°C$

6.6A (a) $HNO_3(aq) + KOH(aq) \longrightarrow KNO_3(aq) + H_2O(l)$

$H^+(aq) + OH^-(aq) \longrightarrow H_2O(l)$

(b) Total mass of solution = (25.0 mL + 50.0 mL) × 1.00 g/mL = 75.0 g

$\Delta T_{\text{soln}} = 27.05°C - 21.90°C = 5.15°C = 5.15$ K

$q_{\text{soln}} = c_{\text{soln}} \times \text{mass}_{\text{soln}} \times \Delta T_{\text{soln}} = -q_{\text{rxn}}$

$= (4.184 \text{ J/g·K})(75.0 \text{ g})(5.15 \text{ K})$

$= 1620 \text{ J} \times 1 \text{ kJ}/1000 \text{ J} = 1.62$ kJ

$q_{\text{rxn}} = -1.62$ kJ

6.6B $2HCl(aq) + Ba(OH)_2(aq) \longrightarrow 2H_2O(l) + BaCl_2(aq)$

$q_{\text{soln}} = -q_{\text{rxn}} = 1.386$ kJ so $q_{\text{rxn}} = -1.386$ kJ

Amount (mol) of HCl = 0.500 mol HCl/L × 0.0500 L = 0.0250 mol HCl

Similarly, we have 0.0250 mol $BaCl_2$.

Finding the limiting reactant from the balanced equation:

$$\text{Amount (mol) of } H_2O = 0.0250 \text{ mol HCl} \times \frac{2 \text{ mol } H_2O}{2 \text{ mol HCl}} = 0.0250 \text{ mol } H_2O$$

$$\text{Amount (mol) of } H_2O = 0.0250 \text{ mol } Ba(OH)_2 \times \frac{2 \text{ mol } H_2O}{1 \text{ mol } Ba(OH)_2} = 0.0500 \text{ mol } H_2O$$

Thus, HCl is limiting.

$$\Delta H \text{ (kJ/mol } H_2O) = \frac{q_{\text{rxn}}}{\text{mol } H_2O} = \frac{-1.386 \text{ kJ}}{0.0250 \text{ mol } H_2O} = -55.4 \text{ kJ/mol } H_2O$$

6.7A $-q_{\text{rxn}} = q_{\text{calorimeter}}$

$$q_{\text{calorimeter}} = -0.8650 \text{ g C} \times \frac{1 \text{ mol C}}{12.01 \text{ g C}} \times \frac{-393.5 \text{ kJ}}{1 \text{ mol C}} = 28.34 \text{ kJ}$$

$q_{\text{calorimeter}} = C_{\text{calorimeter}} \times \Delta T$

$$C_{\text{calorimeter}} = \frac{q_{\text{calorimeter}}}{\Delta T} = \frac{28.34 \text{ kJ}}{3.116 \text{ K}} = 9.09 \text{ kJ/K}$$

6.7B $\Delta T = 26.799°C - 21.862°C = 4.937°C = 4.937$ K

$q_{\text{calorimeter}} = C_{\text{calorimeter}} \times \Delta T = (8.151 \text{ kJ/K})(4.937 \text{ K}) = 40.24$ kJ

$$q \text{ (Calories)} = 40.24 \text{ kJ} \times \frac{1 \text{ kcal}}{4.184 \text{ kJ}} \times \frac{1 \text{ Calorie}}{1 \text{ kcal}} = 9.618 < 10 \text{ Calories}$$

The claim is correct.

6.8A $C_2H_4(g) + H_2(g) \longrightarrow C_2H_6(g) + 137$ kJ

$$\text{Heat (kJ)} = 15.0 \text{ kg} \times \frac{1000 \text{ g}}{1 \text{ kg}} \times \frac{1 \text{ mol } C_2H_6}{30.07 \text{ g } C_2H_6} \times \frac{137 \text{ kJ}}{1 \text{ mol}}$$
$$= 6.83\times10^4 \text{ kJ released}$$

6.8B $N_2(g) + O_2(g) + 180.58 \text{ kJ} \longrightarrow 2NO(g)$

$$\text{Heat (kJ)} = 3.50 \text{ t NO} \times \frac{1000 \text{ kg}}{1 \text{ t}} \times \frac{1000 \text{ g}}{1 \text{ kg}} \times \frac{1 \text{ mol NO}}{30.01 \text{ g NO}} \times \frac{180.58 \text{ kJ}}{2 \text{ mol NO}}$$
$$= 1.05\times10^7 \text{ kJ absorbed}$$

6.9A Reverse equation 1:

$2NO(g) + \frac{3}{2}O_2(g) \longrightarrow N_2O_5(s)$ $\qquad \Delta H = -223.7$ kJ

Reverse equation 2 and multiply by 2:

$2NO_2(g) \longrightarrow 2NO(g) + O_2(g)$ $\qquad \Delta H = -2(-57.1 \text{ kJ}) = 114.2$ kJ

$\cancel{2NO(g)} + \frac{1}{2}\cancel{\frac{3}{2}}O_2(g) + 2NO_2(g) \longrightarrow N_2O_5(s) + \cancel{2NO(g)} + \cancel{O_2(g)}$

$2NO_2(g) + \frac{1}{2}O_2(g) \longrightarrow N_2O_5(s)$ $\qquad \Delta H = -109.5$ kJ

6.9B Reverse equation 1 and divide by 2:

$NH_3(g) \longrightarrow \frac{1}{2}N_2(g) + \frac{3}{2}H_2(g)$ $\qquad \Delta H = -\frac{1}{2}(-91.8 \text{ kJ}) = 45.9$ kJ

Divide equation 2 by 2:

$\frac{1}{2}N_2(g) + 2H_2(g) + \frac{1}{2}Cl_2(g) \longrightarrow NH_4Cl(s)$

$\Delta H = \frac{1}{2}(-628.8 \text{ kJ}) = -314.4$ kJ

Reverse equation 3:

$NH_4Cl(s) \longrightarrow NH_3(g) + HCl(g)$ $\qquad \Delta H = -(-176.2 \text{ kJ}) = 176.2$ kJ

$\cancel{NH_3(g)} + \cancel{\frac{1}{2}N_2(g)} + \frac{1}{2}2H_2(g) + \frac{1}{2}Cl_2(g) + \cancel{NH_4Cl(s)} \longrightarrow \cancel{\frac{1}{2}N_2(g)} + \cancel{\frac{3}{2}H_2(g)} + \cancel{NH_4Cl(s)} + \cancel{NH_3(g)} + HCl(g)$

$\frac{1}{2}H_2(g) + \frac{1}{2}Cl_2(g) \longrightarrow HCl(g)$ $\qquad \Delta H = -92.3$ kJ

6.10A (a) $C(\text{graphite}) + 2H_2(g) + \frac{1}{2}O_2(g) \longrightarrow CH_3OH(l)$ $\qquad \Delta H_f^\circ = -238.6$ kJ

(b) $Ca(s) + \frac{1}{2}O_2(g) \longrightarrow CaO(s)$ $\qquad \Delta H_f^\circ = -635.1$ kJ

(c) $C(\text{graphite}) + \frac{1}{4}S_8(\text{rhombic}) \longrightarrow CS_2(l)$ $\qquad \Delta H_f^\circ = 87.9$ kJ

6.10B (a) $C(\text{graphite}) + \frac{1}{2}H_2(g) + \frac{3}{2}Cl_2(g) \longrightarrow CHCl_3(l)$ $\qquad \Delta H_f^\circ = -132$ kJ

(b) $\frac{1}{2}N_2(g) + 2H_2(g) + \frac{1}{2}Cl_2(g) \longrightarrow NH_4Cl(s)$ $\qquad \Delta H_f^\circ = -314.4$ kJ

(c) $Pb(s) + \frac{1}{8}S_8(\text{rhombic}) + 2O_2(g) \longrightarrow PbSO_4(s)$ $\qquad \Delta H_f^\circ = -918.39$ kJ

6.11A $\Delta H^\circ_{\text{rxn}} = \{4\Delta H_f^\circ[H_3PO_4(l)]\} - \{\Delta H_f^\circ[P_4O_{10}(s)] + 6\Delta H_f^\circ[H_2O(l)]\}$

$= [(4 \text{ mol})(-1271.7 \text{ kJ/mol})] - [(1 \text{ mol})(-2984 \text{ kJ/mol}) + (6 \text{ mol})(-285.840 \text{ kJ/mol})]$

$= -5086.8 \text{ kJ} - (-2984 \text{ kJ} - 1715.04 \text{ kJ})$

$= -387.8$ kJ

6.11B Note that the ΔH_f° of $O_2(g) = 0$.

$\Delta H^\circ_{\text{rxn}} = \{\Delta H_f^\circ[CO_2(g)] + 2\Delta H_f^\circ[H_2O(g)]\} - \{\Delta H_f^\circ[CH_3OH(l)]\}$

ΔH_f° of $CH_3OH(l) = -\Delta H^\circ_{\text{rxn}} + 2\Delta H_f^\circ[H_2O(g)] + \Delta H_f^\circ[CO_2(g)]$

$= 638.6 \text{ kJ} + (2 \text{ mol})(-241.826 \text{ kJ/mol}) + (1 \text{ mol})(-393.5 \text{ kJ/mol})$

$= -238.6$ kJ

PROBLEMS

Problems with **colored** numbers are answered in Appendix E and worked in detail in the Student Solutions Manual. Problem sections match those in the text and give the numbers of relevant sample problems. Most offer Concept Review Questions, Skill-Building Exercises (grouped in pairs covering the same concept), and Problems in Context. The Comprehensive Problems are based on material from any section or previous chapter.

Forms of Energy and Their Interconversion
(Sample Problems 6.1 and 6.2)

Concept Review Questions

6.1 Why do heat (q) and work (w) have positive values when entering a system and negative values when leaving?

6.2 If you feel warm after exercising, have you increased the internal energy of your body? Explain.

6.3 An *adiabatic* process is one that involves no heat transfer. What is the relationship between work and the change in internal energy in an adiabatic process?

6.4 State two ways that you increase the internal energy of your body and two ways that you decrease it.

6.5 Name a common device used to accomplish each change:
(a) Electrical energy to thermal energy
(b) Electrical energy to sound energy
(c) Electrical energy to light energy
(d) Mechanical energy to electrical energy
(e) Chemical energy to electrical energy

6.6 In winter, an electric heater uses a certain amount of electrical energy to heat a room to 20°C. In summer, an air conditioner uses the same amount of electrical energy to cool the room to 20°C. Is the change in internal energy of the heater larger, smaller, or the same as that of the air conditioner? Explain.

6.7 You lift your textbook and drop it onto a desk. Describe the energy transformations (from one form to another) that occur, moving backward in time from a moment after impact.

6.8 Why is the work done when a system expands against a constant external pressure assigned a negative sign?

Skill-Building Exercises (grouped in similar pairs)

6.9 A system receives 425 J of heat from and delivers 425 J of work to its surroundings. What is the change in internal energy of the system (in J)?

6.10 A system releases 255 cal of heat to the surroundings and delivers 428 cal of work. What is the change in internal energy of the system (in cal)?

6.11 What is the change in internal energy (in J) of a system that releases 675 J of thermal energy to its surroundings and has 530 cal of work done on it?

6.12 What is the change in internal energy (in J) of a system that absorbs 0.615 kJ of heat from its surroundings and has 0.247 kcal of work done on it?

6.13 Complete combustion of 2.0 metric tons of coal to gaseous carbon dioxide releases 6.6×10^{10} J of heat. Convert this energy to (a) kilojoules; (b) kilocalories; (c) British thermal units.

6.14 Thermal decomposition of 5.0 metric tons of limestone to lime and carbon dioxide absorbs 9.0×10^{6} kJ of heat. Convert this energy to (a) joules; (b) calories; (c) British thermal units.

6.15 At constant temperature, a sample of helium gas expands from 922 mL to 1.14 L against a pressure of 2.33 atm. Find w (in J) done by the gas (101.3 J = 1 atm·L).

6.16 The external pressure on a gas sample is 2660 mmHg, and the volume changes from 0.88 L to 0.63 L at constant temperature. Find w (in kJ) done on the gas (1 atm·L = 101.3 J).

Problems in Context

6.17 The nutritional calorie (Calorie) is equivalent to 1 kcal. One pound of body fat is equivalent to about 4.1×10^{3} Calories. Express this quantity of energy in joules and kilojoules.

6.18 If an athlete expends 1950 kJ/h, how long does it take her to work off 1.0 lb of body fat? (See Problem 6.17.)

Enthalpy: Changes at Constant Pressure
(Sample Problem 6.3)

Concept Review Questions

6.19 Why is it often more convenient to measure ΔH than ΔE?

6.20 Hot packs used by skiers produce heat via the crystallization of sodium acetate from a concentrated solution. What is the sign of ΔH for this crystallization? Is the reaction exothermic or endothermic?

6.21 Classify the following processes as exothermic or endothermic: (a) freezing of water; (b) boiling of water; (c) digestion of food; (d) a person running; (e) a person growing; (f) wood being chopped; (g) heating with a furnace.

6.22 What are the two main components of the internal energy of a substance? On what are they based?

6.23 For each process, state whether ΔH is less than (more negative), equal to, or greater than ΔE of the system. Explain.
(a) An ideal gas is cooled at constant pressure.
(b) A gas mixture reacts exothermically at fixed volume.
(c) A solid reacts exothermically to yield a mixture of gases in a container of variable volume.

Skill-Building Exercises (grouped in similar pairs)

6.24 Draw an enthalpy diagram for a general exothermic reaction; label the axis, reactants, products, and ΔH with its sign.

6.25 Draw an enthalpy diagram for a general endothermic reaction; label the axis, reactants, products, and ΔH with its sign.

6.26 Write a balanced equation and draw an enthalpy diagram for: (a) combustion of 1 mol of ethane; (b) freezing of liquid water.

6.27 Write a balanced equation and draw an enthalpy diagram for (a) formation of 1 mol of sodium chloride from its elements (heat is released); (b) vaporization of liquid benzene.

6.28 Write a balanced equation and draw an enthalpy diagram for (a) combustion of 1 mol of liquid methanol (CH_3OH); (b) formation of 1 mol of NO_2 from its elements (heat is absorbed).

6.29 Write a balanced equation and draw an enthalpy diagram for (a) sublimation of dry ice [conversion of $CO_2(s)$ directly to $CO_2(g)$]; (b) reaction of 1 mol of SO_2 with O_2.

6.30 The circles represent a phase change at constant temperature:

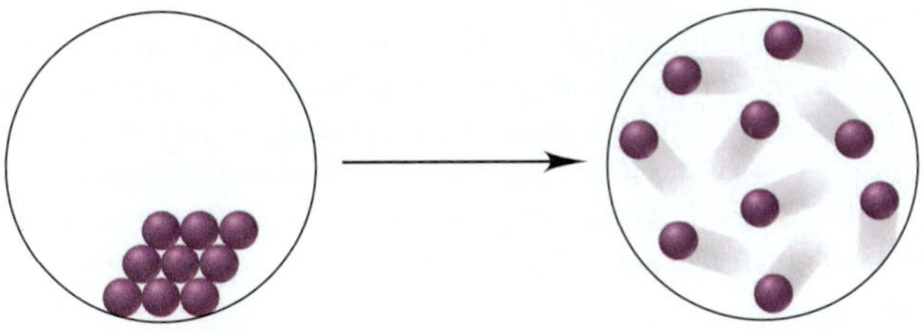

Is the value of each of the following positive (+), negative (−), or zero: (a) q_{sys}; (b) ΔE_{sys}; (c) ΔE_{univ}?

6.31 The scene below represents a physical change taking place in a piston-cylinder assembly:

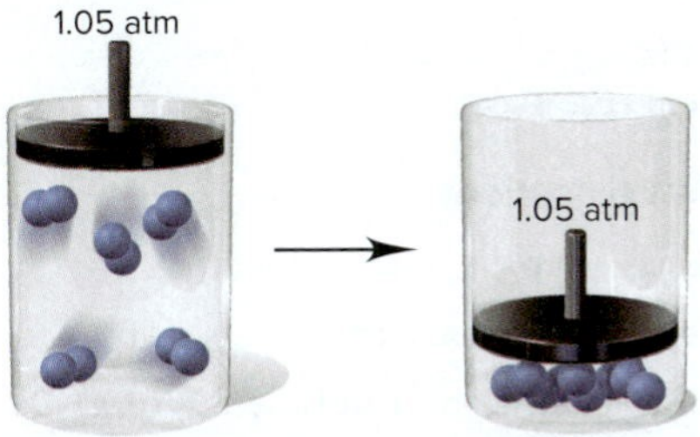

(a) Is w_{sys} +, −, or 0? (b) Is ΔH_{sys} +, −, or 0? (c) Can you determine whether ΔE_{surr} is +, −, or 0? Explain.

Calorimetry: Measuring the Heat of a Chemical or Physical Change

(Sample Problems 6.4 to 6.7)

Concept Review Questions

6.32 Which is larger, the specific heat capacity or the molar heat capacity of a substance? Explain.

6.33 What data do you need to determine the specific heat capacity of a substance?

6.34 Is the specific heat capacity of a substance an intensive or extensive property? Explain.

6.35 Distinguish between specific heat capacity, molar heat capacity, and heat capacity.

6.36 Both a coffee-cup calorimeter and a bomb calorimeter can be used to measure the heat transferred in a reaction. Which measures ΔE and which measures ΔH? Explain.

Skill-Building Exercises (grouped in similar pairs)

6.37 Find q when 22.0 g of water is heated from 25.0°C to 100.°C.

6.38 Calculate q when 0.10 g of ice is cooled from 10.°C to −75°C (c_{ice} = 2.087 J/g·K).

6.39 A 295-g aluminum engine part at an initial temperature of 13.00°C absorbs 75.0 kJ of heat. What is the final temperature of the part (c of Al = 0.900 J/g·K)?

6.40 A 27.7-g sample of the radiator coolant ethylene glycol releases 688 J of heat. What was the initial temperature of the sample if the final temperature is 32.5°C (c of ethylene glycol = 2.42 J/g·K)?

6.41 Two iron bolts of equal mass—one at 100.°C, the other at 55°C—are placed in an insulated container. Assuming the heat capacity of the container is negligible, what is the final temperature inside the container (c of iron = 0.450 J/g·K)?

6.42 One piece of copper jewelry at 105°C has twice the mass of another piece at 45°C. Both are placed in a calorimeter of negligible heat capacity. What is the final temperature inside the calorimeter (c of copper = 0.387 J/g·K)?

6.43 When 155 mL of water at 26°C is mixed with 75 mL of water at 85°C, what is the final temperature? (Assume that no heat is released to the surroundings; d of water = 1.00 g/mL.)

6.44 An unknown volume of water at 18.2°C is added to 24.4 mL of water at 35.0°C. If the final temperature is 23.5°C, what was the unknown volume? (Assume that no heat is released to the surroundings; d of water = 1.00 g/mL.)

6.45 A 455-g piece of copper tubing is heated to 89.5°C and placed in an insulated vessel containing 159 g of water at 22.8°C. Assuming no loss of water and a heat capacity of 10.0 J/K for the vessel, what is the final temperature (c of copper = 0.387 J/g·K)?

6.46 A 30.5-g sample of an alloy at 93.0°C is placed into 50.0 g of water at 22.0°C in an insulated coffee-cup calorimeter with a heat capacity of 9.2 J/K. If the final temperature of the system is 31.1°C, what is the specific heat capacity of the alloy?

6.47 When 25.0 mL of 0.500 M H_2SO_4 is added to 25.0 mL of 1.00 M KOH in a coffee-cup calorimeter at 23.50°C, the temperature rises to 30.17°C. Calculate ΔH in kJ per mole of H_2O formed. (Assume that the total volume is the sum of the volumes and that the density and specific heat capacity of the solution are the same as for water.)

6.48 When 20.0 mL of 0.200 M $AgNO_3$ and 30.0 mL of 0.100 M NaCl, both at 24.72°C, are mixed in a coffee-cup calorimeter, AgCl precipitates and the temperature of the mixture increases to 25.65°C. Calculate ΔH in kJ per mole of AgCl(s) produced. (Assume that the total volume is the sum of the volumes and that the density and specific heat capacity of the solution are the same as for water.)

6.49 When a 2.150-g sample of glucose, $C_6H_{12}O_6$, is burned in a bomb calorimeter with a heat capacity of 6.317 kJ/K, the temperature of the calorimeter increases from 23.446°C to 28.745°C. Calculate ΔE for the combustion of glucose in kJ/mol.

6.50 A chemist places 1.750 g of ethanol, C_2H_6O, in a bomb calorimeter with a heat capacity of 12.05 kJ/K. The sample is burned and the temperature of the calorimeter increases by 4.287°C. Calculate ΔE for the combustion of ethanol in kJ/mol.

Problems in Context

6.51 High-purity benzoic acid (C_6H_5COOH; ΔH for combustion = −3227 kJ/mol) is used to calibrate bomb calorimeters. A 1.221-g sample burns in a calorimeter that has a heat capacity of 6.384 kJ/°C. What is the temperature change?

6.52 Two aircraft rivets, one iron and the other copper, are placed in a calorimeter that has an initial temperature of 20.°C. The data for the rivets are as follows:

	Iron	Copper
Mass (g)	30.0	20.0
Initial T (°C)	0.0	100.0
c (J/g·K)	0.450	0.387

(a) Will heat flow from Fe to Cu or from Cu to Fe?
(b) What other information is needed to correct any measurements in an actual experiment?
(c) What is the maximum final temperature of the system (assuming the heat capacity of the calorimeter is negligible)?

6.53 A chemical engineer burned 1.520 g of a hydrocarbon in the bomb of a calorimeter (see Figure 6.10). The water temperature rose from 20.00°C to 23.55°C. If the calorimeter had a heat capacity of 11.09 kJ/K, what was the heat released (q_V) per gram of hydrocarbon?

Stoichiometry of Thermochemical Equations

(Sample Problem 6.8)

Concept Review Questions

6.54 Does a negative ΔH mean that the heat should be treated as a reactant or as a product?

6.55 Would you expect $O_2(g) \longrightarrow 2O(g)$ to have a positive or a negative ΔH? Explain.

6.56 Is ΔH positive or negative when 1 mol of water vapor condenses to liquid water? Why? How does this value compare with ΔH for the vaporization of 2 mol of liquid water to water vapor?

Skill-Building Exercises (grouped in similar pairs)

6.57 Consider the following balanced thermochemical equation for a reaction sometimes used for H_2S production:

$$\tfrac{1}{8}S_8(s) + H_2(g) \longrightarrow H_2S(g) \quad \Delta H = -20.2 \text{ kJ}$$

(a) Is this an exothermic or endothermic reaction?
(b) What is ΔH for the reverse reaction?
(c) What is ΔH when 2.6 mol of S_8 reacts?
(d) What is ΔH when 25.0 g of S_8 reacts?

6.58 Consider the following balanced thermochemical equation for the decomposition of the mineral magnesite:

$$MgCO_3(s) \longrightarrow MgO(s) + CO_2(g) \quad \Delta H = 117.3 \text{ kJ}$$

(a) Is heat absorbed or released in the reaction?
(b) What is ΔH for the reverse reaction?
(c) What is ΔH when 5.35 mol of CO_2 reacts with excess MgO?
(d) What is ΔH when 35.5 g of CO_2 reacts with excess MgO?

6.59 When 1 mol of NO(g) forms from its elements, 90.29 kJ of heat is absorbed. (a) Write a balanced thermochemical equation. (b) What is ΔH when 3.50 g of NO decomposes to its elements?

6.60 When 1 mol of KBr(s) decomposes to its elements, 394 kJ of heat is absorbed. (a) Write a balanced thermochemical equation. (b) What is ΔH when 10.0 kg of KBr forms from its elements?

Problems in Context

6.61 Liquid hydrogen peroxide, an oxidizing agent in many rocket fuel mixtures, releases oxygen gas on decomposition:

$$2H_2O_2(l) \longrightarrow 2H_2O(l) + O_2(g) \quad \Delta H = -196.1 \text{ kJ}$$

How much heat is released when 652 kg of H_2O_2 decomposes?

6.62 Compounds of boron and hydrogen are remarkable for their unusual bonding (described in Section 14.5) and also for their reactivity. With the more reactive halogens, for example, diborane (B_2H_6) forms trihalides even at low temperatures:

$$B_2H_6(g) + 6Cl_2(g) \longrightarrow 2BCl_3(g) + 6HCl(g) \quad \Delta H = -755.4 \text{ kJ}$$

What is ΔH per kilogram of diborane that reacts?

6.63 Deterioration of buildings, bridges, and other structures through the rusting of iron costs millions of dollars a day. The actual process requires water, but a simplified equation is

$$4Fe(s) + 3O_2(g) \longrightarrow 2Fe_2O_3(s) \quad \Delta H = -1.65\times10^3 \text{ kJ}$$

(a) How much heat is released when 0.250 kg of iron rusts?
(b) How much rust forms when 4.85×10^3 kJ of heat is released?

6.64 A mercury mirror forms inside a test tube as a result of the thermal decomposition of mercury(II) oxide:

$$2HgO(s) \longrightarrow 2Hg(l) + O_2(g) \quad \Delta H = 181.6 \text{ kJ}$$

(a) How much heat is absorbed to decompose 555 g of the oxide?
(b) If 275 kJ of heat is absorbed, how many grams of Hg form?

6.65 Most ethylene (C_2H_4), the starting material for producing polyethylene, comes from petroleum processing. It also occurs naturally as a fruit-ripening hormone and as a component of natural gas. (a) The heat transferred during combustion of C_2H_4 is −1411 kJ/mol. Write a balanced thermochemical equation. (b) How many grams of C_2H_4 must burn to give 70.0 kJ of heat?

6.66 Sucrose ($C_{12}H_{22}O_{11}$, table sugar) is oxidized in the body by O_2 via a complex set of reactions that produces $CO_2(g)$ and $H_2O(g)$ and releases 5.64×10^3 kJ/mol sucrose. (a) Write a balanced thermochemical equation for the overall process. (b) How much heat is released per gram of sucrose oxidized?

Hess's Law: Finding ΔH of Any Reaction

(Sample Problem 6.9)

Concept Review Questions

6.67 Express Hess's law in your own words.

6.68 What is the main application of Hess's law?

6.69 When carbon burns in a deficiency of O_2, a mixture of CO and CO_2 forms. Carbon burns in excess O_2 to form only CO_2, and CO burns in excess O_2 to form only CO_2. Use ΔH values of the latter two reactions (from Appendix B) to calculate ΔH for

$$C(\text{graphite}) + \tfrac{1}{2}O_2(g) \longrightarrow CO(g)$$

Skill-Building Exercises (grouped in similar pairs)

6.70 Calculate ΔH for

$$Ca(s) + \tfrac{1}{2}O_2(g) + CO_2(g) \longrightarrow CaCO_3(s)$$

given the following reactions:

$$Ca(s) + \tfrac{1}{2}O_2(g) \longrightarrow CaO(s) \quad \Delta H = -635.1 \text{ kJ}$$
$$CaCO_3(s) \longrightarrow CaO(s) + CO_2(g) \quad \Delta H = 178.3 \text{ kJ}$$

6.71 Calculate ΔH for

$$2NOCl(g) \longrightarrow N_2(g) + O_2(g) + Cl_2(g)$$

given the following reactions:

$$\tfrac{1}{2}N_2(g) + \tfrac{1}{2}O_2(g) \longrightarrow NO(g) \quad \Delta H = 90.3 \text{ kJ}$$
$$NO(g) + \tfrac{1}{2}Cl_2(g) \longrightarrow NOCl(g) \quad \Delta H = -38.6 \text{ kJ}$$

6.72 Write the balanced overall equation (equation 3) for the following process, calculate $\Delta H_{\text{overall}}$, and match the number of each equation with the letter of the appropriate arrow in Figure P6.72:

(1) $N_2(g) + O_2(g) \longrightarrow 2NO(g) \quad \Delta H = 180.6 \text{ kJ}$
(2) $2NO(g) + O_2(g) \longrightarrow 2NO_2(g) \quad \Delta H = -114.2 \text{ kJ}$
(3) $\Delta H_{\text{overall}} = ?$

6.73 Write the balanced overall equation (equation 3) for the following process, calculate $\Delta H_{\text{overall}}$, and match the number of each equation with the letter of the appropriate arrow in Figure P6.73:

(1) $P_4(s) + 6Cl_2(g) \longrightarrow 4PCl_3(g) \quad \Delta H = -1148 \text{ kJ}$
(2) $4PCl_3(g) + 4Cl_2(g) \longrightarrow 4PCl_5(g) \quad \Delta H = -460 \text{ kJ}$
(3) $\Delta H_{\text{overall}} = ?$

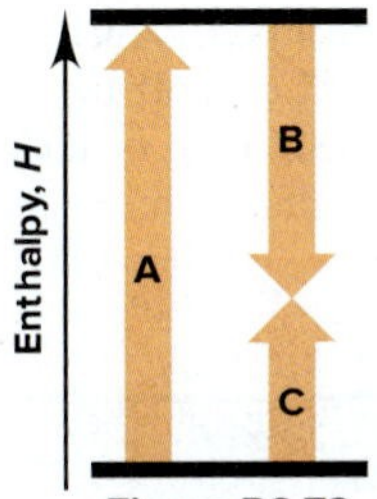

Figure P6.72

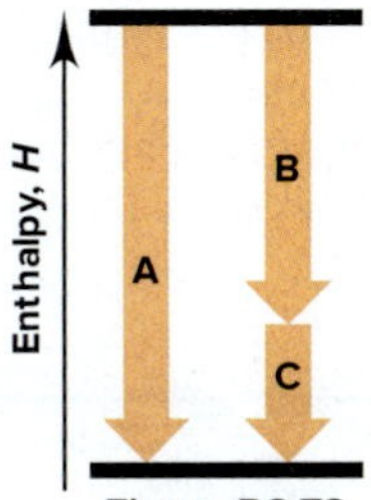

Figure P6.73

6.74 At a given set of conditions, 241.8 kJ of heat is released when 1 mol of $H_2O(g)$ forms from its elements. Under the same conditions, 285.8 kJ is released when 1 mol of $H_2O(l)$ forms from its elements. Find ΔH for the vaporization of water at these conditions.

6.75 When 1 mol of $CS_2(l)$ forms from its elements at 1 atm and 25°C, 89.7 kJ of heat is absorbed, and it takes 27.7 kJ to vaporize 1 mol of the liquid. How much heat is absorbed when 1 mol of $CS_2(g)$ forms from its elements at these conditions?

Problems in Context

6.76 Diamond and graphite are two crystalline forms of carbon. At 1 atm and 25°C, diamond changes to graphite so slowly that the enthalpy change of the process must be obtained indirectly. Using equations from the list below, determine ΔH for

$$\text{C(diamond)} \longrightarrow \text{C(graphite)}$$

(1) $\text{C(diamond)} + O_2(g) \longrightarrow CO_2(g)$ $\quad\Delta H = -395.4$ kJ
(2) $2CO_2(g) \longrightarrow 2CO(g) + O_2(g)$ $\quad\Delta H = 566.0$ kJ
(3) $\text{C(graphite)} + O_2(g) \longrightarrow CO_2(g)$ $\quad\Delta H = -393.5$ kJ
(4) $2CO(g) \longrightarrow \text{C(graphite)} + CO_2(g)$ $\quad\Delta H = -172.5$ kJ

Standard Enthalpies of Reaction (ΔH°_{rxn})

(Sample Problems 6.10 and 6.11)

Concept Review Questions

6.77 What is the difference between the standard enthalpy of formation and the standard enthalpy of reaction?

6.78 How are ΔH°_f values used to calculate ΔH°_{rxn}?

6.79 Make any changes needed in each of the following equations so that the enthalpy change is equal to ΔH°_f for the compound:
(a) $Cl(g) + Na(s) \longrightarrow NaCl(s)$
(b) $H_2O(g) \longrightarrow 2H(g) + \frac{1}{2}O_2(g)$
(c) $\frac{1}{2}N_2(g) + \frac{3}{2}H_2(g) \longrightarrow NH_3(g)$

Skill-Building Exercises (grouped in similar pairs)

6.80 Use Table 6.3 or Appendix B to write a balanced formation equation at standard conditions for each of the following compounds: (a) $CaCl_2$; (b) $NaHCO_3$; (c) CCl_4; (d) HNO_3.

6.81 Use Table 6.3 or Appendix B to write a balanced formation equation at standard conditions for each of the following compounds: (a) HI; (b) SiF_4; (c) O_3; (d) $Ca_3(PO_4)_2$.

6.82 Calculate ΔH°_{rxn} for each of the following:
(a) $2H_2S(g) + 3O_2(g) \longrightarrow 2SO_2(g) + 2H_2O(g)$
(b) $CH_4(g) + Cl_2(g) \longrightarrow CCl_4(l) + HCl(g)$ [unbalanced]

6.83 Calculate ΔH°_{rxn} for each of the following:
(a) $SiO_2(s) + 4HF(g) \longrightarrow SiF_4(g) + 2H_2O(l)$
(b) $C_2H_6(g) + O_2(g) \longrightarrow CO_2(g) + H_2O(g)$ [unbalanced]

6.84 Copper(I) oxide can be oxidized to copper(II) oxide:

$$Cu_2O(s) + \tfrac{1}{2}O_2(g) \longrightarrow 2CuO(s) \qquad \Delta H^\circ_{rxn} = -146.0 \text{ kJ}$$

Given ΔH°_f of $Cu_2O(s) = -168.6$ kJ/mol, find ΔH°_f of $CuO(s)$.

6.85 Acetylene burns in air according to the following equation:

$$C_2H_2(g) + \tfrac{5}{2}O_2(g) \longrightarrow 2CO_2(g) + H_2O(g)$$

$$\Delta H^\circ_{rxn} = -1255.8 \text{ kJ}$$

Given ΔH°_f of $CO_2(g) = -393.5$ kJ/mol and ΔH°_f of $H_2O(g) = -241.8$ kJ/mol, find ΔH°_f of $C_2H_2(g)$.

Problems in Context

6.86 The common lead-acid car battery produces a large burst of current, even at low temperatures, and is rechargeable. The reaction that occurs while recharging a "dead" battery is

$$2PbSO_4(s) + 2H_2O(l) \longrightarrow Pb(s) + PbO_2(s) + 2H_2SO_4(l)$$

(a) Use ΔH°_f values from Appendix B to calculate ΔH°_{rxn}.
(b) Use the following equations to check your answer to part (a):
(1) $Pb(s) + PbO_2(s) + 2SO_3(g) \longrightarrow 2PbSO_4(s)$ $\quad\Delta H^\circ_{rxn} = -768$ kJ
(2) $SO_3(g) + H_2O(l) \longrightarrow H_2SO_4(l)$ $\quad\Delta H^\circ_{rxn} = -132$ kJ

Comprehensive Problems

6.87 Stearic acid ($C_{18}H_{36}O_2$) is a fatty acid, a molecule with a long hydrocarbon chain and an organic acid group (COOH) at the end. It is used to make cosmetics, ointments, soaps, and candles and is found in animal tissue as part of many saturated fats. In fact, when you eat meat, you are ingesting some fats containing stearic acid.
(a) Write a balanced equation for the combustion of stearic acid to gaseous products.
(b) Calculate ΔH°_{rxn} for this combustion (ΔH°_f of $C_{18}H_{36}O_2 = -948$ kJ/mol).
(c) Calculate the heat (q) released in kJ and kcal when 1.00 g of stearic acid is burned completely.
(d) A candy bar contains 11.0 g of fat and provides 100. Cal from fat; is this consistent with your answer for part (c)?

6.88 Diluting sulfuric acid with water is highly exothermic:

$$H_2SO_4(l) \xrightarrow{H_2O} H_2SO_4(aq) + \textit{heat}$$

(a) Use Appendix B to find ΔH°_{rxn} for diluting 1.00 mol of $H_2SO_4(l)$ ($d = 1.83$ g/mL) to 1 L of 1.00 M $H_2SO_4(aq)$ ($d = 1.060$ g/mL).
(b) Suppose you carry out the dilution in a calorimeter. The initial T is 25.0°C, and the specific heat capacity of the final solution is 3.50 J/g·K. What is the final T?
(c) Use the ideas of density and heat capacity to explain why you should add acid to water rather than water to acid.

6.89 A balloonist begins a trip in a helium-filled balloon in early morning when the temperature is 15°C. By mid-afternoon, the temperature is 30.°C. Assuming the pressure remains at 1.00 atm, for each mole of helium, calculate the following:
(a) The initial and final volumes
(b) The change in internal energy, ΔE (*Hint:* Helium behaves like an ideal gas, so $E = \frac{3}{2}nRT$. Be sure the units of R are consistent with those of E.)
(c) The work (w) done by the helium (in J)
(d) The heat (q) transferred (in J)
(e) ΔH for the process (in J)
(f) Explain the relationship between the answers to parts (d) and (e).

6.90 In winemaking, the sugars in grapes undergo *fermentation* by yeast to yield CH_3CH_2OH and CO_2. During cellular *respiration* (combustion), sugar and ethanol yield water vapor and CO_2.
(a) Using $C_6H_{12}O_6$ for sugar, calculate ΔH°_{rxn} of fermentation and of respiration.
(b) Write a combustion reaction for ethanol. Which has a higher ΔH°_{rxn} for combustion per mole of C, sugar or ethanol?

6.91 Three of the reactions that occur when the paraffin of a candle (typical formula $C_{21}H_{44}$) burns are as follows:
(1) Complete combustion forms CO_2 and water vapor.
(2) Incomplete combustion forms CO and water vapor.
(3) Some wax is oxidized to elemental C (soot) and water vapor.
(a) Find ΔH°_{rxn} of each reaction (ΔH°_f of $C_{21}H_{44} = -476$ kJ/mol; use graphite for elemental carbon).
(b) Find q (in kJ) when a 254-g candle burns completely.
(c) Find q (in kJ) when 8.00% by mass of the candle burns incompletely and 5.00% by mass of it undergoes soot formation.

6.92 Ethylene oxide (EO) is prepared by the vapor-phase oxidation of ethylene. Its main uses are in the preparation of the antifreeze ethylene glycol and in the production of poly(ethylene terephthalate), which is used to make beverage bottles and fibers. Pure EO vapor can decompose explosively:

$$\mathrm{H_2C}\overset{\mathrm{O}}{—}\mathrm{CH_2}(g) \longrightarrow \mathrm{CH_4}(g) + \mathrm{CO}(g)$$

Liquid EO has $\Delta H^\circ_f = -77.4$ kJ/mol and ΔH° for its vaporization = 569.4 J/g. (a) Calculate ΔH°_{rxn} for the gas-phase reaction. (b) External heating causes the vapor to decompose at 10 bar and 93°C in a distillation column. What is the final temperature if the average specific heat capacity of the products is 2.5 J/g·°C?

6.93 The following scenes represent a gaseous reaction between compounds of nitrogen *(blue)* and oxygen *(red)* at 298 K:

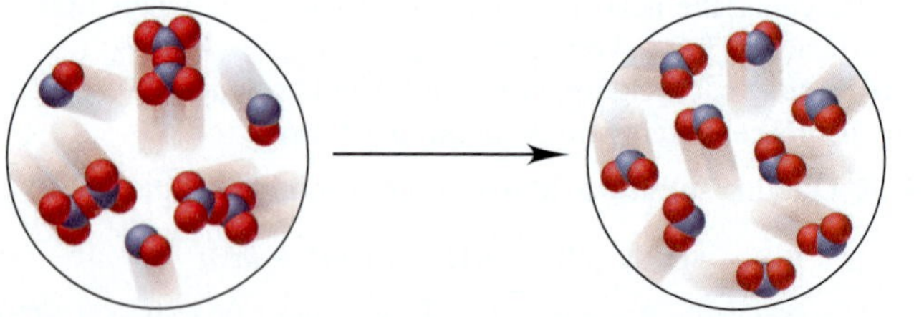

(a) Write a balanced equation and use Appendix B to calculate ΔH°_{rxn}.
(b) If each molecule of product represents 1.50×10^{-2} mol, what quantity of heat (in J) is released or absorbed?

6.94 Isooctane (C_8H_{18}; $d = 0.692$ g/mL) is used as the fuel in a test of a new automobile drive train.
(a) How much energy (in kJ) is released by combustion of 20.4 gal of isooctane to gases ($\Delta H^\circ_{rxn} = -5.44\times10^3$ kJ/mol)?
(b) The energy delivered to the automobile's wheels at 65 mph is 5.5×10^4 kJ/h. Assuming *all* the energy is transferred as work to the wheels, how far (in km) can the vehicle travel on the 20.4 gal of fuel?
(c) If the actual range is 455 mi, explain your answer to (b).

6.95 Four 50.-g samples of different colorless liquids are placed in beakers at $T_{initial} = 25.00$°C. Each liquid is heated until 450. J of heat has been absorbed; T_{final} is shown on each beaker below. Rank the liquids in order of increasing specific heat capacity.

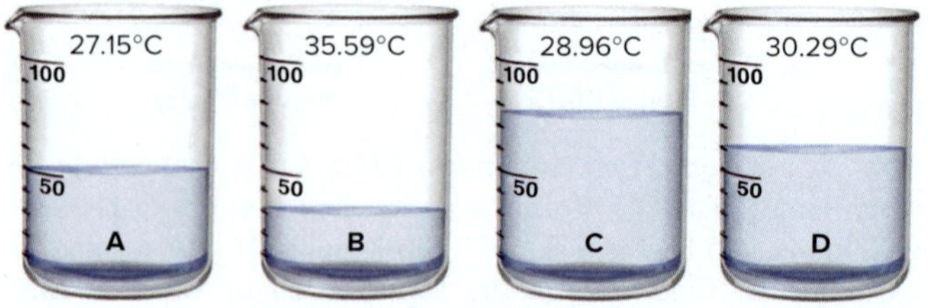

6.96 Reaction of gaseous ClF with F_2 yields liquid ClF_3, an important fluorinating agent. Use the following thermochemical equations to calculate ΔH°_{rxn} for this reaction:

(1) $2ClF(g) + O_2(g) \longrightarrow Cl_2O(g) + OF_2(g)$ $\Delta H^\circ_{rxn} = 167.5$ kJ
(2) $2F_2(g) + O_2(g) \longrightarrow 2OF_2(g)$ $\Delta H^\circ_{rxn} = -43.5$ kJ
(3) $2ClF_3(l) + 2O_2(g) \longrightarrow Cl_2O(g) + 3OF_2(g)$ $\Delta H^\circ_{rxn} = 394.1$ kJ

6.97 Silver bromide is used to coat ordinary black-and-white photographic film, while high-speed film uses silver iodide.
(a) When 50.0 mL of 5.0 g/L $AgNO_3$ is added to a coffee-cup calorimeter containing 50.0 mL of 5.0 g/L NaI, with both solutions at 25°C, what mass of AgI forms?
(b) Use Appendix B to find ΔH°_{rxn}.
(c) What is ΔT_{soln} (assuming the volumes are additive and the solution has the density and specific heat capacity of water)?

6.98 The calorie (4.184 J) is defined as the quantity of energy needed to raise the temperature of 1.00 g of liquid water by 1.00°C. The British thermal unit (Btu) is defined as the quantity of energy needed to raise the temperature of 1.00 lb of liquid water by 1.00°F.
(a) How many joules are in 1.00 Btu (1 lb = 453.6 g; 1.0°C = 1.8°F)?
(b) The *therm* is a unit of energy consumption and is defined as 100,000 Btu. How many joules are in 1.00 therm?
(c) How many moles of methane must be burned to give 1.00 therm of energy? (Assume that water forms as a gas.)
(d) If natural gas costs \$0.66 per therm, what is the cost per mole of methane? (Assume that natural gas is pure methane.)
(e) How much would it cost to warm 318 gal of water in a hot tub from 15.0°C to 42.0°C by burning methane (1 gal = 3.78 L)?

6.99 When organic matter decomposes under oxygen-free (anaerobic) conditions, methane is one of the products. Thus, enormous deposits of natural gas, which is almost entirely methane, serve as a major source of fuel for home and industry.
(a) Known deposits of natural gas can produce 5600 EJ of energy (1 EJ = 10^{18} J). Current total global energy usage is 4.0×10^2 EJ per year. Find the mass (in kg) of known deposits of natural gas (ΔH°_{rxn} for the combustion of $CH_4 = -802$ kJ/mol).
(b) At current rates of usage, for how many years could these deposits supply the world's total energy needs?

(c) What volume (in ft^3) of natural gas, measured at STP, is required to heat 1.00 qt of water from 25.0°C to 100.0°C (d of H_2O = 1.00 g/mL; d of CH_4 at STP = 0.72 g/L)?
(d) The fission of 1 mol of uranium (about 4×10^{-4} ft^3) in a nuclear reactor produces 2×10^{13} J. What volume (in ft^3) of natural gas would produce the same amount of energy?

6.100 A reaction takes place in a steel vessel within a chamber filled with argon gas. Shown below are atomic-scale views of the argon adjacent to the surface of the container wall of the reaction vessel before and after the reaction. Was the reaction exothermic or endothermic? Explain.

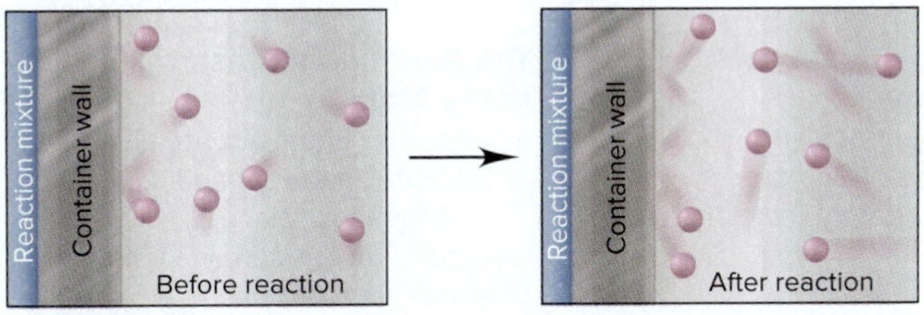

6.101 An aqueous waste stream with a maximum concentration of 0.50 M H_2SO_4 (d = 1.030 g/mL at 25°C) is neutralized by controlled addition of 40% NaOH (d = 1.430 g/L) before it goes to the process sewer and then to the chemical plant waste treatment facility. A safety review finds that the waste stream could meet a small stream of an immiscible organic compound, which could form a flammable vapor in air at 40.°C. The maximum temperature reached by the NaOH solution and the waste stream is 31°C. Could the temperature increase due to the heat transferred by the neutralization cause the organic vapor to explode? Assume that the specific heat capacity of each solution is 4.184 J/g·K.

6.102 Kerosene, a common space-heater fuel, is a mixture of hydrocarbons whose "average" formula is $C_{12}H_{26}$.
(a) Write a balanced equation, using the simplest whole-number coefficients, for the complete combustion of kerosene to gases.
(b) If $\Delta H^\circ_{rxn} = -1.50\times10^4$ kJ for the combustion equation as written in part (a), determine ΔH°_f of kerosene.
(c) Calculate the heat released by combustion of 0.50 gal of kerosene (d of kerosene = 0.749 g/mL).
(d) How many gallons of kerosene must be burned for a kerosene furnace to produce 1250. Btu (1 Btu = 1.055 kJ)?

6.103 Silicon tetrachloride is produced annually on the multikiloton scale and used in making transistor-grade silicon. It can be produced directly from the elements (reaction 1) or, more cheaply, by heating sand and graphite with chlorine gas (reaction 2). If water is present in reaction 2, some tetrachloride may be lost in an unwanted side reaction (reaction 3):

(1) $Si(s) + 2Cl_2(g) \longrightarrow SiCl_4(g)$
(2) $SiO_2(s) + 2C(\text{graphite}) + 2Cl_2(g) \longrightarrow SiCl_4(g) + 2CO(g)$
(3) $SiCl_4(g) + 2H_2O(g) \longrightarrow SiO_2(s) + 4HCl(g)$ $\quad \Delta H^\circ_{rxn} = -139.5$ kJ

(a) Use reaction 3 to calculate the standard enthalpies of reaction of reactions 1 and 2.
(b) What is the standard enthalpy of reaction for a fourth reaction that is the sum of reactions 2 and 3?

6.104 One mole of nitrogen gas confined within a cylinder by a piston is heated from 0°C to 819°C at 1.00 atm.
(a) Calculate the work done by the expanding gas in joules (1 J = 9.87×10^{-3} atm·L). Assume that all the energy is used to do work.
(b) What would be the temperature change if the gas were heated using the same amount of energy in a container of fixed volume? (Assume that the specific heat capacity of N_2 is 1.00 J/g·K.)

6.105 The chemistry of nitrogen oxides is very versatile. Given the following reactions and their standard enthalpy changes,

(1) $NO(g) + NO_2(g) \longrightarrow N_2O_3(g)$ $\quad \Delta H^\circ_{rxn} = -39.8$ kJ
(2) $NO(g) + NO_2(g) + O_2(g) \longrightarrow N_2O_5(g)$ $\quad \Delta H^\circ_{rxn} = -112.5$ kJ
(3) $2NO_2(g) \longrightarrow N_2O_4(g)$ $\quad \Delta H^\circ_{rxn} = -57.2$ kJ
(4) $2NO(g) + O_2(g) \longrightarrow 2NO_2(g)$ $\quad \Delta H^\circ_{rxn} = -114.2$ kJ
(5) $N_2O_5(s) \longrightarrow N_2O_5(g)$ $\quad \Delta H^\circ_{rxn} = 54.1$ kJ

calculate the standard enthalpy of reaction for

$$N_2O_3(g) + N_2O_5(s) \longrightarrow 2N_2O_4(g)$$

6.106 Electric generating plants transport large amounts of hot water through metal pipes, and oxygen dissolved in the water can cause a major corrosion problem. Hydrazine (N_2H_4) added to the water prevents this problem by reacting with the oxygen:

$$N_2H_4(aq) + O_2(g) \longrightarrow N_2(g) + 2H_2O(l)$$

About 4×10^7 kg of hydrazine is produced every year by reacting ammonia with sodium hypochlorite in the *Raschig process:*

$$2NH_3(aq) + NaOCl(aq) \longrightarrow N_2H_4(aq) + NaCl(aq) + H_2O(l) \quad \Delta H^\circ_{rxn} = -151 \text{ kJ}$$

(a) If ΔH°_f of NaOCl(aq) = −346 kJ/mol, find ΔH°_f of $N_2H_4(aq)$.
(b) What is the heat released when aqueous N_2H_4 is added to 5.00×10^3 L of water that is 2.50×10^{-4} M O_2?

6.107 Liquid methanol (CH_3OH) can be used as an alternative fuel by pickup trucks and SUVs. An industrial method for preparing it involves the catalytic hydrogenation of carbon monoxide:

$$CO(g) + 2H_2(g) \xrightarrow{\text{catalyst}} CH_3OH(l)$$

How much heat (in kJ) is released when 15.0 L of CO at 85°C and 112 kPa reacts with 18.5 L of H_2 at 75°C and 744 torr?

6.108 (a) How much heat is released when 25.0 g of methane burns in excess O_2 to form gaseous CO_2 and H_2O?
(b) Calculate the temperature of the product mixture if the methane and air are both at an initial temperature of 0.0°C. Assume a stoichiometric ratio of methane to oxygen from the air, with air being 21% O_2 by volume (c of CO_2 = 57.2 J/mol·K; c of $H_2O(g)$ = 36.0 J/mol·K; c of N_2 = 30.5 J/mol·K).

7 Quantum Theory and Atomic Structure

7.1 The Nature of Light
Wave Nature of Light
Particle Nature of Light

7.2 Atomic Spectra
Line Spectra and the Rydberg Equation
Bohr Model of the Hydrogen Atom
Energy Levels of the Hydrogen Atom

7.3 The Wave-Particle Duality of Matter and Energy
Wave Nature of Electrons and Particle Nature of Photons
Heisenberg's Uncertainty Principle

7.4 The Quantum-Mechanical Model of the Atom
Schrödinger Equation, Atomic Orbital, and Probable Location of the Electron
Quantum Numbers of an Orbital
Quantum Numbers and Energy Levels
Shapes of Atomic Orbitals
The Special Case of the H Atom

Source: © Jeffrey Johnson

Concepts and Skills to Review Before You Study This Chapter

- discovery of the electron and atomic nucleus (Section 2.4)
- major features of atomic structure (Section 2.5)
- changes in the energy state of a system (Section 6.1)

Neon signs, introduced to the United States in 1923, quickly became popular with businesses eager to catch the attention of potential customers. These signs are now everywhere, from the simple "OPEN" in the window of your local pizza place to more elaborate creations in Times Square and Las Vegas. The science behind neon signs (which can involve other elements as well) fascinated scientists in the early 20th century. The fact that neon emitted light was one of several observations that led to a new view of matter and energy that not only explains fireworks, sodium-vapor streetlights, and TV screens, but clarifies the structure of the atom! New ideas about atomic structure were some of many revolutions that took place in Western science (and culture) from around 1890 to 1930. ›

But revolutions in science are not the violent upheavals of political overthrow. Flaws appear in a model due to conflicting evidence, a startling discovery widens the flaws into cracks, and the theoretical structure crumbles. Then, new insight, verified by experiment, builds a model more consistent with reality. So it was when Lavoisier's theory of combustion overthrew the phlogiston model, when Dalton's atomic theory established the idea of individual units of matter, and when Rutherford's model substituted nuclear atoms for "billiard balls" or "plum puddings." You will see this process unfold again as we discuss the development of modern atomic theory.

A Time of Revolution

Here are some of the revolutionary discoveries and events occurring in the Western world during those decades:

1895 Röntgen: discovery of x-rays
1896 Becquerel: discovery of radioactivity
1897 Thomson: discovery of the electron
1900 Freud: theory of the unconscious
1900 Planck: quantum theory
1901 Marconi: invention of the radio
1903 Wright brothers: flight of a plane
1905 Ford: assembly line to build cars
1905 Rutherford: radioactivity explained
1905 Einstein: relativity/photon theories
1906 St. Denis: modern dance
1908 Matisse/Picasso: modern art
1909 Schoenberg/Berg: modern music
1911 Rutherford: nuclear model
1913 Bohr: atomic spectra and new model
1914–1918: World War I
1923 Compton: photon momentum
1924 de Broglie: wave theory of matter
1926 Schrödinger: wave equation
1927 Heisenberg: uncertainty principle

IN THIS CHAPTER . . . *We discuss quantum mechanics, the theory that explains the fundamental nature of energy and matter and accounts for atomic structure.*

- We consider the classical distinction between the wave properties of energy and the particle properties of matter.
- We then examine two observations—blackbody radiation and the photoelectric effect—whose explanations led to a *quantized*, or particulate, model of light.
- We see that light emitted by excited atoms—an *atomic spectrum*—suggests an atom with distinct energy levels, and we apply spectra to chemical analysis.
- We see how wave-particle duality, which shows that matter and energy have similar properties, and the uncertainty principle, which proposes that electron behavior can never be known exactly, led to the current model of the H atom.
- We present quantum numbers, which specify the size, shape, and orientation of atomic orbitals, the regions an electron occupies in the H atom. (In Chapter 8, we'll examine quantum numbers for atoms with more than one electron.)

7.1 THE NATURE OF LIGHT

To understand current atomic theory, you need to know about **electromagnetic radiation** (also called *electromagnetic energy,* or *radiant energy*). Visible light, x-rays, radio waves, and microwaves are some of the types of electromagnetic radiation. All of these consist of energy propagated by electric and magnetic fields that increase and decrease in intensity as they move, wavelike, through space. This *classical wave model* explains why rainbows form, how magnifying glasses work, and many other familiar observations. But it cannot explain observations on the very unfamiliar atomic scale. In this section, we describe some properties of electromagnetic radiation and note how they are distinguished from the properties of matter. But, we will see that some properties blur the distinction between energy and matter, requiring a new model to explain them.

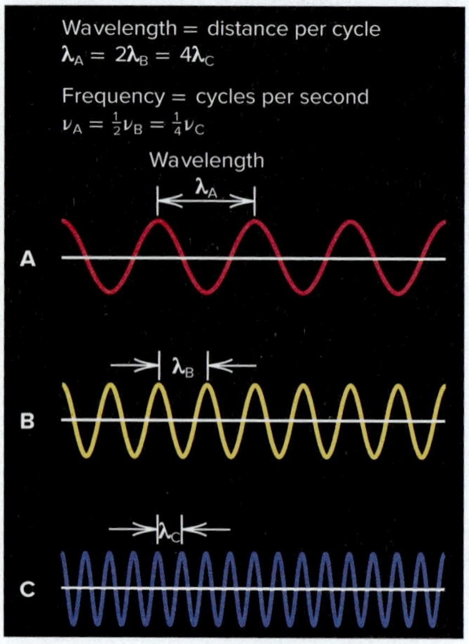

Figure 7.1 The inverse relationship of frequency and wavelength.

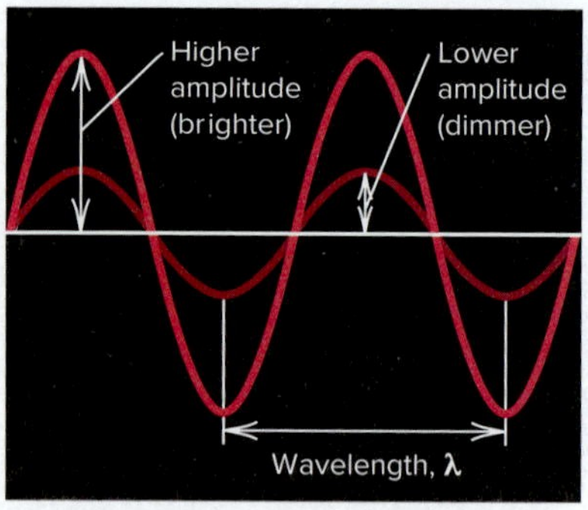

Figure 7.2 Differing amplitude (brightness, or intensity) of a wave.

The Wave Nature of Light

The wave properties of electromagnetic radiation are described by three variables and one constant (Figure 7.1):

1. *Frequency* (ν, Greek *nu*). The **frequency** of electromagnetic radiation is the number of complete waves, or cycles, that pass a given point per second, expressed by the unit 1/second [s^{-1}; also called a *hertz* (Hz)]. An FM radio station's channel is the frequency, in MHz, at which it broadcasts.
2. *Wavelength* (λ, Greek *lambda*). The **wavelength** is the distance between any point on a wave and the corresponding point on the next crest (or trough) of the wave, that is, the distance the wave travels during one cycle. Wavelength has units of length such as meters or, for very short wavelengths, nanometers (nm, 10^{-9} m), picometers (pm, 10^{-12} m), or the non-SI unit angstroms (Å, 10^{-10} m).
3. *Speed.* The speed of a wave is the distance it moves per unit time (meters per second), the product of its frequency (cycles per second) and the wavelength (meters per cycle):

$$\text{Units for speed of wave:}\quad \frac{\cancel{\text{cycles}}}{\text{s}} \times \frac{\text{m}}{\cancel{\text{cycle}}} = \frac{\text{m}}{\text{s}}$$

   In a vacuum, electromagnetic radiation moves at 2.99792458×10^8 m/s (3.00×10^8 m/s to three significant figures), a *physical constant* called the **speed of light (*c*)**:

$$c = \nu \times \lambda \qquad \textbf{(7.1)}$$

   Since the product of ν and λ is a constant, they have an inverse relationship—*radiation with a high frequency has a short wavelength, and radiation with a low frequency has a long wavelength*:

$$\nu\uparrow\lambda\downarrow \quad \text{and} \quad \nu\downarrow\lambda\uparrow$$

   Wave A in Figure 7.1 has a wavelength that is four times longer than that of wave C; since both waves are traveling at the same speed, one wavelength of wave A passes a point in the same amount of time as four wavelengths of wave C—the frequency of wave C is four times that of wave A.
4. *Amplitude.* The **amplitude** of a wave is the height of the crest (or depth of the trough). For an electromagnetic wave, the amplitude is related to the *intensity* of the radiation, or its brightness in the case of visible light. Light of a particular color has a specific frequency (and thus, wavelength), but, as Figure 7.2 shows, its amplitude can vary; the light can be dimmer (lower amplitude, less intense) or brighter (higher amplitude, more intense).

The Electromagnetic Spectrum *All waves of electromagnetic radiation travel at the same speed through a vacuum but differ in frequency and, therefore, wavelength.* The types of radiation are arranged in order of increasing wavelength (decreasing frequency) in the **electromagnetic spectrum** (Figure 7.3). As a *continuum of radiant energy,* the spectrum is broken down into regions according to wavelength and frequency, with each region adjoining the next.

Visible light represents just a small region of the spectrum. We perceive different wavelengths (or frequencies) of visible light as different colors, from violet (λ ≈ 400 nm) to red (λ ≈ 750 nm). Light of a single wavelength is called *monochromatic* (Greek, "one color"), whereas light of many wavelengths is *polychromatic*. White light, composed of all of the colors of visible light, is polychromatic. The region adjacent to visible light on the short-wavelength (high-frequency) end consists of **ultraviolet (UV)** radiation (also called *ultraviolet light*); still shorter wavelengths make up the x-ray and gamma (γ) ray regions. The region adjacent to visible light on the long-wavelength (low-frequency) end consists of **infrared (IR)** radiation; still longer wavelengths make up the microwave and radio wave regions.

Some types of electromagnetic radiation are utilized by familiar devices; for example, long-wavelength, low-frequency radiation is used by microwave ovens, radios, and cell phones. But electromagnetic emissions are everywhere: coming from

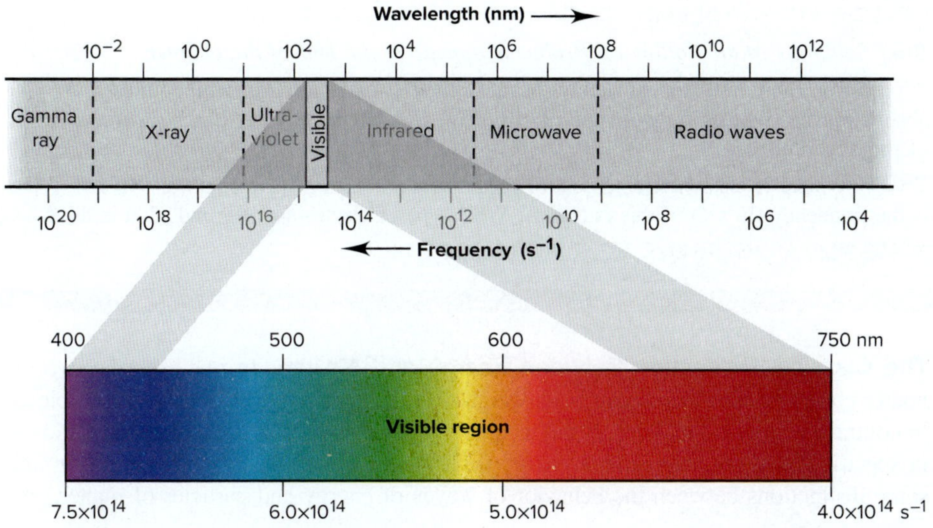

Figure 7.3 **Regions of the electromagnetic spectrum.** The visible region is expanded (and the scale made linear) to show the component colors.

human artifacts such as lightbulbs, x-ray equipment, and car motors, and from natural sources such as the Sun, lightning, radioactivity, and even the glow of fireflies! Our knowledge of the universe comes from radiation that enters our eyes and our light, x-ray, and radio telescopes.

SAMPLE PROBLEM 7.1 Interconverting Wavelength and Frequency

Problem A dental hygienist uses x-rays ($\lambda = 0.50$ Å) to take a series of dental radiographs while the patient listens to a radio station ($\lambda = 325$ cm) and looks out the window at the blue sky ($\lambda = 473$ nm). What is the frequency (in s^{-1}) of the electromagnetic radiation from each source? (Assume that the radiation travels at the speed of light, 3.00×10^8 m/s.)

Plan We are given the wavelengths, so we use Equation 7.1 to find the frequencies. However, we must first convert the wavelengths to meters because c has units of m/s (see the road map).

Road Map

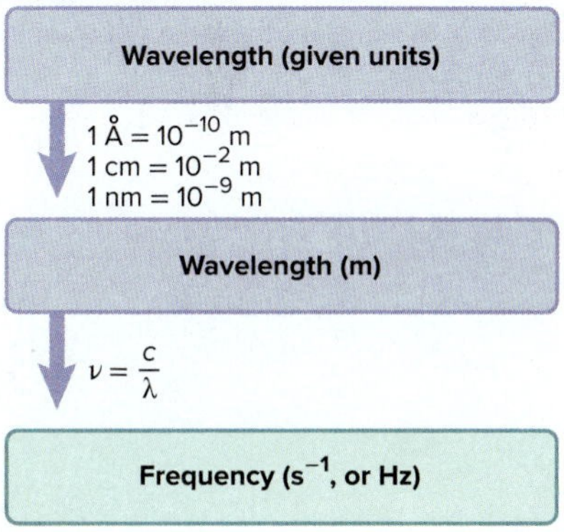

Solution For the x-rays: Converting from angstroms to meters,

$$\lambda = 0.50\ \text{Å} \times \frac{10^{-10}\ \text{m}}{1\ \text{Å}} = 5.0\times10^{-11}\ \text{m}$$

Calculating the frequency:

$$\nu = \frac{c}{\lambda} = \frac{3.00\times10^8\ \text{m/s}}{5.0\times10^{-11}\ \text{m}} = 6.0\times10^{18}\ \text{s}^{-1}$$

For the radio signal: Combining steps to calculate the frequency,

$$\nu = \frac{c}{\lambda} = \frac{3.00\times10^8\ \text{m/s}}{325\ \text{cm} \times \dfrac{10^{-2}\ \text{m}}{1\ \text{cm}}} = 9.23\times10^7\ \text{s}^{-1}$$

For the blue sky: Combining steps to calculate the frequency,

$$\nu = \frac{c}{\lambda} = \frac{3.00\times10^8\ \text{m/s}}{473\ \text{nm} \times \dfrac{10^{-9}\ \text{m}}{1\ \text{nm}}} = 6.34\times10^{14}\ \text{s}^{-1}$$

Check The orders of magnitude are correct for the regions of the electromagnetic spectrum (see Figure 7.3): x-rays (10^{19} to 10^{16} s^{-1}), radio waves (10^9 to 10^4 s^{-1}), and visible light (7.5×10^{14} to 4.0×10^{14} s^{-1}).

Comment Note that the x-ray, with the shortest wavelength, has the highest frequency, while the radio wave, with the longest wavelength, has the lowest frequency. The radio station here is broadcasting at 92.3×10^6 s^{-1}, or 92.3 million Hz (92.3 MHz), about midway in the FM range.

FOLLOW-UP PROBLEMS

Brief Solutions to ***all*** *Follow-up Problems appear at the end of the chapter.*

7.1A Some diamonds appear yellow because they contain nitrogen compounds that absorb purple light of frequency 7.23×10^{14} Hz. Calculate the wavelength (in nm and Å) of the absorbed light.

7.1B A typical television remote control emits radiation with a wavelength of 940 nm. What is the frequency (in s^{-1}) of this radiation? What type of electromagnetic radiation is this?

SOME SIMILAR PROBLEMS 7.7, 7.8, 7.13, and 7.14

Wave theory explains the colors of a rainbow.

Source: © Anthony McAulay/Shutterstock.com

The Classical Distinction Between Energy and Matter In our everyday world, matter comes in chunks you can hold, weigh, and change the quantity of, piece by piece. In contrast, energy is "massless," and its quantity can change continuously. Matter moves in specific paths, whereas radiant energy (light) travels in diffuse waves. Let's examine some distinctions between the behavior of waves of energy and particles of matter.

1. *Refraction and dispersion.* Light of a given wavelength travels at different speeds through various transparent media—vacuum, air, water, quartz, and so forth. Therefore, when a light wave passes from one medium into another, the speed of the wave changes. Figure 7.4A shows the phenomenon known as **refraction.** If the wave strikes the boundary between media, say, between air and water, at an angle other than 90°, the change in speed causes a change in direction, and the wave continues at a different angle. The angle of refraction depends on the two media and the wavelength of the light. In the related process of *dispersion,* white light separates (disperses) into its component colors when it passes through a prism (or other refracting object) because each incoming wave is refracted at a slightly different angle.

Two of many familiar applications of refraction and dispersion involve rainbows and diamonds. You see a rainbow when sunlight entering the near surface of countless droplets is dispersed and reflected off the far surface. Red light is bent least, so it appears higher in the sky and violet lower *(see photo).* A diamond sparkles with various colors because its facets disperse and reflect the incoming light.

Figure 7.4 Different behaviors of waves and particles.

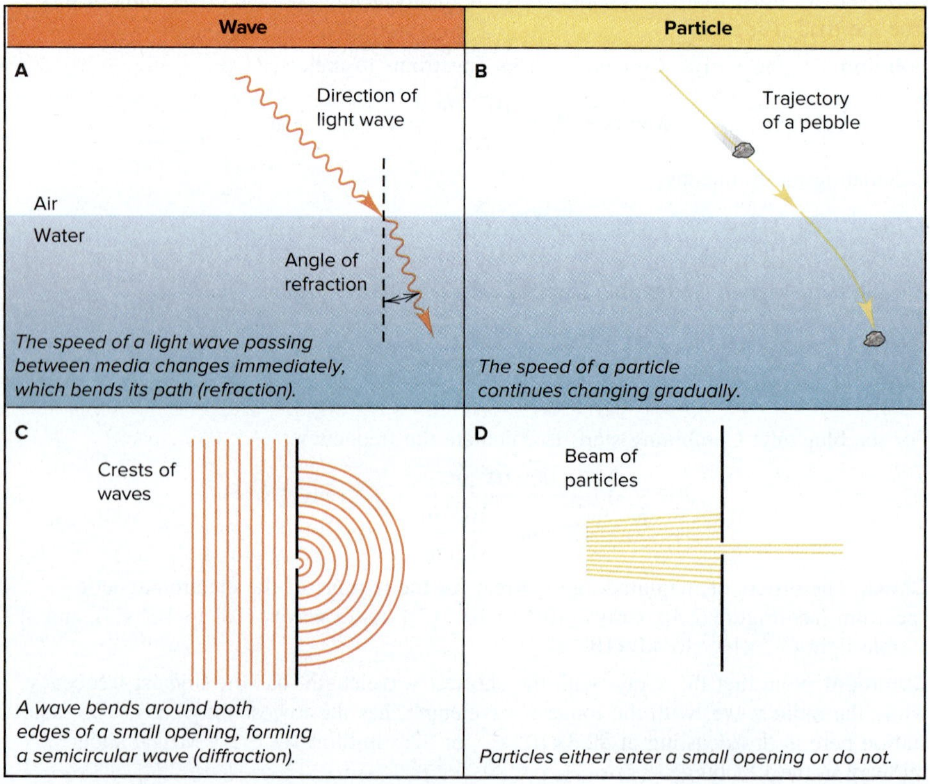

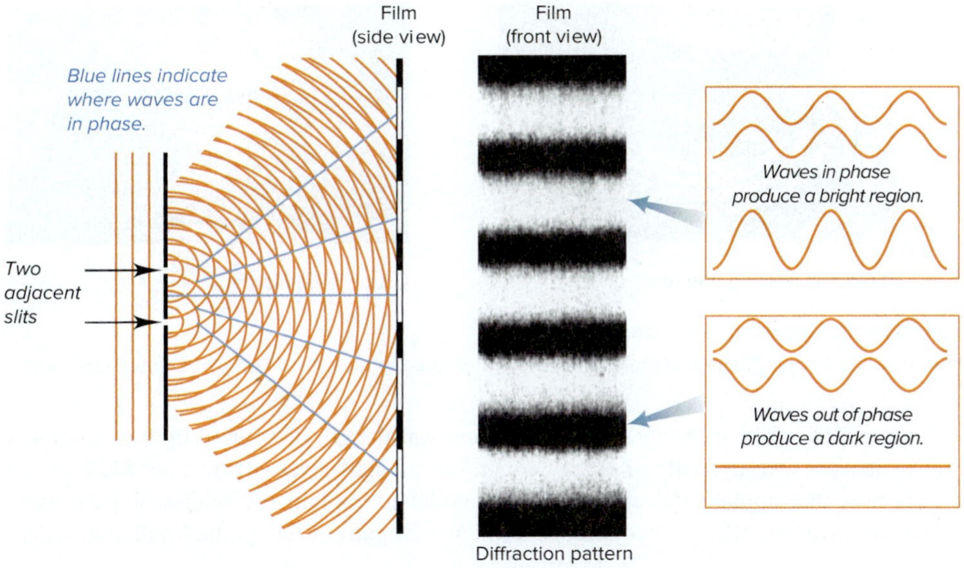

Light waves passing through two slits emerge as circular waves; constructive and destructive interference result in a diffraction pattern.

Constructive and destructive interference occur as water waves pass through two adjacent slits.

Figure 7.5 Formation of a diffraction pattern.

Source: © Berenice Abbott/Science Source

In contrast to a wave of light, a particle of matter, like a pebble, does not undergo refraction. If you throw a pebble through the air into a pond, it continues to slow down gradually along a curved path after entering the water (Figure 7.4B).

2. *Diffraction and interference.* When a wave strikes the edge of an object, it bends around it in a phenomenon called **diffraction.** If the wave passes through a slit about as wide as its wavelength, it bends around both edges of the slit and forms a semicircular wave on the other side of the opening (Figure 7.4C).

In contrast, when you throw a collection of particles, like a handful of sand, at a small opening, some particles hit the edge, while others go through the opening and continue in a narrower group (Figure 7.4D).

When waves of light pass through two adjacent slits, the nearby emerging circular waves interact through the process of *interference.* If the crests of the waves coincide *(in phase),* they interfere *constructively*—the amplitudes add together to form a brighter region. If crests coincide with troughs *(out of phase),* they interfere *destructively*—the amplitudes cancel to form a darker region. The result is a *diffraction pattern* (Figure 7.5).

In contrast, particles passing through adjacent openings continue in straight paths, some colliding and moving at different angles.

The Particle Nature of Light

Three observations involving matter and light confounded physicists at the turn of the 20th century: blackbody radiation, the photoelectric effect, and atomic spectra. Explaining these phenomena required a radically new view of energy. We discuss the first two of them here and the third in Section 7.2.

Blackbody Radiation and the Quantum Theory of Energy The first of the puzzling observations involved the light given off by an object being heated.

1. *Observation: blackbody radiation.* When a solid object is heated to about 1000 K, it begins to emit visible light, as you can see in the red glow of smoldering coal (Figure 7.6A). At about 1500 K, the light is brighter and more orange, like that from an electric heating coil (Figure 7.6B). At temperatures greater than 2000 K, the light is still brighter and whiter, like that emitted by the filament of a lightbulb

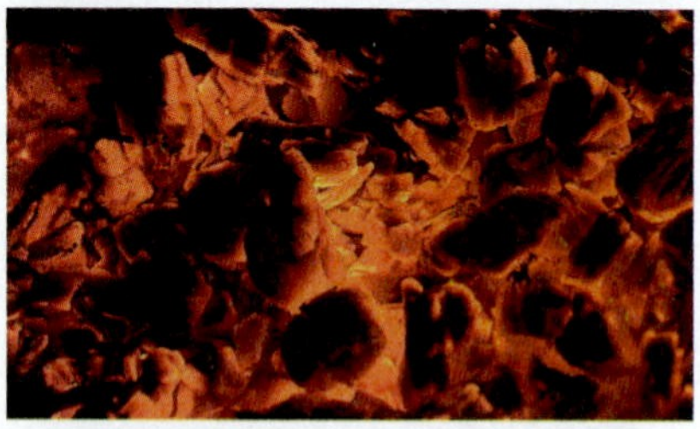

A Smoldering coal

B Electric heating element

C Lightbulb filament

Figure 7.6 Familiar examples of light emission related to blackbody radiation.

Source: (a): © Ravi/Shutterstock.com; (b): © McGraw-Hill Education. Charles Winters/Timeframe Photography, Inc.; (c): © Feng Yu/Shutterstock.com

(Figure 7.6C). These changes in intensity and wavelength of emitted light as an object is heated are characteristic of *blackbody radiation,* light given off by a hot *blackbody.** Attempts that applied the classical wave model to explain the relationship between the energy given off by a hot object and the wavelength of the emitted radiation failed.

2. *Explanation: the quantum theory.* In 1900, the German physicist Max Planck (1858–1947) developed a formula that fit the data perfectly. To find a physical explanation for his formula, Planck assumed, in a self-proclaimed "act of despair," that the hot, glowing object could emit (or absorb) only *certain* quantities of energy:

$$E = nh\nu$$

where E is the energy of the radiation, ν is its frequency, n is a positive integer (1, 2, 3, and so on) called a **quantum number,** and h is **Planck's constant.** With energy in joules (J) and frequency in s^{-1}, h has units of J·s:

$$h = 6.62606876\times10^{-34}\ \text{J·s} = 6.626\times10^{-34}\ \text{J·s} \quad \text{(4 sf)}$$

Since the hot object emits only certain quantities of energy, and the energy must be emitted by the object's atoms, this means that each atom *emits* only certain quantities of energy. It follows, then, that each atom *has* only certain quantities of energy. Thus, the energy of an atom is *quantized:* it occurs in fixed quantities, rather than being continuous. Each change in an atom's energy occurs when the atom absorbs or emits one or more "packets," or definite amounts, of energy. Each energy packet is called a **quantum** ("fixed quantity"; plural, *quanta*). A quantum of energy is equal to $h\nu$. Thus, *an atom changes its energy state by emitting (or absorbing) one or more quanta,* and the energy of the emitted (or absorbed) radiation is equal to the *difference in the atom's energy states:*

$$\Delta E_{\text{atom}} = E_{\text{emitted (or absorbed) radiation}} = \Delta nh\nu$$

Because the atom can change its energy only by integer multiples of $h\nu$, the smallest change occurs when an atom in a given energy state changes to an adjacent state, that is, when $\Delta n = 1$:

$$\Delta E = h\nu \qquad \textbf{(7.2)}$$

Rearranging Equation 7.1,

$$c = \nu \times \lambda \quad \text{or} \quad \frac{c}{\lambda} = \nu$$

and substituting c/λ for frequency, ν, in Equation 7.2 gives the direct relationship between wavelength and energy:

$$\Delta E = h\nu = \frac{hc}{\lambda} \qquad \textbf{(7.3)}$$

These relationships indicate that energy is directly proportional to frequency and inversely proportional to wavelength. Short-wavelength radiation such as x-rays has high frequency and high energy, while long-wavelength radiation such as radio waves has low frequency and low energy.

*A blackbody is an idealized object that absorbs all the radiation incident on it. A hollow cube with a small hole in one wall approximates a blackbody.

The Photoelectric Effect and the Photon Theory of Light Despite the idea of quantization, physicists still pictured energy as traveling in waves. But, the wave model could not explain the second confusing observation, the flow of current when light strikes a metal.

1. *Observation: the* **photoelectric effect.** When monochromatic light of sufficient frequency shines on a metal plate, a current flows (Figure 7.7). It was thought that the current arises because light transfers energy that frees electrons from the metal surface. However, the effect had two confusing features, the presence of a threshold frequency and the absence of a time lag:

- *Presence of a threshold frequency.* For current to flow, the light shining on the metal must have a minimum, or threshold, *frequency,* and different metals have different minimum frequencies. But, the wave theory associates the energy of light with its *amplitude* (intensity), not its frequency (color). Thus, the theory predicts that an electron would break free when it absorbed enough energy from light of *any* color.
- *Absence of a time lag.* Current flows the moment light of the minimum frequency shines on the metal, regardless of the light's intensity. But the wave theory predicts that with dim light there would be a time lag before the current flows, because the electrons would have to absorb enough energy to break free.

2. *Explanation: the photon theory.* Building on Planck's ideas, Albert Einstein (1879–1955) proposed that light itself is particulate, quantized into tiny "bundles" of energy, later called **photons.** Each atom changes its energy by an amount ΔE_{atom} when it absorbs or emits one photon, one "particle" of light, whose energy is related to its *frequency,* not its amplitude:

$$E_{photon} = h\nu = \Delta E_{atom}$$

Let's see how the photon theory explains the two features of the photoelectric effect:

- *Why there is a threshold frequency.* A beam of light consists of an enormous number of photons. The intensity (brightness) is related to the *number* of photons, but *not* to the energy of each. Therefore, a photon of a certain *minimum* energy must be absorbed to free an electron from the surface (see Figure 7.7). Since energy depends on frequency ($h\nu$), the theory predicts a threshold frequency.
- *Why there is no time lag.* An electron breaks free when it absorbs a photon of *enough* energy; it cannot break free by "saving up" energy from several photons, each having less than the minimum energy. The current is weak in dim light because fewer photons of enough energy can free fewer electrons per unit time, but some current flows *as soon as* light of sufficient energy (frequency) strikes the metal. ›

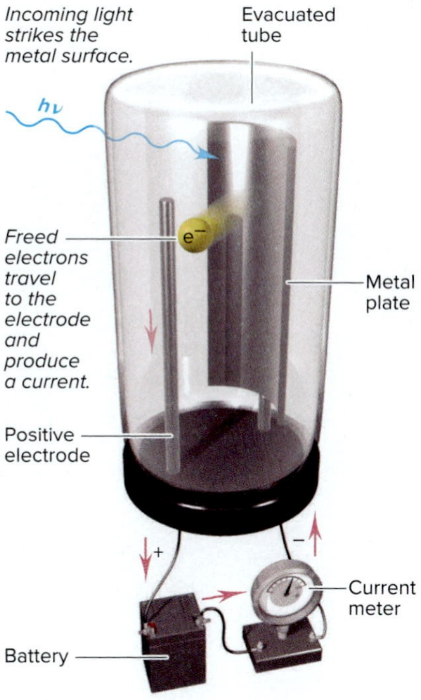

Figure 7.7 The photoelectric effect.

Ping-Pong Photons

This analogy helps explain why light of *insufficient* energy *cannot* free an electron from the metal surface. If one Ping-Pong ball doesn't have enough energy to knock a book off a shelf, neither does a series of Ping-Pong balls, because the book can't save up the energy from the individual impacts. But one baseball traveling at the same speed does have enough energy to move the book.

SAMPLE PROBLEM 7.2 Interconverting Energy, Wavelength, and Frequency

Problem A student uses a microwave oven to heat a meal. The wavelength of the radiation is 1.20 cm. What is the energy of one photon of this microwave radiation?

Plan We know λ in centimeters (1.20 cm) so we convert to meters and then find the energy of one photon with Equation 7.3.

Solution Combining steps to find the energy of a photon:

$$E = \frac{hc}{\lambda} = \frac{(6.626\times10^{-34}\ \text{J·s})(3.00\times10^{8}\ \text{m/s})}{(1.20\ \text{cm})\left(\dfrac{10^{-2}\ \text{m}}{1\ \text{cm}}\right)} = 1.66\times10^{-23}\ \text{J}$$

Check Checking the order of magnitude gives

$$\frac{(10^{-33}\ \text{J·s})(10^{8}\ \text{m/s})}{10^{-2}\ \text{m}} = 10^{-23}\ \text{J}$$

FOLLOW-UP PROBLEMS

7.2A Calculate the wavelength (in nm) and the frequency of a photon with an energy of 8.2×10^{-19} J, the minimum amount of energy required to dislodge an electron from the metal gold.

7.2B Calculate the energy of one photon of **(a)** ultraviolet light ($\lambda = 1\times10^{-8}$ m); **(b)** visible light ($\lambda = 5\times10^{-7}$ m); and **(c)** infrared light ($\lambda = 1\times10^{-4}$ m). What do the answers indicate about the relationship between light's wavelength and energy?

SOME SIMILAR PROBLEMS 7.9–7.12

› Summary of Section 7.1

- Electromagnetic radiation travels in waves characterized by a given wavelength (λ) and frequency (ν).
- Electromagnetic waves travel through a vacuum at the speed of light, c (3.00×10^8 m/s), which equals $\nu \times \lambda$. Therefore, wavelength and frequency are inversely proportional to each other.
- The intensity (brightness) of light is related to the amplitude of its waves.
- The electromagnetic spectrum ranges from very long radio waves to very short gamma rays and includes the visible region between wavelengths 750 nm (red) and 400 nm (violet).
- Refraction (change in a wave's speed when entering a different medium) and diffraction (bend of a wave around an edge of an object) indicate that energy is wavelike, with properties distinct from those of particles of mass.
- Blackbody radiation and the photoelectric effect, however, are consistent with energy occurring in discrete packets, like particles.
- Light exists as photons (quanta) whose energy is directly proportional to the frequency.
- According to quantum theory, an atom has only certain quantities of energy ($E = nh\nu$), and it can change its energy only by absorbing or emitting a photon whose energy equals the change in the atom's energy.

7.2 ATOMIC SPECTRA

The third confusing observation about matter and energy involved the light emitted when an element is vaporized and then excited electrically. In this section, we discuss the nature of that light and see why it created a problem for the existing atomic model and how a new model solved the problem.

Line Spectra and the Rydberg Equation

When light from electrically excited gaseous atoms passes through a slit and is refracted by a prism, it does not create a *continuous spectrum,* or rainbow, as sunlight does. Instead, it creates a **line spectrum,** a series of fine lines at specific wavelengths (specific frequencies) separated by black spaces. Figure 7.8A shows the apparatus for obtaining these spectra and the line spectrum of atomic hydrogen. Figure 7.8B shows that each element produces a unique line spectrum. For example, the brilliant color of a simple neon sign is due to intense orange and red lines in the spectrum of neon.

Features of the Rydberg Equation Spectroscopists studying atomic hydrogen identified several series of spectral lines in different regions of the electromagnetic spectrum. Figure 7.9 shows the spectral lines in the ultraviolet, visible, and infrared regions.

Using data, not theory, the Swedish physicist Johannes Rydberg (1854–1919) developed a relationship, called the *Rydberg equation,* that predicts the position and wavelength of any line in a given series:

$$\frac{1}{\lambda} = R\left(\frac{1}{n_1^2} - \frac{1}{n_2^2}\right) \tag{7.4}$$

where λ is the wavelength of the line, n_1 and n_2 are positive integers with $n_2 > n_1$, and R is the Rydberg constant (1.096776×10^7 m^{-1}). To calculate the wavelengths of the lines in the visible series, $n_1 = 2$:

$$\frac{1}{\lambda} = R\left(\frac{1}{2^2} - \frac{1}{n_2^2}\right), \quad \text{with } n_2 = 3, 4, 5, \ldots$$

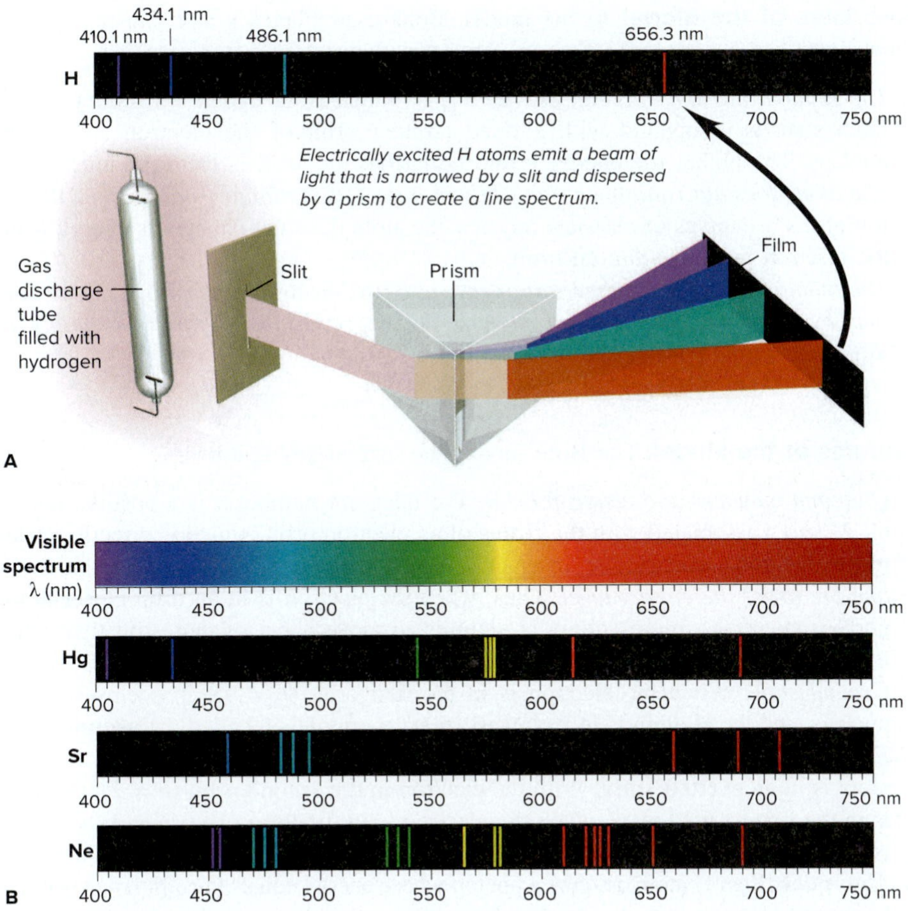

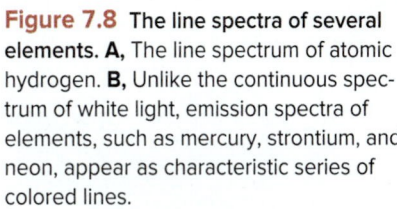

Figure 7.8 **The line spectra of several elements. A,** The line spectrum of atomic hydrogen. **B,** Unlike the continuous spectrum of white light, emission spectra of elements, such as mercury, strontium, and neon, appear as characteristic series of colored lines.

For the ultraviolet series, $n_1 = 1$ ($n_2 = 2, 3, 4, \ldots$), and for the infrared series, $n_1 = 3$ ($n_2 = 4, 5, 6, \ldots$). While the Rydberg equation predicts the wavelengths in the line spectrum of hydrogen, *it does not explain why line spectra occur.*

Problems with Rutherford's Nuclear Model Almost as soon as Rutherford proposed his nuclear model (described in Chapter 2), a major problem arose. A positive nucleus and a negative electron attract each other, and for them to stay apart, the kinetic energy of the electron's motion must counterbalance the potential energy of attraction. However, the laws of classical physics say that a negative particle moving in a curved path around a positive particle *must* emit radiation and thus lose energy. If the orbiting electrons behaved in that way, they would spiral into the nucleus, and all atoms would collapse! The laws of classical physics also say that the frequency of the emitted radiation should change smoothly as the negative particle spirals inward and, thus, create a continuous spectrum, not a line spectrum. The behavior of subatomic particles seemed to violate real-world experience and accepted principles.

The Bohr Model of the Hydrogen Atom

Two years after the nuclear model was proposed, Niels Bohr (1885–1962), a young Danish physicist working in Rutherford's laboratory, suggested a model for the H atom that *did* predict the existence of line spectra.

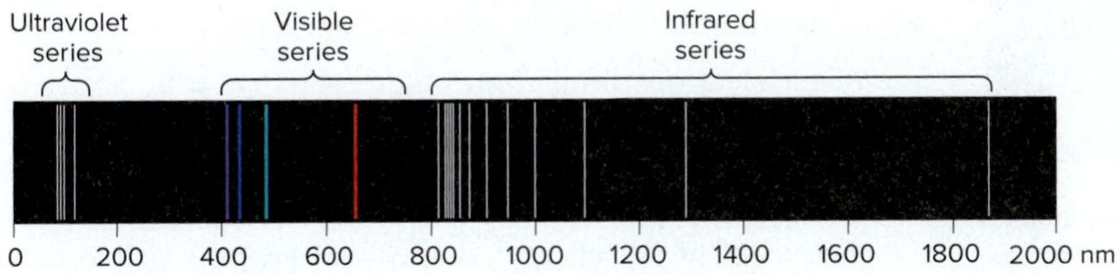

Figure 7.9 **Three series of spectral lines of atomic hydrogen.**

Postulates of the Model In his model, Bohr used Planck's and Einstein's ideas about quantized energy and proposed three postulates:

1. *The H atom has only certain energy levels,* which Bohr called **stationary states.** Each state is associated with a fixed circular orbit of the electron around the nucleus. The higher the energy level, the farther the orbit is from the nucleus.
2. *The atom does **not** radiate energy while in one of its stationary states.* Even though it violates principles of classical physics, the atom does not change its energy while the electron moves *within* an orbit.
3. *The atom changes to another stationary state* (the electron moves to another orbit) *only by absorbing or emitting a photon. The energy of the photon (hν) equals the difference in the energies of the two states:*

$$E_{\text{photon}} = \Delta E_{\text{atom}} = E_{\text{final}} - E_{\text{initial}} = h\nu$$

Features of the Model The Bohr model has several key features:

- *Quantum numbers and electron orbit.* The quantum number n is a positive integer (1, 2, 3, ...) associated with the radius of an electron orbit, which is directly related to the electron's energy: *the lower the n value, the smaller the radius of the orbit, and the lower the energy level.* Thus, when the electron is in an orbit closer to the nucleus (lower n), more energy is required to move it out of that orbit than when it is in an orbit farther from the nucleus (higher n).
- *Ground state.* When the electron is in the first orbit ($n = 1$), it is closest to the nucleus, and the H atom is in its lowest (first) energy level, called the **ground state.**
- *Excited states.* If the electron is in any orbit farther from the nucleus ($n > 1$), the atom is in an **excited state.** With the electron in the second orbit ($n = 2$), the atom is in the first excited state; when the electron is in the third orbit ($n = 3$), the atom is in the second excited state, and so forth.
- *Absorption.* If an H atom *absorbs* a photon whose energy equals the *difference* between lower and higher energy levels, the electron moves to the outer (higher energy) orbit.
- *Emission.* If an H atom in a higher energy level (with its electron in a farther orbit) returns to a lower energy level (electron in a closer orbit), the atom *emits* a photon whose energy equals the difference between the two levels. Figure 7.10 shows a staircase analogy that illustrates absorption and emission. Note that the energy difference between two consecutive orbits decreases as n increases.

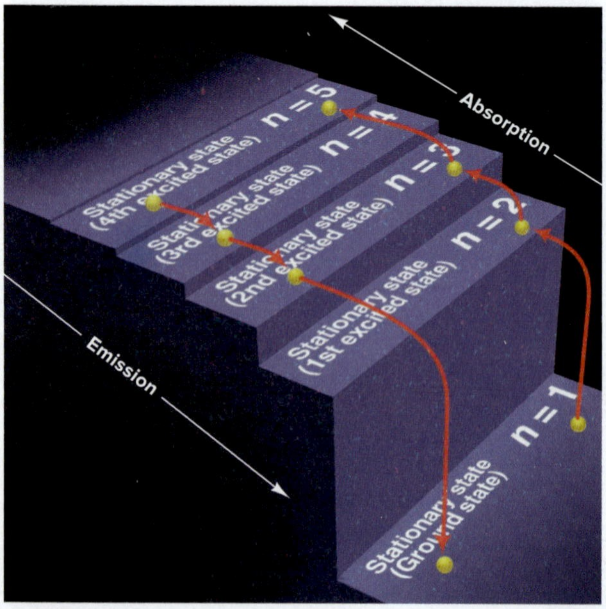

Figure 7.10 **A quantum "staircase" as an analogy for atomic energy levels.** Note that the electron can move up or down one or more steps at a time but cannot lie *between* steps.

How the Model Explains Line Spectra When an electron moves to an orbit closer to the nucleus, the atom's energy changes from a higher state to a lower state, resulting in a loss of a specific quantity of energy; that energy is *emitted* as a photon of a specific wavelength, producing a line in the atom's line spectrum. Since there are only certain energy levels in the atom, only certain quantities of energy (and photons of certain wavelengths) can be emitted as the electron transitions from one orbit to another. *Since an atom's energy is not continuous, but rather has only certain states, an atomic spectrum is not continuous.*

Figure 7.11A shows how Bohr's model accounts for three series of spectral lines of hydrogen. When a sample of gaseous H atoms is excited, the atoms absorb different quantities of energy. Each atom has one electron, but there are so many atoms in the whole sample that all the energy levels (orbits) have electrons. As electrons transition from one energy level to another, the n value of the final energy level determines the region of the electromagnetic spectrum in which the emission line falls.

- When electrons drop from outer orbits ($n = 4, 5, \ldots$) to the $n = 3$ orbit (second excited state), photons with the energy of infrared radiation are emitted, creating the *infrared* series of lines.
- When electrons drop to the $n = 2$ orbit (first excited state), photons of higher energy (shorter wavelength) visible radiation are emitted, creating the *visible* series of lines.

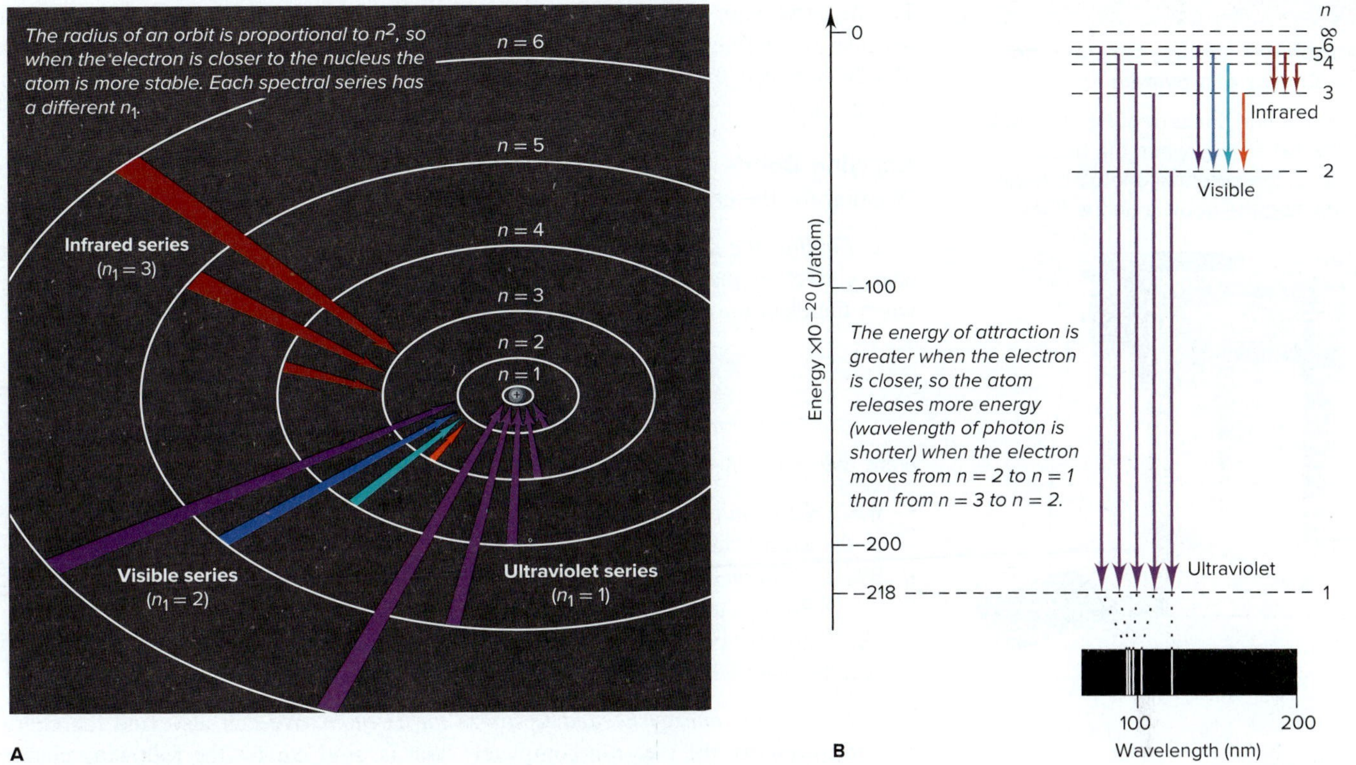

Figure 7.11 **The Bohr explanation of three series of spectral lines emitted by the H atom. A,** In a given series, the outer electron drops to the same inner orbit (the same value of n_1 in the Rydberg equation). **B,** An energy diagram shows how the ultraviolet series arises.

- When electrons drop to the $n = 1$ orbit (ground state), photons of even higher energy, corresponding to ultraviolet radiation, are emitted, resulting in the *ultraviolet* series of lines. Figure 7.11B shows how the specific lines in the ultraviolet series appear.

Limitations of the Model Despite its great success in accounting for the spectral lines of the H atom, the Bohr model failed to predict the spectrum of any other atom. The reason is that it is a *one-electron model:* it works beautifully for the H atom and for other one-electron species, such as He^+ ($Z = 2$), Li^{2+} ($Z = 3$), and Be^{3+} ($Z = 4$). It fails completely for atoms with more than one electron because the electron-electron repulsions and additional nucleus-electron attractions that are present create much more complex interactions. But, even more fundamentally, as we'll see in Section 7.4, *electrons do not move in fixed, defined orbits*. Even though, as a picture of the atom, the Bohr model is incorrect, we still use the terms "ground state" and "excited state" and retain the central idea that *the energy of an atom occurs in discrete levels and it changes when the atom absorbs or emits a photon of specific energy.*

The Energy Levels of the Hydrogen Atom

Bohr's work leads to an equation for calculating the energy levels of an atom:

$$E = -2.18\times10^{-18}\ \text{J}\left(\frac{Z^2}{n^2}\right)$$

where Z is the charge of the nucleus. For the H atom, $Z = 1$, so we have

$$E = -2.18\times10^{-18}\ \text{J}\left(\frac{1^2}{n^2}\right)$$

$$= -2.18\times10^{-18}\ \text{J}\left(\frac{1}{n^2}\right)$$

Therefore, the energy of the ground state ($n = 1$) of the H atom is

$$E = -2.18\times10^{-18}\ \text{J}\left(\frac{1}{1^2}\right)$$

$$= -2.18\times10^{-18}\ \text{J}$$

A Tabletop Analogy for Defining the Energy of a System

If you define the zero point of a book's potential energy when the book is on a table, the potential energy is negative when the book is on the floor.

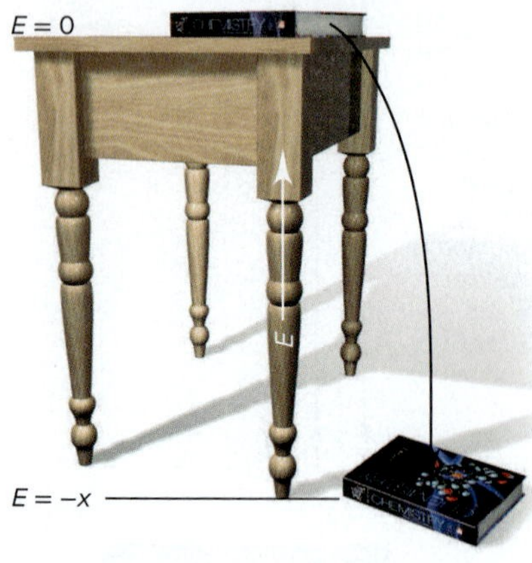

The negative sign for the energy (also used on the axis in Figure 7.11B) appears because we *define* the zero point of the atom's energy when *the electron is completely removed from the nucleus.* Thus, $E = 0$ when $n = \infty$, so $E < 0$ for any smaller *n.* ‹

Applying Bohr's Equation for the Energy Levels of an Atom We can use the equation for the energy levels in several ways:

1. *Finding the difference in energy between two levels.* By subtracting the initial energy level of the atom from its final energy level, we find the change in energy when the electron moves between the two levels:

$$\Delta E = E_{\text{final}} - E_{\text{initial}} = -2.18\times10^{-18}\,\text{J}\left(\frac{1}{n_{\text{final}}^2} - \frac{1}{n_{\text{initial}}^2}\right) \qquad \textbf{(7.5)}$$

Note that, since *n* is in the denominator,

- *When the atom emits energy,* the electron moves closer to the nucleus ($n_{\text{final}} < n_{\text{initial}}$), so the atom's final energy is a *larger* negative number and ΔE is negative.
- *When the atom absorbs energy,* the electron moves away from the nucleus ($n_{\text{final}} > n_{\text{initial}}$), so the atom's final energy is a *smaller* negative number and ΔE is positive. (Analogously, in Chapter 6, you saw that, when the system releases heat, ΔH is negative, and when it absorbs heat, ΔH is positive.)

2. *Finding the energy needed to ionize the H atom.* We can also find the energy needed to remove the electron completely, that is, find ΔE for the following change:

$$H(g) \longrightarrow H^+(g) + e^-$$

We substitute $n_{\text{final}} = \infty$ and $n_{\text{initial}} = 1$ into Equation 7.5 and obtain

$$\Delta E = E_{\text{final}} - E_{\text{initial}} = -2.18\times10^{-18}\,\text{J}\left(\frac{1}{\infty^2} - \frac{1}{1^2}\right)$$
$$= -2.18\times10^{-18}\,\text{J}\,(0 - 1) = 2.18\times10^{-18}\,\text{J}$$

Energy must be *absorbed* to remove the electron from the nucleus, so ΔE is positive.

The *ionization energy* of hydrogen is the energy required to form 1 mol of gaseous H^+ ions from 1 mol of gaseous H atoms. Thus, for 1 mol of H atoms,

$$\Delta E = \left(2.18\times10^{-18}\,\frac{\text{J}}{\text{atom}}\right)\left(6.022\times10^{23}\,\frac{\text{atoms}}{\text{mol}}\right)\left(\frac{1\text{ kJ}}{10^3\text{ J}}\right) = 1.31\times10^3\text{ kJ/mol}$$

Ionization energy is a key atomic property, and we'll return to it in (Chapter 8).

3. *Finding the wavelength of a spectral line.* Once we know ΔE from Equation 7.5, we find the wavelengths of the spectral lines of the H atom from Equation 7.3:

$$\lambda = \frac{hc}{\Delta E}$$

Alternatively, we combine Equation 7.3 with 7.5 and use Bohr's model to derive Rydberg's equation (Equation 7.4):

$$\frac{hc}{\lambda} = \Delta E = -2.18\times10^{-18}\,\text{J}\left(\frac{1}{n_{\text{final}}^2} - \frac{1}{n_{\text{initial}}^2}\right)$$
$$\frac{1}{\lambda} = \frac{-2.18\times10^{-18}\,\text{J}}{hc}\left(\frac{1}{n_{\text{final}}^2} - \frac{1}{n_{\text{initial}}^2}\right)$$
$$= \frac{-2.18\times10^{-18}\,\text{J}}{(6.626\times10^{-34}\,\text{J·s})(3.00\times10^8\,\text{m/s})}\left(\frac{1}{n_{\text{final}}^2} - \frac{1}{n_{\text{initial}}^2}\right)$$
$$= 1.0967\times10^7\,\text{m}^{-1}\left(\frac{1}{n_1^2} - \frac{1}{n_2^2}\right)$$

SAMPLE PROBLEM 7.3 Determining ΔE and λ of an Electron Transition

Problem A hydrogen atom absorbs a photon of UV light (see Figure 7.11), and its electron enters the $n = 4$ energy level. Calculate **(a)** the change in energy of the atom and **(b)** the wavelength (in nm) of the absorbed photon.

Plan **(a)** The H atom absorbs energy, so $E_{final} > E_{initial}$. We are given $n_{final} = 4$, and Figure 7.11 shows that $n_{initial} = 1$ because a UV photon is absorbed. We apply Equation 7.5 to find ΔE. **(b)** Once we know ΔE, we find the wavelength (in m) with Equation 7.3. Then we convert from meters to nanometers.

Solution **(a)** Substituting the known values into Equation 7.5:

$$\Delta E = -2.18\times10^{-18}\text{ J}\left(\frac{1}{n_{\text{final}}^2} - \frac{1}{n_{\text{initial}}^2}\right) = -2.18\times10^{-18}\text{ J}\left(\frac{1}{4^2} - \frac{1}{1^2}\right)$$

$$= -2.18\times10^{-18}\text{ J}\left(\frac{1}{16} - \frac{1}{1}\right) = 2.04\times10^{-18}\text{ J}$$

(b) Using Equation 7.3 to solve for λ:

$$\Delta E = \frac{hc}{\lambda}$$

therefore,

$$\lambda = \frac{hc}{\Delta E} = \frac{(6.626\times10^{-34}\text{ J·s})(3.00\times10^{8}\text{ m/s})}{2.04\times10^{-18}\text{ J}} = 9.74\times10^{-8}\text{ m}$$

Converting m to nm:

$$\lambda = 9.74\times10^{-8}\text{ m} \times \frac{1\text{ nm}}{10^{-9}\text{ m}}$$
$$= 97.4\text{ nm}$$

Check **(a)** The energy change is positive, which is consistent with absorption of a photon. **(b)** The wavelength is within the UV region (about 10–380 nm).

Comment **1.** We could instead use Equation 7.4 to find the wavelength (recall that in this equation, $n_2 > n_1$):

$$\frac{1}{\lambda} = R\left(\frac{1}{n_1^2} - \frac{1}{n_2^2}\right)$$

$$= 1.096776\times10^{7}\text{ m}^{-1}\left(\frac{1}{1^2} - \frac{1}{4^2}\right)$$

$$\frac{1}{\lambda} = 1.028228\times10^{7}\text{ m}^{-1}$$

$$\lambda\text{ (m)} = \frac{1}{1.028228\times10^{7}\text{ m}^{-1}} = 9.73\times10^{-8}\text{ m}$$

$$\lambda\text{ (nm)} = 9.73\times10^{-8}\text{ m} \times \frac{1\text{ nm}}{10^{-9}\text{ m}} = 97.3\text{ nm}$$

2. In Follow-up Problem 7.3A, note that if ΔE is negative (the atom loses energy), we use its absolute value, $|\Delta E|$, because λ must have a positive value.

FOLLOW-UP PROBLEMS

7.3A A hydrogen atom with its electron in the $n = 6$ energy level emits a photon of IR light. Calculate **(a)** the change in energy of the atom and **(b)** the wavelength (in Å) of the photon.

7.3B An electron in the $n = 6$ energy level of an H atom drops to a lower energy level; the atom emits a photon of wavelength 410. nm. **(a)** What is ΔE for this transition in 1 mol of H atoms? **(b)** To what energy level did the electron move?

SOME SIMILAR PROBLEMS 7.23–7.28, 7.31, and 7.32

Spectrometric analysis of the H atom led to the Bohr model, the first step toward our current model of the atom. From its use by 19th-century chemists as a means of identifying elements and compounds, spectrometry has developed into a major tool of modern chemistry (see Tools of the Laboratory at the end of this section).

TOOLS OF THE LABORATORY

Spectrometry in Chemical Analysis

The use of spectral data to identify and quantify substances is essential to modern chemical analysis. The terms *spectroscopy* and **spectrometry** refer to a large group of instrumental techniques that obtain spectra to gather data on a substance's atomic and molecular energy levels.

Types of Spectra

Line spectra occur in two important types:

1. An **emission spectrum,** such as the H atom line spectrum, occurs when atoms in an excited state *emit* photons as they return to a lower energy state. Some elements produce an intense spectral line (or several closely spaced ones) that is evidence of their presence. **Flame tests,** performed by placing a granule of an ionic compound or a drop of its solution in a flame, rely on these intense emissions (Figure B7.1A), and some colors of fireworks are due to the same elements (Figure B7.1B). The colors of sodium-vapor and mercury-vapor streetlamps are due to intense emissions lines in these elements' spectra.

2. An **absorption spectrum** is produced when atoms *absorb* photons of certain wavelengths and become excited. When white light passes through sodium vapor, for example, the absorption spectrum shows dark lines at the same wavelengths as the yellow-orange lines in sodium's emission spectrum (Figure B7.2).

Strontium
Copper
A
B

Figure B7.1 **Flame tests and fireworks. A,** The flame's color is due to intense emission by the element of light of a particular wavelength. **B,** Fireworks display emissions similar to those seen in flame tests.

Source: A(1–2): © McGraw-Hill Education, Stephen Frisch, photographer; B: © Studio Photogram/Alamy RF

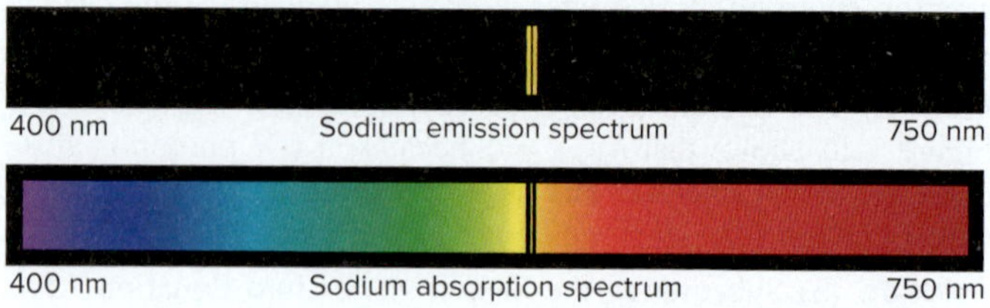

400 nm
Sodium emission spectrum
750 nm
400 nm
Sodium absorption spectrum
750 nm

Figure B7.2 **Emission and absorption spectra of sodium atoms.** The wavelengths of the bright emission lines correspond to those of the dark absorption lines because both are created by the same energy change: $\Delta E_{emission} = -\Delta E_{absorption}$. (Only the two most intense lines in the Na spectra are shown.)

Because elements and compounds have unique line spectra, astronomers can use such spectra to determine the composition of the Sun, planets, stars, comets, and other celestial bodies. For example, the presence of helium (helios, Greek for sun) in the Sun was discovered by examining sunlight with a spectroscope.

Basic Instrumentation

Instruments that are based on absorption spectra are much more common than those based on emission spectra because many substances absorb relatively few wavelengths so their absorption spectra are more characteristic, and absorption is less destructive of fragile molecules.

Design differences depend on the region of the electromagnetic spectrum used to irradiate the sample, but all modern spectrometers have components that perform the same basic functions (Figure B7.3). (We discuss infrared spectroscopy and nuclear magnetic resonance spectroscopy in later chapters.)

Identifying and Quantifying a Substance

In chemical analysis, spectra are used to identify a substance and/or quantify the amount in a sample. Visible light is often used for colored substances, because they absorb only some of the wavelengths. A leaf looks green, for example, because its chlorophyll absorbs red and blue light strongly but green light weakly, so most of the green light is reflected.

Figure B7.3 **Components of a typical spectrometer.**

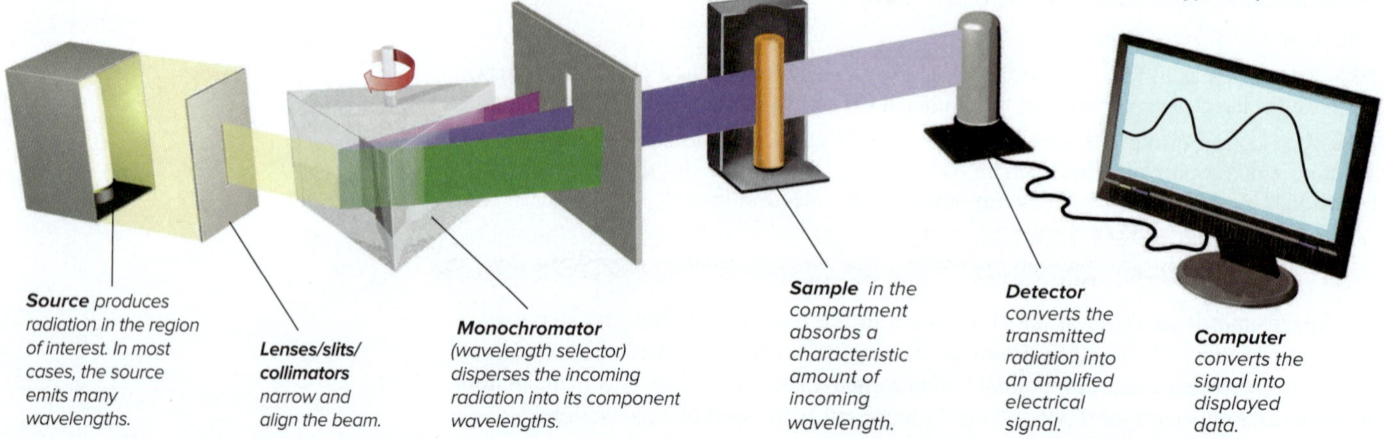

Source produces radiation in the region of interest. In most cases, the source emits many wavelengths.
Lenses/slits/collimators narrow and align the beam.
Monochromator (wavelength selector) disperses the incoming radiation into its component wavelengths.
Sample in the compartment absorbs a characteristic amount of incoming wavelength.
Detector converts the transmitted radiation into an amplified electrical signal.
Computer converts the signal into displayed data.

Figure B7.4 shows the visible absorption spectrum of chlorophyll *a* in ether solution. The shape of the curve and the wavelengths of the major peaks are characteristic of chlorophyll *a*. The curve varies in height because chlorophyll *a* absorbs light of different wavelengths to different extents. The absorptions appear as broad bands, rather than as the distinct lines seen on absorption spectra of elements, because there are greater numbers and types of energy levels within a molecule as well as interactions between molecules and solvent.

In addition to identifying a substance, a spectrometer can be used to measure its concentration because *absorbance*, the quantity of light of a given wavelength absorbed by a substance, is *proportional to the number of molecules*. Suppose you want to determine the concentration of chlorophyll *a* in a leaf extract. You select a strongly absorbed wavelength from the compound's spectrum (such as 663 nm in Figure B7.4A), measure the absorbance of the *unknown* solution, and compare it with the absorbances of a series of solutions of *known* concentration (Figure B7.4B).

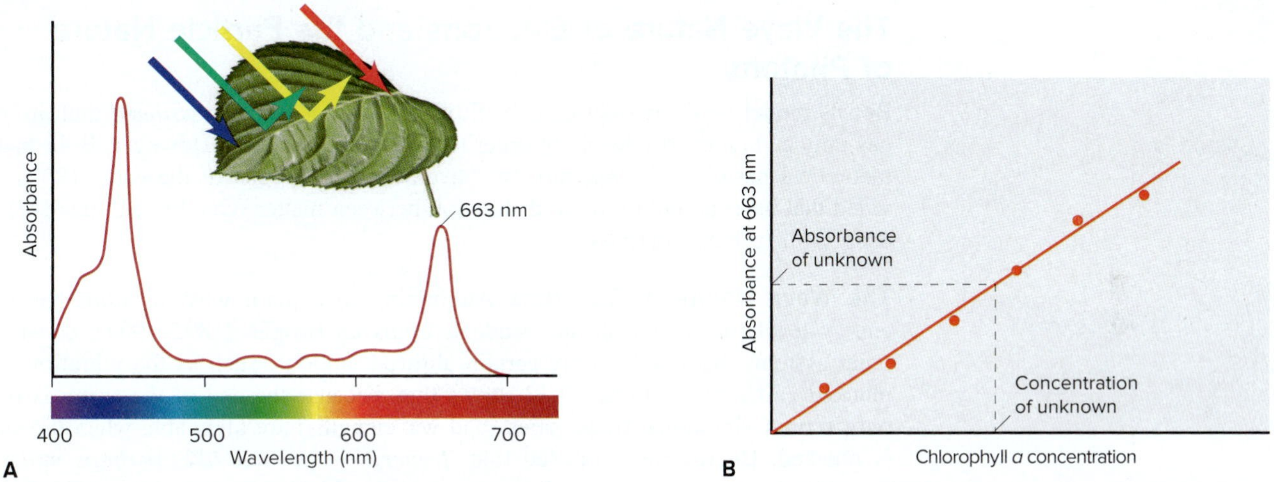

Figure B7.4 **Measuring chlorophyll *a* concentration in leaf extract.** **A,** The spectrometer is set to measure the strong absorption at 663 nm in the chlorophyll *a* spectrum. **B,** The absorbance from the leaf extract is compared to a series of known standards.

Problems

B7.1 The sodium salt of 2-quinizarinsulfonic acid forms a complex with Al^{3+} that absorbs strongly at 560 nm.
(a) Use the data below to draw a plot of absorbance vs. concentration of the complex in solution and find the slope and *y*-intercept:

Concentration (*M*)	Absorbance (560 nm)
1.0×10^{-5}	0.131
1.5×10^{-5}	0.201
2.0×10^{-5}	0.265
2.5×10^{-5}	0.329
3.0×10^{-5}	0.396

(b) When 20.0 mL of a solution of this complex is diluted with water to 150. mL, the measured absorbance is 0.236. Find the concentrations of the diluted solution and of the original solution.

B7.2 In fireworks displays, flame tests, and other emission events, light of a given wavelength may indicate the presence of a particular element or ion. What are the frequency and color of the light associated with each of the following?
(a) Li, $\lambda = 671$ nm
(b) Cs^+, $\lambda = 453$ nm
(c) Na, $\lambda = 589$ nm

› Summary of Section 7.2

- › Unlike sunlight, light emitted by electrically excited atoms of elements and refracted through a prism appears as separate spectral lines.
- › Spectroscopists use an empirical formula (the Rydberg equation) to determine the wavelength of a spectral line. Atomic hydrogen displays several series of spectral lines.
- › To explain the existence of line spectra, Bohr proposed that an electron moves in fixed orbits. It moves from one orbit to another when the atom absorbs or emits a photon whose energy equals the difference in energy levels (orbits).
- › Bohr's model predicts only the spectrum of the H atom and other one-electron species. Despite this, Bohr was correct that an atom's energy is quantized.
- › Spectrometry is an instrumental technique that uses emission and absorption spectra to identify substances and measure their concentrations.

7.3 THE WAVE-PARTICLE DUALITY OF MATTER AND ENERGY

The year 1905 was a busy one for Albert Einstein. He had just presented the photon theory and explained the photoelectric effect. A friend remembered him in his small apartment, rocking his baby in its carriage with one hand, while scribbling with the other ideas for a new branch of physics—the theory of relativity. One of its revelations was that *matter and energy are alternate forms of the same entity.* This idea is embodied in his famous equation $E = mc^2$, identifying the quantity of energy that is equivalent to a given mass. Some results that showed energy to be particle-like had to coexist with others that showed matter to be wavelike. These remarkable ideas are the key to understanding our modern atomic model.

The Wave Nature of Electrons and the Particle Nature of Photons

Bohr's model was a perfect case of fitting theory to data: he *assumed* that an atom has only certain energy levels in order to *explain* line spectra. However, Bohr had no theoretical basis for the assumption. Several breakthroughs in the early 1920s provided that basis and blurred the distinction between matter (chunky and massive) and energy (diffuse and massless).

The Wave Nature of Electrons Attempting to explain why an atom has fixed energy levels, a French physics student, Louis de Broglie (1892–1987), considered other systems that display only certain allowed motions, such as the vibrations of a plucked guitar string. Figure 7.12 shows that, because the end of the string is fixed, only certain vibrational frequencies (and wavelengths) are allowable when the string is plucked. De Broglie proposed that *if energy is particle-like, perhaps matter is wavelike.* He reasoned that *if electrons have wavelike motion* in orbits of fixed radii, they would have only certain allowable frequencies and energies.

Combining the equations for mass-energy equivalence ($E = mc^2$) and energy of a photon ($E = h\nu = hc/\lambda$), de Broglie derived an equation for the wavelength of any particle of mass m—whether planet, baseball, or electron—moving at speed u:

$$\lambda = \frac{h}{mu} \tag{7.6}$$

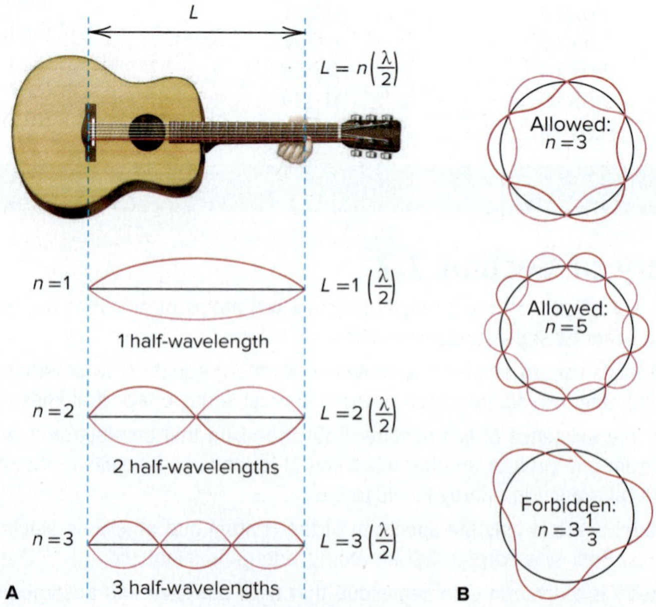

Figure 7.12 Wave motion in restricted systems. A, One half-wavelength ($\lambda/2$) is the "quantum" of the guitar string's vibration. With string length L fixed by a finger on the fret, allowed vibrations occur when L is a whole-number multiple (n) of $\lambda/2$. **B,** In a circular electron orbit, only whole numbers of wavelengths are allowed ($n = 3$ and $n = 5$ are shown). A wave with a fractional number of wavelengths (such as $n = 3\frac{1}{3}$) is "forbidden" because it dies out through overlap of crests and troughs.

According to this equation for the **de Broglie wavelength,** *matter behaves as though it moves in a wave.* An object's wavelength is *inversely* proportional to its mass, so heavy objects such as planets and baseballs have wavelengths *many* orders of magnitude smaller than the objects themselves, too small to be detected, in fact (Table 7.1).

Table 7.1 The de Broglie Wavelengths of Several Objects

Substance	Mass (g)	Speed (m/s)	λ (m)
Slow electron	9×10^{-28}	1.0	7×10^{-4}
Fast electron	9×10^{-28}	5.9×10^{6}	1×10^{-10}
Alpha particle	6.6×10^{-24}	1.5×10^{7}	7×10^{-15}
1-gram mass	1.0	0.01	7×10^{-29}
Baseball	142	40.0	1×10^{-34}
Earth	6.0×10^{27}	3.0×10^{4}	4×10^{-63}

SAMPLE PROBLEM 7.4 Calculating the de Broglie Wavelength of an Electron

Problem Find the de Broglie wavelength of an electron with a speed of 1.00×10^{6} m/s (electron mass $= 9.11\times10^{-31}$ kg; $h = 6.626\times10^{-34}$ kg·m²/s).

Plan We know the speed (1.00×10^{6} m/s) and mass (9.11×10^{-31} kg) of the electron, so we substitute these into Equation 7.6 to find λ.

Solution

$$\lambda = \frac{h}{mu} = \frac{6.626\times10^{-34}\ \text{kg·m}^2/\text{s}}{(9.11\times10^{-31}\ \text{kg})(1.00\times10^{6}\ \text{m/s})} = 7.27\times10^{-10}\ \text{m}$$

Check The order of magnitude and the unit seem correct:

$$\lambda \approx \frac{10^{-33}\ \text{kg·m}^2/\text{s}}{(10^{-30}\ \text{kg})(10^{6}\ \text{m/s})} = 10^{-9}\ \text{m}$$

Comment As you'll see in the upcoming discussion, such fast-moving electrons, with wavelengths in the range of atomic sizes, exhibit remarkable properties.

FOLLOW-UP PROBLEMS

7.4A (**a**) What is the speed of an electron that has a de Broglie wavelength of 100. nm (electron mass $= 9.11\times10^{-31}$ kg)?

(**b**) At what speed would a 45.9-g golf ball need to move to have a de Broglie wavelength of 100. nm?

7.4B Find the de Broglie wavelength of a 39.7-g racquetball traveling at a speed of 55 mi/h.

SOME SIMILAR PROBLEMS 7.39(a), 7.40(a), and 7.41–7.42

If electrons travel in waves, they should exhibit diffraction and interference. A fast-moving electron has a wavelength of about 10^{-10} m, so a beam of such electrons should be diffracted by the spaces between atoms in a crystal—which measure about 10^{-10} m. In 1927, C. Davisson and L. Germer guided a beam of x-rays and then a beam of electrons at a nickel crystal and obtained two diffraction patterns; Figure 7.13 shows such patterns for aluminum. Thus, electrons—particles with mass and charge—create diffraction patterns, just as electromagnetic waves do.

A major application of electrons traveling in waves is the *electron microscope.* Its great advantage over light microscopes is that high-speed electrons have much smaller wavelengths than visible light, which allow much higher resolution. A transmission electron microscope (TEM) focuses a beam of electrons through a lens, and the beam then passes through a thin section of the specimen to a second and then third lens. In this instrument, these "lenses" are electromagnetic fields, which can result in up to 200,000-fold magnification. In a scanning electron microscope (SEM),

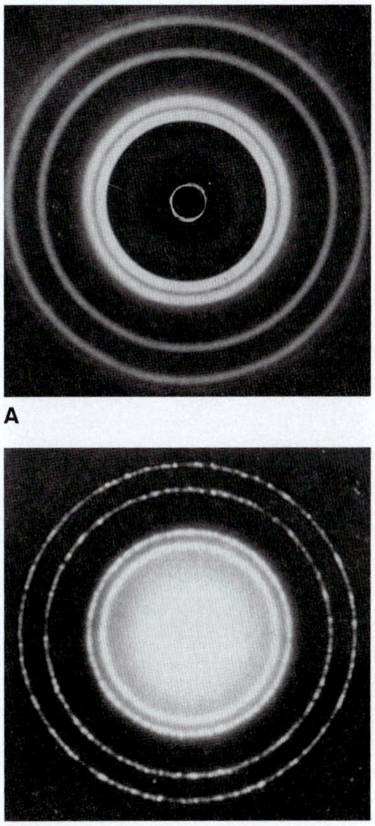

A

B

Figure 7.13 Diffraction patterns of aluminum. **A,** With x-rays. **B,** With electrons.

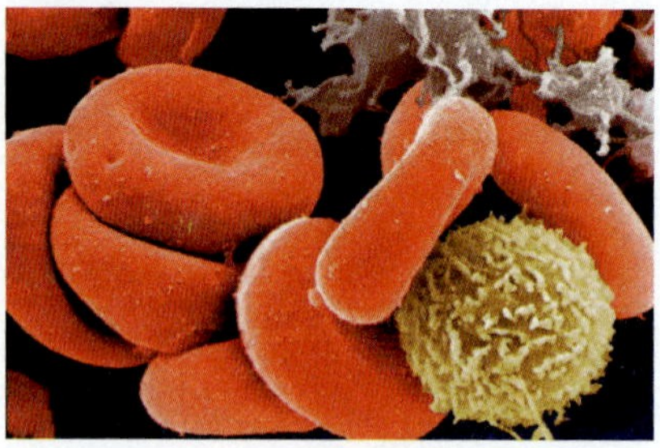

Figure 7.14 False-color scanning electron micrograph of blood cells (×1200).
Source: © Yorgos Nikas/Stone/Getty Images.

the beam scans the specimen, knocking electrons from it that create a current, which generates an image that looks like the object's surface (Figure 7.14).

The Particle Nature of Photons If electrons have properties of energy, do photons have properties of matter? The de Broglie equation suggests that we can calculate the momentum (p), the product of mass and speed, for a photon. Substituting the speed of light (c) for speed u in Equation 7.6 and solving for p gives

$$\lambda = \frac{h}{mc} = \frac{h}{p} \quad \text{and} \quad p = \frac{h}{\lambda}$$

The inverse relationship between p and λ in this equation means that shorter wavelength (higher energy) photons have greater momentum. Thus, a decrease in a photon's momentum should appear as an increase in its wavelength. In 1923, Arthur Compton directed a beam of x-ray photons at graphite and observed an increase in the wavelength of the reflected photons. Thus, just as billiard balls transfer momentum when they collide, the photons transferred momentum to the electrons in the carbon atoms of the graphite. In this experiment, photons behaved as particles.

Wave-Particle Duality Classical experiments had shown matter to be particle-like and energy to be wavelike. But, results on the atomic scale show electrons moving in waves and photons having momentum. Thus, every property of matter was also a property of energy. The truth is that *both* matter and energy show *both* behaviors: each possesses both "faces." In some experiments, we observe one face; in other experiments, we observe the other face. Our everyday distinction between matter and energy is meaningful in the macroscopic world, but *not* in the atomic world. The distinction is in our minds and the limited definitions we have created, not inherent in nature. This dual character of matter and energy is known as the **wave-particle duality**—particles have wavelike properties and energy has particle-like properties. Figure 7.15 summarizes the theories and observations that led to this new understanding.

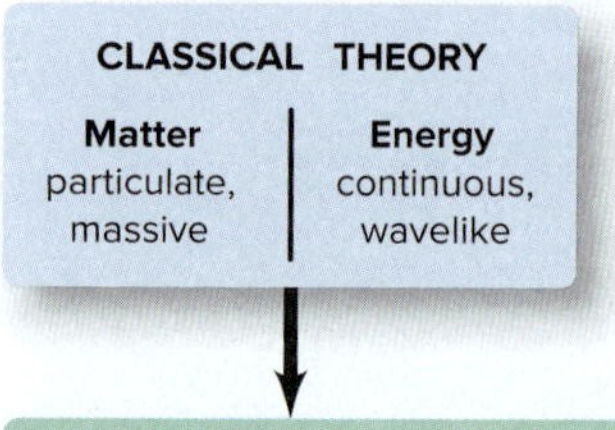
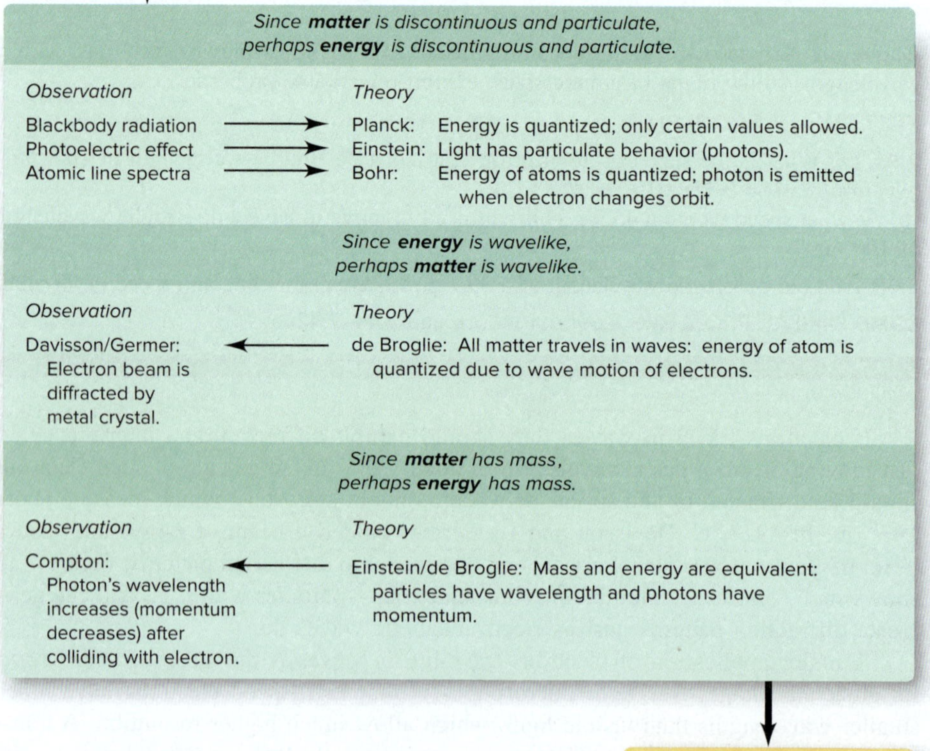
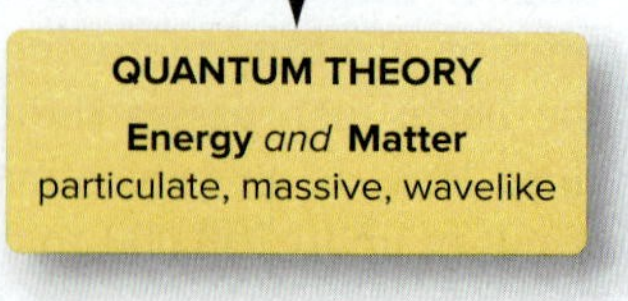

Figure 7.15 Major observations and theories leading from classical theory to quantum theory.

Heisenberg's Uncertainty Principle

In classical physics, a moving particle has a definite location at any instant, whereas a wave is spread out in space. If an electron has the properties of *both* a particle and a wave, can we determine its position in the atom? In 1927, the German physicist Werner Heisenberg (1901–1976) postulated the **uncertainty principle,** which states that it is impossible to know simultaneously the position *and* momentum (mass times speed) of a particle. For a particle with constant mass m, the principle is expressed mathematically as

$$\Delta x \cdot m\Delta u \geq \frac{h}{4\pi} \qquad \textbf{(7.7)}$$

where Δx is the uncertainty in position, Δu is the uncertainty in speed, and h is Planck's constant. The more accurately we know the position of the particle (smaller Δx), the less accurately we know its speed (larger Δu), and vice versa. The best-case scenario is that we know the product of these uncertainties, which is equal to $h/4\pi$.

For a macroscopic object like a baseball, Δx and Δu are insignificant because the mass is enormous compared with $h/4\pi$. Thus, if we know the position and speed of a pitched baseball, we can use the laws of motion to predict its trajectory and whether it will be a ball or a strike. However, using the position and speed of an electron to find its trajectory is a very different proposition, as Sample Problem 7.5 demonstrates.

SAMPLE PROBLEM 7.5 Applying the Uncertainty Principle

Problem An electron moving near an atomic nucleus has a speed of 6×10^6 m/s ± 1%. What is the uncertainty in its position (Δx)?

Plan The uncertainty in the speed (Δu) is given as 1%, so we multiply u (6×10^6 m/s) by 0.01 to calculate the value of Δu, substitute that value into Equation 7.7, and solve for the uncertainty in position (Δx).

Solution Finding the uncertainty in speed, Δu:

$$\Delta u = 1\% \text{ of } u = 0.01(6\times10^6 \text{ m/s}) = 6\times10^4 \text{ m/s}$$

Calculating the uncertainty in position, Δx:

$$\Delta x \cdot m\Delta u \geq \frac{h}{4\pi}$$

Thus,

$$\Delta x \geq \frac{h}{4\pi m\Delta u} \geq \frac{6.626\times10^{-34}\ \text{kg}\cdot\text{m}^2/\text{s}}{4\pi(9.11\times10^{-31}\ \text{kg})(6\times10^4\ \text{m/s})} \geq 1\times10^{-9}\ \text{m}$$

Check Be sure to round off and check the order of magnitude of the answer:

$$\Delta x \geq \frac{10^{-33}\ \text{kg}\cdot\text{m}^2/\text{s}}{(10^1)(10^{-30}\ \text{kg})(10^5\ \text{m/s})} = 10^{-9}\ \text{m}$$

Comment The uncertainty in the electron's position is about 10 times greater than the diameter of the entire atom (10^{-10} m)! Therefore, we have no precise idea where in the atom the electron is located. In Follow-up Problem 7.5A, you'll see whether an umpire has any better idea about the position of a baseball.

FOLLOW-UP PROBLEMS

7.5A How accurately can an umpire know the position of a baseball (mass = 0.142 kg) moving at 100.0 mi/h ± 1.00% (44.7 m/s ± 1.00%)?

7.5B A neutron has a speed of 8×10^7 m/s ± 1%. What is the uncertainty in its position? The mass of a neutron is 1.67×10^{-27} kg.

SOME SIMILAR PROBLEMS 7.39(b) and 7.40(b)

› Summary of Section 7.3

- As a result of Planck's quantum theory and Einstein's theory of relativity, we no longer view matter and energy as distinct entities.
- The de Broglie wavelength is based on the idea that an electron (or any object) has wavelike motion. Allowed atomic energy levels are related to allowed wavelengths of the electron's motion.
- Electrons exhibit diffraction, just as light waves do, and photons exhibit transfer of momentum, just as objects do. This wave-particle duality of matter and energy is observable only on the atomic scale.
- According to the uncertainty principle, we can never know the position and speed of an electron simultaneously.

7.4 THE QUANTUM-MECHANICAL MODEL OF THE ATOM

Acceptance of the dual nature of matter and energy and of the uncertainty principle culminated in the field of **quantum mechanics,** which examines the wave nature of objects on the atomic scale. In 1926, Erwin Schrödinger (1887–1961) derived an equation that is the basis for the *quantum-mechanical model* of the H atom. Key features of the model are described in the following subsections.

The Schrödinger Equation, the Atomic Orbital, and the Probable Location of the Electron

The quantum-mechanical model describes an atom with specific quantities of energy that result from certain allowed values of its electron's wavelike motion. As you'll see, one of the key results of the model is that the electron's position can be known only within a certain probability.

The Schrödinger Equation The **Schrödinger equation** is the mathematical centerpiece of the quantum-mechanical model. Focusing on the wavelike nature of electrons, Schrödinger used a **wave function,** ψ (Greek *psi*, pronounced "sigh"), a mathematical description of the electron's matter-wave in three dimensions, to describe the electron's motion. An example of a wave function for a particle in a one-dimensional system as a function of its position x is shown in Figure 7.16A. Note that the wave function has both positive and negative values. Any point at which the wave function passes through zero is called a **node.** In Figure 7.16A, for example, there are two nodes.

The uncertainty principle says we cannot know *exactly* where the electron is at any moment, but we can know where it *probably* is, that is, where it spends most of its time. We get this information by squaring the wave function, ψ^2. Thus, even though ψ has no physical meaning, ψ^2 does and is called the *probability density*, a measure of the probability of finding an electron in some tiny volume of the atom. Thus, for a one-dimensional system, the probability of finding a particle between x and $x + dx$

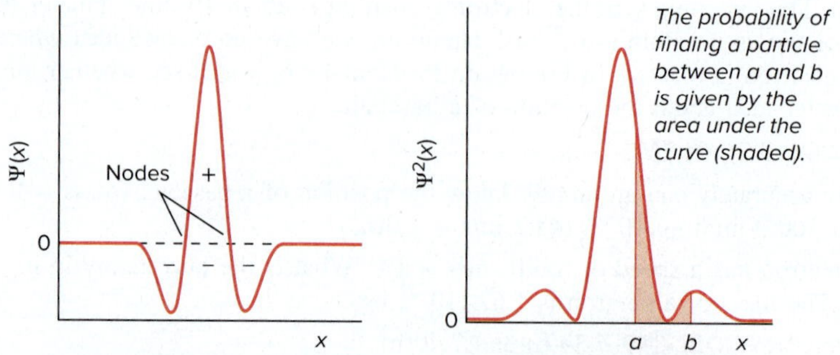

A A wave function for a particle in a one-dimensional system

B The probability density of the wave function

Figure 7.16 A wave function and probability density for a one-dimensional system.

is $\psi^2(x)\,dx$. The probability, P, of finding a particle between $x = a$ and $x = b$ is given by the area under the curve in that region (Figure 7.16B) and can be calculated from the integral of $\psi^2(x)\,dx$:

$$P = \int_a^b \psi^2(x)dx$$

Notice that the probability of finding the particle between $x = a$ and $x = b$ is zero at the nodes of the wave function; since ψ has a value of zero at the nodes, ψ^2 has a value of zero as well.

For a particle of mass m moving in one dimension, the form of the Schrödinger equation is

$$\underbrace{-\frac{h^2}{8\pi^2 m}\frac{d^2\psi(x)}{dx^2}}_{\text{kinetic energy}} + \underbrace{V(x)\psi(x)}_{\text{potential energy}} = \underbrace{E\psi(x)}_{\text{total energy}}$$

where $V(x)$ is the potential energy of the particle and the term $-\frac{h^2}{8\pi^2 m}\frac{d^2\psi(x)}{dx^2}$ represents its kinetic energy. The Schrödinger equation can be represented in simpler form in which the left-hand side of the equation is written as $\hat{\mathscr{H}}\psi$, where $\hat{\mathscr{H}}$ is called the Hamiltonian operator, a mathematical combination of kinetic and potential energy terms that operate on a given wave function to give an energy value:

$$\hat{\mathscr{H}}\Psi = E\Psi$$

The Particle-in-a-Box Model The Schrödinger equation cannot be solved exactly in most situations. But, let's examine a model system for which an exact solution is possible—the particle-in-a-box model (Figure 7.17). In the simplest version of this model, a particle of mass m moves within a one-dimensional box, that is, it moves along a line of length L between two potential energy barriers, which are the ends of the box. Therefore, the potential energy V inside the box is zero because the particle is free to move, but outside the box, V is infinite. The potential energy of the particle is therefore

$$V(x) = 0 \qquad \text{for } 0 \le x \le L$$

$$V(x) = \infty \qquad \text{for } x < 0 \text{ and } x > L$$

where x is the position of the particle along the line.

Inside the box, where the potential energy term $V(x) = 0$, the Schrödinger equation simplifies to

$$-\frac{h^2}{8\pi^2 m}\frac{d^2\psi(x)}{dx^2} = E\psi(x)$$

or

$$\frac{d^2\psi(x)}{dx^2} = -\frac{8\pi^2 mE}{h^2}\psi(x)$$

Thus, the equation describes a function whose second derivative is the function multiplied by a negative constant. The trigonometric sine function is a possible solution to this equation:

$$\psi(x) = A \sin kx$$

where A and k are constants. Substituting the sine function in place of $\psi(x)$ in the simplified Schrödinger equation gives

$$\frac{d^2}{dx^2}[A \sin (kx)] = -\frac{8\pi^2 mE}{h^2}A \sin (kx)$$

Solving for k gives

$$k = \left(\frac{8\pi^2 mE}{h^2}\right)^{1/2}$$

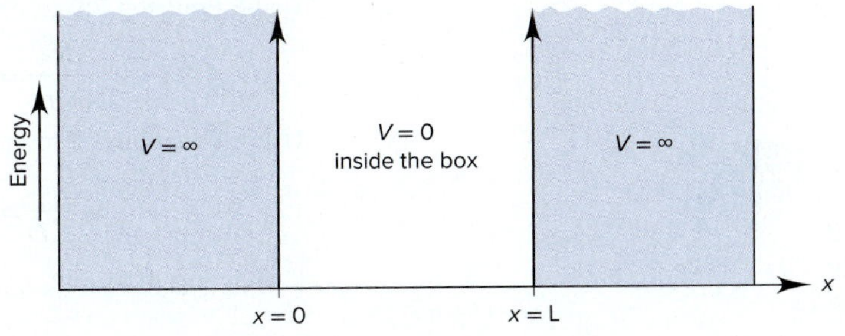

Figure 7.17 The particle in a box model.

so that

$$\psi(x) = A \sin kx = A \sin\left[\left(\frac{8\pi^2 mE}{h^2}\right)^{1/2} x\right]$$

Since the potential energy becomes infinite outside the box, the probability of finding the particle there is 0, that is, $\psi(x) = 0$ at $x = 0$ and $x = L$. Therefore,

$$\psi(L) = A \sin\left[\left(\frac{8\pi^2 mE}{h^2}\right)^{1/2} L\right] = 0$$

Noting that $\sin(x) = 0$ only when x is a multiple of π (that is, $\sin[n\pi] = 0$) means that

$$\left(\frac{8\pi^2 mE}{h^2}\right)^{1/2} L = n\pi \qquad \text{where } n = 1, 2, 3, \ldots$$

Solving for the energy, E_n, of the particle at a particular value of n gives the following:

$$E_n = \frac{n^2h^2}{8mL^2} \qquad n = 1, 2, 3, \ldots \tag{7.8}$$

Notice that since n has only discrete (integer) values, the energy of the particle must also have discrete values; thus, we say that the energy of the particle is *quantized,* and the values of n are called *quantum numbers.* (We discuss quantum numbers further later in the section.) The difference in energy between a level with quantum number n and one with quantum number $n + 1$ is

$$\Delta E = E_{n+1} - E_n = \frac{(n+1)^2h^2}{8mL^2} - \frac{n^2h^2}{8mL^2} = (2n+1)\frac{h^2}{8mL^2} \tag{7.9}$$

Equation 7.9 shows that as the mass m of the particle increases, the difference in energy between levels decreases. Thus, quantization is important only for particles of very small mass, such as electrons.

SAMPLE PROBLEM 7.6 **Determining ΔE and λ of an Electron Transition Using the Particle-in-a-Box Model**

Problem Calculate ΔE and the wavelength (in nm) absorbed when an electron in a one-dimensional box with a length of 120. pm moves from the $n = 3$ to $n = 4$ energy level.

Plan We know n (3), the mass of the electron (9.109×10^{-31} kg), and the length, L, of the box (120. pm). We use Equation 7.9 to calculate ΔE for the transition. The length of the box must be converted to meters. Once we know ΔE, we will use Equation 7.3 to find the wavelength in meters. Then we convert from meters to nanometers.

Solution Converting pm to m:

$$L = (120.\text{ pm})\left(\frac{10^{-12}\text{ m}}{1\text{ pm}}\right) = 1.20\times10^{-10}\text{ m}$$

Using Equation 7.9 to calculate ΔE:

$$\Delta E = (2n+1)\frac{h^2}{8mL^2} = (2(3)+1)\frac{(6.626\times10^{-34}\text{ J·s})^2}{8(9.109\times10^{-31}\text{ kg})(1.20\times10^{-10}\text{ m})^2} = 2.93\times10^{-17}\text{ J}$$

Using Equation 7.3 to calculate λ:

$$\lambda = \frac{hc}{\Delta E} = \frac{(6.626\times10^{-34}\text{ J·s})(3.00\times10^{8}\text{ m/s})}{(2.93\times10^{-17}\text{ J})} = 6.78\times10^{-9}\text{ m}$$

Converting m to nm:

$$\lambda = 6.78\times10^{-9}\text{ m} \times \frac{1\text{ nm}}{10^{-9}\text{ m}} = 6.78\text{ nm}$$

Check ΔE is inversely proportional to the length of the box; since the length unit is quite small, it is reasonable to expect a relatively large value for ΔE and therefore a short wavelength.

FOLLOW-UP PROBLEMS

7.6A Calculate ΔE when an electron in a one-dimensional box with a length of 120. nm moves from the $n = 3$ to $n = 4$ energy level. How does this value compare to that calculated in Sample Problem 7.6?

7.6B A wavelength of 915 nm is required to promote an electron in a one-dimensional box from the $n = 1$ to the $n = 2$ energy level. Calculate the length of the box.

SOME SIMILAR PROBLEMS 7.97 and 7.98

Wave Functions The wave functions that correspond to the energy, E_n, are given by

$$\psi_n(x) = A \sin\left(\frac{n\pi x}{L}\right)$$

Remember that the probability of finding a particle between x and $x + dx$ in the box is the square of the wave function, $\psi^2(x)\,dx$. Also, the probability of finding the particle over all of the possible values of x must be 1, because the particle must be somewhere in the box. Therefore,

$$\int_0^L \psi_n^2(x)dx = A^2 \int_0^L \sin^2\left(\frac{n\pi x}{L}\right)dx = 1$$

The solution to this integral is $L/2$, so $A^2\left(\frac{L}{2}\right) = 1$ and $A = \sqrt{\frac{2}{L}}$. Substituting this expression for A into the wave function gives

$$\psi_n(x) = \sqrt{\frac{2}{L}}\sin\left(\frac{n\pi x}{L}\right) \qquad n = 1, 2, 3, \ldots \tag{7.10}$$

An examination of the wave functions and the probability densities for the first four E_n values are shown in Figure 7.18. Note that

- when $n = 1$, the maximum probability density $\psi^2(x)$ occurs at $L/2$ and there are no nodes;
- when $n = 2$, there are two maxima at $L/4$ and $3L/4$ and there is one node;
- a wave function has $n - 1$ nodes and n maximum probability densities; and
- the number of nodes and the number of maximum probability densities both increase as energy increases.

An interesting point to notice is that the wave functions in Figure 7.18 look very much like the allowed wavelengths for a plucked guitar string (see Figure 7.12).

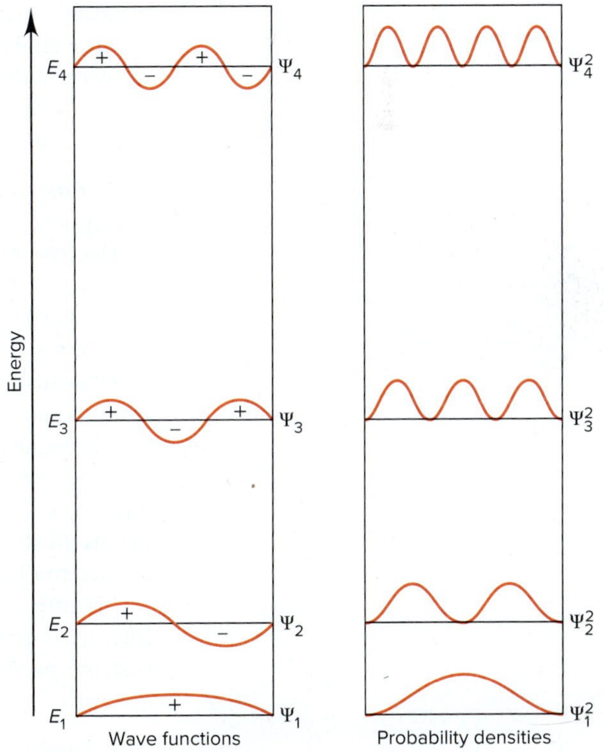

Figure 7.18 Wave functions (ψ) and probability densities (ψ^2) for the first four energy levels of a particle in a one-dimensional box.

The Schrödinger Equation for the Hydrogen Atom The Schrödinger equation can be solved exactly for the H atom because it has only one electron and, thus, no electron-electron interactions. Hydrogen's electron can be thought of as similar to the particle in a box, in that it is confined to a small volume by its attraction to the nucleus. However, unlike the particle in a box, the electron does have potential energy due to that attraction. Moreover, the electron can move in three dimensions, x, y, and z, not just one.

The Schrödinger equation for a single electron moving in three dimensions and attracted to a nucleus is

$$-\frac{h^2}{8\pi^2 m_e}\left[\frac{d^2\psi}{dx^2} + \frac{d^2\psi}{dy^2} + \frac{d^2\psi}{dz^2}\right] + V(x, y, z)\psi = E\psi$$

The potential energy term $V(x, y, z)$ that results from the coulombic attraction between an electron of charge $-e$ separated by a distance r from a proton of charge Ze is expressed as

$$V(r) = -\frac{Ze \times e}{4\pi\varepsilon_0 r} = -\frac{Ze^2}{4\pi\varepsilon_0 r}$$

where ε_0 is a fundamental constant known as the *permittivity of the vacuum.*

For the particle in a one-dimensional box, one quantum number, *n,* is needed to describe a wave function (Equation 7.10). To describe the wave functions of hydrogen's electron moving in *three* dimensions, three quantum numbers are needed. Each set of these three quantum numbers (discussed in more detail later in the section) specifies a wave function for the electron, called an **atomic orbital.** The energies of the atomic orbitals in hydrogen are given by the following equation:

$$E_n = -\frac{Z^2 e^4 m_e}{8h^2 \varepsilon_0{}^2 n^2} \qquad n = 1, 2, 3, \ldots$$

where $Z = 1$ for hydrogen. Substituting known values for e, m_e, h, and ε_0 into this equation gives

$$E_n = -2.18\times10^{-18}\left(\frac{Z^2}{n^2}\right)$$

which is the same relationship obtained by Bohr (Section 7.2)! Like the Bohr model, the quantum-mechanical model based on Schrödinger's equation shows that *the energy of the atom is quantized, that is, it has discrete values.*

The Probable Location of an Electron Each solution of the Schrödinger equation gives an energy state of the atom associated with a given atomic orbital. An important point to keep in mind is that an "orbital" in the quantum-mechanical model bears no resemblance to an "orbit" in the Bohr model: an *orbit* is an electron's actual path around the nucleus, whereas an *orbital* is a mathematical function that describes the electron's matter-wave but has no physical meaning.

Remember that the square of the wave function (ψ^2, probability density) gives a measure of the probability of finding the electron in a particular region of space within the atom. We depict the electron's probable location in several ways. Let's look first at the H atom's *ground state:*

1. *Probability of the electron being in some tiny volume of the atom.* For each energy level, we can create an *electron probability density diagram,* or more simply, an **electron density diagram.** The value of ψ^2 for a given volume is shown with dots: the greater the density of dots, the higher the probability of finding the electron in that volume. Note, that for the ground state of the H atom, *the electron probability density decreases with distance from the nucleus* along a line, r (Figure 7.19A). In other words, the probability of finding the electron decreases with increasing distance from the nucleus.

These diagrams are also called **electron cloud depictions** because, if we *could* take a time-exposure photograph of the electron in wavelike motion around the nucleus, it would appear as a "cloud" of positions. The electron cloud is an *imaginary* picture of the electron changing its position rapidly over time; it does *not* mean that an electron is a diffuse cloud of charge.

Figure 7.19B shows a plot of ψ^2 vs. r. Due to the thickness of the printed line, the curve appears to touch the axis; however, in the blow-up circle, we see that *the probability of the electron being far from the nucleus is very small, but not zero.*

Student Hot Spot

Student data indicate that you may struggle with the concept of probability density. Access the Smartbook to view additional Learning Resources on this topic.

2. *Total probability density at some distance from the nucleus.* To find *radial probability distribution,* that is, the *total* probability of finding the electron at some distance r from the nucleus, we first mentally divide the volume around the nucleus into thin, concentric, spherical layers, like the layers of an onion (shown in cross section

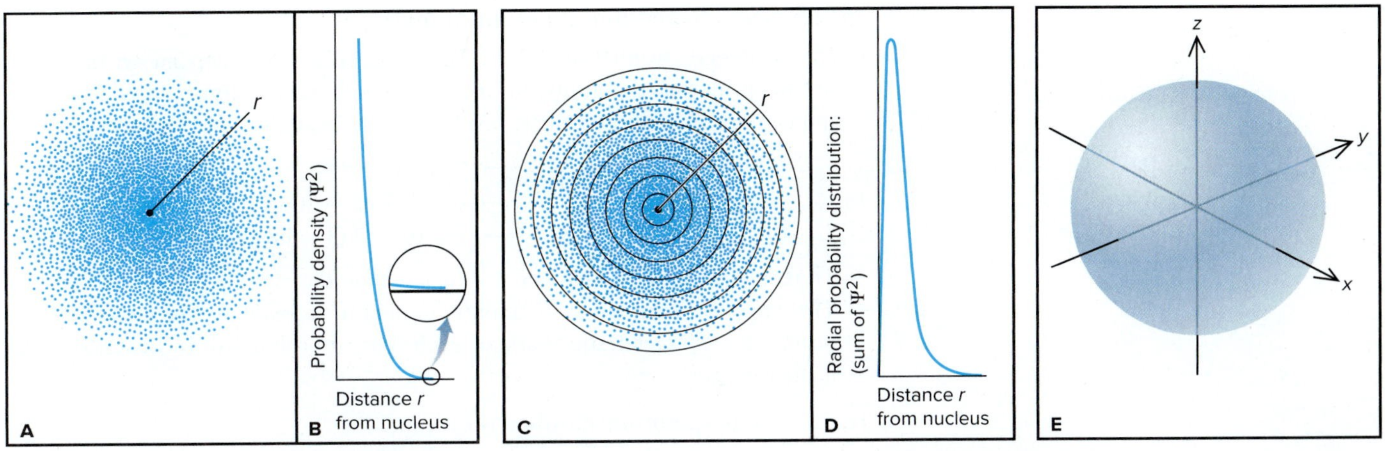

Figure 7.19 **Electron probability density in the ground-state H atom.** **A,** In the electron density diagram, the density of dots represents the probability of the electron being within a tiny volume and decreases with distance, *r*, from the nucleus. **B,** The probability density (ψ^2) decreases with *r* but does not reach zero *(blow-up circle)*. **C,** Counting dots within each layer gives the total probability of the electron being in that layer. **D,** A radial probability distribution plot shows that total electron density peaks *near,* but not *at,* the nucleus. **E,** A 90% probability contour for the ground state of the H atom.

in Figure 7.19C). Then, we find the *sum of* ψ^2 *values* in each layer to see which is most likely to contain the electron.

The falloff in probability density with distance has an important effect. Near the nucleus, *the volume of each layer increases faster than its density of dots decreases.* The result of these opposing effects is that the *total* probability peaks in a layer *near,* but not *at,* the nucleus. For example, the total probability in the second layer is higher than in the first, but this result disappears with greater distance. Figure 7.19D shows this result as a **radial probability distribution plot.** ›

3. *Probability contour and the size of the atom.* How far away from the nucleus can we find the electron? This is the same as asking "How big is the H atom?" Recall from Figure 7.19B that the probability of finding the electron far from the nucleus is not zero. Therefore, we *cannot* assign a definite volume to an atom. However, we can visualize an atom with a 90% **probability contour:** the electron is somewhere within that volume 90% of the time (Figure 7.19E).

As you'll see later in this section, each atomic orbital has a distinctive radial probability distribution and 90% probability contour.

A Radial Probability Distribution of Apples

An analogy might clarify why the curve in the radial probability distribution plot peaks and then falls off. Picture fallen apples around the base of an apple tree: the density of apples is greatest near the trunk and decreases with distance. Divide the ground under the tree into foot-wide concentric rings and collect the apples within each ring. Apple density is greatest in the first ring, but the area of the second ring is larger, and so it contains a greater *total* number of apples. Farther out near the edge of the tree, rings have more area but lower apple "density," so the total number of apples decreases. A plot of "number of apples in each ring" vs. "distance from trunk" shows a peak at some distance close to the trunk, as in Figure 7.19D.

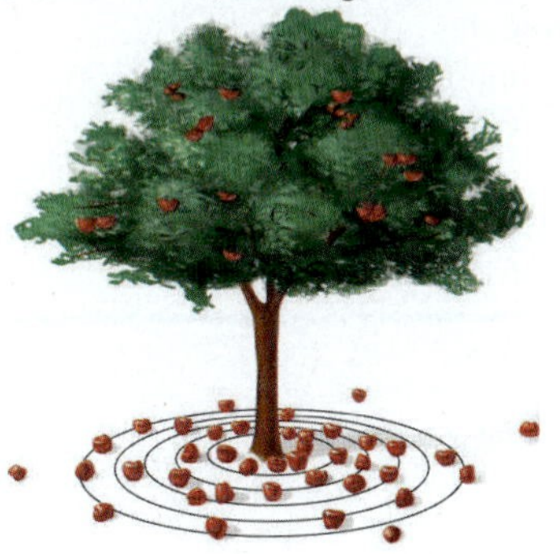

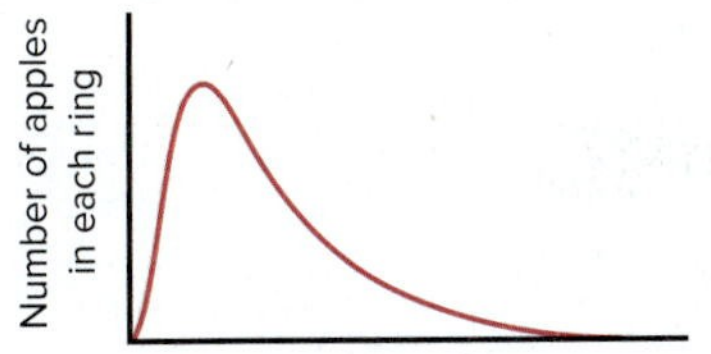

Quantum Numbers of an Atomic Orbital

An atomic orbital is specified by three quantum numbers that are part of the solution of the Schrödinger equation and indicate the size, shape, and orientation in space of the orbital.*

1. The **principal quantum number (*n*):**
   - *n* is a positive integer (1, 2, 3, and so forth).
   - It indicates the relative *size* of the orbital and therefore the relative *distance from the nucleus* of the peak in the radial probability distribution plot. As *n* increases, the orbital becomes larger and the electron spends more time farther from the nucleus.
   - The principal quantum number specifies the *energy level* of the H atom: *the higher the n value, the higher the energy level.* When the electron occupies an orbital with $n = 1$, the H atom is in its ground state and has its lowest energy. When the electron occupies an orbital with $n = 2$ (first excited state), the atom has more energy.

*For ease in discussion, chemists often refer to the size, shape, and orientation of an "atomic orbital," although we really mean the size, shape, and orientation of an "atomic orbital's radial probability distribution."

2. The **angular momentum quantum number (l):**
   - l is an integer from 0 to $n - 1$. Thus, the values of l depend on the value of the principal quantum number, n, which sets a limit on the angular momentum quantum number: n limits l. For example:

$$n = 1 \longrightarrow l = 0 \text{ (1 value)}$$
$$n = 2 \longrightarrow l = 0, 1 \text{ (2 values)}$$
$$n = 3 \longrightarrow l = 0, 1, 2 \text{ (3 values)}$$

     Thus, the *number* of possible l values equals the value of n.
   - The angular momentum quantum number is related to the *shape* of the orbital; the characteristic shape of each type of orbital will be described later in this section.

3. The **magnetic quantum number (m_l):**
   - *m_l is an integer from $-l$ through 0 to $+l$.* The angular momentum quantum number, l, sets a limit on the magnetic quantum number: l limits m_l.
   - The number of m_l values = $2l + 1$ = the number of orbitals for a given l:

$$l = 0 \longrightarrow m_l = 0 \text{ (1 value)}$$
$$l = 1 \longrightarrow m_l = -1, 0, +1 \text{ (3 values)}$$
$$l = 2 \longrightarrow m_l = -2, -1, 0, +1, +2 \text{ (5 values)}$$

     Thus, for a particular value of n, there is only one possible orbital with $l = 0$, but three possible orbitals with $l = 1$, each with its own m_l value.
   - The total number of m_l values for a given n value = n^2 = the total number of orbitals in that energy level.
   - The magnetic quantum number prescribes the three-dimensional *orientation* of the orbital in the space around the nucleus. For example, each of the three orbitals with $l = 1$ has its own orientation.

Table 7.2 presents a summary of these quantum numbers.

Table 7.2 The Hierarchy of Quantum Numbers for Atomic Orbitals

Name, Symbol (Property)	Allowed Values	Quantum Numbers					
Principal, n (size, energy)	Positive integer (1, 2, 3, ...)	1	2		3		
Angular momentum, l (shape)	0 to $n - 1$	0	0	1	0	1	2
Magnetic, m_l (orientation)	$-l, \ldots, 0, \ldots, +l$	0	0	−1 0 +1	0	−1 0 +1	−2 −1 0 +1 +2

SAMPLE PROBLEM 7.7 Determining Quantum Numbers for an Energy Level

Problem What values of the angular momentum (l) and magnetic (m_l) quantum numbers are allowed for a principal quantum number (n) of 3? How many orbitals are there in this energy level?

Plan We determine allowable quantum numbers with the rules from the text: l values are integers from 0 to $n - 1$, and m_l values are integers from $-l$ to 0 to $+l$. One m_l value is assigned to each orbital, so the number of m_l values gives the number of orbitals.

Solution Determining l values: for $n = 3$:

$$l = 0 \text{ to } (3 - 1): l = 0, 1, 2.$$

Determining m_l for each l value:

For $l = 0$, $m_l = 0$

For $l = 1$, $m_l = -1, 0, +1$

For $l = 2$, $m_l = -2, -1, 0, +1, +2$

There are nine m_l values, so there are nine orbitals with $n = 3$.

Check Table 7.2 shows that we are correct. As we saw, the total number of orbitals for a given n value is n^2, and for $n = 3$, $n^2 = 9$.

FOLLOW-UP PROBLEMS

7.7A What are the possible l and m_l values for $n = 4$?

7.7B For what value of the principal quantum number n are there five allowed values of l? What are the allowed m_l values for this n?

SOME SIMILAR PROBLEMS 7.51 and 7.52

Quantum Numbers and Energy Levels

The energy states and orbitals of the atom are described with specific terms and are associated with one or more quantum numbers:

1. *Level.* The atom's energy **levels,** or *shells,* are given by the n value: the smaller the n value, the lower the energy level and the greater the probability that the electron is closer to the nucleus.

2. *Sublevel.* The atom's levels are divided into **sublevels,** or *subshells,* that are given by the l value. Each l value is designated by a letter:

$l = 0$ is an s sublevel.
$l = 1$ is a p sublevel.
$l = 2$ is a d sublevel.
$l = 3$ is an f sublevel.

(The letters derive from names of spectroscopic lines: *s*harp, *p*rincipal, *d*iffuse, and *f*undamental.) Sublevels with l values greater than 3 are designated by consecutive letters after *f*: *g* sublevel, *h* sublevel, and so on. A sublevel is named with its n value and letter designation; for example, the sublevel (subshell) with $n = 2$ and $l = 0$ is called the 2*s* sublevel. We discuss orbital shapes below.

3. *Orbital.* Each combination of n, l, and m_l specifies the size (energy), shape, and spatial orientation of one of the atom's orbitals. We know the quantum numbers of the orbitals in a sublevel from the sublevel name and the quantum-number hierarchy. For example,

- For the 2*s* sublevel: $n = 2$, $l = 0$, $m_l = 0$; ***one*** value of m_l indicates ***one*** orbital in this sublevel.
- For the 3*p* sublevel: $n = 3$, $l = 1$, $m_l = -1, 0, +1$; ***three*** values of m_l indicate ***three*** orbitals in this sublevel, one with $m_l = -1$, one with $m_l = 0$, and one with $m_l = +1$.

Student Hot Spot

Student data indicate that you may struggle with applying quantum numbers. Access the Smartbook to view additional Learning Resources on this topic.

SAMPLE PROBLEM 7.8 **Determining Sublevel Names and Orbital Quantum Numbers**

Problem Give the name, possible magnetic quantum number(s), and number of orbitals for the sublevel that has the given n and l quantum numbers:

(a) $n = 3$, $l = 2$ **(b)** $n = 2$, $l = 0$ **(c)** $n = 5$, $l = 1$ **(d)** $n = 4$, $l = 3$

Plan We name the sublevel (subshell) with the n value and the letter designation of the l value. From the l value, we find the number of possible m_l values, which equals the number of orbitals in that sublevel.

Solution

	n	l	Sublevel Name	Possible m_l Values	No. of Orbitals
(a)	3	2	3*d*	−2, −1, 0, +1, +2	5
(b)	2	0	2*s*	0	1
(c)	5	1	5*p*	−1, 0, +1	3
(d)	4	3	4*f*	−3, −2, −1, 0, +1, +2, +3	7

Check Check the number of orbitals in each sublevel using

$$\text{No. of orbitals} = \text{no. of } m_l \text{ values} = 2l + 1$$

FOLLOW-UP PROBLEMS

7.8A What are the n, l, and possible m_l values for the 2*p* and 5*f* sublevels?

7.8B What are the n, l, and possible m_l values for the 4*d* and 6*s* sublevels? How many orbitals are in these sublevels of an atom?

SOME SIMILAR PROBLEMS 7.49, 7.50, and 7.55–7.58

SAMPLE PROBLEM 7.9 Identifying Incorrect Quantum Numbers

Problem What is wrong with each quantum number designation and/or sublevel name?

	n	l	m_l	Name
(a)	1	1	0	1*p*
(b)	4	3	+1	4*d*
(c)	3	1	−2	3*p*

Solution

(a) A sublevel with $n = 1$ can only have $l = 0$, not $l = 1$. The correct sublevel name is 1*s*.

(b) A sublevel with $l = 3$ is an *f* sublevel, not a *d* sublevel. The name should be 4*f*.

(c) A sublevel with $l = 1$ can have only −1, 0, or +1 for m_l, not −2.

Check Check that l is always less than n, and m_l is always $\geq -l$ and $\leq +l$.

FOLLOW-UP PROBLEMS

7.9A Supply the missing quantum numbers and sublevel names.

	n	l	m_l	Name
(a)	?	?	0	4*p*
(b)	2	1	0	?
(c)	3	2	−2	?
(d)	?	?	?	2*s*

7.9B What is wrong with each quantum number designation and/or sublevel name?

	n	l	m_l	Name
(a)	5	3	4	5*f*
(b)	2	2	1	2*d*
(c)	6	1	−1	6*s*

SOME SIMILAR PROBLEMS 7.59, 7.60, and 7.72

Shapes of Atomic Orbitals

Each sublevel of the H atom consists of a set of orbitals with characteristic shapes. As you'll see in Chapter 8, orbitals for the other atoms have similar shapes.

The *s* Orbital An orbital with $l = 0$ has a *spherical* shape with the nucleus at its center and is called an *s* **orbital.** Because a sphere has only one orientation, an *s* orbital has only one m_l value: for any *s* orbital, $m_l = 0$.

1. *The 1s orbital* holds the electron in the H atom's ground state. *The electron probability density is highest at the nucleus.* Figure 7.20A shows this graphically *(top),* and an electron density *relief map (inset)* depicts the graph's curve in three dimensions. Note the quarter-section of a three-dimensional electron cloud depiction *(middle)* has the darkest shading at the nucleus. For reasons discussed earlier (see Figure 7.19D and the discussion about the radial probability distribution of apples), the radial probability distribution plot *(bottom)* is highest slightly out from the nucleus. Both plots fall off smoothly with distance from the nucleus.

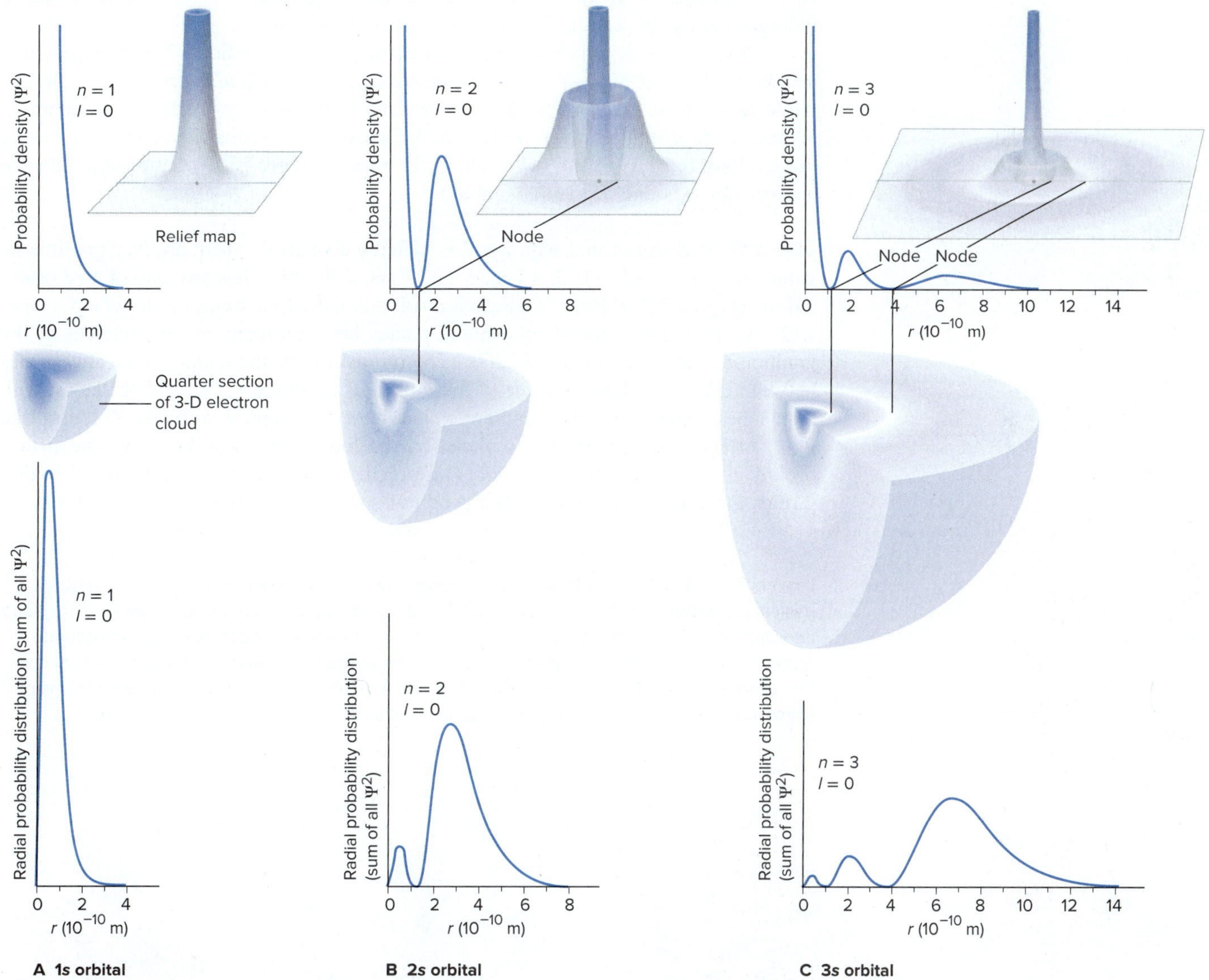

Figure 7.20 **The 1*s*, 2*s*, and 3*s* orbitals.** For each of the *s* orbitals, a plot of probability density vs. distance (*top*, with the relief map, *inset*, showing the plot in three dimensions) lies above a quarter section of an electron cloud depiction of the 90% probability contour *(middle)*, which lies above a radial probability distribution plot *(bottom)*. **A,** The 1*s* orbital. **B,** The 2*s* orbital. **C,** The 3*s* orbital.

2. *The 2s orbital* (Figure 7.20B) has two regions of higher electron density. The radial probability distribution (Figure 7.20B, *bottom*) of the more distant region is *higher* than that of the closer one because the sum of Ψ^2 for it is taken over a much larger volume. Between the two regions is a spherical node, where the probability of finding the electron drops to zero (see Figure 7.16). Because the 2*s* orbital is larger than the 1*s*, an electron in the 2*s* spends more time farther from the nucleus (in the larger of the two regions) than it does when it occupies the 1*s*.

3. *The 3s orbital* (Figure 7.20C) has three regions of high electron density and two nodes. Here again, the highest radial probability is that for the region at the greatest distance from the nucleus. This pattern of more nodes and higher probability with distance from the nucleus continues with the 4*s*, 5*s*, and so forth.

The *p* Orbital An orbital with $l = 1$ is called a ***p* orbital** and has two regions (lobes) of high probability, one on *either side* of the nucleus (Figure 7.21). The *nucleus lies at the nodal plane* of this dumbbell-shaped orbital. Since the maximum value of l is $n - 1$, only levels with $n = 2$ or higher have a *p* orbital: the lowest energy *p* orbital (the one closest to the nucleus) is the 2*p*. *One p* orbital consists of *two* lobes, and the electron spends *equal* time in both. Similar to the pattern for *s* orbitals, a 3*p* orbital is larger than a 2*p*, a 4*p* is larger than a 3*p*, and so forth.

Unlike *s* orbitals, *p* orbitals *have* different spatial orientations. The three possible m_l values of −1, 0, and +1 refer to three *mutually perpendicular* orientations; that is, while identical in size, shape, and energy, the three *p* orbitals differ in orientation. We associate *p* orbitals with the *x*, *y*, and *z* axes: the p_x orbital lies along the *x* axis, the p_y along the *y* axis, and the p_z along the *z* axis. (There is no relationship between a particular axis and a given m_l value.)

The *d* Orbital An orbital with $l = 2$ is called a ***d* orbital.** There are five possible m_l values for $l = 2$: −2, −1, 0, +1, and +2. Thus, a *d* orbital has any one of five orientations (Figure 7.22). Four of the five *d* orbitals have four lobes (a cloverleaf shape) with two mutually perpendicular nodal planes between them and the nucleus at the junction of the lobes (Figure 7.22C). (The orientation of the nodal planes always lies between the orbital lobes.) Three of these orbitals lie in the *xy, xz,* and *yz* planes, with their lobes *between* the axes, and are called the d_{xy}, d_{xz}, and d_{yz} orbitals. A fourth, the $d_{x^2-y^2}$ orbital, also lies in the *xy* plane, but its lobes are *along* the axes. The fifth *d* orbital, the d_{z^2}, has two major lobes *along* the *z* axis, and a donut-shaped region girdles the center. An electron in a *d* orbital spends equal time in all of its lobes.

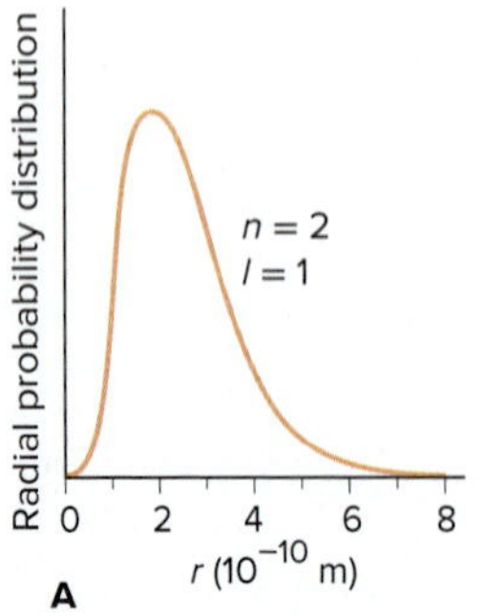

A

Figure 7.21 **The 2*p* orbitals. A,** A radial probability distribution plot of the 2*p* orbital shows a peak much farther from the nucleus than the peak for the 1*s*. **B,** Cross section of an electron cloud depiction of the 90% probability contour of the $2p_z$ orbital shows a nodal plane. **C,** An accurate representation of the $2p_z$ probability contour. **D,** The stylized depiction of the $2p_z$ probability contour used in the text. **E,** The three (stylized) 2*p* orbitals occupy mutually perpendicular regions of space, contributing to the atom's overall spherical shape.

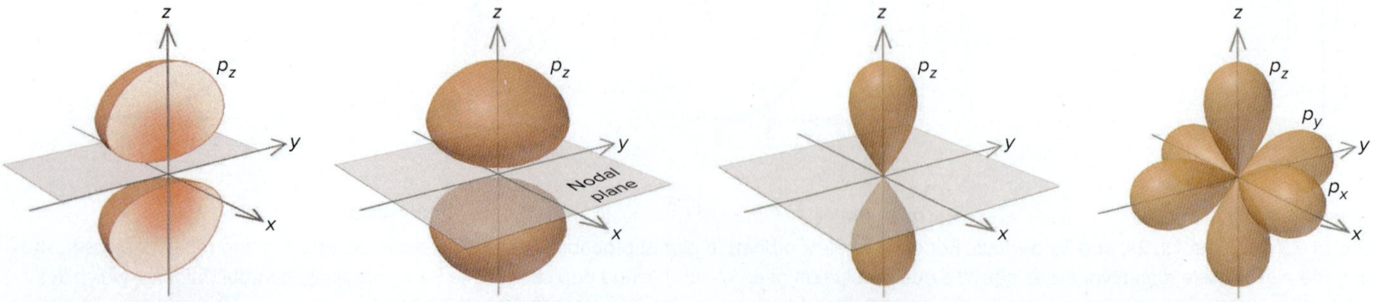

B **Cross section of electron cloud depiction, $2p_z$** **C** **Accurate probability contour, $2p_z$** **D** **Stylized probability contour, $2p_z$** **E** **The three 2*p* orbitals**

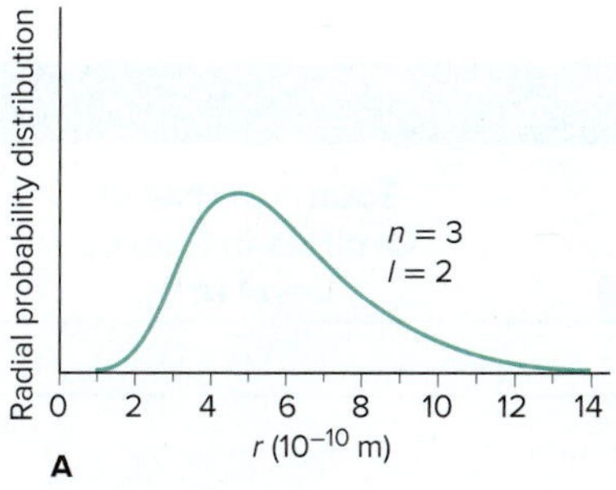

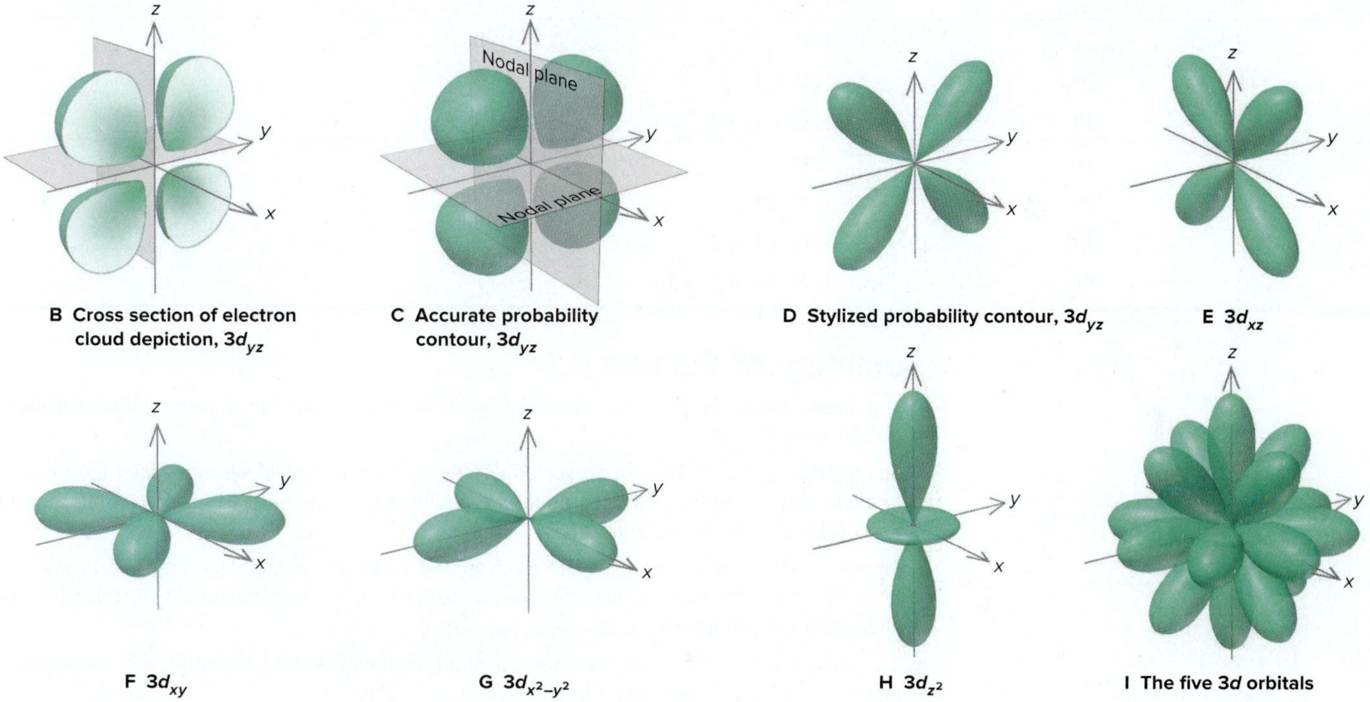

Figure 7.22 The 3*d* orbitals. **A,** A radial probability distribution plot. **B,** Cross section of an electron cloud depiction of the $3d_{yz}$ orbital probability contour shows two mutually perpendicular nodal planes and lobes lying *between* the axes. **C,** An accurate representation of the $3d_{yz}$ orbital probability contour. **D,** The stylized depiction of the $3d_{yz}$ orbital used in the text. **E,** The $3d_{xz}$ orbital. **F,** The $3d_{xy}$ orbital. **G,** The lobes of the $3d_{x^2-y^2}$ orbital lie *along* the *x* and *y* axes. **H,** The $3d_{z^2}$ orbital has two lobes and a central, donut-shaped region. **I,** A composite of the five 3*d* (stylized) orbitals shows how they contribute to the atom's spherical shape.

In keeping with the quantum-number hierarchy, a *d* orbital ($l = 2$) must have a principal quantum number of $n = 3$ or higher, so 3*d* is the lowest energy *d* sublevel. Orbitals in the 4*d* sublevel are larger (extend farther from the nucleus) than those in the 3*d,* and the 5*d* orbitals are larger still.

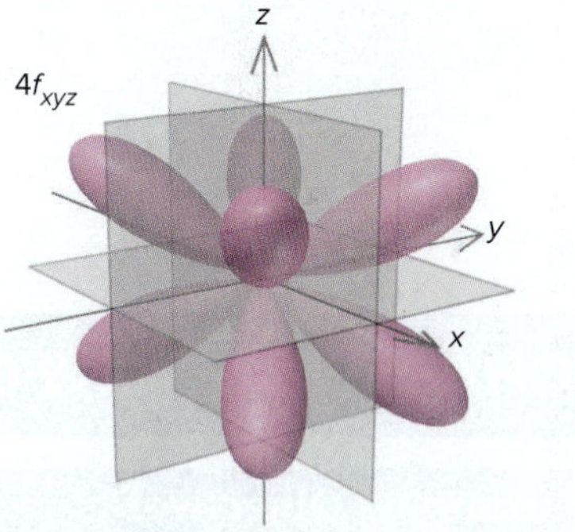

Figure 7.23 The $4f_{xyz}$ orbital, one of the seven 4*f* orbitals.

The *f* Orbital and Orbitals with Higher *l* Values An orbital with $l = 3$ is called an ***f* orbital** and has a principal quantum number of at least $n = 4$. For the $l = 3$ sublevel, $m_l = -3, -2, -1, 0, +1, +2, +3$; seven values of m_l indicate seven *f* orbitals in the sublevel, with seven orientations. Figure 7.23 shows one of the seven *f* orbitals; each *f* orbital has a complex, multilobed shape with several nodal planes.

Orbitals with $l = 4$ are *g* orbitals, but they play no known role in chemical bonding. Table 7.3 summarizes the relationships between the quantum numbers and orbitals for the first four energy levels ($n = 1$, 2, 3, and 4).

The Special Case of Energy Levels in the Hydrogen Atom

With regard to energy levels and sublevels, the H atom is a special case. When an H atom gains energy, its electron occupies an orbital of higher *n* value, which is (on average) farther from the nucleus. But, because it has just one electron, *hydrogen is the only atom whose energy state depends completely on the principal quantum number, n.* As you'll see in Chapter 8, because of additional nucleus-electron attractions and electron-electron repulsions, the energy states of all other atoms depend on the *n and l* values of the occupied orbitals. Thus, *for the H atom only,* all four $n = 2$ orbitals (one 2*s* and three 2*p*) have the same energy, all nine $n = 3$ orbitals (one 3*s,* three 3*p,* and five 3*d*) have the same energy (Figure 7.24), and so forth.

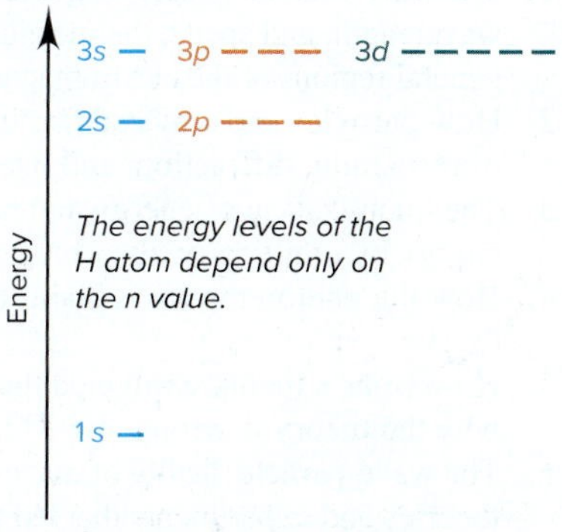

Figure 7.24 Energy levels of the H atom.

Table 7.3 Summary of n, l, and m_l Values for the First Four Principal Energy Levels

n	l	Sublevel Designation	m_l	Number of Orbitals in Sublevel ($2l + 1$)	Total Number of Orbitals in Energy Level (n^2)
1	0	1*s*	0	1	1
2	0	2*s*	0	1	4
	1	2*p*	−1, 0, +1	3	
3	0	3*s*	0	1	9
	1	3*p*	−1, 0, +1	3	
	2	3*d*	−2, −1, 0, +1, +2	5	
4	0	4*s*	0	1	16
	1	4*p*	−1, 0, +1	3	
	2	4*d*	−2, −1, 0, +1, +2	5	
	3	4*f*	−3, −2, −1, 0, +1, +2, +3	7	

› Summary of Section 7.4

- The atomic orbital (ψ, wave function) is a mathematical description of the electron's wavelike behavior in an atom.
- The Schrödinger equation can be solved exactly in a simple model system, called the particle in a box. Applied to the hydrogen atom, the equation converts each allowed wave function to one of the atom's energy states.
- The probability density of finding the electron at a particular location is represented by ψ^2. For a given energy level, an electron density diagram and a radial probability distribution plot show how the electron occupies the space near the nucleus.
- An atomic orbital is described by three quantum numbers: size (n), shape (l), and orientation (m_l): n limits l to n values, and l limits m_l to $2l + 1$ values.
- An energy level has sublevels with the same n value; a sublevel has orbitals with the same n and l values but differing m_l values.
- A sublevel with $l = 0$ has a spherical (*s*) orbital; a sublevel with $l = 1$ has three, two-lobed (*p*) orbitals; a sublevel with $l = 2$ has five multilobed (*d*) orbitals; and a sublevel with $l = 3$ has seven multilobed (*f*) orbitals.
- In the special case of the H atom, the energy levels depend only on the n value.

CHAPTER REVIEW GUIDE

Learning Objectives

Relevant section (§) and/or sample problem (SP) numbers appear in parentheses.

Understand These Concepts

1. The wave characteristics of light (the interrelations of frequency, wavelength, and speed; the meaning of amplitude) and the general regions of the electromagnetic spectrum (§7.1)
2. How particles and waves differ in terms of the phenomena of refraction, diffraction, and interference (§7.1)
3. The quantization of energy and how an atom changes its energy by emitting or absorbing quanta of radiation (§7.1)
4. How the photon theory explains the photoelectric effect (§7.1)
5. How Bohr's theory explained the line spectra of the H atom; why the theory is wrong and which ideas we retain (§7.2)
6. The wave-particle duality of matter and energy and the relevant theories and experiments that led to it (de Broglie wavelength, electron diffraction, and photon momentum) (§7.3)
7. The meaning of the uncertainty principle and how uncertainty limits our knowledge of electron properties (§7.3)
8. How the Schrödinger equation is solved for the particle in a box and for the electron in the hydrogen atom (§7.4)
9. How the Schrödinger equation shows that the energy of a particle is quantized (§7.4)
10. The distinction between ψ (wave function, or atomic orbital) and ψ^2 (probability density) (§7.4)
11. How electron density diagrams and radial probability distribution plots depict the electron's location within the atom (§7.4)
12. The hierarchy of quantum numbers that describe the size and energy (n), shape (l), and orientation (m_l) of an orbital (§7.4)
13. The distinction between energy level (shell), sublevel (subshell), and orbital (§7.4)
14. The shapes of *s*, *p*, *d*, and *f* orbitals (§7.4)

Master These Skills

1. Interconverting wavelength and frequency (SP 7.1)
2. Interconverting energy, wavelength, and frequency (SP 7.2)
3. Finding the energy change and the wavelength of the photon absorbed or emitted when an H atom changes its energy level (SP 7.3)
4. Applying de Broglie's equation to find the wavelength of an electron (SP 7.4)
5. Applying the uncertainty principle to see that the location and speed of a particle cannot be determined simultaneously (SP 7.5)
6. Using the particle-in-a-box model to determine the energy change and wavelength of the photon absorbed or emitted during an electron transition (SP 7.6).
7. Determining quantum numbers and sublevel designations (SPs 7.7–7.9)

Key Terms

Page numbers appear in parentheses.

absorption spectrum (308)
amplitude (296)
angular momentum quantum number (l) (320)
atomic orbital (318)
d orbital (323)
de Broglie wavelength (311)
diffraction (299)
electromagnetic radiation (295)
electromagnetic spectrum (296)
electron cloud depiction (318)
electron density diagram (318)
emission spectrum (308)
excited state (304)
f orbital (325)
flame test (308)
frequency (ν) (296)
ground state (304)
infrared (IR) (296)
level (shell) (321)
line spectrum (302)
magnetic quantum number (m_l) (320)
node (314)
p orbital (323)
photoelectric effect (301)
photon (301)
Planck's constant (h) (300)
principal quantum number (n) (319)
probability contour (319)
quantum (300)
quantum mechanics (314)
quantum number (300)
radial probability distribution plot (319)
refraction (298)
s orbital (323)
Schrödinger equation (314)
spectrometry (308)
speed of light (c) (296)
stationary state (304)
sublevel (subshell) (321)
ultraviolet (UV) (296)
uncertainty principle (313)
wavelength (λ) (296)
wave function (314)
wave-particle duality (312)

Key Equations and Relationships

Page numbers appear in parentheses.

7.1 Relating the speed of light to its frequency and wavelength (296):

$$c = \nu \times \lambda$$

7.2 Relating energy change and frequency (300):

$$\Delta E = h\nu$$

7.3 Relating energy change and wavelength (300):

$$\Delta E = \frac{hc}{\lambda}$$

7.4 Calculating the wavelength of any line in the H atom spectrum (Rydberg equation) (302):

$$\frac{1}{\lambda} = R\left(\frac{1}{n_1^2} - \frac{1}{n_2^2}\right)$$

where n_1 and n_2 are positive integers and $n_2 > n_1$

7.5 Finding the difference between two energy levels in the H atom (306):

$$\Delta E = E_{\text{final}} - E_{\text{initial}} = -2.18\times10^{-18}\text{ J}\left(\frac{1}{n_{\text{final}}^2} - \frac{1}{n_{\text{initial}}^2}\right)$$

7.6 Calculating the wavelength of any moving particle (de Broglie wavelength) (310):

$$\lambda = \frac{h}{mu}$$

7.7 Finding the uncertainty in position or speed of a particle (the uncertainty principle) (313):

$$\Delta x \cdot m\Delta u \geq \frac{h}{4\pi}$$

7.8 Calculating the energy of the particle in a box of length L (316):

$$E_n = \frac{n^2h^2}{8mL^2} \qquad n = 1, 2, 3, \ldots$$

7.9 Calculating the difference between two energy levels for the particle in a box (316):

$$\Delta E = (2n + 1)\frac{h^2}{8mL^2}$$

7.10 Calculating the wave function for the particle in a box (317):

$$\psi_n(x) = \sqrt{\frac{2}{L}}\sin\left(\frac{n\pi x}{L}\right) \qquad n = 1, 2, 3, \ldots$$

BRIEF SOLUTIONS TO FOLLOW-UP PROBLEMS

7.1A $\lambda(\text{nm}) = \frac{c}{\nu} = \frac{3.00\times10^8\text{ m/s}}{7.23\times10^{14}\text{ s}^{-1}} \times \frac{10^9\text{ nm}}{1\text{ m}} = 415\text{ nm}$

$\lambda(\text{Å}) = 415\text{ nm} \times \frac{10^{-9}\text{ m}}{1\text{ nm}} \times \frac{1\text{ Å}}{10^{-10}\text{ m}} = 4150\text{ Å}$

7.1B $\nu = \frac{c}{\lambda} = \frac{3.00\times10^8\text{ m/s}}{940\text{ nm}} \times \frac{10^9\text{ nm}}{1\text{ m}} = 3.2\times10^{14}\text{ s}^{-1}$

This is infrared radiation.

7.2A $\lambda = \frac{hc}{E} = \frac{(6.626\times10^{-34}\text{ J}\cdot\text{s})(3.00\times10^8\text{ m/s})}{8.2\times10^{-19}\text{ J}} \times \frac{10^9\text{ nm}}{1\text{ m}}$

$= 240\text{ nm}$

$\nu = \frac{E}{h} = \frac{8.2\times10^{-19}\text{ J}}{6.626\times10^{-34}\text{ J}\cdot\text{s}} = 1.2\times10^{15}\text{ s}^{-1}$

7.2B (a) UV: $E = hc/\lambda$

$$= \frac{(6.626\times10^{-34}\ \text{J·s})(3.00\times10^{8}\ \text{m/s})}{1\times10^{-8}\ \text{m}}$$

$$= 2\times10^{-17}\ \text{J}$$

(b) Visible: $E = 4\times10^{-19}$ J (c) IR: $E = 2\times10^{-21}$ J

As λ increases, E decreases.

7.3A (a) With $n_{\text{final}} = 3$ for an IR photon:

$$\Delta E = -2.18\times10^{-18}\ \text{J}\left(\frac{1}{n_{\text{final}}^2} - \frac{1}{n_{\text{initial}}^2}\right)$$

$$= -2.18\times10^{-18}\ \text{J}\left(\frac{1}{3^2} - \frac{1}{6^2}\right)$$

$$= -2.18\times10^{-18}\ \text{J}\left(\frac{1}{9} - \frac{1}{36}\right) = -1.82\times10^{-19}\ \text{J}$$

(ΔE is negative since the photon is emitted.)

(b) Use the absolute value of ΔE since wavelength must have a positive value:

$$\lambda = \frac{hc}{|\Delta E|} = \frac{(6.626\times10^{-34}\ \text{J·s})(3.00\times10^{8}\ \text{m/s})}{1.82\times10^{-19}\ \text{J}} \times \frac{1\ \text{Å}}{10^{-10}\ \text{m}}$$

$$= 1.09\times10^{4}\ \text{Å}$$

7.3B

(a) $$\Delta E\ (\text{J/atom}) = \frac{hc}{\lambda}$$

$$= \frac{(6.626\times10^{-34}\ \text{J·s})(3.00\times10^{8}\ \text{m/s})}{410\ \text{nm}} \times \frac{10^{9}\ \text{nm}}{1\ \text{m}}$$

$$= 4.85\times10^{-19}\ \text{J} = -4.85\times10^{-19}\ \text{J}$$

(ΔE is negative since the photon is emitted.)

$$\Delta E\ (\text{kJ/mol}) = \frac{-4.85\times10^{-19}\ \text{J}}{\text{atom}} \times \frac{6.022\times10^{23}\ \text{atoms}}{\text{mol}} \times \frac{1\ \text{kJ}}{10^{3}\ \text{J}}$$

$$= -292\ \text{kJ/mol}$$

(b) $$\Delta E = -2.18\times10^{-18}\ \text{J}\left(\frac{1}{n_{\text{final}}^2} - \frac{1}{n_{\text{initial}}^2}\right)$$

$$\frac{1}{n_{\text{final}}^2} = \frac{\Delta E}{-2.18\times10^{-18}\ \text{J}} + \frac{1}{n_{\text{initial}}^2}$$

$$= \frac{-4.85\times10^{-19}\ \text{J}}{-2.18\times10^{-18}\ \text{J}} + \frac{1}{6^2} = 0.25025$$

$$n_{\text{final}}^2 = \frac{1}{0.25025} = 4$$

$$n_{\text{final}} = 2$$

7.4A (a) $$\lambda = \frac{h}{mu}$$

$$u = \frac{h}{m\lambda} = \frac{6.626\times10^{-34}\ \text{kg·m}^2\text{/s}}{(9.11\times10^{-31}\ \text{kg})\left(100\ \text{nm} \times \frac{1\ \text{m}}{10^{9}\ \text{nm}}\right)}$$

$$= 7.27\times10^{3}\ \text{m/s}$$

(b) $$u = \frac{h}{m\lambda} = \frac{6.626\times10^{-34}\ \text{kg·m}^2\text{/s}}{(0.0459\ \text{kg})\left(100.\ \text{nm} \times \frac{1\ \text{m}}{10^{9}\ \text{nm}}\right)}$$

$$= 1.44\times10^{-25}\ \text{m/s}$$

7.4B $$u\ (\text{m/s}) = \frac{55\ \text{mi}}{1\ \text{h}} \times \frac{1\ \text{h}}{3600\ \text{s}} \times \frac{1.609\ \text{km}}{1\ \text{mi}} \times \frac{10^{3}\ \text{m}}{1\ \text{km}}$$

$$= 25\ \text{m/s}$$

$$\lambda = \frac{h}{mu} = \frac{6.626\times10^{-34}\ \text{kg·m}^2\text{/s}}{(0.0397\ \text{kg})(25\ \text{m/s})}$$

$$= 6.7\times10^{-34}\ \text{m}$$

7.5A $\Delta u = 1.00\%$ of $u = 0.0100 \times 44.7\ \text{m/s} = 0.447\ \text{m/s}$

$$\Delta x \geq \frac{h}{4\pi m \Delta u} \geq \frac{6.626\times10^{-34}\ \text{kg·m}^2\text{/s}}{4\pi(0.142\ \text{kg})(0.447\ \text{m/s})} \geq 8.31\times10^{-34}\ \text{m}$$

7.5B $\Delta u = 1\%$ of $u = 0.01(8\times10^{7}\ \text{m/s}) = 8\times10^{5}\ \text{m/s}$

$$\Delta x \geq \frac{h}{4\pi m \Delta u} \geq \frac{6.626\times10^{-34}\ \text{kg·m}^2\text{/s}}{4\pi(1.67\times10^{-27}\ \text{kg})(8\times10^{5}\ \text{m/s})} \geq 4\times10^{-14}\ \text{m}$$

7.6A $$\Delta E = (2n + 1)\frac{h^2}{8mL^2}$$

$$= (2(3) + 1)\frac{(6.626\times10^{-34}\ \text{J·s})^2}{8(9.109\times10^{-31}\ \text{kg})\left(120\ \text{nm} \times \frac{10^{-9}\ \text{m}}{1\ \text{nm}}\right)^2}$$

$$= 2.93 \times 10^{-23}\ \text{J}$$

The box with a length 120. nm is 1000 times larger than the box with a length 120. pm. Since ΔE is inversely proportional to L^2, ΔE for the larger box is 1×10^{-6} times as large as ΔE for the smaller box.

7.6B $$E = \frac{hc}{\lambda} = \frac{(6.626\times10^{-34}\ \text{J·s})(3.00\times10^{8}\ \text{m/s})}{915\ \text{nm}}\left(\frac{1\ \text{nm}}{10^{-9}\ \text{m}}\right)$$

$$= 2.17\times10^{-19}\ \text{J}$$

$$L(\text{nm}) = \sqrt{\frac{(2n + 1)h^2}{8m\Delta E}}$$

$$= \sqrt{\frac{(2(1) + 1)(6.626\times10^{-34}\ \text{J·s})^2}{8(9.109\times10^{-31}\ \text{kg})(2.17\times10^{-19}\ \text{J})}} \times \frac{1\ \text{nm}}{10^{-9}\ \text{m}}$$

$$= 0.913\ \text{nm}$$

7.7A $n = 4$, so $l = 0, 1, 2, 3$. In addition to the nine m_l values given in Sample Problem 7.6, there are those for $l = 3$: $m_l = -3, -2, -1, 0, +1, +2, +3$.

7.7B For $l = 0, 1, 2, 3, 4$, $n = 5$. In addition to the nine m_l values listed in Sample Problem 7.6 and the seven m_l values listed in Follow-up Problem 7.6A, there are those for $l = 4$: $m_l = -4, -3, -2, -1, 0, +1, +2, +3, +4$.

7.8A For $2p$: $n = 2$, $l = 1$, $m_l = -1, 0, +1$

For $5f$: $n = 5$, $l = 3$, $m_l = -3, -2, -1, 0, +1, +2, +3$

7.8B For $4d$: $n = 4$, $l = 2$, $m_l = -2, -1, 0, +1, +2$; there are five $4d$ orbitals.

For $6s$: $n = 6$, $l = 0$, $m_l = 0$; there is one $6s$ orbital.

7.9A (a) $n = 4$, $l = 1$; (b) name is $2p$; (c) name is $3d$; (d) $n = 2$, $l = 0$, $m_l = 0$

7.9B (a) For $l = 3$, the allowed values for m_l are $-3, -2, -1, 0, +1, +2, +3$, not 4.

(b) For $n = 2$, $l = 0$ or 1 only, not 2; the sublevel is $2p$, since $m_l = 1$.

(c) The value $l = 1$ indicates the p sublevel, not the s; the sublevel name is $6p$.

PROBLEMS

Problems with **colored** numbers are answered in Appendix E and worked in detail in the Student Solutions Manual. Problem sections match those in the text and give the numbers of relevant sample problems. Most offer Concept Review Questions, Skill-Building Exercises (grouped in pairs covering the same concept), and Problems in Context. Comprehensive Problems are based on material from any section or previous chapter.

The Nature of Light

(Sample Problems 7.1 and 7.2)

Concept Review Questions

7.1 In what ways are microwave and ultraviolet radiation the same? In what ways are they different?

7.2 Consider the following types of electromagnetic radiation:
(1) Microwave (2) Ultraviolet (3) Radio waves
(4) Infrared (5) X-ray (6) Visible
(a) Arrange them in order of increasing wavelength.
(b) Arrange them in order of increasing frequency.
(c) Arrange them in order of increasing energy.

7.3 Define each of the following wave phenomena, and give an example of where each occurs: (a) refraction; (b) diffraction; (c) dispersion; (d) interference.

7.4 In the 17th century, Isaac Newton proposed that light was a stream of particles. The wave-particle debate continued for over 250 years until Planck and Einstein presented their ideas. Give two pieces of evidence for the wave model and two for the particle model.

7.5 Portions of electromagnetic waves A, B, and C are represented by the following (not drawn to scale):

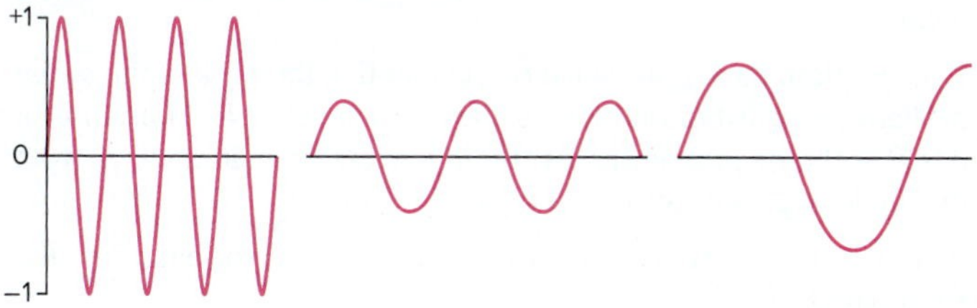

Rank them in order of (a) increasing frequency; (b) increasing energy; (c) increasing amplitude. (d) If wave B just barely fails to cause a current when shining on a metal, is wave A or C more likely to do so? (e) If wave B represents visible radiation, is wave A or C more likely to be IR radiation?

7.6 What new idea about light did Einstein use to explain the photoelectric effect? Why does the photoelectric effect exhibit a threshold frequency but not a time lag?

Skill-Building Exercises (grouped in similar pairs)

7.7 An AM station broadcasts rock music at "950 on your radio dial." Units for AM frequencies are given in kilohertz (kHz). Find the wavelength of the station's radio waves in meters (m), nanometers (nm), and angstroms (Å).

7.8 An FM station broadcasts music at 93.5 MHz (megahertz, or 10^6 Hz). Find the wavelength (in m, nm, and Å) of these waves.

7.9 A radio wave has a frequency of 3.8×10^{10} Hz. What is the energy (in J) of one photon of this radiation?

7.10 An x-ray has a wavelength of 1.3 Å. Calculate the energy (in J) of one photon of this radiation.

7.11 Rank these photons in terms of increasing energy: blue ($\lambda = 453$ nm); red ($\lambda = 660$ nm); yellow ($\lambda = 595$ nm).

7.12 Rank these photons in terms of decreasing energy: IR ($\nu = 6.5\times10^{13}$ s^{-1}); microwave ($\nu = 9.8\times10^{11}$ s^{-1}); UV ($\nu = 8.0\times10^{15}$ s^{-1}).

Problems in Context

7.13 Police often monitor traffic with "K-band" radar guns, which operate in the microwave region at 22.235 GHz (1 GHz = 10^9 Hz). Find the wavelength (in nm and Å) of this radiation.

7.14 Covalent bonds in a molecule absorb radiation in the IR region and vibrate at characteristic frequencies.
(a) The C—O bond absorbs radiation of wavelength 9.6 μm. What frequency (in s^{-1}) corresponds to that wavelength?
(b) The H—Cl bond has a frequency of vibration of 8.652×10^{13} Hz. What wavelength (in μm) corresponds to that frequency?

7.15 Cobalt-60 is a radioactive isotope used to treat cancers. A gamma ray emitted by this isotope has an energy of 1.33 MeV (million electron volts; 1 eV = 1.602×10^{-19} J). What is the frequency (in Hz) and the wavelength (in m) of this gamma ray?

7.16 (a) Ozone formation in the upper atmosphere starts when oxygen molecules absorb UV radiation with wavelengths less than or equal to 242 nm. Find the frequency and energy of the least energetic of these photons. (b) Ozone absorbs radiation with wavelengths in the range 2200–2900 Å, thus protecting organisms from this radiation. Find the frequency and energy of the most energetic of these photons.

Atomic Spectra

(Sample Problem 7.3)

Concept Review Questions

7.17 How is n_1 in the Rydberg equation (Equation 7.4) related to the quantum number n in the Bohr model?

7.18 What key assumption of Bohr's model would a Solar System model of the atom violate? What was the theoretical basis for this assumption?

7.19 Distinguish between an absorption spectrum and an emission spectrum. With which did Bohr work?

7.20 Which of these electron transitions correspond to absorption of energy and which to emission?
(a) $n = 2$ to $n = 4$ (b) $n = 3$ to $n = 1$
(c) $n = 5$ to $n = 2$ (d) $n = 3$ to $n = 4$

7.21 Why couldn't the Bohr model predict spectra for atoms other than hydrogen?

7.22 The H atom and the Be^{3+} ion each have one electron. Would you expect the Bohr model to predict their spectra accurately? Would you expect their spectra to be identical? Explain.

Skill-Building Exercises (grouped in similar pairs)

7.23 Use the Rydberg equation to find the wavelength (in nm) of the photon emitted when an electron in an H atom undergoes a transition from $n = 5$ to $n = 2$.

7.24 Use the Rydberg equation to find the wavelength (in Å) of the photon absorbed when an electron in an H atom undergoes a transition from $n = 1$ to $n = 3$.

7.25 What is the wavelength (in nm) of the least energetic spectral line in the infrared series of the H atom?

7.26 What is the wavelength (in nm) of the least energetic spectral line in the visible series of the H atom?

7.27 Calculate the energy difference (ΔE) for the transition in Problem 7.23 for 1 mol of H atoms.

7.28 Calculate the energy difference (ΔE) for the transition in Problem 7.24 for 1 mol of H atoms.

7.29 Arrange the following H atom electron transitions in order of *increasing* frequency of the photon absorbed or emitted:
(a) $n = 2$ to $n = 4$ (b) $n = 2$ to $n = 1$
(c) $n = 2$ to $n = 5$ (d) $n = 4$ to $n = 3$

7.30 Arrange the following H atom electron transitions in order of *decreasing* wavelength of the photon absorbed or emitted:
(a) $n = 2$ to $n = \infty$ (b) $n = 4$ to $n = 20$
(c) $n = 3$ to $n = 10$ (d) $n = 2$ to $n = 1$

7.31 The electron in a ground-state H atom absorbs a photon of wavelength 97.20 nm. To what energy level does it move?

7.32 An electron in the $n = 5$ level of an H atom emits a photon of wavelength 1281 nm. To what energy level does it move?

Problems in Context

7.33 In addition to continuous radiation, fluorescent lamps emit some visible lines from mercury. A prominent line has a wavelength of 436 nm. What is the energy (in J) of one photon of it?

7.34 A Bohr-model representation of the H atom is shown below with several electron transitions depicted by arrows:

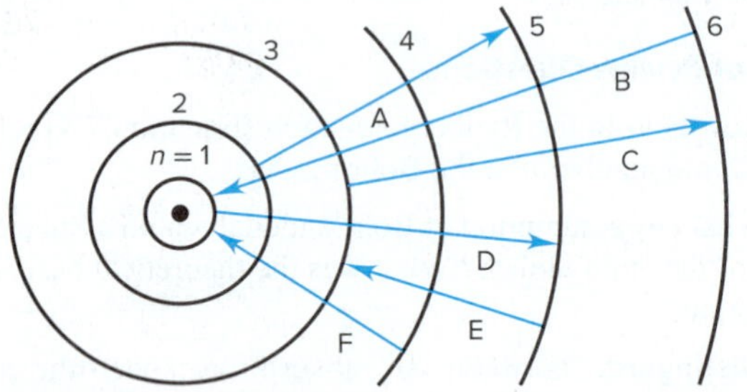

(a) Which transitions are absorptions and which are emissions?
(b) Rank the emissions in terms of increasing energy.
(c) Rank the absorptions in terms of increasing wavelength of light absorbed.

The Wave-Particle Duality of Matter and Energy
(Sample Problems 7.4 and 7.5)

Concept Review Questions

7.35 In what sense is the wave motion of a guitar string analogous to the motion of an electron in an atom?

7.36 What experimental support did de Broglie's concept receive?

7.37 If particles have wavelike motion, why don't we observe that motion in the macroscopic world?

7.38 Why can't we overcome the uncertainty predicted by Heisenberg's principle by building more precise instruments to reduce the error in measurements below the $h/4\pi$ limit?

Skill-Building Exercises (grouped in similar pairs)

7.39 A 232-lb fullback runs 40 yd at 19.8 ± 0.1 mi/h.
(a) What is his de Broglie wavelength (in meters)?
(b) What is the uncertainty in his position?

7.40 An alpha particle (mass = 6.6×10^{-24} g) emitted by a radium isotope travels at $3.4\times10^{7} \pm 0.1\times10^{7}$ mi/h.
(a) What is its de Broglie wavelength (in meters)?
(b) What is the uncertainty in its position?

7.41 How fast must a 56.5-g tennis ball travel to have a de Broglie wavelength equal to that of a photon of green light (5400 Å)?

7.42 How fast must a 142-g baseball travel to have a de Broglie wavelength equal to that of an x-ray photon with $\lambda = 100.$ pm?

7.43 A sodium flame has a characteristic yellow color due to emission of light of wavelength 589 nm. What is the mass equivalence of one photon with this wavelength ($1\text{ J} = 1\text{ kg}\cdot\text{m}^2/\text{s}^2$)?

7.44 A lithium flame has a characteristic red color due to emission of light of wavelength 671 nm. What is the mass equivalence of 1 mol of photons with this wavelength ($1\text{ J} = 1\text{ kg}\cdot\text{m}^2/\text{s}^2$)?

The Quantum-Mechanical Model of the Atom
(Sample Problems 7.7 to 7.9)

Concept Review Questions

7.45 What physical meaning is attributed to ψ^2?

7.46 What does "electron density in a tiny volume of space" mean?

7.47 Explain the significance of the fact that the peak in the radial probability distribution plot for the $n = 1$ level of an H atom is at 0.529 Å. Is the probability of finding an electron at 0.529 Å from the nucleus greater for the 1*s* or the 2*s* orbital?

7.48 What feature of an orbital is related to each of the following?
(a) Principal quantum number (n)
(b) Angular momentum quantum number (l)
(c) Magnetic quantum number (m_l)

Skill-Building Exercises (grouped in similar pairs)

7.49 How many orbitals in an atom can have each of the following designations: (a) 1*s*; (b) 4*d*; (c) 3*p*; (d) $n = 3$?

7.50 How many orbitals in an atom can have each of the following designations: (a) 5*f*; (b) 4*p*; (c) 5*d*; (d) $n = 2$?

7.51 Give all possible m_l values for orbitals that have each of the following: (a) $l = 2$; (b) $n = 1$; (c) $n = 4$, $l = 3$.

7.52 Give all possible m_l values for orbitals that have each of the following: (a) $l = 3$; (b) $n = 2$; (c) $n = 6$, $l = 1$.

7.53 Draw 90% probability contours (with axes) for each of the following orbitals: (a) *s*; (b) p_x.

7.54 Draw 90% probability contours (with axes) for each of the following orbitals: (a) p_z; (b) d_{xy}.

7.55 For each of the following, give the sublevel designation, the allowable m_l values, and the number of orbitals:
(a) $n = 4, l = 2$ (b) $n = 5, l = 1$
(c) $n = 6, l = 3$

7.56 For each of the following, give the sublevel designation, the allowable m_l values, and the number of orbitals:
(a) $n = 2, l = 0$ (b) $n = 3, l = 2$
(c) $n = 5, l = 1$

7.57 For each of the following sublevels, give the n and l values and the number of orbitals: (a) 5*s*; (b) 3*p*; (c) 4*f*.

7.58 For each of the following sublevels, give the n and l values and the number of orbitals: (a) 6*g*; (b) 4*s*; (c) 3*d*.

7.59 Are the following combinations allowed? If not, show two ways to correct them:
(a) $n = 2; l = 0; m_l = -1$ (b) $n = 4; l = 3; m_l = -1$
(c) $n = 3; l = 1; m_l = 0$ (d) $n = 5; l = 2; m_l = +3$

7.60 Are the following combinations allowed? If not, show two ways to correct them:
(a) $n = 1; l = 0; m_l = 0$ (b) $n = 2; l = 2; m_l = +1$
(c) $n = 7; l = 1; m_l = +2$ (d) $n = 3; l = 1; m_l = -2$

Comprehensive Problems

7.61 The orange color of carrots and orange peel is due mostly to β-carotene, an organic compound insoluble in water but soluble in benzene and chloroform. Describe an experiment to determine the concentration of β-carotene in the oil from orange peel.

7.62 The quantum-mechanical treatment of the H atom gives the energy, E, of the electron as a function of n:

$$E = -\frac{h^2}{8\pi^2 m_e a_0^2 n^2} \quad (n = 1, 2, 3, \ldots)$$

where h is Planck's constant, m_e is the electron's mass, and a_0 is 52.92×10^{-12} m.
(a) Write the expression in the form $E = -(\text{constant})(1/n^2)$, evaluate the constant (in J), and compare it with the corresponding expression from Bohr's theory.
(b) Use the expression from part (a) to find ΔE between $n = 2$ and $n = 3$.
(c) Calculate the wavelength of the photon that corresponds to this energy change.

7.63 The photoelectric effect is illustrated in a plot of the kinetic energies of electrons ejected from the surface of potassium metal or silver metal at different frequencies of incident light.

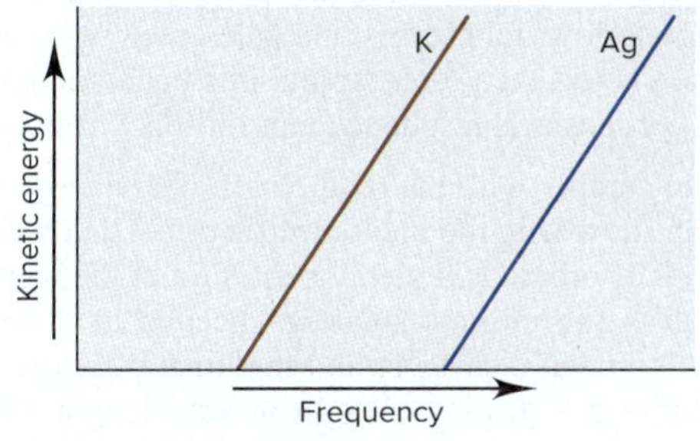

(a) Why don't the lines begin at the origin? (b) Why don't the lines begin at the same point? (c) From which metal will light of shorter wavelength eject an electron? (d) Why are the slopes equal?

7.64 The optic nerve needs a minimum of 2.0×10^{-17} J of energy to trigger a series of impulses that eventually reach the brain. (a) How many photons of red light (700. nm) are needed? (b) How many photons of blue light (475 nm)?

7.65 One reason carbon monoxide (CO) is toxic is that it binds to the blood protein hemoglobin more strongly than oxygen does. The bond between hemoglobin and CO absorbs radiation of 1953 cm^{-1}. (The unit is the reciprocal of the wavelength in centimeters.) Calculate the wavelength (in nm and Å) and the frequency (in Hz) of the absorbed radiation.

7.66 A metal ion M^{n+} has a single electron. The highest energy line in its emission spectrum has a frequency of 2.961×10^{16} Hz. Identify the ion.

7.67 Compare the wavelengths of an electron (mass = 9.11×10^{-31} kg) and a proton (mass = 1.67×10^{-27} kg), each having (a) a speed of 3.4×10^6 m/s; (b) a kinetic energy of 2.7×10^{-15} J.

7.68 Five lines in the H atom spectrum have these wavelengths (in Å): (a) 1212.7; (b) 4340.5; (c) 4861.3; (d) 6562.8; (e) 10,938. Three lines result from transitions to $n_{final} = 2$ (visible series). The other two result from transitions in different series, one with $n_{final} = 1$ and the other with $n_{final} = 3$. Identify $n_{initial}$ for each line.

7.69 In his explanation of the threshold frequency in the photoelectric effect, Einstein reasoned that the absorbed photon must have a minimum energy to dislodge an electron from the metal surface. This energy is called the *work function* (ϕ) of the metal. What is the longest wavelength of radiation (in nm) that could cause the photoelectric effect in each of these metals: (a) calcium, $\phi = 4.60\times10^{-19}$ J; (b) titanium, $\phi = 6.94\times10^{-19}$ J; (c) sodium, $\phi = 4.41\times10^{-19}$ J?

7.70 Refractometry is based on the difference in the speed of light through a substance (v) and through a vacuum (c). In the procedure, light of known wavelength passes through a fixed thickness of the substance at a known temperature. The index of refraction equals c/v. Using yellow light ($\lambda = 589$ nm) at 20°C, for example, the index of refraction of water is 1.33 and that of diamond is 2.42. Calculate the speed of light in (a) water and (b) diamond.

7.71 A laser (*l*ight *a*mplification by *s*timulated *e*mission of *r*adiation) provides nearly monochromatic high-intensity light. Lasers are used in eye surgery, CD/DVD players, basic research, and many other areas. Some dye lasers can be "tuned" to emit a desired wavelength. Fill in the blanks in the following table of the properties of some common types of lasers:

Type	λ (nm)	ν (s^{-1})	*E* (J)	Color
He-Ne	632.8	?	?	?
Ar	?	6.148×10^{14}	?	?
Ar-Kr	?	?	3.499×10^{-19}	?
Dye	663.7	?	?	?

7.72 The following combinations are not allowed. If n and m_l are correct, change the l value to create an allowable combination:
(a) $n = 3; l = 0; m_l = -1$ (b) $n = 3; l = 3; m_l = +1$
(c) $n = 7; l = 2; m_l = +3$ (d) $n = 4; l = 1; m_l = -2$

7.73 A ground-state H atom absorbs a photon of wavelength 94.91 nm, and its electron attains a higher energy level. The

atom then emits two photons: one of wavelength 1281 nm to reach an intermediate energy level, and a second to return to the ground state.
(a) What higher level did the electron reach?
(b) What intermediate level did the electron reach?
(c) What was the wavelength of the second photon emitted?

7.74 Ground-state ionization energies of some one-electron species are

$H = 1.31\times10^3$ kJ/mol $He^+ = 5.24\times10^3$ kJ/mol
$Li^{2+} = 1.18\times10^4$ kJ/mol

(a) Write a general expression for the ionization energy of any one-electron species.
(b) Use your expression to calculate the ionization energy of B^{4+}.
(c) What is the minimum wavelength required to remove the electron from the $n = 3$ level of He^+?
(d) What is the minimum wavelength required to remove the electron from the $n = 2$ level of Be^{3+}?

7.75 Use the relative size of the 3*s* orbital below to answer the following questions about orbitals A–D.

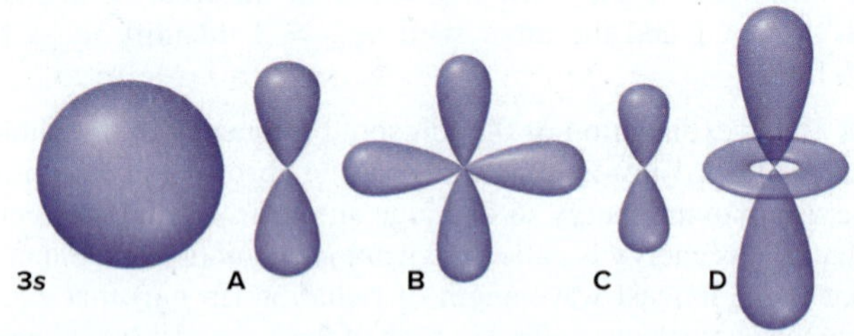

(a) Which orbital has the highest value of n? (b) Which orbital(s) have a value of $l = 1$? $l = 2$? (c) How many other orbitals with the same value of n have the same shape as orbital B? Orbital C? (d) Which orbital has the highest energy? Lowest energy?

7.76 In the course of developing his model, Bohr arrived at the following formula for the radius of the electron's orbit: $r_n = n^2h^2\varepsilon_0/\pi m_e e^2$, where m_e is the electron's mass, e is its charge, and ε_0 is a constant related to charge attraction in a vacuum. Given that $m_e = 9.109\times10^{-31}$ kg, $e = 1.602\times10^{-19}$ C, and $\varepsilon_0 = 8.854\times10^{-12}$ $C^2/J\cdot m$, calculate the following:
(a) The radius of the first ($n = 1$) orbit in the H atom
(b) The radius of the tenth ($n = 10$) orbit in the H atom

7.77 (a) Find the Bohr radius of an electron in the $n = 3$ orbit of an H atom (see Problem 7.76). (b) What is the energy (in J) of the atom in part (a)? (c) What is the energy of an Li^{2+} ion with its electron in the $n = 3$ orbit? (d) Why are the answers to parts (b) and (c) different?

7.78 Enormous numbers of microwave photons are needed to warm macroscopic samples of matter. A portion of soup containing 252 g of water is heated in a microwave oven from 20.°C to 98°C, with radiation of wavelength 1.55×10^{-2} m. How many photons are absorbed by the water in the soup?

7.79 The quantum-mechanical treatment of the hydrogen atom gives this expression for the wave function, ψ, of the 1*s* orbital:

$$\Psi = \frac{1}{\sqrt{\pi}}\left(\frac{1}{a_0}\right)^{3/2} e^{-r/a_0}$$

where r is the distance from the nucleus and a_0 is 52.92 pm. The probability of finding the electron in a tiny volume at distance r from the nucleus is proportional to ψ^2. The total probability of finding the electron at all points at distance r from the nucleus is proportional to $4\pi r^2\psi^2$. Calculate the values (to three significant figures) of ψ, ψ^2, and $4\pi r^2\psi^2$ to fill in the following table and sketch a plot of each set of values versus r. Compare the latter two plots with those in Figure 7.17A.

r (pm)	Ψ ($pm^{-3/2}$)	Ψ^2 (pm^{-3})	$4\pi r^2\Psi^2$ (pm^{-1})
0			
50			
100			
200			

7.80 Lines in one spectral series can overlap lines in another.
(a) Does the range of wavelengths in the $n_1 = 1$ series for the H atom overlap the range in the $n_1 = 2$ series?
(b) Does the range in the $n_1 = 3$ series overlap the range in the $n_1 = 4$ series?
(c) How many lines in the $n_1 = 4$ series lie in the range of the $n_1 = 5$ series?
(d) What does this overlap imply about the H atom line spectrum at longer wavelengths?

7.81 The following values are the only energy levels of a hypothetical one-electron atom:

$E_6 = -2\times10^{-19}$ J $E_5 = -7\times10^{-19}$ J
$E_4 = -11\times10^{-19}$ J $E_3 = -15\times10^{-19}$ J
$E_2 = -17\times10^{-19}$ J $E_1 = -20\times10^{-19}$ J

(a) If the electron were in the $n = 3$ level, what would be the highest frequency (and minimum wavelength) of radiation that could be emitted?
(b) What is the ionization energy (in kJ/mol) of the atom in its ground state?
(c) If the electron were in the $n = 4$ level, what would be the shortest wavelength (in nm) of radiation that could be absorbed without causing ionization?

7.82 Photoelectron spectroscopy applies the principle of the photoelectric effect to study orbital energies of atoms and molecules. High-energy radiation (usually UV or x-ray) is absorbed by a sample and an electron is ejected. The orbital energy can be calculated from the known energy of the radiation and the measured energy of the electron lost. The following energy differences were determined for several electron transitions:

$\Delta E_{2\longrightarrow1} = 4.098\times10^{-17}$ J $\Delta E_{3\longrightarrow1} = 4.854\times10^{-17}$ J
$\Delta E_{5\longrightarrow1} = 5.242\times10^{-17}$ J $\Delta E_{4\longrightarrow2} = 1.024\times10^{-17}$ J

Calculate ΔE and λ of a photon emitted in the following transitions: (a) level 3 → 2; (b) level 4 → 1; (c) level 5 → 4.

7.83 Horticulturists know that, for many plants, dark green leaves are associated with low light levels and pale green with high levels. (a) Use the photon theory to explain this behavior. (b) What change in leaf composition might account for this difference?

7.84 In order to comply with the requirement that energy be conserved, Einstein showed in the photoelectric effect that the energy of a photon ($h\nu$) absorbed by a metal is the sum of the work function (ϕ), which is the minimum energy needed to dislodge an electron from the metal's surface, and the kinetic energy (E_k) of the electron: $h\nu = \phi + E_k$. When light of wavelength 358.1 nm falls on the surface of potassium metal, the speed (u) of the dislodged electron is 6.40×10^5 m/s.
(a) What is E_k ($\frac{1}{2}mu^2$) of the dislodged electron?
(b) What is ϕ (in J) of potassium?

7.85 For any microscope, the size of the smallest observable object is one-half the wavelength of the radiation used. For example, the smallest object observable with 400-nm light is 2×10^{-7} m. (a) What is the smallest observable object for an electron microscope using electrons moving at 5.5×10^4 m/s? (b) At 3.0×10^7 m/s?

7.86 In fireworks, the heat of the reaction of an oxidizing agent, such as $KClO_4$, with an organic compound excites certain salts, which emit specific colors. Strontium salts have an intense emission at 641 nm, and barium salts have one at 493 nm.
(a) What colors do these emissions produce?
(b) What is the energy (in kJ) of these emissions for 5.00 g each of the chloride salts of Sr and Ba? (Assume that all the heat produced is converted to emitted light.)

7.87 Atomic hydrogen produces several series of spectral lines. Each series fits the Rydberg equation with its own particular n_1 value. Calculate the value of n_1 (by trial and error if necessary) that would produce a series of lines in which:
(a) The *highest* energy line has a wavelength of 3282 nm.
(b) The *lowest* energy line has a wavelength of 7460 nm.

7.88 Fish-liver oil is a good source of vitamin A, whose concentration is measured spectrometrically at a wavelength of 329 nm.
(a) Suggest a reason for using this wavelength.
(b) In what region of the spectrum does this wavelength lie?
(c) When 0.1232 g of fish-liver oil is dissolved in 500. mL of solvent, the absorbance is 0.724 units. When 1.67×10^{-3} g of vitamin A is dissolved in 250. mL of solvent, the absorbance is 1.018 units. Calculate the vitamin A concentration in the fish-liver oil.

7.89 Many calculators use photocells as their energy source. Find the maximum wavelength needed to remove an electron from silver ($\phi = 7.59\times10^{-19}$ J). Is silver a good choice for a photocell that uses visible light? [The concept of the work function (ϕ) is explained in Problem 7.69.]

7.90 In a game of "Clue," Ms. White is killed in the conservatory. A spectrometer in each room records who is present to help find the murderer. For example, if someone wearing yellow is in a room, light at 580 nm is reflected. The suspects are Col. Mustard, Prof. Plum, Mr. Green, Ms. Peacock (blue), and Ms. Scarlet. At the time of the murder, the spectrometer in the dining room shows a reflection at 520 nm, those in the lounge and the study record lower frequencies, and the one in the library records the shortest possible wavelength. Who killed Ms. White? Explain.

7.91 Technetium (Tc; $Z = 43$) is a synthetic element used as a radioactive tracer in medical studies. A Tc atom emits a beta particle (electron) with a kinetic energy ($E_k = \frac{1}{2}mv^2$) of 4.71×10^{-15} J. What is the de Broglie wavelength of this electron?

7.92 Electric power is measured in watts (1 W = 1 J/s). About 95% of the power output of an incandescent bulb is converted to heat and 5% to light. If 10% of that light shines on your chemistry textbook, how many photons per second shine on the book from a 75-W bulb? (Assume that the photons have a wavelength of 550 nm.)

7.93 The flame tests for sodium and potassium are based on the emissions at 589 nm and 404 nm, respectively. When both elements are present, the Na^+ emission is so strong that the K^+ emission can be seen only by looking through a cobalt-glass filter.
(a) What are the colors of the Na^+ and K^+ emissions?
(b) What does the cobalt-glass filter do?
(c) Why is $KClO_4$ used as an oxidizing agent in fireworks rather than $NaClO_4$?

7.94 The net change during photosynthesis involves CO_2 and H_2O forming glucose ($C_6H_{12}O_6$) and O_2. Chlorophyll absorbs light in the 600–700 nm region.
(a) Write a balanced thermochemical equation for formation of 1.00 mol of glucose.
(b) What is the minimum number of photons with $\lambda = 680.$ nm needed to form 1.00 mol of glucose?

7.95 Only certain transitions are allowed from one energy level to another. In one-electron species, the change in l for an allowed transition is ±1. For example, a $3p$ electron can move to a $2s$ orbital but not to a $2p$. Thus, in the UV series, where $n_{final} = 1$, allowed transitions can start in a p orbital ($l = 1$) of $n = 2$ or higher, not in an s ($l = 0$) or d ($l = 2$) orbital of $n = 2$ or higher. From what orbital do each of the allowed transitions start for the first four emission lines in the visible series ($n_{final} = 2$)?

7.96 The discharge of phosphate in detergents to the environment has led to imbalances in the life cycle of freshwater lakes. A chemist uses a spectrometric method to measure total phosphate and obtains the following data for known standards:

Absorbance (880 nm)	Concentration (mol/L)
0	0.0×10^{-5}
0.10	2.5×10^{-5}
0.16	3.2×10^{-5}
0.20	4.4×10^{-5}
0.25	5.6×10^{-5}
0.38	8.4×10^{-5}
0.48	10.5×10^{-5}
0.62	13.8×10^{-5}
0.76	17.0×10^{-5}
0.88	19.4×10^{-5}

(a) Draw a curve of absorbance vs. phosphate concentration.
(b) If a sample of lake water has an absorbance of 0.55, what is its phosphate concentration?

7.97 Consider a particle in a one-dimensional box whose length is 0.450 nm. Calculate the wavelength of light required to move the particle from the $n = 2$ to the $n = 3$ energy levels in the box. Assume that the mass of the particle equals the mass of an electron.

7.98 A wavelength of 6545 nm is required to promote an electron in a one-dimensional box from the $n = 2$ to the $n = 3$ energy level. Calculate the length of the box.

7.99 (a) Calculate the difference in energy between the $n = 2$ and $n = 1$ states of an electron in a one-dimensional box with a length of 0.10 nm. (b) Calculate the difference in energy between the $n = 2$ and $n = 1$ states for an oxygen molecule in a one-dimensional box with a length of 10. cm. (c) What do the different values of ΔE for the electron and the O_2 molecule indicate?

7.100 Calculate the energy of the four lowest energy levels of an electron in a one-dimensional box with a length of 1.10 nm.

7.101 Consider three different one-dimensional boxes with the following lengths: 10^{-6} m, 10^{-10} m, and 10^{-4} m. In which box does an electron have the lowest ground state energy? In which does it have the highest ground state energy? Explain your answers.

8 Electron Configuration and Chemical Periodicity

8.1 Characteristics of Many-Electron Atoms
The Electron-Spin Quantum Number
The Exclusion Principle
Electrostatic Effects and Energy-Level Splitting

8.2 The Quantum-Mechanical Model and the Periodic Table
Building Up Period 1
Building Up Period 2
Building Up Period 3
Building Up Period 4
General Principles of Electron Configurations
Transition and Inner Transition Elements
Electron Configurations Within Groups

8.3 Trends in Three Atomic Properties
Atomic Size
Ionization Energy
Electron Affinity

8.4 Atomic Properties and Chemical Reactivity
Trends in Metallic Behavior
Properties of Monatomic Ions

Source: © Mikhail Tolstoy/Alamy RF

Concepts and Skills to Review Before You Study This Chapter

- format of the periodic table (Section 2.6)
- characteristics of metals and nonmetals (Section 2.6)
- attractions, repulsions, and Coulomb's law (Section 2.7)
- characteristics of acids and bases (Section 4.4)
- rules for assigning quantum numbers (Section 7.4)

Every year, we know that spring will follow winter because the seasons are one of the recurring patterns of nature. Day and night, the rhythm of a heartbeat, tides, and the phases of the moon are other examples of nature's patterns that make it possible for us to know what to expect. Elements also exhibit recurring patterns in their properties, patterns that we can use to make predictions about their behavior.

The outpouring of scientific creativity by early 20th-century physicists that led to the new quantum-mechanical model of the atom was preceded by countless hours of laboratory work by 19th-century chemists who were exploring the nature of electrolytes, the kinetic-molecular theory, and chemical thermodynamics. The fields of organic chemistry and biochemistry were born in the 19th century, as were the fertilizer, explosives, glassmaking, soapmaking, bleaching, and dyestuff industries. And, for the first time, chemistry became a subject taught in universities in Europe and America.

Condensed from all these efforts was an enormous body of facts about the elements, which became organized into the periodic table:

- *The original periodic table.* In 1870, the Russian chemist Dmitri Mendeleev arranged the 65 elements known at the time into a table and summarized their behavior in the **periodic law:** when arranged by *atomic mass,* the elements exhibit a periodic recurrence of similar properties. Mendeleev left blank spaces in his table and was even able to *predict* the properties of several elements, for example, germanium, that were not discovered until later.
- *The modern periodic table.* Today's periodic table *(inside front cover)* includes 53 elements not known in 1870 and, most importantly, arranges the elements by *atomic number* (number of protons) *not* atomic mass. This change is based on the work of the British physicist Henry G. J. Moseley, who obtained x-ray spectra of various metals by bombarding the metals with electrons. Moseley found that the largest x-ray peak for each metal was related to the nuclear charge, which increased by 1 for each successive element.

The great test for the new atomic model was to answer one of the central questions in chemistry: *why* do the elements behave as they do? Or, rephrasing to fit the main topic of this chapter, how does the **electron configuration** of an element—*the distribution of electrons within the levels and sublevels of its atoms*—relate to its chemical and physical properties?

IN THIS CHAPTER . . . *We explore recurring patterns of electron distributions in atoms to see how they account for the recurring behavior of the elements.*

- We describe a new quantum number and a restriction on the number of electrons in an orbital, which allows us to define a unique set of quantum numbers for each electron in an atom of any element.
- We explore electrostatic effects that lead to splitting of atomic energy levels into sublevels and give rise to the order in which orbitals fill with electrons. This filling order correlates with the order of elements in the periodic table.
- We examine the reasons for periodic trends in three atomic properties.
- We see how these trends account for chemical reactivity with regard to metallic and redox behavior, oxide acidity, ion formation, and magnetic behavior.

8.1 CHARACTERISTICS OF MANY-ELECTRON ATOMS

The Schrödinger equation (introduced in Chapter 7) does not give *exact* solutions for the energy levels of *many-electron atoms,* those with more than one electron—that is, all atoms except hydrogen. However, unlike the Bohr model, it gives excellent *approximate* solutions. Three additional features become important in many-electron atoms: (1) a fourth quantum number, (2) a limit on the number of electrons in an orbital, and (3) a splitting of energy levels into sublevels.

The Electron-Spin Quantum Number

The three quantum numbers n, l, and m_l describe the size (energy), shape, and orientation in space, respectively, of an atomic orbital. A fourth quantum number describes a property called *spin,* which is a property of the electron and not the orbital.

When a beam of atoms that have one or more lone electrons passes through a nonuniform magnetic field (created by magnet faces with different shapes), it splits into two beams. Half of the electrons are *attracted* by the large external magnetic field, while the other half are *repelled* by it; Figure 8.1 shows this for a beam of H atoms.

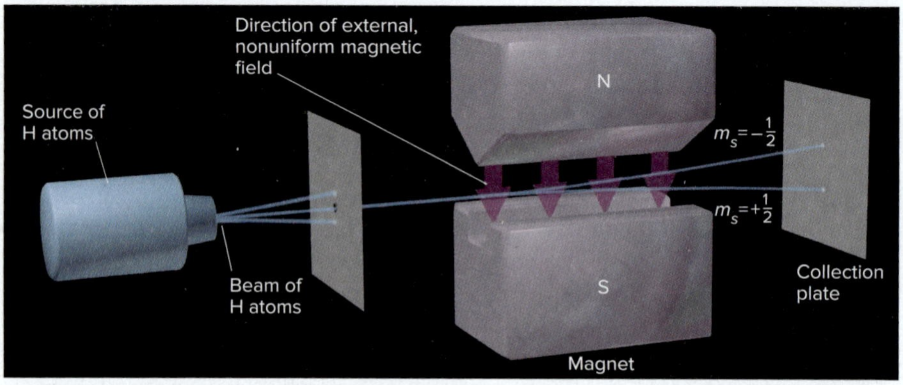

Figure 8.1 The effect of electron spin. A beam of H atoms splits because each atom's electron has one of the two possible values of spin.

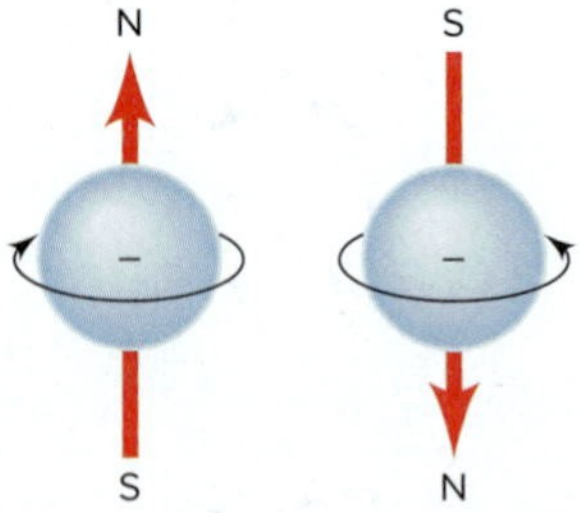

Figure 8.2 The direction of electron spin. The electron spins in one of two opposite directions, generating a magnetic field whose direction depends on the direction of spin.

This observation is explained by the idea that an electron spins on its axis in one of two opposite directions, clockwise or counterclockwise; a spinning charge generates a tiny magnetic field whose direction depends on the direction of the spin (Figure 8.2). Thus, there are two magnetic fields with opposing directions, resulting in the two beams seen in Figure 8.1. Corresponding to the two directions of the electron's magnetic field, the **spin quantum number (m_s)** has two possible values, $+\frac{1}{2}$ or $-\frac{1}{2}$.

Thus, *each electron in an atom is described completely by a set of* ***four*** *quantum numbers: the first three describe its orbital, and the fourth describes its spin.* The quantum numbers are summarized in Table 8.1.

Now we can write a set of four quantum numbers for any electron in the ground state of any atom. For example, the lone electron in hydrogen (H; $Z = 1$), which occupies the 1*s* orbital, has the following quantum numbers: $n = 1$, $l = 0$, $m_l = 0$,

Table 8.1 Summary of Quantum Numbers of Electrons in Atoms

Name	Symbol	Permitted Values	Property
Principal	n	Positive integers (1, 2, 3, . . .)	Orbital energy (size)
Angular momentum	l	Integers from 0 to $n - 1$	Orbital shape (The l values 0, 1, 2, and 3 correspond to *s, p, d,* and *f* orbitals, respectively.)
Magnetic	m_l	Integers from $-l$ to 0 to $+l$	Orbital orientation
Spin	m_s	$+\frac{1}{2}$ or $-\frac{1}{2}$	Direction of e^- spin

and $m_s = +\frac{1}{2}$. (The spin quantum number could just as well be $-\frac{1}{2}$, but by convention, we assign $+\frac{1}{2}$ to the first electron in an orbital.)

The Exclusion Principle

The element after hydrogen is helium (He; $Z = 2$), the first with atoms having more than one electron. The first electron in the He ground state has the same set of quantum numbers as the electron in the H atom, but the second He electron does not. Based on observations of excited states, the Austrian physicist Wolfgang Pauli formulated the **exclusion principle:** *no two electrons in the same atom can have the same four quantum numbers.* Therefore, the second He electron occupies the same 1*s* orbital as the first but has an opposite spin: $n = 1$, $l = 0$, $m_l = 0$, and $m_s = -\frac{1}{2}$. ›

The major consequence of the exclusion principle is that *an atomic orbital can hold a maximum of two electrons, which must have opposing spins.* Two electrons can occupy the same orbital, thus having the same values of *n*, *l* and m_l, but each electron must have a unique value of m_s. Since there are only two values of m_s, a third electron cannot be added to that orbital without repeating an m_s value. We say that the 1*s* orbital in He, with its two electrons, is *filled* and that the electrons have *paired spins.* Thus, a beam of He atoms is not split in an experiment like that in Figure 8.1.

Baseball Quantum Numbers

The unique set of quantum numbers that describes an electron is analogous to the unique location of a box seat at a baseball game. The stadium (atom) is divided into section (*n*, level), box (*l*, sublevel), row (m_l, orbital), and seat (m_s, spin). Only one person (electron) can have this particular set of stadium "quantum numbers." While many people can occupy the same section, box, and row in the stadium, only one person can occupy a particular seat.

Electrostatic Effects and Energy-Level Splitting

Electrostatic effects—attraction of opposite charges and repulsion of like charges—play a major role in determining the energy states of many-electron atoms. Unlike the H atom, in which there is only the attraction between nucleus and electron and the energy state is determined *only* by the *n* value, the energy states of many-electron atoms are also affected by electron-electron repulsions. You'll see shortly how these additional interactions give rise to *the splitting of energy levels into sublevels of differing energies: the energy of an orbital in a many-electron atom depends mostly on its n value (size) and to a lesser extent on its l value (shape).*

Our first encounter with energy-level splitting occurs with lithium (Li; $Z = 3$). The first two electrons of Li fill its 1*s* orbital, so the third Li electron must go into the $n = 2$ level. But, this level has 2*s and* 2*p* sublevels: which does the third electron enter? For reasons we discuss below, the 2*s* is lower in energy than the 2*p*, so the ground state of Li has its third electron in the 2*s*.

This energy difference arises from three factors—*nuclear attraction, electron repulsions,* and *orbital shape* (i.e., radial probability distribution). Their interplay leads to two phenomena—*shielding* and *penetration*—that occur in all atoms *except* hydrogen. [In the following discussion, keep in mind that more energy is needed to remove an electron from a more stable (lower energy) sublevel than from a less stable (higher energy) sublevel.]

The Effect of Nuclear Charge (*Z*) on Sublevel Energy Higher charges interact more strongly than lower charges (Coulomb's law, Section 2.7). Therefore, *a higher nuclear charge (more protons in the nucleus) increases nucleus-electron attractions and, thus, lowers sublevel energy (stabilizes the atom).* We see this effect by comparing the 1*s* sublevel energies of three species with one electron: the H atom ($Z = 1$), the He^+ ion ($Z = 2$), and the Li^{2+} ion ($Z = 3$). Figure 8.3 shows that the 1*s* sublevel in H (one proton attracting one electron) is the least stable (highest energy), so the least energy is needed to remove its electron; and the 1*s* sublevel in Li^{2+} (three protons attracting one electron) is the most stable, so the most energy is needed to remove its electron.

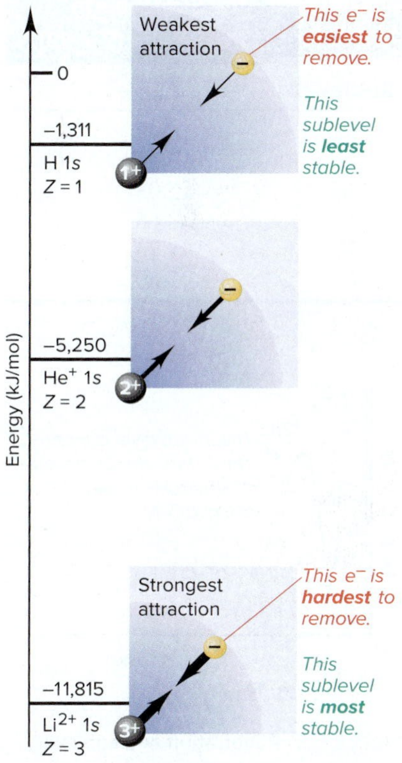

Figure 8.3 The effect of nuclear charge on sublevel energy. Greater nuclear charge lowers sublevel energy (to a more negative number), which makes the electron harder to remove. (The strength of attraction is indicated by the thickness of the black arrows.)

Shielding: The Effect of Electron Repulsions on Sublevel Energy In many-electron atoms, each electron "feels" not only the attraction to the nucleus but also repulsions from other electrons. Repulsions counteract the nuclear attraction somewhat, making each electron easier to remove by, in effect, helping to push it away. We speak of each electron "shielding" the other electrons to some extent from the nuclear charge. **Shielding** (also called *screening*) reduces the full nuclear charge to an **effective nuclear charge (Z_{eff}),** the nuclear charge an electron *actually experiences, and this lower nuclear charge makes the electron easier to remove.*

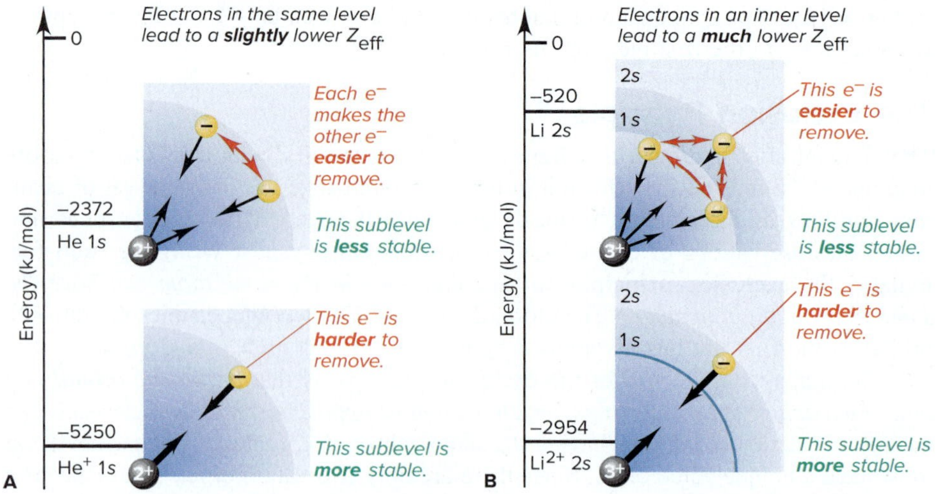

Figure 8.4 Shielding and energy levels. **A,** Within an energy level, each electron shields *(red arrows)* other electrons from the full nuclear charge *(black arrows),* so they experience a lower Z_{eff}. **B,** Inner electrons shield outer electrons *much* more effectively than electrons in the same energy level shield each other.

1. *Shielding by other electrons in a given energy level.* Electrons in the *same* energy level shield each other somewhat. Compare the He atom and He^+ ion: both have a 2+ nuclear charge, but He has two electrons in the 1*s* sublevel and He^+ has only one (Figure 8.4A). It takes less than half as much energy to remove an electron from He (2372 kJ/mol) than from He^+ (5250 kJ/mol) because the second electron in He repels the first, in effect causing a lower Z_{eff}.

2. *Shielding by electrons in inner energy levels.* Because inner electrons spend nearly all their time *between* the outer electrons and the nucleus, they cause a *much* lower Z_{eff} than do electrons in the same level. We can see this by comparing two atomic systems with the same nucleus, one *with* inner electrons and the other *without.* The ground-state Li atom has two inner (1*s*) electrons and one outer (2*s*) electron, while the Li^{2+} ion has only one electron, which occupies the 2*s* orbital in the first excited state (Figure 8.4B). It takes about one-sixth as much energy to remove the 2*s* electron from the Li atom (520 kJ/mol) as it takes to remove it from the Li^{2+} ion (2954 kJ/mol), because the *inner electrons shield very effectively*.

Table 8.2 Values of Z_{eff} for the Sublevels in Potassium ($Z = 19$)

Sublevel	Z_{eff}
1*s*	18.49
2*s*	13.01
2*p*	15.03
3*s*	8.68
3*p*	7.73
4*s*	3.50

Table 8.2 lists the values of Z_{eff} for the electrons in a potassium atom. Although K ($Z = 19$) has a nuclear charge of 19+, all of its electrons have $Z_{eff} < 19$. The two innermost (1*s*) electrons have a slightly lower Z_{eff} of 18.49 because they shield each other. The outermost (4*s*) electron has a significantly lower Z_{eff} of 3.50, as a result of the effective shielding by the 18 inner electrons.

Penetration: The Effect of Orbital Shape on Sublevel Energy To see why the third Li electron occupies the 2*s* sublevel rather than the 2*p*, we have to consider orbital shapes, that is, radial probability distributions (Figure 8.5). A 2*p* orbital *(orange curve)* is slightly closer to the nucleus, on average, than the major portion of the 2*s* orbital *(blue curve).* But a small portion of the 2*s* radial probability distribution peaks within the 1*s* region. Thus, an electron in the 2*s* orbital spends part of its time "penetrating" very close to the nucleus. **Penetration** has two effects:

- It *increases the nuclear attraction* for a 2*s* electron over that for a 2*p* electron.
- It *decreases the shielding* of a 2*s* electron by the 1*s* electrons.

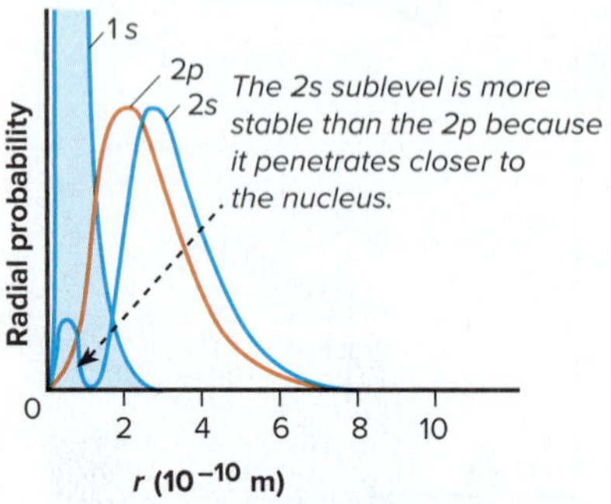

Figure 8.5 Penetration and sublevel energy.

Thus, since it takes more energy to remove a 2*s* electron (520 kJ/mol) than a 2*p* (341 kJ/mol), the 2*s* sublevel is lower in energy than the 2*p*.

Splitting of Levels into Sublevels In general, *penetration and the resulting effects on shielding split an energy level into sublevels of differing energy.* The lower the *l* value of a sublevel, the more its electrons penetrate, and so the greater their attraction to the nucleus. Therefore, *for a given n value, a lower l value indicates a more stable (lower energy) sublevel:*

$$\text{Order of sublevel energies: } s < p < d < f \qquad \textbf{(8.1)}$$

Thus, the 2*s* ($l = 0$) is lower in energy than the 2*p* ($l = 1$), the 3*p* ($l = 1$) is lower than the 3*d* ($l = 2$), and so forth.

Figure 8.6 shows the general energy order of levels (*n* value) and how they are split into sublevels (*l* values) of differing energies. (Compare this with the H atom energy levels in Figure 7.21.) Next, we'll use this energy order to construct a periodic table of ground-state atoms.

› Summary of Section 8.1

- Identifying electrons in many-electron atoms requires four quantum numbers: three (n, l, m_l) describe the orbital, and a fourth (m_s) describes the electron's spin.
- The exclusion principle requires each electron to have a unique set of four quantum numbers; therefore, an orbital can hold no more than two electrons, and their spins must be paired (opposite).
- Electrostatic interactions determine sublevel energies as follows:
  1. Greater nuclear charge lowers sublevel energy, making electrons harder to remove.
  2. Electron-electron repulsions raise sublevel energy, making electrons easier to remove. Repulsions shield electrons from the full nuclear charge, reducing it to an effective nuclear charge, Z_{eff}. Inner electrons shield outer electrons very effectively.
  3. Penetration makes an electron harder to remove because nuclear attraction increases and shielding decreases. As a result, an energy level is split into sublevels with the energy order $s < p < d < f$.

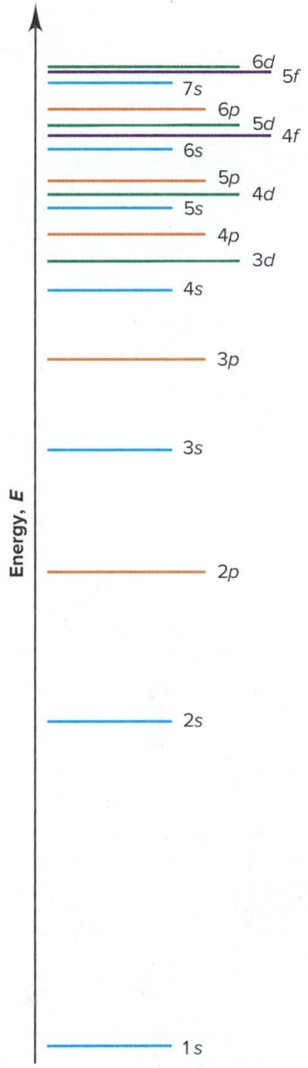

Figure 8.6 Order for filling energy sublevels with electrons. In general, energies of sublevels increase with the principal quantum number *n* ($1 < 2 < 3$, etc.) and the angular momentum quantum number *l* ($s < p < d < f$). As *n* increases, some sublevels overlap; for example, the 4*s* sublevel is lower in energy than the 3*d*. (The color of each line indicates the sublevel type.)

8.2 THE QUANTUM-MECHANICAL MODEL AND THE PERIODIC TABLE

Quantum mechanics provides the theoretical foundation for the experimentally based periodic table. In this section, we fill the table by determining the *ground-state* electron configuration of each element—*the lowest energy distribution of electrons in the sublevels of its atoms*. Note especially the *recurring pattern in electron configurations, which is the basis for recurring patterns in chemical behavior.*

A useful way to determine electron configurations is based on the **aufbau principle** (German *aufbauen,* "to build up"). We start at the beginning of the periodic table and add one proton to the nucleus and one electron to the *lowest energy sublevel available.* (Of course, one or more neutrons are also added to the nucleus.)

There are two common ways to indicate the distribution of electrons in the sublevels:

- *The electron configuration.* This shorthand notation consists of the principal energy level (*n* value), the letter designation of the sublevel (*l* value), and the number of electrons (#) in the sublevel, written as a superscript:

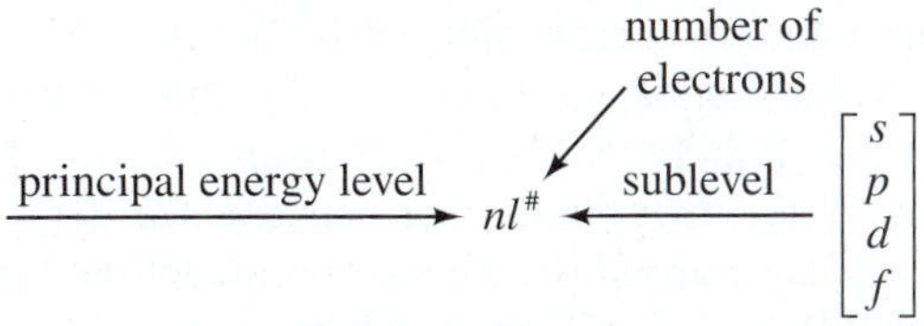

- *The orbital diagram.* An **orbital diagram** consists of a box (or just a line) for each orbital in a given energy level, grouped by sublevel (with *nl* designation shown beneath), with an arrow representing each electron *and* its spin: ↑ is $+\frac{1}{2}$ and ↓ is $-\frac{1}{2}$. (Throughout the text, orbital occupancy is also indicated by color intensity: no color is empty, pale color is half-filled, and full color is filled.)

Building Up Period 1

Let's begin by applying the aufbau principle to Period 1, whose ground-state elements have only the $n = 1$ level and, thus, only the 1*s* sublevel, which consists of only the 1*s* orbital. We'll also assign a set of four quantum numbers to each element's *last added* electron.

1. *Hydrogen.* For the electron in H, as you've seen, the set of quantum numbers is H ($Z = 1$): $n = 1$, $l = 0$, $m_l = 0$, $m_s = +\frac{1}{2}$. The electron configuration (spoken "one-ess-one") and orbital diagram are

H ($Z = 1$) $1s^1$

2. *Helium.* Recall that the first electron in He has the same quantum numbers as the electron in H, but the second He electron has opposing spin (exclusion principle), giving He ($Z = 2$): $n = 1$, $l = 0$, $m_l = 0$, $m_s = -\frac{1}{2}$. The electron configuration (spoken "one-ess-two," *not* "one-ess-squared") and orbital diagram are

He ($Z = 2$) $1s^2$

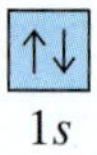

Building Up Period 2

The exclusion principle says an orbital can hold no more than two electrons. Therefore, in He, the $1s$ orbital, the $1s$ sublevel, the $n = 1$ level, and Period 1 are filled. Filling the $n = 2$ level builds up Period 2 and begins with the $2s$ sublevel, which is the next lowest in energy (see Figure 8.6) and consists of only the $2s$ orbital. When the $2s$ sublevel is filled, we proceed to fill the $2p$.

1. *Lithium.* The first two electrons in Li fill the $1s$ sublevel, so the last added Li electron enters the $2s$ sublevel and has quantum numbers $n = 2$, $l = 0$, $m_l = 0$, $m_s = +\frac{1}{2}$. The electron configuration and orbital diagram are

Li ($Z = 3$) $1s^22s^1$

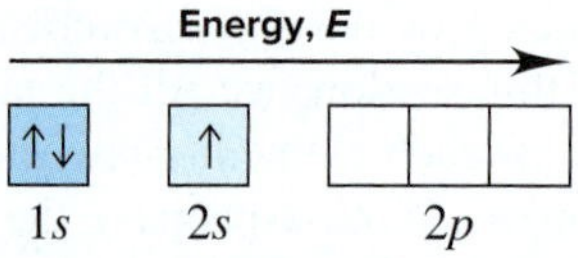

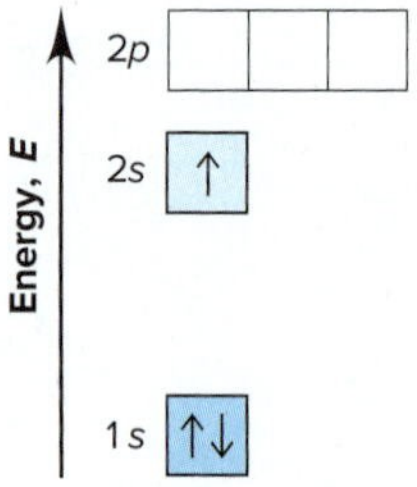

Figure 8.7 A vertical orbital diagram for the Li ground state.

(Note that a complete orbital diagram shows all the orbitals for the given n value, whether or not they are occupied.) To save space on a page, orbital diagrams are written horizontally, with *the sublevel energy increasing left to right.* But Figure 8.7 highlights the energy increase with a vertical orbital diagram for lithium.

2. *Beryllium.* The $2s$ orbital is only half-filled in Li, so the fourth electron of beryllium fills it with the electron's spin paired: $n = 2$, $l = 0$, $m_l = 0$, $m_s = -\frac{1}{2}$.

Be ($Z = 4$) $1s^22s^2$

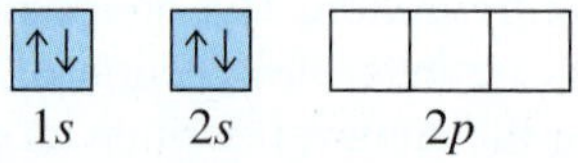

3. *Boron.* The next lowest energy sublevel is the $2p$. A p sublevel has $l = 1$, so the m_l (orientation) value can be -1, 0, or $+1$. The three orbitals in the $2p$ sublevel have *equal energy* (same n and l values), which means that the fifth electron of boron can go into *any one of the 2p orbitals.* For convenience, let's label the boxes from left to right: -1, 0, $+1$. By convention, we start on the left and place the fifth electron in the $m_l = -1$ orbital: $n = 2$, $l = 1$, $m_l = -1$, $m_s = +\frac{1}{2}$.

B ($Z = 5$) $1s^22s^22p^1$

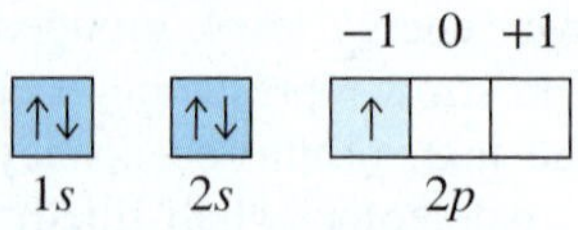

4. *Carbon.* To minimize electron-electron repulsions, the sixth electron of carbon enters one of the *unoccupied* $2p$ orbitals rather than the $2p$ orbital that already holds one electron; by convention, we place it in the $m_l = 0$ orbital. Experiment shows that the spin of this electron is *parallel* to (the same as) the spin of the other $2p$ electron. This fact exemplifies **Hund's rule:** *when orbitals of equal energy are available, the electron configuration of lowest energy has the maximum number of unpaired electrons*

with parallel spins. Thus, the sixth electron of carbon has $n = 2$, $l = 1$, $m_l = 0$ and $m_s = +\frac{1}{2}$.

C ($Z = 6$) $1s^22s^22p^2$

1s	2s	2p		
↑↓	↑↓	↑	↑	

5. *Nitrogen.* Based on Hund's rule, nitrogen's seventh electron enters the last empty $2p$ orbital, with its spin parallel to the other two: $n = 2$, $l = 1$, $m_l = +1$, $m_s = +\frac{1}{2}$.

N ($Z = 7$) $1s^22s^22p^3$

1s	2s	2p		
↑↓	↑↓	↑	↑	↑

6. *Oxygen.* The eighth electron in oxygen must enter one of the three half-filled $2p$ orbitals and "pair up" with (oppose the spin of) the electron already there. We place the electron in the first half-filled $2p$ orbital: $n = 2$, $l = 1$, $m_l = -1$, $m_s = -\frac{1}{2}$.

O ($Z = 8$) $1s^22s^22p^4$

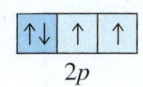

1s	2s	2p		
↑↓	↑↓	↑↓	↑	↑

7. *Fluorine.* Fluorine's ninth electron enters the next of the two remaining half-filled $2p$ orbitals: $n = 2$, $l = 1$, $m_l = 0$, $m_s = -\frac{1}{2}$.

F ($Z = 9$) $1s^22s^22p^5$

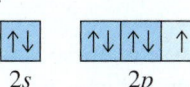

1s	2s	2p		
↑↓	↑↓	↑↓	↑↓	↑

8. *Neon.* Only one half-filled $2p$ orbital remains, so the tenth electron of neon occupies it: $n = 2$, $l = 1$, $m_l = +1$, $m_s = -\frac{1}{2}$. With neon, the $n = 2$ level is filled.

Ne ($Z = 10$) $1s^22s^22p^6$

1s	2s	2p		
↑↓	↑↓	↑↓	↑↓	↑↓

SAMPLE PROBLEM 8.1 Correlating Quantum Numbers and Orbital Diagrams

Problem Use the orbital diagram for fluorine (shown above) to write sets of quantum numbers for the third and eighth electrons of the F atom.

Plan Referring to the orbital diagram, we identify the electron of interest and note its level (n), sublevel (l), orbital (m_l), and spin (m_s).

Solution The orbital diagram of fluorine is shown below with the electrons of interest in red:

1s	2s	2p		
↑↓	↑↓	↑↓	↑↓	↑

The third electron is in the $2s$ orbital. Recall that $l = 0$ for an s orbital. The upward arrow indicates a spin of $+\frac{1}{2}$:

$$n = 2,\ l = 0,\ m_l = 0,\ m_s = +\tfrac{1}{2}$$

The eighth electron is in the first $2p$ orbital ($l = 1$ for a p orbital), which is designated $m_l = -1$, and is represented by a downward arrow because it was the second electron added to that $2p$ orbital:

$$n = 2,\ l = 1,\ m_l = -1,\ m_s = -\tfrac{1}{2}$$

FOLLOW-UP PROBLEMS

Brief Solutions for ***all*** *Follow-up Problems appear at the end of the chapter.*

8.1A Use the periodic table to identify the element with the electron configuration $1s^22s^22p^4$. Write its orbital diagram and the set of quantum numbers for its sixth electron.

8.1B The last electron added to an atom has the following set of quantum numbers: $n = 2$, $l = 1$, $m_l = 0$, $m_s = +\frac{1}{2}$. Identify the element, and write its electron configuration and orbital diagram.

SOME SIMILAR PROBLEMS 8.21(a), 8.22(a), and 8.22(d)

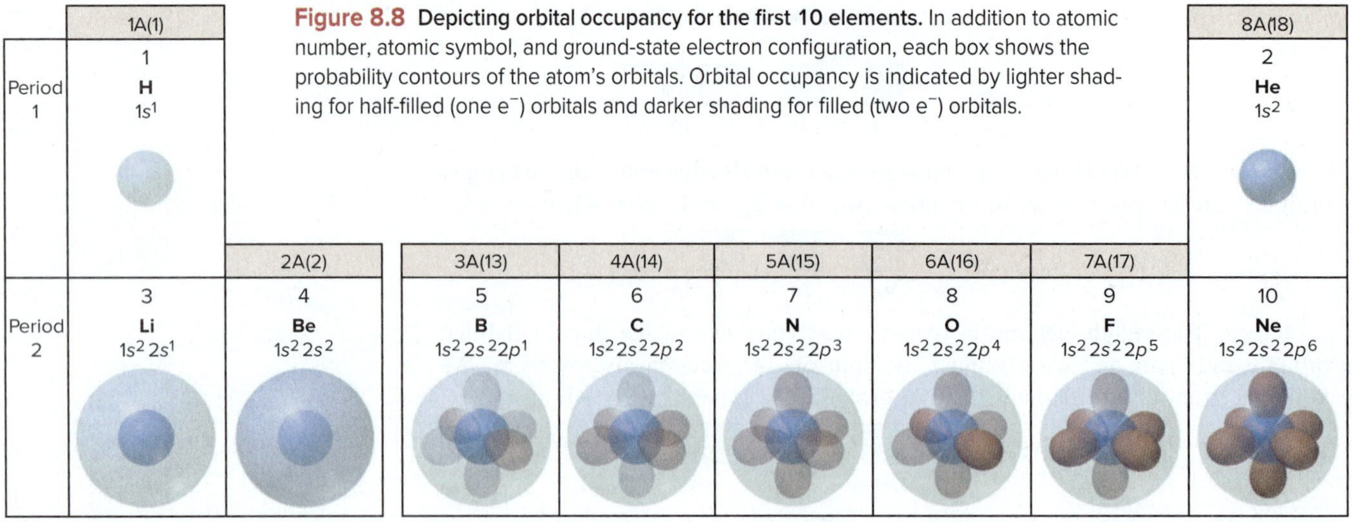

Figure 8.8 Depicting orbital occupancy for the first 10 elements. In addition to atomic number, atomic symbol, and ground-state electron configuration, each box shows the probability contours of the atom's orbitals. Orbital occupancy is indicated by lighter shading for half-filled (one e^-) orbitals and darker shading for filled (two e^-) orbitals.

With our attention on these notations, it's easy to forget that atoms are real objects and electrons occupy volumes with specific shapes and orientations. Figure 8.8 shows orbital contours for the first 10 elements arranged in periodic table format.

At this point, we can make an important correlation: *elements in the same group have similar outer electron configurations and similar patterns of reactivity.* As an example, helium (He) and neon (Ne) in Group 8A(18) both have filled outer sublevels—$1s^2$ for helium and $2s^22p^6$ for neon—and neither element forms compounds. In general, we find that filled outer sublevels make elements much more stable and unreactive.

Building Up Period 3

The Period 3 elements, Na through Ar, lie directly under the Period 2 elements, Li through Ne. That is, even though the $n = 3$ level splits into $3s$, $3p$, and $3d$ sublevels, Period 3 fills only $3s$ and $3p$; as you'll see shortly, the $3d$ is filled in Period 4. Table 8.3

Table 8.3 Partial Orbital Diagrams and Electron Configurations* for the Elements in Period 3

Atomic Number	Element	Partial Orbital Diagram (3s and 3p Sublevels Only): 3s	3p	Full Electron Configuration	Condensed Electron Configuration
11	Na	[↑]	[ ][ ][ ]	$[1s^22s^22p^6]\ 3s^1$	$[Ne]\ 3s^1$
12	Mg	[↑↓]	[ ][ ][ ]	$[1s^22s^22p^6]\ 3s^2$	$[Ne]\ 3s^2$
13	Al	[↑↓]	[↑][ ][ ]	$[1s^22s^22p^6]\ 3s^23p^1$	$[Ne]\ 3s^23p^1$
14	Si	[↑↓]	[↑][↑][ ]	$[1s^22s^22p^6]\ 3s^23p^2$	$[Ne]\ 3s^23p^2$
15	P	[↑↓]	[↑][↑][↑]	$[1s^22s^22p^6]\ 3s^23p^3$	$[Ne]\ 3s^23p^3$
16	S	[↑↓]	[↑↓][↑][↑]	$[1s^22s^22p^6]\ 3s^23p^4$	$[Ne]\ 3s^23p^4$
17	Cl	[↑↓]	[↑↓][↑↓][↑]	$[1s^22s^22p^6]\ 3s^23p^5$	$[Ne]\ 3s^23p^5$
18	Ar	[↑↓]	[↑↓][↑↓][↑↓]	$[1s^22s^22p^6]\ 3s^23p^6$	$[Ne]\ 3s^23p^6$

*Colored type indicates the sublevel to which the last electron has been added.

introduces two ways to present electron distributions more concisely:

- *Partial orbital diagrams* show only the sublevels being filled, here the 3*s* and 3*p*.
- *Condensed electron configurations (rightmost column)* have the element symbol of the noble gas from the previous row in brackets, to represent its configuration, followed by the electron configuration of filled inner sublevels and the energy level being filled. For example, the condensed electron configuration of sulfur is [Ne] $3s^2 3p^4$, where [Ne] stands for $1s^2 2s^2 2p^6$.

In Na (the second alkali metal) and Mg (the second alkaline earth metal), electrons are added to the 3*s* sublevel, which contains only the 3*s* orbital; this is directly comparable to the filling of the 2*s* sublevel in Li and Be in Period 2. Then, in the same way as the 2*p* orbitals of B, C, and N in Period 2 are half-filled, the last electrons added to Al, Si, and P in Period 3 half-fill successive 3*p* orbitals with spins parallel (Hund's rule). The last electrons added to S, Cl, and Ar then successively pair up to fill those 3*p* orbitals, and thus the 3*p* sublevel.

Building Up Period 4: The First Transition Series

Period 4 contains the first series of **transition elements,** those in which *d* orbitals are being filled. Let's examine three factors that affect the filling pattern in a period with a transition series; we focus on Period 4, but similar effects occur in Periods 5–7.

1. *Effects of shielding and penetration on sublevel energy.* The 3*d* sublevel is filled in Period 4, but *the 4s sublevel is filled first.* This switch in filling order is due to shielding and penetration effects. Based on the 3*d* radial probability distribution (see Figure 7.19), a 3*d* electron spends most of the time outside the filled inner $n = 1$ and $n = 2$ levels, so it is shielded very effectively from the nuclear charge. However, the outermost 4*s* electron penetrates close to the nucleus part of the time, so it is subject to a greater attraction. As a result, the 4*s* orbital is slightly *lower* in energy than the 3*d* and fills first. In any period, *the ns sublevel fills before the* $(n - 1)d$ *sublevel.* Other variations in the filling pattern occur at higher values of *n* because sublevel energies become very close together (see Figure 8.6).

2. *Filling the 4s and 3d sublevels.* Table 8.4 on the next page shows the partial orbital diagrams and full and condensed electron configurations for the 18 elements in Period 4. The first two elements, K and Ca, are the next alkali and alkaline earth metals, respectively; the last electron of K half-fills and that of Ca fills the 4*s* sublevel. The last electron of scandium (Sc; $Z = 21$), the first transition element, occupies any one of the five 3*d* orbitals because they are equal in energy; Sc has the electron configuration [Ar] $4s^2 3d^1$. Filling of the 3*d* orbitals proceeds one electron at a time, as with the *p* orbitals, except in two cases, chromium (Cr; $Z = 24$) and copper (Cu; $Z = 29$), discussed next.

3. *Stability of half-filled and filled sublevels.* Vanadium (V; $Z = 23$) has three half-filled *d* orbitals ([Ar] $4s^2 3d^3$). However, the last electron of the next element, Cr, does not enter a fourth empty *d* orbital to give [Ar] $4s^2 3d^4$; instead, Cr has one electron in the 4*s* sublevel and five in the 3*d* sublevel, making both sublevels half-filled: [Ar] $4s^1 3d^5$. In the next element, manganese (Mn; $Z = 25$), the 4*s* sublevel is filled again ([Ar] $4s^2 3d^5$).

Cr ($Z = 24$) [Ar] $4s^1 3d^5$

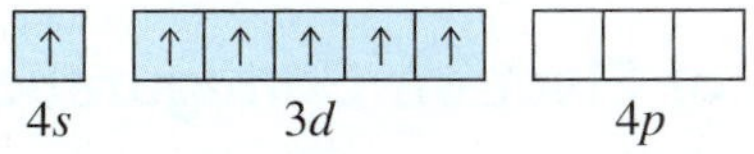

Because it follows nickel (Ni; [Ar] $4s^2 3d^8$), copper would be expected to have the configuration [Ar] $4s^2 3d^9$. Instead, the 4*s* sublevel of Cu is *half-filled* with 1 electron,

Table 8.4 Partial Orbital Diagrams and Electron Configurations* for the Elements in Period 4

Atomic Number	Element	Partial Orbital Diagram (4s, 3d, and 4p Sublevels Only) 4s	3d	4p	Full Electron Configuration	Condensed Electron Configuration
19	K	[↑]	[ ][ ][ ][ ][ ]	[ ][ ][ ]	$1s^22s^22p^63s^23p^64s^1$	[Ar] $4s^1$
20	Ca	[↑↓]	[ ][ ][ ][ ][ ]	[ ][ ][ ]	$1s^22s^22p^63s^23p^64s^2$	[Ar] $4s^2$
21	Sc	[↑↓]	[↑][ ][ ][ ][ ]	[ ][ ][ ]	$1s^22s^22p^63s^23p^64s^23d^1$	[Ar] $4s^23d^1$
22	Ti	[↑↓]	[↑][↑][ ][ ][ ]	[ ][ ][ ]	$1s^22s^22p^63s^23p^64s^23d^2$	[Ar] $4s^23d^2$
23	V	[↑↓]	[↑][↑][↑][ ][ ]	[ ][ ][ ]	$1s^22s^22p^63s^23p^64s^23d^3$	[Ar] $4s^23d^3$
24	Cr	[↑]	[↑][↑][↑][↑][↑]	[ ][ ][ ]	$1s^22s^22p^63s^23p^64s^13d^5$	[Ar] $4s^13d^5$
25	Mn	[↑↓]	[↑][↑][↑][↑][↑]	[ ][ ][ ]	$1s^22s^22p^63s^23p^64s^23d^5$	[Ar] $4s^23d^5$
26	Fe	[↑↓]	[↑↓][↑][↑][↑][↑]	[ ][ ][ ]	$1s^22s^22p^63s^23p^64s^23d^6$	[Ar] $4s^23d^6$
27	Co	[↑↓]	[↑↓][↑↓][↑][↑][↑]	[ ][ ][ ]	$1s^22s^22p^63s^23p^64s^23d^7$	[Ar] $4s^23d^7$
28	Ni	[↑↓]	[↑↓][↑↓][↑↓][↑][↑]	[ ][ ][ ]	$1s^22s^22p^63s^23p^64s^23d^8$	[Ar] $4s^23d^8$
29	Cu	[↑]	[↑↓][↑↓][↑↓][↑↓][↑↓]	[ ][ ][ ]	$1s^22s^22p^63s^23p^64s^13d^{10}$	[Ar] $4s^13d^{10}$
30	Zn	[↑↓]	[↑↓][↑↓][↑↓][↑↓][↑↓]	[ ][ ][ ]	$1s^22s^22p^63s^23p^64s^23d^{10}$	[Ar]$4s^23d^{10}$
31	Ga	[↑↓]	[↑↓][↑↓][↑↓][↑↓][↑↓]	[↑][ ][ ]	$1s^22s^22p^63s^23p^64s^23d^{10}4p^1$	[Ar] $4s^23d^{10}4p^1$
32	Ge	[↑↓]	[↑↓][↑↓][↑↓][↑↓][↑↓]	[↑][↑][ ]	$1s^22s^22p^63s^23p^64s^23d^{10}4p^2$	[Ar] $4s^23d^{10}4p^2$
33	As	[↑↓]	[↑↓][↑↓][↑↓][↑↓][↑↓]	[↑][↑][↑]	$1s^22s^22p^63s^23p^64s^23d^{10}4p^3$	[Ar] $4s^23d^{10}4p^3$
34	Se	[↑↓]	[↑↓][↑↓][↑↓][↑↓][↑↓]	[↑↓][↑][↑]	$1s^22s^22p^63s^23p^64s^23d^{10}4p^4$	[Ar] $4s^23d^{10}4p^4$
35	Br	[↑↓]	[↑↓][↑↓][↑↓][↑↓][↑↓]	[↑↓][↑↓][↑]	$1s^22s^22p^63s^23p^64s^23d^{10}4p^5$	[Ar] $4s^23d^{10}4p^5$
36	Kr	[↑↓]	[↑↓][↑↓][↑↓][↑↓][↑↓]	[↑↓][↑↓][↑↓]	$1s^22s^22p^63s^23p^64s^23d^{10}4p^6$	[Ar] $4s^23d^{10}4p^6$

*Colored type indicates sublevel(s) whose occupancy changes when the last electron is added.

and the 3*d* sublevel is *filled* with 10. From these two exceptions, Cr and Cu, we conclude that *half-filled and filled sublevels are unexpectedly stable (low in energy);* we see this pattern with many other elements.

Cu (Z = 29) [Ar] $4s^13d^{10}$

4s	3d	4p
[↑]	[↑↓][↑↓][↑↓][↑↓][↑↓]	[ ][ ][ ]

With zinc (Zn; Z = 30), the 4*s* sublevel is filled ([Ar] $4s^23d^{10}$), and the first transition series ends. The 4*p* sublevel is filled in the next six elements, and Period 4 ends with the noble gas krypton (Kr; Z = 36).

General Principles of Electron Configurations

Figure 8.9 shows the partial (just the highest energy sublevels being filled) ground-state electron configurations of the 118 known elements. Let's highlight some key relationships among them.

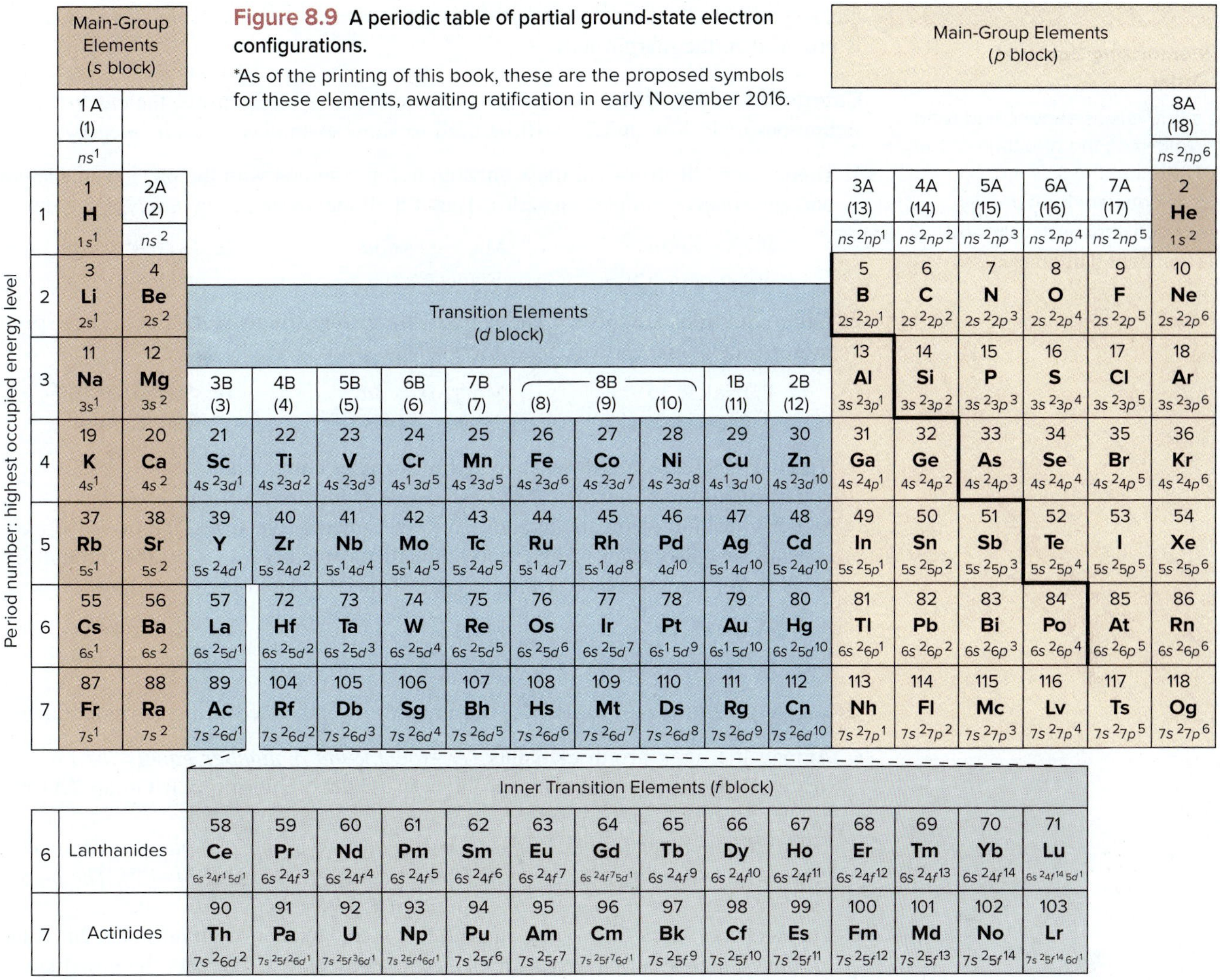

Figure 8.9 A periodic table of partial ground-state electron configurations.

*As of the printing of this book, these are the proposed symbols for these elements, awaiting ratification in early November 2016.

Orbital Filling Order When the elements are "built up" by filling levels and sublevels in order of increasing energy, we obtain the sequence in the periodic table. Reading the table from left to right, like words on a page, gives the energy order of levels and sublevels (Figure 8.10); this is the same energy order illustrated in Figure 8.6. A

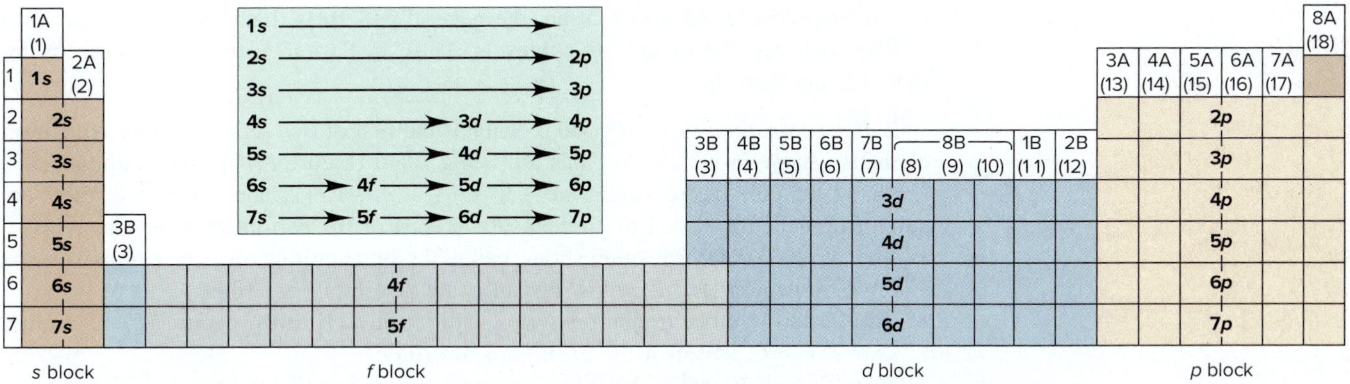

Figure 8.10 Orbital filling and the periodic table. This form of the periodic table shows the sublevel blocks. *Pale green inset:* A summary of the sublevel filling order.

Aid to Memorizing Sublevel Filling Order

List the sublevels as shown, and read from 1*s*, following the direction of the arrows. Note that the

- *n* value is constant horizontally,
- *l* value is constant vertically, and
- *n* + *l* is constant diagonally.

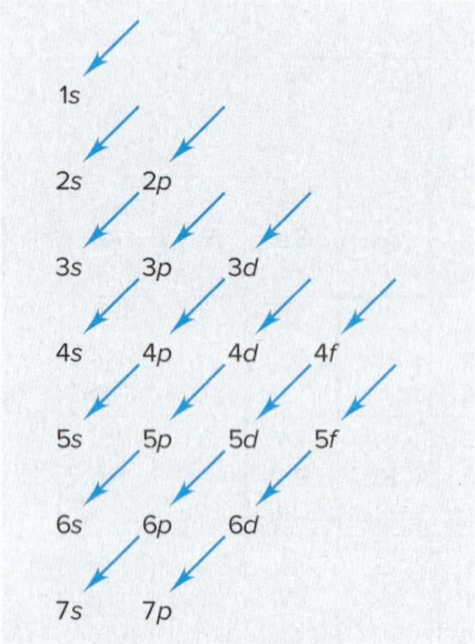

memory aid for recalling sublevel filling order when a periodic table is not available is provided in the margin note. ‹

Categories of Electrons Atoms have three categories of electrons; the electron configurations of P, Mn, and Te will be used to show examples of each category:

1. **Inner (core) electrons** are those an atom has in common with the previous noble gas and any *completed* transition series. They fill all the *lower energy levels* of an atom.

P: [Ne] $3s^23p^3$	Mn: [Ar] $4s^23d^5$	Te: [Kr] $5s^24d^{10}5p^4$
10 inner electrons	18 inner electrons	46 inner electrons

2. **Outer electrons** are those in the *highest energy level* (highest *n* value). They spend most of their time farthest from the nucleus.

P: [Ne] $3s^23p^3$	Mn: [Ar] $4s^23d^5$	Te: [Kr] $5s^24d^{10}5p^4$
5 outer electrons	2 outer electrons	6 outer electrons

3. **Valence electrons** are those involved in forming compounds:
   - For main-group elements, the *valence electrons* ***are*** *the outer electrons.*
   - For transition elements, in addition to the outer *ns* electrons, the $(n - 1)d$ electrons are also valence electrons, though the metals Fe ($Z = 26$) through Zn ($Z = 30$) may use only a few, if any, of their *d* electrons in bonding.

P: [Ne] $3s^23p^3$	Mn: [Ar] $4s^23d^5$	Te: [Kr] $5s^24d^{10}5p^4$
5 valence electrons	7 valence electrons	6 valence electrons

Group and Period Numbers Key information is embedded in the periodic table:

- Among the main-group elements (A groups), *the A number equals the number of outer electrons* (those with the highest *n*); thus, chlorine (Cl; Group 7A) has 7 outer electrons, and so forth.
- *The period number is the n value of the outermost s and p sublevels. The period number − 1 is the n value of the outermost d sublevel (Periods 4 − 7). The period number − 2 is the n value of the outermost f sublevel (Periods 6 and 7).*
- For an energy level, the *n* value squared (n^2) is the number of *orbitals,* and $2n^2$ is the maximum number of *electrons* (or elements). For example, consider the $n = 3$ level:
  Number of orbitals: $n^2 = 3^2 = 9$: one 3*s*, three 3*p*, and five 3*d*.
  Number of electrons: $2n^2 = 2(3^2) = 18$: two 3*s* and six 3*p* electrons for the eight elements of Period 3, and ten 3*d* electrons for the ten transition elements of Period 4.

Intervening Series: Transition and Inner Transition Elements

As Figure 8.10 shows, the *d* block and *f* block occur between the main-group *s* and *p* blocks.

1. *Transition series.* Periods 4, 5, 6, and 7 incorporate the 3*d,* 4*d,* 5*d,* and 6*d* sublevels, respectively. As you've seen, the general pattern is that the $(n - 1)d$ sublevel is filled between the *ns* and *np* sublevels. Thus, in Period 5, the filling order is 5*s,* then 4*d,* and then 5*p.*

2. *Inner transition series.* Period 6 contains the first of two series of **inner transition elements,** those in which *f* orbitals are being filled (Figure 8.10). The *f* orbitals have $l = 3$, so the possible m_l values are −3, −2, −1, 0, +1, +2, and +3; that is, there are seven *f* orbitals, for a total of 14 elements in *each* of the two inner transition series:
- The Period 6 inner transition series, called the **lanthanides** (or **rare earths**), occurs after lanthanum (La; $Z = 57$), and in it the 4*f* orbitals are filled.
- The Period 7 inner transition series, called the **actinides,** occurs after actinium (Ac; $Z = 89$), and in it the 5*f* orbitals are filled.

Thus, in Periods 6 and 7, the filling sequence is *ns*, first of the $(n - 1)d$, all $(n - 2)f$, remainder of the $(n - 1)d$, and *np.* Period 6 ends with the 6*p* sublevel, and Period 7 ends with the 7*p* sublevel.

3. *Irregular filling patterns.* Irregularities in the filling pattern, such as those for Cr and Cu in Period 4, occur in the *d* and *f* blocks because the sublevel energies in these larger atoms differ very little. Even though occasional deviations occur in the *d* block, the sum of *ns* electrons and $(n - 1)d$ electrons always equals the new group number (in parentheses). For instance, despite variations in Group 6B(**6**)—Cr, Mo, W, and Sg—the sum of *ns* and $(n - 1)d$ electrons is 6; for Group 8B(**10**)—Ni, Pd, Pt, and Ds—the sum is 10.

In summary, we will use the following guidelines to write the electron configuration of any element, keeping in mind that deviations do occur:

- Place each successive electron in the next available orbital of lowest possible energy; the order of filling is shown in Figures 8.6 and 8.10.
- Each orbital holds a maximum of two electrons:
  The one *s* orbital in a sublevel can hold two electrons (s^2).
  The three orbitals in a *p* sublevel can hold a total of six electrons (p^6).
  The five orbitals in a *d* sublevel can hold a total of ten electrons (d^{10}).
  The seven orbitals in an *f* sublevel can hold a total of fourteen electrons (f^{14}).
- Electrons will not pair in a set of equal energy orbitals if an empty orbital is available (Hund's rule).

Similar Electron Configurations Within Groups

As shown in Figure 8.9, there is similarity in outer electron configuration within a group of elements. Among the main-group elements (A groups)—the *s*-block and *p*-block elements—outer electron configurations within a group are identical. Some variations do occur in the transition elements (B groups, *d* block) and inner transition elements (*f* block). One of the central points in chemistry is that *similar outer electron configurations correlate with similar chemical behavior.* Here are examples from three of the groups:

- The Group 1A(1) elements, Li, Na, K, Rb, Cs, and Fr, have the outer electron configuration ns^1 (where *n* is the quantum number of the highest energy level). All are highly reactive metals whose atoms lose the outer electron when they form ionic compounds with nonmetals, and all react vigorously with water to displace H_2 (Figure 8.11A).
- The Group 7A(17) elements, F, Cl, Br, I, and At, have the outer electron configuration ns^2np^5. All are reactive nonmetals that occur as diatomic molecules (X_2), and all form ionic compounds with metals (KX, MgX_2) in which their ionic charge is 1− (Figure 8.11B), covalent compounds with hydrogen (HX) that yield acidic solutions in water, and covalent compounds with carbon (CX_4).
- In Group 8A(18), He has the electron configuration ns^2, and all the other elements in the group have the outer configuration ns^2np^6. Consistent with their *filled* energy levels, all the members are very unreactive monatomic gases.

Figure 8.11 **Similar reactivities in a group. A,** Potassium reacting with water. **B,** Chlorine reacting with potassium.

Source: © McGraw-Hill Education/Stephen Frisch, photographer

Student Hot Spot

Student data indicate that you may struggle with writing electron configurations. Access the SmartBook to view additional Learning Resources on this topic.

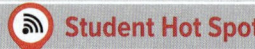

To summarize the major connection between quantum mechanics and chemical periodicity: *sublevels are filled in order of increasing energy, which leads to outer electron configurations that recur periodically, leading to chemical properties that recur periodically.*

SAMPLE PROBLEM 8.2 Determining Electron Configurations

Problem Using only the periodic table (not Figure 8.9 or Table 8.4) and assuming a regular filling pattern, give the full and condensed electron configurations, partial orbital diagram showing valence electrons only, and number of inner electrons for the following elements:

(a) Potassium (K; Z = 19) **(b)** Technetium (Tc; Z = 43) **(c)** Lead (Pb; Z = 82)

Plan The atomic number tells us the number of electrons, and the periodic table shows the order for filling sublevels. In the partial orbital diagrams, we include all electrons added after the previous noble gas *except* those in *filled* inner sublevels. The number of inner electrons is the sum of those in the previous noble gas and in filled *d* and *f* sublevels.

Solution **(a)** For K (Z = 19), the full electron configuration is $1s^22s^22p^63s^23p^64s^1$.

The condensed configuration is [Ar] $4s^1$.

The partial orbital diagram showing valence electrons is

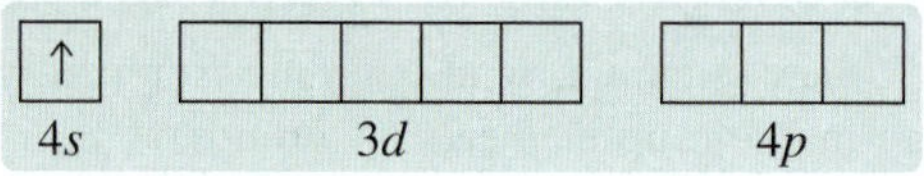

The inner electrons in K are those in the [Ar] part of the condensed electron configuration; there are 18 inner electrons.

(b) For Tc (Z = 43), assuming the expected pattern, the full electron configuration is $1s^22s^22p^63s^23p^64s^23d^{10}4p^65s^24d^5$.

The condensed electron configuration is [Kr] $5s^24d^5$.

The partial orbital diagram showing valence electrons is

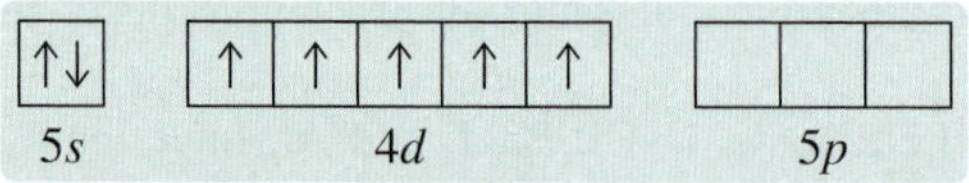

The inner electrons in Tc are those in the [Kr] part of the condensed electron configuration; there are 36 inner electrons.

(c) For Pb (Z = 82), the full electron configuration is $1s^22s^22p^63s^23p^64s^23d^{10}4p^65s^24d^{10}5p^66s^24f^{14}5d^{10}6p^2$.

The condensed electron configuration is [Xe] $6s^24f^{14}5d^{10}6p^2$.

The partial orbital diagram showing valence electrons (no filled inner sublevels) is

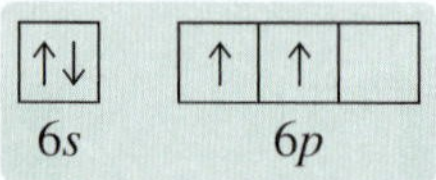

The inner electrons in Pb are the 54 electrons represented by [Xe] plus the 14 electrons in the filled 4*f* sublevel plus the 10 electrons in the filled 5*d* sublevel, for a total of 78 inner electrons.

Check Be sure that the sum of the superscripts (numbers of electrons) in the full electron configuration equals the atomic number and that the number of *valence* electrons in the condensed configuration equals the number of electrons in the partial orbital diagram.

FOLLOW-UP PROBLEMS

8.2A Without referring to Table 8.4 or Figure 8.9, give full and condensed electron configurations, partial orbital diagram showing valence electrons only, and number of inner electrons for the following elements:

(a) Ni (Z = 28) **(b)** Sr (Z = 38) **(c)** Po (Z = 84)

8.2B From each partial orbital diagram (valence electrons only), identify the element and give its full and condensed electron configuration and the number of inner electrons. Use only the periodic table, without referring to Table 8.4 or Figure 8.9.

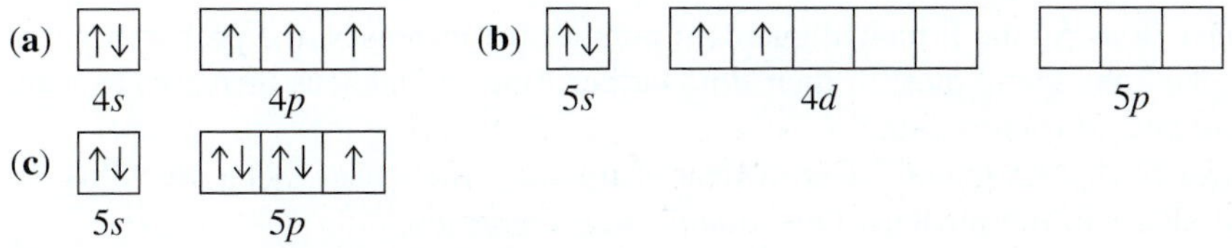

SOME SIMILAR PROBLEMS 8.23–8.36

› Summary of Section 8.2

- In applying the aufbau principle, one electron is added to an atom of each successive element in accord with the exclusion principle (no two electrons can have the same set of quantum numbers) and Hund's rule (orbitals of equal energy become half-filled, with electron spins parallel, before any pairing of spins occurs).
- For the main-group elements, valence electrons (those involved in reactions) are in the outer (highest energy) level only. For transition elements, $(n - 1)d$ electrons are also considered valence electrons.
- Because of shielding of d electrons by electrons in inner sublevels and penetration by the ns electron, the $(n - 1)d$ sublevel fills after the ns and before the np sublevels.
- In Periods 6 and 7, $(n - 2)f$ orbitals fill between the first and second $(n - 1)d$ orbitals.
- The elements of a group have similar outer electron configurations and similar chemical behavior.

8.3 TRENDS IN THREE ATOMIC PROPERTIES

In this section, we focus on three atomic properties directly influenced by electron configuration and effective nuclear charge: atomic size, ionization energy, and electron affinity. Most importantly, these properties are *periodic,* which means they generally exhibit consistent changes, or *trends,* within a group or period.

Trends in Atomic Size

Recall from Chapter 7 that we often represent atoms with spherical contours in which the electrons spend 90% of their time. We *define* **atomic size** (the extent of the spherical contour for a given atom) in terms of how closely one atom lies next to another. But, in practice, as we discuss in Chapter 12, we measure the distance between atomic nuclei in a sample of an element and divide that distance in half. Because atoms do not have hard surfaces, the size of an atom in a given compound depends somewhat on the atoms near it. In other words, *atomic size varies slightly from substance to substance.*

Figure 8.12 shows two common definitions of atomic size:

1. **Metallic radius.** Used mostly for *metals,* it is one-half the shortest distance between nuclei of adjacent, individual atoms in a crystal of the element (Figure 8.12A).
2. **Covalent radius.** Used for elements occurring as molecules, mostly *nonmetals,* it is one-half the shortest distance between nuclei of bonded atoms (Figure 8.12B).

Radii measured for some elements are used to determine the radii of other elements from the distances between the atoms in compounds. For instance, in a carbon-chlorine compound, the distance between nuclei in a C—Cl bond is 177 pm. Using the known covalent radius of Cl (100 pm), we find the covalent radius of C (177 pm − 100 pm = 77 pm) (Figure 8.12C).

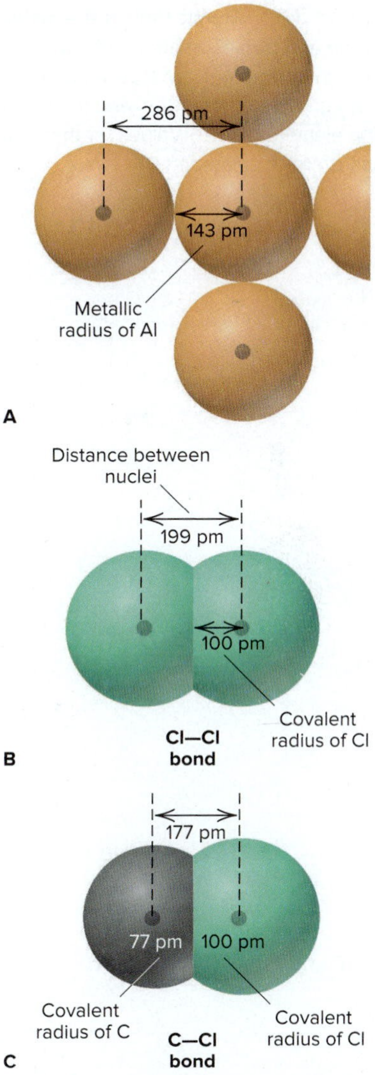

Figure 8.12 Defining atomic size. A, The metallic radius of aluminum. **B,** The covalent radius of chlorine. **C,** Known covalent radii and distances between nuclei can be used to find unknown radii.

Main-Group Elements Figure 8.13 shows the atomic radii of the main-group elements and most of the transition elements. Among the main-group elements, atomic size varies within both groups and periods as a result of two opposing influences:

1. *Changes in n.* As the principal quantum number (n) increases, the probability that outer electrons spend most of their time farther from the nucleus increases as well; thus, *atomic size increases.*
2. *Changes in* Z_{eff}. As the effective nuclear charge (Z_{eff}) increases, outer electrons are pulled closer to the nucleus; thus, *atomic size decreases.*

The net effect of these influences depends on how effectively the inner electrons shield the increasing nuclear charge:

1. *Down a group, n dominates.* As we move down a main group, each member has *one more level of inner electrons that shield the outer electrons very effectively.* Even though additional protons do moderately increase Z_{eff} for the outer electrons, the atoms get larger as a result of the increasing n value:

Atomic radius generally ***increases*** *down a group.*

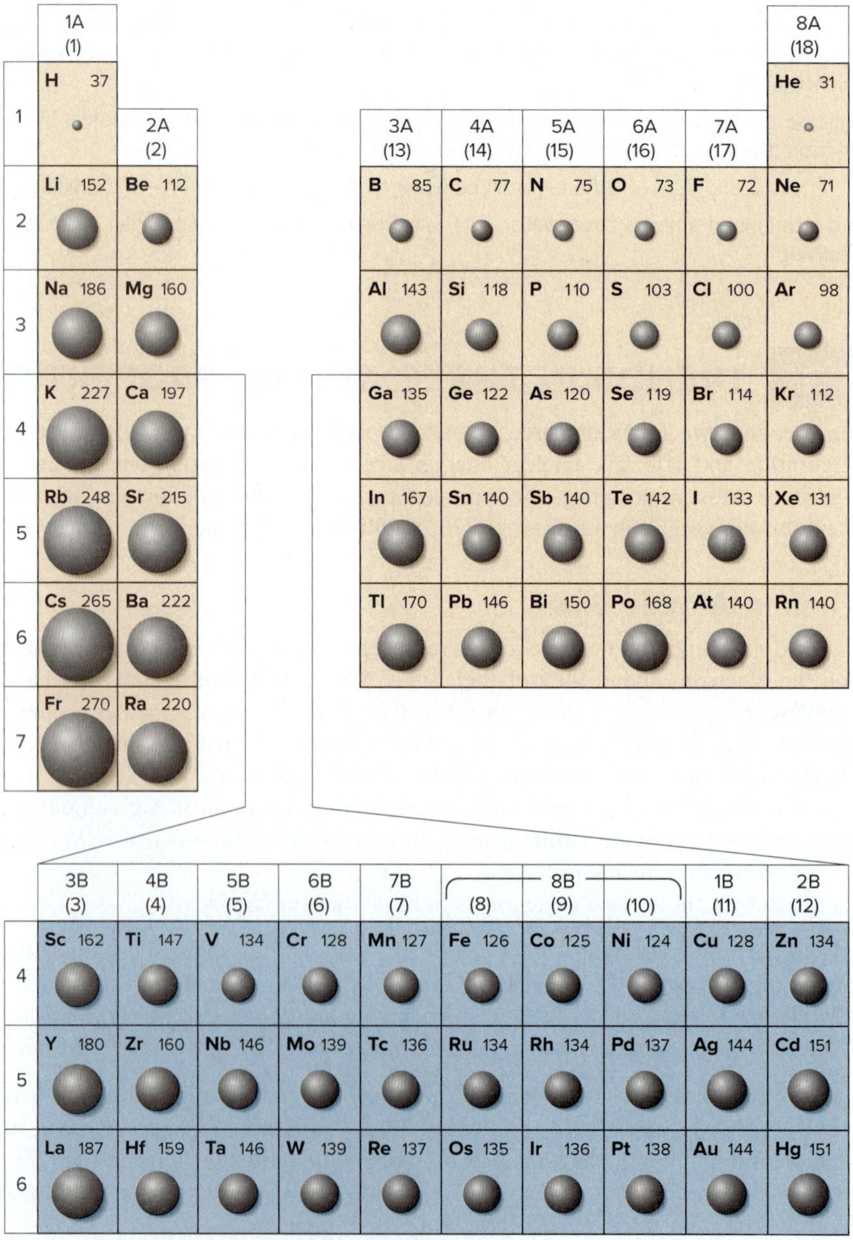

Figure 8.13 Atomic radii of the main-group and transition elements. Atomic radii (in picometers) are shown for the main-group elements *(tan)* and the transition elements *(blue).* (Values for the noble gases are calculated.)

2. *Across a period, Z_{eff} dominates.* Across a period from left to right, electrons are added to the *same* outer level, so the shielding by inner electrons does not change. Despite greater electron repulsions, outer electrons shield each other only slightly, so Z_{eff} rises significantly, and the outer electrons are pulled closer to the nucleus:

*Atomic radius generally **decreases** across a period.*

Transition Elements As Figure 8.13 shows, size trends are *not* as consistent for the transition elements:

1. *Down a transition group, n* increases, but shielding by an additional level of inner electrons results in only a small size increase from Period 4 to 5 and none from 5 to 6.

2. *Across a transition series*, atomic size shrinks through the first two or three elements because of the increasing nuclear charge. But, from then on, *size remains relatively constant* because shielding by the inner *d* electrons counteracts the increase in Z_{eff}. Thus, for example, in Period 4, the third transition element, vanadium (V; $Z = 23$), has the same radius as the last, zinc (Zn; $Z = 30$). This pattern also appears in Periods 5 and 6 in the transition and both inner transition series.

3. *A transition series affects atomic size in neighboring main groups.* Shielding by *d* electrons causes a *major size decrease from Group 2A(2) to Group 3A(13)* in Periods 4 through 6. Because the *np* sublevel has more penetration than the $(n - 1)d$, the first *np* electron [added in Group 3A(13)] "feels" a much greater Z_{eff}, due to all the protons added in the intervening transition elements. The greatest decrease occurs in Period 4: calcium (Ca; $Z = 20$) in Group 2A(2) is nearly 50% larger than gallium (Ga; $Z = 31$) in 3A(13). In fact, *d*-orbital shielding causes gallium to be slightly *smaller* than aluminum (Al; $Z = 13$), the element above it!

SAMPLE PROBLEM 8.3 Ranking Elements by Atomic Size

Problem Using only the periodic table (not Figure 8.13), rank each set of main-group elements in order of *decreasing* atomic size:

(a) Ca, Mg, Sr **(b)** K, Ga, Ca **(c)** Br, Rb, Kr **(d)** Sr, Ca, Rb

Plan To rank the elements by atomic size, we find them in the periodic table. They are main-group elements, so size increases down a group and decreases across a period.

Solution **(a)** Sr > Ca > Mg. These three elements are in Group 2A(2), and size increases down the group.

(b) K > Ca > Ga. These three elements are in Period 4, and size decreases across a period.

(c) Rb > Br > Kr. Rb is largest because it has one more energy level (Period 5) and is farthest to the left. Kr is smaller than Br because Kr is farther to the right in Period 4.

(d) Rb > Sr > Ca. Ca is smallest because it has one fewer energy level. Sr is smaller than Rb because it is farther to the right.

Check From Figure 8.13, we see that the rankings are correct.

FOLLOW-UP PROBLEMS

8.3A Using only the periodic table, rank the elements in each set in order of *increasing* size: **(a)** Se, Br, Cl; **(b)** I, Xe, Ba.

8.3B Using only the periodic table, rank the elements in each set in order of *decreasing* size: **(a)** As, Cs, S; **(b)** F, P, K.

SOME SIMILAR PROBLEMS 8.53 and 8.54

Periodicity of Atomic Size Figure 8.14 on the next page shows the variation in atomic size with atomic number. Note the up-and-down pattern as size drops across a period to the noble gas *(purple)* and then leaps up to the alkali metal *(brown)* that begins the next period. Also note the deviations from the smooth size decrease in each transition *(blue)* and inner transition *(green)* series.

Figure 8.14 Periodicity of atomic radius.

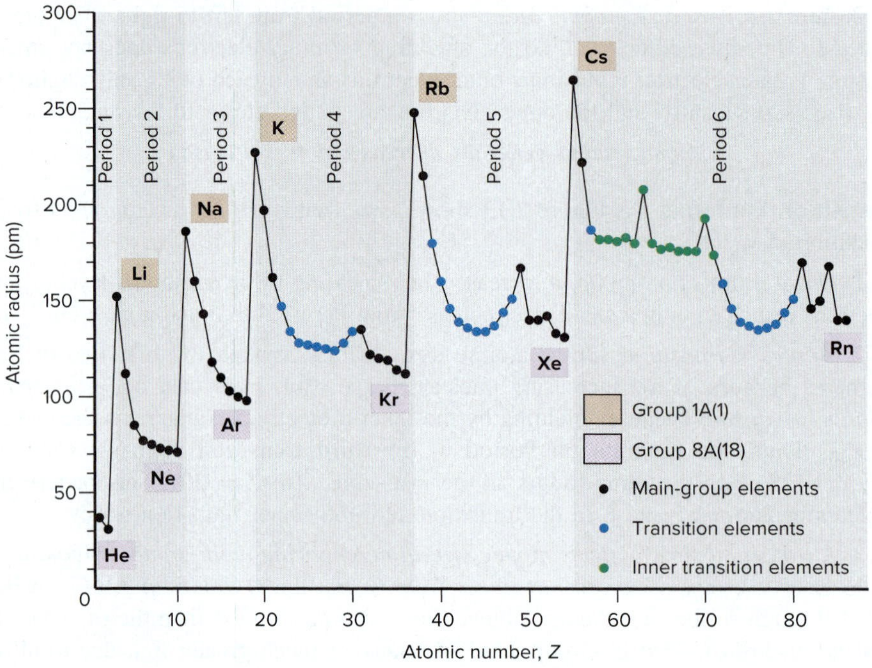

Trends in Ionization Energy

The **ionization energy (IE)** is the energy required for the *complete removal* of 1 mol of electrons from 1 mol of gaseous atoms or ions. Pulling an electron away from a nucleus *requires* energy to overcome electrostatic attraction between the negatively charged electron and the positively charged protons in the nucleus. Because energy flows *into* the system, the ionization energy is always positive (like ΔH of an endothermic reaction). (Chapter 7 viewed the ionization energy of the H atom as the energy difference between $n = 1$ and $n = \infty$, where the electron is completely removed.) The ionization energy is a key factor in an element's reactivity:

- *Atoms with a low IE tend to form cations during reactions.*
- *Atoms with a high IE (except the noble gases) tend to form anions.*

Many-electron atoms can lose more than one electron. The *first* ionization energy (IE_1) removes an outermost electron (one in the highest energy sublevel) from a gaseous atom:

$$\text{Atom}(g) \longrightarrow \text{ion}^+(g) + e^- \qquad \Delta E = IE_1 > 0 \tag{8.2}$$

Example: $Mg(g) \longrightarrow Mg^+(g) + e^-$ $\quad IE_1 = 738$ kJ/mol

The *second* ionization energy (IE_2) removes a second electron. Since this electron is pulled away from a positive ion, more energy is required for its removal and IE_2 is always larger than IE_1:

$$\text{Ion}^+(g) \longrightarrow \text{ion}^{2+}(g) + e^- \qquad \Delta E = IE_2 \text{ (always > } IE_1)$$

Example: $Mg^+(g) \longrightarrow Mg^{2+}(g) + e^-$ $\quad IE_2 = 1450$ kJ/mol

Periodicity of First Ionization Energy Figure 8.15 shows the variation in first ionization energy with atomic number. This up-and-down pattern—IE_1 rising across a period to the noble gas *(purple)* and then dropping down to the next alkali metal *(brown)*—is the *inverse* of the variation in atomic size (Figure 8.14): *as size decreases, it takes more energy to remove an electron because the nucleus is closer to the electron, so IE_1 increases.*

Let's examine the group and period trends and their exceptions:

1. *Down a group.* As we move *down* a main group, the n value increases, so atomic size does as well. As the distance from nucleus to outer electron increases,

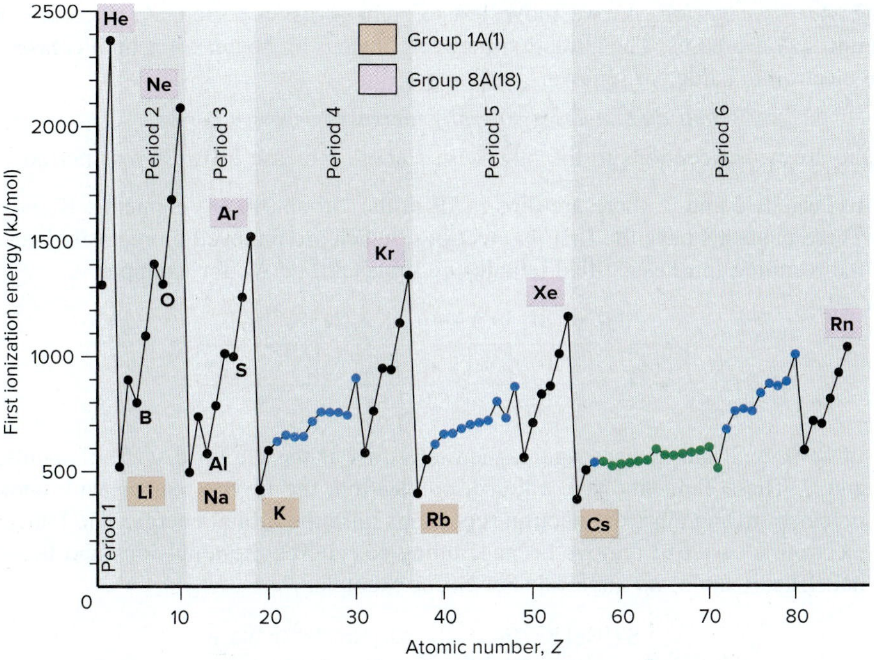

Figure 8.15 Periodicity of first ionization energy (IE_1). This trend is the *inverse* of the trend in atomic size (see Figure 8.14).

the attraction between nucleus and outer electron lessens, so the electron is easier to remove (Figure 8.16):

*Ionization energy generally **decreases** down a group.*

The only significant exception occurs in Group 3A(13): IE_1 decreases from boron (B) to aluminum (Al), but not for the rest of the group. Filling the transition series in Periods 4, 5, and 6 causes a much higher Z_{eff} and an unusually small change in size, so outer electrons in the larger Group 3A members are held tighter.

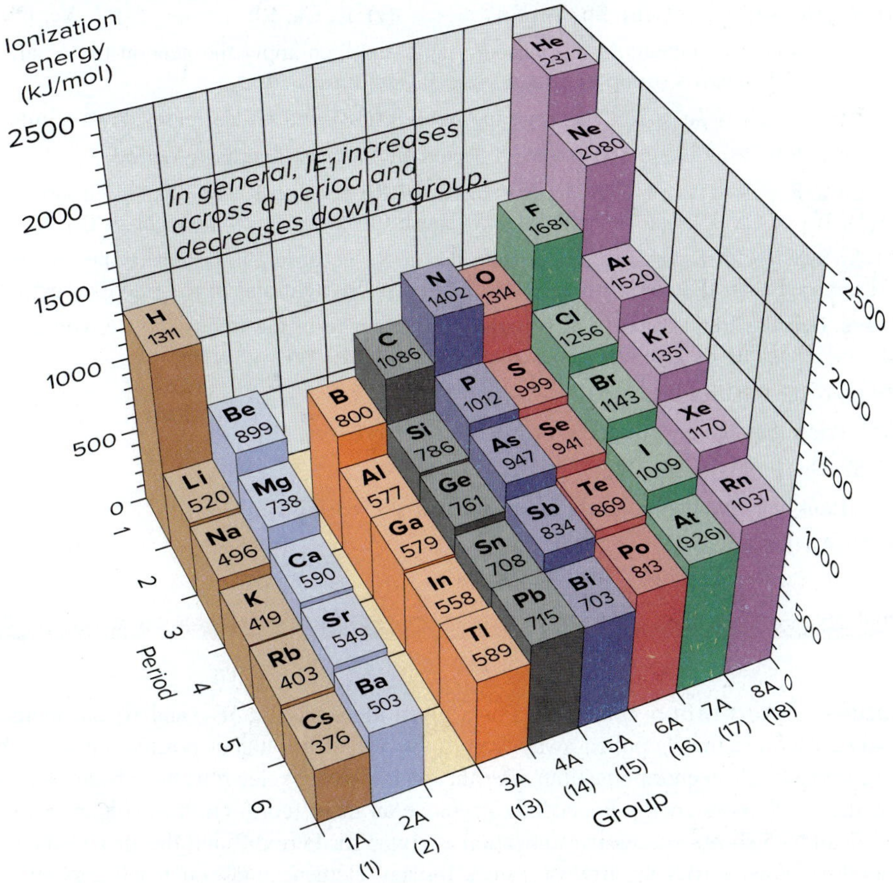

Figure 8.16 First ionization energies of the main-group elements.

2. *Across a period.* As we move left to right across a period, Z_{eff} increases and atomic size decreases. The attraction between nucleus and outer electron increases, so the electron is harder to remove:

Ionization energy generally ***increases*** *across a period.*

There are two exceptions to the otherwise smooth increase in IE_1 across periods:

- In Periods 2 and 3, there are dips in IE at the Group 3A(13) elements, B and Al. These elements have the first *np* electrons, which are removed more easily because the resulting ion has a filled (stable) *ns* sublevel. For Al, for example,

$$\text{Al ([Ne] } 3s^2 3p^1) \longrightarrow \text{Al}^+ \text{ ([Ne] } 3s^2) + e^-$$

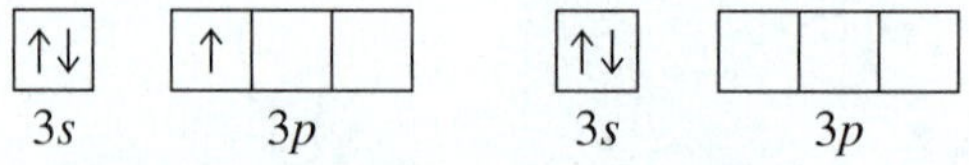

- In Periods 2 and 3, once again, there are dips at the Group 6A(16) elements, O and S. These elements have a fourth *np* electron, the first to pair up with another *np* electron, and electron-electron repulsions raise the orbital energy. The fourth *np* electron is easier to remove because doing so relieves the repulsions and leaves a half-filled (stable) *np* sublevel. For S, for example,

$$\text{S ([Ne] } 3s^2 3p^4) \longrightarrow \text{S}^+ \text{ ([Ne] } 3s^2 3p^3) + e^-$$

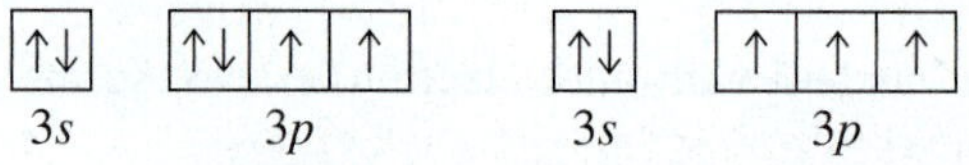

SAMPLE PROBLEM 8.4 **Ranking Elements by First Ionization Energy**

Problem Using the periodic table only, rank the elements in each set in order of *decreasing* IE_1:

(a) Kr, He, Ar **(b)** Sb, Te, Sn **(c)** K, Ca, Rb **(d)** I, Xe, Cs

Plan We find the elements in the periodic table and then apply the general trends of decreasing IE_1 down a group and increasing IE_1 across a period.

Solution (a) He > Ar > Kr. These are in Group 8A(18), and IE_1 decreases down a group.

(b) Te > Sb > Sn. These are in Period 5, and IE_1 increases across a period.

(c) Ca > K > Rb. IE_1 of K is larger than IE_1 of Rb because K is higher in Group 1A(1). IE_1 of Ca is larger than IE_1 of K because Ca is farther to the right in Period 4.

(d) Xe > I > Cs. IE_1 of I is smaller than IE_1 of Xe because I is farther to the left. IE_1 of I is larger than IE_1 of Cs because I is farther to the right and in the previous period.

Check Because trends in IE_1 are generally the opposite of the trends in size, you can rank the elements by size and check that you obtain the reverse order.

FOLLOW-UP PROBLEMS

8.4A Rank the elements in each set in order of *increasing* IE_1:

(a) Sb, Sn, I **(b)** Na, Mg, Cs

8.4B Rank the elements in each set in order of *decreasing* IE_1:

(a) O, As, Rb **(b)** Sn, Cl, Si

SOME SIMILAR PROBLEMS 8.55 and 8.56

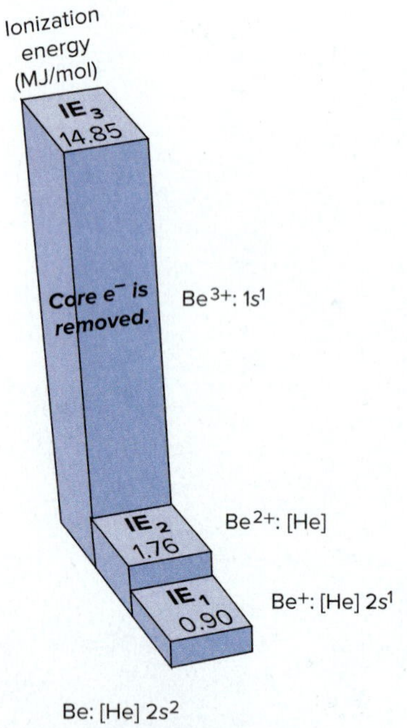

Figure 8.17 The first three ionization energies of beryllium. Beryllium has two valence electrons, so IE_3 is much larger than IE_2.

Successive Ionization Energies For a given element, IE_1, IE_2, and so on, increase because each electron is pulled away from a species with a higher positive charge. This increase includes an enormous jump *after* the outer (valence) electrons have been removed because *much* more energy is needed to remove an inner (core) electron (Figure 8.17).

Table 8.5 shows successive ionization energies for Period 2 and the first element in Period 3. Move across the row of values for any element, and you reach a point that

Table 8.5 Successive Ionization Energies of the Elements Lithium Through Sodium

Z	Element	Number of Valence Electrons	Ionization Energy (MJ/mol)*									
			IE_1	IE_2	IE_3	IE_4	IE_5	IE_6	IE_7	IE_8	IE_9	IE_{10}
3	Li	1	0.52	7.30	11.81							
4	Be	2	0.90	1.76	14.85	21.01						
5	B	3	0.80	2.43	3.66	25.02	32.82					
6	C	4	1.09	2.35	4.62	6.22	37.83	47.28				
7	N	5	1.40	2.86	4.58	7.48	9.44	53.27	64.36			
8	O	6	1.31	3.39	5.30	7.47	10.98	13.33	71.33	84.08		
9	F	7	1.68	3.37	6.05	8.41	11.02	15.16	17.87	92.04	106.43	
10	Ne	8	2.08	3.95	6.12	9.37	12.18	15.24	20.00	23.07	115.38	131.43
11	Na	1	0.50	4.56	6.91	9.54	13.35	16.61	20.11	25.49	28.93	141.37

CORE ELECTRONS

*MJ/mol, or megajoules per mole = 10^3 kJ/mol.

separates relatively low from relatively high IE values *(shaded area)*. For example, follow the values for boron (B): IE_1 (0.80 MJ/mol) is lower than IE_2 (2.43 MJ/mol), which is lower than IE_3 (3.66 MJ/mol), which is *much* lower than IE_4 (25.02 MJ/mol). From this jump, we know that boron has three electrons in the highest energy level ($1s^22s^22p^1$). Because they are so difficult to remove, *core electrons are not involved in reactions.*

Student Hot Spot

Student data indicate that you may struggle with the concept of successive ionization energies. Access the Smartbook to view additional Learning Resources on this topic.

SAMPLE PROBLEM 8.5 Identifying an Element from Its Ionization Energies

Problem Name the Period 3 element with the following ionization energies (kJ/mol), and write its full electron configuration:

IE_1	IE_2	IE_3	IE_4	IE_5	IE_6
1012	1903	2910	4956	6278	22,230

Plan We look for a large jump in the IE values, which occurs after all valence electrons have been removed. Then we refer to the periodic table to find the Period 3 element with this number of valence electrons and write its electron configuration.

Solution The large jump occurs after IE_5, indicating that the element has five valence electrons and, thus, is in Group 5A(15). This Period 3 element is phosphorus (P; $Z = 15$). Its full electron configuration is $1s^22s^22p^63s^23p^3$.

FOLLOW-UP PROBLEMS

8.5A Element Q is in Period 3 and has the following ionization energies (in kJ/mol):

IE_1	IE_2	IE_3	IE_4	IE_5	IE_6
577	1816	2744	11,576	14,829	18,375

Name element Q, and write its full electron configuration.

8.5B Write the condensed electron configurations of the elements Rb, Sr, and Y. Which has the *highest* IE_2? Which has the highest IE_3?

SOME SIMILAR PROBLEMS 8.57–8.60

Trends in Electron Affinity

The **electron affinity (EA)** is the energy change accompanying the *addition* of 1 mol of electrons to 1 mol of gaseous atoms or ions. The *first electron affinity* (EA_1) refers to the formation of 1 mol of monovalent (1−) gaseous anions:

$$\text{Atom}(g) + e^- \longrightarrow \text{ion}^-(g) \qquad \Delta E = EA_1$$

Example: $$F(g) \longrightarrow F^-(g) + e^- \qquad EA_1 = -328 \text{ kJ/mol}$$

1A (1)	2A (2)	3A (13)	4A (14)	5A (15)	6A (16)	7A (17)	8A (18)
H −72.8							He (0.0)
Li −59.6	Be ≤0	B −26.7	C −122	N +7	O −141	F −328	Ne (+29)
Na −52.9	Mg ≤0	Al −42.5	Si −134	P −72.0	S −200	Cl −349	Ar (+35)
K −48.4	Ca −2.37	Ga −28.9	Ge −119	As −78.2	Se −195	Br −325	Kr (+39)
Rb −46.9	Sr −5.03	In −28.9	Sn −107	Sb −103	Te −190	I −295	Xe (+41)
Cs −45.5	Ba −13.95	Tl −19.3	Pb −35.1	Bi −91.3	Po −183	At −270	Rn (+41)

Figure 8.18 Electron affinities of the main-group elements (in kJ/mol). Values for Group 8A(18) are estimates, which is indicated by parentheses.

As with ionization energy, there is a first electron affinity, a second, and so forth. The first electron is *attracted* by the atom's nucleus, so in most cases, *EA_1 is negative* (energy is released), analogous to the negative ΔH for an exothermic reaction.* But, the second electron affinity (EA_2) is always positive because energy must be *absorbed* to overcome electrostatic repulsions and add another electron to a negative ion.

Factors other than Z_{eff} and atomic size affect electron affinities, so trends are not regular, as are those for size and IE_1. The many exceptions arise from changes in sublevel energy and electron-electron repulsion:

- *Down a group.* We might expect a smooth decrease (smaller negative number) down a group because atomic size increases, and the nucleus is farther away from an electron being added. But only Group 1A(1) exhibits this behavior (Figure 8.18).
- *Across a period.* We might expect a regular increase (larger negative number) across a period because atomic size decreases, and higher Z_{eff} should attract the electron being added more strongly. There is an overall left-to-right increase, but it is not at all regular.

Despite these irregularities, relative values of IE and EA show three general behavior patterns:

1. *Reactive nonmetals.* Members of Group 6A(16) and especially Group 7A(17) (the halogens) have high IEs and highly negative (exothermic) EAs: these elements lose electrons with difficulty but attract them strongly. Therefore, *in their ionic compounds, they form negative ions.*
2. *Reactive metals.* Members of Groups 1A(1) and 2A(2) have low IEs and slightly negative (exothermic) EAs: they lose electrons easily but attract them weakly, if at all. Therefore, *in their ionic compounds, they form positive ions.*
3. *Noble gases.* Members of Group 8A(18) have very high IEs and slightly positive (endothermic) EAs: *they tend **not** to lose or gain electrons.* In fact, only the larger members of the group (Kr, Xe, and Rn) form compounds at all.

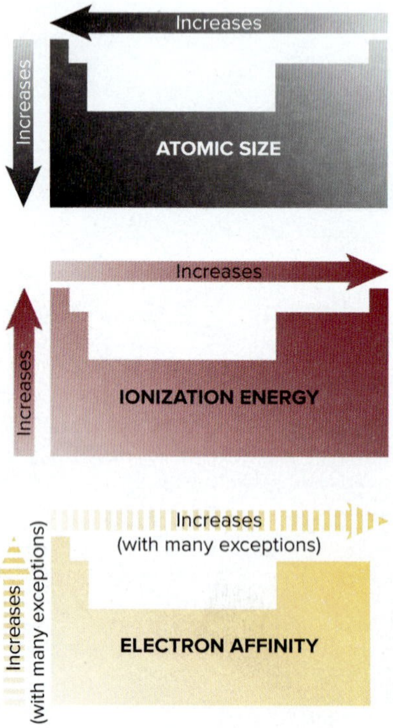

Figure 8.19 Trends in three atomic properties. For electron affinity, the dashed arrows indicate that there are numerous exceptions to the expected trends.

› Summary of Section 8.3

- › Trends in three atomic properties are summarized in Figure 8.19.
- › Atomic size (half the distance between nuclei of adjacent atoms) increases down a main group and decreases across a period. In a transition series, size remains relatively constant.
- › First ionization energy (the energy required to remove a mole of electrons from a mole of gaseous atoms or ions) is inversely related to atomic size: IE_1 decreases down a main group and increases across a period.
- › Successive ionization energies of an element show a very large increase after all valence electrons have been removed, because the first inner (core) electron is in an orbital of much lower energy and so is held very tightly.
- › Electron affinity (the energy involved in adding a mole of electrons to a mole of gaseous atoms or ions) shows many variations from expected trends.
- › Based on the relative sizes of IEs and EAs, Group 1A(1) and 2A(2) elements tend to form cations, and Group 6A(16) and 7A(17) elements tend to form anions in ionic compounds. Group 8A(18) elements are very unreactive.

*Some tables of EA_1 values list them as positive values because that quantity of energy would be *absorbed* to remove an electron from the anion.

8.4 ATOMIC PROPERTIES AND CHEMICAL REACTIVITY

All physical and chemical behaviors of the elements and their compounds are based on electron configuration and effective nuclear charge. In this section, you'll see how atomic properties determine metallic behavior and the properties of ions.

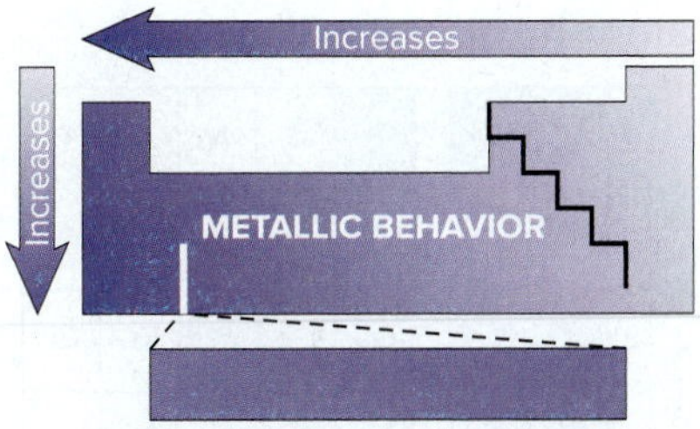

Figure 8.20 **Trends in metallic behavior.** (Hydrogen appears next to helium.)

Trends in Metallic Behavior

The three general classes of elements have distinguishing properties:

- *Metals,* found in the left and lower three-quarters of the periodic table, are typically shiny solids, have moderate to high melting points, are good conductors of heat and electricity, can be machined into wires and sheets, and lose electrons to nonmetals.
- *Nonmetals,* found in the upper right quarter of the table, are typically not shiny, have relatively low melting points, are poor conductors, are mostly crumbly solids or gases, and tend to gain electrons from metals.
- *Metalloids,* found between the other two classes, have intermediate properties.

Thus, *metallic behavior decreases from left to right across a period and increases down a group in the periodic table* (Figure 8.20).

Remember, though, that some elements don't fit these categories: as graphite, nonmetallic carbon is a good electrical conductor; the nonmetal iodine is shiny; metallic gallium melts in your hand; mercury is a liquid; and iron is brittle. Despite such exceptions, in this discussion, we'll make several generalizations about metallic behavior and its applications to redox behavior and the acid-base properties of oxides.

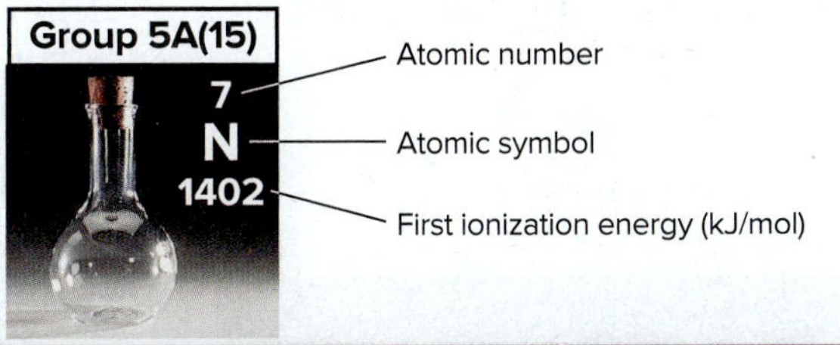

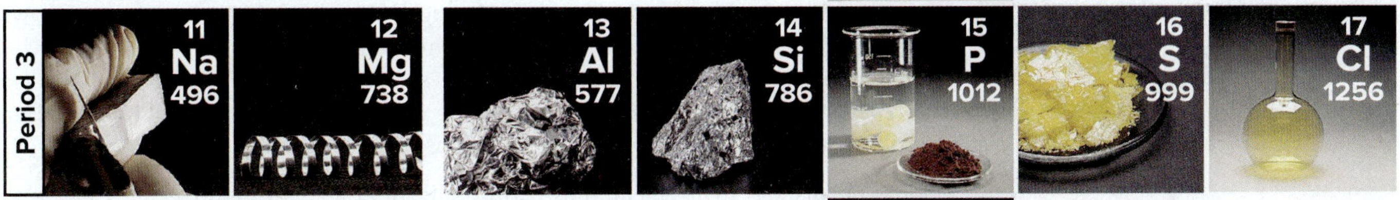

Figure 8.21 **Metallic behavior in Group 5A(15) and Period 3.** Moving down from N to Bi, there is an *increase* in metallic behavior. Moving across from Na to Cl, there is a *decrease* in metallic behavior.

Source: © McGraw-Hill Education/ Stephen Frisch, photographer

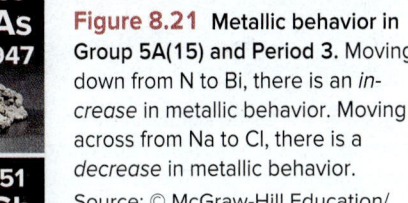

Relative Tendency to Lose or Gain Electrons Metals tend to lose electrons to nonmetals during reactions:

1. *Down a main group.* The increase in metallic behavior down a group is consistent with an increase in size and a decrease in IE and is most obvious in groups with more than one class of element, such as Group 5A(15) (Figure 8.21, *vertical*): *elements at the top can form anions, and those at the bottom can form cations.* Nitrogen (N) is a gaseous nonmetal, and phosphorus (P) is a soft nonmetal; both occur occasionally as 3− anions in their compounds. Arsenic (As) and antimony (Sb) are metalloids, with Sb the more metallic, and neither forms ions readily. Bismuth (Bi) is a typical metal, forming a 3+ cation in its mostly ionic compounds. Groups 3A(13), 4A(14), and 6A(16) show a similar trend. But even in Group 2A(2), which contains only metals, the tendency to form cations increases down the group: beryllium (Be) forms covalent compounds with nonmetals, whereas all compounds of barium (Ba) are ionic.

2. *Across a period.* The decrease in metallic behavior across a period is consistent with a decrease in size, an increase in IE, and a more favorable (more negative) EA. Consider Period 3 (Figure 8.21, *horizontal*): *elements at the left tend to form cations, and those at the right tend to form anions.* Sodium and magnesium are metals that occur as Na^+ and Mg^{2+} in seawater, minerals, and organisms.

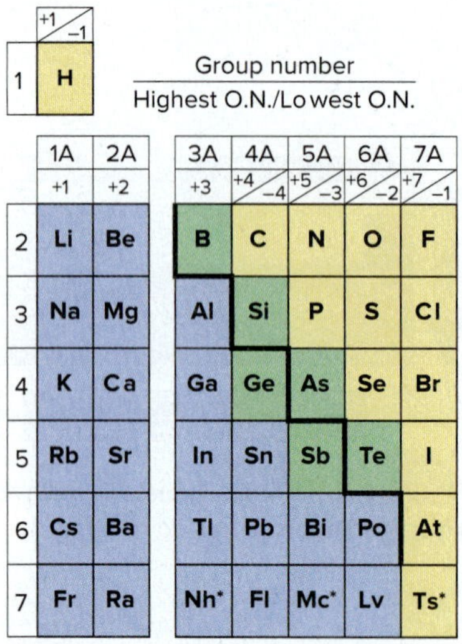

Figure 8.22 **Highest and lowest O.N.s of reactive main-group elements.**

*As of the printing of this book, these are the proposed symbols for these elements, awaiting ratification in early November 2016.

Aluminum is metallic physically and occurs as Al^{3+} in some compounds, but it bonds covalently in most others. Silicon (Si) is a shiny metalloid that does not occur as a monatomic ion. Phosphorus is a white, waxy nonmetal that occurs rarely as P^{3-}, whereas crumbly, yellow sulfur exists as S^{2-} in many compounds, and gaseous, yellow-green chlorine occurs in nature almost always as Cl^-.

Redox Behavior of the Main-Group Elements Closely related to an element's tendency to lose or gain electrons is its redox behavior—that is, whether it behaves as an oxidizing or reducing agent and the associated changes in its oxidation number (O.N.). You can find the highest and lowest oxidation numbers of many main-group elements from the periodic table (Figure 8.22):

- For most elements, the A-group number is the *highest* oxidation number (always positive) of any element in the group. The exceptions are O and F (see Table 4.4).
- For nonmetals and some metalloids, the A-group number minus 8 is the *lowest* oxidation number (always negative) of any element in the group.

For example, the highest oxidation number of S (Group 6A) is +6, as in SF_6, and the lowest is 6 − 8, or −2, as in FeS and other metal sulfides.

Redox behavior is closely related to atomic properties:

- With their low IEs and small EAs, the members of Groups 1A(1) and 2A(2) lose electrons readily, so they are *strong reducing agents* and become oxidized.
- With their high IEs and large EAs, nonmetals in Groups 6A(16) and 7A(17) gain electrons readily, so they are *strong oxidizing agents* and become reduced.

Acid-Base Behavior of Oxides Metals are also distinguished from nonmetals by the acid-base behavior of their oxides in water:

- Most main-group metals *transfer* electrons to oxygen, so their *oxides are ionic. In water, these oxides act as bases,* producing OH^- ions from O^{2-} and reacting with acids. Calcium oxide is an example (Figure 8.23).
- Nonmetals *share* electrons with oxygen, so *nonmetal oxides are covalent. They react with water to form acids,* producing H^+ ions and reacting with bases. Tetraphosphorus decoxide is an example (Figure 8.23).

Figure 8.23 **Oxide acidity.** In water, metal oxides are basic *(left),* and nonmetal oxides are acidic *(right).* The colors are due to the presence of acid-base indicators.

Source: © McGraw-Hill Education/Stephen Frisch, photographer

Some metals and many metalloids form oxides that are **amphoteric:** they can act as acids *or* bases in water.

In Figure 8.24, the acid-base behavior of some common oxides of elements in Group 5A(15) and Period 3 is shown, with a gradient from blue (basic) to red (acidic):

1. *As elements become more metallic down a group (larger size and smaller IE), their oxides become more basic.* In Group 5A(15), dinitrogen pentoxide, N_2O_5, forms the strong acid HNO_3:

$$N_2O_5(s) + H_2O(l) \longrightarrow 2HNO_3(aq)$$

Tetraphosphorus decoxide, P_4O_{10}, forms the weaker acid H_3PO_4:

$$P_4O_{10}(s) + 6H_2O(l) \longrightarrow 4H_3PO_4(aq)$$

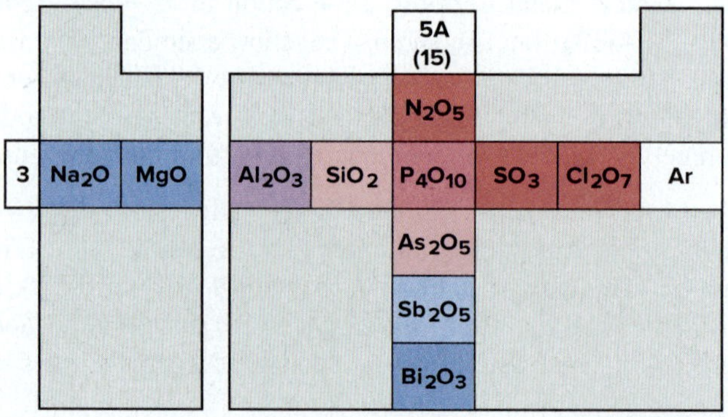

Figure 8.24 Acid-base behavior of some element oxides.

The oxide of the metalloid arsenic is weakly acidic, whereas that of the metalloid antimony is weakly basic. Bismuth, the most metallic of the group, forms a basic oxide that is insoluble in water but reacts with acid to yield a salt and water:

$$Bi_2O_3(s) + 6HNO_3(aq) \longrightarrow 2Bi(NO_3)_3(aq) + 3H_2O(l)$$

2. *As the elements become less metallic across a period (smaller size and higher IE), their oxides become more acidic.* In Period 3, Na_2O and MgO are strongly basic, and amphoteric aluminum oxide (Al_2O_3) reacts with acid or with base:

$$Al_2O_3(s) + 6HCl(aq) \longrightarrow 2AlCl_3(aq) + 3H_2O(l)$$

$$Al_2O_3(s) + 2NaOH(aq) + 3H_2O(l) \longrightarrow 2NaAl(OH)_4(aq)$$

Silicon dioxide is weakly acidic, forming a salt and water with base:

$$SiO_2(s) + 2NaOH(aq) \longrightarrow Na_2SiO_3(aq) + H_2O(l)$$

The common oxides of phosphorus, sulfur, and chlorine form acids of increasing strength: H_3PO_4, H_2SO_4, and $HClO_4$.

Properties of Monatomic Ions

So far we have focused on the reactants—the atoms—in the process of electron loss and gain. Now we focus on the products—the ions—considering their electron configurations, magnetic properties, and sizes.

Electron Configurations of Main-Group Ions Why does an ion have a particular charge: Na^+ not Na^{2+}, or F^- not F^{2-}? Why do some metals form two ions, such as Sn^{2+} and Sn^{4+}? The answer relates to the location of the element in the periodic table and the energy associated with losing or gaining electrons:

1. *Ions with a noble gas configuration.* Atoms of the noble gases have very low reactivity because their highest energy level is *filled* (ns^2np^6). Thus, *when elements at either end of a period form ions, they attain a filled outer level—a noble gas configuration.* These elements lie on either side of Group 8A(18), and their ions are **isoelectronic** (Greek *iso*, "same") with the nearest noble gas (Figure 8.25; see also Figure 2.13).

- Elements in Group 1A(1) *lose one* electron and elements in Group 2A(2) *lose two* electrons and become isoelectronic with the *previous* noble gas. The Na^+ ion, for example, is isoelectronic with neon (Ne):

$$Na(1s^22s^22p^63s^1) \longrightarrow e^- + Na^+([He]2s^22p^6) \text{ [isoelectronic with Ne([He] } 2s^22p^6)]$$

- Elements in Group 6A(16) *gain two* electrons and elements in Group 7A(17) *gain one* electron and become isoelectronic with the *next* noble gas. The Br^- ion, for example, is isoelectronic with krypton (Kr):

$$Br([Ar]\,4s^23d^{10}4p^5) + e^- \longrightarrow Br^-([Ar]\,4s^23d^{10}4p^6)$$
$$\text{[isoelectronic with Kr ([Ar] } 4s^23d^{10}4p^6)]$$

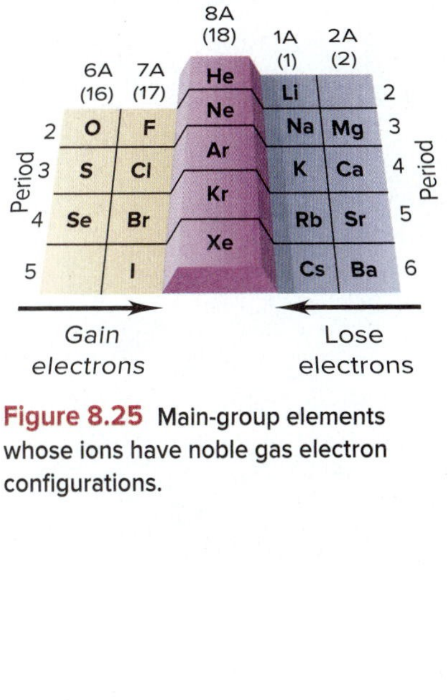

Figure 8.25 Main-group elements whose ions have noble gas electron configurations.

The energy needed to remove electrons from metals or add them to nonmetals determines the charges of the resulting ions:

- *Cations.* Removing another electron from Na^+ or from Mg^{2+} means removing a core electron, which requires too much energy: thus, $NaCl_2$ and MgF_3 do *not* exist.
- *Anions.* Similarly, adding another electron to F^- or to O^{2-} means putting it into the next higher energy level ($n = 3$). With 10 electrons ($1s^22s^22p^6$) acting as inner electrons, the nuclear charge would be shielded very effectively, and adding an outer electron would require too much energy: thus, we never see Na_2F or Mg_3O_2.

2. *Ions without a noble gas configuration.* Except for aluminum, the metals of Groups 3A(13) to 5A(15) do not form ions with noble gas configurations. Instead, they form cations with two different stable configurations:

- *Pseudo–noble gas configuration.* If the metal atom empties its highest energy level, it attains the stability of empty *ns* and *np* sublevels and a filled inner $(n - 1)d$ sublevel. This $(n - 1)d^{10}$ configuration is called a **pseudo–noble gas configuration.**

For example, tin (Sn; $Z = 50$) loses four electrons to form the tin(IV) ion (Sn^{4+}), which has empty 5*s* and 5*p* sublevels and a filled inner 4*d* sublevel:

$$\text{Sn}\ ([\text{Kr}]\ 5s^2 4d^{10} 5p^2) \longrightarrow \text{Sn}^{4+}([\text{Kr}]\ 4d^{10}) + 4e^-$$

- *Inert pair configuration.* Alternatively, the metal atom loses just its *np* electrons and attains a stable configuration with filled *ns* and $(n - 1)d$ sublevels. The retained ns^2 electrons are sometimes called an *inert pair*. For example, in the more common tin(II) ion (Sn^{2+}), the atom loses only the two 5*p* electrons and has filled 5*s* and 4*d* sublevels:

$$\text{Sn}\ ([\text{Kr}]\ 5s^2 4d^{10} 5p^2) \longrightarrow \text{Sn}^{2+}([\text{Kr}]\ 5s^2 4d^{10}) + 2e^-$$

Thallium, lead, and bismuth, the largest and, thus, most metallic members of Groups 3A(13) to 5A(15), form ions that retain the ns^2 pair: Tl^+, Pb^{2+}, and Bi^{3+}.

Once again, energy considerations explain these configurations. It would be energetically impossible for metals in Groups 3A(13) to 5A(15) to achieve noble gas configurations: tin, for example, would have to lose 14 electrons—ten 4*d* in addition to the two 5*p* and two 5*s*—to be isoelectronic with krypton (Kr; $Z = 36$), the previous noble gas.

SAMPLE PROBLEM 8.6 Writing Electron Configurations of Main-Group Ions

Problem Using condensed electron configurations, write equations representing the formation of the ion(s) of the following elements:

(a) Iodine ($Z = 53$) **(b)** Potassium ($Z = 19$) **(c)** Indium ($Z = 49$)

Plan We identify the element's position in the periodic table and recall that

- Ions of elements in Groups 1A(1), 2A(2), 6A(16), and 7A(17) are isoelectronic with the nearest noble gas.
- Metals in Groups 3A(13) to 5A(15) lose the *ns* and *np* electrons or just the *np*.

Solution (a) Iodine is in Group 7A(17), so it gains one electron, and I^- is isoelectronic with xenon:

$$\text{I}\ ([\text{Kr}]\ 5s^2 4d^{10} 5p^5) + e^- \longrightarrow \text{I}^-([\text{Kr}]\ 5s^2 4d^{10} 5p^6) \quad \text{(same as Xe)}$$

(b) Potassium is in Group 1A(1), so it loses one electron; K^+ is isoelectronic with argon:

$$\text{K}([\text{Ar}]4s^1) \longrightarrow \text{K}^+([\text{Ar}]) + e^-$$

(c) Indium is in Group 3A(13), so it loses either three electrons to form In^{3+} (with a pseudo–noble gas configuration) or one to form In^+ (with an inert pair):

$$\text{In}([\text{Kr}]5s^2 4d^{10} 5p^1) \longrightarrow \text{In}^{3+}([\text{Kr}]4d^{10}) + 3e^-$$

$$\text{In}([\text{Kr}]5s^2 4d^{10} 5p^1) \longrightarrow \text{In}^{+}([\text{Kr}]5s^2 4d^{10}) + e^-$$

Check Be sure that the number of electrons in the ion's electron configuration, plus the number gained or lost to form the ion, equals *Z*.

FOLLOW-UP PROBLEMS

8.6A Using condensed electron configurations, write equations representing the formation of the ion(s) of the following elements:

(a) Ba ($Z = 56$) **(b)** O ($Z = 8$) **(c)** Pb ($Z = 82$)

8.6B Using condensed electron configurations, write equations representing the formation of the ion(s) of the following elements:

(a) F ($Z = 9$) **(b)** Tl ($Z = 81$) **(c)** Mg ($Z = 12$)

SOME SIMILAR PROBLEMS 8.75–8.78

Electron Configurations of Transition Metal Ions In contrast to many main-group ions, *transition metal ions rarely attain a noble gas configuration.* Aside from the Period 4 elements scandium, which forms Sc^{3+}, and titanium, which occasionally

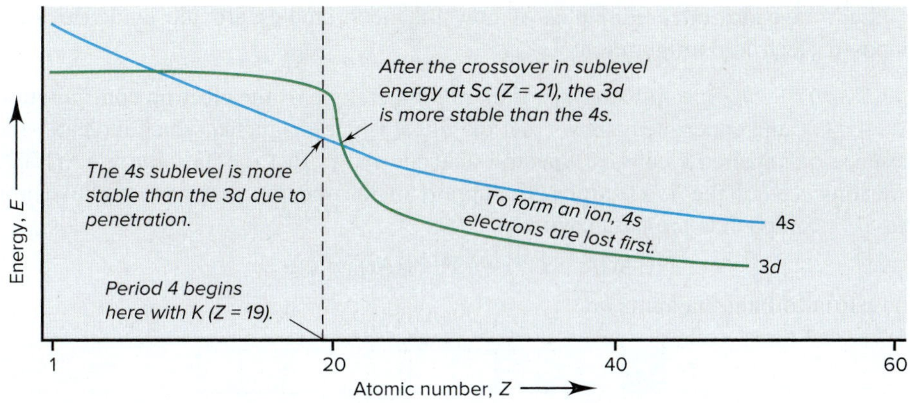

Figure 8.26 The crossover of sublevel energies in Period 4.

forms Ti^{4+}, *a transition element typically forms more than one cation by losing all of its ns and some of its (n − 1)d electrons.*

The reason, once again, is that the energy cost of attaining a noble gas configuration is too high. Let's consider again the filling of Period 4. At the beginning of Period 4 (the same point holds in other periods), penetration makes the 4*s* sublevel *more stable* than the 3*d*. Therefore, the first and second electrons added enter the 4*s*, which is the outer sublevel. But, the 3*d* is an *inner* sublevel, so as it begins to fill, its electrons are not well shielded from the increasing nuclear charge.

A *crossover in sublevel energy* results: the 3*d* becomes *more stable* than the 4*s* in the transition series (Figure 8.26). This crossover has a major effect on the formation of Period 4 transition metal ions: because the 3*d* electrons are held tightly and shield those in the outer sublevel, *the 4s electrons of a transition metal are lost **before** the 3d electrons.* Thus, 4*s* electrons are added before 3*d* electrons to form the *atom* and are lost before them to form the *ion,* the so-called "first-in, first-out" rule.

Ion Formation: A Summary of Electron Loss or Gain The various ways that cations form have one point in common—*outer electrons are removed first.* Here is a summary of the rules for formation of any main-group or transition metal ion:

- Main-group *s*-block metals lose all electrons with the highest *n* value.
- Main-group *p*-block metals lose *np* electrons before *ns* electrons.
- Transition (*d*-block) metals lose *ns* electrons before (*n* − 1)*d* electrons.
- Nonmetals gain electrons in the *p* orbitals of highest *n* value.

Magnetic Properties of Transition Metal Ions We learn a great deal about an element's electron configuration from atomic spectra, and magnetic studies provide additional evidence.

Recall that electron spin generates a tiny magnetic field, which causes a beam of H atoms to split in an external magnetic field (see Figure 8.1). Only a beam of a species (atoms, ions, or molecules) with one or more *unpaired* electrons will split. A beam of silver atoms (Ag; $Z = 47$) was used in the original 1921 experiment:

Ag ($Z = 47$) [Kr] $5s^1 4d^{10}$

5s	4d					5p		
↑	↑↓	↑↓	↑↓	↑↓	↑↓			

Note the unpaired 5*s* electron (recall that this irregularity is due to the stability of filled sublevels). A beam of cadmium atoms (Cd; $Z = 48$) is not split because the 5*s* electrons in this species are *paired* ([Kr] $5s^2 4d^{10}$).

A species with unpaired electrons exhibits **paramagnetism:** it is attracted by an external magnetic field. A species with all of its electrons paired exhibits **diamagnetism:** it is not attracted (and is slightly repelled) by the magnetic field (Figure 8.27). Many transition metals and their compounds are paramagnetic because their atoms and ions have unpaired electrons.

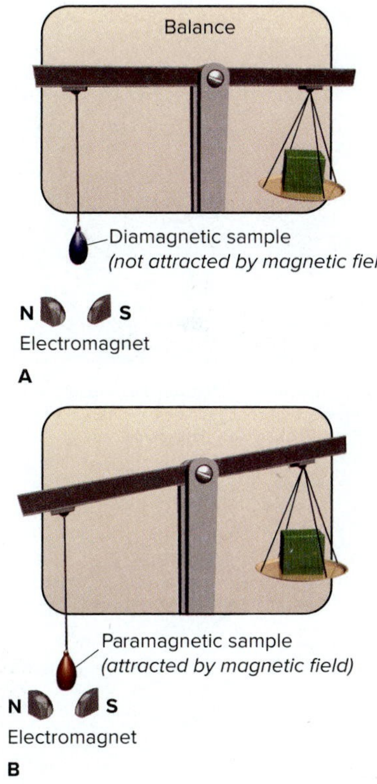

Figure 8.27 Measuring the magnetic behavior of a sample. The substance is weighed with the external magnetic field "off." **A,** If the substance is diamagnetic (has all *paired* electrons), its apparent mass is unaffected (or slightly reduced) with the field "on." **B,** If the substance is paramagnetic (has *unpaired* electrons), its apparent mass increases.

Let's consider three examples of how magnetic studies provide evidence for a proposed electron configuration:

1. *The Ti^{2+} ion.* Spectral analysis of titanium metal yields the electron configuration [Ar] $4s^23d^2$, and experiment shows that the metal is paramagnetic, which indicates the presence of unpaired electrons. Spectral analysis shows that the Ti^{2+} ion is [Ar] $3d^2$, indicating loss of the $4s$ electrons. In support of the spectra, magnetic studies show that Ti^{2+} compounds are paramagnetic:

$$\text{Ti}\,([\text{Ar}]4s^23d^2) \longrightarrow \text{Ti}^{2+}\,([\text{Ar}]\,3d^2) + 2e^-$$

The partial orbital diagrams are

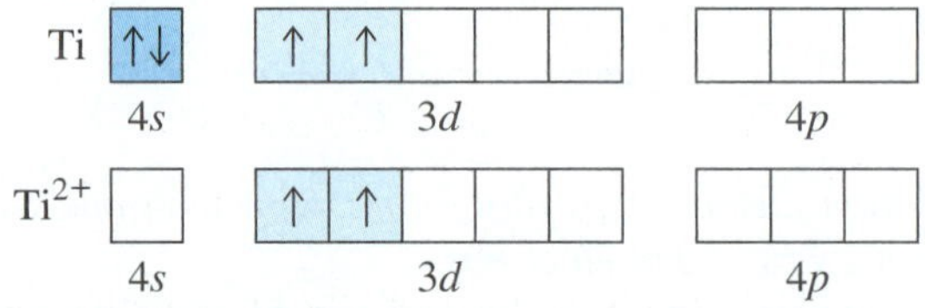

If Ti lost its $3d$ electrons to form Ti^{2+}, its compounds would be diamagnetic.

2. *The Fe^{3+} ion.* An increase in paramagnetism occurs when iron metal (Fe) becomes Fe^{3+} in compounds, which indicates an increase in the number of unpaired electrons. This fact is consistent with Fe losing its $4s$ pair and one electron of a $3d$ pair:

$$\text{Fe}\,([\text{Ar}]4s^23d^6) \longrightarrow \text{Fe}^{3+}\,([\text{Ar}]\,3d^5) + 3e^-$$

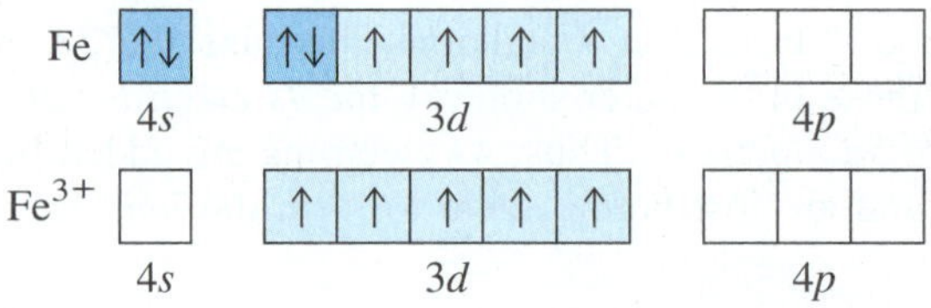

2. *The Cu^+ and Zn^{2+} ions.* Copper (Cu) metal is paramagnetic, but zinc (Zn) is diamagnetic. The Cu^+ and Zn^{2+} ions are diamagnetic, too. These observations are consistent with the ions being isoelectronic, which means $4s$ electrons were lost:

Student Hot Spot

Student data indicate that you may struggle with writing electron configurations of ions. Access the SmartBook to view additional Learning Resources on this topic.

$$\text{Cu}\,([\text{Ar}]4s^13d^{10}) \longrightarrow \text{Cu}^{+}\,([\text{Ar}]\,3d^{10}) + e^-$$

$$\text{Zn}\,([\text{Ar}]4s^23d^{10}) \longrightarrow \text{Zn}^{2+}\,([\text{Ar}]\,3d^{10}) + 2e^-$$

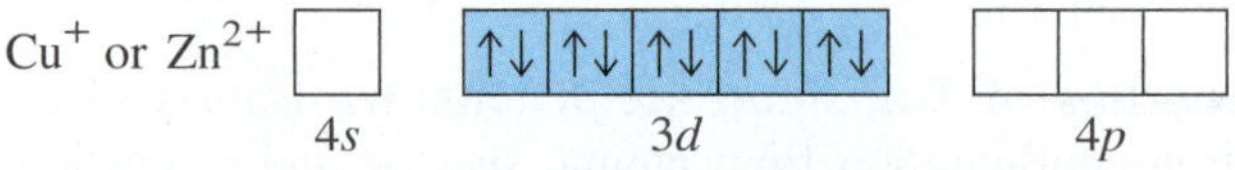

SAMPLE PROBLEM 8.7 Writing Electron Configurations and Predicting Magnetic Behavior of Transition Metal Ions

Problem Use condensed electron configurations to write an equation for the formation of each transition metal ion, and predict whether it is paramagnetic:

(a) Co^{3+} ($Z = 27$) **(b)** Cr^{3+} ($Z = 24$) **(c)** Hg^{2+} ($Z = 80$)

Plan We first write the condensed electron configuration of the atom, recalling the irregularity for Cr. Then we remove electrons, beginning with *ns* electrons, to attain the ion charge. If unpaired electrons are present, the ion is paramagnetic.

Solution

(a) $\text{Co}\,([\text{Ar}]4s^23d^7) \longrightarrow \text{Co}^{3+}\,([\text{Ar}]\,3d^6) + 3e^-$

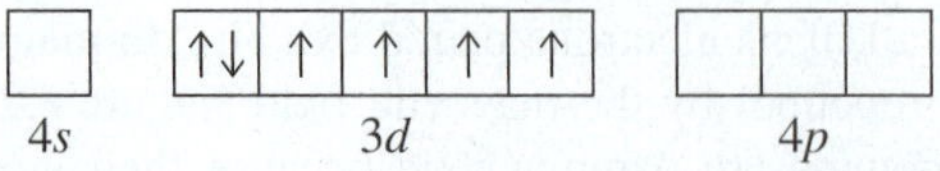

There are four unpaired e^-, so Co^{3+} is paramagnetic.

(b) $Cr([Ar]4s^1 3d^5) \longrightarrow Cr^{3+}([Ar]\,3d^3) + 3e^-$

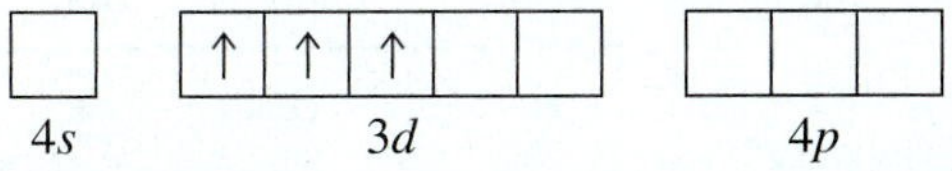

There are three unpaired e^-, so Cr^{3+} is paramagnetic.

(c) $Hg([Xe]6s^2 4f^{14} 5d^{10}) \longrightarrow Hg^{2+}([Xe]\,4f^{14} 5d^{10}) + 2e^-$

The $4f$ and $5d$ sublevels are filled, so there are no unpaired e^-: Hg^{2+} is *not* paramagnetic.

Check We removed the *ns* electrons first, and the sum of the lost electrons and those in the electron configuration of the ion equals Z.

FOLLOW-UP PROBLEMS

8.7A Write the condensed electron configuration of each transition metal ion, and predict whether it is paramagnetic:

(a) V^{3+} ($Z = 23$) **(b)** Ni^{2+} ($Z = 28$) **(c)** La^{3+} ($Z = 57$)

8.7B Write the condensed electron configuration of each transition metal ion, and give the number of unpaired electrons:

(a) Zr^{2+} ($Z = 40$) **(b)** Os^{3+} ($Z = 76$) **(c)** Co^{2+} ($Z = 27$)

SOME SIMILAR PROBLEMS 8.82–8.84

Ionic Size vs. Atomic Size The **ionic radius** is a measure of the size of an ion and is obtained from the distance between the nuclei of adjacent ions in a crystalline ionic compound (Figure 8.28). From the relation between effective nuclear charge (Z_{eff}) and atomic size, we can predict the size of an ion relative to its parent atom:

- *Cations are smaller than parent atoms.* When a cation forms, electrons are *removed from* the outer level. The resulting decrease in shielding and the value of *nl* allows the nucleus to pull the remaining electrons closer.
- *Anions are larger than parent atoms.* When an anion forms, electrons are *added to* the outer level. The increases in shielding and electron repulsions means the electrons occupy more space.

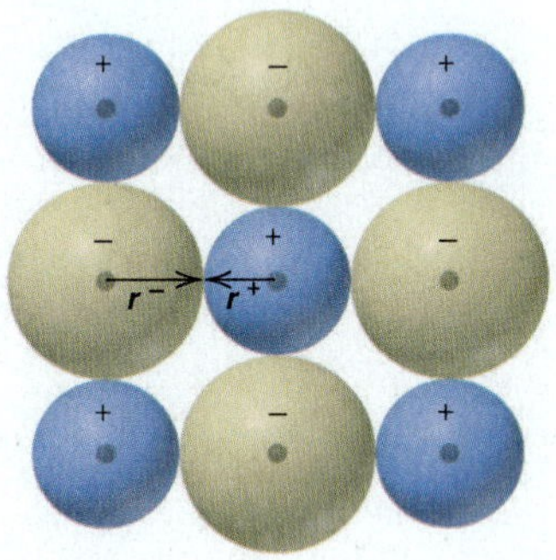

Figure 8.28 Ionic radius. Cation radius (r^+) and anion radius (r^-) together make up the distance between nuclei.

Figure 8.29 on the next page shows the radii of some main-group ions and their parent atoms; we can make the following observations:

1. *Down a group, ionic size increases* because *n* increases.
2. *Across a period,* the pattern is complex. For instance, consider Period 3:
   - *Among cations,* the increase in Z_{eff} from left to right makes Na^+ larger than Mg^{2+}, which is larger than Al^{3+}. Across a period, cation size decreases with increasing charge.
   - *From last cation to first anion,* a great jump in size occurs: we are *adding* electrons rather than removing them, so repulsions increase sharply. For instance, P^{3-} has eight more electrons than Al^{3+}.
   - *Among anions,* the increase in Z_{eff} from left to right makes P^{3-} larger than S^{2-}, which is larger than Cl^-. Across a period, anion size decreases with decreasing charge.
   - *Within an isoelectronic series,* these factors have striking results. Within the dashed outline in Figure 8.29, the ions are isoelectronic with neon. Period 2 anions are much larger than Period 3 cations because the same number of electrons are attracted by an increasing nuclear charge. The size pattern is

$$\begin{array}{ccccccccccc} 3- & > & 2- & > & 1- & > & 1+ & > & 2+ & > & 3+ \\ 7p^+ & & 8p^+ & & 9p^+ & & 11p^+ & & 12p^+ & & 13p^+ \\ 10e^- & & 10e^- & & 10e^- & & 10e^- & & 10e^- & & 10e^- \end{array}$$

Figure 8.29 Ionic vs. atomic radii. Atomic radii *(color)* and ionic radii *(gray)* are given in picometers. Metal atoms *(blue)* form *smaller* positive ions, and nonmetal atoms *(red)* form *larger* negative ions. Ions in the dashed outline are *isoelectronic* with neon.

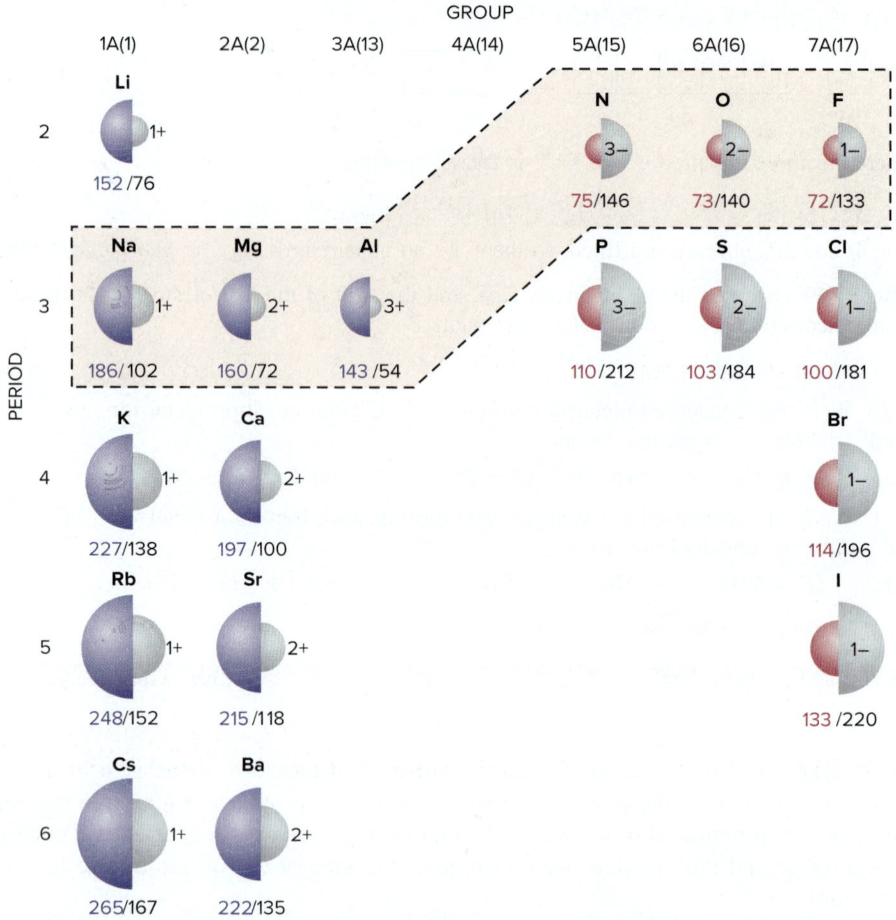

3. *Cation size decreases with increasing charge.* When a metal forms more than one cation, *the greater the ionic charge, the smaller the ionic radius.* With the two ions of iron, for example, Fe^{3+} has one fewer electron than Fe^{2+}, so shielding is reduced somewhat, and the same nucleus is attracting fewer electrons. As a result, Z_{eff} increases, so Fe^{3+} (65 pm) is smaller than Fe^{2+} (78 pm).

SAMPLE PROBLEM 8.8 Ranking Ions by Size

Problem Rank each set of ions in order of *decreasing* size, and explain your ranking:

(a) Ca^{2+}, Sr^{2+}, Mg^{2+} **(b)** K^+, S^{2-}, Cl^- **(c)** Au^+, Au^{3+}

Plan We find the position of each element in the periodic table and apply the ideas presented in the text.

Solution (a) Mg^{2+}, Ca^{2+}, and Sr^{2+} are all from Group 2A(2), so their sizes decrease up the group: $Sr^{2+} > Ca^{2+} > Mg^{2+}$.

(b) The ions K^+, S^{2-}, and Cl^- are isoelectronic. S^{2-} has a lower Z_{eff} than Cl^-, so it is larger. K^+ is a cation and has the highest Z_{eff}, so it is smallest: $S^{2-} > Cl^- > K^+$.

(c) Au^+ has a lower charge than Au^{3+}, so it is larger: $Au^+ > Au^{3+}$.

FOLLOW-UP PROBLEMS

8.8A Rank the ions in each set in order of *increasing* size:

(a) Cl^-, Br^-, F^- **(b)** Na^+, Mg^{2+}, F^- **(c)** Cr^{2+}, Cr^{3+}

8.8B Rank the ions in each set in order of *decreasing* size:

(a) S^{2-}, P^{3-}, Cl^- **(b)** Rb^+, K^+, Cs^+ **(c)** I^-, Ba^{2+}, Cs^+

SOME SIMILAR PROBLEMS 8.87 and 8.88

› Summary of Section 8.4

- Metallic behavior correlates with large atomic size and low ionization energy. Thus, metallic behavior increases down a group and decreases from left to right across a period.
- Elements in Groups 1A(1) and 2A(2) are strong reducing agents; nonmetals in Groups 6A(16) and 7A(17) are strong oxidizing agents.
- Within the main groups, metal oxides are basic and nonmetal oxides acidic. Thus, oxides become more acidic across a period and more basic down a group.
- Many main-group elements form ions that are isoelectronic with the nearest noble gas. Removing (or adding) more electrons than needed to attain the noble gas configuration requires a prohibitive amount of energy.
- Metals in Groups 3A(13) to 5A(15) lose either their *np* electrons or both their *ns* and *np* electrons.
- Transition metals lose *ns* electrons before $(n - 1)d$ electrons and commonly form more than one ion.
- Many transition metals and their compounds are paramagnetic because their atoms (or ions) have unpaired electrons.
- Cations are smaller and anions larger than their parent atoms. Ionic radius increases down a group. Across a period, ionic radii generally decrease, but a large increase occurs from the last cation to the first anion.

CHAPTER REVIEW GUIDE

Learning Objectives

Relevant section (§) and/or sample problem (SP) numbers appear in parentheses.

Understand These Concepts

1. The meaning of the periodic law and the arrangement of elements by atomic number (Introduction)
2. The reason for the spin quantum number and its two possible values (§8.1)
3. How the exclusion principle applies to orbital filling (§8.1)
4. The effects of nuclear charge, shielding, and penetration on the splitting of energy levels; the meaning of effective nuclear charge (§8.1)
5. How the arrangement of the periodic table is based on the order of sublevel energies (§8.2)
6. How sublevels are filled in main-group and transition elements; the importance of Hund's rule (§8.2)
7. The distinction among inner, outer, and valence electrons (§8.2)
8. How outer electron configuration within a group is related to chemical behavior (§8.2)
9. The meaning of atomic radius, ionization energy, and electron affinity (§8.3)
10. How *n* value and effective nuclear charge give rise to the periodic trends of atomic size and ionization energy (§8.3)
11. The importance of core electrons to the pattern of successive ionization energies (§8.3)
12. How atomic properties relate to the tendency to form ions (§8.3)
13. The general properties of metals and nonmetals (§8.4)
14. How vertical and horizontal trends in metallic behavior are related to ion formation and oxide acidity (§8.4)
15. How atomic size and IE relate to redox behavior (§8.4)
16. Why main-group ions are either isoelectronic with the nearest noble gas or have a pseudo–noble gas electron configuration (§8.4)
17. Why transition elements lose *ns* electrons first (§8.4)
18. The origin of paramagnetic and diamagnetic behavior (§8.4)
19. The relation between ionic and atomic size and the trends in ionic size (§8.4)

Master These Skills

1. Correlating an orbital diagram and the set of quantum numbers for any electron in an atom (SP 8.1)
2. Writing full and condensed electron configurations for an element (SP 8.2)
3. Using periodic trends to rank elements by atomic size and first ionization energy (SPs 8.3, 8.4)
4. Identifying an element from its successive ionization energies (SP 8.5)
5. Writing electron configurations of main-group and transition metal ions and predicting magnetic behavior of transition metal ions (SPs 8.6, 8.7)
6. Using periodic trends to rank ions by size (SP 8.8)

Key Terms

Page numbers appear in parentheses.

actinides (342)
amphoteric (354)
atomic size (345)
aufbau principle (335)
covalent radius (345)
diamagnetism (357)
effective nuclear charge (Z_{eff}) (333)
electron affinity (EA) (351)
electron configuration (331)
exclusion principle (333)
Hund's rule (336)
inner (core) electrons (342)
inner transition elements (342)
ionic radius (359)
ionization energy (IE) (348)

isoelectronic (355)
lanthanides (rare earths) (342)
metallic radius (345)
orbital diagram (335)
outer electrons (342)
paramagnetism (357)
penetration (334)
periodic law (331)
pseudo–noble gas configuration (355)
shielding (333)
spin quantum number (m_s) (332)
transition elements (339)
valence electrons (342)

Key Equations and Relationships

Page numbers appear in parentheses.

8.1 Defining the energy order of sublevels in terms of the angular momentum quantum number (l value) (334):

Order of sublevel energies: $s < p < d < f$

8.2 Meaning of the first ionization energy (348):

$$\text{Atom}(g) \longrightarrow \text{ion}^+(g) + e^- \qquad \Delta E = IE_1 > 0$$

BRIEF SOLUTIONS TO FOLLOW-UP PROBLEMS

8.1A The element has eight electrons (two in the 1s orbital, two in the 2s orbital, and four in 2p orbitals), so $Z = 8$: oxygen.

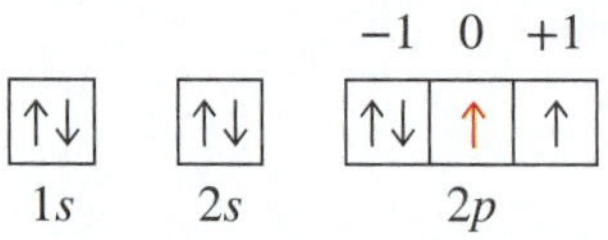

The sixth electron (shown in red) is the first electron in the 2p orbital whose m_l value is 0: $n = 2$, $l = 1$, $m_l = 0$, $+\frac{1}{2}$

8.1B Since $n = 2$ and $l = 1$, the electron is in a 2p orbital; since $m_l = 0$ and $m_s = +\frac{1}{2}$, the electron is the second electron in the 2p sublevel. The 1s and 2s orbitals are full and there are two electrons in the 2p sublevel for a total of 6 electrons. $Z = 6$ and the element is carbon; $1s^22s^22p^2$.

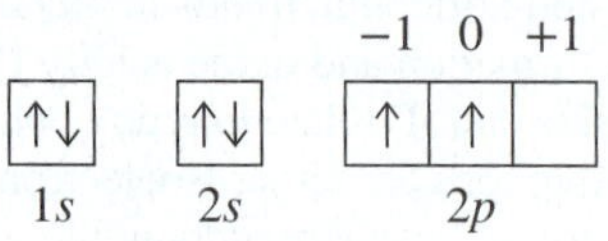

8.2A (a) For Ni, $1s^22s^22p^63s^23p^64s^23d^8$; [Ar] $4s^23d^8$

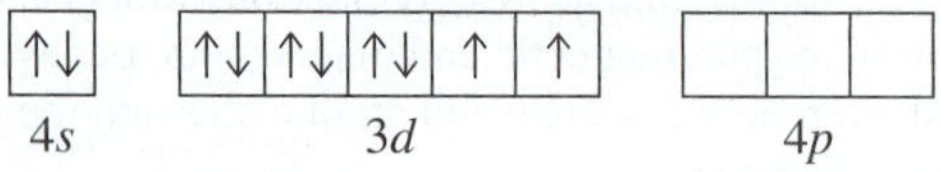

Ni has 18 inner electrons.
(b) For Sr, $1s^22s^22p^63s^23p^64s^23d^{10}4p^65s^2$; [Kr] $5s^2$

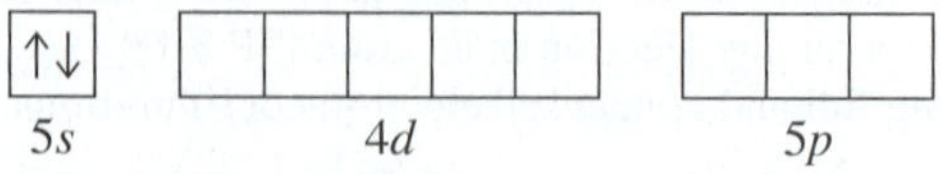

Sr has 36 inner electrons.
(c) For Po, $1s^22s^22p^63s^23p^64s^23d^{10}4p^65s^24d^{10}5p^66s^24f^{14}5d^{10}6p^4$; [Xe] $6s^24f^{14}5d^{10}6p^4$

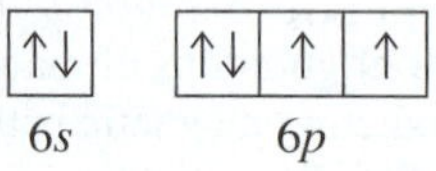

Po has 78 inner electrons.

8.2B (a) As; $1s^22s^22p^63s^23p^64s^23d^{10}4p^3$; [Ar] $4s^23d^{10}4p^3$;
As has 28 inner electrons.
(b) Zr; $1s^22s^22p^63s^23p^64s^23d^{10}4p^65s^24d^2$; [Kr] $5s^24d^2$;
Zr has 36 inner electrons.
(c) I; $1s^22s^22p^63s^23p^64s^23d^{10}4p^65s^24d^{10}5p^5$; [Kr] $5s^24d^{10}5p^5$;
I has 46 inner electrons.

8.3A (a) Cl < Br < Se; (b) Xe < I < Ba

8.3B (a) Cs > As > S; (b) K > P > F

8.4A (a) Sn < Sb < I; (b) Cs < Na < Mg

8.4B (a) O > As > Rb; (b) Cl > Si > Sn

8.5A The large jump occurs after IE_3, indicating that the element has three valence electrons and is in Group 3A(13). Q is aluminum: $1s^22s^22p^63s^23p^1$.

8.5B Rb: [Kr] $5s^1$; Sr: [Kr] $5s^2$; Y: [Kr] $5s^24d^1$
Highest IE_2: Rb, since the second electron removed is a core electron, which is much more difficult to remove. Highest IE_3: Sr, since the third electron removed is a core electron.

8.6A (a) $\text{Ba}([\text{Xe}]\,6s^2) \longrightarrow \text{Ba}^{2+}([\text{Xe}]) + 2e^-$

(b) $\text{O}([\text{He}]\,2s^22p^4) + 2e^- \longrightarrow \text{O}^{2-}([\text{He}]\,2s^22p^6)$ (same as Ne)

(c) $\text{Pb}([\text{Xe}]\,6s^24f^{14}5d^{10}6p^2) \longrightarrow \text{Pb}^{2+}([\text{Xe}]\,6s^24f^{14}5d^{10}) + 2e^-$
$\text{Pb}([\text{Xe}]\,6s^24f^{14}5d^{10}6p^2) \longrightarrow \text{Pb}^{4+}([\text{Xe}]\,4f^{14}5d^{10}) + 4e^-$

8.6B (a) $\text{F}([\text{He}]\,2s^22p^5) + e^- \longrightarrow \text{F}^-([\text{He}]\,2s^22p^6)$ (same as Ne)

(b) $\text{Tl}([\text{Xe}]\,6s^24f^{14}5d^{10}6p^1) \longrightarrow \text{Tl}^{+}([\text{Xe}]\,6s^24f^{14}5d^{10}) + e^-$
$\text{Tl}([\text{Xe}]\,6s^24f^{14}5d^{10}6p^1) \longrightarrow \text{Tl}^{3+}([\text{Xe}]\,4f^{14}5d^{10}) + 3e^-$

(c) $\text{Mg}([\text{Ne}]\,3s^2) \longrightarrow \text{Mg}^{2+}([\text{Ne}]) + 2e^-$

8.7A (a) V: [Ar] $4s^23d^3$; V^{3+}: [Ar] $3d^2$; paramagnetic
(b) Ni: [Ar] $4s^23d^8$; Ni^{2+}: [Ar] $3d^8$; paramagnetic
(c) La: [Xe] $6s^25d^1$; La^{3+}: [Xe]; not paramagnetic (diamagnetic)

8.7B (a) Zr: [Kr] $5s^24d^2$; Zr^{2+}: [Kr] $4d^2$; 2 unpaired electrons
(b) Os: [Xe] $6s^24f^{14}5d^6$; Os^{3+}: [Xe] $4f^{14}5d^5$; 5 unpaired electrons
(c) Co: [Ar] $4s^23d^7$; Co^{2+}: [Ar] $3d^7$; 3 unpaired electrons

8.8A (a) $F^- < Cl^- < Br^-$ Ion size increases down a group.
(b) $Mg^{2+} < Na^+ < F^-$ Across a period, cation size decreases with increasing charge; in a series of isoelectronic ions, anions are larger than the cations.
(c) $Cr^{3+} < Cr^{2+}$ Cation size decreases with increasing charge.

8.8B (a) $P^{3-} > S^{2-} > Cl^-$ Anion size decreases with decreasing charge.
(b) $Cs^+ > Rb^+ > K^+$ Ion size increases down a group.
(c) $I^- > Cs^+ > Ba^{2+}$ Across a period, cation size decreases with increasing charge; in a series of isoelectronic ions, anions are larger than the cations.

PROBLEMS

Problems with **colored** numbers are answered in Appendix E and worked in detail in the Student Solutions Manual. Problem sections match those in the text and give the numbers of relevant sample problems. Most offer Concept Review Questions, Skill-Building Exercises (grouped in pairs covering the same concept), and Problems in Context. Comprehensive Problems are based on material from any section or previous chapter.

Introduction

Concept Review Questions

8.1 What would be your reaction to a claim that a new element had been discovered and it fit between tin (Sn) and antimony (Sb) in the periodic table?

8.2 Based on results of his study of atomic x-ray spectra, Moseley discovered a relationship that replaced atomic mass as the criterion for ordering the elements. By what criterion are the elements now ordered in the periodic table? Give an example of a sequence of element order that was confirmed by Moseley's findings.

Skill-Building Exercises (grouped in similar pairs)

8.3 Before Mendeleev published his periodic table, German scientist Johann Döbereiner grouped elements with similar properties into "triads," in which the unknown properties of one member could be predicted by averaging known values of the properties of the others. To test this idea, predict the values of the following quantities:
(a) The atomic mass of K from the atomic masses of Na and Rb
(b) The melting point of Br_2 from the melting points of Cl_2 (−101.0°C) and I_2 (113.6°C) (actual value = −7.2°C)

8.4 To test Döbereiner's idea (Problem 8.3), predict:
(a) The boiling point of HBr from the boiling points of HCl (−84.9°C) and HI (−35.4°C) (actual value = −67.0°C)
(b) The boiling point of AsH_3 from the boiling points of PH_3 (−87.4°C) and SbH_3 (−17.1°C) (actual value = −55°C)

Characteristics of Many-Electron Atoms

Concept Review Questions

8.5 Summarize the rules for the allowable values of the four quantum numbers of an electron in an atom.

8.6 Which of the quantum numbers relate(s) to the electron only? Which relate(s) to the orbital?

8.7 State the exclusion principle. What does it imply about the number and spin of electrons in an atomic orbital?

8.8 What is the key distinction between sublevel energies in one-electron species, such as the H atom, and those in many-electron species, such as the C atom? What factors lead to this distinction? Would you expect the pattern of sublevel energies in Be^{3+} to be more like that in H or that in C? Explain.

8.9 Define *shielding* and *effective nuclear charge.* What is the connection between the two?

8.10 What is penetration? How is it related to shielding? Use the penetration effect to explain the difference in relative orbital energies of a 3*p* and a 3*d* electron in the same atom.

Skill-Building Exercises (grouped in similar pairs)

8.11 How many electrons in an atom can have each of the following quantum number or sublevel designations?
(a) $n = 2, l = 1$ (b) 3*d* (c) 4*s*

8.12 How many electrons in an atom can have each of the following quantum number or sublevel designations?
(a) $n = 2, l = 1, m_l = 0$ (b) 5*p* (c) $n = 4, l = 3$

8.13 How many electrons in an atom can have each of the following quantum number or sublevel designations?
(a) 4*p* (b) $n = 3, l = 1, m_l = +1$ (c) $n = 5, l = 3$

8.14 How many electrons in an atom can have each of the following quantum number or sublevel designations?
(a) 2*s* (b) $n = 3, l = 2$ (c) 6*d*

The Quantum-Mechanical Model and the Periodic Table

(Sample Problems 8.1 and 8.2)

Concept Review Questions

8.15 State the periodic law, and explain its relation to electron configuration. (Use Na and K in your explanation.)

8.16 State Hund's rule in your own words, and show its application in the orbital diagram of the nitrogen atom.

8.17 How does the aufbau principle, in connection with the periodic law, lead to the format of the periodic table?

8.18 For main-group elements, are outer electron configurations similar or different within a group? Within a period? Explain.

8.19 For which blocks of elements are outer electrons the same as valence electrons? For which elements are *d* electrons often included among the valence electrons?

8.20 What is the electron capacity of the *n*th energy level? What is the capacity of the fourth energy level?

Skill-Building Exercises (grouped in similar pairs)

8.21 Write a full set of possible quantum numbers for each of the following:
(a) The outermost electron in an Rb atom
(b) The electron gained when an S^- ion becomes an S^{2-} ion
(c) The electron lost when an Ag atom ionizes
(d) The electron gained when an F^- ion forms from an F atom

8.22 Write a full set of possible quantum numbers for each of the following:
(a) The outermost electron in an Li atom
(b) The electron gained when a Br atom becomes a Br^- ion
(c) The electron lost when a Cs atom ionizes
(d) The highest energy electron in the ground-state B atom

8.23 Write the full ground-state electron configuration for each:
(a) Rb (b) Ge (c) Ar

8.24 Write the full ground-state electron configuration for each:
(a) Br (b) Mg (c) Se

8.25 Write the full ground-state electron configuration for each:
(a) Cl (b) Si (c) Sr

8.26 Write the full ground-state electron configuration for each:
(a) S (b) Kr (c) Cs

8.27 Draw a partial (valence-level) orbital diagram, and write the condensed ground-state electron configuration for each:
(a) Ti (b) Cl (c) V

8.28 Draw a partial (valence-level) orbital diagram, and write the condensed ground-state electron configuration for each:
(a) Ba (b) Co (c) Ag

8.29 Draw a partial (valence-level) orbital diagram, and write the condensed ground-state electron configuration for each:
(a) Mn (b) P (c) Fe

8.30 Draw a partial (valence-level) orbital diagram, and write the condensed ground-state electron configuration for each:
(a) Ga (b) Zn (c) Sc

8.31 Draw the partial (valence-level) orbital diagram, and write the symbol, group number, and period number of the element:
(a) [He] $2s^22p^4$ (b) [Ne] $3s^23p^3$

8.32 Draw the partial (valence-level) orbital diagram, and write the symbol, group number, and period number of the element:
(a) [Kr] $5s^24d^{10}$ (b) [Ar] $4s^23d^8$

8.33 Draw the partial (valence-level) orbital diagram, and write the symbol, group number, and period number of the element:
(a) [Ne] $3s^23p^5$ (b) [Ar] $4s^23d^{10}4p^3$

8.34 Draw the partial (valence-level) orbital diagram, and write the symbol, group number, and period number of the element:
(a) [Ar] $4s^23d^5$ (b) [Kr] $5s^24d^2$

8.35 From each partial (valence-level) orbital diagram, write the condensed electron configuration and group number:

(a)
(b)
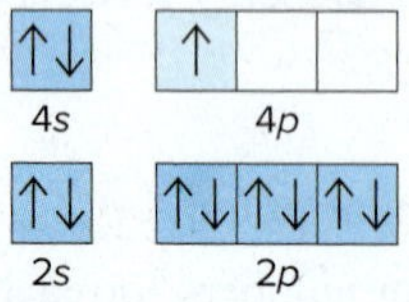

8.36 From each partial (valence-level) orbital diagram, write the condensed electron configuration and group number:

(a)
(b)
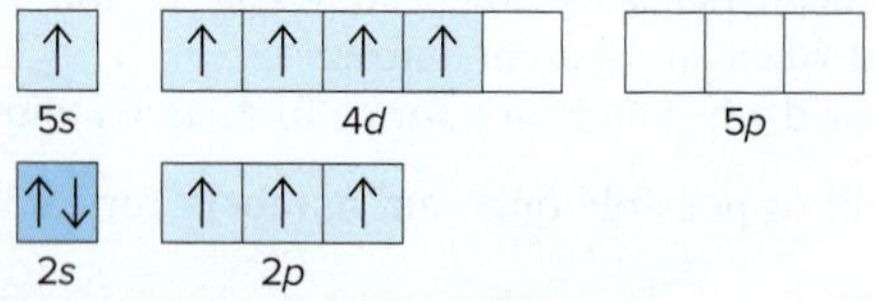

8.37 How many inner, outer, and valence electrons are present in an atom of each of the following elements?
(a) O (b) Sn (c) Ca (d) Fe (e) Se

8.38 How many inner, outer, and valence electrons are present in an atom of each of the following elements?
(a) Br (b) Cs (c) Cr (d) Sr (e) F

8.39 Identify each element below, and give the symbols of the other elements in its group:
(a) [He] $2s^22p^1$ (b) [Ne] $3s^23p^4$ (c) [Xe] $6s^25d^1$

8.40 Identify each element below, and give the symbols of the other elements in its group:
(a) [Ar] $4s^23d^{10}4p^4$ (b) [Xe] $6s^24f^{14}5d^2$ (c) [Ar] $4s^23d^5$

8.41 Identify each element below, and give the symbols of the other elements in its group:
(a) [He] $2s^22p^2$ (b) [Ar] $4s^23d^3$ (c) [Ne] $3s^23p^3$

8.42 Identify each element below, and give the symbols of the other elements in its group:
(a) [Ar] $4s^23d^{10}4p^2$ (b) [Ar] $4s^23d^7$ (c) [Kr] $5s^24d^5$

Problems in Context

8.43 After an atom in its ground state absorbs energy, it exists in an excited state. Spectral lines are produced when the atom returns to its ground state. The yellow-orange line in the sodium spectrum, for example, is produced by the emission of energy when excited sodium atoms return to their ground state. Write the electron configuration and the orbital diagram of the first excited state of sodium. (*Hint:* The outermost electron is excited.)

8.44 One reason spectroscopists study excited states is to gain information about the energies of orbitals that are unoccupied in an atom's ground state. Each of the following electron configurations represents an atom in an excited state. Identify the element, and write its condensed ground-state configuration:
(a) $1s^22s^22p^63s^13p^1$ (b) $1s^22s^22p^63s^23p^44s^1$
(c) $1s^22s^22p^63s^23p^64s^23d^44p^1$ (d) $1s^22s^22p^53s^1$

Trends in Three Atomic Properties
(Sample Problems 8.3 to 8.5)

Concept Review Questions

8.45 If the exact outer limit of an isolated atom cannot be measured, what criterion can we use to determine atomic radii? What is the difference between a covalent radius and a metallic radius?

8.46 Given the following partial (valence-level) electron configurations, (a) identify each element, (b) rank the four elements in order of increasing atomic size, and (c) rank them in order of increasing ionization energy:

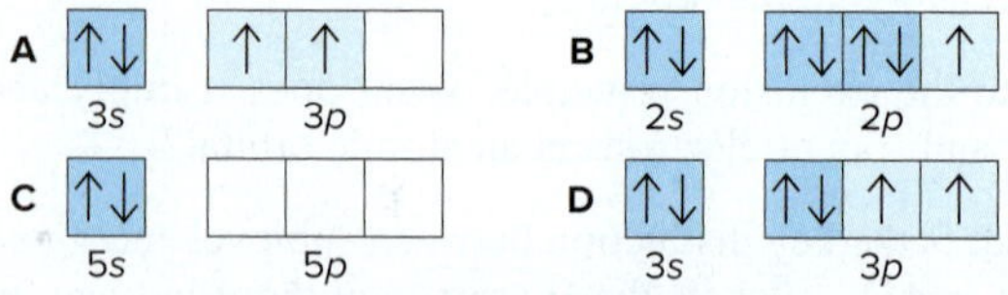

8.47 In what region of the periodic table will you find elements with relatively high IEs? With relatively low IEs?

8.48 (a) Why do successive IEs of a given element always increase? (b) When the difference between successive IEs of a given element is exceptionally large (for example, between IE_1

and IE_2 of K), what do we learn about the element's electron configuration? (c) The bars represent the relative magnitudes of the first five ionization energies of an atom:

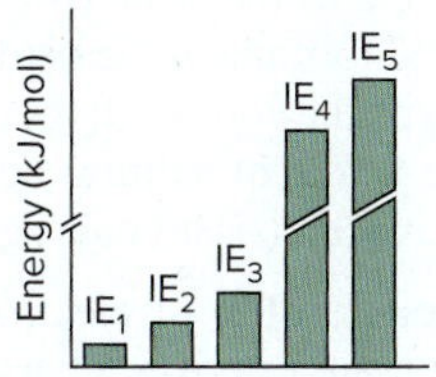

Identify the element and write its complete electron configuration, assuming it comes from (i) Period 2; (ii) Period 3; (iii) Period 4.

8.49 In a plot of IE_1 for the Period 3 elements (see Figure 8.15), why do the values for elements in Groups 3A(13) and 6A(16) drop slightly below the generally increasing trend?

8.50 Which group in the periodic table has elements with high (endothermic) IE_1 and very negative (exothermic) first electron affinities (EA_1)? Give the charge on the ions these atoms form.

8.51 The EA_2 of an oxygen atom is positive, even though its EA_1 is negative. Why does this change of sign occur? Which other elements exhibit a positive EA_2? Explain.

8.52 How does *d*-electron shielding influence atomic size among the Period 4 transition elements?

Skill-Building Exercises (grouped in similar pairs)

8.53 Arrange each set in order of *increasing* atomic size:
(a) Rb, K, Cs (b) C, O, Be
(c) Cl, K, S (d) Mg, K, Ca

8.54 Arrange each set in order of *decreasing* atomic size:
(a) Ge, Pb, Sn (b) Sn, Te, Sr
(c) F, Ne, Na (d) Be, Mg, Na

8.55 Arrange each set of atoms in order of *increasing* IE_1:
(a) Sr, Ca, Ba (b) N, B, Ne
(c) Br, Rb, Se (d) As, Sb, Sn

8.56 Arrange each set of atoms in order of *decreasing* IE_1:
(a) Na, Li, K (b) Be, F, C
(c) Cl, Ar, Na (d) Cl, Br, Se

8.57 Write the full electron configuration of the Period 2 element with the following successive IEs (in kJ/mol):
$IE_1 = 801$ $IE_2 = 2427$ $IE_3 = 3659$
$IE_4 = 25{,}022$ $IE_5 = 32{,}822$

8.58 Write the full electron configuration of the Period 3 element with the following successive IEs (in kJ/mol):
$IE_1 = 738$ $IE_2 = 1450$ $IE_3 = 7732$
$IE_4 = 10{,}539$ $IE_5 = 13{,}628$

8.59 Which element in each of the following sets would you expect to have the *highest* IE_2?
(a) Na, Mg, Al (b) Na, K, Fe
(c) Sc, Be, Mg

8.60 Which element in each of the following sets would you expect to have the *lowest* IE_3?
(a) Na, Mg, Al (b) K, Ca, Sc
(c) Li, Al, B

Atomic Properties and Chemical Reactivity
(Sample Problems 8.6 to 8.8)

Concept Review Questions

8.61 List three ways in which metals and nonmetals differ.

8.62 Summarize the trend in metallic character as a function of position in the periodic table. Is it the same as the trend in atomic size? The trend in ionization energy?

8.63 Explain the relationship between atomic size and reducing strength in Group 1A(1). Explain the relationship between IE and oxidizing strength in Group 7A(17).

8.64 Summarize the acid-base behavior of the main-group metal and nonmetal oxides in water. How does oxide acidity in water change down a group and across a period?

8.65 What ions are possible for the two largest stable elements in Group 4A(14)? How does each arise?

8.66 What is a pseudo–noble gas configuration? Give an example of one ion from Group 3A(13) that has it.

8.67 How are measurements of paramagnetism used to support electron configurations derived spectroscopically? Use Cu(I) and Cu(II) chlorides as examples.

8.68 The charges of a set of isoelectronic ions vary from 3+ to 3−. Place the ions in order of increasing size.

Skill-Building Exercises (grouped in similar pairs)

8.69 Which element would you expect to be *more* metallic?
(a) Ca or Rb (b) Mg or Ra (c) Br or I

8.70 Which element would you expect to be *more* metallic?
(a) S or Cl (b) In or Al (c) As or Br

8.71 Which element would you expect to be *less* metallic?
(a) Sb or As (b) Si or P (c) Be or Na

8.72 Which element would you expect to be *less* metallic?
(a) Cs or Rn (b) Sn or Te (c) Se or Ge

8.73 Does the reaction of a main-group nonmetal oxide in water produce an acidic or a basic solution? Write a balanced equation for the reaction of a Group 6A(16) nonmetal oxide with water.

8.74 Does the reaction of a main-group metal oxide in water produce an acidic or a basic solution? Write a balanced equation for the reaction of a Group 2A(2) oxide with water.

8.75 Write the charge and full ground-state electron configuration of the monatomic ion most likely to be formed by each:
(a) Cl (b) Na (c) Ca

8.76 Write the charge and full ground-state electron configuration of the monatomic ion most likely to be formed by each:
(a) Rb (b) N (c) Br

8.77 Write the charge and full ground-state electron configuration of the monatomic ion most likely to be formed by each:
(a) Al (b) S (c) Sr

8.78 Write the charge and full ground-state electron configuration of the monatomic ion most likely to be formed by each:
(a) P (b) Mg (c) Se

8.79 How many unpaired electrons are present in a ground-state atom from each of the following groups?
(a) 2A(2) (b) 5A(15) (c) 8A(18) (d) 3A(13)

8.80 How many unpaired electrons are present in a ground-state atom from each of the following groups?
(a) 4A(14) (b) 7A(17) (c) 1A(1) (d) 6A(16)

8.81 Which of these atoms are paramagnetic in their ground state?
(a) Ga (b) Si (c) Be (d) Te

8.82 Are compounds of these ground-state ions paramagnetic?
(a) Ti^{2+} (b) Zn^{2+} (c) Ca^{2+} (d) Sn^{2+}

8.83 Write the condensed ground-state electron configurations of these transition metal ions, and state which are paramagnetic:
(a) V^{3+} (b) Cd^{2+} (c) Co^{3+} (d) Ag^{+}

8.84 Write the condensed ground-state electron configurations of these transition metal ions, and state which are paramagnetic:
(a) Mo^{3+} (b) Au^{+} (c) Mn^{2+} (d) Hf^{2+}

8.85 Palladium (Pd; $Z = 46$) is diamagnetic. Draw partial orbital diagrams to show which of the following electron configurations is consistent with this fact:
(a) [Kr] $5s^2 4d^8$ (b) [Kr] $4d^{10}$ (c) [Kr] $5s^1 4d^9$

8.86 Niobium (Nb; $Z = 41$) has an anomalous ground-state electron configuration for a Group 5B(5) element: [Kr] $5s^1 4d^4$. What is the expected electron configuration for elements in this group? Draw partial orbital diagrams to show how paramagnetic measurements could support niobium's actual configuration.

8.87 Rank the ions in each set in order of *increasing* size, and explain your ranking:
(a) Li^+, K^+, Na^+ (b) Se^{2-}, Rb^+, Br^- (c) O^{2-}, F^-, N^{3-}

8.88 Rank the ions in each set in order of *decreasing* size, and explain your ranking:
(a) Se^{2-}, S^{2-}, O^{2-} (b) Te^{2-}, Cs^+, I^- (c) Sr^{2+}, Ba^{2+}, Cs^+

Comprehensive Problems

8.89 Name the element described in each of the following:
(a) Smallest atomic radius in Group 6A(16)
(b) Largest atomic radius in Period 6
(c) Smallest metal in Period 3
(d) Highest IE_1 in Group 4A(14)
(e) Lowest IE_1 in Period 5
(f) Most metallic in Group 5A(15)
(g) Group 3A(13) element that forms the most basic oxide
(h) Period 4 element with highest energy level filled
(i) Condensed ground-state electron configuration of [Ne] $3s^2 3p^2$
(j) Condensed ground-state electron configuration of [Kr] $5s^2 4d^6$
(k) Forms 2+ ion with electron configuration [Ar] $3d^3$
(l) Period 5 element that forms 3+ ion with pseudo–noble gas configuration
(m) Period 4 transition element that forms 3+ diamagnetic ion
(n) Period 4 transition element that forms 2+ ion with a half-filled *d* sublevel
(o) Heaviest lanthanide
(p) Period 3 element whose 2− ion is isoelectronic with Ar
(q) Alkaline earth metal whose cation is isoelectronic with Kr
(r) Group 5A(15) metalloid with the most acidic oxide

8.90 Use electron configurations to account for the stability of the lanthanide ions Ce^{4+} and Eu^{2+}.

8.91 When a nonmetal oxide reacts with water, it forms an oxoacid in which the nonmetal retains the same oxidation number as in the oxide. Give the name and formula of the oxide used to prepare each of these oxoacids: (a) hypochlorous acid; (b) chlorous acid; (c) chloric acid; (d) perchloric acid; (e) sulfuric acid; (f) sulfurous acid; (g) nitric acid; (h) nitrous acid; (i) carbonic acid; (j) phosphoric acid.

8.92 A fundamental relationship of electrostatics states that the energy required to separate opposite charges of magnitudes Q_1 and Q_2 that are a distance d apart is proportional to $\frac{Q_1 \times Q_2}{d}$. Use this relationship and any other relevant factors to explain the following:
(a) The IE_2 of He ($Z = 2$) is *more* than twice the IE_1 of H ($Z = 1$).
(b) The IE_1 of He is *less* than twice the IE_1 of H.

8.93 The energy difference between the 5*d* and 6*s* sublevels in gold accounts for its color. Assuming this energy difference is about 2.7 eV (electron volts; 1 eV = 1.602×10^{-19} J), explain why gold has a warm yellow color.

8.94 Write the formula and name of the compound formed from the following ionic interactions: (a) The 2+ ion and the 1− ion are both isoelectronic with the atoms of a chemically unreactive Period 4 element. (b) The 2+ ion and the 2− ion are both isoelectronic with the Period 3 noble gas. (c) The 2+ ion is the smallest with a filled *d* sublevel; the anion forms from the smallest halogen. (d) The ions form from the largest and smallest ionizable atoms in Period 2.

8.95 The energy changes for many unusual reactions can be determined using Hess's law (Section 6.5).
(a) Calculate ΔE for the conversion of $F^-(g)$ into $F^+(g)$.
(b) Calculate ΔE for the conversion of $Na^+(g)$ into $Na^-(g)$.

8.96 Discuss each conclusion from a study of redox reactions:
(a) The sulfide ion functions only as a reducing agent.
(b) The sulfate ion functions only as an oxidizing agent.
(c) Sulfur dioxide functions as an oxidizing or a reducing agent.

8.97 The hot glowing gases around the Sun, the *corona*, can reach millions of degrees Celsius, temperatures high enough to remove many electrons from gaseous atoms. Iron ions with charges as high as 14+ have been observed in the corona. Which ions from Fe^+ to Fe^{14+} are paramagnetic? Which would be most strongly attracted to a magnetic field?

8.98 There are some exceptions to the trends of first and successive ionization energies. For each of the following pairs, explain which ionization energy would be higher:
(a) IE_1 of Ga or IE_1 of Ge (b) IE_2 of Ga or IE_2 of Ge
(c) IE_3 of Ga or IE_3 of Ge (d) IE_4 of Ga or IE_4 of Ge

8.99 Use Figure 8.16, to find: (a) the longest wavelength of electromagnetic (EM) radiation that can ionize an alkali metal atom; (b) the longest wavelength of EM radiation that can ionize an alkaline earth metal atom; (c) the elements, other than the alkali and alkaline earth metals, that could also be ionized by the radiation in part (b); (d) the region of the EM spectrum in which these photons are found.

8.100 Rubidium and bromine atoms are depicted at right. (a) What monatomic ions do they form? (b) What electronic feature characterizes this pair of ions, and which noble

gas are they related to? (c) Which pair best represents the relative ionic sizes?

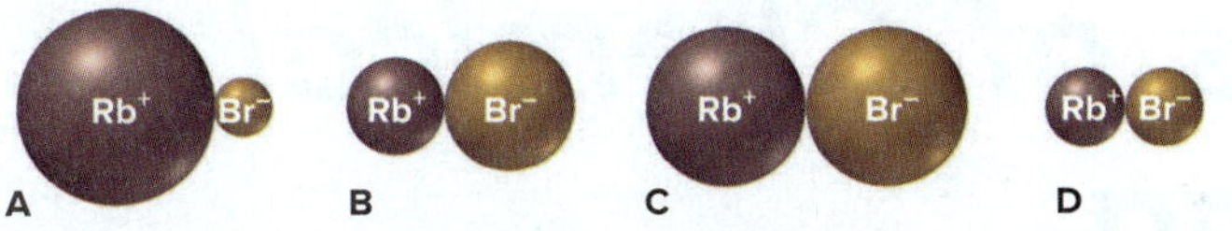

8.101 Partial (valence-level) electron configurations for four different ions are shown below:

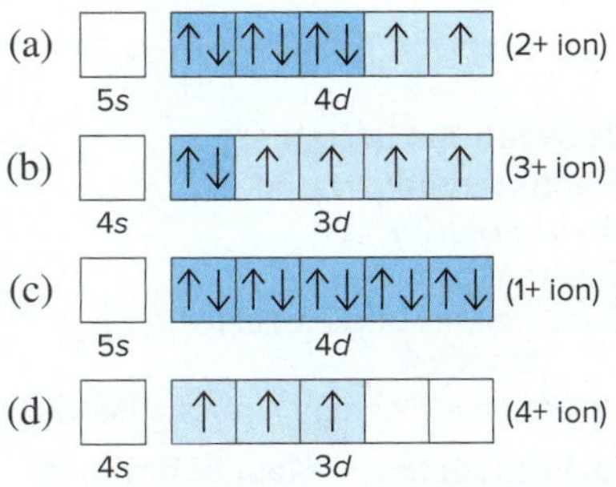

Identify the elements from which the ions are derived, and write the formula of the oxide each ion forms.

8.102 Data from the planet Zog for some main-group elements are shown below (Zoggian units are linearly related to Earth units but are not shown). Radio signals from Zog reveal that balloonium is a monatomic gas with two positive nuclear charges. Use the data to deduce the names that Earthlings give to these elements:

Name	Atomic Radius	IE_1	EA_1
Balloonium	10	339	0
Inertium	24	297	+4.1
Allotropium	34	143	−28.6
Brinium	63	70.9	−7.6
Canium	47	101	−15.3
Fertilium	25	200	0
Liquidium	38	163	−46.4
Utilium	48	82.4	−6.1
Crimsonium	72	78.4	−2.9

9 Models of Chemical Bonding

9.1 Atomic Properties and Chemical Bonds
The Three Ways Elements Combine
Lewis Symbols and the Octet Rule

9.2 The Ionic Bonding Model
Importance of Lattice Energy
Periodic Trends in Lattice Energy
How the Model Explains the Properties of Ionic Compounds

9.3 The Covalent Bonding Model
Formation of a Covalent Bond
Bonding Pairs and Lone Pairs
Bond Order, Energy, and Length
How the Model Explains the Properties of Covalent Substances

9.4 Bond Energy and Chemical Change
Where Does ΔH°_{rxn} Come From?
Using Bond Energies to Calculate ΔH°_{rxn}
Bond Strengths and Heat Released from Fuels and Foods

9.5 Between the Extremes: Electronegativity and Bond Polarity
Electronegativity
Bond Polarity and Partial Ionic Character
Gradation in Bonding Across a Period

9.6 An Introduction to Metallic Bonding
The Electron-Sea Model
How the Model Explains the Properties of Metals

Source: © Chip Clark/Fundamental Photographs, NYC

Concepts and Skills to Review Before You Study This Chapter

- characteristics of ionic and covalent substances; Coulomb's law (Section 2.7)
- polar covalent bonds and the polarity of water (Section 4.1)
- Hess's law, ΔH°_{rxn}, and ΔH°_f (Sections 6.5 and 6.6)
- atomic and ionic electron configurations (Sections 8.2 and 8.4)
- trends in atomic properties and metallic behavior (Sections 8.3 and 8.4)

Two distinctly different elements react to form table salt, a compound that is distinctly different from either reactant! Sodium, like most metals, is shiny, malleable, and able to conduct a current whether molten or solid. Chlorine, like most covalent substances, is low melting (it's a gas at room temperature) and nonconducting. The product of the reaction, sodium chloride, is a hard, brittle, high-melting solid that conducts a current only when molten or dissolved in water, like any other ionic compound. Why do these substances behave so differently? The answer lies in the *type of bonding within each substance*. In Chapter 8, we examined the properties of individual atoms and ions. But the behavior of matter really depends on how those atoms and ions bond.

IN THIS CHAPTER . . . *We examine how atomic properties give rise to three models of chemical bonding—ionic, covalent, and metallic—and how each model explains the behavior of substances.*

- We see how metals and nonmetals combine via three types of bonding and learn how to depict atoms and ions with Lewis symbols.
- We detail the steps in the formation of an ionic solid and focus on the importance of lattice energy.
- Covalent bonding occurs in the vast majority of compounds, so we look at how a bond forms and discuss the relations among bond order, energy, and length.
- We explore the relationship between bond energy and the enthalpy change of a reaction, with a focus on fuels and foods.
- We examine periodic trends in electronegativity and learn its role in the range of bonding, from pure covalent to ionic, and in bond polarity.
- We consider a simple bonding model that explains the properties of metals.

9.1 ATOMIC PROPERTIES AND CHEMICAL BONDS

Just as the electron configuration and the strength of nucleus-electron attractions determine the properties of an atom, the type and strength of chemical bonds determine the properties of a substance.

The Three Ways Elements Combine

Before we examine the types of chemical bonding, we should answer a fundamental question: why do atoms bond at all? In general, *bonding lowers the potential energy between positive and negative particles* (see Figure 1.3C), whether they are oppositely charged ions or nuclei and electrons.

As you saw in Chapter 8, there is, in general, a gradation from more metallic elements to more nonmetallic elements across a period *and* up a group (Figure 9.1, *next page*). Three models of bonding result from the three ways atoms of these two types of elements can combine.

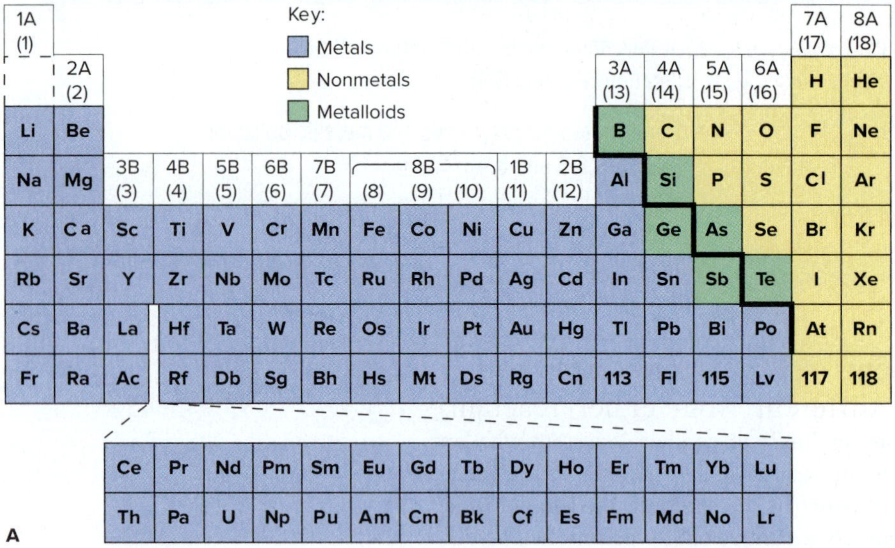

B

PROPERTY	METAL ATOM	NONMETAL ATOM
Atomic size	Larger	Smaller
Z_{eff}	Lower	Higher
IE	Lower	Higher
EA	Less negative	More negative

Figure 9.1 A comparison of metals and nonmetals. **A,** Location within the periodic table. **B,** Relative magnitudes of some atomic properties across a period.

1. *Metal with nonmetal: electron transfer and ionic bonding* (Figure 9.2A). We observe *electron transfer* and **ionic bonding** between atoms with *large differences in their tendencies to lose or gain electrons*. Such differences occur between reactive metals [Groups 1A(1) and 2A(2)] and nonmetals [Group 7A(17) and the top of Group 6A(16)]. A metal atom (low IE) loses its one or two valence electrons, and a nonmetal atom (highly negative EA) gains the electron(s). As the transfer of electron(s) from metal atom to nonmetal atom occurs, each atom forms an ion with a noble gas electron configuration. The electrostatic attractions between these positive and negative ions draw them into a three-dimensional array to form an ionic solid. The chemical formula of an ionic compound is the *empirical formula* because it gives the cation-to-anion ratio.

2. *Nonmetal with nonmetal: electron sharing and covalent bonding* (Figure 9.2B). When two atoms *differ little, or not at all, in their tendencies to lose or gain electrons,* we observe *electron sharing* and **covalent bonding,** which occurs most commonly between nonmetals. Each nonmetal atom holds onto its own electrons tightly (high IE) and attracts other electrons as well (highly negative EA). The nucleus of each atom attracts the valence electrons of the other, which draws the atoms together. The shared electron pair is typically *localized* between the two atoms, linking them in a covalent bond of a particular length and strength. In most cases, separate molecules result when atoms bond covalently. Note that the chemical formula of a covalent substance is the *molecular formula* because it gives the actual numbers of atoms in each molecule.

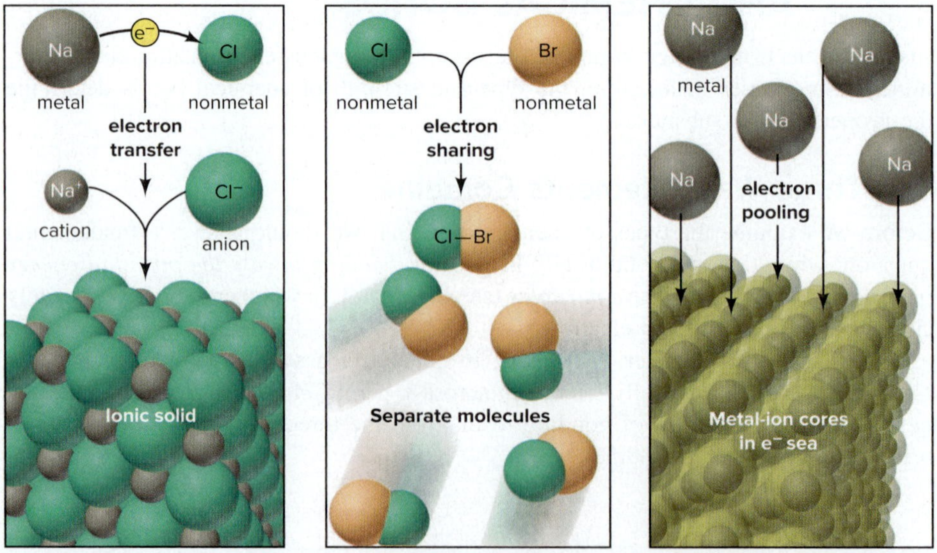

A Ionic bonding   B Covalent bonding   C Metallic bonding

Figure 9.2 Three models of chemical bonding.

3. *Metal with metal: electron pooling and metallic bonding* (Figure 9.2C). Metals share electrons in a different way. Their atoms are relatively large, and the few outer electrons are well shielded by filled inner levels (core electrons). Thus, metals lose outer electrons easily (low IE) and do not gain them readily (slightly negative or positive EA). These properties lead metal atoms to share their valence electrons, but not by covalent bonding. In the simplest model of **metallic bonding,** the enormous number of atoms in a sample of a metal *pool* their valence electrons in a "sea" of electrons that "flows" between and around each metal-ion core (nucleus plus inner electrons), thereby attracting and holding them together. Unlike the localized electrons in covalent bonding, electrons in metallic bonding are *delocalized,* moving freely throughout the entire piece of metal.

In the world of real substances, there are exceptions to these idealized models, so you can't always predict bond type from positions of the elements in the periodic table. As just one example, when the metal beryllium [Group 2A(2)] combines with the nonmetal chlorine [Group 7A(17)], the bonding fits the covalent model better than the ionic model. Just as we see gradations in atomic behavior within a group or period, we see a gradation in bonding from one type to another between atoms from different groups and periods (Figure 9.3).

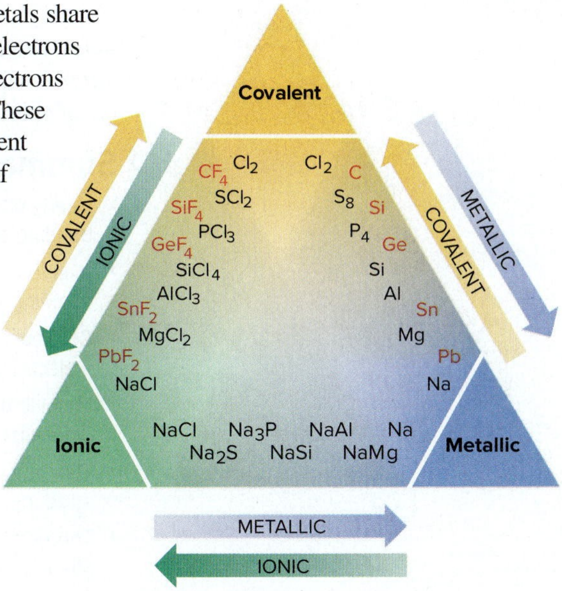

Figure 9.3 Gradations in bond type among Period 3 (*black type*) and Group 4A (*red type*) elements.

Lewis Symbols and the Octet Rule

Before examining each of the models, we consider a method for depicting the valence electrons of interacting atoms that predicts how they bond. In a **Lewis electron-dot symbol** (named for the American chemist G. N. Lewis), the element symbol represents the nucleus *and* inner electrons, and dots around the symbol represent the valence electrons (Figure 9.4). Note that the pattern of dots is the same for elements within a group.

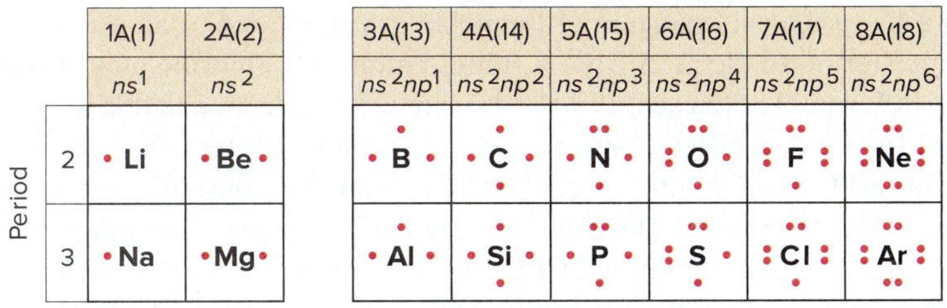

Figure 9.4 Lewis electron-dot symbols for elements in Periods 2 and 3.

We use these steps to write the Lewis symbol for any main-group element:

1. Note its A-group number (1A to 8A), which is the number of valence electrons in the atom and thus the number of valence-electron dots needed in the electron-dot symbol.
2. Place one dot at a time on each of the four sides (top, right, bottom, left) of the element symbol.
3. Keep adding dots, pairing them, until all are used up.

The specific placement of dots is not important; that is, in addition to the one shown in Figure 9.4, the Lewis symbol for nitrogen can *also* be written as

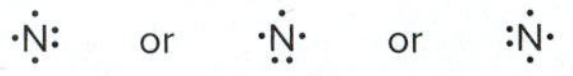

The Lewis symbol provides information about an element's bonding behavior:

- For a *metal,* the *total* number of dots is the number of electrons an atom loses to form a cation; for example, Mg loses 2 to form Mg^{2+}.
- For a *nonmetal,* the number of *unpaired* dots equals either the number of electrons an atom *gains* to form an anion (F gains 1 to form F^-) or the number it *shares* to form covalent bonds.

The Lewis symbol for carbon illustrates the last point. Rather than one pair of dots and two unpaired dots, as its electron configuration seems to call for ([He] $2s^22p^2$), carbon has four unpaired dots because it forms four bonds. Larger nonmetals can form as many bonds as the number of dots in their Lewis symbol (Chapter 10).

In his pioneering studies, Lewis generalized much of bonding behavior into the **octet rule:** *when atoms bond, they lose, gain, or share electrons to attain a filled outer level of eight electrons* (*or two,* for H and Li). The octet rule holds for nearly all of the compounds of Period 2 elements and a large number of others as well.

› Summary of Section 9.1

- Nearly all naturally occurring substances consist of atoms or ions bonded to others. Chemical bonding allows atoms to lower their energy.
- Ionic bonding occurs when metal atoms transfer electrons to nonmetal atoms, and the resulting ions attract each other and form an ionic solid.
- Covalent bonding is most common between nonmetal atoms and usually results in individual molecules. Bonded atoms share one or more pairs of electrons that are localized between them.
- Metallic bonding occurs when many metal atoms pool their valence electrons into a delocalized electron "sea" that holds all the atoms in the sample together.
- The Lewis electron-dot symbol of a main-group atom shows valence electrons as dots surrounding the element symbol.
- The octet rule says that, when bonding, many atoms lose, gain, or share electrons to attain a filled outer level of eight (or two) electrons.

9.2 THE IONIC BONDING MODEL

The central idea of the ionic bonding model is the *transfer of electrons from metal atoms to nonmetal atoms to form ions that attract each other and form a solid compound.* In most cases, for the main groups, the ion that forms has a filled outer level of either two or eight electrons (octet rule), the number in the nearest noble gas. In other words, a metal will lose the number of electrons needed to achieve the configuration of the noble gas that precedes it in the periodic table, while a nonmetal will gain the number of electrons needed to achieve the configuration of the noble gas at the end of its period.

The transfer of an electron from a lithium atom to a fluorine atom is depicted in three ways in Figure 9.5. In each, Li loses its single outer electron and is left with a filled $n = 1$ level (two e^-), while F gains a single electron to fill its $n = 2$ level (eight e^-). In this case, each atom is one electron away from the configuration of its nearest noble gas, so the number of electrons lost by each Li equals the number gained by each F. Therefore, equal numbers of Li^+ and F^- ions form, as the formula LiF indicates. That is, in ionic bonding, *the total number of electrons lost by the metal atom(s) equals the total number of electrons gained by the nonmetal atom(s).*

Figure 9.5 Three ways to depict electron transfer in the formation of Li^+ and F^-. The electron being transferred is shown in red.

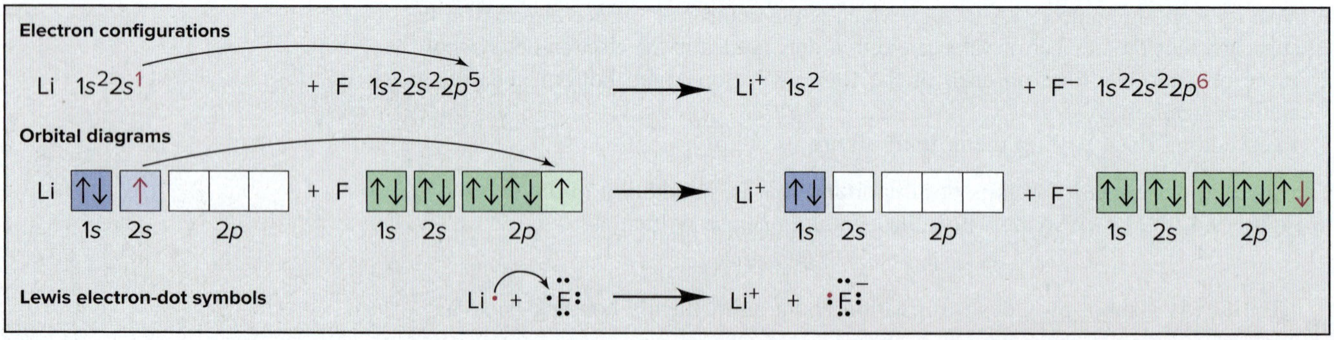

SAMPLE PROBLEM 9.1 Depicting Ion Formation

Problem Use partial orbital diagrams and Lewis symbols to depict the formation of Na^+ and O^{2-} ions from the atoms, and give the formula of the compound formed.

Plan First we draw the orbital diagrams and Lewis symbols for Na and O atoms. To attain filled outer levels (noble gas configurations), Na loses one electron and O gains two. To make the number of electrons lost equal the number gained, two Na atoms are needed for each O atom.

Solution

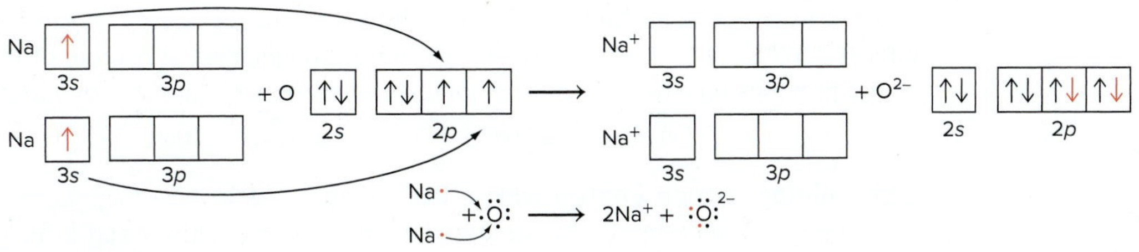

The formula is Na_2O.

FOLLOW-UP PROBLEMS

Brief Solutions for ***all*** *Follow-up Problems appear at the end of the chapter.*

9.1A Use condensed electron configurations, partial orbital diagrams (*ns* and *np* only), and Lewis symbols to depict the formation of Mg^{2+} and Cl^- ions from the atoms, and give the formula of the compound formed.

9.1B Use condensed electron configurations, partial orbital diagrams (*ns* and *np* only), and Lewis symbols to depict the formation of Ca^{2+} and O^{2-} ions, and give the formula of the compound formed.

SOME SIMILAR PROBLEMS 9.20 and 9.21

Why Ionic Compounds Form: The Importance of Lattice Energy

You may be surprised to learn that energy is actually *absorbed* during electron transfer. So why does it occur? And, given this absorption of energy, why do ionic substances exist at all? As you'll see, the answer involves the enormous quantity of energy *released* after electron transfer, as the ions form a solid.

1. *The electron-transfer process.* Consider just the electron-transfer process for the formation of lithium fluoride, which involves a gaseous Li atom losing an electron and a gaseous F atom gaining it:

- The first ionization energy (IE_1) of Li is the energy absorbed when 1 mol of gaseous Li atoms loses 1 mol of valence electrons:

$$Li(g) \longrightarrow Li^+(g) + e^- \qquad IE_1 = 520 \text{ kJ}$$

- The first electron affinity (EA_1) of F is the energy released when 1 mol of gaseous F atoms gains 1 mol of electrons:

$$F(g) + e^- \longrightarrow F^-(g) \qquad EA_1 = -328 \text{ kJ}$$

- Taking the sum shows that electron transfer *by itself* requires energy:

$$Li(g) + F(g) \longrightarrow Li^+(g) + F^-(g) \qquad IE_1 + EA_1 = 192 \text{ kJ}$$

2. *Other steps that absorb energy.* The total energy needed prior to ion formation is added to the sum of IE_1 and EA_1: metallic Li must be made into gaseous Li atoms (161 kJ/mol), and F_2 molecules must be broken into separate F atoms (79.5 kJ/mol).

3. *Steps that release energy.* Despite these endothermic steps, the standard enthalpy of formation (ΔH°_f) of solid LiF is −617 kJ/mol; that is, 617 kJ is *released* when 1 mol of LiF(*s*) forms from its elements. Formation of LiF is typical of reactions between active metals and nonmetals: ionic solids form readily (Figure 9.6).

If the overall reaction releases energy, there must be some step that is exothermic enough to outweigh the endothermic steps. This step involves the *strong attraction between pairs of oppositely charged ions.* When 1 mol of $Li^+(g)$ and 1 mol of $F^-(g)$ form 1 mol of gaseous LiF molecules, a large quantity of heat is released:

$$Li^+(g) + F^-(g) \longrightarrow LiF(g) \qquad \Delta H^\circ = -755 \text{ kJ}$$

But, as you know, under ordinary conditions, LiF does not exist as gaseous molecules: *even more energy is released when the separate gaseous ions coalesce into a crystalline solid* because each ion attracts *several* oppositely charged ions:

$$Li^+(g) + F^-(g) \longrightarrow LiF(s) \qquad \Delta H^\circ = -1050 \text{ kJ}$$

Figure 9.6 **The exothermic formation of sodium bromide. A,** Sodium (in beaker under mineral oil) and bromine. **B,** The reaction is rapid and vigorous.

Source: © McGraw-Hill Education/Stephen Frisch, photographer

The negative of this enthalpy change is 1050 kJ, the lattice energy of LiF. The **lattice energy ($\Delta H^\circ_{\text{lattice}}$)** is the enthalpy change that accompanies the reverse of the process in the previous equation—1 mol of ionic solid separating into gaseous ions:

$$\text{LiF}(s) \longrightarrow \text{Li}^+(g) + \text{F}^-(g) \qquad \Delta H^\circ_{\text{lattice}} = 1050 \text{ kJ}$$

Determining Lattice Energy with a Born-Haber Cycle The magnitude of the lattice energy is a measure of the strength of the ionic interactions and influences macroscopic properties, such as melting point, hardness, and solubility. Despite playing this crucial role in the formation of ionic compounds, lattice energy is usually *not* measured directly. One way to determine it applies Hess's law (see Section 6.5) in a **Born-Haber cycle,** a series of steps from elements to ionic solid for which all the enthalpies* are known except the lattice energy.

Let's go through a Born-Haber cycle for the formation of lithium fluoride to calculate $\Delta H^\circ_{\text{lattice}}$. Figure 9.7 shows two possible paths, a direct combination reaction (ΔH°_f; *black arrow*) and a multistep path *(orange arrows)* in which one step is the unknown $\Delta H^\circ_{\text{lattice}}$. Hess's law tells us both paths involve the same overall enthalpy change:

$$\Delta H^\circ_f \text{ of LiF}(s) = \text{sum of } \Delta H^\circ \text{ values for multistep path}$$

Hess's law lets us choose *hypothetical* steps whose enthalpy changes we can measure, even though *they are* ***not*** *the actual steps that occur when lithium reacts with fluorine.* We identify each ΔH° by its step number in Figure 9.7:

Step 1. From solid Li to gaseous Li atoms. This step, called *atomization,* has the enthalpy change $\Delta H^\circ_{\text{atom}}$. It involves breaking metallic bonds and vaporizing the atoms, so it absorbs energy:

$$\text{Li}(s) \longrightarrow \text{Li}(g) \qquad \Delta H^\circ_{\text{step 1}} = \Delta H^\circ_{\text{atom}} = 161 \text{ kJ}$$

*Strictly speaking, ionization energy (IE) and electron affinity (EA) are internal energy changes (ΔE), not enthalpy changes (ΔH), but in these steps, $\Delta H = \Delta E$ because $\Delta V = 0$ (Section 6.2).

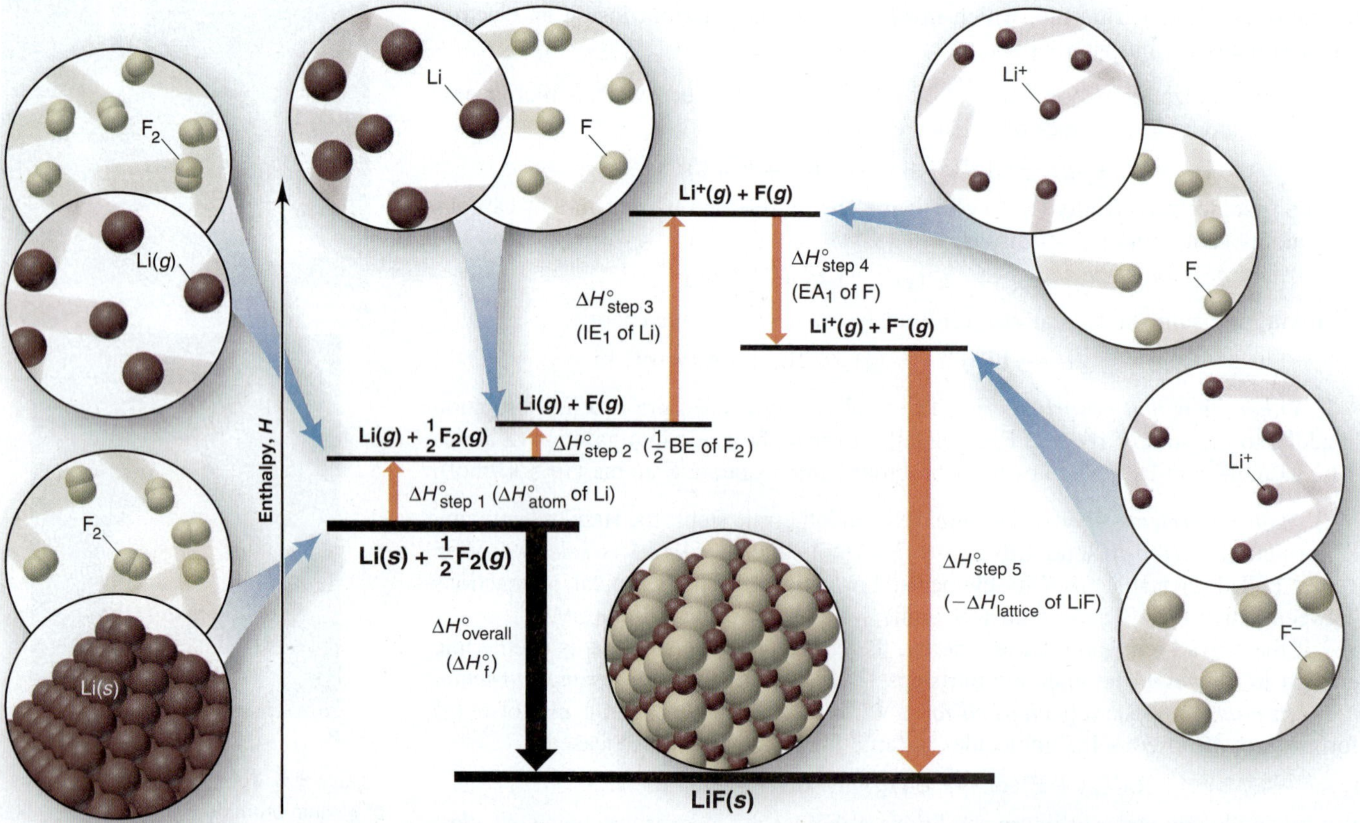

Figure 9.7 **The Born-Haber cycle for lithium fluoride.** The formation of LiF(*s*) from its elements can happen in one combination reaction *(black arrow)* or in five steps *(orange arrows)*. The unknown enthalpy change is $\Delta H^\circ_{\text{step 5}}$ ($-\Delta H^\circ_{\text{lattice}}$ of LiF).

Step 2. From F_2 molecules to F atoms. This step involves breaking a covalent bond, so it absorbs energy; as we discuss later, this is the *bond energy* (BE) of F_2. Since we need 1 mol of F atoms to make 1 mol of LiF, we start with $\frac{1}{2}$ mol of F_2:

$$\tfrac{1}{2}F_2(g) \longrightarrow F(g) \qquad \Delta H^\circ_{\text{step 2}} = \tfrac{1}{2}(\text{BE of } F_2) = \tfrac{1}{2}(159\text{ kJ}) = 79.5\text{ kJ}$$

Step 3. From Li to Li^+. Removing the 2*s* electron from Li absorbs energy:

$$Li(g) \longrightarrow Li^+(g) + e^- \qquad \Delta H^\circ_{\text{step 3}} = IE_1 = 520\text{ kJ}$$

Step 4. From F to F^-. Adding an electron to F releases energy:

$$F(g) + e^- \longrightarrow F^-(g) \qquad \Delta H^\circ_{\text{step 4}} = EA_1 = -328\text{ kJ}$$

Step 5. From gaseous ions to ionic solid. Forming solid LiF from gaseous Li^+ and F^- releases a lot of energy. The enthalpy change for this step is unknown but is, by definition, the negative of the lattice energy:

$$Li^+(g) + F^-(g) \longrightarrow LiF(s) \qquad \Delta H^\circ_{\text{step 5}} = -\Delta H^\circ_{\text{lattice}} \text{ of LiF} = ?$$

The enthalpy change of the combination reaction *(black arrow)* is

$$Li(s) + \tfrac{1}{2}F_2(g) \longrightarrow LiF(s) \qquad \Delta H^\circ_{\text{overall}} = \Delta H^\circ_f = -617\text{ kJ}$$

We set ΔH°_f equal to the sum of the ΔH° values for the steps and solve for $\Delta H^\circ_{\text{lattice}}$:

$$\Delta H^\circ_f = \Delta H^\circ_{\text{step 1}} + \Delta H^\circ_{\text{step 2}} + \Delta H^\circ_{\text{step 3}} + \Delta H^\circ_{\text{step 4}} + (-\Delta H^\circ_{\text{lattice}} \text{ of LiF})$$

Solving for $-\Delta H^\circ_{\text{lattice}}$ of LiF gives

$$\begin{aligned} -\Delta H^\circ_{\text{lattice}} \text{ of LiF} &= \Delta H^\circ_f - (\Delta H^\circ_{\text{step 1}} + \Delta H^\circ_{\text{step 2}} + \Delta H^\circ_{\text{step 3}} + \Delta H^\circ_{\text{step 4}}) \\ &= -617\text{ kJ} - [161\text{ kJ} + 79.5\text{ kJ} + 520\text{ kJ} + (-328\text{ kJ})] \\ &= -1050\text{ kJ} \end{aligned}$$

And changing the sign gives

$$\Delta H^\circ_{\text{lattice}} \text{ of LiF} = -(-1050\text{ kJ}) = 1050\text{ kJ}$$

Note that *the lattice energy is, by far, the largest component of the multistep process.* (Problems 9.30 and 9.31 are among several that focus on Born-Haber cycles.)

The Born-Haber cycle shows that the energy *required* for elements to form ions is *supplied* by the attraction among the ions in the solid. And the "take-home lesson" is that *ionic solids exist* ***only*** *because the lattice energy far exceeds the total energy needed to form the ions.*

Periodic Trends in Lattice Energy

The lattice energy results from electrostatic interactions among ions, so its magnitude depends on ionic size, ionic charge, and the arrangement of the ions in the solid. Therefore, we expect to see periodic trends in lattice energy.

Explaining the Trends with Coulomb's Law Recall from Chapter 2 that **Coulomb's law** states that the electrostatic energy between particles A and B is directly proportional to the product of their charges and inversely proportional to the distance between them:

$$\text{Electrostatic energy} \propto \frac{\text{charge A} \times \text{charge B}}{\text{distance}}$$

Lattice energy is directly proportional to electrostatic energy. In an ionic solid, cations and anions lie as close to each other as possible, so the distance between them is the sum of the ionic radii (see Figure 8.28):

$$\text{Electrostatic energy} \propto \frac{\text{cation charge} \times \text{anion charge}}{\text{cation radius} + \text{anion radius}} \propto \Delta H^\circ_{\text{lattice}} \qquad \textbf{(9.1)}$$

This relationship helps us explain the effects of ionic size and charge on trends in lattice energy (also see Figure 2.12):

1. *Effect of ionic size.* Smaller ions attract each other more strongly than larger ions, because the charges are closer to each other. As we move down a group, ionic

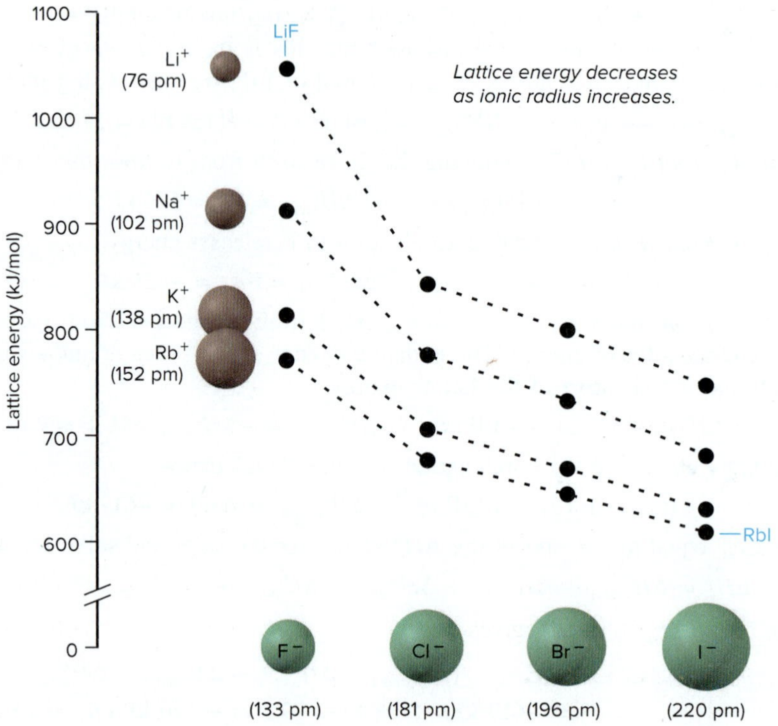

Figure 9.8 Trends in lattice energy. The lattice energies are shown for compounds formed from a given Group 1A(1) cation *(left side)* and one of the Group 7A(17) anions *(bottom)*. LiF (smallest ions) has the highest lattice energy, and RbI (largest ions) has the lowest.

radii increase, so the electrostatic energy between cations and anions decreases; thus, lattice energies should decrease as well:

ionic size increases ↑, lattice energy decreases ↓

Figure 9.8 shows that, for the alkali-metal halides, lattice energy decreases down the group whether the cation remains the same (LiF to LiI) or the anion remains the same (LiF to RbF).

2. *Effect of ionic charge.* Ions with higher charges attract each other more strongly than ions with lower charges; thus, lattice energy increases with increasing ionic charge:

ionic charge increases ↑, lattice energy increases ↑

Across a period, ionic charge changes. For example, lithium fluoride and magnesium oxide have cations and anions of about equal radii (Li^+ = 76 pm and Mg^{2+} = 72 pm; F^- = 133 pm and O^{2-} = 140 pm). The major difference is between singly charged Li^+ and F^- ions and doubly charged Mg^{2+} and O^{2-} ions. The difference in the lattice energies of the two compounds is striking:

$$\Delta H^\circ_{\text{lattice}} \text{ of LiF} = 1050 \text{ kJ/mol} \quad \text{and} \quad \Delta H^\circ_{\text{lattice}} \text{ of MgO} = 3923 \text{ kJ/mol}$$

This nearly fourfold increase in $\Delta H^\circ_{\text{lattice}}$ reflects the fourfold increase in the product of the charges (1 × 1 vs. 2 × 2) in the numerator of Equation 9.1.

Student Hot Spot

Student data indicate that you may struggle with predicting relative lattice energies of ionic compounds. Access the SmartBook to view additional Learning Resources on this topic.

SAMPLE PROBLEM 9.2 Predicting Relative Lattice Energy from Ionic Properties

Problem Use ionic properties to explain which compound in each pair has the larger lattice energy: **(a)** RbI or NaBr; **(b)** KCl or CaS.

Plan To choose the compound with the larger lattice energy, we apply Coulomb's law and periodic trends in ionic radius and charge (see Figure 2.12). We examine the ions in each compound: for ions of similar size, higher charge leads to a larger lattice energy; for ions with the same charge, smaller size leads to larger lattice energy because the ions can get closer together.

Solution **(a)** NaBr. All the ions have single charges, so charge is not a factor. Size increases down a group, so Rb^+ is larger than Na^+, and I^- is larger than Br^-. Therefore, NaBr has the larger lattice energy because it consists of smaller ions.

(b) CaS. Size decreases from left to right across a period, so K^+ is slightly larger than Ca^{2+}, and S^{2-} is slightly larger than Cl^-. However, these small differences are not nearly as important as the charges: Ca^{2+} and S^{2-} have twice the charge of K^+ and Cl^-, so CaS has the larger lattice energy.

Check The actual lattice energies are **(a)** RbI = 598 kJ/mol and NaBr = 719 kJ/mol; **(b)** KCl = 676 kJ/mol and CaS = 3039 kJ/mol.

FOLLOW-UP PROBLEMS

9.2A Use ionic properties to explain which compound has the *smaller* lattice energy: BaF_2 or SrF_2.

9.2B Use ionic properties to arrange the following compounds in order of *increasing* lattice energy: MgF_2, Na_2O, CaO.

SOME SIMILAR PROBLEMS 9.26–9.29

Why Does MgO Exist? We might ask how ionic solids, like MgO, with doubly charged ions, could even form. After all, forming 1 mol of Mg^{2+} involves the sum of the first *and* second ionization energies:

$$Mg(g) \longrightarrow Mg^{2+}(g) + 2e^- \qquad \Delta H° = IE_1 + IE_2 = 738\text{ kJ} + 1450\text{ kJ} = 2188\text{ kJ}$$

And, while forming 1 mol of O^- ions is exothermic (first electron affinity, EA_1), adding a second mole of electrons (second electron affinity, EA_2) is endothermic because the electron is added to a negative ion. The overall formation of O^{2-} ions is endothermic:

$$\begin{array}{llll} O(g) + e^- \longrightarrow O^-(g) & \Delta H° = EA_1 & = & -141\text{ kJ} \\ O^-(g) + e^- \longrightarrow O^{2-}(g) & \Delta H° = EA_2 & = & 878\text{ kJ} \\ \hline O(g) + 2e^- \longrightarrow O^{2-}(g) & \Delta H° = EA_1 + EA_2 & = & 737\text{ kJ} \end{array}$$

There are also the endothermic steps for converting Mg(*s*) to Mg(*g*) (148 kJ/mol) and breaking $\frac{1}{2}$ mol of O_2 molecules into O atoms ($\frac{1}{2}$[498 kJ/mol] = 249 kJ). Thus, the total energy that must be absorbed in the formation of the compound is significant: 2188 + 737 + 148 + 249 kJ = 3322 kJ. Nevertheless, the 2+ and 2− ionic charges make the lattice energy so large ($\Delta H°_{lattice}$ = 3923 kJ/mol) that MgO forms readily whenever Mg burns in air ($\Delta H°_f$ = −601 kJ/mol; Figure 3.7).

How the Model Explains the Properties of Ionic Compounds

The central role of any model is to explain the facts. With atomic-level views, we can see how the ionic bonding model accounts for the properties of ionic solids:

1. *Physical behavior*. As a typical ionic compound, a piece of rock salt (NaCl) is *hard* (does not dent), *rigid* (does not bend), and *brittle* (cracks without deforming). These properties arise from the strong attractive forces that hold the ions *in specific positions*. Moving them out of position requires overcoming these forces, so rock salt does not dent or bend. If enough force *is* applied, ions of like charge are brought next to each other, and repulsions between them crack the sample suddenly (Figure 9.9).

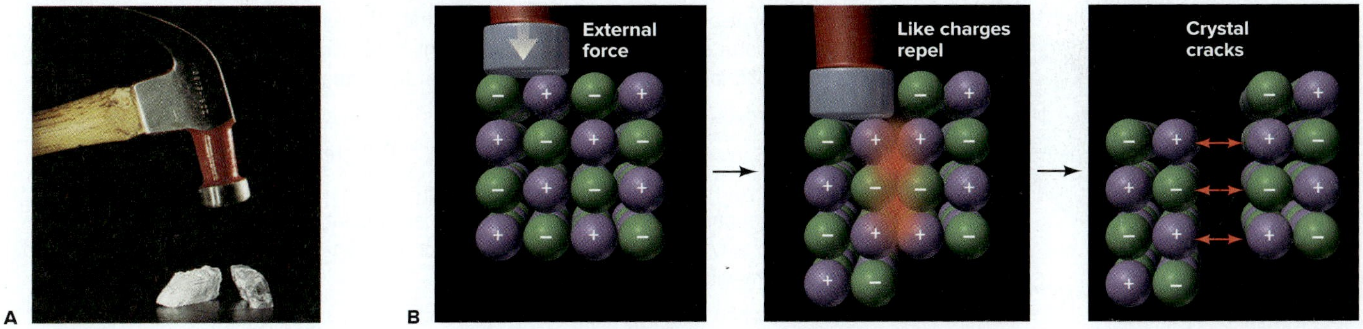

Figure 9.9 **Why ionic compounds crack. A,** Ionic compounds crack when struck with enough force. **B,** When a force moves like charges near each other, repulsions cause a crack.

Source: (A) © McGraw-Hill Education/Stephen Frisch, photographer

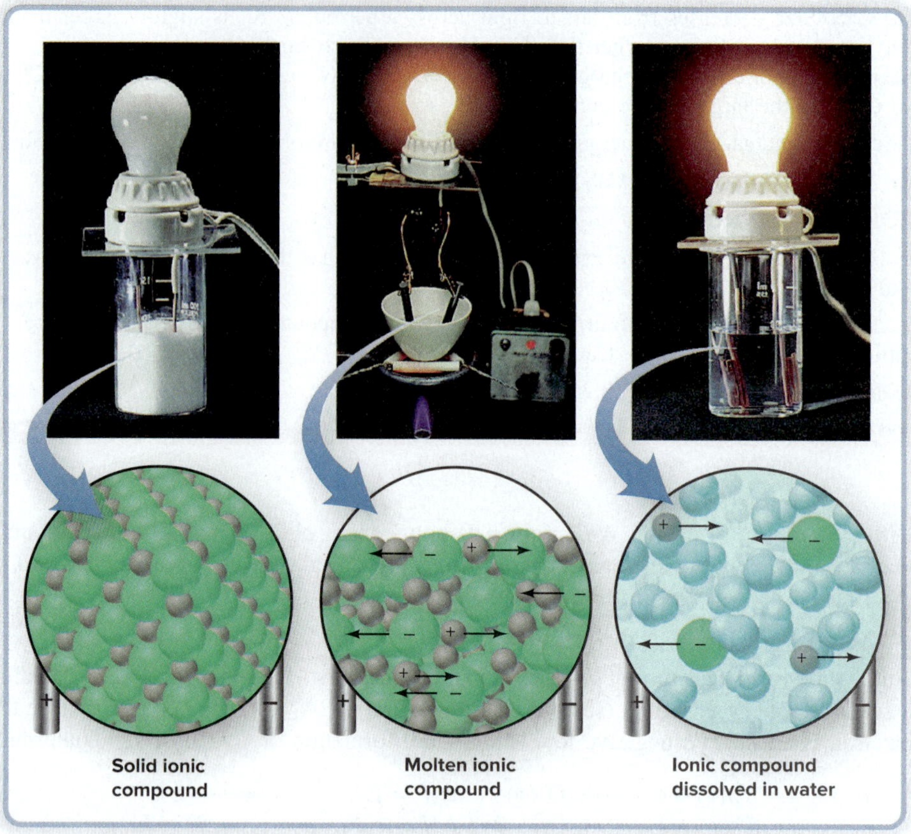

Figure 9.10 Electrical conductance and ion mobility.
Source: © McGraw-Hill Education/Stephen Frisch, photographer

Table 9.1 Melting and Boiling Points of Some Ionic Compounds

Compound	mp (°C)	bp (°C)
CsBr	636	1300
NaI	661	1304
$MgCl_2$	714	1412
KBr	734	1435
$CaCl_2$	782	>1600
NaCl	801	1413
LiF	845	1676
KF	858	1505
MgO	2852	3600

2. *Electrical conductivity.* Ionic compounds typically *do not* conduct electricity in the solid state but *do* conduct when melted or dissolved. According to the model, the solid consists of fixed ions, but when it melts or dissolves, the ions can move and carry a current (Figure 9.10).

3. *Melting and boiling point.* Large amounts of energy are needed to free the ions from their positions and separate them. Thus, we expect ionic compounds to have high melting points and much higher boiling points (Table 9.1). In fact, the interionic attraction is so strong that a vaporized ionic compound consists of **ion pairs,** gaseous ionic molecules, rather than individual ions (Figure 9.11). In their normal state, as you know, *ionic compounds are solid arrays of ions, and no separate molecules exist.*

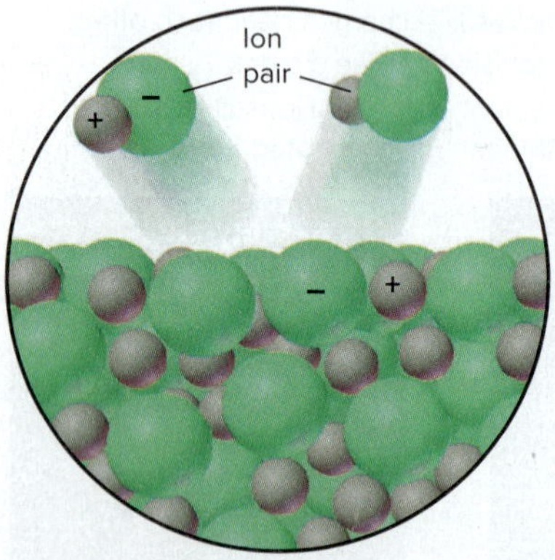

Figure 9.11 Ion pairs formed when an ionic compound vaporizes.

Summary of Section 9.2

- In ionic bonding, a metal transfers electrons to a nonmetal, and the resulting ions attract each other to form a solid.
- Main-group elements often attain a filled outer level (either eight electrons or two electrons) by forming ions that have the electron configuration of the nearest noble gas.
- Ion formation by itself *absorbs* energy, but more than that quantity of energy is *released* when the ions form a solid. The high lattice energy of an ionic solid, the energy required to separate the solid into gaseous ions, is the reason the compound forms.
- The lattice energy is determined by applying Hess's law in a Born-Haber cycle.
- Lattice energies increase with higher ionic charge and decrease with larger ionic radius (Coulomb's law).
- According to the ionic bonding model, the strong electrostatic attractions that keep ions in position explain why ionic solids are hard, conduct a current only when melted or dissolved, and have high melting and boiling points.
- Ion pairs form when an ionic compound vaporizes.

9.3 THE COVALENT BONDING MODEL

Look through the *Handbook of Chemistry and Physics,* and you'll find that the number of covalent compounds dwarfs the number of ionic compounds. Covalent substances range from diatomic hydrogen to biological and synthetic macromolecules with many thousands of atoms and even to some minerals that have covalent bonds throughout the sample. And covalent bonds occur in all polyatomic ions, too. Without doubt, *sharing electrons is the main way that atoms interact.*

The Formation of a Covalent Bond

Why does hydrogen gas consist of H_2 molecules and not separate H atoms? Figure 9.12 plots the potential energy of a system of two isolated H atoms versus the distance between their nuclei (see also Figure 2.14). Let's start at the right end of the curve and move along it to the left, as the atoms get closer:

- *At point 1,* the atoms are far apart, and each acts as though the other were not present.
- *At point 2,* the distance between the atoms has decreased enough for each nucleus to start attracting the other atom's electron, which lowers the potential energy. As the atoms get closer, these attractions increase, but so do repulsions between the nuclei and between the electrons.
- *At point 3* (bottom of the energy "well"), the maximum attraction is achieved in the face of the increasing repulsion, and the system has its minimum energy.
- *At point 4,* if it were reached, the atoms would be too close, and the rise in potential energy from increasing repulsions between the nuclei and between the electrons would push them apart toward point 3 again.

Thus, a **covalent bond** arises from the balance between the nuclei attracting the electrons and electrons and nuclei repelling each other, as shown in the blow-up of the structure at point 3. (We'll return to Figure 9.12 shortly.)

Formation of a covalent bond always results in greater electron density ***between*** *the nuclei.* Figure 9.13 (*on the next page*) depicts this fact with an *electron density contour map* (A), and an *electron density relief map* (B).

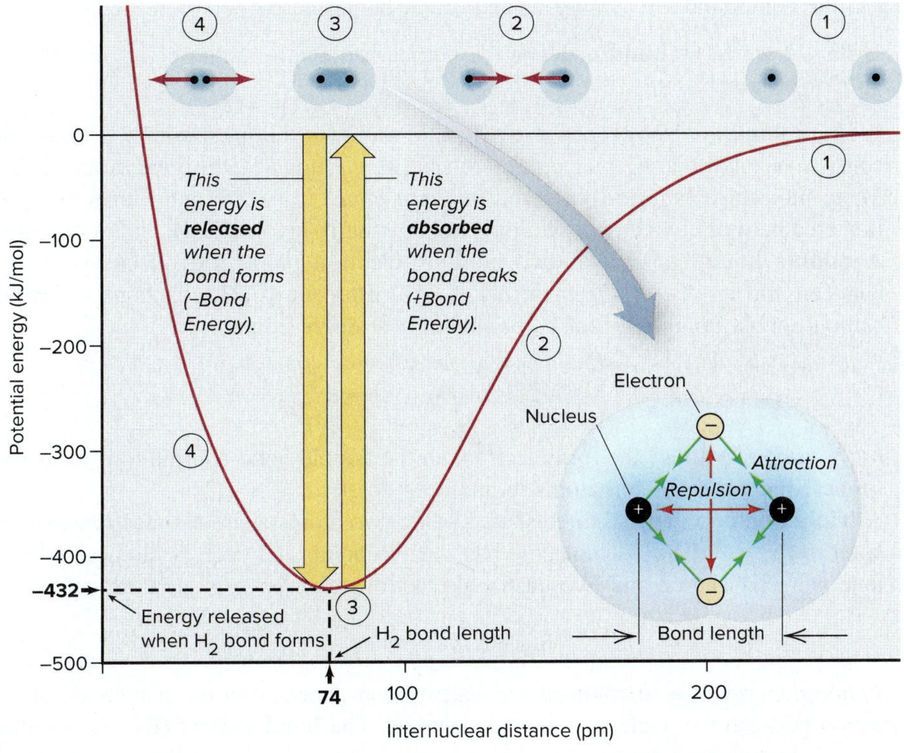

Figure 9.12 **Covalent bond formation in H_2.** The energy difference between points 1 and 3 is the H_2 *bond energy* (432 kJ/mol). The internuclear distance at point 3 is the H_2 *bond length* (74 pm).

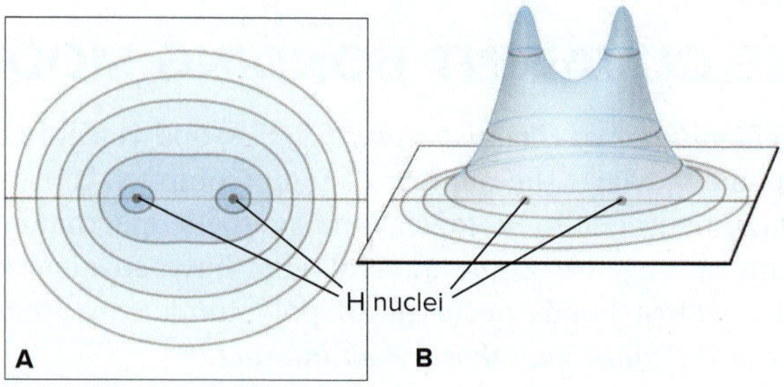

Figure 9.13 **Distribution of electron density in H_2. A,** Electron density *(blue shading)* is high around and between the nuclei. Electron density doubles with each concentric curve. **B,** The highest regions of electron density are shown as peaks.

Bonding Pairs and Lone Pairs

Unlike ionic compounds in which atoms gain or lose enough electrons to obtain an octet of electrons, atoms in covalent compounds share electrons. To achieve a full outer (valence) level of electrons, *each atom in a covalent bond "counts" the shared electrons as belonging entirely to itself.* Thus, the two shared electrons in H_2 simultaneously fill the outer level of *both* H atoms, as clarified by the blue circles added below. The **shared pair,** or **bonding pair,** is represented by a pair of dots or a line:

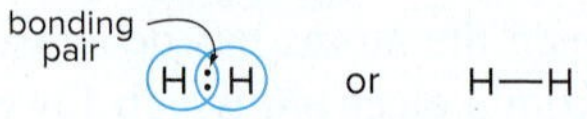

An outer-level electron pair that is *not* involved in bonding is called a **lone pair,** or **unshared pair.** The bonding pair in HF fills the outer level of the H atom *and,* together with three lone pairs, fills the outer level of the F atom as well:

In F_2, the bonding pair and three lone pairs fill the outer level of *each* F atom, so that each has an octet of electrons:

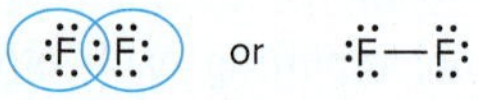

(This text generally shows bonding pairs as lines and lone pairs as dots.)

Properties of a Covalent Bond: Order, Energy, and Length

A covalent bond has three important properties that are closely related to one another and to the compound's reactivity—bond order, bond energy, and bond length.

1. *Bond order.* The **bond order** is the number of electron pairs being shared by a given pair of atoms:

- A **single bond,** as shown above in H_2, HF, or F_2, is the most common bond and consists of one bonding pair of electrons: *a single bond has a bond order of 1.*
- Many molecules (and ions) contain *multiple bonds,* in which more than one pair is shared between two atoms. Multiple bonds usually involve C, O, and/or N atoms. A **double bond** consists of two bonding electron pairs, four electrons shared between two atoms, so *the bond order is 2.* Ethylene (C_2H_4) contains a carbon-carbon double bond and four carbon-hydrogen single bonds:

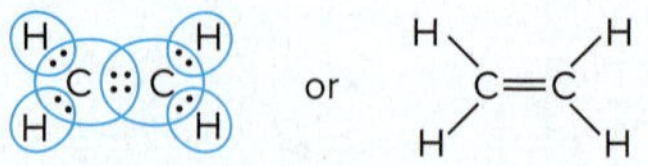

Each carbon "counts" the four electrons in the double bond and the four in its two single bonds to hydrogen atoms to attain an octet.
- A **triple bond** consists of three shared pairs: two atoms share six electrons, so *the bond order is 3.* The N_2 molecule has a triple bond, and each N atom also has a lone pair. Six shared and two unshared electrons give *each* N atom an octet:

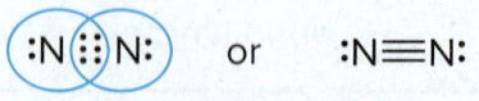

2. *Bond energy.* The strength of a covalent bond depends on the magnitude of the attraction between the nuclei and shared electrons. The **bond energy (BE)** (also called *bond enthalpy* or *bond strength*) is the energy needed to overcome this attraction and

is defined as *the standard enthalpy change for breaking the bond in 1 mol of gaseous molecules.* Bond breakage is an *endothermic* process, so *bond energy is always positive:*

$$\text{A—B}(g) \longrightarrow \text{A}(g) + \text{B}(g) \qquad \Delta H^\circ_{\text{bond breaking}} = \text{BE}_{\text{A—B}} \text{ (always} > 0)$$

The bond energy is the difference in energy between separated and bonded atoms (the potential energy difference between points 1 and 3—the energy "well"—in Figure 9.12). *The same quantity of energy absorbed to break the bond is released when the bond forms.* Bond formation is an *exothermic* process, so *the sign of its enthalpy change is always negative:*

$$\text{A}(g) + \text{B}(g) \longrightarrow \text{A—B}(g) \qquad \Delta H^\circ_{\text{bond forming}} = -\text{BE}_{\text{A—B}} \text{ (always} < 0)$$

Table 9.2 lists the energies of some common bonds. By definition,

- *Stronger bonds have a larger BE because they are lower in energy (have a deeper energy well).*
- *Weaker bonds have a smaller BE because they are higher in energy (have a shallower energy well).*

The energy of a given bond varies slightly from molecule to molecule and even within the same molecule, so each value is an *average* bond energy. In other words, the C—H bond energy value of 413 kJ/mol is the average of the C—H bond energy values in all molecules containing this bond.

3. *Bond length.* A covalent bond has a **bond length,** the distance between the nuclei of the two bonded atoms. In Figure 9.12, bond length is the distance between the nuclei at the point of minimum energy (bottom of the well), and Table 9.2 shows the lengths of some covalent bonds. Like bond energies, these values are *average* bond lengths for a bond in different substances. Bond length is related to the sum of the radii of the bonded atoms. In fact, most atomic radii are calculated from measured bond lengths (see Figure 8.12C). Bond lengths for a series of similar bonds, as in the halogens, increase with atomic size (Figure 9.14).

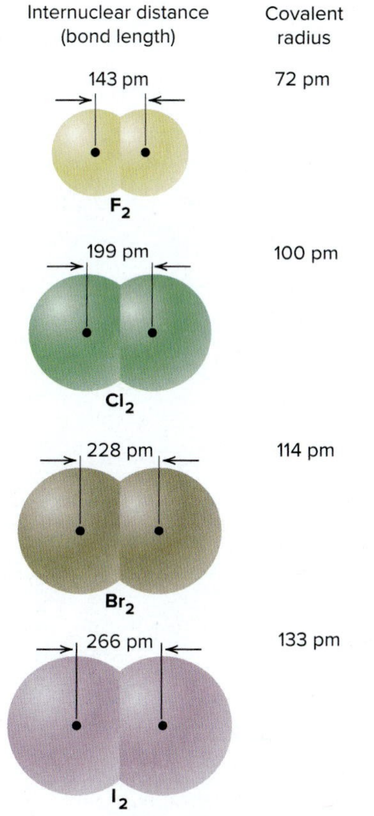

Figure 9.14 Bond length and covalent radius.

Table 9.2 Average Bond Energies (kJ/mol) and Bond Lengths (pm)

Bond	Energy	Length	Bond	Energy	Length	Bond	Energy	Length	Bond	Energy	Length
Single Bonds											
H—H	432	74	N—H	391	101	Si—H	323	148	S—H	347	134
H—F	565	92	N—N	160	146	Si—Si	226	234	S—S	266	204
H—Cl	427	127	N—P	209	177	Si—O	368	161	S—F	327	158
H—Br	363	141	N—O	201	144	Si—S	226	210	S—Cl	271	201
H—I	295	161	N—F	272	139	Si—F	565	156	S—Br	218	225
			N—Cl	200	191	Si—Cl	381	204	S—I	~170	234
C—H	413	109	N—Br	243	214	Si—Br	310	216			
C—C	347	154	N—I	159	222	Si—I	234	240	F—F	159	143
C—Si	301	186							F—Cl	193	166
C—N	305	147	O—H	467	96	P—H	320	142	F—Br	212	178
C—O	358	143	O—P	351	160	P—Si	213	227	F—I	263	187
C—P	264	187	O—O	204	148	P—P	200	221	Cl—Cl	243	199
C—S	259	181	O—S	265	151	P—F	490	156	Cl—Br	215	214
C—F	453	133	O—F	190	142	P—Cl	331	204	Cl—I	208	243
C—Cl	339	177	O—Cl	203	164	P—Br	272	222	Br—Br	193	228
C—Br	276	194	O—Br	234	172	P—I	184	246	Br—I	175	248
C—I	216	213	O—I	234	194				I—I	151	266
Multiple Bonds											
C=C	614	134	N=N	418	122	C≡C	839	121	N≡N	945	110
C=N	615	127	N=O	607	120	C≡N	891	115	N≡O	1020	106
C=O	745	123	O_2	498	121	C≡O	1070	113			
	(799 in CO_2)										

Table 9.3 The Relation of Bond Order, Bond Length, and Bond Energy

Bond	Bond Order	Average Bond Length (pm)	Average Bond Energy (kJ/mol)
C—O	1	143	358
C═O	2	123	745
C≡O	3	113	1070
C—C	1	154	347
C═C	2	134	614
C≡C	3	121	839
N—N	1	146	160
N═N	2	122	418
N≡N	3	110	945

The order, energy, and length of a covalent bond are interrelated. Two nuclei are more strongly attracted to two shared electron pairs than to one, so double-bonded atoms are drawn closer together *and* are more difficult to pull apart than single-bonded atoms. Therefore, *for a given pair of atoms:*

- *a higher bond order results in a smaller bond length.* As bond order increases, bond length decreases.
- *a higher bond order results in a higher bond energy.* As bond order increases, bond energy also increases.

Thus, as Table 9.3 shows, for a given pair of atoms, *a shorter bond is a stronger bond.*

In some cases, we can see a relation among atomic size, bond length, and bond energy by varying one of the atoms involved in a single bond while holding the other constant:

- *Variation within a group.* The trend in carbon-halogen single bond lengths, C—I > C—Br > C—Cl > C—F, parallels the trend in atomic size, I > Br > Cl > F, and is opposite to the trend in bond energy, C—F > C—Cl > C—Br > C—I.
- *Variation within a period.* Looking again at single bonds involving carbon, the trend in bond lengths, C—N > C—O > C—F, parallels the trend in atomic size, N > O > F, and is opposite to the trend in bond energy, C—F > C—O > C—N.

In general, bond length increases (and bond energy decreases) with increasing atomic radii of the atoms in the bond.

Student Hot Spot

Student data indicate that you may struggle with comparing bond length and strength. Access the SmartBook to view additional Learning Resources on this topic.

SAMPLE PROBLEM 9.3 Comparing Bond Length and Bond Strength

Problem Without referring to Table 9.2, rank the bonds in each set in order of *decreasing* bond length and *decreasing* bond strength:
(a) S—F, S—Br, S—Cl **(b)** C═O, C—O, C≡O

Plan **(a)** S is singly bonded to three different halogen atoms, so the bond order is the same. Bond length increases and bond strength decreases as the halogen's atomic radius increases. **(b)** The same two atoms are bonded, but the bond orders differ. In this case, bond strength increases and bond length decreases as bond order increases.

Solution **(a)** Atomic size increases down a group, so F < Cl < Br.

Bond length: S—Br > S—Cl > S—F

Bond strength: S—F > S—Cl > S—Br

(b) By ranking the bond orders, C≡O > C═O > C—O, we obtain

Bond length: C—O > C═O > C≡O

Bond strength: C≡O > C═O > C—O

Check From Table 9.2, we see that the rankings are correct.

Comment Remember that for bonds involving pairs of different atoms, as in part (a), *the relationship between length and strength holds* only *for single bonds* and not in every case, so apply it carefully.

FOLLOW-UP PROBLEMS

9.3A Rank the bonds in each set in order of *decreasing* bond length and *decreasing* bond strength:

(a) C≡N, C≡O, C≡C **(b)** P—I, P—F, P—Br

9.3B Rank the bonds in each set in order of *increasing* bond length and *increasing* bond strength:

(a) Si—F, Si—C, Si—O **(b)** N=N, N—N, N≡N

SOME SIMILAR PROBLEMS 9.39 and 9.40

How the Model Explains the Properties of Covalent Substances

The covalent bonding model proposes that electron sharing between pairs of atoms leads to *strong, localized bonds.* Most, but not all, covalent substances consist of individual molecules. These *molecular* covalent substances have very different *physical* properties than *network* covalent solids do because different types of forces give rise to the two kinds of substances.

1. *Physical properties of molecular covalent substances.* At first glance, the model seems inconsistent with physical properties of covalent substances. Most are gases (such as methane and ammonia), liquids (such as benzene and water), or low-melting solids (such as sulfur and paraffin wax). If covalent bonds are so strong (~200 to 500 kJ/mol), why do covalent substances melt and boil at such low temperatures? To answer this, we'll focus on two different forces:

(1) *strong bonding forces* hold the atoms together *within* the molecule, and

(2) *weak intermolecular forces* act *between* separate molecules in the sample.

It is the weak forces *between* one molecule and the other molecules around it that account for the physical properties of *molecular* covalent substances. For example, look what happens when pentane (C_5H_{12}) boils (Figure 9.15): weak forces *between* pentane molecules are overcome during this process; the strong C—C and C—H covalent bonds *within* each pentane molecule are *not* broken.

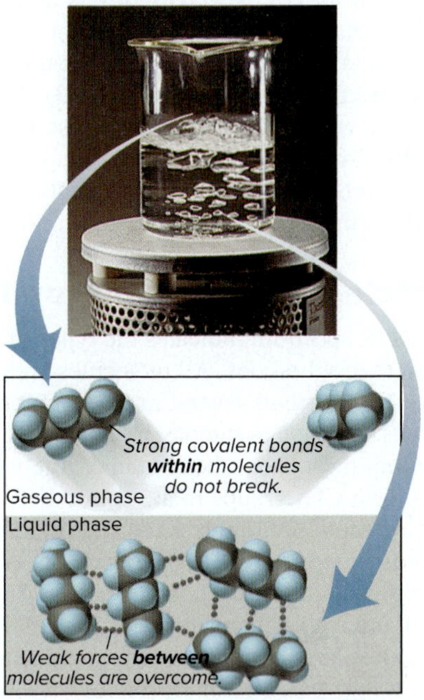

Figure 9.15 Strong forces within molecules and weak forces between them.

Source: © McGraw-Hill Education/Stephen Frisch, photographer

2. *Physical properties of network covalent solids.* Some covalent substances do not consist of separate molecules. Rather, these *network covalent solids* are held together by covalent bonds *between atoms throughout the sample,* and their properties *do* reflect the strength of covalent bonds. Two examples are quartz and diamond (Figure 9.16). Quartz (SiO_2; *top*) has silicon-oxygen covalent bonds in three dimensions; no separate SiO_2 molecules exist. It is very hard and melts at 1550°C. Diamond *(bottom)* has covalent bonds connecting each carbon atom to four others. It is the hardest natural substance known and melts at around 3550°C. Thus, covalent bonds *are* strong, but most covalent substances consist of separate molecules with weak forces between them. (We discuss intermolecular forces in detail in Chapter 12.)

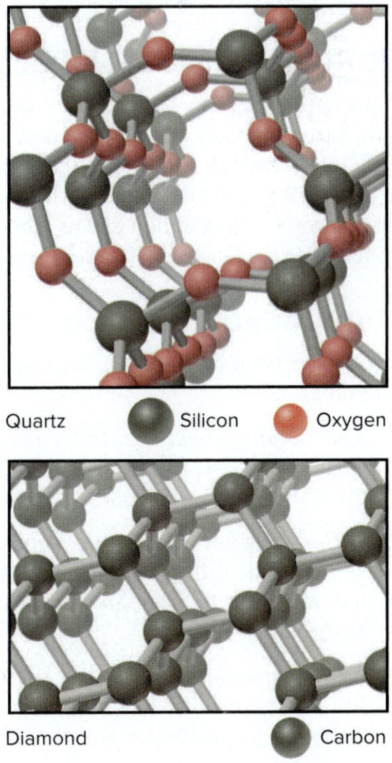

Figure 9.16 Covalent bonds of network covalent solids: quartz and diamond.

3. *Electrical conductivity.* An electric current is carried by either mobile electrons or mobile ions. Most covalent substances are poor electrical conductors, whether melted or dissolved, because their electrons are localized as either shared or unshared pairs and are not mobile, and no ions are present.

The Tools of the Laboratory essay describes a technique used widely for studying the types of bonds in covalent substances.

TOOLS OF THE LABORATORY

Infrared Spectroscopy

Infrared (IR) spectroscopy is an instrumental technique most often used to study covalently bonded molecules. The key components of an IR spectrometer are the same as those of other types of spectrometers (see Figure B7.3). The source emits radiation of many wavelengths, but only those in the IR region are selected. The sample is typically a pure liquid or solid that absorbs varying amounts of different IR wavelengths. An IR spectrum consists of peaks that indicate these various absorptions.

Molecular Vibrations and IR Absorption

An IR spectrum indicates the types of bonds in a molecule based on their vibrations. All molecules undergo motion through space, rotation around several axes, and vibrations between bonded atoms. Consider a sample of ethane gas. The $H_3C—CH_3$ molecules zoom throughout the container, the whole molecule rotates, and its two CH_3 groups rotate about the C—C bond. But let's disregard motion through space and rotation and focus on the motion most important to IR spectroscopy: each pair of atoms is vibrating as though the bonds were springs that stretch and bend. Figure B9.1 depicts the vibrations of diatomic and triatomic molecules; larger molecules vibrate in many more ways.

The energies of IR photons are in the range of these vibrational energies. *Each vibration has a frequency based on the type of motion, masses of the atoms, and strength of the bond.* The frequencies correspond to IR wavelengths between 2.5 and 25 μm. *The energy of these vibrations is quantized.* Just as an atom can absorb a photon whose energy equals the difference between *electron* energy levels, a molecule can absorb an IR photon whose energy equals the difference between *vibrational* energy levels.

IR Radiation and Global Warming

Carbon dioxide, O═C═O, is a linear molecule that bends and stretches symmetrically and asymmetrically when it absorbs IR radiation (Problem B9.2). Sunlight is absorbed by Earth's surface and re-emitted as heat, much of which is IR radiation. Atmospheric CO_2 absorbs this radiation and re-emits it, thus warming the atmosphere (see the Chemical Connections at the end of Section 6.6).

Compound Identification

An IR spectrum can be used to identify a compound for three related reasons:

1. *Each kind of bond absorbs a specific range of wavelengths.* That is, a C—C bond absorbs a different range than does a C═C bond, a C—H bond, a C═O bond, and so forth.
2. *Different types of organic compounds have characteristic spectra.* The different groupings of atoms that define an alcohol, a carboxylic acid, an ether, and so forth (see Chapter 15) absorb differently.
3. *Each compound has a unique spectrum.* The IR spectrum acts like a fingerprint to identify the compound, because the quantity of each wavelength absorbed depends on the detailed molecular structure. For example, no other compound has the IR spectrum of acrylonitrile, a compound used to make plastics (Figure B9.2).

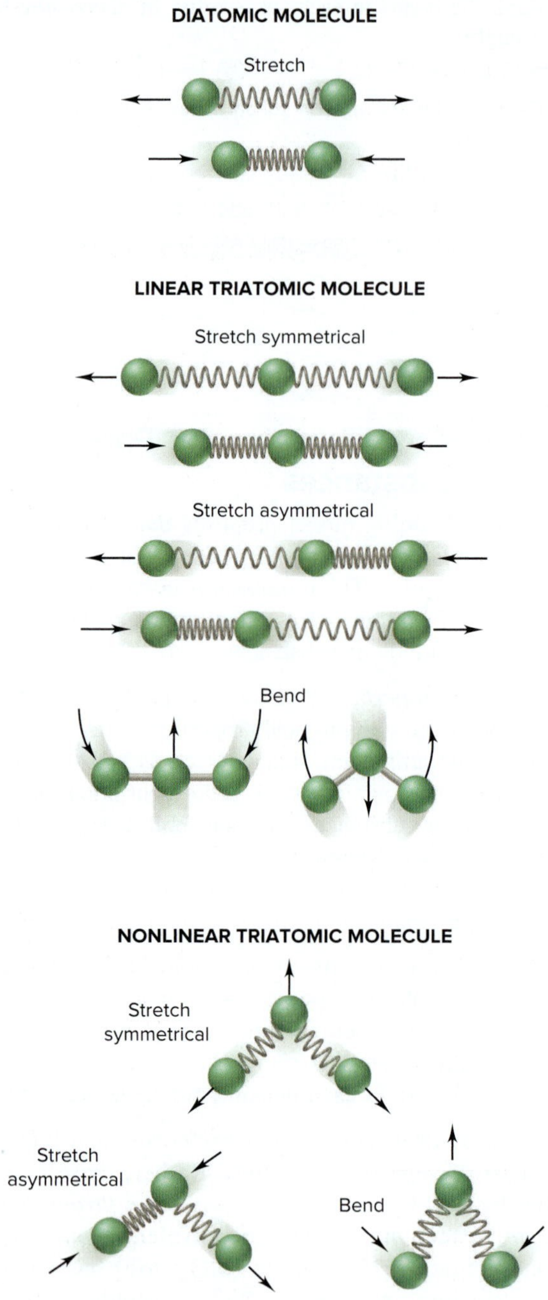

Figure B9.1 Vibrational motions in general diatomic and triatomic molecules.

Recall from Chapter 3 that constitutional (structural) isomers have the same molecular formula but different structures. We expect to see very different IR spectra for the isomers diethyl ether and 2-butanol because their molecular structures are so dissimilar (Figure B9.3). However, as shown in Figure B9.4, even the very similar 1,3-dimethylbenzene and 1,4-dimethylbenzene have different spectra.

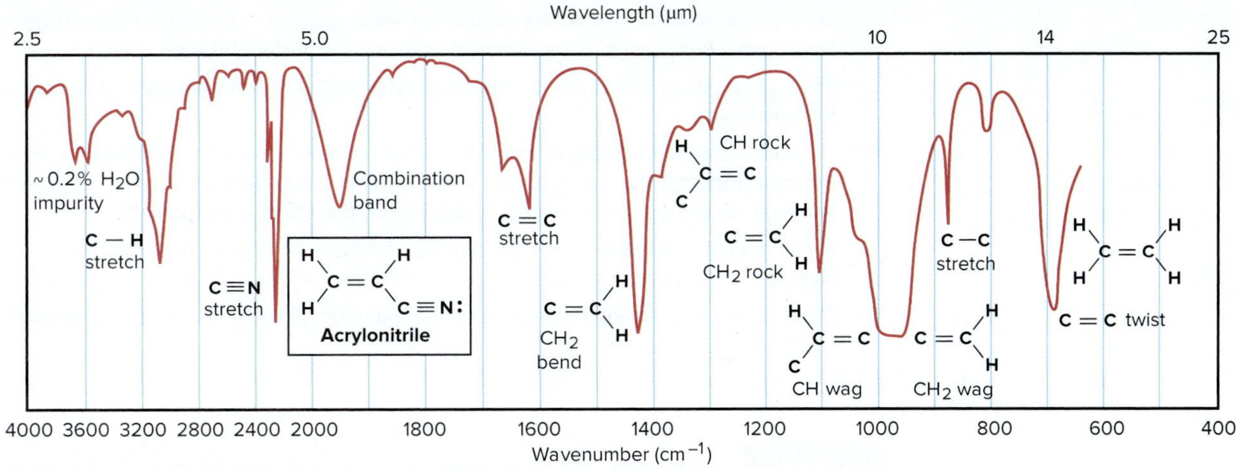

Figure B9.2 The infrared (IR) spectrum of acrylonitrile. In this typical IR spectrum, many absorption bands appear as downward-projecting "peaks" of differing depths and sharpness. Most correspond to a particular type of vibration (stretch, bend, rock, wag, or twist). Some broad peaks (e.g., "combination band") represent several overlapping vibrations. The bottom axis identifies the wavenumber, the inverse of wavelength, which has the unit cm^{-1}. (The scale expands to the right of 2000 cm^{-1}.)

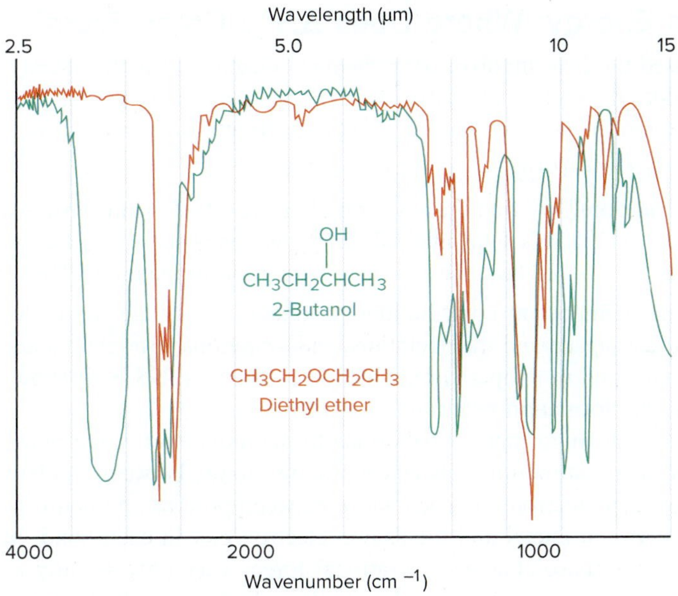

Figure B9.3 The infrared spectra of 2-butanol (*green*) and diethyl ether (*red*).

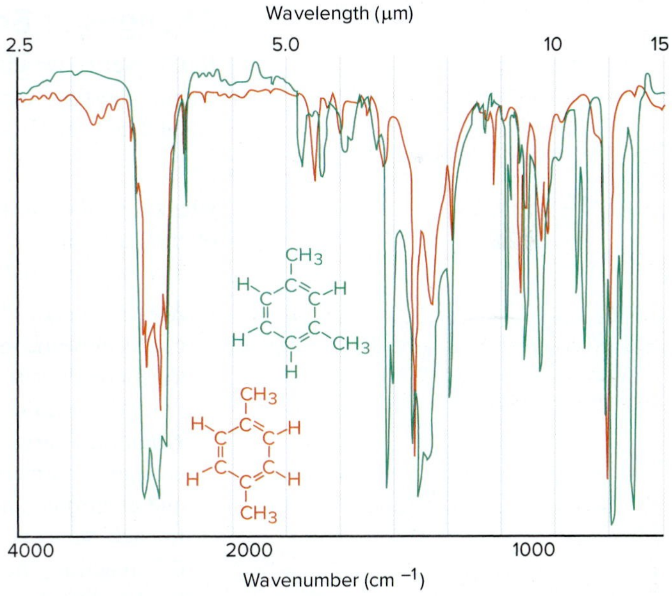

Figure B9.4 The infrared spectra of 1,3-dimethylbenzene (*green*) and 1,4-dimethylbenzene (*red*).

Problems

B9.1 In Figure B9.2, the peak labeled "C=C stretch" occurs at a shorter wavelength (higher wavenumber) than that labeled "C—C stretch," as it does in the IR spectrum of any substance with those bonds. Explain the relative positions of these peaks. In what relative position along the wavelength scale of Figure B9.2 would you expect to find a peak for a C≡C stretch? Explain.

B9.2 The vibrations of a CO_2 molecule are symmetrical stretch, bend, and asymmetrical stretch (Figure B9.1), with frequencies of 4.02×10^{13} s^{-1}, 2.00×10^{13} s^{-1}, and 7.05×10^{13} s^{-1}, respectively.

(a) What wavelengths correspond to these vibrations?

(b) Calculate the energy (in J) of each vibration. Which uses the least energy?

› Summary of Section 9.3

- A shared, localized pair of valence electrons holds the nuclei of two atoms together in a covalent bond, filling each atom's outer level.
- Bond order is the number of shared pairs between two atoms. Bond energy (strength) is the energy absorbed to separate the atoms; the same quantity of energy is released when the bond forms. Bond length is the distance between the nuclei of the two atoms.
- For a given pair of atoms, bond order is directly related to bond energy and inversely related to bond length.
- Molecular covalent substances are soft and low melting because of the weak forces *between* the molecules, not the strong bonding forces *within* them. Network covalent solids are hard and high melting because covalent bonds join all the atoms in the sample.
- Most covalent substances have low electrical conductivity because their electrons are localized and ions are absent.
- Atoms in a covalent bond vibrate, and the energies of these vibrations are studied with IR spectroscopy.

9.4 BOND ENERGY AND CHEMICAL CHANGE

The relative strengths of the bonds in reactants and products determine whether heat is released or absorbed in a chemical reaction. In Chapter 20, you'll see that the change in bond energy is one of two factors determining whether the reaction occurs at all. In this section, we'll discuss the origin of the enthalpy of reaction, use bond energies to calculate it, and look at the energy available from fuels and foods.

Changes in Bond Energy: Where Does ΔH°_{rxn} Come From?

In Chapter 6, we discussed the heat involved in a chemical change but never asked a central question: where does the enthalpy of reaction (ΔH°_{rxn}) come from? For example, when 1 mol of H_2 and 1 mol of F_2 react to form 2 mol of HF at 1 atm and 298 K,

$$H_2(g) + F_2(g) \longrightarrow 2HF(g) + 546\text{ kJ}$$

where does the 546 kJ come from? We find the answer by looking closely at the energies of the molecules involved. A system's total internal energy is composed of its kinetic energy and potential energy. Let's see how these change during the formation of HF:

- *Kinetic energy.* The most important contributions to kinetic energy are the molecules' movements through space and their rotations and vibrations. However, since kinetic energy is proportional to temperature, which is constant at 298 K, it shows no net change during this formation reaction.
- *Potential energy.* The most important contributions to potential energy are phase changes and changes in the attraction between vibrating atoms, between nucleus and electrons (and between electrons) in each atom, between protons and neutrons in each nucleus, and between nuclei and the shared electron pair in each bond. In this reaction, there are no phase changes, vibrational forces vary only slightly as the bonded atoms change, and forces within the atoms and nuclei don't change at all. The only major change in potential energy comes from changes in the attraction between the nuclei and the shared electron pair—the bond energies—of H_2, F_2, and HF.

Thus, our answer to "where does ΔH°_{rxn} come from?" is that it doesn't really "come from" anywhere: *the heat released or absorbed during a chemical change is due to differences between reactant bond energies and product bond energies.*

Using Bond Energies to Calculate ΔH°_{rxn}

Recall that Hess's law (Section 6.5) allows us to think of any reaction as a two-step process, whether or not it actually occurs that way:

1. *A certain quantity of heat is absorbed ($\Delta H^\circ > 0$) to break the reactant bonds and form separate atoms.*
2. *A different quantity of heat is then released ($\Delta H^\circ < 0$) when the atoms form product bonds.*

The sum (symbolized by Σ) of these enthalpy changes is the enthalpy of reaction, ΔH°_{rxn}:

$$\Delta H^\circ_{rxn} = \Sigma\Delta H^\circ_{\text{reactant bonds broken}} + \Sigma\Delta H^\circ_{\text{product bonds formed}} \quad \textbf{(9.2)}$$

- In an exothermic reaction, the energy released during the formation of bonds in the product molecules is *greater* than the energy required to break bonds in the reactant molecules; the magnitude of $\Delta H^\circ_{\text{product bonds formed}}$ is greater than that of $\Delta H^\circ_{\text{reactant bonds broken}}$, so the sum, ΔH°_{rxn}, is *negative* (heat is released).
- In an endothermic reaction, the opposite situation is true: the energy released during bond formation in the product molecules is *smaller* than the energy required to break bonds in the reactant molecules; the magnitude of $\Delta H^\circ_{\text{product bonds formed}}$ is smaller than that of $\Delta H^\circ_{\text{reactant bonds broken}}$, so ΔH°_{rxn} is *positive* (heat is absorbed).

An equivalent form of Equation 9.2 uses bond energies:

$$\Delta H^\circ_{rxn} = \Sigma\text{BE}_{\text{reactant bonds broken}} - \Sigma\text{BE}_{\text{product bonds formed}}$$

(We need the minus sign because all bond energies are positive.)

In reality, only certain bonds break and form during a typical reaction, but we can apply Hess's law and use the following simpler method for calculating ΔH°_{rxn}:

1. Break ***all*** the reactant bonds to obtain individual atoms.
2. Use the atoms to form ***all*** the product bonds.
3. Add the bond energies, with appropriate signs, to obtain the enthalpy of reaction.

(This method assumes reactants and products do not change physical state; additional heat is involved when phase changes occur. We address this topic in Chapter 12.)

Let's use the method to calculate ΔH°_{rxn} for two reactions:

1. *Formation of HF.* When 1 mol of H—H bonds and 1 mol of F—F bonds absorb energy and break, the 2 mol of H atoms and 2 mol of F atoms form 2 mol of H—F bonds, which releases energy (Figure 9.17). We find the bond energy values in Table 9.2 and use a positive sign for bonds broken and a negative sign for bonds formed:

Bonds broken:

$$1 \times \text{H—H} = (1 \text{ mol})(432 \text{ kJ/mol}) = 432 \text{ kJ}$$
$$1 \times \text{F—F} = (1 \text{ mol})(159 \text{ kJ/mol}) = 159 \text{ kJ}$$
$$\Sigma\Delta H^\circ_{\text{reactant bonds broken}} = 591 \text{ kJ}$$

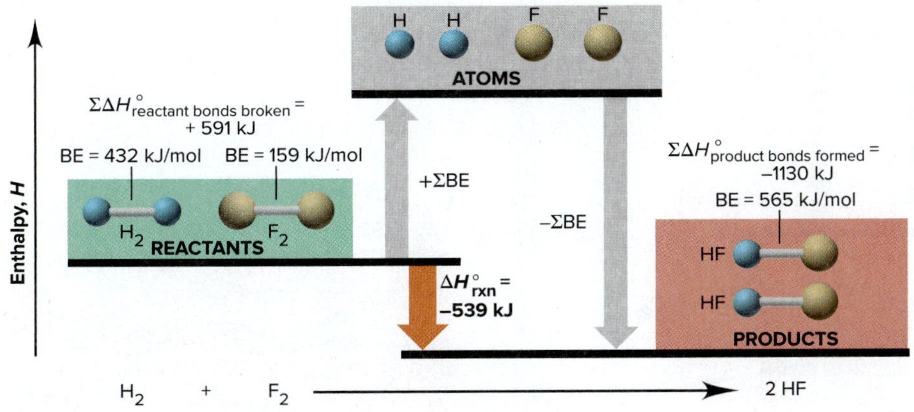

Figure 9.17 Using bond energies to calculate ΔH°_{rxn} for HF formation.

Bonds formed:

$$2 \times \text{H—F} = (2\ \text{mol})(-565\ \text{kJ/mol}) = -1130\ \text{kJ} = \Sigma\Delta H^\circ_{\text{product bonds formed}}$$

Applying Equation 9.2 gives

$$\begin{aligned}\Delta H^\circ_{\text{rxn}} &= \Sigma\Delta H^\circ_{\text{reactant bonds broken}} + \Sigma\Delta H^\circ_{\text{product bonds formed}}\\ &= 591\ \text{kJ} + (-1130\ \text{kJ}) = -539\ \text{kJ}\end{aligned}$$

The small discrepancy between this bond energy value (−539 kJ) and the value from tabulated ΔH° values (−546 kJ) is due to variations in experimental method.

2. *Combustion of* CH_4. For this more complicated reaction, we assume that all the bonds in CH_4 and O_2 break, and the atoms form all the bonds in CO_2 and H_2O (Figure 9.18):

$$CH_4(g) + 2O_2(g) \longrightarrow CO_2(g) + 2H_2O(g)$$

Once again, we use Table 9.2 and appropriate signs for bonds broken and bonds formed:

Bonds broken:

$$\begin{aligned}4 \times \text{C—H} &= (4\ \text{mol})(413\ \text{kJ/mol}) = 1652\ \text{kJ}\\ 2 \times O_2 &= (2\ \text{mol})(498\ \text{kJ/mol}) = 996\ \text{kJ}\\ \Sigma\Delta H^\circ_{\text{reactant bonds broken}} &= 2648\ \text{kJ}\end{aligned}$$

Bonds formed:

$$\begin{aligned}2 \times \text{C=O} &= (2\ \text{mol})(-799\ \text{kJ/mol}) = -1598\ \text{kJ}\\ 4 \times \text{O—H} &= (4\ \text{mol})(-467\ \text{kJ/mol}) = -1868\ \text{kJ}\\ \Sigma\Delta H^\circ_{\text{product bonds formed}} &= -3466\ \text{kJ}\end{aligned}$$

Applying Equation 9.2 gives

$$\begin{aligned}\Delta H^\circ_{\text{rxn}} &= \Sigma\Delta H^\circ_{\text{reactant bonds broken}} + \Sigma\Delta H^\circ_{\text{product bonds formed}}\\ &= 2648\ \text{kJ} + (-3466\ \text{kJ}) = -818\ \text{kJ}\end{aligned}$$

In addition to variations in experimental method, there is a more basic reason for the discrepancy between the $\Delta H^\circ_{\text{rxn}}$ obtained from bond energies (−818 kJ) and the value obtained by calorimetry (−802 kJ; Section 6.3) for the combustion of methane. A bond energy is an *average* value for a given bond in many compounds. The value *in a particular substance* is usually close, but not equal, to the average. For example, the C—H bond energy of 413 kJ/mol is the average for C—H bonds in many molecules. In fact, 415 kJ is actually required to break 1 mol of C—H bonds in methane, or 1660 kJ for 4 mol of these bonds, which gives a $\Delta H^\circ_{\text{rxn}}$ closer to the calorimetric value. Thus, it isn't surprising to find small discrepancies between $\Delta H^\circ_{\text{rxn}}$ values obtained in different ways.

Figure 9.18 Using bond energies to calculate $\Delta H^\circ_{\text{rxn}}$ for the combustion of methane.

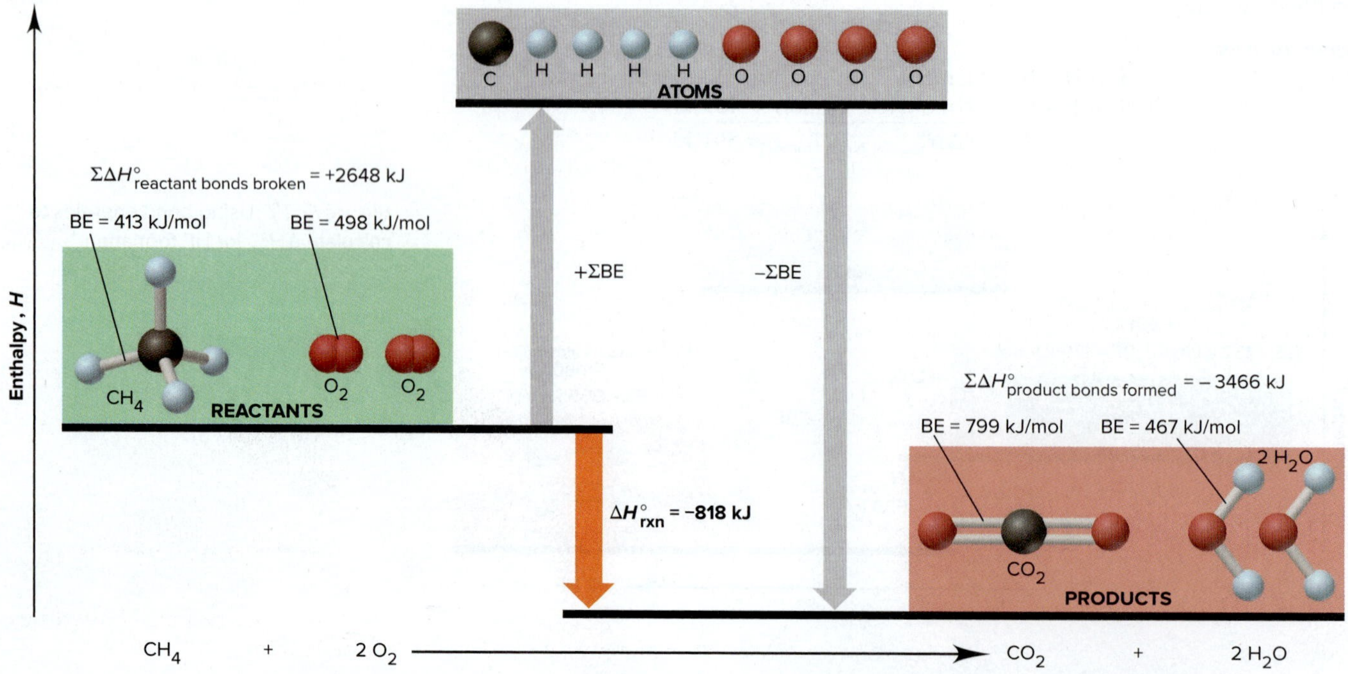

SAMPLE PROBLEM 9.4 Using Bond Energies to Calculate ΔH°_{rxn}

Problem Calculate ΔH°_{rxn} for the chlorination of methane to form chloroform:

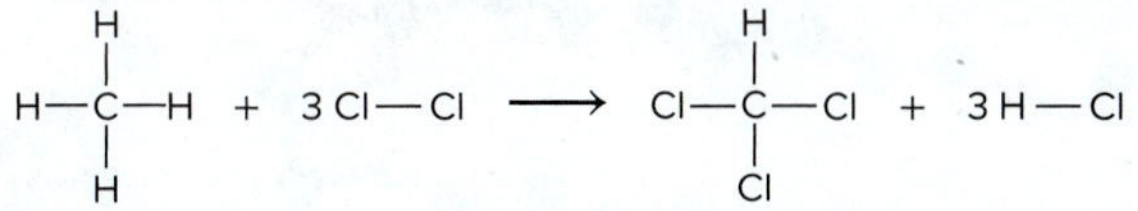

Plan *All* the reactant bonds break, and *all* the product bonds form. We find the bond energies in Table 9.2 and substitute the two sums, with correct signs, into Equation 9.2.

Solution Finding the standard enthalpy changes for bonds broken and for bonds formed: For bonds broken, the bond energy values are positive:

$$4 \times \text{C—H} = (4\ \text{mol})(413\ \text{kJ/mol}) = 1652\ \text{kJ}$$
$$3 \times \text{Cl—Cl} = (3\ \text{mol})(243\ \text{kJ/mol}) = 729\ \text{kJ}$$
$$\Sigma\Delta H^\circ_{\text{bonds broken}} = 2381\ \text{kJ}$$

For bonds formed, the values are negative:

$$3 \times \text{C—Cl} = (3\ \text{mol})(-339\ \text{kJ/mol}) = -1017\ \text{kJ}$$
$$1 \times \text{C—H} = (1\ \text{mol})(-413\ \text{kJ/mol}) = -413\ \text{kJ}$$
$$3 \times \text{H—Cl} = (3\ \text{mol})(-427\ \text{kJ/mol}) = -1281\ \text{kJ}$$
$$\Sigma\Delta H^\circ_{\text{bonds formed}} = -2711\ \text{kJ}$$

Calculating ΔH°_{rxn}:

$$\Delta H^\circ_{rxn} = \Sigma\Delta H^\circ_{\text{bonds broken}} + \Sigma\Delta H^\circ_{\text{bonds formed}} = 2381\ \text{kJ} + (-2711\ \text{kJ}) = -330\ \text{kJ}$$

Check The signs of the enthalpy changes are correct: $\Sigma\Delta H^\circ_{\text{bonds broken}} > 0$, and $\Sigma\Delta H^\circ_{\text{bonds formed}} < 0$. More energy is released than absorbed, so ΔH°_{rxn} is negative:

$$\sim 2400\ \text{kJ} + [\sim(-2700\ \text{kJ})] = -300\ \text{kJ}$$

FOLLOW-UP PROBLEMS

9.4A One of the most important industrial reactions is the formation of ammonia from its elements:

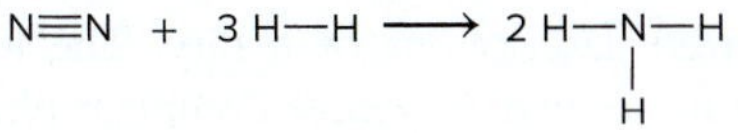

Use bond energies to calculate ΔH°_{rxn}.

9.4B Oxygen difluoride reacts with water to produce hydrofluoric acid:

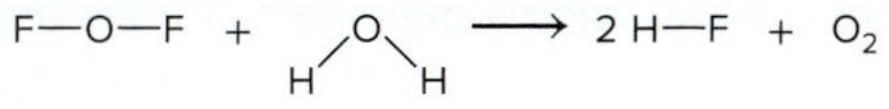

Use bond energies to calculate ΔH°_{rxn}.

SOME SIMILAR PROBLEMS 9.47–9.50

Bond Strengths and the Heat Released from Fuels and Foods

A *fuel* is a material that reacts with atmospheric oxygen to release energy *and* is available at a reasonable cost. The most common fuels for machines are hydrocarbons and coal, and the most common ones for organisms are fats and carbohydrates. All of these fuels contain large organic molecules with many C—C and C—H bonds and fewer C—O and O—H bonds. According to our two-step approach for determining the enthalpy of reaction from bond energies, when the fuel reacts with O_2, all of the bonds break, and the C, H, and O atoms form C═O and O—H bonds in the products (CO_2 and H_2O). Because the combustion is exothermic, the total of the bond energies in the products is *greater* than the total in the reactants. *Weaker bonds (less stable, more reactive) are easier to break than stronger bonds (more stable, less reactive) because they are already higher in energy.* Therefore, the bonds in CO_2 and H_2O are stronger (lower energy, more stable) than those in gasoline (or cooking oil) and O_2 (weaker, higher energy, less stable).

Fuels with more weak bonds yield more energy than fuels with fewer weak bonds. When a hydrocarbon (alkane) burns, C—C and C—H bonds break; when an alcohol burns, C—O and O—H bonds break as well. Table 9.2 shows that the sum for C—C and C—H bonds (760 kJ/mol) is less than the sum for C—O and O—H bonds (825 kJ/mol). Therefore, it takes more energy to break the bonds of a fuel with a lot of C—O and O—H bonds. Thus, in general, *a fuel with fewer bonds to O releases more energy* (Figure 9.19).

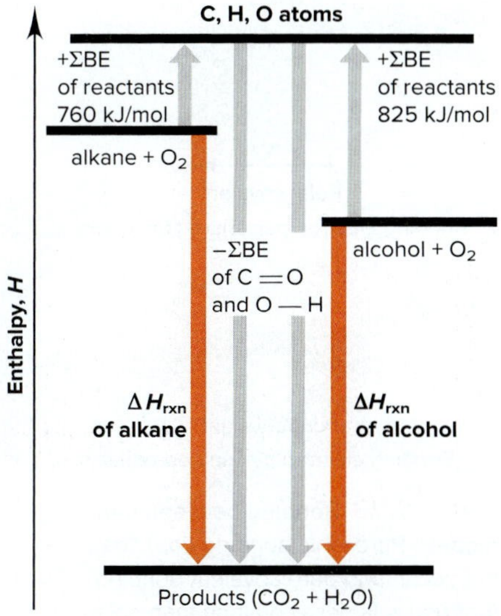

Figure 9.19 Relative bond strength and energy from fuels.

Table 9.4 Enthalpies of Reaction for Combustion of Some Foods

Substance	ΔH_{rxn} (kJ/g)
Fats	
Vegetable oil	–37.0
Margarine	–30.1
Butter	–30.0
Carbohydrates	
Table sugar (sucrose)	–16.2
Brown rice	–14.9
Maple syrup	–10.4

Both fats and carbohydrates serve as high-energy foods and consist of chains or rings of C atoms attached to H atoms, with some C—O, C=O, and O—H bonds *(shown in red below)*:

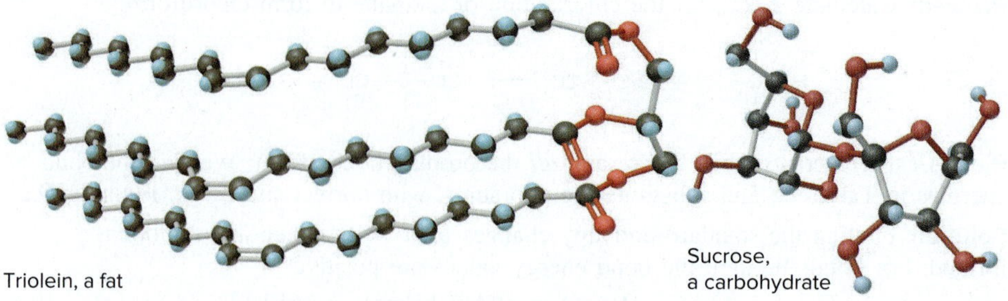

Carbohydrates have fewer chains of C atoms and bonds to H and more bonds to O. Fats "contain more Calories" per gram than carbohydrates because fats have fewer bonds to O (Table 9.4).

› Summary of Section 9.4

- › The only component of internal energy that changes significantly during a reaction is the bond energies of reactants and products, and this change appears as the enthalpy of reaction, ΔH°_{rxn}.
- › A reaction involves breaking reactant bonds and forming product bonds. Applying Hess's law, we use tabulated bond energies to calculate ΔH°_{rxn}.
- › Bonds in fuels are weaker (less stable, higher energy) than bonds in the combustion products. Fuels with more weak bonds release more energy than fuels with fewer.

9.5 BETWEEN THE EXTREMES: ELECTRONEGATIVITY AND BOND POLARITY

Scientific models are idealized descriptions of reality. The ionic and covalent bonding models portray compounds as formed by *either* complete electron transfer *or* complete electron sharing. But, in real substances, most atoms are joined by *polar covalent bonds*—between pure ionic and pure covalent (see Figure 9.20). In this section, we explore the "in-between" nature of these bonds and its importance for the properties of substances.

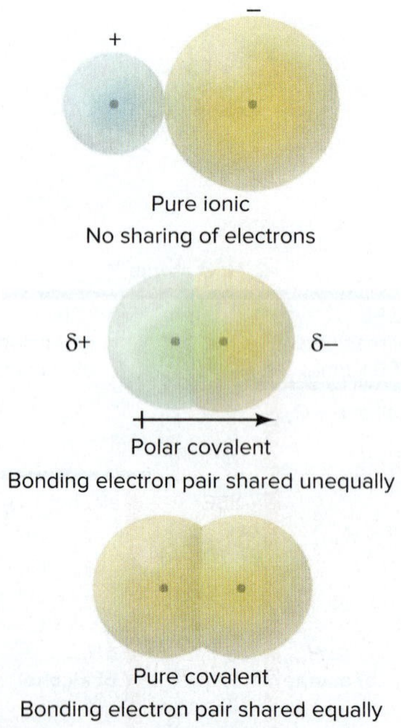

Figure 9.20 Bonding between the models. Pure ionic bonding *(top)* does not occur, and pure covalent bonding *(bottom)* is far less common than polar covalent bonding *(middle)*.

Electronegativity

Electronegativity (EN) is the relative ability of a bonded atom to attract shared electrons.* We might expect, for instance, the H—F bond energy to be the average of an H—H bond (432 kJ/mol) and an F—F bond (159 kJ/mol), or 296 kJ/mol. But, the actual HF bond energy is 565 kJ/mol, or 269 kJ/mol *higher!* To explain this difference, the American chemist Linus Pauling reasoned that, if F attracts the shared electron pair more strongly than H—that is, if F is more *electronegative* than H—the electrons will spend more time closer to F, and this unequal sharing makes the F end of the bond partially negative and the H end partially positive. The *electrostatic attraction* between these partial charges *increases* the energy required to break the bond. From studies with many compounds, Pauling derived a scale of *relative EN values* based on fluorine having the highest EN value, 4.0 (Figure 9.21).

Trends in Electronegativity In general, *electronegativity is inversely related to atomic size* because the nucleus of a smaller atom is closer to the shared pair than the nucleus of a larger atom, so it attracts the electrons more strongly (Figure 9.22):

- *Down a main group,* electronegativity (height of post) decreases as atomic size (hemisphere on top of post) increases.
- *Across a period of main-group elements,* electronegativity increases.
- Nonmetals are *more* electronegative than metals.

*Electronegativity refers to a *bonded* atom attracting a shared pair; electron affinity refers to a gaseous atom gaining an electron to form an anion. Elements with a high EN also have a highly negative EA.

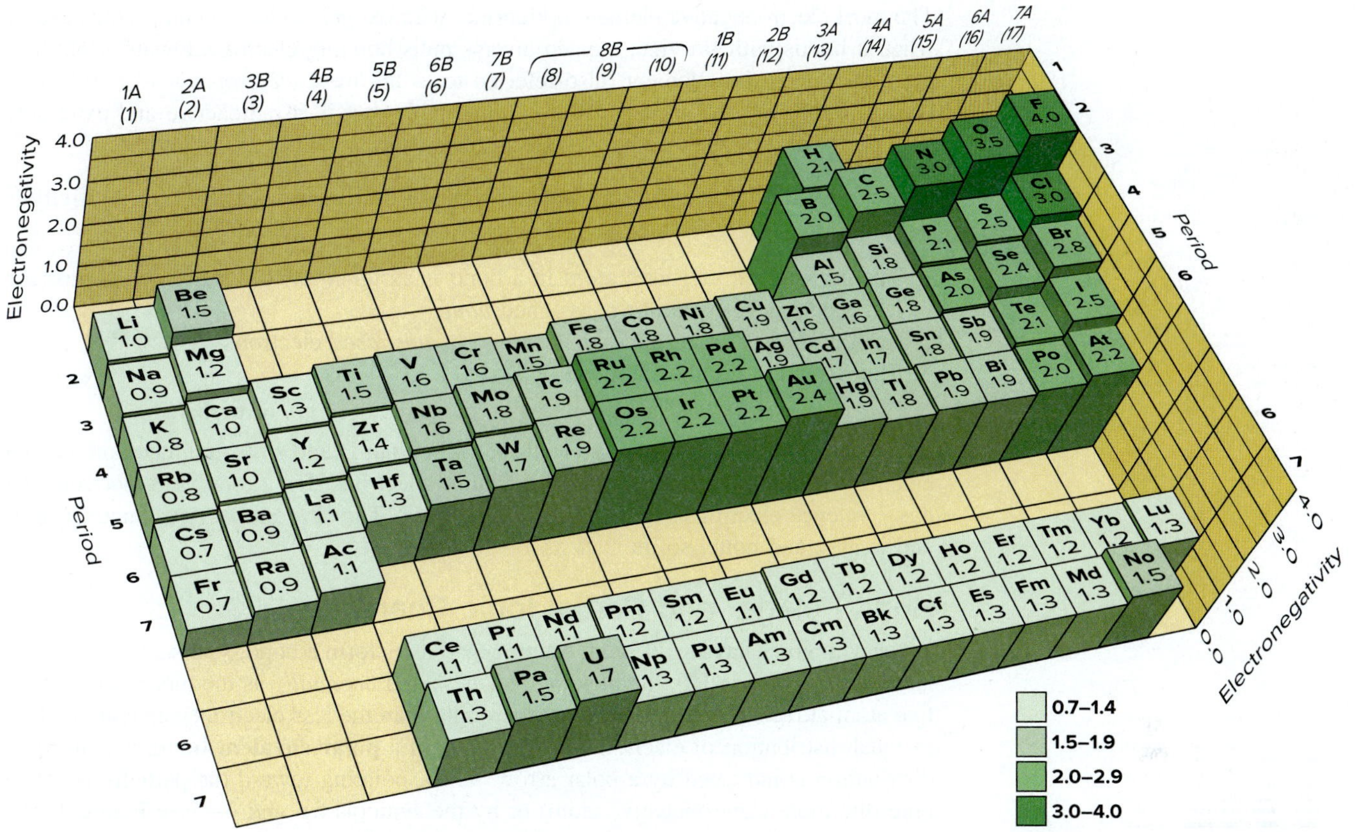

Figure 9.21 **The Pauling electronegativity (EN) scale.** The height of each post is proportional to the EN, which is shown on top. The key has several EN cutoffs. In the main groups, EN *increases* across and *decreases* down. The transition and inner transition elements show little change in EN. Here hydrogen is placed near elements with similar EN values.

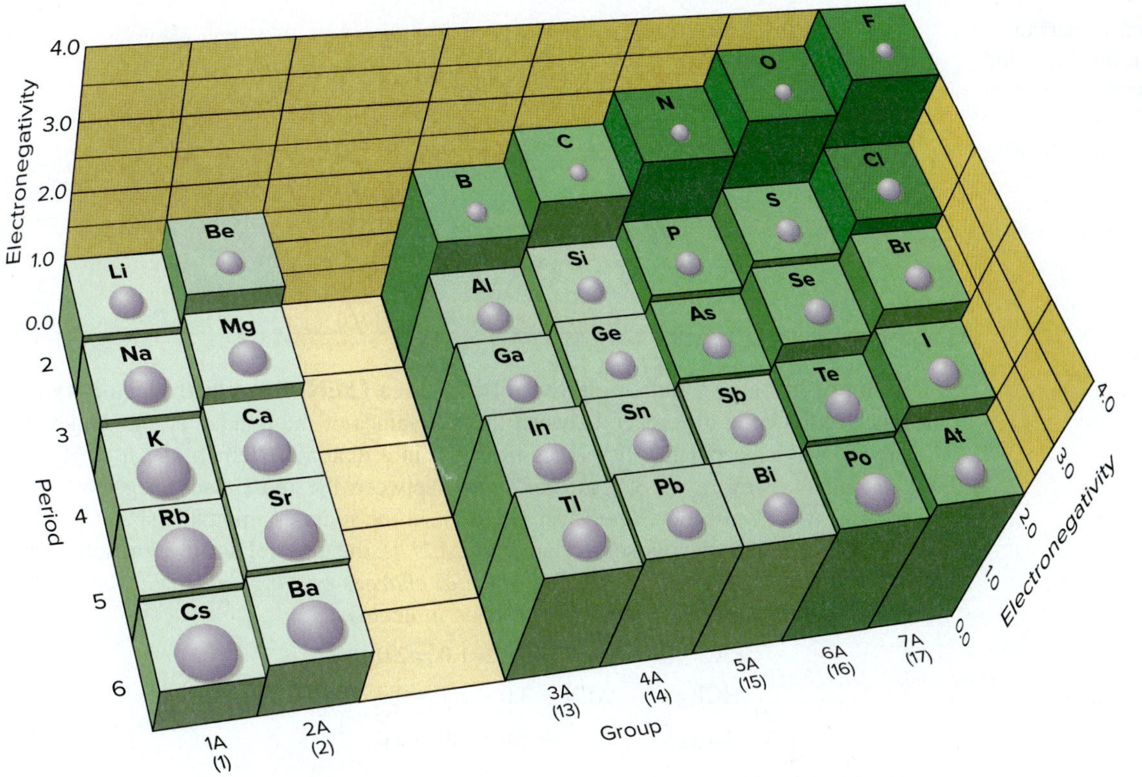

Figure 9.22 Electronegativity and atomic size.

The most electronegative element is fluorine, with oxygen a close second. Thus, except when it bonds with fluorine, oxygen always pulls bonding electrons toward itself. The least electronegative element (also referred to as the most *electropositive*) is francium (Fr), in the lower left corner of the periodic table, but it is radioactive and extremely rare, so for all practical purposes, cesium (Cs) is the most electropositive.*

Electronegativity and Oxidation Number An important use of electronegativity is in determining an atom's oxidation number (O.N.; see Section 4.5):

1. The more electronegative atom in a bond is assigned *all* the *shared* electrons; the less electronegative atom is assigned *none.*
2. Each atom in a bond is assigned *all* of its *unshared* electrons.
3. The oxidation number is given by

$$\text{O.N.} = \text{no. of valence } e^- - (\text{no. of shared } e^- + \text{no. of unshared } e^-)$$

In HCl, for example, Cl is more electronegative than H and so is assigned both shared electrons in the bond and its own 6 unshared electrons, for a total of 8 electrons. Cl has 7 valence electrons so its O.N. is $7 - 8 = -1$. The H atom has 1 valence electron and is assigned none, so its O.N. is $1 - 0 = +1$.

Bond Polarity and Partial Ionic Character

Whenever atoms having different electronegativities form a bond, such as H (EN = 2.1) and F (EN = 4.0) in HF, the bonding pair is shared *unequally* as the more electronegative atom attracts the shared pair more strongly than the less electronegative atom. This unequal distribution of electron density results in a **polar covalent bond.** The unequal distribution is indicated by a polar arrow, $\mapsto$, pointing toward the partially negative pole (the more electronegative atom) or by the symbols δ+ and δ− (see Figure 4.1):

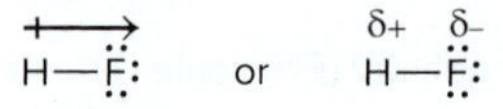

In the H—H and F—F bonds, where the atoms are identical, the bonding pair is shared *equally,* and a **nonpolar covalent bond** results. In Figure 9.23, relief maps show the distribution of electron density in H_2, F_2, and HF.

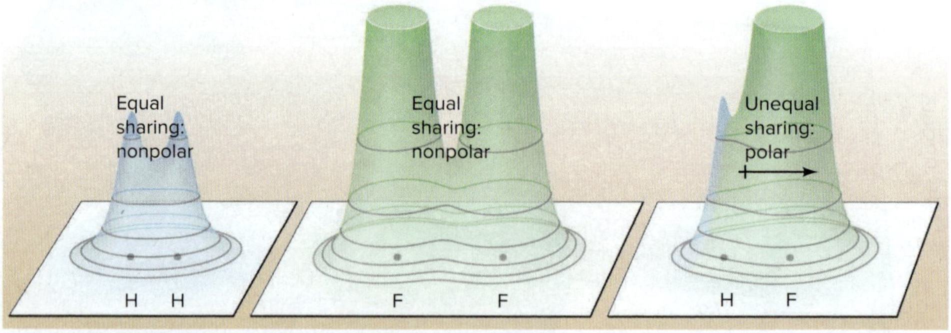

Figure 9.23 Electron density distributions shown as relief maps for H_2, F_2, and HF. In HF, the electron density shifts from H to F. (The electron density peak for F has been cut off to limit the figure height.)

The Importance of Electronegativity Difference (ΔEN) The **electronegativity difference (ΔEN),** the difference between the EN values of the bonded atoms, is directly related to a bond's polarity. It ranges from 0.0 in a diatomic element, such as H_2, O_2, or Cl_2, all the way up to 3.3, the difference between the most electronegative atom, F (4.0), and the most electropositive, Cs (0.7), in the ionic compound CsF.

Another parameter closely related to ΔEN is the **partial ionic character** of a bond: *a greater ΔEN results in larger partial charges and higher partial ionic character.* Consider three Cl-containing gaseous molecules:

$LiCl(g)$:	$\Delta EN = 3.0 - 1.0 = 2.0$	ionic character decreases ↓
$HCl(g)$:	$\Delta EN = 3.0 - 2.1 = 0.9$	
$Cl_2(g)$:	$\Delta EN = 3.0 - 3.0 = 0.0$	

*In 1934, the American physicist Robert S. Mulliken developed electronegativity values based on atomic properties: EN = (IE − EA)/2. By this approach as well, fluorine, with a high ionization energy (IE) and a large negative electron affinity (EA), has a high EN; and cesium, with a low IE and a small EA, has a low EN.

The bond in LiCl has more ionic character than the one in HCl, which has more than the one in Cl_2.

Here are two approaches for quantifying ionic character. Both use arbitrary cutoffs, which is not really consistent with the actual gradation in bonding:

1. *ΔEN range.* This approach divides bonds into mostly ionic, polar covalent, mostly covalent, and nonpolar covalent based on a range of ΔEN values (Figure 9.24).
2. *Percent ionic character.* This approach is based on the behavior of a gaseous diatomic molecule in an electric field. A plot of *percent ionic character* vs. ΔEN for several molecules shows that, as expected, *percent ionic character generally increases with ΔEN* (Figure 9.25A). A value of 50% divides ionic from covalent bonds. Note that a substance like $Cl_2(g)$ has 0% ionic character, but none has 100% ionic character: *electron sharing occurs to some extent in every bond,* even between an ion pair of an alkali halide (Figure 9.25B).

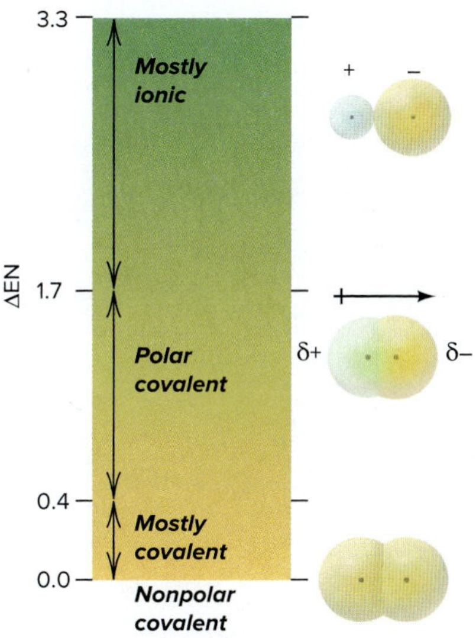

Figure 9.24 **ΔEN ranges for classifying the partial ionic character of bonds.**

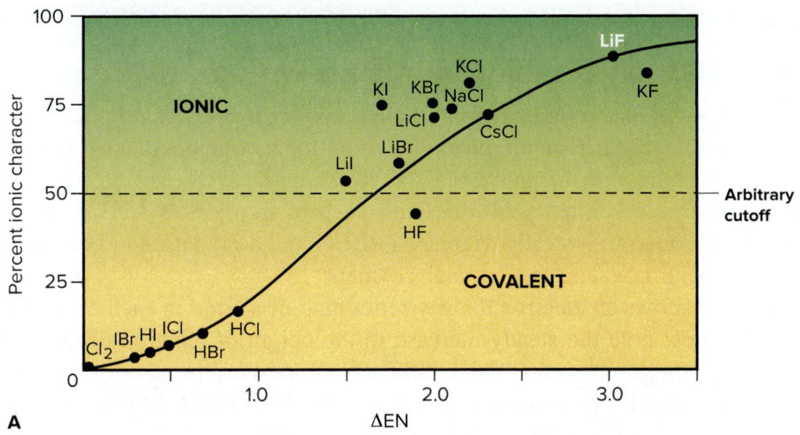

A

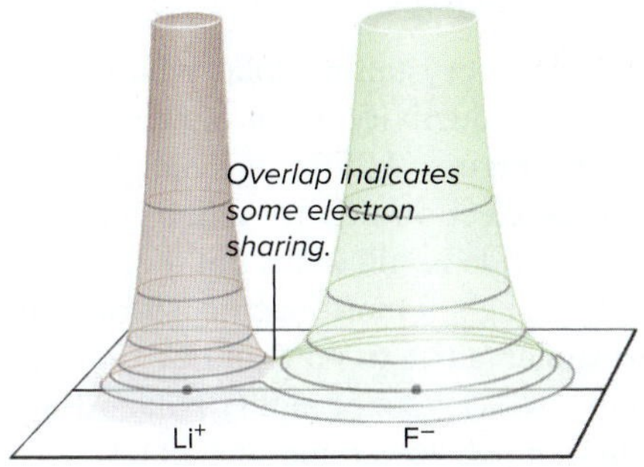

B

Figure 9.25 **Percent ionic character as a function of ΔEN.** **A,** ΔEN correlates with ionic character. **B,** Even in highly ionic LiF (ΔEN = 3.0), some electron sharing occurs between the ions in the gaseous ion pair.

Student Hot Spot

Student data indicate that you may struggle with using electronegativity to determine bond polarity. Access the Smartbook to view additional Learning Resources on this topic.

SAMPLE PROBLEM 9.5 Determining Bond Polarity from EN Values

Problem **(a)** Use a polar arrow to indicate the polarity of each bond: N—H, F—N, I—Cl.
(b) Rank the following bonds in order of *increasing* polarity and *decreasing* percent ionic character: H—N, H—O, H—C.

Plan **(a)** We use Figure 9.21 to find the EN values for the atoms and point the polar arrow toward the more electronegative atom, which is the negative end of the bond.
(b) To rank the bond polarity, we determine ΔEN: the higher the value, the greater the polarity. Percent ionic character is also directly related to ΔEN (and bond polarity); it decreases in the opposite order that polarity increases.

Solution **(a)** The EN of N = 3.0 and the EN of H = 2.1, so $\overleftarrow{\text{N—H}}$.
The EN of F = 4.0 and the EN of N = 3.0, so $\overleftarrow{\text{F—N}}$.
The EN of I = 2.5 and the EN of Cl = 3.0, so $\overrightarrow{\text{I—Cl}}$.
(b) The ΔEN values are 3.0 − 2.1 = 0.9 for H—N, 3.5 − 2.1 = 1.4 for H—O, and 2.5 − 2.1 = 0.4 for H—C.
The order of *increasing* bond polarity is H—C < H—N < H—O.
The order of *decreasing* percent ionic character is H—O > H—N > H—C.

Check In (b), we can check the order of bond polarity using periodic trends. Each bond involves H and a Period 2 atom. Since atomic size decreases and EN increases across a period, the polarity is greatest for the bond to O (farthest to the right in Period 2).

Comment In Chapter 10, you'll see that the bond polarity contributes to the overall polarity of the molecule, which is a major factor in determining behavior.

FOLLOW-UP PROBLEMS

9.5A Arrange each set of bonds in order of *increasing* polarity, and indicate bond polarity with δ+ and δ− symbols:
(a) Cl—F, Br—Cl, Cl—Cl **(b)** P—F, N—O, N—F

9.5B Arrange each set of bonds in order of *decreasing* polarity, and indicate bond polarity with a polar arrow:
(a) Se—Cl, Se—F, Se—Br **(b)** S—B, F—B, Cl—B

SOME SIMILAR PROBLEMS 9.60–9.63, 9.66, and 9.67

The Gradation in Bonding Across a Period

A metal and a nonmetal—elements from the left and right sides of the periodic table—have a relatively large ΔEN and typically form an ionic compound. Two nonmetals—both from the right side of the table—have a small ΔEN and form a covalent compound. When we combine chlorine with each of the Period 3 elements, starting with sodium, we observe a steady decrease in ΔEN and a gradation in bond type from ionic through polar covalent to nonpolar covalent.

Figure 9.26 shows an electron density relief map of a bond in each of the common Period 3 chlorides; note the steady increase in the height of electron density *between* the peaks—the bonding region—which indicates an *increase in electron sharing.* Figure 9.27 shows samples of common Period 3 chlorides—NaCl, $MgCl_2$, $AlCl_3$, $SiCl_4$, PCl_3, and SCl_2, as well as Cl_2—along with the change in ΔEN and two physical properties, melting point and electrical conductivity:

- *NaCl.* Sodium chloride is a white (colorless) crystalline solid with a ΔEN of 2.1, a high melting point, and high electrical conductivity when molten—ionic by any criterion. Nevertheless, just as for LiF (Figure 9.25B), a small but significant amount of electron sharing appears in the electron density relief map for NaCl.
- *$MgCl_2$.* With a ΔEN of 1.8, magnesium chloride is still ionic, but it has a lower melting point and lower conductivity, as well as slightly more electron sharing, than NaCl.
- *$AlCl_3$.* Rather than being a three-dimensional lattice of Al^{3+} and Cl^- ions, aluminum chloride, with a ΔEN value of 1.5, consists of layers of highly polar Al—Cl bonds. Weak forces between layers result in a much lower melting point, and the low conductivity implies few free ions. Electron density between the nuclei is even higher than in $MgCl_2$. $AlCl_3$ has less ionic character than NaCl or $MgCl_2$.
- *$SiCl_4$, PCl_3, SCl_2, and Cl_2.* The trend toward bonding that is more covalent continues through the remaining substances. The three compounds, with polar covalent bonds, occur as separate molecules, which have no conductivity (no ions) and such weak forces *between* them that melting points are below 0°C. In Cl_2, the bond is nonpolar (ΔEN = 0.0). The relief maps show the increasing electron density in the bonding region.

Thus, *as ΔEN decreases, the bond becomes more covalent,* and the character of the substance changes from ionic solid to covalent gas.

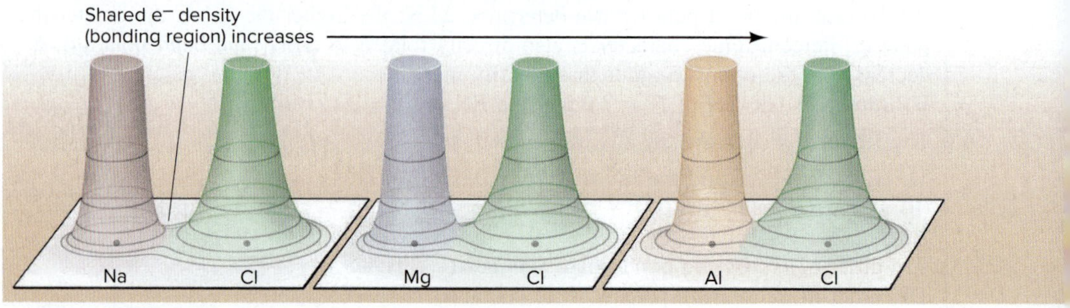

Figure 9.26 Electron density relief maps for bonds of Period 3 chlorides. Note the steady increase in electron sharing from NaCl to Cl_2.

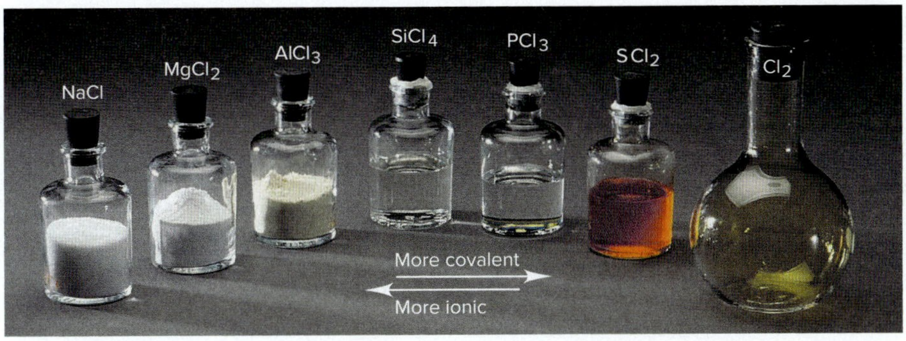

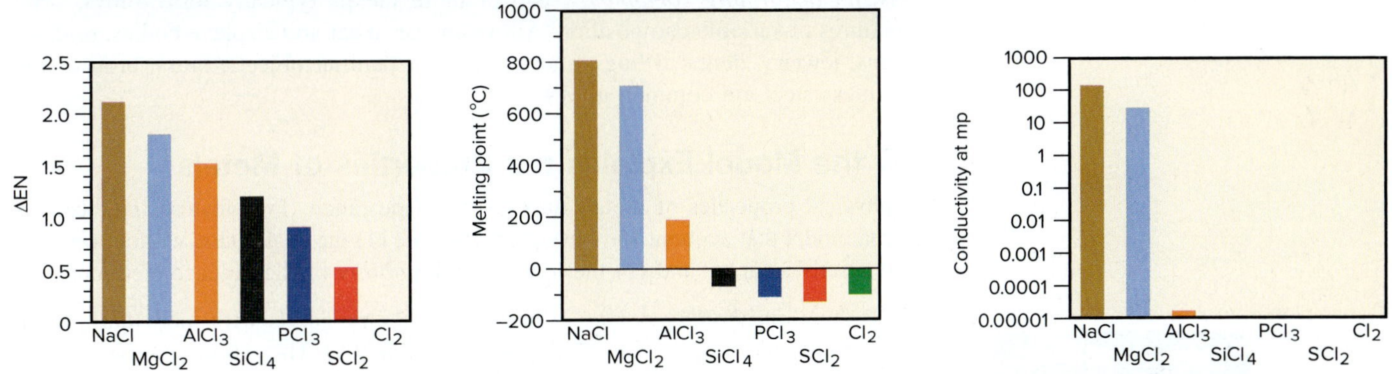

Figure 9.27 **Properties of Period 3 chlorides.** As ΔEN decreases, melting point and electrical conductivity decrease because the bond type changes from ionic to polar covalent to nonpolar covalent.

Source: © McGraw-Hill Education/Stephen Frisch, photographer

› Summary of Section 9.5

- Electronegativity is the ability of a bonded atom to attract shared electrons, which generates opposite partial charges at the ends of the bond and contributes to the bond energy.
- Electronegativity increases across a period and decreases down a group, the reverse of the trends in atomic size.
- The larger the ΔEN for two bonded atoms, the more polar the bond is and the greater its ionic character.
- For Period 3 chlorides, there is a gradation in bond type from ionic to polar covalent to nonpolar covalent.

9.6 AN INTRODUCTION TO METALLIC BONDING

In this section, you'll see how a simple, qualitative model for metallic bonding accounts for the properties of metals; a more detailed model is presented in Chapter 12.

The Electron-Sea Model

Metals can transfer electrons to nonmetals and form ionic solids, such as NaCl. And experiments with metals in the gas phase show that two metal atoms can even share their valence electrons to form gaseous, diatomic molecules, such as Na_2. But what holds the atoms together in a chunk of Na metal? The **electron-sea model** of metallic

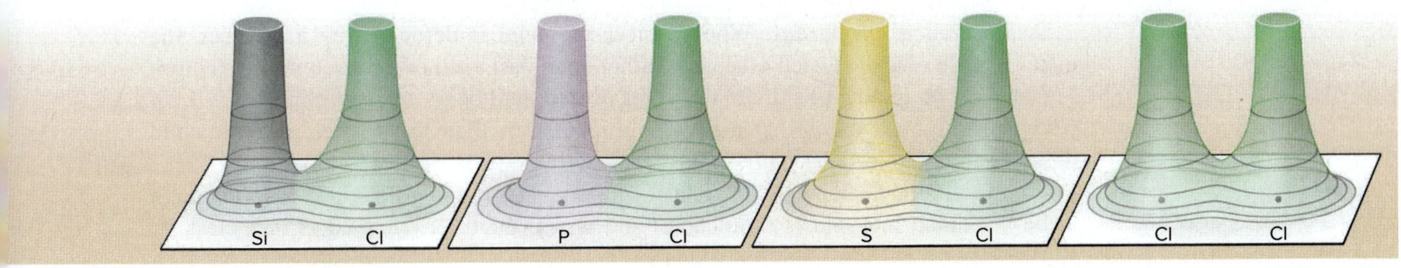

bonding proposes that all the metal atoms in the sample contribute their valence electrons to form a delocalized electron "sea" throughout the piece, with the metal ions (nuclei and core electrons) lying in an orderly array (see Figure 9.2C). *All the atoms in the sample share the electrons,* and the piece is held together by the mutual attraction of the metal cations for the mobile valence electrons. Thus, bonding in metals is fundamentally different than the other two types:

- *In contrast to ionic bonding,* no anions are present, and the metal ions are not held in place as rigidly.
- *In contrast to covalent bonding,* no particular pair of metal atoms is bonded through a localized electron pair.
- *Instead of forming compounds,* two or more metals typically form **alloys,** solid mixtures of variable composition. Alloys appear in car and airplane bodies, bridges, coins, jewelry, dental fillings, and many other familiar objects; brass, bronze, and stainless steel are common alloys.

How the Model Explains the Properties of Metals

The physical properties of metals vary over a wide range. Two features of the electron-sea model that account for these properties are (1) the *regularity,* without rigidity, of the metal-ion array and (2) the *number* and *mobility* of the valence electrons.

Table 9.5 Melting and Boiling Points of Some Metals

Element	mp (°C)	bp (°C)
Lithium (Li)	180	1347
Tin (Sn)	232	2623
Aluminum (Al)	660	2467
Barium (Ba)	727	1850
Silver (Ag)	961	2155
Copper (Cu)	1083	2570
Uranium (U)	1130	3930

1. *Melting and boiling points.* Nearly all metals are solids with moderate to high melting points and much higher boiling points (Table 9.5). These properties are related to the energy of the metallic bonding. Melting points are only moderately high because the cations can move without disrupting their attraction to the surrounding electrons. Boiling points are very high because each cation and its valence electron(s) must break away from the others to vaporize. Gallium provides a striking example: it can melt in your hand (mp 29.8°C) but doesn't boil until over 2400°C.

Periodic trends are consistent with the strength of the bonding:

- *Down a group,* melting points decrease because the larger metal ions have a weaker attraction to the electron sea.
- *Across a period,* melting points increase. Alkaline earth metals [Group 2A(2)] have higher melting points than alkali metals [Group 1A(1)] because their 2+ cations have stronger attractions to twice as many valence electrons (Figure 9.28).

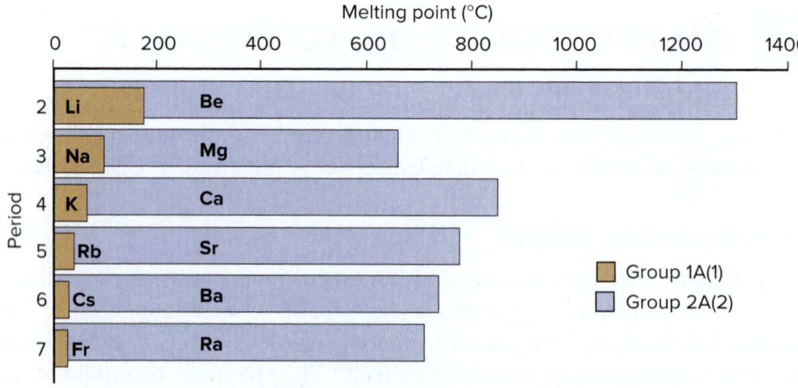

Figure 9.28 Melting points of Group 1A(1) and Group 2A(2) metals.

2. *Mechanical properties.* When a piece of metal is deformed by a hammer, the metal ions do not repel each other, but rather slide past each other through the electron sea and end up in new positions; thus, metals dent and bend, as shown in Figure 9.29. Compare this behavior with the interionic repulsions that occur when an ionic solid is struck (see Figure 9.9).

The common Group 1B(11) metals—copper, silver, and gold—are all soft enough to be machined into sheets (malleable) and wires (ductile), but gold is in a class by

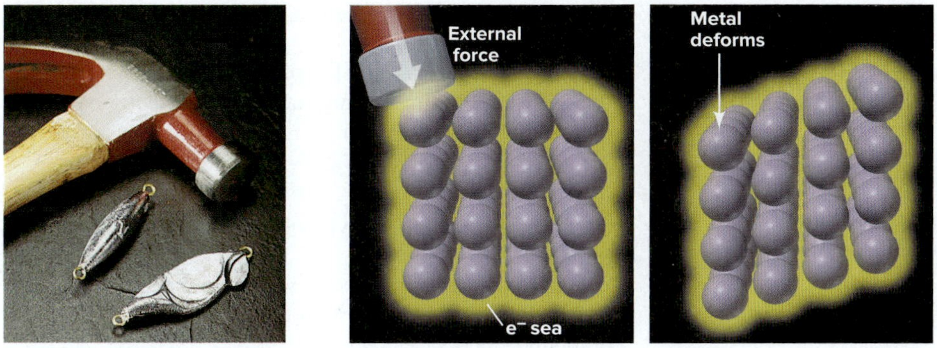

Figure 9.29 Why metals dent and bend rather than crack.
Source: © McGraw-Hill Education/Stephen Frisch, photographer

itself. One gram of gold, about the size of a small ball-bearing, can be drawn into a wire 20 μm thick and 165 m long or hammered into a 1.0-m^2 sheet that is 230 atoms (about 70 nm) thick *(see photo)*!

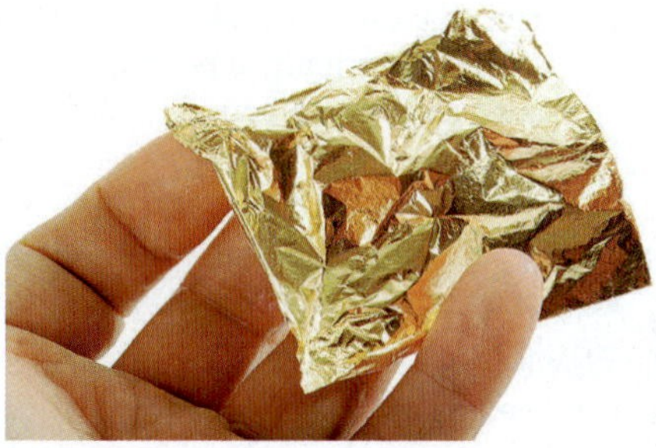

How thin is gold leaf?
Source: © Westend61 GmbH/Alamy RF

3. *Electrical conductivity.* Unlike ionic and covalent substances, metals are good conductors of electricity in both the solid and liquid states because of their mobile electrons. When a piece of metal wire is attached to a battery, electrons flow from the negative terminal into the wire, replacing the electrons that flow from the wire into the positive terminal. Foreign atoms disrupt the array of metal atoms and reduce conductivity. Copper used in electrical wiring is over 99.99% pure because traces of other atoms drastically restrict electron flow.

4. *Thermal conductivity.* Mobile electrons also make metals good conductors of heat. Place your hand on a piece of metal and a piece of wood that are both at room temperature. The metal feels colder because it conducts heat away from your hand much faster than the wood. The mobile, delocalized electrons in the metal disperse the heat from your hand more quickly than the localized electron pairs in the covalent bonds of wood.

› Summary of Section 9.6

- According to the electron-sea model, the valence electrons of the metal atoms in a sample are highly delocalized and attract all the metal cations, holding them together.
- Metals have only moderately high melting points because the metal ions remain attracted to the electrons in the electron sea even if their relative positions change.
- Vaporization involves complete separation of individual cations with their valence electrons from all the others, so metals have very high boiling points.
- Metals can be deformed because the electron sea prevents repulsions among the cations.
- Metals conduct electricity and heat because their electrons are mobile.

CHAPTER REVIEW GUIDE

Learning Objectives

Relevant section (§) and/or sample problem (SP) numbers appear in parentheses.

Understand These Concepts

1. How differences in atomic properties lead to differences in bond type; the basic distinctions among the three types of bonding (§9.1)
2. The essential features of ionic bonding: electron transfer to form ions, and their electrostatic attraction to form a solid (§9.2)
3. How lattice energy is ultimately responsible for the formation of ionic compounds (§9.2)
4. How ionic compound formation is conceptualized as occurring in hypothetical steps (a Born-Haber cycle) to calculate the lattice energy (§9.2)
5. How Coulomb's law explains the periodic trends in lattice energy (§9.2)

6. Why ionic compounds are brittle and high melting and conduct electricity only when molten or dissolved in water (§9.2)
7. How nonmetal atoms form a covalent bond (§9.3)
8. How bonding and lone electron pairs fill the outer (valence) level of each atom in a molecule (§9.3)
9. The interrelationships among bond order, bond length, and bond energy (§9.3)
10. How the distinction between bonding and nonbonding forces explains the properties of covalent molecules and network covalent solids (§9.3)
11. How changes in bond energy account for the enthalpy of reaction (§9.4)
12. How a reaction can be divided conceptually into bond-breaking and bond-forming steps (§9.4)
13. The periodic trends in electronegativity and the inverse relationship of EN values to atomic sizes (§9.5)
14. How bond polarity arises from differences in the electronegativities of the bonded atoms; the direction of bond polarity (§9.5)
15. The relation of partial ionic character to ΔEN and the change from ionic to polar covalent to nonpolar covalent bonding across a period (§9.5)
16. The role of delocalized electrons in metallic bonding (§9.6)
17. How the electron-sea model explains why metals bend, have very high boiling points, and conduct electricity in solid or molten form (§9.6)

Master These Skills

1. Using Lewis electron-dot symbols to depict main-group atoms (§9.1)
2. Depicting the formation of ions with partial orbital diagrams and Lewis symbols, and writing the formula of the ionic compound (SP 9.1)
3. Calculating lattice energy from the enthalpies of the steps to ionic compound formation (§9.2)
4. Ranking lattice energies from ionic properties (SP 9.2)
5. Ranking similar covalent bonds according to their length and strength (SP 9.3)
6. Using bond energies to calculate ΔH°_{rxn} (SP 9.4)
7. Determining bond polarity from EN values (SP 9.5)

Key Terms

Page numbers appear in parentheses.

alloy (396)
bond energy (BE) (380)
bond length (381)
bond order (380)
bonding (shared) pair (380)
Born-Haber cycle (374)
Coulomb's law (375)
covalent bond (379)
covalent bonding (370)
double bond (380)
electron-sea model (395)
electronegativity (EN) (390)
electronegativity difference (ΔEN) (392)
infrared (IR) spectroscopy (384)
ion pairs (378)
ionic bonding (370)
lattice energy ($\Delta H^\circ_{lattice}$) (374)
Lewis electron-dot symbol (371)
lone (unshared) pair (380)
metallic bonding (371)
nonpolar covalent bond (392)
octet rule (372)
partial ionic character (392)
polar covalent bond (392)
single bond (380)
triple bond (380)

Key Equations and Relationships

Page numbers appear in parentheses.

9.1 Relating the energy of attraction to the lattice energy (375):

$$\text{Electrostatic energy} \propto \frac{\text{cation charge} \times \text{anion charge}}{\text{cation radius} + \text{anion radius}} \propto \Delta H^\circ_{lattice}$$

9.2 Calculating enthalpy of reaction from bond enthalpies or bond energies (387):

$$\Delta H^\circ_{rxn} = \Sigma\Delta H^\circ_{\text{reactant bonds broken}} + \Sigma\Delta H^\circ_{\text{product bonds formed}}$$

or

$$\Delta H^\circ_{rxn} = \Sigma BE_{\text{reactant bonds broken}} - \Sigma BE_{\text{product bonds formed}}$$

BRIEF SOLUTIONS TO FOLLOW-UP PROBLEMS

9.1A $Mg([Ne]3s^2) + 2Cl([Ne]3s^23p^5) \longrightarrow Mg^{2+}([Ne]) + 2Cl^-([Ne]3s^23p^6)$

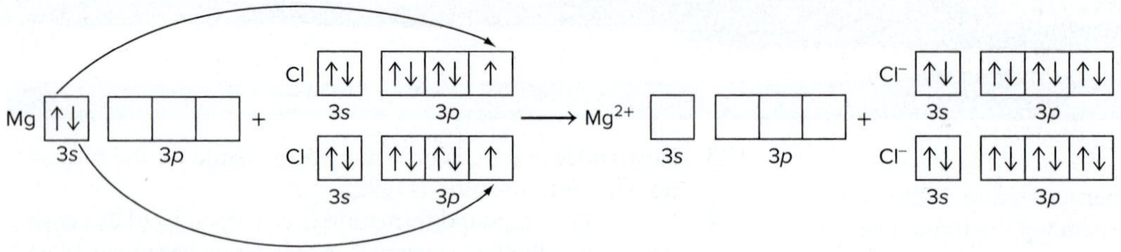

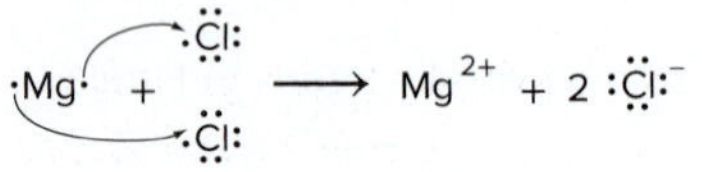

Formula: $MgCl_2$

9.1B $Ca([Ar]4s^2) + O([Ne]2s^22p^4) \longrightarrow Ca^{2+}([Ar]) + O^{2-}([Ne]2s^22p^6)$

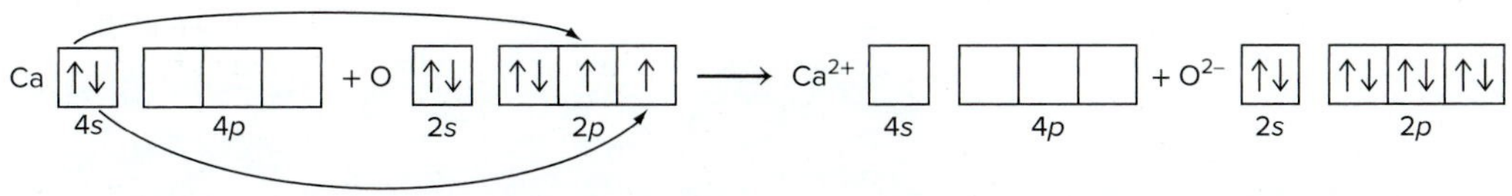

$\cdot Ca \cdot + :\ddot{O}: \longrightarrow Ca^{2+} + :\ddot{O}:^{2-}$ Formula: CaO

9.2A BaF_2. The only difference between these compounds is the size of the cation: the Ba^{2+} ion is larger than the Sr^{2+} ion (ionic size increases down the group), so the lattice energy of BaF_2 is smaller than that of SrF_2.

9.2B $Na_2O < MgF_2 < CaO$. In Na_2O, the ions are Na^+ and O^{2-}; the ions in MgF_2 are Mg^{2+} and F^-. The products of the cation charge and anion charge (Equation 9.1) are the same for these two compounds (1 × 2 and 2 × 1); however, since Na^+ is larger than Mg^{2+}, and O^{2-} is larger than F^-, Na_2O has the smaller lattice energy. In CaO, the ions are Ca^{2+} and O^{2-}. The larger product of cation and anion charge (2 × 2 for CaO) results in a larger lattice energy for CaO. In nearly every case, charge is more important than size.

9.3A (a) All three bonds have a bond order of 3. Since atomic size decreases left to right across a period, C > N > O.

Bond length: C≡C > C≡N > C≡O
Bond strength: C≡O > C≡N > C≡C

(b) All three bonds have a bond order of 1. Since atomic size increases down a group, I > Br > F

Bond length: P—I > P—Br > P—F
Bond strength: P—F > P—Br > P—I

9.3B (a) All three bonds have a bond order of 1. Since atomic size decreases left to right across a period, F < O < C.

Bond length: Si—F < Si—O < Si—C
Bond strength: Si—C < Si—O < Si—F

(b) As bond order decreases, bond length increases and bond strength decreases.

Bond length: N≡N < N═N < N—N
Bond strength: N—N < N═N < N≡N

9.4A N≡N + 3 H—H ⟶ 2 H—N(—H)—H

$$\Sigma\Delta H^\circ_{\text{bonds broken}} = 1\ \text{N≡N} + 3\ \text{H—H} = (1\ \text{mol})(945\ \text{kJ/mol}) + (3\ \text{mol})(432\ \text{kJ/mol}) = 2241\ \text{kJ}$$

$$\Sigma\Delta H^\circ_{\text{bonds formed}} = 6\ \text{N—H} = (6\ \text{mol})(-391\ \text{kJ/mol}) = -2346\ \text{kJ}$$

$$\Delta H^\circ_{\text{rxn}} = \Sigma\Delta H^\circ_{\text{bonds broken}} + \Sigma\Delta H^\circ_{\text{bonds formed}} = 2241\ \text{kJ} + (-2346\ \text{kJ}) = -105\ \text{kJ}$$

9.4B

$$\Sigma\Delta H^\circ_{\text{bonds broken}} = 2\ \text{O—F} + 2\ \text{O—H} = (2\ \text{mol})(190\ \text{kJ/mol}) + (2\ \text{mol})(467\ \text{kJ/mol}) = 1314\ \text{kJ}$$

$$\Sigma\Delta H^\circ_{\text{bonds formed}} = 2\text{H—F} + 1\text{O═O} = (2\ \text{mol})(-565\ \text{kJ/mol}) + (1\ \text{mol})(-498\ \text{kJ/mol}) = -1628\ \text{kJ}$$

$$\Delta H^\circ_{\text{rxn}} = \Sigma\Delta H^\circ_{\text{bonds broken}} + \Sigma\Delta H^\circ_{\text{bonds formed}} = 1314\ \text{kJ} + (-1628\ \text{kJ}) = -314\ \text{kJ}$$

9.5A (a) Cl—F: ΔEN = 4.0 − 3.0 = 1.0;
Br—Cl: ΔEN = 3.0 − 2.8 = 0.2; Cl—Cl: ΔEN = 3.0 − 3.0 = 0

$\text{Cl—Cl} < \overset{\delta+}{\text{Br}}\text{—}\overset{\delta-}{\text{Cl}} < \overset{\delta+}{\text{Cl}}\text{—}\overset{\delta-}{\text{F}}$

(b) P—F: ΔEN = 4.0 − 2.1 = 1.9; N—O: ΔEN = 3.5 − 3.0 = 0.5;
N—F: ΔEN = 4.0 − 3.0 = 1.0

$\overset{\delta+}{\text{N}}\text{—}\overset{\delta-}{\text{O}} < \overset{\delta+}{\text{N}}\text{—}\overset{\delta-}{\text{F}} < \overset{\delta+}{\text{P}}\text{—}\overset{\delta-}{\text{F}}$

9.5B (a) Se—Cl: ΔEN = 3.0 − 2.4 = 0.6;
Se—F: ΔEN = 4.0 − 2.4 = 1.6; Se—Br: ΔEN = 2.8 − 2.4 = 0.4

Se—F > Se—Cl > Se—Br

(b) S—B: ΔEN = 2.5 − 2.0 = 0.5; F—B: ΔEN = 4.0 − 2.0 = 2.0;
Cl—B: ΔEN = 3.0 − 2.0 = 1.0

F—B > Cl—B > S—B

PROBLEMS

Problems with **colored** numbers are answered in Appendix E and worked in detail in the Student Solutions Manual. Problem sections match those in the text and give the numbers of relevant sample problems. Most offer Concept Review Questions, Skill-Building Exercises (grouped in pairs covering the same concept), and Problems in Context. The Comprehensive Problems are based on material from any section or previous chapter.

Atomic Properties and Chemical Bonds

Concept Review Questions

9.1 In general terms, how does each of the following atomic properties influence the metallic character of the main-group elements in a period?
(a) Ionization energy (b) Atomic radius
(c) Number of outer electrons (d) Effective nuclear charge

9.2 Three solids are represented below. What is the predominant type of bonding in each?

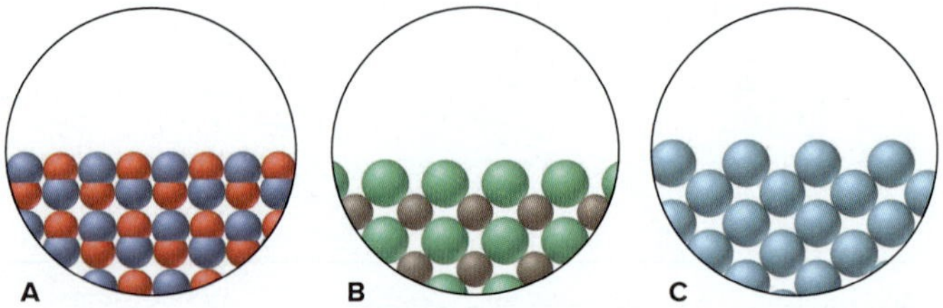

9.3 What is the relationship between the tendency of a main-group element to form a monatomic ion and its position in the periodic table? In what part of the table are the main-group elements that typically form cations? Anions?

Skill-Building Exercises (grouped in similar pairs)

9.4 Which member of each pair is *more* metallic?
(a) Na or Cs (b) Mg or Rb (c) As or N

9.5 Which member of each pair is *less* metallic?
(a) I or O (b) Be or Ba (c) Se or Ge

9.6 State the type of bonding—ionic, covalent, or metallic—you would expect in each:
(a) $CsF(s)$ (b) $N_2(g)$ (c) $Na(s)$

9.7 State the type of bonding—ionic, covalent, or metallic—you would expect in each:
(a) $ICl_3(g)$ (b) $N_2O(g)$ (c) $LiCl(s)$

9.8 State the type of bonding—ionic, covalent, or metallic—you would expect in each:
(a) $O_3(g)$ (b) $MgCl_2(s)$ (c) $BrO_2(g)$

9.9 State the type of bonding—ionic, covalent, or metallic—you would expect in each:
(a) $Cr(s)$ (b) $H_2S(g)$ (c) $CaO(s)$

9.10 Draw a Lewis electron-dot symbol for each atom:
(a) Rb (b) Si (c) I

9.11 Draw a Lew is electron-dot symbol for each atom:
(a) Ba (b) Kr (c) Br

9.12 Draw a Lewis electron-dot symbol for each:
(a) Sr (b) P (c) S

9.13 Draw a Lewis electron-dot symbol for each:
(a) As (b) Se (c) Ga

9.14 Give the group number and condensed electron configuration of an element corresponding to each electron-dot symbol:
(a) $\cdot\dot{\underset{\cdot}{X}}:$ (b) $\dot{X}\cdot$

9.15 Give the group number and condensed electron configuration of an element corresponding to each electron-dot symbol:
(a) $\cdot\dot{\underset{\cdot}{X}}:$ (b) $\cdot\dot{\underset{\cdot}{X}}\cdot$

The Ionic Bonding Model
(Sample Problems 9.1 and 9.2)

Concept Review Questions

9.16 If energy is required to form monatomic ions from metals and nonmetals, why do ionic compounds exist?

9.17 (a) In general, how does the lattice energy of an ionic compound depend on the charges and sizes of the ions?

(b) Ion arrangements of three general salts are represented below. Rank them in order of increasing lattice energy.

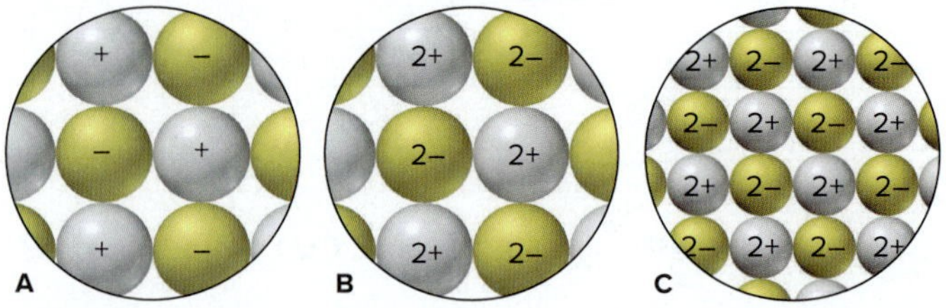

9.18 When gaseous Na^+ and Cl^- ions form ion pairs, 548 kJ/mol of energy is released. Why, then, does NaCl occur as a solid under ordinary conditions?

9.19 To form S^{2-} ions from gaseous sulfur atoms requires 214 kJ/mol, but these ions exist in solids such as K_2S. Explain.

Skill-Building Exercises (grouped in similar pairs)

9.20 Use condensed electron configurations and Lewis electron-dot symbols to depict the ions formed from each of the following atoms, and predict the formula of their compound:
(a) Ba and Cl (b) Sr and O (c) Al and F (d) Rb and O

9.21 Use condensed electron configurations and Lewis electron-dot symbols to depict the ions formed from each of the following atoms, and predict the formula of their compound:
(a) Cs and S (b) O and Ga (c) N and Mg (d) Br and Li

9.22 For each ionic compound formula, identify the main group to which X belongs:
(a) XF_2 (b) MgX (c) X_2SO_4

9.23 For each ionic compound formula, identify the main group to which X belongs:
(a) X_3PO_4 (b) $X_2(SO_4)_3$ (c) $X(NO_3)_2$

9.24 For each ionic compound formula, identify the main group to which X belongs:
(a) X_2O_3 (b) XCO_3 (c) Na_2X

9.25 For each ionic compound formula, identify the main group to which X belongs:
(a) CaX_2 (b) Al_2X_3 (c) XPO_4

9.26 For each pair, choose the compound with the larger lattice energy, and explain your choice:
(a) BaS or CsCl (b) LiCl or CsCl

9.27 For each pair, choose the compound with the larger lattice energy, and explain your choice:
(a) CaO or CaS (b) BaO or SrO

9.28 For each pair, choose the compound with the smaller lattice energy, and explain your choice:
(a) CaS or BaS (b) NaF or MgO

9.29 For each pair, choose the compound with the smaller lattice energy, and explain your choice:
(a) NaF or NaCl (b) K_2O or K_2S

9.30 Use the following to calculate $\Delta H^\circ_{lattice}$ of NaCl:

$Na(s) \longrightarrow Na(g)$	$\Delta H^\circ = 109$ kJ
$Cl_2(g) \longrightarrow 2Cl(g)$	$\Delta H^\circ = 243$ kJ
$Na(g) \longrightarrow Na^+(g) + e^-$	$\Delta H^\circ = 496$ kJ
$Cl(g) + e^- \longrightarrow Cl^-(g)$	$\Delta H^\circ = -349$ kJ
$Na(s) + \frac{1}{2}Cl_2(g) \longrightarrow NaCl(s)$	$\Delta H^\circ = -411$ kJ

Compared with the lattice energy of LiF (1050 kJ/mol), is the magnitude of the value for NaCl what you expected? Explain.

9.31 Use the following to calculate $\Delta H^\circ_{\text{lattice}}$ of MgF_2:

$Mg(s) \longrightarrow Mg(g)$	$\Delta H^\circ = 148$ kJ
$F_2(g) \longrightarrow 2F(g)$	$\Delta H^\circ = 159$ kJ
$Mg(g) \longrightarrow Mg^+(g) + e^-$	$\Delta H^\circ = 738$ kJ
$Mg^+(g) \longrightarrow Mg^{2+}(g) + e^-$	$\Delta H^\circ = 1450$ kJ
$F(g) + e^- \longrightarrow F^-(g)$	$\Delta H^\circ = -328$ kJ
$Mg(s) + F_2(g) \longrightarrow MgF_2(s)$	$\Delta H^\circ = -1123$ kJ

Compared with the lattice energy of LiF (1050 kJ/mol) or the lattice energy you calculated for NaCl in Problem 9.30, does the relative magnitude of the value for MgF_2 surprise you? Explain.

Problems in Context

9.32 Aluminum oxide (Al_2O_3) is a widely used industrial abrasive (known as emery or corundum), for which the specific application depends on the hardness of the crystal. What does this hardness imply about the magnitude of the lattice energy? Would you have predicted from the chemical formula that Al_2O_3 is hard? Explain.

9.33 Born-Haber cycles were used to obtain the first reliable values for electron affinity by considering the EA value as the unknown and using a theoretically calculated value for the lattice energy. Use a Born-Haber cycle for KF and the following values to calculate a value for the electron affinity of fluorine:

$K(s) \longrightarrow K(g)$	$\Delta H^\circ = 90$ kJ
$K(g) \longrightarrow K^+(g) + e^-$	$\Delta H^\circ = 419$ kJ
$F_2(g) \longrightarrow 2F(g)$	$\Delta H^\circ = 159$ kJ
$K(s) + \frac{1}{2}F_2(g) \longrightarrow KF(s)$	$\Delta H^\circ = -569$ kJ
$K^+(g) + F^-(g) \longrightarrow KF(s)$	$\Delta H^\circ = -821$ kJ

The Covalent Bonding Model
(Sample Problem 9.3)

Concept Review Questions

9.34 Describe the interactions that occur between individual chlorine atoms as they approach each other and form Cl_2. What combination of forces gives rise to the energy holding the atoms together and to the final internuclear distance?

9.35 Define *bond energy* using the H—Cl bond as an example. When this bond breaks, is energy absorbed or released? Is the accompanying ΔH value positive or negative? How do the magnitude and sign of this ΔH value relate to the value that accompanies H—Cl bond formation?

9.36 For single bonds between similar types of atoms, how does the strength of the bond relate to the sizes of the atoms? Explain.

9.37 How does the energy of the bond between a given pair of atoms relate to the bond order? Why?

9.38 When liquid benzene (C_6H_6) boils, does the gas consist of molecules, ions, or separate atoms? Explain.

Skill-Building Exercises (grouped in similar pairs)

9.39 Using the periodic table only, arrange the members of each of the following sets in order of increasing bond *strength:*
(a) Br—Br, Cl—Cl, I—I (b) S—H, S—Br, S—Cl
(c) C=N, C—N, C≡N

9.40 Using the periodic table only, arrange the members of each of the following sets in order of increasing bond *length:*
(a) H—F, H—I, H—Cl (b) C—S, C=O, C—O
(c) N—H, N—S, N—O

Problem in Context

9.41 Formic acid (HCOOH; structural formula shown below) is secreted by certain species of ants when they bite.

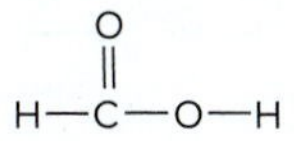

Rank the relative strengths of
(a) the C—O and C=O bonds and
(b) the H—C and O—H bonds. Explain these rankings.

Bond Energy and Chemical Change
(Sample Problem 9.4)

Concept Review Questions

9.42 Write a solution Plan (without actual numbers, but including the bond energies you would use and how you would combine them algebraically) for calculating the total enthalpy change of the following reaction:

$$H_2(g) + O_2(g) \longrightarrow H_2O_2(g) \qquad (\text{H—O—O—H})$$

9.43 The text points out that, for similar types of substances, one with weaker bonds is usually more reactive than one with stronger bonds. Why is this generally true?

9.44 Why is there a discrepancy between an enthalpy of reaction obtained from calorimetry and one obtained from bond energies?

Skill-Building Exercises (grouped in similar pairs)

9.45 Which of the following gases would you expect to have the greater enthalpy of reaction per mole for combustion? Why?

methane or formaldehyde

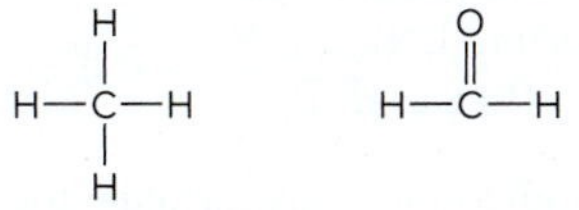

9.46 Which of the following gases would you expect to have the greater enthalpy of reaction per mole for combustion? Why?

ethanol or methanol

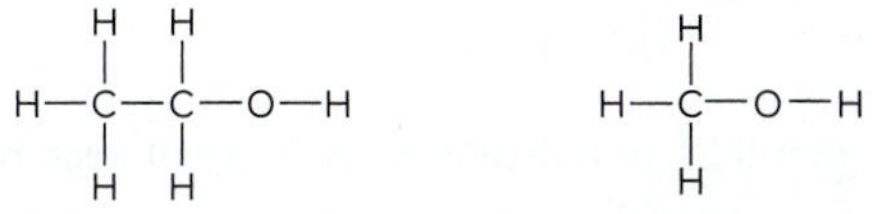

9.47 Use bond energies to calculate the enthalpy of reaction:

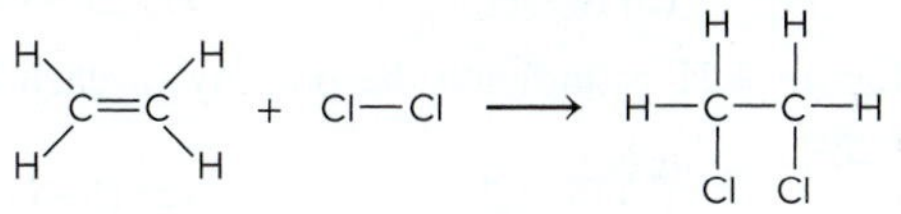

9.48 Use bond energies to calculate the enthalpy of reaction:

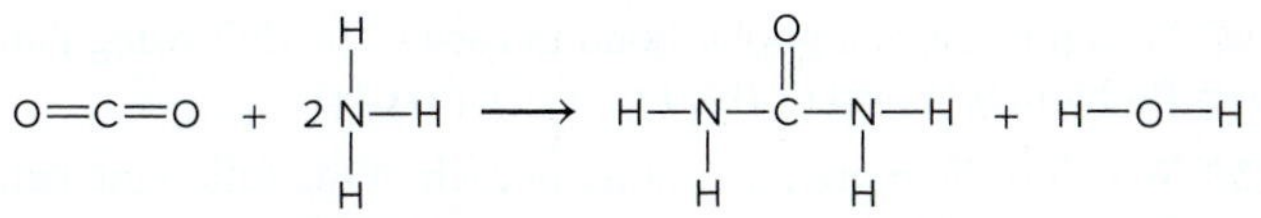

Problems in Context

9.49 An important industrial route to extremely pure acetic acid is the reaction of methanol with carbon monoxide:

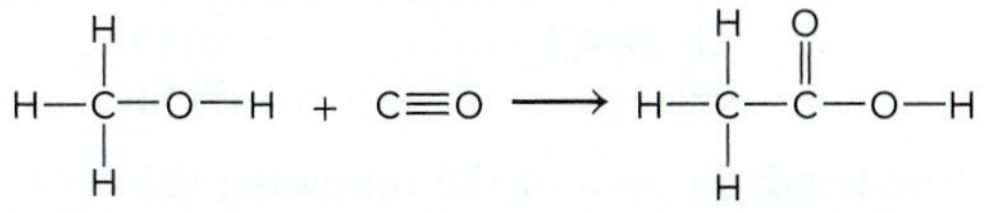

Use bond energies to calculate the enthalpy of reaction.

9.50 Sports trainers treat sprains and soreness with ethyl bromide. It is manufactured by reacting ethylene with hydrogen bromide:

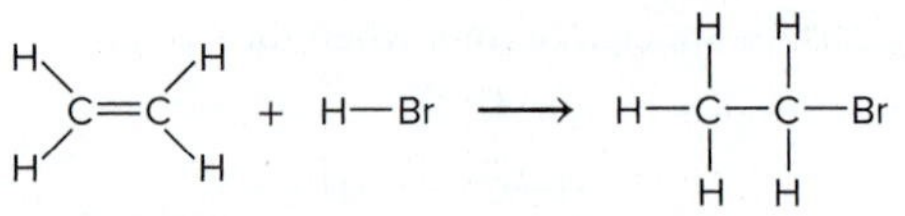

Use bond energies to find the enthalpy of reaction.

Between the Extremes: Electronegativity and Bond Polarity

(Sample Problem 9.5)

Concept Review Questions

9.51 Describe the vertical and horizontal trends in electronegativity (EN) among the main-group elements. According to Pauling's scale, what are the two most electronegative elements? The two least electronegative elements?

9.52 What is the general relationship between IE_1 and EN for the elements? Why?

9.53 Is the H—O bond in water nonpolar covalent, polar covalent, or ionic? Define each term, and explain your choice.

9.54 How does electronegativity differ from electron affinity?

9.55 How is the partial ionic character of a bond in a diatomic molecule related to ΔEN for the bonded atoms? Why?

Skill-Building Exercises (grouped in similar pairs)

9.56 Using the periodic table only, arrange the elements in each set in order of *increasing* EN:
(a) S, O, Si (b) Mg, P, As

9.57 Using the periodic table only, arrange the elements in each set in order of *increasing* EN:
(a) I, Br, N (b) Ca, H, F

9.58 Using the periodic table only, arrange the elements in each set in order of *decreasing* EN:
(a) N, P, Si (b) Ca, Ga, As

9.59 Using the periodic table only, arrange the elements in each set in order of *decreasing* EN:
(a) Br, Cl, P (b) I, F, O

9.60 Use Figure 9.21 to indicate the polarity of each bond with a *polar arrow:*
(a) N—B (b) N—O (c) C—S
(d) S—O (e) N—H (f) Cl—O

9.61 Use Figure 9.21 to indicate the polarity of each bond with *partial charges:*
(a) Br—Cl (b) F—Cl (c) H—O
(d) Se—H (e) As—H (f) S—N

9.62 Which is the more polar bond in each of the following pairs from Problem 9.60: (a) or (b); (c) or (d); (e) or (f)?

9.63 Which is the more polar bond in each of the following pairs from Problem 9.61: (a) or (b); (c) or (d); (e) or (f)?

9.64 Are the bonds in each of the following substances ionic, nonpolar covalent, or polar covalent? Arrange the substances with polar covalent bonds in order of increasing bond polarity:
(a) S_8 (b) RbCl (c) PF_3
(d) SCl_2 (e) F_2 (f) SF_2

9.65 Are the bonds in each of the following substances ionic, nonpolar covalent, or polar covalent? Arrange the substances with polar covalent bonds in order of increasing bond polarity:
(a) KCl (b) P_4 (c) BF_3
(d) SO_2 (e) Br_2 (f) NO_2

9.66 Rank the members of each set of compounds in order of *increasing* ionic character of their bonds. Use *polar arrows* to indicate the bond polarity of each:
(a) HBr, HCl, HI (b) H_2O, CH_4, HF (c) SCl_2, PCl_3, $SiCl_4$

9.67 Rank the members of each set of compounds in order of *decreasing* ionic character of their bonds. Use *partial charges* to indicate the bond polarity of each:
(a) PCl_3, PBr_3, PF_3 (b) BF_3, NF_3, CF_4 (c) SeF_4, TeF_4, BrF_3

Problem in Context

9.68 The energy of the C—C bond is 347 kJ/mol, and that of the Cl—Cl bond is 243 kJ/mol. Which of the following values might you expect for the C—Cl bond energy? Explain.
(a) 590 kJ/mol (sum of the values given)
(b) 104 kJ/mol (difference of the values given)
(c) 295 kJ/mol (average of the values given)
(d) 339 kJ/mol (greater than the average of the values given)

An Introduction to Metallic Bonding

Concept Review Questions

9.69 (a) List four physical characteristics of a solid metal.
(b) List two chemical characteristics of a metallic element.

9.70 Briefly account for the following relative values:
(a) The melting points of Na and K are 89°C and 63°C, respectively.
(b) The melting points of Li and Be are 180°C and 1287°C, respectively.
(c) Li boils more than 1100°C higher than it melts.

9.71 Magnesium metal is easily deformed by an applied force, whereas magnesium fluoride is shattered. Why do these two solids behave so differently?

Comprehensive Problems

9.72 Geologists have a rule of thumb: when molten rock cools and solidifies, crystals of compounds with the smallest lattice energies appear at the bottom of the mass. Suggest a reason for this.

9.73 Acetylene gas (ethyne; HC≡CH) burns in an oxyacetylene torch to produce carbon dioxide and water vapor. The enthalpy of reaction for the combustion of acetylene is 1259 kJ/mol.
(a) Calculate the C≡C bond energy, and compare your value with that in Table 9.2.
(b) When 500.0 g of acetylene burns, how many kilojoules of heat are given off?
(c) How many grams of CO_2 form?
(d) How many liters of O_2 at 298 K and 18.0 atm are consumed?

9.74 Use Lewis electron-dot symbols to represent the formation of
(a) BrF_3 from bromine and fluorine atoms;
(b) AlF_3 from aluminum and fluorine atoms.

9.75 Even though so much energy is required to form a metal cation with a 2+ charge, the alkaline earth metals form halides with the general formula MX_2, rather than MX.
(a) Use the following data to calculate ΔH_f° of MgCl:

$Mg(s) \longrightarrow Mg(g)$ $\Delta H^\circ = 148$ kJ
$Cl_2(g) \longrightarrow 2Cl(g)$ $\Delta H^\circ = 243$ kJ
$Mg(g) \longrightarrow Mg^+(g) + e^-$ $\Delta H^\circ = 738$ kJ
$Cl(g) + e^- \longrightarrow Cl^-(g)$ $\Delta H^\circ = -349$ kJ
$\Delta H^\circ_{lattice}$ of MgCl = 783.5 kJ/mol

(b) Is MgCl favored energetically relative to Mg and Cl_2? Explain.
(c) Use Hess's law to calculate $\Delta H°$ for the conversion of MgCl to $MgCl_2$ and Mg ($\Delta H_f°$ of $MgCl_2 = -641.6$ kJ/mol).
(d) Is MgCl favored energetically relative to $MgCl_2$? Explain.

9.76 Gases react explosively if the heat released when the reaction begins is sufficient to cause more reaction, which leads to a rapid expansion of the gases. Use bond energies to calculate $\Delta H°$ of each of the following reactions, and predict which occurs explosively:
(a) $H_2(g) + Cl_2(g) \longrightarrow 2HCl(g)$
(b) $H_2(g) + I_2(g) \longrightarrow 2HI(g)$
(c) $2H_2(g) + O_2(g) \longrightarrow 2H_2O(g)$

9.77 By using photons of specific wavelengths, chemists can dissociate gaseous HI to produce H atoms with certain speeds. When HI dissociates, the H atoms move away rapidly, whereas the heavier I atoms move more slowly.
(a) What is the longest wavelength (in nm) that can dissociate a molecule of HI?
(b) If a photon of 254 nm is used, what is the excess energy (in J) over that needed for dissociation?
(c) If this excess energy is carried away by the H atom as kinetic energy, what is its speed (in m/s)?

9.78 In developing the concept of electronegativity, Pauling used the term *excess bond energy* for the difference between the actual bond energy of X—Y and the average bond energies of X—X and Y—Y (see the discussion of HF in Section 9.5). Based on the values in Figure 9.21, which of the following substances contain bonds with no excess bond energy?
(a) PH_3 (b) CS_2 (c) BrCl (d) BH_3 (e) Se_8

9.79 Use condensed electron configurations to predict the relative hardnesses and melting points of rubidium ($Z = 37$), vanadium ($Z = 23$), and cadmium ($Z = 48$).

9.80 Without stratospheric ozone (O_3), harmful solar radiation would cause gene alterations. Ozone forms when the bond in O_2 breaks and each O atom reacts with another O_2 molecule. It is destroyed by reaction with the Cl atoms formed when the C—Cl bond in synthetic chemicals breaks. Find the wavelengths of light that can break the C—Cl bond and the bond in O_2.

9.81 "Inert" xenon actually forms many compounds, especially with highly electronegative fluorine. The $\Delta H_f°$ values for xenon difluoride, tetrafluoride, and hexafluoride are −105, −284, and −402 kJ/mol, respectively. Find the average bond energy of the Xe—F bonds in each fluoride.

9.82 The HF bond length is 92 pm, 16% shorter than the sum of the covalent radii of H (37 pm) and F (72 pm). Suggest a reason for this difference. Similar data show that the difference becomes smaller down the group, from HF to HI. Explain.

9.83 There are two main types of covalent bond breakage. In homolytic breakage (as in Table 9.2), each atom in the bond gets one of the shared electrons. In some cases, the electronegativity of adjacent atoms affects the bond energy. In heterolytic breakage, one atom gets both electrons and the other gets none; thus, a cation and an anion form.
(a) Why is the C—C bond in $H_3C—CF_3$ (423 kJ/mol) stronger than that in $H_3C—CH_3$ (376 kJ/mol)?
(b) Use bond energy and any other data to calculate the enthalpy of reaction for the heterolytic cleavage of O_2.

9.84 Find the longest wavelengths of light that can cleave the bonds in elemental nitrogen, oxygen, and fluorine.

9.85 The work function (ϕ) of a metal is the minimum energy needed to remove an electron from its surface.
(a) Is it easier to remove an electron from a gaseous silver atom or from the surface of solid silver ($\phi = 7.59\times10^{-19}$ J; IE = 731 kJ/mol)?
(b) Explain the results in terms of the electron-sea model of metallic bonding.

9.86 Lattice energies can also be calculated for covalent network solids using a Born-Haber cycle, and the network solid silicon dioxide has one of the highest $\Delta H°_{lattice}$ values. Silicon dioxide is found in pure crystalline form as transparent rock quartz. Much harder than glass, this material was once prized for making lenses for optical devices and expensive spectacles. Use Appendix B and the following data to calculate $\Delta H°_{lattice}$ of SiO_2:

$Si(s) \longrightarrow Si(g)$	$\Delta H° = 454$ kJ
$Si(g) \longrightarrow Si^{4+}(g) + 4e^-$	$\Delta H° = 9949$ kJ
$O_2(g) \longrightarrow 2O(g)$	$\Delta H° = 498$ kJ
$O(g) + 2e^- \longrightarrow O^{2-}(g)$	$\Delta H° = 737$ kJ

9.87 The average C—H bond energy in CH_4 is 415 kJ/mol. Use Table 9.2 and the following data to calculate the average C—H bond energy in ethane (C_2H_6; C—C bond), in ethene (C_2H_4; C=C bond), and in ethyne (C_2H_2; C≡C bond):

$C_2H_6(g) + H_2(g) \longrightarrow 2CH_4(g)$	$\Delta H°_{rxn} = -65.07$ kJ/mol
$C_2H_4(g) + 2H_2(g) \longrightarrow 2CH_4(g)$	$\Delta H°_{rxn} = -202.21$ kJ/mol
$C_2H_2(g) + 3H_2(g) \longrightarrow 2CH_4(g)$	$\Delta H°_{rxn} = -376.74$ kJ/mol

9.88 Carbon-carbon bonds form the "backbone" of nearly every organic and biological molecule. The average bond energy of the C—C bond is 347 kJ/mol. Calculate the frequency and wavelength of the least energetic photon that can break an average C—C bond. In what region of the electromagnetic spectrum is this radiation?

9.89 In a future hydrogen-fuel economy, the cheapest source of H_2 will certainly be water. It takes 467 kJ to produce 1 mol of H atoms from water. What is the frequency, wavelength, and minimum energy of a photon that can free an H atom from water?

9.90 Dimethyl ether (CH_3OCH_3) and ethanol (CH_3CH_2OH) are constitutional isomers (see Table 3.3).
(a) Use Table 9.2 to calculate $\Delta H°_{rxn}$ for the formation of each compound as a gas from methane and oxygen; water vapor also forms.
(b) State which reaction is more exothermic.
(c) Calculate $\Delta H°_{rxn}$ for the conversion of ethanol to dimethyl ether.

9.91 Enthalpies of reaction calculated from bond energies and from enthalpies of formation are often, but not always, close to each other.
(a) Industrial ethanol (CH_3CH_2OH) is produced by a catalytic reaction of ethylene ($CH_2=CH_2$) with water at high pressures and temperatures. Calculate $\Delta H°_{rxn}$ for this gas-phase hydration of ethylene to ethanol, using bond energies and then using enthalpies of formation.
(b) Ethylene glycol is produced by the catalytic oxidation of ethylene to ethylene oxide, which then reacts with water to form ethylene glycol:

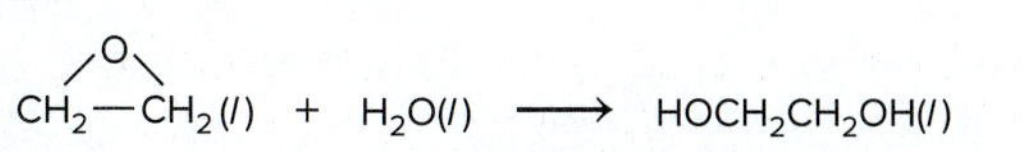

The $\Delta H°_{rxn}$ for this hydrolysis step, based on enthalpies of formation, is −97 kJ/mol. Calculate $\Delta H°_{rxn}$ for the hydrolysis using bond energies.
(c) Why are the two values relatively close for the hydration in part (a) but not close for the hydrolysis in part (b)?

10 The Shapes of Molecules

10.1 Depicting Molecules and Ions with Lewis Structures
Applying the Octet Rule
Resonance
Formal Charge
Exceptions to the Octet Rule

10.2 Valence-Shell Electron-Pair Repulsion (VSEPR) Theory
Electron-Group Arrangements and Molecular Shapes
Molecular Shape with Two Electron Groups
Molecular Shapes with Three Electron Groups
Molecular Shapes with Four Electron Groups
Molecular Shapes with Five Electron Groups
Molecular Shapes with Six Electron Groups
Using VSEPR Theory to Determine Molecular Shape
Molecular Shapes with More Than One Central Atom

10.3 Molecular Shape and Molecular Polarity
Bond Polarity, Bond Angle, and Dipole Moment
Molecular Polarity and Behavior

Source: © Mikhail Rulkov/Shutterstock.com

Concepts and Skills to Review Before You Study This Chapter

- electron configurations of main-group elements (Section 8.2)
- electron-dot symbols (Section 9.1)
- octet rule (Section 9.1)
- bond order, bond length, and bond energy (Sections 9.3 and 9.4)
- polar covalent bonds and bond polarity (Section 9.5)

Very young children often play with a shape-sorting toy that teaches them to match blocks of different shapes to the correct holes in the toy. Chemists, biochemists, and pharmacologists are also interested in shape matching because molecular shape plays a crucial role in the interactions of reactants, in the behavior of synthetic materials, in the life-sustaining processes in cells, in the senses of taste and smell, and in the discovery of new pharmaceutical drugs. All the symbols, lines, and dots you've been seeing in representations of molecules make it easy to forget that every molecule has a characteristic minute architecture. Each atom, bonding pair, and lone pair of electrons has a specific position relative to the others, and the angles and distances between these constituents are determined by the attractive and repulsive forces governing all matter.

IN THIS CHAPTER . . . *We learn how to picture a molecule by writing a two-dimensional structure for it and converting it to a three-dimensional shape, and we examine some effects of molecular shape on physical and biochemical behavior.*

- We learn how to apply the octet rule to convert a molecular formula into a flat structural formula that shows atom attachments and electron-pair locations.
- We see how electron delocalization limits our ability to depict a molecule with a single structural formula and requires us to apply the concept of resonance.
- We introduce bond angle and apply a theory that converts two-dimensional formulas into three-dimensional shapes.
- We describe five basic classes of shapes that many molecules adopt and consider how multiple bonds and lone pairs affect them and how to combine smaller molecular portions into the shapes of more complex molecules.
- We discuss the relation among bond polarity, shape, and molecular polarity and the effect of polarity on behavior.
- We describe a few examples of the influence of shape on biological function.

10.1 DEPICTING MOLECULES AND IONS WITH LEWIS STRUCTURES

The first step toward visualizing a molecule is to convert its molecular formula to its **Lewis structure** (or **Lewis formula***), which shows symbols for the atoms, the bonding electron pairs as lines, and the lone electron pairs that fill each atom's outer level (valence shell) as pairs of dots.

Applying the Octet Rule to Write Lewis Structures

To write a Lewis structure, we decide on the relative placement of the atoms in the molecule or polyatomic ion and then distribute the total number of valence electrons as bonding and lone pairs. In many, but not all, cases, the octet rule (Section 9.1) guides us in distributing the electrons. We begin with species that "obey" the octet rule, in which each atom fills its outer level with eight electrons (or two for hydrogen).

*A Lewis *structure* does ***not*** indicate the three-dimensional shape, so it may be more correct to call it a Lewis *formula*, but we follow convention and use the term *structure*.

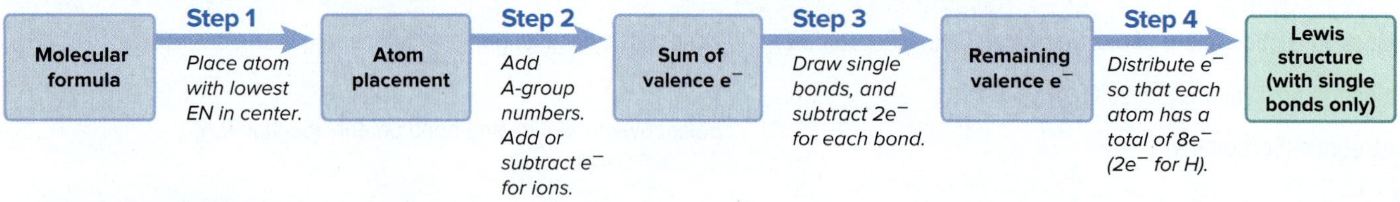

A Species with single bonds only

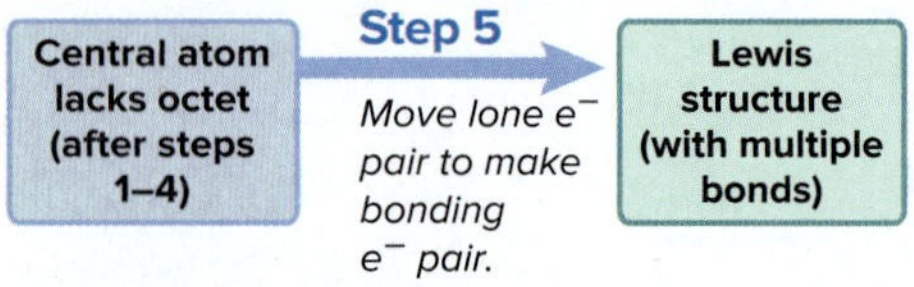

B Species with multiple bonds

Figure 10.1 The steps in converting a molecular formula into a Lewis structure.

Molecules with Single Bonds Figure 10.1A shows the four steps for writing Lewis structures for species with only single bonds. Let's use two species, nitrogen trifluoride, NF_3, and the tetrafluoroborate ion, BF_4^-, to introduce the steps:

Step 1. Place the atoms relative to each other. For compounds with the general molecular formula AB_n, place the atom with the *lower group number* in the center because it needs more electrons to attain an octet; usually, this is also the atom with the *lower electronegativity*. In NF_3, nitrogen [Group 5A(15); EN = 3.0] has a lower group number and lower electronegativity than F [Group 7A(17); EN = 4.0], so it goes in the center with the three F atoms around it. In BF_4^-, boron [Group 3A(13); EN = 2.0] has the lower group number and lower electronegativity, so it goes in the center with the four F atoms around it:

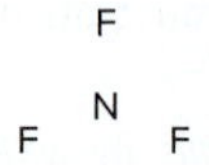

If the atoms have the same group number, as in SO_3, place the atom with the *higher period number* (also the lower EN) in the center. H can form only one bond, so it is *never* a central atom.

Step 2. Determine the total number of valence electrons.

- For molecules, like NF_3, add up the valence electrons of the atoms. (Recall that the number of valence electrons equals the A-group number.) In NF_3, N has five valence electrons, and each F has seven:

$$[1 \times N(5e^-)] + [3 \times F(7e^-)] = 5e^- + 21e^- = 26 \text{ valence } e^-$$

- For polyatomic ions, like BF_4^-, *add* one e$^-$ for each negative charge, or *subtract* one e$^-$ for each positive charge. In the BF_4^- ion, B has three valence electrons, each F atom has seven, and one more electron is included because of the negative charge:

$$[1 \times B(3e^-) + [4 \times F(7e^-)] + [\text{charge } (1e^-)] = 32 \text{ valence } e^-$$

You don't need to keep track of which electrons come from which atoms, because only the total number of valence electrons is important.

Step 3. Draw a single bond from each surrounding atom to the central atom, and subtract 2e$^-$ for each bond from the total number of valence electrons to find the number of e$^-$ remaining:

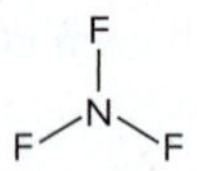

3 N—F bonds × 2e$^-$ = 6e$^-$
26e$^-$ − 6e$^-$ = 20e$^-$ remaining

4 B—F bonds × 2e$^-$ = 8e$^-$
32e$^-$ − 8e$^-$ = 24e$^-$ remaining

Step 4. Distribute the remaining electrons in pairs so that each atom ends up with 8e$^-$ (or 2e$^-$ for H). First, place lone pairs on the *surrounding (more electronegative) atoms* to give each an octet. If any electrons remain, place them around the central atom:

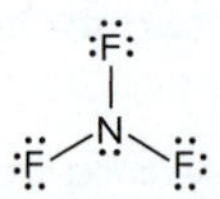

Each F gets three pairs (3 × 6e$^-$ = 18e$^-$)
N gets one pair (2e$^-$)
Total: 18e$^-$ + 2e$^-$ = 20e$^-$
Each atom has an octet (8e$^-$) of electrons.

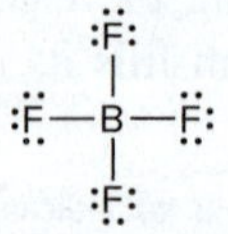

Each F gets three pairs (4 × 6e$^-$ = 24e$^-$)
B does not get any electrons
Total: 24e$^-$
Each atom has an octet (8e$^-$) of electrons.

The Lewis structure for the neutral species NF_3 is complete. Since BF_4^- is a polyatomic ion, we also take the charge into account, as described in step 2, and put the structure in square brackets with the ion charge as a superscript outside the brackets:

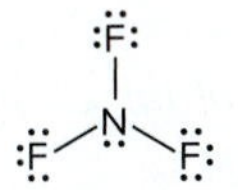

$6e^-$ in three bonds
$+\ 20e^-$ in 10 lone pairs
26 valence e^-

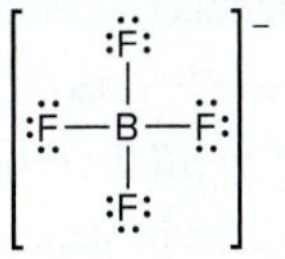

$8e^-$ in four bonds
$+\ 24e^-$ in 12 lone pairs
32 valence e^-

Since Lewis structures do not indicate shape, an equally correct depiction of NF_3 is

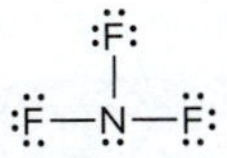

or any other that retains the *same connections among the atoms*—a central N atom connected by single bonds to each of three surrounding F atoms.

Using these four steps, you can write a Lewis structure for any singly bonded species with a central C, N, or O atom, as well as for some species with central atoms from higher periods. In nearly all their compounds,

- Hydrogen atoms form one bond.
- Carbon atoms form four bonds.
- Nitrogen atoms form three bonds.
- Oxygen atoms form two bonds.
- Surrounding halogens form one bond; fluorine is *always* a surrounding atom.

SAMPLE PROBLEM 10.1 Writing Lewis Structures for Species with Single Bonds and One Central Atom

Problem Write a Lewis structure for **(a)** CCl_2F_2; **(b)** PCl_2^-.

Solution **(a)** *Step 1.* Place the atoms relative to each other. In CCl_2F_2, carbon has the lowest group number and EN, so it is the central atom. The halogen atoms surround it, but their specific positions are not important.

(a) *Step 1:*

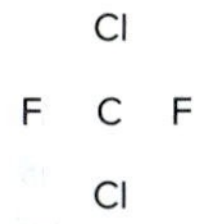

Step 2. Determine the total number of valence electrons (from A-group numbers): C is in Group 4A; F and Cl are in Group 7A. Therefore, we have

$$[1 \times C(4e^-)] + [2 \times F(7e^-)] + [2 \times Cl(7e^-)] = 32 \text{ valence } e^-$$

Step 3. Draw single bonds to the central atom and subtract $2e^-$ for each bond:

$$4 \text{ bonds} \times 2e^- = 8e^- \qquad 32e^- - 8e^- = 24e^- \text{ remaining}$$

Step 3:

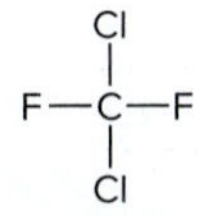

Step 4. Distribute the 24 remaining electrons in pairs, beginning with the surrounding atoms, so that each atom has an octet. Each surrounding fluorine atom gets three pairs. All atoms have an octet of electrons.

Step 4:

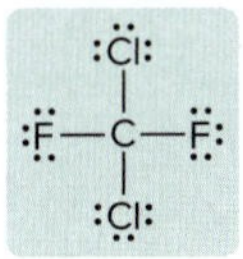

(b) *Step 1.* Place the atoms relative to each other. In PCl_2^-, phosphorus has the lower group number and EN, so it is the central atom. The chlorine atoms surround it.

(b) *Step 1:* Cl P Cl

Step 2. Determine the total number of valence electrons (from A-group numbers and ion charge): P is in Group 5A, and Cl is in Group 7A. Add $1e^-$ to the total number of valence electrons because of the 1− charge:

$$[1 \times P(5e^-)] + [2 \times Cl(7e^-)] + [\text{charge } (1e^-)] = 20 \text{ valence } e^-$$

Step 3. Draw single bonds to the central atom and subtract $2e^-$ for each bond:

$$2 \text{ bonds} \times 2e^- = 4e^- \qquad 20e^- - 4e^- = 16e^- \text{ remaining}$$

Step 3: Cl—P—Cl

Step 4. Distribute the 16 remaining electrons in pairs, beginning with the surrounding atoms, so that each atom has an octet. Each surrounding chlorine atom gets three pairs, and the central phosphorus atom gets two pairs. Each atom has an octet of electrons. Since this species is an ion, place brackets around the Lewis structure and indicate the 1− charge as a superscript negative sign outside the brackets.

Step 4:

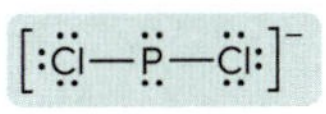

Check Always check that each atom has an octet. Bonding electrons belong to each atom in the bond. In CCl_2F_2, the total number in bonds ($8e^-$) and lone pairs ($24e^-$) equals 32 valence e^-. As expected, C has four bonds and each of the surrounding halogens has one bond. In PCl_2^-, the total number of electrons in bonds ($4e^-$) and lone pairs ($16e^-$) equals 20 valence e^-.

FOLLOW-UP PROBLEMS

*Brief Solutions to **all** Follow-up Problems appear at the end of the chapter.*

10.1A Write a Lewis structure for **(a)** H_2S; **(b)** $AlCl_4^-$; **(c)** $SOCl_2$.

10.1B Write a Lewis structure for **(a)** OF_2; **(b)** CH_2Br_2; **(c)** IBr_2^+.

SOME SIMILAR PROBLEMS 10.5(a), 10.5(b), 10.6(a), 10.6(c), 10.7(a), and 10.8(c)

In molecules with two or more central atoms bonded to each other, it is usually clear which atoms are central and which are surrounding.

SAMPLE PROBLEM 10.2 **Writing Lewis Structures for Molecules with Single Bonds and More Than One Central Atom**

Problem Write the Lewis structure for methanol (molecular formula CH_4O), an important industrial alcohol that can be used as a gasoline alternative in cars.

Step 1:

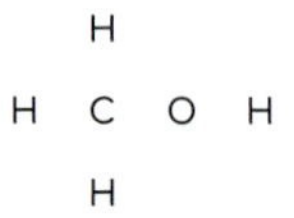

Solution *Step 1.* Place the atoms relative to each other. The H atoms can have only one bond, so C and O must be central and adjacent to each other. C has four bonds and O has two, so we arrange the H atoms accordingly.

Step 2. Find the sum of valence electrons (C is in Group 4A; O is in Group 6A):

$$[1 \times C(4e^-)] + [1 \times O(6e^-)] + [4 \times H(1e^-)] = 14e^-$$

Step 3:

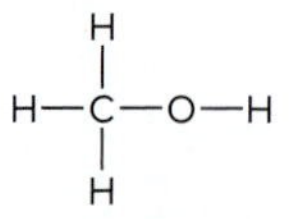

Step 3. Add single bonds and subtract $2e^-$ for each bond:

$$5 \text{ bonds} \times 2e^- = 10e^- \quad 14e^- - 10e^- = 4e^- \text{ remaining}$$

Step 4:

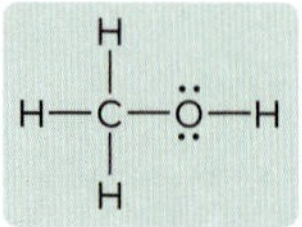

Step 4. Add the remaining four electrons in pairs to fill each valence level. C already has an octet, and each H shares $2e^-$ with the C; so the remaining $4e^-$ form two lone pairs on O to give the Lewis structure for methanol.

Check Each H atom has $2e^-$, and C and O each have $8e^-$. The total number of valence electrons is $14e^-$, which equals $10e^-$ in bonds plus $4e^-$ in two lone pairs. Each H has one bond, C has four, and O has two.

FOLLOW-UP PROBLEMS

10.2A Write a Lewis structure for **(a)** hydroxylamine (NH_3O); **(b)** dimethyl ether (C_2H_6O; no O—H bonds).

10.2B Write a Lewis structure for **(a)** hydrazine (N_2H_4); **(b)** methylamine (CH_3NH_2).

SOME SIMILAR PROBLEMS 10.8(a) and 10.8(b)

Molecules with Multiple Bonds After completing steps 1–4 to write the Lewis structure for a species, we may find that the central atom still does not have an octet. In most of these cases, we add the following step (Figure 10.1B):

Step 5. If a central atom does not end up with an octet, form one or more multiple bonds. Change a lone pair on a surrounding atom into another bonding pair to the central atom, thus forming a double bond. This may be done again if necessary to form a triple bond or to form another double bond.

Let's draw the Lewis structure for the O_2 molecule to illustrate this:

Step 1. Place the atoms relative to each other.

Step 2. Find the sum of valence electrons (O is in Group 6A):

$$[2 \times O(6e^-)] = 12e^-$$

Step 3. Draw single bonds and subtract $2e^-$ for each bond:

$$1 \text{ bond} \times 2e^- = 2e^- \quad \text{so} \quad 12e^- - 2e^- = 10e^- \text{ remaining}$$

Step 4. Add the remaining 10 electrons in pairs; there are enough electrons for the O atom on the left to get three pairs to complete its octet, but the O atom on the right only gets two pairs of electrons and thus does not have an octet. Step 5 is necessary.

Step 5. Change a lone pair on the left O atom to another bonding pair between the two O atoms so that each O atom in the molecule has an octet:

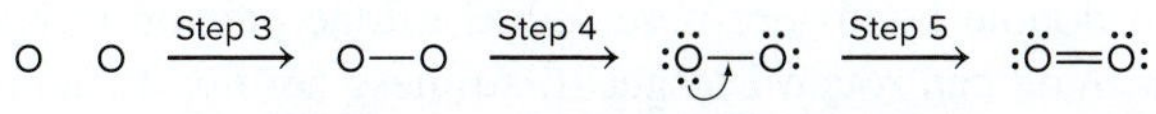

SAMPLE PROBLEM 10.3 Writing Lewis Structures for Molecules with Multiple Bonds

Problem Write Lewis structures for the following:

(a) Ethylene (C_2H_4), the most important reactant in the manufacture of polymers

(b) Nitrogen (N_2), the most abundant atmospheric gas

Solution **(a)** *Step 1.* Place the atoms relative to each other. In C_2H_4, the two C atoms must be bonded to each other since H atoms can have only one bond.
Step 2. Determine the total number of valence electrons (C is in Group 4A; H is in Group 1A):

$$[2 \times C(4e^-)] + [4 \times H(1e^-) = 12 \text{ valence } e^-$$

Step 3. Add single bonds and subtract $2e^-$ for each bond:

$$5 \text{ bonds} \times 2e^- = 10e^- \quad \text{so} \quad 12e^- - 10e^- = 2e^- \text{ remaining}$$

Step 4. Distribute the remaining valence electrons to attain octets.
Step 5. Change a lone pair to a bonding pair. The right C has an octet, but the left C has only $6e^-$, so we change the lone pair on the right C to another bonding pair between the two C atoms.
(b) *Step 1.* Place the atoms relative to each other.
Step 2. Determine the total number of valence electrons (N is in Group 5A):

$$[2 \times N(5e^-)] = 10 \text{ valence } e^-$$

Step 3. Add single bonds and subtract $2e^-$ for each bond:

$$1 \text{ bond} \times 2e^- = 2e^- \quad 10e^- - 2e^- = 8e^- \text{ remaining}$$

Step 4. Distribute the remaining valence electrons to attain octets.
Step 5. Neither N ends up with an octet, so we change a lone pair to a bonding pair. In this case, moving one lone pair to make a double bond still does not give both N atoms an octet, so we also move another lone pair from the other N to make a triple bond.

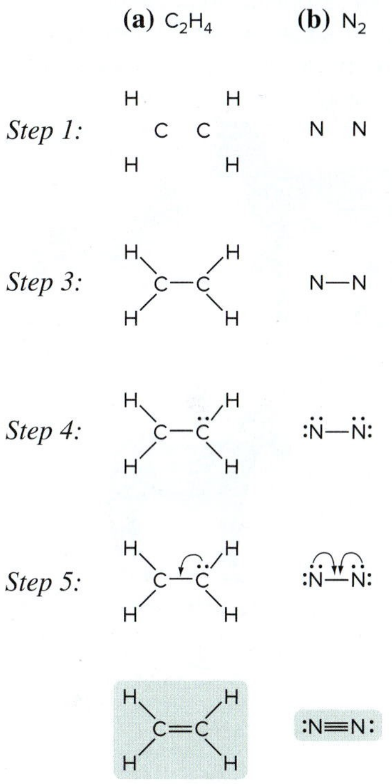

Check **(a)** Each C has four bonds and counts the $4e^-$ in the double bond as part of its own octet. The valence electron total is $12e^-$, all in six bonds. **(b)** Each N counts the $6e^-$ in the triple bond as part of its own octet. The valence electron total is $10e^-$, which equals the electrons in three bonds and two lone pairs.

FOLLOW-UP PROBLEMS

10.3A Write Lewis structures for **(a)** CO (the only common molecule in which C has three bonds); **(b)** HCN; **(c)** CO_2.

10.3B Write Lewis structures for **(a)** NO^+; **(b)** H_2CO; **(c)** N_2H_2.

SOME SIMILAR PROBLEMS 10.5(c), 10.6(b), 10.7(b), and 10.7(c)

Resonance: Delocalized Electron-Pair Bonding

We often find that, for a molecule or polyatomic ion with *double bonds next to single bonds,* we can write more than one Lewis structure. Which, if any, is correct?

The Need for Resonance Structures To understand this issue, consider ozone (O_3), an air pollutant at ground level but an absorber of harmful ultraviolet (UV) radiation in the stratosphere. Since oxygen is in Group 6A(16), there are $[3 \times O(6e^-)] = 18$ valence e^- in the molecule. Four electrons are used in the formation of two single bonds, leaving $18e^- - 4e^- = 14e^-$, enough electrons to give the surrounding O atoms (designated A and C for clarity) an octet of electrons, but not enough to complete the octet of the central O atom (designated B). Applying Step 5 gives two Lewis structures:

A Purple Mule, Not a Blue Horse and a Red Donkey

A mule is a genetic mix, a hybrid, of a horse and a donkey; it is not a horse one instant and a donkey the next. Similarly, the color purple is a mix of red and blue, not red one instant and blue the next. In the same sense, a resonance hybrid is one molecular species, not one resonance form this instant and another resonance form the next. The problem is that we cannot depict the actual species, the hybrid, accurately with a single Lewis structure.

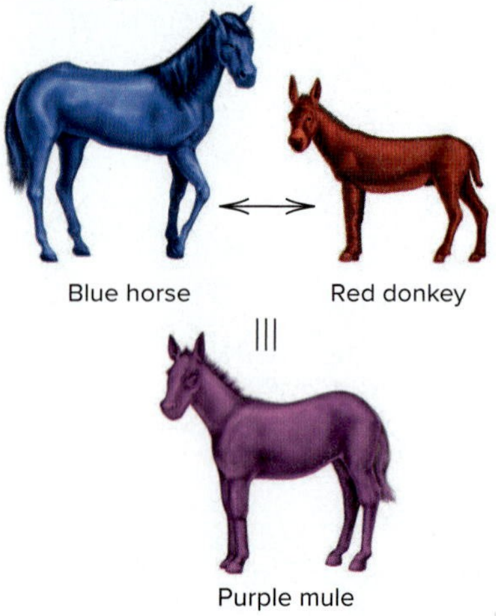

In structure I, a lone pair on oxygen A is changed to another bonding pair so that oxygen B has a double bond to oxygen A and a single bond to oxygen C. In structure II, the single and double bonds are reversed as a lone pair on oxygen C is changed to a bonding pair. You can rotate I to get II, so these are *not* different types of ozone molecules, but different Lewis structures for the *same* molecule.

Lewis structures I and II each have one O=O double bond (bond length = 121 pm) and one O—O single bond (bond length = 148 pm). But, *neither* Lewis structure depicts O_3 accurately, because the two oxygen-oxygen bonds in O_3 are actually identical in length (bond length = 128 pm) and energy. Each bond in O_3 has properties between those of an O—O bond and an O=O bond, making it something like a"one-and-a-half" bond. The molecule is shown more correctly as two Lewis structures, called **resonance structures** (or **resonance forms**), with a two-headed resonance arrow (⟷) between them. Resonance structures *have the same relative placement of atoms but different locations of bonding and lone electron pairs.* You can convert one resonance form to another by moving lone pairs to bonding positions, and vice versa:

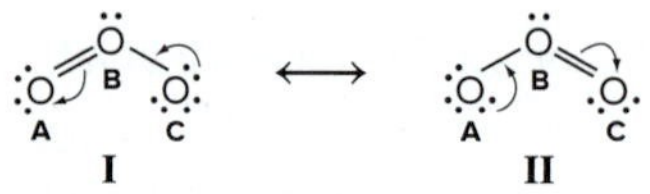

Resonance structures are not real bonding depictions: O_3 does *not* change back and forth quickly from structure I to structure II. The actual molecule is a **resonance hybrid,** an average of the resonance structures. ‹

Electron Delocalization Our need for more than one Lewis structure to depict O_3 is due to **electron-pair delocalization.** In a single, double, or triple bond, each electron pair is *localized* between the bonded atoms. In a resonance hybrid, two of the electron pairs (one bonding and one lone pair) are *delocalized:* their density is "spread" over a few adjacent atoms. (This delocalization involves just a few e^- pairs, so it is *much* less extensive than the electron delocalization in metals that we considered in Section 9.6.)

In O_3, the result is two identical bonds, each consisting of a single bond (the localized pair) and a *partial bond* (the contribution from one of the delocalized pairs). We draw the resonance hybrid with a curved dashed line to show the delocalized pairs:

resonance hybrid 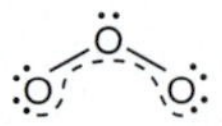

Resonance is very common. For example, benzene (C_6H_6, *shown below*) has two important resonance structures in which alternating single and double bonds have different positions. The actual molecule is an average of the two structures, with six carbon-carbon bonds of equal length and three electron pairs delocalized over all six C atoms. The delocalized pairs are often shown as a dashed circle (or simply a circle):

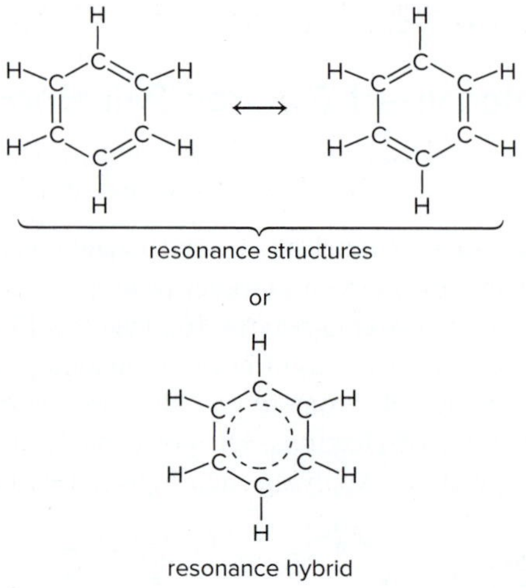

Student Hot Spot

Student data indicate that you may struggle with the concept of resonance. Access the Smartbook to view additional Learning Resources on this topic.

Fractional Bond Orders Partial bonding, as in resonance hybrids, often leads to fractional bond orders. For O_3, which has a total of three bonds (shared electron pairs) between two bonded-atom (O—O) pairs, we have

$$\text{Bond order} = \frac{3 \text{ shared electron pairs}}{2 \text{ bonded-atom (O—O) pairs}} = 1\tfrac{1}{2}$$

The bond order in the benzene ring is 9 shared electron pairs/6 bonded-atom (C—C) pairs, or $1\frac{1}{2}$ also. For the carbonate ion, CO_3^{2-}, three resonance structures can be drawn:

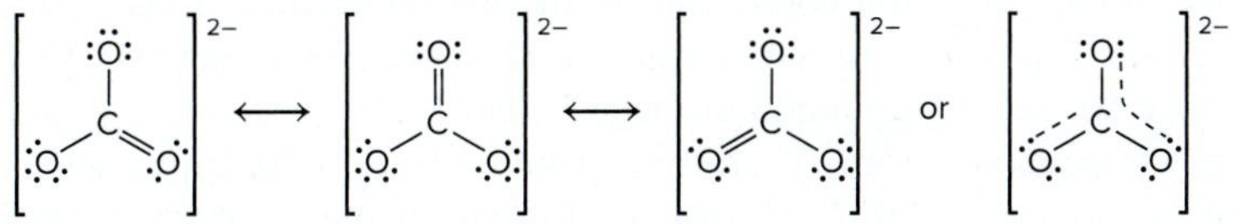

Each has 4 electron pairs shared among 3 bonded-atom (C—O) pairs, so the bond order is $\frac{4}{3}$, or $1\frac{1}{3}$.

Formal Charge: Selecting the More Important Resonance Structure

If one resonance structure "looks" more like the resonance hybrid than the others, it "weights" the average in its favor. One way to select the more important resonance structure is by determining each atom's **formal charge,** the charge it would have *if the bonding electrons were shared equally.* We'll examine this concept and then see how formal charge compares with oxidation number.

Determining Formal Charge An atom's formal charge is its total number of valence electrons minus the number of electrons associated with that atom in the Lewis structure: *all* of its unshared valence electrons and *half* of its shared valence electrons:

Formal charge of atom =
no. of valence e^- − (no. of unshared valence e^- + $\frac{1}{2}$ no. of shared valence e^-) **(10.1)**

If the number of electrons associated with an atom in the Lewis structure equals the number of valence electrons in the atom, its formal charge is 0. If the number of electrons associated with an atom in the Lewis structure is greater than the number of valence electrons in the atom, its formal charge is negative; a positive formal charge results when an atom has fewer valence electrons in the Lewis structure than there are in the atom.

For example, in O_3, the formal charge of oxygen A in resonance form I is

$$6 \text{ valence } e^- - (4 \text{ unshared } e^- + \tfrac{1}{2} \text{ of } 4 \text{ shared } e^-) = 6 - 4 - 2 = 0$$

The formal charges of all the atoms in the two O_3 resonance structures are

$O_A[6 - 4 - \frac{1}{2}(4)] = 0$
$O_B[6 - 2 - \frac{1}{2}(6)] = +1$
$O_C[6 - 6 - \frac{1}{2}(2)] = -1$

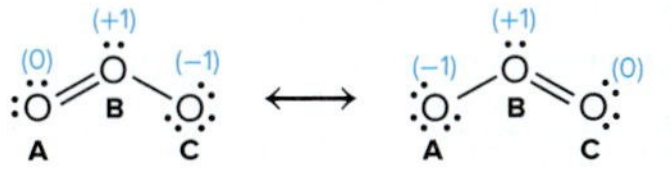

I II

$O_A[6 - 6 - \frac{1}{2}(2)] = -1$
$O_B[6 - 2 - \frac{1}{2}(6)] = +1$
$O_C[6 - 4 - \frac{1}{2}(4)] = 0$

Structures I and II have the same formal charges but on different O atoms, so they contribute equally to the resonance hybrid. *Formal charges must sum to the actual charge on the species:* zero for a molecule or the ionic charge for an ion. In O_3, the formal charges add to 0: (+1) + (−1) + 0 = 0.

Note that, in structure I, instead of oxygen's usual two bonds, O_B has three bonds and O_C has one. Only when an atom has a zero formal charge does it have its usual number of bonds; this observation also applies to C in CO_3^{2-}, N in NO_3^-, and so forth.

Choosing the More Important Resonance Structure Three criteria help us choose the more important resonance structure:

- Smaller formal charges (positive *or* negative) are preferable to larger ones.
- The *same* nonzero formal charges on adjacent atoms are not preferred.
- A more negative formal charge should reside on a more electronegative atom.

Structure I

N $[5 - 6 - \frac{1}{2}(2)] = -2$

C $[4 - 0 - \frac{1}{2}(8)] = 0$

O $[6 - 2 - \frac{1}{2}(6)] = +1$

Structure II

N $[5 - 4 - \frac{1}{2}(4)] = -1$

C $[4 - 0 - \frac{1}{2}(8)] = 0$

O $[6 - 4 - \frac{1}{2}(4)] = 0$

Structure III

N $[5 - 2 - \frac{1}{2}(6)] = 0$

C $[4 - 0 - \frac{1}{2}(8)] = 0$

O $[6 - 6 - \frac{1}{2}(2)] = -1$

Student Hot Spot

Student data indicate that you may struggle with using formal charges to select the more important resonance structure. Access the SmartBook to view additional Learning Resources on this topic.

As in the case of O_3, the resonance structures for CO_3^{2-}, NO_3^-, and benzene all have identical atoms surrounding the central atom(s) and, thus, have identical formal charges and are equally important contributors to the resonance hybrid. But, let's apply these criteria to the cyanate ion, NCO^-, which has two *different* atoms around the central one. Three resonance structures with formal charges are *(see margin)*

Formal charges:
Resonance structures:

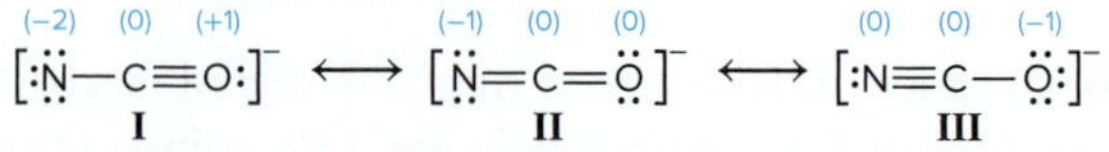

Structure I is *not* an important contributor to the hybrid because it has a larger formal charge on N and a positive formal charge on the more electronegative O. Structures II and III have the same magnitude of charges, but III has a −1 formal charge on O, the more electronegative atom. Therefore, II and III are more important than I, and III is more important than II. Note that the formal charges in each structure sum to −1, the charge of the ion.

Formal Charge vs. Oxidation Number Formal charge (used to examine resonance structures) is *not* the same as oxidation number (used to monitor redox reactions):

- For a *formal charge,* bonding electrons are *shared equally* by the atoms (as if the bonding were *nonpolar covalent*), so each atom has half of them:

$$\text{Formal charge} = \text{valence } e^- - (\text{lone pair } e^- + \tfrac{1}{2} \text{ bonding } e^-)$$

- For an *oxidation number,* bonding electrons are *transferred completely* to the more electronegative atom (as if the bonding were *pure ionic*):

$$\text{Oxidation number} = \text{valence } e^- - (\text{lone pair } e^- + \text{bonding } e^-)$$

Compare the two sets of numbers for the three cyanate ion resonance structures. When determining the oxidation numbers of the atoms in this ion, all electrons in the bonds between N and C are assigned to the more electronegative N atom, while all electrons in the bonds between C and O are assigned to the more electronegative O atom:

Formal charges:
Oxidation numbers:

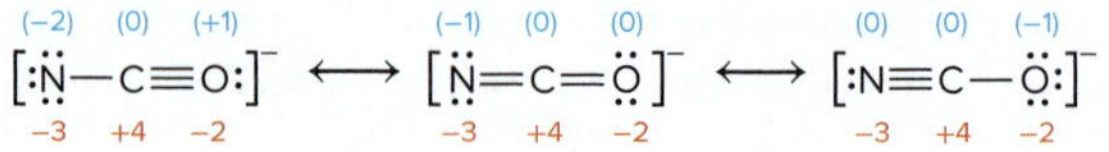

Notice that the oxidation numbers *do not* change from one resonance structure to another (because the electronegativities of the atoms *do not* change), but the formal charges *do* change (because the numbers of bonding and lone pairs *do* change). Neither an atom's formal charge nor its oxidation number represents an actual charge on the atom; both of these types of numbers serve simply to keep track of electrons.

SAMPLE PROBLEM 10.4 Writing Resonance Structures and Assigning Formal Charges

Problem Write resonance structures for the nitrate ion, NO_3^-, assign formal charges to the atoms in each resonance structure, and find the bond order.

Plan We write a Lewis structure, remembering to add $1e^-$ to the total number of valence electrons because of the 1− ionic charge. Then we move lone and bonding pairs to write other resonance structures and connect them with resonance arrows. We use Equation 10.1 to assign formal charges. The bond order is the number of shared electron pairs divided by the number of atom pairs.

Solution *Steps 1 and 2.* Nitrogen has the lower group number and is placed in the center. There are $[1 \times N(5e^-)] + [3 \times O(6e^-)] + [\text{charge } (1e^-)] = 24$ valence e^-.

Steps 3 and 4 give us:

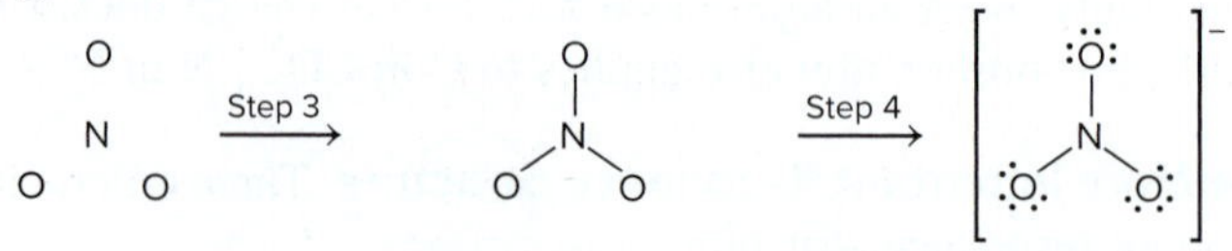

3 bonds × 2e⁻ = 6e⁻ 18e⁻
24e⁻ − 6e⁻ = 18e⁻ distributed

Step 5. Since N has only $6e^-$, we change a lone pair on one of the O atoms to a bonding pair to form a double bond, which gives each atom an octet. All the O atoms are equivalent, however, so we can move a lone pair from any one of the three and obtain three resonance structures:

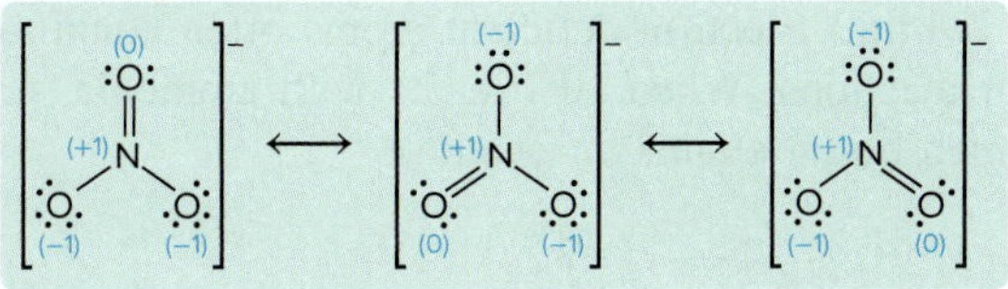

Formal charge of atom = no. of valence e^- − (no. of unshared valence e^- + $\frac{1}{2}$ no. of shared valence e^-)

For N: $[5 - 0 - \frac{1}{2}(8)] = +1$
For O (single-bonded): $[6 - 6 - \frac{1}{2}(2)] = -1$
For O (double-bonded): $[6 - 4 - \frac{1}{2}(4)] = 0$
The bond order is

$$\frac{4 \text{ shared electron pairs}}{3 \text{ bonded-atom (N—O) pairs}} = 1\tfrac{1}{3}$$

Check Each structure has the same relative placement of atoms, an octet around each atom, and $24e^-$ (the sum of the valence electrons and $1e^-$ from the ionic charge, distributed in four bonds and eight lone pairs).

Comment These three resonance structures contribute equally to the resonance hybrid because all of the surrounding atoms are identical and each resonance hybrid has the same set of formal charges.

FOLLOW-UP PROBLEMS

10.4A Write resonance structures for nitromethane, H_3CNO_2 (the H atoms are bonded to C, and the C atom is bonded to N, which is bonded to both O atoms). Assign formal charges to the atoms in each resonance structure.

10.4B Write resonance structures for the thiocyanate ion, SCN^- (C is the central atom). Use formal charges to determine which resonance structure is the most important.

SOME SIMILAR PROBLEMS 10.9–10.16

Lewis Structures for Exceptions to the Octet Rule

The octet rule applies to most molecules (and ions) with Period 2 central atoms, but not every one, and not to many with central atoms from Period 3 and higher. Three important exceptions occur for molecules with (1) electron-deficient atoms, (2) odd-electron atoms, and (3) atoms with expanded valence shells. In discussing these exceptions, you'll also see that formal charge has limitations for selecting the best resonance structure.

Molecules with Electron-Deficient Atoms In gaseous molecules containing either beryllium or boron as the central atom, that atom is often **electron deficient:** it has *fewer* than eight electrons around it (an incomplete octet). The Lewis structures, with formal charges, of gaseous beryllium chloride* and boron trifluoride are

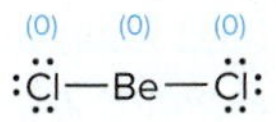

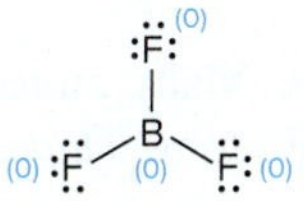

There are only four electrons around Be and only six around B. Surrounding halogen atoms don't form multiple bonds to the central atoms to give them an octet, because the halogens are much more electronegative than either Be or B. Formal charges make the following structures unlikely:

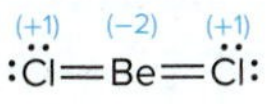

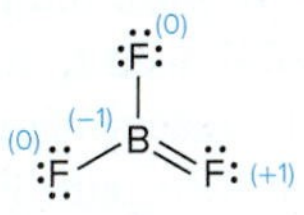

*Even though beryllium is in Group 2A(2), most Be compounds have considerable covalent bonding. For example, molten $BeCl_2$ does not conduct electricity, indicating a lack of ions.

(Some data for BF_3 show a shorter than expected B—F bond. Shorter bonds indicate double-bond character, so the structure with the B=F bond may be a minor contributor to a resonance hybrid.) Electron-deficient atoms often attain an octet by forming additional bonds in reactions. When BF_3 reacts with ammonia, for instance, a compound forms in which boron attains an octet:†

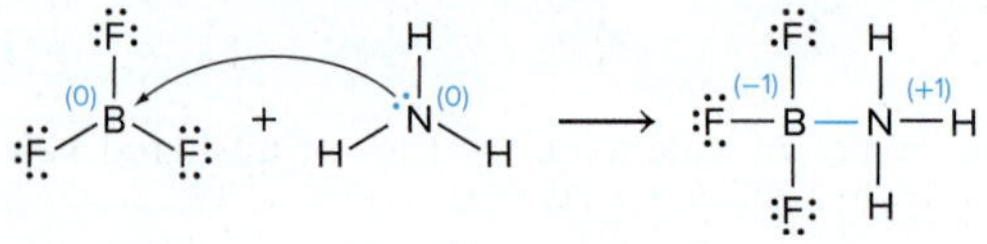

Molecules with Odd-Electron Atoms A few molecules contain a central atom with an odd number of valence electrons, so they cannot have all their electrons in pairs. Most of these molecules have a central atom from an odd-numbered group, such as N [Group 5A(15)] or Cl [Group 7A(17)]. They are called **free radicals,** species that contain a lone (unpaired) electron, which makes them paramagnetic (Section 8.4) and extremely reactive. Free radicals are dangerous because they can bond to an H atom in a biomolecule and extract it, which forms a new free radical. This step repeats and can disrupt genes and membranes. Recent studies suggest that free radicals may be involved in cancer and even aging. Antioxidants like vitamin E interrupt free-radical proliferation.

Consider the free radical nitrogen dioxide, NO_2, a major contributor to urban smog that is formed when the NO in auto exhaust is oxidized. NO_2 has several resonance structures. Two of them differ as to which O atom is doubly bonded to the central atom, as in the case of ozone. Two others have the lone electron residing on the N or on an O, so the resonance hybrid has the lone electron delocalized over these two atoms:

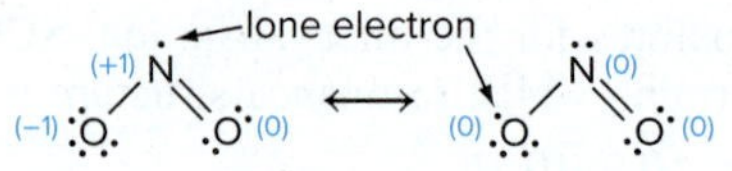

Let's see if formal charge considerations help us decide where the lone electron resides most of the time. The structure with it on the singly bonded O has zero formal charges *(right),* while the structure with it on N *(left)* has some nonzero charges. Thus, based on formal charges, the structure on the right is more important. But, chemical facts suggest otherwise. Free radicals often react with each other to pair their lone electrons. When two NO_2 molecules react, the lone electrons pair up to form the N—N bond in dinitrogen tetroxide (N_2O_4) and each N attains an octet:

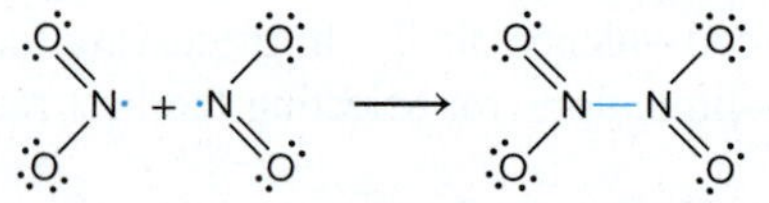

Thus, given the way NO_2 reacts, the lone electron may spend most of its time on N, making *that* resonance structure more important. Apparently, in this case, formal charge is not very useful for picking the most important resonance structure; we'll see other cases below.

Expanded Valence Shells Many molecules (and ions) have more than eight valence electrons around the central atom. *That atom expands its valence shell to form more bonds, which releases energy.* The central atom must be large and have empty orbitals that can hold the additional pairs. Therefore, **expanded valence shells** (expanded octets) occur only with *nonmetals from Period 3 or higher because they have d orbitals available.* Such a central atom may be bonded to more than four atoms, or to four or fewer.

1. *Central atom bonded to more than four atoms.* Phosphorus pentachloride, PCl_5, is a fuming yellow-white solid used to manufacture lacquers and films. It forms when phosphorus trichloride, PCl_3, reacts with chlorine gas. The P in PCl_3 has an octet, but two more bonds to chlorine form and P expands its valence shell to 10 electrons in

†Reactions in which one species "donates" an electron pair to another to form a covalent bond are Lewis acid-base reactions, which we discuss fully in Chapter 18.

PCl_5. Note that when PCl_5 forms, *one* Cl—Cl bond breaks *(left side of the equation),* and two P—Cl bonds form *(right side),* for a net increase of one bond:

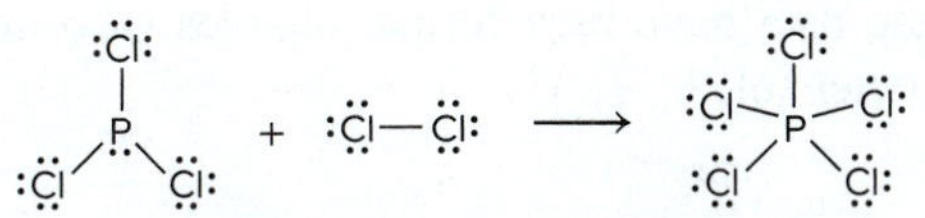

Sulfur hexafluoride, SF_6, is a dense, inert gas used as an electrical insulator. Analogously to PCl_5, it forms when sulfur tetrafluoride, SF_4, reacts with more F_2. The S in SF_4 already has an expanded valence shell of 10 electrons, but two more bonds to F expand the valence shell further to 12 electrons:

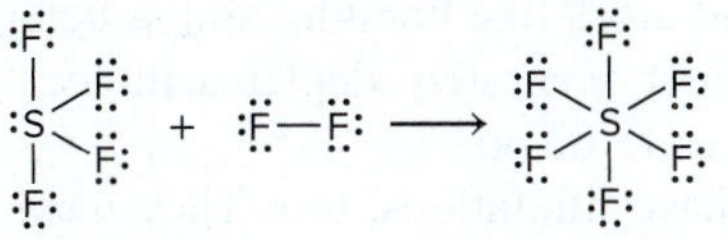

2. *Central atoms bonded to four or fewer atoms.* In some species, a central atom bonded to *four or fewer* atoms has an expanded valence shell. The S in SF_4 is one example; another is the Cl in chlorine trifluoride, ClF_3. There are 28 valence electrons in a ClF_3 molecule, 6 in the three single Cl—F bonds and 20 that complete the octets of the four atoms. The 2 remaining valence electrons are placed on the central Cl, giving it an expanded valence shell of 10 electrons:

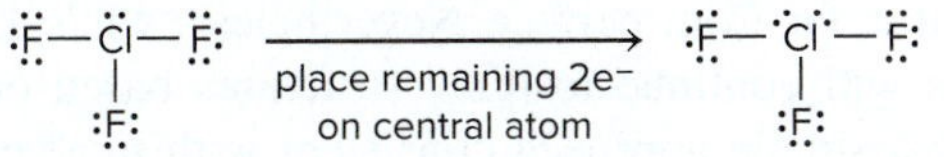

In general, just as in ClF_3, *any valence electrons remaining after completing the octets of all atoms in a species are given to the central atom.*

For some species with central atoms bonded to four or fewer atoms, the concept of formal charge can be applied to draw Lewis structures with expanded valence shells. Some examples follow:

- *Sulfuric acid.* Two resonance structures of H_2SO_4, with formal charges, are

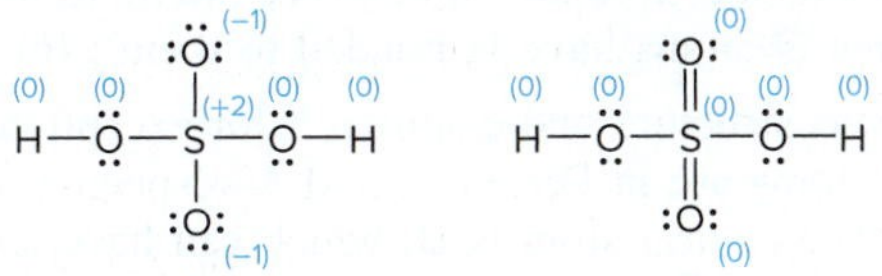

I II

Structure I obeys the octet rule, but it has several nonzero formal charges. In structure II, two lone pairs from two O atoms have been changed to bonding pairs; now sulfur has 12 electrons (six bonds) around it, but all atoms have zero formal charges. Thus, based on the formal charge rules alone, II contributes more than I to the resonance hybrid. More important than whether rules are followed, structure II is consistent with observation. In gaseous H_2SO_4, the two sulfur-oxygen bonds *with* an H atom attached to the O are 157 pm long, whereas the two sulfur-oxygen bonds *without* an H attached to O are 142 pm long. This shorter bond length indicates double-bond character, and other measurements indicate greater electron density in the bonds without the attached H.

- *Sulfate ion.* When sulfuric acid loses two H^+ ions, it forms the sulfate ion, SO_4^{2-}. Measurements indicate that all the bonds in SO_4^{2-} are 149 pm long, between the length of an S=O bond (~142 pm) and that of an S—O bond (~157 pm). Six of the seven resonance structures consistent with these data have an expanded valence shell and zero formal charges on some of the atoms. Two of those six *(left)* and the one that obeys the octet rule *(right)* are

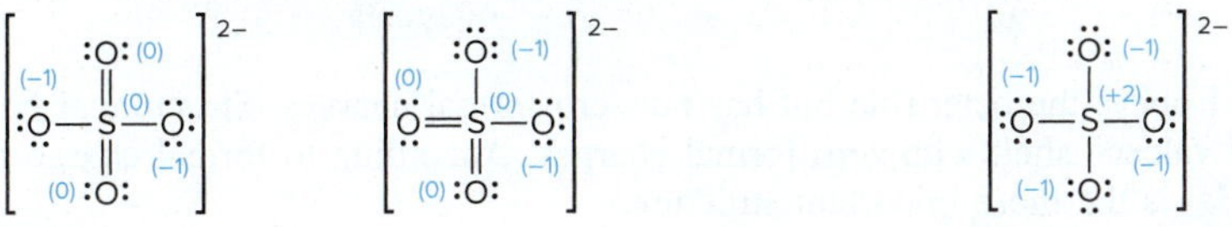

- *Two sulfur oxides*. Measurements show that the sulfur-oxygen bonds in SO_2 and SO_3 are all approximately 142 pm long, indicating S=O bonds. Lewis structures consistent with these data have zero formal charges *(two at left)*, but others *(two at right)* obey the octet rule:

3. *Limitations of Lewis structures and formal charge*. Chemistry has been a central science for well over two centuries, yet controversies often arise over interpretation of data, even in established areas like bonding and structure. We've seen that a single Lewis structure often cannot accurately depict a molecule; in such cases, we need several resonance structures to do so.

Formal charge rules have limitations, too. They were not useful in choosing the correct location for the lone electron in NO_2. And, quantum-mechanical calculations indicate that resonance structures with expanded valence shells that have zero formal charges look *less* like the actual species than structures that follow the octet rule but have higher formal charges. Shorter bonds arise, these findings suggest, not from double-bond character, but because the higher formal charges draw the bonded atoms closer. Such considerations favor the octet-rule structures for H_2SO_4, SO_4^{2-}, SO_2, and SO_3. Thus, formal charge may be a useful tool for selecting the most important resonance structure, but it is far from perfect. Nevertheless, while keeping these contrary findings in mind, we will continue to draw structures based on formal charge rules because they provide a simple approach consistent with most experimental data.

SAMPLE PROBLEM 10.5 Writing Lewis Structures for Octet-Rule Exceptions

Problem Write a Lewis structure and identify the octet-rule exception for **(a)** XeF_4; **(b)** H_3PO_4 (draw two resonance structures and select the more important; all O atoms are bonded to P and three O atoms have H bonded to them); **(c)** $BFCl_2$.

Plan We write each Lewis structure and examine it for exceptions to the octet rule. **(a)** and **(b)** The central atoms are in Periods 5 and 3, respectively, so they can have more than an octet. **(c)** The central atom is B, which can have fewer than an octet of electrons.

Solution **(a)** XeF_4 has a total of 36 valence electrons: 8 form the bonds and 24 complete the octets of the F atoms. The remaining 4 valence electrons are placed on the central atom, resulting in an *expanded valence shell:*

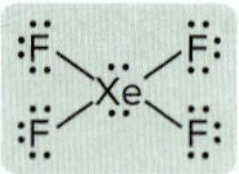

(b) H_3PO_4 has 32 valence electrons: 14 form bonds and 18 complete the octets of the O and P atoms (structure I, shown with formal charges). Another resonance structure (structure II) can be drawn in which a lone pair from the O atom with nonzero formal charge is changed to a bonding pair.

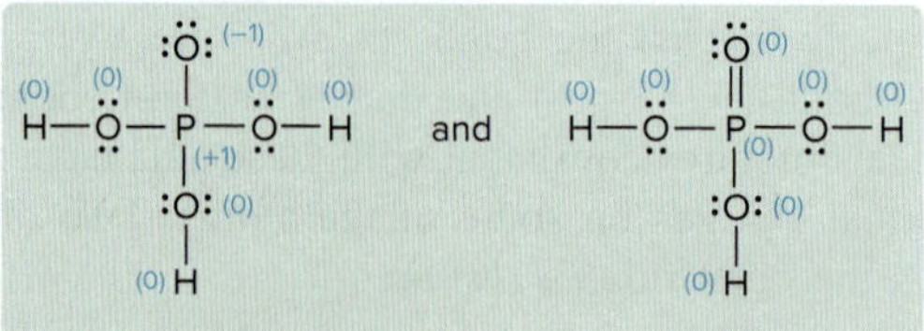

I and II

Structure I obeys the octet rule but has nonzero formal charges. Structure II has an expanded valence shell with zero formal charges. According to formal charge rules, structure II is the more important structure.

(c) $BFCl_2$ has 24 valence electrons: 6 form bonds and 18 complete the octets of the F and Cl atoms. This molecule has an *electron-deficient atom (incomplete octet);* B has only six electrons surrounding it:

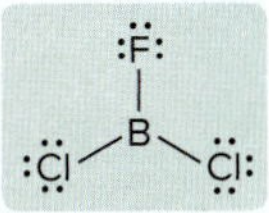

Comment In (b), structure II is consistent with bond-length measurements, which show one shorter (152 pm) phosphorus-oxygen bond and three longer (157 pm) ones. Nevertheless, quantum-mechanical calculations show that structure I may be more important.

FOLLOW-UP PROBLEMS

10.5A Write a Lewis structure with minimal formal charges for **(a)** $POCl_3$; **(b)** ClO_2; **(c)** IBr_4^-.

10.5B Write a Lewis structure with minimal formal charges for **(a)** BeH_2; **(b)** I_3^-; **(c)** XeO_3.

SOME SIMILAR PROBLEMS 10.19–10.22

› Summary of Section 10.1

- A stepwise process converts a molecular formula into a Lewis structure, a two-dimensional representation of a molecule (or ion) that shows the placement of atoms and the distribution of valence electrons among bonding and lone pairs.
- When two or more Lewis structures can be drawn for the same relative placement of atoms, the actual structure is a hybrid of those resonance structures.
- Formal charges can be useful for choosing the more important contributor to the hybrid, but experimental data always determine the choice.
- Molecules with an electron-deficient atom (central Be or B) or an odd-electron atom (free radicals) have less than an octet around the central atom but often attain an octet in reactions.
- In a molecule (or ion) with a central atom from Period 3 or higher, that atom can have more than eight valence electrons because it is larger and has empty *d* orbitals for expanding its valence shell.

10.2 VALENCE-SHELL ELECTRON-PAIR REPULSION (VSEPR) THEORY

Virtually every biochemical process hinges to a great extent on the shapes of interacting molecules. Every medicine you take, odor you smell, or flavor you taste depends on part or all of one molecule fitting together with another. Biologists have found that complex behaviors in many organisms, such as mating, defense, navigation, and feeding, often depend on one molecule's shape matching that of another. In this section, we discuss a model for predicting the shape of a molecule.

To determine the molecular shape, chemists start with the Lewis structure and apply **valence-shell electron-pair repulsion (VSEPR) theory.** Its basic principle is that, *to minimize repulsions, each group of valence electrons around a central atom is located as far as possible from the others.* A "group" of electrons is any number that occupies a localized region around an atom. Each of the following represents a single electron group:

- a single bond
- a double bond
- a triple bond*
- a lone pair of electrons
- one lone electron

Only electron groups around the *central* atom affect shape; electrons on atoms other than the central atom do not. The **molecular shape** is the three-dimensional arrangement of nuclei joined by the bonding groups.

*The two electron pairs in a double bond (or the three pairs in a triple bond) occupy separate orbitals, but they remain near each other and act as one electron group (see Chapter 11).

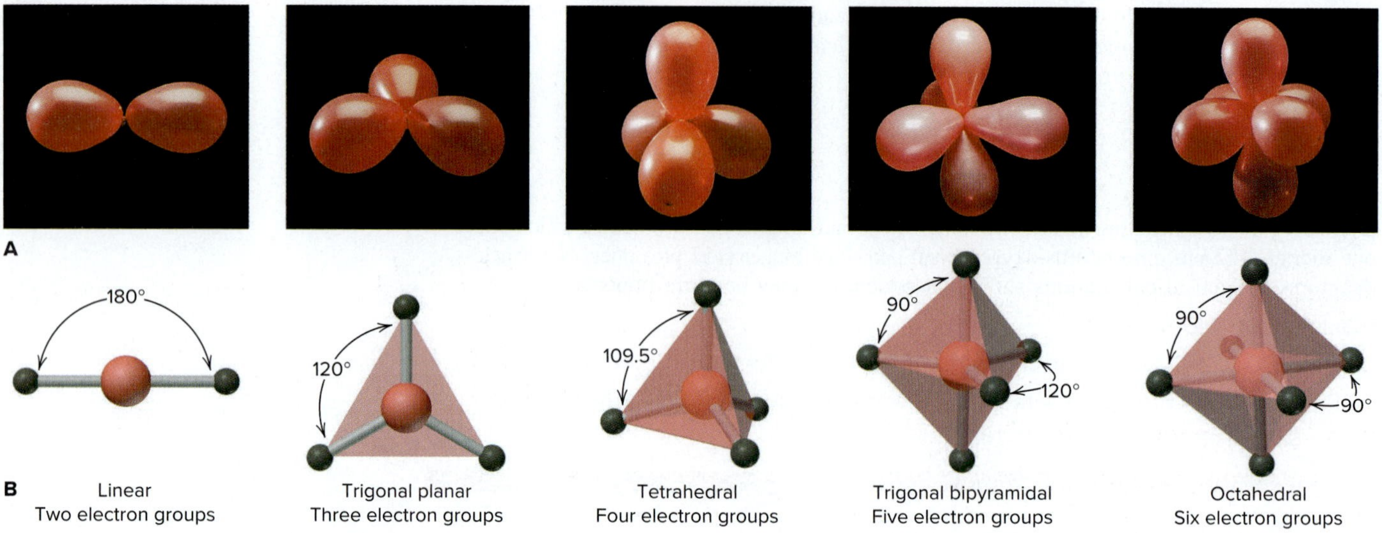

Figure 10.2 Electron-group repulsions and molecular shapes. A, Five geometric orientations arise when each balloon occupies as much space as possible. **B,** Mutually repelling bonding groups *(gray sticks)* attach a surrounding atom *(dark gray)* to the central atom *(red).* The name is the electron-group arrangement.

Source: (A) © McGraw-Hill Education/ Stephen Frisch, photographer

Electron-Group Arrangements and Molecular Shapes

When two, three, four, five, or six objects attached to a central point maximize the space between them, five geometric patterns result, which Figure 10.2A shows with balloons. If the objects are valence-electron groups, repulsions maximize the space each occupies around the central atom, and we obtain the five *electron-group arrangements* seen in the great majority of molecules and polyatomic ions.

Classifying Molecular Shapes *The electron-group arrangement* is defined by the bonding groups (shared electron pairs) *and* nonbonding electron groups (lone electron pairs) around the central atom, but the *molecular shape* is defined by the relative positions of the nuclei, which are connected by the bonding groups only. Figure 10.2B shows the molecular shapes that occur when *all* the surrounding electron groups are *bonding* groups. When some are *nonbonding* groups, different molecular shapes occur. Thus, *the same electron-group arrangement can give rise to different molecular shapes:* some with all bonding groups (as in Figure 10.2B) and others with bonding and nonbonding groups. To classify molecular shapes, we assign each a specific AX_mE_n designation, where m and n are integers. In such a designation,

- A is the central atom;
- X is a surrounding atom; and
- E is a nonbonding valence-electron group (usually a lone pair).

Bond Angle and Deviations from the Ideal Value The **bond angle** is the angle formed by the bonds joining the nuclei of two surrounding atoms to the nucleus of the central atom, which is at the vertex. The angles shown for the shapes in Figure 10.2B are *ideal* bond angles determined by basic geometry alone. We observe them when all the bonding groups are the same and connected to the same type of atom (structure I, *below*). When this is not the case, the real bond angles deviate from the ideal angles. Deviations occur when the bonds are not the same (structure II), the surrounding atoms are not the same (structure III), or one or more of the electron groups are nonbonding groups (structure IV). We'll see examples of these deviations among the compounds discussed in this section.

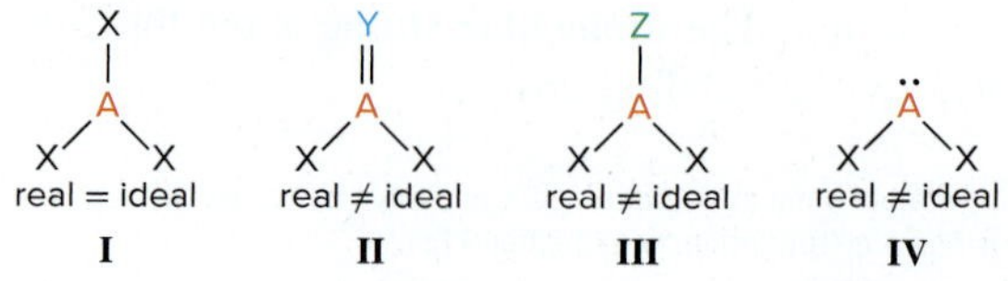

The Molecular Shape with Two Electron Groups (Linear Arrangement)

Two electron groups attached to a central atom point in opposite directions. This **linear arrangement** of electron groups results in a molecule with a **linear shape** and a bond angle of 180°. Figure 10.3 shows the general form *(top)* and shape *(middle)* for VSEPR shape class AX_2 and gives the formulas of some linear molecules.

Gaseous beryllium chloride ($BeCl_2$) is a linear molecule. Recall that in gaseous Be compounds, Be is electron deficient, with two electron pairs around it:

In carbon dioxide, the central C atom forms two double bonds with the O atoms:

Each double bond acts as one electron group and is 180° away from the other. The lone pairs on the O atoms of CO_2 or on the Cl atoms of $BeCl_2$ are *not* involved in the molecular shape: only the two electron groups around the central C or Be atom affect the shape.

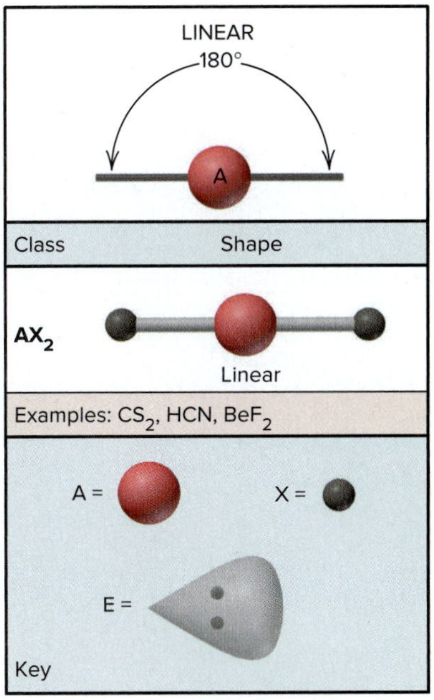

Figure 10.3 **The single molecular shape of the linear electron-group arrangement.** The key *(bottom)* for A, X, and E also refers to Figures 10.4, 10.5, 10.7, and 10.8.

Molecular Shapes with Three Electron Groups (Trigonal Planar Arrangement)

Three electron groups around a central atom point to the corners of an equilateral triangle, which gives the **trigonal planar arrangement** and an ideal bond angle of 120° (Figure 10.4). This arrangement has two molecular shapes—one with all bonding groups and the other with one lone pair. In this case, we can see *the effects of lone pairs and double bonds on bond angles.*

1. *All bonding groups: trigonal planar shape (AX_3).* Boron trifluoride (BF_3), another electron-deficient species, is an example of a trigonal planar molecule. It has six electrons around the central B atom in three single bonds to F atoms. The four nuclei lie in a plane, and each F—B—F angle is 120°:

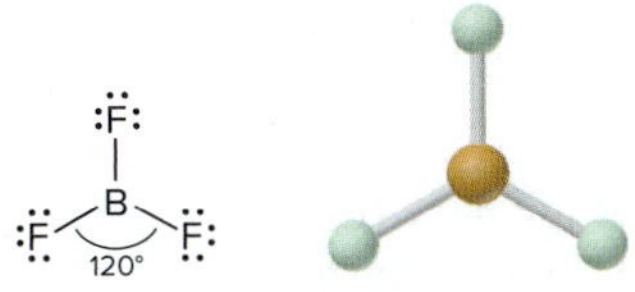

The nitrate ion (NO_3^-) is one of several polyatomic ions with the trigonal planar shape. One of three resonance structures of the nitrate ion (Sample Problem 10.4) is

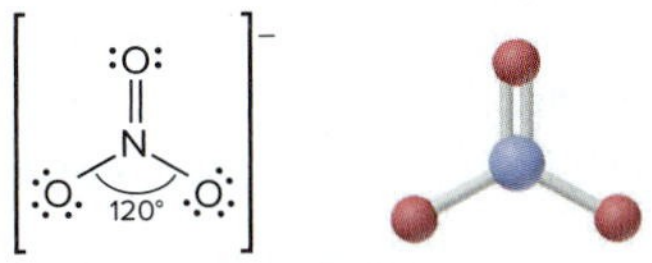

Any of the three resonance structures of NO_3^- can be used to determine its molecular shape. Remember that the resonance hybrid has three identical bonds with bond order $1\frac{1}{3}$, so the ideal bond angle is observed.

2. *One lone pair: bent or V shape (AX_2E).* Gaseous tin(II) chloride is a molecule with a **bent shape,** or **V shape,** with three electron groups in a trigonal plane, including a lone pair at any one of the triangle's corners. A lone pair often has a major effect on bond angle. Because it is held by only one nucleus, a lone pair is less confined than a bonding pair and so exerts stronger repulsions. In general, *a lone pair repels bonding pairs more than bonding pairs repel each other, so it decreases the angle*

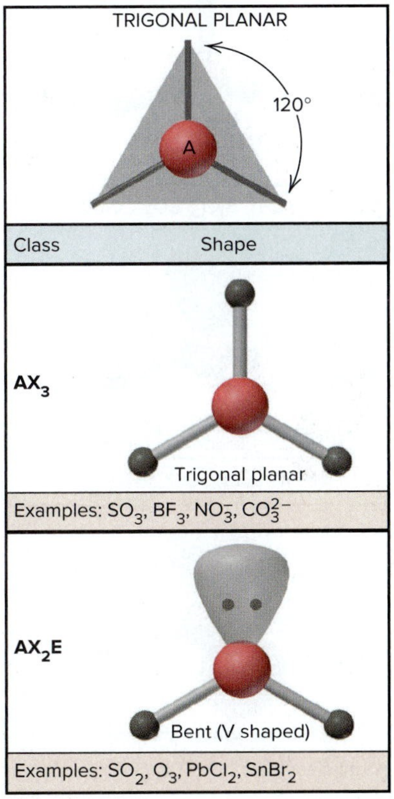

Figure 10.4 **The two molecular shapes of the trigonal planar electron-group arrangement.**

between bonding pairs. Note the 95° bond angle in $SnCl_2$, which is considerably less than the ideal 120°:

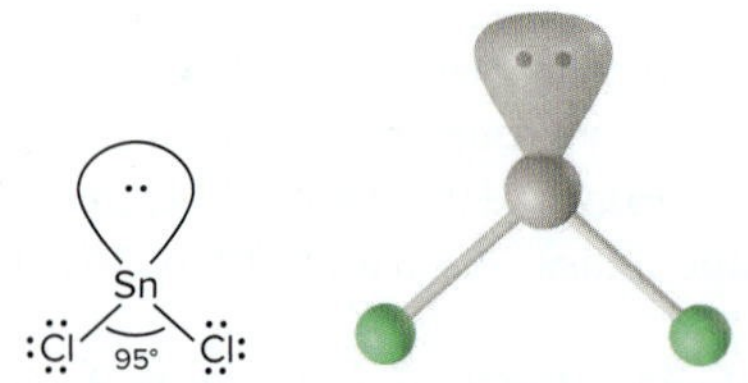

Effect of Double Bonds on Bond Angle When the surrounding atoms and electron groups are not identical, the bond angles are often affected. Consider formaldehyde (CH_2O), whose uses include the manufacture of countertops and the production of methanol. The three electron groups around the central C result in a trigonal planar shape. But there are two types of surrounding atoms (O and H) and two types of electron groups (single and double bonds):

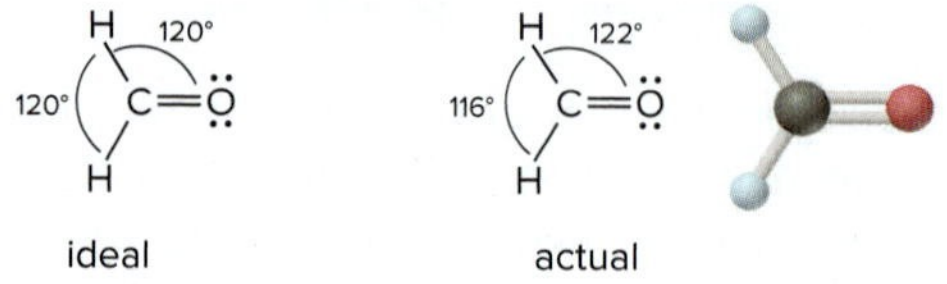

The actual H—C—H bond angle is less than the ideal 120° because *the greater electron density of a double bond repels electrons in single bonds more than the single bonds repel each other.*

Molecular Shapes with Four Electron Groups (Tetrahedral Arrangement)

Shapes based on two or three electron groups lie in a plane, but four electron groups require three dimensions to maximize separation. Consider methane, whose Lewis structure *(below, left)* shows four bonds pointing to the corners of a square, which suggests 90° bond angles. But, *Lewis structures do **not** depict shape.* In three dimensions, the four electron groups lie at the corners of a *tetrahedron,* a polyhedron with four faces made of equilateral triangles, giving bond angles of 109.5° (Figure 10.5).

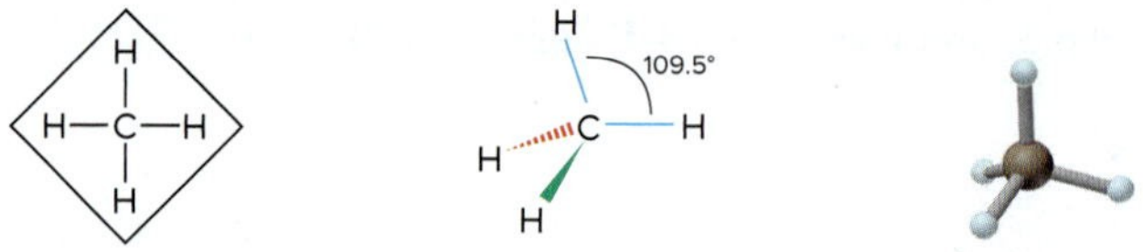

A *perspective drawing* for methane *(above, middle)* indicates depth by using solid and dashed wedges for bonds out of the plane of the page. The normal bond lines *(blue)* are in the plane of the page; the solid wedge *(green)* is the bond from the C atom in the plane of the page to the H above that plane; and the dashed wedge *(red)* is the bond from the C to the H below the plane of the page. The ball-and-stick model *(above, right)* shows the tetrahedral shape more clearly.

All molecules or ions with four electron groups around a central atom adopt the **tetrahedral arrangement.** There are three shapes with this arrangement:

1. *All bonding groups: tetrahedral shape (AX_4).* Methane has a tetrahedral shape, a very common geometry in organic molecules. In Sample Problem 10.1, we drew the Lewis structure for CCl_2F_2, without regard to relative placement of the four halogen atoms around the carbon atom. Because Lewis structures are flat, it may seem like you can write two different ones for CCl_2F_2, but they represent the same molecule, as a twist of the wrist reveals (Figure 10.6).

2. *One lone pair: trigonal pyramidal shape (AX_3E).* Ammonia (NH_3) is an example of a molecule with a **trigonal pyramidal shape,** a tetrahedron with one vertex "missing." Stronger repulsions by the lone pair make the H—N—H bond angle slightly less than the ideal 109.5°. The lone pair forces the N—H pairs closer to each other, and

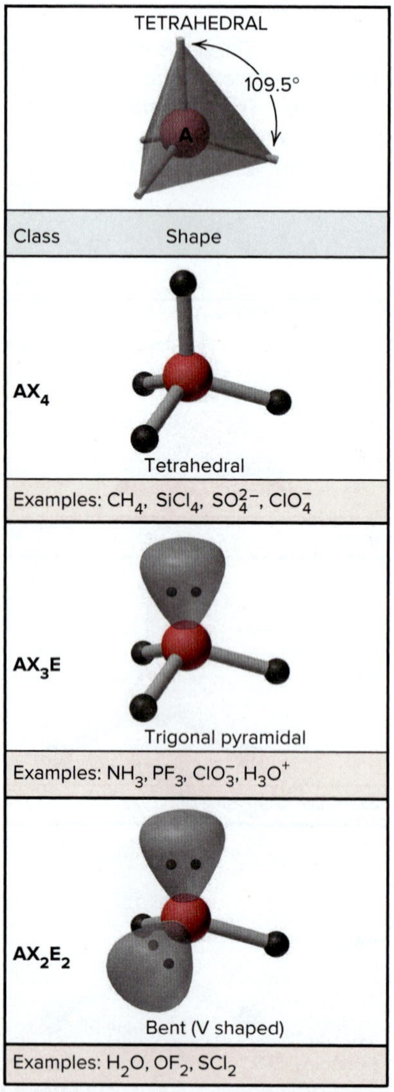

Figure 10.5 The three molecular shapes of the tetrahedral electron-group arrangement.

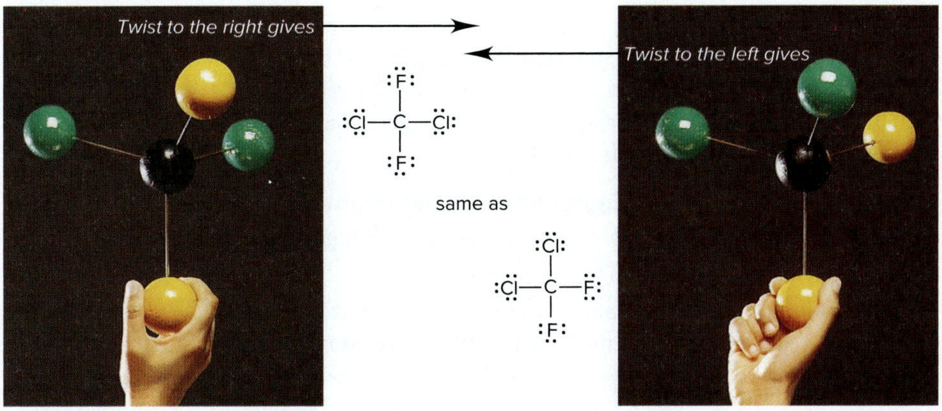

Figure 10.6 Lewis structures do not indicate molecular shape. In this model, Cl is green and F is yellow.

Source: © McGraw-Hill Education/Stephen Frisch, photographer

the bond angle is 107.3°. NF_3, whose Lewis structure we drew at the beginning of Section 10.1, also has a trigonal pyramidal shape.

Picturing shapes is a great way to visualize a reaction. For instance, when ammonia reacts with an acid, the lone pair on N forms a bond to the H^+ and yields the ammonium ion (NH_4^+), one of many tetrahedral polyatomic ions. As the lone pair becomes a bonding pair, the H—N—H angle expands from 107.3° to 109.5°:

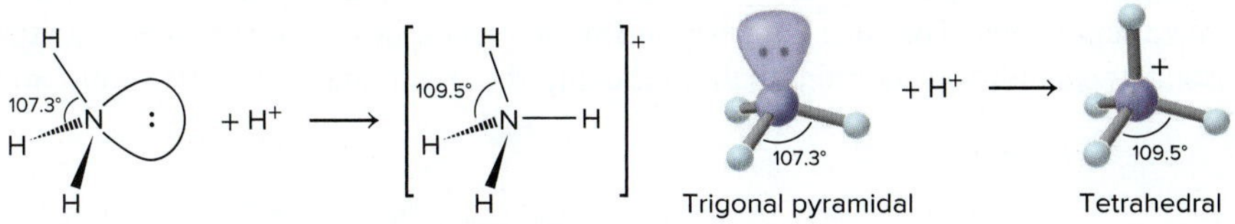

3. *Two lone pairs: bent or V shape (AX_2E_2).* Water is the most important V-shaped molecule with the tetrahedral arrangement. [Note that, in the trigonal planar arrangement, the V shape has two bonding groups and *one* lone pair (AX_2E), and its ideal bond angle is 120°, not 109.5°.] Repulsions from two lone pairs are greater than from one, and the H—O—H bond angle is 104.5°, less than the H—N—H angle in NH_3:

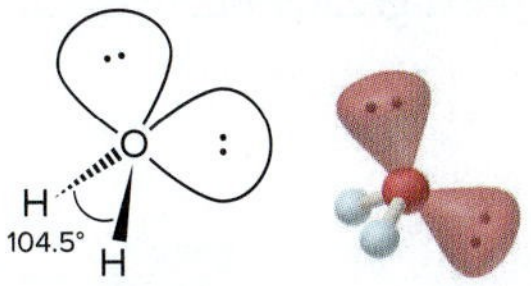

Thus, for similar molecules within a given electron-group arrangement, electron-electron repulsions cause deviations from ideal bond angles in the following order:

Lone pair–lone pair > lone pair–bonding pair > bonding pair–bonding pair **(10.2)**

Molecular Shapes with Five Electron Groups (Trigonal Bipyramidal Arrangement)

All molecules with five or six electron groups have a central atom from Period 3 or higher because only those atoms have *d* orbitals available to expand the valence shell.

Relative Positions of Electron Groups Five mutually repelling electron groups form the **trigonal bipyramidal arrangement,** in which two trigonal pyramids share a common base (Figure 10.7, *next page*). This is the only case in which *there are two different positions for electron groups and two ideal bond angles:*

- Three **equatorial groups** lie in a trigonal plane that includes the central atom; a 120° bond angle separates the equatorial groups.
- Two **axial groups** lie above and below this plane; a 90° angle separates axial from equatorial groups.

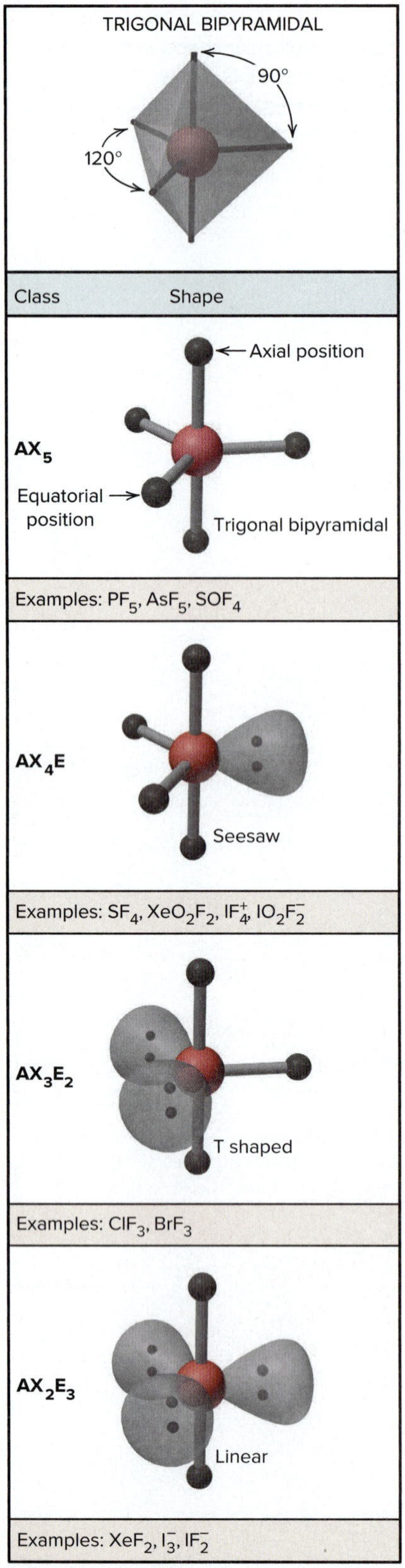

Figure 10.7 The four molecular shapes of the trigonal bipyramidal electron-group arrangement.

Two factors come into play:

- The greater the bond angle, the weaker the repulsions, so *equatorial-equatorial (120°) repulsions are weaker than axial-equatorial (90°) repulsions.*
- The stronger repulsions from lone pairs means that, when possible, *lone pairs occupy equatorial positions.*

Shapes for the Trigonal Bipyramidal Arrangement The tendency for lone pairs to occupy equatorial positions, and thus minimize stronger axial-equatorial repulsions, governs three of the four shapes for this arrangement.

1. *All bonding groups: trigonal bipyramidal shape (AX_5).* Phosphorus pentachloride (PCl_5) has a trigonal bipyramidal shape. With five identical surrounding atoms, the bond angles are ideal:

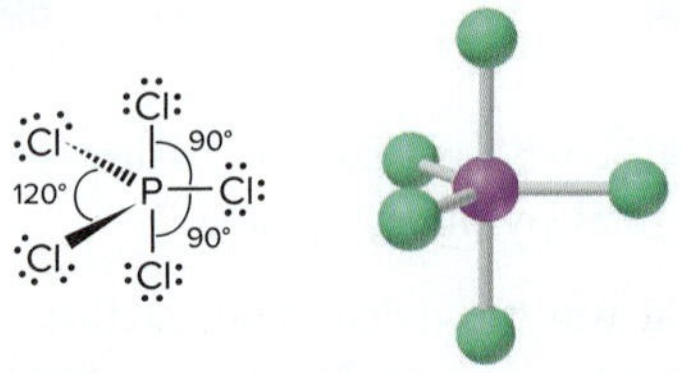

2. *One lone pair: seesaw shape (AX_4E).* Sulfur tetrafluoride (SF_4), a strong fluorinating agent, has the **seesaw shape;** in Figure 10.7, the "seesaw" is tipped on an end. This is the first example of a *lone pair occupying an equatorial position* to minimize repulsions. The lone pair, which can be placed in any of the three equatorial positions, repels all four bonding pairs, reducing the bond angles to 101.5° and 86.8°:

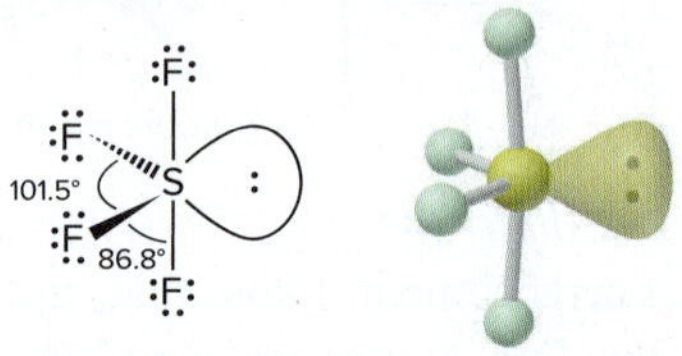

3. *Two lone pairs: T shape (AX_3E_2).* Bromine trifluoride (BrF_3), one of many compounds with fluorine bonded to a larger halogen, has a **T shape.** Since both lone pairs occupy equatorial positions, we see a greater decrease in the axial-equatorial bond angle, bringing it down to 86.2°:

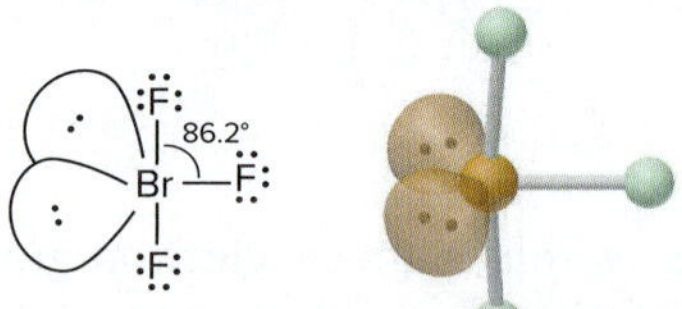

4. *Three lone pairs: linear shape (AX_2E_3).* The triiodide ion (I_3^-), which forms when I_2 dissolves in aqueous I^- solution, is linear. With three equatorial lone pairs and two axial bonding pairs, the three nuclei form a straight line and a 180° X—A—X bond angle:

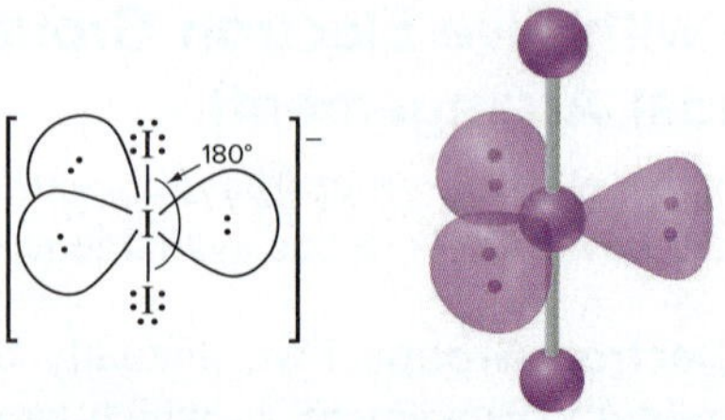

Molecular Shapes with Six Electron Groups (Octahedral Arrangement)

Six electron groups form the **octahedral arrangement.** An *octahedron* is a polyhedron with eight equilateral triangles for faces and six identical vertices (Figure 10.8). Each of the six groups points to a corner, giving 90° ideal bond angles.

1. *All bonding groups: octahedral shape (AX_6).* When seesaw-shaped SF_4 reacts with more F_2, the central S atom expands its valence shell further to form octahedral sulfur hexafluoride (SF_6):

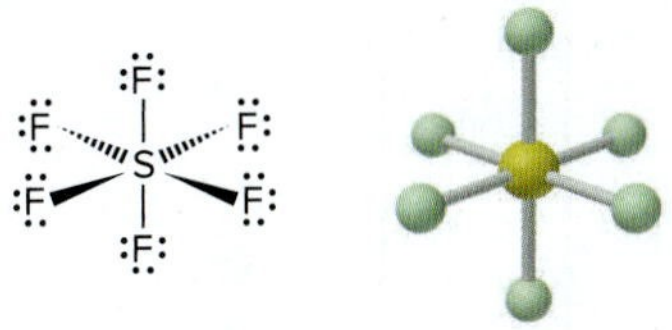

2. *One lone pair: square pyramidal shape (AX_5E).* Iodine pentafluoride (IF_5) has a **square pyramidal shape.** Note that it makes no difference where the one lone pair resides because all the ideal bond angles are 90°. The lone pair reduces the bond angles to 81.9°:

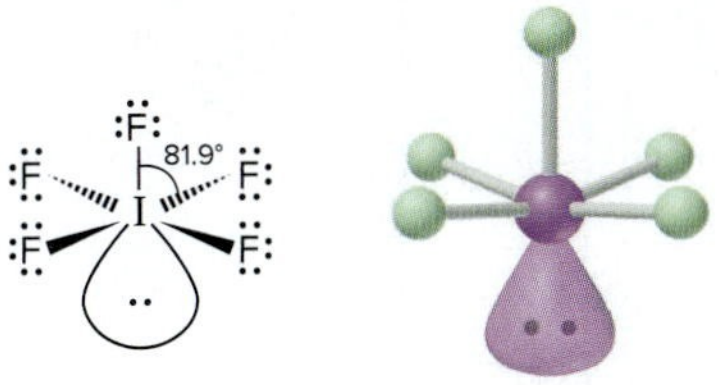

3. *Two lone pairs: square planar shape (AX_4E_2).* Xenon tetrafluoride (XeF_4) has a **square planar shape.** To avoid the stronger lone pair–lone pair repulsions that would result if the two lone pairs were adjacent to each other, they lie *opposite* each other:

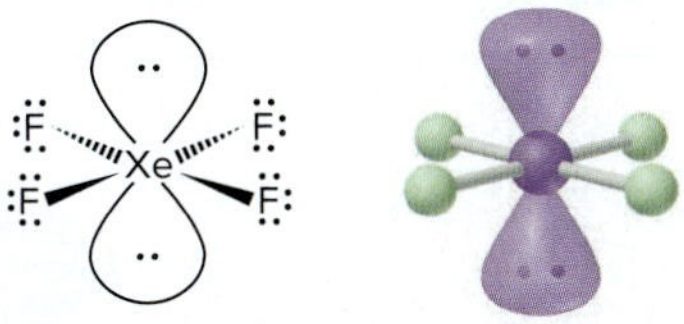

Figure 10.9 displays the shapes possible for elements in different periods, and Figure 10.10 on the next page summarizes the molecular shapes we've discussed.

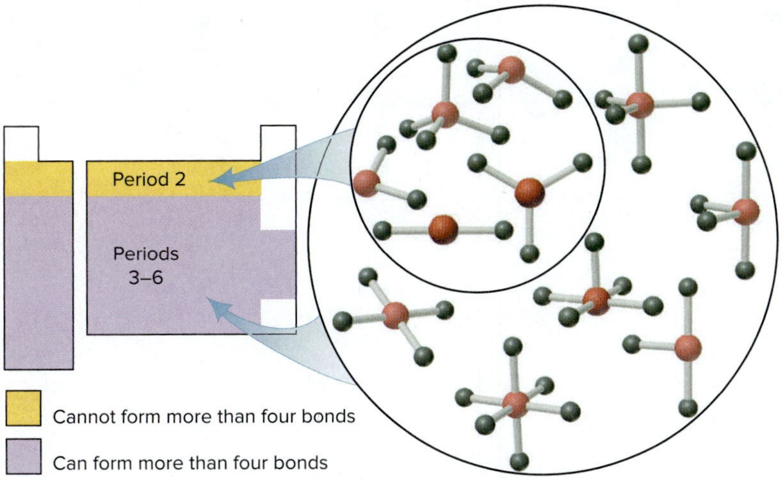

Figure 10.9 Molecular shapes for central atoms in Period 2 and in higher periods.

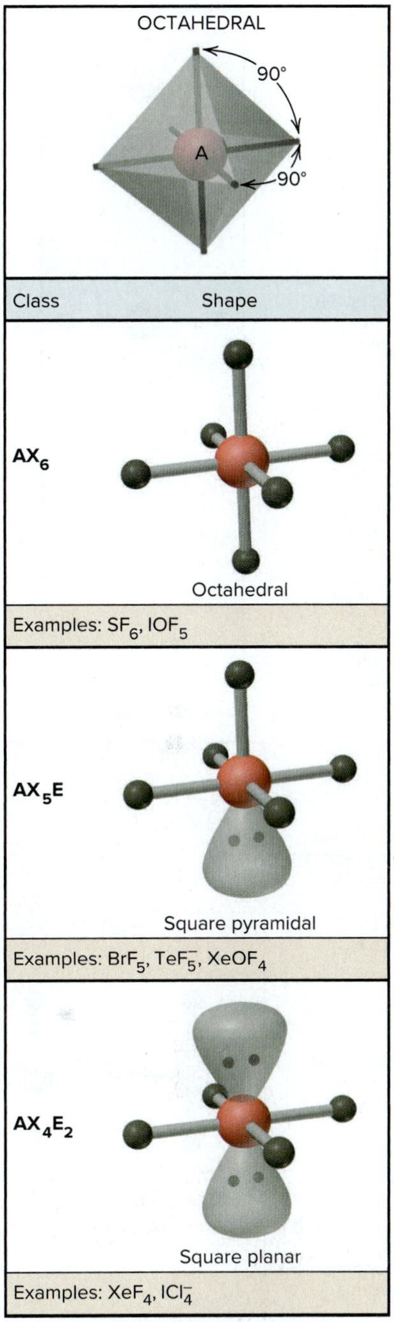

Figure 10.8 The three molecular shapes of the octahedral electron-group arrangement.

Using VSEPR Theory to Determine Molecular Shape

Let's apply a stepwise method for using VSEPR theory to determine a molecular shape from a molecular formula (Figure 10.11, *next page*):

Step 1. Write the Lewis structure from the molecular formula (Figure 10.1) to see the relative placement of atoms and the number of electron groups.

Step 2. Assign one of the five electron-group arrangements (Figure 10.2) by counting *all* electron groups (bonding plus nonbonding) around the central atom.

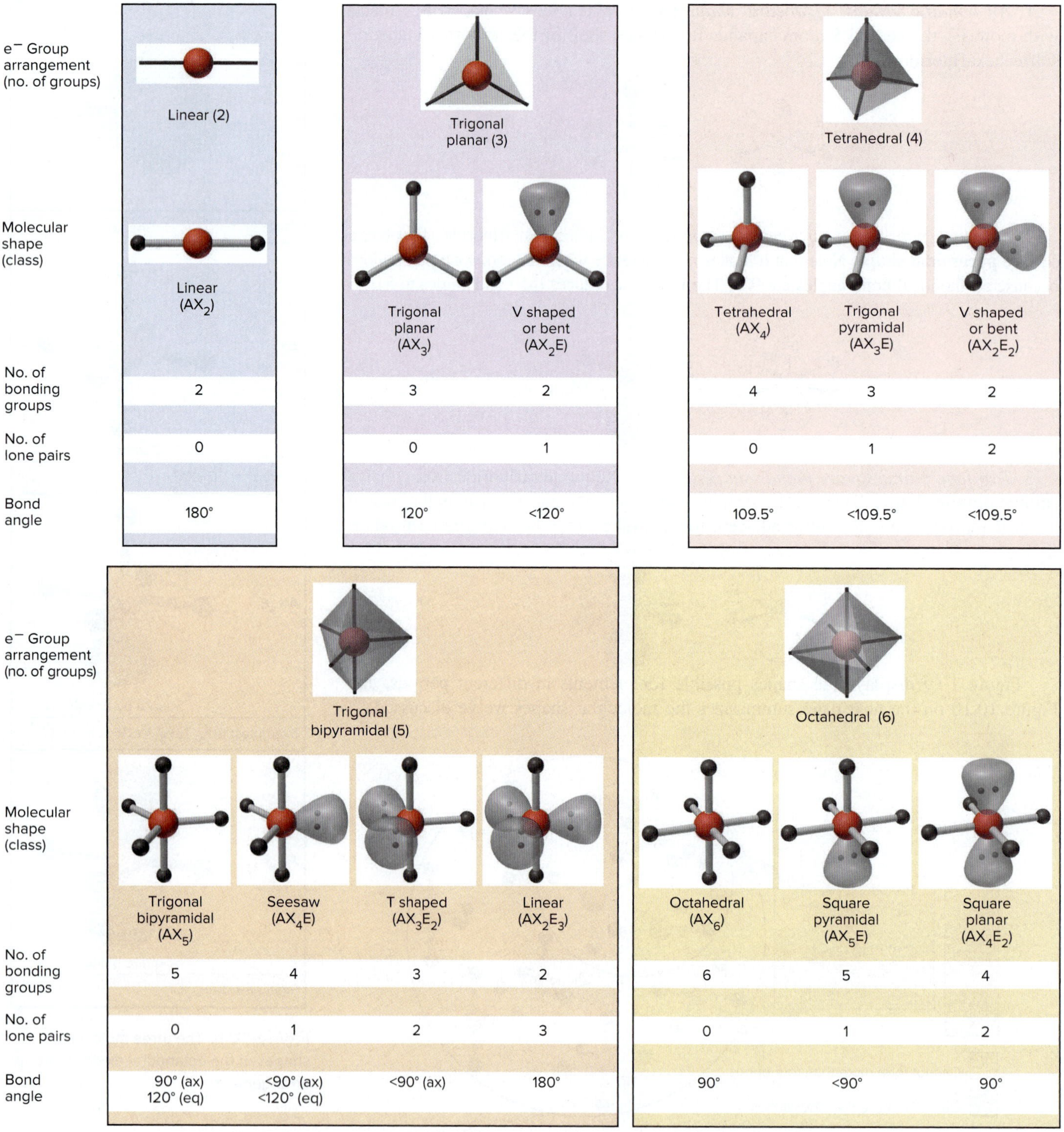

Figure 10.10 A summary of common molecular shapes with two to six electron groups around a central atom.

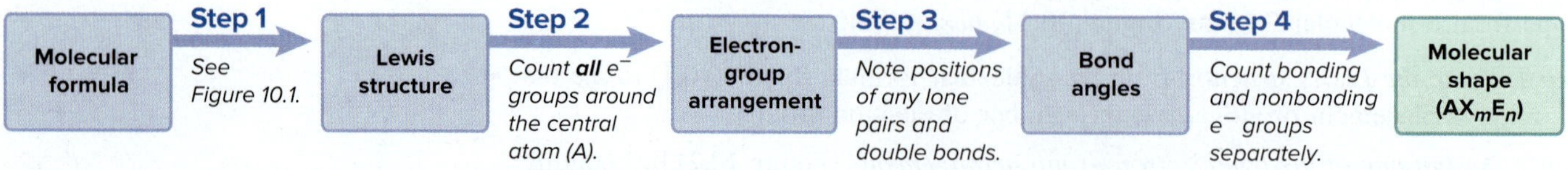

Figure 10.11 The four steps in converting a molecular formula to a molecular shape.

Step 3. Predict the ideal bond angle from the electron-group arrangement and *the effect of any deviation* caused by lone pairs or double bonds.

Step 4. Draw and name the molecular shape by counting bonding groups and nonbonding groups separately.

The next two sample problems apply these steps.

SAMPLE PROBLEM 10.6 Examining Shapes with Two, Three, or Four Electron Groups

Problem Draw the molecular shapes and predict the bond angles (relative to the ideal angles) of **(a)** PF_3 and **(b)** $COCl_2$.

Solution (a) For PF_3.

Step 1. Write the Lewis structure from the formula *(see below).*

Step 2. Assign the electron-group arrangement: Three bonding groups and one lone pair give four electron groups around P and the *tetrahedral arrangement.*

Step 3. Predict the bond angle: The ideal bond angle in a tetrahedron is 109.5°. There is one lone pair, so the actual bond angle will be less than 109.5°.

Step 4. Draw and name the molecular shape: With one lone pair, PF_3 has a trigonal pyramidal shape (AX_3E):

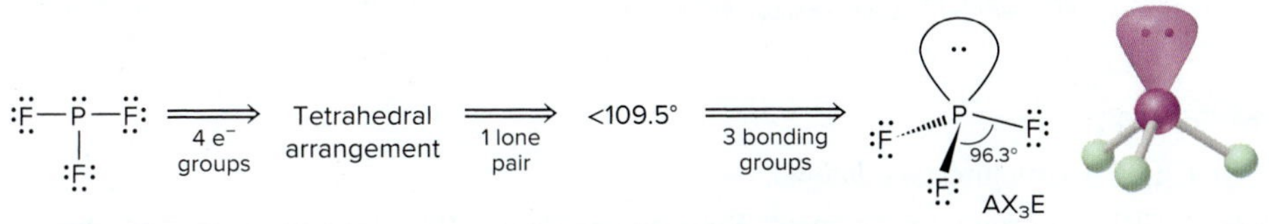

(b) For $COCl_2$.

Step 1. Write the Lewis structure from the formula *(see below).*

Step 2. Assign the electron-group arrangement: Two single bonds and one double bond give three electron groups around C and the *trigonal planar arrangement.*

Step 3. Predict the bond angles: The ideal bond angle in the trigonal planar arrangement is 120°, but the double bond between C and O will compress the Cl—C—Cl angle to less than 120°.

Step 4. Draw and name the molecular shape: With three electron groups and no lone pairs, $COCl_2$ has a trigonal planar shape (AX_3):

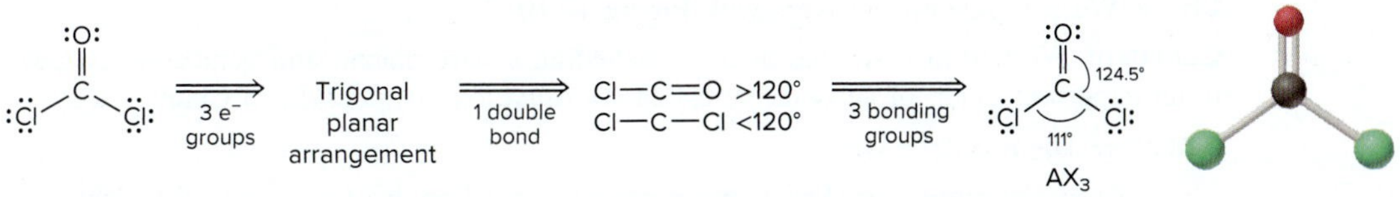

Check We compare the answers with the general information in Figure 10.10.

Comment Be sure the Lewis structure is correct because it determines the other steps. Notice that the electron-group arrangement and the molecular geometry are the same for $COCl_2$ since there are no lone pairs on the central C atom; because the central P atom in PF_3 has a lone pair, the electron-group arrangement and molecular geometry are different for this molecule.

FOLLOW-UP PROBLEMS

10.6A Draw the molecular shapes and predict the bond angles (relative to the ideal angles) of **(a)** CS_2; **(b)** $PbCl_2$; **(c)** CBr_4; **(d)** SF_2.

10.6B Name the electron-group arrangements, draw and name the molecular shapes, and predict the bond angles (relative to the ideal angles) of **(a)** BrO_2^-; **(b)** AsH_3; **(c)** N_3^-; **(d)** SeO_3.

SOME SIMILAR PROBLEMS 10.34–10.37

SAMPLE PROBLEM 10.7 **Examining Shapes with Five or Six Electron Groups**

Problem Draw the molecular shapes and predict the bond angles (relative to the ideal angles) of **(a)** SbF_5 and **(b)** BrF_5.

Plan We proceed as in Sample Problem 10.6, being sure to minimize the number of axial-equatorial repulsions.

Solution **(a)** For SbF_5.

Step 1. Lewis structure *(see below).*

Step 2. Electron-group arrangement: With five electron groups (five bonding pairs), this molecule has the *trigonal bipyramidal* arrangement.

Step 3. Bond angles: All the groups and surrounding atoms are identical, so the bond angles are ideal: 120° between equatorial groups and 90° between axial and equatorial groups.

Step 4. Molecular shape: Five electron groups and no lone pairs give the trigonal bipyramidal shape (AX_5):

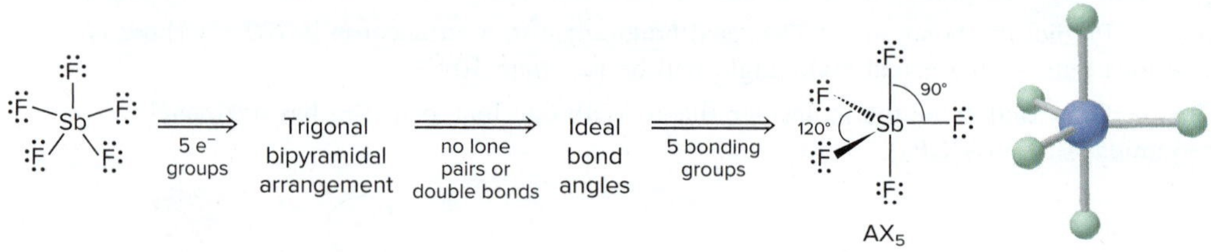

(b) For BrF_5.

Step 1. Lewis structure *(see below).*

Step 2. Electron-group arrangement: Six electron groups (five bonding pairs and one lone pair) give the *octahedral* arrangement.

Step 3. Bond angles: The lone pair will make all bond angles less than the ideal 90°.

Step 4. Molecular shape: With one lone pair, BrF_5 has the square pyramidal shape (AX_5E):

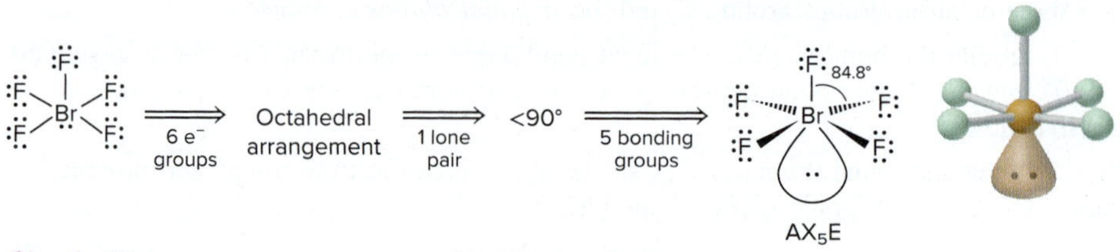

Check We compare our answers with Figure 10.10.

Comment We will also see the linear, tetrahedral, square planar, and octahedral shapes in an important group of substances, called *coordination compounds,* in Chapter 23.

FOLLOW-UP PROBLEMS

10.7A Draw the molecular shapes and predict the bond angles (relative to the ideal angles) of **(a)** ICl_2^-; **(b)** ClF_3; **(c)** SOF_4.

10.7B Draw the molecular shapes and predict the bond angles (relative to the ideal angles) of **(a)** BrF_4^-; **(b)** ClF_4^+; **(c)** PCl_6^-.

SOME SIMILAR PROBLEMS 10.40 and 10.41

Student Hot Spot

Student data indicate that you may struggle with determining molecular shapes and bond angles. Access the Smartbook to view additional Learning Resources on this topic.

Molecular Shapes with More Than One Central Atom

The shapes of molecules with more than one central atom are composites of the shapes around each of the atoms. Here are two examples:

1. Ethane (CH_3CH_3; molecular formula C_2H_6) is a component of natural gas. With four bonding groups and no lone pairs around the two central carbons, ethane is shaped like two overlapping tetrahedra (Figure 10.12A).
2. Ethanol (CH_3CH_2OH; molecular formula C_2H_6O), the intoxicating substance in beer, wine, and whiskey, has three central atoms (Figure 10.12B). The —CH_3 group is tetrahedrally shaped, and the —CH_2— group has four bonding groups around its

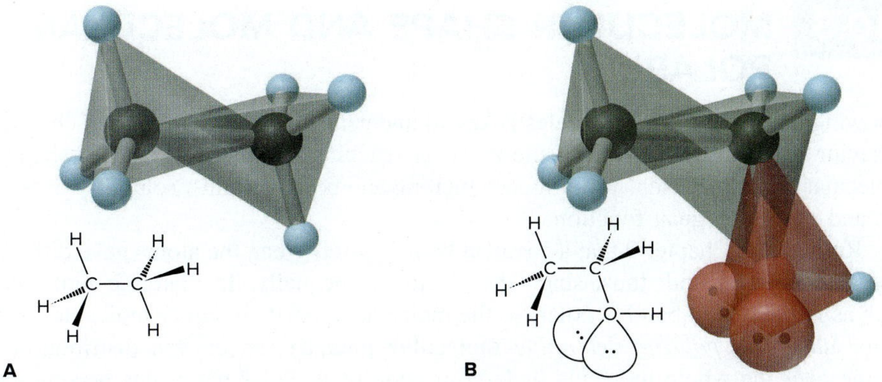

Figure 10.12 The tetrahedral shapes around the central atoms and the overall shapes of ethane (A) and ethanol (B).

central C atom, so it is also tetrahedrally shaped. The O atom has two bonding groups and two lone pairs around it, so the —OH group has a V shape (AX_2E_2).

SAMPLE PROBLEM 10.8 Predicting Molecular Shapes with More Than One Central Atom

Problem Determine the shape around each central atom in acetone, $(CH_3)_2CO$.

Plan There are three central C atoms, two of which are in CH_3— groups. We determine the shape around one central atom at a time.

Solution *Step 1.* Lewis structure *(see below).*

Step 2. Electron-group arrangement: Each CH_3— group has four electron groups around its central C, so its electron-group arrangement is *tetrahedral*. The third C atom has three electron groups around it, so it has the *trigonal planar arrangement.*

Step 3. Bond angles: The H—C—H angle in CH_3— should be near the ideal 109.5°. The C=O double bond will compress the C—C—C angle to less than the ideal 120°.

Step 4. Shapes around central atoms: With four electron groups and no lone pairs, the shape around C in each CH_3— is tetrahedral (AX_4). With three electron groups and no lone pairs, the shape around the middle C is trigonal planar (AX_3):

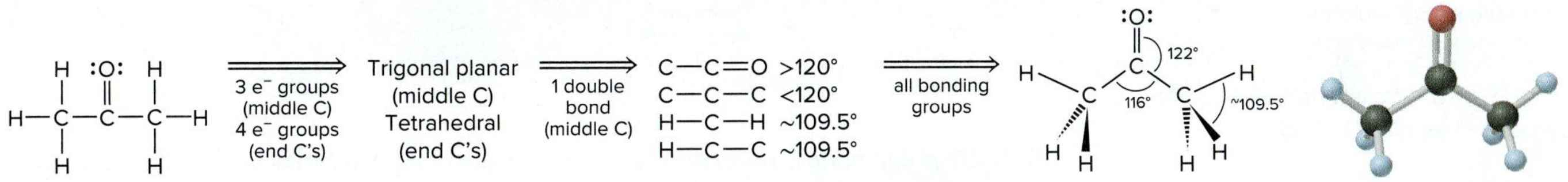

FOLLOW-UP PROBLEMS

10.8A Determine the shape around each central atom and predict any deviations from ideal bond angles in the following: **(a)** H_2SO_4; **(b)** propyne (C_3H_4; there is one C≡C bond); **(c)** S_2F_2.

10.8B Determine the shape around each central atom and predict any deviations from ideal bond angles in the following: **(a)** CH_3NH_2; **(b)** C_2Cl_4; **(c)** Cl_2O_7 ($O_3ClOClO_3$).

SOME SIMILAR PROBLEMS 10.42–10.45

› Summary of Section 10.2

› VSEPR theory proposes that each electron group (single bond, multiple bond, lone pair, or lone electron) around a central atom remains as far from the others as possible.

› Five electron-group arrangements are possible when two, three, four, five, or six electron groups surround a central atom. Each arrangement is associated with one or more molecular shapes, depending on the numbers of bonding and lone pairs.

› Ideal bond angles are based on the regular geometric arrangements. Deviations from them occur when surrounding atoms and/or electron groups are not identical.

› Lone pairs and double bonds exert stronger repulsions on other electron groups than single bonds do.

› Shapes of larger molecules are composites of the shapes around each central atom.

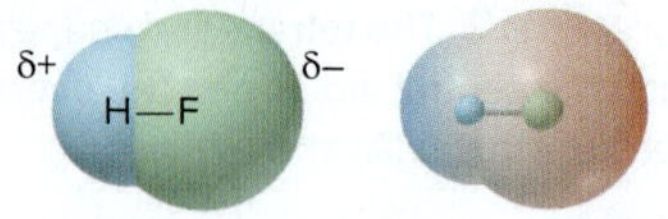

Space-filling model of HF *Electron density model of HF*

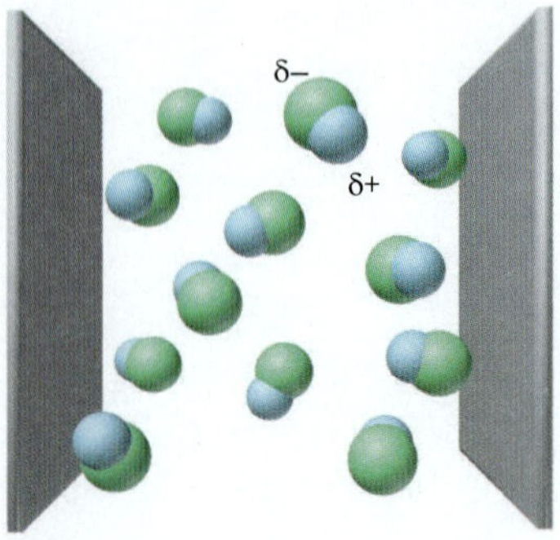

Electric field off: *HF molecules are oriented randomly.*

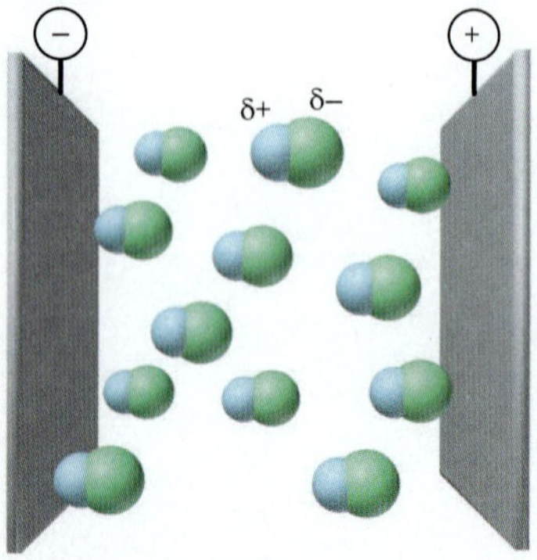

Electric field on: *HF molecules are oriented with their partially charged ends toward the oppositely charged plates.*

Figure 10.13 The orientation of polar molecules in an electric field.

10.3 MOLECULAR SHAPE AND MOLECULAR POLARITY

Knowing the shape of its molecules is key to understanding the physical and chemical behavior of a substance. One of the most far-reaching effects of molecular shape is molecular polarity, which can influence melting and boiling points, solubility, reactivity, and even biological function.

Recall from Chapter 9 that a covalent bond is *polar* when the atoms have different electronegativities and, thus, share the electrons unequally. In diatomic molecules, such as HF, the only bond is polar, so the molecule is polar. In larger molecules, *both shape and bond polarity determine* **molecular polarity,** an uneven distribution of charge over the whole molecule or large portion of it. Polar molecules become oriented in an electric field with their partially charged ends pointing toward the oppositely charged plates (Figure 10.13). **Dipole moment (μ)** is a measure of molecular polarity, given in the unit called a *debye* (D),* which is derived from SI units of charge (coulomb, C) and length (m): $1\ D = 3.34{\times}10^{-30}\ C{\cdot}m$.

Bond Polarity, Bond Angle, and Dipole Moment

The presence of polar bonds does not *always* result in a polar molecule; we must also consider shape and the atoms surrounding the central atom. Here are three cases:

1. *CO_2: polar bonds, nonpolar molecule.* In carbon dioxide, the electronegativity difference between C (EN = 2.5) and O (EN = 3.5) makes each C═O bond polar, with electron density pulled from C toward O. But CO_2 is linear, so the bonds point 180° from each other. Since the two bonds have polarities that are equal in magnitude but opposite in direction, their polarities cancel, resulting in *no net dipole moment* for the molecule (μ = 0 D), just as two numbers of equal magnitude but opposite sign (for example, +1 and −1) add to 0. The electron density model shows regions of high negative charge *(red)* distributed equally on either side of the central region of high positive charge *(blue):*

2. *H_2O: polar bonds, polar molecule.* Water also has two polar bonds, but it *is* polar (μ = 1.85 D). In each O—H bond, electron density is pulled from H (EN = 2.1) toward O (EN = 3.5). Bond polarities do *not* cancel because the molecule is V shaped (see also Figure 4.1). The bond polarities are partially reinforced, making the O end partially negative and the other end (the region between the H atoms) partially positive:

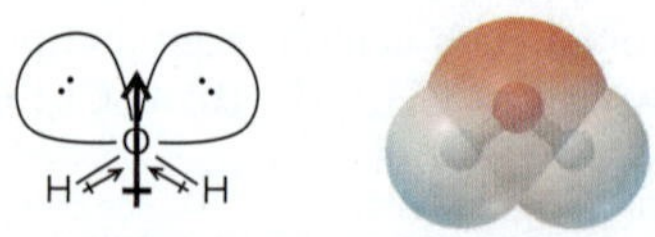

(The molecular polarity of water has some amazing effects, from determining the composition of the oceans to supporting life itself, as you'll see in Chapter 12.)

3. *Same shapes, different polarities.* When different molecules have the same shape, the identities of the surrounding atoms affect polarity. Carbon tetrachloride (CCl_4) and chloroform ($CHCl_3$) are tetrahedral molecules with very different polarities. In CCl_4, all the surrounding atoms are Cl atoms. Each C—Cl bond is equally polar (ΔEN = 0.5), but the molecule is nonpolar (μ = 0 D) because the bond polarities cancel each other. In $CHCl_3$, an H replaces one Cl. The polarity of the H—C bond differs in magnitude (ΔEN = 0.4) and direction from that of the three C—Cl

*The unit is named for the Dutch-American scientist Peter Debye (1884–1966), who won the Nobel Prize in chemistry in 1936 for contributions to the fields of molecular structure and solution behavior.

bonds; thus, it disrupts the balance and gives chloroform a significant dipole moment (μ = 1.01 D):

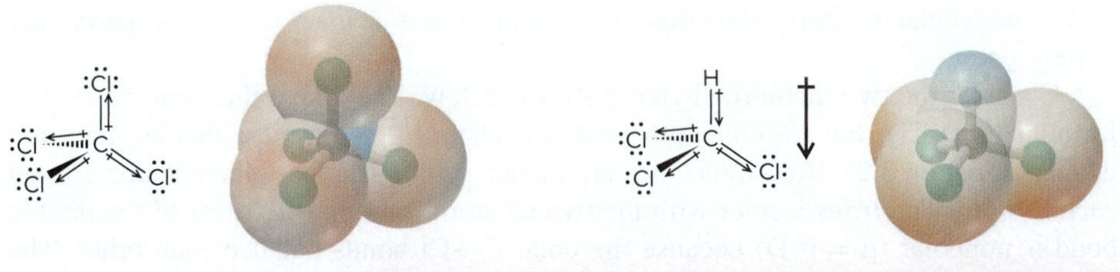

Student Hot Spot

Student data indicate that you may struggle with determining molecular polarity. Access the SmartBook to view additional Learning Resources on this topic.

SAMPLE PROBLEM 10.9 Predicting the Polarity of Molecules

Problem For each of the following, use the molecular shape and EN values and trends (Figure 9.21) to predict the direction of bond and molecular polarity, if present: **(a)** ammonia, NH_3; **(b)** boron trifluoride, BF_3; **(c)** carbonyl sulfide, COS (atom sequence SCO).

Plan We draw and name the molecular shape and point a polar arrow toward the atom with higher EN in each bond. If the bond polarities balance one another, the molecule is nonpolar; if they reinforce each other, we show the direction of the molecular polarity.

Solution **(a)** For NH_3. The molecular shape is trigonal pyramidal. N (EN = 3.0) is more electronegative than H (EN = 2.1), so the bond polarities point toward N and partially reinforce each other; thus the molecular polarity points toward N:

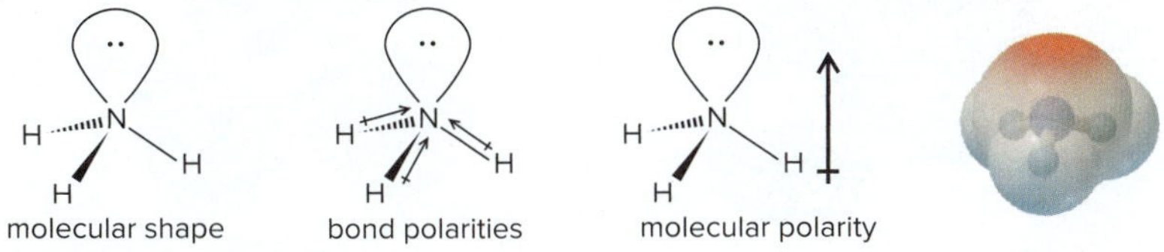

Therefore, ammonia is polar.

(b) For BF_3. The molecular shape is trigonal planar. F (EN = 4.0) is farther to the right in Period 2 than B (EN = 2.0), so it is more electronegative; thus, each bond polarity points toward F. However, the bond angle is 120°, so the three bond polarities balance each other, and BF_3 has no molecular polarity:

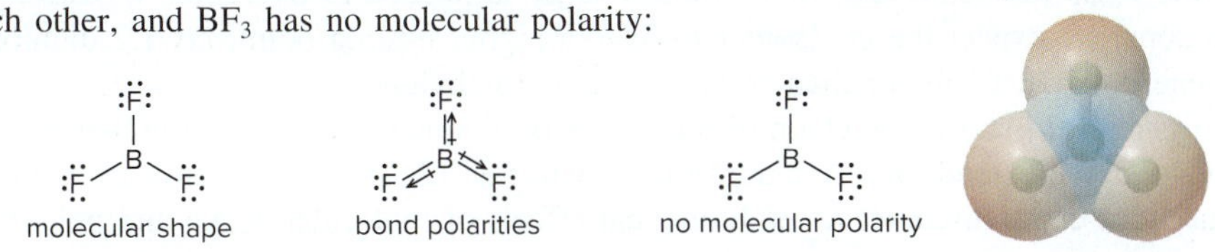

Therefore, boron trifluoride is nonpolar.

(c) For COS. The molecular shape is linear. With C and S having the same EN, the C=S bond is nonpolar, but the C=O bond is quite polar (ΔEN = 1.0), so there is a net molecular polarity toward the O:

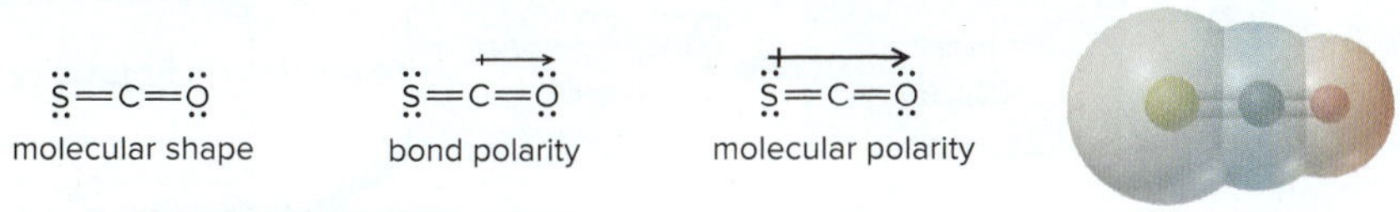

Therefore, carbonyl sulfide is polar.

Check The electron density models confirm our conclusions. Note that, in (b), the negative *(red)* regions surround the central B *(blue)* symmetrically.

FOLLOW-UP PROBLEMS

10.9A Show the bond polarities and molecular polarity, if any, for each of the following: **(a)** dichloromethane (CH_2Cl_2); **(b)** iodine oxide pentafluoride (IOF_5); **(c)** iodine pentafluoride (IF_5).

10.9B Show the bond polarities and molecular polarity, if any, for each of the following: **(a)** xenon tetrafluoride (XeF_4); **(b)** chlorine trifluoride (ClF_3); **(c)** sulfur monoxide tetrafluoride (SOF_4).

SOME SIMILAR PROBLEMS 10.55–10.58

The Effect of Molecular Polarity on Behavior

Earlier we mentioned that molecular polarity influences physical behavior. Let's see how a molecular property like dipole moment affects a macroscopic property like boiling point.

Consider the two dichloroethylenes shown below. They have the *same* molecular formula ($C_2H_2Cl_2$) but *different* physical and chemical properties; that is, they are isomers (Section 3.2). Both molecules are planar, with a trigonal planar shape around each C atom. The *trans* isomer with the two Cl atoms on opposite sides of the double bond is nonpolar (μ = 0 D) because the polar C—Cl bonds balance each other. The *cis* isomer with the Cl atoms on the same side of the double bond is polar (μ = 1.90 D) because the bond polarities partially reinforce each other, with the molecular polarity pointing between the Cl atoms.

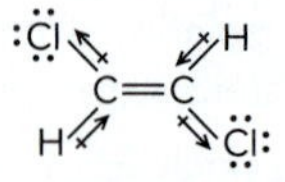

trans

μ = 0 D
Boiling point = 47.5°C

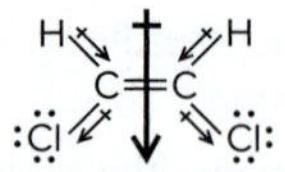

cis

μ = 1.90 D
Boiling point = 60.3°C

A liquid boils when it forms bubbles against the atmospheric pressure. To enter the bubble, molecules in the liquid must overcome the weak attractive forces *between* them. Because of their polarity, the *cis* molecules attract each other more strongly than the *trans* molecules do. Since more energy is needed to overcome the stronger attractions, we expect the *cis* isomer to have a higher boiling point: *cis*-1,2-dichloroethylene boils 13°C higher than *trans*-1,2-dichloroethylene.

Figure 10.14 shows the chain of influences of atomic properties on the behavior of substances. We extend these relationships in Chapter 12, and the upcoming Chemical Connections essay discusses some biological effects of molecular shape and polarity.

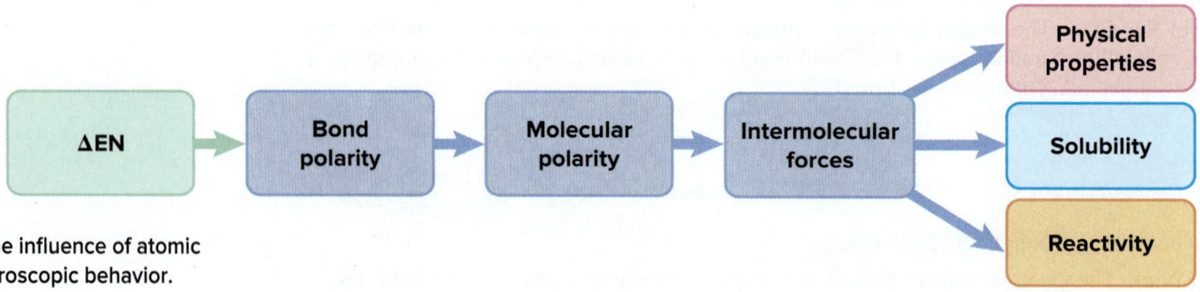

Figure 10.14 The influence of atomic properties on macroscopic behavior.

› Summary of Section 10.3

- Bond polarity and molecular shape determine molecular polarity, which is measured as a dipole moment.
- A molecule with polar bonds is not necessarily a polar molecule. When bond polarities cancel each other, the molecule is nonpolar; when they reinforce each other, the molecule is polar.
- Molecular shape and polarity can affect physical properties, such as boiling point, and they play a central role in biological function.

CHEMICAL CONNECTIONS TO SENSORY PHYSIOLOGY

Molecular Shape, Biological Receptors, and the Sense of Smell

A biological cell can be thought of as a membrane-bound sack filled with an aqueous fluid containing many molecules of various shapes. Many complex processes begin when a molecular "key" fits into a molecular "lock" with a complementary shape. Typically, the key is a small molecule circulating in cellular or other bodily fluid, and the lock, the *biological receptor,* is a large molecule often embedded in a membrane. The *receptor site* is a small region of the receptor with a shape that matches the molecular key. Thousands of molecules per second collide with the receptor site, but when one with the correct shape does, the receptor site "grabs" it through intermolecular attractions, and the biological response begins.

Molecular Shape and the Sense of Smell

Molecular shapes fitting together is crucial to the sense of smell (olfaction). To have an odor, a substance must be a gas or a volatile liquid or solid and must be at least slightly soluble in the aqueous film of the nasal passages. And the odorous molecule, or a portion of it, must fit into one of the receptor sites on nerve endings in this area. When this happens, nerve impulses travel to the brain, which interprets them as a specific odor.

Over half a century ago, a new theory proposed that molecular shape (and sometimes polarity), but *not* composition, determines odor. It stated that any molecule producing one of seven primary odors—camphor-like, musky, floral, minty, ethereal, pungent, and putrid—matches a receptor site of a particular shape. Figure B10.1 shows three of the seven sites holding a molecule with that odor.

Several predictions of the theory proved to be correct:

- If different substances fit a given receptor, they have the same odor. The four molecules in Figure B10.2 fit the camphor-like receptor and smell like moth repellent.
- If different parts of a molecule fit different receptors, it has a mixed odor. Portions of benzaldehyde fit the camphor-like, floral, and minty receptors, and it smells like almonds; other molecules with this odor fit the same three receptors.

But, other predictions were not confirmed. For example, an odor predicted from the shape was often not the odor smelled. The reason is that molecules may not have the same shape in solution at the receptor as they have in the gas phase.

Evidence from the 1990s suggests that the original model is far too simple. The 2004 Nobel Prize in medicine or physiology was given to Richard Axel and Linda B. Buck for showing that olfaction involves about 1000 different receptors and molecules fitting various combinations of them produce the more than 10,000 odors humans can distinguish.

Thus, although the process is much more complex than originally thought, the central idea that odor depends on molecular shape is valid and is being actively researched in the food, cosmetics, and insecticide industries.

Biological Significance of Molecular Shape

Countless other examples show that no chemical property is more crucial to living systems than molecular shape. Here are just a few of the many biochemical processes controlled by one molecule fitting into a receptor site on another:

- Enzymes are proteins that bind cellular reactants and speed their reaction. An early step in energy metabolism involves glucose binding to the "active" site of the enzyme hexokinase (Figure B10.3, *next page*).
- Nerve impulses are transmitted when small molecules released from one nerve fit into receptors on the next. Mind-altering drugs act by chemically mimicking the molecular fit at such nerve receptors in the brain.
- One type of immune response is triggered when a molecule on a bacterial surface binds to the receptors on "killer" cells in the bloodstream.
- Hormones regulate processes by fitting into and activating specific receptors on target tissues and organs.
- Genes function when certain nucleic acid molecules fit into specific regions of others.

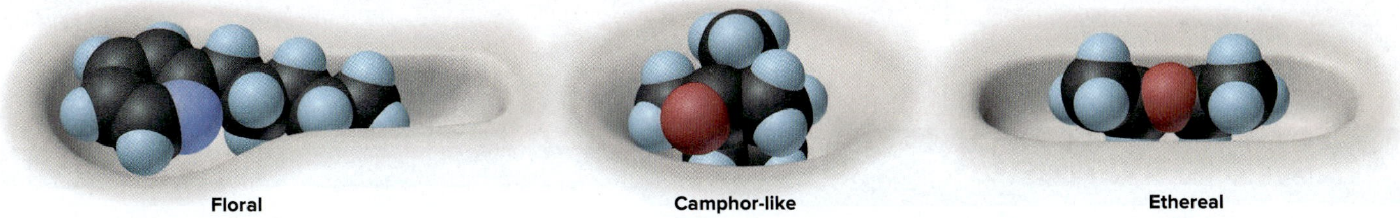

Figure B10.1 Shapes of some olfactory receptor sites.

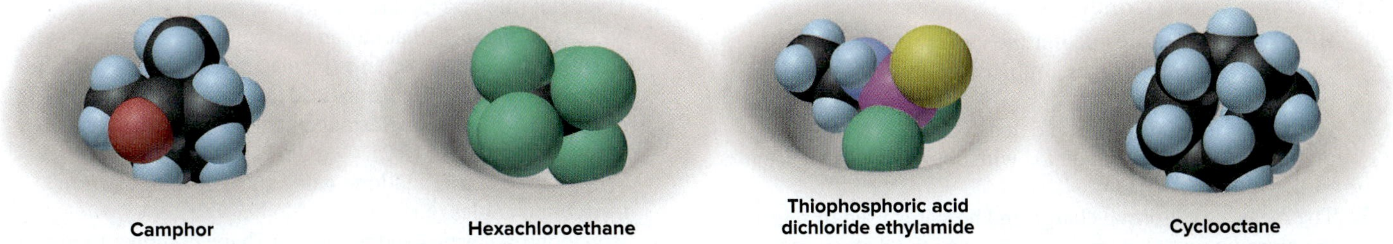

Figure B10.2 Different molecules with the same odor.

(continued)

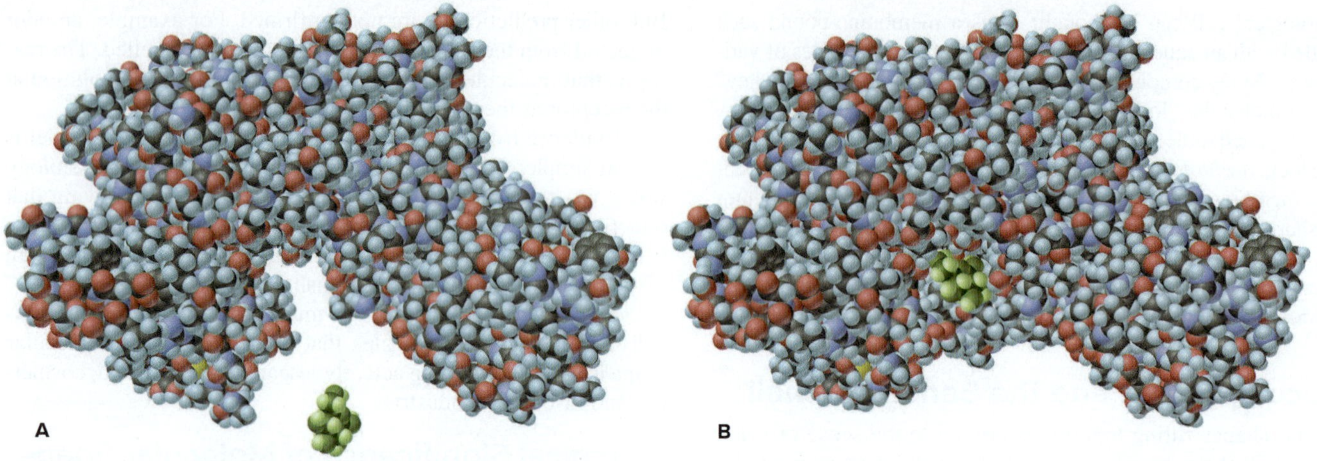

Figure B10.3 **Molecular shape and enzyme action. A,** A small sugar molecule *(bottom)* is shown near a specific region of an enzyme molecule. **B,** When the sugar lands in that region, the reaction begins.

Problems

B10.1 As you'll learn in Chapter 11, groups joined by a single bond rotate freely around the bond, but those joined by double bonds don't. Peptide bonds make up a major portion of protein chains. Determine the molecular shapes around the central C and N atoms in the two resonance forms of the peptide bond:

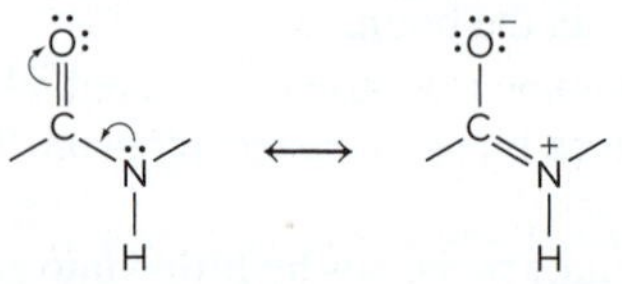

B10.2 Lewis structures of mescaline, a hallucinogenic compound in peyote cactus, and dopamine, a neurotransmitter in the mammalian brain, are shown below. Suggest a reason for mescaline's ability to disrupt nerve impulses.

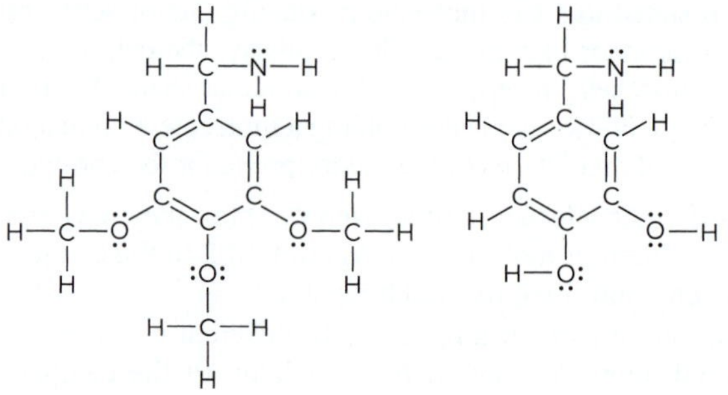

CHAPTER REVIEW GUIDE

Learning Objectives

Relevant section (§) and/or sample problem (SP) numbers appear in parentheses.

Understand These Concepts

1. How Lewis structures depict the atoms, bonding pairs, and lone electron pairs in a molecule or polyatomic ion (§10.1)
2. How resonance and electron delocalization explain bond properties in many compounds with double bonds adjacent to single bonds (§10.1)
3. The meaning of formal charge and how it is used to select the more important resonance structure; the difference between formal charge and oxidation number (§10.1)
4. The octet rule and its three major exceptions—molecules with a central atom that has an electron deficiency, an odd number of electrons, or an expanded valence shell (§10.1)
5. How electron-group repulsions lead to molecular shapes (§10.2)
6. The five electron-group arrangements and their associated molecular shapes (§10.2)
7. Why double bonds and lone pairs cause deviations from ideal bond angles (§10.2)
8. How bond polarities and molecular shape combine to give a molecule polarity (§10.3)

Learning Objectives (continued)

Relevant section (§) and/or sample problem (SP) numbers appear in parentheses.

Master These Skills

1. Using a stepwise method for writing a Lewis structure from a molecular formula (SPs 10.1–10.3, 10.5)
2. Writing resonance structures for molecules and ions (SP 10.4)
3. Calculating the formal charge of any atom in a molecule or ion (SPs 10.4, 10.5)
4. Predicting molecular shapes from Lewis structures (SPs 10.6–10.8)
5. Using molecular shape and electronegativity values to predict the polarity of a molecule (SP 10.9)

Key Terms

Page numbers appear in parentheses.

axial group (421)
bent shape (V shape) (419)
bond angle (418)
dipole moment (μ) (428)
electron deficient (413)
electron-pair delocalization (410)
equatorial group (421)
expanded valence shell (414)
formal charge (411)
free radical (414)
Lewis structure (Lewis formula) (405)
linear arrangement (419)
linear shape (419)
molecular polarity (428)
molecular shape (417)
octahedral arrangement (422)
resonance hybrid (410)
resonance structure (resonance form) (410)
seesaw shape (422)
square planar shape (423)
square pyramidal shape (423)
T shape (422)
tetrahedral arrangement (420)
trigonal bipyramidal arrangement (421)
trigonal planar arrangement (419)
trigonal pyramidal shape (420)
valence-shell electron-pair repulsion (VSEPR) theory (417)

Key Equations and Relationships

Page numbers appear in parentheses.

10.1 Calculating the formal charge on an atom (411):

Formal charge of atom
$$= \text{no. of valence } e^- - (\text{no. of unshared valence } e^- + \tfrac{1}{2}\text{ no. of shared valence } e^-)$$

10.2 Ranking the effect of electron-pair repulsions on bond angle (421):

Lone pair–lone pair > lone pair–bonding pair > bonding pair–bonding pair

BRIEF SOLUTIONS TO FOLLOW-UP PROBLEMS

10.1A (a) H_2S has $[2 \times H(1e^-)] + [1 \times S(6e^-)] = 8$ valence e^-. Draw two bonds: 2 bonds $\times\ 2e^- = 4e^-$, with $8e^- - 4e^- = 4e^-$ remaining to complete the octet of S.

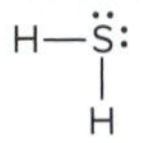

(b) $AlCl_4^-$ has $[1 \times Al(3e^-)] + [4 \times Cl(7e^-)] + [\text{charge } 1e^-] = 32$ valence e^-. Draw four bonds: 4 bonds $\times\ 2e^- = 8e^-$, with $32e^- - 8e^- = 24e^-$ remaining to complete the octets of the Cl atoms (Al already has an octet). Since this is an ion, place the structure in square brackets with the charge as a superscript.

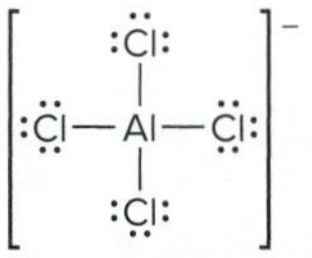

(c) $SOCl_2$ has $[1 \times S(6e^-)] + [1 \times O(6e^-)] + [2 \times Cl(7e^-)] = 26$ valence e^-. Place S in the center since it has the lowest electronegativity value. Draw three bonds: 3 bonds $\times\ 2e^- = 6e^-$, with $26e^- - 6e^- = 20e^-$ remaining to complete the octets of all atoms.

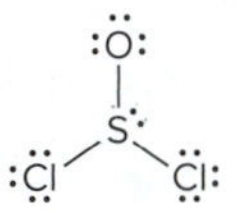

10.1B (a) OF_2 has $[1 \times O(6e^-)] + [2 \times F(7e^-)] = 20$ valence e^-. Draw two bonds: 2 bonds $\times\ 2e^- = 4e^-$, with $20e^- - 4e^- = 16e^-$ remaining to complete the octets of the O and F atoms.

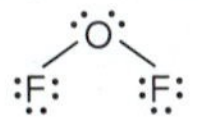

(b) CH_2Br_2 has $[1 \times C(4e^-)] + [2 \times H(1e^-)] + [2 \times Br(7e^-)] = 20$ valence e^-. Place C in the center since it has a lower electronegativity than Br, and H cannot be a central atom. Draw four bonds: 4 bonds $\times\ 2e^- = 8e^-$, with $20e^- - 8e^- = 12e^-$ remaining to complete the octets of the Br atoms.

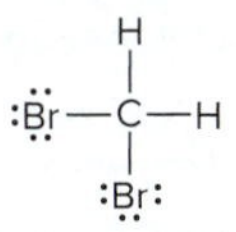

(c) IBr_2^+ has $[1 \times I(7e^-)] + [2 \times Br(7e^-)] - [\text{charge } 1e^-] = 20$ valence e^-. Draw two bonds: 2 bonds $\times\ 2e^- = 4e^-$, with $20e^- - 4e^- = 16e^-$ remaining to complete the octets of all atoms. Since this is an ion, place the structure in square brackets with the charge as a superscript.

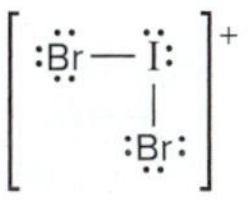

BRIEF SOLUTIONS TO FOLLOW-UP PROBLEMS (continued)

10.2A (a) H atoms can have only one bond; O should have two bonds and N should have three bonds, so the atoms are arranged as shown below. NH_3O has $[1 \times N(5e^-)] + [3 \times H(1e^-)] + [1 \times O(6e^-)] = 14$ valence e^-. Draw four bonds: 4 bonds $\times$ $2e^- = 8e^-$, with $14e^- - 8e^- = 6e^-$ remaining to complete the octets of the O and N atoms.

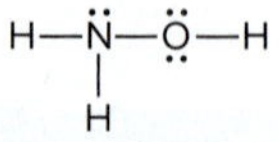

(b) H atoms can have only one bond; O should have two bonds and C should have four bonds; there are no O—H bonds, so the atoms are arranged as shown. C_2H_6O has $[2 \times C(4e^-)] + [6 \times H(1e^-)] + [1 \times O(6e^-)] = 20$ valence e^-. Draw eight bonds: 8 bonds $\times$ $2e^- = 16e^-$, with $20e^- - 16e^- = 4e^-$ remaining to complete the octet of the O atom (the C atoms already have octets).

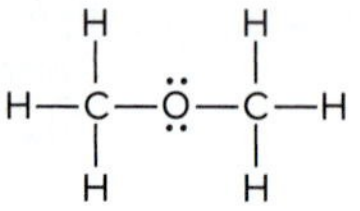

10.2B (a) H atoms can have only one bond, so the atoms are arranged as shown below. N_2H_4 has $[2 \times N(5e^-)] + [4 \times H(1e^-)] = 14$ valence e^-. Draw five bonds: 5 bonds $\times$ $2e^- = 10e^-$, with $14e^- - 10e^- = 4e^-$ remaining to complete the octets of the N atoms.

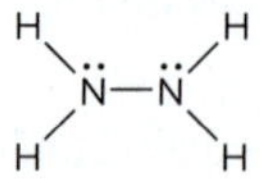

(b) H atoms can have only one bond; N should have three bonds and C should have four bonds, so the atoms are arranged as shown below. CH_3NH_2 has $[1 \times C(4e^-)] + [5 \times H(1e^-)] + [1 \times N(5e^-)] = 14$ valence e^-. Draw six bonds: 6 bonds $\times$ $2e^- = 12e^-$, with $14e^- - 12e^- = 2e^-$ remaining to complete the octet of the N atom (the C atom already has an octet).

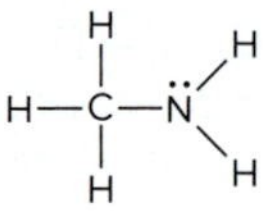

10.3A (a) Using the 10 valence e^- in CO gives a Lewis structure in which C does not have an octet; two lone pairs on O are changed to bonding pairs:

:C≡O:

(b) Using the 10 valence e^- in HCN gives a Lewis structure in which C does not have an octet; two lone pairs on N are changed to bonding pairs:

H—C≡N:

(c) Using the 16 valence e^- in CO_2 gives a Lewis structure in which C does not have an octet; one lone pair on each O atom is changed to a bonding pair:

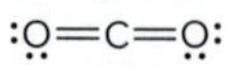

10.3B (a) Using the 10 valence e^- in NO^+ gives a Lewis structure in which an atom does not have an octet; two lone pairs are changed to bonding pairs:

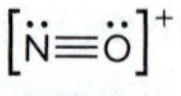

(b) Using the 12 valence e^- in H_2CO gives a Lewis structure in which C does not have an octet; one lone pair on O is changed to a bonding pair:

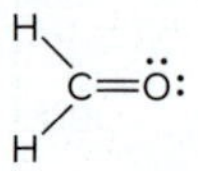

(c) Using the 12 valence e^- in N_2H_2 gives a Lewis structure in which one N atom does not have an octet; one lone pair on the other N atom is changed to a bonding pair:

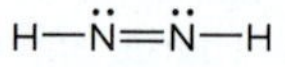

10.4A After distributing the 24 valence e^-, the N atom does not have an octet. A lone pair on either O atom is changed to a bonding pair to give N an octet.

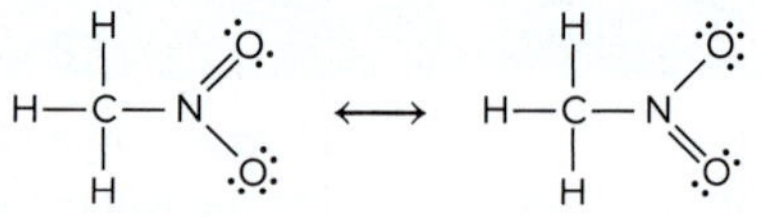

Formal charge of atom = no. of valence e^- − (no. of unshared valence e^- + $\frac{1}{2}$ no. of shared valence e^-)

N $[5 - 0 - \frac{1}{2}(8)] = +1$; C $[4 - 0 - \frac{1}{2}(8)] = 0$; H $[1 - 0 - \frac{1}{2}(2)] = 0$;
O (single-bonded) $[6 - 6 - \frac{1}{2}(2)] = -1$;
O (double-bonded) $[6 - 4 - \frac{1}{2}(4)] = 0$

10.4B After distributing the 16 valence e^-, C has only two bonds (four e^-); lone pairs on either S, N, or both are changed to bonding pairs. Formal charges:

Structure I	Structure II	Structure III
S $[6 - 4 - \frac{1}{2}(4)] = 0$	S $[6 - 6 - \frac{1}{2}(2)] = -1$	S $[6 - 2 - \frac{1}{2}(6)] = +1$
C $[4 - 0 - \frac{1}{2}(8)] = 0$	C $[4 - 0 - \frac{1}{2}(8)] = 0$	C $[4 - 0 - \frac{1}{2}(8)] = 0$
N $[5 - 4 - \frac{1}{2}(4)] = -1$	N $[5 - 2 - \frac{1}{2}(6)] = 0$	N $[5 - 6 - \frac{1}{2}(2)] = -2$

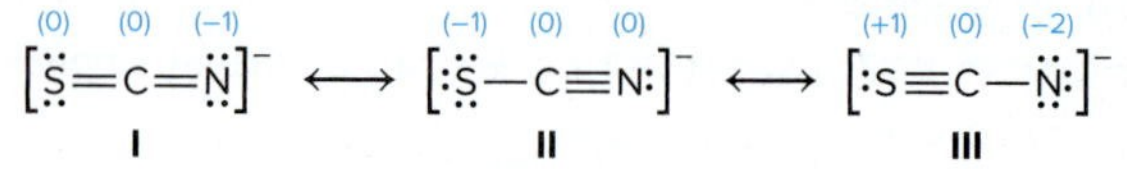

Structure I is the most important resonance structure; the formal charges are lower than those in structure III, and the negative formal charge is on the more electronegative N atom, while the negative formal charge in structure II is on the less electronegative S atom.

10.5A (a) A structure in which all atoms have an octet can be drawn, but minimal formal charges are achieved by changing a lone pair on O to a bonding pair, giving P an expanded octet.

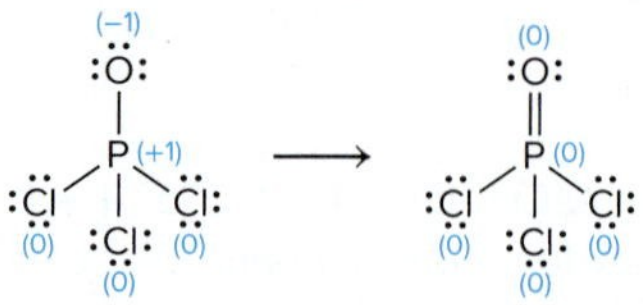

(b) A structure in which all atoms have an octet can be drawn, but minimal formal charges are achieved by changing a lone pair on each O to a bonding pair and giving Cl the unpaired electron.

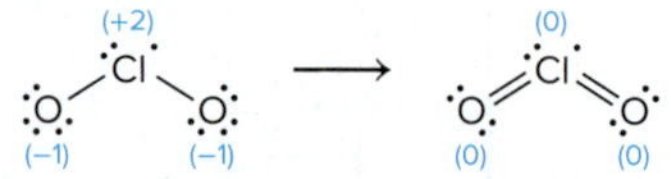

(c) The ion has 36 valence electrons, 32 of which are used in the four single bonds and completing the octets of the four Br atoms. The remaining 4 valence electrons are placed on the central I atom, resulting in an expanded valence shell.

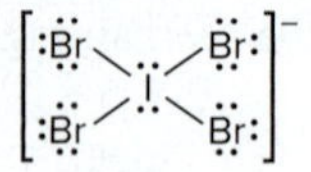

10.5B (a) Be has only four electrons surrounding it so the molecule is electron deficient:

H—Be—H

(b) The ion has 22 valence electrons, 20 of which are used in the two single bonds and completing the octets of the I atoms. The remaining 2 valence electrons are placed on the central I atom, resulting in an expanded valence shell.

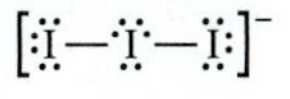

(c) A structure in which all atoms have an octet can be drawn, but minimal formal charges are achieved by changing a lone pair on each O to a bonding pair, giving Xe an expanded octet.

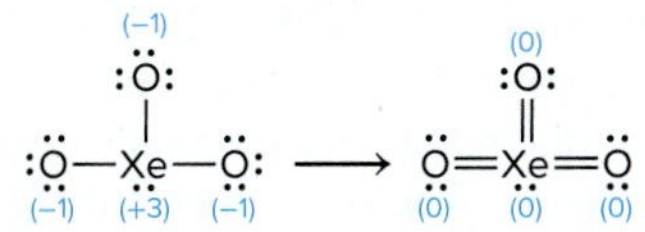

10.6A (a) There are two electron groups with a linear electron-group arrangement. There are no lone pairs, so the molecular shape is linear, with a bond angle of 180°.

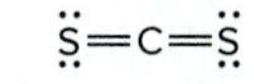

(b) There are three electron groups (two bonding pairs and one lone pair) with a trigonal planar electron-group arrangement. Since there is one lone pair, the molecular shape is bent (V shaped), with a bond angle < 120°.

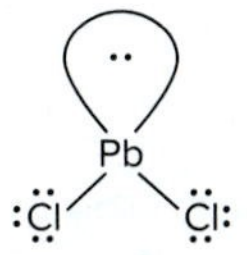

(c) There are four electron groups with a tetrahedral electron-group arrangement. Since there are no lone pairs, the molecular shape is tetrahedral with bond angles of 109.5°.

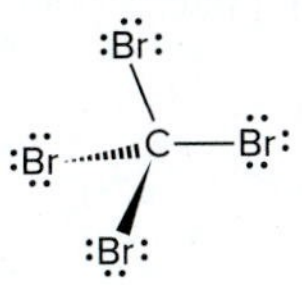

(d) There are four electron groups with a tetrahedral electron-group arrangement. Since there are two lone pairs, the molecular shape is bent (V shaped) with a bond angle < 109.5°.

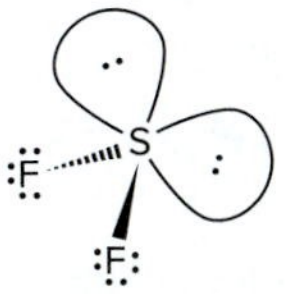

10.6B (a) There are four electron groups with a tetrahedral electron-group arrangement. Since there are two lone pairs, the molecular shape is bent (V shaped) with a bond angle < 109.5°

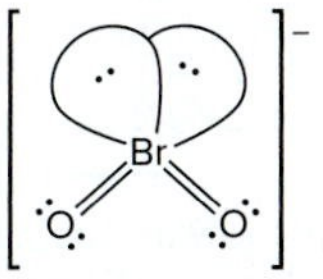

(b) There are four electron groups with a tetrahedral electron-group arrangement. Since there is one lone pair, the molecular shape is trigonal pyramidal with bond angles < 109.5°.

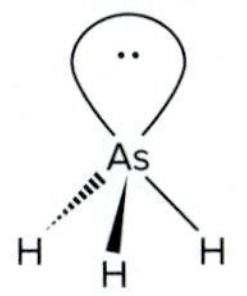

(c) There are two electron groups with a linear electron-group arrangement. There are no lone pairs, so the molecular shape is linear, with a bond angle of 180°.

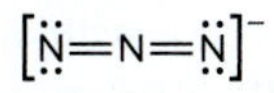

(d) There are three electron groups with a trigonal planar electron-group arrangement. There are no lone pairs, so the molecular shape is trigonal planar, with bond angles of 120°.

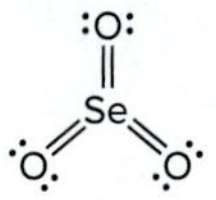

10.7A (a) There are five electron groups with a trigonal bipyramidal electron-group arrangement. Since there are three lone pairs, the molecular shape is linear with a bond angle of 180°. The three lone pairs occupy equatorial positions to minimize repulsions.

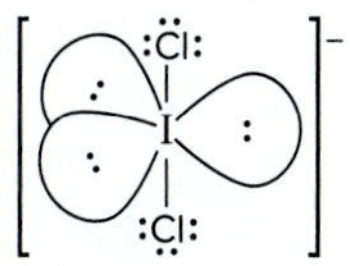

(b) There are five electron groups with a trigonal bipyramidal electron-group arrangement. Since there are two lone pairs, the molecular shape is T shaped with bond angles < 90°. The two lone pairs occupy equatorial positions to minimize repulsions.

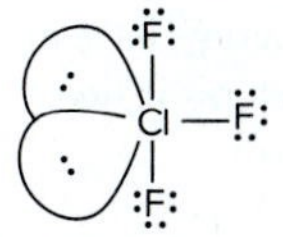

(c) There are five electron groups with a trigonal bipyramidal electron-group arrangement. Since there are no lone pairs, the molecular shape is trigonal bipyramidal. The double bond reduces the bond angles: F_{eq}—S—F_{eq} angle < 120°; F_{ax}—S—F_{eq} angle < 90°.

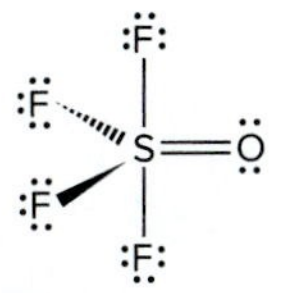

10.7B (a) There are six electron groups with an octahedral electron-group arrangement. Since there are two lone pairs, the molecular shape is square planar with bond angles of 90°.

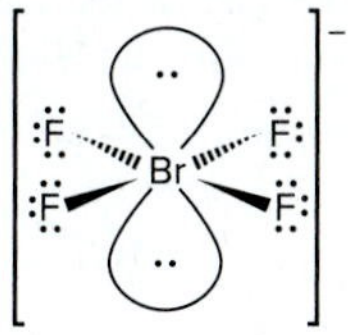

(b) There are five electron groups with a trigonal bipyramidal electron-group arrangement. Since there is one lone pair, the molecular shape is seesaw. The lone pair reduces the bond angles: F_{eq}—Cl—F_{eq} angle < 120°; F_{ax}—Cl—F_{eq} angle < 90°. The lone pair occupies an equatorial position to minimize repulsions.

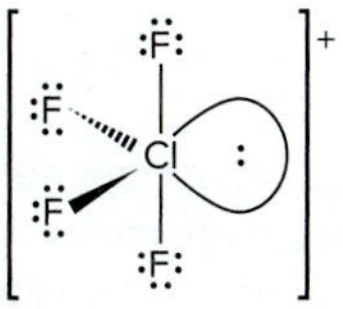

BRIEF SOLUTIONS TO FOLLOW-UP PROBLEMS (continued)

(c) There are six electron groups with an octahedral electron-group arrangement. Since there are no lone pairs, the molecular shape is octahedral with bond angles of 90°.

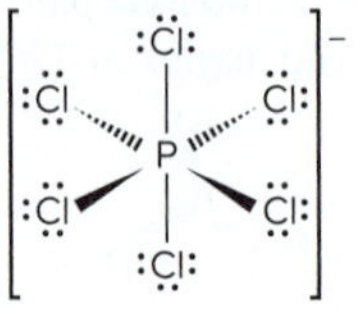

10.8A (a) S has four electron groups, and the shape around it is tetrahedral; double bonds compress the O—S—O angle to <109.5°. The O atoms have four electron groups, two of which are lone pairs, so the shape around each O in an —OH is bent (V shaped); the lone pairs compress the H—O—S angle to <109.5°.

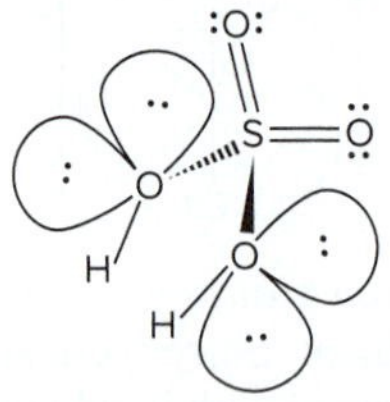

(b) The C in the CH_3— group has four electron groups, all of which are bonding pairs, so the shape around C in CH_3— is tetrahedral, with angles ~109.5°; the other C atoms have two electron groups and the shape around each of these atoms is linear, with a bond angle of 180°.

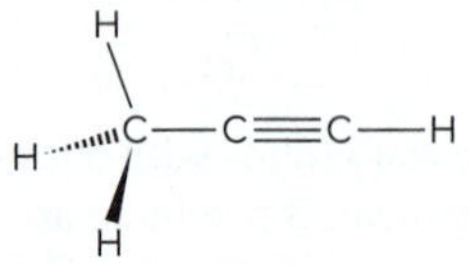

(c) Each S atom has four electron groups (tetrahedral), two of which are lone pairs so the shape around each S is bent (V shaped); F—S—S angle < 109.5°.

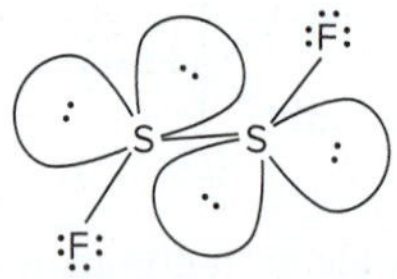

10.8B (a) C, with four bonding pairs, has a tetrahedral shape around it, with H—C—H and H—C—N angles ~109.5°; N has four electron groups, one of which is a lone pair, so the shape around N is trigonal pyramidal; the lone pair compresses the H—N—H angle to <109.5°.

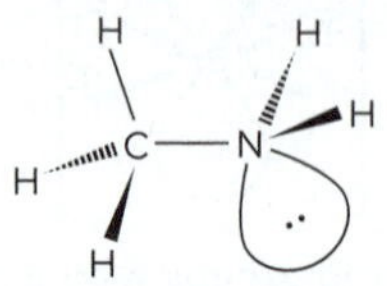

(b) Each C has three electron groups; there are no lone pairs, so the shape around each C is trigonal planar; the double bond compresses each Cl—C—Cl angle to <120°.

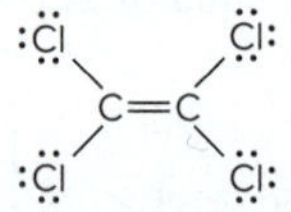

(c) Each Cl, with four electron groups and no lone pairs, has a tetrahedral shape around it, with angles ~109.5°; the central O atom has two bonding pairs and two lone pairs, and thus the shape around this O is bent (V shaped) with the lone pairs compressing the Cl—O—Cl angle to <109.5°.

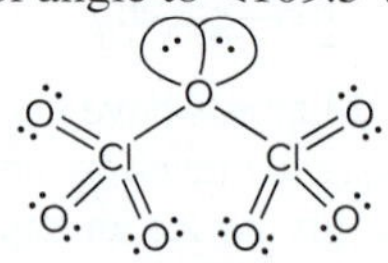

10.9A (a) In the C—H bonds (ΔEN = 0.4), the bond polarity points toward the more electronegative C atom; in the C—Cl bonds (ΔEN = 0.5), the bond polarity points toward the more electronegative Cl atoms. The polar bonds reinforce each other, with the molecular polarity pointing between the Cl atoms.

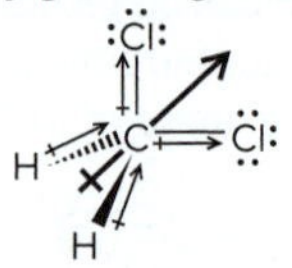

(b) The bond polarity points toward the more electronegative F or O atom. Four of the I—F bonds balance each other but the fifth I—F bond is not balanced by the O—I bond. Since the I—F bond is more polar (ΔEN = 1.5) than the O—I bond (ΔEN = 1.0), the molecular polarity points toward the F atom.

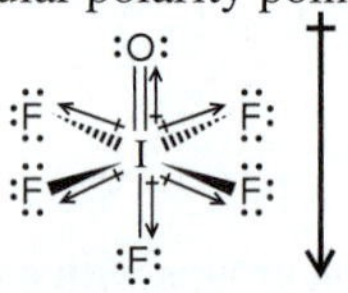

(c) The bond polarity points toward the more electronegative F atom. Four of the I—F bonds balance each other but the fifth I—F bond is not balanced. The molecular polarity points toward the F atom.

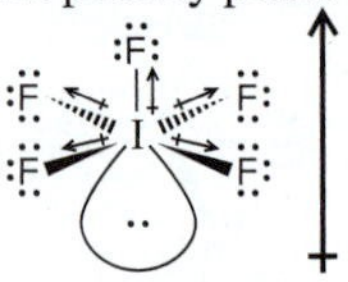

10.9B (a) The bond polarity points toward the more electronegative F atom. The four Xe—F bonds balance each other, so there is no molecular polarity; this is a nonpolar molecule.

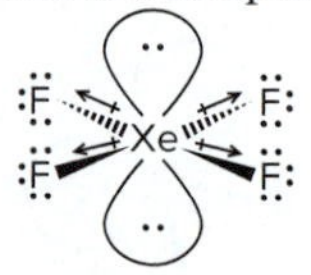

(b) The bond polarity points toward the more electronegative F atom. Two of the Cl—F bonds balance each other, but the third Cl—F bond is not balanced. The molecular polarity points toward the F atom.

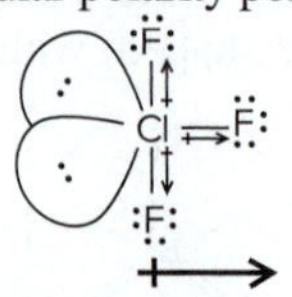

(c) The bond polarity points toward the more electronegative F or O atom. The two axial S—F bonds balance each other, but the two equatorial S—F bonds (ΔEN = 1.5) are not balanced by the double bond between S and O (ΔEN = 1.0). Since the S—F bonds are more polar, the molecular polarity points toward the F atoms.

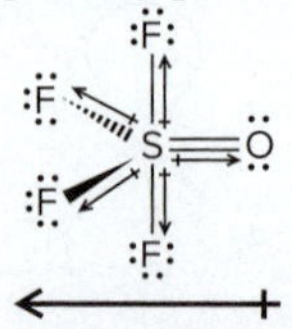

PROBLEMS

Problems with **colored** numbers are answered in Appendix E and worked in detail in the Student Solutions Manual. Problem sections match those in the text and give the numbers of relevant sample problems. Most offer Concept Review Questions, Skill-Building Exercises (grouped in pairs covering the same concept), and Problems in Context. The Comprehensive Problems are based on material from any section or previous chapter.

Depicting Molecules and Ions with Lewis Structures

(Sample Problems 10.1 to 10.5)

Concept Review Questions

10.1 Which of these atoms *cannot* serve as a central atom in a Lewis structure: (a) O; (b) He; (c) F; (d) H; (e) P? Explain.

10.2 When is a resonance hybrid needed to adequately depict the bonding in a molecule? Using NO_2 as an example, explain how a resonance hybrid is consistent with the actual bond length, bond strength, and bond order.

10.3 In which of these structures does X obey the octet rule?

(a) (b) (c) (d) (e) (f) (g) (h)

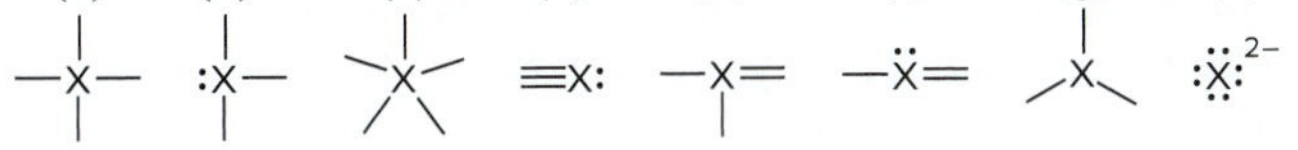

10.4 What is required for an atom to expand its valence shell? Which of the following atoms can expand its valence shell: F, S, H, Al, Se, Cl?

Skill-Building Exercises (grouped in similar pairs)

10.5 Draw a Lewis structure for (a) SiF_4; (b) $SeCl_2$; (c) COF_2 (C is the central atom).

10.6 Draw a Lewis structure for (a) PH_4^+; (b) C_2F_4; (c) SbH_3.

10.7 Draw a Lewis structure for (a) PF_3; (b) H_2CO_3 (both H atoms are attached to O atoms); (c) CS_2.

10.8 Draw a Lewis structure for (a) CH_4S; (b) S_2Cl_2; (c) $CHCl_3$.

10.9 Draw Lewis structures of all the important resonance forms of (a) NO_2^+; (b) NO_2F (N is central).

10.10 Draw Lewis structures of all the important resonance forms of (a) HNO_3 ($HONO_2$); (b) $HAsO_4^{2-}$ ($HOAsO_3^{2-}$).

10.11 Draw Lewis structures of all the important resonance forms of (a) N_3^-; (b) NO_2^-.

10.12 Draw Lewis structures of all the important resonance forms of (a) HCO_2^- (H is attached to C); (b) $HBrO_4$ ($HOBrO_3$).

10.13 Draw the Lewis structure with lowest formal charges, and determine the charge of each atom in (a) IF_5; (b) AlH_4^-.

10.14 Draw the Lewis structure with lowest formal charges, and determine the charge of each atom in (a) OCS; (b) NO.

10.15 Draw the Lewis structure with lowest formal charges, and determine the charge of each atom in (a) CN^-; (b) ClO^-.

10.16 Draw the Lewis structure with lowest formal charges, and determine the charge of each atom in (a) ClF_2^+; (b) ClNO.

10.17 Draw a Lewis structure for a resonance form of each ion with the lowest possible formal charges, show the charges, and give oxidation numbers of the atoms: (a) BrO_3^-; (b) SO_3^{2-}.

10.18 Draw a Lewis structure for a resonance form of each ion with the lowest possible formal charges, show the charges, and give oxidation numbers of the atoms: (a) AsO_4^{3-}; (b) ClO_2^-.

10.19 These species do not obey the octet rule. Draw a Lewis structure for each, and state the type of octet-rule exception:
(a) BH_3 (b) AsF_4^- (c) $SeCl_4$

10.20 These species do not obey the octet rule. Draw a Lewis structure for each, and state the type of octet-rule exception:
(a) PF_6^- (b) ClO_3 (c) H_3PO_3 (one P—H bond)

10.21 These species do not obey the octet rule. Draw a Lewis structure for each, and state the type of octet-rule exception:
(a) BrF_3 (b) ICl_2^- (c) BeF_2

10.22 These species do not obey the octet rule. Draw a Lewis structure for each, and state the type of octet-rule exception:
(a) O_3^- (b) XeF_2 (c) SbF_4^-

Problems in Context

10.23 Molten beryllium chloride reacts with chloride ion from molten NaCl to form the $BeCl_4^{2-}$ ion, in which the Be atom attains an octet. Show the net ionic reaction with Lewis structures.

10.24 Despite many attempts, the perbromate ion (BrO_4^-) was not prepared in the laboratory until about 1970. (In fact, articles were published explaining theoretically why it could never be prepared!) Draw a Lewis structure for BrO_4^- in which all atoms have lowest formal charges.

10.25 Cryolite (Na_3AlF_6) is an indispensable component in the electrochemical production of aluminum. Draw a Lewis structure for the AlF_6^{3-} ion.

10.26 Phosgene is a colorless, highly toxic gas that was employed against troops in World War I and is used today as a key reactant in organic syntheses. From the following resonance structures, select the one with the lowest formal charges:

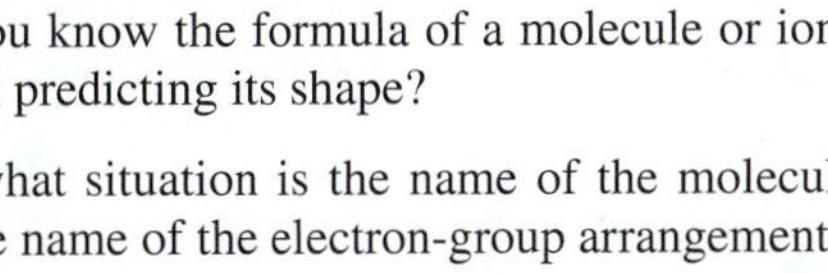

A B C

Valence-Shell Electron-Pair Repulsion (VSEPR) Theory

(Sample Problems 10.6 to 10.8)

Concept Review Questions

10.27 If you know the formula of a molecule or ion, what is the first step in predicting its shape?

10.28 In what situation is the name of the molecular shape the same as the name of the electron-group arrangement?

10.29 Which of the following numbers of electron groups can give rise to a bent (V shaped) molecule: two, three, four, five, six? Draw an example for each case, showing the shape classification (AX_mE_n) and the ideal bond angle.

10.30 Name all the molecular shapes that have a tetrahedral electron-group arrangement.

10.31 Consider the following molecular shapes. (a) Which has the most electron pairs (both bonding and lone pairs) around the central atom? (b) Which has the most lone pairs around the central atom? (c) Do any have only bonding pairs around the central atom?

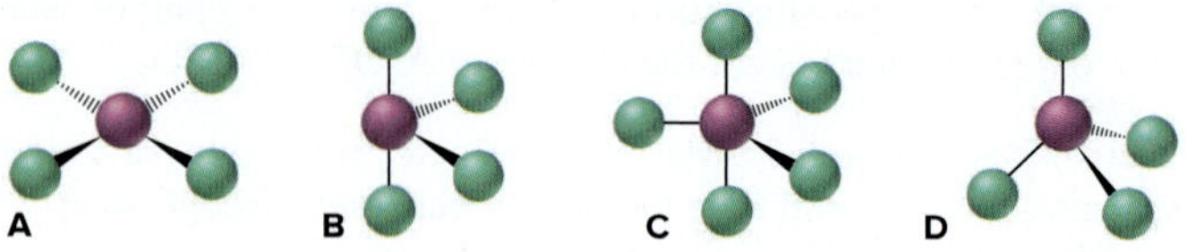

10.32 Use wedge-bond perspective drawings (if necessary) to sketch the atom positions in a general molecule of formula (not shape class) AX_n that has each of the following shapes:
(a) V shaped (b) trigonal planar (c) trigonal bipyramidal
(d) T shaped (e) trigonal pyramidal (f) square pyramidal

10.33 What would you expect to be the electron-group arrangement around atom A in each of the following cases? For each arrangement, give the ideal bond angle and the direction of any expected deviation:

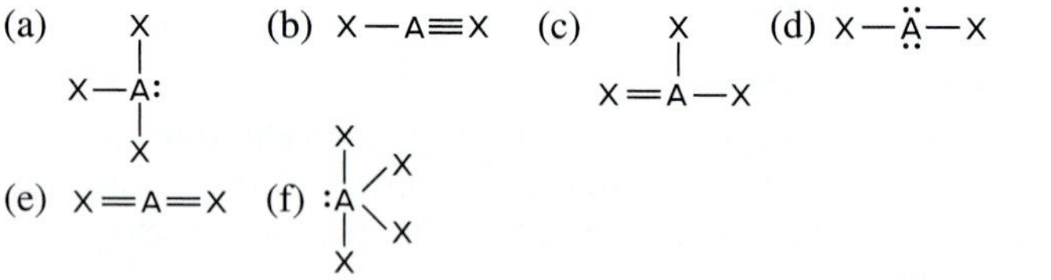

Skill-Building Exercises (grouped in similar pairs)

10.34 Determine the electron-group arrangement, molecular shape, and ideal bond angle(s) for each of the following:
(a) O_3 (b) H_3O^+ (c) NF_3

10.35 Determine the electron-group arrangement, molecular shape, and ideal bond angle(s) for each of the following:
(a) SO_4^{2-} (b) NO_2^- (c) PH_3

10.36 Determine the electron-group arrangement, molecular shape, and ideal bond angle(s) for each of the following:
(a) CO_3^{2-} (b) SO_2 (c) CF_4

10.37 Determine the electron-group arrangement, molecular shape, and ideal bond angle(s) for each of the following:
(a) SO_3 (b) N_2O (N is central) (c) CH_2Cl_2

10.38 Name the shape and give the AX_mE_n classification and ideal bond angle(s) for each of the following general molecules:

10.39 Name the shape and give the AX_mE_n classification and ideal bond angle(s) for each of the following general molecules:

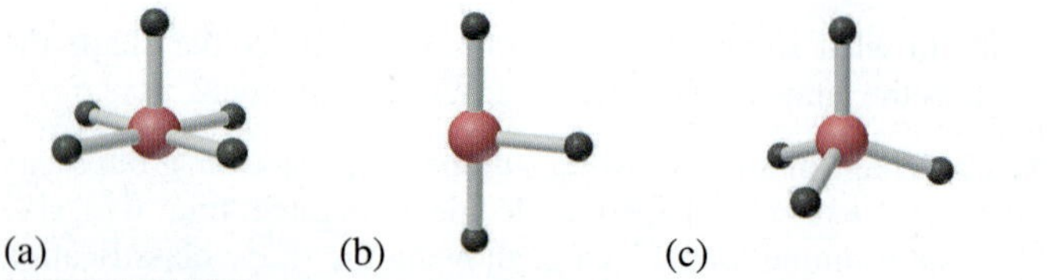

10.40 Determine the shape, ideal bond angle(s), and the direction of any deviation from those angles for each of the following:
(a) ClO_2^- (b) PF_5 (c) SeF_4 (d) KrF_2

10.41 Determine the shape, ideal bond angle(s), and the direction of any deviation from those angles for each of the following:
(a) ClO_3^- (b) IF_4^- (c) $SeOF_2$ (d) TeF_5^-

10.42 Determine the shape around each central atom in each molecule, and explain any deviation from ideal bond angles:
(a) CH_3OH (b) N_2O_4 (O_2NNO_2)

10.43 Determine the shape around each central atom in each molecule, and explain any deviation from ideal bond angles:
(a) H_3PO_4 (no H—P bond) (b) CH_3—O—CH_2CH_3

10.44 Determine the shape around each central atom in each molecule, and explain any deviation from ideal bond angles:
(a) CH_3COOH (b) H_2O_2

10.45 Determine the shape around each central atom in each molecule, and explain any deviation from ideal bond angles:
(a) H_2SO_3 (no H—S bond) (b) N_2O_3 ($ONNO_2$)

10.46 Arrange the following AF_n species in order of *increasing* F—A—F bond angles: BF_3, BeF_2, CF_4, NF_3, OF_2.

10.47 Arrange the following ACl_n species in order of *decreasing* Cl—A—Cl bond angles: SCl_2, OCl_2, PCl_3, $SiCl_4$, $SiCl_6^{2-}$.

10.48 State an ideal value for each of the bond angles in each molecule, and note where you expect deviations:

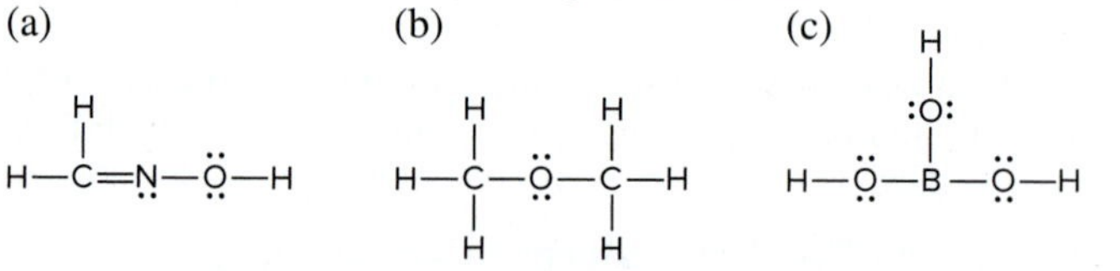

10.49 State an ideal value for each of the bond angles in each molecule, and note where you expect deviations:

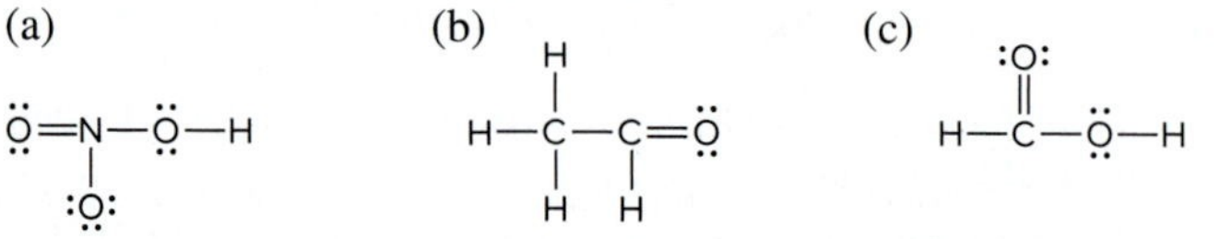

Problems in Context

10.50 Because both tin and carbon are members of Group 4A(14), they form structurally similar compounds. But tin exhibits a greater variety of structures because it forms several ionic species. Predict the shapes and ideal bond angles, including any deviations:
(a) $Sn(CH_3)_2$ (b) $SnCl_3^-$ (c) $Sn(CH_3)_4$ (d) SnF_5^- (e) SnF_6^{2-}

10.51 In the gas phase, phosphorus pentachloride exists as separate molecules. In the solid phase, however, the compound is composed of alternating PCl_4^+ and PCl_6^- ions. What change(s) in molecular shape occur(s) as PCl_5 solidifies? How does the Cl—P—Cl angle change?

Molecular Shape and Molecular Polarity

(Sample Problem 10.9)

Concept Review Questions

10.52 For molecules of general formula AX_n (where $n > 2$), how do you determine if a particular molecule is polar?

10.53 How can a molecule with polar covalent bonds not be polar? Give an example.

10.54 Explain in general why the shape of a biomolecule is important to its function.

Skill-Building Exercises (grouped in similar pairs)

10.55 Consider the molecules SCl_2, F_2, CS_2, CF_4, and BrCl.
(a) Which has bonds that are the most polar?
(b) Which molecules have a dipole moment?

10.56 Consider the molecules BF_3, PF_3, BrF_3, SF_4, and SF_6.
(a) Which has bonds that are the most polar?
(b) Which molecules have a dipole moment?

10.57 Which molecule in each pair has the greater dipole moment? Give the reason for your choice.
(a) SO_2 or SO_3 (b) ICl or IF
(c) SiF_4 or SF_4 (d) H_2O or H_2S

10.58 Which molecule in each pair has the greater dipole moment? Give the reason for your choice.
(a) ClO_2 or SO_2 (b) HBr or HCl
(c) $BeCl_2$ or SCl_2 (d) AsF_3 or AsF_5

Problems in Context

10.59 There are three different dichloroethylenes (molecular formula $C_2H_2Cl_2$), which we can designate X, Y, and Z. Compound X has no dipole moment, but compound Z does. Compounds X and Z each combine with hydrogen to give the same product:

$$C_2H_2Cl_2\ (\text{X or Z}) + H_2 \longrightarrow ClCH_2{-}CH_2Cl$$

What are the structures of X, Y, and Z? Would you expect compound Y to have a dipole moment?

10.60 Dinitrogen difluoride, N_2F_2, is the only stable, simple inorganic molecule with an N=N bond. It occurs in two forms: *cis* (both F atoms on the same side of the N=N bond) and *trans* (the F atoms on opposite sides of the N=N bond).
(a) Draw the molecular shapes of the two forms of N_2F_2.
(b) Predict the direction of the molecular polarity, if any, of each form.

Comprehensive Problems

10.61 In addition to ammonia, nitrogen forms three other hydrides: hydrazine (N_2H_4), diazene (N_2H_2), and tetrazene (N_4H_4).
(a) Use Lewis structures to compare the strength, length, and order of the nitrogen-nitrogen bonds in hydrazine, diazene, and N_2.
(b) Tetrazene (atom sequence H_2NNNNH_2) decomposes above 0°C to hydrazine and nitrogen gas. Draw a Lewis structure for tetrazene, and calculate ΔH°_{rxn} for this decomposition.

10.62 Draw a Lewis structure for each species: (a) PF_5; (b) CCl_4; (c) H_3O^+; (d) ICl_3; (e) BeH_2; (f) PH_2^-; (g) $GeBr_4$; (h) CH_3^-; (i) BCl_3; (j) BrF_4^+; (k) XeO_3; (l) TeF_4.

10.63 Give the molecular shape of each species in Problem 10.62.

10.64 Consider the following reaction of silicon tetrafluoride:

$$SiF_4 + F^- \longrightarrow SiF_5^-$$

(a) Which depiction below best illustrates the change in molecular shape around Si? (b) Give the name and AX_mE_n designation of each shape in the depiction chosen in part (a).

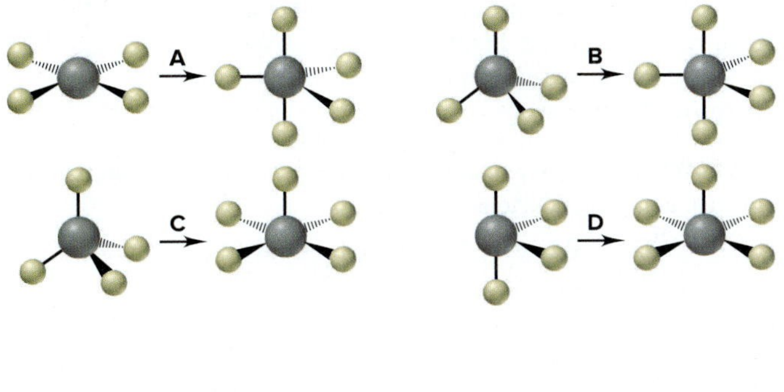

10.65 Both aluminum and iodine form chlorides, Al_2Cl_6 and I_2Cl_6, with "bridging" Cl atoms. The Lewis structures are

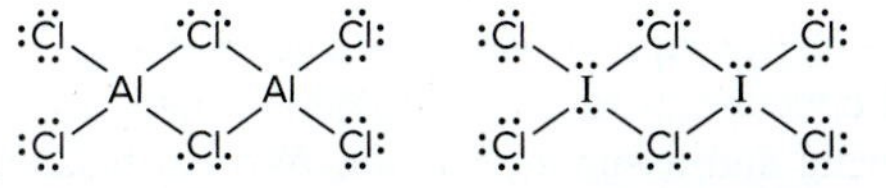

(a) What is the formal charge on each atom in molecule?
(b) Which of these molecules has a planar shape? Explain.

10.66 The VSEPR model was developed before any xenon compounds had been prepared. Thus, these compounds provided an excellent test of the model's predictive power. What would you have predicted for the shapes of XeF_2, XeF_4, and XeF_6?

10.67 When SO_3 gains two electrons, SO_3^{2-} forms. (a) Which depiction best illustrates the change in molecular shape around S? (b) Does molecular polarity change during this reaction?

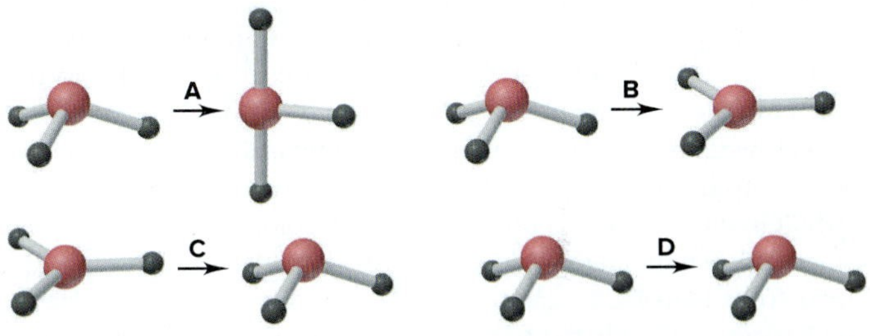

10.68 The actual bond angle in NO_2 is 134.3°, and in NO_2^- it is 115.4°, although the ideal bond angle is 120° for both. Explain.

10.69 "Inert" xenon actually forms several compounds, especially with the highly electronegative elements oxygen and fluorine. The simple fluorides XeF_2, XeF_4, and XeF_6 are all formed by direct reaction of the elements. As you might expect from the size of the xenon atom, the Xe—F bond is not a strong one. Calculate the Xe—F bond energy in XeF_6, given that the enthalpy of formation is −402 kJ/mol.

10.70 Propylene oxide is used to make many products, including plastics such as polyurethane. One method for synthesizing it involves oxidizing propene with hydrogen peroxide:

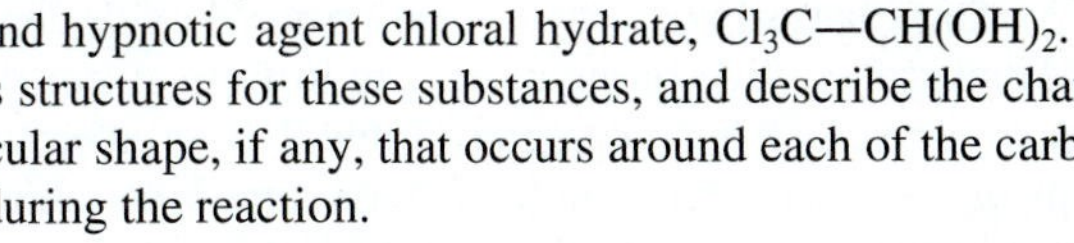

(a) What is the molecular shape and ideal bond angle around each carbon atom in propylene oxide?
(b) Predict any deviation from the ideal for the actual C—C—C bond angles (assume the three atoms in the ring form an equilateral triangle).

10.71 Chloral, $Cl_3C{-}CH{=}O$, reacts with water to form the sedative and hypnotic agent chloral hydrate, $Cl_3C{-}CH(OH)_2$. Draw Lewis structures for these substances, and describe the change in molecular shape, if any, that occurs around each of the carbon atoms during the reaction.

10.72 Like several other bonds, carbon-oxygen bonds have lengths and strengths that depend on the bond order. Draw Lewis structures for the following species, and arrange them in order of increasing carbon-oxygen bond length and then by increasing carbon-oxygen bond strength: (a) CO; (b) CO_3^{2-}; (c) H_2CO; (d) CH_4O; (e) HCO_3^- (H attached to O).

10.73 In the 1980s, there was an international agreement to destroy all stockpiles of mustard gas, $ClCH_2CH_2SCH_2CH_2Cl$. When this substance contacts the moisture in eyes, nasal passages, and skin, the —OH groups of water replace the Cl atoms and create high local concentrations of hydrochloric acid, which cause severe blistering and tissue destruction. Write a balanced equation for this reaction, and calculate ΔH°_{rxn}.

10.74 The four bonds of carbon tetrachloride (CCl_4) are polar, but the molecule is nonpolar because the bond polarity is canceled by the symmetric tetrahedral shape. When other atoms substitute for some of the Cl atoms, the symmetry is broken and the molecule becomes polar. Use Figure 9.21 to rank the following molecules from the least polar to the most polar: CH_2Br_2, CF_2Cl_2, CH_2F_2, CH_2Cl_2, CBr_4, CF_2Br_2.

10.75 Ethanol (CH_3CH_2OH) is being used as a gasoline additive or alternative in many parts of the world.
(a) Use bond energies to find ΔH°_{rxn} for the combustion of gaseous ethanol. (Assume H_2O forms as a gas.)
(b) In its standard state at 25°C, ethanol is a liquid. Its vaporization requires 40.5 kJ/mol. Correct the value from part (a) to find the enthalpy of reaction for the combustion of liquid ethanol.
(c) How does the value from part (b) compare with the value you calculate from standard enthalpies of formation (Appendix B)?
(d) "Greener" methods produce ethanol from corn and other plant material, but the main industrial method involves hydrating ethylene from petroleum. Use Lewis structures and bond energies to calculate ΔH°_{rxn} for the formation of gaseous ethanol from ethylene gas with water vapor.

10.76 In each of the following compounds, the C atoms form a single ring. Draw a Lewis structure for each molecule, identify cases for which resonance exists, and determine the carbon-carbon bond order(s): (a) C_3H_4; (b) C_3H_6; (c) C_4H_6; (d) C_4H_4; (e) C_6H_6.

10.77 An experiment requires 50.0 mL of 0.040 *M* NaOH for the titration of 1.00 mmol of acid. Mass analysis of the acid shows 2.24% hydrogen, 26.7% carbon, and 71.1% oxygen. Draw the Lewis structure of the acid.

10.78 A gaseous compound has a composition by mass of 24.8% carbon, 2.08% hydrogen, and 73.1% chlorine. At STP, the gas has a density of 4.3 g/L. Draw a Lewis structure that fits these facts. Would another structure be equally satisfactory? Explain.

10.79 Perchlorates are powerful oxidizing agents used in fireworks, flares, and the booster rockets of space shuttles. Lewis structures for the perchlorate ion (ClO_4^-) can be drawn with all single bonds or with one, two, or three double bonds. Draw each of these possible resonance forms, use formal charges to determine the more important structure, and calculate its average bond order.

10.80 Methane burns in oxygen to form carbon dioxide and water vapor. Hydrogen sulfide burns in oxygen to form sulfur dioxide and water vapor. Use bond energies (Table 9.2) to determine the enthalpy of each reaction per mole of O_2 (assume Lewis structures with zero formal charges; BE of S=O is 552 kJ/mol).

10.81 Use Lewis structures to determine which *two* of the following are unstable: (a) SF_2; (b) SF_3; (c) SF_4; (d) SF_5; (e) SF_6.

10.82 A major short-lived, neutral species in flames is OH.
(a) What is unusual about the electronic structure of OH?
(b) Use the standard enthalpy of formation of OH(*g*) (39.0 kJ/mol) and bond energies to calculate the O—H bond energy in OH(*g*).
(c) From the average value for the O—H bond energy in Table 9.2 and your value for the O—H bond energy in OH(*g*), find the energy needed to break the first O—H bond in water.

10.83 Pure HN_3 (atom sequence HNNN) is explosive. In aqueous solution, it is a weak acid that yields the azide ion, N_3^-. Draw resonance structures to explain why the nitrogen-nitrogen bond lengths are equal in N_3^- but unequal in HN_3.

10.84 Except for nitrogen, the elements of Group 5A(15) all form pentafluorides, and most form pentachlorides. The chlorine atoms of PCl_5 can be replaced with fluorine atoms one at a time to give, successively, PCl_4F, PCl_3F_2, … , PF_5. (a) Given the sizes of F and Cl, would you expect the first two F substitutions to be at axial or equatorial positions? Explain. (b) Which of the five fluorine-containing molecules have no dipole moment?

10.85 Dinitrogen monoxide (N_2O) supports combustion in a manner similar to oxygen, with the nitrogen atoms forming N_2. Draw three resonance structures for N_2O (one N is central), and use formal charges to decide the relative importance of each. What correlation can you suggest between the more important structure and the observation that N_2O supports combustion?

10.86 Oxalic acid ($H_2C_2O_4$) is found in toxic concentrations in rhubarb leaves. The acid forms two ions, $HC_2O_4^-$ and $C_2O_4^{2-}$, by the sequential loss of H^+ ions. Draw Lewis structures for the three species, and comment on the relative lengths and strengths of their carbon-oxygen bonds. The connections among the atoms are shown below with single bonds only.

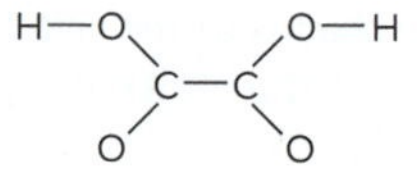

10.87 The Murchison meteorite that landed in Australia in 1969 contained 92 different amino acids, including 21 found in Earth organisms. A skeleton structure (single bonds only) of one of these extraterrestrial amino acids is shown below.

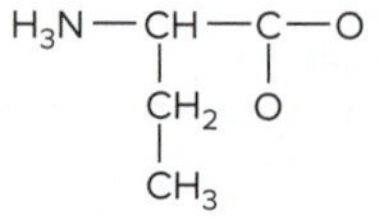

Draw a Lewis structure, and identify any atoms having a nonzero formal charge.

10.88 Hydrazine (N_2H_4) is used as a rocket fuel because it reacts very exothermically with oxygen to form nitrogen gas and water vapor. The heat released and the increase in number of moles of gas provide thrust. Calculate the enthalpy of reaction.

10.89 A student isolates a product with the molecular shape shown at right (F is orange). (a) If the species is a neutral compound, can the black sphere represent selenium (Se)? (b) If the species is an anion, can the black sphere represent N? (c) If the black sphere represents Br, what is the charge of the species?

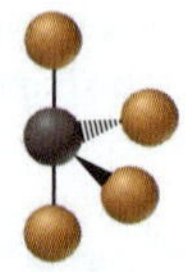

10.90 When gaseous sulfur trioxide is dissolved in concentrated sulfuric acid, disulfuric acid forms:

$$SO_3(g) + H_2SO_4(l) \longrightarrow H_2S_2O_7(l)$$

Use bond energies (Table 9.2) to determine ΔH°_{rxn}. (The S atoms in $H_2S_2O_7$ are bonded through an O atom. Assume Lewis structures with zero formal charges; BE of S═O is 552 kJ/mol.)

10.91 A molecule of formula AY_3 is found experimentally to be polar. Which molecular shapes are possible and which are impossible for AY_3?

10.92 Consider the following molecular shapes:

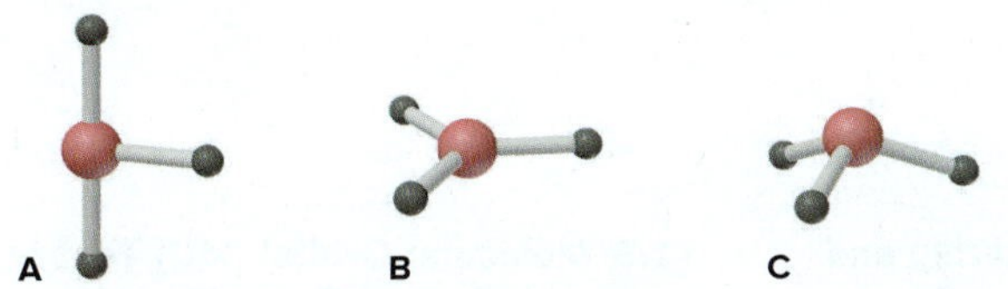

(a) Match each shape with one of the following species: XeF_3^+, $SbBr_3$, $GaCl_3$.
(b) Which, if any, is polar?
(c) Which has the most valence electrons around the central atom?

10.93 Hydrogen cyanide can be catalytically reduced with hydrogen to form methylamine. Use Lewis structures and bond energies to determine ΔH°_{rxn} for

$$HCN(g) + 2H_2(g) \longrightarrow CH_3NH_2(g)$$

10.94 Ethylene, C_2H_4, and tetrafluoroethylene, C_2F_4, are used to make the polymers polyethylene and polytetrafluoroethylene (Teflon), respectively.
(a) Draw the Lewis structures for C_2H_4 and C_2F_4, and give the ideal H—C—H and F—C—F bond angles.
(b) The actual H—C—H and F—C—F bond angles are 117.4° and 112.4°, respectively. Explain these deviations.

10.95 Using bond lengths in Table 9.2 and assuming ideal bond angles, calculate each of the following distances:
(a) Between H atoms in C_2H_2
(b) Between F atoms in SF_6 (two answers)
(c) Between equatorial F atoms in PF_5

10.96 Phosphorus pentachloride, a key industrial compound with annual world production of about 2×10^7 kg, is used to make other compounds. It reacts with sulfur dioxide to produce phosphorus oxychloride ($POCl_3$) and thionyl chloride ($SOCl_2$). Draw a Lewis structure and name the molecular shape of each of these products.

11 Theories of Covalent Bonding

11.1 Valence Bond (VB) Theory and Orbital Hybridization
Central Themes of VB Theory
Types of Hybrid Orbitals

11.2 Modes of Orbital Overlap and the Types of Covalent Bonds
Orbital Overlap in Single and Multiple Bonds
Orbital Overlap and Rotation Within a Molecule

11.3 Molecular Orbital (MO) Theory and Electron Delocalization
Central Themes of MO Theory
Homonuclear Diatomic Molecules of Period 2 Elements
Two Heteronuclear Diatomic Molecules: HF and NO
Two Polyatomic Molecules: Benzene and Ozone

Source: © Richard Megna/Fundamental Photographs, NYC

Concepts and Skills to Review Before You Study This Chapter

- atomic orbital shapes (Section 7.4)
- exclusion principle (Section 8.1)
- Hund's rule (Section 8.2)
- Lewis structures (Section 10.1)
- resonance in covalent bonding (Section 10.1)
- molecular shapes (Section 10.2)
- molecular polarity (Section 10.3)

All scientific models have limitations because they are simplifications of reality. The VSEPR theory, covered in Chapter 10, accounts for molecular shapes, but it doesn't explain how they arise from atomic orbitals. After all, the orbitals described in Chapter 7 aren't oriented toward the corners of, say, a trigonal bipyramid, like the one for phosphorus in PF_5. Moreover, knowing the shape doesn't help us explain the magnetic and spectral properties of a molecule; for example, the Lewis structure of O_2 that we draw does not account for the presence of unpaired electrons that cause liquid oxygen to be attracted to the poles of a magnet *(photo)*. We need an understanding of orbitals and energy levels to explain that.

The two theories presented in this chapter focus on different questions and complement each other. Don't be concerned that we need several models to explain a complex phenomenon like bonding. In every science, one model may account for some aspect of a topic better than another, and several are employed to explain a wider range of phenomena.

IN THIS CHAPTER . . . *We introduce two theories of bonding in molecules that are based on the interactions of the orbitals of their atoms.*

- We discuss valence bond (VB) theory, which rationalizes molecular shapes through interactions of atomic orbitals during bonding to form hybrid orbitals.
- We apply VB theory to sigma and pi bonds, the two types of covalent bonds.
- We see why parts of molecules don't rotate freely around multiple bonds, which has major effects on reactivity, physical properties, and biological behavior.
- We discuss molecular orbital (MO) theory, which explains molecular energy levels and properties based on the formation of orbitals that spread over a whole molecule.
- We apply MO theory to understand key properties of some diatomic and simple polyatomic molecules.

11.1 VALENCE BOND (VB) THEORY AND ORBITAL HYBRIDIZATION

What *is* a covalent bond, and what characteristic gives it strength? And how can we explain *molecular* shapes based on the interactions of *atomic* orbitals? One very useful approach for answering these questions is based on quantum mechanics (Chapter 7) and is called **valence bond (VB) theory.**

The Central Themes of VB Theory

The basic principle of VB theory is that *a covalent bond forms when orbitals of two atoms overlap and a pair of electrons occupy the overlap region.* In the terminology of quantum mechanics, overlap of the two orbitals means their wave functions are *in phase* (constructive interference; see Figure 7.5), so the amplitude between the nuclei increases. The central themes of VB theory derive from this principle:

1. *Opposing spins of the electron pair.* As the exclusion principle (Section 8.1) requires, the space formed by the overlapping orbitals *has a maximum capacity for two electrons that have opposite (paired) spins.* In the simplest case, a molecule of H_2 forms when the 1*s* orbitals of two H atoms overlap, and the electrons, with their spins paired, spend more time in the overlap region (up and down arrows in Figure 11.1A).

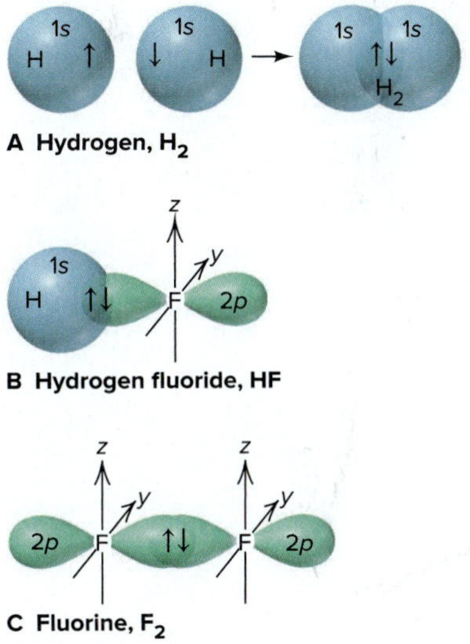

Figure 11.1 Orbital overlap and spin pairing in three diatomic molecules. (In **B** and **C,** the $2p_x$ orbital is shown involved in the bonding; the other two 2*p* orbitals of F are omitted for clarity.)

2. *Maximum overlap of bonding orbitals.* Bond strength depends on the attraction between nuclei and shared electrons, so *the greater the orbital overlap, the closer the nuclei are to the electrons, and the stronger the bond.* Extent of overlap depends on orbital shape and direction. An *s* orbital is spherical, but *p* and *d* orbitals have specified directions. Thus, a *p* or *d* orbital involved in a bond is oriented so as to *maximize overlap.* In HF, for example, the 1*s* orbital of H overlaps a half-filled 2*p* orbital of F *along its long axis* (Figure 11.1B). In F_2, the two half-filled 2*p* orbitals interact end to end, that is, *along their long axes* (Figure 11.1C).

3. *Hybridization of atomic orbitals.* To account for the bonding in diatomic molecules like HF and F_2, we picture direct overlap of *s* and/or *p* orbitals of isolated atoms. But how can we account for the shape of a molecule like methane from the shapes and orientations of C and H atomic orbitals? A C atom ([He] $2s^22p^2$) has two valence electrons in the spherical 2*s* orbital and one valence electron in two of the three mutually perpendicular 2*p* orbitals. If the two half-filled *p* orbitals overlapped the 1*s* orbitals of two H atoms, *two* C—H bonds would form with a 90° H—C—H bond angle. But methane has the formula CH_4, not CH_2, and its bond angles are 109.5°.

To explain such facts, Linus Pauling proposed that, during bonding, *the valence orbitals in the isolated atoms become* ***new*** orbitals *in the molecule.* Quantum-mechanical calculations show that if we mathematically "mix" certain combinations of orbitals, we form new ones whose spatial orientations *do* match the observed molecular shapes. The process of orbital mixing is called **hybridization,** and the new atomic orbitals are called **hybrid orbitals.**

4. *Features of hybrid orbitals.* Here are some central points about the hybrid orbitals that form during bonding:
- The *number* of hybrid orbitals formed *equals* the number of atomic orbitals mixed.
- The *type* of hybrid orbitals formed *varies* with the types of atomic orbitals mixed.
- The *shape* and *orientation* of a hybrid orbital *maximize* its overlap with the orbital of the other atom in the bond.

It is useful to think of hybridization as a process in which atomic orbitals mix, hybrid orbitals form, they overlap other orbitals, and electrons enter the overlap region with opposing spins, thus forming stable bonds. In truth, hybridization is a mathematical concept that helps us explain the molecular shapes we observe.

Types of Hybrid Orbitals

We postulate the type of hybrid orbitals in a molecule *after* we observe its shape. Note that the orientations of the five types of hybrid orbitals we discuss next correspond to the five electron-group arrangements in VSEPR theory (see Figure 10.2).

***sp* Hybridization** When two electron groups surround the central atom, we observe a linear shape, as in $BeCl_2$. Beryllium's two valence electrons are paired in the 2*s* orbital ([He] $2s^2$) in the ground state; with no unpaired electrons, Be cannot form bonds. The beryllium atom can promote an electron from the 2*s* orbital to an empty 2*p* orbital, resulting in two unpaired electrons that could participate in two bonds with Cl atoms to form $BeCl_2$:

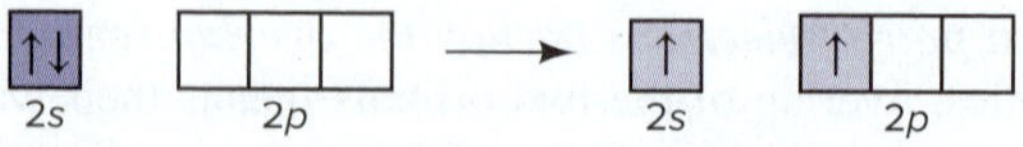

However, this would result in two nonidentical bonds: the 2*s* orbital of Be would overlap with a 3*p* orbital of one Cl atom to form one bond, while the 2*p* orbital of Be would overlap with a 3*p* orbital of a second Cl atom to form the second bond. But experimentally, we know that the two Be—Cl bonds are identical in length and energy. The concept of hybridization offers an explanation for the linear shape and identical bonds in $BeCl_2$.

1. *Orbitals mixed and orbitals formed.* VB theory proposes that two *nonequivalent* orbitals of a central atom, one *s* and one *p*, mix and form two *equivalent* ***sp* hybrid orbitals** that are oriented 180° apart (Figure 11.2A). The shape of these hybrid orbitals,

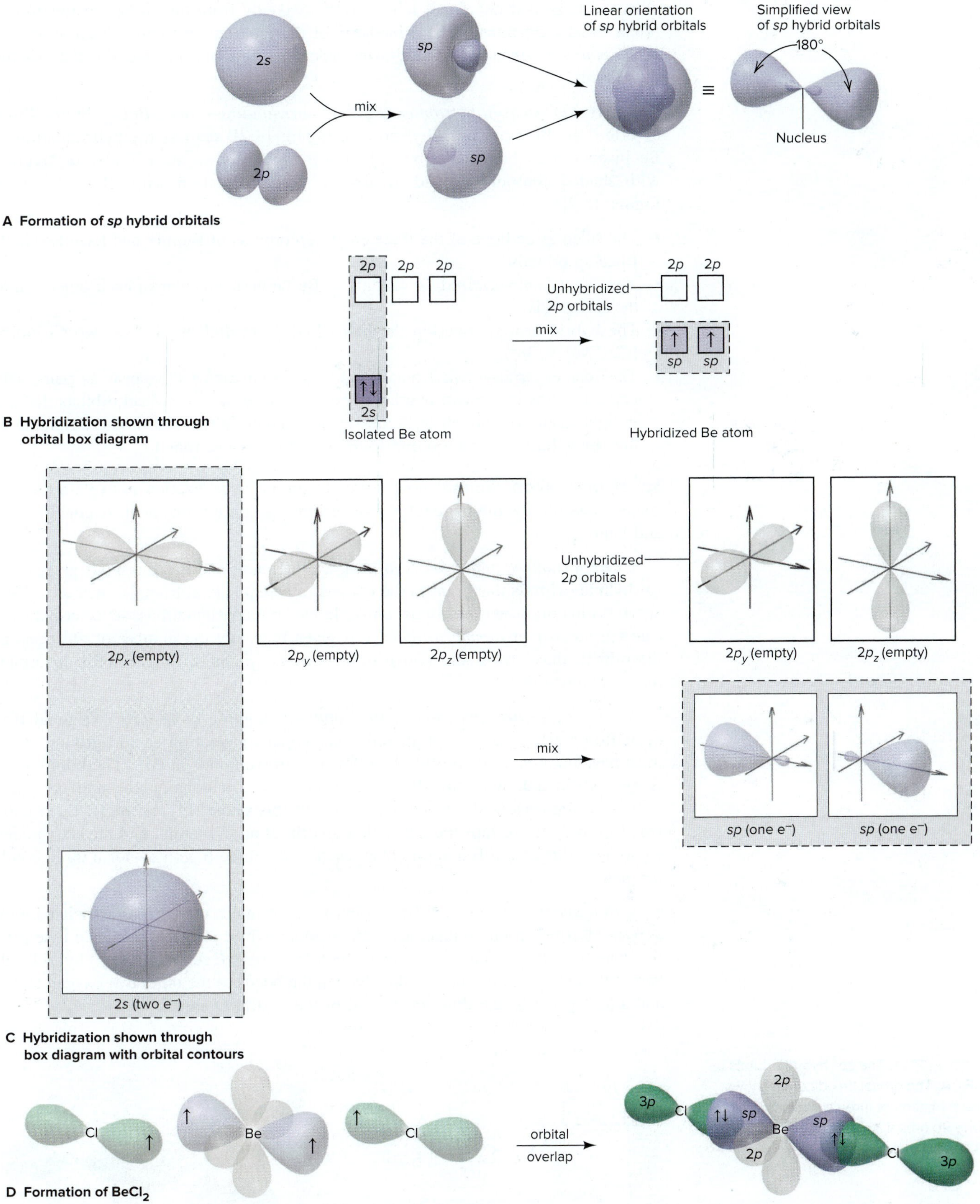

Figure 11.2 **Formation and orientation of *sp* hybrid orbitals and the bonding in $BeCl_2$. A,** One 2*s* and one 2*p* atomic orbital mix to form two *sp* hybrid orbitals. (The simplified hybrid orbitals at far right are used elsewhere, often even without the small lobe.) **B,** The orbital box diagram for hybridization of Be, drawn vertically. (Full color, light shading, or no shading indicates orbital occupancy.) **C,** The orbital box diagram with orbital contours. **D,** Overlap of Be and Cl orbitals to form $BeCl_2$. (Simplified hybrid orbitals are shown for Be; only the 3*p* orbital of Cl that is involved in bonding is shown.)

with one large and one small lobe, differs markedly from the shapes of the atomic orbitals that were mixed. The orientation of these orbitals increases electron density *in the bonding direction* and minimizes repulsions between the electrons that occupy the orbitals.

2. *Overlap of orbitals from central and surrounding atoms: $BeCl_2$.* In beryllium chloride, the Be atom is *sp* hybridized. Figure 11.2B depicts the hybridization of Be in an orbital box diagram, and Figure 11.2C shows an orbital box diagram with shaded contours instead of arrows. Bond formation with Cl is shown in Figure 11.2D:

- The filled 2*s* and one of the three empty 2*p* orbitals of Be mix and form two half-filled *sp* orbitals.
- Two empty unhybridized 2*p* orbitals of Be lie perpendicular to each other and to the *sp* hybrids.
- The hybrid orbitals overlap the half-filled 3*p* orbital in each of two Cl atoms (Cl: [Ne] $3s^23p^5$).
- The four electrons—two from Be and one from each Cl—appear as pairs with opposite spins in the two overlap regions. (The 3*p* and *sp* hybrid orbitals that are in light shades on the left in Figure 11.2D become fully colored on the right, after the bonds form and each orbital is filled with two electrons.)

sp^2 Hybridization We use another type of orbital hybridization to explain the two shapes possible for the trigonal planar electron-group arrangement, trigonal planar and bent.

1. *Orbitals mixed and orbitals formed.* Mixing one *s* and two *p* orbitals gives three **sp^2 hybrid orbitals** that point to the corners of an equilateral triangle, their axes 120° apart. (In hybrid-orbital notations, unlike in electron configurations, superscripts indicate the number of *atomic orbitals* of a given type, *not* the number of *electrons* in the orbital: thus, one *s* and two *p* orbitals give s^1p^2, or sp^2.) The third 2*p* orbital remains unhybridized.

2. *Overlap of orbitals from central and surrounding atoms: BF_3* (trigonal planar). Boron ([He] $2s^22p^1$), with only one unpaired electron in the ground state, must have three half-filled orbitals to form the three bonds in BF_3. The boron atom is sp^2 hybridized, with the three sp^2 orbitals in a trigonal plane and the third 2*p* orbital unhybridized and perpendicular to this plane (Figure 11.3). Each half-filled sp^2 orbital overlaps the half-filled 2*p* orbital of an F atom, and the six valence electrons—three from B and one from each of the three F atoms—form three bonding pairs.

3. *Placement of lone pairs.* To account for other molecular shapes associated with a given electron-group arrangement, one or more hybrid orbitals contains a lone pair. In ozone (O_3), for example, the shape is bent because the central O is sp^2 hybridized; two of the sp^2 hybrid orbitals are involved in the bonds to the other two oxygen atoms and a lone pair fills the third sp^2 orbital of the central O atom.

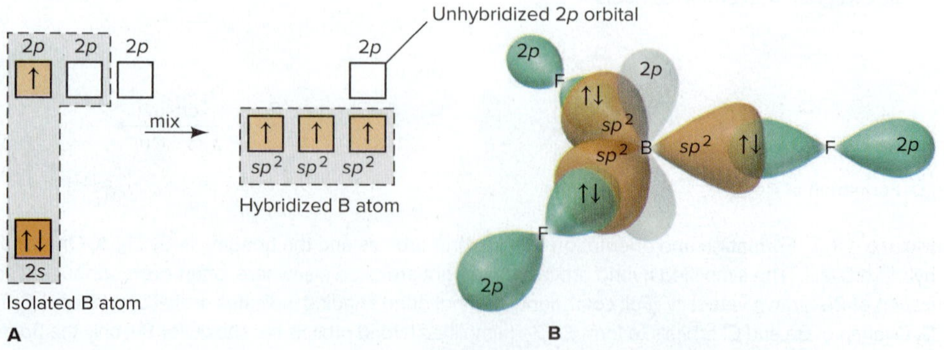

Figure 11.3 **The sp^2 hybrid orbitals in BF_3. A,** The orbital box diagram shows the formation of three sp^2 hybrid orbitals. One 2*p* orbital is unhybridized and empty. **B,** Contour depiction of BF_3.

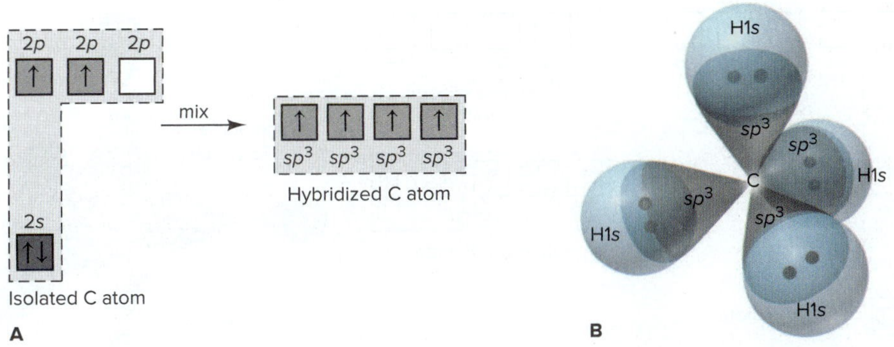

Figure 11.4 **The** ***sp***^{3} **hybrid orbitals in** CH_4**. A,** The orbital box diagram shows formation of four sp^3 hybrids. **B,** Contour depiction of CH_4, with electron pairs shown as dots.

sp^3 Hybridization Let's return to the question posed earlier about the shape of methane. The type of hybridization that accounts for methane's shape applies to any species with a tetrahedral electron-group arrangement.

1. *Orbitals mixed and orbitals formed.* Mixing one *s* and three *p* orbitals gives four ***sp***^{3} **hybrid orbitals** that point to the corners of a tetrahedron.

2. *Overlap of orbitals from central and surrounding atoms:* CH_4. Carbon ([He] $2s^22p^2$), with only two unpaired electrons in the ground state, must have four half-filled orbitals to form the four bonds in CH_4. The C atom in methane is sp^3 hybridized. Its four valence electrons half-fill the four sp^3 hybrids, which overlap the half-filled 1*s* orbitals of four H atoms to form four C—H bonds (Figure 11.4).

3. *Placement of lone pairs.* The trigonal pyramidal shape of NH_3 arises because the three half-filled sp^3 hybrid orbitals form the three N—H bonds and a lone pair fills the fourth sp^3 orbital The bent shape of H_2O arises because lone pairs fill any two of the sp^3 orbitals of O (Figure 11.5).

sp^3d Hybridization Molecules with shapes due to the trigonal bipyramidal electron-group arrangement (trigonal bipyramidal, seesaw, T shaped, and linear) must have central atoms from Period 3 or higher, because *d* orbitals, as well as *s* and *p* orbitals, are mixed to form the hybrid orbitals.

1. *Orbitals mixed and orbitals formed.* Mixing one 3*s*, the three 3*p*, and one of the five 3*d* orbitals gives five ***sp***^{3}***d*** **hybrid orbitals,** which point to the corners of a trigonal bipyramid.

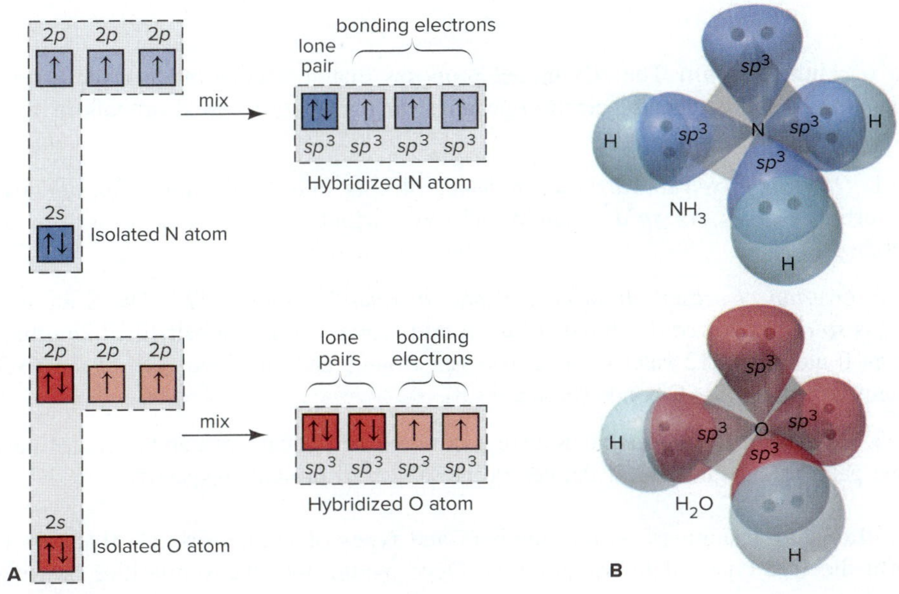

Figure 11.5 **The** ***sp***^{3} **hybrid orbitals in** NH_3 **and** H_2O**. A,** The orbital box diagrams show sp^3 hybridization, with lone pairs filling one (in NH_3) or two (in H_2O) hybrid orbitals. **B,** Contour depictions of NH_3 and H_2O.

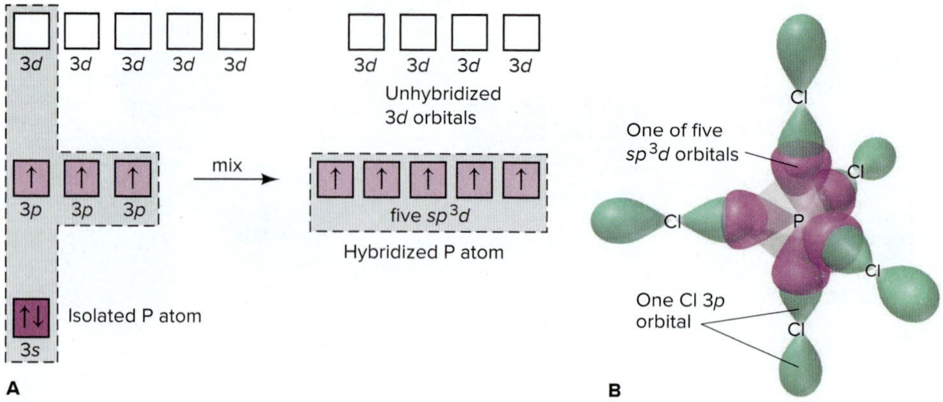

Figure 11.6 **The sp^3d hybrid orbitals in PCl_5. A,** The orbital box diagram shows the formation of five half-filled sp^3d orbitals. Four 3*d* orbitals are unhybridized and empty. **B,** Contour depiction of PCl_5. (For clarity, the unhybridized 3*d* orbitals of P, the other two 3*p* orbitals of Cl, and the five bonding P—Cl pairs are not shown.)

2. *Overlap of orbitals from central and surrounding atoms: PCl_5.* The P atom in PCl_5 is sp^3d hybridized. Each hybrid orbital overlaps a 3*p* orbital of a Cl atom, and ten valence electrons—five from P and one from each of the five Cl atoms—form five P—Cl bonds (Figure 11.6).

3. *Placement of lone pairs.* Seesaw, T-shaped, and linear molecules have lone pairs in one, two, or three of the central atom's sp^3d orbitals, respectively. Figure 11.7 shows the bonding for the T-shaped molecule ClF_3.

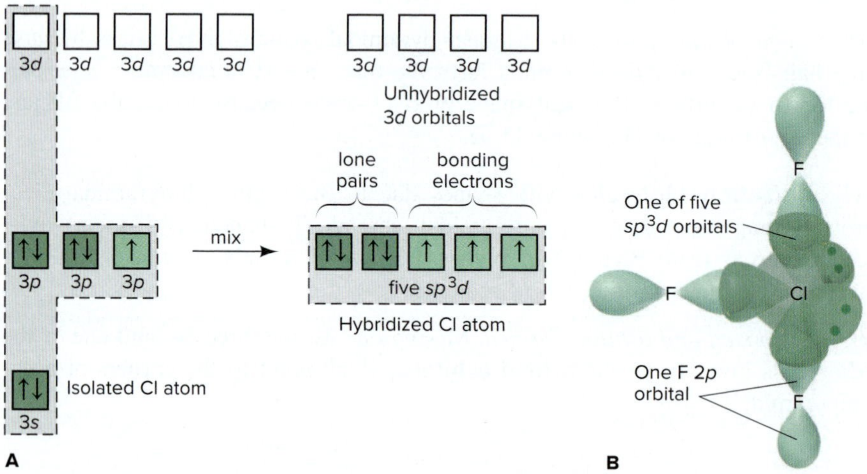

Figure 11.7 **The sp^3d hybrid orbitals in ClF_3. A,** The orbital box diagram shows sp^3d hybridization. **B,** Contour depiction of ClF_3, with lone pairs filling two hybrid orbitals. (For clarity, the unhybridized 3*d* orbitals of Cl, the other two 2*p* orbitals of F, and the three bonding Cl—F pairs are not shown.)

sp^3d^2 Hybridization The VB model proposes that molecules with shapes corresponding to the octahedral electron-group arrangement use two *d* orbitals to form hybrids.

1. *Orbitals mixed and orbitals formed.* Mixing one 3*s*, the three 3*p*, and two 3*d* orbitals gives six **sp^3d^2 hybrid orbitals,** which point to the corners of an octahedron.

2. *Overlap of orbitals from central and surrounding atoms: SF_6.* The S atom in SF_6 is sp^3d^2 hybridized. Each half-filled hybrid orbital overlaps a half-filled 2*p* orbital of an F atom, and 12 valence electrons—six from S and one from each of the six F atoms—form six S—F bonds (Figure 11.8).

3. *Placement of lone pairs.* Square pyramidal and square planar molecules have lone pairs in one and two of the central atom's sp^3d^2 orbitals, respectively.

Table 11.1 summarizes the numbers and types of atomic orbitals that mix to form the five types of hybrid orbitals. Once again, note the similarities between

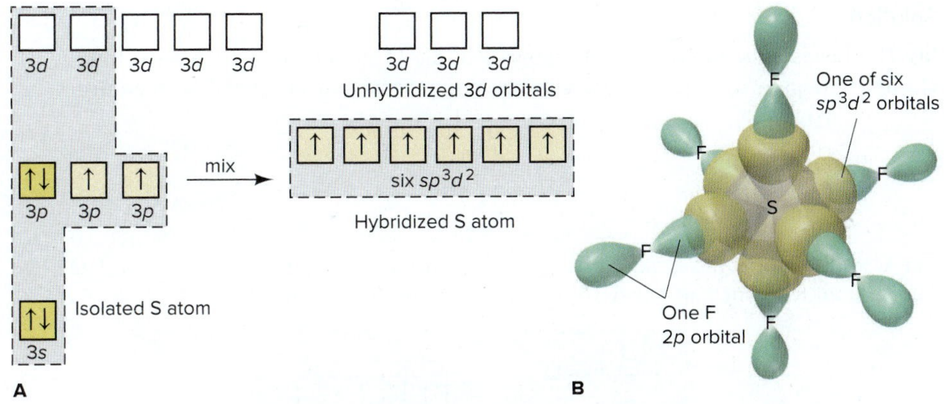

Figure 11.8 **The sp^3d^2 hybrid orbitals in SF_6. A,** The orbital box diagram shows formation of six half-filled sp^3d^2 orbitals; three 3*d* orbitals remain unhybridized and empty. **B,** Contour depiction of SF_6.

Table 11.1 Composition and Orientation of Hybrid Orbitals

	Linear	Trigonal Planar	Tetrahedral	Trigonal Bipyramidal	Octahedral
Atomic orbitals mixed	one *s* one *p*	one *s* two *p*	one *s* three *p*	one *s* three *p* one *d*	one *s* three *p* two *d*
Hybrid orbitals formed	two *sp*	three sp^2	four sp^3	five sp^3d	six sp^3d^2
Unhybridized orbitals remaining	two *p*	one *p*	none	four *d*	three *d*
Orientation of hybrid orbitals	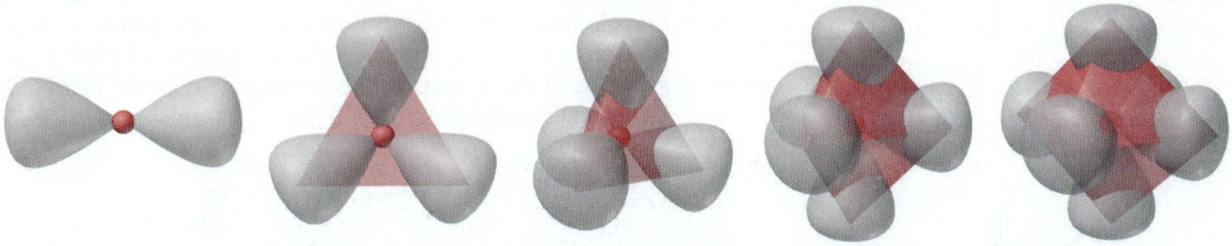				

hybrid-orbital orientations and the shapes predicted by VSEPR theory. Figure 11.9 shows three conceptual steps from a molecular formula to the hybrid orbitals in the molecule, and Sample Problem 11.1 focuses on the last step of that process.

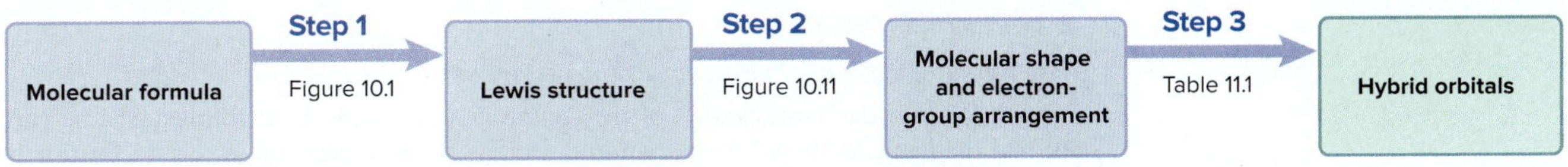

Figure 11.9 **From molecular formula to hybrid orbitals.** (See Figures 10.1 and 10.11.)

SAMPLE PROBLEM 11.1 Postulating Hybrid Orbitals in a Molecule

Problem Use partial orbital diagrams to describe how mixing the atomic orbitals of the central atom(s) leads to the hybrid orbitals in each of the following:

(a) Methanol, CH_3OH **(b)** Sulfur tetrafluoride, SF_4

Plan We use the molecular formula to draw the Lewis structure and determine the electron-group arrangement of each central atom. Then we use Table 11.1 to postulate the type of hybrid orbitals. We write the partial orbital diagram for each central atom before and after the orbitals are hybridized.

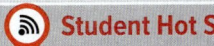

Student Hot Spot

Student data indicate that you may struggle with the concept of orbital hybridization. Access the SmartBook to view additional Learning Resources on this topic.

Solution

(a) The Lewis structure of CH_3OH shows that the C and O atoms each have four electron groups, with a tetrahedral electron-group arrangement around both atoms:

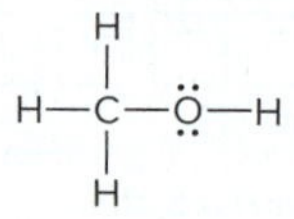

The C and O atoms have each mixed one 2*s* and three 2*p* to become sp^3 hybridized. The C atom has four half-filled sp^3 orbitals to form the four bonds:

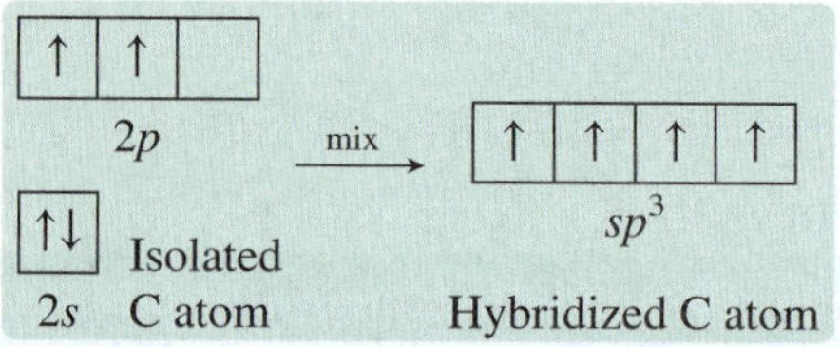

The O atom has two half-filled sp^3 orbitals that are used to form two bonds and two sp^3 orbitals filled with lone pairs:

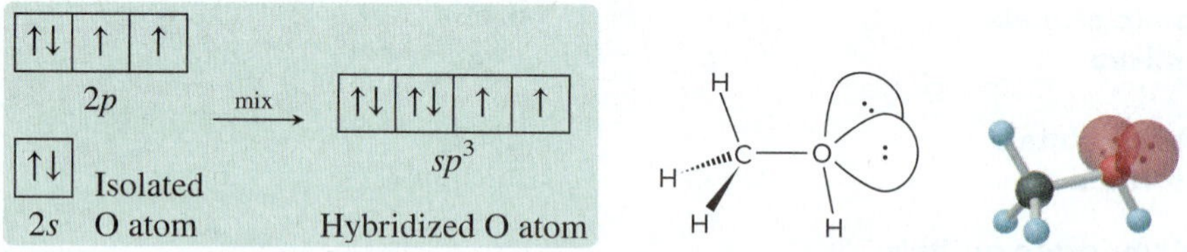

(b) The Lewis structure of SF_4 shows that S has four bonds and one lone pair, for a total of five electron groups. The electron-group arrangement is trigonal bipyramidal, so the central S atom is sp^3d hybridized, which means one 3*s*, three 3*p*, and one 3*d* orbital were mixed. One of the hybrid orbitals is filled with a lone pair, and four are half-filled. Four unhybridized 3*d* orbitals remain empty:

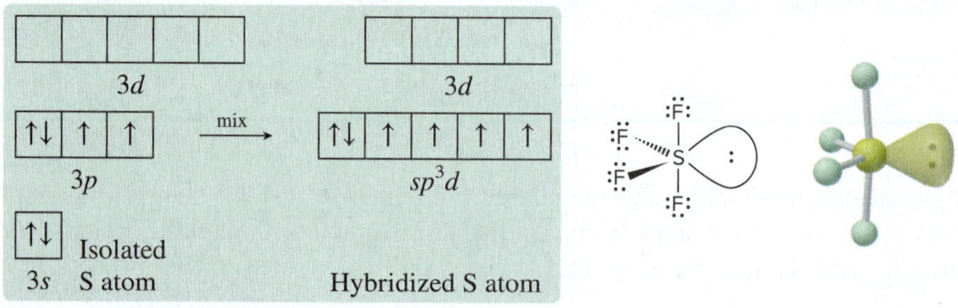

FOLLOW-UP PROBLEMS

Brief Solutions for ***all*** *Follow-up Problems appear at the end of the chapter.*

11.1A What is the hybridization of the central atom in each of the following? Use partial orbital diagrams to show how the atomic orbitals of the central atom mix to form hybrid orbitals in **(a)** beryllium fluoride, BeF_2; **(b)** silicon tetrachloride, $SiCl_4$; **(c)** xenon tetrafluoride, XeF_4.

11.1B What is the hybridization of the central atom in each of the following? Use partial orbital diagrams to show how the atomic orbitals of the central atom mix to form hybrid orbitals in **(a)** nitrogen dioxide, NO_2; **(b)** phosphorus trichloride, PCl_3; **(c)** bromine pentafluoride, BrF_5.

SOME SIMILAR PROBLEMS 11.7–11.12 and 11.15–11.18

Limitations to the Concept of Hybridization We rely on the VSEPR and VB theories to explain an observed molecular shape. In some cases, however, the theories may not be consistent with other findings.

1. *Hybridization does not apply: large nonmetal hydrides.* Consider the Lewis structure and bond angle of H_2S:

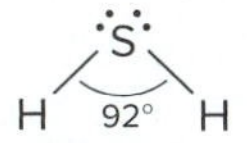

Based on VSEPR theory, we would predict that, as in H_2O, the four electron groups around H_2S point to the corners of a tetrahedron, and the two lone pairs compress the H—S—H bond angle below the ideal 109.5°. Based on VB theory, we would propose that the 3*s* and 3*p* orbitals of the S atom mix and form four sp^3 hybrids, two of which are filled with lone pairs, while the other two overlap 1*s* orbitals of two H atoms and are filled with bonding pairs.

But observation does *not* support these arguments. The bond angle of 92° is close to the 90° angle between *unhybridized p* orbitals. Similar angles occur in the hydrides of other large nonmetals of Groups 5A(15) and 6A(16). Why apply a theory if the facts don't warrant it? Real factors—bond length, atomic size, and electrostatic repulsions—influence shape. Larger atoms form longer bonds to H, which decreases repulsions; thus, overlap of *unhybridized* orbitals adequately explains these shapes.

2. *d-Orbital hybridization is less important: shapes with expanded valence shells.* Quantum-mechanical calculations show that *d* orbitals have such high energies that they do not hybridize effectively with the *much* more stable *s* and *p* orbitals of a given *n* value. Thus, for example, some have proposed that SF_6 is most stable when the bonding orbitals of the central S use a combination of *sp* hybrid orbitals and unhybridized 3*p* orbitals instead of sp^3d^2 hybrid orbitals. Others prefer explanations that involve molecular orbitals or even ionic structures. These topics are beyond the scope of this text, and while they are being actively debated, we will continue to use the traditional, though limited, approach of including *d*-orbital hybridization for molecules with expanded valence shells.

› Summary of Section 11.1

- › VB theory explains that a covalent bond forms when two atomic orbitals overlap and two electrons with paired (opposite) spins occupy the overlap region.
- › To explain molecular shape, VB theory proposes that, during bonding, atomic orbitals mix to form hybrid orbitals whose shape and orientation differ from those of the mixed atomic orbitals. This process gives rise to greater orbital overlap and, thus, stronger bonds.
- › Based on an observed molecular shape (and the related electron-group arrangement), we postulate the type of hybrid orbital that accounts for that shape. In many cases, especially for molecules with larger central atoms, the concept of hybridization has major limitations.

11.2 MODES OF ORBITAL OVERLAP AND THE TYPES OF COVALENT BONDS

Orbitals can overlap by two *modes*—end to end or side to side—which gives rise to two types of covalent bonds—*sigma* and *pi*. We'll use VB theory to describe the two types here; as you'll see, they are described by molecular orbital theory as well.

Orbital Overlap in Single and Multiple Bonds

Ethane (C_2H_6), ethylene (C_2H_4), and acetylene (C_2H_2) have different shapes. Ethane is tetrahedral at both carbons with bond angles near the ideal 109.5°. Ethylene is trigonal planar at both carbons with bond angles near the ideal 120°. Acetylene is linear with bond angles of 180°:

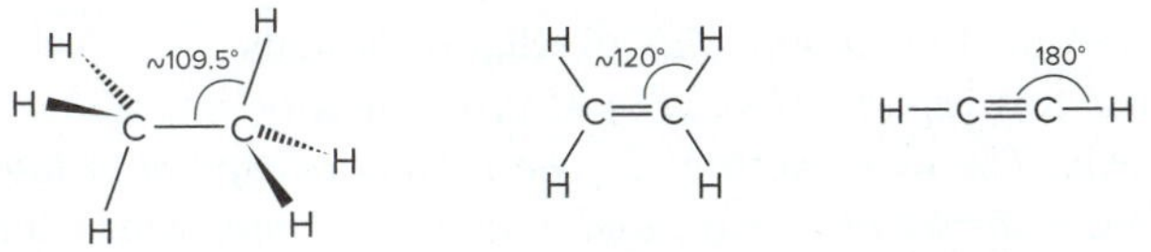

ethane ethylene acetylene

In these molecules, two modes of orbital overlap result in two types of bonds.

End-to-End Overlap and Sigma (σ) Bonding Both C atoms of ethane are sp^3 hybridized; the four valence electrons of C half-fill the four sp^3 orbitals:

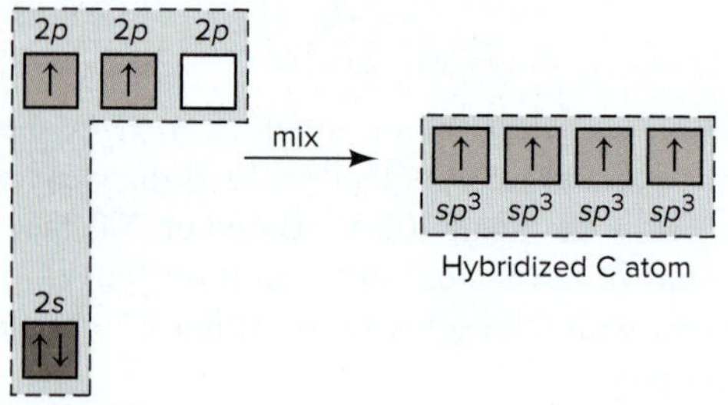

The C—C bond arises from overlap of the end of one sp^3 orbital on one C atom with the end of one sp^3 orbital on the other C atom (Figure 11.10). Each of the six C—H bonds is formed by the overlap of the end of a sp^3 hybrid orbital on the C atom with the 1*s* orbital of a hydrogen atom. Each bond in ethane is a **sigma (σ) bond:**

- A sigma bond is formed by *end-to-end* overlap of orbitals.
- A sigma bond has its *highest electron density along the bond axis* and is shaped like an ellipse rotated about its long axis (like a football).
- *All single bonds are σ bonds.*

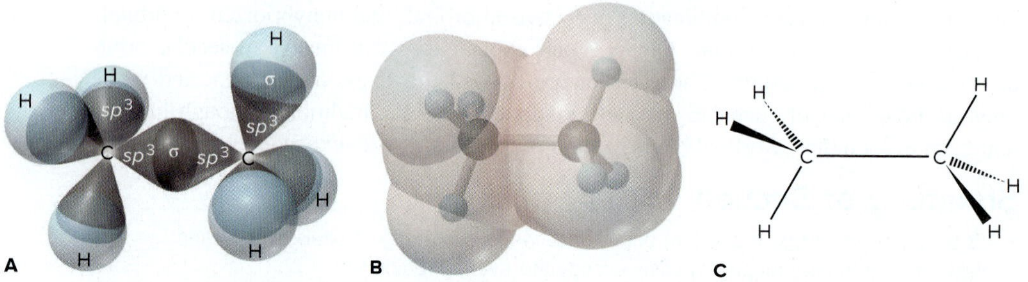

Figure 11.10 **The σ bonds in ethane (C_2H_6). A,** Depiction using atomic contours. **B,** An electron density model shows very slightly positive *(blue)* and negative *(red)* regions. **C,** Wedge-bond perspective drawing of the ethane molecule.

Side-to-Side Overlap and Pi (π) Bonding For this type of bonding, we'll examine ethylene and acetylene:

1. *In ethylene,* each C atom is sp^2 hybridized. The four valence electrons of C half-fill the three sp^2 orbitals *and* the unhybridized 2*p* orbital:

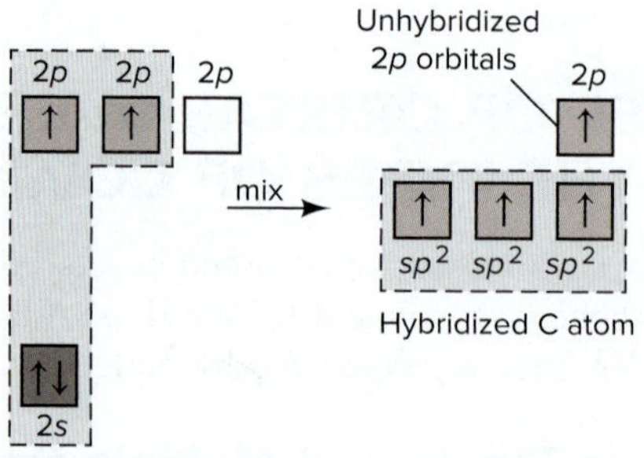

The unhybridized 2*p* orbital lies perpendicular to the sp^2 plane (Figure 11.11). Two sp^2 orbitals of each C form C—H σ bonds. The third sp^2 orbital forms a σ bond with the other C. With the σ-bonded C atoms near each other, a second bond is formed between the two C atoms when their half-filled unhybridized 2*p* orbitals overlap to form a **pi (π) bond:**

- A pi bond is formed by *side to side* overlap of orbitals.
- A pi bond has *two regions (lobes) of electron density,* one above and one below the σ-bond axis. *The two electrons in one π bond occupy both lobes.*
- *A double bond consists of one σ bond and one π bond,* which increases electron density between the nuclei (Figure 11.11D). The two electron pairs act as one electron group because each pair occupies a different orbital, which reduces repulsions.

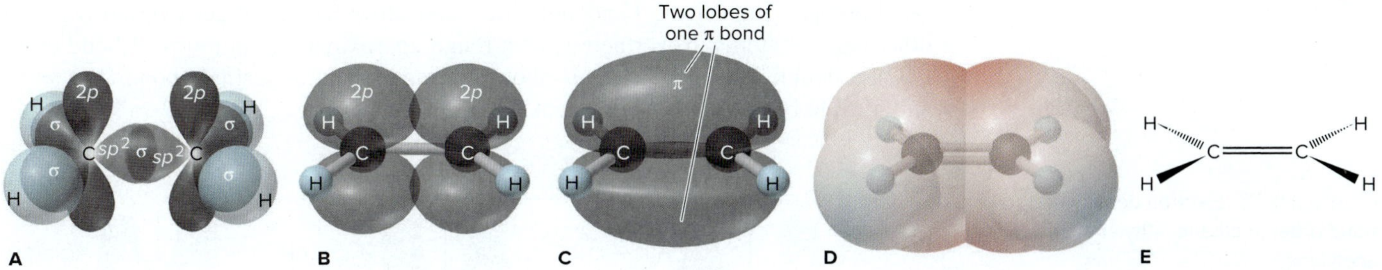

Figure 11.11 **The σ and π bonds in ethylene (C_2H_4). A,** The C—C σ bond and the four C—H σ bonds are shown with the unhybridized 2*p* orbitals. **B,** An accurate depiction of the 2*p* orbitals shows the side-to-side overlap; σ bonds are shown with a ball-and-stick model. **C,** Two overlapping regions comprise *one* π bond, which is occupied by two electrons. **D,** With four electrons (one σ bond and one π bond) between the C atoms, electron density *(red)* is higher there. **E,** Wedge-bond perspective drawing of the ethylene molecule.

2. *In acetylene,* each C atom is *sp* hybridized, and its four valence electrons half-fill the two *sp* hybrids *and* the two unhybridized 2*p* orbitals:

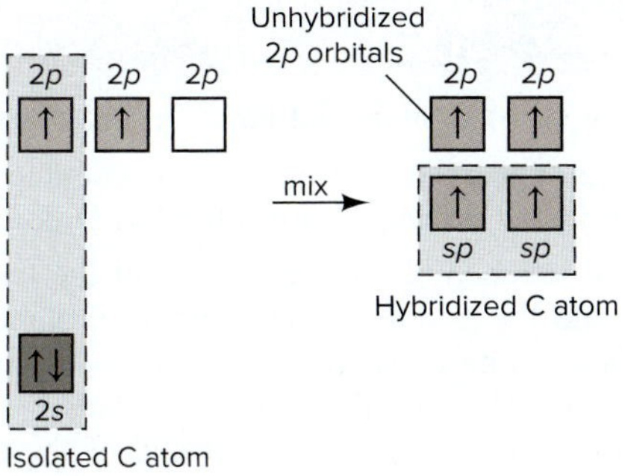

Each C forms a C—H σ bond with one *sp* orbital and a C—C σ bond with the other (Figure 11.12). Side-to-side overlap of one pair of unhybridized 2*p* orbitals gives one π bond, with electron density above and below the σ bond. Side-to-side overlap of the other pair of unhybridized 2*p* orbitals gives another π bond, 90° away from the first, with electron density in front and back of the σ bond. The result is a *cylindrically symmetrical* H—C≡C—H molecule. Note the greater electron density between the C atoms created by the six bonding electrons.

- A triple bond *consists of one* σ *and two* π *bonds.*

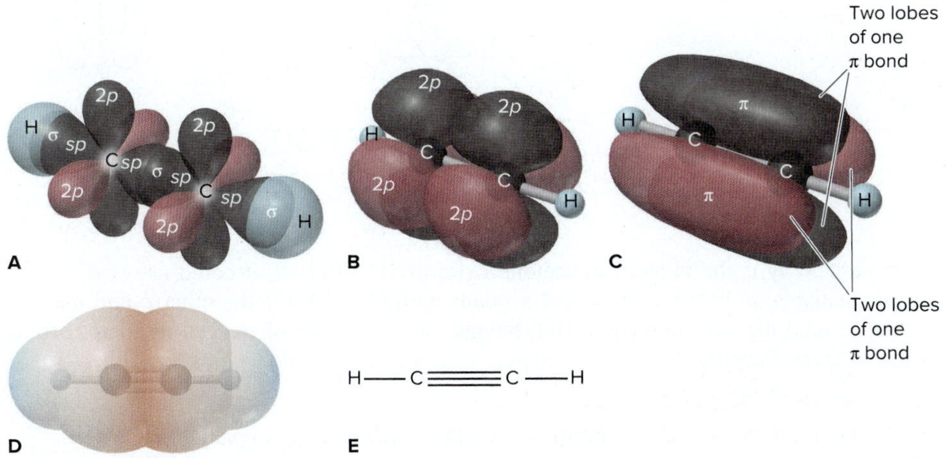

Figure 11.12 **The σ bonds and π bonds in acetylene (C_2H_2). A,** A contour depiction shows the C—C σ bond, the two C—H σ bonds, and two unhybridized 2*p* orbitals (one pair is in *red* for clarity). **B,** The 2*p* orbitals (accurate form) overlap side to side; σ bonds are shown with a ball-and-stick model. **C,** Overlapping regions of two π bonds, one black and the other red. **D,** The molecule has cylindrical symmetry. Six electrons (one σ bond and two π bonds) create even higher electron density *(red)* between the C atoms. **E,** Bond-line drawing of the acetylene molecule.

Mode of Overlap, Bond Strength, and Bond Order Because orbitals overlap less side to side than they do end to end, a π bond is weaker than a σ bond; thus, for carbon-carbon bonds, a double bond is less than twice as strong as a single bond (Table 9.2). Figure 11.13 on the next page shows electron density relief maps of the three types of carbon-carbon bonds; note the increasing electron density between the nuclei from single to double to triple bond.

Lone-pair repulsions, bond polarities, and other factors affect overlap between other pairs of atoms. Nevertheless, as a rough approximation, in terms of bond order (BO), a double bond (BO = 2) is about twice as strong as a single bond (BO = 1), and a triple bond (BO = 3) is about three times as strong.

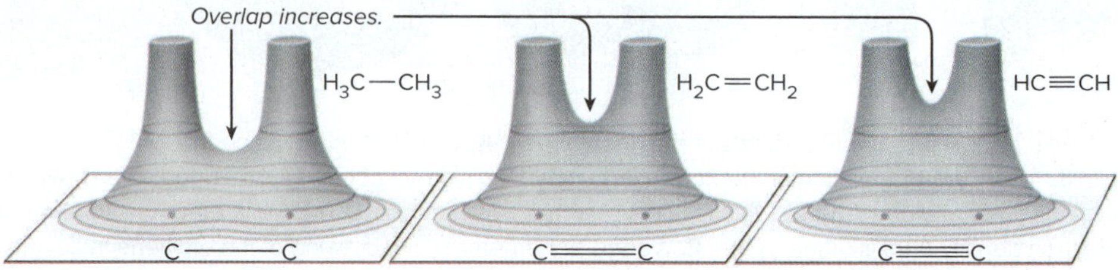

Figure 11.13 Electron density and bond order in ethane, ethylene, and acetylene.

SAMPLE PROBLEM 11.2 Describing the Types of Orbitals and Bonds in Molecules

Problem Describe the types of orbitals and bonds in acetone, $(CH_3)_2CO$.

Plan We use the shape around each central atom to postulate the hybrid orbitals involved and use unhybridized orbitals to form the C═O bond.

Solution The shapes are tetrahedral around each C of the two CH_3 (methyl) groups and trigonal planar around the middle C (see Sample Problem 10.8). Thus, the middle C has three sp^2 orbitals and one unhybridized p orbital. Each of the two methyl C atoms has four sp^3 orbitals. Three of these form σ bonds with the 1s orbitals of H atoms; the fourth forms a σ bond with an sp^2 orbital of the middle C. Thus, two of the three sp^2 orbitals of the middle C form σ bonds to the other two C atoms.

The O atom is also sp^2 hybridized and has an unhybridized p orbital that can form a π bond. Two of the O atom's sp^2 orbitals hold lone pairs, and the third forms a σ bond with the third sp^2 orbital of the middle C atom. The unhybridized, half-filled 2p orbitals of C and O form a π bond. The σ and π bonds constitute the C═O bond:

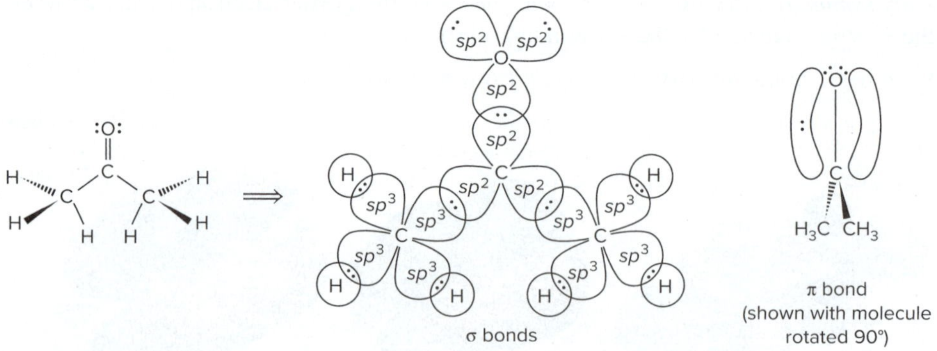

Comment Why is the O atom in acetone hybridized? After all, it could use two perpendicular p orbitals for the σ and π bonds with C and leave the other p and the s orbital to hold the two lone pairs. But, having each lone pair of oxygen in an sp^2 orbital oriented away from the C═O bond lowers electron-electron repulsions.

FOLLOW-UP PROBLEMS

11.2A Describe the types of orbitals and bonds in **(a)** hydrogen cyanide, HCN; **(b)** carbon dioxide, CO_2.

11.2B Describe the types of orbitals and bonds in **(a)** carbon monoxide; **(b)** urea, $H_2NC(═O)NH_2$.

SOME SIMILAR PROBLEMS 11.21–11.24

Orbital Overlap and Rotation Within a Molecule

The type of overlap—end to end or side to side—affects rotation around the bond:

- *Sigma bond.* A σ *bond allows free rotation* because the extent of overlap is not affected. If you could hold one CH_3 group of ethane, the other CH_3 could spin without affecting the overlap of the C═C σ bond (see Figure 11.10).
- *Pi bond.* A π *bond restricts rotation* because *p* orbitals must be parallel to each other to overlap most effectively. Holding one CH_2 group in ethylene and trying to spin the other decreases the side-to-side overlap and breaks the π bond. For this reason, distinct *cis* and *trans* structures exist for compounds such as 1,2-dichloroethylene (Section 10.3). As Figure 11.14 shows, the π bond allows two *different* arrangements of atoms around the C atoms, which has a major effect on molecular polarity (see Section 10.3). Rotation around a triple bond is not meaningful: each triple-bonded C atom is bonded to one other group in a linear arrangement, so there can be no difference in the relative positions of attached groups.

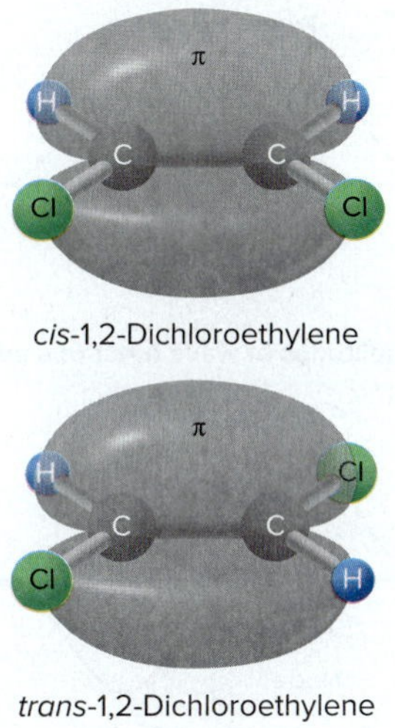

Figure 11.14 Restricted rotation around a π bond. *Cis*- and *trans*-1,2-dichloroethylene are different molecules because the π bond restricts rotation.

› Summary of Section 11.2

› End-to-end overlap of atomic orbitals forms a σ bond, which allows free rotation of the bonded parts of the molecule.

› A multiple bond consists of a σ bond and either one π bond (double bond) or two π bonds (triple bond). Multiple bonds have greater electron density between the nuclei than single bonds do and, thus, higher bond energies.

› Side-to-side overlap of orbitals in a π bond restricts rotation.

11.3 MOLECULAR ORBITAL (MO) THEORY AND ELECTRON DELOCALIZATION

Scientists choose the model that best answers the question they are posing: VSEPR theory for one about molecular shape, or VB theory for one about orbital overlap. But neither model adequately explains magnetic and spectral properties, and both understate the importance of electron delocalization. To deal with these phenomena, which involve molecular energy levels, chemists apply **molecular orbital (MO) theory.** The MO model is a quantum-mechanical treatment for molecules similar to the one for atoms (Chapter 8): just as an atom has atomic orbitals (AOs) of given energies and shapes that are occupied by the atom's electrons, a molecule has **molecular orbitals (MOs)** of given energies and shapes that are occupied by the molecule's electrons.

There is a key distinction between VB and MO theories:

- VB theory pictures a molecule as a group of atoms bonded through overlapping of valence-shell atomic and/or hybrid orbitals occupied by *localized* electrons.
- MO theory pictures a molecule as a collection of nuclei with orbitals that extend over the whole molecule and are occupied by *delocalized* electrons.

Despite its usefulness, MO theory has the drawback that MOs are more difficult to visualize than the shapes of VSEPR theory or the hybrid AOs of VB theory.

The Central Themes of MO Theory

Several key ideas of MO theory appear in its description of H_2 and other simple species: how MOs form, what their energies and shapes are, and how they fill with electrons.

Formation of Molecular Orbitals Just as the Schrödinger equation gives only approximate solutions for any atom with more than one electron (Chapter 8), determining the MOs of even the simplest molecule, H_2, requires an approximation. The approximation is to mathematically *combine* (add or subtract) AOs (atomic wave

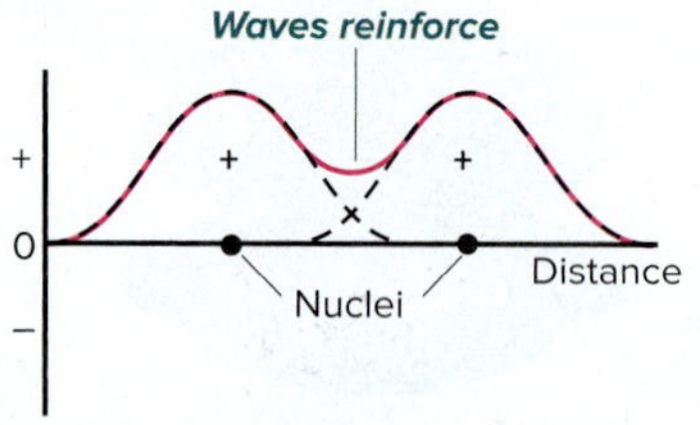

A Amplitudes of wave functions added

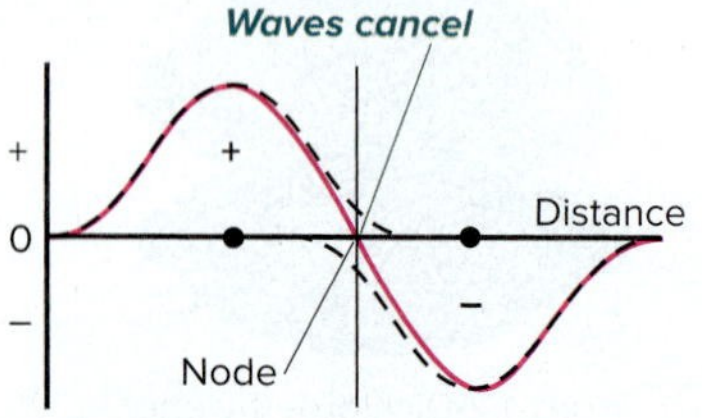

B Amplitudes of wave functions subtracted

Figure 11.15 An analogy between light waves and atomic wave functions.

functions) of nearby atoms to form MOs (molecular wave functions). Thus, when two H nuclei lie near each other, their AOs overlap and combine in two ways:

- *Adding the wave functions together.* This combination forms a **bonding MO,** which has *a region of high electron density between the nuclei.* Additive overlap is analogous to light waves reinforcing each other, which makes the amplitude higher and the light brighter. For electron waves, the overlap *increases* the probability that the electrons are between the nuclei (Figure 11.15A).
- *Subtracting the wave functions from each other.* This combination forms an **antibonding MO,** which has *a node, a region of zero electron density, between the nuclei* (Figure 11.15B). Subtractive overlap is analogous to light waves canceling each other, causing the light to disappear. With electron waves, subtractive overlap means the probability that the electrons lie between the nuclei *decreases* to zero.

The two possible combinations for hydrogen atoms H_A and H_B are

AO of H_A + AO of H_B = bonding MO of H_2 (more e^- density between nuclei)

AO of H_A − AO of H_B = antibonding MO of H_2 (less e^- density between nuclei)

Notice that *the number of AOs combined always equals the number of MOs formed:* two H atomic orbitals combine to form two H_2 molecular orbitals.

Shape and Energy of H_2 Molecular Orbitals Bonding and antibonding MOs have different shapes and energies. Figure 11.16 shows these orbitals for H_2:

- *Bonding MO.* A bonding MO is *lower in energy* than the AOs that form it. Because the electron density is spread mostly *between* the nuclei, nuclear repulsions decrease while nucleus-electron attractions increase. Moreover, two electrons in this MO can delocalize their charges over a larger volume than they could in nearby, separate AOs, which lowers electron repulsions. Because of these electrostatic effects, when electrons occupy this MO, the H_2 molecule is *more stable* than the separate H atoms.
- *Antibonding MO.* An antibonding MO is *higher in energy* than the AOs that form it. With most of the electron density *outside* the internuclear region and *a node between the nuclei,* nuclear repulsions increase. Therefore, when electrons occupy this MO, the H_2 molecule is *less stable* than the separate H atoms.

Both the bonding and antibonding MOs of H_2 are **sigma (σ) MOs** because they are cylindrically symmetrical about an imaginary line between the nuclei. The bonding MO is denoted by σ_{1s}; that is, a σ MO derived from 1*s* AOs. Antibonding orbitals are denoted with a superscript star: the antibonding MO derived from 1*s* AOs is σ^*_{1s} (spoken "sigma, one ess, star").

For AOs to interact enough to form MOs, they must be similar in *energy* and *orientation.* The 1*s* orbitals of two H atoms have identical energy and orientation, so they interact strongly. We'll revisit this requirement for molecules composed of many-electron atoms.

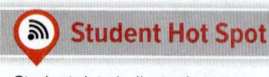

Student Hot Spot

Student data indicate that you may struggle with the concept of molecular orbitals. Access the SmartBook to view additional Learning Resources on this topic.

Electrons in Molecular Orbitals Several aspects of MO theory—filling of MOs, energy-level diagrams, electron configuration, and bond order—relate to earlier ideas:

1. *Filling MOs with electrons.* Electrons enter MOs just as they do AOs:

- MOs are filled in order of increasing energy (aufbau principle).

Figure 11.16 Contours and energies of H_2 bonding and antibonding MOs.

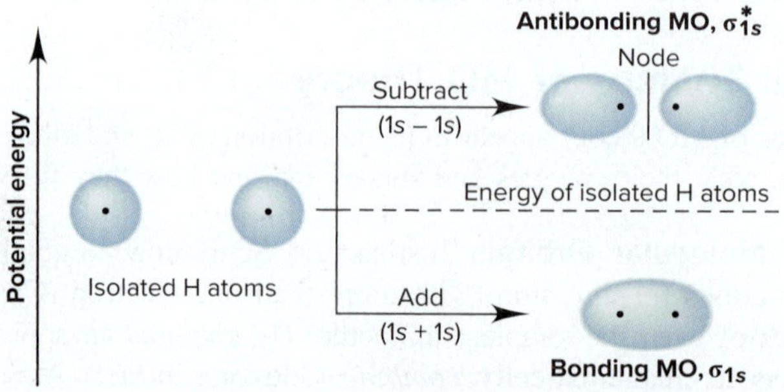

- An MO can hold a maximum of two electrons with opposite spins (exclusion principle).
- Orbitals of equal energy are half-filled, with electron spins parallel, before any of them are filled (Hund's rule).

2. *MO energy-level diagrams.* A **molecular orbital (MO) diagram** shows the relative energy and number of electrons for each MO, as well as for the AOs from which they are formed. In the MO diagram for H_2 (Figure 11.17), two electrons, one from the AO of each H, fill the lower energy bonding MO, while the higher energy antibonding MO remains empty.

3. *Electron configuration.* Just as we can write the electron configuration for an atom, we can write one for a molecule. The symbol of each occupied MO is written in parentheses, with the number of electrons in it as a superscript outside: the electron configuration of H_2 is $(\sigma_{1s})^2$.

4. *Bond order.* In a Lewis structure, bond order is the number of electron pairs per atom-to-atom linkage. The **MO bond order** is the number of electrons in bonding MOs minus the number in antibonding MOs, multiplied by $\frac{1}{2}$:

$$\text{Bond order} = \tfrac{1}{2}[(\text{no. of } e^- \text{ in bonding MOs}) - (\text{no. of } e^- \text{ in antibonding MOs})] \quad \mathbf{(11.1)}$$

Here are three key points about MO bond order:

- Bond order > 0: the molecule is more stable than the separate atoms, so it *will* form. For H_2, the bond order is $\frac{1}{2}(2 - 0) = 1$.
- Bond order = 0: the molecule is as stable as the separate atoms, so it will *not* form (occurs when equal numbers of electrons occupy bonding and antibonding MOs).
- In general, the *higher* the bond order, the *stronger* the bond is.

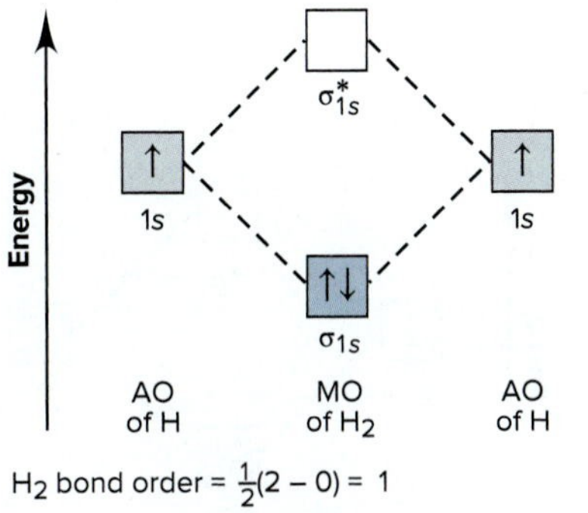

Figure 11.17 MO diagram for H_2. The vertical placement of the boxes indicates relative energies. Orbital occupancy is shown with arrows and color (darker color = full occupancy, paler color = half-filled, no color = empty).

Do He_2^+ and He_2 Exist? One of the early triumphs of MO theory was its ability to *predict* the existence of He_2^+, the helium molecule-ion, which consists of two He nuclei and three electrons. Let's use MO diagrams to see why He_2^+ exists but He_2 doesn't:

- In He_2^+, the 1*s* AOs form MOs (Figure 11.18A). The three electrons are distributed as a pair in the σ_{1s} MO and a lone electron in the σ^*_{1s} MO. The bond order is $\frac{1}{2}(2 - 1) = \frac{1}{2}$. Thus, He_2^+ has a relatively weak bond, but it should exist. Indeed, this species has been observed frequently when He atoms collide with He^+ ions. Its electron configuration is $(\sigma_{1s})^2\,(\sigma^*_{1s})^1$.
- In He_2, with two electrons in the σ_{1s} MO and two electrons in the σ^*_{1s} MO, both the bonding and antibonding orbitals are filled (Figure 11.18B). Stabilization from the electron pair in the bonding MO is canceled by destabilization from the electron pair in the antibonding MO. Because the bond order is zero $[\frac{1}{2}(2 - 2) = 0]$, we predict, and experiment has so far confirmed, that a covalent He_2 molecule does not exist.

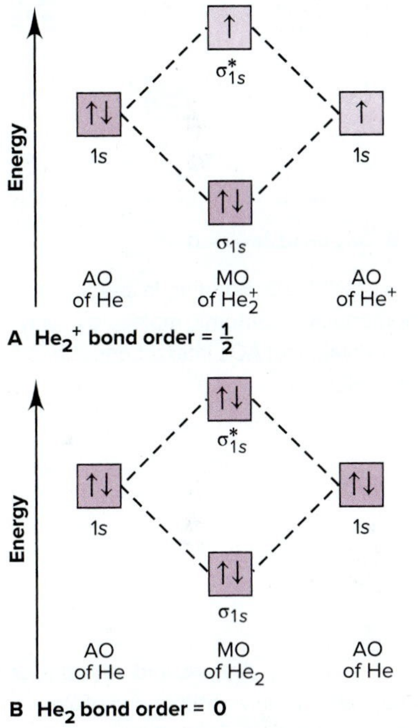

Figure 11.18 MO diagrams for He_2^+ and He_2.

SAMPLE PROBLEM 11.3 Predicting Stability of Species Using MO Diagrams

Problem Use an MO diagram to find the bond order and predict whether H_2^+ exists. If it exists, write its electron configuration.

Plan Since the 1*s* AOs form the MOs, the MO diagram is similar to the one for H_2. We find the number of electrons and distribute them one at a time to the MOs in order of increasing energy. We obtain the bond order with Equation 11.1 and write the electron configuration as described in the text.

Solution H_2 has two e^-, so H_2^+ has only one, which enters the bonding MO *(see diagram in margin)*. The bond order is $\frac{1}{2}(1 - 0) = \frac{1}{2}$, so we predict that H_2^+ exists. The electron configuration is $(\sigma_{1s})^1$.

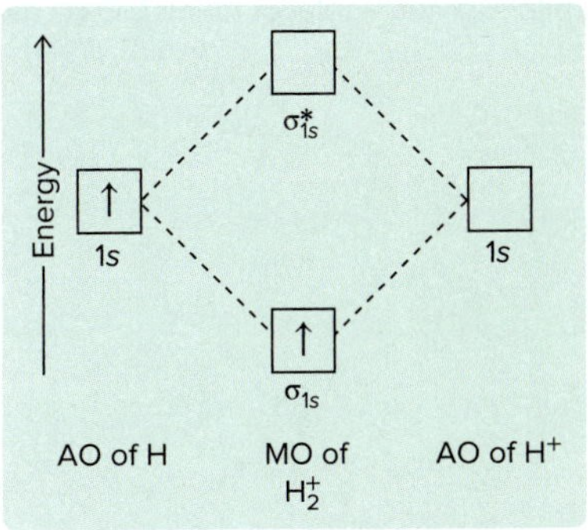

Check The number of electrons in the MOs equals the number of electrons in the AOs.

Comment This species has been detected spectroscopically in the material around stars.

FOLLOW-UP PROBLEMS

11.3A Use an MO diagram to find the bond order and predict whether two H^- ions could form H_2^{2-}. If this species exists, write its electron configuration.

11.3B Use an MO diagram to find the bond order and predict whether two He^+ ions could form He_2^{2+}. If this species exists, write its electron configuration.

A SIMILAR PROBLEM 11.55

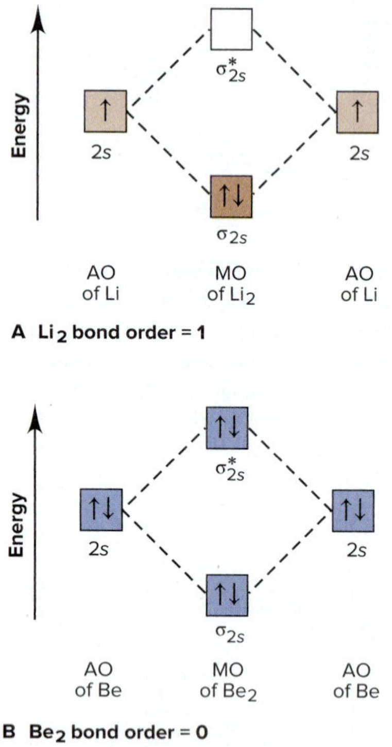

Figure 11.19 **Bonding in *s*-block homonuclear diatomic molecules.** Only outer (valence) AOs interact enough to form MOs.

Homonuclear Diatomic Molecules of Period 2 Elements

Homonuclear diatomic molecules are composed of two identical atoms. In addition to H_2 from Period 1, you're familiar with N_2, O_2, and F_2 from Period 2 as the elemental forms under standard conditions. Others in Period 2—Li_2, Be_2, B_2, C_2, and Ne_2—are observed, if at all, at high temperatures. We'll divide them into molecules from the *s* block, Groups 1A(1) and 2A(2), and those from the *p* block, Groups 3A(13) through 8A(18).

Bonding in the s-Block Homonuclear Diatomic Molecules Both Li and Be occur as metals under normal conditions, but MO theory can examine their stability as diatomic gases, dilithium (Li_2) and diberyllium (Be_2).

These atoms have electrons in inner (1*s*) and outer (2*s*) AOs, but we ignore the inner ones because, in general, *only outer (valence) AOs interact enough to form MOs.* Like those formed from 1*s* AOs, these 2*s* AOs form σ MOs, cylindrically symmetrical around the internuclear axis.

- In Li_2, the two valence electrons fill the bonding (σ_{2s}) MO, with opposing spins, leaving the antibonding (σ^*_{2s}) MO empty (Figure 11.19A). The bond order is $\frac{1}{2}(2 - 0) = 1$. In fact, Li_2 *has* been observed; the electron configuration is $(\sigma_{2s})^2$.
- In Be_2, the four valence electrons fill the σ_{2s} and σ^*_{2s} MOs (Figure 11.19B), giving an orbital occupancy similar to that in He_2. The bond order is $\frac{1}{2}(2 - 2) = 0$, and the ground state of Be_2 has never been observed.

Shape and Energy of MOs from Atomic *p*-Orbital Combinations In boron, atomic 2*p* orbitals are occupied. Recall that *p* orbitals can overlap by two modes, which correspond to two ways their wave functions combine (Figure 11.20):

- End-to-end combination gives a pair of σ MOs, σ_{2p} and σ^*_{2p}.
- Side-to-side combination gives a pair of **pi (π) MOs,** π_{2p} and π^*_{2p}.

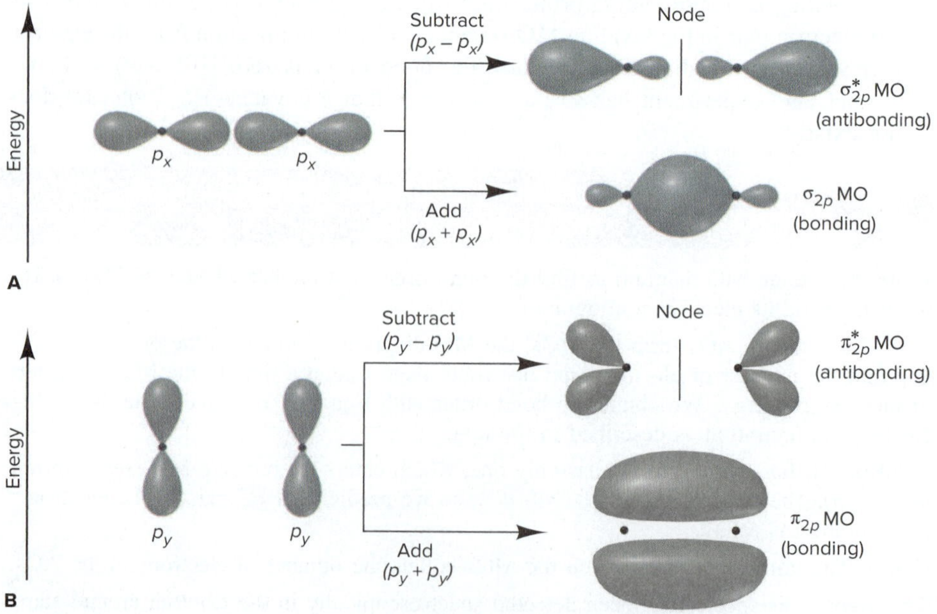

Figure 11.20 **Shapes and energies of σ and π MOs from combinations of 2*p* atomic orbitals. A,** The *p* orbitals lying along the internuclear axis (designated p_x) overlap end to end and form σ_{2p} and σ^*_{2p} MOs. **B,** The *p* orbitals perpendicular to the internuclear axis overlap side to side and form two π MOs. (The p_y interactions, shown here, are the same as for the p_z orbitals, giving a total of four π MOs.)

Despite the different shapes, MOs derived from p orbitals are like those from s orbitals. Bonding MOs have most of the electron density *between* the nuclei, and antibonding MOs have most of it *outside* the internuclear region, with a node between the nuclei.

The order of energy levels for MOs, whether bonding or antibonding, is based on the order of AO energy levels *and* on the mode of the p-orbital overlap:

- MOs formed from 2s orbitals are *lower in energy* than MOs formed from 2p orbitals because 2s AOs are lower in energy than 2p AOs.
- Bonding MOs are *lower in energy* than antibonding MOs: σ_{2p} is lower in energy than σ^*_{2p} and π_{2p} is lower than π^*_{2p}.
- Atomic p orbitals overlap more extensively end to end than side to side. Thus, the σ_{2p} MO is usually lower in energy than the π_{2p} MO. We also find that the destabilizing effect of the σ^*_{2p} MO is greater than that of the π^*_{2p} MO.

Thus, the energy order for MOs derived from 2p orbitals is typically

$$\sigma_{2p} < \pi_{2p} < \pi^*_{2p} < \sigma^*_{2p}$$

Each atom has three mutually perpendicular 2p orbitals. When the six p orbitals in two atoms combine, the two that interact end to end form one σ and one σ* MO, and the two pairs of orbitals that interact side to side form two π MOs and two π* MOs. Placing these orientations within the energy order gives the *expected* MO diagram for the p-block Period 2 homonuclear diatomic molecules (Figure 11.21A).

Recall that only AOs of similar energy interact enough to form MOs. This fact leads to two energy orders:

1. ***Without** mixing of s and p orbitals: O_2, F_2, and Ne_2.* The order in Figure 11.21A assumes that s and p AOs are so different in energy that they do not interact; we say the orbitals do not *mix*. Lying at the right of Period 2, O, F, and Ne are relatively small. Thus, as electrons start to pair up in the half-filled 2p orbitals, strong repulsions raise the energy of the 2p orbitals high enough above the 2s orbitals to prevent orbital mixing. Thus the energy order of the MOs derived from the 2p orbitals is as expected for O_2, F_2, and Ne_2:

$$\sigma_{2p} < \pi_{2p} < \pi^*_{2p} < \sigma^*_{2p}$$

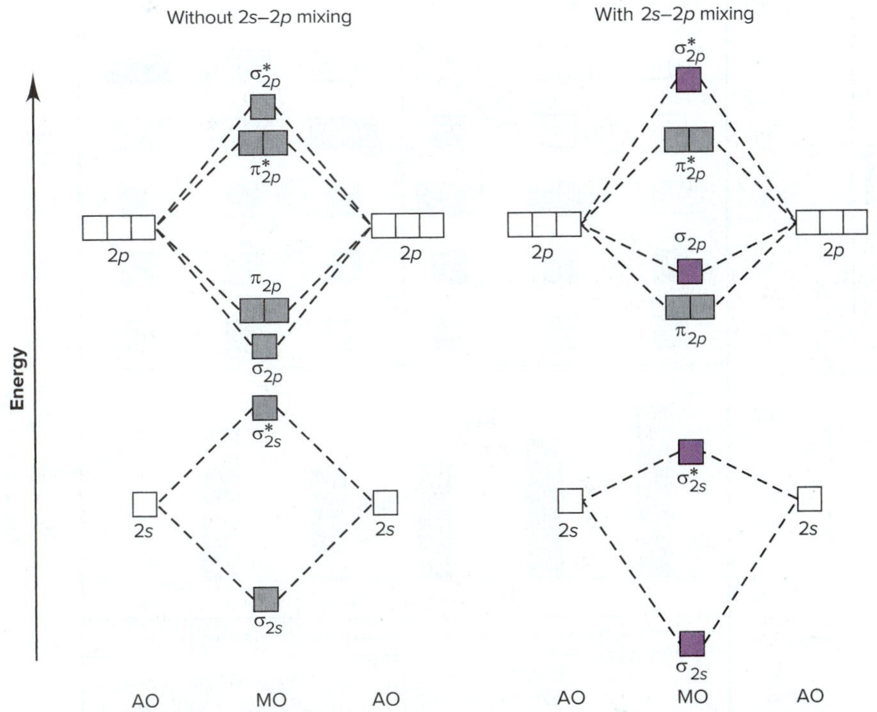

A MO energy levels for O_2, F_2, and Ne_2

B MO energy levels for B_2, C_2, and N_2

Figure 11.21 Relative MO energy levels for Period 2 homonuclear diatomic molecules. (For clarity, the MOs affected by mixing are shown in *purple*.)

2. ***With** mixing of s and p orbitals: B_2, C_2, and N_2*. B, C, and N atoms are relatively large, with $2p$ AOs only half-filled, so repulsions are weaker. As a result, orbital energies are close enough for some mixing to occur between the $2s$ of one atom and the end-on $2p$ of the other. The effect is to *lower* the energy of the σ_{2s} and σ^*_{2s} MOs and *raise* the energy of the σ_{2p} and σ^*_{2p} MOs; the π MOs are not affected. The MO diagram for B_2, C_2, and N_2 reflects this mixing (Figure 11.21B). The only qualitative difference from the MO diagram for O_2, F_2, and Ne_2 is the *reversal in energy order* of the σ_{2p} and π_{2p} MOs:

$$\pi_{2p} < \sigma_{2p} < \pi^*_{2p} < \sigma^*_{2p}$$

Bonding in the *p*-Block Homonuclear Diatomic Molecules Figure 11.22 shows the MOs, electron occupancy, and some properties of B_2 through Ne_2. Note that

1. *Higher* bond order correlates with *greater* bond energy and *shorter* bond length.

2. Orbital occupancy correlates with magnetic properties. Recall from Chapter 8 that if a substance has unpaired electrons, it is *paramagnetic,* and if all electrons are paired, the substance is *diamagnetic;* the same distinction applies to molecules. Let's examine the MO occupancy and some properties of these molecules:

- *B_2*. The B_2 molecule has six outer electrons: four fill the σ_{2s} and σ^*_{2s} MOs. The remaining two electrons occupy the two π_{2p} MOs, one in each orbital, in keeping with Hund's rule. With four electrons in bonding MOs and two in antibonding MOs, the bond order of B_2 is $\frac{1}{2}(4 - 2) = 1$. As expected from the two lone electrons, B_2 is paramagnetic.
- *C_2*. Two additional electrons in C_2 fill the two π_{2p} MOs. With two more bonding electrons than B_2, the bond order of C_2 is 2 and the bond is stronger and shorter. But with all the electrons paired, C_2 is diamagnetic.
- *N_2*. Two more electrons in N_2 fill the σ_{2p} MO, so the molecule is also diamagnetic. The bond order of 3 is consistent with the triple bond in the Lewis structure and with a stronger, shorter bond.

Figure 11.22 **MO occupancy and some properties of B_2 through Ne_2.** Energy order and occupancy of MOs are above a bar graph showing bond energy and length, bond orders, magnetic properties, and outer (valence) electron configurations.

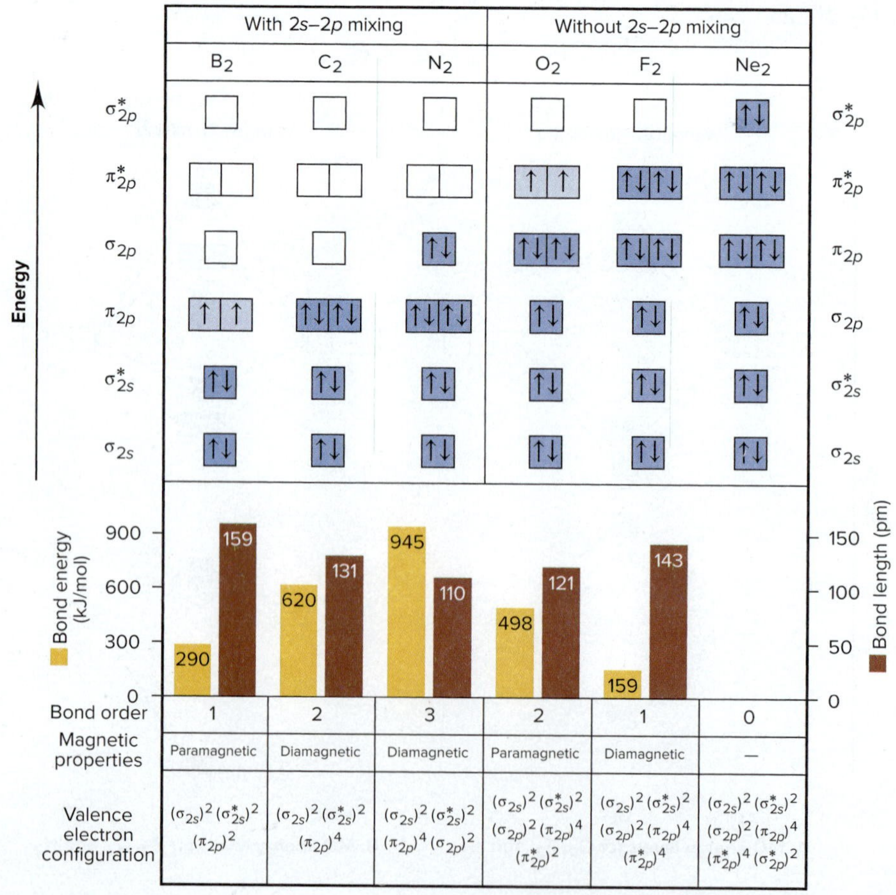

- *O_2*. With this molecule, we see the power of MO theory over VB theory and others based on electrons in localized orbitals. It is impossible to write one Lewis structure consistent with the fact that O_2 has a double bond and is paramagnetic. We can write one structure with a double bond and paired electrons, and another with a single bond and two unpaired electrons:

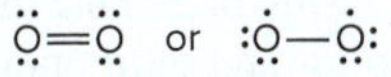

MO theory resolves this paradox beautifully: with eight electrons in bonding MOs and four in antibonding MOs, the bond order is $\frac{1}{2}(8 - 4) = 2$, and the molecule is paramagnetic because *one* electron occupies each of *two* π^*_{2p} MOs and these two electrons have unpaired (parallel) spins. A thin stream of liquid O_2 will remain suspended between the poles of a powerful magnet *(see the chapter-opening photo)*. As expected, the bond is weaker and longer than the one in N_2.
- *F_2*. Two more electrons in F_2 fill the π^*_{2p} orbitals, so it is diamagnetic. The bond has an order of 1 and is weaker and longer than the one in O_2. Note that this bond is shorter than the single bond in B_2 but only about half as strong. We might have expected it to be stronger because F is smaller than B. But 18 electrons in the smaller volume of F_2 cause greater repulsions than the 10 electrons in B_2, making the F_2 bond weaker.
- *Ne_2*. The final member of the Period 2 series doesn't exist for the same reason He_2 doesn't: all the MOs are filled, which gives a bond order of zero.

SAMPLE PROBLEM 11.4 Using MO Theory to Explain Bond Properties

Problem Explain the following data with diagrams showing the occupancy of MOs:

	N_2	N_2^+	O_2	O_2^+
Bond energy (kJ/mol)	945	841	498	623
Bond length (pm)	110	112	121	112

Plan The data show that removing an electron from each parent molecule has opposite effects: N_2^+ has a weaker, longer bond than N_2, but O_2^+ has a stronger, shorter bond than O_2. We determine the valence electrons in each species, draw the sequence of MO energy levels (showing orbital mixing in N_2 but not in O_2), and fill them with electrons. To explain the data, we calculate bond orders, which relate directly to bond energy and inversely to bond length.

Solution Determining the valence electrons:

N has 5 valence e^-, so N_2 has 10 and N_2^+ has 9.

O has 6 valence e^-, so O_2 has 12 and O_2^+ has 11.

Drawing and filling the MO diagrams:

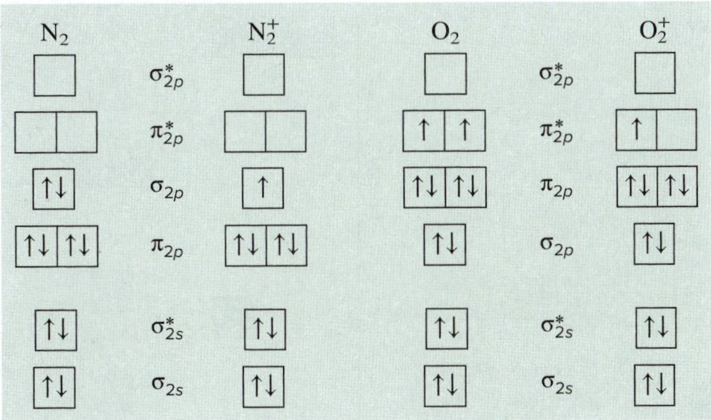

Calculating bond orders:

$\frac{1}{2}(8 - 2) = 3$ $\quad$ $\frac{1}{2}(7 - 2) = 2.5$ $\quad$ $\frac{1}{2}(8 - 4) = 2$ $\quad$ $\frac{1}{2}(8 - 3) = 2.5$

Explaining the data:

1. When N_2 becomes N_2^+, a *bonding* electron is removed, so the bond order decreases. Thus, N_2^+ has a weaker, longer bond than N_2.
2. When O_2 becomes O_2^+, an *antibonding* electron is removed, so the bond order increases. Thus, O_2^+ has a stronger, shorter bond than O_2.

Check The answers make sense in terms of bond order, bond energy, and bond length. Check that the total number of bonding and antibonding electrons equals the total number of valence electrons for each species.

FOLLOW-UP PROBLEMS

11.4A Determine the bond orders for these other dinitrogen species: N_2^{2+}, N_2^{2-}, and N_2^{2-}. List the species in order of decreasing bond energy and in order of decreasing bond length.

11.4A Determine the bond orders for the following difluorine species: F_2^{2-}, F_2^-, F_2, F_2^+, and F_2^{2+}. List the species that exist in order of increasing bond energy and in order of increasing bond length.

SOME SIMILAR PROBLEMS 11.36 and 11.37

Student Hot Spot

Student data indicate that you may struggle with drawing molecular orbital diagrams. Access the Smartbook to view additional Learning Resources on this topic.

Two Heteronuclear Diatomic Molecules: HF and NO

Heteronuclear diatomic molecules have asymmetric MO diagrams because the AOs of the *different* atoms have unequal energies. Atoms with greater effective nuclear charge (Z_{eff}) pull their electrons closer, so they have more stable (lower energy) AOs (Section 8.1) and higher EN values (Section 9.5). Let's examine the bonding in HF and NO.

Bonding in HF To form the MOs in HF, we decide which AOs will combine. The high Z_{eff} of F means that its electrons are held so tightly that the 1*s*, 2*s*, and 2*p* orbitals have lower energy than the 1*s* of H. The half-filled 2*p* orbital of F interacts with the 1*s* of H through end-on overlap, which forms σ and σ* MOs. The two filled 2*p* orbitals of F are called **nonbonding MOs.** Because they are not involved in bonding, they have the same energy as the isolated AOs (Figure 11.23A).

The bonding MO of HF is closer in energy to the AOs of F, so the F 2*p* orbital contributes more to the HF bond than the H 1*s* does. In polar covalent molecules, *bonding MOs are closer in energy to the AOs of the more electronegative atom.* In

Figure 11.23 The MO diagrams for (A) HF and (B) NO.

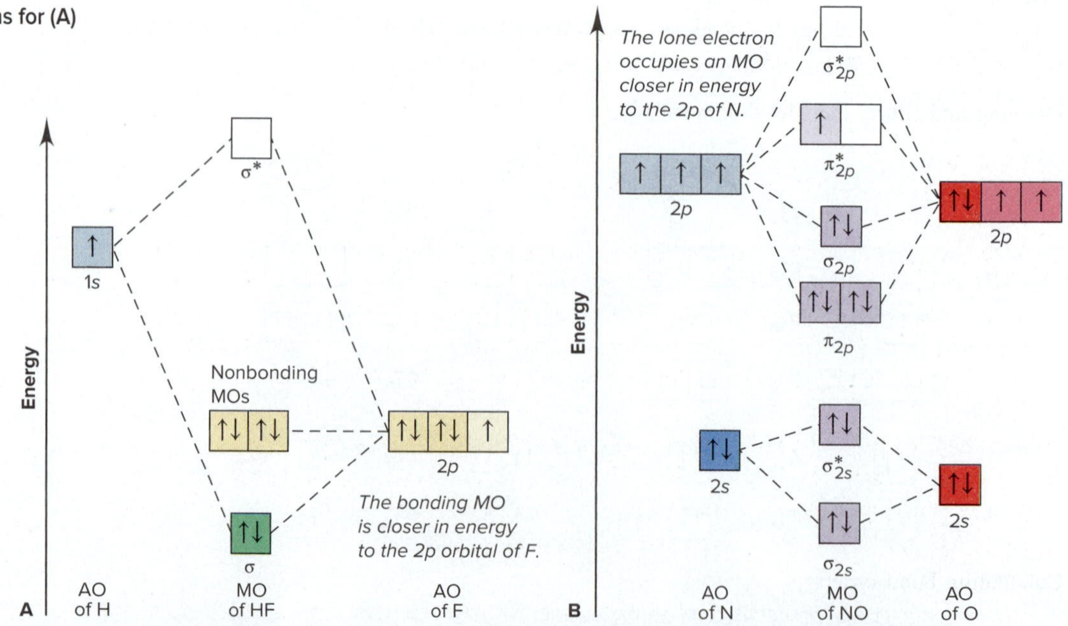

effect, fluorine's greater electronegativity lowers the energy of the bonding MO and draws the bonding electrons closer.

Bonding in NO Nitrogen monoxide (nitric oxide) is highly reactive because of its lone electron. Two possible Lewis structures for NO, with formal charges (Section 10.1), are

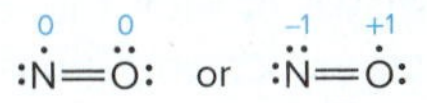

I II

Both structures show a double bond, but the *measured* bond energy suggests a bond order *higher* than 2. Moreover, it is not clear where the lone electron resides, although the lower formal charges for structure I suggest that it is on the N.

The MO diagram is asymmetric, with the AOs of the more electronegative O lower in energy (Figure 11.23B). When the 11 valence electrons fill the MOs of NO, the lone electron occupies one of the π^*_{2p} orbitals. Eight bonding and three antibonding electrons give a bond order of $\frac{1}{2}(8 - 3) = 2.5$, more in keeping with the data than either Lewis structure. Bonding electrons lie in MOs closer in energy to the AOs of the O atom. The 2*p* orbitals of N contribute more to the orbital that holds the lone electron, so it spends more time closer to N. (This diagram shows the energy order *with* mixing of the 2*s* and 2*p* orbitals, as in Figure 11.21, but the result is the same as the order without mixing.)

Two Polyatomic Molecules: Benzene and Ozone

The orbital shapes and MO diagrams for polyatomic molecules are too complex for detailed treatment here. But we'll briefly discuss how the model eliminates the need for resonance forms and helps explain the effects of the absorption of energy.

Recall that we cannot draw one Lewis structure for either benzene or ozone, so we draw separate forms of a resonance hybrid. The VB model also uses resonance because it relies on *localized* bonds. In contrast, MO theory pictures a structure of delocalized σ and π MOs. Figure 11.24 shows the lowest energy π bonding MOs in benzene and ozone. The extended areas of electron density allow delocalization of one electron pair in the π bonding MO over the entire molecule, thus eliminating the need for separate resonance forms:

- *In benzene*, the upper and lower hexagonal lobes of this MO lie above and below the plane of the six C nuclei.
- *In ozone*, the two lobes of this MO extend over and under the three O nuclei.

The full MO diagrams for these molecules help us rationalize how bonding electrons in O_3 become excited and occupy empty antibonding orbitals when the molecule absorbs UV radiation in the stratosphere, and why the UV spectrum of benzene has its characteristic absorption bands.

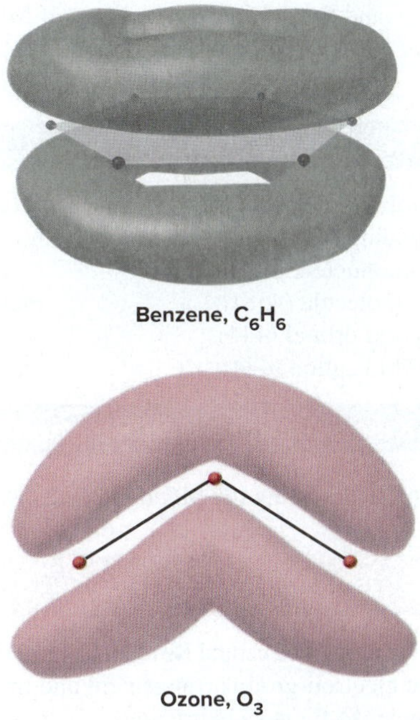

Figure 11.24 The lowest energy π bonding MOs in benzene and ozone.

› Summary of Section 11.3

- › Molecular orbital (MO) theory treats a molecule as a collection of nuclei with MOs delocalized over the entire structure.
- › Atomic orbitals of comparable energy can be added or subtracted to obtain bonding or antibonding MOs, respectively.
- › Bonding MOs, whether σ or π, have most of the electron density between the nuclei and are lower in energy than the AOs that combine to form them; most of the electron density in antibonding MOs does not lie between the nuclei, so these MOs are higher in energy.
- › MOs are filled in order of their energy with paired electrons having opposite spins.
- › MO diagrams show energy levels and orbital occupancy. Diagrams for the Period 2 homonuclear diatomic molecules explain bond energy, bond length, and magnetic behavior.
- › In heteronuclear diatomic molecules, the more electronegative atom contributes more to the bonding MOs.
- › MO theory eliminates the need for resonance forms to depict polyatomic molecules.

CHAPTER REVIEW GUIDE

Learning Objectives

Relevant section (§) and/or sample problem (SP) numbers appear in parentheses.

Understand These Concepts

1. The main ideas of valence bond theory—orbital overlap, opposing electron spins, and hybridization—as a means of rationalizing molecular shapes (§11.1)
2. How orbitals mix to form hybrid orbitals with different spatial orientations (§11.1)
3. The distinction between end-to-end and side-to-side overlap and the origin of sigma (σ) and pi (π) bonds in simple molecules (§11.2)
4. How the two modes of orbital overlap lead to single, double, and triple bonds (§11.2)
5. Why π bonding restricts rotation around double bonds (§11.2)
6. The distinction between the localized bonding of VB theory and the delocalized bonding of MO theory (§11.3)
7. How addition or subtraction of AOs forms bonding or antibonding MOs (§11.3)
8. The shapes of MOs formed from combinations of two *s* orbitals and combinations of two *p* orbitals (§11.3)
9. How MO bond order predicts the stability of molecular species (§11.3)
10. How MO theory explains the bonding and magnetic properties of homonuclear and heteronuclear diatomic molecules of Period 2 elements (§11.3)

Master These Skills

1. Using a partial orbital diagram and molecular shape to postulate the hybrid orbitals used by a central atom (SP 11.1)
2. Describing the types of orbitals and bonds in a molecule (SP 11.2)
3. Using MO diagrams to find the bond order and predict the stability of molecular species and writing the electron configuration of a species that exists (SP 11.3)
4. Explaining bond properties with MO theory (SP 11.4)

Key Terms

Page numbers appear in parentheses.

antibonding MO (456)
bonding MO (456)
homonuclear diatomic molecule (458)
hybrid orbital (444)
hybridization (444)
MO bond order (457)
molecular orbital (MO) diagram (457)
molecular orbital (MO) theory (455)
molecular orbital (MO) (455)
nonbonding MO (462)
pi (π) bond (452)
pi (π) MO (458)
sigma (σ) bond (452)
sigma (σ) MO (456)
sp hybrid orbital (444)
sp^2 hybrid orbital (446)
sp^3 hybrid orbital (447)
sp^3d hybrid orbital (447)
sp^3d^2 hybrid orbital (448)
valence bond (VB) theory (443)

Key Equations and Relationships

Page number appears in parentheses.

11.1 Calculating the MO bond order (457): Bond order $= \frac{1}{2}$[(no. of e^- in bonding MOs) − (no. of e^- in antibonding MOs)]

BRIEF SOLUTIONS TO FOLLOW-UP PROBLEMS

11.1A (a) The central Be atom has two electron groups (two bonds); the electron-group arrangement and the molecular shape are linear, so Be is *sp* hybridized:

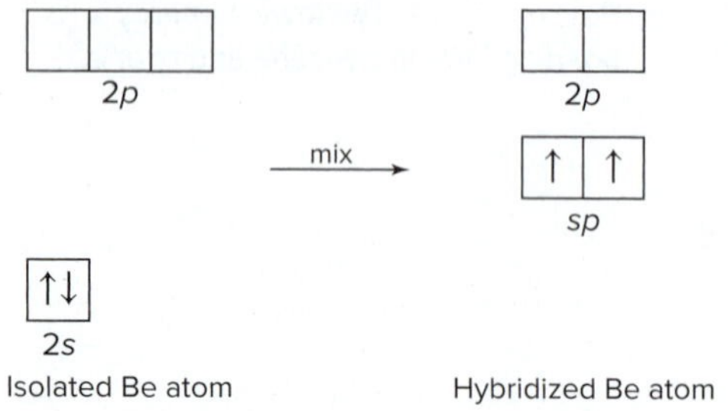

(b) The central Si atom has four electron groups (four bonds); the electron-group arrangement and the molecular shape are tetrahedral, so Si is sp^3 hybridized:

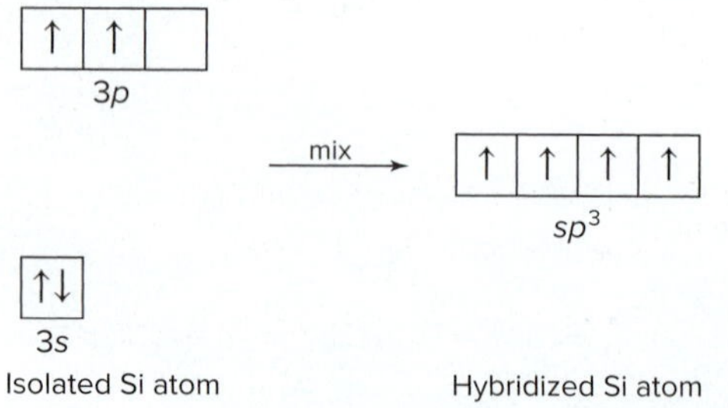

(c) The central Xe atom has six electron groups (four bonds and two lone pairs); the electron-group arrangement is octahedral and the molecular shape is square planar, so Xe is sp^3d^2 hybridized:

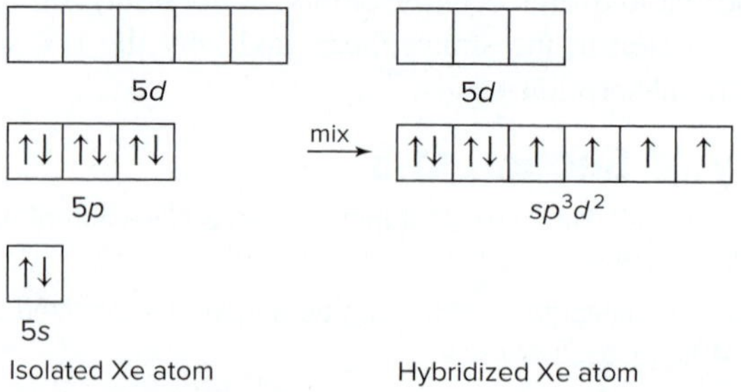

11.1B (a) The central N atom has three electron groups (a single bond, a double bond, and one unpaired electron); the electron-group arrangement is trigonal planar and the molecular shape is bent, so N is sp^2 hybridized:

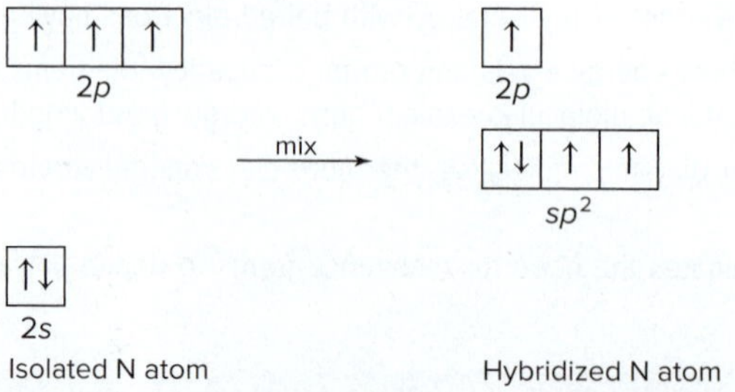

(b) The central P atom has four electron groups (three bonds and one lone pair); the electron-group arrangement is tetrahedral and the molecular shape is trigonal pyramidal, so P is sp^3 hybridized:

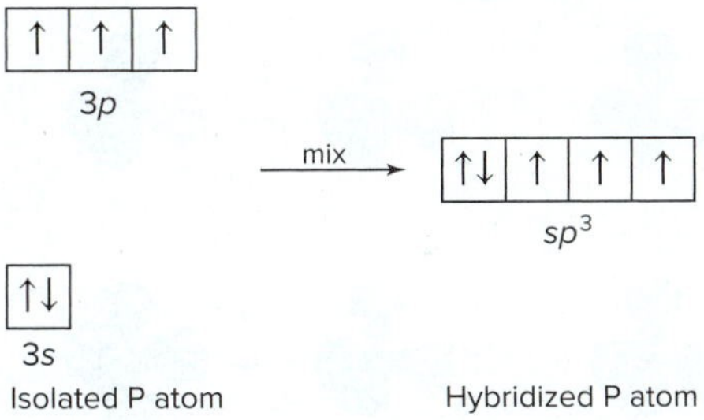

(c) The central Br atom has six electron groups (five bonds and one lone pair); the electron-group arrangement is octahedral and the molecular shape is square pyramidal, so Br is sp^3d^2 hybridized:

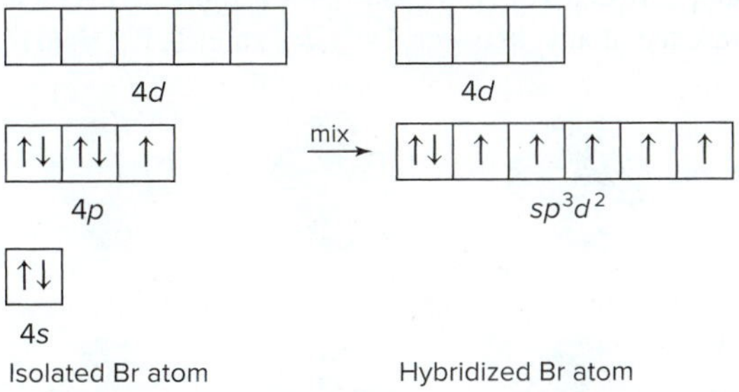

11.2A (a) H—C≡N:

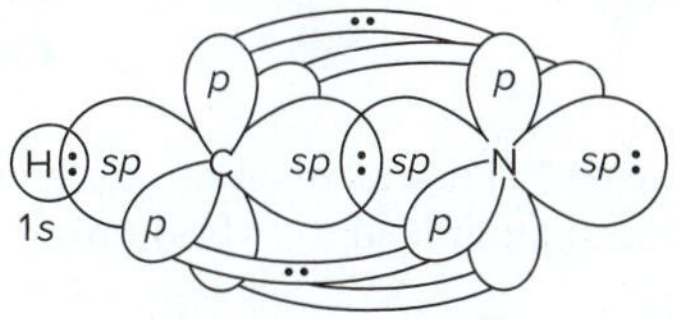

HCN is linear (C has two electron groups), so C is *sp* hybridized. N is also *sp* hybridized. One *sp* of C overlaps the 1*s* of H to form a σ bond. The other *sp* of C overlaps one *sp* of N to form a σ bond. The other *sp* of N holds a lone pair. Two unhybridized *p* orbitals of N and two of C overlap to form two π bonds.

(b) Ö=C=Ö

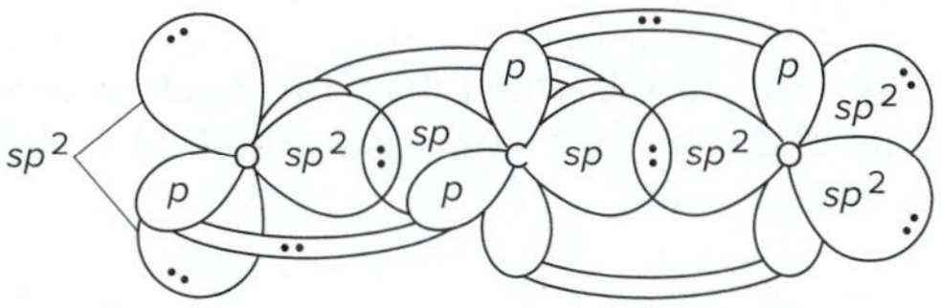

CO_2 is linear (C has two electron groups), so C is *sp* hybridized. Both O atoms are sp^2 hybridized. Each *sp* of C overlaps one sp^2 of an O to form a σ bond, which gives a total of two σ bonds. Each of the two unhybridized *p* orbitals of C forms a π bond with the unhybridized *p* of one of the two O atoms. Two sp^2 orbitals of each O hold lone pairs.

11.2B (a) :C≡O:

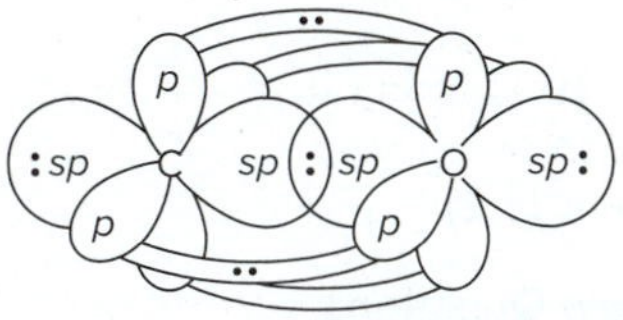

Both C and O are *sp* hybridized. One *sp* of C overlaps one *sp* of O to form one σ bond. The other *sp* of both C and O holds a lone pair. Two unhybridized *p* orbitals of C and two of O overlap to form two π bonds.

(b)

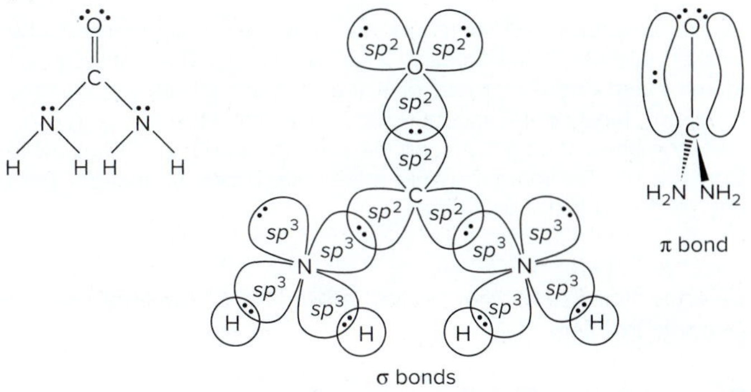

The C and O are sp^2 hybridized, and each N is sp^3 hybridized. The sp^3 orbitals of each N form three σ bonds, two with the 1*s* of two H atoms and one with an sp^2 of the C. A lone pair fills the fourth sp^3 of each N. The third sp^2 of the C forms a σ bond to one sp^2 of the O, and lone pairs fill the other sp^2 orbitals of the O. Unhybridized *p* orbitals on C and O form a π bond.

11.3A Each H^- ion has two electrons so H_2^{2-} has four electrons. Two electrons are in the σ_{1s} MO and two electrons are in the σ^*_{1s} MO. Does not exist: bond order = $\frac{1}{2}(2-2) = 0$.

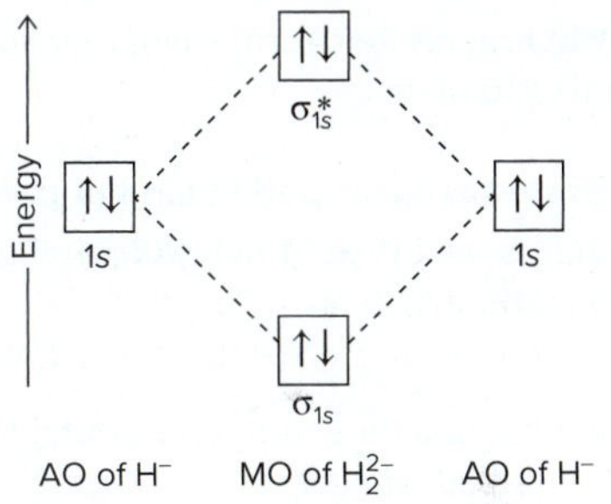

11.3B Each He^+ ion has one electron so He_2^{2+} has two electrons. Both electrons are in the σ_{1s} MO. Does exist: bond order = $\frac{1}{2}(2-0) = 1$; $(\sigma_{1s})^2$.

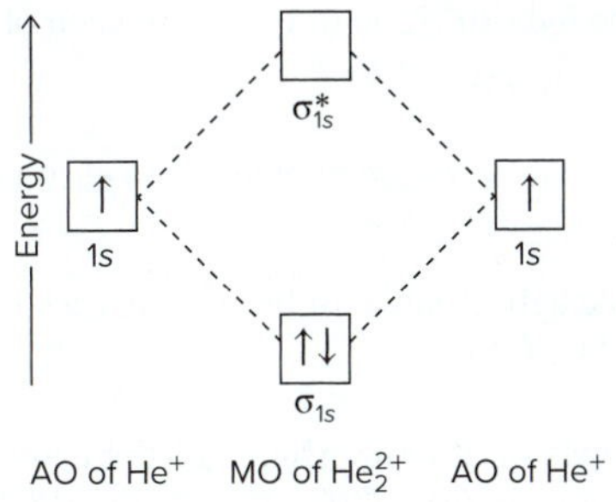

11.4A Bond orders: $N_2^{2+} = 2$; $N_2^- = 2.5$; $N_2^{2-} = 2$

Bond energy: $N_2^- > N_2^{2+} = N_2^{2-}$

Bond length: $N_2^{2+} = N_2^{2-} > N_2^-$

11.4B Bond orders: $F_2^{2-} = 0$; $F_2^- = 0.5$; $F_2 = 1$; $F_2^+ = 1.5$; $F_2^{2+} = 2$

Bond energy: F_2^{2-} does not exist; $F_2^- < F_2 < F_2^+ < F_2^{2+}$

Bond length: $F_2^{2+} < F_2^+ < F_2 < F_2^-$; F_2^{2-} does not exist

PROBLEMS

Problems with **colored** numbers are answered in Appendix E and worked in detail in the Student Solutions Manual. Problem sections match those in the text and give the numbers of relevant sample problems. Most offer Concept Review Questions, Skill-Building Exercises (grouped in pairs covering the same concept), and Problems in Context. The Comprehensive Problems are based on material from any section or previous chapter.

Valence Bond (VB) Theory and Orbital Hybridization

(Sample Problem 11.1)

Concept Review Questions

11.1 What type of central-atom orbital hybridization corresponds to each electron-group arrangement: (a) trigonal planar; (b) octahedral; (c) linear; (d) tetrahedral; (e) trigonal bipyramidal?

11.2 What is the orbital hybridization of a central atom that has one lone pair and bonds to: (a) two other atoms; (b) three other atoms; (c) four other atoms; (d) five other atoms?

11.3 How do carbon and silicon differ with regard to the *types* of orbitals available for hybridization? Explain.

11.4 How many hybrid orbitals form when four atomic orbitals of a central atom mix? Explain.

Skill-Building Exercises (grouped in similar pairs)

11.5 Give the number and type of hybrid orbital that forms when each set of atomic orbitals mixes:
(a) two *d*, one *s*, and three *p* (b) three *p* and one *s*

11.6 Give the number and type of hybrid orbital that forms when each set of atomic orbitals mixes:
(a) one *p* and one *s* (b) three *p*, one *d*, and one *s*

11.7 What is the hybridization of nitrogen in each of the following: (a) NO; (b) NO_2; (c) NO_2^-?

11.8 What is the hybridization of carbon in each of the following: (a) CO_3^{2-}; (b) $C_2O_4^{2-}$; (c) NCO^-?

11.9 What is the hybridization of chlorine in each of the following: (a) ClO_2; (b) ClO_3^-; (c) ClO_4^-?

11.10 What is the hybridization of bromine in each of the following: (a) BrF_3; (b) BrO_2^-; (c) BrF_5?

11.11 Which types of atomic orbitals of the central atom mix to form hybrid orbitals in (a) $SiClH_3$; (b) CS_2; (c) SCl_3F; (d) NF_3?

11.12 Which types of atomic orbitals of the central atom mix to form hybrid orbitals in (a) Cl_2O; (b) $BrCl_3$; (c) PF_5; (d) SO_3^{2-}?

11.13 Phosphine (PH_3) reacts with borane (BH_3) as follows:

$$PH_3 + BH_3 \longrightarrow H_3P\text{—}BH_3$$

(a) Which of the accompanying illustrations in the next column depicts the change, if any, in the orbital hybridization of P during this reaction? (b) Which depicts the change, if any, in the orbital hybridization of B?

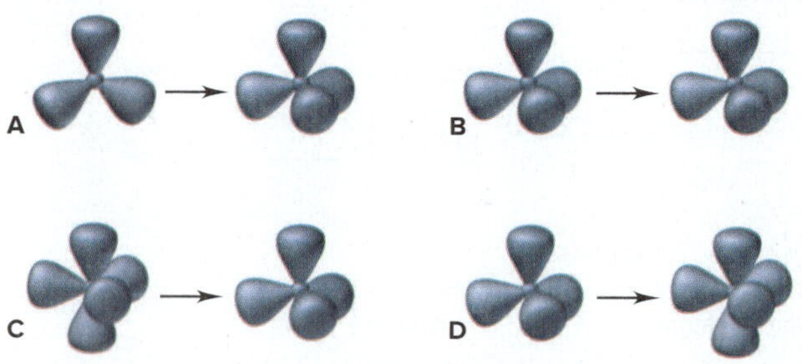

11.14 The illustrations below depict differences in orbital hybridization of some tellurium (Te) fluorides. (a) Which depicts the difference, if any, between TeF_6 *(left)* and TeF_5^- *(right)*? (b) Which depicts the difference, if any, between TeF_4 *(left)* and TeF_6 *(right)*?

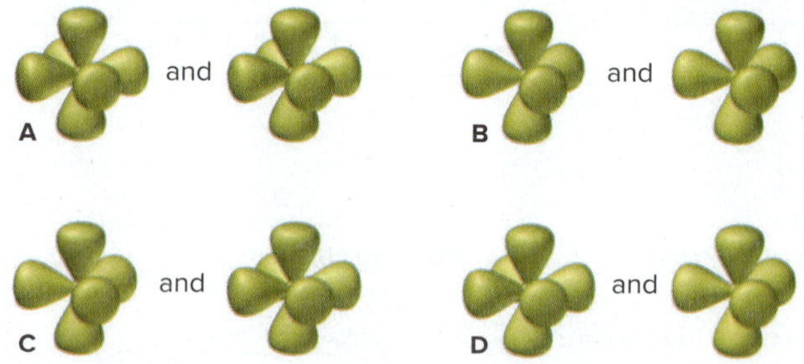

11.15 Use partial orbital diagrams to show how the atomic orbitals of the central atom lead to hybrid orbitals in (a) $GeCl_4$; (b) BCl_3; (c) CH_3^+.

11.16 Use partial orbital diagrams to show how the atomic orbitals of the central atom lead to hybrid orbitals in (a) BF_4^-; (b) PO_4^{3-}; (c) SO_3.

11.17 Use partial orbital diagrams to show how the atomic orbitals of the central atom lead to hybrid orbitals in (a) $SeCl_2$; (b) H_3O^+; (c) IF_4^-.

11.18 Use partial orbital diagrams to show how the atomic orbitals of the central atom lead to hybrid orbitals in (a) $AsCl_3$; (b) $SnCl_2$; (c) PF_6^-.

Problems in Context

11.19 Methyl isocyanate, $CH_3\text{—}\ddot{N}\text{=}C\text{=}\ddot{O}\text{:}$, is an intermediate in the manufacture of many pesticides. In 1984, a leak from a manufacturing plant resulted in the death of more than 2000 people in Bhopal, India. What are the hybridizations of the N atom and the two C atoms in methyl isocyanate? Sketch the molecular shape.

Modes of Orbital Overlap and the Types of Covalent Bonds

(Sample Problem 11.2)

Concept Review Questions

11.20 Are these statements true or false? Correct any false ones.
(a) Two σ bonds comprise a double bond.
(b) A triple bond consists of one π bond and two σ bonds.
(c) Bonds formed from atomic *s* orbitals are always σ bonds.

(d) A π bond restricts rotation about the σ-bond axis.
(e) A π bond consists of two pairs of electrons.
(f) End-to-end overlap results in a bond with electron density above and below the bond axis.

Skill-Building Exercises (grouped in similar pairs)

11.21 Identify the hybrid orbitals used by the central atom and the type(s) of bonds formed in (a) NO_3^-; (b) CS_2; (c) CH_2O.

11.22 Identify the hybrid orbitals used by the central atom and the type(s) of bonds formed in (a) O_3; (b) I_3^-; (c) $COCl_2$ (C is central).

11.23 Identify the hybrid orbitals used by the central atom(s) and the type(s) of bonds formed in (a) FNO; (b) C_2F_4; (c) $(CN)_2$.

11.24 Identify the hybrid orbitals used by the central atom(s) and the type(s) of bonds formed in (a) BrF_3; (b) $CH_3C{\equiv}CH$; (c) SO_2.

Problem in Context

11.25 2-Butene ($CH_3CH{=}CHCH_3$) is a starting material in the manufacture of lubricating oils and many other compounds. Draw two different structures for 2-butene, indicating the σ and π bonds in each.

Molecular Orbital (MO) Theory and Electron Delocalization

(Sample Problems 11.3 and 11.4)

Concept Review Questions

11.26 Two *p* orbitals from one atom and two *p* orbitals from another atom are combined to form molecular orbitals for the joined atoms. How many MOs will result from this combination? Explain.

11.27 Certain atomic orbitals on two atoms were combined to form the following MOs. Name the atomic orbitals used and the MOs formed, and explain which MO has higher energy:

11.28 How do the bonding and antibonding MOs formed from a given pair of AOs compare to each other with respect to (a) energy; (b) presence of nodes; (c) internuclear electron density?

11.29 Antibonding MOs always have at least one node. Can a bonding MO have a node? If so, draw an example.

Skill-Building Exercises (grouped in similar pairs)

11.30 How many electrons does it take to fill (a) a σ bonding MO; (b) a π antibonding MO; (c) the MOs formed from combination of the 1*s* orbitals of two atoms?

11.31 How many electrons does it take to fill (a) the MOs formed from combination of the 2*p* orbitals of two atoms; (b) a σ^*_{2p} MO; (c) the MOs formed from combination of the 2*s* orbitals of two atoms?

11.32 The molecular orbitals depicted are derived from 2*p* atomic orbitals in F_2^+. (a) Give the orbital designations. (b) Which is occupied by at least one electron in F_2^+? (c) Which is occupied by only one electron in F_2^+?

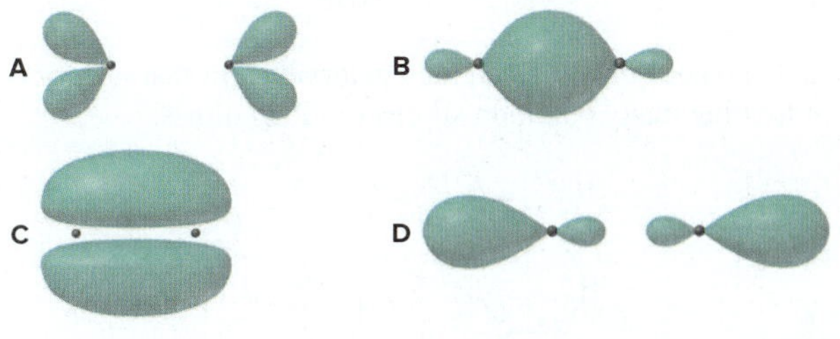

11.33 The molecular orbitals depicted below are derived from $n = 2$ atomic orbitals. (a) Give the orbital designations. (b) Which is highest in energy? (c) Lowest in energy? (d) Rank the MOs in order of increasing energy for B_2.

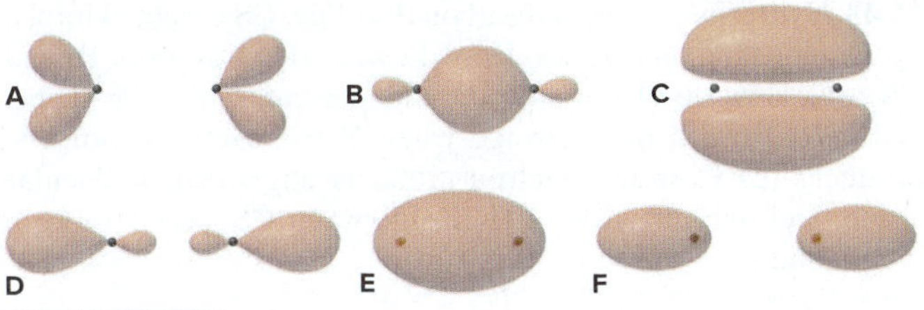

11.34 Use an MO diagram and the bond order you obtain from it to answer these questions: (a) Is Be_2^+ stable? (b) Is Be_2^+ diamagnetic? (c) What is the outer (valence) electron configuration of Be_2^+?

11.35 Use an MO diagram and the bond order you obtain from it to answer these questions: (a) Is O_2^- stable? (b) Is O_2^- paramagnetic? (c) What is the outer (valence) electron configuration of O_2^-?

11.36 Use MO diagrams to rank C_2^-, C_2, and C_2^+ in order of (a) increasing bond energy; (b) increasing bond length.

11.37 Use MO diagrams to rank B_2^+, B_2, and B_2^- in order of (a) decreasing bond energy; (b) decreasing bond length.

Comprehensive Problems

11.38 Predict the shape, state the hybridization of the central atom, and give the ideal bond angle(s) and any expected deviations for: (a) BrO_3^- (b) $AsCl_4^-$ (c) SeO_4^{2-} (d) BiF_5^{2-} (e) SbF_4^+ (f) AlF_6^{3-} (g) IF_4^+

11.39 Butadiene *(right)* is a colorless gas used to make synthetic rubber and many other compounds. (a) How many σ bonds and π bonds does the molecule have? (b) Are *cis-trans* arrangements about the double bonds possible? Explain.

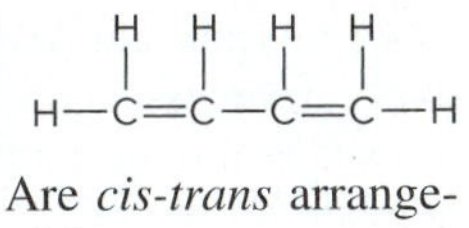

11.40 Epinephrine (or adrenaline; *below*) is a naturally occurring hormone that is also manufactured commercially for use as a heart stimulant, a nasal decongestant, and a glaucoma treatment.

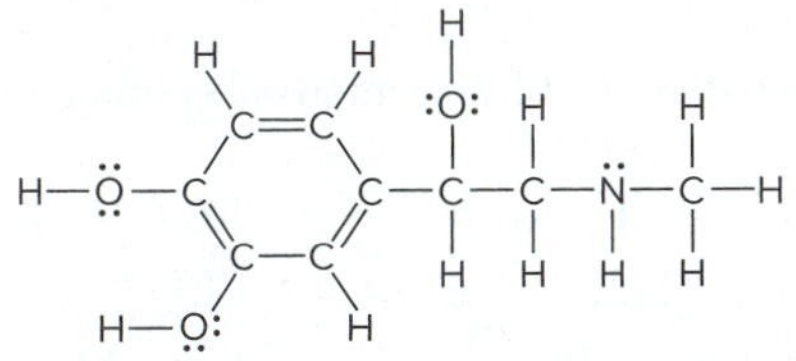

(a) What is the hybridization of each C, O, and N atom? (b) How many σ bonds does the molecule have? (c) How many π electrons are delocalized in the ring?

11.41 Use partial orbital diagrams to show how the atomic orbitals of the central atom lead to the hybrid orbitals in:
(a) IF_2^- (b) ICl_3 (c) $XeOF_4$ (d) BHF_2

11.42 Isoniazid *(below)* is an antibacterial agent that is very useful in treating many common strains of tuberculosis.

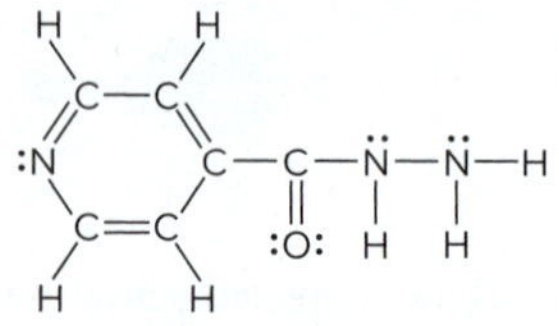

(a) How many σ bonds are in the molecule? (b) What is the hybridization of each C and N atom?

11.43 Hydrazine, N_2H_4, and carbon disulfide, CS_2, react to form a cyclic molecule *(below)*. (a) Draw Lewis structures for N_2H_4 and CS_2. (b) How do electron-group arrangement, molecular shape, and hybridization of N change when N_2H_4 reacts to form the product? (c) How do electron-group arrangement, molecular shape, and hybridization of C change when CS_2 reacts to form the product?

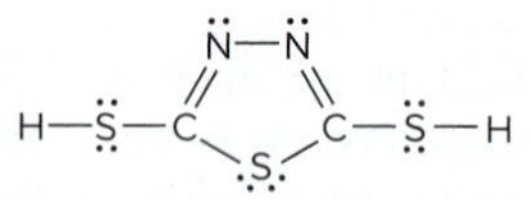

11.44 In each of the following equations, what hybridization change, if any, occurs for the underlined atom?
(a) $\underline{B}F_3 + NaF \longrightarrow Na^+BF_4^-$
(b) $\underline{P}Cl_3 + Cl_2 \longrightarrow PCl_5$
(c) $H\underline{C}{\equiv}CH + H_2 \longrightarrow H_2C{=}CH_2$
(d) $\underline{Si}F_4 + 2F^- \longrightarrow SiF_6^{2-}$
(e) $\underline{S}O_2 + \frac{1}{2}O_2 \longrightarrow SO_3$

11.45 The ionosphere lies about 100 km above Earth's surface. This layer of the atmosphere consists mostly of NO, O_2, and N_2, and photoionization creates NO^+, O_2^+, and N_2^+. (a) Use MO theory to compare the bond orders of the molecules and ions. (b) Does the magnetic behavior of each species change when its ion forms?

11.46 Glyphosate *(below)* is a common herbicide that is relatively harmless to animals but deadly to most plants. Describe the shape around and the hybridization of the P, N, and three numbered C atoms.

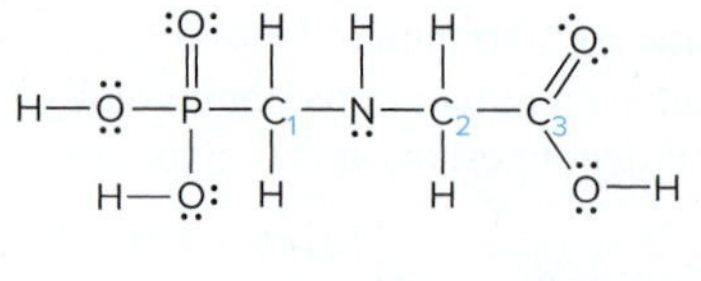

11.47 Tryptophan is one of the amino acids found in proteins:

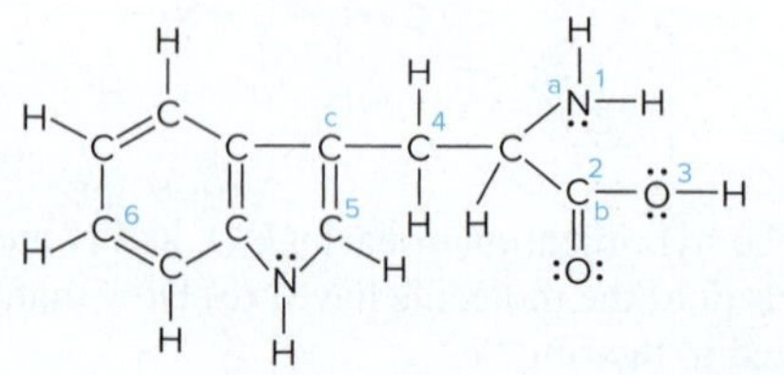

(a) What is the hybridization of each of the numbered C, N, and O atoms? (b) How many σ bonds are present in tryptophan? (c) Predict the bond angles at points a, b, and c.

11.48 Some species consisting of just two oxygen atoms are the oxygen molecule, O_2; the peroxide ion, O_2^{2-}; the superoxide ion, O_2^-; and the dioxygenyl ion, O_2^+. Draw an MO diagram for each, rank the species in order of increasing bond length, and find the number of unpaired electrons in each.

11.49 Molecular nitrogen, carbon monoxide, and cyanide ion are isoelectronic. (a) Draw an MO diagram for each. (b) CO and CN^- are toxic. What property may explain why N_2 isn't?

11.50 Government agencies that deal with health issues are concerned that the American diet contains too much meat, and numerous recommendations have been made urging people to consume more fruit and vegetables. One of the richest sources of vegetable protein is soy, available in many forms. One form is soybean curd, or tofu, which is a staple of many Asian diets. Chemists have isolated an anticancer agent called *genistein* from tofu, which may explain the much lower incidence of cancer among people in the Far East. A valid Lewis structure for genistein is

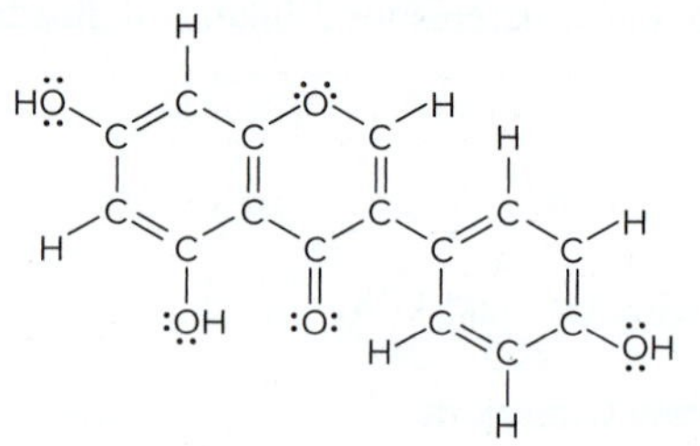

(a) Is the hybridization of each C in the right-hand ring the same? Explain. (b) Is the hybridization of the O atom in the center ring the same as that of the O atoms in the OH groups? Explain. (c) How many carbon-oxygen σ bonds are there? How many carbon-oxygen π bonds? (d) Do all the lone pairs on oxygens occupy the same type of hybrid orbital? Explain.

11.51 An organic chemist synthesizes the molecule below:

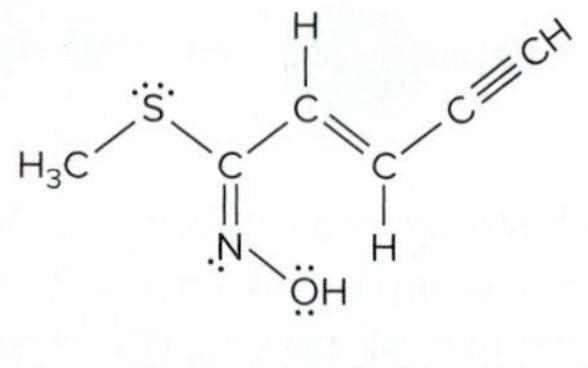

(a) Which of the orientations of hybrid orbitals shown below are present in the molecule? (b) Are there any present that are not shown below? If so, what are they? (c) How many of each type of hybrid orbital are present?

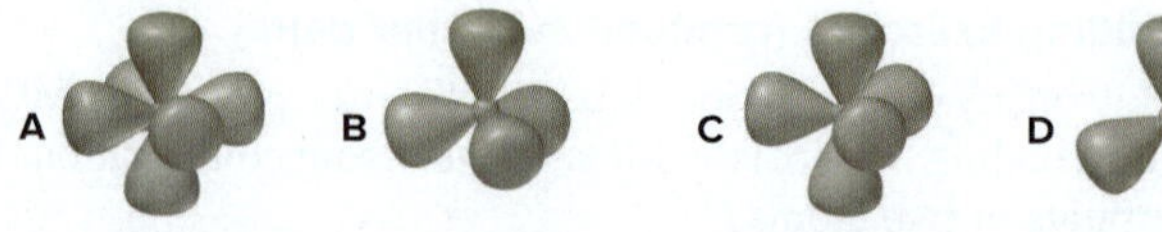

11.52 Simple proteins consist of amino acids linked together in a long chain; a small portion of such a chain is

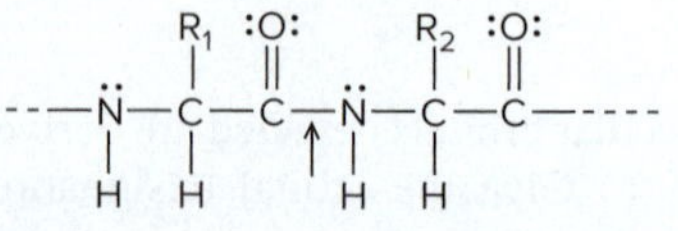

Experiment shows that rotation about the C—N bond (indicated by the arrow) is somewhat restricted. Explain with resonance structures, and show the types of bonding involved.

11.53 Sulfur forms oxides, oxoanions, and halides. What is the hybridization of the central S in SO_2, SO_3, SO_3^{2-}, SCl_4, SCl_6, and S_2Cl_2 (atom sequence Cl—S—S—Cl)?

11.54 The compound 2,6-dimethylpyrazine *(below)* gives chocolate its odor and is used in flavorings. (a) Which atomic orbitals mix to form the hybrid orbitals of N? (b) In what type of hybrid orbital do the lone pairs of N reside? (c) Is C in CH_3 hybridized the same as any C in the ring? Explain.

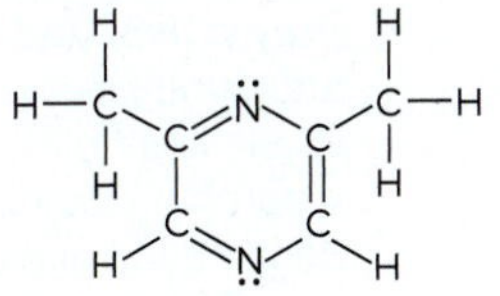

11.55 Use an MO diagram to find the bond order and predict whether H_2^- exists.

11.56 Acetylsalicylic acid (aspirin), the most widely used medicine in the world, has the Lewis structure shown. (a) What is the hybridization of each C and each O atom? (b) How many localized π bonds are present? (c) How many C atoms have a trigonal planar shape around them? A tetrahedral shape?

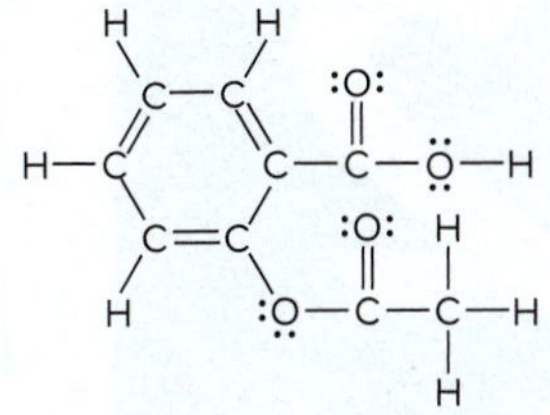

11.57 Linoleic acid is an essential fatty acid found in many vegetable oils, such as soy, peanut, and cottonseed. A key structural feature of the molecule is the *cis* orientation around its two double bonds, where R_1 and R_2 represent two different groups that form the rest of the molecule.

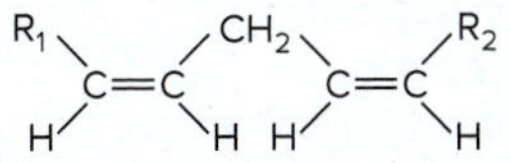

(a) How many different compounds are possible, changing only the *cis-trans* arrangements around these two double bonds?
(b) How many are possible for a similar compound with three double bonds?

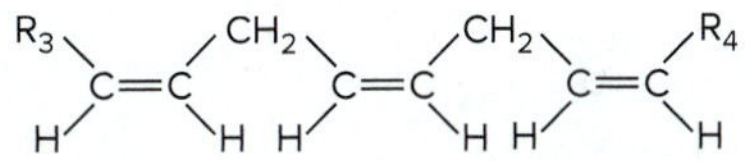

12 Intermolecular Forces: Liquids, Solids, and Phase Changes

12.1 An Overview of Physical States and Phase Changes

12.2 Quantitative Aspects of Phase Changes
Heat Involved in Phase Changes
Equilibrium Nature of Phase Changes
Phase Diagrams

12.3 Types of Intermolecular Forces
How Close Can Molecules Approach Each Other?
Ion-Dipole Forces
Dipole-Dipole Forces
The Hydrogen Bond
Polarizability and Induced Dipole Forces
Dispersion (London) Forces

12.4 Properties of the Liquid State
Surface Tension
Capillarity
Viscosity

12.5 The Uniqueness of Water
Solvent Properties
Thermal Properties
Surface Properties
Unusual Density of Solid Water

12.6 The Solid State: Structure, Properties, and Bonding
Structural Features
Crystalline Solids
Amorphous Solids
Bonding in Solids: Molecular Orbital Band Theory

12.7 Advanced Materials
Electronic Materials
Liquid Crystals
Ceramic Materials
Polymeric Materials
Nanotechnology

Source: © Franck Charton/hemis.fr/Getty Images

Concepts and Skills to Review Before You Study This Chapter

- properties of gases, liquids, and solids (Section 5.1)
- kinetic-molecular theory of gases (Section 5.5)
- kinetic and potential energy (Section 6.1)
- enthalpy change, heat capacity, and Hess's law (Sections 6.2, 6.3, and 6.5)
- diffraction of light (Section 7.1)
- Coulomb's law (Section 9.2)
- chemical bonding models (Chapter 9)
- molecular polarity (Section 10.3)
- molecular orbitals in diatomic molecules (Section 11.3)

All the matter in and around you occurs in one or more of the three physical states—gas, liquid, or solid. Under different conditions, many substances can occur in any of the states. We're all familiar with the states of water, all of which can be seen in the photo of a hot spring in Yellowstone National Park in winter. We inhale and exhale gaseous water; drink, excrete, and wash with liquid water; and shovel, slide on, or cool our drinks with solid water. And, given the right conditions, other solids (gold jewelry, glass), liquids (gasoline, antifreeze), and gases (air, CO_2 bubbles) can exist in one or both of the other states.

The three states were introduced in Chapter 1, and their properties were compared when we examined gases in Chapter 5. You saw there that the particles in gases are, on average, very far apart. Now, we examine liquids and solids, which are called *condensed states* because *their particles are very close together.*

IN THIS CHAPTER . . . *We explore the interplay of forces that give rise to the three states of matter and their changes, with special attention to liquids and solids.*

- We use kinetic-molecular theory to see how relative magnitudes of potential and kinetic energy account for the behavior of gases, liquids, and solids and how temperature is related to phase changes.
- We calculate the heat associated with warming or cooling a phase and with a phase change.
- We highlight the effects of temperature and pressure on phases and their changes with phase diagrams.
- We describe the types and relative strengths of the intermolecular forces that give rise to the phases and phase changes of pure substances.
- We see how intermolecular forces underlie the properties of liquids.
- As a special case, we trace the unique and vital physical properties of water to the electron configurations of the atoms making up its molecules.
- We discuss the properties of solids, emphasizing the relation between type of bonding and predominant intermolecular force, and we examine crystal structures and two of the methods used to study them.
- We introduce key features of several kinds of advanced materials—semiconductors, liquid crystals, ceramics, polymers, and nanostructures.

12.1 AN OVERVIEW OF PHYSICAL STATES AND PHASE CHANGES

Each physical state is called a **phase,** a physically distinct, homogeneous part of a system. The water in a closed container constitutes one phase; the water vapor above the liquid is a second phase; add some ice, and there are three phases.

In this section, you'll see that interactions between the potential energy and the kinetic energy of the particles give rise to the properties of each phase:

- The *potential energy* in the form of **intermolecular forces** (or, more generally, *interparticle forces*) tends to draw the molecules together. According to Coulomb's law, the electrostatic potential energy depends on the charges of the particles and the distances between them (Section 9.2). We'll examine the various types of intermolecular forces in Section 12.3.

- The *kinetic energy* associated with the random motion of the molecules tends to disperse them. It is related to their average speed and is proportional to the absolute temperature (Section 5.5).

These interactions also explain **phase changes**—liquid to solid, solid to gas, and so forth, or even one solid form to another.

A Kinetic-Molecular View of the Three States Imagine yourself among the particles in any of the three states of water. Look closely and you'll discover two types of electrostatic forces at work:

1. *Intra*molecular (bonding) forces exist *within* each molecule. The *chemical* behavior of the three states is identical because all of them consist of the same bent, polar H—O—H molecules, each held together by identical covalent bonding forces.
2. *Inter*molecular (nonbonding) forces exist *between* the molecules. The *physical* behavior of the states is different because the strengths of these forces differ from state to state.

Whether a substance occurs as a gas, liquid, or solid depends on the interplay of the potential and kinetic energy:

- *In a gas,* the potential energy (energy of attraction) is small relative to the kinetic energy (energy of motion); thus, on average, the particles are far apart. This large distance has several macroscopic consequences: a gas fills its container, is highly compressible, and flows easily through another gas (Table 12.1).
- *In a liquid,* attractions are stronger because the particles are touching, but they have enough kinetic energy to move randomly around each other. Thus, a liquid conforms to the shape of its container but has a surface, it resists an applied force and thus compresses very slightly, and it flows, but *much* more slowly than a gas.
- *In a solid,* the attractions dominate the motion so much that the particles are fixed in position relative to one another, just jiggling in place. Thus, a solid has its own shape, compresses even less than a liquid, and does not flow significantly.

Table 12.1 A Macroscopic Comparison of Gases, Liquids, and Solids

State	Shape and Volume	Compressibility	Ability to Flow
Gas	Conforms to shape and volume of container	High	High
Liquid	Conforms to shape of container; volume limited by surface	Very low	Moderate
Solid	Maintains its own shape and volume	Almost none	Almost none

Relative ability to flow influences the degree of mixing and, thus, composition in nature.

Source: © Christopher Meder Photography/Shutterstock.com

The environment provides a perfect demonstration of these differences in ability to flow *(see photo)*. Atmospheric gases mix so well that the lowest 80 km of air has a uniform composition. Much less mixing in the oceans allows the composition at various depths to support different species. And rocks intermingle so little that adjacent strata remain separated for millions of years.

Types of Phase Changes and Their Enthalpies When we consider phase changes, understanding the effect of temperature is critical:

- As *temperature increases,* the average kinetic energy does too, so the particles move faster and overcome attractions more easily.
- As *temperature decreases,* the average kinetic energy does too, so particles move more slowly and attractions can pull them together more easily.

Pressure also affects phase changes, most dramatically when gases are involved and to a lesser extent with liquids. Each phase change has a name and an associated enthalpy change:

1. *Gas to liquid, and vice versa.* As the temperature drops (and/or the pressure rises), the molecules in the gas phase come together and form a liquid in the

process of **condensation;** the opposite process, changing from a liquid to a gas, is **vaporization.**

2. *Liquid to solid, and vice versa.* As the temperature drops further, the particles move more slowly and become fixed in position in the process of **freezing;** the opposite change is called **melting,** or **fusion.** (For most substances, an increase in pressure supports freezing as well.) In common speech, *freezing* implies low temperature because we think of water. But, for example, molten metals freeze (solidify) at much higher temperatures, which gives them many medical, industrial, and artistic applications, such as gold dental crowns, steel auto bodies, and bronze statues.

3. *Gas to solid, and vice versa.* All three states of water are familiar because they are stable under ordinary conditions. Carbon dioxide, on the other hand, is familiar as a gas and a solid (dry ice), but liquid CO_2 occurs only at pressures of 5.1 atm or greater. At ordinary conditions, solid CO_2 changes directly to a gas, a process called **sublimation.** Freeze-dried foods are prepared by freezing the food and subliming the water. The opposite process, changing from a gas directly into a solid, is called **deposition**—ice crystals form on a cold window from the deposition of water vapor (Figure 12.1A).

The accompanying enthalpy changes are either exothermic or endothermic:

- *Exothermic changes.* As the molecules of a gas attract each other into a liquid, and then become fixed in a solid, the system of particles *loses* energy, which is released as heat. Thus, *condensing, freezing, and deposition are exothermic changes.*
- *Endothermic changes.* Heat must be absorbed by the system to overcome the attractive forces that keep the particles fixed in place in a solid or near each other in a liquid. Thus, *melting, vaporizing, and subliming are endothermic changes.*

Sweating has a cooling effect because heat from your body vaporizes the water. To achieve this cooling, cats lick themselves and dogs pant (Figure 12.1B).

For a pure substance, each phase change is accompanied by a standard enthalpy change, given in units of *kilojoules per mole* (measured at 1 atm and the temperature of the change). For vaporization, it is the **heat** (or *enthalpy*) **of vaporization (ΔH°_{vap})**, and for fusion (melting), it is the **heat** (or *enthalpy*) **of fusion (ΔH°_{fus})**. In the case of water, we have

$$H_2O(l) \longrightarrow H_2O(g) \qquad \Delta H = \Delta H^\circ_{vap} = 40.7 \text{ kJ/mol (at 100°C)}$$
$$H_2O(s) \longrightarrow H_2O(l) \qquad \Delta H = \Delta H^\circ_{fus} = 6.02 \text{ kJ/mol (at 0°C)}$$

The reverse processes, condensing and freezing, have enthalpy changes of the *same magnitude but opposite sign:*

$$H_2O(g) \longrightarrow H_2O(l) \qquad \Delta H = -\Delta H^\circ_{vap} = -40.7 \text{ kJ/mol}$$
$$H_2O(l) \longrightarrow H_2O(s) \qquad \Delta H = -\Delta H^\circ_{fus} = -6.02 \text{ kJ/mol}$$

Water behaves typically in that it takes much less energy to melt the solid than to vaporize the liquid: $\Delta H^\circ_{fus} < \Delta H^\circ_{vap}$. That is, it takes less energy to reduce the intermolecular forces enough for the molecules to move out of their fixed positions (melt a solid) than to separate them completely (vaporize a liquid) (Figure 12.2).

A Deposition

B Vaporization

Figure 12.1 Two familiar phase changes.

Source: (A) © dinadesign/Shutterstock.com; (B) © Jill Birschbach Photo Services

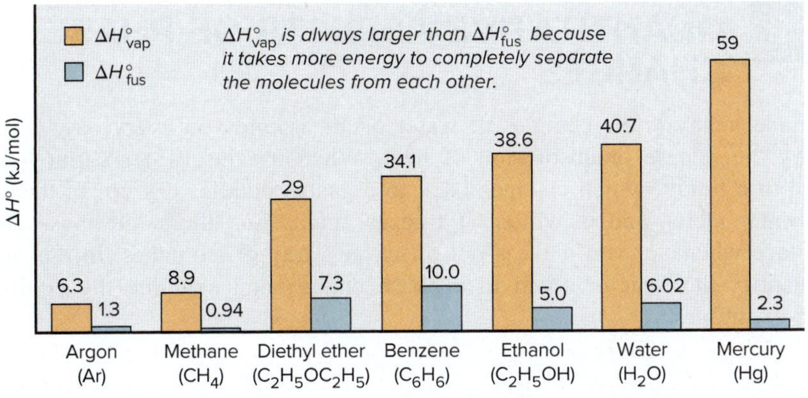

Figure 12.2 Heats of vaporization and fusion for several common substances.

The **heat** (or *enthalpy*) **of sublimation** (ΔH°_{subl}) is the enthalpy change when 1 mol of a substance sublimes, and the negative of this value is the change when 1 mol of the substance deposits. Since sublimation can be considered a combination of melting and vaporizing steps, Hess's law (Section 6.5) says that the heat of sublimation equals the sum of the heats of fusion and vaporization:

$$\begin{array}{lll} \text{Solid} \longrightarrow \text{liquid} & & \Delta H^\circ_{fus} \\ \text{Liquid} \longrightarrow \text{gas} & & \Delta H^\circ_{vap} \\ \hline \text{Solid} \longrightarrow \text{gas} & & \Delta H^\circ_{subl} \end{array}$$

Figure 12.3 summarizes the phase changes and their enthalpy changes.

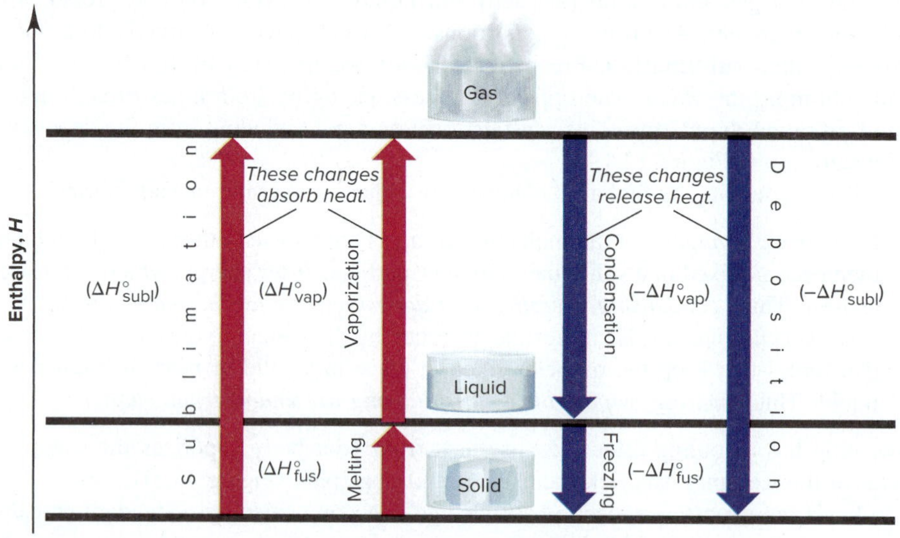

Figure 12.3 Phase changes and their enthalpy changes. Fusion (or melting), vaporization, and sublimation are endothermic changes (positive ΔH°), whereas freezing, condensation, and deposition are exothermic changes (negative ΔH°).

› Summary of Section 12.1

- Because of the relative magnitudes of intermolecular forces (potential energy) and average speed (kinetic energy), the particles in a gas are far apart and moving randomly, those in a liquid are in contact and moving relative to each other, and those in a solid are in contact and in fixed positions. These molecular-level differences account for macroscopic differences in shape, compressibility, and ability to flow.
- When a solid becomes a liquid (melting, or fusion), a liquid becomes a gas (vaporization), or a solid becomes a gas (sublimation), energy is absorbed to overcome intermolecular forces and increase the average distance between particles. As particles come closer together in the reverse changes (freezing, condensation, and deposition), energy is released. Each phase change is associated with a given enthalpy change under specified conditions.

12.2 QUANTITATIVE ASPECTS OF PHASE CHANGES

Of course, many phase changes of water occur around you every day, accompanied by the release or absorption of heat. When it rains, water vapor has condensed to a liquid, which changes back to a gas as puddles dry up. In the spring, solid water melts, and in winter, it freezes again. And the same phase changes take place whenever you make a pot of tea or a tray of ice cubes. In this section, we quantify the heat involved in a phase change and examine the equilibrium nature of the process.

Heat Involved in Phase Changes

We apply a kinetic-molecular approach to phase changes with a **heating-cooling curve,** which shows the changes in temperature of a sample when heat is absorbed or released at a constant rate. Let's examine what happens when 2.50 mol of gaseous water in a closed container undergoes a change from 130°C to −40°C at a constant pressure of 1 atm. We'll divide this process into five heat-releasing (exothermic) stages (Figure 12.4).

Stage 1. Gaseous water cools. Water molecules zoom chaotically at a range of speeds, smashing into each other and the container walls. At the starting temperature, the most probable speed of the molecules, and thus their average kinetic energy (E_k), is high enough to overcome the potential energy (E_p) of attractions. As the temperature falls, the average E_k decreases and attractions become more important. The change is

$$H_2O(g)\ [130°C] \longrightarrow H_2O(g)\ [100°C]$$

The heat lost in the cooling process (q) is the product of the amount (number of moles, n) of water, the molar heat capacity of *gaseous* water, $C_{m(water,\ g)}$ (J/mol·K or J/mol·°C), and the temperature change during this step, ΔT ($T_{final} - T_{initial}$):

$$q = n \times C_{m(water,\ g)} \times \Delta T = (2.50\ mol)\ (33.1\ J/mol{\cdot}°C)(100°C - 130°C)$$
$$= -2482\ J = -2.48\ kJ$$

The negative sign indicates that heat is released. (For purposes of canceling, the units for molar heat capacity, C_m, include °C rather than K, but this doesn't affect the magnitude of C_m because 1°C and 1 K represent the same temperature increment.)

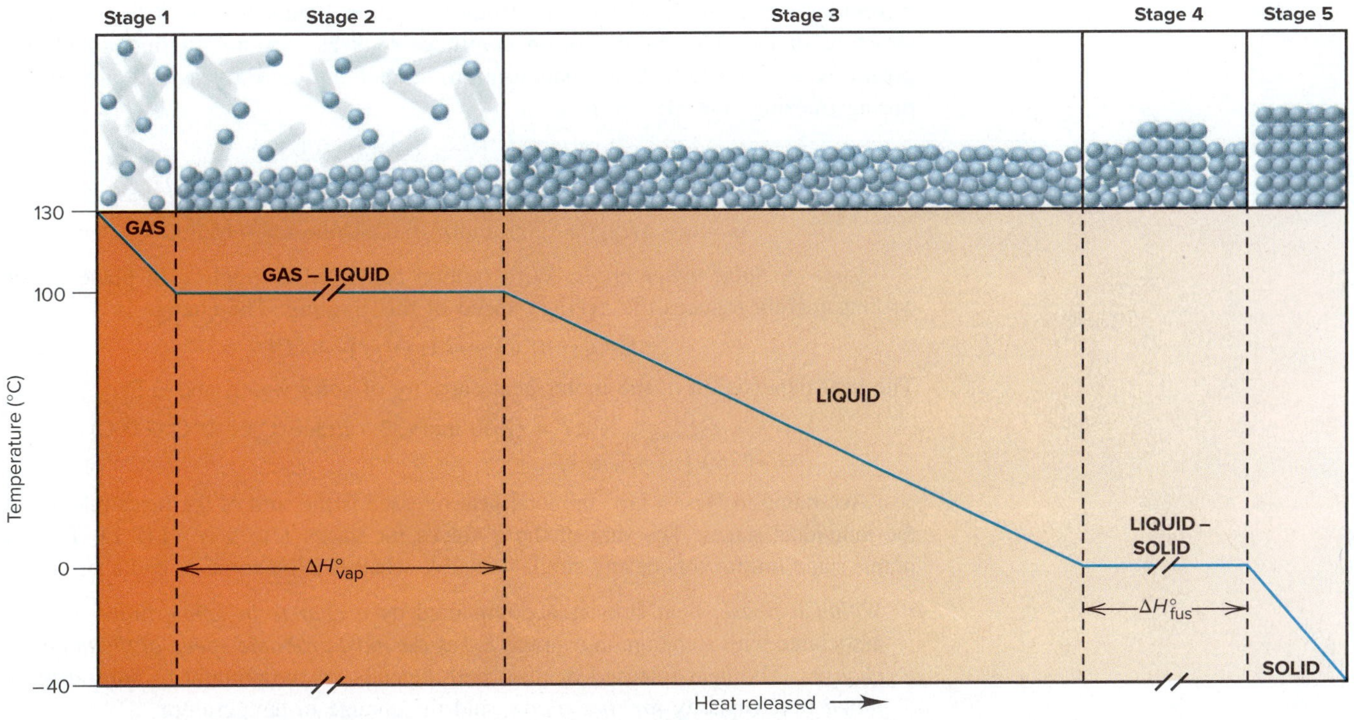

Figure 12.4 A heating-cooling curve for the conversion of gaseous water to ice. A plot of temperature vs. heat released as gaseous water changes to ice is shown, with a molecular-level depiction for each stage. The slopes of the lines in stages 1, 3, and 5 reflect the molar heat capacities of the phases. Although not drawn to scale, the line in stage 2 is longer than the line in stage 4 because $\Delta H°_{vap}$ of water is greater than $\Delta H°_{fus}$. A plot of temperature vs. heat absorbed starting at −40°C would have the same steps but in reverse order.

Stage 2. Gaseous water condenses. At the condensation point, intermolecular attractions cause the slowest of the molecules to aggregate into microdroplets and then a bulk liquid. Note that, during the phase change,

- The temperature of the sample, and thus its average E_k, is constant. At the same temperature, molecules move farther between collisions in a gas than in a liquid, but their *average* speed is the same.
- Releasing heat from the sample decreases the average E_p as the molecules approach and attract each other.

Thus, at 100°C, gaseous and liquid water have the same average E_k, but the liquid has lower average E_p. The change is

$$H_2O(g)\ [100°C] \longrightarrow H_2O(l)\ [100°C]$$

The heat is the amount (n) of water times the negative of the heat of vaporization ($-\Delta H^\circ_{vap}$):

$$q = n(-\Delta H^\circ_{vap}) = (2.50\ \text{mol})(-40.7\ \text{kJ/mol}) = -102\ \text{kJ}$$

This stage contributes the *greatest portion of the total heat released* because of the large decrease in E_p as the molecules become so much closer in the liquid than they were in the gas.

Stage 3. Liquid water cools. The molecules in the liquid state continue to lose heat, and the loss appears as a decrease in temperature, that is, as a decrease in the most probable molecular speed and, thus, the average E_k. The temperature decreases as long as the sample remains liquid. The change is

$$H_2O(l)\ [100°C] \longrightarrow H_2O(l)\ [0°C]$$

The heat depends on amount (n), the molar heat capacity of *liquid* water, and ΔT:

$$q = n \times C_{m(\text{water},\ l)} \times \Delta T = (2.50\ \text{mol})(75.4\ \text{J/mol·°C})(0°C - 100°C)$$
$$= -18{,}850\ \text{J} = -18.8\ \text{kJ}$$

Stage 4. Liquid water freezes. At 0°C, the sample loses E_p as increasing intermolecular attractions cause the molecules to align themselves into the crystalline structure of ice. Molecular motion continues only as random jiggling about fixed positions. As we saw during condensation, temperature and average E_k are constant during freezing. The change is

$$H_2O(l)\ [0°C] \longrightarrow H_2O(s)\ [0°C]$$

The heat is equal to n times the negative of the heat of fusion ($-\Delta H^\circ_{fus}$):

$$q = n(-\Delta H^\circ_{fus}) = (2.50\ \text{mol})(-6.02\ \text{kJ/mol}) = -15.0\ \text{kJ}$$

Stage 5. Solid water cools. With motion restricted to jiggling in place, further cooling merely reduces the average speed of this jiggling. The change is

$$H_2O(s)\ [0°C] \longrightarrow H_2O(s)\ [-40°C]$$

The heat depends on n, the molar heat capacity of *solid* water, and ΔT:

$$q = n \times C_{m(\text{water},\ s)} \times \Delta T = (2.50\ \text{mol})(37.6\ \text{J/mol·°C})(-40°C - 0°C)$$
$$= -3760\ \text{J} = -3.76\ \text{kJ}$$

According to Hess's law, the total heat released is the sum of the heats released for the individual stages. The sum of the q values for stages 1 to 5 is -142 kJ. Two key points stand out for this or any similar process, whether exothermic or endothermic:

- *Within a phase,* heat flow is accompanied by *a change in temperature,* which is associated with a change in average E_k, as *the most probable speed of the molecules changes.* The heat released or absorbed depends on the amount of substance, the molar heat capacity *for that phase,* and the change in temperature.
- *During a phase change,* heat flow occurs at a *constant temperature,* which is associated with a change in average E_p, as *the average distance between molecules changes.* Both phases are present and (as you'll see) are in equilibrium during the change. The heat released or absorbed depends on the amount of substance and the enthalpy change for that phase change.

Let's work a molecular-scene sample problem to make sure these ideas are clear.

Student Hot Spot

Student data indicate that you may struggle with the quantitative aspects of phase changes.. Access the Smartbook to view additional Learning Resources on this topic.

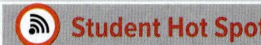

SAMPLE PROBLEM 12.1 Finding the Heat of a Phase Change Depicted by Molecular Scenes

Problem The scenes below represent a phase change of water. Use values for molar heat capacities and heats of phase changes from the text discussion to find the heat (in kJ) released or absorbed when 24.3 g of H_2O undergoes this change.

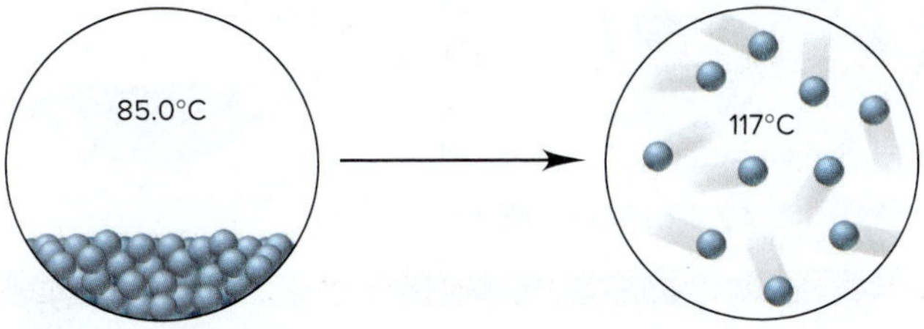

Plan From the molecular scenes, values given in the text discussion, and the given mass (24.3 g) of water, we have to find the heat that accompanies this change. The scenes show a disorderly, condensed phase at 85.0°C changing to separate molecules at 117°C. Thus, the phase change they depict is vaporization, an endothermic process. The values for molar heat capacities and heats of phase changes are per mole, so we first convert the mass (g) of water to amount (mol). There are three stages: (1) heating the liquid from 85.0°C to 100.°C; (2) converting liquid water at 100.°C to gaseous water at 100.°C; and (3) heating the gas from 100.°C to 117°C *(see margin)*. We add the values of q for these stages to obtain the total heat.

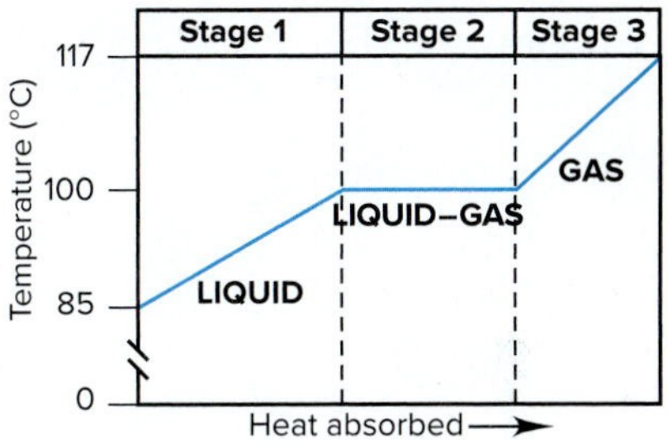

Solution Converting from mass (g) of H_2O to amount (mol):

$$\text{Amount (mol) of } H_2O = 24.3 \text{ g } H_2O \times \frac{1 \text{ mol}}{18.02 \text{ g } H_2O} = 1.35 \text{ mol}$$

Finding the heat accompanying stage 1, $H_2O(l)$ [85.0°C] $\longrightarrow$ $H_2O(l)$ [100.°C]:

$$q = n \times C_{m(\text{water}, l)} \times \Delta T = (1.35 \text{ mol})(75.4 \text{ J/mol·°C})(100.\text{°C} - 85.0\text{°C})$$
$$= 1527 \text{ J} = 1.53 \text{ kJ}$$

Finding the heat accompanying stage 2, $H_2O(l)$ [100.°C] $\longrightarrow$ $H_2O(g)$ [100.°C]:

$$q = n(\Delta H^\circ_{vap}) = (1.35 \text{ mol})(40.7 \text{ kJ/mol}) = 54.9 \text{ kJ}$$

Finding the heat accompanying stage 3, $H_2O(g)$ [100.°C] $\longrightarrow$ $H_2O(g)$ [117°C]:

$$q = n \times C_{m(\text{water}, g)} \times \Delta T = (1.35 \text{ mol})(33.1 \text{ J/mol·°C})(117\text{°C} - 100.\text{°C})$$
$$= 759.6 \text{ J} = 0.760 \text{ kJ}$$

Adding the three heats to find the total heat for the process:

$$\text{Total heat (kJ)} = 1.53 \text{ kJ} + 54.9 \text{ kJ} + 0.760 \text{ kJ} = 57.2 \text{ kJ}$$

Check The heat should have a positive value because it is absorbed. Be sure to round to check each value of q; for example, in stage 1, 1.35 mol × 75 J/mol·°C × 15°C = 1500 J. Note that the phase change itself (stage 2) requires the most energy and, thus, dominates the final answer. The ΔH°_{vap} units include kJ, whereas the molar heat capacity units include J, which is a thousandth as large.

FOLLOW-UP PROBLEMS

Brief Solutions for ***all*** *Follow-up Problems appear at the end of the chapter.*

12.1A The scenes below represent a phase change of water. Use values for molar heat capacities and heats of phase changes from the text discussion to find the heat (in kJ) released or absorbed when 2.25 mol of H_2O undergoes this change.

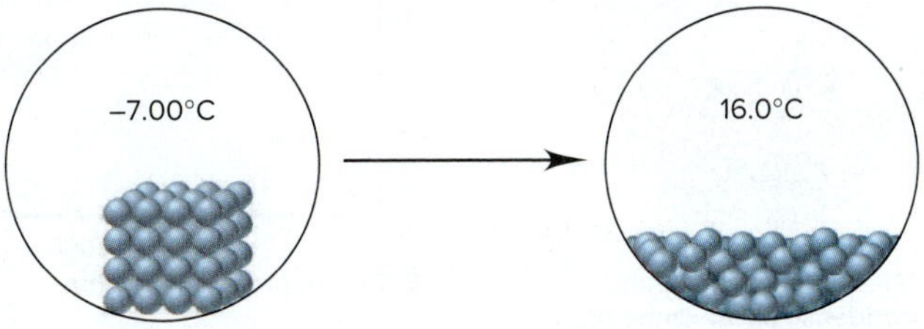

12.1B The scenes below represent a phase change of bromine, Br_2. Use the following data to find the heat (in kJ) released or absorbed when 47.94 g of bromine undergoes this change: $C_{m(liquid)} = 75.7$ J/mol·°C; $C_{m(gas)} = 36.0$ J/mol·°C; $\Delta H^\circ_{fus} = 10.6$ kJ/mol; $\Delta H^\circ_{vap} = 29.6$ kJ/mol; mp = −7.25°C; bp = 59.5°C.

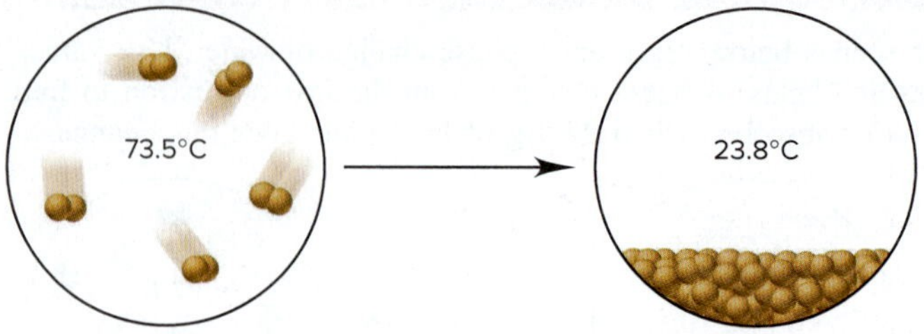

SOME SIMILAR PROBLEMS 12.19 and 12.20

The Equilibrium Nature of Phase Changes

In everyday experience, phase changes take place in *open* containers—the outdoors, a pot on a stove, the freezer compartment of a refrigerator—so they are not reversible. But, in a *closed* container, *phase changes* ***are*** *reversible and reach equilibrium,* just as chemical changes do. In this discussion, we examine the three phase equilibria.

Liquid-Gas Equilibria Vaporization and condensation are familiar events. Let's see how these processes differ in open and closed systems of a liquid in a flask:

1. *Open system: nonequilibrium process.* Picture an *open* flask containing a pure liquid at constant temperature. Within their range of speeds, some molecules at the surface have a high enough E_k to overcome attractions and vaporize. Nearby molecules fill the gap, and with heat supplied by the constant-temperature surroundings, the process continues until the entire liquid phase is gone.

2. *Closed system: equilibrium process.* Now picture a *closed* flask at constant temperature and assume a vacuum exists above a liquid in the flask (Figure 12.5A). Two processes take place: Some molecules at the surface have a high enough E_k to *vaporize*. After a short time, molecules in the vapor collide with the surface, and the slower ones are attracted strongly enough to *condense*.

At first, these two processes occur at different rates. The number of molecules in a given surface area is constant, so the number of molecules leaving the surface per unit time—the rate of vaporization—is also constant, and the pressure increases. With time, the number of molecules colliding with and entering the surface—the rate of condensation—increases as the vapor becomes more populated, so the increase in pressure slows. Eventually, the rate of condensation equals the rate of vaporization; from this time onward, *the pressure is constant* (Figure 12.5B).

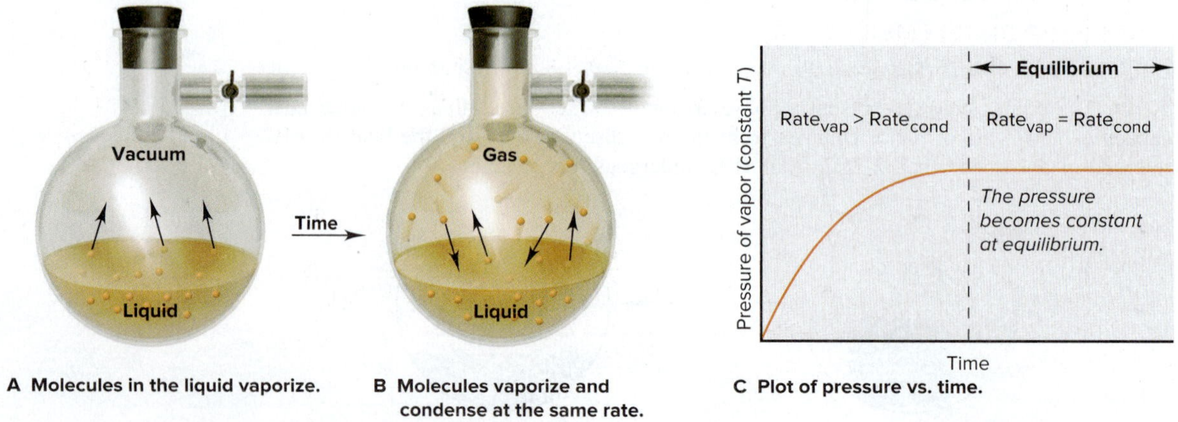

Figure 12.5 Liquid-gas equilibrium. **A,** Molecules leave the surface at a constant rate and the pressure rises. **B,** At equilibrium, the same number of molecules leave and enter the liquid in a given time. **C,** Pressure increases until, at equilibrium, it is constant.

Macroscopically, the situation at this point seems static, but at the molecular level, molecules are entering and leaving the liquid at equal rates. The system has reached a state of *dynamic equilibrium:*

$$\text{liquid} \rightleftharpoons \text{gas}$$

The pressure exerted by the vapor at equilibrium is called the *equilibrium vapor pressure,* or just the **vapor pressure,** of the liquid at that temperature. Figure 12.5C depicts the entire process graphically. If we started with a larger flask, the number of molecules in the vapor would be greater at equilibrium. But, as long as the temperature is constant and some liquid is present, the vapor pressure will be the same.

3. *Disturbing a system at equilibrium.* Let's see what happens if we alter certain conditions, called "disturbing" the system:

- *Decrease in pressure.* Suppose we pump some vapor out of the flask, immediately lowering the pressure. (In a cylinder fitted with a piston, we lower the pressure by moving the piston outward, thus increasing the volume.) The rate of condensation temporarily falls below the rate of vaporization (the forward process is faster) because fewer molecules enter the liquid than leave it. The pressure rises until, after a short time, the condensation rate increases enough for equilibrium to be reached again.
- *Increase in pressure.* Suppose we pump more vapor in (or move the piston inward, thus decreasing the volume), thereby immediately raising the pressure. The rate of condensation temporarily exceeds the rate of vaporization because more molecules enter the liquid than leave it (the reverse process is faster). Soon, however, the condensation rate decreases until the pressure again reaches the equilibrium value.

This general behavior of a liquid and its vapor is seen in any system at equilibrium: *when a system at equilibrium is disturbed, it counteracts the disturbance until it reestablishes equilibrium.* We'll return to this key idea often in later chapters.

The Effects of Temperature and Intermolecular Forces on Vapor Pressure The vapor pressure is affected by two factors—a change in temperature and a change in the gas itself, that is, in the type and/or strength of intermolecular forces:

1. *Effect of temperature.* Temperature has a major effect on vapor pressure because it changes the fraction of molecules moving fast enough to escape the liquid and, by the same token, the fraction moving slowly enough to be recaptured. In Figure 12.6, we see the familiar skewed bell-shaped curve of the distribution of molecular speeds (see also Figure 5.20). At the higher temperature, T_2, more molecules have enough energy to leave the surface. Thus, in general, *the higher the temperature is, the higher the vapor pressure:*

$$\text{higher } T \Longrightarrow \text{higher } P$$

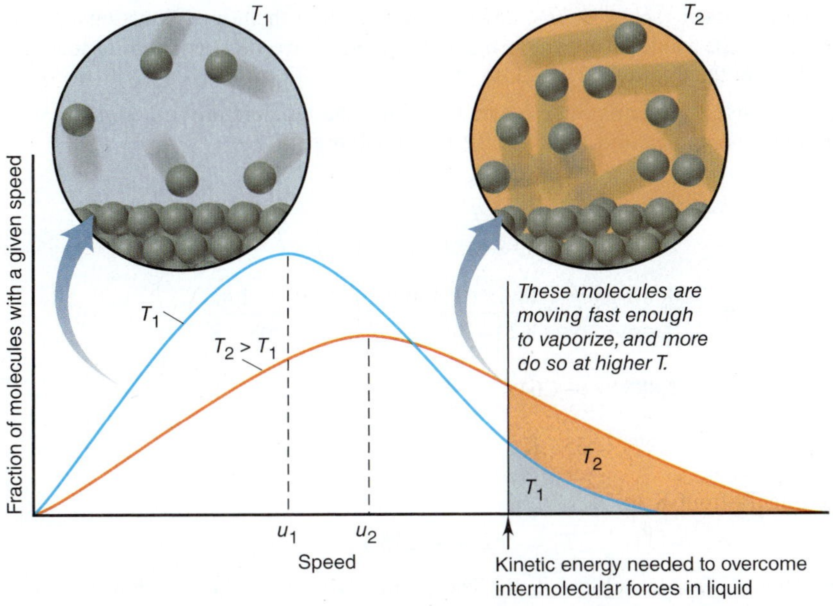

Figure 12.6 The effect of temperature on the distribution of molecular speeds.

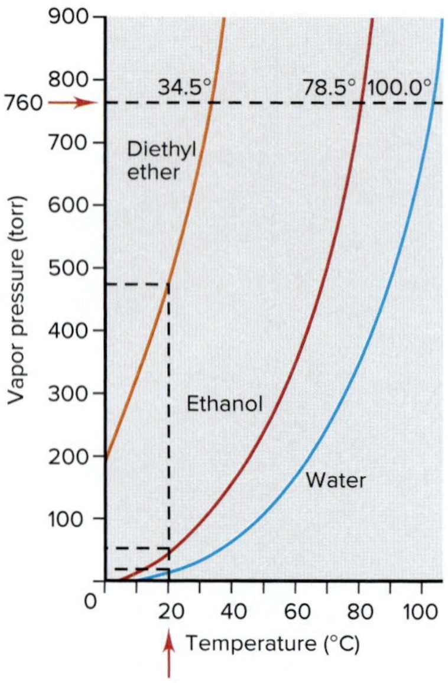

Figure 12.7 Vapor pressure as a function of temperature and intermolecular forces.

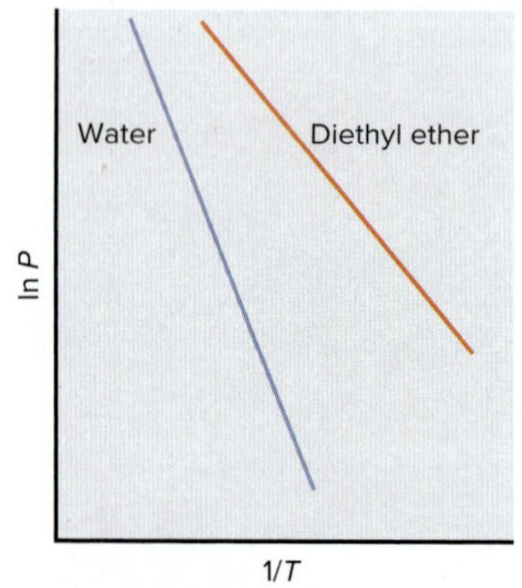

Figure 12.8 Linear plots of the relationship between vapor pressure and temperature. The slope is steeper for water because its ΔH_{vap} is greater.

2. *Effect of intermolecular forces.* At a given T, all substances have the same average E_k. Therefore, molecules with weaker intermolecular forces are held less tightly at the surface and vaporize more easily. In general, *the weaker the intermolecular forces are, the higher the vapor pressure:*

$$\text{weaker forces} \Longrightarrow \text{higher } P$$

Figure 12.7 shows the vapor pressure of three liquids as a function of temperature:

- The effect of temperature is seen in the steeper rise as the temperature increases.
- The effect of intermolecular forces is seen in the values of the vapor pressure, the short, horizontal dashed lines intersecting the vertical (pressure) axis at a given temperature *(vertical dashed line at 20°C)*: the intermolecular forces in diethyl ether (highest vapor pressure) are weaker than those in ethanol, which are weaker than those in water (lowest vapor pressure).

Quantifying the Effect of Temperature The nonlinear relationship between vapor pressure and temperature is converted to a linear one with the **Clausius-Clapeyron equation:**

$$\ln P = \frac{-\Delta H_{vap}}{R}\left(\frac{1}{T}\right) + C$$

$$y \quad = \quad m \quad\quad x \quad + b$$

where ln P is the natural logarithm of the vapor pressure, ΔH_{vap} is the heat of vaporization, R is the universal gas constant (8.314 J/mol·K), T is the absolute temperature, and C is a constant (not related to heat capacity). The equation is often used to find the heat of vaporization. The equation for a straight line is shown under it in blue, with $y = \ln P$, $x = 1/T$, m (the slope) $= -\Delta H_{vap}/R$, and b (the y-axis intercept) $= C$. A plot of ln P vs. $1/T$ gives a straight line, as shown for diethyl ether and water in Figure 12.8.

A two-point version of the Clausius-Clapeyron equation allows us to calculate ΔH_{vap} if vapor pressures at two temperatures are known:

$$\ln \frac{P_2}{P_1} = \frac{-\Delta H_{vap}}{R}\left(\frac{1}{T_2} - \frac{1}{T_1}\right) \quad \textbf{(12.1)}$$

If ΔH_{vap} and P_1 at T_1 are known, we can also calculate the vapor pressure (P_2) at any other temperature (T_2) or the temperature at any other pressure.

SAMPLE PROBLEM 12.2 Applying the Clausius-Clapeyron Equation

Problem The vapor pressure of ethanol is 115 torr at 34.9°C. If ΔH_{vap} of ethanol is 38.6 kJ/mol, calculate the temperature (in °C) when the vapor pressure is 760 torr.

Plan We are given ΔH_{vap}, P_1, P_2, and T_1 and substitute them into Equation 12.1 to solve for T_2. The value of R here is 8.314 J/mol·K, so we must convert T_1 to K to obtain T_2, and then convert T_2 back to °C.

Solution Convert the units of ΔH_{vap} to J/mol and the temperature to kelvins. Substituting the values into Equation 12.1 and solving for T_2:

$$\ln \frac{P_2}{P_1} = \frac{-\Delta H_{vap}}{R}\left(\frac{1}{T_2} - \frac{1}{T_1}\right)$$

$$T_1 = 34.9°\text{C} + 273.15 = 308.0 \text{ K}$$

$$\ln \frac{760 \text{ torr}}{115 \text{ torr}} = \left(-\frac{38.6\times10^3 \text{ J/mol}}{8.314 \text{ J/mol·K}}\right)\left(\frac{1}{T_2} - \frac{1}{308.0 \text{ K}}\right)$$

$$1.888 = (-4.64\times10^3 \text{ K})\left[\frac{1}{T_2} - (3.247\times10^{-3} \text{ K}^{-1})\right]$$

$$T_2 = 352 \text{ K}$$

Converting T_2 from K to °C:

$$T_2 = 352 \text{ K} - 273.15 = 79°\text{C}$$

Check Round off to check the math. The change is in the right direction: higher P should occur at higher T. As we discuss next, a substance has a vapor pressure of

760 torr at its *normal boiling point.* Checking the *CRC Handbook of Chemistry and Physics* shows that the boiling point of ethanol is 78.5°C, very close to our answer.

FOLLOW-UP PROBLEMS

12.2A At 34.1°C, the vapor pressure of water is 40.1 torr. What is the vapor pressure at 85.5°C? The ΔH_{vap} of water is 40.7 kJ/mol.

12.2B Acetone is essential as an industrial solvent and a starting reactant in the manufacturing of countless products, from films to synthetic fibers. The vapor pressure of acetone is 24.50 kPa at 20.2°C and 10.00 kPa at 0.95°C. What is ΔH_{vap} of acetone?

SOME SIMILAR PROBLEMS 12.21–12.24

Vapor Pressure and Boiling Point Let's discuss what is happening when a liquid boils and then see the effect of pressure on boiling point.

1. *How a liquid boils.* In an *open* container, the weight of the atmosphere bears down on a liquid surface. As the temperature rises, molecules move more quickly throughout the liquid. At some temperature, the average E_k of the molecules in the liquid is great enough for them to form bubbles of vapor *in the interior,* and the liquid boils. At any lower temperature, the bubbles collapse as soon as they start to form because the external pressure is greater than the vapor pressure inside the bubbles. Thus, the **boiling point** *is the temperature at which the vapor pressure inside bubbles in the liquid equals the external pressure,* which is usually that of the atmosphere. As in condensation and freezing, once boiling begins, the temperature of the liquid remains constant until all of the liquid is gone.

2. *Effect of pressure on boiling point.* The boiling point of a liquid varies with elevation. At high elevations, a lower atmospheric pressure is exerted on the liquid surface, so molecules in the interior need less kinetic energy to form bubbles. Therefore, in mountainous regions, food takes *more* time to cook because the boiling point is lower and the boiling liquid is not as hot; for instance, in Boulder, Colorado (elevation 5430 ft, or 1655 m), water boils at 94°C. On the other hand, in a pressure cooker, food takes *less* time to cook because the boiling point is higher at the higher pressure. Thus, *the boiling point is directly proportional to the applied pressure:*

$$\text{higher } P \text{ (applied)} \Longrightarrow \text{higher boiling point}$$

The *normal boiling point* is observed at standard atmospheric pressure (760 torr, or 101.3 kPa; *long, horizontal dashed line* in Figure 12.7).

Solid-Liquid Equilibria The particles in a crystal are continually jiggling about their fixed positions. As the temperature rises, the particles jiggle more rapidly, until some have enough kinetic energy to break free of their positions. At this point, melting begins. As more molecules enter the liquid (molten) phase, some collide with the solid and become fixed in position again. Because the phases remain in contact, a dynamic equilibrium is established when the melting rate equals the freezing rate. The temperature at which this occurs is called the **melting point.** The temperature remains fixed at the melting point until all the solid melts.

Because liquids and solids are nearly incompressible, pressure has little effect on the rates of melting and freezing: a plot of pressure vs. temperature for a solid-liquid phase change is typically a *nearly* vertical straight line.

Solid-Gas Equilibria Sublimation is not very familiar because solids have *much* lower vapor pressures than liquids. A substance sublimes rather than melts because the intermolecular attractions are not great enough to keep the molecules near each other when they leave the solid state. Some solids *do* have high enough vapor pressures to sublime at ordinary conditions, including dry ice (carbon dioxide), iodine (Figure 12.9), and moth repellants, all of which consist of nonpolar molecules with weak intermolecular forces.

The plot of pressure vs. temperature for a solid-gas phase change reflects the large effect of temperature on vapor pressure; thus, it resembles the liquid-gas curve in rising steeply with higher temperatures.

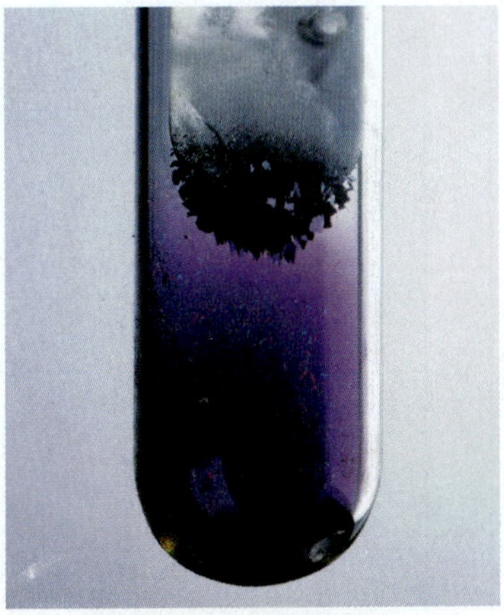

Figure 12.9 Iodine subliming. As the solid sublimes, I_2 vapor deposits on a cold surface (water-filled inner test tube).

Source: © McGraw-Hill Education/ Richard Megna, photographer

Phase Diagrams: Effect of Pressure and Temperature on Physical State

The **phase diagram** of a substance combines the liquid-gas, solid-liquid, and solid-gas curves and gives the conditions of temperature and pressure at which each phase is stable and at which phase changes occur.

The Phase Diagram for Carbon Dioxide and Most Substances The diagram for CO_2, which is typical of most substances, has four general features (Figure 12.10):

1. *Regions of the diagram.* Each region presents the conditions of pressure and temperature at which the phase is stable. If another phase is placed under those conditions, it will change to the stable phase. In general, the solid is stable at low temperature and high pressure, the gas at high temperature and low pressure, and the liquid at intermediate conditions.

2. *Lines between regions.* The lines are the phase-transition curves discussed earlier. Any point along a line shows the pressure and temperature at which the phases are in equilibrium. The solid-liquid line has a slightly *positive* slope (slants to the *right* with increasing pressure) because, for most substances, the solid is more dense than the liquid: an increase in pressure converts the liquid to the solid. (Water is *the* major exception.)

3. *The triple point.* The three phase-transition curves meet at the **triple point,** at which all three phases are in equilibrium. As strange as it sounds, at the triple point in Figure 12.10, CO_2 is subliming and depositing, melting and freezing, and vaporizing and condensing simultaneously! Substances with several solid and/or liquid forms can have more than one triple point.

The CO_2 phase diagram shows why dry ice (solid CO_2) doesn't melt under ordinary conditions. The triple-point pressure is 5.1 atm, so liquid CO_2 doesn't occur at 1 atm because it is not stable. The horizontal dashed line at 1.0 atm crosses the solid-gas line, so when solid CO_2 is heated, it sublimes at −78°C rather than melts. If our normal atmospheric pressure were 5.2 atm, liquid CO_2 *would* occur.

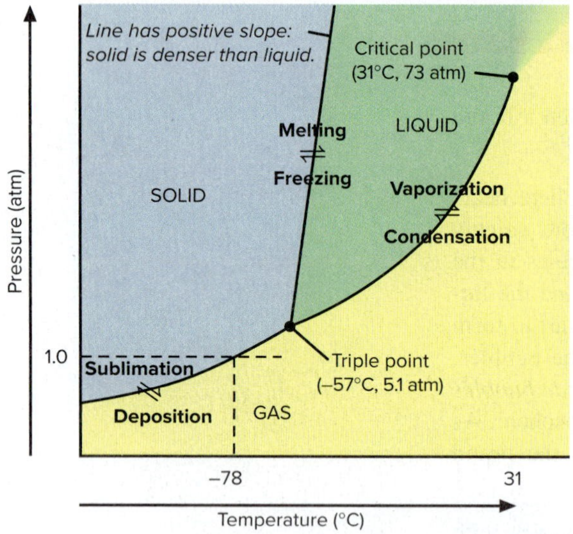

Figure 12.10 **Phase diagram for CO_2.** (In both this figure and Figure 12.11, the slope of the solid-liquid line is exaggerated and the axes are not linear.)

4. *The critical point.* Heat a liquid in a closed container and its density decreases. At the same time, more of the liquid vaporizes, so the density of the vapor increases. At the **critical point,** the two densities become equal and the phase boundary disappears. The temperature at the critical point is the *critical temperature* (T_c), and the pressure is the *critical pressure* (P_c). The average E_k is so high at this point that the vapor cannot be condensed at any pressure. The two most common gases in air have critical temperatures far below room temperature: O_2 cannot be condensed above −119°C, and N_2 cannot be condensed above −147°C.

Beyond the critical temperature, a *supercritical fluid* (SCF) exists rather than separate liquid and gaseous phases. An SCF expands and contracts like a gas and has unusual solvent properties. Supercritical CO_2 is used to extract caffeine from coffee beans, nicotine from tobacco, and fats from potato chips, and acts as a dry cleaning agent. Lower the pressure, and the SCF disperses as a harmless gas. Supercritical H_2O dissolves nonpolar substances, even though liquid water cannot. Studies are under way to use supercritical H_2O for removing nonpolar organic toxins, such as PCBs, from industrial waste.

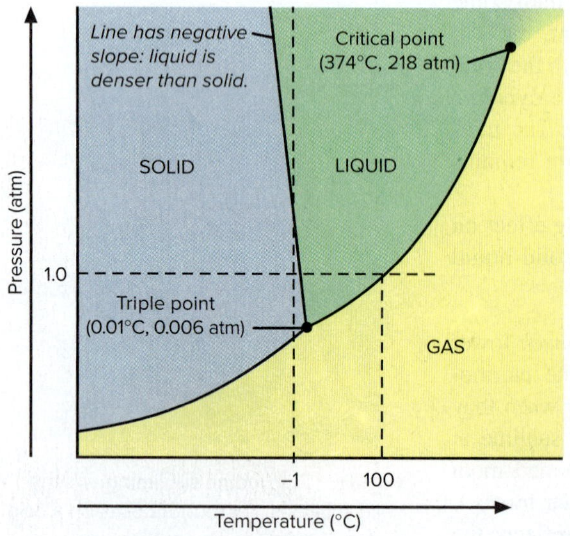

Figure 12.11 **Phase diagram for H_2O.**

The Solid-Liquid Line for Water The phase diagram for water differs from others in one major respect that reveals a key property. Unlike almost any other substance, the solid form is *less dense* than the liquid; that is, *water expands upon freezing.* Thus, the solid-liquid line has a *negative* slope (slants to the *left* with increasing pressure): an increase in pressure converts the solid to the liquid, and the higher the pressure, the lower the temperature at which water freezes (Figure 12.11). The vertical dashed line at −1°C crosses the solid-liquid line, which means that ice melts with only an increase in pressure.

The triple point of water occurs at low pressure (0.006 atm). Therefore, when solid water is heated at 1.0 atm *(horizontal dashed line)*, the solid-liquid line is crossed at 0°C, the normal melting point. Thus, ice melts rather than sublimes. The horizontal dashed line crosses the liquid-gas curve at 100°C, the normal boiling point.

SAMPLE PROBLEM 12.3 Using a Phase Diagram to Predict Phase Changes

Problem Use the phase diagram for carbon *(shown at right)* to describe the phase changes that a sample of carbon undergoes during the following:

(a) The sample is heated at 10^2 bar (99 atm) from 1000 K to 5000 K.

(b) The sample is then compressed at 5000 K to 10^6 bar (990,000 atm).

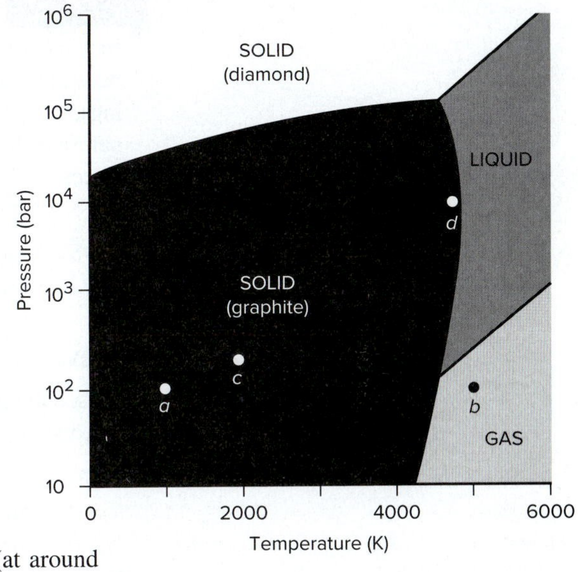

Plan We have to describe the phase changes due to heating at constant pressure (10^2 bar) and then compressing at constant temperature (5000 K).

(a) Using the phase diagram, we find the starting conditions (10^2 bar and 1000 K), the point labeled *a*, and draw a horizontal line from there to point *b*, at the ending conditions (10^2 bar and 5000 K). We note any phase-transition curves that the line crosses.

(b) Then, from the new starting conditions at point *b* (10^2 bar and 5000 K), we draw a vertical line to the ending pressure (10^6 bar) and note any phase-transition curves that this line crosses.

Solution **(a)** At 10^2 bar and 1000 K, the sample is in the form of solid graphite. As it is heated, the graphite sublimes and becomes carbon gas (at around 4400 K).

(b) As it is compressed at 5000 K, the gas first condenses to liquid carbon (at around 300 bar), which then solidifies to diamond (at around 3×10^5 bar).

Check **(a)** From the phase diagram, the first triple point occurs above 10^2 bar, so graphite will vaporize before it liquefies. **(b)** The liquid carbon will solidify to diamond because graphite does not exist at 5000 K under any pressure.

Comment Synthetic diamonds are made by compressing graphite at approximately 7×10^4 bar and 2300 K. The process occurs at a commercially useful rate when carried out in molten nickel. Traces of nickel remain in the diamonds, which are used for grinding tools and are not of gem quality.

FOLLOW-UP PROBLEMS

12.3A Describe the phase changes that a sample of carbon undergoes when it is heated at 3×10^2 bar from 2000 K (point *c* on the phase diagram in the sample problem) to 6000 K.

12.3B Describe the phase changes that a sample of carbon undergoes when it is compressed at 4600 K from 10^4 bar (point *d* on the phase diagram in the sample problem) to 10^6 bar.

SOME SIMILAR PROBLEMS 12.27 and 12.28

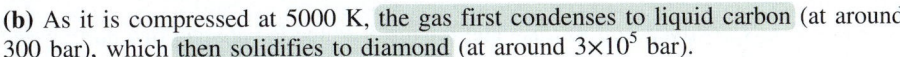

› Summary of Section 12.2

- A heating-cooling curve depicts the change in temperature when a substance absorbs or releases heat at a constant rate. Within a phase, temperature (and average E_k) changes. During a phase change, temperature (and average E_k) is constant, but E_p changes. The total enthalpy change for the system is found using Hess's law.
- In a closed container, the liquid and gas phases of a substance reach equilibrium. The vapor pressure, the pressure of the gas at equilibrium, *increases* with temperature and *decreases* with the strength of the intermolecular forces.
- The Clausius-Clapeyron equation relates the vapor pressure to the temperature and is often used to find ΔH_{vap}.

› A liquid in an open container boils when the vapor pressure inside bubbles forming in the liquid equals the external pressure.

› Solid-liquid equilibrium occurs at the melting point. Some solids sublime because they have very weak intermolecular forces.

› The phase diagram of a substance shows the phase that is stable at any *P* and *T*, the conditions at which phase changes occur, and the conditions at the critical point and the triple point. Water differs from most substances in that its solid phase is less dense than its liquid phase, so its solid-liquid line has a negative slope.

12.3 TYPES OF INTERMOLECULAR FORCES

In Chapter 9, we saw that bonding (*intra*molecular) forces are due to the attraction between cations and anions (ionic bonding), nuclei and electron pairs (covalent bonding), or metal cations and delocalized electrons (metallic bonding). But the physical nature of the phases and their changes are due primarily to *inter*molecular (nonbonding) forces, which arise from the attraction between molecules with partial charges or between ions and molecules. Coulomb's law explains the relative strength of these forces:

- *Bonding forces are relatively strong* because larger charges are closer together.
- *Intermolecular forces are relatively weak* because smaller charges are farther apart.

How Close Can Molecules Approach Each Other?

To see the minimum distance *between* molecules, consider solid Cl_2. When we measure the distances between two Cl nuclei, we obtain two different values (Figure 12.12A):

- *Bond length and covalent radius.* The shorter distance, called the *bond length,* is between *two bonded Cl atoms in the* ***same*** *molecule.* One-half this distance is the *covalent radius.*
- *Van der Waals distance and radius.* The longer distance is between *two nonbonded Cl atoms in* ***adjacent*** *molecules.* It is called the *van der Waals (VDW) distance.* At this distance, intermolecular attractions balance electron-cloud repulsions; thus, the VDW distance is as close as one Cl_2 molecule can approach another. The **van der Waals radius** is one-half the closest distance between nuclei of identical *nonbonded* atoms. The VDW radius of an atom is *always larger than its covalent radius.* Like covalent radii, VDW radii decrease across a period and increase down a group (Figure 12.12B).

As we discuss intermolecular forces (also called *van der Waals forces*), consult Table 12.2, which compares them with bonding forces.

Figure 12.12 Covalent and van der Waals radii and their periodic trends. A, The van der Waals (VDW) radius is one-half the closest distance between nuclei of adjacent *nonbonded* atoms ($\frac{1}{2}$ × VDW distance). B, Like covalent radii *(blue quarter circles and numbers),* VDW radii *(red quarter-circles and numbers)* increase down a group and decrease across a period.

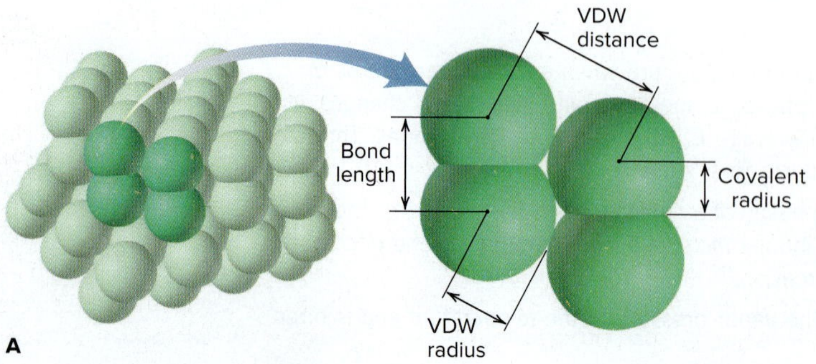

A

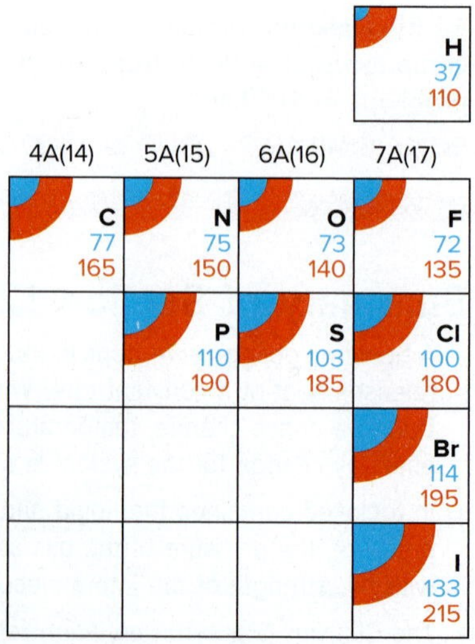

B

Table 12.2 Comparison of Bonding and Nonbonding (Intermolecular) Forces

Force	Model	Basis of Attraction	Energy (kJ/mol)	Example
Bonding				
Ionic	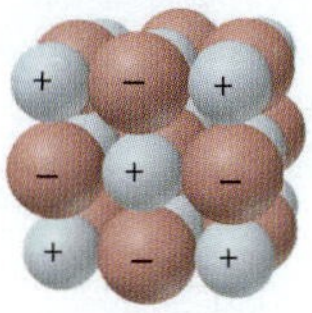	Cation–anion	400–4000	NaCl
Covalent	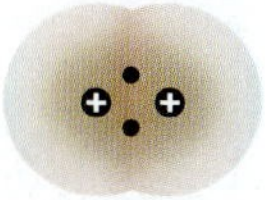	Nuclei–shared e^- pair	150–1100	H—H
Metallic	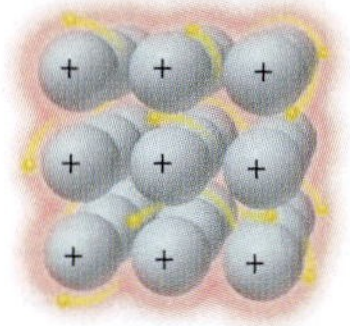	Cations–delocalized electrons	75–1000	Fe
Nonbonding (Intermolecular)				
Ion-dipole		Ion charge–dipole charge	40–600	$Na^+\cdots OH_2$
H bond	$\overset{\delta-}{—A}—\overset{\delta+}{H}\cdots\cdots:\overset{\delta-}{B}—$	Polar bond to H–dipole charge (high EN of N, O, F)	10–40	:Ö(H)—H···:Ö(H)—H
Dipole-dipole		Dipole charges	5–25	I—Cl···I—Cl
Ion–induced dipole		Ion charge–polarizable e^- cloud	3–15	$Fe^{2+}\cdots O_2$
Dipole–induced dipole		Dipole charge–polarizable e^- cloud	2–10	H—Cl···Cl—Cl
Dispersion (London)	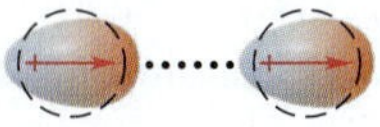	Polarizable e^- clouds	0.05–40	F—F···F—F

Ion-Dipole Forces

When an ion and a nearby polar molecule (dipole) attract each other, an **ion-dipole force** results. The most important example takes place when an ionic compound dissolves in water. As you'll see in Chapter 13, one main reason the ions become separated is because the attractions between the ions and the oppositely charged poles of the H_2O molecules are stronger than the attractions between the ions themselves.

Dipole-Dipole Forces

In Figure 10.13, an external electric field orients gaseous polar molecules. The polar molecules in liquids and solids lie near each other, and their partial charges act as tiny electric fields and give rise to **dipole-dipole forces:** the positive pole *(blue)* of one molecule attracts the negative pole *(red)* of another (Figure 12.13). The orientation is more orderly in a solid than in a liquid because the average kinetic energy of the molecules is lower.

These forces depend on the magnitude of the molecular dipole moment. For compounds of similar molar mass, the greater the molecular dipole moment, the greater the dipole-dipole forces, so the more energy it takes to separate the molecules; thus, the boiling point is higher. Methyl chloride, for instance, has a smaller dipole

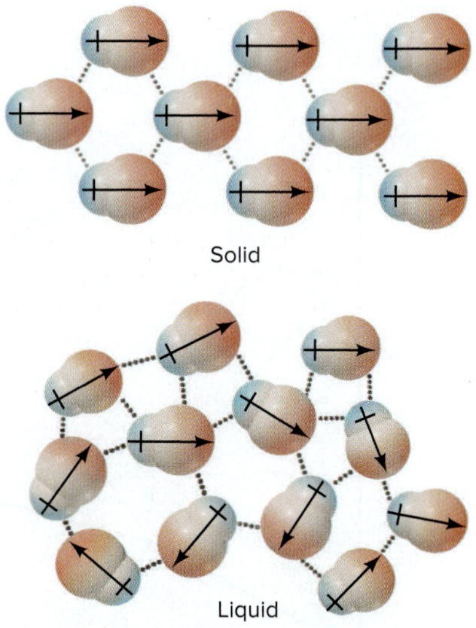

Figure 12.13 Polar molecules and dipole-dipole forces. (Spaces between the molecules are exaggerated.)

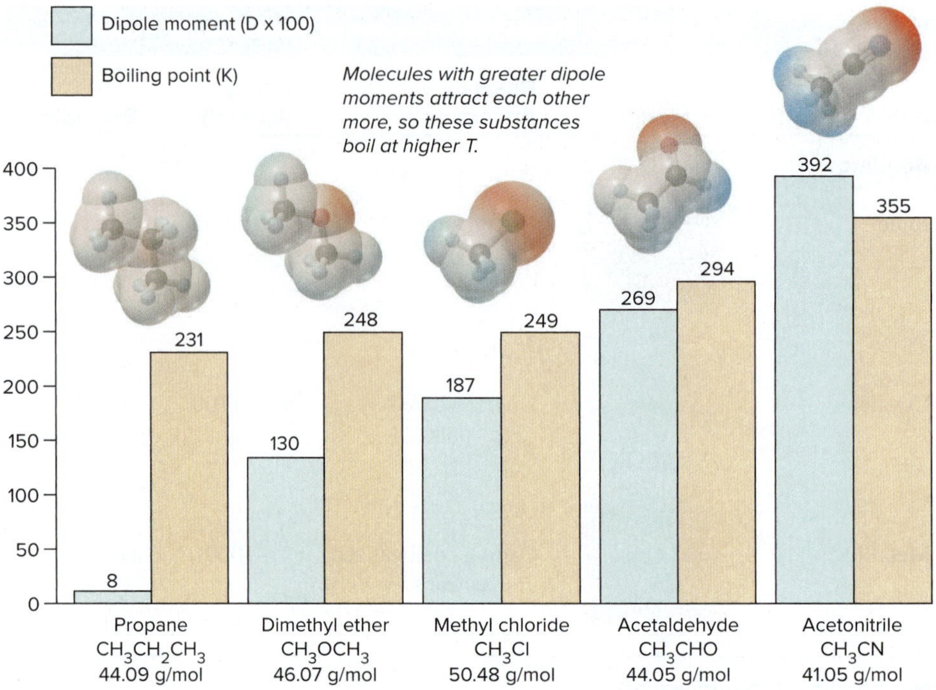

Figure 12.14 Dipole moment and boiling point.

moment than acetaldehyde and boils at a lower temperature (Figure 12.14). Dipole-dipole forces do not exist for nonpolar molecules.

The Hydrogen Bond

A special type of dipole-dipole force arises between molecules that have *an H atom bonded to a small, highly electronegative atom with lone electron pairs, specifically N, O, or F.* The H—N, H—O, and H—F bonds are very polar. When the partially positive H of one molecule is attracted to the partially negative lone pair on the N, O, or F of another molecule, a **hydrogen bond (H bond)** forms. Thus, the atom sequence of an H bond (dotted line) is —B:···H—A—, where A *and* B are N, O, or F. Some examples are

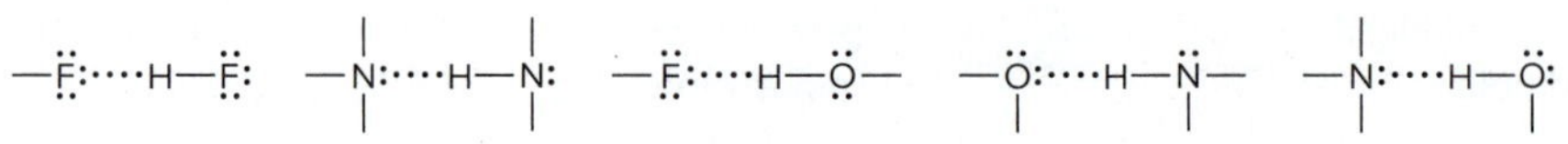

The first two are found in pure samples of HF and NH_3, respectively.

The small sizes of N, O, and F are essential to H bonding for two reasons:

1. The atoms are so electronegative that their covalently bonded H is highly positive.
2. The lone pair on the N, O, or F of the other molecule can come close to the H.

The Significance of Hydrogen Bonding Hydrogen bonding has a profound impact in many systems. We'll examine one effect on physical properties and preview its importance in biological systems, a topic we address in Chapters 13 and 15.

Figure 12.15 shows the effect of H bonding on the boiling points of the binary hydrides of Groups 4A(14) through 7A(17). For reasons we'll discuss shortly, boiling points rise with molar mass, as the Group 4A(14) hydrides show. However, the first member in each of the other groups—NH_3, H_2O, and HF—deviates enormously from this expected trend. Within samples of these substances, the molecules form strong H bonds, so it takes more energy for the molecules to separate and enter the gas phase. For example, on the basis of molar mass alone, we would

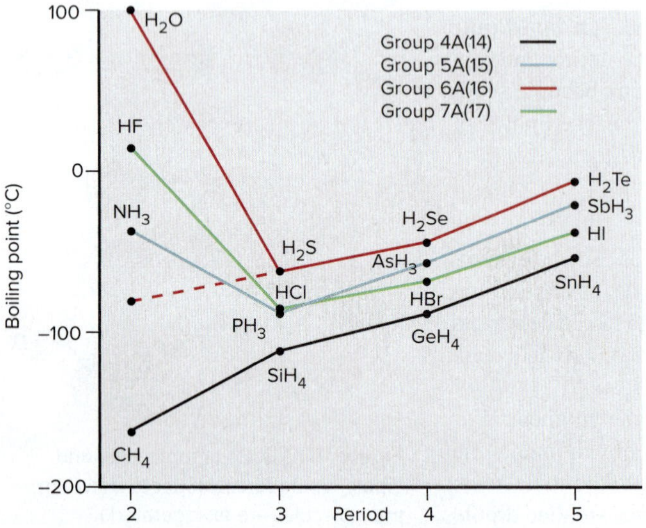

Figure 12.15 Hydrogen bonding and boiling point. NH_3, H_2O, and HF have exceptionally high boiling points because they form H bonds.

expect water to boil about 200°C lower than it actually does *(red dashed line).* (In Section 12.5 we'll discuss the effects that H bonds in water have in nature.)

The significance of hydrogen bonding in biological systems cannot be emphasized too strongly. In Chapter 13, you'll see that it is a key feature in the structure and function of the biological macromolecules. As such, it is responsible for the action of many *enzymes*—the proteins that speed metabolic reactions—and for the functioning of genes, not to mention the strength of wood and texture of cotton.

SAMPLE PROBLEM 12.4 Drawing Hydrogen Bonds Between Molecules of a Substance

Problem Which of these substances exhibits H bonding? Draw examples of the H bonds between two molecules of each substance that does.

(a) C_2H_6 **(b)** CH_3OH **(c)** $CH_3\overset{\overset{\large O}{\|}}{C}—NH_2$

Plan If the molecule does *not* contain N, O, or F, it cannot form H bonds. If it contains any of these atoms covalently bonded to H, we draw two molecules in the —B:····H—A— pattern.

Solution (a) For C_2H_6. No N, O, or F, so no H bonds can form.

(b) For CH_3OH. The H covalently bonded to the O in one molecule forms an H bond to the lone pair on the O of an adjacent molecule:

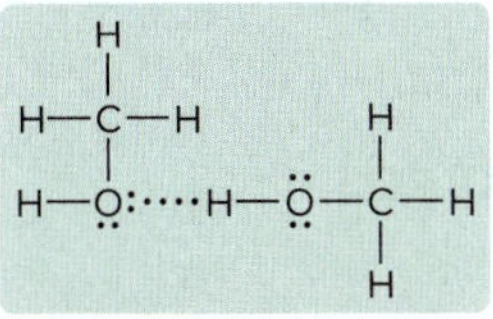

(c) For $CH_3\overset{\overset{\large O}{\|}}{C}—NH_2$. Two of these molecules can form one H bond between an H bonded to N and the O, or they can form two such H bonds:

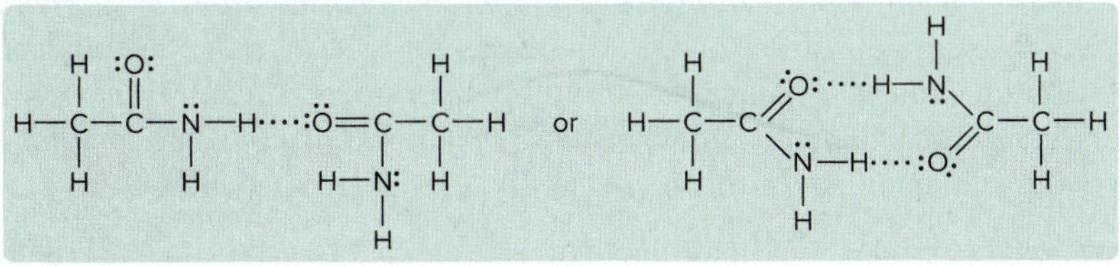

A third possibility (not shown) could be between an H attached to N in one molecule and the lone pair of N in another molecule.

Check The —B:····H—A— sequence (with A and B either N, O, or F) is present.

Comment Note that H covalently bonded to C *does not form H bonds* because carbon is not electronegative enough to make the C—H bond sufficiently polar.

FOLLOW-UP PROBLEMS

12.4A Which of these substances exhibits H bonding? Draw one possible example of the H bond(s) between two molecules of each substance that does.

(a) $CH_3\overset{\overset{\large O}{\|}}{C}—OH$ **(b)** CH_3CH_2OH **(c)** $CH_3\overset{\overset{\large O}{\|}}{C}CH_3$

12.4B Which of these substances exhibits H bonding? Draw one possible example of the H bond(s) between two molecules of each substance that does.

(a) $H_2C{=}O$ **(b)** H_2NOH **(c)** $O{=}CHCH_2OH$

SOME SIMILAR PROBLEMS 12.43 and 12.44

Polarizability and Induced Dipole Forces

Even though electrons are attracted to nuclei and localized in bonding and lone pairs, we often picture them as "clouds" of negative charge because they are in constant motion. A nearby electric field can *induce* a distortion in the cloud, pulling

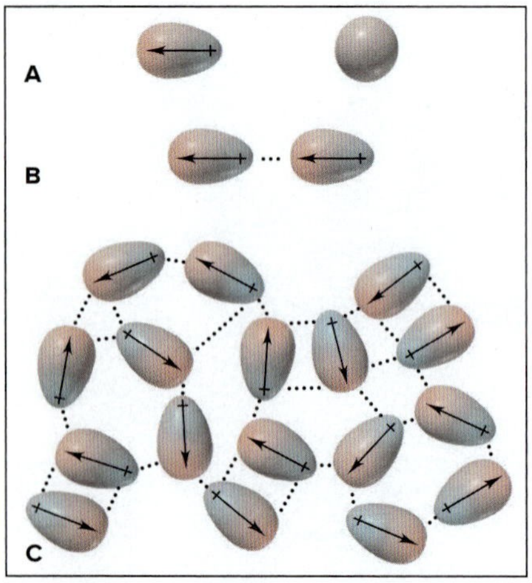

Figure 12.16 Dispersion forces among nonpolar particles. A, When atoms are far apart, an instantaneous dipole in one atom *(left)* doesn't influence another. **B,** When atoms are close together, the instantaneous dipole in one atom induces a dipole in the other. **C,** The process occurs through the sample.

electron density toward a positive pole of a field or pushing it away from a negative one:

- *For a nonpolar molecule,* the distortion induces a temporary dipole moment.
- *For a polar molecule,* the distortion induces an increase in the already existing dipole moment.

In addition to charged plates connected to a battery, the source of the electric field can be the charge of an ion or the partial charges of a polar molecule.

How easily the electron cloud of an atom (or ion) can be distorted is called its **polarizability.** Smaller particles are less polarizable than larger ones because their electrons are closer to the nucleus and therefore held more tightly. Thus, we observe several trends:

- Polarizability *increases down a group* because atomic size increases and larger electron clouds are easier to distort.
- Polarizability *decreases across a period* because increasing Z_{eff} makes the atoms smaller and holds the electrons more tightly.
- Cations are *less* polarizable than their parent atoms because they are smaller; anions are *more* polarizable because they are larger.

Ion–induced dipole and dipole–induced dipole forces are the two types of charge-induced dipole forces; they are most important in solution, so we'll focus on them in Chapter 13. Nevertheless, *polarizability affects all intermolecular forces.*

Dispersion (London) Forces

So far, we've discussed forces that depend on the existing charge of an ion or a polar molecule. But what forces cause nonpolar substances like octane, chlorine, and argon to condense and solidify? As you'll see, polarizability plays the central role in the most universal intermolecular force.

The intermolecular force responsible for the condensed states of nonpolar substances is the **dispersion force** (or **London force,** for Fritz London, the physicist who explained its quantum-mechanical basis). *Dispersion forces are present between all particles (atoms, ions, and molecules) because they result from the motion of electrons in atoms.* Let's examine key aspects of this force:

1. *Origin.* Picture one atom in a sample of, say, argon gas. Over time, its 18 electrons are distributed uniformly, so the atom is nonpolar. But at any instant, there may be more electrons on one side of the nucleus than the other, which gives the atom an *instantaneous dipole.* When a pair of argon atoms is far apart, they don't influence each other, but when close together, *the instantaneous dipole in one atom induces a dipole in its neighbor,* and they attract each other. This process spreads to other atoms and throughout the sample. At low temperatures, these attractions keep the atoms together (Figure 12.16). Thus, dispersion forces are *instantaneous dipole–induced dipole forces.*

2. *Universal presence.* While they are the *only* force existing between nonpolar particles, *dispersion forces contribute to the energy of attraction in all substances* because they exist between *all* particles. In fact, except for the forces between small, highly polar molecules or between molecules forming H bonds, the *dispersion force is the dominant intermolecular force.* Calculations show, for example, that 85% of the attraction between HCl molecules is due to dispersion forces and only 15% to dipole-dipole forces. Even for water, 75% of the attraction comes from H bonds and 25% from dispersion forces.

3. *Relative strength.* The relative strength of dispersion forces depends on the polarizability of the particles, so they are weak for small particles, like H_2 and He, but stronger for larger particles, like I_2 and Xe. *Polarizability depends on the number of electrons, which correlates closely with molar mass* because heavier particles are either larger atoms or molecules with more atoms and, thus, more electrons. For this reason, as molar mass increases down the Group 4A(14) hydrides (see Figure 12.15) or down the halogens or the noble gases, dispersion forces increase and so do boiling points (Figure 12.17).

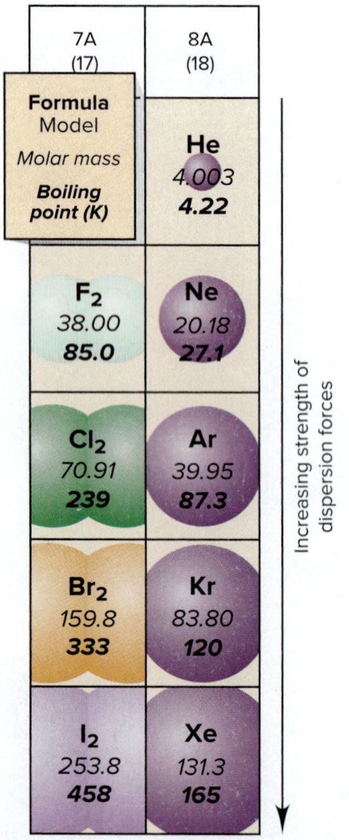

Figure 12.17 Molar mass and trends in boiling point.

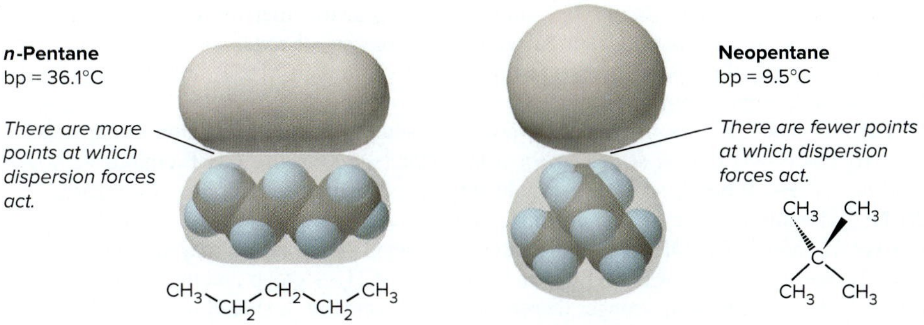

Figure 12.18 Molecular shape, intermolecular contact, and boiling point.

4. *Effect of molecular shape.* For a pair of nonpolar substances with the same molar mass, a molecular shape that has more area over which electron clouds can be distorted allows stronger attractions. For example, the two five-carbon alkanes, *n*-pentane and neopentane (2,2-dimethylpropane) are isomers—same molecular formula (C_5H_{12}) but different structures and properties. *n*-Pentane is more cylindrical and neopentane more spherical (Figure 12.18). Thus, two *n*-pentane molecules make more contact than do two neopentane molecules, so dispersion forces act at more points, and *n*-pentane has a higher boiling point.

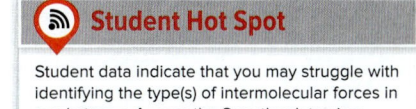

Student Hot Spot

Student data indicate that you may struggle with identifying the type(s) of intermolecular forces in a substance. Access the Smartbook to view additional Learning Resources on this topic.

Figure 12.19 shows how to decide what intermolecular forces are present in a sample.

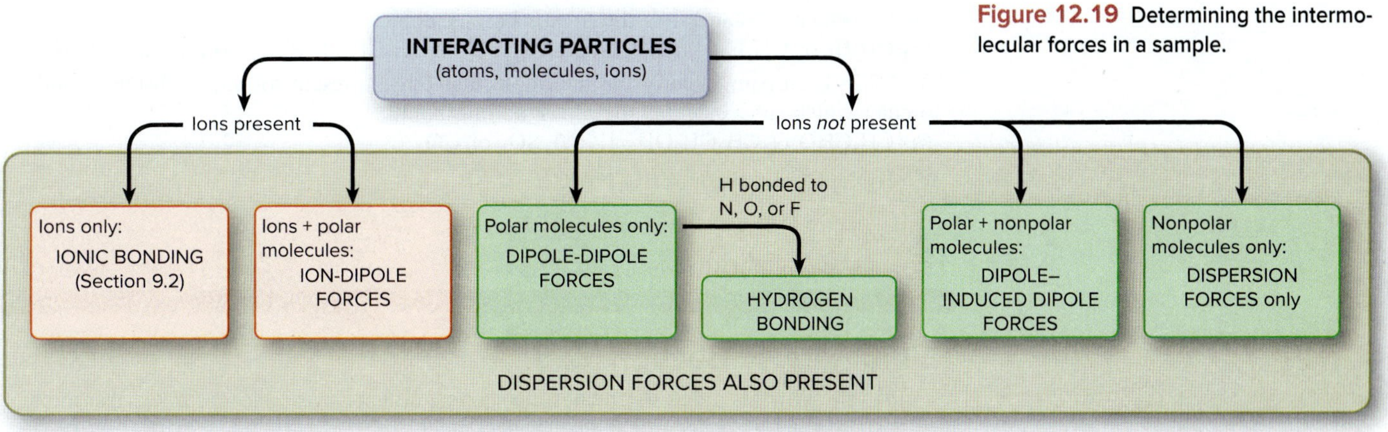

Figure 12.19 Determining the intermolecular forces in a sample.

SAMPLE PROBLEM 12.5 Identifying the Types of Intermolecular Forces

Problem For each substance, identify the key bonding and/or intermolecular force(s), and predict which substance of the pair has the higher boiling point:

(a) $MgCl_2$ or PCl_3

(b) CH_3NH_2 or CH_3F

(c) CH_3OH or CH_3CH_2OH

(d) Hexane ($CH_3CH_2CH_2CH_2CH_2CH_3$) or 2,2-dimethylbutane ($CH_3C(CH_3)_2CH_2CH_3$)

Plan We examine the formulas and structures for key differences between members of the pair: Are ions present? Are molecules polar or nonpolar? Is N, O, or F bonded to H? Do the molecules have different masses or shapes?
To rank boiling points, we consult Figure 12.19 and Table 12.2. Remember that

- Bonding forces are stronger than intermolecular forces.
- Hydrogen bonding is a strong type of dipole-dipole force.
- Dispersion forces are decisive when the difference is molar mass or molecular shape.

(b)
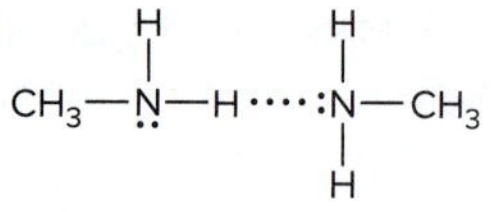

(c)
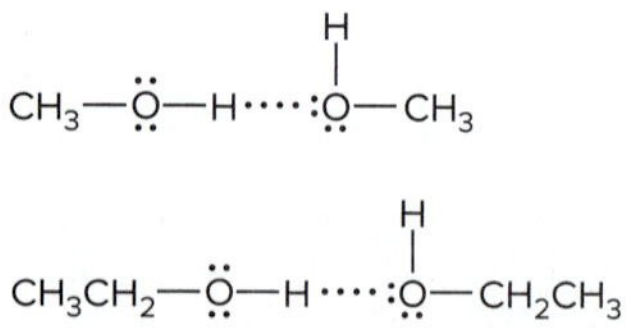

(d)
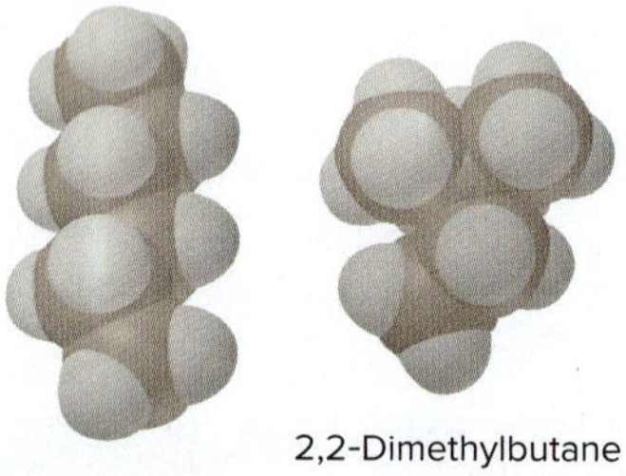

Hexane

2,2-Dimethylbutane

Solution **(a)** $MgCl_2$ consists of Mg^{2+} and Cl^- ions held together by ionic bonding forces; PCl_3, with a trigonal pyramidal geometry, consists of polar molecules, so intermolecular dipole-dipole forces are present. The forces in $MgCl_2$ are stronger, so it should have a higher boiling point.

(b) CH_3NH_2 and CH_3F both consist of polar molecules of about the same molar mass. CH_3NH_2 has N—H bonds, so it can form H bonds *(see margin).* CH_3F contains a C—F bond but no H—F bond, so dipole-dipole forces occur but not H bonds. Therefore, CH_3NH_2 should have the higher boiling point.

(c) CH_3OH and CH_3CH_2OH molecules both contain an O—H bond, so they can form H bonds *(see margin).* CH_3CH_2OH has an additional —CH_2— group and thus a larger molar mass, which correlates with stronger dispersion forces; therefore, it should have a higher boiling point.

(d) Hexane and 2,2-dimethylbutane are nonpolar molecules of the same molar mass but different molecular shapes *(see margin).* Cylindrical hexane molecules make more intermolecular contact than more compact 2,2-dimethylbutane molecules do, so hexane should have stronger dispersion forces and a higher boiling point.

Check The actual boiling points show that our predictions are correct:
(a) $MgCl_2$ (1412°C) and PCl_3 (76°C)
(b) CH_3NH_2 (–6.3°C) and CH_3F (–78.4°C)
(c) CH_3OH (64.7°C) and CH_3CH_2OH (78.5°C)
(d) Hexane (69°C) and 2,2-dimethylbutane (49.7°C)

Comment Dispersion forces are *always* present, but in parts (a) and (b), they are much less significant than the other forces that occur.

FOLLOW-UP PROBLEMS

12.5A In each pair, identify the intermolecular forces present for each substance, and predict which substance has the *higher* boiling point:
(a) CH_3Br or CH_3F **(b)** $CH_3CH_2CH_2OH$ or $CH_3CH_2OCH_3$ **(c)** C_2H_6 or C_3H_8

12.5B In each pair, identify the intermolecular forces present for each substance, and predict which substance has the *lower* boiling point:
(a) CH_3CHO or CH_3CH_2OH **(b)** SO_2 or CO_2
(c) $H_2N\text{—}\underset{\underset{O}{\|}}{C}CH_2CH_3$ or $(CH_3)_2N\text{—}\underset{\underset{O}{\|}}{C}H$

SOME SIMILAR PROBLEMS 12.49–12.54

› Summary of Section 12.3

- The van der Waals radius determines the shortest distance over which intermolecular forces operate; it is always larger than the covalent radius.
- Intermolecular forces are much weaker than bonding (intramolecular) forces.
- Ion-dipole forces occur between ions and polar molecules.
- Dipole-dipole forces occur between oppositely charged poles on polar molecules.
- Hydrogen bonding, a special type of dipole-dipole force, occurs when H bonded to N, O, or F is attracted to the lone pair of N, O, or F in another molecule.
- Electron clouds can be distorted (polarized) in an electric field.
- Ion– and dipole–induced dipole forces arise between a charge and the dipole it induces in another molecule.
- Dispersion (London) forces are instantaneous dipole–induced dipole forces that occur among all particles and increase with number of electrons (molar mass). Molecular shape determines the extent of contact between molecules and can be a factor in the strength of dispersion forces.

12.4 PROPERTIES OF THE LIQUID STATE

Of the three states, only liquids combine the ability to flow with the effects of strong intermolecular forces. We understand this state least at the molecular level. Because of the *random* arrangement of the particles in a gas, any region of the sample is virtually identical to any other. And, different regions of a crystalline solid are identical because of the *orderly* arrangement of the particles (Section 12.6). Liquids, however, have regions that are orderly one moment and random the next. Nevertheless, many macroscopic properties, such as surface tension, capillarity, and viscosity, are well understood.

Surface Tension

Intermolecular forces have different effects on a molecule at the surface than on one in the interior (Figure 12.20):

- An interior molecule is attracted by others on all sides.
- A surface molecule is only attracted by others below and to the sides, so it experiences a *net attraction downward.*

Therefore, to increase attractions and become more stable, a surface molecule tends to move into the interior. For this reason, *a liquid surface has the fewest molecules and, thus, the smallest area possible.* In effect, the surface behaves like a "taut skin" covering the interior.

The only way to increase the surface area is for molecules to move up by breaking attractions in the interior, which requires energy. The **surface tension** is the energy required to increase the surface area by a given amount and has units of J/m^2. This property is dependent on the intermolecular forces present and the temperature:

- In general, *the stronger the intermoleular forces between particles, the more energy it takes to increase the surface area, so the greater the surface tension* (Table 12.3). Water has a high surface tension because its molecules form multiple H bonds. *Surfactants* (*surf*ace-*act*ive age*nts*), such as soaps, petroleum recovery agents, and fat emulsifiers, decrease the surface tension of water by congregating at the surface and disrupting the H bonds.
- Surface tension *decreases with increasing temperature.* For example, the surface tension of water is 7.3×10^{-2} J/m^2 at 20°C, 6.8×10^{-2} J/m^2 at 50°C, and 6.1×10^{-2} J/m^2 at 90°C. At higher temperatures, the liquid molecules have increased kinetic energy with which to break attractions to molecules in the interior. Hot water is more effective than cold water at cleaning greasy dishes as the lower surface tension of hot water allows it to "wet" the surface of the dish to a greater extent. The surfactants in the soap further decrease water's surface tension.

Figure 12.20 The molecular basis of surface tension.

Table 12.3 Surface Tension, Viscosity, and Forces Between Particles

Substance	Formula	Surface Tension (J/m^2) at 20°C	Viscosity ($N\cdot s/m^2$) at 20°C	Major Force(s)
Diethyl ether	$CH_3CH_2OCH_2CH_3$	1.7×10^{-2}	0.240×10^{-3}	Dipole-dipole; dispersion
Ethanol	CH_3CH_2OH	2.3×10^{-2}	1.20×10^{-3}	H bonding
Butanol	$CH_3CH_2CH_2CH_2OH$	2.5×10^{-2}	2.95×10^{-3}	H bonding; dispersion
Water	H_2O	7.3×10^{-2}	1.00×10^{-3}	H bonding
Mercury	Hg	48×10^{-2}	1.55×10^{-3}	Metallic bonding

Surface tension is the cause of many familiar phenomena. Bubbles are round because this shape minimizes the surface area around the gas, and surface tension is the reason some insects can walk on water *(see photo).*

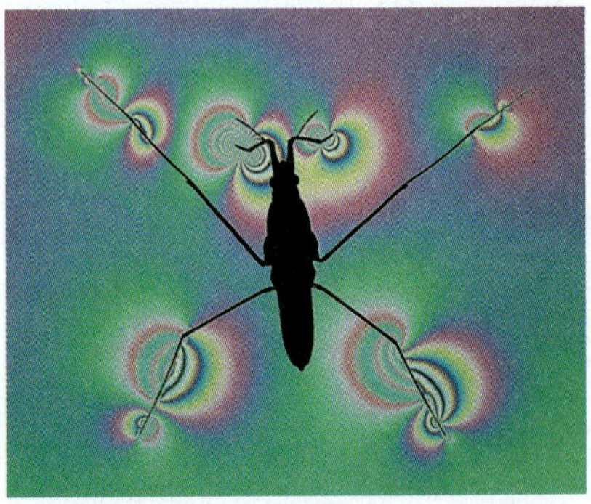

Because of its widespread legs, a water strider doesn't exert enough pressure to exceed the surface tension.

Source: © Nuridsany & Perennou/Science Source

Capillarity

The rising of a liquid against the pull of gravity through a narrow space, such as a thin tube, is called *capillary action,* or **capillarity.** Capillarity results from a competition between the intermolecular forces within the liquid (cohesive forces) and those between the liquid and the tube walls (adhesive forces). Let's look at the difference between the capillarities of water and mercury in glass:

1. *Water in glass.* When you place a narrow glass tube in water, why does the liquid rise up the tube and form a concave meniscus? Glass is mostly silicon dioxide (SiO_2), so water molecules form adhesive H-bonding forces with the O atoms of the glass. As a result, a thin film of water creeps up the wall. At the same time, cohesive H-bonding forces between water molecules, which give rise to surface tension, make the surface taut. These adhesive and cohesive forces combine to raise the

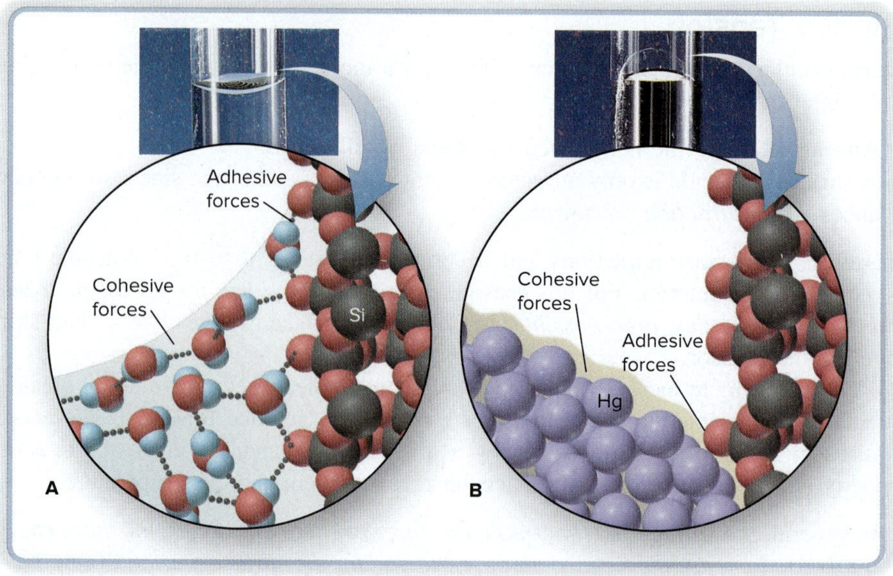

Figure 12.21 **Capillary action and the shape of the water or mercury meniscus in glass.** **A,** Water displays a concave meniscus. **B,** Mercury displays a convex meniscus.

Source: © McGraw-Hill Education/Stephen Frisch, photographer

water level and produce the concave meniscus (Figure 12.21A). The liquid rises until gravity pulling down is balanced by adhesive forces pulling up.

2. *Mercury in glass.* When you place a glass tube in a dish of mercury, why does the liquid drop below the level in the dish and form a convex meniscus? The cohesive forces among the mercury atoms are metallic bonds, so they are *much* stronger than the mostly dispersion adhesive forces between mercury and glass. As a result, the liquid pulls away from the walls. At the same time, the surface atoms are being pulled toward the interior by mercury's high surface tension, so the level drops. These combined forces produce the convex meniscus seen in a laboratory barometer (Figure 12.21B).

Capillarity plays a key role in many everyday events. The adhesive (dipole–induced dipole) forces between water and a nonpolar surface are much weaker than the cohesive (H-bond) forces within water. As a result, water pulls away from a nonpolar surface and forms beaded droplets, as on a freshly waxed car or on the waxy coating of a leaf after a rainfall. Even more familiar, after a shower, capillary action draws water away from your body through the closely spaced fibers in the cotton towel, which is made of cellulose molecules, so the water molecules also form H bonds with the —OH groups of cellulose.

Viscosity

Viscosity is the resistance of a fluid to flow, and it results from intermolecular attractions that impede the movement of molecules around and past each other. Both gases and liquids flow, but liquid viscosities are *much* higher because the much shorter distances between the particles of a liquid result in many more points for intermolecular forces to act. Intermolecular forces, temperature, and molecular shape influence viscosity:

- In general, *the stronger the intermolecular forces between particles, the higher the viscosity* (Table 12.3) since stronger attractions prevent molecules from moving past each other freely.
- *Viscosity decreases with increasing temperature* (Table 12.4). Faster moving molecules overcome intermolecular forces more easily, so the resistance to flow decreases with increasing temperature. Next time you heat cooking oil, watch the oil flow more easily and spread out in the pan as it warms.
- *Viscosity is affected by molecular shape.* Small, spherical molecules make little contact and pour easily, like buckshot from a bowl. Long molecules make more contact and become entangled and pour slowly, like cooked spaghetti from a bowl. Thus, given the same types of intermolecular forces, liquids consisting of longer molecules have higher viscosities. As shown in Table 12.3, the viscosity of the longer butanol molecule is greater than that of the shorter ethanol molecule.

Table 12.4 Viscosity of Water at Several Temperatures

Temperature (°C)	Viscosity ($N \cdot s/m^2$)*
20	1.00×10^{-3}
40	0.65×10^{-3}
60	0.47×10^{-3}
80	0.35×10^{-3}

*The units of viscosity are newton-seconds per square meter.

To protect engine parts during long drives, motor oils contain *polymeric viscosity improvers*. As the oil warms, these additive molecules change from compact spheres to long strands that become tangled with the long hydrocarbon molecules of the oil. Greater dispersion forces increase the viscosity to compensate for the thinning of the oil due to heating.

› Summary of Section 12.4

- › Surface tension is a measure of the energy required to increase a liquid's surface area. Greater intermolecular forces within a liquid create higher surface tension.
- › Capillarity, the rising of a liquid through a narrow space, occurs when the forces between a liquid and a surface (adhesive) are greater than those in the liquid (cohesive).
- › Viscosity, the resistance to flow, depends on molecular shape and decreases with temperature. Stronger intermolecular forces create higher viscosity.

12.5 THE UNIQUENESS OF WATER

Water is absolutely amazing stuff, with some of the most unusual properties of any substance, but it is so familiar we take it for granted. Like any substance, its properties arise inevitably from those of its atoms. Each O and H atom attains a filled outer level by sharing electrons in single bonds. With two bonding pairs and two lone pairs around O and a large electronegativity difference in each O—H bond, the H_2O molecule is bent and highly polar. This arrangement is crucial because it allows each molecule to engage in four H bonds with its neighbors (Figure 12.22). From these basic atomic and molecular facts emerges some unique and remarkable macroscopic behavior.

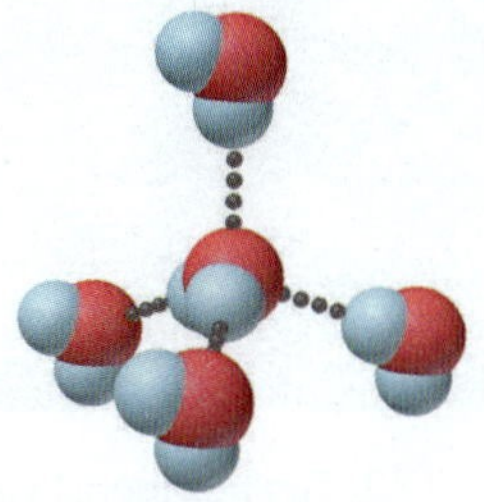

Figure 12.22 H-bonding ability of water. One H_2O molecule can form four H bonds to other molecules, resulting in a tetrahedral arrangement.

Solvent Properties of Water

The *great solvent power* of water results from its polarity and H-bonding ability:

- It dissolves ionic compounds through ion-dipole forces that separate the ions from the solid and keep them in solution (see Figure 4.2).
- It dissolves polar nonionic substances, such as ethanol (CH_3CH_2OH) and glucose ($C_6H_{12}O_6$), by H bonding.
- It dissolves nonpolar atmospheric gases to a limited extent through dipole–induced dipole and dispersion forces.

Water is the environmental and biological solvent, forming the complex solutions we know as oceans, lakes, and cellular fluid. Aquatic animals could not survive without dissolved O_2, nor could aquatic plants without dissolved CO_2. Tiny marine animals form coral reefs made of carbonates from dissolved CO_2 and HCO_3^-. Life began in a "primordial soup," an aqueous mixture of simple molecules from which emerged larger molecules capable of self-sustaining reactions. From a chemical point of view, all organisms, from bacteria to humans, are highly organized systems of membranes enclosing and compartmentalizing complex aqueous solutions.

Thermal Properties of Water

When a substance is heated, some of the added energy increases average molecular speed, some increases molecular vibration and rotation, and some is used to overcome intermolecular forces.

1. *Specific heat capacity.* Because water has so many strong H bonds, its *specific heat capacity* is higher than that of any common liquid. With oceans covering 70% of Earth's surface, daytime energy from the Sun causes relatively small changes in temperature, allowing life to survive. On the waterless, airless Moon, temperatures range from 100°C to −150°C during a complete lunar day. Even in Earth's deserts, day-night temperature differences of 40°C are common.

2. *Heat of vaporization.* Numerous strong H bonds give water a very *high heat of vaporization*. Two examples show why this is crucial. The average adult has 40 kg of

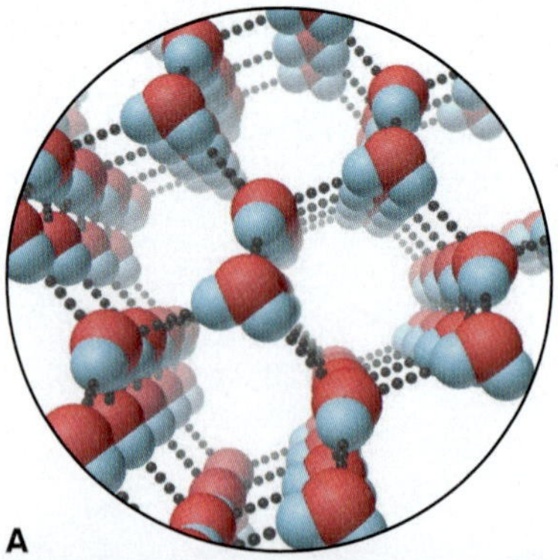

A

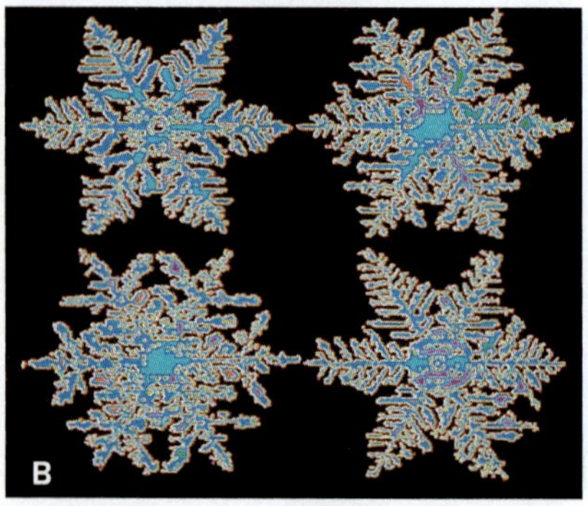

B

Figure 12.23 The hexagonal structure of ice. A, The open, hexagonal molecular structure of ice. **B,** The beauty of six-pointed snowflakes reflects this hexagonal structure.

Source: (B) © Scott Camazine/Science Source

body water and generates about 10,000 kJ of heat a day from metabolism. If this heat were used to increase the average E_k of water molecules in the body, the rise in body temperature of tens of degrees would mean immediate death. Instead, the heat is converted to E_p as it breaks H bonds and evaporates sweat, resulting in a stable body temperature and minimal loss of body fluid. On a planetary scale, the Sun's energy vaporizes ocean water in warm latitudes, and the potential energy is released as heat to warm cooler regions when the vapor condenses to rain. This global-scale cycling of water powers many weather patterns.

Surface Properties of Water

Hydrogen bonding is also responsible for water's *high surface tension* and *high capillarity*. Except for some molten metals and salts, water has the highest surface tension of any liquid. It keeps plant debris resting on a pond surface, providing shelter and nutrients for fish and insects. High capillarity means water rises through the tiny spaces between soil particles, so plant roots can absorb deep groundwater during dry periods.

The Unusual Density of Solid Water

In the solid state, the tetrahedral arrangement of H-bonded water molecules (see Figure 12.22) leads to the hexagonal, *open structure* of ice (Figure 12.23A), and the symmetrical beauty of snowflakes (Figure 12.23B) reflects this hexagonal organization. The large spaces within ice make *the solid less dense than the liquid* and explain the negative slope of the solid-liquid line in the phase diagram for water (see Figure 12.11). As pressure is applied, some H bonds break, so the crystal structure is disrupted, and the ice liquefies. When ice melts at 0°C, the loosened molecules pack much more closely, filling spaces in the collapsing solid structure. As a result, liquid water is most dense (1.000 g/mL) at around 4°C (3.98°C). With more heating, the density decreases through normal thermal expansion. This behavior has major effects in nature:

- *Surface ice of lakes.* When the surface of a lake freezes in winter, the ice floats. If the solid were denser than the liquid, as is true for nearly every other substance, the surface water would freeze and sink until the entire lake was solid. Aquatic life would not survive from year to year.
- *Nutrient turnover.* As lake water becomes colder in early winter, it becomes more dense *before* it freezes. Similarly, in spring, less dense ice thaws to form more dense water *before* the water expands. During both of these seasonal density changes, the top layer of water reaches maximum density first and sinks. The next layer of water rises because it is slightly less dense, reaches 4°C, and likewise sinks. This alternation of sinking and rising distributes nutrients and dissolved oxygen.
- *Soil formation.* When rain fills crevices in rocks and freezes, an outward force is applied that is relieved when the ice melts. In time, this repeated freeze-thaw stress cracks the rock. Over eons, this effect helps produce sand and soil.

The far-reaching consequences of the properties of water illustrate chemistry's central theme: the macroscopic world we know is the cumulative outcome of the atomic world we seek to know (Figure 12.24).

› Summary of Section 12.5

- The atomic properties of H and O result in water's bent molecular shape, polarity, and H-bonding ability.
- These properties give water the ability to dissolve many ionic and polar compounds.
- Water's high specific heat capacity and heat of vaporization give Earth and its organisms a narrow temperature range.
- Water's high surface tension and capillarity are essential to plants and animals.
- Because water expands on freezing, lake life survives in winter, nutrients mix from seasonal density changes, and soil forms through freeze-thaw stress on rocks.

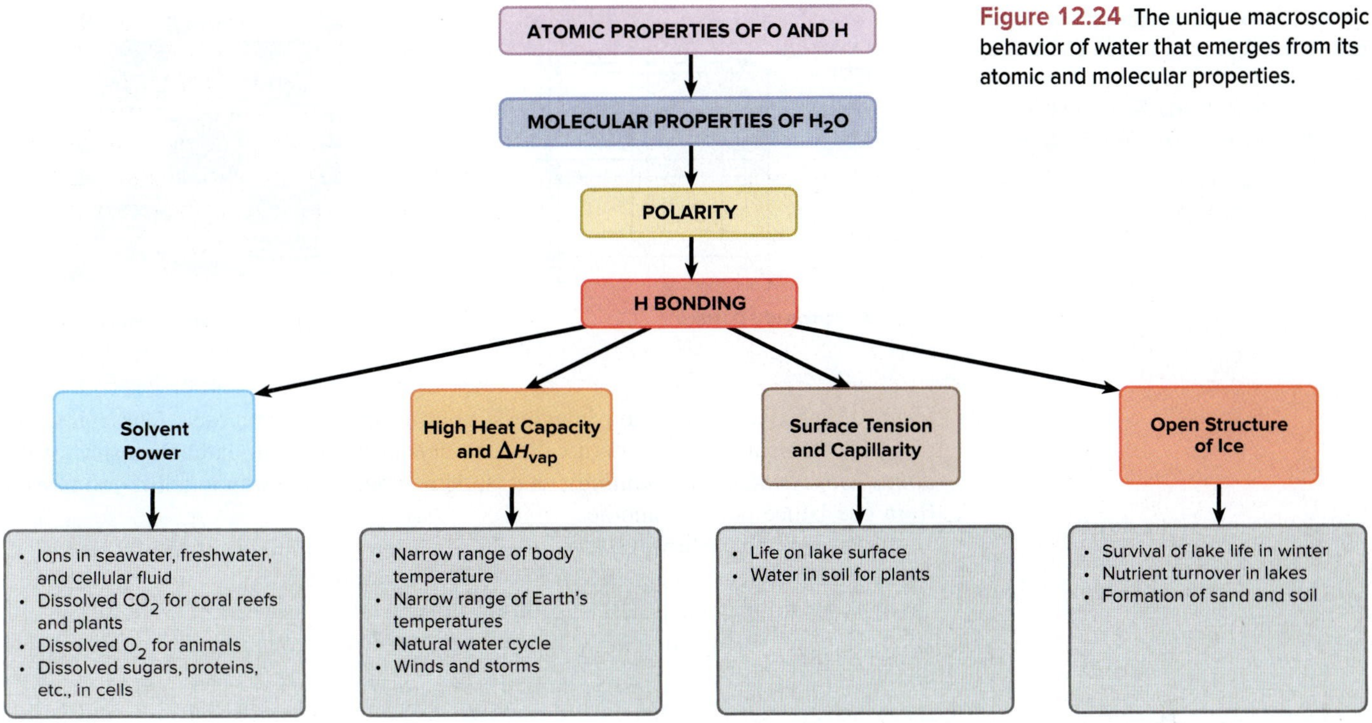

Figure 12.24 The unique macroscopic behavior of water that emerges from its atomic and molecular properties.

12.6 THE SOLID STATE: STRUCTURE, PROPERTIES, AND BONDING

Stroll through a museum's mineral collection, and you'll be struck by the variety and beauty of crystalline solids. In this section, we discuss the structural features of these and other solids and the intermolecular forces that create them. We also consider the main bonding model that explains many properties of solids.

Structural Features of Solids

We can divide solids into two broad categories:

- **Crystalline solids** have well-defined shapes because their particles—atoms, molecules, or ions—occur in an orderly arrangement (Figure 12.25).
- **Amorphous solids** have poorly defined shapes because their particles do not have an orderly arrangement throughout a sample.

The Crystal Lattice and the Unit Cell The particles in a crystal are packed tightly in an orderly, three-dimensional array. As the simplest case, consider the particles as

Figure 12.25 The beauty of crystalline solids. **A,** Wulfenite. **B,** Barite. **C,** Beryl (emerald). **D,** Quartz (amethyst).

Source: (A) © Matteo Chinellato/Shutterstock.com (B) © Albert Russ/Shutterstock.com (C) © Gabbro/Alamy (D) © Walter Geiersperger/Corbis

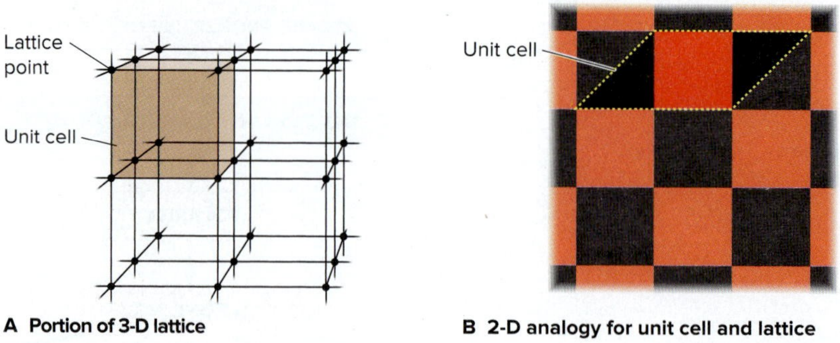

Figure 12.26 The crystal lattice and the unit cell. A, A small portion of a lattice is shown as points connected by lines, with a unit cell *(colored).* **B,** A checkerboard as a two-dimensional analogy for a lattice.

identical spherical atoms, and imagine a point at the center of each. The collection of points forms a regular pattern called the crystal **lattice.** The lattice consists of *all points with identical surroundings;* that is, there would be no way to tell if you moved from one lattice point to another.

Figure 12.26A shows a portion of a lattice and the **unit cell,** the *smallest* portion that yields the crystal if it is repeated in all directions. A two-dimensional analogy for a unit cell and the resulting crystal lattice is a checkerboard (Figure 12.26B), a section of tiled floor, or any pattern that is constructed from a repeating unit.

There are 7 crystal systems and 14 types of unit cells that occur in nature, but we will be concerned primarily with the *cubic system.* The solid states of a majority of metallic elements, some covalent compounds, and many ionic compounds occur as cubic lattices. A key parameter of any lattice is the **coordination number,** the number of *nearest* neighbors of a particle in a crystal. There are three types of cubic unit cells:

1. In the **simple cubic unit cell** (Figure 12.27A), the centers of eight identical particles define the corners of a cube (shown in the expanded view, *top row*). The particles touch along the cube edges (see the space-filling view, *second row*), but they do not touch diagonally along the cube faces or through its center. An expanded portion of the crystal *(third row)* shows that the coordination number of each particle is 6: four in its own layer, one in the layer above, and one in the layer below.
2. In the **body-centered cubic unit cell** (Figure 12.27B), identical particles lie at each corner *and* in the center of the cube. Those at the corners do not touch each other, but they all touch the one in the center. Each particle is surrounded by eight nearest neighbors, four above and four below, so the coordination number is 8.
3. In the **face-centered cubic unit cell** (Figure 12.27C), identical particles lie at each corner *and* in the center of each face but not in the center of the cube. Particles at the corners touch those in the faces but not each other. The coordination number is 12.

How many particles make up a unit cell? For particles of the same size, *the higher the coordination number, the greater the number of particles in a given volume.* Since one unit cell touches another, with no gaps, a particle at a corner or face is *shared* by adjacent cells. In the cubic unit cells, the particle at each corner is part of eight adjacent cells (Figure 12.27, *third row*), so one-eighth of each particle belongs to each cell *(bottom row).* There are eight corners in a cube, so

- A simple cubic unit cell contains $8 \times \frac{1}{8}$ particle = 1 particle.
- A body-centered cubic unit cell contains $8 \times \frac{1}{8}$ particle = 1 particle plus 1 particle in the center, for a total of 2 particles.
- A face-centered cubic unit cell contains $8 \times \frac{1}{8}$ particle = 1 particle plus one-half particle in each of the six faces $6 \times \frac{1}{2}$ particle = 3 particles, for a total of 4 particles.

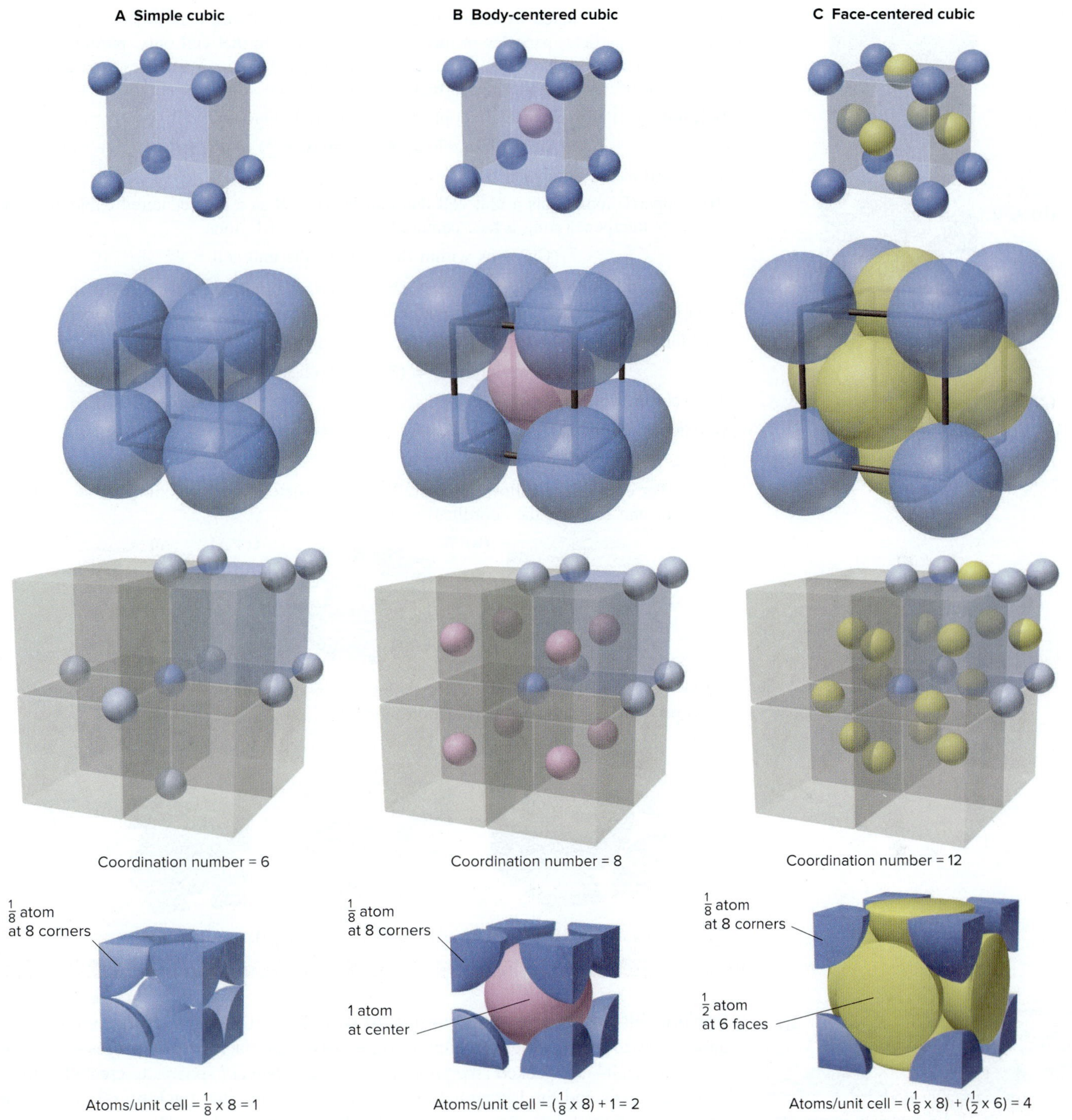

Figure 12.27 **The three cubic unit cells. A,** Simple cubic unit cell. **B,** Body-centered cubic unit cell. **C,** Face-centered cubic unit cell. *Top row:* Cubic arrangements of atoms in expanded view. *Second row:* Space-filling view of these cubic arrangements. All atoms are identical but, for clarity, corner atoms are blue, body-centered atoms pink, and face-centered atoms yellow. *Third row:* A unit cell *(shaded blue)* in an expanded portion of the crystal. The number of nearest neighbors around one particle *(dark blue in center)* is the coordination number. *Bottom row:* The total numbers of atoms in the actual unit cells. The simple cubic has one atom, the body-centered has two, and the face-centered has four.

SAMPLE PROBLEM 12.6 Determining the Number of Particles per Unit Cell and the Coordination Number

Problem For each of the crystalline solids shown in the margin on the next page, determine the number of atoms (or ions) per unit cell and the coordination number.

Plan To determine the number of particles (atoms or ions) in a unit cell, we count the number of particles at the corners, in the faces, and within the body of the cell. The

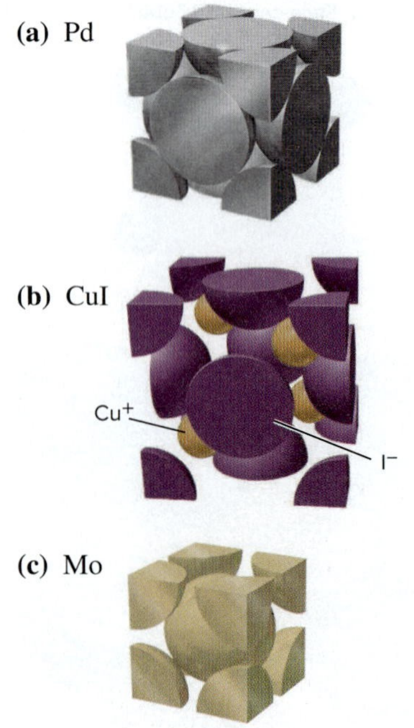

eight corners are shared by eight other cells for a total of 8 corners × $\frac{1}{8}$ particle per corner, or 1 particle; particles in the body of a cell are in that cell only; particles in the faces are shared by two cells, 6 faces × $\frac{1}{2}$ particle per face, or 3 particles. We count the number of nearest neighboring particles to obtain the coordination number.

Solution **(a)** Palladium metal adopts a face-centered cubic unit cell:

$$\text{Atoms/unit cell} = (8 \times \tfrac{1}{8}) + (6 \times \tfrac{1}{2}) = \text{4 atoms}$$

The coordination number is 12.

(b) Copper(I) iodide has a unit cell that can be viewed as a face-centered cubic array of Cu^+ ions interpenetrating a face-centered cubic array of I^- ions:

$$Cu^+ \text{ ions within the body of the unit cell} = 4$$

$$I^- \text{ ions/unit cell} = (8 \times \tfrac{1}{8}) + (6 \times \tfrac{1}{2}) = 4\ Cu^+ \text{ and } 4\ I^- \text{ ions}$$

Each ion is surrounded by four of the oppositely charged ions: coordination number is 4.

(c) Molybdenum metal adopts a body-centered cubic cell:

$$\text{Atoms/unit cell} = (8 \times \tfrac{1}{8}) + (1 \times 1) = \text{2 atoms}$$

The coordination number is 8.

Check Using Figure 12.27, we see that the values are correct.

FOLLOW-UP PROBLEMS

12.6A For each of the following crystalline solids, determine the number of atoms (or ions) per unit cell and the coordination number:

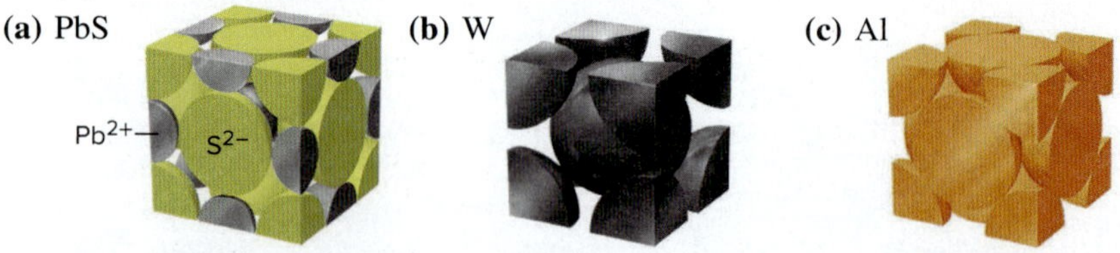

12.6B For each of the following crystalline solids, determine the number of atoms (or ions) per unit cell and the coordination number:

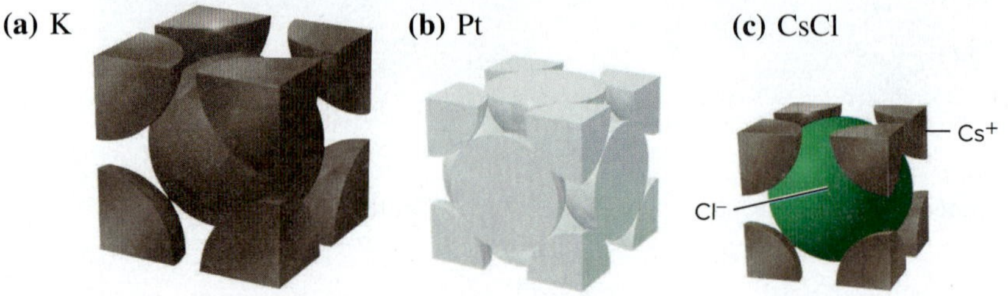

SOME SIMILAR PROBLEMS 12.87, 12.96, 12.97, 12.98(a), and 12.99(a)

The efficient packing of fruit.

Source: © Bryan Busovicki/Shutterstock.com

Packing Efficiency and the Creation of Unit Cells Unit cells result from the ways atoms pack together, which are similar to the ways that macroscopic spheres—marbles, golf balls, fruit—are packed *(see photo)*. Let's pack *identical* spheres to create the three cubic unit cells and the hexagonal unit cell and determine the **packing efficiency,** the percentage of the total volume of the unit cell occupied by the spheres themselves:

1. *The simple cubic unit cell.* When we arrange the first layer of spheres in vertical and horizontal rows, large diamond-shaped spaces are formed (Figure 12.28A, *cutaway portion*). If we place the next layer of spheres *directly above* the first, we obtain an arrangement based on the *simple* cubic unit cell (Figure 12.28B). The spheres occupy only 52% of the unit-cell volume, so 48% is empty space between them. This is a very inefficient way to pack spheres, so neither fruit nor atoms are typically packed this way.

2. *The body-centered cubic unit cell.* Rather than placing the second layer (colored *green* for clarity) directly above the first, we use space more efficiently by placing its spheres over the diamond-shaped spaces in the first layer (Figure 12.28C). Then we pack the third layer onto the diamond-shaped spaces in the second, which makes the first and third layers line up vertically. This arrangement is based on the *body-centered* cubic unit cell, and its packing efficiency, at 68%, is much higher than that of the

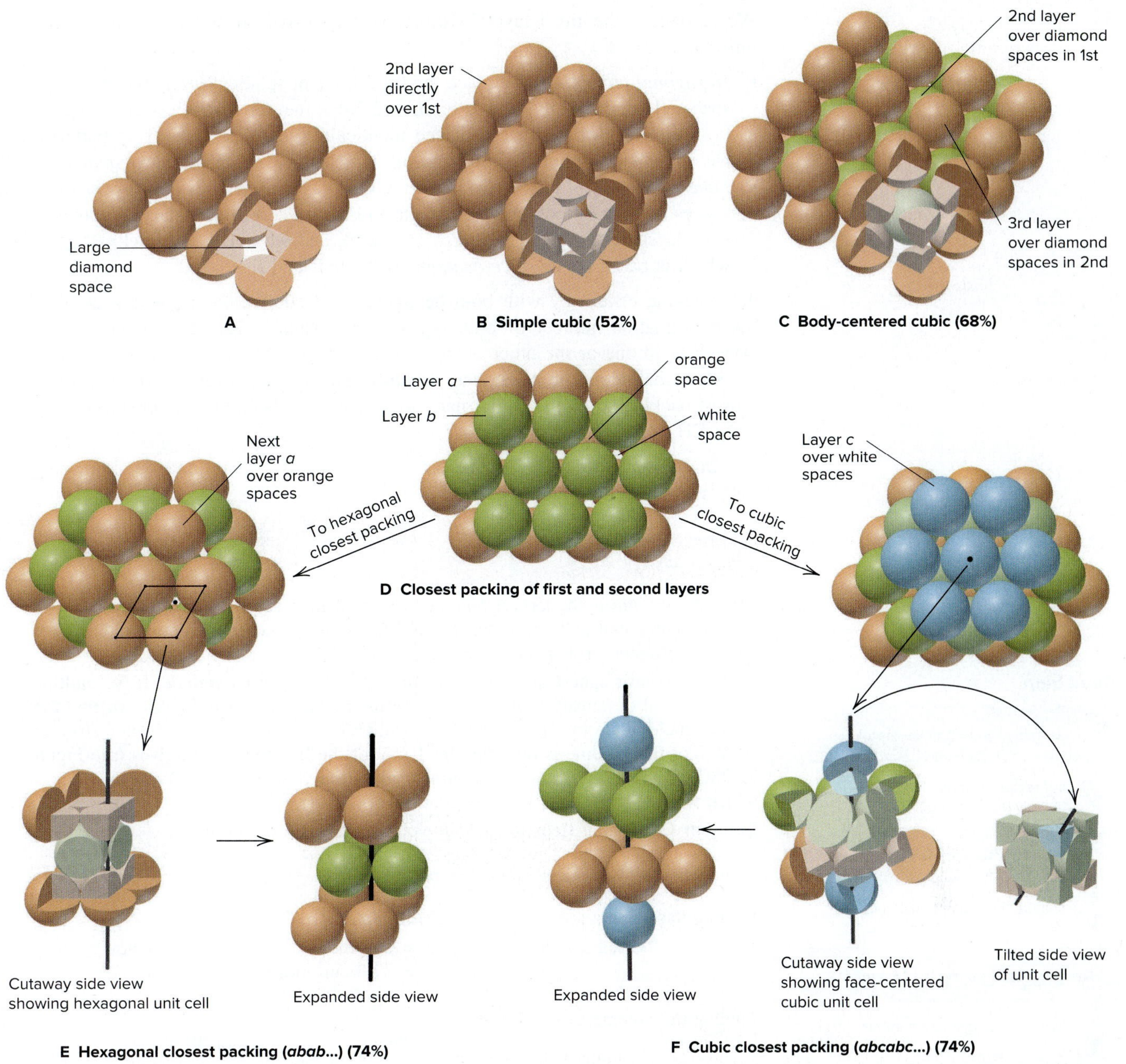

Figure 12.28 **Packing spheres to obtain three cubic and hexagonal unit cells. A,** In the first layer, each sphere lies next to another horizontally and vertically; note the large diamond-shaped spaces *(see cutaway).* **B,** If the spheres in the next layer lie directly over those in the first, the packing is based on the simple cubic unit cell *(pale orange cube, lower right corner).* **C,** If the spheres in the next layer lie in the diamond-shaped spaces of the first layer, the packing is based on the *body-centered cubic* unit cell *(lower right corner).* **D,** The closest possible packing of the first layer *(layer a, orange)* is obtained by shifting every other row in part A to obtain smaller triangular spaces. The spheres of the second layer *(layer b, green)* are placed above these spaces; note the orange and white spaces that result. **E,** When the third layer *(next layer a, orange)* is placed over the orange spaces, we obtain an *abab*. . . pattern and the hexagonal unit cell. **F,** When the third layer *(layer c, blue)* covers the white spaces, we get an *abcabc*. . . pattern and the face-centered cubic unit cell.

simple cubic unit cell. Several metallic elements, including chromium, iron, and all the Group 1A(1) elements, have a crystal structure based on this unit cell.

3. *The hexagonal and face-centered cubic unit cells*. Spheres are packed most efficiently in these cells. First, in the bottom layer (labeled *a, orange*), we shift every other row laterally so that the large diamond-shaped spaces become smaller triangular spaces. Then we place the second layer *(b, green)* over these spaces (Figure 12.28D).

In layer *b,* notice that some spaces are orange because they lie above *spheres* in layer *a,* whereas other spaces are white because they lie above *spaces* in layer *a.*

We can place the third layer in either of two ways, giving rise to two different unit cells:

- *Hexagonal unit cell.* If we place the third layer of spheres *(orange)* over the orange spaces (look down and left to Figure 12.28E), they lie directly over the spheres in layer *a*. Every other layer is placed identically (an *abab*. . . layering pattern), and we obtain **hexagonal closest packing,** which is based on the *hexagonal* unit cell.
- *Face-centered unit cell.* If we place the third layer of spheres *(blue)* over the white spaces in layer *b* (look down and right to Figure 12.28F), the placement is different from layers *a* and *b* (an *abcabc*. . . pattern), and we obtain **cubic closest packing,** which is based on the *face-centered cubic* unit cell.

The packing efficiency with both hexagonal and cubic closest packing is 74%, and the coordination number of both types of unit cells is 12. Most metallic elements crystallize in one or the other of these arrangements. Magnesium, titanium, and zinc are some that adopt the hexagonal structure; nickel, copper, and lead adopt the cubic closest packing structure, as do many ionic compounds and other substances, such as CO_2, CH_4, and most noble gases.

In Sample Problem 12.7, we use the density of a metal and our knowledge of unit cells to find an atomic radius.

SAMPLE PROBLEM 12.7 Determining Atomic Radius

Problem Barium is the largest nonradioactive alkaline earth metal. It has a body-centered cubic unit cell and a density of 3.62 g/cm^3. What is the atomic radius of barium? (Volume of a sphere: $V = \frac{4}{3}\pi r^3$.)

Road Map

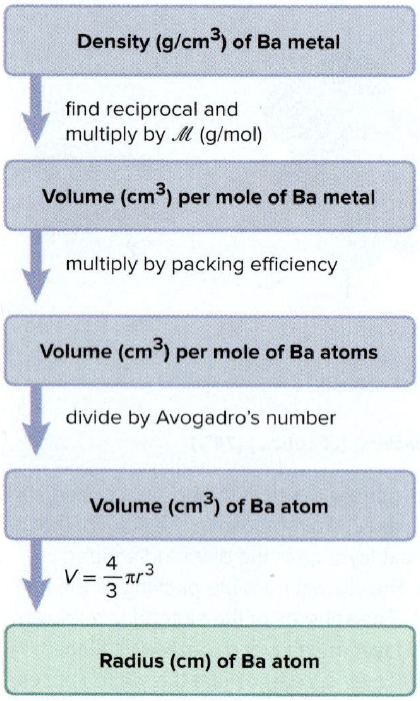

Plan An atom is spherical, so we can find its radius from its volume. If we multiply the reciprocal of density (volume/mass) by the molar mass (mass/mole), we find the volume/mole of Ba metal. The metal crystallizes in the body-centered cubic structure, so 68% of this volume is occupied by 1 mol of the Ba atoms themselves (see Figure 12.28C). Dividing by Avogadro's number gives the volume of one Ba atom, from which we find the radius. (See the road map.)

Solution Combining steps to find the volume of 1 mol of Ba metal:

$$\text{Volume/mole of Ba metal} = \frac{1}{\text{density}} \times \mathcal{M} = \frac{1\ \text{cm}^3}{3.62\ \text{g Ba}} \times \frac{137.3\ \text{g Ba}}{1\ \text{mole Ba}} = 37.9\ \text{cm}^3/\text{mol Ba}$$

Finding the volume (cm^3) of 1 mol of Ba *atoms:*

$$\begin{aligned}\text{Volume/mole of Ba atoms} &= \text{cm}^3/\text{mol Ba} \times \text{packing efficiency}\\ &= 37.9\ \text{cm}^3/\text{mol Ba} \times 0.68\\ &= 26\ \text{cm}^3/\text{mol Ba atoms}\end{aligned}$$

Finding the volume of one Ba atom:

$$\begin{aligned}\text{Volume of Ba atom} &= \frac{26\ \text{cm}^3}{1\ \text{mol Ba atoms}} \times \frac{1\ \text{mol Ba atoms}}{6.022\times10^{23}\ \text{Ba atoms}}\\ &= 4.3\times10^{-23}\ \text{cm}^3/\text{Ba atom}\end{aligned}$$

Finding the atomic radius of Ba from the volume of a sphere:

$$V \text{ of Ba atom} = \frac{4}{3}\pi r^3 \quad \text{so} \quad r^3 = \frac{3V}{4\pi}$$

Thus,

$$r = \sqrt[3]{\frac{3V}{4\pi}} = \sqrt[3]{\frac{3(4.3\times10^{-23}\ \text{cm}^3)}{4 \times 3.14}} = 2.2\times10^{-8}\ \text{cm}$$

Check The order of magnitude is correct for an atom ($\sim 10^{-8}$ cm $\approx 10^{-10}$ m). The actual value for barium is, in fact, 2.22×10^{-8} cm (see Figure 8.13).

FOLLOW-UP PROBLEMS

12.7A Cobalt has a face-centered cubic unit cell and an atomic radius of 125 pm. Calculate the density of cobalt metal.

12.7B Iron crystallizes in a body-centered cubic structure. The volume of one Fe atom is 8.38×10^{-24} cm^3, and the density of Fe is 7.874 g/cm^3. Calculate an approximate value for Avogadro's number.

SOME SIMILAR PROBLEMS 12.88 and 12.89

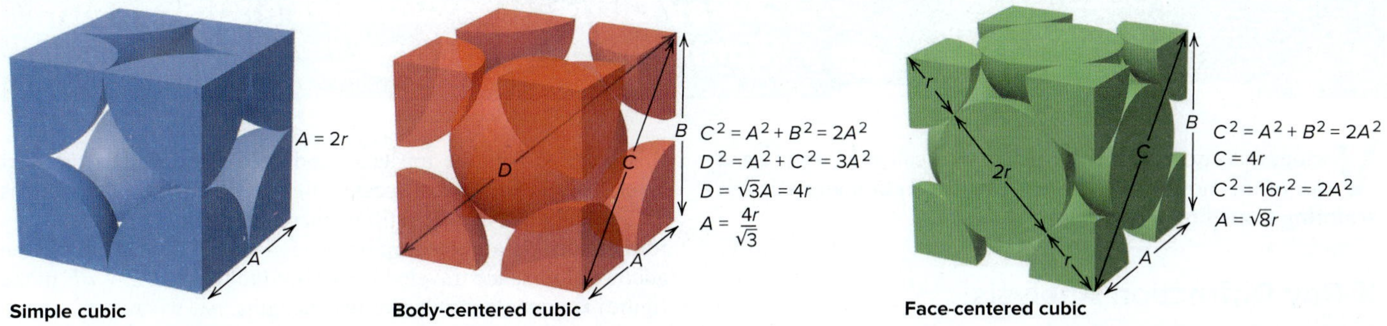

Figure 12.29 Edge length and atomic (ionic) radius in the three cubic unit cells.

Our understanding of crystal structures comes from our ability to "see" them. Two techniques for visualizing crystal structures are described in Tools of the Laboratory on the next page.

The edge length of the unit cell is obtained from x-ray crystallography. We can then apply the Pythagorean theorem and some arithmetic to find atomic (or ionic) radii in the three cubic unit cells (Figure 12.29). Sample Problem 12.8 applies this approach.

SAMPLE PROBLEM 12.8 Determining Atomic Radius from the Unit Cell

Problem Copper adopts cubic closest packing, and the edge length of the unit cell is 361.5 pm. What is the atomic radius of copper?

Plan Cubic closest packing gives a face-centered cubic unit cell, and we know the edge length. From Figure 12.29, with $A = 361.5$ pm, we solve for r.

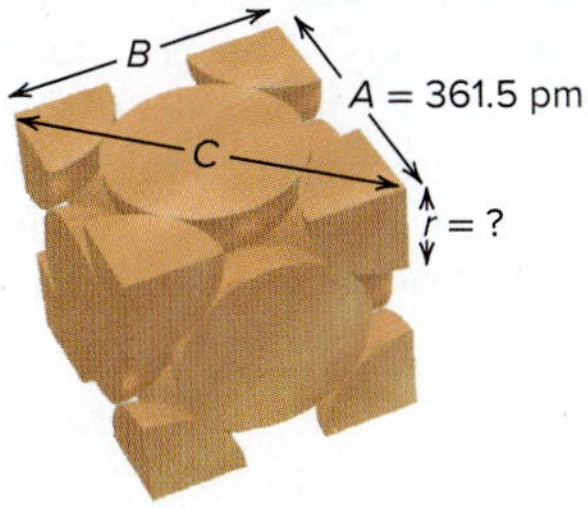

Solution Using the Pythagorean theorem to find C, the diagonal of the cell's face:

$$C = \sqrt{A^2 + B^2}$$

The unit cell is a cube, so $A = B$. Therefore,

$$C = \sqrt{2A^2} = \sqrt{2(361.5 \text{ pm})^2} = 511.2 \text{ pm}$$

Finding r: $C = 4r$. Therefore,

$$r = 511.2 \text{ pm}/4 = 127.8 \text{ pm.}$$

We can combine these steps by using the equation presented in Figure 12.29:

$$A = \sqrt{8}r$$

$$r = 361.5 \text{ pm}/\sqrt{8} = 127.8 \text{ pm}$$

Check Rounding and quickly checking the math gives

$$C = \sqrt{2(4\times10^2 \text{ pm})^2} = \sqrt{2(16\times10^4 \text{ pm}^2)}$$

or ~500–600 pm; thus, $r \approx$ 125–150 pm. The actual value for copper is 128 pm (see Figure 8.13).

FOLLOW-UP PROBLEMS

12.8A Iron crystallizes in a body-centered cubic structure. If the atomic radius of Fe is 126 pm, find the edge length (in nm) of the unit cell.

12.8B Aluminum crystallizes in a cubic closest packed structure. If the edge length of the unit cell is 0.405 nm, find the atomic radius (in pm).

SOME SIMILAR PROBLEMS 12.92 and 12.93

TOOLS OF THE LABORATORY

X-Ray Diffraction Analysis and Scanning Tunneling Microscopy

Various tools exist for measuring atomic-scale dimensions, and two of the most powerful are **x-ray diffraction analysis** and **scanning tunneling microscopy.**

X-Ray Diffraction Analysis

In Chapter 7, we saw how diffraction patterns of bright and dark regions appear when light passes through slits spaced as far apart as the wavelength (see Figure 7.5). X-ray wavelengths are about the same size as the spaces between layers of spheres in a crystal, so we use the spaces as "slits" to diffract x-rays.

As one example, let's use this approach to measure the distance (d) between two adjacent layers of atoms in a simplified lattice (Figure B12.1). Two waves impinge on the crystal at an angle θ and are diffracted at that angle by the layers. When the first wave strikes the top layer and the second strikes the next layer, they are *in phase* (peaks aligned with peaks and troughs with troughs). If they are still in phase after being diffracted, they form a spot on a photographic plate. This will occur only if the additional distance traveled by the second wave ($DE + EF$ in the figure) is a whole number of wavelengths, $n\lambda$, where n is a positive integer referred to as the *order* of the diffraction. From trigonometry, we find that

$$n\lambda = 2d \sin \theta$$

where θ is the angle of incoming light and d, the unknown, is the distance between layers. This is the *Bragg equation,* named for W. H. Bragg and his son W. L. Bragg, who shared the Nobel Prize in physics in 1915 for their work on crystal structure analysis.

Rotating the crystal changes θ and produces different sets of spots. Eventually, the complete diffraction pattern reveals distances and angles in the lattice (Figure B12.2). X-ray diffraction analysis is used in many studies, but its greatest recent impact is uncovering the structures of DNA and proteins.

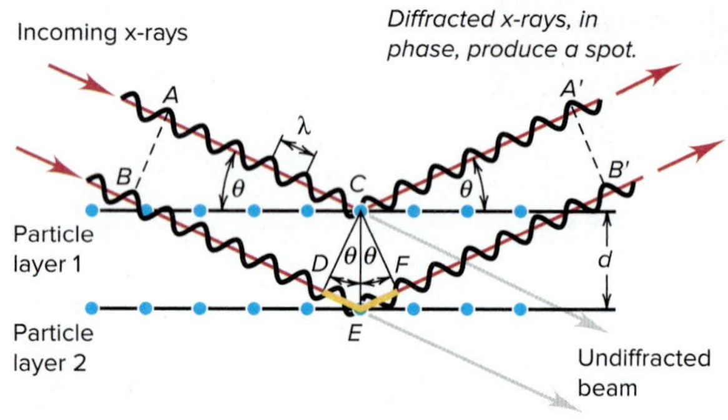

Figure B12.1 **Diffraction of x-rays by crystal planes.**

Scanning Tunneling Microscopy

This technique, invented by Gerd Binnig and Heinrich Rohrer, who won the Nobel Prize in physics in 1986, is used to observe surfaces. It is based on the idea that an electron has a small, but finite, probability of being so far from its nucleus that it can move ("tunnel") closer to another atom.

In practice, an extremely sharp tungsten-tipped probe, the source of the tunneling electrons, is placed very close to (about 0.5 nm from) the surface under study. A small applied electric

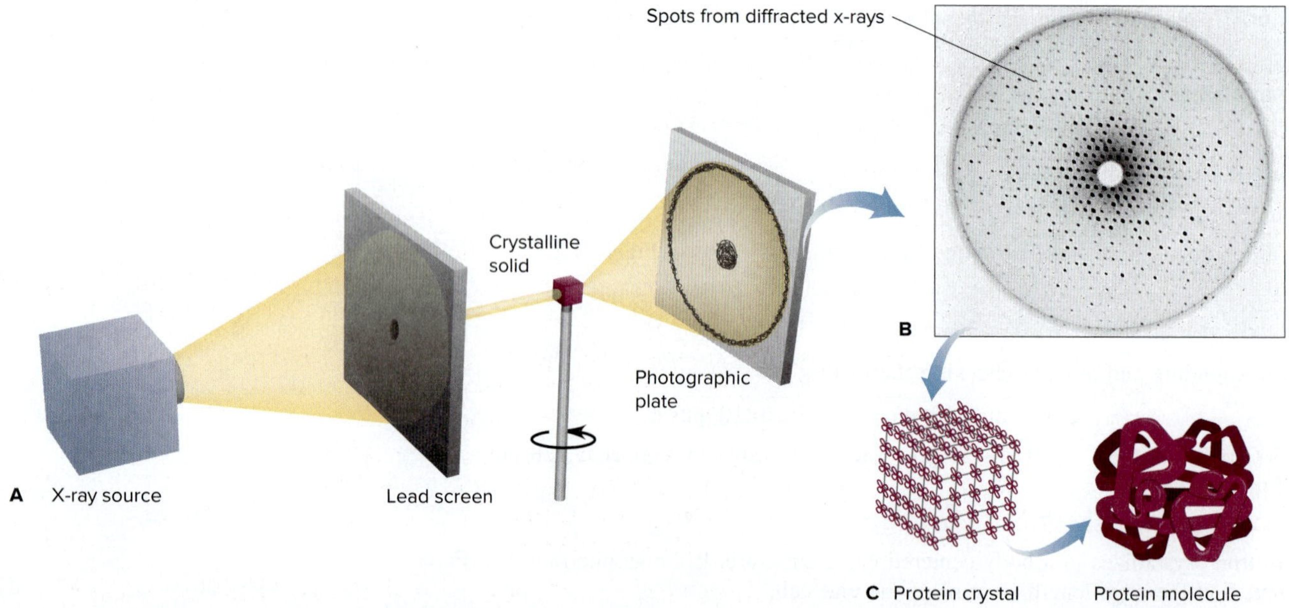

Figure B12.2 **Formation of an x-ray diffraction pattern of the protein hemoglobin. A,** A sample of crystalline protein is rotated to obtain different angles of incoming and diffracted x-rays. **B,** The diffraction pattern is a complex series of spots. **C,** Computerized analysis provides a picture of the molecule.

potential increases the probability that the electrons will tunnel across this minute gap. The probe moves tiny distances up and down to maintain the constant current across the gap, thus following the surface at the atomic scale. This movement is monitored electronically, and after many scans, a three-dimensional map of the surface is obtained. The method has revealed images of atoms and molecules coated on surfaces and is being used to study the nature of defects and many other surface features (Figure B12.3).

Problems

B12.1 X-rays of wavelength $\lambda = 0.709\times10^{-10}$ m undergo a first-order ($n = 1$) diffraction of 11.6° when aimed at crystalline nickel. What is the spacing between layers of Ni atoms?

B12.2 A first-order ($n = 1$) diffraction of x-rays aimed at a crystal of NaCl occurs at 15.9°.

(a) What is the wavelength (in pm) of the x-rays?

(b) At what angle would a second-order ($n = 2$) diffraction appear?

Figure B12.3 A scanning tunneling micrograph of cesium atoms (red) on gallium arsenide.

Source: Courtesy National Institute of Standards and Technology

Types and Properties of Crystalline Solids

The five most important types of solids are defined by the type(s) of particle(s) in the crystal (Table 12.5). We'll highlight interparticle forces and physical properties.

Table 12.5 Characteristics of the Major Types of Crystalline Solids

Type	Particle(s)	Interparticle Forces	Physical Properties	Examples [mp, °C]
Atomic	Atoms	Dispersion	Soft, very low mp, poor thermal and electrical conductors	Elements in Group 8A(18) (Ne [−249] to Rn [−71])
Molecular	Molecules	Dispersion, dipole-dipole, H bonds	Fairly soft, low to moderate mp, poor thermal and electrical conductors	*Nonpolar** O_2 [−219], C_4H_{10} [−138] Cl_2 [−101], C_6H_{14} [−95], P_4 [44.1] *Polar* SO_2 [−73], $CHCl_3$ [−64], HNO_3 [−42], H_2O [0.0], CH_3COOH [17]
Ionic	Positive and negative ions	Ion-ion attraction	Hard and brittle, high mp, good thermal and electrical conductors when molten	NaCl [801] CaF_2 [1423] MgO [2852]
Metallic	Atoms	Metallic bond	Soft to hard, low to very high mp, excellent thermal and electrical conductors, malleable and ductile	Na [97.8] Zn [420] Fe [1535]
Network covalent	Atoms	Covalent bond	Very hard, very high mp, usually poor thermal and electrical conductors	SiO_2 (quartz) [1610] C (diamond) [~4000]

*Nonpolar molecular solids are arranged in order of increasing molar mass. Note the correlation with increasing melting point (mp).

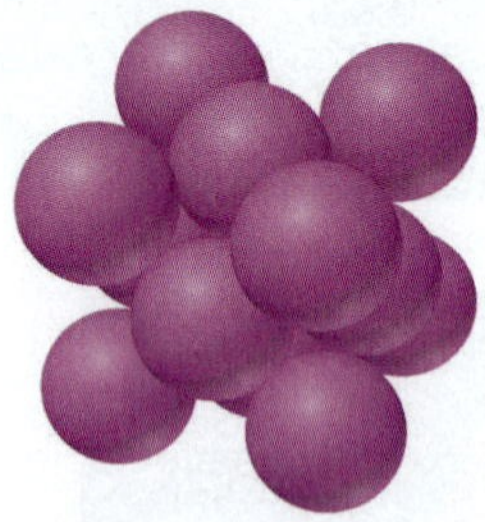

Figure 12.30 Cubic closest packing of argon (face-centered cubic unit cell).

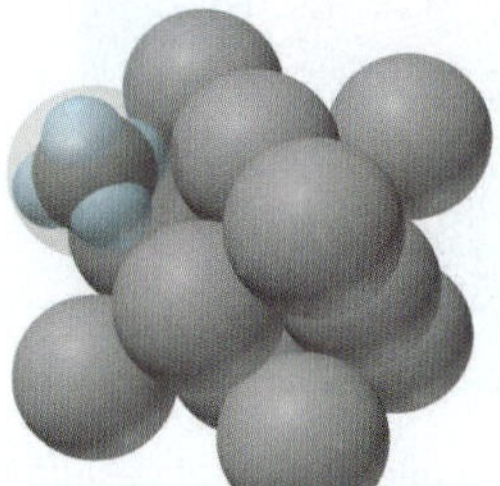

Figure 12.31 Cubic closest packing (face-centered unit cell) of CH_4.

Atomic Solids Individual atoms held together only by *dispersion forces* form an **atomic solid** and the noble gases [Group 8A(18)] are the only substances that form such solids. The very weak forces among the atoms mean melting and boiling points and heats of vaporization and fusion are all very low, rising smoothly with increasing molar mass. Argon crystallizes in a cubic closest packed structure (Figure 12.30), as do the other noble gases.

Molecular Solids In the many thousands of **molecular solids,** individual molecules occupy the lattice points. Various combinations of dipole-dipole, dispersion, and H-bonding forces account for a wide range of physical properties. Dispersion forces in nonpolar substances lead to melting points that generally increase with molar mass (Table 12.5). Among polar molecules, dipole-dipole forces and, where possible, H bonding occur. Most molecular solids have much higher melting points than atomic solids (noble gases) but much lower melting points than other types of solids. Methane crystallizes in a face-centered cubic structure with the center of each carbon as the lattice point (Figure 12.31).

Ionic Solids To maximize attractions in a binary **ionic solid,** cations are surrounded by as many anions as possible, and vice versa, with *the smaller of the ions (usually the cation) lying in the spaces (holes) formed by the packing of the larger (usually the anion).* The unit cell is the smallest portion that maintains the composition; that is, *the unit cell has the same cation/anion ratio as the empirical formula.*

As a result of cation-anion contact, the interparticle forces (ionic bonds) are *much* stronger than the van der Waals forces in atomic or molecular solids. The properties of ionic solids are a direct consequence of the *fixed ion positions* and *very strong attractive forces,* which create a high lattice energy. Thus, ionic solids typically have high melting points and low electrical conductivities. When a large quantity of heat is supplied and the ions gain enough kinetic energy to break free of their positions, the solid melts and the mobile ions conduct a current. Ionic compounds are hard because only a strong external force can change the relative positions of the huge number of ions attracting one another. If enough force *is* applied to move them, ions of like charge are brought near each other, and their repulsions crack the crystal (see Figure 9.9).

Ionic solids often adopt cubic closest packed crystal structures. Let's consider two examples with a 1/1 ratio of ions and then two with a 2/1 (or 1/2) ratio:

1. The *sodium chloride structure* is found in many ionic compounds, including most alkali [Group 1A(1)] halides and hydrides, alkaline earth [Group 2A(2)] oxides and sulfides, and several transition metal oxides and sulfides. To visualize this structure, imagine Cl^- anions and Na^+ cations in separate face-centered cubic arrays. Now picture the two arrays penetrating each other, the smaller Na^+ ions ending up in the holes between the larger Cl^- ions (Figure 12.32A). Thus, each Na^+ is surrounded by six Cl^-, and vice versa (coordination number = 6). Figure 12.32B is a space-filling depiction of the unit cell: four Cl^- $[(8 \times \frac{1}{8}) + (6 \times \frac{1}{2}) = 4\ Cl^-]$ and four Na^+ $[(12 \times \frac{1}{4}) + 1$ in the center $= 4\ Na^+]$, give a 1/1 ion ratio.

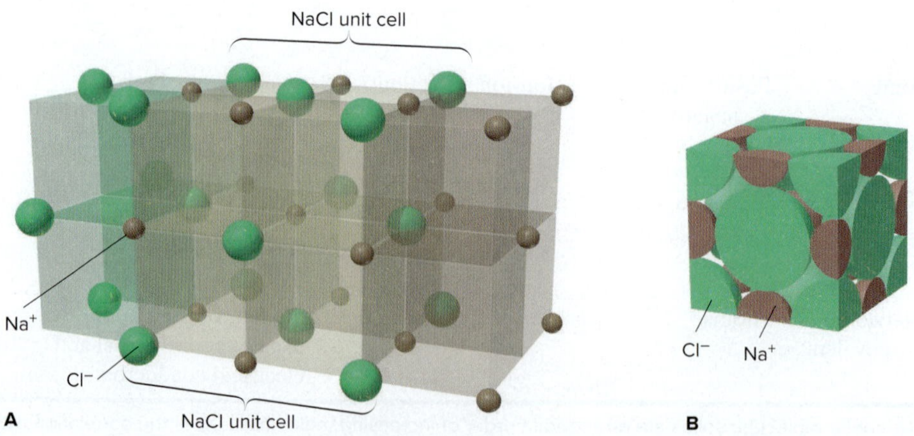

Figure 12.32 The sodium chloride structure. **A,** Expanded view. **B,** Space-filling depiction of the NaCl unit cell.

2. The *zinc blende (ZnS) structure* can be pictured as two face-centered cubic arrays, one of Zn^{2+} ions *(gray)* and the other of S^{2-} ions *(yellow),* penetrating each other such that each ion is tetrahedrally surrounded by four of the other ions (coordination number = 4) (Figure 12.33). There are four $[(8 \times \frac{1}{8}) + (6 \times \frac{1}{2}) = 4]$ S^{2-} ions and four Zn^{2+} ions for a 1/1 ion ratio. Many other compounds, including AgI, CdS, and the copper(I) halides, adopt the zinc blende structure.

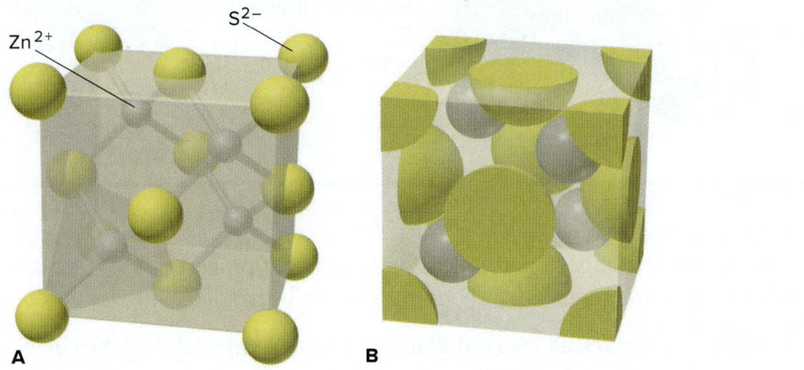

Figure 12.33 **The zinc blende structure. A,** Expanded view (with bonds shown for clarity). **B,** The unit cell is expanded a bit to show interior ions.

3. The *fluorite* (CaF_2) *structure* is common among salts with a 1/2 cation/anion ratio that have relatively large cations and small anions, such as SrF_2 and $BaCl_2$. In CaF_2, the unit cell is a face-centered cubic array of Ca^{2+} ions with F^- ions occupying *all* eight available holes (Figure 12.34). This results in a Ca^{2+}/F^- ratio of 4/8, or 1/2.

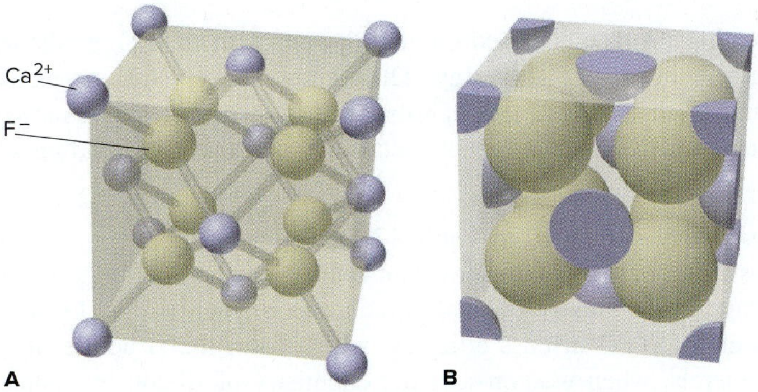

Figure 12.34 **The fluorite structure. A,** Expanded view (with bonds shown for clarity). **B,** The unit cell is expanded a bit to show interior ions.

4. The *antifluorite structure* is common in compounds with a 2/1 cation/anion ratio and a relatively large anion (such as K_2S). The ion arrangement is the opposite of that in the fluorite structure: the cations occupy all eight holes formed by the cubic closest packing of the anions.

Metallic Solids Most metallic elements crystallize in one of the two closest packed structures (Figure 12.35). In contrast to the weak dispersion forces in atomic solids, powerful metallic bonding forces hold atoms together in **metallic solids.** The properties of metals—high electrical and thermal conductivity, luster, and malleability—result from their delocalized electrons (Section 9.6). Melting points and hardnesses are related to the packing efficiency and number of valence electrons. For example, Group 2A metals are harder and melt higher than Group 1A metals (see Figure 9.28), because the 2A metals have twice as many delocalized valence electrons.

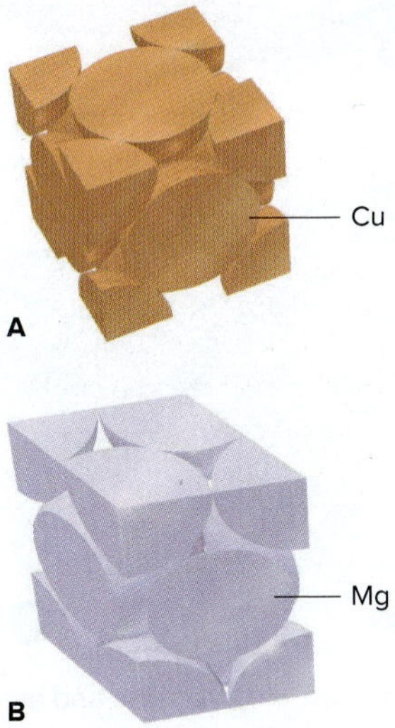

Figure 12.35 **Crystal structures of metals. A,** Copper adopts cubic closest packing. **B,** Magnesium adopts hexagonal closest packing.

Network Covalent Solids Strong covalent bonds link the atoms together in a **network covalent solid;** thus, separate particles are not present. These substances adopt a variety of crystal structures depending on the details of their bonding.

As a consequence of strong bonding, all network covalent solids have extremely high melting and boiling points, but their conductivity and hardness vary. Two examples

Table 12.6 **Comparison of the Properties of Diamond and Graphite**

Property	Graphite	Diamond
Density (g/cm^3)	2.27	3.51
Hardness	<1 (very soft)	10 (hardest)
Melting point (K)	4100	4100
Color	Shiny black	Colorless transparent
Electrical conductivity	High (along sheet)	None
ΔH°_{rxn} for combustion (kJ/mol)	−393.5	−395.4
ΔH°_f (kJ/mol)	0 (standard state)	1.90

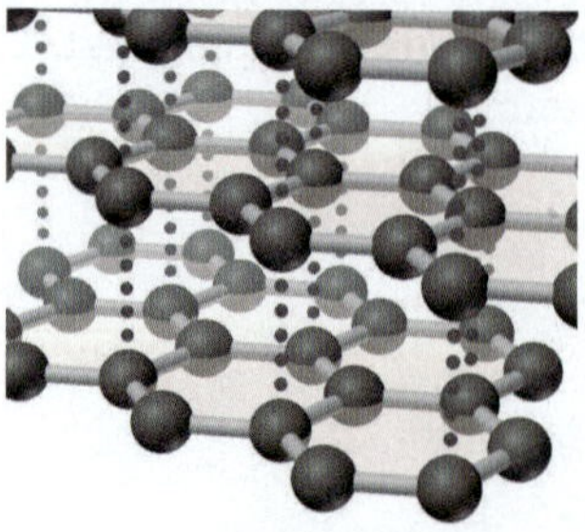

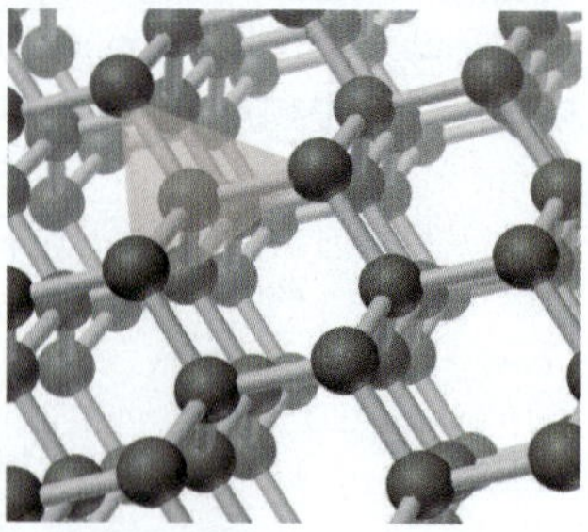

with the same composition but strikingly different properties are the two common crystalline forms of elemental carbon, graphite and diamond:

- *Graphite* occurs as stacked flat sheets of hexagonal carbon rings with a strong σ-bond framework and delocalized π bonds, reminiscent of benzene; the arrangement looks like chicken wire or honeycomb. Whereas the π-bonding electrons of benzene are delocalized over one ring, those of graphite are delocalized over the entire sheet. Thus, graphite conducts electricity well—it is a common electrode material—but only in the plane of the sheets. The sheets interact via dispersion forces, and impurities, such as O_2, between the sheets allow them to slide past each other, which explains why graphite is soft and used as a lubricant.
- *Diamond* adopts a face-centered cubic unit cell, with each C tetrahedrally bonded to four others in an endless array. Throughout the crystal, strong single bonds make diamond the hardest natural substance known. Like most network covalent solids, diamond does not conduct electricity because the bonding electrons are localized.

These properties are compared in Table 12.6.

The most important network covalent solids are the *silicates,* which consist of extended arrays of covalently bonded silicon and oxygen atoms. Quartz (SiO_2) is a common example. We'll discuss silicates, which form the structure of clays, rocks, and many minerals, when we consider the chemistry of silicon in Chapter 14.

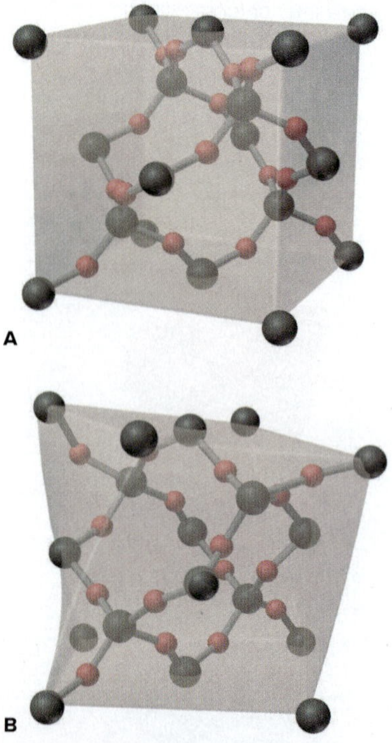

Figure 12.36 **Crystalline and amorphous silicon dioxide. A,** Cristobalite, a crystalline form of silica (SiO_2), shows cubic closest packing. **B,** Quartz glass is amorphous with a generally disordered structure.

Amorphous Solids

Amorphous solids are noncrystalline. Many have small, somewhat ordered regions interspersed among large disordered regions. Charcoal, rubber, and glass are examples.

The process that forms quartz glass is typical of many amorphous solids. Crystalline quartz (SiO_2), which adopts cubic closest packing, is melted and the viscous liquid is cooled rapidly to prevent it from recrystallizing. The chains of Si and O atoms cannot orient themselves quickly enough, so they solidify in a distorted jumble containing gaps and misaligned rows (Figure 12.36). The absence of regularity confers some properties of a liquid; in fact, glasses are sometimes referred to as *supercooled liquids.*

Bonding in Solids: Molecular Orbital Band Theory

Chapter 9 introduced a qualitative model of metallic bonding, with metal ions submerged in a "sea" of delocalized valence electrons. Molecular orbital (MO) theory offers a more quantitative, and therefore more useful, model called **band theory.** We'll focus on bonding in metals and the resulting conductivity of metals, metalloids, and nonmetals.

Formation of Valence and Conduction Bands Recall from Section 11.3 that when two atoms form a diatomic molecule, their atomic orbitals (AOs) combine to form an equal number of molecular orbitals (MOs). Figure 12.37 shows the formation of MOs in lithium. To form Li_2, four valence orbitals (one 2*s* and three 2*p*) of each Li atom combine to form eight MOs, four bonding and four antibonding. The order of MOs shows mixing of the 2*s* and 2*p* AOs (Figure 11.21B). Two more Li atoms form Li_4, a slightly larger aggregate, with 16 delocalized MOs. As more Li atoms join the cluster, more MOs are created, and their energy levels lie closer and closer together. Extending this process to 7 g (1 mol) of lithium results in 6×10^{23} Li atoms (Li_{N_A}) combining to form an extremely large number (4 × Avogadro's number) of delocalized MOs. *The energies of the MOs are so close that they form a continuum, or band, of MOs.*

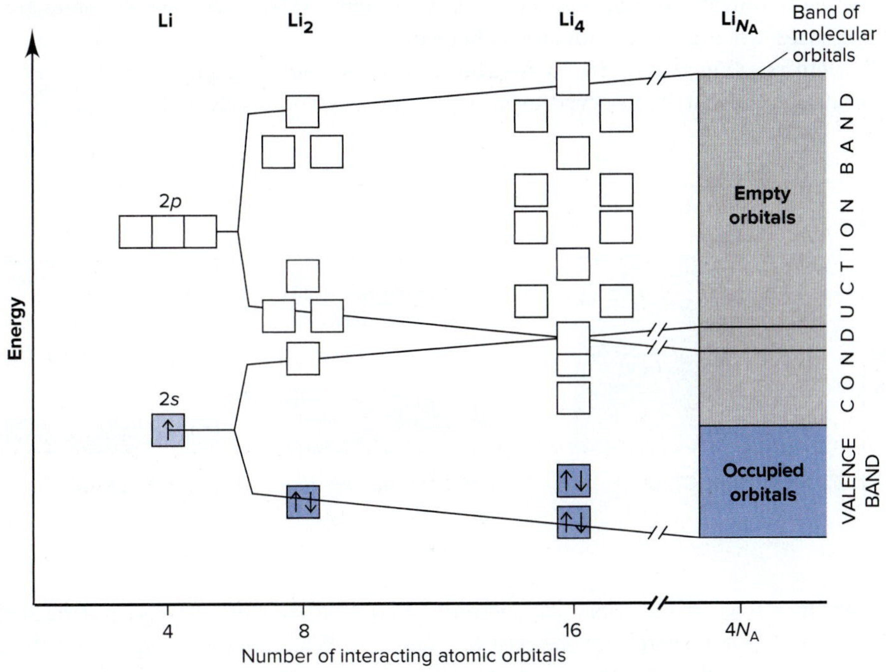

Figure 12.37 The band of molecular orbitals in lithium metal.

The lower energy MOs are occupied by the $2s^1$ valence electrons and make up the **valence band.** The empty MOs that are higher in energy make up the **conduction band.** In Li metal, the valence band is derived from the 2*s* AOs, and the conduction band is derived from 2*s* and mostly 2*p* AOs. In Li_2, two valence electrons fill the lowest energy bonding MO and leave the antibonding MO empty. In Li metal, 1 mol of valence electrons fills the valence band and leaves the conduction band empty.

How Band Theory Explains Metallic Properties The key to understanding the properties of metals is that *the valence and conduction bands are contiguous,* that is, the highest level of one touches the lowest of the other. This means that, given an infinitesimal quantity of energy, electrons jump from the filled valence band to the unfilled conduction band: the electrons are completely delocalized and free to move.

- *Electrical conductivity.* Metals conduct electricity so well because an applied field easily excites the highest energy valence electrons into empty conduction orbitals, allowing them to move through the sample.
- *Luster.* With so many closely spaced levels available, electrons absorb and release photons of many frequencies as they move between the valence and conduction bands.
- *Malleability.* Under an applied force, layers of positive metal ions move past each other, always protected from mutual repulsions by the delocalized electrons.
- *Thermal conductivity.* When a metal wire is heated, the highest energy electrons are excited and their extra energy is transferred as kinetic energy along the wire's length.

Conductivity of Solids and the Size of the Energy Gap Like metal atoms, large numbers of nonmetal and metalloid atoms can form bands of MOs. Band theory explains differences in electrical conductivity and the effect of temperature among these three classes of substances in terms of the presence of an energy gap between their valence and conduction bands (Figure 12.38):

1. *Conductors (metals).* The valence and conduction bands of a **conductor** have *no energy gap* between them, so electrons flow when a tiny electrical potential difference is applied. When the temperature is raised, greater random motion of the atoms hinders electron movement: conductivity *decreases* when a metal is heated.
2. *Semiconductors (metalloids).* In a **semiconductor,** a *small energy gap* exists between the valence and conduction bands. Thermally excited electrons can cross the gap, allowing a small current to flow: in contrast to a conductor, conductivity *increases* when a semiconductor is heated.
3. *Insulators (nonmetals).* In an **insulator,** a *large energy gap* exists between the bands: no current is observed even when the substance is heated.

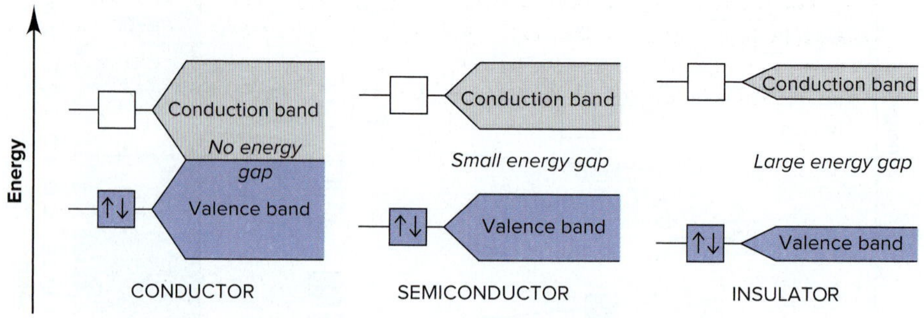

Figure 12.38 Electrical conductivity in a conductor, semiconductor, and insulator.

Another type of electrical conductivity, called **superconductivity,** has been the focus of intensive research for the past few decades. When metals conduct at ordinary temperatures, moving electrons collide with vibrating atoms; the reduction in their flow appears as resistive heating and represents a loss of energy.

For many years, to conduct with no energy loss—to superconduct—required minimizing atom movement by cooling with liquid helium (bp = 4 K; price ≈ $11/L). Then, in 1986, certain ionic oxides that superconduct near the boiling point of liquid nitrogen (bp = 77 K; price = $0.06/L) were prepared. Like metal conductors, oxide superconductors, such as $YBa_2Cu_3O_7$, have no gap between bands. In 1989, oxides with Bi and Tl instead of Y and Ba were synthesized and found to superconduct at 125 K; and in 1993, an oxide with Hg, Ba, and Ca, in addition to Cu and O, superconducted at 133 K. Such materials could transmit electricity with no loss of energy, allowing power plants to be located far from cities. They could be part of ultrasmall microchips for ultrafast computers, electromagnets to levitate superfast trains, and inexpensive medical diagnostic equipment with superb image clarity. However, the oxides are brittle and not easy to machine, and when warmed, the superconductivity may disappear and not return on cooling. Addressing these and related problems will involve chemists, physicists, and engineers for many years.

› Summary of Section 12.6

- Particles in crystalline solids lie at points that form a structure of repeating unit cells.
- The three types of unit cells of the cubic system are simple, body-centered, and face-centered. The highest packing efficiency occurs with cubic (face-centered) and hexagonal closest packing.

- Bond angles and distances in a crystal are determined with x-ray diffraction analysis and scanning tunneling microscopy. These data are used to determine atomic radii.
- Atomic solids [Group 8A(18)] adopt cubic closest packing, with atoms held together by weak dispersion forces.
- Molecular solids have molecules at the lattice points and often adopt cubic closest packing. Combinations of intermolecular forces (dispersion, dipole-dipole, and H bonding) result in physical properties that vary greatly.
- Ionic solids crystallize with one ion filling holes in the cubic closest packed array of the other. High melting points and hardness and low conductivity arise from strong ionic attractions.
- Most metals have a closest packed structure. Their physical properties result from the high packing efficiency and the presence of delocalized electrons.
- Atoms of network covalent solids are covalently bonded throughout the sample, so these substances have very high melting and boiling points.
- Amorphous solids have little regularity in their structure.
- Band theory proposes that atomic orbitals of many atoms combine to form a continuum, or band, of molecular orbitals. Metals are electrical conductors because electrons move freely from the filled (valence) band to the empty (conduction) band. Insulators have a large energy gap between the two portions, and semiconductors have a small gap, which can be bridged by heating.

12.7 ADVANCED MATERIALS

In the last few decades, the exciting field of materials science has applied concepts from solid-state chemistry, physics, and engineering. Objects that were once considered futuristic fantasies have become realities: powerful, ultrafast computers smaller than a cell phone; cars powered by sunlight and made of nonmetallic parts stronger than steel and lighter than aluminum; ultrasmall machines constructed by manipulating individual atoms and molecules. In this section, we briefly discuss some of these remarkable materials.

Electronic Materials

The ideal of a perfectly ordered crystal is attainable only if the crystal is grown very slowly under carefully controlled conditions. When crystals form more rapidly, **crystal defects** inevitably result. Planes of particles are misaligned, particles are out of place or missing entirely, and foreign particles replace those that belong in the lattice. Although defects usually weaken a substance, they can be introduced intentionally to improve materials, giving them increased strength, hardness, or conductivity. Centuries of metal-working exemplify the traditional importance of crystal defects, which is the basis of modern electronic materials.

Introducing Crystal Defects in Welding and Alloying In the process of *welding* two metals together, *vacancies* form near the surface when atoms vaporize, and then these vacancies move deeper as atoms from lower rows rise to fill the gaps. Welding causes the two types of metal atoms to intermingle and fill each other's vacancies. Metal *alloying* introduces several kinds of crystal defects, as when some atoms of a second metal occupy lattice sites of the first. Often, the alloy is harder than the pure metal; an example is brass, an alloy of copper with zinc. One reason the welded metals are stronger and the alloy is harder is that the second metal contributes additional valence electrons for metallic bonding.

Doped Semiconductors By controlling the number of valence electrons through the creation of specific types of crystal defects, chemists and engineers can greatly increase the conductivity of a semiconductor. For example, pure silicon [Si; Group 4A(14)], conducts poorly at room temperature because an energy gap separates its filled valence band from its conduction band (Figure 12.39A). Silicon's conductivity

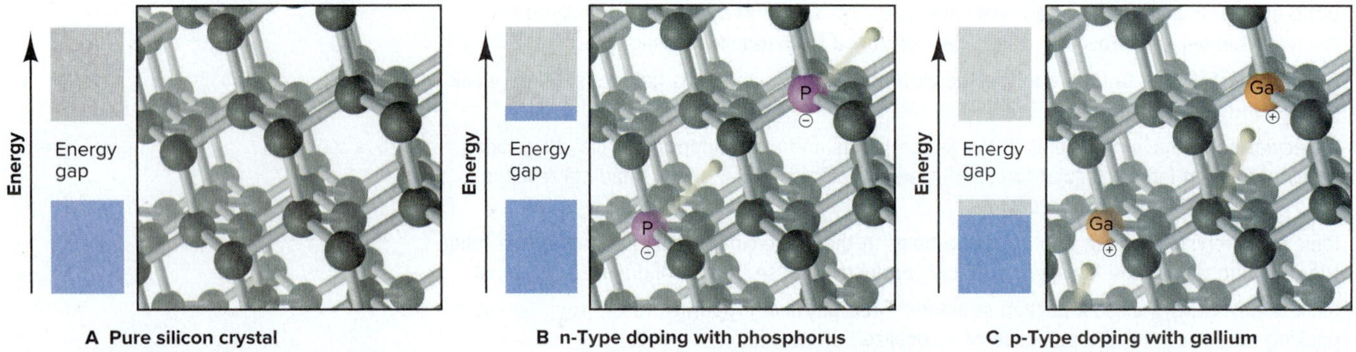

Figure 12.39 Crystal structures and band representations of doped semiconductors. A, Pure silicon has an energy gap between its valence and conduction bands, which keeps conductivity low at room temperature. **B,** Doping silicon with phosphorus *(purple)* adds additional valence electrons. **C,** Doping silicon with gallium *(orange)* removes electrons from the valence band and introduces positive holes.

can be greatly enhanced by **doping,** adding small amounts of other elements to increase or decrease the number of valence electrons in the bands.

- *Doping to create an n-type semiconductor: increasing the number of valence electrons.* When Si is doped with phosphorus [or another Group 5A(15) element], P atoms occupy some of the lattice sites. Since P has one more valence electron than Si, this additional electron must enter an empty orbital in the conduction band, thus bridging the energy gap and increasing conductivity. Such doping creates an *n-type semiconductor,* so called because extra *n*egative charges (electrons) are present (Figure 12.39B).
- *Doping to create a p-type semiconductor: decreasing the number of valence electrons.* When Si is doped with gallium [or another Group 3A(13) element], Ga atoms occupy some sites (Figure 12.39C). Since Ga has one fewer valence electron than Si, some of the orbitals in the valence band are empty, which creates a positive site. Si electrons can migrate to these empty orbitals, thereby increasing conductivity. Such doping creates a *p-type semiconductor,* so called because the empty orbitals act as *p*ositive holes.

In contact with each other, an n-type and a p-type semiconductor form a *p-n junction.* When the negative terminal of a battery is connected to the n-type portion and the positive terminal to the p-type portion, electrons *(yellow, negative spheres)* flow freely in the n-to-p direction, which has the simultaneous effect of moving holes *(white, positive spheres)* in the p-to-n direction (Figure 12.40A). No current flows if the terminals are reversed (Figure 12.40B). Such unidirectional current flow makes a p-n junction act as a *rectifier,* a device that converts alternating current into direct current. A p-n junction in a modern integrated circuit can be made smaller than a square 10 μm on a side.

A modern computer chip the size of a nickel may incorporate millions of p-n junctions in the form of *transistors.* One of the most common types, an n-p-n transistor, is made by sandwiching a p-type portion between two n-type portions to form adjacent p-n junctions. The current flowing through one junction controls the current flowing through the other and results in an amplified signal.

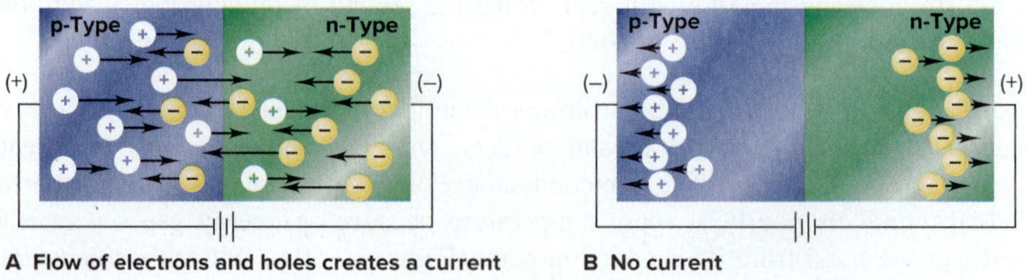

Figure 12.40 The p-n junction. Placing a p-type semiconductor adjacent to an n-type creates a p-n junction.

One of the more common types of *solar cell* is, in essence, a p-n junction with an n-type surface exposed to a light source. Light provides the energy to free electrons from the n-type region and accelerate them through an external circuit into the p-type region, thus producing a current.

Liquid Crystals

Incorporated in the membrane of every cell in your body and in the display of every electronic device are **liquid crystals.** These materials flow like liquids but, like crystalline solids, pack with a high degree of order at the molecular level.

Ordering of the Particles Let's examine how particles are ordered in the three physical states and how this affects their properties. Both gases and liquids are *isotropic*—their physical properties are the same in every direction within the phase. For example, the viscosity of a gas or a liquid is the same regardless of direction. Glasses and other amorphous solids are also isotropic because they have no regular lattice structure.

In contrast, crystalline solids have a high degree of order among their particles. The properties of a crystal *do* depend on direction, so a crystal is *anisotropic.* The facets in a cut diamond, for instance, arise because the crystal cracks in one direction more easily than in another. Liquid crystals are anisotropic with respect to several physical properties, including electrical and optical properties, which differ with direction through the phase.

Molecular Characteristics Liquid crystal phases consist of individual molecules with two characteristics: (1) a long, cylindrical shape and (2) a structure that fosters intermolecular attractions but inhibits perfect crystalline packing. Molecules that form liquid crystal phases have rodlike shapes and certain groups—in the cases shown in Figure 12.41, flat, benzene-like ring systems—that keep the molecules extended. Many of these molecules also have polarity associated with the long molecular axis. A strong electric field can orient these polar molecules like compass needles in a magnetic field.

The viscosity of a liquid crystal phase is lowest in the direction parallel to the long axis. Like moistened microscope slides, it is easier for the molecules to slide

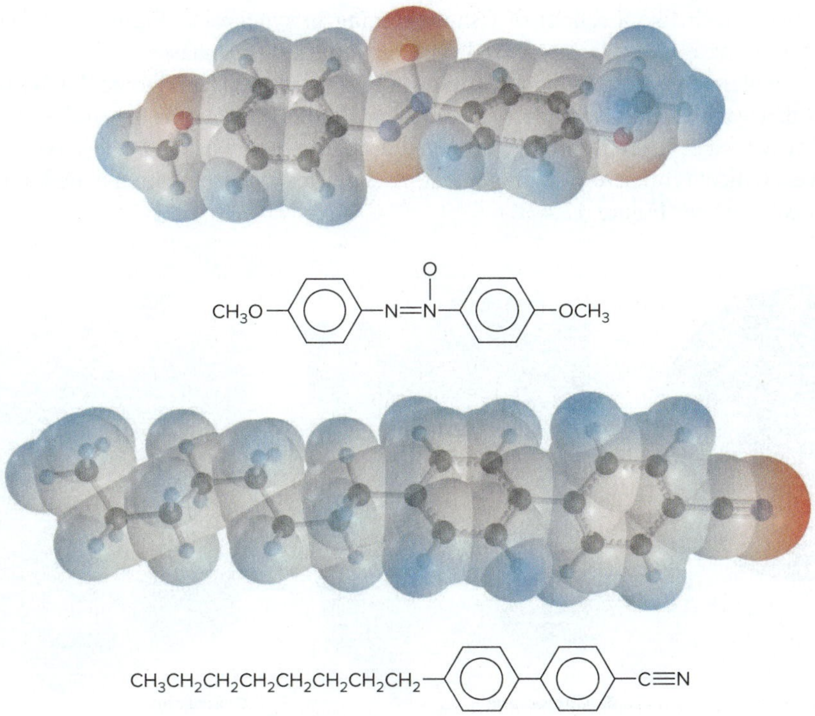

Figure 12.41 **Structures of two typical molecules that form liquid crystal phases.** Note the long, extended shapes and the regions of high *(red)* and low *(blue)* electron density.

along each other (because the total attractive force remains the same), than it is for them to pull apart from each other sideways. As a result, the molecules tend to align while the phase flows.

Controlling Conditions to Form Phases Liquid crystal phases can arise in two general ways, and, sometimes, either way can occur in the same substance:

1. *Thermotropic phases* develop as a result of a change in temperature. As a crystalline solid is heated, the molecules leave their lattice sites, but the intermolecular interactions are still strong enough to keep the molecules aligned with each other along their long axes. Like any other phase, the liquid crystal phase has sharp transition temperatures, but they occur over a relatively small temperature range. Further heating provides enough kinetic energy for the molecules to become disordered, as in a normal liquid. The typical range for liquid crystal phases of pure substances is from <1°C to as much as 10°C, but mixing phases of two or more substances can greatly extend this range. For this reason, the liquid crystal phases used within display devices, as well as those within cell membranes, consist of mixtures of molecules.

2. *Lyotropic phases* occur in solution as the result of changes in concentration, but the conditions for forming such a phase vary for different substances. For example, when purified, some biomolecules that exist in cell membranes form lyotropic phases in water at the temperature that occurs within the organism. At the other extreme, Kevlar, a fiber used in bulletproof vests and high-performance sports equipment, forms a lyotropic phase in concentrated H_2SO_4 solution.

In some cases, a substance that forms a given liquid crystal phase under one set of conditions forms other phases under different conditions. Thus, a given thermotropic liquid crystal substance can pass from disordered liquid through a series of distinct liquid crystal phases to an ordered crystal through a decrease in temperature. A lyotropic substance can undergo similar changes through an increase in concentration.

Types of Order in Liquid Crystal Phases Molecules that form liquid crystal phases can exhibit various types of order. Three common types are nematic, cholesteric, and smectic:

1. In a *nematic phase,* the molecules lie in the same direction but their ends are not aligned, much like a school of fish swimming in synchrony (Figure 12.42A). The nematic phase is the least ordered type of liquid crystal phases.
2. In a *cholesteric phase,* which is somewhat more ordered, the molecules lie in layers that each exhibit nematic-type ordering. Rather than lying in parallel fashion, however, each layer is rotated by a fixed angle with respect to the next layer to give a helical (corkscrew) arrangement. A cholesteric phase is often called a *twisted nematic phase* (Figure 12.42B).

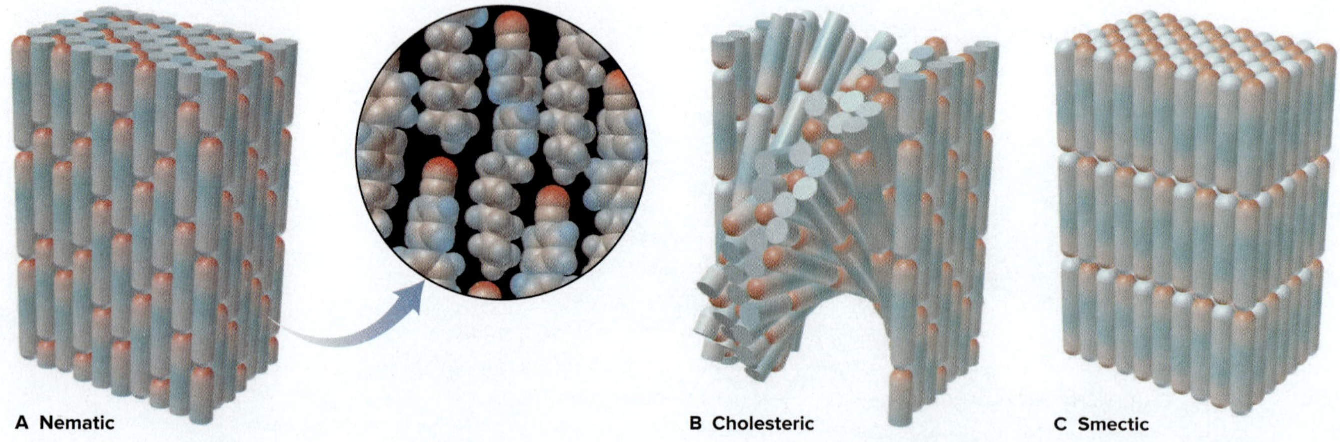

Figure 12.42 The three common types of ordering in liquid crystal phases.

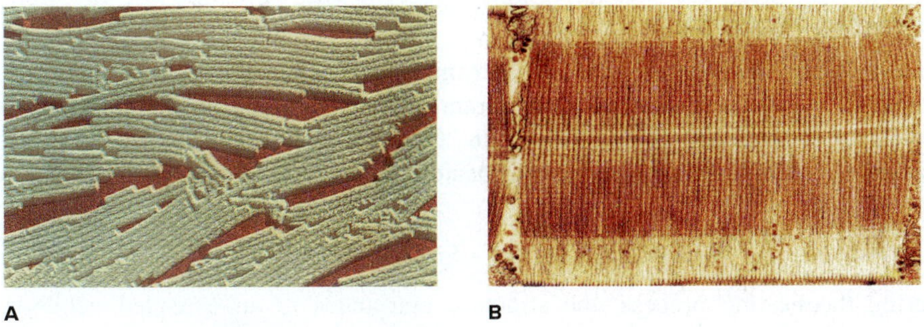

Figure 12.43 Liquid crystal–type phases in biological systems. A, Nematic arrays of tobacco mosaic virus particles within the fluid of a tobacco leaf. **B,** The smectic-like arrangement of actin and myosin protein filaments in voluntary muscle cells.

Source: (A) © J. R. Factor/Science Source (B) © C. F. Armstrong/Science Source

3. In a *smectic phase,* which is the most ordered, the molecules lie parallel to each other, *with* their ends aligned, in layers that are stacked directly over each other (Figure 12.42C). The long molecular axis has a well-defined angle (shown in the figure as 90°) with respect to the plane of the layer. The molecules in Figure 12.41 form nematic or smectic phases. Liquid crystal–type phases appear in many biological systems (Figure 12.43).

Applications of Liquid Crystals The ability to control the orientation of the molecules in a liquid crystal allows us to produce materials with high strength or unique optical properties:

1. *High-strength applications* involve the use of extremely long molecules called *polymers.* While in a thermotropic liquid crystal phase and during their flow through the processing equipment, these molecules become highly aligned, like the fibers in wood. Cooling solidifies them into fibers, rods, and sheets that can be shaped into materials with superior mechanical properties. Sporting equipment, supersonic aircraft parts, and the sails used in the America's Cup races are fabricated from these polymeric materials. (We discuss the structure and physical behavior of polymers later in this section and their synthesis in Chapter 15.)

2. *Optical applications* include the liquid crystal displays (LCDs) used in countless devices, such as watches, calculators, cell phones, and tablets. All depend on *changes in molecular orientation in an electric field.* Figure 12.44 shows a small portion of a wristwatch LCD. Layers of nematic phases are sandwiched between thin glass plates that incorporate transparent electrodes. The long molecular axes lie parallel to the plane of the plates. The distance between plates (6–8 μm) allows the molecular axes within each succeeding layer to twist just enough for the molecular orientation at the bottom plate to be 90° from that at the top plate. Above and below this "sandwich" are thin polarizing filters that allow light waves oriented in only one direction to pass through. The filters are placed in a "crossed" arrangement, so that light passing through the top filter must twist 90° to pass through the bottom filter. This whole grouping of filters, plates, and liquid crystal phase lies on a mirror.

A current generated by the watch battery controls the orientation of the molecules. With the current on in one region of the display, the molecules become oriented *toward* the field, and thus block the light from passing through to the bottom filter, so that region appears dark. With the current off in another region, light passes through the molecules and bottom filter to the mirror and back again, so that region appears bright.

Cholesteric liquid crystals are used in applications that involve color changes with temperature. The twisted molecular orientation "unwinds" with heating, and the extent of the unwinding determines the color. Liquid crystal thermometers include a mixture of substances to widen their range of temperatures. Newer uses include "mapping" the area of a tumor, detecting faulty connections in electronic circuit boards, and nondestructive testing of materials under stress.

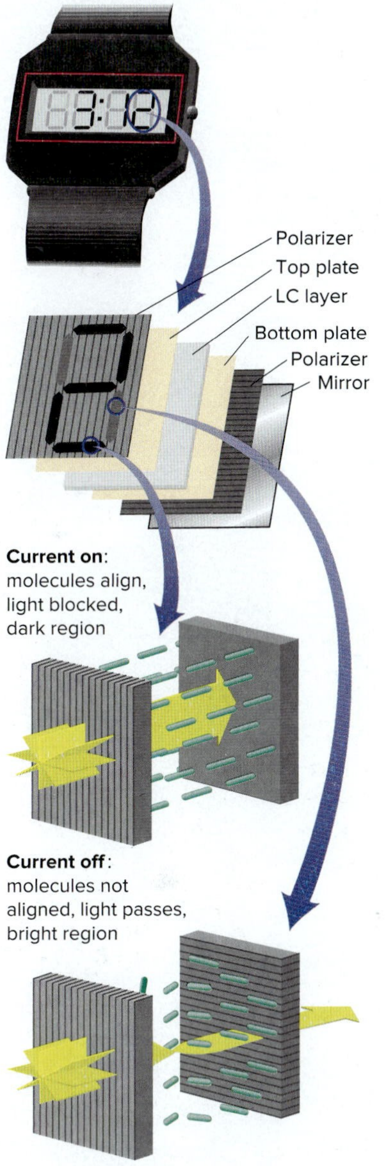

Figure 12.44 A liquid crystal display (LCD). A close-up of the "2" on a wristwatch LCD reveals two polarizers sandwiching two glass plates, which sandwich a liquid crystal (LC) layer, all lying on a mirror. When light waves enter the first polarizer, waves oriented in one direction emerge to enter the LC layer. Enlarging a dark region *(top blow-up)* shows the LC molecules aligned, keeping light waves from passing through; thus, the viewer sees no light. Enlarging a bright region *(bottom blow-up)* shows the LC molecules rotating the plane of the light waves, which can then pass through and reach the viewer.

Ceramic Materials

First developed by Stone Age people, **ceramics** are nonmetallic, nonpolymeric solids hardened by high temperature. Clay ceramics consist of silicate microcrystals suspended in a glassy cementing medium. In "firing" a ceramic pot, for example, a kiln heats the object, made of an aluminosilicate clay, such as kaolinite, to 1500°C and the clay loses water:

$$Si_2Al_2O_5(OH)_4(s) \longrightarrow Si_2Al_2O_7(s) + 2H_2O(g)$$

During the heating process, the structure rearranges to an extended network of Si-centered and Al-centered tetrahedra of O atoms (Section 14.6).

Bricks, porcelain, glazes, and other clay ceramics are hard and resistant to heat and chemicals. Today's high-tech ceramics have these characteristics in addition to superior electrical and magnetic properties (Table 12.7). As just one example, consider the unusual electrical behavior of certain zinc oxide (ZnO) composites. Ordinarily a semiconductor, ZnO can be doped so that it becomes a conductor. Imbedding particles of the doped oxide into an insulating ceramic produces a variable resistor: at low voltage, the material conducts poorly, but at high voltage, it conducts well. The changeover voltage can be "preset" by controlling the size of the ZnO particles and the thickness of the insulating medium.

Table 12.7 Some Uses of Modern Ceramics and Ceramic Mixtures

Ceramic	Applications
SiC, Si_3N_4, TiB_2, Al_2O_3	Whiskers (fibers) to strengthen Al and other ceramics
Si_3N_4	Car engine parts; turbine rotors for "turbo" cars; electronic sensor units
Si_3N_4, BN, Al_2O_3	Support or layering materials (as insulators) in electronic microchips
SiC, Si_3N_4, TiB_2, ZrO_2, Al_2O_3, BN	Cutting tools, edge sharpeners (as coatings and whole devices), scissors, surgical tools, industrial "diamond"
BN, SiC	Armor-plating reinforcement fibers (as in Kevlar composites)
ZrO_2, Al_2O_3	Surgical implants (hip and knee joints)

Preparing Modern Ceramics Among the important modern ceramics are silicon carbide (SiC) and nitride (Si_3N_4), boron nitride (BN), and the superconducting oxides. They are prepared by standard chemical methods that involve driving off a volatile component during the reaction:

1. *SiC ceramics* are made from compounds used in silicone polymer manufacture (we'll discuss these polymers in Section 14.6):

$$n(CH_3)_2SiCl_2(l) + 2nNa(s) \longrightarrow 2nNaCl(s) + [(CH_3)_2Si]_n(s)$$

This product is heated to 800°C to form the ceramic:

$$[(CH_3)_2Si]_n \longrightarrow nCH_4(g) + nH_2(g) + nSiC(s)$$

Silicon carbide can also be prepared by the direct reaction of silicon and graphite under vacuum:

$$Si(s) + C(\text{graphite}) \xrightarrow{\sim 1500°C} SiC(s)$$

The nitride is also prepared by reaction of the elements:

$$3Si(s) + 2N_2(g) \xrightarrow{>1300°C} Si_3N_4(s)$$

2. *BN ceramics* are produced through the reaction of boron trichloride or boric acid with ammonia:

$$B(OH)_3(s) + 3NH_3(g) \longrightarrow B(NH_2)_3(s) + 3H_2O(g)$$

Heat drives off some of the bound nitrogen as NH_3 to yield the ceramic:

$$B(NH_2)_3(s) \xrightarrow{\Delta} 2NH_3(g) + BN(s)$$

3. One type of *superconducting oxide* is made by heating a mixture of barium carbonate with copper and yttrium oxides, followed by further heating in the presence of O_2:

$$4BaCO_3(s) + 6CuO(s) + Y_2O_3(s) \xrightarrow{\Delta} 2YBa_2Cu_3O_{6.5}(s) + 4CO_2(g)$$
$$YBa_2Cu_3O_{6.5}(s) + \tfrac{1}{4}O_2(g) \xrightarrow{\Delta} YBa_2Cu_3O_7(s)$$

Ceramic Structures and Uses Structures of several ceramic materials are shown in Figure 12.45.

1. *Silicon carbide* has a diamond-like structure (Figure 12.45A). Network covalent bonding gives this material great strength. It is made into thin fibers, called *whiskers,* to reinforce other ceramics and prevent cracking in a composite structure, much like steel rods reinforcing concrete.

2. *Silicon nitride* is virtually inert chemically, retains its strength and wear resistance for extended periods above 1000°C, is dense and hard, and acts as an electrical insulator. Many automakers are testing it in high-efficiency car and truck engines because of its low weight, tolerance of high operating temperatures, and little need for lubrication.

3. *BN ceramics* exist in two structures, analogous to the common crystalline forms of carbon. In the graphite-like form, BN has extraordinary properties as an electrical insulator. At high temperature and very high pressure (1800°C and 8.5×10^4 atm), BN converts to a diamond-like structure (Figure 12.45B), which is extremely hard and durable. Both forms are virtually invisible to radar.

4. *Superconducting oxides* often contain copper in an unusual oxidation state. In $YBa_2Cu_3O_7$ (Figure 12.45C), for instance, assuming oxidation states of +3 for Y, +2 for Ba, and −2 for O, the three Cu atoms have a total oxidation state of +7. This is allocated as $Cu(II)_2Cu(III)$, with one Cu in the unusual +3 state. X-ray diffraction analysis indicates that a distortion in the structure makes four of the oxide ions unusually close to the Y^{3+} ion, which aligns the Cu ions into chains within the crystal. It is thought that a specific half-filled 3*d* orbital in Cu oriented toward a neighboring O^{2-} ion may be associated with superconductivity.

Because of brittleness, it is difficult to fashion these ceramics into wires, but methods for making films and ribbons have been developed. The brittleness arises from the strength of the ionic-covalent bonding and the resulting inability to deform. Under stress, a microfine defect widens until the material cracks. One new method forms defect-free superconducting oxide ceramics using controlled packing and heat treatment of extremely small, uniform oxide particles coated with organic polymers. Another method is aimed at arresting a widening crack. The ceramics are embedded with zirconia (ZrO_2), whose crystal structure expands up to 5% under mechanical stress. When an advancing crack reaches them, the zirconia particles pinch it shut. A third method prepares precisely grown crystals of Al_2O_3 and $GdAlO_3$, which become entangled during the solidification process. The resulting material bends without cracking above 1800 K.

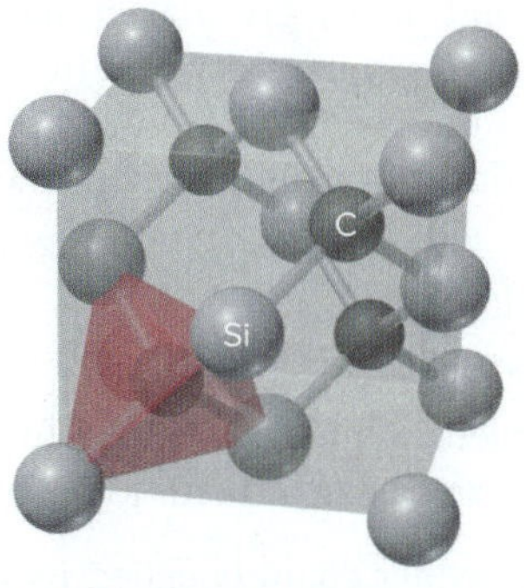

A Silicon carbide

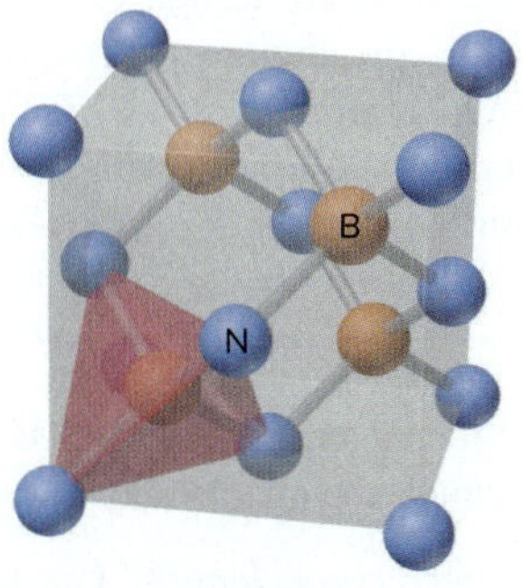

B Cubic BN (borazon)

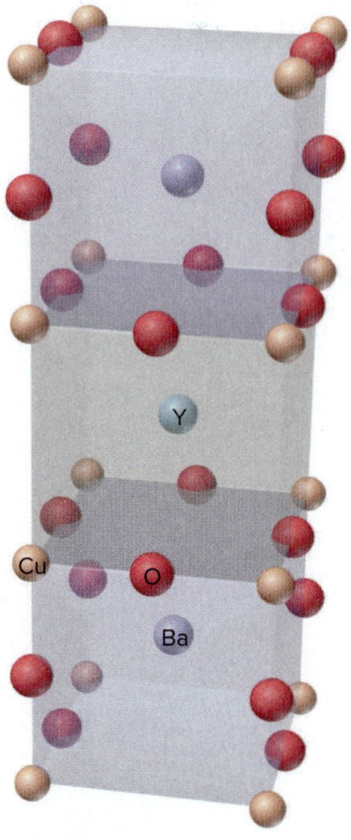

C $YBa_2Cu_3O_7$

Figure 12.45 Expanded view of the atom arrangements in some ceramic materials.

Polymeric Materials

A **polymer** (Greek, "many parts") is an extremely large molecule, or **macromolecule,** consisting of a covalently linked chain of smaller molecules, called **monomers** (Greek, "one part"). The monomer is the *repeat unit* of the polymer, and a polymer may have hundreds to hundreds of thousands of repeat units. Many types of monomers give polymers the complete repertoire of intermolecular forces.

- *Synthetic* polymers are created in the laboratory.
- *Natural* polymers (or *biopolymers*) are created within organisms.

In this section, we examine the physical nature of synthetic polymers and explore the role of intermolecular forces in their properties and uses. In Chapter 15, we'll examine the types of monomers, the preparation of synthetic polymers, and the monomers, structures, and functions of some biopolymers.

Dimensions of a Polymer Chain Polymers differ from smaller molecules in terms of molar mass, chain length, shape, and size. We'll discuss these, using polyethylene, by far the most common synthetic polymer, as an example.

1. *Polymer mass.* The molar mass of a polymer chain ($\mathcal{M}_{\text{polymer}}$, in g/mol, often referred to as the *molecular weight*) depends on the molar mass of the repeat unit ($\mathcal{M}_{\text{repeat}}$) and the **degree of polymerization (*n*),** the number of repeat units in the chain:

$$\mathcal{M}_{\text{polymer}} = \mathcal{M}_{\text{repeat}} \times n$$

For example, the molar mass of the ethylene repeat unit ($—CH_2—CH_2—$) is 28 g/mol. If an individual polyethylene chain in a plastic grocery bag has a degree of polymerization of 7100, the molar mass of that chain is

$$\mathcal{M}_{\text{polymer}} = \mathcal{M}_{\text{repeat}} \times n = (28 \text{ g/mol})\,(7.1\times10^3) = 2.0\times10^5 \text{ g/mol}$$

Table 12.8 shows some other examples.

Table 12.8 Molar Masses of Some Common Polymers

Name	$\mathcal{M}_{\text{polymer}}$ (g/mol)	*n*	Uses
Acrylates	2×10^5	2×10^3	Rugs, carpets
Polyamide (nylons)	1.5×10^4	1.2×10^2	Tires, fishing line
Polycarbonate	1×10^5	4×10^2	Compact discs
Polyethylene	3×10^5	1×10^4	Grocery bags
Polyethylene (ultrahigh molecular weight)	5×10^6	2×10^5	Hip joints
Poly(ethylene terephthalate)	2×10^4	1×10^2	Soda bottles
Polystyrene	3×10^5	3×10^3	Packing, coffee cups
Poly(vinyl chloride)	1×10^5	1.5×10^3	Plumbing

However, even though the monomer repeat unit in any given chain has the same molar mass, the degree of polymerization often varies considerably from chain to chain. Thus, *a sample of a synthetic polymer has a distribution of chain lengths and molar masses.* Polymer chemists use various definitions of the *average* molar mass, and a common one is the *number-average molar mass,* $\mathcal{M}_n$:

$$\mathcal{M}_n = \frac{\text{total mass of all chains}}{\text{number of moles of chains}}$$

If the number-average molar mass of the polyethylene in grocery bags is, for example, 1.6×10^5 g/mol, the chains may vary in molar mass from about 7.0×10^4 to 3.0×10^5 g/mol.

2. *Polymer chain length.* The long axis of a polymer chain is called its *backbone.* The length of an *extended* backbone is simply the number of repeat units (degree of polymerization, *n*) times the length of each repeat unit (l_0). For instance, the length

of an ethylene repeat unit is about 250 pm, so the extended length of our grocery-bag polyethylene chain is

$$\text{Length of extended chain} = n \times l_0 = (7.1\times10^3)\ (2.5\times10^2\ \text{pm}) = 1.8\times10^6\ \text{pm}$$

Comparing this length with the thickness of the chain, which is about 40 pm, shows the threadlike nature of the extended chain (length about 50,000 times width).

3. *Coiled shape and size.* A polymer molecule, however, whether pure or in solution, doesn't exist as an extended chain. In principle, the shape of the molecule arises from free rotation around all the single bonds. Thus, as each repeat unit rotates randomly, the chain continuously changes direction, turning back on itself many times and eventually arriving at the **random coil** shape that most polymers adopt (Figure 12.46). In reality, rotation is not completely free because intermolecular forces between chain portions, between different chains, and/or between chain and solvent have significant effects on the actual shape of a polymer chain.

The size of the coiled chain is expressed by its **radius of gyration, R_g,** the average distance from the center of mass of the molecule to the outer edge of the coil (Figure 12.46). Even though R_g is reported as a single value for a given polymer, it represents an average value for many chains. The mathematical expression for the radius of gyration includes the length of each repeat unit and the degree of polymerization:*

$$R_g = \sqrt{\frac{nl_0^2}{6}}$$

As expected, the radius of gyration increases with the degree of polymerization, and thus with the molar mass as well. Light-scattering measurements correlate with these calculated results, so for many polymers the radius of gyration can be determined experimentally.

*The mathematical derivation of R_g is beyond the scope of this text, but it is analogous to the two-dimensional "walk of the drunken sailor." With each step, the sailor stumbles in random directions and, given enough time, ends up very close to the starting position. The radius of gyration quantifies how far the end of the polymer chain (the sailor) has gone from the origin.

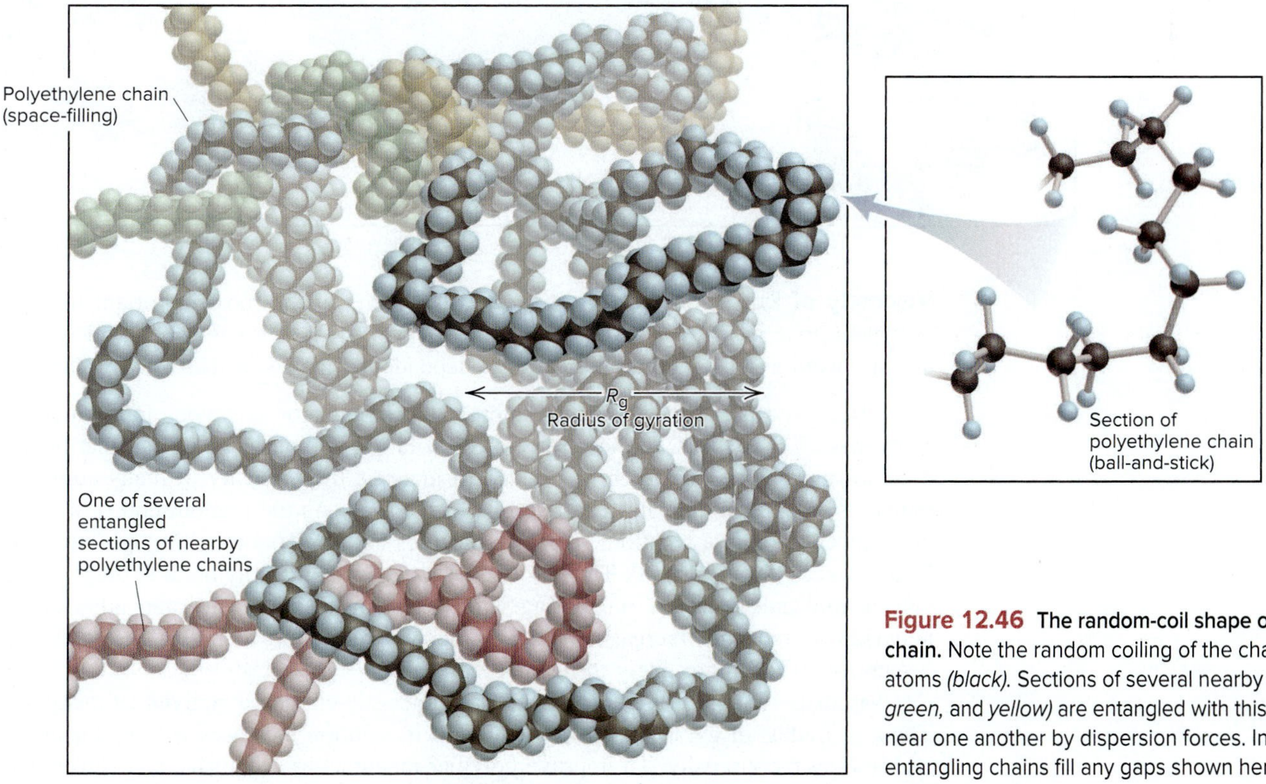

Figure 12.46 The random-coil shape of a polymer chain. Note the random coiling of the chain's carbon atoms *(black)*. Sections of several nearby chains *(red, green,* and *yellow)* are entangled with this chain, kept near one another by dispersion forces. In reality, entangling chains fill any gaps shown here.

For our grocery-bag polyethylene chain, we have

$$R_g = \sqrt{\frac{nl_0^2}{6}} = \sqrt{\frac{(7.1\times10^3)(2.5\times10^2\ \text{pm})^2}{6}} = 8.6\times10^3\ \text{pm}$$

Doubling the radius gives a diameter of 1.7×10^4 pm for the coiled chain, less than one-hundredth the length of the extended chain!

Polymer Crystallinity A sample of a given polymer is not just a disorderly jumble of chains. If the molecular structure allows neighboring chains to pack together and if the chemical groups lead to favorable dipole-dipole, H bonding, and/or dispersion forces, portions of the chains can align regularly and exhibit crystallinity.

However, the crystallinity of a polymer is very different from the crystal structures we discussed earlier. In those structures, the orderly array extends over many atoms or molecules, and the unit cell includes at least one atom or molecule. In contrast, the orderly regions of a polymer rarely involve even one whole molecule (Figure 12.47). At best, a polymer is *semicrystalline,* because parts of the molecule align with parts of neighboring molecules (or with other parts of the same chain), while most of the chain remains as a random coil.

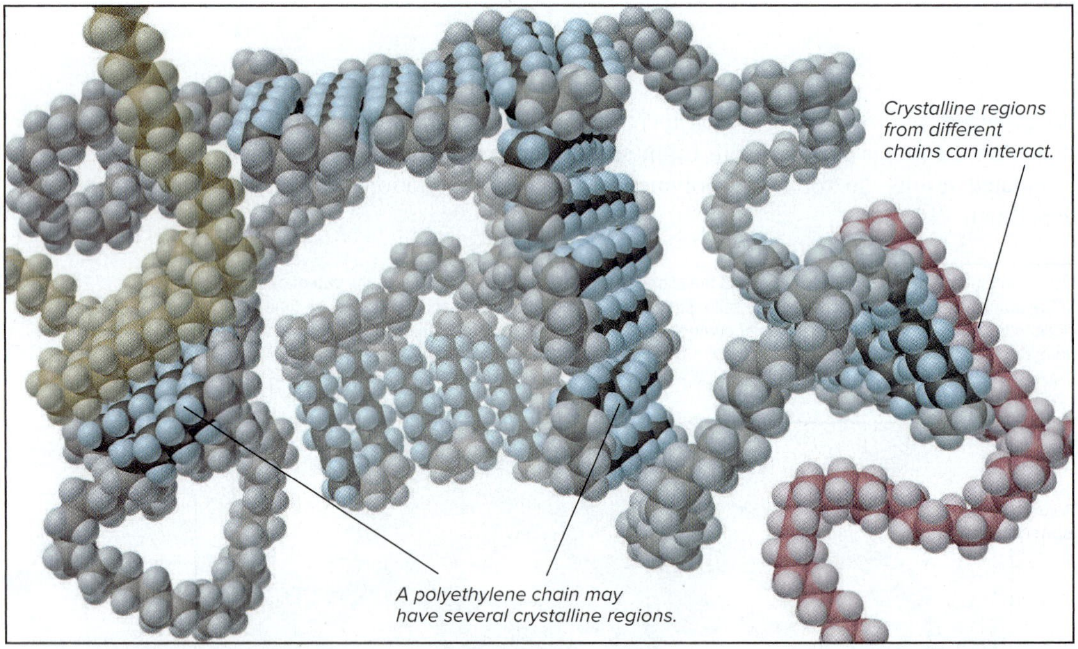
Crystalline regions from different chains can interact.
A polyethylene chain may have several crystalline regions.

Figure 12.47 The semicrystallinity of a polymer chain. Several crystalline regions *(darker color)* have randomly coiled regions between them. Crystalline regions from nearby chains *(red* and *yellow)* align with those in the main chain.

Viscosity of Pure and Dissolved Polymers Some of the most important uses of polymers arise from their ability to change the viscosity of a solvent in which they are dissolved and to undergo temperature-dependent changes in their own viscosity.

1. *Dissolved polymers.* When an appreciable amount of polymer (approximately 5–15 mass %) dissolves, the viscosity of the solution is much higher than that of the pure solvent. In fact, polymers are added to increase the viscosity of many common materials, such as motor oil, paint, and salad dressing. As the random coil of a polymer moves through a solution, solvent molecules are attracted to it through intermolecular forces (Figure 12.48). Thus, as the solution flows, the polymer coil drags along solvent molecules that are interacting with other solvent molecules and other coils, and flow is reduced. Increasing the polymer concentration increases the viscosity because coils are more likely to become entangled.

Viscosity is a property of a particular polymer-solvent pair at a given temperature. The size (radius of gyration) of a random coil in solution increases with molar mass, and so does the viscosity. To improve polymer manufacture, chemists have developed

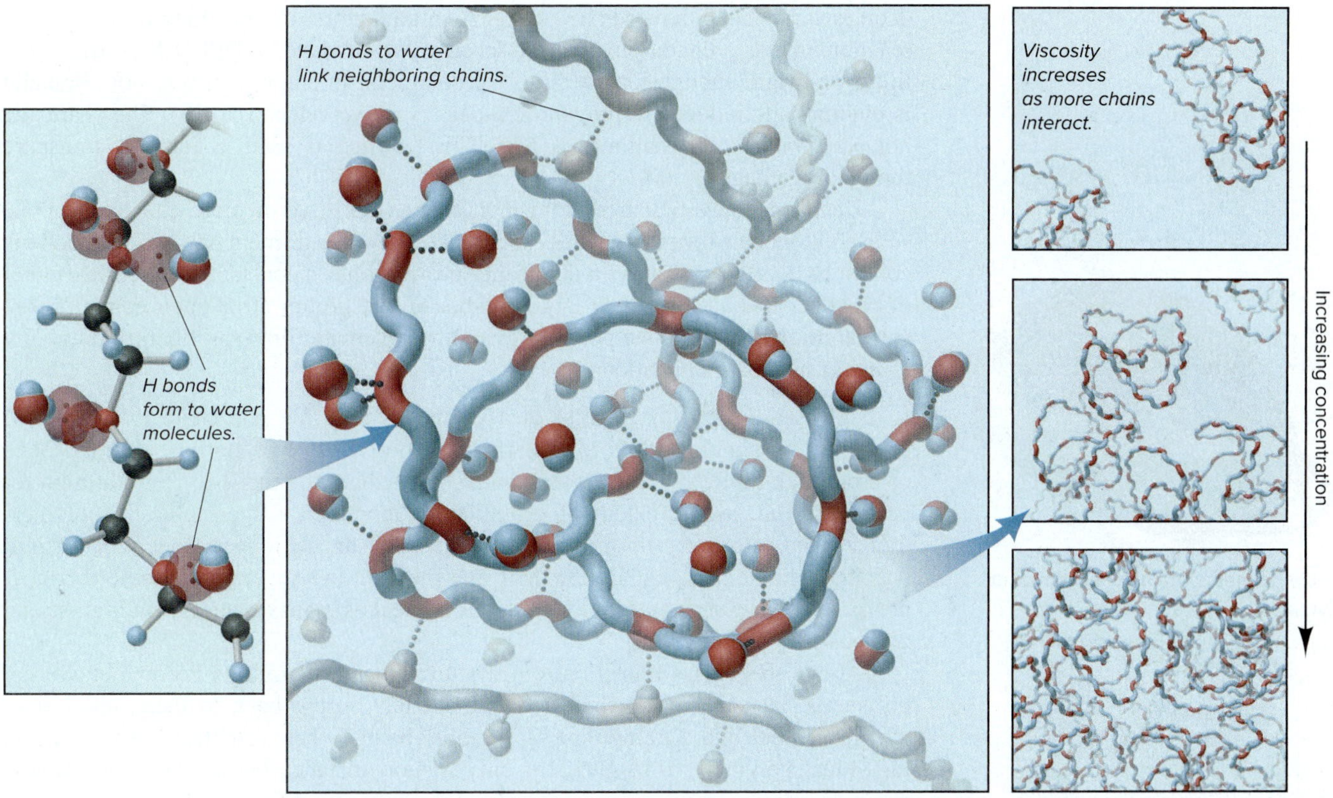

Figure 12.48 **The viscosity of a polymer in aqueous solution.** A section of a poly(ethylene oxide) chain *(left panel)* forms H bonds between the lone pairs of the chain's O atoms and the H atoms of water molecules. A polymer chain *(blue and red coiled rod, center panel)* forms many H bonds with water molecules that allow the chain to interact with other chains. As the concentration of polymer increases *(three right panels),* the viscosity of the solution increases.

equations to predict the viscosity of polymers of different molar masses in a variety of solvents (see Problem 12.147).

2. *Pure polymers.* Intermolecular forces also play a major role in the flow of a pure polymer. At temperatures high enough to melt them, many polymers exist as viscous liquids, flowing more like honey than water. The forces between the chains, as well as the entangling of chains, hinder the flow. As the temperature decreases, the intermolecular attractions exert a greater effect, and the eventual result is a rigid solid. If the chains don't crystallize, the resulting material is called a *polymer glass.* The transition from a liquid to a glass occurs over a narrow temperature range (10–20°C) for a given polymer; the temperature at the midpoint of the range is called the *glass transition temperature,* T_g. Like window glass, many polymer glasses are transparent, including polystyrene in drinking cups and polycarbonate in eyeglass lenses.

Plasticity of Polymers The flow-related properties of polymers give rise to their familiar **plastic** mechanical behavior. The word *plastic* refers to a material that, when deformed, retains its new shape; in contrast, when an *elastic* object is deformed, it returns to its original shape. Many polymers can be deformed (stretched, bent, twisted) when warm and retain their deformed shape when cooled. In this way, they are made into milk bottles, car parts, and countless other everyday objects.

Molecular Architecture of Polymers A polymer's architecture—its overall spatial layout and molecular structure—is crucial to its properties. In addition to linear chains, more complex architectures arise through branching and crosslinking:

1. **Branches** are smaller chains appended to a polymer backbone. As the number of branches increases, the chains cannot pack as well, so the degree of crystallinity

decreases and the polymer is less rigid. A small amount of branching occurs as a side reaction in the preparation of high-density polyethylene (HDPE). It is still largely linear and rigid enough for use in milk containers. In contrast, much more branching is intentionally induced to prepare low-density polyethylene (LDPE). The chains cannot pack well, so crystallinity is low. This polymer is used in flexible, transparent food storage bags.

Dendrimers are prepared from monomers with three or more attachment points, so each monomer forms branches. In essence, then, dendrimers have no backbone and consist only of branches. As a dendrimer grows, it has a constantly increasing number of branches and an incredibly large number of end groups at its outer edge. Chemists use dendrimers to bind one polymer to another to create films and fibers and to deliver drugs in medical applications.

Bicycle helmet containing a thermoset polymer.

Source: © Ingram Publishing/Alamy RF

2. **Crosslinks** can be thought of as branches that link one chain to another. The extent of crosslinking leads to remarkable differences in properties. A small degree of crosslinking often yields a *thermoplastic* polymer, one that still flows at high temperatures. But, as the extent of crosslinking increases, a thermoplastic polymer is transformed into a *thermoset* polymer, one that can no longer flow because it has become a single network. Below their glass transition temperatures, some thermosets are extremely rigid and strong, making them ideal as matrix materials in high-strength composites *(see photo)*.

Above their glass transition temperatures, many thermosets become **elastomers,** polymers that can be stretched and immediately spring back to their initial shapes when released, like a trampoline or a rubber band. When you stretch a rubber band, individual polymer chains flow for only a short distance before the connectivity of the network returns them to their original positions. Table 12.9 is a list of some elastomers.

Table 12.9 Some Common Elastomers

Name	T_g (°C)	Uses
Poly(dimethyl siloxane)	−123	Breast implants
Polybutadiene	−106	Rubber bands
Polyisoprene	−65	Surgical gloves
Polychloroprene (neoprene)	−43	Footwear; medical tubing

Impact of Monomer Sequence Differences in monomer sequences influence polymer properties as well. A *homopolymer* consists of one type of monomer (A—A—A—A—A—···), whereas a **copolymer** consists of two or more types. The simplest copolymer is called an *AB block copolymer* because a chain (block) of monomer A and a chain of monomer B are linked at one point:

··· —A—A—A—A—A—B—B—B—B—B— ···

If the intermolecular forces between the A and B portions of the chain are weaker than those between different regions within each portion, the A and B portions form their own random coils. This ability makes AB block copolymers ideal *adhesives* for joining two polymer surfaces covalently. An ABA block copolymer has A chains linked at each end of a B chain:

··· —A—A—A—B— $(B)_n$—B—A—A—A— ···

Some of these block copolymers act as *thermoplastic elastomers,* materials shaped at high temperature that become elastomers at room temperature; many are used in the modern footwear industry.

Silicone polymers are described in Chapter 14, and organic reactions that form polymer chains from monomers are examined in Chapter 15.

Nanotechnology: Designing Materials Atom by Atom

Nanotechnology is the science and engineering of nanoscale systems, whose sizes range from 1 to 100 nm. (Recall that 1 nm is 10^{-9} m—about one-billionth the width of your dorm-room desk or one-millionth the width of your pen tip. A nanometer is to a meter as a marble is to Earth!) Nanotechnology joins scientists from physics, materials science, chemistry, biology, environmental science, medicine, and many branches of engineering.

Nanoscale materials behave neither like atoms, which are smaller (about 1×10^1 nm), nor like crystals, which are larger (at least 1×10^5 nm). The different behavior is due to a surface area that is similar to their interior volume—for example, a 5-nm particle has about half of its atoms on its surface. Cut a piece of ordinary aluminum foil into pieces just large enough to see with a light microscope and they still behave like the original piece. But, if you could cut the original piece into 20–30-nm pieces, they would explode. Indeed, aluminum nanoparticles are used in fireworks and propellants.

As so often happens in science, a technological advance allowed this new field to blossom. Nanotechnologists can use the scanning tunneling microscope (see Tools of the Laboratory in Section 12.6) and the similar scanning probe and atomic force microscopes to examine the chemical and physical properties of nanostructures and build them up, in many cases, atom by atom.

Two features of nanoscale engineering occur routinely in nature:

1. *Self-assembly* is the ability of smaller, simpler parts to organize themselves into a larger, more complex whole. On the molecular scale, self-assembly refers to atoms or small molecules aggregating through intermolecular forces. Oppositely charged regions on two such particles make contact to form a larger particle, which in turn aggregates with another to form a still larger one.
2. *Controlled orientation* is the positioning of two molecules near each other long enough for intermolecular forces to result in a change, often a bond breaking or forming. All enzymes function this way.

Because this field is changing so fast, we can only provide a glimpse of a few exciting research directions in nanotechnology.

Nanoscale Optical Materials *Quantum dots* are nanoparticles of a semiconducting material, such as gallium arsenide (GaAs) or gallium selenide (GaSe), that are smaller than 10 nm. Rather than having a band of energy levels, the dot has discrete energy levels, much like a single atom. As in atoms, the energy levels can be studied with spectroscopy; unlike atoms, quantum dots can be modified chemically. When high-energy (short-wavelength) UV light irradiates a suspension of quantum dots, they become excited and emit radiation of lower energy (longer wavelength). Most importantly, the emitted wavelengths depend on the size of the dots: the larger the particles, the longer the wavelength emitted. Thus, the color of the emitted light can be "tuned" by varying the dimensions of the nanoparticles (Figure 12.49). One use for quantum dots is the imaging of specific cells. The dot is encapsulated in a coating and then bonded to a particular protein used by a target organ. The cells in the target incorporate the protein, along with its attached quantum dot, and the location of the protein is viewed spectroscopically.

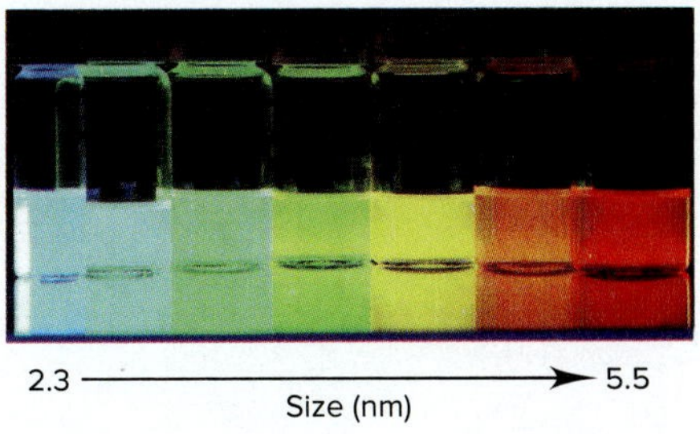

Figure 12.49 **The colors of quantum dots.** The smaller the dots, the shorter the emitted wavelengths.
Source: Courtesy of Benoit Dubertret

Nanostructured Materials *Nanostructuring,* the construction of bulk materials from nanoscale building blocks, increases strength, ductility, plasticity, and many other properties. The resulting *nanocomposites* behave as liquid magnets, ductile cements, conducting elastomers, and many other unique materials.

One example mimics a familiar natural material. Bones are natural nanocomposites of hydroxyapatite (a type of calcium phosphate) and other minerals. Synthetic nanocrystalline hydroxyapatite has weight-bearing properties identical to bone and can be formed into a framework into which natural bone tissue can grow to heal a fracture. Another material, called a *ferrofluid,* consists of magnetic nanoparticles (usually

Figure 12.50 The magnetic behavior of a ferrofluid. Nanoparticles of magnetite (Fe_3O_4) dispersed in a viscous fluid are suspended between the poles of a magnet.

Source: © & by permission from Ferrotec (USA) Corporation. May not be reproduced without written permission.

Figure 12.51 Driving a nanocar.

Source: T. Sasaki/Rice University (Image courtesy of Rice University Office of Media Relations and Information)

magnetite, Fe_3O_4) dispersed in a viscous fluid. When placed in a magnetic field, the particles align to conform to the shape of the field (Figure 12.50). Ferrofluids find uses in automobile shock absorbers, magnetic plastics, audio speakers, computer hard drives, high-vacuum valves, and many other devices.

High-Surface-Area Materials Carbon nanotubes have remarkable mechanical and electrical properties. But, a possible future application for them is in storing H_2 gas for hydrogen-powered vehicles. Currently, H_2 is stored in metal cylinders at high pressures (>70 atm). Porous solids, such as carbon nanotubes, can store the same amount of gas at much lower pressures because they have surface areas of 4500 m^2/g—about four football fields per gram of material! Thus, a medium-sized container of nanotubes could store a large amount of H_2 fuel under conditions safe enough for a family car.

Other high-surface-area materials being developed are porous membranes for water purification or batteries and multilayer films that incorporate photosynthetic molecules for high-efficiency solar cells. A very exciting development is a molecule-specific biosensor, which can detect as little as 10^{-14} mol of DNA using color changes in gold nanoparticles. Such biosensors could circulate freely in the bloodstream to measure levels of specific disease-related molecules, deliver drugs to individual cells, and even alter individual genes.

Nanomachines In nature, a virus is a marvel of nanoscale engineering, designed for delivering genetic material to infect a host cell. For example, the T4 virus latches onto a bacterial cell with its tail fibers, bores through the bacterial membrane with its end plate, and then injects the payload of DNA, packaged in its head, into the bacterium. These complex functions are performed by a biological "machine" measuring 60 by 200 nm. Researchers are exploring ways for viruses to deliver medicinal agents to specific cells. In other feats of nanoscale engineering, researchers have created nanovalves, nanopropellers, and even a nanocar, complete with a chassis, axles, and buckyball wheels, that is only 4 nm wide and is "driven" on a gold surface under the direction of an atomic force microscope (Figure 12.51).

› Summary of Section 12.7

- Doping increases the conductivity of semiconductors and is essential to modern electronic materials. Doping silicon with Group 5A(15) atoms introduces negative sites (creating an n-type semiconductor) by adding valence electrons to the conduction band, whereas doping with Group 3A(13) atoms adds positive holes (for a p-type semiconductor) by emptying some orbitals in the valence band. Placing these two types of semiconductors in contact with one another forms a p-n junction. Sandwiching a p-type portion between two n-type portions forms a transistor.
- Liquid crystal phases flow like liquids but have molecules ordered like crystalline solids. Typically, the molecules have rodlike shapes, and their intermolecular forces keep them aligned. Thermotropic phases are prepared by heating the solid; lyotropic phases form when the solvent concentration is varied. The nematic, cholesteric, and smectic phases of liquid crystals differ in their molecular order. Liquid crystal applications depend on controlling the orientation of the molecules.
- Ceramics are very resistant to heat and chemicals. Most are network covalent solids formed at high temperature from simple reactants. They add lightweight strength to other materials.
- Polymers are extremely large molecules that adopt the shape of a random coil, as a result of intermolecular forces. A polymer sample has an average molar mass because it consists of chains with a range of lengths. The high viscosity of a polymer arises from attractions between chains or, in the case of a dissolved polymer, between chains and solvent. By varying the degrees of branching, crosslinking, and ordering (crystallinity), chemists tailor polymers with specific properties.
- Nanoscale materials can be made through construction processes involving self-assembly and controlled orientation of molecules.

CHAPTER REVIEW GUIDE

Learning Objectives

Relevant section (§) and/or sample problem (SP) numbers appear in parentheses.

Understand These Concepts

1. How the interplay between kinetic and potential energy underlies the properties of the three states of matter and their phase changes (§12.1)
2. The processes involved, both within a phase and during a phase change, when heat is added or removed from a pure substance (§12.2)
3. The meaning of vapor pressure and how phase changes are dynamic equilibrium processes (§12.2)
4. How temperature and intermolecular forces influence vapor pressure (§12.2)
5. The relation between vapor pressure and boiling point (§12.2)
6. How a phase diagram shows the phases of a substance at differing conditions of pressure and temperature (§12.2)
7. The distinction between bonding and intermolecular forces on the basis of Coulomb's law and the meaning of the van der Waals radius of an atom (§12.3)
8. The types and relative strengths of intermolecular forces acting in a substance (dipole-dipole, H bonding, dispersion), the impact of H bonding on physical properties, and the meaning of polarizability (§12.3)
9. The meanings of surface tension, capillarity, and viscosity and how intermolecular forces influence their magnitudes (§12.4)
10. How the important macroscopic properties of water arise from atomic and molecular properties (§12.5)
11. The structure of a crystal lattice and the characteristics of the three types of cubic unit cells (§12.6)
12. How packing of spheres gives rise to the hexagonal and cubic unit cells (§12.6)
13. Types of crystalline solids and how their intermolecular forces give rise to their properties (§12.6)
14. How band theory accounts for the properties of metals and the relative conductivities of metals, nonmetals, and metalloids (§12.6)
15. The structures, properties, and functions of modern materials (doped semiconductors, liquid crystals, ceramics, polymers, and nanostructures) (§12.7)

Master These Skills

1. Calculating the overall enthalpy change when heat is gained or lost by a pure substance (§12.2 and SP 12.1)
2. Using the Clausius-Clapeyron equation to examine the relationship between vapor pressure and temperature (SP 12.2)
3. Using a phase diagram to predict the phase changes of a substance (§12.2 and SP 12.3)
4. Determining whether a substance can form H bonds and drawing the H-bonded structures (SP 12.4)
5. Predicting the types and relative strength of the bonding and intermolecular forces acting within a substance from its structure (SP 12.5)
6. Finding the number of particles per unit cell and the coordination number of a crystalline solid (§12.6 and SP 12.6)
7. Calculating atomic radius from the density and/or crystal structure of an element (SPs 12.7, 12.8)

Key Terms

Page numbers appear in parentheses.

amorphous solid (495)
atomic solid (504)
band theory (506)
body-centered cubic unit cell (496)
boiling point (481)
branch (519)
capillarity (491)
ceramic (514)
Clausius-Clapeyron equation (480)
condensation (473)
conduction band (507)
conductor (508)
coordination number (496)
copolymer (520)
critical point (482)
crosslink (520)
crystal defect (509)
crystalline solid (495)
cubic closest packing (500)
degree of polymerization (n) (516)
deposition (473)
dipole-dipole force (485)
dispersion (London) force (488)
doping (510)
elastomer (520)
face-centered cubic unit cell (496)
freezing (473)
fusion (473)
heat of fusion (ΔH°_{fus}) (473)
heat of sublimation (ΔH°_{subl}) (474)
heat of vaporization (ΔH°_{vap}) (473)
heating-cooling curve (475)
hexagonal closest packing (500)
hydrogen bond (H bond) (486)
insulator (508)
intermolecular forces (471)
ion-dipole force (485)
ionic solid (504)
lattice (496)
liquid crystal (511)
London force (488)
macromolecule (516)
melting point (481)
melting (fusion) (473)
metallic solid (505)
molecular solid (504)
monomer (516)
nanotechnology (521)
network covalent solid (505)
packing efficiency (498)
phase (471)
phase change (472)
phase diagram (482)
plastic (519)
polarizability (488)
polymer (516)
radius of gyration (R_g) (517)
random coil (517)
scanning tunneling microscopy (502)
semiconductor (508)
simple cubic unit cell (496)
sublimation (473)
superconductivity (508)
surface tension (491)
triple point (482)
unit cell (496)
valence band (507)
van der Waals radius (484)
vapor pressure (479)
vaporization (473)
viscosity (492)
x-ray diffraction analysis (502)

Key Equations and Relationships

Page number appears in parentheses.

12.1 Using the vapor pressure at one temperature to find the vapor pressure at another temperature (two-point form of the Clausius-Clapeyron equation) (480):

$$\ln \frac{P_2}{P_1} = \frac{-\Delta H_{vap}}{R}\left(\frac{1}{T_2} - \frac{1}{T_1}\right)$$

BRIEF SOLUTIONS TO FOLLOW-UP PROBLEMS

12.1A The scenes represent solid water at −7.00°C melting to liquid water at 16.0°C, so heat is absorbed by the system (q is positive). There are three stages:

Stage 1, increasing the temperature of ice to the melting point:
$H_2O(s)$ [−7.00°C] ⟶ $H_2O(s)$ [0.00°C]:
$q = n \times C_{m(\text{water}, s)} \times \Delta T = (2.25\ \text{mol})(37.6\ \text{J/mol·°C})[0°\text{C} - (-7.00°\text{C})]$
$= 592\ \text{J} = 0.592\ \text{kJ}$

Stage 2, melting the ice: $H_2O(s)$ [0.00°C] ⟶ $H_2O(l)$ [0.00°C]:
$q = n(\Delta H^\circ_{\text{fus}}) = (2.25\ \text{mol})(6.02\ \text{kJ/mol}) = 13.5\ \text{kJ}$

Stage 3, increasing the temperature of the water:
$H_2O(l)$ [0.00°C] ⟶ $H_2O(l)$ [16.0°C]:
$q = n \times C_{m(\text{water}, l)} \times \Delta T = (2.25\ \text{mol})(75.4\ \text{J/mol·°C})(16.0°\text{C} - 0°\text{C})$
$= 2714\ \text{J} = 2.71\ \text{kJ}$
Total heat (kJ) = 0.592 kJ + 13.5 kJ + 2.71 kJ = 16.8 kJ

12.1B The scenes represent gaseous bromine at 73.5°C condensing to liquid bromine at 23.8°C, so heat is released by the system (q is negative). First convert mass to moles:

$$\text{Amount (mol) } Br_2 = 47.94\ \text{g } Br_2 \times \frac{1\ \text{mol } Br_2}{159.8\ \text{g } Br_2} = 0.3000\ \text{mol } Br_2$$

There are three stages:
Stage 1, cooling the gas: $Br_2(g)$ [73.5°C] ⟶ $Br_2(g)$ [59.5°C]:
$q = n \times C_{m(Br_2, g)} \times \Delta T$
$= (0.3000\ \text{mol})(36.0\ \text{J/mol·°C})(59.5°\text{C} - 73.5°\text{C})$
$= -151\ \text{J} = -0.151\ \text{kJ}$
Stage 2, condensing the gas to liquid at the boiling point:
$Br_2(g)$ [59.5°C] ⟶ $Br_2(l)$ [59.5°C]:
$q = n(-\Delta H^\circ_{\text{vap}}) = (0.3000\ \text{mol})(-29.6\ \text{kJ/mol}) = -8.88\ \text{kJ}$
Stage 3, cooling the liquid: $Br_2(l)$ [59.5°C] ⟶ $Br_2(l)$ [23.8°C] :
$q = n \times C_{m(Br_2, l)} \times \Delta T = (0.3000\ \text{mol})(75.7\ \text{J/mol·°C})(23.8°\text{C} - 59.5°\text{C})$
$= -811\ \text{J} = -0.811\ \text{kJ}$
Total heat (kJ) = −0.151 kJ + (−8.88 kJ) + (−0.811 kJ)
= −9.84 kJ

12.2A $\ln \dfrac{P_2}{P_1} = \dfrac{-\Delta H_{\text{vap}}}{R}\left(\dfrac{1}{T_2} - \dfrac{1}{T_1}\right)$

$$\ln \frac{P_2}{P_1} = \left(\frac{-40.7\times10^3\ \text{J/mol}}{8.314\ \text{J/mol·K}}\right) \times \left(\frac{1}{273.15 + 85.5\ \text{K}} - \frac{1}{273.15 + 34.1\ \text{K}}\right)$$

$$\ln \frac{P_2}{P_1} = (-4.90\times10^3\ \text{K})(-4.66\times10^{-4}\ \text{K}^{-1}) = 2.28$$

$$\frac{P_2}{P_1} = 9.8$$

$$\frac{P_2}{40.1\ \text{torr}} = 9.8; \qquad P_2 = 40.1\ \text{torr} \times 9.8 = 390\ \text{torr}$$

12.2B $\ln \dfrac{P_2}{P_1} = \dfrac{-\Delta H_{\text{vap}}}{R}\left(\dfrac{1}{T_2} - \dfrac{1}{T_1}\right)$

$$\ln \frac{24.50\ \text{kPa}}{10.00\ \text{kPa}} = \frac{-\Delta H_{\text{vap}}}{8.314\ \text{J/K·mol}} \times \left(\frac{1}{273.15 + 20.2\ \text{K}} - \frac{1}{273.15 + 0.95\ \text{K}}\right)$$

$$0.8961 = \frac{-\Delta H_{\text{vap}}}{8.314\ \text{J/K·mol}} \times (-0.0002394\ \text{K}^{-1})$$

$$\Delta H_{\text{vap}} = 3.11\times10^4\ \text{J/mol} = 31.1\ \text{kJ/mol}$$

12.3A At the starting conditions (3×10^2 bar and 2000 K; point c), the carbon sample is in the form of graphite. As the sample is heated at constant pressure, it melts at around 4500 K, and then vaporizes at around 5000 K.

12.3B At the starting conditions (4600 K and 10^4 bar; point d), the carbon sample is in the form of graphite. As the sample is compressed at constant temperature, it melts at around 4×10^4 bar and then solidifies to diamond at around 2×10^5 bar.

12.4A

(a)
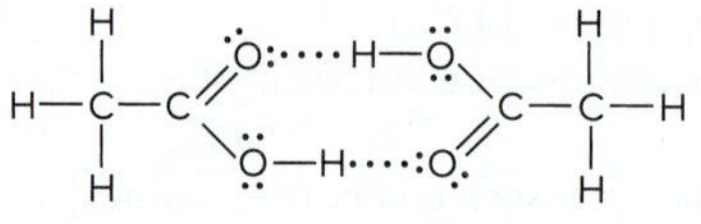

(b)
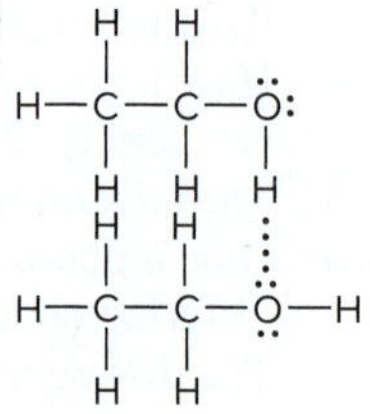

(c) No H bonding (H is not bonded to O in this molecule.)

12.4B (a) No H bonding (H is not bonded to O in this molecule.)

(b)
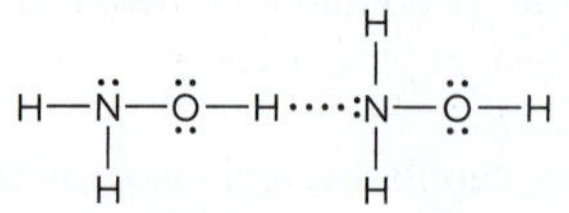

(c)
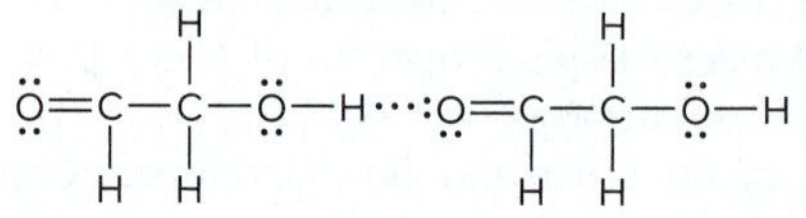

12.5A (a) Both molecules are polar (tetrahedral molecular geometry). Dipole-dipole and dispersion forces are present for both; CH_3Br has the higher boiling point because it has the larger molar mass.
(b) H bonding in $CH_3CH_2CH_2OH$, dipole-dipole and dispersion in both; $CH_3CH_2CH_2OH$ has the higher boiling point due to the H bonding.
(c) Both are nonpolar molecules. Dispersion forces; C_3H_8 has the higher boiling point because it has the larger molar mass.

12.5B (a) Dispersion and dipole-dipole forces in both, H bonding in CH_3CH_2OH; CH_3CHO has the lower boiling point since it does not have H bonding.
(b) SO_2 molecules, with a bent molecular geometry, are polar and have dispersion and dipole-dipole forces; CO_2, which is a linear molecule, has only dispersion forces. With the weaker forces, CO_2 has the lower boiling point.
(c) Dispersion and dipole-dipole forces in both molecules; H bonding in $H_2N—C(=O)CH_2CH_3$; $(CH_3)_2N—C(=O)H$ has the lower boiling point since it does not have H bonding.

12.6A (a) PbS: Pb^{2+} ions/unit cell = S^2 ions/unit cell = $(8 \times \frac{1}{8}) + (6 \times \frac{1}{2}) = 4$ Pb^{2+} and 4 S^{2-} ions; coordination number of each ion is 6.
(b) W: Atoms/unit cell = $(8 \times \frac{1}{8}) + 1 = 2$; coordination number is 8.
(c) Al: Atoms/unit cell = $(8 \times \frac{1}{8}) + (6 \times \frac{1}{2}) = 4$; coordination number is 12.

12.6B (a) K: Atoms/unit cell = $(8 \times \frac{1}{8}) + 1 = 2$; coordination number is 8.
(b) Pt: Atoms/unit cell = $(8 \times \frac{1}{8}) + (6 \times \frac{1}{2}) = 4$; coordination number is 12.
(c) CsCl: Cs^+ ions/unit cell = $(8 \times \frac{1}{8}) = 1$; Cl^- ions/unit cell = 1; coordination number of each ion is 8.

12.7A The packing efficiency in the face-centered cubic cell is 74%.

$$V \text{ of Co atom} = \frac{4}{3}\pi r^3 = \frac{4}{3}(3.14)(125\text{ pm})^3 = \frac{8.18\times10^6\text{ pm}^3}{\text{atom}}$$

$$V \text{ of 1 mol of Co atoms (cm}^3) = \frac{8.18\times10^6\text{ pm}^3}{\text{atom}} \times \frac{6.022\times10^{23}\text{ atoms}}{\text{mol atoms}} \times \frac{1\text{ cm}^3}{10^{30}\text{ pm}^3} = 4.93\text{ cm}^3/\text{mol atoms}$$

$$V \text{ of 1 mol of Co metal (cm}^3) = \frac{4.93\text{ cm}^3/\text{mol atoms}}{0.74} = 6.66\text{ cm}^3/\text{mol Co metal}$$

$$d \text{ of Co metal} = \frac{58.93\text{ g}}{\text{mol}} \times \frac{1\text{ mol}}{6.66\text{ cm}^3} = 8.85\text{ g/cm}^3$$

12.7B The packing efficiency in the body-centered cubic cell is 68%.

$$\text{Avogadro's no.} = \frac{1\text{ cm}^3}{7.874\text{ g Fe}} \times \frac{55.85\text{ g Fe}}{1\text{ mol Fe}} \times 0.68 \times \frac{1\text{ Fe atom}}{8.38\times10^{-24}\text{ cm}^3} = 5.8\times10^{23}\text{ Fe atoms/mol Fe}$$

12.8A From Figure 12.29,

$$A\text{ (nm)} = \frac{4r}{\sqrt{3}} = \frac{4(126\text{ pm})}{\sqrt{3}} = 291\text{ pm} \times \frac{1\text{ nm}}{1000\text{ pm}} = 0.291\text{ nm}$$

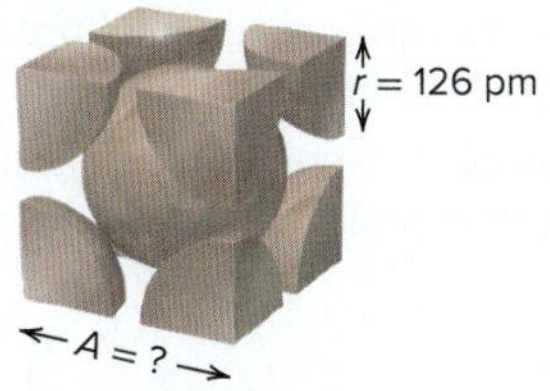

12.8B $A = \sqrt{8}r$
$r = A/\sqrt{8} = 0.405\text{ nm}/\sqrt{8} = 0.143\text{ nm} = 143\text{ pm}$

PROBLEMS

Problems with **colored** numbers are answered in Appendix E and worked in detail in the Student Solutions Manual. Problem sections match those in the text and give the numbers of relevant sample problems. Most offer Concept Review Questions, Skill-Building Exercises (grouped in pairs covering the same concept), and Problems in Context. The Comprehensive Problems are based on material from any section or previous chapter.

An Overview of Physical States and Phase Changes

Concept Review Questions

12.1 How does the energy of attraction between particles compare with their energy of motion in a gas and in a solid? As part of your answer, identify two macroscopic properties that differ between a gas and a solid.

12.2 (a) Why are gases more easily compressed than liquids?
(b) Why do liquids have a greater ability to flow than solids?

12.3 What type of forces, intramolecular or intermolecular:
(a) Prevent ice cubes from adopting the shape of their container?
(b) Are overcome when ice melts?
(c) Are overcome when liquid water is vaporized?
(d) Are overcome when gaseous water is converted to hydrogen gas and oxygen gas?

12.4 (a) Why is the heat of fusion (ΔH_{fus}) of a substance smaller than its heat of vaporization (ΔH_{vap})?
(b) Why is the heat of sublimation (ΔH_{subl}) of a substance greater than its ΔH_{vap}?
(c) At a given temperature and pressure, how does the magnitude of the heat of vaporization of a substance compare with that of its heat of condensation?

Skill-Building Exercises (grouped in similar pairs)

12.5 Which forces are intramolecular and which intermolecular?
(a) Those preventing oil from evaporating at room temperature
(b) Those preventing butter from melting in a refrigerator
(c) Those allowing silver to tarnish
(d) Those preventing O_2 in air from forming O atoms

12.6 Which forces are intramolecular and which intermolecular?
(a) Those allowing fog to form on a cool, humid evening
(b) Those allowing water to form when H_2 is sparked
(c) Those allowing liquid benzene to crystallize when cooled
(d) Those responsible for the low boiling point of hexane

12.7 Name the phase change in each of these events: (a) Dew appears on a lawn in the morning. (b) Icicles change into liquid water. (c) Wet clothes dry on a summer day.

12.8 Name the phase change in each of these events: (a) A diamond film forms on a surface from gaseous carbon atoms in a vacuum. (b) Mothballs in a bureau drawer disappear over time. (c) Molten iron from a blast furnace is cast into ingots ("pigs").

Problems in Context

12.9 Liquid propane, a widely used fuel, is produced by compressing gaseous propane. During the process, approximately 15 kJ of energy is released for each mole of gas liquefied. Where does this energy come from?

12.10 Many heat-sensitive and oxygen-sensitive solids, such as camphor, are purified by warming under vacuum. The solid vaporizes directly, and the vapor crystallizes on a cool surface. What phase changes are involved in this method?

Quantitative Aspects of Phase Changes

(Sample Problems 12.1–12.3)

Concept Review Questions

12.11 Describe the changes (if any) in potential energy and in kinetic energy among the molecules when gaseous PCl_3 condenses to a liquid at a fixed temperature.

12.12 When benzene is at its melting point, two processes occur simultaneously and balance each other. Describe these processes on the macroscopic and molecular levels.

12.13 Liquid hexane (bp = 69°C) is placed in a closed container at room temperature. At first, the pressure of the vapor phase increases, but after a short time, it stops changing. Why?

12.14 Explain the effect of strong intermolecular forces on each of these parameters: (a) critical temperature; (b) boiling point; (c) vapor pressure; (d) heat of vaporization.

12.15 Match each numbered point in the phase diagram for compound Q with the correct molecular scene below:

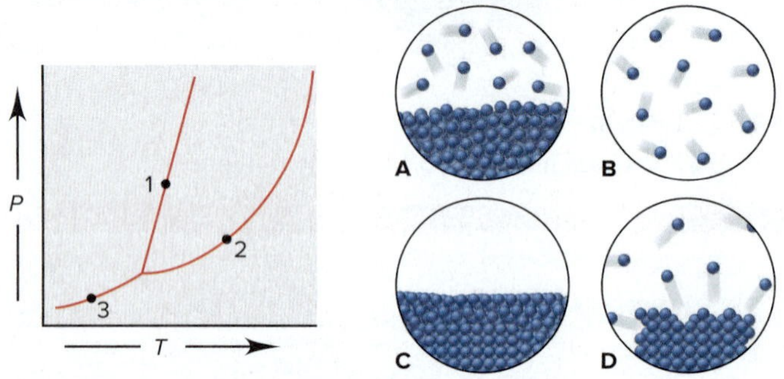

12.16 A liquid is in equilibrium with its vapor in a closed vessel at a fixed temperature. The vessel is connected by a stopcock to an evacuated vessel. When the stopcock is opened, will the final pressure of the vapor be different from the initial value if (a) some liquid remains; (b) all the liquid is first removed? Explain.

12.17 The phase diagram for substance A has a solid-liquid line with a positive slope, and that for substance B has a solid-liquid line with a negative slope. What macroscopic property can distinguish A from B?

12.18 Why does water vapor at 100°C cause a more severe burn than liquid water at 100°C?

Skill-Building Exercises (grouped in similar pairs)

12.19 From the data below, calculate the total heat (in J) needed to convert 22.00 g of ice at −6.00°C to liquid water at 0.500°C:

mp at 1 atm:	0.0°C	ΔH°_{fus}:	6.02 kJ/mol
c_{liquid}:	4.21 J/g·°C	c_{solid}:	2.09 J/g·°C

12.20 From the data below, calculate the total heat (in J) needed to convert 0.333 mol of gaseous ethanol at 300°C and 1 atm to liquid ethanol at 25.0°C and 1 atm:

bp at 1 atm:	78.5°C	ΔH°_{vap}:	40.5 kJ/mol
c_{gas}:	1.43 J/g·°C	c_{liquid}:	2.45 J/g·°C

12.21 A liquid has a ΔH°_{vap} of 35.5 kJ/mol and a boiling point of 122°C at 1.00 atm. What is its vapor pressure at 113°C?

12.22 Diethyl ether has a ΔH°_{vap} of 29.1 kJ/mol and a vapor pressure of 0.703 atm at 25.0°C. What is its vapor pressure at 95.0°C?

12.23 What is the ΔH°_{vap} of a liquid that has a vapor pressure of 621 torr at 85.2°C and a boiling point of 95.6°C at 1 atm?

12.24 Methane (CH_4) has a boiling point of −164°C at 1 atm and a vapor pressure of 42.8 atm at −100°C. What is the heat of vaporization of CH_4?

12.25 Use these data to draw a qualitative phase diagram for ethylene (C_2H_4). Is $C_2H_4(s)$ more or less dense than $C_2H_4(l)$?

bp at 1 atm:	−103.7°C
mp at 1 atm:	−169.16°C
Critical point:	9.9°C and 50.5 atm
Triple point:	−169.17°C and 1.20×10^{-3} atm

12.26 Use these data to draw a qualitative phase diagram for H_2. Does H_2 sublime at 0.05 atm? Explain.

mp at 1 atm:	13.96 K
bp at 1 atm:	20.39 K
Triple point:	13.95 K and 0.07 atm
Critical point:	33.2 K and 13.0 atm
Vapor pressure of solid at 10 K:	0.001 atm

12.27 The phase diagram for sulfur is shown below.
(a) Give a set of conditions under which it is possible to sublime the rhombic form of solid sulfur.
(b) Describe the phase changes that a sample of sulfur undergoes at 1 atm when it is heated from 90°C to 450°C.

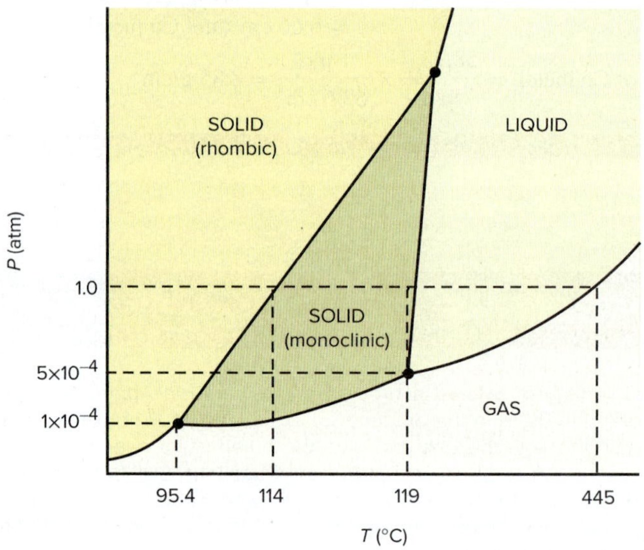

12.28 The phase diagram for xenon is shown below.
(a) What phase is xenon in at room temperature and pressure?
(b) Describe the phase changes that a sample of xenon undergoes at −115°C as it is compressed from 0.5 atm to 25 atm. (The critical pressure of xenon is 58 atm.)

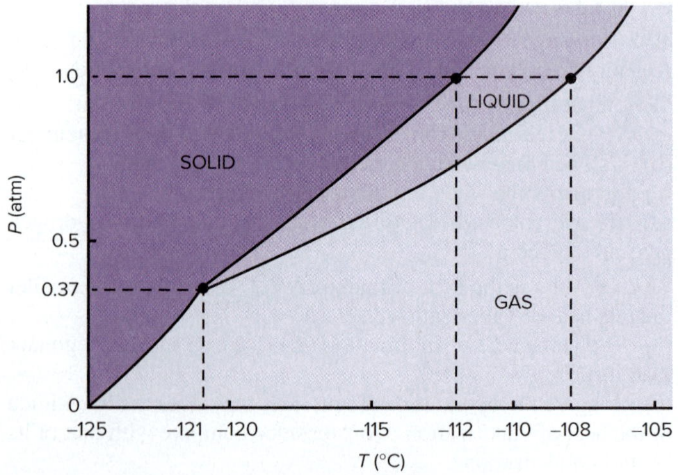

Problems in Context

12.29 Sulfur dioxide is produced in enormous amounts for sulfuric acid production. It melts at −73°C and boils at −10.°C. Its $\Delta H^\circ_{\text{fus}}$ is 8.619 kJ/mol, and its $\Delta H^\circ_{\text{vap}}$ is 25.73 kJ/mol. The specific heat capacities of the liquid and gas are 0.995 J/g·K and 0.622 J/g·K, respectively. How much heat is required to convert 2.500 kg of solid SO_2 at the melting point to a gas at 60.°C?

12.30 Butane is a common fuel used in cigarette lighters and camping stoves. Normally supplied in metal containers under pressure, the fuel exists as a mixture of liquid and gas, so high temperatures may cause the container to explode. At 25.0°C, the vapor pressure of butane is 2.3 atm. What is the pressure in the container at 135°C ($\Delta H^\circ_{\text{vap}}$ = 24.3 kJ/mol)?

12.31 Use Figure 12.10 to answer the following:
(a) Carbon dioxide is sold in steel cylinders under a pressure of approximately 20 atm. Is there liquid CO_2 in the cylinder at room temperature (~20°C)? At 40°C? At −40°C? At −120°C?
(b) Carbon dioxide is also sold as solid chunks, called *dry ice,* in insulated containers. If the chunks are warmed by leaving them in an open container at room temperature, will they melt?
(c) If a container is nearly filled with dry ice and then sealed and warmed to room temperature, will the dry ice melt?
(d) If dry ice is compressed at a temperature below its triple point, will it melt?

Types of Intermolecular Forces
(Sample Problems 12.4 and 12.5)

Concept Review Questions

12.32 Why are covalent bonds typically much stronger than intermolecular forces?

12.33 (a) Name the type of force depicted in each scene below.
(b) Rank the forces in order of increasing strength.

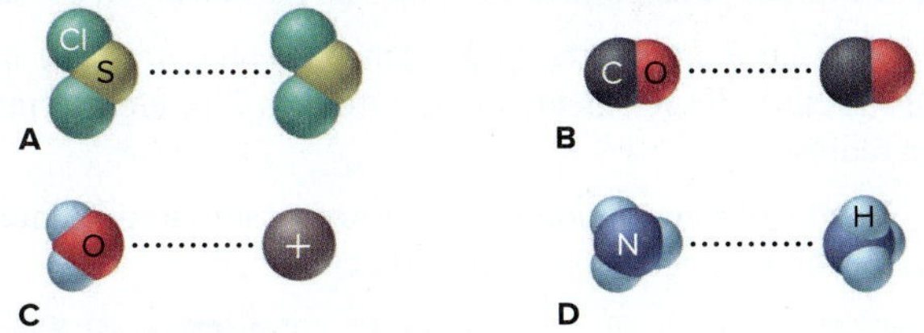

12.34 Oxygen and selenium are members of Group 6A(16). Water forms H bonds, but H_2Se does not. Explain.

12.35 In solid I_2, is the distance between the two I nuclei of one I_2 molecule longer or shorter than the distance between two I nuclei of adjacent I_2 molecules? Explain.

12.36 Polar molecules exhibit dipole-dipole forces. Do they also exhibit dispersion forces? Explain.

12.37 Distinguish between *polarizability* and *polarity*. How does each influence intermolecular forces?

12.38 How can one nonpolar molecule induce a dipole in a nearby nonpolar molecule?

Skill-Building Exercises (grouped in similar pairs)

12.39 What is the strongest interparticle force in each substance?
(a) CH_3OH (b) CCl_4 (c) Cl_2

12.40 What is the strongest interparticle force in each substance?
(a) H_3PO_4 (b) SO_2 (c) $MgCl_2$

12.41 What is the strongest interparticle force in each substance?
(a) CH_3Cl (b) CH_3CH_3 (c) NH_3

12.42 What is the strongest interparticle force in each substance?
(a) Kr (b) BrF (c) H_2SO_4

12.43 Which member of each pair of compounds forms intermolecular H bonds? Draw the H-bonded structures in each case:
(a) CH_3CHCH_3 (with OH on the central C) or CH_3SCH_3 (b) HF or HBr

12.44 Which member of each pair of compounds forms intermolecular H bonds? Draw the H-bonded structures in each case:
(a) $(CH_3)_2NH$ or $(CH_3)_3N$ (b) $HOCH_2CH_2OH$ or FCH_2CH_2F

12.45 Which forces oppose vaporization of each substance?
(a) Hexane (b) Water (c) $SiCl_4$

12.46 Which forces oppose vaporization of each substance?
(a) Br_2 (b) SbH_3 (c) CH_3NH_2

12.47 Which species in each pair has the greater polarizability? Explain.
(a) Br^- or I^- (b) $CH_2{=}CH_2$ or $CH_3{-}CH_3$ (c) H_2O or H_2Se

12.48 Which species in each pair has the greater polarizability? Explain.
(a) Ca^{2+} or Ca (b) CH_3CH_3 or $CH_3CH_2CH_3$ (c) CCl_4 or CF_4

12.49 Which liquid in each pair has the *higher* vapor pressure at a given temperature? Explain.
(a) C_2H_6 or C_4H_{10} (b) CH_3CH_2OH or CH_3CH_2F
(c) NH_3 or PH_3

12.50 Which liquid in each pair has the *lower* vapor pressure at a given temperature? Explain.
(a) $HOCH_2CH_2OH$ or $CH_3CH_2CH_2OH$
(b) CH_3COOH or $(CH_3)_2C{=}O$ (c) HF or HCl

12.51 Which substance has the *lower* boiling point? Explain.
(a) LiCl or HCl (b) NH_3 or PH_3 (c) Xe or I_2

12.52 Which substance has the *higher* boiling point? Explain.
(a) CH_3CH_2OH or $CH_3CH_2CH_3$ (b) NO or N_2
(c) H_2S or H_2Te

12.53 Which substance has the *lower* boiling point? Explain.
(a) $CH_3CH_2CH_2CH_3$ or

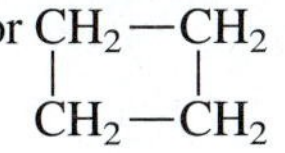

(b) NaBr or PBr_3
(c) H_2O or HBr

12.54 Which substance has the *higher* boiling point? Explain.
(a) CH_3OH or CH_3CH_3 (b) FNO or ClNO
(c)

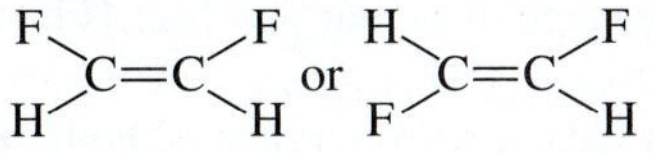

Problems in Context

12.55 For pairs of molecules in the gas phase, average H-bond dissociation energies are 17 kJ/mol for NH_3, 22 kJ/mol for H_2O, and 29 kJ/mol for HF. Explain this increase in H-bond strength.

12.56 Dispersion forces are the only intermolecular forces present in motor oil, yet it has a high boiling point. Explain.

12.57 Why does the antifreeze ingredient ethylene glycol ($HOCH_2CH_2OH$; $\mathcal{M}$ = 62.07g/mol) have a boiling point of 197.6°C, whereas propanol ($CH_3CH_2CH_2OH$; $\mathcal{M}$ = 60.09 g/mol), a compound with a similar molar mass, has a boiling point of only 97.4°C?

Properties of the Liquid State

Concept Review Questions

12.58 Before the phenomenon of surface tension was understood, physicists described the surface of water as being covered with a "skin." What causes this skinlike phenomenon?

12.59 Small, equal-sized drops of oil, water, and mercury lie on a waxed floor. How does each liquid behave? Explain.

12.60 Why does an aqueous solution of ethanol (CH_3CH_2OH) have a lower surface tension than water?

12.61 Why are units of energy per area (J/m^2) used for surface tension values?

12.62 Does the *strength* of the intermolecular forces in a liquid change as the liquid is heated? Explain. Why does a liquid's viscosity decrease with rising temperature?

Skill-Building Exercises (grouped in similar pairs)

12.63 Rank the following in order of *increasing* surface tension at a given temperature, and explain your ranking:
(a) $CH_3CH_2CH_2OH$ (b) $HOCH_2CH(OH)CH_2OH$
(c) $HOCH_2CH_2OH$

12.64 Rank the following in order of *decreasing* surface tension at a given temperature, and explain your ranking:
(a) CH_3OH (b) CH_3CH_3 (c) $H_2C{=}O$

12.65 Rank the compounds in Problem 12.63 in order of *decreasing* viscosity at a given temperature; explain your ranking.

12.66 Rank the compounds in Problem 12.64 in order of *increasing* viscosity at a given temperature; explain your ranking.

Problems in Context

12.67 Soil vapor extraction (SVE) is used to remove volatile organic pollutants, such as chlorinated solvents, from soil at hazardous waste sites. Vent wells are drilled, and a vacuum pump is applied to the subsurface. (a) How does this method remove pollutants? (b) Why does heating combined with SVE speed the process?

12.68 Use Figure 12.2, to answer the following: (a) Does it take more heat to melt 12.0 g of CH_4 or 12.0 g of Hg? (b) Does it take more heat to vaporize 12.0 g of CH_4 or 12.0 g of Hg? (c) What is the principal intermolecular force in each sample?

12.69 Pentanol ($C_5H_{11}OH$; $\mathcal{M} = 88.15$ g/mol) has nearly the same molar mass as hexane (C_6H_{14}; $\mathcal{M} = 86.17$ g/mol) but is more than 12 times as viscous at 20°C. Explain.

The Uniqueness of Water

Concept Review Questions

12.70 For what types of substances is water a good solvent? For what types is it a poor solvent? Explain.

12.71 A water molecule can engage in as many as four H bonds. Explain.

12.72 Warm-blooded animals have a narrow range of body temperature because their bodies have a high water content. Explain.

12.73 What property of water keeps plant debris on the surface of lakes and ponds? What is the ecological significance of this?

12.74 A drooping plant can be made to stand upright by watering the ground around it. Explain.

12.75 Describe the molecular basis of the property of water responsible for the presence of ice on the surface of a frozen lake.

12.76 Describe in molecular terms what occurs when ice melts.

The Solid State: Structure, Properties, and Bonding

(Sample Problems 12.6–12.8)

Concept Review Questions

12.77 What is the difference between an amorphous solid and a crystalline solid on the macroscopic and molecular levels? Give an example of each.

12.78 How are a solid's unit cell and crystal structure related?

12.79 For structures consisting of identical atoms, how many atoms are contained in a simple cubic, a body-centered cubic, and a face-centered cubic unit cell? Explain how you obtained the values.

12.80 An element has a crystal structure in which the width of the cubic unit cell equals the diameter of an atom. What type of unit cell does it have?

12.81 What specific difference in the positioning of spheres gives a crystal structure based on the face-centered cubic unit cell less empty space than one based on the body-centered cubic unit cell?

12.82 Both solid Kr and solid Cu consist of individual atoms. Why do their physical properties differ so much?

12.83 What is the energy gap in band theory? Compare its size in superconductors, conductors, semiconductors, and insulators.

12.84 Predict the effect (if any) of an increase in temperature on the electrical conductivity of (a) a conductor; (b) a semiconductor; (c) an insulator.

12.85 Besides the type of unit cell, what information is needed to find the density of a solid consisting of identical atoms?

Skill-Building Exercises (grouped in similar pairs)

12.86 What type of unit cell does each metal use in its crystal lattice? (The number of atoms per unit cell is given in parentheses.)
(a) Ni (4) (b) Cr (2) (c) Ca (4)

12.87 What is the number of atoms per unit cell for each metal?
(a) Polonium, Po (b) Manganese, Mn (c) Silver, Ag

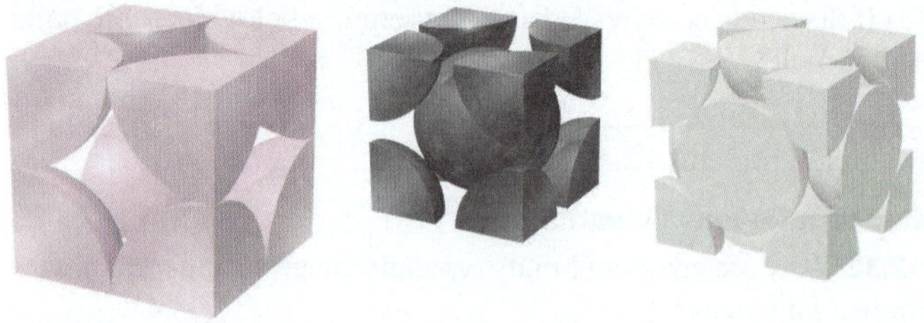

12.88 Calcium crystallizes in a cubic closest packed structure. If the atomic radius of calcium is 197 pm, find the density of the solid.

12.89 Chromium adopts the body-centered cubic unit cell in its crystal structure. If the density of chromium is 7.14 g/cm^3, find its atomic radius.

12.90 When cadmium oxide reacts to form cadmium selenide, a change in unit cell occurs, as depicted below:

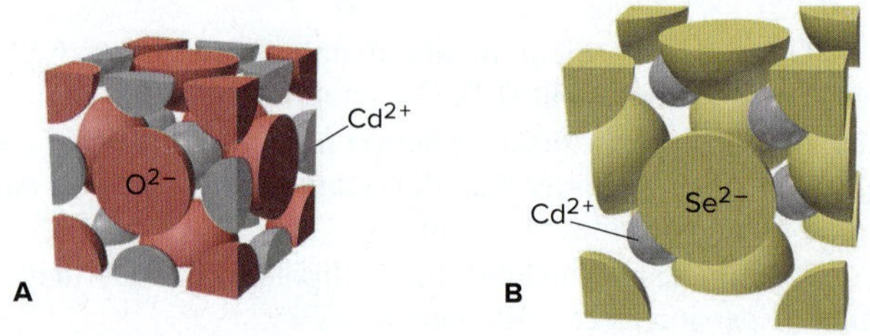

(a) What is the change in unit cell?
(b) Does the coordination number of cadmium change? Explain.

12.91 As molten iron cools to 1674 K, it adopts one type of cubic unit cell; then, as the temperature drops below 1181 K, it changes to another, as depicted below:

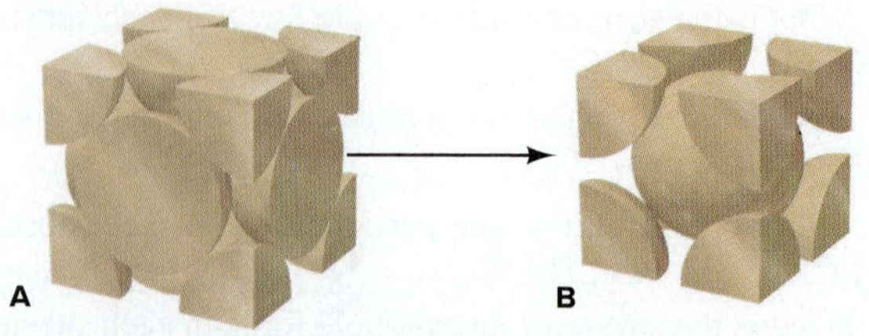

(a) What is the change in unit cell?
(b) Which crystal structure has the greater packing efficiency?

12.92 Potassium adopts the body-centered cubic unit cell in its crystal structure. If the atomic radius of potassium is 227 pm, find the edge length of the unit cell.

12.93 Lead adopts the face-centered cubic unit cell in its crystal structure. If the edge length of the unit cell is 495 pm, find the atomic radius of lead.

12.94 Of the five major types of crystalline solid, which does each of the following form, and why: (a) Ni; (b) F_2; (c) CH_3OH; (d) Sn; (e) Si; (f) Xe?

12.95 Of the five major types of crystalline solid, which does each of the following form, and why: (a) SiC; (b) Na_2SO_4; (c) SF_6; (d) cholesterol ($C_{27}H_{45}OH$); (e) KCl; (f) BN?

12.96 Zinc oxide adopts the zinc blende crystal structure (Figure P12.96). How many Zn^{2+} ions are in the ZnO unit cell?

12.97 Calcium sulfide adopts the sodium chloride crystal structure (Figure P12.97). How many S^{2-} ions are in the CaS unit cell?

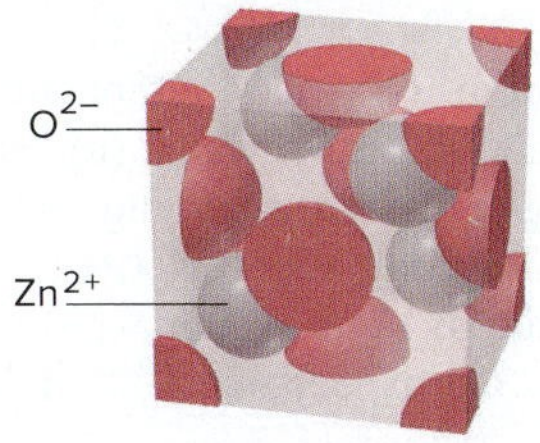

Figure P12.96

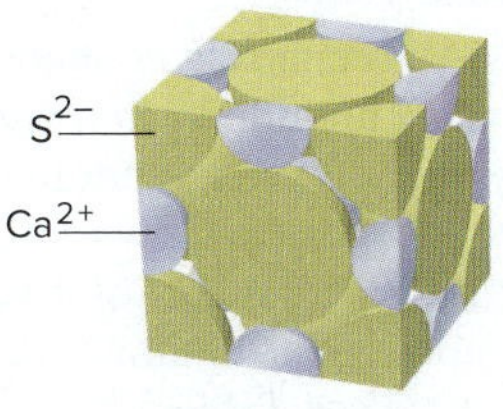

Figure P12.97

12.98 Zinc selenide (ZnSe) crystallizes in the zinc blende structure (see Figure P12.96) and has a density of 5.42 g/cm^3.
(a) How many Zn and Se ions are in each unit cell?
(b) What is the mass of a unit cell?
(c) What is the volume of a unit cell?
(d) What is the edge length of a unit cell?

12.99 An element crystallizes in a face-centered cubic lattice, and it has a density of 1.45 g/cm^3. The edge of its unit cell is 4.52×10^{-8} cm.
(a) How many atoms are in each unit cell?
(b) What is the volume of a unit cell?
(c) What is the mass of a unit cell?
(d) Calculate an approximate atomic mass for the element.

12.100 Classify each of the following as a conductor, insulator, or semiconductor: (a) phosphorus; (b) mercury; (c) germanium.

12.101 Classify each of the following as a conductor, insulator, or semiconductor: (a) carbon (graphite); (b) sulfur; (c) platinum.

12.102 Predict the effect (if any) of an increase in temperature on the electrical conductivity of (a) antimony; (b) tellurium; (c) bismuth.

12.103 Predict the effect (if any) of a decrease in temperature on the electrical conductivity of (a) silicon; (b) lead; (c) germanium.

Problems in Context

12.104 Polonium, the Period 6 member of Group 6A(16), is a rare radioactive metal that is the only element with a crystal structure based on the simple cubic unit cell. If its density is 9.142 g/cm^3, calculate an approximate atomic radius for polonium.

12.105 The coinage metals—copper, silver, and gold—crystallize in a cubic closest packed structure. Use the density of copper (8.95 g/cm^3) and its molar mass (63.55 g/mol) to calculate an approximate atomic radius for copper.

12.106 One of the most important enzymes in the world—nitrogenase, the plant protein that catalyzes nitrogen fixation—contains active clusters of iron, sulfur, and molybdenum atoms. Crystalline molybdenum (Mo) has a body-centered cubic unit cell (d of Mo = 10.28 g/cm^3). (a) Determine the edge length of the unit cell. (b) Calculate the atomic radius of Mo.

12.107 Tantalum (Ta; d = 16.634 g/cm^3 and $\mathcal{M}$ = 180.9479 g/mol) has a body-centered cubic structure with a unit-cell edge length of 3.3058 Å. Use these data to calculate Avogadro's number.

Advanced Materials

Concept Review Questions

12.108 When tin is added to copper, the resulting alloy (bronze) is much harder than copper. Explain.

12.109 In the process of doping a semiconductor, certain impurities are added to increase the electrical conductivity. Explain this process for an n-type and a p-type semiconductor.

12.110 State two molecular characteristics of substances that typically form liquid crystals. How is each related to function?

12.111 Distinguish between isotropic and anisotropic substances. To which category do liquid crystals belong?

12.112 How are the properties of high-tech ceramics the same as those of traditional clay ceramics, and how are they different? Refer to specific substances in your answer.

12.113 Why is the average molar mass of a polymer sample different from the molar mass of an individual chain?

12.114 How does the random coil shape relate to the radius of gyration of a polymer chain?

12.115 What factor(s) influence the viscosity of a polymer solution? What factor(s) influence the viscosity of a molten polymer? What is a polymer glass?

12.116 Use an example to show how branching and crosslinking can affect the physical behavior of a polymer.

Skill-Building Exercises (grouped in similar pairs)

12.117 Silicon and germanium are both semiconducting elements from Group 4A(14) that can be doped to improve their conductivity. Would each of the following form an n-type or a p-type semiconductor: (a) Ge doped with P; (b) Si doped with In?

12.118 Would each of the following form an n-type or a p-type semiconductor: (a) Ge doped with As; (b) Si doped with B?

12.119 The repeat unit of the polystyrene of a coffee cup has the formula $C_6H_5CHCH_2$. If the molar mass of the polymer is 3.5×10^5 g/mol, what is the degree of polymerization?

12.120 The monomer of poly(vinyl chloride) has the formula C_2H_3Cl. If there are 1565 repeat units in a single chain of the polymer, what is the molecular mass (in amu) of that chain?

12.121 The polypropylene (repeat unit CH_3CHCH_2) in a plastic toy has a molar mass of 2.8×10^5 g/mol, and the length of a repeat unit is 0.252 pm. Calculate the radius of gyration.

12.122 The polymer that is used to make 2-L soda bottles [poly(ethylene terephthalate)] has a repeat unit with molecular formula $C_{10}H_8O_4$ and a length of 1.075 nm. Calculate the radius of gyration of a chain with a molar mass of 2.30×10^4 g/mol.

Comprehensive Problems

12.123 A 0.75-L bottle is cleaned, dried, and closed in a room where the air is 22°C and has 44% relative humidity (that is, the water vapor in the air is 0.44 of the equilibrium vapor pressure at 22°C). The bottle is then brought outside and stored at 0.0°C.
(a) What mass of liquid water condenses inside the bottle?
(b) Would liquid water condense at 10°C? (See Table 5.2.)

12.124 In an experiment, 5.00 L of N_2 is saturated with water vapor at 22°C and then compressed to half its volume at constant T. (a) What is the partial pressure of H_2O in the compressed gas mixture? (b) What mass of water vapor condenses to liquid?

12.125 Barium is the largest nonradioactive alkaline earth metal. It has a body-centered cubic unit cell and a density of 3.62 g/cm^3. What is the atomic radius of barium? (Volume of a sphere: $V = \frac{4}{3}\pi r^3$.)

12.126 Two important characteristics used to evaluate the risk of fire or explosion are a compound's *lower flammable limit* (LFL) and *flash point*. The LFL is the minimum percentage by volume in air that is ignitable. Below that, the mixture is too "lean" to burn. The flash point is the temperature at which the air over a confined liquid becomes ignitable. *n*-Hexane boils at 68.7°C at 1 atm. At 20.0°C, its vapor pressure is 121 mmHg. The LFL of *n*-hexane is 1.1%. Calculate the flash point of *n*-hexane.

12.127 Bismuth is used to calibrate instruments employed in high-pressure studies because it has several well-characterized crystalline phases. Its phase diagram *(right)* shows the liquid phase and five solid phases that are stable above 1 katm (1000 atm) and up to 300°C.

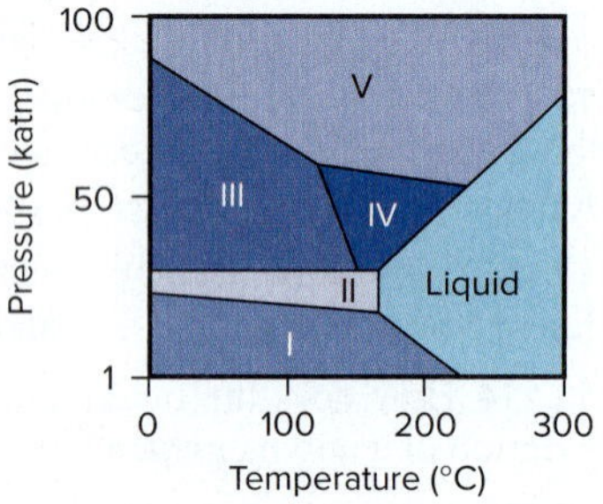

(a) Which solid phases are stable at 25°C? (b) Which phase is stable at 50 katm and 175°C? (c) As the pressure is reduced from 100 to 1 katm at 200°C, what phase transitions does bismuth undergo? (d) What phases are present at each of the triple points?

12.128 In making computer chips, a 4.00-kg cylindrical ingot of ultrapure n-type doped silicon that is 5.20 inches in diameter is sliced into wafers 1.12×10^{-4} m thick.
(a) Assuming no waste, how many wafers can be made?
(b) What is the mass of a wafer (d of Si = 2.34 g/cm^3; V of a cylinder = $\pi r^2 h$)?
(c) A key step in making p-n junctions for the chip is chemical removal of the oxide layer on the wafer through treatment with gaseous HF. Write a balanced equation for this reaction.
(d) If 0.750% of the Si atoms are removed during the treatment in part (c), how many moles of HF are required per wafer, assuming 100% reaction yield?

12.129 Methyl salicylate, $C_8H_8O_3$, the odorous constituent of oil of wintergreen, has a vapor pressure of 1.00 torr at 54.3°C and 10.0 torr at 95.3°C. (a) What is its vapor pressure at 25°C? (b) What is the minimum number of liters of air that must pass over a sample of the compound at 25°C to vaporize 1.0 mg of it?

12.130 Mercury (Hg) vapor is toxic and readily absorbed from the lungs. At 20.°C, mercury (ΔH_{vap} = 59.1 kJ/mol) has a vapor pressure of 1.20×10^{-3} torr, which is high enough to be hazardous. To reduce the danger to workers in processing plants, Hg is cooled to lower its vapor pressure. At what temperature would the vapor pressure of Hg be at the safer level of 5.0×10^{-5} torr?

12.131 Polytetrafluoroethylene (Teflon) has a repeat unit with the formula $F_2C—CF_2$. A sample of the polymer consists of fractions with the following distribution of chains:

Fraction	Average Number of Repeat Units	Amount (mol) of Polymer
1	273	0.10
2	330	0.40
3	368	1.00
4	483	0.70
5	525	0.30
6	575	0.10

(a) Determine the molar mass of each fraction.
(b) Determine the number-average molar mass of the sample.
(c) Another type of average molar mass of a polymer sample is called the *weight-average molar mass*, $\mathcal{M}_w$:

$$\mathcal{M}_w = \frac{\Sigma(\mathcal{M} \text{ of fraction} \times \text{mass of fraction})}{\text{total mass of all fractions}}$$

Calculate the weight-average molar mass of the sample of polytetrafluoroethylene.

12.132 A greenhouse contains 256 m^3 of air at a temperature of 26°C, and a humidifier in it vaporizes 4.20 L of water. (a) What is the pressure of water vapor in the greenhouse, assuming that none escapes and that the air was originally completely dry (d of H_2O = 1.00 g/mL)? (b) What total volume of liquid water would have to be vaporized to saturate the air (that is, achieve 100% relative humidity)? (See Table 5.2.)

12.133 Like most transition metals, tantalum (Ta) exhibits several oxidation states. Give the formula of each tantalum compound whose unit cell is depicted below:

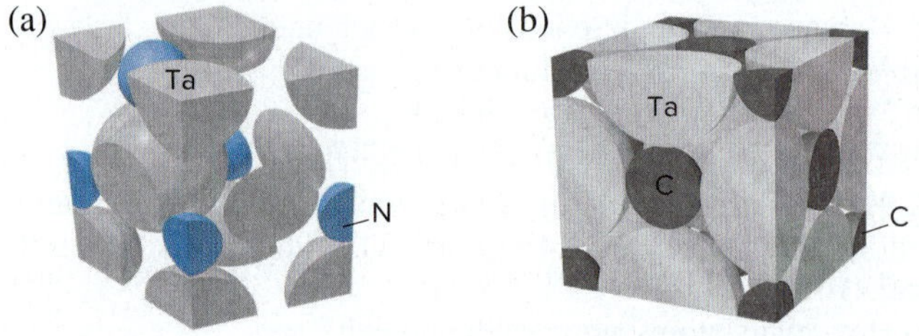

12.134 KF has the same type of crystal structure as NaCl. The unit cell of KF has an edge length of 5.39 Å. Find the density of KF.

12.135 Furfural, which is prepared from corncobs, is an important solvent in synthetic rubber manufacturing, and it is reduced to furfuryl alcohol, which is used to make polymer resins. Furfural can also be oxidized to 2-furoic acid.

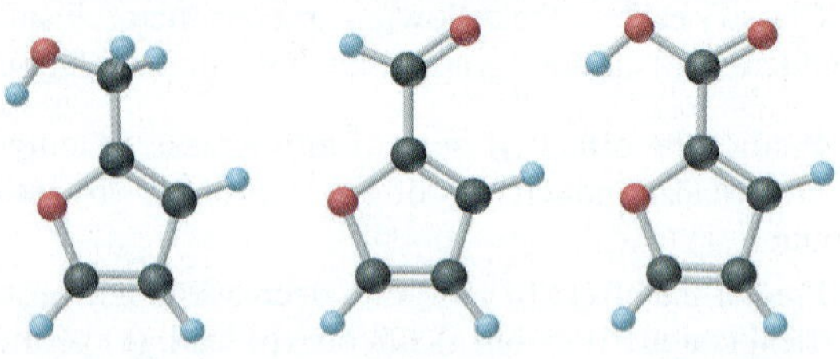

furfuryl alcohol furfural 2-furoic acid

(a) Which of these compounds can form H bonds? Draw structures in each case.
(b) The molecules of some substances can form an "internal" H bond, that is, an H bond *within* a molecule. This takes the form of a polygon with atoms as corners and bonds as sides and an H bond as one of the sides. Which of these molecules is (are) likely to form a stable internal H bond? Draw the structure. (*Hint:* Structures with 5 or 6 atoms as corners are most stable.)

12.136 On a humid day in New Orleans, the temperature is 22.0°C, and the partial pressure of water vapor in the air is 31.0 torr. The 9000-ton air-conditioning system in the Louisiana Superdome maintains the inside air temperature at the same 22.0°C, but produces a partial pressure of water vapor of 10.0 torr. The volume of air in the dome is 2.4×10^6 m^3, and the total pressure is 1.0 atm both inside and outside the dome.
(a) What mass of water (in metric tons) must be removed every time the inside air is completely replaced with outside air? (*Hint:* How many moles of gas are in the dome? How many moles of water vapor? How many moles of dry air? How many moles of outside air must be added to the air in the dome to simulate the composition of outside air?)
(b) Find the heat released when this mass of water condenses.

12.137 The boiling point of amphetamine, $C_9H_{13}N$, is 201°C at 760 torr and 83°C at 13 torr. What is the concentration (in g/m^3) of amphetamine when it is in contact with 20.°C air?

12.138 Diamond has a face-centered cubic unit cell, with four more C atoms in tetrahedral holes within the cell. Densities of diamonds vary from 3.01 g/cm^3 to 3.52 g/cm^3 because C atoms are missing from some holes. (a) Calculate the unit-cell edge length of the densest diamond. (b) Assuming the cell dimensions are fixed, how many C atoms are in the unit cell of the diamond with the lowest density?

12.139 Is it possible for a salt of formula AB_3 to have a face-centered cubic unit cell of anions with cations in all eight of the available holes? Explain.

12.140 The density of solid gallium at its melting point is 5.9 g/cm^3, whereas that of liquid gallium is 6.1 g/cm^3. Is the temperature at the triple point higher or lower than the normal melting point? Is the slope of the solid-liquid line for gallium positive or negative?

12.141 A 4.7-L sealed bottle containing 0.33 g of liquid ethanol, C_2H_6O, is placed in a refrigerator and reaches equilibrium with its vapor at −11°C. (a) What mass of ethanol is present in the vapor? (b) When the container is removed and warmed to room temperature, 20.°C, will all the ethanol vaporize? (c) How much liquid ethanol would be present at 0.0°C? The vapor pressure of ethanol is 10. torr at −2.3°C and 40. torr at 19°C.

12.142 Substance A has the following properties.

mp at 1 atm:	−20.°C	bp at 1 atm:	85°C
ΔH_{fus}:	180. J/g	ΔH_{vap}:	500. J/g
c_{solid}:	1.0 J/g·°C	c_{liquid}:	2.5 J/g·°C
c_{gas}:	0.5 J/g·°C		

At 1 atm, a 25-g sample of A is heated from −40.°C to 100.°C at a constant rate of 450. J/min. (a) How many minutes does it take to heat the sample to its melting point? (b) How many minutes does it take to melt the sample? (c) Perform any other necessary calculations, and draw a curve of temperature vs. time for the entire heating process.

12.143 An aerospace manufacturer is building a prototype experimental aircraft that cannot be detected by radar. Boron nitride is chosen for incorporation into the body parts, and the boric acid/ammonia method is used to prepare the ceramic material. Given 85.5% and 86.8% yields for the two reaction steps, how much boron nitride can be prepared from 1.00 metric ton of boric acid and 12.5 m^3 of ammonia at 275 K and 3.07×10^3 kPa? Assume that ammonia does not behave ideally under these conditions and is recycled completely in the reaction process.

12.144 The ball-and-stick models below represent three compounds with the same molecular formula, $C_4H_8O_2$:

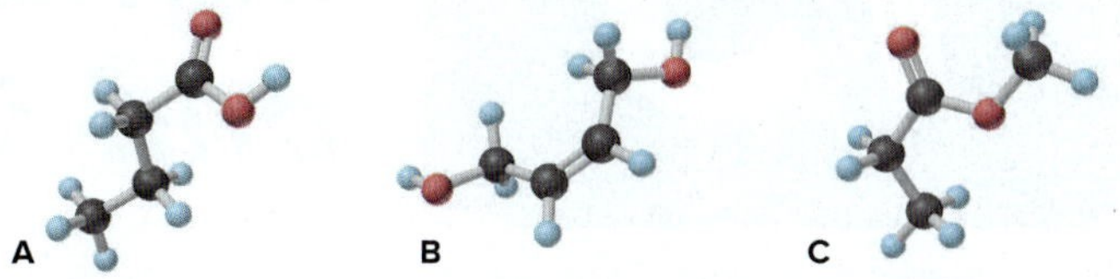

(a) Which compound(s) can form intermolecular H bonds?
(b) Which has the highest viscosity?

12.145 The ΔH_f° of gaseous dimethyl ether (CH_3OCH_3) is −185.4 kJ/mol; the vapor pressure is 1.00 atm at −23.7°C and 0.526 atm at −37.8°C. (a) Calculate ΔH_{vap}° of dimethyl ether. (b) Calculate ΔH_f° of liquid dimethyl ether.

12.146 The crystal structure of sodium is based on the body-centered cubic unit cell. What is the mass of one unit cell of Na?

12.147 The *intrinsic viscosity* of a polymer solute in a solvent, $[\eta]_{solvent}$, is the portion of the total viscosity due to the solute and is related to solute shape. It has also been found to relate to intermolecular interactions between solvent and polymer: higher $[\eta]_{solvent}$ means stronger interaction. The $[\eta]_{solvent}$ values of polymers in solution are given by the Mark-Houwink equation, $[\eta]_{solvent} = K\mathcal{M}^a$, where $\mathcal{M}$ is the molar mass of the polymer and K and a are constants specific to the polymer and solvent. Use the data below for substances at 25°C to answer the following questions:

Polymer	Solvent	*K* (mL/g)	*a*
Polystyrene	Benzene	9.5×10^{-3}	0.74
	Cyclohexane	8.1×10^{-2}	0.50
Polyisobutylene	Benzene	8.3×10^{-2}	0.50
	Cyclohexane	2.6×10^{-1}	0.70

(a) A polystyrene sample has a molar mass of 104,160 g/mol. Calculate the intrinsic viscosity of this polymer in benzene and in cyclohexane. Which solvent has stronger interactions with the polymer?
(b) A different polystyrene sample has a molar mass of 52,000 g/mol. Calculate its $[\eta]_{benzene}$. Given a polymer standard of known $\mathcal{M}$, how could you use its measured $[\eta]$ in a given solvent to determine the molar mass of any sample of that polymer?
(c) Compare $[\eta]$ values of a polyisobutylene sample [repeat unit $(CH_3)_2CCH_2$] with a molar mass of 104,160 g/mol with those of the polystyrene in part (a). What does this suggest about the solvent-polymer interactions of the two samples?

12.148 One way of purifying gaseous H_2 is to pass it under high pressure through the holes of a metal's crystal structure. Palladium, which adopts a cubic closest packed structure, absorbs more H_2 than any other element and is one of the metals used for this purpose. How the metal and H_2 interact is unclear, but it is estimated that the density of absorbed H_2 approaches that of liquid hydrogen (70.8 g/L). What volume (in L) of gaseous H_2 (at STP) can be packed into the spaces of 1 dm^3 of palladium metal?

13

The Properties of Mixtures: Solutions and Colloids

13.1 Types of Solutions: Intermolecular Forces and Solubility
Intermolecular Forces in Solution
Liquid Solutions and Molecular Polarity
Gas and Solid Solutions

13.2 Intermolecular Forces and Biological Macromolecules
Structures of Proteins
Dual Polarity in Soaps, Membranes, and Antibiotics
Structure of DNA

13.3 Why Substances Dissolve: Breaking Down the Solution Process
Heat of Solution and Its Components
Heat of Hydration: Dissolving Ionic Solids in Water
Solution Process and Entropy Change

13.4 Solubility as an Equilibrium Process
Effect of Temperature on Solubility
Effect of Pressure on Solubility

13.5 Concentration Terms
Molarity and Molality
Parts of Solute by Parts of Solution
Interconverting Concentration Terms

13.6 Colligative Properties of Solutions
Nonvolatile Nonelectrolyte Solutions
Using Colligative Properties to Find Solute Molar Mass
Volatile Nonelectrolyte Solutions
Strong Electrolyte Solutions
Applications of Colligative Properties

13.7 Structure and Properties of Colloids

Source: © amnat11/Shutterstock.com

Concepts and Skills to Review Before You Study This Chapter

- separation of mixtures (Section 2.9)
- calculations involving mass percent (Section 3.1) and molarity (Section 4.1)
- electrolytes; water as solvent (Sections 4.1 and 12.5)
- mole fraction and Dalton's law (Section 5.4)
- intermolecular forces and polarizability (Section 12.3)
- monomers and polymers (Section 12.7)
- equilibrium nature of phase changes and vapor pressure of liquids; phase diagrams (Section 12.2)

Virtually all the gases, liquids, and solids in the real world are mixtures—two or more substances mixed together physically, not combined chemically. Synthetic mixtures usually contain only a dozen or so components; for example, the soda you may be drinking while reading your chemistry text contains water, carbon dioxide, sugar, caffeine, caramel color, phosphoric and citric acids, and other flavorings. Natural mixtures, such as seawater and soil, often contain over 50 components. Living mixtures, such as trees and students, are the most complex—even a simple bacterial cell contains nearly 6000 different compounds (Table 13.1).

Table 13.1 Approximate Composition of a Bacterium

Substance	Mass % of Cell	Number of Types	Number of Molecules
Water	~70	1	5×10^{10}
Ions	1	20	?
Sugars*	3	200	3×10^{8}
Amino acids*	0.4	100	5×10^{7}
Lipids*	2	50	3×10^{7}
Nucleotides*	0.4	200	1×10^{7}
Other small molecules	0.2	~200	?
Macromolecules (proteins, nucleic acids, polysaccharides)	23	~5000	6×10^{6}

*Includes precursors and metabolites.

Recall from Chapter 2 that a mixture has two defining characteristics: *its composition is variable,* and *it retains some properties of its components.* We focus here on two common types of mixtures—solutions and colloids—whose main differences relate to particle size and number of phases:

- A *solution* is a *homogeneous* mixture; that is, it exists as one phase. In a solution, the particles are individual atoms, ions, or small molecules.
- A *colloid* is a type of *heterogeneous* mixture. A heterogeneous mixture has two or more phases. They may be visibly distinct, like pebbles in concrete, or not, like the much smaller particles in the colloids smoke and milk. In a colloid, the particles are typically macromolecules or aggregations of small molecules that are dispersed so finely they don't settle out.

IN THIS CHAPTER . . . *We focus on how intermolecular forces and other energy considerations affect a solute dissolving in a solvent, how to calculate concentration, and how solutions differ from pure substances. We also briefly consider the behavior and applications of colloids.*

- We survey intermolecular forces between solute and solvent and find that substances with similar types of forces form a solution.
- We see how the dual polarity of some organic molecules gives rise to these same intermolecular forces, and we find that they determine the structures and functions of soaps, antibiotics, biological macromolecules, and cell membranes.
- We use a stepwise cycle to see why a substance dissolves and examine the heat involved and the dispersal of matter that occurs when a solution forms. To understand the latter factor, we introduce the concept of entropy.
- We examine the equilibrium nature of solubility and see how temperature and pressure affect it.
- We define various solution concentration units and see how to interconvert them mathematically.
- We see why the physical properties of solutions are different from those of pure substances and learn how to apply those differences.
- We investigate colloids and apply solution and colloid chemistry to the purification of water.

13.1 TYPES OF SOLUTIONS: INTERMOLECULAR FORCES AND SOLUBILITY

A **solute** dissolves in a **solvent** to form a solution. In general, *the solvent is the most abundant component,* but in some cases, the substances are **miscible**—soluble in each other in any proportion—so the terms "solute" and "solvent" lose their meaning. *The physical state of the solvent usually determines the physical state of the solution.* Solutions can be gaseous, liquid, or solid, but we focus mostly on liquid solutions because they are by far the most important.

The **solubility (*S*)** of a solute is the maximum amount that dissolves in a fixed quantity of a given solvent at a given temperature, when an excess of the solute is present. Different solutes have different solubilities:

- Sodium chloride (NaCl), $S = 39.12$ g/100. mL water at 100.°C
- Silver chloride (AgCl), $S = 0.0021$ g/100. mL water at 100.°C

Solubility is a *quantitative* term, but *dilute* and *concentrated* are qualitative, referring to the *relative* amounts of dissolved solute:

- The NaCl solution above is concentrated (a relatively large amount of solute dissolved in a given quantity of solvent).
- The AgCl solution is dilute (a relatively small amount of solute in a given quantity of solvent).

A given solute may dissolve in one solvent and not another. The explanation lies in the relative strengths of the intermolecular forces within both solute and solvent and between them. The useful rule-of-thumb **"like dissolves like"** says that *substances with similar types of intermolecular forces dissolve in each other.* Thus, by knowing the forces, we can often predict whether a solute will dissolve in a solvent.

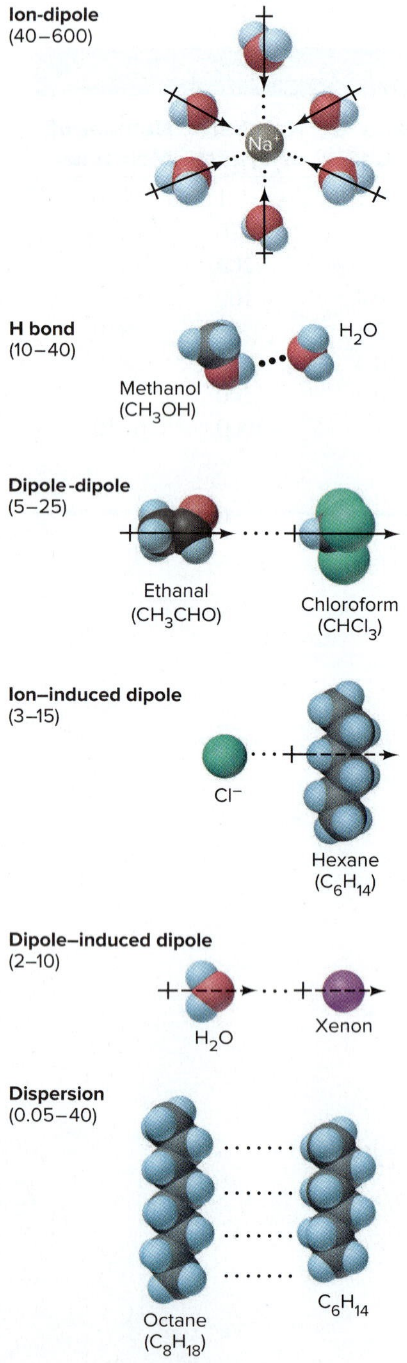

Figure 13.1 Types of intermolecular forces in solutions. Forces are listed in order of decreasing strength (values are in kJ/mol), with an example of each.

Intermolecular Forces in Solution

All the intermolecular forces we discussed for pure substances also occur in solutions (Figure 13.1; also see Section 12.3):

1. *Ion-dipole forces* (attractions between ions and polar molecules) are the principal force involved when an ionic compound dissolves in water. Two events occur simultaneously:
   - *Forces compete.* When a soluble salt is added to water, each type of ion attracts the oppositely charged pole of a water molecule. These attractions between ions and water compete with and overcome attractions between the ions, and the crystal structure breaks down.
   - *Hydration shells form.* As an ion separates from the crystal structure, water molecules cluster around it in **hydration shells.** The number of water molecules in the innermost shell depends on the ion's size: four fit tetrahedrally around small ions like Li^+, while the larger Na^+ and F^- have six water molecules surrounding them octahedrally (Figure 13.2). In the innermost shell, normal hydrogen bonding between water molecules is disrupted to form the ion-dipole forces. But these water molecules are H bonded to others in the next shell, and those are H bonded to others still farther away.
2. *Hydrogen bonding* (attractions between molecules with an H atom bonded to N, O, or F) is the principal force in solutions of polar, O- and N-containing organic and biological compounds, such as alcohols, amines, and amino acids.
3. *Dipole-dipole forces* (attractions between polar molecules), in the absence of H bonding, allow polar molecules like propanal (CH_3CH_2CHO) to dissolve in polar solvents like dichloromethane (CH_2Cl_2).
4. **Ion–induced dipole forces,** one type of *charge-induced dipole force,* rely on polarizability. They arise when an ion's charge distorts the electron cloud of a nearby nonpolar molecule, giving it a temporary dipole moment. This type of force initiates the binding of the Fe^{2+} ion in hemoglobin to an O_2 molecule that enters a red blood cell.

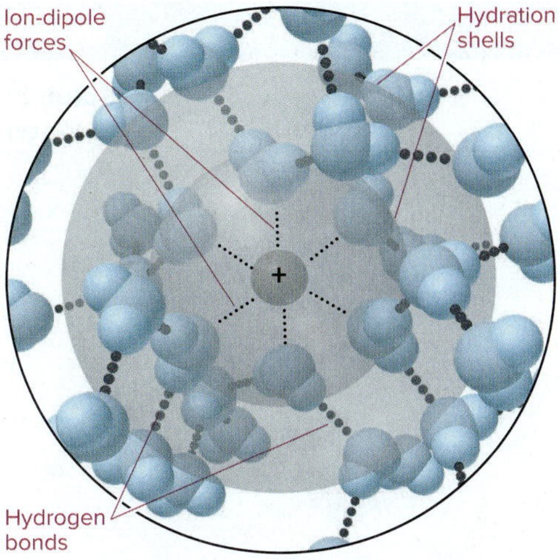

Figure 13.2 Hydration shells around an Na^+ ion. Ion-dipole forces orient water molecules around an ion. In the innermost shell here, six water molecules surround the cation octahedrally.

5. **Dipole–induced dipole forces,** also based on polarizability, arise when a polar molecule distorts the electron cloud of a nonpolar molecule. They are weaker than ion–induced dipole forces because the charge of each pole is less than an ion's (Coulomb's law). The solubility in water of atmospheric O_2, N_2, and noble gases, while limited, is due in part to these forces. Paint thinners and grease solvents also rely on them.
6. *Dispersion forces* contribute to the solubility of all solutes in all solvents, but they are the *principal* intermolecular force in solutions of nonpolar substances, such as petroleum and gasoline.

The same forces maintain the shapes of biological macromolecules (Section 13.2).

Liquid Solutions and the Role of Molecular Polarity

From cytoplasm to tree sap, gasoline to cleaning fluid, iced tea to urine, liquid solutions are very familiar. Water is the most prominent solvent, but there are many other liquid solvents, with polarities from very polar to nonpolar.

Applying the Like-Dissolves-Like Rule The like-dissolves-like rule says that when the forces *within* the solute are *similar* to those *within* the solvent, the forces can *replace* each other and a solution forms. Thus,

- *Salts are soluble in water* because the ion-dipole attractions between ion and water are similar in strength to the strong attractions between the ions and the strong H bonds between water molecules, so they *can* replace each other.
- *Salts are insoluble in hexane (C_6H_{14})* because the ion–induced dipole forces between ion and nonpolar hexane are very weak and *cannot* replace the strong attractions between the ions.
- *Oil is insoluble in water* because the weak dipole–induced dipole forces between oil and water molecules *cannot* replace the strong H bonds between water molecules or the extensive dispersion forces within the oil.
- *Oil is soluble in hexane* because dispersion forces in one *can* replace the similar dispersion forces in the other.

Dual Polarity and Effects on Solubility To examine these ideas further, let's compare the solubilities of a series of alcohols in water and in hexane ($CH_3CH_2CH_2CH_2CH_2CH_3$), two solvents with very different intermolecular forces; polar water molecules exhibit H bonds, while nonpolar hexane molecules have dispersion forces. Alcohols are organic compounds that have a dual polarity, a polar hydroxyl (—OH) group bonded to a nonpolar hydrocarbon group:

- The —OH portion interacts through strong H bonds with water and through weak dipole–induced dipole forces with hexane.
- The hydrocarbon portion interacts through dispersion forces with hexane and through very weak dipole–induced dipole forces with water.

Table 13.2 Solubility* of a Series of Alcohols in Water and in Hexane

Alcohol	Model	Solubility in Water	Solubility in Hexane
CH_3OH (methanol)		∞	1.2
CH_3CH_2OH (ethanol)		∞	∞
$CH_3(CH_2)_2OH$ (1-propanol)	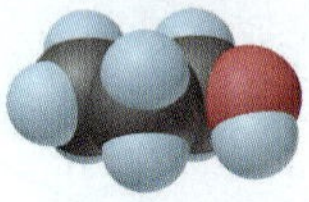	∞	∞
$CH_3(CH_2)_3OH$ (1-butanol)	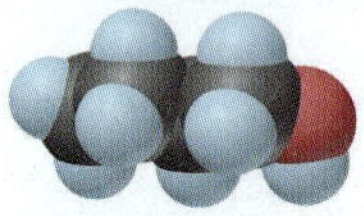	1.1	∞
$CH_3(CH_2)_4OH$ (1-pentanol)	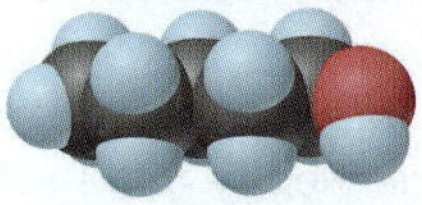	0.30	∞
$CH_3(CH_2)_5OH$ (1-hexanol)		0.058	∞

*Expressed in mol alcohol/1000 g solvent at 20°C.

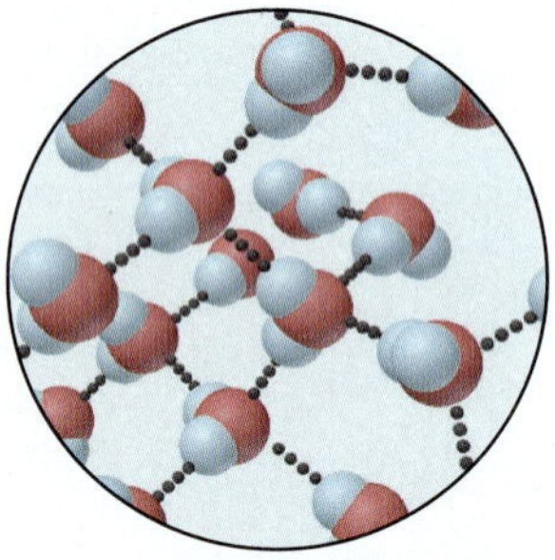

In pure water, H bonds link the molecules.

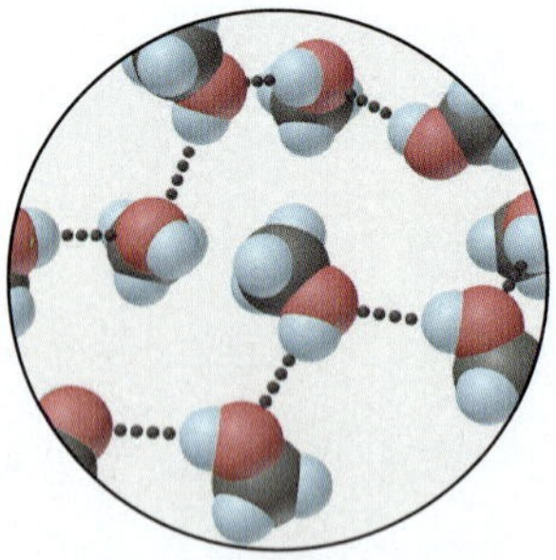

In pure methanol, H bonds link the molecules.

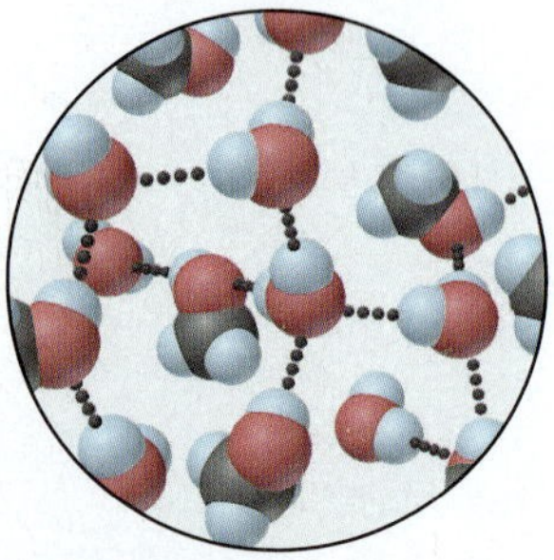

In a solution of water and methanol, H bonds link the two types of molecules.

Figure 13.3 Like dissolves like: solubility of methanol in water.

The general formula for an alcohol is $CH_3(CH_2)_nOH$, and we'll look at straight-chain examples with one to six carbons (n = 0 to 5):

1. *Solubility in water is high for smaller alcohols.* From the models in Table 13.2, we see that the —OH group is a relatively large portion of the alcohols with one to three carbons (n = 0 to 2). These molecules interact with each other through H bonding, just as water molecules do. When they mix with water, H bonding *within* solute and *within* solvent is replaced by H bonding *between* solute and solvent (Figure 13.3). As a result, these smaller alcohols are miscible with water.

2. *Solubility in water is low for larger alcohols.* Solubility decreases dramatically for alcohols larger than three carbons ($n > 2$); in fact, those with chains longer than six carbons ($n > 5$) are insoluble in water. For larger alcohols to dissolve, the nonpolar chains have to move among the water molecules, replacing the strong H-bond attractions between water molecules with the weak attractions of the chains for water. While the —OH portion of such an alcohol forms H bonds to water, these cannot make up for all the other H bonds between water molecules that have to break to make room for the long hydrocarbon portion.

Table 13.2 shows that the opposite trend occurs with hexane:

1. *Solubility in hexane is low for the smallest alcohol.* For alcohols in hexane, in addition to dispersion forces, weak dipole–induced dipole forces exist between the —OH of methanol (CH_3OH) and hexane. These cannot replace the strong H bonding between CH_3OH molecules, so solubility is relatively low.

2. *Solubility in hexane is high for larger alcohols.* In any larger alcohol ($n > 0$), dispersion forces between the hydrocarbon portion and hexane *can* replace dispersion forces between hexane molecules. With only weak forces within the solvent to be replaced, even ethanol, with a two-carbon chain, has enough dispersion forces between it and hexane to be miscible.

Many organic molecules have polar and nonpolar portions, which determine their solubility. For example, carboxylic acids and amines behave like alcohols: methanoic acid (HCOOH, formic acid) and methanamine (CH_3NH_2) are miscible with water and slightly soluble in hexane, whereas hexanoic acid [$CH_3(CH_2)_4COOH$] and 1-hexanamine [$CH_3(CH_2)_5NH_2$] are slightly soluble in water and very soluble in hexane.

SAMPLE PROBLEM 13.1 Predicting Relative Solubilities

Problem Predict which solvent will dissolve more of the given solute:
(a) Sodium chloride in methanol (CH_3OH) or in 1-propanol ($CH_3CH_2CH_2OH$)
(b) Ethylene glycol ($HOCH_2CH_2OH$) in hexane ($CH_3CH_2CH_2CH_2CH_2CH_3$) or in water
(c) Diethyl ether ($CH_3CH_2OCH_2CH_3$) in water or in ethanol (CH_3CH_2OH)

Plan We examine the formulas of solute and solvent to determine the forces in and between solute and solvent. A solute is more soluble in a solvent whose intermolecular forces are similar to, and therefore can replace, its own.

Solution **(a)** Methanol. NaCl is ionic, so it dissolves through ion-dipole forces. Both methanol and 1-propanol have a polar —OH group, but the hydrocarbon portion of each alcohol interacts only weakly with the ions and 1-propanol has a longer hydrocarbon portion than methanol.
(b) Water. Ethylene glycol molecules have two —OH groups, so they interact with each other through H bonding. H bonds formed with H_2O can replace these H bonds between solute molecules better than dipole–induced dipole forces with hexane can.
(c) Ethanol. Diethyl ether molecules interact through dipole-dipole and dispersion forces. They can form H bonds to H_2O or to ethanol. But ethanol can also interact with the ether effectively through dispersion forces because it has a hydrocarbon chain.

FOLLOW-UP PROBLEMS

Brief Solutions for ***all*** *Follow-up Problems appear at the end of the chapter.*

13.1A State which solute is more soluble in the given solvent and which forces are most important: **(a)** 1-butanol ($CH_3CH_2CH_2CH_2OH$) or 1,4-butanediol ($HOCH_2CH_2CH_2CH_2OH$) in water; **(b)** chloroform ($CHCl_3$) or carbon tetrachloride (CCl_4) in water.

13.1B State which solvent can dissolve more of the given solute and which forces are most important: **(a)** chloromethane (CH_3Cl) in chloroform ($CHCl_3$) or in methanol (CH_3OH); **(b)** pentanol ($C_5H_{11}OH$) in water or in hexane (C_6H_{14}).

SOME SIMILAR PROBLEMS 13.13 and 13.14

Gas-Liquid Solutions A substance with very weak intermolecular attractions should have a low boiling point and, thus, would be a gas under ordinary conditions. Also, it would not be very soluble in water because of weak solute-solvent forces. Thus, for nonpolar or slightly polar gases, boiling point generally correlates with solubility in water (Table 13.3). A higher boiling point is an indication of stronger intermolecular forces, which result in a greater solubility in water.

The small amount of a nonpolar gas that *does* dissolve may be vital. At 25°C and 1 atm, the solubility of O_2 is only 3.2 mL/100. mL of water, but aquatic animal life requires it. At times, the solubility of a nonpolar gas may *seem* high because it is also *reacting* with the solvent. Oxygen seems more soluble in blood than in water because it bonds chemically to hemoglobin in red blood cells. Carbon dioxide, which is essential for aquatic plants and coral-reef growth, seems very soluble in water (~81 mL of CO_2/100. mL of H_2O at 25°C and 1 atm) because it is dissolving *and* reacting:

$$CO_2(g) + H_2O(l) \rightleftharpoons H^+(aq) + HCO_3^-(aq)$$

Table 13.3 Correlation Between Boiling Point and Solubility in Water

Gas	Solubility (mol/L)*	bp (K)
He	4.2×10^{-4}	4.2
Ne	6.6×10^{-4}	27.1
N_2	10.4×10^{-4}	77.4
CO	15.6×10^{-4}	81.6
O_2	21.8×10^{-4}	90.2
NO	32.7×10^{-4}	121.4

*At 273 K and 1 atm.

Gas Solutions and Solid Solutions

Gas solutions and solid solutions also have vital importance and numerous applications.

Gas-Gas Solutions *All gases are miscible with each other.* Air is the classic example of a gaseous solution, consisting of about 18 gases in widely differing proportions.

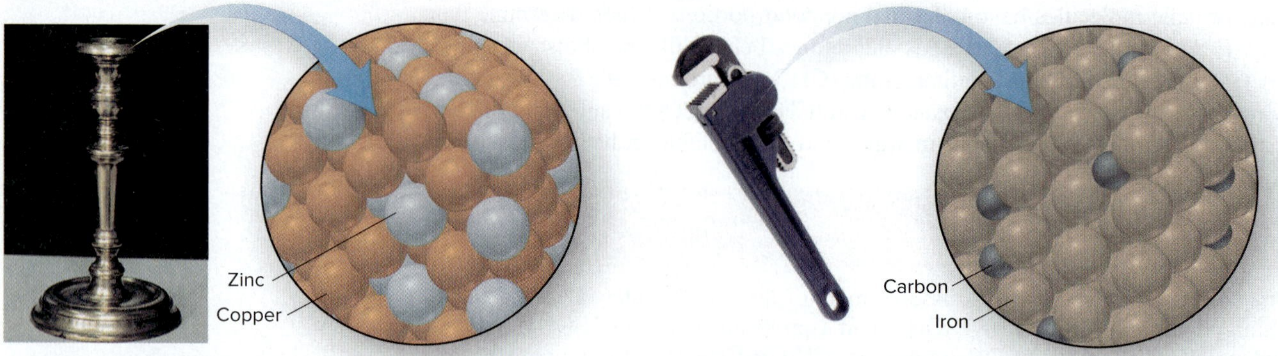

A Brass, a substitutional alloy

B Carbon steel, an interstitial alloy

Figure 13.4 The arrangement of atoms in two types of alloys.

Source: (A) © Ruth Melnick; (B) © Ingram Publishing/age fotostock RF

Anesthetic gas proportions are finely adjusted to the needs of the patient and the length of the surgical procedure. The proportions of many industrial gas mixtures, such as CO/H_2 in syngas production or N_2/H_2 in ammonia production, are controlled to optimize product yield under varying conditions of temperature and pressure.

Gas-Solid Solutions *When a gas dissolves in a solid, it occupies the spaces between the closely packed particles.* Hydrogen gas can be purified by passing an impure sample through palladium. Only H_2 molecules are small enough to fit between the Pd atoms, where they form Pd—H bonds. The H atoms move from one Pd atom to another and emerge from the metal as H_2 molecules (see Figure 14.2).

The ability of gases to dissolve in a solid also has disadvantages. The electrical conductivity of copper is greatly reduced by the presence of O_2, which dissolves into the crystal structure and reacts to form copper(I) oxide. High-conductivity copper is prepared by melting and recasting the metal in an O_2-free atmosphere.

Solid-Solid Solutions Solids diffuse so little that their mixtures are usually heterogeneous. Some solid-solid solutions can be formed by melting the solids and then mixing them and allowing them to freeze. Many **alloys,** mixtures of elements that have a metallic character, are solid-solid solutions (although several have microscopic heterogeneous regions). Alloys generally fall into one of two categories:

- In a *substitutional alloy* like brass (Figure 13.4A), atoms of zinc *replace* atoms of the main element, copper, at some sites in the cubic closest packed array. This occurs when the atoms of the elements in the alloy are similar in size.
- In an *interstitial alloy* like carbon steel (Figure 13.4B), atoms of carbon (a nonmetal is typical in this type of alloy) *fill some spaces (interstices)* between atoms of the main element, iron, in the body-centered array.

Waxes are also solid-solid solutions. Most are amorphous solids with some small regions of crystalline regularity. A *wax* is defined as a solid of biological origin that is insoluble in water and soluble in nonpolar solvents. Beeswax, which bees secrete to build their combs, is a homogeneous mixture of fatty acids, long-chain carboxylic acids, and hydrocarbons in which some of the molecules are more than 40 carbon atoms long. Carnauba wax, from a South American palm, is a mixture of compounds, each consisting of a fatty acid bound to a long-chain alcohol. It is hard but forms a thick gel in nonpolar solvents, so it is perfect for waxing cars.

› Summary of Section 13.1

- › A solution is a homogeneous mixture of a solute dissolved in a solvent through the action of intermolecular forces.
- › Ion-dipole, ion–induced dipole, and dipole–induced dipole forces occur in solutions, in addition to all the intermolecular forces that also occur in pure substances.
- › If similar intermolecular forces occur in solute and solvent, they replace each other when the substances mix and a solution is likely to form (the like-dissolves-like rule).
- › When ionic compounds dissolve in water, the ions become surrounded by hydration shells of H-bonded water molecules.

› Solubility of organic molecules in various solvents depends on the relative sizes of their polar and nonpolar portions.
› The solubility of nonpolar gases in water is low because of weak intermolecular forces. Gases are miscible with one another and dissolve in solids by fitting into spaces in the crystal structure.
› Solid-solid solutions include alloys (some of which are formed by mixing molten components) and waxes.

13.2 INTERMOLECULAR FORCES AND BIOLOGICAL MACROMOLECULES

We discuss the shapes of proteins, nucleic acids, and cell membranes, as well as the functions of soaps and antibiotics, in this chapter on solutions because they depend on intermolecular forces too. These shapes and functions are explained by two ideas:

- Polar and ionic groups attract water, but nonpolar groups do not.
- Just as separate molecules attract each other, so do distant groups on the same molecule.

The Structures of Proteins

Proteins are very large molecules (called polymers) formed by linking together many smaller molecules called **amino acids;** about 20 different amino acids occur in proteins, which range in size from about 50 amino acids ($\mathcal{M} \approx 5\times10^3$ g/mol) to several thousand ($\mathcal{M} \approx 5\times10^5$ g/mol). Proteins with a few types of amino acids in repeating patterns have extended helical or sheetlike shapes and give structure to hair, skin, and so forth. Proteins with many types of amino acids have complex, globular shapes and function as antibodies, enzymes, and so forth. Let's see how intermolecular forces among amino acids influence the shapes of globular proteins.

The Polarity of Amino-Acid Side Chains In a cell, a free amino acid has four groups bonded to one C, which is called the α-carbon (Figure 13.5): charged carboxyl ($—COO^-$) and amine ($—NH_3^+$) groups, an H atom, and a *side chain,* represented by an R, which ranges from another H atom, to a few C atoms, to a two-ringed C_9H_8N group. In a protein, the carboxyl group of one amino acid is linked covalently to the amine group of the next by a *peptide* bond. (We discuss amino acid structures and peptide bond formation in Chapter 15.) Thus, as Figure 13.6 shows, the *backbone* of a protein is a polypeptide chain: an *α-carbon connected through a peptide bond (orange screen) to the next α-carbon,* and so forth. The various side chains *(gray screens)* dangle off the α-carbons on alternate sides of the chain.

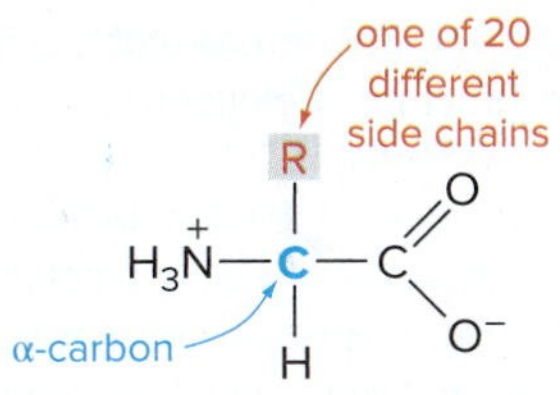

Figure 13.5 The charged form of an amino acid under physiological conditions.

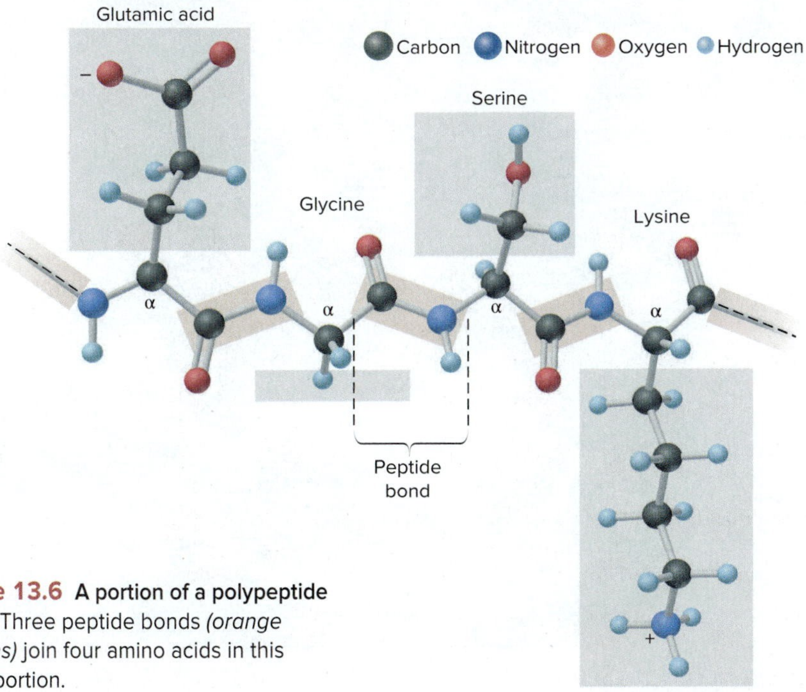

Figure 13.6 A portion of a polypeptide chain. Three peptide bonds *(orange screens)* join four amino acids in this chain portion.

Amino acids can be classified by the polarity or charge of their side chains: nonpolar, polar, and ionic. A few examples are

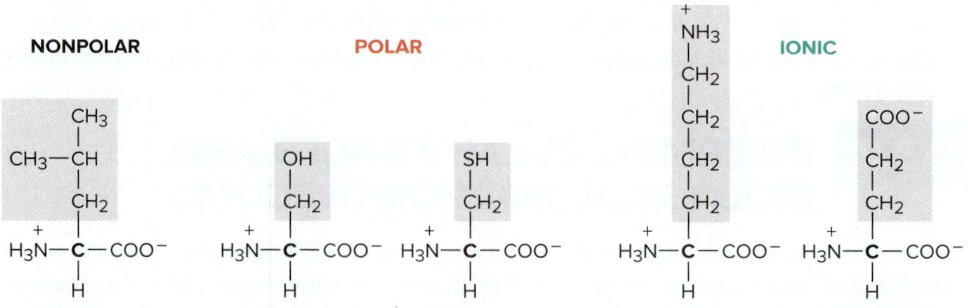

Intermolecular Forces and Protein Shape The same forces that act between separate molecules are responsible for a protein's shape, because *distant groups on the protein chain end up near each other as the chain bends.* Figure 13.7 depicts the forces within a small portion of a protein and between the protein and the aqueous medium of the cell. In general order of importance, they are

- Covalent peptide bonds create the backbone (polypeptide chain).
- Helical and sheetlike segments arise from *H bonds between the C═O of one peptide bond and the N—H of another.*
- Polar and ionic side chains protrude into the surrounding cell fluid, interacting with water through ion-dipole forces and H bonds.
- Nonpolar side chains interact through dispersion forces within the nonaqueous protein interior.

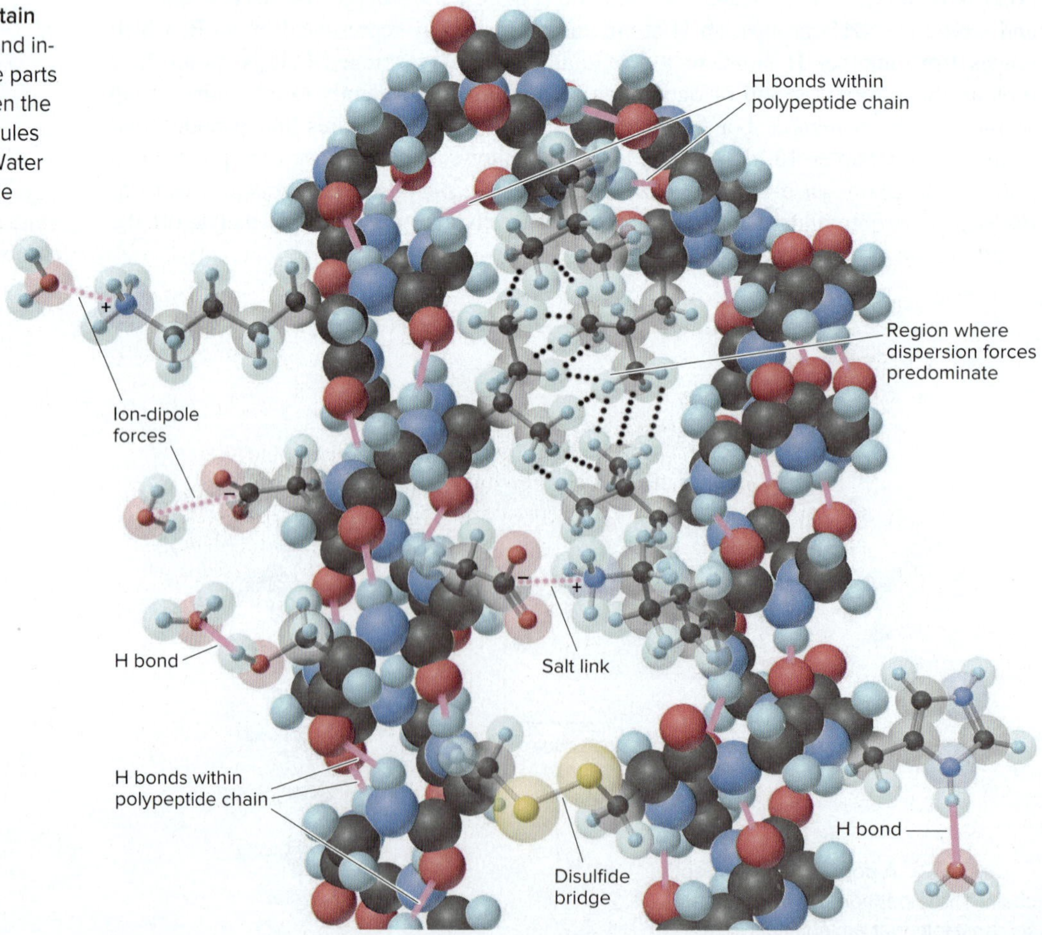

Figure 13.7 The forces that maintain protein structure. Covalent, ionic, and intermolecular forces act between the parts of a portion of a protein and between the protein and surrounding H_2O molecules to determine the protein's shape. (Water molecules and some amino-acid side chains are shown as ball-and-stick models within space-filling contours.)

- The —SH ends of two cysteine side chains form a covalent —S—S— bond, a *disulfide bridge,* between distant parts of the chain that creates a loop.
- Oppositely charged ends of ionic side chains, $—COO^-$ and $—NH_3^+$ groups, that lie near each other form an electrostatic *salt link* (or *ion pair*) that creates bend in the protein chain.
- Other H bonds between side chains keep distant chain portions near each other.

Thus, *soluble proteins have polar-ionic exteriors and nonpolar interiors.* As we emphasize in Chapter 15, the amino acid *sequence* of a protein determines its *shape,* which determines its *function.*

Dual Polarity in Soaps, Membranes, and Antibiotics

Dual polarity, which we noted as a key factor in the solubility of alcohols, also helps explain how soaps, cell membranes, and antibiotics function.

Action of Soaps A **soap** is the salt formed when a strong base (a metal hydroxide) reacts with a *fatty acid,* a carboxylic acid with a long hydrocarbon chain. A typical soap molecule is made up of a nonpolar "tail" 15–19 carbons long and a polar-ionic "head" consisting of a $—COO^-$ group and the cation of the strong base. The cation greatly influences a soap's properties. Lithium soaps are hard and high melting and used in car lubricants. Potassium soaps are low melting and used in liquid form. Several sodium soaps, including sodium stearate, $CH_3(CH_2)_{16}COONa$, are components of common bar soaps:

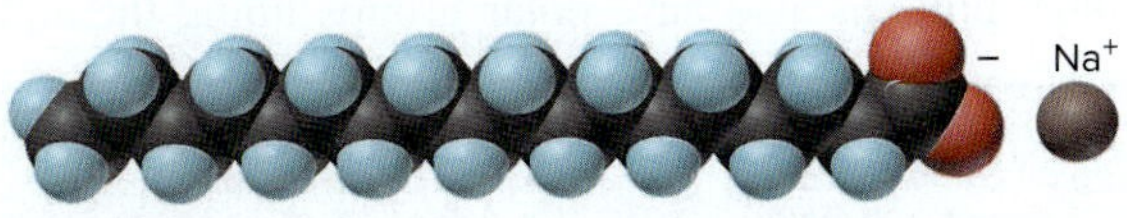
–
Na^+

Sodium stearate, $C_{17}H_{35}COONa$

When grease on your hands or clothes is immersed in soapy water, the soap molecules' nonpolar tails interact with the nonpolar grease molecules through dispersion forces, while the polar-ionic heads attract water molecules through ion-dipole forces and H bonds. Tiny aggregates of grease molecules, embedded with soap molecules whose polar-ionic heads stick into the water, are flushed away by added water (Figure 13.8).

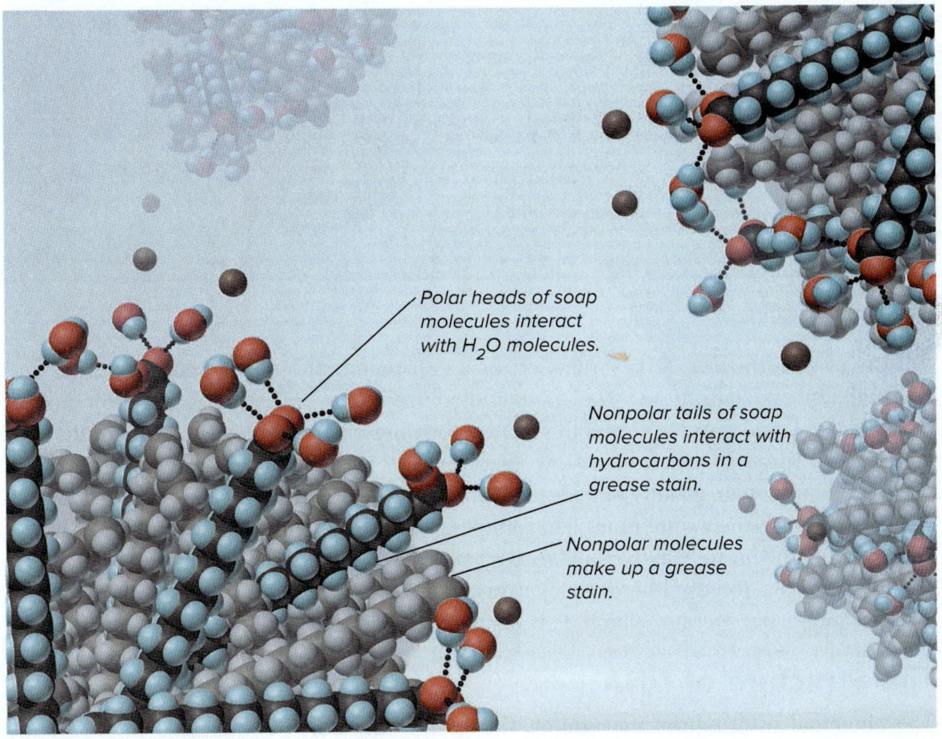
Polar heads of soap molecules interact with H_2O molecules.
Nonpolar tails of soap molecules interact with hydrocarbons in a grease stain.
Nonpolar molecules make up a grease stain.

Figure 13.8 The cleaning ability of a soap depends on the dual polarity of its molecules.

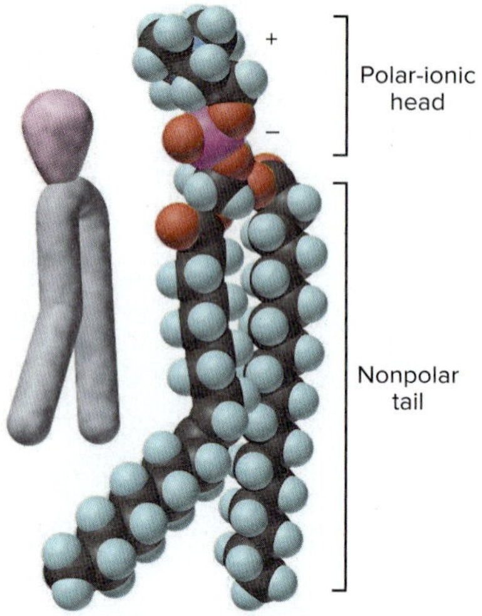
+
−
Polar-ionic head
Nonpolar tail

Figure 13.9 A membrane phospholipid. Lecithin (phosphatidylcholine), a phospholipid, is shown as a space-filling model and as a simplified purple-and-gray shape.

Lipid Bilayers and the Structure of the Cell Membrane The most abundant molecules in cell membranes are *phospholipids*. Like soaps, they have a dual polarity—a nonpolar tail consists of two fatty acid chains, and an organophosphate group is the polar-ionic head (Figure 13.9). Remarkably, phospholipids self-assemble in water into a sheetlike double layer called a **lipid bilayer,** with the tails of the two layers touching and the heads in the water. In the laboratory, bilayers form spherical vesicles that trap water inside. These structures are favored energetically because of their intermolecular forces:

- Ion-dipole forces occur between polar heads and water inside and outside.
- Dispersion forces occur between nonpolar tails within the bilayer interior.
- Minimal contact exists between nonpolar tails and water.

A typical animal cell membrane consists of a phospholipid bilayer with proteins partially embedded in it; a small portion of such a membrane appears in Figure 13.10. Membrane proteins, which play countless essential roles, differ fundamentally from soluble proteins in terms of their dual polarity:

- Soluble proteins have polar exteriors and nonpolar interiors. They form ion-dipole and H-bonding forces between water and *polar groups on the exterior* and dispersion forces between *nonpolar groups in the interior* (see Figure 13.7).
- Membrane proteins have exteriors that are partially polar *(red)* and partially nonpolar *(blue)*. They have polar groups on the *exterior portion that juts into the aqueous surroundings* and nonpolar groups on *the exterior portion embedded in the membrane*. These nonpolar groups form dispersion forces with the phospholipid tails of the bilayer. Channel proteins also have polar groups lining the aqueous channel.

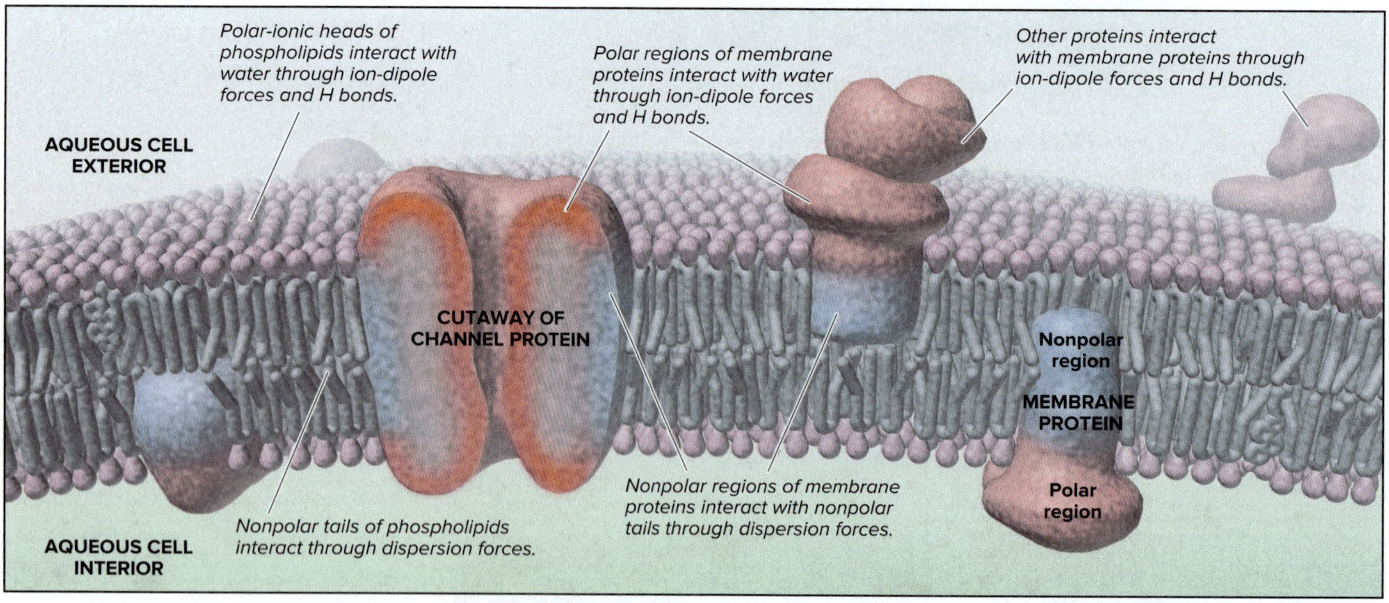
Polar-ionic heads of phospholipids interact with water through ion-dipole forces and H bonds.
Polar regions of membrane proteins interact with water through ion-dipole forces and H bonds.
Other proteins interact with membrane proteins through ion-dipole forces and H bonds.
AQUEOUS CELL EXTERIOR
CUTAWAY OF CHANNEL PROTEIN
Nonpolar region
MEMBRANE PROTEIN
Polar region
Nonpolar regions of membrane proteins interact with nonpolar tails through dispersion forces.
Nonpolar tails of phospholipids interact through dispersion forces.
AQUEOUS CELL INTERIOR

Figure 13.10 Intermolecular forces and cell membrane structure.

Action of Antibiotics A key function of a cell membrane is to balance internal and external ion concentrations: Na^+ is excluded from the cell, and K^+ is kept inside. Gramicidin A and similar antibiotics act by forming channels in the cell membrane of a bacterium through which ions flow (Figure 13.11). Two helical gramicidin A molecules, their nonpolar groups outside and polar groups inside, lie end to end to form a channel through the membrane. The nonpolar outside stabilizes the molecule in the cell membrane through dispersion forces, and the polar inside passes the ions along using ion-dipole forces, like a "bucket brigade." Over 10^7 ions pass through each of these channels per second, which disrupts the vital ion balance, and the bacterium dies.

The Structure of DNA

The chemical information that guides the design and construction, and therefore the function, of all proteins is contained in **nucleic acids,** unbranched polymers made up

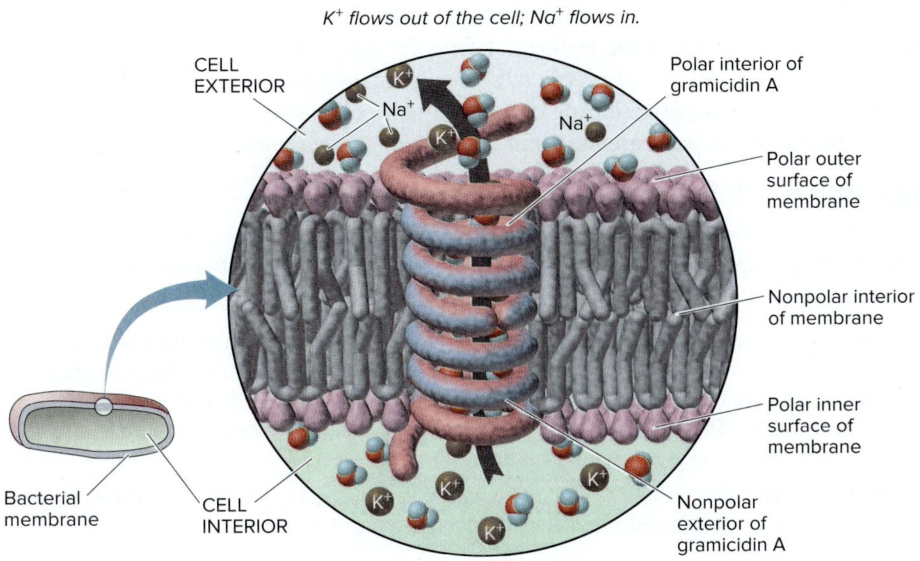

Figure 13.11 **The mode of action of the antibiotic gramicidin A.** The K^+ ions are shown leaving the cell. At the same time (not shown), Na^+ ions enter the cell.

of smaller units (monomers) called **mononucleotides.** Each mononucleotide consists of an N-containing base, a sugar, and a phosphate group (Figure 13.12). In DNA *(**d**eoxyribo**n**ucleic **a**cid),* the sugar is *2-deoxyribose,* in which —H substitutes for —OH on the second C atom of the five-C sugar ribose.

The repeating pattern of the DNA chain is *sugar linked to phosphate linked to sugar linked to phosphate,* and so on. Attached to each sugar is one of four N-containing bases, flat ring structures that dangle off the polynucleotide chain, similar to the way amino-acid side chains dangle off the polypeptide chain.

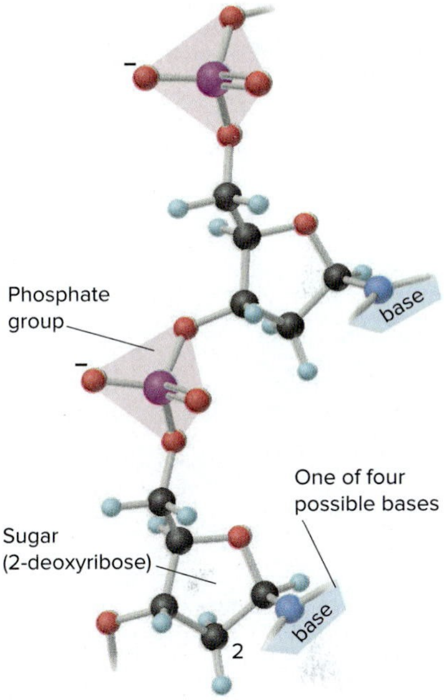

Figure 13.12 **A short portion of the polynucleotide chain of DNA.**

Intermolecular Forces and the Double Helix DNA exists as two chains wrapped around each other in a **double helix** that is stabilized by intermolecular forces (Figure 13.13):

- *On the more polar exterior,* negatively charged sugar-phosphate groups interact with the aqueous surroundings via ion-dipole forces and H bonds.
- *In the less polar interior,* flat, N-containing bases stack above each other and interact by dispersion forces.
- *Bases form specific interchain H bonds;* that is, each base in one chain is always H bonded with its complementary base in the other chain. Thus, *the base sequence of one chain is the H-bonded complement of the base sequence of the other.*

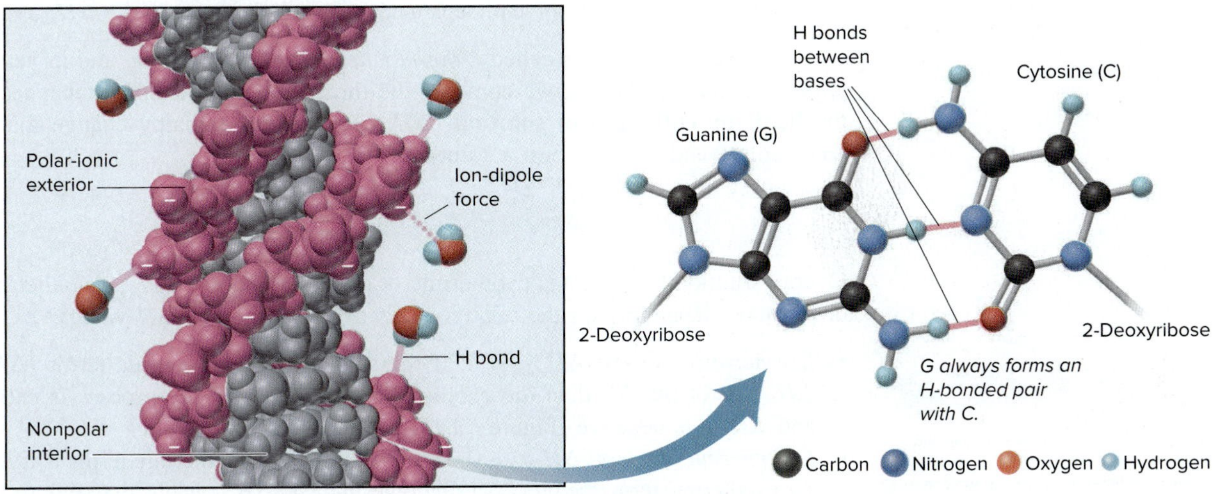

Figure 13.13 **The double helix of DNA.** A segment of DNA *(left)* has its polar-ionic sugar-phosphate portion *(pink)* facing the water and the nonpolar bases *(gray)* stacking in the interior. The expanded portion *(right)* shows an H-bonded pair of the bases guanine and cytosine.

A DNA molecule contains millions of H bonds linking bases in these prescribed pairs. The *total* energy of the H bonds keeps the chains together, but each H bond is weak enough (around 5% of a typical covalent single bond) that a few at a time can break as the chains separate during crucial cellular processes. (In Chapter 15, we'll see how H-bonded base pairs are essential for protein synthesis and DNA replication.)

› Summary of Section 13.2

- In soluble proteins, polar and ionic amino-acid side chains on the exterior interact with surrounding water, and nonpolar side chains in the interior interact with each other.
- With polar-ionic heads *and* nonpolar tails, a soap dissolves grease *and* interacts with water.
- Like soaps, phospholipids have dual polarity. They assemble into a water-impermeable lipid bilayer. In a cell membrane, the embedded portions of membrane proteins have exterior nonpolar side chains that interact with the nonpolar tails in the lipid bilayer through dispersion forces. Some antibiotics form channels with nonpolar exteriors and polar interiors that shuttle ions through the cell membrane.
- DNA forms a double helix with a sugar-phosphate, polar-ionic exterior. In the interior, N-containing bases H bond in specific pairs and stack through dispersion forces.

13.3 WHY SUBSTANCES DISSOLVE: BREAKING DOWN THE SOLUTION PROCESS

The qualitative *macroscopic* rule "like dissolves like" is based on *molecular* interactions between the solute and the solvent. To see *why* like dissolves like, we'll break down the solution process conceptually into steps and examine each of them quantitatively.

The Heat of Solution and Its Components

Before a solution forms, solute particles (ions or molecules) are attracting each other, as are solvent particles (molecules). For one to dissolve in the other, three steps must take place, each accompanied by an enthalpy change:

Step 1. Solute particles separate from each other. This step involves overcoming intermolecular (or ionic) attractions, so it is *endothermic:*

$$\text{Solute (aggregated)} + heat \longrightarrow \text{solute (separated)} \qquad \Delta H_{\text{solute}} > 0$$

Step 2. Solvent particles separate from each other. This step also involves overcoming attractions, so it is *endothermic,* too:

$$\text{Solvent (aggregated)} + heat \longrightarrow \text{solvent (separated)} \qquad \Delta H_{\text{solvent}} > 0$$

Step 3. Solute and solvent particles mix and form a solution. The different particles attract each other and come together, so this step is *exothermic:*

$$\text{Solute (separated)} + \text{solvent (separated)} \longrightarrow \text{solution} + heat \qquad \Delta H_{\text{mix}} < 0$$

The overall process is called a *thermochemical solution cycle,* and in yet another application of Hess's law, we combine the three individual enthalpy changes to find the **heat (or enthalpy) of solution (ΔH_{soln})**, the total enthalpy change that occurs when solute and solvent form a solution:

$$\Delta H_{\text{soln}} = \Delta H_{\text{solute}} + \Delta H_{\text{solvent}} + \Delta H_{\text{mix}} \qquad \textbf{(13.1)}$$

Overall solution formation is exothermic or endothermic, and ΔH_{soln} is either negative or positive, depending on the relative sizes of the individual ΔH values:

- *Exothermic process:* $\Delta H_{\text{soln}} < 0$. If the sum of the endothermic terms ($\Delta H_{\text{solute}} + \Delta H_{\text{solvent}}$) is *smaller* than the exothermic term (ΔH_{mix}), the process is exothermic and ΔH_{soln} is negative (Figure 13.14A).
- *Endothermic process:* $\Delta H_{\text{soln}} > 0$. If the sum of the endothermic terms is *larger* than the exothermic term, the process is endothermic and ΔH_{soln} is positive (Figure 13.14B). If ΔH_{soln} is highly positive, the solute may not dissolve significantly in that solvent.

Student Hot Spot

Student data indicate that you may struggle with the concept of energy changes and the solution process. Access the Smartbook to view additional Learning Resources on this topic.

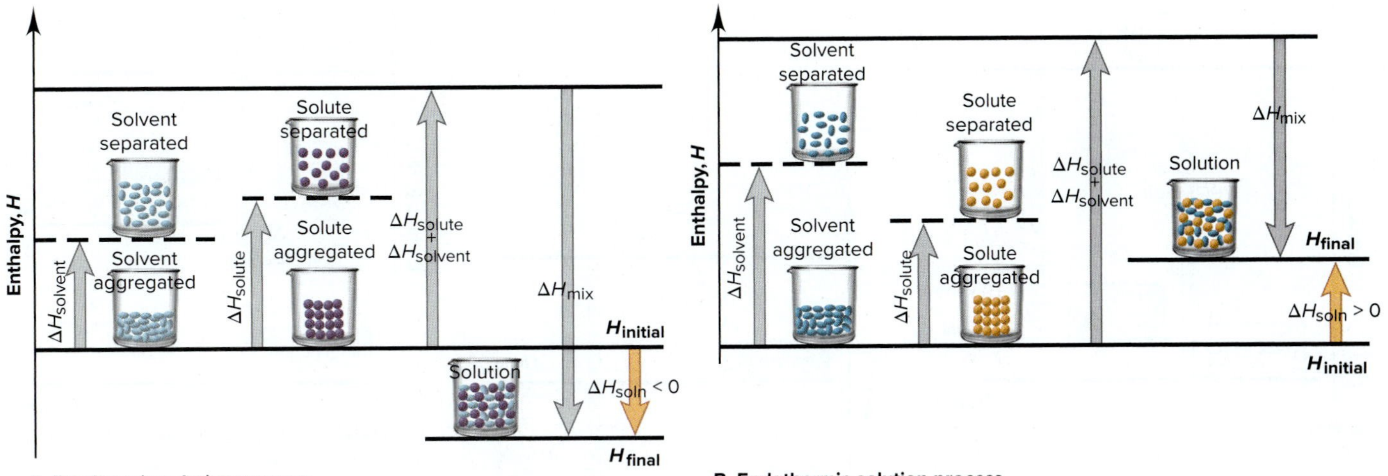

A Exothermic solution process   **B** Endothermic solution process

Figure 13.14 Enthalpy components of the heat of solution. **A,** ΔH_{mix} is larger than the sum of ΔH_{solute} and $\Delta H_{solvent}$, so ΔH_{soln} is negative. **B,** ΔH_{mix} is smaller than the sum of ΔH_{solute} and $\Delta H_{solvent}$, so ΔH_{soln} is positive.

The Heat of Hydration: Dissolving Ionic Solids in Water

The $\Delta H_{solvent}$ and ΔH_{mix} components of the solution cycle are difficult to measure individually. Combined, they equal the enthalpy change for **solvation,** the process of surrounding a solute particle with solvent particles:

$$\Delta H_{solvation} = \Delta H_{solvent} + \Delta H_{mix}$$

Solvation in water is called **hydration.** Thus, enthalpy changes for separating the water molecules ($\Delta H_{solvent}$) and mixing the separated solute with them (ΔH_{mix}) are combined into the **heat** (or enthalpy) **of hydration (ΔH_{hydr}).** In water, Equation 13.1 becomes

$$\Delta H_{soln} = \Delta H_{solute} + \Delta H_{hydr}$$

The heat of hydration is a key factor in dissolving an ionic solid. Breaking the H bonds in water is more than compensated for by forming the stronger ion-dipole forces, so hydration of an ion is *always* exothermic. The ΔH_{hydr} of an ion is defined as the enthalpy change for the hydration of 1 mol of separated (gaseous) ions:

$$M^+(g)\ [\text{or } X^-(g)] \xrightarrow{H_2O} M^+(aq)\ [\text{or } X^-(aq)] \qquad \Delta H_{\text{hydr of the ion}}\ (\text{always} <0)$$

Importance of Charge Density Heats of hydration exhibit trends based on the ion's **charge density,** the ratio of its charge to its volume. In general, the higher the charge density, the more negative ΔH_{hydr} is. Coulomb's law says that the higher the charge of an ion and the smaller its radius, the closer it gets to the oppositely charged pole of an H_2O molecule (see Figure 2.12), and the stronger the attraction. Thus,

- A 2+ ion attracts H_2O molecules more strongly than a 1+ ion of similar size.
- A small 1+ ion attracts H_2O molecules more strongly than a large 1+ ion.

Periodic trends in ΔH_{hydr} values are based on trends in charge density (Table 13.4):

- *Down a group,* the charge stays the same and the size increases; thus, the charge densities decrease, as do the ΔH_{hydr} values.
- *Across a period,* say, from Group 1A(1) to Group 2A(2), the 2A ion has a smaller radius *and* a higher charge, so its charge density and ΔH_{hydr} are greater.

Components of Aqueous Heats of Solution To separate an ionic solute, MX, into gaseous ions requires a lot of energy (ΔH_{solute}); recall from Chapter 9 that this is the lattice energy, and it is highly positive:

$$MX(s) \longrightarrow M^+(g) + X^-(g) \qquad \Delta H_{solute}\ (\text{always} > 0) = \Delta H_{lattice}$$

Thus, for ionic compounds in water, the heat of solution is the lattice energy (always positive) plus the combined heats of hydration of the ions (always negative):

$$\Delta H_{soln} = \Delta H_{lattice} + \Delta H_{\text{hydr of the ions}} \qquad (13.2)$$

Table 13.4 Trends in Ionic Heats of Hydration

Ion	Ionic Radius (pm)	ΔH_{hydr} (kJ/mol)
Group 1A(1)		
Na^+	102	−410
K^+	138	−336
Rb^+	152	−315
Cs^+	167	−282
Group 2A(2)		
Mg^{2+}	72	−1903
Ca^{2+}	100	−1591
Sr^{2+}	118	−1424
Ba^{2+}	135	−1317
Group 7A(17)		
F^-	133	−431
Cl^-	181	−313
Br^-	196	−284
I^-	220	−247

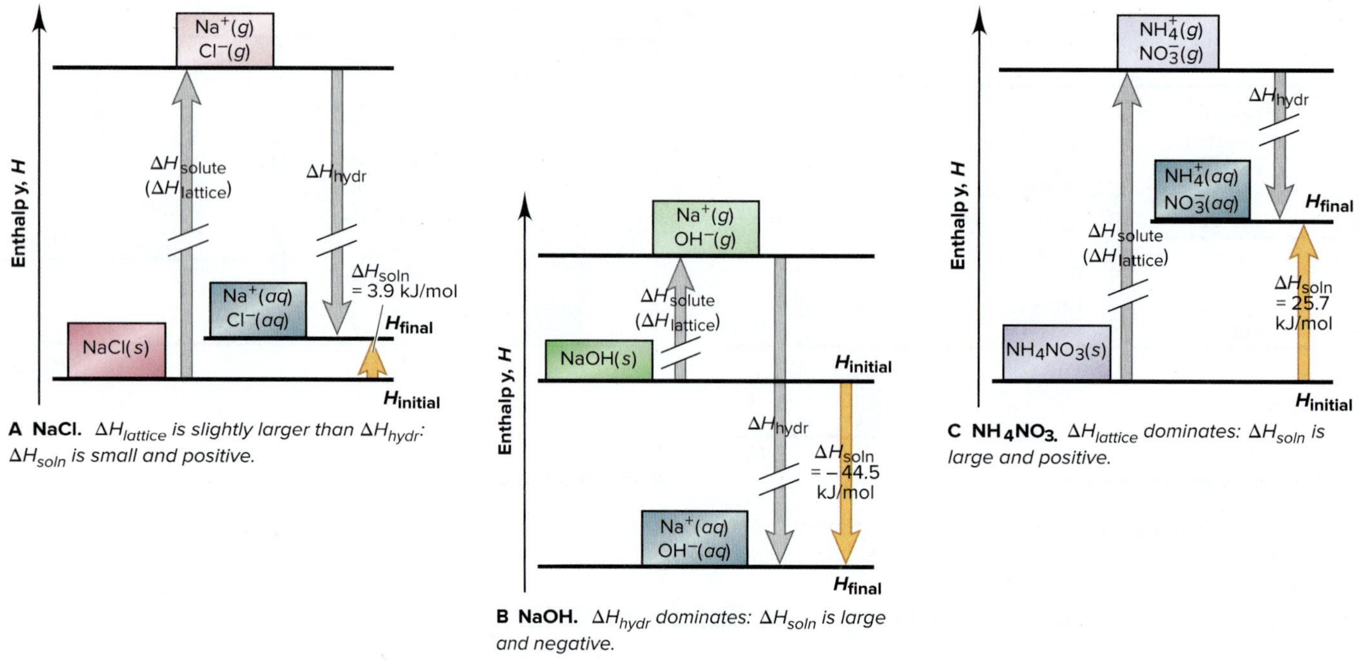

Figure 13.15 Enthalpy diagrams for three ionic compounds dissolving in water.

Once again, the sizes of the individual terms determine the sign of ΔH_{soln}.

Figure 13.15 shows qualitative enthalpy diagrams for three ionic solutes dissolving in water:

- *NaCl.* Sodium chloride has a small positive ΔH_{soln} (3.9 kJ/mol) because its lattice energy is only slightly greater than the combined ionic heats of hydration: if you dissolve NaCl in water in a flask, you don't feel the small temperature change.
- *NaOH.* Sodium hydroxide has a large negative ΔH_{soln} (–44.5 kJ/mol) because its lattice energy is much *smaller* than the combined ionic heats of hydration: if you dissolve NaOH in water, the flask feels hot.
- *NH_4NO_3.* Ammonium nitrate has a large positive ΔH_{soln} (25.7 kJ/mol) because its lattice energy is much *larger* than the combined ionic heats of hydration: if you dissolve NH_4NO_3 in water, the flask feels cold.

Hot and cold "packs" consist of a thick outer pouch of water and a thin inner pouch of a salt. A squeeze breaks the inner pouch, and the salt dissolves. Most hot packs use anhydrous $CaCl_2$ ($\Delta H_{soln} = -82.8$ kJ/mol). In Japan, some soup is sold in double-walled cans with a salt in a packet immersed in water between the walls. Open the can and the packet breaks, the salt dissolves, and the soup quickly warms to about 90°C. Cold packs use NH_4NO_3 (ΔH_{soln} = 25.7 kJ/mol). A cold pack can keep the solution at 0°C for about half an hour, long enough to soothe a sprain.

SAMPLE PROBLEM 13.2 Calculating an Aqueous Ionic Heat of Solution

Problem With secondary applications ranging from sedative to fire retardant, calcium bromide is used primarily in concentrated solution as an industrial drilling fluid.
(a) Use Table 13.4 and the lattice energy (2132 kJ/mol) to find the heat of solution (kJ/mol) of calcium bromide.
(b) Draw an enthalpy diagram for this solution process.

(a) Calculating the heat of solution of $CaBr_2$.

Plan We are given, or can look up, the individual enthalpy components for a salt dissolving in water and have to determine their signs to calculate the overall heat of solution (ΔH_{soln}). The components are the lattice energy (the heat absorbed when the solid separates into gaseous ions) and the heat of hydration for each ion (the heat released when the ion becomes hydrated). The lattice energy is always positive, so $\Delta H_{lattice}$ = 2132 kJ/mol. Heats of hydration are always negative, so from Table 13.4, ΔH_{hydr} of Ca^{2+} = −1591 kJ/mol and ΔH_{hydr} of Br^- = −284 kJ/mol. We use Equation 13.2, noting that there are 2 mol of Br^-, to obtain ΔH_{soln}.

Solution

$$\begin{aligned}\Delta H_{soln} &= \Delta H_{lattice} + \Delta H_{hydr\ of\ the\ ions}\\ &= \Delta H_{lattice} + \Delta H_{hydr}\text{ of }Ca^{2+} + 2(\Delta H_{hydr}\text{ of }Br^-)\\ &= 2132\text{ kJ/mol} + (-1591\text{ kJ/mol}) + 2(-284\text{ kJ/mol})\\ &= -27\text{ kJ/mol}\end{aligned}$$

Check Rounding to check the math gives

$$2100\text{ kJ/mol} - 1600\text{ kJ/mol} - 560\text{ kJ/mol} = -60\text{ kJ/mol}$$

This small negative sum indicates that our answer is correct.

(b) Drawing an enthalpy diagram for the process of dissolving $CaBr_2$.

Plan Along a vertical enthalpy axis, the lattice energy (endothermic) is represented by an upward arrow leading from solid salt to gaseous ions. The hydration of the ions (exothermic) is represented by a downward arrow from gaseous to hydrated ions. From part (a), ΔH_{soln} is small and negative, so the downward arrowhead is slightly below the tail of the upward arrow.

Solution The enthalpy diagram is shown in the margin.

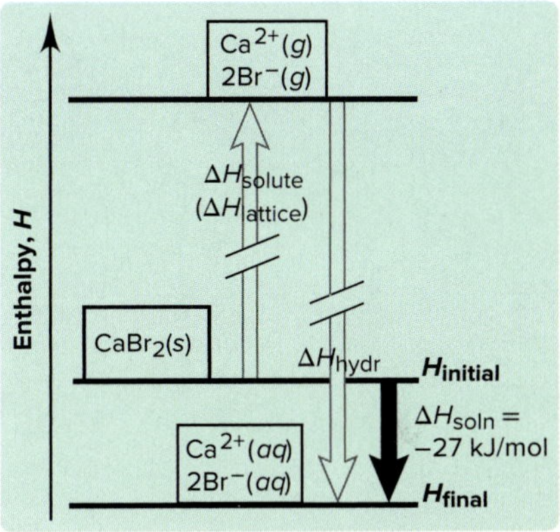

FOLLOW-UP PROBLEMS

13.2A Use the following data to find the combined heat of hydration for the ions in KNO_3: ΔH_{soln} = 34.89 kJ/mol and $\Delta H_{lattice}$ = 685 kJ/mol.

13.2B Use the following data to find the heat of hydration of CN^-: ΔH_{soln} of NaCN = 1.21 kJ/mol, $\Delta H_{lattice}$ of NaCN = 766 kJ/mol, and ΔH_{hydr} of Na^+ = −410. kJ/mol.

SOME SIMILAR PROBLEMS 13.30, 13.31, 13.36, and 13.37

The Solution Process and the Change in Entropy

The heat of solution (ΔH_{soln}) is one of two factors that determine whether a solute dissolves. The other factor is the natural tendency of a system of particles to spread out, which results in the system's kinetic energy becoming more dispersed or more widely distributed. A thermodynamic variable called **entropy (S)** is directly related to the number of ways a system can distribute its energy, which involves the freedom of motion of the particles.

Let's see what it means for a system to "distribute its energy." We'll first compare the three physical states and then compare solute and solvent with solution.

Entropy and the Three Physical States The states of matter differ significantly in their entropy.

- In a solid, the particles are fixed in their positions with little freedom of motion. In a liquid, they can move around each other and so have greater freedom of motion. And in a gas, the particles have little restriction and much more freedom of motion.
- The more freedom of motion the particles have, the more ways they can distribute their kinetic energy; thus, a liquid has higher entropy than a solid, and a gas has higher entropy than a liquid:

$$S_{gas} > S_{liquid} > S_{solid}$$

- Thus, there is a *change* in entropy (ΔS) associated with a phase change, and it can be positive or negative. For example, when a liquid vaporizes, the change in entropy ($\Delta S_{vap} = S_{gas} - S_{liquid}$) is positive ($\Delta S_{vap} > 0$, increase in entropy) since $S_{gas} > S_{liquid}$; when a liquid freezes (fusion), the change in entropy ($\Delta S_{fus} = S_{solid} - S_{liquid}$) is negative ($\Delta S_{fus} < 0$, decrease in entropy) since $S_{solid} < S_{liquid}$.

Entropy and the Formation of Solutions The formation of solutions also involves a change in entropy. *A solution usually has higher entropy than the pure solute and pure solvent* because the number of ways to distribute the energy is related to the number of interactions between different molecules. There are far more interactions possible when solute and solvent are mixed than when they are pure; thus,

$$S_{soln} > (S_{solute} + S_{solvent}) \qquad \text{or} \qquad \Delta S_{soln} > 0$$

You know from everyday experience that solutions form naturally, but pure solutes and solvents don't: you've seen sugar dissolve in water, but you've never seen a sugar solution separate into pure sugar and water. In Chapter 20, we'll see that energy is needed to reverse the natural tendency of systems to distribute their energy—to get "mixed up." Water treatment plants, oil refineries, steel mills, and many other industrial facilities expend a lot of energy to separate mixtures into pure components.

Enthalpy vs. Entropy Changes in Solution Formation Solution formation involves the interplay of two factors: systems change toward a state of *lower enthalpy* and *higher entropy,* so the relative sizes of ΔH_{soln} and ΔS_{soln} determine whether a solution forms. Let's consider three solute-solvent pairs to see which factor dominates in each case:

1. *NaCl in hexane*. Given their very different intermolecular forces, we predict that sodium chloride does *not* dissolve in hexane (C_6H_{14}). An enthalpy diagram (Figure 13.16A) shows that separating the nonpolar solvent is easy because the dispersion forces are weak ($\Delta H_{solvent} \geq 0$), but separating the solute requires supplying the very large $\Delta H_{lattice}$ ($\Delta H_{solute} \gg 0$). Mixing releases little heat because ion–induced dipole forces between Na^+ (or Cl^-) and hexane are weak ($\Delta H_{mix} \leq 0$). Because the sum of the endothermic terms is *much* larger than the exothermic term, $\Delta H_{soln} \gg 0$. *A solution does not form because the entropy increase from mixing solute and solvent would be much smaller than the enthalpy increase required to separate the solute:* $\Delta S_{mix} \ll \Delta H_{solute}$.

2. *Octane in hexane*. We predict that octane (C_8H_{18}) is soluble in hexane because both are held together by dispersion forces of similar strength; in fact, these two substances are miscible. That is, both ΔH_{solute} and $\Delta H_{solvent}$ are around zero. The similar forces mean that ΔH_{mix} is also around zero. And a lot of heat is not released; in fact, ΔH_{soln} is around zero (Figure 13.16B). So why does a solution form so readily? *With no enthalpy change driving the process, octane dissolves in hexane because the entropy increases greatly when the pure substances mix:* $\Delta S_{mix} \gg \Delta H_{soln}$.

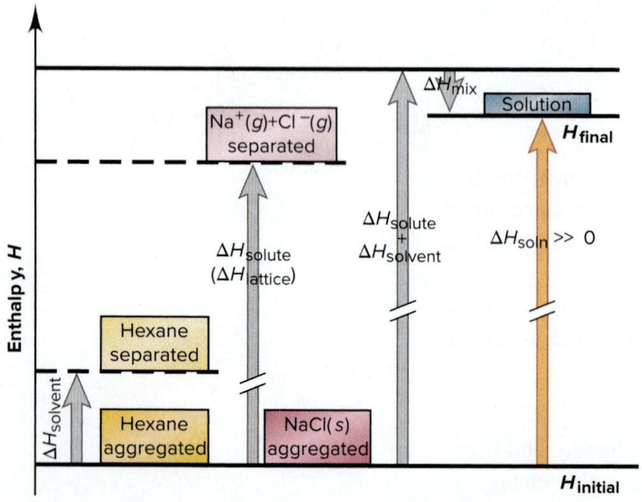

A NaCl. ΔH_{mix} *is* ***much*** *smaller than* ΔH_{solute}*;* ΔH_{soln} *is so much larger than the entropy increase due to mixing that NaCl does* ***not*** *dissolve.*

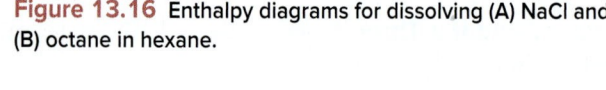

Figure 13.16 Enthalpy diagrams for dissolving (A) NaCl and (B) octane in hexane.

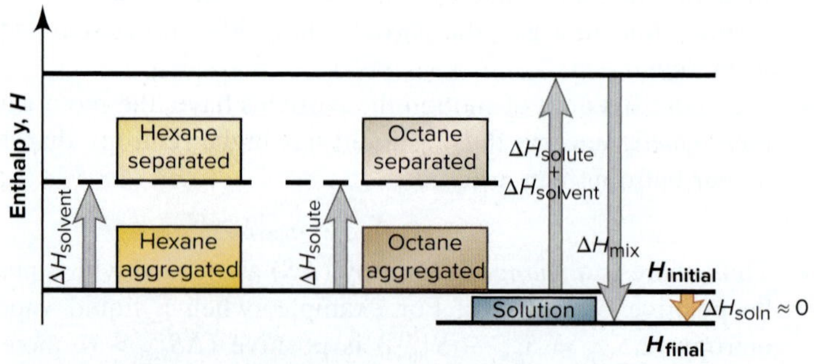

B Octane. ΔH_{soln} *is very small, but the entropy increase due to mixing is large, so octane* ***does*** *dissolve.*

3. *NH_4NO_3 in water.* A large enough increase in entropy can sometimes cause a solution to form even when the enthalpy increase is large ($\Delta H_{soln} \gg 0$). As we saw previously, in Figure 13.15C, when ammonium nitrate dissolves in water, the process is highly endothermic; that is, $\Delta H_{lattice} \gg \Delta H_{hydr\ of\ the\ ions}$. Nevertheless, *the increase in entropy that occurs when the crystal breaks down and the ions mix with water molecules is greater than the increase in enthalpy: $\Delta S_{soln} > \Delta H_{soln}$.*

In Chapter 20, we'll return in depth to the relation between enthalpy and entropy to understand physical and chemical systems.

Figure 13.17 Equilibrium in a saturated solution. At some temperature, the number of solute particles dissolving *(white arrows)* per unit time equals the number recrystallizing *(black arrows).*

› Summary of Section 13.3

- In a thermochemical solution cycle, the heat of solution is the sum of the endothermic separations of solute and of solvent and the exothermic mixing of their particles.
- In water, solvation (surrounding solute particles with solvent) is called hydration. For ions, heats of hydration depend on the ion's charge density but are always negative because ion-dipole forces are strong. Charge density exhibits periodic trends.
- Systems naturally increase their entropy (distribute their energy in more ways). A gas has higher entropy than a liquid, which has higher entropy than a solid, and a solution has higher entropy than the pure solute and solvent.
- Relative sizes of the enthalpy and entropy changes determine solution formation. A substance with a positive ΔH_{soln} dissolves *only* if ΔS_{soln} is larger than ΔH_{soln}.

13.4 SOLUBILITY AS AN EQUILIBRIUM PROCESS

When an excess amount of solid is added to a solvent, particles leave the crystal, are surrounded by solvent, and move away. Some dissolved solute particles collide with undissolved solute and recrystallize, but, as long as the rate of dissolving is greater than the rate of recrystallizing, the concentration rises. At a given temperature, when solid is dissolving at the same rate as dissolved particles are recrystallizing, the concentration remains constant and *undissolved solute is in equilibrium with dissolved solute:*

$$\text{Solute (undissolved)} \rightleftharpoons \text{solute (dissolved)}$$

Figure 13.17 shows an ionic solid in equilibrium with dissolved cations and anions. (We learned about the concept of equilibrium in Section 4.7 and saw how it occurs between phases in Section 12.2.)

Three terms express the extent of this solution process:

- A **saturated solution** is *at* equilibrium and contains the maximum amount of dissolved solute at a given temperature in the presence of undissolved solute. Therefore, if you filter off the solution and add more solute, the added solute doesn't dissolve.
- An **unsaturated solution** contains *less* than the equilibrium concentration of dissolved solute; add more solute, and more will dissolve until the solution is saturated.
- A **supersaturated solution** contains *more* than the equilibrium concentration and is unstable relative to the saturated solution. You can often prepare a supersaturated solution if the solute is more soluble at higher temperature. While heating, dissolve more than the amount required for saturation at some lower temperature, and then slowly cool the solution. If the excess solute remains dissolved, the solution is supersaturated. Add a "seed" crystal of solute or tap the container, and the excess solute crystallizes immediately, leaving behind a saturated solution (Figure 13.18).

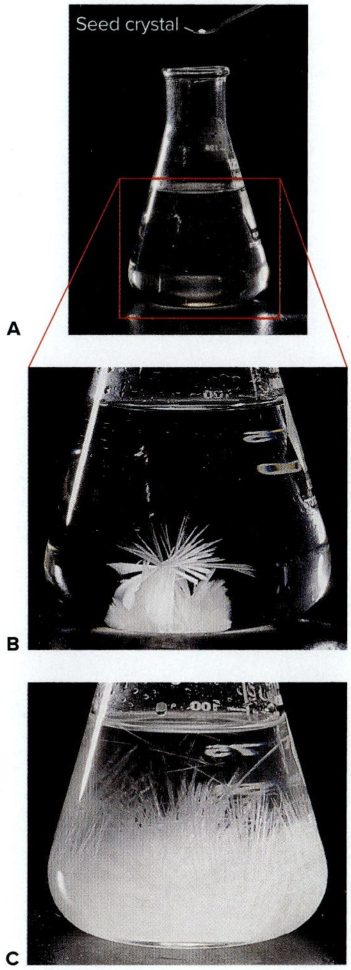

Figure 13.18 Sodium acetate crystallizing from a supersaturated solution. When a seed crystal of sodium acetate is added to a supersaturated solution of the compound **(A),** solute begins to crystallize **(B)** and continues until the remaining solution is saturated **(C).**

Source: A–C: © McGraw-Hill Education/ Stephen Frisch, photographer

Effect of Temperature on Solubility

You know that more sugar dissolves in hot tea than in iced tea; in fact, temperature affects the solubility of most substances. Let's examine the effects of temperature on the solubility of solids and gases.

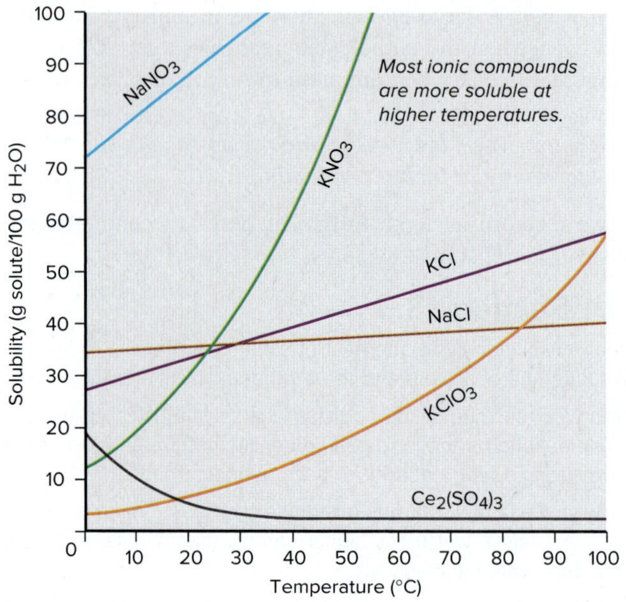

Figure 13.19 Relation between solubility and temperature for several ionic compounds.

Temperature and the Solubility of Solids in Water Like sugar, *most solids are more soluble at higher temperatures* (Figure 13.19). Note that cerium sulfate is one of several exceptions, most of which are other sulfates. Some salts have higher solubility up to a certain temperature and then lower solubility at still higher temperatures.

The effect of temperature on solubility is a complex phenomenon, and the sign of ΔH_{soln} does not reflect that complexity. Tabulated ΔH_{soln} values give the enthalpy change for making a solution at the standard state of 1 *M*; but in order to understand the effect of temperature, we need to know the sign of the enthalpy change very close to the point of saturation, which may differ from the sign of the tabulated value. For example, tables give a negative ΔH_{soln} for NaOH and a positive one for NH_4NO_3, yet both compounds are more soluble at higher temperatures. Thus, even though the effect of temperature indicates the equilibrium nature of solubility, no single measure can predict the effect for a given solute.

Temperature and the Solubility of Gases in Water The effect of temperature on *gas* solubility is much more predictable. When a solid dissolves in a liquid, the solute particles must separate, so $\Delta H_{solute} > 0$. In contrast, gas particles are already separated, so $\Delta H_{solute} \approx 0$. Because hydration is exothermic ($\Delta H_{hydr} < 0$), the sum of these two terms is negative. Thus, $\Delta H_{soln} < 0$ for all gases in water:

$$\text{Solute}(g) + \text{water}(l) \rightleftharpoons \text{saturated solution}(aq) + \text{heat}$$

Thus, *gas solubility in water* ***decreases*** *with rising temperature (addition of heat).* Gases have weak intermolecular forces with water. When the temperature rises, the average kinetic energy increases, allowing the gas particles to easily overcome these forces and re-enter the gas phase.

This behavior leads to *thermal pollution.* Many electric power plants use large amounts of water from a nearby river or lake for cooling, and the warmed water is returned to the source. The metabolic rates of fish and other aquatic animals increase in this warmer water, increasing their need for O_2. But the concentration of dissolved O_2 is lower in warm water, so they become "oxygen deprived." Also, the less dense warm water floats and prevents O_2 from reaching the cooler water below. Thus, even creatures at deeper levels become oxygen deprived. Farther from the plant, the water temperature and O_2 solubility return to normal. To mitigate the problem, cooling towers lower the temperature of the water before it exits the plant (Figure 13.20); nuclear power plants use a similar approach (Section 24.7).

Figure 13.20 Preventing thermal pollution with cooling towers.

Source: © Kristin Smith/Shutterstock.com

Effect of Pressure on Solubility

Pressure has little effect on the solubility of liquids and solids because they are almost incompressible. But it has a *major* effect on the solubility of gases. Consider a piston-cylinder assembly with a gas above a saturated aqueous solution of the gas (Figure 13.21A). At equilibrium, at a given pressure, the same number of gas molecules enter and leave the solution per unit time:

$$\text{Gas} + \text{solvent} \rightleftharpoons \text{saturated solution}$$

Push down on the piston, and you disturb the equilibrium: gas volume decreases, so gas pressure (and concentration) increases, and gas particles collide with the liquid surface more often. Thus, more particles enter than leave the solution per unit time (Figure 13.21B). More gas dissolves to reduce this disturbance (a shift to the right in the preceding equation) until the system re-establishes equilibrium (Figure 13.21C).

The relation between gas pressure and solubility has many familiar applications. In a closed can of cola, dissolved CO_2 is in equilibrium with 4 atm of CO_2 in the small volume above the solution. Open the can and the dissolved CO_2 bubbles out of solution until the drink goes "flat" because its CO_2 concentration reaches equilibrium with CO_2 in air ($P_{CO_2} = 4\times10^{-4}$ atm). In a different situation *(see photo)*, scuba divers who breathe compressed air have a high concentration of N_2 dissolved in their blood. If they ascend too quickly, they may suffer decompression sickness (the "bends"), in which lower external pressure allows the dissolved N_2 to form bubbles in the blood that block capillaries.

What happens to gases dissolved in the blood of a scuba diver?

Source: © Peter Scoones/Science Source

Henry's law expresses the quantitative relationship between gas pressure and solubility: *the solubility of a gas (S_{gas}) is directly proportional to the partial pressure of the gas (P_{gas}) above the solution:*

$$S_{\text{gas}} = k_{\text{H}} \times P_{\text{gas}} \qquad \textbf{(13.3)}$$

where k_H is the *Henry's law constant* and is specific for a given gas-solvent combination at a given temperature. With S_{gas} in mol/L and P_{gas} in atm, the units of k_H are mol/L·atm, or $\text{mol}\cdot\text{L}^{-1}\cdot\text{atm}^{-1}$.

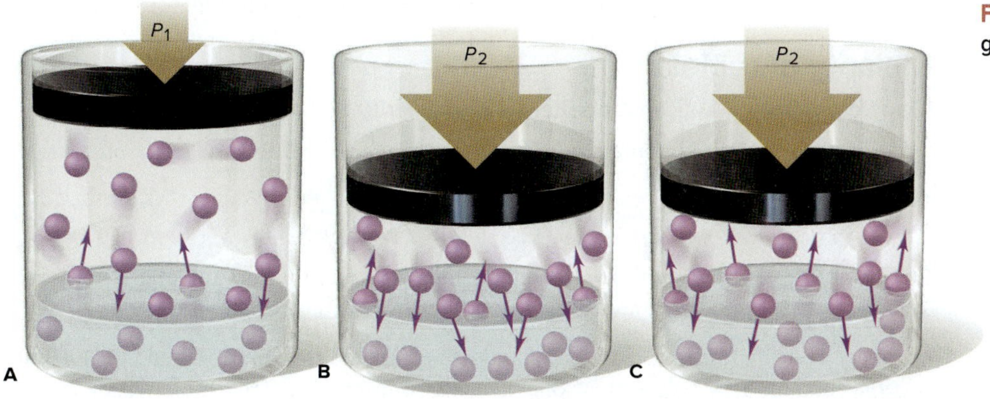

Figure 13.21 The effect of pressure on gas solubility.

SAMPLE PROBLEM 13.3 Using Henry's Law to Calculate Gas Solubility

Problem The partial pressure of carbon dioxide gas inside a bottle of cola is 4 atm at 25°C. What mass of CO_2 is dissolved in a 0.5-L bottle of cola? The Henry's law constant for CO_2 in water is 3.3×10^{-2} mol/L·atm at 25°C.

Plan We know P_{CO_2} (4 atm) and the value of k_H (3.3×10^{-2} mol/L·atm), so we substitute them into Equation 13.3 to find S_{CO_2} in mol/L. Then we multiply the solubility by the volume of cola and molar mass of CO_2 to obtain the mass of dissolved CO_2.

Solution $S_{CO_2} = k_H \times P_{CO_2} = (3.3\times10^{-2}\text{ mol/L·atm})(4\text{ atm}) = 0.1\text{ mol/L}$

$$\text{Mass (g) } CO_2 = \frac{0.1\text{ mol}}{\text{L}} \times 0.5\text{ L} \times \frac{44.01\text{ g } CO_2}{1\text{ mol } CO_2} = 2\text{ g } CO_2$$

Check The units are correct. We rounded to one significant figure to match the number of significant figures in the given values for the pressure and the volume of cola.

FOLLOW-UP PROBLEMS

13.3A If air contains 78% N_2 by volume, what is the solubility of N_2 in water at 25°C and 1 atm (k_H for N_2 in H_2O at 25°C = 7×10^{-4} mol/L·atm)?

13.3B Dinitrogen monoxide (N_2O), commonly known as *laughing gas,* is mixed with oxygen for use as an analgesic during dental surgery. A mixture of N_2O and O_2 is 40.% N_2O by volume and has a total pressure of 1.2 atm; the solubility of N_2O in water at 25°C is 1.2×10^{-2} mol/L. Calculate k_H for N_2O in H_2O at this temperature.

SOME SIMILAR PROBLEMS 13.46 and 13.47

› Summary of Section 13.4

- A solution that contains the maximum amount of dissolved solute in the presence of excess undissolved solute is saturated. A saturated solution is in equilibrium with excess solute, because solute particles are entering and leaving the solution at the same rate. An unsaturated solution contains less than this amount.
- Most solids are more soluble at higher temperatures.
- All gases have a negative ΔH_{soln} in water, so heating lowers gas solubility in water.
- Henry's law says that the solubility of a gas is directly proportional to its partial pressure above the solution.

13.5 CONCENTRATION TERMS

Concentration is the *proportion* of a substance in a mixture, so it is an *intensive* property (like density and temperature), one that does not depend on the quantity of mixture: 1.0 L or 1.0 mL of 0.1 *M* NaCl have the same concentration. Concentration is a ratio of quantities (Table 13.5), most often solute to *solution,* but sometimes solute to *solvent.* Both parts of the ratio can be given in units of mass, volume, or amount (mol), and chemists express concentration by several terms, including molarity, molality, and various expressions of "parts of solute per part by solution."

Molarity and Molality

Two very common concentration terms are molarity and molality:

1. Molarity (*M*) is the *number of moles of solute dissolved in 1 L of solution:*

$$\text{Molarity } (M) = \frac{\text{amount (mol) of solute}}{\text{volume (L) of solution}} \qquad \textbf{(13.4)}$$

In Chapter 4, we used molarity to convert liters of solution into amount of dissolved solute. Molarity has two drawbacks that affect its use in precise work:

- *Effect of temperature.* A liquid expands when heated, so a unit volume of hot solution contains less solute than one of cold solution; thus, the molarity of the hot and cold solutions is different.

Table 13.5 Concentration Definitions

Concentration Term	Ratio
Molarity (M)	$\frac{\text{amount (mol) of solute}}{\text{volume (L) of solution}}$
Molality (m)	$\frac{\text{amount (mol) of solute}}{\text{mass (kg) of solvent}}$
Parts by mass	$\frac{\text{mass of solute}}{\text{mass of solution}}$
Parts by volume	$\frac{\text{volume of solute}}{\text{volume of solution}}$
Mole fraction (X)	$\frac{\text{amount (mol) of solute}}{\text{amount (mol) of solute} + \text{amount (mol) of solvent}}$

- *Effect of mixing*. Because of solute-solvent interactions that are difficult to predict, *volumes may not be additive:* adding 500. mL of one solution to 500. mL of another may not give 1000. mL of final solution.

2. **Molality (m)** does not contain volume in its ratio; it is *the number of moles of solute dissolved in 1000 g (1 kg) of solvent:*

$$\text{Molality } (m) = \frac{\text{amount (mol) of solute}}{\text{mass (kg) of solvent}} \quad (13.5)$$

Note that molality includes the quantity of *solvent,* not of solution. For precise work, molality has two advantages over molarity:

- *Effect of temperature*. Molal solutions are based on *masses* of components, not *volume*. Since mass does not change with temperature, neither does molality.
- *Effect of mixing*. Unlike volumes, masses *are* additive: adding 500. g of one solution to 500. g of another *does* give 1000. g of final solution.

For these reasons, molality is the preferred term when temperature, and hence density, may change, as in a study of physical properties. Note that, in the case of water, 1 L has a mass of 1 kg, so *molality and molarity are nearly the same for dilute aqueous solutions.*

SAMPLE PROBLEM 13.4 Calculating Molality

Problem What is the molality of a solution prepared by dissolving 32.0 g of $CaCl_2$ in 271 g of water?

Plan To use Equation 13.5, we convert the mass of $CaCl_2$ (32.0 g) to amount (mol) using the molar mass (g/mol) and then divide by the mass of water (271 g), being sure to convert from grams to kilograms (see the road map).

Solution Converting from grams of solute to moles:

$$\text{Amount (mol) of } CaCl_2 = 32.0 \text{ g } CaCl_2 \times \frac{1 \text{ mol } CaCl_2}{110.98 \text{ g } CaCl_2} = 0.288 \text{ mol } CaCl_2$$

Finding the molality:

$$\text{Molality} = \frac{\text{mol solute}}{\text{kg solvent}} = \frac{0.288 \text{ mol } CaCl_2}{271 \text{ g} \times \frac{1 \text{ kg}}{10^3 \text{ g}}}$$

$$= 1.06\ m\ CaCl_2$$

Check The answer seems reasonable: the given amount (mol) of $CaCl_2$ and mass (kg) of H_2O are about the same, so their ratio is about 1.

Road Map

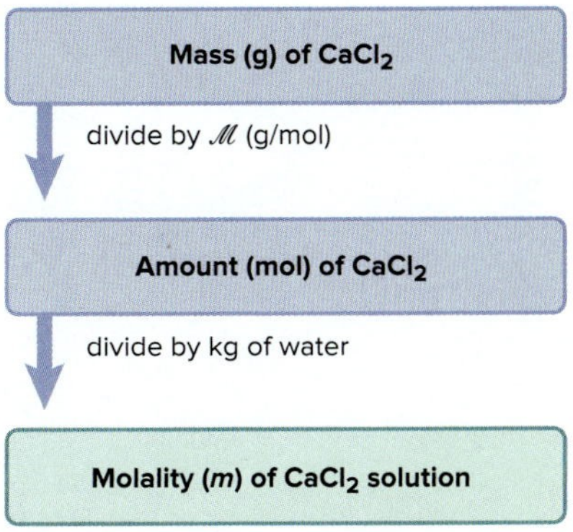

FOLLOW-UP PROBLEMS

13.4A How many grams of glucose ($C_6H_{12}O_6$) must be dissolved in 563 g of ethanol (C_2H_5OH) to prepare a 2.40×10^{-2} *m* solution?

13.4B What is the molality of a solution made by dissolving 15.20 g of I_2 in 1.33 mol of diethyl ether, $(CH_3CH_2)_2O$?

SOME SIMILAR PROBLEMS 13.64 and 13.65

Parts of Solute by Parts of Solution

Several concentration terms relate the parts of solute (or solvent) to the parts of *solution*. Both can be expressed in units of mass, volume, or amount (mol).

Parts by Mass The most common of these terms is **mass percent [% (w/w)],** which you encountered in Chapter 3 as the fraction by mass of an element in a compound. The word *percent* means "per hundred," so with respect to solution concentration, *mass percent* means mass of solute dissolved in 100. parts by mass of solution:

$$\text{Mass percent} = \frac{\text{mass of solute}}{\text{mass of solute} + \text{mass of solvent}} \times 100 = \frac{\text{mass of solute}}{\text{mass of solution}} \times 100 \quad \textbf{(13.6)}$$

Mass percent values appear commonly on jars of solid laboratory chemicals to indicate impurities.

Two very similar terms are parts per million (ppm) by mass and parts per billion (ppb) by mass, or grams of solute per million or per billion grams of solution: in Equation 13.6, you multiply by 10^6 or by 10^9, respectively, instead of by 100, to obtain these much smaller concentrations. Environmental toxicologists use these units to measure pollutants. For example, TCDD (*t*etra*c*hloro*d*ibenzo*d*ioxin) *(see margin)*, a byproduct of paper bleaching, is unsafe at soil levels above 1 ppb. From normal contact with air, water, and soil, North Americans have an average of 0.01 ppb TCDD in their tissues.

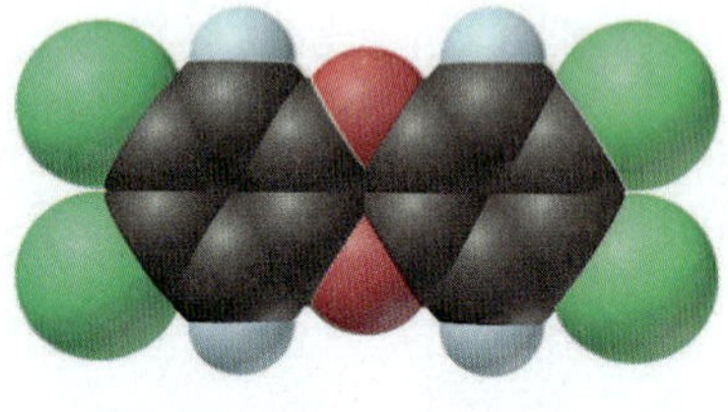

Is this deadly toxin in your body?

Parts by Volume The most common parts-by-volume term is **volume percent [% (v/v)],** the volume of solute in 100. volumes of solution:

$$\text{Volume percent} = \frac{\text{volume of solute}}{\text{volume of solution}} \times 100 \quad \textbf{(13.7)}$$

For example, rubbing alcohol is an aqueous solution of isopropanol (a three-carbon alcohol) that contains 70 volumes of alcohol in 100. volumes of solution, or 70% (v/v).

Parts by volume is often used to express tiny concentrations of liquids or gases. Minor atmospheric components occur in parts per million by volume (ppmv). For example, about 0.05 ppmv of the toxic gas carbon monoxide (CO) is in clean air, 1000 times as much (50 ppmv of CO) in air over urban traffic, and 10,000 times as much (500 ppmv of CO) in cigarette smoke. *Pheromones* are compounds secreted by members of a species to signal food, danger, sexual readiness, and so forth. Many organisms, including dogs and monkeys, release pheromones, and researchers suspect that humans do as well. Some insect pheromones, such as the sexual attractant of the gypsy moth, are active at a few hundred molecules per milliliter of air, 100 parts per quadrillion by volume.

A concentration term often used in health-related facilities for aqueous solutions is % (w/v), a ratio of solute *weight* (actually mass) to solution *volume*. Thus, a 1.5% (w/v) NaCl solution contains 1.5 g of NaCl per 100. mL of *solution*.

Parts by Mole: Mole Fraction The **mole fraction (*X*)** of a solute is the ratio of the number of moles of solute to the total number of moles (solute plus solvent):

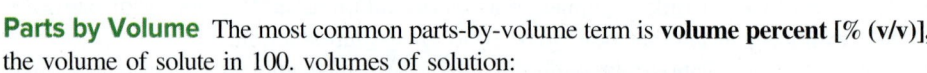

$$\text{Mole fraction } (X) = \frac{\text{amount (mol) of solute}}{\text{amount (mol) of solute} + \text{amount (mol) of solvent}} \quad \textbf{(13.8)}$$

Put another way, the mole fraction gives the proportion of solute (or solvent) particles in solution. The *mole percent* is the mole fraction expressed as a percentage:

$$\text{Mole percent (mol \%)} = \text{mole fraction} \times 100$$

SAMPLE PROBLEM 13.5 **Expressing Concentrations in Parts by Mass, Parts by Volume, and Mole Fraction**

Problem **(a)** Find the concentration of calcium ion (in ppm) in a 3.50-g pill that contains 40.5 mg of Ca^{2+}.
(b) The label on a 0.750-L bottle of Italian chianti says "11.5% alcohol by volume." How many liters of alcohol does the wine contain?
(c) A sample of rubbing alcohol contains 142 g of isopropyl alcohol (C_3H_7OH) and 58.0 g of water. What are the mole fractions of alcohol and water?

Plan **(a)** We know the mass of Ca^{2+} (40.5 mg) and the mass of the pill (3.50 g). We convert the mass of Ca^{2+} from mg to g, find the mass ratio of Ca^{2+} to pill, and multiply by 10^6 to obtain the concentration in ppm. **(b)** We know the volume % (11.5%, or 11.5 parts by volume of alcohol to 100. parts of chianti) and the total volume (0.750 L), so we use Equation 13.7 to find the number of liters of alcohol. **(c)** We know the mass and formula of each component, so we convert masses to amounts (mol) and apply Equation 13.8 to find the mole fractions.

Solution **(a)** Finding parts per million by mass of Ca^{2+}. Combining the steps, we have

$$\text{ppm } Ca^{2+} = \frac{\text{mass of } Ca^{2+}}{\text{mass of pill}} \times 10^6 = \frac{40.5 \text{ mg } Ca^{2+} \times \dfrac{1 \text{ g}}{10^3 \text{ mg}}}{3.50 \text{ g}} \times 10^6$$

$$= 1.16 \times 10^4 \text{ ppm } Ca^{2+}$$

(b) Finding volume (L) of alcohol:

$$\text{Volume percent} = \frac{\text{volume of solute}}{\text{volume of solution}} \times 100$$

$$\text{Volume of alcohol} = \text{volume of solution} \times \frac{\text{volume percent}}{100}$$

$$\text{Volume (L) of alcohol} = 0.750 \text{ L chianti} \times \frac{11.5 \text{ L alcohol}}{100. \text{ L chianti}}$$

$$= 0.0862 \text{ L}$$

(c) Finding mole fractions. Converting from mass (g) to amount (mol):

$$\text{Amount (mol) of } C_3H_7OH = 142 \text{ g } C_3H_7OH \times \frac{1 \text{ mol } C_3H_7OH}{60.09 \text{ g } C_3H_7OH} = 2.36 \text{ mol } C_3H_7OH$$

$$\text{Amount (mol) of } H_2O = 58.0 \text{ g } H_2O \times \frac{1 \text{ mol } H_2O}{18.02 \text{ g } H_2O} = 3.22 \text{ mol } H_2O$$

Calculating mole fractions:

$$X_{C_3H_7OH} = \frac{\text{moles of } C_3H_7OH}{\text{total moles}} = \frac{2.36 \text{ mol}}{2.36 \text{ mol} + 3.22 \text{ mol}}$$

$$= 0.423$$

$$X_{H_2O} = \frac{\text{moles of } H_2O}{\text{total moles}} = \frac{3.22 \text{ mol}}{2.36 \text{ mol} + 3.22 \text{ mol}}$$

$$= 0.577$$

Check **(a)** The mass ratio is about 0.04 g/4 g = 10^{-2}, and $10^{-2} \times 10^6 = 10^4$ ppm, so it seems correct. **(b)** The volume % is a bit more than 10%, so the volume of alcohol should be a bit more than 75 mL (0.075 L). **(c)** Always check that the *mole fractions add up to 1:* 0.423 + 0.577 = 1.000.

FOLLOW-UP PROBLEMS

13.5A An alcohol solution contains 35.0 g of 1-propanol (C_3H_7OH) and 150. g of ethanol (C_2H_5OH). Calculate the mass percent and the mole fraction of each alcohol.

13.5B A sample of gasoline contains 1.87 g of ethanol (C_2H_5OH), 27.4 g of 2,2,4-trimethylpentane (iso-octane, C_8H_{18}), and 4.10 g of heptane (C_7H_{16}). Calculate the mass percent and the mole percent of each component.

SOME SIMILAR PROBLEMS 13.70 and 13.71

Interconverting Concentration Terms

All the terms we just discussed represent different ways of expressing concentration, so they are interconvertible. Keep these points in mind:

- To convert a term based on amount to one based on mass, you need the molar mass. These conversions are similar to the mass-mole conversions you've done earlier.
- To convert a term based on mass to one based on volume, you need the solution *density*. Given the mass of solution, the density (mass/volume) gives the volume, or vice versa.
- Molality includes quantity of *solvent;* the other terms include quantity of *solution.*

Student Hot Spot

Student data indicate that you may struggle with converting from one concentration term to another. Access the SmartBook to view additional Learning Resources on this topic.

SAMPLE PROBLEM 13.6 Interconverting Concentration Terms

Problem Hydrogen peroxide is a powerful oxidizing agent; it is used in concentrated solution in rocket fuel and in dilute solution in hair bleach. An aqueous solution of H_2O_2 is 30.0% by mass and has a density of 1.11 g/mL. Calculate the **(a)** molality, **(b)** mole fraction of H_2O_2, and **(c)** molarity.

Plan We know the mass % (30.0) and the density (1.11 g/mL). **(a)** For molality, we need the amount (mol) of solute and the mass (kg) of *solvent.* If we assume 100.0 g of solution, the mass % equals the grams of H_2O_2, which we subtract from 100.0 g to obtain the grams of solvent. To find molality, we convert grams of H_2O_2 to moles and divide by mass of solvent (converting g to kg). **(b)** To find the mole fraction, we use the moles of H_2O_2 [from part (a)] and convert the grams of H_2O to moles. Then we divide the moles of H_2O_2 by the total moles. **(c)** To find molarity, we assume 100.0 g of solution and use the solution density to find the volume. Then we divide the moles of H_2O_2 [from part (a)] by *solution* volume (in L).

Solution (a) From mass % to molality:

Finding mass of solvent (assume a 100.0-g sample of solution; 30.0% of 100 g = 30.0 g of H_2O_2):

$$\text{Mass (g) of } H_2O = 100.0 \text{ g solution} - 30.0 \text{ g } H_2O_2 = 70.0 \text{ g } H_2O$$

Converting from grams of H_2O_2 to moles:

$$\text{Amount (mol) of } H_2O_2 = 30.0 \text{ g } H_2O_2 \times \frac{1 \text{ mol } H_2O_2}{34.02 \text{ g } H_2O_2} = 0.882 \text{ mol } H_2O_2$$

Calculating molality:

$$\text{Molality of } H_2O_2 = \frac{\text{mol } H_2O_2}{\text{kg } H_2O} = \frac{0.882 \text{ mol } H_2O_2}{70.0 \text{ g} \times \frac{1 \text{ kg}}{10^3 \text{ g}}} = 12.6\ m\ H_2O_2$$

(b) From mass % to mole fraction:

$$\text{Amount (mol) of } H_2O_2 = 0.882 \text{ mol } H_2O_2 \text{ [from part (a)]}$$

$$\text{Amount (mol) of } H_2O = 70.0 \text{ g } H_2O \times \frac{1 \text{ mol } H_2O}{18.02 \text{ g } H_2O} = 3.88 \text{ mol } H_2O$$

$$X_{H_2O_2} = \frac{\text{mol } H_2O_2}{\text{mol } H_2O_2 + \text{mol } H_2O} = \frac{0.882 \text{ mol}}{0.882 \text{ mol} + 3.88 \text{ mol}} = 0.185$$

(c) From mass % and density to molarity:

Using density to convert from solution mass to volume:

$$\text{Volume (mL) of solution} = 100.0 \text{ g} \times \frac{1 \text{ mL}}{1.11 \text{ g}} = 90.1 \text{ mL}$$

Calculating molarity:

$$\text{Molarity} = \frac{\text{mol } H_2O_2}{\text{L soln}} = \frac{0.882 \text{ mol } H_2O_2}{90.1 \text{ mL} \times \frac{1 \text{ L soln}}{10^3 \text{ mL}}} = 9.79\ M\ H_2O_2$$

Check Rounding shows the sizes of the answers to be reasonable: **(a)** The ratio ~0.9 mol/0.07 kg is greater than 10. **(b)** ~0.9 mol H_2O_2/(1 mol + 4 mol) ≈ 0.2. **(c)** The ratio of moles to liters (0.9/0.09) is around 10.

FOLLOW-UP PROBLEMS

13.6A Concentrated hydrochloric acid is 11.8 *M* HCl and has a density of 1.190 g/mL. Calculate the mass %, molality, and mole fraction of HCl.

13.6B Concentrated aqueous solutions of calcium bromide are used in drilling oil wells. If a solution of $CaBr_2$ is 5.44 *m* and has a density of 1.70 g/mL, find the mass %, molarity, and mole percent of $CaBr_2$.

SOME SIMILAR PROBLEMS 13.72–13.75

› Summary of Section 13.5

- The concentration of a solution is independent of the quantity of solution and can be expressed as molarity (mol solute/L solution), molality (mol solute/kg solvent), parts by mass (mass solute/mass solution), parts by volume (volume solute/volume solution), or mole fraction [mol solute/(mol solute + mol solvent)] (Table 13.5).
- Molality is based on mass, so it is independent of temperature; the mole fraction gives the proportion of dissolved particles.
- If the quantities of solute and solution as well as the solution density are known, the various ways of expressing concentration are interconvertible.

13.6 COLLIGATIVE PROPERTIES OF SOLUTIONS

The presence of solute gives a solution different physical properties than the pure solvent has. But, in the case of four important properties, it is the *number* of solute particles, *not* their chemical identity, that makes the difference. These **colligative properties** (*colligative* means "collective") are vapor pressure lowering, boiling point elevation, freezing point depression, and osmotic pressure. Most of these effects are small, but they have many applications, including some that are vital to organisms.

We predict the magnitude of a colligative property from the solute formula, which shows the number of particles in solution and is closely related to our classification of solutes (Chapter 4) by their ability to conduct an electric current (Figure 13.22):

1. *Electrolytes.* An aqueous solution of an **electrolyte** conducts a current because the solute separates into ions as it dissolves.
   - *Strong electrolytes*—soluble salts, strong acids, and strong bases—dissociate completely, so their solutions conduct well.
   - *Weak electrolytes*—weak acids and weak bases—dissociate very little, so their solutions conduct poorly.
2. *Nonelectrolytes.* Compounds such as sugar and alcohol do not dissociate into ions at all. They are **nonelectrolytes** because their solutions do not conduct a current.

Thus, we can predict that

- *For nonelectrolytes,* 1 mol of compound yields 1 mol of particles when it dissolves in solution. For example, 0.35 *M* glucose contains 0.35 mol of solute particles per liter.
- *For strong electrolytes,* 1 mol of compound dissolves to yield the amount (mol) of ions shown in the formula unit: 0.4 *M* Na_2SO_4 has 0.8 mol of Na^+ ions and 0.4 mol of SO_4^{2-} ions, or 1.2 mol of particles, per liter (see Sample Problem 4.2).
- *For weak electrolytes,* the calculation is complicated because the solution reaches equilibrium; we examine these systems in Chapters 18 and 19.

In this section, we discuss colligative properties of three types of solutions—those of nonvolatile nonelectrolytes, volatile nonelectrolytes, and strong electrolytes.

Figure 13.22 Conductivity of three types of electrolyte solutions. A, For strong electrolytes, the number of particles equals the number of ions in a formula unit. **B,** Weak electrolytes form few ions. **C,** For nonelectrolytes, the number of particles equals the number of molecules.

Source: A–C: © McGraw-Hill Education/ Stephen Frisch, photographer

Nonvolatile Nonelectrolyte Solutions

We start with solutions of *nonvolatile nonelectrolytes,* because they provide the clearest examples of the colligative properties. These solutions contain solutes that are not ionic and thus do not dissociate, and they have negligible vapor pressure at the boiling point of the solvent; sucrose (table sugar) dissolved in water is an example.

Vapor Pressure Lowering *The vapor pressure of a nonvolatile nonelectrolyte solution is always lower than the vapor pressure of the pure solvent.* The difference in vapor pressures is the **vapor pressure lowering (ΔP).**

1. *Why the vapor pressure of a solution is lower.* The fundamental reason for this lowering involves entropy, specifically the relative change in entropy that accompanies vaporization of solvent versus vaporization of the solution. A liquid vaporizes because a gas has higher entropy. In a closed container, vaporization continues until the numbers of particles leaving and entering the liquid phase per unit time are equal, that is, when the system reaches equilibrium. But, as we said, the entropy of a solution is already higher than that of a pure solvent, so fewer solvent particles need to vaporize to reach the same entropy. With fewer particles in the gas phase, the vapor above a solution has lower pressure (Figure 13.23).

Figure 13.23 Effect of solute on the vapor pressure of solution.

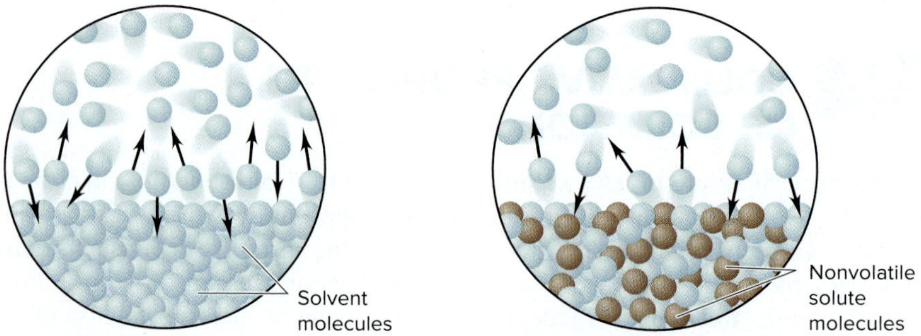

A Pure solvent. *Equilibrium between a liquid and its vapor occurs when equal numbers of molecules vaporize and condense in a given time.*

B Solvent and dissolved solute. *Solute molecules increase the entropy, so equilibrium occurs at a lower vapor pressure.*

2. *Quantifying vapor pressure lowering.* **Raoult's law** says that the vapor pressure of solvent above a solution ($P_{solvent}$) equals the mole fraction of solvent ($X_{solvent}$) times the vapor pressure of the pure solvent ($P^{\circ}_{solvent}$):

$$P_{solvent} = X_{solvent} \times P^{\circ}_{solvent} \quad \textbf{(13.9)}$$

Since $X_{solvent}$ is less than 1 in a solution, $P_{solvent}$ is less than $P^{\circ}_{solvent}$. An **ideal solution** is one that follows Raoult's law at any concentration. However, just as real gases deviate from ideality, so do real solutions. In practice, Raoult's law works reasonably well for *dilute* solutions and becomes exact at infinite dilution.

Let's see how the *amount* of dissolved solute affects ΔP. The solution consists of solvent and solute, so the sum of their mole fractions equals 1:

$$X_{solvent} + X_{solute} = 1; \quad \text{thus} \quad X_{solvent} = 1 - X_{solute}$$

From Raoult's law, we have

$$P_{solvent} = X_{solvent} \times P^{\circ}_{solvent} = (1 - X_{solute}) \times P^{\circ}_{solvent}$$

Multiplying through on the right side gives

$$P_{solvent} = P^{\circ}_{solvent} - (X_{solute} \times P^{\circ}_{solvent})$$

Rearranging and introducing ΔP gives

$$P^{\circ}_{solvent} - P_{solvent} = \Delta P = X_{solute} \times P^{\circ}_{solvent} \quad \textbf{(13.10)}$$

Thus, ΔP equals the mole fraction of solute times the vapor pressure of the pure solvent—a relationship applied in the next sample problem.

SAMPLE PROBLEM 13.7 Using Raoult's Law to Find ΔP

Problem Find the vapor pressure lowering, ΔP, when 10.0 mL of glycerol ($C_3H_8O_3$) is added to 500. mL of water at 50.°C. At this temperature, the vapor pressure of pure water is 92.5 torr and its density is 0.988 g/mL. The density of glycerol is 1.26 g/mL.

Plan To calculate ΔP, we use Equation 13.10. We are given the vapor pressure of pure water ($P^\circ_{H_2O}$ = 92.5 torr), so we just need the mole fraction of glycerol, $X_{glycerol}$. We convert the given volume of glycerol (10.0 mL) to mass using the given density (1.26 g/mL), find the molar mass from the formula, and convert mass (g) to amount (mol). The same procedure gives amount of H_2O. From these amounts, we find $X_{glycerol}$ and ΔP.

Road Map

- Volume (mL) of glycerol (or H_2O)
  - multiply by density (g/mL)
- Mass (g) of glycerol (or H_2O)
  - divide by $\mathcal{M}$ (g/mol)
- Amount (mol) of glycerol (or H_2O)
  - divide by total number of moles
- Mole fraction (X) of glycerol
  - multiply by $P^\circ_{H_2O}$
- Vapor pressure lowering (ΔP)

Solution Calculating the amount (mol) of glycerol and of water:

$$\text{Amount (mol) of glycerol} = 10.0\ \text{mL glycerol} \times \frac{1.26\ \text{g glycerol}}{1\ \text{mL glycerol}} \times \frac{1\ \text{mol glycerol}}{92.09\ \text{g glycerol}} = 0.137\ \text{mol glycerol}$$

$$\text{Amount (mol) of } H_2O = 500.\ \text{mL } H_2O \times \frac{0.988\ \text{g } H_2O}{1\ \text{mL } H_2O} \times \frac{1\ \text{mol } H_2O}{18.02\ \text{g } H_2O} = 27.4\ \text{mol } H_2O$$

Calculating the mole fraction of glycerol:

$$X_{glycerol} = \frac{0.137\ \text{mol}}{0.137\ \text{mol} + 27.4\ \text{mol}} = 0.00498$$

Finding the vapor pressure lowering:

$$\Delta P = X_{glycerol} = P^\circ_{H_2O} = 0.00498 \times 92.5\ \text{torr} = 0.461\ \text{torr}$$

Check The amount of each component seems correct: for glycerol, ~10 mL × 1.25 g/mL ÷ 100 g/mol = 0.125 mol; for H_2O, ~500 mL × 1 g/mL ÷ 20 g/mol = 25 mol. The small ΔP is reasonable because the mole fraction of solute is small.

Comment 1. We can also find X_{H_2O} (1 – 0.00498 = 0.99502) and use Equation 13.9 to calculate P_{H_2O} (92.0 torr). Then $P^\circ_{H_2O} - P_{H_2O}$ gives ΔP (with fewer significant figures).

2. The calculation assumes that glycerol is nonvolatile. At 1 atm, glycerol boils at 290.0°C, so the vapor pressure of glycerol at 50°C is negligible.

3. We have assumed the solution is close to ideal and that Raoult's law holds.

FOLLOW-UP PROBLEMS

13.7A Calculate the vapor pressure lowering of a solution of 2.00 g of aspirin ($\mathcal{M}$ = 180.15 g/mol) in 50.0 g of methanol (CH_3OH) at 21.2°C. Methanol has a vapor pressure of 101 torr at this temperature.

13.7B Because menthol ($C_{10}H_{20}O$) has a minty taste and soothes throat soreness, it is added to many medicinal products. Find the vapor pressure (in torr) of a solution that contains 6.49 g of menthol in 25.0 g of ethanol (C_2H_5OH) at 22°C. Ethanol has a vapor pressure of 6.87 kPa at this temperature (101.325 kPa = 1 atm).

SOME SIMILAR PROBLEMS 13.96 and 13.97

Boiling Point Elevation *A solution boils at a higher temperature than the pure solvent.* This colligative property results from the vapor pressure lowering:

1. *Why a solution boils at a higher T.* Recall that the boiling point, T_b, of a liquid is the temperature at which its vapor pressure equals the external pressure, P_{ext}. But, the vapor pressure of a solution is always lower than that of the pure solvent. Therefore, it is lower than P_{ext} at T_b of the solvent, so the solution doesn't boil. Thus, **boiling point elevation (ΔT_b)** results because a higher temperature is needed to raise the solution's vapor pressure to equal P_{ext}.

We superimpose a phase diagram for the solution on one for the solvent to see ΔT_b (Figure 13.24, *next page*): the gas-liquid line for the solution lies *below* the line for the solvent at any T and to the right of it at any P, and the line crosses 1 atm (P_{ext}, or P_{atm}) at a higher T.

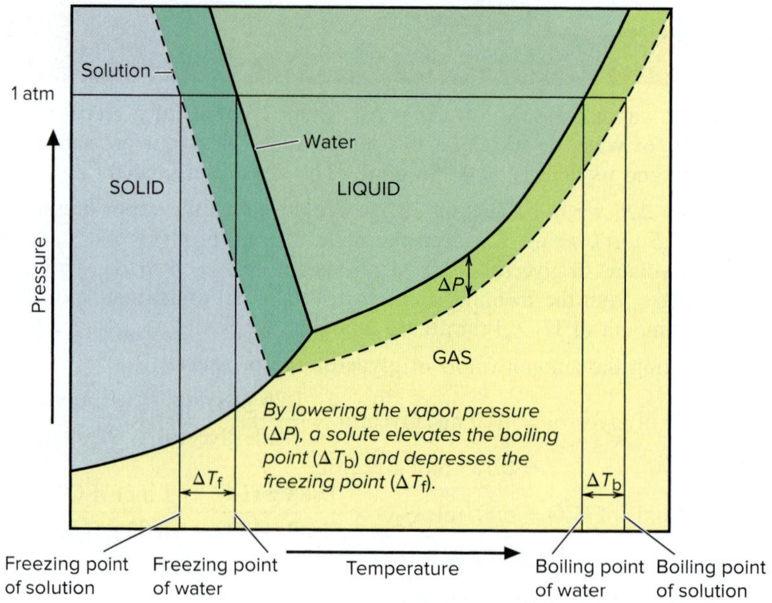

Figure 13.24 Boiling and freezing points of solvent and solution. Phase diagrams of an aqueous solution *(dashed lines)* and of pure water *(solid lines)*. (The slope of the solid-liquid line and the size of ΔP are exaggerated.)

2. *Quantifying boiling point elevation.* Like vapor pressure lowering, boiling point elevation is proportional to the concentration of solute:

$$\Delta T_b \propto m \quad \text{or} \quad \Delta T_b = K_b m \tag{13.11}$$

where m is the solution molality and K_b is the *molal boiling point elevation constant.* Since $\Delta T_b > 0$, we subtract the lower solvent T_b from the higher solution T_b:

$$\Delta T_b = T_{b(\text{solution})} - T_{b(\text{solvent})}$$

Molality is used to determine the boiling point elevation because it relates to mole fraction and, thus, to particles of solute, and it is not affected by temperature.

The constant K_b has units of degrees Celsius per molal unit (°C/m) and is specific for a given solvent (Table 13.6). The K_b for water is 0.512°C/m, so ΔT_b for aqueous solutions is quite small. For example, if you dissolve 1.00 mol of glucose (180. g; 1.00 mol of particles) or 0.500 mol of NaCl (29.2 g; also 1.00 mol of particles) in 1.00 kg of water at 1 atm, both solutions will boil at 100.512°C instead of 100.000°C.

Table 13.6 Molal Boiling Point Elevation and Freezing Point Depression Constants of Several Solvents

Solvent	Boiling Point (°C)*	K_b (°C/m)	Freezing Point (°C)	K_f (°C/m)
Acetic acid	117.9	3.07	16.6	3.90
Benzene	80.1	2.53	5.56	4.90
Carbon disulfide	46.2	2.34	−111.5	3.83
Carbon tetrachloride	76.5	5.03	−23	30.
Chloroform	61.7	3.63	−63.5	4.70
Diethyl ether	34.5	2.02	−116.2	1.79
Ethanol	78.5	1.22	−117.3	1.99
Water	100.0	0.512	0.0	1.86

*At 1 atm

Freezing Point Depression *A solution freezes at a lower temperature than the pure solvent,* and this colligative property also results from vapor pressure lowering:

1. *Why a solution freezes at a lower T.* Only solvent vaporizes from a solution, so solute molecules are left behind. Similarly, *only solvent freezes,* again leaving solute molecules behind. The freezing point of a solution is the temperature at which its vapor pressure equals that of the pure solvent, that is, when solid solvent and liquid

solution are in equilibrium. **Freezing point depression (ΔT_f)** occurs because the vapor pressure of the solution is always lower than that of the solvent, so the solution freezes at a lower temperature; that is, only at a lower temperature will solvent particles leave and enter the solid at the same rate. In Figure 13.24, the solid-liquid line for the solution is to the left of the pure solvent line at 1 atm, or at any pressure.

2. *Quantifying freezing point depression.* Like ΔT_b, the freezing point depression is proportional to the molal concentration of solute:

$$\Delta T_f \propto m \qquad \text{or} \qquad \Delta T_f = K_f m \qquad \textbf{(13.12)}$$

where K_f is the *molal freezing point depression constant,* which also has units of °C/m (see Table 13.6). Also, like ΔT_b, $\Delta T_f > 0$, but in this case, we subtract the lower solution T_f from the higher solvent T_f:

$$\Delta T_f = T_{f(solvent)} - T_{f(solution)}$$

Here, too, the effect in aqueous solution is small because K_f for water is just 1.86°C/m. Thus, at 1 atm, 1 m glucose, 0.5 m NaCl, and 0.33 m K_2SO_4, all solutions with 1 mol of particles per kilogram of water, freeze at –1.86°C instead of at 0.00°C.

Student Hot Spot

Student data indicate that you may struggle with colligative property calculations. Access the Smartbook to view additional Learning Resources on this topic.

SAMPLE PROBLEM 13.8 Determining Boiling and Freezing Points of a Solution

Problem You add 1.00 kg of ethylene glycol ($C_2H_6O_2$) antifreeze to 4450 g of water in your car's radiator. What are the boiling and freezing points of the solution?

Plan To find the boiling and freezing points, we need ΔT_b and ΔT_f. We first find the molality by converting mass of solute (1.00 kg) to amount (mol) and dividing by mass of solvent (4450 g, converted to kg). Then we calculate ΔT_b and ΔT_f from Equations 13.11 and 13.12 (using constants from Table 13.6). We add ΔT_b to the solvent boiling point and subtract ΔT_f from its freezing point. The road map shows the steps.

Road Map

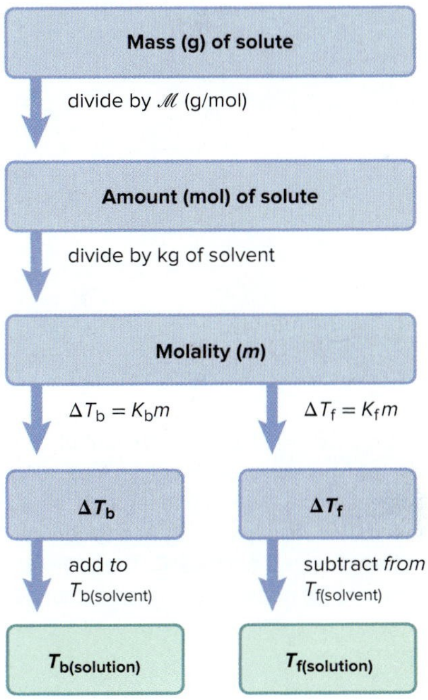

Solution Calculating the molality:

$$\text{Amount (mol) of } C_2H_6O_2 = 1.00 \text{ kg } C_2H_6O_2 \times \frac{10^3 \text{ g}}{1 \text{ kg}} \times \frac{1 \text{ mol } C_2H_6O_2}{62.07 \text{ g } C_2H_6O_2}$$

$$= 16.1 \text{ mol } C_2H_6O_2$$

$$\text{Molality} = \frac{\text{mol solute}}{\text{kg solvent}} = \frac{16.1 \text{ mol } C_2H_6O_2}{4450 \text{ g } H_2O \times \dfrac{1 \text{ kg}}{10^3 \text{ g}}} = 3.62\ m\ C_2H_6O_2$$

Finding the boiling point elevation and $T_{b(solution)}$, with $K_b = 0.512°C/m$:

$$\Delta T_b = \frac{0.512°C}{m} \times 3.62\ m = 1.85°C$$

$$T_{b(solution)} = T_{b(solvent)} + \Delta T_b = 100.00°C + 1.85°C = 101.85°C$$

Finding the freezing point depression and $T_{f(solution)}$, with $K_f = 1.86°C/m$:

$$\Delta T_f = \frac{1.86°C}{m} \times 3.62\ m = 6.73°C$$

$$T_{f(solution)} = T_{f(solvent)} - \Delta T_f = 0.00°C - 6.73°C = -6.73°C$$

Check The changes in boiling and freezing points should be in the same proportion as the constants used. That is, $\Delta T_b/\Delta T_f$ should equal K_b/K_f: 1.85/6.73 = 0.275 = 0.512/1.86.

Comment These answers are approximate because the concentration far exceeds that of a *dilute* solution, for which Raoult's law is most accurate.

FOLLOW-UP PROBLEMS

13.8A Carbon disulfide (CS_2) has the unusual ability to dissolve several nonmetals as well as rubbers and resins. What are the boiling and freezing points of a solution consisting of 8.44 g of phosphorus (P_4) dissolved in 60.0 g of CS_2 (see Table 13.6)?

13.8B What is the lowest molality of ethylene glycol ($C_2H_6O_2$) solution that will protect your car's coolant from freezing at 0.00°F? To make a solution of this molality, what mass of ethylene glycol must be added to 4.00 kg of water? (Assume that the solution is ideal.)

SOME SIMILAR PROBLEMS 13.98–13.103

Osmotic Pressure Another colligative property is observed when solutions of higher and lower concentrations are separated by a **semipermeable membrane,** one that allows solvent, but *not* solute, to pass through. The phenomenon is called **osmosis:** a net flow of solvent into the more concentrated solution causes a pressure difference known as *osmotic pressure*. Many organisms regulate internal concentrations by osmosis.

1. *Why osmotic pressure arises*. Consider a simple apparatus in which a semipermeable membrane lies at the curve of a U tube and separates an aqueous sugar solution from pure water. Water molecules pass in *either* direction, but the larger sugar molecules do not. Because sugar molecules are on the solution side of the membrane, fewer water molecules touch that side, so fewer leave the solution than enter it in a given time (Figure 13.25A). As a result, the solution maintains higher entropy. This *net flow of water into the solution* increases its volume and thus decreases its concentration.

As the height of the solution column rises and the height of the water column falls, the difference in the weights of the columns produces a pressure difference that resists more water entering and pushes some water back through the membrane. When water is being pushed out of the solution at the same rate it is entering, the system is at equilibrium (Figure 13.25B). The pressure difference at this point is the **osmotic pressure (Π),** which is the same as the pressure that must be applied to *prevent* net movement of water from solvent to solution (or from lower to higher concentration, Figure 13.25C).

2. *Quantifying osmotic pressure*. The osmotic pressure is proportional to the number of solute particles in a given solution *volume,* that is, to the molarity (M):

$$\Pi \propto \frac{n_{\text{solute}}}{V_{\text{soln}}} \quad \text{or} \quad \Pi \propto M$$

The proportionality constant is R times the absolute temperature T. Thus,

$$\Pi = \frac{n_{\text{solute}}}{V_{\text{soln}}} RT = MRT \tag{13.13}$$

The similarity of Equation 13.13 to the ideal gas law ($P = nRT/V$) is not surprising, because both relate the pressure of a system to its concentration and temperature.

The Underlying Theme of Colligative Properties A common theme runs through the explanations of the four colligative properties of nonvolatile solutes. Each property arises because solute particles cannot move between two phases:

- They cannot enter the gas phase, which leads to *vapor pressure lowering* and *boiling point elevation*.

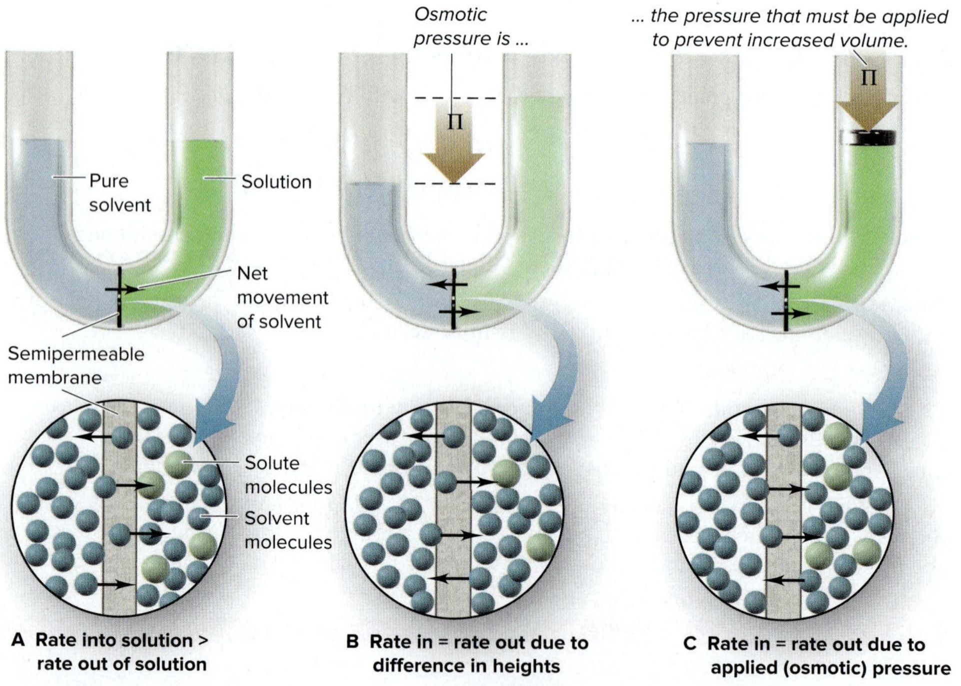

Figure 13.25 **The development of osmotic pressure. A,** In a given time, more solvent enters the solution through the membrane than leaves. **B,** At equilibrium, solvent flow is equalized. **C,** The *osmotic pressure* (Π) *prevents* the volume change.

- They cannot enter the solid phase, which leads to *freezing point depression.*
- They cannot cross a semipermeable membrane, which leads to *osmotic pressure.*

In each situation, the presence of solute decreases the mole fraction of solvent, which lowers the number of solvent particles leaving the solution per unit time. This lowering maintains higher entropy and requires a new balance in numbers of particles moving between two phases per unit time, and this new balance results in the measured colligative property.

Using Colligative Properties to Find Solute Molar Mass

Each colligative property is proportional to solute concentration. Thus, by measuring the property—lower freezing point, higher boiling point, and so forth—we determine the amount (mol) of solute particles and, given the mass of solute, the molar mass.

In principle, any of the colligative properties can be used, but generally only freezing point depression and osmotic pressure are widely used. Osmotic pressure creates the largest changes and, thus, the most precise measurements. Polymer chemists and biochemists estimate molar masses as great as 10^5 g/mol by measuring osmotic pressure. Because only a tiny fraction of a mole of a macromolecular solute dissolves, the change in the other colligative properties would be too small.

SAMPLE PROBLEM 13.9 Determining Molar Mass from Colligative Properties

Problem Biochemists have discovered more than 400 mutant varieties of hemoglobin, the blood protein that binds O_2 and carries it to the body's cells. A physician dissolves 21.5 mg of one variety in water to make 1.50 mL of solution at 5.0°C. She measures an osmotic pressure of 3.61 torr. What is the molar mass of the protein?

Plan We know the osmotic pressure (Π = 3.61 torr), R, and T (5.0°C). We convert Π from torr to atm, and T from °C to K, and then use Equation 13.13 to solve for molarity (M). Then we calculate the amount (mol) of hemoglobin from the known volume (1.50 mL) and use the known mass (21.5 mg) to find $\mathcal{M}$ (see the road map).

Road Map

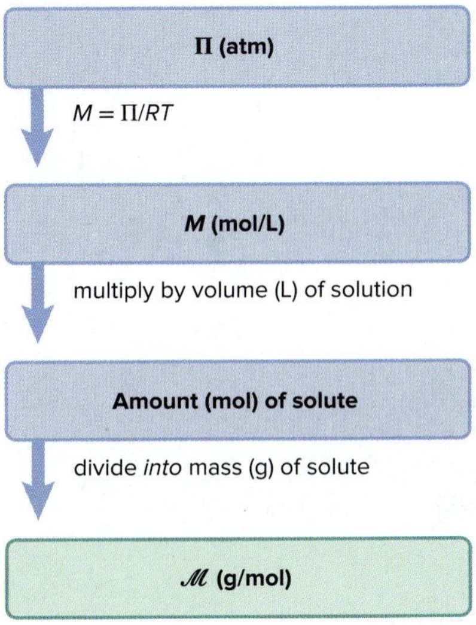

Solution Combining unit conversion steps and solving for molarity:

$$M = \frac{\Pi}{RT} = \frac{\dfrac{3.61\ \text{torr}}{760\ \text{torr/1 atm}}}{\left(0.0821\ \dfrac{\text{atm·L}}{\text{mol·K}}\right)(273.15\ \text{K} + 5.0)} = 2.08\times10^{-4}\ M$$

Finding amount (mol) of solute (after changing mL to L):

$$\text{Amount (mol) of solute} = M \times V = \frac{2.08\times10^{-4}\ \text{mol}}{1\ \text{L soln}} \times 0.00150\ \text{L soln} = 3.12\times10^{-7}\ \text{mol}$$

Calculating molar mass of hemoglobin (after changing mg to g):

$$\mathcal{M} = \frac{0.0215\ \text{g}}{3.12\times10^{-7}\ \text{mol}} = 6.89\times10^4\ \text{g/mol}$$

Check The small osmotic pressure implies a very low molarity. Hemoglobin is a biopolymer, so we expect a small number of moles [($\sim$2×10^{-4} mol/L) (1.5×10^{-3} L) = 3×10^{-7} mol] and a high $\mathcal{M}$ ($\sim$21×10^{-3} g/3×10^{-7} mol = 7×10^4 g/mol). Mammalian hemoglobin has a molar mass of 64,500 g/mol.

FOLLOW-UP PROBLEMS

13.9A Pepsin is an enzyme in the intestines of mammals that breaks down proteins into amino acids. A 12.0-mL aqueous solution contains 0.200 g of pepsin and has an osmotic pressure of 8.98 torr at 27.0°C. What is the molar mass of pepsin?

13.9B Naphthalene is the main ingredient in some mothballs. The freezing point of a solution made by dissolving 7.01 g of naphthalene in 200. g of benzene is 4.20°C. What is the molar mass of naphthalene?

SOME SIMILAR PROBLEMS 13.106 and 13.107

Volatile Nonelectrolyte Solutions

What is the effect on vapor pressure when the solute *is* volatile, that is, when the vapor consists of solute *and* solvent molecules? From Raoult's law (Equation 13.9),

$$P_{\text{solvent}} = X_{\text{solvent}} \times P^{\circ}_{\text{solvent}} \quad \text{and} \quad P_{\text{solute}} = X_{\text{solute}} \times P^{\circ}_{\text{solute}}$$

where X_{solvent} and X_{solute} are the mole fractions in the *liquid* phase. According to Dalton's law (Section 5.4), the total vapor pressure is the sum of the partial vapor pressures:

$$P_{\text{total}} = P_{\text{solvent}} + P_{\text{solute}} = (X_{\text{solvent}} \times P^{\circ}_{\text{solvent}}) + (X_{\text{solute}} \times P^{\circ}_{\text{solute}})$$

Thus, just as a nonvolatile solute lowers the vapor pressure of the solvent by making the solvent's mole fraction less than 1, *the presence of each volatile component lowers the vapor pressure of the other* by making each mole fraction less than 1.

Let's examine this idea with a solution of benzene (C_6H_6) and toluene (C_7H_8), which are miscible. The mole fractions of the two solution components in this example are equal: $X_{\text{ben}} = X_{\text{tol}} = 0.500$. At 25°C, the vapor pressures of the pure substances are 95.1 torr for benzene (P°_{ben}) and 28.4 torr for toluene (P°_{tol}). Note that benzene is *more volatile* than toluene. We find the partial pressures from Raoult's law:

$$P_{\text{ben}} = X_{\text{ben}} \times P^{\circ}_{\text{ben}} = 0.500 \times 95.1 \text{ torr} = 47.6 \text{ torr}$$

$$P_{\text{tol}} = X_{\text{tol}} \times P^{\circ}_{\text{tol}} = 0.500 \times 28.4 \text{ torr} = 14.2 \text{ torr}$$

Thus, the presence of benzene in the liquid lowers the vapor pressure of toluene, and vice versa.

Now let's calculate the mole fraction of each substance *in the vapor* by applying Dalton's law. Recall from Section 5.4 that $X_A = P_A/P_{\text{total}}$. Therefore, for benzene and toluene in the vapor,

$$X_{\text{ben}} = \frac{P_{\text{ben}}}{P_{\text{total}}} = \frac{47.6 \text{ torr}}{47.6 \text{ torr} + 14.2 \text{ torr}} = 0.770$$

$$X_{\text{tol}} = \frac{P_{\text{tol}}}{P_{\text{total}}} = \frac{14.2 \text{ torr}}{47.6 \text{ torr} + 14.2 \text{ torr}} = 0.230$$

The key point is that *the vapor has a higher mole fraction of the* ***more*** *volatile component.* Through a single vaporization-condensation step, a 50/50 *liquid* ratio of benzene to toluene created a 77/23 *vapor* ratio. Condense this vapor into a separate container, and the new *liquid* would have this 77/23 composition, and the new *vapor* above it would be enriched still further in the more volatile benzene.

In the laboratory method of **fractional distillation,** a solution of two or more volatile components is attached to a *fractionating column* packed with glass beads (or ceramic chunks), which increase surface area; the column is connected to a condenser and collection flask. As the solution is heated and the vapor mixture meets the beads, numerous vaporization-condensation steps enrich the mixture until the vapor leaving the column, and thus the liquid finally collected, consists only of the most volatile component. (As we'll see shortly, this method is essential in the oil refining industry.)

Strong Electrolyte Solutions

For strong electrolyte solutions, the solute formula tells us the number of particles for determining colligative properties. For instance, the boiling point elevation (ΔT_b) of 0.050 *m* NaCl should be $2 \times \Delta T_b$ of 0.050 *m* glucose ($C_6H_{12}O_6$), because NaCl dissociates into two particles per formula unit. Thus, we use a multiplying factor called the *van't Hoff factor (i),* named after the Dutch chemist Jacobus van't Hoff (1852–1911):

$$i = \frac{\text{measured value for electrolyte solution}}{\text{expected value for nonelectrolyte solution}}$$

To calculate colligative properties for strong electrolyte solutions, we include *i*:

For vapor pressure lowering: $\Delta P = i(X_{\text{solute}} \times P^{\circ}_{\text{solvent}})$

For boiling point elevation: $\Delta T_b = i(K_b m)$

For freezing point depression: $\Delta T_f = i(K_f m)$

For osmotic pressure: $\Pi = i(MRT)$

Nonideal Solutions and Ionic Atmospheres *If* strong electrolyte solutions behaved ideally, the factor i would be the amount (mol) of particles in solution divided by the amount (mol) of dissolved solute; that is, i would be 2 for KBr, 3 for $Mg(NO_3)_2$, and so forth. However, *most strong electrolyte solutions are* ***not*** *ideal,* and the measured value of i is typically *lower* than the value expected from the formula. For example, for the boiling point elevation of 0.050 m NaCl, the expected value is 2.0, but, from experiment, we have

$$i = \frac{\Delta T_b \text{ of } 0.050\ m \text{ NaCl}}{\Delta T_b \text{ of } 0.050\ m \text{ glucose}} = \frac{0.049°C}{0.026°C} = 1.9$$

It seems as though the ions are not behaving independently, even though other evidence indicates that soluble salts dissociate completely. One clue to understanding the results is that multiply-charged ions cause a larger deviation (Figure 13.26).

To explain this nonideal behavior, we picture positive ions clustered, on average, near negative ions, and vice versa, to form an **ionic atmosphere** of net opposite charge (Figure 13.27). In effect, each type of ion acts "tied up," so its actual concentration seems *lower*. The *effective* concentration is the *stoichiometric* concentration, which is based on the formula, multiplied by i. The greater the charge, the stronger the attractions, which explains the larger deviation for salts with multiply-charged ions.

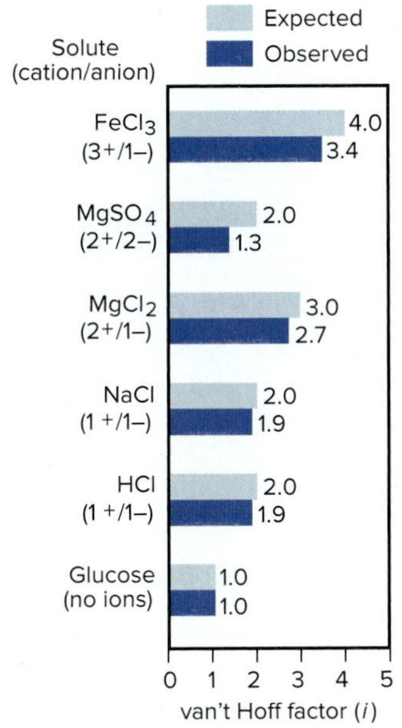

Figure 13.26 Nonideal behavior of strong electrolyte solutions. Van't Hoff factors (i) for 0.050 m solutions show the largest deviation for salts with multiply-charged ions. Glucose (a nonelectrolyte) behaves as expected.

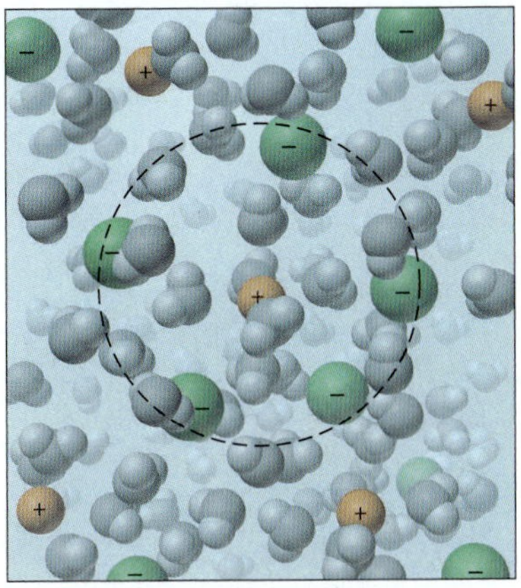

Figure 13.27 An ionic atmosphere model for nonideal behavior of electrolyte solutions.

Comparing Real Solutions and Real Gases At ordinary conditions and concentrations, particles are *much* closer together in liquids than in gases. And, as a result, nonideal behavior is much more common for solutions, and the observed deviations are much larger. Nevertheless, the two systems have similarities:

- Gases display nearly ideal behavior at low pressures because the distances between particles are large. Similarly, the van't Hoff factor (i) approaches the ideal value as the solution becomes more dilute, that is, as the distance between ions increases.
- Attractions between particles cause deviations from the expected pressure in gases and from the expected size of a colligative property in ionic solutions.
- For both real gases and real solutions, we use empirically determined numbers (van der Waals constants or van't Hoff factors) to transform theories (the ideal gas law or Raoult's law) into more useful relations.

SAMPLE PROBLEM 13.10 Depicting Strong Electrolyte Solutions

Problem A 0.952-g sample of magnesium chloride dissolves in 100. g of water in a flask.

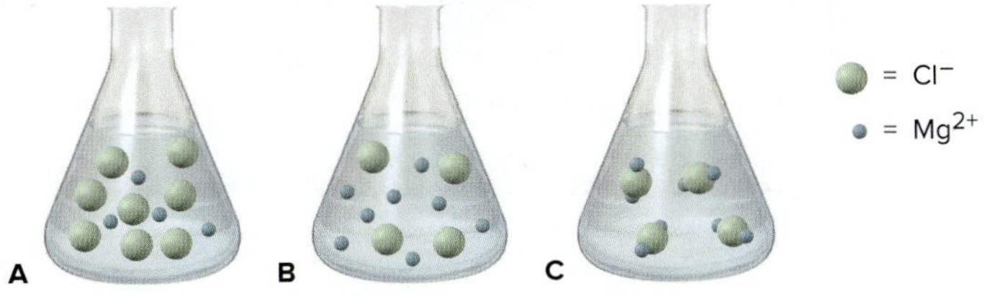

(a) Which scene depicts the solution best?
(b) What is the amount (mol) represented by each green sphere?
(c) Assuming the solution is ideal, what is its freezing point (at 1 atm)?

Plan (a) We find the numbers of cations and anions per formula unit from the name and compare it with the three scenes. **(b)** We convert the given mass to amount (mol), use the answer from part (a) to find moles of chloride ions *(green spheres),* and divide by the number of green spheres to get moles/sphere. **(c)** We find the molality (m) from amount (mol) of solute divided by the given mass of water (changed to kg). We multiply K_f for water (Table 13.6) by m to get ΔT_f, and then subtract that from 0.000°C to get the solution freezing point.

Solution (a) The formula is $MgCl_2$; only scene A has 1 Mg^{2+} for every 2 Cl^-.

(b) $$\text{Amount (mol) of } MgCl_2 = \frac{0.952 \text{ g } MgCl_2}{95.21 \text{ g/mol } MgCl_2} = 0.0100 \text{ mol } MgCl_2$$

Therefore, $$\text{Amount (mol) of } Cl^- = 0.0100 \text{ mol } MgCl_2 \times \frac{2\ Cl^-}{1\ MgCl_2} = 0.0200 \text{ mol } Cl^-$$

$$\text{Moles/sphere} = \frac{0.0200 \text{ mol } Cl^-}{8 \text{ spheres}} = 2.50\times10^{-3} \text{ mol/sphere}$$

(c) $$\text{Molality}(m) = \frac{\text{mol of solute}}{\text{kg of solvent}} = \frac{0.0100 \text{ mol } MgCl_2}{100.\text{ g} \times \frac{1 \text{ kg}}{10^3 \text{ g}}} = 0.100\ m\ MgCl_2$$

Assuming an ideal solution, $i = 3$ for $MgCl_2$ (3 ions per formula unit), so we have

$$\Delta T_f = i(K_f m) = 3(1.86°\text{C}/m \times 0.100\ m) = 0.558°\text{C}$$

and $$T_f = 0.000°\text{C} - 0.558°\text{C} = -0.558°\text{C}$$

Check Let's quickly check the ΔT_f in part (c): We have 0.01 mol dissolved in 0.1 kg, or 0.1 *m*. Then, rounding K_f, we have about $3(2°\text{C}/m \times 0.1\ m) = 0.6°\text{C}$.

FOLLOW-UP PROBLEMS

13.10A A solution is made by dissolving 31.2 g of potassium phosphate in 85.0 g of water in a beaker. **(a)** Which scene depicts the solution best? **(b)** Assuming ideal behavior, what is the boiling point of the solution (at 1 atm)?

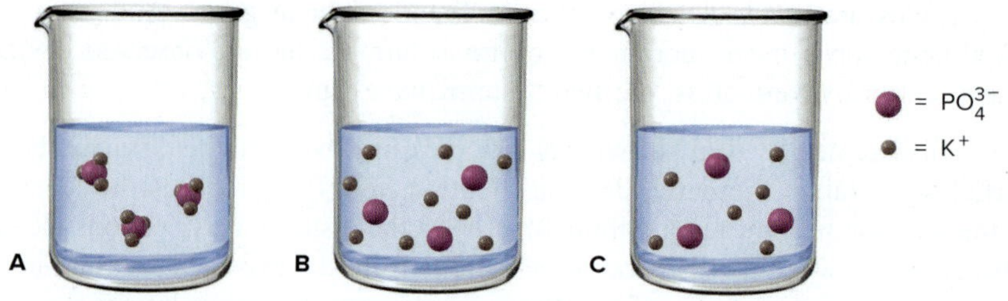

13.10B The $MgCl_2$ solution in the sample problem has a density of 1.006 g/mL at 20.0°C. **(a)** What is its osmotic pressure? **(b)** A U tube, with semipermeable membrane, is filled with this solution in the left arm and a glucose solution of equal molarity in the right. After time, which scene depicts the U tube best?

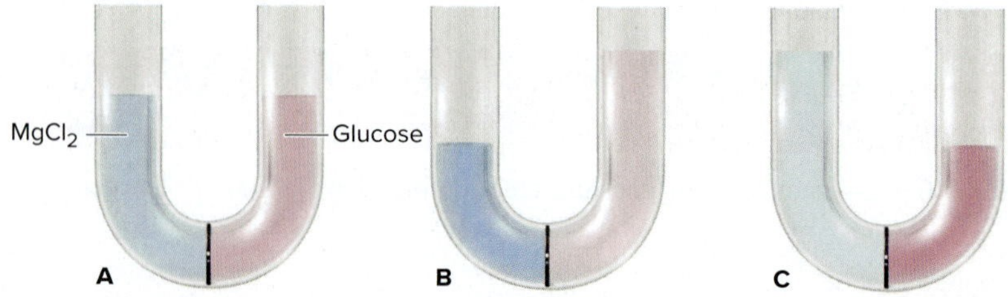

SOME SIMILAR PROBLEMS 13.108 and 13.109

Applications of Colligative Properties

Two colligative properties, freezing point depression (or boiling point elevation) and osmotic pressure, have key applications that involve all three types of solutes we've discussed, nonvolatile and volatile nonelectrolytes and strong electrolytes.

Uses of Freezing Point Depression Applications of this property appear in everyday life, nature, and industry:

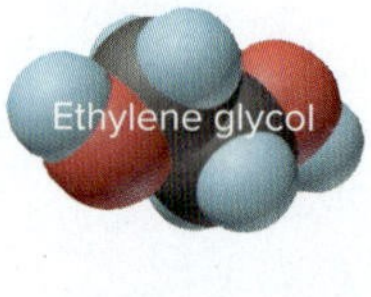

- *Plane de-icer and car antifreeze.* The main ingredient in plane de-icer and car antifreeze, ethylene glycol ($C_2H_6O_2$), lowers the freezing point of water in winter and raises its boiling point in summer. Due to extensive H bonding, ethylene glycol is miscible with water and nonvolatile at 100°C.

- *Biological antifreeze.* Structurally similar to ethylene glycol and also miscible with water, glycerol ($C_3H_8O_3$) is produced by some fish and insects, including the common housefly, to lower the freezing point of their blood and thus allow them to survive winters.
- *De-icing sidewalks and roads.* NaCl, or a mixture of it with $CaCl_2$, is used to melt ice on roads. A small amount dissolves in the ice by lowering its freezing point and melting it. More salt dissolves, more ice melts, and so forth. An advantage of $CaCl_2$ is that it has a highly negative ΔH_{soln}, so heat is released when it dissolves, which melts more ice.
- *Refining petroleum and silicon.* In the oil refining industry, countless vaporization-condensation steps within a 30-m fractionating tower separate the hundreds of volatile compounds in crude oil into a few "fractions" based on boiling point range (Figure 13.28). Impure silicon is refined by continuous melting and refreezing into a sample pure enough for use in computer chips (Chapter 22).

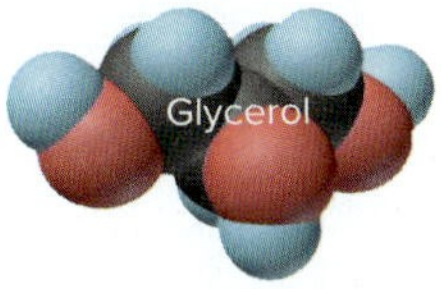

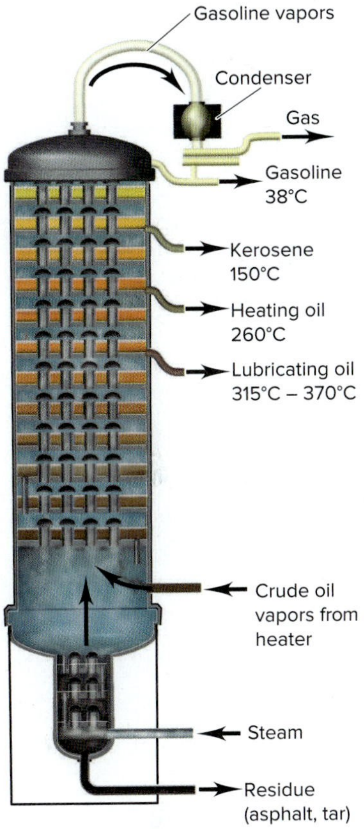

Figure 13.28 **Fractional distillation in petroleum refining.**

Uses of Osmotic Pressure Of the four colligative properties, this one has the most applications vital to organisms.

- *Tonicity and cell shape.* The term *tonicity* refers to the tone, or firmness, of a cell. Placing a cell in an *isotonic* solution, one that has the same concentration of particles as the cell fluid, maintains the cell's normal shape because water enters and leaves the cell at the same rate (Figure 13.29A). A *hypotonic* solution has a lower concentration of particles, so water enters the cell faster than it leaves, causing the cell to burst (Figure 13.29B). In contrast, a *hypertonic* solution has a higher concentration of particles, so the cell shrinks because water leaves faster than it enters (Figure 13.29C). To maintain cell shape, contact-lens rinse is isotonic saline (0.15 *M* NaCl), as are solutions used to deliver nutrients or drugs intravenously.
- *Absorption of water by trees.* Tree sap is a more concentrated solution than surrounding groundwater, so water passes through root membranes into the tree, creating an osmotic pressure that can exceed 20 atm in the tallest trees.
- *Regulation of water volume in the body.* Of the four major biological cations—Na^+, K^+, Mg^{2+}, and Ca^{2+}—sodium ion has the primary role in regulating water volume. It accounts for over 90 mol % of cations *outside* a cell: high Na^+ draws water out of the cell and low Na^+ leaves more inside. And the primary role of the kidneys is to regulate Na^+ concentration.
- *Food preservation.* Before refrigeration, salt was valued as a preservative. Packed onto food, salt causes microbes on the surface to shrivel as they lose water. In fact, it was so highly prized for this purpose that Roman soldiers were paid in salt, from which comes the word *salary*. Salt is also used to draw water out of cucumbers to make pickles.

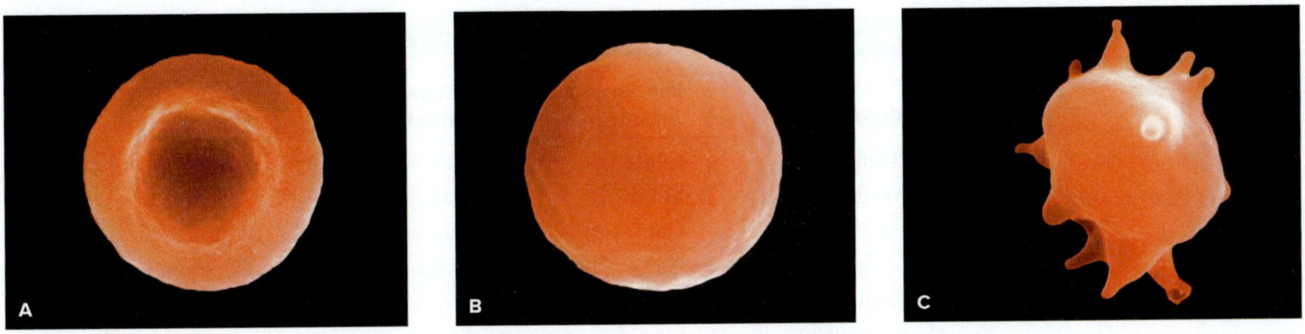

Figure 13.29 **Osmotic pressure and cell shape.** Isotonic **(A)**, hypotonic **(B)**, and hypertonic **(C)** solutions influence the shape of a red blood cell.

Source: A–C: © David M. Phillips/Science Source

› Summary of Section 13.6

- Colligative properties arise from the number, not the type, of dissolved solute particles.
- Compared to pure solvent, a solution has higher entropy, which results in lower vapor pressure (Raoult's law), elevated boiling point, and depressed freezing point. A difference in concentration of a solute between two solutions gives rise to osmotic pressure.
- Colligative properties are used to determine solute molar mass; osmotic pressure gives the most precise measurements.
- When solute *and* solvent are volatile, each lowers the vapor pressure of the other, with the vapor pressure of the more volatile component greater. When the vapor is condensed, the new solution is richer in that component than the original solution.
- Calculating colligative properties of strong electrolyte solutions requires a factor (i) that adjusts for the number of ions per formula unit. These solutions exhibit nonideal behavior because charge attractions effectively reduce the concentration of ions.

13.7 THE STRUCTURE AND PROPERTIES OF COLLOIDS

Particle size plays a defining role in three types of mixtures:

- *Suspensions*. Stir a handful of fine sand into a glass of water, and the particles are suspended at first but gradually settle to the bottom. Sand in water is a **suspension,** a *heterogeneous* mixture containing particles large enough (>1000 nm) to be visibly distinct from the surrounding fluid.
- *Solutions*. In contrast, stirring sugar into water forms a solution, a *homogeneous* mixture in which the particles are invisible, individual sugar molecules (around 1 nm) distributed evenly throughout the surrounding fluid.
- *Colloids*. Between these extremes is a large group of mixtures called *colloidal dispersions,* or **colloids,** in which a dispersed (solute-like) substance is distributed throughout a dispersing (solvent-like) substance. The particles are larger than simple molecules but too small to settle out.

In this section, we examine the classification and key features of colloids.

1. *Particle size and surface area*. Colloidal particles range in diameter from 1 to 1000 nm (10^{-9} to 10^{-6} m). Each particle may be a single macromolecule (natural or synthetic) or an aggregate of many atoms, ions, or molecules. As a result of the small particle size, a colloid has a *very large total surface area*. A cube with 1-cm sides has a total surface area of 6 cm^2. If we divide it equally into 10^{12} cubes, each cube is the size of a large colloidal particle and the total surface area is 60,000 cm^2, or 6 m^2. The enormous surface area of a colloid attracts other particles through various intermolecular forces.

2. *Classification of colloids*. Colloids are commonly classified by the physical state of the dispersed and dispersing substances (Table 13.7). Many familiar commercial products and natural objects are colloids. Whipped cream and shaving cream are

Table 13.7 Types of Colloids

Colloid Type	Dispersed Substance	Dispersing Substance	Example(s)
Aerosol	Liquid	Gas	Fog
Aerosol	Solid	Gas	Smoke
Foam	Gas	Liquid	Whipped cream
Solid foam	Gas	Solid	Marshmallow
Emulsion	Liquid	Liquid	Milk
Solid emulsion	Liquid	Solid	Butter
Sol	Solid	Liquid	Paint, cell fluid
Solid sol	Solid	Solid	Opal

A

B

Figure 13.30 Light scattering and the Tyndall effect. A, The narrow, barely visible light beam that passes through a solution *(left)*, is scattered and broadened by passing through a colloid *(right)*. **B,** Sunlight is scattered by dust in air.

Source: (A) © McGraw-Hill Education/Charles Winters/Timeframe Photography, Inc. (B) © Corbis Royalty-Free RF

foams, a gas dispersed in a liquid. Styrofoam is a *solid foam,* a gas dispersed in a solid. Most biological fluids are aqueous *sols,* solids dispersed in water: proteins and nucleic acids are the colloidal-size particles dispersed in an aqueous fluid of ions and small molecules within a cell. Soaps and detergents work by forming an *emulsion,* a liquid dispersed in another liquid. Other emulsions are mayonnaise and hand cream. Bile salts convert fats to an emulsion in the watery fluid of the small intestine.

3. *Tyndall effect and Brownian motion*. Light passing through a colloid is scattered randomly because the dispersed particles have sizes similar to wavelengths of visible light (400 to 750 nm). The scattered light beam appears broader than one passing through a solution, an example of the **Tyndall effect** (Figure 13.30). Dust scatters sunlight shining through it, as does mist with headlight beams.

Under low magnification, colloidal particles exhibit *Brownian motion,* an erratic change of speed and direction. Brownian motion results from collisions of the particles with molecules of the dispersing medium, and Einstein's explanation of it in 1905 led many to accept the molecular nature of matter.

4. *Stabilizing and destabilizing colloids*. Why *don't* colloidal particles aggregate and settle out? Colloidal particles dispersed in water have charged surfaces that stabilize the colloid through ion-dipole forces. Molecules with dual polarities, like lipids and soaps, form spherical *micelles,* with the charged heads on the exterior and hydrocarbon tails in the interior (Section 13.2). Aqueous proteins mimic this arrangement, with charged amino acid groups facing the water and nonpolar groups buried within the molecule. Oily particles can be dispersed in water by adding ions, which are adsorbed onto their surfaces. Repulsions between the ions on the oil and partial charges on water molecules prevent the particles from aggregating.

Despite these forces, various methods that coagulate the particles destabilize the colloid. Heating makes the particles collide more often and with enough force to coalesce and settle out. Addition of an electrolyte solution containing oppositely charged ions neutralizes the surface charges, so the particles coagulate and settle. In smokestack gases from a coal-burning power plant, ions become adsorbed onto uncharged colloidal particles, which are then attracted to the charged plates of a device called a *Cotrell precipitator* installed in the stack. At the mouths of rivers, where salt concentrations increase near an ocean or sea, colloidal clay particles coalesce into muddy deltas, like those of the Mississippi and the Nile (Figure 13.31). Thus, the city of New Orleans and the ancient Egyptian empire were made possible by large-scale colloid chemistry.

Figure 13.31 The Nile delta (reddish-brown area).

Source: © Earth Satellite Corporation/SPL/Science Source

The upcoming Chemical Connections essay applies solution and colloid chemistry to the purification of water for residential and industrial use.

› Summary of Section 13.7

- Particles in a colloid are smaller than those in a suspension and larger than those in a solution.
- Colloids are classified by the physical states of the dispersed and dispersing substances and involve many combinations of gas, liquid, and/or solid.
- Colloids have extremely large surface areas, scatter incoming light (Tyndall effect), and exhibit random (Brownian) motion.
- Colloidal particles in water are stabilized by charged surfaces that keep them dispersed, but they can be coagulated by heating or by the addition of ions.
- Solution behavior and colloid chemistry are applied to water treatment and purification.

CHEMICAL CONNECTIONS TO ENVIRONMENTAL ENGINEERING

Solutions and Colloids in Water Purification

Most water destined for human use comes from lakes, rivers, reservoirs, or groundwater. Present in this essential resource may be soluble toxic organic compounds and high concentrations of NO_3^- and Fe^{3+}, colloidal clay and microbes, and suspended debris. Let's see how water is treated to remove these dissolved, dispersed, and suspended particles.

Water Treatment Plants

Treating water involves several steps (Figure B13.1):

Step 1. Screening and settling. As water enters the facility, screens remove debris, and settling removes sand and other particles.

Step 2. Coagulating. This step and the next two remove colloids. These particles have negative surfaces that repel each other. Added aluminum sulfate [cake alum; $Al_2(SO_4)_3$] or iron(III) chloride ($FeCl_3$), which supply Al^{3+} or Fe^{3+} ions that neutralize the charges, coagulates the particles through intermolecular forces.

Step 3. Flocculating and sedimenting. Mixing water and flocculating agents in large basins causes a fluffy *floc* to form. Added cationic polymers form long-chain bridges between floc particles, which grow bigger and flow into other basins, where they form a sediment and are removed. Some plants use *dissolved air flotation* (DAF) instead: bubbles forced through the water attach to the floc, and the floating mass is skimmed.

Step 4. Filtering. Various filters remove remaining particles. In *slow sand filters,* the water passes through sand and/or gravel of increasing particle size. In *rapid sand filters,* the sand is backwashed with water, and the colloidal mass is removed. Membrane filters *(not shown)* with pore sizes of 0.1–10 μm are thin tubes bundled together inside a vessel. The water is forced into these tubes, and the colloid-free filtrate is collected from a large central tube. Filtration is very effective at removing microorganisms resistant to disinfectants.

Step 5. Disinfecting. Water sources often contain harmful microorganisms that are killed by one of three agents:

- Chlorine, as aqueous bleach (ClO^-) or Cl_2, is most common, but carcinogenic chlorinated organic compounds can form.
- UV light emitted by high-intensity fluorescent tubes disinfects by disrupting microorganisms' DNA.
- Ozone (O_3) gas is a powerful oxidizing agent.

Sodium fluoride (NaF) to prevent tooth decay and phosphate salts to prevent leaching of lead from pipes may then be added.

Step 6 (not shown). Adsorbing onto granular activated carbon (GAC). Petroleum and other organic contaminants are removed by adsorption. GAC is a highly porous agent formed by "activating" wood, coal, or coconut shells with steam: 1 kg of GAC has a surface area of 275 acres!

Water Softening via Ion Exchange

Water with large amounts of 2+ ions, such as Ca^{2+} and Mg^{2+}, is called **hard water.** Combined with fatty-acid anions in soap, these cations form solid deposits on clothes, washing machines, and sinks:

$$\underset{}{Ca^{2+}(aq)} + \underset{\text{soap}}{2C_{17}H_{35}COONa(aq)} \longrightarrow \underset{\text{insoluble deposit}}{(C_{17}H_{35}COO)_2Ca(s)} + 2Na^+(aq)$$

When a large amount of HCO_3^- is present, the cations form *scale,* a carbonate deposit in boilers and hot-water pipes that interferes with the transfer of heat:

$$Ca^{2+}(aq) + 2HCO_3^-(aq) \longrightarrow CaCO_3(s) + CO_2(g) + H_2O(l)$$

Removing hard-water cations, called **water softening,** is done by exchanging Na^+ ions for Ca^{2+} and Mg^{2+} ions. A home system for **ion exchange** contains an insoluble polymer *resin* with bonded

Figure B13.1 The typical steps in municipal water treatment.

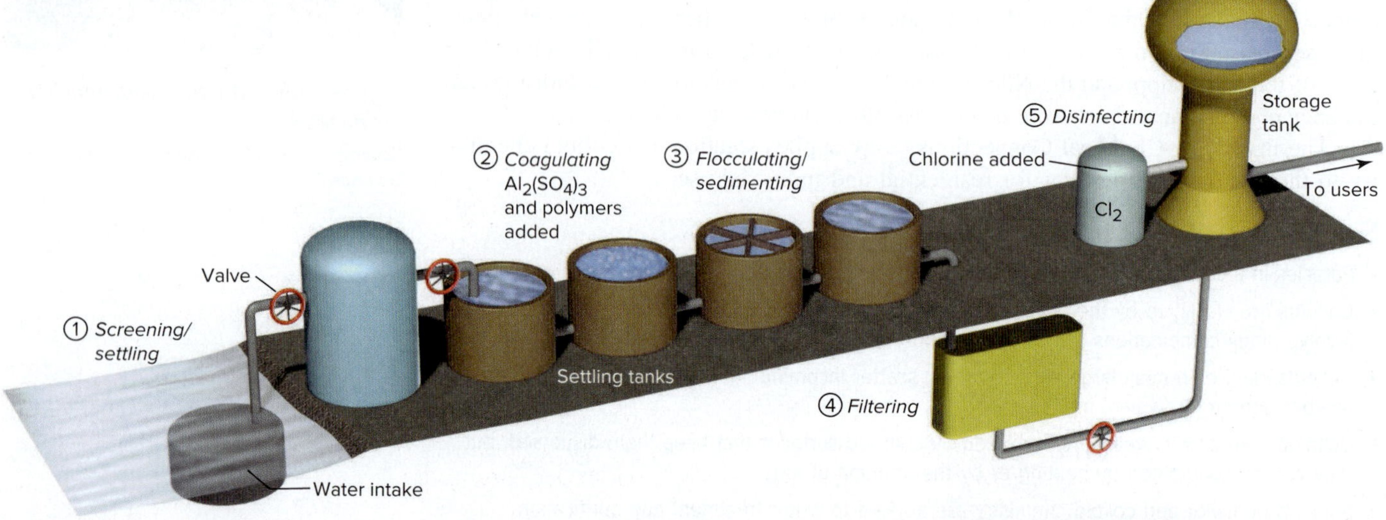

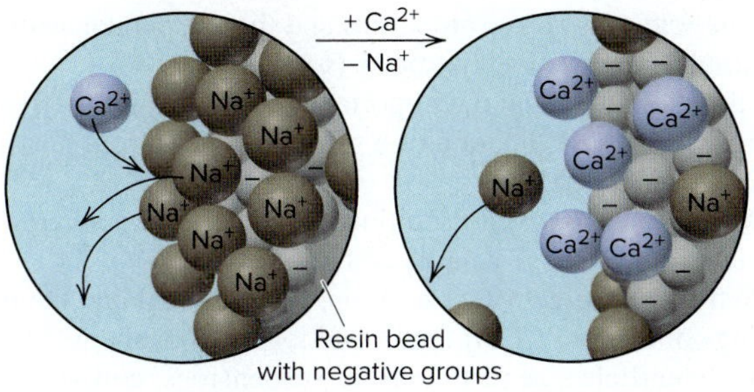

Figure B13.2 Ion exchange to remove hard-water cations.

anionic groups, such as —SO_3^- or —COO^-, and Na^+ ions for charge balance (Figure B13.2). The hard-water cations displace the Na^+ ions and bind to the anionic groups. When all resin sites are occupied, the resin is regenerated with concentrated Na^+ solution that exchanges Na^+ ions for bound Ca^{2+} and Mg^{2+}.

Membrane Processes and Reverse Osmosis

Membranes with 0.0001–0.01 μm pores can remove unwanted ions from water. Recall that solutions of different concentrations separated by a semipermeable membrane create osmotic pressure. In **reverse osmosis,** a pressure *greater* than the osmotic pressure is *applied* to the more concentrated solution to force water back through the membrane and filter out ions. In homes, toxic *heavy-metal ions,* such as Pb^{2+}, Cd^{2+}, and Hg^{2+}, are removed this way. On a large scale, reverse osmosis is used for **desalination,** which can convert seawater (40,000 ppm of ions) to drinking water (400 ppm) (Figure B13.3).

Wastewater Treatment

Wastewater, used domestic or industrial water, is treated in several ways before being returned to a natural source:

- In *primary* treatment, the water enters a settling basin to remove particles.
- In *biological* treatment, bacteria metabolize organic compounds and are then removed by settling.
- In *advanced* treatment, a process is tailored to remove a specific pollutant. For example, ammonia, which causes excessive growth of plants and algae, is removed in two steps:
  1. *Nitrification.* Certain bacteria oxidize ammonia (electron donor) with O_2 (electron acceptor) to form nitrate ion:
  $$NH_4^+ + 2O_2 \longrightarrow NO_3^- + 2H^+ + H_2O$$
  2. *Denitrification.* Other bacteria oxidize an added compound like methanol (CH_3OH) using the NO_3^-:
  $$5CH_3OH + 6NO_3^- \longrightarrow 3N_2 + 5CO_2 + 7H_2O + 6OH^-$$

  Thus, the process converts NH_3 in wastewater to N_2, which is released to the atmosphere.

Problems

B13.1 Briefly answer each of the following:
(a) Why is cake alum [$Al_2(SO_4)_3$] added during water purification?
(b) Why is water that contains large amounts of Ca^{2+} and Mg^{2+} difficult to use for cleaning?
(c) What is the meaning of "reverse" in reverse osmosis?
(d) Why might a water treatment plant use ozone as a disinfectant instead of chlorine?
(e) How does passing a saturated NaCl solution through a "spent" ion-exchange resin regenerate the resin?

B13.2 Wastewater discharged into a stream by a sugar refinery contains 3.55 g of sucrose ($C_{12}H_{22}O_{11}$) per liter. A government-sponsored study is testing the feasibility of removing the sugar by reverse osmosis. What pressure must be applied to the wastewater solution at 20.°C to produce pure water?

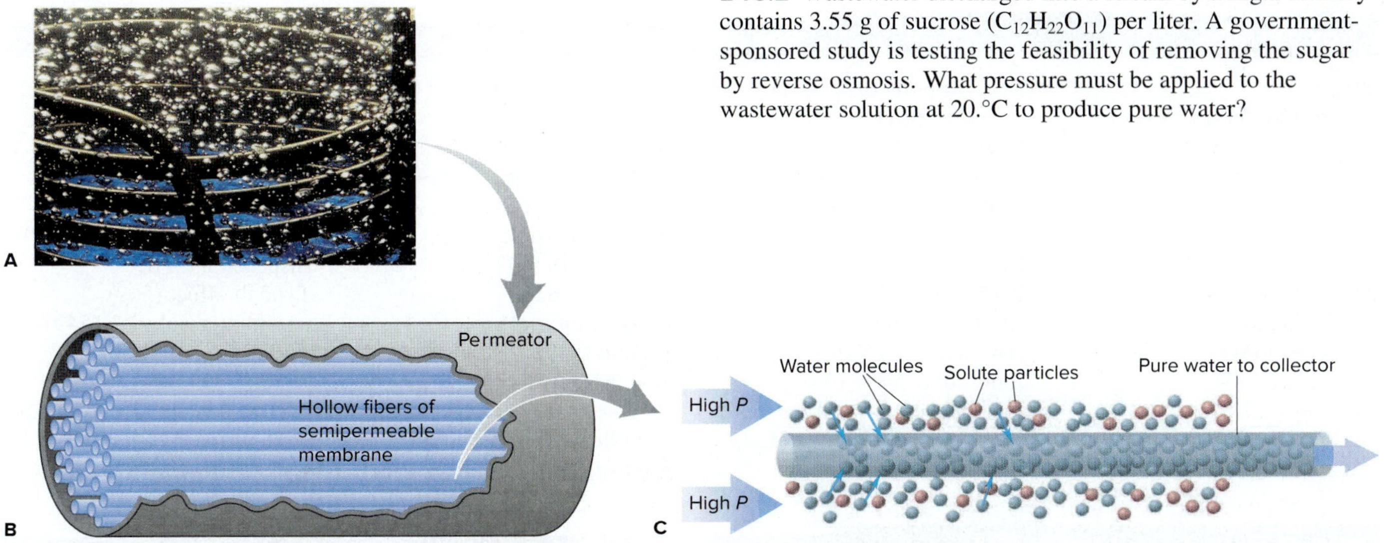

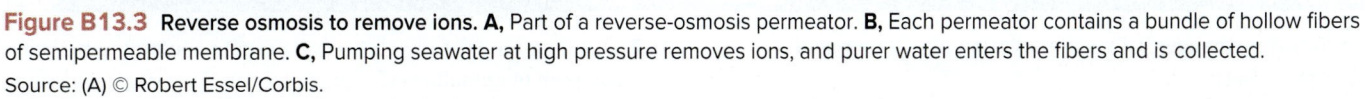

Figure B13.3 Reverse osmosis to remove ions. **A,** Part of a reverse-osmosis permeator. **B,** Each permeator contains a bundle of hollow fibers of semipermeable membrane. **C,** Pumping seawater at high pressure removes ions, and purer water enters the fibers and is collected.
Source: (A) © Robert Essel/Corbis.

CHAPTER REVIEW GUIDE

Learning Objectives

Relevant section (§) and/or sample problem (SP) numbers appear in parentheses.

Understand These Concepts

1. The quantitative meaning of solubility (§13.1)
2. The major types of intermolecular forces in solution and their relative strengths (§13.1)
3. How the like-dissolves-like rule depends on intermolecular forces (§13.1)
4. Why gases have relatively low solubilities in water (§13.1)
5. General characteristics of solutions formed by various combinations of gases, liquids, and solids (§13.1)
6. How intermolecular forces stabilize the structures of proteins, the cell membrane, and DNA (§13.2)
7. The enthalpy components of a solution cycle and their effect on ΔH_{soln} (§13.3)
8. The dependence of ΔH_{hydr} on ionic charge density and the factors that determine whether ionic solution processes are exothermic or endothermic (§13.3)
9. The meaning of entropy and how the balance between the change in enthalpy and the change in entropy governs the solution process (§13.3)
10. The distinctions among saturated, unsaturated, and supersaturated solutions, and the equilibrium nature of a saturated solution (§13.4)
11. The relation between temperature and the solubility of solids (§13.4)
12. Why the solubility of gases in water decreases with a rise in temperature (§13.4)
13. The effect of gas pressure on solubility and its quantitative expression as Henry's law (§13.4)
14. The meaning of molarity, molality, mole fraction, and parts by mass or by volume of a solution, and how to convert among them (§13.5)
15. The distinction between electrolytes and nonelectrolytes in solution (§13.6)
16. The four colligative properties and their dependence on number of dissolved particles (§13.6)
17. Ideal solutions and the importance of Raoult's law (§13.6)
18. How the phase diagram of a solution differs from that of the pure solvent (§13.6)
19. Why the vapor over a solution of a volatile nonelectrolyte is richer in the more volatile component (§13.6)
20. Why strong electrolyte solutions are not ideal and the meanings of the van't Hoff factor and ionic atmosphere (§13.6)
21. How particle size distinguishes suspensions, colloids, and solutions (§13.7)
22. How colloidal behavior is demonstrated by the Tyndall effect and Brownian motion (§13.7)

Master These Skills

1. Predicting relative solubilities from intermolecular forces (SP 13.1)
2. Calculating the heat of solution for an ionic compound (SP 13.2)
3. Using Henry's law to calculate the solubility of a gas (SP 13.3)
4. Expressing concentration in terms of molality, parts by mass, parts by volume, and mole fraction (SPs 13.4, 13.5)
5. Interconverting among the various terms for expressing concentration (SP 13.6)
6. Using Raoult's law to calculate the vapor pressure lowering of a solution (SP 13.7)
7. Determining boiling and freezing points of a solution (SP 13.8)
8. Using a colligative property to calculate the molar mass of a solute (SP 13.9)
9. Calculating the composition of vapor over a solution of volatile nonelectrolyte (§13.6)
10. Calculating the van't Hoff factor (i) from the magnitude of a colligative property (§13.6)
11. Using a depiction to determine colligative properties (SP 13.10)

Key Terms

Page numbers appear in parentheses.

alloy (538)
amino acid (539)
boiling point elevation (ΔT_b) (559)
charge density (545)
colligative property (557)
colloid (568)
desalination (571)
dipole–induced dipole force (535)
double helix (543)
electrolyte (557)
entropy (S) (547)
fractional distillation (564)
freezing point depression (ΔT_f) (561)
hard water (570)
heat of hydration (ΔH_{hydr}) (545)
heat of solution (ΔH_{soln}) (544)
Henry's law (551)
hydration (545)
hydration shell (534)
ideal solution (558)
ion exchange (570)
ionic atmosphere (565)
ion–induced dipole force (534)
like-dissolves-like rule (534)
lipid bilayer (542)
mass percent [% (w/w)] (554)
miscible (534)
molality (m) (553)
mole fraction (X) (554)
mononucleotide (543)
nonelectrolyte (557)
nucleic acid (542)
osmosis (562)
osmotic pressure (Π) (562)
protein (539)
Raoult's law (558)
reverse osmosis (571)
saturated solution (549)
semipermeable membrane (562)
soap (541)
solubility (S) (534)
solute (534)
solvation (545)
solvent (534)
supersaturated solution (549)
suspension (568)
Tyndall effect (569)
unsaturated solution (549)
vapor pressure lowering (ΔP) (558)
volume percent [% (v/v)] (554)
wastewater (571)
water softening (570)

Key Equations and Relationships

Page numbers appear in parentheses.

13.1 Dividing the general heat of solution into component enthalpies (544):

$$\Delta H_{soln} = \Delta H_{solute} + \Delta H_{solvent} + \Delta H_{mix}$$

13.2 Dividing the heat of solution of an ionic compound in water into component enthalpies (545):

$$\Delta H_{soln} = \Delta H_{lattice} + \Delta H_{hydr\ of\ the\ ions}$$

13.3 Relating gas solubility to its partial pressure (Henry's law) (551):

$$S_{gas} = k_H \times P_{gas}$$

13.4 Defining concentration in terms of molarity (552):

$$\text{Molarity } (M) = \frac{\text{amount (mol) of solute}}{\text{volume (L) of solution}}$$

13.5 Defining concentration in terms of molality (553):

$$\text{Molality } (m) = \frac{\text{amount (mol) of solute}}{\text{mass (kg) of solvent}}$$

13.6 Defining concentration in terms of mass percent (554):

$$\text{Mass percent } [\% \text{ (w/w)}] = \frac{\text{mass of solute}}{\text{mass of solution}} \times 100$$

13.7 Defining concentration in terms of volume percent (554):

$$\text{Volume percent } [\% \text{ (v/v)}] = \frac{\text{volume of solute}}{\text{volume of solution}} \times 100$$

13.8 Defining concentration in terms of mole fraction (554):

Mole fraction (X)

$$= \frac{\text{amount (mol) of solute}}{\text{amount (mol) of solute + amount (mol) of solvent}}$$

13.9 Expressing the relationship between the vapor pressure of solvent above a solution and its mole fraction in the solution (Raoult's law) (558):

$$P_{solvent} = X_{solvent} \times P^{\circ}_{solvent}$$

13.10 Calculating the vapor pressure lowering due to solute (558):

$$\Delta P = X_{solute} \times P^{\circ}_{solvent}$$

13.11 Calculating the boiling point elevation of a solution (560):

$$\Delta T_b = K_b m$$

13.12 Calculating the freezing point depression of a solution (561):

$$\Delta T_f = K_f m$$

13.13 Calculating the osmotic pressure of a solution (562):

$$\Pi = \frac{n_{solute}}{V_{soln}} RT = MRT$$

BRIEF SOLUTIONS TO FOLLOW-UP PROBLEMS

13.1A (a) 1-Butanol has one —OH group/molecule, while 1,4-butanediol has two —OH groups/molecule. 1,4-Butanediol is more soluble in water because it can form more H bonds.
(b) Chloroform is more soluble in water because of dipole-dipole forces between the polar $CHCl_3$ molecules and water. The forces between nonpolar CCl_4 molecules and water are weaker dipole–induced dipole forces, which do not effectively replace H bonds between water molecules.

13.1B (a) Chloroform dissolves more chloromethane due to similar dipole-dipole forces between the polar molecules of these two substances. CH_3Cl molecules do not exhibit H bonding and so do not effectively replace H bonds between methanol molecules.
(b) Hexane dissolves more pentanol due to dispersion forces between the hydrocarbon chains in each molecule.

13.2A From Equation 13.2, we have

$$\Delta H_{soln} \text{ of } KNO_3 = \Delta H_{lattice} \text{ of } KNO_3 + (\Delta H_{hydr} \text{ of } K^+ + \Delta H_{hydr} \text{ of } NO_3^-)$$

$$34.89 \text{ kJ/mol} = 685 \text{ kJ/mol} + (\Delta H_{hydr} \text{ of } K^+ + \Delta H_{hydr} \text{ of } NO_3^-)$$

$$\Delta H_{hydr} \text{ of } K^+ + \Delta H_{hydr} \text{ of } NO_3^- = 34.89 \text{ kJ/mol} - 685 \text{ kJ/mol} = -650. \text{ kJ/mol}$$

13.2B From Equation 13.2, we have

$$\Delta H_{soln} \text{ of NaCN} = \Delta H_{lattice} \text{ of NaCN} + (\Delta H_{hydr} \text{ of } Na^+ + \Delta H_{hydr} \text{ of } CN^-)$$

$$1.21 \text{ kJ/mol} = 766 \text{ kJ/mol} + (-410. \text{ kJ/mol} + \Delta H_{hydr} \text{ of } CN^-)$$

$$\Delta H_{hydr} \text{ of } CN^- = 1.21 \text{ kJ/mol} - 766 \text{ kJ/mol} + 410. \text{ kJ/mol} = -355 \text{ kJ/mol}$$

13.3A The partial pressure of N_2 in air is the volume percent divided by 100 times the total pressure (Dalton's law, Section 5.4): $P_{N_2} = 0.78 \times 1$ atm $= 0.78$ atm.

$$S_{gas} = k_H \times P_{gas}$$

$$S_{N_2} = (7\times10^{-4} \text{ mol/L·atm})(0.78 \text{ atm}) = 5\times10^{-4} \text{ mol/L}$$

13.3B In a mixture of gases, the volume percent of a gas divided by 100 times the total pressure equals the gas's partial pressure (Dalton's law, Section 5.4):

$P_{gas} = 0.40 \times 1.2$ atm $= 0.48$ atm.

$$k_H = \frac{S_{gas}}{P_{gas}} = \frac{1.2\times10^{-2} \text{ mol/L}}{0.48 \text{ atm}} = 2.5\times10^{-2} \text{ mol/L·atm}$$

13.4A Convert mass (g) of ethanol to kg, multiply by the molality to obtain amount (mol) of glucose, and then multiply amount (mol) of glucose by the molar mass to obtain mass of glucose.

Amount (mol) of glucose

$$= 563 \text{ g ethanol} \times \frac{1 \text{ kg}}{10^3 \text{ g}} \times \frac{2.40\times10^{-2} \text{ mol glucose}}{1 \text{ kg ethanol}} = 1.35\times10^{-2} \text{ mol glucose}$$

$$\text{Mass (g) glucose} = 1.35\times10^{-2} \text{ mol } C_6H_{12}O_6 \times \frac{180.16 \text{ g } C_6H_{12}O_6}{1 \text{ mol } C_6H_{12}O_6} = 2.43 \text{ g glucose}$$

13.4B Convert mass (g) of I_2 to amount (mol) and amount (mol) of $(CH_3CH_2)_2O$ to mass (kg). Then divide moles of I_2 by kg of $(CH_3CH_2)_2O$.

$$\text{Amount (mol) of } I_2 = 15.20 \text{ g } I_2 \times \frac{1 \text{ mol } I_2}{253.8 \text{ g } I_2} = 5.989\times10^{-2} \text{ mol } I_2$$

Mass (kg) of $(CH_3CH_2)_2O$

$$= 1.33 \text{ mol } (CH_3CH_2)_2O \times \frac{74.12 \text{ g } (CH_3CH_2)_2O}{1 \text{ mol } (CH_3CH_2)_2O} \times \frac{1 \text{ kg}}{10^3 \text{ g}} = 9.86\times10^{-2} \text{ kg } (CH_3CH_2)_2O$$

$$\text{Molality } (m) = \frac{5.989\times10^{-2} \text{ mol}}{9.86\times10^{-2} \text{ kg}} = 0.607\ m$$

BRIEF SOLUTIONS TO FOLLOW-UP PROBLEMS (continued)

13.5A $\text{Mass \%} = \dfrac{\text{mass of solute}}{\text{mass of solution}} \times 100$

$$\text{Mass \% } C_3H_7OH = \frac{35.0\ \text{g}}{35.0\ \text{g} + 150.\ \text{g}} \times 100 = 18.9\ \text{mass \%}$$

$$\text{Mass \% } C_2H_5OH = 100.0 - 18.9 = 81.1\ \text{mass \%}$$

$$X_{C_3H_7OH} = \frac{35.0\ \text{g}\ C_3H_7OH \times \dfrac{1\ \text{mol}\ C_3H_7OH}{60.09\ \text{g}\ C_3H_7OH}}{\left(35.0\ \text{g}\ C_3H_7OH \times \dfrac{1\ \text{mol}\ C_3H_7OH}{60.09\ \text{g}\ C_3H_7OH}\right) + \left(150.\ \text{g}\ C_2H_5OH \times \dfrac{1\ \text{mol}\ C_2H_5OH}{46.07\ \text{g}\ C_2H_5OH}\right)} = 0.152$$

$$X_{C_2H_5OH} = 1.000 - 0.152 = 0.848$$

13.5B Total mass (g) = 1.87 g + 27.4 g + 4.10 g = 33.4 g

$$\text{Mass \%} = \frac{\text{mass of solute}}{\text{mass of solution}} \times 100$$

$$\text{Mass \% } C_2H_5OH = \frac{1.87\ \text{g}}{33.4\ \text{g}} \times 100 = 5.60\%$$

Mass % C_8H_{18} = 82.0%; mass % C_7H_{16} = 12.3%

$$\text{Amount (mol) of } C_2H_5OH = 1.87\ \text{g}\ C_2H_5OH \times \frac{1\ \text{mol}\ C_2H_5OH}{46.07\ \text{g}\ C_2H_5OH} = 0.0406\ \text{mol}\ C_2H_5OH$$

Amount (mol) of C_8H_{18} = 0.240 mol C_8H_{18}
Amount (mol) of C_7H_{16} = 0.0409 mol C_7H_{16}
Total amount (mol) = 0.0406 mol + 0.240 mol + 0.0409 mol = 0.322 mol

$$\text{Mole \% } C_2H_5OH = \frac{\text{mol } C_2H_5OH}{\text{total mol}} \times 100$$

$$\text{Mole \% } C_2H_5OH = \frac{0.0406\ \text{mol}}{0.322\ \text{mol}} \times 100 = 12.6\%$$

Mole % C_8H_{18} = 74.5%; mole % C_7H_{16} = 12.7%
(Slight differences from 100% are due to rounding.)

13.6A There are 11.8 moles of HCl in 1.00 L of solution:

$$\text{Mass (g) of HCl} = 11.8\ \text{mol HCl} \times \frac{36.46\ \text{g HCl}}{1\ \text{mol HCl}} = 430.\ \text{g HCl}$$

$$\text{Mass (g) of soln} = 1\ \text{L soln} \times \frac{10^3\ \text{mL soln}}{1\ \text{L soln}} \times \frac{1.190\ \text{g soln}}{1\ \text{mL soln}} = 1190.\ \text{g soln}$$

$$\text{Mass \% HCl} = \frac{\text{mass of HCl}}{\text{mass of soln}} \times 100 = \frac{430.\ \text{g}}{1190.\ \text{g}} \times 100 = 36.1\ \text{mass \% HCl}$$

$$\text{Mass (kg) of } H_2O = (1190.\ \text{g soln} - 430.\ \text{g HCl}) \times \frac{1\ \text{kg}}{1000\ \text{g}} = 0.760\ \text{kg}\ H_2O$$

$$\text{Molality of HCl} = \frac{\text{mol HCl}}{\text{kg water}} = \frac{11.8\ \text{mol HCl}}{0.760\ \text{kg}\ H_2O} = 15.5\ m\ \text{HCl}$$

$$\text{Amount (mol) of } H_2O = 760\ \text{g}\ H_2O \times \frac{1\ \text{mol}\ H_2O}{18.02\ \text{g}\ H_2O} = 42.2\ \text{mol}\ H_2O$$

$$X_{HCl} = \frac{\text{mol HCl}}{\text{mol HCl} + \text{mol } H_2O} = \frac{11.8\ \text{mol}}{11.8\ \text{mol} + 42.2\ \text{mol}} = 0.219$$

13.6B There are 5.44 moles of $CaBr_2$ in 1 kg (1000 g) of H_2O:

$$\text{Mass (g) of } CaBr_2 = 5.44\ \text{mol}\ CaBr_2 \times \frac{199.88\ \text{g}\ CaBr_2}{1\ \text{mol}\ CaBr_2} = 1087\ \text{g}\ CaBr_2$$

Mass (g) of soln = 1087 g $CaBr_2$ + 1000 g H_2O = 2087 g soln

$$\text{Mass \% } CaBr_2 = \frac{\text{mass of } CaBr_2}{\text{mass of soln}} \times 100 = \frac{1087\ \text{g}}{2087\ \text{g}} \times 100 = 52.1\ \text{mass \% } CaBr_2$$

$$\text{Volume(L) of soln} = 2087\ \text{g soln} \times \frac{1\ \text{mL soln}}{1.70\ \text{g soln}} \times \frac{1\ \text{L}}{10^3\ \text{mL}} = 1.228\ \text{L soln}$$

$$\text{Molarity } (M) = \frac{\text{amount (mol) } CaBr_2}{\text{L soln}} = \frac{5.44\ \text{mol}\ CaBr_2}{1.228\ \text{L}} = 4.43\ M$$

$$\text{Amount (mol) of } H_2O = 1000\ \text{g}\ H_2O \times \frac{1\ \text{mol}\ H_2O}{18.02\ \text{g}\ H_2O} = 55.5\ \text{mol}\ H_2O$$

$$\text{Mole \%} = \frac{\text{amount (mol) } CaBr_2}{\text{amount (mol) } CaBr_2 + \text{amount (mol) } H_2O} \times 100 = \frac{5.44\ \text{mol}\ CaBr_2}{5.44\ \text{mol}\ CaBr_2 + 55.5\ \text{mol}\ H_2O} \times 100 = 8.93\%$$

13.7A Find the mole fraction of aspirin and use Equation 13.10:

$$\text{Amount (mol) of aspirin} = 2.00\ \text{g aspirin} \times \frac{1\ \text{mol aspirin}}{180.15\ \text{g aspirin}} = 0.0111\ \text{mol aspirin}$$

$$\text{Amount (mol) of } CH_3OH = 50.0\ \text{g}\ CH_3OH \times \frac{1\ \text{mol}\ CH_3OH}{32.04\ \text{g}\ CH_3OH} = 1.56\ \text{mol}\ CH_3OH$$

$$X_{\text{aspirin}} = \frac{\text{mol aspirin}}{\text{mol aspirin} + \text{mol } CH_3OH} = \frac{0.0111\ \text{mol}}{0.0111\ \text{mol} + 1.56\ \text{mol}} = 0.00707$$

$$\Delta P = X_{\text{aspirin}} \times P^\circ_{\text{methanol}} = 0.00707 \times 101\ \text{torr} = 0.714\ \text{torr}$$

13.7B Find mole fraction of ethanol, apply Equation 13.9, and convert kPa to torr:

$$\text{Amount (mol) of menthol} = 6.49 \text{ g menthol} \times \frac{1 \text{ mol menthol}}{156.26 \text{ g menthol}} = 0.0415 \text{ mol menthol}$$

$$\text{Amount (mol) of ethanol} = 25.0 \text{ g ethanol} \times \frac{1 \text{ mol ethanol}}{46.07 \text{ g ethanol}} = 0.543 \text{ mol ethanol}$$

$$X_{\text{ethanol}} = \frac{\text{mol of ethanol}}{\text{mol of ethanol + mol of menthol}} = \frac{0.543 \text{ mol}}{0.543 \text{ mol} + 0.0415 \text{ mol}} = 0.929$$

$$P_{\text{ethanol}} \text{ (kPa)} = X_{\text{ethanol}} \times P^{\circ}_{\text{ethanol}} = 0.929 \times 6.87 \text{ kPa} = 6.38 \text{ kPa}$$

$$P_{\text{ethanol}} \text{ (torr)} = 6.38 \text{ kPa} \times \frac{760 \text{ torr}}{101.325 \text{ kPa}} = 47.9 \text{ torr}$$

13.8A Find molality and then apply Equations 13.11 and 13.12:

$$\text{Amount (mol) of } P_4 = 8.44 \text{ g } P_4 \times \frac{1 \text{ mol } P_4}{123.88 \text{ g } P_4} = 0.0681 \text{ mol } P_4$$

$$\text{Molality } (m) = \frac{\text{amount (mol) } P_4}{\text{mass (kg) } CS_2} = \frac{0.0681 \text{ mol } P_4}{60.0 \text{ g } CS_2 \times \frac{1 \text{ kg}}{10^3 \text{ g}}} = 1.14\ m$$

$$\Delta T_b = K_b m = 2.34°\text{C}/m \times 1.14\ m = 2.67°\text{C}$$

$$T_{b(\text{solution})} = T_{b(\text{solvent})} + \Delta T_b = 46.2°\text{C} + 2.67°\text{C} = 48.9°\text{C}$$

$$\Delta T_f = K_f m = 3.83°\text{C}/m \times 1.14\ m = 4.37°\text{C}$$

$$T_{f(\text{solution})} = T_{f(\text{solvent})} - \Delta T_f = -111.5°\text{C} - 4.37°\text{C} = -115.9°\text{C}$$

13.8B Calculate ΔT_f in °C and use Equation 13.12:

$$\Delta T_f = T_{f(\text{solvent})} - T_{f(\text{solution})} = 32.00°\text{F} - 0.00°\text{F} = 32.00°\text{F} \times \frac{5°\text{C}}{9°\text{F}} = 17.78°\text{C}$$

$$\text{Molality of } C_2H_6O_2 = \frac{\Delta T_f}{K_f} = \frac{17.78°\text{ C}}{1.86°\text{ C}/m} = 9.56\ m$$

$$\text{Mass of } C_2H_6O_2 = 4.00 \text{ kg } H_2O \times \frac{9.56 \text{ mol } C_2H_6O_2}{1 \text{ kg } H_2O} \times \frac{62.07 \text{ g } C_2H_6O_2}{1 \text{ mol } C_2H_6O_2} = 2370 \text{ g } C_2H_6O_2$$

13.9A Find the molarity with Equation 13.13, use V to find the amount (mol), and then use mass to find the molar mass:

$$\text{Molarity } (M) = \frac{\Pi}{RT} = \frac{8.98 \text{ torr} \times \frac{1 \text{ atm}}{760 \text{ torr}}}{\left(0.0821 \frac{\text{atm·L}}{\text{mol·K}}\right)(273.15 \text{ K} + 27.0)} = 4.79\times10^{-4} \text{ mol/L}$$

$$\text{Amount (mol)} = M \times V = 4.79\times10^{-4} \text{ mol/L} \times 0.0120 \text{ L} = 5.75\times10^{-6} \text{ mol}$$

$$\mathcal{M} = \frac{0.200 \text{ g}}{5.75\times10^{-6} \text{ mol}} = 3.48\times10^{4} \text{ g/mol}$$

13.9B Find the molality with Equation 13.12, use the mass of solvent to find the amount (mol) of napthalene, and then use mass to find the molar mass:

$$\text{Molality} = \frac{\Delta T_f}{K_f} = \frac{5.56°\text{ C} - 4.20°\text{ C}}{4.90°\text{ C}/m} = 0.278\ m$$

Amount (mol) of naphthalene

$$= 0.200 \text{ kg benzene} \times \frac{0.278 \text{ mol naphthalene}}{1 \text{ kg benzene}} = 0.0556 \text{ mol}$$

$$\mathcal{M} = \frac{7.01 \text{ g}}{0.0556 \text{ mol}} = 126 \text{ g/mol}$$

13.10A (a) Scene B shows separate ions in a 3 K^+/1 PO_4^{3-} ratio.

(b) Find the moles of K_3PO_4 and then the molality of the solution; use Equation 13.11 to find the boiling point:

$$\text{Amount (mol) of } K_3PO_4 = 31.2 \text{ g } K_3PO_4 \times \frac{1 \text{ mol } K_3PO_4}{212.27 \text{ g } K_3PO_4} = 0.147 \text{ mol } K_3PO_4$$

$$\text{Molality } (m) = \frac{\text{amount (mol) of } K_3PO_4}{\text{mass (kg) of } H_2O} = \frac{0.147 \text{ mol}}{0.0850 \text{ kg}} = 1.73\ m$$

With $i = 4$, $\Delta T_b = iK_b m = 4 \times 0.512°\text{C}/m \times 1.73\ m = 3.54°\text{C}$

$$T_{b(\text{solution})} = T_{b(\text{solvent})} + \Delta T_b = 100.00°\text{C} + 3.54°\text{C} = 103.54°\text{C}$$

13.10B (a) Find the molarity of the solution and use Equation 13.13; assume $i = 3$:

$$\text{Mass of solution} = 100.\text{ g water} + 0.952 \text{ g } MgCl_2 = 100.952 \text{ g}$$

$$\text{Volume of solution} = 100.952 \text{ g} \times \frac{1 \text{ mL}}{1.006 \text{ g}} \times \frac{1 \text{ L}}{10^3 \text{ mL}} = 0.1003 \text{ L}$$

$$\text{Amount (mol) of } MgCl_2 = 0.952 \text{ g } MgCl_2 \times \frac{1 \text{ mol } MgCl_2}{95.21 \text{ g } MgCl_2} = 0.0100 \text{ mol } MgCl_2$$

$$\text{Molarity} = \frac{0.0100 \text{ mol } MgCl_2}{0.1003 \text{ L soln}} = 9.97\times10^{-2}\ M$$

$$\text{Osmotic pressure } (\Pi) = i(MRT) = 3(9.97\times10^{-2} \text{ mol/L})\left(0.0821 \frac{\text{atm·L}}{\text{mol·K}}\right)(293 \text{ K}) = 7.19 \text{ atm}$$

(b) Scene C. There is net flow of water from the glucose solution into the $MgCl_2$ solution since there are fewer particles of solute in the glucose solution ($i = 1$ for glucose and $i = 3$ for $MgCl_2$).

PROBLEMS

Problems with **colored** numbers are answered in Appendix E and worked in detail in the Student Solutions Manual. Problem sections match those in the text and give the numbers of relevant sample problems. Most offer Concept Review Questions, Skill-Building Exercises (grouped in pairs covering the same concept), and Problems in Context. The Comprehensive Problems are based on material from any section or previous chapter.

Types of Solutions: Intermolecular Forces and Solubility

(Sample Problem 13.1)

Concept Review Questions

13.1 Describe how properties of seawater illustrate the two characteristics that define mixtures.

13.2 What types of intermolecular forces give rise to hydration shells in an aqueous solution of sodium chloride?

13.3 Acetic acid is miscible with water. Would you expect carboxylic acids with the general formula $CH_3(CH_2)_nCOOH$ to become more or less water soluble as *n* increases? Explain.

13.4 Which would you expect to be more effective as a soap, sodium acetate or sodium stearate? Explain.

13.5 Hexane and methanol are miscible as gases but only slightly soluble in each other as liquids. Explain.

13.6 Hydrogen chloride (HCl) gas is much more soluble than propane gas (C_3H_8) in water, even though HCl has a lower boiling point. Explain.

Skill-Building Exercises (grouped in similar pairs)

13.7 Which gives the more concentrated solution, (a) KNO_3 in H_2O or (b) KNO_3 in carbon tetrachloride (CCl_4)? Explain.

13.8 Which gives the more concentrated solution, stearic acid [$CH_3(CH_2)_{16}COOH$] in (a) H_2O or (b) CCl_4? Explain.

13.9 What is the strongest type of intermolecular force between solute and solvent in each solution?

(a) $CsCl(s)$ in $H_2O(l)$ (b) $CH_3\overset{O}{\overset{\|}{C}}CH_3(l)$ in $H_2O(l)$
(c) $CH_3OH(l)$ in $CCl_4(l)$

13.10 What is the strongest type of intermolecular force between solute and solvent in each solution?
(a) $Cu(s)$ in $Ag(s)$ (b) $CH_3Cl(g)$ in $CH_3OCH_3(g)$
(c) $CH_3CH_3(g)$ in $CH_3CH_2CH_2NH_2(l)$

13.11 What is the strongest type of intermolecular force between solute and solvent in each solution?
(a) $CH_3OCH_3(g)$ in $H_2O(l)$ (b) $Ne(g)$ in $H_2O(l)$
(c) $N_2(g)$ in $C_4H_{10}(g)$

13.12 What is the strongest type of intermolecular force between solute and solvent in each solution?
(a) $C_6H_{14}(l)$ in $C_8H_{18}(l)$ (b) $H_2C{=}O(g)$ in $CH_3OH(l)$
(c) $Br_2(l)$ in $CCl_4(l)$

13.13 Which member of each pair is more soluble in diethyl ether? Why?

(a) $NaCl(s)$ or $HCl(g)$ (b) $H_2O(l)$ or $CH_3\overset{O}{\overset{\|}{C}}H(l)$
(c) $MgBr_2(s)$ or $CH_3CH_2MgBr(s)$

13.14 Which member of each pair is more soluble in water? Why?
(a) $CH_3CH_2OCH_2CH_3(l)$ or $CH_3CH_2OCH_3(g)$
(b) $CH_2Cl_2(l)$ or $CCl_4(l)$
(c)

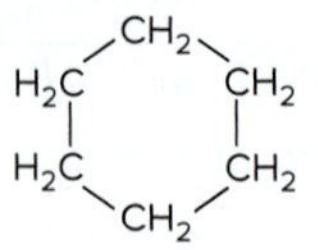

or

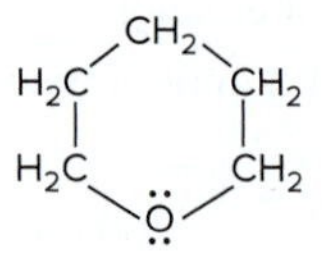

cyclohexane tetrahydropyran

Problems in Context

13.15 A dictionary definition of *homogeneous* is "uniform in composition throughout." River water is a mixture of dissolved compounds, such as calcium bicarbonate, and suspended soil particles. Is river water homogeneous? Explain.

13.16 Gluconic acid is a derivative of glucose used in cleaners and in the dairy and brewing industries. Caproic acid is a carboxylic acid used in the flavoring industry. Although both are six-carbon acids *(see structures below)*, gluconic acid is soluble in water and nearly insoluble in hexane, whereas caproic acid has the opposite solubility behavior. Explain.

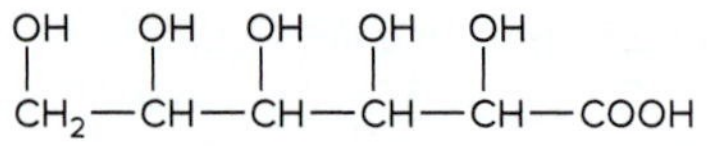

gluconic acid

$CH_3—CH_2—CH_2—CH_2—CH_2—COOH$

caproic acid

Intermolecular Forces and Biological Macromolecules

Concept Review Questions

13.17 Name three intermolecular forces that stabilize the shape of a soluble, globular protein, and explain how they act.

13.18 Name three intermolecular forces that stabilize the structure of DNA, and explain how they act.

13.19 How can relatively weak H bonds hold the double helix together yet allow DNA to function?

13.20 Is sodium propanoate (the sodium salt of propanoic acid) as effective a soap as sodium stearate (the sodium salt of stearic acid)? Explain.

13.21 What intermolecular forces stabilize a lipid bilayer?

13.22 In what way do proteins embedded in a membrane differ structurally from soluble proteins?

13.23 Histones are proteins that control gene function by attaching through salt links to exterior regions of DNA. Name an amino acid whose side chain is often found on the exterior of histones.

Why Substances Dissolve: Understanding the Solution Process

(Sample Problem 13.2)

Concept Review Questions

13.24 What is the relationship between solvation and hydration?

13.25 For a general solvent, which enthalpy terms in the thermochemical solution cycle are combined to obtain $\Delta H_{solvation}$?

13.26 (a) What is the charge density of an ion, and what two properties of an ion affect it?
(b) Arrange the following in order of increasing charge density:

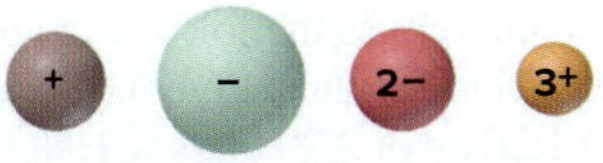

(c) How do the two properties in part (a) affect the ionic heat of hydration, ΔH_{hydr}?

13.27 For ΔH_{soln} to be very small, what quantities must be nearly equal in magnitude? Will their signs be the same or opposite?

13.28 Water is added to a flask containing solid NH_4Cl. As the salt dissolves, the solution becomes colder.
(a) Is the dissolving of NH_4Cl exothermic or endothermic?
(b) Is the magnitude of $\Delta H_{lattice}$ for NH_4Cl larger or smaller than the combined ΔH_{hydr} values of its ions? Explain.
(c) Given the answer to part (a), why does NH_4Cl dissolve in water?

13.29 An ionic compound has a highly negative ΔH_{soln} in water. Would you expect it to be very soluble or nearly insoluble in water? Explain in terms of enthalpy and entropy changes.

Skill-Building Exercises (grouped in similar pairs)

13.30 Sketch an enthalpy diagram for the process of dissolving $KCl(s)$ in H_2O (endothermic).

13.31 Sketch an enthalpy diagram for the process of dissolving $NaI(s)$ in H_2O (exothermic).

13.32 Which ion in each pair has greater charge density? Explain.

(a) Na^+ or Cs^+	(b) Sr^{2+} or Rb^+	(c) Na^+ or Cl^-
(d) O^{2-} or F^-	(e) OH^- or SH^-	(f) Mg^{2+} or Ba^{2+}
(g) Mg^{2+} or Na^+	(h) NO_3^- or CO_3^{2-}	

13.33 Which ion has the lower ratio of charge to volume? Explain.

(a) Br^- or I^-	(b) Sc^{3+} or Ca^{2+}	(c) Br^- or K^+
(d) S^{2-} or Cl^-	(e) Sc^{3+} or Al^{3+}	(f) SO_4^{2-} or ClO_4^-
(g) Fe^{3+} or Fe^{2+}	(h) Ca^{2+} or K^+	

13.34 Which ion of each pair in Problem 13.32 has the *larger* ΔH_{hydr}?

13.35 Which ion of each pair in Problem 13.33 has the *smaller* ΔH_{hydr}?

13.36 (a) Use the following data to calculate the combined heat of hydration for the ions in potassium bromate ($KBrO_3$):

$$\Delta H_{lattice} = 745 \text{ kJ/mol} \qquad \Delta H_{soln} = 41.1 \text{ kJ/mol}$$

(b) Which ion contributes more to the answer for part (a)? Why?

13.37 (a) Use the following data to calculate the combined heat of hydration for the ions in sodium acetate ($NaC_2H_3O_2$):

$$\Delta H_{lattice} = 763 \text{ kJ/mol} \qquad \Delta H_{soln} = 17.3 \text{ kJ/mol}$$

(b) Which ion contributes more to the answer for part (a)? Why?

13.38 State whether the entropy of the system increases or decreases in each of the following processes:
(a) Gasoline burns in a car engine.
(b) Gold is extracted and purified from its ore.
(c) Ethanol (CH_3CH_2OH) dissolves in 1-propanol ($CH_3CH_2CH_2OH$).

13.39 State whether the entropy of the system increases or decreases in each of the following processes:
(a) Pure gases are mixed to prepare an anesthetic.
(b) Electronic-grade silicon is prepared from sand.
(c) Dry ice (solid CO_2) sublimes.

Problems in Context

13.40 Besides being used in black-and-white film, silver nitrate ($AgNO_3$) is used similarly in forensic science. The NaCl left behind in the sweat of a fingerprint is treated with $AgNO_3$ solution to form AgCl. This precipitate is developed to show the black-and-white fingerprint pattern. Given that $\Delta H_{lattice} = 822$ kJ/mol and $\Delta H_{hydr} = -799$ kJ/mol for $AgNO_3$, calculate its ΔH_{soln}.

Solubility as an Equilibrium Process

(Sample Problem 13.3)

Concept Review Questions

13.41 You are given a bottle of solid X and three aqueous solutions of X—one saturated, one unsaturated, and one supersaturated. How would you determine which solution is which?

13.42 Potassium permanganate ($KMnO_4$) has a solubility of 6.4 g/100 g of H_2O at 20°C and a curve of solubility vs. temperature that slopes upward to the right. How would you prepare a supersaturated solution of $KMnO_4$?

13.43 Why does the solubility of any gas in water decrease with rising temperature?

Skill-Building Exercises (grouped in similar pairs)

13.44 For a saturated aqueous solution of each of the following at 20°C and 1 atm, will the solubility increase, decrease, or stay the same when the indicated change occurs?
(a) $O_2(g)$, increase P (b) $N_2(g)$, increase V

13.45 For a saturated aqueous solution of each of the following at 20°C and 1 atm, will the solubility increase, decrease, or stay the same when the indicated change occurs?
(a) $He(g)$, decrease T (b) $RbI(s)$, increase P

13.46 The Henry's law constant (k_H) for O_2 in water at 20°C is 1.28×10^{-3} mol/L·atm.
(a) How many grams of O_2 will dissolve in 2.50 L of H_2O that is in contact with pure O_2 at 1.00 atm?
(b) How many grams of O_2 will dissolve in 2.50 L of H_2O that is in contact with air, where the partial pressure of O_2 is 0.209 atm?

13.47 Argon makes up 0.93% by volume of air. Calculate its solubility (mol/L) in water at 20°C and 1.0 atm. The Henry's law constant for Ar under these conditions is 1.5×10^{-3} mol/L·atm.

Problems in Context

13.48 Caffeine is about 10 times as soluble in hot water as in cold water. A chemist puts a hot-water extract of caffeine into an ice bath, and some caffeine crystallizes. Is the remaining solution saturated, unsaturated, or supersaturated?

13.49 The partial pressure of CO_2 gas above the liquid in a bottle of champagne at 20°C is 5.5 atm. What is the solubility of CO_2 in champagne? Assume Henry's law constant is the same for champagne as for water: at 20°C, $k_H = 3.7\times10^{-2}$ mol/L·atm.

13.50 Respiratory problems are treated with devices that deliver air with a higher partial pressure of O_2 than normal air. Why?

Concentration Terms

(Sample Problems 13.4 to 13.6)

Concept Review Questions

13.51 Explain the difference between molarity and molality. Under what circumstances would molality be a more accurate measure of the concentration of a prepared solution than molarity? Why?

13.52 Which way of expressing concentration includes (a) volume of solution; (b) mass of solution; (c) mass of solvent?

13.53 A solute has a solubility in water of 21 g/kg water. Is this value the same as 21 g/kg solution? Explain.

13.54 You want to convert among molarity, molality, and mole fraction of a solution. You know the masses of solute and solvent and the volume of solution. Is this enough information to carry out all the conversions? Explain.

13.55 When a solution is heated, which ways of expressing concentration change in value? Which remain unchanged? Explain.

Skill-Building Exercises (grouped in similar pairs)

13.56 Calculate the molarity of each aqueous solution:
(a) 32.3 g of table sugar ($C_{12}H_{22}O_{11}$) in 100. mL of solution
(b) 5.80 g of $LiNO_3$ in 505 mL of solution

13.57 Calculate the molarity of each aqueous solution:
(a) 0.82 g of ethanol (C_2H_5OH) in 10.5 mL of solution
(b) 1.27 g of gaseous NH_3 in 33.5 mL of solution

13.58 Calculate the molarity of each aqueous solution:
(a) 78.0 mL of 0.240 *M* NaOH diluted to 0.250 L with water
(b) 38.5 mL of 1.2 *M* HNO_3 diluted to 0.130 L with water

13.59 Calculate the molarity of each aqueous solution:
(a) 25.5 mL of 6.25 *M* HCl diluted to 0.500 L with water
(b) 8.25 mL of 2.00×10^{-2} *M* KI diluted to 12.0 mL with water

13.60 How would you prepare the following aqueous solutions?
(a) 365 mL of 8.55×10^{-2} *M* KH_2PO_4 from solid KH_2PO_4
(b) 465 mL of 0.335 *M* NaOH from 1.25 *M* NaOH

13.61 How would you prepare the following aqueous solutions?
(a) 2.5 L of 0.65 *M* NaCl from solid NaCl
(b) 15.5 L of 0.3 *M* urea [$(NH_2)_2C{=}O$] from 2.1 *M* urea

13.62 How would you prepare the following aqueous solutions?
(a) 1.40 L of 0.288 *M* KBr from solid KBr
(b) 255 mL of 0.0856 *M* $LiNO_3$ from 0.264 *M* $LiNO_3$

13.63 How would you prepare the following aqueous solutions?
(a) 57.5 mL of 1.53×10^{-3} *M* $Cr(NO_3)_3$ from solid $Cr(NO_3)_3$
(b) 5.8×10^3 m^3 of 1.45 *M* NH_4NO_3 from 2.50 *M* NH_4NO_3

13.64 Calculate the molality of the following:
(a) A solution containing 85.4 g of glycine (NH_2CH_2COOH) dissolved in 1.270 kg of H_2O
(b) A solution containing 8.59 g of glycerol ($C_3H_8O_3$) in 77.0 g of ethanol (C_2H_5OH)

13.65 Calculate the molality of the following:
(a) A solution containing 174 g of HCl in 757 g of H_2O
(b) A solution containing 16.5 g of naphthalene ($C_{10}H_8$) in 53.3 g of benzene (C_6H_6)

13.66 What is the molality of a solution consisting of 44.0 mL of benzene (C_6H_6; $d = 0.877$ g/mL) in 167 mL of hexane (C_6H_{14}; $d = 0.660$ g/mL)?

13.67 What is the molality of a solution consisting of 2.66 mL of carbon tetrachloride (CCl_4; $d = 1.59$ g/mL) in 76.5 mL of methylene chloride (CH_2Cl_2; $d = 1.33$ g/mL)?

13.68 How would you prepare the following aqueous solutions?
(a) 3.10×10^2 g of 0.125 *m* ethylene glycol ($C_2H_6O_2$) from ethylene glycol and water (b) 1.20 kg of 2.20 mass % HNO_3 from 52.0 mass % HNO_3

13.69 How would you prepare the following aqueous solutions?
(a) 1.50 kg of 0.0355 *m* ethanol (C_2H_5OH) from ethanol and water
(b) 445 g of 13.0 mass % HCl from 34.1 mass % HCl

13.70 A solution contains 0.35 mol of isopropanol (C_3H_7OH) dissolved in 0.85 mol of water.
(a) What is the mole fraction of isopropanol?
(b) The mass percent?
(c) The molality?

13.71 A solution contains 0.100 mol of NaCl dissolved in 8.60 mol of water.
(a) What is the mole fraction of NaCl?
(b) The mass percent?
(c) The molality?

13.72 What mass of cesium bromide must be added to 0.500 L of water ($d = 1.00$ g/mL) to produce a 0.400 *m* solution? What are the mole fraction and the mass percent of CsBr?

13.73 What are the mole fraction and the mass percent of a solution made by dissolving 0.30 g of KI in 0.400 L of water ($d = 1.00$ g/mL)?

13.74 Calculate the molality, molarity, and mole fraction of NH_3 in ordinary household ammonia, which is an 8.00 mass % aqueous solution ($d = 0.9651$ g/mL).

13.75 Calculate the molality, molarity, and mole fraction of $FeCl_3$ in a 28.8 mass % aqueous solution ($d = 1.280$ g/mL).

Problems in Context

13.76 Wastewater from a cement factory contains 0.25 g of Ca^{2+} ions and 0.056 g of Mg^{2+} ions per 100.0 L of solution. The solution density is 1.001 g/mL. Calculate the Ca^{2+} and Mg^{2+} concentrations in ppm (by mass).

13.77 An automobile antifreeze mixture is made by mixing equal volumes of ethylene glycol ($d = 1.114$ g/mL; $\mathcal{M} = 62.07$ g/mol) and water ($d = 1.00$ g/mL) at 20°C. The density of the mixture is 1.070 g/mL. Express the concentration of ethylene glycol as:
(a) Volume percent (b) Mass percent (c) Molarity
(d) Molality (e) Mole fraction

Colligative Properties of Solutions

(Sample Problems 13.7 to 13.10)

Concept Review Questions

13.78 The chemical formula of a solute does *not* affect the extent of the solution's colligative properties. What characteristic of a solute *does* affect these properties? Name a physical property of a solution that *is* affected by the chemical formula of the solute.

13.79 What is a nonvolatile nonelectrolyte? Why is using this type of solute the simplest way to demonstrate colligative properties?

13.80 In what sense is a strong electrolyte "strong"? What property of a substance makes it a strong electrolyte?

13.81 Express Raoult's law in words. Is Raoult's law valid for a solution of a volatile solute? Explain.

13.82 What are the most important differences between the phase diagram of a pure solvent and the phase diagram of a solution of that solvent?

13.83 Is the composition of the vapor at the top of a fractionating column different from the composition at the bottom? Explain.

13.84 Is the boiling point of 0.01 *m* KF(*aq*) higher or lower than that of 0.01 *m* glucose(*aq*)? Explain.

13.85 Which aqueous solution has a boiling point closer to its predicted value, 0.050 *m* NaF or 0.50 *m* KCl? Explain.

13.86 Which aqueous solution has a freezing point closer to its predicted value, 0.01 *m* NaBr or 0.01 *m* $MgCl_2$? Explain.

13.87 The freezing point depression constants of the solvents cyclohexane and naphthalene are 20.1°C/*m* and 6.94°C/*m*, respectively. Which solvent would give a more accurate result if you are using freezing point depression to determine the molar mass of a substance that is soluble in either one? Why?

Skill-Building Exercises (grouped in similar pairs)

13.88 Classify each substance as a strong electrolyte, weak electrolyte, or nonelectrolyte:
(a) Hydrogen chloride (HCl) (b) Potassium nitrate (KNO_3)
(c) Glucose ($C_6H_{12}O_6$) (d) Ammonia (NH_3)

13.89 Classify each substance as a strong electrolyte, weak electrolyte, or nonelectrolyte:
(a) Sodium permanganate ($NaMnO_4$)
(b) Acetic acid (CH_3COOH)
(c) Methanol (CH_3OH)
(d) Calcium acetate [$Ca(C_2H_3O_2)_2$]

13.90 How many moles of solute particles are present in 1 L of each of the following aqueous solutions?
(a) 0.3 *M* KBr (b) 0.065 *M* HNO_3
(c) 10^{-4} *M* $KHSO_4$ (d) 0.06 *M* ethanol (C_2H_5OH)

13.91 How many moles of solute particles are present in 1 mL of each of the following aqueous solutions?
(a) 0.02 *M* $CuSO_4$ (b) 0.004 *M* $Ba(OH)_2$
(c) 0.08 *M* pyridine (C_5H_5N) (d) 0.05 *M* $(NH_4)_2CO_3$

13.92 Which solution has the lower freezing point?
(a) 11.0 g of CH_3OH in 100. g of H_2O *or*
22.0 g of CH_3CH_2OH in 200. g of H_2O
(b) 20.0 g of H_2O in 1.00 kg of CH_3OH *or*
20.0 g of CH_3CH_2OH in 1.00 kg of CH_3OH

13.93 Which solution has the higher boiling point?
(a) 38.0 g of $C_3H_8O_3$ in 250. g of ethanol *or*
38.0 g of $C_2H_6O_2$ in 250. g of ethanol
(b) 15 g of $C_2H_6O_2$ in 0.50 kg of H_2O *or*
15 g of NaCl in 0.50 kg of H_2O

13.94 Rank the following aqueous solutions in order of increasing (a) osmotic pressure; (b) boiling point; (c) freezing point; (d) vapor pressure at 50°C:
(I) 0.100 *m* $NaNO_3$
(II) 0.100 *m* glucose
(III) 0.100 *m* $CaCl_2$

13.95 Rank the following aqueous solutions in order of decreasing (a) osmotic pressure; (b) boiling point; (c) freezing point; (d) vapor pressure at 298 K:
(I) 0.04 *m* urea [$(NH_2)_2C{=}O$]
(II) 0.01 *m* $AgNO_3$
(III) 0.03 *m* $CuSO_4$

13.96 Calculate the vapor pressure of a solution of 34.0 g of glycerol ($C_3H_8O_3$) in 500.0 g of water at 25°C. The vapor pressure of water at 25°C is 23.76 torr. (Assume ideal behavior.)

13.97 Calculate the vapor pressure of a solution of 0.39 mol of cholesterol in 5.4 mol of toluene at 32°C. Pure toluene has a vapor pressure of 41 torr at 32°C. (Assume ideal behavior.)

13.98 What is the freezing point of 0.251 *m* urea in water?

13.99 What is the boiling point of 0.200 *m* lactose in water?

13.100 The boiling point of ethanol (C_2H_5OH) is 78.5°C. What is the boiling point of a solution of 6.4 g of vanillin ($\mathcal{M}$ = 152.14 g/mol) in 50.0 g of ethanol (K_b of ethanol = 1.22°C/*m*)?

13.101 The freezing point of benzene is 5.5°C. What is the freezing point of a solution of 5.00 g of naphthalene ($C_{10}H_8$) in 444 g of benzene (K_f of benzene = 4.90°C/*m*)?

13.102 What is the minimum mass of ethylene glycol ($C_2H_6O_2$) that must be dissolved in 14.5 kg of water to prevent the solution from freezing at −12.0°F? (Assume ideal behavior.)

13.103 What is the minimum mass of glycerol ($C_3H_8O_3$) that must be dissolved in 11.0 mg of water to prevent the solution from freezing at −15°C? (Assume ideal behavior.)

13.104 A small protein has a molar mass of 1.50×10^4 g/mol. What is the osmotic pressure exerted at 24.0°C by 25.0 mL of an aqueous solution that contains 37.5 mg of the protein?

13.105 At 37°C, 0.30 *M* sucrose has about the same osmotic pressure as blood. What is the osmotic pressure of blood?

13.106 A solution made by dissolving 31.7 g of an unknown compound in 150. g of water freezes at −1.15°C. What is the molar mass of the compound?

13.107 A 125-mL sample of an aqueous solution of the protein ovalbumin from chicken egg white contains 1.31 g of the dissolved protein and has an osmotic pressure of 4.32 torr at 25°C. What is the molar mass of ovalbumin?

13.108 Assuming ideal behavior, find the freezing point of a solution made by dissolving 13.2 g of ammonium phosphate in 45.0 g of water.

13.109 Assuming ideal behavior, find the boiling point of a solution made by dissolving 32.8 g of calcium nitrate in 108 g of water.

13.110 Calculate the molality and van't Hoff factor (*i*) for the following aqueous solutions:
(a) 1.00 mass % NaCl, freezing point = −0.593°C
(b) 0.500 mass % CH_3COOH, freezing point = −0.159°C

13.111 Calculate the molality and van't Hoff factor (*i*) for the following aqueous solutions:
(a) 0.500 mass % KCl, freezing point = −0.234°C
(b) 1.00 mass % H_2SO_4, freezing point = −0.423°C

Problems in Context

13.112 In a study designed to prepare new gasoline-resistant coatings, a polymer chemist dissolves 6.053 g of poly(vinyl alcohol) in enough water to make 100.0 mL of solution. At 25°C, the osmotic pressure of this solution is 0.272 atm. What is the molar mass of the polymer sample?

13.113 The U.S. Food and Drug Administration lists dichloromethane (CH_2Cl_2) and carbon tetrachloride (CCl_4) among the many cancer-causing chlorinated organic compounds. What are the partial pressures of these substances in the vapor above a solution of 1.60 mol of CH_2Cl_2 and 1.10 mol of CCl_4 at 23.5°C? The vapor pressures of pure CH_2Cl_2 and CCl_4 at 23.5°C are 352 torr and 118 torr, respectively. (Assume ideal behavior.)

The Structure and Properties of Colloids

Concept Review Questions

13.114 Is the fluid inside a bacterial cell considered a solution, a colloid, or both? Explain.

13.115 What type of colloid is each of the following?
(a) Milk (b) Fog (c) Shaving cream

13.116 What is Brownian motion, and what causes it?

13.117 In a movie theater, you can see the beam of projected light. What phenomenon does this exemplify? Why does it occur?

13.118 Why don't soap micelles coagulate and form large globules? Is soap more effective in freshwater or in seawater? Why?

Comprehensive Problems

13.119 The three aqueous ionic solutions represented below have total volumes of 25. mL for A, 50. mL for B, and 100. mL for C. If each sphere represents 0.010 mol of ions, calculate:
(a) the total molarity of ions for each solution;
(b) the highest molarity of solute;
(c) the lowest molality of solute (assuming the solution densities are equal);
(d) the highest osmotic pressure (assuming ideal behavior).

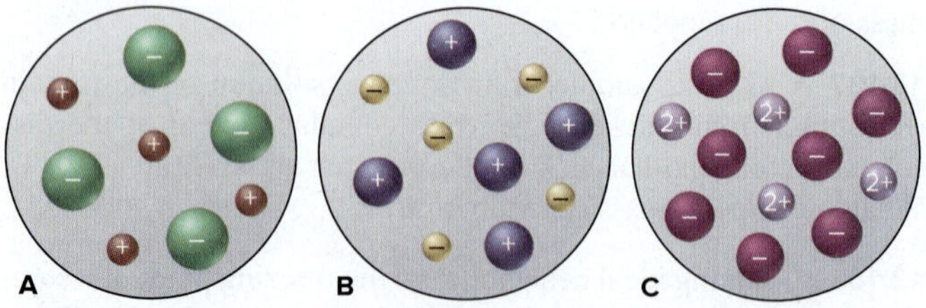

13.120 An aqueous solution is 10.% glucose by mass (d = 1.039 g/mL at 20°C). Calculate its freezing point, boiling point at 1 atm, and osmotic pressure.

13.121 Because zinc has nearly the same atomic radius as copper (d = 8.95 g/cm^3), zinc atoms substitute for some copper atoms in the many types of brass. Calculate the density of the brass with (a) 10.0 atom % Zn and (b) 38.0 atom % Zn.

13.122 Gold occurs in seawater at an average concentration of 1.1×10^{-2} ppb. How many liters of seawater must be processed to recover 1 troy ounce of gold, assuming 81.5% efficiency (d of seawater = 1.025 g/mL; 1 troy ounce = 31.1 g)?

13.123 Use atomic properties to explain why xenon is 11 times as soluble as helium in water at 0°C on a mole basis.

13.124 Which of the following best represents a molecular-scale view of an ionic compound in aqueous solution? Explain.

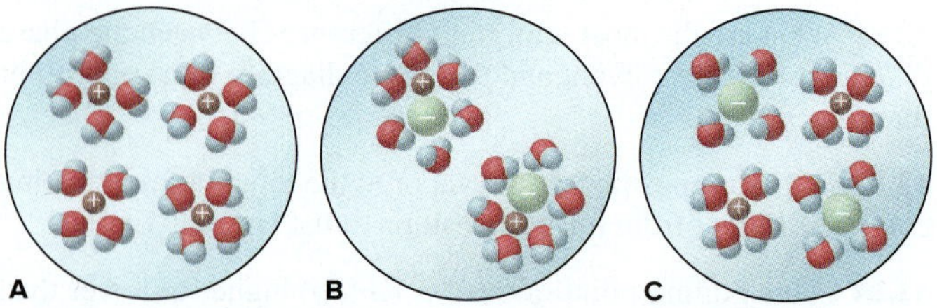

13.125 Four 0.50 m aqueous solutions are depicted below. Assume that the solutions behave ideally.
(a) Which has the highest boiling point?
(b) Which has the lowest freezing point?
(c) Can you determine which one has the highest osmotic pressure? Explain.

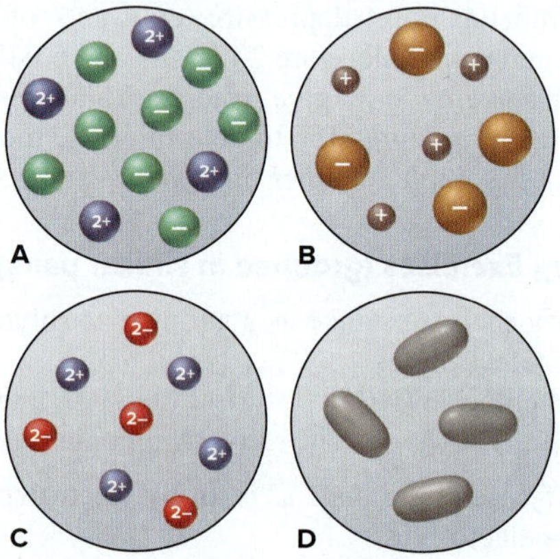

13.126 Thermal pollution from industrial wastewater causes the temperature of river or lake water to rise, which can affect fish survival as the concentration of dissolved O_2 decreases. Use the following data to find the molarity of O_2 at each temperature (assume the solution density is the same as water).

Temperature (°C)	Solubility of O_2 (mg/kg H_2O)	Density of H_2O (g/mL)
0.0	14.5	0.99987
20.0	9.07	0.99823
40.0	6.44	0.99224

13.127 Pyridine *(right)* is an essential portion of many biologically active compounds, such as nicotine and vitamin B_6. Like ammonia, it has a nitrogen with a lone pair, which makes it act as a weak base. Because it is miscible in a wide range of solvents, from water to benzene, pyridine is one of the most important bases and solvents in organic syntheses. Account for its solubility behavior in terms of intermolecular forces.

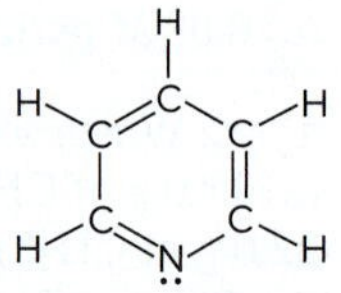

13.128 A chemist is studying small organic compounds to evaluate their potential for use as an antifreeze. When 0.243 g of a compound is dissolved in 25.0 mL of water, the freezing point of the solution is −0.201°C.
(a) Calculate the molar mass of the compound (d of water = 1.00 g/mL).
(b) Analysis shows that the compound is 53.31 mass % C and 11.18 mass % H, the remainder being O. Determine the empirical and molecular formulas of the compound.
(c) Draw a Lewis structure for a compound with this formula that forms H bonds and another for one that does not.

13.129 Air in a smoky bar contains 4.0×10^{-6} mol/L of CO. What mass of CO is inhaled by a bartender who respires at a rate of 11 L/min during an 8.0-h shift?

13.130 Is 50% by mass of methanol dissolved in ethanol different from 50% by mass of ethanol dissolved in methanol? Explain.

13.131 Three gaseous mixtures of N_2 *(blue)*, Cl_2 *(green)*, and Ne *(purple)* are depicted below.
(a) Which has the smallest mole fraction of N_2?
(b) Which have the same mole fraction of Ne?
(c) Rank all three in order of increasing mole fraction of Cl_2.

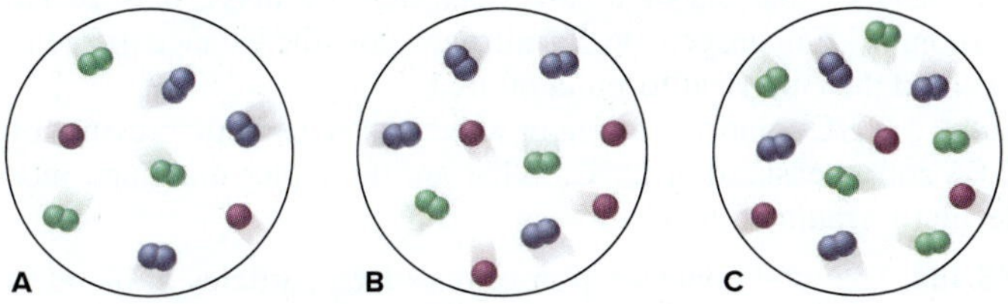

13.132 A water treatment plant needs to attain a fluoride concentration of 4.50×10^{-5} *M* in the drinking water it produces.
(a) What mass of NaF must be added to 5000. L of water in a blending tank?
(b) What mass per day of fluoride is ingested by a person who drinks 2.0 L of this water?

13.133 Four U tubes each have distilled water in the right arm, a solution in the left arm, and a semipermeable membrane between the arms.
(a) If the solute is KCl, which solution is most concentrated?
(b) If each solute is different but all the solutions have the same molarity, which contains the smallest number of dissolved ions?

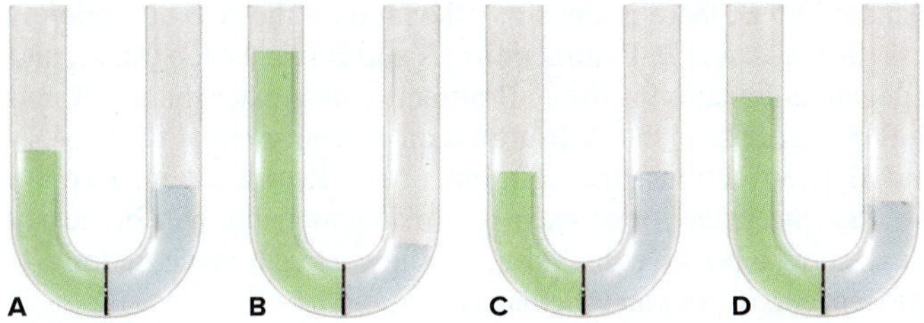

13.134 β-Pinene ($C_{10}H_{16}$) and α-terpineol ($C_{10}H_{18}O$) are used in cosmetics to provide a "fresh pine" scent. At 367 K, the pure substances have vapor pressures of 100.3 torr and 9.8 torr, respectively. What is the composition of the vapor (in terms of mole fractions) above a solution containing equal masses of these compounds at 367 K? (Assume ideal behavior.)

13.135 A solution of 1.50 g of solute dissolved in 25.0 mL of H_2O at 25°C has a boiling point of 100.45°C.
(a) What is the molar mass of the solute if it is a nonvolatile nonelectrolyte and the solution behaves ideally (*d* of H_2O at 25°C = 0.997 g/mL)?
(b) Conductivity measurements show that the solute is ionic with general formula AB_2 or A_2B. What is the molar mass if the solution behaves ideally?
(c) Analysis indicates that the solute has an empirical formula of CaN_2O_6. Explain the difference between the actual formula mass and that calculated from the boiling point elevation.
(d) Find the van't Hoff factor (*i*) for this solution.

13.136 A pharmaceutical preparation made with ethanol (C_2H_5OH) is contaminated with methanol (CH_3OH). A sample of vapor above the liquid mixture contains a 97/1 mass ratio of C_2H_5OH to CH_3OH. What is the mass ratio of these alcohols in the liquid mixture? At the temperature of the liquid mixture, the vapor pressures of C_2H_5OH and CH_3OH are 60.5 torr and 126.0 torr, respectively.

13.137 Water treatment plants commonly use chlorination to destroy bacteria. A byproduct is chloroform ($CHCl_3$), a suspected carcinogen, produced when HOCl, formed by reaction of Cl_2 and water, reacts with dissolved organic matter. The United States, Canada, and the World Health Organization have set a limit of 100. ppb of $CHCl_3$ in drinking water. Convert this concentration into molarity, molality, mole fraction, and mass percent.

13.138 A saturated Na_2CO_3 solution is prepared, and a small excess of solid is present (white pile in beaker). A seed crystal of $Na_2\,{}^{14}CO_3$ (${}^{14}C$ is a radioactive isotope of ${}^{12}C$) is added (small red piece), and the radioactivity is measured over time.
(a) Would you expect radioactivity in the solution? Explain.
(b) Would you expect radioactivity in all the solid or just in the seed crystal? Explain.

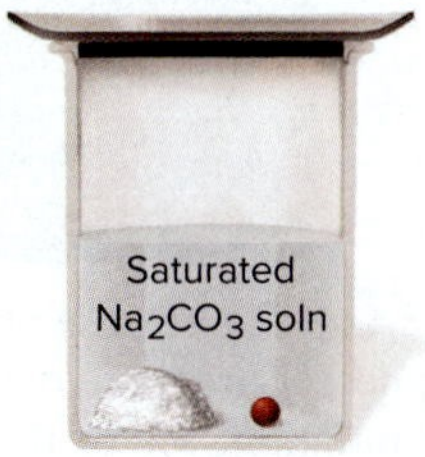

13.139 A biochemical engineer isolates a bacterial gene fragment and dissolves a 10.0-mg sample in enough water to make 30.0 mL of solution. The osmotic pressure of the solution is 0.340 torr at 25°C.
(a) What is the molar mass of the gene fragment?
(b) If the solution density is 0.997 g/mL, how large is the freezing point depression for this solution (K_f of water = 1.86°C/*m*)?

13.140 A river is contaminated with 0.65 mg/L of dichloroethylene ($C_2H_2Cl_2$). What is the concentration (in ng/L) of dichloroethylene at 21°C in the air breathed by a person swimming in the river (k_H for $C_2H_2Cl_2$ in water is 0.033 mol/L·atm)?

13.141 At an air-water interface, fatty acids such as oleic acid lie in a one-molecule-thick layer (a monolayer), with the heads in the water and the tails perpendicular in the air. When 2.50 mg of oleic acid is placed on a water surface, it forms a circular monolayer 38.6 cm in diameter. Find the surface area (in cm^2) occupied by one molecule ($\mathcal{M}$ of oleic acid = 283 g/mol).

13.142 A simple device used for estimating the concentration of total dissolved solids in an aqueous solution works by measuring the electrical conductivity of the solution. The method assumes that equal concentrations of different solids give approximately the same conductivity, and that the conductivity is proportional to concentration. The table below gives some actual electrical conductivities (in arbitrary units) for solutions of selected solids at the indicated concentrations (in ppm by mass):

	Conductivity		
Sample	**0 ppm**	**5.00×10^3 ppm**	**10.00×10^3 ppm**
$CaCl_2$	0.0	8.0	16.0
K_2CO_3	0.0	7.0	14.0
Na_2SO_4	0.0	6.0	11.0
Seawater (dilute)	0.0	8.0	15.0
Sucrose ($C_{12}H_{22}O_{11}$)	0.0	0.0	0.0
Urea [$(NH_2)_2C{=}O$]	0.0	0.0	0.0

(a) How reliable are these measurements for estimating concentrations of dissolved solids?
(b) For what types of substances might this method have a large error? Why?
(c) Based on this method, an aqueous $CaCl_2$ solution has a conductivity of 14.0 units. Calculate its mole fraction and molality.

13.143 Two beakers are placed in a closed container *(left)*. One beaker contains water, the other a concentrated aqueous sugar solution. With time, the solution volume increases and the water volume decreases *(right)*. Explain on the molecular level.

13.144 The release of volatile organic compounds into the atmosphere is regulated to limit ozone formation. In a laboratory simulation, 5% of the ethanol in a liquid detergent is released. Thus, a "down-the-drain" factor of 0.05 is used to estimate ethanol emissions from the detergent. The k_H values for ethanol and 2-butoxyethanol ($C_4H_9OCH_2CH_2OH$) are 5×10^{-6} atm·m^3/mol and 1.6×10^{-6} atm·m^3/mol, respectively.
(a) Estimate a "down-the-drain" factor for 2-butoxyethanol in the detergent.
(b) What is the k_H for ethanol in units of L·atm/mol?
(c) Is the value found in part (b) consistent with a value given as 0.64 Pa·m^3/mol?

13.145 Although other solvents are available, dichloromethane (CH_2Cl_2) is still often used to "decaffeinate" drinks because the solubility of caffeine in CH_2Cl_2 is 8.35 times that in water.
(a) A 100.0-mL sample of cola containing 10.0 mg of caffeine is extracted with 60.0 mL of CH_2Cl_2. What mass of caffeine remains in the aqueous phase?
(b) A second identical cola sample is extracted with two successive 30.0-mL portions of CH_2Cl_2. What mass of caffeine remains in the aqueous phase after each extraction?
(c) Which approach extracts more caffeine?

13.146 How would you prepare 250. g of 0.150 *m* aqueous $NaHCO_3$?

13.147 Tartaric acid occurs in crystalline residues found in wine vats. It is used in baking powders and as an additive in foods. It contains 32.3% by mass carbon and 3.97% by mass hydrogen; the balance is oxygen. When 0.981 g of tartaric acid is dissolved in 11.23 g of water, the solution freezes at –1.26°C. Find the empirical and molecular formulas of tartaric acid.

13.148 Methanol (CH_3OH) and ethanol (C_2H_5OH) are miscible because the major intermolecular force for each is H bonding. In some methanol-ethanol solutions, the mole fraction of methanol is higher, but the mass percent of ethanol is higher. What is the range of mole fraction of methanol for these solutions?

13.149 A solution of 5.0 g of benzoic acid (C_6H_5COOH) in 100.0 g of carbon tetrachloride has a boiling point of 77.5°C.
(a) Calculate the molar mass of benzoic acid in the solution.
(b) Suggest a reason for the difference between the molar mass based on the formula and that found in part (a). (*Hint:* Consider intermolecular forces in this compound.)

13.150 Derive a general equation that expresses the relationship between the molarity and the molality of a solution. Why are the numerical values of these two terms approximately equal for very dilute aqueous solutions?

13.151 A florist prepares a solution of nitrogen-phosphorus fertilizer by dissolving 5.66 g of NH_4NO_3 and 4.42 g of $(NH_4)_3PO_4$ in enough water to make 20.0 L of solution. What are the molarities of NH_4^+ and of PO_4^{3-} in the solution?

13.152 Suppose coal-fired power plants used water in scrubbers to remove SO_2 from smokestack gases (see Chemical Connections, Section 6.6).
(a) If the partial pressure of SO_2 in the stack gases is 2.0×10^{-3} atm, what is the solubility of SO_2 in the scrubber liquid (k_H for SO_2 in water is 1.23 mol/L·atm at 200.°C)?
(b) From your answer to part (a), why are basic solutions, such as limewater slurries [$Ca(OH)_2$], used in scrubbers?

13.153 Urea is a white crystalline solid used as a fertilizer, in the pharmaceutical industry, and in the manufacture of certain polymer resins. Analysis of urea reveals that, by mass, it is 20.1% carbon, 6.7% hydrogen, 46.5% nitrogen, and the balance oxygen.
(a) Find the empirical formula of urea.
(b) A 5.0 g/L solution of urea in water has an osmotic pressure of 2.04 atm, measured at 25°C. What are the molar mass and molecular formula of urea?

13.154 The total concentration of dissolved particles in blood is 0.30 *M*. An intravenous (IV) solution must be isotonic with blood, which means it must have the same concentration.
(a) To relieve dehydration, a patient is given 100. mL/h of IV glucose ($C_6H_{12}O_6$) for 2.5 h. What mass (g) of glucose did she receive?
(b) If isotonic saline (NaCl) is used, what is the molarity of the solution?
(c) If the patient is given 150. mL/h of IV saline for 1.5 h, how many grams of NaCl did she receive?

13.155 Deviations from Raoult's law lead to the formation of *azeotropes,* constant boiling mixtures that cannot be separated by distillation, making industrial separations difficult. For components A and B, there is a positive deviation if the A-B attraction is less than A-A and B-B attractions (A and B reject each other), and a negative deviation if the A-B attraction is greater than A-A and B-B attractions. If the A-B attraction is nearly equal to the A-A and B-B attractions, the solution obeys Raoult's law. Explain whether the behavior of each pair of components will be nearly ideal, show a positive deviation, or show a negative deviation:
(a) Benzene (C_6H_6) and methanol
(b) Water and ethyl acetate
(c) Hexane and heptane
(d) Methanol and water
(e) Water and hydrochloric acid

13.156 Acrylic acid (CH_2=CHCOOH) is a monomer used to make superabsorbent polymers and various compounds for paint and adhesive production. At 1 atm, it boils at 141.5°C but is prone to polymerization. Its vapor pressure at 25°C is 4.1 mbar. What pressure (in mmHg) is needed to distill the pure acid at 65°C?

13.157 To effectively stop polymerization, certain inhibitors require the presence of a small amount of O_2. At equilibrium with 1 atm of air, the concentration of O_2 dissolved in the monomer acrylic acid (CH_2=CHCOOH) is 1.64×10^{-3} *M*.
(a) What is k_H (mol/L·atm) for O_2 in acrylic acid?
(b) If 0.005 atm of O_2 is sufficient to stop polymerization, what is the molarity of O_2?
(c) What is the mole fraction?
(d) What is the concentration in ppm? (Pure acrylic acid is 14.6 *M*; P_{O_2} in air is 0.2095 atm.)

13.158 Volatile organic solvents have been implicated in adverse health effects observed in industrial workers. Greener methods are phasing these solvents out. Rank the solvents in Table 13.6 in terms of increasing volatility.

13.159 At ordinary temperatures, water is a poor solvent for organic substances. But at high pressure and above 200°C, water develops many properties of organic solvents. Find the minimum pressure needed to maintain water as a liquid at 200.°C (ΔH_{vap} = 40.7 kJ/mol at 100°C and 1.00 atm; assume that this value remains constant with temperature).

13.160 In ice-cream making, the ingredients are kept below 0.0°C in an ice-salt bath.
(a) Assuming that NaCl dissolves completely and forms an ideal solution, what mass of it is needed to lower the melting point of 5.5 kg of ice to –5.0°C?
(b) Given the same assumptions as in part (a), what mass of $CaCl_2$ is needed?

13.161 Perfluorocarbons (PFCs), hydrocarbons with all H atoms replaced by F atoms, have very weak cohesive forces. One interesting consequence of this property is that a live mouse can breathe while submerged in O_2-saturated PFCs.
(a) At 298 K, perfluorohexane (C_6F_{14}, $\mathscr{M}$ = 338 g/mol and d = 1.674 g/mL) in equilibrium with 101,325 Pa of O_2 has a mole fraction of O_2 of 4.28×10^{-3}. What is k_H in mol/L·atm?
(b) According to one source, k_H for O_2 in water at 25°C is 756.7 L·atm/mol. What is the solubility of O_2 in water at 25°C in ppm?
(c) Rank in descending order the k_H for O_2 in water, ethanol, C_6F_{14}, and C_6H_{14}. Explain your ranking.

13.162 The solubility of N_2 in blood is a serious problem for divers breathing compressed air (78% N_2 by volume) at depths greater than 50 ft.
(a) What is the molarity of N_2 in blood at 1.00 atm?
(b) What is the molarity of N_2 in blood at a depth of 50. ft?
(c) Find the volume (in mL) of N_2, measured at 25°C and 1.00 atm, released per liter of blood when a diver at a depth of 50. ft rises to the surface (k_H for N_2 in water at 25°C is 7.0×10^{-4} mol/L·atm and at 37°C is 6.2×10^{-4} mol/L·atm; assume d of water is 1.00 g/mL).

13.163 Figure 12.11 shows the phase changes of pure water. Consider how the diagram would change if air were present at 1 atm and dissolved in the water.
(a) Would the three phases of water still attain equilibrium at some temperature? Explain.
(b) In principle, would that temperature be higher, lower, or the same as the triple point for pure water? Explain.
(c) Would ice sublime at a few degrees below the freezing point under this pressure? Explain.
(d) Would the liquid have the same vapor pressure as that shown in Figure 12.7 at 100°C? At 120°C?

13.164 KNO_3, $KClO_3$, KCl, and NaCl are recrystallized as follows:
Step 1. A saturated aqueous solution of the compound is prepared at 50°C.
Step 2. The mixture is filtered to remove undissolved compound.
Step 3. The filtrate is cooled to 0°C.
Step 4. The crystals that form are filtered, dried, and weighed.
(a) Which compound has the highest percent recovery and which the lowest (see Figure 13.19)?
(b) Starting with 100. g of each compound, how many grams of each can be recovered?

13.165 Eighty proof whiskey is 40% ethanol (C_2H_5OH) by volume. A man has 7.0 L of blood and drinks 28 mL of the whiskey, of which 22% of the ethanol goes into his blood.
(a) What concentration (in g/mL) of ethanol is in his blood (d of ethanol = 0.789 g/mL)?
(b) What volume (in mL) of whiskey would raise his blood alcohol level to 8.0×10^{-4} g/mL, the level at which a person is considered intoxicated?

13.166 Soft drinks are canned under 4 atm of CO_2 and release CO_2 when the can is opened.
(a) How many moles of CO_2 are dissolved in 355 mL of soda in a can before it is opened?
(b) After the soda has gone flat?
(c) What volume (in L) would the released CO_2 occupy at 1.00 atm and 25°C (k_H for CO_2 at 25°C is 3.3×10^{-2} mol/L·atm; P_{CO_2} in air is 4×10^{-4} atm)?

13.167 Gaseous O_2 in equilibrium with O_2 dissolved in water at 283 K is depicted at right.
(a) Which scene below represents the system at 298 K?
(b) Which scene represents the system when the pressure of O_2 is increased by half?

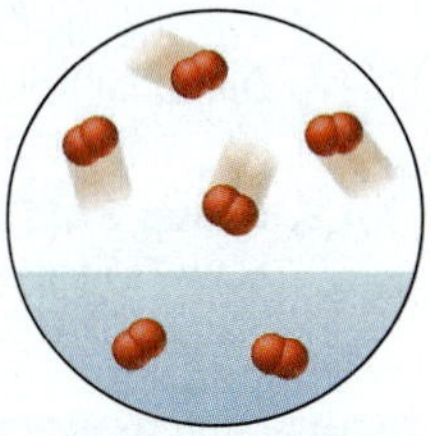

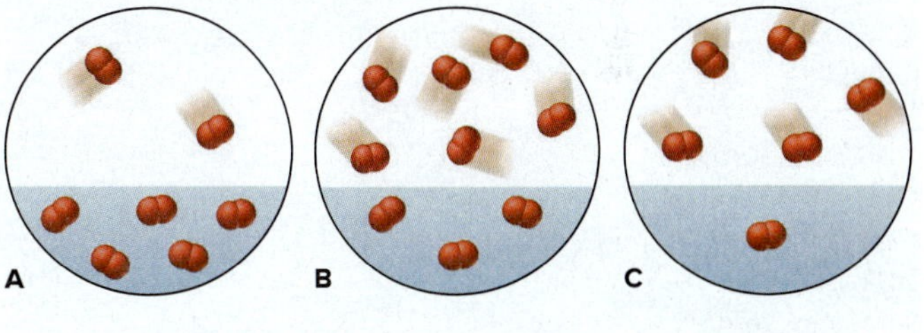

14 Periodic Patterns in the Main-Group Elements

14.1 Hydrogen, the Simplest Atom
Where Hydrogen Fits in the Periodic Table
Highlights of Hydrogen Chemistry

14.2 Trends Across the Periodic Table: The Period 2 Elements

14.3 Group 1A(1): The Alkali Metals
Why the Alkali Metals Are Unusual Physically
Why the Alkali Metals Are So Reactive

14.4 Group 2A(2): The Alkaline Earth Metals
How the Alkaline Earth and Alkali Metals Compare Physically
How the Alkaline Earth and Alkali Metals Compare Chemically
Diagonal Relationships: Lithium and Magnesium

14.5 Group 3A(13): The Boron Family
How Transition Elements Influence This Group's Properties
Features That First Appear in This Group's Chemical Properties
Highlights of Boron Chemistry
Diagonal Relationships: Beryllium and Aluminum

14.6 Group 4A(14): The Carbon Family
How Type of Bonding Affects Physical Properties
How Bonding Changes in This Group's Compounds
Highlights of Carbon Chemistry
Highlights of Silicon Chemistry
Diagonal Relationships: Boron and Silicon

14.7 Group 5A(15): The Nitrogen Family
The Wide Range of Physical Behavior
Patterns in Chemical Behavior
Highlights of Nitrogen Chemistry
Highlights of Phosphorus Chemistry

14.8 Group 6A(16): The Oxygen Family
How the Oxygen and Nitrogen Families Compare Physically
How the Oxygen and Nitrogen Families Compare Chemically
Highlights of Oxygen Chemistry
Highlights of Sulfur Chemistry

14.9 Group 7A(17): The Halogens
Physical Behavior of the Halogens
Why the Halogens Are So Reactive
Highlights of Halogen Chemistry

14.10 Group 8A(18): The Noble Gases
How the Noble Gases and Alkali Metals Contrast Physically
How Noble Gases Can Form Compounds

Source: NASA

Concepts and Skills to Review Before You Study This Chapter

- acids, bases, and salts (Section 4.4)
- redox behavior and oxidation states (Section 4.5)
- electron configurations (Section 8.2)
- trends in atomic size, ionization energy, metallic behavior, and electronegativity (Sections 8.3, 8.4, and 9.5)
- trends in element properties and type of bonding (Sections 8.4 and 9.5)
- models of ionic, covalent, and metallic bonding (Sections 9.2, 9.3, 9.6, and 12.6)
- resonance and formal charge (Section 10.1)
- molecular shape and polarity (Sections 10.2 and 10.3)
- orbital hybridization and modes of orbital overlap (Sections 11.1 and 11.2)
- phase changes and phase diagrams, intermolecular forces, and crystalline solids (Sections 12.2, 12.3, and 12.6)

Recurring patterns appear throughout nature, helping us make sense of the diversity in the world around us. Physical patterns we observe in animals and plants give rise to systems of biological classification; patterns within our bodies, such as heartbeats, allow us to monitor health and disease; and astronomical patterns allow us to predict phases of the Moon, eclipses, and comet appearances. For example, astronomers know from studying the periodic pattern of its orbit that Halley's Comet *(see photo)*, last seen in 1986, will appear next in 2061. As you'll see in the upcoming discussions, patterns also appear in the atomic, chemical, and physical properties of the elements, which help us make sense of their behavior.

IN THIS CHAPTER . . . *We apply general ideas of bonding, structure, and reactivity (from Chapters 7–12) to the main-group elements and see how their behavior correlates with their position in the periodic table.*

- We begin with hydrogen, the simplest element, considering its position in the periodic table and examining the three types of hydrides.
- We then survey Period 2 as an example of the general changes in chemical and physical properties from left to right *across* the periodic table.
- We go on to discuss each of the eight families of main-group elements by exploring vertical trends in physical and chemical properties. In the process, we highlight some of the most important elements—boron, carbon, silicon, nitrogen, phosphorus, oxygen, sulfur, and the halogens.

14.1 HYDROGEN, THE SIMPLEST ATOM

A hydrogen atom consists of a nucleus with a single positive charge, surrounded by a single electron. Perhaps because of this simple structure, hydrogen may be the most important element of all. In the Sun, hydrogen (H) nuclei combine to form helium (He) nuclei in a process that provides nearly all Earth's energy. About 90% of all the atoms in the universe are H atoms, so it is the most abundant element by far. On Earth, only tiny amounts of the free, diatomic element occur naturally, but hydrogen is abundant in combination with oxygen in water. With a simple structure and low molar mass, nonpolar H_2 is a colorless, odorless gas with extremely weak dispersion forces that result in very low melting (−259°C) and boiling points (−253°C).

Where Hydrogen Fits in the Periodic Table

Hydrogen has no perfectly suitable position in the periodic table (Figure 14.1). Depending on the property, hydrogen may fit better in Group 1A(1), 4A(14), or 7A(17):

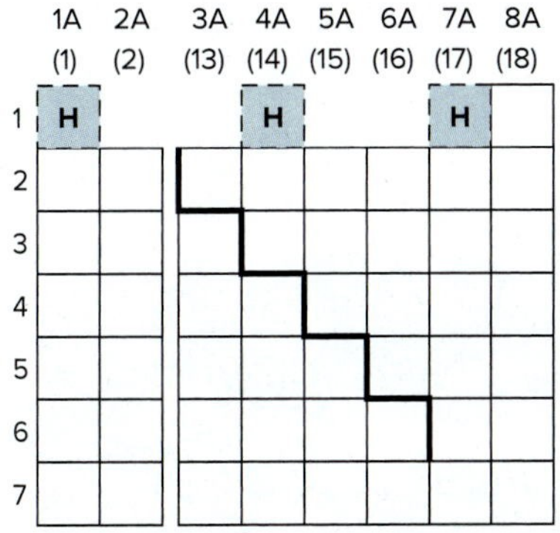

Figure 14.1 Where does hydrogen belong?

- *Like the Group 1A(1) elements,* hydrogen has an outer electron configuration of ns^1 and most commonly a +1 oxidation state. However, unlike the alkali metals, hydrogen *shares* its single valence electron with nonmetals rather than transferring it to them.

Moreover, hydrogen has a much higher ionization energy (IE = 1311 kJ/mol) and electronegativity (EN = 2.1) than any of the alkali metals. By comparison, lithium has an IE of only 520 kJ/mol and an EN of 1.0, the highest of the alkali metals.

- *Like the Group 4A(14) elements,* hydrogen's valence level is half-filled, but with only one electron, and it has an ionization energy, electron affinity, electronegativity, and bond energy similar to the values for Group 4A(14).
- *Like the Group 7A(17) elements,* hydrogen occurs as diatomic molecules and fills its outer level either by electron sharing or by gaining one electron from a metal to form a 1− ion (hydride, H^-). However, while the monatomic halide ions (X^-) are common and stable, H^- is rare and reactive. Moreover, hydrogen has a lower electronegativity (EN = 2.1) than any of the halogens (whose ENs range from 4.0 to 2.2), and it lacks their three valence electron pairs.

Hydrogen's unique behavior arises from its tiny size. It has a high IE because its electron is very close to the nucleus, with no inner electrons to shield it from the positive charge. And it has a low EN (for a nonmetal) because it has only one proton to attract bonding electrons. In this chapter, H will be discussed as part of either Group 1A(1) or 7A(17) depending on the property being considered.

Highlights of Hydrogen Chemistry

In Chapters 12 and 13, we discussed the vital importance of H bonding, and in Chapter 22, we'll look at hydrogen's many uses in industry. Elemental hydrogen is very reactive and combines with nearly every other element. It forms three types of hydrides—ionic, covalent, and metallic.

Ionic (Saltlike) Hydrides With very reactive metals, such as those in Group 1A(1) and the larger members of Group 2A(2) (Ca, Sr, and Ba), hydrogen forms *saltlike hydrides*—white, crystalline solids composed of the metal cation and the hydride ion:

$$2Li(s) + H_2(g) \longrightarrow 2LiH(s)$$
$$Ca(s) + H_2(g) \longrightarrow CaH_2(s)$$

In water, H^- is a strong base that pulls H^+ from surrounding H_2O molecules to form H_2 and OH^-:

$$NaH(s) + H_2O(l) \longrightarrow Na^+(aq) + OH^-(aq) + H_2(g)$$

The hydride ion is also a powerful reducing agent; for example, it reduces Ti(IV) to the free metal:

$$TiCl_4(l) + 4LiH(s) \longrightarrow Ti(s) + 4LiCl(s) + 2H_2(g)$$

Covalent (Molecular) Hydrides Hydrogen reacts with nonmetals to form many *covalent hydrides,* such as CH_4, PH_3, H_2S, and HCl. Most are gases, but many hydrides of boron and carbon are liquids or solids that consist of much larger molecules. In most covalent hydrides, hydrogen has an oxidation number of +1 because the other nonmetal has a higher electronegativity.

Conditions for preparing the covalent hydrides depend on the reactivity of the other nonmetal. For example, with stable, triple-bonded N_2, hydrogen reacts at high temperatures (~400°C) and pressures (~250 atm), and the reaction needs a catalyst to proceed at any practical speed:

$$N_2(g) + 3H_2(g) \xrightarrow{\text{catalyst}} 2NH_3(g) \qquad \Delta H^\circ_{rxn} = -91.8\ \text{kJ}$$

Industrial facilities throughout the world use this reaction to produce millions of tons of ammonia each year for fertilizers, explosives, and synthetic fibers. On the other hand, hydrogen combines rapidly with reactive, single-bonded F_2, even at extremely low temperatures (−196°C):

$$F_2(g) + H_2(g) \longrightarrow 2HF(g) \qquad \Delta H^\circ_{rxn} = -546\ \text{kJ}$$

Metallic (Interstitial) Hydrides Many transition elements form *metallic (interstitial) hydrides,* in which H_2 molecules (and H atoms) occupy the holes in the metal's crystal structure (Figure 14.2). Thus, such hydrides are *not* compounds but gas-solid solutions.

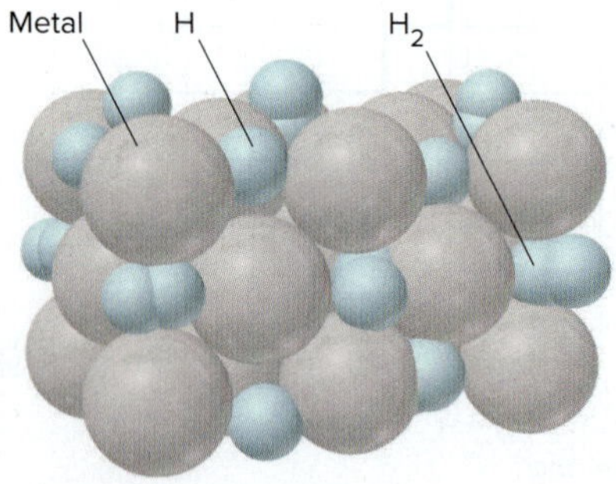

Figure 14.2 A metallic (interstitial) hydride.

Also, unlike ionic and covalent hydrides, interstitial hydrides, such as $TiH_{1.7}$, typically do not have a specific stoichiometric formula because the metal can incorporate variable amounts of hydrogen, depending on the pressure and temperature. Metallic hydrides cannot serve as "storage containers" for hydrogen fuel in cars; the metals that store the most hydrogen are expensive and heavy and release hydrogen only at very high temperatures. But other systems, including carbon nanotubes, are being studied for this role.

14.2 TRENDS ACROSS THE PERIODIC TABLE: THE PERIOD 2 ELEMENTS

Table 14.1 on pages 588–589 presents the main horizontal trends in atomic properties and the physical and chemical properties that emerge from them for the Period 2 elements, lithium through neon. In general, *these trends apply to the other periods as well.* Note the following points:

- Electrons fill the one *ns* and the three *np* orbitals according to Pauli's exclusion principle and Hund's rule.
- As a result of increasing nuclear charge and the addition of electrons to orbitals of the same energy level (same *n* value), atomic size generally decreases, whereas first ionization energy and electronegativity generally increase (see bar graphs).
- Metallic character decreases with increasing nuclear charge as elements change from metals to metalloids to nonmetals.
- Reactivity is highest at the left and right ends of the period, except for the inert noble gas, because members of Groups 1A(1) and 7A(17) are only one electron away from attaining a filled outer level.
- Bonding between atoms of an element changes from metallic, to covalent in networks, to covalent in individual molecules, to none (noble gases exist as separate atoms). As expected, physical properties, such as melting point, change abruptly at the network/molecule boundary, which occurs between carbon (solid) and nitrogen (gas).
- Bonding between each element and an active nonmetal changes from ionic, to polar covalent, to covalent. Bonding between each element and an active metal changes from metallic to polar covalent to ionic (see Figure 9.2).
- The acid-base behavior of the common element oxide in water changes from basic to amphoteric to acidic as the bond between the element and oxygen changes from ionic to covalent (see the similar trend for the Period 3 elements in Figure 8.24).
- Reducing strength decreases through the metals, and oxidizing strength increases through the nonmetals. In Period 2, oxidation numbers (O.N.s) equal the A-group number for Li and Be and the A-group number minus 8 for O and F. Boron has several O.N.s, Ne has none, and C and N show all the O.N.s that are possible for their groups.

The Anomalous Behavior of Period 2 Members One point that is not made in Table 14.1 is the *anomalous (unrepresentative) behavior specific for the Period 2 elements within their groups*. This behavior arises from the relatively *small atomic size* of these elements and the *small number of orbitals* in their outer energy level.

1. *Anomalous properties of lithium.* In Group 1A(1), Li is the only member that forms a simple oxide and nitride, Li_2O and Li_3N, on reaction with O_2 and N_2 in air and the only member that forms molecular compounds with organic halides:

$$2Li(s) + CH_3CH_2Cl(g) \longrightarrow CH_3CH_2Li(s) + LiCl(s)$$

Because of its small size, Li^+ has a relatively high charge density. Therefore, it can deform nearby electron clouds to a much greater extent than the other Group 1A(1) ions can, which increases orbital overlap and gives many lithium salts significant covalent character. Thus, LiCl, LiBr, and LiI are much more soluble in polar organic solvents, such as ethanol and acetone, than are the halides of Na and K, because the bond polarity of a lithium halide allows it to interact with these solvents through

Table 14.1 **Trends in Atomic, Physical, and Chemical Properties of the Period 2 Elements**

Group: Element/Atomic No.:	1A(1) Lithium (Li) $Z = 3$	2A(2) Beryllium (Be) $Z = 4$	3A(13) Boron (B) $Z = 5$	4A(14) Carbon (C) $Z = 6$
Atomic Properties				
Condensed electron configuration; partial orbital diagram	[He] $2s^1$ 2s: [↑] 2p: [ ][ ][ ]	[He] $2s^2$ 2s: [↑↓] 2p: [ ][ ][ ]	[He] $2s^2 2p^1$ 2s: [↑↓] 2p: [↑][ ][ ]	[He] $2s^2 2p^2$ 2s: [↑↓] 2p: [↑][↑][ ]
Physical Properties				
Appearance				
Metallic character	Metal	Metal	Metalloid	Nonmetal
Hardness	Soft	Hard	Very hard	Graphite: soft Diamond: extremely hard
Melting point/ boiling point	Low mp for a metal	High mp	Extremely high mp	Extremely high mp
Chemical Properties				
General reactivity	Reactive	Low reactivity at room temperature	Low reactivity at room temperature	Low reactivity at room temperature; graphite more reactive
Bonding among atoms of element	Metallic	Metallic	Network covalent	Network covalent
Bonding with nonmetals	Ionic	Polar covalent	Polar covalent	Covalent (π bonds common)
Bonding with metals	Metallic	Metallic	Polar covalent	Polar covalent
Acid-base behavior of common oxide	Strongly basic	Amphoteric	Very weakly acidic	Very weakly acidic
Redox behavior (O.N.)	Strong reducing agent (+1)	Moderately strong reducing agent (+2)	Complex hydrides good reducing agents (+3, −3)	Every oxidation state from +4 to −4
Relevance/Uses of Element and Compounds				
	Li soaps for auto grease; thermonuclear bombs; high-voltage, low-weight batteries; treatment of bipolar disorders (Li_2CO_3)	Rocket nose cones; alloys for springs and gears; nuclear reactor parts; x-ray tubes	Cleaning agent (borax); eyewash, antiseptic (boric acid); armor (B_4C); borosilicate glass; plant nutrient	Graphite: lubricant, structural fiber Diamond: jewelry, cutting tools, protective films Limestone ($CaCO_3$) Organic compounds: drugs, fuels, textiles, biomolecules, etc.

Source: All photos © McGraw-Hill Education/Stephen Frisch, photographer

	5A(15) Nitrogen (N) $Z = 7$	**6A(16) Oxygen (O) $Z = 8$**	**7A(17) Fluorine (F) $Z = 9$**	**8A(18) Neon (Ne) $Z = 10$**
	[He] $2s^22p^3$ ↑↓ | ↑ ↑ ↑ 2s 2p	[He] $2s^22p^4$ ↑↓ | ↑↓ ↑ ↑ 2s 2p	[He] $2s^22p^5$ ↑↓ | ↑↓ ↑↓ ↑ 2s 2p	[He] $2s^22p^6$ ↑↓ | ↑↓ ↑↓ ↑↓ 2s 2p
			No sample available	
	Nonmetal	Nonmetal	Nonmetal	Nonmetal
	—	—	—	—
	Very low mp and bp	Very low mp and bp	Very low mp and bp	Extremely low mp and bp
	Inactive at room temperature	Very reactive	Extremely reactive	Chemically inert
	Covalent N_2 molecules	Covalent O_2 (or O_3) molecules	Covalent F_2 molecules	None; separate atoms
	Covalent (π bonds common)	Covalent (π bonds common)	Covalent	None
	Ionic/polar covalent; anions with active metals	Ionic	Ionic	None
	Strongly acidic (NO_2)	—	Acidic	None
	Every oxidation state from +5 to −3	O_2 (and O_3) very strong oxidizing agents (−2)	Strongest oxidizing agent (−1)	None
	Component of proteins, nucleic acids; ammonia for fertilizers, explosives; oxides involved in manufacturing and air pollution (smog, acid rain)	Component of biological macromolecules; final oxidizer in residential, industrial, and biological energy production	Manufacture of coatings (Teflon); glass etching (HF); refrigerants involved in ozone depletion (CFCs); dental protection (NaF, SnF_2)	Electrified gas in advertising signs

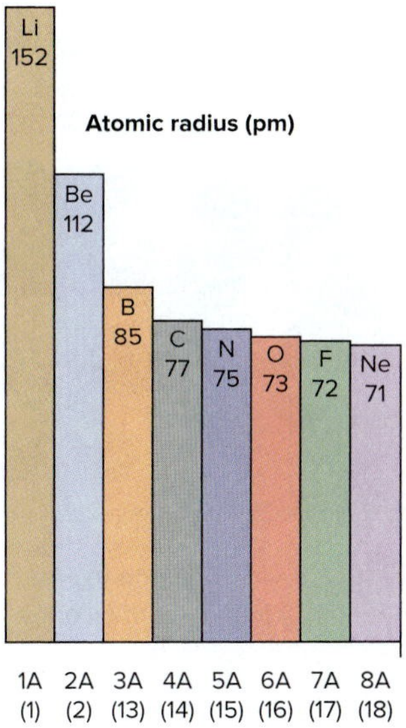

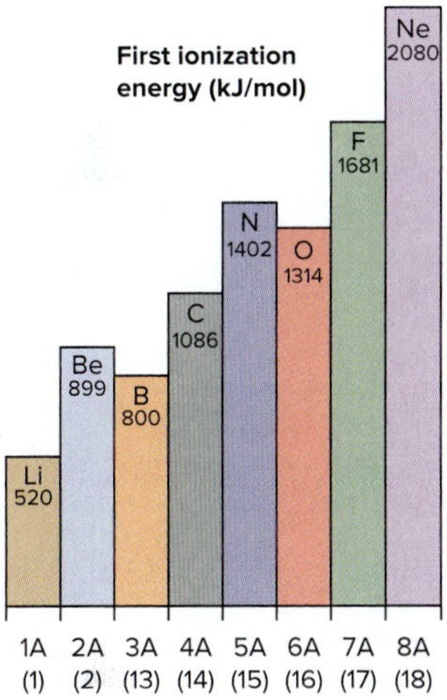

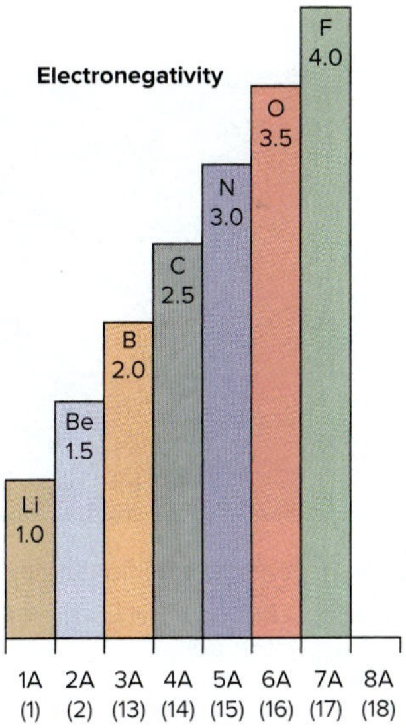

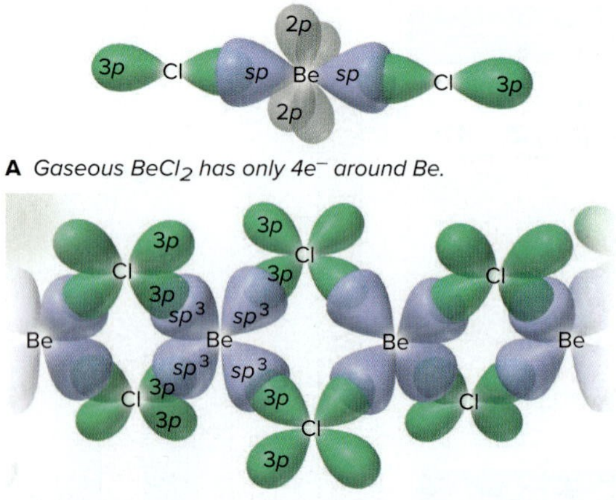

A *Gaseous $BeCl_2$ has only $4e^-$ around Be.*

B *Solid $BeCl_2$ has an octet around Be.*

Figure 14.3 **Overcoming electron deficiency in beryllium chloride.**

dipole-dipole forces. The small, highly positive Li^+ makes Li salts much less soluble in water than those of Na and K.

2. *Anomalous properties of beryllium.* Because of the extremely high charge density of Be^{2+}, the discrete ion does not exist, and *all Be compounds exhibit covalent bonding*. With only two valence electrons, Be does not attain an octet in its simple gaseous compounds (Section 10.1), but this electron deficiency is overcome as the gas condenses.

Consider beryllium chloride ($BeCl_2$). At temperatures greater than 900°C, it consists of linear molecules in which two *sp* hybrid orbitals hold four electrons around the central Be (Figure 14.3A). As the temperature decreases, the molecules bond together, solidifying in long chains with each Cl bridging two Be atoms. Each Be is sp^3 hybridized and has attained an octet (Figure 14.3B).

3. *Anomalous properties of other Period 2 elements.* Boron is the only member of its group to form complex families of compounds with metals and with hydrogen (boranes). Carbon shows extremely unusual behavior: carbon atoms bond to other carbons (and to a small number of other elements) extensively and diversely, giving rise to countless organic compounds. Triple-bonded, unreactive, gaseous nitrogen is very different from its reactive, solid family members. Oxygen, the only gas in its group, is much more reactive than sulfur and the other members. Fluorine is so electronegative that it reacts violently with water, and it is the only member of its group that forms a weak hydrohalic acid, HF.

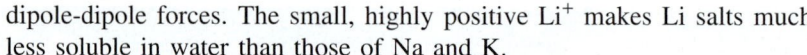

14.3 GROUP 1A(1): THE ALKALI METALS

The first group of elements in the periodic table is named for the alkaline (basic) nature of their oxides and for the basic solutions the elements form in water. Group 1A(1) provides the best example of regular trends with no significant exceptions. All the elements in the group—lithium (Li), sodium (Na), potassium (K), rubidium (Rb), cesium (Cs), and rare, radioactive francium (Fr)*—are very reactive metals. The Family Portrait of Group 1A(1)is the first in a series that provides an overview of each of the main groups, summarizing key atomic, physical, and chemical properties.

Why the Alkali Metals Are Unusual Physically

Alkali metals have some properties that are unique for metals:

- They are unusually soft and can be easily cut with a knife. Na has the consistency of cold butter, and K can be squeezed like clay.
- Alkali metals have lower melting and boiling points than any other group of metals. Li is the only member that melts above 100°C, and Cs melts only a few degrees above room temperature.
- They have lower densities than most metals. Li floats on lightweight mineral oil *(see photo)*.

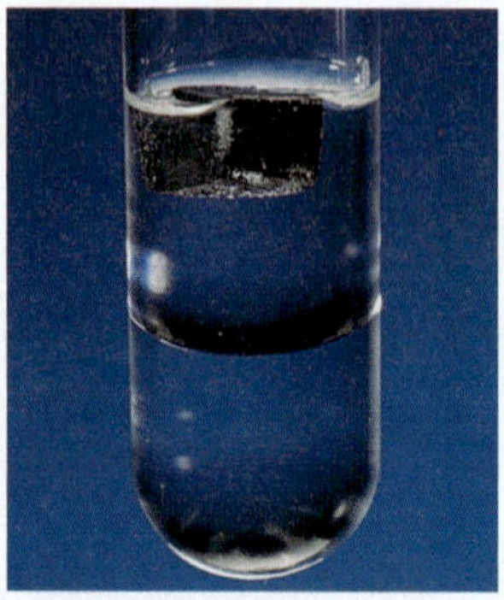

Lithium floating in oil floating on water.

Source: © McGraw-Hill Education. Stephen Frisch, photographer

The unusual physical behavior of these metals can be traced to the largest atomic size in their respective periods and to the ns^1 valence electron configuration. Because the single valence electron is relatively far from the nucleus, only weak attractions exist in the solid between the delocalized electrons and the metal-ion cores. Such weak metallic bonding means that the alkali metal crystal structure can be easily deformed or broken down, which results in a soft consistency and low melting point. The low densities of the alkali metals result from their having the lowest molar masses and largest atomic radii (and, thus, volumes) in their periods.

*Francium is so rare (estimates indicate only 15 g of the element in the top kilometer of Earth's crust) that its properties are largely unknown. Therefore, the discussion of this group mentions it only occasionally.

KEY ATOMIC PROPERTIES, PHYSICAL PROPERTIES, AND REACTIONS

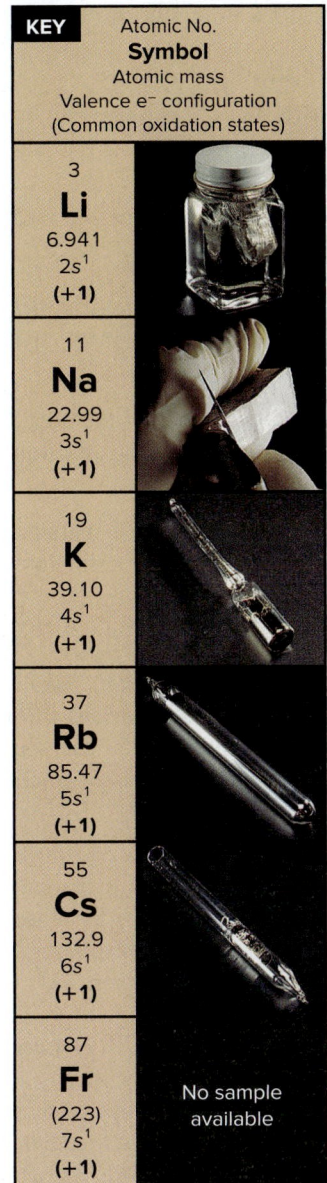

KEY
Atomic No. **Symbol** Atomic mass Valence e⁻ configuration (Common oxidation states)
3 **Li** 6.941 $2s^1$ (+1)
11 **Na** 22.99 $3s^1$ (+1)
19 **K** 39.10 $4s^1$ (+1)
37 **Rb** 85.47 $5s^1$ (+1)
55 **Cs** 132.9 $6s^1$ (+1)
87 **Fr** (223) $7s^1$ (+1)

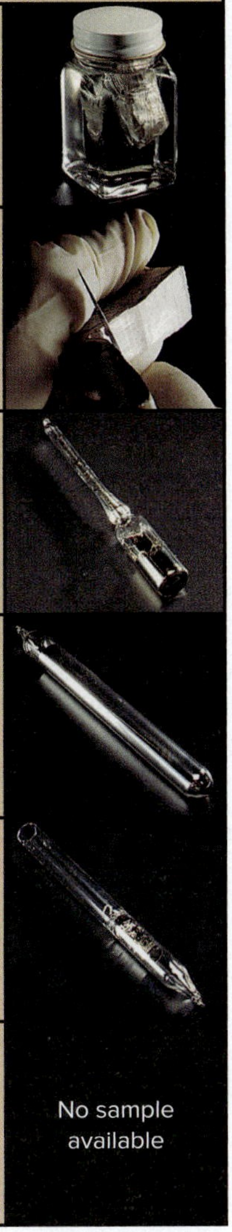

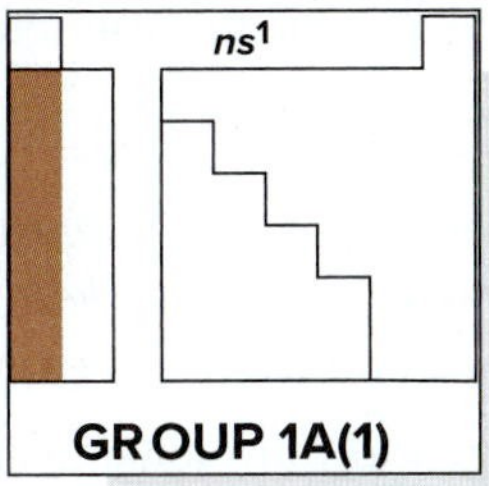

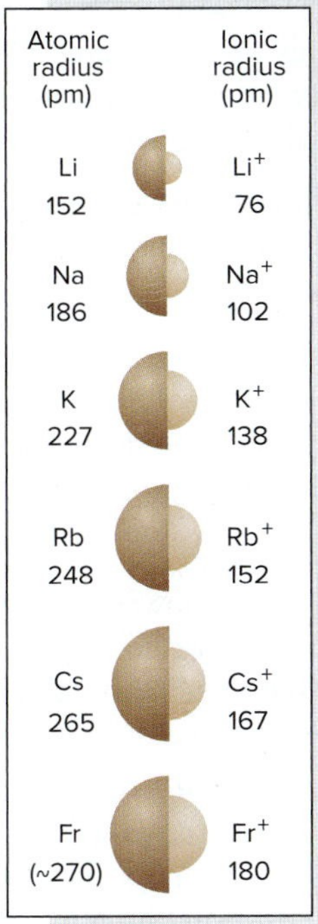

Atomic Properties

Group electron configuration is ns^1. All members have the +1 oxidation state and form an E^+ ion. Atoms have the largest size and lowest IE and EN in their periods. Down the group, atomic and ionic size increase, while IE and EN decrease.

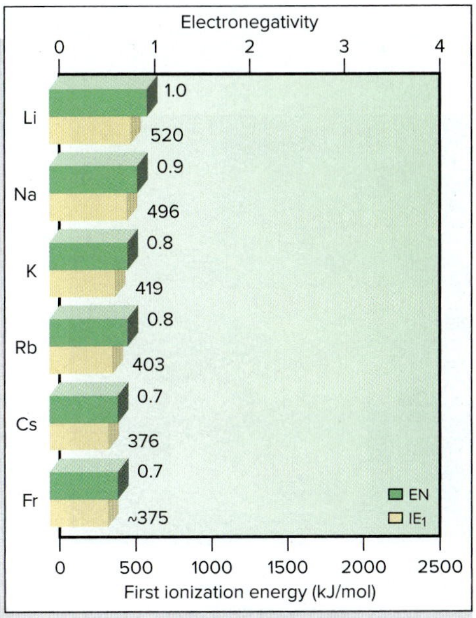

Physical Properties

Metallic bonding is relatively weak because there is only one valence electron. Therefore, these metals are soft with relatively low melting and boiling points. These values decrease down the group because larger atom cores attract delocalized electrons less strongly. Large atomic size and low atomic mass result in low density; thus, density generally increases down the group because mass increases more than size.

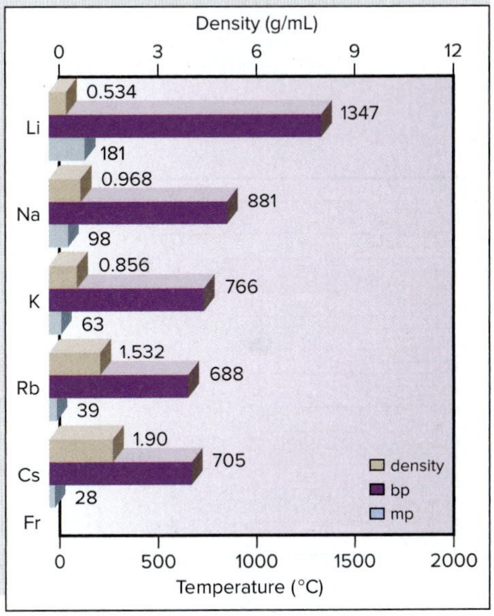

Source: (c) McGraw-Hill Education. Stephen Frisch, photographer

Reactions

1. The alkali metals reduce H in H_2O from the +1 to the 0 oxidation state (E represents any element in the group):

$$2E(s) + 2H_2O(l) \longrightarrow 2E^+(aq) + 2OH^-(aq) + H_2(g)$$

The reaction becomes more vigorous down the group.

2. The alkali metals reduce oxygen, but the product depends on the metal. Li forms the oxide, Li_2O; Na forms the peroxide (O.N. of O = −1), Na_2O_2; K, Rb, and Cs form the superoxide (O.N. of O = $-\frac{1}{2}$), EO_2:

$$4Li(s) + O_2(g) \longrightarrow 2Li_2O(s)$$
$$2Na(s) + O_2(g) \longrightarrow Na_2O_2(s)$$
$$K(s) + O_2(g) \longrightarrow KO_2(s)$$

In emergency breathing units, KO_2 reacts with H_2O and CO_2 in exhaled air to release O_2 (Section 22.4).

3. The alkali metals reduce hydrogen to form ionic (saltlike) hydrides:

$$2E(s) + H_2(g) \longrightarrow 2EH(s)$$

NaH is an industrial base and reducing agent that is used to prepare other reducing agents, such as $NaBH_4$.

4. The alkali metals reduce halogens to form ionic halides:

$$2E(s) + X_2 \longrightarrow 2EX(s) \qquad (X = F, Cl, Br, I)$$

Potassium reacting with water.

Source: © McGraw-Hill Education/Stephen Frisch, photographer

Why the Alkali Metals Are So Reactive

The alkali metals are extremely reactive elements. They are *powerful reducing agents,* readily losing an electron and always occurring in nature as 1+ cations rather than as free metals. Some examples of their reactivity follow:

- The alkali metals (E)* reduce halogens to form ionic solids in highly exothermic reactions:

$$2E(s) + X_2 \longrightarrow 2EX(s) \qquad (X = F, Cl, Br, I)$$

- They reduce hydrogen in water, reacting vigorously (Rb and Cs explosively) to form H_2 and a metal hydroxide solution *(see photo):*

$$2E(s) + 2H_2O(l) \longrightarrow 2E^+(aq) + 2OH^-(aq) + H_2(g)$$

- They reduce molecular hydrogen to form ionic hydrides:

$$2E(s) + H_2(g) \longrightarrow 2EH(s)$$

- They reduce O_2 in air, and thus tarnish rapidly. Because of this reactivity, Na and K are usually kept under mineral oil (an unreactive liquid) in the laboratory, and Rb and Cs are handled with gloves under an inert argon atmosphere.

The ns^1 configuration, which is the basis for the physical properties of the alkali metals, is also the basis of their reactivity, as shown in the steps for the reaction between an alkali metal and a nonmetal:

1. *Atomization: the solid metal separates into gaseous atoms.* The weak metallic bonding leads to *low values for* ΔH_{atom} (the heat needed to convert the solid into individual gaseous atoms), which decrease down the group:

$$E(s) \longrightarrow E(g) \qquad \Delta H_{atom}\ (Li > Na > K > Rb > Cs)$$

2. *Ionization: the metal atom transfers its outer electron to the nonmetal.* Alkali metals have *low ionization energies* (the lowest in their periods) and form *cations with small radii* since a great decrease in size occurs when the outer electron is lost: the volume of the Li^+ is about 13% the volume of Li! Thus, Group 1A(1) ions are small spheres with considerable charge density.

3. *Lattice formation: the resulting cations and anions attract each other to form an ionic solid.* Group 1A(1) salts have *high lattice energies,* which easily overcome the endothermic atomization and ionization steps, because the small cations lie close to the anions. For a given anion, the trend in lattice energy is the inverse of the trend in cation size: *as cation radius increases, lattice energy decreases.* The Group 1A(1) and 2A(2) chlorides exemplify this steady decrease in lattice energy (Figure 14.4).

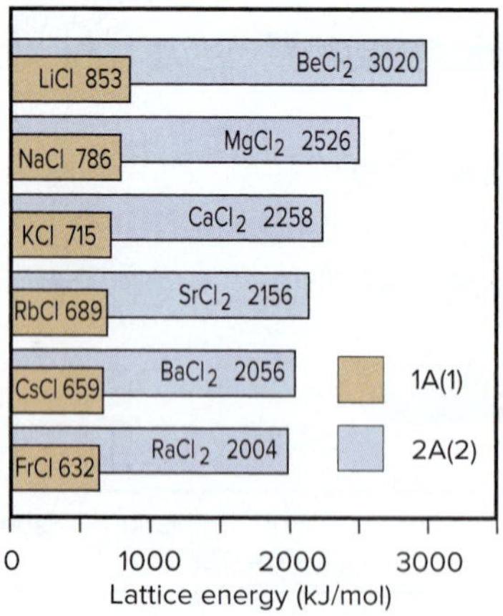

Figure 14.4 Lattice energies of the Group 1A(1) and 2A(2) chlorides.

Despite these strong ionic attractions in the solid, *nearly all Group 1A(1) salts are water soluble.* The attraction between the ions and water molecules creates a highly exothermic heat of hydration (ΔH_{hydr}), and a large increase in entropy occurs when ions in the organized crystal become dispersed and hydrated in solution; together, these factors outweigh the high lattice energy.

The magnitude of the hydration energy *decreases* as ionic size increases:

$$E^+(g) \longrightarrow E^+(aq) \qquad \Delta H = -\Delta H_{hydr}\ (Li^+ > Na^+ > K^+ > Rb^+ > Cs^+)$$

Interestingly, the *smaller* ions form larger *hydrated ions.* This size trend is key to the function of nerves, kidneys, and cell membranes because the *sizes* of $Na^+(aq)$ and $K^+(aq)$, the most common cations in cell fluids, influence their movement into and out of cells.

14.4 GROUP 2A(2): THE ALKALINE EARTH METALS

The Group 2A(2) elements are called *alkaline earth metals* because their oxides give basic (alkaline) solutions and melt at such high temperatures that they remained as solids ("earths") in the alchemists' fires. The group is a fascinating collection of

*Throughout the chapter, we use E to represent any element in a group other than in Group 7A(17), for which the halogens are represented as X.

elements: rare beryllium (Be), common magnesium (Mg) and calcium (Ca), less familiar strontium (Sr) and barium (Ba), and radioactive radium (Ra). The Group 2A(2) Family Portrait *(next page)* presents an overview of these elements.

How the Alkaline Earth and Alkali Metals Compare Physically

In general terms, the elements in Groups 1A(1) and 2A(2) behave as close cousins physically, with the differences due to the change in outer electron configuration from ns^1 to ns^2. Two electrons from each alkaline earth atom and one more proton in each nucleus strengthen metallic bonding. The following changes result:

- Melting and boiling points are much higher for Group 2A(2) elements; in fact, they melt at around the same temperatures as the Group 1A(1) elements boil.
- Compared to many transition metals, the alkaline earths are soft and lightweight, but they are harder and denser than the alkali metals.

How the Alkaline Earth and Alkali Metals Compare Chemically

The second valence electron in an alkaline earth metal lies in the same sublevel as the first and thus it is not shielded very well from the additional nuclear charge, so Z_{eff} is greater. As a result, Group 2A(2) elements have smaller atomic radii and higher ionization energies than Group 1A(1) elements. Despite the higher IEs, *all the alkaline earths (except Be) occur as 2+ cations in ionic compounds.* (As we said, Be behaves anomolously because so much energy is needed to remove two electrons from this tiny atom that it never forms discrete Be^{2+} ions, and so its bonds are polar covalent.)

Some important chemical properties of Group 2A(2) elements are the following:

1. *Reducing strength.* Like the alkali metals, the alkaline earth metals are *strong reducing agents:*

- Each reduces O_2 in air to form the oxide (Ba also forms the peroxide, BaO_2).
- Except for Be and Mg, which form adherent oxide coatings, each reduces H_2O at room temperature to form H_2.
- Except for Be, each reduces the halogens, N_2, and H_2 to form ionic compounds.

2. *Basicity of oxides.* The oxides are strongly basic (except for amphoteric BeO) and react with acidic oxides to form salts, such as sulfites and carbonates; for example,

$$SrO(s) + CO_2(g) \longrightarrow SrCO_3(s)$$

Natural carbonates, such as limestone and marble, are major structural materials and the commercial sources for most alkaline earth compounds. Calcium carbonate is heated to obtain calcium oxide (lime); this important industrial compound has essential roles in steelmaking, water treatment, and smokestack scrubbing and is used to make glass, whiten paper, and neutralize acidic soil.

3. *Lattice energies and solubilities.* Lattice energy plays a role in the reactivity of the Group 2A(2) elements and the solubility of their salts in water:

- The elements are reactive because the high lattice energies of their compounds more than compensate for the large total IE required to form 2+ cations (Section 9.2). Because the cations are smaller and doubly charged, their charge densities and lattice energies are much higher than those for salts of Group 1A(1) (see Figure 14.4).
- High lattice energy also leads to lower solubility of Group 2A(2) salts in water. Higher charge density increases heat of hydration, but it increases lattice energy even more. Thus, unlike the corresponding Group 1A(1) compounds, most Group 2A(2) fluorides, carbonates, phosphates, and sulfates have very low solubility.
- Although solubility is limited for many Group 2A(2) compounds, the ion-dipole attractions between 2+ ions and water molecules are so strong that many slightly soluble salts of these elements crystallize as hydrates; two examples are Epsom salt, $MgSO_4{\cdot}7H_2O$, used as an aqueous soaking solution for treating inflammation, and gypsum, $CaSO_4{\cdot}2H_2O$, used as the bonding material between the paper sheets in wallboard and as the cement in surgical casts.

KEY ATOMIC PROPERTIES, PHYSICAL PROPERTIES, AND REACTIONS

KEY
Atomic No. / **Symbol** / Atomic mass / Valence e^- configuration / (Common oxidation states)
4 / **Be** / 9.012 / $2s^2$ / **(+2)**
12 / **Mg** / 24.30 / $3s^2$ / **(+2)**
20 / **Ca** / 40.08 / $4s^2$ / **(+2)**
38 / **Sr** / 87.62 / $5s^2$ / **(+2)**
56 / **Ba** / 137.3 / $6s^2$ / **(+2)**
88 / **Ra** / (226) / $7s^2$ / **(+2)**

Source: © McGraw-Hill Education/Stephen Frisch, photographer

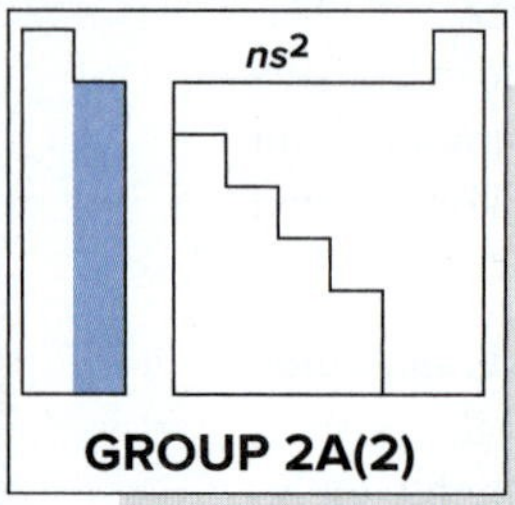

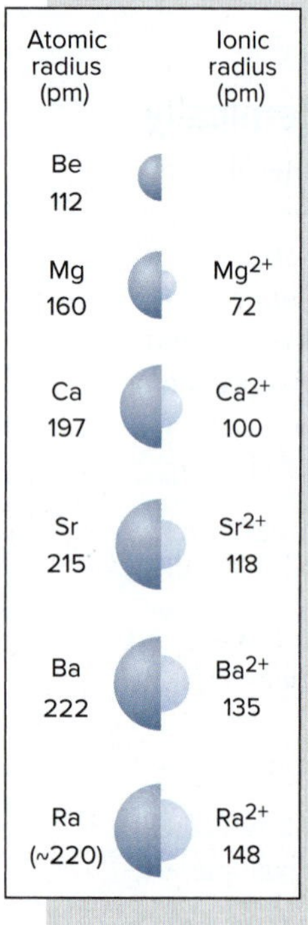

Atomic Properties

Group electron configuration is ns^2 (filled ns sublevel). All members have the +2 oxidation state and, except for Be, form compounds with an E^{2+} ion. Atomic and ionic sizes increase down the group but are smaller than for the corresponding 1A(1) elements. IE and EN decrease down the group but are higher than for the corresponding 1A(1) elements.

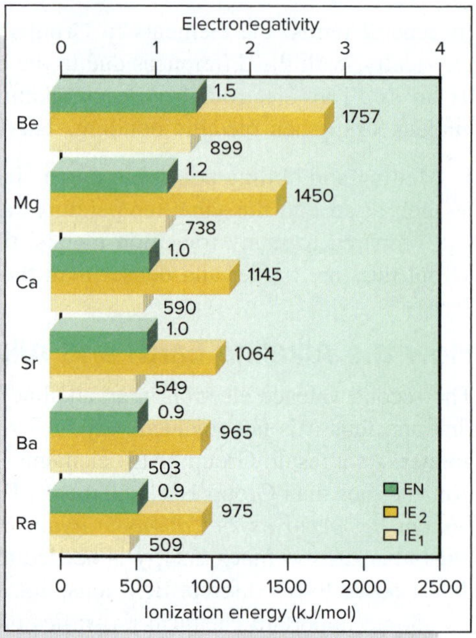

Physical Properties

Metallic bonding involves two valence electrons. These metals are still relatively soft but are much harder than the 1A(1) metals. Melting and boiling points generally decrease and densities generally increase down the group. These values are much higher than for 1A(1) elements, and the trend is not as regular.

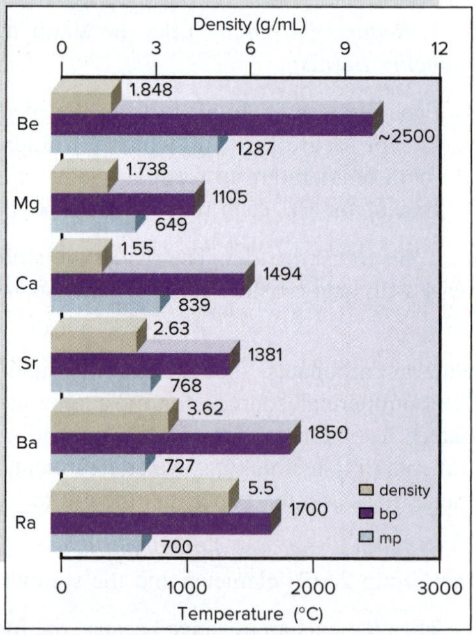

Reactions

1. The metals reduce O_2 to form the oxides:

$$2E(s) + O_2(g) \longrightarrow 2EO(s)$$

Ba also forms the peroxide, BaO_2.

2. The larger metals reduce water to form hydrogen gas:

$$E(s) + 2H_2O(l) \longrightarrow E^{2+}(aq) + 2OH^-(aq) + H_2(g) \qquad (E = Ca, Sr, Ba)$$

Be and Mg form an oxide coating that allows only slight reaction.

3. The metals reduce halogens to form ionic halides:

$$E(s) + X_2 \longrightarrow EX_2(s) \qquad [X = F \text{ (not with Be)}, Cl, Br, I]$$

4. Most of the elements reduce hydrogen to form ionic hydrides:

$$E(s) + H_2(g) \longrightarrow EH_2(s) \qquad (E = \text{all except Be})$$

5. The elements reduce nitrogen to form ionic nitrides:

$$3E(s) + N_2(g) \longrightarrow E_3N_2(s)$$

6. Except for amphoteric BeO, the element oxides are basic:

$$EO(s) + H_2O(l) \longrightarrow E^{2+}(aq) + 2OH^-(aq)$$

$Ca(OH)_2$ is a component of cement and mortar.

7. All carbonates undergo thermal decomposition to the oxide:

$$ECO_3(s) \xrightarrow{\Delta} EO(s) + CO_2(g)$$

This reaction is used to produce CaO (lime) in huge amounts from naturally occurring limestone.

Diagonal Relationships: Lithium and Magnesium

One of the clearest ways to see how atomic properties influence chemical behavior is to look at three **diagonal relationships,** similarities between a Period 2 element and one diagonally down and to the right in Period 3.

The first of these occurs between Li and Mg, which have similar atomic and ionic sizes (Figure 14.5). Note that *moving one period down increases atomic (or ionic) size and moving one group to the right decreases it.* The Li radius is 152 pm and that of Mg is 160 pm; the Li^+ radius is 76 pm and that of Mg^{2+} is 72 pm. From similar atomic properties emerge similar chemical properties. Both elements form nitrides with N_2, hydroxides and carbonates that decompose easily with heat, organic compounds with a polar covalent metal-carbon bond, and salts with similar solubilities. We'll discuss the relationships between Be and Al and between B and Si in upcoming sections.

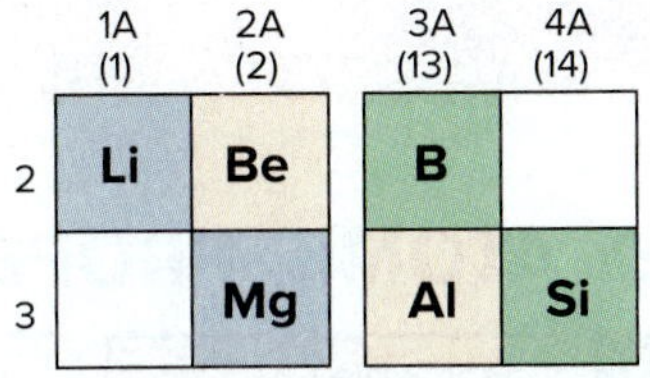

Figure 14.5 Three diagonal relationships in the periodic table.

14.5 GROUP 3A(13): THE BORON FAMILY

The third family of main-group elements contains both familiar and unusual members, which engage in some exotic bonding and have strange physical properties. Boron (B) heads the family, but, as we said, its properties are not representative. Metallic aluminum (Al) has properties more typical of the group, but its great abundance and importance contrast with the rareness of gallium (Ga), indium (In), thallium (Tl), and recently synthesized nihonium (Nh). The atomic, physical, and chemical properties of these elements are summarized in the Group 3A(13) Family Portrait *(next page).*

How the Transition Elements Influence This Group's Properties

If you look only at the main groups, Group 3A(13), the first of the *p* block, seems to be just one group away from Group 2A(2). In Period 4 and higher, however, 10 transition elements (*d* block) separate these groups (see Figure 8.10). And an additional 14 inner transition elements (*f* block) appear in Periods 6 and 7. Thus, the heavier Group 3A(13) members have nuclei with many more protons, but since *d* and *f* electrons penetrate very little (Section 8.1), the outer (*s* and *p*) electrons of Ga, In, and Tl are poorly shielded from the much higher positive charge. As a result, *these elements have greater Z_{eff} than the two lighter members,* and this stronger nuclear pull explains why Ga, In, and Tl have smaller atomic radii and larger ionization energies and electronegativities than expected. This effect is observed in higher groups, too.

Physical properties are influenced by the type of bonding. Boron is a network covalent metalloid—black, hard, and very high melting. The other group members are metals—shiny and relatively soft and low melting. Aluminum's low density and three valence electrons make it an exceptional conductor: for a given mass, aluminum conducts a current twice as effectively as copper. Gallium's metallic bonding gives it the largest liquid temperature range of any element: it melts at skin temperature *(see photo in the Family Portrait)* but does not boil until 2403°C. The bonding is too weak to keep the Ga atoms fixed when the solid is warmed, but strong enough to keep them from escaping the molten metal until it is very hot.

Features That First Appear in This Group's Chemical Properties

Looking down Group 3A(13), we see a wide range of chemical behavior:

- Boron, the anomalous member from Period 2, is the only metalloid. It is much less reactive at room temperature than the other members and forms covalent bonds exclusively.
- Although aluminum acts like a metal physically, its halides exist in the gas phase as covalent *dimers*—molecules formed by joining two identical smaller molecules (Figure 14.6)—and its oxide is amphoteric rather than basic.
- Most of the other Group 3A(13) compounds are ionic. However, because the cations of this group are smaller and triply charged, they polarize an anion more effectively than do Group 2A(2) cations, and so their compounds are more covalent.

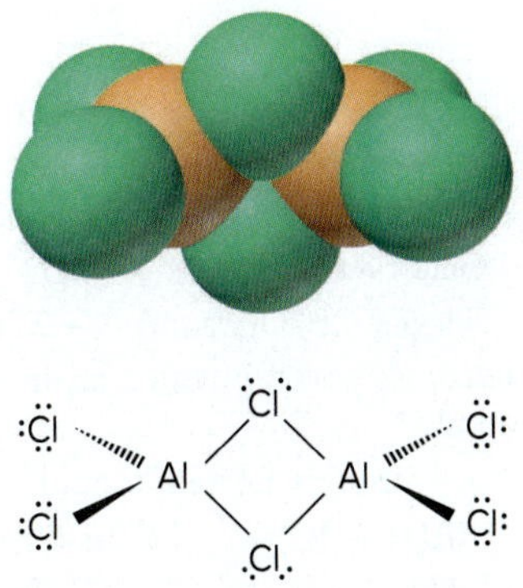

Figure 14.6 The dimeric structure of gaseous aluminum chloride.

Group 3A(13): The Boron Family

KEY ATOMIC PROPERTIES, PHYSICAL PROPERTIES, AND REACTIONS

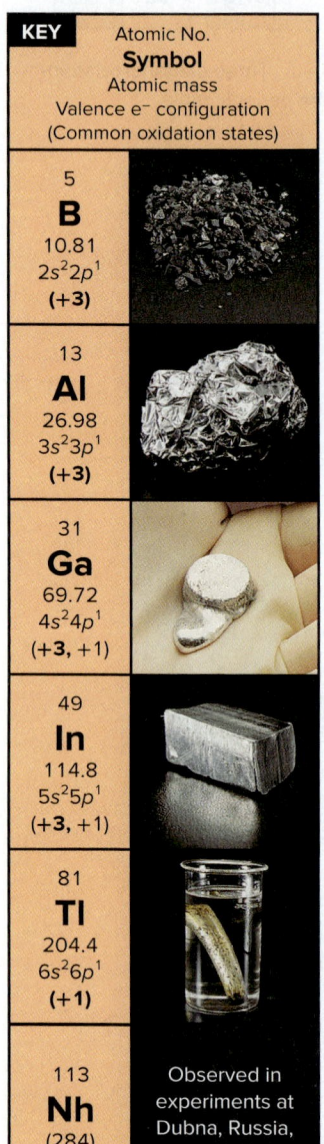

KEY
Atomic No.
Symbol
Atomic mass
Valence e^- configuration
(Common oxidation states)

5
B
10.81
$2s^22p^1$
(+3)

13
Al
26.98
$3s^23p^1$
(+3)

31
Ga
69.72
$4s^24p^1$
(+3, +1)

49
In
114.8
$5s^25p^1$
(+3, +1)

81
Tl
204.4
$6s^26p^1$
(+1)

113
Nh
(284)
$7s^27p^1$

Observed in experiments at Dubna, Russia, in 2003

Source: © McGraw-Hill Education/Stephen Frisch, photographer

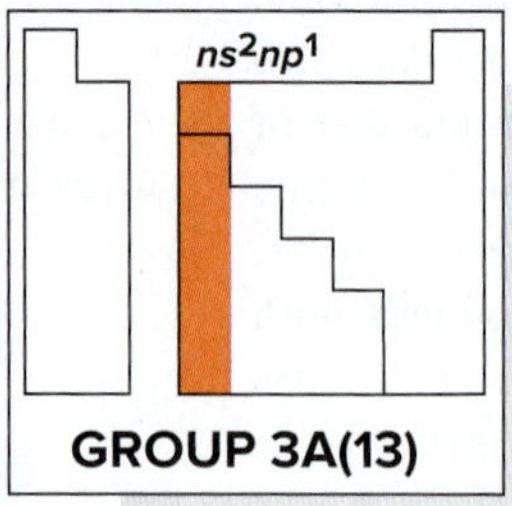

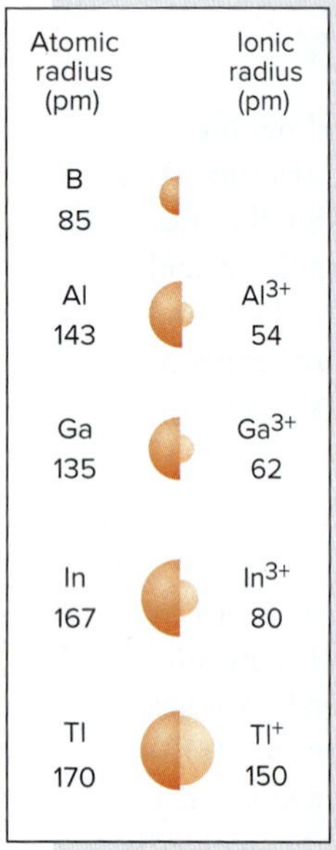

Atomic Properties

Group electron configuration is ns^2np^1. All except Tl commonly display the +3 oxidation state. The +1 state becomes more common down the group. Atomic size is smaller and EN is higher than for 2A(2) elements; IE is lower, however, because it is easier to remove an electron from the higher energy *p* sublevel. Atomic size, IE, and EN do not change as expected down the group because there are intervening transition and inner transition elements.

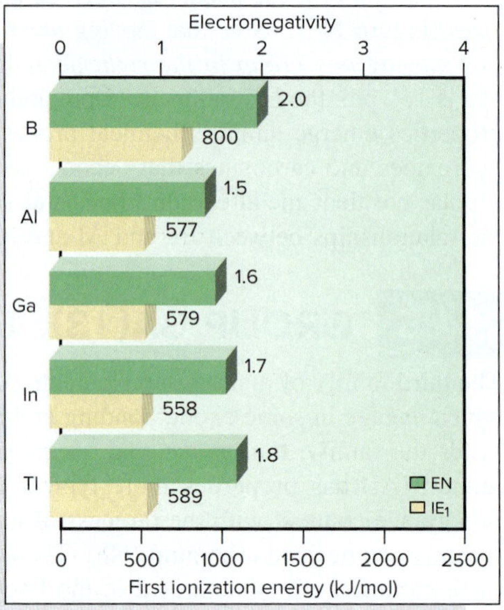

Physical Properties

Bonding changes from network covalent in B to metallic in the rest of the group. Thus, B has a much higher melting point than the others, but there is no overall trend. Boiling points decrease down the group. Densities increase down the group.

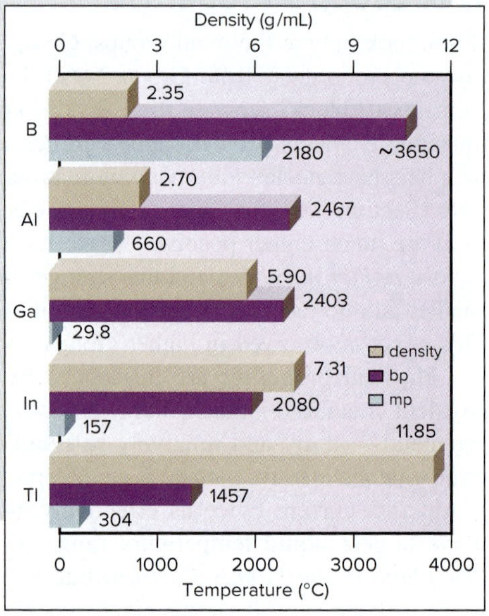

Reactions

1. The elements react sluggishly, if at all, with water:

$$2\text{Ga}(s) + 6\text{H}_2\text{O}(hot) \longrightarrow 2\text{Ga}^{3+}(aq) + 6\text{OH}^-(aq) + 3\text{H}_2(g)$$

$$2\text{Tl}(s) + 2\text{H}_2\text{O}(steam) \longrightarrow 2\text{Tl}^+(aq) + 2\text{OH}^-(aq) + \text{H}_2(g)$$

Al becomes covered with a layer of Al_2O_3 that prevents further reaction.

2. When strongly heated in pure O_2, all members form oxides:

$$4\text{E}(s) + 3\text{O}_2(g) \xrightarrow{\Delta} 2\text{E}_2\text{O}_3(s) \qquad (\text{E} = \text{B, Al, Ga, In})$$

$$4\text{Tl}(s) + \text{O}_2(g) \xrightarrow{\Delta} 2\text{Tl}_2\text{O}(s)$$

Oxide acidity decreases down the group: B_2O_3 (weakly acidic) > Al_2O_3 > Ga_2O_3 > In_2O_3 > Tl_2O (strongly basic), and the +1 oxide is more basic than the +3 oxide.

3. All members reduce halogens (X_2):

$$2\text{E}(s) + 3\text{X}_2 \longrightarrow 2\text{EX}_3 \qquad (\text{E} = \text{B, Al, Ga, In})$$

$$2\text{Tl}(s) + \text{X}_2 \longrightarrow 2\text{TlX}(s)$$

The BX_3 compounds are volatile and covalent. Trihalides of Al, Ga, and In are (mostly) ionic solids.

The redox behavior of Group 3A(13) exhibits three features that appear first here, but occur in Groups 4A(14) to 6A(16) as well:

1. *Presence of multiple oxidation states.* Larger members of these groups also have an important oxidation state of +1, *two lower than the A-group number.* The lower state occurs when the atoms lose their *np* electrons only, not their two *ns* electrons. This fact is often called the *inert-pair effect* (Section 8.4).

2. *Increasing stability of the lower oxidation state.* For these groups, *the lower state becomes more stable going down the group.* In Group 3A(13), for instance, all members exhibit the +3 state, but the +1 state first appears with some compounds of gallium and becomes the only important state of thallium.

3. *Increasing metallic behavior and basicity of oxides.* In general, *oxides of the element in the lower oxidation state are more basic.* Thus, for example, in Group 3A(13), In_2O is more basic than In_2O_3. The reason is that *an element acts more like a metal in its lower state.* In this example, the lower charge of In^+ does not polarize the O^{2-} ion as much as the higher charge of In^{3+} does, so the In-to-O bonding is more ionic and the O^{2-} ion is more available to act as a base.

Highlights of Boron Chemistry

Like the other Period 2 elements, the chemical behavior of boron is strikingly different from that of the other members of its group. *All boron compounds are covalent,* and unlike the other Group 3A(13) members, boron forms network covalent compounds or large molecules with metals, H, O, N, and C. The unifying feature of many boron compounds is the element's *electron deficiency.* Boron adopts two strategies to fill its outer level: accepting a bonding pair from an electron-rich atom and forming bridge bonds with an electron-poor atom.

Accepting a Bonding Pair from an Electron-Rich Atom In gaseous boron trihalides (BX_3), the B atom is electron deficient, with only six electrons around it (Section 10.1). To attain an octet, the B atom accepts a lone pair (*blue*) from an electron-rich atom and forms a covalent bond:

$$BF_3(g) + :NH_3(g) \longrightarrow F_3B{-}NH_3(g)$$

(Reactions in which one reactant accepts an electron pair from another to form a covalent bond are very common and are known as *Lewis acid-base reactions.* We'll discuss them in Chapters 18 and 23 and see examples of them throughout the rest of the text.)

Similarly, B has only six electrons in boric acid, $B(OH)_3$ (sometimes written as H_3BO_3). In water, the acid itself does not release a proton. Rather, it accepts an electron pair from the O in H_2O, forming a fourth bond and releasing an H^+ ion:

$$B(OH)_3(s) + H_2O(l) \rightleftharpoons B(OH)_4^-(aq) + H^+(aq)$$

Boron's outer shell is filled in the wide variety of borate salts, such as the mineral borax (sodium borate), $Na_2[B_4O_5(OH)_4]{\cdot}8H_2O$, used for decades as a household cleaning agent. Strong heating of boric acid (or borate salts) drives off water molecules and gives molten boron oxide:

$$2B(OH)_3(s) \xrightarrow{\Delta} B_2O_3(l) + 3H_2O(g)$$

When mixed with silica (SiO_2), this molten oxide forms borosilicate glass. Its high transparency and small change in size when heated or cooled make borosilicate glass useful in cookware and in lab glassware *(see photo).*

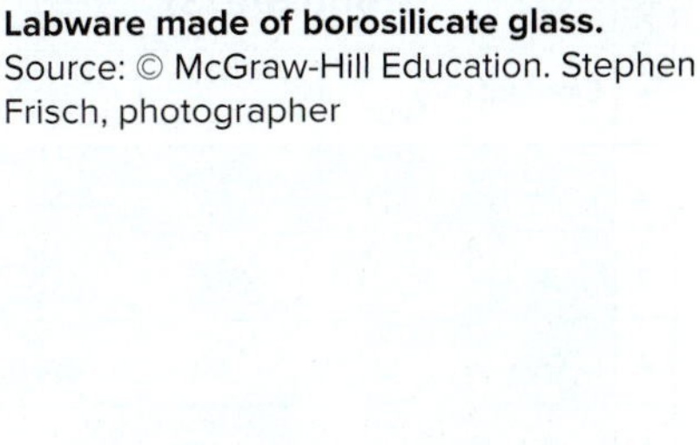

Labware made of borosilicate glass.
Source: © McGraw-Hill Education. Stephen Frisch, photographer

Forming a Bridge Bond with an Electron-Poor Atom In elemental boron and its many hydrides (boranes), there is no electron-rich atom to supply boron with electrons. In these substances, boron attains an octet through some unusual bonding. In diborane (B_2H_6) and many larger boranes, for example, two types of B—H bonds exist. The first type is a normal electron-pair bond. The valence bond picture in Figure 14.7 *(next page)* shows an sp^3 orbital of B overlapping a 1*s* orbital of H in each of the four terminal B—H bonds, using two of the three electrons in the valence level of each B atom.

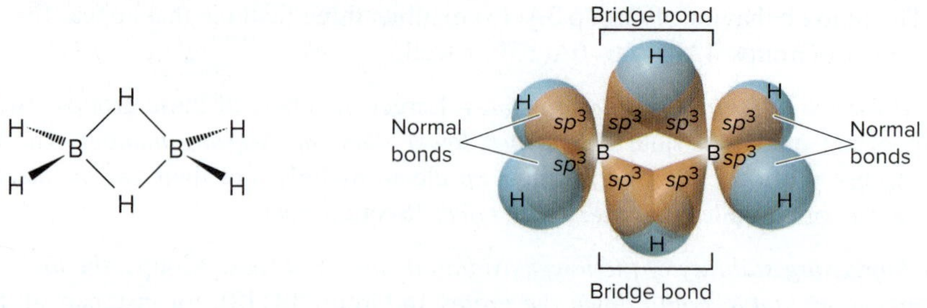

Figure 14.7 The two types of covalent bonding in diborane.

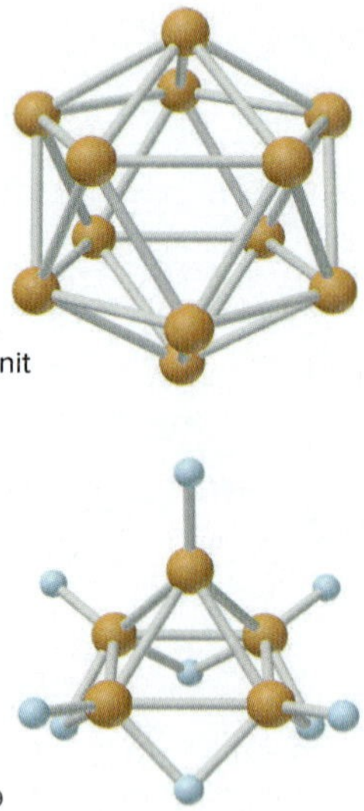

Figure 14.8 The boron icosahedron and one of the boranes.

The other type of bond is a hydride **bridge bond** (or *three-center, two-electron bond*), in which *each B—H—B grouping is held together by only two electrons.* Two sp^3 orbitals, one from *each* B, overlap an H 1*s* orbital between them. Two electrons move through this extended bonding orbital—one from one of the B atoms and the other from the H atom—and join the two B atoms via the H-atom bridge. Notice that *each B atom is surrounded by eight electrons:* four from the two normal B—H bonds and four from the two B—H—B bridge bonds with a tetrahedral arrangement around each B atom. In many boranes and in elemental boron (Figure 14.8), one B atom bridges two others in a three-center, two-electron B—B—B bond.

Diagonal Relationships: Beryllium and Aluminum

Beryllium in Group 2A(2) and aluminum in Group 3A(13) are another pair of diagonally related elements. Both form oxoanions in strong base: beryllate, $Be(OH)_4^{2-}$, and aluminate, $Al(OH)_4^-$. Both have bridge bonds in their hydrides and chlorides. Both form oxide coatings impervious to reaction with water (which explains aluminum's great use as a structural metal), and both oxides are amphoteric, extremely hard, and high melting. Although the atomic and ionic sizes of these elements differ, the small, highly charged Be^{2+} and Al^{3+} ions polarize nearby electron clouds strongly. Therefore, some Al compounds and all Be compounds have significant covalent character.

14.6 GROUP 4A(14): THE CARBON FAMILY

The whole range of behavior occurs in Group 4A(14): nonmetallic carbon (C) leads off, followed by the metalloids silicon (Si) and germanium (Ge), then metallic tin (Sn) and lead (Pb), and ending with flerovium (Fl), which was synthesized in 1998. Information about the compounds of C and of Si fills libraries: organic chemistry, most polymer chemistry, and biochemistry are based on carbon, whereas geochemistry and some essential polymer and electronic technologies are based on silicon. The Group 4A(14) Family Portrait summarizes atomic, physical, and chemical properties.

How Type of Bonding Affects Physical Properties

The elements of Group 4A(14) and their neighbors in Groups 3A(13) and 5A(15) illustrate how some physical properties depend on the type of bonding in an element (Table 14.2). Within Group 4A(14), the large decrease in melting point

Table 14.2 Bond Type and the Melting Process in Groups 3A(13) to 5A(15)

Period	Group 3A(13) Element	Bond Type	Melting Point (°C)	ΔH_{fus} (kJ/mol)	Group 4A(14) Element	Bond Type	Melting Point (°C)	ΔH_{fus} (kJ/mol)	Group 5A(15) Element	Bond Type	Melting Point (°C)	ΔH_{fus} (kJ/mol)
2	B	Covalent network	2180	23.6	C	Covalent network	4100	Very high	N	Covalent molecule	−210	0.7
3	Al	Metallic	660	10.5	Si	Covalent network	1420	50.6	P	Covalent molecule	44.1	2.5
4	Ga	Metallic	30	5.6	Ge	Covalent network	945	36.8	As	Covalent network	816	27.7
5	In	Metallic	157	3.3	Sn	Metallic	232	7.1	Sb	Covalent network	631	20.0
6	Tl	Metallic	304	4.3	Pb	Metallic	327	4.8	Bi	Metallic	271	10.5

Key: Metallic; Covalent network; Covalent molecule; Metal; Metalloid; Nonmetal

KEY ATOMIC PROPERTIES, PHYSICAL PROPERTIES, AND REACTIONS

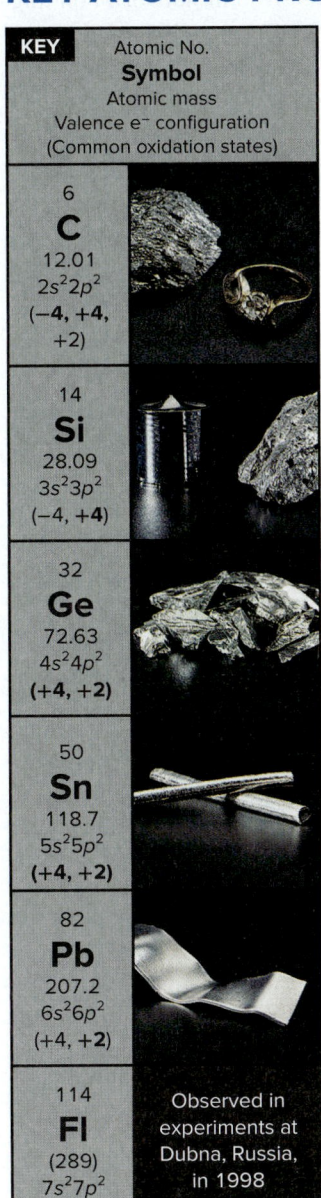

KEY
Atomic No.
Symbol
Atomic mass
Valence e⁻ configuration
(Common oxidation states)

6
C
12.01
$2s^22p^2$
(**−4**, **+4**, +2)

14
Si
28.09
$3s^23p^2$
(−4, **+4**)

32
Ge
72.63
$4s^24p^2$
(**+4**, **+2**)

50
Sn
118.7
$5s^25p^2$
(**+4**, **+2**)

82
Pb
207.2
$6s^26p^2$
(+4, **+2**)

114
Fl
(289)
$7s^27p^2$
Observed in experiments at Dubna, Russia, in 1998

Source: © McGraw-Hill Education/Stephen Frisch, photographer

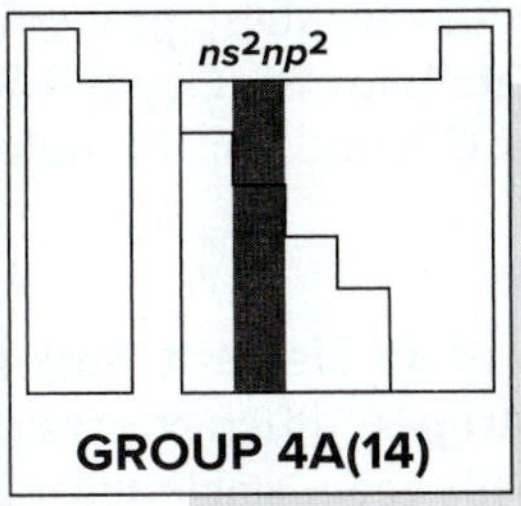

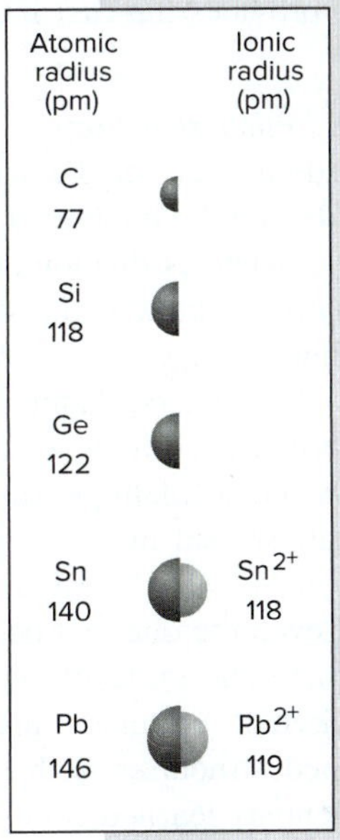

Atomic Properties

Group electron configuration is ns^2np^2. Down the group, the number of oxidation states decreases, and the lower (+2) state becomes more common. Down the group, size increases. Because transition and inner transition elements intervene, IE and EN do not decrease smoothly.

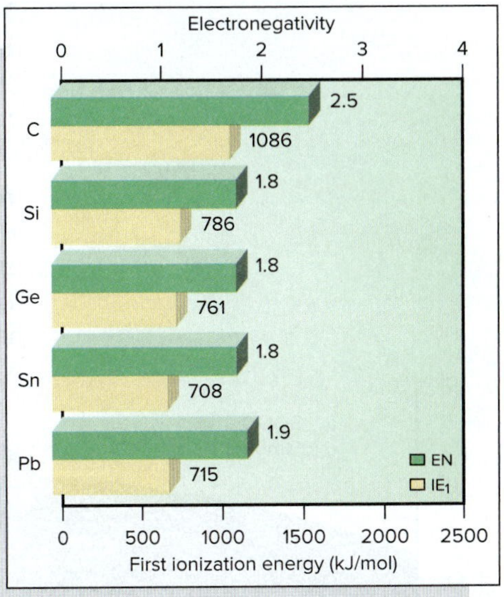

Physical Properties

Trends in properties, such as decreasing hardness and melting point, are due to changes in types of bonding within the solid: covalent network in C, Si, and Ge; metallic in Sn and Pb *(see text)*. Down the group, density increases because of several factors, including differences in crystal packing.

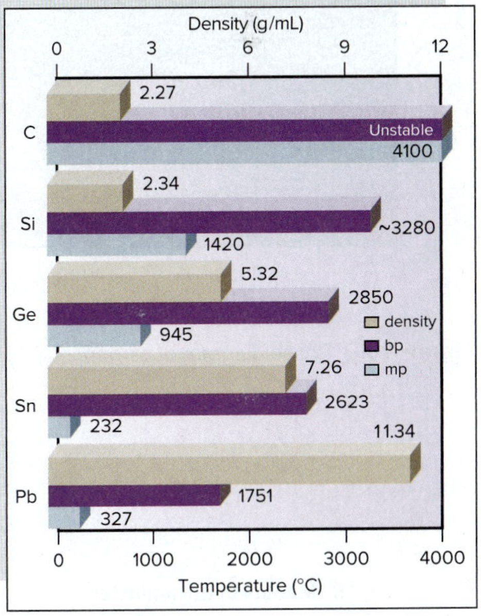

Reactions

1. The elements are oxidized by halogens:

$$E(s) + 2X_2 \longrightarrow EX_4 \qquad (E = C, Si, Ge)$$

The +2 halides are more stable for tin and lead, SnX_2 and PbX_2.

2. The elements are oxidized by O_2:

$$E(s) + O_2(g) \longrightarrow EO_2 \qquad (E = C, Si, Ge, Sn)$$

Pb forms the +2 oxide, PbO. Oxides become more basic down the group. The reaction of CO_2 and H_2O provides the weak acidity of natural unpolluted waters:

$$CO_2(g) + H_2O(l) \rightleftharpoons [H_2CO_3(aq)] \rightleftharpoons H^+(aq) + HCO_3^-(aq)$$

3. Hydrocarbons react with O_2 to form CO_2 and H_2O. The reaction for methane is adapted to yield heat or electricity:

$$CH_4(g) + 2O_2(g) \longrightarrow CO_2(g) + 2H_2O(g)$$

4. Silica is reduced to form elemental silicon:

$$SiO_2(s) + 2C(s) \longrightarrow Si(s) + 2CO(g)$$

This crude silicon is made ultrapure through zone refining for use in the manufacture of computer chips.

between the network covalent solids C and Si is due to longer, weaker bonds in the Si structure; the large decrease between Ge and Sn is due to the change from covalent network to metallic bonding. Similarly, considering horizontal trends, the large increases in melting point and ΔH_{fus} across a period between Al and Si and between Ga and Ge reflect the change from metallic to covalent network bonding. Note the abrupt rises in the values for these properties from metallic Al, Ga, and Sn to the network covalent metalloids Si, Ge, and Sb, and note the abrupt drops from the covalent networks of C and Si to the individual molecules of N and P in Group 5A(15).

Allotropism: Different Forms of an Element Striking variations in physical properties often appear among **allotropes,** different crystalline or molecular forms of a substance. One allotrope is usually more stable than another at a particular pressure and temperature. Group 4A(14) provides the first important examples of allotropism, in the forms of carbon and tin.

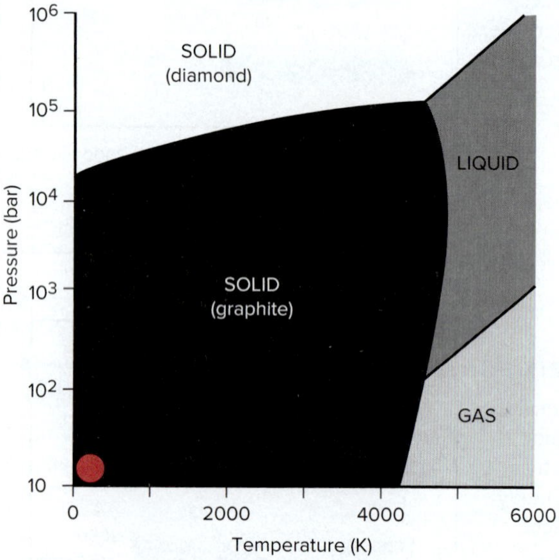

Figure 14.9 Phase diagram of carbon.

1. *Allotropes and other structures of carbon.* It is difficult to imagine two substances made entirely of the same atom that are more different than graphite and diamond. Graphite is a black electrical conductor that is soft and "greasy," whereas diamond is a colorless electrical insulator that is extremely hard. Graphite is the standard state of carbon, the more stable form at ordinary temperature and pressure (Figure 14.9, *red dot*). Fortunately for jewelry owners, diamond changes to graphite at a negligible rate under normal conditions.

In the mid-1980s, a new allotrope was discovered. Mass spectra of soot showed a soccer ball–shaped molecule of formula C_{60} (Figure 14.10A). The molecule has also been found in geological samples formed by meteorite impacts, even the one that occurred around the time the dinosaurs became extinct. The molecule was dubbed *buckminsterfullerene* (and called a "buckyball") after the architect-engineer R. Buckminster Fuller, who designed structures with similar shapes. Excitement rose in 1990, when scientists learned how to prepare enough C_{60} to study its behavior and applications. Since then, metal atoms have been incorporated in the balls and many different groups (fluorine, hydroxyl groups, sugars, etc.) have been attached, resulting in compounds with a range of useful properties.

In 1991, scientists passed an electric discharge through graphite rods and obtained extremely thin (~1 nm in diameter) graphite-like tubes with fullerene ends called *nanotubes* (Figure 14.10B; the model shows concentric nanotubes, colored differently, without the fullerene ends). Rigid and, on a mass basis, stronger than steel along their

Figure 14.10 Models of buckminsterfullerene (A), a carbon nanotube (B), and graphene (C).

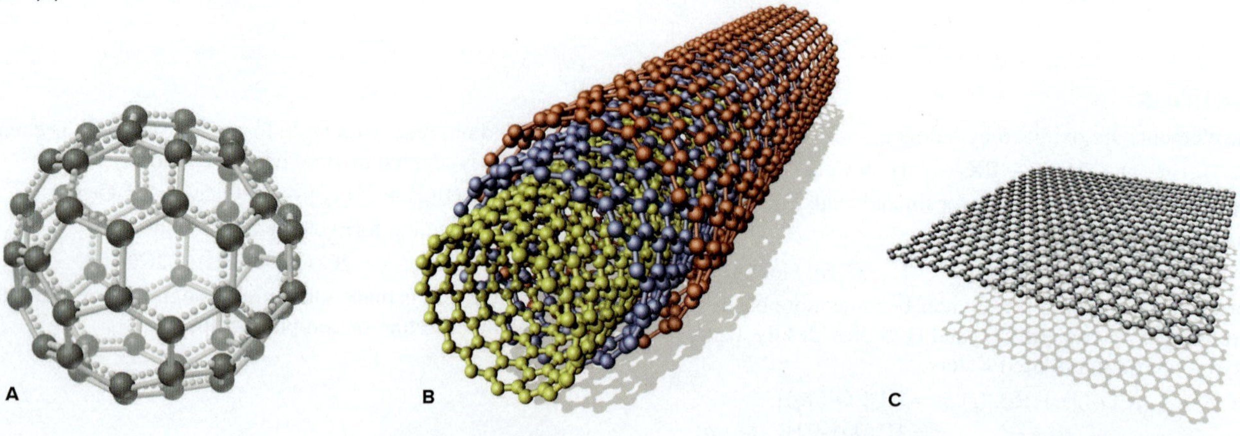

A B C

long axis, they also conduct electricity along this axis because of the delocalized electrons. With applications in nanoscale electronics, energy storage, catalysis, polymers, and medicine, nanotube chemistry is a major area of materials research. Then, in 2010, the Nobel Prize in physics was awarded for studies of a new form of carbon called *graphene*, which exists as extended sheets only one atom thick but has remarkable conductivity and strength (Figure 14.10C).

2. *Allotropes of tin.* Tin has two allotropes. White β-tin is stable at room temperature and above, whereas gray α-tin is the more stable form below 13°C (56°F). When white tin is kept for long periods at a low temperature, some converts to microcrystals of gray tin. The random formation and growth of these regions of gray tin, which has a different crystal structure, weaken the metal and make it crumble. In the unheated cathedrals of medieval northern Europe, tin pipes of magnificent organs were sometimes destroyed by the "tin disease" caused by this allotropic transition.

How Bonding Changes in This Group's Compounds

The Group 4A(14) elements display a wide range of chemical behavior, from the covalent compounds of carbon to the ionic compounds of lead, and the features we saw first in Group 3A(13) appear here as well.

1. *Multiple oxidation states.* All the members have at least two important states (+4 and +2), and carbon has three (+4, +2, and −4).

2. *Increasing stability of lower oxidation state.* Compounds with Si in the +4 state are much more stable than those with Si in the +2 state, whereas compounds with Pb in the +2 state are more stable than those with Pb in the +4 state.

3. *Increasing metallic behavior, ionic bonding, and oxide basicity.* Carbon's intermediate EN of 2.5 ensures that it always bonds covalently, but the larger members form bonds with increasing ionic character. With nonmetals, Si and Ge form strong polar covalent bonds, such as the Si—O bond, one of the strongest Period 3 bonds (BE = 368 kJ/mol) and responsible for the stability of Earth's solid surface. Although individual ions of Sn or Pb rarely exist, their bonding with a nonmetal has considerable ionic character. And the bonding becomes more ionic with the metal in the lower oxidation state. Thus, $SnCl_2$ and $PbCl_2$ are white, high-melting, water-soluble crystals—typical properties of a salt (Figure 14.11)—whereas $SnCl_4$ is a volatile, nonpolar liquid, and $PbCl_4$ is a thermally unstable oil. Similarly, SnO and PbO are more basic than SnO_2 and PbO_2: because the +2 metals are less able to polarize the oxide ion, the E-to-O bonding is more ionic.

Figure 14.11 Saltlike +2 chlorides and oily +4 chlorides show the greater metallic character of tin and lead in their lower oxidation state.

Source: © McGraw-Hill Education/Stephen Frisch, photographer

Highlights of Carbon Chemistry

Like the other Period 2 elements, carbon is an anomaly in its group; indeed, it may be an anomaly in the entire periodic table. Carbon forms bonds with the smaller Group 1A(1) and 2A(2) metals, many transition metals, the halogens, and many other metalloids and nonmetals. In addition to the three common oxidation states, carbon exhibits all the others possible for its group, from +4 through −4.

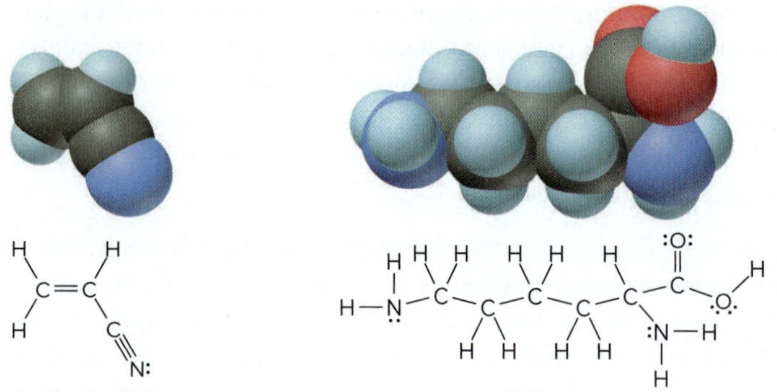

A **Acrylonitrile** **B** **Lysine**

Figure 14.12 **Two of the several million known organic compounds of carbon.** **A,** Acrylonitrile, a precursor of acrylic fibers. **B,** Lysine, one of about 20 amino acids that occur in proteins.

Organic Compounds Two major properties of carbon give rise to the enormous field of organic chemistry.

1. Carbon forms covalent bonds with other carbon atoms—a process known as *catenation.* As a result of its small size and its capacity for four bonds, carbon can form chains, branches, and rings that lead to myriad structures. Add a lot of H, some O and N, a bit of S, P, halogens, and a few metals, and you have the whole organic world! Figure 14.12 shows two of the several million organic compounds known.

Halogenated organic compounds have major polymer applications, such as poly (vinyl chloride) in plumbing and Teflon in cookware. Some are very long-lived in the environment. PCBs, previously used in electrical equipment, are now banned because they are carcinogenic (Figure 14.13). And Freons, used as cleaners of electronic parts and coolants in air conditioners, are responsible for a severe reduction in Earth's protective ozone layer (Chapter 16).

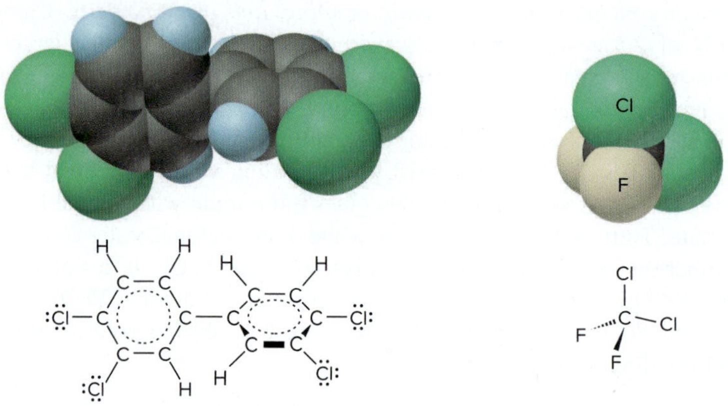

A **One of many PCBs** **B** **Freon-12**

Figure 14.13 **Two important halogenated organic compounds.** **A,** A typical PCB (one of the polychlorinated biphenyls). **B,** Freon-12 (CCl_2F_2), a chlorofluorocarbon.

2. Carbon has the ability to form multiple bonds. Multiple bonds are common in carbon structures because the C—C bond is short enough for side-to-side overlap of two half-filled 2*p* orbitals to form π bonds. (In Chapter 15, we discuss how the properties of carbon give rise to the diverse structures and reactivities of organic compounds.)

Because the other Group 4A(4) members are larger, E—E bonds become longer and weaker, so catenation and multiple bonding are much less important down the group.

Inorganic Compounds In contrast to its organic compounds, carbon's inorganic compounds are simple.

1. *Carbonates.* Metal carbonates are the main mineral form. Marble, limestone, chalk, coral, and several other types are found in enormous deposits throughout the world. Many of these compounds are remnants of fossilized marine organisms. Carbonates are essential to many industries, and they occur in several common antacids because they react with the HCl in stomach acid:

$$CaCO_3(s) + 2HCl(aq) \longrightarrow CaCl_2(aq) + CO_2(g) + H_2O(l)$$

Identical net ionic reactions with sulfuric and nitric acids protect lakes bounded by limestone deposits from the harmful effects of acid rain.

2. *Oxides.* Unlike the other Group 4A(4) members, which form only solid network covalent or ionic oxides, carbon forms two common gaseous oxides, CO_2 and CO.

- Carbon dioxide is essential to all life: it is the primary source of carbon in plants, through photosynthesis, and in animals who eat the plants. Its aqueous solution is the cause of mild acidity in natural waters. However, its atmospheric buildup from deforestation and excessive use of fossil fuels is severely affecting the global climate through warming.
- Carbon monoxide forms when carbon or its compounds burn in an inadequate supply of O_2:

$$2C(s) + O_2(g) \longrightarrow 2CO(g)$$

It is a key component of syngas fuels (see Chemical Connections, Chapter 6) and is widely used in the production of methanol, formaldehyde, and other major industrial compounds. CO binds strongly to many transition metals. When inhaled in cigarette smoke or polluted air, it enters the blood and binds strongly to the Fe(II) in hemoglobin, preventing the normal binding of O_2, and to other iron-containing proteins. The cyanide ion (CN^-) is *isoelectronic* with CO:

$[:C{\equiv}N:]^-$ same electronic structure as $:C{\equiv}O:$

Cyanide binds to many of the same iron-containing proteins and is also toxic.

Highlights of Silicon Chemistry

To a great extent, the chemistry of silicon is the chemistry of the *silicon-oxygen bond.* Just as carbon forms lengthy C—C chains, the —Si—O— grouping repeats itself almost endlessly in a wide variety of **silicates,** the most important minerals on the planet, and in **silicones,** synthetic polymers that have many applications:

1. *Silicate minerals.* From common sand and clay to semiprecious amethyst and carnelian, silicate minerals are the dominant form of matter in the nonliving world. Oxygen, the most abundant element on Earth, and silicon, the next most abundant, account for four of every five atoms on the surface of the planet!

The silicate building unit is the *orthosilicate grouping,* —SiO_4—, a tetrahedral arrangement of four O atoms around the central Si. Several minerals contain SiO_4^{4-} ions or small groups of them linked together. The gemstone zircon ($ZrSiO_4$) contains one unit; hemimorphite [$Zn_4(OH)_2Si_2O_7{\cdot}H_2O$] contains two units linked through an oxygen corner of each one; and beryl ($Be_3Al_2Si_6O_{18}$), the major source of beryllium, contains six units joined in a cyclic ion (Figure 14.14).

In extended structures, one of the O atoms links the next Si—O group to form chains, a second one forms crosslinks to neighboring chains to form sheets, and the

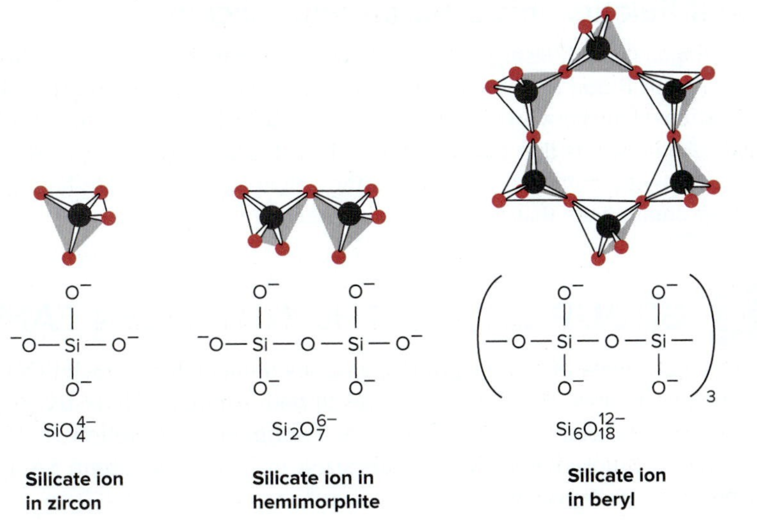

Figure 14.14 Structures of the silicate anions in some minerals.

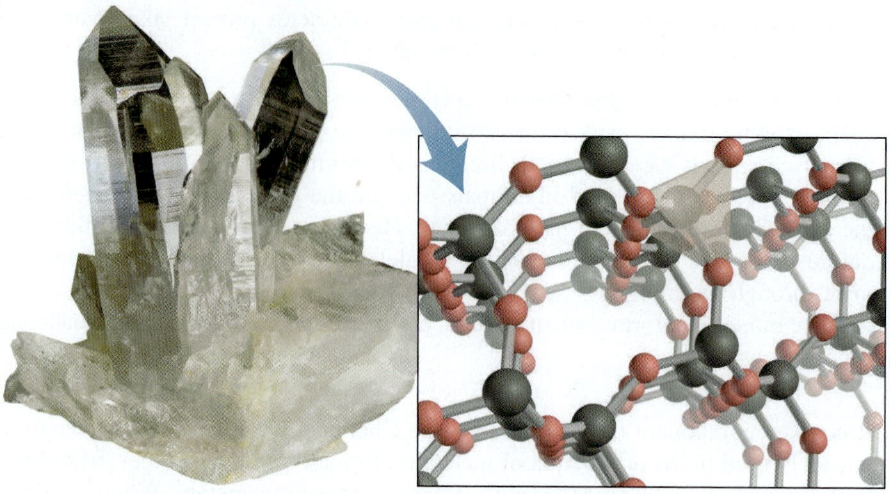

Figure 14.15 **Quartz is a three-dimensional framework silicate.**
Source: © Albert Russ/Shutterstock.com

third forms more crosslinks to create three-dimensional frameworks. Chains of silicate groups compose the asbestos minerals, sheets give rise to talc and mica, and frameworks occur in feldspar and quartz (Figure 14.15).

2. *Silicone polymers.* Unlike the naturally occurring silicates, silicone polymers are manufactured substances that consist of alternating Si and O atoms with two organic groups also bonded to each Si atom in a very long Si—O chain, as in *poly(dimethyl siloxane):*

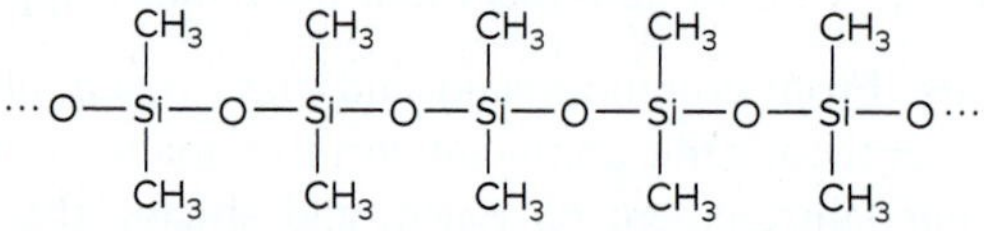

Silicones have properties of both plastics and minerals. The organic groups give them the flexibility and weak intermolecular forces between chains that are characteristic of a plastic, while the O—Si—O backbone confers the thermal stability and nonflammability of a mineral. Structural categories similar to those of the silicates can be created by adding various reactants to form silicone chains, sheets, and frameworks. Chains are oily liquids used as lubricants and as components of car polish and makeup. Sheets are components of gaskets, space suits, and contact lenses. Frameworks have uses as laminates on circuit boards, in nonstick cookware, and in artificial skin and bone.

Diagonal Relationships: Boron and Silicon

Our final diagonal relationship occurs between the semiconducting metalloids boron and silicon. Both B and Si and their mineral oxoanions—borates and silicates—occur in extended covalent networks. Both boric acid [$B(OH)_3$] and silicic acid [$Si(OH)_4$] are weakly acidic solids that occur as layers held together by widespread H bonding. Their hydrides—the compact boranes and the extended silanes—are both flammable, low-melting compounds that act as reducing agents.

14.7 GROUP 5A(15): THE NITROGEN FAMILY

The first two elements of Group 5A(15), gaseous nonmetallic nitrogen (N) and solid nonmetallic phosphorus (P), play major roles in both nature and industry. Below these nonmetals are two metalloids, arsenic (As) and antimony (Sb), followed by the metal bismuth (Bi), and moscovium (Mc), synthesized in 2003. The Group 5A(15) Family Portrait provides an overview.

KEY ATOMIC PROPERTIES, PHYSICAL PROPERTIES, AND REACTIONS

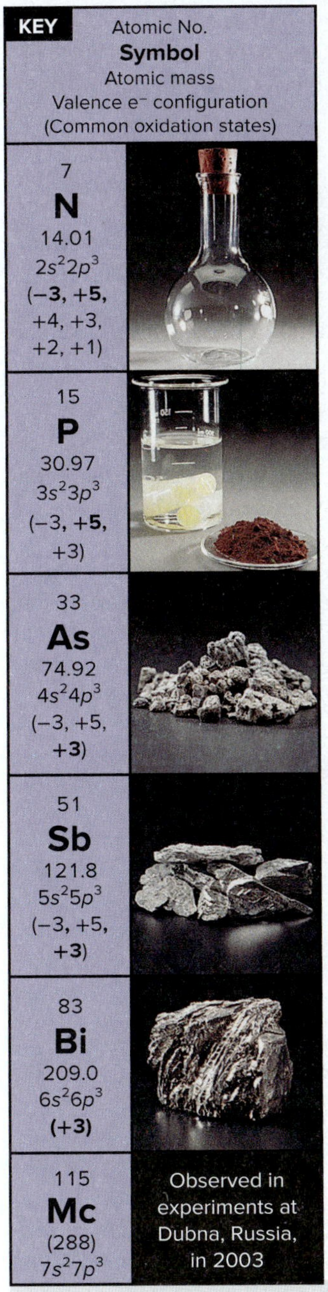

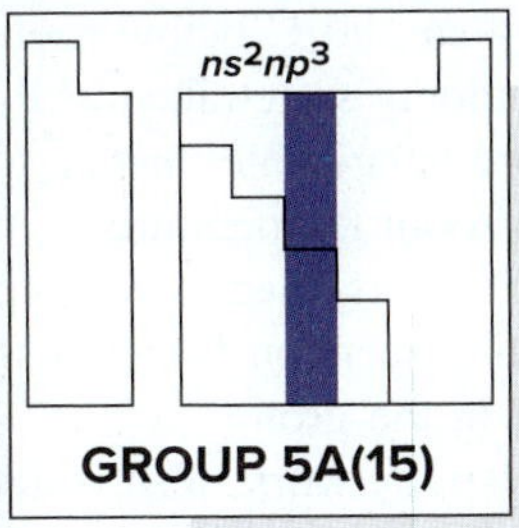

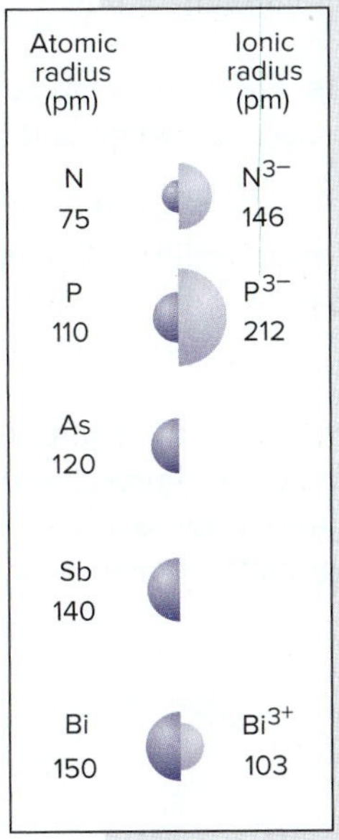

Atomic Properties

Group electron configuration is ns^2np^3. The *np* sublevel is half-filled, with each *p* orbital containing one electron (parallel spin). The number of oxidation states decreases down the group, and the lower (+3) state becomes more common. Atomic properties follow generally expected trends. The large (~50%) increase in size from N to P correlates with the much lower IE and EN of P.

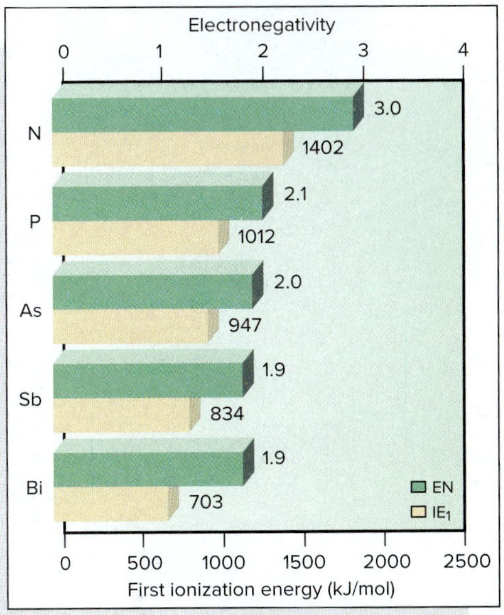

Physical Properties

Physical properties reflect the change from individual molecules (N, P) to network covalent solid (As, Sb) to metal (Bi). Thus, melting points increase and then decrease. Large atomic size and low atomic mass result in low density. Because mass increases more than size down the group, the density of the elements as solids increases. The dramatic increase in density from P to As is due to the intervening transition elements.

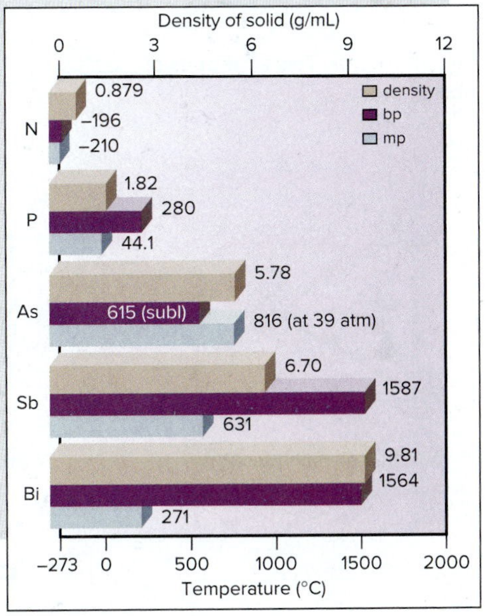

Source: © McGraw-Hill Education/Stephen Frisch, photographer

Reactions

1. Nitrogen is "fixed" industrially in the Haber process:

$$N_2(g) + 3H_2(g) \rightleftharpoons 2NH_3(g)$$

Further reactions convert NH_3 to NO, NO_2, and HNO_3 *(see text)*. Hydrides of some other group members are formed from reaction in water (or with H_3O^+) of a metal phosphide, arsenide, and so forth:

$$Ca_3P_2(s) + 6H_2O(l) \longrightarrow 2PH_3(g) + 3Ca(OH)_2(aq)$$

2. Halides are formed by direct combination of the elements:

$$2E(s) + 3X_2 \longrightarrow 2EX_3 \qquad (E = \text{all except N})$$

$$EX_3 + X_2 \longrightarrow EX_5$$ (E = all except N and Bi with X = F and Cl, but no $BiCl_5$; E = P for X = Br)

3. Oxoacids are formed from the halides in a reaction with water that is common to many nonmetal halides:

$$EX_3 + 3H_2O(l) \longrightarrow H_3EO_3(aq) + 3HX(aq)$$
(E = all except N)

$$EX_5 + 4H_2O(l) \longrightarrow H_3EO_4(aq) + 5HX(aq)$$
(E = all except N and Bi)

Note that the oxidation number of E does *not* change.

The Wide Range of Physical Behavior

Group 5A(15) displays the widest range of physical behavior we've seen so far because of large changes in bonding and intermolecular forces:

- *Nitrogen* is a gas consisting of N_2 molecules with such weak intermolecular forces that the element *boils* more than 200°C below room temperature.
- *Phosphorus* exists most commonly as tetrahedral P_4 molecules in the solid phase. Because P is heavier and more polarizable than N, it has stronger dispersion forces and melts about 25°C above room temperature.
- *Arsenic* consists of extended sheets. Each As atom bonds to three others in the sheet, and each sheet exhibits dispersion forces with adjacent sheets, which gives As the highest melting point in the group.
- *Antimony* has a similar covalent network, also resulting in a high melting point.
- *Bismuth* has metallic bonding and thus a lower melting point than As and Sb.

Two Allotropes of Phosphorus Phosphorus has several allotropes, which have very different properties. Two major ones are white and red phosphorus:

- *White phosphorus* consists of tetrahedral molecules (Figure 14.16A). It is a low-melting, whitish, waxy solid that is soluble in nonpolar solvents, such as CS_2. Each P atom uses its half-filled 3*p* orbitals to bond to the other three; with a small bond angle (60°), and thus weak P—P bonds that are easily broken, it is highly reactive (Figure 14.16B).
- *Red phosphorus* is formed by heating the white form in the absence of air. One of the P—P bonds in each tetrahedron breaks, and those 3*p* orbitals overlap with others to form chains of P_4 units (Figure 14.16C). The chains make the red allotrope much less reactive, high melting, and insoluble.

Figure 14.16 Two allotropes of phosphorus.

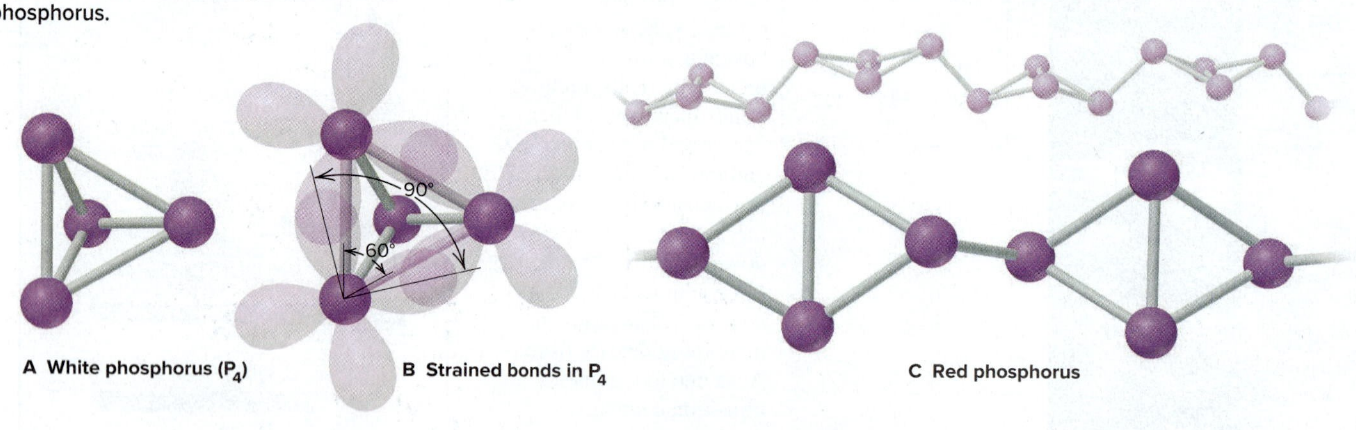

Patterns in Chemical Behavior

Many of the patterns we saw in Group 4A(14) appear here in the change from nonmetallic N to metallic Bi. The great majority of Group 5A(15) compounds have *covalent bonds*. Whereas N can form no more than four bonds, the next three members can form more by expanding their valence levels and using empty *d* orbitals.

1. *Formation of ions.* For a Group 5A(15) element to form an ion with a noble gas electron configuration, it must *gain* three electrons, the last two in endothermic steps. Nevertheless, the enormous lattice energy released when such highly charged anions attract cations drives their formation. However, the 3− anion of N occurs only in compounds with active metals, such as Li_3N and Mg_3N_2 (and the 3− anion of P may occur in $Na_3P_3{\cdot}5H_2O$). Metallic Bi mostly bonds covalently but exists as a cation in a few compounds, such as BiF_3 and $Bi(NO_3)_3{\cdot}5H_2O$, through *loss* of its three *p* valence electrons.

2. *Oxidation states and basicity of oxides.* As in Groups 3A(13) and 4A(14), the lower oxidation state becomes more prominent down the group: N exhibits every state

possible for a Group 5A(15) element, from +5 to −3; only the +5 and +3 states are common for P, As, and Sb; and +3 is the only common state of Bi. The oxides change from acidic to amphoteric to basic, reflecting the increase in the metallic character of the elements. In addition, the lower oxide of an element is more basic than the higher oxide, reflecting the greater ionic character of its E-to-O bonding.

3. *Formation of hydrides.* All the Group 5A(15) elements form gaseous hydrides of formula EH_3. Except for NH_3, these are extremely reactive and poisonous and are synthesized by reaction of a metal phosphide, arsenide, and so forth, which acts as a strong base in water or aqueous acid. For example,

$$Ca_3As_2(s) + 6H_2O(l) \longrightarrow 2AsH_3(g) + 3Ca(OH)_2(aq)$$

Ammonia is made industrially by direct combination of the elements at high pressure and moderately high temperature:

$$N_2(g) + 3H_2(g) \rightleftharpoons 2NH_3(g)$$

Nitrogen forms a second hydride, hydrazine, N_2H_4. Like NH_3, hydrazine is a weak base; it is used to make antituberculin drugs, plant growth regulators, and fungicides.

Molecular properties of the Group 5A(15) hydrides reveal some interesting bonding and structural patterns:

- Despite its much lower molar mass, NH_3 melts and boils at higher temperatures than the other 5A(15) hydrides, as a result of its *H bonding* (see Figure 12.15).
- Bond angles decrease from 107.3° for NH_3 to around 90° for the other hydrides, which suggests that the larger atoms use unhybridized *p* orbitals.
- E—H bond lengths increase down the group, so bond strength and thermal stability decrease: AsH_3 decomposes at 250°C, SbH_3 at 20°C, and BiH_3 at −45°C.

We'll see these features—H bonding for the smallest member, change in bond angles, change in bond energies—in the hydrides of Group 6A(16) as well.

4. *Types and properties of halides.* The Group 5A(15) elements all form trihalides (EX_3). All except nitrogen form pentafluorides (EF_5), but only a few other pentahalides (PCl_5, PBr_5, $AsCl_5$, and $SbCl_5$) are known. Nitrogen cannot form pentahalides because it cannot expand its valence level. Most trihalides are prepared by direct combination:

$$P_4(s) + 6Cl_2(g) \longrightarrow 4PCl_3(l)$$

The pentahalides form with excess halogen:

$$PCl_3(l) + Cl_2(g) \longrightarrow PCl_5(s)$$

As with the hydrides, thermal stability of the halides decreases as the E—X bond becomes longer; we see this trend easiest if we change the halogen. Among the nitrogen halides, for example, NF_3 is a stable, rather unreactive gas. NCl_3 is explosive and reacts rapidly with water. (The chemist who first prepared it lost three fingers and an eye!) NBr_3 can only be made below −87°C. NI_3 has never been prepared, but an ammoniated product ($NI_3 \cdot NH_3$) explodes at the slightest touch. Other Group 5A(15) members show less drastic trends, but stability decreases as the halogen gets larger.

5. *Reaction of halides in water.* In a reaction pattern *typical of many nonmetal halides,* each 5A(15) halide reacts with water to yield the hydrogen halide and the oxoacid, in which E has the *same* O.N. as in the original halide. For example, PX_5 (O.N. of P = +5) produces phosphoric acid (O.N. of P = +5) and HX:

$$PCl_5(s) + 4H_2O(l) \longrightarrow H_3PO_4(l) + 5HCl(g)$$

Highlights of Nitrogen Chemistry

The most striking highlight of nitrogen chemistry is the inertness of N_2. Even with four-fifths of the atmosphere being N_2 and the other fifth nearly all O_2, it takes the searing temperature caused by lightning to form significant amounts of nitrogen oxides. Although N_2 is inert at moderate temperatures, it reacts at high temperatures with H_2, Li, Group 2A(2) metals, B, Al, C, Si, Ge, and many transition elements. In fact, nearly every element forms bonds to N. Here we focus on the oxides and the oxoacids and their salts.

Nitrogen Oxides Nitrogen is remarkable for having six stable oxides, each with a *positive* enthalpy of formation because of the great strength of the N≡N bond (BE = 945 kJ/mol). Their structures and some properties are shown in Table 14.3. Unlike the hydrides and halides of nitrogen, the oxides are planar. Nitrogen displays all its positive oxidation states in these compounds, and in N_2O and N_2O_3, the two N atoms have different states. Let's highlight the three most important:

1. *Dinitrogen monoxide* (N_2O; also called *nitrous oxide*) is the dental anesthetic "laughing gas" and the propellant in canned whipped cream. It is a linear molecule with an electronic structure described by three resonance forms (note formal charges):

$$\overset{0}{:\text{N}}\equiv\overset{+1}{\text{N}}-\overset{-1}{\ddot{\underset{\cdot\cdot}{\text{O}}}}: \longleftrightarrow \overset{-1}{:\ddot{\text{N}}}=\overset{+1}{\text{N}}=\overset{0}{\ddot{\text{O}}}: \longleftrightarrow \overset{-2}{:\ddot{\underset{\cdot\cdot}{\text{N}}}}-\overset{+1}{\text{N}}\equiv\overset{+1}{\text{O}}:$$

most important ... least important

2. *Nitrogen monoxide* (NO; also called *nitric oxide*) is a molecule with an odd electron and with biochemical functions ranging from neurotransmission to control of blood flow. In Section 11.3, we used MO theory to explain its bonding. The commercial preparation of NO occurs as a first step in the production of nitric acid:

$$4NH_3(g) + 5O_2(g) \longrightarrow 4NO(g) + 6H_2O(g)$$

Nitrogen monoxide is also produced whenever air is heated to high temperatures, as in a car engine or by lightning during a thunderstorm:

$$N_2(g) + O_2(g) \xrightarrow{\text{high } T} 2NO(g)$$

Heating converts NO to two other oxides:

$$3NO(g) \xrightarrow{\Delta} N_2O(g) + NO_2(g)$$

Table 14.3 Structures and Properties of the Nitrogen Oxides

Formula	Name	Space-Filling Model	Lewis Structure	Oxidation State of N	ΔH_f° (kJ/mol) at 298 K	Comment
N_2O	Dinitrogen monoxide (nitrous oxide)		$:N\equiv N-\ddot{O}:$	+1 (0, +2)	82.0	Colorless gas; used as dental anesthetic ("laughing gas") and aerosol propellant
NO	Nitrogen monoxide (nitric oxide)		$:\dot{N}=\ddot{O}:$	+2	90.3	Colorless, paramagnetic gas; biochemical messenger; air pollutant
N_2O_3	Dinitrogen trioxide	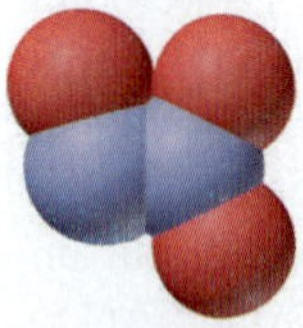	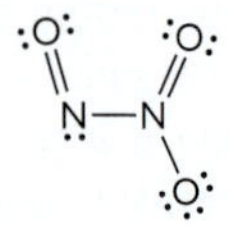	+3 (+2, +4)	83.7	Reddish-brown gas (reversibly dissociates to NO and NO_2)
NO_2	Nitrogen dioxide		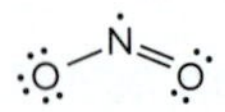	+4	33.2	Orange-brown, paramagnetic gas formed during HNO_3 manufacture; poisonous air pollutant
N_2O_4	Dinitrogen tetroxide	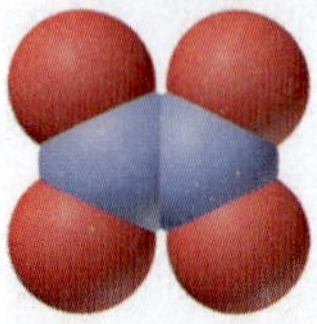	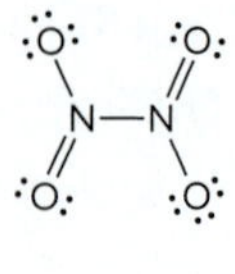	+4	9.16	Colorless to yellow liquid (reversibly dissociates to NO_2)
N_2O_5	Dinitrogen pentoxide	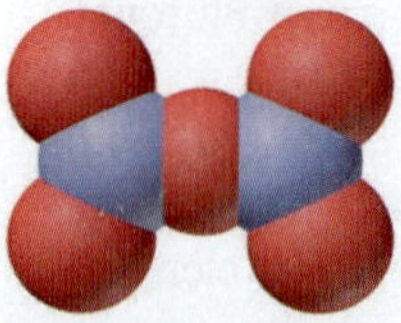	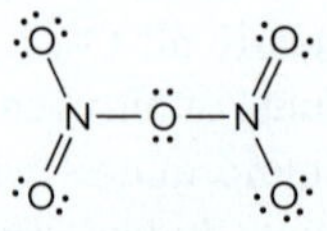	+5	11.3	Colorless, volatile solid consisting of NO_2^+ and NO_3^-; gas consists of N_2O_5 molecules

This type of redox reaction is called a **disproportionation,** in which *one substance acts as both the oxidizing and reducing agents*. In the process, an atom with an intermediate oxidation state in the reactant occurs in both lower and higher states in the products: the oxidation state of N in NO (+2) is intermediate between that in N_2O (+1) and that in NO_2 (+4).

3. *Nitrogen dioxide* (NO_2), a brown poisonous gas, forms to a small extent when NO reacts with additional oxygen:

$$2NO(g) + O_2(g) \rightleftharpoons 2NO_2(g)$$

Like NO, NO_2 is a molecule with an odd electron, which is more localized on the N atom in this species. Thus, NO_2 dimerizes reversibly to *dinitrogen tetroxide:*

$$O_2N\cdot(g) + \cdot NO_2(g) \rightleftharpoons O_2N{-}NO_2(g) \quad (\text{or } N_2O_4)$$

Thunderstorms form NO and NO_2 and carry them down to the soil, where they act as natural fertilizers. In urban traffic, however, their formation leads to *photochemical smog (see photo)* in a series of reactions also involving sunlight, ozone (O_3), unburned gasoline, and various other species.

Photochemical smog over Los Angeles, California.

Source: © Daniel Stein/Getty Images RF

Nitrogen Oxoacids and Oxoanions There are two common oxoacids of nitrogen (Figure 14.17):

1. *Nitric acid* (HNO_3) is produced in the *Ostwald process;* we have already seen the first two steps—the oxidations of NH_3 to NO and of NO to NO_2. The final step is a disproportionation, as the oxidation numbers show:

$$3\overset{+4}{N}O_2(g) + H_2O(l) \longrightarrow 2H\overset{+5}{N}O_3(aq) + \overset{+2}{N}O(g)$$

The NO is recycled to make more NO_2.

In nitric acid, as in all oxoacids, *the acidic H is attached to one of the O atoms.* When the proton is lost, the trigonal planar nitrate ion is formed (Figure 14.17A). In the laboratory, nitric acid is used as a strong oxidizing acid. The products of its reactions with metals vary with the metal's reactivity and the acid's concentration. In the following examples, notice that *the NO_3^- ion is the oxidizing agent.* Nitrate ion that is not reduced is a spectator ion and does not appear in the net ionic equations.

- With an active metal, such as Al, and dilute nitric acid, N is reduced from the +5 state all the way to the −3 state in the ammonium ion, NH_4^+:

$$8Al(s) + 30HNO_3(aq;\ 1\ M) \longrightarrow 8Al(NO_3)_3(aq) + 3NH_4NO_3(aq) + 9H_2O(l)$$
$$8Al(s) + 30H^+(aq) + 3NO_3^-(aq) \longrightarrow 8Al^{3+}(aq) + 3NH_4^+(aq) + 9H_2O(l)$$

- With a less reactive metal, such as Cu, and more concentrated acid, N is reduced to the +2 state in NO:

$$3Cu(s) + 8HNO_3(aq;\ 3\text{ to }6\ M) \longrightarrow 3Cu(NO_3)_2(aq) + 4H_2O(l) + 2NO(g)$$
$$3Cu(s) + 8H^+(aq) + 2NO_3^-(aq) \longrightarrow 3Cu^{2+}(aq) + 4H_2O(l) + 2NO(g)$$

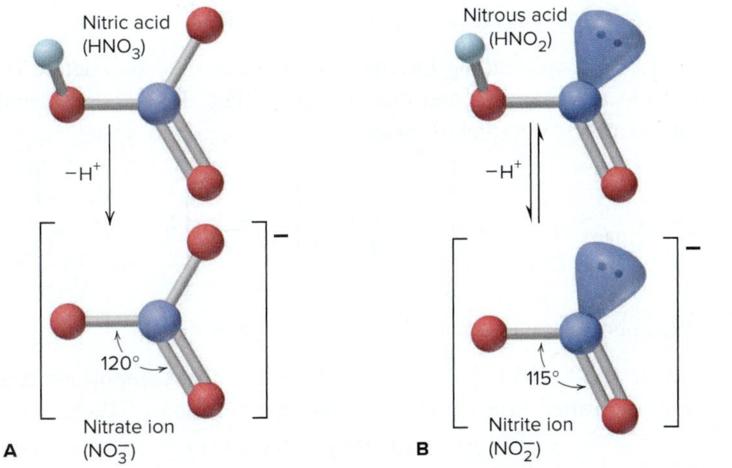

Figure 14.17 The structures of nitric and nitrous acids and their oxoanions.

- With still more concentrated acid, N is reduced only to the +4 state in NO_2:

$$Cu(s) + 4HNO_3(aq;\ 12\ M) \longrightarrow Cu(NO_3)_2(aq) + 2H_2O(l) + 2NO_2(g)$$
$$Cu(s) + 4H^+(aq) + 2NO_3^-(aq) \longrightarrow Cu^{2+}(aq) + 2H_2O(l) + 2NO_2(g)$$

Nitrates form when HNO_3 reacts with metals or with their hydroxides, oxides, or carbonates. *All nitrates are soluble in water.*

2. *Nitrous acid* (HNO_2) is a much weaker acid that forms when metal nitrites are treated with a strong acid:

$$NaNO_2(aq) + HCl(aq) \longrightarrow HNO_2(aq) + NaCl(aq)$$

This acid forms the planar nitrite ion (Figure 14.17B) in which nitrogen's lone pair reduces the ideal 120° bond angle to 115°.

These two acids reveal a *general pattern in relative acid strength among oxoacids: the more O atoms bonded to the central nonmetal, the stronger the acid.* The O atoms pull electron density from the N atom, which in turn pulls electron density from the O of the O—H bond, facilitating the release of the H^+ ion. The O atoms also stabilize the resulting oxoanion by delocalizing its negative charge. The same pattern occurs in the oxoacids of sulfur and the halogens; we'll discuss the pattern quantitatively in Chapter 18.

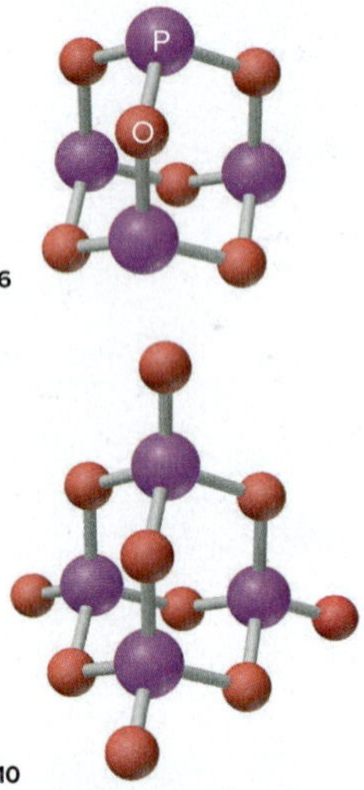

A P_4O_6

B P_4O_{10}

Figure 14.18 Important oxides of phosphorus.

Highlights of Phosphorus Chemistry

Like nitrogen, phosphorus forms important oxides (although not as many) and oxoacids. Here we focus on these compounds as well as on some other important phosphorus compounds.

Phosphorus Oxides Phosphorus forms two important oxides:

1. *Tetraphosphorus hexoxide* (P_4O_6) has P in its +3 oxidation state. It forms when P_4 reacts with limited oxygen:

$$P_4(s) + 3O_2(g) \longrightarrow P_4O_6(s)$$

P_4O_6 has the same tetrahedral orientation of the P atoms as in P_4, with an O atom between each pair of P atoms (Figure 14.18A).

2. *Tetraphosphorus decoxide* (P_4O_{10}) has P in the +5 oxidation state. Commonly known as "phosphorus pentoxide" from the empirical formula (P_2O_5), it forms when P_4 burns in excess O_2:

$$P_4(s) + 5O_2(g) \longrightarrow P_4O_{10}(s)$$

Its structure is that of P_4O_6 with another O atom bonded to each of the four corner P atoms (Figure 14.18B). P_4O_{10} is a powerful drying agent.

Phosphorus Oxoacids and Oxoanions The two common phosphorus oxoacids are phosphorous acid (note the different spelling) and phosphoric acid.

1. *Phosphorous acid* (H_3PO_3) is formed when P_4O_6 reacts with water:

$$P_4O_6(s) + 6H_2O(l) \longrightarrow 4H_3PO_3(l)$$

The formula H_3PO_3 is misleading because the acid has only two acidic H atoms; the third is bonded to the central P and does not dissociate. Phosphorous acid is a weak acid in water but reacts completely in two steps with excess strong base:

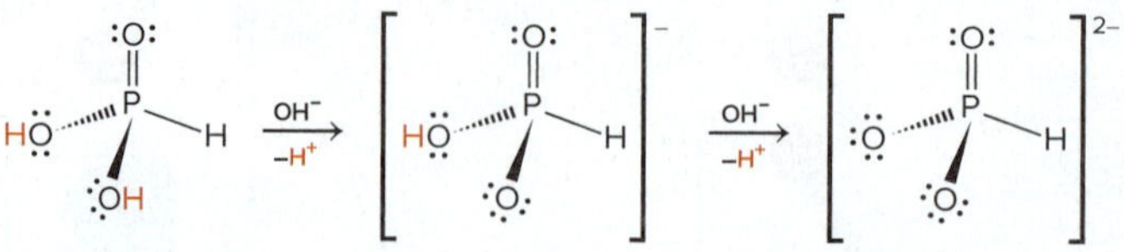

Salts of phosphorous acid contain the phosphite ion, HPO_3^{2-}.

2. *Phosphoric acid* (H_3PO_4), one of the "top-10" most important compounds in manufacturing, is formed in a vigorous exothermic reaction of P_4O_{10} with water:

$$P_4O_{10}(s) + 6H_2O(l) \longrightarrow 4H_3PO_4(l)$$

The presence of many H bonds makes pure H_3PO_4 syrupy, more than 75 times as viscous as water. The laboratory-grade concentrated acid is an 85% by mass aqueous solution. H_3PO_4 is a weak triprotic acid; in water, it loses one proton:

$$H_3PO_4(l) + H_2O(l) \rightleftharpoons H_2PO_4^-(aq) + H_3O^+(aq)$$

In excess strong base, however, the three protons dissociate completely in three steps to give the three phosphate oxoanions:

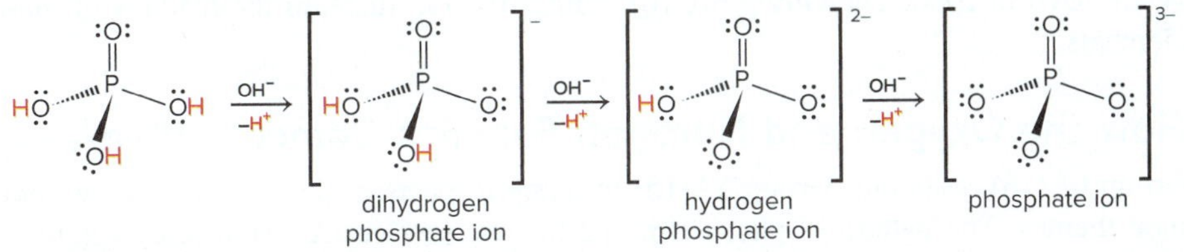

Phosphoric acid has a central role in fertilizer production and is also added to soft drinks for tartness. The various phosphate salts also have numerous essential applications. Na_3PO_4 is a paint stripper and grease remover. K_3PO_4 is used to stabilize latex for synthetic rubber, and K_2HPO_4 is a radiator corrosion inhibitor. Ammonium phosphates are used as fertilizers and as flame retardants on curtains, and calcium phosphates are used in baking powders and toothpastes, as mineral supplements in livestock feed, and as fertilizers.

Polyphosphates When they are heated, hydrogen phosphates lose water and form P—O—P linkages in compounds called *polyphosphates.* A reaction in which an H_2O molecule is lost for every pair of groups that join is called a **dehydration-condensation;** this type of reaction occurs frequently in the formation of polyoxoanion chains and other polymeric structures, both synthetic and natural. For example, sodium diphosphate, $Na_4P_2O_7$, is prepared by heating sodium hydrogen phosphate:

$$2Na_2HPO_4(s) \xrightarrow{\Delta} Na_4P_2O_7(s) + H_2O(g)$$

The diphosphate ion, $P_2O_7^{4-}$, the smallest of the polyphosphates, consists of two PO_4 units linked through a common oxygen corner (Figure 14.19A). Its reaction with water, the reverse of the previous reaction, generates heat:

$$P_2O_7^{4-}(aq) + H_2O(l) \longrightarrow 2HPO_4^-(aq) + heat$$

A similar process is put to vital use by organisms, when a third PO_4 unit linked to diphosphate creates the triphosphate grouping, part of the all-important high-energy biomolecule adenosine triphosphate (ATP). In Chapters 20 and 21, we discuss the central role of ATP in biological energy production. Extended polyphosphate chains consist of many tetrahedral PO_4 units (Figure 14.19B) and are structurally similar to silicate chains.

Phosphorus Compounds with Sulfur and Nitrogen Phosphorus forms many sulfides and nitrides. P_4S_3 is used in "strike anywhere" match heads, and P_4S_{10} is used in the manufacture of organophosphorus pesticides, such as malathion. Compounds of phosphorus and nitrogen called *polyphosphazenes* have properties similar to those of silicones. The —$(R_2)P{=}N$— unit is isoelectronic with the silicone unit, —$(R_2)Si{-}O$—. Sheets, films, fibers, and foams of polyphosphazene are water repellent, flame resistant, solvent resistant, and flexible at low temperatures—perfect for the gaskets and O-rings in spacecraft and polar vehicles.

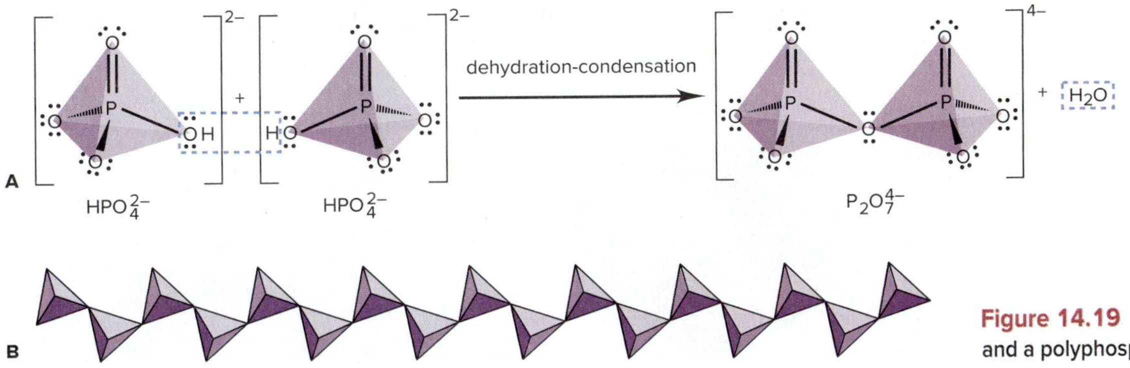

Figure 14.19 The diphosphate ion (A) and a polyphosphate chain (B).

14.8 GROUP 6A(16): THE OXYGEN FAMILY

The first two members of this group—gaseous nonmetallic oxygen (O) and solid nonmetallic sulfur (S)—are among the most important elements in industry, the environment, and living things. Two metalloids, selenium (Se) and tellurium (Te), appear next, followed by the radioactive metal polonium (Po), and finally livermorium (Lv), synthesized in 2000. The Group 6A(16) Family Portrait displays the features of these elements.

How the Oxygen and Nitrogen Families Compare Physically

Group 6A(16) resembles Group 5A(15) in many respects, so let's look at some common themes. The pattern of physical properties we saw in 5A(15) appears again:

- *Oxygen,* like nitrogen, occurs as a low-boiling diatomic gas.
- *Sulfur,* like phosphorus, occurs as a polyatomic molecular solid.
- *Selenium,* like arsenic, commonly occurs as a gray metalloid.
- *Tellurium,* like antimony, is slightly more metallic but still displays network covalent bonding.
- *Polonium,* like bismuth, has a metallic crystal structure.

As in Group 5A(15), electrical conductivities increase steadily down Group 6A(16) as bonding changes from nonmetal molecules (insulators) to metalloid networks (semiconductors) to a metallic solid (conductor).

Allotropism in the Oxygen Family Allotropism is even more common in Group 6A(16) than in Group 5A(15).

1. *Oxygen.* Oxygen has two allotropes: life-giving dioxygen (O_2), and poisonous triatomic ozone (O_3). Dioxygen is colorless, odorless, paramagnetic, and thermally stable. In contrast, ozone is bluish, has a pungent odor, is diamagnetic, and decomposes in heat and especially in ultraviolet (UV) light:

$$2O_3(g) \xrightarrow{\text{UV}} 3O_2(g)$$

This ability to absorb high-energy photons makes stratospheric ozone vital to life. A thinning of the ozone layer, observed above the North Pole and especially the South Pole, means that more UV light is reaching Earth's surface, with potentially hazardous effects. (We'll discuss the chemical causes of ozone depletion in Chapter 16.)

2. *Sulfur.* Sulfur is the allotrope "champion" of the periodic table, with more than 10 forms. The S atom's ability to bond to other S atoms (catenate) creates rings and chains, many with S—S bond lengths that range from 180 pm to 260 pm and bond angles ranging from 90° to 180°. The most stable allotrope is orthorhombic α-S_8, which consists of a crown-shaped ring of eight atoms called *cyclo-*S_8 (Figure 14.20); all other S allotropes eventually revert to this one.

3. *Selenium.* Selenium also has several allotropes, some consisting of crown-shaped Se_8 molecules. Gray Se is composed of layers of helical chains. When molten glass, cadmium sulfide, and gray Se are mixed and heated in the absence of air, a ruby-red glass forms, which is still used in traffic lights. The ability of gray Se to

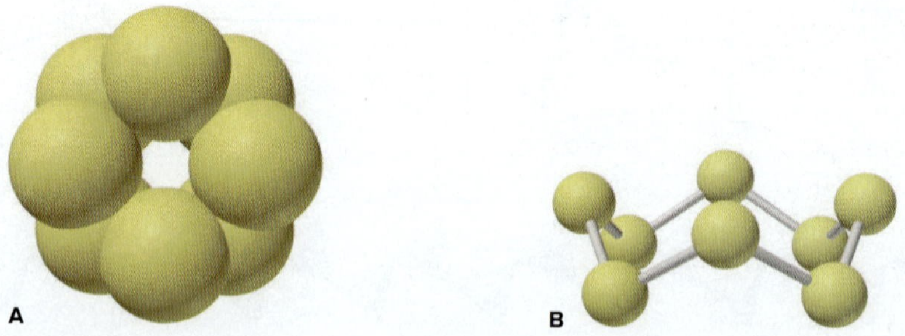

Figure 14.20 **The cyclo-S_8 molecule.** **A,** Top view of a space-filling model. **B,** Side view of a ball-and-stick model; note the crownlike shape.

KEY ATOMIC PROPERTIES, PHYSICAL PROPERTIES, AND REACTIONS

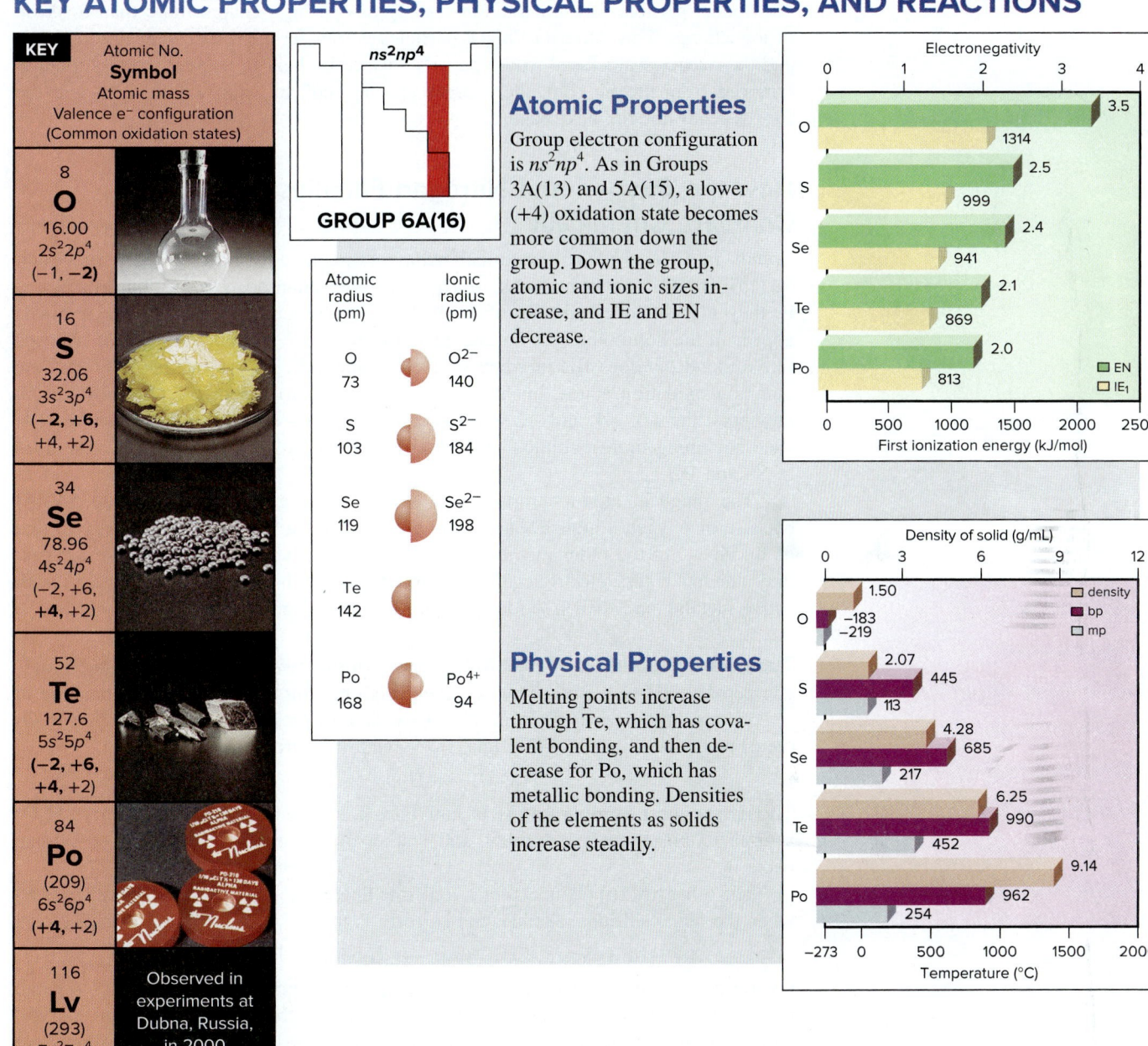

KEY
Atomic No.
Symbol
Atomic mass
Valence e⁻ configuration
(Common oxidation states)

8
O
16.00
$2s^22p^4$
(−1, **−2**)

16
S
32.06
$3s^23p^4$
(**−2, +6,** +4, +2)

34
Se
78.96
$4s^24p^4$
(−2, +6, **+4,** +2)

52
Te
127.6
$5s^25p^4$
(**−2, +6,** **+4,** +2)

84
Po
(209)
$6s^26p^4$
(**+4,** +2)

116
Lv
(293)
$7s^27p^4$
Observed in experiments at Dubna, Russia, in 2000

Source: © McGraw-Hill Education/Stephen Frisch, photographer

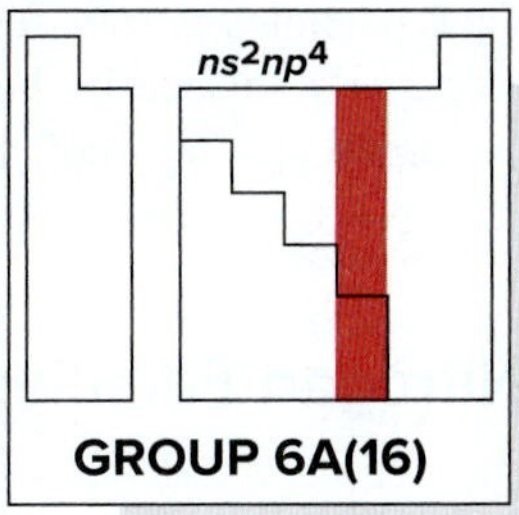

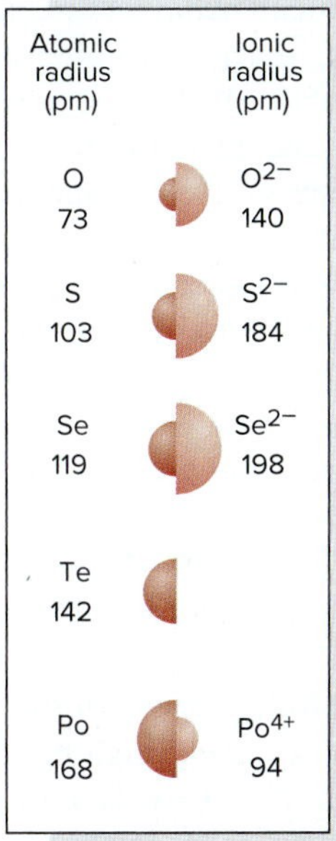

Atomic Properties

Group electron configuration is ns^2np^4. As in Groups 3A(13) and 5A(15), a lower (+4) oxidation state becomes more common down the group. Down the group, atomic and ionic sizes increase, and IE and EN decrease.

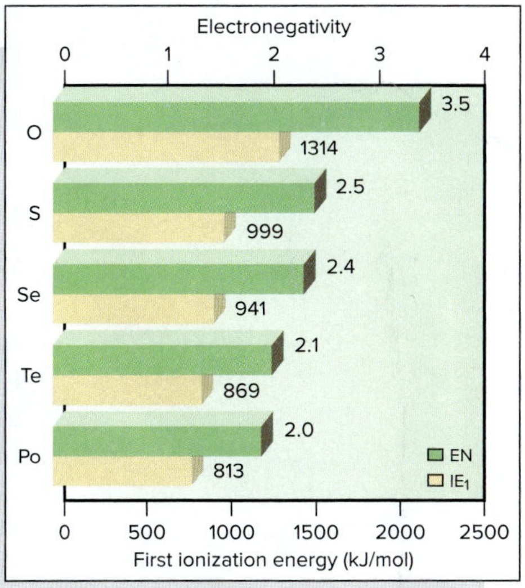

Physical Properties

Melting points increase through Te, which has covalent bonding, and then decrease for Po, which has metallic bonding. Densities of the elements as solids increase steadily.

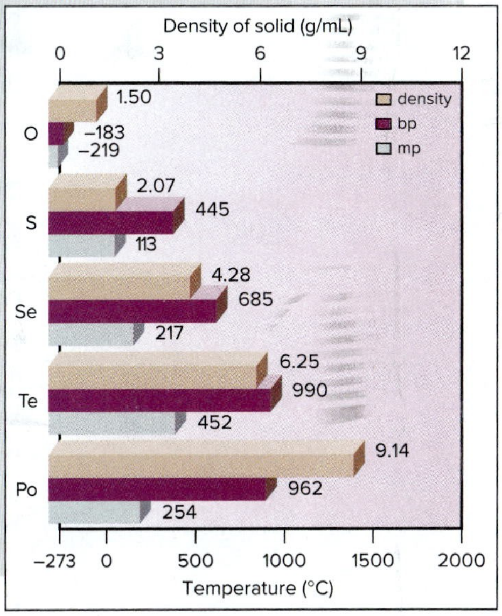

Reactions

1. Halides are formed by direct combination:

$$E(s) + X_2(g) \longrightarrow \text{various halides} \qquad (E = S, Se, Te; X = F, Cl)$$

2. The other elements in the group are oxidized by O_2:

$$E(s) + O_2(g) \longrightarrow EO_2 \qquad (E = S, Se, Te, Po)$$

SO_2 is oxidized further, and the product is used in the final step of H_2SO_4 manufacture *(see text):*

$$2SO_2(g) + O_2(g) \longrightarrow 2SO_3(g)$$

3. The thiosulfate ion is formed when an alkali metal sulfite reacts with sulfur, as in the preparation of "hypo" (photographers' developing solution):

$$S_8(s) + 8Na_2SO_3(s) \longrightarrow 8Na_2S_2O_3(aq)$$

conduct an electric current when illuminated is applied in photocopying. A film of amorphous Se is deposited on an aluminum drum and electrostatically charged. Exposure to a document produces an "image" of low and high positive charges corresponding to the document's bright and dark areas. Negatively charged black, dry ink (toner) particles are attracted to the regions of high charge more than to those of low charge. This pattern of black particles is transferred electrostatically to paper, and the particles are fused to the paper's surface by heat or solvent. Excess toner is removed from the Se film, the charges are "erased" by exposure to light, and the film is ready for the next page.

How the Oxygen and Nitrogen Families Compare Chemically

Trends in Group 6A(16) chemical behavior are also similar to those in Group 5A(15). O and S occur as anions much more often than do N and P, but like N and P, they also bond covalently with almost every other nonmetal. Covalent bonds appear in the compounds of Se and Te (as in those of As and Sb), and Po behaves like a metal (as does Bi) in some saltlike compounds. In contrast to N, O has few common oxidation states, but the earlier pattern returns with the other Group 6A members: the +6, +4, and −2 states occur most often, with the lower positive (+4) state becoming more common in Te and Po [as the lower positive (+3) state does in Sb and Bi].

The range in atomic properties is wider in this group than in Group 5A(15) because of oxygen's high EN (3.5) and great oxidizing strength, second only to that of fluorine. But the other members of Group 6A(16) behave very little like oxygen: they are much less electronegative, form anions much less often (S^{2-} occurs with active metals), and their hydrides exhibit no H bonding.

Types and Properties of Hydrides Oxygen forms two hydrides, water and hydrogen peroxide (H_2O_2). Both have relatively high boiling points and viscosities due to H bonding. In peroxides, O is in the −1 oxidation state, midway between that in O_2 (zero) and that in oxides (−2); thus, H_2O_2 readily disproportionates:

$$H_2O_2(l) \longrightarrow H_2O(l) + \tfrac{1}{2}O_2(g)$$

Though its most familiar use is in hair bleach and disinfectants, much more H_2O_2 is used to bleach paper, textiles, and leather and in sewage treatment to oxidize bacteria.

The other 6A(16) elements form foul-smelling, poisonous, gaseous hydrides (H_2E) when the metal sulfide, selenide, and so forth is treated with acid. For example,

$$FeSe(s) + 2HCl(aq) \longrightarrow H_2Se(g) + FeCl_2(aq)$$

Hydrogen sulfide also forms naturally in swamps from the breakdown of organic matter. It is as toxic as HCN, and even worse, it anesthetizes your olfactory nerves, so that as its concentration increases, you smell it less! And the other hydrides are about 100 times *more* toxic.

In their bonding and thermal stability, these Group 6A(16) hydrides have several features in common with those of Group 5A(15):

- Only water and H_2O_2 can form H bonds, so these substances melt and boil much higher than the other H_2E compounds (see Figure 12.15).
- Bond angles drop from the nearly tetrahedral 104.5° for H_2O to around 90° for the larger 6A(16) hydrides, suggesting that the central atom uses unhybridized *p* orbitals.
- E—H bond length increases, so bond energy decreases down the group. Thus, H_2Te decomposes above 0°C, and H_2Po can be made only in extreme cold because thermal energy from radioactive Po decomposes it. Another result of longer (weaker) bonds is that the 6A(16) hydrides are acids in water, and their acidity increases from H_2S to H_2Po.

Types and Properties of Halides Except for O, the Group 6A(16) elements form a wide range of halides, whose structure and reactivity patterns depend on the *sizes of the central atom and the surrounding halogens:*

- Sulfur forms many fluorides, a few chlorides, and one bromide, but no stable iodides.
- As the central atom becomes larger, the halides become more stable. Thus, tetrachlorides and tetrabromides of Se, Te, and Po are known, as are tetraiodides of Te and Po. Hexafluorides are known only for S, Se, and Te.

The inverse relationship between bond length and bond strength that we've seen previously does not account for this pattern. Rather, it is based on the effect of electron repulsions due to crowding of lone pairs and halogen (X) atoms around the central Group 6A(16) atom. With S, the larger X atoms would be too crowded, which explains why sulfur iodides do not occur. With increasing size of E, and therefore increasing length of E—X bonds, however, lone pairs and X atoms do not crowd each other as much, so a greater number of stable halides form.

Highlights of Oxygen Chemistry: Range of Oxide Properties

Oxygen is the most abundant element on Earth's surface, occurring both as the free element and in innumerable oxides, silicates, carbonates, and phosphates, as well as in water. Virtually all free O_2 has been formed for billions of years by photosynthetic algae and multicellular plants in an overall equation that only looks simple:

$$n\text{H}_2\text{O}(l) + n\text{CO}_2(g) \xrightarrow{\text{light}} n\text{O}_2(g) + (\text{CH}_2\text{O})_n \text{ (carbohydrates)}$$

The reverse process occurs during combustion and respiration. Through these O_2-forming and O_2-utilizing processes, the 1.5×10^9 km^3 of water on Earth is, on average, used and remade every 2 million years!

Every element (except He, Ne, and Ar) forms at least one oxide, many by direct combination. A spectrum of properties characterizes these compounds. Some oxides are gases that condense at very low temperatures, such as CO (bp = −192°C); others are solids that melt at extremely high temperatures, such as BeO (mp = 2530°C). Oxides cover the full range of conductivity: insulators (MgO), semiconductors (NiO), conductors (ReO_3), and superconductors ($YBa_2Cu_3O_7$). They may be thermally stable (CaO) or unstable (HgO), as well as chemically reactive (Li_2O) or inert (Fe_2O_3).

Another useful way to classify element oxides is by their acid-base properties. The oxides of Group 6A(16) exhibit expected trends in acidity, with SO_3 [the higher (+6) oxide] the most acidic and PoO_2 [the lower (+4) oxide] the most basic.

Highlights of Sulfur Chemistry

Like phosphorus, sulfur forms two common oxides and two oxoacids, one of which is essential to many industries. There are also several important metal sulfides.

Sulfur Oxides Sulfur forms two important oxides.

1. *Sulfur dioxide* (SO_2) has S in its +4 oxidation state. It is a colorless, choking gas that forms when S, H_2S, or a metal sulfide burns in air:

$$2\text{H}_2\text{S}(g) + 3\text{O}_2(g) \longrightarrow 2\text{H}_2\text{O}(g) + 2\text{SO}_2(g)$$
$$4\text{FeS}_2(s) + 11\text{O}_2(g) \longrightarrow 2\text{Fe}_2\text{O}_3(s) + 8\text{SO}_2(g)$$

2. *Sulfur trioxide* (SO_3), which has S in the +6 oxidation state, is produced when SO_2 reacts in O_2. A catalyst (Chapter 16) must be used to speed up this very slow reaction. For the production of sulfuric acid, a vanadium(V) oxide catalyst is used:

$$\text{SO}_2(g) + \tfrac{1}{2}\text{O}_2(g) \xrightarrow{\text{V}_2\text{O}_5/\text{K}_2\text{O catalyst}} \text{SO}_3(g)$$

Sulfur Oxoacids Sulfur forms two important oxoacids.

1. *Sulfurous acid* (H_2SO_3), formed when SO_2 dissolves in water, exists in equilibrium with hydrated SO_2 rather than as stable H_2SO_3 molecules:

$$\text{SO}_2(aq) + \text{H}_2\text{O}(l) \rightleftharpoons [\text{H}_2\text{SO}_3(aq)] \rightleftharpoons \text{H}^+(aq) + \text{HSO}_3^-(aq)$$

Sulfurous acid is weak and has two acidic protons, forming the hydrogen sulfite (bisulfite, HSO_3^-) and sulfite (SO_3^{2-}) ions with strong base. Because the S in SO_3^{2-} is in the +4 state and is easily oxidized to the +6 state, sulfites are good reducing agents and preserve foods and wine by eliminating undesirable products of air oxidation.

2. *Sulfuric acid* (H_2SO_4) is produced when SO_2 is oxidized catalytically to SO_3, which is then absorbed into concentrated H_2SO_4 and treated with H_2O:

$$SO_3 \text{ (in concentrated } H_2SO_4) + H_2O(l) \longrightarrow H_2SO_4(l)$$

With more than 60 million tons produced each year in the United States alone, H_2SO_4 ranks first among all industrial chemicals: it is vital to fertilizer production; metal, pigment, and textile processing; and soap and detergent manufacturing. (The production of H_2SO_4 will be discussed in detail in Chapter 22.)

Concentrated laboratory-grade sulfuric acid is a viscous, colorless liquid that is 98% H_2SO_4 by mass. Like other strong acids, H_2SO_4 dissociates completely in water, forming the hydrogen sulfate (or bisulfate) ion, a much weaker acid:

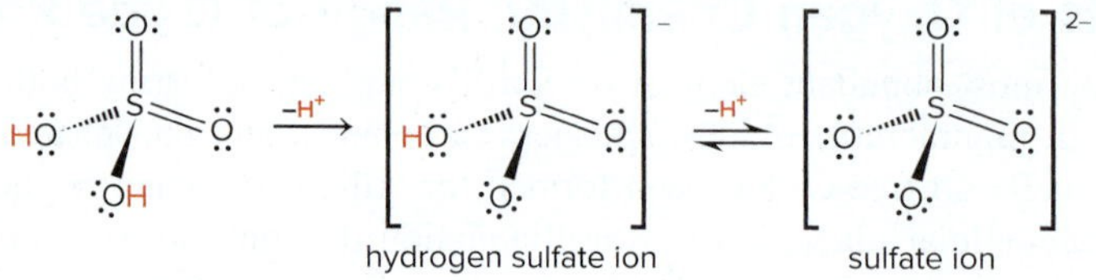

Most common hydrogen sulfates and sulfates are water soluble, but those of the larger Group 2A(2) members (Ca^{2+}, Sr^{2+}, Ba^{2+}, and Ra^{2+}), Ag^+, and Pb^{2+} are not.

Concentrated sulfuric acid is an excellent dehydrating agent. Its loosely held proton transfers to water in a highly exothermic formation of hydronium (H_3O^+) ions. This process can occur even when the reacting substance contains no free water. For example, H_2SO_4 dehydrates wood, natural fibers, and many other organic substances, such as table sugar $(CH_2O)_n$, by removing the components of water from the molecular structure, leaving behind a carbonaceous mass (Figure 14.21).

Figure 14.21 The dehydration of sugar by sulfuric acid.
Source: © McGraw-Hill Education/Stephen Frisch, photographer

Industrial sources produce the sulfur oxides that lead to acid rain.
Source: © Steve Cole/Getty Images RF

Sulfuric acid is one of the components of acid rain. Enormous amounts of SO_2 are emitted by coal-burning power plants, petroleum refineries, and metal-ore smelters *(see photo)*. In contact with H_2O, this SO_2 and its oxidation product, SO_3, form H_2SO_3 and H_2SO_4 in the atmosphere, which then fall in rain, snow, and dust on animals, plants, buildings, and lakes (see Chemical Connections, Chapter 19).

Metal Sulfides Many metals combine directly with S to form *metal sulfides*. Sulfide ores are mined for the extraction of many metals, including copper, zinc, lead, and silver. Aside from the sulfides of Groups 1A(1) and 2A(2), most metal sulfides do not have discrete S^{2-} ions. Several transition metals, such as chromium, iron, and nickel, form covalent, alloy-like, nonstoichiometric compounds with S, such as $Cr_{0.88}S$ and $Fe_{0.86}S$. Some important minerals contain S_2^{2-} ions; an example is iron pyrite, or "fool's gold" (FeS_2). We discuss the metallurgy of ores in Chapter 22.

14.9 GROUP 7A(17): THE HALOGENS

The halogens, the last elements of great reactivity, begin with fluorine (F), the strongest electron "grabber" of all. Chlorine (Cl, the most important halogen industrially), bromine (Br), and iodine (I) also form compounds with most elements, and even rare, radioactive astatine (At) is reactive; little is known of tennessine (Ts, first synthesized in 2010). The key features of this group are presented in the Group 7A(17) Family Portrait *(next page)*.

Physical Behavior of the Halogens

As expected from the increase in molar mass, the halogens display regular trends in their physical properties: melting and boiling points and heats of fusion and vaporization *increase* down Group 7A(17). The halogens exist as diatomic molecules that interact through dispersion forces, which *increase* in strength as the atoms become larger and, thus, more easily polarized. Even their color darkens with molar mass: F_2 is a very pale yellow gas, Cl_2 a yellow-green gas, Br_2 a brown-orange liquid, and I_2 a purple-black solid.

Why the Halogens Are So Reactive

The Group 7A(17) elements react with most metals and nonmetals to form many ionic and covalent compounds: metal and nonmetal halides, halogen oxides, and oxoacids. The main reason for halogen reactivity is the same as for alkali metal reactivity—an electron configuration one electron away from that of a noble gas. Whereas a Group 1A(1) metal atom must lose one electron to attain a filled outer level, *a Group 7A(17) nonmetal atom must gain one electron to fill its outer level.* It accomplishes this filling in either of two ways:

1. Gaining an electron from a metal atom, thus forming a negative ion and an ionic bond with the positive metal ion.
2. Sharing an electron pair with a nonmetal atom, thus forming a covalent bond.

Electronegativity and Bond Properties The halogens display the largest range in electronegativity of any group, but all are electronegative enough to behave as nonmetals. Down the group, reactivity reflects the decrease in electronegativity: F_2 is the most reactive and I_2 the least. The exceptional reactivity of elemental F_2 is also related to the weakness of the F—F bond. Although bond energy generally decreases as atomic size increases down the group (Figure 14.22), F_2 deviates from this trend. The short F—F bond is weaker than expected because lone pairs on each small F atom repel those on the other. As a result of these factors, F_2 reacts with every element (except He, Ne, and Ar), in many cases, explosively.

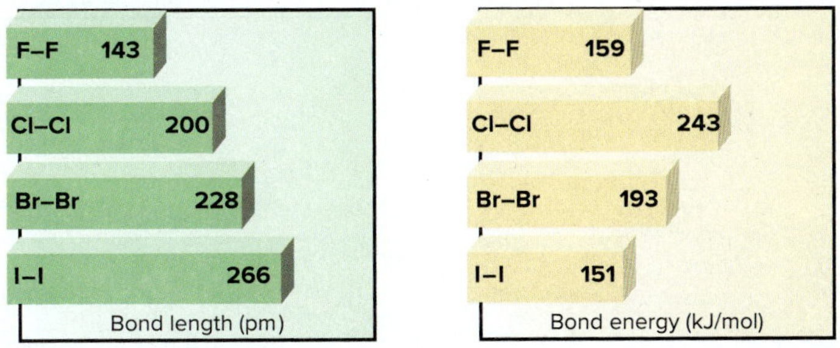

Figure 14.22 Bond energies and bond lengths of the halogens.

Redox Behavior The halogens act as *oxidizing agents* in the majority of their reactions, and halogens higher in the group can oxidize halide ions lower down:

$$F_2(g) + 2X^-(aq) \longrightarrow 2F^-(aq) + X_2(aq) \qquad (X = Cl, Br, I)$$

Thus, the oxidizing ability of X_2 increases *up* the group: the higher the EN, the more strongly each X atom pulls electrons away. Similarly, the reducing ability of

FAMILY PORTRAIT

Group 7A(17): The Halogens

KEY ATOMIC PROPERTIES, PHYSICAL PROPERTIES, AND REACTIONS

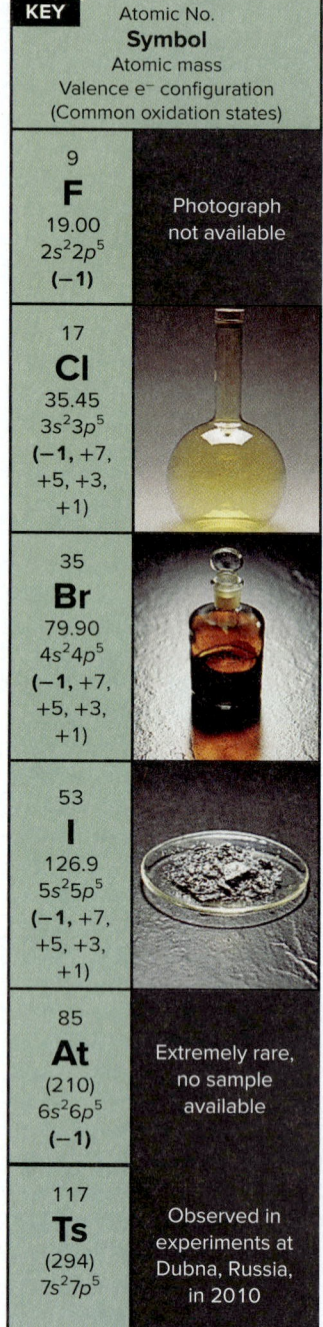

KEY: Atomic No. / Symbol / Atomic mass / Valence e⁻ configuration / (Common oxidation states)	
9 **F** 19.00 $2s^22p^5$ **(−1)**	Photograph not available
17 **Cl** 35.45 $3s^23p^5$ (**−1**, +7, +5, +3, +1)	
35 **Br** 79.90 $4s^24p^5$ (**−1**, +7, +5, +3, +1)	
53 **I** 126.9 $5s^25p^5$ (**−1**, +7, +5, +3, +1)	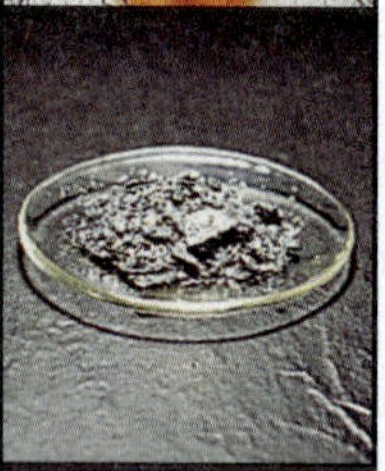
85 **At** (210) $6s^26p^5$ **(−1)**	Extremely rare, no sample available
117 **Ts** (294) $7s^27p^5$	Observed in experiments at Dubna, Russia, in 2010

Source: © McGraw-Hill Education/Stephen Frisch, photographer

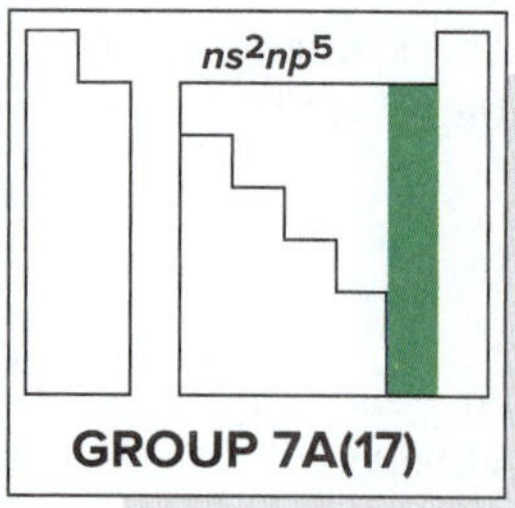

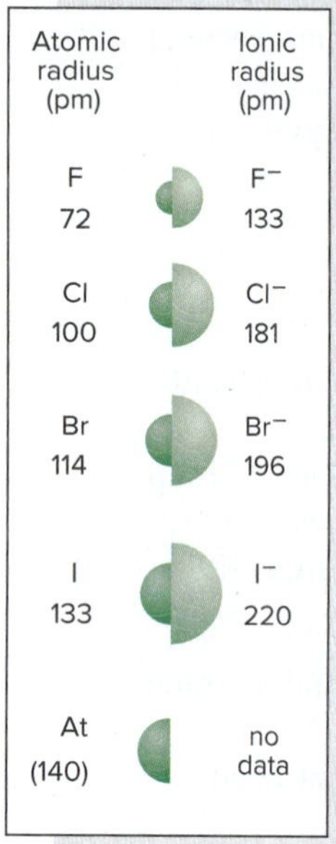

Atomic Properties

Group electron configuration is ns^2np^5; elements lack one electron to complete their outer level. The −1 oxidation state is the most common for all members. Except for F, the halogens exhibit all odd-numbered states (+7 through −1). Down the group, atomic and ionic sizes increase steadily, as IE and EN decrease.

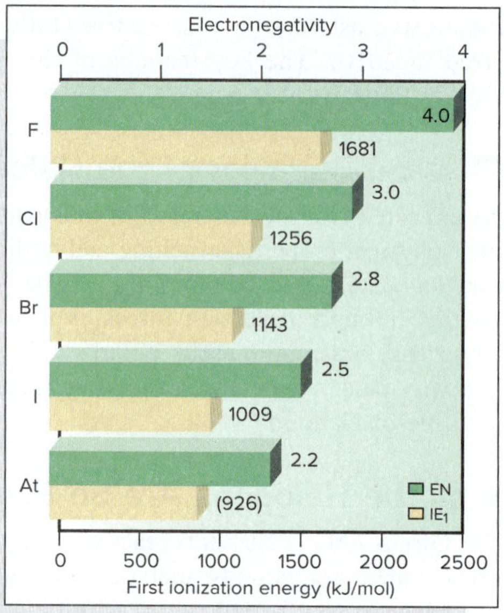

Physical Properties

Down the group, melting and boiling points increase smoothly as a result of stronger dispersion forces between larger molecules. The densities of the elements as liquids (at the given temperatures) increase steadily with molar mass.

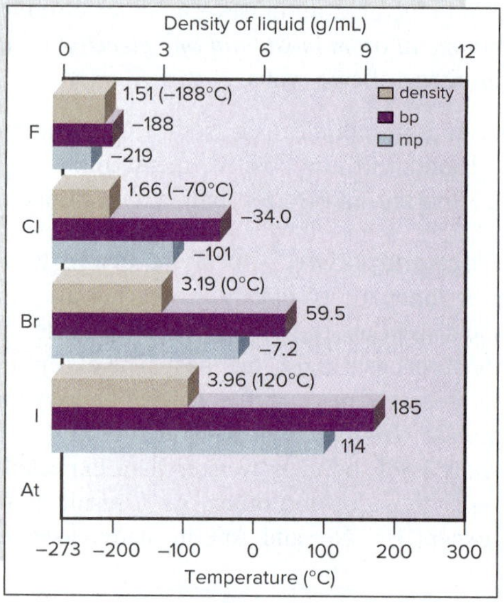

Reactions

1. The halogens (X_2) oxidize many metals and nonmetals. The reaction with hydrogen, although not used commercially for HX production (except for high-purity HCl), is characteristic of these strong oxidizing agents:

$$X_2 + H_2(g) \longrightarrow 2HX(g)$$

2. The halogens undergo disproportionation in water:

$$X_2 + H_2O(l) \rightleftharpoons HX(aq) + HXO(aq) \qquad (X = Cl, Br, I)$$

In aqueous base, the reaction goes to completion to form hypohalites and, at higher temperatures, halates, for example:

$$3Cl_2(g) + 6OH^-(aq) \xrightarrow{\Delta} ClO_3^-(aq) + 5Cl^-(aq) + 3H_2O(l)$$

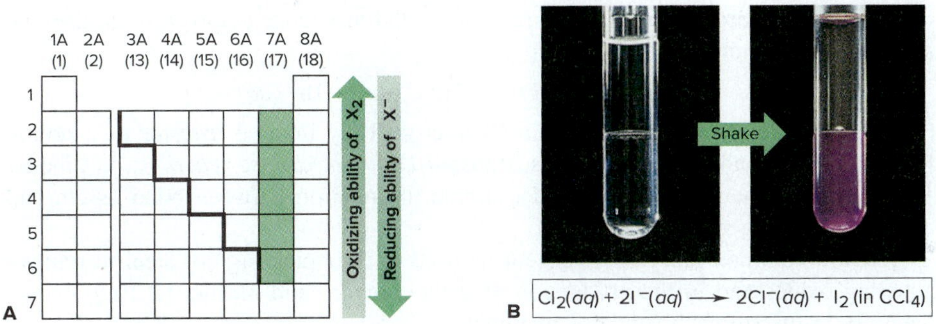

Figure 14.23 The relative oxidizing ability of the halogens.

Source: B: © McGraw-Hill Education/ Stephen Frisch, photographer

X^- increases *down* the group: the larger the ion, the more easily it gives up its electron (Figure 14.23A). Aqueous Cl_2 added to a solution of I^- (Figure 14.23B, *top layer*) oxidizes the I^- to I_2, which dissolves in the CCl_4 solvent *(bottom layer)* to give a purple solution.

The halogens undergo some important aqueous redox chemistry. Fluorine is such a powerful oxidizing agent that it reacts vigorously with water, oxidizing the O to produce O_2, some O_3, and HFO (hypofluorous acid). The other halogens undergo disproportionations (note the oxidation numbers):

$$\overset{0}{X_2} + H_2O(l) \rightleftharpoons \overset{-1}{HX}(aq) + \overset{+1}{HXO}(aq) \qquad (X = Cl, Br, I)$$

At equilibrium, very little product is present unless excess OH^- ion is added, which reacts with the HX and HXO and drives the reaction to completion:

$$X_2 + 2OH^-(aq) \longrightarrow X^-(aq) + XO^-(aq) + H_2O(l)$$

When X is Cl, the product mixture acts as a bleach: household bleach is a dilute solution of sodium hypochlorite (NaClO). Heating causes XO^- to disproportionate further, creating oxoanions with X in a higher oxidation state:

$$3\overset{+1}{X}O^-(aq) \xrightarrow{\Delta} 2\overset{-1}{X}^-(aq) + \overset{+5}{X}O_3^-(aq)$$

Highlights of Halogen Chemistry

Now, let's examine the compounds the halogens form with hydrogen and with each other, as well as their oxides, oxoanions, and oxoacids.

The Hydrogen Halides The halogens form gaseous hydrogen halides (HX) through direct combination with H_2 or through the action of a concentrated acid on the metal halide (a nonoxidizing acid is used for HBr and HI):

$$CaF_2(s) + H_2SO_4(l) \longrightarrow CaSO_4(s) + 2HF(g)$$
$$3NaBr(s) + H_3PO_4(l) \longrightarrow Na_3PO_4(s) + 3HBr(g)$$

Commercially, most HCl is formed as a byproduct in the chlorination of hydrocarbons for plastics production:

$$CH_2{=}CH_2(g) + Cl_2(g) \longrightarrow ClCH_2CH_2Cl(l) \xrightarrow{500^\circ C} \underset{\text{vinyl chloride}}{CH_2{=}CHCl(g)} + HCl(g)$$

In this case, the vinyl chloride reacts in a separate process to form poly(vinyl chloride), or PVC, a polymer used extensively in pipes for plumbing and other purposes.

In water, gaseous HX molecules form a *hydrohalic acid.* Only HF, with its relatively short, strong bond, forms a weak acid:

$$HF(g) + H_2O(l) \rightleftharpoons H_3O^+(aq) + F^-(aq)$$

HF has many uses, including the synthesis of cryolite (Na_3AlF_6) for aluminum production (Chapter 22), of fluorocarbons for refrigeration, and of NaF for water fluoridation. HF is also used in nuclear fuel processing and for glass etching.

The other hydrohalic acids are strong acids and dissociate completely to form the stoichiometric amount of H_3O^+ ions:

$$HBr(g) + H_2O(l) \longrightarrow H_3O^+(aq) + Br^-(aq)$$

(We saw reactions similar to these in Chapter 4. They involve *transfer* of a proton from acid to H_2O and are classified as *Brønsted-Lowry acid-base reactions*. In Chapter 18, we discuss them thoroughly and examine the relation between bond length and acidity of the larger HX molecules.)

HCl, a common laboratory reagent, is used in the "pickling" of steel to remove adhering oxides and in the production of syrups, rayon, and plastic. HCl(*aq*) occurs naturally in the stomach fluids of mammals.

Interhalogen Compounds: The "Halogen Halides" Halogens react exothermically with one another to form many **interhalogen compounds.** The simplest are diatomic molecules, such as ClF and BrCl. Every binary combination of the four common halogens is known. The more electronegative halogen is in the −1 oxidation state, and the less electronegative is in the +1 state. Interhalogens of general formula XY_n (n = 3, 5, 7) form through a variety of reactions, including direct reaction of the elements. In every case, the central atom has the lower *electronegativity* and a positive oxidation state.

Some interhalogens are used commercially as powerful *fluorinating agents,* which react with metals, nonmetals, and oxides—even wood and asbestos:

$$Sn(s) + ClF_3(l) \longrightarrow SnF_2(s) + ClF(g)$$
$$P_4(s) + 5ClF_3(l) \longrightarrow 4PF_3(g) + 3ClF(g) + Cl_2(g)$$
$$2B_2O_3(s) + 4BrF_3(l) \longrightarrow 4BF_3(g) + 2Br_2(l) + 3O_2(g)$$

Their reactions with water are nearly explosive and yield HF and the *oxoacid with the central halogen in the same oxidation state,* for example:

$$3H_2O(l) + \overset{+5}{Br}F_5(l) \longrightarrow 5HF(g) + H\overset{+5}{Br}O_3(aq)$$

The Oddness and Evenness of Oxidation States *Almost all stable molecules have paired electrons,* either as bonding or lone pairs. Therefore, *when bonds form or break, two electrons are involved, so the oxidation state changes by 2.* For this reason, odd-numbered groups exhibit odd-numbered oxidation states and even-numbered groups exhibit even-numbered states.

1. *Odd-numbered oxidation states.* Consider the interhalogens. Four general formulas are XY, XY_3, XY_5, and XY_7; examples are shown in Figure 14.24. With Y in the −1 state, X must be in the +1, +3, +5, or +7 state, respectively. The −1 state arises when Y fills its valence level; the +7 state arises when the central halogen (X) is completely oxidized, that is, when all seven valence electrons have shifted away from it to the more electronegative Y atoms around it.

Let's examine the iodine fluorides to see why the oxidation states jump by two units. When I_2 reacts with F_2, IF forms (note the oxidation number of I):

$$I_2 + F_2 \longrightarrow 2\overset{+1}{I}F$$

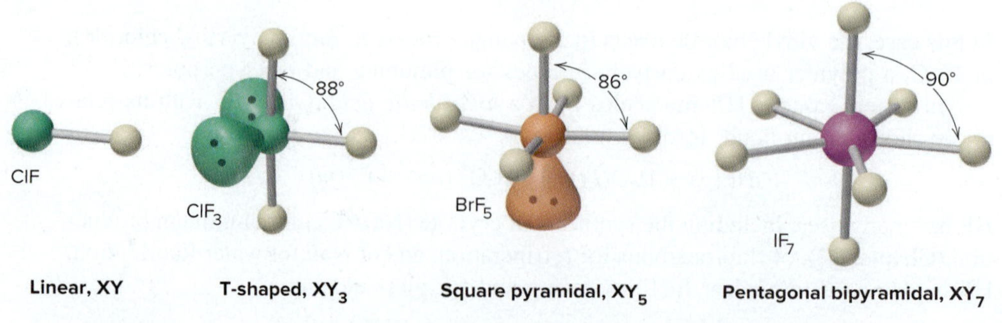

Figure 14.24 Molecular shapes of the main types of interhalogen compounds.

In IF_3, I uses *two* more valence electrons to form *two* more bonds:

$$\overset{+1}{\text{IF}} + \text{F}_2 \longrightarrow \overset{+3}{\text{IF}_3}$$

Otherwise, an unstable lone-electron species containing two fluorines would form. When IF_3 reacts with more fluorine, another jump of two units occurs and the pentafluoride forms:

$$\overset{+3}{\text{IF}_3} + \text{F}_2 \longrightarrow \overset{+5}{\text{IF}_5}$$

With still more fluorine, the heptafluoride forms:

$$\overset{+5}{\text{IF}_5} + \text{F}_2 \longrightarrow \overset{+7}{\text{IF}_7}$$

2. *Even-numbered oxidation states.* An element in an even-numbered group, such as sulfur in Group 6A(16), shows the same tendency to have paired electrons in its compounds. Elemental sulfur (oxidation number, O.N. = 0) gains or shares two electrons to complete its shell (O.N. = −2). It uses two electrons to react with fluorine, for example, to form SF_2 (O.N. = +2), two more electrons to form SF_4 (O.N. = +4), and two more to form SF_6 (O.N. = +6).

Thus, an element with one even state typically has all even states, and an element with one odd state typically has all odd states. To reiterate the main point, *successive oxidation states differ by two units because stable molecules have electrons in pairs around their atoms.*

Halogen Oxides, Oxoacids, and Oxoanions The Group 7A(17) elements form many oxides that are *powerful oxidizing agents and acids in water.* Dichlorine monoxide (Cl_2O) and especially chlorine dioxide (ClO_2) are used to bleach paper (Figure 14.25). ClO_2 is unstable to heat and shock, so it is prepared on site, and more than 100,000 tons are used annually:

$$2\text{NaClO}_3(s) + \text{SO}_2(g) + \text{H}_2\text{SO}_4(aq) \longrightarrow 2\text{ClO}_2(g) + 2\text{NaHSO}_4(aq)$$

The dioxide has an unpaired electron and Cl in the unusual +4 oxidation state.

Chlorine is in its highest (+7) oxidation state in dichlorine heptoxide, Cl_2O_7, which is a symmetrical molecule formed when two $HClO_4$ ($HO—ClO_3$) molecules undergo a dehydration-condensation reaction:

$$\text{O}_3\text{Cl—O}\boxed{\text{H + HO}}\text{—ClO}_3 \longrightarrow \text{O}_3\text{Cl—O—ClO}_3(l) + \text{H}_2\text{O}(l)$$

The halogen oxoacids and oxoanions are produced by reaction of the halogens and their oxides with water. Most of the oxoacids are stable only in solution. Table 14.4 *(on the next page)* shows ball-and-stick models of the acids in which each atom has its lowest formal charge; note the formulas, which emphasize that the H is bonded to O. The hypohalites (XO^-), halites (XO_2^-), and halates (XO_3^-) are oxidizing agents formed by aqueous disproportionation reactions [see the Group 7A(17) Family Portrait, reaction 2]. You may have heated solid alkali chlorates in the laboratory to form small amounts of O_2:

$$2\text{MClO}_3(s) \xrightarrow{\Delta} 2\text{MCl}(s) + 3\text{O}_2(g)$$

Potassium chlorate is the oxidizer in "safety" matches.

Several perhalates (XO_4^-) are also strong oxidizing agents. Thousands of tons of perchlorates are made each year for use in explosives and fireworks. Ammonium

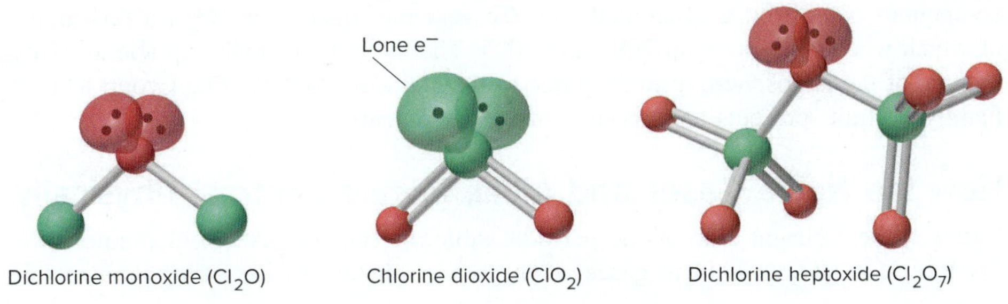

Figure 14.25 **Chlorine oxides.** Structures show each central Cl atom with its lowest formal charge.

Table 14.4 The Known Halogen Oxoacids*

Central Atom	Hypohalous Acid (HOX)	Halous Acid (HOXO)	Halic Acid ($HOXO_2$)	Perhalic Acid ($HOXO_3$)
	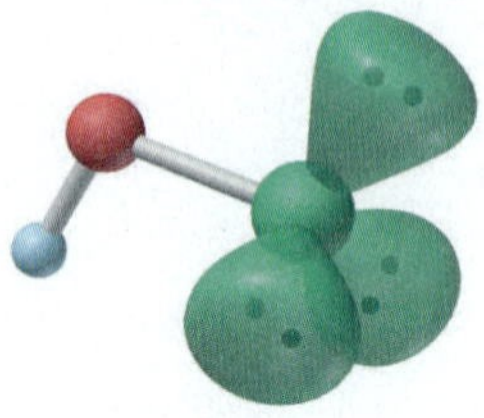	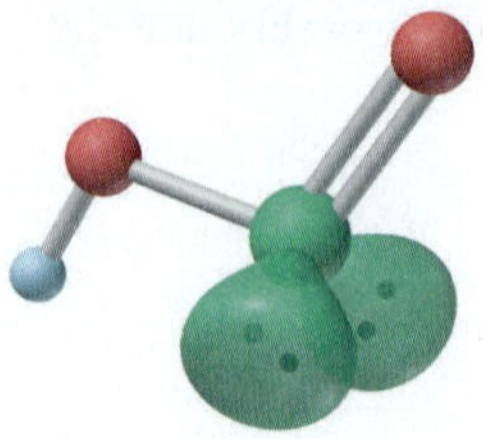	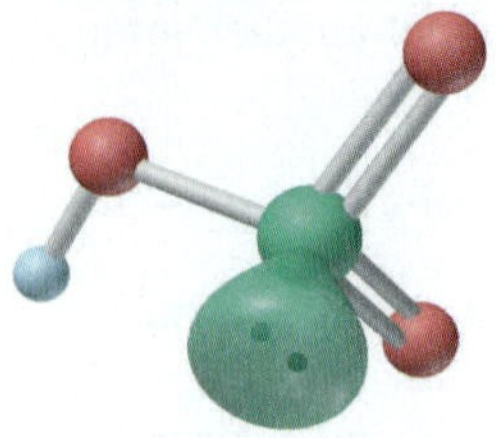	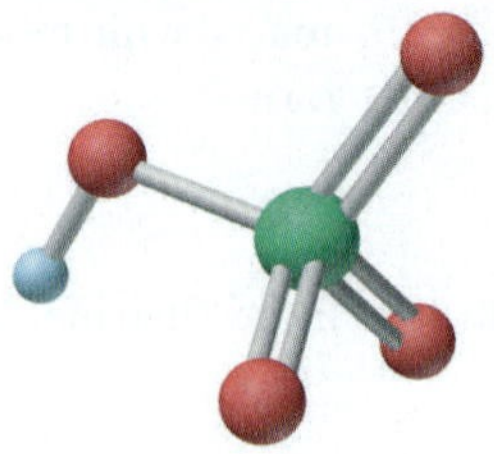
Fluorine	HOF	—	—	—
Chlorine	HOCl	HOClO	$HOClO_2$	$HOClO_3$
Bromine	HOBr	HOBrO	$HOBrO_2$	$HOBrO_3$
Iodine	HOI	—	$HOIO_2$	$HOIO_3$, $(HO)_5IO$
Oxoanion	Hypohalite	Halite	Halate	Perhalate

*Lone pairs are shown only on the halogen atom, and each atom has its lowest formal charge.

perchlorate, prepared from sodium perchlorate, was the oxidizing agent for the aluminum powder in the solid-fuel booster rockets of space shuttles; each launch used more than 700 tons of NH_4ClO_4:

$$10Al(s) + 6NH_4ClO_4(s) \longrightarrow 4Al_2O_3(s) + 12H_2O(g) + 3N_2(g) + 2AlCl_3(g)$$

The relative strengths of the halogen oxoacids depend on two factors:

1. *Electronegativity of the halogen.* Among oxoacids with the halogen in the same oxidation state, such as the halic acids, HXO_3 (or $HOXO_2$), acid strength decreases as the halogen's EN decreases:

$$HOClO_2 > HOBrO_2 > HOIO_2$$

The more electronegative the halogen, the more electron density it removes from the O—H bond, and the more easily the proton is lost.

2. *Oxidation state of the halogen.* Among oxoacids of a given halogen, such as chlorine, acid strength decreases as the oxidation state of the halogen decreases:

$$HOClO_3 > HOClO_2 > HOClO > HOCl$$

The higher the oxidation state (number of attached O atoms) of the halogen, the more electron density it pulls from the O—H bond. We consider these trends quantitatively in Chapter 18.

14.10 GROUP 8A(18): THE NOBLE GASES

The last main group consists of individual atoms too "noble" to interact much with others. The Group 8A(18) elements display regular trends in physical properties and very low, if any, reactivity. The group consists of the following elements: helium (He), the second most abundant element in the universe; neon (Ne); argon (Ar); krypton (Kr), xenon (Xe), and radioactive radon (Rn), the only members for which compounds have been well studied; and the very rare oganesson (Og), a few atoms of which were synthesized in 2002 and 2005. The noble gases make up about 1% by volume of the atmosphere, primarily due to the abundance of Ar. The Group 8A(18) Family Portrait presents an overview of these elements.

How the Noble Gases and Alkali Metals Contrast Physically

Lying at the far right side of the periodic table, the Group 8A(18) elements come as close to behaving as ideal gases as any other substances. Like the alkali metals

KEY ATOMIC AND PHYSICAL PROPERTIES

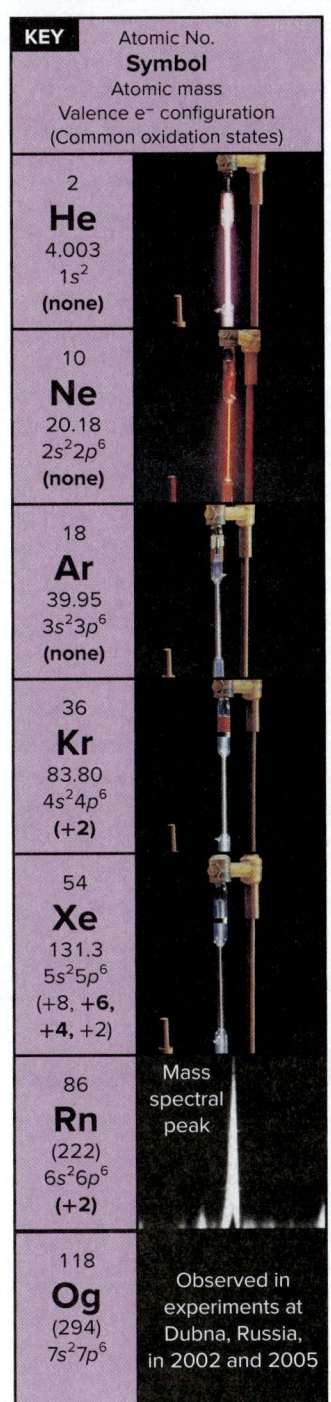

KEY
Atomic No.
Symbol
Atomic mass
Valence e⁻ configuration
(Common oxidation states)

Atomic No.	Symbol	Atomic mass	Valence e^- configuration	Common oxidation states
2	**He**	4.003	$1s^2$	**(none)**
10	**Ne**	20.18	$2s^22p^6$	**(none)**
18	**Ar**	39.95	$3s^23p^6$	**(none)**
36	**Kr**	83.80	$4s^24p^6$	**(+2)**
54	**Xe**	131.3	$5s^25p^6$	(+8, **+6**, **+4**, +2)
86	**Rn**	(222)	$6s^26p^6$	**(+2)**
118	**Og**	(294)	$7s^27p^6$	

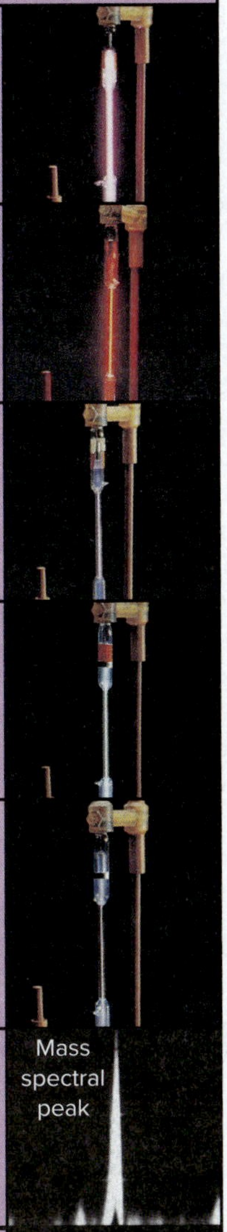

Og: Observed in experiments at Dubna, Russia, in 2002 and 2005

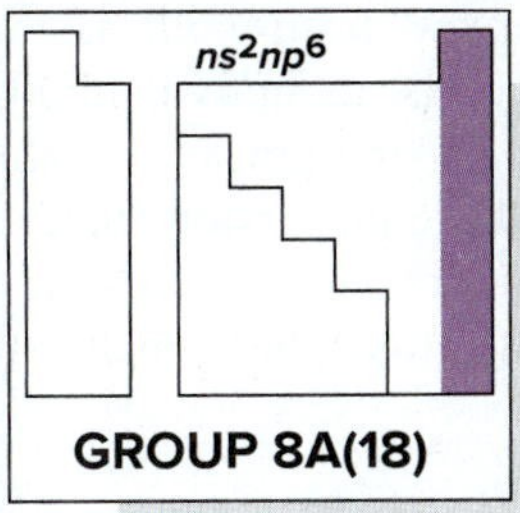

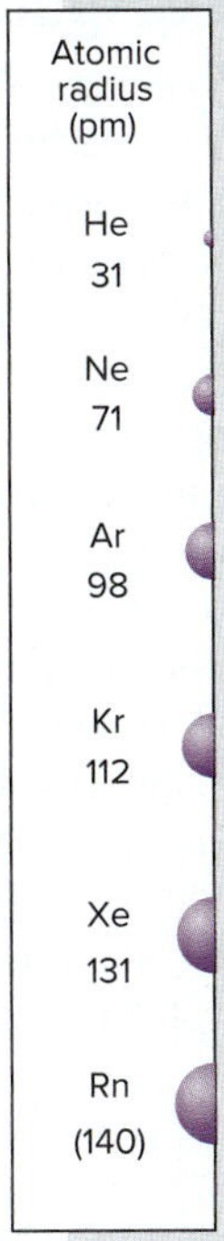

Atomic Properties

Group electron configuration is $1s^2$ for He and ns^2np^6 for the others. The valence shell is filled. Only Kr, Xe, and Rn are known to form compounds. The more reactive Xe exhibits all even oxidation states (+2 to +8). This group contains the smallest atoms with the highest IEs in their periods. Down the group, atomic size increases and IE decreases steadily. (EN values are given only for Kr and Xe.)

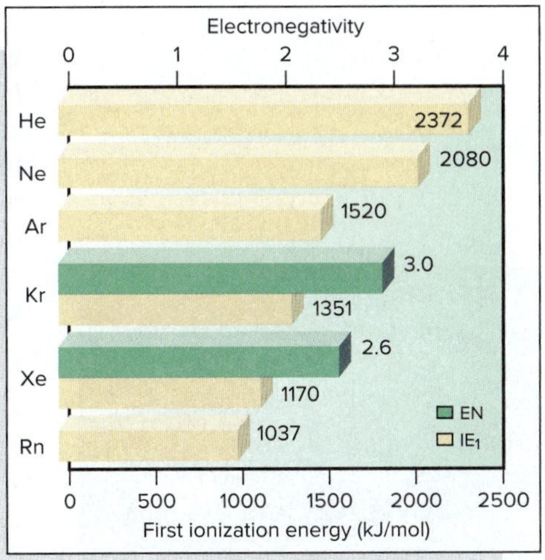

Physical Properties

Melting and boiling points of these gaseous elements are extremely low but increase down the group because of stronger dispersion forces. Note the extremely small liquid ranges. Densities (at STP) increase steadily, as expected.

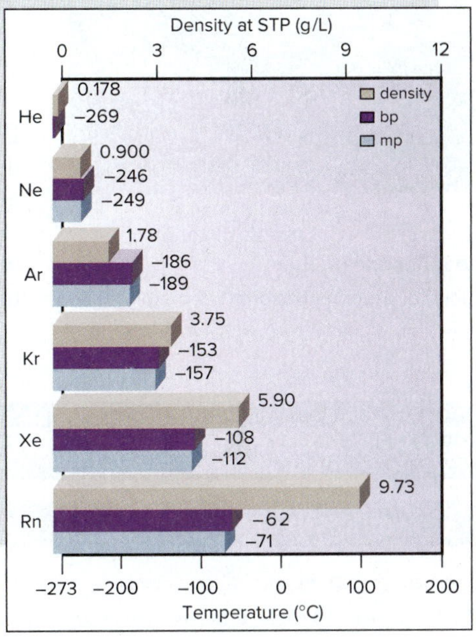

Source: © McGraw-Hill Education/Stephen Frisch, photographer

at the other end of the periodic table, the noble gases display regular trends in physical properties. However, whereas melting and boiling points and heats of fusion and vaporization *decrease* down Group 1A(1), these properties *increase* down Group 8A(18). The reason for these opposite trends is the different bonding interactions in the elements. The alkali metals consist of atoms held together by metallic bonding, which *decreases* in strength as the atoms become larger. The noble gases, on the other hand, exist as individual atoms that interact through dispersion forces, which *increase* in strength as the atoms become larger and more easily polarized. Nevertheless, these dispersion forces are so weak that only at very low temperatures do the elements condense and solidify. For example, even argon, which makes up almost 1% of the atmosphere, melts at −189°C and boils only about three degrees higher.

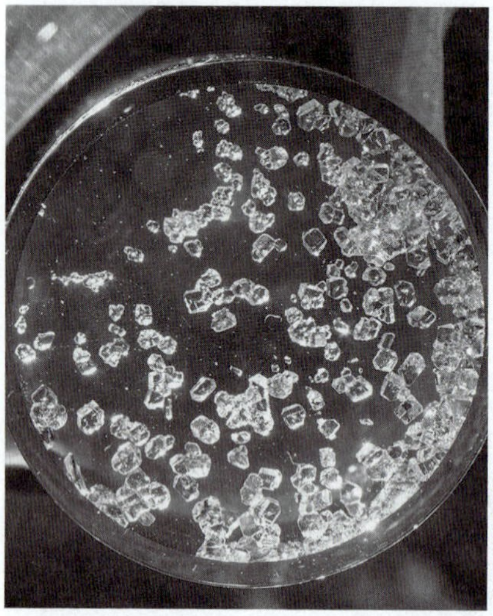

Figure 14.26 Crystals of xenon tetrafluoride (XeF_4).
Source: Argonne National Laboratory

How Noble Gases Can Form Compounds

Following their discovery in the late 19th century, these elements were considered, and even formerly named, the "inert" gases. Early 20th-century atomic theory and, more important, all experiments supported this idea. Then, in 1962, all this changed when the first noble gas compound was prepared. How, with filled outer levels and extremely high ionization energies, *can* noble gases react?

The discovery of noble gas reactivity is a classic example of clear thinking in the face of an unexpected event. A young inorganic chemist named Neil Bartlett was studying platinum fluorides, known to be strong oxidizing agents. When he accidentally exposed PtF_6 to air, its deep-red color lightened slightly, and analysis showed that the PtF_6 had oxidized O_2 to form the ionic compound $[O_2]^+[PtF_6]^-$. Knowing that the ionization energy of the oxygen molecule ($O_2 \longrightarrow O_2^+ + e^-$; IE = 1175 kJ/mol) is very close to IE_1 of xenon (1170 kJ/mol), Bartlett reasoned that PtF_6 might be able to oxidize xenon. Shortly thereafter, he prepared $XePtF_6$, an orange-yellow solid. Within a few months, the white crystalline XeF_2 and XeF_4 (Figure 14.26) were also prepared. In addition to its +2 and +4 oxidation states, Xe has the +6 state in several compounds, such as XeF_6, and the +8 state in the unstable oxide, XeO_4. A few compounds of Kr and Rn have also been made.

CHAPTER REVIEW GUIDE

Learning Objectives

Relevant section (§) numbers appear in parentheses.

Understand These Concepts

Note: Many characteristic reactions appear in the "Reactions" section of each group's Family Portrait.

1. How hydrogen is similar to, yet different from, alkali metals and halogens; the differences between ionic, covalent, and metallic hydrides (§14.1)
2. Key horizontal trends in atomic properties, types of bonding, oxide acid-base properties, and redox behavior as the elements change from metals to nonmetals (§14.2)
3. How small atomic size and limited number of valence orbitals account for the anomalous behavior of the Period 2 member of each group (§14.2)
4. How the ns^1 configuration accounts for the physical and chemical properties of the alkali metals (§14.3)
5. How the ns^2 configuration accounts for the key differences between Groups 1A(1) and 2A(2) (§14.4)
6. The basis of the three important diagonal relationships (Li/Mg, Be/Al, B/Si) (§14.4–14.6)
7. How the presence of inner $(n-1)d$ electrons affects properties in Group 3A(13) (§14.5)
8. Patterns among larger members of Groups 3A(13) to 6A(16): two common oxidation states (inert-pair effect), lower state more important going down the group, and more basic oxide with the element in the lower state (§14.5–14.8)
9. How boron attains an octet of electrons (§14.5)
10. The effect of bonding on the physical behavior of Groups 4A(14) to 6A(16) (§14.6–14.8)
11. Allotropism in carbon, phosphorus, and sulfur (§14.6–14.8)
12. How atomic properties lead to catenation and multiple bonds in organic compounds (§14.6)
13. Structures and properties of the silicates and silicones (§14.6)
14. Patterns of behavior among hydrides and halides of Groups 5A(15) and 6A(16) (§14.7, §14.8)
15. The meaning of disproportionation (§14.7)

16. Structure and chemistry of the nitrogen oxides and oxoacids (§14.7)
17. Structure and chemistry of the phosphorus oxides and oxoacids (§14.7)
18. Dehydration-condensation reactions and polyphosphate structures (§14.7)
19. Structure and chemistry of the sulfur oxides and oxoacids (§14.8)
20. How the ns^2np^5 configuration accounts for halogen reactivity with metals (§14.9)
21. Why the oxidation states of an element change by two units (§14.9)
22. Structure and chemistry of the halogen oxides and oxoacids (§14.9)
23. How the ns^2np^6 configuration accounts for the relative inertness of the noble gases (§14.10)

Key Terms

Page numbers appear in parentheses.

allotrope (600)
bridge bond (598)
diagonal relationship (595)
disproportionation reaction (609)
dehydration-condensation reaction (611)
interhalogen compound (620)
silicate (603)
silicone (603)

PROBLEMS

Problems with **colored** numbers are answered in Appendix E and worked in detail in the Student Solutions Manual. Problem sections match those in the text. Most offer Concept Review Questions, Skill-Building Exercises (grouped in pairs covering the same concept), and Problems in Context. The Comprehensive Problems are based on material from any section or previous chapter.

Hydrogen, the Simplest Atom

Concept Review Questions

14.1 Hydrogen has only one proton, but its IE_1 is much greater than that of lithium, which has three protons. Explain.

14.2 Sketch a periodic table, and label the areas containing elements that give rise to the three types of hydrides.

Skill-Building Exercises (grouped in similar pairs)

14.3 Draw Lewis structures for the following compounds, and predict which member of each pair will form H bonds:
(a) NF_3 or NH_3 (b) CH_3OCH_3 or CH_3CH_2OH

14.4 Draw Lewis structures for the following compounds, and predict which member of each pair will form H bonds:
(a) NH_3 or AsH_3 (b) CH_4 or H_2O

14.5 Complete and balance the following equations:
(a) An active metal reacting with acid,

$$Al(s) + HCl(aq) \longrightarrow$$

(b) A saltlike (alkali metal) hydride reacting with water,

$$LiH(s) + H_2O(l) \longrightarrow$$

14.6 Complete and balance the following equations:
(a) A saltlike (alkaline earth metal) hydride reacting with water,

$$CaH_2(s) + H_2O(l) \longrightarrow$$

(b) Reduction of a metal halide by hydrogen to form a metal,

$$PdCl_2(aq) + H_2(g) \longrightarrow$$

Problems in Context

14.7 Compounds such as $NaBH_4$, $Al(BH_4)_3$, and $LiAlH_4$ are complex hydrides used as reducing agents in many syntheses.
(a) Give the oxidation state of each element in these compounds.
(b) Write a Lewis structure for the polyatomic anion in $NaBH_4$, and predict its shape.

14.8 Unlike the F^- ion, which has an ionic radius close to 133 pm in all alkali metal fluorides, the ionic radius of H^- varies from 137 pm in LiH to 152 pm in CsH. Suggest an explanation for the large variability in the size of H^- but not F^-.

Trends Across the Periodic Table: The Period 2 Elements

Concept Review Questions

14.9 How does the maximum oxidation number vary across a period in the main groups? Is the pattern in Period 2 different?

14.10 What correlation, if any, exists for the Period 2 elements between group number and the number of covalent bonds the element typically forms? How is the correlation different for elements in Periods 3 to 6?

14.11 Each of the chemically active Period 2 elements forms stable compounds in which it has bonds to fluorine.
(a) What are the names and formulas of these compounds?
(b) Does ΔEN increase or decrease left to right across the period?
(c) Does percent ionic character increase or decrease left to right?
(d) Draw Lewis structures for these compounds.

14.12 Period 6 contains the first series of inner transition elements.
(a) How many elements belong to Period 6? How many metals?
(b) It contains no metalloids. Where is the metal/nonmetal boundary in Period 6?

14.13 An element forms an oxide, E_2O_3, and a fluoride, EF_3.
(a) Of which two groups might E be a member?
(b) How does the group to which E belongs affect the properties of the oxide and the fluoride?

14.14 Fluorine lies between oxygen and neon in Period 2. Whereas atomic sizes and ionization energies of these three elements change smoothly, their electronegativities display a dramatic change. What is this change, and how do their electron configurations explain it?

Group 1A(1): The Alkali Metals

Concept Review Questions

14.15 Lithium salts are often much less soluble in water than the corresponding salts of other alkali metals. For example, at 18°C, the concentration of a saturated LiF solution is 1.0×10^{-2} *M*, whereas that of a saturated KF solution is 1.6 *M*. How can you explain this behavior?

14.16 The alkali metals play virtually the same general chemical role in all their reactions.
(a) What is this role?
(b) How is it based on atomic properties?
(c) Using sodium, write two balanced equations that illustrate this role.

14.17 How do atomic properties account for the low densities of the Group 1A(1) elements?

Skill-Building Exercises (grouped in similar pairs)

14.18 Each of the following properties shows a regular trend in Group 1A(1). Predict whether each increases or decreases *down* the group:
(a) Density (b) Ionic size (c) E—E bond energy
(d) IE_1 (e) Magnitude of ΔH_{hydr} of E^+ ion

14.19 Each of the following properties shows a regular trend in Group 1A(1). Predict whether each increases or decreases *up* the group:
(a) Melting point (b) E—E bond length (c) Hardness
(d) Molar volume (e) Lattice energy of EBr

14.20 Write a balanced equation for the formation from its elements of sodium peroxide, an industrial bleach.

14.21 Write a balanced equation for the formation of rubidium bromide through reaction of a strong acid and a strong base.

Problems in Context

14.22 Although the alkali metal halides can be prepared directly from the elements, the far less expensive industrial route is treatment of the carbonate or hydroxide with aqueous hydrohalic acid (HX) followed by recrystallization. Balance the reaction between potassium carbonate and aqueous hydriodic acid.

14.23 Lithium forms several useful organolithium compounds. Calculate the mass percent of Li in the following:
(a) Lithium stearate ($C_{17}H_{35}COOLi$), a water-resistant grease used in cars because it does not harden at cold temperatures
(b) Butyllithium (LiC_4H_9), a reagent in organic syntheses

Group 2A(2): The Alkaline Earth Metals

Concept Review Questions

14.24 How do Groups 1A(1) and 2A(2) compare with respect to reaction of the metals with water?

14.25 Alkaline earth metals are involved in two key diagonal relationships in the periodic table.
(a) Give the two pairs of elements in these diagonal relationships.
(b) For each pair, cite two similarities that demonstrate the relationship.
(c) Why are the members of each pair so similar in behavior?

14.26 The melting points of alkaline earth metals are many times higher than those of the alkali metals. Explain this difference on the basis of atomic properties. Name three other physical properties for which Group 2A(2) metals have higher values than the corresponding 1A(1) metals.

Skill-Building Exercises (grouped in similar pairs)

14.27 Write a balanced equation for each reaction:
(a) "Slaking" of lime (treatment with water)
(b) Combustion of calcium in air

14.28 Write a balanced equation for each reaction:
(a) Thermal decomposition of witherite (barium carbonate)
(b) Neutralization of stomach acid (HCl) by milk of magnesia (magnesium hydroxide)

Problems in Context

14.29 Lime (CaO) is one of the most abundantly produced chemicals in the world. Write balanced equations for these reactions:
(a) The preparation of lime from natural sources
(b) The use of slaked lime to remove SO_2 from flue gases
(c) The reaction of lime with arsenic acid (H_3AsO_4) to manufacture the insecticide calcium arsenate
(d) The regeneration of NaOH in the paper industry by reaction of lime with aqueous sodium carbonate

14.30 In some reactions, Be behaves like a typical alkaline earth metal; in others, it does not. Complete and balance the following:
(a) $BeO(s) + H_2O(l) \longrightarrow$
(b) $BeCl_2(l) + Cl^-(l;\text{ from molten NaCl}) \longrightarrow$
In which reaction does Be behave like the other Group 2A(2) members?

Group 3A(13): The Boron Family

Concept Review Questions

14.31 How do the transition metals in Period 4 affect the pattern of ionization energies in Group 3A(13)? How does this pattern compare with that in Group 3B(3)?

14.32 How do the acidities of aqueous solutions of Tl_2O and Tl_2O_3 compare with each other? Explain.

14.33 Despite the expected decrease in atomic size, there is an unexpected drop in the first ionization energy between Groups 2A(2) and 3A(13) in Periods 2 through 4. Explain this pattern in terms of electron configurations and orbital energies.

14.34 Many compounds of Group 3A(13) elements have chemical behavior that reflects an electron deficiency.
(a) What is the meaning of *electron deficiency?*
(b) Give two reactions that illustrate this behavior.

14.35 Boron's chemistry is not typical of its group.
(a) Cite three ways in which boron and its compounds differ significantly from the other 3A(13) members and their compounds.
(b) What is the reason for these differences?

Skill-Building Exercises (grouped in similar pairs)

14.36 Rank the following oxides in order of increasing *acidity* of their aqueous solutions: Ga_2O_3, Al_2O_3, In_2O_3.

14.37 Rank the following hydroxides in order of increasing *basicity* of their aqueous solutions: $Al(OH)_3$, $B(OH)_3$, $In(OH)_3$.

14.38 Thallium forms the compound TlI_3. What is the apparent oxidation state of Tl in this compound? Given that the anion is I_3^-, what is the actual oxidation state of Tl? Draw the shape of the anion, giving its VSEPR class and bond angles. Propose a reason why the compound does not exist as $(Tl^{3+})(I^-)_3$.

14.39 Very stable dihalides of the Group 3A(13) metals are known. What is the apparent oxidation state of Ga in $GaCl_2$? Given that $GaCl_2$ consists of a Ga^+ cation and a $GaCl_4^-$ anion, what are the actual oxidation states of Ga? Draw the shape of the anion, giving its VSEPR class and bond angles.

Problems in Context

14.40 Give the name and symbol or formula of a Group 3A(13) element or compound that fits each description or use:
(a) Component of heat-resistant (Pyrex-type) glass
(b) Largest temperature range for liquid state of an element
(c) Elemental substance with three-center, two-electron bonds
(d) Metal protected from oxidation by adherent oxide coat
(e) Toxic metal that lies between two other toxic metals

14.41 Indium (In) reacts with HCl to form a diamagnetic solid with the formula $InCl_2$.
(a) Write condensed electron configurations for In, In^+, In^{2+}, and In^{3+}.
(b) Which of these species is (are) diamagnetic and which paramagnetic?
(c) What is the apparent oxidation state of In in $InCl_2$?
(d) Given your answers to parts (b) and (c), explain how $InCl_2$ can be diamagnetic.

14.42 Use VSEPR theory to draw structures, with ideal bond angles, for boric acid and the anion it forms when it reacts with water.

14.43 Boron nitride (BN) has a structure similar to graphite, but is a white insulator rather than a black conductor. It is synthesized by heating diboron trioxide with ammonia at about 1000°C.
(a) Write a balanced equation for the formation of BN; water also forms.
(b) Calculate ΔH°_{rxn} for the production of BN (ΔH°_f of BN is −254 kJ/mol).
(c) Boron is obtained from the mineral borax, $Na_2B_4O_7 \cdot 10H_2O$. How much borax is needed to produce 1.0 kg of BN, assuming 72% yield?

Group 4A(14): The Carbon Family

Concept Review Questions

14.44 How does the basicity of SnO_2 in water compare with that of CO_2? Explain.

14.45 Nearly every compound of silicon has the element in the +4 oxidation state. In contrast, most compounds of lead have the element in the +2 state.
(a) What general observation do these facts illustrate?
(b) Explain in terms of atomic and molecular properties.
(c) Give an analogous example from Group 3A(13).

14.46 The sum of IE_1 through IE_4 for Group 4A(14) elements shows a decrease from C to Si, a slight increase from Si to Ge, a decrease from Ge to Sn, and an increase from Sn to Pb.
(a) What is the expected trend for IEs down a group?
(b) Suggest a reason for the deviations in Group 4A(14).
(c) Which group might show even greater deviations?

14.47 Give explanations for the large drops in melting point from C to Si and from Ge to Sn.

14.48 What is an allotrope? Name two Group 4A(14) elements that exhibit allotropism, and identify two of their allotropes.

14.49 Even though EN values vary relatively little down Group 4A(14), the elements change from nonmetal to metal. Explain.

14.50 How do atomic properties account for the enormous number of carbon compounds? Why don't other Group 4A(14) elements behave similarly?

Skill-Building Exercises (grouped in similar pairs)

14.51 Draw a Lewis structure for each species:
(a) The cyclic silicate ion $Si_4O_{12}^{8-}$
(b) A cyclic hydrocarbon with formula C_4H_8

14.52 Draw a Lewis structure for each species:
(a) The cyclic silicate ion $Si_6O_{18}^{12-}$
(b) A cyclic hydrocarbon with formula C_6H_{12}

Problems in Context

14.53 Zeolite A, $Na_{12}[(AlO_2)_{12}(SiO_2)_{12}] \cdot 27H_2O$, is used to soften water because it replaces Ca^{2+} and Mg^{2+} dissolved in the water with Na^+. Hard water from a certain source is 4.5×10^{-3} *M* Ca^{2+} and 9.2×10^{-4} *M* Mg^{2+}, and a pipe delivers 25,000 L of this hard water per day. What mass (in kg) of zeolite A is needed to soften a week's supply of the water? (Assume zeolite A loses its capacity to exchange ions when 85 mol % of its Na^+ has been lost.)

14.54 Give the name and symbol or formula of a Group 4A(14) element or compound that fits each description or use:
(a) Hardest known natural substance
(b) Medicinal antacid
(c) Atmospheric gas implicated in the greenhouse effect
(d) Gas that binds to Fe(II) in blood
(e) Element used in the manufacture of computer chips

14.55 One similarity between B and Si is the explosive combustion of their hydrides in air. Write balanced equations for the combustion of B_2H_6 and of Si_4H_{10}.

Group 5A(15): The Nitrogen Family

Concept Review Questions

14.56 Which Group 5A(15) elements form trihalides? Pentahalides? Explain.

14.57 As you move down Group 5A(15), the melting points of the elements increase and then decrease. Explain.

14.58 (a) What is the range of oxidation states shown by the elements of Group 5A(15) as you move down the group?
(b) How does this range illustrate the general rule for the range of oxidation states in groups on the right side of the periodic table?

14.59 Bismuth(V) compounds are such powerful oxidizing agents that they have not been prepared in pure form. How is this fact consistent with the location of Bi in the periodic table?

14.60 Rank the following oxides in order of increasing acidity in water: Sb_2O_3, Bi_2O_3, P_4O_{10}, Sb_2O_5.

Skill-Building Exercises (grouped in similar pairs)

14.61 Assuming acid strength relates directly to electronegativity of the central atom, rank H_3PO_4, HNO_3, and H_3AsO_4 in order of *increasing* acid strength.

14.62 Assuming acid strength relates directly to number of O atoms bonded to the central atom, rank $H_2N_2O_2$ [or $(HON)_2$], HNO_3 (or $HONO_2$), and HNO_2 (or HONO) in order of *decreasing* acid strength.

14.63 Complete and balance the following:
(a) $As(s) + \text{excess } O_2(g) \longrightarrow$
(b) $Bi(s) + \text{excess } F_2(g) \longrightarrow$
(c) $Ca_3As_2(s) + H_2O(l) \longrightarrow$

14.64 Complete and balance the following:
(a) Excess $Sb(s) + Br_2(l) \longrightarrow$
(b) $HNO_3(aq) + MgCO_3(s) \longrightarrow$
(c) $K_2HPO_4(s) \xrightarrow{\Delta}$

14.65 Complete and balance the following:
(a) $N_2(g) + Al(s) \xrightarrow{\Delta}$ (b) $PF_5(g) + H_2O(l) \longrightarrow$

14.66 Complete and balance the following:
(a) $AsCl_3(l) + H_2O(l) \longrightarrow$ (b) $Sb_2O_3(s) + NaOH(aq) \longrightarrow$

14.67 Based on the relative sizes of F and Cl, predict the structure of PF_2Cl_3.

14.68 Use the VSEPR model to predict the structure of the cyclic ion $P_3O_9^{3-}$.

Problems in Context

14.69 The pentafluorides of the larger members of Group 5A(15) have been prepared, but N can have only eight electrons. A claim has been made that, at low temperatures, a compound with the empirical formula NF_5 forms. Draw a possible Lewis structure for this compound. (*Hint:* NF_5 is ionic.)

14.70 Give the name and symbol or formula of a Group 5A(15) element or compound that fits each description or use:
(a) Hydride that exhibits hydrogen bonding
(b) Compound used in "strike-anywhere" match heads
(c) Oxide used as a laboratory drying agent
(d) Molecule with an odd-electron atom (two examples)
(e) Compound used as an additive in soft drinks

14.71 In addition to those in Table 14.3, other less stable nitrogen oxides exist. Draw a Lewis structure for each of the following:
(a) N_2O_2, a dimer of nitrogen monoxide with an N—N bond
(b) N_2O_2, a dimer of nitrogen monoxide with no N—N bond
(c) N_2O_3 with no N—N bond
(d) NO^+ and NO_3^-, products of the ionization of liquid N_2O_4

14.72 Nitrous oxide (N_2O), the "laughing gas" used as an anesthetic by dentists, is made by thermal decomposition of solid NH_4NO_3. Write a balanced equation for this reaction. What are the oxidation states of N in NH_4NO_3 and in N_2O?

14.73 Write balanced equations for the thermal decomposition of potassium nitrate (O_2 is also formed in both cases):
(a) At low temperature to the nitrite
(b) At high temperature to the metal oxide and nitrogen

Group 6A(16): The Oxygen Family

Concept Review Questions

14.74 Rank the following in order of increasing electrical conductivity, and explain your ranking: Po, S, Se.

14.75 The oxygen and nitrogen families have some obvious similarities and differences.
(a) State two general physical similarities between Group 5A(15) and 6A(16) elements.
(b) State two general chemical similarities between Group 5A(15) and 6A(16) elements.
(c) State two chemical similarities between P and S.
(d) State two physical similarities between N and O.
(e) State two chemical differences between N and O.

14.76 A molecular property of the Group 6A(16) hydrides changes abruptly down the group. This change has been explained in terms of a change in orbital hybridization.
(a) Between what periods does the change occur?
(b) What is the change in the molecular property?
(c) What is the change in hybridization?
(d) What other group displays a similar change?

Skill-Building Exercises (grouped in similar pairs)

14.77 Complete and balance the following:
(a) $NaHSO_4(aq) + NaOH(aq) \longrightarrow$
(b) $S_8(s) + \text{excess } F_2(g) \longrightarrow$
(c) $FeS(s) + HCl(aq) \longrightarrow$
(d) $Te(s) + I_2(s) \longrightarrow$

14.78 Complete and balance the following:
(a) $H_2S(g) + O_2(g) \longrightarrow$
(b) $SO_3(g) + H_2O(l) \longrightarrow$
(c) $SF_4(g) + H_2O(l) \longrightarrow$
(d) $Al_2Se_3(s) + H_2O(l) \longrightarrow$

14.79 Is each oxide basic, acidic, or amphoteric in water?
(a) SeO_2 (b) N_2O_3 (c) K_2O
(d) BeO (e) BaO

14.80 Is each oxide basic, acidic, or amphoteric in water?
(a) MgO (b) N_2O_5 (c) CaO
(d) CO_2 (e) TeO_2

14.81 Rank the following hydrides in order of *increasing* acid strength: H_2S, H_2O, H_2Te.

14.82 Rank the following species in order of *decreasing* acid strength: H_2SO_4, H_2SO_3, HSO_3^-.

Problems in Context

14.83 Selenium tetrafluoride reacts with fluorine to form selenium hexafluoride.
(a) Draw Lewis structures for both selenium fluorides, and predict any deviations from ideal bond angles.
(b) Describe the change in orbital hybridization of the central Se during the reaction.

14.84 Give the name and symbol or formula of a Group 6A(16) element or compound that fits each description or use:
(a) Unstable allotrope of oxygen
(b) Oxide having sulfur with the same O.N. as in sulfuric acid
(c) Air pollutant produced by burning sulfur-containing coal
(d) Powerful dehydrating agent
(e) Compound used in solution in the photographic process

14.85 Give the oxidation state of sulfur in each substance:
(a) S_8 (b) SF_4 (c) SF_6
(d) H_2S (e) FeS_2 (f) H_2SO_4
(g) $Na_2S_2O_3 \cdot 5H_2O$

14.86 Disulfur decafluoride is intermediate in reactivity between SF_4 and SF_6. It disproportionates at 150°C to these monosulfur fluorides. Write a balanced equation for this reaction, and give the oxidation state of S in each compound.

Group 7A(17): The Halogens

Concept Review Questions

14.87 (a) Give the physical state and color of each halogen at STP. (b) Explain the change in physical state down Group 7A(17) in terms of molecular properties.

14.88 (a) What are the common oxidation states of the halogens? (b) Give an explanation based on electron configuration for the range and values of the oxidation states of chlorine.

(c) Why is fluorine an exception to the pattern of oxidation states found for the other group members?

14.89 How many electrons does a halogen atom need to complete its octet? Give examples of the ways a Cl atom can do so.

14.90 Select the stronger bond in each pair:
(a) Cl—Cl or Br—Br
(b) Br—Br or I—I
(c) F—F or Cl—Cl. Why doesn't the F—F bond strength follow the group trend?

14.91 In addition to interhalogen compounds, many interhalogen ions exist. Would you expect interhalogen ions with a 1+ or a 1− charge to have an even or odd number of atoms? Explain.

14.92 (a) A halogen (X_2) disproportionates in base in several steps to X^- and XO_3^-. Write the overall equation for the disproportionation of Br_2 in excess OH^- to Br^- and BrO_3^-.
(b) Write a balanced equation for the reaction of ClF_5 with aqueous base (*Hint:* Add base to the reaction of BrF_5 shown in Section 14.9).

Skill-Building Exercises (grouped in similar pairs)

14.93 Complete and balance the following equations. If no reaction occurs, write NR:
(a) $Rb(s) + Br_2(l) \longrightarrow$ (b) $I_2(s) + H_2O(l) \longrightarrow$
(c) $Br_2(l) + I^-(aq) \longrightarrow$ (d) $CaF_2(s) + H_2SO_4(l) \longrightarrow$

14.94 Complete and balance the following equations. If no reaction occurs, write NR:
(a) $H_3PO_4(l) + NaI(s) \longrightarrow$ (b) $Cl_2(g) + I^-(aq) \longrightarrow$
(c) $Br_2(l) + Cl^-(aq) \longrightarrow$ (d) $ClF(g) + F_2(g) \longrightarrow$

14.95 Rank the following acids in order of *increasing* acid strength: HClO, $HClO_2$, HBrO, HIO.

14.96 Rank the following acids in order of *decreasing* acid strength: $HBrO_3$, $HBrO_4$, HIO_3, $HClO_4$.

Problems in Context

14.97 Give the name and symbol or formula of a Group 7A(17) element or compound that fits each description or use:
(a) Used in etching glass
(b) Compound used in household bleach
(c) Weakest hydrohalic acid
(d) Element that is a liquid at room temperature
(e) Organic chloride used to make PVC

14.98 An industrial chemist treats solid NaCl with concentrated H_2SO_4 and obtains gaseous HCl and $NaHSO_4$. When she substitutes solid NaI for NaCl, she obtains gaseous H_2S, solid I_2, and S_8, but no HI.
(a) What type of reaction did the H_2SO_4 undergo with NaI?
(b) Why does NaI, but not NaCl, cause this type of reaction?
(c) To produce HI(g) by reacting NaI with an acid, how does the acid have to differ from sulfuric acid?

14.99 Rank the halogens Cl_2, Br_2, and I_2 in order of increasing oxidizing strength based on their products with the metal rhenium (Re): $ReCl_6$, $ReBr_5$, ReI_4. Explain your ranking.

Group 8A(18): The Noble Gases

14.100 Which noble gas is the most abundant in the universe? In Earth's atmosphere?

14.101 What oxidation states does Xe show in its compounds?

14.102 Why do the noble gases have such low boiling points?

14.103 Explain why Xe, and to a limited extent Kr, form compounds, whereas He, Ne, and Ar do not.

14.104 (a) Why do stable xenon fluorides have an even number of F atoms?
(b) Why do the ionic species XeF_3^+ and XeF_7^- have odd numbers of F atoms?
(c) Predict the shape of XeF_3^+.

Comprehensive Problems

14.105 Xenon tetrafluoride reacts with antimony pentafluoride to form the ionic complex $[XeF_3]^+[SbF_6]^-$.
(a) Which depiction shows the molecular shapes of the reactants and product?
(b) How, if at all, does the hybridization of xenon change in the reaction?

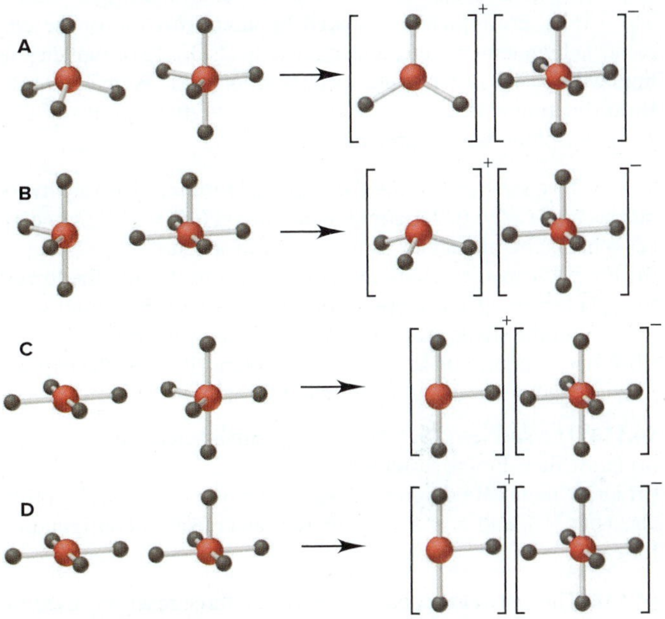

14.106 Given the following information,

$$H^+(g) + H_2O(g) \longrightarrow H_3O^+(g) \quad \Delta H = -720 \text{ kJ}$$
$$H^+(g) + H_2O(l) \longrightarrow H_3O^+(aq) \quad \Delta H = -1090 \text{ kJ}$$
$$H_2O(l) \longrightarrow H_2O(g) \quad \Delta H = 40.7 \text{ kJ}$$

calculate the heat of solution of the hydronium ion:

$$H_3O^+(g) \xrightarrow{H_2O} H_3O^+(aq)$$

14.107 The electronic transition in Na from $3p^1$ to $3s^1$ gives rise to a bright yellow-orange emission at 589.2 nm. What is the energy of this transition?

14.108 Unlike other Group 2A(2) metals, beryllium reacts like aluminum and zinc with concentrated aqueous base to release hydrogen gas and form oxoanions of formula $M(OH)_4^{n-}$. Write equations for the reactions of these three metals with NaOH.

14.109 The interhalogen IF undergoes the reaction depicted below (I is purple and F is green):

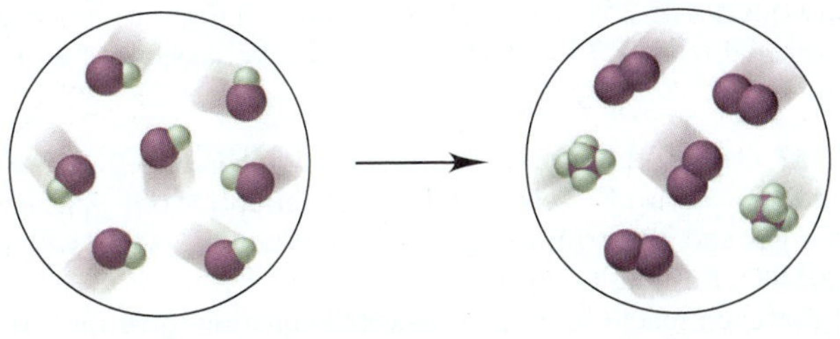

(a) Write the balanced equation. (b) Name the interhalogen product. (c) What type of reaction is shown? (d) If each molecule of IF represents 2.50×10^{-3} mol, what mass of each product forms?

14.110 The main reason alkali metal dihalides (MX_2) do *not* form is the high IE_2 of the metal.
(a) Why is IE_2 so high for alkali metals?
(b) The IE_2 for Cs is 2255 kJ/mol, low enough for CsF_2 to form exothermically ($\Delta H_f^\circ = -125$ kJ/mol). This compound cannot be synthesized, however, because CsF forms with a much greater release of heat ($\Delta H_f^\circ = -530$ kJ/mol). Thus, the breakdown of CsF_2 to CsF happens readily. Write the equation for this breakdown, and calculate the enthalpy of reaction per mole of CsF.

14.111 Semiconductors made from elements in Groups 3A(13) and 5A(15) are typically prepared by direct reaction of the elements at high temperature. An engineer treats 32.5 g of molten gallium with 20.4 L of white phosphorus vapor at 515 K and 195 kPa. If purification losses are 7.2% by mass, how many grams of gallium phosphide will be prepared?

14.112 Two substances with the empirical formula HNO are hyponitrous acid ($\mathcal{M} = 62.04$ g/mol) and nitroxyl ($\mathcal{M} = 31.02$ g/mol).
(a) What is the molecular formula of each species?
(b) For each species, draw the Lewis structure having the lowest formal charges. (*Hint:* Hyponitrous acid has an N=N bond.)
(c) Predict the shape around the N atoms of each species.
(d) When hyponitrous acid loses two protons, it forms the hyponitrite ion. Draw *cis* and *trans* forms of this ion.

14.113 The species CO, CN^-, and C_2^{2-} are isoelectronic.
(a) Draw their Lewis structures.
(b) Draw their MO diagrams (assume mixing of 2*s* and 2*p* orbitals, as in N_2), and give the bond order and electron configuration for each.

14.114 The Ostwald process is a series of three reactions used for the industrial production of nitric acid from ammonia.
(a) Write a series of balanced equations for the Ostwald process.
(b) If NO is *not* recycled, how many moles of NH_3 are consumed per mole of HNO_3 produced?
(c) In a typical industrial unit, the process is very efficient, with a 96% yield for the first step. Assuming 100% yields for the subsequent steps, what volume of concentrated aqueous nitric acid (60.% by mass; $d = 1.37$ g/mL) can be prepared for each cubic meter of a gas mixture that is 90.% air and 10.% NH_3 by volume at the industrial conditions of 5.0 atm and 850.°C?

14.115 All common plant fertilizers contain nitrogen compounds. Determine the mass % of N in each compound:
(a) Ammonia
(b) Ammonium nitrate
(c) Ammonium hydrogen phosphate

14.116 Producer gas is a fuel formed by passing air over red-hot coke (amorphous carbon). What mass of a producer gas that consists of 25% CO, 5.0% CO_2, and 70.% N_2 by mass can be formed from 1.75 metric tons (t) of coke, assuming an 87% yield?

14.117 Gaseous F_2 reacts with water to form HF and O_2. In NaOH solution, F_2 forms F^-, water, and oxygen difluoride (OF_2), a highly toxic gas and powerful oxidizing agent. The OF_2 reacts with excess OH^-, forming O_2, water, and F^-.
(a) For each reaction, write a balanced equation, give the oxidation state of O in all compounds, and identify the oxidizing and reducing agents.
(b) Draw a Lewis structure for OF_2, and predict its shape.

14.118 What is a disproportionation reaction, and which of the following fit the description?
(a) $I_2(s) + KI(aq) \longrightarrow KI_3(aq)$
(b) $2ClO_2(g) + H_2O(l) \longrightarrow HClO_3(aq) + HClO_2(aq)$
(c) $Cl_2(g) + 2NaOH(aq) \longrightarrow NaCl(aq) + NaClO(aq) + H_2O(l)$
(d) $NH_4NO_2(s) \longrightarrow N_2(g) + 2H_2O(g)$
(e) $3MnO_4^{2-}(aq) + 2H_2O(l) \longrightarrow$
$2MnO_4^-(aq) + MnO_2(s) + 4OH^-(aq)$
(f) $3AuCl(s) \longrightarrow AuCl_3(s) + 2Au(s)$

14.119 Explain the following observations:
(a) In reactions with Cl_2, phosphorus forms PCl_5 in addition to the expected PCl_3, but nitrogen forms only NCl_3.
(b) Carbon tetrachloride is unreactive toward water, but silicon tetrachloride reacts rapidly and completely. (To give what?)
(c) The sulfur-oxygen bond in SO_4^{2-} is shorter than expected for an S—O single bond.
(d) Chlorine forms ClF_3 and ClF_5, but ClF_4 is unknown.

14.120 Which group(s) of the periodic table is (are) described by each of the following general statements?
(a) The elements form compounds of VSEPR class AX_3E.
(b) The free elements are strong oxidizing agents and form monatomic ions and oxoanions.
(c) The atoms form compounds by combining with two other atoms that donate one electron each.
(d) The free elements are strong reducing agents, show only one nonzero oxidation state, and form mainly ionic compounds.
(e) The elements can form stable compounds with only three bonds, but as a central atom, they can accept a pair of electrons from a fourth atom without expanding their valence shell.
(f) Only larger members of the group are chemically active.

14.121 Diiodine pentoxide (I_2O_5) was discovered by Joseph Gay-Lussac in 1813, but its structure was unknown until 1970! Like Cl_2O_7, it can be prepared by the dehydration-condensation of the corresponding oxoacid.
(a) Name the precursor oxoacid, write a reaction for formation of the oxide, and draw a likely Lewis structure.
(b) Data show that the bonds to the terminal O are shorter than the bonds to the bridging O. Why?
(c) I_2O_5 is one of the few chemicals that can oxidize CO rapidly and completely; elemental iodine forms in the process. Write a balanced equation for this reaction.

14.122 Bromine monofluoride (BrF) disproportionates to bromine gas and bromine trifluoride or pentafluoride. Use the following to find ΔH_{rxn}° for the decomposition of BrF to its elements:

$$3BrF(g) \longrightarrow Br_2(g) + BrF_3(l) \qquad \Delta H_{rxn} = -125.3 \text{ kJ}$$
$$5BrF(g) \longrightarrow 2Br_2(g) + BrF_5(l) \qquad \Delta H_{rxn} = -166.1 \text{ kJ}$$
$$BrF_3(l) + F_2(g) \longrightarrow BrF_5(l) \qquad \Delta H_{rxn} = -158.0 \text{ kJ}$$

14.123 White phosphorus is prepared by heating phosphate rock [principally $Ca_3(PO_4)_2$] with sand and coke:

$$Ca_3(PO_4)_2(s) + SiO_2(s) + C(s) \longrightarrow CaSiO_3(s) + CO(g) + P_4(g) \text{ [unbalanced]}$$

How many kilograms of phosphate rock are needed to produce 315 mol of P_4, assuming that the conversion is 90.% efficient?

14.124 Element E forms an oxide of general structure A and a chloride of general structure B, shown at right. For the anion EF_5^-, what is (a) the molecular shape; (b) the hybridization of E; (c) the O.N. of E?

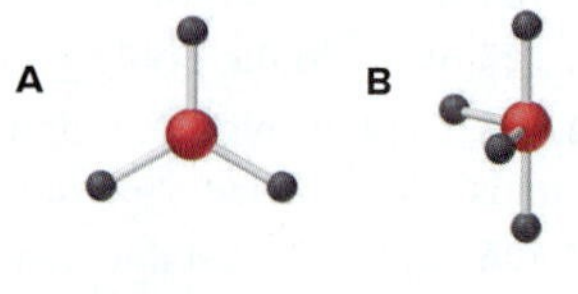

14.125 From its formula, one might expect CO to be quite polar, but its dipole moment is actually small (0.11 D).
(a) Draw the Lewis structure for CO.
(b) Calculate the formal charges.
(c) Based on your answers to parts (a) and (b), explain why the dipole moment is so small.

14.126 When an alkaline earth carbonate is heated, it releases CO_2, leaving the metal oxide. The temperature at which each Group 2A(2) carbonate yields a CO_2 partial pressure of 1 atm is listed below:

Carbonate	Temperature (°C)
$MgCO_3$	542
$CaCO_3$	882
$SrCO_3$	1155
$BaCO_3$	1360

(a) Suggest a reason for this trend.
(b) Mixtures of $CaCO_3$ and MgO are used to absorb dissolved silicates from boiler water. How would you prepare a mixture of $CaCO_3$ and MgO from dolomite, which contains $CaCO_3$ and $MgCO_3$?

14.127 The bond angles in the nitrite ion, nitrogen dioxide, and the nitronium ion (NO_2^+) are 115°, 134°, and 180°, respectively. Explain these values using Lewis structures and VSEPR theory.

14.128 A common method for producing a gaseous hydride is to treat a salt containing the anion of the volatile hydride with a strong acid.
(a) Write an equation for each of the following examples: (1) the production of HF from CaF_2; (2) the production of HCl from NaCl; (3) the production of H_2S from FeS.
(b) In some cases, even as weak an acid as water can be used for this preparation if the anion of the salt has a sufficiently strong attraction for protons. An example is the production of PH_3 from Ca_3P_2 and water. Write the equation for this reaction.
(c) By analogy, predict the products and write the equation for the reaction of Al_4C_3 with water.

14.129 Chlorine trifluoride was formerly used in the production of uranium hexafluoride for the U.S. nuclear industry:

$$U(s) + 3ClF_3(l) \longrightarrow UF_6(l) + 3ClF(g)$$

How many grams of UF_6 can form from 1.00 metric ton of uranium ore that is 1.55% by mass uranium and 12.75 L of chlorine trifluoride (d = 1.88 g/mL)?

14.130 Chlorine is used to make bleach solutions containing 5.25% NaClO (by mass). Assuming 100% yield in the reaction producing NaClO from Cl_2, how many liters of $Cl_2(g)$ at STP will be needed to make 1000. L of bleach solution (d = 1.07 g/mL)?

14.131 The triatomic molecular ion H_3^+ was first detected and characterized by J. J. Thomson using mass spectrometry. Use the bond energy of H_2 (432 kJ/mol) and the proton affinity of H_2 ($H_2 + H^+ \longrightarrow H_3^+$; $\Delta H = -337$ kJ/mol) to calculate the enthalpy of reaction for $H + H + H^+ \longrightarrow H_3^+$.

14.132 An atomic hydrogen torch is used for cutting and welding thick sheets of metal. When H_2 passes through an electric arc, the molecules decompose into atoms, which react with O_2. Temperatures over 5000°C are reached, which can melt all metals. Write equations for the breakdown of H_2 to H atoms and for the subsequent overall reaction of the H atoms with oxygen. Use Appendix B to find the standard enthalpy of each reaction per mole of product.

14.133 Which of the following oxygen ions are paramagnetic: O^+, O^-, O^{2-}, O^{2+}?

14.134 Copper(II) hydrogen arsenite ($CuHAsO_3$) is a green pigment once used in wallpaper. In damp conditions, mold metabolizes this compound to trimethylarsine [$(CH_3)_3As$], a highly toxic gas.
(a) Calculate the mass percent of As in each compound.
(b) How much $CuHAsO_3$ must react to reach a toxic level in a room that measures 12.35 m × 7.52 m × 2.98 m (arsenic is toxic at 0.50 mg/m^3)?

14.135 Hydrogen peroxide can act as either an oxidizing agent or a reducing agent.
(a) When H_2O_2 is treated with aqueous KI, I_2 forms. In which role is H_2O_2 acting? What oxygen-containing product is formed?
(b) When H_2O_2 is treated with aqueous $KMnO_4$, the purple color of MnO_4^- disappears and a gas forms. In which role is H_2O_2 acting? What is the oxygen-containing product formed?

15 Organic Compounds and the Atomic Properties of Carbon

15.1 The Special Nature of Carbon and the Characteristics of Organic Molecules
- Structural Complexity of Organic Molecules
- Chemical Diversity of Organic Molecules

15.2 The Structures and Classes of Hydrocarbons
- Carbon Skeletons and Hydrogen Skins
- Alkanes
- Dispersion Forces and the Physical Properties of Alkanes
- Constitutional Isomerism
- Chiral Molecules and Optical Isomerism
- Alkenes
- Geometric (*cis-trans*) Isomerism
- Alkynes
- Aromatic Hydrocarbons
- Catenated Inorganic Hydrides

15.3 Some Important Classes of Organic Reactions
- Types of Organic Reactions
- Organic Redox Reactions

15.4 Properties and Reactivities of Common Functional Groups
- Groups with Only Single Bonds
- Groups with Double Bonds
- Groups with Both Single and Double Bonds
- Groups with Triple Bonds

15.5 The Monomer-Polymer Theme I: Synthetic Macromolecules
- Addition Polymers
- Condensation Polymers

15.6 The Monomer-Polymer Theme II: Biological Macromolecules
- Sugars and Polysaccharides
- Amino Acids and Proteins
- Nucleotides and Nucleic Acids

Source: © McGraw-Hill Education/ Mark A. Dierker, photographer

Concepts and Skills to Review Before You Study This Chapter

- naming straight-chain alkanes (Section 2.8)
- constitutional isomerism (Section 3.2)
- ΔEN and bond polarity (Section 9.5)
- resonance structures (Section 10.1)
- VSEPR theory (Section 10.2)
- orbital hybridization (Section 11.1)
- σ and π bonding (Section 11.2)
- intermolecular forces and synthetic and biological macromolecules (Sections 12.3, 12.7, and 13.2)
- properties of the Period 2 elements (Section 14.2)
- properties of the Group 4A(14) elements (Section 14.6)

If you eat a snack during a study break, chances are you will consume organic compounds from containers composed of more organic compounds! Milk, apples, and cookies contain carbohydrates, fats, and proteins, and the plastic milk container and plate are a combination of polymeric organic compounds. In fact, except for a few inorganic salts and ever-present water, nearly everything you put into or on your body—food, medicine, cosmetics, and clothing—consists of organic compounds. Organic fuels warm our homes, cook our meals, and power our vehicles. Major industries are devoted to producing organic compounds, including plastics, pharmaceuticals, and insecticides.

What *is* an organic compound? According to the dictionary, it is "a compound of carbon," but that definition is too general because carbonates, cyanides, carbides, cyanates, and so forth, also contain carbon but are classified as inorganic. Here is a more specific definition: all **organic compounds** contain carbon, nearly always bonded to one or more other carbons and to hydrogen, and often to other elements.

In the early 19th century, organic compounds were usually obtained from living things, so they were thought to possess a spiritual "vital force," which made them impossible to synthesize and fundamentally different from inorganic compounds. But, in 1828, Friedrich Wöhler destroyed this notion when he obtained urea by heating ammonium cyanate. Today, we know that *the same chemical principles govern organic and inorganic systems* because the behavior of a compound—no matter how amazing—can be explained by understanding the properties of its elements.

IN THIS CHAPTER . . . *We see that the structures and reactivities of organic molecules emerge naturally from the properties of their component atoms.*

- We begin by reviewing the atomic properties of carbon and seeing how they lead to the complex structures and reactivity of organic molecules.
- We focus on drawing and naming hydrocarbons as a prelude to naming other types of organic compounds.
- We classify the main types of organic reactions in terms of bond order and apply them to the functional groups that characterize families of organic compounds.
- We examine the giant organic molecules of commerce and life—synthetic and natural polymers.

15.1 THE SPECIAL NATURE OF CARBON AND THE CHARACTERISTICS OF ORGANIC MOLECULES

Although there is nothing mystical about organic molecules, their indispensable role in biology and industry leads us to wonder if carbon is somehow special. Of course, each element has specific properties, but the atomic properties of carbon do give it bonding capabilities beyond those of any element, and this exceptional behavior leads to the two characteristics of organic molecules—structural complexity and chemical diversity.

The Structural Complexity of Organic Molecules

Most organic molecules have more complex structures than most inorganic molecules. A quick review of carbon's atomic properties and bonding behavior shows why.

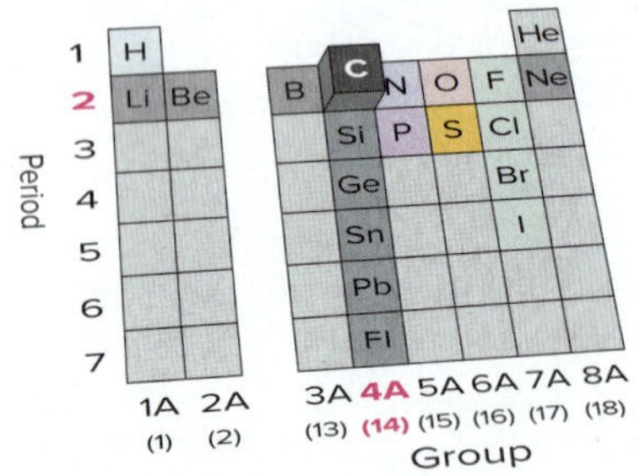

Figure 15.1 The position of carbon in the periodic table.

1. *Electron configuration, electronegativity, and bonding.* Carbon forms covalent, rather than ionic, bonds in all its elemental forms and compounds. This bonding behavior is the result of carbon's electron configuration and electronegativity:

- Carbon's ground-state electron configuration of [He] $2s^22p^2$—four electrons more than He and four fewer than Ne—means that the formation of carbon ions is energetically impossible under ordinary conditions. The loss of four e^- to form the C^{4+} cation requires the sum of IE_1 through IE_4 (>14,000 kJ/mol!); the gain of four e^- to form the C^{4-} anion requires the sum of EA_1 through EA_4, the last three steps of which are endothermic.
- Lying at the center of Period 2, carbon has an electronegativity (EN = 2.5) that is midway between that of the most metallic element (Li, EN = 1.0) and the most nonmetallic active element (F, EN = 4.0) (Figure 15.1).

2. *Catenation, bond properties, and molecular shape.* The *number* and *strength* of carbon's bonds lead to its property of **catenation,** the ability to bond to one or more other carbon atoms, which results in a multitude of chemically and thermally stable chain, ring, and branched compounds:

- Through the process of orbital hybridization (Section 11.1), carbon forms four bonds in virtually all its compounds, and they point in as many as four different directions.
- The small size of carbon allows close approach to another atom and thus greater orbital overlap, so carbon forms relatively short, strong bonds.
- The C—C bond is short enough to allow side-to-side overlap of half-filled, unhybridized *p* orbitals and the formation of multiple bonds. These restrict rotation of attached groups (see Figure 11.14), leading to additional structures.

3. *Molecular stability.* Although silicon and several other elements also catenate, none forms chains as stable as those of carbon. Atomic and bonding properties explain why carbon chains are so stable and, therefore, so common:

- *Atomic size and bond strength.* As atomic size increases down Group 4A(14), bonds between identical atoms become longer and weaker. Thus, a C—C bond (347 kJ/mol) is much stronger than an Si—Si bond (226 kJ/mol).
- *Relative enthalpies of reaction.* A C—C bond (347 kJ/mol), a C—O bond (358 kJ/mol), and a C—Cl bond (339 kJ/mol) have nearly the same energy, so relatively little heat is released when a C chain reacts and one bond replaces the other. In contrast, an Si—O bond (368 kJ/mol) or an Si—Cl bond (381 kJ/mol) is much stronger than an Si—Si bond (226 kJ/mol), so the large quantity of heat released when an Si chain reacts favors reactivity.
- *Orbitals available for reaction.* Unlike C, Si has low-energy *d* orbitals that can be attacked (occupied) by the lone pairs of incoming reactants. Thus, for example, ethane (CH_3—CH_3) is stable in water and reacts in air only when sparked, whereas disilane (SiH_3—SiH_3) breaks down in water and ignites spontaneously in air.

The Chemical Diversity of Organic Molecules

In addition to their complex structures, organic compounds are remarkable for their sheer number and diverse chemical behavior. Several million organic compounds are known, and thousands more are discovered or synthesized each year. This diversity is also founded on atomic behavior and is due to three interrelated factors—bonding to heteroatoms, variations in reactivity, and the occurrence of functional groups.

1. *Bonding to heteroatoms.* Many organic compounds contain **heteroatoms,** atoms other than C or H. The most common heteroatoms are N and O, but S, P, and the halogens often occur, and organic compounds with other elements are known as well. Figure 15.2 shows that 23 different molecular structures are possible from various arrangements of four C atoms singly bonded to each other, just one O atom (either singly or doubly bonded), and the necessary number of H atoms.

2. *Electron density and reactivity.* Most reactions start—that is, a new bond begins to form—*when a region of high electron density on one molecule meets a region of low electron density on another.* These regions may be due to the properties of a multiple bond or a carbon-heteroatom bond. Consider the reactivities of four bonds commonly found in organic molecules:

- *The C—C bond.* When C is singly bonded to another C, as occurs in portions of nearly every organic molecule, the EN values are equal and the bond is nonpolar. Therefore, in most cases, *C—C bonds are unreactive.*
- *The C—H bond.* This bond, which also occurs in nearly every organic molecule, is short (109 pm) and, with EN values of 2.1 for H and 2.5, very nearly nonpolar. Thus, *C—H bonds are largely unreactive.*
- *The C—O bond.* This bond, which occurs in many types of organic molecules, is highly polar (ΔEN = 1.0), with the O end electron rich and the C end electron poor. As a result of this imbalance in electron density, *the C—O bond is reactive* (easy to break), and, given appropriate conditions, a reaction will occur there.
- *Bonds to other heteroatoms.* Even when a carbon-heteroatom bond has a small ΔEN, such as that for C—Br (ΔEN = 0.3), or none at all, as for C—S (ΔEN = 0), the heteroatoms are generally large, so their bonds to carbon are long, weak, and thus *reactive.*

3. *Importance of functional groups.* The central feature of most organic molecules is the **functional group,** a specific combination of bonded atoms that reacts in a *characteristic* way no matter what molecule it occurs in. In nearly every case, *the reaction of an organic compound takes place at the functional group.* (In fact, as you'll see in Section 15.3, we often write a general symbol for the remainder of the molecule because it stays the same while the functional group reacts.) Functional groups vary from carbon-carbon multiple bonds to several combinations of carbon-heteroatom bonds, and each has its own pattern of reactivity. A particular bond may be a functional group itself or *part* of one or more functional groups. For example, the C—O bond occurs in four functional groups. We will discuss the reactivity of three of these groups in this chapter:

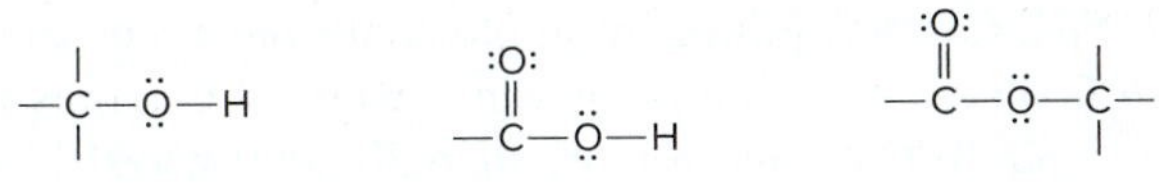

alcohol group　　carboxylic acid group　　ester group

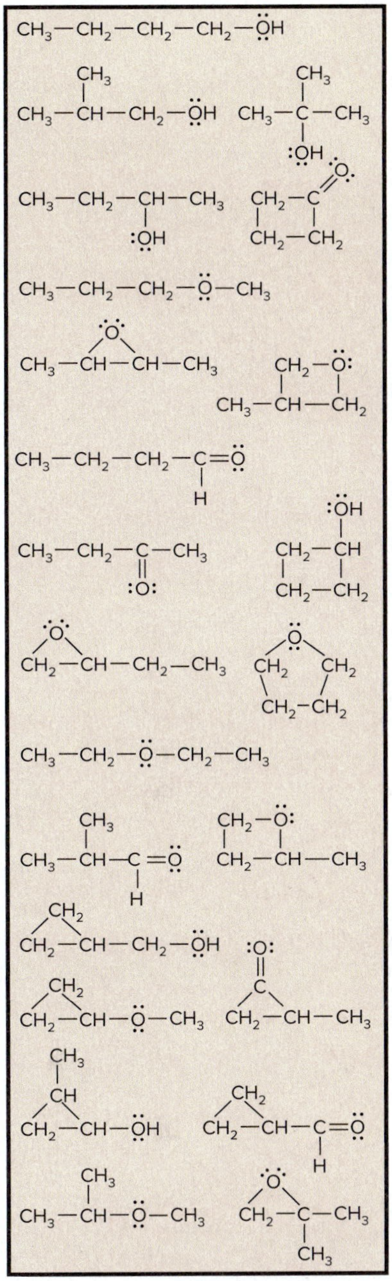

Figure 15.2 Heteroatoms and different bonding arrangements lead to great chemical diversity among organic molecules.

› Summary of Section 15.1

- › Carbon's small size, intermediate electronegativity, four valence electrons, and ability to form multiple bonds result in the structural complexity of organic compounds.
- › These factors lead to carbon's ability to catenate, which creates chains, branches, and rings of C atoms. Small size and the absence of *d* orbitals in the valence level lead to strong, chemically resistant bonds that point in as many as four directions from each C.
- › Carbon's ability to bond to many other elements, especially O and N, creates polar, reactive bonds, which leads to the chemical diversity of organic compounds.
- › Most organic compounds contain functional groups, specific combinations of bonded atoms that react in characteristic ways.

15.2 THE STRUCTURES AND CLASSES OF HYDROCARBONS

Let's make an anatomical analogy between an organic molecule and an animal. The carbon-carbon bonds form the skeleton: the longest continual chain is the backbone, and any branches are the limbs. Covering the skeleton is a skin of hydrogen atoms, with functional groups protruding at specific locations, like chemical hands ready to grab an incoming reactant. In this section, we "dissect" one group of compounds down to their skeletons and see how to draw and name them.

Hydrocarbons, the simplest type of organic compound, contain only H and C atoms. Some common fuels, such as natural gas and gasoline, are hydrocarbon mixtures. Some hydrocarbons, such as ethylene, acetylene, and benzene, are important *feedstocks,* that is, precursor reactants used to make other compounds.

Carbon Skeletons and Hydrogen Skins

We'll begin by examining the possible bonding arrangements of C atoms only (leaving off the H atoms for now) in simple skeletons without multiple bonds or rings. To distinguish different skeletons, focus on the *arrangement* of C atoms (that is, the successive linkages of one to another) and keep in mind that *groups joined by a single (sigma) bond are relatively free to rotate* (Section 11.2).

Number of C Atoms and Number of Arrangements One, two, or three C atoms can be arranged in only one way. Whether you draw three C atoms in a line or with a bend, the arrangement is the same. Four C atoms, however, have two possible arrangements—a four-C chain or a three-C chain with a one-C branch at the central C:

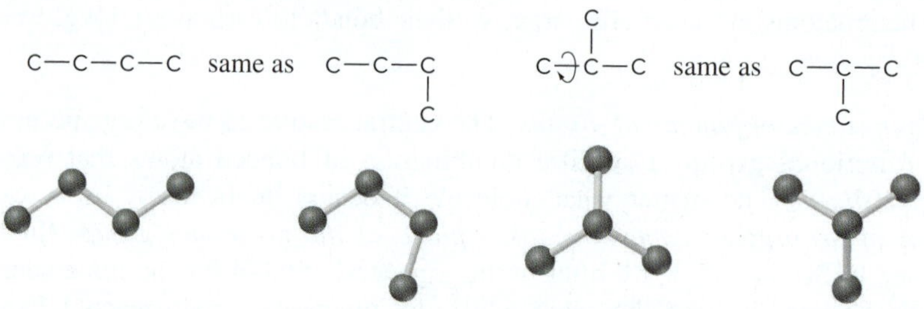

When we show the chains with bends due to the tetrahedral C (as in the ball-and-stick models above), the different arrangements are especially clear. Notice that if the branch is added to either end of the three-C chain, it is simply a bend in a four-C chain, *not* a different arrangement. Similarly, if the branch points down instead of up, it represents the same arrangement because single-bonded groups rotate.

As the total number of C atoms increases, the number of different arrangements increases as well. Five C atoms joined by single bonds have 3 possible arrangements (Figure 15.3A); 6 C atoms can be arranged in 5 ways, 7 C atoms in 9 ways, 10 C atoms in 75 ways, and 20 C atoms in more than 300,000 ways! If we include multiple bonds and rings, the number of arrangements increases further. For example,

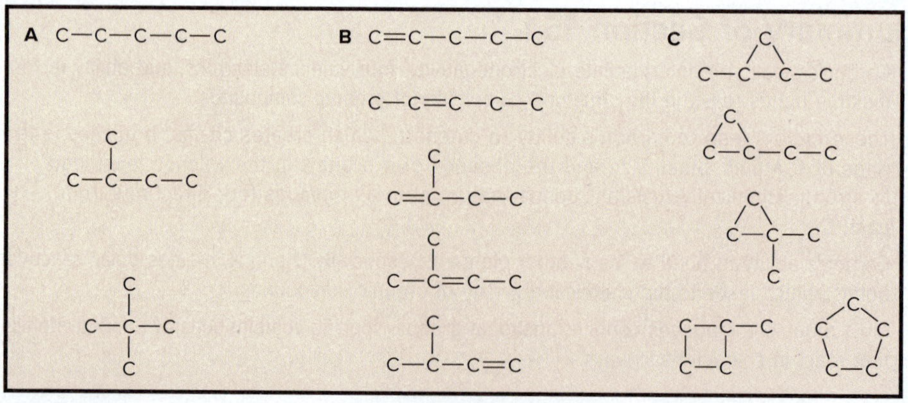

Figure 15.3 Some five-carbon skeletons.

including just one C=C bond in the five-C skeletons creates 5 more arrangements (Figure 15.3B), and including one ring creates 5 more (Figure 15.3C).

Drawing Hydrocarbons When determining the number of different skeletons for a given number of C atoms, remember that

- Each C atom can form four single bonds, two single and one double bond, or one single and one triple bond:

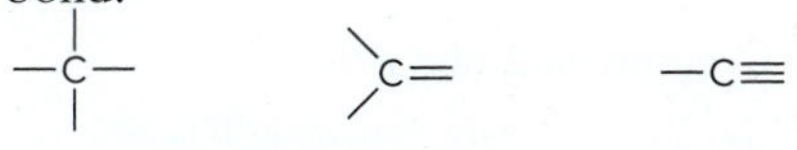

- A straight chain or a bent chain of C atoms represent the same skeleton.
- For a single-bonded arrangement, a branch pointing down is the same as one pointing up.
- A double bond *restricts rotation,* so different groups joined by a double bond can result in different arrangements.

If we put a hydrogen "skin" on a carbon skeleton, we obtain a hydrocarbon. Figure 15.4 shows that a skeleton has the correct number of H atoms *when each C has a total of four bonds.* Sample Problem 15.1 provides practice drawing hydrocarbons.

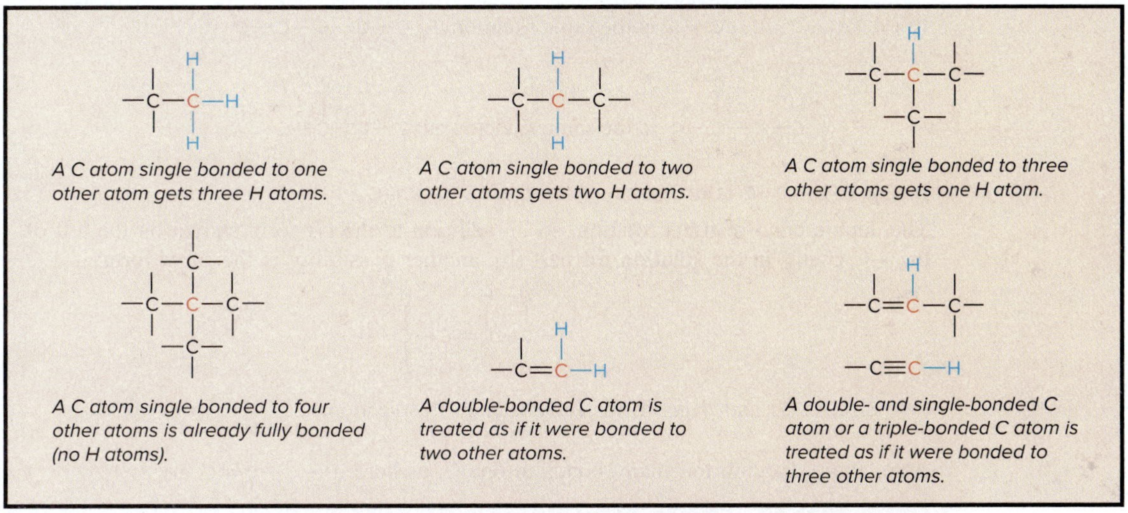
A C atom single bonded to one other atom gets three H atoms.
A C atom single bonded to two other atoms gets two H atoms.
A C atom single bonded to three other atoms gets one H atom.
A C atom single bonded to four other atoms is already fully bonded (no H atoms).
A double-bonded C atom is treated as if it were bonded to two other atoms.
A double- and single-bonded C atom or a triple-bonded C atom is treated as if it were bonded to three other atoms.

Figure 15.4 Adding the H-atom skin to the C-atom skeleton.

SAMPLE PROBLEM 15.1 Drawing Hydrocarbons

Problem Draw structures that have different atom arrangements for hydrocarbons with:

(a) Six C atoms, no multiple bonds, and no rings
(b) Four C atoms, one double bond, and no rings
(c) Four C atoms, no multiple bonds, and one ring

Plan In each case, we draw the longest carbon chain first and then work down to smaller chains with branches at different points along them. The process typically involves trial and error. Then, we add H atoms to give each C a total of four bonds.

Solution **(a)** Compounds with six C atoms:

6-C chain: 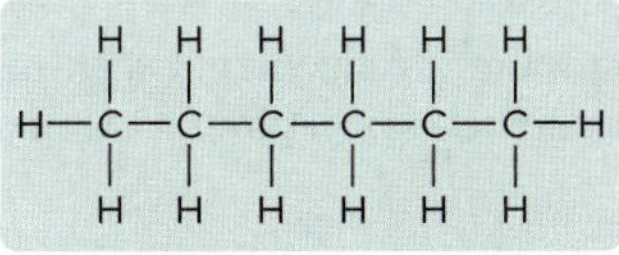

5-C chains and 1 branch: 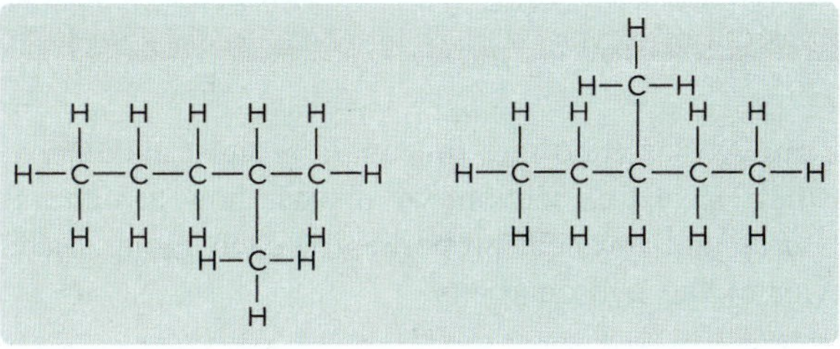

4-C chains and 2 branches: 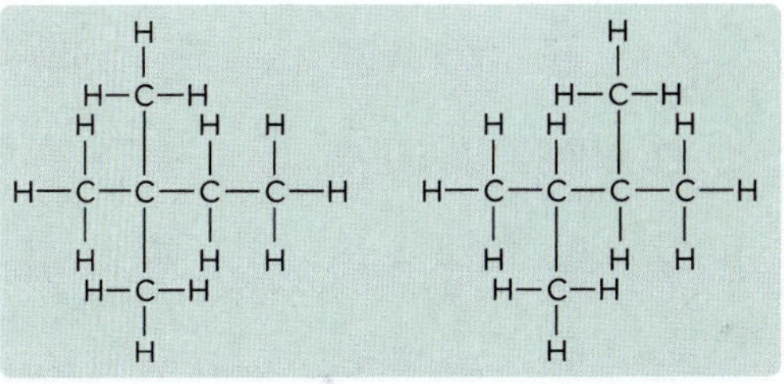

(b) Compounds with four C atoms and one double bond:

4-C chains 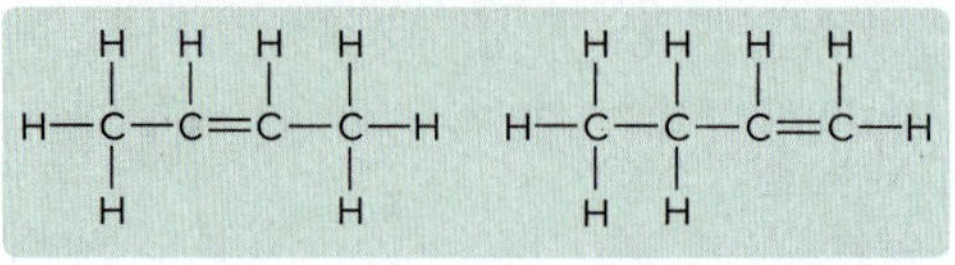 3-C chain and 1 branch: 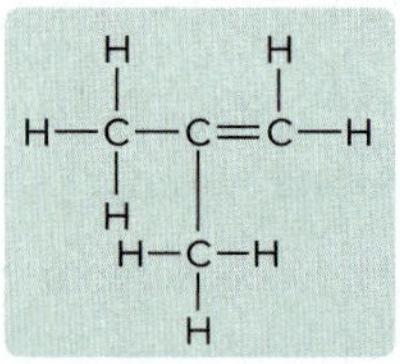

(c) Compounds with four C atoms and one ring:

4-C ring: 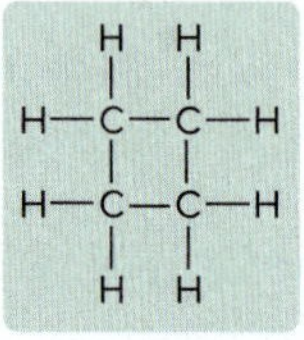 3-C ring: 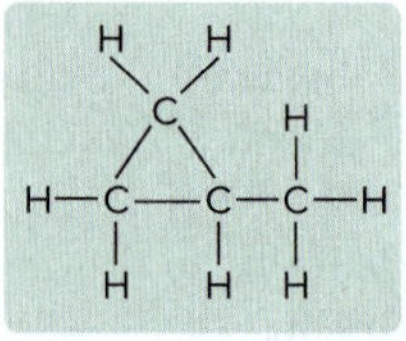

Check Be sure each skeleton has the correct number of C atoms, multiple bonds, and/or rings, and no arrangements are repeated or omitted; remember a double bond counts as two bonds.

Comment Avoid some *common mistakes:*

In (a): 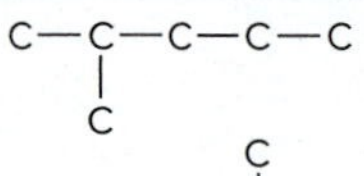 is the same skeleton as 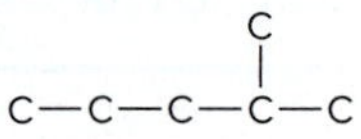

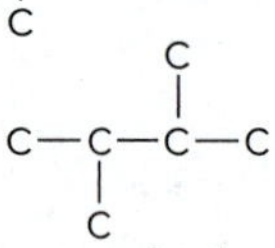 is the same skeleton as 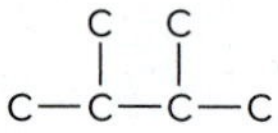

In (b): C—C—C═C is the same skeleton as C═C—C—C

The double bond restricts rotation, so, in addition to the *cis* form shown on the left of the 4-C chains in the solution for part (b), another possibility is the *trans* form:

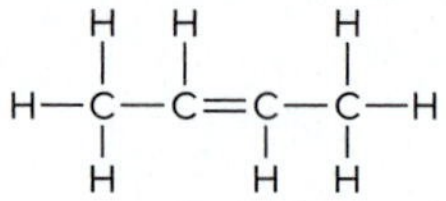

(We discuss *cis* and *trans* forms fully later in this section.)

Also, avoid drawing too many bonds to one C, as here: 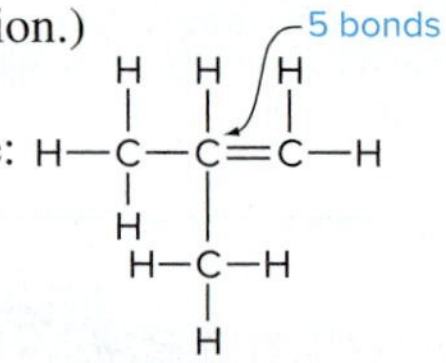

In (c): there would be too many bonds to one C in 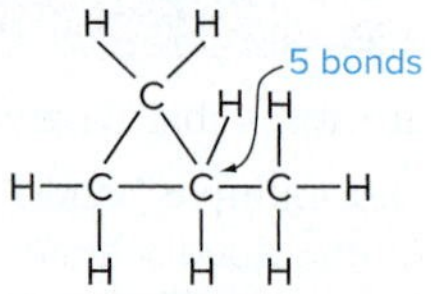

FOLLOW-UP PROBLEMS

*Brief Solutions to **all** Follow-up Problems appear at the end of the chapter.*

15.1A Draw all hydrocarbons that have different atom arrangements with **(a)** seven C atoms, no multiple bonds, and no rings (nine arrangements); **(b)** five C atoms, one triple bond, and no rings (three arrangements).

15.1B Draw all hydrocarbons that have different atom arrangements with **(a)** four C atoms, one ring, and one double bond (four arrangements); **(b)** four C atoms, no rings, and two double bonds (two arrangements).

SOME SIMILAR PROBLEMS 15.13–15.16

Hydrocarbons can be classified into four main groups: alkanes, alkenes, alkynes, and aromatics. In the rest of this section, we discuss how to name them, as well as some structural features and physical properties of each group. Later, we'll discuss the chemical behavior of the hydrocarbons.

Alkanes: Hydrocarbons with Only Single Bonds

Hydrocarbons that contain only single bonds are **alkanes** and have the following general features:

- Alkanes have the general formula $\mathbf{C_nH_{2n+2}}$, where n is a positive integer. For example, if $n = 5$, the formula is $C_5H_{[(2\times5)+2]}$, or C_5H_{12}.
- The alkanes comprise a **homologous series,** one in which each member differs from the next by a —CH_2— (methylene) group.
- In an alkane, each C is sp^3 hybridized.
- Because each C is bonded to the *maximum number (4) of other atoms* (C or H), alkanes are referred to as **saturated hydrocarbons.**

Naming Alkanes You learned the names of the 10 smallest straight-chain alkanes in Section 2.8. Here we discuss the rules for naming any alkane and, by extension, other organic compounds as well. The key point is that *each chain, branch, or ring has a name based on the **number** of C atoms.* The name of a compound has three portions:

PREFIX + ROOT + SUFFIX

- *Root:* The root tells the number of C atoms in the longest *continuous* chain in the molecule. The roots for the ten smallest alkanes are listed in Table 15.1. As you can see, there are special roots for compounds with chains of one to four C atoms; roots of longer chains are based on Greek numbers.
- *Suffix:* The suffix tells the *type of organic compound* that is being named; that is, it identifies the key functional group the molecule possesses. The suffix is placed *after* the root.
- *Prefix:* Each prefix identifies a *group attached to the main chain* and the number of the carbon atom to which it is attached. Prefixes identifying hydrocarbon branches are the same as root names (Table 15.1) but have *-yl* as their ending; for example, a two-carbon branch is an ethyl group. Each prefix is placed *before* the root.

Table 15.1 Numerical Roots for Carbon Chains and Branches

Roots	Number of C Atoms
meth-	1
eth-	2
prop-	3
but-	4
pent-	5
hex-	6
hept-	7
oct-	8
non-	9
dec-	10

For example, in the name 2-methylbutane, *2-methyl-* is the prefix, *-but-* is the root, and *-ane* is the suffix, with each portion indicating information about the molecular structure:

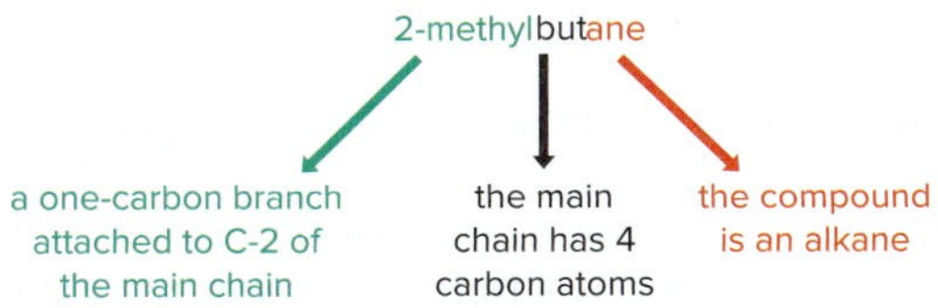

To obtain the systematic name of a compound,

1. Name the longest chain (ROOT).
2. Add the compound type (SUFFIX).
3. Name any branches (PREFIX).

Table 15.2 on the next page presents the rules for naming any organic compound and applies them to an alkane component of gasoline. Other organic compounds are named with a variety of other prefixes and suffixes (see Table 15.5). In addition to these *systematic* names, as you'll see, some *common* names are still in use.

Depicting Alkanes with Formulas and Models Chemists have several ways to depict organic compounds. Expanded, condensed, and carbon-skeleton formulas are easy to draw; ball-and-stick and space-filling models show the actual shapes and bond angles:

- The *expanded formula* is a Lewis structure, so it shows each atom and bond.
- One type of *condensed formula* groups each C atom with its H atoms; not all bonds are shown.
- The *carbon-skeleton formula* shows only carbon-carbon bonds and appears as a zig-zag line, with one or more branches as needed. *Each end or bend of a line or*

Table 15.2 Rules for Naming an Organic Compound

1. Naming the longest chain (root)
   (a) Find the longest *continuous* chain of C atoms.
   (b) Select the root that corresponds to the number of C atoms in this chain.

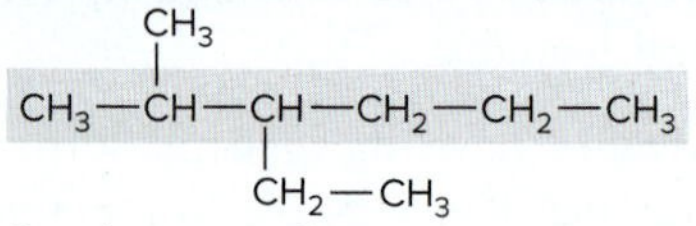

6 carbons ⟹ hex-

2. Naming the compound type (suffix)
   (a) For alkanes, add the suffix *-ane* to the chain root. (Other suffixes appear in Table 15.5 with their functional group and compound type.)
   (b) If the chain forms a ring, the name is preceded by *cyclo-*.

hex- + -ane ⟹ hexane

3. Naming the branches (prefixes) (If the compound has no branches, the name consists only of the root and suffix.)
   (a) Each branch name consists of a root (indicating the number of C atoms) and the ending *-yl* to signify that it is not part of the main chain.

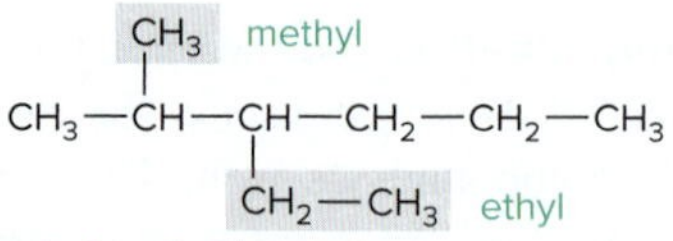

ethylmethylhexane

   (b) Branch names precede the chain name. When two or more branches are present, their names appear in *alphabetical* order.
   (c) To specify where a branch occurs along the chain, number the main-chain C atoms consecutively, starting at the end *closer* to the first branch, to achieve the *lowest* numbers for all the branches. Precede each branch name with the number of the main-chain C to which that branch is attached.

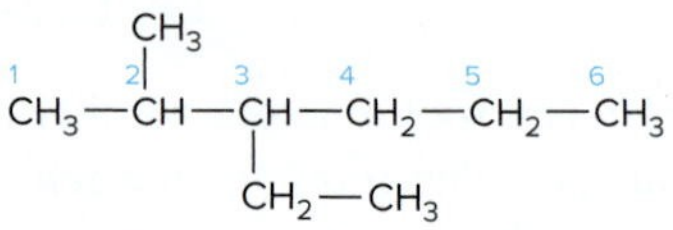

3-ethyl-2-methylhexane

branch represents a C atom attached to the number of H atoms that gives it four bonds:

propane 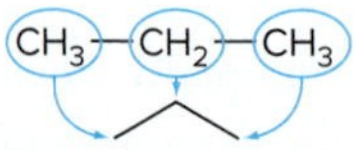

2,3-dimethylbutane 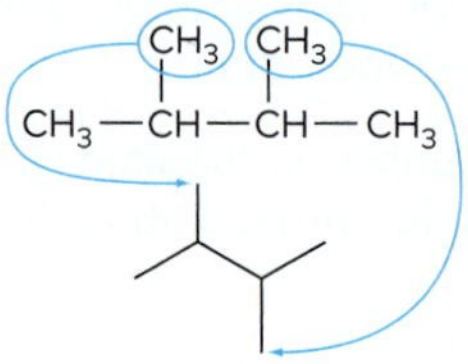

Figure 15.5 shows the formulas (and models) of the compound named in Table 15.2.

Cyclic Hydrocarbons A **cyclic hydrocarbon** contains one or more rings. When a straight-chain alkane (C_nH_{2n+2}) forms a ring, two H atoms are lost as the two ends of the chain form a C—C bond. Thus, *cycloalkanes* have the general formula C_nH_{2n}. Cyclic hydrocarbons are often drawn with carbon-skeleton formulas (Figure 15.6, *top row*). Except for three-carbon rings, *cycloalkanes are nonplanar,* as the ball-and-stick and space-filling models show. This structural feature arises from the tetrahedral shape around each C atom and the need to minimize electron repulsions between adjacent H atoms. The most stable form of cyclohexane is called the *chair conformation* (see the ball-and-stick model in Figure 15.6D).

Figure 15.5 Ways of depicting the alkane 3-ethyl-2-methylhexane.

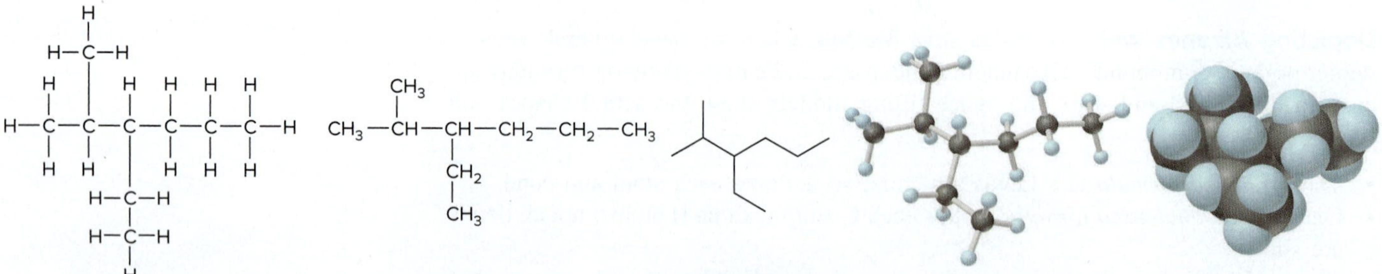

Expanded formula | Condensed formula | Carbon-skeleton formula | Ball-and-stick model | Space-filling model

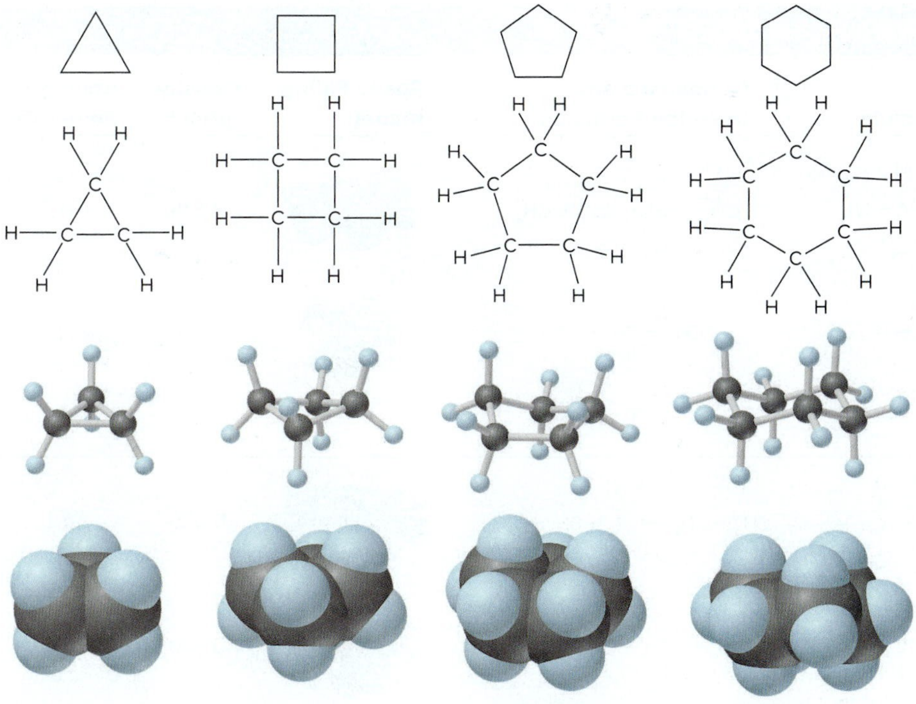

A Cyclopropane B Cyclobutane C Cyclopentane D Cyclohexane

Figure 15.6 Depicting cycloalkanes.

Dispersion Forces and the Physical Properties of Alkanes

Because alkanes are nearly nonpolar, we expect their physical properties and solubility behavior to be determined by dispersion forces (Section 12.3). A particularly clear example of this effect occurs among the unbranched alkanes (*n*-alkanes). Boiling points increase steadily with chain length (Figure 15.7): the longer the chain, the greater the molar mass and the greater the intermolecular contact, the stronger the dispersion forces, and, thus, the higher the boiling point. Pentane (five C atoms) is the smallest *n*-alkane that exists as a liquid at room temperature.

The solubility of alkanes, and of all hydrocarbons, is easy to predict from the like-dissolves-like rule (Section 13.1). Alkanes are miscible in each other and in other nonpolar solvents, such as benzene, but are nearly insoluble in water. The solubility of pentane in water, for example, is only 0.36 g/L at room temperature.

Constitutional Isomerism

Recall from Section 3.2 that two or more compounds that have the same molecular formula but different properties are called **isomers.** Those with *different arrangements*

Figure 15.7 Formulas, molar masses ($\mathcal{M}$, in g/mol), structures, and boiling points (°C, at 1 atm pressure) of the first 10 unbranched alkanes.

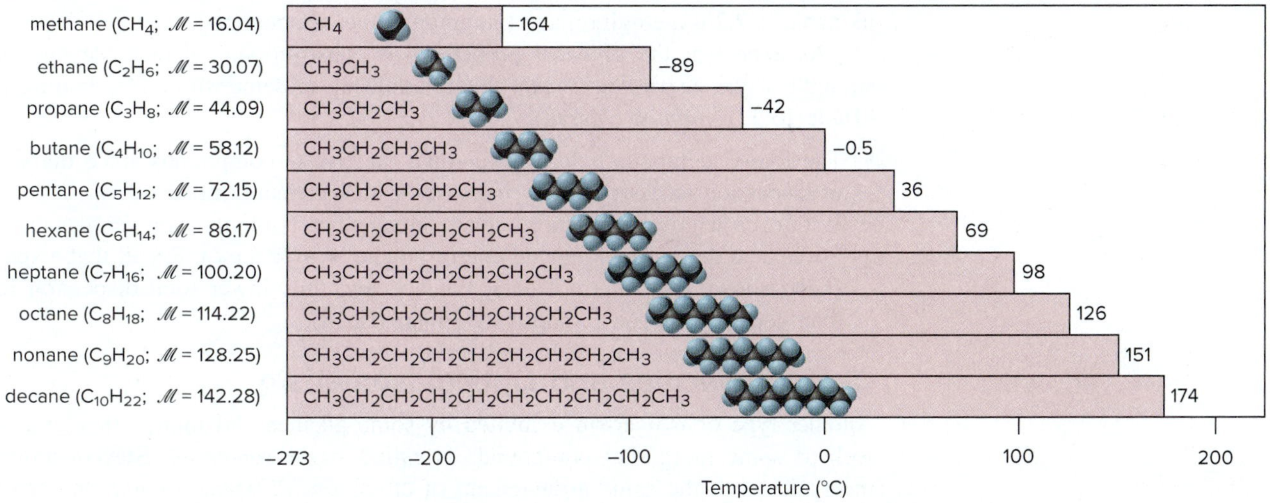

Table 15.3 The Constitutional Isomers of C_4H_{10} and C_5H_{12}

Systematic Name (Common Name)	Expanded Formula	Condensed and Skeleton Formulas	Space-Filling Model	Density (g/mL)	Boiling Point (°C)
Butane (*n*-butane)	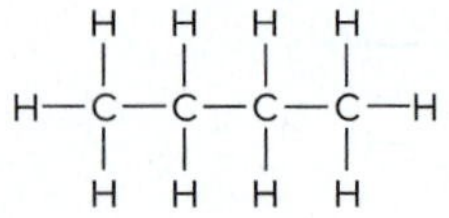	$CH_3-CH_2-CH_2-CH_3$ 	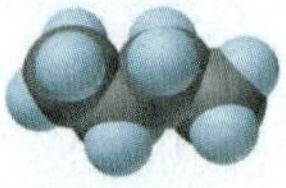	0.579	−0.5
2-Methylpropane (isobutane)	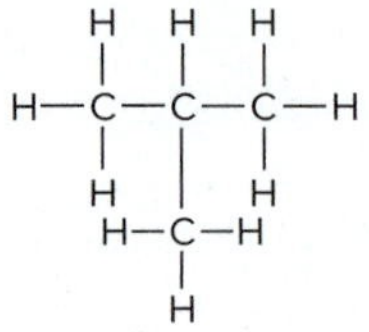	$CH_3-\underset{CH_3}{CH}-CH_3$ 	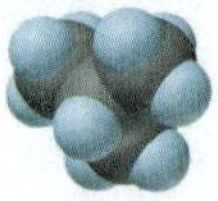	0.549	−11.6
Pentane (*n*-pentane)	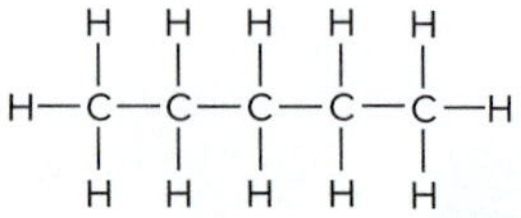	$CH_3-CH_2-CH_2-CH_2-CH_3$ 	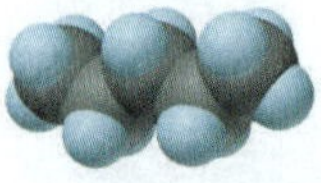	0.626	36.1
2-Methylbutane (isopentane)	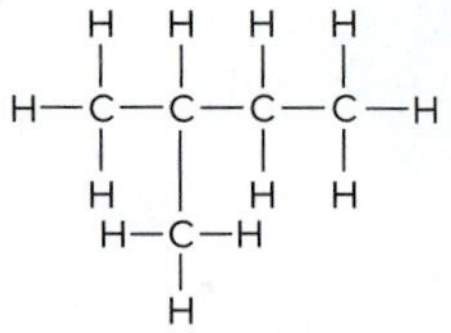	$CH_3-\underset{CH_3}{CH}-CH_2-CH_3$ 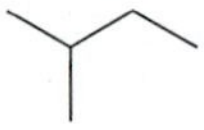	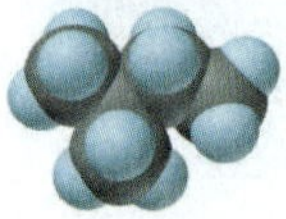	0.620	27.8
2,2-Dimethylpropane (neopentane)	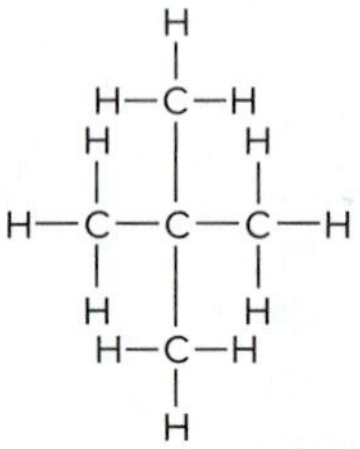	$CH_3-\overset{CH_3}{\underset{CH_3}{C}}-CH_3$ 	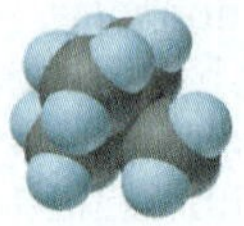	0.614	9.5

of bonded atoms are **constitutional** (or **structural**) **isomers;** alkanes with the same number of C atoms but different skeletons are examples. The smallest alkane to exhibit constitutional isomerism has four C atoms, C_4H_{10}; its two isomers are shown in the top of Table 15.3. The unbranched isomer is butane (common name, *n*-butane; *n*- stands for "normal," or having a straight chain), and the other is 2-methylpropane (common name, *iso*butane). Similarly, the bottom three isomers in the table have the formula C_5H_{12}. The unbranched isomer is pentane (common name, *n*-pentane); the one with a methyl branch at C-2 of a four-C chain is 2-methylbutane (common name, *iso*pentane). The third isomer has two methyl branches on C-2 of a three-C chain, so its name is 2,2-dimethylpropane (common name, *neo*pentane).

As expected, the physical properties of constitutional isomers depend on the strength of the dispersion forces; this relationship is demonstrated by boiling points (Table 15.3, *rightmost column*).

- The four-C compounds boil lower than the five-C compounds since the strength of dispersion forces generally increases with increasing molar mass.
- Within each group, the more spherical member (isobutane or neopentane) boils lower than the more elongated one (*n*-butane or *n*-pentane). Recall that a spherical shape allows less intermolecular contact, and thus lower total dispersion forces, than does an elongated shape (see Figure 12.18).

Chiral Molecules and Optical Isomerism

Another type of isomerism exhibited by some alkanes and many other organic (as well as some inorganic) compounds is called *stereoisomerism.* **Stereoisomers** are molecules with the same arrangement of atoms *but different orientations of groups in space. Optical isomerism* is one type of stereoisomerism: *when two objects are* *mirror images of each other and cannot be superimposed, they are* **optical isomers,** also called *enantiomers.* Your right hand is an optical isomer of your left. Look at your right hand in a mirror: the *image* is identical to your left hand (Figure 15.8). No matter how you twist your arms around, however, your hands cannot lie on top of each other with your palms facing in the same direction and be superimposed.

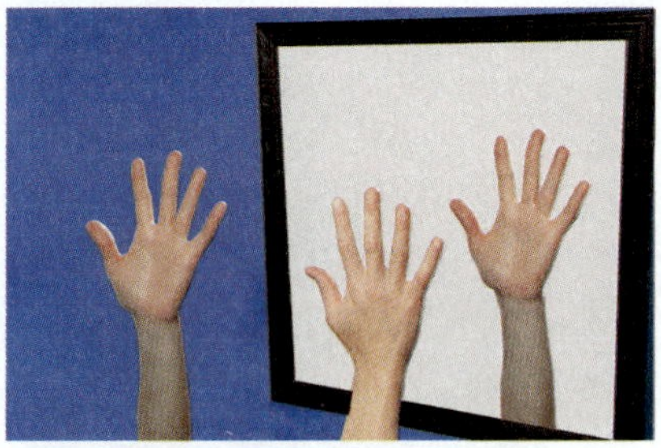

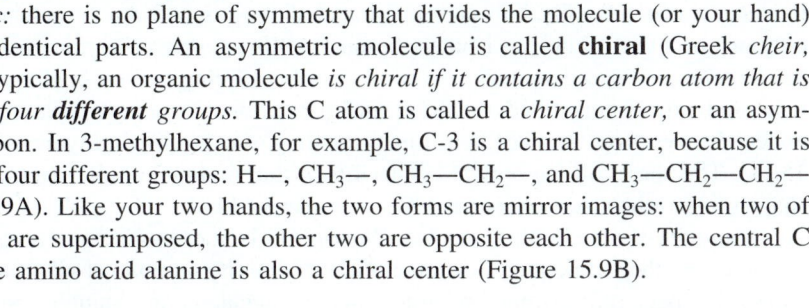

Figure 15.8 An analogy for optical isomers.

Source: © McGraw-Hill Education/Jill Braaten, photographer

Asymmetry and Chirality Optical isomers are not superimposable because each is *asymmetric:* there is no plane of symmetry that divides the molecule (or your hand) into two identical parts. An asymmetric molecule is called **chiral** (Greek *cheir,* "hand"). Typically, an organic molecule *is chiral if it contains a carbon atom that is bonded to four* ***different*** *groups.* This C atom is called a *chiral center,* or an asymmetric carbon. In 3-methylhexane, for example, C-3 is a chiral center, because it is bonded to four different groups: H—, CH_3—, CH_3—CH_2—, and CH_3—CH_2—CH_2— (Figure 15.9A). Like your two hands, the two forms are mirror images: when two of the groups are superimposed, the other two are opposite each other. The central C atom in the amino acid alanine is also a chiral center (Figure 15.9B).

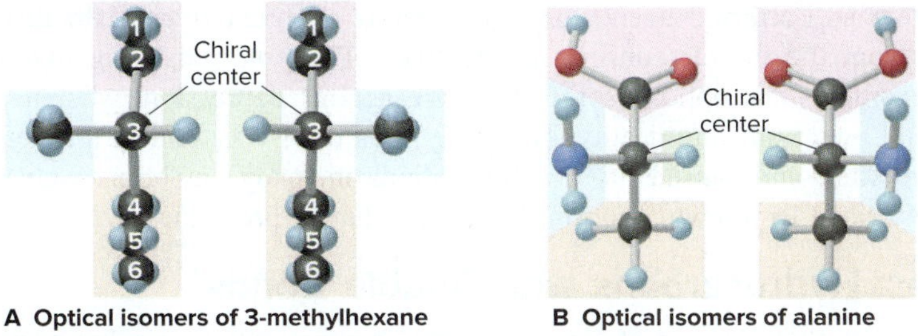

Figure 15.9 Two chiral molecules.

Properties of Optical Isomers Unlike constitutional isomers, which have different physical properties, optical isomers are identical in all but two respects:

1. In their physical properties, *optical isomers differ only in the direction that each isomer rotates the plane of polarized light.* A **polarimeter** is used to measure the angle that the plane is rotated (Figure 15.10). A beam of light consists of waves that oscillate in all planes. A polarizing filter blocks all waves except those in one plane, so the light emerging through the filter is *plane-polarized.* An optical isomer is **optically active** because it rotates the plane of this polarized light. (Liquid crystal displays incorporate polarizing filters and optically active compounds; see Figure 12.44.) The *dextrorotatory* isomer (designated *d* or +) rotates the plane of light clockwise; the *levorotatory* isomer (designated *l* or −) is the mirror image of the *d* isomer and rotates the plane counterclockwise. An equimolar mixture of the two isomers (called a *racemic mixture*) does not rotate the plane of light because the dextrorotation cancels the levorotation. The *specific rotation* is a characteristic, measurable property of an optical isomer at a certain temperature and concentration and with a particular wavelength of light.

2. In their chemical properties, *optical isomers differ only in a chiral (asymmetric) chemical environment,* one that distinguishes "right-handed" from "left-handed" molecules. As an analogy, your right hand fits well in your right glove but not in your left glove. Typically, one isomer of an optically active reactant is added to a mixture

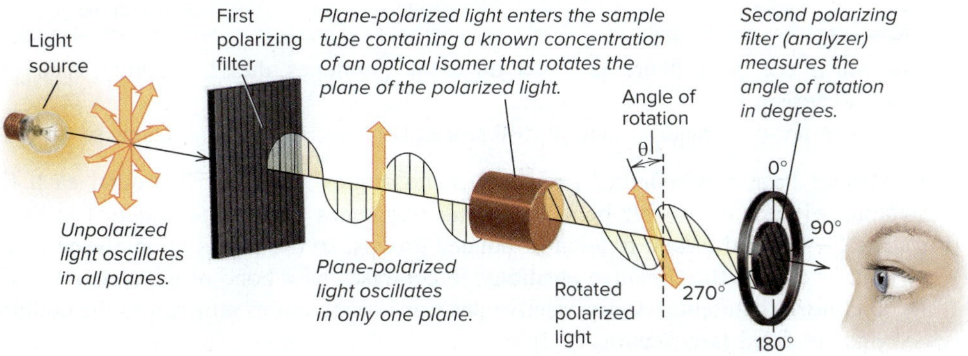

Figure 15.10 The rotation of plane-polarized light by an optically active substance.

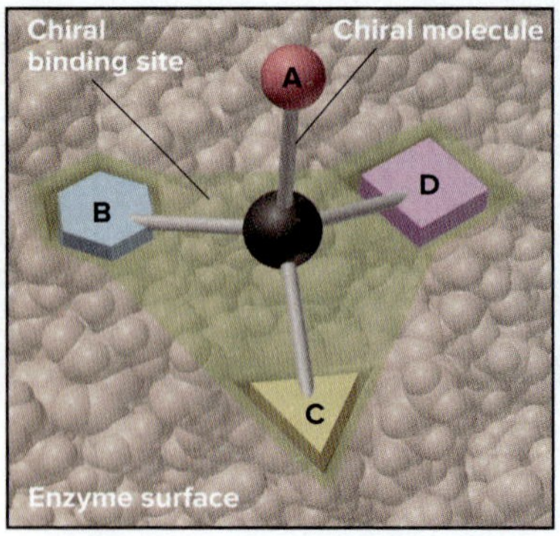

Figure 15.11 The binding site of an enzyme.

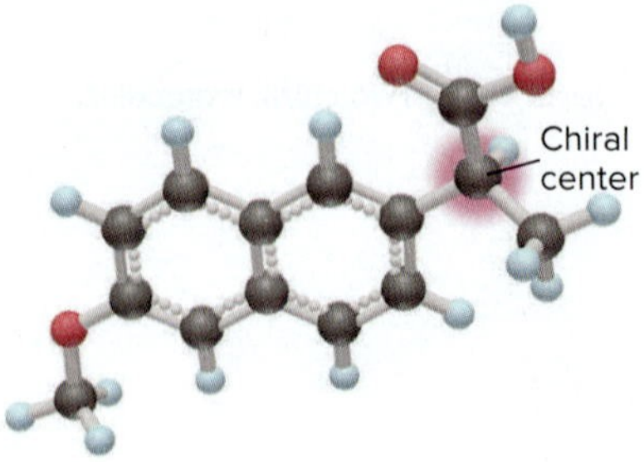

Naproxen

of optical isomers of another compound. The products of the reaction have different properties and can be separated.

The Role of Optical Isomerism in Organisms and Medicines Optical isomerism plays a vital role in living cells. Nearly all carbohydrates and amino acids are optically active, but only one of the isomers is biologically usable. For example, *d*-glucose is metabolized for energy, but *l*-glucose is not, and is excreted unused. Similarly, *l*-alanine is incorporated naturally into proteins, but *d*-alanine is not. An organism can utilize only one of a pair of optical isomers because of its enzymes (Section 16.7). Enzymes are proteins that speed virtually every reaction in a living cell by binding to the reactants and influencing bond breakage and formation. An enzyme distinguishes one optical isomer from another because its binding site is chiral (asymmetric) (Figure 15.11). The shape of one optical isomer fits at the binding site, but the mirror image shape of the other isomer does not fit, so it cannot bind.

Many drugs are chiral molecules. One optical isomer has certain biological activity, and the other has either a different type of activity or none at all. Naproxen *(see margin)*, a pain reliever and anti-inflammatory agent, is an example: one isomer is active as an anti-arthritic agent, and the other is a potent liver toxin that must be removed from the mixture during manufacture. The notorious drug thalidomide is another example: one optical isomer is active against depression, whereas the other causes fetal mutations and deaths. Tragically, this drug was sold in the 1950s as the racemic mixture and caused limb malformations in many newborns whose mothers had it prescribed to relieve "morning sickness" during pregnancy.

Alkenes: Hydrocarbons with Double Bonds

A hydrocarbon that contains at least one C═C bond is called an **alkene,** and has the following features:

- With two H atoms removed to make the double bond, alkenes have the general formula $\mathbf{C_nH_{2n}}$. The five carbon alkene ($n = 5$), for example, has a formula of $C_5H_{(2\times5)}$, or C_5H_{10}.
- The double-bonded C atoms are sp^2 hybridized.
- Because their carbon atoms are bonded to fewer than the maximum of four atoms each, alkenes are considered **unsaturated hydrocarbons.**

Alkene names differ from those of alkanes in two respects:

1. The main chain (root) *must* contain both C atoms of the double bond, even if it is not the longest chain. The chain is numbered from the end *closer* to the C═C bond, and the position of the bond is indicated by the number of the *first* C atom in it.
2. The suffix for alkenes is *-ene*.

For example, the three four-C alkenes (C_4H_8), two unbranched and one branched, whose structures were drawn in Sample Problem 15.1b, are named as follows:

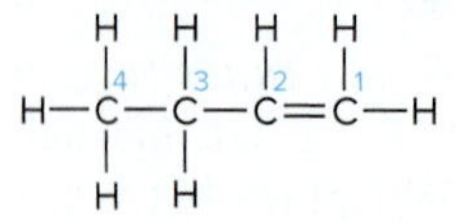

1-butene
(C═C bond between C-1 and C-2)

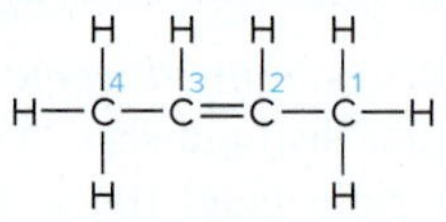

2-butene
(C═C bond between C-2 and C-3)

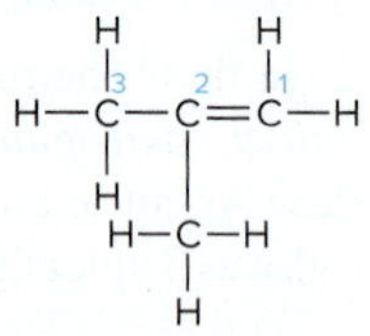

2-methylpropene
(methyl branch on C-2)

(As you'll see next, there are two different 2-butenes due to another type of stereoisomerism.)

There are two major structural differences between alkanes and alkenes.

- Alkanes have a *tetrahedral* geometry (bond angles of ∼109.5°) around each C atom, whereas the double-bonded C atoms in alkenes are *trigonal planar* (∼120°).
- The C—C bond *allows* rotation of bonded groups, so the atoms in an alkane continually change their relative positions; in contrast, the π bond of the alkene C═C bond *restricts* rotation, so the relative positions of the atoms attached to the double bond are fixed (see Section 11.2).

Table 15.4 The Geometric Isomers of 2-Butene

Systematic Name	Condensed and Skeleton Formulas	Space-Filling Model	Density (g/mL)	Boiling Point (°C)
cis-2-Butene	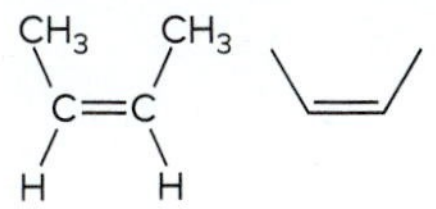	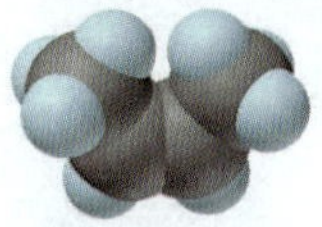	0.621	3.7
trans-2-Butene	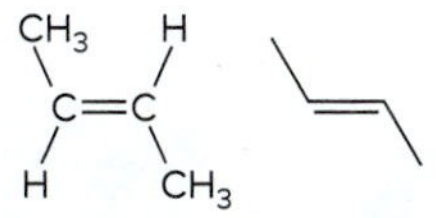		0.604	0.9

Restricted Rotation and Geometric *(cis-trans)* Isomerism

This rotational restriction of the π bond in the C═C bond leads to another type of stereoisomerism. **Geometric isomers** (also called ***cis-trans* isomers**) have different orientations of groups around a double bond (or similar structural feature). Table 15.4 shows the two geometric isomers of 2-butene (see Comment, Sample Problem 15.1):

- One isomer, *cis*-2-butene, has the CH_3 groups on the *same* side of the C═C bond; in general, the *cis* isomer has the *larger portions of the main chain* (in this case, two CH_3 groups) *on the same side* of the double bond.
- The other isomer, *trans*-2-butene, has the CH_3 on *opposite* sides of the C═C bond; the *trans* isomer generally has the *larger portions of the main chain on the opposite sides* of the double bond.

For a molecule to have geometric isomers, *each C atom in the C═C bond must also be bonded to two different groups.*

Like structural isomers, geometric isomers have different physical properties. Note in Table 15.4 that the two 2-butenes differ in molecular shape *and* physical properties. The *cis* isomer has a bend in the chain that the *trans* isomer lacks. In Chapters 10 and 12, you saw how such a difference affects molecular polarity and physical properties, which arise from differing strengths of intermolecular attractions.

Geometric Isomers and the Chemistry of Vision Differences in geometric isomers have profound effects in biological systems. For example, the first step in the sequence that allows us to see relies on the different shapes of two geometric isomers. *Retinal,* a 20-C compound consisting of a 15-C chain, including a ring, with five C═C bonds and five 1-C branches, is part of the molecule responsible for receiving light energy that is then converted to electrical signals transmitted to the brain. There are two biologically occurring isomers of retinal:

- The all-*trans* isomer has a *trans* orientation around all five double bonds and is elongated.
- The 11-*cis* isomer has a *cis* orientation around the C═C bond between C-11 and C-12 and, thus, is bent.

Certain cells of the retina are densely packed with *rhodopsin,* a large molecule consisting of a protein covalently bonded to 11-*cis* retinal. The initial chemical event in vision occurs when rhodopsin absorbs a photon of visible light. The energy range of visible photons (165–293 kJ/mol) encompasses the energy needed to break a C═C π bond (about 250 kJ/mol). Within a few millionths of a second after rhodopsin absorbs a photon, the 11-*cis* π bond of retinal breaks, the intact σ bond between C-11 and C-12 rotates, and the π bond re-forms to produce all-*trans* retinal (Figure 15.12, *next page*).

This rapid, and quite sizeable, change in retinal's shape causes the attached protein to change its shape as well, triggering a flow of ions into the retina's cells and initiating electrical impulses, which the optic nerve conducts to the brain. Meanwhile, the free all-*trans* retinal diffuses away from the protein and is changed back to the

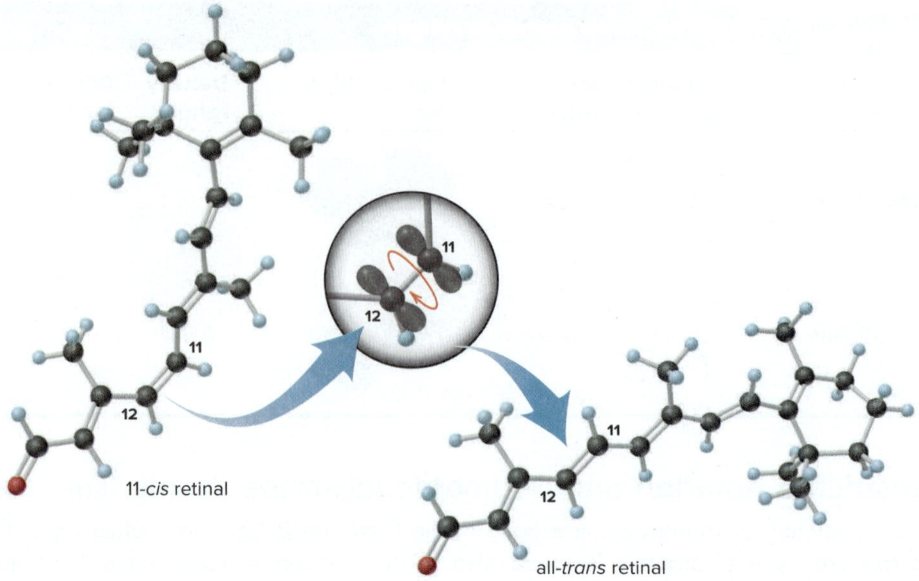

Figure 15.12 The initial chemical event in vision and the change in the shape of retinal.

cis form, which then binds to the protein portion again. Because of the speed and efficiency with which light causes such a large structural change in retinal, evolutionary selection has made it the photon absorber in organisms as different as purple bacteria, mollusks, insects, and vertebrates.

Alkynes: Hydrocarbons with Triple Bonds

Hydrocarbons that contain at least one C≡C bond are called **alkynes,** with the following features:

- Because they have two H atoms fewer than alkenes, their general formula is $\mathbf{C_nH_{2n-2}}$. The alkyne with five carbon atoms has the formula $C_5H_{[(2\times5)-2]}$, or C_5H_8.
- A carbon involved in a C≡C bond can bond to only one other atom, so the geometry around each C atom is linear (180°): each C is *sp* hybridized.
- Alkynes are named in the same way as alkenes, except that the suffix is *-yne.*

Because of their localized π electrons, C═C and C≡C bonds are electron rich and act as functional groups. Thus, alkenes and alkynes are much more reactive than alkanes, as we'll discuss in Section 15.4.

SAMPLE PROBLEM 15.2 Naming Hydrocarbons and Understanding Chirality and Geometric Isomerism

Problem Give the systematic name for each of the following, identify the chiral center in part (d), and draw two geometric isomers for the alkene in part (e):

(a)
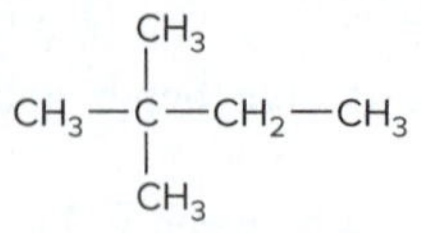

(b)
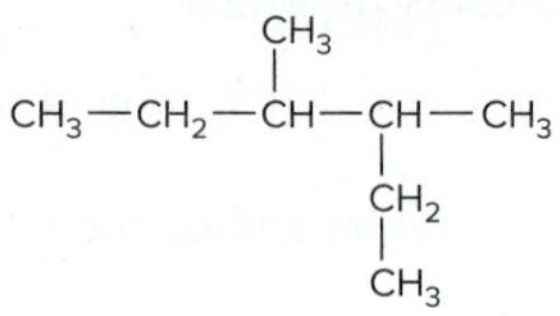

(c)
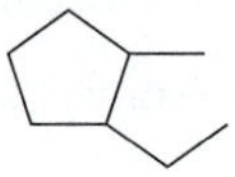

(d)
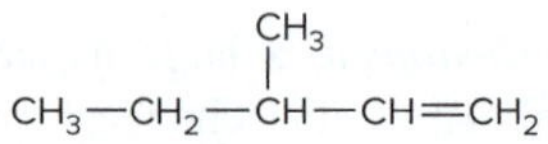

(e)
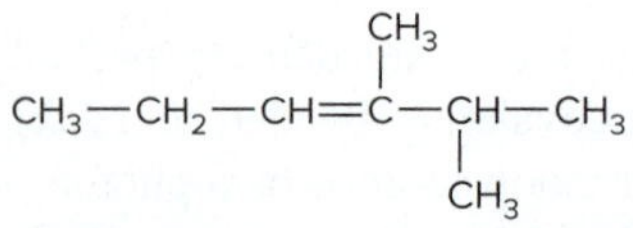

Plan For **(a)** to **(c),** we refer to Table 15.2. We first name the longest chain (*root-*) and add the suffix *-ane* because there are only single bonds. Then we find the *lowest* branch numbers by counting C atoms from the end *closer* to a branch. Finally, we name each branch (*root-* + *-yl*) and put these names alphabetically before the root name. For **(c)**, we add *cyclo-* before the root. For **(d)** and **(e)**, we number the longest chain that *includes* the multiple bond starting at the end closer to it. For **(d)**, the chiral center is the C atom bonded to four different groups. In **(e)**, the *cis* isomer has larger groups on the same side of the double bond, and the *trans* isomer has them on opposite sides.

Solution

(a)

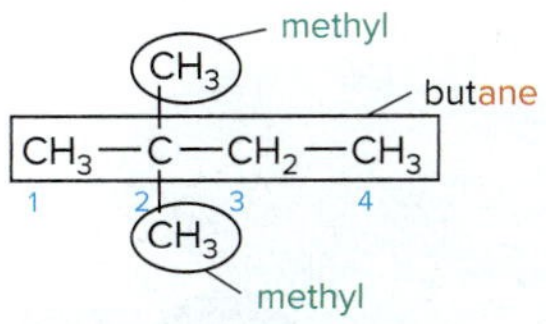

2,2-dimethylbutane

When a type of branch appears more than once, we group the branch numbers and indicate the number of branches with a prefix, as in 2,2-*di*methyl.

(b)

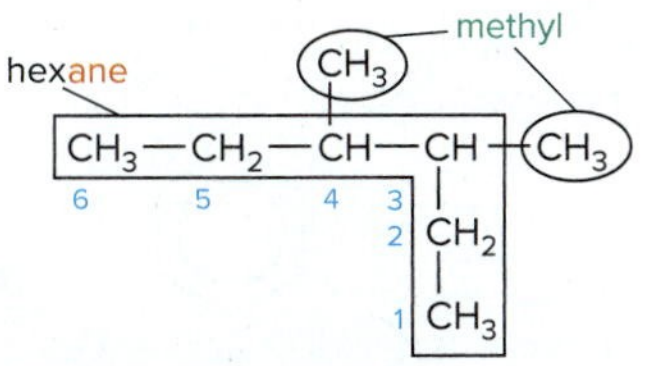

3,4-dimethylhexane

In this case, we can number the chain from either end because the branches are the same and are attached to the two central C atoms.

(c)

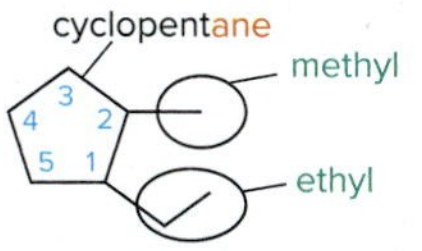

1-ethyl-2-methylcyclopentane

We number the ring C atoms so that a branch is attached to C-1.

(d)

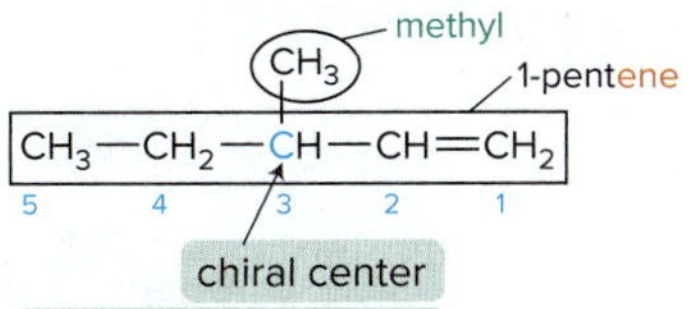

3-methyl-1-pentene

(e)

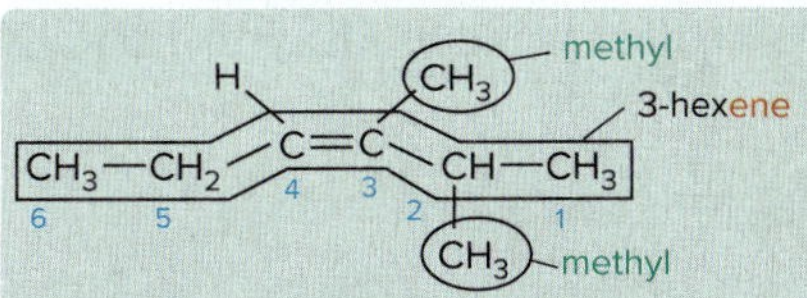

cis-2,3-dimethyl-3-hexene

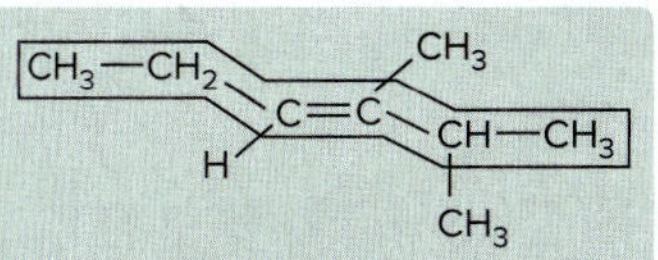

trans-2,3-dimethyl-3-hexene

Check A good check (and excellent practice) is to reverse the process by drawing structures for the names to see if you come up with the structures given in the problem.

Comment **1.** In (b), C-3 and C-4 are chiral centers, as are C-1 and C-2 in (c). However, in (b) the molecule is not chiral: it has a plane of symmetry between C-3 and C-4, so each half of the molecule rotates light in opposite directions. **2.** Avoid these common mistakes: In (b), 2-ethyl-3-methylpentane would be wrong: the longest chain is *hexane.* In (c), 1-methyl-2-ethylcyclopentane would be wrong: the branch names should appear *alphabetically.*

FOLLOW-UP PROBLEMS

15.2A Give the systematic name for each of the following, name the geometric isomer in (c), and identify the chiral centers in part (d):

(a)

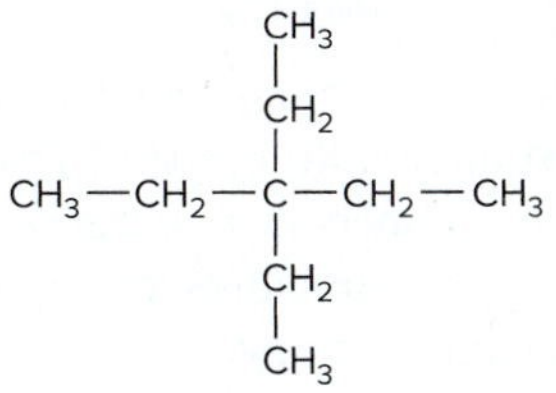

(b)

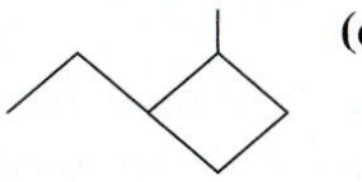

(c)

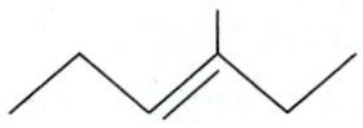

(d)

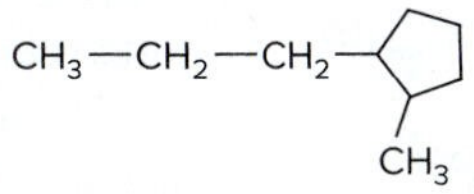

15.2B Draw condensed formulas for the following compounds: **(a)** 3-ethyl-3-methyloctane; **(b)** 1-ethyl-3-propylcyclohexane (also draw a carbon-skeleton formula for this compound); **(c)** 3,3-diethyl-1-hexyne; **(d)** *trans*-3-methyl-3-heptene.

SOME SIMILAR PROBLEMS 15.19–15.30

Aromatic Hydrocarbons: Cyclic Molecules with Delocalized π Electrons

Unlike most cycloalkanes, **aromatic hydrocarbons** are planar molecules, usually with one or more rings of six C atoms, and are often drawn with alternating single and double bonds. As you learned for benzene (Section 10.1), however, all the ring bonds are identical, with values of length and strength *between* those of a C—C and a C═C bond. To indicate this, benzene is also shown as a resonance hybrid, with a circle (or dashed circle) representing the delocalized character of the π electrons (Figure 15.13A). An orbital picture shows the two lobes of the delocalized π cloud above and below the hexagonal plane of the σ-bonded C atoms (Figure 15.13B, *next page*).

The systematic naming of aromatic compounds in which benzene is the main structure is straightforward, because attached groups, or *substituents,* are named as prefixes. However, many common names are still in use. For example, benzene with one methyl group attached is systematically named *methylbenzene* but is better known

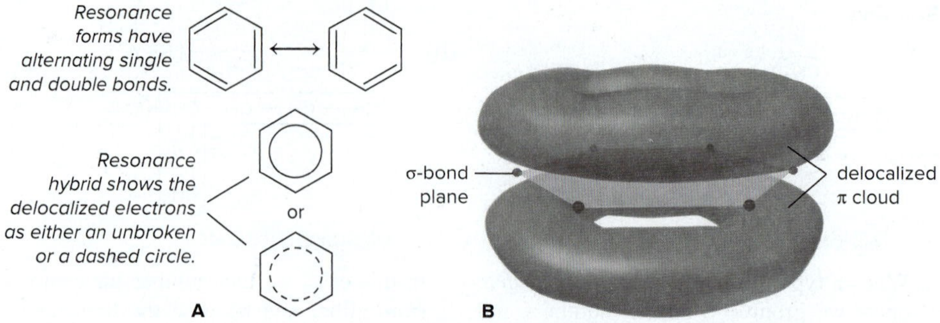
Resonance forms have alternating single and double bonds.
Resonance hybrid shows the delocalized electrons as either an unbroken or a dashed circle.
or
A
σ-bond plane
delocalized π cloud
B

Figure 15.13 Representations of benzene.

by its common name, *toluene.* With only one substituent present, as in toluene, we do not number the ring C atoms; when two or more groups are attached, however, we number in such a way that one of the groups is attached to ring C-1:

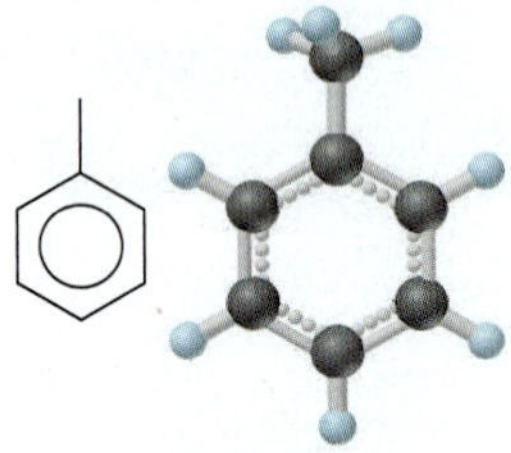

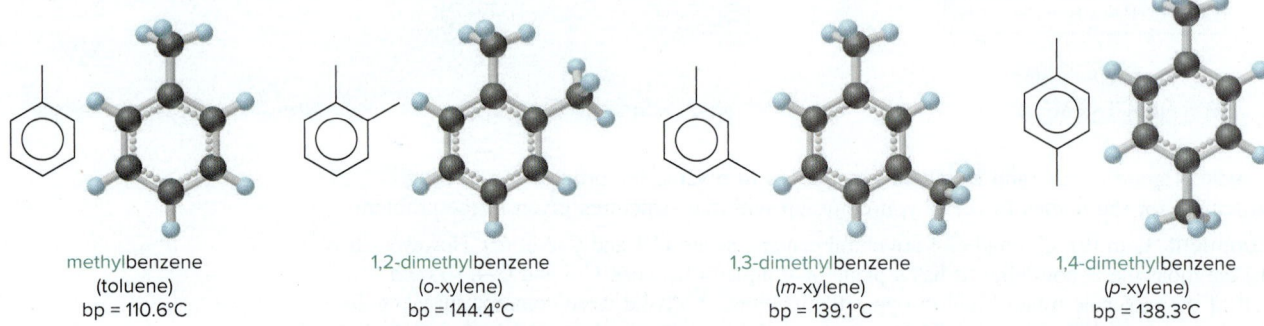

methylbenzene
(toluene)
bp = 110.6°C

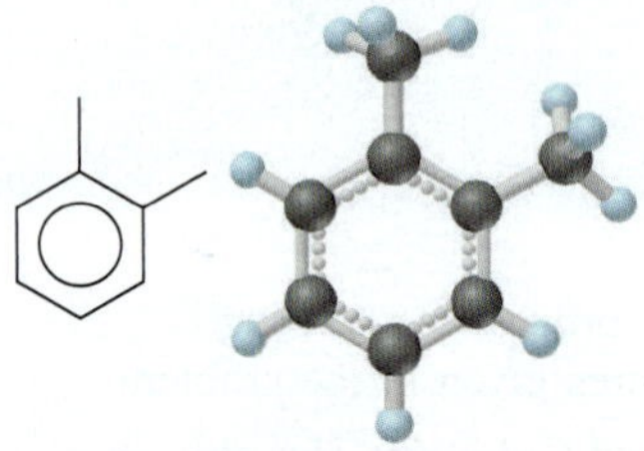

1,2-dimethylbenzene
(*o*-xylene)
bp = 144.4°C

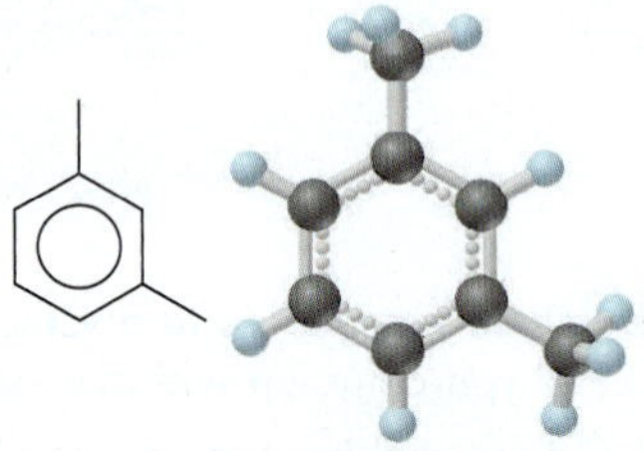

1,3-dimethylbenzene
(*m*-xylene)
bp = 139.1°C

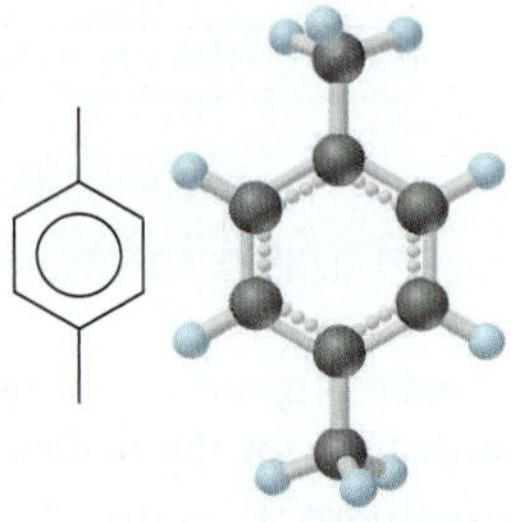

1,4-dimethylbenzene
(*p*-xylene)
bp = 138.3°C

In common names, the positions of two groups are indicated by *o*- (ortho) for groups on adjacent ring C atoms, *m*- (meta) for groups separated by one ring C atom, and *p*- (para) for groups on opposite ring C atoms. The dimethylbenzenes (commonly known as *xylenes*) are important solvents and feedstocks for polyester fibers and dyes. Benzene and many other aromatic hydrocarbons have been shown to have carcinogenic (cancer-causing) activity.

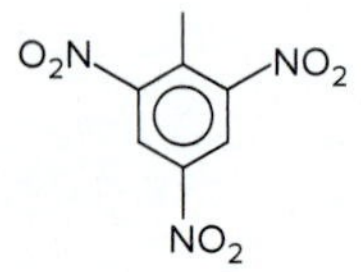
O_2N
NO_2
NO_2

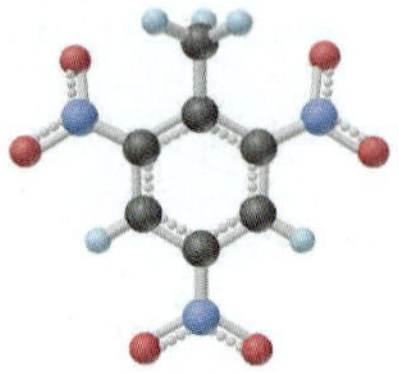

2,4,6-trinitromethylbenze
(trinitrotoluene, TNT)

The number of isomers increases with more than two attached groups. For example, there are six isomers for a compound with one methyl and three nitro ($—NO_2$) groups attached to a benzene ring; the explosive TNT is one of the isomers *(see margin).*

One of the most important methods for determining the structures of organic molecules is discussed in the upcoming Tools of the Laboratory essay.

Variations on a Theme: Catenated Inorganic Hydrides

In brief discussions throughout this chapter, called Variations on a Theme, we examine similarities between organic and inorganic compounds. Although no other element has as many different hydrides as carbon, catenated hydrides occur for several other elements, and many ring, chain, and cage structures are known.

- *Boranes.* One such group of inorganic compounds is the boron hydrides, or boranes (Section 14.5). Although their varied shapes rival those of the hydrocarbons, the weakness of their unusual bridge bonds renders most of them thermally and chemically unstable.
- *Silanes.* An obvious structural similarity exists between alkanes and the silicon hydrides, or silanes. Silanes even have an analogous general formula (Si_nH_{2n+2}). Branched silanes are known, but no cyclic or unsaturated (containing an Si═Si bond) compounds had been prepared until very recently. Unlike alkanes, silanes are unstable thermally and ignite spontaneously in air.
- *Polysulfanes.* Sulfur's ability to catenate is second only to carbon's, and many chains and rings occur among its allotropes (Section 14.8). A number of sulfur hydrides, or polysulfanes, are known. However, these molecules are unbranched chains with H atoms only at the ends ($H—S_n—H$). Like the silanes, the polysulfanes are oxidized easily and decompose readily, yielding sulfur's only stable hydride, H_2S, and its most stable allotrope, cyclo-S_8.

Nuclear Magnetic Resonance (NMR) Spectroscopy

In addition to mass spectrometry (Chapter 2) and infrared (IR) spectroscopy (Chapter 9), one of the most useful tools for analyzing organic and biochemical structures is **nuclear magnetic resonance (NMR) spectroscopy,** which measures the molecular environments of certain nuclei in a molecule.

Like electrons, several types of nuclei, such as ^{13}C, ^{19}F, ^{31}P, and ^{1}H, act as if they spin in either of two directions, each of which creates a tiny magnetic field. In this discussion, we focus primarily on ^{1}H-NMR spectroscopy, which measures changes in the nuclei of the most common isotope of hydrogen. Oriented randomly, the magnetic fields of all the ^{1}H nuclei in a sample of compound, when placed in a strong external magnetic field ($\boldsymbol{B_0}$), become aligned either *with* the external field (parallel) or *against* it (antiparallel). Most nuclei adopt the parallel orientation, which is slightly lower in energy. The energy difference (ΔE) between the two energy states (spin states) lies in the radio-frequency (rf) region of the electromagnetic spectrum (Figure B15.1).

When an ^{1}H *(blue arrow)* in the lower energy (parallel) spin state absorbs a photon in the radio-frequency region with an energy equal to ΔE, it "flips," in a process called *resonance,* to the higher energy (antiparallel) spin state. The system then re-emits that energy, which is detected by the rf receiver of the ^{1}H-NMR spectrometer. The ΔE between the two states depends on the *actual* magnetic field acting on each ^{1}H nucleus, which is affected by the tiny magnetic fields of the *electrons* of atoms adjacent to that nucleus. Thus, the ΔE required for resonance of each ^{1}H nucleus depends on its specific molecular environment—the C atoms, electronegative atoms, multiple bonds, and aromatic rings around it. ^{1}H nuclei in different molecular environments produce different peaks in the ^{1}H-NMR spectrum.

An ^{1}H-NMR spectrum, which is unique for each compound, is a series of peaks that represents the resonance as a function of the changing magnetic field. The *chemical shift* of the ^{1}H nuclei in a given environment is where a peak appears. Chemical shifts are shown relative to that of an added standard, tetramethylsilane [$(CH_3)_4Si$, or TMS]. TMS has 12 ^{1}H nuclei bonded to four C atoms that are bonded to one Si atom in a tetrahedral arrangement, so all 12 are in identical environments and produce only one peak.

Figure B15.2 shows the ^{1}H-NMR spectrum of acetone. The six ^{1}H nuclei of acetone have identical environments: all six are bonded to two C atoms that are each bonded to the C atom involved in the C=O bond. So one peak is produced, but at a different position from the TMS peak. The spectrum of dimethoxymethane in Figure B15.3 shows *two* peaks in addition to the TMS peak since the ^{1}H nuclei have two different evironments. The taller peak is due to the six ^{1}H nuclei in the two CH_3 groups, and the shorter peak is due to the two ^{1}H nuclei in the CH_2 group. The area under each peak (given as a number of chart-paper grid spaces) is proportional to *the number of* ^{1}*H nuclei in a given environment*. Note that the area ratio is $20.3/6.8 \approx 3/1$, the same as the ratio of six nuclei in the CH_3 groups to two in the CH_2 group. Thus, by analyzing the chemical shifts and peak areas, the chemist learns the type and number of hydrogen atoms in the compound.

(continued)

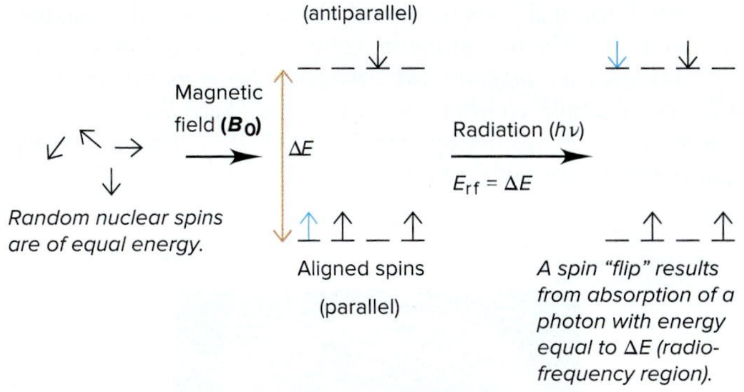

Figure B15.1 The basis of ^{1}H spin resonance.

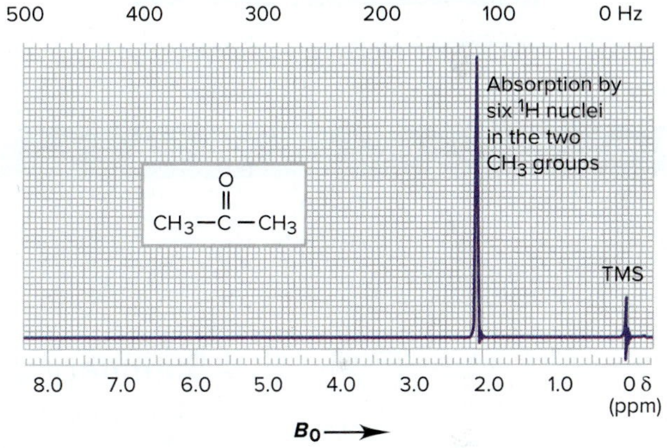

Figure B15.2 The ^{1}H-NMR spectrum of acetone.

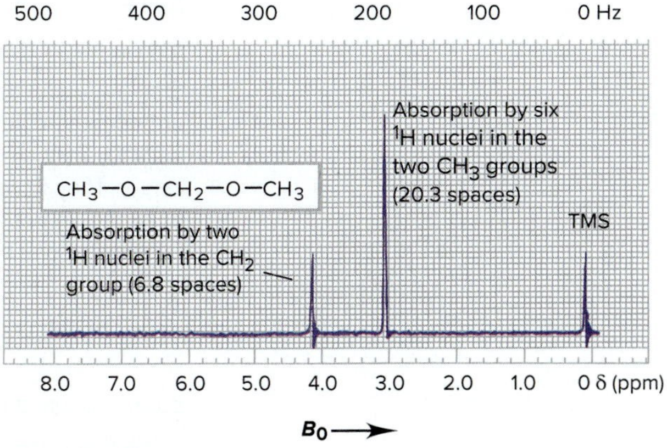

Figure B15.3 The ^{1}H-NMR spectrum of dimethoxymethane.

TOOLS OF THE LABORATORY

Nuclear Magnetic Resonance (NMR) Spectroscopy (Continued)

NMR has many applications in biochemistry and medicine. Applying the principle to the imaging of organs and other body parts is known as *computer-aided magnetic resonance imaging*, or *MRI*. For example, an MRI scan of a person's head (Figure B15.4) can reveal a brain tumor by mapping levels of metabolic activity in different regions of the brain.

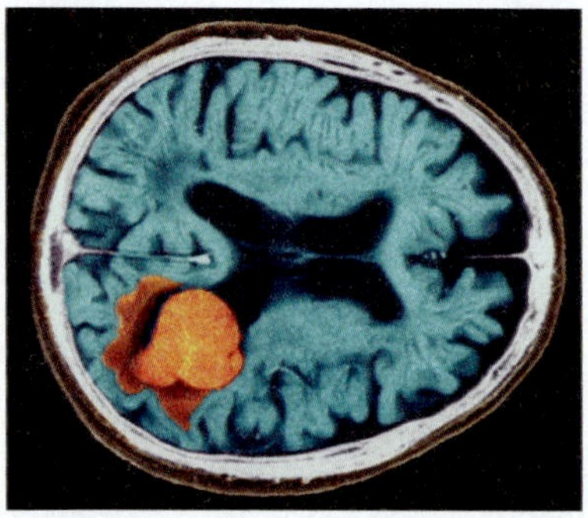

Figure B15.4 An MRI scan showing a brain tumor.

Source: © Zephyr/Science Source

Problems

B15.1 How many peaks appear in the NMR spectrum of each isomer of C_4H_{10} and of C_5H_{12} shown in Table 15.3?

B15.2 Radio-frequency radiation is used in NMR since ΔE, the energy difference between the two spin states of the ^1H nucleus, is very small. Calculate the ΔE (in J/mol) that corresponds to a frequency of 200. MHz.

B15.3 Consider three isomers of $C_4H_8Br_2$:

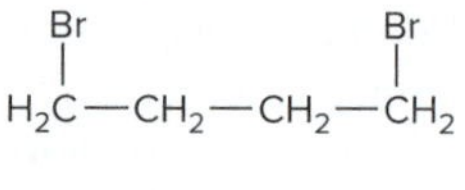

A

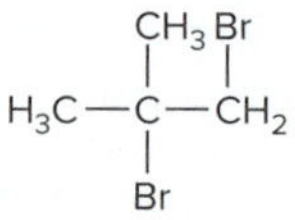

B

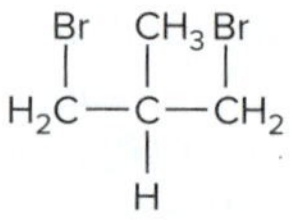

C

Which of these isomers is characterized by a ^1H-NMR spectrum that consists of two peaks with a 3/1 area ratio?

› Summary of Section 15.2

- Hydrocarbons contain only C and H atoms, so their physical properties depend on the strength of their dispersion forces.
- Names of organic compounds consist of a root for the longest chain, a prefix for any attached group, and a suffix for the type of compound.
- Alkanes (C_nH_{2n+2}) have only single bonds. Cycloalkanes (C_nH_{2n}) have ring structures that are typically nonplanar. Alkenes (C_nH_{2n}) have at least one C═C bond. Alkynes (C_nH_{2n-2}) have at least one C≡C bond. Aromatic hydrocarbons have at least one planar ring with delocalized π electrons.
- Isomers are compounds with the same molecular formula but different properties.
- Constitutional (structural) isomers have different arrangements of atoms.
- Stereoisomers (optical and geometric) have the same arrangement of atoms, but their atoms are oriented differently in space. Optical isomers cannot be superimposed on each other because they are asymmetric, with four different groups bonded to the C that is the chiral center. They have identical physical and chemical properties except in their rotation of plane-polarized light, their reactivity with chiral reactants, and their biological activity. Geometric *(cis-trans)* isomers have groups oriented differently around a C═C bond, which restricts rotation.
- Light converts a *cis* isomer of retinal to the all-*trans* form, which initiates the process of vision.
- Boron, silicon, and sulfur also form catenated hydrides, but these are unstable.
- ^1H-NMR spectroscopy indicates the relative numbers of H atoms in the various environments within an organic molecule.

15.3 SOME IMPORTANT CLASSES OF ORGANIC REACTIONS

In Chapter 4, we classified chemical reactions based on the chemical process involved (precipitation, acid-base, or redox) and then briefly included a classification based on the number of reactants and products (combination, decomposition, or displacement). We take a similar approach here with organic reactions.

To depict reactions more generally, it's common practice to use R— to signify a nonreacting organic group attached to one of the atoms; you can usually picture R— as an **alkyl group,** a saturated hydrocarbon chain with one bond available. Thus, R—CH_2—Br has an alkyl group attached to a CH_2 group bonded to a Br atom; R—CH=CH_2 is an alkene with an alkyl group attached to one of the carbons in the double bond; and so forth. (Often, when more than one R group is present, we write R, R′, R″, and so forth, to indicate that these groups may be different.)

Types of Organic Reactions

Three important organic reaction types are *addition, elimination,* and *substitution* reactions, which can be identified by comparing the *number of bonds to C* in reactants and products.

Addition Reactions An **addition reaction** occurs when an unsaturated reactant becomes a saturated product:

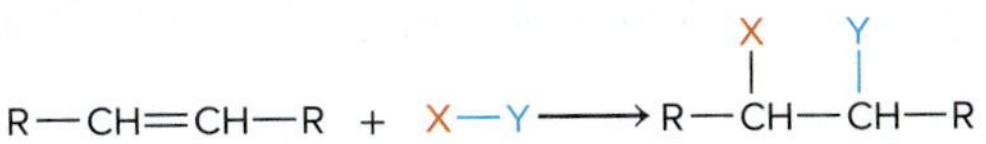

Note the C atoms are bonded to *more* atoms in the product than in the reactant.

The C=C and C≡C bonds and the C=O bond commonly undergo addition reactions. In each case, the π bond breaks, leaving the σ bond intact. In the product, the two C atoms (or C and O) form two additional σ bonds. Let's examine the standard enthalpy of reaction (ΔH°_{rxn}) for a typical addition reaction to see why these reactions occur. Consider the reaction between ethene (common name, ethylene) and HCl:

$$CH_2{=}CH_2 \quad + \quad H{-}Cl \quad \longrightarrow \quad H{-}CH_2{-}CH_2{-}Cl$$

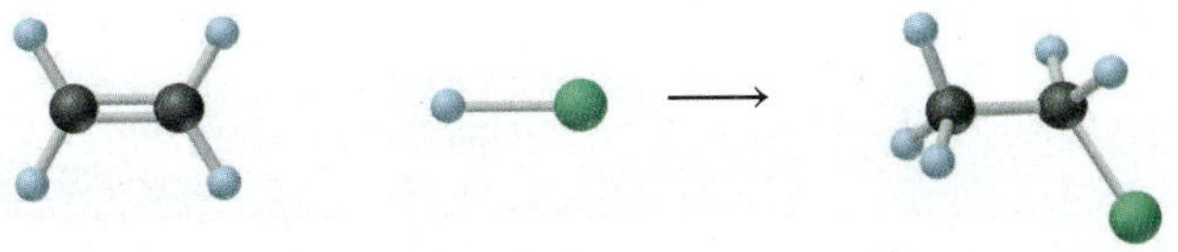

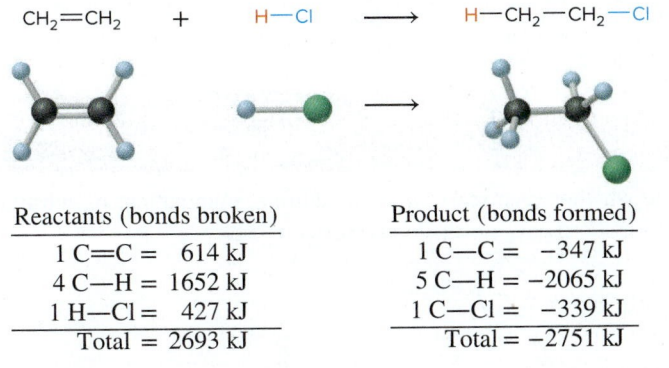

Reactants (bonds broken)
1 C=C = 614 kJ
4 C—H = 1652 kJ
1 H—Cl = 427 kJ
Total = 2693 kJ

Product (bonds formed)
1 C—C = −347 kJ
5 C—H = −2065 kJ
1 C—Cl = −339 kJ
Total = −2751 kJ

$$\Delta H^\circ_{rxn} = \Sigma\Delta H^\circ_{bonds\ broken} + \Sigma\Delta H^\circ_{bonds\ formed} = 2693\ kJ + (-2751\ kJ) = -58\ kJ$$

The reaction is exothermic. By looking at the *net* change in bonds, we see that the driving force for many additions is the formation of two σ bonds (in this case, C—H and C—Cl) from one σ bond (in this case, H—Cl) and one relatively weak π bond. An addition reaction is the basis of a color test for the presence of C=C bonds (Figure 15.14).

A B

Figure 15.14 A color test for C=C bonds. A, Br_2 (in pipet) reacts with a compound that has a C=C bond (in beaker), and its orange-brown color disappears:

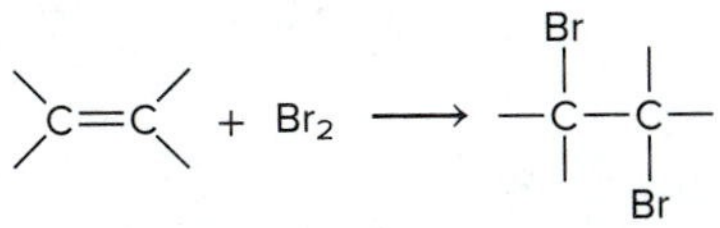

B, Here the compound in the beaker has no C=C bond, so the Br_2 does not react.

Source: © McGraw-Hill Education/Richard Megna, photographer

Elimination Reactions **Elimination reactions** are the opposite of addition reactions. They occur when a saturated reactant becomes an unsaturated product:

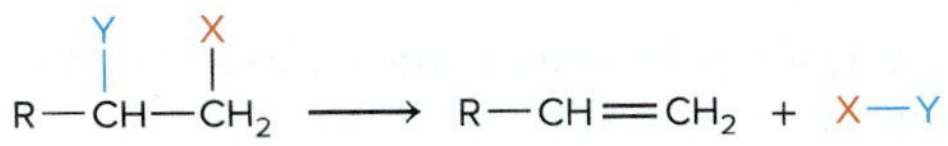

Note that the C atoms are bonded to *fewer* atoms in the product than in the reactant. A pair of halogen atoms, an H atom and a halogen atom, or an H atom and an —OH group are typically eliminated, but C atoms are not. Thus, the driving force for many elimination reactions is the formation of a small, stable molecule, such as HCl(*g*) or H_2O, which increases the entropy of the system (Section 13.3):

$$CH_3-\underset{\underset{\text{OH}}{|}}{CH}-\underset{\underset{\text{H}}{|}}{CH_2} \xrightarrow{H_2SO_4} CH_3-CH=CH_2 \;+\; H-OH$$

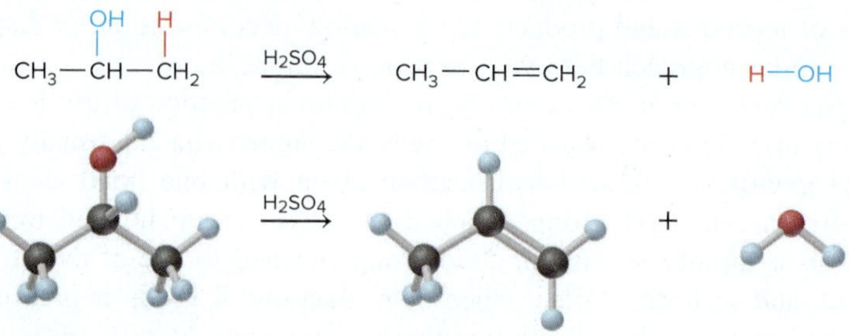

Substitution Reactions A **substitution reaction** occurs when an atom (or group) in one reactant substitutes for one attached to a carbon in the other reactant:

$$R-\underset{|}{\overset{|}{C}}-X \;+\; :Y \longrightarrow R-\underset{|}{\overset{|}{C}}-Y \;+\; :X$$

Note that the C atom is bonded to the *same number* of atoms in the product as in the reactant. The C atom may be saturated or unsaturated, and X and Y can be many different atoms, but generally *not* C. The main flavor ingredient in banana oil, for instance, forms through a substitution reaction; note that the O substitutes for the Cl:

$$CH_3-\overset{\overset{\text{O}}{\|}}{C}-Cl \;+\; HO-CH_2-CH_2-\underset{\underset{CH_3}{|}}{CH}-CH_3 \longrightarrow CH_3-\overset{\overset{\text{O}}{\|}}{C}-O-CH_2-CH_2-\underset{\underset{CH_3}{|}}{CH}-CH_3 \;+\; H-Cl$$

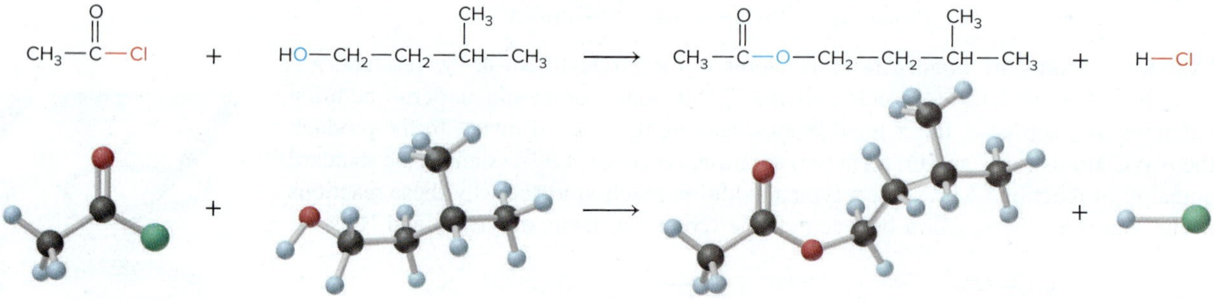

SAMPLE PROBLEM 15.3   Recognizing the Type of Organic Reaction

Problem State whether each reaction is an addition, elimination, or substitution:

(a) $CH_3-CH_2-CH_2-Br \longrightarrow CH_3-CH=CH_2 \;+\; HBr$

(b) $+\; H_2 \longrightarrow$ 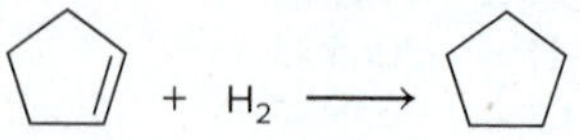

(c) $CH_3-\overset{\overset{\text{O}}{\|}}{C}-Br \;+\; CH_3-CH_2-OH \longrightarrow CH_3-\overset{\overset{\text{O}}{\|}}{C}-O-CH_2-CH_3 \;+\; HBr$ 

Plan We determine the type of reaction by looking for any change in the number of atoms bonded to C:

• More atoms bonded to C is an *addition*.
• Fewer atoms bonded to C is an *elimination*.
• Same number of atoms bonded to C is a *substitution*.

Solution (a) Elimination: two bonds in the reactant, C—H and C—Br, are absent in the product, so fewer atoms are bonded to C.

(b) Addition: two more C—H bonds have formed in the product, so more atoms are bonded to C.

(c) Substitution: the reactant C—Br bond becomes a C—O bond in the product, so the same number of atoms are bonded to C.

FOLLOW-UP PROBLEMS

15.3A State whether each reaction is an addition, elimination, or substitution:

(a) $CH_3—CH_2—Br + NaOH \longrightarrow CH_3—CH_2—OH + NaBr$

(b) $CH_3—O—\overset{\overset{\displaystyle O}{\|}}{C}—\bigcirc—Cl + H_2O \xrightarrow{H^+} HO—\overset{\overset{\displaystyle O}{\|}}{C}—\bigcirc—Cl + CH_3—OH$ 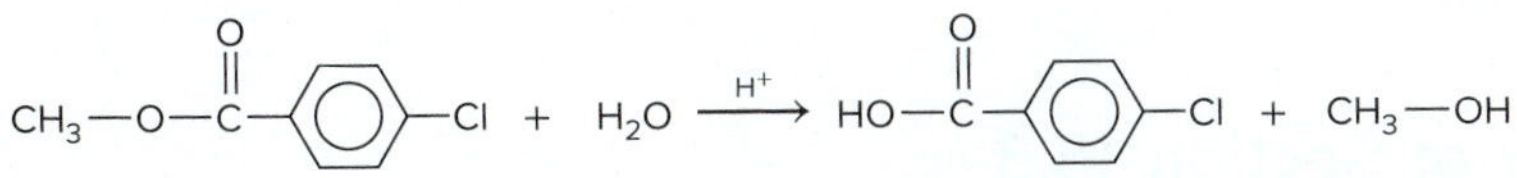

(c) $Cl—\overset{\overset{\displaystyle Cl}{|}}{\underset{\underset{\displaystyle Cl}{|}}{C}}—\overset{\overset{\displaystyle O}{\|}}{C}H + H_2O \xrightarrow{H^+} Cl—\overset{\overset{\displaystyle Cl}{|}}{\underset{\underset{\displaystyle Cl}{|}}{C}}—\overset{\overset{\displaystyle OH}{|}}{\underset{\underset{\displaystyle OH}{|}}{C}}H$ 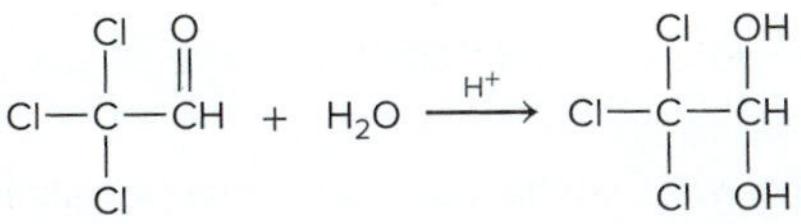

15.3B Write a balanced equation for each of the following:

(a) An addition reaction between 2-butene and Cl_2

(b) A substitution reaction between $CH_3—CH_2—CH_2—Br$ and OH^-

(c) The elimination of H_2O from $(CH_3)_3C—OH$

SOME SIMILAR PROBLEMS 15.40–15.43

The Redox Process in Organic Reactions

An important process in many organic reactions is *oxidation-reduction*. Even though a redox reaction always involves both an oxidation and a reduction, organic chemists typically *focus on the organic reactant only*. Instead of monitoring the change in oxidation numbers of the various C atoms, organic chemists usually note the movement of electron density around a C atom by counting the number of bonds to more electronegative atoms (usually O) or to less electronegative atoms (usually H). A more electronegative atom takes some electron density from the C, whereas a less electronegative atom gives some electron density to the C. Therefore,

- When a C atom in the organic reactant forms more bonds to O or fewer bonds to H, thus losing some electron density, the reactant is oxidized, and the reaction is called an *oxidation*.
- When a C atom in the organic reactant forms fewer bonds to O or more bonds to H, thus gaining some electron density, the reactant is reduced, and the reaction is called a *reduction*.

The most dramatic redox reactions are combustion reactions. Virtually all organic compounds contain C and H atoms and burn in excess O_2 to form CO_2 and H_2O. For ethane, the reaction is

$$2CH_3—CH_3 + 7O_2 \longrightarrow 4CO_2 + 6H_2O$$

Obviously, when ethane is converted to CO_2 and H_2O, each of its C atoms has more bonds to O and fewer bonds to H. Thus, ethane is oxidized, and this reaction is referred to as an *oxidation,* even though O_2 is reduced as well.

Most oxidations do not involve a breakup of the entire molecule, however. When 2-propanol reacts with potassium dichromate in acidic solution (a common oxidizing agent in organic reactions), it forms 2-propanone (acetone):

$$CH_3—\overset{\overset{\displaystyle |}{\underset{\underset{\displaystyle OH}{|}}{C}}H}{}—CH_3 \xrightarrow[H_2SO_4]{K_2Cr_2O_7} CH_3—\overset{\overset{\displaystyle |}{\underset{\underset{\displaystyle O}{\|}}{C}}}{}—CH_3$$

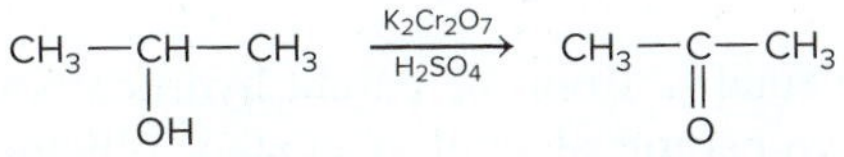

2-propanol   2-propanone

Note that C-2 has one fewer bond to H and one more bond to O in 2-propanone than it does in 2-propanol. Thus, 2-propanol is oxidized, so this is an *oxidation*. Don't forget, however, that the dichromate ion is reduced at the same time:

$$Cr_2O_7^{2-} + 14H^+ + 6e^- \longrightarrow 2Cr^{3+} + 7H_2O$$

The reaction between H_2 and an alkene is both an addition and a *reduction:*

$$CH_2{=}CH_2 + H_2 \xrightarrow{Pd} CH_3—CH_3$$

Note that each C has more bonds to H in ethane than it has in ethene, so the ethene is reduced. The H_2 is oxidized in the process, and the palladium shown over the arrow acts as a catalyst to speed up the reaction. (We discuss catalysts in Section 16.7.)

› Summary of Section 15.3

› In an addition reaction, a π bond breaks, and the two C atoms (or one C and one O) are bonded to more atoms in the product.

› In an elimination reaction, a π bond forms, with the two C atoms bonded to fewer atoms in the product.

› In a substitution reaction, one atom bonded to C is replaced by another, but the total number of atoms bonded to the C does not change.

› In an organic redox reaction, the organic reactant is oxidized if at least one of its C atoms forms more bonds to O atoms (or fewer bonds to H atoms), and it is reduced if one of its C atoms forms more bonds to H (or fewer bonds to O).

Methanol (methyl alcohol)
Byproduct in coal gasification; de-icing agent; gasoline substitute; precursor of organic compounds

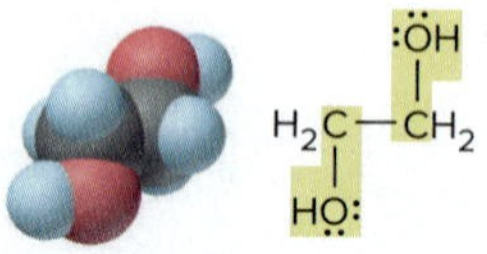

1,2-Ethanediol (ethylene glycol)
Main component of auto antifreeze

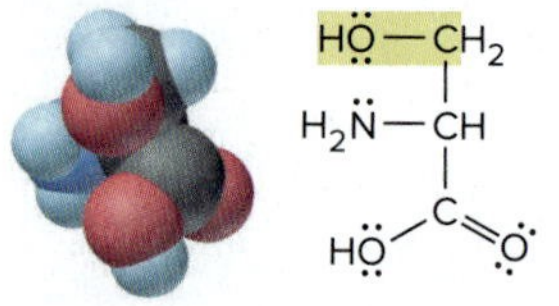

Serine
Amino acid found in most proteins

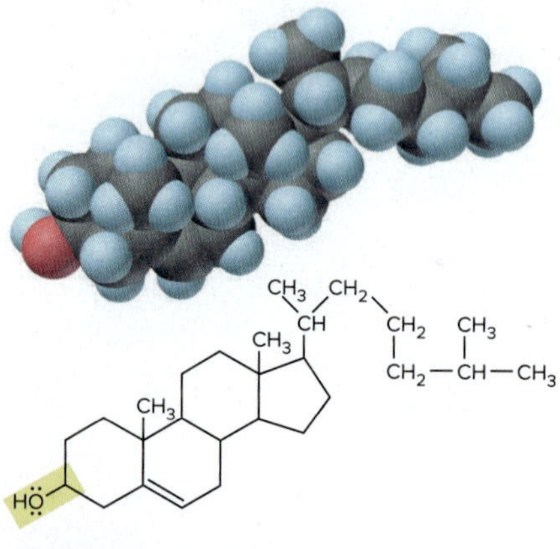

Cholesterol
Major sterol in animals; essential for cell membranes; precursor of steroid hormones

Figure 15.15 Some molecules with the alcohol functional group.

15.4 PROPERTIES AND REACTIVITIES OF COMMON FUNCTIONAL GROUPS

A central organizing principle of organic reactions is that *the distribution of electron density in a functional group is a key to the reactivity of the compound.* The electron density can be high, as in the C═C and C≡C bonds, or it can be low at one end of a bond and high at the other, as in the C—Cl and C—O bonds. Such electron-rich or polar bonds enhance the opposite charge in a polar bond of the other reactant. As a result, the reactants attract each other and begin a sequence of bond-forming and bond-breaking steps that lead to a product. Thus, *the intermolecular forces that affect physical properties and solubility also affect reactivity.* Table 15.5 lists some of the important functional groups in organic compounds.

When we classify functional groups by bond order, they follow certain patterns of reactivity:

- Functional groups with only single bonds undergo elimination or substitution.
- Functional groups with double or triple bonds undergo addition.
- Functional groups with both single and double bonds undergo substitution.

In this section, we see how to name compounds that contain each functional group, examine some of their properties, and summarize their common reactions.

Functional Groups with Only Single Bonds

The most common functional groups with only single bonds are alcohols, haloalkanes, and amines.

Alcohols The **alcohol** functional group consists of a carbon bonded to an OH group, $-\overset{|}{\underset{|}{C}}-\ddot{\underset{\cdot\cdot}{O}}-H$, and the general formula of an alcohol is R—OH. Alcohols are named by dropping the final *-e* from the parent hydrocarbon name and adding the suffix *-ol.* Thus, the two-carbon alcohol is ethanol (ethan- + -ol). The common name of an alcohol is the hydrocarbon *root-* + *-yl,* followed by "alcohol"; thus, the common name of ethanol is ethyl alcohol. (This substance, obtained from fermented grain, has been consumed by people as an intoxicant in beverages since ancient times; today, it is recognized as the most abused drug in the world.) Alcohols are common organic reactants, and the functional group occurs in many biomolecules, including carbohydrates, sterols, and some amino acids. Figure 15.15 shows the names, structures, and uses of some important compounds that contain the alcohol group.

Table 15.5 Important Functional Groups in Organic Compounds

Functional Group	Compound Type	Prefix or Suffix of Name	Example: Lewis Structure	Example: Ball-and-Stick Model	Example: Systematic Name (Common Name)
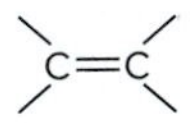	alkene	-ene	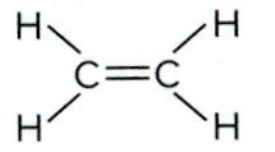	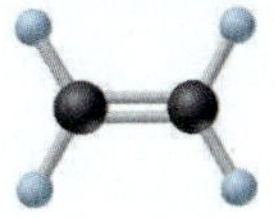	ethene (ethylene)
	alkyne	-yne	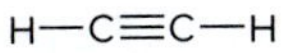		ethyne (acetylene)
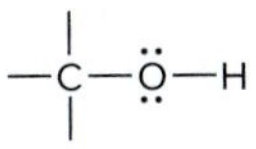	alcohol	-ol	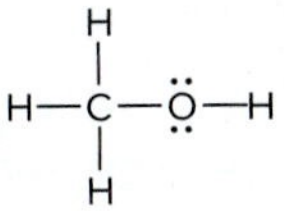		methanol (methyl alcohol)
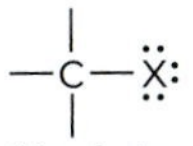 (X = halogen)	haloalkane	halo-	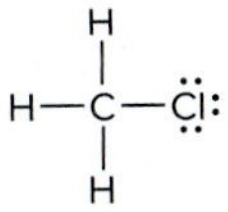	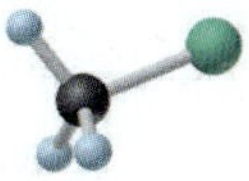	chloromethane (methyl chloride)
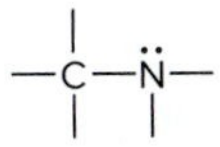	amine	-amine	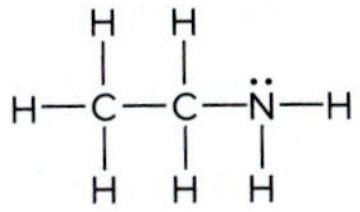	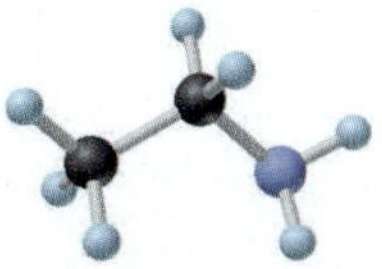	ethanamine (ethylamine)
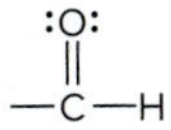	aldehyde	-al	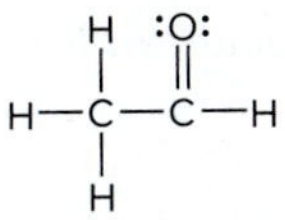	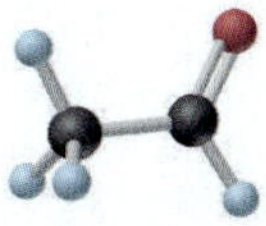	ethanal (acetaldehyde)
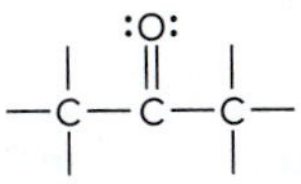	ketone	-one	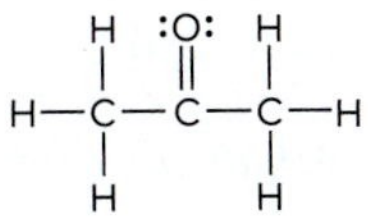	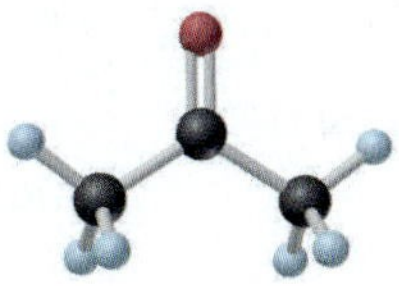	2-propanone (acetone)
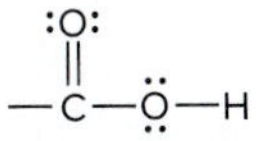	carboxylic acid	-oic acid	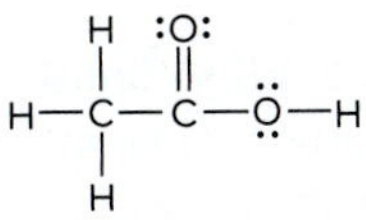	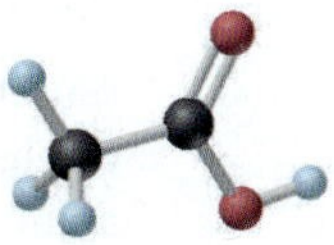	ethanoic acid (acetic acid)
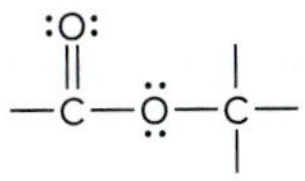	ester	-oate	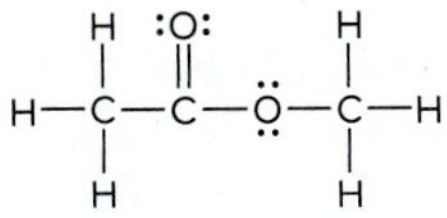	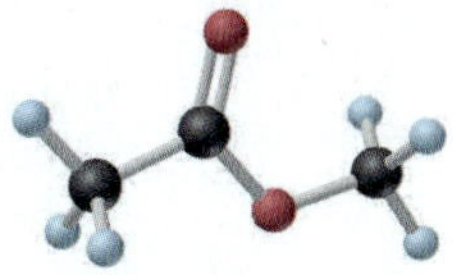	methyl ethanoate (methyl acetate)
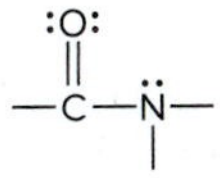	amide	-amide	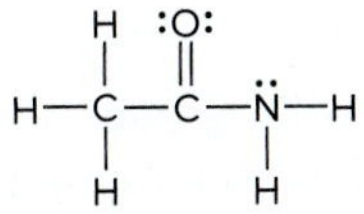	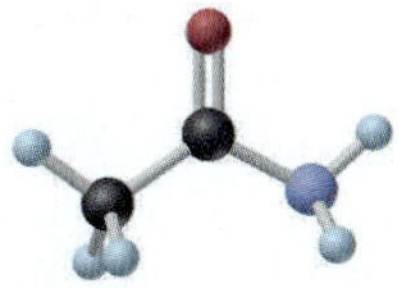	ethanamide (acetamide)
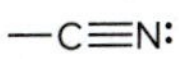	nitrile	-nitrile	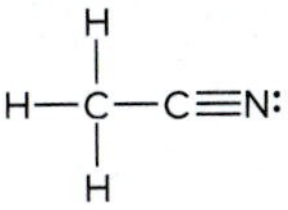	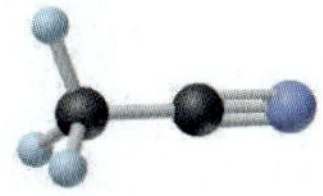	ethanenitrile (acetonitrile, methyl cyanide)

The physical properties of the smaller alcohols are similar to those of water. They have high melting and boiling points as a result of hydrogen bonding, and they dissolve polar molecules and some salts.

Alcohols undergo elimination and substitution reactions:

- *Elimination* of H and OH, called *dehydration,* requires acid and forms an alkene:

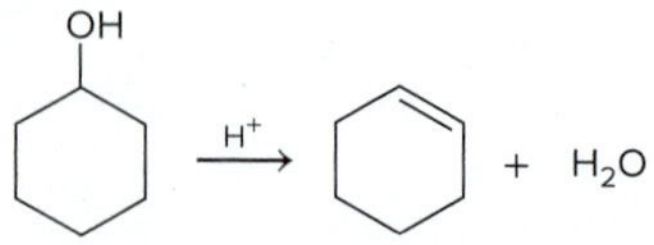

cyclohexanol cyclohexene

- *Elimination* of two H atoms is an *oxidation* that requires an inorganic oxidizing agent, such as potassium dichromate ($K_2Cr_2O_7$) in aqueous H_2SO_4. The product has a C=O group:

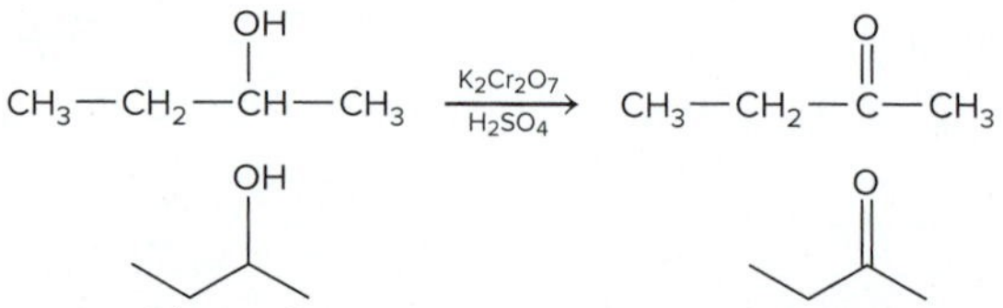

2-butanol 2-butanone

- Alcohols with an OH group at the end of the chain (R—CH_2—OH) can be oxidized further to acids (but this step is *not* an elimination). Wine turns sour, for example, when the ethanol in contact with air is oxidized to acetic acid:

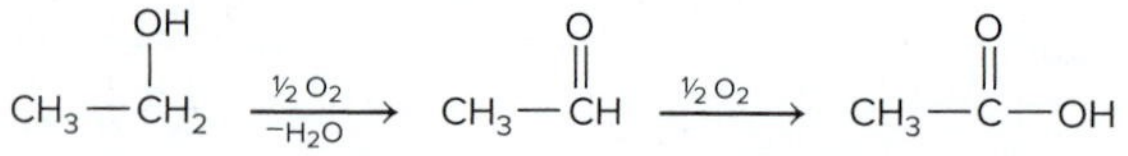

- *Substitution* yields products with other single-bonded functional groups. Reactions of hydrohalic acids with many alcohols give haloalkanes:

$$R_2CH—OH + HBr \longrightarrow R_2CH—Br + HOH$$

As you'll see below, *the C atom undergoing the change in a substitution is bonded to a more electronegative element,* which makes it partially positive and, thus, a target for a negatively charged or electron-rich group of an incoming reactant.

Haloalkanes A *halogen* atom (X) bonded to a carbon forms the **haloalkane** functional group, —C—$\ddot{\text{X}}$:, found in compounds with the general formula R—X. Haloalkanes (common name, **alkyl halides**) are named by identifying the halogen using a prefix on the hydrocarbon name and numbering the C atom to which the halogen is attached, as in bromomethane, 2-chloropropane, or 1,3-diiodohexane.

Like alcohols, haloalkanes undergo substitution and elimination reactions.

- Just as many alcohols undergo substitution to form alkyl halides when treated with halide ions in acid, many haloalkanes undergo substitution in base to form alcohols. For example, OH^- attacks the positive C end of the C—X bond and displaces X^-:

$$CH_3—CH_2—CH_2—CH_2—Br + OH^- \longrightarrow CH_3—CH_2—CH_2—CH_2—OH + Br^-$$

1-bromobutane 1-butanol

  Substitutions by groups such as —CN, —SH, —OR, and —NH_2 allow chemists to convert haloalkanes to a host of other types of compounds.

- Just as addition of HX *to* an alkene produces haloalkanes, elimination of HX *from* a haloalkane by reaction with a strong base, such as potassium ethoxide, produces an alkene:

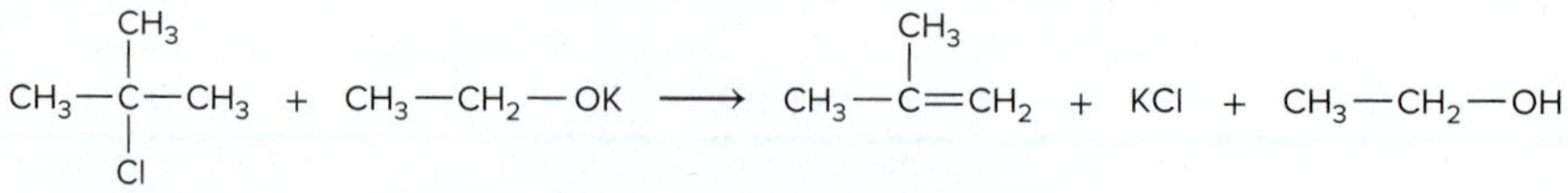

2-chloro-2-methylpropane potassium ethoxide 2-methylpropene

Haloalkanes have many important uses, but some, like the chlorofluorocarbons, cause serious environmental problems (Chapter 16 Chemical Connections). Also, some halogenated *aromatic* hydrocarbons (*aryl* halides) are carcinogenic in mammals, have severe neurological effects in humans, and, to make matters worse, are very stable in the environment. For example, the polychlorinated biphenyls (PCBs) (Figure 15.16), used as insulating fluids in electrical transformers, have accumulated for decades in rivers and lakes and have been incorporated into the food chain. PCBs in natural waters pose health risks and present an enormous cleanup problem.

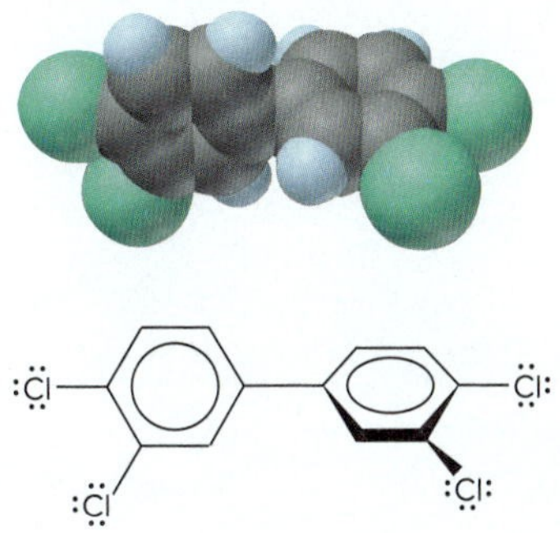

Figure 15.16 A tetrachlorobiphenyl, one of 209 polychlorinated biphenyls (PCBs).

Amines The **amine** functional group is $-\overset{|}{\underset{|}{C}}-\underset{|}{\ddot{N}}-$. Chemists classify amines as derivatives of ammonia, with R groups in place of one or more of the H atoms in NH_3:

- *Primary* (1°) amines are RNH_2.
- *Secondary* (2°) amines are R_2NH.
- *Tertiary* (3°) amines are R_3N.

The R groups in secondary and tertiary amines can be the same or different. Like ammonia, amines have trigonal pyramidal shapes and a lone pair of electrons on a partially negative N atom, which is the key to amine reactivity (Figure 15.17).

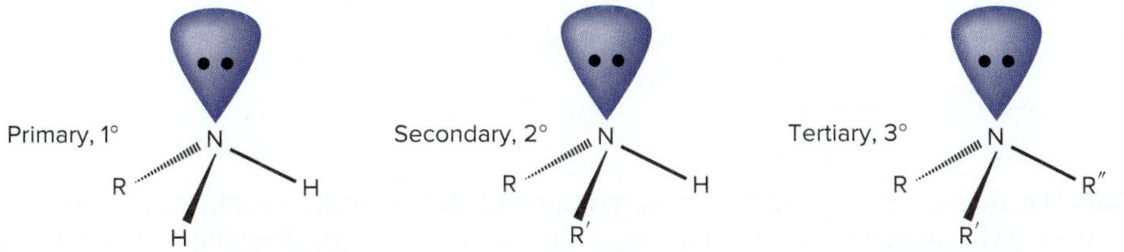

Figure 15.17 General structures of amines.

Systematic names drop the final *-e* of the alkane and add the suffix *-amine,* as in ethanamine. However, there is still wide usage of common names, in which the suffix *-amine* follows the name of the alkyl group; thus, methylamine has one methyl group attached to N, diethylamine has two ethyl groups attached, and so forth. Figure 15.18 shows that the amine functional group occurs in many biomolecules.

Figure 15.18 Some biomolecules with the amine functional group.

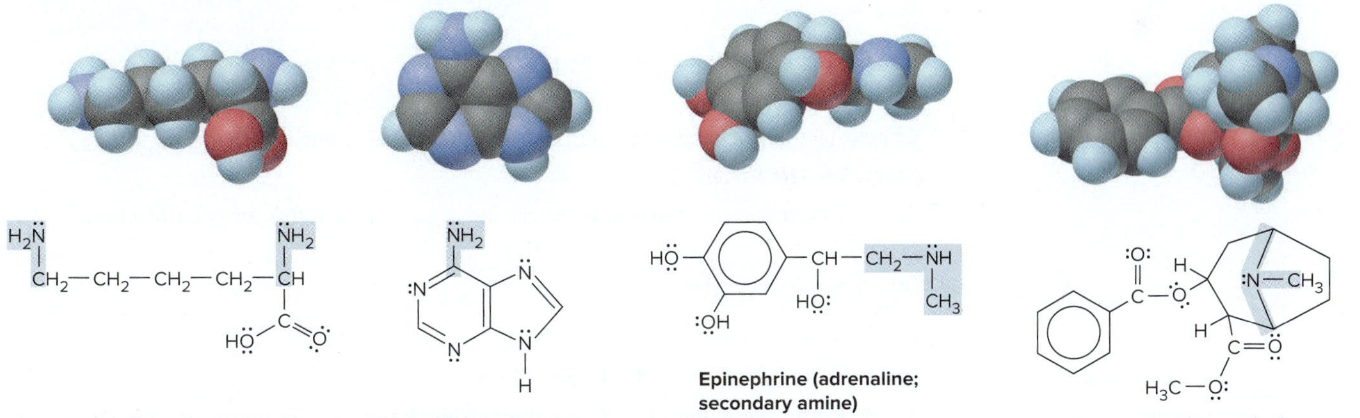

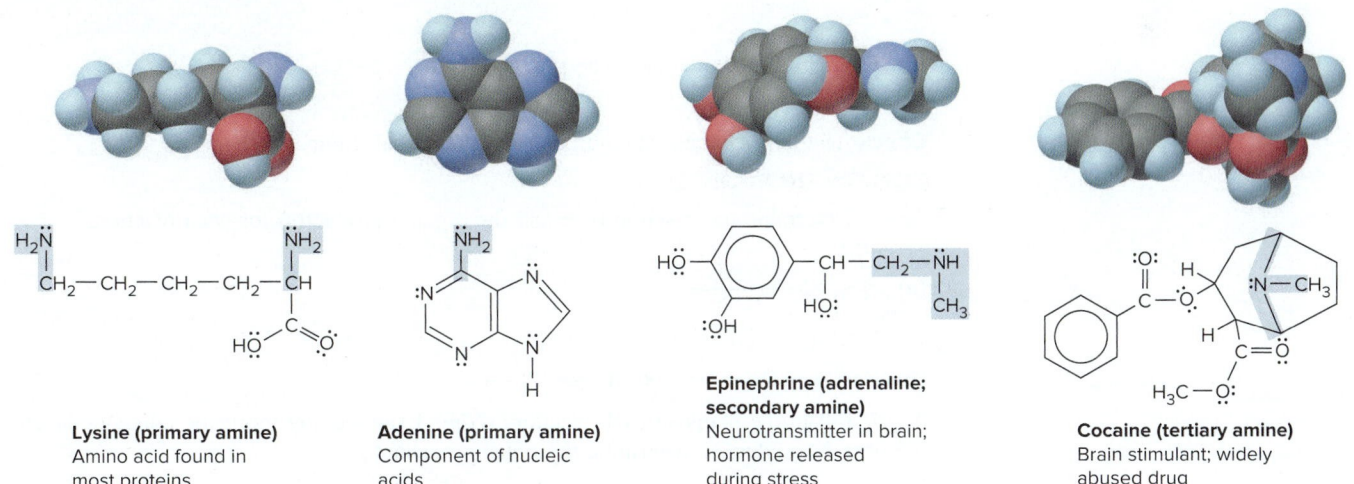

Lysine (primary amine) Amino acid found in most proteins

Adenine (primary amine) Component of nucleic acids

Epinephrine (adrenaline; secondary amine) Neurotransmitter in brain; hormone released during stress

Cocaine (tertiary amine) Brain stimulant; widely abused drug

Primary and secondary amines can form H bonds, so they have higher melting and boiling points than hydrocarbons and alkyl halides of similar molar mass. For example, dimethylamine ($\mathcal{M}$ = 45.09 g/mol) boils 45°C higher than ethyl fluoride ($\mathcal{M}$ = 48.06 g/mol). Trimethylamine has a greater molar mass than dimethylamine, but it melts more than 20°C *lower* because its N does not have an attached H atom that could participate in H bonding.

Amines of low molar mass are fishy smelling, water soluble, and weakly basic because of the lone pair on N. The reaction with water proceeds only slightly to the right to reach equilibrium:

$$CH_3{-}\ddot{N}H_2 + H_2O \rightleftharpoons CH_3{-}\overset{+}{N}H_3 + OH^-$$

The lone pair is also the reason amines undergo substitution reactions: the lone pair attacks the partially positive C in an alkyl halide to displace X^- and form a larger amine:

$$\underset{\text{ethylamine}}{2\,CH_3{-}CH_2{-}\ddot{N}H_2} + \underset{\text{chloroethane}}{CH_3{-}CH_2{-}Cl} \longrightarrow \underset{\text{diethylamine}}{CH_3{-}CH_2{-}\ddot{N}H{-}CH_2{-}CH_3} + \underset{\text{ethylammonium chloride}}{CH_3{-}CH_2{-}\overset{+}{N}H_3Cl^-}$$

(Two molecules of amine are needed: one attacks the chloroethane, and the other binds the released H^+ to prevent it from remaining on the diethylamine product.)

SAMPLE PROBLEM 15.4 Predicting the Reactions of Alcohols, Alkyl Halides, and Amines

Problem Determine the reaction type and the organic product(s) for each reaction:

(a) $CH_3{-}CH_2{-}CH_2{-}I + NaOH \longrightarrow$

(b) $CH_3{-}CH_2{-}Br + 2\,CH_3{-}CH_2{-}CH_2{-}NH_2 \longrightarrow$

(c) $CH_3{-}\underset{OH}{\underset{|}{C}H}{-}CH_3 \xrightarrow[H_2SO_4]{Cr_2O_7^{2-}}$

Plan We first determine the functional group(s) of the organic reactant(s) and then examine any inorganic reactant(s) to decide on the reaction type, keeping in mind that, in general, these functional groups undergo substitution or elimination. **(a)** The reactant is a haloalkane, so the OH^- of the inorganic base NaOH substitutes for I^-. **(b)** The reactants are an amine and an alkyl halide, so the N of the amine substitutes for the Br. **(c)** The organic reactant is an alcohol, the inorganic reactants form a strong oxidizing agent, and the alcohol group undergoes elimination to form a C=O group.

Solution **(a)** Substitution: The product is $CH_3{-}CH_2{-}CH_2{-}OH$ + NaI

(b) Substitution: The products are $CH_3{-}CH_2{-}CH_2{-}\underset{CH_2-CH_3}{\underset{|}{N}H} + CH_3{-}CH_2{-}CH_2{-}\overset{+}{N}H_3Br^-$

(c) Elimination (oxidation): The product is $CH_3{-}\underset{O}{\underset{\|}{C}}{-}CH_3$

Check The only changes should be at the functional group.

FOLLOW-UP PROBLEMS

15.4A Determine the reaction type and the organic product(s) for each reaction:

(a) 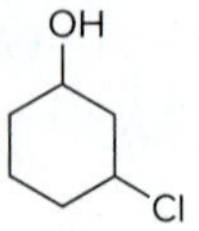 $\xrightarrow[H_2SO_4]{Cr_2O_7^{2-}}$

(b) $Br{-}CH_2{-}CH_2{-}CH_2{-}Br + NaCN \longrightarrow$

15.4B Fill in the blank in each reaction. (*Hint:* Examine any inorganic compounds and the organic product to determine the organic reactant.)

(a) ____ $+ CH_3{-}ONa \longrightarrow CH_3{-}CH{=}\overset{CH_3}{\overset{|}{C}}{-}CH_3 + NaCl + CH_3{-}OH$

(b) ____ $\xrightarrow[H_2SO_4]{Cr_2O_7^{2-}} CH_3{-}CH_2{-}\overset{O}{\overset{\|}{C}}{-}OH$

SOME SIMILAR PROBLEMS 15.65–15.74

Variations on a Theme: Inorganic Compounds with Single Bonds to O, X, or N

The —OH group occurs frequently in inorganic compounds. All oxoacids contain at least one —OH, usually bonded to a relatively electronegative nonmetal atom, which in most cases is bonded to other O atoms. Oxoacids are acidic in water because these additional O atoms pull electron density from the central nonmetal, which pulls electron density from the O—H bond, releasing an H^+ ion and stabilizing the oxoanion through resonance. Alcohols are *not* acidic in water because they lack the additional O atoms and the electronegative nonmetal.

Halides of nearly every nonmetal are known, and many undergo substitution reactions in base. As in the case of a haloalkane, the process involves an attack on the partially positive central atom by OH^-:

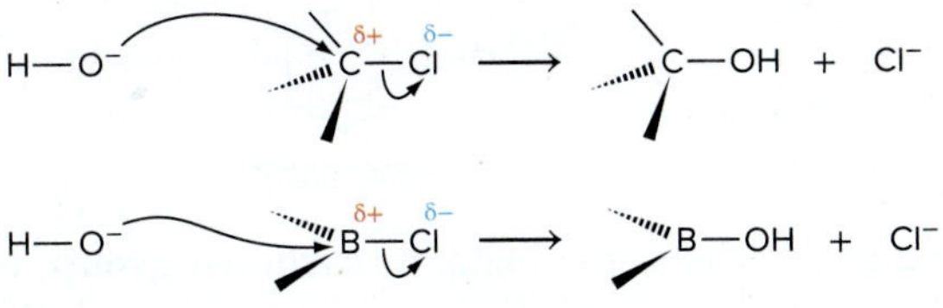

Thus, haloalkanes undergo the same general reaction as other nonmetal halides, such as BCl_3, SiF_4, and PCl_5.

The bonds between nitrogen and larger nonmetals, such as Si, P, and S, have significant double-bond character, which affects structure and reactivity. For example, $(SiH_3)_3N$, trisilylamine, the Si analog of trimethylamine, is trigonal planar, rather than trigonal pyramidal, because of the formation of a double bond between N and Si. Because the lone pair on N is delocalized in this π bond, trisilylamine is not basic.

Functional Groups with Double Bonds

The most important functional groups with double bonds are the C=C group of alkenes and the C=O group of aldehydes and ketones. Both appear in many organic and biological molecules. *Their most common reaction type is addition.*

Alkenes The C=C bond is the essential portion of the *alkene* functional group, 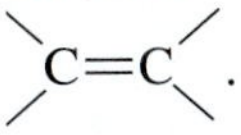. Although they undergo elimination to alkynes, *alkenes typically undergo addition.* The electron-rich double bond is readily attracted to the partially positive H atoms of hydronium ions and hydrohalic acids, yielding alcohols and alkyl halides, respectively:

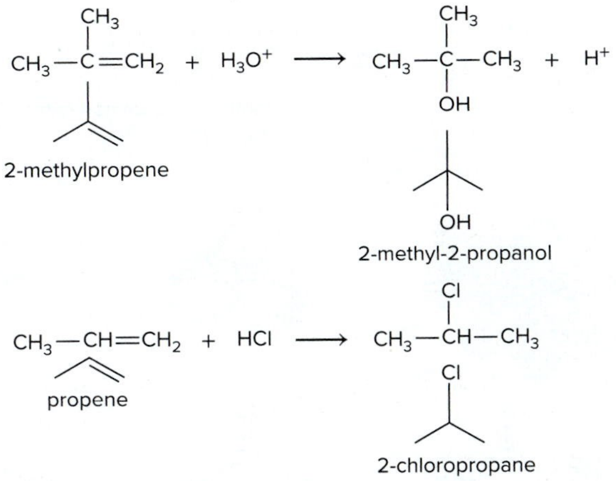

Comparing the Reactivities of Alkenes and Aromatic Compounds The *localized* unsaturation of alkenes is very different from the *delocalized* unsaturation of aromatic compounds. That is, despite the way we depict its resonance forms, benzene does *not* have double bonds and so, for example, it does not decolorize Br_2 (see Figure 15.14).

In general, aromatic rings are much *less* reactive than alkenes because of their delocalized π electrons. For example, let's compare the enthalpies of reaction for the

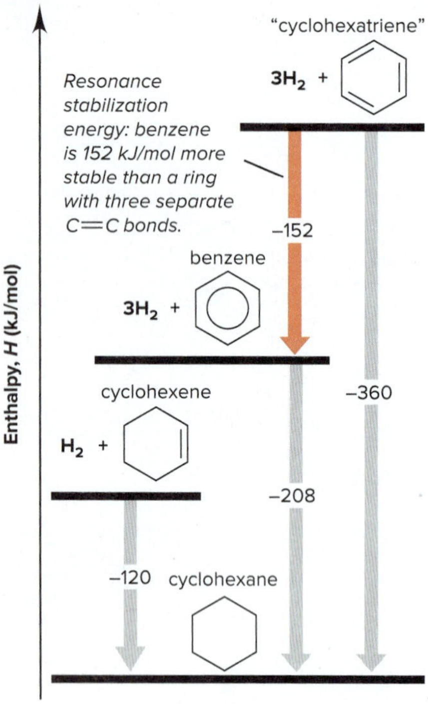

Figure 15.19 The stability of benzene.

addition of H_2. Hydrogenation of cyclohexene, with one C=C bond, has a ΔH°_{rxn} of −120 kJ/mol. For the imaginary molecule "cyclohexatriene," that is, the structure with three C=C bonds, the hypothetical ΔH°_{rxn} for hydrogenation would be three times as much, or −360 kJ/mol. Hydrogenation of benzene has a ΔH°_{rxn} of −208 kJ/mol. Thus, the ΔH°_{rxn} for hydrogenation of benzene is 152 kJ/mol *less* than that for "cyclohexatriene." This energy difference is attributed to the aromatic resonance stabilization of benzene (Figure 15.19).

This extra energy needed to break up the delocalized π system means *addition* reactions with benzene occur rarely. But, benzene does undergo many *substitution* reactions in which the delocalization is retained when an H atom attached to a ring C is replaced by another group:

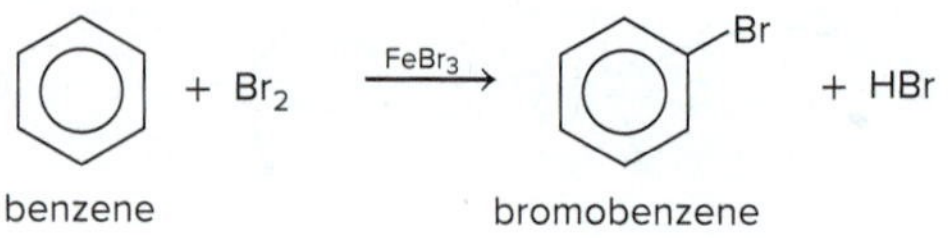

Aldehydes and Ketones The C=O bond, or **carbonyl group,** is one of the most chemically versatile.

- In the **aldehyde** functional group, the carbonyl C is bonded to H (and typically to another C), so it occurs *at the end of a chain,* R—C(=:O:)—H. Aldehyde names drop the final *-e* from the alkane name and add *-al;* thus, the three-C aldehyde is propanal.
- In the **ketone** functional group, the carbonyl C is bonded to two other C atoms, —C—C(=:O:)—C—, so it occurs *within the chain.* Ketones, R—C(=O)—R′, are named by numbering the carbonyl C, dropping the final *-e* from the alkane name, and adding *-one*. For example, the unbranched, five-C ketone with the carbonyl C as C-2 in the chain is named 2-pentanone. Figure 15.20 shows some common carbonyl compounds.

Aldehydes and ketones are formed by the oxidation of alcohols:

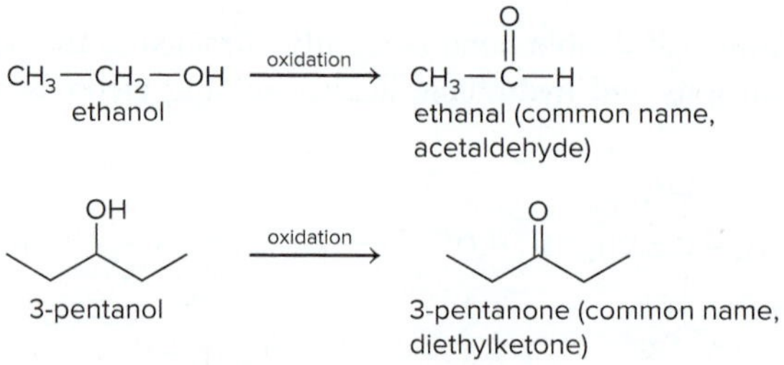

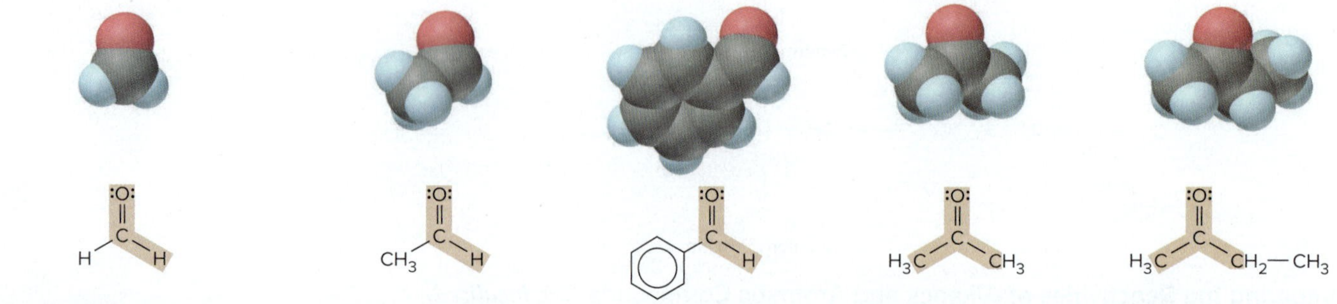

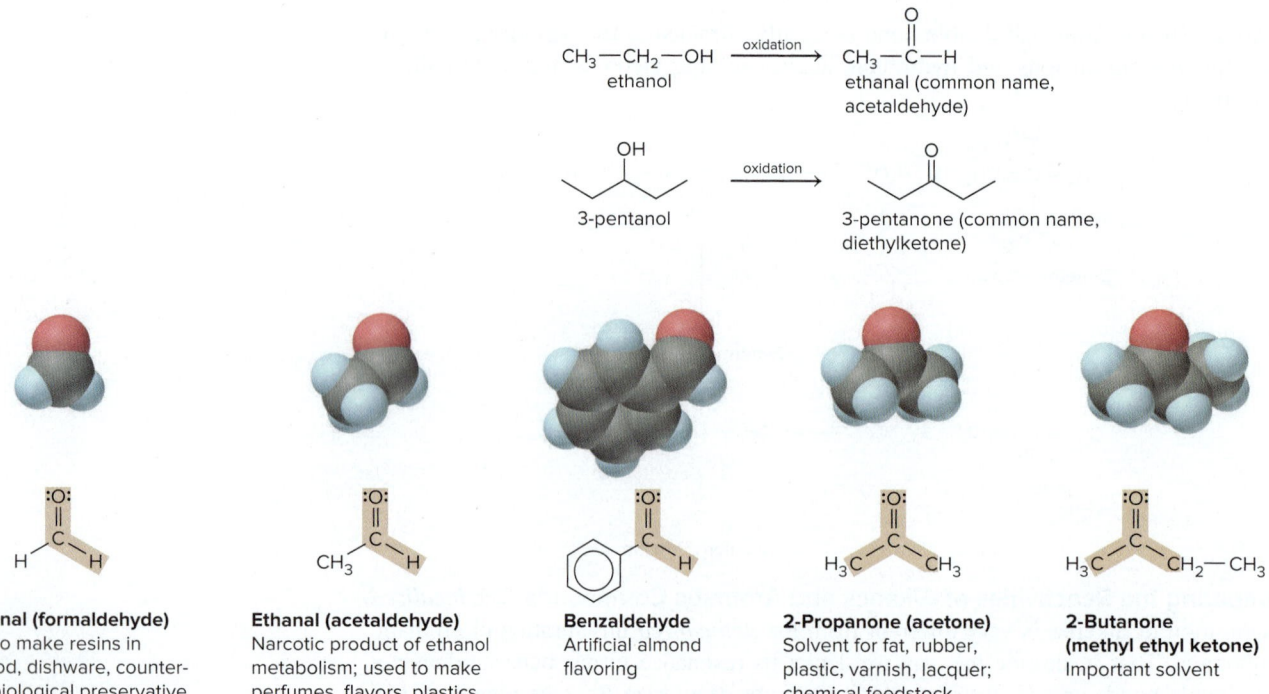

Methanal (formaldehyde) Used to make resins in plywood, dishware, countertops; biological preservative

Ethanal (acetaldehyde) Narcotic product of ethanol metabolism; used to make perfumes, flavors, plastics, other chemicals

Benzaldehyde Artificial almond flavoring

2-Propanone (acetone) Solvent for fat, rubber, plastic, varnish, lacquer; chemical feedstock

2-Butanone (methyl ethyl ketone) Important solvent

Figure 15.20 Some common aldehydes and ketones.

Conversely, as a result of their unsaturation, carbonyl compounds can undergo *addition* and be reduced to alcohols:

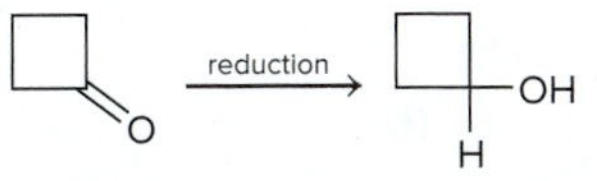

cyclobutanone cyclobutanol

Like the C=C bond, the C=O bond is *electron rich;* unlike the C=C bond, it is *highly polar* (ΔEN = 1.0). Figure 15.21 depicts this polarity with an orbital contour model (Figure 15.21A) and a charged resonance form (Figure 15.21B).

As a result of the bond polarity, addition often occurs with an electron-rich group bonding to the carbonyl C and an electron-poor group bonding to the carbonyl O. **Organometallic compounds,** which have a metal atom (usually Li or Mg) attached to an R group through a polar covalent bond, take part in this type of reaction. In a two-step sequence, they convert carbonyl compounds to alcohols with *different carbon skeletons:*

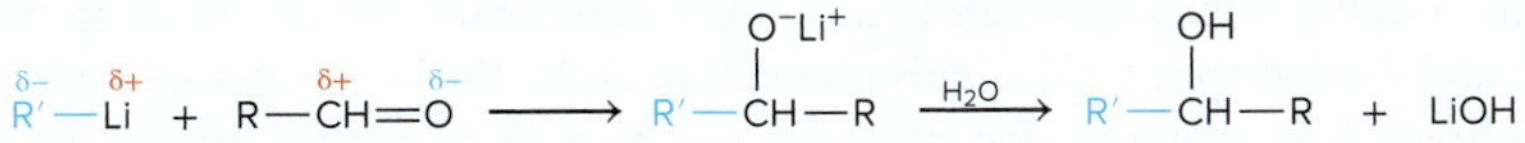

In the following example, the electron-rich C bonded to Li in ethyllithium, CH_3CH_2—Li, attacks the electron-poor carbonyl C of 2-propanone, adding the ethyl group to that C; at the same time, the Li adds to the carbonyl O. Treating the mixture with water forms the C—OH group in the product, 2-methyl-2-butanol:

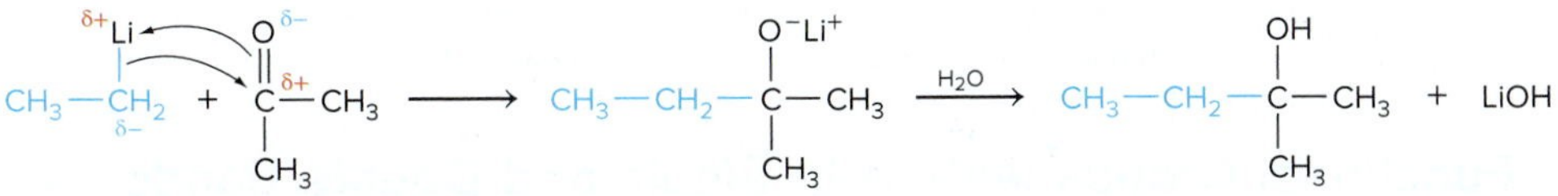

Note that the product skeleton combines the two reactant skeletons. The field of *organic synthesis* often employs organometallic compounds to create new compounds with different skeletons.

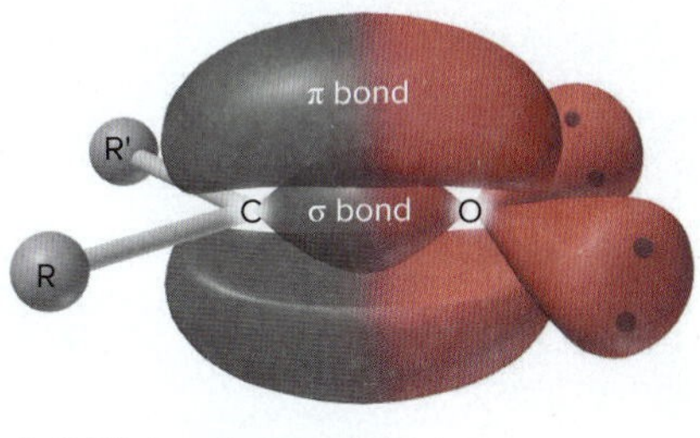

A Orbital contour model

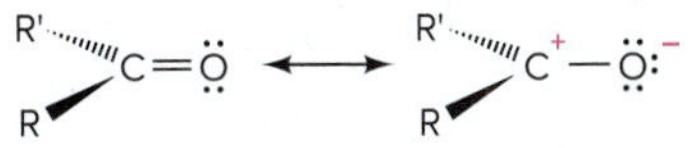

B Charged resonance form

Figure 15.21 The polar carbonyl group.

SAMPLE PROBLEM 15.5 Predicting the Steps in a Reaction Sequence

Problem Fill in the blanks in the following reaction sequence:

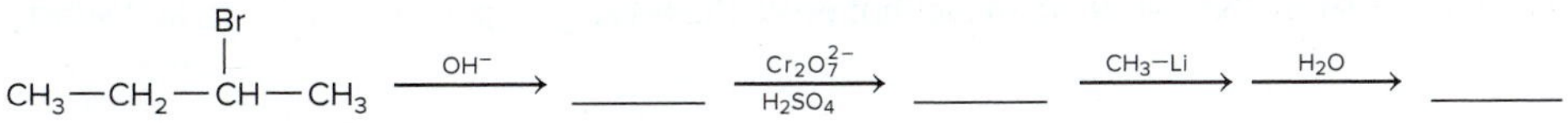

Plan For each step, we examine the functional group of the reactant and the reagent above the yield arrow to decide on the most likely product.

Solution The sequence starts with an alkyl halide reacting with OH^-. Substitution gives an alcohol. Oxidation of this alcohol with acidic dichromate gives a ketone. Finally, a two-step reaction of a ketone with CH_3—Li and then water forms an alcohol with a carbon skeleton that has the CH_3— group attached to the carbonyl C:

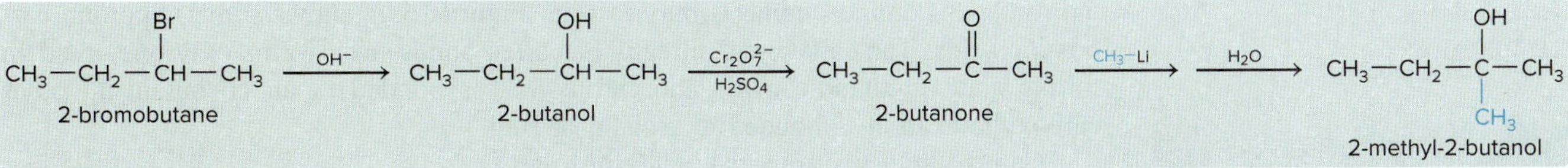

Check In this case, make sure that the first two reactions alter the functional group only and that the final steps change the C skeleton.

FOLLOW-UP PROBLEMS

15.5A Determine the organic product of each of the following reactions:

(a) 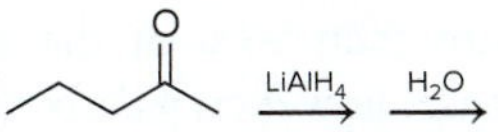

(*Hint:* Lithium aluminum hydride, $LiAlH_4$, is an important reducing agent in organic reactions; the second step of the reaction requires H_2O.)

(b) 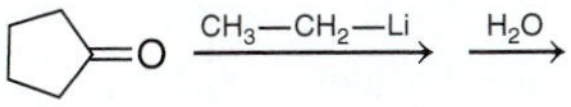

15.5B Choose reactants that will yield the following products:

(a) 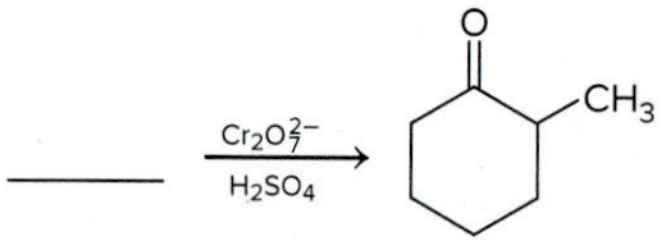

(b) 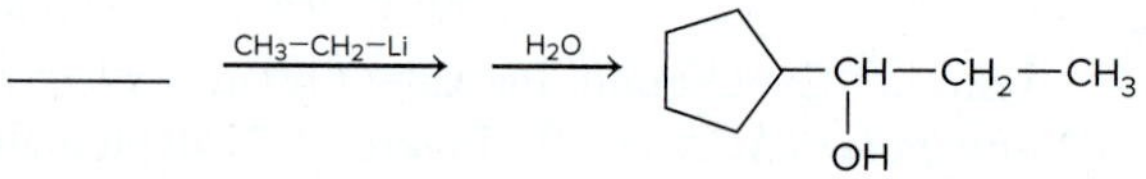

SOME SIMILAR PROBLEMS 15.65–15.74

Variations on a Theme: Inorganic Compounds with Double Bonds Homonuclear (same kind of atom) double bonds are rare for atoms other than C. However, we've seen many double bonds between O and other nonmetals, for example, in the oxides of S, N, and the halogens. Like carbonyl compounds, these substances undergo addition reactions. For example, the partially negative O of water attacks the partially positive S of SO_3 to form sulfuric acid:

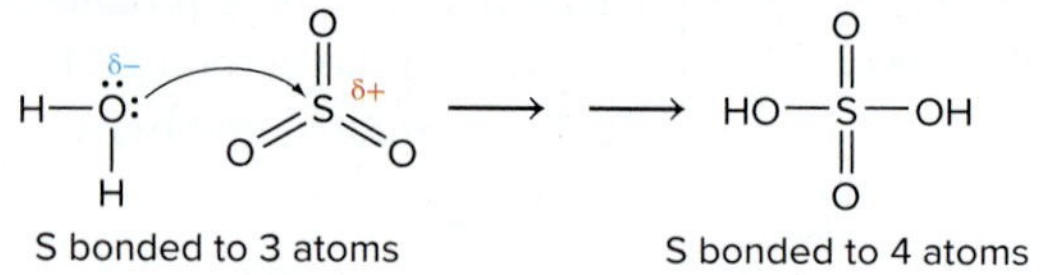

Functional Groups with Both Single and Double Bonds

A family of three functional groups contains C double bonded to O (a carbonyl group) *and* single bonded to O or N. The parent of the family is the **carboxylic acid** group, —C(=O)—OH, also called the *carboxyl group* and written —COOH. *The most important reaction type of this family is substitution from one member to another.* Substitution for the —OH by the —OR of an alcohol gives the **ester** group, —C(=O)—O—R, and substitution for the —OH by the —N(—)— of an amine gives the **amide** group, —C(=O)—N(—)—.

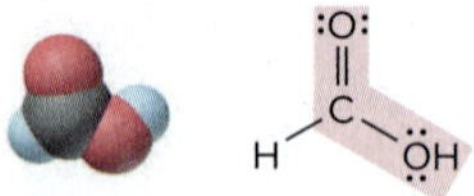

Methanoic acid (formic acid)
An irritating component of ant and bee stings

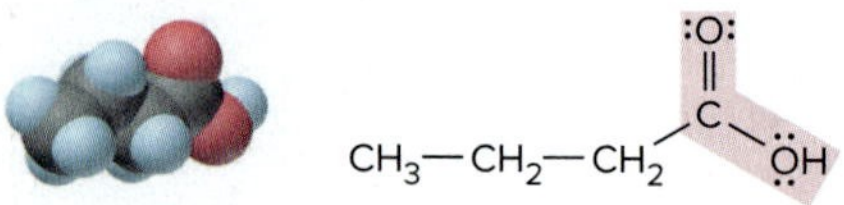

Butanoic acid (butyric acid)
Odor of rancid butter; suspected component of monkey sex attractant

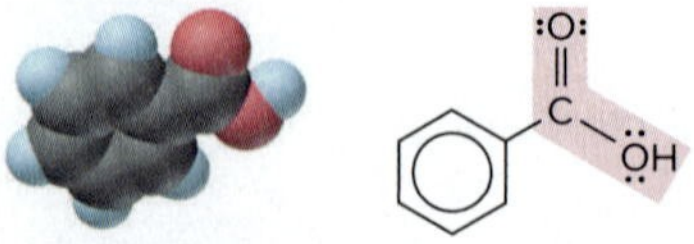

Benzoic acid
Calorimetric standard; used in preserving food, dyeing fabric, curing tobacco

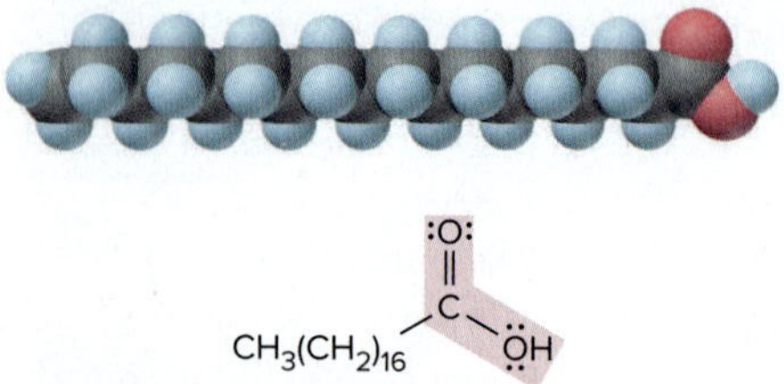

Octadecanoic acid (stearic acid)
Found in animal fats; used in making candles and soaps

Figure 15.22 Some molecules with the carboxylic acid functional group.

Carboxylic Acids Carboxylic acids, R—C(=O)—OH, are named by dropping the *-e* from the alkane name and adding *-oic acid;* however, many common names are used. For example, the four-C acid is butanoic acid (the carboxyl C is counted when choosing the root); its common name is butyric acid. Figure 15.22 shows some important carboxylic acids. The carboxyl C already has three bonds, so it forms only one other. In formic acid (methanoic acid), the carboxyl C is bonded to an H, but in all other carboxylic acids it is bonded to a chain or ring.

Carboxylic acids are weak acids in water:

$$CH_3—C(=O)—OH(l) + H_2O(l) \rightleftharpoons CH_3—C(=O)—O^-(aq) + H_3O^+(aq)$$

ethanoic acid (acetic acid)

At equilibrium in a solution of typical concentration, more than 99% of the acid molecules are undissociated at any given moment. In strong base, however, a carboxylic acid reacts completely to form a salt and water:

$$CH_3—C(=O)—OH(l) + NaOH(aq) \longrightarrow CH_3—C(=O)—O^-(aq) + Na^+(aq) + H_2O(l)$$

The anion of the salt is the *carboxylate ion,* named by dropping *-oic acid* and adding *-oate;* the sodium salt of butanoic acid, for instance, is sodium butanoate.

Carboxylic acids with long hydrocarbon chains are **fatty acids,** an essential group of compounds found in all cells. Animal fatty acids have saturated chains (see stearic acid, Figure 15.22, *bottom*), whereas many fatty acids from plants are unsaturated, usually with the C═C bonds in the *cis* configuration. The double bonds make them much easier to metabolize. Nearly all fatty acid skeletons have an even number of C atoms—16 and 18 carbons are very common—because cells use two-carbon units in synthesizing them. Fatty acid salts, usually with a cation from Group 1A(1) or 2A(2), are soaps (Section 13.2).

Substitution of carboxylic acids and other members of this family occurs through a two-step sequence: *addition plus elimination equals substitution.* Addition to the trigonal planar C atom gives an unstable tetrahedral intermediate, which immediately undergoes elimination to revert to a trigonal planar product (in this case, X is OH):

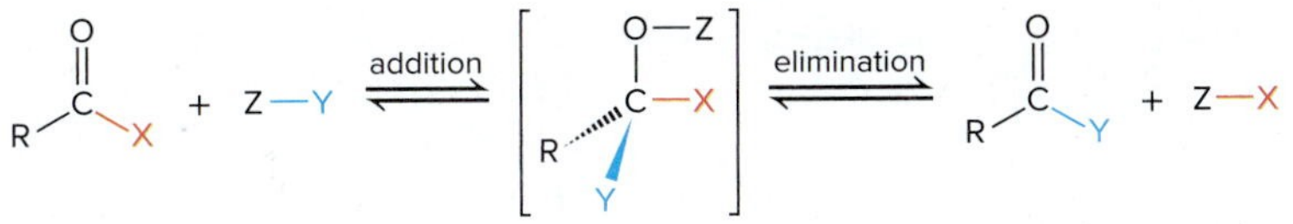

Strong heating of a carboxylic acid forms an **acid anhydride** through a type of substitution called a *dehydration-condensation reaction* (Section 14.7). Viewing this reaction as similar to the one above, but where X is OH, Z is H, and Y is the OOC—R group, we see the result is that two molecules condense into one with loss of a water molecule:

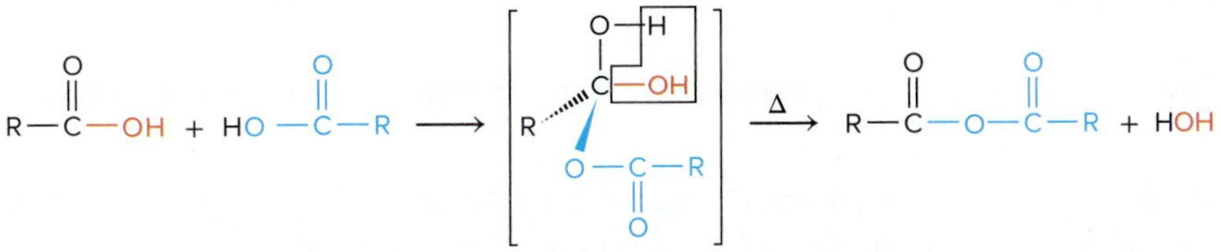

Esters An ester, $\mathrm{R{-}\overset{O}{\overset{\|}{C}}{-}O{-}R}$, is formed from an alcohol and a carboxylic acid. The first part of an ester name designates the alcohol portion and the second the acid portion (named in the same way as the carboxylate ion). For example, the ester formed between ethanol and ethanoic acid is ethyl ethanoate (common name, ethyl acetate), a solvent for nail polish and model glue.

The ester group occurs commonly in **lipids,** a large group of fatty biological substances. Most dietary fats are *triglycerides,* esters that are composed of three fatty acids linked to the alcohol 1,2,3-trihydroxypropane (common name, glycerol) and that function as energy stores. Some important lipids are shown in Figure 15.23; lecithin is one of several phospholipids that make up the lipid bilayer in all cell membranes (Section 13.2).

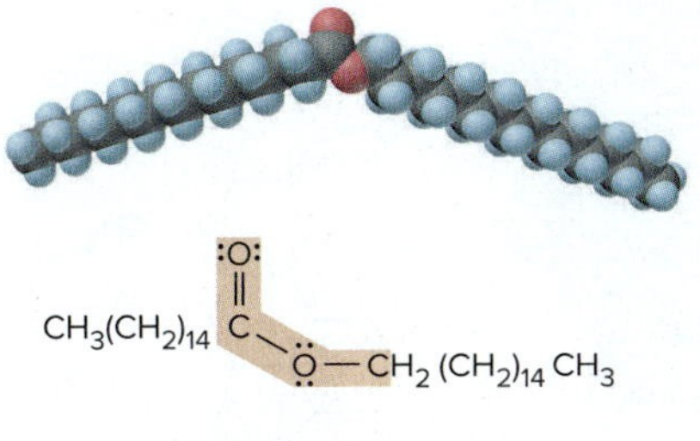

Cetyl palmitate The most common lipid in whale blubber

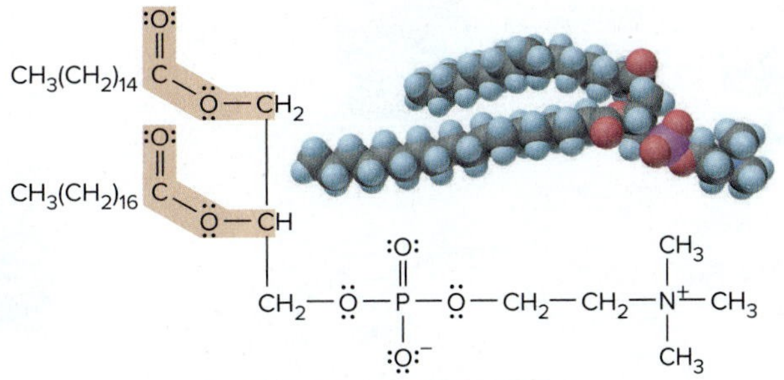

Lecithin Phospholipid found in all cell membranes

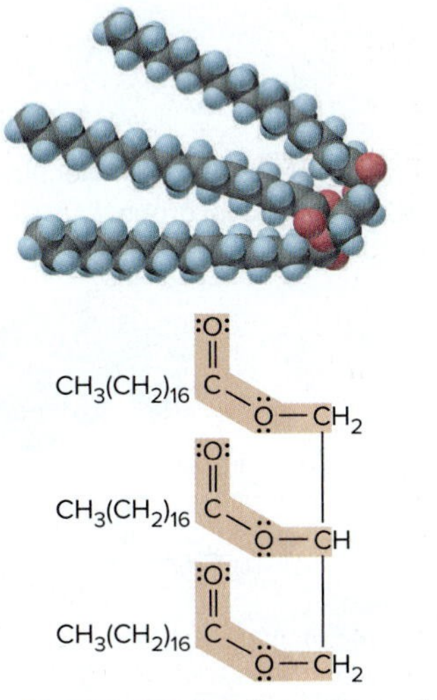

Tristearin Typical dietary fat used as an energy store in animals

Figure 15.23 Some lipid molecules with the ester functional group.

Esters, like acid anhydrides, form through a dehydration-condensation reaction; in this case, it is called an *esterification:*

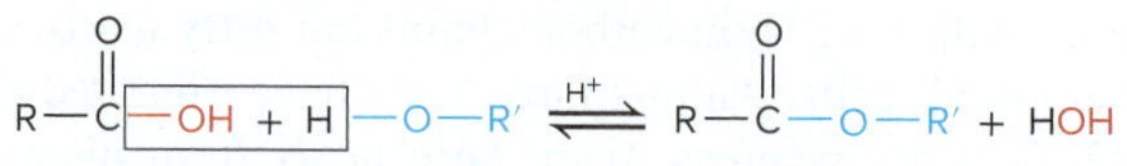

Carboxylic acids and esters have some familiar and distinctive odors.

Source: © Jill Braaten

Note that the esterification reaction is reversible. The opposite of dehydration-condensation is **hydrolysis,** in which the O atom of water is attracted to the partially positive C atom of the ester, cleaving (lysing) the ester molecule into two parts. One part bonds to water's —OH, and the other part to water's other H. In the process of soap manufacture, or *saponification* (Latin *sapon,* "soap"), used since ancient times, the ester bonds in animal or vegetable fats are hydrolyzed with a strong base:

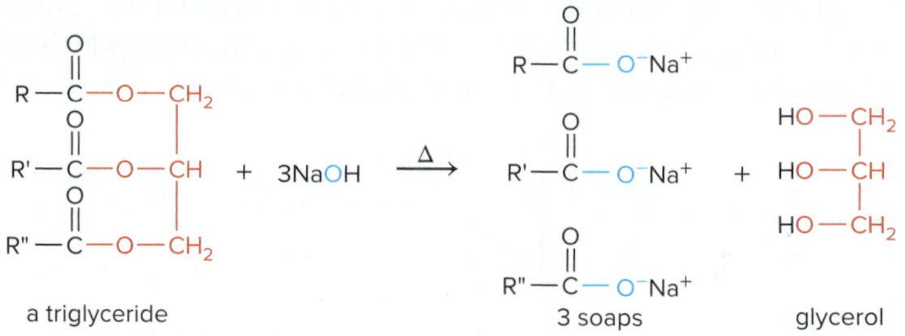

While carboxylic acids have pungent, vinegary, or cheesy odors, esters have pleasant odors. For example, ethyl butanoate has the scent of pineapple, and pentylpentanoate has the aroma of apples *(see photo).* Naturally occurring and synthetic esters are used to add fruity, floral, and herbal odors to foods, cosmetics, household deodorizers, and medicines.

Amides The product of a substitution between an amine (or NH_3) and an ester is an amide, $R{-}\overset{O}{\overset{\|}{C}}{-}\overset{|}{N}{-}$. The partially negative N of the amine is attracted to the partially positive C of the ester, an alcohol (ROH) is lost, and an amide forms:

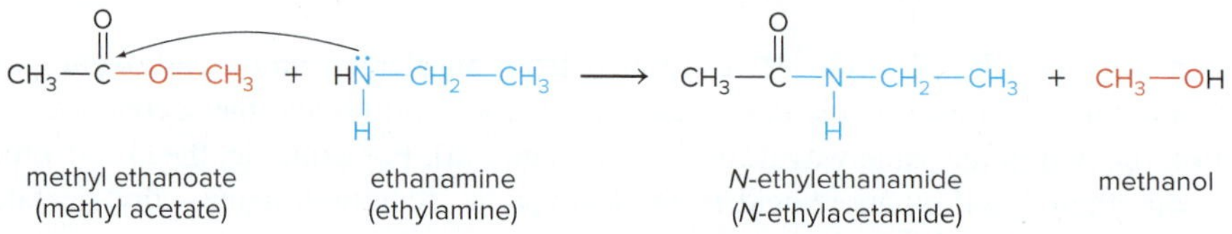

Amides are named by denoting the amine portion with *N*- (note that it is italic) and replacing *-oic acid* from the parent carboxylic acid with *-amide*. In the amide produced in the preceding reaction, the ethyl group comes from the amine, and the acid portion comes from ethanoic acid (acetic acid). Some amides are shown in Figure 15.24. The most important example of the amide group is the *peptide bond* (discussed in Sections 13.2 and 15.6), which links amino acids in a protein.

Figure 15.24 Some molecules with the amide functional group.

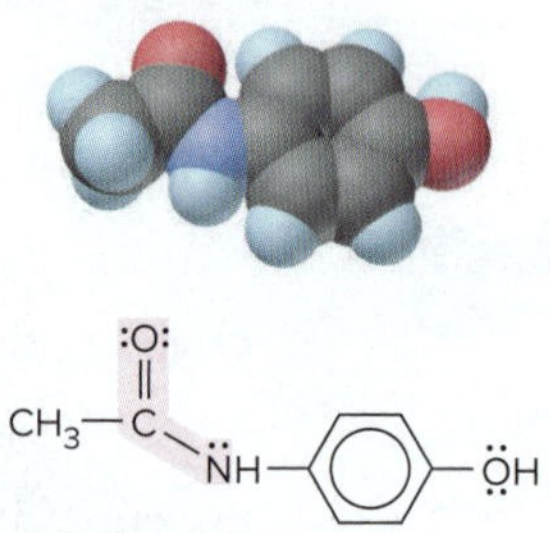

Acetaminophen Active ingredient in nonaspirin pain relievers; used to make dyes and photographic chemicals

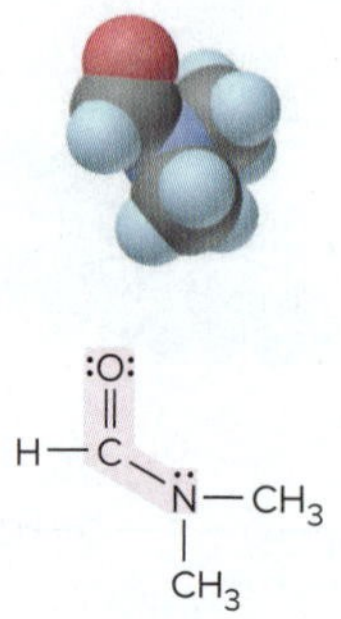

***N,N*-Dimethylmethanamide (dimethylformamide)** Major organic solvent; used in production of synthetic fibers

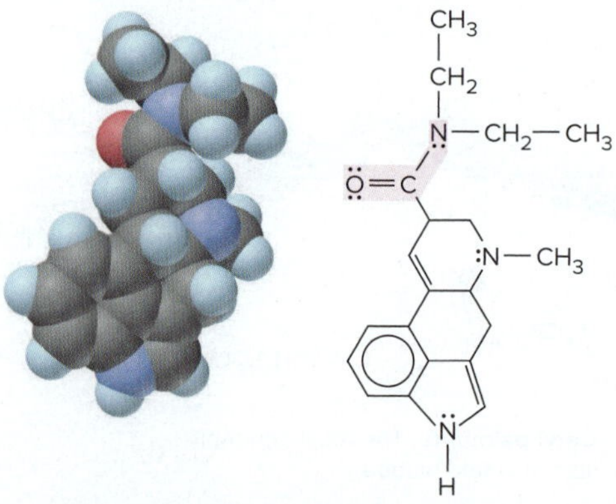

Lysergic acid diethylamide (LSD-25) A potent hallucinogen

An amide is hydrolyzed in hot water (or base) to a carboxylic acid (or carboxylate ion) and an amine. Thus, even though amides are not normally formed in the following way, they can be viewed as the result of a reversible dehydration-condensation:

$$R-\overset{\overset{\displaystyle O}{\|}}{C}\boxed{-OH + H-}N-R' \rightleftharpoons R-\overset{\overset{\displaystyle O}{\|}}{C}-\overset{\overset{\displaystyle H}{|}}{N}-R' + HOH$$

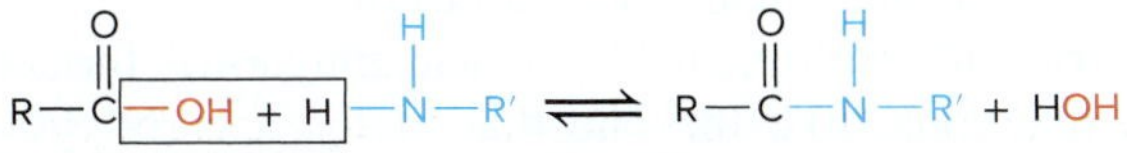

Reduction Reactions Compounds in the carboxylic acid family also undergo reduction to form other functional groups. For example, certain inorganic *reducing agents* (such as lithium aluminum hydride, $LiAlH_4$, or sodium borohydride, $NaBH_4$) convert acids or esters to alcohols and convert amides to amines:

$$R-\overset{\overset{\displaystyle O}{\|}}{C}-OH \ (or\ R-\overset{\overset{\displaystyle O}{\|}}{C}-O-R') \xrightarrow{reduction} R-CH_2-OH + HOH\ (or\ R'-OH)$$

$$R-\overset{\overset{\displaystyle O}{\|}}{C}-NH-R' \xrightarrow{reduction} R-CH_2-NH-R' + H_2O$$

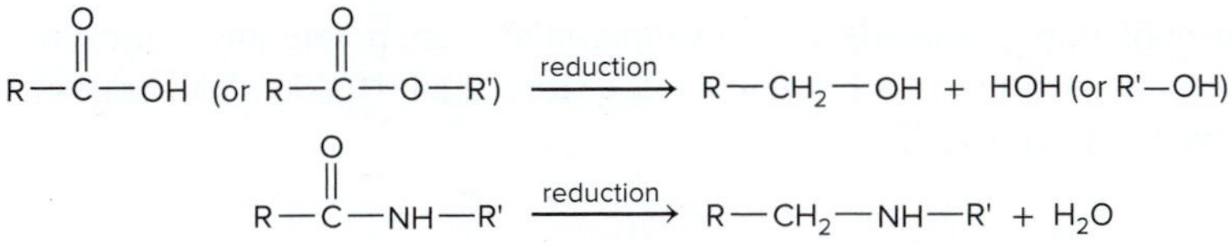

SAMPLE PROBLEM 15.6	Predicting Reactions of the Carboxylic Acid Family

Problem Predict the organic product(s) of the following reactions:

(a) $CH_3-CH_2-CH_2-\overset{\overset{\displaystyle O}{\|}}{C}-OH + CH_3-\overset{\overset{\displaystyle OH}{|}}{CH}-CH_3 \underset{}{\overset{H^+}{\rightleftharpoons}}$

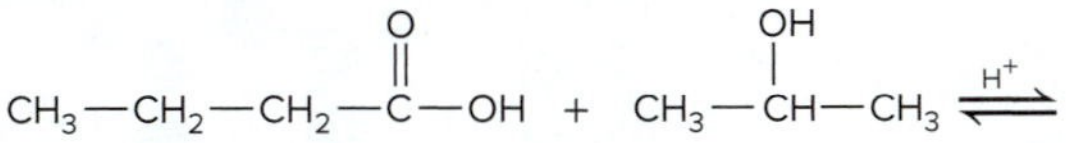

(b) $CH_3-\overset{\overset{\displaystyle CH_3}{|}}{CH}-CH_2-CH_2-\overset{\overset{\displaystyle O}{\|}}{C}-NH-CH_2-CH_3 \xrightarrow[H_2O]{NaOH}$

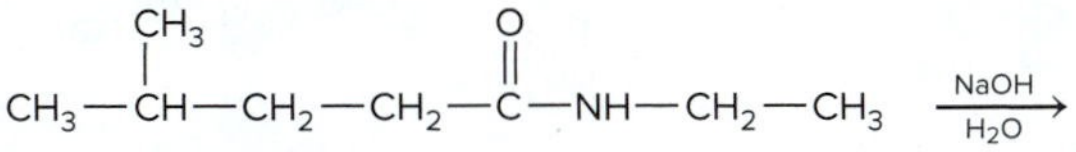

Plan (a) A carboxylic acid and an alcohol react, so the reaction must be a substitution to form an ester and water. (b) An amide reacts with OH^-, so it is hydrolyzed to an amine and a sodium carboxylate.

Solution (a) Formation of an ester:

$$CH_3-CH_2-CH_2-\overset{\overset{\displaystyle O}{\|}}{C}-O-\overset{\overset{\displaystyle CH_3}{|}}{CH}-CH_3 + H_2O$$

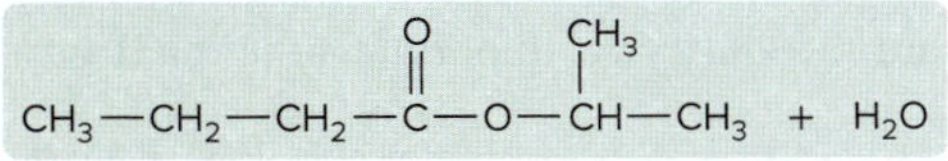

(b) Basic hydrolysis of an amide:

$$CH_3-\overset{\overset{\displaystyle CH_3}{|}}{CH}-CH_2-CH_2-\overset{\overset{\displaystyle O}{\|}}{C}-O^- + Na^+ + H_2N-CH_2-CH_3$$

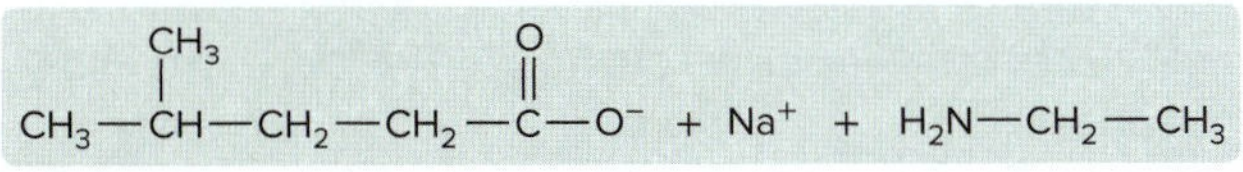

Check Note that in part (b), the carboxylate ion forms, rather than the acid, because the aqueous NaOH that is present reacts with any carboxylic acid as it forms.

FOLLOW-UP PROBLEMS

15.6A Predict the organic product(s) of the following reactions:

(a) $CH_3-(CH_2)_{14}-\overset{\overset{\displaystyle O}{\|}}{C}-O-CH_2-(CH_2)_{14}-CH_3 \xrightarrow[H_2O]{NaOH}$

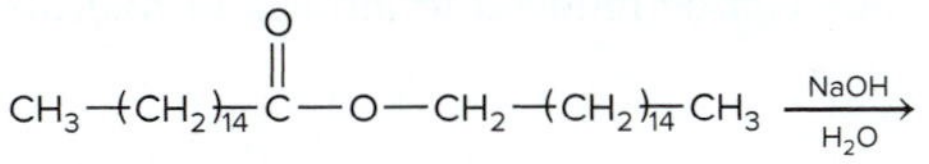

(b) $CH_3-\langle\bigcirc\rangle-\overset{\overset{\displaystyle O}{\|}}{C}-NH-\langle\bigcirc\rangle-CH_3 \xrightarrow{LiAlH_4} \xrightarrow{H_2O}$

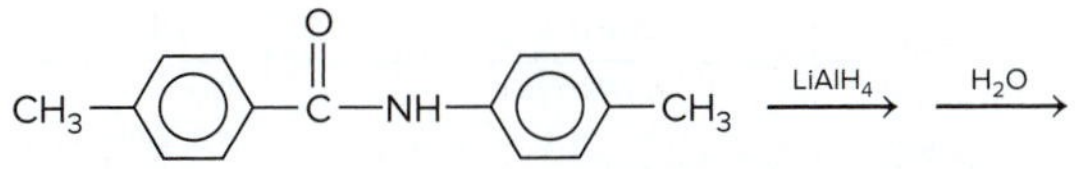

15.6B Fill in the blanks in the following reactions:

(a) ＿＿＿ $+ CH_3-OH \overset{H^+}{\rightleftharpoons} \langle\bigcirc\rangle-CH_2-\overset{\overset{\displaystyle O}{\|}}{C}-O-CH_3 + H_2O$

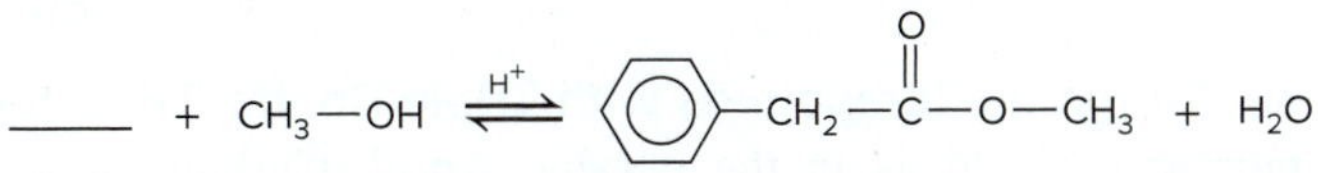

(b) ＿＿＿ $+$ ＿＿＿ $\longrightarrow CH_3-CH_2-CH_2-\overset{\overset{\displaystyle O}{\|}}{C}-NH-CH_2-CH_3 + CH_3-OH$

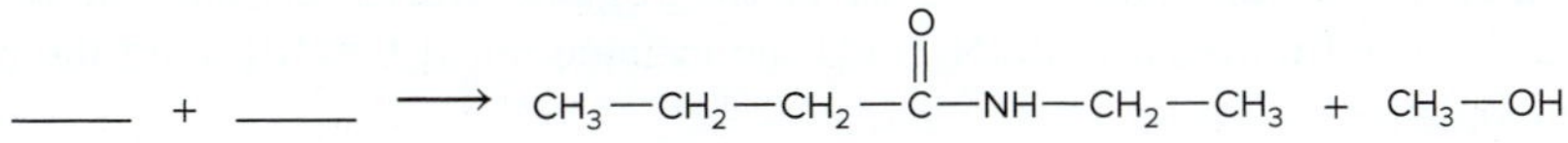

SOME SIMILAR PROBLEMS 15.67–15.74

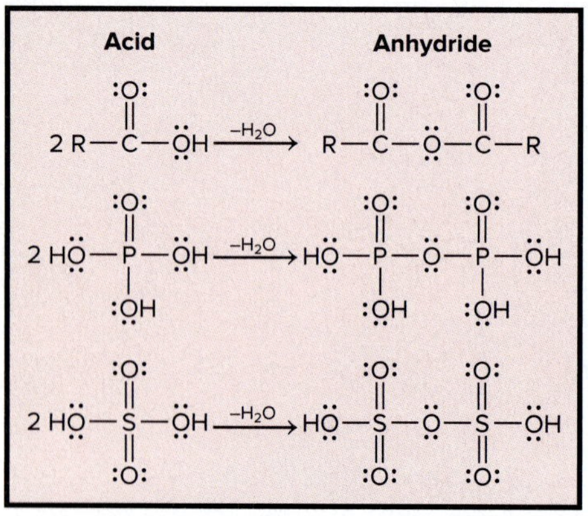

Acid	Anhydride
$2 \; R-\overset{\overset{\displaystyle :O:}{\|}}{C}-\ddot{\overset{..}{O}}H \xrightarrow{-H_2O}$	$R-\overset{\overset{\displaystyle :O:}{\|}}{C}-\ddot{\overset{..}{O}}-\overset{\overset{\displaystyle :O:}{\|}}{C}-R$
$2 \; HO-\overset{\overset{\displaystyle :O:}{\|}}{\underset{\underset{\displaystyle :\ddot{O}H}{\|}}{P}}-\ddot{\overset{..}{O}}H \xrightarrow{-H_2O}$	$H\ddot{O}-\overset{\overset{\displaystyle :O:}{\|}}{\underset{\underset{\displaystyle :\ddot{O}H}{\|}}{P}}-\ddot{\overset{..}{O}}-\overset{\overset{\displaystyle :O:}{\|}}{\underset{\underset{\displaystyle :\ddot{O}H}{\|}}{P}}-\ddot{\overset{..}{O}}H$
$2 \; H\ddot{O}-\overset{\overset{\displaystyle :O:}{\|}}{\underset{\underset{\displaystyle :O:}{\|}}{S}}-\ddot{\overset{..}{O}}H \xrightarrow{-H_2O}$	$H\ddot{O}-\overset{\overset{\displaystyle :O:}{\|}}{\underset{\underset{\displaystyle :O:}{\|}}{S}}-\ddot{\overset{..}{O}}-\overset{\overset{\displaystyle :O:}{\|}}{\underset{\underset{\displaystyle :O:}{\|}}{S}}-\ddot{\overset{..}{O}}H$

Figure 15.25 The formation of carboxylic, phosphoric, and sulfuric acid anhydrides.

Variations on a Theme: Oxoacids, Esters, and Amides of Other Nonmetals A nonmetal that has both double and single bonds to oxygens occurs in most inorganic oxoacids, such as phosphoric, sulfuric, and chlorous acids. Those with additional O atoms are stronger acids than carboxylic acids.

Diphosphoric and disulfuric acids are acid anhydrides formed by dehydration-condensation reactions, like the one that yields a carboxylic acid anhydride (Figure 15.25). Inorganic oxoacids form esters and amides that are part of many biological molecules. We already saw that certain lipids are phosphate esters (see Figure 15.23). The first compound formed when glucose is digested is a phosphate ester (Figure 15.26A); a similar phosphate ester is a major structural feature of nucleic acids, as we'll see shortly. Amides of organic sulfur-containing oxoacids, called *sulfonamides,* are potent antibiotics; the simplest of these is depicted in Figure 15.26B. More than 10,000 different sulfonamides have been synthesized.

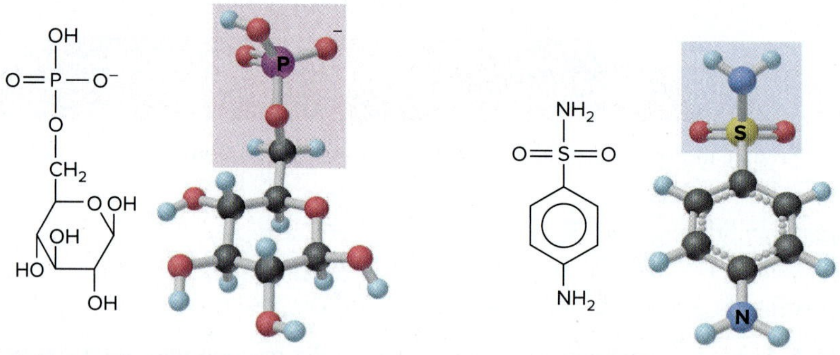

Figure 15.26 A phosphate ester and a sulfonamide.

A Glucose-6-phosphate   **B Sulfanilamide**

Functional Groups with Triple Bonds

Alkynes and nitriles are the only two important functional groups with triple bonds.

Alkynes Alkynes, with an electron-rich —C≡C— group, undergo addition (by H_2O, H_2, HX, X_2, and so forth) to form double-bonded or saturated compounds:

$$CH_3-C{\equiv}CH \xrightarrow{H_2} CH_3-CH{=}CH_2 \xrightarrow{H_2} CH_3-CH_2-CH_3$$
$$\text{propyne} \qquad\qquad \text{propene} \qquad\qquad \text{propane}$$

Nitriles Nitriles (R—C≡N) contain the **nitrile** group (—C≡N:) and are made by substituting a CN^- (cyanide) ion for X^- in a reaction with an alkyl halide:

$$CH_3-CH_2-Cl + NaCN \longrightarrow CH_3-CH_2-C{\equiv}N + NaCl$$

This reaction is useful because it *increases the chain by one C atom.* Nitriles are versatile because once they are formed, they can be reduced to amines or hydrolyzed to carboxylic acids:

$$CH_3-CH_2-CH_2-NH_2 \xleftarrow{\text{reduction}} CH_3-CH_2-C{\equiv}N \xrightarrow[\text{hydrolysis}]{H_3O^+,\ H_2O} CH_3-CH_2-\overset{\overset{\displaystyle O}{\|}}{C}-OH + NH_4^+$$

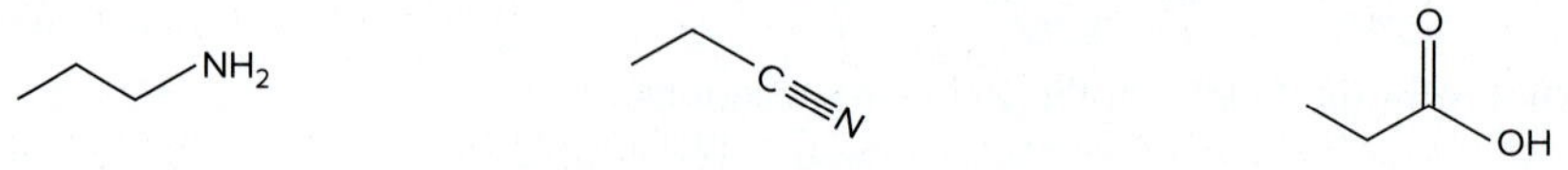

Variations on a Theme: Inorganic Compounds with Triple Bonds Triple bonds are as scarce in the inorganic world as in the organic world. Carbon monoxide (:C≡O:), elemental nitrogen (:N≡N:), and the cyanide ion ([:C≡N:]⁻) are the only common examples.

You've seen quite a few functional groups by this time, and it is especially important that you can recognize them in a complex organic molecule. Sample Problem 15.7 provides some practice.

SAMPLE PROBLEM 15.7 Recognizing Functional Groups

Problem Circle and name the functional groups in the following molecules:

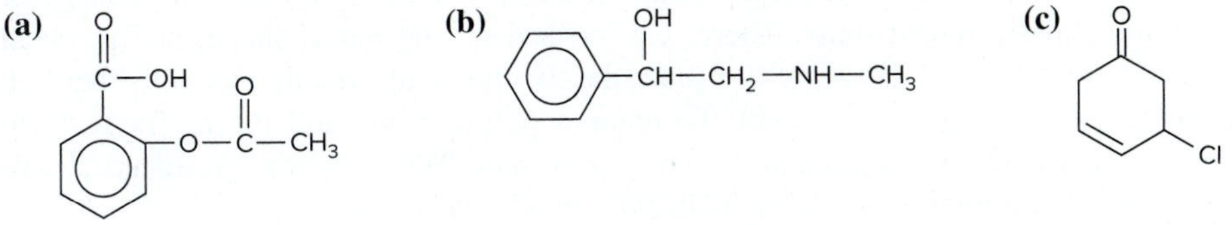

Plan We use Table 15.5 to identify the various functional groups.

Solution

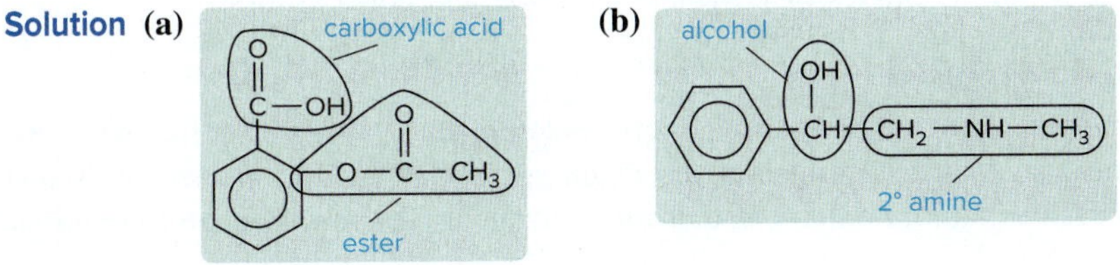

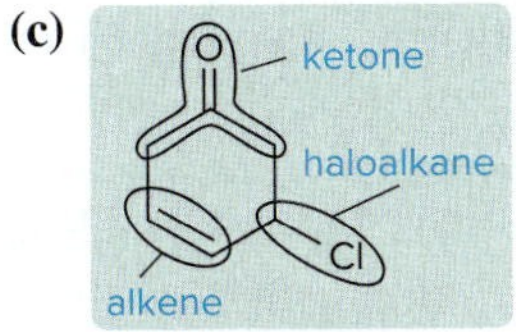

FOLLOW-UP PROBLEMS

15.7A Circle and name the functional groups:

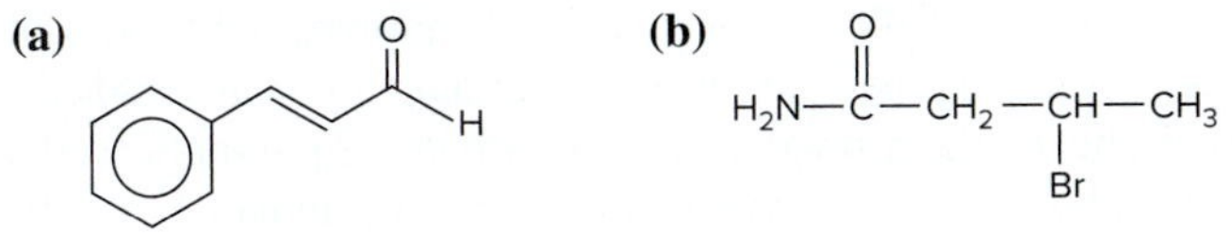

15.7B Circle and name the functional groups:

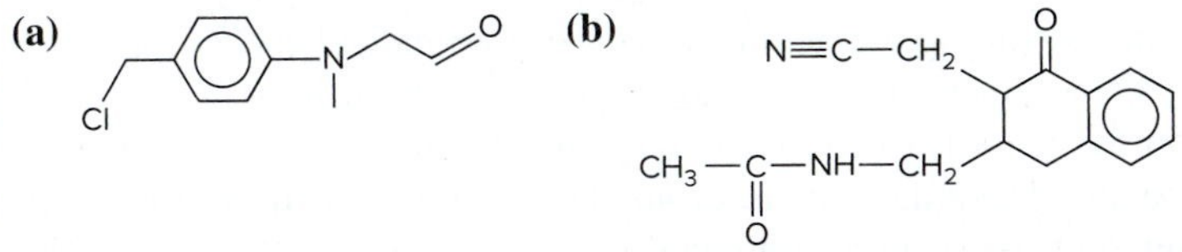

SOME SIMILAR PROBLEMS 15.59 and 15.60

› Summary of Section 15.4

- Organic reactions are initiated when regions of high and low electron density in reactant molecules attract each other.
- Functional groups containing only single bonds—alcohols, haloalkanes, and amines—undergo substitution and elimination reactions.
- Functional groups with double or triple bonds—alkenes, aldehydes, ketones, alkynes, and nitriles—mostly undergo addition reactions.
- Aromatic compounds typically undergo substitution, rather than addition, because delocalization of the π electrons stabilizes the ring.
- Functional groups with both double and single bonds—carboxylic acids, esters, and amides—undergo substitution (addition plus elimination) reactions.
- Many reactions change one functional group to another, but some, including reactions with organometallic compounds and with the cyanide ion, change the carbon skeleton.

15.5 THE MONOMER-POLYMER THEME I: SYNTHETIC MACROMOLECULES

In Chapter 12, you saw that polymers are extremely large molecules that consist of many monomeric repeat units. There, we focused on the mass, shape, and physical properties of polymers. Now we'll discuss the two types of organic reactions that link monomers covalently into a chain. To name a polymer, we add the prefix *poly-* to the monomer name, as in *polyethylene* or *polystyrene*. When the monomer has a two-word name, parentheses are used, as in *poly(vinyl chloride)*.

The two major types of reaction processes that form synthetic polymers lend their names to the resulting classes of polymer—addition and condensation.

Addition Polymers

Addition polymers form when monomers undergo an addition reaction with one another. These substances are also called *chain-reaction* (or *chain-growth*) *polymers* because as each monomer adds to the chain, it forms a new reactive site to continue the process. The monomers of most addition polymers have the $\rangle C{=}C\langle$ grouping. As you can see from Table 15.6, the remarkably different physical behaviors of an acrylic sweater, a plastic grocery bag, and a bowling ball result from the different groups that are attached to the double-bonded C atoms of the monomers.

The *free-radical polymerization* of ethene (ethylene, $CH_2{=}CH_2$) to polyethylene is a simple example of the addition process (Figure 15.27). The monomer reacts to form a *free radical,* a species with an unpaired electron, which forms a covalent bond with an electron from another monomer:

Step 1. The process begins when an *initiator,* usually a peroxide, generates a free radical.
Step 2. The free radical attacks the π bond of an ethylene molecule, forming a σ bond with one of the *p* electrons and leaving the other unpaired, creating a new free radical.
Step 3. This new free radical attacks the π bond of another ethylene, joining it to the chain end, and the backbone of the polymer grows one unit longer.
Step 4. This process stops when two free radicals form a covalent bond or when a very stable free radical is formed by addition of an *inhibitor* molecule.

Recent progress in controlling the high reactivity of free-radical species promises an even wider range of polymers. In one method, polymerization is initiated by the formation of a cation (or anion) instead of a free radical. The cationic (or anionic) reactive end of the chain attacks the π bond of another monomer to form a new cationic (or anionic) end, and the process continues.

The most important polymerization reactions take place under relatively mild conditions through the use of catalysts that incorporate transition metals. In 1963, Karl Ziegler and Giulio Natta received the Nobel Prize in chemistry for developing *Ziegler-Natta catalysts,* which employ an organoaluminum compound, such as $Al(C_2H_5)_3$, and the tetrachloride of titanium or vanadium. Today, chemists use organometallic catalysts that are *stereoselective* to create polymers whose repeat units have groups spatially oriented in particular ways. Use of these catalysts with varying conditions and reagents allows polyethylene chains with molar masses of 10^4 to 10^5 g/mol to be made.

Similar methods are used to make polypropylenes, $\text{+}CH_2—\underset{\displaystyle CH_3}{\underset{|}{CH}}\text{+}_n$, that have all the CH_3 groups of the repeat units oriented either on one side of the chain or on alternating sides. The different orientations lead to different packing efficiencies of the chains and, thus, different degrees of crystallinity, which lead to differences in such physical properties as density, rigidity, and elasticity (Section 12.7).

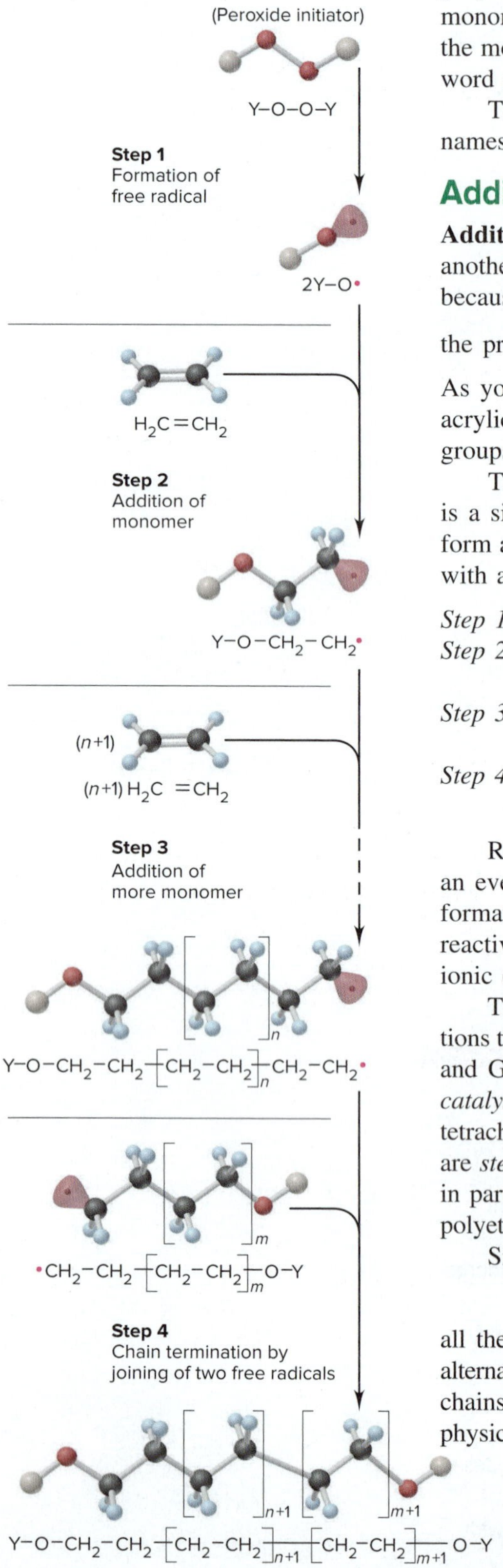

Figure 15.27 Steps in the free-radical polymerization of ethylene.

Table 15.6 Some Major Addition Polymers

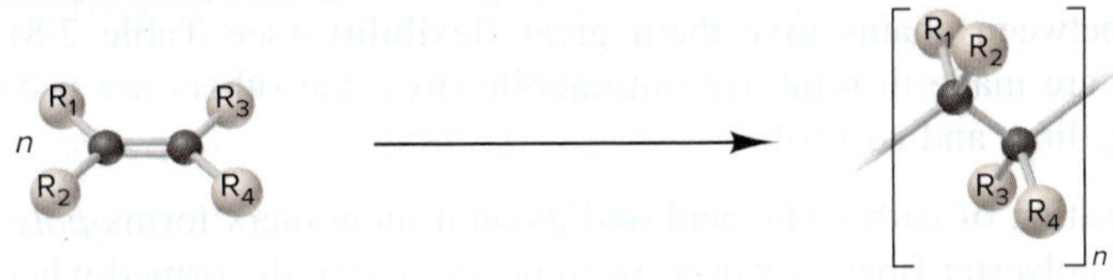

Monomer	Polymer	Applications
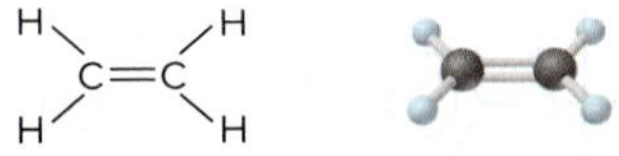	polyethylene ($R_1 = R_2 = R_3 = R_4 = H$)	Plastic bags; bottles; toys
	polytetrafluoroethylene ($R_1 = R_2 = R_3 = R_4 = F$)	Cooking utensils (e.g., Teflon)
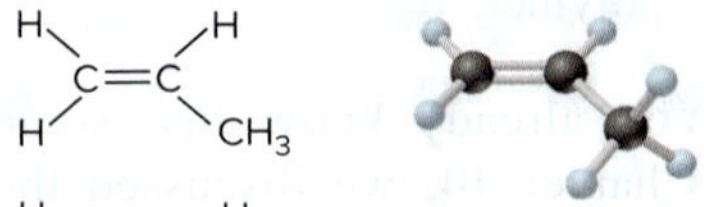	polypropylene ($R_1 = R_2 = R_3 = H$; $R_4 = CH_3$)	Carpeting (indoor-outdoor); bottles
	poly(vinyl chloride) ($R_1 = R_2 = R_3 = H$; $R_4 = Cl$)	Plastic wrap; garden hose; indoor plumbing
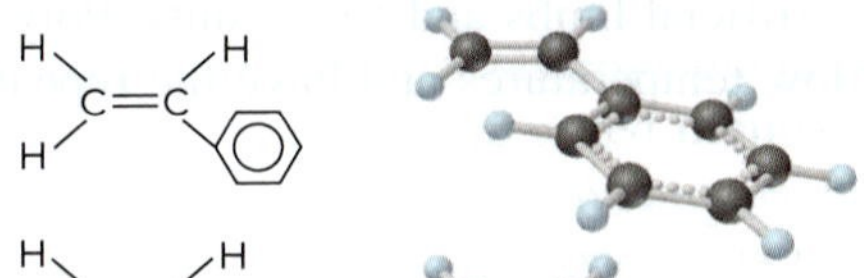	polystyrene ($R_1 = R_2 = R_3 = H$; $R_4 = C_6H_5$)	Insulation; furniture; packing materials
	polyacrylonitrile ($R_1 = R_2 = R_3 = H$; $R_4 = CN$)	Yarns (e.g., Orlon, Acrilan); fabrics; wigs
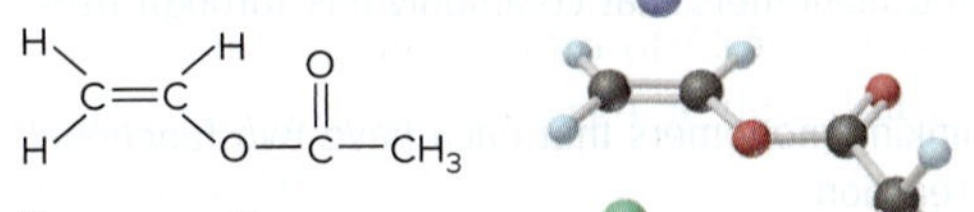	poly(vinyl acetate) ($R_1 = R_2 = R_3 = H$; $R_4 = O\overset{O}{\overset{\Vert}{C}}CH_3$)	Adhesives; paints; textile coatings; computer disks
	poly(vinylidene chloride) ($R_1 = R_2 = H$; $R_3 = R_4 = Cl$)	Food wrap (e.g., Saran)
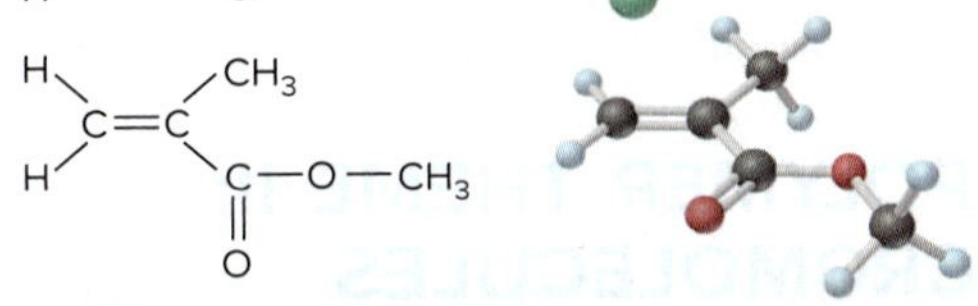	poly(methyl methacrylate) ($R_1 = R_2 = H$; $R_3 = CH_3$; $R_4 = \overset{O}{\overset{\Vert}{C}}OCH_3$)	Glass substitute (e.g., Lucite, Plexiglas); bowling balls; paint

Condensation Polymers

The monomers of **condensation polymers** must have *two functional groups;* we can designate such a monomer as A—**R**—B (where A and B may or may not be the same, and **R** is the rest of the molecule). Most commonly, the monomers link when an A group on one undergoes a *dehydration-condensation reaction* with a B group on another:

$$\tfrac{1}{2}n\text{H—A—R—B—OH} + \tfrac{1}{2}n\text{H—A—R—B—OH} \xrightarrow{-(n-1)\text{HOH}} \text{H}\!+\!\text{A—R—B}\!+_{n}\!\text{OH}$$

Many condensation polymers are *copolymers,* those consisting of two or more different repeat units (Section 12.7). Two major types are polyamides and polyesters.

1. *Polyamides.* Condensation of carboxylic acid and amine monomers forms *polyamides (nylons).* One of the most common is *nylon-66,* manufactured by mixing equimolar amounts of a six-C diamine (1,6-diaminohexane) and a six-C diacid (1,6-hexanedioic acid). The basic amine reacts with the acid to form a "nylon salt." Heating drives off water and forms the amide bonds:

$$n\text{HO}-\overset{O}{\overset{\Vert}{C}}-(CH_2)_4-\overset{O}{\overset{\Vert}{C}}-\text{OH} + n\text{H}_2\text{N}-(CH_2)_6-\text{NH}_2 \xrightarrow[-(2n-1)H_2O]{\Delta} \text{HO}\left[\overset{O}{\overset{\Vert}{C}}-(CH_2)_4-\overset{O}{\overset{\Vert}{C}}-\text{NH}-(CH_2)_6-\text{NH}\right]_n\text{H}$$

signs; $[H_2O]$ increases, so it has a plus sign. Another check is to use Equation 16.2, with A = H_2, $a = 2$; B = O_2, $b = 1$; and C = H_2O, $c = 2$:

$$\text{Rate} = -\frac{1}{a}\frac{d[\text{A}]}{dt} = -\frac{1}{b}\frac{d[\text{B}]}{dt} = \frac{1}{c}\frac{d[\text{C}]}{dt}$$

or

$$\text{Rate} = -\frac{1}{2}\frac{d[\text{H}_2]}{dt} = -\frac{d[\text{O}_2]}{dt} = \frac{1}{2}\frac{d[\text{H}_2\text{O}]}{dt}$$

(b) Given the rate expression, it makes sense that the numerical value of the rate of $[H_2O]$ increase is twice that of $[O_2]$ decrease.

Comment Thinking through this type of problem at the molecular level is the best approach, but use Equation 16.2 to confirm your answer.

FOLLOW-UP PROBLEMS

*Brief Solutions for **all** Follow-up Problems appear at the end of the chapter.*

16.1A **(a)** Balance the following equation and express the rate in terms of the change in concentration with time for each substance:

$$\text{NO}(g) + \text{O}_2(g) \longrightarrow \text{N}_2\text{O}_3(g)$$

(b) How fast is $[O_2]$ decreasing when [NO] is decreasing at a rate of 1.60×10^{-4} mol/L·s?

16.1B For a particular reaction, the reaction rate in terms of the change in concentration with time for each substance is

$$\text{Rate} = -\frac{1}{4}\frac{d[\text{NH}_3]}{dt} = -\frac{1}{5}\frac{d[\text{O}_2]}{dt} = \frac{1}{4}\frac{d[\text{NO}]}{dt} = \frac{1}{6}\frac{d[\text{H}_2\text{O}]}{dt}$$

(a) Write a balanced equation for this gaseous reaction.
(b) When $[H_2O]$ is increasing at a rate of 2.52×10^{-2} mol/L·s, how fast is $[O_2]$ decreasing?

SOME SIMILAR PROBLEMS 16.14–16.19

› Summary of Section 16.2

- The average reaction rate is the change in reactant (or product) concentration over a change in time, Δt. The rate slows as the reaction proceeds because reactants are used up.
- The instantaneous rate at time t is the slope of the tangent to a curve that plots concentration vs. time, and equals $-d[\text{A}]/dt$.
- The initial rate, the instantaneous rate at $t = 0$, occurs when reactants have just been mixed and before any product accumulates.
- The expression for a reaction rate and its numerical value depend on which reaction component is being referenced.

16.3 THE RATE LAW AND ITS COMPONENTS

The centerpiece of any kinetic study of a reaction is the **rate law (*or* rate equation),** which expresses the rate as a function of concentrations and temperature. The rate law is based on experiment, so any hypothesis about how the reaction occurs on the molecular level must conform to it.

In this discussion, we generally consider reactions for which the products do not appear in the rate law, so the rate depends only on *reactant* concentrations and temperature. For a general reaction occurring at a fixed temperature,

$$a\text{A} + b\text{B} + \cdots \longrightarrow c\text{C} + d\text{D} + \cdots$$

the rate law is

$$\text{Rate} = k[\text{A}]^m[\text{B}]^n \ldots \qquad \textbf{(16.3)}$$

The term k is a proportionality constant, called the **rate constant,** that is specific for a given reaction at a given temperature and does *not* change as the reaction proceeds. (As we'll see in Section 16.5, k *does* change with temperature.) The exponents m and n, called the **reaction orders,** define how the rate is affected by reactant concentration; we'll see how to determine them shortly. Two key points to remember are

- *The balancing coefficients a and b in the reaction equation are* ***not*** *necessarily related in any way to the reaction orders m and n.*
- *The components of the rate law—rate, reaction orders, and rate constant—**must** be found by experiment.*

In the remainder of this section, we'll find the components of the rate law by measuring concentrations to determine the *initial rate,* using initial rates to determine the *reaction orders,* and using these values to calculate the *rate constant.* With the rate law for a reaction, we can predict the rate for any initial concentrations.

Some Laboratory Methods for Determining the Initial Rate

We determine an initial rate from a plot of concentration vs. time, so we need a quick, accurate method for measuring concentration as a reaction proceeds. Let's briefly discuss three common approaches.

1. *Spectrometric methods* measure the concentration of a component that absorbs (or emits) characteristic wavelengths of light. For example, in the reaction of NO and O_3, only NO_2 has a color (an indication that NO_2 absorbs some wavelength of visible light):

$$NO(g,\text{ colorless}) + O_3(g,\text{ colorless}) \longrightarrow O_2(g,\text{ colorless}) + NO_2(g,\text{ brown})$$

Known amounts of reactants are injected into a tube of known volume within a spectrometer (see Tools of the Laboratory, end of Section 7.2), which is set to measure the wavelength and intensity of the color. The rate of NO_2 formation is proportional to the increase in that intensity over time.

2. *Conductometric methods* rely on the change in electrical conductivity of the reaction solution when nonionic reactants form ionic products, or vice versa. Consider the substitution reaction between a haloalkane, such as 2-bromo-2-methylpropane, and water:

$$(CH_3)_3C\text{—}Br(l) + H_2O(l) \longrightarrow (CH_3)_3C\text{—}OH(l) + H^+(aq) + Br^-(aq)$$

The HBr that forms is a strong acid, so it dissociates completely in the water. As time passes, more ions form, so the conductivity of the reaction mixture increases.

3. *Manometric methods* employ a manometer attached to a reaction vessel of fixed volume and temperature. The manometer measures the pressure change in the vessel due to a reaction that involves a change in the number of moles of gas. Consider the decomposition reaction of hydrogen peroxide:

$$2H_2O_2(aq) \longrightarrow 2H_2O(l) + O_2(g)$$

The rate is directly proportional to the increase in pressure as O_2 gas forms.

Determining Reaction Orders

With the initial rate in hand, we can determine reaction orders. Let's first discuss what reaction orders are and then see how to determine them by controlling reactant concentrations.

First, Second, and Zero Orders A reaction has an *individual* order "with respect to" or "in" each reactant, and an *overall* order, the sum of the individual orders.

Consider first the simplest case, a reaction with only one reactant, A:

$$A \longrightarrow \text{products}$$

- *First order.* If the rate doubles when [A] doubles, the rate depends on [A] raised to the first power, $[A]^1$ (the superscript 1 is generally omitted). Thus, the reaction is *first order* in (or with respect to) A and *first order* overall:

$$\text{Rate} = k[A]^1 = k[A]$$

- *Second order.* If the rate quadruples when [A] doubles, the rate depends on [A] squared, $[A]^2$. In this case, the reaction is *second order* in A and *second order* overall:

$$\text{Rate} = k[A]^2$$

- *Zero order*. If the rate does not change when [A] doubles, the rate does *not* depend on [A], but we express this fact mathematically by saying that the rate depends on [A] raised to the zero power, $[A]^0$. The reaction is *zero order* in A and *zero order* overall:

$$\text{Rate} = k[\text{A}]^0 = k(1) = k$$

Figure 16.7 shows plots of [A] vs. time for first-, second-, and zero-order reactions. (In all cases, the value of k was assumed to be the same.) Notice that

- The decrease in [A] doesn't change as time goes on for a zero-order reaction; the rate is constant.
- The decrease slows down as time goes on for a first-order reaction.
- The decrease slows even more for a second-order reaction.

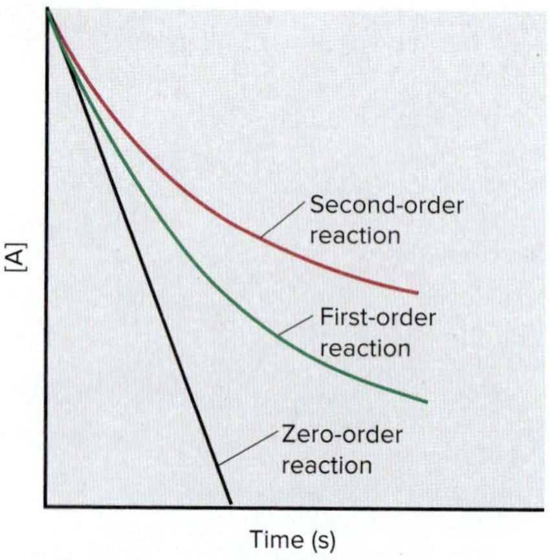

Figure 16.7 Plots of reactant concentration, [A], vs. time for first-, second-, and zero-order reactions.

These results are reflected in Figure 16.8, which shows plots of rate vs. [A] for the same reaction orders. Notice that

- The plot is a horizontal line for the zero-order reaction because the rate doesn't change regardless of the value of [A].
- The plot is an upward-sloping *line* for the first-order reaction because the rate is directly proportional to [A].
- The plot is an upward-sloping *curve* for the second-order reaction because the rate increases exponentially with [A].

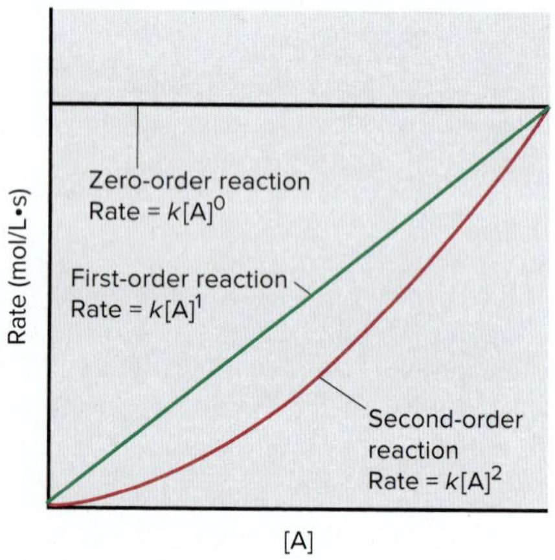

Figure 16.8 Plots of rate vs. reactant concentration, [A], for first-, second-, and zero-order reactions.

Let's look at some examples of observed rate laws and note the reaction orders.

1. For the reaction between nitrogen monoxide and ozone,

$$\text{NO}(g) + \text{O}_3(g) \longrightarrow \text{NO}_2(g) + \text{O}_2(g)$$

the rate law is

$$\text{Rate} = k[\text{NO}][\text{O}_3]$$

This reaction is first order with respect to NO and first order with respect to O_3, so it is second order overall (1 + 1 = 2).

2. For the hydrolysis of 2-bromo-2-methylpropane,

$$(\text{CH}_3)_3\text{C—Br}(l) + \text{H}_2\text{O}(l) \longrightarrow (\text{CH}_3)_3\text{C—OH}(l) + \text{H}^+(aq) + \text{Br}^-(aq)$$

the rate law is

$$\text{Rate} = k[(\text{CH}_3)_3\text{CBr}]$$

This reaction is first order in 2-bromo-2-methylpropane and zero order with respect to H_2O, *despite its coefficient of 1 in the balanced equation.* If we want to note that water is a reactant, we can write

$$\text{Rate} = k[(\text{CH}_3)_3\text{CBr}][\text{H}_2\text{O}]^0$$

Overall, this is a first-order reaction (1 + 0 = 1).

3. Finally, for the reaction between NO and hydrogen gas,

$$2\text{NO}(g) + 2\text{H}_2(g) \longrightarrow \text{N}_2(g) + 2\text{H}_2\text{O}(g)$$

the rate law is

$$\text{Rate} = k[\text{NO}]^2[\text{H}_2]$$

This reaction is second order in NO. And even though *H_2 has a coefficient of 2 in the balanced equation,* the reaction is *first order* in H_2. It is third order overall (2 + 1 = 3).

These examples reiterate an important point: *reaction orders* ***cannot*** *be deduced from the balanced equation but* ***must*** *be determined from experimental data.*

The larger the reaction order for a reactant, the greater the effect the concentration of that reactant has on the rate. For example, consider the result of doubling both [NO] and $[H_2]$ in the third reaction discussed above:

$$\text{Rate} = k[\text{NO}]^2[\text{H}_2] = k(2)^2(2) = k \times 4 \times 2 = k \times 8, \text{ or } 8k$$

Since the change in [NO] is squared, doubling [NO] results in the rate increasing by a factor of 4, while doubling $[H_2]$ increases the rate only by a factor of 2; the net effect of the change in both concentrations is that the rate increases by a factor of 8.

Other Reaction Orders Although usually positive integers or zero, reaction orders can also be fractional or negative.

1. For the reaction

$$CHCl_3(g) + Cl_2(g) \longrightarrow CCl_4(g) + HCl(g)$$

a fractional order appears in the rate law:

$$\text{Rate} = k[CHCl_3][Cl_2]^{1/2}$$

This *fractional reaction order* means that if, for example, $[Cl_2]$ increases by a factor of 4, the rate increases by a factor of 2, the square root of the change in $[Cl_2]$, or $(4)^{1/2}$. The overall order of this reaction is $1 + \frac{1}{2} = \frac{3}{2}$.

2. A *negative reaction order* means the rate *decreases* when the concentration of that component increases. Negative orders are often seen when the rate law includes products. For example, for the atmospheric reaction

$$2O_3(g) \rightleftharpoons 3O_2(g)$$

the rate law is

$$\text{Rate} = k[O_3]^2[O_2]^{-1} = k\frac{[O_3]^2}{[O_2]}$$

If $[O_2]$ doubles, the reaction proceeds half as fast. This reaction is second order in O_3 and negative first order in O_2, so it is first order overall $[2 + (-1) = 1]$.

SAMPLE PROBLEM 16.2 Determining Reaction Orders from Rate Laws

Problem For each of the following reactions, use the given rate law to determine the reaction order with respect to each reactant and the overall order; for the reaction in (a), determine the factor by which the rate changes if [NO] is tripled and $[O_2]$ is doubled.

(a) $2NO(g) + O_2(g) \longrightarrow 2NO_2(g)$; rate $= k[NO]^2[O_2]$
(b) $CH_3CHO(g) \longrightarrow CH_4(g) + CO(g)$; rate $= k[CH_3CHO]^{3/2}$
(c) $H_2O_2(aq) + 3I^-(aq) + 2H^+(aq) \longrightarrow I_3^-(aq) + 2H_2O(l)$; rate $= k[H_2O_2][I^-]$

Plan We inspect the exponents in the rate law, *not* the coefficients of the balanced equation, to find the individual orders, and then take their sum to find the overall reaction order. To find the change in rate for the reaction in (a), we substitute 3 for [NO] in the rate law (since that concentration changes by a factor of 3) and 2 for $[O_2]$ (since that concentration changes by a factor of 2), and solve for the rate.

Solution (a) The exponent of [NO] is 2, so the reaction is second order with respect to NO, first order with respect to O_2, and third order overall.

Finding the change in rate:

$$\text{Rate} = k[NO]^2[O_2] = k(3)^2(2) = k \times 18 = 18k$$

The rate increases by a factor of 18.

(b) The reaction is $\frac{3}{2}$ order in CH_3CHO and $\frac{3}{2}$ order overall.

(c) The reaction is first order in H_2O_2, first order in I^-, and second order overall.

The reactant H^+ does not appear in the rate law, so the reaction is zero order in H^+.

Check Be sure that each reactant has an order and that the sum of the individual orders gives the overall order. For the rate change in (a), be sure that the 3 used to indicate a tripled concentration is squared since the reaction is second order in NO.

FOLLOW-UP PROBLEMS

16.2A Experiment shows that the reaction

$$6I^-(aq) + BrO_3^-(aq) + 6H^+(aq) \longrightarrow 3I_2(aq) + Br^-(aq) + 3H_2O(l)$$

obeys this rate law: rate $= k[I^-][BrO_3^-][H^+]^2$. **(a)** What is the reaction order in each reactant and the overall reaction order? **(b)** By what factor does the rate change if $[I^-]$ and $[BrO_3^-]$ are tripled and $[H^+]$ is doubled?

16.2B The rate law for the reaction

$$2ClO_2(aq) + 2OH^-(aq) \longrightarrow ClO_3^-(aq) + ClO_2^-(aq) + H_2O(l)$$

is rate = $k[ClO_2]^2[OH^-]$. **(a)** What is the reaction order in each reactant and the overall reaction order? **(b)** By what factor does the rate change if $[ClO_2]$ is halved and $[OH^-]$ is doubled?

SOME SIMILAR PROBLEMS 16.26–16.33

Determining Reaction Orders by Changing Reactant Concentrations Now let's see how reaction orders are found *before* the rate law is known. Before looking at a real reaction, we'll go through the process for substances A and B in this reaction:

$$A + 2B \longrightarrow C + D$$

The rate law, expressed in general terms, is

$$\text{Rate} = k[A]^m[B]^n$$

To find the values of m and n, *we run a series of experiments in which one reactant concentration changes while the other is kept constant, and we measure the effect on the initial rate in each case.* Table 16.1 shows the results.

Table 16.1 Initial Rates for the Reaction Between A and B

Experiment	Initial Rate (mol/L·s)	Initial [A] (mol/L)	Initial [B] (mol/L)
1	1.75×10^{-3}	2.50×10^{-2}	3.00×10^{-2}
2	3.50×10^{-3}	5.00×10^{-2}	3.00×10^{-2}
3	3.50×10^{-3}	2.50×10^{-2}	6.00×10^{-2}
4	7.00×10^{-3}	5.00×10^{-2}	6.00×10^{-2}

1. *Finding m, the order with respect to A.* By comparing experiments 1 and 2, in which [A] doubles and [B] is constant, we can obtain m. First, we take the ratio of the general rate laws for these two experiments:

$$\frac{\text{Rate 2}}{\text{Rate 1}} = \frac{k[A]_2^m[B]_2^n}{k[A]_1^m[B]_1^n}$$

where $[A]_2$ is the concentration of A in experiment 2, $[B]_1$ is the concentration of B in experiment 1, and so forth. Because k is a constant and [B] does not change between these two experiments, those quantities cancel:

$$\frac{\text{Rate 2}}{\text{Rate 1}} = \frac{\cancel{k}[A]_2^m\cancel{[B]_2^n}}{\cancel{k}[A]_1^m\cancel{[B]_1^n}} = \frac{[A]_2^m}{[A]_1^m} = \left(\frac{[A]_2}{[A]_1}\right)^m$$

Substituting the values from Table 16.1, we have

$$\frac{3.50\times10^{-3}\text{ mol/L·s}}{1.75\times10^{-3}\text{ mol/L·s}} = \left(\frac{5.00\times10^{-2}\text{ mol/L}}{2.50\times10^{-2}\text{ mol/L}}\right)^m$$

Dividing, we obtain

$$2.00 = (2.00)^m \qquad \text{so} \qquad m = 1$$

Thus, the reaction is first order in A, because when [A] doubles, the rate doubles.

2. *Finding n, the order with respect to B.* To find n, we compare experiments 3 and 1 in which [A] is held constant and [B] doubles:

$$\frac{\text{Rate 3}}{\text{Rate 1}} = \frac{k[A]_3^m[B]_3^n}{k[A]_1^m[B]_1^n}$$

As before, k is a constant, and in this pair of experiments, [A] does not change, so those quantities cancel, and we have

$$\frac{\text{Rate 3}}{\text{Rate 1}} = \frac{\cancel{k}\cancel{[A]_3^m}[B]_3^n}{\cancel{k}\cancel{[A]_1^m}[B]_1^n} = \frac{[B]_3^n}{[B]_1^n} = \left(\frac{[B]_3}{[B]_1}\right)^n$$

The actual values give

$$\frac{3.50\times10^{-3}\ \text{mol/L·s}}{1.75\times10^{-3}\ \text{mol/L·s}} = \left(\frac{6.00\times10^{-2}\ \text{mol/L}}{3.00\times10^{-2}\ \text{mol/L}}\right)^n$$

Dividing, we obtain

$$2.00 = (2.00)^n \qquad \text{so} \qquad n = 1$$

Thus, the reaction is also first order in B because when [B] doubles, the rate doubles. We can check this conclusion from experiments 1 and 4: when *both* [A] and [B] double, the rate should quadruple, and it does. Thus, the rate law, with m and n equal to 1, is

$$\text{Rate} = k[\text{A}][\text{B}]$$

Note, especially, that while the order with respect to B is 1, the coefficient of B in the balanced equation is 2. Thus, as we said earlier, *reaction orders must be determined from experiment*, not from the balanced equation.

Next, let's go through this process for a real reaction, the one between oxygen and nitrogen monoxide, a key step in the formation of acid rain and in the industrial production of nitric acid:

$$O_2(g) + 2NO(g) \longrightarrow 2NO_2(g)$$

The general rate law is

$$\text{Rate} = k[O_2]^m[NO]^n$$

Table 16.2 shows experiments that change one reactant concentration while keeping the other constant.

Table 16.2 Initial Rates for the Reaction Between O_2 and NO

Experiment	Initial Rate (mol/L·s)	Initial Reactant Concentrations (mol/L) $[O_2]$	[NO]
1	3.21×10^{-3}	1.10×10^{-2}	1.30×10^{-2}
2	6.40×10^{-3}	2.20×10^{-2}	1.30×10^{-2}
3	12.8×10^{-3}	1.10×10^{-2}	2.60×10^{-2}
4	9.60×10^{-3}	3.30×10^{-2}	1.30×10^{-2}
5	28.8×10^{-3}	1.10×10^{-2}	3.90×10^{-2}

1. *Finding m, the order with respect to O_2.* If we compare experiments 1 and 2, in which $[O_2]$ doubles and [NO] is constant, we see the effect of doubling $[O_2]$ on the rate. First, we take the ratio of their rate laws:

$$\frac{\text{Rate 2}}{\text{Rate 1}} = \frac{k[O_2]_2^m\,[NO]_2^n}{k[O_2]_1^m\,[NO]_1^n}$$

As before, the constant quantities—k and [NO]—cancel:

$$\frac{\text{Rate 2}}{\text{Rate 1}} = \frac{\cancel{k}\,[O_2]_2^m\,\cancel{[NO]_2^n}}{\cancel{k}\,[O_2]_1^m\,\cancel{[NO]_1^n}} = \frac{[O_2]_2^m}{[O_2]_1^m} = \left(\frac{[O_2]_2}{[O_2]_1}\right)^m$$

Substituting the values from Table 16.2, we obtain

$$\frac{6.40\times10^{-3}\ \text{mol/L·s}}{3.21\times10^{-3}\ \text{mol/L·s}} = \left(\frac{2.20\times10^{-2}\ \text{mol/L}}{1.10\times10^{-2}\ \text{mol/L}}\right)^m$$

Dividing, we obtain

$$1.99 = (2.00)^m$$

Rounding to one significant figure gives

$$2 = 2^m, \qquad \text{so} \qquad m = 1$$

Sometimes, the exponent is not as easy to find by inspection as it is here. In those cases, we solve for m with an equation of the form $a = b^m$, which in this case gives

$$\log a = m \log b \qquad \text{or} \qquad m = \frac{\log a}{\log b} = \frac{\log 1.99}{\log 2.00} = 0.993$$

which rounds to 1. Thus, the reaction is first order in O_2: when $[O_2]$ doubles, the rate doubles.

2. *Finding n, the order with respect to NO.* We compare experiments 3 and 1, in which $[O_2]$ is held constant and [NO] is doubled, to find the order of NO:

$$\frac{\text{Rate 3}}{\text{Rate 1}} = \frac{k[O_2]_3^m\,[\text{NO}]_3^n}{k[O_2]_1^m\,[\text{NO}]_1^n}$$

Canceling the constant k and the unchanging $[O_2]$, we have

$$\frac{\text{Rate 3}}{\text{Rate 1}} = \frac{\cancel{k}\,\cancel{[O_2]_3^m}\,[\text{NO}]_3^n}{\cancel{k}\,\cancel{[O_2]_1^m}\,[\text{NO}]_1^n} = \frac{[\text{NO}]_3^n}{[\text{NO}]_1^n} = \left(\frac{[\text{NO}]_3}{[\text{NO}]_1}\right)^n$$

The actual values give

$$\frac{12.8\times10^{-3}\ \text{mol/L·s}}{3.21\times10^{-3}\ \text{mol/L·s}} = \left(\frac{2.60\times10^{-2}\ \text{mol/L}}{1.30\times10^{-2}\ \text{mol/L}}\right)^n$$

Dividing, we obtain

$$3.99 = (2.00)^n$$

Solving for n:

$$n = \frac{\log 3.99}{\log 2.00} = 2.00 \quad (\text{or } 2)$$

The reaction is second order in NO: when [NO] doubles, the rate quadruples. Thus, the actual rate law is

$$\text{Rate} = k[O_2][\text{NO}]^2$$

In this case, the reaction orders happen to be the same as the equation coefficients; nevertheless, they must *always* be determined by experiment.

Student Hot Spot

Student data indicate that you may struggle with determining reaction orders by changing reactant concentrations. Access the Smartbook to view additional Learning Resources on this topic.

Determining the Rate Constant

Let's find the rate constant for the reaction of O_2 and NO. With the rate, reactant concentrations, and reaction orders known, the sole remaining unknown in the rate law is the rate constant, k. We can use data from *any* of the experiments in Table 16.2 to solve for k. From experiment 1, for instance, we have

$$\text{Rate} = k[O_2]_1[\text{NO}]_1^2$$

$$k = \frac{\text{rate 1}}{[O_2]_1\,[\text{NO}]_1^2} = \frac{3.21\times10^{-3}\ \text{mol/L·s}}{(1.10\times10^{-2}\ \text{mol/L})\,(1.30\times10^{-2}\ \text{mol/L})^2}$$

$$= \frac{3.21\times10^{-3}\ \text{mol/L·s}}{1.86\times10^{-6}\ \text{mol}^3/\text{L}^3} = 1.73\times10^3\ \text{L}^2/\text{mol}^2\text{·s}$$

Always check that the values of k for a series of experiments are constant within experimental error. To three significant figures, the average value of k for the five experiments in Table 16.2 is 1.72×10^3 $\text{L}^2/\text{mol}^2\text{·s}$.

With concentrations in mol/L and the reaction rate in units of mol/L·time, the units for k depend on the order of the reaction and, of course, the time unit. For this reaction, the units for k have to be $\text{L}^2/\text{mol}^2\text{·s}$ to give a rate with units of mol/L·s:

$$\text{Rate} = k[O_2][\text{NO}]^2$$

$$\frac{\text{mol}}{\text{L·s}} = \frac{\text{L}^2}{\text{mol}^2\text{·s}} \times \frac{\text{mol}}{\text{L}} \times \left(\frac{\text{mol}}{\text{L}}\right)^2$$

The rate constant will *always* have these units for an overall third-order reaction with the time unit of seconds. Table 16.3 shows the units of k for common integer overall orders, but you can always determine the units mathematically.

Figure 16.9 summarizes the steps for studying the kinetics of a reaction. The next two sample problems offer practice in applying this approach; the first is based on data and the second on molecular scenes.

Table 16.3 Units of the Rate Constant k for Several Overall Reaction Orders

Overall Reaction Order	Units of k (t in seconds)
0	mol/L·s (or mol L^{-1} s^{-1})
1	1/s (or s^{-1})
2	L/mol·s (or L mol^{-1} s^{-1})
3	$\text{L}^2/\text{mol}^2\text{·s}$ (or L^2 mol^{-2} s^{-1})

General formula:

$$\text{Units of } k = \frac{\left(\frac{\text{L}}{\text{mol}}\right)^{\text{order}-1}}{\text{unit of } t}$$

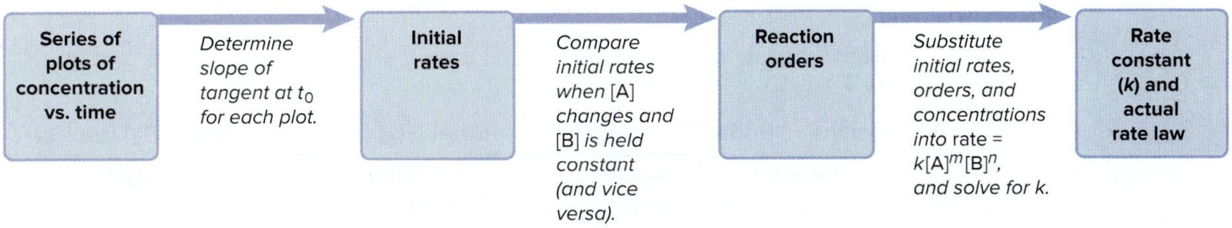

Figure 16.9 Information sequence to determine the kinetic parameters of a reaction.

SAMPLE PROBLEM 16.3 Determining Reaction Orders and Rate Constants from Rate Data

Problem Many gaseous reactions occur in car engines and exhaust systems. One of these is

$$NO_2(g) + CO(g) \longrightarrow NO(g) + CO_2(g) \qquad \text{Rate} = k[NO_2]^m[CO]^n$$

(a) Use the following data to determine the individual and overall reaction orders.
(b) Calculate k using the data from experiment 1.

Experiment	Initial Rate (mol/L·s)	Initial [NO_2] (mol/L)	Initial [CO] (mol/L)
1	0.0050	0.10	0.10
2	0.080	0.40	0.10
3	0.0050	0.10	0.20

Plan **(a)** We need to solve the general rate law for m and for n and then add those orders to get the overall order. To solve for each exponent, we proceed as in the text, taking the ratio of the rate laws for two experiments in which only the reactant in question changes. **(b)** We substitute the values for the rate and the reactant concentrations from experiment 1 into the rate law and solve for k.

Solution **(a)** Calculating m in $[NO_2]^m$: We take the ratio of the rate laws for experiments 1 and 2, in which $[NO_2]$ varies but [CO] is constant:

$$\frac{\text{Rate 2}}{\text{Rate 1}} = \frac{k[NO_2]_2^m[CO]_2^n}{k[NO_2]_1^m[CO]_1^n} = \left(\frac{[NO_2]_2}{[NO_2]_1}\right)^m \quad \text{or} \quad \frac{0.080\ \text{mol/L·s}}{0.0050\ \text{mol/L·s}} = \left(\frac{0.40\ \text{mol/L}}{0.10\ \text{mol/L}}\right)^m$$

This gives $16 = (4.0)^m$, so we have $m = \log 16/\log 4.0 = 2.0$. The reaction is second order in NO_2.

Calculating n in $[CO]^n$: We take the ratio of the rate laws for experiments 1 and 3, in which [CO] varies but $[NO_2]$ is constant:

$$\frac{\text{Rate 3}}{\text{Rate 1}} = \frac{k[NO_2]_3^2[CO]_3^n}{k[NO_2]_1^2[CO]_1^n} = \left(\frac{[CO]_3}{[CO]_1}\right)^n \quad \text{or} \quad \frac{0.0050\ \text{mol/L·s}}{0.0050\ \text{mol/L·s}} = \left(\frac{0.20\ \text{mol/L}}{0.10\ \text{mol/L}}\right)^n$$

We have $1.0 = (2.0)^n$, so $n = 0$. The rate does not change when [CO] varies, so the reaction is zero order in CO.

Therefore, the rate law is

$$\text{Rate} = k[NO_2]^2[CO]^0 = k[NO_2]^2(1) = k[NO_2]^2$$

The reaction is second order overall.

(b) From experiment 1, rate = 0.0050 mol/L·s and $[NO_2]$ = 0.10 mol/L. Calculating k:

$$\text{Rate} = k[NO_2]^2$$

$$k = \frac{\text{Rate}}{[NO_2]^2} = \frac{0.0050\ \text{mol/L·s}}{(0.10\ \text{mol/L})^2}$$

$$= 0.50\ \text{L/mol·s}$$

Check A good check is to reason through the orders. If $m = 1$, quadrupling $[NO_2]$ would quadruple the rate; but the rate *more* than quadruples, so $m > 1$. If $m = 2$, quadrupling $[NO_2]$ would increase the rate by a factor of 16 (4^2). The ratio of rates is 0.080/0.005 = 16, so $m = 2$. In contrast, increasing [CO] has no effect on the rate, which can happen only if $[CO]^n = 1$, so $n = 0$. The calculated value for the rate constant can be verified using the data from experiments 2 and 3.

FOLLOW-UP PROBLEMS

16.3A Find the rate law, the individual and overall reaction orders, and the average value of *k* for the reaction $H_2 + I_2 \longrightarrow 2HI$, using the following data at 450°C:

Experiment	Initial Rate (mol/L·s)	Initial $[H_2]$ (mol/L)	Initial $[I_2]$ (mol/L)
1	1.9×10^{-23}	0.0113	0.0011
2	1.1×10^{-22}	0.0220	0.0033
3	9.3×10^{-23}	0.0550	0.0011
4	1.9×10^{-22}	0.0220	0.0056

16.3B For the reaction $H_2SeO_3(aq) + 6I^-(aq) + 4H^+(aq) \longrightarrow Se(s) + 2I_3^-(aq) + 3H_2O(l)$ at 0°C, the following data were obtained:

Experiment	Initial Rate (mol/L·s)	Initial $[H_2SeO_3]$ (mol/L)	Initial $[I^-]$ (mol/L)	Initial $[H^+]$ (mol/L)
1	9.85×10^{-7}	2.5×10^{-3}	1.5×10^{-2}	1.5×10^{-2}
2	7.88×10^{-6}	2.5×10^{-3}	3.0×10^{-2}	1.5×10^{-2}
3	3.94×10^{-6}	1.0×10^{-2}	1.5×10^{-2}	1.5×10^{-2}
4	3.15×10^{-5}	2.5×10^{-3}	3.0×10^{-2}	3.0×10^{-2}

(a) Find the rate law for the reaction.
(b) Find the value of *k* (using the data in experiment 1).

SOME SIMILAR PROBLEMS 16.34, 16.35, and 16.38

SAMPLE PROBLEM 16.4 Determining Reaction Orders from Molecular Scenes

Problem At a particular temperature and volume, two gases, A *(red)* and B *(blue)*, react. The following molecular scenes represent starting mixtures for four experiments:

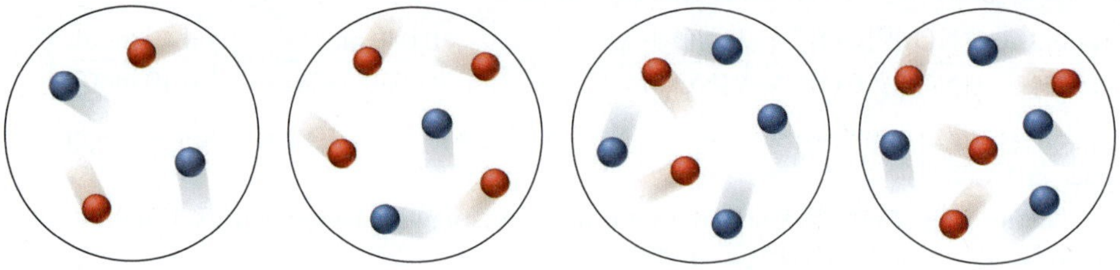

Expt no.:	1	2	3	4
Initial rate (mol/L·s):	0.50×10^{-4}	1.0×10^{-4}	2.0×10^{-4}	?

(a) What is the reaction order with respect to A? With respect to B? The overall order?
(b) Write the rate law for the reaction.
(c) Predict the initial rate of Expt 4.

Plan **(a)** As before, we find the individual reaction orders by seeing how a change in each reactant changes the rate. In this case, however, instead of using concentration data, we count numbers of particles. The sum of the individual orders is the overall order. **(b)** To write the rate law, we use the orders from part (a) as exponents in the general rate law. **(c)** Using the results from Expts 1 through 3 and the rate law from part (b), we find the unknown initial rate of Expt 4.

Solution **(a)** The rate law is rate $= k[A]^m[B]^n$. For reactant A *(red)* in Expts 1 and 2 (*k* and number of B particles are constant):

$$\frac{\text{Rate 2}}{\text{Rate 1}} = \frac{k[A]_2^m[B]_2^n}{k[A]_1^m[B]_1^n} = \left(\frac{[A]_2}{[A]_1}\right)^m \quad \text{or} \quad \frac{1.0\times10^{-4}\ \text{mol/L·s}}{0.50\times10^{-4}\ \text{mol/L·s}} = \left(\frac{4\ \text{particles}}{2\ \text{particles}}\right)^m$$

Thus, $2 = (2)^m$ so the order with respect to A is 1. For reactant B *(blue)* in Expts 1 and 3 (*k* and and number of A particles are constant):

$$\frac{\text{Rate 3}}{\text{Rate 1}} = \frac{k[A]_3^m[B]_3^n}{k[A]_1^m[B]_1^n} = \left(\frac{[B]_3}{[B]_1}\right)^n \quad \text{or} \quad \frac{2.0\times10^{-4}\ \text{mol/L·s}}{0.50\times10^{-4}\ \text{mol/L·s}} = \left(\frac{4\ \text{particles}}{2\ \text{particles}}\right)^n$$

Thus, $4 = (2)^n$ so the order with respect to B is 2. The overall order is $1 + 2 = 3$.

(b) Writing the rate law: The general rate law is rate = $k[A]^m[B]^n$, so we have

$$\text{Rate} = k[A][B]^2$$

(c) Finding the initial rate of Expt 4: There are several possibilities, but let's compare Expts 3 and 4, in which the number of particles of A doubles (from 2 to 4) and the number of particles of B doesn't change. Since the rate law shows that the reaction is first order in A, the initial rate in Expt 4 should be double the initial rate in Expt 3, or 4.0×10^{-4} mol/L·s.

Check A good check is to compare other pairs of experiments. **(a)** Comparing Expts 2 and 3 shows that the number of B doubles, which causes the rate to quadruple, and the number of A decreases by half, which causes the rate to halve; so the overall rate change should double (from 1.0×10^{-4} mol/L·s to 2.0×10^{-4} mol/L·s), which it does. **(c)** Comparing Expts 2 and 4, in which the number of A is constant and the number of B doubles, the rate should quadruple, which means the initial rate of Expt 4 would be 4.0×10^{-4} mol/L·s, as we found.

FOLLOW-UP PROBLEMS

16.4A The molecular scenes below show three experiments at a given temperature and volume involving reactants X *(black)* and Y *(green):*

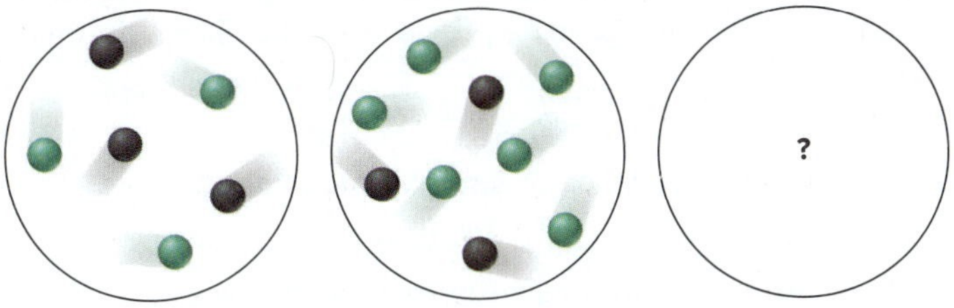

Expt no.:	1	2	3
Initial rate (mol/L·s):	0.25×10^{-5}	?	1.0×10^{-5}

If the rate law for the reaction is rate = $k[X]^2$: **(a)** What is the initial rate of Expt 2? **(b)** Draw a scene for Expt 3 that involves a single change of the scene for Expt 1.

16.4B The scenes below show mixtures of reactants A *(blue)* and B *(yellow)* at a given temperature and volume:

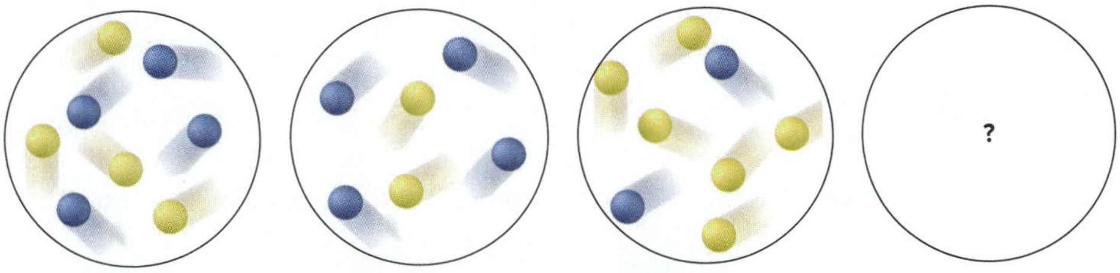

Expt no.:	1	2	3	4
Initial rate (mol/L·s):	1.6×10^{-3}	1.6×10^{-3}	8.0×10^{-4}	3.2×10^{-3}

(a) Write the rate law for the reaction.
(b) How many particles of A should be shown in the scene for Expt 4?

SOME SIMILAR PROBLEMS 16.93, 16.122(a), and 16.122(b)

› Summary of Section 16.3

- An experimentally determined rate law shows how the rate of a reaction depends on concentration. Considering only initial rates (that is, no products), the expression for a general rate law is rate = $k[A]^m[B]^n \cdots$. This reaction is mth order with respect to A and nth order with respect to B; the overall reaction order is $m + n$.
- With an accurate method for obtaining initial rates, reaction orders are determined experimentally by varying the concentration of one reactant at a time to see its effect on the rate.
- By substituting the known rate, concentrations, and reaction orders into the rate law, we solve for the rate constant, k.

16.4 INTEGRATED RATE LAWS: CONCENTRATION CHANGES OVER TIME

The rate laws we've developed so far tell us the rate or concentration at a given instant, allowing us to answer questions such as "How fast is the reaction proceeding at the moment *y* moles per liter of A are mixed with *z* moles per liter of B?" and "What is [B], when [A] is *x* moles per liter?" By employing different forms of the rate laws, called **integrated rate laws,** we can include time as a variable and answer questions such as "How long will it take to use up *x* moles per liter of A?" and "What is [A] after *y* minutes of reaction?"

Integrated Rate Laws for First-, Second-, and Zero-Order Reactions

Let's examine the integrated rate laws for reactions that are first, second, or zero order in reactant A:

1. *First-order reactions.* As we've seen, for a general first-order reaction, the rate is the negative of the change in [A] divided by the change in time:

$$\text{Rate} = -\frac{d[\text{A}]}{dt}$$

The rate can also be expressed in terms of the rate law:

$$\text{Rate} = k[\text{A}]$$

Setting these two expressions equal to each other gives

$$-\frac{d[\text{A}]}{dt} = k[\text{A}] \qquad \text{or} \qquad \frac{1}{[\text{A}]} = d[\text{A}] = -kdt$$

Since the instantaneous rate is the derivative of concentration with respect to time, we find the dependence of [A] on time by integrating this equation from an initial concentration $[\text{A}]_0$ at $t = 0$ to any later concentration $[\text{A}]t$ at time = t:

$$\int_{[\text{A}]_0}^{[\text{A}]_t} \frac{1}{[\text{A}]}\, d[\text{A}] = -k \int_0^t dt$$

to obtain

$$\ln[\text{A}]_t - \ln[\text{A}]_0 = -kt$$

The integrated rate law for first-order reactions can also be written as:

$$\ln\frac{[\text{A}]_0}{[\text{A}]_t} = kt \qquad \text{(first-order reaction; rate} = k[\text{A}]) \qquad \textbf{(16.4)}$$

where ln is the natural logarithm, $[\text{A}]_0$ is the concentration of A at $t = 0$, and $[\text{A}]_t$ is the concentration of A at any time t.

2. *Second-order reactions with a single reactant, A.* For a general second-order reaction involving only *one* reactant [A], setting the two rate expressions equal to each other gives

$$\text{Rate} = -\frac{d[\text{A}]}{dt} = k[\text{A}]^2 \qquad \text{or} \qquad \frac{1}{[\text{A}]^2}\, d[\text{A}] = -kdt$$

Integrating this equation from an initial concentration of $[\text{A}]_0$ at $t = 0$ to a final concentration $[\text{A}]_t$ at time = t gives the integrated rate law for a second-order reaction involving one reactant:

$$\int_{[\text{A}]_0}^{[\text{A}]_t} \frac{1}{[\text{A}]^2}\, d[\text{A}] = -k \int_0^t dt$$

$$-\frac{1}{[\text{A}]_t} - \left(-\frac{1}{[\text{A}]_0}\right) = -kt$$

which is more conveniently written as:

$$\frac{1}{[A]_t} - \frac{1}{[A]_0} = kt \qquad \text{(second-order reaction; rate} = k[A]^2\text{)} \qquad \textbf{(16.5)}$$

3. *Second-order reactions with two reactants, A and B*. For a general overall second-order reaction which is first-order in *two* reactants A and B, the expression including time is complex. The rate expression for such a reaction is

$$\text{Rate} = -\frac{d[A]}{dt} = -\frac{d[B]}{dt} = k[A][B]$$

Integrating this rate law, assuming $[A]_0 < [B]_0$ and stoichiometric coefficients of 1 for both reactants, gives the integrated rate law:

$$\frac{1}{[B]_0 - [A]_0} \ln \frac{[A]_0 [B]_t}{[B]_0 [A]_t} = kt$$

We can simplify this relationship by choosing initial concentrations that make the reaction *appear* to be first order. We carry out the reaction with [A] much greater than [B] (that is, $[A]_0 >> [B]_0$); then $[A]_0$ is so large that relatively little of this reactant is consumed as the reaction proceeds, which means that $[A]_t \approx [A]_0$. And, since $[B]_0$ is much smaller than $[A]_0$, $[B]_0 - [A]_0 \approx -[A]_0$. The integrated rate law then simplifies to

$$\frac{1}{-[A]_0} \ln \frac{[A]_0 [B]_t}{[B]_0 [A]_0} = kt \qquad \text{or} \qquad \ln \frac{[B]_0}{[B]_t} = k[A]_0 t$$

This is the integrated rate law for a first-order reaction in reactant B in which the observed rate constant $k' = k[A]_0$:

$$\ln \frac{[B]_0}{[B]_t} = k't \qquad \text{where} \qquad k' = k[A]_0$$

The second-order reaction now behaves as a first-order reaction due to a large excess of one reactant, so it is called a **pseudo–first-order** reaction, and the observed rate constant is the pseudo–first-order rate constant that depends on the initial concentration of A.

4. *Zero-order reactions*. For a general zero-order reaction, setting the two rate expressions equal to each other gives

$$\text{Rate} = -\frac{d[A]}{dt} = k[A]^0 = k \qquad \text{or} \qquad -d[A] = kdt$$

Integrating over time gives the integrated rate law for a zero-order reaction:

$$[A]_t - [A]_0 = -kt \qquad \text{(zero-order reaction; rate} = k[A]^0 = k\text{)} \qquad \textbf{(16.6)}$$

Sample Problem 16.5 shows one way integrated rate laws are applied.

SAMPLE PROBLEM 16.5 **Determining the Reactant Concentration After a Given Time**

Student Hot Spot

Student data indicate that you may struggle with applying an integrated rate law. Access the Smartbook to view additional Learning Resources on this topic.

Problem At 1000°C, cyclobutane (C_4H_8) decomposes in a first-order reaction, with the very high rate constant of 87 s^{-1}, to two molecules of ethylene (C_2H_4).

(a) The initial C_4H_8 concentration is 2.00 *M*. What is the concentration after 0.010 s?

(b) How long will it take for 70.0% of the C_4H_8 to decompose?

Plan **(a)** We must find the concentration of cyclobutane at time *t*, $[C_4H_8]_t$. We are told that this is a first-order reaction, so we use the integrated first-order rate law:

$$\ln \frac{[C_4H_8]_0}{[C_4H_8]_t} = kt$$

We know *k* (87 s^{-1}), *t* (0.010 s), and $[C_4H_8]_0$ (2.00 *M*), so we can solve for $[C_4H_8]_t$.

(b) If 70.0% of the C_4H_8 has decomposed, 30.0% of it remains ($[C_4H_8]_t$). We know k ($87\ s^{-1}$), $[C_4H_8]_0$ (2.00 *M*), and $[C_4H_8]_t$ ($0.300 \times 2.00\ M = 0.600\ M$), so we can solve for t.

Solution (a) Substituting the data into the integrated rate law:

$$\ln\frac{2.00\ \text{mol/L}}{[C_4H_8]_t} = (87\ s^{-1})\ (0.010\ s) = 0.87$$

Taking the antilog of both sides:

$$\frac{2.00\ \text{mol/L}}{[C_4H_8]_t} = e^{0.87} = 2.4$$

Solving for $[C_4H_8]_t$:

$$[C_4H_8]_t = \frac{2.00\ \text{mol/L}}{2.4} = 0.83\ \text{mol/L}$$

(b) Substituting the data into the integrated rate law:

$$\ln\frac{2.00\ \text{mol/L}}{0.600\ \text{mol/L}} = (87\ s^{-1})\ (t)$$

$$t = 0.014\ s$$

Check The concentration remaining after 0.010 s (0.83 mol/L) is less than the starting concentration (2.00 mol/L), which makes sense. Raising e to an exponent slightly less than 1 should give a number (2.4) slightly less than the value of e (2.718). Moreover, the final result makes sense: a high rate constant indicates a fast reaction, so it's not surprising that so much decomposes in such a short time.

Comment Be sure to note that $[A]_t$ is the concentration of reactant A that *remains* at time t, not the amount of A that has reacted.

FOLLOW-UP PROBLEMS

16.5A At 25°C, hydrogen iodide breaks down very slowly to hydrogen and iodine: rate = $k[HI]^2$. The rate constant at 25°C is 2.4×10^{-21} L/mol·s. If 0.0100 mol of HI(*g*) is placed in a 1.0-L container, how long will it take for [HI] to reach 0.00900 mol/L (10.0% reacted)?

16.5B Hydrogen peroxide (H_2O_2) decomposes to water and oxygen in a first-order reaction. **(a)** If 1.28 *M* H_2O_2 changes to 0.85 *M* in 10.0 min at a given temperature, what is the rate constant for the decomposition reaction? **(b)** How long will it take for 25% of an initial amount of H_2O_2 to decompose? (*Hint:* Assume that $[H_2O_2]_0 = 1.0\ M$.)

SOME SIMILAR PROBLEMS 16.41 and 16.42

Determining Reaction Orders from an Integrated Rate Law

In Sample Problem 16.3, we found the reaction orders using rate data. If rate data are not available, we rearrange the integrated rate law into an equation for a straight line, $y = mx + b$, where m is the slope and b is the y-axis intercept. We then use a graphical method to find the order:

- For a *first-order reaction,* we have

$$\ln\frac{[A]_0}{[A]_t} = kt$$

  From Appendix A, we know that $\ln\frac{a}{b} = \ln a - \ln b$, so we have

$$\ln[A]_0 - \ln[A]_t = kt$$

  Rearranging gives

$$\ln[A]_t = -kt + \ln[A]_0$$
$$y = mx + b$$

  Therefore, a plot of $\ln[A]_t$ vs. t gives a straight line with slope $= -k$ and y-intercept $= \ln[A]_0$ (Figure 16.10A).

- For a *second-order reaction* with one reactant, we have

$$\frac{1}{[A]_t} - \frac{1}{[A]_0} = kt$$

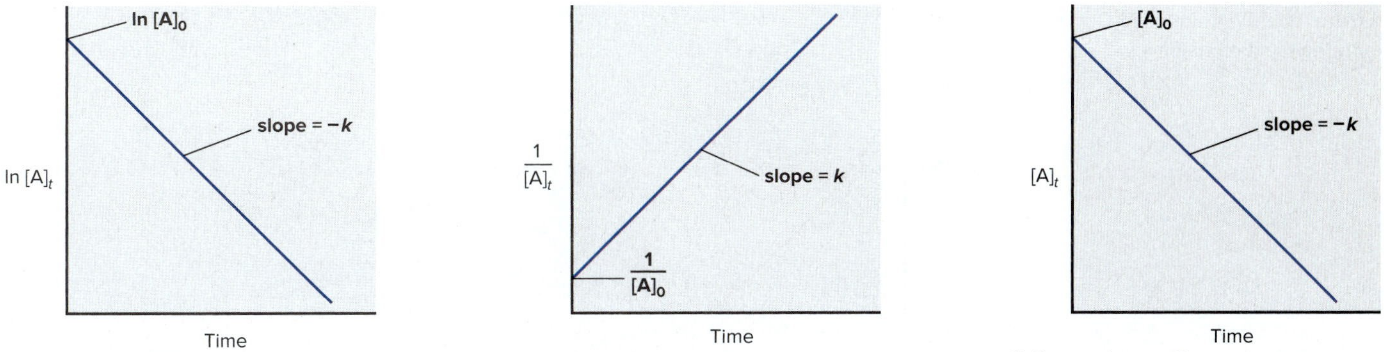

A First-order reaction **B** Second-order reaction **C** Zero-order reaction

Figure 16.10 Graphical method for finding the reaction order from the integrated rate law.

Rearranging gives

$$\frac{1}{[A]_t} = kt + \frac{1}{[A]_0}$$
$$y = mx + b$$

In this case, a plot of $1/[A]_t$ vs. t gives a straight line with slope = k and y-intercept = $1/[A]_0$ (Figure 16.10B). Note that for a second-order reaction with two reactants, in which one reactant is present in much higher concentration than the other (pseudo–first-order), we obtain results similar to those for a first-order reaction.

- For a *zero-order reaction,* we have

$$[A]_t - [A]_0 = -kt$$

Rearranging gives

$$[A]_t = -kt + [A]_0$$
$$y = mx + b$$

A plot of $[A]_t$ vs. t gives a straight line with slope = $-k$ and y-intercept = $[A]_0$ (Figure 16.10C).

Some trial-and-error graphical plotting is required to find the reaction order from the concentration and time data:

- If you obtain a straight line when you plot ln [reactant] vs. t, the reaction is *first order* with respect to that reactant.
- If you obtain a straight line when you plot 1/[reactant] vs. t, the reaction is *second order* with respect to that reactant.
- If you obtain a straight line when you plot [reactant] vs. t, the reaction is *zero order* with respect to that reactant.

In Figure 16.11 we use this approach to determine the order for the decomposition of N_2O_5. When we plot the data from each column in the table vs. time, we find that the plot of ln $[N_2O_5]$ vs. t *is* linear (part B), while the plots of $[N_2O_5]$ vs. t (part A) and of $1/[N_2O_5]$ vs. t (part C) are *not;* therefore, the decomposition is first order in N_2O_5.

Figure 16.11 **Graphical determination of the reaction order for the decomposition of N_2O_5.** The time and concentration data in the table are used to obtain the three plots, **A, B,** and **C.**

Time (min)	$[N_2O_5]$	ln $[N_2O_5]$	$1/[N_2O_5]$
0	0.0165	−4.104	60.6
10	0.0124	−4.390	80.6
20	0.0093	−4.68	1.1×10^2
30	0.0071	−4.95	1.4×10^2
40	0.0053	−5.24	1.9×10^2
50	0.0039	−5.55	2.6×10^2
60	0.0029	−5.84	3.4×10^2

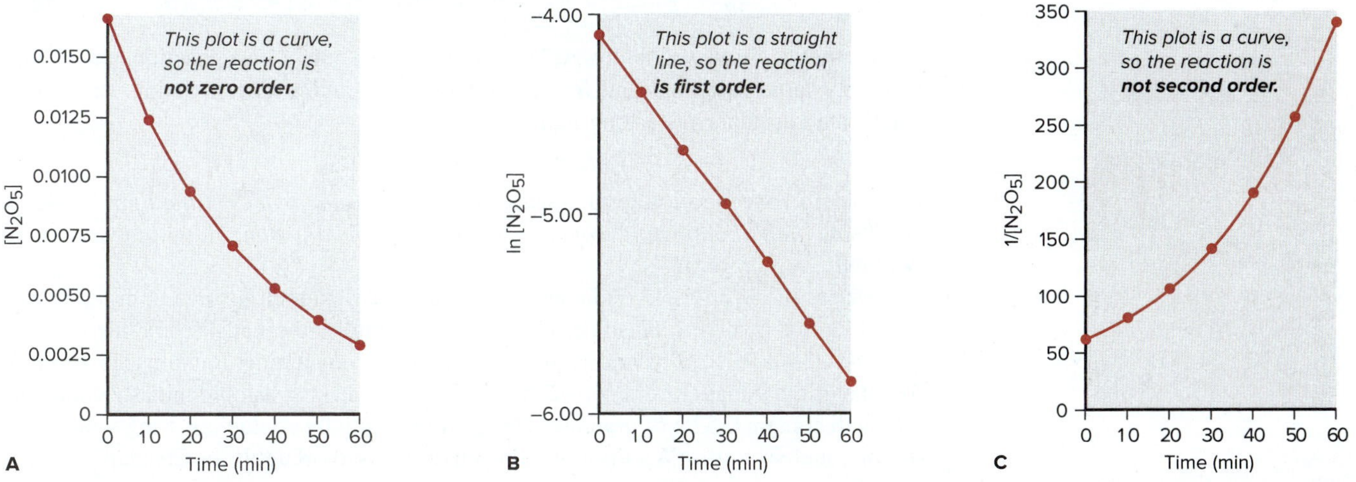

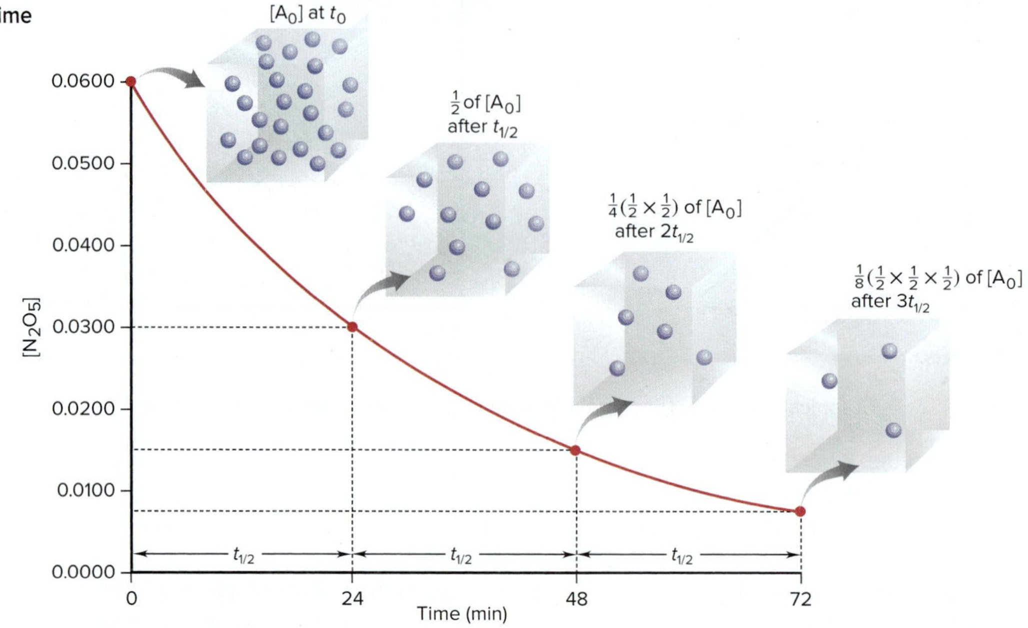

Figure 16.12 A plot of $[N_2O_5]$ vs. time for three reaction half-lives.

Reaction Half-Life

The **half-life ($t_{1/2}$)** of a reaction is the time it takes for the reactant concentration to reach *half its initial value.* A half-life has time units appropriate for the specific reaction and is characteristic of that reaction at a given temperature. For example, at 45°C, the half-life for the decomposition of N_2O_5, which we know is first order, is 24.0 min. Therefore, if we start with, say, 0.0600 mol/L of N_2O_5 at 45°C, 0.0300 mol/L will have reacted after 24 min (one half-life), and 0.0300 mol/L will remain; after 48 min (two half-lives), 0.0150 mol/L will remain; after 72 min (three half-lives), 0.0075 mol/L will remain, and so forth (Figure 16.12). The mathematical expression for the half-life depends on the overall order of the reaction.

First-Order Reactions We can derive an expression for the half-life of a first-order reaction from the integrated rate law, which is

$$\ln \frac{[A]_0}{[A]_t} = kt$$

By definition, after one half-life, $t = t_{1/2}$, and $[A]_t = \frac{1}{2}[A]_0$. Substituting and canceling $[A]_0$ gives

$$\ln \frac{\cancel{[A]_0}}{\frac{1}{2}\cancel{[A]_0}} = kt_{1/2} \qquad \text{or} \qquad \ln 2 = kt_{1/2}$$

Then, solving for $t_{1/2}$, we have

$$t_{1/2} = \frac{\ln 2}{k} = \frac{0.693}{k} \qquad \text{(first-order process; rate} = k[A]) \qquad \textbf{(16.7)}$$

Notice that *$t_{1/2}$ and k are inversely proportional.* Thus, a fast reaction, one with a relatively large rate constant, has a short half-life, and a slow reaction, one with a small rate constant, has a long half-life:

$$k\uparrow,\ t_{1/2}\downarrow \qquad \text{and} \qquad k\downarrow,\ t_{1/2}\uparrow$$

Because no concentration term appears, *for a first-order reaction, the time it takes to reach one-half the starting concentration is a constant and, thus, independent of reactant concentration.*

Decay of an unstable, radioactive nucleus is an example of a first-order process that does not involve a *chemical* change. For example, the half-life for the decay of uranium-235 is 7.1×10^8 years. Thus, a sample of ore containing uranium-235 will have half the original mass of uranium-235 after 7.1×10^8 years: a sample initially containing 1 kg will contain 0.5 kg of uranium-235, a sample initially containing 1 mg will contain 0.5 mg, and so forth. (We discuss radioactive decay thoroughly in Chapter 24.)

The next two sample problems show some ways to use half-life in calculations: the first is based on molecular scenes.

SAMPLE PROBLEM 16.6 Using Molecular Scenes to Find Quantities at Various Times

Problem Substance A *(green)* decomposes to two other substances, B *(blue)* and C *(yellow),* in a first-order gaseous reaction. The molecular scenes below show a portion of the reaction mixture at two different times:

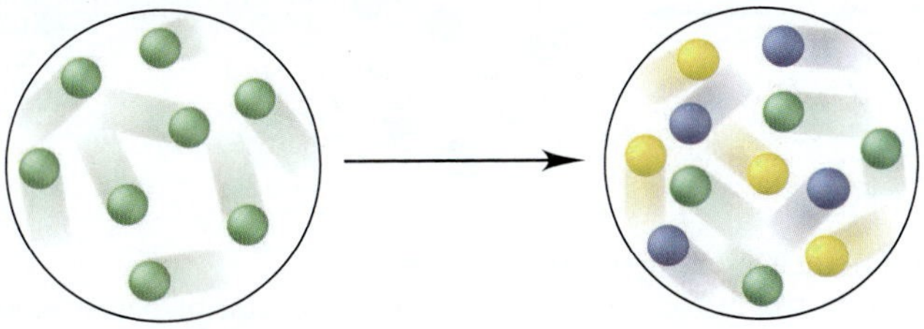

$t = 0.0$ s $\qquad$ $t = 30.0$ s

(a) Draw a similar molecular scene of the reaction mixture at $t = 60.0$ s.
(b) Find the rate constant for the reaction.
(c) If the total pressure (P_{total}) of the mixture is 5.00 atm at 90.0 s, what is the partial pressure of substance B (P_B)?

Plan We are shown molecular scenes of a reaction at two times, with various numbers of reactant and product particles, and have to predict quantities at two later times.
(a) We count the number of A particles and see that A has decreased by half after 30.0 s; thus, the half-life is 30.0 s. The time $t = 60.0$ s represents two half-lives, so the number of A will decrease by half again, and each A forms one B and one C. **(b)** We substitute the value of the half-life in Equation 16.7 to find k. **(c)** First, we find the numbers of particles at $t = 90.0$ s, which represents three half-lives. To find P_B, we multiply the mole fraction of B, X_B, by P_{total} (5.00 atm) (see Equation 5.12). To find X_B, we know that the number of particles is equivalent to the number of moles, so we divide the number of B particles by the total number of particles.

Solution **(a)** The number of A particles decreased from 8 to 4 in one half-life (30.0 s), so after two half-lives (60.0 s), the number of A particles will be 2 (1/2 of 4). Each A decomposes to 1 B and 1 C, so 6 (8 − 2) particles of A form 6 of B and 6 of C *(see margin)*.

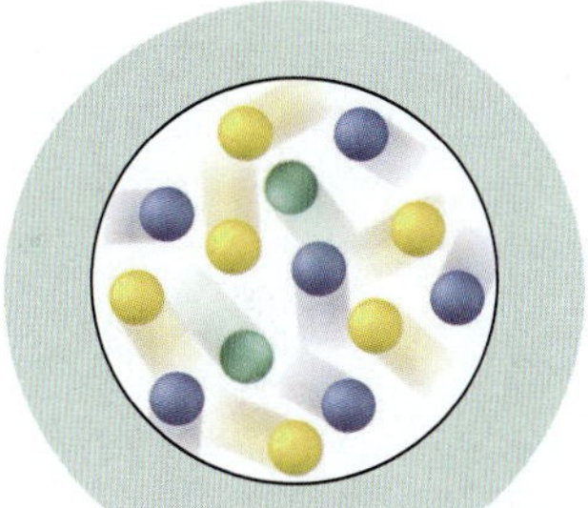

$t = 60.0$ s

(b) Finding the rate constant, k:

$$t_{1/2} = \frac{0.693}{k} \quad \text{so} \quad k = \frac{0.693}{t_{1/2}} = \frac{0.693}{30.0\text{ s}} = 2.31\times10^{-2}\text{ s}^{-1}$$

(c) Finding the number of particles after 90.0 s: After a third half-life, there will be 1 A, 7 B, and 7 C particles.

Finding the mole fraction of B, X_B:

$$X_B = \frac{7\text{ B particles}}{1 + 7 + 7\text{ total particles}} = \frac{7}{15} = 0.467$$

Finding the partial pressure of B, P_B:

$$P_B = X_B \times P_{\text{total}} = 0.467 \times 5.00\text{ atm} = 2.33\text{ atm}$$

Check For (b), rounding gives 0.7/30, which is a bit over 0.02, so the answer seems correct. For (c), X_B is almost 0.5, so P_B is a bit less than half of 5 atm, or <2.5 atm.

FOLLOW-UP PROBLEMS

16.6A Substance X *(black)* changes to substance Y *(red)* in a first-order gaseous reaction. The scenes below represent the reaction mixture in a cubic container at two different times:

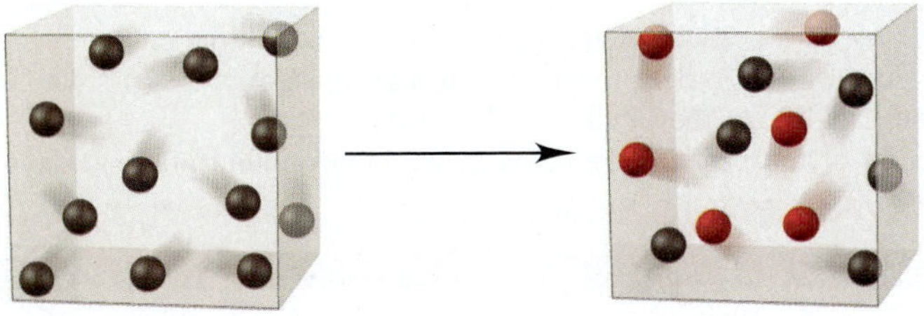

$t = 0.0$ min $\qquad$ $t = 2.5$ min

(a) Draw a scene that represents the mixture at 5.0 min.
(b) If each particle in the container actually represents 0.20 mol of particles and the volume of the cubic container is 0.50 L, what is the molarity of X at 10.0 min?

16.6B Substance A *(red)* changes to substance B *(blue)* in a first-order gaseous reaction. The scenes below represent the reaction mixture at three different times:

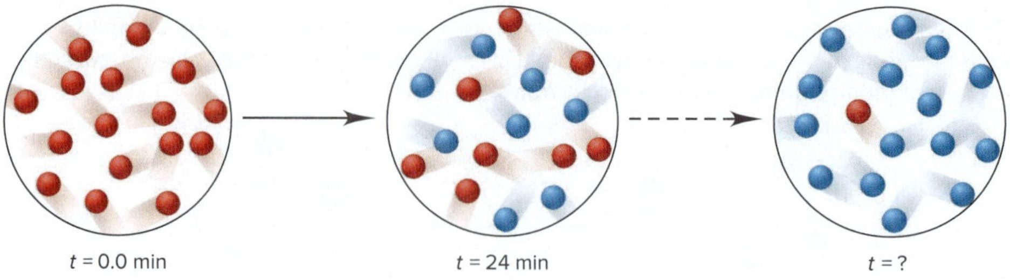

(a) What is the time at which the third scene occurs?
(b) If each particle in the circle actually represents 0.10 mol of particles and the volume of the container is 0.25 L, what is the molarity of A at 72 min?

SOME SIMILAR PROBLEMS 16.43 and 16.44

SAMPLE PROBLEM 16.7 Determining the Half-Life of a First-Order Reaction

Problem Cyclopropane is the smallest cyclic hydrocarbon. Because its 60° bond angles reduce orbital overlap, its bonds are weak. As a result, it is thermally unstable and rearranges to propene at 1000°C via the following first-order reaction:

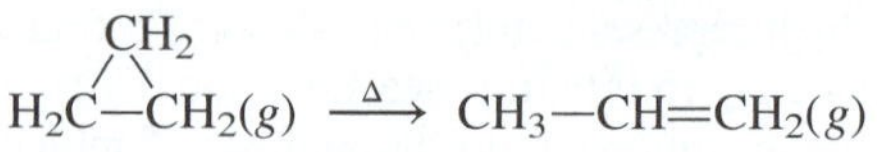

The rate constant is $9.2\ s^{-1}$.
(a) What is the half-life of the reaction?
(b) How long does it take for the concentration of cyclopropane to reach one-quarter of the initial value?

Plan (a) The cyclopropane rearrangement is first order, so to find $t_{1/2}$ we use Equation 16.7 and substitute for k ($9.2\ s^{-1}$). **(b)** Each half-life decreases the concentration to one-half of its initial value, so two half-lives decrease it to one-quarter of its initial value.

Solution (a) Solving for $t_{1/2}$:

$$t_{1/2} = \frac{\ln 2}{k} = \frac{0.693}{9.2\ s^{-1}}$$
$$= 0.075\ s$$

It takes 0.075 s for half the cyclopropane to form propene at this temperature.

(b) Finding the time to reach one-quarter of the initial concentration:

$$\text{Time} = 2(t_{1/2}) = 2(0.075\ s)$$
$$= 0.15\ s$$

Check For part (a), rounding gives $0.7/9\ s^{-1} = 0.08$ s, so the answer seems correct.

FOLLOW-UP PROBLEMS

16.7A Iodine-123 is used to study thyroid gland function. This radioactive isotope breaks down in a first-order process with a half-life of 13.1 h. What is the rate constant for the process?

16.7B The half-life of a pesticide determines its persistence in the environment. A common pesticide degrades in a first-order process with a rate constant of $9\times10^{-2}\ day^{-1}$.
(a) What is the half-life of the breakdown reaction?
(b) What fraction of the pesticide remains in the environment after 40 days?

SOME SIMILAR PROBLEMS 16.45 and 16.46

Second-Order Reactions In contrast to the half-life of a first-order reaction, the half-life of a simple second-order reaction *does* depend on reactant concentration:

$$t_{1/2} = \frac{1}{k[\mathrm{A}]_0} \qquad \text{(second-order process; rate} = k[\mathrm{A}]^2\text{)}$$

Note that here *the half-life is* ***inversely*** *proportional to the initial reactant concentration.* This relationship means that a second-order reaction with a high initial reactant concentration has a *shorter* half-life, and one with a low initial reactant concentration has a *longer* half-life. It also means that, for a particular reaction, each successive half-life is double the preceding one, since [A] is halved during each half-life. (Note, however, that since a second-order reaction with two reactants, in which one reactant is present in much higher concentration than the other, follows first-order kinetics, its half-life behaves like the half-life of a first-order reaction.)

Zero-Order Reactions In contrast to the half-life of a second-order reaction, *the half-life of a zero-order reaction is* ***directly*** *proportional to the initial reactant concentration:*

$$t_{1/2} = \frac{[\mathrm{A}]_0}{2k} \qquad \text{(zero-order process; rate} = k\text{)}$$

Thus, if a zero-order reaction begins with a high reactant concentration, it has a longer half-life than if it begins with a low reactant concentration.

Table 16.4 summarizes the features of zero-, first-, and second-order reactions.

Table 16.4 An Overview of Zero-Order, First-Order, and Simple Second-Order Reactions

	Zero Order	First Order	Second Order
Rate law	rate $= k$	rate $= k[\mathrm{A}]$	rate $= k[\mathrm{A}]^2$
Units for k	mol/L·s	1/s	L/mol·s
Half-life	$\frac{[\mathrm{A}]_0}{2k}$	$\frac{\ln 2}{k}$	$\frac{1}{k[A]_0}$
Integrated rate law in straight-line form	$[\mathrm{A}]_t = -kt + [\mathrm{A}]_0$	$\ln [\mathrm{A}]_t = -kt + \ln [\mathrm{A}]_0$	$1/[\mathrm{A}]_t = kt + 1/[\mathrm{A}]_0$
Plot for straight line	$[\mathrm{A}]_t$ vs. t	$\ln [\mathrm{A}]_t$ vs. t	$1/[\mathrm{A}]_t$ vs. t
Slope, y-intercept	$-k$, $[\mathrm{A}]_0$	$-k$, $\ln [\mathrm{A}]_0$	k, $1/[\mathrm{A}]_0$

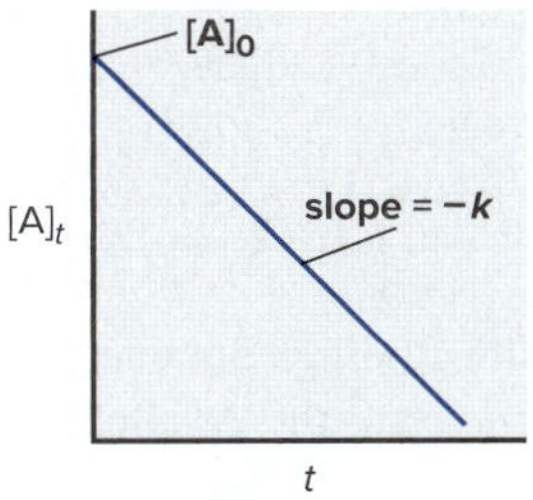

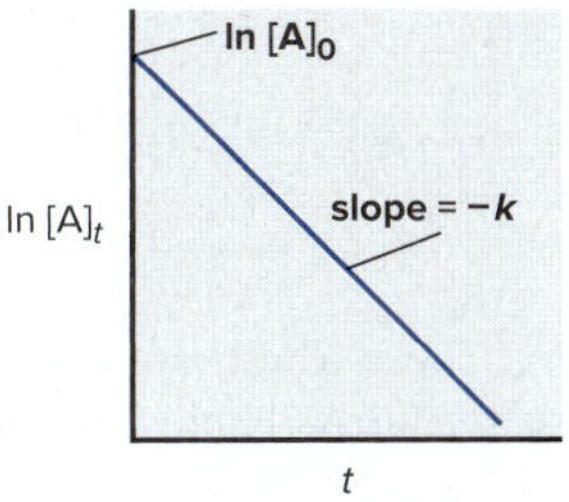

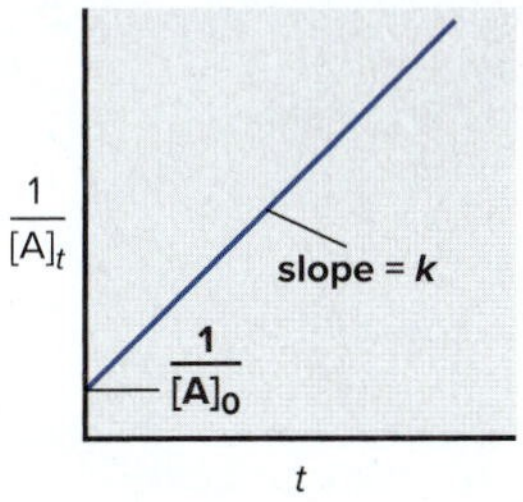

› Summary of Section 16.4

- Integrated rate laws are used to find either the time needed to reach a certain concentration of reactant or the concentration present after a given time.
- Rearrangements of the integrated rate laws that give equations that graph as a straight line allow us to determine reaction orders and rate constants graphically.
- The half-life is the time needed for the reactant concentration to reach half its initial value; for first-order reactions, the half-life is constant, that is, independent of concentration.

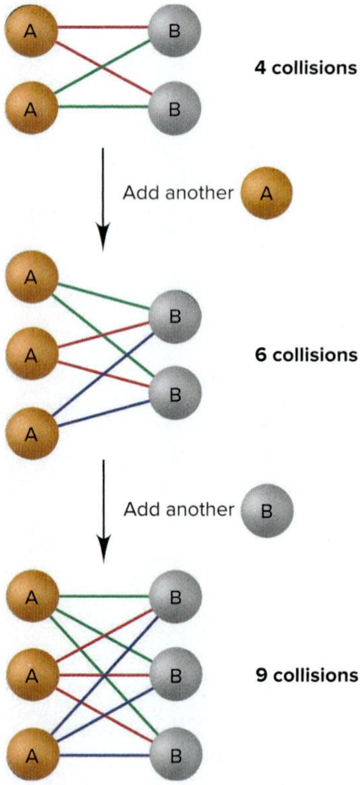

Figure 16.13 The number of possible collisions is the product, not the sum, of reactant concentrations.

16.5 THEORIES OF CHEMICAL KINETICS

As was pointed out in Section 16.1, concentration and temperature have major effects on reaction rate. Chemists employ two models—*collision theory* and *transition state theory*—to explain these effects.

Collision Theory: Basis of the Rate Law

The basic tenet of **collision theory** is that particles—atoms, molecules, or ions—must collide to react. But number of collisions can't be the only factor determining rate, or all reactions would be over in an instant. For example, at 1 atm and 20°C, the N_2 and O_2 molecules in 1 mL of air experience about 10^{27} collisions per second. If all that was needed for a reaction to occur was an N_2 molecule colliding with an O_2 molecule, our atmosphere would consist of almost all NO; in fact, only traces are present. Thus, this theory also relies on the concepts of collision energy and molecular structure to explain the effects of concentration and temperature on rate.

Why Concentrations Are Multiplied in the Rate Law Particles must collide to react, so the collision frequency, the number of collisions per unit time, provides an *upper limit* on how fast a reaction can take place. In its basic form, collision theory deals with one-step reactions, those in which two particles collide and form products: $A + B \longrightarrow$ products. Suppose we have only two particles of A and two of B confined in a vessel. Figure 16.13 shows that four A-B collisions are possible. The laws of probability tell us that the number of collisions depends on the *product* of the numbers of reactant particles, not their sum. Thus, when we add another particle of A, six A-B collisions (3×2) are possible, not just five $(3 + 2)$. Similarly, when we add another particle of B, nine A-B collisions (3×3) are possible, not just six $(3 + 3)$. Thus, collision theory explains why we *multiply* the concentrations in the rate law to obtain the observed rate.

The Effect of Temperature on the Rate Constant and the Rate Temperature typically has a dramatic effect on reaction rate: for many reactions near room temperature, an increase of 10 K (10°C) doubles or triples the rate. Figure 16.14A shows kinetic data for an organic reaction—hydrolysis, or reaction with water, of the organic compound ethyl acetate. To understand the effect of temperature, we measure concentrations and times for the reaction run at different temperatures. Solving each rate expression for k and plotting the results (Figure 16.14B), we find that *k increases exponentially as T increases.*

Figure 16.14 Increase of the rate constant with temperature for the hydrolysis of an ester.

Expt	[Ester]	$[H_2O]$	T (K)	Rate (mol/L•s)	k (L/mol•s)
1	0.100	0.200	288	1.04×10^{-3}	0.0521
2	0.100	0.200	298	2.02×10^{-3}	0.101
3	0.100	0.200	308	3.68×10^{-3}	0.184
4	0.100	0.200	318	6.64×10^{-3}	0.332

A

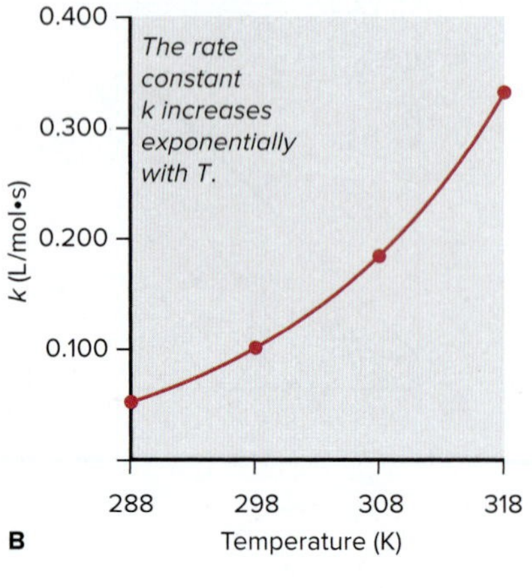

B

These results are consistent with findings obtained in 1889 by the Swedish chemist Svante Arrhenius. In its modern form, the **Arrhenius equation** is

$$k = Ae^{-E_a/RT} \quad \textbf{(16.8)}$$

where k is the rate constant, e is the base of natural logarithms, T is the absolute temperature, and R is the universal gas constant. (We'll focus on the term E_a in the next subsection and on the term A a bit later.) Note the relationship between k and T, especially that T is in the denominator of a negative exponent. Thus, *as T increases, the value of the negative exponent becomes smaller, which means k becomes larger, so the rate increases:*

$$\text{Higher } T \Longrightarrow \text{larger } k \Longrightarrow \text{increased rate}$$

The Central Importance of Activation Energy The effect of temperature on k is closely related to the **activation energy (E_a)** of a reaction, an energy *threshold* that the colliding molecules must exceed in order to react. As an analogy, in order to succeed at the high jump, an athlete must exert at least enough energy to get over the bar. Similarly, if reactant molecules collide with a certain minimum energy, they reach an *activated state,* from which they can change to products (Figure 16.15); collisions that occur with an energy below this minimum leave the reactants unchanged.

As you can see from Figure 16.15, a reversible reaction has two activation energies. The activation energy for the forward reaction, $E_{a(fwd)}$, is the energy difference between the activated state and the reactants; the activation energy for the reverse reaction, $E_{a(rev)}$, is the energy difference between the activated state and the products. The reaction represented in the diagram is exothermic ($\Delta H_{rxn} < 0$) in the forward direction. Thus, the products are at a lower energy than the reactants, and $E_{a(fwd)}$ is less than $E_{a(rev)}$. This difference equals the enthalpy of reaction, ΔH_{rxn}:

$$\Delta H_{rxn} = E_{a(fwd)} - E_{a(rev)} \quad \textbf{(16.9)}$$

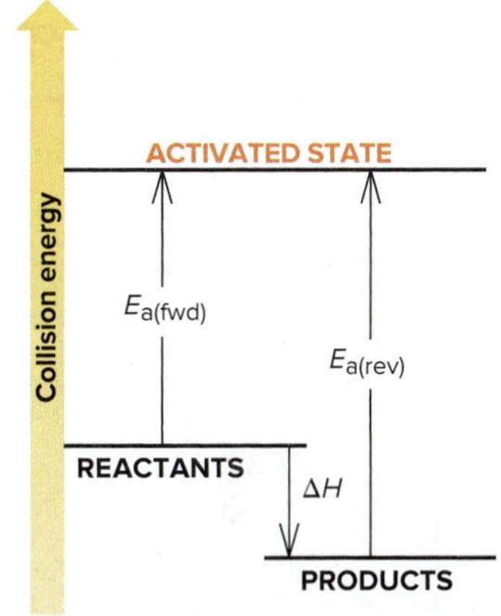

Figure 16.15 Energy-level diagram for a reaction. Like a high jumper with enough energy to go over the bar, molecules must collide with enough energy, E_a, to reach an activated state. (This reaction is reversible and is exothermic in the forward direction.)
Source: © Sipa via AP Images

The Effect of Temperature on Collision Energy As was mentioned in Section 16.1, a rise in temperature has two effects on moving particles: it causes a higher collision *frequency* and a higher collision *energy*. Let's see how each affects rate.

- *Collision frequency*. If particles move faster, they collide more often. Calculations show that a 10 K rise in temperature from, say, 288 K to 298 K, increases the average molecular speed by 2%, which would lead to, at most, a 4% increase in rate. Thus, higher collision frequency cannot possibly account for the doubling or tripling of rates observed with a 10 K rise. Indeed, the effect of temperature on collision *frequency* is only a minor factor.
- *Collision energy*. On the other hand, the effect of temperature on collision *energy* is a major factor. At a given temperature, the fraction f of collisions with energy equal to or greater than E_a is

$$f = e^{-E_a/RT}$$

where e is the base of natural logarithms, T is the absolute temperature, and R is the universal gas constant. The right side of this expression appears in the Arrhenius equation (Equation 16.8), which shows that a rise in T causes a larger k. We now see why—*a rise in temperature increases the kinetic energy of the reactant particles and enlarges the fraction of collisions with enough energy to exceed* E_a (Figure 16.16). It is important to note that increasing the temperature does not change the *value* of E_a—only the fraction of collisions with enough energy to get over this energy barrier.

Table 16.5 *(next page)* shows that the magnitudes of *both* E_a *and* T *affect the size of this fraction* for the hydrolysis of ethyl acetate that we discussed above:

- In the top portion of the table, temperature is held constant, and the fraction of sufficiently energetic collisions shrinks several orders of magnitude with each 25-kJ/mol increase in E_a. (To extend the high jump analogy, as the height of the bar is raised, a smaller fraction of the athletes have enough energy to jump over it.)
- In the bottom portion, E_a is held constant, and the fraction nearly doubles for each 10 K (10°C) rise in temperature.

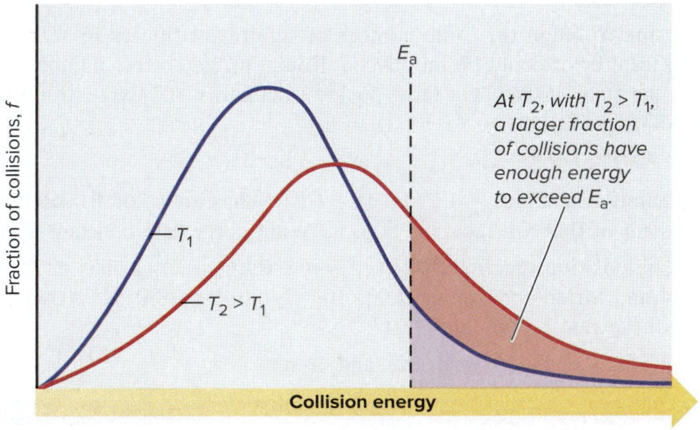

Figure 16.16 **The effect of temperature on the distribution of collision energies.**

Table 16.5 The Effect of E_a and T on the Fraction (f) of Collisions with Sufficient Energy to Allow Reaction

E_a (kJ/mol)	f (at T = 298 K)
50	1.70×10^{-9}
75	7.03×10^{-14}
100	2.90×10^{-18}
T	**f (at E_a = 50 kJ/mol)**
25°C (298 K)	1.70×10^{-9}
35°C (308 K)	3.29×10^{-9}
45°C (318 K)	6.12×10^{-9}

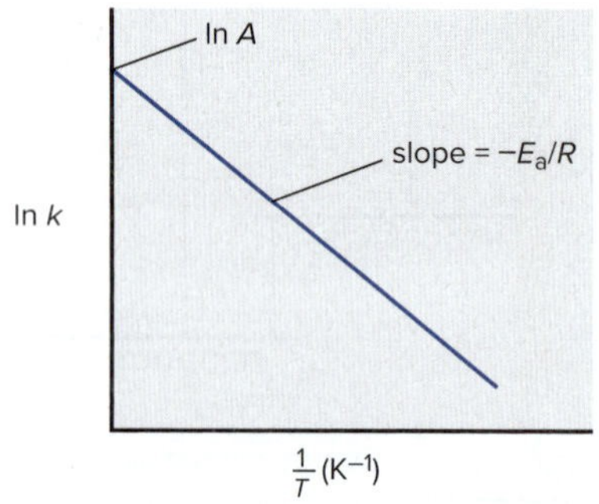

Figure 16.17 Graphical determination of the activation energy.

Therefore, *the smaller the activation energy (or the higher the temperature), the larger the fraction of sufficiently energetic collisions, the larger the value of k, and the higher the reaction rate:*

Smaller E_a (or higher T) $\Longrightarrow$ larger f $\Longrightarrow$ larger k $\Longrightarrow$ higher rate

Calculating the Activation Energy We can calculate E_a from the Arrhenius equation by taking the natural logarithm of both sides of the equation, which recasts it into the form of an equation for a straight line:

$$k = Ae^{-E_a/RT}$$

$$\ln k = \ln A - \frac{E_a}{R}\left(\frac{1}{T}\right)$$

$$y = b + mx$$

A plot of ln k vs. $1/T$ gives a straight line whose slope is $-E_a/R$ and whose y-intercept is ln A (Figure 16.17). We know the constant R, so we can determine E_a graphically from a series of k values at different temperatures.

We can find E_a in another way if we know the rate constants at two temperatures, T_2 and T_1:

$$\ln k_2 = \ln A - \frac{E_a}{R}\left(\frac{1}{T_2}\right) \qquad \ln k_1 = \ln A - \frac{E_a}{R}\left(\frac{1}{T_1}\right)$$

When we subtract ln k_1 from ln k_2, the term ln A drops out and the other terms can be rearranged to give

$$\ln\frac{k_2}{k_1} = \frac{E_a}{R}\left(\frac{1}{T_1} - \frac{1}{T_2}\right) \qquad (16.10)$$

Then, we can solve for E_a, as in the next sample problem.

SAMPLE PROBLEM 16.8 Determining the Energy of Activation

Problem The decomposition of hydrogen iodide $2HI(g) \longrightarrow H_2(g) + I_2(g)$, has rate constants of 9.51×10^{-9} L/mol·s at 500. K and 1.10×10^{-5} L/mol·s at 600. K. Find E_a.

Plan We are given the rate constants, k_1 and k_2, at two temperatures, T_1 and T_2, so we substitute into Equation 16.10 and solve for E_a.

Solution Rearranging Equation 16.10 to solve for E_a:

$$\ln\frac{k_2}{k_1} = \frac{E_a}{R}\left(\frac{1}{T_1} - \frac{1}{T_2}\right)$$

$$E_a = R\left(\ln\frac{k_2}{k_1}\right)\left(\frac{1}{T_1} - \frac{1}{T_2}\right)^{-1}$$

$$= (8.314\text{ J/mol·K})\left(\ln\frac{1.10\times10^{-5}\text{ L/mol·s}}{9.51\times10^{-9}\text{ L/mol·s}}\right)\left(\frac{1}{500.\text{ K}} - \frac{1}{600.\text{ K}}\right)^{-1}$$

$$= 1.76\times10^{5}\text{ J/mol} = 1.76\times10^{2}\text{ kJ/mol}$$

Comment Be sure to retain the same number of significant figures in $1/T$ as you have in T, or a significant error could be introduced. Round to the correct number of significant figures only at the final answer. On most pocket calculators, the expression $(1/T_1 - 1/T_2)$ is entered as follows: $(T_1)(1/x) - (T_2)(1/x) =$.

FOLLOW-UP PROBLEMS

16.8A The reaction $2NOCl(g) \longrightarrow 2NO(g) + Cl_2(g)$ has an E_a of 1.00×10^{2} kJ/mol and a rate constant of 0.286 L/mol·s at 500. K. What is the rate constant at 490. K?

16.8B The decomposition reaction $2NO_2(g) \longrightarrow 2NO(g) + O_2(g)$ has an E_a of 1.14×10^{5} J/mol and a rate constant of 7.02×10^{-3} L/mol·s at 500. K. At what temperature will the rate be twice as fast?

SOME SIMILAR PROBLEMS 16.61, 16.62, and 16.65

The Effect of Molecular Structure on Rate At ordinary temperatures, the enormous number of collisions per second between reactant particles is reduced by six or more orders of magnitude by counting only those with enough energy to react. And, even this tiny fraction of all collisions is typically much larger than the number of **effective collisions,** those that actually lead to product because *the atoms that become bonded in the product make contact.* Thus, to be effective, a collision must have enough energy *and* the appropriate *molecular orientation.*

In the Arrhenius equation, molecular orientation is contained in the term A:

$$k = Ae^{-E_a/RT}$$

This term is the **frequency factor,** the product of the collision frequency Z and an *orientation probability factor, p,* which is specific for each reaction:

$$A = pZ$$

The factor p is related to the structural complexity of the reactants. You can think of p as the ratio of effectively oriented collisions to all possible collisions. Figure 16.18 shows a few of the possible collision orientations for the following reaction:

$$NO(g) + NO_3(g) \longrightarrow 2NO_2(g)$$

Of the five orientations shown, only one has the effective orientation, with the N of NO making contact with an O of NO_3 to form a new nitrogen-oxygen bond. Actually, the p value for this reaction is 0.006: only 6 collisions in every 1000 (1 in 167) have the correct orientation.

The more complex the molecular structure, the smaller the p value is. Individual atoms are spherical, so reactions between them have p values near 1: as long as reacting atoms collide with enough energy, the product can form. At the other extreme are biochemical reactions in which two small molecules (or portions of larger molecules) react only when they collide with enough energy on a specific tiny region of a giant protein. The p value for these reactions is often much less than 10^{-6}, fewer than one in a million. The fact that countless such biochemical reactions are occurring in you right now attests to the astounding number of collisions per second!

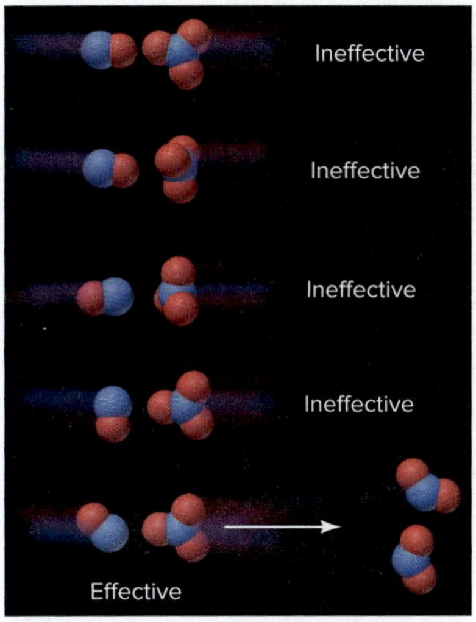

Figure 16.18 The importance of molecular orientation to an effective collision. In the one effective orientation *(bottom),* contact occurs between the atoms that become bonded in the product.

Transition State Theory: What the Activation Energy Is Used For

Collision theory explains the importance of effective collisions, and **transition state theory** focuses on the high-energy species that exists at the moment of an effective collision when reactants are becoming products.

Visualizing the Transition State As two molecules approach each other, repulsions between their electron clouds continually increase, so they slow down as some of their kinetic energy is converted to potential energy. If they collide, but the energy of the collision is *less* than the activation energy, the molecules bounce off each other and no reaction occurs.

However, in a tiny fraction of collisions in which the molecules are moving fast enough, *their kinetic energies push them together with enough force to overcome the repulsions and surpass the activation energy.* And, in an even tinier fraction of these sufficiently energetic collisions, the molecules are oriented effectively. In those cases, nuclei in one molecule attract electrons in the other, atomic orbitals overlap, electron densities shift, and some bonds lengthen and weaken while others shorten and strengthen. At some point during this smooth transformation, *a species with partial bonds exists* that is neither reactant nor product. This very unstable species, called the **transition state** (or **activated complex**) exists only at the instant of highest potential energy. Thus, *the activation energy of a reaction is used to reach the transition state.*

Transition states cannot be isolated, but the work of Ahmed H. Zewail, who received the 1999 Nobel Prize in chemistry, greatly expanded our knowledge of them. Using lasers pulsing at the time scale of bond vibrations (10^{-15} s), his team observed transition states forming and decomposing. A few, such as the one that forms as methyl bromide reacts with hydroxide ion, have been well studied:

$$BrCH_3 + OH^- \longrightarrow Br^- + CH_3OH$$

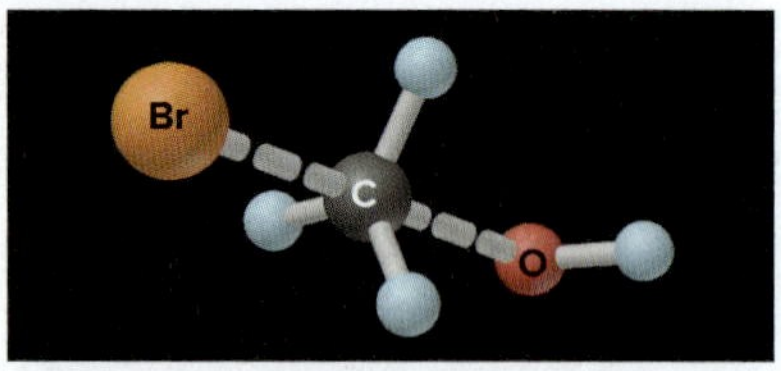

Figure 16.19 **The transition state of the reaction between $BrCH_3$ and OH^-.** Note the partial *(dashed)* Br—C and C—O bonds and the trigonal bipyramidal shape.

The electronegative bromine makes the carbon in $BrCH_3$ partially positive, and the carbon attracts the negatively charged oxygen in OH^-. As a C—O bond begins to form, the Br—C bond begins to weaken. In the transition state (Figure 16.19), C is surrounded by five atoms (trigonal bipyramidal; Section 10.2), which never occurs in carbon's stable compounds. This high-energy species has three normal C—H bonds and two partial bonds, one from C to O and the other from Br to C.

Reaching the transition state does not guarantee that a reaction will proceed to products because *a transition state can change in either direction.* In this case, if the C—O bond continues to strengthen, products form; but, if the Br—C bond becomes stronger again, the transition state reverts to reactants.

Depicting the Change with a Reaction Energy Diagram A useful way to depict these events is with a **reaction energy diagram,** which plots how potential energy changes as the reaction proceeds from reactants to products (the *reaction progress*). The diagram shows the relative energy levels of reactants, products, and transition state, as well as the forward and reverse activation energies and the enthalpy of reaction.

The bottom of Figure 16.20 shows the diagram for the reaction of $BrCH_3$ and OH^-. This reaction is exothermic, so reactants are higher in energy than products, which means $E_{a(fwd)}$ is less than $E_{a(rev)}$. Above the diagram are molecular-scale views at various points during the reaction and corresponding structural formulas. At the top of the figure

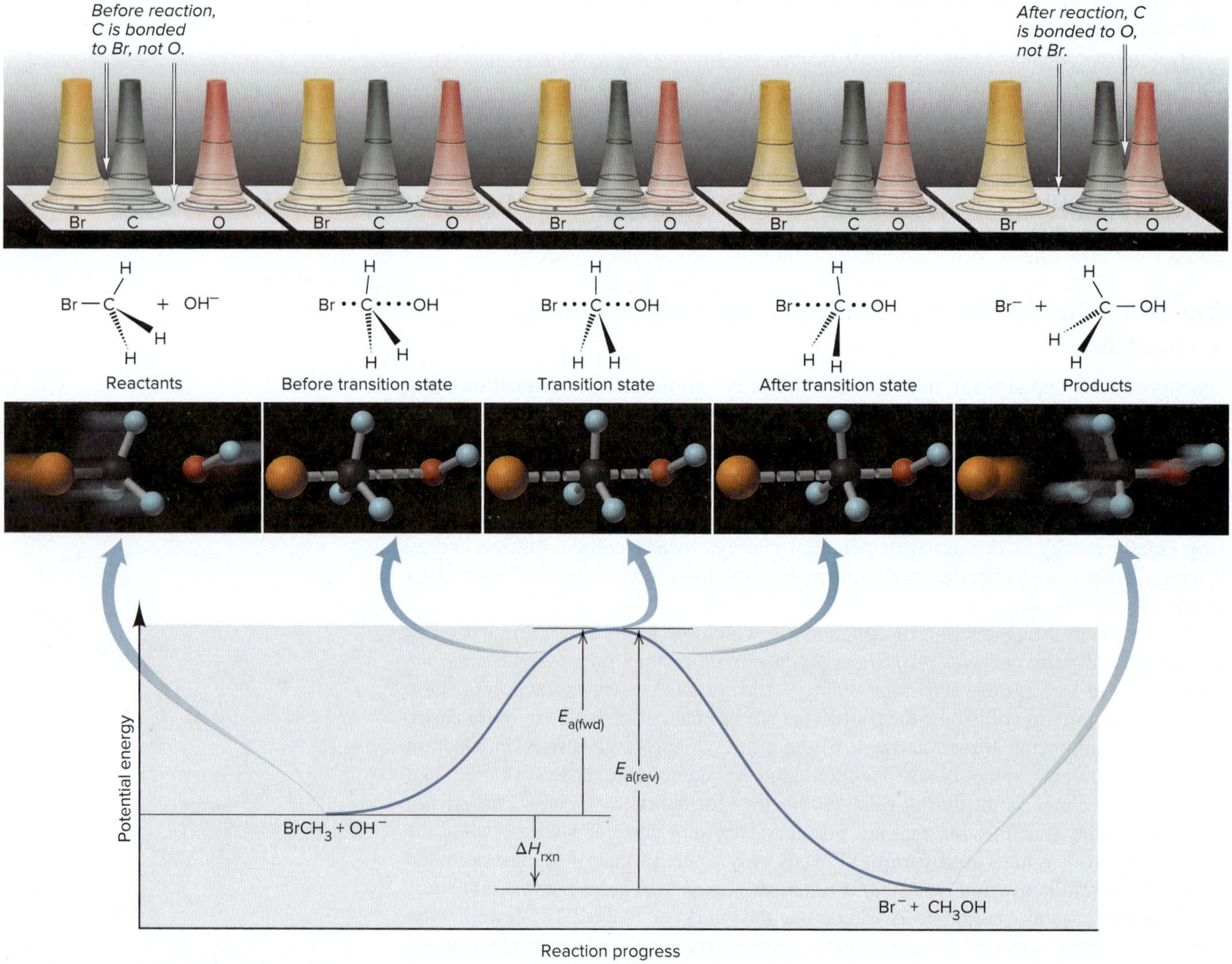

Figure 16.20 **Depicting the reaction between $BrCH_3$ and OH^-.** In order *(bottom to top)* are the reaction energy diagram with blow-up arrows at five points during the reaction, molecular-scale views, structural formulas, and electron density relief maps. Note the gradual, simultaneous C—O bond forming and Br—C bond breaking.

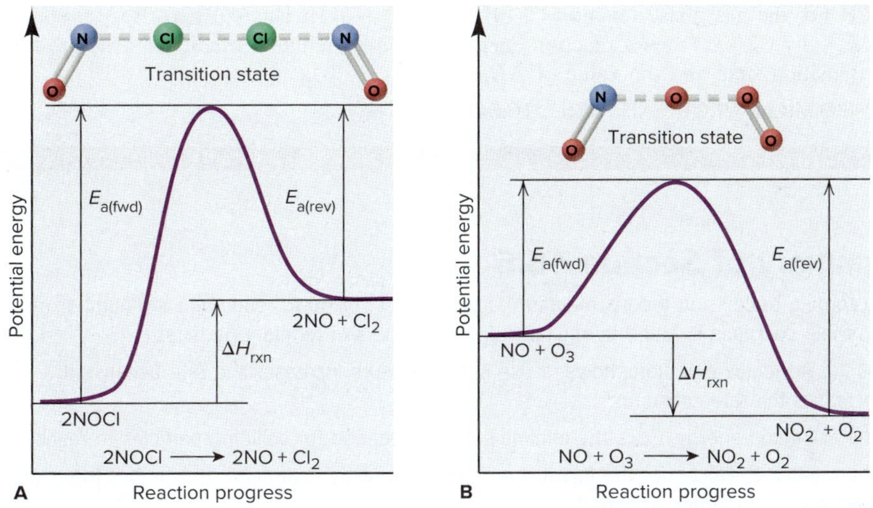

Figure 16.21 Reaction energy diagrams and possible transition states for two reactions.

are electron density relief maps of the Br, C, and O atoms: note the gradual change in electron density from C overlapping Br *(left)* to C overlapping O *(right)*.

Transition state theory proposes that *every reaction (or every step in an overall reaction) goes through its own transition state.* Figure 16.21 presents reaction energy diagrams for two gas-phase reactions. The reaction shown in Figure 16.21A is endothermic, while the reaction shown in Figure 16.21B is exothermic. In each case, the structure of the transition state is predicted from the orientations of the reactant atoms that must become bonded in the product.

SAMPLE PROBLEM 16.9 Drawing Reaction Energy Diagrams and Transition States

Problem A key reaction in the upper atmosphere is

$$O_3(g) + O(g) \longrightarrow 2O_2(g)$$

The $E_{a(fwd)}$ is 19 kJ, and the ΔH_{rxn} for the reaction as written is −392 kJ. Draw a reaction energy diagram, predict a structure for the transition state, and calculate $E_{a(rev)}$.

Plan The reaction is highly exothermic ($\Delta H_{rxn} = -392$ kJ), so the products are much lower in energy than the reactants. The small $E_{a(fwd)}$ (19 kJ) means that the energy of the reactants lies slightly below that of the transition state. We use Equation 16.9 to calculate $E_{a(rev)}$. To predict the transition state, we sketch the species and note that one of the bonds in O_3 weakens, and this partially bonded O begins forming a bond to the separate O atom.

Solution Solving for $E_{a(rev)}$:

$$\Delta H_{rxn} = E_{a(fwd)} - E_{a(rev)}$$

So, $$E_{a(rev)} = E_{a(fwd)} - \Delta H_{rxn} = 19 \text{ kJ} - (-392 \text{ kJ}) = 411 \text{ kJ}$$

The reaction energy diagram (not drawn to scale), with transition state, is shown in the margin.

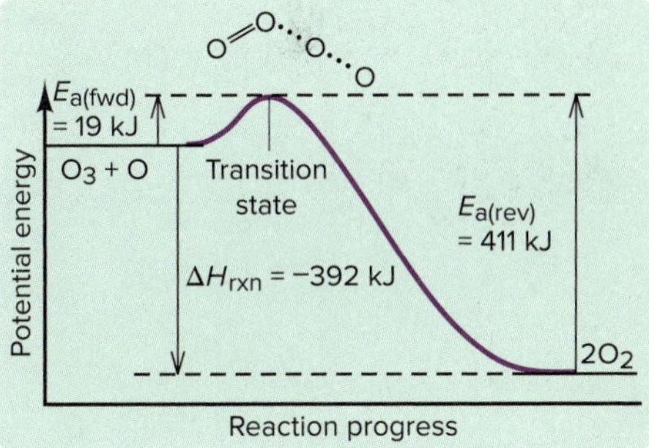

Check Rounding to find $E_{a(rev)}$ gives ~20 + 390 = 410.

FOLLOW-UP PROBLEMS

16.9A The following reaction energy diagram depicts another key atmospheric reaction. Label the axes, identify $E_{a(fwd)}$, $E_{a(rev)}$, and ΔH_{rxn}, draw and label the transition state, and calculate $E_{a(rev)}$ for the reaction.

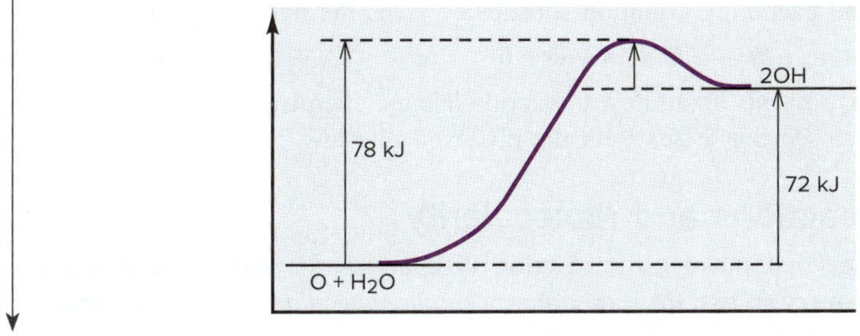

16.9B For the gas phase reaction $Cl(g) + HBr(g) \longrightarrow HCl(g) + Br(g)$, $E_{a(fwd)} = 7$ kJ and $E_{a(rev)} = 72$ kJ. Draw a reaction energy diagram; include a predicted structure for the transition state and the value of ΔH_{rxn}.

SOME SIMILAR PROBLEMS 16.63, 16.64, and 16.66

› Summary of Section 16.5

- › According to collision theory, reactant particles must collide to react, and the number of possible collisions is found by multiplying the numbers of reactant particles.
- › As the Arrhenius equation shows, a rise in temperature increases the rate because it increases the rate constant.
- › The activation energy, E_a, is the minimum energy needed for colliding particles to react.
- › The relative E_a values for the forward and reverse reactions depend on whether the overall reaction is exothermic or endothermic.
- › At higher temperatures, more collisions have enough energy to exceed E_a.
- › E_a can be determined graphically from k values obtained at different T values.
- › Molecules must collide with an effective orientation for them to react, so structural complexity decreases rate.
- › Transition state theory focuses on the change of kinetic energy to potential energy as reactant particles collide and form an unstable transition state.
- › Given a sufficiently energetic collision and an effective molecular orientation, the reactant species form the transition state, which either continues toward product(s) or reverts to reactant(s).
- › A reaction energy diagram depicts the changing potential energy throughout a reaction's progress from reactants through transition states to products.

16.6 REACTION MECHANISMS: THE STEPS FROM REACTANT TO PRODUCT

You can't understand how a car works by examining the body, wheels, and dashboard, or even the engine as a whole. You need to look inside the engine to see how its parts fit together and function. Similarly, by examining the overall balanced equation, we can't know how a reaction "works." We must look "inside the reaction" to see how reactants change into products, step by step.

The steps for most reactions can be described by a **reaction mechanism,** a sequence of single reaction steps that sum to the overall equation. For example, a possible mechanism for the overall reaction

$$2A + B \longrightarrow E + F$$

might involve these three simpler steps:

(1) $A + B \longrightarrow C$

(2) $C + A \longrightarrow D$

(3) $D \longrightarrow E + F$

Adding the steps and canceling common substances gives the overall equation:

$$A + B + \cancel{C} + A + \cancel{D} \longrightarrow \cancel{C} + \cancel{D} + E + F \qquad \text{or} \qquad 2A + B \longrightarrow E + F$$

A mechanism is a hypothesis about how a reaction occurs; chemists *propose* a mechanism and then *test* to see that it fits with the observed rate law.

Elementary Reactions and Molecularity

The individual steps that make up a reaction mechanism are called **elementary reactions** (or **elementary steps**). Each describes a *single molecular event*—one particle decomposing, two particles combining, and so forth.

An elementary step is characterized by its **molecularity,** the number of *reactant* particles in the step. Consider the mechanism for the breakdown of ozone in the stratosphere. The overall equation is

$$2O_3(g) \longrightarrow 3O_2(g)$$

A two-step mechanism has been proposed:

(1) $O_3(g) \longrightarrow O_2(g) + O(g)$

(2) $O_3(g) + O(g) \longrightarrow 2O_2(g)$

- The first step is a **unimolecular reaction,** one that involves the decomposition or rearrangement of a single particle (O_3).
- The second step is a **bimolecular reaction,** one in which two particles (O_3 and O) collide and react.

Some *termolecular* elementary steps occur, but they are extremely rare because the probability of three particles colliding simultaneously with enough energy and an effective orientation is very small. Higher molecularities are not known. Therefore, in general, *we propose unimolecular and/or bimolecular reactions as reasonable steps in a mechanism.*

Because an elementary reaction occurs in one step, its rate law, unlike that for an overall reaction, *can* be deduced from the reaction stoichiometry: *reaction order equals molecularity*. Therefore, *only* for an elementary step, *we use the equation coefficients as the reaction orders in the rate law* (Table 16.6). For example, in step 2 of the proposed mechanism for the decomposition of O_3, one O_3 molecule collides with one O atom; therefore, this step is first-order in O_3 and first-order in O and the rate law for this step is rate = $k[O_3][O]$.

Table 16.6 Rate Laws for General Elementary Steps

Elementary Step	Molecularity	Rate Law
$A \longrightarrow$ product	Unimolecular	Rate = $k[A]$
$2A \longrightarrow$ product	Bimolecular	Rate = $k[A]^2$
$A + B \longrightarrow$ product	Bimolecular	Rate = $k[A][B]$
$2A + B \longrightarrow$ product	Termolecular	Rate = $k[A]^2[B]$

SAMPLE PROBLEM 16.10 Determining Molecularities and Rate Laws for Elementary Steps

Problem The following elementary steps are proposed for a reaction mechanism:

(1) $Cl_2(g) \rightleftharpoons 2Cl(g)$

(2) $Cl(g) + CHCl_3(g) \longrightarrow HCl(g) + CCl_3(g)$

(3) $Cl(g) + CCl_3(g) \longrightarrow CCl_4(g)$

(a) Write the balanced equation for the overall reaction.

(b) Determine the molecularity of each step.

(c) Write the rate law for each step.

Plan We find the overall equation by adding the elementary steps. The molecularity of each step equals the total number of *reactant* particles. We write the rate law for each step using the equation coefficients as reaction orders.

Solution (a) Writing the overall balanced equation:

$$Cl_2(g) \rightleftharpoons 2Cl(g)$$
$$Cl(g) + CHCl_3(g) \longrightarrow HCl(g) + CCl_3(g)$$
$$Cl(g) + CCl_3(g) \longrightarrow CCl_4(g)$$

$$Cl_2(g) + \cancel{Cl(g)} + CHCl_3(g) + \cancel{Cl(g)} + \cancel{CCl_3(g)} \longrightarrow \cancel{2Cl(g)} + HCl(g) + \cancel{CCl_3(g)} + CCl_4(g)$$

$$Cl_2(g) + CHCl_3(g) \longrightarrow HCl(g) + CCl_4(g)$$

(b) Determining the molecularity of each step: The first step has one reactant, Cl_2, so it is unimolecular. The second and third steps each have two reactants (Cl and $CHCl_3$; Cl and CCl_3), so they are bimolecular.

(c) Writing the rate laws for the elementary steps:

(1) $\text{Rate}_1 = k_1[Cl_2]$

(2) $\text{Rate}_2 = k_2[Cl][CHCl_3]$

(3) $\text{Rate}_3 = k_3[Cl][CCl_3]$

Check In part (a), be sure the equation is balanced; in part (c), be sure the substances in brackets are the *reactants* of each elementary step.

FOLLOW-UP PROBLEMS

16.10A The mechanism below is proposed for the decomposition of hydrogen peroxide:

(1) $H_2O_2(aq) \longrightarrow 2OH(aq)$
(2) $H_2O_2(aq) + OH(aq) \longrightarrow H_2O(l) + HO_2(aq)$
(3) $HO_2(aq) + OH(aq) \longrightarrow H_2O(l) + O_2(g)$

(a) Write the balanced equation for the overall reaction. **(b)** Determine the molecularity of each step. **(c)** Write the rate law for each step.

16.10B The following elementary steps are proposed for a reaction mechanism:

(1) $2NO(g) \rightleftharpoons N_2O_2(g)$
(2) $N_2O_2(g) + H_2(g) \longrightarrow N_2O(g) + H_2O(g)$
(3) $N_2O(g) + H_2(g) \longrightarrow N_2(g) + H_2O(g)$

(a) Write the balanced equation for the overall reaction. **(b)** Determine the molecularity of each step. **(c)** Write the rate law for each step.

SOME SIMILAR PROBLEMS 16.74(a), 16.74(c), 16.75(a), and 16.75(c)

The Rate-Determining Step of a Reaction Mechanism

All the elementary steps in a mechanism have their own rates. However, one step is usually *much* slower than the others. This step, called the **rate-determining step (*or* rate-limiting step),** limits how fast the overall reaction proceeds. Therefore, *the rate law of the rate-determining step becomes the rate law for the overall reaction.* ›

A Rate-Determining Step for Traffic Flow

Imagine driving home on a wide street that goes over a toll bridge. Traffic flows smoothly until the road narrows at the approach to the toll booth. Past the bridge, traffic returns to normal, and you proceed a couple of miles further to your destination. Getting through the bottleneck to pay the toll takes longer than the rest of the trip combined and, therefore, largely determines the time for the whole trip.

Consider the reaction between nitrogen dioxide and carbon monoxide:

$$NO_2(g) + CO(g) \longrightarrow NO(g) + CO_2(g)$$

If this reaction were an elementary step—that is, if the mechanism consisted of only one step—we could immediately write the overall rate law as

$$\text{Rate} = k[NO_2][CO]$$

But, as you saw in Sample Problem 16.3, experimental data show the rate law is

$$\text{Rate} = k[NO_2]^2$$

Thus, the overall reaction cannot be elementary.

A proposed two-step mechanism is

(1) $NO_2(g) + NO_2(g) \longrightarrow NO_3(g) + NO(g)$ [slow; rate determining]

(2) $NO_3(g) + CO(g) \longrightarrow NO_2(g) + CO_2(g)$ [fast]

In this mechanism, NO_3 functions as a **reaction intermediate,** a substance formed in one step of the mechanism and used up in a subsequent step during the reaction. Even though it does not appear in the overall balanced equation, a reaction intermediate is essential for the reaction to occur. Intermediates are less stable than the reactants and products, but unlike *much* less stable transition states, they have normal bonds and can sometimes be isolated.

Rate laws for the two elementary steps listed above are

(1) $\text{Rate}_1 = k_1[NO_2][NO_2] = k_1[NO_2]^2$
(2) $\text{Rate}_2 = k_2[NO_3][CO]$

Three key points to notice about this mechanism are

- If $k_1 = k$, the rate law for the rate-determining step (step 1) becomes identical to the observed rate law.

- Because the first step is slow, $[NO_3]$ is low. As soon as any NO_3 forms, it is consumed by the fast second step, so the reaction takes as long as the first step does.
- CO does not appear in the rate law (reaction order = 0) because it takes part in the mechanism *after* the rate-determining step and so its concentration does not affect the rate.

Correlating the Mechanism with the Rate Law

Coming up with a reasonable reaction mechanism is a classic demonstration of the scientific method. Using observations and data from rate experiments, we hypothesize the individual steps and then test our hypothesis with further evidence. If the evidence supports our mechanism, we can accept it; if not, we must propose a different one. We can never *prove* that a mechanism represents the *actual* chemical change, only that it is consistent with the data.

A valid mechanism must meet three criteria:

1. *The elementary steps must add up to the overall balanced equation.*
2. *The elementary steps must be reasonable.* They should generally involve one reactant particle (unimolecular) or two (bimolecular).
3. *The mechanism must correlate with the rate law,* not the other way around.

Mechanisms with a Slow Initial Step The reaction between NO_2 and CO that we considered earlier has a mechanism with a slow initial step; that is, the first step is rate-determining. The reaction between nitrogen dioxide and fluorine is another example of such a reaction:

$$2NO_2(g) + F_2(g) \longrightarrow 2NO_2F(g)$$

The experimental rate law is first order in NO_2 and in F_2,

$$\text{Rate} = k[NO_2][F_2]$$

and the accepted mechanism is

(1) $NO_2(g) + F_2(g) \longrightarrow NO_2F(g) + F(g)$ [slow; rate determining]
(2) $NO_2(g) + F(g) \longrightarrow NO_2F(g)$ [fast]

Note that the free fluorine atom is a reaction intermediate.

Let's see how this mechanism meets the three criteria.

1. The elementary reactions sum to the balanced equation:

$$NO_2(g) + NO_2(g) + F_2(g) + \cancel{F(g)} \longrightarrow NO_2F(g) + NO_2F(g) + \cancel{F(g)}$$

or

$$2NO_2(g) + F_2(g) \longrightarrow 2NO_2F(g)$$

2. Both steps are bimolecular and, thus, reasonable.
3. To determine whether the mechanism is consistent with the observed rate law, we first write the rate laws for the elementary steps:
   (1) $\text{Rate}_1 = k_1[NO_2][F_2]$
   (2) $\text{Rate}_2 = k_2[NO_2][F]$
   Step 1 is the rate-determining step, and with $k_1 = k$, it is the same as the overall rate law, so the third criterion is met.

Note that the second molecule of NO_2 is involved *after* the rate-determining step, so it does not appear in the overall rate law. Thus, as in the mechanism for the reaction of NO_2 and CO, *the overall rate law includes all the reactants involved in the rate-determining step.*

Mechanisms with a Fast Initial Step If the rate-limiting step in a mechanism is *not* the initial step, the product of the fast initial step builds up and starts reverting to reactant. With time, this *fast, reversible step reaches equilibrium,* as product changes to reactant as fast as it forms. As you'll see, this situation allows us to fit the mechanism to the overall rate law.

Consider the oxidation of nitrogen monoxide:

$$2NO(g) + O_2(g) \longrightarrow 2NO_2(g)$$

The observed rate law is

$$\text{Rate} = k[NO]^2[O_2]$$

and a proposed mechanism is

(1) $NO(g) + O_2(g) \rightleftharpoons NO_3(g)$ [fast, reversible]

(2) $NO_3(g) + NO(g) \longrightarrow 2NO_2(g)$ [slow; rate determining]

Let's go through the three criteria to see if this mechanism is valid.

1. With cancellation of the reaction intermediate, NO_3, the sum of the steps gives the overall equation, so the first criterion is met:

$$NO(g) + O_2(g) + \cancel{NO_3(g)} + NO(g) \longrightarrow \cancel{NO_3(g)} + 2NO_2(g)$$

or

$$2NO(g) + O_2(g) \longrightarrow 2NO_2(g)$$

2. Both steps are bimolecular, so the second criterion is met.
3. To see whether the third criterion—that the mechanism is consistent with the observed rate law—is met, we first write rate laws for the elementary steps:

(1) $\text{Rate}_{1(\text{fwd})} = k_1[NO][O_2]$

$\text{Rate}_{1(\text{rev})} = k_{-1}[NO_3]$

where k_{-1} is the rate constant and NO_3 is the reactant for the reverse reaction.

(2) $\text{Rate}_2 = k_2[NO_3][NO]$

Next, we show that the rate law for the rate-determining step (step 2) gives the overall rate law. As written, it does not, because it contains the intermediate NO_3, and *an overall rate law includes only reactants (and products).* We eliminate $[NO_3]$ from the rate law for step 2 by expressing it in terms of reactants, as follows.

- Step 1 reaches equilibrium when the forward and reverse rates are equal:

$$\text{Rate}_{1(\text{fwd})} = \text{Rate}_{1(\text{rev})} \quad \text{or} \quad k_1[NO][O_2] = k_{-1}[NO_3]$$

- To express $[NO_3]$ in terms of reactants, we isolate it algebraically:

$$[NO_3] = \frac{k_1}{k_{-1}}[NO][O_2]$$

- Then, substituting for $[NO_3]$ in the rate law for the slow step, step 2, we obtain

$$\text{Rate}_2 = k_2[NO_3][NO] = k_2\underbrace{\left(\frac{k_1}{k_{-1}}[NO][O_2]\right)}_{[NO_3]}[NO] = \frac{k_2k_1}{k_{-1}}[NO]^2[O_2]$$

With $k = \frac{k_2k_1}{k_{-1}}$, this rate law is identical to the overall rate law.

To summarize, we assess the validity of a mechanism with a fast initial step by

1. Writing rate laws for the fast step (both directions) and for the slow step.
2. Expressing [intermediate] in terms of [reactant] by setting the forward rate law of the reversible step equal to the reverse rate law, and solving for [intermediate].
3. Substituting the expression for [intermediate] into the rate law for the slow step to obtain the overall rate law.

Student Hot Spot

Student data indicate that you may struggle with the concept of reaction mechanisms. Access the Smartbook to view additional Learning Resources on this topic.

An important point to note is that, *for any mechanism, only reactants involved up to and including the slow (rate-determining) step appear in the overall rate law.*

SAMPLE PROBLEM 16.11 Identifying Intermediates and Correlating Rate Laws and Reaction Mechanisms

Problem Consider the reaction mechanism in Sample Problem 16.10.
(a) Identify the intermediates in the mechanism.
(b) Assuming that step 2 is the slow (rate-determining) step, show that the mechanism is consistent with the observed rate law: $\text{Rate} = k[Cl_2]^{1/2}[CHCl_3]$.

Plan **(a)** We write each elementary step and sum to get the overall reaction. Intermediates are substances that are produced in one elementary step and consumed as reactants in a subsequent elementary step. **(b)** We use the reactants in the slow step to write its rate law; if an intermediate is present in the rate law, we eliminate it by expressing it in terms of reactants.

Solution (a) Determining the intermediates:

$$\begin{array}{rll} Cl_2(g) \rightleftharpoons 2Cl(g) & & [\text{fast}] \\ Cl(g) + CHCl_3(g) \longrightarrow HCl(g) + CCl_3(g) & & [\text{slow}] \\ Cl(g) + CCl_3(g) \longrightarrow CCl_4(g) & & [\text{fast}] \end{array}$$

$$Cl_2(g) + \cancel{Cl(g)} + CHCl_3(g) + \cancel{Cl(g)} + \cancel{CCl_3(g)} \longrightarrow 2\cancel{Cl(g)} + HCl(g) + \cancel{CCl_3(g)} + CCl_4(g)$$

$$Cl_2(g) + CHCl_3(g) \longrightarrow HCl(g) + CCl_4(g)$$

Cl and CCl_3 are formed and used up, so the intermediates are $Cl(g)$ and $CCl_3(g)$.

(b) Correlating the reaction mechanism and the rate law:

Step 2 is the rate-determining step:

$$Cl(g) + CHCl_3(g) \longrightarrow HCl(g) + CCl_3(g) \quad [\text{slow}]$$

$$\text{Rate} = k_2[Cl][CHCl_3]$$

Since it is an intermediate, [Cl] must be expressed in terms of reactants:

From step 1: $\text{Rate}_{1(\text{fwd})} = \text{Rate}_{1(\text{rev})}$ or $k_1[Cl_2] = k_{-1}[Cl]^2$

Thus,

$$[Cl]^2 = \frac{k_1}{k_{-1}}[Cl_2] \quad \text{and} \quad [Cl] = \sqrt{\frac{k_1}{k_{-1}}[Cl_2]} = \sqrt{\frac{k_1}{k_{-1}}}[Cl_2]^{1/2}$$

Substituting for [Cl] in the rate law for the slow step:

$$\text{Rate} = k_2\,[Cl]\,[CHCl_3] = k_2\sqrt{\frac{k_1}{k_{-1}}}[Cl_2]^{1/2}[CHCl_3]$$

With $k = k_2\sqrt{\frac{k_1}{k_{-1}}}$, this rate law is consistent with the observed rate law.

Check Be sure that the substances identified as intermediates are not included in the overall balanced equation. In part (b), be sure that the rate-determining step is used to write the rate law and that no intermediates are included in the rate law.

FOLLOW-UP PROBLEMS

16.11A Examine the reaction mechanism in Follow-up Problem 16.10A. **(a)** Identify the intermediates in the mechanism. **(b)** The observed rate law for the decomposition of hydrogen peroxide is Rate = $k[H_2O_2]$. For the reaction mechanism to be consistent with this rate law, which step must be the rate-determining step?

16.11B The observed rate law for the overall reaction whose mechanism is shown in Follow-up Problem 16.10B is Rate = $k[NO]^2[H_2]$. **(a)** Identify the intermediates in this mechanism. **(b)** Assuming that step 2 is the slow (rate-determining) step, show that this mechanism is consistent with the observed rate law.

SOME SIMILAR PROBLEMS 16.74(b), 16.74(d), 16.74(e), 16.75(b), and 16.75(d)

Using the Steady-State Approximation Another approach to determining the rate law from a mechanism is to use the **steady-state approximation,** which assumes that the concentration of an intermediate is essentially constant because it is produced in one step of the mechanism at the same rate it is consumed in another step. Let's apply the steady-state approximation to the oxidation of nitrogen monoxide that we previously examined,

$$2NO(g) + O_2(g) \longrightarrow 2NO_2(g)$$

with the following proposed mechanism:

(1) $NO(g) + O_2(g) \rightleftharpoons NO_3(g)$ [fast, reversible]

(2) $NO_3(g) + NO(g) \longrightarrow 2NO_2(g)$ [slow; rate determining]

The rate of NO_2 production is given by the second step of the mechanism, with k_2 as the rate constant:

$$\text{Rate of } NO_2 \text{ production} = k_2[NO_3][NO]$$

We must eliminate the NO_3 intermediate from this expression. First, we write the expression for the *net* rate of formation of NO_3. It is produced in the forward reaction of step 1 at a rate of $k_1[NO][O_2]$, consumed in the reverse reaction of step 1 at a rate of $k_{-1}[NO_3]$, and consumed in step 2 at a rate of $k_2[NO_3][NO]$. Therefore, the net rate of NO_3 formation is the difference between the rate of production and the rates of consumption:

$$\text{Net rate of NO}_3\text{ formation} = \underbrace{k_1[\text{NO}][\text{O}_2]}_{\substack{\text{production in}\\ \text{step 1}\\ \text{(forward}\\ \text{reaction)}}} - \underbrace{k_{-1}[\text{NO}_3]}_{\substack{\text{consumption}\\ \text{in step 1}\\ \text{(reverse}\\ \text{reaction)}}} - \underbrace{k_2[\text{NO}_3][\text{NO}]}_{\substack{\text{consumption}\\ \text{in step 2}}}$$

At this point, we make the steady-state approximation that $[NO_3]$ remains constant during the reaction, so that its net rate of formation is zero and the previous equation becomes

$$\text{Net rate of NO}_3\text{ formation} = 0 = k_1[\text{NO}][\text{O}_2] - k_{-1}[\text{NO}_3] - k_2[\text{NO}_3][\text{NO}]$$

We can rearrange this equation to obtain an expression for $[NO_3]$:

$$k_1[\text{NO}][\text{O}_2] = k_{-1}[\text{NO}_3] + k_2[\text{NO}_3][\text{NO}]$$

$$k_1[\text{NO}][\text{O}_2] = [\text{NO}_3](k_{-1} + k_2[\text{NO}])$$

$$\frac{k_1[\text{NO}][\text{O}_2]}{k_{-1} + k_2[\text{NO}]} = [\text{NO}_3]$$

Substituting this expression for $[NO_3]$ into the rate law for the production of NO_2 gives

$$\text{Rate of NO}_2\text{ production} = k_2[\text{NO}_3][\text{NO}] = \frac{k_1k_2[\text{NO}]^2[\text{O}_2]}{k_{-1} + k_2[\text{NO}]}$$

At first, this expression seems to differ from the observed rate law, which is rate = $k[NO]^2[O_2]$. But, recall that step 2 is slow relative to step 1, which means that $k_2 << k_{-1}$, so $k_2[NO] << k_{-1}$ and can be ignored. Thus, we have

$$\text{Rate of NO}_2\text{ production} = \frac{k_1k_2}{k_{-1}}[\text{NO}]^2[\text{O}_2] = k[\text{NO}]^2[\text{O}_2]$$

As you can see, with $k = \dfrac{k_1k_2}{k_{-1}}$, the expression becomes the same as the observed rate law. The steady-state approximation method is also useful with mechanisms in which there is no single rate-determining step.

Depicting a Multistep Mechanism with a Reaction Energy Diagram Figure 16.22 shows reaction energy diagrams for two reactions, each of which has a two-step mechanism, shown below the diagram. The reaction of NO_2 and F_2 (part A) starts with a slow step, and the reaction of NO and O_2 (part B) starts with a fast step. Both overall reactions are exothermic, so the product is lower in energy than the reactants. Note these key points:

- *Each step in the mechanism has its own peak with the transition state at the top.*
- The intermediates (F in part A and NO_3 in part B) are reactive, unstable species, so they are higher in energy than the reactants or product.
- The slow (rate-determining) step (step 1 in A and step 2 in B) has a *larger* E_a than the other step.

› Summary of Section 16.6

- › The mechanisms of most common reactions consist of two or more elementary steps, each of which describes a single molecular event.
- › The molecularity of an elementary step equals the number of reactant particles and is the same as the total reaction order of the step. Only unimolecular and bimolecular steps are reasonable.
- › The rate-determining (slowest) step in a mechanism determines how fast the overall reaction occurs, and its rate law is equivalent to the overall rate law.

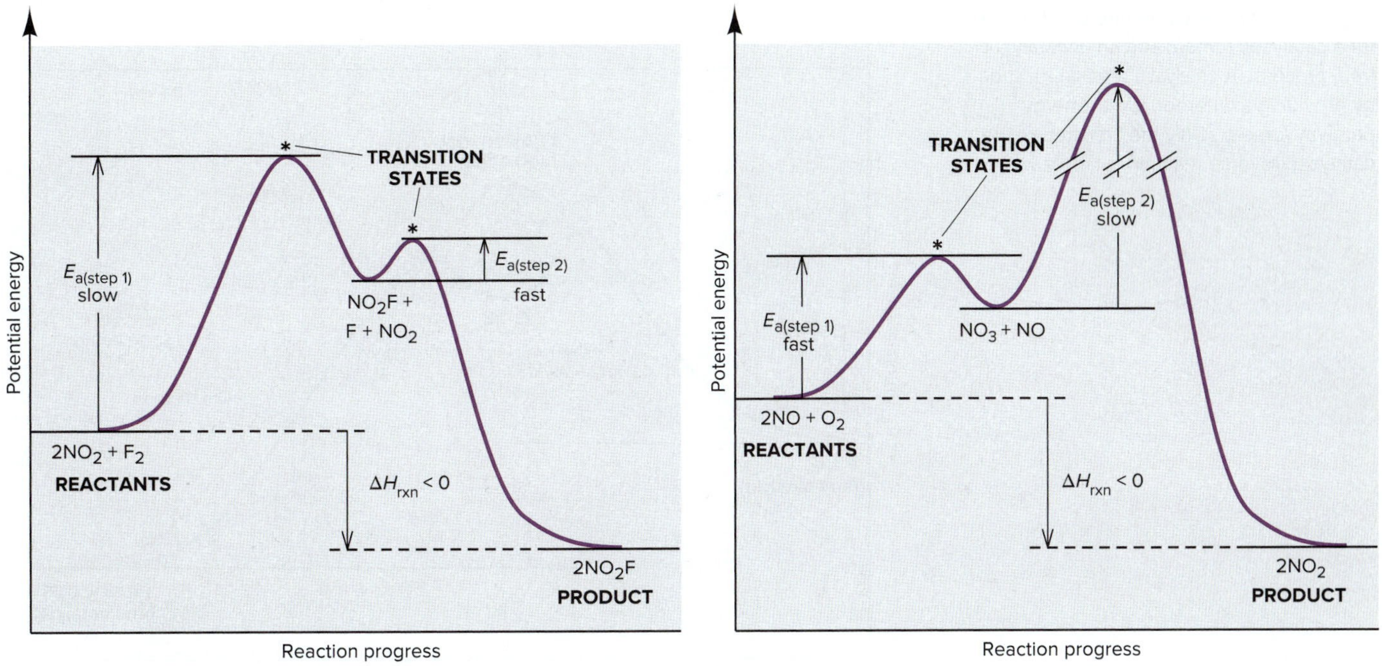

A

(1) $NO_2(g) + F_2(g) \longrightarrow NO_2F(g) + F(g)$ [slow]

(2) $NO_2(g) + F(g) \longrightarrow NO_2F(g)$ [fast]

B

(1) $NO(g) + O_2(g) \rightleftharpoons NO_3(g)$ [fast]

(2) $NO_3(g) + NO(g) \longrightarrow 2NO_2(g)$ [slow]

Figure 16.22 Reaction energy diagrams for the two-step reaction of (A) NO_2 and F_2 and of (B) NO and O_2.

- Reaction intermediates are species that form in one step and react in a later one.
- For a mechanism to be valid, (1) the elementary steps must add up to the overall balanced equation, (2) the steps must be reasonable, and (3) the mechanism must correlate with the rate law.
- If a mechanism begins with a slow step, only those reactants involved in the slow step appear in the overall rate law.
- If a mechanism begins with a fast step, the product of the fast step accumulates as an intermediate, and the step reaches equilibrium. To show that the mechanism is valid, we express [intermediate] in terms of [reactant]. In some cases, we use the steady-state approximation, which assumes that [intermediate] is constant, to see if the mechanism is consistent with the rate law.
- Only reactants involved in steps up to and including the slow (rate-determining) step appear in the overall rate law.
- Each step in a mechanism has its own transition state, which appears at the top of a peak in the reaction energy diagram. A slower step has a higher peak (larger E_a).

16.7 CATALYSIS: SPEEDING UP A REACTION

Increasing the rate of a reaction has countless applications, in both engineering and biology. Higher temperature can speed up a reaction, but energy for industrial processes is costly and many organic and biological substances are heat sensitive. More commonly, by far, a reaction is accelerated by a **catalyst,** a substance that increases the reaction rate *without* being consumed. Thus, only a small, nonstoichiometric amount of the catalyst is required to speed the reaction. Despite this, catalysts are employed in so many processes that several million tons are produced annually in the United States alone. Nature is the master designer and user of catalysts: every organism relies on protein catalysts, known as *enzymes,* to speed up life-sustaining reactions, and even the simplest bacterium employs thousands of them.

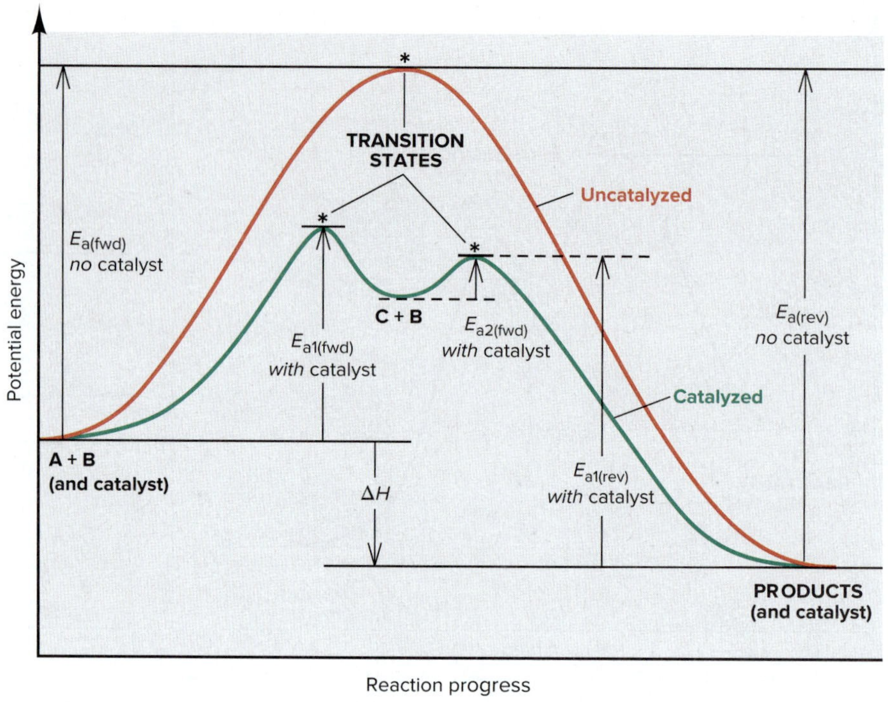

Figure 16.23 Reaction energy diagram for a catalyzed (*green*) and an uncatalyzed (*red*) process. A catalyst speeds a reaction by providing a different, lower energy pathway *(green)*. (Only the first step of the catalyzed reverse reaction is labeled.)

The Basis of Catalytic Action

Each catalyst has its own specific way of functioning, but in general, *a catalyst provides a different reaction mechanism with a lower activation energy, which in turn makes the rate constant larger and, thus, the reaction rate higher:*

$$\text{Catalyst} \Longrightarrow \text{lower } E_a \Longrightarrow \text{larger } k \Longrightarrow \text{higher rate}$$

Consider a general *uncatalyzed* reaction that proceeds by a one-step mechanism involving a bimolecular collision between the reactants A and B (Figure 16.23). The activation energy is relatively large, so the rate is relatively low:

$$A + B \longrightarrow \text{product} \qquad [\text{larger } E_a \Longrightarrow \text{lower rate}]$$

In the *catalyzed* reaction, reactant A interacts with the catalyst in one step to form the intermediate C, and then C reacts with B in a second step to form product and regenerate the catalyst:

$$(1)\ A + \text{catalyst} \longrightarrow C \qquad [\text{smaller } E_a \Longrightarrow \text{higher rate}]$$

$$(2)\ C + B \longrightarrow \text{product} + \text{catalyst} \qquad [\text{smaller } E_a \Longrightarrow \text{higher rate}]$$

Three points to note in Figure 16.23 are

- A catalyst speeds up the forward *and* reverse reactions. Thus, a reaction has the *same yield* with or without a catalyst, but the product forms *faster*.
- A catalyst causes a *lower total activation energy* by providing a *different mechanism* for the reaction. The total of the activation energies for both steps of the catalyzed pathway $[E_{a1(fwd)} + E_{a2(fwd)}]$ is less than the forward activation energy of the uncatalyzed pathway.
- The catalyst is *not* consumed, but rather used and then regenerated.

Homogeneous Catalysis

Chemists classify catalysts based on whether or not they act in the same phase as the reactants and products. A **homogeneous catalyst** exists in solution with the reaction mixture, so it must be a gas, liquid, or soluble solid.

A thoroughly studied example of homogeneous catalysis in the gas phase was formerly used in sulfuric acid manufacture. The key step, the oxidation of sulfur dioxide to sulfur trioxide, occurs so slowly that it is not economical:

$$SO_2(g) + \tfrac{1}{2}O_2(g) \longrightarrow SO_3(g)$$

In the presence of nitrogen monoxide, however, the overall reaction rate speeds up dramatically:

$$\begin{array}{rl} NO(g) + \frac{1}{2}O_2(g) \longrightarrow & NO_2(g) \\ NO_2(g) + SO_2(g) \longrightarrow & NO(g) + SO_3(g) \\ \hline \cancel{NO(g)} + \frac{1}{2}O_2(g) + \cancel{NO_2(g)} + SO_2(g) \longrightarrow & \cancel{NO_2(g)} + \cancel{NO(g)} + SO_3(g) \\ \text{or} \quad SO_2(g) + \frac{1}{2}O_2(g) \longrightarrow & SO_3(g) \end{array}$$

Note that NO and NO_2 cancel to give the overall reaction:

- NO_2 acts as an intermediate (formed in the first step and then consumed in the second step) and
- NO acts as a catalyst (used in the first step and then regenerated in the second step).

Another well-studied example of homogeneous catalysis involves the decomposition of hydrogen peroxide in aqueous solution:

$$2H_2O_2(aq) \longrightarrow 2H_2O(l) + O_2(g)$$

Commercial H_2O_2 decomposes in light and in the presence of the small amounts of ions dissolved from glass, but it is quite stable in dark plastic containers. Many other substances speed its decomposition, including bromide ion, Br^- (Figure 16.24). The catalyzed process is thought to occur in two steps, both of which are oxidation-reduction reactions:

$$\begin{array}{rl} 2Br^-(aq) + H_2O_2(aq) + 2H^+(aq) \longrightarrow & Br_2(aq) + 2H_2O(l) \\ Br_2(aq) + H_2O_2(aq) \longrightarrow & 2Br^-(aq) + 2H^+(aq) + O_2(g) \\ \hline \cancel{2Br^-(aq)} + H_2O_2(aq) + \cancel{2H^+(aq)} + \cancel{Br_2(aq)} + H_2O_2(aq) \longrightarrow & \\ & \cancel{Br_2(aq)} + 2H_2O(l) + \cancel{2Br^-(aq)} + \cancel{2H^+(aq)} + O_2(g) \\ \text{or} \quad 2H_2O_2(aq) \longrightarrow & 2H_2O(l) + O_2(g) \end{array}$$

In this case, Br_2, Br^-, and H^+ cancel to give the overall balanced equation:

- Br^- (in the presence of H^+) is the catalyst, and
- Br_2 (which is orange in color) is the intermediate.

Heterogeneous Catalysis

A **heterogeneous catalyst** speeds up a reaction in a different phase. Most often solids interacting with gaseous or liquid reactants, these catalysts have enormous surface areas, sometimes as much as 500 m^2/g. Very early in the reaction, the rate depends on reactant concentration, but almost immediately, the reaction becomes zero order: the rate-determining step occurs on the catalyst's surface, so once the reactant covers it, adding more reactant cannot increase the rate further.

A very important organic example of heterogeneous catalysis is used to speed up **hydrogenation,** the addition of H_2 to C=C bonds to form C—C bonds. The petroleum, plastics, and food industries employ this process on an enormous scale. The simplest hydrogenation converts ethylene (ethene) to ethane:

$$H_2C{=}CH_2(g) + H_2(g) \longrightarrow H_3C{-}CH_3(g)$$

In the absence of a catalyst, the reaction is very slow. But, at high H_2 pressure (high $[H_2]$) and in the presence of finely divided Ni, Pd, or Pt metal, it is rapid even at ordinary temperatures. The Group 8B(10) metals catalyze by *chemically adsorbing the reactants onto their surface* (Figure 16.25, *next page*). In the rate-determining step, the adsorbed H_2 splits into two H atoms that become weakly bonded to the catalyst's surface (catM):

$$H{-}H(g) + 2\text{catM}(s) \longrightarrow 2\text{catM}{-}H \text{ (H atoms bound to metal surface)}$$

Then, C_2H_4 adsorbs and reacts with the H atoms, one at a time, to form C_2H_6. Thus, the catalyst acts by lowering the activation energy of the slow step (H—H bond breakage) as part of a different mechanism.

A solid mixture of transition metals and their oxides forms a heterogeneous catalyst in your car's exhaust system. The catalytic converter speeds both the oxidation of toxic CO and unburned gasoline to CO_2 and H_2O *and* the reduction of the pollutant

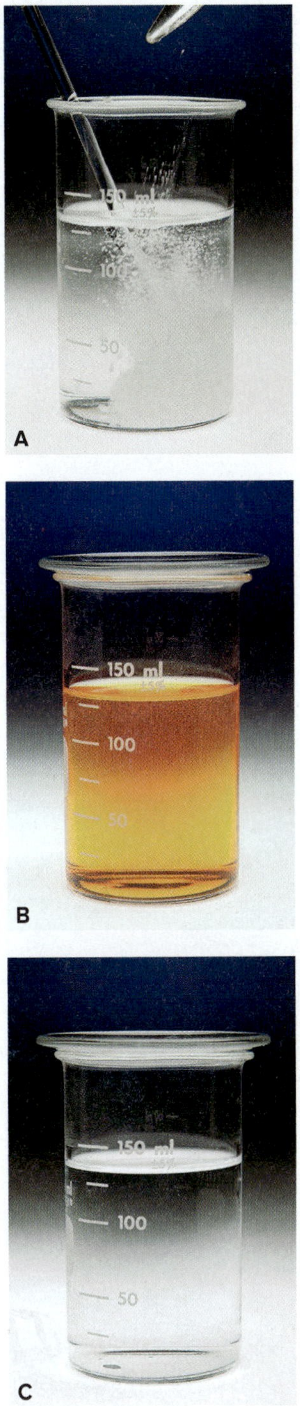

Figure 16.24 The catalyzed decomposition of H_2O_2. **A,** Solid NaBr is dissolved in an aqueous solution of H_2O_2. **B,** Bubbles of O_2 gas form quickly as $Br^-(aq)$ catalyzes the decomposition of H_2O_2. Note the intermediate Br_2 turns the solution orange. **C,** The solution becomes colorless when the Br_2 is consumed in the final step of the mechanism.

Source: © McGraw-Hill Education/Charles Winters/Timeframe Photography, Inc.

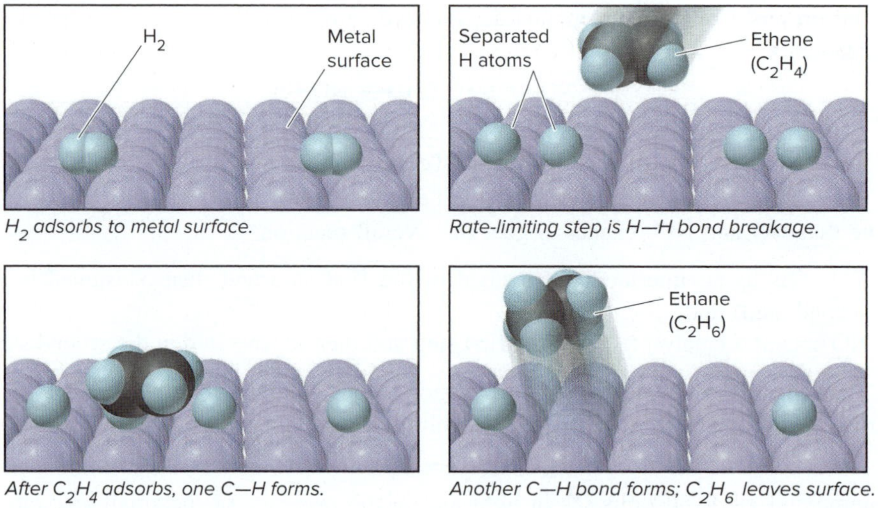
H$_2$
Metal surface
H$_2$ adsorbs to metal surface.
Separated H atoms
Ethene (C_2H_4)
Rate-limiting step is H—H bond breakage.
After C_2H_4 adsorbs, one C—H forms.
Ethane (C_2H_6)
Another C—H bond forms; C_2H_6 leaves surface.

Figure 16.25 The metal-catalyzed hydrogenation of ethene.

NO to N_2. As in the hydrogenation mechanism, the catalyst adsorbs the molecules, which weakens and splits their bonds, thus allowing the atoms to form new bonds more quickly. The process is extremely efficient: for example, an NO molecule is split into catalyst-bound N and O atoms in less than 2×10^{-12} s.

Kinetics and Function of Biological Catalysts

Whereas most industrial chemical reactions occur under extreme conditions with high concentrations, thousands of complex reactions occur in every living cell in dilute solution at ordinary temperatures and pressures. Moreover, the rate of each reaction responds to the rates of other reactions, chemical signals from other cells, and environmental stress. In this marvelous chemical harmony, each rate is controlled by an **enzyme,** a protein catalyst whose function has been perfected through evolution.

Enzymes are typically globular proteins with complex shapes (Section 15.6) and molar masses ranging from about 15,000 to 1,000,000 g/mol. A small part of an enzyme's surface—like a hollow carved into a mountainside—is the **active site,** a region whose shape results from the amino-acid side chains involved in catalyzing the reaction. Often far apart in the sequence of the polypeptide chain, these groups lie near each other because of the chain's three-dimensional folding. When the reactant molecules, called **substrates,** collide effectively with the active site, they become attached through *intermolecular forces,* and the chemical change begins.

Characteristics of Enzyme Action The catalytic behavior of enzymes has several common features:

1. *Catalytic activity.* An enzyme behaves like both types of catalyst. Enzymes are typically *much* larger than the substrate, and many enzymes are embedded within cell membranes. Thus, like a heterogeneous catalyst, an enzyme provides a surface on which one reactant is temporarily immobilized to wait until the other reactant lands nearby. But, like many homogeneous catalysts, the active site groups interact with the substrate(s) in multistep sequences in the presence of solvent and other species.
2. *Catalytic efficiency.* Enzymes are incredibly *efficient* in terms of the number of reactions catalyzed per unit time. For example, for the hydrolysis of urea,

$$(NH_2)_2C{=}O(aq) + 2H_2O(l) + H^+(aq) \longrightarrow 2NH_4^+(aq) + HCO_3^-(aq)$$

the rate constant is 3×10^{-10} s^{-1} for the uncatalyzed reaction in water at room temperature. Under the same conditions in the presence of the enzyme *urease* (pronounced "YUR-ee-ase"), the rate constant is 3×10^{4} s^{-1}, a 10^{14}-fold increase.

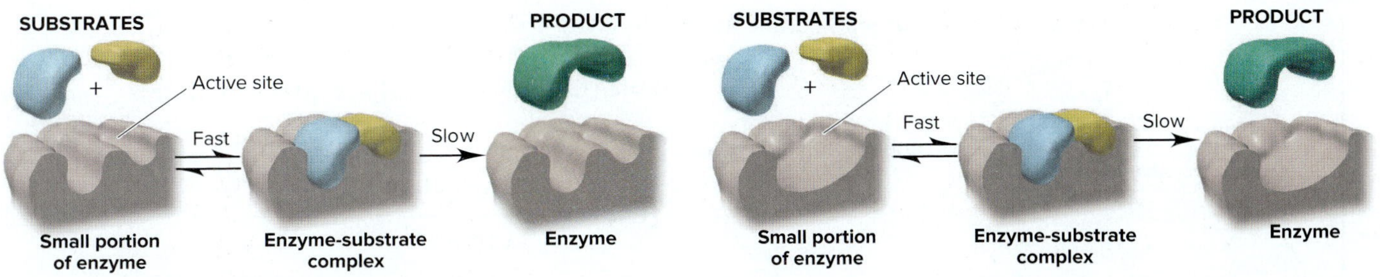

A Lock-and-key model: *fixed shape of active site matches shape of substrate(s).*

B Induced-fit model: *active site changes shape to bind substrate(s) more effectively.*

Figure 16.26 Two models of enzyme action.

Enzymes typically increase rates by 10^8 to 10^{20} times, values that industrial chemists who use catalysts can only dream of!

3. *Catalytic specificity*. As a result of the particular groups at the active site, enzymes are also highly *specific:* each enzyme generally catalyzes only one reaction. Urease catalyzes *only* the hydrolysis of urea, and no other enzyme does this.

Models of Enzyme Action Two models of enzyme action have been proposed.

1. According to the older **lock-and-key model** (Figure 16.26A), when the "key" (substrate) fits the "lock" (active site), the chemical change begins.
2. According to the more current **induced-fit model** (Figure 16.26B), the substrate induces the active site to adopt a perfect fit. X-ray crystallographic and spectroscopic methods show that, in most cases, *the enzyme changes shape when the substrate lands at the active site.* Rather than a rigid key in a lock, we picture a flexible hand in a glove; the "glove" (active site) does not attain its functional shape until the "hand" (substrate) moves into it.

Kinetics of Enzyme Action Understanding the kinetics of enzyme catalysis relies on the steady-state approximation (Section 16.6) in a mechanism proposed by Leonor Michaelis (1875–1949) and Maud Menten (1879–1960) in 1913:

(1) $E + S \rightleftharpoons ES$ [fast, reversible]

(2) $ES \longrightarrow E + P$ [slow; rate-determining]

In the first step, substrate (S) and enzyme (E) form an intermediate **enzyme-substrate complex (ES),** which breaks down to form product (P) and enzyme in the second step. Thus, the rate is proportional to the concentration of ES, that is, of reactant bound to catalyst:

$$\text{Rate} = k_2[\text{ES}]$$

where k_2 is the rate constant of step 2. Let's use the steady-state approximation to eliminate the intermediate concentration, [ES], from this rate law. ES is produced in the forward reaction of step 1 (with rate constant = k_1), consumed in the reverse reaction of step 1 (with rate constant = k_{-1}), and consumed in step 2. From the approximation, the net rate of ES production is zero:

$$\frac{d[\text{ES}]}{dt} = 0 = k_1[\text{E}][\text{S}] - k_{-1}[\text{ES}] - k_2[\text{ES}]$$

The concentration of free enzyme, [E], is difficult to measure, but we can determine it by noting that the initial, measured amount of enzyme, E_0, must equal the sum of E and ES. Therefore, $[\text{E}]_0 = [\text{E}] + [\text{ES}]$, and solving for [E] gives

$$[\text{E}] = [\text{E}]_0 - [\text{ES}]$$

Substituting this concentration into the expression for the net rate of ES production gives

$$\frac{d[\text{ES}]}{dt} = 0 = k_1([\text{E}]_0 - [\text{ES}])[\text{S}] - k_{-1}[\text{ES}] - k_2[\text{ES}]$$

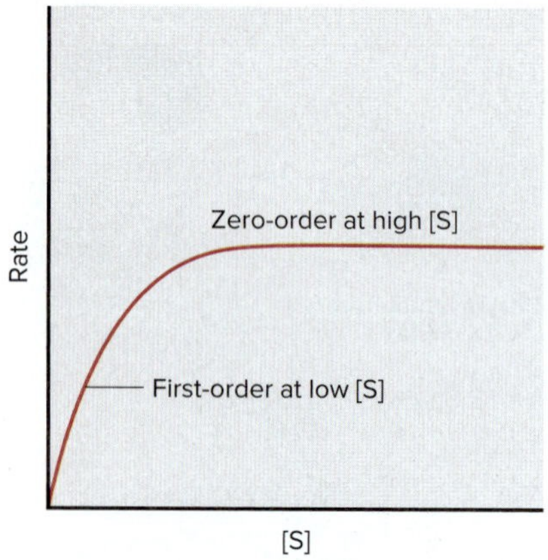

Figure 16.27 Plot of the initial rate of an enzyme-catalyzed reaction versus substrate concentration

Carrying out the arithmetic gives

$$\frac{d[\text{ES}]}{dt} = 0 = k_1[\text{E}]_0[\text{S}] - k_1[\text{ES}][\text{S}] - k_{-1}[\text{ES}] - k_2[\text{ES}]$$

Solving this equation for [ES] gives

$$[\text{ES}] = \frac{k_1[\text{E}]_0[\text{S}]}{k_{-1} + k_2 + k_1[\text{S}]} = \frac{[\text{E}]_0[\text{S}]}{K_\text{M} + [\text{S}]}$$

where $K_\text{M} = \dfrac{k_{-1} + k_2}{k_1}$ and is referred to as the **Michaelis constant.** Substituting the expression for [ES] into the rate law for the formation of product gives the **Michaelis-Menten equation** for enzyme kinetics:

$$\text{Rate} = k_2[\text{ES}] = \frac{k_2[\text{E}]_0[\text{S}]}{K_\text{M} + [\text{S}]} \qquad \textbf{(16.11)}$$

The Michaelis-Menten equation shows that the rate of reaction is always first order in the initial concentration of enzyme, $[\text{E}]_0$. Figure 16.27 shows the relationship between rate and substrate concentration. Two portions of the plot are important:

- When [S] is low, such that $[\text{S}] << K_\text{M}$, Equation 16.11 simplifies to

$$\text{Rate} = \frac{k_2}{K_\text{M}}[\text{E}]_0[\text{S}]$$

  and the rate is first order in [S], as shown by the rising, linear portion of the plot. As more S is added, more ES forms, so more P forms as well.
- When [S] is high, such that $[\text{S}] >> K_\text{M}$, Equation 16.11 simplifies to

$$\text{Rate} = k_2[\text{E}]_0$$

  and the rate becomes independent of substrate concentration; that is, the reaction is zero order in [S]. When all the enzyme molecules are bound to substrate, [ES] is as high as possible, so adding more S has no effect on rate, as shown by the horizontal leveling off of the plot.

By taking the inverse of both sides of Equation 16.11, we get

$$\frac{1}{\text{rate}} = \frac{K_\text{M}}{k_2[\text{E}]_0[\text{S}]} + \frac{1}{k_2[\text{E}]_0}$$

$$y = mx + b$$

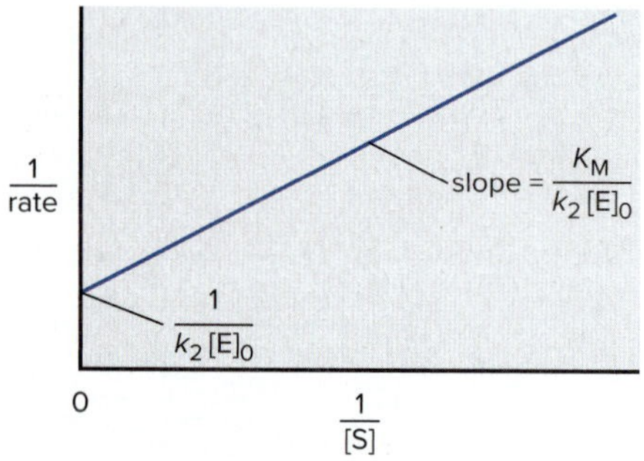

Figure 16.28 Graphical determination of the Michaelis constant.

Figure 16.28 shows that a plot of 1/rate vs. 1/[S] gives a straight line with the slope, $m = \dfrac{K_\text{M}}{k_2[\text{E}]_0}$ and y-intercept, $b = \dfrac{1}{k_2[\text{E}]_0}$, from which we can solve for the Michaelis constant, K_M.

Mechanisms of Enzyme Action Enzymes employ a variety of reaction mechanisms. In some cases, the active site groups bring specific atoms of the substrate close together. In others, the groups bound to the substrate move apart, stretching the bond to be broken. Many *hydrolases* have acidic groups that provide H^+ ions to speed bond cleavage. Two examples are lysozyme, an enzyme found in tears, which hydrolyzes a polysaccharide in bacterial cell walls to protect the eyes from infection, and chymotrypsin, an enzyme found in the small intestine, which hydrolyzes proteins during digestion.

No matter what their mode of action, *all enzymes catalyze by stabilizing the reaction's transition state.* For instance, in the lysozyme-catalyzed reaction, the transition state is a portion of the polysaccharide whose bonds have been twisted and stretched by the active site groups until it fits the active site well. Stabilizing the transition state lowers the activation energy and thus increases the rate.

Like enzyme-catalyzed reactions, the process of stratospheric ozone depletion, discussed in the following Chemical Connections essay, includes both homogeneous and heterogeneous catalysis.

Depletion of Earth's Ozone Layer

The stratospheric ozone layer absorbs UV radiation that is emitted by the Sun and has wavelengths between 280 and 320 nm. If it reaches Earth's suface, this radiation has enough energy to damage genes by breaking bonds in DNA. Thus, depletion of stratospheric ozone increases human health risks, particularly the risks of skin cancer and cataracts (clouding of the eye's lens). The radiation may also damage plant life, especially forms at the bottom of the food chain. Both homogeneous and heterogeneous catalysts play roles in the depletion of ozone.

Before 1976, stratospheric ozone concentration varied seasonally but remained nearly constant from year to year as a result of a series of atmospheric reactions:

$$O_2 \xrightarrow{UV} 2O \quad \text{[dissociation of } O_2 \text{ by UV]}$$
$$O + O_2 \longrightarrow O_3 \quad \text{[ozone formation]}$$
$$O_3 + O \longrightarrow 2O_2 \quad \text{[ozone breakdown]}$$

Then, in 1985, British scientists reported that a severe reduction of ozone, an *ozone hole,* appeared in the stratosphere over the Antarctic when the Sun ended the long winter darkness. Subsequent research by Paul J. Crutzen, Mario J. Molina, and F. Sherwood Rowland, for which they received the Nobel Prize in chemistry in 1995, revealed that industrial chlorofluorocarbons (CFCs) were lowering $[O_3]$ in the stratosphere by catalyzing the reaction in which ozone is broken down.

Widely used as aerosol propellants, foaming agents, and air-conditioning coolants, large quantities of CFCs were released for years. Unreactive in the troposphere, CFC molecules gradually reach the stratosphere, where UV radiation splits them:

$$CF_2Cl_2 \xrightarrow{UV} \cdot CF_2Cl + Cl\cdot$$

(The dots are lone electrons resulting from C—Cl bond cleavage.)

Like many species with a lone electron (free radicals), atomic Cl is very reactive. It reacts with stratospheric O_3 to produce the intermediate chlorine monoxide ($ClO\cdot$), which then reacts with a free O atom to regenerate a Cl atom:

$$O_3 + Cl\cdot \longrightarrow ClO\cdot + O_2$$
$$ClO\cdot + O \longrightarrow Cl\cdot + O_2$$

The sum of these steps is the ozone breakdown reaction:

$$O_3 + \cancel{Cl\cdot} + \cancel{ClO\cdot} + O \longrightarrow \cancel{ClO\cdot} + O_2 + \cancel{Cl\cdot} + O_2$$

or

$$O_3 + O \longrightarrow 2O_2$$

Studies finding high $[ClO\cdot]$ over Antarctica supported this mechanism. Figure B16.1 shows the ozone hole expanding over the years in which CFCs accumulated.

Figure B16.2 shows the reaction energy diagram for the process. Note that *chlorine acts as a homogeneous catalyst:* it exists in the same phase as the reactants, lowers the total activation energy via a different mechanism, and is regenerated. Each Cl atom has a stratospheric half-life of about 2 years, during which it speeds the breakdown of about 100,000 ozone molecules.

Later studies showed that the ozone hole enlarges by heterogeneous catalysis. Stratospheric clouds provide a surface for reactions that convert inactive chlorine compounds, such as HCl and chlorine nitrate ($ClONO_2$), to substances, such as Cl_2, that are cleaved by UV radiation to Cl atoms. Fine particles in the stratosphere act in the same way: dust from the 1991 eruption of Mt. Pinatubo reduced stratospheric ozone for 2 years. The Montreal Protocol of 1987 and later amendments curtailed CFC production and proposed substituting hydrocarbon propellants, such as isobutane, $(CH_3)_3CH$. Nevertheless, because of the long lifetimes of CFCs, complete recovery of the ozone layer may take another century! The good news is that tropospheric halogen levels have begun to fall.

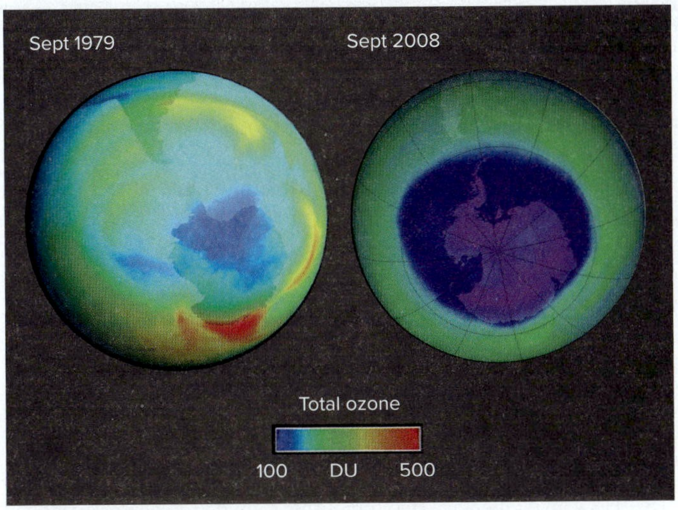

Figure B16.1 Satellite images of the increasing size of the Antarctic ozone hole (*purple*).

Source: NASA image courtesy Ozone Hole Watch

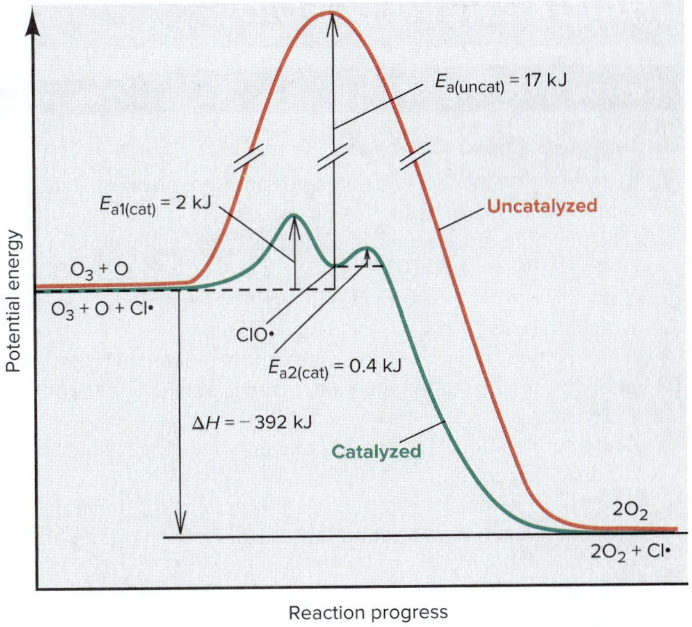

Figure B16.2 Reaction energy diagram for breakdown of O_3 by Cl atoms. (Not drawn to scale.)

(continued)

Problems

B16.1 The catalytic destruction of ozone occurs via a two-step mechanism, where X can be one of several species:

(1) $X + O_3 \longrightarrow XO + O_2$ [slow]

(2) $XO + O \longrightarrow X + O_2$ [fast]

(a) Write the overall reaction and the rate law for each step.

(b) X acts as ______, and XO acts as ______.

B16.2 Aircraft in the stratosphere release NO, which catalyzes ozone breakdown by the mechanism shown in Problem B16.1.

(a) Write the mechanism.

(b) What is the rate of O_3 depletion (k for the rate-determining step is 6×10^{-15} cm^3/molecule·s) when $[O_3]$ is 5×10^{12} molecules/cm^3 and [NO] is 1.0×10^9 molecules/cm^3?

B16.3 In the mechanism for O_3 breakdown in Problem B16.1:

(a) Which would you expect to have a higher value for the orientation probability factor, p, step 1 with NO or step 1 with Cl? Explain.

(b) Draw a possible transition state for the step with Cl.

› Summary of Section 16.7

- A catalyst increases the rate of a reaction without being consumed. It accomplishes this by providing another mechanism that has a lower activation energy.
- Homogeneous catalysts function in the same phase as the reactants. Heterogeneous catalysts act in a different phase from the reactants.
- The hydrogenation of carbon-carbon double bonds takes place on a solid metal catalyst, which speeds up the rate-determining step, the breakage of the H_2 bond.
- Enzymes are protein catalysts of high efficiency and specificity that act by stabilizing the transition state and, thus, lowering the activation energy. The rate of an enzyme-catalyzed reaction is first order in substrate concentration, [S], at low [S] and zero order at high [S].
- Chlorine atoms from CFC molecules speed the breakdown of stratospheric O_3 in a process that is affected by both homogeneous and heterogeneous catalysis.

CHAPTER REVIEW GUIDE

Learning Objectives

Relevant section (§) and/or sample problem (SP) numbers appear in parentheses.

Understand These Concepts

1. How reaction rate depends on concentration, physical state, and temperature (§16.1)
2. The meaning of *reaction rate* in terms of changing concentrations over time (§16.2)
3. How the rate can be expressed in terms of reactant or product concentrations (§16.2)
4. The distinction between average and instantaneous rate and why the instantaneous rate changes during a reaction (§16.2)
5. The interpretation of a reaction rate in terms of reactant and product concentrations (§16.2)
6. The experimental basis of the rate law and the information needed to determine it—initial rate data, reaction orders, and rate constant (§16.3)
7. The importance of reaction order in determining the rate (§16.3)
8. How reaction order is determined from initial rates at different concentrations (§16.3)
9. How integrated rate laws show the dependence of concentration on time (§16.4)
10. What reaction half-life means and why it is constant for a first-order reaction (§16.4)
11. Why concentrations are multiplied in the rate law (§16.5)
12. Activation energy and the effect of temperature on the rate constant (Arrhenius equation) (§16.5)
13. How temperature affects rate by influencing collision energy and, thus, the fraction of collisions with energy exceeding the activation energy (§16.5)
14. Why molecular orientation and complexity influence the number of effective collisions and the rate (§16.5)
15. The transition state as the species that exists only momentarily between reactants and products and whose formation requires the activation energy (§16.5)
16. How an elementary step represents a single molecular event and its molecularity equals the number of colliding particles (§16.6)
17. How a reaction mechanism consists of several elementary steps, with the slowest step determining the overall rate (§16.6)

18. The criteria for a valid reaction mechanism (§16.6)
19. How the steady-state approximation can be used to determine a rate law from a reaction mechanism (16.6)
20. How a catalyst speeds a reaction by lowering the activation energy (§16.7)
21. The distinction between homogeneous and heterogeneous catalysis (§16.7)
22. How the steady-state approximation can be used to understand enzyme kinetics (16.7)

Master These Skills

1. Calculating the instantaneous rate from the slope of a tangent to a concentration vs. time plot (§16.2)
2. Expressing reaction rate in terms of changes in concentration with time (SP 16.1)
3. Determining reaction order from a known rate law (SP 16.2)
4. Determining reaction orders from rate data or molecular scenes (SPs 16.3, 16.4)
5. Calculating the rate constant and its units (§16.3 and SP 16.3)
6. Using an integrated rate law to find concentration at a given time or the time to reach a given concentration (SP 16.5)
7. Determining reaction order graphically with a rearranged integrated rate law (§16.4)
8. Determining the half-life of a reaction (SPs 16.6, 16.7)
9. Using a form of the Arrhenius equation to calculate the activation energy (SP 16.8)
10. Using reaction energy diagrams to depict the energy changes during a reaction (SP 16.9)
11. Predicting a transition state for a simple reaction (SP 16.9)
12. Determining the molecularity and rate law for an elementary step (SP 16.10)
13. Constructing a mechanism with either a slow or a fast initial step (§16.6)
14. Identifying intermediates and correlating rate laws and reaction mechanisms (SP 16.11)

Key Terms

Page numbers appear in parentheses.

activation energy (E_a) (716)
active site (732)
Arrhenius equation (716)
average rate (695)
bimolecular reaction (723)
catalyst (729)
chemical kinetics (691)
collision theory (716)
effective collision (719)
elementary reaction (elementary step) (722)
enzyme (732)
enzyme-substrate complex (ES) (733)
frequency factor (719)
half-life ($t_{1/2}$) (712)
heterogeneous catalyst (731)
homogeneous catalyst (730)
hydrogenation (731)
induced-fit model (733)
initial rate (696)
instantaneous rate (695)
integrated rate law (708)
lock-and-key model (733)
Michaelis constant (734)
Michaelis-Menten equation (734)
molecularity (723)
pseudo-first-order (709)
rate constant (698)
rate-determining (rate-limiting) step (724)
rate law (rate equation) (698)
reaction energy diagram (720)
reaction intermediate (724)
reaction mechanism (722)
reaction orders (699)
reaction rate (691)
steady-state approximation (727)
substrate (732)
transition state (activated complex) (719)
transition state theory (719)
unimolecular reaction (723)

Key Equations and Relationships

Page numbers appear in parentheses.

16.1 Expressing reaction rate in terms of reactant A (694):

$$\text{Rate} = -\frac{\Delta[\text{A}]}{\Delta t}$$

16.2 Expressing the rate of a general reaction (697):

$$\text{Rate} = -\frac{1}{a}\frac{d[\text{A}]}{dt} = -\frac{1}{b}\frac{d[\text{B}]}{dt} = \frac{1}{c}\frac{d[\text{C}]}{dt} = \frac{1}{d}\frac{d[\text{D}]}{dt}$$

16.3 Writing a general rate law (in which products do not appear) (698):

$$\text{Rate} = k[\text{A}]^m[\text{B}]^n \cdots$$

16.4 Calculating the time to reach a given [A] for a first-order reaction (rate = k[A]) (708):

$$\ln\frac{[\text{A}]_0}{[\text{A}]_t} = kt$$

16.5 Calculating the time to reach a given [A] for a simple second-order reaction (rate = $k[\text{A}]^2$) (709):

$$\frac{1}{[\text{A}]_t} - \frac{1}{[\text{A}]_0} = kt$$

16.6 Calculating the time to reach a given [A] for a zero-order reaction (rate = k) (709):

$$[\text{A}]_t - [\text{A}]_0 = -kt$$

16.7 Finding the half-life of a first-order process (712):

$$t_{1/2} = \frac{\ln 2}{k} = \frac{0.693}{k}$$

16.8 Relating the rate constant to the temperature (Arrhenius equation) (716):

$$k = Ae^{-E_a/RT}$$

16.9 Relating the heat of reaction to the forward and reverse activation energies (717):

$$\Delta H_{\text{rxn}} = E_{\text{a(fwd)}} - E_{\text{a(rev)}}$$

16.10 Calculating the activation energy (rearranged form of Arrhenius equation) (718):

$$\ln\frac{k_2}{k_1} = \frac{E_a}{R}\left(\frac{1}{T_1} - \frac{1}{T_2}\right)$$

16.11 Relating the rate of enzyme catalysis to substrate concentration (Michaelis-Menten equation) (734):

$$\text{Rate} = k_2\,[\text{ES}] = \frac{k_2[\text{E}]_0\,[\text{S}]}{K_\text{M} + [\text{S}]}$$

BRIEF SOLUTIONS TO FOLLOW-UP PROBLEMS

16.1A (a) $4NO(g) + O_2(g) \longrightarrow 2N_2O_3(g)$

$$\text{Rate} = -\frac{d[O_2]}{dt} = -\frac{1}{4}\frac{d[NO]}{dt} = \frac{1}{2}\frac{d[N_2O_3]}{dt}$$

(b) $$-\frac{d[O_2]}{dt} = -\frac{1}{4}\frac{d[NO]}{dt} = -\frac{1}{4}(-1.60\times10^{-4}\ \text{mol/L·s})$$
$$= 4.00\times10^{-5}\ \text{mol/L·s}$$

16.1B (a) Since the changes in concentrations of NH_3 and O_2 are negative, those substances are the reactants:

(b) $$-\frac{1}{5}\frac{d[O_2]}{dt} = \frac{1}{6}\frac{d[H_2O]}{dt}$$
$$-\frac{d[O_2]}{dt} = \frac{5}{6}\frac{d[H_2O]}{dt} = \frac{5}{6}(2.52\times10^{-2}\ \text{mol/L·s})$$
$$= 2.10\times10^{-2}\ \text{mol/L·s}$$

16.2A (a) First order in I^-, first order in BrO_3^-, second order in H^+; fourth order overall
(b) Rate $= k[I^-][BrO_3^-][H^+]^2 = k(3)(3)(2)^2 = 36k$; the rate increases by a factor of 36.

16.2B (a) Second order in ClO_2, first order in OH^-; third order overall
(b) Rate $= k[ClO_2]^2[OH^-] = k(0.5)^2(2) = 0.5$; the rate decreases by a factor of ½.

16.3A Rate $= k[H_2]^m[I_2]^n$. From Experiments 1 and 3:

$$\frac{\text{Rate 3}}{\text{Rate 1}} = \frac{k[H_2]_3^m[I_2]_3^n}{k[H_2]_1^m[I_2]_1^n} = \left(\frac{[H_2]_3}{[H_2]_1}\right)^m$$

or $$\frac{9.3\times10^{-23}\ \text{mol/L·s}}{1.9\times10^{-23}\ \text{mol/L·s}} = \left(\frac{0.0550\ \text{mol/L}}{0.0113\ \text{mol/L}}\right)^m$$

This gives $4.9 = (4.9)^m$, so $m = 1$. From Experiments 2 and 4:

$$\frac{\text{Rate 4}}{\text{Rate 2}} = \frac{k[H_2]_4^m[I_2]_4^n}{k[H_2]_2^m[I_2]_2^n} = \left(\frac{[I_2]_4}{[I_2]_2}\right)^n$$

or $$\frac{1.9\times10^{-22}\ \text{mol/L·s}}{1.1\times10^{-22}\ \text{mol/L·s}} = \left(\frac{0.0056\ \text{mol/L}}{0.0033\ \text{mol/L}}\right)^n$$

This gives $1.7 = (1.7)^n$, so $n = 1$. Therefore, rate $= k[H_2][I_2]$; second order overall. From Experiment 1:

$$k = \frac{\text{rate}}{[H_2][I_2]} = \frac{1.9\times10^{-23}\ \text{mol/L·s}}{(0.0113\ \text{mol/L})(0.0011\ \text{mol/L})}$$
$$= 1.5\times10^{-18}\ \text{L/mol·s}$$

The average value of k from all four experiments is 1.5×10^{-18} L/mol·s.

16.3B (a) Rate $= k[H_2SeO_3]^m[I^-]^n[H^+]^p$. From Experiments 1 and 3:

$$\frac{\text{Rate 3}}{\text{Rate 1}} = \left(\frac{[H_2SeO_3]_3}{[H_2SeO_3]_1}\right)^m$$

or $$\frac{3.94\times10^{-6}\ \text{mol/L·s}}{9.85\times10^{-7}\ \text{mol/L·s}} = \left(\frac{1.0\times10^{-2}\ \text{mol/L}}{2.5\times10^{-3}\ \text{mol/L}}\right)^m$$

This gives $4 = (4)^m$, so $m = 1$. From Experiments 1 and 2:

$$\frac{\text{Rate 2}}{\text{Rate 1}} = \left(\frac{[I^-]_2}{[I^-]_1}\right)^n$$

or $$\frac{7.88\times10^{-6}\ \text{mol/L·s}}{9.85\times10^{-7}\ \text{mol/L·s}} = \left(\frac{3.0\times10^{-2}\ \text{mol/L}}{1.5\times10^{-2}\ \text{mol/L}}\right)^n$$

This gives $8 = (2)^n$, so $n = 3$. From Experiments 2 and 4:

$$\frac{\text{Rate 4}}{\text{Rate 2}} = \left(\frac{[H^+]_4}{[H^+]_2}\right)^p$$

or $$\frac{3.15\times10^{-5}\ \text{mol/L·s}}{7.88\times10^{-6}\ \text{mol/L·s}} = \left(\frac{3.0\times10^{-2}\ \text{mol/L}}{1.5\times10^{-2}\ \text{mol/L}}\right)^p$$

This gives $4 = (2)^p$, so $p = 2$. Therefore, rate $= k[H_2SeO_3][I^-]^3[H^+]^2$.
(b) From Experiment 1:

$$k = \frac{\text{rate}}{[H_2SeO_3][I^-]^3[H^+]^2}$$
$$= \frac{9.85\times10^{-7}\ \text{mol/L·s}}{(2.5\times10^{-3}\ \text{mol/L})(1.5\times10^{-2}\ \text{mol/L})^3(1.5\times10^{-2}\ \text{mol/L})^2}$$
$$= 5.2\times10^5\ \text{L}^5/\text{mol}^5\text{·s}$$

16.4A (a) The rate law shows the reaction is zero order in Y, so the rate is not affected by doubling Y. The reaction is second order in X; since the amount of X does not change, the rate does not change either: rate of Expt 2 = rate of Expt 1 = 0.25×10^{-5} mol/L·s
(b) The rate of Expt 3 is four times that of Expt 1, so [X] doubles.

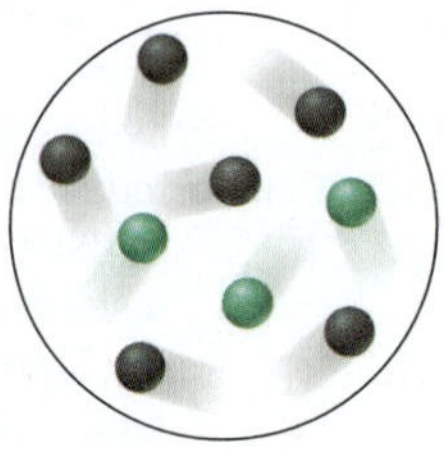

16.4B (a) The rate law is rate $= k[A]^m[B]^n$. For reactant B in Expts 1 and 2:

$$\frac{\text{Rate 2}}{\text{Rate 1}} = \frac{k[A]_2^m[B]_2^n}{k[A]_1^m[B]_1^n} = \left(\frac{[B]_2}{[B]_1}\right)^n$$

or $$\frac{1.6\times10^{-3}\ \text{mol/L·s}}{1.6\times10^{-3}\ \text{mol/L·s}} = \left(\frac{2\ \text{particles}}{4\ \text{particles}}\right)^n$$

Thus, $1 = (0.5)^n$ so $n = 0$. For reactant A in Expts 1 and 3:

$$\frac{\text{Rate 3}}{\text{Rate 1}} = \frac{k[A]_3^m[B]_3^n}{k[A]_1^m[B]_1^n} = \left(\frac{[A]_3}{[A]_1}\right)^m$$

or $$\frac{8.0\times10^{-4}\ \text{mol/L·s}}{1.6\times10^{-3}\ \text{mol/L·s}} = \left(\frac{2\ \text{particles}}{4\ \text{particles}}\right)^m$$

Thus, $0.5 = (0.5)^m$ so $m = 1$. Therefore, rate $= k[A][B]^0$ or rate $= k[A]$.
(b) The rate for Expt 4 is twice as great as the rate for Expt 1; since [A] has a reaction order of 1, twice as many A particles should appear in the scene for Expt 4: 2×4 particles = 8 particles.

16.5A The reaction is second order in HI.

$$\frac{1}{[A]_t} - \frac{1}{[A]_0} = kt$$

$$\frac{1}{0.00900\ \text{mol/L}} - \frac{1}{0.0100\ \text{mol/L}} = (2.4\times10^{-21}\ \text{L/mol·s})(t)$$

$t = 4.6\times10^{21}$ s (or 1.5×10^{14} yr)

16.5B (a) $\ln \frac{[H_2O_2]_0}{[H_2O_2]_t} = kt$

$$\ln \frac{1.28 \text{ mol/L}}{0.85 \text{ mol/L}} = k(10.0 \text{ min})$$

$$k = 0.041/\text{min}$$

(b) If 25% of the H_2O_2 decomposes, 75% remains. Assuming an initial concentration of 1.0 M means that the concentration is 0.75 M at time t:

$$\ln \frac{[H_2O_2]_0}{[H_2O_2]_t} = kt$$

$$\ln \frac{1.0 \text{ mol/L}}{0.75 \text{ mol/L}} = (0.041/\text{min})(t)$$

$$t = 7.0 \text{ min}$$

16.6A (a) After 2.5 min, the number of particles of X has decreased from 12 to 6, so $t_{1/2} =$ 2.5 min. At 5.0 min, which is two half-lives, 3 particles of X will remain and an additional 3 particles of Y will be formed, for a total of 9 particles of Y.

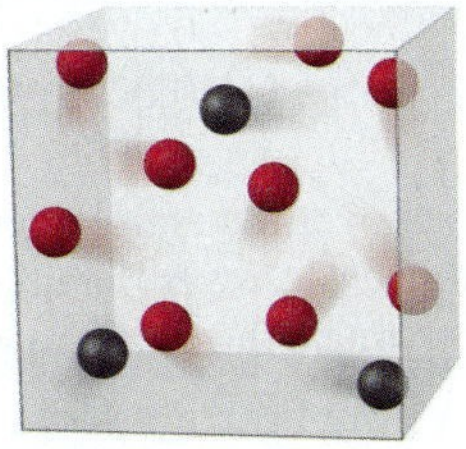

(b) At 10.0 min (four half-lives), there are 0.75 particles of X $(12 \times \frac{1}{2} \times \frac{1}{2} \times \frac{1}{2} \times \frac{1}{2})$.

$$\text{Amount (mol)} = 0.75 \text{ particles} \times \frac{0.20 \text{ mol X}}{1 \text{ particle}} = 0.15 \text{ mol X}$$

$$M = \frac{0.15 \text{ mol X}}{0.50 \text{ L}} = 0.30\ M$$

16.6B (a) The half-life is 24 min; the third scene represents four half-lives (1 particle is $\frac{1}{16}$ of the initial amount; $\frac{1}{2} \times \frac{1}{2} \times \frac{1}{2} \times \frac{1}{2} = \frac{1}{16}$). Four half-lives is 4×24 min = 96 min.

(b) At 72 min (three half-lives), there are 2 particles of A.

$$\text{Amount (mol)} = 2 \text{ particles} \times \frac{0.10 \text{ mol A}}{1 \text{ particle}} = 0.20 \text{ mol A}$$

$$M = \frac{0.20 \text{ mol A}}{0.25 \text{ L}} = 0.80\ M$$

16.7A $k = \frac{\ln 2}{t_{1/2}} = \frac{0.693}{13.1 \text{ h}} = 5.29\times10^{-2} \text{ h}^{-1}$

16.7B (a) $t_{1/2} = \frac{\ln 2}{k} = \frac{0.693}{9\times10^{-2} \text{ day}^{-1}} = 8 \text{ days}$

(b) 40 days is five half-lives; $\frac{1}{2} \times \frac{1}{2} \times \frac{1}{2} \times \frac{1}{2} \times \frac{1}{2} = (\frac{1}{2})^5 = \frac{1}{32}$ remains in the environment.

16.8A $\ln \frac{k_2}{k_1} = \frac{E_a}{R}\left(\frac{1}{T_1} - \frac{1}{T_2}\right)$

$$\ln \frac{0.286 \text{ L/mol·s}}{k_1} = \frac{1.00\times10^5 \text{J/mol}}{8.314 \text{ J/mol·K}} \times \left(\frac{1}{490.\text{ K}} - \frac{1}{500.\text{ K}}\right) = 0.491$$

$$\frac{0.286 \text{ L/mol·s}}{k_1} = e^{0.491} = 1.63$$

$$k_1 = 0.175 \text{ L/mol·s}$$

16.8B If the reaction rate is twice as fast, $k_2 = k_1 \times 2 = 7.02\times10^{-3}$ L/mol·s $\times 2 = 1.40\times10^{-2}$ L/mol·s.

$$\ln \frac{k_2}{k_1} = \frac{E_a}{R}\left(\frac{1}{T_1} - \frac{1}{T_2}\right)$$

$$\ln \frac{1.40\times10^{-2} \text{ L/mol·s}}{7.02\times10^{-3} \text{ L/mol·s}} = \frac{1.14\times10^5 \text{ J/mol}}{8.314 \text{ J/mol·K}} \times \left(\frac{1}{500.\text{ K}} - \frac{1}{T_2}\right)$$

$$0.69029 = 13712 \text{ K} \times \left(0.00200 \text{ K}^{-1} - \frac{1}{T_2}\right)$$

$$5.0342\times10^{-5} \text{ K}^{-1} = \left(0.00200 \text{ K}^{-1} - \frac{1}{T_2}\right)$$

$$0.00195 \text{ K}^{-1} = \frac{1}{T_2}$$

$$T_2 = 513 \text{ K}$$

16.9A $E_{a(rev)} = E_{a(fwd)} - \Delta H_{rxn} = 78 \text{ kJ} - 72 \text{ kJ} = 6 \text{ kJ}$

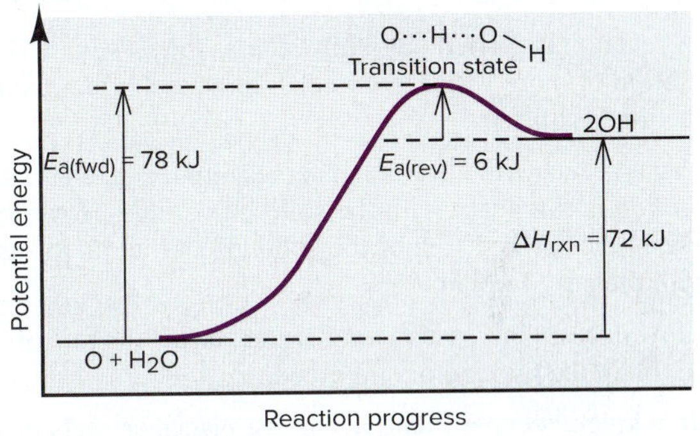

16.9B $\Delta H_{rxn} = E_{a(fwd)} - E_{a(rev)} = 7 \text{ kJ} - 72 \text{ kJ} = -65 \text{ kJ}$

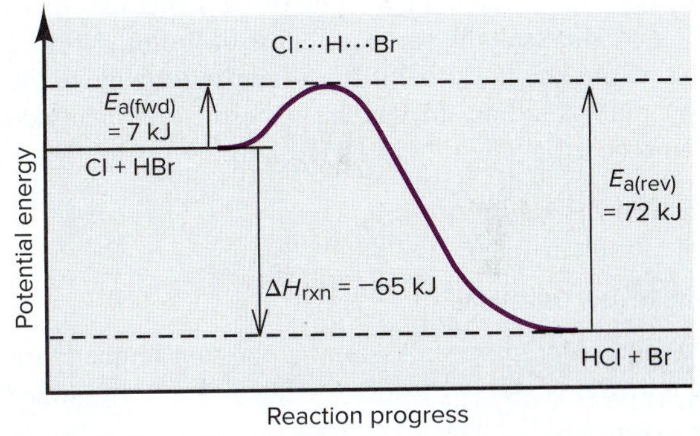

16.10A (a) Balanced equation: $2H_2O_2(aq) \longrightarrow 2H_2O(l) + O_2(g)$

(b) Step 1 is unimolecular; steps 2 and 3 are bimolecular.

(c) $\text{Rate}_1 = k_1[H_2O_2]$; $\text{Rate}_2 = k_2[H_2O_2][OH]$; $\text{Rate}_3 = k_3[HO_2][OH]$

16.10B (a) Balanced equation:

$2NO(g) + 2H_2(g) \longrightarrow N_2(g) + 2H_2O(g)$

(b) All steps are bimolecular.

(c) $\text{Rate}_1 = k_1[NO]^2$; $\text{Rate}_2 = k_2[N_2O_2][H_2]$; $\text{Rate}_3 = k_3[N_2O][H_2]$

16.11A (a) The intermediates are OH and HO_2.

(b) The reaction is first order in H_2O_2; step 1 must be the rate-determining step.

16.11B (a) The intermediates are N_2O_2 and N_2O.

(b) From step 2, rate $= k_2[N_2O_2][H_2]$; from step 1, $k_1[NO]^2 = k_{-1}[N_2O_2]$ or $[N_2O_2] = \frac{k_1}{k_{-1}}[NO]^2$. Substituting into the rate law from step 2 gives

$$\text{rate} = k_2\,[N_2O_2]\,[H_2] = k_2 \frac{k_1}{k_{-1}}[NO]^2\,[H_2], \text{ or rate} = k[NO]^2\,[H_2].$$

PROBLEMS

Problems with **colored** numbers are answered in Appendix E and worked in detail in the Student Solutions Manual. Problem sections match those in the text and give the numbers of relevant sample problems. Most offer Concept Review Questions, Skill-Building Exercises (grouped in pairs covering the same concept), and Problems in Context. The Comprehensive Problems are based on material from any section or previous chapter.

Focusing on Reaction Rate

Concept Review Questions

16.1 What variable of a chemical reaction is measured over time to obtain the reaction rate?

16.2 How does an increase in pressure affect the rate of a gas-phase reaction? Explain.

16.3 A reaction is carried out with water as the solvent. How does the addition of more water to the reaction vessel affect the rate of the reaction? Explain.

16.4 A gas reacts with a solid that is present in large chunks. Then the reaction is run again with the solid pulverized. How does the increase in the surface area of the solid affect the rate of its reaction with the gas? Explain.

16.5 How does an increase in temperature affect the rate of a reaction? Explain the two factors involved.

16.6 In a kinetics experiment, a chemist places crystals of iodine in a closed reaction vessel, introduces a given quantity of H_2 gas, and obtains data to calculate the rate of HI formation. In a second experiment, she uses the same amounts of iodine and hydrogen, but first warms the flask to 130°C, a temperature above the sublimation point of iodine. In which of these experiments does the reaction proceed at a higher rate? Explain.

Expressing the Reaction Rate

(Sample Problem 16.1)

Concept Review Questions

16.7 Define *reaction rate*. Assuming constant temperature and a closed reaction vessel, why does the rate change with time?

16.8 (a) What is the difference between an average rate and an instantaneous rate? (b) What is the difference between an initial rate and an instantaneous rate?

16.9 Give two reasons to measure initial rates in a kinetics study.

16.10 For the reaction $A(g) \longrightarrow B(g)$, sketch two curves on the same set of axes that show
(a) The formation of product as a function of time
(b) The consumption of reactant as a function of time

16.11 For the reaction $C(g) \longrightarrow D(g)$, [C] vs. time is plotted:

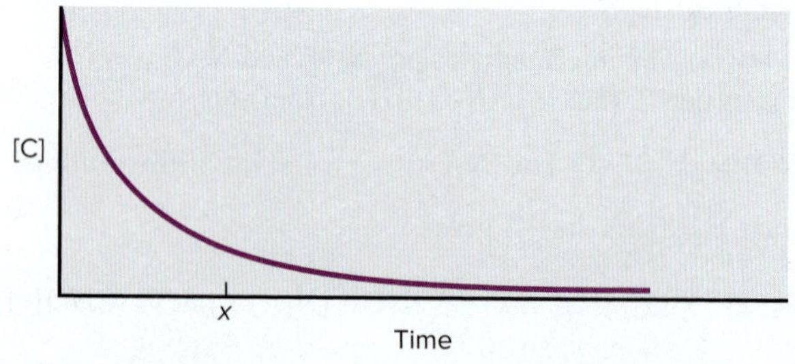

How do you determine each of the following?
(a) The average rate over the entire experiment
(b) The reaction rate at time x
(c) The initial reaction rate
(d) Would the values in parts (a), (b), and (c) be different if you plotted [D] vs. time? Explain.

Skill-Building Exercises (grouped in similar pairs)

16.12 The compound AX_2 decomposes according to the equation $2AX_2(g) \rightarrow 2AX(g) + X_2(g)$. In one experiment, $[AX_2]$ was measured at various times and these data were obtained:

Time (s)	$[AX_2]$ (mol/L)
0.0	0.0500
2.0	0.0448
6.0	0.0300
8.0	0.0249
10.0	0.0209
20.0	0.0088

(a) Find the average rate over the entire experiment.
(b) Is the initial rate higher or lower than the rate in part (a)? Use graphical methods to estimate the initial rate.

16.13 (a) Use the data from Problem 16.12 to calculate the average rate from 8.0 to 20.0 s.
(b) Is the rate at exactly 5.0 s higher or lower than the rate in part (a)? Use graphical methods to estimate the rate at 5.0 s.

16.14 Express the rate of this reaction in terms of the change in concentration of each of the reactants and products:

$$2A(g) \longrightarrow B(g) + C(g)$$

When [C] is increasing at 2 mol/L·s, how fast is [A] decreasing?

16.15 Express the rate of this reaction in terms of the change in concentration of each of the reactants and products:

$$D(g) \longrightarrow \tfrac{3}{2}E(g) + \tfrac{5}{2}F(g)$$

When [E] is increasing at 0.25 mol/L·s, how fast is [F] increasing?

16.16 Express the rate of this reaction in terms of the change in concentration of each of the reactants and products:

$$A(g) + 2B(g) \longrightarrow C(g)$$

When [B] is decreasing at 0.5 mol/L·s, how fast is [A] decreasing?

16.17 Express the rate of this reaction in terms of the change in concentration of each of the reactants and products:

$$2D(g) + 3E(g) + F(g) \longrightarrow 2G(g) + H(g)$$

When [D] is decreasing at 0.1 mol/L·s, how fast is [H] increasing?

16.18 Reaction rate is expressed in terms of changes in concentration of reactants and products. Write a balanced equation for the reaction with this rate expression:

$$\text{Rate} = -\frac{1}{2}\frac{d[N_2O_5]}{dt} = \frac{1}{4}\frac{d[NO_2]}{dt} = \frac{d[O_2]}{dt}$$

16.19 Reaction rate is expressed in terms of changes in concentration of reactants and products. Write a balanced equation for the reaction with this rate expression:

$$\text{Rate} = -\frac{d[CH_4]}{dt} = -\frac{1}{2}\frac{d[O_2]}{dt} = \frac{1}{2}\frac{d[H_2O]}{dt} = \frac{d[CO_2]}{dt}$$

Problems in Context

16.20 The decomposition of NOBr is studied manometrically because the number of moles of gas changes; it cannot be studied colorimetrically because both NOBr and Br_2 are reddish brown:

$$2NOBr(g) \longrightarrow 2NO(g) + Br_2(g)$$

Use the data below to answer the following:
(a) Determine the average rate over the entire experiment.
(b) Determine the average rate between 2.00 and 4.00 s.
(c) Use graphical methods to estimate the initial reaction rate.
(d) Use graphical methods to estimate the rate at 7.00 s.
(e) At what time does the instantaneous rate equal the average rate over the entire experiment?

Time (s)	[NOBr] (mol/L)
0.00	0.0100
2.00	0.0071
4.00	0.0055
6.00	0.0045
8.00	0.0038
10.00	0.0033

16.21 The formation of ammonia is one of the most important processes in the chemical industry:

$$N_2(g) + 3H_2(g) \longrightarrow 2NH_3(g)$$

Express the rate in terms of changes in $[N_2]$, $[H_2]$, and $[NH_3]$.

16.22 Although the depletion of stratospheric ozone threatens life on Earth today, its accumulation was one of the crucial processes that allowed life to develop in prehistoric times:

$$3O_2(g) \longrightarrow 2O_3(g)$$

(a) Express the reaction rate in terms of $[O_2]$ and $[O_3]$.
(b) At a given instant, the reaction rate in terms of $[O_2]$ is 2.17×10^{-5} mol/L·s. What is it in terms of $[O_3]$?

The Rate Law and Its Components

(Sample Problems 16.2 to 16.4)

Concept Review Questions

16.23 The rate law for the general reaction

$$aA + bB + \cdots \longrightarrow cC + dD + \cdots$$

is rate $= k[A]^m[B]^n \cdots$. (a) Explain the meaning of k. (b) Explain the meanings of m and n. Does $m = a$ and $n = b$? Explain. (c) If the reaction is first order in A and second order in B, and time is measured in minutes (min), what are the units for k?

16.24 You are studying the reaction

$$A_2(g) + B_2(g) \longrightarrow 2AB(g)$$

to determine its rate law. Assuming that you have a valid experimental procedure for obtaining $[A_2]$ and $[B_2]$ at various times, explain how you determine (a) the initial rate, (b) the reaction orders, and (c) the rate constant.

16.25 By what factor does the rate change in each of the following cases (assuming constant temperature)?
(a) A reaction is first order in reactant A, and [A] is doubled.
(b) A reaction is second order in reactant B, and [B] is halved.
(c) A reaction is second order in reactant C, and [C] is tripled.

Skill-Building Exercises (grouped in similar pairs)

16.26 Give the individual reaction orders for all substances and the overall reaction order from the following rate law:

$$\text{Rate} = k[BrO_3^-][Br^-][H^+]^2$$

16.27 Give the individual reaction orders for all substances and the overall reaction order from the following rate law:

$$\text{Rate} = k\frac{[O_3]^2}{[O_2]}$$

16.28 By what factor does the rate in Problem 16.26 change if each of the following changes occurs: (a) $[BrO_3^-]$ is doubled; (b) $[Br^-]$ is halved; (c) $[H^+]$ is quadrupled?

16.29 By what factor does the rate in Problem 16.27 change if each of the following changes occurs: (a) $[O_3]$ is doubled; (b) $[O_2]$ is doubled; (c) $[O_2]$ is halved?

16.30 Give the individual reaction orders for all substances and the overall reaction order from this rate law:

$$\text{Rate} = k[NO_2]^2[Cl_2]$$

16.31 Give the individual reaction orders for all substances and the overall reaction order from this rate law:

$$\text{Rate} = k\frac{[HNO_2]^4}{[NO]^2}$$

16.32 By what factor does the rate in Problem 16.30 change if each of the following changes occurs: (a) $[NO_2]$ is tripled; (b) $[NO_2]$ and $[Cl_2]$ are doubled; (c) $[Cl_2]$ is halved?

16.33 By what factor does the rate in Problem 16.31 change if each of the following changes occurs: (a) $[HNO_2]$ is doubled; (b) [NO] is doubled; (c) $[HNO_2]$ is halved?

16.34 For the reaction

$$4A(g) + 3B(g) \longrightarrow 2C(g)$$

the following data were obtained at constant temperature:

Experiment	Initial Rate (mol/L·min)	Initial [A] (mol/L)	Initial [B] (mol/L)
1	5.00	0.100	0.100
2	45.0	0.300	0.100
3	10.0	0.100	0.200
4	90.0	0.300	0.200

(a) What is the order with respect to each reactant? (b) Write the rate law. (c) Calculate k (using the data from Expt 1).

16.35 For the reaction

$$A(g) + B(g) + C(g) \longrightarrow D(g)$$

the following data were obtained at constant temperature:

Expt	Initial Rate (mol/L·s)	Initial [A] (mol/L)	Initial [B] (mol/L)	Initial [C] (mol/L)
1	6.25×10^{-3}	0.0500	0.0500	0.0100
2	1.25×10^{-2}	0.1000	0.0500	0.0100
3	5.00×10^{-2}	0.1000	0.1000	0.0100
4	6.25×10^{-3}	0.0500	0.0500	0.0200

(a) What is the order with respect to each reactant? (b) Write the rate law. (c) Calculate k (using the data from Expt 1).

16.36 Without consulting Table 16.3, give the units of the rate constants for reactions with the following overall orders: (a) first order; (b) second order; (c) third order; (d) $\frac{5}{2}$ order.

16.37 Give the overall reaction order that corresponds to a rate constant with each of the following units: (a) mol/L·s; (b) yr^{-1}; (c) $(mol/L)^{1/2} \cdot s^{-1}$; (d) $(mol/L)^{-5/2} \cdot min^{-1}$.

Problems in Context

16.38 Phosgene is a toxic gas prepared by the reaction of carbon monoxide with chlorine:

$$CO(g) + Cl_2(g) \longrightarrow COCl_2(g)$$

These data were obtained in a kinetics study of its formation:

Experiment	Initial Rate (mol/L·s)	Initial [CO] (mol/L)	Initial [Cl_2] (mol/L)
1	1.29×10^{-29}	1.00	0.100
2	1.33×10^{-30}	0.100	0.100
3	1.30×10^{-29}	0.100	1.00
4	1.32×10^{-31}	0.100	0.0100

(a) Write the rate law for the formation of phosgene.
(b) Calculate the average value of the rate constant.

Integrated Rate Laws: Concentration Changes over Time
(Sample Problems 16.5 to 16.7)

Concept Review Questions

16.39 How are integrated rate laws used to determine reaction order? What is the reaction order in each of these cases?
(a) A plot of the natural logarithm of [reactant] vs. time is linear.
(b) A plot of the inverse of [reactant] vs. time is linear.
(c) [reactant] vs. time is linear.

16.40 Define the *half-life* of a reaction. Explain on the molecular level why the half-life of a first-order reaction is constant.

Skill-Building Exercises (grouped in similar pairs)

16.41 For the simple decomposition reaction

$$AB(g) \longrightarrow A(g) + B(g)$$

rate = $k[AB]^2$ and $k = 0.2$ L/mol·s. How long will it take for [AB] to reach one-third of its initial concentration of 1.50 *M*?

16.42 For the reaction in Problem 16.41, what is [AB] after 10.0 s?

16.43 The first-order rate constant for the reaction A *(green)* → B *(yellow)* is 0.063 min^{-1}. The scenes below represent the reaction mixture at two different times:

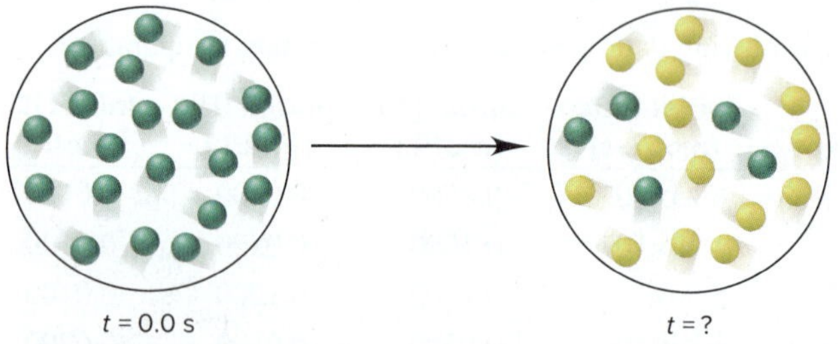

What is the time at which the second scene occurs?

16.44 The molecular scenes below represent the first-order reaction in which cyclopropane *(red)* is converted to propene *(green)*:

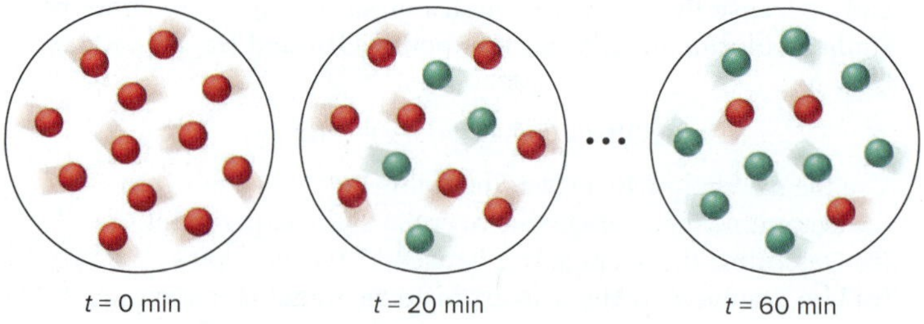

Determine (a) the half-life and (b) the rate constant.

16.45 In a first-order decomposition reaction, 50.0% of a compound decomposes in 10.5 min. (a) What is the rate constant of the reaction? (b) How long does it take for 75.0% of the compound to decompose?

16.46 A decomposition reaction has a rate constant of 0.0012 yr^{-1}. (a) What is the half-life of the reaction? (b) How long does it take for [reactant] to reach 12.5% of its original value?

Problems in Context

16.47 In a study of ammonia production, an industrial chemist discovers that the compound decomposes to its elements N_2 and H_2 in a first-order process. She collects the following data:

Time (s)	0	1.000	2.000
[NH_3](mol/L)	4.000	3.986	3.974

(a) Use graphical methods to determine the rate constant.
(b) What is the half-life for ammonia decomposition?

Theories of Chemical Kinetics
(Sample Problems 16.8 and 16.9)

Concept Review Questions

16.48 What is the central idea of collision theory? How does this model explain the effect of concentration on reaction rate?

16.49 Is collision frequency the only factor affecting rate? Explain.

16.50 Arrhenius proposed that each reaction has an energy threshold that must be reached for the particles to react. The kinetic theory of gases proposes that the average kinetic energy of the particles is proportional to the absolute temperature. How do these concepts relate to the effect of temperature on rate?

16.51 Use the exponential term in the Arrhenius equation to explain how temperature affects reaction rate.

16.52 How is the activation energy determined from the Arrhenius equation?

16.53 (a) Graph the relationship between *k* (vertical axis) and *T* (horizontal axis). (b) Graph the relationship between ln *k* (vertical axis) and 1/*T* (horizontal axis). How is the activation energy determined from this graph?

16.54 (a) For a reaction with a given E_a, how does an increase in *T* affect the rate? (b) For a reaction at a given *T*, how does a decrease in E_a affect the rate?

16.55 In the reaction $AB + CD \rightleftharpoons EF$, 4×10^{-5} mol of AB molecules collide with 4×10^{-5} mol of CD molecules. Will 4×10^{-5} mol of EF form? Explain.

16.56 Assuming the activation energies are equal, which of the following reactions will proceed at a higher rate at 50°C? Explain.

$$NH_3(g) + HCl(g) \longrightarrow NH_4Cl(s)$$
$$N(CH_3)_3(g) + HCl(g) \longrightarrow (CH_3)_3NHCl(s)$$

Skill-Building Exercises (grouped in similar pairs)

16.57 For the reaction $A(g) + B(g) \longrightarrow AB(g)$, how many unique collisions between A and B are possible if there are four particles of A and three particles of B present in the vessel?

16.58 For the reaction $A(g) + B(g) \longrightarrow AB(g)$, how many unique collisions between A and B are possible if 1.01 mol of A(*g*) and 2.12 mol of B(*g*) are present in the vessel?

16.59 At 25°C, what is the fraction of collisions with energy equal to or greater than an activation energy of 100. kJ/mol?

16.60 If the temperature in Problem 16.59 is increased to 50.°C, by what factor does the fraction of collisions with energy equal to or greater than the activation energy change?

16.61 The rate constant of a reaction is 4.7×10^{-3} s^{-1} at 25°C, and the activation energy is 33.6 kJ/mol. What is *k* at 75°C?

16.62 The rate constant of a reaction is 4.50×10^{-5} L/mol·s at 195°C and 3.20×10^{-3} L/mol·s at 258°C. What is the activation energy of the reaction?

16.63 For the reaction $ABC + D \rightleftharpoons AB + CD$, $\Delta H^\circ_{rxn} = -55$ kJ/mol and $E_{a(fwd)} = 215$ kJ/mol. Assuming a one-step reaction, (a) draw a reaction energy diagram; (b) calculate $E_{a(rev)}$; and (c) sketch a possible transition state if ABC is V shaped.

16.64 For the reaction $A_2 + B_2 \longrightarrow 2AB$, $E_{a(fwd)} = 125$ kJ/mol and $E_{a(rev)} = 85$ kJ/mol. Assuming the reaction occurs in one step, (a) draw a reaction energy diagram; (b) calculate ΔH°_{rxn}; and (c) sketch a possible transition state.

Problems in Context

16.65 Understanding the high-temperature formation and breakdown of the nitrogen oxides is essential for controlling the pollutants generated from power plants and cars. The first-order breakdown of dinitrogen monoxide to its elements has rate constants of 0.76/s at 727°C and 0.87/s at 757°C. What is the activation energy of this reaction?

16.66 Aqua regia, a mixture of HCl and HNO_3, has been used since alchemical times to dissolve many metals, including gold. Its orange color is due to the presence of nitrosyl chloride. Consider this one-step reaction for the formation of this compound:

$$NO(g) + Cl_2(g) \longrightarrow NOCl(g) + Cl(g) \qquad \Delta H^\circ = 83 \text{ kJ}$$

(a) Draw a reaction energy diagram, given $E_{a(fwd)} = 86$ kJ/mol.
(b) Calculate $E_{a(rev)}$.
(c) Sketch a possible transition state for the reaction. (*Note:* The atom sequence of nitrosyl chloride is Cl—N—O.)

Reaction Mechanisms: The Steps from Reactant to Product

(Sample Problems 16.10 and 16.11)

Concept Review Questions

16.67 Is the rate of an overall reaction lower, higher, or equal to the average rate of the individual steps? Explain.

16.68 Explain why the coefficients of an elementary step equal the reaction orders of its rate law but those of an overall reaction do not.

16.69 Is it possible for more than one mechanism to be consistent with the rate law of a given reaction? Explain.

16.70 What is the difference between a reaction intermediate and a transition state?

16.71 Why is a bimolecular step more reasonable physically than a termolecular step?

16.72 If a slow step precedes a fast step in a two-step mechanism, do the substances in the fast step appear in the overall rate law? Explain.

16.73 If a fast step precedes a slow step in a two-step mechanism, how is the fast step affected? How is this effect used to determine the validity of the mechanism?

Skill-Building Exercises (grouped in similar pairs)

16.74 The proposed mechanism for a reaction is
(1) $A(g) + B(g) \rightleftharpoons X(g)$ [fast]
(2) $X(g) + C(g) \longrightarrow Y(g)$ [slow]
(3) $Y(g) \longrightarrow D(g)$ [fast]
(a) What is the overall equation?
(b) Identify the intermediate(s), if any.
(c) What are the molecularity and the rate law for each step?
(d) Is the mechanism consistent with the actual rate law: Rate = *k*[A][B][C]?
(e) Is the following one-step mechanism equally valid? $A(g) + B(g) + C(g) \longrightarrow D(g)$?

16.75 Consider the following mechanism:
(1) $ClO^-(aq) + H_2O(l) \rightleftharpoons HClO(aq) + OH^-(aq)$ [fast]
(2) $I^-(aq) + HClO(aq) \longrightarrow HIO(aq) + Cl^-(aq)$ [slow]
(3) $OH^-(aq) + HIO(aq) \longrightarrow H_2O(l) + IO^-(aq)$ [fast]
(a) What is the overall equation?
(b) Identify the intermediate(s), if any.
(c) What are the molecularity and the rate law for each step?
(d) Is the mechanism consistent with the actual rate law: Rate = $k[ClO^-][I^-]$?

Problems in Context

16.76 In a study of nitrosyl halides, a chemist proposes the following mechanism for the synthesis of nitrosyl bromide:

$$NO(g) + Br_2(g) \rightleftharpoons NOBr_2(g) \qquad \text{[fast]}$$
$$NOBr_2(g) + NO(g) \longrightarrow 2NOBr(g) \qquad \text{[slow]}$$

If the rate law is rate = $k[NO]^2[Br_2]$, is the proposed mechanism valid? If so, show that it satisfies the three criteria for validity.

16.77 The rate law for $2NO(g) + O_2(g) \longrightarrow 2NO_2(g)$ is rate = $k[NO]^2[O_2]$. In addition to the mechanism in the text (Section 16.6), the following ones have been proposed:

I	$2NO(g) + O_2(g) \longrightarrow 2NO_2(g)$	
II	$2NO(g) \rightleftharpoons N_2O_2(g)$	[fast]
	$N_2O_2(g) + O_2(g) \longrightarrow 2NO_2(g)$	[slow]
III	$2NO(g) \rightleftharpoons N_2(g) + O_2(g)$	[fast]
	$N_2(g) + 2O_2(g) \longrightarrow 2NO_2(g)$	[slow]

(a) Which of these mechanisms is consistent with the rate law?
(b) Which of these mechanisms is most reasonable? Why?

Catalysis: Speeding Up a Reaction

Concept Review Questions

16.78 Consider the reaction $N_2O(g) \xrightarrow{Au} N_2(g) + \frac{1}{2}O_2(g)$
(a) Does the gold catalyst (Au, above the arrow) act as a homogeneous or a heterogeneous catalyst?
(b) On the same set of axes, sketch the reaction energy diagrams for the catalyzed and the uncatalyzed reaction.

16.79 Does a catalyst increase reaction rate by the same means as a rise in temperature does? Explain.

16.80 In a classroom demonstration, hydrogen gas and oxygen gas are mixed in a balloon. The mixture is stable under normal conditions, but if a spark is applied to it or some powdered metal is added, the mixture explodes. (a) Is the spark acting as a catalyst? Explain. (b) Is the metal acting as a catalyst? Explain.

16.81 A principle of green chemistry is that the energy needs of industrial processes should have minimal environmental impact. How can the use of catalysts lead to "greener" technologies?

16.82 Enzymes are remarkably efficient catalysts that can increase reaction rates by as many as 20 orders of magnitude.
(a) How does an enzyme affect the transition state of a reaction, and how does this effect increase the reaction rate?
(b) What characteristics of enzymes give them this effectiveness as catalysts?

Comprehensive Problems

16.83 Experiments show that each of the following redox reactions is second order overall:

Reaction 1: $NO_2(g) + CO(g) \longrightarrow NO(g) + CO_2(g)$
Reaction 2: $NO(g) + O_3(g) \longrightarrow NO_2(g) + O_2(g)$

(a) When $[NO_2]$ in reaction 1 is doubled, the rate quadruples. Write the rate law for this reaction.
(b) When [NO] in reaction 2 is doubled, the rate doubles. Write the rate law for this reaction.
(c) In each reaction, the initial concentrations of the reactants are equal. For each reaction, what is the ratio of the initial rate to the rate when the reaction is 50% complete?
(d) In reaction 1, the initial $[NO_2]$ is twice the initial [CO]. What is the ratio of the initial rate to the rate at 50% completion?
(e) In reaction 2, the initial [NO] is twice the initial $[O_3]$. What is the ratio of the initial rate to the rate at 50% completion?

16.84 Consider the following reaction energy diagram:

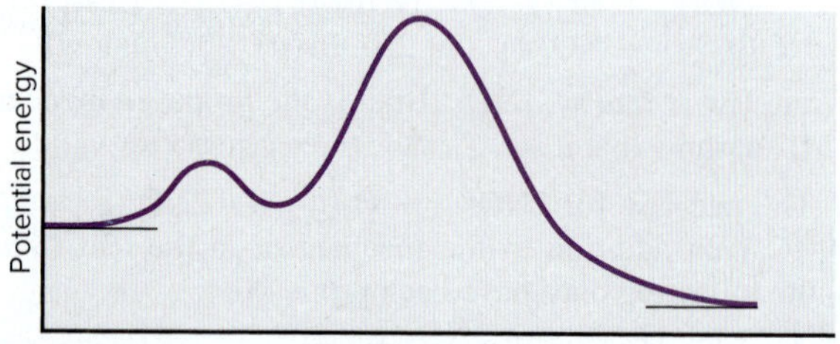

(a) How many elementary steps are in the reaction mechanism?
(b) Which step is rate-limiting?
(c) Is the overall reaction exothermic or endothermic?

16.85 Reactions between certain haloalkanes (alkyl halides) and water produce alcohols. Consider the overall reaction for *t*-butyl bromide (2-bromo-2-methylpropane):

$$(CH_3)_3CBr(aq) + H_2O(l) \longrightarrow (CH_3)_3COH(aq) + H^+(aq) + Br^-(aq)$$

The experimental rate law is rate $= k[(CH_3)_3CBr]$. The accepted mechanism for the reaction is
(1) $(CH_3)_3C{-}Br(aq) \longrightarrow (CH_3)_3C^+(aq) + Br^-(aq)$ [slow]
(2) $(CH_3)_3C^+(aq) + H_2O(l) \longrightarrow (CH_3)_3C{-}OH_2^+(aq)$ [fast]
(3) $(CH_3)_3C{-}OH_2^+(aq) \longrightarrow H^+(aq) + (CH_3)_3C{-}OH(aq)$ [fast]
(a) Why doesn't H_2O appear in the rate law?
(b) Write rate laws for the elementary steps.
(c) What reaction intermediates appear in the mechanism?
(d) Show that the mechanism is consistent with the experimental rate law.

16.86 Archaeologists can determine the age of an artifact made of wood or bone by measuring the amount of the radioactive isotope ^{14}C present in the object. The amount of this isotope decreases in a first-order process. If 15.5% of the original amount of ^{14}C is present in a wooden tool at the time of analysis, what is the age of the tool? The half-life of ^{14}C is 5730 yr.

16.87 A slightly bruised apple will rot extensively in about 4 days at room temperature (20°C). If it is kept in the refrigerator at 0°C, the same extent of rotting takes about 16 days. What is the activation energy for the rotting reaction?

16.88 Benzoyl peroxide, the substance most widely used against acne, has a half-life of $9.8{\times}10^3$ days when refrigerated. How long will it take to lose 5% of its potency (95% remaining)?

16.89 The rate law for the reaction

$$NO_2(g) + CO(g) \longrightarrow NO(g) + CO_2(g)$$

is rate $= k[NO_2]^2$; one possible mechanism is shown in Section 16.6.
(a) Draw a reaction energy diagram for that mechanism, given that $\Delta H^\circ_{overall} = -226$ kJ/mol.
(b) Consider the following alternative mechanism:
(1) $2NO_2(g) \longrightarrow N_2(g) + 2O_2(g)$ [slow]
(2) $2CO(g) + O_2(g) \longrightarrow 2CO_2(g)$ [fast]
(3) $N_2(g) + O_2(g) \longrightarrow 2NO(g)$ [fast]
Is the alternative mechanism consistent with the rate law? Is one mechanism more reasonable physically? Explain.

16.90 Consider the following general reaction and data:

$$2A + 2B + C \longrightarrow D + 3E$$

Expt	Initial Rate (mol/L·s)	Initial [A] (mol/L)	Initial [B] (mol/L)	Initial [C] (mol/L)
1	$6.0{\times}10^{-6}$	0.024	0.085	0.032
2	$9.6{\times}10^{-5}$	0.096	0.085	0.032
3	$1.5{\times}10^{-5}$	0.024	0.034	0.080
4	$1.5{\times}10^{-6}$	0.012	0.170	0.032

(a) What is the reaction order with respect to each reactant?
(b) Calculate the rate constant.
(c) Write the rate law for this reaction.
(d) Express the rate in terms of changes in concentration with time for each of the components.

16.91 In acidic solution, the breakdown of sucrose into glucose and fructose has this rate law: rate $= k[H^+]$[sucrose]. The initial rate of

sucrose breakdown is measured in a solution that is 0.01 M H^+, 1.0 M sucrose, 0.1 M fructose, and 0.1 M glucose. How does the rate change if
(a) [sucrose] is changed to 2.5 M?
(b) [sucrose], [fructose], and [glucose] are all changed to 0.5 M?
(c) $[H^+]$ is changed to 0.0001 M?
(d) [sucrose] and $[H^+]$ are both changed to 0.1 M?

16.92 The citric acid cycle is the central reaction sequence in the cellular metabolism of humans and many other organisms. One of the key steps is catalyzed by the enzyme isocitrate dehydrogenase and the oxidizing agent NAD^+. In yeast, the reaction is eleventh order:

$$\text{Rate} = k[\text{enzyme}]\,[\text{isocitrate}]^4\,[\text{AMP}]^2\,[\text{NAD}^+]^m\,[\text{Mg}^{2+}]^2$$

What is the order with respect to NAD^+?

16.93 The following molecular scenes represent starting mixtures I and II for the reaction of A *(black)* with B *(orange):*

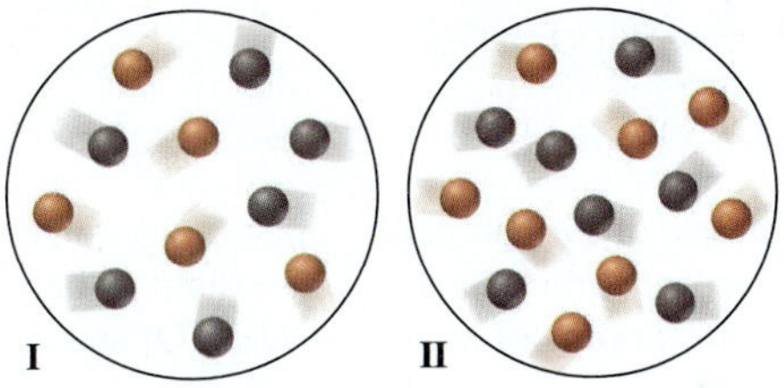

Each sphere represents 0.010 mol, and the volume is 0.50 L. If the reaction is first order in A and first order in B and the initial rate for I is 8.3×10^{-4} mol/L·min, what is the initial rate for II?

16.94 Experiment shows that the rate of formation of carbon tetrachloride from chloroform,

$$CHCl_3(g) + Cl_2(g) \longrightarrow CCl_4(g) + HCl(g)$$

is first order in $CHCl_3$, $\frac{1}{2}$ order in Cl_2, and $\frac{3}{2}$ order overall. Show that the following mechanism is consistent with the rate law:
(1) $Cl_2(g) \rightleftharpoons 2Cl(g)$ [fast]
(2) $Cl(g) + CHCl_3(g) \longrightarrow HCl(g) + CCl_3(g)$ [slow]
(3) $CCl_3(g) + Cl(g) \longrightarrow CCl_4(g)$ [fast]

16.95 A biochemist studying the breakdown of the insecticide DDT finds that it decomposes by a first-order reaction with a half-life of 12 yr. How long does it take DDT in a soil sample to decrease from 275 ppbm to 10. ppbm (parts per billion by mass)?

16.96 Insulin is a polypeptide hormone that is released into the blood from the pancreas and stimulates fat and muscle to take up glucose; the insulin is used up in a first-order process. In a certain patient, this process has a half-life of 3.5 min. To maintain an adequate blood concentration of insulin, it must be replenished in a time interval equal to $1/k$. How long is the time interval for this patient?

16.97 For the reaction $A(g) + B(g) \longrightarrow AB(g)$, the rate is 0.20 mol/L·s, when $[A]_0 = [B]_0 = 1.0$ mol/L. If the reaction is first order in B and second order in A, what is the rate when $[A]_0 = 2.0$ mol/L and $[B]_0 = 3.0$ mol/L?

16.98 The acid-catalyzed hydrolysis of sucrose occurs by the following overall reaction, whose kinetic data are given below:

$$\underset{\text{sucrose}}{C_{12}H_{22}O_{11}(s)} + H_2O(l) \longrightarrow \underset{\text{glucose}}{C_6H_{12}O_6(aq)} + \underset{\text{fructose}}{C_6H_{12}O_6(aq)}$$

[Sucrose] (mol/L)	Time (h)
0.501	0.00
0.451	0.50
0.404	1.00
0.363	1.50
0.267	3.00

(a) Determine the rate constant and the half-life of the reaction.
(b) How long does it take to hydrolyze 75% of the sucrose?
(c) Some studies have shown that this reaction is actually second order overall but appears to follow pseudo–first-order kinetics. Suggest a reason for this apparent first-order behavior.

16.99 At body temperature (37°C), the rate constant of an enzyme-catalyzed decomposition is 2.3×10^{14} times that of the uncatalyzed reaction. If the frequency factor, A, is the same for both processes, by how much does the enzyme lower the E_a?

16.100 Is each of these statements true? If not, explain why.
(a) At a given T, all molecules have the same kinetic energy.
(b) Halving the P for a gaseous reaction doubles the rate.
(c) A higher activation energy gives a lower reaction rate.
(d) A temperature rise of 10°C doubles the rate of any reaction.
(e) If reactant molecules collide with greater energy than the activation energy, they change into product molecules.
(f) The activation energy of a reaction depends on temperature.
(g) The rate of a reaction increases as the reaction proceeds.
(h) Activation energy depends on collision frequency.
(i) A catalyst increases the rate by increasing collision frequency.
(j) Exothermic reactions are faster than endothermic reactions.
(k) Temperature has no effect on the frequency factor (A).
(l) The activation energy of a reaction is lowered by a catalyst.
(m) For most reactions, ΔH_{rxn} is lowered by a catalyst.
(n) The orientation probability factor (p) is near 1 for reactions between single atoms.
(o) The initial rate of a reaction is its maximum rate.
(p) A bimolecular reaction is generally twice as fast as a unimolecular reaction.
(q) The molecularity of an elementary reaction is proportional to the molecular complexity of the reactant(s).

16.101 For the decomposition of gaseous dinitrogen pentoxide,

$$2N_2O_5(g) \longrightarrow 4NO_2(g) + O_2(g)$$

the rate constant is $k = 2.8\times10^{-3}\ s^{-1}$ at 60°C. The initial concentration of N_2O_5 is 1.58 mol/L. (a) What is $[N_2O_5]$ after 5.00 min? (b) What fraction of the N_2O_5 has decomposed after 5.00 min?

16.102 Even when a mechanism is consistent with the rate law, later work may show it to be incorrect. For example, the reaction between hydrogen and iodine has this rate law: rate $= k[H_2][I_2]$. The long-accepted mechanism had a single bimolecular step; that is, the overall reaction was thought to be elementary:

$$H_2(g) + I_2(g) \longrightarrow 2HI(g)$$

In the 1960s, however, spectroscopic evidence showed the presence of free I atoms during the reaction. Kineticists have since proposed a three-step mechanism:
(1) $I_2(g) \rightleftharpoons 2I(g)$ [fast]
(2) $H_2(g) + I(g) \rightleftharpoons H_2I(g)$ [fast]
(3) $H_2I(g) + I(g) \longrightarrow 2HI(g)$ [slow]
Show that this mechanism is consistent with the rate law.

16.103 Suggest an experimental method for measuring the change in concentration with time for each of the following reactions:

(a) $CH_3CH_2Br(l) + H_2O(l) \longrightarrow CH_3CH_2OH(l) + HBr(aq)$

(b) $2NO(g) + Cl_2(g) \longrightarrow 2NOCl(g)$

16.104 An atmospheric chemist fills a container with gaseous N_2O_5 to a pressure of 125 kPa, and the gas decomposes to NO_2 and O_2. What is the partial pressure of NO_2, P_{NO2} (in kPa), when the total pressure is 178 kPa?

16.105 Many drugs decompose in blood by a first-order process. (a) Two tablets of aspirin supply 0.60 g of the active compound. After 30 min, this compound reaches a maximum concentration of 2 mg/100 mL of blood. If the half-life for its breakdown is 90 min, what is its concentration (in mg/100 mL) 2.5 h after it reaches its maximum concentration? (b) For the decomposition of an antibiotic in a person with a normal temperature (98.6°F), $k = 3.1\times10^{-5}\ s^{-1}$; for a person with a fever (temperature of 101.9°F), $k = 3.9\times10^{-5}\ s^{-1}$. If the person with the fever must take another pill when $\frac{2}{3}$ of the first pill has decomposed, how many hours should she wait to take a second pill? A third pill? (Assume that the pill is effective immediately.) (c) Calculate E_a for decomposition of the antibiotic in part (b).

16.106 While developing a catalytic process to make ethylene glycol from synthesis gas ($CO + H_2$), a chemical engineer finds the rate to be fourth order with respect to gas pressure. The uncertainty in the pressure reading is 5%. When the catalyst is modified, the rate increases by 10%. If you were the company patent attorney, would you file for a patent on this catalyst modification? Explain.

16.107 Iodide ion reacts with chloromethane to displace chloride ion in a common organic substitution reaction:

$$I^- + CH_3Cl \longrightarrow CH_3I + Cl^-$$

(a) Draw a wedge-bond structure of chloroform, and indicate the most effective direction of I^- attack.

(b) The analogous reaction with 2-chlorobutane results in a major change in specific rotation as measured by polarimetry. Explain, showing a wedge-bond structure of the product.

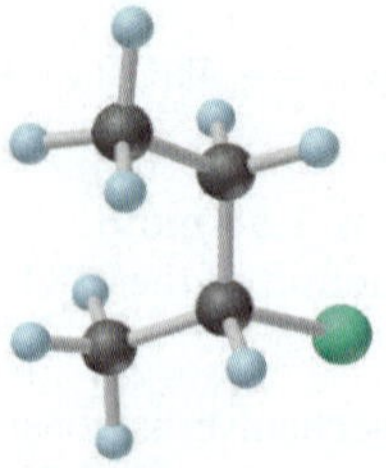

(c) Under different conditions, 2-chlorobutane loses Cl^- in a rate-determining step to form a planar intermediate. This cationic species reacts with HI and then loses H^+ to form a product that exhibits no optical activity. Explain, showing a wedge-bond structure.

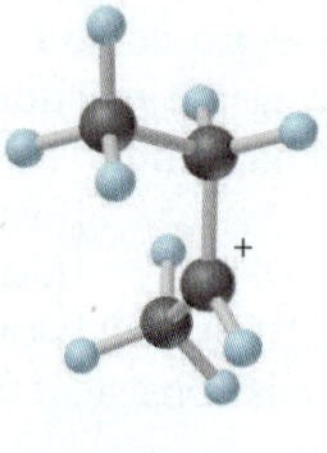

16.108 Assume that water boils at 100.0°C in Houston (near sea level), and at 90.0°C in Cripple Creek, Colorado (near 9500 ft). If it takes 4.8 min to cook an egg in Cripple Creek and 4.5 min in Houston, what is E_a for this process?

16.109 Sulfonation of benzene has the following mechanism:

(1) $2H_2SO_4 \longrightarrow H_3O^+ + HSO_4^- + SO_3$ [fast]
(2) $SO_3 + C_6H_6 \longrightarrow H(C_6H_5^+)SO_3^-$ [slow]
(3) $H(C_6H_5^+)SO_3^- + HSO_4^- \longrightarrow C_6H_5SO_3^- + H_2SO_4$ [fast]
(4) $C_6H_5SO_3^- + H_3O^+ \longrightarrow C_6H_5SO_3H + H_2O$ [fast]

(a) Write an overall equation for the reaction. (b) Write the overall rate law in terms of the initial rate of the reaction.

16.110 In the lower troposphere, ozone is one of the components of photochemical smog. It is generated in air when nitrogen dioxide, formed by the oxidation of nitrogen monoxide from car exhaust, reacts by the following mechanism:

(1) $NO_2(g) \xrightarrow[h\nu]{k_1} NO(g) + O(g)$

(2) $O(g) + O_2(g) \xrightarrow[h\nu]{k_2} O_3(g)$

Assuming the rate of formation of atomic oxygen in step 1 equals the rate of its consumption in step 2, use the data below to calculate (a) the concentration of atomic oxygen [O] and (b) the rate of ozone formation.

$k_1 = 6.0\times10^{-3}\ s^{-1}$ $[NO_2] = 4.0\times10^{-9}\ M$
$k_2 = 1.0\times10^{6}$ L/mol·s $[O_2] = 1.0\times10^{-2}\ M$

16.111 Chlorine is commonly used to disinfect drinking water, and inactivation of pathogens by chlorine follows first-order kinetics. The following data are for *E. coli* inactivation:

Contact Time (min)	Percent (%) Inactivation
0.00	0.0
0.50	68.3
1.00	90.0
1.50	96.8
2.00	99.0
2.50	99.7
3.00	99.9

(a) Determine the first-order inactivation constant, *k*. [*Hint:* % inactivation = $100 \times (1 - [A]_t/[A]_0)$.]
(b) How much contact time is required for 95% inactivation?

16.112 The overall equation and rate law for the gas-phase decomposition of dinitrogen pentoxide are

$$2N_2O_5(g) \longrightarrow 4NO_2(g) + O_2(g) \qquad \text{rate} = k[N_2O_5]$$

Which of the following can be considered valid mechanisms for the reaction?

I One-step collision
II $2N_2O_5(g) \longrightarrow 2NO_3(g) + 2NO_2(g)$ [slow]
$2NO_3(g) \longrightarrow 2NO_2(g) + 2O(g)$ [fast]
$2O(g) \longrightarrow O_2(g)$ [fast]
III $N_2O_5(g) \rightleftharpoons NO_3(g) + NO_2(g)$ [fast]
$NO_2(g) + N_2O_5(g) \longrightarrow 3NO_2(g) + O(g)$ [slow]
$NO_3(g) + O(g) \longrightarrow NO_2(g) + O_2(g)$ [fast]

IV $2N_2O_5(g) \rightleftharpoons 2NO_2(g) + N_2O_3(g) + 3O(g)$ [fast]
$N_2O_3(g) + O(g) \longrightarrow 2NO_2(g)$ [slow]
$2O(g) \longrightarrow O_2(g)$ [fast]
V $2N_2O_5(g) \longrightarrow N_4O_{10}(g)$ [slow]
$N_4O_{10}(g) \longrightarrow 4NO_2(g) + O_2(g)$ [fast]

16.113 Nitrification is a biological process for removing NH_3 from wastewater as NH_4^+:

$$NH_4^+ + 2O_2 \longrightarrow NO_3^- + 2H^+ + H_2O$$

The first-order rate constant is given as

$$k_1 = 0.47e^{0.095(T - 15°C)}$$

where k_1 is in day^{-1} and T is in °C.
(a) If the initial concentration of NH_3 is 3.0 mol/m^3, how long will it take to reduce the concentration to 0.35 mol/m^3 in the spring (T = 20°C)?
(b) In the winter (T = 10°C)?
(c) Using your answer to part (a), what is the rate of O_2 consumption?

16.114 Carbon disulfide, a poisonous, flammable liquid, is an excellent solvent for phosphorus, sulfur, and some other nonmetals. A kinetic study of its gaseous decomposition gave these data:

Experiment	Initial Rate (mol/L·s)	Initial [CS_2] (mol/L)
1	12.7×10^{-7}	0.100
2	2.2×10^{-7}	0.080
3	1.5×10^{-7}	0.055
4	1.2×10^{-7}	0.044

(a) Write the rate law for the decomposition of CS_2.
(b) Calculate the average value of the rate constant.

16.115 Like any catalyst, palladium, platinum, or nickel catalyzes both directions of a reaction: addition of hydrogen to (hydrogenation) and its elimination from (dehydrogenation) carbon double bonds.
(a) Which variable determines whether an alkene will be hydrogenated or dehydrogenated?
(b) Which reaction requires a higher temperature?
(c) How can all-*trans* fats arise during hydrogenation of fats that contain some double bonds with a *cis* orientation?

16.116 In a *clock reaction,* a dramatic color change occurs at a time determined by concentration and temperature. Consider the iodine clock reaction, whose overall equation is

$$2I^-(aq) + S_2O_8^{2-}(aq) \longrightarrow I_2(aq) + 2SO_4^{2-}(aq)$$

As I_2 forms, it is immediately consumed by its reaction with a fixed amount of added $S_2O_3^{2-}$:

$$I_2(aq) + 2S_2O_3^{2-}(aq) \longrightarrow 2I^-(aq) + S_4O_6^{2-}(aq)$$

Once the $S_2O_3^{2-}$ is consumed, the excess I_2 forms a blue-black product with starch present in solution:

$$I_2 + \text{starch} \longrightarrow \text{starch}{\cdot}I_2 \text{ (blue-black)}$$

The rate of the reaction is also influenced by the total concentration of ions, so KCl and $(NH_4)_2SO_4$ are added to maintain a constant value. Use the data below, obtained at 23°C, to determine:
(a) The average rate for each trial
(b) The order with respect to each reactant
(c) The rate constant
(d) The rate law for the overall reaction

	Expt 1	Expt 2	Expt 3
0.200 *M* KI (mL)	10.0	20.0	20.0
0.100 *M* $Na_2S_2O_8$ (mL)	20.0	20.0	10.0
0.0050 *M* $Na_2S_2O_3$ (mL)	10.0	10.0	10.0
0.200 *M* KCl (mL)	10.0	0.0	0.0
0.100 *M* $(NH_4)_2SO_4$ (mL)	0.0	0.0	10.0
Time to color (s)	29.0	14.5	14.5

16.117 Heat transfer to and from a reaction flask is often a critical factor in controlling reaction rate. The heat transferred (q) depends on a heat transfer coefficient (h) for the flask material, the temperature difference (ΔT) across the flask wall, and the commonly "wetted" area (A) of the flask and bath: $q = hA\Delta T$. When an exothermic reaction is run at a given T, there is a bath temperature at which the reaction can no longer be controlled, and the reaction "runs away" suddenly. A similar problem is often seen when a reaction is "scaled up" from, say, a half-filled small flask to a half-filled large flask. Explain these behaviors.

16.118 The growth of *Pseudomonas* bacteria is modeled as a first-order process with $k = 0.035$ min^{-1} at 37°C. The initial *Pseudomonas* population density is 1.0×10^3 cells/L. (a) What is the population density after 2 h? (b) What is the time required for the population to go from 1.0×10^3 to 2.0×10^3 cells/L?

16.119 Consider the following organic reaction, in which one halogen replaces another in a haloalkane:

$$CH_3CH_2Br + KI \longrightarrow CH_3CH_2I + KBr$$

In acetone, this particular reaction goes to completion because KI is soluble in acetone but KBr is not. In the mechanism, I^- approaches the carbon *opposite* to the Br (see Figure 16.20, with I^- instead of OH^-). After Br^- has been replaced by I^- and precipitates as KBr, other I^- ions react with the ethyl iodide by the same mechanism.
(a) If we designate the carbon bonded to the halogen as C-1, what is the shape around C-1 and the hybridization of C-1 in ethyl iodide?
(b) In the transition state, one of the two lobes of the unhybridized $2p$ orbital of C-1 overlaps a p orbital of I, while the other lobe overlaps a p orbital of Br. What is the shape around C-1 and the hybridization of this carbon in the transition state?
(c) The deuterated reactant, CH_3CHDBr (where D is deuterium, ^2H), has two optical isomers because C-1 is chiral. If the reaction is run with one of the isomers, the ethyl iodide is *not* optically active. Explain.

16.120 Another radioisotope of iodine, ^{131}I, is also used to study thyroid function (see Follow-up Problem 16.7A). A patient is given a sample that is 1.7×10^{-4} *M* ^{131}I. If the half-life is 8.04 days, what fraction of the radioactivity remains after 30. days?

16.121 The effect of substrate concentration on the first-order growth rate of a microbial population follows the Monod equation:

$$\mu = \frac{\mu_{max}S}{K_s + S}$$

where μ is the first-order growth rate (s^{-1}), μ_{max} is the maximum growth rate (s^{-1}), S is the substrate concentration (kg/m^3), and K_s

is the value of S that gives one-half of the maximum growth rate (in kg/m^3). For $\mu_{max} = 1.5\times10^{-4}\ s^{-1}$ and $K_s = 0.03\ kg/m^3$:

(a) Plot μ vs. S for S between 0.0 and 1.0 kg/m^3.
(b) The initial population density is 5.0×10^3 cells/m^3. What is the density after 1.0 h, if the initial S is 0.30 kg/m^3?
(c) What is it if the initial S is 0.70 kg/m^3?

16.122 The scenes depict four initial reaction mixtures for the reaction of A *(blue)* and B *(yellow)*, with and without a solid present *(gray cubes)*. The initial rate, $-d[A]/dt$ (in mol/L·s), is shown, with each sphere representing 0.010 mol and the container volume at 0.50 L.

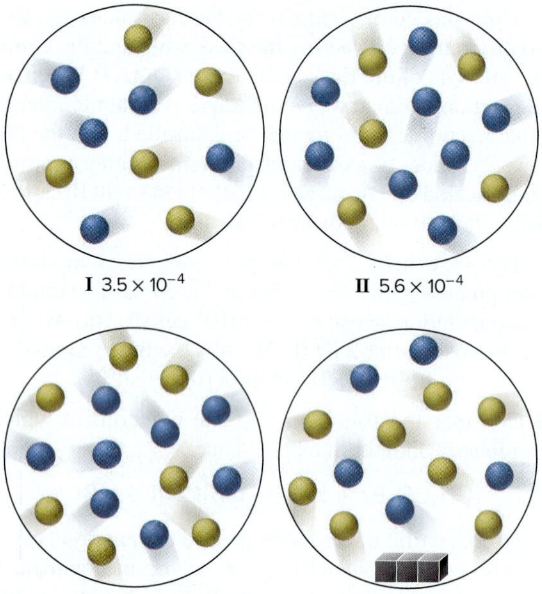

I 3.5×10^{-4} II 5.6×10^{-4} III 5.6×10^{-4} IV 4.9×10^{-4}

(a) What is the rate law in the absence of a catalyst?
(b) What is the overall reaction order?
(c) Find the rate constant.
(d) Do the gray cubes have a catalytic effect? Explain.

16.123 The mathematics of the first-order rate law can be applied to any situation in which a quantity decreases by a constant fraction per unit of time (or unit of any other variable).
(a) As light moves through a solution, its intensity decreases per unit distance traveled in the solution. Show that

$$\ln\left(\frac{\text{intensity of light leaving the solution}}{\text{intensity of light entering the solution}}\right) = -\text{fraction of light removed per unit of length} \times \text{distance traveled in solution}$$

(b) The value of your savings declines under conditions of constant inflation. Show that

$$\ln\left(\frac{\text{value remaining}}{\text{initial value}}\right) = -\text{fraction lost per unit of time} \times \text{savings time interval}$$

16.124 Figure 16.25 shows key steps in the metal-catalyzed (M) hydrogenation of ethene:

$$C_2H_4(g) + H_2(g) \longrightarrow C_2H_6(g)$$

Use the following symbols to write a mechanism that gives the overall equation:

$H_2(ads)$	adsorbed hydrogen molecules
M—H	hydrogen atoms bonded to metal atoms
$C_2H_4(ads)$	adsorbed ethene molecules
$C_2H_5(ads)$	adsorbed ethyl radicals

16.125 Human liver enzymes catalyze the degradation of ingested toxins. By what factor is the rate of a detoxification changed if an enzyme lowers the E_a by 5 kJ/mol at 37°C?

16.126 Acetone is one of the most important solvents in organic chemistry, used to dissolve everything from fats and waxes to airplane glue and nail polish. At high temperatures, it decomposes in a first-order process to methane and ketene ($CH_2{=}C{=}O$). At 600°C, the rate constant is $8.7\times10^{-3}\ s^{-1}$.
(a) What is the half-life of the reaction?
(b) How long does it take for 40.% of a sample of acetone to decompose?
(c) How long does it take for 90.% of a sample of acetone to decompose?

16.127 A *(green)*, B *(blue)*, and C *(red)* are structural isomers. The molecular filmstrip depicts them undergoing a chemical change as time proceeds.

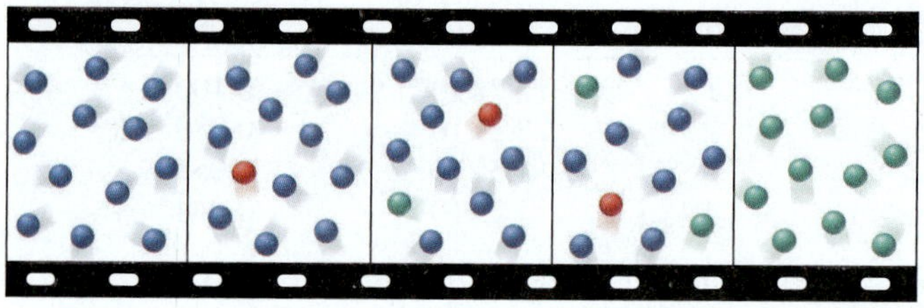

1 $t = 0$ 2 3 4 5 $t = \infty$

(a) Write a mechanism for the reaction.
(b) What role does C play?

16.128 The reaction $A + B \longrightarrow C$ is first order in A, first order in B, and second order overall with a rate constant of $0.0350\ M^{-1}s^{-1}$. What is the concentration of C after 1.5 min if $[A]_0 = 2.00\ M$ and $[B]_0 = 1.00\ M$?

16.129 The rate law for the reaction $HOCl + NH_3 \longrightarrow NH_2Cl + H_2O$ is rate $= k[HOCl][NH_3]$ with $k = 5.1\times10^6$ L/mol·s at 25°C. The reaction becomes pseudo–first-order in NH_3 by using a large excess of HOCl. How long does it take for 30% of the NH_3 to react if the initial concentration of HOCl is $2\times10^{-3}\ M$?

16.130 In the lower atmosphere, one of the mechanisms proposed for the decomposition of ozone to produce oxygen, $2O_3 \longrightarrow 3O_2$, is
(1) $O_3 \rightleftharpoons O_2 + O$
(2) $O + O_3 \longrightarrow 2O_2$
(a) Use the steady-state approximation to write the rate law, assuming that step 2 is the rate-determining step.
(b) The experimentally determined rate law is rate $= k\dfrac{[O_3]^2}{[O_2]}$. How can the rate law derived in (a) be rewritten to explain the known rate law?

16.131 Consider the following mechanism for the decomposition of NO_2Cl to NO_2 and Cl_2:
(1) $NO_2Cl \rightleftharpoons NO_2 + Cl$
(2) $NO_2Cl + Cl \longrightarrow NO_2 + Cl_2$

(a) Use the steady-state approximation to express the rate of Cl_2 production.
(b) How is the rate law simplified if step 1 is the rate-determining step?
(c) How is the rate law simplified if step 2 is the rate-determining step?

16.132 The rate equation for an enzyme-catalyzed reaction has the following constants: $K_M = 3\times10^{-2}\ M$ and $k_2 = 4\times10^4\ s^{-1}$.
(a) Calculate the maximum rate when the enzyme concentration is 3×10^{-6} mol/L.
(b) What is the substrate concentration if the rate is 25% of that calculated in (a)?

16.133 Urea, $(NH_2)_2CO$, is used as a fertilizer because many soil bacteria contain the enzyme urease, which catalyzes the hydrolysis of urea into carbon dioxide and ammonia. Under a specific set of conditions, the Michaelis constant, K_M, for urease is $2\times10^{-3}\ M$ and $k_2 = 2.5\times10^4\ s^{-1}$. (a) Calculate the initial rate when [urea] = $2.5\times10^{-5}\ M$ and [urease] = $4.0\times10^{-6}\ M$. (b) What is the maximum rate of reaction?

17 Equilibrium: The Extent of Chemical Reactions

17.1 The Equilibrium State and the Equilibrium Constant

17.2 The Reaction Quotient and the Equilibrium Constant
- The Changing Value of the Reaction Quotient
- Writing the Reaction Quotient (Q)

17.3 Expressing Equilibria with Pressure Terms: Relation Between K_c and K_p

17.4 Comparing Q and K to Determine Reaction Direction

17.5 How to Solve Equilibrium Problems
- Using Quantities to Find K
- Using K to Find Quantities
- Problems Involving Mixtures of Reactants and Products

17.6 Reaction Conditions and Equilibrium: Le Châtelier's Principle
- Change in Concentration
- Change in Pressure (Volume)
- Change in Temperature
- Lack of Effect of a Catalyst
- Synthesis of Ammonia

Source: © hxdbzxy/Shutterstock.com

Concepts and Skills to Review Before You Study This Chapter

- reversibility of reactions (Section 4.7)
- equilibrium vapor pressure (Section 12.2)
- equilibrium nature of a saturated solution (Section 13.4)
- dependence of rate on concentration (Sections 16.2 and 16.5)
- rate laws for elementary reactions (Section 16.6)
- function of a catalyst (Section 16.7)

Just as reactions vary greatly in their speed, they also vary greatly in their extent. Indeed, kinetics and equilibrium apply to different aspects of a reaction:

- Kinetics applies to the *speed* (or rate) of a reaction, the concentration of reactant that disappears (or of product that appears) per unit time.
- Equilibrium applies to the *extent* of a reaction, the concentrations of reactant and product present after an unlimited time, that is, when no further change is occurring.

As you'll see, at equilibrium, no further *net* change occurs because the forward and reverse reactions reach a *balance*. Like traffic going back and forth across a bridge or people going up and down an escalator, the change in one direction is balanced by the change in the other. A fast reaction may go almost completely, just partially, or only slightly toward products before this balance is reached. Consider acid dissociation in water. In 1 M HCl, virtually all the hydrogen chloride molecules are dissociated into ions, whereas fewer than 1% of the acetic acid molecules are dissociated in 1 M CH_3COOH. Yet both reactions take less than a second. Similarly, some slow reactions yield a large amount of product, whereas others yield very little. After a few years, a steel water tank will start to rust, and it will do so completely given enough time. But, no matter how long you wait, the water inside the tank will not decompose to hydrogen and oxygen.

Knowing the extent of a given reaction is crucial. How much product—medicine, polymer, or fuel—can you obtain from a particular reaction mixture? How can you adjust conditions to obtain more? If a reaction is slow but eventually has a good yield, will a catalyst speed it up enough to make it useful?

IN THIS CHAPTER . . . *We consider the principles of equilibrium in systems of gases and/or pure liquids and solids.*

- We examine the equilibrium state at the macroscopic and molecular levels and see that equilibrium occurs when the forward and reverse reaction rates are equal.
- We contrast the reaction quotient, Q, which changes as the reaction proceeds, with the equilibrium constant, K, which applies to the reaction at equilibrium.
- We express the equilibrium condition in terms of concentrations or pressures and see how the two expressions are related.
- We compare values of Q and K to determine the direction in which a system must proceed to reach equilibrium.
- We develop a systematic approach to solving a variety of equilibrium problems.
- We explore Le Châtelier's principle, which explains how a change in conditions—in concentration, pressure, or temperature or by addition of a catalyst—affects the equilibrium state, and we apply this principle to a major industrial process.
- We introduce equilibrium concepts operating in metabolic pathways.

17.1 THE EQUILIBRIUM STATE AND THE EQUILIBRIUM CONSTANT

Experimental results from countless reactions have shown that, *given sufficient time, the concentrations of reactants and products attain certain values that no longer change.* This apparent cessation of chemical change occurs because *all reactions are reversible and reach a state of equilibrium.* Let's examine a chemical system at the macroscopic and molecular levels to see how equilibrium arises and then consider some quantitative aspects of the process.

1. *A macroscopic view of equilibrium.* The system we'll consider is the reversible gaseous reaction between colorless dinitrogen tetroxide and brown nitrogen dioxide:

$$N_2O_4(g;\ \text{colorless}) \rightleftharpoons 2NO_2(g;\ \text{brown})$$

As soon as we introduce some liquid N_2O_4 (bp = 21°C) into a sealed container kept at 200°C (473 K), it vaporizes, and the gas begins to turn pale brown as NO_2 is produced. As time passes, the brown darkens, until, after more time, the color stops changing. The first three photos in Figure 17.1 show the color change, and the last photo shows no further change.

2. *A molecular view of equilibrium.* On the molecular level, as shown in the blow-up circles of Figure 17.1, a dynamic scene unfolds:

- The N_2O_4 molecules fly wildly throughout the container, a few splitting into two NO_2 molecules.
- As time passes, more N_2O_4 molecules decompose and the concentration of NO_2 rises.
- As the number of N_2O_4 molecules decreases, N_2O_4 decomposition slows down (recall that reaction rate decreases with decreasing reactant concentration for most reactions).
- At the same time, as the number of NO_2 molecules increases, more collide, and formation of N_2O_4 (the reverse reaction) speeds up.
- Eventually, the system reaches equilibrium: N_2O_4 molecules are decomposing into NO_2 molecules just as fast as NO_2 molecules are forming N_2O_4.

Thus, at equilibrium, *reactant and product concentrations are constant because a change in one direction is balanced by a change in the other as the forward and reverse rates become equal:*

$$\text{At equilibrium: rate}_{\text{fwd}} = \text{rate}_{\text{rev}} \qquad \textbf{(17.1)}$$

3. *A quantitiative view of equilibrium: a constant ratio of constants.* Let's see how reactant and product concentrations affect this process. At a particular temperature, when the system reaches equilibrium, we have

$$\text{rate}_{\text{fwd}} = \text{rate}_{\text{rev}}$$

In this reaction system, both forward and reverse reactions are elementary steps (Section 16.6), so we can write their rate laws directly from the balanced equation:

$$k_{\text{fwd}}[N_2O_4]_{\text{eq}} = k_{\text{rev}}[NO_2]^2_{\text{eq}}$$

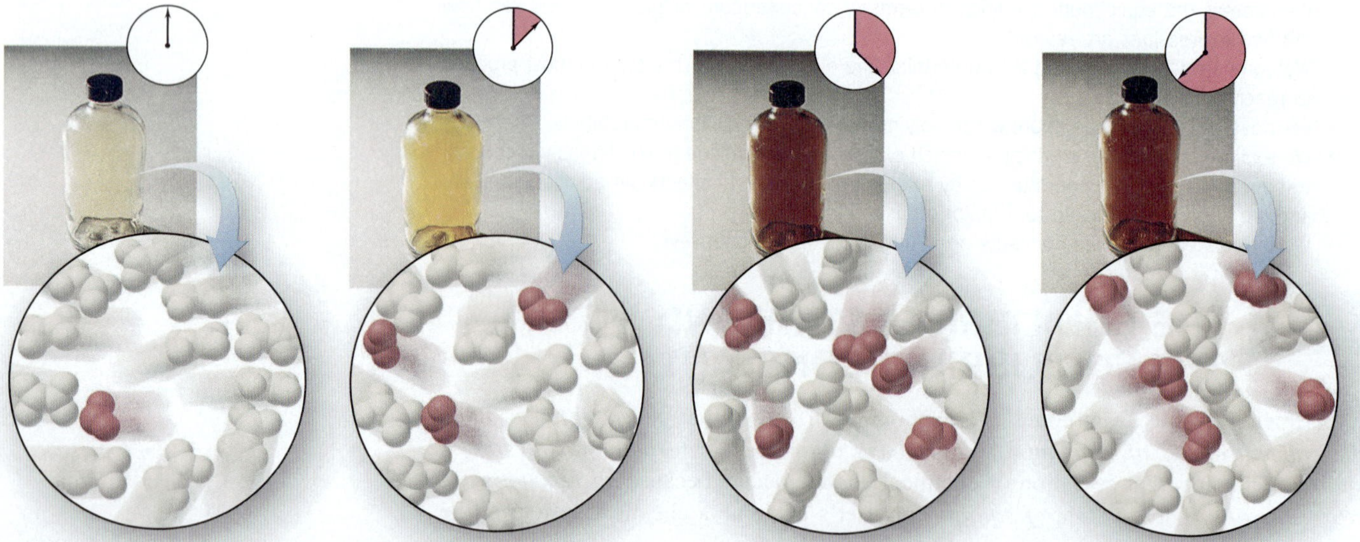

Figure 17.1 **Reaching equilibrium on the macroscopic and molecular levels.**
Source: © McGraw-Hill Education/Stephen Frisch, photographer

The reaction mixture consists mostly of colorless N_2O_4.

As N_2O_4 decomposes to NO_2, the mixture becomes pale brown.

The brown color deepens as more N_2O_4 decomposes and more NO_2 is produced, until the color reaches a maximum at equilibrium.

The reaction continues in both directions at equal rates, so the concentrations of NO_2 and N_2O_4, and therefore the color, no longer change.

where k_{fwd} and k_{rev} are the forward and reverse rate constants, respectively, and the subscript "eq" refers to concentrations at equilibrium. By rearranging, we set the ratio of the rate constants equal to the ratio of the concentration terms:

$$\frac{k_{fwd}}{k_{rev}} = \frac{[NO_2]^2_{eq}}{[N_2O_4]_{eq}}$$

The ratio of rate constants creates a new constant called the **equilibrium constant (*K*):**

$$K = \frac{k_{fwd}}{k_{rev}} = \frac{[NO_2]^2_{eq}}{[N_2O_4]_{eq}} \qquad \textbf{(17.2)}$$

Thus, *K is a number equal to a particular ratio of equilibrium concentrations of product(s) and reactant(s) at a particular temperature.*

4. *K as a measure of reaction extent.* The *magnitude* of *K* is an indication of *how far a reaction proceeds toward product at a given temperature*. Different reactions, even at the same temperature, have a wide range of concentrations at equilibrium—from almost all reactant to almost all product—and so, they have a wide range of equilibrium constants. Here are three examples of different magnitudes of *K*:

- *Small K* (Figure 17.2A). If a reaction yields little product before reaching equilibrium, the ratio of product concentration to reactant concentration is small, and the reaction has a small *K*; if *K* is very small, we may say there is "no reaction." For example, there is "no reaction" between nitrogen and oxygen at 1000 K:*

$$N_2(g) + O_2(g) \rightleftharpoons 2NO(g) \qquad K = 1\times10^{-30}$$

- *Large K* (Figure 17.2B). Conversely, if a reaction reaches equilibrium with little reactant remaining, its ratio of product concentration to reactant concentration is large, and it has a large *K*; if *K* is very large, we say the reaction "goes to completion." The oxidation of carbon monoxide "goes to completion" at 1000 K:

$$2CO(g) + O_2(g) \rightleftharpoons 2CO_2(g) \qquad K = 2.2\times10^{22}$$

- *Intermediate K* (Figure 17.2C). When significant amounts of both reactant and product are present at equilibrium, *K* has an intermediate value, as when bromine monochloride breaks down to its elements at 1000 K:

$$2BrCl(g) \rightleftharpoons Br_2(g) + Cl_2(g) \qquad K = 5$$

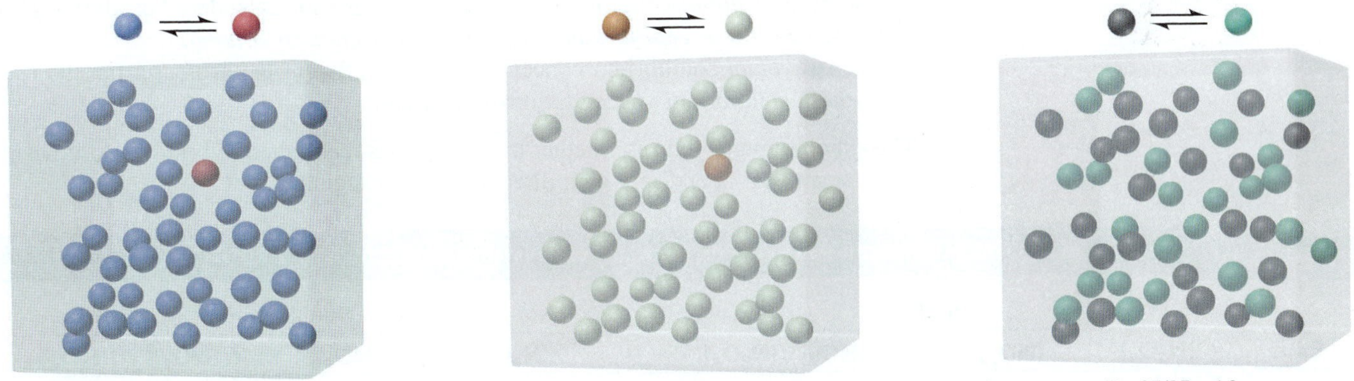

A $K = 1/49 = 0.020$ B $K = 49/1 = 49$ C $K = 25/25 = 1.0$

Figure 17.2 The range of equilibrium constants.

› Summary of Section 17.1

- › Kinetics and equilibrium are distinct aspects of a reaction system, and the rate and extent of a reaction are not necessarily related.
- › When the forward and reverse reactions occur at the same rate, concentrations no longer change and the system has reached equilibrium.
- › The equilibrium constant (*K*) is a number based on a particular ratio of product and reactant concentrations: *K* is small if a high concentration of reactant(s) is present at equilibrium, and it is large if a high concentration of product(s) is present at equilibrium.

*Note that the equilibrium constant is represented by a capital italic *K*, whereas the temperature unit, the kelvin, is a capital roman K. Also, since the kelvin is a unit, it always follows a number.

17.2 THE REACTION QUOTIENT AND THE EQUILIBRIUM CONSTANT

We have introduced the equilibrium constant in terms of a ratio of rate constants, but the original research on chemical equilibrium was developed many years before the principles of kinetics. In 1864, two Norwegian chemists, Cato Guldberg and Peter Waage, observed that *at a given temperature, a chemical system reaches a state in which a particular ratio of reactant and product concentrations has a constant value.* This is a statement of the **law of chemical equilibrium,** or the **law of mass action.**

The Changing Value of the Reaction Quotient

The particular ratio of concentration terms that we write for a given reaction is called the **reaction quotient (*Q*)** (also known as the *mass-action expression*). For the reversible breakdown of N_2O_4 to NO_2, the reaction quotient is

$$N_2O_4(g) \rightleftharpoons 2NO_2(g) \qquad Q = \frac{[NO_2]^2}{[N_2O_4]}$$

As the reaction proceeds toward equilibrium, the concentrations of reactants and products change continually, and so does their ratio, which is the value of Q: at a given temperature, at the beginning of the reaction, the concentrations have initial values, and Q has an initial value; a moment later, the concentrations have slightly different values, and so does Q; after another moment, the concentrations and the value of Q change further. These changes continue, *until the system reaches equilibrium.* At that point, reactant and product concentrations have their equilibrium values and no longer change. Thus, the value of Q no longer changes and equals K at that temperature:

$$\text{At equilibrium: } Q = K \qquad \textbf{(17.3)}$$

In formulating the law of mass action, Guldberg and Waage found that, *for a particular system and temperature, the same equilibrium state is attained regardless of starting concentrations.* For example, for the N_2O_4-NO_2 system at 200°C (473 K), the data from four experiments appear in Table 17.1. Two essential points stand out:

- The ratio of *initial* concentrations *(fourth column)* varies widely but always gives the same ratio of *equilibrium* concentrations *(rightmost column).*
- The *individual* equilibrium concentrations are different in each case, but the *ratio* of these equilibrium concentrations is constant.

Thus, monitoring Q tells us whether the system has reached equilibrium or, if it hasn't, how far away it is and in which direction it is changing.

Table 17.1 Initial and Equilibrium Concentration Ratios for the N_2O_4-NO_2 System at 200°C (473 K)

	Initial			Equilibrium		
Expt	$[N_2O_4]$	$[NO_2]$	Q, $[NO_2]^2/[N_2O_4]$	$[N_2O_4]_{eq}$	$[NO_2]_{eq}$	K, $[NO_2]^2_{eq}/[N_2O_4]_{eq}$
1	0.1000	0.0000	$\frac{(0.0000)^2}{(0.1000)} = 0.0000$	0.00357	0.193	$\frac{(0.193)^2}{(0.00357)} = 10.4$
2	0.0000	0.1000	$\frac{(0.1000)^2}{(0.0000)} = \infty$	0.000924	0.0982	$\frac{(0.0982)^2}{(0.000924)} = 10.4$
3	0.0500	0.0500	$\frac{(0.0500)^2}{(0.0500)} = 0.0500$	0.00204	0.146	$\frac{(0.146)^2}{(0.00204)} = 10.4$
4	0.0750	0.0250	$\frac{(0.0250)^2}{(0.0750)} = 0.00833$	0.00275	0.170	$\frac{(0.170)^2}{(0.00275)} = 10.5$

The curves in Figure 17.3 show experiment 1 in Table 17.1. Note that $[N_2O_4]$ and $[NO_2]$ change smoothly during the course of the reaction (as indicated by the changing brown color at the top), and so does the value of Q. Once the system reaches equilibrium (constant brown color), the concentrations no longer change and Q equals K. In other words, for any given chemical system, *K is a special value of Q that occurs when the reactant and product concentrations have their equilibrium values.*

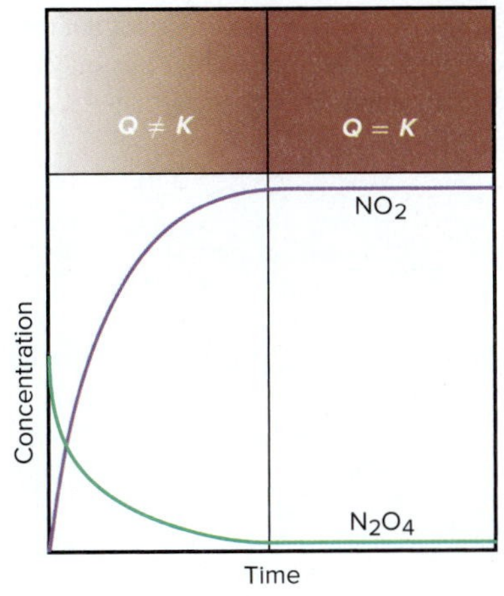

Figure 17.3 The change in Q during the N_2O_4-NO_2 reaction. Based on data from Experiment 1 in Table 17.1.

Writing the Reaction Quotient in Its Various Forms

The reaction quotient *must* be written directly from the balanced equation. In contrast, as discussed in Chapter 16, the rate law for an overall reaction *cannot* be written from the balanced equation, but must be determined from rate data.

Constructing the Reaction Quotient The most common form of the reaction quotient shows reactant and product terms as molar concentrations, which are designated by square brackets, [ ]. Thus, from now on, we designate the *reaction quotient based on concentrations* as Q_c. (We also designate the *equilibrium constant based on concentrations* as K_c.) For the general equation

$$aA + bB \rightleftharpoons cC + dD$$

where a, b, c, and d are the stoichiometric coefficients, the reaction quotient is

$$Q_c = \frac{[C]^c[D]^d}{[A]^a[B]^b} \quad (17.4)$$

Thus, *Q is a ratio consisting of product concentration terms multiplied together and divided by reactant concentration terms multiplied together, with each term raised to the power of its balancing coefficient.*

Two steps are needed to construct a reaction quotient for any chemical system:

1. *Start with the balanced equation.* For example, for the formation of ammonia from its elements, the balanced equation (with colored coefficients for reference) is

$$N_2(g) + 3H_2(g) \rightleftharpoons 2NH_3(g)$$

2. *Arrange the terms and exponents.* Place the product terms in the numerator and the reactant terms in the denominator, multiplied by each other, and raise each term to the power of its balancing coefficient (colored as in the balanced equation):

$$Q_c = \frac{[NH_3]^2}{[N_2][H_2]^3}$$

Why Q and K Have No Units In this text (and most others), *the values of Q and K are unitless numbers.* As we will discuss in Chapter 20, the reason is that Q and K are based on thermodynamic parameters. That is, each term in Q or K is actually an *activity,* or active concentration, the *ratio* of the concentration (or pressure) of a substance to its reference standard-state concentration (or pressure). Recall from Section 6.6 that the standard states are 1 M for a substance in solution, 1 atm for a gas, and the pure substance itself for a liquid or solid. The terms in Q or K thus become unitless as follows:

- For a concentration of, say, 1.20 M NaCl, we have

$$[NaCl] = \frac{1.20\ \cancel{M}\ \text{(measured quantity)}}{1\ \cancel{M}\ \text{(standard-state quantity)}} = 1.20$$

- For a pressure of, say, 0.53 atm N_2, we have

$$P_{N_2} = \frac{0.53\ \cancel{\text{atm}}\ \text{(measured quantity)}}{1\ \cancel{\text{atm}}\ \text{(standard-state quantity)}} = 0.53$$

- For a pure solid or a pure liquid, the concentration is the density of the substance, which when divided by its density, must equal 1.

Since each term used to write Q or K is a unitless ratio, Q and K are unitless also.

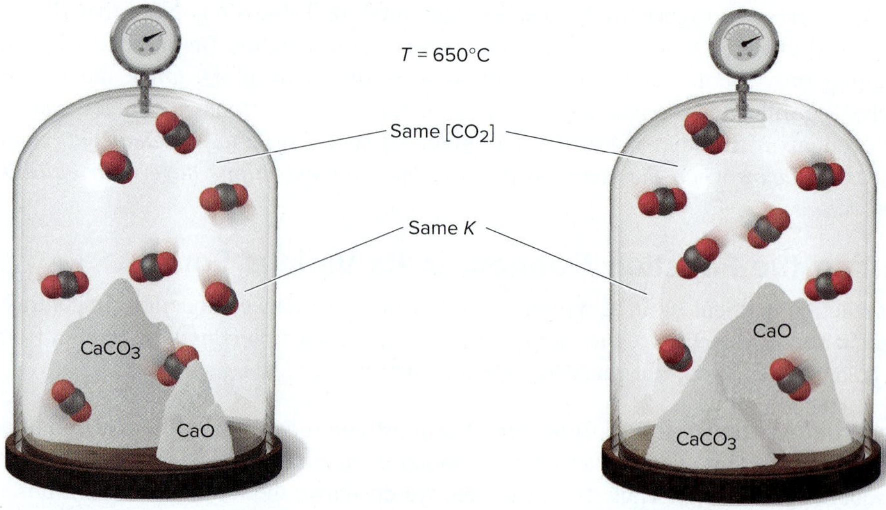

Figure 17.4 The reaction quotient for a heterogeneous system depends only on concentrations that change.

Form of Q for a Reaction Involving Pure Liquids or Solids The formation of gaseous NH_3 from its elements in the gas phase is an example of a *homogeneous* equilibrium system in which all the components are in the same phase. When the components are in different phases, the equilibrium system is *heterogeneous*.

Consider the decomposition of limestone to lime and carbon dioxide:

$$CaCO_3(s) \rightleftharpoons CaO(s) + CO_2(g)$$

Using the rules for writing the reaction quotient, we would obtain

$$Q_c = \frac{[CaO][CO_2]}{[CaCO_3]}$$

However, as the preceding subsection about Q and K being unitless numbers stated, each term in the reaction quotient is actually the ratio of the measured quantity to the standard-state quantity and *for all pure solids and liquids, this ratio is 1.* Thus, we have

$$Q_c = \frac{(1)[CO_2]}{(1)} = [CO_2]$$

This means that as long as some CaO and some $CaCO_3$ are present, only the CO_2 concentration affects the value of Q_c (Figure 17.4).

It makes sense that *we eliminate the terms for pure solids and liquids in the expressions for Q and K* because we are concerned only with concentrations that *change* as they approach equilibrium, and the concentrations, like the densities, of pure solids and liquids are constant.

SAMPLE PROBLEM 17.1 Writing the Reaction Quotient from the Balanced Equation

Problem Write the reaction quotient, Q_c, for each of the following reactions:

(a) $N_2O_5(g) \rightleftharpoons NO_2(g) + O_2(g)$

(b) $Na(s) + H_2O(l) \rightleftharpoons NaOH(aq) + H_2(g)$

(c) $C_3H_8(g) + O_2(g) \rightleftharpoons CO_2(g) + H_2O(g)$

Plan We balance the equation and then construct the reaction quotient using Equation 17.4. Remember that concentration terms for pure liquids and solids are *not* included in the reaction quotient.

Solution (a) From the balanced equation, $2N_2O_5(g) \rightleftharpoons 4NO_2(g) + O_2(g)$, we have

$$Q_c = \frac{[NO_2]^4[O_2]}{[N_2O_5]^2}$$

(b) The solid Na and pure liquid H_2O are not included in the reaction quotient. $NaOH(aq)$ is a solution in the liquid phase, but it is included in the reaction quotient since it is not a pure liquid. From the balanced equation, $2Na(s) + 2H_2O(l) \rightleftharpoons 2NaOH(aq) + H_2(g)$, we have

$$Q_c = [NaOH]^2[H_2]$$

(c) From the balanced equation, $C_3H_8(g) + 5O_2(g) \rightleftharpoons 3CO_2(g) + 4H_2O(g)$, we have

$$Q_c = \frac{[CO_2]^3[H_2O]^4}{[C_3H_8][O_2]^5}$$

Check Be sure that the exponents in the reaction quotient are the same as the balancing coefficients. A good check is to reverse the process and see if you obtain the balanced equation, except for any substances that are pure liquids or solids: change the numerator of the expression to products, the denominator to reactants, and the exponents to coefficients.

FOLLOW-UP PROBLEMS

Brief Solutions for ***all*** *Follow-up Problems appear at the end of the chapter.*

17.1A Write the reaction quotient, Q_c, for each of the following reactions:
(a) $NH_3(g) + O_2(g) \rightleftharpoons NO(g) + H_2O(g)$
(b) $NO_2(g) + H_2(g) \rightleftharpoons NH_3(g) + H_2O(l)$
(c) $KClO_3(s) \rightleftharpoons KCl(s) + O_2(g)$

17.1B Write the reaction quotient, Q_c, for each of the following reactions:
(a) $CH_4(g) + CO_2(g) \rightleftharpoons CO(g) + H_2(g)$
(b) $H_2S(g) + SO_2(g) \rightleftharpoons S(s) + H_2O(g)$
(c) $HCN(aq) + NaOH(aq) \rightleftharpoons NaCN(aq) + H_2O(l)$

SOME SIMILAR PROBLEMS 17.12–17.15 and 17.18–17.21

Form of Q for a Forward and a Reverse Reaction The form of the reaction quotient depends on the *direction* in which the balanced equation is written. Consider, for example, the oxidation of sulfur dioxide to sulfur trioxide. This reaction is a key step in acid-rain formation and sulfuric acid production. The balanced equation is

$$2SO_2(g) + O_2(g) \rightleftharpoons 2SO_3(g)$$

The reaction quotient for this equation *as written* is

$$Q_{c(fwd)} = \frac{[SO_3]^2}{[SO_2]^2[O_2]}$$

If we had written the reverse reaction, the decomposition of sulfur trioxide,

$$2SO_3(g) \rightleftharpoons 2SO_2(g) + O_2(g)$$

the reaction quotient would be the *reciprocal* of $Q_{c(fwd)}$:

$$Q_{c(rev)} = \frac{[SO_2]^2[O_2]}{[SO_3]^2} = \frac{1}{Q_{c(fwd)}}$$

Thus, *a reaction quotient (or equilibrium constant) for a forward reaction is the* ***reciprocal*** *of the reaction quotient (or equilibrium constant) for the reverse reaction:*

$$Q_{c(fwd)} = \frac{1}{Q_{c(rev)}} \quad \text{and} \quad K_{c(fwd)} = \frac{1}{K_{c(rev)}} \qquad (17.5)$$

The K_c values for the forward and reverse reactions shown above at 1000 K are

$$K_{c(fwd)} = \frac{[SO_3]^2}{[SO_2]^2[O_2]} = 261 \quad \text{and} \quad K_{c(rev)} = \frac{[SO_2]^2[O_2]}{[SO_3]^2} = \frac{1}{K_{c(fwd)}} = \frac{1}{261} = 3.83\times10^{-3}$$

These values make sense: if the forward reaction goes far to the right (high K_c), yielding a large concentration of the product SO_3 at equilibrium, the reverse reaction does not go far to the right (low K_c), again resulting in a large concentration of unreacted SO_3 when equilibrium is reached.

Form of Q for a Reaction with Coefficients Multiplied by a Common Factor Multiplying all the coefficients of the equation by some factor also changes the form of Q. For example, multiplying all the coefficients in the equation for the formation of SO_3 by $\frac{1}{2}$ gives

$$SO_2(g) + \tfrac{1}{2}O_2(g) \rightleftharpoons SO_3(g)$$

For this equation, the reaction quotient is

$$Q'_{c(fwd)} = \frac{[SO_3]}{[SO_2][O_2]^{1/2}}$$

Notice that Q_c for the halved equation equals Q_c for the original equation raised to the $\frac{1}{2}$ power:

$$Q'_{c(fwd)} = Q_{c(fwd)}^{1/2} = \left(\frac{[SO_3]^2}{[SO_2]^2[O_2]}\right)^{1/2} = \frac{[SO_3]}{[SO_2][O_2]^{1/2}}$$

Once again, the same property holds for the equilibrium constants. Relating the halved reaction to the original, we have

$$K'_{c(fwd)} = K_{c(fwd)}^{1/2} = (261)^{1/2} = 16.2$$

It may seem that we have changed the extent of the reaction, as indicated by a different K, just by changing the coefficients of the equation, but that can't be true. *A particular K has meaning only in relation to a particular equation.* If $K_{c(fwd)}$ and $K'_{c(fwd)}$ relate to different equations, we cannot compare them directly.

In general, *if all the coefficients of the balanced equation are multiplied by some factor, that factor becomes the exponent for relating the reaction quotients and the equilibrium constants.* For a multiplying factor n, which we can write as

$$n(aA + bB \rightleftharpoons cC + dD)$$

the reaction quotient and equilibrium constant are

$$Q' = Q^n = \left(\frac{[C]^c[D]^d}{[A]^a[B]^b}\right)^n \quad \text{and} \quad K' = K^n \tag{17.6}$$

Form of Q for an Overall Reaction We follow the same procedure for writing a reaction quotient whether an equation represents an individual reaction step or an overall multistep reaction. *If an overall reaction is the **sum** of two or more reactions, the overall reaction quotient (or equilibrium constant) is the **product** of the reaction quotients (or equilibrium constants) for the steps:*

$$Q_{overall} = Q_1 \times Q_2 \times Q_3 \times \cdots$$

and

$$K_{overall} = K_1 \times K_2 \times K_3 \times \cdots \tag{17.7}$$

Sample Problem 17.2 demonstrates calculations involving K values for reactions that are reversed or multiplied by a factor, and for an overall reaction.

SAMPLE PROBLEM 17.2 Finding K for Reactions Multiplied by a Common Factor or Reversed and for an Overall Reaction

Problem A chemist wants to find K_c for the following reaction at 700 K:

$$2NH_3(g) + 3I_2(g) \rightleftharpoons 6HI(g) + N_2(g) \qquad K_c = ?$$

Use the following data at 700 K to find the unknown K_c:

(1) $N_2(g) + 3H_2(g) \rightleftharpoons 2NH_3(g) \qquad K_{c1} = 0.343$

(2) $H_2(g) + I_2(g) \rightleftharpoons 2HI(g) \qquad K_{c2} = 54$

Plan We must manipulate reactions (1) and (2) so that they add together to give the target reaction. Reaction (1) must be reversed; the new equilibrium constant (K'_{c1}) is the reciprocal of K_{c1}. Reaction (2) must be multiplied by 3; the new equilibrium constant (K'_{c2}) is K_{c2} raised to the third power. We then add the two manipulated reactions and multiply their K_c values to obtain the overall K_c.

Solution Reaction (1) must be reversed since $2NH_3$ is a reactant in the target reaction; the K_c value of the reversed reaction is the reciprocal of the original K_c:

$$(1)\ 2NH_3(g) \rightleftharpoons N_2(g) + 3H_2(g) \qquad K'_{c1} = \frac{1}{0.343} = 2.92$$

Reaction (2) must be multiplied by 3 (in the target reaction I_2 has a coefficient of 3 and HI has a coefficient of 6) so the new K_c value equals the original K_c value raised to the third power:

$$(2)\ 3H_2(g) + 3I_2(g) \rightleftharpoons 6HI(g) \qquad K'_{c2} = 54^3 = 1.6\times10^5$$

Adding the two reactions (cancelling common terms):

$$\begin{array}{lrcl} (1) & 2NH_3(g) & \rightleftharpoons & N_2(g) + \cancel{3H_2(g)} \qquad K'_{c1} = 2.92 \\ (2) & \cancel{3H_2(g)} + 3I_2(g) & \rightleftharpoons & 6HI(g) \qquad K'_{c2} = 1.6\times10^5 \\ \hline & 2NH_3(g) + 3I_2(g) & \rightleftharpoons & N_2(g) + 6HI(g) \end{array}$$

Multiplying the individual reaction quotients and calculating the overall K_c:

$$Q'_{c1} \times Q'_{c2} = \frac{[N_2]\cancel{[H_2]^3}}{[NH_3]^2} \times \frac{[HI]^6}{\cancel{[H_2]^3}[I_2]^3} = \frac{[N_2][HI]^6}{[NH_3]^2[I_2]^3} = Q_{c(overall)}$$

$$K_{c(overall)} = K'_{c1} \times K'_{c2} = (2.92)(1.6\times10^5) = 4.7\times10^5$$

Check Round off and check the calculation of the overall K_c value:

$$K_c \approx (3)(2\times10^5) = 6\times10^5$$

FOLLOW-UP PROBLEMS

17.2A Find the K_c value of the following reaction at 1123 K

$$C(s) + CO_2(g) + 2Cl_2(g) \rightleftharpoons 2COCl_2(g)$$

using the following data at the same temperature:

(1) $C(s) + CO_2(g) \rightleftharpoons 2CO(g)$ $\quad K_{c1} = 1.4\times10^{12}$

(2) $CO(g) + Cl_2(g) \rightleftharpoons COCl_2(g)$ $\quad K_{c2} = 0.55$

17.2B The reaction $2NO(g) + Br_2(g) \rightleftharpoons 2NOBr(g)$ has a K_c value of 1.3×10^{-2} at 1000 K. Calculate K_c for the following reactions:

(a) $4NOBr(g) \rightleftharpoons 4NO(g) + 2Br_2(g)$ **(b)** $\frac{1}{2}NO(g) + \frac{1}{4}Br_2(g) \rightleftharpoons \frac{1}{2}NOBr(g)$

SOME SIMILAR PROBLEMS 17.16, 17.17, 17.23, and 17.24

Table 17.2 summarizes the ways of writing Q and calculating K.

Table 17.2 Ways of Expressing Q and Calculating K

Form of Chemical Equations	Form of Q	Value of K
Reference reaction (ref): $A \rightleftharpoons B$	$Q_{(ref)} = \frac{[B]}{[A]}$	$K_{(ref)} = \frac{[B]_{eq}}{[A]_{eq}}$
Reverse reaction: $B \rightleftharpoons A$	$Q = \frac{1}{Q_{(ref)}} = \frac{[A]}{[B]}$	$K = \frac{1}{K_{(ref)}}$
Coefficients multiplied by n	$Q = Q^n_{(ref)}$	$K = K^n_{(ref)}$
Reaction as sum of two steps:		
(1) $A \rightleftharpoons C$	$Q_1 = \frac{[C]}{[A]}$; $Q_2 = \frac{[B]}{[C]}$	
(2) $C \rightleftharpoons B$	$Q_{overall} = Q_1 \times Q_2 = Q_{(ref)}$ $= \frac{\cancel{[C]}}{[A]} \times \frac{[B]}{\cancel{[C]}} = \frac{[B]}{[A]}$	$K_{overall} = K_1 \times K_2$ $= K_{(ref)}$
Reaction with pure solid or liquid component, such as $A(s)$	$Q = Q_{(ref)}(1) = [B]$	$K = K_{(ref)}(1) = [B]$

› Summary of Section 17.2

- The reaction quotient based on concentration, Q_c, is a particular ratio of product to reactant concentrations. The value of Q_c changes as the reaction proceeds. When the system reaches equilibrium at a particular temperature, $Q_c = K_c$.
- The concentration terms for pure liquids or solids do not appear in Q because they are equal to 1.
- The *form* of Q is based on the balanced equation exactly as written, so it changes if the equation is reversed or multiplied by some factor, and K changes accordingly.
- If a reaction is the sum of two or more steps, the overall Q (or K) is the product of the individual Q's (or K's).
- Three criteria define a system at equilibrium:
  1. Reactant and product concentrations are constant over time.
  2. The opposing reaction rates are equal: $\text{rate}_{\text{fwd}} = \text{rate}_{\text{rev}}$.
  3. The reaction quotient equals the equilibrium constant: $Q = K$.

17.3 EXPRESSING EQUILIBRIA WITH PRESSURE TERMS: RELATION BETWEEN K_c AND K_p

It is easier to measure gas *pressure* than concentration, so when a reaction involves gases, we often express the reaction quotient in terms of partial pressures instead of concentrations. As long as the gases behave nearly ideally during the experiment, the ideal gas law (Section 5.3) allows us to relate pressure (P) to molar concentration (n/V):

$$PV = nRT, \qquad \text{so} \qquad P = \frac{n}{V}RT \qquad \text{or} \qquad \frac{P}{RT} = \frac{n}{V}$$

Thus, at constant T, *pressure is directly proportional to molar concentration.* For example, in the reaction between gaseous NO and O_2,

$$2NO(g) + O_2(g) \rightleftharpoons 2NO_2(g)$$

the *reaction quotient based on pressures,* Q_p, is

$$Q_p = \frac{P_{NO_2}^2}{P_{NO}^2 \times P_{O_2}}$$

The equilibrium constant obtained when all components are present at their equilibrium partial pressures is designated K_p, the *equilibrium constant based on pressures.* In many cases, K_p has a value different from K_c. But if you know one of these values, the *change in amount (mol) of gas,* Δn_{gas}, from the balanced equation allows you to calculate the other. Let's see how this conversion works for the oxidation of NO:

$$2NO(g) + O_2(g) \rightleftharpoons 2NO_2(g)$$

As the balanced equation shows,

$$3 \text{ mol } (2 \text{ mol} + 1 \text{ mol}) \text{ gaseous reactants} \rightleftharpoons 2 \text{ mol gaseous products}$$

With Δ meaning "final *minus* initial" (products *minus* reactants), we have

$$\Delta n_{gas} = \text{moles of gaseous product} - \text{moles of gaseous reactant} = 2 - 3 = -1$$

Keep this value of Δn_{gas} in mind because it appears in the conversion that follows.

The reaction quotient based on concentrations is

$$Q_c = \frac{[NO_2]^2}{[NO]^2\,[O_2]}$$

Rearranging the ideal gas law to $n/V = P/RT$, we write the terms in square brackets as n/V and convert them to partial pressures, P; then we collect the RT terms and cancel:

$$Q_c = \frac{\dfrac{n_{NO_2}^2}{V^2}}{\dfrac{n_{NO}^2}{V^2} \times \dfrac{n_{O_2}}{V}} = \frac{\dfrac{P_{NO_2}^2}{(RT)^2}}{\dfrac{P_{NO}^2}{(RT)^2} \times \dfrac{P_{O_2}}{RT}} = \frac{P_{NO_2}^2}{P_{NO}^2 \times P_{O_2}} \times \frac{\dfrac{1}{\cancel{(RT)^2}}}{\dfrac{1}{\cancel{(RT)^2}} \times \dfrac{1}{RT}}$$

$$= \frac{P_{NO_2}^2}{P_{NO}^2 \times P_{O_2}} \times RT$$

Note that the far right side of the previous expression is Q_p multiplied by RT. Thus,

$$Q_c = Q_p(RT)$$

Also, at equilibrium, $K_c = K_p(RT)$; and, solving for K_p gives

$$K_p = \frac{K_c}{RT} = K_c(RT)^{-1}$$

Note especially that *the exponent of the RT term equals the change in the amount (mol) of gas (Δn_{gas}) from the balanced equation,* −1 in this case. Thus, in general, we have

$$K_p = K_c(RT)^{\Delta n_{gas}} \quad \textbf{(17.8)}$$

Based on Equation 17.8, *if the amount (mol) of gas does not change in the reaction,* $\Delta n_{gas} = 0$, and the RT term drops out (i.e., equals 1); then, $K_p = K_c$. (In calculations, be sure the units for R are consistent with units for pressure.)

SAMPLE PROBLEM 17.3 Converting Between K_c and K_p

Problem Ammonium hydrosulfide decomposes to ammonia and hydrogen sulfide gases. Find K_c for the decomposition reaction:

$$NH_4SH(s) \rightleftharpoons NH_3(g) + H_2S(g) \qquad K_p = 0.19 \text{ (at 218°C)}$$

Plan We know K_p (0.19), so to convert between K_p and K_c, we must first determine Δn_{gas} from the balanced equation. Then we rearrange Equation 17.8. If we assume that pressure is measured in atmospheres, $R = 0.0821$ atm·L/mol·K; the given temperature must be converted to kelvins.

Solution Determining Δn_{gas}: There are 2 moles of gaseous product (1 mol NH_3 and 1 mol H_2S) and no gaseous reactant, so $\Delta n_{gas} = 2 - 0 = 2$. (Note that only moles of *gaseous* reactants and products are used to determine Δn_{gas}.)

Converting T from °C to K: T (K) = 218°C + 273.15 = 491 K

Rearranging Equation 17.8 and calculating K_c:

$$K_p = K_c(RT)^{\Delta n_{gas}} \qquad \text{or} \qquad K_p = K_c(RT)^2$$

so

$$K_c = K_p(RT)^{-2} = (0.19)(0.0821 \times 491)^{-2} = 1.2 \times 10^{-4}$$

Check Rounding gives

$$(0.2)(0.1 \times 500)^{-2} = (0.2)(4 \times 10^{-4}) = 0.8 \times 10^{-4}$$

FOLLOW-UP PROBLEMS

17.3A Calculate K_p for the following reaction:

$$PCl_3(g) + Cl_2(g) \rightleftharpoons PCl_5(g) \qquad K_c = 1.67 \text{ (at 500. K)}$$

17.3B Calculate K_c for the following reaction:

$$CS_2(g) + 4H_2(g) \rightleftharpoons CH_4(g) + 2H_2S(g) \qquad K_p = 3.0 \times 10^{-5} \text{ (at 900.°C)}$$

SOME SIMILAR PROBLEMS 17.30–17.33

› Summary of Section 17.3

- › For a gaseous reaction, the reaction quotient and the equilibrium constant can be expressed in terms of partial pressures (Q_p and K_p).
- › If you know K_p, you can find K_c, and vice versa: $K_p = K_c(RT)^{\Delta n_{gas}}$.

17.4 COMPARING Q AND K TO DETERMINE REACTION DIRECTION

Suppose you have a mixture of reactants and products and you know K at the temperature of the reaction. By comparing the value of Q with the known value of K, you can tell whether the reaction has reached equilibrium or, if not, in which direction it is progressing. The value of K remains constant at a particular temperature, but as we've seen, the value of Q varies as the concentrations of reactants and products change. Since Q has product terms in the numerator and reactant terms in the denominator, *more product makes Q larger, and more reactant makes Q smaller:*

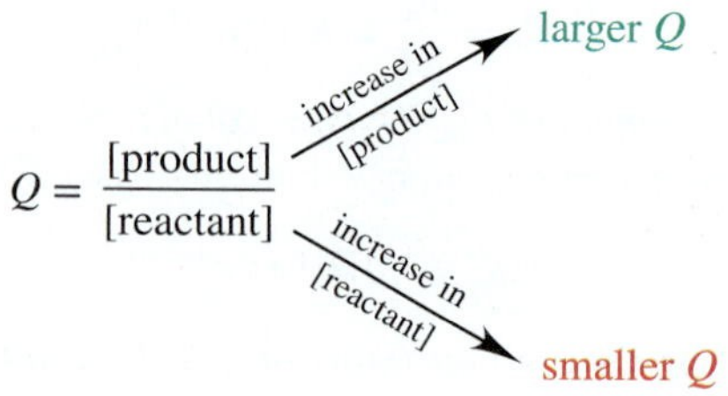

There are three possibilities for the relative sizes of Q and K (Figure 17.5):

- $Q < K$. If Q is smaller than K, the denominator (reactants) is large relative to the numerator (products). For Q to equal K, the reactants must decrease and the products increase. The reaction is not at equilibrium and will progress to the right, toward products, until $Q = K$:

$$\text{If } Q < K, \text{ reactants} \longrightarrow \text{products}$$

- $Q > K$. If Q is larger than K, the numerator (products) will decrease and the denominator (reactants) will increase. The reaction is not at equilibrium and will progress to the left, toward reactants, until $Q = K$:

$$\text{If } Q > K, \text{ reactants} \longleftarrow \text{products}$$

- $Q = K$. This situation occurs when the reactant and product terms equal their equilibrium values. The reaction is at equilibrium, and no further net change takes place:

$$\text{If } Q = K, \text{ reactants} \rightleftharpoons \text{products}$$

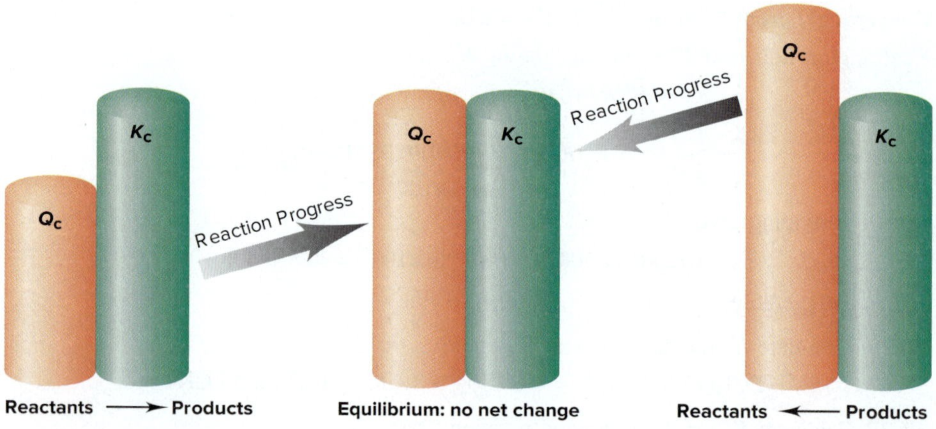

Figure 17.5 Reaction direction and the relative sizes of Q and K. When Q_c is smaller or larger than K_c, the reaction continues until $Q_c = K_c$. Note that K_c remains the same throughout.

Sample Problem 17.4 relies on molecular scenes to determine reaction direction, and Sample Problem 17.5 uses concentration data.

SAMPLE PROBLEM 17.4 Using Molecular Scenes to Determine Reaction Direction

Problem For the reaction $A(g) \rightleftharpoons B(g)$, the equilibrium mixture at 175°C is $[A] = 2.8\times10^{-4}\ M$ and $[B] = 1.2\times10^{-4}\ M$. The molecular scenes below represent four different mixtures for the reaction (A is *red;* B is *blue*). For each mixture, does the reaction progress to the right or to the left or neither to reach equilibrium?

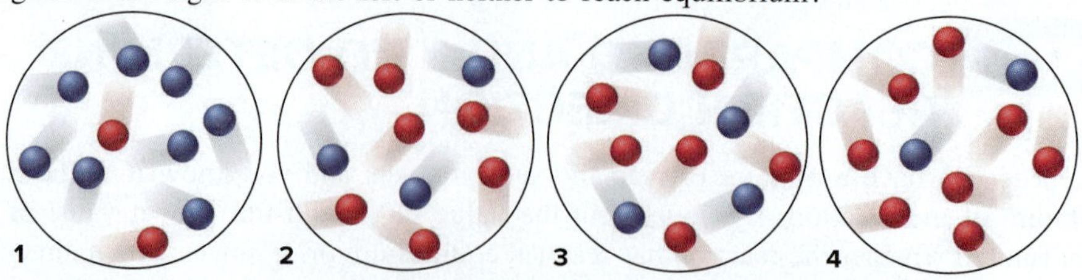

Plan We must compare Q_c with K_c to determine the reaction direction, so we first use the given equilibrium concentrations to find K_c. Then we count spheres and calculate Q_c for each mixture. If $Q_c < K_c$, the reaction progresses to the right (reactants to products); if $Q_c > K_c$, it progresses to the left (products to reactants); and if $Q_c = K_c$, there is no further net change.

Solution Writing the reaction quotient and using the data to find K_c:

$$Q_c = \frac{[B]}{[A]} = \frac{1.2\times10^{-4}}{2.8\times10^{-4}} = 0.43 = K_c$$

Counting red (A) and blue (B) spheres to calculate Q_c for each mixture:

1. $Q_c = 8/2 = 4.0$ 2. $Q_c = 3/7 = 0.43$ 3. $Q_c = 4/6 = 0.67$ 4. $Q_c = 2/8 = 0.25$

Comparing Q_c with K_c to determine reaction direction:

1. $Q_c > K_c$: left 2. $Q_c = K_c$: no net change 3. $Q_c > K_c$: left 4. $Q_c < K_c$: right

Check Making an error in the calculation for K_c would lead to incorrect conclusions throughout, so check that step: the exponents are the same, and 1.2/2.8 is a bit less than 0.5, as is the calculated K_c. You can check the final answers by inspection; for example, for the number of B (8) in mixture 1 to equal the number at equilibrium (3), more B must change to A, so the reaction must proceed to the left.

FOLLOW-UP PROBLEMS

17.4A At 338 K, the reaction $X(g) \rightleftharpoons Y(g)$ has a K_c of 1.4. The scenes below represent different mixtures at 338 K, with X *orange* and Y *green*. In which direction does the reaction proceed (if there is any net change) for each mixture to reach equilibrium?

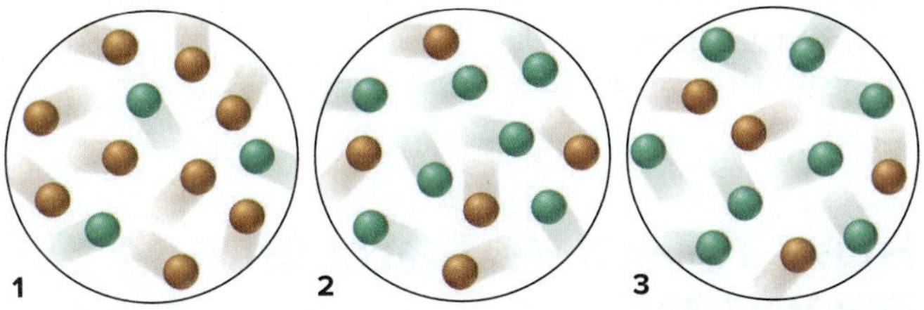

17.4B Scene 1 below represents the reaction $2C(g) \rightleftharpoons D(g)$ at equilibrium at 25°C (C is *yellow* and D is *blue*). Scenes 2 and 3 represent two different mixtures of C and D at 25°C. Calculate K_c for the reaction, and determine in which direction the reaction proceeds (if there is any net change) for the mixtures in scenes 2 and 3.

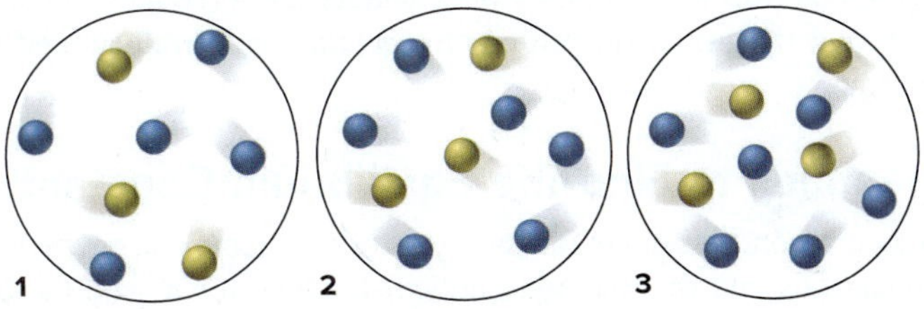

A SIMILAR PROBLEM 17.35

SAMPLE PROBLEM 17.5 Using Concentrations to Determine Reaction Direction

Problem For the reaction $N_2O_4(g) \rightleftharpoons 2NO_2(g)$, $K_c = 0.21$ at 100°C. At a point during the reaction, $[N_2O_4] = 0.12\ M$ and $[NO_2] = 0.55\ M$. Is the reaction at equilibrium at that time? If not, in which direction is it progressing?

Plan We write Q_c, find its value by substituting the given concentrations, and compare its value with the given K_c.

Solution Writing the reaction quotient and solving for Q_c:

$$Q_c = \frac{[NO_2]^2}{[N_2O_4]} = \frac{(0.55)^2}{0.12} = 2.5$$

With $Q_c > K_c$ (2.5 > 0.21), the reaction is not at equilibrium and will proceed to the left until $Q_c = K_c$.

Check With $[NO_2] > [N_2O_4]$, we expect to obtain a value for Q_c that is greater than 0.21. If $Q_c > K_c$, the numerator will decrease and the denominator will increase until $Q_c = K_c$; that is, this reaction will proceed toward reactants.

FOLLOW-UP PROBLEMS

17.5A Chloromethane forms by the reaction

$$CH_4(g) + Cl_2(g) \rightleftharpoons CH_3Cl(g) + HCl(g)$$

At 1500 K, $K_p = 1.6\times10^4$. In the reaction mixture, $P_{CH_4} = 0.13$ atm, $P_{Cl_2} = 0.035$ atm, $P_{CH_3Cl} = 0.24$ atm, and $P_{HCl} = 0.47$ atm. Is CH_3Cl or CH_4 forming?

17.5B Sulfur trioxide is produced by the reaction of sulfur dioxide with oxygen:

$$2SO_2(g) + O_2(g) \rightleftharpoons 2SO_3(g) \qquad K_c = 4.2\times10^{-2} \text{ (at 727°C)}$$

Suppose 3.4 mol of SO_2, 1.5 mol of O_2, and 1.2 mol of SO_3 are placed in a 2.0-L flask. Is the system at equilibrium? If not, in which direction will the reaction proceed?

SOME SIMILAR PROBLEMS 17.36–17.38

› Summary of Section 17.4

- › We compare the values of Q and K to determine the direction in which a reaction will proceed toward equilibrium.
  - If $Q_c < K_c$, more product forms.
  - If $Q_c > K_c$, more reactant forms.
  - If $Q_c = K_c$, there is no further net change.

17.5 HOW TO SOLVE EQUILIBRIUM PROBLEMS

Many kinds of equilibrium problems arise in the real world—and on chemistry exams—but we can group most of them into two types:

1. We know the equilibrium quantities of reactants and products and solve for K.
2. We know K and the initial quantities of reactants and products and solve for the equilibrium quantities.

Using Quantities to Find the Equilibrium Constant

There are two common variations on this type of problem: one involves substituting equilibrium quantities into the reaction quotient to solve for K, and the other requires first finding some of the quantities.

Substituting Equilibrium Quantities into Q to Find K In this type of problem, we use given equilibrium quantities to calculate K. Suppose, for example, that equal amounts of gaseous hydrogen and iodine are injected into a 1.50-L flask at a fixed temperature. In time, the following equilibrium is attained:

$$H_2(g) + I_2(g) \rightleftharpoons 2HI(g)$$

At equilibrium, the flask contains 1.80 mol of H_2, 1.80 mol of I_2, and 0.520 mol of HI. We calculate K_c by finding the concentrations from the amounts and flask volume and substituting into the reaction quotient from the balanced equation:

$$Q_c = \frac{[HI]^2}{[H_2][I_2]}$$

We divide each amount (mol) by the volume (L) to find each concentration (mol/L):

$$[H_2] = \frac{1.80 \text{ mol}}{1.50 \text{ L}} = 1.20\ M$$

Similarly, $[I_2]$ = 1.20 *M*, and [HI] = 0.347 *M*. Substituting these values into the expression for Q_c gives K_c:

$$K_c = \frac{(0.347)^2}{(1.20)(1.20)} = 8.36 \times 10^{-2}$$

Using a Reaction Table to Find Equilibrium Quantities and *K* When some quantities are not given, we determine them from the reaction stoichiometry and then find *K*. In the following example, note the use of the *reaction table.*

In a study of carbon oxidation, an evacuated vessel containing a small amount of powdered graphite is heated to 1080 K. Gaseous CO_2 is added to a pressure of 0.458 atm, and CO forms. At equilibrium, the total pressure is 0.757 atm. Calculate K_p.

As always, we start by writing the balanced equation and the reaction quotient:

$$CO_2(g) + C(\text{graphite}) \rightleftharpoons 2CO(g)$$

The data are given in atmospheres and we must find K_p, so we write the expression for Q_p; note that it does *not* include a term for the solid, C(graphite):

$$Q_p = \frac{P^2_{CO}}{P_{CO_2}}$$

We are given the initial P_{CO_2} and the P_{total} at equilibrium. To find K_p, we must find the equilibrium pressures of CO_2 and CO and then substitute them into Q_p.

Let's think through what happened in the vessel. An unknown portion of the CO_2 reacted with graphite to form an unknown amount of CO. We already know the *relative* amounts of CO_2 and CO from the balanced equation: for each mole of CO_2 that reacts, 2 mol of CO forms, which means that when x atm of CO_2 reacts, $2x$ atm of CO forms:

$$x \text{ atm } CO_2 \longrightarrow 2x \text{ atm CO}$$

The pressure of CO_2 at equilibrium, $P_{CO_2(eq)}$, is the initial pressure, $P_{CO_2(init)}$, *minus* x, the CO_2 that reacts (the change in P_{CO_2} due to the reaction):

$$P_{CO_2(init)} - x = P_{CO_2(eq)} = 0.458 - x$$

Similarly, the pressure of CO at equilibrium, $P_{CO(eq)}$, is the initial pressure, $P_{CO(init)}$, *plus* $2x$, the CO that forms (the change in P_{CO} due to the reaction). $P_{CO(init)}$ is zero, so

$$P_{CO(init)} + 2x = 0 + 2x = 2x = P_{CO(eq)}$$

We'll summarize this information in a reaction table, similar to those in Chapter 3, but for "Final" quantities, we use those at equilibrium. The table shows the balanced equation at the top, and

- the *initial* quantities (concentrations or pressures) of reactants and products,
- the *changes* in these quantities during the reaction, and
- the *equilibrium* quantities.

Pressure (atm)	**$CO_2(g)$**	**+**	**C(graphite)**	**$\rightleftharpoons$**	**$2CO(g)$**
Initial	0.458		—		0
Change	$-x$		—		$+2x$
Equilibrium	$0.458 - x$		—		$2x$

We treat each column like a list of numbers to add together: *the initial quantity plus the change in that quantity gives the equilibrium quantity*. Note that we *only* consider substances whose concentrations change; thus, the column for C(graphite) is blank.

To solve for K_p, we substitute equilibrium values into Q_p, so we first have to find x. To do this, we use the other piece of data given, $P_{total(eq)}$. According to Dalton's law of partial pressures and using the equilibrium quantities from the reaction table,

$$P_{total(eq)} = 0.757 \text{ atm} = P_{CO_2(eq)} + P_{CO(eq)} = (0.458 \text{ atm} - x) + 2x$$

Thus,

$$0.757 \text{ atm} = 0.458 \text{ atm} + x \qquad \text{and} \qquad x = 0.299 \text{ atm}$$

With x known, we determine the equilibrium partial pressures:

$$P_{CO_2(eq)} = 0.458\text{ atm} - x = 0.458\text{ atm} - 0.299\text{ atm} = 0.159\text{ atm}$$

$$P_{CO(eq)} = 2x = 2(0.299\text{ atm}) = 0.598\text{ atm}$$

Then we substitute them into Q_p to find K_p:

$$Q_p = \frac{P^2_{CO(eq)}}{P_{CO_2(eq)}} = \frac{(0.598)^2}{0.159} = 2.25 = K_p$$

(From now on, the subscripts "init" and "eq" appear only when it is not clear whether a quantity is an initial or equilibrium value.)

Student Hot Spot

Student data indicate that you may struggle with constructing a reaction table in equilibrium calculations. Access the Smartbook to view additional Learning Resources on this topic.

SAMPLE PROBLEM 17.6 Calculating K_c from Concentration Data

Problem In order to study hydrogen halide decomposition, a researcher fills an evacuated 2.00-L flask with 0.200 mol of HI gas and allows the reaction to proceed at 453°C:

$$2HI(g) \rightleftharpoons H_2(g) + I_2(g)$$

At equilibrium, [HI] = 0.078 M. Calculate K_c.

Plan To calculate K_c, we need the equilibrium concentrations. We can find the initial [HI] from the amount (0.200 mol) and the flask volume (2.00 L), and we are given [HI] at equilibrium (0.078 M). From the balanced equation, when $2x$ mol of HI reacts, x mol of H_2 and x mol of I_2 form. We set up a reaction table, use the known [HI] at equilibrium to solve for x (the change in $[H_2]$ or $[I_2]$), and substitute the concentrations into Q_c.

Solution Calculating initial [HI]:

$$[HI] = \frac{0.200\text{ mol}}{2.00\text{ L}} = 0.100\ M$$

Setting up the reaction table, with $x = [H_2]$ or $[I_2]$ that forms and $2x$ = [HI] that reacts:

Concentration (M)	$2HI(g)$	$\rightleftharpoons$	$H_2(g)$	+	$I_2(g)$
Initial	0.100		0		0
Change	$-2x$		$+x$		$+x$
Equilibrium	$0.100 - 2x$		x		x

Solving for x, using the known [HI] at equilibrium:

$$[HI] = 0.100\ M - 2x = 0.078\ M \quad \text{so} \quad x = 0.011\ M$$

Therefore, the equilibrium concentrations are

$$[H_2] = [I_2] = 0.011\ M \quad \text{and we were given} \quad [HI] = 0.078\ M$$

Substituting into the reaction quotient:

$$Q_c = \frac{[H_2][I_2]}{[HI]^2}$$

Thus,

$$K_c = \frac{(0.011)(0.011)}{0.078^2} = 0.020$$

Check Rounding gives $\sim 0.01^2/0.08^2 = 0.02$. Because the initial [HI] of 0.100 M fell slightly at equilibrium to 0.078 M, relatively little product formed; so we expect $K_c < 1$.

FOLLOW-UP PROBLEMS

17.6A The atmospheric oxidation of nitrogen monoxide, $2NO(g) + O_2(g) \rightleftharpoons 2NO_2(g)$, was studied at 184°C with initial pressures of 1.000 atm of NO and 1.000 atm of O_2. At equilibrium, $P_{O_2} = 0.506$ atm. Calculate K_p.

17.6B To study the reaction between ammonia and oxygen,

$$4NH_3(g) + 7O_2(g) \rightleftharpoons 2N_2O_4(g) + 6H_2O(g)$$

a flask was filled with 2.40 M NH_3 and 2.40 M O_2 at a particular temperature; the reaction proceeded, and at equilibrium $[N_2O_4] = 0.134\ M$. Calculate K_c.

SOME SIMILAR PROBLEMS 17.52, 17.53, 17.55, and 17.56

Using the Equilibrium Constant to Find Quantities

The type of problem that involves finding equilibrium quantities also has several variations. Sample Problem 17.7 is one variation, in which we know K and some of the equilibrium concentrations and must find another equilibrium concentration.

SAMPLE PROBLEM 17.7 Determining Equilibrium Concentrations from K_c

Problem In a study of the conversion of methane to other fuels, a chemical engineer mixes gaseous CH_4 and H_2O in a 0.32-L flask at 1200 K. At equilibrium, the flask contains 0.028 mol of CO, 0.084 mol of H_2, and 0.045 mol of CH_4. What is $[H_2O]$ at equilibrium ($K_c = 0.26$ for this process at 1200 K)?

Plan First, we write the balanced equation and the reaction quotient. We calculate the equilibrium concentrations from the given numbers of moles and the flask volume (0.32 L). Substituting these concentrations into Q_c and setting its value equal to the given K_c (0.26), we solve for the unknown equilibrium concentration, $[H_2O]$.

Solution Writing the balanced equation and reaction quotient:

$$CH_4(g) + H_2O(g) \rightleftharpoons CO(g) + 3H_2(g) \qquad Q_c = \frac{[CO][H_2]^3}{[CH_4][H_2O]}$$

Determining the equilibrium concentrations:

$$[CH_4] = \frac{0.045\ \text{mol}}{0.32\ \text{L}} = 0.14\ M$$

Similarly, $[CO] = 0.088\ M$ and $[H_2] = 0.26\ M$.
Calculating $[H_2O]$ at equilibrium: Since $Q_c = K_c$, rearranging gives

$$[H_2O] = \frac{[CO][H_2]^3}{[CH_4]\,K_c} = \frac{(0.088)(0.26)^3}{(0.14)(0.26)} = 0.042\ M$$

Check Always check by substituting the concentrations into Q_c to confirm that the result is equal to K_c:

$$Q_c = \frac{[CO][H_2]^3}{[CH_4][H_2O]} = \frac{(0.088)(0.26)^3}{(0.14)(0.042)} = 0.26 = K_c$$

FOLLOW-UP PROBLEMS

17.7A Nitrogen monoxide decomposes by the following equation: $2NO(g) \rightleftharpoons N_2(g) + O_2(g)$; $K_c = 2.3\times10^{30}$ at 298 K. In the atmosphere, $P_{O_2} = 0.209$ atm and $P_{N_2} = 0.781$ atm. What is the equilibrium partial pressure of NO in the air we breathe? (*Hint:* You need K_p to find the partial pressure.)

17.7B Phosphine decomposes by the following equation:

$$2PH_3(g) \rightleftharpoons P_2(g) + 3H_2(g) \qquad K_p = 19.6\ (\text{at } 750.\ \text{K})$$

In a reaction mixture at equilibrium, $P_{PH_3} = 0.112$ atm and $P_{P_2} = 0.215$ atm. What is the equilibrium partial pressure of H_2?

SOME SIMILAR PROBLEMS 17.46 and 17.47

In a somewhat more involved variation, we know K and *initial* quantities and must find *equilibrium* quantities, for which we use a reaction table. Sample Problem 17.8 focuses on this type of situation.

SAMPLE PROBLEM 17.8 Determining Equilibrium Concentrations from Initial Concentrations and K_c

Problem Fuel engineers use the extent of the change from CO and H_2O to CO_2 and H_2 to regulate the proportions of synthetic fuel mixtures. If 0.250 mol of CO gas and 0.250 mol of H_2O gas are placed in a 125-mL flask at 900 K, what is the composition of the equilibrium mixture? At this temperature, K_c is 1.56.

Plan We have to find the "composition" of the equilibrium mixture, in other words, the equilibrium concentrations. As always, we write the balanced equation and use it to write the reaction quotient. We find the initial [CO] and [H_2O] from the amounts (0.250 mol of each) and volume (0.125 L), use the balanced equation to define x and set up a reaction table, substitute into Q_c, and solve for x, from which we calculate the concentrations.

Solution Writing the balanced equation and reaction quotient:

$$CO(g) + H_2O(g) \rightleftharpoons CO_2(g) + H_2(g) \qquad Q_c = \frac{[CO_2][H_2]}{[CO][H_2O]}$$

Calculating initial reactant concentrations:

$$[CO] = [H_2O] = \frac{0.250 \text{ mol}}{0.125 \text{ L}} = 2.00 \, M$$

Setting up the reaction table, with x = [CO] and [H_2O] that react (note the 1/1 molar ratio among all reactants and products):

Concentration (*M*)	$CO(g)$	+	$H_2O(g)$	$\rightleftharpoons$	$CO_2(g)$	+	$H_2(g)$
Initial	2.00		2.00		0		0
Change	$-x$		$-x$		$+x$		$+x$
Equilibrium	$2.00 - x$		$2.00 - x$		x		x

Substituting into the reaction quotient and solving for x:

$$Q_c = \frac{[CO_2][H_2]}{[CO][H_2O]} = \frac{(x)(x)}{(2.00 - x)(2.00 - x)} = \frac{x^2}{(2.00 - x)^2}$$

At equilibrium, we have

$$Q_c = K_c = 1.56 = \frac{x^2}{(2.00 - x)^2}$$

We can apply the following math shortcut in this case *but not in general:* Because the right side of the preceding equation is a perfect square, we take the square root of both sides:

$$\sqrt{1.56} = \frac{x}{2.00 - x} = \pm 1.25$$

A positive number (1.56) has a positive *and* a negative root, but in this case, only the positive root has any chemical meaning, so we ignore the negative root:*

$$1.25 = \frac{x}{2.00 - x} \quad \text{or} \quad 2.50 - 1.25x = x$$

So $\quad 2.50 = 2.25\,x \quad$ therefore $\quad x = 1.11\,M$

Calculating equilibrium concentrations:

$$[CO] = [H_2O] = 2.00\,M - x = 2.00\,M - 1.11\,M = 0.89\,M$$
$$[CO_2] = [H_2] = x = 1.11\,M$$

Check Given the intermediate size of K_c (1.56), it makes sense that the changes in concentration are moderate. It's a good idea to check that the sign of x in the reaction table is correct—only reactants were initially present, so x has a negative sign for reactants and a positive sign for products. Also check that the equilibrium concentrations give the known K_c: $\dfrac{(1.11)(1.11)}{(0.89)(0.89)} = 1.56$.

*The negative root gives $-1.25 = \dfrac{x}{2.00 - x}$, or $-2.50 + 1.25x = x$.

So $\quad 2.50 = 0.25x \quad$ and $\quad x = 10.\,M$

This value has no chemical meaning because we started with 2.00 *M* as the concentration of each reactant, so it is impossible for x to be 10. *M*. Moreover, the square root of an equilibrium constant is another equilibrium constant, which cannot have a negative value.

FOLLOW-UP PROBLEMS

17.8A The decomposition of HI at low temperature was studied by injecting 2.50 mol of HI into a 10.32-L vessel at 25°C. What is $[H_2]$ at equilibrium for the reaction $2HI(g) \rightleftharpoons H_2(g) + I_2(g)$; $K_c = 1.26\times10^{-3}$?

17.8B The following reaction has a K_c value of 0.18 at 500. K:

$$Cl_2O(g) + H_2O(g) \rightleftharpoons 2HOCl(g)$$

If 6.15 mol of Cl_2O gas and 6.15 mol of H_2O gas are added to a 5.00-L flask at 500. K, calculate the equilibrium concentrations of Cl_2O, H_2O, and HOCl.

SOME SIMILAR PROBLEMS 17.48, 17.54, and 17.57

Using the Quadratic Formula to Find the Unknown The shortcut we used to simplify the math in Sample Problem 17.8 is a special case that occurred because we started with equal concentrations of the reactants, but typically, we start with different concentrations of reactants. Suppose, for example, we start with 2.00 *M* CO and 1.00 *M* H_2O. Now, the reaction table becomes

Concentration (*M*)	$CO(g)$	+	$H_2O(g)$	$\rightleftharpoons$	$CO_2(g)$	+	$H_2(g)$
Initial	2.00		1.00		0		0
Change	$-x$		$-x$		$+x$		$+x$
Equilibrium	$2.00 - x$		$1.00 - x$		x		x

Substituting these values into Q_c, we obtain

$$Q_c = \frac{[CO_2][H_2]}{[CO][H_2O]} = \frac{(x)(x)}{(2.00 - x)(1.00 - x)} = \frac{x^2}{x^2 - 3.00x + 2.00}$$

At equilibrium, using the value of K_c from Sample Problem 17.8, we have

$$1.56 = \frac{x^2}{x^2 - 3.00x + 2.00}$$

To solve for x in this case, we rearrange the previous expression into the form of a *quadratic equation:*

$$a\,x^2 + b\,x + c = 0$$
$$0.56x^2 - 4.68x + 3.12 = 0$$

where $a = 0.56$, $b = -4.68$, and $c = 3.12$. Then we find x with the quadratic formula (Appendix A):

$$x = \frac{-b \pm \sqrt{b^2 - 4ac}}{2a}$$

The $\pm$ sign means that we obtain two possible values for x:

$$x = \frac{4.68 \pm \sqrt{(-4.68)^2 - 4(0.56)(3.12)}}{2(0.56)}$$

$$x = 7.6\ M \quad \text{and} \quad x = 0.73\ M$$

But only one of the values makes sense chemically. In this case, addition of the square root term gave the larger value, which would yield negative concentrations at equilibrium (for example, 2.00 *M* − 7.6 *M* = −5.6 *M*); these have no meaning. Therefore, $x = 0.73\ M$, and we have

$$[CO] = 2.00\ M - x = 2.00\ M - 0.73\ M = 1.27\ M$$
$$[H_2O] = 1.00\ M - x = 0.27\ M$$
$$[CO_2] = [H_2] = x = 0.73\ M$$

Checking to see if these values give the known K_c, we have

$$K_c = \frac{(0.73)(0.73)}{(1.27)(0.27)} = 1.6 \text{ (within rounding of 1.56)}$$

A Simplifying Assumption for Finding the Unknown In many cases, we can use chemical "common sense" to make an assumption that simplifies the math by avoiding the need to use the quadratic formula to find x. In general,

- *if a reaction has a relatively small K and a relatively large initial reactant concentration, the concentration change (x) can often be neglected.*

This assumption does not mean that $x = 0$, because then there would be no reaction. It means that if a reaction starts with a high $[\text{reactant}]_{\text{init}}$ and proceeds very little to reach equilibrium (small K), the reactant concentration at equilibrium, $[\text{reactant}]_{\text{eq}}$, will be nearly the same as $[\text{reactant}]_{\text{init}}$.

Here's an everyday analogy for this assumption. On a bathroom scale, you weigh 158 lb. Take off your wristwatch, and you still weigh 158 lb. Within the accuracy of the scale, the weight of the watch is so small compared with your body weight that it can be neglected:

$$\text{Initial body weight} - \text{weight of watch} = \text{final body weight} \approx \text{initial body weight}$$

Thus, let's say the initial concentration of A is 0.500 M and, because of a small K_c, the concentration of A that reacts is 0.002 M. We can then assume that

$$0.500\ M - 0.002\ M = 0.498\ M \approx 0.500\ M$$

that is,

$$[\text{A}]_{\text{init}} - [\text{A}]_{\text{reacting}} = [\text{A}]_{\text{eq}} \approx [\text{A}]_{\text{init}} \quad \textbf{(17.9)}$$

To justify the assumption that x is negligible, we make sure the error introduced is not significant. One common criterion for "significant" is the 5% rule:

- *if the assumption results in a change that is less than 5% of the initial concentration, the error is not significant, and the assumption is justified.*

Let's see how making this assumption simplifies the math and if it is justified in the cases of two different $[\text{reactant}]_{\text{init}}$ values.

Student Hot Spot

Student data indicate that you may struggle with applying the simplifying assumption in equilibrium problems. Access the Smartbook to view additional Learning Resources on this topic.

SAMPLE PROBLEM 17.9 Making a Simplifying Assumption to Calculate Equilibrium Concentrations

Problem Phosgene is a potent chemical warfare agent that is now outlawed by international agreement. It decomposes by the reaction

$$COCl_2(g) \rightleftharpoons CO(g) + Cl_2(g) \qquad K_c = 8.3\times10^{-4} \text{ (at 360°C)}$$

Calculate [CO], $[Cl_2]$, and $[COCl_2]$ when each of the following amounts of phosgene decomposes and reaches equilibrium in a 10.0-L flask:

(a) 5.00 mol of $COCl_2$ **(b)** 0.100 mol of $COCl_2$

Plan We know from the balanced equation that when x mol of $COCl_2$ decomposes, x mol of CO and x mol of Cl_2 form. We use the volume (10.0 L) to convert amount (5.00 mol or 0.100 mol) to molar concentration, define x and set up the reaction table, and substitute the values into Q_c. Before using the quadratic formula, we assume that x is negligibly small. After solving for x, we check the assumption and find the equilibrium concentrations. If the assumption is not justified, we use the quadratic formula to find x.

Solution **(a)** For 5.00 mol of $COCl_2$. Writing the reaction quotient:

$$Q_c = \frac{[CO][Cl_2]}{[COCl_2]}$$

Calculating the initial reactant concentration, $[COCl_2]_{\text{init}}$:

$$[COCl_2]_{\text{init}} = \frac{5.00\text{ mol}}{10.0\text{ L}} = 0.500\ M$$

Setting up the reaction table, with x equal to $[COCl_2]_{reacting}$:

Concentration (*M*)	$COCl_2(g)$	$\rightleftharpoons$	$CO(g)$	+	$Cl_2(g)$
Initial	0.500		0		0
Change	$-x$		$+x$		$+x$
Equilibrium	$0.500 - x$		x		x

If we use the equilibrium values in Q_c with the given K_c, we obtain

$$Q_c = \frac{[CO][Cl_2]}{[COCl_2]} = \frac{x^2}{0.500 - x} = K_c = 8.3\times10^{-4}$$

Because K_c is small, the reaction does not proceed very far to the right, so let's assume that x ($[COCl_2]_{reacting}$) can be neglected. In other words, we assume that the equilibrium concentration is nearly the same as the initial concentration, 0.500 *M*:

$$[COCl_2]_{init} - [COCl_2]_{reacting} = [COCl_2]_{eq}$$
$$0.500\ M - x \approx 0.500\ M$$

Using this assumption, we substitute and solve for x:

$$K_c = 8.3\times10^{-4} \approx \frac{x^2}{0.500}$$
$$x^2 \approx (8.3\times10^{-4})(0.500) \quad \text{so} \quad x \approx 2.0\times10^{-2}$$

Checking the assumption by seeing if the error is <5%:

$$\frac{[\text{Change}]}{[\text{Initial}]} \times 100 = \frac{2.0\times10^{-2}}{0.500} \times 100 = 4\% \ (<5\%, \text{ so the assumption is justified})$$

Solving for the equilibrium concentrations:

$$[CO] = [Cl_2] = x = 2.0\times10^{-2}\ M$$
$$[COCl_2] = 0.500\ M - x = 0.480\ M$$

(b) For 0.100 mol of $COCl_2$. The calculation in this case is the same as the calculation in part (a), except that $[COCl_2]_{init}$ = 0.100 mol/10.0 L = 0.0100 *M*. Thus, at equilibrium,

$$Q_c = \frac{[CO][Cl_2]}{[COCl_2]} = \frac{x^2}{0.0100 - x} = K_c = 8.3\times10^{-4}$$

Making the assumption that $0.0100\ M - x \approx 0.0100\ M$ and solving for x:

$$K_c = 8.3\times10^{-4} \approx \frac{x^2}{0.0100} \quad \text{so} \quad x \approx 2.9\times10^{-3}$$

Checking the assumption:

$$\frac{2.9\times10^{-3}}{0.0100} \times 100 = 29\% \ (>5\%, \text{ so the assumption is } \textit{not} \text{ justified})$$

Rearranging Q_c and its value at equilibrium into the form of a quadratic equation gives

$$8.3\times10^{-4} = \frac{x^2}{0.0100 - x} \quad \text{or} \quad x^2 = (8.3\times10^{-6}) - (8.3\times10^{-4})x$$

and so

$$x^2 + (8.3\times10^{-4})x - (8.3\times10^{-6}) = 0$$

Solving this equation with the quadratic formula shows that

$$x = \frac{-8.3\times10^{-4} \pm \sqrt{(8.3\times10^{-4})^2 - 4(1)(-8.3\times10^{-6})}}{2(1)}$$

The only meaningful value of x is 2.5×10^{-3}.

Solving for the equilibrium concentrations:

$$[CO] = [Cl_2] = x = 2.5\times10^{-3}\ M$$
$$[COCl_2] = 1.00\times10^{-2}\ M - x = 7.5\times10^{-3}\ M$$

Check Once again, use the calculated values to be sure you obtain the given K_c.

Comment The main point is that the simplifying assumption was justified at the high $[COCl_2]_{init}$ in part (a) but *not* at the low $[COCl_2]_{init}$ in part (b). The amount of $[COCl_2]_{reacting}$

is small in (b), but since $[COCl_2]_{init}$ is also small, $[COCl_2]_{reacting}$ is a high percentage of $[COCl_2]_{init}$ and cannot be neglected without introducing significant error.

FOLLOW-UP PROBLEMS

17.9A In a study of the effect of temperature on halogen decomposition, 0.50 mol of I_2 was heated in a 2.5-L vessel, and the following reaction occurred: $I_2(g) \rightleftharpoons 2I(g)$
(a) Calculate $[I_2]$ and $[I]$ at equilibrium at 600 K; $K_c = 2.94\times10^{-10}$.
(b) Calculate $[I_2]$ and $[I]$ at equilibrium at 2000 K; $K_c = 0.209$.

17.9B Phosphorus pentachloride decomposes as follows:

$$PCl_5(g) \rightleftharpoons PCl_3(g) + Cl_2(g) \qquad K_p = 3.4\times10^{-4} \text{ (at 400. K)}$$

Calculate the equilibrium partial pressure of PCl_5 if the reaction begins with:
(a) a PCl_5 partial pressure of 0.18 atm;
(b) a PCl_5 partial pressure of 0.025 atm.

SOME SIMILAR PROBLEMS 17.49–17.51

Predicting When the Assumption Will Be Justified To summarize, we assume that x ($[A]_{reacting}$) can be neglected if K_c is relatively small and/or $[A]_{init}$ is relatively large. The same holds for K_p and $P_{A(init)}$. But *how* small or large must these variables be? Here's a benchmark for deciding when to make the assumption:

- If $\dfrac{[A]_{init}}{K_c} > 400$, the assumption is justified: neglecting x introduces an error <5%.
- If $\dfrac{[A]_{init}}{K_c} < 400$, the assumption is *not* justified; neglecting x introduces an error >5%, so we solve a quadratic equation to find x.

For example, using the values from Sample Problem 17.9, we have

Part (a): For $[A]_{init} = 0.500\ M$, $\dfrac{0.500}{8.3\times10^{-4}} = 6.0\times10^2$, which is greater than 400.

Part (b): For $[A]_{init} = 0.0100\ M$, $\dfrac{0.0100}{8.3\times10^{-4}} = 12$, which is less than 400.

We will make a similar assumption in many problems in Chapters 18 and 19.

Problems Involving Mixtures of Reactants and Products

In the problems so far, the reaction *had* to go toward products because it started with only reactants. And therefore, in the reaction tables, we knew that the unknown change in reactant concentration had a negative sign ($-x$) and the change in product concentration had a positive sign ($+x$). If, however, we start with a *mixture* of reactants and products, the reaction direction is not obvious. In those cases, we apply the idea from Section 17.4 and first *compare the values of Q and K to find the direction in which the reaction is proceeding to reach equilibrium.* (To focus on this idea, Sample Problem 17.10 uses concentrations that avoid the need for the quadratic formula.)

Student Hot Spot

Student data indicate that you may struggle with this type of equilibrium calculation. Access the Smartbook to view additional Learning Resources on this topic.

SAMPLE PROBLEM 17.10 Predicting Reaction Direction and Calculating Equilibrium Concentrations

Problem The research and development unit of a chemical company is studying the reaction of CH_4 and H_2S, two components of natural gas:

$$CH_4(g) + 2H_2S(g) \rightleftharpoons CS_2(g) + 4H_2(g)$$

In one experiment, 1.00 mol of CH_4, 1.00 mol of CS_2, 2.00 mol of H_2S, and 2.00 mol of H_2 are mixed in a 250.-mL vessel at 960°C. At this temperature, $K_c = 0.036$.
(a) In which direction will the reaction proceed to reach equilibrium?
(b) If $[CH_4] = 5.56\ M$ at equilibrium, what are the equilibrium concentrations of the other substances?

Plan **(a)** To find the direction, we convert the given initial amounts and volume (0.250 L) to concentrations, calculate Q_c, and compare it with K_c. **(b)** Based on this information, we determine the sign of each concentration change for the reaction table and then use the known $[CH_4]$ at equilibrium (5.56 *M*) to determine *x* and the other equilibrium concentrations.

Solution **(a)** Calculating the initial concentrations:

$$[CH_4] = \frac{1.00 \text{ mol}}{0.250 \text{ L}} = 4.00\ M$$

Similarly, $[H_2S] = 8.00\ M$, $[CS_2] = 4.00\ M$, and $[H_2] = 8.00\ M$.

Calculating the value of Q_c:

$$Q_c = \frac{[CS_2][H_2]^4}{[CH_4][H_2S]^2} = \frac{(4.00)(8.00)^4}{(4.00)(8.00)^2} = 64.0$$

Comparing Q_c and K_c: $Q_c > K_c$ (64.0 > 0.036), so the reaction proceeds to the left. Therefore, concentrations of reactants increase (+*x*) and those of products decrease (–*x*).

(b) Setting up a reaction table, with $x = [CS_2]$ that reacts, which equals $[CH_4]$ that forms:

Concentration (*M*)	$CH_4(g)$	+	$2H_2S(g)$	$\rightleftharpoons$	$CS_2(g)$	+	$4H_2(g)$
Initial	4.00		8.00		4.00		8.00
Change	$+x$		$+2x$		$-x$		$-4x$
Equilibrium	$4.00 + x$		$8.00 + 2x$		$4.00 - x$		$8.00 - 4x$

Solving for *x*: At equilibrium,

$$[CH_4] = 5.56\ M = 4.00\ M + x \qquad \text{so} \qquad x = 1.56\ M$$

Thus,

$$[H_2S] = 8.00\ M + 2x = 8.00\ M + 2(1.56\ M) = 11.12\ M$$

$$[CS_2] = 4.00\ M - x = 4.00\ M - 1.56\ M = 2.44\ M$$

$$[H_2] = 8.00\ M - 4x = 8.00\ M - 4(1.56\ M) = 1.76\ M$$

Check The comparison of Q_c and K_c showed the reaction proceeding to the left. The given data from part (b) confirm this because $[CH_4]$ increases from 4.00 *M* to 5.56 *M* during the reaction. As always, you should check that the concentrations give the known K_c:

$$\frac{(2.44)(1.76)^4}{(5.56)(11.12)^2} = 0.0341, \text{ which is close to } 0.036$$

FOLLOW-UP PROBLEMS

17.10A An inorganic chemist studying the reactions of phosphorus halides mixes 0.1050 mol of PCl_5 with 0.0450 mol of Cl_2 and 0.0450 mol of PCl_3 in a 0.5000-L flask at 250°C:

$$PCl_5(g) \rightleftharpoons PCl_3(g) + Cl_2(g) \qquad K_c = 4.2\times10^{-2}$$

(a) In which direction will the reaction proceed?
(b) If $[PCl_5] = 0.2065\ M$ at equilibrium, what are the equilibrium concentrations of the other components?

17.10B A chemist studying the production of nitrogen monoxide (NO) in the atmosphere adds 0.500 atm of N_2, 0.500 atm of O_2, and 0.750 atm of NO to a sealed container at 2500°C:

$$N_2(g) + O_2(g) \rightleftharpoons 2NO(g) \qquad K_p = 8.44\times10^{-3}$$

Determine the direction in which the reaction proceeds, and calculate the equilibrium pressures of all three gases.

SOME SIMILAR PROBLEMS 17.94 and 17.102

Figure 17.6 on the next page groups the steps for solving equilibrium problems in which you know *K* and some initial quantities and must find the equilibrium quantities.

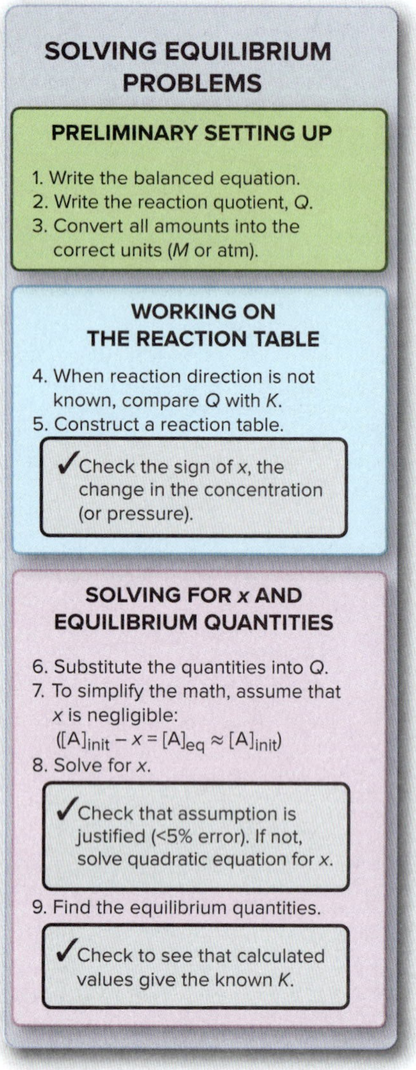
SOLVING EQUILIBRIUM PROBLEMS
PRELIMINARY SETTING UP
1. Write the balanced equation.
2. Write the reaction quotient, Q.
3. Convert all amounts into the correct units (M or atm).
WORKING ON THE REACTION TABLE
4. When reaction direction is not known, compare Q with K.
5. Construct a reaction table.
✓Check the sign of x, the change in the concentration (or pressure).
SOLVING FOR x AND EQUILIBRIUM QUANTITIES
6. Substitute the quantities into Q.
7. To simplify the math, assume that x is negligible: ($[A]_{init} - x = [A]_{eq} \approx [A]_{init}$)
8. Solve for x.
✓Check that assumption is justified (<5% error). If not, solve quadratic equation for x.
9. Find the equilibrium quantities.
✓Check to see that calculated values give the known K.

Figure 17.6 Steps in solving equilibrium problems.

Le Châtelier's principle applies to many systems.

Source: © AfriPics.com/Alamy

› Summary of Section 17.5

- In equilibrium problems, we typically use quantities (concentrations or pressures) of reactants and products to find K, or we use K to find quantities.
- Reaction tables summarize the initial quantities, their changes during the reaction, and the equilibrium quantities.
- To simplify calculations, we assume that if K is small and the initial quantity of reactant is large, the unknown change in reactant (x) can be neglected. If this assumption is not justified (that is, if the resulting error is greater than 5%), we use the quadratic formula to find x.
- For reactions that start with a mixture of reactants and products, we first determine reaction direction by comparing Q and K to decide on the sign of x.

17.6 REACTION CONDITIONS AND EQUILIBRIUM: LE CHÂTELIER'S PRINCIPLE

Change conditions so that a system is no longer at equilibrium, and it has the remarkable ability to adjust itself and reattain equilibrium. This phenomenon is described by **Le Châtelier's principle:** when a chemical system at equilibrium is disturbed, it reattains equilibrium by undergoing a net reaction that reduces the effect of the disturbance.

Two phrases in this statement *need* further explanation:

1. How is a system "disturbed"? At equilibrium, Q equals K. The system is disturbed when a change in conditions forces it temporarily out of equilibrium ($Q \neq K$). Three common disturbances are a change in concentration, a change in pressure (caused by a change in volume), or a change in temperature. We'll discuss each below.

2. What does a "net reaction" mean? This phrase refers to a shift in the *equilibrium position* to the right or left. The *equilibrium position* is defined by the specific equilibrium concentrations (or pressures). *Concentrations (or pressures) change in a way that reduces the effect of the change in conditions, and the system attains a new equilibrium position:*

- a shift to the right is a net reaction that converts reactant to product until equilibrium is reattained;
- a shift to the left is a net reaction that converts product to reactant until equilibrium is reattained.

For the remainder of this section, we'll examine a system at equilibrium to see how it responds to changes in concentration, pressure (volume), or temperature; then, we'll see what happens when we add a catalyst. As our example, we use the gaseous reaction between phosphorus trichloride and chlorine to produce phosphorus pentachloride:

$$PCl_3(g) + Cl_2(g) \rightleftharpoons PCl_5(g)$$

Le Châtelier's principle holds for many systems at equilibrium in both the natural and social sciences—the effects of disease or drought on populations of predators and prey on the African savannah *(see photo)*, a change in the cost of petroleum based on a balance of supply and demand, and even the formation of certain elements in the core of a star.

The Effect of a Change in Concentration

When a system at equilibrium is disturbed by a change in concentration of one of the components, it reacts in the direction that reduces the change:

- If the concentration of A is increased, the system reacts to consume some of that component.
- If the concentration of A is decreased, the system reacts to produce some of it.

Only components that appear in Q can have an effect, so changes in the amounts of pure liquids and solids cannot.

A Qualitative View of a Concentration Change At 523 K, the PCl_3-Cl_2-PCl_5 system reaches equilibrium when

$$PCl_3(g) + Cl_2(g) \rightleftharpoons PCl_5(g) \qquad Q_c = \frac{[PCl_5]}{[PCl_3][Cl_2]} = 24.0 = K_c$$

Starting with Q_c equal to K_c, let's think through some changes in concentration:

1. *Adding a reactant.* What happens if we disturb the system by adding some Cl_2 gas? To reduce this disturbance, the system will consume some of the added Cl_2 by shifting toward product. With regard to the reaction quotient, when the $[Cl_2]$ term increases, the value of Q_c decreases; thus, $Q_c < K_c$. As some of the added Cl_2 reacts with some of the PCl_3 to form more PCl_5, the denominator becomes smaller again and the numerator larger, until eventually $Q_c = K_c$ again. Notice the changes in the new equilibrium concentrations: $[Cl_2]$ and $[PCl_5]$ are higher than in the original equilibrium position, and $[PCl_3]$ is lower. Nevertheless, the ratio of values gives the same K_c. Thus, *the equilibrium position shifts to the right when a component on the left is added:*

$$PCl_3 + Cl_2(\text{added}) \longrightarrow PCl_5$$

2. *Removing a reactant.* What happens if we disturb the system by removing some PCl_3? To reduce this disturbance, the system will replace the PCl_3 by consuming some PCl_5 and proceeding toward reactants. With regard to Q_c, when the $[PCl_3]$ term decreases, Q_c increases, so $Q_c > K_c$. As some PCl_5 decomposes to PCl_3 and Cl_2, the numerator decreases and the denominator increases until $Q_c = K_c$ again. Once again, the new and old equilibrium concentrations are different, but the value of K_c is not. Thus, *the equilibrium position shifts to the left when a component on the left is removed:*

$$PCl_3(\text{removed}) + Cl_2 \longleftarrow PCl_5$$

3. *Adding or removing a product.* The same points we just made for a reactant hold for a product. If we add PCl_5, the equilibrium position shifts to the left to consume PCl_5; if we remove some PCl_5, the equilibrium position shifts to the right to produce PCl_5.

In other words, no matter how the disturbance in concentration comes about, *the system reacts to consume some of the added substance or produce some of the removed substance* to make $Q_c = K_c$ again (Figure 17.7):

- The equilibrium position shifts to the *right* if a reactant is added or a product is removed: [reactant] increases or [product] decreases.
- The equilibrium position shifts to the *left* if a reactant is removed or a product is added: [reactant] decreases or [product] increases.

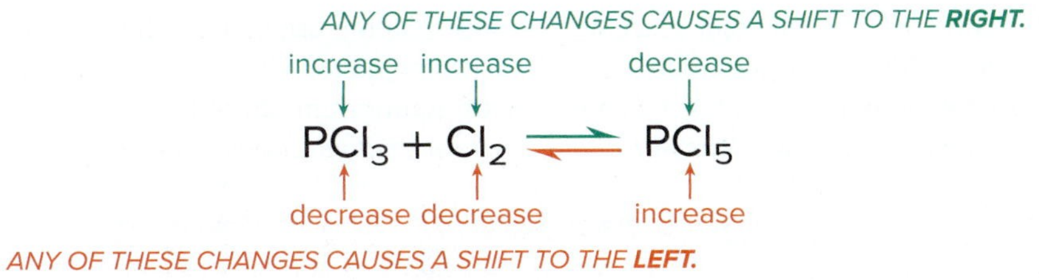
ANY OF THESE CHANGES CAUSES A SHIFT TO THE RIGHT.
increase
increase
decrease
PCl3 + Cl2 ⇌ PCl5
decrease
decrease
increase
ANY OF THESE CHANGES CAUSES A SHIFT TO THE LEFT.

Figure 17.7 The effect of a change in concentration on a system at equilibrium.

A Quantitative View of a Concentration Change As you saw, the system "reduces the effect of the disturbance." But the effect is not completely eliminated, as we can see from a quantitative comparison of original and new equilibrium positions.

Consider what happens when we add Cl_2 to a system whose original equilibrium position was established with $[PCl_3] = 0.200\ M$, $[Cl_2] = 0.125\ M$, and $[PCl_5] = 0.600\ M$. That is,

$$Q_c = \frac{[PCl_5]}{[PCl_3][Cl_2]} = \frac{0.600}{(0.200)(0.125)} = 24.0 = K_c$$

Table 17.3 The Effect of Added Cl_2 on the PCl_3-Cl_2-PCl_5 System

Concentration (*M*)	$PCl_3(g)$	+	$Cl_2(g)$	$\rightleftharpoons$	$PCl_5(g)$
Original equilibrium	0.200		0.125		0.600
Disturbance			+0.075		
New initial	0.200		0.200		0.600
Change	$-x$		$-x$		$+x$
New equilibrium	$0.200 - x$		$0.200 - x$		$0.600 + x$ (0.637)*

*Experimentally determined value.

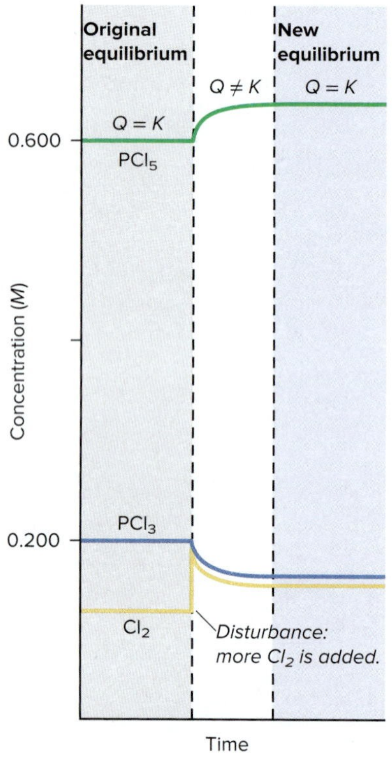

Figure 17.8 The effect of added Cl_2 on the PCl_3-Cl_2-PCl_5 system. The original equilibrium concentrations are shown at left *(gray region)*. When Cl_2 *(yellow curve)* is added, its concentration increases instantly *(vertical part of yellow curve)* and then falls gradually as it reacts with PCl_3 to form more PCl_5. Soon, equilibrium is re-established at new concentrations *(blue region)* but with the same *K*.

Now we add enough Cl_2 to increase its concentration by 0.075 *M* to a new $[Cl_2]_{init}$ of 0.200 *M*; the system is no longer at equilibrium since $Q_c < K_c$:

$$Q_c = \frac{[PCl_5]}{[PCl_3][Cl_2]} = \frac{0.600}{(0.200)(0.200)} = 15.0 < K_c$$

The reaction proceeds, and the system comes to a new equilibrium position. From Le Châtelier's principle, we predict that adding more reactant will shift the equilibrium position to the right. Experimental measurement shows that the new $[PCl_5]_{eq}$ is 0.637 *M*.

Table 17.3 shows a reaction table for the entire process: the original equilibrium position, the disturbance, the (new) initial concentrations, the direction of x (the change needed to reattain equilibrium), and the new equilibrium position. Figure 17.8 depicts the process.

Let's determine the new equilibrium concentrations. From Table 17.3,

$$[PCl_5] = 0.600\ M + x = 0.637\ M \qquad \text{so} \qquad x = 0.037\ M$$

Thus, $[PCl_3] = [Cl_2] = 0.200\ M - x = 0.200\ M - 0.037\ M = 0.163\ M$

Therefore, at equilibrium,

$$K_{c(\text{original})} = \frac{0.600}{(0.200)(0.125)} = 24.0$$

$$K_{c(\text{new})} = \frac{0.637}{(0.163)(0.163)} = 24.0$$

There are several key points to notice about the new equilibrium concentrations:

- As we predicted, $[PCl_5]$ (0.637 *M*) is higher than the original concentration (0.600 *M*).
- $[Cl_2]$ (0.163 *M*) is higher than the *original* equilibrium concentration (0.125 *M*), but lower than the new initial concentration (0.200 *M*); thus, the disturbance is *reduced but not eliminated.*
- $[PCl_3]$ (0.163 *M*), the other reactant, is lower than the original equilibrium concentration (0.200 *M*) because some reacted with the added Cl_2.
- Most importantly, although the position of equilibrium shifted to the right, *at a given temperature, K_c does **not** change with a change in concentration.*

SAMPLE PROBLEM 17.11 Predicting the Effect of a Change in Concentration on the Equilibrium Position

Problem To improve air quality and obtain a useful product, chemists often remove sulfur from coal and natural gas by treating the contaminant hydrogen sulfide with O_2:

$$2H_2S(g) + O_2(g) \rightleftharpoons 2S(s) + 2H_2O(g)$$

What happens to

(a) $[H_2O]$ if O_2 is added?
(b) $[H_2S]$ if O_2 is added?
(c) $[O_2]$ if H_2S is removed?
(d) $[H_2S]$ if sulfur is added?

Plan We write the reaction quotient to see how Q_c is affected by each disturbance, relative to K_c. In each case, the effect tells us the direction in which the reaction proceeds for the system to reattain equilibrium and how each concentration changes.

Solution Writing the reaction quotient (solid sulfur is not included): $Q_c = \dfrac{[H_2O]^2}{[H_2S]^2[O_2]}$

(a) When O_2 is added, the denominator of Q_c increases, so $Q_c < K_c$. The reaction proceeds to the right until $Q_c = K_c$ again, so $[H_2O]$ increases.

(b) As in part (a), when O_2 is added, $Q_c < K_c$. Some H_2S reacts with the added O_2 as the reaction proceeds to the right, so $[H_2S]$ decreases.

(c) When H_2S is removed, the denominator of Q_c decreases, so $Q_c > K_c$. As the reaction proceeds to the left to re-form H_2S, more O_2 forms as well, so $[O_2]$ increases.

(d) The *concentration* of solid S does not change, so it does not appear in the reaction quotient. As long as some S is present, adding more S has no effect, so $[H_2S]$ is unchanged (but see Comment 2 below).

Check Apply Le Châtelier's principle to see that the reaction proceeds in the direction that lowers the increased concentration or raises the decreased concentration.

Comment **1.** As you know, sulfur exists most commonly as S_8. How would this change in formula affect the answers? The balanced equation and Q_c would be

$$8H_2S(g) + 4O_2(g) \rightleftharpoons S_8(s) + 8H_2O(g) \qquad Q_c = \frac{[H_2O]^8}{[H_2S]^8[O_2]^4}$$

The value of K_c is different for this equation, but the changes described in the problem have the same effects. Thus, shifts in equilibrium position predicted by Le Châtelier's principle are not affected by a change in the balancing coefficients.

2. In part (d), you saw that adding a solid has no effect on the concentrations of other components: because *the* ***concentration*** *of the solid cannot change,* it does not appear in Q. But *the* ***amount*** *of solid can change.* Adding H_2S shifts the reaction to the right, so more S forms.

FOLLOW-UP PROBLEMS

17.11A In a study of glass etching, a chemist examines the reaction between sand (SiO_2) and hydrogen fluoride at 150°C:

$$SiO_2(s) + 4HF(g) \rightleftharpoons SiF_4(g) + 2H_2O(g)$$

Predict the effect on $[SiF_4]$ when
(a) $H_2O(g)$ is removed.
(b) some liquid water is added.
(c) HF is removed.
(d) some sand is removed.

17.11B Water gas, a fuel mixture of carbon monoxide and hydrogen, is produced by passing steam over hot carbon:

$$C(s) + H_2O(g) \rightleftharpoons CO(g) + H_2(g)$$

For each of the following changes, what happens to [CO]?
(a) Carbon is added.
(b) $H_2O(g)$ is removed.
(c) $H_2(g)$ is removed.
(d) $H_2O(g)$ is added.

SOME SIMILAR PROBLEMS 17.64 and 17.65

The Effect of a Change in Pressure (Volume)

Changes in pressure can have a large effect on equilibrium systems containing gaseous components. (A change in pressure has a negligible effect on liquids and solids because they are nearly incompressible.) Pressure changes can occur in three ways:

- *Changing the concentration of a gaseous component.* We just considered the effect of changing the concentration of a component, and that reasoning applies here.
- *Adding an inert gas (one that does not take part in the reaction).* As long as the volume of the system is constant, adding an inert gas has no effect on the equilibrium position because *all concentrations, and thus partial pressures, remain the same.* Moreover, the inert gas does not appear in Q, so it cannot have an effect.
- *Changing the volume of the reaction vessel.* This change can cause a large shift in equilibrium position, but only for reactions in which the number of moles of gas, n_{gas}, changes.

Let's consider the two possible situations for the third way: changing the volume of the reaction vessel.

1. *Reactions in which n_{gas} changes.* Suppose the PCl_3-Cl_2-PCl_5 system is in a piston-cylinder assembly. We press down on the piston to halve the volume, so the gas pressure doubles (recall that pressure and volume are inversely proportional—Boyle's law). To reduce this disturbance, the system responds by *reducing the number of gas molecules* in order to reduce the pressure. And the only way to do that is through a net reaction toward the side with *fewer moles of gas,* in this case, toward product:

$$PCl_3(g) + Cl_2(g) \longrightarrow PCl_5(g)$$
$$\text{2 mol gas} \longrightarrow \text{1 mol gas}$$

Recall that $Q_c = \dfrac{[PCl_5]}{[PCl_3][Cl_2]}$ for this reaction. When the volume is halved, the concentrations double, but the denominator of Q_c is the product of two concentrations, so it quadruples while the numerator only doubles. Thus, Q_c becomes less than K_c. As a result, the system forms more PCl_5 and a new equilibrium position is reached.

Thus, for a system that consists of gases at equilibrium, in which the amount (mol) of gas, n_{gas}, changes during the reaction (Figure 17.9):

- If the volume becomes smaller (pressure is higher), the reaction shifts so that the total number of gas molecules decreases.
- If the volume becomes larger (pressure is lower), the reaction shifts so that the total number of gas molecules increases.

2. *Reactions in which n_{gas} does **not** change.* For the formation of hydrogen iodide from its elements, we have the same amount (mol) of gas on both sides:

$$H_2(g) + I_2(g) \rightleftharpoons 2HI(g)$$
$$\text{2 mol gas} \longrightarrow \text{2 mol gas}$$

Therefore, Q_c has the same number of terms in the numerator and denominator:

$$Q_c = \frac{[HI]^2}{[H_2][I_2]} = \frac{[HI][HI]}{[H_2][I_2]}$$

Because a change in volume has the same effect on the numerator and denominator, *there is **no** effect on the equilibrium position.*

In terms of the equilibrium constant, *a change in volume is, in effect, a change in concentration:* a decrease in volume raises the concentration, and vice versa. Therefore, like other changes in concentration, *a change in pressure due to a change in volume does **not** alter K_c.*

Student Hot Spot

Student data indicate that you may struggle with using Le Chatelier's Principle to predict a change in equilibrium. Access the Smartbook to view additional Learning Resources on this topic.

Figure 17.9 The effect of a change in pressure (volume) on a system at equilibrium.

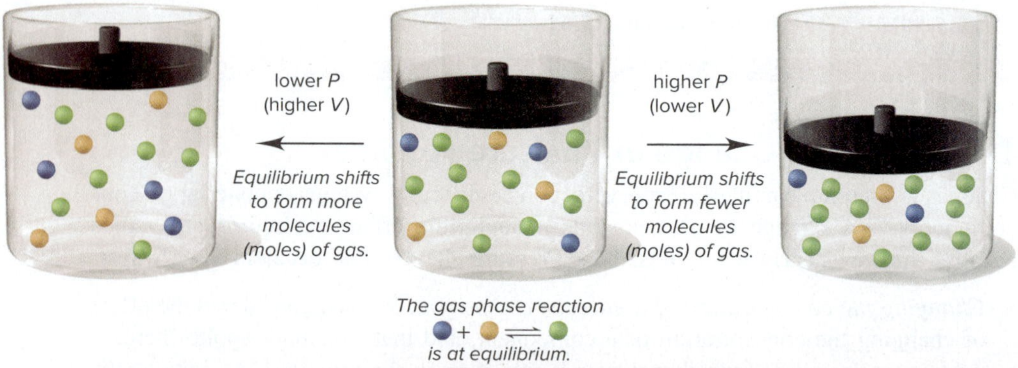
lower P
(higher V)
Equilibrium shifts
to form more
molecules
(moles) of gas.
higher P
(lower V)
Equilibrium shifts
to form fewer
molecules
(moles) of gas.
The gas phase reaction
is at equilibrium.

SAMPLE PROBLEM 17.12 Predicting the Effect of a Change in Volume (Pressure) on the Equilibrium Position

Problem How would you change the volume for each of the following reactions to *increase* the yield of the product(s)?

(a) $CaCO_3(s) \rightleftharpoons CaO(s) + CO_2(g)$ **(b)** $S(s) + 3F_2(g) \rightleftharpoons SF_6(g)$
(c) $Cl_2(g) + I_2(g) \rightleftharpoons 2ICl(g)$

Plan Whenever gases are present, a change in volume causes a change in concentration. For reactions in which the number of moles of gas changes, if the volume decreases (pressure increases), the equilibrium position shifts to lower the pressure by reducing the number of moles of gas. A volume increase (pressure decrease) has the opposite effect.

Solution **(a)** The only gas is the product CO_2. To make the system produce more molecules of gas, that is, more CO_2, we increase the volume (decrease the pressure).

(b) With 3 mol of gas on the left and only 1 mol on the right, we decrease the volume (increase the pressure) to form fewer molecules of gas and, thus, more SF_6.

(c) The number of moles of gas is the same on both sides of the equation, so a change in volume (pressure) will have no effect on the yield of ICl.

Check Let's predict the relative values of Q_c and K_c.
(a) $Q_c = [CO_2]$, so increasing the volume will make $Q_c < K_c$, and the system will yield more CO_2.
(b) $Q_c = [SF_6]/[F_2]^3$. Lowering the volume increases $[F_2]$ and $[SF_6]$ proportionately, but Q_c decreases because of the exponent 3 in the denominator. To make $Q_c = K_c$ again, $[SF_6]$ must increase.
(c) $Q_c = [ICl]^2/[Cl_2][I_2]$. A change in volume (pressure) affects the numerator (2 mol) and denominator (2 mol) equally, so it will have no effect.

FOLLOW-UP PROBLEMS

17.12A Would you increase or decrease the pressure (via a volume change) of each of the following reaction mixtures to *decrease* the yield of product(s)?
(a) $2SO_2(g) + O_2(g) \rightleftharpoons 2SO_3(g)$
(b) $4NH_3(g) + 5O_2(g) \rightleftharpoons 4NO(g) + 6H_2O(g)$
(c) $CaC_2O_4(s) \rightleftharpoons CaCO_3(s) + CO(g)$

17.12B Predict the effect of an increase in container volume (decrease in P) on the yield of products in each of the following:
(a) $CH_4(g) + CO_2(g) \rightleftharpoons 2CO(g) + 2H_2(g)$
(b) $NO(g) + CO_2(g) \rightleftharpoons NO_2(g) + CO(g)$
(c) $2H_2S(g) + SO_2(g) \rightleftharpoons 3S(s) + 2H_2O(g)$

SOME SIMILAR PROBLEMS 17.66–17.71

The Effect of a Change in Temperature

Of the three types of disturbances that may occur—a change in concentration, pressure, or temperature—*only temperature changes alter the value of K.* To see why, let's focus on the sign of ΔH°_{rxn}:

$$PCl_3(g) + Cl_2(g) \rightleftharpoons PCl_5(g) \qquad \Delta H^\circ_{rxn} = -111 \text{ kJ}$$

The forward reaction is exothermic (releases heat; $\Delta H^\circ_{rxn} < 0$), so the reverse reaction is endothermic (absorbs heat; $\Delta H^\circ_{rxn} > 0$):

$$PCl_3(g) + Cl_2(g) \longrightarrow PCl_5(g) + \textbf{heat} \text{ (exothermic)}$$
$$PCl_3(g) + Cl_2(g) \longleftarrow PCl_5(g) + \textbf{heat} \text{ (endothermic)}$$

If we consider *heat as a component of the equilibrium system,* a rise in temperature occurs when heat is "added" to the system and a drop in temperature occurs when heat is "removed" from the system. As with a change in any other component, the system shifts to reduce the effect of the change. Therefore, *a temperature increase (adding heat) favors the endothermic (heat-absorbing) direction, and a temperature decrease (removing heat) favors the exothermic (heat-releasing) direction.*

If we start with the system at equilibrium, Q_c equals K_c. Increase the temperature, and the system absorbs the added heat by decomposing some PCl_5 to PCl_3 and Cl_2. The denominator of Q_c becomes larger and the numerator smaller, so the system reaches a new equilibrium position at a smaller ratio of concentration terms, that is, a lower K_c. Similarly, if the temperature drops, the system releases more heat by forming more PCl_5 from some PCl_3 and Cl_2. The numerator of Q_c becomes larger, the denominator smaller, and the new equilibrium position has a higher K_c. Thus,

- *A temperature rise will increase K_c for a system with a positive ΔH°_{rxn}.*
- *A temperature rise will decrease K_c for a system with a negative ΔH°_{rxn}.*

Let's review these ideas with a sample problem.

SAMPLE PROBLEM 17.13 Predicting the Effect of a Change in Temperature on the Equilibrium Position

Problem How does an *increase* in temperature affect the equilibrium concentration of the underlined substance and K for each of the following reactions?

(a) $CaO(s) + H_2O(l) \rightleftharpoons \underline{Ca(OH)_2(aq)}$ $\Delta H° = -82$ kJ
(b) $CaCO_3(s) \rightleftharpoons CaO(s) + \underline{CO_2(g)}$ $\Delta H° = 178$ kJ
(c) $\underline{SO_2(g)} \rightleftharpoons S(s) + O_2(g)$ $\Delta H° = 297$ kJ

Plan We write each equation to show heat as a reactant or product. The temperature increases when we add heat, so the system shifts to absorb the heat; that is, the endothermic reaction occurs. Thus, K will increase if the forward reaction is endothermic and decrease if it is exothermic.

Solution **(a)** $CaO(s) + H_2O(l) \rightleftharpoons Ca(OH)_2(aq) + \textbf{\textit{heat}}$

Adding heat shifts the system to the left: $[Ca(OH)_2]$ and K will decrease.

(b) $CaCO_3(s) + \textbf{\textit{heat}} \rightleftharpoons CaO(s) + CO_2(g)$

Adding heat shifts the system to the right: $[CO_2]$ and K will increase.

(c) $SO_2(g) + \textbf{\textit{heat}} \rightleftharpoons S(s) + O_2(g)$

Adding heat shifts the system to the right: $[SO_2]$ will decrease and K will increase.

Check Check your answers by reasoning through a *decrease* in temperature: heat is removed and the exothermic direction is favored. All the answers should be opposite.

Comment Note that, as part (a) shows, these ideas hold for solutions as well.

FOLLOW-UP PROBLEMS

17.13A How does a *decrease* in T affect the partial pressure of the underlined substance and the value of K for each of the following reactions?

(a) $C(\text{graphite}) + \underline{2H_2(g)} \rightleftharpoons CH_4(g)$ $\Delta H° = -75$ kJ
(b) $\underline{N_2(g)} + O_2(g) \rightleftharpoons 2NO(g)$ $\Delta H° = 181$ kJ
(c) $P_4(s) + 10Cl_2(g) \rightleftharpoons \underline{4PCl_5(g)}$ $\Delta H° = -1528$ kJ

17.13B Should T be increased or decreased to yield more product(s) in each of the following reactions? What effect does the change in T have on the value of K for each reaction?

(a) $2Cl_2(g) + 2H_2O(g) \rightleftharpoons 4HCl(g) + O_2(g)$ $\Delta H° = 114$ kJ
(b) $2NO(g) + Cl_2(g) \rightleftharpoons 2NOCl(g)$ $\Delta H° = -77.2$ kJ
(c) $2HgO(s) \rightleftharpoons 2Hg(l) + O_2(g)$ $\Delta H° = 182$ kJ

SOME SIMILAR PROBLEMS 17.72 and 17.73

The van't Hoff Equation: The Effect of *T* on *K* The *van't Hoff equation* shows quantitatively how the equilibrium constant is affected by changes in temperature:

$$\ln \frac{K_2}{K_1} = -\frac{\Delta H°_{\text{rxn}}}{R}\left(\frac{1}{T_2} - \frac{1}{T_1}\right) \tag{17.10}$$

where K_1 is the equilibrium constant at T_1, K_2 is the equilibrium constant at T_2, and R is the universal gas constant (8.314 J/mol·K). If we know $\Delta H°_{\text{rxn}}$ and K at one temperature, the van't Hoff equation allows us to find K at any other temperature (or to find $\Delta H°_{\text{rxn}}$, given the two K's at two T's).

Equation 17.10 confirms the qualitative prediction from Le Châtelier's principle: for a temperature rise, we have

$$T_2 > T_1 \quad \text{and} \quad 1/T_2 < 1/T_1$$

so

$$1/T_2 - 1/T_1 < 0$$

Therefore,

- For an endothermic reaction ($\Delta H°_{\text{rxn}} > 0$), the $-(\Delta H°_{\text{rxn}}/R)$ term in Equation 17.10 is < 0. With $1/T_2 - 1/T_1 < 0$, the right side of the equation is > 0. Thus, $\ln (K_2/K_1) > 0$, so $K_2 > K_1$. An increase in temperature results in a shift to the right and an increase in K.

- For an exothermic reaction ($\Delta H^\circ_{rxn} < 0$), the $-(\Delta H^\circ_{rxn}/R)$ term in Equation 17.10 is > 0. With $1/T_2 - 1/T_1 < 0$, the right side of the equation is < 0. Thus, $\ln (K_2/K_1) < 0$, so $K_2 < K_1$. An increase in temperature results in a shift to the left and a decrease in K.

Here's a typical problem that requires the van't Hoff equation. Many coal gasification processes begin with the formation of syngas from carbon and steam:

$$C(s) + H_2O(g) \rightleftharpoons CO(g) + H_2(g) \qquad \Delta H^\circ_{rxn} = 131 \text{ kJ/mol}$$

An engineer knows that K_p is only 9.36×10^{-17} at 25°C and therefore wants to find a temperature that allows a much higher yield. Calculate K_p at 700.°C.

$$\ln \frac{K_2}{K_1} = -\frac{\Delta H^\circ_{rxn}}{R}\left(\frac{1}{T_2} - \frac{1}{T_1}\right)$$

The temperatures must be in kelvins, and the units of ΔH° and R must be made consistent:

$$\ln\left(\frac{K_{p2}}{9.36\times10^{-17}}\right) = -\frac{131\times10^3 \text{ J/mol}}{8.314 \text{ J/mol·K}}\left(\frac{1}{973 \text{ K}} - \frac{1}{298 \text{ K}}\right)$$

$$\frac{K_{p2}}{9.36\times10^{-17}} = 8.51\times10^{15}$$

$$K_{p2} = 0.797 \qquad \text{(a much higher yield)}$$

(For further practice with the van't Hoff equation, see Problems 17.74 and 17.75.)

The Lack of Effect of a Catalyst

Let's briefly consider what effect, if any, adding a catalyst will have on the system. Recall from Chapter 16 that a catalyst speeds up a reaction by lowering the activation energy, thereby increasing the forward *and* reverse rates to the same extent. Thus, *a catalyst shortens the time it takes to reach equilibrium but has* ***no*** *effect on the equilibrium position.* That is, if we add a catalyst to a mixture of PCl_3 and Cl_2 at 523 K, the system attains the *same* equilibrium concentrations of PCl_3, Cl_2, and PCl_5 *more quickly* than it does without the catalyst. As you'll see in a moment, however, catalysts play key roles in optimizing reaction systems.

Table 17.4 summarizes the effects of changing conditions. Many changes alter the equilibrium *position*, but only temperature changes alter the equilibrium *constant*. Sample Problem 17.14 shows how to visualize equilibrium at the molecular level.

Table 17.4 Effects of Various Disturbances on a System at Equilibrium

Disturbance	Effect on Equilibrium Position	Effect on Value of K
Concentration		
Increase [reactant]	Toward formation of product	None
Decrease [reactant]	Toward formation of reactant	None
Increase [product]	Toward formation of reactant	None
Decrease [product]	Toward formation of product	None
Pressure		
Increase P (decrease V)	Toward formation of fewer moles of gas	None
Decrease P (increase V)	Toward formation of more moles of gas	None
Increase P (add inert gas, no change in V)	None; concentrations unchanged	None
Temperature		
Increase T	Toward absorption of heat	Increases if $\Delta H^\circ_{rxn} > 0$ Decreases if $\Delta H^\circ_{rxn} < 0$
Decrease T	Toward release of heat	Increases if $\Delta H^\circ_{rxn} < 0$ Decreases if $\Delta H^\circ_{rxn} > 0$
Catalyst added	None; forward and reverse rates increase equally, and equilibrium is reached sooner	None

SAMPLE PROBLEM 17.14 Determining Equilibrium Parameters from Molecular Scenes

Problem For the reaction,

$$X(g) + Y_2(g) \rightleftharpoons XY(g) + Y(g) \qquad \Delta H > 0$$

the following molecular scenes depict different reaction mixtures (X is *green,* Y is *purple*):

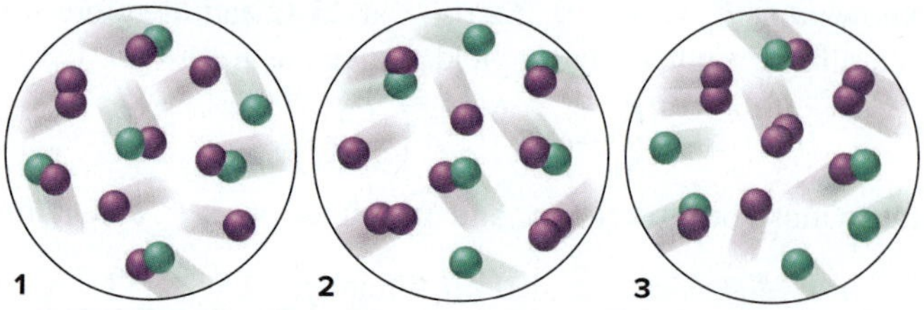

(a) If $K = 2$ at the temperature of the reaction, which scene represents the mixture at equilibrium? **(b)** Will the reaction mixtures in the other two scenes proceed toward reactants or toward products to reach equilibrium? **(c)** For the mixture at equilibrium, how will a rise in temperature affect $[Y_2]$?

Plan **(a)** We are given the balanced equation and K and must choose the scene that represents the mixture at equilibrium. We write Q, and for each scene, count particles and find the value of Q. Whichever scene gives a Q equal to K (that is, equal to 2) represents the mixture at equilibrium. **(b)** For each of the other two reaction mixtures, we compare the value of Q with 2. If $Q > K$, the numerator (product side) is too high, so the reaction proceeds toward reactants; if $Q < K$, the reaction proceeds toward products. **(c)** We are given that $\Delta H > 0$, so we must see whether a rise in T increases or decreases $[Y_2]$, one of the reactants.

Solution **(a)** For the reaction, we have $Q = \frac{[XY][Y]}{[X][Y_2]}$. Thus,

$$\text{scene 1}: Q = \frac{5 \times 3}{1 \times 1} = 15 \quad \text{scene 2}: Q = \frac{4 \times 2}{2 \times 2} = 2 \quad \text{scene 3}: Q = \frac{3 \times 1}{3 \times 3} = \frac{1}{3}$$

For scene 2, $Q = K$, so scene 2 represents the mixture at equilibrium.

(b) For scene 1, Q (15) $> K$ (2), so the reaction proceeds toward reactants.

For scene 3, Q ($\frac{1}{3}$) $< K$ (2), so the reaction proceeds toward products.

(c) The reaction is endothermic, so heat acts as a reactant:

$$X(g) + Y_2(g) + \textit{heat} \rightleftharpoons XY(g) + Y(g)$$

Therefore, adding heat shifts the reaction to the right, so $[Y_2]$ decreases.

Check **(a)** Remember that quantities in the numerator (or denominator) of Q are multiplied, not added. For example, the denominator for scene 1 is $1 \times 1 = 1$, not $1 + 1 = 2$.

(c) A good check is to imagine that $\Delta H < 0$ and see if you get the opposite result:

$$X(g) + Y_2(g) \rightleftharpoons XY(g) + Y(g) + \textit{heat}$$

If $\Delta H < 0$, adding heat would shift the reaction to the left and increase $[Y_2]$.

FOLLOW-UP PROBLEMS

17.14A For the reaction $C_2(g) + D_2(g) \rightleftharpoons 2CD(g)$, for which $\Delta H < 0$, these molecular scenes depict different reaction mixtures (C is *red,* D is *blue*):

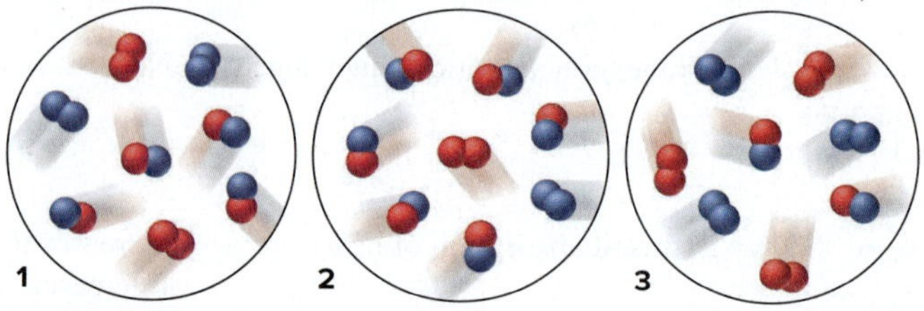

At equilibrium

(a) Calculate the value of K_p. **(b)** In which direction will the reaction proceed for the mixtures *not* at equilibrium? **(c)** For the mixture at equilibrium, what effect will a rise in T have on the total amount (mol) of gas (increase, decrease, no effect)? Explain.

17.14B For the reaction $A(g) + B(g) \rightleftharpoons AB(g)$, these molecular scenes depict reaction mixtures at two different temperatures (A is *red*, B is *green*):

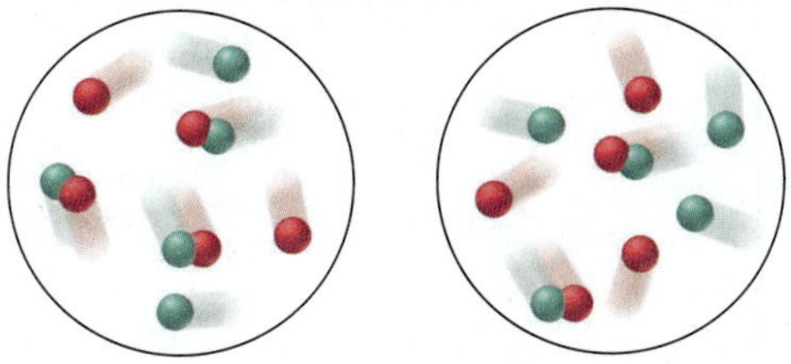

At equilibrium at T_1 At equilibrium at T_2

(a) Calculate the value of K at T_1. **(b)** If $T_2 < T_1$, what is the sign of ΔH for the reaction? **(c)** Calculate K at T_2.

SOME SIMILAR PROBLEMS 17.60 and 17.81

Applying Le Châtelier's Principle to the Synthesis of Ammonia

Le Châtelier's principle has countless applications in natural systems and in the chemical industry. As a case study, we'll look at the synthesis of ammonia, which, on a mole basis, is produced industrially in greater amount than any other compound. Then, a Chemical Connections essay applies the principle to cellular metabolism.

Even though four of every five molecules in the atmosphere are N_2, the supply of *usable* nitrogen is limited because the strong triple bond in N_2 lowers its reactivity. Thus, the N atom is very difficult to "fix," that is, to combine with other atoms into useful compounds. Natural nitrogen fixation occurs through the fine-tuned activity of enzymes found in bacteria that live on plant roots, or through the brute force of lightning. But, nearly 13% of nitrogen fixation is done industrially via the **Haber process:**

$$N_2(g) + 3H_2(g) \rightleftharpoons 2NH_3(g) \qquad \Delta H^\circ_{rxn} = -91.8 \text{ kJ}$$

Developed by the German chemist Fritz Haber in 1913 and first used in a plant making 12,000 tons of ammonia a year, the process now yields over 110 million tons a year. Over 80% of this is used as fertilizer, with most of the remainder used to make explosives and nylons and other polymers, and smaller amounts going into production of refrigerants, rubber stabilizers, household cleaners, and pharmaceuticals.

Optimizing Reaction Conditions: Yield vs. Rate The Haber process applies equilibrium *and* kinetics principles to achieve a compromise that makes the process economical. From the balanced equation, we see three ways to maximize NH_3 yield:

1. *Decrease [NH_3].* Removing NH_3 as it forms will make the system shift toward producing more to reattain equilibrium.
2. *Decrease volume (increase pressure).* Because 4 mol of reactant gases react to form 2 mol of product gas, decreasing the volume will shift the system toward making fewer moles of gas.
3. *Decrease temperature.* Because the formation of NH_3 is exothermic, decreasing the temperature (removing heat) will shift the equilibrium position toward product, thereby increasing K_c (Table 17.5).

Therefore, the conditions for maximizing the yield of product are continual removal of NH_3, high pressure, and low temperature. Figure 17.10 on the next page shows the percent yield of NH_3 at various combinations of pressure and temperature. Note the almost complete conversion (98.3%) to product at 1000 atm and 473 K (200.°C).

Although the *yield* is favored at this relatively low temperature, the *rate* of formation is so low that the process is uneconomical. The higher temperatures that would increase reaction rate would also result in a lower yield. In practice, a compromise optimizes yield *and* rate. High pressure and continuous removal are used to increase yield, but the temperature is raised to a moderate level and *a catalyst is used to increase the rate.* Achieving the same rate without a catalyst would require much higher temperatures and, thus, result in a much lower yield.

Table 17.5 Effect of Temperature on K_c for Ammonia Synthesis

T (K)	K_c
200.	7.17×10^{15}
300.	2.69×10^{8}
400.	3.94×10^{4}
500.	1.72×10^{2}
600.	4.53×10^{0}
700.	2.96×10^{-1}
800.	3.96×10^{-2}

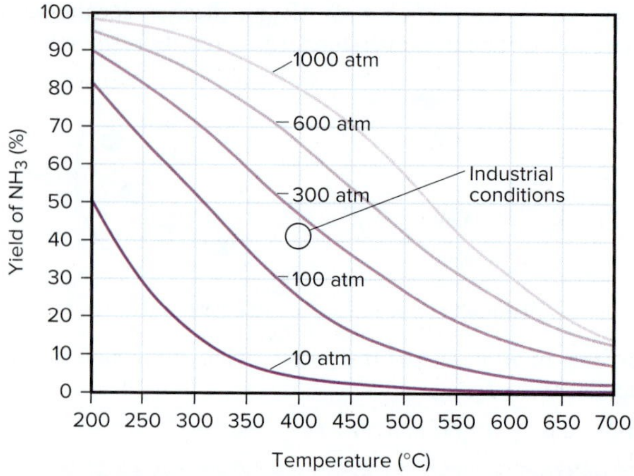

Figure 17.10 Percent yield of ammonia vs. temperature at five different pressures. At very high *P* and low *T (top left),* the yield is high, but the rate is low. Industrial conditions *(circle)* are between 200 and 300 atm at about 400°C.

The Industrial Process Stages in the industrial production of NH_3 are shown in Figure 17.11. To extend equipment life and minimize cost, modern plants operate at about 200 to 300 atm and around 673 K (400.°C). The stoichiometric ratio of reactant gases ($N_2/H_2 = 1/3$ by volume) is injected into the heated, pressurized reaction chamber. Some of the needed heat is supplied by ΔH°_{rxn}. The gases flow over catalyst beds that consist of 5-mm to 10-mm chunks of iron crystals embedded in a fused mixture of MgO, Al_2O_3, and SiO_2. The emerging equilibrium mixture contains about 35% NH_3 and is cooled by refrigeration until the NH_3 (bp −33.4°C) condenses; it is then removed and stored. Because N_2 and H_2 have much lower boiling points, they are recycled as gases by pumping them back into the reaction chamber.

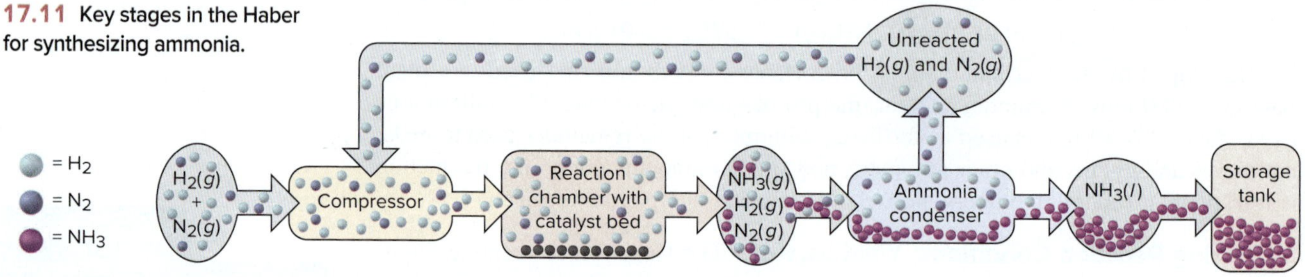

Figure 17.11 Key stages in the Haber process for synthesizing ammonia.

› Summary of Section 17.6

- If a system at equilibrium is disturbed, it undergoes a net reaction that reduces the disturbance and returns the system to equilibrium.
- Changes in concentration cause a net reaction to consume the added component or to produce the removed component.
- For a reaction that involves a change in amount (mol) of gas, an increase in pressure (decrease in volume) causes a net reaction toward fewer moles of gas, and a decrease in pressure causes the opposite change.
- Although the equilibrium position changes as a result of a concentration or volume change, *K* does not.
- A temperature change affects *K*: higher *T* increases *K* for an endothermic reaction (positive ΔH°_{rxn}) and decreases *K* for an exothermic reaction (negative ΔH°_{rxn}).
- A catalyst causes the system to reach equilibrium more quickly by speeding forward and reverse reactions equally, but it does not affect the equilibrium position.
- Ammonia is produced in a process favored by high pressure, low temperature, and continual removal of product. To make the process economical, intermediate temperature and pressure and a catalyst are used.
- A metabolic pathway is a cellular reaction sequence with each step shifted completely toward product. Its overall yield is controlled by a feedback system that inhibits the activity of certain key enzymes.

CHEMICAL CONNECTIONS TO CELLULAR METABOLISM

Design and Control of a Metabolic Pathway

From the simplest bacterium to the most specialized neuron, every cell performs thousands of reactions that allow it to grow and reproduce, feed and excrete, move and communicate. Taken together, these myriad feats of breakdown, synthesis, and energy flow constitute the cell's *metabolism* and are organized into reaction sequences called **metabolic pathways.**

Continual Shift Toward Product

In principle, each step in a metabolic pathway is a reversible reaction catalyzed by a specific enzyme (Section 16.7). However, equilibrium is never reached in a pathway, because the product of each reaction becomes the reactant of the next. Consider the five-step pathway by which the amino acid threonine is converted into the amino acid isoleucine (Figure B17.1). Threonine, supplied from a different region of the cell, forms ketobutyrate through the catalytic action of enzyme 1. The equilibrium position of reaction 1 shifts to the right because the product, ketobutyrate, is the reactant in reaction 2. Similarly, reaction 2 shifts to the right as its product, acetohydroxybutyrate, is used up in reaction 3. In this way, each subsequent reaction shifts the previous reaction to the right. The final product, isoleucine, is removed to make proteins elsewhere in the cell. Thus, the entire pathway operates in one direction.

Creation of a Steady State

This continuous shift in equilibrium position has two consequences for metabolic pathways:

1. *Each step proceeds with nearly 100% yield:* Virtually every molecule of threonine that enters this region of the cell eventually changes to ketobutyrate, every molecule of ketobutyrate to the next product, and so on.
2. *Reactant and product concentrations remain nearly constant,* because they reach a *steady state*.
   - In an equilibrium system, equal rates in *opposing* directions create constant concentrations of reactants and products.
   - In a steady-state system, the rates of reactions in *one* direction—into, through, and out of the system—create constant concentrations of intermediates.

Ketobutyrate, for example, is formed in reaction 1 as fast as it is consumed in reaction 2, so its concentration is constant. (A steady-state amount of water results if you fill a sink and then adjust the flow of faucet and drain so that water enters as fast as it leaves.)

Regulation by Feedback Inhibition

Recall from Chapter 16 that substrate concentrations are *much* higher than enzyme concentrations. If the active sites on all the enzyme molecules were always occupied, all cellular reactions would occur at their maximum rates. This might be ideal for an industrial process, but an organism needs to control amounts carefully. To regulate product formation, certain key steps are catalyzed by *regulatory enzymes,* which contain an *inhibitor site* (Figure B17.2A) in addition to an active site: when the inhibitor site is occupied, the shape of the active site is deformed and the reaction is not catalyzed (Figure B17.2B).

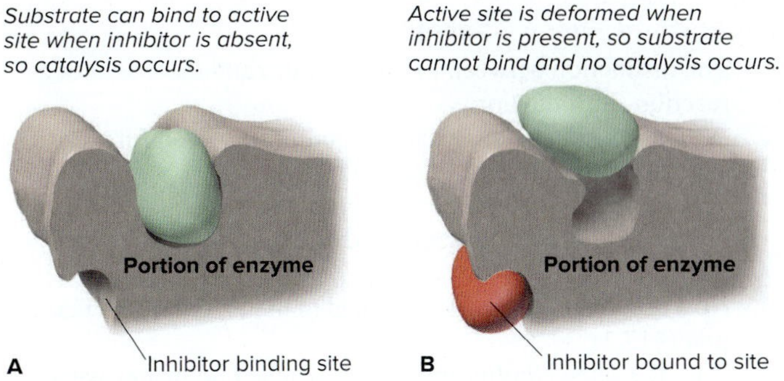

Figure B17.2 Effect of inhibitor binding on shape of active site.

In the simplest case of metabolic regulation (Figure B17.1), *the final product of a pathway is the inhibitor, and the regulatory enzyme catalyzes the first step.* Suppose, for instance, that a cell is temporarily making less protein, so isoleucine is not being removed as quickly. As its concentration rises, isoleucine lands more often on the inhibitor site of threonine dehydratase, the first enzyme in the pathway, thereby inhibiting its own production. This process is called *end-product feedback inhibition.* More complex pathways have more elaborate regulatory schemes.

Problem

B17.1 Many metabolites are products in branched pathways. In the one below, letters are compounds, and numbers are enzymes:

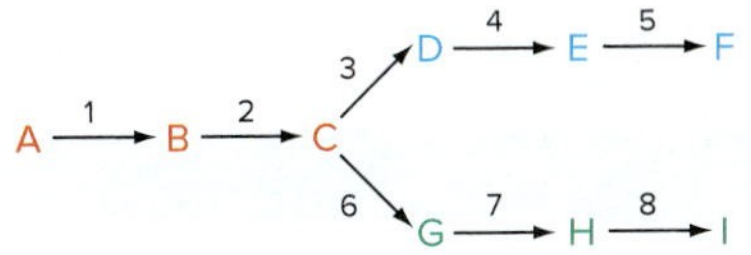

One method of control of these pathways occurs through inhibition of the first enzyme specific for a branch. (a) Which enzyme is inhibited by F? (b) Which enzyme is inhibited by I? (c) What is the disadvantage if F inhibited enzyme 1? (d) What would be the disadvantage if F inhibited enzyme 6?

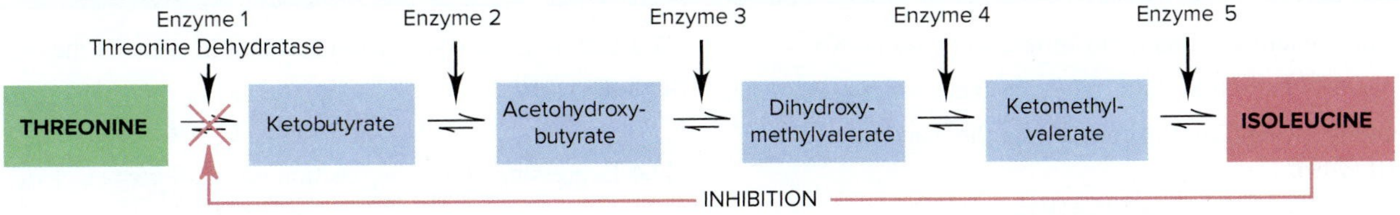

Figure B17.1 **The biosynthesis of isoleucine from threonine.** Isoleucine is synthesized from threonine in a sequence of five enzyme-catalyzed reactions. Once enough isoleucine is present, its concentration builds up and inhibits threonine dehydratase, the first enzyme in the pathway.

CHAPTER REVIEW GUIDE

Learning Objectives

Relevant section (§) and/or sample problem (SP) numbers appear in parentheses.

Understand These Concepts

1. The distinction between the speed (rate) and the extent of a reaction (Introduction)
2. Why a system attains dynamic equilibrium when forward and reverse reaction rates are equal (§17.1)
3. The equilibrium constant as a number that is equal to a particular ratio of rate constants and of concentration terms (§17.1)
4. How the magnitude of K is related to the extent of the reaction (§17.1)
5. Why the same equilibrium state is reached no matter what the starting concentrations of the reacting system (§17.2)
6. How the reaction quotient (Q) changes continuously until the system reaches equilibrium, at which point $Q = K$ (§17.2)
7. Why the form of Q is based exactly on the balanced equation *as written* (§17.2)
8. Why terms for pure solids and liquids do not appear in Q (§17.2)
9. How the *sum* of reaction steps gives the overall reaction, and the *product* of Q's (or K's) gives the overall Q (or K) (§17.2)
10. How the interconversion of K_c and K_p is based on the ideal gas law and Δn_{gas} (§17.3)
11. How the reaction direction depends on the relative values of Q and K (§17.4)
12. How a reaction table is used to find an unknown quantity (concentration or pressure) (§17.5)
13. How assuming that the change in [reactant] is relatively small simplifies finding equilibrium quantities (§17.5)
14. How Le Châtelier's principle explains the effects of a change in concentration, pressure (volume), or temperature on a system at equilibrium and on K (§17.6)
15. Why a change in temperature *does* affect K (§17.6)
16. Why the addition of a catalyst does *not* affect K (§17.6)
17. How adjusting reaction conditions and using a catalyst optimizes the synthesis of ammonia (§17.6)

Master These Skills

1. Writing a reaction quotient (Q) from a balanced equation (SP 17.1)
2. Writing Q and finding K for a reaction multiplied by a common factor or reversed and for an overall reaction (SP 17.2)
3. Writing Q for heterogeneous equilibria (§17.2)
4. Converting between K_c and K_p (SP 17.3)
5. Comparing Q and K to determine reaction direction (SPs 17.4, 17.5)
6. Substituting quantities (concentrations or pressures) into Q to find K (§17.5)
7. Using a reaction table to determine quantities and find K (SP 17.6)
8. Finding one equilibrium quantity from other equilibrium quantities and K (SP 17.7)
9. Finding an equilibrium quantity from initial quantities and K (SP 17.8)
10. Solving a quadratic equation for an unknown equilibrium quantity (§17.5)
11. Assuming that the change in [reactant] is relatively small to find equilibrium quantities and checking the assumption (SP 17.9)
12. Comparing the values of Q and K to find reaction direction and x, the unknown change in a quantity (SP 17.10)
13. Using the relative values of Q and K to predict the effect of a change in concentration on the equilibrium position and on K (SP 17.11)
14. Using Le Châtelier's principle and Δn_{gas} to predict the effect of a change in pressure (volume) on the equilibrium position (SP 17.12)
15. Using Le Châtelier's principle and $\Delta H°$ to predict the effect of a change in temperature on the equilibrium position and on K (SP 17.13)
16. Using the van't Hoff equation to calculate K at one temperature given K at another temperature (§17.6)
17. Using molecular scenes to find equilibrium parameters (SP 17.14)

Key Terms

Page numbers appear in parentheses.

equilibrium constant (K) (749)
Haber process (779)
law of chemical equilibrium (law of mass action) (750)
Le Châtelier's principle (770)
metabolic pathway (781)
reaction quotient (Q) (750)

Key Equations and Relationships

Page numbers appear in parentheses.

17.1 Defining equilibrium in terms of reaction rates (748):

$$\text{At equilibrium: rate}_{fwd} = \text{rate}_{rev}$$

17.2 Defining the equilibrium constant for the reaction $A \rightleftharpoons 2B$ (749):

$$K = \frac{k_{fwd}}{k_{rev}} = \frac{[B]^2_{eq}}{[A]_{eq}}$$

17.3 Defining the equilibrium constant in terms of the reaction quotient (750):

$$\text{At equilibrium: } Q = K$$

17.4 Expressing Q_c for the reaction $aA + bB \rightleftharpoons cC + dD$ (751):

$$Q_c = \frac{[C]^c[D]^d}{[A]^a[B]^b}$$

17.5 Finding K of a reaction from K of the reverse reaction (753):

$$K_{fwd} = \frac{1}{K_{rev}}$$

17.6 Finding Q and K of a reaction multiplied by a factor n (754):

$$Q' = Q^n = \left(\frac{[C]^c[D]^d}{[A]^a[B]^b}\right)^n \quad \text{and} \quad K' = K^n$$

17.7 Finding the overall K for a reaction sequence (754):

$$K_{overall} = K_1 \times K_2 \times K_3 \times \cdots$$

17.8 Relating K based on pressures to K based on concentrations (757):

$$K_p = K_c(RT)^{\Delta n_{gas}}$$

17.9 Assuming that ignoring the concentration that reacts introduces no significant error (766):

$$[A]_{init} - [A]_{reacting} = [A]_{eq} \approx [A]_{init}$$

17.10 Finding K at one temperature given K at another (van't Hoff equation) (776):

$$\ln \frac{K_2}{K_1} = -\frac{\Delta H^\circ_{rxn}}{R}\left(\frac{1}{T_2} - \frac{1}{T_1}\right)$$

BRIEF SOLUTIONS TO FOLLOW-UP PROBLEMS

17.1A (a) $Q_c = \dfrac{[NO]^4[H_2O]^6}{[NH_3]^4[O_2]^5}$

(b) $Q_c = \dfrac{[NH_3]^2}{[NO_2]^2[H_2]^7}$ (Liquid H_2O is not included.)

(c) $Q_c = [O_2]^3$ (Solid $KClO_3$ and KCl are not included.)

17.1B (a) $Q_c = \dfrac{[CO]^2[H_2]^2}{[CH_4][CO_2]}$

(b) $Q_c = \dfrac{[H_2O]^2}{[H_2S]^2[SO_2]}$ (Solid S is not included.)

(c) $Q_c = \dfrac{[NaCN]}{[HCN][NaOH]}$ (Liquid H_2O is not included.)

17.2A

(1) $C(s) + CO_2(g) \rightleftharpoons \cancel{2CO(g)}$ $\quad K_{c1} = 1.4\times10^{12}$

Multiply by 2

(2) $\cancel{2CO(g)} + 2Cl_2(g) \rightleftharpoons 2COCl_2(g)$ $\quad K_{c2} = (0.55)^2 = 0.30$

$C(s) + CO_2(g) + 2Cl_2(g) \rightleftharpoons 2COCl_2(g)$

$K_{c(overall)} = K_{c1} \times K_{c2} = (1.4\times10^{12})\ (0.30) = 4.2\times10^{11}$

17.2B (a) $K_c = \left(\dfrac{1}{K_{c(ref)}}\right)^2 = \left(\dfrac{1}{1.3\times10^{-2}}\right)^2 = 5.9\times10^3$

(b) $K_c = K^{1/4}_{c(ref)} = (1.3\times10^{-2})^{1/4} = 0.34$

17.3A $\Delta n_{gas} = 1 - 2 = -1$

$K_p = K_c(RT)^{-1} = 1.67\left(0.0821\ \dfrac{atm\cdot L}{mol\cdot K} \times 500.\ K\right)^{-1} = 4.07\times10^{-2}$

17.3B $\Delta n_{gas} = 3 - 5 = -2$ and $K_p = K_c(RT)^{-2}$

$K_c = \dfrac{K_p}{(RT)^{-2}} = K_p(RT)^2 = 3.0\times10^{-5}\left(0.0821\ \dfrac{atm\cdot L}{mol\cdot K} \times 1173\ K\right)^2 = 0.28$

17.4A $K_c = \dfrac{[Y]}{[X]} = 1.4$

Scene 1: $Q_c = \dfrac{3}{9} = 0.33$; $Q_c < K_c$, right

Scene 2: $Q_c = \dfrac{7}{5} = 1.4$; $Q_c = K_c$, no net change

Scene 3: $Q_c = \dfrac{8}{4} = 2.0$; $Q_c > K_c$, left

17.4B $K_c = \dfrac{[D]}{[C]^2} = \dfrac{5}{3^2} = 0.56$

Scene 2: $Q_c = \dfrac{6}{3^2} = 0.67$; $Q_c > K_c$, left

Scene 3: $Q_c = \dfrac{7}{4^2} = 0.44$; $Q_c < K_c$, right

17.5A $Q_p = \dfrac{(P_{CH_3Cl})(P_{HCl})}{(P_{CH_4})(P_{Cl_2})} = \dfrac{(0.24)(0.47)}{(0.13)(0.035)} = 25$

$Q_p < K_p$, so the reaction progresses to the right and CH_3Cl is forming.

17.5B $M_{SO_2} = \dfrac{3.4\ mol}{2.0\ L} = 1.7\ M$; $\quad M_{O_2} = \dfrac{1.5\ mol}{2.0\ L} = 0.75\ M$

$M_{SO_3} = \dfrac{1.2\ mol}{2.0\ L} = 0.60\ M$

$Q_c = \dfrac{[SO_3]^2}{[SO_2]^2[O_2]} = \dfrac{[0.60]^2}{[1.7]^2[0.75]} = 0.17$

$Q_c > K_c$, so the system is not at equilibrium, and the reaction will proceed toward the left.

17.6A

Pressure (atm)	$2NO(g)$	+	$O_2(g)$	$\rightleftharpoons$	$2NO_2(g)$
Initial	1.000		1.000		0
Change	$-2x$		$-x$		$+2x$
Equilibrium	$1.000 - 2x$		$1.000 - x$		$2x$

$P_{O_2} = 1.000\ atm - x = 0.506\ atm$, so $x = 0.494$ atm

$P_{NO_2} = 2x = 2(0.494\ atm) = 0.988$ atm

$P_{NO} = 1.000\ atm - 2x = 1.000\ atm - 2(0.494\ atm) = 0.012$ atm

$K_p = \dfrac{P^2_{NO_2}}{(P^2_{NO})(P_{O_2})} = \dfrac{0.988^2}{(0.012)^2\ (0.506)} = 1.3\times10^4$

17.6B

Concentration (M)	$4NH_3(g)$	+	$7O_2(g)$	$\rightleftharpoons$	$2N_2O_4(g)$	+	$6H_2O(g)$
Initial	2.40		2.40		0		0
Change	$-4x$		$-7x$		$+2x$		$+6x$
Equilibrium	$2.40 - 4x$		$2.40 - 7x$		$2x$		$6x$

$[N_2O_4] = 2x = 0.134\ M$, so $x = 0.0670\ M$

$[NH_3] = 2.40 - 4x = 2.40 - 4(0.0670) = 2.13\ M$

$[O_2] = 2.40 - 7x = 2.40 - 7(0.0670) = 1.93\ M$

$[H_2O] = 6x = 6(0.0670) = 0.402\ M$

$K_c = \dfrac{[N_2O_4]^2\ [H_2O]^6}{[NH_3]^4\ [O_2]^7} = \dfrac{(0.134)^2\ (0.402)^6}{(2.13)^4\ (1.93)^7} = 3.69\times10^{-8}$

BRIEF SOLUTIONS TO FOLLOW-UP PROBLEMS (continued)

17.7A Since $\Delta n_{gas} = 2 - 2 = 0$,

$$K_p = K_c = 2.3\times10^{30} = \frac{(P_{N_2})(P_{O_2})}{P_{NO}^2}$$

$$P_{NO} = \sqrt{\frac{(P_{N_2})(P_{O_2})}{2.3\times10^{30}}} = \sqrt{\frac{(0.781)(0.209)}{2.3\times10^{30}}} = 2.7\times10^{-16}\text{ atm}$$

17.7B $K_p = 19.6 = \frac{(P_{P_2})(P_{H_2}^3)}{P_{PH_3}^2}$

$$P_{H_2} = \sqrt[3]{\frac{(P_{PH_3}^2)(K_p)}{(P_{P_2})}} = \sqrt[3]{\frac{(0.112)^2(19.6)}{0.215}} = 1.05\text{ atm}$$

17.8A $M_{HI} = \frac{2.50\text{ mol}}{10.32\text{ L}} = 0.242\ M$

Concentration (*M*)	$2HI(g)$	⇌ $H_2(g)$	+ $I_2(g)$
Initial	0.242	0	0
Change	$-2x$	$+x$	$+x$
Equilibrium	$0.242 - 2x$	x	x

$$K_c = 1.26\times10^{-3} = \frac{[H_2][I_2]}{[HI]^2} = \frac{x^2}{(0.242 - 2x)^2}$$

Taking the square root of both sides, ignoring the negative root, and solving gives $x = [H_2] = 8.02\times10^{-3}\ M$.

17.8B $M_{Cl_2O} = M_{H_2O} = \frac{6.15\text{ mol}}{5.00\text{ L}} = 1.23\ M$

Concentration (*M*)	$Cl_2O(g)$	+ $H_2O(g)$	⇌ $2HOCl(g)$
Initial	1.23	1.23	0
Change	$-x$	$-x$	$+2x$
Equilibrium	$1.23 - x$	$1.23 - x$	$2x$

$$K_c = 0.18 = \frac{[HOCl]^2}{[Cl_2O][H_2O]} = \frac{(2x)^2}{(1.23 - x)^2}$$

Taking the square root of both sides, ignoring the negative root, and solving gives $x = 0.21\ M$.
$[H_2O] = [Cl_2O] = 1.23 - x = 1.23 - 0.21 = 1.02\ M$
$[HClO] = 2x = 2(0.21) = 0.42\ M$

17.9A (a) $M_{I_2} = \frac{0.50\text{ mol}}{2.5\text{ L}} = 0.20\ M$

Concentration (*M*)	$I_2(g)$	⇌ $2I(g)$
Initial	0.20	0
Change	$-x$	$+2x$
Equilibrium	$0.20 - x$	$2x$

Assuming that $0.20\ M - x \approx 0.20\ M$:

$$K_c = 2.94\times10^{-10} = \frac{[I]^2}{[I_2]} = \frac{(2x)^2}{0.20 - x} \approx \frac{4x^2}{0.20} \qquad x \approx 3.8\times10^{-6}\ M$$

Error $= (3.8\times10^{-6}/0.20)\times 100 = 1.9\times10^{-3}\%$, so assumption is justified; therefore, at equilibrium, $[I_2] = 0.20\ M$ and $[I] = 2x = 2(3.8\times10^{-6}) = 7.6\times10^{-6}\ M$.
(b) Based on the same reaction table and assumption that $0.20\ M - x \approx 0.20\ M$:

$$K_c = 0.209 = \frac{[I]^2}{[I_2]} = \frac{(2x)^2}{0.20 - x} \approx \frac{4x^2}{0.20} \qquad x \approx 0.10\ M$$

Error is $(0.10/0.20)\times 100 = 50\%$, so assumption is *not* justified. Solve the quadratic equation:

$$4x^2 + 0.209x - 0.042 = 0, \quad \text{so} \quad x = 0.080\ M$$

Therefore, at equilibrium, $[I_2] = 0.20 - x = 0.20 - 0.080 = 0.12\ M$; and $[I] = 2x = 2(0.080) = 0.16\ M$.

17.9B

(a)

Pressure (atm)	$PCl_5(g)$	⇌ $PCl_3(g)$	+ $Cl_2(g)$
Initial	0.18	0	0
Change	$-x$	$+x$	$+x$
Equilibrium	$0.18 - x$	x	x

Assuming that $0.18\text{ atm} - x \approx 0.18$ atm:

$$K_p = 3.4\times10^{-4} = \frac{[PCl_3][Cl_2]}{[PCl_5]} = \frac{x^2}{0.18 - x} \approx \frac{x^2}{0.18}$$

$x \approx 7.8\times10^{-3}$ atm; error $= (7.8\times10^{-3}/0.18)\times 100 = 4.3\%$, so assumption is justified; at equilibrium, $P_{PCl_5} = 0.18 - x = 0.18 - (7.8\times10^{-3}) = 0.17$ atm.

(b)

Pressure (atm)	$PCl_5(g)$	⇌ $PCl_3(g)$	+ $Cl_2(g)$
Initial	0.025	0	0
Change	$-x$	$+x$	$+x$
Equilibrium	$0.025 - x$	x	x

Assuming that $0.025\text{ atm} - x \approx 0.025$ atm:

$$K_p = 3.4\times10^{-4} = \frac{[PCl_3][Cl_2]}{[PCl_5]} = \frac{x^2}{0.025 - x} \approx \frac{x^2}{0.025}$$

$x \approx 2.9\times10^{-3}$ atm; error $= (2.9\times10^{-3}/0.025)\times 100 = 12\%$, so assumption is *not* justified. Solving the quadratic equation:

$$x^2 + 3.4\times10^{-4}x - 8.5\times10^{-6} = 0, \quad \text{so} \quad x = 0.0028$$

Therefore, at equilibrium, $P_{PCl_5} = 0.025 - x = 0.025 - 0.0028 = 0.022$ atm

17.10A (a) $M_{PCl_5} = 0.1050\text{ mol}/0.5000\text{ L} = 0.2100\ M$; similarly, $M_{Cl_2} = M_{PCl_3} = 0.0900\ M$

$$Q_c = \frac{[PCl_3][Cl_2]}{[PCl_5]} = \frac{[0.0900][0.0900]}{[0.2100]} = 3.86\times10^{-2}$$

$Q_c < K_c$, so reaction proceeds to the right.

(b)

Concentration (*M*)	$PCl_5(g)$	⇌ $PCl_3(g)$	+ $Cl_2(g)$
Initial	0.2100	0.0900	0.0900
Change	$-x$	$+x$	$+x$
Equilibrium	$0.2100 - x$	$0.0900 + x$	$0.0900 + x$

$[PCl_5] = 0.2100\ M - x = 0.2065\ M$, so $x = 0.0035\ M$
$[Cl_2] = [PCl_3] = 0.0900\ M + x = 0.0900\ M + 0.0035\ M = 0.0935\ M$

17.10B $Q_p = \frac{[NO]^2}{[N_2][O_2]} = \frac{[0.750]^2}{[0.500][0.500]} = 2.25$

$Q_p > K_p$, so reaction proceeds to the left.

Pressure (atm)	$N_2(g)$	+ $O_2(g)$	⇌ $2NO(g)$
Initial	0.500	0.500	0.750
Change	$+x$	$+x$	$-2x$
Equilibrium	$0.500 + x$	$0.500 + x$	$0.750 - 2x$

$$K_p = 8.44\times10^{-3} = \frac{P_{NO}^2}{(P_{N_2})(P_{O_2})} = \frac{(0.750 - 2x)^2}{(0.500 + x)^2}$$

Taking the square root of both sides and solving gives $x = 0.337$, so

$$P_{N_2} = P_{O_2} = 0.500 + x = 0.500 + 0.337 = 0.837 \text{ atm}$$
$$P_{NO} = 0.750 - 2x = 0.750 - 2(0.337) = 0.076 \text{ atm}$$

17.11A (a) $[SiF_4]$ increases; (b) decreases since adding $H_2O(l)$ above the boiling point of water results in an increase in the concentration of $H_2O(g)$; (c) decreases; (d) no effect.

17.11B (a) [CO] does not change; (b) decreases; (c) increases; (d) increases.

17.12A (a) Decrease P; (b) increase P; (c) increase P.

17.12B (a) More CO and H_2; (b) no change; (c) less S and H_2O.

17.13A (a) P_{H_2} will decrease; K will increase. (b) P_{N_2} will increase; K will decrease. (c) P_{PCl_5} will increase; K will increase.

17.13B (a) Increase T, and K increases; (b) decrease T, and K increases; (c) increase T, and K increases.

17.14A (a) Since $P = \frac{n}{V}RT$ and, in this case, V, R, and T cancel,

$$K_p = \frac{n_{CD}^2}{n_{C_2} \times n_{D_2}} = \frac{4^2}{2 \times 2} = 4$$

(b) Scene 2: $Q_p = \frac{(6)^2}{(1)(1)} = 36 > K_p\ (4)$; to the left;

Scene 3: $Q_p = \frac{(2)^2}{(3)(3)} = 0.44 < K_p\ (4)$; to the right.

(c) There are 2 mol of gas on each side of the balanced equation, so increasing T has no effect on total moles of gas.

17.14B (a) $K = \frac{[AB]}{[A][B]} = \frac{3}{2 \times 2} = 0.75$

(b) $\Delta H > 0$

(c) $K = \frac{2}{3 \times 3} = 0.22$

PROBLEMS

Problems with **colored** numbers are answered in Appendix E and worked in detail in the Student Solutions Manual. Problem sections match those in the text and give the numbers of relevant sample problems. Most offer Concept Review Questions, Skill-Building Exercises (grouped in pairs covering the same concept), and Problems in Context. The Comprehensive Problems are based on material from any section or previous chapter.

The Equilibrium State and the Equilibrium Constant

Concept Review Questions

17.1 A change in reaction conditions increases the rate of a certain forward reaction more than that of the reverse reaction. What is the effect on the equilibrium constant and on the concentrations of reactants and products at equilibrium?

17.2 When a chemical company employs a new reaction to manufacture a product, the chemists consider its rate (kinetics) and yield (equilibrium). How do each of these affect the usefulness of a manufacturing process?

17.3 If there is no change in concentrations, why is the equilibrium state considered dynamic?

17.4 Is K very large or very small for a reaction that goes essentially to completion? Explain.

17.5 White phosphorus, P_4, is produced by the reduction of phosphate rock, $Ca_3(PO_4)_2$. If exposed to oxygen, the waxy, white solid smokes, bursts into flames, and releases a large quantity of heat:

$$P_4(g) + 5O_2(g) \rightleftharpoons P_4O_{10}(s) + \textit{heat}$$

Does this reaction have a large or small equilibrium constant? Explain.

The Reaction Quotient and the Equilibrium Constant
(Sample Problems 17.1 to 17.2)

Concept Review Questions

17.6 For a given reaction at a given temperature, the value of K is constant. Is the value of Q also constant? Explain.

17.7 In a study of the thermal decomposition of lithium peroxide,

$$2Li_2O_2(s) \rightleftharpoons 2Li_2O(s) + O_2(g)$$

a chemist finds that, as long as some Li_2O_2 is present at the end of the experiment, the amount of O_2 obtained in a given container at a given T is the same. Explain.

17.8 In a study of the formation of HI from its elements,

$$H_2(g) + I_2(g) \rightleftharpoons 2HI(g)$$

equal amounts of H_2 and I_2 were placed in a container, which was then sealed and heated.
(a) On one set of axes, sketch concentration vs. time curves for H_2 and HI, and explain how Q changes as a function of time.
(b) Is the value of Q different if $[I_2]$ is plotted instead of $[H_2]$?

17.9 Explain the difference between a heterogeneous and a homogeneous equilibrium. Give an example of each.

17.10 Does Q for the formation of 1 mol of NO from its elements differ from Q for the decomposition of 1 mol of NO to its elements? Explain and give the relationship between the two Q's.

17.11 Does Q for the formation of 1 mol of NH_3 from H_2 and N_2 differ from Q for the formation of NH_3 from H_2 and 1 mol of N_2? Explain and give the relationship between the two Q's.

Skill-Building Exercises (grouped in similar pairs)

17.12 Balance each reaction and write its reaction quotient, Q_c:
(a) $NO(g) + O_2(g) \rightleftharpoons N_2O_3(g)$
(b) $SF_6(g) + SO_3(g) \rightleftharpoons SO_2F_2(g)$
(c) $SClF_5(g) + H_2(g) \rightleftharpoons S_2F_{10}(g) + HCl(g)$

17.13 Balance each reaction and write its reaction quotient, Q_c:
(a) $C_2H_6(g) + O_2(g) \rightleftharpoons CO_2(g) + H_2O(g)$
(b) $CH_4(g) + F_2(g) \rightleftharpoons CF_4(g) + HF(g)$
(c) $SO_3(g) \rightleftharpoons SO_2(g) + O_2(g)$

17.14 Balance each reaction and write its reaction quotient, Q_c:
(a) $NO_2Cl(g) \rightleftharpoons NO_2(g) + Cl_2(g)$
(b) $POCl_3(g) \rightleftharpoons PCl_3(g) + O_2(g)$
(c) $NH_3(g) + O_2(g) \rightleftharpoons N_2(g) + H_2O(g)$

17.15 Balance each reaction and write its reaction quotient, Q_c:
(a) $O_2(g) \rightleftharpoons O_3(g)$
(b) $NO(g) + O_3(g) \rightleftharpoons NO_2(g) + O_2(g)$
(c) $N_2O(g) + H_2(g) \rightleftharpoons NH_3(g) + H_2O(g)$

17.16 At a particular temperature, $K_c = 1.6\times10^{-2}$ for

$$2H_2S(g) \rightleftharpoons 2H_2(g) + S_2(g)$$

Calculate K_c for each of the following reactions:
(a) $\frac{1}{2}S_2(g) + H_2(g) \rightleftharpoons H_2S(g)$
(b) $5H_2S(g) \rightleftharpoons 5H_2(g) + \frac{5}{2}S_2(g)$

17.17 At a particular temperature, $K_c = 6.5\times10^{2}$ for

$$2NO(g) + 2H_2(g) \rightleftharpoons N_2(g) + 2H_2O(g)$$

Calculate K_c for each of the following reactions:
(a) $NO(g) + H_2(g) \rightleftharpoons \frac{1}{2}N_2(g) + H_2O(g)$
(b) $2N_2(g) + 4H_2O(g) \rightleftharpoons 4NO(g) + 4H_2(g)$

17.18 Balance each of the following examples of heterogeneous equilibria and write each reaction quotient, Q_c:
(a) $Na_2O_2(s) + CO_2(g) \rightleftharpoons Na_2CO_3(s) + O_2(g)$
(b) $H_2O(l) \rightleftharpoons H_2O(g)$
(c) $NH_4Cl(s) \rightleftharpoons NH_3(g) + HCl(g)$

17.19 Balance each of the following examples of heterogeneous equilibria and write each reaction quotient, Q_c:
(a) $H_2O(l) + SO_3(g) \rightleftharpoons H_2SO_4(aq)$
(b) $KNO_3(s) \rightleftharpoons KNO_2(s) + O_2(g)$
(c) $S_8(s) + F_2(g) \rightleftharpoons SF_6(g)$

17.20 Balance each of the following examples of heterogeneous equilibria and write each reaction quotient, Q_c:
(a) $NaHCO_3(s) \rightleftharpoons Na_2CO_3(s) + CO_2(g) + H_2O(g)$
(b) $SnO_2(s) + H_2(g) \rightleftharpoons Sn(s) + H_2O(g)$
(c) $H_2SO_4(l) + SO_3(g) \rightleftharpoons H_2S_2O_7(l)$

17.21 Balance each of the following examples of heterogeneous equilibria and write each reaction quotient, Q_c:
(a) $Al(s) + NaOH(aq) + H_2O(l) \rightleftharpoons Na[Al(OH)_4](aq) + H_2(g)$
(b) $CO_2(s) \rightleftharpoons CO_2(g)$
(c) $N_2O_5(s) \rightleftharpoons NO_2(g) + O_2(g)$

Problems in Context

17.22 Write Q_c for each of the following:
(a) Hydrogen chloride gas reacts with oxygen gas to produce chlorine gas and water vapor.
(b) Solid diarsenic trioxide reacts with fluorine gas to produce liquid arsenic pentafluoride and oxygen gas.
(c) Gaseous sulfur tetrafluoride reacts with liquid water to produce gaseous sulfur dioxide and hydrogen fluoride gas.
(d) Solid molybdenum(VI) oxide reacts with gaseous xenon difluoride to form liquid molybdenum(VI) fluoride, xenon gas, and oxygen gas.

17.23 The interhalogen ClF_3 is prepared via a two-step fluorination of chlorine gas:

$$Cl_2(g) + F_2(g) \rightleftharpoons ClF(g)$$
$$ClF(g) + F_2(g) \rightleftharpoons ClF_3(g)$$

(a) Balance each step and write the overall equation.
(b) Show that the overall Q_c equals the product of the Q_c's for the individual steps.

17.24 As an EPA scientist studying catalytic converters and urban smog, you want to find K_c for the following reaction:

$$2NO_2(g) \rightleftharpoons N_2(g) + 2O_2(g) \qquad K_c = ?$$

Use the following data to find the unknown K_c:

$$\tfrac{1}{2}N_2(g) + \tfrac{1}{2}O_2(g) \rightleftharpoons NO(g) \qquad K_c = 4.8\times10^{-10}$$
$$2NO_2(g) \rightleftharpoons 2NO(g) + O_2(g) \qquad K_c = 1.1\times10^{-5}$$

Expressing Equilibria with Pressure Terms: Relation Between K_c and K_p

(Sample Problem 17.3)

Concept Review Questions

17.25 Guldberg and Waage proposed the definition of the equilibrium constant as a certain ratio of *concentrations*. What relationship allows us to use a particular ratio of *partial pressures* (for a gaseous reaction) to express an equilibrium constant? Explain.

17.26 When are K_c and K_p equal, and when are they not?

17.27 A certain reaction at equilibrium has more moles of gaseous products than of gaseous reactants.
(a) Is K_c larger or smaller than K_p?
(b) Write a statement about the relative sizes of K_c and K_p for any gaseous equilibrium.

Skill-Building Exercises (grouped in similar pairs)

17.28 Determine Δn_{gas} for each of the following reactions:
(a) $2KClO_3(s) \rightleftharpoons 2KCl(s) + 3O_2(g)$
(b) $2PbO(s) + O_2(g) \rightleftharpoons 2PbO_2(s)$
(c) $I_2(s) + 3XeF_2(s) \rightleftharpoons 2IF_3(s) + 3Xe(g)$

17.29 Determine Δn_{gas} for each of the following reactions:
(a) $MgCO_3(s) \rightleftharpoons MgO(s) + CO_2(g)$
(b) $2H_2(g) + O_2(g) \rightleftharpoons 2H_2O(l)$
(c) $HNO_3(l) + ClF(g) \rightleftharpoons ClONO_2(g) + HF(g)$

17.30 Calculate K_c for each of the following equilibria:
(a) $CO(g) + Cl_2(g) \rightleftharpoons COCl_2(g)$; $K_p = 3.9\times10^{-2}$ at 1000. K
(b) $S_2(g) + C(s) \rightleftharpoons CS_2(g)$; $K_p = 28.5$ at 500. K

17.31 Calculate K_c for each of the following equilibria:
(a) $H_2(g) + I_2(g) \rightleftharpoons 2HI(g)$; $K_p = 49$ at 730. K
(b) $2SO_2(g) + O_2(g) \rightleftharpoons 2SO_3(g)$; $K_p = 2.5\times10^{10}$ at 500. K

17.32 Calculate K_p for each of the following equilibria:
(a) $N_2O_4(g) \rightleftharpoons 2NO_2(g)$; $K_c = 6.1\times10^{-3}$ at 298 K
(b) $N_2(g) + 3H_2(g) \rightleftharpoons 2NH_3(g)$; $K_c = 2.4\times10^{-3}$ at 1000. K

17.33 Calculate K_p for each of the following equilibria:
(a) $H_2(g) + CO_2(g) \rightleftharpoons H_2O(g) + CO(g)$; $K_c = 0.77$ at 1020. K
(b) $3O_2(g) \rightleftharpoons 2O_3(g)$; $K_c = 1.8\times10^{-56}$ at 570. K

Comparing Q and K to Determine Reaction Direction

(Sample Problems 17.4 and 17.5)

Concept Review Questions

17.34 When the numerical value of Q is less than that of K, in which direction does the reaction proceed to reach equilibrium? Explain.

17.35 The following molecular scenes depict the aqueous reaction $2D \rightleftharpoons E$, with D *red* and E *blue*. Each sphere represents 0.0100 mol, but the volume is 1.00 L in scene A, whereas in scenes B and C, it is 0.500 L.

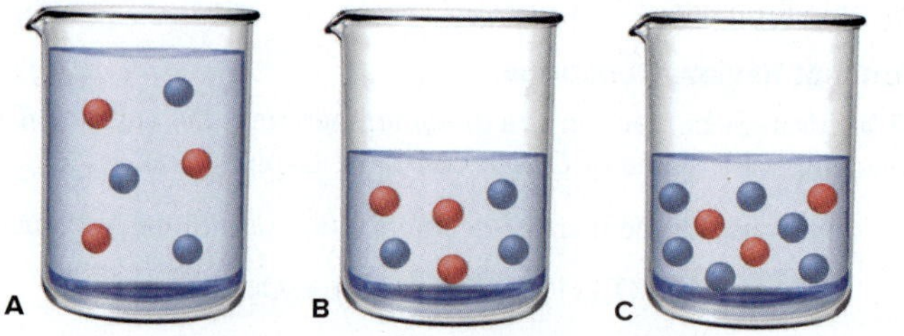

(a) If the reaction in scene A is at equilibrium, calculate K_c.

(b) Are the reactions in scenes B and C at equilibrium? Which, if either, is not, and in which direction will it proceed?

Skill-Building Exercises (grouped in similar pairs)

17.36 At 425°C, $K_p = 4.18\times10^{-9}$ for the reaction

$$2HBr(g) \rightleftharpoons H_2(g) + Br_2(g)$$

In one experiment, 0.20 atm of $HBr(g)$, 0.010 atm of $H_2(g)$, and 0.010 atm of $Br_2(g)$ are introduced into a container. Is the reaction at equilibrium? If not, in which direction will it proceed?

17.37 At 100°C, $K_p = 60.6$ for the reaction

$$2NOBr(g) \rightleftharpoons 2NO(g) + Br_2(g)$$

In a given experiment, 0.10 atm of each component is placed in a container. Is the system at equilibrium? If not, in which direction will the reaction proceed?

Problems in Context

17.38 The water-gas shift reaction plays a central role in the chemical methods for obtaining cleaner fuels from coal:

$$CO(g) + H_2O(g) \rightleftharpoons CO_2(g) + H_2(g)$$

At a given temperature, $K_p = 2.7$. If 0.13 mol of CO, 0.56 mol of H_2O, 0.62 mol of CO_2, and 0.43 mol of H_2 are put in a 2.0-L flask, in which direction does the reaction proceed?

How to Solve Equilibrium Problems

(Sample Problems 17.6 to 17.10)

Concept Review Questions

17.39 In the 1980s, CFC-11 was one of the most heavily produced chlorofluorocarbons. The last step in its formation is

$$CCl_4(g) + HF(g) \rightleftharpoons CFCl_3(g) + HCl(g)$$

If you start the reaction with equal concentrations of CCl_4 and HF, you obtain equal concentrations of $CFCl_3$ and HCl at equilibrium. Are the final concentrations of $CFCl_3$ and HCl equal if you start with unequal concentrations of CCl_4 and HF? Explain.

17.40 For a problem involving the catalyzed reaction of methane and steam, the following reaction table was prepared:

Pressure (atm)	$CH_4(g)$ +	$2H_2O(g)$ $\rightleftharpoons$	$CO_2(g)$ +	$4H_2(g)$
Initial	0.30	0.40	0	0
Change	$-x$	$-2x$	$+x$	$+4x$
Equilibrium	$0.30 - x$	$0.40 - 2x$	x	$4x$

Explain the entries in the "Change" and "Equilibrium" rows.

17.41 (a) What is the basis of the approximation that avoids the need to use the quadratic formula to find an equilibrium concentration?

(b) When should this approximation *not* be made?

Skill-Building Exercises (grouped in similar pairs)

17.42 In an experiment to study the formation of HI(*g*),

$$H_2(g) + I_2(g) \rightleftharpoons 2HI(g)$$

$H_2(g)$ and $I_2(g)$ were placed in a sealed container at a certain temperature. At equilibrium, $[H_2] = 6.50\times10^{-5}$ *M*, $[I_2] = 1.06\times10^{-3}$ *M*, and $[HI] = 1.87\times10^{-3}$ *M*. Calculate K_c for the reaction at this temperature.

17.43 Gaseous ammonia was introduced into a sealed container and heated to a certain temperature:

$$2NH_3(g) \rightleftharpoons N_2(g) + 3H_2(g)$$

At equilibrium, $[NH_3] = 0.0225$ *M*, $[N_2] = 0.114$ *M*, and $[H_2] = 0.342$ *M*. Calculate K_c for the reaction at this temperature.

17.44 Gaseous PCl_5 decomposes according to the reaction

$$PCl_5(g) \rightleftharpoons PCl_3(g) + Cl_2(g)$$

In one experiment, 0.15 mol of $PCl_5(g)$ was introduced into a 2.0-L container. Construct the reaction table for this process.

17.45 Hydrogen fluoride, HF, can be made by the reaction

$$H_2(g) + F_2(g) \rightleftharpoons 2HF(g)$$

In one experiment, 0.10 mol of $H_2(g)$ and 0.050 mol of $F_2(g)$ are added to a 0.50-L flask. Write a reaction table for this process.

17.46 For the following reaction, $K_p = 6.5\times10^4$ at 308 K:

$$2NO(g) + Cl_2(g) \rightleftharpoons 2NOCl(g)$$

At equilibrium, $P_{NO} = 0.35$ atm and $P_{Cl_2} = 0.10$ atm. What is the equilibrium partial pressure of NOCl(*g*)?

17.47 For the following reaction, $K_p = 0.262$ at 1000°C:

$$C(s) + 2H_2(g) \rightleftharpoons CH_4(g)$$

At equilibrium, P_{H_2} is 1.22 atm. What is the equilibrium partial pressure of $CH_4(g)$?

17.48 Ammonium hydrogen sulfide decomposes according to the following reaction, for which $K_p = 0.11$ at 250°C:

$$NH_4HS(s) \rightleftharpoons H_2S(g) + NH_3(g)$$

If 55.0 g of $NH_4HS(s)$ is placed in a sealed 5.0-L container, what is the partial pressure of $NH_3(g)$ at equilibrium?

17.49 Hydrogen sulfide decomposes according to the following reaction, for which $K_c = 9.30\times10^{-8}$ at 700°C:

$$2H_2S(g) \rightleftharpoons 2H_2(g) + S_2(g)$$

If 0.45 mol of H_2S is placed in a 3.0-L container, what is the equilibrium concentration of $H_2(g)$ at 700°C?

17.50 Even at high *T*, the formation of NO is not favored:

$$N_2(g) + O_2(g) \rightleftharpoons 2NO(g) \qquad K_c = 4.10\times10^{-4} \text{ at } 2000°C$$

What is [NO] when a mixture of 0.20 mol of $N_2(g)$ and 0.15 mol of $O_2(g)$ reach equilibrium in a 1.0-L container at 2000°C?

17.51 Nitrogen dioxide decomposes according to the reaction

$$2NO_2(g) \rightleftharpoons 2NO(g) + O_2(g)$$

where $K_p = 4.48\times10^{-13}$ at a certain temperature. If 0.75 atm of NO_2 is added to a container and allowed to come to equilibrium, what are the equilibrium partial pressures of NO(*g*) and $O_2(g)$?

17.52 Hydrogen iodide decomposes according to the reaction

$$2HI(g) \rightleftharpoons H_2(g) + I_2(g)$$

A sealed 1.50-L container initially holds 0.00623 mol of H_2, 0.00414 mol of I_2, and 0.0244 mol of HI at 703 K. When equilibrium is reached, the concentration of $H_2(g)$ is 0.00467 *M*. What are the concentrations of HI(*g*) and $I_2(g)$?

17.53 Compound A decomposes according to the equation

$$A(g) \rightleftharpoons 2B(g) + C(g)$$

A sealed 1.00-L container initially contains 1.75×10^{-3} mol of A(*g*), 1.25×10^{-3} mol of B(*g*), and 6.50×10^{-4} mol of C(*g*) at 100°C. At equilibrium, [A] is 2.15×10^{-3} *M*. Find [B] and [C].

Problems in Context

17.54 In an analysis of interhalogen reactivity, 0.500 mol of ICl was placed in a 5.00-L flask, where it decomposed at a high *T*:

$$2ICl(g) \rightleftharpoons I_2(g) + Cl_2(g)$$

Calculate the equilibrium concentrations of I_2, Cl_2, and ICl (K_c = 0.110 at this temperature).

17.55 A toxicologist studying mustard gas, $S(CH_2CH_2Cl)_2$, a blistering agent, prepares a mixture of 0.675 *M* SCl_2 and 0.973 *M* C_2H_4 and allows it to react at room temperature (20.0°C):

$$SCl_2(g) + 2C_2H_4(g) \rightleftharpoons S(CH_2CH_2Cl)_2(g)$$

At equilibrium, $[S(CH_2CH_2Cl)_2] = 0.350$ *M*. Calculate K_p.

17.56 The first step in HNO_3 production is the catalyzed oxidation of NH_3. Without a catalyst, a different reaction predominates:

$$4NH_3(g) + 3O_2(g) \rightleftharpoons 2N_2(g) + 6H_2O(g)$$

When 0.0150 mol of $NH_3(g)$ and 0.0150 mol of $O_2(g)$ are placed in a 1.00-L container at a certain temperature, the N_2 concentration at equilibrium is 1.96×10^{-3} *M*. Calculate K_c.

17.57 A key step in the extraction of iron from its ore is

$$FeO(s) + CO(g) \rightleftharpoons Fe(s) + CO_2(g) \quad K_p = 0.403 \text{ at } 1000°C$$

This step occurs in the 700°C to 1200°C zone within a blast furnace. What are the equilibrium partial pressures of $CO(g)$ and $CO_2(g)$ when 1.00 atm of $CO(g)$ and excess $FeO(s)$ react in a sealed container at 1000°C?

Reaction Conditions and Equilibrium: Le Châtelier's Principle

(Sample Problems 17.11 to 17.14)

Concept Review Questions

17.58 What does "disturbance" mean in Le Châtelier's principle?

17.59 What is the difference between the equilibrium position and the equilibrium constant of a reaction? Which changes as a result of a change in reactant concentration?

17.60 Scenes A, B, and C below depict the following reaction at three temperatures:

$$NH_4Cl(s) \rightleftharpoons NH_3(g) + HCl(g) \qquad \Delta H^\circ_{rxn} = 176 \text{ kJ}$$

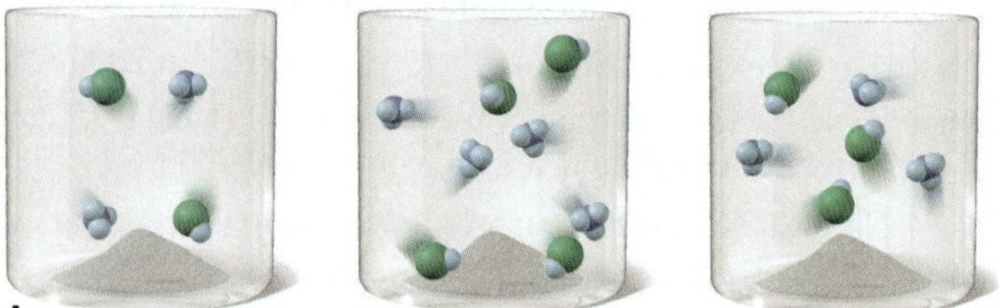

A B C

(a) Which best represents the reaction mixture at the highest temperature? Explain. (b) Which best represents the reaction mixture at the lowest temperature? Explain.

17.61 What is implied by the word "constant" in the term *equilibrium constant?* Give two reaction parameters that can be changed without changing the value of an equilibrium constant.

17.62 Le Châtelier's principle is related ultimately to the rates of the forward and reverse steps in a reaction. Explain (a) why an increase in reactant concentration shifts the equilibrium position to the right but does not change *K*; (b) why a decrease in *V* shifts the equilibrium position toward fewer moles of gas but does not change *K*; (c) why a rise in *T* shifts the equilibrium position of an exothermic reaction toward reactants and also changes *K*; and (d) why a rise in temperature of an endothermic reaction from T_1 to T_2 results in K_2 being larger than K_1.

17.63 An equilibrium mixture of two solids and a gas, in the reaction $XY(s) \rightleftharpoons X(g) + Y(s)$, is depicted at right (X is *green* and Y is *black*). Does scene A, B, or C best represent the system at equilibrium after two formula units of $Y(s)$ is added? Explain.

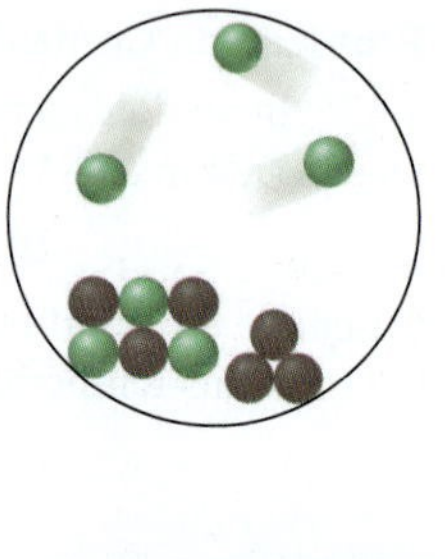

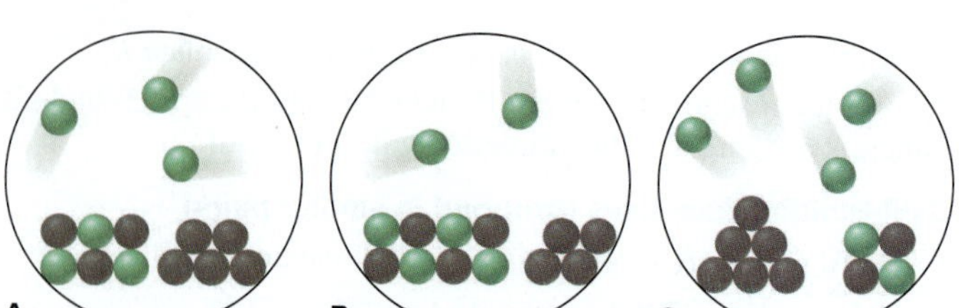

A B C

Skill-Building Exercises (grouped in similar pairs)

17.64 Consider this equilibrium system:

$$CO(g) + Fe_3O_4(s) \rightleftharpoons CO_2(g) + 3FeO(s)$$

How does the equilibrium position shift as a result of each of the following disturbances? (a) CO is added. (b) CO_2 is removed by adding solid NaOH. (c) Additional $Fe_3O_4(s)$ is added to the system. (d) Dry ice is added at constant temperature.

17.65 Sodium bicarbonate undergoes thermal decomposition according to the reaction

$$2NaHCO_3(s) \rightleftharpoons Na_2CO_3(s) + CO_2(g) + H_2O(g)$$

How does the equilibrium position shift as a result of each of the following disturbances? (a) 0.20 atm of argon gas is added. (b) $NaHCO_3(s)$ is added. (c) $Mg(ClO_4)_2(s)$ is added as a drying agent to remove H_2O. (d) Dry ice is added at constant *T*.

17.66 Predict the effect of *increasing* the container volume on the amounts of each reactant and product in the following reactions:
(a) $F_2(g) \rightleftharpoons 2F(g)$
(b) $2CH_4(g) \rightleftharpoons C_2H_2(g) + 3H_2(g)$

17.67 Predict the effect of *increasing* the container volume on the amounts of each reactant and product in the following reactions:
(a) $CH_3OH(l) \rightleftharpoons CH_3OH(g)$
(b) $CH_4(g) + NH_3(g) \rightleftharpoons HCN(g) + 3H_2(g)$

17.68 Predict the effect of *decreasing* the container volume on the amounts of each reactant and product in the following reactions:
(a) $H_2(g) + Cl_2(g) \rightleftharpoons 2HCl(g)$
(b) $2H_2(g) + O_2(g) \rightleftharpoons 2H_2O(l)$

17.69 Predict the effect of *decreasing* the container volume on the amounts of each reactant and product in the following reactions:
(a) $C_3H_8(g) + 5O_2(g) \rightleftharpoons 3CO_2(g) + 4H_2O(l)$
(b) $4NH_3(g) + 3O_2(g) \rightleftharpoons 2N_2(g) + 6H_2O(g)$

17.70 How would you adjust the *volume* of the container in order to maximize product yield in each of the following reactions?
(a) $Fe_3O_4(s) + 4H_2(g) \rightleftharpoons 3Fe(s) + 4H_2O(g)$
(b) $2C(s) + O_2(g) \rightleftharpoons 2CO(g)$

17.71 How would you adjust the *volume* of the container in order to maximize product yield in each of the following reactions?
(a) $Na_2O_2(s) \rightleftharpoons 2Na(l) + O_2(g)$
(b) $C_2H_2(g) + 2H_2(g) \rightleftharpoons C_2H_6(g)$

17.72 Predict the effect of *increasing* the temperature on the amount(s) of product(s) in the following reactions:
(a) $CO(g) + 2H_2(g) \rightleftharpoons CH_3OH(g) \qquad \Delta H^\circ_{rxn} = -90.7 \text{ kJ}$
(b) $C(s) + H_2O(g) \rightleftharpoons CO(g) + H_2(g) \qquad \Delta H^\circ_{rxn} = 131 \text{ kJ}$
(c) $2NO_2(g) \rightleftharpoons 2NO(g) + O_2(g)$ (endothermic)
(d) $2C(s) + O_2(g) \rightleftharpoons 2CO(g)$ (exothermic)

17.73 Predict the effect of *decreasing* the temperature on the amount(s) of reactant(s) in the following reactions:
(a) $C_2H_2(g) + H_2O(g) \rightleftharpoons CH_3CHO(g) \qquad \Delta H^\circ_{rxn} = -151 \text{ kJ}$
(b) $CH_3CH_2OH(l) + O_2(g) \rightleftharpoons CH_3CO_2H(l) + H_2O(g)$
$\Delta H^\circ_{rxn} = -451 \text{ kJ}$

(c) $2C_2H_4(g) + O_2(g) \rightleftharpoons 2CH_3CHO(g)$ (exothermic)
(d) $N_2O_4(g) \rightleftharpoons 2NO_2(g)$ (endothermic)

17.74 The molecule D_2 (where D, deuterium, is ^{2}H) undergoes a reaction with ordinary H_2 that leads to isotopic equilibrium:

$$D_2(g) + H_2(g) \rightleftharpoons 2DH(g) \qquad K_p = 1.80 \text{ at } 298 \text{ K}$$

If ΔH°_{rxn} is 0.32 kJ/mol DH, calculate K_p at 500. K.

17.75 The formation of methanol is important to the processing of new fuels. At 298 K, $K_p = 2.25\times10^4$ for the reaction

$$CO(g) + 2H_2(g) \rightleftharpoons CH_3OH(l)$$

If $\Delta H^\circ_{rxn} = -128$ kJ/mol CH_3OH, calculate K_p at 0°C.

Problems in Context

17.76 The minerals hematite (Fe_2O_3) and magnetite (Fe_3O_4) exist in equilibrium with atmospheric oxygen:

$$4Fe_3O_4(s) + O_2(g) \rightleftharpoons 6Fe_2O_3(s) \quad K_p = 2.5\times10^{87} \text{ at } 298 \text{ K}$$

(a) Determine P_{O_2} at equilibrium. (b) Given that P_{O_2} in air is 0.21 atm, in which direction will the reaction proceed to reach equilibrium? (c) Calculate K_c at 298 K.

17.77 The oxidation of SO_2 is the key step in H_2SO_4 production:

$$SO_2(g) + \tfrac{1}{2}O_2(g) \rightleftharpoons SO_3(g) \qquad \Delta H^\circ_{rxn} = -99.2 \text{ kJ}$$

(a) What qualitative combination of T and P maximizes SO_3 yield?
(b) How does addition of O_2 affect Q? K?
(c) Why is catalysis used for this reaction?

17.78 A mixture of 3.00 volumes of H_2 and 1.00 volume of N_2 reacts at 344°C to form ammonia. The equilibrium mixture at 110. atm contains 41.49% NH_3 by volume. Calculate K_p for the reaction, assuming that the gases behave ideally.

17.79 You are a member of a research team of chemists discussing plans for a plant to produce ammonia:

$$N_2(g) + 3H_2(g) \rightleftharpoons 2NH_3(g)$$

(a) The plant will operate at close to 700 K, at which K_p is 1.00×10^{-4}, and employs the stoichiometric 1/3 ratio of N_2/H_2. At equilibrium, the partial pressure of NH_3 is 50. atm. Calculate the partial pressures of each reactant and P_{total}.
(b) One member of the team suggests the following: since the partial pressure of H_2 is cubed in the reaction quotient, the plant could produce the same amount of NH_3 if the reactants were in a 1/6 ratio of N_2/H_2 and could do so at a lower pressure, which would cut operating costs. Calculate the partial pressure of each reactant and P_{total} under these conditions, assuming an unchanged partial pressure of 50. atm for NH_3. Is the suggestion valid?

Comprehensive Problems

17.80 For the following equilibrium system, which of the changes will form more $CaCO_3$?

$$CO_2(g) + Ca(OH)_2(s) \rightleftharpoons CaCO_3(s) + H_2O(l) \qquad \Delta H^\circ = -113 \text{ kJ}$$

(a) Decrease temperature at constant pressure (no phase change)
(b) Increase volume at constant temperature
(c) Increase partial pressure of CO_2
(d) Remove one-half of the initial $CaCO_3$

17.81 The "filmstrip" represents five molecular scenes of a gaseous mixture as it reaches equilibrium over time:

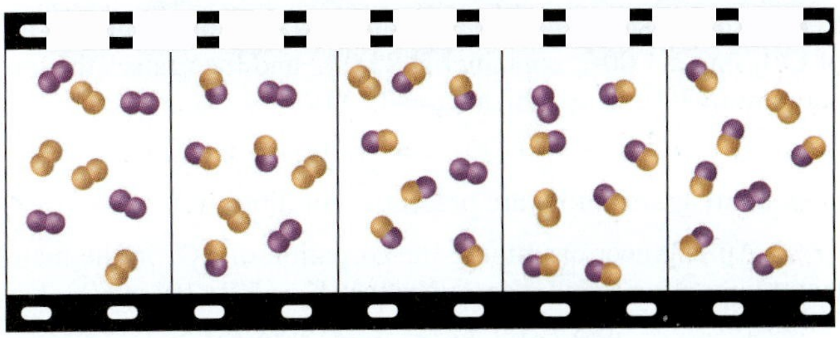

X is *purple* and Y is *orange*: $X_2(g) + Y_2(g) \rightleftharpoons 2XY(g)$.
(a) Write the reaction quotient, Q, for this reaction.
(b) If each particle represents 0.1 mol, find Q for each scene.
(c) If $K > 1$, is time progressing to the right or to the left? Explain.
(d) Calculate K at this temperature.
(e) If $\Delta H^\circ_{rxn} < 0$, which scene, if any, best represents the mixture at a higher temperature? Explain.
(f) Which scene, if any, best represents the mixture at a higher pressure (lower volume)? Explain.

17.82 Ammonium carbamate (NH_2COONH_4) is a salt of carbamic acid that is found in the blood and urine of mammals. At 250.°C, $K_c = 1.58\times10^{-8}$ for the following equilibrium:

$$NH_2COONH_4(s) \rightleftharpoons 2NH_3(g) + CO_2(g)$$

If 7.80 g of NH_2COONH_4 is put into a 0.500-L evacuated container, what is the total pressure at equilibrium?

17.83 Isolation of Group 8B(10) elements, used as industrial catalysts, involves a series of steps. For nickel, the sulfide ore is roasted in air: $Ni_3S_2(s) + O_2(g) \rightleftharpoons NiO(s) + SO_2(g)$. The metal oxide is reduced by the H_2 in water gas ($CO + H_2$) to impure Ni: $NiO(s) + H_2(g) \rightleftharpoons Ni(s) + H_2O(g)$. The CO in water gas then reacts with the metal in the Mond process to form gaseous nickel carbonyl, $Ni(s) + CO(g) \rightleftharpoons Ni(CO)_4(g)$, which is subsequently decomposed to the metal. (a) Balance each of the three steps, and obtain an overall balanced equation for the conversion of Ni_3S_2 to $Ni(CO)_4$. (b) Show that the overall Q_c is the product of the Q_c's for the individual reactions.

17.84 Consider the formation of ammonia in two experiments.
(a) To a 1.00-L container at 727°C, 1.30 mol of N_2 and 1.65 mol of H_2 are added. At equilibrium, 0.100 mol of NH_3 is present. Calculate the equilibrium concentrations of N_2 and H_2, and find K_c for the reaction:

$$2NH_3(g) \rightleftharpoons N_2(g) + 3H_2(g)$$

(b) In a different 1.00-L container at the same temperature, equilibrium is established with 8.34×10^{-2} mol of NH_3, 1.50 mol of N_2, and 1.25 mol of H_2 present. Calculate K_c for the reaction:

$$NH_3(g) \rightleftharpoons \tfrac{1}{2}N_2(g) + \tfrac{3}{2}H_2(g)$$

(c) What is the relationship between the K_c values in parts (a) and (b)? Why aren't these values the same?

17.85 An important industrial source of ethanol is the reaction, catalyzed by H_3PO_4, of steam with ethylene derived from oil:

$$C_2H_4(g) + H_2O(g) \rightleftharpoons C_2H_5OH(g)$$
$$\Delta H^\circ_{rxn} = -47.8 \text{ kJ} \qquad K_c = 9\times10^3 \text{ at } 600. \text{ K}$$

(a) At equilibrium, $P_{C_2H_5OH} = 200.$ atm and $P_{H_2O} = 400.$ atm. Calculate $P_{C_2H_4}$. (b) Is the highest yield of ethanol obtained at high or low P? High or low T? (c) Calculate K_c at 450. K. (d) In NH_3 manufacture, the yield is increased by condensing the NH_3 to a liquid and removing it. Would condensing the C_2H_5OH have the same effect in ethanol production? Explain.

17.86 An industrial chemist introduces 2.0 atm of H_2 and 2.0 atm of CO_2 into a 1.00-L container at 25.0°C and then raises the temperature to 700.°C, at which $K_c = 0.534$:

$$H_2(g) + CO_2(g) \rightleftharpoons H_2O(g) + CO(g)$$

How many grams of H_2 are present at equilibrium?

17.87 An engineer examining the oxidation of SO_2 in the manufacture of sulfuric acid determines that $K_c = 1.7\times10^8$ at 600. K:

$$2SO_2(g) + O_2(g) \rightleftharpoons 2SO_3(g)$$

(a) At equilibrium, $P_{SO_3} = 300.$ atm and $P_{O_2} = 100.$ atm. Calculate P_{SO_2}. (b) The engineer places a mixture of 0.0040 mol of $SO_2(g)$ and 0.0028 mol of $O_2(g)$ in a 1.0-L container and raises the temperature to 1000 K. At equilibrium, 0.0020 mol of $SO_3(g)$ is present. Calculate K_c and P_{SO_2} for this reaction at 1000. K.

17.88 Phosgene ($COCl_2$) is a toxic substance that forms readily from carbon monoxide and chlorine at elevated temperatures:

$$CO(g) + Cl_2(g) \rightleftharpoons COCl_2(g)$$

If 0.350 mol of each reactant is placed in a 0.500-L flask at 600 K, what are the concentrations of all three substances at equilibrium ($K_c = 4.95$ at this temperature)?

17.89 When 0.100 mol of $CaCO_3(s)$ and 0.100 mol of $CaO(s)$ are placed in an evacuated sealed 10.0-L container and heated to 385 K, $P_{CO_2} = 0.220$ atm after equilibrium is established:

$$CaCO_3(s) \rightleftharpoons CaO(s) + CO_2(g)$$

An additional 0.300 atm of $CO_2(g)$ is pumped in. What is the total mass (in g) of $CaCO_3$ after equilibrium is re-established?

17.90 Use each of the following reaction quotients to write the balanced equation:

(a) $Q = \dfrac{[CO_2]^2[H_2O]^2}{[C_2H_4][O_2]^3}$ (b) $Q = \dfrac{[NH_3]^4[O_2]^7}{[NO_2]^4[H_2O]^6}$

17.91 Hydrogenation of carbon-carbon π bonds is important in the petroleum and food industries. The conversion of acetylene to ethylene is a simple example of the process:

$$C_2H_2(g) + H_2(g) \rightleftharpoons C_2H_4(g)$$

The calculated K_c at 2000. K is 2.9×10^8. But the process is run at lower temperatures with the aid of a catalyst to prevent decomposition. Use $\Delta H°$ values to calculate K_c at 300. K.

17.92 For the reaction $M_2 + N_2 \rightleftharpoons 2MN$, scene A represents the mixture at equilibrium, with M *black* and N *orange*. If each molecule represents 0.10 mol and the volume is 1.0 L, how many moles of each substance will be present in scene B when that mixture reaches equilibrium?

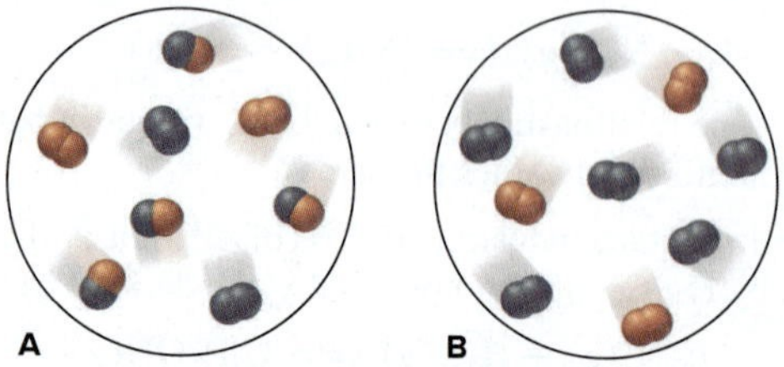

A B

17.93 Highly toxic disulfur decafluoride decomposes by a free-radical process: $S_2F_{10}(g) \rightleftharpoons SF_4(g) + SF_6(g)$. In a study of the decomposition, S_2F_{10} was placed in a 2.0-L flask and heated to 100°C; $[S_2F_{10}]$ was 0.50 *M* at equilibrium. More S_2F_{10} was added, and when equilibrium was reattained, $[S_2F_{10}]$ was 2.5 *M*. How did $[SF_4]$ and $[SF_6]$ change from the original to the new equilibrium position after the addition of more S_2F_{10}?

17.94 A study of the water-gas shift reaction (see Problem 17.38) was made in which equilibrium was reached with $[CO] = [H_2O] = [H_2] = 0.10$ *M* and $[CO_2] = 0.40$ *M*. After 0.60 mol of H_2 is added to the 2.0-L container and equilibrium is re-established, what are the new concentrations of all the components?

17.95 A gaseous mixture of 10.0 volumes of CO_2, 1.00 volume of unreacted O_2, and 50.0 volumes of unreacted N_2 leaves an engine at 4.0 atm and 800. K. Assuming that the mixture reaches equilibrium, what are (a) the partial pressure and (b) the concentration (in picograms per liter, pg/L) of CO in this exhaust gas?

$$2CO_2(g) \rightleftharpoons 2CO(g) + O_2(g) \qquad K_p = 1.4\times10^{-28} \text{ at 800. K}$$

(The actual concentration of CO in exhaust gas is much higher because the gases do *not* reach equilibrium in the short transit time through the engine and exhaust system.)

17.96 When ammonia is made industrially, the mixture of N_2, H_2, and NH_3 that emerges from the reaction chamber is far from equilibrium. Why does the plant supervisor use reaction conditions that produce less than the maximum yield of ammonia?

17.97 The following reaction can be used to make H_2 for the synthesis of ammonia from the greenhouse gases carbon dioxide and methane:

$$CH_4(g) + CO_2(g) \rightleftharpoons 2CO(g) + 2H_2(g)$$

(a) What is the percent yield of H_2 when an equimolar mixture of CH_4 and CO_2 with a total pressure of 20.0 atm reaches equilibrium at 1200. K, at which $K_p = 3.548\times10^6$?

(b) What is the percent yield of H_2 for this system at 1300. K, at which $K_p = 2.626\times10^7$?

(c) Use the van't Hoff equation to find $\Delta H°_{rxn}$.

17.98 The methane used to obtain H_2 for NH_3 manufacture is impure and usually contains other hydrocarbons, such as propane, C_3H_8. Imagine the reaction of propane occurring in two steps:

$$C_3H_8(g) + 3H_2O(g) \rightleftharpoons 3CO(g) + 7H_2(g) \qquad K_p = 8.175\times10^{15} \text{ at 1200. K}$$

$$CO(g) + H_2O(g) \rightleftharpoons CO_2(g) + H_2(g) \qquad K_p = 0.6944 \text{ at 1200. K}$$

(a) Write the overall equation for the reaction of propane and steam to produce carbon dioxide and hydrogen.

(b) Calculate K_p for the overall process at 1200. K.

(c) When 1.00 volume of C_3H_8 and 4.00 volumes of H_2O, each at 1200. K and 5.0 atm, are mixed in a container, what is the final pressure? Assume the total volume remains constant, that the reaction is essentially complete, and that the gases behave ideally.

(d) What percentage of the C_3H_8 remains unreacted?

17.99 Using CH_4 and steam as a source of H_2 for NH_3 synthesis requires high temperatures. Rather than burning CH_4 separately to heat the mixture, it is more efficient to inject some O_2 into the reaction mixture. All of the H_2 is thus released for the synthesis, and the heat of reaction for the combustion of CH_4 helps maintain the required temperature. Imagine the reaction occurring in two steps:

$$2CH_4(g) + O_2(g) \rightleftharpoons 2CO(g) + 4H_2(g) \qquad K_p = 9.34\times10^{28} \text{ at 1000. K}$$

$$CO(g) + H_2O(g) \rightleftharpoons CO_2(g) + H_2(g) \qquad K_p = 1.374 \text{ at 1000. K}$$

(a) Write the overall equation for the reaction of methane, steam, and oxygen to form carbon dioxide and hydrogen.

(b) What is K_p for the overall reaction?

(c) What is K_c for the overall reaction?

(d) A mixture of 2.0 mol of CH_4, 1.0 mol of O_2, and 2.0 mol of steam with a total pressure of 30. atm reacts at 1000. K at constant volume. Assuming that the reaction is complete and the ideal gas law is a valid approximation, what is the final pressure?

17.100 One mechanism for the synthesis of ammonia proposes that N_2 and H_2 molecules catalytically dissociate into atoms:

$$N_2(g) \rightleftharpoons 2N(g) \qquad \log K_p = -43.10$$
$$H_2(g) \rightleftharpoons 2H(g) \qquad \log K_p = -17.30$$

(a) Find the partial pressure of N in N_2 at 1000. K and 200. atm.

(b) Find the partial pressure of H in H_2 at 1000. K and 600. atm.

(c) How many N atoms and H atoms are present per liter?

(d) Based on these answers, which of the following is a more reasonable step to continue the mechanism after the catalytic dissociation? Explain.

$$N(g) + H(g) \longrightarrow NH(g)$$
$$N_2(g) + H(g) \longrightarrow NH(g) + N(g)$$

17.101 The molecular scenes below depict the reaction $Y \rightleftharpoons 2Z$ at four different times, out of sequence, as it reaches equilibrium. Each sphere (Y is *red* and Z is *green*) represents 0.025 mol, and the volume is 0.40 L. (a) Which scene(s) represent(s) equilibrium? (b) List the scenes in the correct sequence. (c) Calculate K_c.

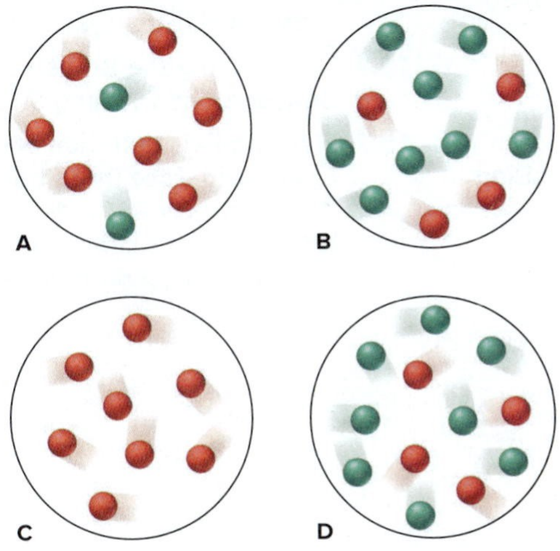

17.102 For the equilibrium

$$H_2S(g) \rightleftharpoons 2H_2(g) + S_2(g) \qquad K_c = 9.0\times10^{-8} \text{ at } 700°C$$

the initial concentrations of the three gases are 0.300 M H_2S, 0.300 M H_2, and 0.150 M S_2. Determine the equilibrium concentrations of the gases.

17.103 The two most abundant atmospheric gases react to a tiny extent at 298 K in the presence of a catalyst:

$$N_2(g) + O_2(g) \rightleftharpoons 2NO(g) \qquad K_p = 4.35\times10^{-31}$$

(a) What are the equilibrium pressures of the three gases when the atmospheric partial pressures of O_2 (0.210 atm) and of N_2 (0.780 atm) are put into an evacuated 1.00-L flask at 298 K with the catalyst? (b) What is P_{total} in the container? (c) Find K_c at 298 K.

17.104 The oxidation of nitrogen monoxide is favored at 457 K:

$$2NO(g) + O_2(g) \rightleftharpoons 2NO_2(g) \qquad K_p = 1.3\times10^4$$

(a) Calculate K_c at 457 K. (b) Find $\Delta H°_{rxn}$ from standard heats of formation. (c) At what temperature does $K_c = 6.4\times10^9$?

17.105 The kinetics and equilibrium of the decomposition of hydrogen iodide have been studied extensively:

$$2HI(g) \rightleftharpoons H_2(g) + I_2(g)$$

(a) At 298 K, $K_c = 1.26\times10^{-3}$ for this reaction. Calculate K_p.

(b) Calculate K_c for the *formation* of HI at 298 K.

(c) Calculate $\Delta H°_{rxn}$ for HI decomposition from $\Delta H°_f$ values.

(d) At 729 K, $K_c = 2.0\times10^{-2}$ for HI decomposition. Calculate ΔH_{rxn} for this reaction from the van't Hoff equation.

17.106 Isopentyl alcohol reacts with pure acetic acid to form isopentyl acetate, the essence of banana oil:

$$C_5H_{11}OH + CH_3COOH \rightleftharpoons CH_3COOC_5H_{11} + H_2O$$

A student adds a drying agent to remove H_2O and thus increase the yield of banana oil. Is this approach reasonable? Explain.

17.107 Isomers Q (*blue*) and R (*yellow*) interconvert. They are depicted in an equilibrium mixture in scene A. Scene B represents the mixture after addition of more Q. How many molecules of each isomer are present when the mixture in scene B attains equilibrium again?

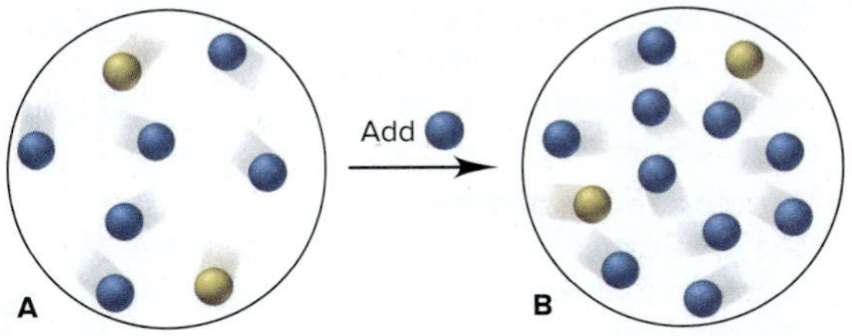

17.108 Glauber's salt, $Na_2SO_4 \cdot 10H_2O$, was used by J. R. Glauber in the 17th century as a medicinal agent. At 25°C, $K_p = 4.08\times10^{-25}$ for the loss of waters of hydration from Glauber's salt:

$$Na_2SO_4 \cdot 10H_2O(s) \rightleftharpoons Na_2SO_4(s) + 10H_2O(g)$$

(a) What is the vapor pressure of water at 25°C in a closed container holding a sample of $Na_2SO_4 \cdot 10H_2O(s)$?

(b) How do the following changes affect the ratio (higher, lower, same) of hydrated form to anhydrous form for the system above?

(1) Add more $Na_2SO_4(s)$

(2) Reduce the container volume

(3) Add more water vapor

(4) Add N_2 gas

17.109 In a study of synthetic fuels, 0.100 mol of CO and 0.100 mol of water vapor are added to a 20.00-L container at 900.°C, and they react to form CO_2 and H_2. At equilibrium, [CO] is 2.24×10^{-3} M. (a) Calculate K_c at this temperature. (b) Calculate P_{total} in the flask at equilibrium. (c) How many moles of CO must be added to double this pressure? (d) After P_{total} is doubled and the system reattains equilibrium, what is $[CO]_{eq}$?

17.110 Synthetic diamonds are made under conditions of high temperature (2000 K) and high pressure (10^{10} Pa; 10^5 atm) in the presence of catalysts. Carbon's phase diagram is useful for finding the conditions for formation of natural and synthetic diamonds. Along the diamond-graphite line, the two allotropes are in equilibrium. (a) At point A, what is the sign of ΔH for the formation of diamond from graphite? Explain. (b) Which allotrope is denser? Explain.

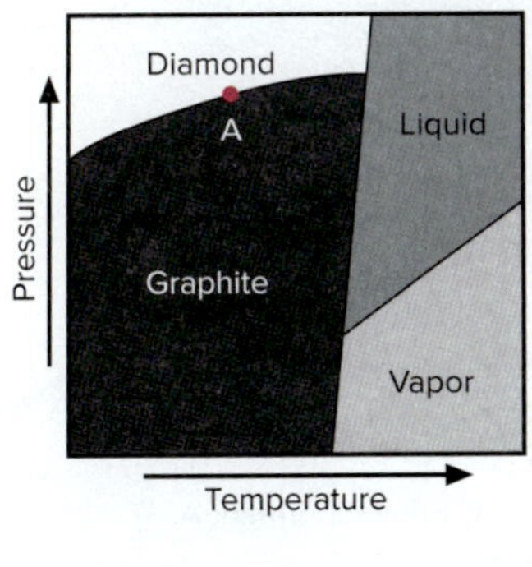

18 Acid-Base Equilibria

18.1 Acids and Bases in Water
Arrhenius Acid-Base Definition
Acid-Dissociation Constant (K_a)
Relative Strengths of Acids and Bases

18.2 Autoionization of Water and the pH Scale
Autoionization and K_w
The pH Scale

18.3 Proton Transfer and the Brønsted-Lowry Acid-Base Definition
Conjugate Acid-Base Pairs
Net Direction of Acid-Base Reactions

18.4 Solving Problems Involving Weak-Acid Equilibria
Finding K_a Given Concentrations
Finding Concentrations Given K_a
Extent of Acid Dissociation
Polyprotic Acids

18.5 Molecular Properties and Acid Strength
Nonmetal Hydrides
Oxoacids
Hydrated Metal Ions

18.6 Weak Bases and Their Relation to Weak Acids
Ammonia and the Amines
Anions of Weak Acids
Relation Between K_a and K_b

18.7 Acid-Base Properties of Salt Solutions
Salts That Yield Neutral Solutions
Salts That Yield Acidic Solutions
Salts That Yield Basic Solutions
Salts of Weakly Acidic Cations and Weakly Basic Anions
Salts of Amphiprotic Anions

18.8 Generalizing the Brønsted-Lowry Concept: The Leveling Effect

18.9 Electron-Pair Donation and the Lewis Acid-Base Definition
Molecules as Lewis Acids
Metal Cations as Lewis Acids
Overview of Acid-Base Definitions

Source: © McGraw-Hill Education/Mark A. Dierker, photographer

Concepts and Skills to Review Before You Study This Chapter

- role of water as solvent (Section 4.1)
- writing ionic equations (Section 4.2)
- acids, bases, and acid-base reactions (Section 4.4)
- proton transfer in acid-base reactions (Section 4.4)
- properties of an equilibrium constant (Section 17.2)
- solving equilibrium problems (Section 17.5)

It's lunch time: how about a meal of citric and ascorbic acids (lemons and oranges), oxalic acid (spinach), folic acid (broccoli), and lactic acid (yogurt), all washed down with some phosphoric acid (soft drink)—if you eat too much, you may need some magnesium or aluminum hydroxide base (antacid)! Acids and bases are in many common consumer products (Table 18.1) and are indispensable in academic and industrial research.

Acids gives substances such as lemon juice and vinegar a sour taste. In fact, sourness was a defining property of an acid since the 17th century: an acid was any substance that had a sour taste; reacted with active metals, such as aluminum and zinc, to produce hydrogen gas; and turned certain organic compounds specific colors. (We discuss *indicators* in this chapter and Chapter 19.) Similarly, a base was any substance that had a bitter taste and slippery feel and turned the same organic compounds different colors. Moreover, it was known that *when an acid and a base react, each cancels the properties of the other in a process called neutralization.* Although these early definitions described distinctive properties, they gave way to others based on molecular behavior. As science progresses, limited definitions are replaced by broader ones that explain more phenomena.

Table 18.1 Some Common Acids and Bases and Their Household Uses

Substance	Use
Acids	
Acetic acid, CH_3COOH	Flavoring, preservative (vinegar)
Citric acid, $H_3C_6H_5O_7$	Flavoring (lemon juice)
Ascorbic acid, $H_2C_6H_6O_6$	Vitamin C; nutritional supplement
Aluminum salts, $NaAl(SO_4)_2{\cdot}12H_2O$	In baking powder, with sodium hydrogen carbonate
Bases	
Sodium hydroxide (lye), $NaOH$	Oven and drain cleaners
Ammonia, NH_3	Household cleaner
Sodium carbonate, Na_2CO_3	Water softener, grease remover
Sodium hydrogen carbonate, $NaHCO_3$	Fire extinguisher, rising agent in cake mixes (baking soda), mild antacid
Sodium phosphate, Na_3PO_4	Cleaner for surfaces before painting or wallpapering

Source: © McGraw-Hill Education/Stephen Frisch, photographer

IN THIS CHAPTER . . . *We develop three definitions of acids and bases that explain an expanded range of reactions, and we apply equilibrium principles to understand acid-base behavior.*

- We begin with the *Arrhenius* acid-base definition, which relies on formulas and behavior in water.
- We examine acid dissociation to see how variation in acid strength is expressed by a new equilibrium constant.
- We introduce the pH scale to measure the acidity of aqueous solutions.
- We discuss proton transfer in the *Brønsted-Lowry* acid-base definition, which expands the meaning of "base", along with the scope of acid-base reactions.
- We apply a systematic approach to solving acid-base equilibrium problems.
- We examine the molecular structures of acids to rationalize their relative strengths.

In a dilute solution of a weak acid, *the great majority of HA molecules are undissociated.* Thus, $[H_3O^+] = [A^-] << [HA]_{init}$, and $[HA]_{eq} \approx [HA]_{init}$, so K_c is very small. Hydrocyanic acid is an example of a weak acid:

$$HCN(aq) + H_2O(l) \rightleftharpoons H_3O^+(aq) + CN^-(aq)$$

$$Q_c = \frac{[H_3O^+][CN^-]}{[HCN][H_2O]} \quad \text{(at equilibrium, } Q_c = K_c << 1)$$

(As in Chapter 17, brackets with no subscript mean molar concentration *at equilibrium;* that is, [X] means $[X]_{eq}$. In this chapter, we are dealing with systems *at equilibrium,* so instead of writing Q and stating that Q equals K at equilibrium, we'll express K directly as a collection of equilibrium concentration terms.)

The difference in $[H_3O^+]$ causes a much higher rate for the reaction of a strong acid with an active metal like zinc than for the same reaction of a weak acid (Figure 18.2):

$$Zn(s) + 2H_3O^+(aq) \longrightarrow Zn^{2+}(aq) + 2H_2O(l) + H_2(g)$$

In a strong acid, with its much higher $[H_3O^+]$, zinc reacts rapidly, forming bubbles of H_2 vigorously. In a weak acid, $[H_3O^+]$ is much lower, so zinc reacts slowly.

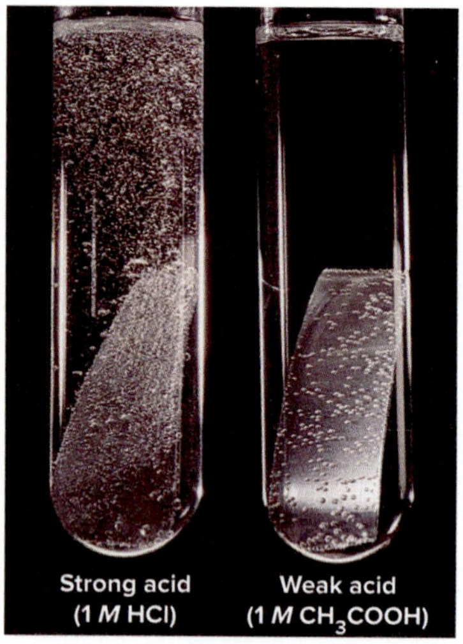

Figure 18.2 Reaction of zinc with a strong acid *(left)* and a weak acid *(right).*
Source: © McGraw-Hill Education/Stephen Frisch, photographer

The Meaning of K_a We write a *specific* equilibrium constant for acid dissociation that includes only the species whose concentrations change to any significant extent. For the dissociation of a general *weak acid,* HA,

$$HA(aq) + H_2O(l) \rightleftharpoons H_3O^+(aq) + A^-(aq)$$

the equilibrium expression is

$$K_c = \frac{[H_3O^+][A^-]}{[HA][H_2O]}$$

The concentration of water $\left([H_2O] = \frac{1000\ g}{18.02\ g/mol} = 55.5\ M\right)$ is typically several orders of magnitude larger than [HA]. Therefore, $[H_2O]$ is essentially constant when HA dissociates, and so H_2O can be considered a pure liquid. As we discussed in Section 17.2, the concentration terms for pure liquids and solids are equal to 1 and do not appear in the equilibrium expression. Thus, we can define a new equilibrium constant, the **acid-dissociation constant (*or* acid-ionization constant), K_a:**

$$K_c \times 1 = K_a = \frac{[H_3O^+][A^-]}{[HA]} \quad \mathbf{(18.1)}$$

Like any equilibrium constant, K_a is a number whose magnitude is temperature dependent and tells how far to the right the reaction has proceeded to reach equilibrium. Thus, *the stronger the acid, the higher $[H_3O^+]$ is at equilibrium, and the larger the value of K_a:*

Stronger acid $\Longrightarrow$ higher $[H_3O^+]$ $\Longrightarrow$ larger K_a

The Range of K_a Values Acid-dissociation constants of weak acids range over many orders of magnitude. Some benchmark K_a values for typical weak acids in Table 18.2 give an idea of the fraction of HA molecules that dissociate into ions.

Table 18.2 Magnitude of K_a Values and Percent Dissociation for Weak Acids

Magnitude of K_a	% Dissociation in a 1 *M* Solution of HA	Specific Example (K_a Value, % Dissociation)
Relatively high K_a (~10^{-2})	~10%	1 *M* chlorous acid ($HClO_2$) ($K_a = 1.1\times10^{-2}$, 10.%)
Moderate K_a (~10^{-5})	~0.3%	1 *M* acetic acid (CH_3COOH) ($K_a = 1.8\times10^{-5}$, 0.42%)
Relatively low K_a (~10^{-10})	~0.001%	1 *M* hydrocyanic acid (HCN) ($K_a = 6.2\times10^{-10}$, 0.0025%)

Thus, for solutions of the same initial HA concentration, *the smaller the K_a, the lower the percent dissociation of HA:*

Weaker acid $\Longrightarrow$ lower % dissociation of HA $\Longrightarrow$ smaller K_a

Table 18.3 K_a **Values for Some Monoprotic Acids at 25°C**

Name (Formula)*	Lewis Structure*	Dissociation Reaction	K_a
Chlorous acid ($HClO_2$)	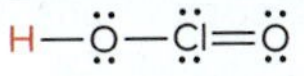	$HClO_2(aq) + H_2O(l) \rightleftharpoons H_3O^+(aq) + ClO_2^-(aq)$	1.1×10^{-2}
Nitrous acid (HNO_2)	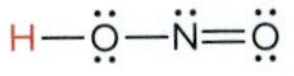	$HNO_2(aq) + H_2O(l) \rightleftharpoons H_3O^+(aq) + NO_2^-(aq)$	7.1×10^{-4}
Hydrofluoric acid (HF)	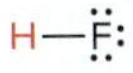	$HF(aq) + H_2O(l) \rightleftharpoons H_3O^+(aq) + F^-(aq)$	6.8×10^{-4}
Formic acid (HCOOH)	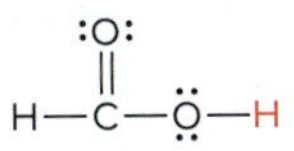	$HCOOH(aq) + H_2O(l) \rightleftharpoons H_3O^+(aq) + HCOO^-(aq)$	1.8×10^{-4}
Acetic acid (CH_3COOH)	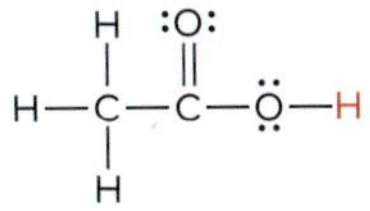	$CH_3COOH(aq) + H_2O(l) \rightleftharpoons H_3O^+(aq) + CH_3COO^-(aq)$	1.8×10^{-5}
Propanoic acid (CH_3CH_2COOH)	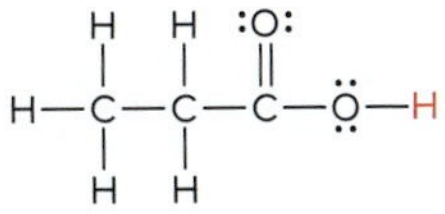	$CH_3CH_2COOH(aq) + H_2O(l) \rightleftharpoons H_3O^+(aq) + CH_3CH_2COO^-(aq)$	1.3×10^{-5}
Hypochlorous acid (HClO)	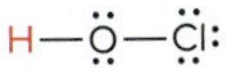	$HClO(aq) + H_2O(l) \rightleftharpoons H_3O^+(aq) + ClO^-(aq)$	2.9×10^{-8}
Hydrocyanic acid (HCN)	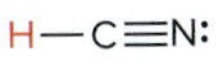	$HCN(aq) + H_2O(l) \rightleftharpoons H_3O^+(aq) + CN^-(aq)$	6.2×10^{-10}

*Red type indicates the ionizable proton; all atoms have zero formal charge.

Table 18.3 lists K_a values of some weak *monoprotic* acids, those with one ionizable proton. (A more extensive list is in Appendix C.) Note that the ionizable proton in organic acids is bound to the O in —COOH; H atoms bonded to C do *not* ionize.

Classifying the Relative Strengths of Acids and Bases

Using a table of K_a values is the surest way to quantify strengths of weak acids, but you can classify acids and bases qualitatively as strong or weak from their formulas:

- *Strong acids.* Two types of strong acids, with examples *you should memorize,* are
  1. The hydrohalic acids HCl, HBr, and HI
  2. Oxoacids in which the number of O atoms exceeds the number of ionizable protons by two or more, such as HNO_3, H_2SO_4, and $HClO_4$; for example, in the case of H_2SO_4, 4 O's − 2 H's = 2
- *Weak acids.* There are many *more* weak acids than strong ones. Four types are
  1. The hydrohalic acid HF
  2. Acids in which H is not bonded to O or to a halogen, such as HCN and H_2S
  3. Oxoacids in which the number of O atoms equals or exceeds by one the number of ionizable protons, such as HClO, HNO_2, and H_3PO_4
  4. Carboxylic acids (general formula RCOOH, with the ionizable proton shown in red), such as CH_3COOH and C_6H_5COOH
- *Strong bases.* Water-soluble compounds containing O^{2-} or OH^- ions are strong bases. The cations are usually those of the most active metals:
  1. M_2O or MOH, where M = Group 1A(1) metal (Li, Na, K, Rb, Cs)
  2. MO or $M(OH)_2$, where M = Group 2A(2) metal (Ca, Sr, Ba) [MgO_2 and $Mg(OH)_2$ are only slightly soluble in water, but the soluble portion dissociates completely.]
- *Weak bases.* Many compounds with an electron-rich nitrogen atom are weak bases (none is an Arrhenius base). The common structural feature is an N atom with a lone electron pair (shown in blue):
  1. Ammonia ($\ddot{N}H_3$)
  2. Amines (general formula $R\ddot{N}H_2$, $R_2\ddot{N}H$, or $R_3\ddot{N}$), such as $CH_3CH_2\ddot{N}H_2$, $(CH_3)_2\ddot{N}H$, and $(C_3H_7)_3\ddot{N}$

SAMPLE PROBLEM 18.1 Classifying Acid and Base Strength from the Chemical Formula

Problem Classify each of the following compounds as a strong acid, weak acid, strong base, or weak base, and write the K_a expression for any weak acid:
(a) KOH **(b)** $(CH_3)_2CHCOOH$ **(c)** H_2SeO_4 **(d)** $(CH_3)_2CHNH_2$

Plan We examine the formula and classify each compound as acid or base, using the text descriptions. Particular points to note for acids are the numbers of O atoms relative to ionizable H atoms and the presence of the —COOH group. For bases, we note the nature of the cation or the presence of an N atom that has a lone pair.

Solution (a) Strong base: KOH is one of the Group 1A(1) hydroxides.

(b) Weak acid: $(CH_3)_2CHCOOH$ is a carboxylic acid, as indicated by the —COOH group; the dissociation reaction is:

$$(CH_3)_2CHCOOH(aq) + H_2O(l) \rightleftharpoons H_3O^+(aq) + (CH_3)_2CHCOO^-(aq)$$

and

$$K_a = \frac{[H_3O^+][(CH_3)_2CHCOO^-]}{[(CH_3)_2CHCOOH]}$$

(c) Strong acid: H_2SeO_4 is an oxoacid in which the number of O atoms exceeds the number of ionizable protons by 2.
(d) Weak base: $(CH_3)_2CHNH_2$ has a lone electron pair on the N and is an amine.

FOLLOW-UP PROBLEMS

Brief Solutions for ***all*** *Follow-up Problems appear at the end of the chapter.*

18.1A Which member of each pair is the stronger acid or base?
(a) HClO or $HClO_3$ **(b)** HCl or CH_3COOH **(c)** NaOH or CH_3NH_2

18.1B Classify each of the following compounds as a strong acid, weak acid, strong base, or weak base, and write the K_a expression for any weak acid:
(a) $(CH_3)_3N$ **(b)** HI **(c)** HBrO **(d)** $Ca(OH)_2$

SOME SIMILAR PROBLEMS 18.15–18.18

› Summary of Section 18.1

- In aqueous solution, water binds the proton released from an acid to form a hydrated species represented by $H_3O^+(aq)$.
- By the Arrhenius definition, acids contain H and yield H_3O^+ in water, bases contain OH and yield OH^- in water, and an acid-base reaction (neutralization) is the reaction of H^+ and OH^- to form H_2O.
- Acid strength depends on $[H_3O^+]$ relative to [HA] in aqueous solution. Strong acids dissociate completely and weak acids slightly.
- The extent of dissociation is expressed by the acid-dissociation constant, K_a. Most weak acids have K_a values ranging from about 10^{-2} to 10^{-10}.
- Many acids and bases can be classified as strong or weak based on their formulas.

18.2 AUTOIONIZATION OF WATER AND THE pH SCALE

Before we discuss the next major acid-base definition, let's examine a crucial property of water that enables us to quantify $[H_3O^+]$: *water dissociates very slightly into ions in an equilibrium process known as* **autoionization** (or *self-ionization*):

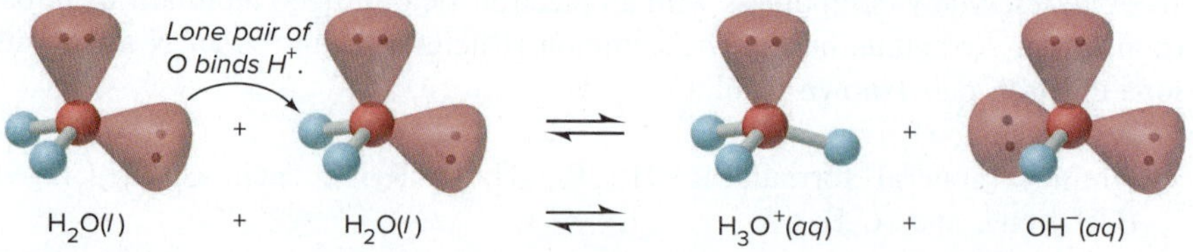

The Equilibrium Nature of Autoionization: The Ion-Product Constant for Water (K_w)

Like any equilibrium process, the autoionization of water is described quantitatively by an equilibrium constant:

$$K_c = \frac{[H_3O^+][OH^-]}{[H_2O]^2}$$

Because the concentration of H_2O (55.5 *M*) remains essentially constant, it is considered a pure liquid and is eliminated from the equilibrium expression. Thus, we obtain the **ion-product constant for water, K_w:**

$$K_c \times (1)^2 = K_w = [H_3O^+][OH] = 1.0{\times}10^{14} \text{ (at 25°C)} \qquad \textbf{(18.2)}$$

Notice that *one H_3O^+ ion and one OH^- ion form for each H_2O molecule that dissociates.* Therefore, in pure water, we find that

$$[H_3O^+] = [OH^-] = \sqrt{1.0{\times}10^{-14}} = 1.0{\times}10^{-7}\ M \text{ (at 25°C)}$$

Since pure water has a concentration of about 55.5 *M*, these equilibrium concentrations are attained when only 1 in 555 million water molecules dissociates reversibly into ions!

Autoionization of water affects aqueous acid-base chemistry in two major ways:

1. *A change in $[H_3O^+]$ causes an inverse change in $[OH^-]$,* and vice versa:

Higher $[H_3O^+]$ ⟹ lower $[OH^-]$ and Higher $[OH^-]$ ⟹ lower $[H_3O^+]$

Recall from Le Châtelier's principle (Section 17.6) that a change in concentration shifts the equilibrium position but does *not* change the equilibrium constant. Therefore, if some acid is added, $[H_3O^+]$ increases and $[OH^-]$ decreases as the autoionization reaction proceeds to the left and the ions react to form water; similarly, if some base is added, $[OH^-]$ increases and $[H_3O^+]$ decreases. In both cases, as long as the temperature is constant, the value of K_w is constant.

2. *Both ions are present in all aqueous systems.* Thus, all acidic solutions contain a low $[OH^-]$, and all basic solutions contain a low $[H_3O^+]$. The equilibrium nature of autoionization allows us to define "acidic" and "basic" solutions in terms of relative magnitudes of $[H_3O^+]$ and $[OH^-]$:

In an *acidic* solution, $[H_3O^+] > [OH^-]$

In a *neutral* solution, $[H_3O^+] = [OH^-]$

In a *basic* solution, $[H_3O^+] < [OH^-]$

Figure 18.3 summarizes these relationships. Moreover, if you know the value of K_w at a particular temperature and the concentration of one of the two ions, you can find the concentration of the other:

$$[H_3O^+] = \frac{K_w}{[OH^-]} \quad \text{or} \quad [OH^-] = \frac{K_w}{[H_3O^+]}$$

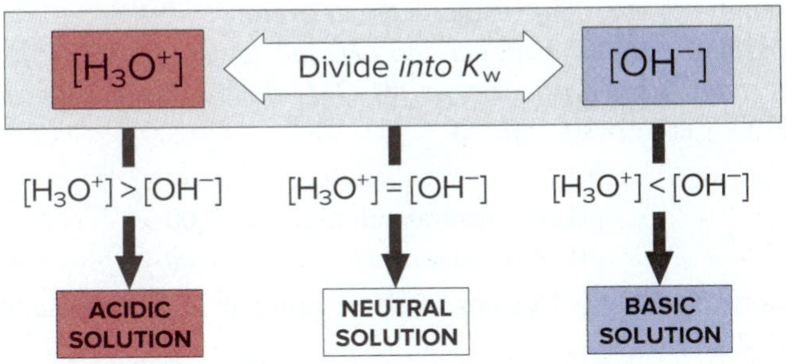

Figure 18.3 The relationship between $[H_3O^+]$ and $[OH^-]$ and the relative acidity of solutions.

SAMPLE PROBLEM 18.2 Calculating $[H_3O^+]$ or $[OH^-]$ in Aqueous Solution

Problem A research chemist adds a measured amount of HCl gas to pure water at 25°C and obtains a solution with $[H_3O^+] = 3.0\times10^{-4}\ M$. Calculate $[OH^-]$. Is the solution neutral, acidic, or basic?

Plan We use the known value of K_w at 25°C (1.0×10^{-14}) and the given $[H_3O^+]$ ($3.0\times10^{-4}\ M$) to solve for $[OH^-]$. Then we compare $[H_3O^+]$ with $[OH^-]$ to determine whether the solution is acidic, basic, or neutral (see Figure 18.3).

Solution Calculating $[OH^-]$:

$$[OH^-] = \frac{K_w}{[H_3O^+]} = \frac{1.0\times10^{-14}}{3.0\times10^{-4}} = 3.3\times10^{-11}\ M$$

Because $[H_3O^+] > [OH^-]$, the solution is acidic.

Check It makes sense that adding an acid to water results in an acidic solution. Also, since $[H_3O^+]$ is greater than $10^{-7}\ M$, $[OH^-]$ must be less than $10^{-7}\ M$ to give a constant K_w.

FOLLOW-UP PROBLEMS

18.2A Calculate $[H_3O^+]$ in a solution that has $[OH^-] = 6.7\times10^{-2}\ M$ at 25°C. Is the solution neutral, acidic, or basic?

18.2B An aqueous solution of window cleaner has $[H_3O^+] = 1.8\times10^{-10}\ M$ at 25°C. Calculate $[OH^-]$. Is the solution neutral, acidic, or basic?

SOME SIMILAR PROBLEMS 18.27–18.30

Expressing the Hydronium Ion Concentration: The pH Scale

In aqueous solutions, $[H_3O^+]$ can vary from about 10 *M* to $10^{-15}\ M$. To handle numbers with negative exponents more conveniently in calculations, we convert them to positive numbers using a numerical system called a *p-scale,* the negative of the common (base-10) logarithm of the number. Applying this numerical system to $[H_3O^+]$ gives **pH,** the negative of the common logarithm of $[H^+]$ (or $[H_3O^+]$):

$$pH = -\log\ [H_3O^+] \qquad (18.3)$$

What is the pH of a $10^{-12}\ M\ H_3O^+$ solution?

$$pH = -\log\ [H_3O^+] = -\log\ 10^{-12} = (-1)(-12) = 12$$

Similarly, a $10^{-3}\ M\ H_3O^+$ solution has a pH of 3, and a $5.4\times10^{-4}\ M\ H_3O^+$ solution has a pH of 3.27:

$$pH = -\log\ [H_3O^+] = (-1)(\log 5.4 + \log 10^{-4}) = 3.27$$

As with any measurement, the number of significant figures in a pH value reflects the precision with which the concentration is known. However, a pH value is a logarithm, so the number of significant figures in the concentration equals the number of digits *to the right of the decimal point in the pH value* (see Appendix A). In the preceding example, $5.4\times10^{-4}\ M$ has two significant figures, so its negative logarithm, 3.27, has two digits to the right of the decimal point.

Note in particular that *the higher the pH, the lower the $[H_3O^+]$*. Therefore, *an acidic solution has a lower pH (higher $[H_3O^+]$) than a basic solution.* At 25°C in pure water, $[H_3O^+]$ is $1.0\times10^{-7}\ M$, so

pH of an acidic solution	<	7.00
pH of a neutral solution	=	7.00
pH of a basic solution	>	7.00

Figure 18.4 shows that the pH values of some familiar aqueous solutions fall within a range of 0 to 14.

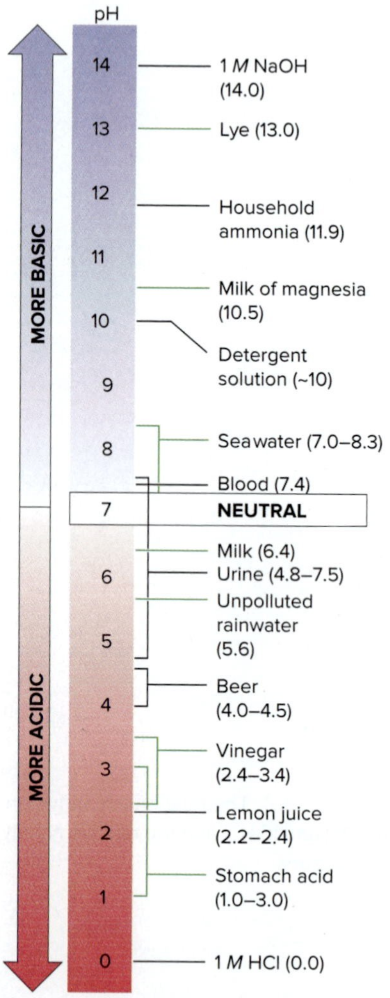

Figure 18.4 The pH values of some familiar aqueous solutions.

Table 18.4 The Relationship Between K_a and pK_a

Acid Name (Formula)	K_a at 25°C	pK_a
Hydrogen sulfate ion (HSO_4^-)	1.0×10^{-2}	1.99
Nitrous acid (HNO_2)	7.1×10^{-4}	3.15
Acetic acid (CH_3COOH)	1.8×10^{-5}	4.75
Hypobromous acid (HBrO)	2.3×10^{-9}	8.64
Phenol (C_6H_5OH)	1.0×10^{-10}	10.00

ACID STRENGTH

Because the pH scale is logarithmic, a solution of pH 1.0 has an $[H_3O^+]$ that is 10 times higher than that of a pH 2.0 solution, 100 times higher than that of a pH 3.0 solution, and so forth. To find the $[H_3O^+]$ from the pH, you perform the opposite arithmetic process; that is, you find the negative antilog of pH:

$$[H_3O^+] = 10^{-pH}$$

A p-scale is used to express other quantities as well:

1. Hydroxide ion concentration can be expressed as pOH:

$$pOH = -\log [OH^-]$$

Acidic solutions have a higher pOH (lower $[OH^-]$) than basic solutions.

2. Equilibrium constants can be expressed as pK:

$$pK = -\log K$$

Specifically for weak acids:

$$pK_a = -\log K_a \text{ and } K_a = 10^{-pK_a}$$

A low pK corresponds to a high K. So
- a reaction that reaches equilibrium with mostly products present (proceeds far to the right) has a low pK (high K);
- a reaction that has mostly reactants present at equilibrium has a high pK (low K).

Table 18.4 shows this relationship for aqueous equilibria of some weak acids.

The Relationships Among pH, pOH, and pK_w Taking the negative log of both sides of the K_w expression gives a useful relationship among pK_w, pH, and pOH:

$$K_w = [H_3O^+][OH^-] = 1.0\times10^{-14} \text{ (at 25°C)}$$

$$-\log K_w = (-\log [H_3O^+]) + (-\log [OH^-]) = -\log (1.0\times10^{-14})$$

$$pK_w = pH + pOH = 14.00 \quad \text{(at 25°C)} \qquad \textbf{(18.4)}$$

Note these important points:

1. The sum of pH and pOH is pK_w for any aqueous solution at any temperature, and pK_w equals 14.00 at 25°C.
2. Because K_w is constant, pH, pOH, $[H_3O^+]$, and $[OH^-]$ are interrelated:
   - $[H_3O^+]$ and $[OH^-]$ change in opposite directions.
   - pH and pOH also change in opposite directions.
   - At 25°C, the product of $[H_3O^+]$ and $[OH^-]$ is 1.0×10^{-14}, and the sum of pH and pOH is 14.00 (Figure 18.5, *next page*).

Calculating pH for Strong Acids and Bases Since strong acids and bases dissociate completely, calculating pH, $[H_3O^+]$, $[OH^-]$, and pOH for these substances is straightforward:

- The equilibrium concentration of H_3O^+ is equal to the initial concentration of a strong acid: 0.200 M HCl dissociates to produce 0.200 M H_3O^+.
- The equilibrium concentration of OH^- is equal to the initial concentration of a Group 1A(1) hydroxide: 0.200 M KOH dissociates to produce 0.200 M OH^-.

Figure 18.5 The relationships among $[H_3O^+]$, pH, $[OH^-]$, and pOH.

MORE BASIC ↑ / MORE ACIDIC ↓

	$[H_3O^+]$	pH	$[OH^-]$	pOH
	1.0×10^{-15}	15.00	1.0×10^{1}	−1.00
	1.0×10^{-14}	14.00	1.0×10^{0}	0.00
	1.0×10^{-13}	13.00	1.0×10^{-1}	1.00
BASIC	1.0×10^{-12}	12.00	1.0×10^{-2}	2.00
	1.0×10^{-11}	11.00	1.0×10^{-3}	3.00
	1.0×10^{-10}	10.00	1.0×10^{-4}	4.00
	1.0×10^{-9}	9.00	1.0×10^{-5}	5.00
	1.0×10^{-8}	8.00	1.0×10^{-6}	6.00
NEUTRAL	1.0×10^{-7}	7.00	1.0×10^{-7}	7.00
	1.0×10^{-6}	6.00	1.0×10^{-8}	8.00
	1.0×10^{-5}	5.00	1.0×10^{-9}	9.00
	1.0×10^{-4}	4.00	1.0×10^{-10}	10.00
	1.0×10^{-3}	3.00	1.0×10^{-11}	11.00
ACIDIC	1.0×10^{-2}	2.00	1.0×10^{-12}	12.00
	1.0×10^{-1}	1.00	1.0×10^{-13}	13.00
	1.0×10^{0}	0.00	1.0×10^{-14}	14.00
	1.0×10^{1}	−1.00	1.0×10^{-15}	15.00

- The equilibrium concentration of OH^- is equal to twice the initial concentration of a Group 2A(2) hydroxide: 0.200 *M* $Ba(OH)_2$ dissociates to produce 2(0.200 *M*) = 0.400 *M* OH^- since there are two moles of OH^- in every mole of $Ba(OH)_2$.

SAMPLE PROBLEM 18.3 Calculating $[H_3O^+]$, pH, $[OH^-]$, and pOH for Strong Acids and Bases

Problem Calculate $[H_3O^+]$, pH, $[OH^-]$, and pOH for each solution at 25°C:
(a) 0.30 *M* HNO_3, used for etching copper metal
(b) 0.0042 *M* $Ca(OH)_2$, used in leather tanning to remove hair from hides

Plan We know that HNO_3 is a strong acid and dissociates completely; thus, $[H_3O^+] = [HNO_3]_{init}$. Similarly, $Ca(OH)_2$ is a strong base and dissociates completely; the molar ratio is 1 mol $Ca(OH)_2$/2 mol OH^- so $[OH^-] = 2[Ca(OH)_2]_{init}$. We use these concentrations and the value of K_w at 25°C (1.0×10^{-14}) to find $[OH^-]$ or $[H_3O^+]$, which we then use to calculate pH and pOH.

Solution **(a)** For 0.30 *M* HNO_3:

$$[H_3O^+] = 0.30\ M$$

$$pH = -\log [H_3O^+] = -\log 0.30 = 0.52$$

$$[OH^-] = \frac{K_w}{[H_3O^+]} = \frac{1.0\times10^{-14}}{0.30} = 3.3\times10^{-14}\ M$$

$$pOH = -\log [OH^-] = -\log (3.3\times10^{-14}) = 13.48$$

(b) For 0.0042 *M* $Ca(OH)_2$: $Ca(OH)_2(aq) \longrightarrow Ca^{2+}(aq) + 2OH^-(aq)$

$$[OH^-] = 2(0.0042\ M) = 0.0084\ M$$

$$pOH = -\log [OH^-] = -\log (0.0084) = 2.08$$

$$[H_3O^+] = \frac{K_w}{[OH^-]} = \frac{1.0\times10^{-14}}{0.0084} = 1.2\times10^{-12}\ M$$

$$pH = -\log [H_3O^+] = -\log (1.2\times10^{-12}) = 11.92$$

Check The strong acid has pH < 7 and the strong base has a pH > 7, as expected. In each case, pH + pOH = 14, so the arithmetic seems correct.

FOLLOW-UP PROBLEMS

18.3A Sodium hydroxide is used to clear clogged drains. A solution of NaOH has a pOH of 4.48 at 25°C. What are its pH, $[OH^-]$, and $[H_3O^+]$?

18.3B Hydrochloric acid is sold commercially under the name *muriatic acid* and is used to clean concrete masonry. A dilute solution of HCl has a pH of 2.28 at 25°C. Calculate its pOH, $[H_3O^+]$, and $[OH^-]$.

SOME SIMILAR PROBLEMS 18.23–18.30

Measuring pH In the laboratory, pH values are usually obtained in two ways:

1. **Acid-base indicators** are organic molecules whose colors depend on the acidity of the solution in which they are dissolved. A pH can be estimated quickly with *pH paper,* a paper strip impregnated with one or a mixture of indicators. A drop of solution is placed on the strip, and the color is compared with a chart (Figure 18.6A).

2. A *pH meter* measures $[H_3O^+]$ by means of two electrodes immersed in the test solution. One electrode supplies a reference system; the other consists of a very thin glass membrane that separates a known internal $[H_3O^+]$ from the unknown external $[H_3O^+]$. The difference in $[H_3O^+]$ creates a voltage difference across the membrane, which is displayed in pH units (Figure 18.6B). We examine this device in Chapter 21.

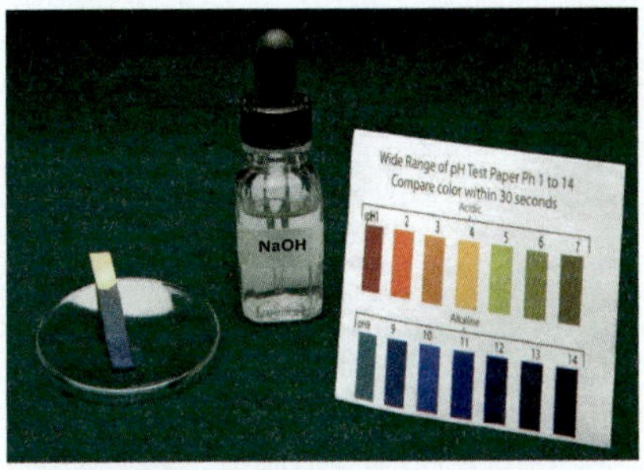

A

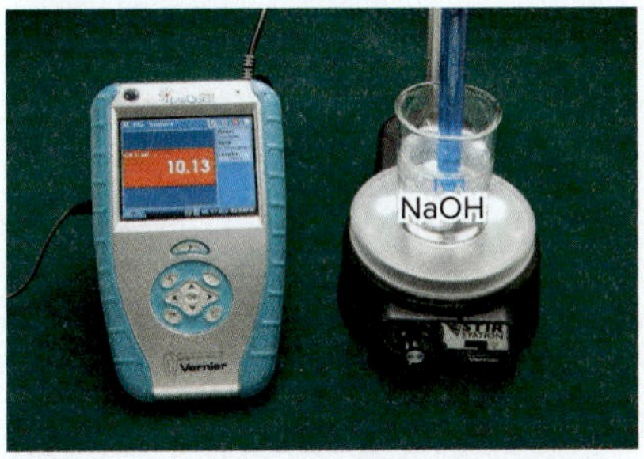

B

Figure 18.6 Methods for measuring the pH of an aqueous solution. **A,** pH paper. **B,** pH meter. Solution is 10^{-4} *M* NaOH.

Source: © McGraw-Hill Education/Charles Winters/Timeframe Photography, Inc.

› Summary of Section 18.2

- › Pure water has a low conductivity because it autoionizes to a small extent in a process whose equilibrium constant is the ion-product constant for water, K_w (1.0×10^{-14} at 25°C).
- › $[H_3O^+]$ and $[OH^-]$ are inversely related: in acidic solution, $[H_3O^+]$ is greater than $[OH^-]$; the reverse is true in basic solution; and the two are equal in neutral solution.
- › To express small values of $[H_3O^+]$, we use the pH scale: $pH = -\log [H_3O^+]$. Similarly, $pOH = -\log [OH^-]$, and $pK = -\log K$.
- › A high pH corresponds to a low $[H_3O^+]$. In acidic solutions, pH < 7.00; in basic solutions, pH > 7.00; and in neutral solutions, pH = 7.00. The sum of pH and pOH equals pK_w (14.00 at 25°C).
- › A pH is typically measured with either an acid-base indicator or a pH meter.

18.3 PROTON TRANSFER AND THE BRØNSTED-LOWRY ACID-BASE DEFINITION

Earlier we noted a key limitation of the Arrhenius definition: many substances that yield OH^- ions in water do not contain OH in their formulas. Examples include ammonia, the amines, and many salts of weak acids, such as NaF. Another limitation is that water had to be the solvent for acid-base reactions. In the early 20th century, J. N. Brønsted and T. M. Lowry suggested definitions that remove these limitations. (We introduced some of these ideas in Section 4.4.)

According to the **Brønsted-Lowry acid-base definition,**

- *An acid is a* **proton donor,** *any species that donates an* H^+ *ion.* An acid must contain H in its formula; HNO_3 and $H_2PO_4^-$ are two of many examples. All Arrhenius acids are Brønsted-Lowry acids.
- *A base is a* **proton acceptor,** *any species that accepts an* H^+ *ion.* A base must contain a lone pair of electrons to bind H^+; a few examples are NH_3, CO_3^{2-}, and F^-, as well as OH^- itself. Brønsted-Lowry bases are not Arrhenius bases, but all Arrhenius bases contain the Brønsted-Lowry base OH^-.

From this perspective, an acid-base reaction occurs when *one species donates a proton and another species simultaneously accepts it: an acid-base reaction is thus a proton-transfer process.* Acid-base reactions can occur between gases, in nonaqueous solutions, and in heterogeneous mixtures, as well as in aqueous solutions.

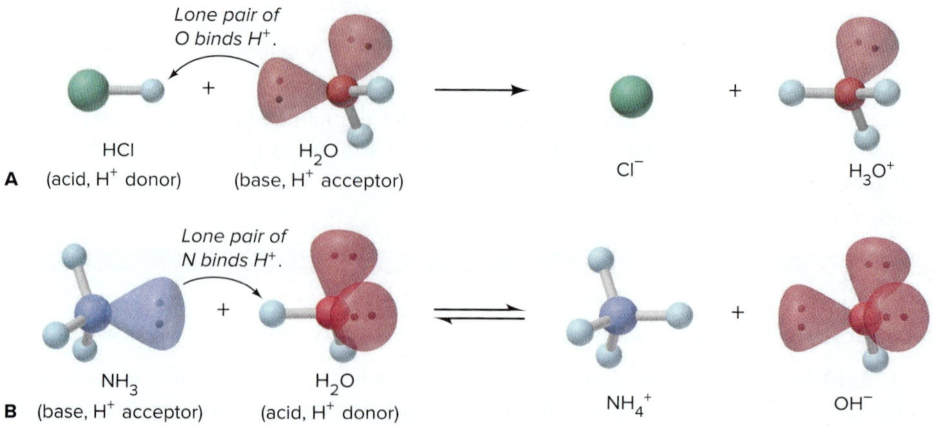

Figure 18.7 Dissolving of an acid or base in water as a Brønsted-Lowry acid-base reaction. A, The acid HCl dissolving in the base water. **B,** The base NH_3 dissolving in the acid water.

According to this definition, an acid-base reaction occurs even when an acid (or a base) just dissolves in water, because water acts as the proton acceptor (or donor):

1. *Acid donates a proton to water* (Figure 18.7A). When HCl dissolves in water, an H^+ ion (a proton) is transferred from HCl to H_2O, where it becomes attached to a lone pair of electrons on the O atom, forming H_3O^+. Thus, HCl (the acid) has *donated* the H^+, and H_2O (the base) has *accepted* it:

$$HCl(g) + H_2\ddot{\underset{\cdot\cdot}{O}}(l) \longrightarrow Cl^-(aq) + H_3\ddot{O}^+(aq)$$

2. *Base accepts a proton from water* (Figure 18.7B). When ammonia dissolves in water, an H^+ from H_2O is transferred to the lone pair of N, forming NH_4^+, and the H_2O becomes an OH^- ion:

$$\ddot{N}H_3(aq) + H_2O(l) \rightleftharpoons NH_4^+(aq) + OH^-(aq)$$

In this case, H_2O (the acid) has *donated* the H^+, and NH_3 (the base) has *accepted* it.

Note that H_2O is *amphiprotic:* it acts as a base (accepts an H^+) in one case and as an acid (donates an H^+) in the other. Many other species are amphiprotic as well.

Conjugate Acid-Base Pairs

The Brønsted-Lowry definition provides a new way to look at acid-base reactions because it focuses on the reactants *and* the products as acids and bases. For example, let's examine the reaction between hydrogen sulfide and ammonia:

$$H_2S + NH_3 \rightleftharpoons HS^- + NH_4^+$$

In the forward reaction, H_2S acts as an acid by donating an H^+ to NH_3, which acts as a base by accepting it. In the reverse reaction, the ammonium ion, NH_4^+, acts as an acid by donating an H^+ to the hydrogen sulfide ion, HS^-, which acts as a base. Notice that the acid, H_2S, becomes a base, HS^-, and the base, NH_3, becomes an acid, NH_4^+.

In Brønsted-Lowry terminology, H_2S and HS^- are a **conjugate acid-base pair:** HS^- is the conjugate base of the acid H_2S. Similarly, NH_3 and NH_4^+ are a conjugate acid-base pair: NH_4^+ is the conjugate acid of the base NH_3. *Every acid has a conjugate base, and every base has a conjugate acid.* For any conjugate acid-base pair,

- The conjugate base has one *fewer* H and one *more* negative charge than the acid.
- The conjugate acid has one *more* H and one *fewer* negative charge than the base.

A Brønsted-Lowry acid-base reaction occurs when *an acid and a base react to form their conjugate base and conjugate acid, respectively:*

$$\text{acid}_1 + \text{base}_2 \rightleftharpoons \text{base}_1 + \text{acid}_2$$

Table 18.5 shows some Brønsted-Lowry acid-base reactions. Note these points:

- Each reaction has an acid and a base as reactants *and* as products, comprising two conjugate acid-base pairs.
- Acids and bases can be neutral molecules, cations, or anions.
- The same species can be an acid or a base (amphiprotic), depending on the other species reacting. Water behaves this way in reactions 1 and 4 in Table 18.5 and HPO_4^{2-} does so in reactions 4 and 6.

Table 18.5 The Conjugate Pairs in Some Acid-Base Reactions

Conjugate Pair (Acid / Base)

	Acid	+	Base	⇌	Base	+	Acid
					Conjugate Pair (Base / Acid)		
Reaction 1	HF	+	H_2O	⇌	F^-	+	H_3O^+
Reaction 2	$HCOOH$	+	CN^-	⇌	$HCOO^-$	+	HCN
Reaction 3	NH_4^+	+	CO_3^{2-}	⇌	NH_3	+	HCO_3^-
Reaction 4	$H_2PO_4^-$	+	OH^-	⇌	HPO_4^{2-}	+	H_2O
Reaction 5	H_2SO_4	+	$N_2H_5^+$	⇌	HSO_4^-	+	$N_2H_6^{2+}$
Reaction 6	HPO_4^{2-}	+	SO_3^{2-}	⇌	PO_4^{3-}	+	HSO_3^-

SAMPLE PROBLEM 18.4 Identifying Conjugate Acid-Base Pairs

Problem The following reactions are important environmental processes. Identify the conjugate acid-base pairs.

(a) $H_2PO_4^-(aq) + CO_3^{2-}(aq) \rightleftharpoons HCO_3^-(aq) + HPO_4^{2-}(aq)$

(b) $H_2O(l) + SO_3^{2-}(aq) \rightleftharpoons OH^-(aq) + HSO_3^-(aq)$

Plan To find the conjugate pairs, we find the species that donated an H^+ (acid) and the species that accepted it (base). The acid (or base) on the left becomes its conjugate base (or conjugate acid) on the right. Remember, the conjugate acid has one more H and one fewer negative charge than its conjugate base.

Solution (a) $H_2PO_4^-$ has one more H^+ than HPO_4^{2-}; CO_3^{2-} has one fewer H^+ than HCO_3^-. Therefore, $H_2PO_4^-$ and HCO_3^- are the acids, and HPO_4^{2-} and CO_3^{2-} are the bases. The conjugate acid-base pairs are $H_2PO_4^-/HPO_4^{2-}$ and HCO_3^-/CO_3^{2-}.

(b) H_2O has one more H^+ than OH^-; SO_3^{2-} has one fewer H^+ than HSO_3^-. The acids are H_2O and HSO_3^-; the bases are OH^- and SO_3^{2-}. The conjugate acid-base pairs are H_2O/OH^- and HSO_3^-/SO_3^{2-}.

FOLLOW-UP PROBLEMS

18.4A Identify the conjugate acid-base pairs:

(a) $CH_3COOH(aq) + H_2O(l) \rightleftharpoons CH_3COO^-(aq) + H_3O^+(aq)$

(b) $H_2O(l) + F^-(aq) \rightleftharpoons OH^-(aq) + HF(aq)$

18.4B Give the formula of each of the following:

(a) The conjugate acid of HSO_3^-
(b) The conjugate base of $C_5H_5NH^+$
(c) The conjugate acid of CO_3^{2-}
(d) The conjugate base of HCN

SOME SIMILAR PROBLEMS 18.43–18.52

Student Hot Spot

Student data indicate that you may struggle with the identification of conjugate acid-base pairs. Access the Smartbook to view additional Learning Resources on this topic.

Relative Acid-Base Strength and the Net Direction of Reaction

The *net* direction of an acid-base reaction depends on relative acid and base strengths: *A reaction proceeds to the greater extent in the direction in which a stronger acid and stronger base form a weaker acid and weaker base.*

Competition for the Proton The net direction of the reaction of H_2S and NH_3 is to the right ($K_c > 1$) because H_2S is a stronger acid than NH_4^+, the other acid present, and NH_3 is a stronger base than HS^-, the other base:

$$\underset{\text{stronger acid}}{H_2S} + \underset{\text{stronger base}}{NH_3} \rightleftharpoons \underset{\text{weaker base}}{HS^-} + \underset{\text{weaker acid}}{NH_4^+}$$

stronger acid + stronger base ⟶ weaker base + weaker acid

You might think of the process as *a competition for the proton between the two bases,* NH_3 and HS^-, in which NH_3 wins.

Plan A stronger acid and base yield a weaker acid and base, so we have to determine the relative acid strengths of HX and HY in order to choose the correct molecular scene. The concentrations of the acid solutions are equal, so we can pick the stronger acid directly from the pH values of the two acid solutions. Because the stronger acid reacts to a greater extent, fewer molecules of it will be in the scene than molecules of the weaker acid.

Solution The HX solution has a lower pH (2.88) than the HY solution (3.52), so we know right away that HX is the stronger acid and X^- is the weaker base and that HY is the weaker acid and Y^- is the stronger base. Therefore, the reaction of HX and Y^- has $K_c > 1$, which means the equilibrium mixture will have more HY than HX. Scene 1 has equal numbers of HX and HY, which would occur if the acids were of equal strength, and scene 2 shows fewer HY than HX, which would occur if HY were stronger. Therefore, only scene 3 is consistent with the relative acid strengths.

FOLLOW-UP PROBLEMS

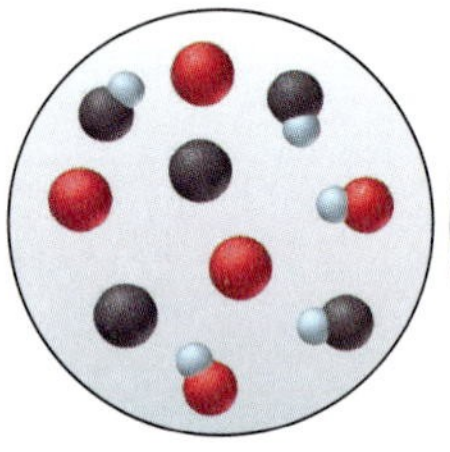

18.6A

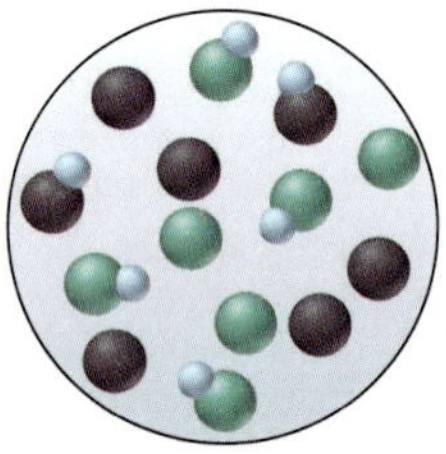

18.6B

18.6A The left-hand scene in the margin represents the equilibrium mixture after 0.10 *M* solutions of HA *(blue and red)* and B^- *(black)* react: Does this reaction have a K_c greater or less than 1? Which acid is stronger, HA or HB?

18.6B The right-hand scene depicts an aqueous solution of two conjugate acid-base pairs: HC/C^- and HD/D^-. HD is a stronger acid than HC. What colors represent the base C^- and the base D^-? Does the reaction between HC and D^- have a K_c greater or less than 1?

A SIMILAR PROBLEM 18.39

› Summary of Section 18.3

- The Brønsted-Lowry acid-base definition does not require that bases contain OH in their formula or that acid-base reactions occur in aqueous solution.
- An acid is a species that donates a proton and a base is one that accepts it, so an acid-base reaction is a proton-transfer process.
- When an acid donates a proton, it becomes the conjugate base; when a base accepts a proton, it becomes the conjugate acid. In an acid-base reaction, acids and bases form their conjugates. A stronger acid has a weaker conjugate base, and vice versa.
- An acid-base reaction proceeds to the greater extent ($K > 1$) in the direction in which a stronger acid and base form a weaker base and acid.

18.4 SOLVING PROBLEMS INVOLVING WEAK-ACID EQUILIBRIA

Just as you saw in Chapter 17 for equilibrium problems in general, there are two types of equilibrium problems involving weak acids and their conjugate bases:

1. Given equilibrium concentrations, find K_a.
2. Given K_a and some concentrations, find other equilibrium concentrations.

For all of these problems, we'll apply the same problem-solving approach, notation system, and assumptions:

- *The problem-solving approach.* Start with what is given in the problem and move toward what you want to find. Make a habit of applying the following steps:
  1. Write the balanced equation and K_a expression; these tell you what to find.
  2. Define x as the unknown change in concentration that occurs during the reaction. Frequently, $x = [HA]_{dissoc}$, the concentration of HA that dissociates, which, based on certain assumptions, also equals $[H_3O^+]$ and $[A^-]$ at equilibrium.
  3. Construct a reaction table (for most problems) that incorporates x.
  4. Make assumptions (usually that x is very small relative to the initial concentration) that simplify the calculations.
  5. Substitute the values into the K_a expression, and solve for x.
  6. Check that the assumptions are justified with the 5% test first used in Sample Problem 17.9. If they are not justified, use the quadratic formula to find x.

- *The notation system.* As always, molar concentration is indicated with brackets. A subscript indicates where the species comes from or when it occurs in the reaction process. For example, $[H_3O^+]_{\text{from HA}}$ is the molar concentration of H_3O^+ that comes from the dissociation of HA; $[HA]_{\text{init}}$ is the initial molar concentration of HA, that is, before dissociation; $[HA]_{\text{dissoc}}$ is the molar concentration of HA that dissociates; and so forth. A bracketed formula with *no* subscript represents the molar concentration of the species *at equilibrium.*
- *The assumptions.* We make two assumptions to simplify the arithmetic:
  1. $[H_3O^+]$ from the autoionization of water is negligible. It is so much smaller than the $[H_3O^+]$ from the dissociation of HA that we can neglect it in these problems:

     $$[H_3O^+] = [H_3O^+]_{\text{from HA}} + [H_3O^+]_{\text{from } H_2O} \approx [H_3O^+]_{\text{from HA}}$$

     Note that each molecule of HA that dissociates forms one H_3O^+ and one A^-:

     $$[HA]_{\text{dissoc}} = [H_3O^+] = [A^-]$$

  2. A weak acid has a small K_a. Therefore, it dissociates to such a small extent that we can neglect the change in its concentration to find its equilibrium concentration:

     $$[HA] = [HA]_{\text{init}} - [HA]_{\text{dissoc}} \approx [HA]_{\text{init}}$$

Finding K_a Given Concentrations

This type of problem involves finding K_a of a weak acid from the concentration of one of the species in solution, usually $[H_3O^+]$ from a given pH:

$$HA(aq) + H_2O(l) \rightleftharpoons H_3O^+(aq) + A^-(aq)$$

$$K_a = \frac{[H_3O^+][A^-]}{[HA]}$$

You prepare an aqueous solution of HA and measure its pH. Thus, you know $[HA]_{\text{init}}$, can calculate $[H_3O^+]$ from the pH, and then determine $[A^-]$ and [HA] at equilibrium. You substitute these values into the K_a expression and solve for K_a. Let's go through the approach in a sample problem.

SAMPLE PROBLEM 18.7 Finding K_a of a Weak Acid from the Solution pH

Problem Phenylacetic acid ($C_6H_5CH_2COOH$, simplified here to HPAc; *see model*) builds up in the blood of persons with phenylketonuria, an inherited disorder that, if untreated, causes mental retardation and death. A study of the acid shows that the pH of 0.12 *M* HPAc is 2.62. What is the K_a of phenylacetic acid?

Plan We are given $[HPAc]_{\text{init}}$ (0.12 *M*) and the pH (2.62) and must find K_a. As always, we first write the equation for HPAc dissociation and the expression for K_a to see which values we need. We set up a reaction table and use the given pH to find $[H_3O^+]$, which equals $[PAc^-]$ and $[HPAc]_{\text{dissoc}}$ (we assume that $[H_3O^+]_{\text{from } H_2O}$ is negligible). To find [HPAc], we assume that, because it is a weak acid, very little dissociates, so $[HPAc]_{\text{init}} - [HPAc]_{\text{dissoc}} = [HPAc] \approx [HPAc]_{\text{init}}$. We make these assumptions, substitute the equilibrium values, solve for K_a, and then check the assumptions using the 5% rule (Sample Problem 17.9).

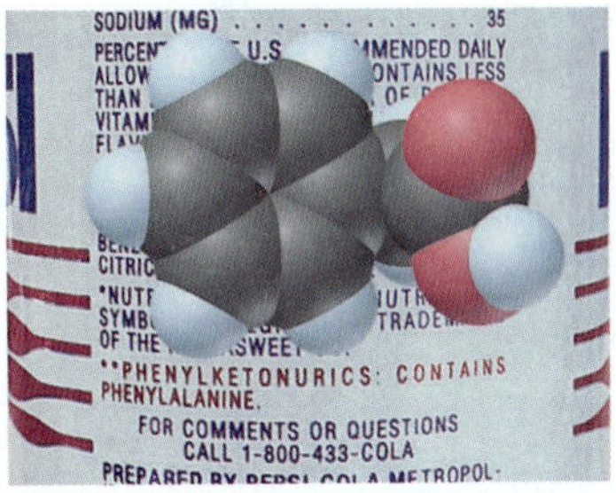

Phenylalanine, one of the amino acids that make up aspartame, is metabolized to phenylacetic acid (model).

Solution Writing the dissociation equation and K_a expression:

$$HPAc(aq) + H_2O(l) \rightleftharpoons H_3O^+(aq) + PAc^-(aq) \qquad K_a = \frac{[H_3O^+][PAc^-]}{[HPAc]}$$

Setting up a reaction table, with $x = [HPAc]_{\text{dissoc}} = [H_3O^+]_{\text{from HPAc}} = [PAc^-]$:

Concentration (*M*)	$HPAc(aq)$	+	$H_2O(l)$	$\rightleftharpoons$	$H_3O^+(aq)$	+	$PAc^-(aq)$
Initial	0.12		—		0		0
Change	$-x$		—		$+x$		$+x$
Equilibrium	$0.12 - x$		—		x		x

Calculating $[H_3O^+]$:

$$[H_3O^+] = 10^{-pH} = 10^{-2.62} = 2.4{\times}10^{-3}\ M$$

Making the assumptions:

1. The calculated $[H_3O^+]$ ($2.4{\times}10^{-3}\ M$) >> $[H_3O^+]_{\text{from } H_2O}$ ($1.0{\times}10^{-7}\ M$), so we assume that $[H_3O^+] \approx [H_3O^+]_{\text{from HPAc}} = [PAc^-] = x$ (the change in [HPAc], or $[HPAc]_{dissoc}$).
2. HPAc is a weak acid, so we assume that $[HPAc] = 0.12\ M - x \approx 0.12\ M$.

Solving for the equilibrium concentrations:

$$x \approx [H_3O^+] = [PAc^-] = 2.4{\times}10^{-3}\ M$$

$$[HPAc] = 0.12\ M - x = 0.12\ M - (2.4{\times}10^{-3}\ M) \approx 0.12\ M \text{ (to 2 sf)}$$

Substituting these values into K_a:

$$K_a = \frac{[H_3O^+][PAc^-]}{[HPAc]} \approx \frac{(2.4{\times}10^{-3})(2.4{\times}10^{-3})}{0.12} = 4.8{\times}10^{-5}$$

Checking the assumptions by finding the percent error in concentration:

1. For $[H_3O^+]_{\text{from } H_2O}$: $\dfrac{1{\times}10^{-7}\ M}{2.4{\times}10^{-3}\ M} \times 100 = 4{\times}10^{-3}\%$ (<5%; assumption is justified.)

2. For $[HPAc]_{dissoc}$: $\dfrac{2.4{\times}10^{-3}\ M}{0.12\ M} \times 100 = 2.0\%$ (<5%; assumption is justified.)

Check The $[H_3O^+]$ makes sense: pH 2.62 should give $[H_3O^+]$ between 10^{-2} and 10^{-3} *M*. The K_a calculation also seems in the correct range: $(10^{-3})^2/10^{-1} = 10^{-5}$, and this value seems reasonable for a weak acid.

FOLLOW-UP PROBLEMS

18.7A The conjugate acid of ammonia is the weak acid NH_4^+. If a 0.2 *M* NH_4Cl solution has a pH of 5.0, what is the K_a of NH_4^+?

18.7B Over a million tons of acrylic acid ($H_2C{=}CHCOOH$) are produced each year for manufacturing plastics, adhesives, and paint. A 0.30 *M* solution of the acid has a pH of 2.43. What is K_a for this acid?

SOME SIMILAR PROBLEMS 18.64, 18.65, 18.72, and 18.73

Finding Concentrations Given K_a

The second type of equilibrium problem gives some concentration data and K_a and asks for the equilibrium concentration of some component. Such problems are very similar to those we solved in Chapter 17 in which a substance with a given initial concentration reacted to an unknown extent (see Sample Problems 17.8 to 17.10). We will use a reaction table in these problems to find the values, and, as we just found, $[H_3O^+]_{\text{from } H_2O}$ is so small relative to $[H_3O^+]_{\text{from HA}}$ that we will neglect it and enter the initial $[H_3O^+]$ in all reaction tables as zero.

SAMPLE PROBLEM 18.8 Determining Concentration and pH from K_a and Initial [HA]

Problem Propanoic acid (CH_3CH_2COOH, which we simplify to HPr) is a carboxylic acid whose salts are used to retard mold growth in foods. What are the $[H_3O^+]$ and the pH of 0.10 *M* HPr ($K_a = 1.3{\times}10^{-5}$)?

Plan We know the initial concentration (0.10 *M*) and K_a ($1.3{\times}10^{-5}$) of HPr, and we need to find $[H_3O^+]$ and pH. First, we write the balanced equation and the expression for K_a. We know $[HPr]_{init}$ but not [HPr] (that is, the concentration at equilibrium). If we let $x = [HPr]_{dissoc}$, x is also $[H_3O^+]_{\text{from HPr}}$ and $[Pr^-]$ because each HPr dissociates into one H_3O^+ and one Pr^-. With this information, we set up a reaction table. We assume that, because HPr has a small K_a, it dissociates very little. After solving for x, which is $[H_3O^+]$, we check the assumption. Then we use the value for $[H_3O^+]$ to find the pH.

Solution Writing the balanced equation and expression for K_a:

$$HPr(aq) + H_2O(l) \rightleftharpoons H_3O^+(aq) + Pr^-(aq) \qquad K_a = \frac{[H_3O^+][Pr^-]}{[HPr]} = 1.3\times10^{-5}$$

Setting up a reaction table, with $x = [HPr]_{dissoc} = [H_3O^+]_{from\ HPr} = [Pr^-] = [H_3O^+]$:

Concentration (M)	$HPr(aq)$	+	$H_2O(l)$	$\rightleftharpoons$	$H_3O^+(aq)$	+	$Pr^-(aq)$
Initial	0.10		—		0		0
Change	$-x$		—		$+x$		$+x$
Equilibrium	$0.10 - x$		—		x		x

Making the assumption: K_a is small, so x is small compared with $[HPr]_{init}$; therefore, $[HPr]_{init} - x = [HPr] \approx [HPr]_{init}$, or $0.10\ M - x \approx 0.10\ M$. Substituting into the K_a expression and solving for x:

$$K_a = \frac{[H_3O^+][Pr^-]}{[HPr]} = 1.3\times10^{-5} \approx \frac{(x)(x)}{0.10}$$

$$x \approx \sqrt{(0.10)(1.3\times10^{-5})} = 1.1\times10^{-3}\ M = [H_3O^+]$$

Checking the assumption for $[HPr]_{dissoc}$:

$$\frac{[H_3O^+]}{[HPr]_{init}} \times 100 = \frac{1.1\times10^{-3}\ M}{0.10\ M} \times 100 = 1.1\% \ (<5\%;\ \text{assumption is justified.})$$

Finding the pH:

$$pH = -\log [H_3O^+] = -\log (1.1\times10^{-3}) = 2.96$$

Check The $[H_3O^+]$ and pH seem reasonable for a dilute solution of a weak acid with a moderate K_a. By reversing the calculation, we can check the math: $(1.1\times10^{-3})^2/0.10 = 1.2\times10^{-5}$, which is within rounding of the given K_a.

Comment In Chapter 17 we introduced a benchmark, aside from the 5% rule, to see if the assumption is justified (see the discussion following Sample Problem 17.9):

- If $\frac{[HA]_{init}}{K_a} > 400$, the assumption is justified: neglecting x introduces an error <5%.
- If $\frac{[HA]_{init}}{K_a} < 400$, the assumption is *not* justified; neglecting x introduces an error >5%, so we solve a quadratic equation to find x.

In this sample problem, we have $\frac{0.10}{1.3\times10^{-5}} = 7.7 \times 10^3$, which is greater than 400. The alternative situation occurs in the next follow-up problem.

FOLLOW-UP PROBLEMS

18.8A Cyanic acid (HOCN) is an extremely acrid, unstable substance. What are the $[H_3O^+]$ and the pH of 0.10 M HOCN ($K_a = 3.5\times10^{-4}$)?

18.8B Benzoic acid (C_6H_5COOH) is used as a food preservative. What are the $[H_3O^+]$ and the pH of 0.25 M C_6H_5COOH ($pK_a = 4.20$)?

SOME SIMILAR PROBLEMS 18.66–18.69 and 18.74–18.77

Student Hot Spot

Student data indicate that you may struggle with calculating the pH of a weak acid. Access the Smartbook to view additional Learning Resources on this topic.

The Effect of Concentration on the Extent of Acid Dissociation

If we repeat the calculation in Sample Problem 18.8, but start with a lower [HPr], we observe a very interesting fact about the extent of dissociation of a weak acid. Suppose the initial concentration of HPr is one-tenth as much, 0.010 M, rather than

0.10 *M*. After filling in the reaction table and making the same assumptions, we find that

$$x = [H_3O^+] = [HPr]_{dissoc} = 3.6\times10^{-4}\ M$$

Now, let's compare the percentages of HPr molecules dissociated at the two different initial acid concentrations, using the relationship

$$\text{Percent HA dissociated} = \frac{[HA]_{dissoc}}{[HA]_{init}} \times 100 \qquad \textbf{(18.5)}$$

Case 1: $[HPr]_{init} = 0.10\ M$

$$\text{Percent HPr dissociated} = \frac{1.1\times10^{-3}\ M}{1.0\times10^{-1}\ M} \times 100 = 1.1\%$$

Case 2: $[HPr]_{init} = 0.010\ M$

$$\text{Percent HPr dissociated} = \frac{3.6\times10^{-4}\ M}{1.0\times10^{-2}\ M} \times 100 = 3.6\%$$

As the initial acid concentration decreases, the percent dissociation of the acid increases. Don't confuse the *concentration* of HA dissociated with the *percent* HA dissociated. The concentration, $[HA]_{dissoc}$, is lower in the diluted HA solution because the actual *number* of dissociated HA molecules is smaller. It is the *fraction* (or the *percent*) of dissociated HA molecules that increases with dilution.

This phenomenon is analogous to a change in container volume (pressure) for a reaction involving gases at equilibrium (Section 17.6). In the case of gases, an increase in volume shifts the equilibrium position to favor more moles of gas. In the case of HA dissociation, a lower HA concentration, which is the same as an increase in volume, shifts the equilibrium position to favor more moles of ions.

SAMPLE PROBLEM 18.9 Finding the Percent Dissociation of a Weak Acid

Problem In 2011, researchers showed that hypochlorous acid (HClO) generated by white blood cells kills bacteria. Calculate the percent dissociation of **(a)** 0.40 *M* HClO; **(b)** 0.035 *M* HClO ($K_a = 2.9\times10^{-8}$).

Plan We know the K_a of HClO and need $[HClO]_{dissoc}$ to find the percent dissociation at two different initial concentrations. We write the balanced equation and the expression for K_a and then set up a reaction table, with $x = [HClO]_{dissoc} = [ClO^-] = [H_3O^+]$. We assume that because HClO has a small K_a, it dissociates very little. Once $[HClO]_{dissoc}$ is known, we use Equation 18.5 to find the percent dissociation and check the assumption.

Solution (a) Writing the balanced equation and the expression for K_a:

$$HClO(aq) + H_2O(l) \rightleftharpoons H_3O^+(aq) + ClO^-(aq) \qquad K_a = \frac{[ClO^-][H_3O^+]}{[HClO]} = 2.9\times10^{-8}$$

Setting up a reaction table with $x = [HClO]_{dissoc} = [ClO^-] = [H_3O^+]$:

Concentration (*M*)	$HClO(aq)$	+	$H_2O(l)$	$\rightleftharpoons$	$H_3O^+(aq)$	+	$ClO^-(aq)$
Initial	0.40		—		0		0
Change	$-x$		—		$+x$		$+x$
Equilibrium	$0.40 - x$		—		x		x

Making the assumption: K_a is small, so x is small compared with $[HClO]_{init}$; therefore, $[HClO]_{init} - x \approx [HClO]_{init}$, or $0.40\ M - x \approx 0.40\ M$. Substituting into the K_a expression and solving for x:

$$K_a = \frac{[ClO^-][H_3O^+]}{[HClO]} = 2.9\times10^{-8} \approx \frac{(x)(x)}{0.40}$$

Thus,

$$x^2 = (0.40)(2.9\times10^{-8})$$
$$x = \sqrt{(0.40)(2.9\times10^{-8})} = 1.1\times10^{-4}\ M = [\text{HClO}]_{\text{dissoc}}$$

Finding the percent dissociation:

$$\text{Percent dissociation} = \frac{[\text{HClO}]_{\text{dissoc}}}{[\text{HClO}]_{\text{init}}} \times 100 = \frac{1.1\times10^{-4}\ M}{0.40\ M} \times 100 = 0.028\%$$

Since the percent dissociation is <5%, the assumption is justified.

(b) Performing the same calculations using $[\text{HClO}]_{\text{init}} = 0.035\ M$:

$$K_a = \frac{[\text{ClO}^-][\text{H}_3\text{O}^+]}{[\text{HClO}]} = 2.9\times10^{-8} \approx \frac{(x)(x)}{0.035}$$
$$x = \sqrt{(0.035)(2.9\times10^{-8})} = 3.2\times10^{-5}\ M = [\text{HClO}]_{\text{dissoc}}$$

Finding the percent dissociation:

$$\text{Percent dissociation} = \frac{[\text{HClO}]_{\text{dissoc}}}{[\text{HClO}]_{\text{init}}} \times 100 = \frac{3.2\times10^{-5}\ M}{0.035\ M} \times 100 = 0.091\%$$

Since the percent dissociation is <5%, the assumption is justified.

Check The percent dissociation is very small, as we expect for an acid with such a low K_a. Note, however, that the percent dissociation is larger for the lower initial concentration, as we also expect.

FOLLOW-UP PROBLEMS

18.9A Calculate the percent dissociation of 0.75 *M* HCN, an extremely poisonous acid that can be obtained from the pits of fruits such as cherries and apples ($K_a = 6.2\times10^{-10}$).

18.9B A weak acid is 3.16% dissociated in a 1.5 *M* solution. What is the K_a of the acid?

SOME SIMILAR PROBLEMS 18.70, 18.71, 18.78, and 18.79

The Behavior of Polyprotic Acids

Acids with more than one ionizable proton are **polyprotic acids.** In solution, each dissociation step has a different K_a. For example, phosphoric acid is a triprotic acid (three ionizable protons), so it has three K_a values:

$$\text{H}_3\text{PO}_4(aq) + \text{H}_2\text{O}(l) \rightleftharpoons \text{H}_2\text{PO}_4^-(aq) + \text{H}_3\text{O}^+(aq)$$
$$K_{a1} = \frac{[\text{H}_2\text{PO}_4^-][\text{H}_3\text{O}^+]}{[\text{H}_3\text{PO}_4]} = 7.2\times10^{-3}$$
$$\text{H}_2\text{PO}_4^-(aq) + \text{H}_2\text{O}(l) \rightleftharpoons \text{HPO}_4^{2-}(aq) + \text{H}_3\text{O}^+(aq)$$
$$K_{a2} = \frac{[\text{HPO}_4^{2-}][\text{H}_3\text{O}^+]}{[\text{H}_2\text{PO}_4^-]} = 6.3\times10^{-8}$$
$$\text{HPO}_4^{2-}(aq) + \text{H}_2\text{O}(l) \rightleftharpoons \text{PO}_4^{3-}(aq) + \text{H}_3\text{O}^+(aq)$$
$$K_{a3} = \frac{[\text{PO}_4^{3-}][\text{H}_3\text{O}^+]}{[\text{HPO}_4^{2-}]} = 4.2\times10^{-13}$$

The relative K_a values show that H_3PO_4 is a much stronger acid than $H_2PO_4^-$, which is much stronger than HPO_4^{2-}.

Table 18.6 *(next page)* lists some common polyprotic acids and their K_a values. (More are listed in Appendix C.) Note that the general pattern seen for H_3PO_4 occurs for all polyprotic acids:

$$K_{a1} >> K_{a2} >> K_{a3}$$

This trend occurs because it is more difficult for the positively charged H^+ ion to leave a singly charged anion (such as $H_2PO_4^-$) than to leave a neutral molecule (such as H_3PO_4), and more difficult still for it to leave a doubly charged anion (such as HPO_4^{2-}). Successive K_a values typically differ by several orders of magnitude. This fact simplifies calculations because *we usually neglect the H_3O^+ coming from the subsequent dissociations.*

Table 18.6 Successive K_a Values for Some Polyprotic Acids at 25°C

Name (Formula)	Lewis Structure*	K_{a1}	K_{a2}	K_{a3}
Oxalic acid ($H_2C_2O_4$)	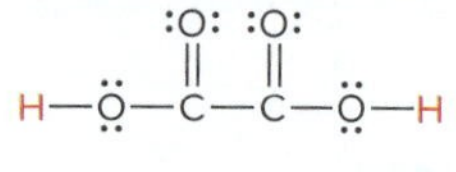	5.6×10^{-2}	5.4×10^{-5}	
Sulfurous acid (H_2SO_3)	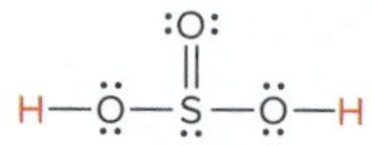	1.4×10^{-2}	6.5×10^{-8}	
Phosphoric acid (H_3PO_4)	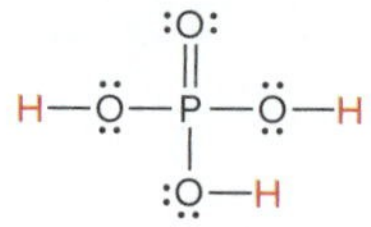	7.2×10^{-3}	6.3×10^{-8}	4.2×10^{-13}
Arsenic acid (H_3AsO_4)	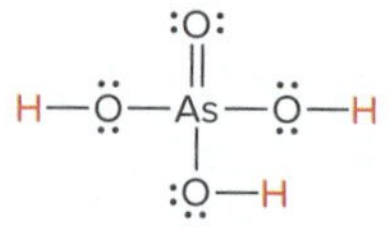	6×10^{-3}	1.1×10^{-7}	3×10^{-12}
Carbonic acid (H_2CO_3)	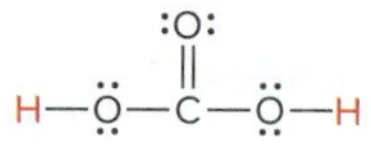	4.5×10^{-7}	4.7×10^{-11}	
Hydrosulfuric acid (H_2S)	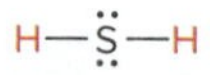	9×10^{-8}	1×10^{-17}	

*Red type indicates the ionizable protons.

SAMPLE PROBLEM 18.10 Calculating Equilibrium Concentrations for a Polyprotic Acid

Problem Ascorbic acid ($H_2C_6H_6O_6$; represented as H_2Asc for this problem), known as vitamin C, is a diprotic acid ($K_{a1} = 1.0\times10^{-5}$ and $K_{a2} = 5\times10^{-12}$) found in citrus fruit. Calculate $[HAsc^-]$, $[Asc^{2-}]$, and the pH of 0.050 *M* H_2Asc.

Plan We know the initial concentration (0.050 *M*) and both K_a's for H_2Asc, and we have to calculate the equilibrium concentrations of all species and convert $[H_3O^+]$ to pH. We first write the equations and K_a expressions. Because $K_{a1} >> K_{a2}$, we can assume that the first dissociation produces almost all the H_3O^+: $[H_3O^+]_{\text{from } H_2Asc} >> [H_3O^+]_{\text{from } HAsc^-}$. Also, because K_{a1} is small, the amount of H_2Asc that dissociates can be neglected. We set up a reaction table for the first dissociation, with $x = [H_2Asc]_{dissoc}$, and then we solve for $[H_3O^+]$ and $[HAsc^-]$. Because the second dissociation occurs to a much lesser extent, we can substitute values from the first dissociation directly to find $[Asc^{2-}]$ from the second.

Solution Writing the equations and K_a expressions:

$$H_2Asc(aq) + H_2O(l) \rightleftharpoons HAsc^-(aq) + H_3O^+(aq)$$

$$K_{a1} = \frac{[HAsc^-][H_3O^+]}{[H_2Asc]} = 1.0\times10^{-5}$$

$$HAsc^-(aq) + H_2O(l) \rightleftharpoons Asc^{2-}(aq) + H_3O^+(aq)$$

$$K_{a2} = \frac{[Asc^{2-}][H_3O^+]}{[HAsc^-]} = 5\times10^{-12}$$

Setting up a reaction table with $x = [H_2Asc]_{dissoc} = [HAsc^-] \approx [H_3O^+]$:

Concentration (*M*)	$H_2Asc(aq)$	+	$H_2O(l)$	$\rightleftharpoons$	$H_3O^+(aq)$	+	$HAsc^-(aq)$
Initial	0.050		—		0		0
Change	$-x$		—		$+x$		$+x$
Equilibrium	$0.050 - x$		—		x		x

Making the assumptions:

1. Because $K_{a2} << K_{a1}$, $[H_3O^+]_{\text{from } HAsc^-} << [H_3O^+]_{\text{from } H_2Asc}$. Therefore,

$$[H_3O^+]_{\text{from } H_2Asc} \approx [H_3O^+]$$

2. Because K_{a1} is small, $[H_2Asc]_{init} - x = [H_2Asc] \approx [H_2Asc]_{init}$. Thus,

$$[H_2Asc] = 0.050\ M - x \approx 0.050\ M$$

Substituting into the expression for K_{a1} and solving for x:

$$K_{a1} = \frac{[H_3O^+][HAsc^-]}{[H_2Asc]} = 1.0\times10^{-5} = \frac{x^2}{0.050 - x} \approx \frac{x^2}{0.050}$$

$$x = [HAsc^-] \approx [H_3O^+] \approx 7.1\times10^{-4}\ M$$

$$pH = -\log [H_3O^+] = -\log (7.1\times10^{-4}) = 3.15$$

Checking the assumptions:

1. $[H_3O^+]_{\text{from } HAsc^-} << [H_3O^+]_{\text{from } H_2Asc}$: For any second dissociation that does occur,

$$K_{a2} = \frac{[H_3O^+][Asc^{2-}]}{[HAsc^-]} = 5\times10^{-12} = \frac{(x)(x)}{7.1\times10^{-4}}$$

$$x = [H_3O^+]_{\text{from } HAsc^-} = 6\times10^{-8}\ M$$

This is even less than $[H_3O^+]_{\text{from } H_2O}$, so the assumption is justified.

2. $[H_2Asc]_{dissoc} << [H_2Asc]_{init}$: $\frac{7.1\times10^{-4}\ M}{0.050\ M} \times 100 = 1.4\%$ (<5%; assumption is justified).

Also, note that $\frac{[H_2Asc]_{init}}{K_{a1}} = \frac{0.050}{1.0\times10^{-5}} = 5000 > 400$

Using the equilibrium concentrations from the first dissociation to calculate $[Asc^{2-}]$:

$$K_{a2} = \frac{[H_3O^+][Asc^{2-}]}{[HAsc^-]} \quad \text{and} \quad [Asc^{2-}] = \frac{(K_{a2})[HAsc^-]}{[H_3O^+]}$$

$$[Asc^{2-}] = \frac{(5\times10^{-12})(7.1\times10^{-4})}{7.1\times10^{-4}} = 5\times10^{-12}\ M$$

Check $K_{a1} >> K_{a2}$, so it makes sense that $[HAsc^-] >> [Asc^{2-}]$ because Asc^{2-} is produced only in the second (much weaker) dissociation. Both K_a's are small, so all concentrations except $[H_2Asc]$ should be much lower than the original 0.050 *M*.

FOLLOW-UP PROBLEMS

18.10A Oxalic acid (HOOC—COOH, or $H_2C_2O_4$) is the simplest diprotic carboxylic acid. Its commercial uses include bleaching straw and leather and removing rust and ink stains. Calculate the equilibrium values of $[H_2C_2O_4]$, $[HC_2O_4^-]$, and $[C_2O_4^{2-}]$, and find the pH of a 0.150 *M* $H_2C_2O_4$ solution. Use K_a values from Appendix C. (*Hint:* First check whether you need the quadratic equation to find *x*.)

18.10B Carbonic acid (H_2CO_3) plays a role in blood chemistry, cave formation, and ocean acidification. Using its K_a values from Appendix C, calculate the equilibrium values of $[H_2CO_3]$, $[HCO_3^-]$, and $[CO_3^{2-}]$, and find the pH of a 0.075 *M* H_2CO_3 solution.

SOME SIMILAR PROBLEMS 18.80 and 18.81

› Summary of Section 18.4

- Two types of weak-acid equilibrium problems involve finding K_a from a given concentration and finding a concentration from a given K_a.
- We simplify the arithmetic by assuming (1) that $[H_3O^+]_{\text{from } H_2O}$ is much smaller than $[H_3O^+]_{\text{from } HA}$ and can be neglected and (2) that weak acids dissociate so little that $[HA]_{init} \approx [HA]$ at equilibrium.
- The *fraction* of weak acid molecules that dissociate is greater in a more dilute solution, even though the total $[H_3O^+]$ is lower.
- Polyprotic acids have more than one ionizable proton, but we assume that the first dissociation provides virtually all the H_3O^+.

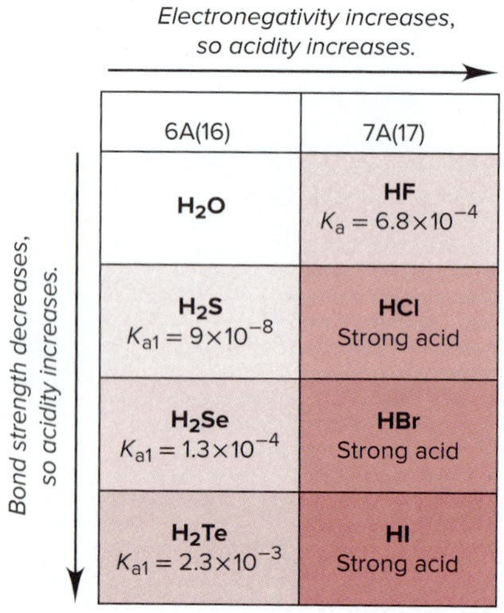

Figure 18.9 The effect of atomic and molecular properties on nonmetal hydride acidity.

18.5 MOLECULAR PROPERTIES AND ACID STRENGTH

The strength of an acid depends on its ability to donate a proton, which depends in turn on the strength of the bond to the acidic proton. In this section, we apply trends in atomic and bond properties to determine the trends in acid strength of nonmetal hydrides and oxoacids and then discuss the acidity of hydrated metal ions.

Acid Strength of Nonmetal Hydrides

Two factors determine how easily a proton is released from a nonmetal hydride:

- The electronegativity of the central nonmetal (E)
- The strength of the E—H bond

Figure 18.9 displays two periodic trends among the nonmetal hydrides:

1. *Across a period, acid strength increases.* The electronegativity of the nonmetal E determines this horizontal trend. From left to right, as E becomes more electronegative, it withdraws electron density from H, and the E—H bond becomes more polar. As a result, H^+ is pulled away more easily by an O atom of a water molecule. In aqueous solution, the hydrides of Groups 3A(13) to 5A(15) do not behave as acids, but an increase in acid strength is seen from Group 6A(16) to 7A(17).

2. *Down a group, acid strength increases.* E—H bond strength determines this vertical trend. As E becomes larger, the E—H bond becomes longer and weaker, so H^+ comes off more easily.* For example, hydrohalic acid strength increases down Group 7A(17):

Acid strength:	HF <<	HCl <	HBr <	HI
Bond length (pm):	92	127	141	161
Bond energy (kJ/mol):	565	427	363	295

(This trend is not seen in aqueous solution, where HCl, HBr, and HI are all equally strong; we discuss how it *is* observed in Section 18.8.)

Acid Strength of Oxoacids

All oxoacids have the acidic H atom bonded to an O atom, so bond length is not involved. As we mentioned for the halogen oxoacids (Section 14.9), two other factors determine the acid strength of oxoacids:

- The electronegativity of the central nonmetal (E)
- The number of O atoms around E (related to the oxidation number, O.N., of E)

Figure 18.10 summarizes these trends:

1. *For oxoacids with the* ***same*** *number of O atoms, acid strength increases with the electronegativity of E.* Consider the hypohalous acids (HOE, where E is a halogen atom). The more electronegative E is, the more polar the O—H bond becomes and the more easily H^+ is lost (Figure 18.10A). Electronegativity (EN) decreases down a group, as does acid strength:

K_a of HOCl = 2.9×10^{-8} $\quad$ K_a of HOBr = 2.3×10^{-9} $\quad$ K_a of HOI = 2.3×10^{-11}
EN of Cl = 3.5 $\quad$ EN of Br = 2.8 $\quad$ EN of I = 2.5

Similarly, in Group 6A(16), H_2SO_4 (EN of S = 2.5) is stronger than H_2SeO_4 (EN of Se = 2.4); in Group 5A(15), H_3PO_4 (EN of P = 2.1) is stronger than H_3AsO_4 (EN of As = 2.0), and so forth.

2. *For oxoacids with* ***different*** *numbers of O atoms, acid strength increases with the number of O atoms (or with the O.N. of the central nonmetal).* The electronegative O atoms pull electron density away from E, which makes the O—H bond more polar. The

*Actually, bond energy refers to bond breakage that forms an H atom, whereas acidity refers to bond breakage that forms an H^+ ion. Although these two types of bond breakage are not the same, the trends in bond energy and acid strength are opposites.

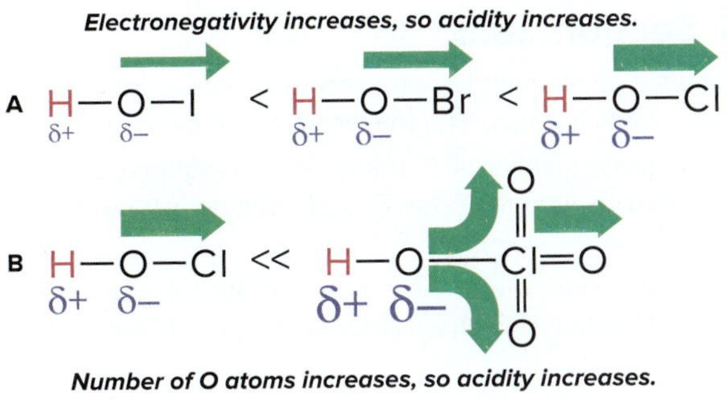

Figure 18.10 The relative strengths of oxoacids. A, Cl withdraws electron density (thickness of green arrow) from the O—H bond most effectively, making that bond most polar (relative size of δ symbols). B, Additional O atoms pull more electron density from the O—H bond.

more O atoms present, the greater the shift in electron density, and the more easily the H^+ ion comes off (Figure 18.10B). Therefore, the chlorine oxoacids ($HOClO_n$, with n from 0 to 3) increase in strength with the number of O atoms (and the O.N. of Cl):

$$K_a \text{ of } HO\overset{+1}{Cl} = 2.9\times10^{-8} \qquad K_a \text{ of } HO\overset{+3}{Cl}O = 1.12\times10^{-2} \qquad K_a \text{ of } HO\overset{+5}{Cl}O_2 \approx 1 \qquad K_a \text{ of } HO\overset{+7}{Cl}O_3 > 10^7$$

Similarly, HNO_3 is stronger than HNO_2, H_2SO_4 is stronger than H_2SO_3, and so forth.

Acidity of Hydrated Metal Ions

The aqueous solutions of certain metal ions are acidic because the *hydrated* metal ion transfers an H^+ ion to water. Consider a general metal nitrate, $M(NO_3)_n$, as it dissolves in water. The ions separate and the metal ion becomes bonded to some number of H_2O molecules. This equation shows the hydration of the cation (M^{n+}) using H_2O molecules and "(aq)"; hydration of the anion (NO_3^-) is indicated with just "(aq)":

$$M(NO_3)_n(s) + xH_2O(l) \longrightarrow M(H_2O)_x^{n+}(aq) + nNO_3^-(aq)$$

If the metal ion, M^{n+}, is *small and highly charged,* its high charge density withdraws sufficient electron density from the O—H bonds of the bound water molecules for an H^+ to be released. Thus, the hydrated cation, $M(H_2O)_x^{n+}$, is a typical Brønsted-Lowry acid. The bound H_2O that releases the H^+ becomes a bound OH^- ion:

$$M(H_2O)_x^{n+}(aq) + H_2O(l) \rightleftharpoons M(H_2O)_{x-1}OH^{(n-1)+}(aq) + H_3O^+(aq)$$

Salts of most M^{2+} and M^{3+} ions yield acidic aqueous solutions. The K_a values for some acidic hydrated metal ions appear in Appendix C.

Consider the small, highly charged Al^{3+} ion. When an aluminum salt, such as $Al(NO_3)_3$, dissolves in water, the following steps occur:

$$Al(NO_3)_3(s) + 6H_2O(l) \longrightarrow Al(H_2O)_6^{3+}(aq) + 3NO_3^-(aq) \qquad \text{[dissolution and hydration]}$$

$$Al(H_2O)_6^{3+}(aq) + H_2O(l) \rightleftharpoons Al(H_2O)_5OH^{2+}(aq) + H_3O^+(aq) \qquad K_a = 1\times10^{-5} \qquad \text{[dissociation of weak acid]}$$

Note the formulas of the hydrated metal ions in the last step. When H^+ is released, the number of bound H_2O molecules decreases by 1 (from 6 to 5) and the number of bound OH^- ions increases by 1 (from 0 to 1), which reduces the ion's positive charge by 1 (from 3 to 2) (Figure 18.11). This pattern of changes in the formula of the hydrated metal ion before and after it loses a proton occurs with any highly charged metal ion in water.

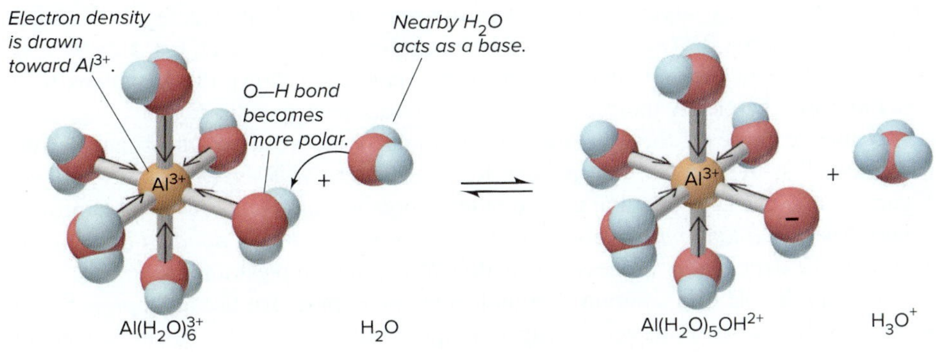

Figure 18.11 The acidic behavior of the hydrated Al^{3+} ion. The hydrated Al^{3+} ion is small and highly charged and pulls electron density from the O—H bonds, so an H^+ ion can be transferred to a nearby water molecule.

› Summary of Section 18.5

- For nonmetal hydrides, acid strength increases across a period, with the electronegativity of the nonmetal (E), and down a group, with the length of the E—H bond.
- For oxoacids with the same number of O atoms, acid strength increases with electronegativity of E; for oxoacids with the same E, acid strength increases with number of O atoms (or O.N. of E).
- Small, highly charged metal ions are acidic in water because they withdraw electron density from the O—H bonds of bound H_2O molecules, releasing an H^+ ion to the solution.

18.6 WEAK BASES AND THEIR RELATION TO WEAK ACIDS

The Brønsted-Lowry concept expands the definition of a base to encompass a host of species that the Arrhenius definition excludes: *to accept a proton, a base needs only a lone electron pair.*

Let's examine the equilibrium system of a weak base (B) as it dissolves in water: B accepts a proton from H_2O, which acts as an acid, leaving behind an OH^- ion:

$$B(aq) + H_2O(l) \rightleftharpoons BH^+(aq) + OH^-(aq)$$

This general reaction for a base in water is described by the equilibrium expression

$$K_c = \frac{[BH^+][OH^-]}{[B][H_2O]}$$

Based on our earlier reasoning, we eliminate $[H_2O]$ from the equilibrium expression and obtain the **base-dissociation constant** (or **base-ionization constant**), K_b:

$$K_b = \frac{[BH^+][OH^-]}{[B]} \quad \textbf{(18.6)}$$

Despite the name "base-dissociation constant," *no base dissociates in the process.*

Base-dissociation constants can be expressed as pK_b values:

$$pK_b = -\log K_b \qquad \text{and} \qquad 10^{-pK_b} = K_b$$

The larger the value of K_b, the smaller the value of pK_b and the stronger the base, producing a higher $[OH^-]$ at equilibrium:

$$\text{stronger base} \Longrightarrow \text{higher } [OH^-] \Longrightarrow \text{larger } K_b \text{ (lower } pK_b)$$

In aqueous solution, the two large classes of weak bases are (1) the molecular species ammonia and the amines and (2) the anions of weak acids.

Molecules as Weak Bases: Ammonia and the Amines

Ammonia is the simplest N-containing compound that acts as a weak base in water:

$$NH_3(aq) + H_2O(l) \rightleftharpoons NH_4^+(aq) + OH^-(aq) \qquad K_b = 1.76\times10^{-5} \text{ (at 25°C)}$$

Despite labels on reagent bottles that read "ammonium hydroxide," an aqueous solution of ammonia consists largely of water and *unprotonated* NH_3 molecules, as its small K_b indicates. In a 1.0 *M* NH_3 solution, for example, $[OH^-] = [NH_4^+] = 4.2\times10^{-3}$ *M*, so about 99.6% of the NH_3 is not protonated. Table 18.7 shows the K_b values for a few molecular bases. (A more extensive list appears in Appendix C.)

If one or more of the H atoms in ammonia is replaced by an organic group (designated as R), an *amine* results: RNH_2, R_2NH, or R_3N (Section 15.4; see Figure 15.17). The key structural feature of these organic compounds, as in all Brønsted-Lowry bases, is *a lone pair of electrons that can bind the proton donated by the acid.* Figure 18.12 depicts this process for methylamine, the simplest amine.

To find the pH of a solution of a molecular weak base, we use an approach very similar to that for a weak acid: write the equilibrium expression, set up a reaction

Table 18.7 K_b **Values for Some Molecular (Amine) Bases at 25°C**

Name (Formula)	Lewis Structure*	Reaction	K_b	pK_b
Diethylamine [$(CH_3CH_2)_2NH$]	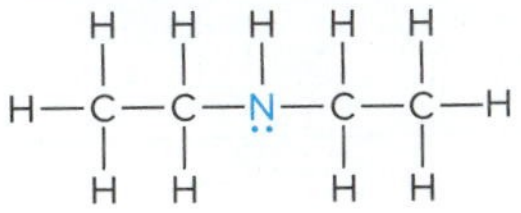	$(CH_3CH_2)_2NH(aq) + H_2O(l) \rightleftharpoons (CH_3CH_2)_2NH_2^+(aq) + OH^-(aq)$	8.6×10^{-4}	3.07
Dimethylamine [$(CH_3)_2NH$]	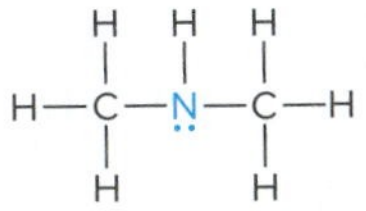	$(CH_3)_2NH(aq) + H_2O(l) \rightleftharpoons (CH_3)_2NH_2^+(aq) + OH^-(aq)$	5.9×10^{-4}	3.23
Methylamine (CH_3NH_2)	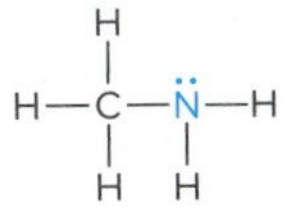	$CH_3NH_2(aq) + H_2O(l) \rightleftharpoons CH_3NH_3^+(aq) + OH^-(aq)$	4.4×10^{-4}	3.36
Ammonia (NH_3)	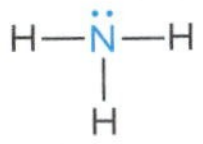	$NH_3(aq) + H_2O(l) \rightleftharpoons NH_4^+(aq) + OH^-(aq)$	1.76×10^{-5}	4.75
Pyridine (C_5H_5N)		$C_5H_5N(aq) + H_2O(l) \rightleftharpoons C_5H_5NH^+(aq) + OH^-(aq)$	1.7×10^{-9}	8.77
Aniline ($C_6H_5NH_2$)	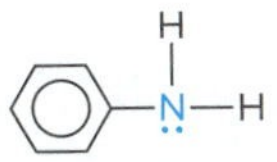	$C_6H_5NH_2(aq) + H_2O(l) \rightleftharpoons C_6H_5NH_3^+(aq) + OH^-(aq)$	4.0×10^{-10}	9.40

*Blue type indicates the basic nitrogen and its lone pair.

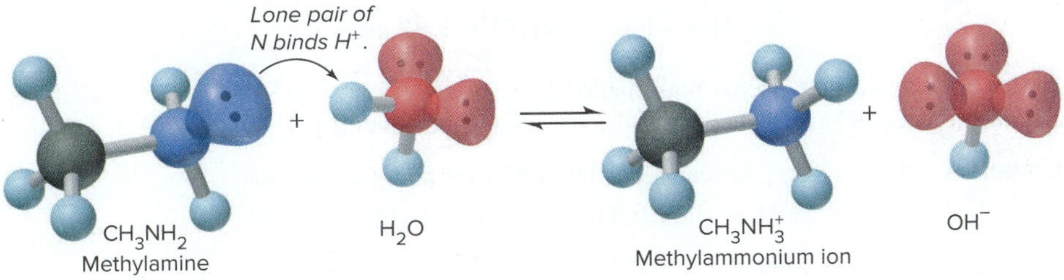

Figure 18.12 Abstraction of a proton from water by the base methylamine.

table to find $[B]_{reacting}$, make the usual assumptions, and then solve for $[OH^-]$. The only additional step is to convert $[OH^-]$ to $[H_3O^+]$ in order to calculate pH.

SAMPLE PROBLEM 18.11 Determining pH from K_b and Initial [B]

Problem Dimethylamine, $(CH_3)_2NH$ *(see the space-filling model),* a key intermediate in detergent manufacture, has a K_b of 5.9×10^{-4}. What is the pH of 1.5 *M* $(CH_3)_2NH$?

Dimethylamine

Plan We know the initial concentration (1.5 *M*) and K_b (5.9×10^{-4}) of $(CH_3)_2NH$ and have to find the pH. The amine reacts with water to form OH^-, so we have to find $[OH^-]$ and then calculate $[H_3O^+]$ and pH. We first write the balanced equation and K_b expression. Because $K_b >> K_w$, the $[OH^-]$ from the autoionization of water is negligible, so we disregard it and assume that all the $[OH^-]$ comes from the base reacting with water. Because K_b is small, we assume that the amount of amine reacting, $[(CH_3)_2NH]_{reacting}$, can be neglected. We set up a reaction table, make the assumption, and solve for *x*. Then we check the assumption and convert $[OH^-]$ to $[H_3O^+]$ using K_w; finally, we calculate pH.

Solution Writing the balanced equation and K_b expression:

$$(CH_3)_2NH(aq) + H_2O(l) \rightleftharpoons (CH_3)_2NH_2^+(aq) + OH^-(aq)$$

$$K_b = \frac{[(CH_3)_2NH_2^+][OH^-]}{[(CH_3)_2NH]}$$

Setting up the reaction table, with $x = [(CH_3)_2NH]_{reacting} = [(CH_3)_2NH_2^+] = [OH^-]$:

Concentration (*M*)	$(CH_3)_2NH(aq)$	+ $H_2O(l)$	$\rightleftharpoons$ $(CH_3)_2NH_2^+(aq)$	+ $OH^-(aq)$
Initial	1.5	—	0	0
Change	$-x$	—	$+x$	$+x$
Equilibrium	$1.5 - x$	—	x	x

Making the assumption: K_b is small, so

$$[(CH_3)_2NH]_{init} - [(CH_3)_2NH]_{reacting} = [(CH_3)_2NH] \approx [(CH_3)_2NH]_{init}$$

Thus, $1.5\ M - x \approx 1.5\ M$.

Substituting into the K_b expression and solving for x:

$$K_b = \frac{[(CH_3)_2NH_2^+][OH^-]}{[(CH_3)_2NH]} = 5.9\times10^{-4} \approx \frac{x^2}{1.5}$$

$$x = [OH^-] \approx 3.0\times10^{-2}\ M$$

Checking the assumption:

$$\frac{3.0\times10^{-2}\ M}{1.5\ M} \times 100 = 2.0\%\ (< 5\%;\ \text{assumption is justified.})$$

Note that the Comment in Sample Problem 18.8 applies in these cases as well:

$$\frac{[B]_{init}}{K_b} = \frac{1.5}{5.9\times10^{-4}} = 2.5\times10^3 > 400$$

Calculating pH:

$$[H_3O^+] = \frac{K_w}{[OH^-]} = \frac{1.0\times10^{-14}}{3.0\times10^{-2}} = 3.3\times10^{-13}\ M$$

$$pH = -\log(3.3\times10^{-13}) = 12.48$$

Check The value of x seems reasonable: $\sqrt{(\sim6\times10^{-4})(1.5)} = \sqrt{9\times10^{-4}} = 3\times10^{-2}$. Because $(CH_3)_2NH$ is a weak base, the pH should be several pH units above 7.

Comment Alternatively, we can find the pOH first and then the pH:

$$pOH = -\log[OH^-] = -\log(3.0\times10^{-2}) = 1.52$$

$$pH = 14 - pOH = 14 - 1.52 = 12.48$$

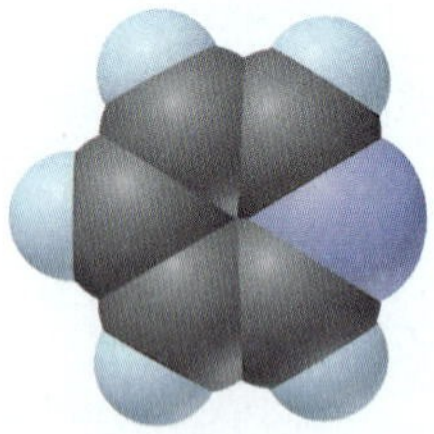

Pyridine

FOLLOW-UP PROBLEMS

18.11A Pyridine (C_5H_5N, *see the space-filling model*) serves as a solvent *and* a base in many organic syntheses. It has a pK_b of 8.77. What is the pH of 0.10 *M* pyridine?

18.11B The weak base amphetamine, $C_6H_5CH_2CH(CH_3)NH_2$, is a stimulant used to treat narcolepsy and attention deficit disorder. What is the pH of 0.075 *M* amphetamine ($K_b = 6.3\times10^{-5}$)?

SOME SIMILAR PROBLEMS 18.103–18.106

Anions of Weak Acids as Weak Bases

The other large group of Brønsted-Lowry bases consists of anions of weak acids:*

$$A^-(aq) + H_2O(l) \rightleftharpoons HA(aq) + OH^-(aq) \qquad K_b = \frac{[HA][OH^-]}{[A^-]}$$

For example, F^-, the anion of the weak acid HF, is a weak base:

$$F^-(aq) + H_2O(l) \rightleftharpoons HF(aq) + OH^-(aq) \qquad K_b = \frac{[HF][OH^-]}{[F^-]}$$

*This equation and equilibrium expression are sometimes referred to as a *hydrolysis reaction* and a *hydrolysis constant*, K_h, because water is dissociated (hydrolyzed). Actually, except for the charge on the base, this process is the same as the proton-abstraction process by ammonia and amines. That is, K_h is just another symbol for K_b, so we'll use K_b throughout.

Why is a solution of HA acidic and a solution of A^- basic? We'll find the answer from relative concentrations of species in 1 *M* HF and in 1 *M* NaF:

1. *The acidity of HA(aq).* HF is a weak acid, so the equilibrium position of the acid dissolving in water lies far to the left:

$$HF(aq) + H_2O(l) \rightleftharpoons H_3O^+(aq) + F^-(aq)$$

Water also contributes H_3O^+ and OH^-, but their concentrations are extremely small:

$$2H_2O(l) \rightleftharpoons H_3O^+(aq) + OH^-(aq)$$

Of all the species present—HF, H_2O, H_3O^+, F^-, and OH^-—the two that can influence the acidity of the solution are H_3O^+, predominantly from HF, and OH^- from water. Thus, the HF solution is acidic because $[H_3O^+]_{\text{from HF}} >> [OH^-]_{\text{from } H_2O}$.

2. *The basicity of $A^-(aq)$.* Now, consider the species present in 1 *M* NaF. The salt dissociates completely to yield 1 *M* Na^+ and 1 *M* F^-. The Na^+ behaves as a spectator ion, while some F^- reacts as a weak base to produce small amounts of HF and OH^-:

$$F^-(aq) + H_2O(l) \rightleftharpoons HF(aq) + OH^-(aq)$$

As before, water dissociation contributes minute amounts of H_3O^+ and OH^-. Thus, in addition to the Na^+ ion, the species present are the same as in the HF solution: HF, H_2O, H_3O^+, F^-, and OH^-. The two species that affect the acidity are OH^-, predominantly from F^- reacting with water, and H_3O^+ from water. In this case, $[OH^-]_{\text{from } F^-} >> [H_3O^+]_{\text{from } H_2O}$, so the solution is basic.

To summarize,

- In an HA solution, $[HA] >> [A^-]$ and $[H_3O^+]_{\text{from HA}} >> [OH^-]_{\text{from } H_2O}$, so the solution is acidic.
- In an A^- solution, $[A^-] >> [HA]$ and $[OH^-]_{\text{from } A^-} >> [H_3O^+]_{\text{from } H_2O}$, so the solution is basic.

The Relation Between K_a and K_b of a Conjugate Acid-Base Pair

A key relationship exists between the K_a of HA and the K_b of its conjugate base, A^-, which we can see by writing the two reactions as a reaction sequence and adding them:

$$\begin{array}{rcl} \cancel{HA} + H_2O & \rightleftharpoons & H_3O^+ + \cancel{A^-} \\ \cancel{A^-} + H_2O & \rightleftharpoons & \cancel{HA} + OH^- \\ \hline 2H_2O & \rightleftharpoons & H_3O^+ + OH^- \end{array}$$

The sum of the two dissociation reactions is the autoionization of water. Recall from Chapter 17 that, for a reaction that is the *sum* of two or more reactions, the overall equilibrium constant is the *product* of the individual equilibrium constants. Therefore, writing the equilibrium expressions for each reaction gives

$$\frac{[H_3O^+]\cancel{[A^-]}}{\cancel{[HA]}} \times \frac{\cancel{[HA]}[OH^-]}{\cancel{[A^-]}} = [H_3O^+][OH^-]$$

or

$$K_a \text{ (of HA)} \times K_b \text{ (of } A^-) = K_w \qquad \textbf{(18.7)}$$

This relationship allows us to find K_a of the acid in a conjugate pair given K_b of the base, and vice versa. Reference tables typically have K_a and K_b values for *molecular species only*. The K_b for F^- and the K_a for $CH_3NH_3^+$, for example, do not appear in standard tables, but you can calculate them by looking up the value for the molecular conjugate species and relating it to K_w. To find the K_b value for F^-, for instance, we look up the K_a value for HF and apply Equation 18.7:

$$K_a \text{ of HF} = 6.8{\times}10^{-4} \text{ (from Appendix C)}$$

So, we have

$$K_a \text{ of HF} \times K_b \text{ of } F^- = K_w$$

or

$$K_b \text{ of } F^- = \frac{K_w}{K_a \text{ of HF}} = \frac{1.0{\times}10^{-14}}{6.8{\times}10^{-4}} = 1.5{\times}10^{-11}$$

SAMPLE PROBLEM 18.12 Determining the pH of a Solution of A^-

Problem Sodium acetate (CH_3COONa, represented by NaAc for this problem) is used in textile dyeing. What is the pH of 0.25 *M* NaAc at 25°C? The K_a of acetic acid (HAc) is 1.8×10^{-5}.

Plan We know the initial concentration of Ac^- (0.25 *M*) and the K_a of HAc (1.8×10^{-5}), and we have to find the pH of the Ac^- solution, knowing that Ac^- acts as a weak base in water. We write the base-dissociation equation and K_b expression. If we can find $[OH^-]$, we can use K_w to find $[H_3O^+]$ and convert it to pH. To solve for $[OH^-]$, we need the K_b of Ac^-, which we obtain from the K_a of its conjugate acid HAc by applying Equation 18.7. We set up a reaction table to find $[OH^-]$ and make the usual assumption that K_b is small, so $[Ac^-]_{init} \approx [Ac^-]$. Notice that Na^+ is a spectator ion and is not included in the base-dissociation equation.

Solution Writing the base-dissociation equation and K_b expression:

$$Ac^-(aq) + H_2O(l) \rightleftharpoons HAc(aq) + OH^-(aq) \qquad K_b = \frac{[HAc][OH^-]}{[Ac^-]}$$

Setting up the reaction table, with $x = [Ac^-]_{reacting} = [HAc] = [OH^-]$:

Concentration (*M*)	$Ac^-(aq)$	+	$H_2O(l)$	$\rightleftharpoons$	$HAc(aq)$	+	$OH^-(aq)$
Initial	0.25		—		0		0
Change	$-x$		—		$+x$		$+x$
Equilibrium	$0.25 - x$		—		x		x

Solving for K_b of Ac^-:

$$K_b = \frac{K_w}{K_a} = \frac{1.0\times10^{-14}}{1.8\times10^{-5}} = 5.6\times10^{-10}$$

Making the assumption: Because K_b is small, $0.25\ M - x \approx 0.25\ M$.

Substituting into the expression for K_b and solving for x:

$$K_b = \frac{[HAc][OH^-]}{[Ac^-]} = 5.6\times10^{-10} \approx \frac{x^2}{0.25} \qquad x = [OH^-] \approx 1.2\times10^{-5}\ M$$

Checking the assumption:

$$\frac{1.2\times10^{-5}\ M}{0.25\ M} \times 100 = 4.8\times10^{-3}\%\ (<5\%;\ \text{assumption is justified.})$$

Also note that

$$\frac{0.25}{5.6\times10^{-10}} = 4.5\times10^{8} > 400$$

Solving for pH:

$$[H_3O^+] = \frac{K_w}{[OH^-]} = \frac{1.0\times10^{-14}}{1.2\times10^{-5}} = 8.3\times10^{-10}\ M$$

$$pH = -\log(8.3\times10^{-10}) = 9.08$$

Check The K_b calculation seems reasonable: $\sim10\times10^{-15}/2\times10^{-5} = 5\times10^{-10}$. Because Ac^- is a weak base, $[OH^-] > [H_3O^+]$; thus, pH > 7, which makes sense.

FOLLOW-UP PROBLEMS

18.12A Sodium hypochlorite (NaClO) is the active ingredient in household laundry bleach. What is the pH of 0.20 *M* NaClO?

18.12B Sodium nitrite ($NaNO_2$) is added to cured meats such as bacon to prevent growth of bacteria. What is the pH of 0.80 *M* $NaNO_2$?

SOME SIMILAR PROBLEMS 18.111–18.114

Student Hot Spot

Student data indicate that you may struggle with pH calculations involving weakly basic anions. Access the Smartbook to view additional Learning Resources on this topic.

› Summary of Section 18.6

- › The extent to which a weak base accepts a proton from water to form OH^- is expressed by a base-dissociation constant, K_b.
- › Brønsted-Lowry bases include NH_3 and amines and the anions of weak acids. All produce basic solutions by accepting H^+ from water, which yields OH^-, thus making $[H_3O^+] < [OH^-]$.
- › A solution of HA is acidic because $[HA] >> [A^-]$, so $[H_3O^+] > [OH^-]$. A solution of A^- is basic because $[A^-] >> [HA]$, so $[OH^-] > [H_3O^+]$.
- › By multiplying the expressions for K_a of HA and K_b of A^-, we obtain K_w. This relationship allows us to calculate either K_a of BH^+ or K_b of A^-.

18.7 ACID-BASE PROPERTIES OF SALT SOLUTIONS

In many cases, when a salt dissolves, one or both of its ions may react with water and affect the pH of the solution. You've seen that cations of weak bases (such as NH_4^+) are acidic, anions of weak acids (such as CN^-) are basic, and small, highly charged metal cations (such as Al^{3+}) are acidic. In addition, certain ions (such as $H_2PO_4^-$ and HCO_3^-) can act as an acid or a base. In this section, we classify the acid-base behavior of the various types of salt solutions.

Salts That Yield Neutral Solutions

A salt consisting of the anion of a strong acid and the cation of a strong base yields a neutral solution because the ions do not react with water. When a strong acid such as HNO_3 dissolves, the reaction goes essentially to completion because *the anion of a strong acid is a much weaker base than water.* The anion, in this case NO_3^-, is hydrated, but *it does not react with water:*

$$HNO_3(l) + H_2O(l) \longrightarrow NO_3^-(aq) + H_3O^+(aq) \qquad \text{[dissolution and hydration]}$$

Similarly, a strong base, such as NaOH, dissolves completely. The cation, in this case Na^+, is hydrated, but *it is not small and charged enough to react with water:*

$$NaOH(s) \xrightarrow{H_2O} Na^+(aq) + OH^-(aq) \qquad \text{[dissolution and hydration]}$$

The ions that do not react with water are:

- the anions of the strong hydrohalic acids: Cl^-, Br^-, I^-;
- the ions of strong oxoacids, such as NO_3^- and ClO_4^-;
- the Group 1A(1) ions (cations of strong bases); and
- Ca^{2+}, Sr^{2+}, and Ba^{2+} in Group 2A(2) (cations of strong bases).

Salts containing only these anions and cations yield neutral solutions. The salts $LiNO_3$ and $CaCl_2$ are two examples.

Salts That Yield Acidic Solutions

There are two types of salts that yield acidic solutions; in both types, the cation is responsible for the acidity:

1. *A salt consisting of the cation of a weak base and the anion of a strong acid yields an acidic solution because the cation acts as a weak acid,* and the anion does not react. For example, NH_4Cl yields an acidic solution because NH_4^+, the cation of the weak base NH_3, is a weak acid; Cl^-, the anion of the strong HCl, does not react:

$$NH_4Cl(s) \xrightarrow{H_2O} NH_4^+(aq) + Cl^-(aq) \qquad \text{[dissolution and hydration]}$$
$$NH_4^+(aq) + H_2O(l) \rightleftharpoons NH_3(aq) + H_3O^+(aq) \qquad \text{[dissociation of weak acid]}$$

2. *A salt consisting of a small, highly charged metal cation and the anion of a strong acid yields an acidic solution because the cation acts as a weak acid (see Figure 18.11),* and the anion does not react. For example, $Fe(NO_3)_3$ yields an acidic

solution because the hydrated Fe^{3+} ion is a weak acid; NO_3^-, the anion of the strong HNO_3, does not react:

$$Fe(NO_3)_3(s) + 6H_2O(l) \xrightarrow{H_2O} Fe(H_2O)_6^{3+}(aq) + 3NO_3^-(aq) \quad \text{[dissolution and hydration]}$$

$$Fe(H_2O)_6^{3+}(aq) + H_2O(l) \rightleftharpoons Fe(H_2O)_5OH^{2+}(aq) + H_3O^+(aq) \quad \text{[dissociation of weak acid]}$$

Salts That Yield Basic Solutions

A salt consisting of the anion of a weak acid and the cation of a strong base yields a basic solution because the anion acts as a weak base, and the cation does not react. Sodium acetate, for example, yields a basic solution because the CH_3COO^- ion, the anion of the weak acid CH_3COOH, acts as a weak base; Na^+, the cation of the strong base NaOH, does not react:

$$CH_3COONa(s) \xrightarrow{H_2O} Na^+(aq) + CH_3COO^-(aq) \quad \text{[dissolution and hydration]}$$

$$CH_3COO^-(aq) + H_2O(l) \rightleftharpoons CH_3COOH(aq) + OH^-(aq) \quad \text{[reaction of weak base]}$$

SAMPLE PROBLEM 18.13 Predicting Relative Acidity of Salt Solutions from Reactions of the Ions with Water

Problem Predict whether aqueous solutions of the following salts are acidic, basic, or neutral, and write an equation for the reaction of any ion with water:
(a) Potassium perchlorate, $KClO_4$ **(b)** Sodium benzoate, C_6H_5COONa
(c) Chromium(III) nitrate, $Cr(NO_3)_3$

Plan The formula shows the cation and the anion. Depending on an ion's ability to react with water, the solution will be neutral (strong-acid anion and strong-base cation), acidic (weak-base cation or highly charged metal cation with strong-acid anion), or basic (weak-acid anion and strong-base cation).

Solution (a) Neutral. The ions are K^+ and ClO_4^-. The K^+ is from the strong base KOH, and the ClO_4^- is from the strong acid $HClO_4$. Neither ion reacts with water.
(b) Basic. The ions are Na^+ and $C_6H_5COO^-$. The Na^+ is the cation of the strong base NaOH, so it does not react with water. The benzoate ion, $C_6H_5COO^-$, is the anion of the weak acid benzoic acid, so it reacts with water to produce OH^- ion:

$$C_6H_5COO^-(aq) + H_2O(l) \rightleftharpoons C_6H_5COOH(aq) + OH^-(aq)$$

(c) Acidic. The ions are Cr^{3+} and NO_3^-. The NO_3^- is the anion of the strong acid HNO_3, so it does not react with water. The Cr^{3+} ion is small and highly charged, so the hydrated ion, $Cr(H_2O)_6^{3+}$, reacts with water to produce H_3O^+:

$$Cr(H_2O)_6^{3+}(aq) + H_2O(l) \rightleftharpoons Cr(H_2O)_5OH^{2+}(aq) + H_3O^+(aq)$$

FOLLOW-UP PROBLEMS

18.13A Write equations to predict whether solutions of the following salts are acidic, basic, or neutral: **(a)** $KClO_2$; **(b)** $CH_3NH_3NO_3$; **(c)** RbBr.

18.13B Write equations to predict whether solutions of the following salts are acidic, basic, or neutral: **(a)** $FeBr_3$; **(b)** $Ca(NO_2)_2$; **(c)** $C_6H_5NH_3I$.

SOME SIMILAR PROBLEMS 18.120, 18.121(a), 18.122, 18.123(a), 18.123(b), 18.124, 18.125(c), 18.126(c), and 18.127(c)

Student Hot Spot

Student data indicate that you may struggle with predicting the pH of a salt solution. Access the Smartbook to view additional Learning Resources on this topic.

Salts of Weakly Acidic Cations and Weakly Basic Anions

If a salt consists of a cation that is a weak acid *and* an anion that is a weak base, the overall acidity of the solution depends on the relative acid strength (K_a) and base strength (K_b) of the separated ions. Consider a solution of ammonium cyanide, NH_4CN, and the reactions that occur between the separated ions and water. Ammonium ion is the conjugate acid of a weak base, so it is a weak acid:

$$NH_4^+(aq) + H_2O(l) \rightleftharpoons NH_3(aq) + H_3O^+(aq)$$

Cyanide ion is the anion of a weak acid, so it is a weak base:

$$CN^-(aq) + H_2O(l) \rightleftharpoons HCN(aq) + OH^-(aq)$$

The reaction that goes farther to the right determines the pH of the solution, so we compare the K_a of NH_4^+ with the K_b of CN^-. Only molecular compounds are listed in K_a and K_b tables, so we use Equation 18.7 to calculate these values for the two ions:

$$K_a \text{ of } NH_4^+ = \frac{K_w}{K_b \text{ of } NH_3} = \frac{1.0\times10^{-14}}{1.76\times10^{-5}} = 5.7\times10^{-10}$$

$$K_b \text{ of } CN^- = \frac{K_w}{K_a \text{ of } HCN} = \frac{1.0\times10^{-14}}{6.2\times10^{-10}} = 1.6\times10^{-5}$$

Because K_b of CN^- > K_a of NH_4^+, we know that CN^- is a stronger base than NH_4^+ is an acid. Thus, more OH^- than H_3O^+ is produced, and the NH_4CN solution is basic.

Salts of Amphiprotic Anions

The only salts left to consider are those in which the cation comes from a strong base and therefore does not react with water, and the anion comes from a polyprotic acid with one or more ionizable protons still attached. These anions are amphiprotic—they can act as an acid and release a proton *to* water or as a base and abstract a proton *from* water. As in the previous case, to determine the overall acidity of their solutions, we compare the magnitudes of K_a and K_b, but here we compare the K_a and K_b of the *same* species, the anion.

For example, Na_2HPO_4 consists of Na^+, the cation of a strong base, which does not react with water, and HPO_4^{2-}, the second anion of the weak polyprotic acid H_3PO_4. In water, the salt undergoes three steps:

1. $Na_2HPO_4(s) \xrightarrow{H_2O} 2Na^+(aq) + HPO_4^{2-}(aq)$ [dissolution and hydration]
2. $HPO_4^{2-}(aq) + H_2O(l) \rightleftharpoons PO_4^{3-}(aq) + H_3O^+(aq)$ [acting as a weak acid]
3. $HPO_4^{2-}(aq) + H_2O(l) \rightleftharpoons H_2PO_4^-(aq) + OH^-(aq)$ [acting as a weak base]

We must decide whether step 2 or step 3 goes farther to the right. Appendix C lists the K_a of HPO_4^{2-} as 4.2×10^{-13}, but the K_b of HPO_4^{2-} is not given, so we find it from the K_a of its conjugate acid, $H_2PO_4^-$, using Equation 18.7:

$$K_b \text{ of } HPO_4^{2-} = \frac{K_w}{K_a \text{ of } H_2PO_4^-} = \frac{1.0\times10^{-14}}{6.3\times10^{-8}} = 1.6\times10^{-7}$$

Because K_b (1.6×10^{-7}) > K_a (4.2×10^{-13}), HPO_4^{2-} is stronger as a base than as an acid, so a solution of Na_2HPO_4 is basic.

Table 18.8 displays the acid-base behavior of the various types of salts in water.

Table 18.8 The Acid-Base Behavior of Salts in Water

Nature of Ions	Examples	Ion(s) That React(s) with Water: Example(s)	pH
Neutral			
Cation of strong base Anion of strong acid	NaCl, KBr, $Ba(NO_3)_2$	None	7.0
Acidic			
Cation of weak base Anion of strong acid	NH_4Cl, NH_4NO_3, CH_3NH_3Br	Cation: $NH_4^+ + H_2O \rightleftharpoons NH_3 + H_3O^+$	<7.0
Small, highly charged metal cation Anion of strong acid	$Al(NO_3)_3$, $CrBr_3$, $FeCl_3$	Cation: $Al(H_2O)_6^{3+} + H_2O \rightleftharpoons Al(H_2O)_5OH^{2+} + H_3O^+$	<7.0
Basic			
Cation of strong base Anion of weak acid	KNO_2, NaClO, Na_2CO_3	Anion: $NO_2^- + H_2O \rightleftharpoons HNO_2 + OH^-$	>7.0
Acidic/Basic			
Cation of weak base or small, highly charged metal cation Anion of weak acid	NH_4ClO_2, NH_4CN, $Pb(CH_3COO)_2$	Cation *and* anion: $NH_4^+ + H_2O \rightleftharpoons NH_3 + H_3O^+$ $ClO_2^- + H_2O \rightleftharpoons HClO_2 + OH^-$	<7.0 if $K_{a(cation)} > K_{b(anion)}$ >7.0 if $K_{b(anion)} > K_{a(cation)}$
Cation of strong base Anion of polyprotic acid	NaH_2PO_4, $KHCO_3$, $NaHSO_3$	Anion: $HSO_3^- + H_2O \rightleftharpoons SO_3^{2-} + H_3O^+$ $HSO_3^- + H_2O \rightleftharpoons H_2SO_3 + OH^-$	<7.0 if $K_{a(anion)} > K_{b(anion)}$ >7.0 if $K_{b(anion)} > K_{a(anion)}$

SAMPLE PROBLEM 18.14 **Predicting the Relative Acidity of a Salt Solution from K_a and K_b of the Ions**

Problem Determine whether aqueous solutions of the following salts are acidic, basic, or neutral at 25°C:

(a) zinc formate, $Zn(HCOO)_2$ **(b)** potassium hydrogen sulfite, $KHSO_3$

Plan (a) The formula consists of the small, highly charged, and therefore weakly acidic, Zn^{2+} cation and the weakly basic $HCOO^-$ anion of the weak acid HCOOH. To determine the relative acidity of the solution, we write equations that show the reactions of the ions with water, and then find K_a of Zn^{2+} (in Appendix C) and calculate K_b of $HCOO^-$ (from K_a of HCOOH in Appendix C) to see which ion reacts with water to a greater extent.

(b) K^+ is from the strong base KOH and does not react with water; the anion, HSO_3^-, is the first anion of the weak polyprotic acid H_2SO_3. We write the reactions of this anion acting as a weak acid and as a weak base in water; we find its K_a (in Appendix C) and calculate its K_b (from the K_a of its conjugate acid, H_2SO_3, in Appendix C) to determine if the ion is a stronger acid or a stronger base.

Solution (a) Writing the reactions with water:

$$Zn(H_2O)_6^{2+}(aq) + H_2O(l) \rightleftharpoons Zn(H_2O)_5OH^+(aq) + H_3O^+(aq)$$
$$HCOO^-(aq) + H_2O(l) \rightleftharpoons HCOOH(aq) + OH^-(aq)$$

Obtaining K_a and K_b of the ions: From Appendix C, the K_a of $Zn(H_2O)_6^{2+}(aq)$ is 1×10^{-9}. We obtain K_a of HCOOH and solve for K_b of $HCOO^-$:

$$K_b \text{ of } HCOO^- = \frac{K_w}{K_a \text{ of } HCOOH} = \frac{1.0\times10^{-14}}{1.8\times10^{-4}} = 5.6\times10^{-11}$$

K_a of $Zn(H_2O)_6^{2+} > K_b$ of $HCOO^-$, so the solution is acidic.

(b) Writing the reactions with water:

$$HSO_3^-(aq) + H_2O(l) \rightleftharpoons SO_3^{2-}(aq) + H_3O^+(aq) \quad \text{(acid dissociation)}$$
$$HSO_3^-(aq) + H_2O(l) \rightleftharpoons H_2SO_3(aq) + OH^-(aq) \quad \text{(base reaction)}$$

Obtaining K_a and K_b of HSO_3^-: From Appendix C, K_a is 6.5×10^{-8}. We obtain K_a of H_2SO_3 and solve for K_b of HSO_3^-:

$$K_b \text{ of } HSO_3^- = \frac{K_w}{K_a \text{ of } H_2SO_3} = \frac{1.0\times10^{-14}}{1.4\times10^{-2}} = 7.1\times10^{-13}$$

$K_a > K_b$, so the solution is acidic.

FOLLOW-UP PROBLEMS

18.14A Determine whether solutions of the following salts are acidic, basic, or neutral at 25°C: **(a)** $Cu(CH_3COO)_2$; **(b)** NH_4F; **(c)** $KHC_6H_6O_6$.

18.14B Determine whether solutions of the following salts are acidic, basic, or neutral at 25°C: **(a)** $NaHCO_3$; **(b)** $C_6H_5NH_3NO_2$; **(c)** NaH_2PO_4.

SOME SIMILAR PROBLEMS 18.121(b), 18.121(c), 18.123(c), 18.125(a), 18.125(b), 18.126(a), 18.127(a), and 18.127(b)

› Summary of Section 18.7

- Salts that yield a neutral solution consist of ions that do not react with water.
- Salts that yield an acidic solution contain an unreactive anion (from a strong acid) and a cation that releases a proton to water.
- Salts that yield a basic solution contain an unreactive cation (from a strong base) and an anion that accepts a proton from water.
- If both cation and anion react with water, the ion that reacts to the greater extent (higher *K*) determines the acidity or basicity of the salt solution.
- If the anion is amphiprotic (from a polyprotic acid), the strength of the anion as an acid (K_a) or as a base (K_b) determines the acidity of the salt solution.

18.8 GENERALIZING THE BRØNSTED-LOWRY CONCEPT: THE LEVELING EFFECT

In general, in solution, *an acid yields the cation and a base yields the anion that would be produced by autoionization of the solvent.* For example, in H_2O, all Brønsted-Lowry acids yield H_3O^+ and all Brønsted-Lowry bases yield OH^-, which are the ions that form when water autoionizes.

This information lets us examine a key question: Why are all strong acids and strong bases *equally* strong in water? The answer is that *in water, the strongest acid possible is H_3O^+ and the strongest base possible is OH^-:*

- *For strong acids.* The moment we put gaseous HCl in water, it reacts with the base H_2O to form H_3O^+. The same holds for any strong acid because it dissociates *completely* to form H_3O^+. Thus, we are actually observing the acid strength of H_3O^+.
- *For strong bases.* A strong base, such as $Ba(OH)_2$, dissociates completely in water to yield OH^-. Even strong bases that do not contain hydroxide ions in the solid, such as K_2O, do so. The oxide ion, which is a stronger base than OH^-, immediately takes a proton from water to form OH^-:

$$O^{2-}(aq) + H_2O(l) \longrightarrow 2OH^-(aq)$$

Thus, water exerts a **leveling effect** on any strong acid or base by reacting with it to form the products of water's autoionization. Acting as a base, water levels the strength of all strong acids by making them appear equally strong, and acting as an acid, it levels the strength of all strong bases.

Therefore, to rank the relative strengths of strong acids, we must dissolve them in a solvent that is a *weaker* base than water, so it accepts their protons less readily. For example, the hydrohalic acids increase in strength as the halogen becomes larger, as a result of the longer, weaker H—X bond (see Figure 18.9). But, in water, HF is weak, and HCl, HBr, and HI appear equally strong because they dissociate completely. When we dissolve them in pure acetic acid, however, *the acetic acid acts as the base* and accepts a proton:

$$\underset{\text{acid}}{HCl(g)} + \underset{\text{base}}{CH_3COOH(l)} \rightleftharpoons \underset{\text{base}}{Cl^-(acet)} + \underset{\text{acid}}{CH_3COOH_2^+(acet)}$$

$$HBr(g) + CH_3COOH(l) \rightleftharpoons Br^-(acet) + CH_3COOH_2^+(acet)$$

$$HI(g) + CH_3COOH(l) \rightleftharpoons I^-(acet) + CH_3COOH_2^+(acet)$$

[The use of "(*acet*)" instead of "(*aq*)" indicates solvation by CH_3COOH.] Because acetic acid is a *weaker base* than water, the three acids donate their protons to *different* extents. Measurements show that HI protonates the solvent to a greater extent than HBr, and HBr does so more than HCl; that is, in pure acetic acid, $K_{HI} > K_{HBr} > K_{HCl}$. Similarly, the relative strengths of strong bases are determined in a solvent that is a weaker acid than H_2O, such as pure liquid NH_3.

› Summary of Section 18.8

- › Strong acids (or strong bases) dissociate completely to yield H_3O^+ (or OH^-) in water; in effect, water equalizes (levels) their strengths.
- › Strong acids show differences in strength when dissolved in a weaker base than water, such as acetic acid.

18.9 ELECTRON-PAIR DONATION AND THE LEWIS ACID-BASE DEFINITION

The final acid-base concept we consider was developed by Gilbert N. Lewis, whose contribution to understanding valence electron pairs in bonding we discussed in Chapter 9. Whereas the Brønsted-Lowry concept focuses on the proton in defining a species as an acid or a base, the Lewis concept highlights the role of the *electron pair*. The **Lewis acid-base definition** holds that

- A *base* is any species that *donates* an electron pair to form a bond.
- An *acid* is any species that *accepts* an electron pair to form a bond.

The Lewis definition, like the Brønsted-Lowry definition, requires that a base have an electron pair to donate, so it does not expand the classes of bases. However, *it greatly expands the classes of acids*. Many species, such as CO_2 and Cu^{2+}, that do not contain H in their formula (and thus cannot be Brønsted-Lowry acids) are Lewis acids because they accept an electron pair in reactions. In fact, the proton itself is a Lewis acid because it accepts the electron pair donated by a base:

$$\text{B:} + \text{H}^+ \rightleftharpoons \text{B—H}^+$$

Thus, *all Brønsted-Lowry acids donate* H^+, *a Lewis acid.*

The product of a Lewis acid-base reaction is an **adduct,** *a single species that contains a* ***new*** *covalent bond:*

$$\text{A} + \text{:B} \rightleftharpoons \text{A—B (adduct)}$$

Thus, the Lewis concept radically broadens the idea of an acid-base reaction:

- To Arrhenius, it is the formation of H_2O from H^+ and OH^-.
- To Brønsted and Lowry, it is H^+ transfer from a stronger acid to a stronger base to form a weaker base and weaker acid.
- To Lewis, it is *the donation and acceptance of an electron pair to form a covalent bond in an adduct.*

By definition, then,

- *A Lewis base must have a lone pair of electrons to donate.*
- *A Lewis acid must have a vacant orbital* (or the ability to rearrange its bonds to form one) to accept a lone electron pair and form a new bond.

In this section, we discuss molecules and positive metal ions that act as Lewis acids.

Molecules as Lewis Acids

Many molecules act as Lewis acids. In every case, the atom accepting the electron pair has *low electron density* due to either an electron deficiency or a polar multiple bond.

Lewis Acids with Electron-Deficient Atoms The most important of the *electron-deficient* Lewis acids are compounds of the Group 3A(13) elements boron and aluminum. Recall from Chapters 10 and 14 that these compounds have fewer than eight electrons around the central atom, so they react to complete that atom's octet. For example, boron trifluoride accepts an electron pair from ammonia to form a covalent bond:

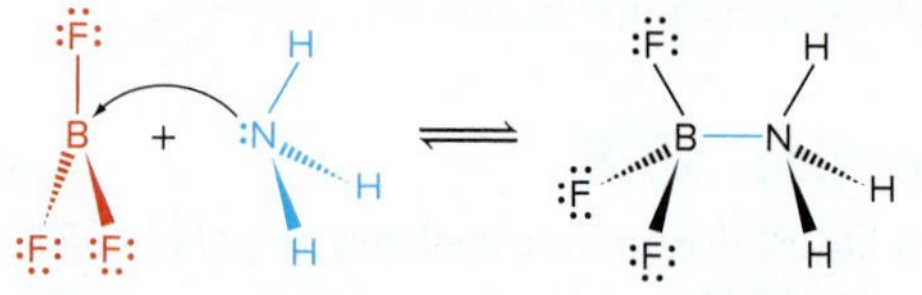

Aluminum chloride, a salt, dissolves freely in relatively nonpolar diethyl ether when the ether's O atom donates an electron pair to Al to form a covalent bond:

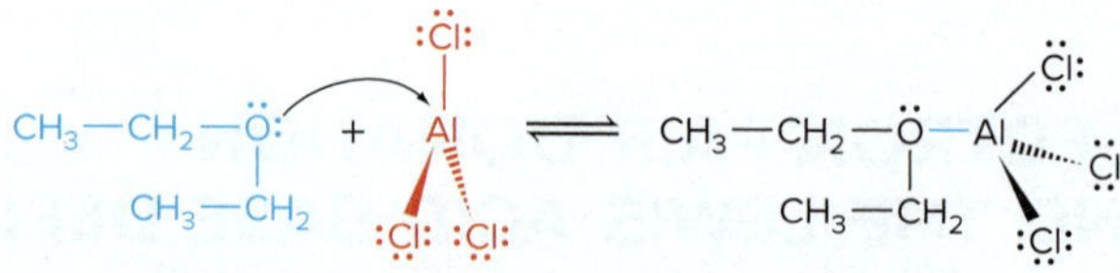

This acidic behavior of boron and aluminum halides is often used in organic syntheses. For example, a methyl group is added to the benzene ring by the action of CH_3Cl

in the presence of $AlCl_3$. The Lewis acid $AlCl_3$ abstracts the Lewis base Cl^- from CH_3Cl to form an ionic adduct that has a reactive CH_3^+ group, which then attacks the electron-rich benzene ring:

$$\underset{\text{base}}{CH_3Cl} + \underset{\text{acid}}{AlCl_3} \rightleftharpoons \underset{\text{adduct}}{[CH_3]^+[Cl{-}AlCl_3]^-}$$

$$\underset{\text{benzene}}{C_6H_6} + [CH_3]^+[Cl{-}AlCl_3]^- \rightleftharpoons \underset{\text{toluene}}{C_6H_5CH_3} + AlCl_3 + HCl$$

Lewis Acids with Polar Multiple Bonds Molecules with a polar double bond also function as Lewis acids. An electron pair on the Lewis base approaches the partially positive end of the double bond to form the new bond in the adduct, as the π bond breaks. For example, consider the reaction of SO_2 in water. The electronegative O atoms in SO_2 make the central S partially positive. The O atom of water donates a lone pair to the S, thus forming an S—O bond and breaking one of the π bonds. Then, a proton is transferred from water to the O that was part of the π bond. The resulting adduct is sulfurous acid:

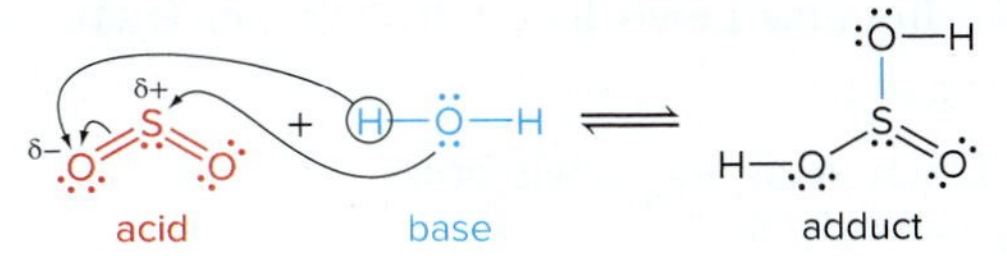

The analogous formation of a carbonate from a metal oxide and carbon dioxide occurs in a nonaqueous system. The O^{2-} ion (shown coming from CaO) donates an electron pair to the partially positive C in CO_2, a π bond breaks, and the CO_3^{2-} ion (shown as part of $CaCO_3$) is the adduct:

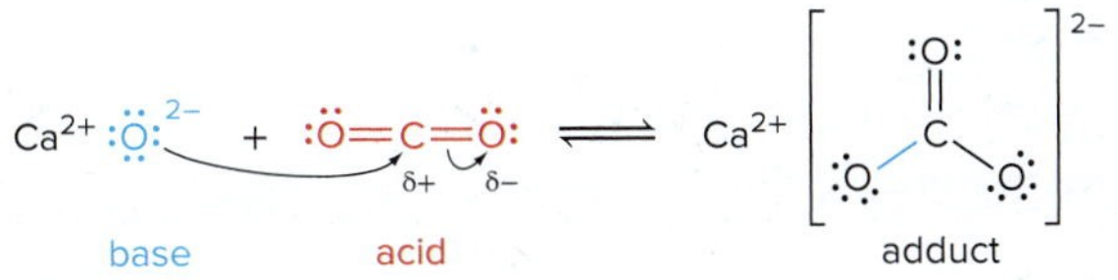

Metal Cations as Lewis Acids

In the Lewis sense, hydration of a metal ion is itself an acid-base reaction. When electron pairs on the O atoms of H_2O molecules form covalent bonds, the hydrated cation is the adduct; thus, *a metal ion acts as a Lewis acid when it dissolves in water:*

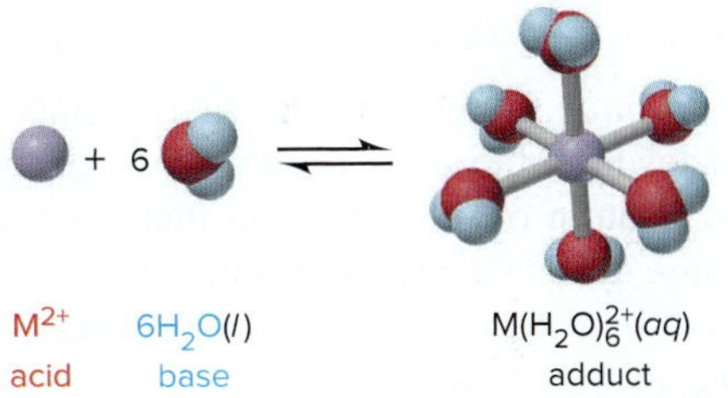

Ammonia is a stronger Lewis base than water because, when it is added to the aqueous solution of the hydrated cation, it displaces H_2O, with $K >> 1$:

$$\underset{\text{hydrated adduct}}{Ni(H_2O)_6^{2+}(aq)} + \underset{\text{base}}{6NH_3(aq)} \rightleftharpoons \underset{\text{ammoniated adduct}}{Ni(NH_3)_6^{2+}(aq)} + 6H_2O(l)$$

We discuss the equilibrium nature of these acid-base reactions in greater detail in Chapter 19, and we investigate the structures of these ions in Chapter 23.

Many biomolecules with central metal ions are Lewis adducts. Most often, O and N atoms of organic groups donate their lone pairs as the Lewis bases. Chlorophyll is a Lewis adduct of a central Mg^{2+} and the four N atoms of a tetrapyrrole (porphin) ring system (Figure 18.13). Vitamin B_{12} has a similar structure with a central Co^{3+}, and so does heme with a central Fe^{2+}. Several other metal ions, such as Zn^{2+}, Mo^{2+}, and Cu^{2+}, are bound at the active sites of enzymes and function as Lewis acids in the catalytic process (see the Chemical Connections at the end of Section 23.4).

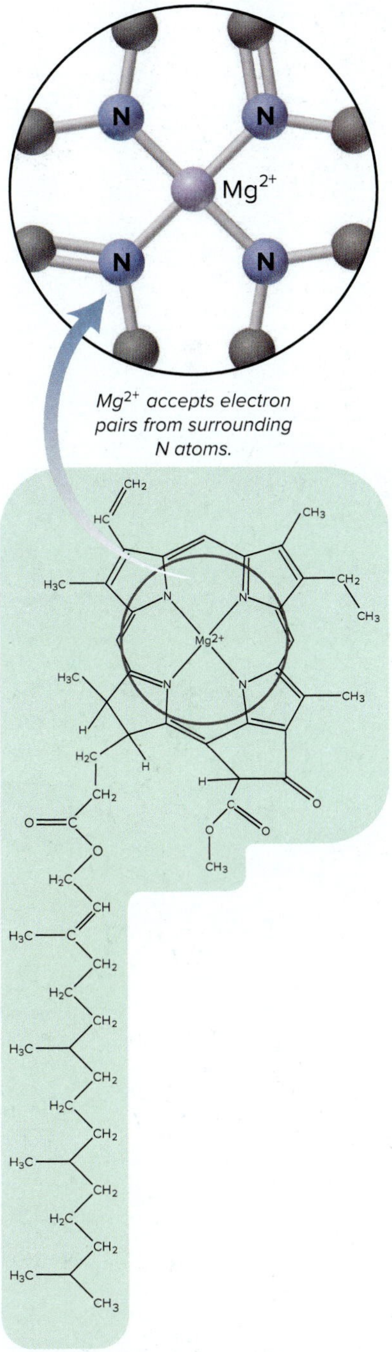

Figure 18.13 The Mg^{2+} ion as a Lewis acid in chlorophyll.

SAMPLE PROBLEM 18.15 **Identifying Lewis Acids and Bases**

Problem Identify the Lewis acids and Lewis bases in the following reactions:

(a) $H^+ + OH^- \rightleftharpoons H_2O$ **(b)** $Cl^- + BCl_3 \rightleftharpoons BCl_4^-$ **(c)** $K^+ + 6H_2O \rightleftharpoons K(H_2O)_6^+$

Plan We examine the formulas to see which species accepts the electron pair (Lewis acid) and which donates it (Lewis base) in forming the adduct. The Lewis base must have a lone pair of electrons.

Solution **(a)** The H^+ ion accepts an electron pair from the OH^- ion in forming a bond. H^+ is the Lewis acid and OH^- is the Lewis base.

(b) The Cl^- ion has four lone pairs and uses one to form a new bond to the central B. BCl_3 is the Lewis acid and Cl^- is the Lewis base.

(c) The K^+ ion does not have any valence electrons to provide, so the bond is formed when electron pairs from O atoms of water enter empty orbitals on K^+. K^+ is the Lewis acid and H_2O is the Lewis base.

Check The Lewis acids (H^+, BCl_3, and K^+) each have an unfilled valence shell that can accept an electron pair from the Lewis base (OH^-, Cl^-, or H_2O).

FOLLOW-UP PROBLEMS

18.15A Identify the Lewis acids and Lewis bases:
(a) $OH^- + Al(OH)_3 \rightleftharpoons Al(OH)_4^-$
(b) $SO_3 + H_2O \rightleftharpoons H_2SO_4$
(c) $Co^{3+} + 6NH_3 \rightleftharpoons Co(NH_3)_6^{3+}$

18.15B Identify the Lewis acids and Lewis bases:
(a) $B(OH)_3 + H_2O \rightleftharpoons B(OH)_4^- + H^+$
(b) $Cd^{2+} + 4I^- \rightleftharpoons CdI_4^{2-}$
(c) $2F^- + SiF_4 \rightleftharpoons SiF_6^{2-}$

SOME SIMILAR PROBLEMS 18.142 and 18.143

An Overview of Acid-Base Definitions

From a broader chemical perspective, the diversity of acid-base reactions takes on more unity. Chemists see a common theme in reactions as diverse as a standardized base being used to analyze an unknown fatty acid, baking soda being used in bread-making, and even oxygen binding to hemoglobin in a blood cell. Let's see how the three acid-base definitions fit together.

1. The *Arrhenius definition,* which was the first attempt at describing acids and bases on the molecular level, is the most limited and narrow of the three. It applies only to species with an H atom or an OH group that is released as an ion when the species dissolves in water. Because relatively few species have these structural prerequisites, Arrhenius acid-base reactions are relatively few in number, and all occur in H_2O and result in the formation of H_2O.

2. The *Brønsted-Lowry definition* sees acid-base reactions as proton-transfer processes that need not occur in water. A Brønsted-Lowry acid, like an Arrhenius acid, must have an H, but a Brønsted-Lowry base is any species with an electron pair available to accept a proton. This definition includes many more species as bases, including OH^-. It defines the acid-base reaction in terms of conjugate acid-base pairs, with an acid and a base on both sides of the reaction. The system reaches equilibrium based on the relative strengths of the acid, the base, and their conjugates.

3. The *Lewis definition* has the widest scope and includes the other two. The defining event of a Lewis acid-base reaction is the donation and acceptance of an electron pair to form a new covalent bond in an adduct. Lewis bases still must have an electron pair to donate, but Lewis acids—the electron-pair acceptors—include many species not encompassed by the other definitions, including molecules with electron-deficient atoms or with polar double bonds, metal ions, and even H^+ itself.

› Summary of Section 18.9

- › The Lewis acid-base definition focuses on the donation or acceptance of an electron pair to form a new covalent bond in an adduct, the product of an acid-base reaction. Lewis bases donate the electron pair, and Lewis acids accept it.
- › Thus, many species that do not contain H are Lewis acids. Molecules with polar double bonds act as Lewis acids, as do those with electron-deficient atoms.
- › Metal ions act as Lewis acids when they dissolve in water, which acts as a Lewis base, to form a hydrated cation as the adduct.
- › Many metal ions function as Lewis acids in biomolecules.

CHAPTER REVIEW GUIDE

Learning Objectives

Relevant section (§) and/or sample problem (SP) numbers appear in parentheses.

Understand These Concepts

1. Why the proton is bonded to a water molecule, as H_3O^+, in all aqueous acid-base systems (§18.1)
2. The Arrhenius definitions of an acid and a base (§18.1)
3. Why all reactions of a strong acid and a strong base have the same ΔH°_{rxn} (§18.1)
4. How the strength of an acid (or base) relates to the extent of its dissociation into ions in water (§18.1)
5. How relative acid strength is expressed by the acid-dissociation constant K_a (§18.1)
6. How the autoionization of water is expressed by K_w (§18.2)
7. Why $[H_3O^+]$ is inversely related to $[OH^-]$ in any aqueous solution (§18.2)
8. How the relative magnitudes of $[H_3O^+]$ and $[OH^-]$ define whether a solution is acidic, basic, or neutral (§18.2)
9. The Brønsted-Lowry definitions of an acid and a base and how an acid-base reaction can be viewed as a proton-transfer process (§18.3)
10. How water acts as a base (or as an acid) when an acid (or a base) dissolves in it (§18.3)
11. How a conjugate acid-base pair differs by one proton (§18.3)
12. How a Brønsted-Lowry acid-base reaction involves two conjugate acid-base pairs (§18.3)
13. Why a stronger acid and base react ($K_c > 1$) to form a weaker base and acid (§18.3)
14. How the percent dissociation of a weak acid increases as its concentration decreases (§18.4)
15. How a polyprotic acid dissociates in two or more steps and why only the first step supplies significant $[H_3O^+]$ (§18.4)
16. The effects of electronegativity, bond polarity, and bond strength on acid strength (§18.5)
17. Why aqueous solutions of small, highly charged metal ions are acidic (§18.5)
18. How weak bases in water accept a proton rather than dissociate; the meaning of K_b and pK_b (§18.6)
19. How ammonia, amines, and weak-acid anions act as weak bases in water (§18.6)
20. Why relative concentrations of HA and A^- determine the acidity or basicity of their solutions (§18.6)
21. The relationship of the K_a and K_b of a conjugate acid-base pair to K_w (§18.6)
22. The various combinations of cations and anions that lead to acidic, basic, or neutral salt solutions (§18.7)
23. Why the strengths of strong acids are leveled in water but differ in a less basic solvent (§18.8)
24. The Lewis definitions of an acid and a base and how a Lewis acid-base reaction involves the donation and acceptance of an electron pair to form a covalent bond (§18.9)
25. How molecules with electron-deficient atoms, molecules with polar multiple bonds, and metal cations act as Lewis acids (§18.9)

Master These Skills

1. Classifying strong and weak acids and bases from their formulas (SP 18.1)
2. Using K_w to calculate $[H_3O^+]$ or $[OH^-]$ in an aqueous solution (SP 18.2)
3. Using p-scales to express $[H_3O^+]$, $[OH^-]$, and K (§18.2)
4. Calculating $[H_3O^+]$, pH, $[OH^-]$, and pOH (SP 18.3)
5. Identifying conjugate acid-base pairs (SP 18.4)
6. Using relative acid strengths to predict the net direction of an acid-base reaction (SPs 18.5, 18.6)
7. Calculating K_a of a weak acid from pH (SP 18.7)
8. Calculating $[H_3O^+]$ (and, thus, pH) from K_a and $[HA]_{init}$ (SP 18.8)
9. Applying the quadratic equation to find a concentration (Follow-Up Problem 18.8A)
10. Calculating the percent dissociation of a weak acid (§18.4 and SP 18.9)
11. Calculating $[H_3O^+]$ and other concentrations for a polyprotic acid (SP 18.10)
12. Predicting relative acid strengths of nonmetal hydrides and oxoacids (§18.5)
13. Calculating pH from K_b and $[B]_{init}$ (SP 18.11)
14. Finding K_b of A^- from K_a of HA and K_w (§18.6 and SP 18.12)
15. Calculating pH from K_b of A^- and $[A^-]_{init}$ (SP 18.12)
16. Predicting the relative acidity of a salt solution from the nature of the cation and anion (SPs 18.13, 18.14)
17. Identifying Lewis acids and bases (SP 18.15)

Key Terms

Page numbers appear in parentheses.

acid-base indicator (803)
acid-dissociation (acid-ionization) constant (K_a) (796)
adduct (828)
Arrhenius acid-base definition (794)
autoionization (798)
base-dissociation (base-ionization) constant (K_b) (818)
Brønsted-Lowry acid-base definition (803)
conjugate acid-base pair (804)
hydronium ion, H_3O^+ (794)
ion-product constant for water (K_w) (799)
leveling effect (827)
Lewis acid-base definition (828)
neutralization (794)
pH (800)
polyprotic acid (813)
proton acceptor (803)
proton donor (803)

Key Equations and Relationships

Page numbers appear in parentheses.

18.1 Defining the acid-dissociation constant (796):

$$K_a = \frac{[H_3O^+][A^-]}{[HA]}$$

18.2 Defining the ion-product constant for water (799):

$$K_w = [H_3O^+][OH^-] = 1.0\times10^{-14} \text{ (at 25°C)}$$

18.3 Defining pH (800):

$$pH = -\log [H_3O^+]$$

18.4 Relating pK_w to pH and pOH (801):

$$pK_w = pH + pOH = 14.00 \text{ (at 25°C)}$$

18.5 Finding the percent dissociation of HA (812):

$$\text{Percent HA dissociated} = \frac{[HA]_{dissoc}}{[HA]_{init}} \times 100$$

18.6 Defining the base-dissociation constant (818):

$$K_b = \frac{[BH^+][OH^-]}{[B]}$$

18.7 Expressing the relationship among K_a, K_b, and K_w (821):

$$K_a \times K_b = K_w$$

BRIEF SOLUTIONS TO FOLLOW-UP PROBLEMS

18.1A (a) $HClO_3$; number of O atoms exceeds number of H atoms by two.
(b) HCl; one of the strong hydrohalic acids.
(c) NaOH; one of the Group 1A(1) hydroxides.

18.1B (a) Weak base; this is an amine with a lone electron pair on the N.
(b) Strong acid; one of the strong hydrohalic acids.
(c) Weak acid; number of O atoms equals the number of H atoms.

$$HBrO(aq) + H_2O(l) \rightleftharpoons H_3O^+(aq) + BrO^-(aq)$$

$$K_a = \frac{[H_3O^+][BrO^-]}{[HBrO]}$$

(d) Strong base; Group 2A(2) metal hydroxide with Ca as the metal.

18.2A $[H_3O^+] = \frac{K_w}{[OH^-]} = \frac{1.0\times10^{-14}}{6.7\times10^{-2}} = 1.5\times10^{-13}\ M$; basic

18.2B $[OH^-] = \frac{K_w}{[H_3O^+]} = \frac{1.0\times10^{-14}}{1.8\times10^{-10}} = 5.6\times10^{-5}\ M$; basic

18.3A $pH = 14.00 - pOH = 14.00 - 4.48 = 9.52$
$[OH^-] = 10^{-pOH} = 10^{-4.48} = 3.3\times10^{-5}\ M$

$$[H_3O^+] = \frac{K_w}{[OH^-]} = \frac{1.0\times10^{-14}}{3.3\times10^{-5}} = 3.0\times10^{-10}\ M$$

18.3B $pOH = 14.00 - pH = 14.00 - 2.28 = 11.72$
$[H_3O^+] = 10^{-pH} = 10^{-2.28} = 5.2\times10^{-3}\ M$

$$[OH^-] = \frac{K_w}{[H_3O^+]} = \frac{1.0\times10^{-14}}{5.2\times10^{-3}} = 1.9\times10^{-12}\ M$$

18.4A (a) CH_3COOH/CH_3COO^- and H_3O^+/H_2O
(b) H_2O/OH^- and HF/F^-

18.4B (a) H_2SO_3 (b) C_5H_5N (c) HCO_3^- (d) CN^-

18.5A (a) $H_2SO_3(aq) + CO_3^{2-}(aq) \rightleftharpoons HSO_3^-(aq) + HCO_3^-(aq)$
stronger acid + stronger base ⟶ weaker base + weaker acid

(b) $HCN(aq) + F^-(aq) \rightleftharpoons CN^-(aq) + HF(aq)$
weaker acid + weaker base ⟵ stronger base + stronger acid

18.5B (a) $NH_3(g) + H_2O(l) \rightleftharpoons NH_4^+(aq) + OH^-(aq)$
(b) $NH_3(g) + H_3O^+(aq;\text{ from HCl}) \longrightarrow NH_4^+(aq) + H_2O(l)$
(c) $NH_4^+(aq) + OH^-(aq;\text{ from NaOH}) \longrightarrow NH_3(g) + H_2O(l)$

18.6A There are more HB molecules than HA, so $K_c > 1$ and HA is the stronger acid.

18.6B Since HD is the stronger acid, there should be more D^- particles than C^- particles: thus, D^- is black (5 particles) and C^- is green (3 particles).
$K_c < 1$ for $HC(aq) + D^-(aq) \rightleftharpoons HD(aq) + C^-(aq)$

18.7A $NH_4^+(aq) + H_2O(l) \rightleftharpoons NH_3(aq) + H_3O^+(aq)$
$[H_3O^+] = 10^{-pH} = 10^{-5.0} = 1\times10^{-5}\ M = [NH_3]$
And, $[NH_4^+] = 0.2\ M - (1\times10^{-5}\ M) \approx 0.2\ M$

$$K_a = \frac{[H_3O^+][NH_3]}{[NH_4^+]} \approx \frac{(1\times10^{-5})^2}{0.2} = 5\times10^{-10}$$

18.7B $CH_2CHCOOH(aq) + H_2O(l) \rightleftharpoons CH_2CHCOO^-(aq) + H_3O^+(aq)$
$[H_3O^+] = 10^{-pH} = 10^{-2.43} = 0.0037\ M = [CH_2CHCOO^-]$

$$K_a = \frac{[H_3O^+][CH_2CHCOO^-]}{[CH_2CHCOOH]} \approx \frac{(0.0037)^2}{0.30} = 4.6\times10^{-5}$$

18.8A $\frac{[HOCN]_{init}}{K_a} = \frac{0.10}{3.5\times10^{-4}} = 286 < 400$
You must a solve a quadratic equation:

$$K_a = \frac{[H_3O^+][OCN^-]}{[HOCN]} = 3.5\times10^{-4} = \frac{(x)(x)}{0.10 - x}$$

$x^2 + (3.5\times10^{-4})x - (3.5\times10^{-5}) = 0$
$x = [H_3O^+] = 5.7\times10^{-3}\ M$
$pH = -\log(5.7\times10^{-3}) = 2.24$

18.8B $K_a = 10^{-pKa} = 10^{-4.20} = 6.3\times10^{-5}$

$$\frac{[C_6H_5COOH]_{init}}{K_a} = \frac{0.25}{6.3\times10^{-5}} = 3968 > 400$$

You may assume that $0.25 - x = 0.25\ M$:

$$K_a = \frac{[H_3O^+][C_6H_5COO^-]}{[C_6H_5COOH]} = 6.3\times10^{-5} = \frac{(x)(x)}{0.25 - x} \approx \frac{x^2}{0.25}$$

$x = [H_3O^+] = 4.0\times10^{-3}\ M$

$pH = -\log [H_3O^+] = -\log [4.0\times10^{-3}] = 2.40$

18.9A $\frac{[HCN]_{init}}{K_a} = \frac{0.75}{6.2\times10^{-10}} = 1.2\times10^9 > 400$

You may assume that $0.75 - x = 0.75$:

$$K_a = \frac{[H_3O^+][CN^-]}{[HCN]} = 6.2\times10^{-10} = \frac{(x)(x)}{0.75 - x} \approx \frac{x^2}{0.75}$$

$x = 2.2\times10^{-5}\ M = [HCN]_{dissoc}$

$$\text{Percent dissociation} = \frac{[HCN]_{dissoc}}{[HCN]_{init}} \times 100 = \frac{2.2\times10^{-5}}{0.75} \times 100 = 0.0029\%$$

18.9B $3.16\% = \frac{[HA]_{dissoc}}{[HA]_{init}} \times 100 = \frac{x}{1.5} \times 100$

$x = 0.047\ M = [HA]_{dissoc} = [H_3O^+] = [A^-]$

$$K_a = \frac{[H_3O^+][A^-]}{[HA]} = \frac{(0.047)(0.047)}{1.5 - 0.047} = 1.5\times10^{-3}$$

18.10A $\frac{[H_2C_2O_4]_{init}}{K_{a1}} = \frac{0.150}{5.6\times10^{-2}} = 2.7 < 400$

You must solve a quadratic equation:

$$K_{a1} = \frac{[H_3O^+][HC_2O_4^-]}{[H_2C_2O_4]} = 5.6\times10^{-2} = \frac{(x)(x)}{0.150 - x}$$

$x^2 + (5.6\times10^{-2})x - (8.4\times10^{-3}) = 0$

$x = [H_3O^+] = 0.068\ M$; $pH = -\log (0.068) = 1.17$

$x = [HC_2O_4^-] = 0.068\ M$; $[H_2C_2O_4] = 0.150\ M - x = 0.082\ M$

$$K_{a2} = \frac{[H_3O^+][C_2O_4^{2-}]}{[HC_2O_4^-]}$$

$$[C_2O_4^{2-}] = \frac{K_{a2}[HC_2O_4^-]}{[H_3O^+]} = \frac{(5.4\times10^{-5})(0.068)}{0.068} = 5.4\times10^{-5}\ M$$

18.10B $\frac{[H_2CO_3]_{init}}{K_{a1}} = \frac{0.075}{4.5\times10^{-7}} = 1.7\times10^5 > 400$

You may assume that $0.075 - x = 0.075\ M$:

$$K_{a1} = \frac{[H_3O^+][HCO_3^-]}{[H_2CO_3]} = 4.5\times10^{-7} = \frac{(x)(x)}{0.075 - x} \approx \frac{x^2}{0.075}$$

$x = [H_3O^+] = 1.8\times10^{-4}\ M$; $pH = 3.74$

$x = [HCO_3^-] = 1.8\times10^{-4}\ M$; $[H_2CO_3] = 0.075\ M - x = 0.075\ M$

$$K_{a2} = \frac{[H_3O^+][CO_3^{2-}]}{[HCO_3^-]}$$

$$[CO_3^{2-}] = \frac{K_{a2}[HCO_3^-]}{[H_3O^+]} = \frac{(4.7\times10^{-11})(1.8\times10^{-4})}{1.8\times10^{-4}} = 4.7\times10^{-11}\ M$$

18.11A $K_b = 10^{-pK_b} = 10^{-8.77} = 1.7\times10^{-9}$

$$\frac{[C_5H_5N]_{init}}{K_b} = \frac{0.10}{1.7\times10^{-9}} = 5.9\times10^7 > 400$$

You may assume that $0.10 - x = 0.10\ M$:

$$K_b = \frac{[C_5H_5NH^+][OH^-]}{[C_5H_5N]} = 1.7\times10^{-9} = \frac{(x)(x)}{0.10 - x} \approx \frac{x^2}{0.10}$$

$x = [OH^-] = 1.3\times10^{-5}\ M$;

$$[H_3O^+] = \frac{K_w}{[OH^-]} = \frac{1.0\times10^{-14}}{1.3\times10^{-5}} = 7.7\times10^{-10}\ M$$

$pH = -\log [H_3O^+] = -\log (7.7\times10^{-10}) = 9.11$

18.11B $\frac{[C_6H_5CH_2CH(CH_3)NH_2]_{init}}{K_b} = \frac{0.075}{6.3\times10^{-5}} = 1200 > 400$

You may assume that $0.075 - x = 0.075\ M$:

$$K_b = \frac{[C_6H_5CH_2CH(CH_3)NH_3^+][OH^-]}{[C_6H_5CH_2CH(CH_3)NH_2]} = 6.3\times10^{-5} = \frac{(x)(x)}{0.075 - x} \approx \frac{x^2}{0.075}$$

$x = [OH^-] = 2.2\times10^{-3}\ M$;

$$[H_3O^+] = \frac{K_w}{[OH^-]} = \frac{1.0\times10^{-14}}{2.2\times10^{-3}} = 4.5\times10^{-12}\ M$$

$pH = -\log [H_3O^+] = -\log (4.5\times10^{-12}) = 11.35$

18.12A $K_b \text{ of } ClO^- = \frac{K_w}{K_a \text{ of } HClO} = \frac{1.0\times10^{-14}}{2.9\times10^{-8}} = 3.4\times10^{-7}$

Assuming $0.20\ M - x \approx 0.20\ M$,

$$K_b = \frac{[HClO][OH^-]}{[ClO^-]} = 3.4\times10^{-7} = \frac{(x)(x)}{0.20 - x} \approx \frac{x^2}{0.20}$$

$x = [OH^-] = 2.6\times10^{-4}\ M$; $[H_3O^+] = 3.8\times10^{-11}\ M$; $pH = 10.42$

18.12B $K_b \text{ of } NO_2^- = \frac{K_w}{K_a \text{ of } HNO_2} = \frac{1.0\times10^{-14}}{7.1\times10^{-4}} = 1.4\times10^{-11}$

Assuming $0.80\ M - x \approx 0.80\ M$,

$$K_b = \frac{[HNO_2][OH^-]}{[NO_2^-]} = 1.4\times10^{-11} = \frac{(x)(x)}{0.80 - x} \approx \frac{x^2}{0.80}$$

$x = [OH^-] = 3.3\times10^{-6}\ M$; $[H_3O^+] = 3.0\times10^{-9}\ M$; $pH = 8.52$

18.13A (a) Basic: K^+ is from the strong base KOH, so it does not react with water; ClO_2^- is the anion of the weak acid $HClO_2$ and reacts with water to produce OH^-:

$$ClO_2^-(aq) + H_2O(l) \rightleftharpoons HClO(aq) + OH^-(aq)$$

(b) Acidic: NO_3^- is from the strong acid HNO_3, so it does not react with water; $CH_3NH_3^+$ is the cation of the weak base CH_3NH_2 and reacts with water to produce H_3O^+:

$$CH_3NH_3^+(aq) + H_2O(l) \rightleftharpoons CH_3NH_2(aq) + H_3O^+(aq)$$

(c) Neutral: Rb^+ is from the strong base RbOH; Br^- is from the strong acid HBr. Neither ion reacts with water.

18.13B (a) Acidic: Br^- is from the strong acid HBr, so it does not react with water; the small, highly charged Fe^{3+} ion forms $Fe(H_2O)_6^{3+}$, which reacts with water to produce H_3O^+:

$$Fe(H_2O)_6^{3+}(aq) + H_2O(l) \rightleftharpoons Fe(H_2O)_5OH^{2+}(aq) + H_3O^+(aq)$$

(b) Basic: Ca^{2+} is from the strong base $Ca(OH)_2$, so it does not react with water; NO_2^- is the anion of the weak acid HNO_2 and reacts with water to produce OH^-:

$$NO_2^-(aq) + H_2O(l) \rightleftharpoons HNO_2(aq) + OH^-(aq)$$

(c) Acidic: I^- is from the strong acid HI, so it does not react with water; $C_6H_5NH_3^+$ is the cation of the weak base $C_6H_5NH_2$ and reacts with water to produce H_3O^+:

$$C_6H_5NH_3^+(aq) + H_2O(l) \rightleftharpoons C_6H_5NH_2(aq) + H_3O^+(aq)$$

18.14A (a) From Appendix C, K_a of $Cu(H_2O)_6^{2+} = 3\times10^{-8}$

$$K_b \text{ of } CH_3COO^- = \frac{K_w}{K_a \text{ of } CH_3COOH} = \frac{1.0\times10^{-14}}{1.8\times10^{-5}} = 5.6\times10^{-10}$$

Since $K_a > K_b$, $Cu(CH_3COO)_2(aq)$ is acidic.

(b) $K_a \text{ of } NH_4^+ = \frac{K_w}{K_b \text{ of } NH_3} = \frac{1.0\times10^{-14}}{1.76\times10^{-5}} = 5.7\times10^{-10}$

$$K_b \text{ of } F^- = \frac{K_w}{K_a \text{ of } HF} = \frac{1.0\times10^{-14}}{6.8\times10^{-4}} = 1.5\times10^{-11}$$

Because $K_a > K_b$, $NH_4F(aq)$ is acidic.

BRIEF SOLUTIONS TO FOLLOW-UP PROBLEMS (continued)

(c) From Appendix C, K_a of $HC_6H_6O_6^- = 5\times10^{-12}$

$$K_b \text{ of } HC_6H_6O_6^- = \frac{K_w}{K_a \text{ of } H_2C_6H_6O_6} = \frac{1.0\times10^{-14}}{1.0\times10^{-5}} = 1.0\times10^{-9}$$

Because $K_b > K_a$, $KHC_6H_6O_6(aq)$ is basic.

18.14B (a) From Appendix C, K_a of $HCO_3^- = 4.7\times10^{-11}$

$$K_b \text{ of } HCO_3^- = \frac{K_w}{K_a \text{ of } H_2CO_3} = \frac{1.0\times10^{-14}}{4.5\times10^{-7}} = 2.2\times10^{-8}$$

Because $K_b > K_a$, $NaHCO_3(aq)$ is basic.

(b) $$K_a \text{ of } C_6H_5NH_3^+ = \frac{K_w}{K_b \text{ of } C_6H_5NH_2} = \frac{1.0\times10^{-14}}{4.0\times10^{-10}} = 2.5\times10^{-5}$$

$$K_b \text{ of } NO_2^- = \frac{K_w}{K_a \text{ of } HNO_2} = \frac{1.0\times10^{-14}}{7.1\times10^{-4}} = 1.4\times10^{-11}$$

Because $K_a > K_b$, $C_6H_5NH_3NO_2(aq)$ is acidic.

(c) From Appendix C, K_a of $H_2PO_4^- = 6.3\times10^{-8}$

$$K_b \text{ of } H_2PO_4^- = \frac{K_w}{K_a \text{ of } H_3PO_4} = \frac{1.0\times10^{-14}}{7.2\times10^{-3}} = 1.4\times10^{-12}$$

Because $K_a > K_b$, $NaH_2PO_4(aq)$ is acidic.

18.15A (a) OH^- is the Lewis base; $Al(OH)_3$ is the Lewis acid.
(b) H_2O is the Lewis base; SO_3 is the Lewis acid.
(c) NH_3 is the Lewis base; Co^{3+} is the Lewis acid.

18.15B (a) H_2O is the Lewis base; $B(OH)_3$ is the Lewis acid.
(b) I^- is the Lewis base; Cd^{2+} is the Lewis acid.
(c) F^- is the Lewis base; SiF_4 is the Lewis acid.

PROBLEMS

Problems with **colored** numbers are answered in Appendix E and worked in detail in the Student Solutions Manual. Problem sections match those in the text and give the numbers of relevant sample problems. Most offer Concept Review Questions, Skill-Building Exercises (grouped in pairs covering the same concept), and Problems in Context. The Comprehensive Problems are based on material from any section or previous chapter.

Note: Unless stated otherwise, all problems refer to aqueous solutions at 298 K (25°C).

Acids and Bases in Water

(Sample Problem 18.1)

Concept Review Questions

18.1 What is the role of water in the Arrhenius acid-base definition?

18.2 What do Arrhenius acids have in common? What do Arrhenius bases have in common? Explain neutralization in terms of the Arrhenius acid-base definition. What data led Arrhenius to propose this idea of neutralization?

18.3 Why is the Arrhenius acid-base definition too limited? Give an example for which the Arrhenius definition does not apply.

18.4 What do "strong" and "weak" mean for acids and bases? The K_a values of weak acids vary over more than 10 orders of magnitude. What do the acids have in common that makes them "weak"?

Skill-Building Exercises (grouped in similar pairs)

18.5 Which of the following are Arrhenius acids?
(a) H_2O (b) $Ca(OH)_2$ (c) H_3PO_3 (d) HI

18.6 Which of the following are Arrhenius acids?
(a) $NaHSO_4$ (b) CH_4 (c) NaH (d) H_3N

18.7 Which of the following are Arrhenius bases?
(a) H_3AsO_4 (b) $Ba(OH)_2$ (c) HClO (d) KOH

18.8 Which of the following are Arrhenius bases?
(a) CH_3COOH (b) HOH (c) CH_3OH (d) H_2NNH_2

18.9 Write the K_a expression for each of the following in water:
(a) HCN (b) HCO_3^- (c) HCOOH

18.10 Write the K_a expression for each of the following in water:
(a) $CH_3NH_3^+$ (b) HClO (c) H_2S

18.11 Write the K_a expression for each of the following in water:
(a) HNO_2 (b) CH_3COOH (c) $HBrO_2$

18.12 Write the K_a expression for each of the following in water:
(a) $H_2PO_4^-$ (b) H_3PO_2 (c) HSO_4^-

18.13 Use Appendix C to rank the following in order of *increasing* acid strength: HIO_3, HI, CH_3COOH, HF.

18.14 Use Appendix C to rank the following in order of *decreasing* acid strength: HClO, HCl, HCN, HNO_2.

18.15 Classify each as a strong or weak acid or base:
(a) H_3AsO_4 (b) $Sr(OH)_2$ (c) HIO (d) $HClO_4$

18.16 Classify each as a strong or weak acid or base:
(a) CH_3NH_2 (b) K_2O (c) HI (d) HCOOH

18.17 Classify each as a strong or weak acid or base:
(a) RbOH (b) HBr (c) H_2Te (d) HClO

18.18 Classify each as a strong or weak acid or base:
(a) $HOCH_2CH_2NH_2$ (b) H_2SeO_4 (c) HS^- (d) $B(OH)_3$

Autoionization of Water and the pH Scale

(Sample Problems 18.2 and 18.3)

Concept Review Questions

18.19 What is an autoionization reaction? Write equations for the autoionization reactions of H_2O and of H_2SO_4.

18.20 What is the difference between K_c and K_w for the autoionization of water?

18.21 (a) What is the change in pH when $[OH^-]$ increases by a factor of 10? (b) What is the change in $[H_3O^+]$ when the pH decreases by 3 units?

18.22 Which solution has the higher pH? Explain.
(a) A 0.1 M solution of an acid with $K_a = 1\times10^{-4}$ or one with $K_a = 4\times10^{-5}$
(b) A 0.1 M solution of an acid with $pK_a = 3.0$ or one with $pK_a = 3.5$
(c) A 0.1 M solution or a 0.01 M solution of a weak acid
(d) A 0.1 M solution of a weak acid or a 0.1 M solution of a strong acid
(e) A 0.1 M solution of an acid or a 0.01 M solution of a base
(f) A solution of pOH 6.0 or one of pOH 8.0

Skill-Building Exercises (grouped in similar pairs)

18.23 (a) What is the pH of 0.0111 M NaOH? Is the solution neutral, acidic, or basic? (b) What is the pOH of 1.35×10^{-3} M HCl? Is the solution neutral, acidic, or basic?

18.24 (a) What is the pH of 0.0333 M HNO_3? Is the solution neutral, acidic, or basic? (b) What is the pOH of 0.0347 M KOH? Is the solution neutral, acidic, or basic?

18.25 (a) What is the pH of 6.14×10^{-3} M HI? Is the solution neutral, acidic, or basic? (b) What is the pOH of 2.55 M $Ba(OH)_2$? Is the solution neutral, acidic, or basic?

18.26 (a) What is the pH of 7.52×10^{-4} M CsOH? Is the solution neutral, acidic, or basic? (b) What is the pOH of 1.59×10^{-3} M $HClO_4$? Is the solution neutral, acidic, or basic?

18.27 (a) What are $[H_3O^+]$, $[OH^-]$, and pOH in a solution with a pH of 9.85? (b) What are $[H_3O^+]$, $[OH^-]$, and pH in a solution with a pOH of 9.43?

18.28 (a) What are $[H_3O^+]$, $[OH^-]$, and pOH in a solution with a pH of 3.47? (b) What are $[H_3O^+]$, $[OH^-]$, and pH in a solution with a pOH of 4.33?

18.29 (a) What are $[H_3O^+]$, $[OH^-]$, and pOH in a solution with a pH of 4.77? (b) What are $[H_3O^+]$, $[OH^-]$, and pH in a solution with a pOH of 5.65?

18.30 (a) What are $[H_3O^+]$, $[OH^-]$, and pOH in a solution with a pH of 8.97? (b) What are $[H_3O^+]$, $[OH^-]$, and pH in a solution with a pOH of 11.27?

18.31 How many moles of H_3O^+ or OH^- must you add to a liter of strong acid solution to adjust its pH from 3.15 to 3.65? Assume a negligible volume change.

18.32 How many moles of H_3O^+ or OH^- must you add to a liter of strong base solution to adjust its pH from 9.33 to 9.07? Assume a negligible volume change.

18.33 How many moles of H_3O^+ or OH^- must you add to 5.6 L of strong acid solution to adjust its pH from 4.52 to 5.25? Assume a negligible volume change.

18.34 How many moles of H_3O^+ or OH^- must you add to 87.5 mL of strong base solution to adjust its pH from 8.92 to 8.45? Assume a negligible volume change.

Problems in Context

18.35 The two molecular scenes shown depict the relative concentrations of H_3O^+ *(purple)* in solutions of the same volume (with counter ions and solvent molecules omitted for clarity). If the pH in scene A is 4.8, what is the pH in scene B?

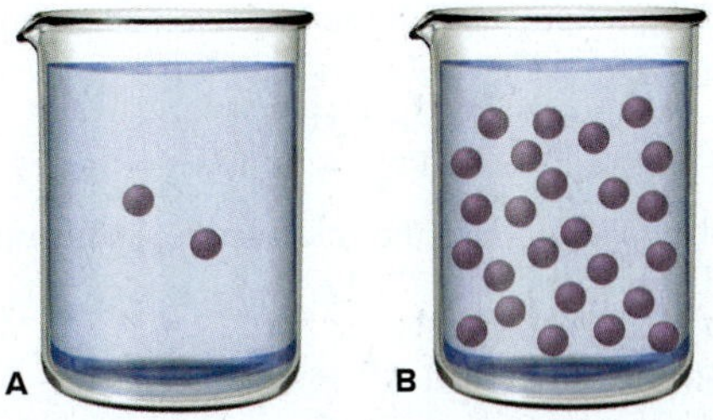

18.36 Like any equilibrium constant, K_w changes with temperature. (a) Given that autoionization is endothermic, how does K_w change with rising T? Explain with a reaction that includes heat as reactant or product. (b) In many medical applications, the value of K_w at 37°C (body temperature) may be more appropriate than the value at 25°C, 1.0×10^{-14}. The pH of pure water at 37°C is 6.80. Calculate K_w, pOH, and $[OH^-]$ at this temperature.

Proton Transfer and the Brønsted-Lowry Acid-Base Definition
(Sample Problems 18.4 to 18.6)

Concept Review Questions

18.37 How are the Arrhenius and Brønsted-Lowry acid-base definitions different? How are they similar? Name two Brønsted-Lowry bases that are not Arrhenius bases. Can you do the same for acids? Explain.

18.38 What is a conjugate acid-base pair? What is the relationship between the two members of the pair?

18.39 (a) A Brønsted-Lowry acid-base reaction proceeds in the net direction in which a stronger acid and stronger base form a weaker acid and weaker base. Explain.
(b) The molecular scene at right depicts an aqueous solution of two conjugate acid-base pairs: HA/A^- and HB/B^-. The base in the first pair is represented by red spheres, and the base in the second pair by green spheres; solvent molecules are omitted for clarity. Which is the stronger acid? The stronger base? Explain.

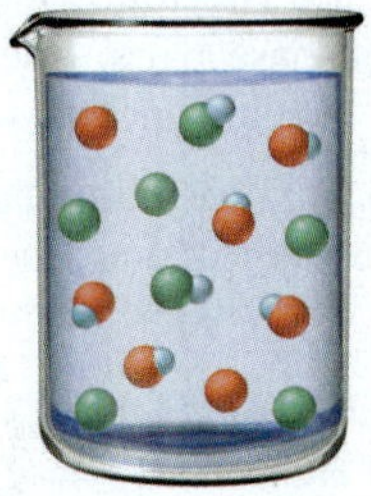

18.40 What is an amphiprotic species? Name one and write balanced equations that show why it is amphiprotic.

Skill-Building Exercises (grouped in similar pairs)

18.41 Write balanced equations and K_a expressions for these Brønsted-Lowry acids in water:
(a) H_3PO_4 (b) C_6H_5COOH (c) HSO_4^-

18.42 Write balanced equations and K_a expressions for these Brønsted-Lowry acids in water:
(a) HCOOH (b) $HClO_3$ (c) $H_2AsO_4^-$

18.43 Give the formula of the conjugate base:
(a) HCl (b) H_2CO_3 (c) H_2O

18.44 Give the formula of the conjugate base:
(a) HPO_4^{2-} (b) NH_4^+ (c) HS^-

18.45 Give the formula of the conjugate acid:
(a) NH_3 (b) NH_2^- (c) nicotine, $C_{10}H_{14}N_2$

18.46 Give the formula of the conjugate acid:
(a) O^{2-} (b) SO_4^{2-} (c) H_2O

18.47 In each equation, label the acids, bases, and conjugate pairs:
(a) $HCl + H_2O \rightleftharpoons Cl^- + H_3O^+$
(b) $HClO_4 + H_2SO_4 \rightleftharpoons ClO_4^- + H_3SO_4^+$
(c) $HPO_4^{2-} + H_2SO_4 \rightleftharpoons H_2PO_4^- + HSO_4^-$

18.48 In each equation, label the acids, bases, and conjugate pairs:
(a) $NH_3 + HNO_3 \rightleftharpoons NH_4^+ + NO_3^-$
(b) $O^{2-} + H_2O \rightleftharpoons OH^- + OH^-$
(c) $NH_4^+ + BrO_3^- \rightleftharpoons NH_3 + HBrO_3$

18.49 In each equation, label the acids, bases, and conjugate pairs:
(a) $NH_3 + H_3PO_4 \rightleftharpoons NH_4^+ + H_2PO_4^-$
(b) $CH_3O^- + NH_3 \rightleftharpoons CH_3OH + NH_2^-$
(c) $HPO_4^{2-} + HSO_4^- \rightleftharpoons H_2PO_4^- + SO_4^{2-}$

18.50 In each equation, label the acids, bases, and conjugate pairs:
(a) $NH_4^+ + CN^- \rightleftharpoons NH_3 + HCN$
(b) $H_2O + HS^- \rightleftharpoons OH^- + H_2S$
(c) $HSO_3^- + CH_3NH_2 \rightleftharpoons SO_3^{2-} + CH_3NH_3^+$

18.51 Write balanced net ionic equations for the following reactions, and label the conjugate acid-base pairs:
(a) $NaOH(aq) + NaH_2PO_4(aq) \rightleftharpoons H_2O(l) + Na_2HPO_4(aq)$
(b) $KHSO_4(aq) + K_2CO_3(aq) \rightleftharpoons K_2SO_4(aq) + KHCO_3(aq)$

18.52 Write balanced net ionic equations for the following reactions, and label the conjugate acid-base pairs:
(a) $HNO_3(aq) + Li_2CO_3(aq) \rightleftharpoons LiNO_3(aq) + LiHCO_3(aq)$
(b) $2NH_4Cl(aq) + Ba(OH)_2(aq) \rightleftharpoons 2H_2O(l) + BaCl_2(aq) + 2NH_3(aq)$

18.53 The following aqueous species constitute two conjugate acid-base pairs. Use them to write one acid-base reaction with $K_c > 1$ and another with $K_c < 1$: HS^-, Cl^-, HCl, H_2S.

18.54 The following aqueous species constitute two conjugate acid-base pairs. Use them to write one acid-base reaction with $K_c > 1$ and another with $K_c < 1$: NO_3^-, F^-, HF, HNO_3.

18.55 Use Figure 18.8 to determine whether $K_c > 1$ for
(a) $HCl + NH_3 \rightleftharpoons NH_4^+ + Cl^-$
(b) $H_2SO_3 + NH_3 \rightleftharpoons HSO_3^- + NH_4^+$

18.56 Use Figure 18.8 to determine whether $K_c > 1$ for
(a) $OH^- + HS^- \rightleftharpoons H_2O + S^{2-}$
(b) $HCN + HCO_3^- \rightleftharpoons H_2CO_3 + CN^-$

18.57 Use Figure 18.8 to determine whether $K_c < 1$ for
(a) $NH_4^+ + HPO_4^{2-} \rightleftharpoons NH_3 + H_2PO_4^-$
(b) $HSO_3^- + HS^- \rightleftharpoons H_2SO_3 + S^{2-}$

18.58 Use Figure 18.8 to determine whether $K_c < 1$ for
(a) $H_2PO_4^- + F^- \rightleftharpoons HPO_4^{2-} + HF$
(b) $CH_3COO^- + HSO_4^- \rightleftharpoons CH_3COOH + SO_4^{2-}$

Solving Problems Involving Weak-Acid Equilibria
(Sample Problems 18.7 to 18.10)

Concept Review Questions

18.59 In each of the following cases, is the concentration of acid before and after dissociation nearly the same or very different? Explain your reasoning: (a) a concentrated solution of a strong acid; (b) a concentrated solution of a weak acid; (c) a dilute solution of a weak acid; (d) a dilute solution of a strong acid.

18.60 A sample of 0.0001 *M* HCl has $[H_3O^+]$ close to that of a sample of 0.1 *M* CH_3COOH. Are acetic acid and hydrochloric acid equally strong in these samples? Explain.

18.61 A 0.15 *M* solution of HA *(blue and green)* is 33% dissociated. Which of the scenes below represents a sample of that solution after it is diluted with water? (An actual weak acid does not have a percent dissociation this high.)

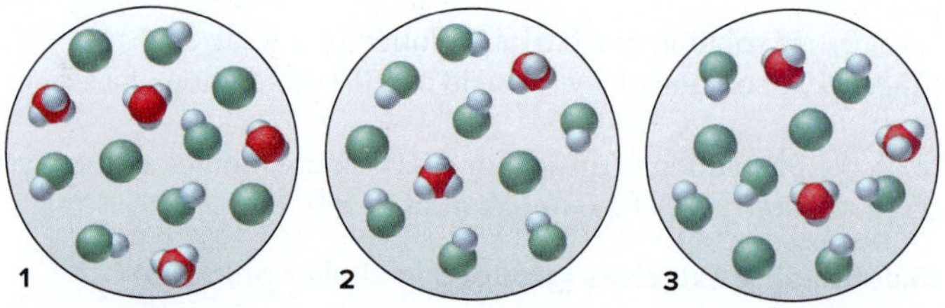

18.62 In which of the following solutions will $[H_3O^+]$ be approximately equal to $[CH_3COO^-]$: (a) 0.1 *M* CH_3COOH; (b) 1×10^{-7} *M* CH_3COOH; (c) a solution containing both 0.1 *M* CH_3COOH and 0.1 *M* CH_3COONa? Explain.

18.63 Why do successive K_a's decrease for all polyprotic acids?

Skill-Building Exercises (grouped in similar pairs)

18.64 A 0.15 *M* solution of butanoic acid, $CH_3CH_2CH_2COOH$, contains 1.51×10^{-3} *M* H_3O^+. What is the K_a of butanoic acid?

18.65 A 0.035 *M* solution of a weak acid (HA) has a pH of 4.88. What is the K_a of the acid?

18.66 Nitrous acid, HNO_2, has a K_a of 7.1×10^{-4}. What are $[H_3O^+]$, $[NO_2^-]$, and $[OH^-]$ in 0.60 *M* HNO_2?

18.67 Hydrofluoric acid, HF, has a K_a of 6.8×10^{-4}. What are $[H_3O^+]$, $[F^-]$, and $[OH^-]$ in 0.75 *M* HF?

18.68 Chloroacetic acid, $ClCH_2COOH$, has a pK_a of 2.87. What are $[H_3O^+]$, pH, $[ClCH_2COO^-]$, and $[ClCH_2COOH]$ in 1.25 *M* $ClCH_2COOH$?

18.69 Hypochlorous acid, HClO, has a pK_a of 7.54. What are $[H_3O^+]$, pH, $[ClO^-]$, and [HClO] in 0.115 *M* HClO?

18.70 In a 0.20 *M* solution, a weak acid is 3.0% dissociated.
(a) Calculate the $[H_3O^+]$, pH, $[OH^-]$, and pOH of the solution.
(b) Calculate K_a of the acid.

18.71 In a 0.735 *M* solution, a weak acid is 12.5% dissociated.
(a) Calculate the $[H_3O^+]$, pH, $[OH^-]$, and pOH of the solution.
(b) Calculate K_a of the acid.

18.72 A 0.250-mol sample of HX is dissolved in enough H_2O to form 655 mL of solution. If the pH of the solution is 3.54, what is the K_a of HX?

18.73 A 4.85×10^{-3}-mol sample of HY is dissolved in enough H_2O to form 0.095 L of solution. If the pH of the solution is 2.68, what is the K_a of HY?

18.74 The weak acid HZ has a K_a of 2.55×10^{-4}.
(a) Calculate the pH of 0.075 *M* HZ.
(b) Calculate the pOH of 0.045 *M* HZ.

18.75 The weak acid HQ has a pK_a of 4.89.
(a) Calculate the $[H_3O^+]$ of 3.5×10^{-2} *M* HQ.
(b) Calculate the $[OH^-]$ of 0.65 *M* HQ.

18.76 (a) Calculate the pH of 0.175 *M* HY, if $K_a = 1.50\times10^{-4}$.
(b) Calculate the pOH of 0.175 *M* HX, if $K_a = 2.00\times10^{-2}$.

18.77 (a) Calculate the pH of 0.55 *M* HCN, if $K_a = 6.2\times10^{-10}$.
(b) Calculate the pOH of 0.044 *M* HIO_3, if $K_a = 0.16$.

18.78 Use Appendix C to calculate the percent dissociation of 0.55 M benzoic acid, C_6H_5COOH.

18.79 Use Appendix C to calculate the percent dissociation of 0.050 M CH_3COOH.

18.80 Use Appendix C to calculate [H_2S], [HS^-], [S^{2-}], [H_3O^+], pH, [OH^-], and pOH in a 0.10 M solution of the diprotic acid hydrosulfuric acid.

18.81 Use Appendix C to calculate [$H_2C_3H_2O_4$], [$HC_3H_2O_4^-$], [$C_3H_2O_4^{2-}$], [H_3O^+], pH, [OH^-], and pOH in a 0.200 M solution of the diprotic acid malonic acid.

Problems in Context

18.82 Acetylsalicylic acid (aspirin), $HC_9H_7O_4$, is the most widely used pain reliever and fever reducer. Find the pH of 0.018 M aqueous aspirin at body temperature (K_a at 37°C = 3.6×10^{-4}).

18.83 Formic acid, HCOOH, the simplest carboxylic acid, is used in the textile and rubber industries and is secreted as a defense by many species of ants (family *Formicidae*). Calculate the percent dissociation of 0.75 M HCOOH.

Molecular Properties and Acid Strength

Concept Review Questions

18.84 Across a period, how does the electronegativity of a nonmetal affect the acidity of its binary hydride?

18.85 How does the atomic size of a nonmetal affect the acidity of its binary hydride?

18.86 A strong acid has a weak bond to its acidic proton, whereas a weak acid has a strong bond to its acidic proton. Explain.

18.87 Perchloric acid, $HClO_4$, is the strongest of the halogen oxoacids, and hypoiodous acid, HIO, is the weakest. What two factors govern this difference in acid strength?

Skill-Building Exercises (grouped in similar pairs)

18.88 Choose the *stronger* acid in each of the following pairs:
(a) H_2SeO_3 or H_2SeO_4 (b) H_3PO_4 or H_3AsO_4 (c) H_2S or H_2Te

18.89 Choose the *weaker* acid in each of the following pairs:
(a) HBr or H_2Se (b) $HClO_4$ or H_2SO_4 (c) H_2SO_3 or H_2SO_4

18.90 Choose the *stronger* acid in each of the following pairs:
(a) H_2Se or H_3As (b) $B(OH)_3$ or $Al(OH)_3$ (c) $HBrO_2$ or HBrO

18.91 Choose the *weaker* acid in each of the following pairs:
(a) HI or HBr (b) H_3AsO_4 or H_2SeO_4 (c) HNO_3 or HNO_2

18.92 Use Appendix C to choose the solution with the *lower* pH:
(a) 0.5 M $CuBr_2$ or 0.5 M $AlBr_3$
(b) 0.3 M $ZnCl_2$ or 0.3 M $SnCl_2$

18.93 Use Appendix C to choose the solution with the *lower* pH:
(a) 0.1 M $FeCl_3$ or 0.1 M $AlCl_3$
(b) 0.1 M $BeCl_2$ or 0.1 M $CaCl_2$

18.94 Use Appendix C to choose the solution with the *higher* pH:
(a) 0.2 M $Ni(NO_3)_2$ or 0.2 M $Co(NO_3)_2$
(b) 0.35 M $Al(NO_3)_3$ or 0.35 M $Cr(NO_3)_3$

18.95 Use Appendix C to choose the solution with the *higher* pH:
(a) 0.1 M $NiCl_2$ or 0.1 M NaCl
(b) 0.1 M $Sn(NO_3)_2$ or 0.1 M $Co(NO_3)_2$

Weak Bases and Their Relation to Weak Acids

(Sample Problems 18.11 and 18.12)

Concept Review Questions

18.96 What is the key structural feature of all Brønsted-Lowry bases? How does this feature function in an acid-base reaction?

18.97 Why are most anions basic in H_2O? Give formulas of four anions that are not basic.

18.98 Except for the Na^+ spectator ion, aqueous solutions of CH_3COOH and CH_3COONa contain the same species. (a) What are the species (other than H_2O)? (b) Why is 0.1 M CH_3COOH acidic and 0.1 M CH_3COONa basic?

Skill-Building Exercises (grouped in similar pairs)

18.99 Write balanced equations and K_b expressions for these Brønsted-Lowry bases in water:
(a) Pyridine, C_5H_5N (b) CO_3^{2-}

18.100 Write balanced equations and K_b expressions for these Brønsted-Lowry bases in water:
(a) Benzoate ion, $C_6H_5COO^-$ (b) $(CH_3)_3N$

18.101 Write balanced equations and K_b expressions for these Brønsted-Lowry bases in water:
(a) Hydroxylamine, HO—NH_2 (b) HPO_4^{2-}

18.102 Write balanced equations and K_b expressions for these Brønsted-Lowry bases in water:
(a) Guanidine, $(H_2N)_2C{=}NH$ (the double-bonded N is more basic)
(b) Acetylide ion, $HC{\equiv}C^-$

18.103 What is the pH of 0.070 M dimethylamine?

18.104 What is the pH of 0.12 M diethylamine?

18.105 What is the pH of 0.25 M ethanolamine?

18.106 What is the pH of 0.26 M aniline?

18.107 (a) What is the K_b of the acetate ion?
(b) What is the K_a of the anilinium ion, $C_6H_5NH_3^+$?

18.108 (a) What is the K_b of the benzoate ion, $C_6H_5COO^-$?
(b) What is the K_a of the 2-hydroxyethylammonium ion, $HOCH_2CH_2NH_3^+$ (pK_b of $HOCH_2CH_2NH_2$ = 4.49)?

18.109 (a) What is the pK_b of ClO_2^-?
(b) What is the pK_a of the dimethylammonium ion, $(CH_3)_2NH_2^+$?

18.110 (a) What is the pK_b of NO_2^-?
(b) What is the pK_a of the hydrazinium ion, $H_2N{-}NH_3^+$ (K_b of hydrazine = 8.5×10^{-7})?

18.111 (a) What is the pH of 0.150 M KCN?
(b) What is the pH of 0.40 M triethylammonium chloride, $(CH_3CH_2)_3NHCl$?

18.112 (a) What is the pH of 0.100 M sodium phenolate, C_6H_5ONa, the sodium salt of phenol?
(b) What is the pH of 0.15 M methylammonium bromide, CH_3NH_3Br (K_b of CH_3NH_2 = 4.4×10^{-4})?

18.113 (a) What is the pH of 0.65 M potassium formate, HCOOK?
(b) What is the pH of 0.85 M NH_4Br?

18.114 (a) What is the pH of 0.75 M NaF?
(b) What is the pH of 0.88 M pyridinium chloride, C_5H_5NHCl?

Problems in Context

18.115 Sodium hypochlorite solution, sold as chlorine bleach, is potentially dangerous because of the basicity of ClO^-, the active bleaching ingredient. What is $[OH^-]$ in an aqueous solution that is 6.5% NaClO by mass? What is the pH of the solution? (Assume that d of the solution is 1.0 g/mL.)

18.116 Codeine ($C_{18}H_{21}NO_3$) is a narcotic pain reliever that forms a salt with HCl. What is the pH of 0.050 M codeine hydrochloride (pK_b of codeine = 5.80)?

Acid-Base Properties of Salt Solutions

(Sample Problems 18.13 and 18.14)

Concept Review Questions

18.117 What determines whether an aqueous solution of a salt will be acidic, basic, or neutral? Give an example of each type of salt.

18.118 Why is aqueous NaF basic but aqueous NaCl neutral?

18.119 The NH_4^+ ion forms acidic solutions, and the CH_3COO^- ion forms basic solutions. However, a solution of ammonium acetate is almost neutral. Do all of the ammonium salts of weak acids form neutral solutions? Explain your answer.

Skill-Building Exercises (grouped in similar pairs)

18.120 Explain with equations and calculations, when necessary, whether an aqueous solution of each of these salts is acidic, basic, or neutral: (a) KBr; (b) NH_4I; (c) KCN.

18.121 Explain with equations and calculations, when necessary, whether an aqueous solution of each of these salts is acidic, basic, or neutral: (a) $SnCl_2$; (b) NaHS; (c) $Zn(CH_3COO)_2$.

18.122 Explain with equations and calculations, when necessary, whether an aqueous solution of each of these salts is acidic, basic, or neutral: (a) Na_2CO_3; (b) $CaCl_2$; (c) $Cu(NO_3)_2$.

18.123 Explain with equations and calculations, when necessary, whether an aqueous solution of each of these salts is acidic, basic, or neutral: (a) CH_3NH_3Cl; (b) $LiClO_4$; (c) CoF_2.

18.124 Explain with equations and calculations, when necessary, whether an aqueous solution of each of these salts is acidic, basic, or neutral: (a) $SrBr_2$; (b) $Ba(CH_3COO)_2$; (c) $(CH_3)_2NH_2Br$.

18.125 Explain with equations and calculations, when necessary, whether an aqueous solution of each of these salts is acidic, basic, or neutral: (a) $Fe(HCOO)_3$; (b) $KHCO_3$; (c) K_2S.

18.126 Explain with equations and calculations, when necessary, whether an aqueous solution of each of these salts is acidic, basic, or neutral: (a) $(NH_4)_3PO_4$; (b) Na_2SO_4; (c) LiClO.

18.127 Explain with equations and calculations, when necessary, whether an aqueous solution of each of these salts is acidic, basic, or neutral: (a) $Pb(CH_3COO)_2$; (b) $Cr(NO_2)_3$; (c) CsI.

18.128 Rank the following salts in order of *increasing* pH of their 0.1 M aqueous solutions:
(a) KNO_3, K_2SO_3, K_2S, $Fe(NO_3)_2$
(b) NH_4NO_3, $NaHSO_4$, $NaHCO_3$, Na_2CO_3

18.129 Rank the following salts in order of *decreasing* pH of their 0.1 M aqueous solutions:
(a) $FeCl_2$, $FeCl_3$, $MgCl_2$, $KClO_2$
(b) NH_4Br, $NaBrO_2$, NaBr, $NaClO_2$

Generalizing the Brønsted-Lowry Concept: The Leveling Effect

Concept Review Questions

18.130 Methoxide ion, CH_3O^-, and amide ion, NH_2^-, are very strong bases that are "leveled" by water. What does this mean? Write the reactions that occur in the leveling process. What species do the two leveled solutions have in common?

18.131 Explain the differing extents of dissociation of H_2SO_4 in CH_3COOH, H_2O, and NH_3.

18.132 In H_2O, HF is weak and the other hydrohalic acids are equally strong. In NH_3, however, all the hydrohalic acids are equally strong. Explain.

Electron-Pair Donation and the Lewis Acid-Base Definition

(Sample Problem 18.15)

Concept Review Questions

18.133 What feature must a molecule or ion have in order to act as a Lewis base? A Lewis acid? Explain the roles of these features.

18.134 How do Lewis acids differ from Brønsted-Lowry acids? How are they similar? Do Lewis bases differ from Brønsted-Lowry bases? Explain.

18.135 (a) Is a weak Brønsted-Lowry base necessarily a weak Lewis base? Explain with an example.
(b) Identify the Lewis bases in the following reaction:

$$Cu(H_2O)_4^{2+}(aq) + 4CN^-(aq) \rightleftharpoons Cu(CN)_4^{2-}(aq) + 4H_2O(l)$$

(c) Given that $K_c > 1$ for the reaction in part (b), which Lewis base is stronger?

18.136 In which of the three acid-base concepts can water be a product of an acid-base reaction? In which is it the only product?

18.137 (a) Give an example of a *substance* that is a base in two of the three acid-base definitions, but not in the third.
(b) Give an example of a *substance* that is an acid in one of the three acid-base definitions, but not in the other two.

Skill-Building Exercises (grouped in similar pairs)

18.138 Which are Lewis acids and which are Lewis bases?
(a) Cu^{2+} (b) Cl^- (c) $SnCl_2$ (d) OF_2

18.139 Which are Lewis acids and which are Lewis bases?
(a) Na^+ (b) NH_3 (c) CN^- (d) BF_3

18.140 Which are Lewis acids and which are Lewis bases?
(a) BF_3 (b) S^{2-} (c) SO_3^{2-} (d) SO_3

18.141 Which are Lewis acids and which are Lewis bases?
(a) Mg^{2+} (b) OH^- (c) SiF_4 (d) $BeCl_2$

18.142 Identify the Lewis acid and Lewis base in each reaction:
(a) $Na^+ + 6H_2O \rightleftharpoons Na(H_2O)_6^+$ (b) $CO_2 + H_2O \rightleftharpoons H_2CO_3$
(c) $F^- + BF_3 \rightleftharpoons BF_4^-$

18.143 Identify the Lewis acid and Lewis base in each reaction:
(a) $Fe^{3+} + 2H_2O \rightleftharpoons FeOH^{2+} + H_3O^+$
(b) $H_2O + H^- \rightleftharpoons OH^- + H_2$
(c) $4CO + Ni \rightleftharpoons Ni(CO)_4$

18.144 Classify the following as Arrhenius, Brønsted-Lowry, or Lewis acid-base reactions. A reaction may fit all, two, one, or none of the categories:
(a) $Ag^+ + 2NH_3 \rightleftharpoons Ag(NH_3)_2^+$
(b) $H_2SO_4 + NH_3 \rightleftharpoons HSO_4^- + NH_4^+$
(c) $2HCl \rightleftharpoons H_2 + Cl_2$
(d) $AlCl_3 + Cl^- \rightleftharpoons AlCl_4^-$

18.145 Classify the following as Arrhenius, Brønsted-Lowry, or Lewis acid-base reactions. A reaction may fit all, two, one, or none of the categories:
(a) $Cu^{2+} + 4Cl^- \rightleftharpoons CuCl_4^{2-}$
(b) $Al(OH)_3 + 3HNO_3 \rightleftharpoons Al^{3+} + 3H_2O + 3NO_3^-$
(c) $N_2 + 3H_2 \rightleftharpoons 2NH_3$
(d) $CN^- + H_2O \rightleftharpoons HCN + OH^-$

Comprehensive Problems

18.146 Chloral ($Cl_3C—CH=O$) forms a monohydrate, chloral hydrate, the sleep-inducing depressant called "knockout drops" in old movies. (a) Write two possible structures for chloral hydrate, one involving hydrogen bonding and one that is a Lewis adduct. (b) What spectroscopic method could be used to identify the real structure? Explain.

18.147 In humans, blood pH is maintained within a narrow range: *acidosis* occurs if the blood pH is below 7.35, and *alkalosis* occurs if the pH is above 7.45. Given that the pK_w of blood is 13.63 at 37°C (body temperature), what is the normal range of $[H_3O^+]$ and of $[OH^-]$ in blood?

18.148 The disinfectant phenol, C_6H_5OH, has a pK_a of 10.0 in water, but 14.4 in methanol.
(a) Why are the values different?
(b) Is methanol a stronger or weaker base than water?
(c) Write the dissociation reaction of phenol in methanol.
(d) Write an expression for the autoionization constant of methanol.

18.149 When carbon dioxide dissolves in water, it undergoes a multistep equilibrium process, with $K_{overall} = 4.5\times10^{-7}$, which is simplified to the following:

$$CO_2(g) + H_2O(l) \rightleftharpoons H_2CO_3(aq)$$
$$H_2CO_3(aq) + H_2O(l) \rightleftharpoons HCO_3^-(aq) + H_3O^+(aq)$$

(a) Classify each step as a Lewis or a Brønsted-Lowry reaction.
(b) What is the pH of nonpolluted rainwater in equilibrium with clean air (P_{CO_2} in clean air = 4×10^{-4} atm; Henry's law constant for CO_2 at 25°C is 0.033 mol/L·atm)?
(c) What is $[CO_3^{2-}]$ in rainwater (K_a of $HCO_3^- = 4.7\times10^{-11}$)?
(d) If the partial pressure of CO_2 in clean air doubles in the next few decades, what will the pH of rainwater become?

18.150 Seashells are mostly calcium carbonate, which reacts with H_3O^+ according to the equation

$$CaCO_3(s) + H_3O^+(aq) \rightleftharpoons Ca^{2+}(aq) + HCO_3^-(aq) + H_2O(l)$$

If K_w increases at higher pressure, will seashells dissolve more rapidly near the surface of the ocean or at great depths? Explain.

18.151 Many molecules with central atoms from Period 3 or higher take part in Lewis acid-base reactions in which the central atom expands its valence shell. $SnCl_4$ reacts with $(CH_3)_3N$ as follows:

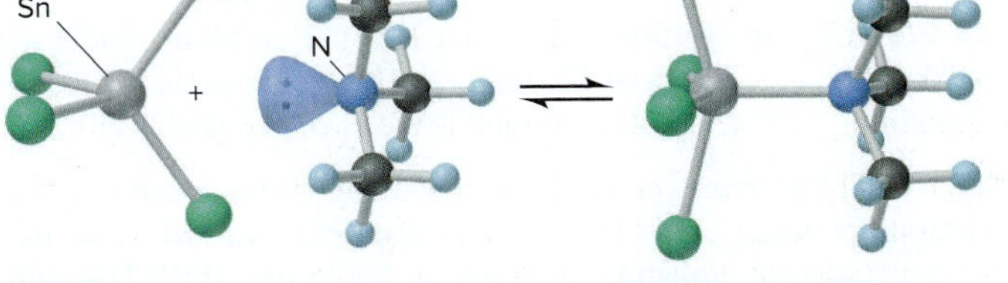

(a) Identify the Lewis acid and the Lewis base in the reaction.
(b) Give the *nl* designation of the sublevel of the central atom in the acid before it accepts the lone pair.

18.152 A chemist makes four successive ten-fold dilutions of 1.0×10^{-5} *M* HCl. Calculate the pH of the original solution and of each diluted solution (through 1.0×10^{-9} *M* HCl).

18.153 Chlorobenzene, C_6H_5Cl, is a key intermediate in the manufacture of dyes and pesticides. It is made by the chlorination of benzene, catalyzed by $FeCl_3$, in this series of steps:
(1) $Cl_2 + FeCl_3 \rightleftharpoons FeCl_5$ (or $Cl^+FeCl_4^-$)
(2) $C_6H_6 + Cl^+FeCl_4^- \rightleftharpoons C_6H_6Cl^+ + FeCl_4^-$
(3) $C_6H_6Cl^+ \rightleftharpoons C_6H_5Cl + H^+$
(4) $H^+ + FeCl_4^- \rightleftharpoons HCl + FeCl_3$
(a) Which of the step(s) is (are) Lewis acid-base reactions?
(b) Identify the Lewis acids and bases in each of those steps.

18.154 The beakers shown below contain 0.300 L of aqueous solutions of a moderately weak acid HY. Each particle represents 0.010 mol; solvent molecules are omitted for clarity. (a) The reaction in beaker A is at equilibrium. Calculate Q for the reactions in beakers B, C, and D to determine which, if any, is also at equilibrium. (b) For any not at equilibrium, in which direction does the reaction proceed? (c) Does dilution affect the extent of dissociation of a weak acid? Explain.

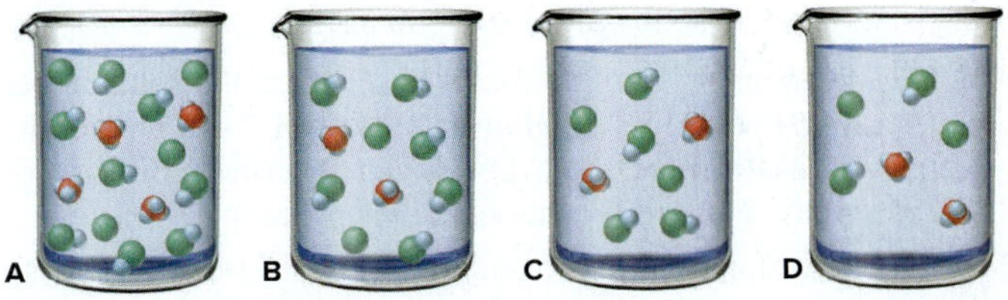

18.155 The strength of an acid or base is related to its strength as an electrolyte. (a) Is the electrical conductivity of 0.1 *M* HCl higher, lower, or the same as that of 0.1 *M* CH_3COOH? Explain. (b) Is the electrical conductivity of 1×10^{-7} *M* HCl higher, lower, or the same as that of 1×10^{-7} *M* CH_3COOH? Explain.

18.156 Esters, RCOOR′, are formed by the reaction of carboxylic acids, RCOOH, and alcohols, R′OH, where R and R′ are hydrocarbon groups. Many esters are responsible for the odors of fruit and, thus, have important uses in the food and cosmetics industries. The first two steps in the mechanism of ester formation are

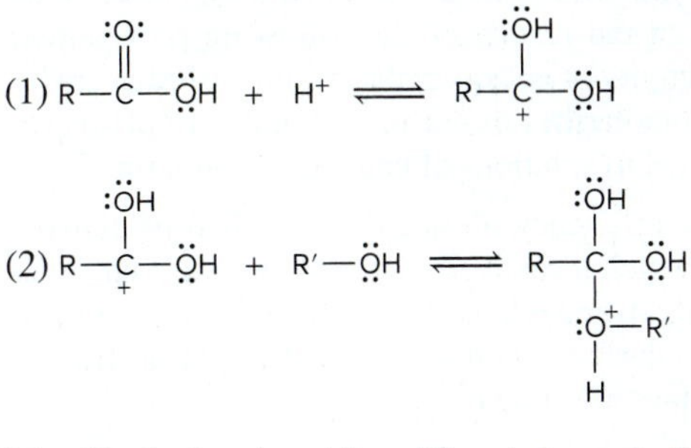

Identify the Lewis acids and Lewis bases in these two steps.

18.157 Three beakers contain 100. mL of 0.10 *M* HCl, $HClO_2$, and HClO, respectively. (a) Find the pH of each. (b) Describe quantitatively how to make the pH equal in the beakers through the addition of water only.

18.158 Human urine has a normal pH of 6.2. If a person eliminates an average of 1250. mL of urine per day, how many H^+ ions are eliminated per week?

18.159 Liquid ammonia autoionizes like water:

$$2NH_3(l) \longrightarrow NH_4^+(am) + NH_2^-(am)$$

where (*am*) represents solvation by NH_3.
(a) Write the ion-product constant expression, K_{am}.
(b) What are the strongest acid and base that can exist in $NH_3(l)$?
(c) HNO_3 and HCOOH are leveled in $NH_3(l)$. Explain with equations.
(d) At the boiling point of ammonia (−33°C), $K_{am} = 5.1{\times}10^{-27}$. Calculate $[NH_4^+]$ at this temperature.
(e) Pure sulfuric acid also autoionizes. Write the ion-product constant expression, K_{sulf}, and find the concentration of the conjugate base at 20°C ($K_{sulf} = 2.7{\times}10^{-4}$ at 20°C).

18.160 Thiamine hydrochloride ($C_{12}H_{18}ON_4SCl_2$) is a water-soluble form of thiamine (vitamin B_1; $K_a = 3.37{\times}10^{-7}$). How many grams of the hydrochloride must be dissolved in 10.00 mL of water to give a pH of 3.50?

18.161 Tris(hydroxymethyl)aminomethane, known as TRIS or THAM, is a water-soluble base used in synthesizing surfactants and pharmaceuticals, as an emulsifying agent in cosmetics, and in cleaning mixtures for textiles and leather. In biomedical research, solutions of TRIS are used to maintain nearly constant pH for the study of enzymes and other cellular components. Given that the pK_b is 5.91, calculate the pH of 0.075 *M* TRIS.

18.162 When an Fe^{3+} salt is dissolved in water, the solution becomes acidic due to formation of $Fe(H_2O)_5OH^{2+}$ and H_3O^+. The overall process involves both Lewis and Brønsted-Lowry acid-base reactions. Write the equations for the process.

18.163 What is the pH of a vinegar with 5.0% (w/v) acetic acid in water?

18.164 The scene below represents a sample of a weak acid HB (*blue and purple*) dissolved in water. Draw a scene that represents the same volume after the solution has been diluted with water.

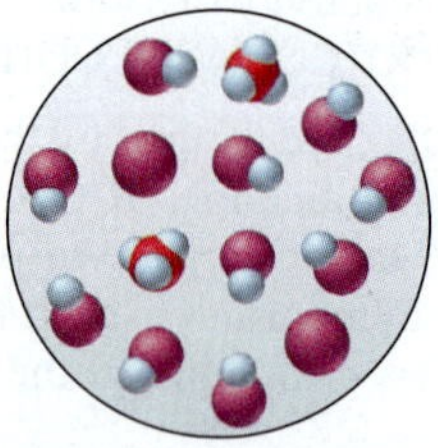

18.165 How would you differentiate between a strong and a weak monoprotic acid from the results of the following procedures? (a) Electrical conductivity of an equimolar solution of each acid is measured. (b) Equal molarities of each are tested with pH paper. (c) Zinc metal is added to solutions of equal concentration.

18.166 The catalytic efficiency of an enzyme, called its *activity*, indicates the rate at which it catalyzes the reaction. Most enzymes have optimum activity over a relatively narrow pH range, which is related to the pH of the local cellular fluid. The pH profiles of three digestive enzymes are shown.

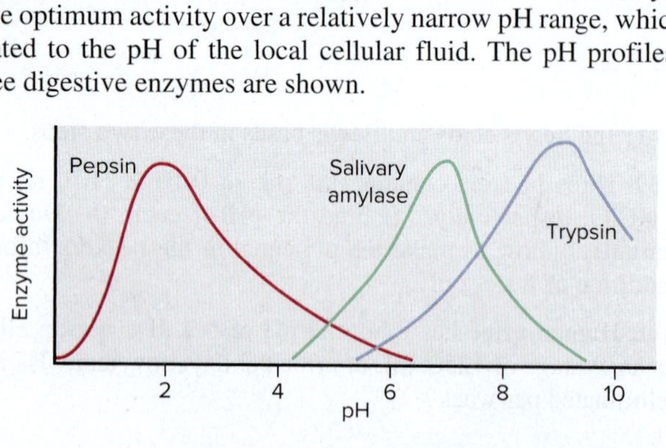

Salivary amylase begins digestion of starches in the mouth and has optimum activity at a pH of 6.8; pepsin begins protein digestion in the stomach and has optimum activity at a pH of 2.0; and trypsin, released in pancreatic juices, continues protein digestion in the small intestine and has optimum activity at a pH of 9.5. Calculate $[H_3O^+]$ in the local cellular fluid for each enzyme.

18.167 Acetic acid has a K_a of $1.8{\times}10^{-5}$, and ammonia has a K_b of $1.8{\times}10^{-5}$. Find $[H_3O^+]$, $[OH^-]$, pH, and pOH for (a) 0.240 *M* acetic acid and (b) 0.240 *M* ammonia.

18.168 The uses of sodium phosphate include clarifying crude sugar, manufacturing paper, removing boiler scale, and washing concrete. What is the pH of a solution containing 33 g of Na_3PO_4 per liter? What is $[OH^-]$ of this solution?

18.169 The Group 5A(15) hydrides react with boron trihalides in a reversible Lewis acid-base reaction. When 0.15 mol of $PH_3BCl_3(s)$ is introduced into a 3.0-L container at a certain temperature, $8.4{\times}10^{-3}$ mol of PH_3 is present at equilibrium:

$$PH_3BCl_3(s) \rightleftharpoons PH_3(g) + BCl_3(g).$$

(a) Find K_c for the reaction at this temperature.
(b) Draw a Lewis structure for the reactant.

18.170 A 1.000 *m* solution of chloroacetic acid ($ClCH_2COOH$) freezes at −1.93°C. Find the K_a of chloroacetic acid. (Assume that the molarities equal the molalities.)

18.171 Sodium stearate ($C_{17}H_{35}COONa$) is a major component of bar soap. The K_a of the stearic acid is $1.3{\times}10^{-5}$. What is the pH of 10.0 mL of a solution containing 0.42 g of sodium stearate?

18.172 Calcium propionate [$Ca(CH_3CH_2COO)_2$; calcium propanoate] is a mold inhibitor used in food, tobacco, and pharmaceuticals. (a) Use balanced equations to show whether aqueous calcium propionate is acidic, basic, or neutral. (b) Use Appendix C to find the resulting pH when 8.75 g of $Ca(CH_3CH_2COO)_2$ dissolves in enough water to give 0.500 L of solution.

18.173 A site in Pennsylvania receives a total annual deposition of 2.688 g/m^2 of sulfate from fertilizer and acid rain. The mass ratio of ammonium sulfate/ammonium bisulfate/sulfuric acid is 3.0/5.5/1.0. (a) How much acid, expressed as kilograms (kg) of sulfuric acid, is deposited over an area of 10. km^2? (b) How many pounds of $CaCO_3$ are needed to neutralize this acid? (c) If 10. km^2 is the area of an unpolluted lake 3 m deep and there is no loss of acid, what pH will the lake water attain by the end of the year? (Assume constant volume.)

18.174 (a) If $K_w = 1.139{\times}10^{-15}$ at 0°C and $5.474{\times}10^{-14}$ at 50°C, find $[H_3O^+]$ and pH of water at 0°C and 50°C.
(b) The autoionization constant for heavy water (deuterium oxide, D_2O) is $3.64{\times}10^{-16}$ at 0°C and $7.89{\times}10^{-15}$ at 50°C. Find $[D_3O^+]$ and pD of heavy water at 0°C and 50°C.
(c) Suggest a reason for these differences.

18.175 HX ($\mathcal{M}$ = 150. g/mol) and HY ($\mathcal{M}$ = 50.0 g/mol) are weak acids. A solution of 12.0 g/L of HX has the same pH as one containing 6.00 g/L of HY. Which is the stronger acid? Why?

18.176 The beakers on the facing page depict the aqueous dissociations of weak acids HA (*blue and green*) and HB (*blue and yellow*); solvent molecules are omitted for clarity. If the HA solution is 0.50 L, and the HB solution is 0.25 L, and each particle represents 0.010 mol, find the K_a of each acid. Which acid, if either, is stronger?

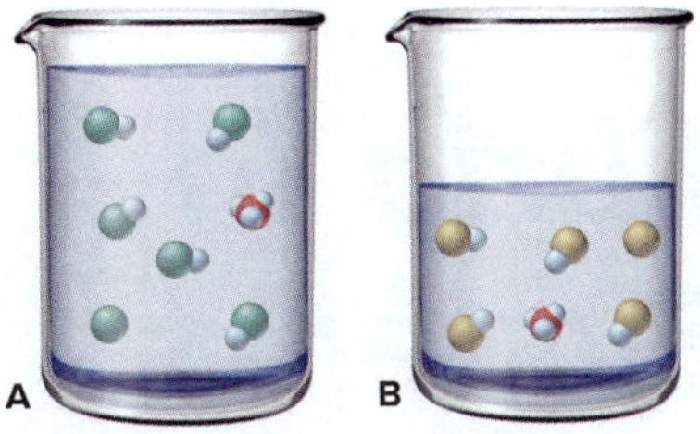

18.177 In his acid-base studies, Arrhenius discovered an important fact involving reactions like the following:

$$KOH(aq) + HNO_3(aq) \longrightarrow ?$$
$$NaOH(aq) + HCl(aq) \longrightarrow ?$$

(a) Complete the reactions and use the data for the individual ions in Appendix B to calculate each ΔH°_{rxn}.
(b) Explain your results and use them to predict ΔH°_{rxn} for

$$KOH(aq) + HCl(aq) \longrightarrow ?$$

18.178 Putrescine [$NH_2(CH_2)_4NH_2$], found in rotting animal tissue, is now known to be in all cells and essential for normal and abnormal (cancerous) growth. It also plays a key role in the formation of GABA, a neurotransmitter. A 0.10 *M* aqueous solution of putrescine has $[OH^-] = 2.1\times10^{-3}$. What is the K_b?

18.179 The molecular scene depicts the relative concentrations of H_3O^+ *(purple)* and OH^- *(green)* in an aqueous solution at 25°C. (Counter ions and solvent molecules are omitted for clarity.) (a) Calculate the pH. (b) How many H_3O^+ ions would you have to draw for every OH^- ion to depict a solution of pH 4?

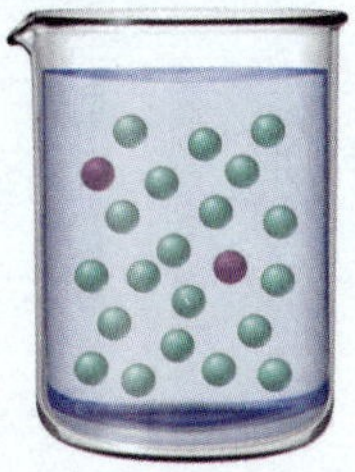

18.180 Polymers are not very soluble in water, but their solubility increases if they have charged groups. (a) Casein, a milk protein, contains many $—COO^-$ groups on its side chains. How does the solubility of casein vary with pH? (b) Histones are proteins essential to the function of DNA. They are weakly basic due to the presence of side chains with $—NH_2$ and $=NH$ groups. How does the solubility of a histone vary with pH?

18.181 Hemoglobin (Hb) transports oxygen in the blood:

$$HbH^+(aq) + O_2(aq) + H_2O(l) \longrightarrow HbO_2(aq) + H_3O^+(aq)$$

In blood, $[H_3O^+]$ is held nearly constant at 4×10^{-8} *M*.
(a) How does the equilibrium position change in the lungs?
(b) How does it change in O_2-deficient cells?
(c) Excessive vomiting may lead to metabolic *alkalosis,* in which $[H_3O^+]$ in blood *decreases.* How does this condition affect the ability of Hb to transport O_2?
(d) Diabetes mellitus may lead to metabolic *acidosis,* in which $[H_3O^+]$ in blood *increases.* How does this condition affect the ability of Hb to transport O_2?

18.182 Nitrogen is discharged from wastewater treatment facilities into rivers and streams, usually as NH_3 and NH_4^+:

$$NH_3(aq) + H_2O(l) \rightleftharpoons NH_4^+(aq) + OH^-(aq) \qquad K_b = 1.76\times10^{-5}$$

One strategy for removing nitrogen is to raise the pH and "strip" the NH_3 from solution by bubbling air through the water. (a) At pH 7.00, what fraction of the total nitrogen in solution is NH_3, defined as $[NH_3]/([NH_3] + [NH_4^+])$? (b) What is the fraction at pH 10.00? (c) Explain the basis of ammonia stripping.

18.183 A solution of propanoic acid (CH_3CH_2COOH), made by dissolving 7.500 g in sufficient water to make 100.0 mL, has a freezing point of −1.890°C.
(a) Calculate the molarity of the solution.
(b) Calculate the molarity of the propanoate ion. (Assume the molarity of the solution equals the molality.)
(c) Calculate the percent dissociation of propanoic acid.

18.184 The antimalarial properties of quinine ($C_{20}H_{24}N_2O_2$) saved thousands of lives during construction of the Panama Canal. This substance is a classic example of the medicinal wealth that tropical forests hold. Both N atoms are basic, but the N (colored) of the 3° amine group is far more basic ($pK_b = 5.1$) than the N within the aromatic ring system ($pK_b = 9.7$).

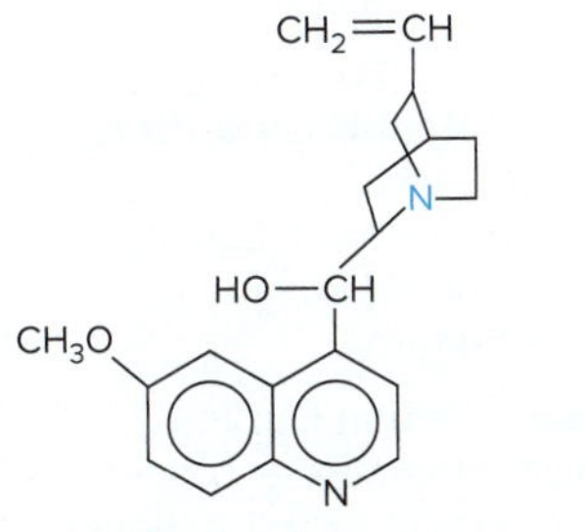

(a) A saturated solution of quinine in water is only 1.6×10^{-3} *M*. What is the pH of this solution?
(b) Show that the aromatic N contributes negligibly to the pH of the solution.
(c) Because of its low solubility, quinine is given as the salt quinine hydrochloride ($C_{20}H_{24}N_2O_2{\cdot}HCl$), which is 120 times more soluble than quinine. What is the pH of 0.33 *M* quinine hydrochloride?
(d) An antimalarial concentration in water is 1.5% quinine hydrochloride by mass (d = 1.0 g/mL). What is the pH?

18.185 Drinking water is often disinfected with Cl_2, which hydrolyzes to form HClO, a weak acid but powerful disinfectant:

$$Cl_2(aq) + 2H_2O(l) \longrightarrow HClO(aq) + H_3O^+(aq) + Cl^-(aq)$$

The fraction of HClO in solution is defined as

$$\frac{[HClO]}{[HClO] + [ClO^-]}$$

(a) What is the fraction of HClO at pH 7.00 (K_a of HClO = 2.9×10^{-8})?
(b) What is the fraction at pH 10.00?

18.186 The following scenes represent three weak acids HA (where A = X, Y, or Z) dissolved in water (H_2O is not shown):

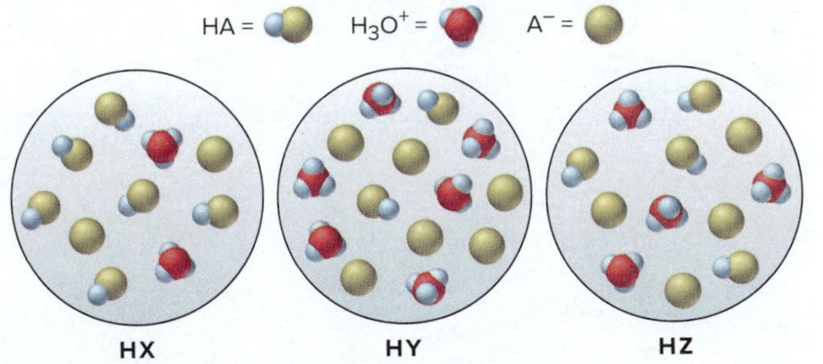

(a) Rank the acids in order of increasing K_a. (b) Rank the acids in order of increasing pK_a. (c) Rank the conjugate bases in order of increasing pK_b. (d) What is the percent dissociation of HX? (e) If equimolar amounts of the sodium salts of the acids (NaX, NaY, and NaZ) were dissolved in water, which solution would have the highest pOH? The lowest pH?

19 Ionic Equilibria in Aqueous Systems

19.1 Equilibria of Acid-Base Buffers
Common-Ion Effect
Henderson-Hasselbalch Equation
Buffer Capacity and Range
Preparing a Buffer

19.2 Acid-Base Titration Curves
Strong Acid–Strong Base Titrations
Weak Acid–Strong Base Titrations
Weak Base–Strong Acid Titrations
Acid-Base Indicators
Polyprotic Acid Titrations
Amino Acids as Polyprotic Acids

19.3 Equilibria of Slightly Soluble Ionic Compounds
Solubility-Product Constant (K_{sp})
Calculations Involving K_{sp}
Effect of a Common Ion
Effect of pH
Formation of a Limestone Cave
Predicting Precipitate Formation: Q_{sp} vs. K_{sp}
Selective Precipitation and Simultaneous Equilibria

19.4 Equilibria Involving Complex Ions
Formation of Complex Ions
Complex Ions and Solubility
Amphoteric Hydroxides

Source: © Joe Scherschel/Getty Images

Concepts and Skills to Review Before You Study This Chapter

- solubility rules for ionic compounds (Section 4.3)
- equilibrium nature of a saturated solution (Section 13.4)
- effect of concentration on equilibrium position (Section 17.6)
- conjugate acid-base pairs (Section 18.3)
- calculations for weak-acid and weak-base equilibria (Sections 18.4 and 18.6)
- acid-base properties of salt solutions (Section 18.7)
- Lewis acids and bases (Section 18.9)

Continuing our exploration of the universal nature of equilibrium, we now examine aqueous ionic systems. Acid-base and carbonate solubility equilibria are responsible for limestone caves and formations such as those found at Mammoth Hot Springs in Yellowstone National Part *(see photo)*. Organisms survive by maintaining cellular pH within narrow limits through complex carbonate and phosphate equilibria. In soils, equilibria involving clays control the availability of ionic nutrients for plants. In industrial settings, the principles of equilibrium govern the softening of water and the purification of products by precipitation of unwanted ions. And they even explain how the weak acids in wine and vinegar influence the delicate taste of a fine sauce.

IN THIS CHAPTER . . . *We define and quantify three aqueous ionic equilibrium systems—acid-base buffers, slightly soluble salts, and complex ions.*

- We introduce the common-ion effect to explain how buffers work, and see why buffers are important and how to prepare them.
- We describe several types of acid-base titrations and explore the role buffers play in them.
- We see that slightly soluble salts have their own equilibrium constant and learn how a common ion and pH influence the solubilities of these compounds.
- We use concepts of aqueous ionic equilibria to understand how a cave forms.
- We investigate how complex ions form and change from one type to another.

19.1 EQUILIBRIA OF ACID-BASE BUFFERS

Why do some lakes become acidic when showered by acid rain, while others remain unaffected? How does blood maintain a constant pH in contact with countless cellular acid-base reactions? How can a chemist sustain a nearly constant $[H_3O^+]$ in reactions that consume or produce H_3O^+ or OH^-? The answer in each case depends on the action of a buffer, and in this section we discuss how buffers work and how to prepare them.

What a Buffer Is and How It Works: The Common-Ion Effect

In everyday language, a *buffer* is something that lessens the impact of an external force. An **acid-base buffer** is a solution that *lessens the impact on its pH of the addition of acid or base*. Add a small amount of H_3O^+ or OH^- to an *un*buffered solution, and the pH changes by several units; thus, $[H_3O^+]$ changes by *several orders of magnitude* (Figure 19.1).

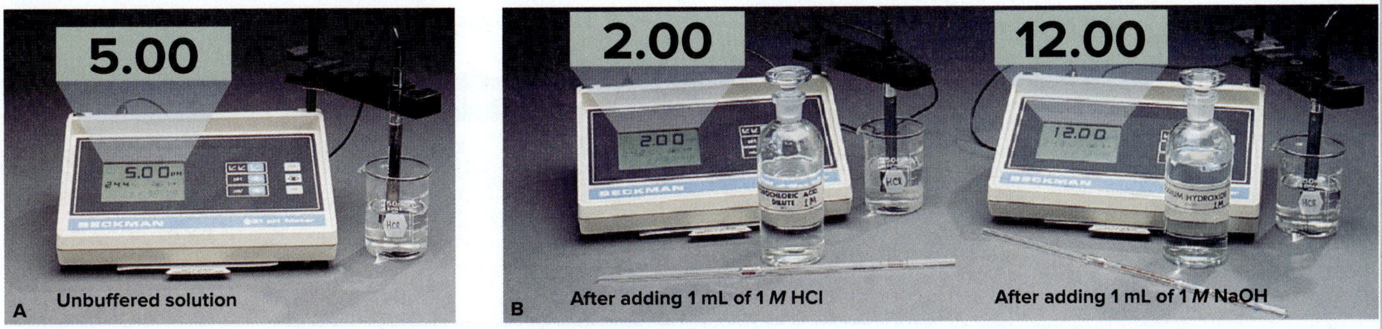

Figure 19.1 **The effect of adding acid or base to an unbuffered solution. A,** A 100-mL sample of dilute HCl is adjusted to pH 5.00 *(left)*. **B,** Adding 1 mL of strong acid *(left)* or of strong base *(right)* changes the pH by several units.

Source: © McGraw-Hill Education/Stephen Frisch, photographer

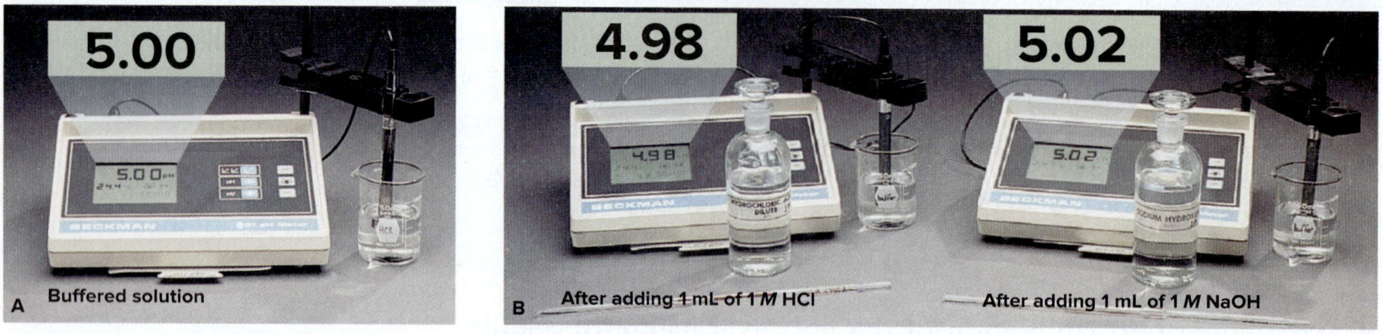

Figure 19.2 The effect of adding acid or base to a buffered solution. A, A 100-mL sample of an acetate buffer, which was made by mixing 1 *M* CH_3COOH with 1 *M* CH_3COONa, is adjusted to pH 5.00. **B,** Adding 1 mL of strong acid *(left)* or of strong base *(right)* changes the pH negligibly.

Source: © McGraw-Hill Education/Stephen Frisch, photographer

The same addition of strong acid or strong base to a buffered solution causes only a minor change in pH (Figure 19.2). To withstand these additions, a buffer must have an acidic component that reacts with the added OH^- *and* a basic component that reacts with the added H_3O^+. However, these components can't be any acid and base because they would neutralize each other.

- Most often, *the components of a buffer are a conjugate acid-base pair* (weak acid and conjugate base or weak base and conjugate acid).

The buffer in Figure 19.2, for example, is a mixture of acetic acid (CH_3COOH) and acetate ion (CH_3COO^-).

Presence of a Common Ion Buffers work through the **common-ion effect.** When you dissolve acetic acid in water, the acid dissociates slightly:

$$CH_3COOH(aq) + H_2O(l) \rightleftharpoons H_3O^+(aq) + CH_3COO^-(aq)$$

What happens if you now introduce acetate ion by adding the soluble salt sodium acetate? First, we need to recognize that, as a strong electrolyte, sodium acetate dissociates completely in water:

$$CH_3COONa(aq) \longrightarrow Na^+(aq) + CH_3COO^-(aq)$$

The sodium ion does not interact with water (Section 18.7) and thus is a spectator ion. From Le Châtelier's principle (Section 17.6), we know that adding CH_3COO^- ion will shift the equilibrium position to the left; thus, $[H_3O^+]$ decreases, in effect lowering the extent of acid dissociation:

$$CH_3COOH(aq) + H_2O(l) \rightleftharpoons H_3O^+(aq) + CH_3COO^-(aq;\ \text{added})$$

We get the same result when we add acetic acid to a sodium acetate solution instead of water. The acetate ion already present suppresses the acid from dissociating as much as it does in water, thus keeping the $[H_3O^+]$ lower (and pH higher). In either case, the effect is less acid dissociation. Acetate ion is called *the common ion* because it is "common" to both the acetic acid and sodium acetate solutions. *The common-ion effect occurs when a given ion is added to an equilibrium mixture that already contains that ion, and the position of equilibrium shifts away from forming it.*

Table 19.1 shows that the percent dissociation (and the $[H_3O^+]$) of an acetic acid solution decreases as the concentration of acetate ion (supplied by dissolving sodium acetate) increases. Thus, the *common ion, A^-, suppresses the dissociation of HA,* which makes the solution less acidic (higher pH).

Table 19.1 The Effect of Added Acetate Ion on the Dissociation of Acetic Acid

$[CH_3COOH]_{init}$	$[CH_3COO^-]_{added}$	% Dissociation*	H_3O^+	pH
0.10	0.00	1.3	1.3×10^{-3}	2.89
0.10	0.050	0.036	3.6×10^{-5}	4.44
0.10	0.10	0.018	1.8×10^{-5}	4.74
0.10	0.15	0.012	1.2×10^{-5}	4.92

$$\text{*\% Dissociation} = \frac{[CH_3COOH]_{dissoc}}{[CH_3COOH]_{init}} \times 100$$

Relative Concentrations of Buffer Components A buffer works because a *large amount of the acidic component (HA) of the buffer consumes small amounts of added OH^- and a large amount of the basic component (A^-) consumes small amounts of added H_3O^+.* Consider what happens in a solution containing high [CH_3COOH] and high [CH_3COO^-] when we add small amounts of strong acid or base. The equilibrium expression for HA dissociation is

$$K_a = \frac{[CH_3COO^-][H_3O^+]}{[CH_3COOH]}$$

Solving for [H_3O^+] gives

$$[H_3O^+] = K_a \times \frac{[CH_3COOH]}{[CH_3COO^-]}$$

Since K_a is constant, *the [H_3O^+] of the solution depends on the buffer-component concentration ratio,* $\frac{[CH_3COOH]}{[CH_3COO^-]}$:

- If the ratio [HA]/[A^-] goes up (more acid and less conjugate base), then [H_3O^+] goes up.
- If the ratio [HA]/[A^-] goes down (less acid and more conjugate base), then [H_3O^+] goes down.

Let's track this ratio as we add strong acid or strong base to a buffer in which [HA] = [A^-] (Figure 19.3, *middle*):

1. *Strong acid.* When we add a small amount of strong acid, the H_3O^+ ions react with an *equal (stoichiometric) amount* of acetate ion, the base component in the buffer, to form more acetic acid:

$$H_3O^+(aq;\ \text{added}) + CH_3COO^-(aq;\ \text{from buffer}) \longrightarrow CH_3COOH(aq) + H_2O(l)$$

As a result, [CH_3COO^-] goes down by that small amount and [CH_3COOH] goes up by that amount, which increases the buffer-component concentration ratio (Figure 19.3, *left*). The [H_3O^+] increases *very* slightly.

2. *Strong base.* The addition of a small amount of strong base produces the opposite result. The OH^- ions react with an *equal (stoichiometric) amount* of CH_3COOH, the acid component in the buffer, to form that much more CH_3COO^- (Figure 19.3, *right*):

$$CH_3COOH(aq;\ \text{from buffer}) + OH^-(aq;\ \text{added}) \longrightarrow CH_3COO^-(aq) + H_2O(l)$$

This time, the buffer-component concentration ratio decreases, which decreases [H_3O^+] *very* slightly.

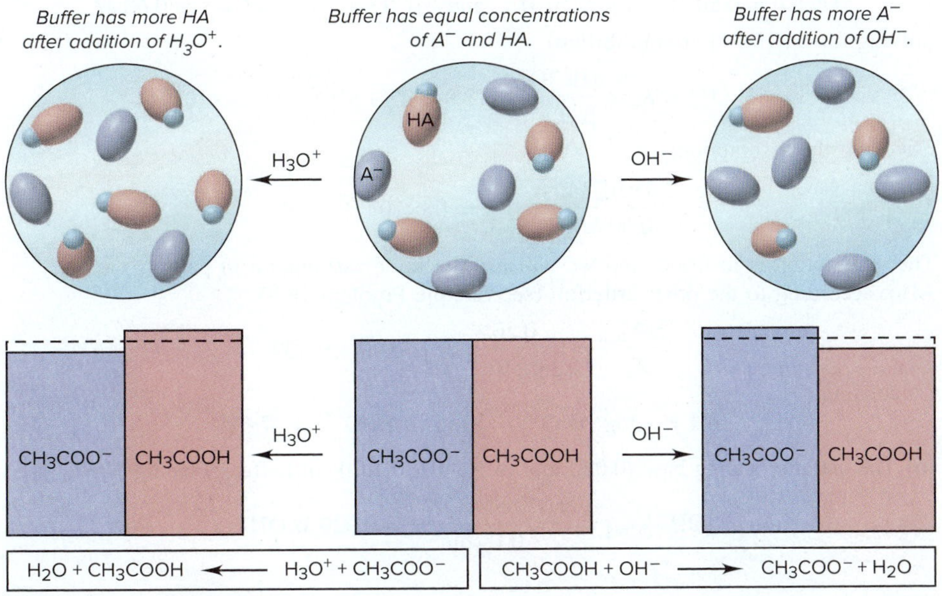

Figure 19.3 How a buffer works. The relative concentrations of the buffer components, acetic acid (CH_3COOH, *red*) and acetate ion (CH_3COO^-, *blue*) are indicated by the heights of the bars.

Thus, the buffer components consume *nearly all* the added H_3O^+ or OH^-. To reiterate, as long as the amount of added H_3O^+ or OH^- is small compared with the amounts of the buffer components, *the conversion of one component into the other produces a small change in the buffer-component concentration ratio and, consequently, a small change in $[H_3O^+]$ and in pH.* Sample Problem 19.1 demonstrates how small these pH changes typically are. Note that the latter two parts of the problem combine a stoichiometry portion, like the problems in Chapter 4, and a weak-acid dissociation portion, like those in Chapter 18.

SAMPLE PROBLEM 19.1 Calculating the Effect of Added H_3O^+ or OH^- on Buffer pH

Problem Calculate the pH: **(a)** of a buffer solution consisting of 0.50 *M* HClO ($K_a = 2.9\times10^{-8}$) and 0.60 *M* NaClO; **(b)** after adding 0.020 mol of solid NaOH to 1.0 L of the buffer solution in part (a); **(c)** after adding 30.0 mL of 1.5 *M* HCl to 0.50 L of the buffer solution in part (a). [In part (b), assume that the addition of NaOH causes negligible volume change.]

Plan For each part, we know, or can find, $[HClO]_{init}$ and $[ClO^-]_{init}$. We know the K_a of HClO (2.9×10^{-8}) and need to find $[H_3O^+]$ at equilibrium and convert it to pH. **(a)** We use the given concentrations of buffer components (0.50 *M* HClO and 0.60 *M* ClO^-) as the initial values. As in earlier problems, we assume that *x*, the [HClO] that dissociates, which equals $[H_3O^+]$, is so small relative to $[HClO]_{init}$ that it can be neglected. We set up a reaction table, solve for *x*, and check the assumption. **(b)** and **(c)** We assume that the added OH^- or H_3O^+ reacts completely with the buffer components to yield new $[HClO]_{init}$ and $[ClO^-]_{init}$, and then the acid dissociates to an unknown extent. We set up two reaction tables. The first summarizes the stoichiometry of adding strong base or acid. The second summarizes the dissociation of the new $[HClO]_{init}$, and we proceed as in part (a) to find the new $[H_3O^+]$. Since there is a volume change due to the added HCl in part (c), we must calculate the new initial concentrations of the buffer components.

Solution **(a)** The original pH: $[H_3O^+]$ in the original buffer. Setting up a reaction table with $x = [HClO]_{dissoc} = [H_3O^+]$ (as in Chapter 18, we assume that $[H_3O^+]$ from H_2O is negligible and disregard it):

Concentration (*M*)	$HClO(aq)$	+	$H_2O(l)$	$\rightleftharpoons$	$ClO^-(aq)$	+	$H_3O^+(aq)$
Initial	0.50		—		0.60		0
Change	$-x$		—		$+x$		$+x$
Equilibrium	$0.50 - x$		—		$0.60 + x$		x

Making the assumption and finding the equilibrium [HClO] and $[ClO^-]$: With K_a small, *x* is small, so we assume

$$[HClO] = 0.50\ M - x \approx 0.50\ M \quad \text{and} \quad [ClO^-] = 0.60\ M + x \approx 0.60\ M$$

Solving for *x* ($[H_3O^+]$ at equilibrium):

$$x = [H_3O^+] = K_a \times \frac{[HClO]}{[ClO^-]} \approx (2.9\times10^{-8}) \times \frac{0.50}{0.60} = 2.4\times10^{-8}\ M$$

Checking the assumption:

$$\frac{2.4\times10^{-8}\ M}{0.50\ M} \times 100 = 4.8\times10^{-6}\% < 5\%$$

The assumption is justified, and we will use the same assumption in parts (b) and (c). Also, according to the other criterion (see Sample Problem 18.8),

$$\frac{[HA]_{init}}{K_a} = \frac{0.50}{2.9\times10^{-8}} = 1.7\times10^{7} > 400$$

Calculating pH:

$$pH = -\log [H_3O^+] = -\log (2.4\times10^{-8}) = 7.62$$

(b) The pH after adding base (0.020 mol of NaOH to 1.0 L of buffer). Finding $[OH^-]_{added}$:

$$[OH^-]_{added} = \frac{0.020\ \text{mol}\ OH^-}{1.0\ \text{L soln}} = 0.020\ M\ OH^-$$

Setting up a reaction table for the *stoichiometry* of adding OH^- to HClO (notice that the reaction between weak acid and strong base goes to completion):

Concentration (*M*)	HClO(*aq*)	+	OH^-(*aq*)	⟶	ClO^-(*aq*)	+	H_2O(*l*)
Initial	0.50		0.020		0.60		—
Change	−0.020		−0.020		+0.020		—
Final	0.48		0		0.62		—

Setting up a reaction table for the *acid dissociation,* using these new initial concentrations. As in part (a), $x = [\text{HClO}]_{\text{dissoc}} = [\text{H}_3\text{O}^+]$:

Concentration (*M*)	HClO(*aq*)	+	H_2O(*l*)	⇌	ClO^-(*aq*)	+	H_3O^+(*aq*)
Initial	0.48		—		0.62		0
Change	−*x*		—		+*x*		+*x*
Equilibrium	0.48 − *x*		—		0.62 + *x*		*x*

Making the assumption that x is small, and solving for x:

$$[\text{HClO}] = 0.48\ M - x \approx 0.48\ M \quad \text{and} \quad [\text{ClO}^-] = 0.62\ M + x \approx 0.62\ M$$

$$x = [\text{H}_3\text{O}^+] = K_a \times \frac{[\text{HClO}]}{[\text{ClO}^-]} \approx (2.9\times10^{-8}) \times \frac{0.48}{0.62} = 2.2\times10^{-8}\ M$$

Calculating the pH:

$$\text{pH} = -\log\,[\text{H}_3\text{O}^+] = -\log\,(2.2\times10^{-8}) = \boxed{7.66}$$

The addition of strong base increased the concentration of the basic buffer component at the expense of the acidic buffer component. Note especially that the pH *increased only slightly,* from 7.62 to 7.66.

(c) The pH after adding acid (30.0 mL of 1.5 *M* HCl to 0.50 L of buffer). Use Equation 4.2 to find the new $[\text{HClO}]_{\text{init}}$ and $[\text{ClO}^-]_{\text{init}}$ and $[\text{H}_3\text{O}^+]_{\text{added}}$ after combining the HCl and buffer solutions; the new volume is 0.50 L + 0.0300 L = 0.53 L:

$$[\text{HClO}]_{\text{init}} = \frac{M_{\text{conc}} \times V_{\text{conc}}}{V_{\text{dil}}} = \frac{0.50\ M \times 0.50\ \text{L}}{0.53\ \text{L}} = 0.47\ M\ \text{HClO}$$

$$[\text{ClO}^-]_{\text{init}} = \frac{M_{\text{conc}} \times V_{\text{conc}}}{V_{\text{dil}}} = \frac{0.60\ M \times 0.50\ \text{L}}{0.53\ \text{L}} = 0.57\ M\ \text{ClO}^-$$

$$[\text{H}_3\text{O}^+]_{\text{added}} = \frac{M_{\text{conc}} \times V_{\text{conc}}}{V_{\text{dil}}} = \frac{1.5\ M \times 0.0300\ \text{L}}{0.53\ \text{L}} = 0.085\ M\ \text{H}_3\text{O}^+$$

Now we proceed as in part (b), by first setting up a reaction table for the *stoichiometry* of adding H_3O^+ to ClO^- (notice that the reaction between strong acid and weak base goes to completion):

Concentration (*M*)	ClO^-(*aq*)	+	H_3O^+(*aq*)	⟶	HClO(*aq*)	+	H_2O(*l*)
Initial	0.57		0.085		0.47		—
Change	−0.085		−0.085		+0.085		—
Final	0.48		0		0.56		—

The reaction table for the *acid dissociation,* with $x = [\text{HClO}]_{\text{dissoc}} = [\text{H}_3\text{O}^+]$ is

Concentration (*M*)	HClO(*aq*)	+	H_2O(*l*)	⇌	ClO^-(*aq*)	+	H_3O^+(*aq*)
Initial	0.56		—		0.48		0
Change	− *x*		—		+*x*		+*x*
Equilibrium	0.56 − *x*		—		0.48 + *x*		*x*

Making the assumption that x is small, and solving for x:

$$[\text{HClO}] = 0.56\ M - x \approx 0.56\ M \quad \text{and} \quad [\text{ClO}^-] = 0.48\ M + x \approx 0.48\ M$$

$$x = [\text{H}_3\text{O}^+] = K_a \times \frac{[\text{HClO}]}{[\text{ClO}^-]} \approx (2.9\times10^{-8}) \times \frac{0.56}{0.48} = 3.4\times10^{-8}\ M$$

Calculating the pH:

$$\text{pH} = -\log\,[\text{H}_3\text{O}^+] = -\log\,(3.4\times10^{-8}) = \boxed{7.47}$$

The addition of strong acid increased the concentration of the acidic buffer component at the expense of the basic buffer component and *lowered* the pH only slightly, from 7.62 to 7.47.

Check The changes in [HClO] and $[ClO^-]$ occur in opposite directions in parts (b) and (c), which makes sense. The pH increased when base was added in (b) and decreased when acid was added in (c).

Comment In part (a), we justified our assumption that x can be neglected. Therefore, in parts (b) and (c), we could have used the "Final" values from the last line of the stoichiometry reaction tables directly for the ratio of buffer components; that would have allowed us to dispense with a reaction table for the acid dissociation. In subsequent problems in this chapter, we will follow this more straightforward approach.

FOLLOW-UP PROBLEMS

Brief Solutions for ***all*** *Follow-up Problems appear at the end of the chapter.*

19.1A Calculate the pH of a buffer consisting of 0.50 *M* HF and 0.45 *M* KF **(a)** before and **(b)** after addition of 0.40 g of NaOH to 1.0 L of the buffer (K_a of HF = 6.8×10^{-4}).

19.1B Calculate the pH of a buffer consisting of 0.25 *M* $(CH_3)_2NH_2Cl$ and 0.30 *M* $(CH_3)_2NH$ **(a)** before and **(b)** after addition of 50.0 mL of 0.75 *M* HCl to 1.0 L of the buffer [pK_b of $(CH_3)_2NH$ = 3.23].

SOME SIMILAR PROBLEMS 19.12–19.21 and 19.28–19.31

Figure 19.4 summarizes the changes in concentrations of buffer components upon addition of strong acid or strong base.

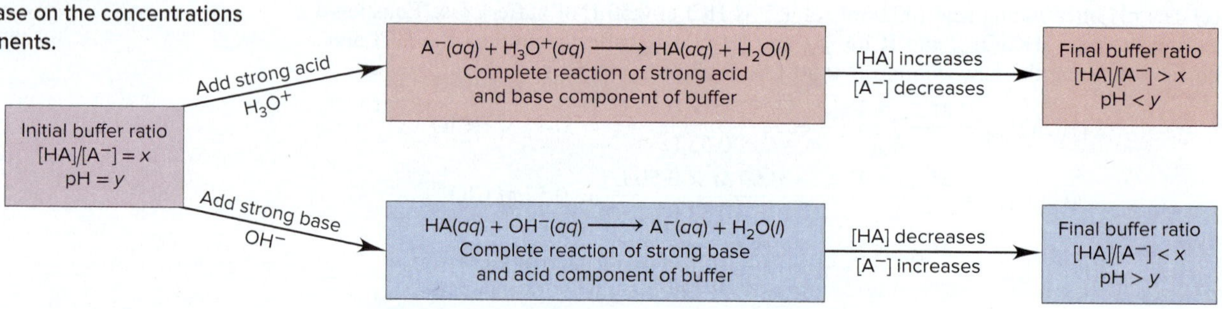

Figure 19.4 Summary of the effect of added acid or base on the concentrations of buffer components.

The Henderson-Hasselbalch Equation

For any weak acid, HA, the dissociation equation and K_a expression are

$$HA + H_2O \rightleftharpoons H_3O^+ + A^-$$

$$K_a = \frac{[H_3O^+][A^-]}{[HA]}$$

The key variable that determines $[H_3O^+]$ is the concentration *ratio* of acid species to base species, so, as before, rearranging to isolate $[H_3O^+]$ gives

$$[H_3O^+] = K_a \times \frac{[HA]}{[A^-]}$$

Taking the negative common (base 10) logarithm of both sides gives

$$\underbrace{-\log[H_3O^+]} = \underbrace{-\log K_a} - \log\left(\frac{[HA]}{[A^-]}\right)$$

which, from definitions, is $\text{pH} = pK_a - \log\frac{[HA]}{[A^-]}$

Then, because of the nature of logarithms, when we invert the buffer-component concentration ratio, the sign of the logarithm changes, to give $\text{pH} = pK_a + \log\left(\frac{[A^-]}{[HA]}\right)$.

Generalizing the previous equation for any conjugate acid-base pair gives the **Henderson-Hasselbalch equation:**

$$pH = pK_a + \log\left(\frac{[\text{base}]}{[\text{acid}]}\right) \quad \textbf{(19.1)}$$

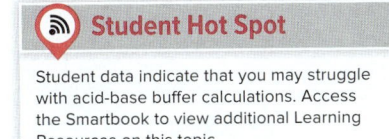

This relationship allows us to solve directly for pH instead of having to calculate $[H_3O^+]$ first. For instance, using the Henderson-Hasselbalch equation in part (b) of Sample Problem 19.1, the pH of the buffer after the addition of NaOH is

$$pH = pK_a + \log\left(\frac{[ClO^-]}{[HClO]}\right) = 7.54 + \log\left(\frac{0.62}{0.48}\right) = 7.65$$

Student Hot Spot

Student data indicate that you may struggle with acid-base buffer calculations. Access the Smartbook to view additional Learning Resources on this topic.

Buffer Capacity and Buffer Range

Let's consider two key aspects of a buffer—its capacity and the closely related range.

Buffer Capacity **Buffer capacity** is a measure of the "strength" of the buffer, its ability to maintain the pH following addition of strong acid or base. Capacity depends ultimately on component concentrations, both the absolute and relative concentrations:

1. In terms of *absolute* concentrations, *the more concentrated the buffer components, the greater the capacity.* Thus, for a given amount of added H_3O^+ or OH^-, the pH of a higher capacity buffer changes less than the pH of a lower capacity buffer (Figure 19.5). Note that *buffer pH is independent of buffer capacity.* A buffer made of equal volumes of 1.0 M CH_3COOH and 1.0 M CH_3COO^- has the same pH (4.74) as a buffer made of equal volumes of 0.10 M CH_3COOH and 0.10 M CH_3COO^-, but the more concentrated buffer has a greater capacity.

2. In terms of *relative* concentrations, *the closer the component concentrations are to each other, the greater the capacity.* As a buffer functions, the concentration of one component increases relative to the other. Because the concentration ratio determines the pH, the less the ratio changes, the less the pH changes. Let's compare the percent change in component concentration ratio for a buffer at two different initial ratios:

- Add 0.010 mol of OH^- to 1.00 L of buffer with initial concentrations $[HA] = [A^-] = 1.000\ M$: $[A^-]$ becomes 1.010 M and [HA] becomes 0.990 M,

$$\frac{[A^-]_{\text{init}}}{[HA]_{\text{init}}} = \frac{1.000\ M}{1.000\ M} = 1.000 \qquad \frac{[A^-]_{\text{final}}}{[HA]_{\text{final}}} = \frac{1.010\ M}{1.990\ M} = 1.02$$

$$\text{Percent change} = \frac{1.02 - 1.000}{1.000} \times 100 = 2\%$$

- Add 0.010 mol of OH^- to 1.00 L of buffer with initial concentrations $[HA] = 0.250\ M$ and $[A^-] = 1.750\ M$: $[A^-]$ becomes 1.760 M and [HA] becomes 0.240 M,

$$\frac{[A^-]_{\text{init}}}{[HA]_{\text{init}}} = \frac{1.750\ M}{0.250\ M} = 7.000 \qquad \frac{[A^-]_{\text{final}}}{[HA]_{\text{final}}} = \frac{1.760\ M}{0.240\ M} = 7.33$$

$$\text{Percent change} = \frac{7.33 - 7.00}{7.00} \times 100 = 4.7\%$$

Note that the change in the concentration ratio is more than twice as large when the initial component concentrations are different than when they are the same. Thus, *a buffer has the highest capacity when the component concentrations are equal.* In the Henderson-Hasselbalch equation, when $[A^-] = [HA]$, their ratio is 1, the log term is 0, and so $pH = pK_a$:

$$pH = pK_a + \log\left(\frac{[A^-]}{[HA]}\right) = pK_a + \log 1 = pK_a + 0 = pK_a$$

Therefore, *a buffer whose pH is equal to or near the pK_a of its acid component has the highest capacity* for a given concentration.

Buffer Range **Buffer range** is the pH range over which the buffer is effective and is also related to the *relative* buffer-component concentrations. The further the concentration ratio is from 1, the less effective the buffer (the lower the buffer capacity). In practice, if the $[A^-]/[HA]$ ratio is greater than 10 or less than 0.1—that is, if one component concentration is more than 10 times the other—buffering action is poor.

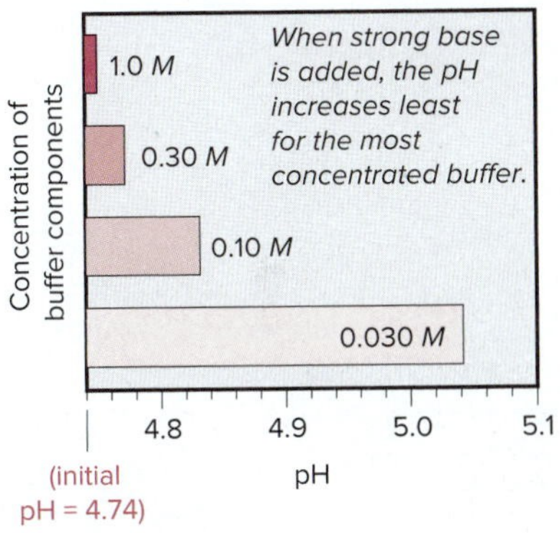

Figure 19.5 The relation between buffer capacity and pH change. The bars indicate the final pH values, after strong base was added, for four CH_3COOH/CH_3COO^- buffers with the same initial pH (4.74) and different component concentrations.

Consider the two extreme concentration ratios for effective buffering:

- For a buffer with $[A^-]/[HA] = 10$: $pH = pK_a + \log\left(\frac{10}{1}\right) = pK_a + 1$.
- For a buffer with $[A^-]/[HA] = 0.1$: $pH = pK_a + \log\left(\frac{1}{10}\right) = pK_a - 1$.

Therefore:

- *Buffers have a usable range within ±1 pH unit of the pK_a of the acid component:* $pH = pK_a \pm 1$.

For example, a buffer consisting of CH_3COOH ($pK_a = 4.74$) and CH_3COO^- has a usable pH range of 4.74 ± 1, or from about 3.7 to 5.7.

SAMPLE PROBLEM 19.2 Using Molecular Scenes to Examine Buffers

Problem The molecular scenes below represent equal-volume samples of four HA/A^- buffers. (HA is *blue and green*, and A^- is *green*; other ions and water are not shown.)

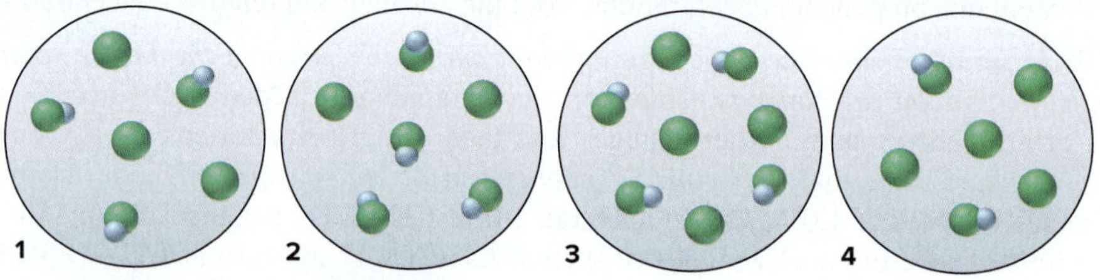

(a) Which buffer has the highest pH?
(b) Which buffer has the greatest capacity?
(c) Should you add a small amount of concentrated strong acid or strong base to convert sample 1 to sample 2 (assuming no volume change)?

Plan The molecular scenes show varying numbers of weak acid molecules (HA) and the conjugate base (A^-). Because the volumes are equal, the scenes represent molarities as well as numbers. **(a)** As the pH rises, more HA loses its H^+ and becomes A^-, so the $[A^-]/[HA]$ ratio will increase. We examine the scenes to see which has the highest ratio. **(b)** Buffer capacity depends on buffer-component concentration *and* ratio. We examine the scenes to see which has a high concentration and a ratio close to 1. **(c)** Adding strong acid converts some A^- to HA, and adding strong base does the opposite. Comparing the $[A^-]/[HA]$ ratios in samples 1 and 2 tells which to add.

Solution **(a)** The $[A^-]/[HA]$ ratios are as follows: For sample 1, $[A^-]/[HA] = 3/3 = 1$. Similarly, sample 2 = 0.5; sample 3 = 1; sample 4 = 2. Sample 4 has the highest pH because it has the highest $[A^-]/[HA]$ ratio.
(b) Samples 1 and 3 have a $[A^-]/[HA]$ ratio of 1, but sample 3 has the greater capacity because it has a higher concentration.
(c) Sample 2 has a lower $[A^-]/[HA]$ ratio than sample 1, so you would add strong acid to sample 1 to convert some A^- to HA.

FOLLOW-UP PROBLEMS

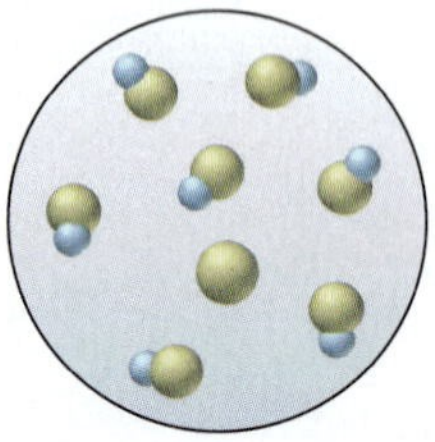

19.2A The molecular scene *(see margin)* shows a sample of an HB/B^- buffer. (HB is *blue and yellow,* and B^- is *yellow;* other ions and water are not shown.)
(a) Would you add a small amount of concentrated strong acid or strong base to increase the buffer capacity?
(b) Assuming no volume change, draw a scene that represents the buffer with the highest possible capacity after the addition in part (a).

19.2B The molecular scenes below show three samples of a buffer consisting of HA (pK_a = 4.2) and A^-. (HA is *blue and red,* and A^- is *red;* other ions and water are not shown.)

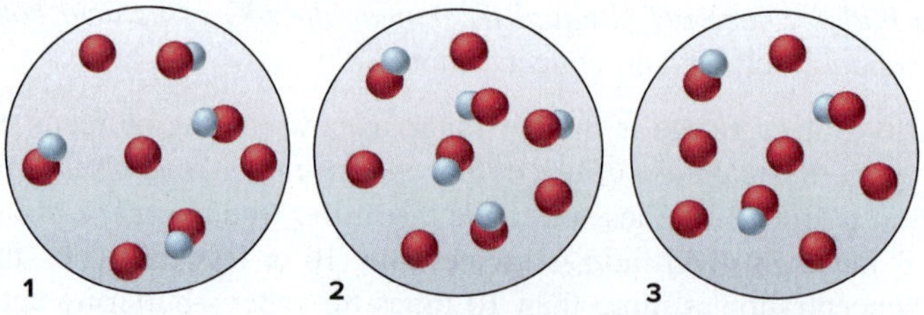

(a) Which buffer has pH > pK_a?
(b) Which buffer can react with the largest amount of added strong base?
(c) Is this conjugate acid-base pair suitable for preparing a buffer with a pH of 6.1?

SIMILAR PROBLEMS 19.10 and 19.11

Preparing a Buffer

Even though chemical supply-houses offer buffers in a variety of pH values and concentrations, you may have to prepare a specific one, for example, in an environmental or biomedical application. Several steps are required to prepare a buffer:

1. *Choose the conjugate acid-base pair.* Deciding on the chemical composition is based to a large extent on the desired pH. Suppose that, for a biochemical experiment, you need a buffer whose pH is 3.90. Therefore, to maximize the capacity, the pK_a of the acid component should be close to 3.90; or $K_a = 10^{-3.90} = 1.3\times10^{-4}$. A table of K_a values (see Appendix C) shows that lactic acid (pK_a = 3.86), glycolic acid (pK_a = 3.83), and formic acid (pK_a = 3.74) are possibilities. To avoid substances that are common in biochemical systems, you choose formic acid, HCOOH, and formate ion, $HCOO^-$, supplied by a soluble salt, such as sodium formate, HCOONa, as the basic component.

2. *Calculate the ratio of buffer component concentrations.* To find the ratio $[A^-]/[HA]$ that gives the desired pH, you use Equation 19.1:

$$\text{pH} = \text{p}K_a + \log\left(\frac{[A^-]}{[HA]}\right) \quad \text{or} \quad 3.90 = 3.74 + \log\left(\frac{[HCOO^-]}{[HCOOH]}\right)$$

$$\log\left(\frac{[HCOO^-]}{[HCOOH]}\right) = 0.16 \quad \text{so} \quad \left(\frac{[HCOO^-]}{[HCOOH]}\right) = 10^{0.16} = 1.4$$

Thus, for every 1.0 mol of HCOOH in a given volume of solution, you need 1.4 mol of HCOONa.

3. *Determine the buffer concentration.* For most laboratory-scale applications, concentrations of about 0.5 *M* are suitable, but the decision is often based on availability of stock solutions. Suppose you have a large stock of 0.40 *M* HCOOH and you need approximately 1.0 L of final buffer. First, you find the amount (mol) of sodium formate that will give the needed 1.4/1.0 ratio, and then convert to mass (g):

$$\text{Amount (mol) of HCOOH} = 1.0\text{ L soln} \times \frac{0.40\text{ mol HCOOH}}{1.0\text{ L soln}} = 0.40\text{ mol HCOOH}$$

$$\text{Amount (mol) of HCOONa} = 0.40\text{ mol HCOOH} \times \frac{1.4\text{ mol HCOONa}}{1.0\text{ mol HCOOH}} = 0.56\text{ mol HCOONa}$$

$$\text{Mass (g) of HCOONa} = 0.56\text{ mol HCOONa} \times \frac{68.01\text{ g HCOONa}}{1\text{ mol HCOONa}} = 38\text{ g HCOONa}$$

4. *Mix the solution and correct the pH.* You dissolve 38 g of sodium formate in the stock 0.40 *M* HCOOH to a total volume of 1.0 L. Because of nonideal behavior (Section 13.6), the buffer may vary slightly from the desired pH, so you add strong acid or strong base dropwise, while monitoring the solution with a pH meter.

SAMPLE PROBLEM 19.3 Preparing a Buffer

Problem An environmental chemist needs a carbonate buffer of pH 10.00 to study the effects of acid rain on limestone-rich soils. How many grams of Na_2CO_3 must she add to 1.5 L of 0.20 *M* $NaHCO_3$ to make the buffer (K_a of $HCO_3^- = 4.7\times10^{-11}$)?

Plan The conjugate pair is HCO_3^- (acid) and CO_3^{2-} (base), and we know the buffer volume (1.5 L) and the concentration (0.20 *M*) of HCO_3^-, so we need to find the buffer-component concentration ratio that gives a pH of 10.00 and the mass of Na_2CO_3 to dissolve. We convert K_a to pK_a and use Equation 19.1 to find the ratio $[CO_3^{2-}]/[HCO_3^-]$ that gives a pH of 10.00. Multiplying the given molarity of HCO_3^- by the volume of solution gives the amount (mol) of HCO_3^- and the ratio of $[CO_3^{2-}]/[HCO_3^-]$ gives the amount (mol) of CO_3^{2-} needed, which we convert to mass (g) of Na_2CO_3.

Solution Calculating pK_a:

$$pK_a = -\log K_a = -\log (4.7\times10^{-11}) = 10.33$$

Solving for $[CO_3^{2-}]/[HCO_3^-]$:

$$pH = pK_a + \log\left(\frac{[CO_3^{2-}]}{[HCO_3^-]}\right) \qquad 10.00 = 10.33 + \log\left(\frac{[CO_3^{2-}]}{[HCO_3^-]}\right)$$

$$-0.33 = \log\left(\frac{[CO_3^{2-}]}{[HCO_3^-]}\right) \quad \text{and} \quad \left(\frac{[CO_3^{2-}]}{[HCO_3^-]}\right) = 10^{-0.33} = 0.47$$

Calculating the amount (mol) of CO_3^{2-} needed for the given volume of solution:

$$\text{Amount (mol) of } HCO_3^- = 1.5 \text{ L soln} \times \frac{0.20 \text{ mol } HCO_3^-}{1.0 \text{ L soln}} = 0.30 \text{ mol } HCO_3^-$$

$$\text{Amount (mol) of } CO_3^{2-} = 0.30 \text{ mol } HCO_3^- \times \frac{0.47 \text{ mol } CO_3^{2-}}{1.0 \text{ mol } HCO_3^-} = 0.14 \text{ mol } CO_3^{2-}$$

Calculating the mass (g) of Na_2CO_3 needed:

$$\text{Mass (g) of } Na_2CO_3 = 0.14 \text{ mol } Na_2CO_3 \times \frac{105.99 \text{ g } Na_2CO_3}{1 \text{ mol } Na_2CO_3} = 15 \text{ g } Na_2CO_3$$

The buffer is prepared by dissolving 15 g of Na_2CO_3 into about 1.3 L of 0.20 *M* $NaHCO_3$ and adding more 0.20 *M* $NaHCO_3$ to make 1.5 L. Then a pH meter is used to adjust the pH to 10.00 by dropwise addition of concentrated strong acid or base.

Check For a useful buffer range, the concentration of the acidic component, $[HCO_3^-]$ in this case, must be within a factor of 10 of the concentration of the basic component, $[CO_3^{2-}]$. And we have 0.30 mol of HCO_3^-, and 0.14 mol of CO_3^{2-}; 0.30/0.14 = 2.1. Make sure the relative amounts of components are reasonable: we want a pH below the pK_a of HCO_3^- (10.33), so we want more of the acidic than the basic species.

FOLLOW-UP PROBLEMS

19.3A How would you prepare a benzoic acid/benzoate buffer with a pH of 4.25, starting with 5.0 L of 0.050 *M* sodium benzoate (C_6H_5COONa) solution and adding the acidic component (K_a of benzoic acid (C_6H_5COOH) = 6.3×10^{-5})?

19.3B What is the component concentration ratio, $[NH_3]/[NH_4^+]$ of a buffer that has a pH of 9.18? What mass of NH_4Cl must be added to 750 mL of 0.15 *M* NH_3 to prepare the buffer with this pH (K_b of NH_3 = 1.76×10^{-5})?

SOME SIMILAR PROBLEMS 19.24–19.27

Another way to prepare a buffer is to form one of the components by *partial neutralization* of the other. For example, you can prepare an $HCOOH/HCOO^-$ buffer by mixing aqueous solutions of HCOOH and NaOH. As OH^- reacts with HCOOH, neutralization of some of the HCOOH produces the $HCOO^-$ needed:

$$\text{HCOOH (HA total)} + OH^-\text{(amt added)} \longrightarrow$$
$$\text{HCOOH (HA total} - OH^- \text{ amt added)} + HCOO^-(OH^- \text{ amt added}) + H_2O$$

This method is based on the same chemical process that occurs when a weak acid is titrated with a strong base, as you'll see in Section 19.2.

› Summary of Section 19.1

- The pH of a buffered solution changes much less than the pH of an unbuffered solution when H_3O^+ or OH^- is added.
- A buffer consists of a weak acid and its conjugate base (or a weak base and its conjugate acid). To be effective, the amounts of the components must be *much* greater than the amount of H_3O^+ or OH^- added.
- The buffer-component concentration ratio determines the pH; the ratio and the pH are related by the Henderson-Hasselbalch equation.

- When H_3O^+ or OH^- is added to a buffer, one component reacts to form the other; thus, $[H_3O^+]$ (and pH) changes only slightly.
- A concentrated (higher capacity) buffer undergoes smaller changes in pH than a dilute buffer. When the buffer pH equals the pK_a of the acid component, the buffer has its highest capacity.
- A buffer has an effective pH range of $pK_a \pm 1$ pH unit.
- To prepare a buffer, choose the conjugate acid-base pair, calculate the ratio of components, determine the buffer concentration, and adjust the final solution to the desired pH.

19.2 ACID-BASE TITRATION CURVES

In Chapter 4, we used titrations to quantify acid-base (and redox) reactions. In this section, we focus on the **acid-base titration curve,** a plot of pH vs. volume of titrant added. We discuss curves for strong acid–strong base, weak acid–strong base, weak base–strong acid, and polyprotic acid–strong base titrations. Running a titration is an exercise for the lab, but understanding the role of acid-base indicators and seeing how salt solutions (Section 18.7) and buffers are involved apply key principles of acid-base equilibria.

Strong Acid–Strong Base Titration Curves

Figure 19.6 shows a typical curve for the titration of a strong acid with a strong base, the data used to construct it, and molecular views of the key species in solution at various points during the titration.

Volume of NaOH added (mL)	pH
00.00	1.00
10.00	1.22
20.00	1.48
30.00	1.85
35.00	2.18
39.00	2.89
39.50	3.20
39.75	3.50
39.90	3.90
39.95	4.20
39.99	4.90
40.00	7.00
40.01	9.10
40.05	9.80
40.10	10.10
40.25	10.50
40.50	10.79
41.00	11.09
45.00	11.76
50.00	12.05
60.00	12.30
70.00	12.43
80.00	12.52

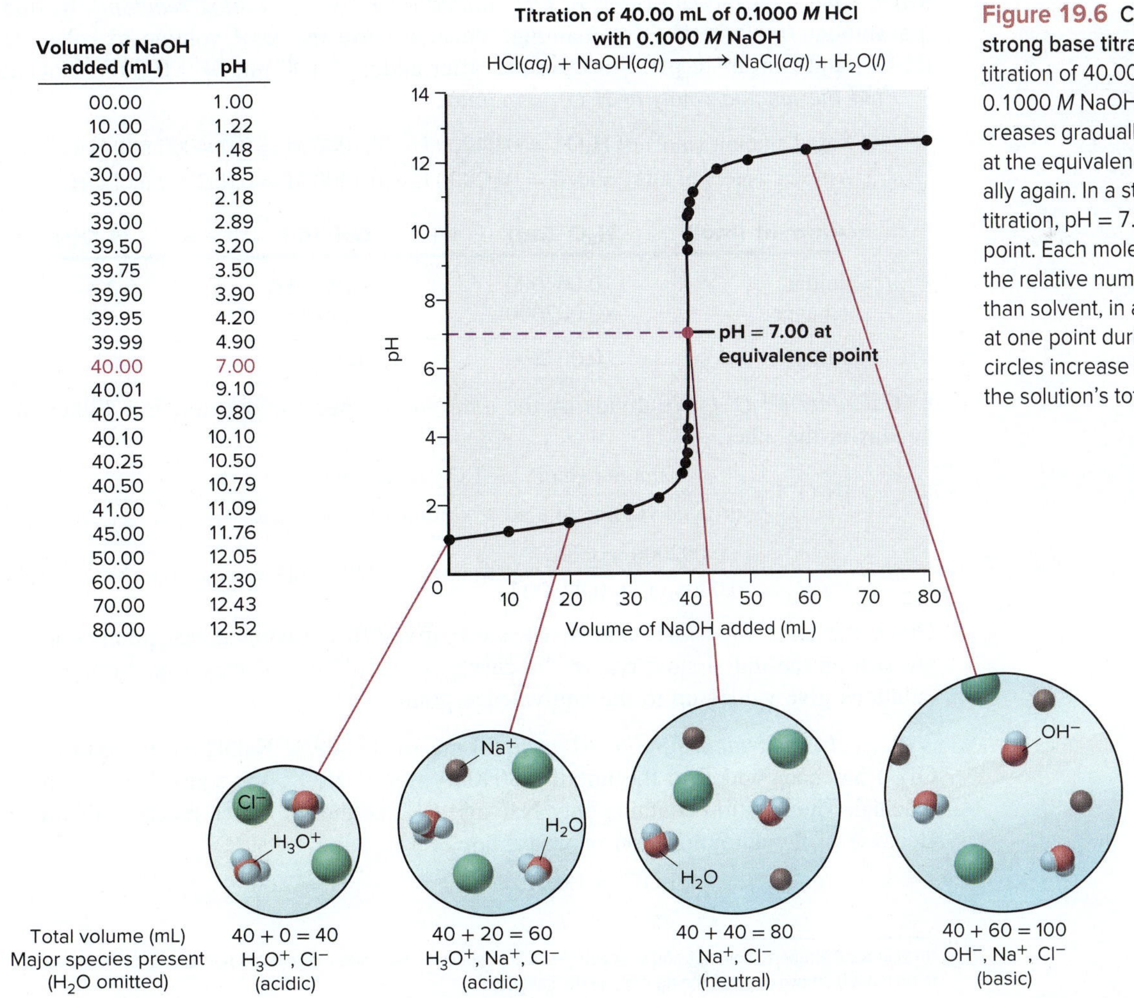

Figure 19.6 **Curve for a strong acid–strong base titration.** Data *(at left)* for the titration of 40.00 mL of 0.1000 *M* HCl with 0.1000 *M* NaOH. In the titration, pH increases gradually, then rapidly near and at the equivalence point, and then gradually again. In a strong acid–strong base titration, pH = 7.00 at the equivalence point. Each molecular-scale view shows the relative numbers of species, other than solvent, in a portion of the solution at one point during the titration. The circles increase in size in proportion to the solution's total volume.

Features of the Curve There are three distinct regions of the titration curve, which correspond to three major changes in slope:

1. *The pH starts out low,* reflecting the high $[H_3O^+]$ of the strong acid, and increases slowly as acid is gradually neutralized by the added base.
2. *The pH rises 6 to 8 units very rapidly.* This steep increase begins when the amount (mol) of OH^- added nearly equals the amount (mol) of H_3O^+ originally present in the acid. One or two more drops of base neutralize the remaining tiny excess of acid and introduce a tiny excess of base.

   The **equivalence point** occurs when the *number of moles of added* OH^- *equals the number of moles of* H_3O^+ originally present. *The solution consists of the anion of the strong acid and the cation of the strong base.* Recall from Chapter 18 that *these ions do not react with water, so the solution is neutral: pH = 7.00.*
3. *The pH increases slowly* beyond the steep rise as more base is added.

Calculating the pH During This Titration By knowing the chemical species present during the titration, we can calculate the pH at various points along the way:

1. *Initial solution of strong HA.* In Figure 19.6, 40.00 mL of 0.1000 *M* HCl is titrated with 0.1000 *M* NaOH. Because a strong acid is completely dissociated, $[HCl] = [H_3O^+] = 0.1000\ M$. Therefore, the initial pH is*

$$pH = -\log [H_3O^+] = -\log (0.1000) = 1.00$$

2. *Before the equivalence point.* As we start adding base, some acid is neutralized and the volume of solution increases. To find the pH at various points up to the equivalence point, we find the *initial* amount (mol) of H_3O^+ and amount (mol) of added OH^-; next, we set up a reaction table and subtract the amount of H_3O^+ *reacted, which equals the amount (mol) of* OH^- *added* (OH^- *is the limiting reactant),* to find the amount (mol) of H_3O^+ remaining. Then, we use the *total* volume to calculate $[H_3O^+]$ and convert to pH. For example, after adding 20.00 mL of 0.1000 *M* NaOH:

- *Find the amount (mol) of* H_3O^+ *remaining.*

$$\text{Initial amount (mol) of } H_3O^+ = 0.04000\ L \times 0.1000\ M = 0.004000 \text{ mol } H_3O^+$$
$$\text{Amount (mol) of } OH^- \text{ added} = 0.02000\ L \times 0.1000\ M = 0.002000 \text{ mol } OH^-$$

Amount (mol)	$H_3O^+(aq)$	+	$OH^-(aq)$	$\longrightarrow$	$H_2O(l)$
Initial	0.004000		0.002000		—
Change	−0.002000		−0.002000		—
Final	0.002000		0		—

- *Calculate* $[H_3O^+]$. We divide by the *total volume* because one solution dilutes the ions in the other:

$$[H_3O^+] = \frac{\text{amount (mol) of } H_3O^+ \text{ remaining}}{\text{original volume of acid} + \text{volume of added base}}$$

$$= \frac{0.002000 \text{ mol } H_3O^+}{0.04000\ L + 0.02000\ L} = 0.03333\ M \qquad pH = -\log (0.03333) = 1.48$$

Given the amount of OH^- added, we are halfway to the equivalence point; but we are still on the initial slow rise of the curve, so the pH is still very low. Similar calculations give values up to the equivalence point.

3. *At the equivalence point.* After 40.00 mL of 0.1000 *M* NaOH (0.004000 mol of OH^-) has been added to the initial 0.004000 mol of H_3O^+, the equivalence point is reached. The solution contains only Na^+ and Cl^-, neither of which reacts with water. Because of the autoionization of water, however,

$$[H_3O^+] = 1.0\times10^{-7}\ M \qquad pH = 7.00$$

*In acid-base titrations, volumes and concentrations are usually known to four significant figures, but pH is reported to no more than two digits to the right of the decimal point.

4. *After the equivalence point.* From the equivalence point on, the pH calculation is based on the amount (mol) of *excess* OH^- present. Find the amount (mol) of added OH^- and set up a reaction table with the *original* amount of HCl; subtract the amount of OH^- *reacted, which equals the amount (mol) of initial H_3O^+ (H_3O^+ is now the limiting reactant),* to find the amount (mol) of OH^- remaining. Then, we use the *total* volume to calculate $[OH^-]$ and convert to pOH and then to pH. For example, after adding 50.00 mL of NaOH to the 40.00 mL of HCl:

- *Find the amount (mol) of OH^- remaining.*

Total amount (mol) of OH^- added $= 0.05000\ L \times 0.1000\ M = 0.005000$ mol OH^-

Amount (mol)	$H_3O^+(aq)$	+	$OH^-(aq)$	⟶	$H_2O(l)$
Initial	0.004000		0.005000		—
Change	−0.004000		−0.004000		—
Final	0		0.001000		—

- *Calculate $[OH^-]$.*

$$[OH^-] = \frac{\text{amount (mol) of } OH^- \text{ remaining}}{\text{original volume of acid + volume of added base}}$$

$$= \frac{0.001000 \text{ mol } OH^-}{0.04000\ L + 0.05000\ L} = 0.01111\ M \qquad pOH = -\log(0.01111) = 1.95$$

$$pH = pK_w - pOH = 14.00 - 1.95 = 12.05$$

Weak Acid–Strong Base Titration Curves

Figure 19.7 shows a curve for the titration of a weak acid with strong base: 40.00 mL of 0.1000 M propanoic acid ($K_a = 1.3\times10^{-5}$) titrated with 0.1000 M NaOH. (We abbreviate the acid, CH_3CH_2COOH, as HPr and the conjugate base, $CH_3CH_2COO^-$, as Pr^-.)

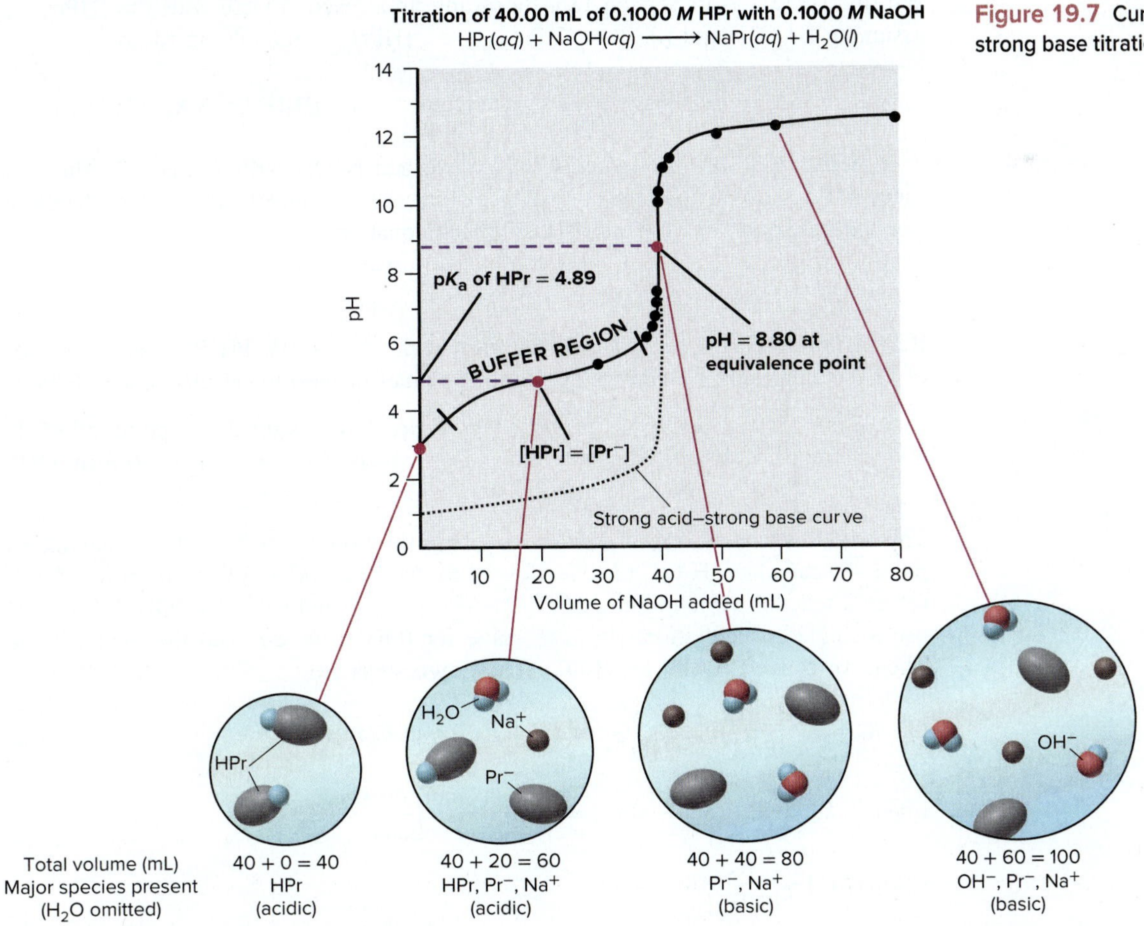

Figure 19.7 Curve for a weak acid–strong base titration.

Features of the Curve The dotted portion of the curve in Figure 19.7 corresponds to the bottom half of the strong acid–strong base curve (Figure 19.6). There are four key points to note for the weak acid curve (the first three differ from the points about the strong acid curve):

1. *The initial pH is higher.* Because the weak acid (HPr) dissociates slightly, much less H_3O^+ is present than with the strong acid.
2. *The curve rises gradually in the so-called buffer region before the steep rise to the equivalence point.* As HPr reacts with strong base, more Pr^- forms, which creates an HPr/Pr^- buffer. At the midpoint of the buffer region, half the initial HPr has reacted (that is, half of the OH^- needed to reach the equivalence point has been added), so $[HPr] = [Pr^-]$, or $[Pr^-]/[HPr] = 1$. Therefore, at this point, *the pH equals the pK_a*:

$$\text{pH} = \text{p}K_a + \log\left(\frac{[\text{Pr}^-]}{[\text{HPr}]}\right) = \text{p}K_a + \log 1 = \text{p}K_a + 0 = \text{p}K_a$$

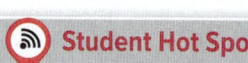

   The pH observed at this point, halfway through the titration, is used to estimate the pK_a of an unknown acid.
3. *The pH at the equivalence point is above 7.00.* The solution contains the strong-base cation Na^+, which does not react with water, and Pr^-, the conjugate base of HPr, which, as a weak base, accepts a proton from H_2O and yields OH^-.
4. *The pH increases slowly* beyond the equivalence point as excess OH^- is added.

Student Hot Spot

Student data indicate that you may struggle with the concept of acid-base titration curves. Access the Smartbook to view additional Learning Resources on this topic.

Calculating the pH During This Titration During the weak acid–strong base titration we must take into account the partial dissociation of the weak acid, the presence of the buffer, and the reaction of the conjugate base with water. Thus, each of four regions of the curve requires its own calculation to find $[H_3O^+]$:

1. *Initial HA solution.* Before base is added, a weak acid dissociates:

$$\text{HPr}(aq) + \text{H}_2\text{O}(l) \rightleftharpoons \text{H}_3\text{O}^+(aq) + \text{Pr}^-(aq)$$

We find $[H_3O^+]$ as in Section 18.4 by setting up a reaction table with $x = [\text{HPr}]_{\text{dissoc}}$, assuming $[H_3O^+]$ (and $[Pr^-]$) $= [\text{HPr}]_{\text{dissoc}} << [\text{HPr}]_{\text{init}}$, and solving for x:

$$K_a = \frac{[\text{H}_3\text{O}^+][\text{Pr}^-]}{[\text{HPr}]} \approx \frac{x^2}{[\text{HPr}]_{\text{init}}} \quad \text{therefore,} \quad x = [\text{H}_3\text{O}^+] \approx \sqrt{K_a \times [\text{HPr}]_{\text{init}}}$$

2. *Solution of HA and added base.* As we add NaOH, HPr forms Pr^-. Thus, for much of the titration up to the equivalence point, we have an HPr/Pr^- buffer. Therefore, we find pH from the Henderson-Hasselbalch equation:

$$\text{pH} = \text{p}K_a + \log\frac{[\text{Pr}^-]}{[\text{HPr}]}$$

Because *volumes cancel in the concentration ratio,* that is, $[Pr^-]/[HPr]$ = amount (mol) of Pr^-/amount (mol) of HPr, we don't need volumes or need to calculate concentrations.

3. *Equivalent amounts of HA and added base.* At the equivalence point, all of the HPr has reacted, so the solution contains Pr^-, which reacts with water to form OH^-:

$$\text{Pr}^-(aq) + \text{H}_2\text{O}(l) \rightleftharpoons \text{HPr}(aq) + \text{OH}^-(aq)$$

This explains why, in a weak acid–strong base titration, pH > 7.00 at the equivalence point. To calculate $[H_3O^+]$ (see Section 18.6), we first find K_b of Pr^- from K_a of HPr, set up a reaction table (assume $[\text{Pr}^-] >> [\text{Pr}^-]_{\text{reacting}}$), and solve for $[OH^-]$. Since we use a single concentration, $[Pr^-]$, to solve for $[OH^-]$, we *do* need the total volume. Then, we convert $[OH^-]$ to $[H_3O^+]$. These two steps are

$$(1)\ K_b = \frac{[\text{OH}^-][\text{HPr}]}{[\text{Pr}^-]} \approx \frac{x^2}{[\text{Pr}^-]}, \text{ or } x = [\text{OH}^-] \approx \sqrt{K_b \times [\text{Pr}^-]},$$

$$\text{where} \quad K_b = \frac{K_w}{K_a} \quad \text{and} \quad [\text{Pr}^-] = \frac{\text{amount (mol) of HPr}_{\text{init}}}{\text{total volume}}$$

$$(2)\ [\text{H}_3\text{O}^+] = \frac{K_w}{[\text{OH}^-]}$$

4. *Solution of excess base.* Beyond the equivalence point, as in the strong acid–strong base titration, we are adding excess OH^-; the very small amount of OH^- produced by the reaction of Pr^- with water is negligible compared to the amount of OH^- from the excess NaOH:

$$[H_3O^+] = \frac{K_w}{[OH^-]} \qquad \text{where} \qquad [OH^-] = \frac{\text{amount (mol) of excess } OH^-}{\text{total volume}}$$

Sample Problem 19.4 shows the overall approach.

SAMPLE PROBLEM 19.4 Finding the pH During a Weak Acid–Strong Base Titration

Problem Calculate the pH during the titration of 40.00 mL of 0.1000 *M* propanoic acid (HPr; $K_a = 1.3\times10^{-5}$) after each of the following additions of 0.1000 *M* NaOH:
(a) 0.00 mL **(b)** 30.00 mL **(c)** 40.00 mL **(d)** 50.00 mL

Plan **(a)** 0.00 mL: No base has been added yet, so this is a weak-acid dissociation. We calculate the pH as we did in Section 18.4. **(b)** 30.00 mL: We find the amount (mol) of Pr^- and of HPr, and substitute into the Henderson-Hasselbalch equation to solve for pH. **(c)** 40.00 mL: The amount (mol) of NaOH added equals the initial amount (mol) of HPr, so a solution of Na^+ and Pr^- exists. We calculate the pH as we did in Section 18.6, except that we need *total* volume to find $[Pr^-]$. **(d)** 50.00 mL: We calculate the amount (mol) of excess OH^- in the total volume, convert to $[H_3O^+]$ and then to pH.

Solution **(a)** 0.00 mL of 0.1000 *M* NaOH added. Following the approach used in Sample Problem 18.8 and just described in the text, we obtain

$$HPr(aq) + H_2O(l) \rightleftharpoons H_3O^+(aq) + Pr^-(aq) \qquad K_a = \frac{[H_3O^+][Pr^-]}{[HPr]} = 1.3\times10^{-5}$$

Concentration (*M*)	HPr(*aq*)	+	$H_2O(l)$	⇌	$H_3O^+(aq)$	+	$Pr^-(aq)$
Initial	0.1000		—		0		0
Change	$-x$		—		$+x$		$+x$
Equilibrium	$0.1000 - x$		—		x		x

$$K_a = \frac{[H_3O^+][Pr^-]}{[HPr]} = 1.3\times10^{-5} \approx \frac{(x)(x)}{0.1000}$$

$$[H_3O^+] = x \approx \sqrt{K_a \times [HPr]_{init}} = \sqrt{(1.3\times10^{-5})(0.1000)} = 1.1\times10^{-3}\ M$$

$$pH = 2.96$$

(b) 30.00 mL of 0.1000 *M* NaOH added. Calculating the initial amounts of HPr and OH^-:

Initial amount (mol) of HPr = 0.04000 L × 0.1000 *M* = 0.004000 mol HPr
Amount (mol) of NaOH added = 0.03000 L × 0.1000 *M* = 0.003000 mol OH^-

For every mole of NaOH, 1 mol of Pr^- forms, so we have this stoichiometry reaction table:

Amount (mol)	HPr(*aq*)	+	$OH^-(aq)$	⟶	$Pr^-(aq)$	+	$H_2O(l)$
Initial	0.004000		0.003000		0		—
Change	−0.003000		−0.003000		+0.003000		—
Final	0.001000		0		0.003000		—

The last line of the table gives the new initial amounts of HPr and Pr^- that react to attain a new equilibrium. With *x* very small, we assume that the $[HPr]/[Pr^-]$ ratio at equilibrium is essentially equal to the ratio of these new initial amounts (see Comment in Sample Problem 19.1).

Calculating pK_a:

$$pK_a = -\log K_a = -\log (1.3\times10^{-5}) = 4.89$$

Solving for pH using the Henderson-Hasselbalch equation:

$$pH = pK_a + \log\frac{[Pr^-]}{[HPr]} = 4.89 + \log\frac{0.003000}{0.001000}$$

$$pH = 5.37$$

(c) 40.00 mL of 0.1000 M NaOH added. Calculating the initial amounts of HPr and OH^-:

$$\text{Initial amount (mol) of HPr} = 0.04000 \text{ L} \times 0.1000\ M = 0.004000 \text{ mol HPr}$$
$$\text{Amount (mol) of NaOH added} = 0.04000 \text{ L} \times 0.1000\ M = 0.004000 \text{ mol } OH^-$$

We have this stoichiometry reaction table:

Amount (mol)	**HPr(*aq*)**	+	**OH^-(*aq*)**	⟶	**Pr^-(*aq*)**	+	**H_2O(*l*)**
Initial	0.004000		0.004000		0		—
Change	0.004000		−0.004000		+0.004000		—
Final	0		0		0.004000		—

Calculating $[Pr^-]$ after all HPr has reacted:

$$[Pr^-] = \frac{0.004000 \text{ mol}}{0.04000 \text{ L} + 0.04000 \text{ L}} = 0.05000\ M$$

Setting up a reaction table for the base dissociation, with $x = [Pr^-]_{reacting} = [OH^-]$ and assuming $[Pr^-]_{reacting} << [Pr^-]_{init}$, calculating K_b, and solving for x (see Sample Problem 18.12 for similar steps):

Concentration, *M*	**Pr^-(*aq*)**	+	**H_2O(*l*)**	⇌	**HPr(*aq*)**	+	**OH^-**
Initial	0.05000		—		0		0
Change	$-x$		—		$+x$		$+x$
Equilibrium	$0.05000 - x$		—		x		x

$$K_b = \frac{K_w}{K_a} = \frac{1.0\times10^{-14}}{1.3\times10^{-5}} = 7.7\times10^{-10}$$

$$K_b = \frac{[HPr][OH^-]}{[Pr^-]} = 7.7\times10^{-10} \approx \frac{(x)(x)}{0.05000}$$

$$[OH^-] = x \approx \sqrt{K_b \times [Pr^-]_{init}} = \sqrt{(7.7\times10^{-10})(0.05000)} = 6.2\times10^{-6}\ M$$

$$[H_3O^+] = \frac{K_w}{[OH^-]} = \frac{1.0\times10^{-14}}{6.2\times10^{-6}} = 1.6\times10^{-9}\ M$$

$$\text{pH} = 8.80$$

(d) 50.00 mL of 0.1000 M NaOH added. Calculating the initial amounts of HPr and OH^-:

$$\text{Initial amount (mol) of HPr} = 0.04000 \text{ L} \times 0.1000\ M = 0.004000 \text{ mol HPr}$$
$$\text{Amount (mol) of NaOH added} = 0.05000 \text{ L} \times 0.1000\ M = 0.005000 \text{ mol } OH^-$$

We have this stoichiometry reaction table:

Amount (mol)	**HPr(*aq*)**	+	**OH^-(*aq*)**	⟶	**Pr^-(*aq*)**	+	**H_2O(*l*)**
Initial	0.004000		0.005000		0		—
Change	−0.004000		−0.004000		+0.004000		—
Final	0		0.001000		0.004000		—

$$[OH^-] = \frac{\text{amount (mol) of excess } OH^-}{\text{total volume}} = \frac{0.001000 \text{ mol}}{0.04000 \text{ L} + 0.05000 \text{ L}} = 0.01111\ M$$

$$[H_3O^+] = \frac{K_w}{[OH^-]} = \frac{1.0\times10^{-14}}{0.01111} = 9.0\times10^{-13}\ M$$

$$\text{pH} = 12.05$$

Check As expected, the pH increases through the four regions of the titration. Be sure to round off and check the arithmetic along the way.

FOLLOW-UP PROBLEMS

19.4A A chemist titrates 20.00 mL of 0.2000 M HBrO ($K_a = 2.3\times10^{-9}$) with 0.1000 M NaOH. What is the pH:

(a) before any base is added?
(b) when $[HBrO] = [BrO^-]$?
(c) at the equivalence point?

(d) when the amount (mol) of OH^- added is twice the amount of HBrO present initially?
(e) Sketch the titration curve, and label the pK_a and the equivalence point.

19.4B For the titration of 30.00 mL of 0.1000 *M* benzoic acid (C_6H_5COOH; $K_a = 6.3\times10^{-5}$) with 0.1500 *M* NaOH, calculate:
(a) the initial pH, before the titration begins;
(b) the pH after the addition of 12.00 mL of base;
(c) the pH at the equivalence point; and
(d) the pH after addition of 22.00 mL of base.

SOME SIMILAR PROBLEMS 19.51, 19.53(a), and 19.54(a)

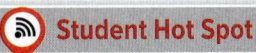

Student Hot Spot

Student data indicate that you may struggle with the calculation of pH during a weak acid-strong base titration. Access the Smartbook to view additional Learning Resources on this topic.

Weak Base–Strong Acid Titration Curves

The opposite of a weak acid–strong base titration is the titration of a weak base (NH_3) with a strong acid (HCl). This titration curve, shown in Figure 19.8, has the *same shape as the weak acid–strong base curve, but it is inverted.*

Thus, we have the same features as the weak acid–strong base curve, but *the pH decreases* throughout the process:

1. *The initial weak-base solution has a pH well above 7.00* since NH_3 in water produces OH^-:

$$NH_3(aq) + H_2O(l) \rightleftharpoons NH_4^+(aq) + OH^-(aq)$$

$[H_3O^+]$ can be calculated from the following relationships (see Sample Problem 18.11):

$$K_b = \frac{[NH_4^+][OH^-]}{[NH_3]} \approx \frac{x^2}{[NH_3]_{init}} \quad \text{therefore,} \quad x = [OH^-] \approx \sqrt{K_b \times [NH_3]_{init}}$$

$$[H_3O^+] \approx \frac{K_w}{[OH^-]}$$

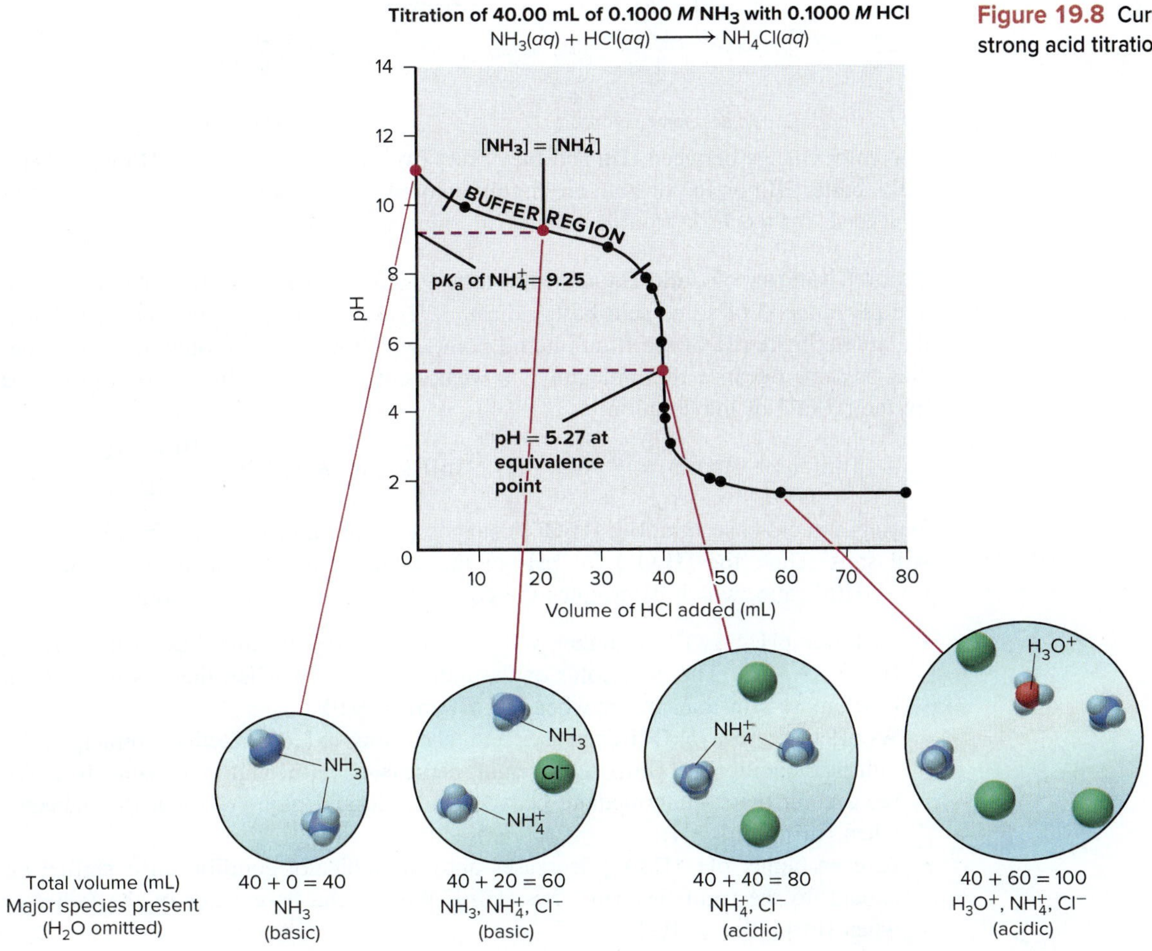

Figure 19.8 Curve for a weak base–strong acid titration.

2. *The pH decreases gradually in the buffer region,* where significant amounts of NH_3 and its conjugate acid, NH_4^+, are present. At the midpoint of this region, *the pH equals the* pK_a of NH_4^+:

$$pH = pK_a + \log\left(\frac{[NH_3]}{[NH_4^+]}\right) = pK_a + \log 1 = pK_a + 0 = pK_a$$

3. *The curve drops steeply at the equivalence point.* All the NH_3 has reacted with added HCl, and the solution contains only NH_4^+ and Cl^-. Note that *the pH at the equivalence point is below 7.00* because Cl^- does not react with water and NH_4^+ is acidic:

$$NH_4^+(aq) + H_2O(l) \rightleftharpoons NH_3(aq) + H_3O^+(aq)$$

The pH at the equivalence point is calculated with this relationship:

$$K_a = \frac{[NH_3][H_3O^+]}{[NH_4^+]} \approx \frac{x^2}{[NH_4^+]} \quad \text{therefore,} \quad x = [H_3O^+] \approx \sqrt{K_a \times [NH_4^+]}$$

4. *The pH decreases slowly* beyond the equivalence point as excess H_3O^+ is added.

$$[H_3O^+] = \frac{\text{amount (mol) of excess } H_3O^+}{\text{total volume}}$$

Monitoring pH with Acid-Base Indicators

An *acid-base indicator* is a weak organic acid (denoted as HIn) whose color (A) differs from the color (B) of its conjugate base (In^-) because the two species have slightly different structures. For example, the acid form of the very commonly used indicator phenolphthalein is colorless, while the conjugate base form is pink:

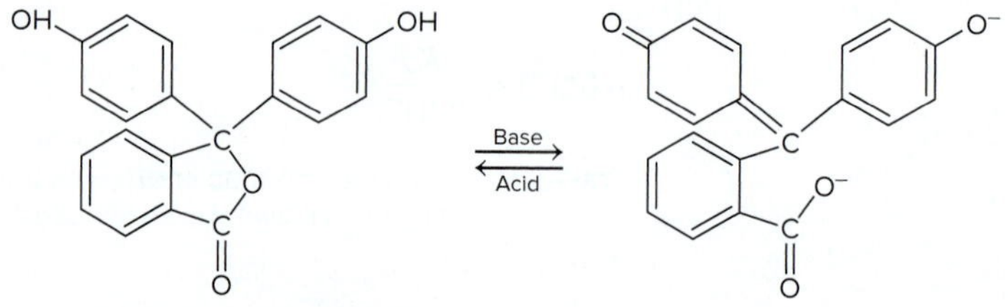

HIn acid form (colorless) In^- base form (pink)

The color change between HIn and In^- occurs over a specific, narrow pH range. Typically, either HIn or In^- or both are highly colored, so only a tiny amount of indicator is needed, far too little to affect the pH during the titration.

Color Changes of Acid-Base Indicators Figure 19.9 shows the color change(s) and pH range(s) of some acid-base indicators. To select an indicator, you must know the approximate pH of the titration end point, which means you know the ionic species present. Because the indicator is a weak acid, the $[In^-]/[HIn]$ ratio is governed by the $[H_3O^+]$ of the solution:

$$\underset{\text{color A}}{HIn(aq)} + H_2O(l) \rightleftharpoons H_3O^+(aq) + \underset{\text{color B}}{In^-(aq)} \qquad K_a \text{ of HIn} = \frac{[H_3O^+][In^-]}{[HIn]}$$

During an acid-base titration, $[H_3O^+]$ changes and the indicator equilibrium reaction will shift. Thus, the $[H_3O^+]$ (or pH) of the solution being titrated determines the $[In^-]/[HIn]$ ratio, which determines the color of the indicator in the solution:

- At lower pH, $[H_3O^+]$ is higher and the indicator equilibrium lies to the left, so $[HIn] >> [In^-]$. The acid color of the indicator is seen when there is at least ten times more HIn than In^- or when $[In^-]/[HIn] \leq 1/10$.
- As pH increases, $[H_3O^+]$ decreases and, according to Le Châtelier's principle, the indicator equilibrium shifts to the right, decreasing [HIn] and increasing $[In^-]$. We see a color that is intermediate between the acid and base colors of the indicator when $[In^-]/[HIn] \approx 1$.
- At even higher pH, $[H_3O^+]$ decreases until the indicator equilibrium is shifted far enough to the right that $[In^-] >> [HIn]$. We see the base color of the indicator when $[In^-]/[HIn] \geq 10/1$.

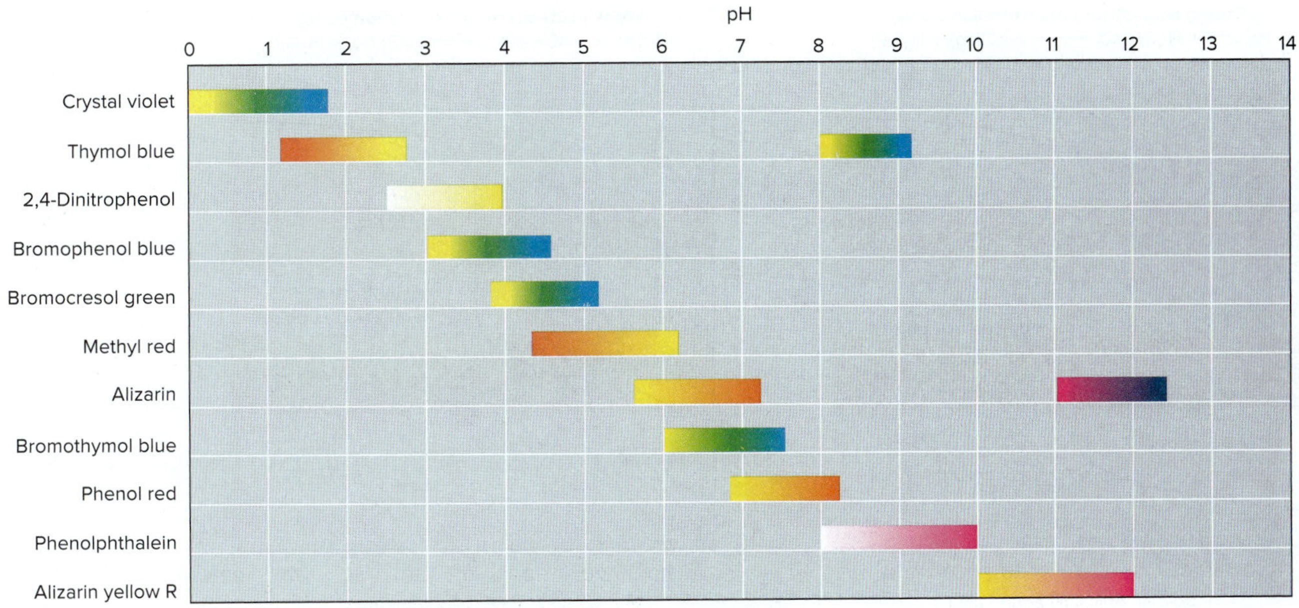

Figure 19.9 Colors and approximate pH ranges of some common acid-base indicators.

The Henderson-Hasselbalch equation relates the ratio of $[In^-]/[HIn]$ to the pK_a of the indicator and the pH of the solution:

$$\text{pH} = \text{p}K_a + \log\left(\frac{[\text{In}^-]}{[\text{HIn}]}\right)$$

- When $\text{pH} = \text{p}K_a - 1$, $\dfrac{[\text{In}^-]}{[\text{HIn}]} = 1/10$ and the acid (HIn) color is seen.
- When $\text{pH} = \text{p}K_a$, $\dfrac{[\text{In}^-]}{[\text{HIn}]} = 1$ and the acid (HIn) and base (In^-) colors merge into an intermediate hue.
- When $\text{pH} = \text{p}K_a + 1$, $\dfrac{[\text{In}^-]}{[\text{HIn}]} = 10/1$ and the base (In^-) color is seen.

Thus, an indicator has a *color range* equal to a 10^2-fold range in the $[In^-]/[HIn]$ ratio (from 1/10, or 0.1, to 10/1, or 10): an *indicator changes color over a range of about 2 pH units approximately centered around its* pK_a*:*

$$\text{pH of indicator color change} \approx \text{p}K_a \text{ of indicator} \pm 1$$

For example, bromothymol blue ($pK_a = 7.0$) has a pH range of about 6.0 to 7.6. As Figure 19.10 shows, it is yellow below a pH of 6.0 *(left),* blue above a pH of 7.6 *(right),* and greenish in between *(center).*

Figure 19.10 The color change of the indicator bromothymol blue.

Source: © McGraw-Hill Education/Stephen Frisch, photographer

Choosing an Acid-Base Indicator During an acid-base titration, we are generally interested in the *equivalence point* of the titration, which must be distinguished from the *end point*:

- The equivalence point occurs when the number of moles of added OH^- (or H_3O^+) equals the number of moles of H_3O^+ (or OH^-) originally present in the flask.
- The **end point** occurs when the indicator, which we added before the titration began, changes color. *We choose an indicator with a color change close to the pH of the equivalence point* because we want the *visible* end point to signal the *invisible* equivalence point.

Figure 19.11 on the next page shows titration curves for the three titrations we examined earlier in this section, with the color change of two indicators, methyl red and phenolphthalein, superimposed on the curves. Figure 19.11A shows that both indicators are suitable for a strong acid-strong base titration. Methyl red changes from red at pH 4.2 to yellow at pH 6.3, and phenolphthalein changes from colorless at pH 8.3 to pink at pH 10.0. Neither change occurs at the equivalence point (pH = 7.00), but both occur on the

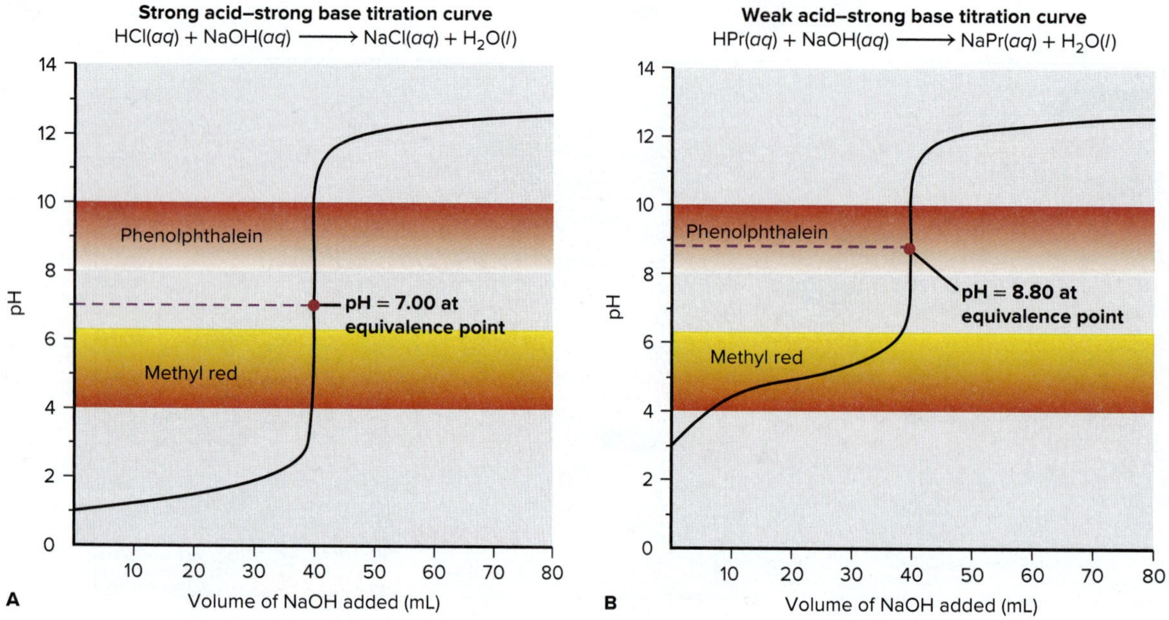

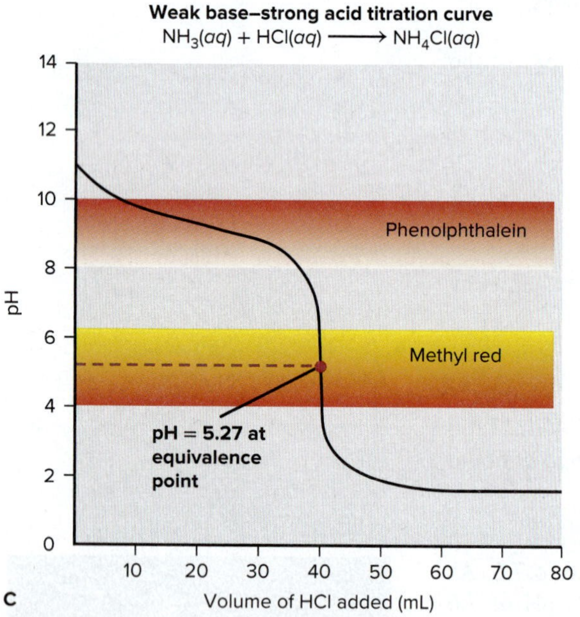

Figure 19.11 Acid-base indicator selection for different acid-base titrations.

vertical portion of the curve. Therefore, when methyl red turns yellow or phenolphthalein pink, we are within a drop or two of the equivalence point.

As shown in Figure 19.11B, there is a more limited choice of indicator for the weak acid-strong base titration because the steep rise occurs within a smaller pH range. Phenolphthalein will work because it changes color within this range. But methyl red will not because its color change occurs significantly before the equivalence point and requires about 30 mL of titrant, rather than just a drop or two. The opposite is true for the weak base-strong acid titration shown in Figure 19.11C. Phenolphthalein changes color too slowly and before the equivalence point is reached, but methyl red's change occurs on the steep portion of the curve and straddles the equivalence point.

Titration Curves for Polyprotic Acids

Except for sulfuric acid, the common polyprotic acids are all weak. The successive K_a values for a polyprotic acid differ by several orders of magnitude, as these values for sulfurous acid show:

$$H_2SO_3(aq) + H_2O(l) \rightleftharpoons HSO_3^-(aq) + H_3O^+(aq)$$
$$K_{a1} = 1.4\times10^{-2} \quad \text{and} \quad pK_{a1} = 1.85$$

$$HSO_3^-(aq) + H_2O(l) \rightleftharpoons SO_3^{2-}(aq) + H_3O^+(aq)$$
$$K_{a2} = 6.5\times10^{-8} \quad \text{and} \quad pK_{a2} = 7.19$$

In a titration of a diprotic acid such as H_2SO_3, two OH^- ions are required to react with the two H^+ ions of each acid molecule. Figure 19.12 shows the titration curve for sulfurous acid with strong base. Because of the large difference in K_a values, each mole of H^+ is titrated separately, so H_2SO_3 molecules lose one H^+ before any HSO_3^- ions do:

$$H_2SO_3 \xrightarrow{1\text{ mol }OH^-} HSO_3^- \xrightarrow{1\text{ mol }OH^-} SO_3^{2-}$$

Several features of the curve are

- The same amount of base (0.004000 mol OH^-) is required per mole of H^+.
- There are two equivalence points and two buffer regions. The pH at the midpoint of each buffer region is equal to the pK_a of the acidic species present then.
- The pH of the first equivalence point is below 7.00, because HSO_3^- is a stronger acid than it is a base (K_a of $HSO_3^- = 6.5\times10^{-8}$; K_b of $HSO_3^- = 7.1\times10^{-13}$).
- The pH of the second equivalence point is above 7.00, because SO_3^{2-} acts as a weak base and accepts a proton from H_2O to yield OH^-.

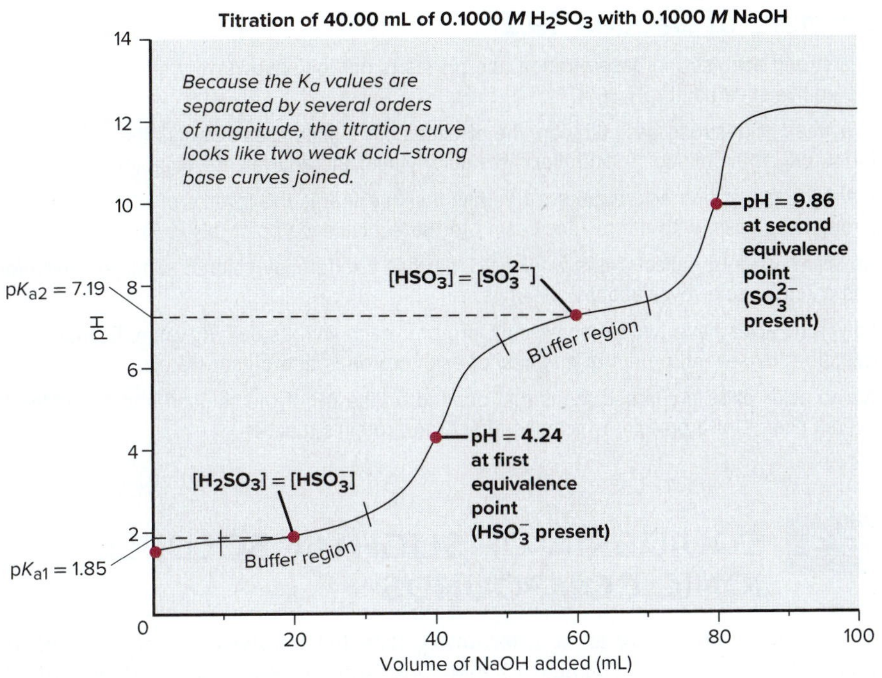

Figure 19.12 Curve for the titration of a weak polyprotic acid.

Amino Acids as Biological Polyprotic Acids

Amino acids have the general formula NH_2—CH(R)—COOH, where R can be one of about 20 different groups (Sections 13.2 and 15.6). In essence, amino acids contain a weak base (—NH_2) and a weak acid (—COOH) on the same molecule. Both the amino and carboxylic acid groups are protonated at low pH: $^{+}NH_3$—CH(R)—COOH. Thus, in this form, the amino acid behaves like a polyprotic acid. For glycine, the simplest amino acid (R = H), the dissociation reactions and pK_a values are

$$^{+}NH_3CH_2COOH(aq) + H_2O(l) \rightleftharpoons {}^{+}NH_3CH_2COO^-(aq) + H_3O^+(aq) \qquad pK_{a1} = 2.35$$

$$^{+}NH_3CH_2COO^-(aq) + H_2O(l) \rightleftharpoons NH_2CH_2COO^-(aq) + H_3O^+(aq) \qquad pK_{a2} = 9.78$$

These values show that the —COOH group is *much* more acidic than the $-NH_3^+$ group. As we saw with H_2SO_3, the acidic protons *(black circles)* are titrated separately, so all the —COOH protons are removed before any $-NH_3^+$ protons are:

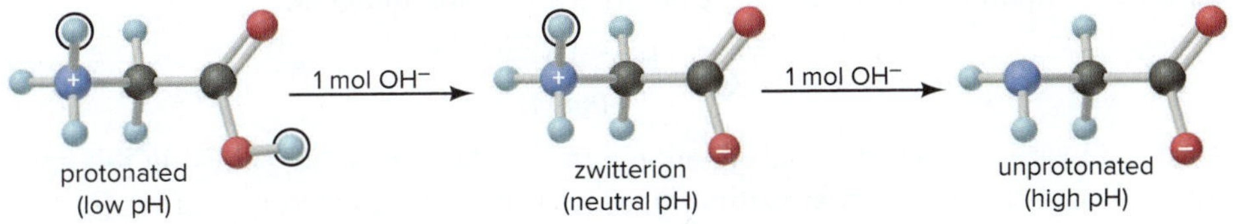

Thus, at physiological pH (~7), which is between the two pK_a values, *an amino acid exists as a zwitterion* (German *zwitter,* "double"), a species with opposite charges on the same molecule; the zwitterion of glycine is $^{+}NH_3CH_2COO^-$.

Of the 20 different R groups of amino acids, several have *additional* —COO^- or $-NH_3^+$ groups at pH 7 (see Figure 15.30). When amino acids link to form a protein, their charged R groups give the protein its overall charge, which is often related to its function. A widely studied example of this relationship occurs in the hereditary disease sickle cell anemia. Normal red blood cells contain hemoglobin molecules that have two glutamic acid R groups (—$CH_2CH_2COO^-$), each providing a negative charge at a critical region. Abnormal hemoglobin molecules in sickle cell anemia have two valine R groups (—CH_3), which are uncharged, at the same region. This change in just 2 of hemoglobin's 574 amino acids lowers charge repulsions between the hemoglobin molecules. As a result, they clump together in fiber-like structures, which leads to the sickle shape of the red blood cells (Figure 19.13). The misshapen cells block capillaries, and the painful course of sickle cell anemia usually ends in early death.

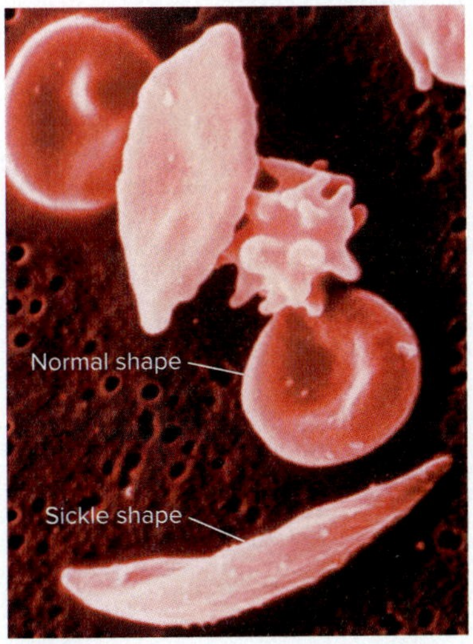

Figure 19.13 Abnormal shape of red blood cells in sickle cell anemia.

Source: © Jackie Lewin, Royal Free Hospital/Science Source

› Summary of Section 19.2

- › In a strong acid–strong base titration, the pH starts out low, rises slowly, then shoots up near the equivalence point (pH = 7).
- › In a weak acid–strong base titration, the pH starts out higher, rises slowly in the buffer region (pH = pK_a at the midpoint), and then rises quickly near the equivalence point (pH > 7).
- › A weak base–strong acid titration curve has a shape that is the inverse of the weak acid–strong base case, with the pH decreasing to the equivalence point (pH < 7).
- › An acid-base (pH) indicator is a weak acid that has a differently colored conjugate base form and changes color over about 2 pH units.
- › Polyprotic acids have two or more acidic protons, each with its own K_a value. Because the K_a's differ by several orders of magnitude, each proton is titrated separately.
- › Amino acids exist in charged forms that depend on the pH of the solution and determine the overall charge of a protein, which can affect the protein's function.

19.3 EQUILIBRIA OF SLIGHTLY SOLUBLE IONIC COMPOUNDS

In this section, we explore an equilibrium system that involves the solubility of ionic compounds. Recall from Chapter 13 that most solutes, even those called "soluble," have a limited solubility in a particular solvent. In a saturated solution at a particular temperature, equilibrium exists between dissolved and undissolved solute. Slightly soluble ionic compounds, which we've been calling "insoluble," reach equilibrium with very little solute dissolved. In this introductory discussion, we will *assume* that, as with a soluble ionic compound, the small amount of a slightly soluble ionic compound that does dissolve dissociates completely into ions.

The Ion-Product Expression (Q_{sp}) and the Solubility-Product Constant (K_{sp})

For a slightly soluble ionic compound, *equilibrium exists between solid solute and aqueous ions.* Thus, for example, for a saturated solution of lead(II) fluoride, we have

$$PbF_2(s) \rightleftharpoons Pb^{2+}(aq) + 2F^-(aq)$$

As for any equilibrium system, we can write a reaction quotient:

$$Q_c = \frac{[Pb^{2+}][F^-]^2}{[PbF_2]}$$

And, as in earlier cases, the concentration term for the pure solid, $[PbF_2]$ in this case, equals 1, and so we obtain a relationship called the *ion-product expression,* Q_{sp}:

$$Q_c \times 1 = Q_{sp} = [Pb^{2+}][F^-]^2$$

When the solution is saturated, the numerical value of Q_{sp} attains a constant value called the **solubility-product constant, K_{sp}.** The K_{sp} for PbF_2 at 25°C, for example, is 3.6×10^{-8}. Like other equilibrium constants, K_{sp} depends *only* on the temperature, not on the individual ion concentrations.

The form of Q_{sp} is identical to that of other reaction quotients: each concentration is raised to an exponent equal to the coefficient for that species in the balanced equation, which in this case also *equals the subscript of each ion in the compound's formula.* At saturation, the concentration terms have their equilibrium values. Thus, in general, for a slightly soluble ionic compound, M_pX_q, composed of the ions M^{n+} and X^{z-}, the ion-product expression at equilibrium is

$$Q_{sp} = [M^{n+}]^p[X^{z-}]^q = K_{sp} \qquad \textbf{(19.2)}$$

We'll write Q_{sp} directly using the symbol K_{sp}.

SAMPLE PROBLEM 19.5 Writing Ion-Product Expressions

Problem Write the ion-product expression at equilibrium for each compound: **(a)** magnesium carbonate; **(b)** iron(II) hydroxide; **(c)** calcium phosphate; **(d)** silver sulfide.

Plan We write an equation for a saturated solution and then write the ion-product expression at equilibrium, K_{sp} (Equation 19.2).

Solution **(a)** Magnesium carbonate:

$$MgCO_3(s) \rightleftharpoons Mg^{2+}(aq) + CO_3^{2-}(aq) \qquad K_{sp} = [Mg^{2+}][CO_3^{2-}]$$

(b) Iron(II) hydroxide:

$$Fe(OH)_2(s) \rightleftharpoons Fe^{2+}(aq) + 2OH^-(aq) \qquad K_{sp} = [Fe^{2+}][OH^-]^2$$

(c) Calcium phosphate:

$$Ca_3(PO_4)_2(s) \rightleftharpoons 3Ca^{2+}(aq) + 2PO_4^{3-}(aq) \qquad K_{sp} = [Ca^{2+}]^3[PO_4^{3-}]^2$$

(d) Silver sulfide:

$$Ag_2S(s) \rightleftharpoons 2Ag^+(aq) + S^{2-}(aq) \qquad K_{sp} = [Ag^+]^2[S^{2-}]$$

Check You can check by using the exponents as subscripts to see if you obtain the formula of the compound from K_{sp}.

Comment The S^{2-} ion in Ag_2S in part (d) is so basic that it reacts completely with water to form the hydrogen sulfide ion (HS^-) and OH^-:

$$S^{2-}(aq) + H_2O(l) \longrightarrow HS^-(aq) + OH^-(aq)$$

We more correctly write the ion-product expression for Ag_2S (and other metal sulfides) by replacing $[S^{2-}]$ with $[HS^-][OH^-]$:

$$K_{sp} = [Ag^+]^2[HS^-][OH^-]$$

However, as a simplification, we will present the ion-product expression for a metal sulfide in terms of $[S^{2-}]$, as shown in part (d) of this sample problem.

FOLLOW-UP PROBLEMS

19.5A Write the ion-product expression at equilibrium for each compound: **(a)** calcium sulfate; **(b)** chromium(III) carbonate; **(c)** magnesium hydroxide; **(d)** aluminum hydroxide.

19.5B Write the formula and give the name of the compound having each ion-product expression: **(a)** $K_{sp} = [Pb^{2+}][CrO_4^{2-}]$; **(b)** $K_{sp} = [Fe^{2+}][S^{2-}]$; **(c)** $K_{sp} = [Sr^{2+}][F^-]^2$; **(d)** $K_{sp} = [Cu^{2+}]^3[PO_4^{3-}]^2$.

SOME SIMILAR PROBLEMS 19.67–19.70

Calculations Involving the Solubility-Product Constant

In Chapters 17 and 18, we considered two types of equilibrium problems. In one type, we use concentrations (or other quantities) to find K, and in the other, we use K to find concentrations. Here we encounter the same two types.

The Problem with Assuming Complete Dissociation Before we focus on calculations, let's address a complication that affects accuracy and results in approximate answers. Our assumption that the small dissolved amount of these compounds dissociates completely into separate ions is an oversimplification. Many slightly soluble salts have polar covalent metal-nonmetal bonds (Section 9.5), and partially dissociated or even undissociated species occur in solution. Here are two of many examples:

- With slightly soluble lead(II) chloride, the aqueous solution contains not only the separate $Pb^{2+}(aq)$ and $Cl^-(aq)$ ions we expect from complete dissociation, but also undissociated $PbCl_2(aq)$ molecules and $PbCl^+(aq)$ ions.
- In an aqueous solution of $CaSO_4$, undissociated ion pairs, $Ca^{2+}SO_4^{2-}(aq)$, are present.

These partly dissociated and undissociated species increase the *apparent* solubility of a slightly soluble salt above the value obtained by assuming complete dissociation.

Table 19.2 Solubility-Product Constants (K_{sp}) of Selected Ionic Compounds at 25°C

Name, Formula	K_{sp}
Aluminum hydroxide, $Al(OH)_3$	3×10^{-34}
Cobalt(II) carbonate, $CoCO_3$	1.0×10^{-10}
Iron(II) hydroxide, $Fe(OH)_2$	4.1×10^{-15}
Lead(II) fluoride, PbF_2	3.6×10^{-8}
Lead(II) sulfate, $PbSO_4$	1.6×10^{-8}
Mercury(I) iodide, Hg_2I_2	4.7×10^{-29}
Silver sulfide, Ag_2S	8×10^{-48}
Zinc iodate, $Zn(IO_3)_2$	3.9×10^{-6}

More advanced courses discuss these factors, but we simply mention them in the Comments of several sample problems. Thus, it is best to view the specific results of the calculations as first approximations.

The Meaning of K_{sp} and How to Determine It from Solubility Values The K_{sp} value indicates *how far the dissolution proceeds at equilibrium (saturation).* Table 19.2 presents a few K_{sp} values; Appendix C lists many more. Note that all the values are low, but they range over many orders of magnitude. In Sample Problem 19.6, we find the K_{sp} from the solubility of a compound.

SAMPLE PROBLEM 19.6 Determining K_{sp} from Solubility

Problem **(a)** Lead(II) sulfate ($PbSO_4$) is a key component in lead-acid car batteries. Its solubility in water at 25°C is 4.25×10^{-3} g/100. mL solution. What is the K_{sp} of $PbSO_4$?
(b) When lead(II) fluoride (PbF_2) is shaken with pure water at 25°C, the solubility is found to be 0.64 g/L. Calculate the K_{sp} of PbF_2.

Plan We are given solubilities in various units and must find K_{sp} values. As always, we write the dissolution equation and ion-product expression for each compound, which show the number of moles of each ion. We use the molar mass to convert the solubility of the compound from the given mass units to *molar solubility* (molarity), then use that value to find the molarity of each ion, and substitute into the ion-product expression to calculate K_{sp}.

Solution **(a)** For $PbSO_4$. Writing the equation and ion-product (K_{sp}) expression:

$$PbSO_4(s) \rightleftharpoons Pb^{2+}(aq) + SO_4^{2-}(aq) \qquad K_{sp} = [Pb^{2+}][SO_4^{2-}]$$

Converting solubility to molar solubility:

$$\text{Molar solubility of } PbSO_4 = \frac{0.00425 \text{ g } PbSO_4}{100. \text{ mL soln}} \times \frac{1000 \text{ mL}}{1 \text{ L}} \times \frac{1 \text{ mol } PbSO_4}{303.3 \text{ g } PbSO_4} = 1.40 \times 10^{-4}\ M\ PbSO_4$$

Determining molarities of the ions: Because 1 mol of Pb^{2+} and 1 mol of SO_4^{2-} form when 1 mol of $PbSO_4$ dissolves, $[Pb^{2+}] = [SO_4^{2-}] = 1.40 \times 10^{-4}\ M$.

Substituting these values into the ion-product expression to calculate K_{sp}:

$$K_{sp} = [Pb^{2+}][SO_4^{2-}] = (1.40 \times 10^{-4})^2 = 1.96 \times 10^{-8}$$

(b) For PbF_2. Writing the equation and K_{sp} expression:

$$PbF_2(s) \rightleftharpoons Pb^{2+}(aq) + 2F^-(aq) \qquad K_{sp} = [Pb^{2+}][F^-]^2$$

Converting solubility to molar solubility:

$$\text{Molar solubility of } PbF_2 = \frac{0.64 \text{ g } PbF_2}{1 \text{ L soln}} \times \frac{1 \text{ mol } PbF_2}{245.2 \text{ g } PbF_2} = 2.6 \times 10^{-3}\ M\ PbF_2$$

Determining molarities of the ions: Since 1 mol of Pb^{2+} and 2 mol of F^- form when 1 mol of PbF_2 dissolves, we have

$$[Pb^{2+}] = 2.6 \times 10^{-3}\ M \qquad \text{and} \qquad [F^-] = 2(2.6 \times 10^{-3}\ M) = 5.2 \times 10^{-3}\ M$$

Substituting these values into the ion-product expression to calculate K_{sp}:

$$K_{sp} = [Pb^{2+}][F^-]^2 = (2.6 \times 10^{-3})(5.2 \times 10^{-3})^2 = 7.0 \times 10^{-8}$$

Check The low solubilities are consistent with K_{sp} values being small. **(a)** The molar solubility seems about right: $\sim \dfrac{4 \times 10^{-2} \text{ g/L}}{3 \times 10^2 \text{ g/mol}} \approx 1.3 \times 10^{-4}\ M$. Squaring this number gives 1.7×10^{-8}, close to the calculated K_{sp}. **(b)** Let's check the math in the final step as follows: $\sim(3 \times 10^{-3})(5 \times 10^{-3})^2 = 7.5 \times 10^{-8}$, close to the calculated K_{sp}.

Comment **1.** In part (b), the formula PbF_2 means that $[F^-]$ is twice $[Pb^{2+}]$. We follow the ion-product expression exactly and square this value of $[F^-]$.

2. The tabulated K_{sp} values for these compounds (Table 19.2) are lower than our calculated values. For PbF_2, for instance, the tabulated value is 3.6×10^{-8}, but we calculated 7.0×10^{-8} from solubility data. The discrepancy arises because we assumed that PbF_2 in solution dissociates completely to Pb^{2+} and F^-. Here is an example of the complication pointed out earlier. Actually, about a third of the PbF_2 dissolves as $PbF^+(aq)$ and a small amount dissolves as undissociated $PbF_2(aq)$. The solubility given in the problem statement (0.64 g/L) is determined experimentally and includes these other species, which we did *not* include in our calculation.

FOLLOW-UP PROBLEMS

19.6A When fluorite (CaF_2; *see photo*) is pulverized and shaken in water at 18°C, 10.0 mL of the solution contains 1.5×10^{-4} g of solute. Find the K_{sp} of CaF_2 at 18°C.

19.6B At 20°C, 3.2×10^{-4} g of silver phosphate (Ag_3PO_4) is soluble in 50. mL of solution. Find the K_{sp} of Ag_3PO_4 at 20°C.

SOME SIMILAR PROBLEMS 19.71–19.74

A sample of fluorite.

Source: © Joel Arem/Science Source

Determining Solubility from K_{sp} The reverse of Sample Problem 19.6 involves finding the solubility of a compound based on its formula and K_{sp} value. We'll use an approach similar to the one we used for weak acids in Sample Problem 18.8: we define the unknown amount dissolved—the molar solubility—as S, include ion concentrations in terms of this unknown in a reaction table, and solve for S.

SAMPLE PROBLEM 19.7 Determining Solubility from K_{sp}

Problem Calcium hydroxide (slaked lime) is a major component of mortar, plaster, and cement, and solutions of $Ca(OH)_2$ are used in industry as a strong, inexpensive base. Calculate the molar solubility of $Ca(OH)_2$ in water given that the K_{sp} is 6.5×10^{-6}.

Plan We write the dissolution equation and the ion-product expression. We know K_{sp} (6.5×10^{-6}), so to find molar solubility (S), we set up a reaction table that expresses $[Ca^{2+}]$ and $[OH^-]$ in terms of S, substitute into the ion-product expression, and solve for S.

Solution Writing the equation and ion-product expression:

$$Ca(OH)_2(s) \rightleftharpoons Ca^{2+}(aq) + 2OH^-(aq) \qquad K_{sp} = [Ca^{2+}][OH^-]^2 = 6.5\times10^{-6}$$

Setting up a reaction table, with S = molar solubility:

Concentration (M)	$Ca(OH)_2(s)$	$\rightleftharpoons$	$Ca^{2+}(aq)$	+	$2OH^-(aq)$
Initial	—		0		0
Change	—		$+S$		$+2S$
Equilibrium	—		S		$2S$

Substituting into the ion-product expression and solving for S:

$$K_{sp} = [Ca^{2+}][OH^-]^2 = (S)(2S)^2 = (S)(4S^2) = 4S^3 = 6.5\times10^{-6}$$

$$S = \sqrt[3]{\frac{6.5\times10^{-6}}{4}} = 1.2\times10^{-2}\ M$$

Check We expect a low solubility from a slightly soluble salt. If we reverse the calculation, we should obtain the given K_{sp}: $4(1.2\times10^{-2})^3 = 6.9\times10^{-6}$, which is close to 6.5×10^{-6}.

Comment 1. Note that we did not double and *then* square $[OH^-]$. $2S$ *is* the $[OH^-]$, so we just squared it, as the ion-product expression required.

2. Once again, we assumed that the solid dissociates completely. Actually, the solubility is increased to about $2.0\times10^{-2}\ M$ by the presence of $CaOH^+(aq)$ formed in the reaction $Ca(OH)_2(s) \rightleftharpoons CaOH^+(aq) + OH^-(aq)$. Our calculated answer is only approximate because we did not take this other species into account.

FOLLOW-UP PROBLEMS

19.7A Milk of magnesia, a suspension of $Mg(OH)_2$ in water, relieves indigestion by neutralizing stomach acid. What is the molar solubility of $Mg(OH)_2$ ($K_{sp} = 6.3\times10^{-10}$) in water?

19.7B Calcium phosphate, $Ca_3(PO_4)_2$, is used as a dietary supplement to treat low blood calcium. What is the molar solubility of $Ca_3(PO_4)_2$ ($K_{sp} = 1.2\times10^{-29}$) in water?

SOME SIMILAR PROBLEMS 19.75(a) and 19.76(a)

Table 19.3 Relationship Between K_{sp} and Solubility at 25°C

No. of Ions	Formula	Cation/Anion	K_{sp}	Solubility (*M*)
2	$MgCO_3$	1/1	3.5×10^{-8}	1.9×10^{-4}
2	$PbSO_4$	1/1	1.6×10^{-8}	1.3×10^{-4}
2	$BaCrO_4$	1/1	2.1×10^{-10}	1.4×10^{-5}
3	$Ca(OH)_2$	1/2	6.5×10^{-6}	1.2×10^{-2}
3	BaF_2	1/2	1.5×10^{-6}	7.2×10^{-3}
3	CaF_2	1/2	3.2×10^{-11}	2.0×10^{-4}
3	Ag_2CrO_4	2/1	2.6×10^{-12}	8.7×10^{-5}

Using K_{sp} Values to Compare Solubilities As long as we compare compounds with the *same total number of ions* in their formulas, K_{sp} values indicate *relative* solubility: *the higher the K_{sp}, the greater the solubility* (Table 19.3). Note that for compounds that form three ions, the relationship holds whether the cation/anion ratio is 1/2 or 2/1, because the mathematical expression containing S is the same ($4S^3$) in the calculation (see Sample Problem 19.7).

Effect of a Common Ion on Solubility

From Le Châtelier's principle (Section 17.6), we know that *adding a common ion decreases the solubility of a slightly soluble ionic compound.* Consider a saturated solution of lead(II) chromate:

$$PbCrO_4(s) \rightleftharpoons Pb^{2+}(aq) + CrO_4^{2-}(aq) \qquad K_{sp} = [Pb^{2+}][CrO_4^{2-}] = 2.3\times10^{-13}$$

At a given temperature, K_{sp} depends on the product of the ion concentrations. If the concentration of either ion goes up, the other goes down to maintain K_{sp}. Suppose we add Na_2CrO_4, a soluble salt, to the saturated $PbCrO_4$ solution. The concentration of the common ion, CrO_4^{2-}, increases, and some of it combines with Pb^{2+} ion to form more solid $PbCrO_4$ (Figure 19.14). That is, the equilibrium position shifts to the left:

$$PbCrO_4(s) \leftleftharpoons Pb^{2+}(aq) + CrO_4^{2-}(aq;\text{ added})$$

As a result of the addition, $[Pb^{2+}]$ is lower. And, since $[Pb^{2+}]$ defines the solubility of $PbCrO_4$, in effect, the solubility of $PbCrO_4$ has decreased. Note that we would get the same result if Na_2CrO_4 solution were the solvent; that is, $PbCrO_4$ is more soluble in water than in aqueous Na_2CrO_4.

Lead(II) chromate, a slightly soluble salt, forms a saturated solution:
PbCrO4(s) ⇌ Pb2+(aq) + CrO42−(aq)
When Na2CrO4 solution is added, the amount of PbCrO4(s) increases:
PbCrO4(s) ⇌ Pb2+(aq) + CrO42−(aq; added)

Figure 19.14 The effect of a common ion on solubility.

Source: © McGraw-Hill Education/Stephen Frisch, photographer

SAMPLE PROBLEM 19.8 Calculating the Effect of a Common Ion on Solubility

Problem In Sample Problem 19.7, we calculated the solubility of $Ca(OH)_2$ in water. What is its solubility in 0.10 *M* $Ca(NO_3)_2$ (K_{sp} of $Ca(OH)_2 = 6.5\times10^{-6}$)?

Plan Addition of Ca^{2+}, the common ion, should lower the solubility. We write the equation and ion-product expression and set up a reaction table, with $[Ca^{2+}]_{init}$ reflecting the 0.10 *M* $Ca(NO_3)_2$ and *S* equal to $[Ca^{2+}]_{from\ Ca(OH)_2}$. To simplify the math, we assume that, because K_{sp} is low, *S* is so small relative to $[Ca^{2+}]_{init}$ that it can be neglected. Then we solve for *S* and check the assumption.

Solution $Ca(NO_3)_2$ is a soluble salt that dissociates completely in water; in a solution of 0.10 *M* $Ca(NO_3)_2$, $[Ca^{2+}] = 0.10\ M$, and NO_3^- is a spectator ion that does not affect the solubility of $Ca(OH)_2$.

Writing the equation and ion-product expression for $Ca(OH)_2$:

$$Ca(OH)_2(s) \rightleftharpoons Ca^{2+}(aq) + 2OH^-(aq) \qquad K_{sp} = [Ca^{2+}][OH^-]^2 = 6.5\times10^{-6}$$

Setting up the reaction table, with $S = [Ca^{2+}]_{from\ Ca(OH)_2}$:

Concentration (*M*)	$Ca(OH)_2(s)$	$\rightleftharpoons$	$Ca^{2+}(aq)$	+	$2OH^-(aq)$
Initial	—		0.10 (from $Ca(NO_3)_2$)		0
Change	—		$+S$ (from $Ca(OH)_2$)		$+2S$
Equilibrium	—		$0.10 + S$		$2S$

Making the assumption: K_{sp} is small, so $S << 0.10\ M$; thus, $0.10\ M + S \approx 0.10\ M$. Substituting into the ion-product expression and solving for *S*:

$$K_{sp} = [Ca^{2+}][OH^-]^2 = 6.5\times10^{-6} \approx (0.10)(2S)^2$$

Therefore, $$4S^2 \approx \frac{6.5\times10^{-6}}{0.10} \quad \text{so} \quad S \approx \sqrt{\frac{6.5\times10^{-5}}{4}} = 4.0\times10^{-3}\ M$$

Checking the assumption by comparing the magnitude of *S* to 0.10 *M*:

$$\frac{4.0\times10^{-3}\ M}{0.10\ M} \times 100 = 4.0\% < 5\%$$

Check In Sample Problem 19.7, the solubility of $Ca(OH)_2$ was 0.012 *M*; here it is 0.0040 *M*, one-third as much. As expected, the solubility *decreased* in the presence of added Ca^{2+}, the common ion.

FOLLOW-UP PROBLEMS

19.8A To improve the x-ray image used to diagnose an intestinal disorder, a patient drinks an aqueous suspension of $BaSO_4$, because Ba^{2+} is opaque to x-rays *(photo)*. However, Ba^{2+} is also toxic; thus, $[Ba^{2+}]$ is lowered by adding dilute Na_2SO_4. What is the solubility of $BaSO_4$ ($K_{sp} = 1.1\times10^{-10}$) in **(a)** water and **(b)** 0.10 *M* Na_2SO_4?

19.8B Calculate the solubility of CaF_2 ($K_{sp} = 3.2\times10^{-11}$) in **(a)** water, **(b)** 0.20 *M* $CaCl_2$, and **(c)** 0.20 *M* NiF_2.

SOME SIMILAR PROBLEMS 19.75(b), 19.76(b), 19.77, and 19.78

Student Hot Spot

Student data indicate that you may struggle with solubility calculations involving a common ion. Access the Smartbook to view additional Learning Resources on this topic.

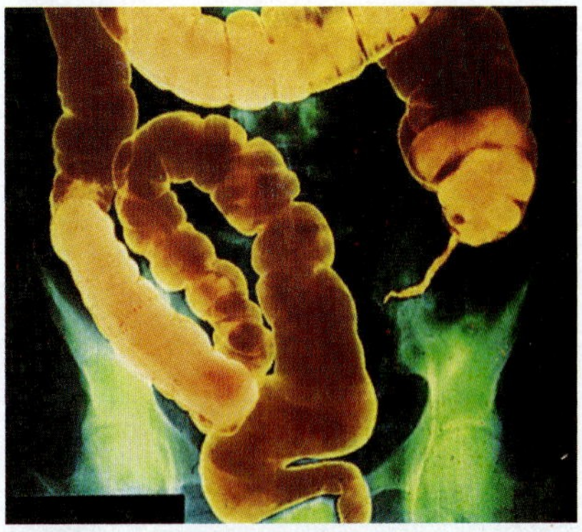

$BaSO_4$ imaging of a large intestine.
Source: © CNRI/Science Source

Effect of pH on Solubility

If a slightly soluble ionic compound contains the anion of a weak acid, *addition of H_3O^+ (from a strong acid) increases the compound's solubility.* Once again, Le Châtelier's principle explains why. Consider a saturated solution of calcium carbonate:

$$CaCO_3(s) \rightleftharpoons Ca^{2+}(aq) + CO_3^{2-}(aq)$$

Adding strong acid introduces H_3O^+, which reacts with the anion of a weak acid, CO_3^{2-}, to form the anion of another weak acid, HCO_3^-:

$$CO_3^{2-}(aq) + H_3O^+(aq) \longrightarrow HCO_3^-(aq) + H_2O(l)$$

If enough H_3O^+ is added, carbonic acid forms, which decomposes to H_2O and CO_2, and the gas escapes the container:

$$HCO_3^-(aq) + H_3O^+(aq) \longrightarrow [H_2CO_3(aq)] + H_2O(l) \longrightarrow CO_2(g) + 2H_2O(l)$$

The net effect of adding H_3O^+ is a shift in the equilibrium position to the right, and more $CaCO_3$ dissolves:

$$CaCO_3(s) \rightleftharpoons Ca^{2+} + CO_3^{2-} \xrightarrow{H_3O^+} HCO_3^- \xrightarrow{H_3O^+} [H_2CO_3] \longrightarrow CO_2(g) + H_2O + Ca^{2+}$$

In fact, this example illustrates a qualitative field test for carbonate minerals because the CO_2 bubbles vigorously (Figure 19.15).

Figure 19.15 Test for the presence of a carbonate. When a carbonate mineral is treated with HCl, bubbles of CO_2 form.

Source: © McGraw-Hill Education/Stephen Frisch, photographer

In contrast, adding H_3O^+ to a saturated solution of a slightly soluble ionic compound that contains a strong-acid anion, such as AgCl, has no effect on its solubility:

$$AgCl(s) \rightleftharpoons Ag^+(aq) + Cl^-(aq)$$

The Cl^- ion can coexist with high $[H_3O^+]$, so the equilibrium position is not affected.

SAMPLE PROBLEM 19.9 Predicting the Effect on Solubility of Adding Strong Acid

Problem Write balanced equations to explain whether addition of H_3O^+ from a strong acid affects the solubility of each ionic compound:
(a) Lead(II) bromide **(b)** Copper(II) hydroxide **(c)** Iron(II) sulfide

Plan We write the balanced dissolution equation and note the anion:

- Weak-acid anions react with H_3O^+ and increase solubility when a strong acid is added.
- Strong-acid anions do not react with H_3O^+, so addition of a strong acid has no effect.

Solution **(a)** $PbBr_2(s) \rightleftharpoons Pb^{2+}(aq) + 2Br^-(aq)$

No effect. Br^- is the anion of HBr, a strong acid, so it does not react with H_3O^+.

(b) $Cu(OH)_2(s) \rightleftharpoons Cu^2(aq) + 2OH^-(aq)$

Increases solubility. OH^- is the anion of H_2O, a very weak acid, so it reacts with the added H_3O^+:

$$OH^-(aq) + H_3O^+(aq) \longrightarrow 2H_2O(l)$$

As OH^- is removed from solution by this reaction, the solubility equilibrium shifts to the right and additional $Cu(OH)_2$ dissolves.

(c) $FeS(s) \rightleftharpoons Fe^{2+}(aq) + S^{2-}(aq)$

Increases solubility. S^{2-} is the anion of the weak acid HS^-, so it reacts with the added H_3O^+, removing it from solution; the solubility equilibrium shifts to the right and additional FeS dissolves:

$$S^{2-}(aq) + H_3O^+(aq) \longrightarrow HS^-(aq) + H_2O(l)$$

FOLLOW-UP PROBLEMS

19.9A Write balanced equations to show how addition of $HNO_3(aq)$ affects the solubility of: **(a)** calcium fluoride; **(b)** iron(III) hydroxide; **(c)** silver iodide.

19.9B Write balanced equations to show how addition of $HBr(aq)$ affects the solubility of: **(a)** silver cyanide; **(b)** copper(I) chloride; **(c)** magnesium phosphate.

SOME SIMILAR PROBLEMS 19.83–19.86

Applying Ionic Equilibria to the Formation of a Limestone Cave

Figure 19.16 Limestone cave in Nerja, Málaga, Spain.

Source: © Ruth Melnick

Limestone caves and the remarkable structures within them provide striking evidence of the results of aqueous ionic equilibria involving carbonate rocks and the carbon dioxide and water that have flowed through them for many hundreds of millennia (Figure 19.16).

The Cave-Forming Reactions Limestone is mostly calcium carbonate ($CaCO_3$; $K_{sp} = 3.3\times10^{-9}$). Two key reactions help us understand how limestone caves form:

1. Gaseous CO_2 in air is in equilibrium with aqueous CO_2 in natural waters:

$$CO_2(g) \xrightleftharpoons{H_2O(l)} CO_2(aq) \qquad \textbf{(equation 1)}$$

The concentration of aqueous CO_2 is proportional to the partial pressure of $CO_2(g)$ in contact with the water (Henry's law; Section 13.4):

$$[CO_2(aq)] \propto P_{CO_2}$$

Due to the continual release of CO_2 from within Earth (outgassing), P_{CO_2} in soil-trapped air is *higher* than P_{CO_2} in the atmosphere.

2. The reaction of CO_2 with water produces H_3O^+:

$$CO_2(aq) + 2H_2O(l) \rightleftharpoons H_3O^+(aq) + HCO_3^-(aq)$$

Thus, since $CaCO_3$ contains the anion of a weak acid, this formation of H_3O^+ increases the solubility of $CaCO_3$:

$$CaCO_3(s) + CO_2(aq) + H_2O(l) \rightleftharpoons Ca^{2+}(aq) + 2HCO_3^-(aq) \qquad \textbf{(equation 2)}$$

The Cave-Forming Process Here is an overview of the process that forms most limestone caves:

1. As surface water trickles through cracks in the ground, it meets soil-trapped air with a high P_{CO_2}. As a result, $[CO_2(aq)]$ increases (equation 1 shifts to the right), and the solution becomes more acidic.
2. When this CO_2-rich water contacts $CaCO_3$, more $CaCO_3$ dissolves (equation 2 shifts to the right). As a result, more rock is carved out, more water flows in, and so on. Centuries pass, and a cave slowly begins to form.
3. Some of the aqueous solution, dilute $Ca(HCO_3)_2$, passes through the ceiling of the growing cave. As it drips, it meets air, which has a lower P_{CO_2} than the soil, so some $CO_2(aq)$ comes out of solution (equation 1 shifts to the left).
4. Consequently, some $CaCO_3$ precipitates on the ceiling and on the floor below, where the drops land (equation 2 shifts to the left). After many decades, the ceiling bears a *stalactite*, while a spike, called a *stalagmite*, grows up from the floor. Eventually, they meet to form a column of precipitated limestone.

The same chemical process can lead to different shapes. Standing pools of $Ca(HCO_3)_2$ solution form limestone "lily pads" or "corals." Cascades of solution form delicate limestone "draperies" on a cave wall, with fabulous colors arising from trace metal ions, such as iron (reddish brown) or copper (bluish green).

Predicting the Formation of a Precipitate: Q_{sp} vs. K_{sp}

As in Section 17.4, here we compare the values of Q_{sp} and K_{sp} to see if a reaction has reached equilibrium and, if not, in which net direction it will progress until it does. Using solutions of *soluble* salts that contain the ions of *slightly* soluble salts, we can calculate the ion concentrations and predict the result when we mix the solutions:

- If $Q_{sp} = K_{sp}$, the solution is saturated and no change will occur.
- If $Q_{sp} > K_{sp}$, a precipitate will form until the remaining solution is saturated.
- If $Q_{sp} < K_{sp}$, no precipitate will form because the solution is unsaturated.

SAMPLE PROBLEM 19.10 Predicting Whether a Precipitate Will Form

Problem A common laboratory method for preparing a precipitate is to mix solutions containing the component ions. Does a precipitate form when 0.100 L of 0.30 *M* $Ca(NO_3)_2$ is mixed with 0.200 L of 0.060 *M* NaF?

Plan First, we decide which slightly soluble salt could form, look up its K_{sp} value in Appendix C, and write a dissolution equation and ion-product expression. To see

Precipitation of CaF_2.
Source: © McGraw-Hill Education/ Stephen Frisch, photographer

whether mixing these solutions forms the precipitate, we find the initial ion concentrations by calculating the amount (mol) of each ion from its concentration and volume, and dividing by the *total* volume because each solution dilutes the other. Finally, we substitute these concentrations to calculate Q_{sp}, and compare Q_{sp} with K_{sp}.

Solution The ions present are Ca^{2+}, Na^+, F^-, and NO_3^-. All sodium and all nitrate salts are soluble (Table 4.1), so the only possible compound that could precipitate is CaF_2 ($K_{sp} = 3.2\times10^{-11}$). Writing the equation and ion-product expression:

$$CaF_2(s) \rightleftharpoons Ca^{2+}(aq) + 2F^-(aq) \qquad Q_{sp} = [Ca^{2+}][F^-]^2$$

Calculating the ion concentrations:

$$\text{Amount (mol) of } Ca^{2+} = 0.30\ M\ Ca^{2+} \times 0.100\ \text{L} = 0.030\ \text{mol } Ca^{2+}$$

$$[Ca^{2+}]_{init} = \frac{0.030\ \text{mol } Ca^{2+}}{0.100\ \text{L} + 0.200\ \text{L}} = 0.10\ M\ Ca^{2+}$$

$$\text{Amount (mol) of } F^- = 0.060\ M\ F^- \times 0.200\ \text{L} = 0.012\ \text{mol } F^-$$

$$[F^-]_{init} = \frac{0.012\ \text{mol } F^-}{0.100\ \text{L} + 0.200\ \text{L}} = 0.040\ M\ F^-$$

Substituting into the ion-product expression and comparing Q_{sp} with K_{sp}:

$$Q_{sp} = [Ca^{2+}]_{init}[F^-]^2_{init} = (0.10)(0.040)^2 = 1.6\times10^{-4}$$

Because $Q_{sp} > K_{sp}$, CaF_2 will precipitate until $Q_{sp} = 3.2\times10^{-11}$.

Check Make sure you round off and quickly check the math. For example, $Q_{sp} = (1\times10^{-1})(4\times10^{-2})^2 = 1.6\times10^{-4}$. With K_{sp} so low, CaF_2 must have a low solubility, and given the sizable concentrations being mixed, we would expect CaF_2 to precipitate.

FOLLOW-UP PROBLEMS

19.10A As a result of mineral erosion and biological activity, phosphate ion is common in natural waters, where it often precipitates as insoluble salts such as $Ca_3(PO_4)_2$. If $[Ca^{2+}]_{init} = [PO_4^{3-}]_{init} = 1.0\times10^{-9}\ M$ in a given river, will $Ca_3(PO_4)_2$ ($K_{sp} = 1.2\times10^{-29}$) precipitate?

19.10B Addition of a soluble sulfide compound such as sodium sulfide is an effective way to remove metals such as lead from wastewater. Does a precipitate form when 0.500 L of 0.10 M Na_2S is added to 25 L of wastewater containing 0.015 g Pb^{2+}/L?

SOME SIMILAR PROBLEMS 19.87–19.90

SAMPLE PROBLEM 19.11 Using Molecular Scenes to Predict Whether a Precipitate Will Form

Problem These four scenes represent solutions of silver *(gray)* and carbonate *(black and red)* ions above solid silver carbonate. (The solid, other ions, and water are not shown.)

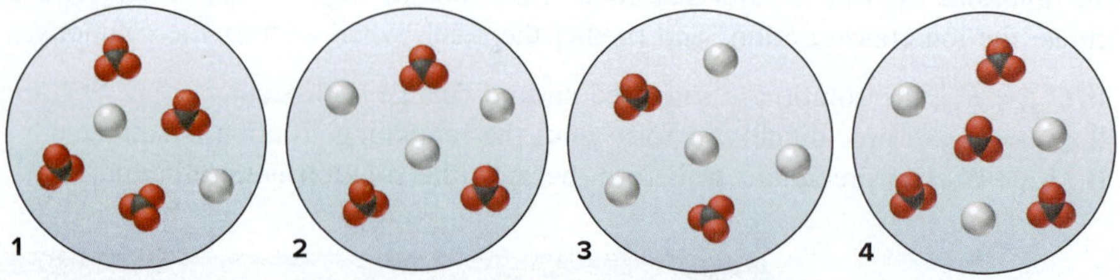

(a) Which scene best represents the solution in equilibrium with the solid?
(b) In which, if any, other scene(s) will additional solid silver carbonate form?
(c) Explain how, if at all, addition of a small volume of concentrated strong acid affects the $[Ag^+]$ in scene 4 and the mass of solid present.

Plan (a) The solution of silver and carbonate ions in equilibrium with the solid (Ag_2CO_3) should have the same relative numbers of cations and anions as in the formula. We examine the scenes to see which has a ratio of 2 Ag^+ to 1 CO_3^{2-}. **(b)** A solid forms if the value of Q_{sp} exceeds that of K_{sp}. We write the dissolution equation and Q_{sp} expression. Then we count ions to calculate Q_{sp} for each scene and see which Q_{sp} value, if any, exceeds the value for the scene identified in part (a). **(c)** The CO_3^{2-} ion reacts with added H_3O^+, so adding strong acid will shift the equilibrium to the right. We write the equations and determine how a shift to the right affects $[Ag^+]$ and the mass of solid Ag_2CO_3.

Solution (a) Scene 3 is the only one with an Ag^+/CO_3^{2-} ratio of 2/1, as in the solid's formula.

(b) Calculating the ion products:

$$Ag_2CO_3(s) \rightleftharpoons 2Ag^+(aq) + CO_3^{2-}(aq) \qquad Q_{sp} = [Ag^+]^2[CO_3^{2-}]$$

Scene 1: $Q_{sp} = (2)^2(4) = 16$ Scene 2: $Q_{sp} = (3)^2(3) = 27$
Scene 3: $Q_{sp} = (4)^2(2) = 32$ Scene 4: $Q_{sp} = (3)^2(4) = 36$

Therefore, for scene 3, $K_{sp} = 32$; the Q_{sp} value for scene 4 is the only other one that equals or exceeds 32, so a precipitate of Ag_2CO_3 will form there.

(c) Writing the equations:

$$(1) \quad Ag_2CO_3(s) \rightleftharpoons 2Ag^+(aq) + CO_3^{2-}(aq)$$

$$(2) \quad CO_3^{2-}(aq) + 2H_3O^+(aq) \longrightarrow [H_2CO_3(aq)] + 2H_2O(l) \longrightarrow 3H_2O(l) + CO_2(g)$$

The CO_2 leaves as a gas, so adding H_3O^+ shifts the equilibrium position of reaction 2 to the right. This change lowers the $[CO_3^{2-}]$ in reaction 1, thereby causing more CO_3^{2-} to form. As a result, more solid dissolves, which means that the $[Ag^+]$ increases and the mass of Ag_2CO_3 decreases.

Check (a) In scene 1, the formula has two CO_3^{2-} per formula unit, not two Ag^+.
(b) Even though scene 4 has fewer Ag^+ ions than scene 3, its Q_{sp} value is higher and exceeds the K_{sp}.

FOLLOW-UP PROBLEMS

19.11A The following scenes represent solutions of nickel(II) *(gray)* and hydroxide *(red and blue)* ions that are above solid nickel(II) hydroxide. (The solid, other ions, and water are not shown.)

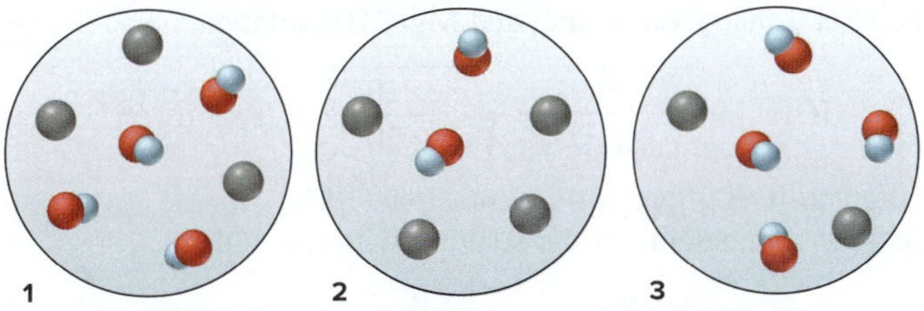

(a) Which scene best depicts the solution at equilibrium with the solid?
(b) In which, if any, other scene(s) will additional solid form?
(c) Will addition of a small amount of concentrated strong acid or strong base affect the mass of solid present in any scene? Explain.

19.11B The following scenes represent solutions of lead(II) *(black)* and chloride *(green)* ions that are above solid lead(II) chloride. (The solid, other ions, and water are not shown.)

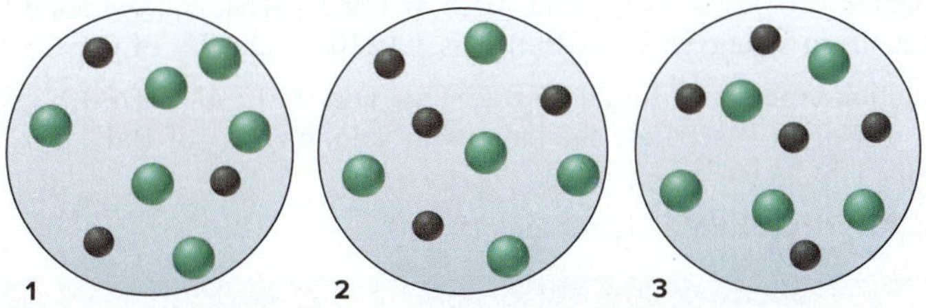

(a) Which scene best represents the solution in equilibrium with the solid?
(b) In which, if any, other scene(s) will additional solid form?
(c) Explain how, if at all, addition of a small amount of concentrated $HCl(aq)$ will affect the mass of solid present in any scene.

SOME SIMILAR PROBLEMS 19.121 and 19.147

Separating Ions by Selective Precipitation and Simultaneous Equilibria

Let's consider two ways to separate one ion from another by reacting them with a given precipitating ion to form compounds with different solubilities.

Selective Precipitation In the process of **selective precipitation,** a solution of one precipitating ion is added to a solution of two ionic compounds until the Q_{sp} of the *more soluble* compound is almost equal to its K_{sp}. This method ensures that the K_{sp} of the *less soluble* compound is exceeded as much as possible. As a result, the maximum amount of the less soluble compound precipitates, but none of the more soluble compound does.

SAMPLE PROBLEM 19.12 Separating Ions by Selective Precipitation

Problem A solution consists of 0.20 M $MgCl_2$ and 0.10 M $CuCl_2$. Calculate the $[OH^-]$ needed to separate the metal ions. The K_{sp} of $Mg(OH)_2$ is 6.3×10^{-10}, and the K_{sp} of $Cu(OH)_2$ is 2.2×10^{-20}.

Plan Because both compounds have a 1/2 ratio of cation/anion (see the discussion on page 868), when we compare their K_{sp} values, we find that $Mg(OH)_2$ is about 10^{10} times more soluble than $Cu(OH)_2$; thus, $Cu(OH)_2$ precipitates first. We write the dissolution equations and ion-product expressions. We are given both cation concentrations, so we solve for the $[OH^-]$ that gives a saturated solution of $Mg(OH)_2$, because this $[OH^-]$ will precipitate the most Cu^{2+}. Then, we calculate the $[Cu^{2+}]$ remaining to see if the separation was accomplished.

Solution Writing the equations and ion-product expressions:

$$Mg(OH)_2(s) \rightleftharpoons Mg^{2+}(aq) + 2OH^-(aq) \qquad K_{sp} = [Mg^{2+}][OH^-]^2 = 6.3\times10^{-10}$$
$$Cu(OH)_2(s) \rightleftharpoons Cu^{2+}(aq) + 2OH^-(aq) \qquad K_{sp} = [Cu^{2+}][OH^-]^2 = 2.2\times10^{-20}$$

Calculating the $[OH^-]$ that gives a saturated $Mg(OH)_2$ solution:

$$[OH^-] = \sqrt{\frac{K_{sp}}{[Mg^{2+}]}} = \sqrt{\frac{6.3\times10^{-10}}{0.20}} = 5.6\times10^{-5}\ M$$

This is the maximum $[OH^-]$ that will *not* precipitate Mg^{2+} ion.
Calculating the $[Cu^{2+}]$ remaining in the solution with this $[OH^-]$:

$$[Cu^{2+}] = \frac{K_{sp}}{[OH^-]^2} = \frac{2.2\times10^{-20}}{(5.6\times10^{-5})^2} = 7.0\times10^{-12}\ M$$

Since the initial $[Cu^{2+}]$ is 0.10 M, virtually all the Cu^{2+} ion is precipitated.

Check Rounding, we find that $[OH^-]$ seems right: $\sim\sqrt{(6\times10^{-10})/0.2} = 5\times10^{-5}$. The $[Cu^{2+}]$ remaining also seems correct: $(200\times10^{-22})/(5\times10^{-5})^2 = 8\times10^{-12}$.

FOLLOW-UP PROBLEMS

19.12A A solution is 0.050 M $BaCl_2$ and 0.025 M $CaCl_2$. What concentration of SO_4^{2-} will separate the cations in solution? K_{sp} of $BaSO_4$ is 1.1×10^{-10}, and K_{sp} of $CaSO_4$ is 2.4×10^{-5}.

19.12B A solution consists of 0.020 M $Ba(NO_3)_2$ and 0.015 M $Ca(NO_3)_2$. What concentration of NaF will separate the cations in solution? K_{sp} of BaF_2 is 1.5×10^{-6}, and K_{sp} of CaF_2 is 3.2×10^{-11}.

SOME SIMILAR PROBLEMS 19.92 and 19.140

Simultaneous Equilibria Sometimes two or more equilibrium systems are controlled simultaneously to separate one metal ion from another as their sulfides, using S^{2-} as the precipitating ion. We control the $[S^{2-}]$ in order to exceed the K_{sp} value of one metal sulfide but not the other, and we do so by shifting H_2S dissociation through adjustments of $[H_3O^+]$:

$$H_2S(aq) + 2H_2O(aq) \rightleftharpoons 2H_3O^+(aq) + S^{2-}(aq)$$

- If we add strong acid, the high $[H_3O^+]$ shifts H_2S dissociation to the left, which decreases $[S^{2-}]$, so the *less* soluble sulfide precipitates.
- If we add strong base, the low $[H_3O^+]$ shifts H_2S dissociation to the right, which increases $[S^{2-}]$, so the *more* soluble sulfide precipitates.

Thus, we shift one equilibrium system (H_2S dissociation) by adjusting a second (H_2O ionization) to control a third (metal sulfide solubility).

As the upcoming Chemical Connections essay demonstrates, the principles of ionic equilibria often help us understand the chemical basis of complex environmental problems and may provide ways to solve them.

› Summary of Section 19.3

- › Only as a first approximation does the dissolved portion of a slightly soluble salt dissociate completely into ions.
- › In a saturated solution, dissolved ions and the undissolved solid salt are in equilibrium. The product of the ion concentrations, each raised to the power of its subscript in the formula, has a constant value ($Q_{sp} = K_{sp}$).
- › The value of K_{sp} can be obtained from the solubility, and vice versa.
- › Adding a common ion lowers a compound's solubility.
- › Adding H_3O^+ (lowering pH) increases an ionic compound's solubility if the anion of the compound is also that of a weak acid.
- › Limestone caves result from shifts in the $CaCO_3/CO_2$ equilibrium system.
- › When two solutions, each containing one of the ions of a slightly soluble ionic compound, are mixed, an ionic solid forms if $Q_{sp} > K_{sp}$.
- › Ions are precipitated selectively from a solution of two ionic compounds by adding a precipitating ion until the K_{sp} of one compound is exceeded as much as possible without exceeding the K_{sp} of the other. An extension of this approach uses simultaneous control of three equilibrium systems to separate metal ions as their sulfides.
- › Lakes bounded by limestone-rich soils form buffer systems that prevent acidification.

CHEMICAL CONNECTIONS TO ENVIRONMENTAL SCIENCE

The Acid-Rain Problem

Acid rain, the deposition of acids in wet form as rain, snow, or fog or in dry form as solid particles, is a global environmental problem. It has been observed in the United States, Canada, Mexico, the Amazon basin, Europe, Russia, many parts of Asia, and even at the North and South Poles. Let's see how it arises and how to prevent some of its harmful effects.

Origins of Acid Rain

The strong acids H_2SO_4 and HNO_3 cause the greatest concern, so let's see how they form (Figure B19.1, on the next page):

1. *Sulfuric acid.* Sulfur dioxide (SO_2), produced mostly by the burning of high-sulfur coal, forms sulfurous acid (H_2SO_3) in contact with water. The atmospheric pollutants hydrogen peroxide (H_2O_2) and ozone (O_3) dissolve in the water in clouds and oxidize the sulfurous acid to sulfuric acid:

$$H_2O_2(aq) + H_2SO_3(aq) \longrightarrow H_2SO_4(aq) + H_2O(l)$$

Alternatively, SO_2 is oxidized by atmospheric hydroxyl radicals (HO·) to sulfur trioxide (SO_3), which forms H_2SO_4 with water.

2. *Nitric acid.* Nitrogen oxides (NO_x) form from N_2 and O_2 during high-temperature combustion in car and truck engines and electric power plants. NO then forms NO_2 and HNO_3 in a process that creates smog. At night, NO_x is converted to N_2O_5, which reacts with water to form HNO_3:

$$N_2O_5(g) + H_2O(l) \longrightarrow 2HNO_3(aq)$$

Ammonium salts, deposited as NH_4HSO_4 or NH_4NO_3, produce HNO_3 in soil through biochemical oxidation.

The pH of Acid Rain

Normal rainwater is weakly acidic from the reaction of atmospheric CO_2 with water (see Problem 19.141):

$$CO_2(g) + 2H_2O(l) \rightleftharpoons H_3O^+(aq) + HCO_3^-(aq)$$

Yet, as long ago as 1984, rainfall in the United States had already reached an average pH of 4.2, an increase of about 25 times as much H_3O^+. Rain in Wheeling, West Virginia, has had a pH of 1.8, lower than the pH of lemon juice. And rain in industrial parts of Sweden once had a pH of 2.7, about the same as vinegar.

Effects of Acid Rain

Some fish and shellfish die at pH values between 4.5 and 5.0. The young of most species are generally more vulnerable. At pH 5, most fish eggs cannot hatch. With tens of thousands of rivers and lakes around the world becoming or already acidified, the loss of freshwater fish became a major concern long ago.

(continued)

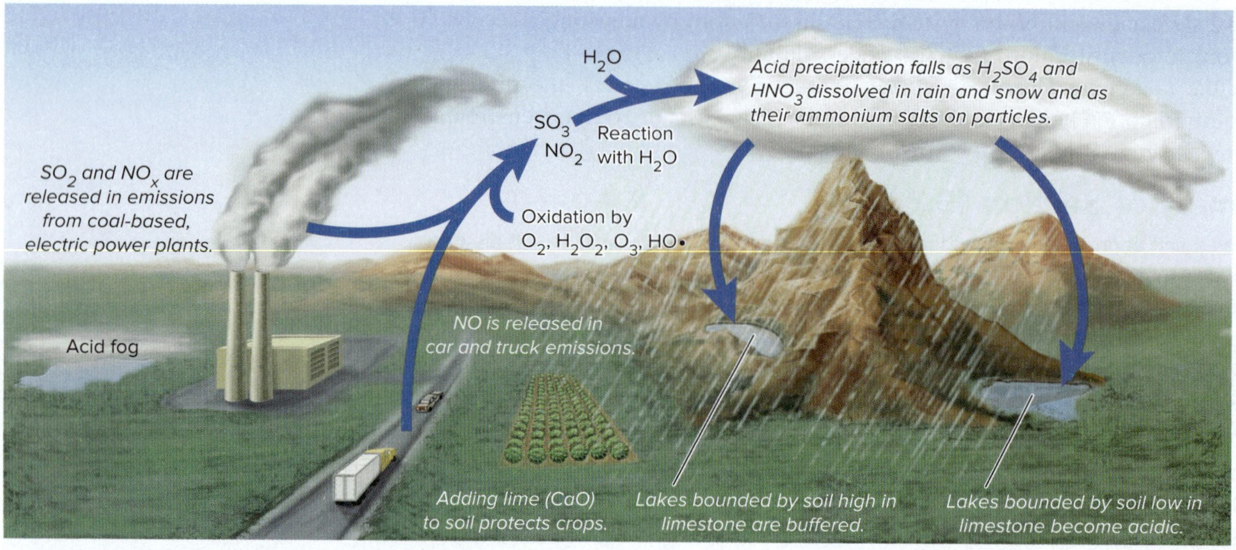

Figure B19.1 Formation of acidic precipitation. A complex interplay of human activities, atmospheric chemistry, and environmental distribution leads to acidic precipitation and its harmful effects.

In addition, acres of forest have been harmed by the acid, which removes nutrients and releases toxic substances from the soil. The aluminosilicates that make up most soils are extremely insoluble in water. Through a complex series of simultaneous equilibria, contact with water at pH < 5 causes these materials to release some bound Al^{3+}, which is toxic to fish and plants. At the same time, acid rain dissolves Ca^{2+} and Mg^{2+} ions from the soil, which are nutrients for plants and animals.

Marble and limestone (both primarily $CaCO_3$) in buildings and monuments react with sulfuric acid to form gypsum ($CaSO_4 \cdot 2H_2O$), which flakes off. Ironically, the same process that destroys these structures rescues lakes that are bounded by limestone-rich soil. Limestone dissolves sufficiently in lake water to form an HCO_3^-/CO_3^{2-} buffer capable of absorbing the incoming H_3O^+ and maintaining a mildly basic pH:

$$CO_3^{2-}(aq) + H_3O^+(aq) \rightleftharpoons HCO_3^-(aq) + H_2O(l)$$

Acidified lakes and rivers that are in contact with granite and other weathering-resistant bedrock can be remediated by *liming* (treating with limestone). This approach is expensive and is only a stopgap measure because the lakes are acidic again within several years.

Preventing Acid Rain

Effective prevention of acid rain has to address the sources of the sulfur and nitrogen pollutants:

1. *Sulfur pollutants.* As was pointed out earlier (Chemical Connections, Chapter 6), the principal means of minimizing sulfur dioxide release is by "scrubbing" power plant emissions with limestone in both dry and wet form. Another method reduces SO_2 with methane, coal, or H_2S, and the mixture is converted catalytically to sulfur, which is sold as a byproduct:

$$16H_2S(g) + 8SO_2(g) \xrightarrow{\text{catalyst}} 3S_8(s) + 16H_2O(l)$$

Although low-sulfur coal is rare and expensive to mine, coal can be converted into gaseous and liquid low-sulfur fuels. The sulfur is removed (as H_2S) in an acid-gas scrubber after gasification.

2. *Nitrogen pollutants.* Through the use of a catalytic converter in an auto exhaust system, NO_x species are reduced to N_2. In power plant emissions, NO_x is decreased by adjusting conditions and by treating hot stack gases with ammonia in the presence of a heterogeneous catalyst:

$$4NO(g) + 4NH_3(g) + O_2(g) \xrightarrow{\text{catalyst}} 4N_2(g) + 6H_2O(g)$$

Reducing power plant emissions in North America and Europe has increased the pH of rainfall and some surface waters. Lime (CaO) is routinely used to react with acid rain falling on croplands. Further progress is expected under current legislation, but, in much of eastern North America and northern Europe, additional measures are needed for full recovery.

Problems

B19.1 An environmental technician collects a sample of rainwater. Back in her lab, the pH meter isn't working, so she uses indicator solutions to estimate the pH. A piece of litmus paper turns red, indicating acidity, so she divides the sample into thirds and obtains the following results: thymol blue turns yellow; bromphenol blue turns green; and methyl red turns red. Estimate the pH of the rainwater.

B19.2 A lake that has a surface area of 10.0 acres (1 acre = 4.840×10^3 yd^2) receives 1.00 in of rain of pH 4.20. (Assume that the acidity of the rain is due to a strong, monoprotic acid.) (a) How many moles of H_3O^+ are in the rain falling on the lake? (b) If the lake is unbuffered and its average depth is 10.0 ft before the rain, find the pH after the rain has been mixed with lake water. (Assume the initial pH is 7.00, and ignore runoff from the surrounding land.) (c) If the lake contains hydrogen carbonate ions (HCO_3^-), what mass of HCO_3^- will neutralize the acid in the rain?

19.4 EQUILIBRIA INVOLVING COMPLEX IONS

A third kind of aqueous ionic equilibrium involves a type of ion we mentioned briefly in Section 18.9. A *simple ion,* such as Na^+ or CH_3COO^-, consists of one or a few bonded atoms, with an excess or deficit of electrons. A **complex ion** consists of a central metal ion covalently bonded to two or more anions or molecules, called **ligands.** Hydroxide, chloride, and cyanide ions are examples of ionic ligands; water, carbon monoxide, and ammonia are some molecular ligands. In the complex ion $Cr(NH_3)_6^{3+}$, for example, the central Cr^{3+} is surrounded by six NH_3 ligands (Figure 19.17). Hydrated metal ions are complex ions with water molecules as ligands (Section 18.5). In Chapter 23, we discuss the transition metals and the structures and properties of the numerous complex ions they form. Here, we focus on equilibria of hydrated ions with ligands other than water.

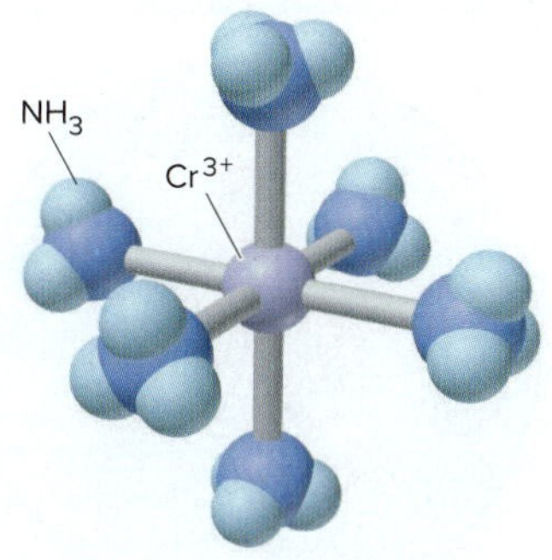

Figure 19.17 $Cr(NH_3)_6^{3+}$, a typical complex ion.

Formation of Complex Ions

When a salt dissolves in water, a complex ion forms, with water as ligands around the metal ion. In many cases, when we treat this hydrated cation with a solution of another ligand, the bound water molecules are replaced by the other ligand. For example, a hydrated M^{2+} ion, $M(H_2O)_4^{2+}$, forms the ammoniated ion, $M(NH_3)_4^{2+}$, in aqueous NH_3:

$$M(H_2O)_4^{2+}(aq) + 4NH_3(aq) \rightleftharpoons M(NH_3)_4^{2+}(aq) + 4H_2O(l)$$

At equilibrium, this system is expressed by a ratio of concentration terms whose form follows that of any other equilibrium expression:

$$K_c = \frac{[M(NH_3)_4^{2+}][H_2O]^4}{[M(H_2O)_4^{2+}][NH_3]^4}$$

Once again, because $[H_2O]$ is essentially constant, water is treated as a pure liquid, with its concentration term equal to 1. Thus, we obtain the expression for a new equilibrium constant called the **formation constant, K_f:**

$$\frac{K_c}{(1)^4} = K_f = \frac{[M(NH_3)_4^{2+}]}{[M(H_2O)_4^{2+}][NH_3]^4}$$

At the molecular level, depicted in Figure 19.18 *(next page)*, the actual process is stepwise, with ammonia molecules replacing water molecules one at a time. This process yields a series of intermediate species, each with its own formation constant:

$$M(H_2O)_4^{2+}(aq) + NH_3(aq) \rightleftharpoons M(H_2O)_3(NH_3)^{2+}(aq) + H_2O(l)$$

$$K_{f1} = \frac{[M(H_2O)_3(NH_3)^{2+}]}{[M(H_2O)_4^{2+}][NH_3]}$$

$$M(H_2O)_3(NH_3)^{2+}(aq) + NH_3(aq) \rightleftharpoons M(H_2O)_2(NH_3)_2^{2+}(aq) + H_2O(l)$$

$$K_{f2} = \frac{[M(H_2O)_2(NH_3)_2^{2+}]}{[M(H_2O)_3(NH_3)^{2+}][NH_3]}$$

$$M(H_2O)_2(NH_3)_2^{2+}(aq) + NH_3(aq) \rightleftharpoons M(H_2O)(NH_3)_3^{2+}(aq) + H_2O(l)$$

$$K_{f3} = \frac{[M(H_2O)(NH_3)_3^{2+}]}{[M(H_2O)_2(NH_3)_2^{2+}][NH_3]}$$

$$M(H_2O)(NH_3)_3^{2+}(aq) + NH_3(aq) \rightleftharpoons M(NH_3)_4^{2+}(aq) + H_2O(l)$$

$$K_{f4} = \frac{[M(NH_3)_4^{2+}]}{[M(H_2O)(NH_3)_3^{2+}][NH_3]}$$

The *sum* of the equations gives the overall equation, so the *product* of the individual formation constants gives the overall formation constant:

$$K_f = K_{f1} \times K_{f2} \times K_{f3} \times K_{f4}$$

Recall that *all complex ions are Lewis adducts* (Section 18.9). The metal ion acts as a Lewis acid (accepts an electron pair), and the ligand acts as a Lewis base (donates an electron pair). In the formation of $M(NH_3)_4^{2+}$, the K_f for each step is

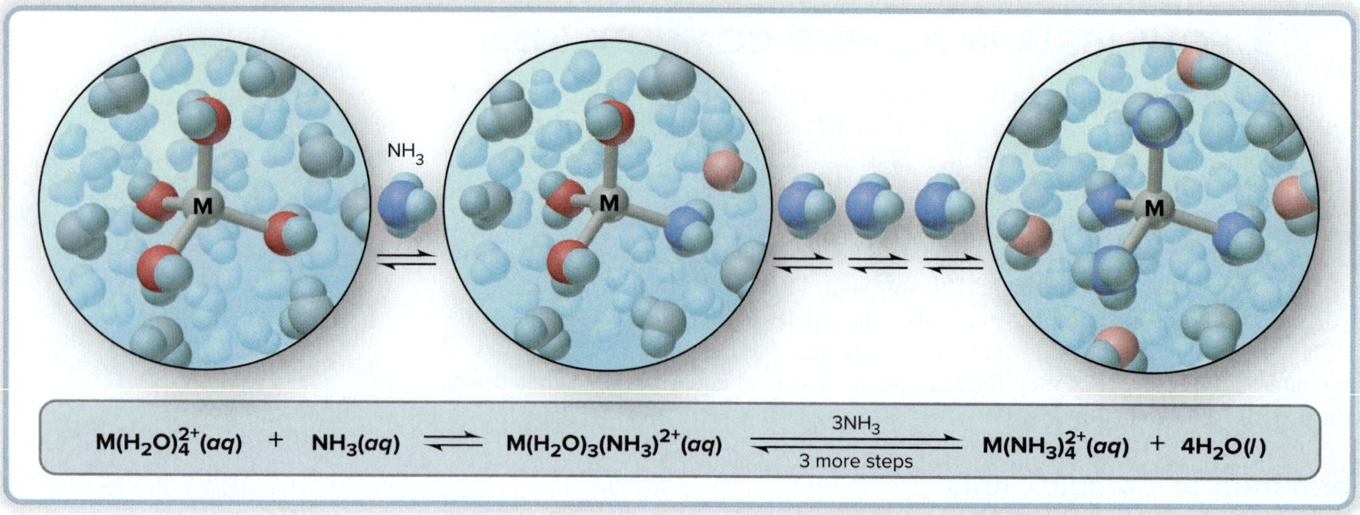

Figure 19.18 The stepwise exchange of NH_3 for H_2O in $M(H_2O)_4^{2+}$. The molecular views show the first exchange and the fully ammoniated ion.

much larger than 1 because ammonia is a stronger Lewis base than water. Therefore, if we add excess ammonia to the $M(H_2O)_4^{2+}$ solution, nearly all the M^{2+} ions exist as $M(NH_3)_4^{2+}(aq)$.

Table 19.4 (and Appendix C) shows K_f values of some complex ions. Notice that they are all of the order 10^6 or greater, which means that these ions form readily from the hydrated ion. Because of this behavior, some uses of complex-ion formation are to retrieve a metal from its ore, eliminate a toxic or unwanted metal ion from a solution, or convert a metal ion to a different form, as Sample Problem 19.13 shows for the zinc ion.

Table 19.4 Formation Constants (K_f) of Some Complex Ions at 25°C

Complex Ion	K_f	Complex Ion	K_f
$Ag(CN)_2^-$	3.0×10^{20}	$Fe(CN)_6^{4-}$	3×10^{35}
$Ag(NH_3)_2^+$	1.7×10^{7}	$Fe(CN)_6^{3-}$	4.0×10^{43}
$Ag(S_2O_3)_2^{3-}$	4.7×10^{13}	$Hg(CN)_4^{2-}$	9.3×10^{38}
AlF_6^{3-}	4×10^{19}	$Ni(NH_3)_6^{2+}$	2.0×10^{8}
$Al(OH)_4^-$	3×10^{33}	$Pb(OH)_3^-$	8×10^{13}
$Be(OH)_4^{2-}$	4×10^{18}	$Sn(OH)_3^-$	3×10^{25}
CdI_4^{2-}	1×10^{6}	$Zn(CN)_4^{2-}$	4.2×10^{19}
$Co(OH)_4^{2-}$	5×10^{9}	$Zn(NH_3)_4^{2+}$	7.8×10^{8}
$Cr(OH)_4^-$	8.0×10^{29}	$Zn(OH)_4^{2-}$	3×10^{15}
$Cu(NH_3)_4^{2+}$	5.6×10^{11}		

SAMPLE PROBLEM 19.13 Calculating the Concentration of a Complex Ion

Problem An industrial chemist converts $Zn(H_2O)_4^{2+}$ to the more stable $Zn(NH_3)_4^{2+}$ ($K_f = 7.8\times10^8$) by mixing 50.0 L of 0.0020 *M* $Zn(H_2O)_4^{2+}$ and 25.0 L of 0.15 *M* NH_3. What is the $[Zn(H_2O)_4^{2+}]$ at equilibrium?

Plan We write the complex-ion formation equation and the K_f expression and use a reaction table to calculate the equilibrium concentrations. To set up the table, we must first find $[Zn(H_2O)_4^{2+}]_{init}$. We are given the individual volumes and molar concentrations, so we find the number of moles and divide by the *total* volume because the solutions are mixed. With the large excess of NH_3 and the high K_f, we assume that almost all the $Zn(H_2O)_4^{2+}$ is converted to $Zn(NH_3)_4^{2+}$. Because $[Zn(H_2O)_4^{2+}]$ at equilibrium is very small, we use x to represent it.

Solution Writing the equation and the K_f expression:

$$Zn(H_2O)_4^{2+}(aq) + 4NH_3(aq) \rightleftharpoons Zn(NH_3)_4^{2+}(aq) + 4H_2O(l)$$

$$K_f = \frac{[Zn(NH_3)_4^{2+}]}{[Zn(H_2O)_4^{2+}][NH_3]^4}$$

Finding the initial reactant concentrations:

$$[Zn(H_2O)_4^{2+}]_{init} = \frac{50.0\text{ L} \times 0.0020\text{ }M}{50.0\text{ L} + 25.0\text{ L}} = 1.3\times10^{-3}\text{ }M$$

$$[NH_3]_{init} = \frac{25.0\text{ L} \times 0.15\text{ }M}{50.0\text{ L} + 25.0\text{ L}} = 5.0\times10^{-2}\text{ }M$$

Setting up a reaction table: We assume that nearly all the $Zn(H_2O)_4^{2+}$ is converted to $Zn(NH_3)_4^{2+}$, so we set up the table with $x = [Zn(H_2O)_4^{2+}]$ at equilibrium. Because 4 mol of NH_3 is needed per mole of $Zn(H_2O)_4^{2+}$, the change in $[NH_3]$ is

$$[NH_3]_{reacted} \approx 4(1.3\times10^{-3}\text{ }M) = 5.2\times10^{-3}\text{ }M \qquad \text{and} \qquad [Zn(NH_3)_4^{2+}] \approx 1.3\times10^{-3}\text{ }M$$

Concentration (*M*)	**$Zn(H_2O)_4^{2+}(aq)$**	**+ $4NH_3(aq)$**	**$\rightleftharpoons$ $Zn(NH_3)_4^{2+}(aq)$**	**+ $4H_2O(l)$**
Initial	1.3×10^{-3}	5.0×10^{-2}	0	—
Change	~(-1.3×10^{-3})	~(-5.2×10^{-3})	~$(+1.3\times10^{-3})$	—
Equilibrium	x	4.5×10^{-2}	1.3×10^{-3}	—

Solving for x, the $[Zn(H_2O)_4^{2+}]$ remaining at equilibrium:

$$K_f = \frac{[Zn(NH_3)_4^{2+}]}{[Zn(H_2O)_4^{2+}][NH_3]^4} = 7.8\times10^8 \approx \frac{1.3\times10^{-3}}{x(4.5\times10^{-2})^4}$$

$$x = [Zn(H_2O)_4^{2+}] \approx 4.1\times10^{-7}\text{ }M$$

Check The K_f is large, so we expect the $[Zn(H_2O)_4^{2+}]$ remaining to be very low.

FOLLOW-UP PROBLEMS

19.13A Cyanide ion is toxic because it forms stable complex ions with the Fe^{3+} ion in certain proteins involved in cellular energy production. To study this effect, a biochemist mixes 25.5 mL of 3.1×10^{-2} *M* $Fe(H_2O)_6^{3+}$ with 35.0 mL of 1.5 *M* NaCN. What is the final $[Fe(H_2O)_6^{3+}]$? The K_f of $Fe(CN)_6^{3-}$ is 4.0×10^{43}.

19.13B The production of aluminum from its ore utilizes a stable complex ion formed from the reaction of $Al(H_2O)_6^{3+}$ with F^-. When 2.4 g of $AlCl_3$ is dissolved in 250 mL of 0.560 *M* NaF, what is the final $[Al(H_2O)_6^{3+}]$? The K_f of AlF_6^{3-} is 4×10^{19}.

SOME SIMILAR PROBLEMS 19.100, 19.101, 19.104, and 19.105

Complex Ions and the Solubility of Precipitates

In Section 19.3, you saw that H_3O^+ increases the solubility of a slightly soluble ionic compound if its anion is that of a weak acid. Similarly, *a ligand increases the solubility of a slightly soluble ionic compound if it forms a complex ion with the compound's cation.* For example, iron(II) sulfide is very slightly soluble:

$$FeS(s) \rightleftharpoons Fe^{2+}(aq) + S^{2-}(aq) \qquad K_{sp} = 8\times10^{-16}$$

When we add some 1.0 *M* NaCN, the CN^- ions act as ligands and react with the small amount of $Fe^{2+}(aq)$ to form the complex ion $Fe(CN)_6^{4-}$:

$$Fe^{2+}(aq) + 6CN^-(aq) \rightleftharpoons Fe(CN)_6^{4-}(aq) \qquad K_f = 3\times10^{35}$$

To see the effect of complex-ion formation on the solubility of FeS, we add the equations and, therefore, multiply their equilibrium constants:

$$FeS(s) + 6CN^-(aq) \rightleftharpoons Fe(CN)_6^{4-}(aq) + S^{2-}(aq)$$

$$K_{overall} = K_{sp} \times K_f = (8\times10^{-16})(3\times10^{35}) = 2\times10^{20}$$

The overall dissociation of FeS into ions increased by more than a factor of 10^{35} in the presence of the ligand.

SAMPLE PROBLEM 19.14 Calculating the Effect of Complex-Ion Formation on Solubility

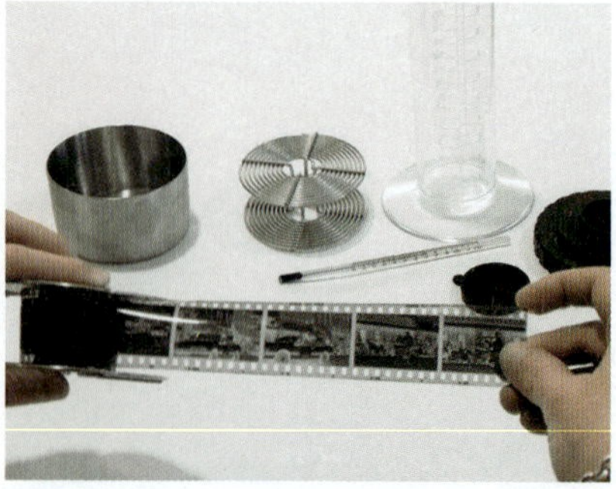

Developing an image in "hypo."

Source: © McGraw-Hill Education/Mark A. Dierker, photographer

Problem In black-and-white film developing *(see photo),* excess AgBr is removed from a film negative with "hypo," an aqueous solution of sodium thiosulfate ($Na_2S_2O_3$), which causes the formation of the complex ion $Ag(S_2O_3)_2^{3-}$. Calculate the solubility of AgBr in **(a)** H_2O; **(b)** 1.0 *M* hypo. K_f of $Ag(S_2O_3)_2^{3-}$ is 4.7×10^{13}, and K_{sp} of AgBr is 5.0×10^{-13}.

Plan **(a)** After writing the dissolution equation and the ion-product expression, we use the given K_{sp} to solve for *S*, the molar solubility of AgBr. **(b)** In hypo, Ag^+ forms a complex ion with $S_2O_3^{2-}$, which shifts the equilibrium and dissolves more AgBr. We write the complex-ion equation and add it to the equation for dissolving AgBr to obtain the overall equation for dissolving AgBr in hypo. We multiply K_{sp} by K_f to find $K_{overall}$. To find the solubility of AgBr in hypo, we set up a reaction table, with $S = [Ag(S_2O_3)_2^{3-}]$, substitute into the expression for $K_{overall}$, and solve for *S*.

Solution **(a)** Solubility in water. Writing the equation for the saturated solution and the ion-product expression:

$$AgBr(s) \rightleftharpoons Ag^+(aq) + Br^-(aq) \qquad K_{sp} = [Ag^+][Br^-]$$

Solving for solubility (*S*) directly from the equation: We know that

$$S = [AgBr]_{dissolved} = [Ag^+] = [Br^-]$$

Thus,

$$K_{sp} = [Ag^+][Br^-] = S^2 = 5.0\times10^{-13}$$

so

$$S = 7.1\times10^{-7}\ M$$

(b) Solubility in 1.0 *M* hypo. Writing the overall equation:

$$AgBr(s) \rightleftharpoons \cancel{Ag^+(aq)} + Br^-(aq)$$
$$\cancel{Ag^+(aq)} + 2S_2O_3^{2-}(aq) \rightleftharpoons Ag(S_2O_3)_2^{3-}(aq)$$
$$AgBr(s) + 2S_2O_3^{2-}(aq) \rightleftharpoons Ag(S_2O_3)_2^{3-}(aq) + Br^-(aq)$$

Calculating $K_{overall}$:

$$K_{overall} = \frac{[Ag(S_2O_3)_2^{3-}][Br^-]}{[S_2O_3^{2-}]^2} = K_{sp} \times K_f = (5.0\times10^{-13})(4.7\times10^{13}) = 24$$

Setting up a reaction table, with $S = [AgBr]_{dissolved} = [Ag(S_2O_3)_2^{3-}]$:

Concentration (*M*)	AgBr(*s*)	+	$2S_2O_3^{2-}(aq)$	$\rightleftharpoons$	$Ag(S_2O_3)_2^{3-}(aq)$	+	$Br^-(aq)$
Initial	—		1.0		0		0
Change	—		$-2S$		$+S$		$+S$
Equilibrium	—		$1.0 - 2S$		S		S

Substituting the values into the expression for $K_{overall}$ and solving for *S*:

$$K_{overall} = \frac{[Ag(S_2O_3)_2^{3-}][Br^-]}{[S_2O_3^{2-}]^2} = \frac{S^2}{(1.0\ M - 2S)^2} = 24$$

Taking the square root of both sides gives

$$\frac{S}{1.0\ M - 2S} = \sqrt{24} = 4.9 \quad \text{so} \quad S = 4.9\ M - 9.8S \quad \text{and} \quad 10.8S = 4.9\ M$$

$$[Ag(S_2O_3)_2^{3-}] = S = 0.45\ M$$

Check **(a)** From the number of ions in the formula of AgBr, we know that $S = \sqrt{K_{sp}}$, so the order of magnitude seems right: $\sim\sqrt{10^{-14}} \approx 10^{-7}$. **(b)** The $K_{overall}$ seems correct: the exponents cancel, and $5 \times 5 = 25$. Most importantly, the answer makes sense because the photographic process requires the remaining AgBr to be washed off the film and the large K_{overall} confirms that. We can check S by rounding and working backward to find K_{overall}: from the reaction table, we find that

$$[(S_2O_3)^{2-}] = 1.0\ M - 2S = 1.0\ M - 2(0.45\ M) = 1.0\ M - 0.90\ M = 0.1\ M$$

so $K_{\text{overall}} \approx (0.45)^2/(0.1)^2 = 20$, within rounding of the calculated value.

FOLLOW-UP PROBLEMS

19.14A How does the solubility of AgBr in 1.0 M NH_3 compare with its solubility in 1.0 M hypo, calculated in part (b) of Sample Problem 19.14? The K_f of $Ag(NH_3)_2^+$ is 1.7×10^7.

19.14B Calculate the solubility of $PbCl_2$ in 0.75 M NaOH. The K_f of $Pb(OH)_3^-$ is 8×10^{13}, and the K_{sp} of $PbCl_2$ is 1.7×10^{-5}.

SOME SIMILAR PROBLEMS 19.102 and 19.103

Complex Ions of Amphoteric Hydroxides

Many of the same metals that form amphoteric oxides (Chapter 8, p. 354) also form slightly soluble *amphoteric hydroxides.* These compounds dissolve very little in water, but they dissolve to a much greater extent in both acidic and basic solutions. Aluminum hydroxide is one of several examples:

$$Al(OH)_3(s) \rightleftharpoons Al^{3+}(aq) + 3OH^-(aq)$$

It is insoluble in water ($K_{sp} = 3\times10^{-34}$), but

- It dissolves in acid because H_3O^+ reacts with the OH^- anion [Sample Problem 19.9(b)],

$$3H_3O^+(aq) + 3OH^-(aq) \longrightarrow 6H_2O(l)$$

giving the overall equation

$$Al(OH)_3(s) + 3H_3O^+(aq) \longrightarrow Al^{3+}(aq) + 6H_2O(l)$$

- It dissolves in base through the formation of a complex ion:

$$Al(OH)_3(s) + OH^-(aq) \longrightarrow Al(OH)_4^-(aq)$$

Let's look more closely at this behavior. When we dissolve a soluble aluminum salt, such as $Al(NO_3)_3$, in water and then slowly add a strong base, a white precipitate first forms and then dissolves as more base is added. What reactions are occurring? The formula for the hydrated Al^{3+} ion is $Al(H_2O)_6^{3+}(aq)$. It acts as a weak polyprotic acid and reacts with added OH^- ions in a stepwise manner. In each step, one of the bound H_2O molecules loses a proton and becomes a bound OH^- ion, so the number of bound H_2O molecules is reduced by 1:

$$Al(H_2O)_6^{3+}(aq) + OH^-(aq) \rightleftharpoons Al(H_2O)_5OH^{2+}(aq) + H_2O(l)$$
$$Al(H_2O)_5OH^{2+}(aq) + OH^-(aq) \rightleftharpoons Al(H_2O)_4(OH)_2^+(aq) + H_2O(l)$$
$$Al(H_2O)_4(OH)_2^+(aq) + OH^-(aq) \rightleftharpoons Al(H_2O)_3(OH)_3(s) + H_2O(l)$$

After three protons have been removed from each $Al(H_2O)_6^{3+}$, the white precipitate has formed, which is the insoluble hydroxide $Al(H_2O)_3(OH)_3(s)$, often written more simply as $Al(OH)_3(s)$. Now you can see that the precipitate actually consists of the hydrated Al^{3+} ion with an H^+ removed from each of three bound H_2O molecules (Figure 19.19, *next page, center*). Addition of H_3O^+ protonates the OH^- ions and re-forms the hydrated Al^{3+} ion (Figure 19.19, *left*).

Further addition of OH^- removes a fourth H^+ and the precipitate dissolves as the soluble ion $Al(H_2O)_2(OH)_4^-(aq)$ forms (Figure 19.19, *right*), which we usually write with the formula $Al(OH)_4^-(aq)$:

$$Al(H_2O)_3(OH)_3(s) + OH^-(aq) \rightleftharpoons Al(H_2O)_2(OH)_4^-(aq) + H_2O(l)$$

In other words, this complex ion is not created by ligands substituting for bound water molecules but through an acid-base reaction in which added OH^- ions titrate bound water molecules.

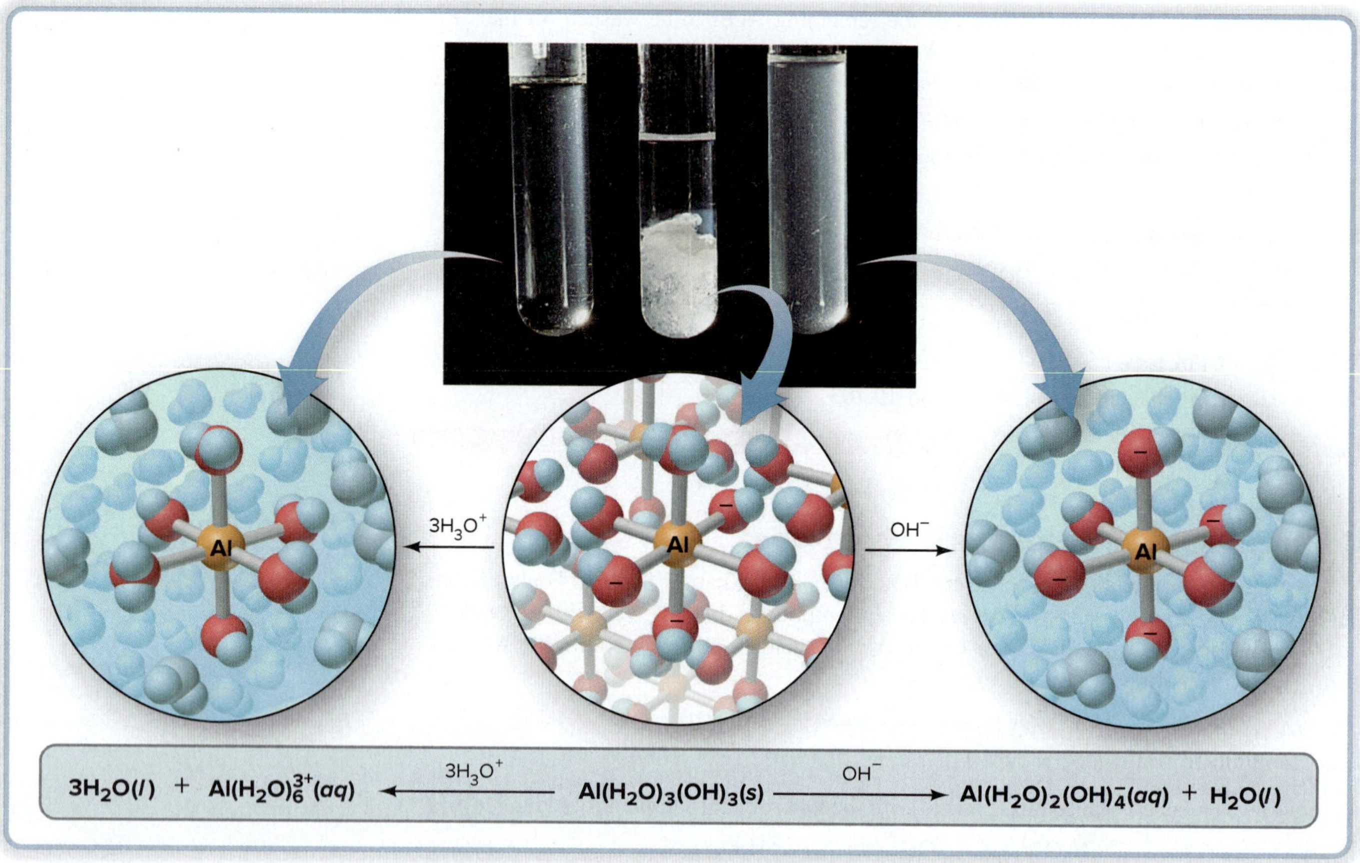

Figure 19.19 The amphoteric behavior of aluminum hydroxide. When solid $Al(OH)_3$ is treated with H_3O^+ *(left)* or with OH^- *(right)*, it dissolves as a result of the formation of soluble complex ions.

Source: © McGraw-Hill Education/Stephen Frisch, photographer

Several other slightly soluble hydroxides, including those of cadmium, chromium(III), cobalt(III), lead(II), tin(II), and zinc, are amphoteric and exhibit similar reactions:

$$Zn(H_2O)_2(OH)_2(s) + OH^-(aq) \rightleftharpoons Zn(H_2O)(OH)_3^-(aq) + H_2O(l)$$

In contrast, the slightly soluble hydroxides of iron(II), iron(III), and calcium dissolve in acid, but do *not* dissolve in base, because the three remaining bound water molecules are not acidic enough to lose any of their protons:

$$Fe(H_2O)_3(OH)_3(s) + 3H_3O^+(aq) \longrightarrow Fe(H_2O)_6^{3+}(aq) + 3H_2O(l)$$
$$Fe(H_2O)_3(OH)_3(s) + OH^-(aq) \longrightarrow \text{no reaction}$$

This difference in solubility in base between $Al(OH)_3$ and $Fe(OH)_3$ is the key to an important separation step in the production of aluminum metal, so we'll consider it again in Section 22.4.

› Summary of Section 19.4

- A complex ion consists of a central metal ion covalently bonded to two or more negatively charged or neutral ligands. Its formation is characterized by a formation constant, K_f.
- A hydrated metal ion is a complex ion with water molecules as ligands. Other ligands (stronger Lewis bases) can displace the water molecules in a stepwise process. In most cases, the K_f value for each step is large, so the fully substituted complex ion forms almost completely in the presence of excess ligand.
- Adding a solution containing a ligand increases the solubility of an ionic precipitate if the cation forms a complex ion with the ligand.
- Amphoteric metal hydroxides dissolve in acid and base due to reactions that involve complex ions.

CHAPTER REVIEW GUIDE

Learning Objectives

Relevant section (§) and/or sample problem (SP) numbers appear in parentheses.

Understand These Concepts

1. How the presence of a common ion suppresses a reaction that forms it (§19.1)
2. Why the concentrations of buffer components must be high to minimize the change in pH due to addition of small amounts of H_3O^+ or OH^- (§19.1)
3. How buffer capacity depends on buffer concentration and on the pK_a of the acid component; why buffer range is within ±1 pH unit of the pK_a (§19.1)
4. Why the shapes of strong acid–strong base, weak acid–strong base, and weak base–strong acid titration curves differ (§19.2)
5. How the pH at the equivalence point is determined by the species present; why the pH at the midpoint of the buffer region equals the pK_a of the acid (§19.2)
6. The nature of an acid-base indicator as a conjugate acid-base pair with differently colored acidic and basic forms (§19.2)
7. The distinction between equivalence point and end point in an acid-base titration (§19.2)
8. How the titration curve of a polyprotic acid has a buffer region and equivalence point for each ionizable proton (§19.2)
9. How a slightly soluble ionic compound reaches equilibrium in water, expressed by an equilibrium (solubility-product) constant, K_{sp} (§19.3)
10. Why incomplete dissociation of an ionic compound means that calculated values for K_{sp} and solubility are approximations (§19.3)
11. Why a common ion in a solution decreases the solubility of its compounds (§19.3)
12. How pH affects the solubility of a compound that contains a weak-acid anion (§19.3)
13. How precipitate formation depends on the relative values of Q_{sp} and K_{sp} (§19.3)
14. How selective precipitation and simultaneous equilibria are used to separate ions (§19.3)
15. How complex-ion formation occurs in steps and is characterized by an overall equilibrium (formation) constant, K_f (§19.4)
16. Why addition of a ligand increases the solubility of a compound whose metal ion forms a complex ion (§19.4)
17. How the aqueous chemistry of amphoteric hydroxides involves precipitation, complex-ion formation, and acid-base equilibria (§19.4)

Master These Skills

1. Using stoichiometry and equilibrium problem-solving techniques to calculate the effect of added H_3O^+ or OH^- on buffer pH (SP 19.1)
2. Using the Henderson-Hasselbalch equation to calculate buffer pH (§19.1)
3. Using molecular scenes to examine buffers (SP 19.2)
4. Choosing the components for a buffer with a given pH and calculating their quantities (SP 19.3)
5. Calculating the pH at any point in an acid-base titration (§19.2 and SP 19.4)
6. Choosing an appropriate indicator based on the pH at various points in a titration (§19.2)
7. Writing K_{sp} expressions for slightly soluble ionic compounds (SP 19.5)
8. Calculating a K_{sp} value from solubility data (SP 19.6)
9. Calculating solubility from a K_{sp} value (SP 19.7)
10. Using K_{sp} values to compare solubilities for compounds with the same total number of ions (§19.3)
11. Calculating the decrease in solubility caused by the presence of a common ion (SP 19.8)
12. Predicting the effect of added H_3O^+ on solubility (SP 19.9)
13. Using ion concentrations to calculate Q_{sp} and compare it with K_{sp} to predict whether a precipitate forms (SPs 19.10, 19.11)
14. Comparing K_{sp} values in order to separate ions by selective precipitation (SP 19.12)
15. Calculating the equilibrium concentration of a hydrated metal ion after addition of excess ligand forms a complex ion (SP 19.13)
16. Using an overall equilibrium constant ($K_{sp} \times K_f$) to calculate the effect of complex-ion formation on solubility (SP 19.14)

Key Terms

Page numbers appear in parentheses.

acid-base buffer (843)
acid-base titration curve (853)
buffer capacity (849)
buffer range (849)
common-ion effect (844)
complex ion (877)
end point (861)
equivalence point (854)
formation constant (K_f) (877)
Henderson-Hasselbalch equation (849)
ligand (877)
selective precipitation (874)
solubility-product constant (K_{sp}) (864)

Key Equations and Relationships

Page numbers appear in parentheses.

19.1 Finding the pH from known concentrations of a conjugate acid-base pair (Henderson-Hasselbalch equation) (849):

$$pH = pK_a + \log\left(\frac{[\text{base}]}{[\text{acid}]}\right)$$

19.2 Defining the equilibrium condition for a saturated solution of a slightly soluble compound, M_pX_q, composed of M^{n+} and X^{z-} ions (864):

$$Q_{sp} = [M^{n+}]^p[X^{z-}]^q = K_{sp}$$

BRIEF SOLUTIONS TO FOLLOW-UP PROBLEMS

19.1A (a) Before addition:
Assuming x is small enough to be neglected,

$[HF] = 0.50\ M$ and $[F^-] = 0.45\ M$

$$[H_3O^+] = K_a \times \frac{[HF]}{[F^-]} \approx (6.8\times10^{-4})\left(\frac{0.50}{0.45}\right) = 7.6\times10^{-4}\ M$$

$\mathrm{pH} = 3.12$

(b) After addition of 0.40 g of NaOH to 1.0 L of buffer: Since 0.40 g of NaOH = 0.010 mol, we have

Concentration (M)	$HF(aq)$	+	$OH^-(aq)$	$\longrightarrow$	$F^-(aq)$	+	$H_2O(l)$
Initial	0.50		0.010		0.45		—
Change	−0.010		−0.010		+0.010		—
Equilibrium	0.49		0		0.46		—

$[HF] = 0.49\ M$ and $[F^-] = 0.46\ M$

$$[H_3O^+] \approx (6.8\times10^{-4})\left(\frac{0.49}{0.46}\right) = 7.2\times10^{-4}\ M;\ \mathrm{pH} = 3.14$$

19.1B (a) Before addition:

K_b of $(CH_3)_2NH = 10^{-pK_b} = 10^{-3.23} = 5.9\times10^{-4}$

$$K_a \text{ of } (CH_3)_2NH_2^+ = \frac{K_w}{K_b \text{ of } (CH_3)_2NH} = \frac{1.0\times10^{-14}}{5.9\times10^{-4}} = 1.7\times10^{-11}$$

$[(CH_3)_2NH_2^+] = 0.25\ M$ and $[(CH_3)_2NH] = 0.30\ M$

$$[H_3O^+] = K_a \times \frac{[(CH_3)_2NH_2^+]}{[(CH_3)_2NH]} \approx (1.7\times10^{-11})\left(\frac{0.25}{0.30}\right)$$
$$= 1.4\times10^{-11}\ M$$

$\mathrm{pH} = 10.85$

(b) After addition of 50.0 mL of 0.75 M HCl to 1.0 L of buffer:

$$[(CH_3)_2NH_2^+]_{init} = \frac{M_{conc} \times V_{conc}}{V_{dil}} = \frac{0.25\ M \times 1.0\ L}{1.05\ L} = 0.24\ M$$

Similarly, $[(CH_3)_2NH] = 0.29\ M$ and $[H_3O^+] = 0.036\ M$

Concentration (M)	$(CH_3)_2NH(aq)$	+	H_3O^+	$\rightleftharpoons$	$(CH_3)_2NH_2^+(aq)$	+	$H_2O(l)$
Initial	0.29		0.036		0.24		—
Change	−0.036		−0.036		+0.036		—
Equilibrium	0.25		0		0.28		—

$$[H_3O^+] \approx (1.7\times10^{-11})\left(\frac{0.28}{0.25}\right) = 1.9\times10^{-11}\ M;\ \mathrm{pH} = 10.72$$

19.2A (a) There are more HB molecules than B^- ions. Strong base would convert HB to B^-, thereby making the ratio closer to 1.

(b) Enough strong base is added to convert three HB molecules to three B^- ions, resulting in a buffer with four acid molecules and four conjugate base ions:

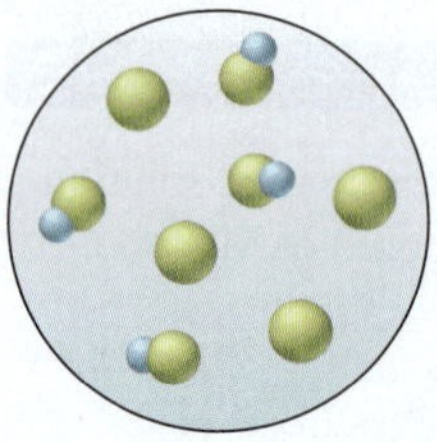

19.2B (a) $\mathrm{pH} = pK_a + \log\left(\frac{A^-}{HA}\right)$, and sample 3 has $\frac{[A^-]}{[HA]} = \frac{6}{3} = 2$; $\frac{[A^-]}{[HA]} > 1$ results in $\mathrm{pH} > pK_a$

(b) Added strong base reacts with the acid HA; sample 2 has the largest concentration of HA.

(c) pH of buffer $= pK_a \pm 1 = 4.2 \pm 1$, so the buffer pH range is 3.2–5.2; the acid-base pair HA/A^- is not suitable for preparing a buffer with a pH of 6.1.

19.3A $pK_a = -\log K_a = -\log(6.3\times10^{-5}) = 4.20$

$$\mathrm{pH} = pK_a + \log\left(\frac{[C_6H_5COO^-]}{[C_6H_5COOH]}\right)$$

$$4.25 = 4.20 + \log\left(\frac{[C_6H_5COO^-]}{[C_6H_5COOH]}\right)$$

$$\left(\frac{[C_6H_5COO^-]}{[C_6H_5COOH]}\right) = 1.1$$

Amount (mol) C_6H_5COOH

$$= 5.0\ \text{L soln} \times \frac{0.050\ \text{mol}\ C_6H_5COO^-}{1\ \text{L soln}} \times \frac{1.0\ \text{mol}\ C_6H_5COOH}{1.1\ \text{mol}\ C_6H_5COO^-}$$
$$= 0.23\ \text{mol}\ C_6H_5COOH$$

Mass (g) of C_6H_5COOH

$$= 0.23\ \text{mol}\ C_6H_5COOH \times \frac{122.12\ \text{g}\ C_6H_5COOH}{1\ \text{mol}\ C_6H_5COOH}$$
$$= 28\ \text{g}\ C_6H_5COOH$$

Dissolve 28 g of C_6H_5COOH into about 4.8 L of 0.050 M C_6H_5COONa and add more 0.050 M C_6H_5COONa for a total volume of 5.0 L. Add strong acid or base dropwise as needed.

19.3B $$K_a \text{ of } NH_4^+ = \frac{K_w}{K_b \text{ of } NH_3} = \frac{1.0\times10^{-14}}{1.76\times10^{-5}} = 5.68\times10^{-10}$$

$pK_a = -\log(5.68\times10^{-10}) = 9.25$

$$\mathrm{pH} = pK_a + \log\left(\frac{[NH_3]}{[NH_4^+]}\right)$$

$$9.18 = 9.25 + \log\left(\frac{[NH_3]}{[NH_4^+]}\right)$$

$$\left(\frac{[NH_3]}{[NH_4^+]}\right) = 0.85 \text{ (buffer-component concentration ratio)}$$

$$\text{Mass of } NH_4Cl = 0.75\ \text{L soln} \times \frac{0.15\ \text{mol}\ NH_3}{1\ \text{L soln}} \times \frac{1\ \text{mol}\ NH_4Cl}{0.85\ \text{mol}\ NH_3} \times \frac{53.49\ \text{g}\ NH_4Cl}{1\ \text{mol}\ NH_4Cl}$$
$$= 7.1\ \text{g}\ NH_4Cl$$

19.4A (a) $HBrO(aq) + H_2O(l) \rightleftharpoons H_3O^+(aq) + Br^-(aq)$

$$K_a = \frac{[H_3O^+][Br^-]}{[HBrO]} = 2.3\times10^{-9} \approx \frac{(x)(x)}{0.2000}$$

$$x \approx [H_3O^+] \approx \sqrt{(2.3\times10^{-9})(0.2000)} = 2.1\times10^{-5}\ M$$

pH = 4.68

(b) When $[HBrO] = [BrO^-]$, $[HBrO]/[BrO^-] = 1$:

$$pH = pK_a + \log\frac{[BrO^-]}{[HBrO]} = 8.64 + \log 1 = 8.64$$

(c) Initial amount (mol) of HBrO = 0.02000 L × 0.2000 M = 0.004000 mol

Amount (mol) of added NaOH at the equivalence point = 0.004000 mol

Volume (L) of added NaOH = 0.004000 mol × 1 L/0.1000 mol = 0.04000 L

From the reaction table, there are 0.004000 mol of BrO^- at the equivalence point.

$$[BrO^-] = \frac{\text{amount (mol) of } BrO^-}{\text{total volume (L)}} = \frac{0.004000 \text{ mol}}{0.02000 \text{ L} + 0.04000 \text{ L}}$$

$$= 0.06667\ M$$

$BrO^-(aq) + H_2O(l) \rightleftharpoons HBrO(aq) + OH^-(aq)$

$$K_b \text{ of } BrO^- = \frac{K_w}{K_a \text{ of HBrO}} = \frac{1.0\times10^{-14}}{2.3\times10^{-9}} = 4.3\times10^{-6}$$

$$K_b = \frac{[HBrO][OH^-]}{[BrO^-]} = 4.3\times10^{-6} = \frac{(x)(x)}{0.06667}$$

$$x = [OH^-] = \sqrt{(4.3\times10^{-6})(0.06667)}$$

$$= 5.3\times10^{-4}\ M$$

$[H_3O^+] = 1.9\times10^{-11}\ M$; pH = 10.72

(d) Amount (mol) of OH^- added = 2 × 0.004000 mol = 0.008000 mol

Volume (L) of OH^- soln = 0.008000 mol × 1 L/0.1000 mol = 0.08000 L

$$[OH^-] = \frac{\text{amount (mol) of } OH^- \text{ unreacted}}{\text{total volume (L)}}$$

$$= \frac{0.008000 \text{ mol} - 0.00400 \text{ mol}}{(0.02000 + 0.08000) \text{ L}} = 0.04000\ M$$

$$[H_3O^+] = \frac{K_w}{[OH^-]} = \frac{1.0\times10^{-14}}{0.04000} = 2.5\times10^{-13}\ M$$

pH = 12.60

(e)

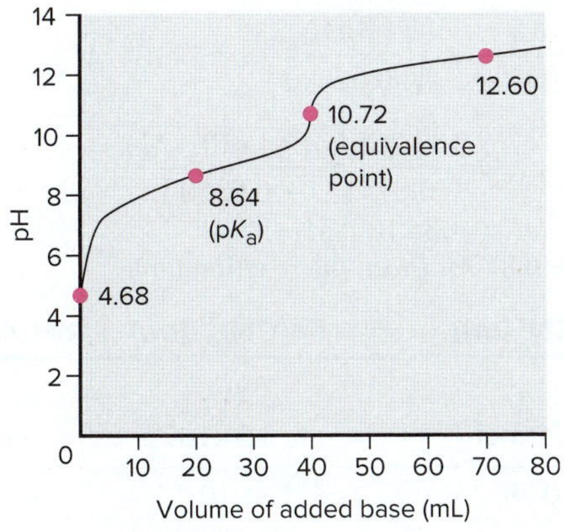

19.4B (a) $C_6H_5COOH(aq) + H_2O(l) \rightleftharpoons H_3O^+(aq) + C_6H_5COO^-(aq)$

$$K_a = \frac{[H_3O^+][C_6H_5COO^-]}{[C_6H_5COOH]} = 6.3\times10^{-5} \approx \frac{(x)(x)}{0.1000}$$

$$x \approx [H_3O^+] \approx \sqrt{(6.3\times10^{-5})(0.1000)} = 2.5\times10^{-3}\ M$$

pH = 2.60

(b) Initial amount (mol) of C_6H_5COOH = 0.03000 L × 0.1000 M = 0.003000 mol

Amount (mol) of added NaOH = 0.01200 L × 0.1500 M = 0.001800 mol

From the reaction table, 0.003000 mol − 0.001800 mol = 0.001200 mol of C_6H_5COOH remains and 0.001800 mol of $C_6H_5COO^-$ is produced:

$$pH = pK_a + \log\frac{[C_6H_5COO^-]}{[C_6H_5COOH]} = 4.20 + \log\frac{0.001800 \text{ mol}}{0.001200 \text{ mol}} = 4.38$$

(c) Initial amount (mol) of C_6H_5COOH = 0.03000 L × 0.1000 M = 0.003000 mol

Amount (mol) of added NaOH at the equivalence point = 0.003000 mol

Volume (L) of added NaOH = 0.003000 mol × 1 L/0.1500 mol = 0.02000 L

From the reaction table, 0.003000 mol of $C_6H_5COO^-$ is present at the equivalence point.

$$[C_6H_5COO^-] = \frac{\text{amount (mol) of } C_6H_5COO^-}{\text{total volume (L)}}$$

$$= \frac{0.003000 \text{ mol}}{0.03000 \text{ L} + 0.02000 \text{ L}} = 0.06000\ M$$

$C_6H_5COO^-(aq) + H_2O(l) \rightleftharpoons C_6H_5COOH(aq) + OH^-(aq)$

$$K_b \text{ of } C_6H_5COO^- = \frac{K_w}{K_a \text{ of } C_6H_5COOH} = \frac{1.0\times10^{-14}}{6.3\times10^{-5}} = 1.6\times10^{-10}$$

$$K_b = \frac{[C_6H_5COOH][OH^-]}{[C_6H_5COO^-]} = 1.6\times10^{-10} = \frac{(x)(x)}{0.06000}$$

$$x = [OH^-] = \sqrt{(1.6\times10^{-10})(0.06000)}$$

$$= 3.1\times10^{-6}\ M$$

$[H_3O^+] = 3.2\times10^{-9}\ M$; pH = 8.49

(d) Amount (mol) of OH^- added = 0.02200 L × 0.1500 M = 0.003300 mol

$$[OH^-] = \frac{\text{amount (mol) of } OH^- \text{ unreacted}}{\text{total volume (L)}}$$

$$= \frac{0.003300 \text{ mol} - 0.003000 \text{ mol}}{(0.03000 + 0.02200) \text{ L}} = 0.00577\ M$$

$$[H_3O^+] = \frac{K_w}{[OH^-]} = 1.73\times10^{-12}\ M$$

pH = 11.76 (pH is shown to two decimal places)

19.5A (a) $K_{sp} = [Ca^{2+}][SO_4^{2-}]$ (b) $K_{sp} = [Cr^{3+}]^2[CO_3^{2-}]^3$

(c) $K_{sp} = [Mg^{2+}][OH^-]^2$ (d) $K_{sp} = [Al^{3+}][OH^-]^3$

19.5B (a) $PbCrO_4$, lead(II) chromate; (b) FeS, iron(II) sulfide; (c) SrF_2, strontium fluoride; (d) $Cu_3(PO_4)_2$, copper(II) phosphate

19.6A $$[CaF_2] = \frac{1.5\times10^{-4} \text{ g } CaF_2}{10.0 \text{ mL soln}} \times \frac{1000 \text{ mL}}{1 \text{ L}} \times \frac{1 \text{ mol } CaF_2}{78.08 \text{ g } CaF_2}$$

$$= 1.9\times10^{-4}\ M$$

$CaF_2(s) \rightleftharpoons Ca^{2+}(aq) + 2F^-(aq)$

$[Ca^{2+}] = 1.9\times10^{-4}\ M$ and $[F^-] = 2(1.9\times10^{-4}\ M) = 3.8\times10^{-4}\ M$

$K_{sp} = [Ca^{2+}][F^-]^2 = (1.9\times10^{-4})(3.8\times10^{-4})^2 = 2.7\times10^{-11}$

BRIEF SOLUTIONS TO FOLLOW-UP PROBLEMS (continued)

19.6B $[Ag_3PO_4] = \frac{3.2\times10^{-4}\text{ g }Ag_3PO_4}{50.\text{ mL soln}} \times \frac{1000\text{ mL}}{1\text{ L}} \times \frac{1\text{ mol }Ag_3PO_4}{418.7\text{ g }Ag_3PO_4}$
$= 1.5\times10^{-5}\ M$

$Ag_3PO_4(s) \rightleftharpoons 3Ag^+(aq) + PO_4^{3-}(aq)$

$[Ag^+] = 3(1.5\times10^{-5}\ M) = 4.5\times10^{-5}\ M$ and $[PO_4^{3-}] = 1.5\times10^{-5}\ M$

$K_{sp} = [Ag^+]^3[PO_4^{3-}] = (4.5\times10^{-5})^3(1.5\times10^{-5}) = 1.4\times10^{-18}$

19.7A From the reaction table, $[Mg^{2+}] = S$ and $[OH^-] = 2S$.
$K_{sp} = [Mg^{2+}][OH^-]^2 = [S][2S]^2 = 4S^3 = 6.3\times10^{-10}$; $S = 5.4\times10^{-4}\ M$

19.7B From the reaction table, $[Ca^{2+}] = 3S$ and $[PO_4^{3-}] = 2S$.
$K_{sp} = [Ca^{2+}]^3[PO_4^{3-}]^2 = (3S)^3(2S)^2 = 27S^3 \times 4S^2 = 108S^5 = 1.2\times10^{-29}$
$S = 6.4\times10^{-7}\ M$

19.8A (a) In water: $K_{sp} = [Ba^{2+}][SO_4^{2-}] = (S)(S) = S^2 = 1.1\times10^{-10}$
$S = 1.0\times10^{-5}\ M$
(b) In 0.10 M Na_2SO_4: $[SO_4^{2-}] = 0.10\ M$ from Na_2SO_4;
$[SO_4^{2-}] = S$ from $BaSO_4$
Assume $S + 0.10\ M \approx 0.10\ M$
$K_{sp} = 1.1\times10^{-10} \approx S \times 0.10$; $S = 1.1\times10^{-9}\ M$
S decreases in presence of the common ion $[SO_4^{2-}]$.

19.8B (a) In water: $K_{sp} = [Ca^{2+}][F^-]^2 = (S)(2S)^2 = 4S^3 = 3.2\times10^{-11}$
$S = 2.0\times10^{-4}\ M$
(b) In 0.20 M $CaCl_2$: $[Ca^{2+}]$ from $CaCl_2 = 0.20\ M$;
$[Ca^{2+}]$ from $CaF_2 = S$
Assume $S + 0.20\ M \approx 0.20\ M$
$K_{sp} = 3.2\times10^{-11} \approx 0.20 \times (2S)^2$; $S = 6.3\times10^{-6}\ M$
(c) In 0.20 M NiF_2: $[F^-]$ from $NiF_2 = 2(0.20) = 0.40\ M$;
$[F^-]$ from $CaF_2 = 2S$
Assume $2S + 0.40\ M \approx 0.40\ M$
$K_{sp} = 3.2\times10^{-11} \approx S \times (0.40)^2$; $S = 2.0\times10^{-10}\ M$

19.9A (a) Increases solubility.

$CaF_2(s) \rightleftharpoons Ca^{2+}(aq) + 2F^-(aq)$

$F^-(aq) + H_3O^+(aq) \longrightarrow HF(aq) + H_2O(l)$

(b) Increases solubility.

$Fe(OH)_3(s) \rightleftharpoons Fe^{3+}(aq) + 3OH^-(aq)$

$OH^-(aq) + H_3O^+(aq) \longrightarrow 2H_2O(l)$

(c) No effect. $I^-(aq)$ is the conjugate base of strong acid HI.

19.9B (a) Increases solubility.

$AgCN(s) \rightleftharpoons Ag^+(aq) + CN^-(aq)$

$CN^-(aq) + H_3O^+(aq) \longrightarrow HCN(aq) + H_2O(l)$

(b) No effect. $Cl^-(aq)$ is the conjugate base of strong acid HCl.
(c) Increases solubility.

$Mg_3(PO_4)_2(s) \rightleftharpoons 3Mg^{2+}(aq) + 2PO_4^{3-}(aq)$

$PO_4^{3-}(aq) + H_3O^+(aq) \longrightarrow HPO_4^{2-}(aq) + H_2O(l)$

$HPO_4^{2-}(aq) + H_3O^+(aq) \longrightarrow H_2PO_4^-(aq) + H_2O(l)$

$H_2PO_4^-(aq) + H_3O^+(aq) \longrightarrow H_3PO_4(aq) + H_2O(l)$

19.10A $Ca_3(PO_4)_2(s) \rightleftharpoons 3Ca^{2+}(aq) + 2PO_4^{3-}(aq)$

$Q_{sp} = [Ca^{2+}]^3[PO_4^{3-}]^2 = (1.0\times10^{-9})^3(1.0\times10^{-9})^2$
$= (1.0\times10^{-9})^5 = 1.0\times10^{-45}$

$Q_{sp} < K_{sp}$, so $Ca_3(PO_4)_2$ will not precipitate.

19.10B $PbS(s) \rightleftharpoons Pb^{2+}(aq) + S^{2-}(aq)$

Amount (mol) of $Pb^{2+} = 25\text{ L} \times \frac{0.015\text{ g }Pb^{2+}}{1.0\text{ L}} \times \frac{1\text{ mol }Pb^{2+}}{207.2\text{ g }Pb^{2+}}$
$= 1.8\times10^{-3}$ mol

$[Pb^{2+}] = \frac{1.8\times10^{-3}\text{ mol }Pb^{2+}}{25\text{ L} + 0.500\text{ L}} = 7.1\times10^{-5}\ M$

Amount (mol) of $S^{2-} = 0.10\ M\ S^{2-} \times 0.500\text{ L} = 0.050$ mol

$[S^{2-}] = \frac{0.050\text{ mol}}{25\text{ L} + 0.500\text{ L}} = 2.0\times10^{-3}\ M$

$Q_{sp} = [Pb^{2+}][S^{2-}] = (7.1\times10^{-5})(2.0\times10^{-3}) = 1.4\times10^{-7}$
$Q_{sp} > K_{sp}$ (which is 3×10^{-25}), so PbS will precipitate.

19.11A (a) Scene 3 has the same relative numbers of ions as in the compound's formula (ratio of 1 Ni^{2+} to every 2 OH^-).
(b) Based on $Ni(OH)_2(s) \rightleftharpoons Ni^{2+}(aq) + 2OH^-(aq)$ and $Q_{sp} = [Ni^{2+}][OH^-]^2$, the ion products are $(3)(4)^2 = 48$ in scene 1; $(4)(2)^2 = 16$ in scene 2; and $(2)(4)^2 = 32 = K_{sp}$ in scene 3. Since Q_{sp} for scene 1 exceeds K_{sp} for scene 3, more solid forms.
(c) Addition of acid will decrease the mass of $Ni(OH)_2(s)$ by reacting with OH^-, thereby causing more solid to dissolve; addition of base will increase the mass of $Ni(OH)_2(s)$ due to the common-ion effect.

19.11B (a) Scene 1 has the same relative numbers of ions as in the compound's formula (ratio of 1 Pb^{2+} to every 2 Cl^-).
(b) Based on $PbCl_2(s) \rightleftharpoons Pb^{2+}(aq) + 2Cl^-(aq)$ and $Q_{sp} = [Pb^{2+}][Cl^-]^2$, the ion products are $(3)(6)^2 = 108 = K_{sp}$ in scene 1; $(4)(5)^2 = 100$ in scene 2; and $(5)(5)^2 = 125$ in scene 3. Since Q_{sp} for scene 3 exceeds K_{sp} for scene 1, more solid forms.
(c) Addition of $HCl(aq)$ will increase the mass of $PbCl_2$ due to the common-ion effect (Cl^- is the common ion).

19.12A Both salts have 1/1 ion ratios, and K_{sp} values show that $CaSO_4$ is more soluble:

$$[SO_4^{2-}] = \frac{K_{sp}}{[Ca^{2+}]} = \frac{2.4\times10^{-5}}{0.025}\ 9.6\times10^{-4}\ M$$

19.12B Both salts have 1/2 ion ratios, and K_{sp} values show that BaF_2 is more soluble:

$$[F^-] = \sqrt{\frac{K_{sp}}{[Ba^{2+}]}} = \sqrt{\frac{1.5\times10^{-6}}{0.020\ M}} = 8.7\times10^{-3}\ M$$

19.13A $[Fe(H_2O)_6^{3+}]_{init} = \frac{(0.0255\text{ L})(3.1\times10^{-2}\ M)}{0.0255\text{ L} + 0.0350\text{ L}}$
$= 1.3\times10^{-2}\ M$

Similarly, $[CN^-]_{init} = 0.87\ M$. From the reaction table,

Concentration (M)	$Fe(H_2O)_6^{3+}(aq)$ +	$6CN^-(aq)$	$\rightleftharpoons$ $Fe(CN)_6^{3-}(aq)$ +	$6H_2O(l)$
Initial	1.3×10^{-2}	0.87	0	—
Change	~(-1.3×10^{-2})	~$(-6\times[1.3\times10^{-2}])$	~$(+1.3\times10^{-2})$	—
Equilibrium	x	0.79	1.3×10^{-2}	—

$$K_f = \frac{[Fe(CN)_6^{3-}]}{[Fe(H_2O)_6^{3+}][CN^-]^6} = 4.0\times10^{43} \approx \frac{1.3\times10^{-2}}{x(0.79)^6}$$

$x = [Fe(H_2O)_6^{3+}] \approx 1.3\times10^{-45}\ M$

19.13B $[Al(H_2O)_6^{3+}]_{init} = \dfrac{2.4\text{ g } AlCl_3 \times \dfrac{1\text{ mol } AlCl_3}{133.33\text{ g } AlCl_3}}{0.250\text{ L}}$

$= 0.072\ M$

$[F^-]_{init} = 0.560\ M$

Concentration (*M*)	$Al(H_2O)_6^{3+}(aq)$	+ $6F^-(aq)$	$\rightleftharpoons$ $AlF_6^{3-}(aq)$	+ $6H_2O(l)$
Initial	0.072	0.560	0	—
Change	~(−0.072)	~(−6×0.072)	~(+0.072)	—
Equilibrium	*x*	0.128	0.072	—

$$K_f = \frac{[AlF_6^{3-}]}{[Al(H_2O)_6^{3+}][F^-]^6} = 4\times10^{19} \approx \frac{0.072}{x(0.128)^6}$$

$x = [Al(H_2O)_6^{3+}] \approx 4\times10^{-16}\ M$

19.14A $AgBr(s) + 2NH_3(aq) \rightleftharpoons Ag(NH_3)_2^+(aq) + Br^-(aq)$

$K_{overall} = K_{sp}$ of AgBr × K_f of $Ag(NH_3)_2^+$
$= 8.5\times10^{-6}$

From the reaction table, $[NH_3] = 1.0 - 2S$ and $[Ag(NH_3)_2^+] = [Br^-] = S$.

$$K_{overall} = \frac{[Ag(NH_3)_2^+][Br^-]}{[NH_3]^2} = 8.5\times10^{-6} = \frac{[S][S]}{(1.0 - 2S)^2}$$

$$\frac{S}{1.0 - 2S} = \sqrt{8.5\times10^{-6}} = 2.9\times10^{-3}$$

$S = [Ag(NH_3)_2^+] = 2.9\times10^{-3}\ M$

Solubility is greater in 1.0 *M* hypo than in 1.0 *M* NH_3.

19.14B $PbCl_2(s) + 3OH^-(aq) \rightleftharpoons Pb(OH)_3^-(aq) + 2Cl^-(aq)$

$K_{overall} = K_{sp} \times K_f = 1\times10^9$

From the reaction table, $[OH^-] = 0.75 - 3S$ and $[Pb(OH)_3^-] = S$; $[Cl^-] = 2S$.

$$K_{overall} = \frac{[Pb(OH)_3^-][Cl^-]^2}{[OH^-]^3} = 1\times10^9 = \frac{(S)(2S)^2}{(0.75 - 3S)^3}$$

$$= \frac{4S^3}{(0.75 - 3S)^3}$$

$$\frac{\sqrt[3]{4}S}{0.75 - 3S} = \sqrt[3]{1\times10^9}$$

$S = [Pb(OH)_3^-] = 0.25\ M$

PROBLEMS

Problems with **colored** numbers are answered in Appendix E and worked in detail in the Student Solutions Manual. Problem sections match those in the text and give the numbers of relevant sample problems. Most offer Concept Review Questions, Skill-Building Exercises (grouped in pairs covering the same concept), and Problems in Context. The Comprehensive Problems are based on material from any section or previous chapter.

Note: Unless stated otherwise, all of the problems for this chapter refer to aqueous solutions at 298 K (25°C).

Equilibria of Acid-Base Buffers

(Sample Problems 19.1 to 19.3)

Concept Review Questions

19.1 What is the purpose of an acid-base buffer?

19.2 How do the acid and base components of a buffer function? Why are they often the conjugate acid-base pair of a weak acid?

19.3 What is the common-ion effect? How is it related to Le Châtelier's principle? Explain with equations that include HF and NaF.

19.4 What is the difference between buffers with high and low capacities? Will adding 0.01 mol of HCl produce a greater pH change in a buffer with a high or a low capacity? Explain.

19.5 Which of these factors influence buffer capacity? How?
(a) The identity of the conjugate acid-base pair
(b) pH of the buffer
(c) Concentration of buffer components
(d) Buffer range
(e) pK_a of the acid component

19.6 What is the relationship between the buffer range and the buffer-component concentration ratio?

19.7 A chemist needs a pH 3.5 buffer. Should she use NaOH with formic acid ($K_a = 1.8\times10^{-4}$) or with acetic acid ($K_a = 1.8\times10^{-5}$)? Why? What is the disadvantage of choosing the other acid? What is the role of the NaOH?

19.8 State and explain the relative change in the pH and in the buffer-component concentration ratio, [NaA]/[HA], for each of the following additions:
(a) Add 0.1 *M* NaOH to the buffer
(b) Add 0.1 *M* HCl to the buffer
(c) Dissolve pure NaA in the buffer
(d) Dissolve pure HA in the buffer

19.9 Does the pH increase or decrease with each of the following additions, and does it do so to a large or small extent in each case?
(a) 5 drops of 0.1 *M* NaOH to 100 mL of 0.5 *M* acetate buffer
(b) 5 drops of 0.1 *M* HCl to 100 mL of 0.5 *M* acetate buffer
(c) 5 drops of 0.1 *M* NaOH to 100 mL of 0.5 *M* HCl
(d) 5 drops of 0.1 *M* NaOH to distilled water

Skill-Building Exercises (grouped in similar pairs)

19.10 The scenes below depict solutions of the same HA/A^- buffer (HA is *red and blue* and A^- is *red*; other ions and water molecules have been omitted for clarity). (a) Which solution has the greatest buffer capacity? (b) Explain how the pH ranges of the buffers compare. (c) Which solution can react with the largest amount of added strong acid?

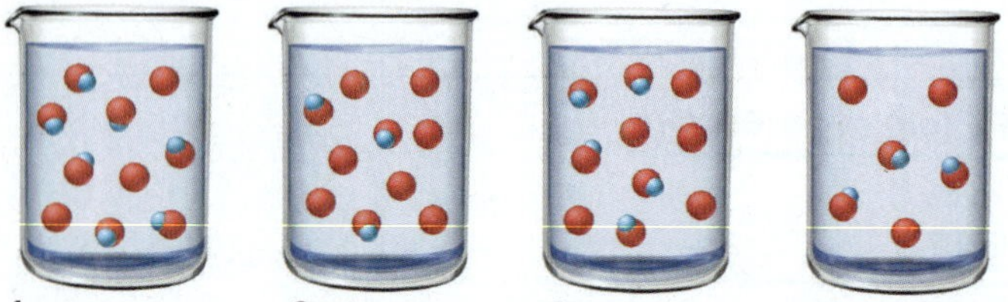

1 2 3 4

19.11 The scenes below show three samples of a buffer consisting of HA and A^- (HA is *green and blue* and A^- is *green*; other ions and water are not shown). (a) Which buffer can react with the largest amount of added strong base? (b) Which buffer has the highest pH? (c) Which buffer has pH = pK_a?

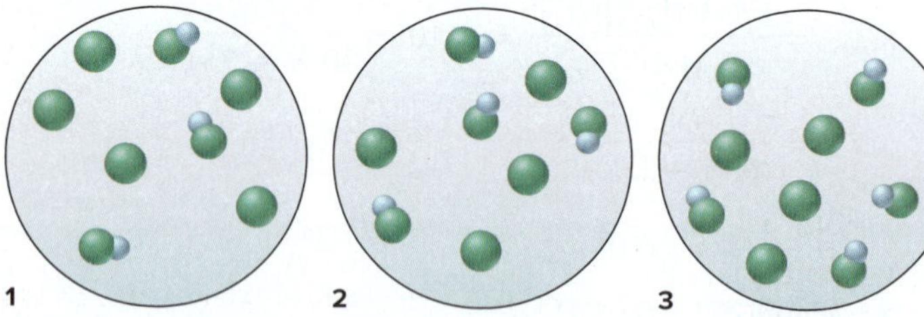

1 2 3

19.12 What are the $[H_3O^+]$ and the pH of a propanoic acid–propanoate buffer that consists of 0.35 *M* CH_3CH_2COONa and 0.15 *M* CH_3CH_2COOH (K_a of propanoic acid = 1.3×10^{-5})?

19.13 What are the $[H_3O^+]$ and the pH of a benzoic acid–benzoate buffer that consists of 0.33 *M* C_6H_5COOH and 0.28 *M* C_6H_5COONa (K_a of benzoic acid = 6.3×10^{-5})?

19.14 What are the $[H_3O^+]$ and the pH of a buffer that consists of 0.55 *M* HNO_2 and 0.75 *M* KNO_2 (K_a of HNO_2 = 7.1×10^{-4})?

19.15 What are the $[H_3O^+]$ and the pH of a buffer that consists of 0.20 *M* HF and 0.25 *M* KF (K_a of HF = 6.8×10^{-4})?

19.16 Find the pH of a buffer that consists of 0.45 *M* HCOOH and 0.63 *M* HCOONa (pK_a of HCOOH = 3.74).

19.17 Find the pH of a buffer that consists of 0.95 *M* HBrO and 0.68 *M* KBrO (pK_a of HBrO = 8.64).

19.18 Find the pH of a buffer that consists of 1.3 *M* sodium phenolate (C_6H_5ONa) and 1.2 *M* phenol (C_6H_5OH) (pK_a of phenol = 10.00).

19.19 Find the pH of a buffer that consists of 0.12 *M* boric acid (H_3BO_3) and 0.82 *M* sodium borate (NaH_2BO_3) (pK_a of boric acid = 9.24).

19.20 Find the pH of a buffer that consists of 0.25 *M* NH_3 and 0.15 *M* NH_4Cl (pK_b of NH_3 = 4.75).

19.21 Find the pH of a buffer that consists of 0.50 *M* methylamine (CH_3NH_2) and 0.60 *M* CH_3NH_3Cl (pK_b of CH_3NH_2 = 3.35).

19.22 A buffer consists of 0.22 *M* $KHCO_3$ and 0.37 *M* K_2CO_3. Carbonic acid is a diprotic acid with $K_{a1} = 4.5\times10^{-7}$ and $K_{a2} = 4.7\times10^{-11}$.
(a) Which K_a value is more important to this buffer?
(b) What is the buffer pH?

19.23 A buffer consists of 0.50 *M* NaH_2PO_4 and 0.40 *M* Na_2HPO_4. Phosphoric acid is a triprotic acid ($K_{a1} = 7.2\times10^{-3}$, $K_{a2} = 6.3\times10^{-8}$, and $K_{a3} = 4.2\times10^{-13}$). (a) Which K_a value is most important to this buffer? (b) What is the buffer pH?

19.24 What is the component concentration ratio, $[Pr^-]/[HPr]$, of a buffer that has a pH of 5.44 (K_a of HPr = 1.3×10^{-5})?

19.25 What is the component concentration ratio, $[NO_2^-]/[HNO_2]$, of a buffer that has a pH of 2.95 (K_a of HNO_2 = 7.1×10^{-4})?

19.26 What is the component concentration ratio, $[BrO^-]/[HBrO]$, of a buffer that has a pH of 7.95 (K_a of HBrO = 2.3×10^{-9})?

19.27 What is the component concentration ratio, $[CH_3COO^-]/[CH_3COOH]$, of a buffer that has a pH of 4.39 (K_a of CH_3COOH = 1.8×10^{-5})?

19.28 A buffer containing 0.2000 *M* of acid, HA, and 0.1500 *M* of its conjugate base, A^-, has a pH of 3.35. What is the pH after 0.0015 mol of NaOH is added to 0.5000 L of this solution?

19.29 A buffer that contains 0.40 *M* of a base, B, and 0.25 *M* of its conjugate acid, BH^+, has a pH of 8.88. What is the pH after 0.0020 mol of HCl is added to 0.25 L of this solution?

19.30 A buffer that contains 0.110 *M* HY and 0.220 *M* Y^- has a pH of 8.77. What is the pH after 0.0015 mol of $Ba(OH)_2$ is added to 0.350 L of this solution?

19.31 A buffer that contains 1.05 *M* B and 0.750 *M* BH^+ has a pH of 9.50. What is the pH after 0.0050 mol of HCl is added to 0.500 L of this solution?

19.32 A buffer is prepared by mixing 204 mL of 0.452 *M* HCl and 0.500 L of 0.400 *M* sodium acetate. (See Appendix C.) (a) What is the pH? (b) How many grams of KOH must be added to 0.500 L of the buffer to change the pH by 0.15 units?

19.33 A buffer is prepared by mixing 50.0 mL of 0.050 *M* sodium bicarbonate and 10.7 mL of 0.10 *M* NaOH. (See Appendix C.) (a) What is the pH? (b) How many grams of HCl must be added to 25.0 mL of the buffer to change the pH by 0.07 units?

19.34 Choose specific acid-base conjugate pairs to make the following buffers: (a) pH ≈ 4.5; (b) pH ≈ 7.0. (See Appendix C.)

19.35 Choose specific acid-base conjugate pairs to make the following buffers: (a) $[H_3O^+] \approx 1\times10^{-9}$ *M*; (b) $[OH^-] \approx 3\times10^{-5}$ *M*. (See Appendix C.)

19.36 Choose specific acid-base conjugate pairs to make the following buffers: (a) pH ≈ 3.5; (b) pH ≈ 5.5. (See Appendix C.)

19.37 Choose specific acid-base conjugate pairs to make the following buffers: (a) $[OH^-] \approx 1\times10^{-6}$ *M*; (b) $[H_3O^+] \approx 4\times10^{-4}$ *M*. (See Appendix C.)

Problems in Context

19.38 An industrial chemist studying bleaching and sterilizing prepares several hypochlorite buffers. Find the pH of
(a) 0.100 *M* HClO and 0.100 *M* NaClO;
(b) 0.100 *M* HClO and 0.150 *M* NaClO;
(c) 0.150 *M* HClO and 0.100 *M* NaClO;
(d) 1.0 L of the solution in part (a) after 0.0050 mol of NaOH has been added.

19.39 Oxoanions of phosphorus are buffer components in blood. For a KH_2PO_4/Na_2HPO_4 buffer with pH = 7.40 (pH of normal arterial blood), what is the buffer-component concentration ratio?

Acid-Base Titration Curves

(Sample Problem 19.4)

Concept Review Questions

19.40 The scenes below depict the relative concentrations of H_3PO_4, $H_2PO_4^-$, and HPO_4^{2-} during a titration with aqueous NaOH, but they are out of order. (Phosphate groups are *purple,* hydrogens are *blue,* and Na^+ ions and water molecules are not shown.) (a) List the scenes in the correct order. (b) What is the pH in the correctly ordered second scene (see Appendix C)? (c) If it requires 10.00 mL of the NaOH solution to reach this scene, how much more is needed to reach the last scene?

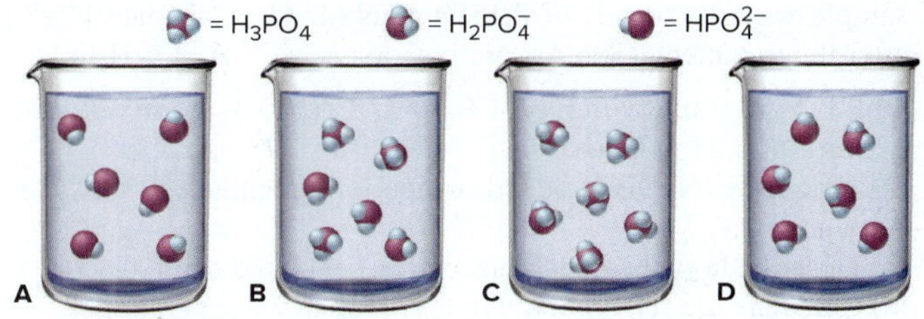

19.41 Explain how *strong acid*–strong base, *weak acid*–strong base, and *weak base*–strong acid titrations using the same concentrations differ in terms of (a) the initial pH and (b) the pH at the equivalence point. (The component in italics is in the flask.)

19.42 What species are in the buffer region of a weak acid–strong base titration? How are they different from the species at the equivalence point? How are they different from the species in the buffer region of a weak base–strong acid titration?

19.43 Why is the midpoint of the buffer region of a weak acid–strong base titration significant?

19.44 How can you estimate the pH range of an indicator's color change? Why do some indicators have two separate pH ranges?

19.45 Why does the color change of an indicator take place over a range of about 2 pH units?

19.46 Why doesn't the addition of an acid-base indicator affect the pH of a test solution?

19.47 What is the difference between the end point of a titration and the equivalence point? Is the equivalence point always reached first? Explain.

19.48 How does the titration curve of a monoprotic acid differ from that of a diprotic acid?

Skill-Building Exercises (grouped in similar pairs)

19.49 Calculate the pH during the titration of 40.00 mL of 0.1000 *M* HCl with 0.1000 *M* NaOH solution after each of the following additions of base:
(a) 0 mL (b) 25.00 mL (c) 39.00 mL (d) 39.90 mL
(e) 40.00 mL (f) 40.10 mL (g) 50.00 mL

19.50 Calculate the pH during the titration of 30.00 mL of 0.1000 *M* KOH with 0.1000 *M* HBr solution after each of the following additions of acid:
(a) 0 mL (b) 15.00 mL (c) 29.00 mL (d) 29.90 mL
(e) 30.00 mL (f) 30.10 mL (g) 40.00 mL

19.51 Find the pH during the titration of 20.00 mL of 0.1000 *M* butanoic acid, $CH_3CH_2CH_2COOH$ ($K_a = 1.54\times10^{-5}$), with 0.1000 *M* NaOH solution after the following additions of titrant:
(a) 0 mL (b) 10.00 mL (c) 15.00 mL (d) 19.00 mL
(e) 19.95 mL (f) 20.00 mL (g) 20.05 mL (h) 25.00 mL

19.52 Find the pH during the titration of 20.00 mL of 0.1000 *M* triethylamine, $(CH_3CH_2)_3N$ ($K_b = 5.2\times10^{-4}$), with 0.1000 *M* HCl solution after the following additions of titrant:
(a) 0 mL (b) 10.00 mL (c) 15.00 mL (d) 19.00 mL
(e) 19.95 mL (f) 20.00 mL (g) 20.05 mL (h) 25.00 mL

19.53 Find the pH of the equivalence point(s) and the volume (mL) of 0.0372 *M* NaOH needed to reach the point(s) in titrations of
(a) 42.2 mL of 0.0520 *M* CH_3COOH
(b) 28.9 mL of 0.0850 *M* H_2SO_3 (two equivalence points)

19.54 Find the pH of the equivalence point(s) and the volume (mL) of 0.0588 *M* KOH needed to reach the point(s) in titrations of
(a) 23.4 mL of 0.0390 *M* HNO_2
(b) 17.3 mL of 0.130 *M* H_2CO_3 (two equivalence points)

19.55 Find the pH of the equivalence point(s) and the volume (mL) of 0.125 *M* HCl needed to reach the point(s) in titrations of
(a) 65.5 mL of 0.234 *M* NH_3
(b) 21.8 mL of 1.11 *M* CH_3NH_2

19.56 Find the pH of the equivalence point(s) and the volume (mL) of 0.447 *M* HNO_3 needed to reach the point(s) in titrations of
(a) 2.65 L of 0.0750 *M* pyridine (C_5H_5N)
(b) 0.188 L of 0.250 *M* ethylenediamine ($H_2NCH_2CH_2NH_2$)

19.57 The indicator cresol red has $K_a = 3.5\times10^{-9}$. Over what approximate pH range does it change color?

19.58 The indicator ethyl red has $K_a = 3.8\times10^{-6}$. Over what approximate pH range does it change color?

19.59 Use Figure 19.9 to find an indicator for these titrations:
(a) 0.10 *M* HCl with 0.10 *M* NaOH
(b) 0.10 *M* HCOOH (see Appendix C) with 0.10 *M* NaOH

19.60 Use Figure 19.9 to find an indicator for these titrations:
(a) 0.10 *M* CH_3NH_2 (see Appendix C) with 0.10 *M* HCl
(b) 0.50 *M* HI with 0.10 *M* KOH

19.61 Use Figure 19.9 to find an indicator for these titrations:
(a) 0.5 *M* $(CH_3)_2NH$ (see Appendix C) with 0.5 *M* HBr
(b) 0.2 *M* KOH with 0.2 *M* HNO_3

19.62 Use Figure 19.9 to find an indicator for these titrations:
(a) 0.25 *M* C_6H_5COOH (see Appendix C) with 0.25 *M* KOH
(b) 0.50 *M* NH_4Cl (see Appendix C) with 0.50 *M* NaOH

Equilibria of Slightly Soluble Ionic Compounds

(Sample Problems 19.5 to 19.12)

Concept Review Questions

19.63 The molar solubility (*S*) of M_2X is 5×10^{-5} *M*. Find *S* of each ion. How do you set up the calculation to find K_{sp}? What assumption must you make about the dissociation of M_2X into ions? Why is the calculated K_{sp} higher than the actual value?

19.64 Why does pH affect the solubility of BaF_2 but not of $BaCl_2$?

19.65 A list of K_{sp} values like that in Appendix C can be used to compare the solubility of silver chloride directly with that of silver bromide but not with that of silver chromate. Explain.

19.66 In a gaseous equilibrium, the reverse reaction occurs when $Q_c > K_c$. What occurs in aqueous solution when $Q_{sp} > K_{sp}$?

Skill-Building Exercises (grouped in similar pairs)

19.67 Write the ion-product expressions for (a) silver carbonate; (b) barium fluoride; (c) copper(II) sulfide.

19.68 Write the ion-product expressions for (a) iron(III) hydroxide; (b) barium phosphate; (c) magnesium fluoride.

19.69 Write the ion-product expressions for (a) calcium chromate; (b) silver cyanide; (c) silver phosphate.

19.70 Write the ion-product expressions for (a) lead(II) iodide; (b) strontium sulfate; (c) chromium(III) hydroxide.

19.71 The solubility of silver carbonate is 0.032 *M* at 20°C. Calculate its K_{sp}.

19.72 The solubility of zinc oxalate is 7.9×10^{-3} *M* at 18°C. Calculate its K_{sp}.

19.73 The solubility of silver dichromate at 15°C is 8.3×10^{-3} g/100. mL solution. Calculate its K_{sp}.

19.74 The solubility of calcium sulfate at 30°C is 0.209 g/100. mL solution. Calculate its K_{sp}.

19.75 Find the molar solubility of $SrCO_3$ ($K_{sp} = 5.4\times10^{-10}$) in (a) pure water and (b) 0.13 *M* $Sr(NO_3)_2$.

19.76 Find the molar solubility of $BaCrO_4$ ($K_{sp} = 2.1\times10^{-10}$) in (a) pure water and (b) 1.5×10^{-3} *M* Na_2CrO_4.

19.77 Calculate the molar solubility of $Ca(IO_3)_2$ in (a) 0.060 *M* $Ca(NO_3)_2$ and (b) 0.060 *M* $NaIO_3$. (See Appendix C.)

19.78 Calculate the molar solubility of Ag_2SO_4 in (a) 0.22 *M* $AgNO_3$ and (b) 0.22 *M* Na_2SO_4. (See Appendix C.)

19.79 Which compound in each pair is more soluble in water?
(a) Magnesium hydroxide or nickel(II) hydroxide
(b) Lead(II) sulfide or copper(II) sulfide
(c) Silver sulfate or magnesium fluoride

19.80 Which compound in each pair is more soluble in water?
(a) Strontium sulfate or barium chromate
(b) Calcium carbonate or copper(II) carbonate
(c) Barium iodate or silver chromate

19.81 Which compound in each pair is more soluble in water?
(a) Barium sulfate or calcium sulfate
(b) Calcium phosphate or magnesium phosphate
(c) Silver chloride or lead(II) sulfate

19.82 Which compound in each pair is more soluble in water?
(a) Manganese(II) hydroxide or calcium iodate
(b) Strontium carbonate or cadmium sulfide
(c) Silver cyanide or copper(I) iodide

19.83 Write equations to show whether the solubility of either of the following is affected by pH: (a) AgCl; (b) $SrCO_3$.

19.84 Write equations to show whether the solubility of either of the following is affected by pH: (a) CuBr; (b) $Ca_3(PO_4)_2$.

19.85 Write equations to show whether the solubility of either of the following is affected by pH: (a) $Fe(OH)_2$; (b) CuS.

19.86 Write equations to show whether the solubility of either of the following is affected by pH: (a) PbI_2; (b) $Hg_2(CN)_2$.

19.87 Does any solid $Cu(OH)_2$ form when 0.075 g of KOH is dissolved in 1.0 L of 1.0×10^{-3} *M* $Cu(NO_3)_2$?

19.88 Does any solid $PbCl_2$ form when 3.5 mg of NaCl is dissolved in 0.250 L of 0.12 *M* $Pb(NO_3)_2$?

19.89 Does any solid $Ba(IO_3)_2$ form when 7.5 mg of $BaCl_2$ is dissolved in 500. mL of 0.023 *M* $NaIO_3$?

19.90 Does any solid Ag_2CrO_4 form when 2.7×10^{-5} g of $AgNO_3$ is dissolved in 15.0 mL of 4.0×10^{-4} *M* K_2CrO_4?

Problems in Context

19.91 When blood is donated, sodium oxalate solution is used to precipitate Ca^{2+}, which triggers clotting. A 104-mL sample of blood contains 9.7×10^{-5} g Ca^{2+}/mL. A technologist treats the sample with 100.0 mL of 0.1550 *M* $Na_2C_2O_4$. Calculate $[Ca^{2+}]$ after the treatment. (See Appendix C for K_{sp} of $CaC_2O_4{\cdot}H_2O$.)

19.92 A 50.0-mL volume of 0.50 *M* $Fe(NO_3)_3$ is mixed with 125 mL of 0.25 *M* $Cd(NO_3)_2$.
(a) If aqueous NaOH is added, which ion precipitates first? (See Appendix C.)
(b) Describe how the metal ions can be separated using NaOH.
(c) Calculate the $[OH^-]$ that will accomplish the separation.

Equilibria Involving Complex Ions

(Sample Problems 19.13 and 19.14)

Concept Review Questions

19.93 How can a metal cation be at the center of a complex anion?

19.94 Write equations to demonstrate the stepwise reaction of $Cd(H_2O)_4^{2+}$ in an aqueous solution of KI to form CdI_4^{2-}. Show that $K_{f(overall)} = K_{f1} \times K_{f2} \times K_{f3} \times K_{f4}$.

19.95 Consider the dissolution of $Zn(OH)_2$ in water:

$$Zn(OH)_2(s) \rightleftharpoons Zn^{2+}(aq) + 2OH^-(aq)$$

Adding aqueous NaOH causes more $Zn(OH)_2$ to dissolve. Does this violate Le Châtelier's principle? Explain.

Skill-Building Exercises (grouped in similar pairs)

19.96 Write a balanced equation for the reaction of $Hg(H_2O)_4^{2+}$ in aqueous KCN.

19.97 Write a balanced equation for the reaction of $Zn(H_2O)_4^{2+}$ in aqueous NaCN.

19.98 Write a balanced equation for the reaction of $Ag(H_2O)_2^{+}$ in aqueous $Na_2S_2O_3$.

19.99 Write a balanced equation for the reaction of $Al(H_2O)_6^{3+}$ in aqueous KF.

19.100 What is $[Ag^+]$ when 25.0 mL each of 0.044 *M* $AgNO_3$ and 0.57 *M* $Na_2S_2O_3$ are mixed? [K_f of $Ag(S_2O_3)_2^{3-} = 4.7\times10^{13}$]?

19.101 Potassium thiocyanate, KSCN, is often used to detect the presence of Fe^{3+} ions in solution through the formation of the red $Fe(H_2O)_5SCN^{2+}$ (or, more simply, $FeSCN^{2+}$; $K_f = 8.9\times10^2$). What is $[Fe^{3+}]$ when 0.50 L each of 0.0015 *M* $Fe(NO_3)_3$ and 0.20 *M* KSCN are mixed?

19.102 Find the solubility of $Cr(OH)_3$ in a buffer of pH 13.0 [K_{sp} of $Cr(OH)_3 = 6.3\times10^{-31}$; K_f of $Cr(OH)_4^- = 8.0\times10^{29}$].

19.103 Find the solubility of AgI in 2.5 *M* NH_3 [K_{sp} of AgI = 8.3×10^{-17}; K_f of $Ag(NH_3)_2^+ = 1.7\times10^7$].

19.104 When 0.84 g of $ZnCl_2$ is dissolved in 245 mL of 0.150 *M* NaCN, what are $[Zn^{2+}]$, $[Zn(CN)_4^{2-}]$, and $[CN^-]$ [K_f of $Zn(CN)_4^{2-} = 4.2\times10^{19}$]?

19.105 When 2.4 g of $Co(NO_3)_2$ is dissolved in 0.350 L of 0.22 *M* KOH, what are $[Co^{2+}]$, $[Co(OH)_4^{2-}]$, and $[OH^-]$ [K_f of $Co(OH)_4^{2-} = 5\times10^9$]?

Comprehensive Problems

19.106 What volumes of 0.200 *M* HCOOH and 2.00 *M* NaOH would make 500. mL of a buffer with the same pH as a buffer made from 475 mL of 0.200 *M* benzoic acid and 25 mL of 2.00 *M* NaOH?

19.107 A microbiologist is preparing a medium on which to culture *E. coli* bacteria. She buffers the medium at pH 7.00 to minimize the effect of acid-producing fermentation. What volumes of equimolar aqueous solutions of K_2HPO_4 and KH_2PO_4 must she combine to make 100. mL of the pH 7.00 buffer?

19.108 As an FDA physiologist, you need 0.700 L of formic acid–formate buffer with a pH of 3.74. (a) What is the required buffer-component concentration ratio? (b) How do you prepare this solution from stock solutions of 1.0 *M* HCOOH and 1.0 *M* NaOH? (c) What is the final concentration of HCOOH in this solution?

19.109 Tris(hydroxymethyl)aminomethane [$(HOCH_2)_3CNH_2$], known as TRIS, is a weak base used in biochemical experiments to make buffer solutions in the pH range of 7 to 9. A certain TRIS buffer has a pH of 8.10 at 25°C and a pH of 7.80 at 37°C. Why does the pH change with temperature?

19.110 Water flowing through pipes of carbon steel must be kept at pH 5 or greater to limit corrosion. If an 8.0×10^3 lb/h water stream contains 10 ppm sulfuric acid and 0.015% acetic acid, how many pounds per hour of sodium acetate trihydrate must be added to maintain that pH?

19.111 Gout is caused by an error in metabolism that leads to a buildup of uric acid in body fluids, which is deposited as slightly soluble sodium urate ($C_5H_3N_4O_3Na$) in the joints. If the extracellular $[Na^+]$ is 0.15 *M* and the solubility of sodium urate is 0.085 g/100. mL, what is the minimum urate ion concentration (abbreviated $[Ur^-]$) that will cause a deposit of sodium urate?

19.112 In the process of cave formation (Section 19.3), the dissolution of CO_2 (equation 1) has a K_{eq} of 3.1×10^{-2}, and the formation of aqueous $Ca(HCO_3)_2$ (equation 2) has a K_{eq} of 1×10^{-12}. The fraction by volume of atmospheric CO_2 is 4×10^{-4}. (a) Find $[CO_2(aq)]$ in equilibrium with atmospheric CO_2. (b) Determine $[Ca^{2+}]$ arising from (equation 2) given current levels of atmospheric CO_2. (c) Calculate $[Ca^{2+}]$ if atmospheric CO_2 doubles.

19.113 Phosphate systems form essential buffers in organisms. Calculate the pH of a buffer made by dissolving 0.80 mol of NaOH in 0.50 L of 1.0 *M* H_3PO_4.

19.114 The solubility of KCl is 3.7 *M* at 20°C. Two beakers each contain 100. mL of saturated KCl solution: 100. mL of 6.0 *M* HCl is added to the first beaker and 100. mL of 12 *M* HCl is added to the second. (a) Find the ion-product constant for KCl at 20°C. (b) What mass, if any, of KCl will precipitate from each beaker?

19.115 It is possible to detect NH_3 gas over 10^{-2} *M* NH_3. To what pH must 0.15 *M* NH_4Cl be raised to form detectable NH_3?

19.116 Manganese(II) sulfide is one of the compounds found in the nodules on the ocean floor that may eventually be a primary source of many transition metals. The solubility of MnS is 4.7×10^{-4} g/100. mL solution. Estimate the K_{sp} of MnS.

19.117 The normal pH of blood is 7.40 ± 0.05 and is controlled in part by the H_2CO_3/HCO_3^- buffer system.
(a) Assuming that the K_a value for carbonic acid at 25°C applies to blood, what is the $[H_2CO_3]/[HCO_3^-]$ ratio in normal blood?
(b) In a condition called *acidosis*, the blood is too acidic. What is the $[H_2CO_3]/[HCO_3^-]$ ratio in a patient whose blood pH is 7.20?

19.118 A bioengineer preparing cells for cloning bathes a small piece of rat epithelial tissue in a TRIS buffer (see Problem 19.109). The buffer is made by dissolving 43.0 g of TRIS ($pK_b = 5.91$) in enough 0.095 *M* HCl to make 1.00 L of solution. What are the molarity of TRIS and the pH of the buffer?

19.119 Sketch a qualitative curve for the titration of ethylenediamine, $H_2NCH_2CH_2NH_2$, with 0.1 *M* HCl.

19.120 Amino acids [general formula $NH_2CH(R)COOH$] can be considered polyprotic acids. In many cases, the R group contains additional amine and carboxyl groups.
(a) Can an amino acid dissolved in pure water have a protonated COOH group and an unprotonated NH_2 group (K_a of COOH group $= 4.47\times10^{-3}$; K_b of NH_2 group $= 6.03\times10^{-5}$)? Use glycine, NH_2CH_2COOH, to explain why.
(b) Calculate $[^+NH_3CH_2COO^-]/[^+NH_3CH_2COOH]$ at pH 5.5.
(c) The R group of lysine is $—CH_2CH_2CH_2CH_2NH_2$ ($pK_b = 3.47$). Draw the structure of lysine at pH 1, physiological pH (~7), and pH 13.
(d) The R group of glutamic acid is $—CH_2CH_2COOH$ ($pK_a = 4.07$). Of the forms of glutamic acid that are shown below, which predominates at (1) pH 1, (2) physiological pH (~7), and (3) pH 13?

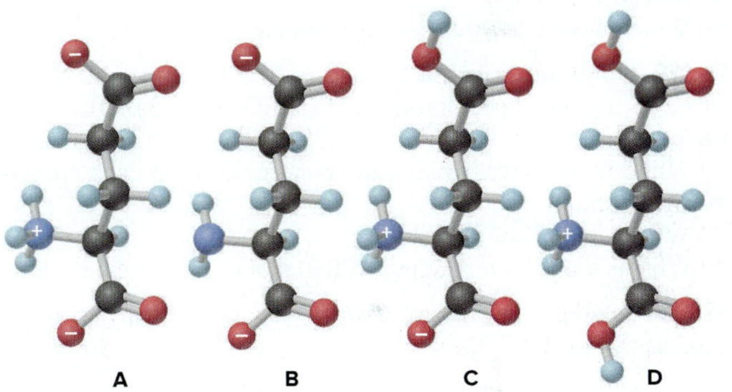

A B C D

19.121 The scene at right depicts a saturated solution of $MCl_2(s)$ in the presence of dilute aqueous NaCl; each sphere represents 1.0×10^{-6} mol of ion, and the volume is 250.0 mL (solid MCl_2 is shown as *green* chunks, M^{2+} is *blue*, and Cl^- is *yellow*; Na^+ ions and water molecules are not shown). (a) Calculate the K_{sp} of MCl_2. (b) If $M(NO_3)_2(s)$ is added, is there an increase, decrease, or no change in the number of Cl^- particles? In the K_{sp}? In the mass of $MCl_2(s)$?

19.122 Tooth enamel consists of hydroxyapatite, $Ca_5(PO_4)_3OH$ ($K_{sp} = 6.8\times10^{-37}$). Fluoride ion added to drinking water reacts with $Ca_5(PO_4)_3OH$ to form the more tooth decay–resistant fluorapatite, $Ca_5(PO_4)_3F$ ($K_{sp} = 1.0\times10^{-60}$). Fluoridated water has dramatically decreased cavities among children. Calculate the solubility of $Ca_5(PO_4)_3OH$ and of $Ca_5(PO_4)_3F$ in water.

19.123 The acid-base indicator ethyl orange turns from red to yellow over the pH range 3.4 to 4.8. Estimate K_a for ethyl orange.

19.124 Use the values obtained in Problem 19.49 to sketch a curve of $[H_3O^+]$ vs. mL of added titrant. Are there advantages or disadvantages to viewing the results in this form? Explain.

19.125 Instrumental acid-base titrations use a pH meter to monitor the changes in pH and volume. The equivalence point is found from the volume at which the curve has the steepest slope.
(a) Use the data in Figure 19.6 to calculate the slope ($\Delta pH/\Delta V$) for all pairs of adjacent points and to calculate the average volume (V_{avg}) for each interval.
(b) Plot $\Delta pH/\Delta V$ vs. V_{avg} to find the steepest slope, and thus the volume at the equivalence point. (For example, the first pair of points gives $\Delta pH = 0.22$, $\Delta V = 10.00$ mL; hence, $\Delta pH/\Delta V = 0.022$ mL^{-1}, and $V_{avg} = 5.00$ mL.)

19.126 What is the pH of a solution of 6.5×10^{-9} mol of $Ca(OH)_2$ in 10.0 L of water [K_{sp} of $Ca(OH)_2 = 6.5\times10^{-6}$]?

19.127 Muscle physiologists study the accumulation of lactic acid [$CH_3CH(OH)COOH$] during exercise. Food chemists study its occurrence in sour milk, beer, wine, and fruit. Industrial microbiologists study its formation by various bacterial species from carbohydrates. A biochemist prepares a lactic acid–lactate buffer by mixing 225 mL of 0.85 *M* lactic acid ($K_a = 1.38\times10^{-4}$) with 435 mL of 0.68 *M* sodium lactate. What is the buffer pH?

19.128 A student wants to dissolve the maximum amount of CaF_2 ($K_{sp} = 3.2\times10^{-11}$) to make 1 L of aqueous solution.
(a) Into which of the following solvents should she dissolve the salt?
(i) Pure water (ii) 0.01 *M* HF
(iii) 0.01 *M* NaOH (iv) 0.01 *M* HCl
(v) 0.01 *M* $Ca(OH)_2$
(b) Which would dissolve the least amount of salt?

19.129 A 500.-mL solution consists of 0.050 mol of solid NaOH and 0.13 mol of hypochlorous acid (HClO; $K_a = 3.0\times10^{-8}$) dissolved in water.
(a) Aside from water, what is the concentration of each species that is present?
(b) What is the pH of the solution?
(c) What is the pH after adding 0.0050 mol of HCl to the flask?

19.130 Calcium ion present in water supplies is easily precipitated as calcite ($CaCO_3$):

$$Ca^{2+}(aq) + CO_3^{2-}(aq) \rightleftharpoons CaCO_3(s)$$

Because the K_{sp} decreases with temperature, heating hard water forms a calcite "scale," which clogs pipes and water heaters. Find the solubility of calcite in water (a) at 10°C ($K_{sp} = 4.4\times10^{-9}$) and (b) at 30°C ($K_{sp} = 3.1\times10^{-9}$).

19.131 Calculate the molar solubility of $Hg_2C_2O_4$ ($K_{sp} = 1.75\times10^{-13}$) in 0.13 *M* $Hg_2(NO_3)_2$.

19.132 Environmental engineers use alkalinity as a measure of the capacity of carbonate buffering systems in water samples:

$$\text{Alkalinity (mol/L)} = [HCO_3^-] + 2[CO_3^{2-}] + [OH^-] - [H^+]$$

Find the alkalinity of a water sample that has a pH of 9.5, 26.0 mg/L CO_3^{2-}, and 65.0 mg/L HCO_3^-.

19.133 Human blood contains one buffer system based on phosphate species and one based on carbonate species. Assuming that blood has a normal pH of 7.4, what are the principal phosphate and carbonate species present? What is the ratio of the two phosphate species? (In the presence of the dissolved ions and other species in blood, K_{a1} of $H_3PO_4 = 1.3\times10^{-2}$, $K_{a2} = 2.3\times10^{-7}$, and $K_{a3} = 6\times10^{-12}$; K_{a1} of $H_2CO_3 = 8\times10^{-7}$ and $K_{a2} = 1.6\times10^{-10}$.)

19.134 Quantitative analysis of Cl^- ion is often performed by a titration with silver nitrate, using sodium chromate as an indicator. As standardized $AgNO_3$ is added, both white AgCl and red Ag_2CrO_4 precipitate, but as long as some Cl^- remains, the Ag_2CrO_4 redissolves as the mixture is stirred. When the red color is permanent, the equivalence point has been reached.
(a) Calculate the equilibrium constant for the reaction

$$2AgCl(s) + CrO_4^{2-}(aq) \rightleftharpoons Ag_2CrO_4(s) + 2Cl^-(aq)$$

(b) Explain why the silver chromate redissolves.
(c) If 25.00 cm^3 of 0.1000 *M* NaCl is mixed with 25.00 cm^3 of 0.1000 *M* $AgNO_3$, what is the concentration of Ag^+ remaining in solution? Is this sufficient to precipitate any silver chromate?

19.135 An ecobotanist separates the components of a tropical bark extract by chromatography. She discovers a large proportion of quinidine, a dextrorotatory isomer of quinine used for control of arrhythmic heartbeat. Quinidine has two basic nitrogens ($K_{b1} = 4.0\times10^{-6}$ and $K_{b2} = 1.0\times10^{-10}$). To measure the concentration of quinidine, she carries out a titration. Because of the low solubility of quinidine, she first protonates both nitrogens with excess HCl and titrates the acidified solution with standardized base. A 33.85-mg sample of quinidine ($\mathcal{M} = 324.41$ g/mol) is acidified with 6.55 mL of 0.150 *M* HCl.
(a) How many milliliters of 0.0133 *M* NaOH are needed to titrate the excess HCl?
(b) How many additional milliliters of titrant are needed to reach the first equivalence point of quinidine dihydrochloride?
(c) What is the pH at the first equivalence point?

19.136 Some kidney stones form by the precipitation of calcium oxalate monohydrate ($CaC_2O_4{\cdot}H_2O$, $K_{sp} = 2.3\times10^{-9}$). The pH of urine varies from 5.5 to 7.0, and the average [Ca^{2+}] in urine is 2.6×10^{-3} *M*.
(a) If the [oxalic acid] in urine is 3.0×10^{-13} *M*, will kidney stones form at pH = 5.5?
(b) Will they form at pH = 7.0?
(c) Vegetarians have a urine pH above 7. Are they more or less likely to form kidney stones?

19.137 A biochemist needs a medium for acid-producing bacteria. The pH of the medium must not change by more than 0.05 pH units for every 0.0010 mol of H_3O^+ generated by the organisms per liter of medium. A buffer consisting of 0.10 *M* HA and 0.10 *M* A^- is included in the medium to control its pH. What volume of this buffer must be included in 1.0 L of medium?

19.138 A 35.00-mL solution of 0.2500 *M* HF is titrated with a standardized 0.1532 *M* solution of NaOH at 25°C.
(a) What is the pH of the HF solution before titrant is added?
(b) How many milliliters of titrant are required to reach the equivalence point?
(c) What is the pH at 0.50 mL before the equivalence point?
(d) What is the pH at the equivalence point?
(e) What is the pH at 0.50 mL after the equivalence point?

19.139 Because of the toxicity of mercury compounds, mercury(I) chloride is used in antibacterial salves. The mercury(I) ion (Hg_2^{2+}) consists of two bonded Hg^+ ions.
(a) What is the empirical formula of mercury(I) chloride?
(b) Calculate [Hg_2^{2+}] in a saturated solution of mercury(I) chloride ($K_{sp} = 1.5\times10^{-18}$).
(c) A seawater sample contains 0.20 lb of NaCl per gallon. Find [Hg_2^{2+}] if the seawater is saturated with mercury(I) chloride.
(d) How many grams of mercury(I) chloride are needed to saturate 4900 km^3 of pure water (the volume of Lake Michigan)?
(e) How many grams of mercury(I) chloride are needed to saturate 4900 km^3 of seawater?

19.140 A 35.0-mL solution of 0.075 M $CaCl_2$ is mixed with 25.0 mL of 0.090 M $BaCl_2$.
(a) If aqueous KF is added, which fluoride precipitates first?
(b) Describe how the metal ions can be separated using KF to form the fluorides.
(c) Calculate the fluoride ion concentration that will accomplish the separation.

19.141 Rainwater is slightly acidic due to dissolved CO_2. Use the following data to calculate the pH of unpolluted rainwater at 25°C: vol % in air of CO_2 = 0.040 vol %; solubility of CO_2 in pure water at 25°C and 1 atm = 88 mL CO_2/100. mL H_2O; K_{a1} of $H_2CO_3 = 4.5\times10^{-7}$.

19.142 Seawater at the surface has a pH of about 8.5.
(a) Which of the following species has the highest concentration at this pH: H_2CO_3; HCO_3^-; CO_3^{2-}? Explain.
(b) What are the concentration ratios $[CO_3^{2-}]/[HCO_3^-]$ and $[HCO_3^-]/[H_2CO_3]$ at this pH?
(c) In the deep sea, light levels are low, and the pH is around 7.5. Suggest a reason for the lower pH at the greater ocean depth. (*Hint:* Consider the presence or absence of plant and animal life and the effects on carbon dioxide concentrations.)

19.143 Ethylenediaminetetraacetic acid (abbreviated H_4EDTA) is a tetraprotic acid. Its salts are used to treat toxic metal poisoning by forming soluble complex ions that are then excreted. Because $EDTA^{4-}$ also binds essential calcium ions, it is often administered as the calcium disodium salt. For example, when $Na_2Ca(EDTA)$ is given to a patient, the $[Ca(EDTA)]^{2-}$ ions react with circulating Pb^{2+} ions and the metal ions are exchanged:

$$[Ca(EDTA)]^{2-}(aq) + Pb^{2+}(aq) \rightleftharpoons [Pb(EDTA)]^{2-}(aq) + Ca^{2+}(aq) \qquad K_c = 2.5\times10^{7}$$

A child has a dangerous blood lead level of 120 μg/100 mL. If the child is administered 100. mL of 0.10 M $Na_2Ca(EDTA)$, what is the final concentration of Pb^{2+} in μg/100 mL blood, assuming that the exchange reaction and excretion process are 100% efficient? (Total blood volume is 1.5 L.)

19.144 Buffers that are based on 3-morpholinopropanesulfonic acid (MOPS) are often used in RNA analysis. The useful pH range of a MOPS buffer is 6.5 to 7.9. Estimate the K_a of MOPS.

19.145 NaCl is purified by adding HCl to a saturated solution of NaCl (317 g/L). Will pure NaCl precipitate when 28.5 mL of 8.65 M HCl is added to 0.100 L of saturated solution?

19.146 Scenes A to D represent tiny portions of 0.10 M aqueous solutions of a weak acid HA (*red and blue;* $K_a = 4.5\times10^{-5}$), its conjugate base A^- *(red),* or a mixture of the two (only these two species are shown):

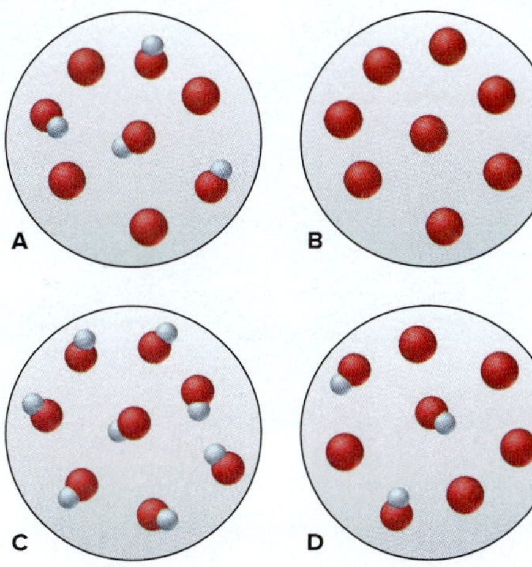

(a) Which scene(s) show(s) a buffer?
(b) What is the pH of each solution?
(c) Arrange the scenes in sequence, assuming that they represent stages in a weak acid–strong base titration.
(d) Which scene represents the titration at its equivalence point?

19.147 Scenes A to C represent aqueous solutions of the slightly soluble salt MZ (only the ions of this salt are shown):

$$MZ(s) \rightleftharpoons M^{2+}(aq) + Z^{2-}(aq)$$

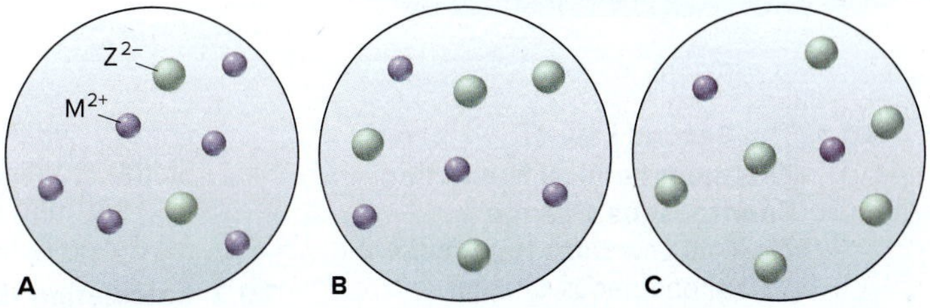

(a) Which scene represents the solution just after solid MZ is stirred thoroughly in distilled water?
(b) If each sphere represents an ion concentration of 2.5×10^{-6} M, what is the K_{sp} of MZ?
(c) Which scene represents the solution after $Na_2Z(aq)$ has been added?
(d) If Z^{2-} is CO_3^{2-}, which scene represents the solution after the pH has been lowered?

19.148 The solubility of Ag(I) in aqueous solutions containing different concentrations of Cl^- is based on the following equilibria:

$$Ag^+(aq) + Cl^-(aq) \rightleftharpoons AgCl(s) \qquad K_{sp} = 1.8\times10^{-10}$$
$$Ag^+(aq) + 2Cl^-(aq) \rightleftharpoons AgCl_2^-(aq) \qquad K_f = 1.8\times10^{5}$$

When solid AgCl is shaken with a solution containing Cl^-, Ag(I) is present as both Ag^+ and $AgCl_2^-$. The solubility of AgCl is the sum of the concentrations of Ag^+ and $AgCl_2^-$.
(a) Show that $[Ag^+]$ in solution is given by

$$[Ag^+] = 1.8\times10^{-10}/[Cl^-]$$

and that $[AgCl_2^-]$ in solution is given by

$$[AgCl_2^-] = (3.2\times10^{-5})([Cl^-])$$

(b) Find the $[Cl^-]$ at which $[Ag^+] = [AgCl_2^-]$.
(c) Explain the shape of a plot of AgCl solubility vs. $[Cl^-]$.
(d) Find the solubility of AgCl at the $[Cl^-]$ of part (b), which is the minimum solubility of AgCl in the presence of Cl^-.

19.149 EDTA binds metal ions to form complex ions (see Problem 19.143), so it is used to determine the concentrations of metal ions in solution:

$$M^{n+}(aq) + EDTA^{4-}(aq) \longrightarrow MEDTA^{n-4}(aq)$$

A 50.0-mL sample of 0.048 M Co^{2+} is titrated with 0.050 M $EDTA^{4-}$. Find $[Co^{2+}]$ and $[EDTA^{4-}]$ after addition of (a) 25.0 mL and (b) 75.0 mL of $EDTA^{4-}$ (log K_f of $CoEDTA^{2-}$ = 16.31).

20 Thermodynamics: Entropy, Free Energy, and Reaction Direction

20.1 The Second Law of Thermodynamics: Predicting Spontaneous Change
- The First Law Does Not Predict Spontaneous Change
- The Sign of ΔH Does Not Predict Spontaneous Change
- Freedom of Motion and Dispersal of Kinetic Energy
- Entropy and the Number of Microstates
- Thermodynamic Treatment of Entropy
- Entropy and the Second Law
- Standard Molar Entropies and the Third Law
- Predicting Relative S° of a System

20.2 Calculating the Change in Entropy of a Reaction
- Standard Entropy of Reaction
- Entropy Changes in the Surroundings
- Entropy Change and the Equilibrium State
- Spontaneous Exothermic and Endothermic Reactions

20.3 Entropy, Free Energy, and Work
- Free Energy Change and Reaction Spontaneity
- Standard Free Energy Changes
- Free Energy and Work
- Temperature and Reaction Spontaneity
- Coupling of Reactions

20.4 Free Energy, Equilibrium, and Reaction Direction

Source: © Luciano Mortula/Shutterstock.com

Concepts and Skills to Review Before You Study This Chapter

- internal energy, heat, and work (Section 6.1)
- state functions (Section 6.1) and standard states (Section 6.6)
- enthalpy, ΔH, and Hess's law (Sections 6.2 and 6.5)
- entropy and solution formation (Section 13.3)
- comparing Q and K to find reaction direction (Section 17.4)

In the last few chapters, we've examined some fundamental questions about change, whether chemical or physical: How fast does it occur? How far does it go toward completion? How is a change affected by concentration and temperature? And we've explored these questions in systems ranging from the stratosphere to a limestone cave to the cells of your body.

But why does the change occur in the first place? Some changes seem to happen *in one direction* but not the other. Copper statues turn green over time as the copper oxidizes *(photo),* but the green oxidation products do not react to re-form the shiny copper. Gasoline burns in air to produce carbon dioxide and water vapor as it runs a car engine, but those products do not react to form the gasoline and oxygen. A cube of sugar dissolves in a cup of coffee after a little stirring, but wait for another millennium and the dissolved sugar won't form the cube again. Chemists call these *spontaneous* changes. Most release energy, but some absorb it.

The principles of thermodynamics, developed almost 200 years ago, explain *why* a spontaneous change occurs. And, as far as we know, they apply to every change in every system in the universe—every working machine, every growing plant, every thinking student, every moving continent, every exploding star.

IN THIS CHAPTER . . . *We discuss* ***why*** *changes occur and how we can utilize them by focusing on the concepts of entropy and free energy and their relation to the direction of a spontaneous change.*

- We discuss the need for a criterion to predict the *direction* of a spontaneous change.
- We review the first law of thermodynamics and see that it accounts for the energy of a change but *not* the direction.
- We see that the sign of the enthalpy change does not predict the direction either.
- We find the criterion for predicting the direction of a spontaneous change in the second law of thermodynamics, which focuses on entropy (S), a state function based on the natural tendency of a system's energy to become dispersed.
- We examine entropy changes for exothermic and endothermic processes.
- We develop the concept of free energy as a simplified criterion for the occurrence of spontaneous change and see how free energy relates to the work a system can do.
- We explore the key relationship between the free energy change of a reaction and its equilibrium constant.

20.1 THE SECOND LAW OF THERMODYNAMICS: PREDICTING SPONTANEOUS CHANGE

A **spontaneous change** of a system is one that occurs under specified conditions without a continuous input of energy from outside the system. The freezing of water, for example, is spontaneous at 1 atm and −5°C. A spontaneous process, such as burning or falling, may need a little "push" to get started—a spark to ignite gasoline vapors in your car's engine, a shove to knock a book off your desk—but once the process begins, it supplies the energy needed for it to continue. In contrast, a *nonspontaneous* change occurs only if the surroundings *continuously* supply the system with an input of energy. Under given conditions, *if a change* **is** *spontaneous in one direction, it is* ***not*** *spontaneous in the other.* A book falls off your desk spontaneously, but does not spontaneously move from floor to desk.

The term *spontaneous* does not mean "instantaneous" nor does it reveal anything about how long a process takes to occur; it means that, given enough time, the process will happen by itself. Many processes are spontaneous but slow—including ripening, rusting, and aging.

Can we predict the direction of a spontaneous change in cases that are not as obvious as burning gasoline or falling books? In this section, we first briefly consider two concepts that do *not* predict the direction of a spontaneous change and then focus on one that does—the concept of entropy. In Section 20.1, we examine several aspects of this central idea. We discuss the meaning of entropy in terms of the relation between the freedom of motion of the particles in a system and the dispersal of its kinetic energy. Then, we show how to obtain the change in a system's entropy through a statistical approach and through the measurement of changes in heat or work. We explore entropy changes during a phase change and during a change in temperature. Finally, you'll see how to compare the relative entropy of systems based on temperature, physical state, whether pure or dissolved, and the atomic or molecular structure.

The First Law of Thermodynamics Does Not Predict Spontaneous Change

Let's see whether energy changes can predict the direction of spontaneous change. Recall from Chapter 6 that the first law of thermodynamics (law of energy conservation) states that the internal energy (E) of a system, the sum of the kinetic and potential energies of its particles, changes when heat (q) and/or work (w) are absorbed or released:

$$\Delta E = q + w$$

Whatever is not part of the system (sys) is part of the surroundings (surr); thus, *the change in energy and, therefore, the heat and/or work absorbed by the system are released by the surroundings,* or vice versa:

$$\Delta E_{\text{sys}} = -\Delta E_{\text{surr}} \quad \text{or} \quad (q + w)_{\text{sys}} = -(q + w)_{\text{surr}}$$

Since the system plus the surroundings is the universe (univ), it follows that *the total energy of the universe is constant, so the change in energy of the universe is zero:**

$$\Delta E_{\text{sys}} + \Delta E_{\text{surr}} = -\Delta E_{\text{surr}} + \Delta E_{\text{surr}} = 0 = \Delta E_{\text{univ}}$$

The first law accounts for the energy, but not the direction, of a process. When gasoline burns in a car engine, the first law states that the potential energy difference between the bonds in the fuel mixture and those in the exhaust gases is converted to the kinetic energy of the moving car and its parts plus the heat released to the environment. But, why doesn't the heat released in the engine convert exhaust fumes back into gasoline and oxygen? When an ice cube melts in your hand, the first law states that energy from your hand is converted to kinetic energy as the solid changes to a liquid. But, why doesn't the pool of water in your cupped hand transfer the heat back to your hand and refreeze? Neither of these events violates the first law—if you measure the work and heat in each case, you find that energy is conserved—but these reverse changes never happen. That is, *the first law by itself does not predict the* ***direction*** *of a spontaneous change.*

The Sign of ΔH Does Not Predict Spontaneous Change

Perhaps the sign of the enthalpy change (ΔH), the heat gained or lost at constant pressure (q_P), is the criterion for spontaneity; in fact, leading scientists thought so through most of the 19th century. If so, we would expect exothermic processes ($\Delta H < 0$) to be spontaneous and endothermic processes ($\Delta H > 0$) to be nonspontaneous. Let's examine some examples to see if this is true.

1. *Spontaneous processes with* $\Delta H < 0$. All freezing and condensing processes are exothermic and spontaneous at certain conditions:

$$H_2O(l) \longrightarrow H_2O(s) \qquad \Delta H^\circ_{\text{rxn}} = -\Delta H^\circ_{\text{fus}} = -6.02 \text{ kJ (1 atm; } T = 0°\text{C)}$$

*Any modern statement of conservation of energy must take into account mass-energy equivalence and the processes in stars, which convert enormous amounts of matter into energy. Thus, the total *mass-energy* of the universe is constant.

The burning of methane and all other combustion reactions are spontaneous and exothermic:

$$CH_4(g) + 2O_2(g) \longrightarrow CO_2(g) + 2H_2O(g) \qquad \Delta H^\circ_{rxn} = -802\ kJ$$

Oxidation of iron and other metals occurs spontaneously and exothermically:

$$2Fe(s) + \tfrac{3}{2}O_2(g) \longrightarrow Fe_2O_3(s) \qquad \Delta H^\circ_{rxn} = -826\ kJ$$

Ionic compounds form spontaneously and exothermically from their elements:

$$Na(s) + \tfrac{1}{2}Cl_2(g) \longrightarrow NaCl(s) \qquad \Delta H^\circ_{rxn} = -411\ kJ$$

2. *Spontaneous processes with* $\Delta H > 0$. In many cases, an exothermic process occurs spontaneously under one set of conditions, whereas the opposite, endothermic, process occurs spontaneously under another set. All melting and vaporizing processes are endothermic *and* spontaneous at certain conditions:

$$H_2O(s) \longrightarrow H_2O(l) \qquad \Delta H^\circ_{rxn} = \Delta H^\circ_{fus} = +6.02\ kJ\ (1\ atm;\ T = 0°C)$$

At ordinary pressure, water vaporizes spontaneously:

$$H_2O(l) \longrightarrow H_2O(g) \qquad \Delta H^\circ_{rxn} = \Delta H^\circ_{vap} = +44.0\ kJ\ (1\ atm;\ T = 100°C)$$

Most soluble salts dissolve endothermically *and* spontaneously:

$$NaCl(s) \xrightarrow{H_2O} Na^+(aq) + Cl^-(aq) \qquad \Delta H^\circ_{soln} = +3.9\ kJ$$

$$RbClO_3(s) \xrightarrow{H_2O} Rb^+(aq) + ClO_3^-(aq) \qquad \Delta H^\circ_{soln} = +47.7\ kJ$$

$$NH_4NO_3(s) \xrightarrow{H_2O} NH_4^+(aq) + NO_3^-(aq) \qquad \Delta H^\circ_{soln} = +25.7\ kJ$$

Even some endothermic reactions are spontaneous:

$$N_2O_5(s) \longrightarrow 2NO_2(g) + \tfrac{1}{2}O_2(g) \qquad \Delta H^\circ_{rxn} = +109.5\ kJ$$

$$Ba(OH)_2{\cdot}8H_2O(s) + 2NH_4NO_3(s) \longrightarrow$$
$$Ba^{2+}(aq) + 2NO_3^-(aq) + 2NH_3(aq) + 10H_2O(l) \qquad \Delta H^\circ_{rxn} = +62.3\ kJ$$

In the latter reaction, ionic solids are mixed (Figure 20.1A) and the waters of hydration released by the $Ba(OH)_2$ solvate the ions. The reaction mixture absorbs heat from the surroundings so quickly that the beaker freezes to a wet board (Figure 20.1B).

Considering just these few examples, we see that, as for the first law, *the sign of* ΔH *by itself does not predict the* ***direction*** *of a spontaneous change.*

Figure 20.1 A spontaneous endothermic reaction.

Source: © McGraw-Hill Education/Stephen Frisch, photographer

Freedom of Particle Motion and Dispersal of Kinetic Energy

When we look closely at the previous examples of spontaneous *endothermic* processes, they have one major feature in common: in every case, the chemical entities—atoms, molecules, or ions—have more freedom of motion *after* the change. Put another way, after the change, the particles have a wider range of energy of motion (kinetic energy); we say that their energy has become more dispersed, distributed, or spread out:

- The phase changes convert a solid, in which motion is restricted, to a liquid, in which particles have more freedom to move around each other, and then to a gas, in which the particles have much greater freedom of motion. Thus, the energy of motion is more dispersed.
- Dissolving a salt changes a crystalline solid and a pure liquid into separate ions and solvent molecules moving and interacting, so their freedom of motion is greater and their energy of motion more dispersed.
- In the chemical reactions, *fewer* moles of crystalline solids produce *more* moles of gases and/or solvated ions, so once again, the freedom of motion of the particles increases and their energy of motion is more dispersed:

	less freedom of particle motion	⟶	more freedom of particle motion
	localized energy of motion	⟶	dispersed energy of motion
Phase change:	solid	⟶	liquid ⟶ gas
Dissolving of salt:	crystalline solid + liquid	⟶	ions in solution
Chemical change:	crystalline solids	⟶	gases + ions in solution

In thermodynamic terms, a change in the freedom of motion of particles in a system, that is, in the dispersal of their energy of motion, is a key factor for predicting the ***direction*** *of a spontaneous process.*

Entropy and the Number of Microstates

Earlier, we discussed the quantized *electronic* energy levels of an atom (Chapter 7) and a molecule (Chapter 11), and mentioned the quantized *kinetic* energy levels—vibrational, rotational, and translational—of a molecule (see Chapter 9, Tools of the Laboratory). Now, we'll see that the energy state of a whole system of atoms or molecules is quantized, too.

Energy Dispersal and the Meaning of Entropy Let's see why freedom of motion and dispersal of energy relate to spontaneous change:

- *Quantization of energy*. Picture a system of, say, 1 mol of N_2 gas and focus on one molecule. At any instant, it is moving through space (translating) at some speed and rotating at some frequency, and its atoms are vibrating at some frequency. In the next instant, the molecule collides with another or with the container, and these motional (kinetic) energy states change to different values. The complete quantum state of the molecule at any instant consists of its electronic states and these translational, rotational, and vibrational states. In this discussion, we focus on the latter three, that is, on the kinetic energy states.
- *Number of microstates*. The energy of all the molecules in a system is similarly quantized. Each quantized state of the system is called a **microstate,** and at any instant, the total energy of the system is dispersed throughout one microstate. In the next instant, it is dispersed throughout a different microstate. The number of microstates possible for a system of 1 mol of molecules is staggering, on the order of $10^{10^{23}}$.
- *Dispersal of energy*. At a given set of conditions, each microstate has the *same* total energy as any other. Therefore, each microstate is equally possible for the system, and the laws of probability say that, over time, all microstates are equally likely. The number of microstates for a system is the number of ways it can disperse (distribute or spread) its kinetic energy among the various motions of all its particles.

In 1877, the Austrian mathematician and physicist Ludwig Boltzmann related the number of microstates (W) to the **entropy (S)** of a system:

$$S = k \ln W \qquad \textbf{(20.1)}$$

where k, the *Boltzmann constant,* is the universal gas constant (R, 8.314 J/mol·K) divided by Avogadro's number (N_A), that is, R/N_A, which equals 1.38×10^{-23} J/K. The term W is the number of microstates, so it has no units; therefore, S has units of joules/kelvin (J/K). Thus,

- A system with fewer microstates (smaller W) has *lower entropy (lower S).*
- A system with more microstates (larger W) has *higher entropy (higher S).*

For our earlier examples of endothermic processes,

	lower entropy (fewer microstates)	⟶	higher entropy (more microstates)
Phase change:	solid	⟶	liquid ⟶ gas
Dissolving of salt:	crystalline solid + liquid	⟶	ions in solution
Chemical change:	crystalline solids	⟶	gases + ions in solution

(Recall from Chapter 13 that entropy is a key factor in the formation of solutions.)

Entropy as a State Function If a change results in a greater number of microstates, there are more ways to disperse the energy of the system and the entropy increases:

$$S_{\text{more microstates}} > S_{\text{fewer microstates}}$$

If a change results in a lower number of microstates, the entropy decreases. Like internal energy (E) and enthalpy (H), *entropy is a state function,* so it depends only on the present state of the system, not on how it arrived at that state (see Chapter 6).

Therefore, the change in entropy of the system (ΔS_{sys}) depends only on the *difference* between its final and initial values:

$$\Delta S_{sys} = S_{final} - S_{initial}$$

As with any state function, $\Delta S_{sys} > 0$ when the value of S increases during a change. For example, for the phase change when dry ice sublimes, the entropy increases since gaseous CO_2 has more microstates and higher entropy than solid CO_2:

$$CO_2(s) \longrightarrow CO_2(g) \qquad \Delta S_{sys} = S_{final} - S_{initial} = S_{gaseous\ CO_2} - S_{solid\ CO_2} > 0$$

And, $\Delta S_{sys} < 0$ when the entropy decreases; when water vapor condenses, for example, the entropy decreases since liquid H_2O has fewer microstates and lower entropy than gaseous H_2O:

$$H_2O(g) \longrightarrow H_2O(l) \qquad \Delta S_{sys} = S_{liquid\ H_2O} - S_{gaseous\ H_2O} < 0$$

As an example of a reaction during which entropy increases, consider the decomposition of dinitrogen tetroxide (N_2O_4, written here as $O_2N{-}NO_2$):

$$O_2N{-}NO_2(g) \longrightarrow 2NO_2(g)$$

When the N—N bond in 1 mol of dinitrogen tetroxide breaks, the 2 mol of NO_2 molecules have many more possible motions; thus, at any instant, the energy of the system is dispersed into any one of a larger number of microstates. Thus, the change in entropy of the system, which is the change in entropy of the reaction (ΔS_{rxn}), goes up:

$$\Delta S_{sys} = \Delta S_{rxn} = S_{final} - S_{initial} = S_{products} - S_{reactants} = 2S_{NO_2} - S_{O_2N{-}NO_2} > 0$$

Quantitative Meaning of an Entropy Change—Counting Microstates We consider two approaches for quantifying a change in entropy. The first is a statistical approach, which we discuss here; the second relies on the measurement of thermodynamic variables, and we discuss it later. The statistical approach is based on *counting the number of microstates* possible for a system. We will consider a system of 1 mole of a gas expanding from 1 L to 2 L and behaving ideally, much as neon does at 298 K:

$$\text{1 mol Ne (initial: 1 L and 298 K)} \longrightarrow \text{1 mol Ne (final: 2 L and 298 K)}$$

Figure 20.2 shows two flasks connected by a stopcock—the right flask is evacuated, and the left flask contains 1 mol of neon. When we open the stopcock, the gas expands until each flask contains 0.5 mol—but *why*? Opening the stopcock increases the volume, which increases the number of translational energy levels the particles can occupy as they move to more locations. Thus, the number of microstates—and the entropy—increases.

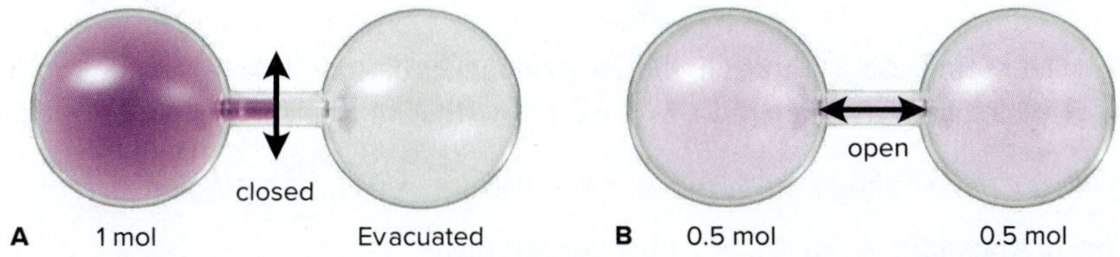

Figure 20.2 Spontaneous expansion of a gas. A, With the stopcock closed, the left flask contains 1 mol of Ne. **B,** With the stopcock open, the gas expands and each flask contains 0.5 mol of Ne.

Figure 20.3 presents this idea with particles on energy levels in a box of changeable volume. When the stopcock opens, there are more energy levels, and they are closer together on average, so more distributions of particles are possible.

In Figure 20.4 *(next page)*, the number of microstates is represented by the placement of particles in the left and/or right flasks:

- *One Ne atom.* At a given instant, a Ne atom in the left flask has its energy in one of some number (W) of microstates. Opening the stopcock increases the volume, which increases the number of possible locations and the number of translational energy levels. Thus, the system has 2^1, or 2, times as many microstates available when the atom moves through both flasks (final state, W_{final}) as when it is confined to the left flask (initial state, $W_{initial}$).
- *Two Ne atoms.* For atoms A and B moving through both flasks, there are 2^2, or 4, times as many microstates as when the atoms were initially in the left flask—some number of microstates with both A and B in the left, that many with A in the left and B in the right or with B in the left and A in the right, and that many with both in the right.

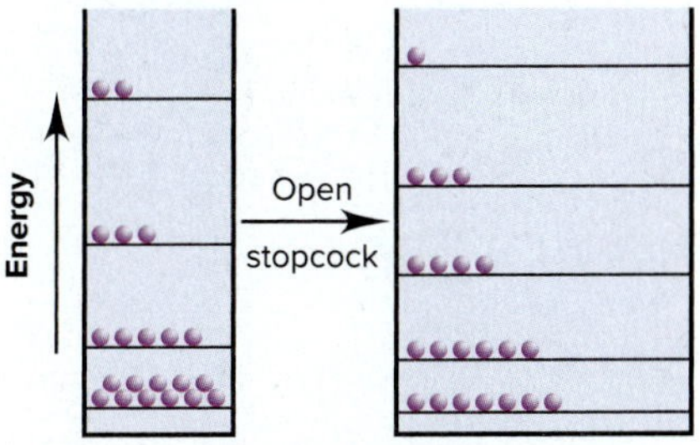

Figure 20.3 The entropy increase due to expansion of a gas. Energy levels are shown as lines in a box of narrow width (*left*). Each distribution of energies for the 21 particles is one microstate. When the stopcock is opened, the box is wider (volume increases, *right*), and the particles have more energy levels available.

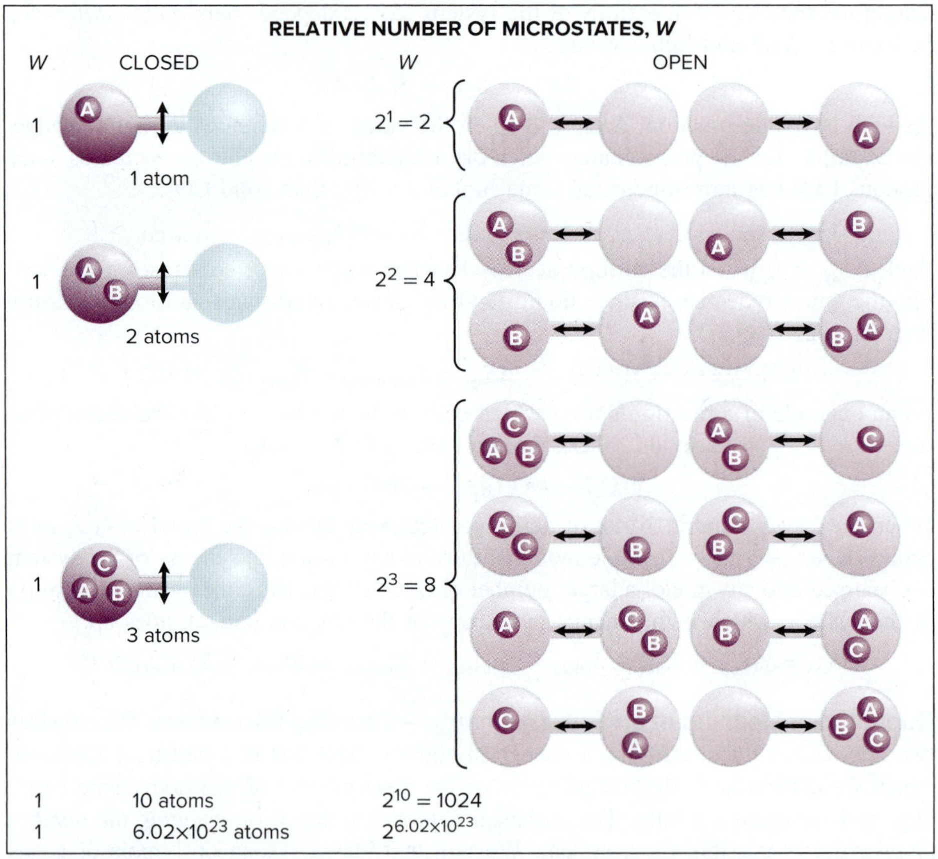

Figure 20.4 Expansion of a gas and the increase in number of microstates. Each set of particle locations represents a different microstate. When the volume increases (stopcock opens), the *relative* number of microstates is 2^n, where *n* is the number of particles.

- *Three Ne atoms.* Add another atom, and there are 2^3, or 8, times as many microstates when the stopcock is open.
- *Ten Ne atoms.* With 10 Ne atoms, there are 2^{10}, or 1024, times as many microstates for the atoms in both flasks as there were for the 10 atoms in the left flask.
- *One mole of Ne atoms.* With 1 mol (N_A) of Ne, there are 2^{N_A} times as many microstates for the atoms in the larger volume (W_{final}) than in the smaller ($W_{initial}$):

$$\frac{W_{final}}{W_{initial}} = 2^{N_A}$$

Now let's find ΔS_{sys} through the Boltzmann equation, $S = k \ln W$. From the properties of logarithms (Appendix A), we know that $\ln A - \ln B = \ln A/B$. Thus,

$$\Delta S_{sys} = S_{final} - S_{initial} = k \ln W_{final} - k \ln W_{initial} = k \ln \frac{W_{final}}{W_{initial}} = k \ln 2^{6.02\times10^{23}}$$

Also, from Appendix A, $\ln A^y = y \ln A$; so we have

$$\begin{aligned}\Delta S_{sys} &= k \ln 2^{6.02\times10^{23}} = k(6.02\times10^{23})(\ln 2)\\ &= (1.38\times10^{-23}\ \text{J/K})(6.02\times10^{23})(0.693) = 5.76\ \text{J/K}\end{aligned}$$

Quantitative Meaning of an Entropy Change—Measuring Thermodynamic Variables

The second approach to quantifying an entropy change measures thermodynamic quantities such as heat and enthalpy that occur during the expansion (or contraction) of gases, phase changes, and the heating of a system.

Entropy Change During an Isothermal Process The great German physicist Rudolf Clausius (1822–1888) defined the entropy change from an initial state to a final state for a reversible process in terms of the heat transfer involved, *dq,* divided by the temperature, *T*:

$$\Delta S = S_{final} - S_{initial} = \int_{initial}^{final} \frac{dq_{rev}}{T}$$

A **reversible process** is one that occurs in such tiny increments that the system remains at equilibrium and the direction of the change can be reversed by an *infinitesimal* reversal of conditions. If the process takes place at constant temperature, it is called an **isothermal process,** and the equation for ΔS can be integrated to obtain

$$\Delta S_{sys} = \frac{q_{rev}}{T} \tag{20.2}$$

We'll approximate a reversible, isothermal expansion by placing 1 mol of Ne gas in a piston-cylinder assembly within a heat reservoir to maintain a constant T of 298 K and start by confining the gas to a volume of 1 L by the "pressure" of a beaker of sand on the piston (Figure 20.5). We remove one grain of sand (an "infinitesimal" decrease in pressure) with a pair of tweezers, and the gas (system) expands a tiny amount, raising the piston and doing work, $-w$. Assuming Ne behaves ideally, it absorbs an equivalent tiny increment of heat, q, from the heat reservoir. With each grain of sand removed, the expanding gas absorbs another tiny increment of heat. This process simulates a reversible expansion because we can reverse it by putting back a grain of sand, which causes the surroundings to do a tiny quantity of work compressing the gas, thus releasing a tiny quantity of heat to the reservoir.

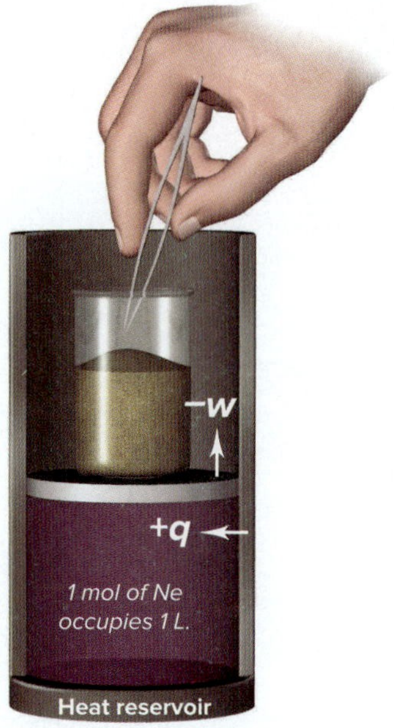

Figure 20.5 Simulating a reversible process.

If we continue this nearly reversible expansion to 2 L and use calculus to integrate the tiny increments of heat together, q_{rev} is 1718 J. From Equation 20.2,

$$\Delta S_{sys} = \frac{q_{rev}}{T} = \frac{1718 \text{ J}}{298 \text{ K}} = 5.76 \text{ J/K}$$

Another way to calculate the entropy change for this reversible, isothermal process is to consider the tiny increments of work done by the gas during expansion:

$$dw = -PdV$$

For an ideal gas, $P = \frac{nRT}{V}$, which replaces P in the work equation to give

$$dw = -\frac{nRT}{V}dV$$

Integrating this relationship from the initial to the final volume gives

$$w = -nRT\int_{V_{initial}}^{V_{final}} \frac{dV}{V} = -nRT \ln\frac{V_{final}}{V_{initial}}$$

For an isothermal process involving an ideal gas, $\Delta E = q + w = 0$ and therefore $q = -w$; thus, the equation above can be written as

$$q_{rev} = nRT \ln\frac{V_{final}}{V_{initial}}$$

When this relationship is combined with Equation 20.2, we obtain

$$\Delta S_{sys} = \frac{q_{rev}}{T} = \frac{nRT \ln(V_{final}/V_{initial})}{T}$$

or

$$\Delta S_{sys} = nR \ln\frac{V_{final}}{V_{initial}} \tag{20.3}$$

For the reversible expansion of 1 mol of neon from 1 L to 2 L shown in Figure 20.5, we have

$$\Delta S_{sys} = nR \ln\frac{V_{final}}{V_{initial}} = (1 \text{ mol})(8.314 \text{ J/mol·K}) \ln\frac{2.0 \text{ L}}{1.0 \text{ L}} = 5.76 \text{ J/K}$$

Note that this value of ΔS_{sys}, obtained from integrating increments of work, is the same as the value we obtained from integrating increments of heat (Equation 20.2) and is also the same as the value we obtained from counting the number of microstates. We would obtain the entropy change accompanying a reversible, isothermal *contraction* of an ideal gas in the same way, but V_{final} would be *smaller* than $V_{initial}$.

Recall that Boyle's law (Equation 5.1) shows the inverse relationship between the volume and pressure of a fixed amount of ideal gas at constant temperature:

$$P_{\text{initial}}V_{\text{initial}} = P_{\text{final}}V_{\text{final}} \quad \text{or} \quad \frac{V_{\text{final}}}{V_{\text{initial}}} = \frac{P_{\text{initial}}}{P_{\text{final}}}$$

Substituting the relationship from Boyle's law into Equation 20.3 expresses the entropy change for an isothermal expansion or contraction of an ideal gas in terms of pressure:

$$\Delta S_{\text{sys}} = nR \ln \frac{P_{\text{initial}}}{P_{\text{final}}} \tag{20.4}$$

SAMPLE PROBLEM 20.1 **Calculating the Change in Entropy During an Isothermal Volume Change of an Ideal Gas**

Problem What is the change in entropy of 75.0 g of neon gas when it undergoes isothermal contraction from 16.2 L to 12.1 L? Assume ideal gas behavior.

Plan We know the mass (75.0 g) and the initial (16.2 L) and final (12.1 L) volumes of the gas and we have to find ΔS of the contraction. To do so, we use Equation 20.3 after converting the mass (g) to amount (mol).

Solution Converting from mass (g) to amount (mol):

$$\text{Amount (mol) of Ne} = 75.0 \text{ g Ne} \times \frac{1 \text{ mol Ne}}{20.18 \text{ g Ne}} = 3.72 \text{ mol Ne}$$

Calculating ΔS:

$$\Delta S_{\text{sys}} = nR \ln \frac{V_{\text{final}}}{V_{\text{initial}}} = 3.72 \text{ mol} \times 8.314 \text{ J/mol·K} \times \ln \frac{12.1 \text{ L}}{16.2 \text{ L}} = -9.02 \text{ J/K}$$

Check Since the gaseous atoms are now in a smaller volume, fewer microstates are possible and thus the entropy should decrease. The calculated entropy change is negative, as expected for a decrease in entropy. Rounding gives 4.0 mol × 8 J/mol·K × ln 0.75 = –9 J/K.

FOLLOW-UP PROBLEMS

*Brief Solutions for **all** Follow-up Problems appear at the end of the chapter.*

20.1A Calculate the entropy change for 200.0 g of N_2 gas that is compressed from a pressure of 0.715 atm to a pressure of 1.16 atm at a constant temperature of 30.0°C. Assume that the gas behaves ideally.

20.1B Calculate the entropy change that occurs when 100. g of CO_2 gas expands from 3.35 L to 4.75 L at a constant temperature of 298 K. Assume ideal gas behavior.

SOME SIMILAR PROBLEMS 20.112 and 20.113

Entropy Change During a Phase Change Equation 20.2 can also be used to calculate the change in entropy for a phase change such as melting (fusion) or vaporization, since these are isothermal processes:

$$\Delta S_{\text{fus}} = \frac{\Delta H_{\text{fus}}}{T_{\text{f}}} \quad \text{and} \quad \Delta S_{\text{vap}} = \frac{\Delta H_{\text{vap}}}{T_{\text{b}}} \tag{20.5}$$

where ΔH_{fus} is the enthalpy of fusion, ΔH_{vap} is the enthalpy of vaporization, and T_{f} and T_{b} are the melting point and boiling point, respectively. Note that we can replace q_{rev} in Equation 20.2 by ΔH in Equation 20.5 because the phase transition occurs at constant pressure (see Equation 6.6).

A very interesting observation made in the late 19th century, known as **Trouton's rule,** shows that most liquids have a ΔS_{vap} value close to 88 J/mol·K (Table 20.1, *next page*). This finding implies that the entropies of most substances in the liquid phase

Table 20.1 Boiling Point (T_b), Molar Enthalpy of Vaporization, (H_{vap}), and Molar Entropy of Vaporization (S_{vap}) of Several Substances

Substance	T_b (K)	H_{vap} (kJ/mol)	S_{vap} (J/mol·K)
Acetone (CH_3COCH_3)	329.4	29.1	88.3
Benzene (C_6H_6)	353.2	31.0	87.7
Carbon tetrachloride (CCl_4)	349.8	32.5	92.9
Diethyl ether $(C_2H_5)_2O$	308.6	27.4	88.8
Ethanol C_2H_5OH	351.4	39.3	111.8
Mercury (Hg)	630	59.0	93.7
Methane (CH_4)	109	9.2	84
Methanol (CH_3OH)	337.8	35.5	105
Water (H_2O)	373.15	40.79	109.4

have a similar number of microstates, as do the substances in the gas phase; thus, the entropy difference between the two phases gives a consistent value. The greatest deviations in Table 20.1 from the ΔS_{vap} value of Trouton's rule occur for water, ethanol (C_2H_5OH), and methanol (CH_3OH), all of which are higher than expected. A common structural feature of these compounds is their ability to form H bonds, and we attribute the deviations to the fact that hydrogen bonding in the liquid phase restricts freedom of molecular motion, so these three liquids would have a lower number of microstates and lower than expected entropies. In the gas phase, however, these three substances have a similar number of microstates as other substances since the molecules are too far apart for hydrogen bonding to affect freedom of motion. Thus, the liquids that can form H bonds have a greater increase in entropy upon vaporizing than liquids that cannot.

Sample Problem 20.2 shows how we find the change in entropy, ΔS, that accompanies a phase change.

SAMPLE PROBLEM 20.2 Calculating the Change in Entropy During a Phase Change

Problem Calculate the change in entropy in J/K for
(a) the vaporization of 10.0 g of water at 100.°C (ΔH_{vap} = 40.7 kJ/mol).
(b) the melting of 10.0 g of ice at 0.0°C (ΔH_{fus} = 6.02 kJ/mol).

Plan We'll combine the Plans for parts (a) and (b) because the steps are similar. We know the mass (10.0 g), the boiling point (100.°C), the melting point (0.0°C), and the values of ΔH_{vap} and ΔH_{fus}, and we have to find ΔS of the phase changes. To do so, we use Equation 20.5, so we find the total ΔH for the amount of substance and divide by the boiling or melting point. We convert the temperatures from °C to K and the mass (g) to amount (mol). Then we multiply the amount by ΔH_{vap} and ΔH_{fus} to find ΔH (in J). Finally, we apply Equation 20.5.

Solution Converting T from °C to K:

$$T_b\,(\text{K}) = 100.°\text{C} + 273.15 = 373\text{ K} \qquad T_f\,(\text{K}) = 0.0°\text{C} = 273.15 = 273.2\text{ K}$$

Converting from mass (g) to amount (mol):

$$\text{Amount (mol) of } H_2O = 10.0\text{ g } H_2O \times \frac{1\text{ mol } H_2O}{18.0\text{ g } H_2O} = 0.556\text{ mol } H_2O$$

Finding ΔH from the molar enthalpies and the amount (mol):
(a) For vaporization:

$$\Delta H(\text{J}) = 0.556\text{ mol } H_2O \times \frac{40.7\text{ kJ}}{1\text{ mol } H_2O} \times \frac{10^3\text{ J}}{1\text{ kJ}} = 22{,}629\text{ J}$$

(b) For melting:

$$\Delta H(\text{J}) = 0.556\text{ mol } H_2O \times \frac{6.02\text{ kJ}}{1\text{ mol } H_2O} \times \frac{10^3\text{ J}}{1\text{ kJ}} = 3{,}347\text{ J}$$

Calculating ΔS:

$$\textbf{(a)}\ \Delta S_{\text{vap}} = \frac{\Delta H_{\text{vap}}}{T_{\text{b}}} = \frac{22{,}629\ \text{J}}{373\ \text{K}} = 60.7\ \text{J/K}$$

$$\textbf{(b)}\ \Delta S_{\text{fus}} = \frac{\Delta H_{\text{fus}}}{T_{\text{f}}} = \frac{3{,}347\ \text{J}}{273.2\ \text{K}} = 12.25\ \text{J/K}$$

Check The units are correct. Since $\Delta H_{\text{fus}} < \Delta H_{\text{vap}}$, $\Delta S_{\text{fus}} < \Delta S_{\text{vap}}$. The arithmetic seems correct: in (a), for example, rounding gives 23,000 J/400 K = 57.5 J/K.

FOLLOW-UP PROBLEMS

20.2A Calculate the entropy change for the vaporization of 34.0 g of acetone, C_3H_6O, at its boiling point of 56.2°C. The heat of vaporization for acetone is 31.3 kJ/mol.

20.2B Calculate the entropy change for the freezing of 112 g of ethanol, C_2H_5OH, at its freezing point of –114°C. The heat of fusion for ethanol is 5.02 kJ/mol.

SOME SIMILAR PROBLEMS 20.114 and 20.115

Entropy Change Due to a Temperature Change We can calculate the change in entropy due to a change in temperature by integrating Equation 20.2 from T_1 to T_2:

$$\Delta S_{\text{sys}} = \int_{T_1}^{T_2} \frac{dq_{\text{rev}}}{T}$$

We saw in Section 6.3 that the quantity of heat absorbed or released by n moles of substance during a temperature change is determined by the molar heat capacity, C_{m}:

$$q = n \times C_{\text{m}} \times \Delta T \qquad \text{or} \qquad dq = n \times C_{\text{m}} \times dT$$

Combining these two equations gives

$$\Delta S_{\text{sys}} = \int_{T_1}^{T_2} \frac{nC_{\text{m}}}{T} dT$$

We remove n from the quantity to be integrated and we'll assume that C_{m} is independent of temperature, so it can be removed, too, and we have

$$\Delta S_{\text{sys}} = nC_{\text{m}} \int_{T_1}^{T_2} \frac{1}{T} dT$$

If the temperature change occurs at constant volume, we use the *constant-volume molar heat capacity*, $C_{V,\text{m}}$, and integrate to obtain

$$\Delta S_{\text{sys}} = nC_{V,\text{m}} \ln \frac{T_2}{T_1} \qquad \textbf{(20.6)}$$

The *constant-pressure molar heat capacity*, $C_{P,\text{m}}$, is used for temperature changes occurring at constant pressure:

$$\Delta S_{\text{sys}} = nC_{P,\text{m}} \ln \frac{T_2}{T_1} \qquad \textbf{(20.7)}$$

We use these quantities to find the change in entropy when the temperature changes.

SAMPLE PROBLEM 20.3 Calculating the Entropy Change Resulting from a Change in Temperature

Problem Calculate the entropy change when a 700.-g sample of mercury is heated from 25°C to 250.°C at constant pressure. $C_{P,\text{m}}$ for mercury = 27.98 J/mol·K.

Plan We know the mass (700. g), the initial (25.0°C) and final (250.°C) temperatures, and the constant-pressure molar heat capacity (27.98 J/mol·K), and we need to find ΔS. We convert the mass (g) to amount (mol) and the temperatures from °C to K. We then use Equation 20.7 to calculate the entropy change.

Solution Converting from mass (g) to amount (mol):

$$\text{Amount (mol) of Hg} = 700.\ \text{g} \times \frac{1\ \text{mol Hg}}{200.6\ \text{g Hg}} = 3.49\ \text{mol Hg}$$

Converting T from °C to K:

$$T_1\ (\text{K}) = 25°\text{C} + 273.15 = 298\ \text{K} \qquad T_2\ (\text{K}) = 250.0°\text{C} + 273.15 = 523\ \text{K}$$

Calculating ΔS:

$$\Delta S_{\text{sys}} = nC_{P,\text{m}} \ln\frac{T_2}{T_1} = 3.49\ \text{mol} \times 27.98\ \text{J/mol·K} \times \ln\frac{523\ \text{K}}{298\ \text{K}} = 54.9\ \text{J/K}$$

Check The units are correct. Rounding gives 3.5 mol × 28 J/mol·K × ln 1.7 = 50 J/K.

FOLLOW-UP PROBLEMS

20.3A Calculate the entropy change when a 375-g sample of nitrogen gas is cooled from 25°C to 0°C at constant volume. $C_{V,\text{m}}$ for nitrogen = 20.8 J/mol·K.

20.3B When 100. g of water at 25°C is heated at constant pressure, ΔS is 70.9 J/K. What is the final temperature of the water (in °C)? $C_{P,\text{m}}$ for water = 75.3 J/mol·K.

SOME SIMILAR PROBLEMS 20.116 and 20.117

Entropy and the Second Law of Thermodynamics

The change in entropy determines the direction of a spontaneous process, but we must consider more than just the entropy change of the *system.* After all, at 20°C, solid water melts spontaneously and ΔS_{sys} goes up, while at −10°C, liquid water freezes spontaneously and ΔS_{sys} goes down. But when we consider *both* the system *and* its surroundings, we find the following:

- *All real processes occur spontaneously in the direction that increases the entropy of the universe (system plus surroundings).* This is one way to state the **second law of thermodynamics.**

The second law says that *either* the entropy change of the system *or* that of the surroundings may be negative. But, for a process to be spontaneous, the *sum* of the two entropy changes must be positive. If the entropy of the system decreases, the entropy of the surroundings must increase even more to offset that decrease, so that the entropy of the universe (system *plus* surroundings) increases. A quantitative statement of the second law, for any real spontaneous process, is

$$\Delta S_{\text{univ}} = \Delta S_{\text{sys}} + \Delta S_{\text{surr}} > 0 \qquad \textbf{(20.8)}$$

(We return to this idea in Section 20.2.)

Standard Molar Entropies and the Third Law

Entropy and enthalpy are state functions, but their values differ in a fundamental way:

- For *enthalpy,* there is no zero point, so we can measure only *changes.*
- For *entropy,* there *is* a zero point, and we can determine *absolute* values by applying the **third law of thermodynamics:** *a perfect crystal has zero entropy at absolute zero:* $S_{\text{sys}} = 0$ at 0 K.

A "perfect" crystal means all the particles are aligned flawlessly. At absolute zero, the particles have minimum energy, so there is only one microstate. Thus, in Equation 20.1,

$$W = 1 \qquad \text{so} \qquad S = k \ln 1 = 0$$

When we warm the crystal to any temperature above 0 K, the total energy increases, so it can be dispersed into more than one microstate. Thus,

$$W > 1 \qquad \text{and} \qquad \ln W > 0 \qquad \text{so} \qquad S > 0$$

To find S of 1 mole of substance at a given temperature, T, we use the equation for the change in entropy due to a change in temperature at constant pressure and the fact that $S = 0$ at 0 K:

$$\Delta S_{sys} = S(T) - S(0\text{ K}) = C_P \int_0^T \frac{1}{T} dT$$

The heat capacity C_P is measured at all temperatures from $T = 0$ K to the temperature of interest, usually 298 K. From a graph of C_P/T vs T, the area under the curve is the entropy of the substance at T. If the substance undergoes a phase change during the heating process from 0 K to T, the entropy change for the phase change must be added to the S value obtained from the graph.

Therefore, S of a substance at a given temperature is an *absolute* value. As with other thermodynamic variables,

- We compare entropy values for substances at the temperature of interest in their *standard states: 1 atm for gases, 1 M for solutions, and the pure substance in its most stable form for solids or liquids.*
- Because entropy is an *extensive* property—one that depends on the amount of substance—we specify the **standard molar entropy ($S°$),** in units of J/mol·K (or J mol^{-1} K^{-1}). ($S°$ values at 298 K for many elements, compounds, and ions appear, with other thermodynamic variables, in Appendix B.)

Predicting Relative $S°$ of a System

Let's see how the standard molar entropy of a substance is affected by several parameters: temperature, physical state, dissolution, and atomic size or molecular complexity. (Unless stated otherwise, the $S°$ values refer to the system at 298 K.)

Temperature Changes Temperature has a direct effect on entropy:

- *For any substance, $S°$ increases as T rises.*

This trend is shown by the entropy values for copper metal at three temperatures:

T (K):	273	295	298
$S°$(J/mol·K):	31.0	32.9	33.2

As heat is absorbed ($q > 0$), temperature, which is a measure of the average kinetic energy of the particles, increases. Recall that the kinetic energies of gas particles are distributed over a range, which becomes wider as T rises (Figure 5.14); liquids and solids behave the same. Thus, at any instant, at a higher temperature, there are more microstates available in which the energy can be dispersed, so the entropy of the substance goes up. Figure 20.6 presents three ways to view the effect of temperature on entropy.

Physical States and Phase Changes In melting or vaporizing, heat is absorbed ($q > 0$). The particles have more freedom of motion, and their energy is more dispersed. Thus,

- *$S°$ increases as the physical state of a substance changes from solid to liquid to gas:*

$$S°_{solid} < S°_{liquid} < S°_{gas}$$

$S°$ (J/mol·K)	Na	H_2O	C(graphite)
$S°$(s or l):	51.4(s)	69.9(l)	5.7(s)
$S°$(g):	153.6	188.7	158.0

Figure 20.7 plots entropy versus T as solid O_2 is heated and changes to liquid and then to gas, with $S°$ values at various points; this behavior is typical of many substances. At the molecular scale, several stages occur:

- Particles in the solid vibrate about their positions but, on average, remain fixed. The energy of the solid is dispersed least, that is, has the fewest available microstates, so the solid has the lowest entropy.
- As T rises, the entropy increases gradually as the particles' kinetic energy increases.

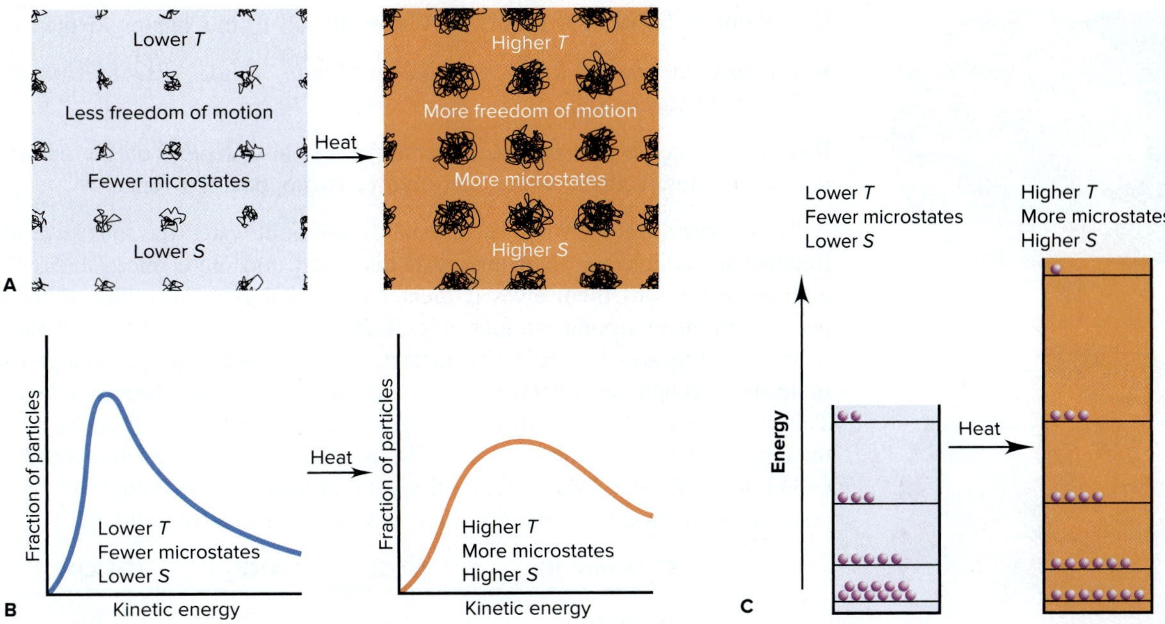

Figure 20.6 Visualizing the effect of temperature on entropy. A, Computer simulations show each particle in a crystal moving about its lattice position. Adding heat increases *T* and the total energy, so the particles have greater freedom of motion, and their energy is more dispersed. Thus, *S* increases. **B,** At any *T*, there is a range of occupied energy levels and, thus, a certain number of microstates. Adding heat increases the total energy *(area under curve),* so the range of occupied energy levels becomes greater, as does the number of microstates (higher *S*). **C,** A system consisting of 21 particles that occupy energy levels *(lines)* in a box whose height represents the total energy. When heat is added, the total energy increases *(box is higher) and* becomes more dispersed *(more lines),* so *S* increases.

- When the solid melts, the particles move much more freely between and around each other, so there is an abrupt increase in entropy (ΔS°_{fus}).
- Further heating of the liquid increases the speed of the particles, and the entropy increases gradually.
- When the liquid vaporizes and becomes a gas, the particles undergo a much larger, abrupt entropy increase (ΔS°_{vap}); *the increase in entropy from liquid to gas is much larger than from solid to liquid:* $\Delta S^\circ_{vap} >> \Delta S^\circ_{fus}$.
- Finally, with further heating of the gas, the entropy increases gradually.

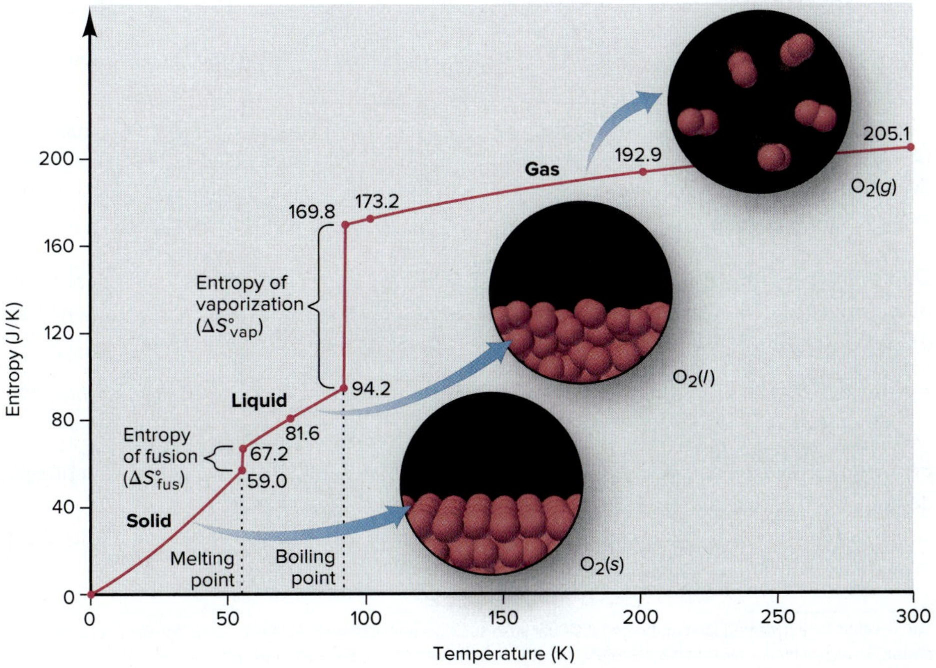

Figure 20.7 The increase in entropy during phase changes from solid to liquid to gas. The data shown are for 1 mol of O_2.

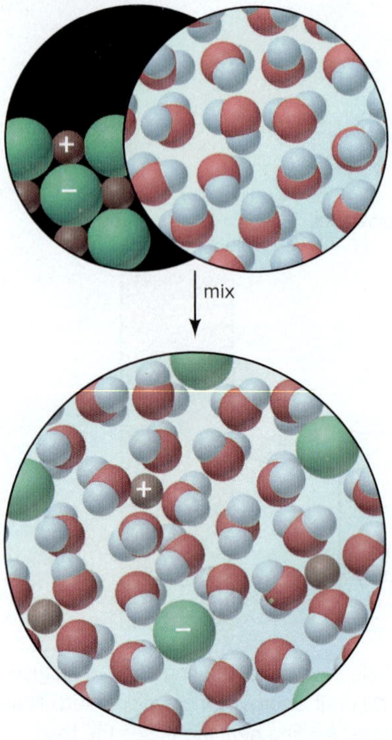
mix

Figure 20.8 The entropy change accompanying the dissolution of a salt. The entropy of a salt solution is usually *greater* than that of the solid and of water, but it is affected by water molecules becoming organized around each ion.

Dissolving a Solid or a Liquid in Water Recall from Chapter 13 that, in general,

- *Entropy increases when a solid or liquid solute dissolves in a solvent:* $S_{soln} > (S_{solute} + S_{solvent})$.

But, when water is the solvent, the entropy may also depend on the nature of solute *and* solvent interactions, which can involve two opposing events:

1. *For ionic solutes,* when the crystal dissolves in water, the ions have much more freedom of motion and their energy is dispersed into more microstates. That is, the entropy of the ions themselves is greater in the solution. However, water molecules become arranged around the ions (Figure 20.8), which limits the molecules' freedom of motion (see also Figure 13.2). In fact, around small, multiply charged ions, H_2O molecules become so organized that their energy of motion becomes *less* dispersed. This negative portion of the total entropy change can lead to *negative* $S°$ values for the ions in solution. In the case of $AlCl_3$, the $Al^{3+}(aq)$ ion has such a negative $S°$ value (−313 J/mol·K) that when $AlCl_3$ dissolves in water, even though $S°$ of $Cl^-(aq)$ is positive, the entropy of aqueous $AlCl_3$ is lower than that of solid $AlCl_3$.*

S° (J/mol·K)	NaCl	$AlCl_3$	CH_3OH
$S°$(*s* or *l*):	72.1(*s*)	167(*s*)	127(*l*)
$S°$(*aq*):	115.1	−148	132

2. *For molecular solutes,* the increase in entropy upon dissolving is typically much smaller than for ionic solutes. After all, for a solid such as glucose, there is no separation into ions, and for a liquid such as methanol or ethanol (Figure 20.9), there is no breakdown of a crystal structure. Furthermore, in these small alcohols, as in pure water, the molecules form many H bonds, so there is relatively little change in their freedom of motion either before or after they are mixed.

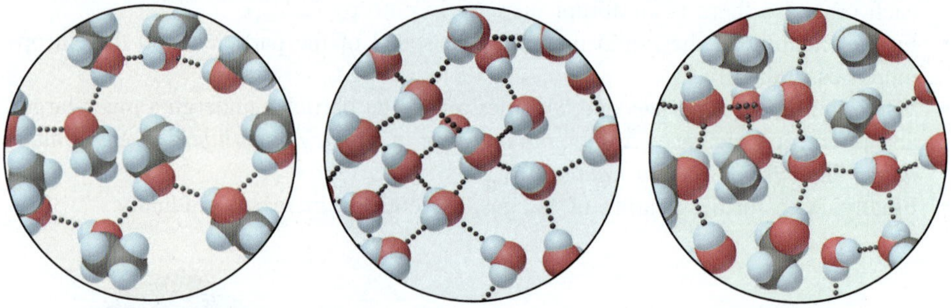

A Ethanol **B** Water **C** Solution of water and ethanol

Figure 20.9 The small increase in entropy when ethanol dissolves in water. In pure ethanol **(A)** and pure water **(B),** molecules form many H bonds to other like molecules. **C,** In solution, these two kinds of molecules form H bonds to each other, so their freedom of motion does not change significantly.

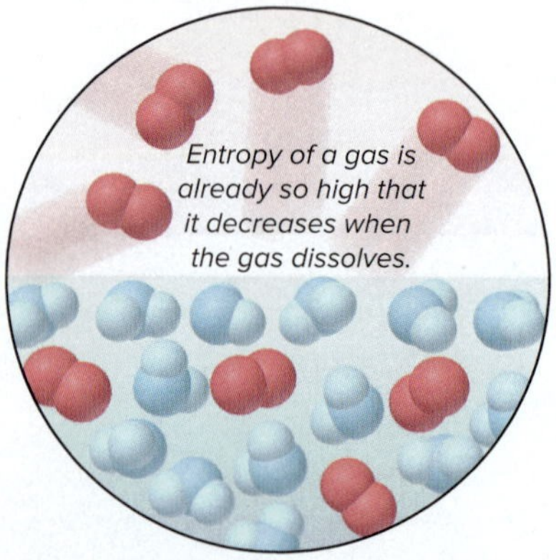
Entropy of a gas is already so high that it decreases when the gas dissolves.

Figure 20.10 The entropy of a gas dissolved in a liquid.

Dissolving a Gas in a Liquid The particles in a gas already have so much freedom of motion—and, thus, such highly dispersed energy—that they lose some when they dissolve in a liquid or solid. Therefore,

- *The entropy of a solution of a gas in a liquid or a solid is always* **less** *than the entropy of the gas itself.*

For instance, when gaseous O_2 [$S°(g)$ = 205.0 J/mol•K] dissolves in water, its entropy decreases considerably [$S°(aq)$ = 110.9 J/mol•K] (Figure 20.10).

When a gas dissolves in another gas, however, the entropy increases as a result of the separation and mixing of the molecules.

*An $S°$ value for a hydrated ion can be negative because such a value is relative to the $S°$ value for the hydrated proton, $H^+(aq)$, which is assigned a value of 0. In other words, $Al^{3+}(aq)$ has a lower entropy than $H^+(aq)$.

Atomic Size and Molecular Complexity Differences in $S°$ values for substances in the same phase are usually based on atomic size and molecular complexity.

1. *Within a periodic group,* energy levels become closer together for heavier atoms, so the number of microstates, and thus the molar entropy, increases:

	Li	Na	K	Rb	Cs
Molar mass (g/mol):	6.941	22.99	39.10	85.47	132.9
$S°(s)$ (J/mol·K):	29.1	51.4	64.7	69.5	85.2

The same trend of increasing entropy holds for similar compounds down a group:

	HF	HCl	HBr	HI
Molar mass (g/mol):	20.01	36.46	80.91	127.9
$S°(g)$ (J/mol·K):	173.7	186.8	198.6	206.3

2. *For different forms of an element (allotropes),* the entropy is *higher* in the form that allows the atoms more freedom of motion. For example, $S°$ of graphite is 5.69 J/mol·K, whereas $S°$ of diamond is 2.44 J/mol·K. In graphite, covalent bonds extend within a two-dimensional sheet, and the sheets move past each other relatively easily; in diamond, covalent bonds extend in three dimensions, allowing the atoms little movement (Table 12.6).

3. *For compounds,* entropy increases with chemical complexity (that is, with number of atoms in the formula), and this trend holds for ionic and covalent substances:

$S°$(J/mol·K)	NaCl	$AlCl_3$	P_4O_{10}	NO	NO_2	N_2O_4
$S°(s)$:	72.1	167	229			
$S°(g)$:				211	240	304

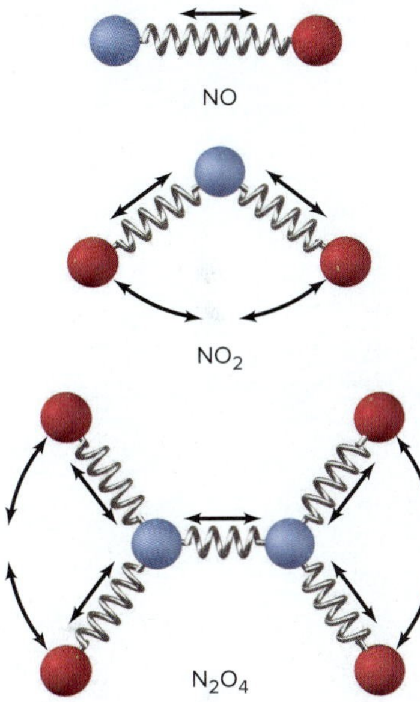

Figure 20.11 Entropy, vibrational motion, and molecular complexity.

The trend is also based on the different motions that are available. Thus, for example, among the three nitrogen oxides listed in the preceding table, the number of different vibrational motions increases with the number of atoms in the molecule (Figure 20.11; see also the Tools of the Laboratory in Chapter 9).

For larger compounds, we also consider the motion of different parts of a molecule. A long hydrocarbon chain can rotate and vibrate in more ways than a short one, so *entropy increases with chain length.* A ring compound, such as cyclopentane (C_5H_{10}), has lower entropy than the chain compound with the same molar mass, pentene (C_5H_{10}), because the ring structure restricts freedom of motion:

	$CH_4(g)$	$C_2H_6(g)$	$C_3H_8(g)$	$C_4H_{10}(g)$	$C_5H_{10}(g)$	C_5H_{10}(cyclo, g)	$C_2H_5OH(l)$
S (J/mol·K):	186	230	270	310	348	293	161

Remember, these trends hold only for *substances in the same physical state.* Gaseous methane (CH_4) has higher entropy than liquid ethanol (C_2H_5OH), even though ethanol molecules are more complex. When gases are compared with liquids, *the effect of physical state dominates the effect of molecular complexity.*

SAMPLE PROBLEM 20.4 Predicting Relative Entropy Values

Problem Select the substance with the higher entropy in each pair, and explain your choice [assume constant temperature, except in part (e)]:
(a) 1 mol of $SO_2(g)$ or 1 mol of $SO_3(g)$
(b) 1 mol of $CO_2(s)$ or 1 mol of $CO_2(g)$
(c) 3 mol of $O_2(g)$ or 2 mol of $O_3(g)$
(d) 1 mol of $KBr(s)$ or 1 mol of $KBr(aq)$
(e) Seawater at 2°C or at 23°C
(f) 1 mol of $CF_4(g)$ or 1 mol of $CCl_4(g)$

Plan In general, particles with more freedom of motion have more microstates in which to disperse their kinetic energy, so they have higher entropy. We know that either raising temperature or having *more* particles increases entropy. We apply the general categories described in the text to choose the member with the higher entropy.

Solution **(a)** 1 mol of $SO_3(g)$. For equal numbers of moles of substances with the same types of atoms in the same physical state, the more atoms in the molecule, the more types of motion available, and thus the higher the entropy.
(b) 1 mol of $CO_2(g)$. For a given substance, entropy increases in the sequence $s < l < g$.
(c) 3 mol of $O_2(g)$. The two samples contain the same number of oxygen atoms but different numbers of molecules. Despite the greater complexity of O_3, the greater number of molecules dominates because there are many more microstates possible for 3 mol of particles than for 2 mol.
(d) 1 mol of $KBr(aq)$. The two samples have the same number of ions, but their motion is more limited and their energy less dispersed in the solid than in the solution.
(e) Seawater at 23°C. Entropy increases with rising temperature.
(f) 1 mol of $CCl_4(g)$. For similar compounds, entropy increases with molar mass.

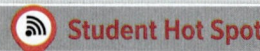

Student Hot Spot

Student data indicate that you may struggle with entropy trends. Access the Smartbook to view additional Learning Resources on this topic.

FOLLOW-UP PROBLEMS

20.4A Select the substance with the *higher* entropy in each pair, and explain your choice (assume 1 mol of each at the same T):
(a) $PCl_3(g)$ or $PCl_5(g)$ **(b)** $CaF_2(s)$ or $BaCl_2(s)$ **(c)** $Br_2(g)$ or $Br_2(l)$

20.4B Select the substance with the *lower* entropy in each pair, and explain your choice (assume 1 mol of each at the same T):
(a) $LiBr(aq)$ or $NaBr(aq)$ **(b)** quartz or glass **(c)**

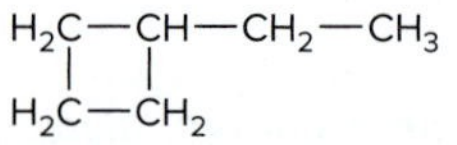

ethylcyclobutane

or

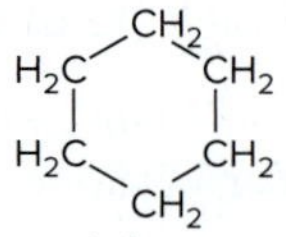

cyclohexane

SOME SIMILAR PROBLEMS 20.22–20.29

› Summary of Section 20.1

- A change is spontaneous under specified conditions if it occurs without a continuous input of energy.
- Neither the first law of thermodynamics nor the sign of ΔH predicts the direction of a spontaneous change.
- Many spontaneous processes involve an increase in the freedom of motion of the system's particles and, thus, in the dispersal of the system's energy of motion.
- Entropy is a state function that measures the extent of energy dispersed into the number of microstates possible for a system. Each microstate consists of the quantized energy levels of the system at a given instant.
- The entropy change that occurs during an isothermal change in gas volume, during a phase change, or during a temperature change can be calculated.
- The second law of thermodynamics states that, in a spontaneous process, the entropy of the universe (system plus surroundings) increases.
- Absolute entropy values can be determined because perfect crystals have zero entropy at 0 K (third law of thermodynamics).
- Standard molar entropy, $S°$ (in J/mol·K), is affected by temperature, phase changes, dissolution, and atomic size or molecular complexity.

20.2 CALCULATING THE CHANGE IN ENTROPY OF A REACTION

Chemists are especially interested in learning how to predict the sign *and* calculate the value of the entropy change that occurs during a reaction.

Entropy Changes in the System: Standard Entropy of Reaction ($\Delta S°_{rxn}$)

The **standard entropy of reaction, $\Delta S°_{rxn}$,** is the entropy change that occurs when all reactants and products are in their standard states.

Predicting the Sign of ΔS°_{rxn} Changes in structure and especially in amount (mol) of a gas help us predict the sign of ΔS°_{rxn}. Because gases have such great freedom of motion and, thus, high molar entropies, *if the number of moles of gas increases, ΔS°_{rxn} is positive; if the number decreases, ΔS°_{rxn} is negative.* Here are a few examples:

- *Increase in amount of gas.* When gaseous H_2 reacts with solid I_2 to form gaseous HI, the total number of moles of *substance* stays the same. Nevertheless, the sign of ΔS°_{rxn} is positive (entropy increases) because the number of moles of *gas* increases:

$$H_2(g) + I_2(s) \longrightarrow 2HI(g) \qquad \Delta S^\circ_{rxn} = S^\circ_{products} - S^\circ_{reactants} > 0$$

- *Decrease in amount of gas.* When ammonia forms from its elements, 4 mol of gas produce 2 mol of gas, so ΔS°_{rxn} is negative (entropy decreases):

$$N_2(g) + 3H_2(g) \rightleftharpoons 2NH_3(g) \qquad \Delta S^\circ_{rxn} = S^\circ_{products} - S^\circ_{reactants} < 0$$

- *No change in amount of gas, but change in structure.* When the amount (mol) of gas doesn't change, we *cannot* predict the sign of ΔS°_{rxn}. But, a change in one of the structures can make it easier to predict the sign of ΔS°_{rxn}. For example, when cyclopropane is heated to 500°C, the ring opens and propene forms. The chain has more freedom of motion than the ring, so ΔS°_{rxn} is positive:

$$\underset{\text{(ring: } H_2C\text{—}CH_2\text{ bridged by }CH_2)}{H_2C\text{—}CH_2(g)} \xrightarrow{\Delta} CH_3\text{—}CH\text{=}CH_2(g) \qquad \Delta S^\circ = S^\circ_{products} - S^\circ_{reactants} > 0$$

Keep in mind, however, that in general *we cannot predict the sign of the entropy change unless the reaction involves a change in number of moles of gas.*

Calculating ΔS°_{rxn} from S° Values By applying Hess's law (Chapter 6), we combined ΔH°_f values to find ΔH°_{rxn}. Similarly, we combine S° values to find the standard entropy of reaction, ΔS°_{rxn}:

$$\Delta S^\circ_{rxn} = \Sigma m S^\circ_{products} - \Sigma n S^\circ_{reactants} \qquad \textbf{(20.9)}$$

where m and n are the amounts (mol) of products and reactants, respectively, given by the coefficients in the balanced equation. For the formation of ammonia, we have

$$\Delta S^\circ_{rxn} = [(2 \text{ mol } NH_3)(S^\circ \text{ of } NH_3)] - [(1 \text{ mol } N_2)(S^\circ \text{ of } N_2) + (3 \text{ mol } H_2)(S^\circ \text{ of } H_2)]$$

From Appendix B, we find the S° values:

$$\Delta S^\circ_{rxn} = [(2 \text{ mol})(193 \text{ J/mol·K})] - [(1 \text{ mol})(191.5 \text{ J/mol·K}) + (3 \text{ mol})(130.6 \text{ J/mol·K})]$$
$$= -197 \text{ J/K}$$

As we predicted above from the decrease in number of moles of gas, $\Delta S^\circ_{rxn} < 0$.

SAMPLE PROBLEM 20.5 Calculating the Standard Entropy of Reaction, ΔS°_{rxn}

Problem Predict the sign of ΔS°_{rxn}, if possible, and calculate its value for the combustion of 1 mol of propane at 25°C:

$$C_3H_8(g) + 5O_2(g) \longrightarrow 3CO_2(g) + 4H_2O(l)$$

Plan We use the change in the number of moles of gas to predict the sign of ΔS°_{rxn}. To find ΔS°_{rxn}, we apply Equation 20.9.

Solution The amount (mol) of gas decreases (6 mol of reactant gas yields 3 mol of product gas), so the entropy should decrease ($\Delta S^\circ_{rxn} < 0$).
Calculating ΔS°_{rxn}. Using Appendix B values,

$$\Delta S^\circ_{rxn} = [(3 \text{ mol } CO_2)(S^\circ \text{ of } CO_2) + (4 \text{ mol } H_2O)(S^\circ \text{ of } H_2O)]$$
$$- [(1 \text{ mol } C_3H_8)(S^\circ \text{ of } C_3H_8) + (5 \text{ mol } O_2)(S^\circ \text{ of } O_2)]$$
$$= [(3 \text{ mol})(213.7 \text{ J/mol·K}) + (4 \text{ mol})(69.940 \text{ J/mol·K})]$$
$$- [(1 \text{ mol})(269.9 \text{ J/mol·K}) + (5 \text{ mol})(205.0 \text{ J/mol·K})]$$
$$= -374.0 \text{ J/K}$$

Check $\Delta S^\circ_{rxn} < 0$, so our prediction is correct. Rounding gives $[3(200) + 4(70)] - [270 + 5(200)] = 880 - 1270 = -390$, close to the calculated value.

Comment Notice that there are 6 mol of reactants and 7 mol of products in the propane combustion reaction; however, we do not count the 4 mol of $H_2O(l)$ when we look for an increase or decrease in mol of gas to predict ΔS°_{rxn}.

FOLLOW-UP PROBLEMS

20.5A Balance each equation, predict the sign of ΔS°_{rxn} if possible, and calculate its value at 298 K:
(a) $NO(g) \longrightarrow N_2O(g) + N_2O_3(g)$ **(b)** $CH_3OH(g) \longrightarrow CO(g) + H_2(g)$

20.5B Balance each equation, predict the sign of ΔS°_{rxn} if possible, and calculate its value at 25°C:
(a) $NaOH(s) + CO_2(g) \longrightarrow Na_2CO_3(s) + H_2O(l)$
(b) $Fe(s) + H_2O(g) \longrightarrow Fe_2O_3(s) + H_2(g)$

SOME SIMILAR PROBLEMS 20.33–20.38

Entropy Changes in the Surroundings: The Other Part of the Total

In the synthesis of ammonia, the combustion of propane, and many other spontaneous reactions, the entropy of the *system* decreases ($\Delta S^\circ_{rxn} < 0$). Remember that the second law dictates that, for a spontaneous process, a ***decrease*** *in the entropy of the system is outweighed by an* ***increase*** *in the entropy of the surroundings.* In this section, we examine the influence of the surroundings—in particular, the addition (or removal) of heat and the temperature at which this heat flow occurs—on the *total* entropy change.

The Role of the Surroundings In essence, the surroundings *add heat to or remove heat from the system.* That is, the surroundings function as an enormous heat source or heat sink, one so large that its temperature remains constant, even though its entropy changes. The surroundings participate in the two types of enthalpy changes as follows:

1. *In an exothermic change, heat released by the system is absorbed by the surroundings.* More heat increases the freedom of motion of the particles and makes the energy more dispersed, so the entropy of the surroundings increases:

   For an exothermic change: $q_{sys} < 0, \quad q_{surr} > 0, \quad \text{and} \quad \Delta S_{surr} > 0$

2. *In an endothermic change, heat absorbed by the system is released by the surroundings.* Less heat reduces the freedom of motion of the particles and makes the energy less dispersed, so the entropy of the surroundings decreases:

   For an endothermic change: $q_{sys} > 0, \quad q_{surr} < 0, \quad \text{and} \quad \Delta S_{surr} < 0$

Temperature at Which Heat Is Transferred The *temperature* of the surroundings when the heat is transferred also affects ΔS_{surr}. Consider the effect of an exothermic reaction at a low or at a high temperature:

- At a low T, such as 20 K, there is little motion of particles in the surroundings because they have relatively low energy. This means there are few energy levels in each microstate and few microstates in which to disperse the energy. Transferring a given quantity of heat to these surroundings causes a relatively large change in how much energy is dispersed.
- At a high T, such as 298 K, the surroundings have a large quantity of energy dispersed among the particles. There are more energy levels in each microstate and a greater number of microstates. Transferring the same given quantity of heat to these surroundings causes a relatively small change in how much energy is dispersed. ‹

A Checkbook Analogy for Heating the Surroundings

If you have \$10 in your checking account, a \$10 deposit represents a 100% increase in your net worth; that is, a given change to a low initial state has a large impact. If, however, you have a \$1000 balance, a \$10 deposit represents only a 1% increase. Thus, the same absolute change to a high initial state has a smaller impact.

In other words, ΔS_{surr} is greater when heat is added at a lower T. Putting these ideas together, *ΔS_{surr} is directly related to an opposite change in the heat of the system (q_{sys}) and inversely related to the temperature at which the heat is transferred:*

$$\Delta S_{surr} = -\frac{q_{sys}}{T}$$

For a process at *constant pressure,* the heat (q_P) is ΔH (Section 6.2), so

$$\Delta S_{surr} = -\frac{\Delta H_{sys}}{T} \quad \textbf{(20.10)}$$

Thus, we find ΔS_{surr} by measuring ΔH_{sys} and T at which the change takes place.

The main point: If a spontaneous reaction has a negative ΔS_{sys} (fewer microstates into which energy is dispersed), ΔS_{surr} must be positive enough (even more microstates into which energy is dispersed) for ΔS_{univ} to be positive (net increase in number of microstates for dispersing the energy).

SAMPLE PROBLEM 20.6 Determining Reaction Spontaneity

Problem At 298 K, the formation of ammonia has a negative ΔS°_{sys}:

$$N_2(g) + 3H_2(g) \longrightarrow 2NH_3(g) \qquad \Delta S^\circ_{sys} = -197 \text{ J/K}$$

Calculate ΔS_{univ}, and state whether the reaction occurs spontaneously at this temperature.

Plan For the reaction to occur spontaneously, $\Delta S_{univ} > 0$, so ΔS_{surr} must be greater than +197 J/K. To find ΔS_{surr}, we need ΔH°_{sys}, which is the same as ΔH°_{rxn}. We use ΔH°_f values from Appendix B to find ΔH°_{rxn}. Then, we divide ΔH°_{rxn} by the given T (298 K) to find ΔS_{surr}. To find ΔS_{univ}, we add the calculated ΔS_{surr} to the given ΔS°_{sys} (−197 J/K).

Solution Calculating ΔH°_{sys}:

$$\begin{aligned}\Delta H^\circ_{sys} &= \Delta H^\circ_{rxn}\\ &= [(2 \text{ mol } NH_3)(\Delta H^\circ_f \text{ of } NH_3)] - [(3 \text{ mol } H_2)(\Delta H^\circ_f \text{ of } H_2) + (1 \text{ mol } N_2)(\Delta H^\circ_f \text{ of } N_2)]\\ &= [(2 \text{ mol } NH_3)(-45.9 \text{ kJ/mol})] - [(3 \text{ mol } H_2)(0 \text{ kJ/mol}) + (1 \text{ mol } N_2)(0 \text{ kJ/mol})]\\ &= -91.8 \text{ kJ}\end{aligned}$$

Calculating ΔS_{surr}:

$$\Delta S_{surr} = -\frac{\Delta H^\circ_{sys}}{T} = -\frac{-91.8 \text{ kJ} \times \frac{1000 \text{ J}}{1 \text{ kJ}}}{298 \text{ K}} = 308 \text{ J/K}$$

Determining ΔS_{univ}:

$$\Delta S_{univ} = \Delta S^\circ_{sys} + \Delta S_{surr} = -197 \text{ J/K} + 308 \text{ J/K} = 111 \text{ J/K}$$

Since $\Delta S_{univ} > 0$, the reaction occurs spontaneously at 298 K *(see margin).*

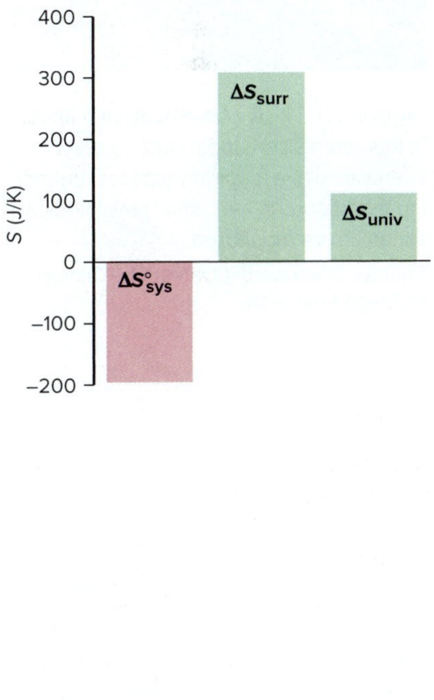

Check Rounding to check the math, we have

$$\begin{aligned}\Delta H^\circ_{rxn} &\approx 2(-45 \text{ kJ}) = -90 \text{ kJ}\\ \Delta S_{surr} &\approx -(-90{,}000 \text{ J})/300 \text{ K} = 300 \text{ J/K}\\ \Delta S_{univ} &\approx -200 \text{ J/K} + 300 \text{ J/K} = 100 \text{ J/K}\end{aligned}$$

Given the negative ΔH°_{rxn}, Le Châtelier's principle says that low T favors NH_3 formation, so the answer is reasonable (see Section 17.6).

Comment **1.** Because ΔH° has units of kJ, and ΔS has units of J/K, don't forget to convert kJ to J, or you'll introduce a large error.

2. This example highlights the distinction between thermodynamics and kinetics. NH_3 forms spontaneously, but so slowly that catalysts are required to achieve a practical rate.

FOLLOW-UP PROBLEMS

20.6A Gaseous phosphorus trichloride forms from the elements through this reaction:

$$P_4(s) + 6Cl_2(g) \longrightarrow 4PCl_3(g)$$

Calculate ΔH°_{rxn}, ΔS_{surr}, and ΔS°_{rxn} to determine if the reaction is spontaneous at 298 K.

20.6B Does the oxidation of FeO(*s*) to $Fe_2O_3(s)$ occur spontaneously at 298 K? (Show the calculation for 1 mol of Fe_2O_3.)

SOME SIMILAR PROBLEMS 20.39–20.42

Do Organisms Obey the Laws of Thermodynamics? Taking the surroundings into account is crucial, not only for determining reaction spontaneity, as in Sample Problem 20.6, but also for understanding the relevance of thermodynamics to biology. Let's examine the first and second laws to see if they apply to living systems.

1. *Do organisms comply with the first law?* The chemical bond energy in food and oxygen is converted into the mechanical energy of jumping, flying, crawling, swimming, and countless other movements; the electrical energy of nerve conduction; the thermal energy of warming the body; and so forth. Many experiments have demonstrated that the total energy is conserved in these situations. Some of the earliest were performed by Lavoisier, who showed that "animal heat" was produced by slow, continual combustion. In experiments with guinea pigs, he invented a calorimeter to measure the heat released from intake of food and O_2 and output of CO_2 and water, and he included respiration in his new theory of combustion. Modern, room-sized calorimeters measure these and other variables to confirm the conservation of energy for an exercising human (Figure 20.12).

Figure 20.12 A whole-body calorimeter. In this room-sized apparatus, a person exercises while respiratory gases, energy input and output, and other physiological variables are monitored.

Source: © Vanderbilt University–Clinical Nutrition Research

2. *Do organisms comply with the second law?* Mature humans are far more complex than the egg and sperm cells from which they develop, and modern organisms are far more complex than the one-celled ancestral specks from which they evolved. Are the growth of an organism and the evolution of life exceptions to the spontaneous tendency of natural processes to increase freedom of motion and disperse energy? For an organism to grow or a species to evolve, many moles of oxygen and nutrients—carbohydrates, proteins, and fats—undergo exothermic reactions to form many *more* moles of gaseous CO_2 and H_2O. Formation of these waste gases and the accompanying release of heat result in an enormous *increase* in the entropy of the surroundings. Thus, the localization of energy and synthesis of macromolecular structures required for the growth and evolution of organisms (system) cause a far greater dispersal of energy and freedom of motion in the environment (surroundings). When system *and* surroundings are considered together, the entropy of the universe, as always, increases.

The Entropy Change and the Equilibrium State

For a process approaching equilibrium, $\Delta S_{univ} > 0$. When the process reaches equilibrium, there is no further *net* change, $\Delta S_{univ} = 0$, because any entropy change in the system is balanced by an opposite entropy change in the surroundings:

$$\text{At equilibrium:} \quad \Delta S_{univ} = \Delta S_{sys} + \Delta S_{surr} = 0 \quad \text{so} \quad \Delta S_{sys} = -\Delta S_{surr}$$

As an example, let's calculate ΔS_{univ} for the vaporization-condensation of 1 mol of water at 100°C (373 K),

$$H_2O(l; 373\ K) \rightleftharpoons H_2O(g; 373\ K)$$

First, we find ΔS°_{sys} for the forward change (vaporization) of 1 mol of water:

$$\Delta S^\circ_{sys} = \Sigma mS^\circ_{products} - \Sigma nS^\circ_{reactants} = S^\circ \text{ of } H_2O(g; 373\ K) - S^\circ \text{ of } H_2O(l; 373\ K)$$
$$= 195.9\ J/K - 86.8\ J/K = 109.1\ J/K$$

As we expect, the entropy of the system increases ($\Delta S^\circ_{sys} > 0$) as the liquid absorbs heat and changes to a gas.

For ΔS_{surr} of the vaporization step, we have

$$\Delta S_{surr} = -\frac{\Delta H^\circ_{sys}}{T}$$

where $\Delta H^\circ_{sys} = \Delta H^\circ_{vap}$ at 373 K = 40.7 kJ/mol = 40.7×10^3 J/mol. For 1 mol of water, we have

$$\Delta S_{surr} = -\frac{\Delta H^\circ_{vap}}{T} = -\frac{40.7\times10^3\ \text{J}}{373\ \text{K}} = -109\ \text{J/K}$$

The surroundings lose heat, and the negative sign means that the entropy of the surroundings decreases. The two entropy changes have the same magnitude but opposite signs, so they cancel:

$$\Delta S_{univ} = 109\ \text{J/K} + (-109\ \text{J/K}) = 0$$

For the reverse change (condensation), ΔS_{univ} also equals zero, but ΔS°_{sys} and ΔS_{surr} have signs opposite those for vaporization.

A similar treatment of a chemical change shows the same result: the entropy change of the forward reaction is *equal in magnitude but opposite in sign* to the entropy change of the reverse reaction. Thus, *when a system reaches equilibrium, neither the forward nor the reverse reaction is spontaneous,* and so there is no net reaction in either direction.

Spontaneous Exothermic and Endothermic Changes

No matter what its *enthalpy* change, a reaction occurs because the total *entropy* of the reacting system *and* its surroundings increases. There are two ways this can happen:

1. *In an exothermic reaction* ($\Delta H_{sys} < 0$), the heat released by the system increases the freedom of motion and dispersal of energy in the surroundings; thus, $\Delta S_{surr} > 0$.

- If the entropy of the products is *more* than that of the reactants ($\Delta S_{sys} > 0$), the total entropy change ($\Delta S_{sys} + \Delta S_{surr}$) will be positive (Figure 20.13A). For example, in the oxidation of glucose, an essential reaction for all higher organisms,

$$C_6H_{12}O_6(s) + 6O_2(g) \longrightarrow 6CO_2(g) + 6H_2O(g) + \textbf{\textit{heat}}$$

6 mol of gas yields 12 mol of gas; thus, $\Delta S_{sys} > 0$, $\Delta S_{surr} > 0$, and $\Delta S_{univ} > 0$.
- If the entropy of the products is *less* than that of the reactants ($\Delta S_{sys} < 0$), the entropy of the surroundings must increase even more ($\Delta S_{surr} >> 0$) to make the total ΔS positive (Figure 20.13B). For example, when calcium oxide and carbon dioxide form calcium carbonate, the amount (mol) of gas decreases from 1 to 0:

$$CaO(s) + CO_2(g) \longrightarrow CaCO_3(s) + \textbf{\textit{heat}}$$

However, even though the system's entropy goes down, the heat released increases the entropy of the surroundings even more; thus, $\Delta S_{sys} < 0$, but $\Delta S_{surr} >> 0$, so $\Delta S_{univ} > 0$.

2. *In an endothermic reaction* ($\Delta H_{sys} > 0$), the heat absorbed by the system decreases molecular freedom of motion and dispersal of energy in the surroundings;

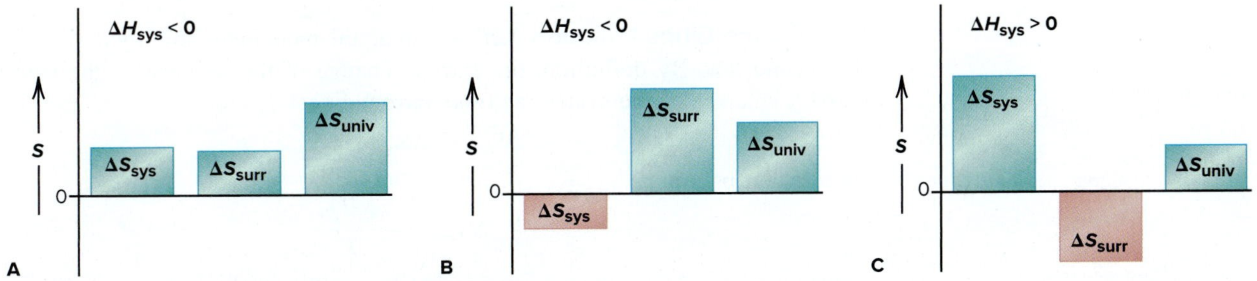

Figure 20.13 Components of ΔS_{univ} for spontaneous reactions. For a reaction to occur spontaneously, ΔS_{univ} must be positive. **A,** An exothermic reaction in which ΔS_{sys} increases; the size of ΔS_{surr} is not important. **B,** An exothermic reaction in which ΔS_{sys} decreases; ΔS_{surr} must be larger than ΔS_{sys}. **C,** An endothermic reaction in which ΔS_{sys} increases; ΔS_{surr} must be smaller than ΔS_{sys}.

so $\Delta S_{surr} < 0$. Thus, the only way an endothermic reaction can occur spontaneously is if ΔS_{sys} is positive and large enough to outweigh the negative ΔS_{surr} (Figure 20.13C).

- In the solution process for many ionic compounds, heat is absorbed to form the solution, so the entropy of the surroundings decreases ($\Delta S_{surr} < 0$). However, when the crystalline solid becomes freely moving ions, the entropy increase is so large ($\Delta S_{sys} >> 0$) that it outweighs the negative ΔS_{surr}. Thus, ΔS_{univ} is positive.
- Spontaneous endothermic reactions are similar. Recall the reaction between barium hydroxide octahydrate and ammonium nitrate (Figure 20.1),

$$\textbf{\textit{heat}} + Ba(OH)_2{\cdot}8H_2O(s) + 2NH_4NO_3(s) \longrightarrow Ba^{2+}(aq) + 2NO_3^-(aq) + 2NH_3(aq) + 10H_2O(l)$$

3 mol of crystalline solids absorb heat from the surroundings ($\Delta S_{surr} < 0$) and yields 15 mol of dissolved ions and molecules, which have much more freedom of motion and, therefore, much greater entropy ($\Delta S_{sys} >> 0$).

› Summary of Section 20.2

- › The standard entropy of reaction, ΔS°_{rxn}, is calculated from S° values.
- › When the amount (mol) of gas increases in a reaction, usually $\Delta S^\circ_{rxn} > 0$.
- › The change in entropy of the surroundings, ΔS_{surr}, is related directly to $-\Delta H^\circ_{sys}$ and inversely to the absolute temperature T at which the change occurs.
- › In a spontaneous change, the entropy of the system can decrease only if the entropy of the surroundings increases even more, so that $\Delta S_{univ} > 0$.
- › The second law is obeyed in living systems when we consider system *plus* surroundings.
- › For a system at equilibrium, $\Delta S_{univ} = 0$, so $\Delta S^\circ_{sys} = -\Delta S_{surr}$.
- › Even if $\Delta S^\circ_{sys} < 0$, an exothermic reaction ($\Delta H^\circ_{rxn} < 0$) is spontaneous ($\Delta S_{univ} > 0$) if $\Delta S_{surr} >> 0$; an endothermic reaction ($\Delta H^\circ_{rxn} > 0$) is spontaneous only if $\Delta S^\circ_{sys} > \Delta S_{surr}$.

20.3 ENTROPY, FREE ENERGY, AND WORK

By measuring *both* ΔS_{sys} *and* ΔS_{surr}, we can predict whether a reaction will be spontaneous at a particular temperature. It would be useful to have *one* criterion for spontaneity that we can determine *by measuring the system only*. The Gibbs free energy, or simply **free energy (*G*)**, combines the system's enthalpy and entropy:

$$G = H - TS$$

One of the greatest but least recognized of American scientists, Josiah Willard Gibbs (1839–1903), established chemical thermodynamics as well as major principles of equilibrium and electrochemistry. Although the great European scientists of his time, James Clerk Maxwell and Henri Le Châtelier, realized Gibbs's achievements, he was not recognized by his American colleagues until nearly 50 years after his death!

Free Energy Change and Reaction Spontaneity

The free energy change (ΔG) is a measure of the spontaneity of a process and of the useful energy available from it.

Deriving the Gibbs Equation Let's examine the meaning of ΔG by deriving it from the second law. By definition, the entropy change of the universe is the sum of the entropy changes of the system and the surroundings:

$$\Delta S_{univ} = \Delta S_{sys} + \Delta S_{surr}$$

At constant pressure,

$$\Delta S_{surr} = -\frac{\Delta H_{sys}}{T}$$

Substituting for ΔS_{surr} gives a relationship that relies solely on the system:

$$\Delta S_{univ} = \Delta S_{sys} - \frac{\Delta H_{sys}}{T}$$

Multiplying both sides by $-T$ and rearranging gives

$$-T\Delta S_{univ} = \Delta H_{sys} - T\Delta S_{sys}$$

Using the Gibbs free energy relationship, $G = H - TS$, we obtain the *Gibbs equation* for the *change* in the free energy of the system (ΔG_{sys}) at constant T and P:

$$\Delta G_{sys} = \Delta H_{sys} - T\Delta S_{sys} \quad \textbf{(20.11)}$$

Combining Equation 20.11 with the one preceding it shows that

$$-T\Delta S_{univ} = \Delta G_{sys} = \Delta H_{sys} - T\Delta S_{sys}$$

Spontaneity and the Sign of ΔG Let's see how the *sign* of ΔG tells if a reaction is spontaneous. According to the second law,

- $\Delta S_{univ} > 0$ for a spontaneous process
- $\Delta S_{univ} < 0$ for a nonspontaneous process
- $\Delta S_{univ} = 0$ for a process at equilibrium

Since the absolute temperature is always positive, for a spontaneous process,

$$T\Delta S_{univ} > 0 \qquad \text{so} \qquad -T\Delta S_{univ} < 0$$

From our derivation above, $\Delta G = -T\Delta S_{univ}$, so we have

- $\Delta G < 0$ for a spontaneous process
- $\Delta G > 0$ for a nonspontaneous process
- $\Delta G = 0$ for a process at equilibrium

Without incorporating any new ideas, we can now predict spontaneity with one variable (ΔG_{sys}) rather than two (ΔS_{sys} and ΔS_{surr}).

Calculating Standard Free Energy Changes

The *sign* of ΔG reveals *whether* a reaction is spontaneous, but the *magnitude* of ΔG tells *how* spontaneous it is. Because free energy (G) combines three state functions, H, S, and T, it is also a state function. As we do with enthalpy, we focus on the free energy *change* (ΔG). As we do with other thermodynamic variables, to compare the free energy changes of different reactions, we calculate the ***standard* free energy change (ΔG°),** which occurs when all components of the system are in their standard states.

Using the Gibbs Equation to Find ΔG° One way to calculate ΔG° is by writing the Gibbs equation (Equation 20.11) at standard-state conditions and using Appendix B to find ΔH°_{sys} and ΔS°_{sys}. Adapting the Gibbs equation, we have

$$\Delta G^\circ_{sys} = \Delta H^\circ_{sys} - T\Delta S^\circ_{sys} \quad \textbf{(20.12)}$$

This important relationship is used to find any one of these three variables, given the other two, as in the following sample problem.

Fireworks explode spontaneously.
Source: © Karl Weatherly/PhotoDisc, RF

SAMPLE PROBLEM 20.7 Calculating ΔG°_{rxn} from Enthalpy and Entropy Values

Problem Potassium chlorate, a common oxidizing agent in fireworks *(see photo)* and matchheads, undergoes a solid-state disproportionation reaction when heated:

$$4\overset{+5}{KClO_3}(s) \xrightarrow{\Delta} 3\overset{+7}{KClO_4}(s) + \overset{-1}{KCl}(s)$$

Use ΔH°_f and S° values to calculate ΔG°_{sys} (which is ΔG°_{rxn}) at 25°C for this reaction.

Plan To solve for ΔG°, we need values from Appendix B. We use ΔH°_f values to calculate ΔH°_{rxn} (ΔH°_{sys}), use S° values to calculate ΔS°_{rxn} (ΔS°_{sys}), and then apply Equation 20.12.

Solution Calculating ΔH°_{sys} from ΔH°_f values (with Equation 6.10):

$$\Delta H^\circ_{sys} = \Delta H^\circ_{rxn} = \Sigma m\Delta H^\circ_{f(products)} - \Sigma n\Delta H^\circ_{f(reactants)}$$
$$= [(3 \text{ mol } KClO_4)(\Delta H^\circ_f \text{ of } KClO_4) + (1 \text{ mol } KCl)(\Delta H^\circ_f \text{ of } KCl)] - [(4 \text{ mol } KClO_3)(\Delta H^\circ_f \text{ of } KClO_3)]$$
$$= [(3 \text{ mol})(-432.75 \text{ kJ/mol}) + (1 \text{ mol})(-436.7 \text{ kJ/mol})] - [(4 \text{ mol})(-397.7 \text{ kJ/mol})]$$
$$= -144.2 \text{ kJ}$$

Calculating ΔS°_{sys} from S° values (with Equation 20.9):

$$\Delta S^\circ_{sys} = \Delta S^\circ_{rxn} = \Sigma mS^\circ_{products} - \Sigma nS^\circ_{reactants}$$
$$= [(3 \text{ mol } KClO_4)(S^\circ \text{ of } KClO_4) + (1 \text{ mol } KCl)(S^\circ \text{ of } KCl)] - [(4 \text{ mol } KClO_3)(S^\circ \text{ of } KClO_3)]$$
$$= [(3 \text{ mol})(151.0 \text{ J/mol·K}) + (1 \text{ mol})(82.6 \text{ J/mol·K})] - [(4 \text{ mol})(143.1 \text{ J/mol·K})]$$
$$= -36.8 \text{ J/K}$$

Calculating ΔG°_{sys} at 298 K:

$$\Delta G^\circ_{sys} = \Delta H^\circ_{sys} - T\Delta S^\circ_{sys} = -144.2 \text{ kJ} - \left[(298 \text{ K})(-36.8 \text{ J/K})\left(\frac{1 \text{ kJ}}{1000 \text{ J}}\right)\right] = -133.2 \text{ kJ}$$

Check Rounding to check the math:

$$\Delta H^\circ \approx [3(-433 \text{ kJ}) + (-440 \text{ kJ})] - [4(-400 \text{ kJ})] = -1740 \text{ kJ} + 1600 \text{ kJ} = -140 \text{ kJ}$$
$$\Delta S^\circ \approx [3(150 \text{ J/K}) + 85 \text{ J/K}] - [4(145 \text{ J/K})] = 535 \text{ J/K} - 580 \text{ J/K} = -45 \text{ J/K}$$
$$\Delta G^\circ \approx -140 \text{ kJ} - 300 \text{ K}(-0.04 \text{ kJ/K}) = -140 \text{ kJ} + 12 \text{ kJ} = -128 \text{ kJ}$$

Comment 1. Recall from Section 20.1 that reaction spontaneity tells nothing about rate. Even though this reaction is spontaneous, the rate is very low in the solid. When $KClO_3$ is heated slightly above its melting point, the ions can move and the reaction occurs readily. **2.** Under *any* conditions, a spontaneous reaction has a negative change in free energy: $\Delta G < 0$. Under standard-state conditions, a spontaneous reaction has a negative *standard* free energy change: $\Delta G^\circ < 0$.

FOLLOW-UP PROBLEMS

20.7A Use ΔH°_{rxn} and ΔS°_{rxn} to calculate ΔG°_{rxn} at 298 K for this reaction:

$$2NO(g) + Cl_2(g) \longrightarrow 2NOCl(g)$$

20.7B Determine the standard free energy change at 298 K for this reaction:

$$4NH_3(g) + 5O_2(g) \longrightarrow 4NO(g) + 6H_2O(g)$$

SOME SIMILAR PROBLEMS 20.53 and 20.54

Using Standard Free Energies of Formation to Find ΔG°_{rxn} Another way to calculate ΔG°_{rxn} is with values for the **standard free energy of formation (ΔG°_f)** of the components. Analogous to the standard enthalpy of formation, ΔH°_f (Section 6.6), ΔG°_f is the free energy change that occurs when 1 mol of a compound is made *from its elements,* with all components in their standard states. Because free energy is a state function, we can apply Hess's law and combine ΔG°_f values of reactants and products to calculate ΔG°_{rxn}, no matter how the reaction takes place:

$$\Delta G^\circ_{rxn} = \Sigma m\Delta G^\circ_{f(products)} - \Sigma n\Delta G^\circ_{f(reactants)} \quad \textbf{(20.13)}$$

ΔG°_f values have properties similar to ΔH°_f values:

- ΔG°_f of an element in its standard state is zero.
- A coefficient in the reaction equation (*m* or *n* in Equation 20.13) multiplies ΔG°_f by that number.
- Reversing a reaction changes the sign of ΔG°_f.

Many ΔG°_f values appear along with those for ΔH°_f and S° in Appendix B.

SAMPLE PROBLEM 20.8 Calculating ΔG°_{rxn} from ΔG°_f Values

Problem Use ΔG°_f values to calculate ΔG°_{rxn} for the reaction in Sample Problem 20.7:

$$4KClO_3(s) \longrightarrow 3KClO_4(s) + KCl(s)$$

Plan We apply Equation 20.13 to calculate ΔG°_{rxn}.

Solution Applying Equation 20.13 with values from Appendix B:

$$\begin{aligned}\Delta G^\circ_{rxn} &= \Sigma m \Delta G^\circ_{f(products)} - \Sigma n \Delta G^\circ_{f(reactants)} \\ &= [(3 \text{ mol } KClO_4)(\Delta G^\circ_f \text{ of } KClO_4) + (1 \text{ mol } KCl)(\Delta G^\circ_f \text{ of } KCl)] \\ &\quad - [(4 \text{ mol } KClO_3)(\Delta G^\circ_f \text{ of } KClO_3)] \\ &= [(3 \text{ mol})(-303.2 \text{ kJ/mol}) + (1 \text{ mol})(-409.2 \text{ kJ/mol})] \\ &\quad - [(4 \text{ mol})(-296.3 \text{ kJ/mol})] \\ &= -134 \text{ kJ}\end{aligned}$$

Check Rounding to check the math:

$$\begin{aligned}\Delta G^\circ_{rxn} &\approx [3(-300 \text{ kJ}) + 1(-400 \text{ kJ})] - [4(-300 \text{ kJ})] \\ &= -1300 \text{ kJ} + 1200 \text{ kJ} = -100 \text{ kJ}\end{aligned}$$

Comment The slight discrepancy between this answer and the one obtained in Sample Problem 20.7 is due to rounding. As you can see, when ΔG°_f values are available, this method is simpler arithmetically than the approach in Sample Problem 20.7.

FOLLOW-UP PROBLEMS

20.8A Use ΔG°_f values to calculate ΔG°_{rxn} at 298 K:
(a) $2NO(g) + Cl_2(g) \longrightarrow 2NOCl(g)$ (from Follow-up Problem 20.7A)
(b) $3H_2(g) + Fe_2O_3(s) \longrightarrow 2Fe(s) + 3H_2O(g)$

20.8B Use ΔG°_f values to calculate ΔG°_{rxn} at 25°C:
(a) $4NH_3(g) + 5O_2(g) \longrightarrow 4NO(g) + 6H_2O(g)$ (from Follow-up Problem 20.7B)
(b) $2C(\text{graphite}) + O_2(g) \longrightarrow 2CO(g)$

SOME SIMILAR PROBLEMS 20.51, 20.52, 20.55(b), and 20.56(b)

The Free Energy Change and the Work a System Can Do

Thermodynamics developed after the invention of the steam engine, a major advance that spawned a new generation of machines. Thus, some of the field's key ideas applied the relationships between the free energy change and the work a system can do:

- ΔG is the *maximum useful work* that can be done ***by*** a system during a *spontaneous* process at constant T and P:

$$\Delta G = w_{max} \qquad \textbf{(20.14)}$$

- ΔG is the *minimum work* that must be done ***to*** a system to make a *nonspontaneous* process occur at constant T and P.

The free energy change is the maximum work the system can *possibly* do. But the work it *actually* does is always less and depends on how the free energy is released. Let's consider the work done by two common systems—a car engine and a battery.

1. *"Useful" work done by a car engine.* When gasoline (represented by octane, C_8H_{18}) is burned in a car engine,

$$C_8H_{18}(l) + \tfrac{25}{2}O_2(g) \longrightarrow 8CO_2(g) + 9H_2O(g)$$

a large amount of energy is released as heat ($\Delta H_{sys} < 0$), and because the number of moles of gas increases, the entropy of the system increases ($\Delta S_{sys} > 0$). Therefore, the reaction is spontaneous ($\Delta G_{sys} < 0$). The free energy released turns the wheels, moves the belts, plays the radio, and so on—all examples of "useful" work. However, the *maximum* work is done by any spontaneous process *only if the free energy is released reversibly,* that is, *in an infinite number of steps.* Of course, in any *real* process, work is performed in a *finite* number of steps, that is, *irreversibly,* so *the maximum work is never done.* Any free energy not used for work is lost to the surroundings as heat. For a car engine, much of the free energy released just warms the engine and the outside air, which increases the freedom of motion of the particles in the universe, in accord with the second law.

2. *"Useful" work done by a battery.* As you'll see in Chapter 21, a battery is essentially a packaged spontaneous redox reaction that releases free energy in the form of an electric current to the surroundings (flashlight, computer, motor, etc.). If we connect the battery terminals to each other through a short piece of wire, the free energy change is released all at once but does no work—it just heats the wire and battery. If we connect the terminals to a motor, a significant portion of the free energy runs the motor, but some is still converted to heat. If we connect the battery to a more "efficient" device, one that discharges the free energy still more slowly, more of the energy does work and less is converted to heat. However, as with all systems, only when the battery discharges infinitely slowly can it do the maximum work.

Efficiency can be defined as the percentage of work output relative to the energy input. The range of efficiencies among common devices is very large: an incandescent bulb converts <7% of incoming electricity to light, with the rest given off as heat. At the other extreme, an electrical generator converts 95% of the incoming mechanical energy to electricity. Here are the efficiencies of some other devices: home oil furnace, 65%; hand-tool motor, 63%; liquid fuel rocket, 50%; car engine, <30%; compact fluorescent bulb, 18%; solar cell, ~15%. Therefore, all engineers must face the fact that *no* ***real*** *process uses all the available free energy to do work because some is always "wasted" as heat.*

Let's summarize the relation between the free energy change of a reaction and the work it can do:

- A spontaneous reaction ($\Delta G_{sys} < 0$) will do work on the surroundings ($-w$). For any real machine, the actual work done is *always less than the maximum* because some of the ΔG is released as heat.
- A nonspontaneous reaction ($\Delta G_{sys} > 0$) will occur only if the surroundings do work on the system ($+w$). For any real machine, the actual work done on the system is *always more than the minimum* because some of the added free energy is wasted as heat.
- A reaction at equilibrium ($\Delta G_{sys} = 0$) can no longer do any work.

The Effect of Temperature on Reaction Spontaneity

In most cases, the enthalpy contribution (ΔH) to the free energy change (ΔG) is much *larger* than the entropy contribution ($T\Delta S$). In fact, the reason most exothermic reactions are spontaneous is that the large negative ΔH makes ΔG negative. However, the *temperature of a reaction influences the magnitude of the $T\Delta S$ term,* so, for many reactions, the overall spontaneity depends on the temperature. From the signs of ΔH and ΔS, we can predict how the temperature affects the sign of ΔG. (The values we'll use below for the thermodynamic variables are standard-state values from Appendix B, but we show them without the degree sign to emphasize that the relationships among ΔG, ΔH, and ΔS are valid at any conditions. Also, we assume that ΔH and ΔS change little with temperature, which is true as long as no phase change occurs.)

Let's examine the four combinations of positive and negative ΔH and ΔS—two that are independent of temperature and two that are dependent on temperature:

- *Temperature-independent cases.* When ΔH and ΔS have *opposite* signs, the reaction occurs spontaneously either at all temperatures or at none (nonspontaneous).
  1. *Reaction is spontaneous at all temperatures:* an exothermic reaction that has an increase in entropy has $\Delta H < 0$, $\Delta S > 0$. Since ΔS is positive, $-T\Delta S$ is negative; thus, both contributions favor a negative ΔG. Most combustion reactions are in this category. The decomposition of hydrogen peroxide, a common disinfectant, is also spontaneous at all temperatures:

$$2H_2O_2(l) \longrightarrow 2H_2O(l) + O_2(g)$$
$$\Delta H = -196 \text{ kJ} \quad \text{and} \quad \Delta S = 125 \text{ J/K}$$

2. *Reaction is nonspontaneous at all temperatures:* an endothermic reaction with a decrease in entropy has $\Delta H > 0$, $\Delta S < 0$. Both contributions oppose spontaneity: ΔH is positive and ΔS is negative, so $-T\Delta S$ is positive; thus, ΔG is always positive. The formation of ozone from oxygen requires a continual energy input, so it is not spontaneous at any temperature:

$$3O_2(g) \longrightarrow 2O_3(g)$$
$$\Delta H = 286 \text{ kJ} \quad \text{and} \quad \Delta S = -137 \text{ J/K}$$

- *Temperature-dependent cases.* When ΔH and ΔS have the *same* sign, the relative magnitudes of $-T\Delta S$ and ΔH determine the sign of ΔG. In these cases, the *direction* of the change in T is crucial.

3. *Reaction becomes spontaneous as temperature increases:* an endothermic reaction with an increase in entropy has $\Delta H > 0$ *and* $\Delta S > 0$. With a positive ΔH, the reaction will occur spontaneously only when $-T\Delta S$ becomes large enough to make ΔG negative, which will happen as the temperature rises. For example,

$$2N_2O(g) + O_2(g) \longrightarrow 4NO(g)$$
$$\Delta H = 197.1 \text{ kJ} \quad \text{and} \quad \Delta S = 198.2 \text{ J/K}$$

The oxidation of N_2O occurs spontaneously at any $T > 994$ K.

4. *Reaction becomes spontaneous as temperature decreases:* an exothermic reaction with a decrease in entropy has $\Delta H < 0$ *and* $\Delta S < 0$. Here, ΔH favors spontaneity, but ΔS does not ($-T\Delta S > 0$). The reaction will occur spontaneously only when $-T\Delta S$ becomes smaller than ΔH, and this happens as the temperature drops. For example,

$$4Fe(s) + 3O_2(g) \longrightarrow 2Fe_2O_3(s)$$
$$\Delta H = -1651 \text{ kJ} \quad \text{and} \quad \Delta S = -549.4 \text{ J/K}$$

The production of iron(III) oxide occurs spontaneously at any $T < 3005$ K.

Table 20.2 summarizes these four possible combinations of ΔH and ΔS, and Sample Problem 20.9 applies them.

Student Hot Spot

Student data indicate that you may struggle with understanding how temperature affects spontaneity. Access the Smartbook to view additional Learning Resources on this topic.

Table 20.2 Reaction Spontaneity and the Signs of ΔH, ΔS, and ΔG

ΔH	ΔS	$-T\Delta S$	ΔG	Description
−	+	−	−	Spontaneous at all T
+	−	+	+	Nonspontaneous at all T
+	+	−	+ or −	Spontaneous at higher T; nonspontaneous at lower T
−	−	+	+ or −	Spontaneous at lower T; nonspontaneous at higher T

SAMPLE PROBLEM 20.9 Using Molecular Scenes to Determine the Signs of ΔH, ΔS, and ΔG

Problem The scenes below represent a familiar phase change for water *(blue spheres)*:

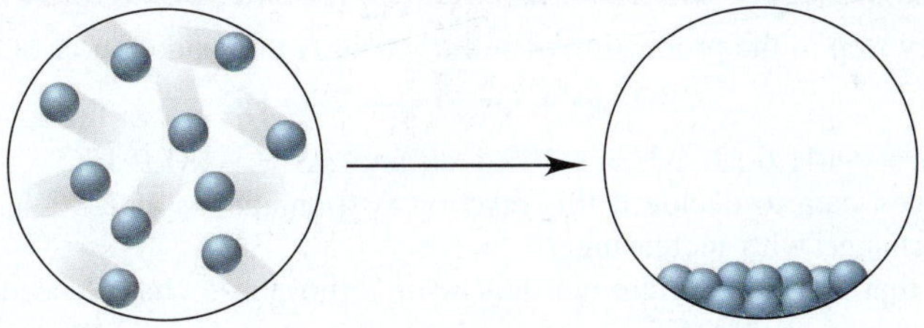

(a) What are the signs of ΔH and ΔS for this process? Explain.
(b) Is the process spontaneous at all T, no T, low T, or high T? Explain.

Plan **(a)** From the scenes, we determine any change in amount of gas and/or any change in the freedom of motion of the particles, which will indicate the sign of ΔS. Also, since the scenes represent a physical change, freedom of particle motion indicates whether heat is absorbed or released, and thus tells us the sign of ΔH. **(b)** The question refers to the sign of ΔG (+ or −) at the different temperature possibilities, so we apply Equation 20.11 and refer to the previous text discussion and Table 20.2.

Solution **(a)** The scenes represent the condensation of water vapor, so the amount of gas decreases dramatically, and the separated molecules give up energy as they come closer together. Therefore, $\Delta S < 0$ and $\Delta H < 0$.

(b) With ΔS negative, the $-T\Delta S$ term is positive. In order for $\Delta G < 0$, the magnitude of T must be small. Therefore, the process is spontaneous at low T.

Check The answer in part (b) seems reasonable based on our analysis in part (a). The answer makes sense because we know from everyday experience that water condenses spontaneously, and it does so at low temperatures.

FOLLOW-UP PROBLEMS

20.9A The scenes below represent the reaction of X_2Y_2 to yield X_2 *(red)* and Y_2 *(blue)*:

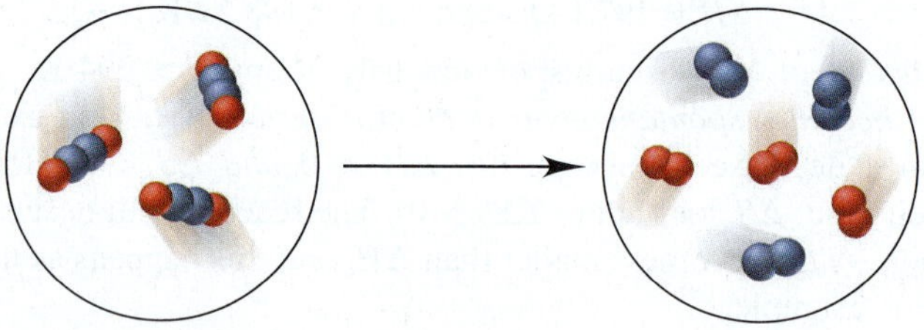

(a) What is the sign of ΔS for the reaction?
(b) If the reaction is spontaneous only above 325°C, what is the sign of ΔH? Explain.

20.9B The scenes below represent the reaction of AB to yield A *(gray)* and B_2 *(red)*:

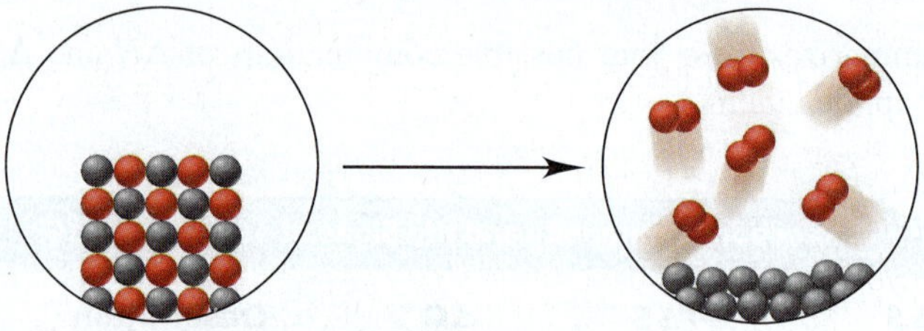

(a) What are the signs of ΔS and ΔH for the reaction?
(b) Is $\Delta G < 0$ at higher T, lower T, all T, or no T? Explain.

SOME SIMILAR PROBLEMS 20.47 and 20.48

As you saw in Sample Problem 20.7, one way to calculate ΔG is from enthalpy and entropy changes. As long as phase changes don't occur, ΔH and ΔS usually change little with temperature, so we use their values at 298 K in the following sample problem 20.10 to examine the effect of T on ΔG and, thus, on reaction spontaneity.

SAMPLE PROBLEM 20.10 Determining the Effect of Temperature on ΔG

Problem A key step in the production of sulfuric acid is the oxidation of $SO_2(g)$ to $SO_3(g)$:

$$2SO_2(g) + O_2(g) \longrightarrow 2SO_3(g)$$

At 298 K, $\Delta G = -141.6$ kJ; $\Delta H = -198.4$ kJ; and $\Delta S = -187.9$ J/K.

(a) Use the given data to decide if this reaction is spontaneous at 25°C, and predict how ΔG will change with increasing T.

(b) Assuming that ΔH and ΔS are constant with T (no phase change occurs), is the reaction spontaneous at 900.°C?

Plan **(a)** We note the sign of ΔG to see if the reaction is spontaneous and the signs of ΔH and ΔS to see the effect of T. **(b)** We use Equation 20.11 to calculate ΔG from the given ΔH and ΔS at the higher T (in K).

Solution **(a)** $\Delta G < 0$, so the reaction is spontaneous at 298 K: SO_2 and O_2 will form SO_3 spontaneously. With $\Delta S < 0$, the term $-T\Delta S > 0$, and this term will become more positive at higher T. Therefore, ΔG will become less negative, and the reaction less spontaneous, with increasing T.

(b) Calculating ΔG at 900.°C ($T = 273 + 900. = 1173$ K):

$$\Delta G = \Delta H - T\Delta S = -198.4\text{ kJ} - [(1173\text{ K})(-187.9\text{ J/K})(1\text{ kJ}/1000\text{ J})] = 22.0\text{ kJ}$$

Since $\Delta G > 0$, the reaction is nonspontaneous at the higher T.

Check The answer in part (b) seems reasonable based on our prediction in part (a). The arithmetic seems correct, with considerable rounding:

$$\Delta G \approx -200\text{ kJ} - [(1200\text{ K})(-200\text{ J/K})(1\text{ kJ}/1000\text{ J})] = +40\text{ kJ}$$

FOLLOW-UP PROBLEMS

20.10A For the following reaction at 298 K, $\Delta S = -308.2$ J/K and $\Delta H = -192.7$ kJ:

$$4NO(g) \longrightarrow N_2O(g) + N_2O_3(g)$$

(a) Is the reaction spontaneous at 298 K?
(b) Would the reaction become more or less spontaneous at higher T?
(c) Assuming that ΔS and ΔH don't change with T, find ΔG at 500.°C.

20.10B A reaction is nonspontaneous at room temperature but *is* spontaneous at −40°C. What can you say about the signs and relative magnitudes of ΔH, ΔS, and $-T\Delta S$?

SOME SIMILAR PROBLEMS 20.57(b) and 20.58(b)

The Temperature at Which a Reaction Becomes Spontaneous As you've just seen, when the signs are the same for ΔH and ΔS of a reaction, it can be nonspontaneous at one temperature and spontaneous at another. The "crossover" temperature occurs when a positive ΔG switches to a negative ΔG because of the magnitude of the $-T\Delta S$ term. We find this temperature by setting ΔG equal to zero and solving for T:

$$\Delta G = \Delta H - T\Delta S = 0$$

Therefore,

$$\Delta H = T\Delta S \quad \text{and} \quad T = \frac{\Delta H}{\Delta S} \qquad \textbf{(20.15)}$$

Consider the reaction of copper(I) oxide with carbon. It does *not* occur at very low temperatures but does at higher temperatures and is used to extract copper from one of its ores:

$$Cu_2O(s) + C(s) \xrightarrow{\Delta} 2Cu(s) + CO(g)$$

We predict that this reaction has a positive ΔS because the number of moles of gas increases; in fact, $\Delta S = 165$ J/K. Furthermore, because the reaction is *non*spontaneous at lower temperatures, it must have a positive ΔH; the actual value is 58.1 kJ. As the $-T\Delta S$ term becomes more negative with higher T, it eventually outweighs the positive ΔH term, so ΔG becomes negative and the reaction occurs spontaneously.

SAMPLE PROBLEM 20.11 Finding the Temperature at Which a Reaction Becomes Spontaneous

Problem At 25°C (298 K), the reduction of copper(I) oxide to copper is nonspontaneous ($\Delta G = 8.9$ kJ). Calculate the temperature at which the reaction becomes spontaneous.

Plan As just discussed, we want the temperature at which ΔG crosses over from a positive to a negative value. We set ΔG equal to zero, and use Equation 20.15 to solve for T, using the values of ΔH (58.1 kJ) and ΔS (165 J/K) given in the text.

Solution From $\Delta G = \Delta H - T\Delta S = 0$, we have

$$T = \frac{\Delta H}{\Delta S} = \frac{58.1\text{ kJ} \times \dfrac{1000\text{ J}}{1\text{ kJ}}}{165\text{ J/K}} = 352\text{ K}$$

Thus, at any temperature above 352 K (79°C), which is a moderate temperature for extracting a metal from its ore, $\Delta G < 0$, so the reaction becomes spontaneous.

Check Rounding to quickly check the math gives

$$T = \frac{60{,}000\ \text{J}}{150\ \text{J/K}} = 400\ \text{K}$$

which is close to the answer.

FOLLOW-UP PROBLEMS

20.11A Find the temperature (in °C) above which the reaction in Follow-up Problem 20.10A is no longer spontaneous.

20.11B Use Appendix B values to find the temperature at which the following reaction becomes spontaneous (assume ΔH and ΔS are constant with T):

$$CaO(s) + CO_2(g) \longrightarrow CaCO_3(s)$$

SOME SIMILAR PROBLEMS 20.59(d), 20.60(d), and 20.61(c)

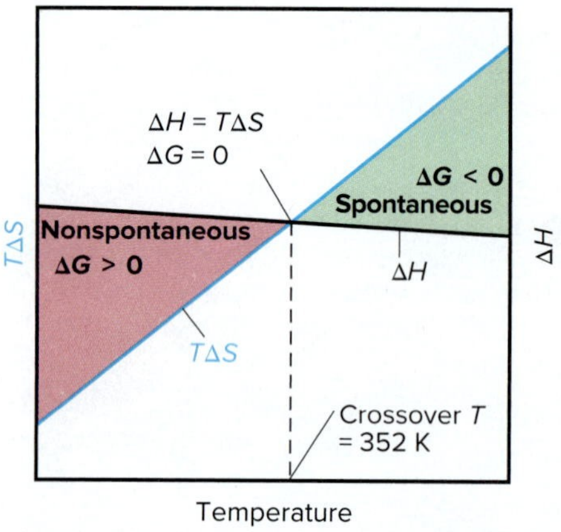

Figure 20.14 The effect of temperature on reaction spontaneity. At low T, $\Delta G > 0$ because ΔH dominates for the reduction of copper(I) oxide to copper. At 352 K, $\Delta H = T\Delta S$, so $\Delta G = 0$. At any higher T, $\Delta G < 0$ because $-T\Delta S$ dominates.

Figure 20.14 shows that the line for $T\Delta S$ rises steadily (and thus the $-T\Delta S$ term becomes more negative) with increasing temperature. This line crosses the relatively constant ΔH line at 352 K for the reduction of copper(I) oxide to copper, as we found in Sample Problem 20.11. At any higher temperature, the $-T\Delta S$ term is greater than the ΔH term, so ΔG is negative.

Coupling of Reactions to Drive a Nonspontaneous Change

In a complex, multistep reaction, we often see a nonspontaneous step driven by a spontaneous step. In such a **coupling of reactions,** *one step supplies enough free energy for the other to occur,* just as burning gasoline supplies enough free energy to move a car.

Look again at the reduction of copper(I) oxide by carbon. In Sample Problem 20.11, we found that the *overall* reaction becomes spontaneous at any temperature above 352 K. Dividing the reaction into two steps, however, we find that even at a higher temperature, say 375 K, copper(I) oxide does not spontaneously decompose to its elements:

$$Cu_2O(s) \longrightarrow 2Cu(s) + \tfrac{1}{2}O_2(g) \qquad \Delta G_{375} = 140.0\ \text{kJ}$$

However, the oxidation of carbon to CO at 375 K is quite spontaneous:

$$C(s) + \tfrac{1}{2}O_2(g) \longrightarrow CO(g) \qquad \Delta G_{375} = -143.8\ \text{kJ}$$

Coupling these reactions means having the carbon in contact with the Cu_2O, which allows the reaction with the larger negative ΔG to "drive" the one with the smaller positive ΔG. Adding the reactions together and canceling the common substance ($\tfrac{1}{2}O_2$) gives an overall reaction with a negative ΔG:

$$Cu_2O(s) + C(s) \longrightarrow 2Cu(s) + CO(g) \qquad \Delta G_{375} = -3.8\ \text{kJ}$$

Many biochemical reactions are also nonspontaneous, including key steps in the syntheses of proteins and nucleic acids, the formation of fatty acids, the maintenance of ion balance, and the breakdown of nutrients. Driving a nonspontaneous step by coupling it to a spontaneous one is a life-sustaining strategy common to all organisms, as you'll see in the upcoming Chemical Connections essay.

› Summary of Section 20.3

- The sign of the free energy change, $\Delta G = \Delta H - T\Delta S$, is directly related to reaction spontaneity: a negative ΔG corresponds to a positive ΔS_{univ}.
- We use the standard free energy of formation (ΔG_f°) to calculate ΔG_{rxn}° at 298 K.
- The maximum work a system can do is never obtained from a real (irreversible) process because some free energy is always converted to heat.
- The magnitude of T influences the spontaneity of a temperature-dependent reaction (same signs of ΔH and ΔS) by affecting the size of $T\Delta S$. For such a reaction, the T at which the reaction becomes spontaneous can be found by setting $\Delta G = 0$.
- A nonspontaneous reaction ($\Delta G > 0$) can be coupled to a more spontaneous one ($\Delta G << 0$) to make it occur. For example, in organisms, the hydrolysis of ATP drives many reactions that have a positive ΔG.

The Universal Role of ATP

Despite their incredible diversity, virtually all organisms use the same amino acids to make their proteins, the same nucleotides to make their nucleic acids, and the same carbohydrate (glucose) to provide energy.

In addition, *all organisms use the same spontaneous reaction to drive a variety of nonspontaneous ones.* This reaction is the hydrolysis of **adenosine triphosphate (ATP)** to adenosine diphosphate (ADP):*

$$ATP^{4-} + H_2O \rightleftharpoons ADP^{3-} + HPO_4^{2-} + H^+ \qquad \Delta G^{\circ\prime} = -30.5\ \text{kJ}$$

In the metabolic breakdown of glucose, for example, the first step, addition of HPO_4^{2-} to glucose, is nonspontaneous:

$$\text{Glucose} + HPO_4^{2-} + H^+ \rightleftharpoons [\text{glucose phosphate}]^- + H_2O \qquad \Delta G^{\circ\prime} = 13.8\ \text{kJ}$$

Coupling this nonspontaneous reaction to ATP hydrolysis makes the overall process spontaneous. If we add the two reactions, HPO_4^{2-}, H^+, and H_2O cancel:

$$\text{Glucose} + ATP^{4-} \rightleftharpoons [\text{glucose phosphate}]^- + ADP^{3-} \qquad \Delta G^{\circ\prime} = -16.7\ \text{kJ}$$

Coupling cannot occur if reactions are physically separated, so these reactions take place on an enzyme (Section 16.7) that simultaneously binds glucose and ATP, and the phosphate group of ATP that will be transferred lies next to the —OH group of glucose that will bind it (Figure B20.1).

The ADP produced in energy-releasing reactions combines with phosphate to regenerate ATP in energy-absorbing reactions catalyzed by other enzymes. Thus, there is a continuous cycling of ATP to ADP and back to ATP again to supply energy to the cells (Figure B20.2).

*In biochemical systems, the standard-state concentration of H^+ is 10^{-7} *M*, not the usual 1 *M*, and the standard free energy change is represented by the symbol $\Delta G^{\circ\prime}$.

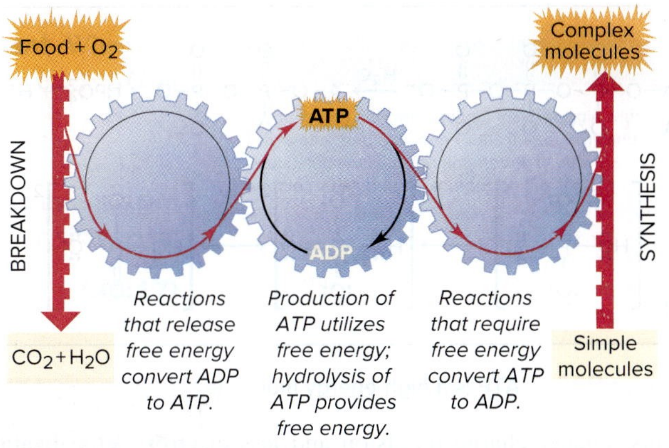
Food + O_2
Complex molecules
ATP
ADP
BREAKDOWN
SYNTHESIS
CO_2+H_2O
Simple molecules
Reactions that release free energy convert ADP to ATP.
Production of ATP utilizes free energy; hydrolysis of ATP provides free energy.
Reactions that require free energy convert ATP to ADP.

Figure B20.2 **The cycling of metabolic free energy.**

Why is ATP a "high-energy" molecule? By examining the phosphate portions of ATP, ADP, and HPO_4^{2-}, we can see two basic chemical reasons why ATP hydrolysis supplies so much free energy (Figure B20.3):

1. *Charge repulsion.* At physiological pH (~7), the triphosphate group of ATP has four negative charges close together. This *high charge repulsion* is reduced in ADP (Figure B20.3A).
2. *Electron delocalization.* Once HPO_4^{2-} is free, there is extensive delocalization and resonance stabilization of the π electrons (Figure B20.3B).

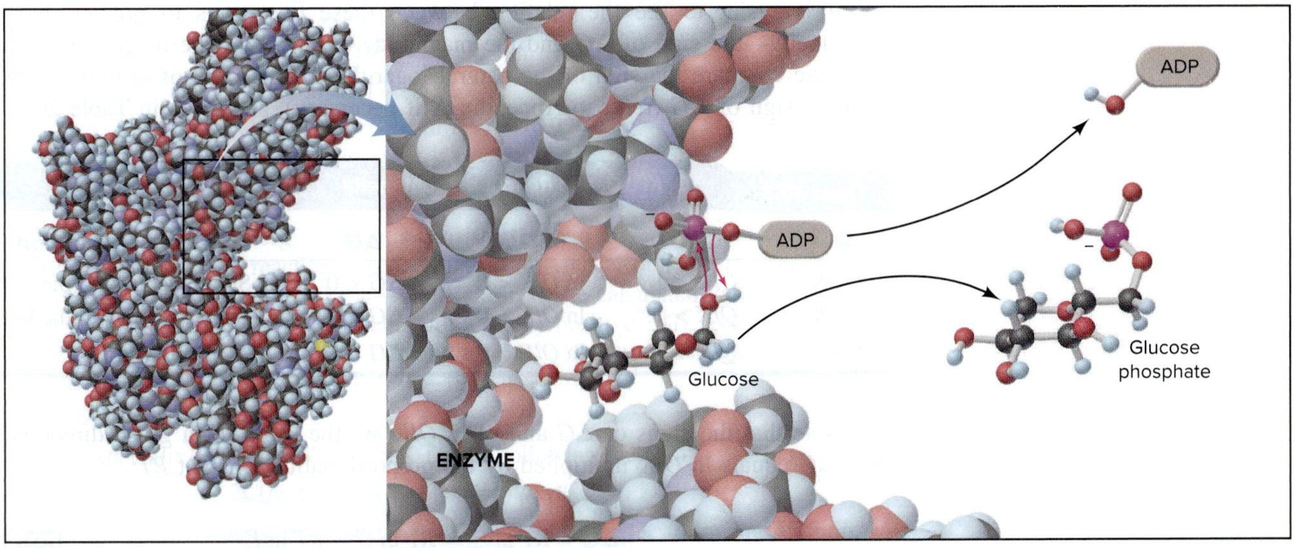
ADP
ADP
Glucose
Glucose phosphate
ENZYME

Figure B20.1 **The coupling of a nonspontaneous reaction to the hydrolysis of ATP.** Glucose lies next to ATP (shown as ADP—O—PO_3H) in the enzyme's active site. ADP (shown as ADP—OH) and glucose phosphate are released.

(continued)

Table 20.4 The Relationship Between $\Delta G°$ and K at 298 K

$\Delta G°$(kJ)	K	Significance
200	9×10^{-36}	Essentially no forward reaction;
100	3×10^{-18}	reverse reaction goes to completion.
50	2×10^{-9}	
10	2×10^{-2}	
1	7×10^{-1}	
0	1	Forward and reverse reactions
−1	1.5	proceed to same extent.
−10	5×10^{1}	
−50	6×10^{8}	
−100	3×10^{17}	Forward reaction goes to completion;
−200	1×10^{35}	essentially no reverse reaction.

FORWARD REACTION (arrow pointing down) · REVERSE REACTION (arrow pointing up)

Table 20.4 shows that, due to their logarithmic relationship, a small change in $\Delta G°$ causes a large change in K. Note just these two examples:

- As $\Delta G°$ becomes more positive, K becomes smaller: if $\Delta G° = +10$ kJ, K is 0.02, so the product terms are $\frac{1}{50}$ the size of the reactant terms (see Figure 17.2A).
- As $\Delta G°$ becomes more negative, K becomes larger: if $\Delta G° = -10$ kJ, K is 50, so the product terms are 50 times larger than the reactant terms (see Figure 17.2B).

Finding the Free Energy Change Under Any Conditions In reality, reactions rarely begin with all components in their standard states. By substituting the relationship between $\Delta G°$ and K (Equation 20.17) into the expression for ΔG (Equation 20.16), we obtain a relationship that applies to *any starting concentrations:*

$$\Delta G = \Delta G° + RT \ln Q \qquad \textbf{(20.18)}$$

Note that, by definition, when the components are in their standard states (1 *M*, 1 atm, etc.), $\Delta G = \Delta G°$:

$$\Delta G = \Delta G° + RT \ln Q = \Delta G° + RT \ln 1$$
$$\Delta G = \Delta G°$$

Sample Problem 20.13 uses molecular scenes to explore these ideas, and Sample Problem 20.14 applies them to an important industrial reaction.

SAMPLE PROBLEM 20.13 Using Molecular Scenes to Find ΔG for a Reaction at Nonstandard Conditions

Problem These molecular scenes represent three mixtures in which A_2 *(black)* and B_2 *(green)* are forming AB. Each molecule represents 0.10 atm. The equation is

$$A_2(g) + B_2(g) \rightleftharpoons 2AB(g) \qquad \Delta G° = -3.4 \text{ kJ/mol}$$

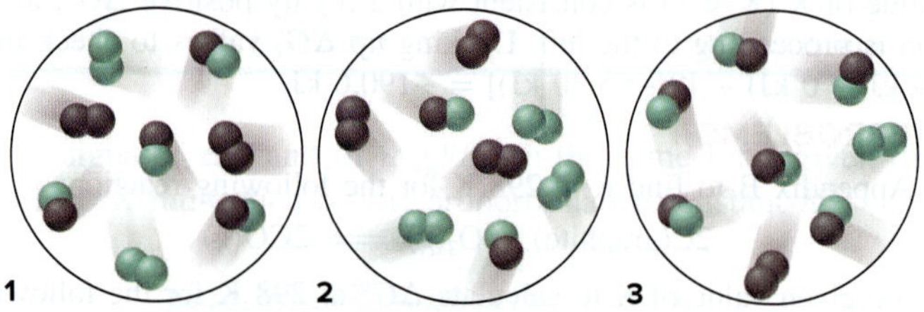

(a) If mixture 1 is at equilibrium, calculate K.
(b) Which mixture has the most negative ΔG, and which has the most positive?

Plan **(a)** Mixture 1 is at equilibrium, so we first write the expression for Q and then find the partial pressure of each substance from the numbers of molecules and calculate K. **(b)** To find ΔG, we apply Equation 20.18. We are given $\Delta G°$ (−3.4 kJ/mol) and know R (8.314 J/mol·K), but we still need to find T. We calculate T from Equation 20.17 using K from part (a), and substitute the partial pressure of each substance (by counting particles) to get Q.

Solution **(a)** Writing the expression for Q and calculating K:

$$A_2(g) + B_2(g) \rightleftharpoons 2AB(g) \qquad Q = \frac{(P_{AB})^2}{P_{A_2} \times P_{B_2}} \qquad K = \frac{(0.40)^2}{(0.20)(0.20)} = 4.0$$

(b) Calculating T from Equation 20.17 for use in Equation 20.18:

$$\Delta G° = -RT \ln K$$

$$T = \frac{\Delta G°}{-R \ln K} = \frac{\dfrac{-3.4\text{ kJ}}{\text{mol}}\left(\dfrac{1000\text{ J}}{1\text{ kJ}}\right)}{-8.314\text{ J/mol·K} \times \ln 4.0} = 295\text{ K}$$

Calculating ΔG from Equation 20.18 for each reaction mixture:

Mixture 1:

$$\Delta G = \Delta G° + RT \ln Q = -3.4\text{ kJ} + RT \ln 4.0$$
$$= -3.4\text{ kJ/mol}\left(\frac{1000\text{ J}}{1\text{ kJ}}\right) + \left(\frac{8.314\text{ J}}{\text{mol·K}}\right)(295\text{ K}) \ln 4.0$$
$$= -3400\text{ J/mol} + 3400\text{ J/mol} = 0.0\text{ J}$$

Mixture 2:

$$\Delta G = -3.4\text{ kJ/mol} + RT \ln \frac{(0.20)^2}{(0.30)(0.30)}$$
$$= -3.4\text{ kJ/mol}\left(\frac{1000\text{ J}}{1\text{ kJ}}\right) + \left(\frac{8.314\text{ J}}{\text{mol·K}}\right)(295\text{ K}) \ln 0.44$$
$$= -5.4\times10^3\text{ J/mol}$$

Mixture 3:

$$\Delta G = -3.4\text{ kJ/mol} + RT \ln \frac{(0.60)^2}{(0.10)(0.10)}$$
$$= -3.4\text{ kJ/mol}\left(\frac{1000\text{ J}}{1\text{ kJ}}\right) + \left(\frac{8.314\text{ J}}{\text{mol·K}}\right)(295\text{ K}) \ln 36$$
$$= 5.4\times10^3\text{ J/mol}$$

Mixture 2 has the most negative ΔG, and mixture 3 has the most positive ΔG.

Check In part (b), round to check the arithmetic; for example, for mixture 3, $\Delta G \approx -3000$ J/mol + (8 J/mol·K)(300 K)4 $\approx$ 7000 J/mol, which is in the ballpark.

Comment **1.** By using the properties of logarithms, we did *not* have to calculate T and ΔG in part (b). For mixture 2, $Q < 1$, so ln Q is negative, which makes ΔG more negative. Also, note that $Q(0.44) < K(4.0)$, which tells us that the reaction is proceeding to the right, so $\Delta G < 0$. For mixture 3, $Q > 1$ (and is greater than it is for mixture 1), so ln Q is positive, which makes ΔG positive. Also, $Q(36) > K(4.0)$, which tells us that the reaction is proceeding to the left, so $\Delta G > 0$.

2. In part (b), the value of zero for ΔG of the equilibrium mixture (mixture 1) makes sense, because a system at equilibrium has released all of its free energy.

3. Note especially that, by definition, when the components are in their standard states, $\Delta G = \Delta G°$:

$$\Delta G = \Delta G° + RT \ln Q = -3.4\text{ kJ/mol} + RT \ln \frac{(1.0)^2}{(1.0)(1.0)}$$
$$= -3.4\text{ kJ/mol} + RT \ln 1.0 = -3.4\text{ kJ/mol}$$

FOLLOW-UP PROBLEMS

20.13A The scenes below depict three mixtures in which A *(orange)* and B *(green)* are forming AB_3 in a reaction for which $\Delta G° = -4.6$ kJ/mol. Assume that each molecule represents 0.10 mol and the volume is 1.0 L.

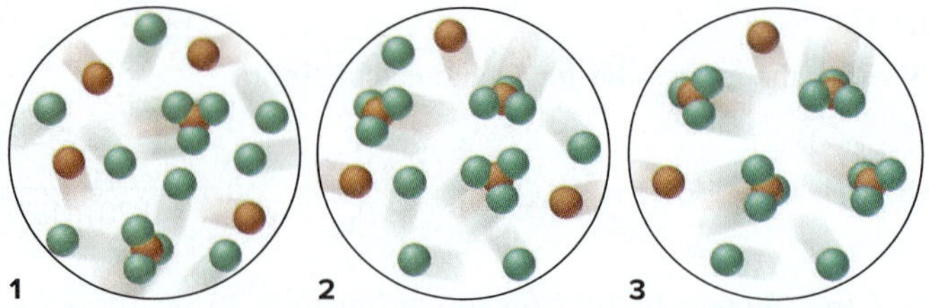

(a) Write a balanced equation for the reaction. **(b)** If $K = 8.0$, which mixture is at equilibrium? **(c)** Rank the three mixtures from the highest (most positive) ΔG to the lowest (most negative) ΔG.

20.13B The scenes below depict mixtures in which X_2 *(tan)* and Y_2 *(blue)* are forming XY_2. Each molecule represents 0.10 mol, and the volume is 0.10 L. The equation is

$$X_2(g) + 2Y_2(g) \rightleftharpoons 2XY_2(g);\ \Delta G° = -1.3 \text{ kJ/mol}$$

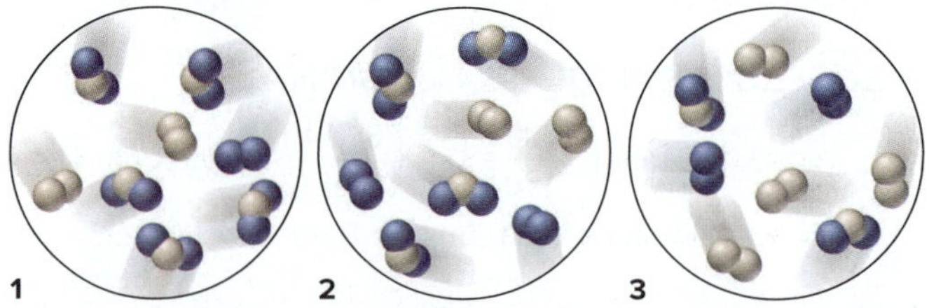

(a) If $K = 2.0$, which mixture is at equilibrium? **(b)** Rank the three mixtures from the lowest (most negative) ΔG to highest (most positive) ΔG. **(c)** What is the sign of ΔG for the change that occurs as each nonequilibrium mixture approaches equilibrium?

SOME SIMILAR PROBLEMS 20.81 and 20.82

Student Hot Spot

Student data indicate that you may struggle with the calculation of ΔG under nonstandard conditions. Access the Smartbook to view additional Learning Resources on this topic.

SAMPLE PROBLEM 20.14 Calculating ΔG at Nonstandard Conditions

Problem The oxidation of $SO_2(g)$, $2SO_2(g) + O_2(g) \longrightarrow 2SO_3(g)$, is the key reaction in the manufacture of sulfuric acid.

(a) Calculate K at 298 K. ($\Delta G°_{298} = -141.6$ kJ/mol of reaction as written.)
(b) A container is filled with 0.500 atm of SO_2, 0.0100 atm of O_2, and 0.100 atm of SO_3 and kept at 25°C. In which direction, if any, will the reaction proceed to reach equilibrium?
(c) Calculate ΔG for the system in part (b).

Plan (a) We know $\Delta G°$, T, and R, so we can calculate K from Equation 20.17 [see Sample Problem 20.12(a)]. **(b)** To determine if a net reaction will occur, we find Q from the given partial pressures and compare it with K from part (a). **(c)** These are *not* standard-state pressures, so we find ΔG at 298 K with Equation 20.18, using the values of $\Delta G°$ (given) and Q [from part (b)].

Solution (a) Calculating K at 298 K:

$$\Delta G° = -RT \ln K \qquad \text{so} \qquad K = e^{-(\Delta G°/RT)}$$

At 298 K, the exponent is

$$-(\Delta G°/RT) = -\left(\frac{-141.6 \text{ kJ/mol} \times \dfrac{1000 \text{ J}}{1 \text{ kJ}}}{8.314 \text{ J/mol·K} \times 298 \text{ K}}\right) = 57.2$$

so

$$K = e^{-(\Delta G°/RT)} = e^{57.2} = 7\times10^{24}$$

(b) Calculating the value of Q:

$$Q = \frac{(P_{SO_3})^2}{(P_{SO_2})^2 \times P_{O_2}} = \frac{0.100^2}{0.500^2 \times 0.0100} = 4.00$$

Because $Q < K$, the denominator will decrease and the numerator will increase—more SO_3 will form—until Q equals K. To reach equilibrium, the reaction will proceed to the right.

(c) Calculating ΔG, the nonstandard free energy change, at 298 K:

$$\begin{aligned}\Delta G_{298} &= \Delta G° + RT \ln Q \\ &= -141.6 \text{ kJ/mol} + \left(8.314 \text{ J/mol·K} \times \frac{1 \text{ kJ}}{1000 \text{ J}} \times 298 \text{ K} \times \ln 4.00\right) \\ &= -138.2 \text{ kJ/mol}\end{aligned}$$

Check Note that in parts (a) and (c), we made the free energy units (kJ) consistent with the units in R (J). Applying the rules for significant figures in addition and subtraction, we retain one digit to the right of the decimal place in part (c).

Comment As with the synthesis of NH_3 (Section 17.6), where the *yield* is high but the *rate* is low at a lower temperature, this process is carried out at higher temperature *with a catalyst* to attain a higher *rate*. We discuss the details of the industrial production of sulfuric acid in Chapter 22.

FOLLOW-UP PROBLEMS

20.14A At 298 K, $\Delta G° = -33.5$ kJ/mol for the formation of chloroethane from ethylene and hydrogen chloride:

$$C_2H_4(g) + HCl(g) \rightleftharpoons C_2H_5Cl(g)$$

(a) Calculate K at 298 K.
(b) Calculate ΔG at 298 K if $[C_2H_5Cl] = 1.5\ M$, $[C_2H_4] = 0.50\ M$, and $[HCl] = 1.0\ M$.

20.14B At 298 K, hypobromous acid (HBrO) dissociates in water with $K_a = 2.3\times10^{-9}$.
(a) Calculate $\Delta G°$ for the dissociation of HBrO.
(b) Calculate ΔG if $[H_3O^+] = 6.0\times10^{-4}\ M$, $[BrO^-] = 0.10\ M$, and $[HBrO] = 0.20\ M$.

SOME SIMILAR PROBLEMS 20.79 and 20.80

Another Look at the Meaning of Spontaneity At this point, we introduce two terms related to *spontaneous* and *nonspontaneous:*

1. *Product-favored reaction.* For the general reaction

$$A \rightleftharpoons B \qquad K = [B]/[A] > 1$$

Therefore, the reaction proceeds largely from left to right (Figure 20.15A, *next page*). From pure A to equilibrium, $Q < K$ and the curved *green* arrow in the figure indicates that the reaction is spontaneous ($\Delta G < 0$). From equilibrium to pure B, the curved *red* arrow indicates that the reaction is nonspontaneous ($\Delta G > 0$). Similarly, from pure B to equilibrium, $Q > K$ and the reaction is also spontaneous ($\Delta G < 0$), but not thereafter. In either case, *free energy decreases until the reaction reaches a minimum at the equilibrium mixture:* $Q = K$ *and* $\Delta G = 0$. For the overall reaction $A \rightleftharpoons B$ (starting with all components in their standard states), $G°_B$ is smaller than $G°_A$, so $\Delta G°$ is negative, which corresponds to $K > 1$. We call this a *product-favored* reaction because in its final state the system contains mostly product.

2. *Reactant-favored reaction.* For the general reaction

$$C \rightleftharpoons D \qquad K = [D]/[C] < 1$$

and this reaction proceeds slightly from left to right (Figure 20.15B). Here, too, whether we start with pure C or pure D, the reaction is spontaneous ($\Delta G < 0$) until equilibrium. In this case, however, the equilibrium mixture contains mostly C (the reactant), so we say the reaction is *reactant favored.* Here, $G°_D$ is *larger* than $G°_C$, so $\Delta G°$ is *positive,* which corresponds to $K < 1$.

Thus, "spontaneous" refers to that *portion* of a reaction in which the free energy decreases—from the starting mixture to the equilibrium mixture. A product-favored

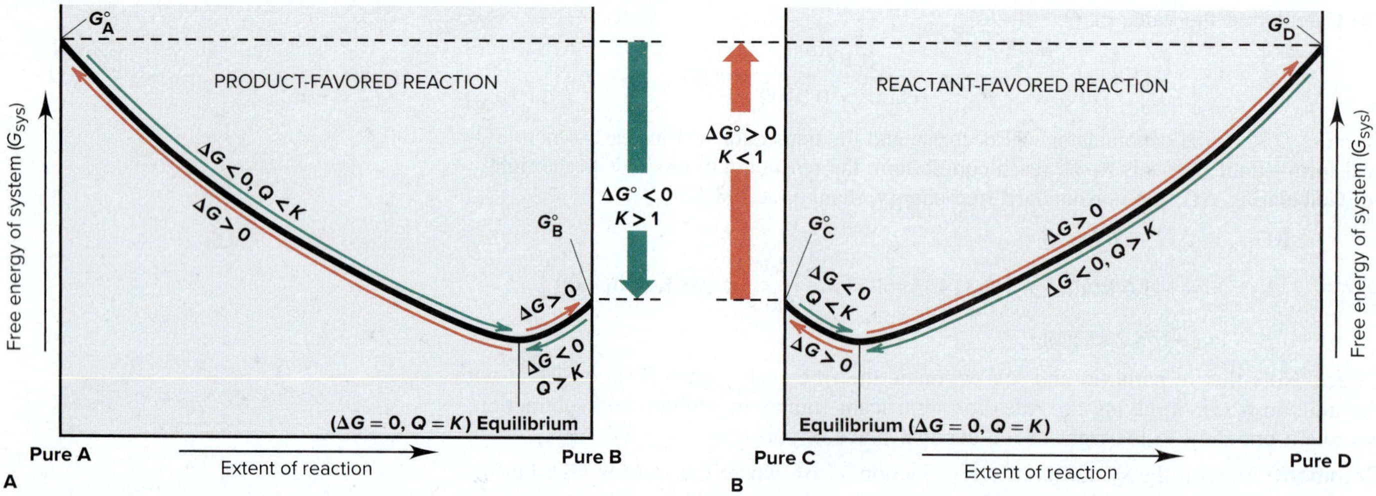

Figure 20.15 **Free energy and the extent of reaction.** Each reaction proceeds spontaneously *(curved green arrows)* from reactants (A or C) or products (B or D) to the equilibrium mixture, at which point $\Delta G = 0$. After that, the reaction is nonspontaneous *(curved red arrows)*. **A,** For the product-favored reaction A $\rightleftharpoons$ B, $G_A^\circ > G_B^\circ$, so $\Delta G^\circ < 0$ and $K > 1$. **B,** For the reactant-favored reaction C $\rightleftharpoons$ D, $G_D^\circ > G_C^\circ$, so $\Delta G^\circ > 0$ and $K < 1$.

reaction goes predominantly, but *not* completely, toward product, and a reactant-favored reaction goes relatively little toward product (see Table 20.4).

› Summary of Section 20.4

- Two ways of predicting reaction spontaneity are from the sign of ΔG or from the value of Q/K. These variables are related to each other by $\Delta G = RT \ln Q/K$. When $Q = K$, $Q/K = 1$ and $\ln Q/K = 0$. Thus, the system is at equilibrium and can release (or absorb) no more free energy.
- When Q is determined with standard-state values, the free energy change is ΔG° and is related to the equilibrium constant: $\Delta G^\circ = -RT \ln K$.
- Any nonequilibrium mixture of reactants and products moves spontaneously ($\Delta G < 0$) toward the equilibrium mixture.
- A product-favored reaction goes predominantly toward product and, thus, has $K > 1$ and $\Delta G^\circ < 0$; a reactant-favored reaction has $K < 1$ and $\Delta G^\circ > 0$.

CHAPTER REVIEW GUIDE

Learning Objectives

Relevant section (§) and/or sample problem (SP) numbers appear in parentheses.

Understand These Concepts

1. How the tendency of a process to occur by itself is distinct from how long it takes to occur (Introduction)
2. The distinction between a spontaneous and a nonspontaneous change (§20.1)
3. Why the first law of thermodynamics and the sign of ΔH° cannot predict the direction of a spontaneous process (§20.1)
4. How the entropy (S) of a system is defined by the number of microstates over which its energy is dispersed (§20.1)
5. How entropy is alternatively defined by the heat absorbed (or released) at constant T in a reversible process (§20.1)
6. The criterion for spontaneity according to the second law of thermodynamics: that a change increases S_{univ} (§20.1)
7. How absolute values of standard molar entropies (S°) can be obtained because the third law of thermodynamics provides a zero point (§20.1)
8. How temperature, physical state, dissolution, atomic size, and molecular complexity influence S° values (§20.1)
9. How ΔS°_{rxn} is based on the difference between the summed S° values for the reactants and those for products (§20.2)
10. How the surroundings add heat to or remove heat from the system and how ΔS_{surr} influences overall ΔS°_{rxn} (§20.2)
11. The relationship between ΔS_{surr} and ΔH_{sys} (§20.2)
12. How reactions proceed spontaneously toward equilibrium ($\Delta S_{univ} > 0$) but proceed no further at equilibrium ($\Delta S_{univ} = 0$) (§20.2)
13. How the free energy change (ΔG) combines a system's entropy and enthalpy changes (§20.3)
14. How the expression for the free energy change is derived from the second law (§20.3)
15. The relationship between ΔG and the maximum work a system can perform and why this quantity of work is never performed in a real process (§20.3)

16. How temperature determines spontaneity for reactions in which ΔS and ΔH have the same sign (§20.3)
17. Why the temperature at which a reaction becomes spontaneous occurs when $\Delta G = 0$ (§20.3)
18. How a spontaneous change can be coupled to a nonspontaneous change to make it occur (§20.3)
19. How ΔG is related to the ratio of Q to K (§20.4)
20. The meaning of $\Delta G°$ and its relation to K (§20.4)
21. The relation of ΔG to $\Delta G°$ and Q (§20.4)
22. Why G decreases, no matter what the starting concentrations, as a reacting system moves toward equilibrium (§20.4)
23. The distinction between product-favored and reactant-favored reactions (§20.4)

Master These Skills

1. Calculating ΔS°_{sys} for an isothermal change in the volume of a gas (SP 20.1)
2. Calculating ΔS°_{sys} for a phase change (SP 20.2)
3. Calculating ΔS°_{sys} for a change in temperature (SP 20.3)
4. Predicting relative $S°$ values of systems (§20.1 and SP 20.4)
5. Calculating ΔS°_{rxn} for a chemical change (SP 20.5)
6. Determining reaction spontaneity from ΔS_{surr} and ΔH°_{sys} (SP 20.6)
7. Calculating ΔG°_{rxn} from ΔH°_f and $S°$ values (SP 20.7)
8. Calculating ΔG°_{rxn} from ΔG°_f values (SP 20.8)
9. Using molecular scenes to determine the signs of ΔH, ΔS, and ΔG (SP 20.9)
10. Determining the effect of temperature on ΔG (SP 20.10)
11. Calculating the temperature at which a reaction becomes spontaneous (§20.3 and SP 20.11)
12. Calculating K from $\Delta G°$, and vice versa (§20.4 and SP 20.12)
13. Using $\Delta G°$ and Q to calculate ΔG at any conditions (SPs 20.13, 20.14)

Key Terms

Page numbers appear in parentheses.

adenosine triphosphate (ATP) (925)
coupling of reactions (924)
entropy (S) (898)
free energy (G) (916)
isothermal process (901)
microstate (898)
reversible process (901)
second law of thermodynamics (905)
spontaneous change (895)
standard entropy of reaction (ΔS°_{rxn}) (910)
standard free energy change ($\Delta G°$) (917)
standard free energy of formation (ΔG°_f) (918)
standard molar entropy ($S°$) (906)
third law of thermodynamics (905)
Trouton's rule (902)

Key Equations and Relationships

Page numbers appear in parentheses.

20.1 Quantifying entropy in terms of the number of microstates (W) over which the energy of a system can be dispersed (898):

$$S = k \ln W$$

20.2 Quantifying the entropy change in terms of heat absorbed (or released) in a reversible process (901):

$$\Delta S_{sys} = \frac{q_{rev}}{T}$$

20.3 Quantifying the entropy change that results from an isothermal volume change of a gas exhibiting ideal behavior (901):

$$\Delta S_{sys} = nR \ln \frac{V_{final}}{V_{initial}}$$

20.4 Quantifying the entropy change that results from an isothermal change in volume (expressed as a change in pressure) of a gas exhibiting ideal behavior (902):

$$\Delta S_{sys} = nR \ln \frac{P_{initial}}{P_{final}}$$

20.5 Quantifying the entropy change that results from a phase change (902):

$$\Delta S_{fus} = \frac{\Delta H_{fus}}{T_f} \quad \text{and} \quad \Delta S_{vap} = \frac{\Delta H_{vap}}{T_b}$$

20.6 Quantifying the entropy change that results from a change in temperature of a system at constant volume (904):

$$\Delta S_{fus} = nC_{V,m} \ln \frac{T_2}{T_1}$$

20.7 Quantifying the entropy change that results from a change in temperature of a system at constant pressure (904):

$$\Delta S_{fus} = nC_{P,m} \ln \frac{T_2}{T_1}$$

20.8 Stating the second law of thermodynamics, for a spontaneous process (905):

$$\Delta S_{univ} = \Delta S_{sys} + \Delta S_{surr} > 0$$

20.9 Calculating the standard entropy of reaction from the standard molar entropies of reactants and products (911):

$$\Delta S^\circ_{rxn} = \Sigma m S^\circ_{products} - \Sigma n S^\circ_{reactants}$$

20.10 Relating the entropy change in the surroundings to the enthalpy change of the system and the temperature (913):

$$\Delta S_{surr} = -\frac{\Delta H_{sys}}{T}$$

20.11 Expressing the free energy change of the system in terms of its component enthalpy and entropy changes (Gibbs equation) (917):

$$\Delta G_{sys} = \Delta H_{sys} - T\Delta S_{sys}$$

20.12 Calculating the standard free energy change from standard enthalpy and entropy changes (917):

$$\Delta G^\circ_{sys} = \Delta H^\circ_{sys} - T\Delta S^\circ_{sys}$$

20.13 Calculating the standard free energy change from the standard free energies of formation (918):

$$\Delta G^\circ_{rxn} = \Sigma m \Delta G^\circ_{f(products)} - \Sigma n \Delta G^\circ_{f(reactants)}$$

20.14 Relating the free energy change to the maximum work a system can do (919):

$$\Delta G = w_{max}$$

20.15 Finding the temperature at which a reaction becomes spontaneous (923):

$$T = \frac{\Delta H}{\Delta S}$$

20.16 Expressing the free energy change in terms of Q and K (926):

$$\Delta G = RT \ln \frac{Q}{K} = RT \ln Q - RT \ln K$$

20.17 Expressing the free energy change with Q at standard-state conditions (927):

$$\Delta G^\circ = -RT \ln K$$

20.18 Expressing the free energy change for nonstandard initial conditions (928):

$$\Delta G = \Delta G^\circ + RT \ln Q$$

BRIEF SOLUTIONS TO FOLLOW-UP PROBLEMS

20.1A Amount (mol) of $N_2 = 200.0 \text{ g } N_2 \times \frac{1 \text{ mol } N_2}{28.02 \text{ g } N_2}$

$= 7.138 \text{ mol } N_2$

$$\Delta S_{sys} = nR \ln \frac{P_{initial}}{P_{final}} = 7.138 \text{ mol} \times 8.314 \text{ J/mol·K} \times \ln \frac{0.715 \text{ atm}}{1.16 \text{ atm}}$$
$= -28.7 \text{ J/K}$

20.1B Amount (mol) of $CO_2 = 100. \text{ g } CO_2 \times \frac{1 \text{ mol } CO_2}{44.01 \text{ g } CO_2}$

$= 2.27 \text{ mol } CO_2$

$$\Delta S_{sys} = nR \ln \frac{V_{final}}{V_{initial}} = 2.27 \text{ mol} \times 8.314 \text{ J/mol·K} \times \ln \frac{4.75 \text{ L}}{3.35 \text{ L}}$$
$= 6.59 \text{ J/K}$

20.2A

$$\Delta H \text{ (J)} = 34.0 \text{ g } C_3H_6O \times \frac{1 \text{ mol } C_3H_6O}{58.08 \text{ g } C_3H_6O} \times \frac{31.3 \text{ kJ}}{1 \text{ mol } C_3H_6O} \times \frac{10^3 \text{ J}}{1 \text{ kJ}}$$
$= 18{,}323 \text{ J}$

$$\Delta S = \frac{\Delta H_{vap}}{T_b} = \frac{18{,}323 \text{ kJ}}{329.4 \text{ K}} = 55.6 \text{ J/K}$$

20.2B

$$\Delta H \text{ (J)} = 112 \text{ g } C_2H_5OH \times \frac{1 \text{ mol } C_2H_5OH}{46.07 \text{ g } C_2H_5OH} \times \frac{-5.02 \text{ kJ}}{1 \text{ mol } C_2H_5OH} \times \frac{10^3 \text{ J}}{1 \text{ kJ}}$$
$= -12{,}204 \text{ J}$

$$\Delta S = \frac{\Delta H_{fus}}{T_f} = \frac{-12{,}204 \text{ kJ}}{159 \text{ K}} = -76.8 \text{ J/K}$$

Note that since the value of 5.02 kJ/mol is ΔH_{fus} for melting solid C_2H_5OH, –5.02 kJ/mol is the value for the opposite process of freezing C_2H_5OH. Energy is required to melt the solid; the same amount of energy is released when the liquid freezes.

20.3A Amount (mol) of $N_2 = 375 \text{ g } N_2 \times \frac{1 \text{ mol } N_2}{28.02 \text{ g } N_2}$

$= 13.4 \text{ mol } N_2$

$$\Delta S_{sys} = nC_{V,m} \ln \frac{T_2}{T_1} = 13.4 \text{ mol} \times 20.8 \text{ J/mol·K} \times \ln \frac{273 \text{ K}}{298 \text{ K}}$$
$= -24.4 \text{ J/K}$

20.3B

Amount (mol) of $H_2O = 100. \text{ g } H_2O \times \frac{1 \text{ mol } H_2O}{18.02 \text{ g } H_2O} = 5.55 \text{ mol } H_2O$

$T_2 = T_1 \times e^{\Delta S_{sys}/nC_{P,m}} = 298 \times e^{(70.9 \text{ J/K})/(5.55 \text{ mol} \times 75.3 \text{ J/mol·K})}$
$T_2 = 353 \text{ K} = 80.°\text{C}$

20.4A (a) $PCl_5(g)$: higher molar mass and more complex molecule; (b) $BaCl_2(s)$: higher molar mass; (c) $Br_2(g)$: gases have more freedom of motion and dispersal of energy than liquids.

20.4B (a) LiBr(*aq*): same number of ions, but lower molar mass of ions; (b) quartz: the particles have less freedom of motion in the crystalline structure; (c) cyclohexane: molecule with no side chain has less freedom of motion.

20.5A (a) $4NO(g) \longrightarrow N_2O(g) + N_2O_3(g)$

$\Delta n_{gas} = -2$, so $\Delta S^\circ_{rxn} < 0$

$\Delta S^\circ_{rxn} = [(1 \text{ mol } N_2O)(219.7 \text{ J/mol·K}) + (1 \text{ mol } N_2O_3)(314.7 \text{ J/mol·K})] - [(4 \text{ mol NO})(210.65 \text{ J/mol·K})]$
$= -308.2 \text{ J/K}$

(b) $CH_3OH(g) \longrightarrow CO(g) + 2H_2(g)$

$\Delta n_{gas} = 2$, so $\Delta S^\circ_{rxn} > 0$

$\Delta S^\circ_{rxn} = [(1 \text{ mol CO})(197.5 \text{ J/mol·K}) + (2 \text{ mol } H_2)(130.6 \text{ J/mol·K})] - [(1 \text{ mol } CH_3OH)(238 \text{ J/mol·K})]$
$= 221 \text{ J/K}$

20.5B (a) $2NaOH(s) + CO_2(g) \longrightarrow Na_2CO_3(s) + H_2O(l)$

$\Delta n_{gas} = -1$, so $\Delta S^\circ_{rxn} < 0$

$\Delta S^\circ_{rxn} = [(1 \text{ mol } H_2O)(69.940 \text{ J/mol·K}) + (1 \text{ mol } Na_2CO_3)(139 \text{ J/mol·K})] - [(1 \text{ mol } CO_2)(213.7 \text{ J/mol·K}) + (2 \text{ mol NaOH})(64.454 \text{ J/mol·K})]$
$= -134 \text{ J/K}$

(b) $2Fe(s) + 3H_2O(g) \longrightarrow Fe_2O_3(s) + 3H_2(g)$

$\Delta n_{gas} = 0$, so cannot predict sign of ΔS°_{rxn}

$\Delta S^\circ_{rxn} = [(1 \text{ mol } Fe_2O_3)(87.400 \text{ J/mol·K}) + (3 \text{ mol } H_2)(130.6 \text{ J/mol·K})] - [(2 \text{ mol Fe})(27.3 \text{ J/mol·K}) + (3 \text{ mol } H_2O)(188.72 \text{ J/mol·K})]$
$= -141.6 \text{ J/K}$

20.6A To find ΔS_{surr}, use ΔH°_f values to calculate ΔH°_{rxn} and divide it by T:

$\Delta H^\circ_{rxn} = [(4 \text{ mol } PCl_3)(-287 \text{ kJ/mol})] - [(1 \text{ mol } P_4)(0 \text{ kJ/mol}) + (6 \text{ mol } Cl_2)(0 \text{ kJ/mol})]$
$= -1148 \text{ kJ}$

$$\Delta S_{surr} = -\frac{\Delta H^\circ_{rxn}}{T} = -\frac{(-1148 \text{ kJ})(1000 \text{ J/1 kJ})}{298 \text{ K}} = 3850 \text{ J/K}$$

To find ΔS_{univ}, use S° values to calculate ΔS°_{rxn} and add it to ΔS_{surr}:

$$\Delta S^\circ_{rxn} = [(4\ \text{mol PCl}_3)(312\ \text{J/mol·K})] - [(1\ \text{mol P}_4)(41.1\ \text{J/mol·K}) + (6\ \text{mol Cl}_2)(223.0\ \text{J/mol·K})] = -131\ \text{J/K}$$

$$\Delta S_{univ} = \Delta S_{surr} + \Delta S^\circ_{rxn} = 3850\ \text{J/K} + (-131\ \text{J/K}) = 3719\ \text{J/K}$$

Reaction is spontaneous at 298 K since ΔS_{univ} is positive.

20.6B $2FeO(s) + \frac{1}{2}O_2(g) \longrightarrow Fe_2O_3(s)$

Use S° values to calculate ΔS°_{rxn} and ΔH°_f values to calculate ΔH°_{rxn}; divide ΔH°_{rxn} by T to obtain ΔS_{surr} and add that value to ΔS°_{rxn} to obtain ΔS_{univ}.

$$\Delta S^\circ_{rxn} = [(1\ \text{mol Fe}_2\text{O}_3)(87.400\ \text{J/mol·K})] - [(2\ \text{mol FeO})(60.75\ \text{J/mol·K}) + (\tfrac{1}{2}\ \text{mol O}_2)(205.0\ \text{J/mol·K})] = -136.6\ \text{J/K}$$

$$\Delta H^\circ_{rxn} = [(1\ \text{mol Fe}_2\text{O}_3)(-825.5\ \text{kJ/mol})] - [(2\ \text{mol FeO})(-272.0\ \text{kJ/mol}) + (\tfrac{1}{2}\ \text{mol O}_2)(0\ \text{kJ/mol})] = -281.5\ \text{kJ}$$

$$\Delta S_{surr} = -\frac{\Delta H^\circ_{sys}}{T} = -\frac{(-281.5\ \text{kJ})(1000\ \text{J/1 kJ})}{298\ \text{K}} = +945\ \text{J/K}$$

$$\Delta S_{univ} = \Delta S^\circ_{rxn} + \Delta S_{surr} = -136.6\ \text{J/K} + 945\ \text{J/K} = 808\ \text{J/K}$$

Reaction is spontaneous at 298 K since ΔS_{univ} is positive.

20.7A Using ΔH°_f and S° values from Appendix B:

$$\Delta H^\circ_{rxn} = [(2\ \text{mol NOCl})(51.71\ \text{kJ/mol})] - [(2\ \text{mol NO})(90.29\ \text{kJ/mol}) + (1\ \text{mol Cl}_2)(0\ \text{kJ/mol})] = -77.16\ \text{kJ}$$

$$\Delta S^\circ_{rxn} = [(2\ \text{mol NOCl})(261.6\ \text{J/mol·K})] - [(2\ \text{mol NO})(210.65\ \text{J/mol·K}) + (1\ \text{mol Cl}_2)(223.0\ \text{J/mol·K})] = -121.1\ \text{J/K}$$

$$\Delta G^\circ_{rxn} = \Delta H^\circ_{rxn} - T\Delta S^\circ_{rxn} = -77.16\ \text{kJ} - [(298\ \text{K})(-121.1\ \text{J/K})(1\ \text{kJ/1000 J})] = -41.1\ \text{kJ}$$

20.7B Using ΔH°_f and S° values from Appendix B:

$$\Delta H^\circ_{rxn} = [(4\ \text{mol NO})(90.29\ \text{kJ/mol}) + (6\ \text{mol H}_2\text{O})(-241.826\ \text{kJ/mol})] - [(4\ \text{mol NH}_3)(-45.9\ \text{kJ/mol}) + (5\ \text{mol O}_2)(0\ \text{kJ/mol})] = -906.2\ \text{kJ}$$

$$\Delta S^\circ_{rxn} = [(4\ \text{mol NO})(210.65\ \text{J/mol·K}) + (6\ \text{mol H}_2\text{O})(188.72\ \text{J/mol·K}] - [(4\ \text{mol NH}_3)(193\ \text{J/mol·K}) + (5\ \text{mol O}_2)(205.0\ \text{J/mol·K})] = 178\ \text{J/K}$$

$$\Delta G^\circ_{rxn} = \Delta H^\circ_{rxn} - T\Delta S^\circ_{rxn} = -906.2\ \text{kJ} - [(298\ \text{K})(178\ \text{J/K})(1\ \text{kJ/1000 J})] = -959\ \text{kJ}$$

20.8A (a) Using ΔG°_f values from Appendix B:

$$\Delta G^\circ_{rxn} = [(2\ \text{mol NOCl})(66.07\ \text{kJ/mol})] - [(2\ \text{mol NO})(86.60\ \text{kJ/mol}) + (1\ \text{mol Cl}_2)(0\ \text{kJ/mol})] = -41.06\ \text{kJ}$$

(b) Using ΔG°_f values from Appendix B:

$$\Delta G^\circ_{rxn} = [(2\ \text{mol Fe})(0\ \text{kJ/mol}) + (3\ \text{mol H}_2\text{O})(-228.60\ \text{kJ/mol})] - [(3\ \text{mol H}_2)(0\ \text{kJ/mol}) + (1\ \text{mol Fe}_2\text{O}_3)(-743.6\ \text{kJ/mol})] = 57.8\ \text{kJ}$$

20.8B (a) Using ΔG°_f values from Appendix B:

$$\Delta G^\circ_{rxn} = [(4\ \text{mol NO})(86.60\ \text{kJ/mol}) + (6\ \text{mol H}_2\text{O})(-228.60\ \text{kJ/mol})] - [(4\ \text{mol NH}_3)(-16\ \text{kJ/mol}) + (5\ \text{mol O}_2)(0\ \text{kJ/mol})] = -961\ \text{kJ}$$

(b) Using ΔG°_f values from Appendix B:

$$\Delta G^\circ_{rxn} = (2\ \text{mol CO})(-137.2\ \text{kJ/mol}) - [(2\ \text{mol C})(0\ \text{kJ/mol}) + (1\ \text{mol O}_2)(0\ \text{kJ/mol})] = -274.4\ \text{kJ}$$

20.9A (a) More moles of gas are present after the reaction, so $\Delta S > 0$. (b) The problem says the reaction is spontaneous ($\Delta G < 0$) only above 325°C, which implies high T. If $\Delta S > 0$, $-T\Delta S < 0$, so ΔG will become negative at higher T only if $\Delta H > 0$.

20.9B (a) A solid forms a gas and a liquid, so $\Delta S > 0$, and a crystalline array breaks down, so $\Delta H > 0$. (b) For the reaction to occur spontaneously ($\Delta G < 0$), $-T\Delta S$ must be greater than ΔH, which would occur only at higher T.

20.10A (a) $\Delta G = \Delta H - T\Delta S$

$$= -192.7\ \text{kJ} - [(298\ \text{K})(-308.2\ \text{J/K})(1\ \text{kJ/1000 J})] = -100.9\ \text{kJ}$$

The reaction is spontaneous at 298 K.

(b) As T increases, $-T\Delta S$ becomes more positive, so the reaction becomes *less* spontaneous.

(c) $\Delta G = \Delta H - T\Delta S$

$$= -192.7\ \text{kJ} - [(500. + 273\ \text{K})(-308.2\ \text{J/K})(1\ \text{kJ/1000 J})] = 45.5\ \text{kJ}$$

20.10B ΔG becomes negative at lower T, so $\Delta H < 0$, $\Delta S < 0$, and $-T\Delta S > 0$. At the lower T, the negative ΔH value becomes larger than the positive $-T\Delta S$ value.

20.11A $\Delta G = 0$ when $T = \dfrac{\Delta H}{\Delta S} = \dfrac{(-192.7\ \text{kJ})(1000\ \text{J/1 kJ})}{-308.2\ \text{J/K}} = 625.2\ \text{K}$

$T\,(°\text{C}) = 625.2\ \text{K} - 273.15 = 352.0°\text{C}$

20.11B

$$\Delta H = [(1\ \text{mol})(\Delta H^\circ_f\ \text{CaCO}_3)] - [(1\ \text{mol})(\Delta H^\circ_f\ \text{CaO}) + (1\ \text{mol})(\Delta H^\circ_f\ \text{CO}_2)]$$
$$= [(1\ \text{mol})(-1206.9\ \text{kJ/mol})] - [(1\ \text{mol})(-635.1\ \text{kJ/mol}) + (1\ \text{mol})(-393.5\ \text{kJ/mol})] = -178.3\ \text{kJ}$$

$$\Delta S = [(1\ \text{mol})(S^\circ\ \text{CaCO}_3)] - [(1\ \text{mol})(S^\circ\ \text{CaO}) + (1\ \text{mol})(S^\circ\ \text{CO}_2)]$$
$$= [(1\ \text{mol})(92.9\ \text{J/mol·K})] - [(1\ \text{mol})(38.2\ \text{J/mol·K}) + (1\ \text{mol})(213.7\ \text{J/mol·K})] = -159.0\ \text{J/K}$$

$$T = \frac{\Delta H}{\Delta S} = \frac{(-178.3\ \text{kJ})(1000\ \text{J/1 kJ})}{-159.0\ \text{J/K}} = 1121\ \text{K}$$

Reaction becomes spontaneous ($\Delta G < 0$) at any $T < 1121\ \text{K} - 273.15 = 848°\text{C}$.

20.12A Using ΔG°_f values from Appendix B,

$$\Delta G^\circ = [(2\ \text{mol CO})(-137.2\ \text{kJ/mol})] - [(2\ \text{mol C})(0\ \text{kJ/mol}) + (1\ \text{mol O}_2)(0\ \text{kJ/mol})] = -274.4\ \text{kJ}$$

$$\ln K = -\frac{\Delta G^\circ}{RT} = -\frac{(-274.4\ \text{kJ})(1000\ \text{J/1 kJ})}{(8.314\ \text{J/mol·K})(298\ \text{K})} = 111$$

$$K = e^{111} = 1.61\times10^{48}$$

BRIEF SOLUTIONS TO FOLLOW-UP PROBLEMS (continued)

20.12B $K = 2.22\times10^{-15}$, so $\ln K = -33.7$

$\Delta G^\circ = -RT \ln K = -(8.314\ \text{J/mol·K})(298\ \text{K})(-33.7)$
$= 8.35\times10^4\ \text{J} = 83.5\ \text{kJ}$

20.13A (a) $A(g) + 3B(g) \rightleftharpoons AB_3(g)$; $\Delta G^\circ = -4.6$ kJ/mol

(b) For mixture 1, $Q_1 = \dfrac{[AB_3]}{[A][B]^3} = \dfrac{0.20}{(0.40)(0.80)^3} = 0.98$

For mixture 2, $Q_2 = \dfrac{(0.30)}{(0.30)(0.50)^3} = 8.0$

For mixture 3, $Q_3 = \dfrac{0.40}{(0.20)(0.20)^3} = 250$

Mixture 2 is at equilibrium.

(c) Since $Q_1 < K$, the reaction is proceeding to the right (it is spontaneous) and so $\Delta G_1 < 0$; $\Delta G_2 = 0$ since this mixture is at equilibrium; $Q_3 > K$, so the reaction is proceeding to the left, not to the right, and $\Delta G_3 > 0$. Therefore, $\Delta G_3 > \Delta G_2 > \Delta G_1$.

20.13B

(a) For mixture 1, $Q_1 = \dfrac{[XY_2]^2}{[X_2][Y_2]^2} = \dfrac{(5.0)^2}{(2.0)(1.0)^2} = 12.5$

For mixture 2, $Q_2 = \dfrac{(4.0)^2}{(2.0)(2.0)^2} = 2.0$

For mixture 3, $Q_3 = \dfrac{(2.0)^2}{(4.0)(2.0)^2} = 0.25$

Mixture 2 is at equilibrium.

(b) Since $Q_1 > K$, the reaction is proceeding to the left, not to the right, and $\Delta G_1 > 0$; $\Delta G_2 = 0$ since this mixture is at equilibrium; $Q_3 < K$, so the reaction is proceeding to the right (it is spontaneous), so $\Delta G_3 < 0$. Therefore, $\Delta G_3 < \Delta G_2 < \Delta G_1$. (c) Any reaction mixture moves spontaneously toward equilibrium, so both changes have a negative ΔG.

20.14A (a) $\Delta G^\circ = -RT \ln K$, so

$$\ln K = -\frac{\Delta G^\circ}{RT} = -\frac{(-33.5\ \text{kJ})(1000\ \text{J/1 kJ})}{(8.314\ \text{J/mol·K})(298\ \text{K})} = 13.521$$

$K = e^{13.521} = 7.45\times10^5$

(b) $Q = \dfrac{[C_2H_5Cl]}{[C_2H_4][HCl]} = \dfrac{1.5}{(0.50)(1.0)} = 3.0$

$\Delta G = \Delta G^\circ + RT \ln Q$
$= -33.5\ \text{kJ} + [(8.314\ \text{J/mol·K})(1\ \text{kJ/1000 J})(298\ \text{K})(\ln 3.0)]$
$= -30.8$ kJ

20.14B (a) $\Delta G^\circ = -RT \ln K$

$= -8.314\ \text{J/mol·K} \times \dfrac{1\ \text{kJ}}{1000\ \text{J}} \times 298\ \text{K}$
$\times \ln (2.3\times10^{-9})$
$= 49$ kJ/mol

(b) $Q = \dfrac{[H_3O^+][BrO^-]}{[HBrO]} = \dfrac{(6.0\times10^{-4})(0.10)}{0.20} = 3.0\times10^{-4}$

$\Delta G = \Delta G^\circ + RT \ln Q$
$= 49\ \text{kJ/mol} + [(8.314\ \text{J/mol·K})(1\ \text{kJ/1000 J})(298\ \text{K})(\ln 3.0\times10^{-4})]$
$= 29$ kJ/mol

PROBLEMS

Problems with **colored** numbers are answered in Appendix E and worked in detail in the Student Solutions Manual. Problem sections match those in the text and give the numbers of relevant sample problems. Most offer Concept Review Questions, Skill-Building Exercises (grouped in pairs covering the same concept), and Problems in Context. The Comprehensive Problems are based on material from any section or previous chapter.

Note: Unless stated otherwise, problems refer to systems at 298 K (25°C). Solving these problems may require values from Appendix B.

The Second Law of Thermodynamics: Predicting Spontaneous Change

(Sample Problem 20.4)

Concept Review Questions

20.1 Distinguish between the terms *spontaneous* and *instantaneous*. Give an example of a process that is spontaneous but very slow, and one that is very fast but not spontaneous.

20.2 Distinguish between the terms *spontaneous* and *nonspontaneous*. Can a nonspontaneous process occur? Explain.

20.3 State the first law of thermodynamics in terms of (a) the energy of the universe; (b) the creation or destruction of energy; (c) the energy change of system and surroundings. Does the first law reveal the direction of spontaneous change? Explain.

20.4 State qualitatively the relationship between entropy and freedom of particle motion. Use this idea to explain why you will probably never (a) be suffocated because all the air near you has moved to the other side of the room; (b) see half the water in your cup of tea freeze while the other half boils.

20.5 Why is ΔS_{vap} of a substance always larger than ΔS_{fus}?

20.6 How does the entropy of the surroundings change during an exothermic reaction? An endothermic reaction? Other than the examples in text, describe a spontaneous endothermic process.

20.7 (a) What is the entropy of a perfect crystal at 0 K?
(b) Does entropy increase or decrease as the temperature rises?
(c) Why is $\Delta H^\circ_f = 0$ but $S^\circ > 0$ for an element?
(d) Why does Appendix B list ΔH°_f values but not ΔS°_f values?

Skill-Building Exercises (grouped in similar pairs)

20.8 Which of these processes are spontaneous? (a) Water evaporates from a puddle. (b) A lion chases an antelope. (c) An isotope undergoes radioactive disintegration.

20.9 Which of these processes are spontaneous? (a) Earth moves around the Sun. (b) A boulder rolls up a hill. (c) Sodium metal and chlorine gas form solid sodium chloride.

20.10 Which of these processes are spontaneous? (a) Methane burns in air. (b) A teaspoonful of sugar dissolves in a cup of hot coffee. (c) A soft-boiled egg becomes raw.

20.11 Which of these processes are spontaneous? (a) A satellite falls to Earth. (b) Water decomposes to H_2 and O_2 at 298 K and 1 atm. (c) Average car prices increase.

20.12 Predict the sign of ΔS_{sys} for each process: (a) A piece of wax melts. (b) Silver chloride precipitates from solution. (c) Dew forms on a lawn in the morning.

20.13 Predict the sign of ΔS_{sys} for each process: (a) Gasoline vapors mix with air in a car engine. (b) Hot air expands. (c) Humidity condenses in cold air.

20.14 Predict the sign of ΔS_{sys} for each process: (a) Alcohol evaporates. (b) A solid explosive converts to a gas. (c) Perfume vapors diffuse through a room.

20.15 Predict the sign of ΔS_{sys} for each process: (a) A pond freezes in winter. (b) Atmospheric CO_2 dissolves in the ocean. (c) An apple tree bears fruit.

20.16 Without using Appendix B, predict the sign of $\Delta S°$ for
(a) $2K(s) + F_2(g) \longrightarrow 2KF(s)$
(b) $NH_3(g) + HBr(g) \longrightarrow NH_4Br(s)$
(c) $NaClO_3(s) \longrightarrow Na^+(aq) + ClO_3^-(aq)$

20.17 Without using Appendix B, predict the sign of $\Delta S°$ for
(a) $H_2S(g) + \frac{1}{2}O_2(g) \longrightarrow \frac{1}{8}S_8(s) + H_2O(g)$
(b) $HCl(aq) + NaOH(aq) \longrightarrow NaCl(aq) + H_2O(l)$
(c) $2NO_2(g) \longrightarrow N_2O_4(g)$

20.18 Without using Appendix B, predict the sign of $\Delta S°$ for
(a) $CaCO_3(s) + 2HCl(aq) \longrightarrow CaCl_2(aq) + H_2O(l) + CO_2(g)$
(b) $2NO(g) + O_2(g) \longrightarrow 2NO_2(g)$
(c) $2KClO_3(s) \longrightarrow 2KCl(s) + 3O_2(g)$

20.19 Without using Appendix B, predict the sign of $\Delta S°$ for
(a) $Ag^+(aq) + Cl^-(aq) \longrightarrow AgCl(s)$
(b) $KBr(s) \longrightarrow KBr(aq)$
(c) $CH_3CH{=}CH_2(g) \longrightarrow$ $H_2C{-}CH_2(g)$ (cyclic, with CH_2 bridging, as drawn)

20.20 Predict the sign of ΔS for each process:
(a) $C_2H_5OH(g)$ (350 K and 500 torr) $\longrightarrow$ $C_2H_5OH(g)$ (350 K and 250 torr)
(b) $N_2(g)$ (298 K and 1 atm) $\longrightarrow$ $N_2(aq)$ (298 K and 1 atm)
(c) $O_2(aq)$ (303 K and 1 atm) $\longrightarrow$ $O_2(g)$ (303 K and 1 atm)

20.21 Predict the sign of ΔS for each process:
(a) $O_2(g)$ (1.0 L at 1 atm) $\longrightarrow$ $O_2(g)$ (0.10 L at 10 atm)
(b) $Cu(s)$ (350°C and 2.5 atm) $\longrightarrow$ $Cu(s)$ (450°C and 2.5 atm)
(c) $Cl_2(g)$ (100°C and 1 atm) $\longrightarrow$ $Cl_2(g)$ (10°C and 1 atm)

20.22 Predict which substance has greater molar entropy. Explain.
(a) Butane $CH_3CH_2CH_2CH_3(g)$ or 2-butene $CH_3CH{=}CHCH_3(g)$
(b) $Ne(g)$ or $Xe(g)$
(c) $CH_4(g)$ or $CCl_4(l)$

20.23 Predict which substance has greater molar entropy. Explain.
(a) $NO_2(g)$ or $N_2O_4(g)$
(b) $CH_3OCH_3(l)$ or $CH_3CH_2OH(l)$
(c) $HCl(g)$ or $HBr(g)$

20.24 Predict which substance has greater molar entropy. Explain.
(a) $CH_3OH(l)$ or $C_2H_5OH(l)$
(b) $KClO_3(s)$ or $KClO_3(aq)$
(c) $Na(s)$ or $K(s)$

20.25 Predict which substance has greater molar entropy. Explain.
(a) $P_4(g)$ or $P_2(g)$
(b) $HNO_3(aq)$ or $HNO_3(l)$
(c) $CuSO_4(s)$ or $CuSO_4{\cdot}5H_2O(s)$

20.26 Without consulting Appendix B, arrange each group in order of *increasing* standard molar entropy ($S°$). Explain.
(a) Graphite, diamond, charcoal
(b) Ice, water vapor, liquid water
(c) O_2, O_3, O atoms

20.27 Without consulting Appendix B, arrange each group in order of *increasing* standard molar entropy ($S°$). Explain.
(a) Glucose ($C_6H_{12}O_6$), sucrose ($C_{12}H_{22}O_{11}$), ribose ($C_5H_{10}O_5$)
(b) $CaCO_3$, $Ca + C + \frac{3}{2}O_2$, $CaO + CO_2$
(c) $SF_6(g)$, $SF_4(g)$, $S_2F_{10}(g)$

20.28 Without consulting Appendix B, arrange each group in order of *decreasing* standard molar entropy ($S°$). Explain.
(a) $ClO_4^-(aq)$, $ClO_2^-(aq)$, $ClO_3^-(aq)$
(b) $NO_2(g)$, $NO(g)$, $N_2(g)$
(c) $Fe_2O_3(s)$, $Al_2O_3(s)$, $Fe_3O_4(s)$

20.29 Without consulting Appendix B, arrange each group in order of *decreasing* standard molar entropy ($S°$). Explain.
(a) Mg metal, Ca metal, Ba metal
(b) Hexane (C_6H_{14}), benzene (C_6H_6), cyclohexane (C_6H_{12})
(c) $PF_2Cl_3(g)$, $PF_5(g)$, $PF_3(g)$

Calculating the Change in Entropy of a Reaction

(Sample Problems 20.5 and 20.6)

Concept Review Questions

20.30 In the reaction depicted in the molecular scenes, X is *red* and Y is *green*.

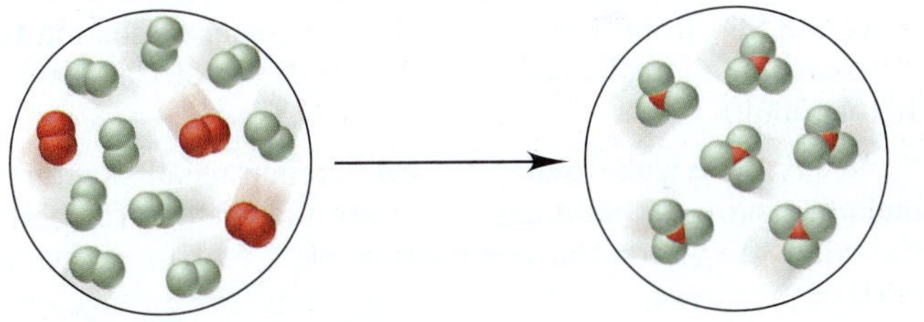

(a) Write a balanced equation.
(b) Determine the sign of ΔS_{rxn}.
(c) Which species has the highest molar entropy?

20.31 Describe the equilibrium condition in terms of the entropy changes of a system and its surroundings. What does this description say about the entropy change of the universe?

20.32 For the reaction $H_2O(g) + Cl_2O(g) \longrightarrow 2HClO(g)$, you know $\Delta S°_{rxn}$ and $S°$ of $HClO(g)$ and of $H_2O(g)$. Write an expression that can be used to determine $S°$ of $Cl_2O(g)$.

Skill-Building Exercises (grouped in similar pairs)

20.33 For each reaction, predict the sign and find the value of $\Delta S°_{rxn}$:
(a) $3NO(g) \longrightarrow N_2O(g) + NO_2(g)$
(b) $3H_2(g) + Fe_2O_3(s) \longrightarrow 2Fe(s) + 3H_2O(g)$
(c) $P_4(s) + 5O_2(g) \longrightarrow P_4O_{10}(s)$

20.34 For each reaction, predict the sign and find the value of ΔS°_{rxn}:
(a) $3NO_2(g) + H_2O(l) \longrightarrow 2HNO_3(l) + NO(g)$
(b) $N_2(g) + 3F_2(g) \longrightarrow 2NF_3(g)$
(c) $C_6H_{12}O_6(s) + 6O_2(g) \longrightarrow 6CO_2(g) + 6H_2O(g)$

20.35 Find ΔS°_{rxn} for the combustion of ethane (C_2H_6) to carbon dioxide and gaseous water. Is the sign of ΔS°_{rxn} as expected?

20.36 Find ΔS°_{rxn} for the combustion of methane to carbon dioxide and liquid water. Is the sign of ΔS°_{rxn} as expected?

20.37 Find ΔS°_{rxn} for the reaction of nitrogen monoxide with hydrogen to form ammonia and water vapor. Is the sign of ΔS°_{rxn} as expected?

20.38 Find ΔS°_{rxn} for the combustion of ammonia to nitrogen dioxide and water vapor. Is the sign of ΔS°_{rxn} as expected?

20.39 (a) Find ΔS°_{rxn} for the formation of $Cu_2O(s)$ from its elements.
(b) Calculate ΔS_{univ}, and state whether the reaction is spontaneous at 298 K.

20.40 (a) Find ΔS°_{rxn} for the formation of $HI(g)$ from its elements.
(b) Calculate ΔS_{univ}, and state whether the reaction is spontaneous at 298 K.

20.41 (a) Find ΔS°_{rxn} for the formation of $CH_3OH(l)$ from its elements.
(b) Calculate ΔS_{univ}, and state whether the reaction is spontaneous at 298 K.

20.42 (a) Find ΔS°_{rxn} for the formation of $N_2O(g)$ from its elements.
(b) Calculate ΔS_{univ}, and state whether the reaction is spontaneous at 298 K.

Problems in Context

20.43 Sulfur dioxide is released in the combustion of coal. Scrubbers use aqueous slurries of calcium hydroxide to remove the SO_2 from flue gases. Write a balanced equation for this reaction and calculate ΔS°_{rxn} at 298 K [S° of $CaSO_3(s)$ = 101.4 J/mol·K].

20.44 Oxyacetylene welding is used to repair metal structures, including bridges, buildings, and even the Statue of Liberty. Calculate ΔS°_{rxn} for the combustion of 1 mol of acetylene (C_2H_2).

Entropy, Free Energy, and Work

(Sample Problems 20.7 to 20.11)

Concept Review Questions

20.45 What is the advantage of calculating free energy changes rather than entropy changes to determine reaction spontaneity?

20.46 Given that $\Delta G_{sys} = -T\Delta S_{univ}$, explain how the sign of ΔG_{sys} correlates with reaction spontaneity.

20.47 (a) Is an endothermic reaction more likely to be spontaneous at higher temperatures or lower temperatures? Explain.
(b) The change depicted below occurs at constant pressure. Explain your answers to each of the following: (1) What is the sign of ΔH_{sys}? (2) What is the sign of ΔS_{sys}? (3) What is the sign of ΔS_{surr}? (4) How does the sign of ΔG_{sys} vary with temperature?

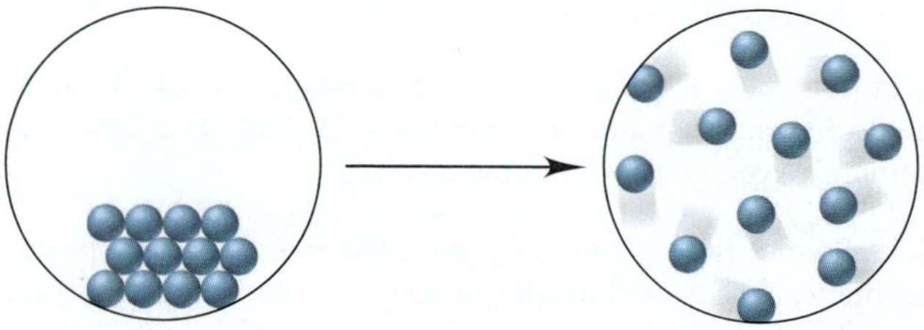

20.48 Explain your answers to each of the following for the change depicted below. (a) What is the sign of ΔH_{sys}? (b) What is the sign of ΔS_{sys}? (c) What is the sign of ΔS_{surr}? (d) How does the sign of ΔG_{sys} vary with temperature?

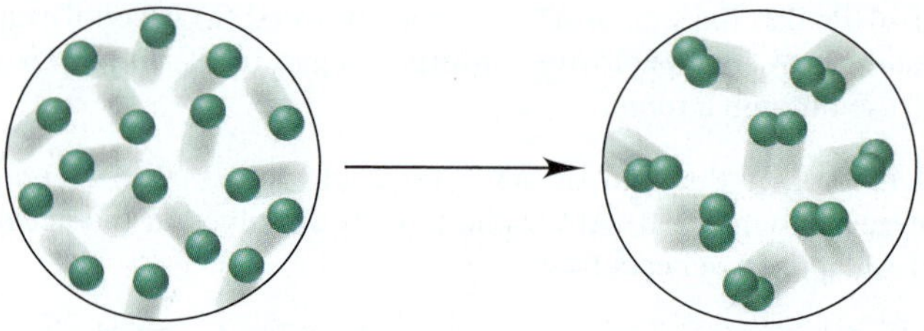

20.49 With its components in their standard states, a certain reaction is spontaneous only at high T. What do you know about the signs of ΔH° and ΔS°? Describe a process for which this is true.

20.50 How can ΔS° for a reaction be relatively independent of T if S° of each reactant and product increases with T?

Skill-Building Exercises (grouped in similar pairs)

20.51 Calculate ΔG° for each reaction using ΔG°_f values:
(a) $2Mg(s) + O_2(g) \longrightarrow 2MgO(s)$
(b) $2CH_3OH(g) + 3O_2(g) \longrightarrow 2CO_2(g) + 4H_2O(g)$
(c) $BaO(s) + CO_2(g) \longrightarrow BaCO_3(s)$

20.52 Calculate ΔG° for each reaction using ΔG°_f values:
(a) $H_2(g) + I_2(s) \longrightarrow 2HI(g)$
(b) $MnO_2(s) + 2CO(g) \longrightarrow Mn(s) + 2CO_2(g)$
(c) $NH_4Cl(s) \longrightarrow NH_3(g) + HCl(g)$

20.53 Find ΔG° for the reactions in Problem 20.51 using ΔH°_f and S° values.

20.54 Find ΔG° for the reactions in Problem 20.52 using ΔH°_f and S° values.

20.55 Consider the oxidation of carbon monoxide:

$$CO(g) + \tfrac{1}{2}O_2(g) \longrightarrow CO_2(g)$$

(a) Predict the signs of ΔS° and ΔH°. Explain.
(b) Calculate ΔG° by two different methods.

20.56 Consider the combustion of butane gas:

$$C_4H_{10}(g) + \tfrac{13}{2}O_2(g) \longrightarrow 4CO_2(g) + 5H_2O(g)$$

(a) Predict the signs of ΔS° and ΔH°. Explain.
(b) Calculate ΔG° by two different methods.

20.57 For the gaseous reaction of xenon and fluorine to form xenon hexafluoride:
(a) Calculate ΔS° at 298 K (ΔH° = −402 kJ/mol and ΔG° = −280. kJ/mol).
(b) Assuming that ΔS° and ΔH° change little with temperature, calculate ΔG° at 500. K.

20.58 For the gaseous reaction of carbon monoxide and chlorine to form phosgene ($COCl_2$):
(a) Calculate $\Delta S°$ at 298 K ($\Delta H° = -220.$ kJ/mol and $\Delta G° = -206$ kJ/mol).
(b) Assuming that $\Delta S°$ and $\Delta H°$ change little with temperature, calculate $\Delta G°$ at 450. K.

20.59 One reaction used to produce small quantities of pure H_2 is

$$CH_3OH(g) \rightleftharpoons CO(g) + 2H_2(g)$$

(a) Determine $\Delta H°$ and $\Delta S°$ for the reaction at 298 K.
(b) Assuming that these values are relatively independent of temperature, calculate $\Delta G°$ at 28°C, 128°C, and 228°C.
(c) What is the significance of the different values of $\Delta G°$?
(d) At what temperature (in K) does the reaction become spontaneous?

20.60 A reaction that occurs in the internal combustion engine is

$$N_2(g) + O_2(g) \rightleftharpoons 2NO(g)$$

(a) Determine $\Delta H°$ and $\Delta S°$ for the reaction at 298 K.
(b) Assuming that these values are relatively independent of temperature, calculate $\Delta G°$ at 100.°C, 2560.°C, and 3540.°C.
(c) What is the significance of the different values of $\Delta G°$?
(d) At what temperature (in K) does the reaction become spontaneous?

Problems in Context

20.61 As a fuel, $H_2(g)$ produces only nonpolluting $H_2O(g)$ when it burns. Moreover, it combines with $O_2(g)$ in a fuel cell (Chapter 21) to provide electrical energy.
(a) Calculate $\Delta H°$, $\Delta S°$, and $\Delta G°$ per mole of H_2 at 298 K.
(b) Is the spontaneity of this reaction dependent on T? Explain.
(c) At what temperature does the reaction become spontaneous?

20.62 The U.S. government requires automobile fuels to contain a renewable component. Fermentation of glucose from corn yields ethanol, which is added to gasoline to fulfill this requirement:

$$C_6H_{12}O_6(s) \longrightarrow 2C_2H_5OH(l) + 2CO_2(g)$$

Calculate $\Delta H°$, $\Delta S°$, and $\Delta G°$ for the reaction at 25°C. Is the spontaneity of this reaction dependent on T? Explain.

Free Energy, Equilibrium, and Reaction Direction

(Sample Problems 20.12 to 20.14)

Concept Review Questions

20.63 (a) If $K << 1$ for a reaction, what do you know about the sign and magnitude of $\Delta G°$? (b) If $\Delta G° << 0$ for a reaction, what do you know about the magnitude of K? Of Q?

20.64 How is the free energy change of a process related to the work that can be obtained from the process? Is this quantity of work obtainable in practice? Explain.

20.65 The scenes and the graph relate to the reaction of $X_2(g)$ *(black)* with $Y_2(g)$ *(orange)* to form $XY(g)$. (a) If reactants and products are in their standard states, what quantity is represented on the graph by x? (b) Which scene represents point 1? Explain. (c) Which scene represents point 2? Explain.

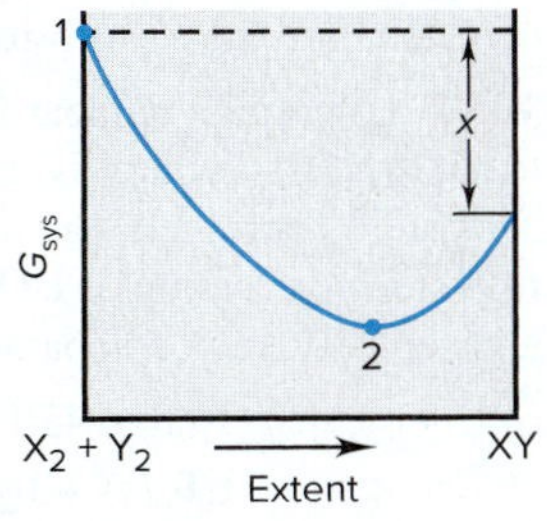

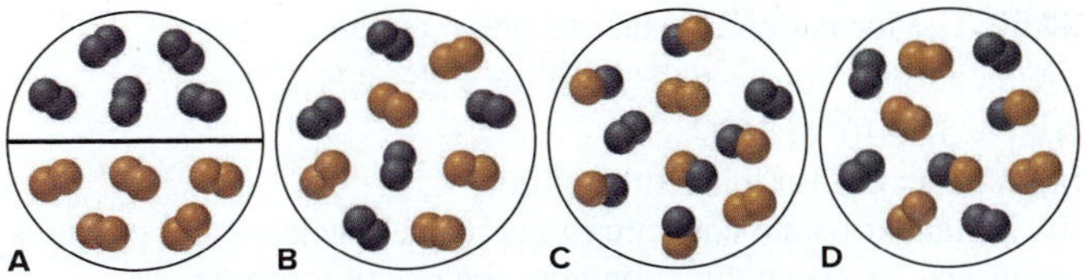

20.66 What is the difference between $\Delta G°$ and ΔG? Under what circumstances does $\Delta G = \Delta G°$?

Skill-Building Exercises (grouped in similar pairs)

20.67 Calculate K at 298 K for each reaction:
(a) $MgCO_3(s) \rightleftharpoons Mg^{2+}(aq) + CO_3^{2-}(aq)$
(b) $H_2(g) + O_2(g) \rightleftharpoons H_2O_2(l)$

20.68 Calculate $\Delta G°$ at 298 K for each reaction:
(a) $2H_2S(g) + 3O_2(g) \rightleftharpoons 2H_2O(g) + 2SO_2(g)$; $K = 6.57\times10^{173}$
(b) $H_2SO_4(l) \rightleftharpoons H_2O(l) + SO_3(g)$; $K = 4.46\times10^{-15}$

20.69 Calculate K at 298 K for each reaction:
(a) $HCN(aq) + NaOH(aq) \rightleftharpoons NaCN(aq) + H_2O(l)$
(b) $SrSO_4(s) \rightleftharpoons Sr^{2+}(aq) + SO_4^{2-}(aq)$

20.70 Calculate $\Delta G°$ at 298 K for each reaction:
(a) $2NO(g) + Cl_2(g) \rightleftharpoons 2NOCl(g)$; $K = 1.58\times10^{7}$
(b) $Cu_2S(s) + O_2(g) \rightleftharpoons 2Cu(s) + SO_2(g)$; $K = 3.25\times10^{37}$

20.71 Use $\Delta H°$ and $\Delta S°$ values for the following process at 1 atm to find the normal boiling point of Br_2:

$$Br_2(l) \rightleftharpoons Br_2(g)$$

20.72 Use $\Delta H°$ and $\Delta S°$ values to find the temperature at which these sulfur allotropes reach equilibrium at 1 atm:

$$S(\text{rhombic}) \rightleftharpoons S(\text{monoclinic})$$

20.73 Use Appendix B to determine the K_{sp} of Ag_2S.

20.74 Use Appendix B to determine the K_{sp} of CaF_2.

20.75 For the reaction $I_2(g) + Cl_2(g) \rightleftharpoons 2ICl(g)$, calculate K_p at 25°C [$\Delta G_f°$ of $ICl(g) = -6.075$ kJ/mol].

20.76 For the reaction $CaCO_3(s) \rightleftharpoons CaO(s) + CO_2(g)$, calculate the equilibrium P_{CO_2} at 25°C.

20.77 The K_{sp} of $PbCl_2$ is 1.7×10^{-5} at 25°C. What is $\Delta G°$? Is it possible to prepare a solution that contains $Pb^{2+}(aq)$ and $Cl^-(aq)$, at their standard-state concentrations?

20.78 The K_{sp} of ZnF_2 is 3.0×10^{-2} at 25°C. What is $\Delta G°$? Is it possible to prepare a solution that contains $Zn^{2+}(aq)$ and $F^-(aq)$ at their standard-state concentrations?

20.79 The equilibrium constant for the reaction

$$2Fe^{3+}(aq) + Hg_2^{2+}(aq) \rightleftharpoons 2Fe^{2+}(aq) + 2Hg^{2+}(aq)$$

is $K_c = 9.1\times10^{-6}$ at 298 K.
(a) What is $\Delta G°$ at this temperature?
(b) If standard-state concentrations of the reactants and products are mixed, in which direction does the reaction proceed?
(c) Calculate ΔG when $[Fe^{3+}] = 0.20\ M$, $[Hg_2^{2+}] = 0.010\ M$, $[Fe^{2+}] = 0.010\ M$, and $[Hg^{2+}] = 0.025\ M$. In which direction will the reaction proceed to achieve equilibrium?

20.80 The formation constant for the reaction

$$Ni^{2+}(aq) + 6NH_3(aq) \rightleftharpoons Ni(NH_3)_6^{2+}(aq)$$

is $K_f = 5.6\times10^8$ at 25°C.
(a) What is $\Delta G°$ at this temperature?
(b) If standard-state concentrations of the reactants and products are mixed, in which direction does the reaction proceed?
(c) Determine ΔG when $[Ni(NH_3)_6^{2+}] = 0.010\ M$, $[Ni^{2+}] = 0.0010\ M$, and $[NH_3] = 0.0050\ M$. In which direction will the reaction proceed to achieve equilibrium?

20.81 The scenes below depict three gaseous mixtures in which A is reacting with itself to form A_2. Assume that each particle represents 0.10 mol and the volume is 0.10 L.

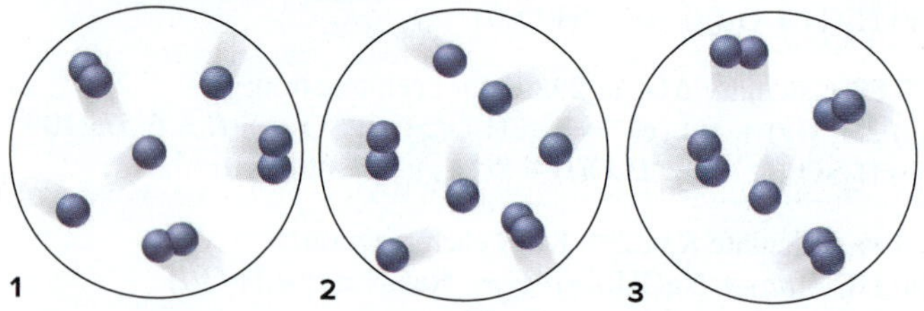

(a) If $K = 0.33$, which mixture is at equilibrium? (b) Rank the mixtures from the most positive ΔG to the most negative ΔG.

20.82 The scenes below depict three gaseous mixtures in which X *(orange)* and Y_2 *(black)* are reacting to form XY and Y. Assume that each gas has a partial pressure of 0.10 atm.

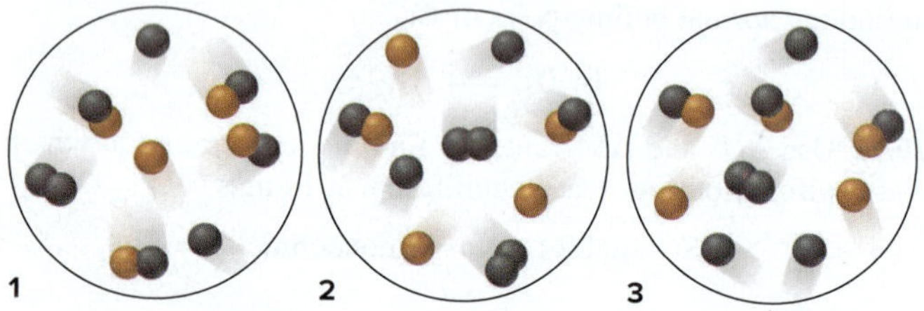

(a) If $K = 4.5$, which mixture is at equilibrium? (b) Rank the mixtures from the most positive ΔG to the most negative ΔG.

Problems in Context

20.83 High levels of ozone (O_3) cause rubber to deteriorate, green plants to turn brown, and many people to have difficulty breathing.
(a) Is the formation of O_3 from O_2 favored at all T, no T, high T, or low T?
(b) Calculate $\Delta G°$ for this reaction at 298 K.
(c) Calculate ΔG at 298 K for this reaction in urban smog where $[O_2] = 0.21$ atm and $[O_3] = 5\times10^{-7}$ atm.

20.84 A $BaSO_4$ slurry is ingested before the gastrointestinal tract is x-rayed because it is opaque to x-rays and defines the contours of the tract. Ba^{2+} ion is toxic, but the compound is nearly insoluble. If $\Delta G°$ at 37°C (body temperature) is 59.1 kJ/mol for the dissolution,

$$BaSO_4(s) \rightleftharpoons Ba^{2+}(aq) + SO_4^{2-}(aq)$$

what is $[Ba^{2+}]$ in the intestinal tract? (Assume that the only source of SO_4^{2-} is the ingested slurry.)

Comprehensive Problems

20.85 According to advertisements, "a diamond is forever."
(a) Calculate $\Delta H°$, $\Delta S°$, and $\Delta G°$ at 298 K for the phase change

$$\text{Diamond} \longrightarrow \text{graphite}$$

(b) Given the conditions under which diamond jewelry is normally kept, argue for and against the statement in the ad.
(c) Given the answers in part (a), what would need to be done to make synthetic diamonds from graphite?
(d) Assuming $\Delta H°$ and $\Delta S°$ do not change with temperature, can graphite be converted to diamond spontaneously at 1 atm?

20.86 Replace each question mark with the correct information:

	ΔS_{rxn}	ΔH_{rxn}	ΔG_{rxn}	Comment
(a)	+	−	−	?
(b)	?	0	−	Spontaneous
(c)	−	+	?	Not spontaneous
(d)	0	?	−	Spontaneous
(e)	?	0	+	?
(f)	+	+	?	$T\Delta S > \Delta H$

20.87 Among the many complex ions of cobalt are the following:

$$Co(NH_3)_6^{3+}(aq) + 3en(aq) \rightleftharpoons Co(en)_3^{3+}(aq) + 6NH_3(aq)$$

where "en" stands for ethylenediamine, $H_2NCH_2CH_2NH_2$. Six Co—N bonds are broken and six Co—N bonds are formed in this reaction, so $\Delta H°_{rxn} \approx 0$; yet $K > 1$. What are the signs of $\Delta S°$ and $\Delta G°$? What drives the reaction?

20.88 What is the change in entropy when 0.200 mol of potassium freezes at 63.7°C ($\Delta H_{fus} = 2.39$ kJ/mol)?

20.89 Is each statement true or false? If false, correct it.
(a) All spontaneous reactions occur quickly.
(b) The reverse of a spontaneous reaction is nonspontaneous.
(c) All spontaneous processes release heat.
(d) The boiling of water at 100°C and 1 atm is spontaneous.
(e) If a process increases the freedom of motion of the particles of a system, the entropy of the system decreases.
(f) The energy of the universe is constant; the entropy of the universe decreases toward a minimum.
(g) All systems disperse their energy spontaneously.
(h) Both ΔS_{sys} and ΔS_{surr} equal zero at equilibrium.

20.90 Hemoglobin carries O_2 from the lungs to tissue cells, where the O_2 is released. The protein is represented as Hb in its unoxygenated form and as Hb•O_2 in its oxygenated form. One reason CO is toxic is that it competes with O_2 in binding to Hb:

$$Hb{\cdot}O_2(aq) + CO(g) \rightleftharpoons Hb{\cdot}CO(aq) + O_2(g)$$

(a) If $\Delta G° \approx -14$ kJ at 37°C (body temperature), what is the ratio of [Hb•CO] to [Hb•O_2] at 37°C with $[O_2] = [CO]$?
(b) How is Le Châtelier's principle used to treat CO poisoning?

20.91 Magnesia (MgO) is used for fire brick, crucibles, and furnace linings because of its high melting point. It is produced by decomposing magnesite ($MgCO_3$) at around 1200°C.
(a) Write a balanced equation for magnesite decomposition.
(b) Use $\Delta H°$ and $S°$ values to find $\Delta G°$ at 298 K.
(c) Assuming that $\Delta H°$ and $S°$ do not change with temperature, find the minimum temperature at which the reaction is spontaneous.
(d) Calculate the equilibrium P_{CO_2} above $MgCO_3$ at 298 K.
(e) Calculate the equilibrium P_{CO_2} above $MgCO_3$ at 1200 K.

20.92 To prepare nuclear fuel, U_3O_8 ("yellow cake") is converted to $UO_2(NO_3)_2$, which is then converted to UO_3 and finally UO_2. The fuel is enriched (the proportion of the ^{235}U is increased) by a two-step conversion of UO_2 into UF_6, a volatile solid, followed by a gaseous-diffusion separation of the ^{235}U and ^{238}U isotopes:

$$UO_2(s) + 4HF(g) \longrightarrow UF_4(s) + 2H_2O(g)$$
$$UF_4(s) + F_2(g) \longrightarrow UF_6(s)$$

Calculate $\Delta G°$ for the overall process at 85°C:

	$\Delta H_f°$ (kJ/mol)	$S°$(J/mol·K)	$\Delta G_f°$ (kJ/mol)
$UO_2(s)$	−1085	77.0	−1032
$UF_4(s)$	−1921	152	−1830
$UF_6(s)$	−2197	225	−2068

20.93 Methanol, a major industrial feedstock, is made by several catalyzed reactions, such as $CO(g) + 2H_2(g) \longrightarrow CH_3OH(l)$.
(a) Show that this reaction is thermodynamically feasible.
(b) Is it favored at low or at high temperatures?
(c) One concern about using CH_3OH as an auto fuel is its oxidation in air to yield formaldehyde, $CH_2O(g)$, which poses a health hazard. Calculate $\Delta G°$ at 100.°C for this oxidation.

20.94 (a) Write a balanced equation for the gaseous reaction between N_2O_5 and F_2 to form NF_3 and O_2. (b) Determine $\Delta G°_{rxn}$. (c) Find ΔG_{rxn} at 298 K if $P_{N_2O_5} = P_{F_2} = 0.20$ atm, $P_{NF_3} = 0.25$ atm, and $P_{O_2} = 0.50$ atm.

20.95 Consider the following reaction:

$$2NOBr(g) \rightleftharpoons 2NO(g) + Br_2(g) \qquad K = 0.42 \text{ at } 373 \text{ K}$$

Given that $S°$ of $NOBr(g)$ = 272.6 J/mol·K and that $\Delta S°_{rxn}$ and $\Delta H°_{rxn}$ are constant with temperature, find
(a) $\Delta S°_{rxn}$ at 298 K
(b) $\Delta G°_{rxn}$ at 373 K
(c) $\Delta H°_{rxn}$ at 373 K
(d) $\Delta H_f°$ of NOBr at 298 K
(e) $\Delta G°_{rxn}$ at 298 K
(f) $\Delta G_f°$ of NOBr at 298 K

20.96 Hydrogenation is the addition of H_2 to double (or triple) carbon-carbon bonds. Peanut butter and most commercial baked goods include hydrogenated oils. Find $\Delta H°$, $\Delta S°$, and $\Delta G°$ for the hydrogenation of ethene (C_2H_4) to ethane (C_2H_6) at 25°C.

20.97 Styrene is produced by catalytic dehydrogenation of ethylbenzene at high temperature in the presence of superheated steam. (a) Find $\Delta H°_{rxn}$, $\Delta G°_{rxn}$, and $\Delta S°_{rxn}$, given these data at 298 K:

Compound	$\Delta H_f°$ (kJ/mol)	$\Delta G_f°$ (kJ/mol)	$S°$ (J/mol·K)
Ethylbenzene, $C_6H_5—CH_2CH_3$	−12.5	119.7	255
Styrene, $C_6H_5—CH{=}CH_2$	103.8	202.5	238

(b) At what temperature is the reaction spontaneous?
(c) What are $\Delta G°_{rxn}$ and K at 600.°C?
(d) With 5.0 parts steam to 1.0 part ethylbenzene in the reactant mixture and the total pressure kept constant at 1.3 atm, what is ΔG at 50.% conversion, that is, when 50.% of the ethylbenzene has reacted?

20.98 Propylene (propene; $CH_3CH{=}CH_2$) is used to produce polypropylene and many other chemicals. Although most propylene is obtained from the cracking of petroleum, about 2% is produced by catalytic dehydrogenation of propane ($CH_3CH_2CH_3$):

$$CH_3CH_2CH_3 \xrightarrow{Pt/Al_2O_3} CH_3CH{=}CH_2 + H_2$$

Because this reaction is endothermic, heaters are placed between the reactor vessels to maintain the required temperature.
(a) If the molar entropy, $S°$, of propylene is 267.1 J/mol·K, find its entropy of formation, $S_f°$.
(b) Find $\Delta G_f°$ of propylene ($\Delta H_f°$ for propylene = 20.4 kJ/mol).
(c) Calculate $\Delta H°_{rxn}$ and $\Delta G°_{rxn}$ for the dehydrogenation.
(d) What is the theoretical yield of propylene at 580°C if the initial pressure of propane is 1.00 atm?
(e) Would the yield change if the reactor walls were permeable to H_2? Explain.
(f) At what temperature is the dehydrogenation spontaneous, with all substances in the standard state?

Note: Problems 20.99 and 20.100 relate to the thermodynamics of adenosine triphosphate (ATP). Refer to the Chemical Connections essay at the end of Section 20.3.

20.99 Find K for (a) the hydrolysis of ATP, (b) the dehydration-condensation to form glucose phosphate, and (c) the coupled reaction between ATP and glucose. (d) How does each K change when T changes from 25°C to 37°C?

20.100 Energy from ATP hydrolysis drives many nonspontaneous cell reactions:

$$ATP^{4-}(aq) + H_2O(l) \rightleftharpoons ADP^{3-}(aq) + HPO_4^{2-}(aq) + H^+(aq)$$
$$\Delta G°' = -30.5 \text{ kJ}$$

Energy for the reverse process comes ultimately from glucose metabolism:

$$C_6H_{12}O_6(s) + 6O_2(g) \longrightarrow 6CO_2(g) + 6H_2O(l)$$

(a) Find K for the hydrolysis of ATP at 37°C.
(b) Find $\Delta G°'_{rxn}$ for metabolism of 1 mol of glucose.
(c) How many moles of ATP can be produced by metabolism of 1 mol of glucose?
(d) If 36 mol of ATP is formed, what is the actual yield?

20.101 From the following reaction and data, find (a) $S°$ of $SOCl_2$ and (b) the T at which the reaction becomes nonspontaneous:

$$SO_3(g) + SCl_2(l) \longrightarrow SOCl_2(l) + SO_2(g) \qquad \Delta G°_{rxn} = -75.2 \text{ kJ}$$

	$SO_3(g)$	$SCl_2(l)$	$SOCl_2(l)$	$SO_2(g)$
$\Delta H_f°$ (kJ/mol)	−396	−50.0	−245.6	−296.8
$S°$ (J/mol·K)	256.7	184	—	248.1

20.102 Write equations for the oxidation of Fe and of Al. Use $\Delta G_f°$ to determine whether either process is spontaneous at 25°C.

20.103 The molecular scene depicts a gaseous equilibrium mixture at 460°C for the reaction of H_2 *(blue)* and I_2 *(purple)* to form HI. Each molecule represents 0.010 mol and the container volume is 1.0 L. (a) Is $K_c > 1$, = 1, or < 1? (b) Is $K_p > K_c$, = K_c, or < K_c? (c) Calculate $\Delta G°_{rxn}$. (d) How would the value of $\Delta G°_{rxn}$ change if the purple molecules represented H_2 and the blue I_2? Explain.

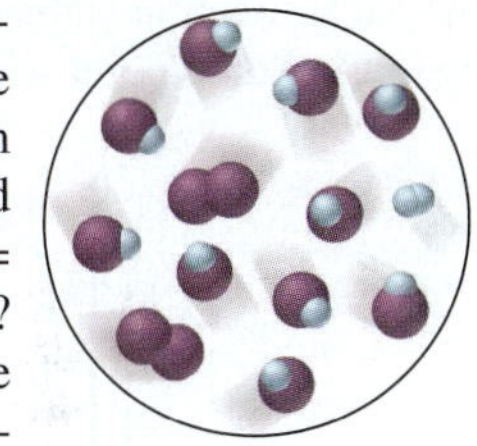

20.104 A key step in the metabolism of glucose for energy is the isomerization of glucose-6-phosphate (G6P) to fructose-6-phosphate (F6P): G6P $\rightleftharpoons$ F6P; $K = 0.510$ at 298 K.
(a) Calculate $\Delta G°$ at 298 K.
(b) Calculate ΔG when Q, the [F6P]/[G6P] ratio, equals 10.0.
(c) Calculate ΔG when $Q = 0.100$.
(d) Calculate Q if $\Delta G = -2.50$ kJ/mol.

20.105 A chemical reaction, such as HI forming from its elements, can reach equilibrium at many temperatures. In contrast, a phase change, such as ice melting, is in equilibrium at a given pressure and temperature. Each of the graphs below depicts G_{sys} vs. extent of change. (a) Which graph depicts how G_{sys} changes for the formation of HI? Explain. (b) Which graph depicts how G_{sys} changes as ice melts at 1°C and 1 atm? Explain.

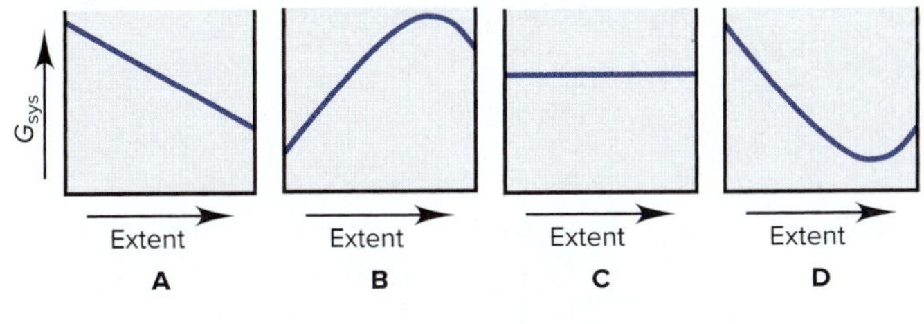

20.106 When heated, the DNA double helix separates into two random coil single strands. When cooled, the random coils re-form the double helix: double helix $\rightleftharpoons$ 2 random coils.
(a) What is the sign of ΔS for the forward process? Why?
(b) Energy must be added to break H bonds and overcome dispersion forces between the strands. What is the sign of ΔG for the forward process when $T\Delta S$ is smaller than ΔH?
(c) Write an expression for T in terms of ΔH and ΔS when the reaction is at equilibrium. (This temperature is called the *melting temperature* of the nucleic acid.)

20.107 In the process of respiration, glucose is oxidized completely. In fermentation, O_2 is absent and glucose is broken down to ethanol and CO_2. Ethanol is oxidized to CO_2 and H_2O.
(a) Balance the following equations for these processes:

Respiration: $C_6H_{12}O_6(s) + O_2(g) \longrightarrow CO_2(g) + H_2O(l)$

Fermentation: $C_6H_{12}O_6(s) \longrightarrow C_2H_5OH(l) + CO_2(g)$

Ethanol oxidation: $C_2H_5OH(l) + O_2(g) \longrightarrow CO_2(g) + H_2O(l)$

(b) Calculate ΔG°_{rxn} for respiration of 1.00 g of glucose.
(c) Calculate ΔG°_{rxn} for fermentation of 1.00 g of glucose.
(d) Calculate ΔG°_{rxn} for oxidation of the ethanol from part (c).

20.108 Consider the formation of ammonia:

$$N_2(g) + 3H_2(g) \rightleftharpoons 2NH_3(g)$$

(a) Assuming that ΔH° and ΔS° are constant with temperature, find the temperature at which $K_p = 1.00$.
(b) Find K_p at 400.°C, a typical temperature for NH_3 production.
(c) Given the lower K_p at the higher temperature, why are these conditions used industrially?

20.109 Kyanite, sillimanite, and andalusite all have the formula Al_2SiO_5. Each is stable under different conditions (see the graph below). At the point where the three phases intersect:

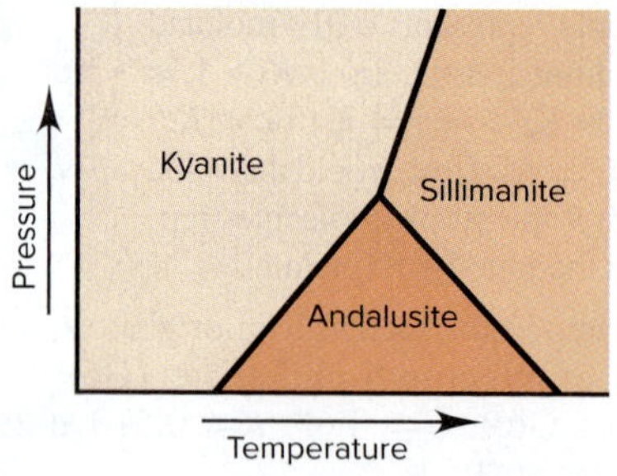

(a) Which mineral, if any, has the lowest free energy?
(b) Which mineral, if any, has the lowest enthalpy?
(c) Which mineral, if any, has the highest entropy?
(d) Which mineral, if any, has the lowest density?

20.110 Acetylene is produced commercially by the partial oxidation of methane. At 1500°C and pressures of 1–10 bar, the yield of acetylene is about 20%. The major side product is carbon monoxide, and some soot and carbon dioxide also form.
(a) At what temperature is the desired reaction spontaneous:

$$2CH_4 + \tfrac{1}{2}O_2 \longrightarrow C_2H_2 + 2H_2 + H_2O$$

(b) Acetylene can also be made by the reaction of its elements, carbon (graphite) and hydrogen. At what temperature is this formation reaction spontaneous?
(c) Why must this reaction mixture be immediately cooled?

20.111 Synthesis gas, a mixture that includes the fuels CO and H_2, is used to produce liquid hydrocarbons and methanol. It is made at pressures up to 100 atm by oxidation of methane followed by the steam re-forming and water-gas shift reactions. Because the process is exothermic, temperatures reach 950–1100°C, and the conditions are such that the amounts of H_2, CO, CO_2, CH_4, and H_2O leaving the reactor are close to the equilibrium amounts for the steam re-forming and water-gas shift reactions:

$$CH_4(g) + H_2O(g) \rightleftharpoons CO(g) + 3H_2(g) \quad \text{(steam re-forming)}$$

$$CO(g) + H_2O(g) \rightleftharpoons CO_2(g) + H_2(g) \quad \text{(water-gas shift)}$$

(a) At 1000.°C, what are ΔG° and ΔH° for the steam re-forming reaction and for the water-gas shift reaction?
(b) By doubling the steam re-forming step and adding it to the water-gas shift step, we obtain the following combined reaction:

$$2CH_4(g) + 3H_2O(g) \rightleftharpoons CO_2(g) + CO(g) + 7H_2(g)$$

Is this reaction spontaneous at 1000.°C in the standard state?
(c) Is it spontaneous at 98 atm and 50.% conversion (when 50.% of the starting materials have reacted)?
(d) Is it spontaneous at 98 atm and 90.% conversion?

20.112 Calculate the change in entropy when 10.0 g of CO_2 isothermally expands from a volume of 6.15 L to 11.5 L. Assume that the gas behaves ideally.

20.113 Calculate the change in entropy when the pressure of a 46.0 g sample of oxygen gas is increased from 625 mmHg to 1.24 atm at a constant temperature of 25°C. Assume ideal behavior.

20.114 Iron melts at 1535°C and its enthalpy of fusion is 13.8 kJ/mol. Calculate the entropy of fusion per mole of iron.

20.115 Calculate the entropy change for the vaporization of 25.0 g of hexane, C_6H_{14}, at its boiling point of 68.8°C. The enthalpy of vaporization of hexane is 28.8 kJ/mol.

20.116 Calculate the entropy change of a 235 g sample of copper that is heated at constant pressure from 32.0°C to 125°C. $C_{P,m}$ for copper = 24.5 J/mol·K.

20.117 The temperature of a 625 g piece of granite is increased from 18.0°C to 60.0°C. What is its change in entropy? The specific heat capacity of granite is 0.790 J/g·K.

20.118 The boiling point of chloroform, $CHCl_3$, is 61.2°C. Use Trouton's rule to estimate the enthalpy of vaporization of chloroform.

21

Electrochemistry: Chemical Change and Electrical Work

21.1 Redox Reactions and Electrochemical Cells
Oxidation-Reduction Review
Half-Reaction Method for Balancing Redox Reactions
Electrochemical Cells

21.2 Voltaic Cells: Using Spontaneous Reactions to Generate Electrical Energy
Cell Construction and Operation
Cell Notation
Why Does the Cell Work?

21.3 Cell Potential: Output of a Voltaic Cell
Standard Cell Potential (E°_{cell})
Strengths of Oxidizing and Reducing Agents
Writing Spontaneous Redox Reactions
Explaining the Activity Series

21.4 Free Energy and Electrical Work
E°_{cell} and the Equilibrium Constant
Effect of Concentration on E_{cell}
Following Changes in E_{cell}
Concentration Cells

21.5 Electrochemical Processes in Batteries
Primary (Nonrechargeable) Batteries
Secondary (Rechargeable) Batteries
Fuel Cells

21.6 Corrosion: An Environmental Voltaic Cell
Corrosion of Iron
Protecting Against Corrosion

21.7 Electrolytic Cells: Using Electrical Energy to Drive Nonspontaneous Reactions
Cell Construction and Operation
Predicting Electrolysis Products
Stoichiometry of Electrolysis

Source: © Griffin Technology

Concepts and Skills to Review Before You Study This Chapter

- redox terminology (Section 4.5)
- activity series of the metals (Section 4.6)
- trends in ionization energy (Section 8.3) and electronegativity (Section 9.5)
- free energy, work, and equilibrium (Sections 20.3 and 20.4)
- Q vs. K (Section 17.4) and ΔG vs. $\Delta G°$ (Section 20.4)

Thermodynamics has countless applications. Some are probably at your fingertips right now: a laptop computer or tablet, MP3 player, or smart phone. These are a few of the objects you use every day that rely on a major field in applied thermodynamics called **electrochemistry,** the study of the relationship between chemical change and electrical work. This field relies on **electrochemical cells,** systems that incorporate a redox reaction to produce or utilize electrical energy.

IN THIS CHAPTER . . . *We examine the essential features of electrochemical cells and quantify the relationship between free energy and electrical work.*

- We review oxidation-reduction (redox) reactions and describe a method for balancing redox reactions that take place in electrochemical cells.
- We highlight key differences between the two types of electrochemical cells.
- We focus on voltaic cells, which *release* free energy to do electrical work, and see how they operate as a combination of two half-cells.
- We calculate the voltage, the cell's electrical potential, which is the sum of the voltages of the half-cells.
- We use half-cell voltages to determine the relative strengths of redox couples and to write spontaneous redox equations.
- We examine the interdependence of the standard free energy change, the standard cell potential, and the equilibrium constant.
- We follow the change in voltage as a voltaic cell operates.
- We see how differences in concentration between the solutions in the half-cells can be harnessed to generate electricity.
- We analyze the makeup and operation of some major types of batteries.
- We describe the problem of corrosion, which has key similarities to the operation of voltaic cells.
- We examine electrolytic cells, which *absorb* free energy from an external source to do electrical work. We predict the products that form under different conditions and quantify the relation between charge and amount of substance.
- We explore the redox system that generates energy in living cells.

21.1 REDOX REACTIONS AND ELECTROCHEMICAL CELLS

To begin, we review the oxidation-reduction process that we covered in Chapter 4 and describe the *half-reaction method* of balancing redox reactions. Then we see how redox reactions are used in the two types of electrochemical cells.

A Quick Review of Oxidation-Reduction Concepts

All electrochemical processes involve the *movement of electrons from one chemical species to another* in an oxidation-reduction (redox) reaction. In any redox process:

- *Oxidation* is the loss of electrons, and *reduction* is the gain of electrons. Oxidation and reduction occur *simultaneously.*

- The *oxidizing agent* does the oxidizing by *taking electrons* from the substance being oxidized. The *reducing agent* does the reducing by *giving electrons* to the substance being reduced. Therefore:

  The oxidizing agent is reduced, and the reducing agent is oxidized.

- The oxidized substance ends up with a *higher* (more positive or less negative) oxidation number (O.N.), and the reduced substance ends up with a *lower* (less positive or more negative) O.N.
- The total number of electrons gained by the atoms/ions of the oxidizing agent equals the total number lost by the atoms/ions of the reducing agent.

Figure 21.1 presents these ideas for the aqueous reaction between zinc metal and H^+ from a strong acid. Be sure you can identify the oxidation and reduction parts of a redox process. You may want to review the full discussion in Section 4.5.

PROCESS	$\overset{0}{Zn}(s) + 2\overset{+1}{H^+}(aq) \longrightarrow \overset{+2}{Zn^{2+}}(aq) + \overset{0}{H_2}(g)$
OXIDATION • One reactant loses electrons. • Reducing agent is oxidized. • Oxidation number increases.	Zinc **loses** electrons. Zinc is the reducing agent and becomes **oxidized.** The oxidation number of Zn **increases** from 0 to +2.
REDUCTION • Other reactant gains electrons. • Oxidizing agent is reduced. • Oxidation number decreases.	Hydrogen ion **gains** electrons. Hydrogen ion is the oxidizing agent and becomes **reduced.** The oxidation number of H **decreases** from +1 to 0.

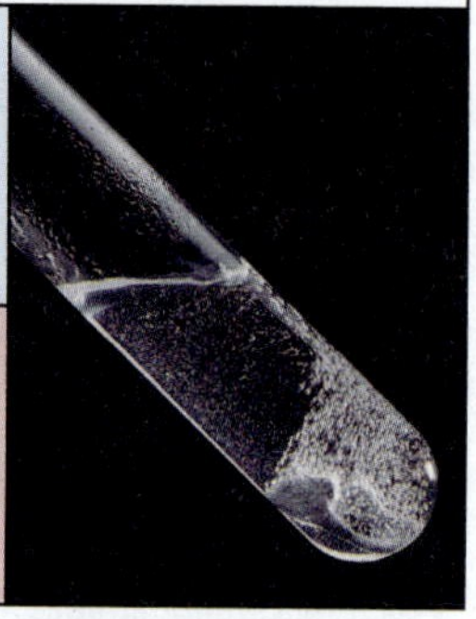

Figure 21.1 A summary of redox terminology, as applied to the reaction of zinc with hydrogen ion.

Source: © McGraw-Hill Education/Stephen Frisch, photographer

Half-Reaction Method for Balancing Redox Reactions

In Chapter 4, you learned how to determine the oxidation number of each element in a compound, how to identify oxidizing and reducing agents, and how to decide if a chemical change is a redox reaction. Now, let's see how to balance redox reactions. The **half-reaction method** is especially useful for studying electrochemistry:

- It *divides the overall redox reaction* into oxidation and reduction *half-reactions,* which, as you'll see, reflect their actual physical separation in electrochemical cells.
- It applies well to reactions in acidic or basic solution, which are common in cells.
- It (usually) does *not* require assigning O.N.s. (In cases where the half-reactions are not obvious, we assign O.N.s to determine which atoms undergo a change and write half-reactions with the species that contain those atoms.)

Steps in the Half-Reaction Method The balancing process begins with a "skeleton" ionic reaction that consists of only species that are oxidized and reduced. Here are the steps in the half-reaction method:

Step 1. Divide the skeleton reaction into two half-reactions. Each half-reaction contains the oxidized and reduced forms of one of the species: *if the oxidized form of a species is on the left side, the reduced form must be on the right, and vice versa.*

Step 2. Balance the atoms and charges in each half-reaction.
- Atoms are balanced *in this order:* atoms other than O and H, then O, then H.
- Charge is balanced by *adding* electrons (e^-) *to the left side in the reduction half-reaction* because the reactant gains them and *to the right side in the oxidation half-reaction* because the reactant loses them.

Step 3. If necessary, multiply one or both half-reactions by an integer so that

$$\text{number of } e^- \text{ gained in reduction} = \text{number of } e^- \text{ lost in oxidation}$$

Step 4. Add the balanced half-reactions, and include states of matter.

Step 5. Check that the atoms and charges are balanced.

Let's balance a redox reaction that occurs in acidic solution and then see how to balance one in basic solution in Sample Problem 21.1.

Balancing Redox Reactions in Acidic Solution For a reaction in acidic solution, H_2O molecules and H^+ ions are present for the balancing. We've usually used H_3O^+ to represent the proton in water, but we use H^+ in this chapter because it makes the balancing simpler. To make that point, we'll also balance this first reaction with H_3O^+ to show that the only difference is in the number of water molecules.

The reaction between dichromate ion and iodide ion to form chromium(III) ion and solid iodine occurs in acidic solution *(see photo)*. The skeleton ionic reaction is

$$Cr_2O_7^{2-}(aq) + I^-(aq) \longrightarrow Cr^{3+}(aq) + I_2(s) \qquad \text{[acidic solution]}$$

Dichromate ion *(left)* and iodide ion *(center)* form chromium(III) ion and solid iodine *(right)*.

Source: © McGraw-Hill Education/Stephen Frisch, photographer

Step 1. Divide the reaction into half-reactions. Each half-reaction contains the oxidized and reduced forms of one species:

$$Cr_2O_7^{2-} \longrightarrow Cr^{3+}$$
$$I^- \longrightarrow I_2$$

Step 2. Balance atoms and charges in each half-reaction. We use H_2O to balance O, H^+ to balance H, and e^- to balance charges.

For the $Cr_2O_7^{2-}/Cr^{3+}$ half-reaction:

a. *Balance atoms other than O and H.* We balance the two Cr in $Cr_2O_7^{2-}$ on the left with a coefficient of 2 on the right:

$$Cr_2O_7^{2-} \longrightarrow 2Cr^{3+}$$

b. *Balance O atoms by adding H_2O molecules.* Each H_2O has one O atom, so we add seven H_2O on the right to balance the seven O in $Cr_2O_7^{2-}$:

$$Cr_2O_7^{2-} \longrightarrow 2Cr^{3+} + 7H_2O$$

c. *Balance H atoms by adding H^+ ions.* Each H_2O contains two H, and we added seven H_2O, so we add 14 H^+ ions on the left:

$$14H^+ + Cr_2O_7^{2-} \longrightarrow 2Cr^{3+} + 7H_2O$$

d. *Balance charge by adding electrons.* Each H^+ ion has a 1+ charge, and 14 H^+ plus the 2− of $Cr_2O_7^{2-}$ gives 12+ on the left. Two Cr^{3+} give 6+ on the right. There is an excess of 6+ on the left, so we add six e^- on the left:

$$6e^- + 14H^+ + Cr_2O_7^{2-} \longrightarrow 2Cr^{3+} + 7H_2O \qquad \text{[reduction]}$$

This half-reaction is balanced. We note that

- it is the *reduction* because electrons appear on the *left, as a reactant;*
- $Cr_2O_7^{2-}$ gains electrons (is reduced), so $Cr_2O_7^{2-}$ is the *oxidizing agent;*
- the O.N. of Cr decreases from +6 on the left to +3 on the right.

For the I^-/I_2 half-reaction:

a. *Balance atoms other than O and H.* Two I atoms on the right require a coefficient of 2 on the left:

$$2I^- \longrightarrow I_2$$

b. *Balance O atoms with H_2O.* Not needed; there are no O atoms.
c. *Balance H atoms with H^+.* Not needed; there are no H atoms.
d. *Balance charge with e^-.* To balance the 2− on the left, we add two e^- on the right:

$$2I^- \longrightarrow I_2 + 2e^- \qquad \text{[oxidation]}$$

This half-reaction is balanced. We see that

- it is the *oxidation* because electrons appear on the *right, as a product;*
- the reactant I^- loses electrons (is oxidized), so I^- is the *reducing agent;*
- the O.N. of I increases from −1 to 0.

Step 3. Multiply each half-reaction, if necessary, by an integer so that the number of e^- lost in the oxidation equals the number of e^- gained in the reduction. Two e^- are lost in the oxidation and six e^- are gained in the reduction, so we multiply the oxidation by 3:

$$3(2I^- \longrightarrow I_2 + 2e^-)$$
$$6I^- \longrightarrow 3I_2 + 6e^-$$

Step 4. Add the half-reactions, canceling species that appear on both sides, and include states of matter. In this example, only the electrons cancel:

$$6e^- + 14H^+ + Cr_2O_7^{2-} \longrightarrow 2Cr^{3+} + 7H_2O$$
$$6I^- \longrightarrow 3I_2 + 6e^-$$
$$6I^-(aq) + 14H^+(aq) + Cr_2O_7^{2-}(aq) \longrightarrow 3I_2(s) + 7H_2O(l) + 2Cr^{3+}(aq)$$

Step 5. Check that atoms and charges balance:

$$\text{Reactants (6I, 14H, 2Cr, 7O; 6+)} \longrightarrow \text{products (6I, 14H, 2Cr, 7O; 6+)}$$

Using H_3O^+, Instead of H^+, to Balance H Now let's see what happens if we use H_3O^+ to supply H atoms, instead of H^+. In step 2 for the reduction half-reaction, balancing Cr atoms and balancing O atoms with H_2O are the same as before, so we have

$$Cr_2O_7^{2-} \longrightarrow 2Cr^{3+} + 7H_2O$$

Now we use H_3O^+ to balance H atoms. Because each H_3O^+ is an H^+ bonded to an H_2O, we balance the 14 H on the right with 14 H_3O^+ on the left and immediately add 14 more H_2O on the right:

$$14H_3O^+ + Cr_2O_7^{2-} \longrightarrow 2Cr^{3+} + 7H_2O + 14H_2O$$

Taking the sum of the H_2O molecules on the right gives

$$14H_3O^+ + Cr_2O_7^{2-} \longrightarrow 2Cr^{3+} + 21H_2O$$

None of this affects the redox change, so balancing the charge still requires six e^- on the left, and we obtain the balanced reduction half-reaction:

$$6e^- + 14H_3O^+ + Cr_2O_7^{2-} \longrightarrow 2Cr^{3+} + 21H_2O \qquad \text{[reduction]}$$

Adding the balanced oxidation half-reaction gives the balanced redox equation:

$$6I^-(aq) + 14H_3O^+(aq) + Cr_2O_7^{2-}(aq) \longrightarrow 3I_2(s) + 21H_2O(l) + 2Cr^{3+}(aq)$$

Note that, as mentioned earlier, the only difference is the number of H_2O molecules: 21 H_2O instead of 7 H_2O. So, even though this approach uses H_3O^+, which shows more accurately how H^+ exists, only the number of water molecules changes and the balancing involves more steps. Therefore, we'll continue to use H^+ to balance H atoms in these equations.

Balancing Redox Reactions in Basic Solution In acidic solution, H_2O molecules and H^+ ions are available to balance a redox reaction, but in basic solution, H_2O molecules and OH^- ions are available.

We need only one additional step to balance a redox reaction that takes place in basic solution. It appears after we balance the half-reactions *as if they occur in acidic solution* and are combined (step 4). At this point, *we add one* ***OH^- to both sides of the equation*** *for every H^+ present.* (This step is labeled "4 Basic" in Sample Problem 21.1.) The OH^- ions added on the side with H^+ ions combine with them to form H_2O molecules, while the OH^- on the other side remain in the equation, and then excess H_2O is canceled.

Student Hot Spot

Student data indicate that you may struggle with balancing redox reactions in basic solution. Access the Smartbook to view additional Learning Resources on this topic.

SAMPLE PROBLEM 21.1 Balancing a Redox Reaction in Basic Solution

Problem Permanganate ion reacts in basic solution with oxalate ion to form carbonate ion and solid manganese dioxide. Balance the skeleton ionic equation for the reaction between $NaMnO_4$ and $Na_2C_2O_4$ in basic solution:

$$MnO_4^-(aq) + C_2O_4^{2-}(aq) \longrightarrow MnO_2(s) + CO_3^{2-}(aq) \qquad \text{[basic solution]}$$

Plan We follow the numbered steps as described in text, and proceed through step 4 as if this reaction occurs in acidic solution. Then, we add the appropriate number of OH^- ions and cancel excess H_2O molecules (see "4 Basic").

Solution

1. Divide into half-reactions.

$$MnO_4^- \longrightarrow MnO_2 \qquad\qquad C_2O_4^{2-} \longrightarrow CO_3^{2-}$$

2. Balance.

a. Atoms other than O and H.
Not needed.

a. Atoms other than O and H,
$$C_2O_4^{2-} \longrightarrow 2CO_3^{2-}$$

b. O atoms with H_2O,
$$MnO_4^- \longrightarrow MnO_2 + 2H_2O$$

b. O atoms with H_2O,
$$2H_2O + C_2O_4^{2-} \longrightarrow 2CO_3^{2-}$$

c. H atoms with H^+,
$$4H^+ + MnO_4^- \longrightarrow MnO_2 + 2H_2O$$

c. H atoms with H^+,
$$2H_2O + C_2O_4^{2-} \longrightarrow 2CO_3^{2-} + 4H^+$$

d. Charge with e^-,
$$3e^- + 4H^+ + MnO_4^- \longrightarrow MnO_2 + 2H_2O$$
[reduction]

d. Charge with e^-,
$$2H_2O + C_2O_4^{2-} \longrightarrow 2CO_3^{2-} + 4H^+ + 2e^-$$
[oxidation]

3. Multiply each half-reaction, if necessary, by some integer to make e^- lost equal e^- gained.

$$2(3e^- + 4H^+ + MnO_4^- \longrightarrow MnO_2 + 2H_2O) \qquad 3(2H_2O + C_2O_4^{2-} \longrightarrow 2CO_3^{2-} + 4H^+ + 2e^-)$$
$$6e^- + 8H^+ + 2MnO_4^- \longrightarrow 2MnO_2 + 4H_2O \qquad 6H_2O + 3C_2O_4^{2-} \longrightarrow 6CO_3^{2-} + 12H^+ + 6e^-$$

4. Add half-reactions, and cancel species appearing on both sides.
The six e^- cancel, eight H^+ cancel to leave four H^+ on the right, and four H_2O cancel to leave two H_2O on the left:

$$\cancel{6e^-} + \cancel{8H^+} + 2MnO_4^- \longrightarrow 2MnO_2 + \cancel{4H_2O}$$
$$2\,\cancel{6}H_2O + 3C_2O_4^{2-} \longrightarrow 6CO_3^{2-} + 4\,\cancel{12}H^+ + \cancel{6e^-}$$
$$\overline{2MnO_4^- + 2H_2O + 3C_2O_4^{2-} \longrightarrow 2MnO_2 + 6CO_3^{2-} + 4H^+}$$

4 **Basic.** Add OH^- to both sides to neutralize the H^+ present, and cancel excess H_2O. Adding four OH^- to both sides forms four H_2O on the right. Two of those cancel the two H_2O on the left and leave two H_2O on the right:

$$2MnO_4^- + 2H_2O + 3C_2O_4^{2-} + 4OH^- \longrightarrow 2MnO_2 + 6CO_3^{2-} + [4H^+ + 4OH^-]$$
$$2MnO_4^- + \cancel{2H_2O} + 3C_2O_4^{2-} + 4OH^- \longrightarrow 2MnO_2 + 6CO_3^{2-} + 2\,\cancel{4}H_2O$$

Including states of matter gives the final balanced equation:

$$2MnO_4^-(aq) + 3C_2O_4^{2-}(aq) + 4OH^-(aq) \longrightarrow 2MnO_2(s) + 6CO_3^{2-}(aq) + 2H_2O(l)$$

5. Check that atoms and charges balance.

$$(2Mn, 24O, 6C, 4H; 12-) \longrightarrow (2Mn, 24O, 6C, 4H; 12-)$$

Comment As a final step, let's see how to obtain the balanced *molecular* equation for this reaction. We note the amount (mol) of each anion in the balanced ionic equation and add the correct amount (mol) of spectator ions (in this case, Na^+, as given in the problem statement) to obtain neutral compounds. The balanced molecular equation is

$$2NaMnO_4(aq) + 3Na_2C_2O_4(aq) + 4NaOH(aq) \longrightarrow$$
$$2MnO_2(s) + 6Na_2CO_3(aq) + 2H_2O(l)$$

FOLLOW-UP PROBLEMS

*Brief Solutions to **all** Follow-up Problems appear at the end of the chapter.*

21.1A Write a balanced molecular equation for the reaction between $KMnO_4$ and KI in basic solution. The skeleton ionic reaction is

$$MnO_4^-(aq) + I^-(aq) \longrightarrow MnO_4^{2-}(aq) + IO_3^-(aq)$$

21.1B Write a balanced molecular equation for the reaction between chromium(III) hydroxide and sodium iodate in basic solution. The skeleton ionic reaction is

$$Cr(OH)_3(s) + IO_3^-(aq) \longrightarrow CrO_4^{2-}(aq) + I^-(aq)$$

SOME SIMILAR PROBLEMS 21.12–21.19

An Overview of Electrochemical Cells

The fundamental difference between the two types of electrochemical cells is based on whether the overall redox reaction in the cell is spontaneous (free energy is released) or nonspontaneous (free energy is absorbed):

1. A **voltaic cell** (also called a **galvanic cell**) uses a *spontaneous* redox reaction ($\Delta G < 0$) to generate electrical energy. In the cell reaction, some of the difference in free energy between higher energy reactants and lower energy products is converted into electrical energy, which operates the load—flashlight, MP3 player, car starter motor, and so forth. Thus, *the system does work on the surroundings (load).* All batteries contain voltaic cells.
2. An **electrolytic cell** uses electrical energy to drive a *nonspontaneous* redox reaction ($\Delta G > 0$). In the cell reaction, an external power source supplies free energy to convert lower energy reactants into higher energy products. Thus, *the surroundings (power supply) do work on the system.* Electroplating and the recovery of metals from ores utilize electrolytic cells.

The two types of cell have several similarities. Two **electrodes,** which conduct the electricity between cell and surroundings, are dipped into an **electrolyte,** a mixture of ions (usually in aqueous solution) that are involved in the reaction or that carry the charge (Figure 21.2). An electrode is identified as either **anode** or **cathode** depending on the half-reaction that takes place there:

- *The oxidation half-reaction occurs at the anode.* Electrons lost by the substance being oxidized (reducing agent) *leave the oxidation half-cell* at the anode.
- *The reduction half-reaction occurs at the cathode.* Electrons gained by the substance being reduced (oxidizing agent) *enter the reduction half-cell* at the cathode.

Note that, for reasons we discuss shortly, *the relative charges of the electrodes are* ***opposite in sign*** *in the two types of cell.*

Which Half-Reaction Occurs at Which Electrode?

Here are some memory aids to help you remember which half-reaction occurs at which electrode:

1. The words *anode* and *oxidation* start with vowels; the words *cathode* and *reduction* start with consonants.
2. Alphabetically, the *A* in anode comes before the *C* in cathode, and the *O* in oxidation comes before the *R* in reduction.
3. Look at the first syllables and use your imagination:
   ANode, OXidation; REDuction, CAThode ⟹ AN OX and a RED CAT

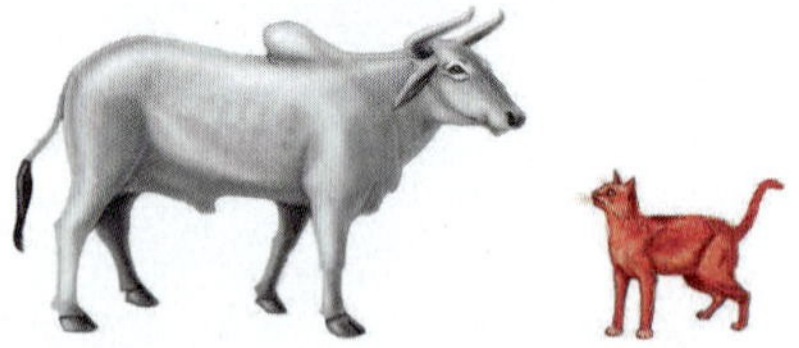

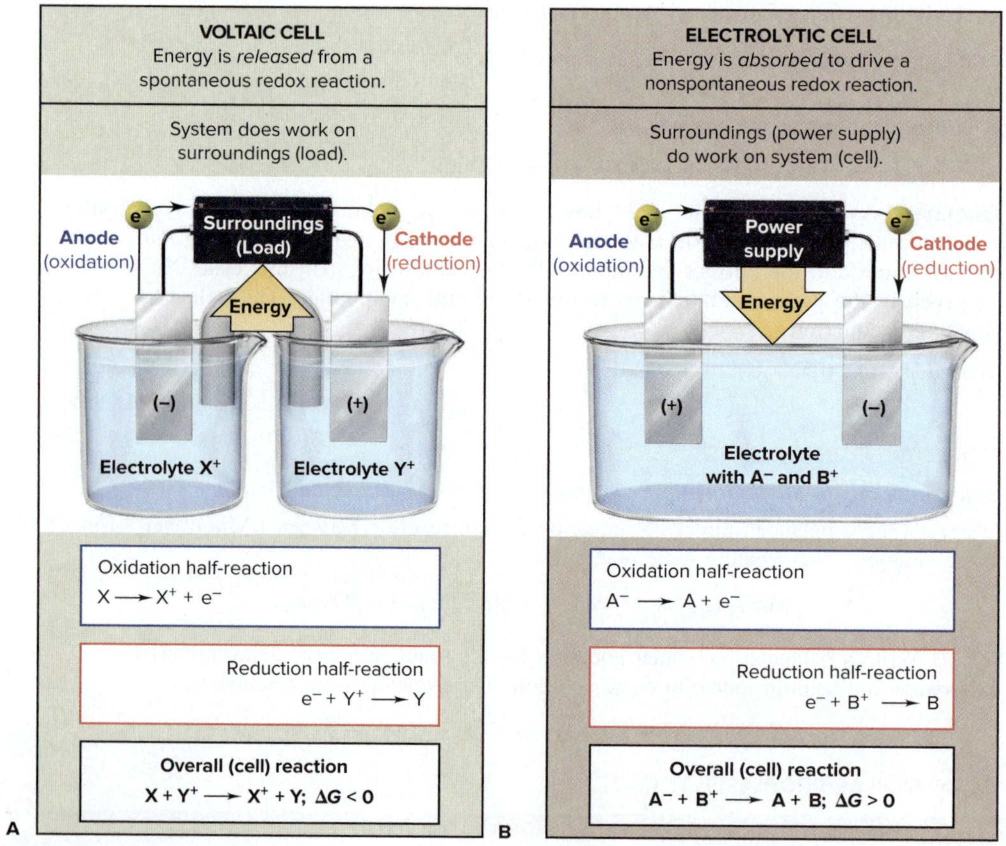

Figure 21.2 General characteristics of (A) voltaic cells and (B) electrolytic cells.

› Summary of Section 21.1

- An oxidation-reduction (redox) reaction involves the transfer of electrons from a reducing agent to an oxidizing agent.
- The half-reaction method of balancing divides the overall reaction into half-reactions that are balanced separately and then recombined.
- There are two types of electrochemical cells. In a voltaic cell, a spontaneous reaction generates electricity and does work on the surroundings. In an electrolytic cell, the surroundings supply electricity that does work to drive a nonspontaneous reaction.
- In both types of cell, two electrodes dip into electrolyte solutions; oxidation occurs at the anode, and reduction occurs at the cathode.

21.2 VOLTAIC CELLS: USING SPONTANEOUS REACTIONS TO GENERATE ELECTRICAL ENERGY

When you put a strip of zinc metal in a solution of Cu^{2+} ion, the blue color of the solution fades and a brown-black crust of copper metal forms on the zinc strip (Figure 21.3). During this spontaneous reaction, Cu^{2+} ion (blue in color) is reduced to Cu metal, while Zn metal is oxidized to Zn^{2+} ion (colorless). The overall reaction consists of two half-reactions:

$$Cu^{2+}(aq) + 2e^- \longrightarrow Cu(s) \qquad \text{[reduction]}$$

$$Zn(s) \longrightarrow Zn^{2+}(aq) + 2e^- \qquad \text{[oxidation]}$$

$$Zn(s) + Cu^{2+}(aq) \longrightarrow Zn^{2+}(aq) + Cu(s) \qquad \text{[overall reaction]}$$

Let's examine this spontaneous reaction as the basis of a voltaic cell.

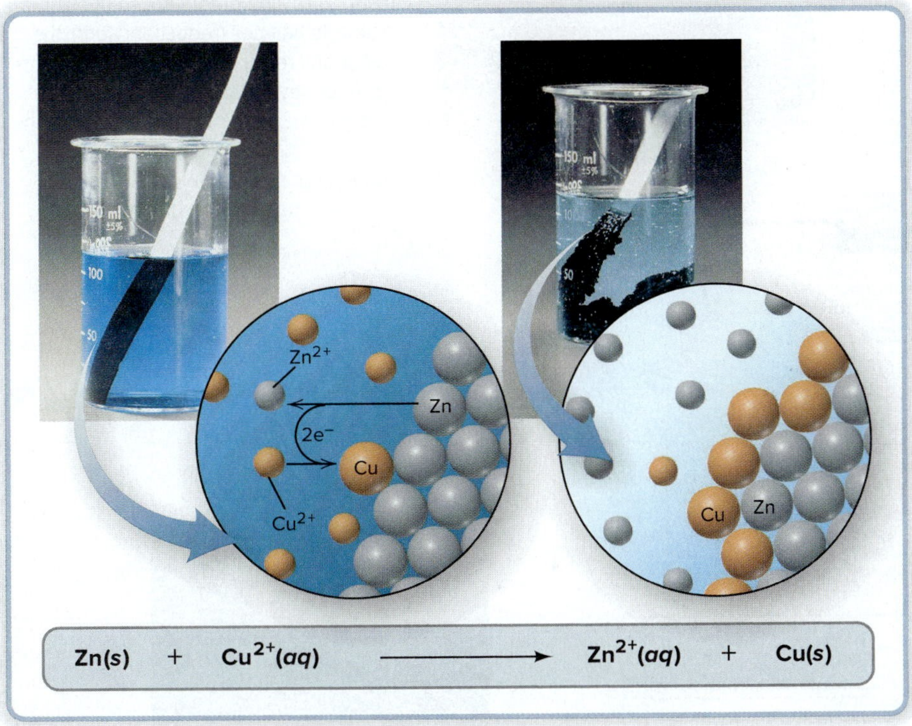

Figure 21.3 The spontaneous reaction between zinc and copper(II) ion. When zinc metal is placed in a solution of Cu^{2+} ion *(left)*, zinc is oxidized to Zn^{2+}, and Cu^{2+} is reduced to copper metal *(right)*. (The very finely divided Cu appears black.)

Source: © McGraw-Hill Education/Stephen Frisch, photographer

Construction and Operation of a Voltaic Cell

Electrons are being transferred in the Zn/Cu^{2+} reaction, but the system does not generate electrical energy because the oxidizing agent (Cu^{2+}) and the reducing agent (Zn) are in the *same* beaker. If, however, we physically separate the half-reactions and connect them by an external circuit, the electrons lost by the zinc produce an electric current as they travel through the circuit toward the copper ions that gain them.

Basis of the Voltaic Cell: Separation of Half-Reactions In any voltaic cell, the components of each half-reaction are placed in a separate container, or **half-cell,** which consists of one electrode dipping into an electrolyte solution (Figure 21.4A). The two half-cells are joined by the external circuit, and a voltmeter measures the voltage generated. A switch (not shown) closes (completes) or opens (breaks) the circuit. Here are some key points about the half-cells of the Zn/Cu^{2+} voltaic cell:

1. *Oxidation half-cell (anode compartment, always shown on the left).* The anode compartment consists of a bar of zinc (the anode) immersed in a Zn^{2+} electrolyte (such as aqueous zinc sulfate, $ZnSO_4$). Zinc metal is the reactant in the oxidation half-reaction, and the bar loses electrons *and* conducts them *out* of this half-cell.
2. *Reduction half-cell (cathode compartment, always shown on the right).* The cathode compartment consists of a bar of copper (the cathode) immersed in a Cu^{2+} electrolyte [such as aqueous copper(II) sulfate, $CuSO_4$]. Copper is the product in the reduction half-reaction, and the bar conducts electrons *into* its half-cell, where Cu^{2+} is reduced.

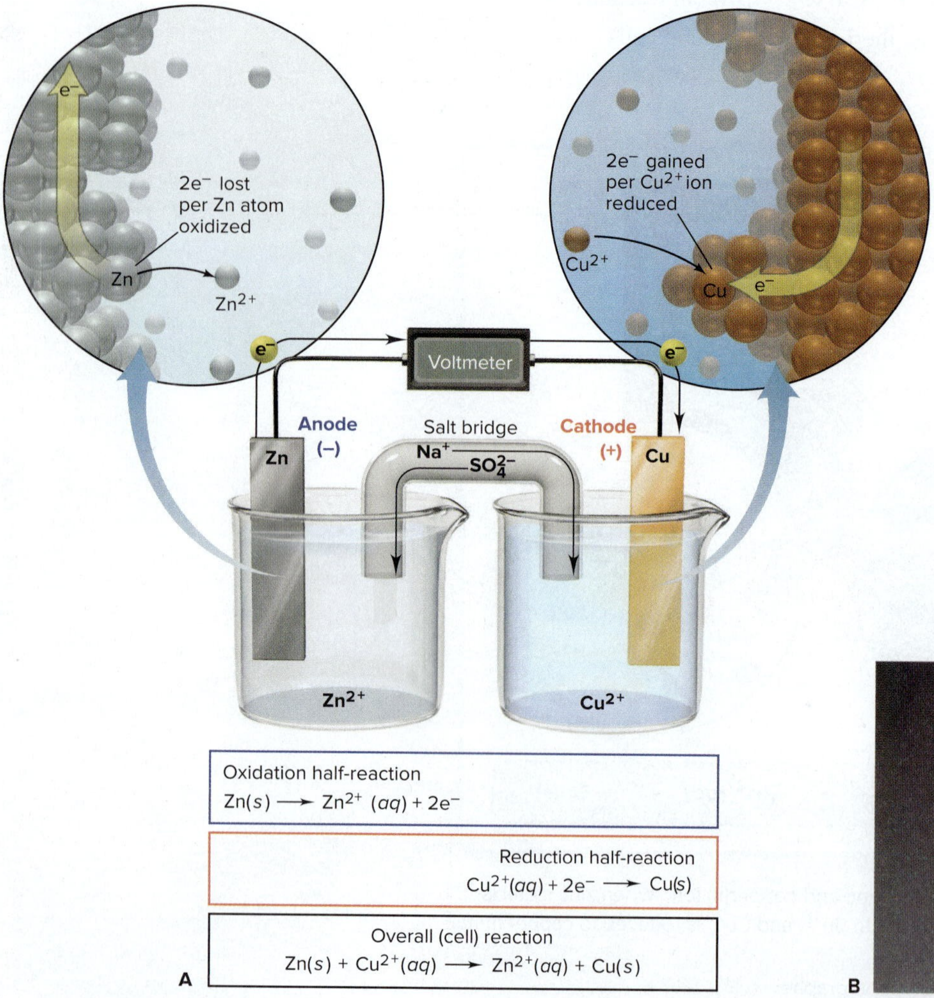

A

B

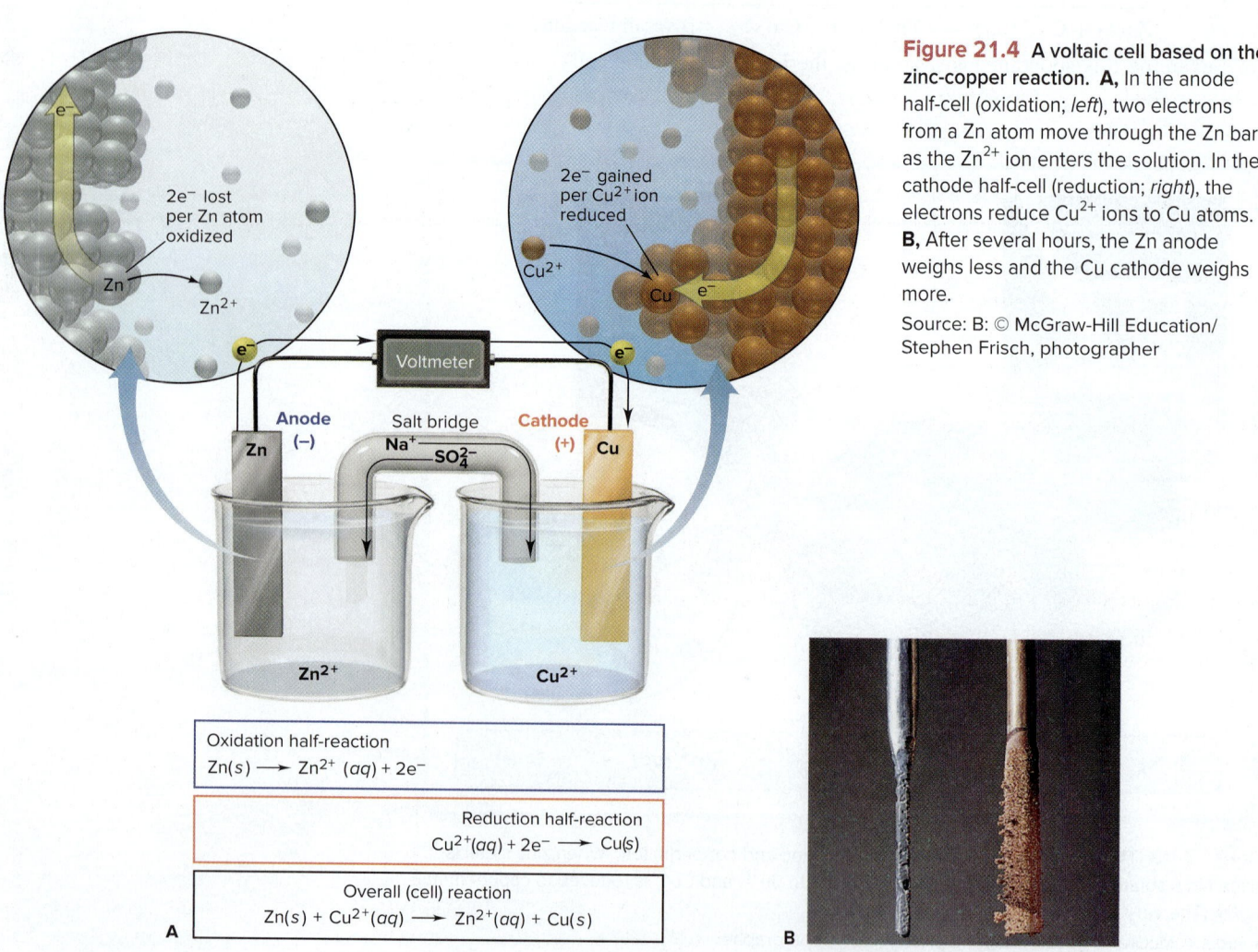

Figure 21.4 **A voltaic cell based on the zinc-copper reaction. A,** In the anode half-cell (oxidation; *left*), two electrons from a Zn atom move through the Zn bar as the Zn^{2+} ion enters the solution. In the cathode half-cell (reduction; *right*), the electrons reduce Cu^{2+} ions to Cu atoms. **B,** After several hours, the Zn anode weighs less and the Cu cathode weighs more.

Source: B: © McGraw-Hill Education/ Stephen Frisch, photographer

Charges of the Electrodes The charges of the electrodes are determined by the *source of electrons* and the *direction of electron flow* through the circuit. In this cell, Zn atoms are oxidized at the anode to Zn^{2+} ions and are the source of electrons. The Zn^{2+} ions enter the half-cell's electrolyte, while the electrons move through the bar and into the wire. *The electrons flow left to right* through the wire to the cathode, where Cu^{2+} ions in this half-cell's electrolyte are reduced to Cu atoms. As the cell operates, electrons are continuously generated at the anode and consumed at the cathode. Therefore, the anode has an excess of electrons and the cathode has a deficit of electrons: *in any* ***voltaic*** *cell, the anode is negative relative to the cathode.*

Completing the Circuit with a Salt Bridge A cell cannot operate unless the circuit is complete. The oxidation half-cell initially contains a neutral solution of Zn^{2+} and SO_4^{2-} ions, but as Zn atoms in the bar lose electrons, the Zn^{2+} ions that form enter the solution and would give it a net positive charge. Similarly, in the reduction half-cell, the initial neutral solution of Cu^{2+} and SO_4^{2-} ions would develop a net negative charge as Cu^{2+} ions leave the solution and form Cu atoms. If allowed to occur, such a charge imbalance between the half-cells would stop cell operation.

This situation is avoided by use of a **salt bridge.** It joins the half-cells and acts like a "liquid wire," allowing ions to flow through both compartments and complete the circuit while keeping the electrolyte solutions neutral. The salt bridge is an inverted U tube containing nonreacting ions, such as Na^+ and SO_4^{2-}, dissolved in a gel, which does not flow out of the tube but allows ions to diffuse into or out of the half-cells:

- *Maintaining a neutral reduction half-cell (right; cathode compartment).* As Cu^{2+} ions change to Cu atoms, Na^+ ions move from the salt bridge into the electrolyte solution (and some SO_4^{2-} ions move from the solution into the salt bridge).
- *Maintaining a neutral oxidation half-cell (left; anode compartment).* As Zn atoms change to Zn^{2+} ions, SO_4^{2-} ions move from the salt bridge into the electrolyte solution (and some Zn^{2+} ions move from the solution into the salt bridge).

Thus, the wire and the salt bridge complete the circuit:

- *Electrons move left to right* through the wire.
- *Anions move right to left* through the salt bridge into the anode half-cell.
- *Cations move left to right* through the salt bridge into the cathode half-cell.

Active vs. Inactive Electrodes The electrodes in the Zn/Cu^{2+} cell are *active* because the metals themselves are components of the half-reactions. As the cell operates, the mass of the zinc bar gradually decreases, as the $[Zn^{2+}]$ in the anode half-cell increases. At the same time, the mass of the copper bar increases as the $[Cu^{2+}]$ in the cathode half-cell decreases and the ions form atoms that "plate out" on the electrode (Figure 21.4B).

In many cases, however, there are no reaction components that can be physically used as an electrode, so *inactive* electrodes are used. Most commonly, inactive electrodes are rods of *graphite* or *platinum.* In a voltaic cell based on the following half-reactions, for instance, the reacting species cannot be made into electrodes:

$$2I^-(aq) \longrightarrow I_2(s) + 2e^- \quad \text{[anode; oxidation]}$$

$$MnO_4^-(aq) + 8H^+(aq) + 5e^- \longrightarrow Mn^{2+}(aq) + 4H_2O(l) \quad \text{[cathode; reduction]}$$

Each half-cell consists of an inactive electrode immersed in an electrolyte that contains *all the reactant species involved in that half-reaction* (Figure 21.5):

- In the *anode half-cell,* I^- ions are oxidized to solid I_2, and the released electrons flow into the graphite (C) electrode and through the wire.
- From the wire, the electrons enter the graphite electrode in the *cathode half-cell* and reduce MnO_4^- ions to Mn^{2+} ions. (A KNO_3 salt bridge is used.)

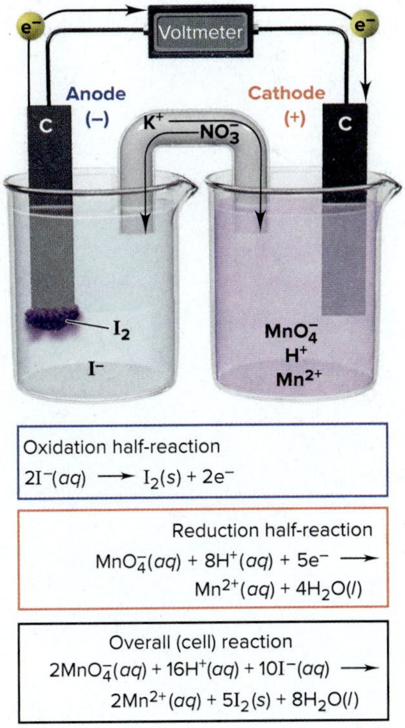

Figure 21.5 A voltaic cell using inactive electrodes.

Diagram of a Voltaic Cell As Figures 21.4A and 21.5 show, there are certain consistent features in the *diagram* of any voltaic cell:

- Components of the half-cells include electrode materials, electrolyte ions, and other species involved in the reaction.
- Electrode name (anode or cathode) and charge are shown. By convention, the anode compartment appears *on the left.*
- Each half-reaction and the overall cell reaction are given.
- Direction of electron flow in the external circuit is from left to right.
- Nature of ions and direction of ion flow in the salt bridge are shown, with cations moving right and anions moving left.

Notation for a Voltaic Cell

A useful shorthand notation describes the components of a voltaic cell. The notation for the Zn/Cu^{2+} cell is

$$\mathrm{Zn}(s)\,|\,\mathrm{Zn}^{2+}(aq)\,\|\,\mathrm{Cu}^{2+}(aq)\,|\,\mathrm{Cu}(s)$$

Key parts of the notation are as follows:

- The components of the anode compartment (oxidation half-cell) are written *to the left* of the components of the cathode compartment (reduction half-cell).
- A double vertical line indicates the half-cells are physically separated.
- A single vertical line represents a phase boundary. For example, $\mathrm{Zn}(s)\,|\,\mathrm{Zn}^{2+}(aq)$ indicates that *solid* Zn is a *different* phase from *aqueous* Zn^{2+}.
- A comma separates the half-cell components that are in the *same* phase. For example, the notation for the voltaic cell shown in Figure 21.5 is

  $$\text{graphite}\,|\,\mathrm{I}^-(aq)\,|\,\mathrm{I}_2(s)\,\|\,\mathrm{MnO}_4^-(aq),\ \mathrm{H}^+(aq),\ \mathrm{Mn}^{2+}(aq)\,|\,\text{graphite}$$

  That is, in the cathode compartment, MnO_4^-, H^+, and Mn^{2+} ions are in an aqueous solution with solid graphite immersed in it.
- If needed, the concentrations of dissolved components are given in parentheses; for example, if the concentrations of Zn^{2+} and Cu^{2+} are 1 *M*, we write

  $$\mathrm{Zn}(s)\,|\,\mathrm{Zn}^{2+}(1M)\,\|\,\mathrm{Cu}^{2+}(1M)\,|\,\mathrm{Cu}(s)$$
- Half-cell components usually appear in the same order as in the half-reaction, and electrodes appear at the far left (anode) and far right (cathode) of the notation.
- Ions in the salt bridge are not part of the reaction so they are not in the notation.

SAMPLE PROBLEM 21.2 Describing a Voltaic Cell with a Diagram and Notation

Problem Draw a diagram, show balanced half-cell and cell equations, and write the notation for a voltaic cell that consists of one half-cell with a Cr bar in a $Cr(NO_3)_3$ solution, another half-cell with an Ag bar in an $AgNO_3$ solution, and a KNO_3 salt bridge. Measurement indicates that the Cr electrode is negative relative to the Ag electrode.

Plan From the given contents of the half-cells, we write the half-reactions. To determine which is the anode compartment (oxidation) and which is the cathode (reduction), we note the relative electrode charges (which are based on the direction of the spontaneous redox reaction). Electrons are released into the anode during oxidation, so it has a negative charge. We are told that Cr is negative, so it is the anode and Ag is the cathode.

Solution Writing the balanced half-reactions. The Ag half-reaction consumes e^-:

$$\mathrm{Ag}^+(aq) + \mathrm{e}^- \longrightarrow \mathrm{Ag}(s) \qquad \text{[reduction; cathode]}$$

The Cr half-reaction releases e^-:

$$\mathrm{Cr}(s) \longrightarrow \mathrm{Cr}^{3+}(aq) + 3\mathrm{e}^- \qquad \text{[oxidation; anode]}$$

Writing the balanced overall cell equation. We triple the reduction half-reaction to balance e^- and combine the half-reactions:

$$\mathrm{Cr}(s) + 3\mathrm{Ag}^+(aq) \longrightarrow \mathrm{Cr}^{3+}(aq) + 3\mathrm{Ag}(s)$$

Determining direction of electron and ion flow. The released e^- in the Cr electrode (negative) flow through the external circuit to the Ag electrode (positive). As Cr^{3+} ions enter the anode electrolyte, NO_3^- ions enter from the salt bridge to maintain neutrality. As Ag^+ ions leave the cathode electrolyte and plate out on the Ag electrode, K^+ ions enter from the salt bridge to maintain neutrality. The cell diagram is shown in the margin. Writing the cell notation with the oxidation components on the left and the reduction components on the right:

$$Cr(s) \mid Cr^{3+}(aq) \parallel Ag^+(aq) \mid Ag(s)$$

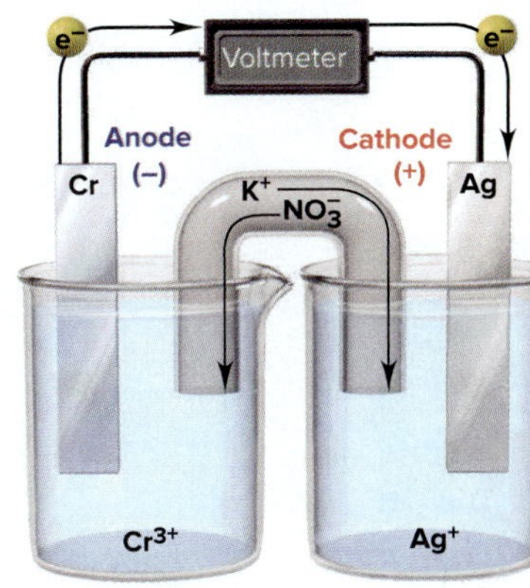

Check Always be sure that the half-reactions and the cell reaction are balanced, the half-cells contain *all* components of the half-reactions, and the electron and ion flow are shown. You should be able to write the half-reactions from the cell notation as a check.

Comment The diagram of a voltaic cell relies on the *direction of the spontaneous reaction* to give the oxidation (anode; negative) and reduction (cathode; positive) half-reactions.

FOLLOW-UP PROBLEMS

21.2A In one half-cell of a voltaic cell, a graphite rod dips into an acidic solution of $K_2Cr_2O_7$ and $Cr(NO_3)_3$; in the other half-cell, a tin bar dips into a $Sn(NO_3)_2$ solution. A KNO_3 salt bridge joins the half-cells. The tin electrode is negative relative to the graphite. Diagram the cell, and write the balanced half-cell and cell equations and the cell notation.

21.2B In one half-cell of a voltaic cell, a graphite rod dips into an acidic solution of $KClO_3$ and KCl; in the other half-cell, a nickel bar dips into a $Ni(NO_3)_2$ solution. A KNO_3 salt bridge joins the half-cells, and the graphite is positive relative to the nickel. Diagram the cell, and write balanced half-cell and cell equations and the cell notation.

SOME SIMILAR PROBLEMS 21.29–21.34

Why Does a Voltaic Cell Work?

What principles explain *how* the Zn/Cu^{2+} cell reaction takes place and *why* electrons flow in the direction they do? Let's examine what is happening when the switch is open and no reaction is occurring. In each half-cell, the metal electrode is in equilibrium with the metal ions in the electrolyte and the electrons within the metal:

$$Zn(s) \rightleftharpoons Zn^{2+}(aq) + 2e^- \text{ (in Zn metal)}$$
$$Cu(s) \rightleftharpoons Cu^{2+}(aq) + 2e^- \text{ (in Cu metal)}$$

Given the direction of the overall spontaneous reaction, Zn gives up its electrons more easily than Cu does; that is, Zn is a stronger reducing agent. Therefore, the equilibrium position of the Zn half-reaction lies farther to the right. We can think of the electrons in the Zn electrode as being subject to a greater "pressure" than those in the Cu electrode, a greater potential energy (referred to as *electrical potential*) ready to "push" them through the circuit. Close the switch, and electrons flow from the Zn to the Cu electrode to equalize this difference in electrical potential. The flow disturbs the equilibrium at each electrode. The Zn half-reaction shifts to the right to restore the electrons flowing out, and the Cu half-reaction shifts to the left to remove the electrons flowing in. Thus, *the spontaneous reaction occurs as a result of the different abilities of these metals to give up their electrons.* ›

Electron Flow and Water Flow

Consider this analogy between electron "pressure" and water pressure. A U tube (cell) is separated into two arms (two half-cells) by a stopcock (switch), and the two arms contain water at different heights (the half-cells have different electrical potentials). Open the stopcock (close the switch), and the different heights (potential difference) become equal as water flows (electrons flow and current is generated).

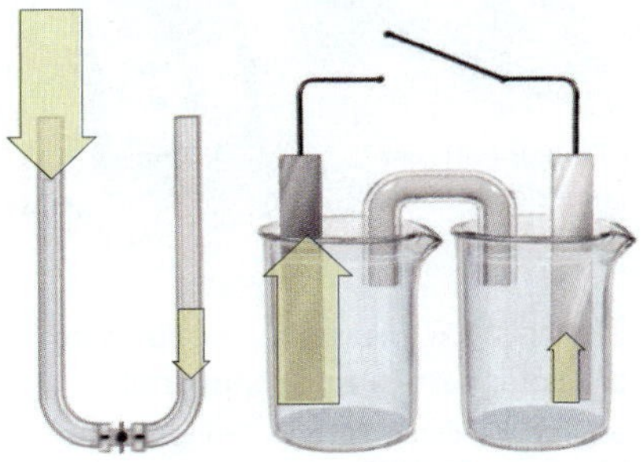

› Summary of Section 21.2

- A voltaic cell consists of oxidation (anode) and reduction (cathode) half-cells, connected by a wire to conduct electrons and a salt bridge to maintain charge neutrality.
- Electrons move from the anode (left) to the cathode (right), while cations move from the salt bridge into the cathode half-cell and anions move from the salt bridge into the anode half-cell.
- The cell notation shows the species and their phases in each half-cell, as well as the direction of current flow.
- A voltaic cell operates because species in the two half-cells differ in the tendency to lose electrons (strength as a reducing agent).

21.3 CELL POTENTIAL: OUTPUT OF A VOLTAIC CELL

A voltaic cell converts the free energy change of a spontaneous reaction into the kinetic energy of electrons moving through an external circuit (electrical energy). This electrical energy is proportional to the *difference in electrical potential between the two electrodes,* which is called the **cell potential (E_{cell})**, the **voltage** of the cell, or the **electromotive force (emf).**

Electrons flow spontaneously from the negative to the positive electrode, that is, toward the electrode with the more positive electrical potential. Thus, when the cell operates *spontaneously,* there is a *positive* cell potential:

$$E_{cell} > 0 \text{ for a spontaneous process} \quad \textbf{(21.1)}$$

- $E_{cell} > 0$ for a spontaneous reaction. The more positive E_{cell} is, the more work the cell can do, and the farther the reaction proceeds to the right as written.
- $E_{cell} < 0$ for a *nonspontaneous* cell reaction.
- $E_{cell} = 0$ when the reaction has reached equilibrium, so the cell can do no more work. (We return to these ideas in Section 21.4.)

Table 21.1 Voltages of Some Voltaic Cells

Voltaic Cell	Voltage (V)
Common alkaline flashlight battery	1.5
Lead-acid car battery (6 cells ≈ 12 V)	2.1
Calculator battery (mercury)	1.3
Lithium-ion laptop battery	3.7
Electric eel (~5000 cells in 6-ft eel = 750 V)	0.15
Nerve of giant squid (across cell membrane)	0.070

Standard Cell Potential (E°_{cell})

The SI unit of electrical potential is the **volt (V),** and the SI unit of electrical charge is the **coulomb (C).** By definition, if two electrodes differ by 1 volt of electrical potential, 1 joule of energy is released (that is, 1 joule of work can be done) for each coulomb of charge that moves between the electrodes. That is,

$$1 \text{ V} = 1 \text{ J/C} \quad \textbf{(21.2)}$$

Table 21.1 lists the voltages of some commercial and natural voltaic cells.

The *measured* cell potential is affected by changes in concentration as the reaction proceeds and by energy losses from heating the cell and external circuit. Therefore, as with other thermodynamic quantities, to compare potentials of different cells, we obtain a **standard cell potential (E°_{cell})**, which is measured at a specified temperature (usually 298 K) with no current flowing* and *all components in their standard states:* 1 atm for gases, 1 *M* for solutions, the pure solid for electrodes. For example, the zinc-copper cell produces 1.10 V when it operates at 298 K with $[Zn^{2+}] = [Cu^{2+}] = 1\ M$ (Figure 21.6):

$$Zn(s) + Cu^{2+}(aq; 1\ M) \longrightarrow Zn^{2+}(aq; 1\ M) + Cu(s) \qquad E_{cell} = E^{\circ}_{cell} = 1.10 \text{ V}$$

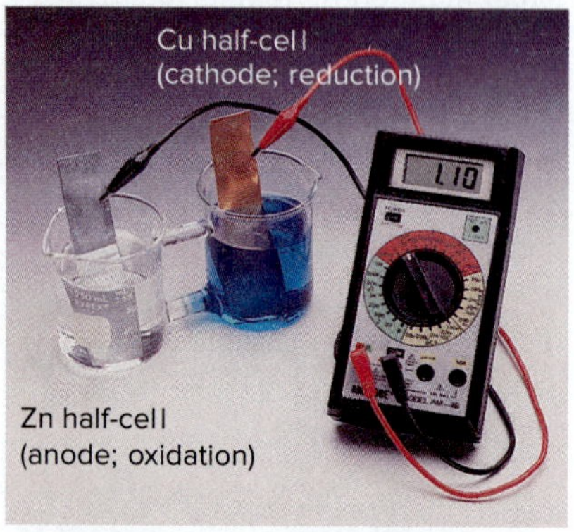

Figure 21.6 Measuring the standard cell potential of a zinc-copper cell.

Source: © Richard Megna/Fundamental Photographs, NYC

Standard Electrode (Half-Cell) Potential ($E^{\circ}_{half\text{-}cell}$) Just as each half-reaction makes up part of the overall reaction, each half-cell (electrode) potential makes up a part of the overall cell potential. The **standard electrode potential ($E^{\circ}_{half\text{-}cell}$)** (or **standard half-cell potential**) is the electrical potential of a given half-reaction (half-cell compartment) with all components in their standard states.

By convention (and as reference tables list them), *standard electrode potentials refer to the half-reactions written as* ***reductions.*** (A list of these potentials, like that in Appendix D, is called an *emf series* or a *table of standard electrode potentials.*) For example, for the zinc-copper reaction, both the zinc half-reaction (E°_{zinc}, anode compartment) and the copper half-reaction (E°_{copper}, cathode compartment) are written as reductions:

$$Zn^{2+}(aq) + 2e^- \longrightarrow Zn(s) \qquad E^{\circ}_{zinc}\ (E^{\circ}_{anode}) \qquad \text{[reduction]}$$
$$Cu^{2+}(aq) + 2e^- \longrightarrow Cu(s) \qquad E^{\circ}_{copper}\ (E^{\circ}_{cathode}) \qquad \text{[reduction]}$$

But, since the overall cell reaction involves the *oxidation* of zinc, not the *reduction* of Zn^{2+}, we can show what is actually occurring by reversing the zinc (anode) half-reaction:

$$Zn(s) \longrightarrow Zn^{2+}(aq) + 2e^- \qquad \text{[oxidation]}$$
$$Cu^{2+}(aq) + 2e^- \longrightarrow Cu(s) \qquad \text{[reduction]}$$

*The tiny current required to operate modern digital voltmeters makes a negligible difference in the value of E°_{cell}

Calculating E°_{cell} from $E^\circ_{half\text{-}cell}$ The overall redox reaction for the zinc-copper cell is the sum of its half-reactions:

$$Zn(s) + Cu^{2+}(aq) \longrightarrow Zn^{2+}(aq) + Cu(s)$$

Because electrons flow spontaneously from the negative to the positive electrode, the copper electrode (cathode) has a more positive $E^\circ_{half\text{-}cell}$ than the zinc electrode (anode). As you'll see shortly, to obtain a positive E°_{cell} for this spontaneous redox reaction, we subtract E°_{zinc} from E°_{copper}:

$$E^\circ_{cell} = E^\circ_{copper} - E^\circ_{zinc}$$

We can generalize this result for any voltaic cell: *the standard cell potential is the standard electrode potential of the cathode (reduction) half-cell* ***minus*** *the standard electrode potential of the anode (oxidation) half-cell:*

$$E^\circ_{cell} = E^\circ_{cathode(reduction)} - E^\circ_{anode(oxidation)} \quad \textbf{(21.3)}$$

For a *spontaneous* reaction at standard conditions, this calculation gives $E^\circ_{cell} > 0$.

SAMPLE PROBLEM 21.3 Using $E^\circ_{half\text{-}cell}$ Values to Find E°_{cell}

Problem Balance the following skeleton ionic reaction, and calculate E°_{cell} to decide whether the reaction is spontaneous:

$$Mn^{2+}(aq) + Br_2(l) \longrightarrow MnO_4^-(aq) + Br^-(aq) \quad \text{[acidic solution]}$$

Plan We balance the skeleton reaction (see Sample Problem 21.1). Then, we look up the $E^\circ_{half\text{-}cell}$ values in Appendix D and use Equation 21.3 to find E°_{cell}. If E°_{cell} is positive, the reaction is spontaneous.

Solution Balancing the skeleton reaction:

$$2[Mn^{2+} + 4H_2O \longrightarrow MnO_4^- + 8H^+ + 5e^-] \quad \text{[oxidation]}$$

$$5[Br_2 + 2e^- \longrightarrow 2Br^-] \quad \text{[reduction]}$$

$$2Mn^{2+}(aq) + 5Br_2(l) + 8H_2O(l) \longrightarrow 2MnO_4^-(aq) + 10Br^-(aq) + 16H^+(aq) \quad \text{[overall]}$$

Using $E^\circ_{half\text{-}cell}$ values to find E°_{cell}:

$$E^\circ_{cell} = E^\circ_{cathode(reduction)} - E^\circ_{anode(oxidation)} = 1.07\text{ V} - 1.51\text{ V}$$
$$= -0.44\text{ V; not spontaneous}$$

Check Rounding to check the calculation: 1.00 V − 1.50 V = −0.50 V. A negative E°_{cell} indicates a nonspontaneous reaction.

FOLLOW-UP PROBLEMS

21.3A Balance the following skeleton ionic reaction, and calculate E°_{cell} to decide whether the reaction is spontaneous:

$$Ag(s) + Cu^{2+}(aq) \longrightarrow Ag^+(aq) + Cu(s)$$

21.3B Balance the following skeleton ionic reaction, and calculate E°_{cell} to decide whether the reaction is spontaneous:

$$Cl_2(g) + Fe^{2+}(aq) \longrightarrow Cl^-(aq) + Fe^{3+}(aq)$$

SOME SIMILAR PROBLEMS 21.42–21.45

Determining $E^\circ_{half\text{-}cell}$ with the Standard Hydrogen Electrode To compare half-cell potentials, we need to know the portion of E°_{cell} contributed by each half-cell. But, how can we find individual half-cell potentials if we can measure only the potential of the overall cell? We can do so because half-cell potentials, such as E°_{zinc} and E°_{copper}, are measured *relative* to a standard reference half-cell, *which has a standard electrode potential defined as zero* ($E^\circ_{reference} \equiv 0.00$ V). To find an unknown standard electrode potential ($E^\circ_{unknown}$), we construct a voltaic cell consisting of this reference half-cell and the unknown half-cell. Since $E^\circ_{reference}$ is zero, the overall E°_{cell} gives $E^\circ_{unknown}$.

The **standard reference half-cell** is a **standard hydrogen electrode,** which consists of a platinum electrode that has H_2 gas at 1 atm bubbling through it and is immersed in 1 *M* strong acid, which means 1 *M* $H^+(aq)$ [or $H_3O^+(aq)$]. Thus, the reference half-reaction is

$$2H^+(aq;\, 1\,M) + 2e^- \rightleftharpoons H_2(g;\, 1\text{ atm}) \qquad E^\circ_{\text{reference}} = 0.00\text{ V}$$

Depending on the unknown half-cell, the reference half-cell can be the anode or the cathode:

- When H_2 is oxidized, the reference half-cell is the anode, and so *reduction* occurs at the unknown half-cell:

$$E^\circ_{\text{cell}} = E^\circ_{\text{cathode}} - E^\circ_{\text{anode}} = E^\circ_{\text{unknown}} - E^\circ_{\text{reference}} = E^\circ_{\text{unknown}} - 0.00\text{ V} = E^\circ_{\text{unknown}}$$

- When H^+ is reduced, the reference half-cell is the cathode, and so *oxidation* occurs at the unknown half-cell:

$$E^\circ_{\text{cell}} = E^\circ_{\text{cathode}} - E^\circ_{\text{anode}} = E^\circ_{\text{reference}} - E^\circ_{\text{unknown}} = 0.00\text{ V} - E^\circ_{\text{unknown}} = -E^\circ_{\text{unknown}}$$

Figure 21.7 shows a voltaic cell that has the Zn/Zn^{2+} half-reaction in one compartment and the H^+/H_2 (or H_3O^+/H_2) half-reaction in the other. The zinc electrode is negative relative to the hydrogen electrode, so we know that the zinc is being oxidized and is the anode. The measured E°_{cell} is +0.76 V (the reaction is spontaneous), and we use this value to find the unknown standard electrode potential, E°_{zinc}:

$$2H^+(aq) + 2e^- \longrightarrow H_2(g) \qquad E^\circ_{\text{reference}} = 0.00\text{ V [cathode; reduction]}$$

$$Zn(s) \longrightarrow Zn^{2+}(aq) + 2e^- \qquad E^\circ_{\text{zinc}} = ?\text{ V} \quad \text{[anode; oxidation]}$$

$$Zn(s) + 2H^+(aq) \longrightarrow Zn^{2+}(aq) + H_2(g) \qquad E^\circ_{\text{cell}} = 0.76\text{ V}$$

$$E^\circ_{\text{cell}} = E^\circ_{\text{cathode}} - E^\circ_{\text{anode}} = E^\circ_{\text{reference}} - E^\circ_{\text{zinc}}$$

Solving for E°_{zinc}, the standard half-cell potential for $Zn^{2+} + 2e^- \longrightarrow Zn$, gives

$$E^\circ_{\text{zinc}} = E^\circ_{\text{reference}} - E^\circ_{\text{cell}} = 0.00\text{ V} - 0.76\text{ V} = -0.76\text{ V}$$

Now we can return to the zinc-copper cell and use the measured E°_{cell} (1.10 V) and the value we just found for E°_{zinc} to calculate E°_{copper}:

$$E^\circ_{\text{cell}} = E^\circ_{\text{cathode}} - E^\circ_{\text{anode}} = E^\circ_{\text{copper}} - E^\circ_{\text{zinc}}$$

and,

$$E^\circ_{\text{copper}} = E^\circ_{\text{cell}} + E^\circ_{\text{zinc}} = 1.10\text{ V} + (-0.76\text{ V}) = 0.34\text{ V}$$

By continuing this process of constructing cells with one known and one unknown electrode potential, we find other standard electrode potentials.

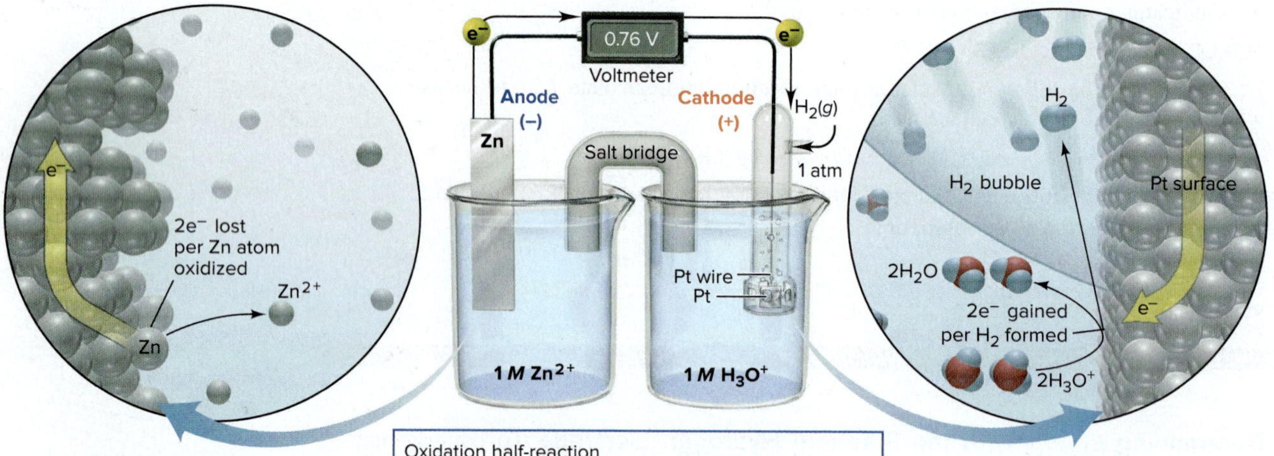

Oxidation half-reaction
$Zn(s) \longrightarrow Zn^{2+}(aq) + 2e^-$

Reduction half-reaction
$2H_3O^+(aq) + 2e^- \longrightarrow H_2(g) + 2H_2O(l)$

Overall (cell) reaction
$Zn(s) + 2H_3O^+(aq) \longrightarrow Zn^{2+}(aq) + H_2(g) + 2H_2O(l)$

Figure 21.7 Determining an unknown $E^\circ_{\text{half-cell}}$ with the standard reference (hydrogen) electrode. The magnified view of the hydrogen half-reaction *(right)* shows two H_3O^+ ions being reduced to two H_2O molecules and an H_2 molecule, which enters the H_2 bubble.

SAMPLE PROBLEM 21.4 Calculating an Unknown $E^\circ_{\text{half-cell}}$ from E°_{cell}

Problem A voltaic cell houses the reaction between aqueous bromine and zinc metal:

$$Br_2(aq) + Zn(s) \longrightarrow Zn^{2+}(aq) + 2Br^-(aq) \qquad E^\circ_{\text{cell}} = 1.83\text{ V}$$

Calculate E°_{bromine}, given $E^\circ_{\text{zinc}} = -0.76$ V.

Plan E°_{cell} is positive, so the reaction is spontaneous as written. By dividing the reaction into half-reactions, we see that Br_2 is reduced and Zn is oxidized; thus, the zinc half-cell contains the anode. We use Equation 21.3 and the known E°_{zinc} to find E°_{unknown} (E°_{bromine}).

Solution Dividing the reaction into half-reactions:

$$Br_2(aq) + 2e^- \longrightarrow 2Br^-(aq) \qquad E^\circ_{\text{unknown}} = E^\circ_{\text{bromine}} = ?\text{ V}$$
$$Zn(s) \longrightarrow Zn^{2+}(aq) + 2e^- \qquad E^\circ_{\text{zinc}} = -0.76\text{ V}$$

Calculating E°_{bromine}:

$$E^\circ_{\text{cell}} = E^\circ_{\text{cathode}} - E^\circ_{\text{anode}} = E^\circ_{\text{bromine}} - E^\circ_{\text{zinc}}$$
$$E^\circ_{\text{bromine}} = E^\circ_{\text{cell}} + E^\circ_{\text{zinc}} = 1.83\text{ V} + (-0.76\text{ V}) = 1.07\text{ V}$$

Check A good check is to make sure that calculating $E^\circ_{\text{bromine}} - E^\circ_{\text{zinc}}$ gives E°_{cell}: 1.07 V − (−0.76 V) = 1.83 V.

Comment Keep in mind that, whichever half-cell potential is unknown, reduction is the cathode half-reaction and oxidation is the anode half-reaction. *Always* subtract E°_{anode} from E°_{cathode} to get E°_{cell}. (We use E°_{bromine} in Follow-up Problem 21.4A.)

FOLLOW-UP PROBLEMS

21.4A A voltaic cell based on the reaction between aqueous Br_2 and vanadium(III) ions has $E^\circ_{\text{cell}} = 0.73$ V:

$$Br_2(aq) + 2V^{3+}(aq) + 2H_2O(l) \longrightarrow 2VO^{2+}(aq) + 4H^+(aq) + 2Br^-(aq)$$

What is $E^\circ_{\text{vanadium}}$, the standard electrode potential for the reduction of VO^{2+} to V^{3+}?

21.4B A voltaic cell based on the reaction between vanadium(III) and nitrate ions has $E^\circ_{\text{cell}} = 0.62$ V:

$$3V^{3+}(aq) + NO_3^-(aq) + H_2O(l) \longrightarrow 3VO^{2+}(aq) + NO(g) + 2H^+(aq)$$

Use the value for $E^\circ_{\text{vanadium}}$ you obtained in Follow-up Problem 21.4A to find E°_{nitrate}, the standard electrode potential for the reduction of NO_3^- to NO.

SOME SIMILAR PROBLEMS 21.38 and 21.39

Relative Strengths of Oxidizing and Reducing Agents

We learn the relative strengths of oxidizing and reducing agents from measuring cell potentials. Three oxidizing agents we just discussed are Cu^{2+}, H^+, and Zn^{2+}. Let's rank, from high *(top)* to low *(bottom)*, their relative oxidizing strengths by writing each half-reaction as a reduction (gain of electrons), with its corresponding standard electrode potential:

$$Cu^{2+}(aq) + 2e^- \longrightarrow Cu(s) \qquad E^\circ = 0.34\text{ V}$$
$$2H^+(aq) + 2e^- \longrightarrow H_2(g) \qquad E^\circ = 0.00\text{ V}$$
$$Zn^{2+}(aq) + 2e^- \longrightarrow Zn(s) \qquad E^\circ = -0.76\text{ V}$$

The more positive the E° value, the more readily the reaction (as written) occurs; thus, Cu^{2+} gains two e^- more readily than H^+, which gains them more readily than Zn^{2+}:

- Strength as an oxidizing agent *decreases* top to bottom: $Cu^{2+} > H^+ > Zn^{2+}$.
- Strength as a reducing agent *increases* top to bottom: $Zn > H_2 > Cu$.

By continuing this process with other half-cells, we create a list of reduction half-reactions in *decreasing* order of standard electrode potential (from most positive

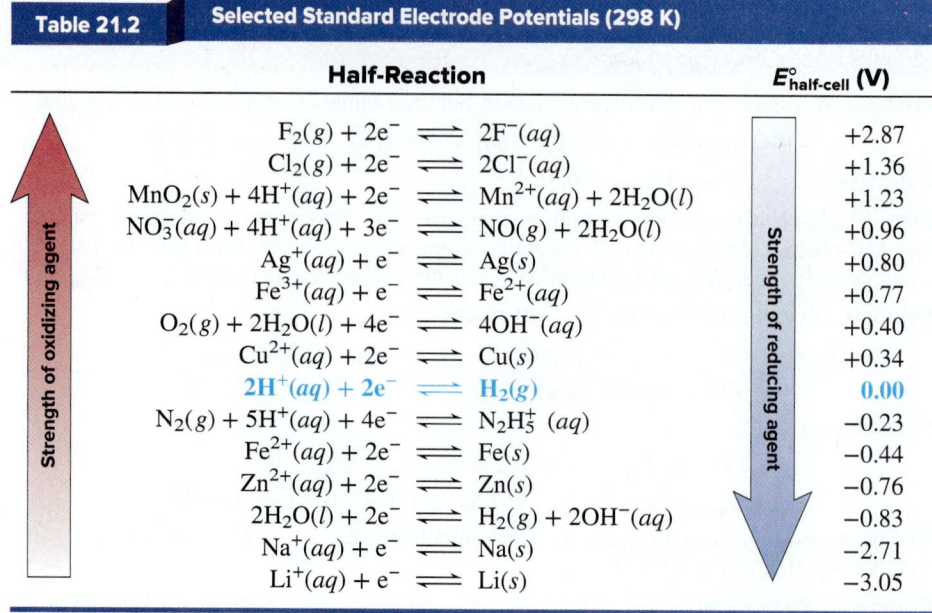

Table 21.2 Selected Standard Electrode Potentials (298 K)

(Strength of oxidizing agent increases going up; strength of reducing agent increases going down.)

Half-Reaction	$E^\circ_{half\text{-}cell}$ (V)
$F_2(g) + 2e^- \rightleftharpoons 2F^-(aq)$	+2.87
$Cl_2(g) + 2e^- \rightleftharpoons 2Cl^-(aq)$	+1.36
$MnO_2(s) + 4H^+(aq) + 2e^- \rightleftharpoons Mn^{2+}(aq) + 2H_2O(l)$	+1.23
$NO_3^-(aq) + 4H^+(aq) + 3e^- \rightleftharpoons NO(g) + 2H_2O(l)$	+0.96
$Ag^+(aq) + e^- \rightleftharpoons Ag(s)$	+0.80
$Fe^{3+}(aq) + e^- \rightleftharpoons Fe^{2+}(aq)$	+0.77
$O_2(g) + 2H_2O(l) + 4e^- \rightleftharpoons 4OH^-(aq)$	+0.40
$Cu^{2+}(aq) + 2e^- \rightleftharpoons Cu(s)$	+0.34
$2H^+(aq) + 2e^- \rightleftharpoons H_2(g)$	0.00
$N_2(g) + 5H^+(aq) + 4e^- \rightleftharpoons N_2H_5^+(aq)$	−0.23
$Fe^{2+}(aq) + 2e^- \rightleftharpoons Fe(s)$	−0.44
$Zn^{2+}(aq) + 2e^- \rightleftharpoons Zn(s)$	−0.76
$2H_2O(l) + 2e^- \rightleftharpoons H_2(g) + 2OH^-(aq)$	−0.83
$Na^+(aq) + e^- \rightleftharpoons Na(s)$	−2.71
$Li^+(aq) + e^- \rightleftharpoons Li(s)$	−3.05

to most negative), as in Appendix D; a few examples are presented in Table 21.2. There are several key points to keep in mind:

- All values are relative to the standard hydrogen (reference) electrode.
- Since the half-reactions are written as *reductions, reactants are oxidizing agents* and *products are reducing agents.*
- The more positive the $E^\circ_{half\text{-}cell}$, the more readily the half-reaction occurs, as written.
- Half-reactions are shown with an equilibrium arrow because each can occur as a reduction (at the cathode) or an oxidation (at the anode), depending on the $E^\circ_{half\text{-}cell}$ of the other half-reaction. In calculations, *always* use $E^\circ_{half\text{-}cell}$ for the half-reaction *written as a reduction.*
- As Appendix D (or Table 21.2) is arranged, the strength of the oxidizing agent (reactant) *increases going up (bottom to top),* and the strength of the reducing agent (product) *increases going down (top to bottom).*

Thus, $F_2(g)$ is the strongest oxidizing agent (has the largest positive E°), which means $F^-(aq)$ is the weakest reducing agent. Similarly, $Li^+(aq)$ is the weakest oxidizing agent (has the most negative E°), which means $Li(s)$ is the strongest reducing agent: a *strong oxidizing agent forms a weak reducing agent,* and vice versa. Rely on your knowledge of the redox behavior of the elements (see Section 8.4) if you forget the ranking in the table:

- F_2 is very electronegative and occurs naturally as F^-, so it is easily reduced (gains electrons) and must be a strong oxidizing agent (high, positive E°).
- Li metal has a low ionization energy and occurs naturally as Li^+, so it is easily oxidized (loses electrons) and must be a strong reducing agent (low, negative E°).

Using $E^\circ_{half\text{-}cell}$ Values to Write Spontaneous Redox Reactions

Every redox reaction is the sum of two half-reactions, so there is a reducing agent and an oxidizing agent on each side. In the zinc-copper reaction, for instance, Zn and Cu are the reducing agents, and Cu^{2+} and Zn^{2+} are the oxidizing agents. The stronger oxidizing and reducing agents react spontaneously to form the weaker oxidizing and reducing agents:

$$\underset{\text{stronger reducing agent}}{Zn(s)} + \underset{\text{stronger oxidizing agent}}{Cu^{2+}(aq)} \longrightarrow \underset{\text{weaker oxidizing agent}}{Zn^{2+}(aq)} + \underset{\text{weaker reducing agent}}{Cu(s)}$$

Based on the order of the $E°$ values in Appendix D, and as we just saw for the Cu^{2+}/Cu, H^+/H_2, and Zn^{2+}/Zn redox pairs (or redox *couples*), *the stronger oxidizing agent (species on the left) has a half-reaction with a larger (more positive or less negative) $E°$ value, and the stronger reducing agent (species on the right) has a half-reaction with a smaller (less positive or more negative) $E°$ value.* Therefore, we can use Appendix D to choose a redox reaction for constructing a voltaic cell.

Writing a Spontaneous Reaction *with* Appendix D A spontaneous reaction ($E°_{cell} > 0$) will occur between an oxidizing agent and any reducing agent that lies *below* it in the list of standard electrode potentials in Appendix D. In other words:

- *For a spontaneous reaction to occur, the half-reaction higher in the list proceeds at the cathode as written, and the half-reaction lower in the list proceeds at the anode in reverse.*

This pairing ensures that the stronger oxidizing agent (higher on the left) and stronger reducing agent (lower on the right) will be the reactants. For example, two half-reactions in the order they appear in Appendix D are

$$Cl_2(g) + 2e^- \longrightarrow 2Cl^-(aq) \qquad E°_{chlorine} = 1.36\ V$$
$$Ni^{2+}(aq) + 2e^- \longrightarrow Ni(s) \qquad E°_{nickel} = -0.25\ V$$

Here are the steps we follow to write a spontaneous reaction between these reactants:

- Reverse the nickel half-reaction (it is lower in the list):

$$Ni(s) \longrightarrow Ni^{2+}(aq) + 2e^- \qquad E°_{nickel} = -0.25\ V$$

  Note, however, that *we do **not** reverse the sign of $E°_{half\text{-}cell}$* because the arithmetic operation of subtraction in Equation 21.3 (cathode – anode) will do that.
- Make sure that e^- lost equals e^- gained; if not, multiply one or both half-reactions by a coefficient to accomplish this. In this case, both e^- lost and e^- gained are $2e^-$, so we skip this step.
- Add the half-reactions to get the balanced equation:

$$Cl_2(g) + Ni(s) \longrightarrow Ni^{2+}(aq) + 2Cl^-(aq)$$

- Apply Equation 21.3 to find $E°_{cell}$:

$$E°_{cell} = E°_{cathode} - E°_{anode} = E°_{chlorine} - E°_{nickel} = 1.36\ V - (-0.25\ V) = 1.61\ V$$

Writing a Spontaneous Reaction *Without* Appendix D Even when a list like Appendix D is not available, we can write a spontaneous redox reaction from a given pair of half-reactions. For example, here are two half-reactions:

$$Sn^{2+}(aq) + 2e^- \longrightarrow Sn(s) \qquad E°_{tin} = -0.14\ V$$
$$Ag^+(aq) + e^- \longrightarrow Ag(s) \qquad E°_{silver} = 0.80\ V$$

Two steps are required:

1. Reverse one of the half-reactions into an oxidation step so that the difference of the electrode potentials (cathode *minus* anode) gives a *positive* $E°_{cell}$. (Remember that when we reverse the half-reaction, we do *not* reverse the sign of $E°_{half\text{-}cell}$.)
2. Multiply by one or more coefficients to make e^- lost equal e^- gained, add the rearranged half-reactions to get a balanced overall equation, and cancel species common to both sides.

We want the reactants to be the stronger oxidizing and reducing agents:

- The larger (more positive) $E°$ value for the silver half-reaction means that Ag^+ is a stronger oxidizing agent (gains electrons more readily) than Sn^{2+}.
- The smaller (more negative) $E°$ value for the tin half-reaction means that Sn is a stronger reducing agent (loses electrons more readily) than Ag.

For step 1, we reverse the tin half-reaction (but *not* the sign of E°_{tin}):

$$Sn(s) \longrightarrow Sn^{2+}(aq) + 2e^- \qquad E^\circ_{tin} = -0.14\ V$$

because when we subtract $E^\circ_{half\text{-}cell}$ of the tin half-reaction (anode, oxidation) from $E^\circ_{half\text{-}cell}$ of the silver half-reaction (cathode, reduction), we get a positive E°_{cell}:

$$0.80\ V - (-0.14\ V) = 0.94\ V$$

For step 2, the number of electrons lost in the oxidation must equal the number gained in the reduction, so we double the silver (reduction) half-reaction. Adding the half-reactions and applying Equation 21.3 gives

$$2Ag^+(aq) + 2e^- \longrightarrow 2Ag(s) \qquad E^\circ_{silver} = 0.80\ V \quad [reduction]$$

$$Sn(s) \longrightarrow Sn^{2+}(aq) + 2e^- \qquad E^\circ_{tin} = -0.14\ V \quad [oxidation]$$

$$Sn(s) + 2Ag^+(aq) \longrightarrow Sn^{2+}(aq) + 2Ag(s)$$

$$E^\circ_{cell} = E^\circ_{cathode} - E^\circ_{anode} = E^\circ_{silver} - E^\circ_{tin} = 0.94\ V$$

A *very important point* to note is that, when we double the coefficients of the silver half-reaction, we do *not* double its $E^\circ_{half\text{-}cell}$. *Changing the coefficients of a half-reaction does* **not** *change its* $E^\circ_{half\text{-}cell}$ because a standard electrode potential is an *intensive* property, one that does *not* depend on the amount of substance. Let's see why. The potential is the *ratio* of energy to charge. When we change the coefficients to change the amounts, the energy *and* the charge change proportionately, so their ratio stays the same. (Similarly, density does not change with the amount of substance because the mass *and* the volume change proportionately.)

Student Hot Spot

Student data indicate that you may struggle with using electrode potentials to determine spontaneity. Access the Smartbook to view additional Learning Resources on this topic.

SAMPLE PROBLEM 21.5 Writing Spontaneous Redox Reactions and Ranking Oxidizing and Reducing Agents by Strength

Problem **(a)** Combine pairs of the balanced half-reactions (1), (2), and (3) into three spontaneous reactions A, B, and C, and calculate E°_{cell} for each. **(b)** Rank the relative strengths of the oxidizing and reducing agents.

(1) $NO_3^-(aq) + 4H^+(aq) + 3e^- \longrightarrow NO(g) + 2H_2O(l) \qquad E^\circ = 0.96\ V$

(2) $N_2(g) + 5H^+(aq) + 4e^- \longrightarrow N_2H_5^+(aq) \qquad E^\circ = -0.23\ V$

(3) $MnO_2(s) + 4H^+(aq) + 2e^- \longrightarrow Mn^{2+}(aq) + 2H_2O(l) \qquad E^\circ = 1.23\ V$

Plan **(a)** To write spontaneous redox reactions, we combine the possible pairs of half-reactions: (1) and (2) give reaction A, (1) and (3) give B, and (2) and (3) give C. They are all written as reductions, so the oxidizing agents appear as reactants and the reducing agents appear as products. For each pair, we reverse the half-reaction that has the smaller (less positive or more negative) E° value to an oxidation (without changing the sign of E°) to obtain a positive E°_{cell}. We make e^- lost equal e^- gained (without changing the E° value), add the half-reactions together, and then apply Equation 21.3 to find E°_{cell}. **(b)** Because each reaction is spontaneous as written, the stronger oxidizing and reducing agents are the reactants. To obtain the overall ranking, we first rank the relative strengths within each equation and then compare them.

Solution **(a)** Combining half-reactions (1) and (2) gives equation A. The E° value for half-reaction (1) is larger (more positive) than that for (2), so we reverse (2) to obtain a positive E°_{cell}:

(1) $NO_3^-(aq) + 4H^+(aq) + 3e^- \longrightarrow NO(g) + 2H_2O(l) \qquad E^\circ = 0.96\ V$

(rev 2) $N_2H_5^+(aq) \longrightarrow N_2(g) + 5H^+(aq) + 4e^- \qquad E^\circ = -0.23\ V$

To make e^- lost equal e^- gained, we multiply (1) by four and the reversed (2) by three; then add half-reactions and cancel appropriate numbers of common species (H^+ and e^-):

$$4NO_3^-(aq) + 16H^+(aq) + 12e^- \longrightarrow 4NO(g) + 8H_2O(l) \qquad E^\circ = 0.96\ V$$

$$3N_2H_5^+(aq) \longrightarrow 3N_2(g) + 15H^+(aq) + 12e^- \qquad E^\circ = -0.23\ V$$

$$(A)\ 3N_2H_5^+(aq) + 4NO_3^-(aq) + H^+(aq) \longrightarrow 3N_2(g) + 4NO(g) + 8H_2O(l)$$

$$E^\circ_{cell} = 0.96\ V - (-0.23\ V) = 1.19\ V$$

Combining half-reactions (1) and (3) gives equation B. Half-reaction (1) has a smaller $E°$, so it is reversed:

(rev 1) $NO(g) + 2H_2O(l) \longrightarrow NO_3^-(aq) + 4H^+(aq) + 3e^-$ $E° = 0.96$ V

(3) $MnO_2(s) + 4H^+(aq) + 2e^- \longrightarrow Mn^{2+}(aq) + 2H_2O(l)$ $E° = 1.23$ V

We multiply reversed (1) by two and (3) by three, then add and cancel:

$$2NO(g) + 4H_2O(l) \longrightarrow 2NO_3^-(aq) + 8H^+(aq) + 6e^- \quad E° = 0.96 \text{ V}$$
$$3MnO_2(s) + 12H^+(aq) + 6e^- \longrightarrow 3Mn^{2+}(aq) + 6H_2O(l) \quad E° = 1.23 \text{ V}$$

$$\text{(B) } 3MnO_2(s) + 4H^+(aq) + 2NO(g) \longrightarrow 3Mn^{2+}(aq) + 2H_2O(l) + 2NO_3^-(aq)$$
$$E°_{cell} = 1.23 \text{ V} - 0.96 \text{ V} = 0.27 \text{ V}$$

Combining half-reactions (2) and (3) gives equation C. Half-reaction (2) has a smaller $E°$, so it is reversed:

(rev 2) $N_2H_5^+(aq) \longrightarrow N_2(g) + 5H^+(aq) + 4e^-$ $E° = -0.23$ V

(3) $MnO_2(s) + 4H^+(aq) + 2e^- \longrightarrow Mn^{2+}(aq) + 2H_2O(l)$ $E° = 1.23$ V

We multiply reaction (3) by two, add the half-reactions, and cancel:

$$N_2H_5^+(aq) \longrightarrow N_2(g) + 5H^+(aq) + 4e^- \quad E° = -0.23 \text{ V}$$
$$2MnO_2(s) + 8H^+(aq) + 4e^- \longrightarrow 2Mn^{2+}(aq) + 4H_2O(l) \quad E° = 1.23 \text{ V}$$

$$\text{(C) } N_2H_5^+(aq) + 2MnO_2(s) + 3H^+(aq) \longrightarrow N_2(g) + 2Mn^{2+}(aq) + 4H_2O(l)$$
$$E°_{cell} = 1.23 \text{ V} - (-0.23 \text{ V}) = 1.46 \text{ V}$$

(b) Ranking the oxidizing and reducing agents within each reaction:

Equation A	Oxidizing agents: $NO_3^- > N_2$	Reducing agents: $N_2H_5^+ > NO$
Equation B	Oxidizing agents: $MnO_2 > NO_3^-$	Reducing agents: $NO > Mn^{2+}$
Equation C	Oxidizing agents: $MnO_2 > N_2$	Reducing agents: $N_2H_5^+ > Mn^{2+}$

Determining the overall ranking of oxidizing and reducing agents. Comparing the relative strengths from the three balanced equations gives

Oxidizing agents: $MnO_2 > NO_3^- > N_2$
Reducing agents: $N_2H_5^+ > NO > Mn^{2+}$

Check As always, check that atoms and charges balance on both sides of each equation. A good way to check the ranking and equations is to list the given half-reactions in order of decreasing $E°$ value:

$$MnO_2(s) + 4H^+(aq) + 2e^- \longrightarrow Mn^{2+}(aq) + 2H_2O(l) \quad E° = 1.23 \text{ V}$$
$$NO_3^-(aq) + 4H^+(aq) + 3e^- \longrightarrow NO(g) + 2H_2O(l) \quad E° = 0.96 \text{ V}$$
$$N_2(g) + 5H^+(aq) + 4e^- \longrightarrow N_2H_5^+(aq) \quad E° = -0.23 \text{ V}$$

Then the oxidizing agents (reactants) decrease in strength going down the list, so the reducing agents (products) decrease in strength going up. Moreover, each of the equations for the spontaneous reactions (A, B, and C) should combine a reactant with a product that is lower down on this list.

FOLLOW-UP PROBLEMS

21.5A (a) Combine pairs of the balanced half-reactions (1), (2), and (3) into three spontaneous reactions A, B, and C, and calculate $E°_{cell}$ for each:

(1) $Ag_2O(s) + H_2O(l) + 2e^- \longrightarrow 2Ag(s) + 2OH^-(aq)$ $E° = 0.34$ V

(2) $BrO_3^-(aq) + 3H_2O(l) + 6e^- \longrightarrow Br^-(aq) + 6OH^-(aq)$ $E° = 0.61$ V

(3) $Zn(OH)_2(s) + 2e^- \longrightarrow Zn(s) + 2OH^-(aq)$ $E° = -1.25$ V

(b) Rank the relative strengths of the oxidizing and reducing agents.

21.5B Is the following reaction spontaneous as written (see Appendix D)?

$$3Fe^{2+}(aq) \longrightarrow Fe(s) + 2Fe^{3+}(aq)$$

If not, write the equation for the spontaneous reaction, calculate $E°_{cell}$, and rank the three species of iron in order of decreasing strength as reducing agents.

SOME SIMILAR PROBLEMS 21.46–21.49

Explaining the Activity Series of the Metals

In Chapter 4, we discussed the activity series of the metals (see Figure 4.24), which ranks metals by their ability to "displace" one another from aqueous solution. Now you'll see *why* this displacement occurs, as well as why many, but not all, metals react with acid to form H_2, and why a few metals form H_2 even in water.

1. *Metals that can displace H_2 from acid.* The standard hydrogen half-reaction represents the reduction of H^+ ions from an acid to H_2:

$$2H^+(aq) + 2e^- \longrightarrow H_2(g) \qquad E^\circ = 0.00\ \text{V}$$

To see which metals reduce H^+ (referred to as "displacing H_2") from acids, choose a metal, write its half-reaction as an oxidation, combine this half-reaction with the hydrogen half-reaction, and see if E°_{cell} is positive. We find that the metals Li through Pb, those that lie *below* the standard hydrogen (reference) half-reaction in Appendix D, give a positive E°_{cell} when reducing H^+. Iron, for example, reduces H^+ from an acid to H_2:

$$\text{Fe}(s) \longrightarrow \text{Fe}^{2+}(aq) + 2e^- \qquad E^\circ = -0.44\ \text{V} \qquad \text{[anode; oxidation]}$$
$$2H^+(aq) + 2e^- \longrightarrow H_2(g) \qquad E^\circ = 0.00\ \text{V} \qquad \text{[cathode; reduction]}$$
$$\text{Fe}(s) + 2H^+(aq) \longrightarrow H_2(g) + \text{Fe}^{2+}(aq) \qquad E^\circ_{\text{cell}} = 0.00\ \text{V} - (-0.44\ \text{V}) = 0.44\ \text{V}$$

The lower the metal in the list, the stronger it is as a reducing agent; therefore, *if E°_{cell} for the reduction of H^+ is more positive with metal A than with metal B, metal A is a stronger reducing agent than metal B and a more* ***active*** *metal.*

2. *Metals that cannot displace H_2 from acid.* For metals that are *above* the standard hydrogen (reference) half-reaction, E°_{cell} is negative when we reverse the metal half-reaction, so the reaction does not occur. For example, the coinage metals—copper, silver, and gold, which are in Group 1B(11)—are not strong enough reducing agents to reduce H^+ from acids:

$$\text{Ag}(s) \longrightarrow \text{Ag}^+(aq) + e^- \qquad E^\circ = 0.80\ \text{V} \qquad \text{[anode; oxidation]}$$
$$2H^+(aq) + 2e^- \longrightarrow H_2(g) \qquad E^\circ = 0.00\ \text{V} \qquad \text{[cathode; reduction]}$$
$$2\text{Ag}(s) + 2H^+(aq) \longrightarrow 2\text{Ag}^+(aq) + H_2(g) \qquad E^\circ_{\text{cell}} = 0.00\ \text{V} - 0.80\ \text{V} = -0.80\ \text{V}$$

The *higher* the metal in the list, the *more negative* is its E°_{cell} for the reduction of H^+ to H_2, the *lower* is its reducing strength, and the *less active* it is. Thus, gold is less active than silver, which is less active than copper.

3. *Metals that can displace H_2 from water. Metals active enough to reduce H_2O lie below that half-reaction:*

$$2H_2O(l) + 2e^- \longrightarrow H_2(g) + 2OH^-(aq) \qquad E^\circ = -0.83\ \text{V}$$

For example, consider the reaction of sodium in water (with the Na^+/Na half-reaction reversed and doubled):

$$2\text{Na}(s) \longrightarrow 2\text{Na}^+(aq) + 2e^- \qquad E^\circ = -2.71\ \text{V} \qquad \text{[anode; oxidation]}$$
$$2H_2O(l) + 2e^- \longrightarrow H_2(g) + 2OH^-(aq) \qquad E^\circ = -0.83\ \text{V} \qquad \text{[cathode; reduction]}$$
$$2\text{Na}(s) + 2H_2O(l) \longrightarrow 2\text{Na}^+(aq) + H_2(g) + 2OH^-(aq)$$
$$E^\circ_{\text{cell}} = -0.83\ \text{V} - (-2.71\ \text{V}) = 1.88\ \text{V}$$

The alkali metals [Group 1A(1)] and the larger alkaline earth metals [Group 2A(2)] can reduce water (displace H_2 from H_2O). Figure 21.8 shows calcium reducing H_2O to H_2.

Figure 21.8 The reaction of calcium in water.

Source: © McGraw-Hill Education/Stephen Frisch, photographer

4. *Metals that displace other metals from solution.* We can also predict whether one metal can reduce the aqueous ion of another metal. Any metal that is lower in the list in Appendix D can reduce the ion of a metal that is higher up, and thus displace that metal from solution. For example, zinc can displace iron from solution:

$$\text{Zn}(s) \longrightarrow \text{Zn}^{2+}(aq) + 2e^- \qquad E^\circ = -0.76\ \text{V} \qquad \text{[anode; oxidation]}$$
$$\text{Fe}^{2+}(aq) + 2e^- \longrightarrow \text{Fe}(s) \qquad E^\circ = -0.44\ \text{V} \qquad \text{[cathode; reduction]}$$
$$\text{Zn}(s) + \text{Fe}^{2+}(aq) \longrightarrow \text{Zn}^{2+}(aq) + \text{Fe}(s) \qquad E^\circ_{\text{cell}} = -0.44\ \text{V} - (-0.76\ \text{V}) = 0.32\ \text{V}$$

This particular reaction has tremendous economic importance in protecting iron from rusting, as we'll discuss in Section 21.6.

A common incident involving the reducing power of metals occurs when you bite down with a filled tooth on a scrap of aluminum foil left on a piece of food. The foil acts as an active anode ($E^\circ_{\text{aluminum}} = -1.66$ V), saliva as the electrolyte, and the filling (usually a silver/tin/mercury alloy) as an inactive cathode at which O_2 is reduced ($E^\circ_{O_2} = 1.23$ V) to water. The circuit between the foil and the filling creates a current that is sensed as pain by the nerve of the tooth (Figure 21.9).

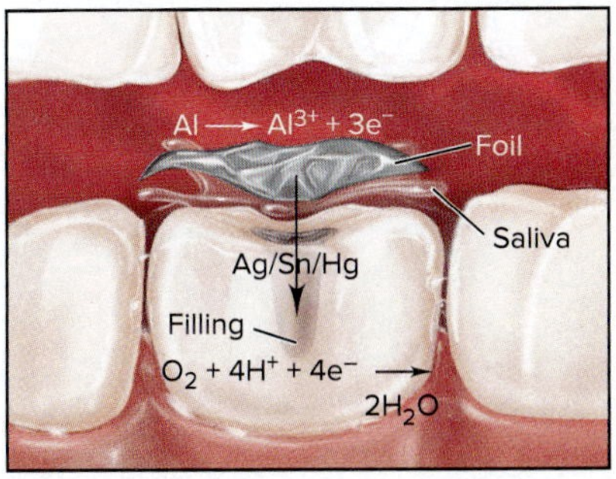

Figure 21.9 A dental "voltaic cell."

› Summary of Section 21.3

- The output of a voltaic cell is the cell potential (E_{cell}), measured in volts (1 V = 1 J/C).
- With all substances in their standard states, the output is the standard cell potential (E°_{cell}).
- $E^\circ_{\text{cell}} > 0$ for a spontaneous reaction at standard-state conditions.
- By convention, a standard electrode potential ($E^\circ_{\text{half-cell}}$) refers to the *reduction* half-reaction.
- E°_{cell} equals $E^\circ_{\text{half-cell}}$ of the cathode *minus* $E^\circ_{\text{half-cell}}$ of the anode.
- A standard hydrogen (reference) electrode ($E^\circ_{\text{reference}} = 0$ V) is used to measure $E^\circ_{\text{half-cell}}$ values and to rank oxidizing (or reducing) agents.
- Spontaneous redox reactions combine stronger oxidizing and reducing agents to form weaker reducing and oxidizing agents, respectively.
- A metal can reduce another species (H^+, H_2O, or an ion of another metal) if E°_{cell} for the overall reaction is positive.

21.4 FREE ENERGY AND ELECTRICAL WORK

Following up on our discussion in Chapter 20, in this section, we examine the relationship of useful work, free energy, and the equilibrium constant in the context of electrochemical cells and see the effect of concentration on cell potential.

Standard Cell Potential and the Equilibrium Constant

The signs of ΔG and E_{cell} are *opposite* for a spontaneous reaction: a *negative* free energy change ($\Delta G < 0$; Section 20.3) and a *positive* cell potential ($E_{\text{cell}} > 0$). These two indicators of spontaneity are proportional to each other:

$$\Delta G \propto -E_{\text{cell}}$$

Let's determine this proportionality constant. The electrical work done (w, in joules) is the product of the potential (E_{cell}, in volts) and the charge that flows (in coulombs). Since E_{cell} is measured with no current flowing and, thus, no energy lost as heat, it is the maximum voltage possible, and, thus, determines the maximum work possible (w_{max}).* Work done *by* the cell *on* the surroundings has a negative sign:

$$w_{\text{max}} = -E_{\text{cell}} \times \text{charge}$$

The maximum work done *on* the surroundings is equal to ΔG (Equation 20.9):

$$w_{\text{max}} = \Delta G = -E_{\text{cell}} \times \text{charge}$$

The charge that flows through the cell equals the amount (mol) of electrons transferred (n) times the charge of 1 mol of electrons (which has the symbol F):

$$\text{Charge} = \text{amount (mol) of } e^- \times \frac{\text{charge}}{\text{mol } e^-} \qquad \text{or} \qquad \text{charge} = nF$$

The charge of 1 mol of electrons is the **Faraday constant (*F*),** named for Michael Faraday, the 19th-century British scientist who pioneered the study of electrochemistry:

$$F = \frac{96{,}485 \text{ C}}{\text{mol } e^-}$$

*Recall from Chapter 20 that only a reversible process can do maximum work. For no current to flow and the process to be reversible, E_{cell} must be opposed by an equal potential in the measuring circuit: if the opposing potential is infinitesimally smaller, the cell reaction goes forward; if it is infinitesimally larger, the reaction goes backward.

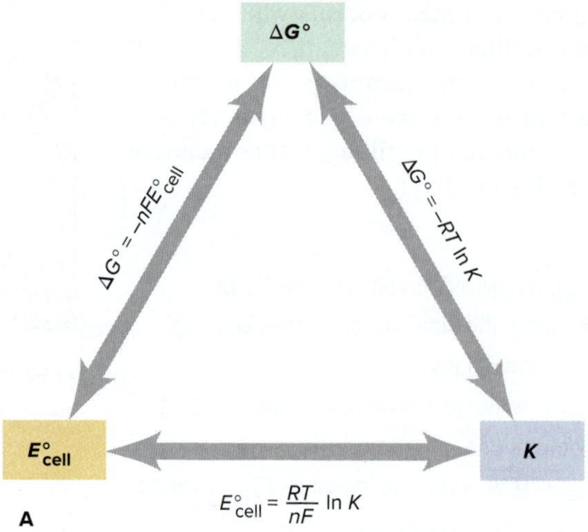

A

B

Reaction Parameters at the Standard State			
ΔG°	K	E°_{cell}	Reaction at standard-state conditions
<0	>1	>0	Spontaneous
0	1	0	At equilibrium
>0	<1	<0	Nonspontaneous

Figure 21.10 **The interrelationship of ΔG°, E°_{cell}, and K. A,** Any one parameter can be used to find the other two. **B,** The signs of ΔG°, K, and E°_{cell} determine reaction direction.

Because 1 V = 1 J/C, we have 1 C = 1 J/V, and

$$F = 9.65\times10^4 \frac{\text{J}}{\text{V}\cdot\text{mol e}^-} \quad (3\text{ sf}) \qquad \textbf{(21.4)}$$

Substituting for charge, the proportionality constant is nF:

$$\Delta G = -nFE_{cell} \qquad \textbf{(21.5)}$$

And, when all components are in their standard states, we have

$$\Delta G^\circ = -nFE^\circ_{cell} \qquad \textbf{(21.6)}$$

Using this relationship, we can relate the standard cell potential to the equilibrium constant of the redox reaction. Recall from Equation 20.12 that

$$\Delta G^\circ = -RT \ln K$$

Substituting for ΔG° from Equation 21.6 gives

$$-nFE^\circ_{cell} = -RT \ln K$$

Solving for E°_{cell} gives

$$E^\circ_{cell} = \frac{RT}{nF} \ln K \qquad \textbf{(21.7)}$$

Figure 21.10 summarizes the interconnections among the standard free energy change, the equilibrium constant, and the standard cell potential. In Chapter 20, we determined K from ΔG°, which we found either from ΔH° and ΔS° values or from ΔG°_f values. Now, for redox reactions, we have a direct *experimental* method for determining K *and* ΔG°: measure E°_{cell}.

In calculations, we adjust Equation 21.7 as follows:

- Substitute 8.314 J/(mol rxn·K) for the constant R.
- Substitute 9.65×10^4 J/(V·mol e$^-$) for the constant F.
- Substitute 298.15 K for T, keeping in mind that the cell can run at other temperatures.
- Multiply by 2.303 to convert the natural logarithm to the common (base-10) logarithm. This conversion shows that a *10-fold change in K makes E°_{cell} change by 1.*

Thus, when n moles of e$^-$ are transferred per mole of reaction, based on the balanced equation, we have

$$E^\circ_{cell} = \frac{RT}{nF} \ln K = 2.303 \times \frac{8.314 \frac{\text{J}}{\text{mol rxn}\cdot\text{K}} \times 298.15\text{ K}}{\frac{n\text{ mol e}^-}{\text{mol rxn}}\left(9.65\times10^4 \frac{\text{J}}{\text{V}\cdot\text{mol e}^-}\right)} \log K$$

And this becomes

$$E^\circ_{cell} = \frac{0.0592\text{ V}}{n} \log K \quad \text{or} \quad \log K = \frac{nE^\circ_{cell}}{0.0592\text{ V}} \quad (\text{at } 298.15\text{ K}) \qquad \textbf{(21.8)}$$

SAMPLE PROBLEM 21.6 Calculating K and $\Delta G°$ from $E°_{cell}$

Problem Silver occurs in trace amounts in some ores of lead, and lead can displace silver from solution:

$$Pb(s) + 2Ag^+(aq) \longrightarrow Pb^{2+}(aq) + 2Ag(s)$$

As a consequence, silver is a valuable byproduct in the industrial extraction of lead from its ores. Calculate K and $\Delta G°$ at 298.15 K for this reaction.

Plan We divide the spontaneous redox reaction into half-reactions and use values from Appendix D to calculate $E°_{cell}$. Then, we substitute this result into Equations 21.7 and 21.8 to find K and into Equation 21.6 to find $\Delta G°$.

Solution Writing the half-reactions with their $E°$ values:

$$(1)\quad Ag^+(aq) + e^- \longrightarrow Ag(s) \qquad E° = 0.80\text{ V}$$
$$(2)\quad Pb^{2+}(aq) + 2e^- \longrightarrow Pb(s) \qquad E° = -0.13\text{ V}$$

Calculating $E°_{cell}$: We double half-reaction (1) to obtain a gain of 2 mol of electrons, reverse (2) since this is the oxidation half-reaction, add the half-reactions, and subtract $E°_{lead}$ from $E°_{silver}$:

$$2Ag^+(aq) + 2e^- \longrightarrow 2Ag(s) \qquad E° = 0.80\text{ V}$$
$$Pb(s) \longrightarrow Pb^{2+}(aq) + 2e^- \qquad E° = -0.13\text{ V}$$
$$Pb(s) + 2Ag^+(aq) \longrightarrow Pb^{2+}(aq) + 2Ag(s) \qquad E°_{cell} = 0.80\text{ V} - (-0.13\text{ V}) = 0.93\text{ V}$$

Calculating K with Equations 21.7 and 21.8: The adjusted half-reactions show that 2 mol of e^- are transferred per mole of reaction as written, so $n = 2$. Then, performing the substitutions for R and F that we just discussed, changing to common logarithms, and running the cell at 25°C (298.15 K), we have

$$E°_{cell} = \frac{RT}{nF}\ln K = \frac{0.0592\text{ V}}{n}\log K = \frac{0.0592\text{ V}}{2}\log K = 0.93\text{ V}$$

So, $$\log K = \frac{0.93\text{ V} \times 2}{0.0592\text{ V}} = 31.42 \quad\text{and}\quad K = 10^{31.42} = 2.6\times10^{31}$$

Calculating $\Delta G°$ (Equation 21.6):

$$\Delta G° = -nFE°_{cell} = -\frac{2\text{ mol }e^-}{\text{mol rxn}} \times \frac{96.5\text{ kJ}}{\text{V·mol }e^-} \times 0.93\text{ V} = -1.8\times10^2\text{ kJ/mol rxn}$$

Check The three variables are consistent with the reaction being spontaneous at standard-state conditions: $E°_{cell} > 0$, $\Delta G° < 0$, and $K > 1$. Be sure to round and check the order of magnitude: to find $\Delta G°$, for instance, $\Delta G° \approx -2 \times 100 \times 1 = -200$, so the overall math seems right. Another check would be to obtain $\Delta G°$ directly from its relation with K:

$$\begin{aligned}\Delta G° &= -RT\ln K = -8.314\text{ J/mol rxn·K} \times 298.15\text{ K} \times \ln(2.6\times10^{31})\\ &= -1.8\times10^5\text{ J/mol rxn} = -1.8\times10^2\text{ kJ/mol rxn}\end{aligned}$$

FOLLOW-UP PROBLEMS

21.6A Calculate K and $\Delta G°$ for the following reaction at 298.15 K:

$$2MnO_4^-(aq) + 4H_2O(l) + 3Cu(s) \longrightarrow 2MnO_2(s) + 8OH^-(aq) + 3Cu^{2+}(aq)$$

21.6B When cadmium metal reduces Cu^{2+} in solution, Cd^{2+} forms in addition to copper metal. Given that $\Delta G° = -143$ kJ/mol rxn, calculate K at 25°C. What is $E°_{cell}$ of a voltaic cell that uses this reaction?

SOME SIMILAR PROBLEMS 21.56–21.67

The Effect of Concentration on Cell Potential

So far, we've considered cells at standard-state conditions, but most cells don't start with those concentrations, and even if they did, concentrations change as the cell operates. Moreover, in all batteries, reactant concentrations are far from the standard state.

To determine E_{cell}, the cell potential under *nonstandard* conditions, we'll derive an expression for the relation between E_{cell} and concentration based on the relation between ΔG and concentration. Recall from Chapter 20 (Equation 20.13) that

$$\Delta G = \Delta G^\circ + RT \ln Q$$

ΔG is related to E_{cell}, and ΔG° is related to E°_{cell} (Equations 21.5 and 21.6); substituting, we get

$$-nFE_{cell} = -nFE^\circ_{cell} + RT \ln Q$$

Dividing both sides by $-nF$, we obtain the equation developed by the great German chemist Walther Hermann Nernst when he was only 25 years old. (In his career, which culminated in the 1920 Nobel Prize in chemistry, he also formulated the third law of thermodynamics and established the concept of the solubility product.) The **Nernst equation** says that E_{cell} depends on E°_{cell} *and* a term for the potential at any ratio of concentrations:

$$E_{cell} = E^\circ_{cell} - \frac{RT}{nF} \ln Q \qquad \textbf{(21.9)}$$

How do changes in Q affect cell potential? From Equation 21.9, we see that

- When $Q < 1$ and thus [reactant] > [product], $\ln Q < 0$, so $E_{cell} > E^\circ_{cell}$.
- When $Q = 1$ and thus [reactant] = [product], $\ln Q = 0$, so $E_{cell} = E^\circ_{cell}$.
- When $Q > 1$ and thus [reactant] < [product], $\ln Q > 0$, so $E_{cell} < E^\circ_{cell}$.

As before, to obtain a form we can use for calculations, we substitute known values of R and F, operate the cell at 298.15 K, and convert to the common (base-10) logarithm:

$$E_{cell} = E^\circ_{cell} - \frac{RT}{nF} \ln Q$$

$$= E^\circ_{cell} - 2.303 \times \frac{8.314 \dfrac{\cancel{J}}{\cancel{\text{mol rxn}}\cdot\cancel{K}} \times 298.15\ \cancel{K}}{\dfrac{n\ \cancel{\text{mol e}^-}}{\cancel{\text{mol rxn}}}\left(9.65\times10^4 \dfrac{\cancel{J}}{\text{V}\cdot\cancel{\text{mol e}^-}}\right)} \log Q$$

We obtain

$$E_{cell} = E^\circ_{cell} - \frac{0.0592\ \text{V}}{n} \log Q \quad (\text{at } 298.15\ \text{K}) \qquad \textbf{(21.10)}$$

Remember that the expression for Q *contains only those species with concentrations (and/or pressures) that can vary;* thus, pure liquids do not appear nor do solids, even when they are the electrodes. For example, in the reaction between cadmium and silver ion, the Cd and Ag electrodes do not appear in the expression for Q:

$$\text{Cd}(s) + 2\text{Ag}^+(aq) \longrightarrow \text{Cd}^{2+}(aq) + 2\text{Ag}(s) \qquad Q = \frac{[\text{Cd}^{2+}]}{[\text{Ag}^+]^2}$$

Student Hot Spot

Student data indicate that you may struggle with calculations involving the Nernst equation. Access the Smartbook to view additional Learning Resources on this topic.

SAMPLE PROBLEM 21.7 Using the Nernst Equation to Calculate E_{cell}

Problem In a test of a new reference electrode, a chemist constructs a voltaic cell consisting of a Zn/Zn^{2+} half-cell and an H_2/H^+ half-cell under the following conditions:

$$[\text{Zn}^{2+}] = 0.010\ M \qquad [\text{H}^+] = 2.5\ M \qquad P_{\text{H}_2} = 0.30\ \text{atm}$$

Calculate E_{cell} at 298.15 K.

Plan To apply the Nernst equation and determine E_{cell}, we must know E°_{cell} and Q. We write the spontaneous reaction and calculate E°_{cell} from standard electrode potentials (Appendix D). Then we substitute into Equation 21.10.

Solution Determining the cell reaction and E°_{cell}:

$$2H^+(aq) + 2e^- \longrightarrow H_2(g) \qquad E^\circ = 0.00\text{ V}$$
$$Zn(s) \longrightarrow Zn^{2+}(aq) + 2e^- \qquad E^\circ = -0.76\text{ V}$$
$$2H^+(aq) + Zn(s) \longrightarrow H_2(g) + Zn^{2+}(aq) \qquad E^\circ_{cell} = 0.00\text{ V} - (-0.76\text{ V}) = 0.76\text{ V}$$

Calculating Q:

$$Q = \frac{P_{H_2} \times [Zn^{2+}]}{[H^+]^2} = \frac{0.30 \times 0.010}{2.5^2} = 4.8\times10^{-4}$$

Solving for E_{cell} at 25°C (298.15 K), with $n = 2$:

$$E_{cell} = E^\circ_{cell} - \frac{0.0592\text{ V}}{n}\log Q$$
$$= 0.76\text{ V} - \frac{0.0592\text{ V}}{2}\log(4.8\times10^{-4}) = 0.76\text{ V} - (-0.098\text{ V}) = 0.86\text{ V}$$

Check After you check the arithmetic, reason through the answer: $E_{cell} > E^\circ_{cell}$ (0.86 > 0.76) because the log Q term was negative, which is consistent with $Q < 1$.

FOLLOW-UP PROBLEMS

21.7A A voltaic cell consists of a Cr/Cr^{3+} half-cell and an Sn/Sn^{2+} half-cell, with $[Sn^{2+}] = 0.20\ M$ and $[Cr^{3+}] = 1.60\ M$. Use Appendix D to calculate E_{cell} at 298.15 K.

21.7B Consider a voltaic cell based on the following reaction:

$$Fe(s) + Cu^{2+}(aq) \longrightarrow Fe^{2+}(aq) + Cu(s)$$

If $[Cu^{2+}] = 0.30\ M$, what must $[Fe^{2+}]$ be to increase E_{cell} by 0.25 V above E°_{cell} at 25°C?

SOME SIMILAR PROBLEMS 21.68–21.71

Following Changes in Potential During Cell Operation

As with any voltaic cell, the potential of a zinc-copper cell changes during cell operation because the concentrations of the components do. Because both of the other components are solids, the only variables are $[Cu^{2+}]$ and $[Zn^{2+}]$:

$$Zn(s) + Cu^{2+}(aq) \longrightarrow Zn^{2+}(aq) + Cu(s) \qquad Q = \frac{[Zn^{2+}]}{[Cu^{2+}]}$$

In this section, we follow the potential as the zinc-copper cell operates.

1. *Starting point of cell operation.* The positive E°_{cell} (1.10 V) means that this reaction proceeds *spontaneously* to the right at standard-state conditions, $[Zn^{2+}] = [Cu^{2+}] = 1\ M$ ($Q = 1$). But, if we start the cell when $[Zn^{2+}] < [Cu^{2+}]$ ($Q < 1$), for example, when $[Zn^{2+}] = 1.0\times10^{-4}\ M$ and $[Cu^{2+}] = 2.0\ M$, the cell potential starts out *higher* than the standard cell potential, because $\log Q < 0$:

$$E_{cell} = E^\circ_{cell} - \frac{0.0592\text{ V}}{2}\log\frac{[Zn^{2+}]}{[Cu^{2+}]} = 1.10\text{ V} - \left(\frac{0.0592\text{ V}}{2}\log\frac{1.0\times10^{-4}}{2.0}\right)$$
$$= 1.10\text{ V} - \left[\frac{0.0592\text{ V}}{2}(-4.30)\right] = 1.10\text{ V} + 0.127\text{ V} = 1.23\text{ V}$$

2. *Key stages during cell operation.* Using Equation 21.10, we identify four key stages of operation. Figure 21.11A on the next page shows the first three. The main point is that *as the cell operates, its potential decreases.* As product concentration ($[P] = [Zn^{2+}]$) increases and reactant concentration ($[R] = [Cu^{2+}]$) decreases, Q becomes larger, the term $[(0.0592\text{ V}/n)\log Q]$ becomes less negative (more positive), and E_{cell} decreases:

Stage 1. $E_{cell} > E^\circ_{cell}$ when $Q < 1$: when the cell begins operation, $[Cu^{2+}] > [Zn^{2+}]$, so $(0.0592\text{ V}/n)\log Q < 0$ and $E_{cell} > E^\circ_{cell}$.

Stage 2. $E_{cell} = E^\circ_{cell}$ when $Q = 1$: at the point when $[Cu^{2+}] = [Zn^{2+}]$, $Q = 1$, so $(0.0592\text{ V}/n)\log Q = 0$ and $E_{cell} = E^\circ_{cell}$.

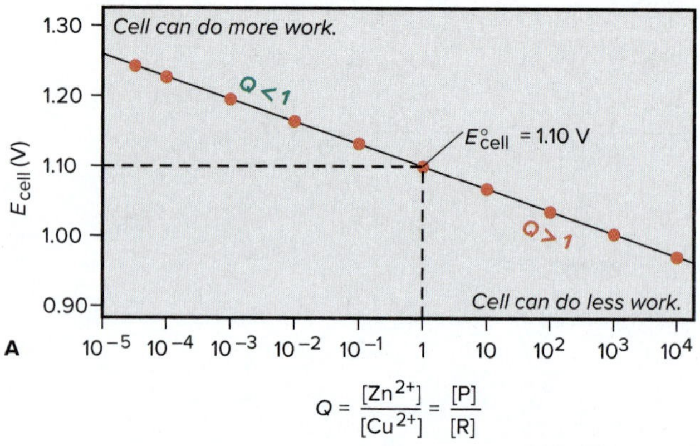

A

Changes in E_{cell} and Concentration

Stage in cell operation	Q	Relative [P] and [R]	$\frac{0.0592\ V}{n} \log Q$
1. $E > E^\circ$	<1	[P] < [R]	<0
2. $E = E^\circ$	=1	[P] = [R]	=0
3. $E < E^\circ$	>1	[P] > [R]	>0
4. $E = 0$	$=K$	[P] ≫ [R]	$=E^\circ$

B

Figure 21.11 The relation between E_{cell} and log Q for the zinc-copper cell. A, A plot of E_{cell} vs. Q (on a logarithmic scale) decreases linearly. When $Q < 1$ *(left)*, the cell does relatively more work. When $Q = 1$, $E_{cell} = E^\circ_{cell}$. When $Q > 1$ (right), the cell does relatively less work. **B,** A summary of the changes in E_{cell} as this or any voltaic cell operates. ($[Zn^{2+}]$ is [P] for [product], and $[Cu^{2+}]$ is [R] for [reactant].)

Stage 3. $E_{cell} < E^\circ_{cell}$ when $Q > 1$: as the $[Zn^{2+}]/[Cu^{2+}]$ ratio continues to increase, $(0.0592\ V/n) \log Q > 0$, so $E_{cell} < E^\circ_{cell}$.

Stage 4. $E_{cell} = 0$ when $Q = K$: eventually, $(0.0592\ V/n) \log Q$ becomes so large that it equals E°_{cell}, which means that E_{cell} is zero. This occurs *when the system reaches* ***equilibrium:*** *no more free energy is released, so the cell can do no more work.* At this point, we say that a battery is "dead."

Figure 21.11B summarizes these four key stages during operation of a voltaic cell.

3. *Q/K and the work the cell can do.* At equilibrium, Equation 21.10 becomes

$$0 = E^\circ_{cell} - \frac{0.0592\ V}{n} \log K, \quad \text{which rearranges to} \quad E^\circ_{cell} = \frac{0.0592\ V}{n} \log K$$

Note that this result is identical to Equation 21.8, which we obtained from ΔG°. Solving for K of the zinc-copper cell ($E^\circ_{cell} = 1.10$ V),

$$\log K = \frac{2 \times E^\circ_{cell}}{0.0592\ V} \quad \text{so} \quad K = 10^{(2\times1.10\ V)/0.0592\ V} = 10^{37.16} = 1.4\times10^{37}$$

As you can see, this cell does work until $[Zn^{2+}]/[Cu^{2+}]$ is *very* high.

The three relations between initial Q/K and E_{cell} are

- If $Q/K < 1$, E_{cell} is positive for the reaction *as written*. The smaller Q/K is, the greater the value of E_{cell}, and the more electrical work the cell can do.
- If $Q/K = 1$, $E_{cell} = 0$. The cell is at equilibrium and can no longer do work.
- If $Q/K > 1$, E_{cell} is negative for the reaction as written. The reverse reaction will take place, and the cell will do work until Q/K equals 1 at equilibrium.

Concentration Cells

If you mix a concentrated solution of a substance with a dilute solution of it, the final solution has an intermediate concentration. A **concentration cell** employs this simple, spontaneous change to generate electrical energy: different concentrations of species involved in the same half-reaction are in separate half-cells, and the cell does work until the concentrations become equal.

Finding E_{cell} for a Concentration Cell Suppose a voltaic cell has the Cu/Cu^{2+} half-reaction in both compartments. The cell reaction is the sum of identical half-reactions, written in opposite directions. The *standard* cell potential, E°_{cell}, is zero because the *standard* electrode potentials are both based on 1 *M* Cu^{2+}, so they cancel. As we said, however, *in a concentration cell, the concentrations are* ***different.*** Thus, even though

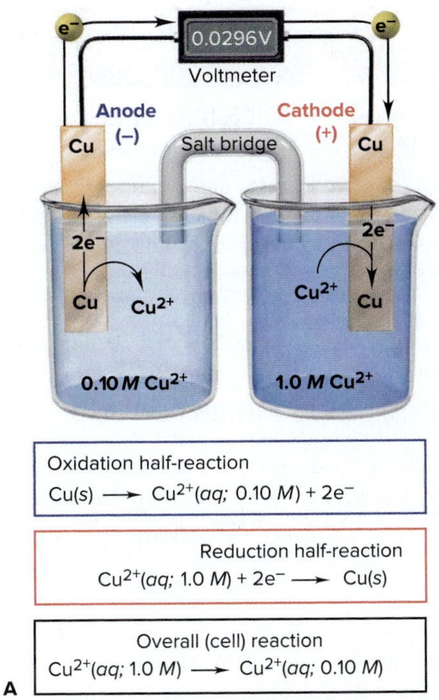

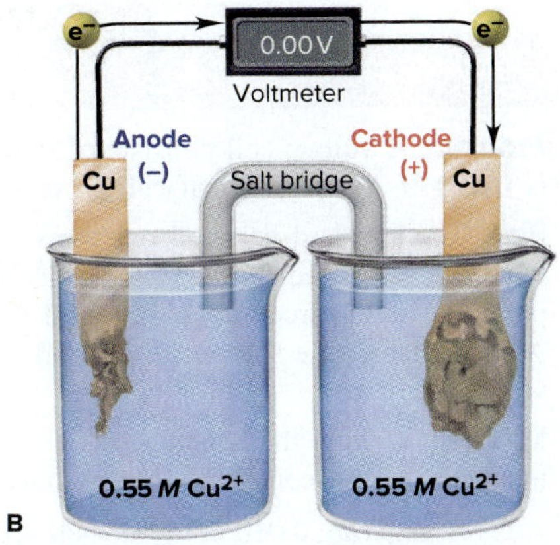

Figure 21.12 **A concentration cell based on the Cu/Cu^{2+} half-reaction.** **A,** $E_{cell} > 0$ as long as the half-cell *concentrations* are different (lighter and darker colors). **B,** Half-cell concentrations are equal (same color), and the sizes of the electrodes (exaggerated for clarity) are different.

E°_{cell} is still zero, the *nonstandard* cell potential, E_{cell}, depends on the *ratio of concentrations,* so it is *not* zero.

For the final concentrations to be equal, a concentration cell must have the dilute solution in the anode compartment and the concentrated solution in the cathode compartment. For example, let's use 0.10 M Cu^{2+} in the anode half-cell and 1.0 M Cu^{2+}, a 10-fold higher concentration, in the cathode half-cell (Figure 21.12A):

$$Cu(s) \longrightarrow Cu^{2+}(aq;\ 0.10\ M) + 2e^- \qquad \text{[anode; oxidation]}$$

$$Cu^{2+}(aq;\ 1.0\ M) + 2e^- \longrightarrow Cu(s) \qquad \text{[cathode; reduction]}$$

The overall cell reaction is the sum of the half-reactions:

$$Cu^{2+}(aq;\ 1.0\ M) \longrightarrow Cu^{2+}(aq;\ 0.10\ M) \qquad E_{cell} = ?$$

The cell potential at the initial concentrations of 0.10 M (dilute) and 1.0 M (concentrated), with $n = 2$, is obtained from the Nernst equation:

$$E_{cell} = E^\circ_{cell} - \frac{0.0592\ \text{V}}{2} \log \frac{[Cu^{2+}]_{dil}}{[Cu^{2+}]_{conc}} = 0\ \text{V} - \left(\frac{0.0592\ \text{V}}{2} \log \frac{0.10\ M}{1.0\ M}\right)$$

$$= 0\ \text{V} - \left[\frac{0.0592\ \text{V}}{2}(-1.00)\right] = 0.0296\ \text{V}$$

Since E°_{cell} is zero, E_{cell} depends entirely on the term $[(0.0592\ \text{V}/n) \log Q]$.

How a Concentration Cell Works Let's see what is happening as this concentration cell operates:

- *In the anode (dilute) half-cell,* Cu atoms in the electrode give up electrons and the resulting Cu^{2+} ions enter the solution and make it *more* concentrated.
- *In the cathode (concentrated) half-cell,* Cu^{2+} ions gain the electrons and the resulting Cu atoms plate out on the electrode, which makes that solution *less* concentrated.

As in any voltaic cell, E_{cell} decreases until equilibrium is attained, which happens when $[Cu^{2+}]$ is the same in both half-cells (Figure 21.12B). The same final concentration would result if we mixed the two solutions, but no electrical work would be done.

SAMPLE PROBLEM 21.8 Calculating the Potential of a Concentration Cell

Problem A voltaic cell consists of two Ag/Ag^+ half-cells. In half-cell A, the electrolyte is 0.010 *M* $AgNO_3$; in half-cell B, it is 4.0×10^{-4} *M* $AgNO_3$. What is E_{cell} at 298.15 K?

Plan The standard half-cell reactions are identical, so E°_{cell} is zero, and we find E_{cell} from the Nernst equation. Half-cell A has a higher $[Ag^+]$, so Ag^+ ions are reduced and plate out on electrode A. In half-cell B, Ag atoms of the electrode are oxidized and Ag^+ ions enter the solution. As in all voltaic cells, reduction occurs at the cathode, so it is positive.

Solution Writing the spontaneous reaction: The $[Ag^+]$ decreases in half-cell A and increases in half-cell B, so the spontaneous reaction is

$$Ag^+(aq; 0.010\ M)\ [\text{half-cell A}] \longrightarrow Ag^+(aq; 4.0\times10^{-4}\ M)\ [\text{half-cell B}]$$

Calculating E_{cell}, with $n = 1$:

$$E_{cell} = E^\circ_{cell} - \frac{0.0592\ V}{1}\log\frac{[Ag^+]_{dil}}{[Ag^+]_{conc}} = 0\ V - \left(0.0592\ V \log\frac{4.0\times10^{-4}}{0.010}\right)$$

$$= 0.083\ V$$

FOLLOW-UP PROBLEMS

21.8A A voltaic cell consists of two Ni/Ni^{2+} half-cells. In half-cell A, the electrolyte is 0.015 *M* $Ni(NO_3)_2$; in half-cell B, it is 0.40 *M* $Ni(NO_3)_2$. Calculate E_{cell} at 298.15 K.

21.8B A voltaic cell is built using two Au/Au^{3+} half-cells. In half-cell A, $[Au^{3+}] = 7.0\times10^{-4}$ *M*, and in half-cell B, $[Au^{3+}] = 2.5\times10^{-2}$ *M*. What is E_{cell}, and which electrode is negative?

SOME SIMILAR PROBLEMS 21.72 and 21.73

Applications of Concentration Cells The principle of a concentration cell has many applications. Here, we'll discuss three:

1. *Measuring pH.* The most important laboratory application of this principle is in measuring $[H^+]$. If we construct a concentration cell in which the cathode compartment is the standard hydrogen electrode and the anode compartment has the same apparatus dipping into a solution of unknown $[H^+]$, the half-reactions and overall reaction are

$$\begin{array}{llr} H_2(g;\ 1\ atm) \longrightarrow 2H^+(aq;\ \text{unknown}) + 2e^- & & [\text{anode; oxidation}] \\ 2H^+(aq;\ 1\ M) + 2e^- \longrightarrow H_2(g;\ 1\ atm) & & [\text{cathode; reduction}] \\ \hline 2H^+(aq;\ 1\ M) \longrightarrow 2H^+(aq;\ \text{unknown}) & E_{cell} = ? & \end{array}$$

E°_{cell} is zero, but E_{cell} is *not* because the half-cells differ in $[H^+]$. From the Nernst equation, with $n = 2$, we have

$$E_{cell} = E^\circ_{cell} - \frac{0.0592\ V}{2}\log\frac{[H^+]^2_{unknown}}{[H^+]^2_{standard}}$$

Substituting 1 *M* for $[H^+]_{standard}$ and 0 V for E°_{cell} gives

$$E_{cell} = 0\ V - \frac{0.0592\ V}{2}\log\frac{[H^+]^2_{unknown}}{1^2} = -\frac{0.0592\ V}{2}\log\ [H^+]^2_{unknown}$$

Because $\log x^2 = 2 \log x$ (see Appendix A), we obtain

$$E_{cell} = -\left[\frac{0.0592\ V}{2}(2\log\ [H^+]_{unknown})\right] = -0.0592\ V \times \log\ [H^+]_{unknown}$$

Substituting $-\log\ [H^+] = pH$, we have

$$E_{cell} = 0.0592\ V \times pH$$

Thus, by measuring E_{cell}, we can find the pH.

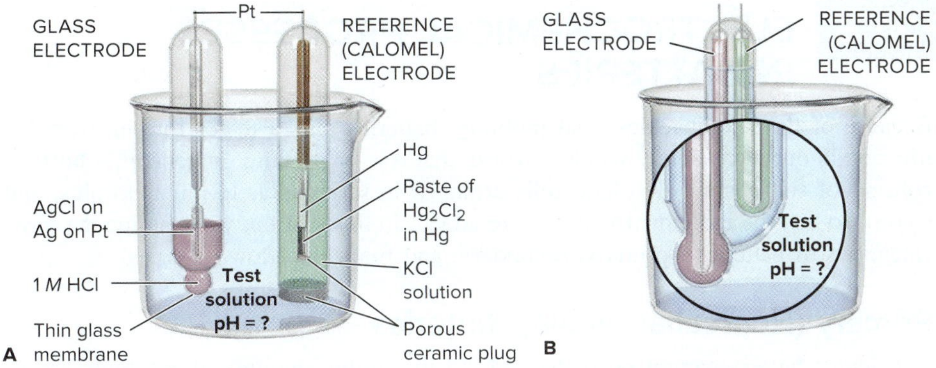

Figure 21.13 Laboratory measurement of pH. A, An older style pH meter includes a glass electrode *(left)* and a reference calomel electrode *(right)*. **B,** Modern pH meters use a combination electrode.

For routine lab measurement of pH, a concentration cell made of two hydrogen electrodes is too bulky and difficult to maintain. Instead, we use a pH meter (Figure 21.13A). In a common, but older, design, two separate electrodes dip into the solution being tested:

- The *glass electrode* consists of an Ag/AgCl half-reaction immersed in HCl solution (usually 1.000 *M*) and enclosed by a thin (~0.05 mm) membrane made of a glass that is very sensitive to H^+ ions.
- The *reference electrode,* usually a *saturated calomel electrode,* consists of a platinum wire immersed in calomel (Hg_2Cl_2) paste, liquid Hg, and saturated KCl solution.

The glass electrode monitors the solution's $[H^+]$ relative to its own fixed internal $[H^+]$, and the instrument converts the potential difference between the glass and reference electrodes into a measure of pH. In modern instruments, a *combination* electrode houses both electrodes in one tube (Figure 21.13B).

2. *Measuring ions selectively*. The pH electrode is one type of *ion-selective (or ion-specific) electrode*. These electrodes are designed with specialized membranes to selectively measure certain ion concentrations in a mixture of many ions, as in natural waters and soils. Biologists can implant a tiny ion-selective electrode in a single cell to study ion channels and receptors (Figure 21.14). Recent advances allow measurement in the femtomolar (10^{-15} *M*) range. Table 21.3 lists a few of the ions studied.

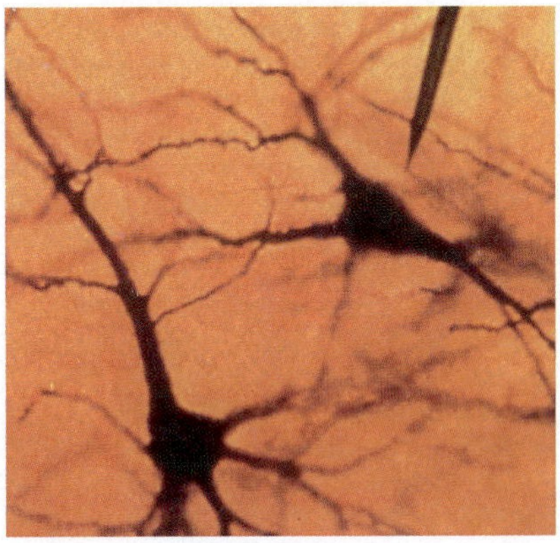

Figure 21.14 Microanalysis. A microelectrode records electrical impulses of a single neuron in a monkey's visual cortex.

Source: Photo © F. W. Goro

3. *Concentration cells in nerves*. A nerve cell membrane is embedded with enzyme "gates" that use the energy of one-third of the body's ATP to create an ion gradient of low $[Na^+]$ and high $[K^+]$ inside and high $[Na^+]$ and low $[K^+]$ outside. As a result of these differences, the outside of a nerve cell is more positive than the inside. (The 1997 Nobel Prize in chemistry was shared by Jens C. Skou for elucidating this mechanism.) When the nerve membrane is stimulated, an electrical impulse is created as Na^+ ions rush in spontaneously and the inside becomes more positive than the outside. This event is followed by K^+ ions spontaneously rushing out, and the outside becomes more positive again; the whole process takes about 0.002 s! These large changes in charge in one membrane region stimulate the neighboring region and the electrical impulse moves along the cell.

Table 21.3 Some Ions Measured with Ion-Specific Electrodes

Species Detected	Typical Sample
NH_3/NH_4^+	Industrial wastewater, seawater
CO_2/HCO_3^-	Blood, groundwater
F^-	Drinking water, urine, soil, industrial stack gases
Br^-	Grain, plant tissue
I^-	Milk, pharmaceuticals
NO_3^-	Soil, fertilizer, drinking water
K^+	Blood serum, soil, wine
H^+	Laboratory solutions, soil, natural waters

› Summary of Section 21.4

- › A spontaneous process has a negative ΔG and a positive E_{cell}: $\Delta G = -nFE_{cell}$. The ΔG of the cell reaction represents the maximum electrical work the voltaic cell can do.
- › The standard free energy change, $\Delta G°$, is related to $E°_{cell}$ and to K.
- › For nonstandard conditions, the Nernst equation shows that E_{cell} depends on $E°_{cell}$ and a correction term based on Q. E_{cell} is high when Q is small (high [reactant]), and it decreases as the cell operates. At equilibrium, ΔG and E_{cell} are zero, which means that $Q = K$.
- › Concentration cells have identical half-reactions, but solutions of differing concentration. They generate electrical energy as the concentrations become equal.
- › Ion-specific electrodes, such as the pH electrode, measure the concentration of one species.
- › The principle of the concentration cell—spontaneous movement of ions "down" a concentration gradient—creates an electrical impulse in a nerve cell.

21.5 ELECTROCHEMICAL PROCESSES IN BATTERIES

Because of their compactness and mobility, batteries play a major role in everyday life, and in our increasingly wireless world, that role is growing. In general, a **battery** consists of self-contained voltaic cells arranged in series (plus-to-minus-to-plus, and so on), so that the individual voltages are added. In this section, we examine the three categories of batteries—primary, secondary, and fuel cells (flow batteries).

Primary (Nonrechargeable) Batteries

A *primary battery* cannot be recharged, so the initial amounts of reactants are as far from equilibrium as is practical. This type of battery is discarded when the cell reaction has reached equilibrium, that is, when the battery is "dead." We'll discuss the alkaline battery, mercury and silver "button" batteries, and the primary lithium battery.

Alkaline Battery Invented in the 1860s, the common dry cell was a familiar item into the 1970s, but has now been replaced by the ubiquitous alkaline battery. The electrode materials are zinc and manganese dioxide, and the electrolyte is a basic paste of KOH and water (Figure 21.15). The half-reactions are

Anode (oxidation): $Zn(s) + 2OH^-(aq) \longrightarrow ZnO(s) + H_2O(l) + 2e^-$

Cathode (reduction): $MnO_2(s) + 2H_2O(l) + 2e^- \longrightarrow Mn(OH)_2(s) + 2OH^-(aq)$

Overall (cell) reaction:

$$Zn(s) + MnO_2(s) + H_2O(l) \longrightarrow ZnO(s) + Mn(OH)_2(s) \qquad E_{cell} = 1.5\ V$$

The alkaline battery powers portable radios, toys, flashlights, and so on, is safe, and comes in many sizes. It has no voltage drop, a long shelf life, and reliable performance in terms of power capability and stored energy.

Figure 21.15 Alkaline battery.
Source: © Jill Braaten

Mercury and Silver (Button) Batteries Both mercury and silver batteries use a zinc container as the anode (reducing agent) in a basic medium. The mercury battery employs HgO as the oxidizing agent, the silver uses Ag_2O, and both have a steel can around the cathode. The solid reactants are compacted with KOH and separated with moist paper. The half-reactions are

Anode (oxidation): $Zn(s) + 2OH^-(aq) \longrightarrow ZnO(s) + H_2O(l) + 2e^-$

Cathode (reduction) (mercury): $HgO(s) + H_2O(l) + 2e^- \longrightarrow Hg(l) + 2OH^-(aq)$

Cathode (reduction) (silver): $Ag_2O(s) + H_2O(l) + 2e^- \longrightarrow 2Ag(s) + 2OH^-(aq)$

Overall (cell) reaction (mercury):

$$Zn(s) + HgO(s) \longrightarrow ZnO(s) + Hg(l) \qquad E_{cell} = 1.3\ V$$

Overall (cell) reaction (silver):

$$Zn(s) + Ag_2O(s) \longrightarrow ZnO(s) + 2Ag(s) \qquad E_{cell} = 1.6\ V$$

Both cells are manufactured as button-sized batteries. The mercury cell, valued for its long life and stability, has been used in calculators, watches, and cameras; due to the toxicity of discarded mercury, the sale of mercury batteries is banned in many countries. The silver cell (Figure 21.16) is used in watches, cameras, heart pacemakers, and hearing aids because of its very steady output; its disadvantage is the high cost of silver.

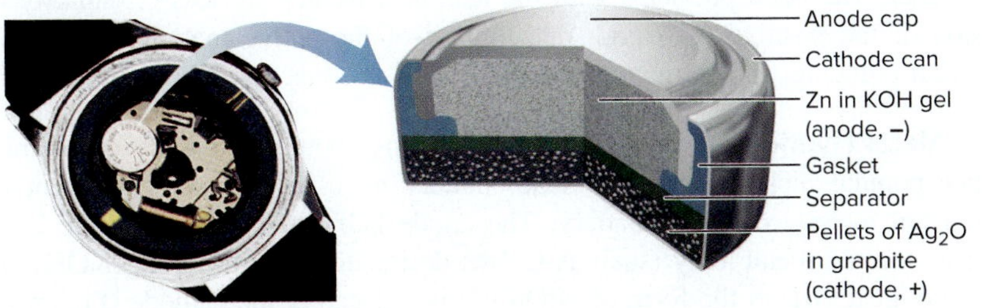

Figure 21.16 **Silver button battery.**
Source: © McGraw-Hill Education/Pat Watson, photographer

Primary Lithium Batteries The primary lithium battery is also widely used in watches and implanted medical products, and more recently in numerous remote-control devices. It offers an extremely high energy/mass ratio, producing 1 mol of e^- (1 *F*) from less than 7 g of metal ($\mathscr{M}$ of Li = 6.941 g/mol). The anode is lithium foil in a nonaqueous electrolyte. The cathode is one of several metal oxides in which lithium ions lie between oxide layers. Some pacemakers have a silver vanadium oxide (SVO; $AgV_2O_{5.5}$) cathode and can provide power (1.5 to 3.5 V) for several years, but at a low rate because energy storage is limited (Figure 21.17). The half-reactions are

Anode (oxidation): $3.5Li(s) \longrightarrow 3.5Li^+ + 3.5e^-$

Cathode (reduction): $AgV_2O_{5.5} + 3.5Li^+ + 3.5e^- \longrightarrow Li_{3.5}AgV_2O_{5.5}$

Overall (cell) reaction: $AgV_2O_{5.5} + 3.5Li(s) \longrightarrow Li_{3.5}AgV_2O_{5.5}$

Secondary (Rechargeable) Batteries

In contrast to a primary battery, a *secondary battery* is *rechargeable;* when it runs down, *electrical energy is supplied to reverse the cell reaction* and form more reactant. In other words, in a secondary battery, the voltaic cells are periodically converted to electrolytic cells to restore the starting *nonequilibrium* concentrations of the cell components. We'll discuss the common car battery, the nickel–metal hydride battery, and the lithium-ion battery, a secondary lithium battery.

Lead-Acid Battery A typical lead-acid car battery has six cells connected in series, each of which delivers about 2.1 V for a total of about 12 V. Each cell contains two lead grids packed with high-surface-area (spongy) Pb in the anode and high-surface-area PbO_2 in the cathode. The grids are immersed in a solution of ~4.5 *M* H_2SO_4. Fiberglass sheets between the grids prevent shorting due to physical contact (Figure 21.18, *next page*).

1. *Discharging.* When the cell discharges as a voltaic cell, it generates electrical energy:

Anode (oxidation): $Pb(s) + HSO_4^-(aq) \longrightarrow PbSO_4(s) + H^+ + 2e^-$

Cathode (reduction):

$$PbO_2(s) + 3H^+(aq) + HSO_4^-(aq) + 2e^- \longrightarrow PbSO_4(s) + 2H_2O(l)$$

Both half-reactions form Pb^{2+} ions, one through oxidation of Pb, the other through reduction of PbO_2. The Pb^{2+} forms $PbSO_4(s)$ at both electrodes by reacting with HSO_4^-.

Overall (cell) reaction (discharge):

$$PbO_2(s) + Pb(s) + 2H_2SO_4(aq) \longrightarrow 2PbSO_4(s) + 2H_2O(l) \qquad E_{cell} = 2.1 \text{ V}$$

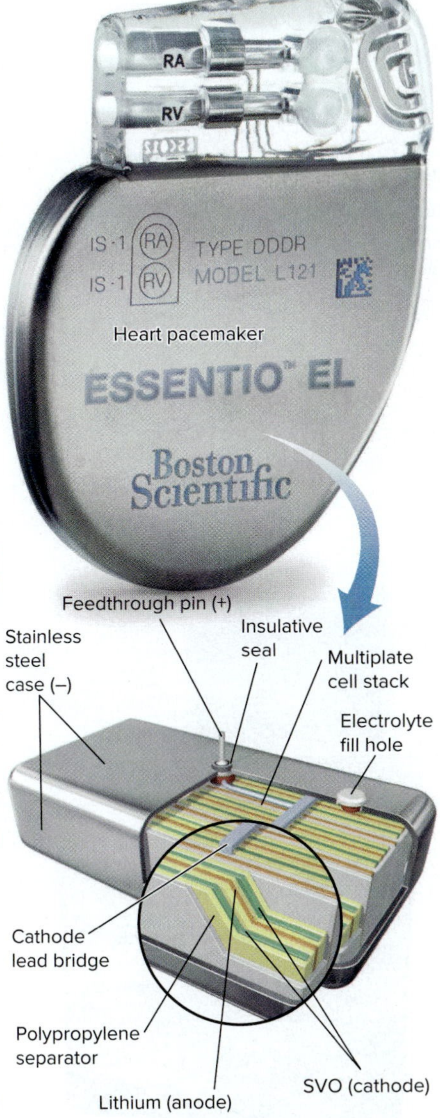

Figure 21.17 **Primary lithium battery.**
Source: © 2015 Boston Scientific Corporation or its affiliates. All rights reserved.

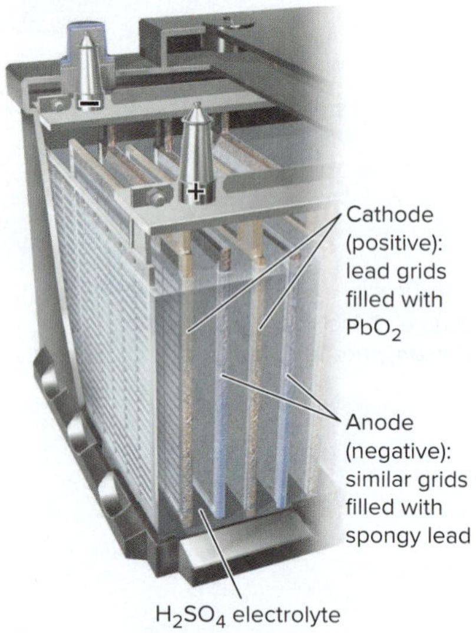

Figure 21.18 Lead-acid battery.

2. *Recharging.* When the cell recharges as an electrolytic cell, it uses electrical energy and the half-cell and overall reactions are reversed.

Overall (cell) reaction (recharge):

$$2PbSO_4(s) + 2H_2O(l) \longrightarrow PbO_2(s) + Pb(s) + 2H_2SO_4(aq)$$

Car and truck owners have relied on the lead-acid battery for over a century to provide the large burst of current needed to start the engine—and to do so for years, in hot and cold weather. The main problems with the lead-acid battery are loss of capacity due to corrosion of the positive (Pb) grid, detachment of the active material due to normal mechanical bumping, and formation of large $PbSO_4$ crystals that hinder recharging.

Nickel–Metal Hydride (Ni-MH) Battery Concerns about the toxicity of cadmium in the once popular nickel-cadmium (nicad) battery have led to its replacement by the nickel–metal hydride (Ni-MH) battery. The anode half-reaction oxidizes the hydrogen absorbed within a metal alloy (such as $LaNi_5$; designated M) in a basic (KOH) electrolyte, while nickel(III) in the form of NiO(OH) is reduced at the cathode (Figure 21.19):

Anode (oxidation): $MH(s) + OH^-(aq) \longrightarrow M(s) + H_2O(l) + e^-$

Cathode (reduction): $NiO(OH)(s) + H_2O(l) + e^- \longrightarrow Ni(OH)_2(s) + OH^-(aq)$

Overall (cell) reaction: $MH(s) + NiO(OH)(s) \longrightarrow M(s) + Ni(OH)_2(s)$

$E_{cell} = 1.4\ V$

The cell reaction is reversed during recharging. The Ni-MH battery is common in cordless razors, camera flash units, and power tools. It is lightweight, has high power, and is nontoxic, but it discharges significantly during storage.

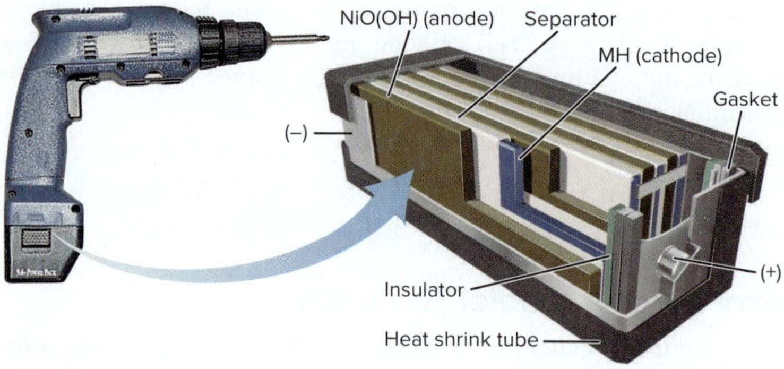

Figure 21.19 Nickel–metal hydride battery.

Source: © McGraw-Hill Education/Stephen Frisch, photographer

Lithium-Ion Battery The secondary lithium-ion battery has an anode of Li atoms that lie between sheets of graphite (designated Li_xC_6). The cathode is a lithium metal oxide, such as $LiMn_2O_4$ or $LiCoO_2$, and a typical electrolyte is 1 *M* $LiPF_6$ in an organic solvent, such as dimethyl carbonate mixed with methylethyl carbonate. Electrons flow through the circuit, while solvated Li^+ ions flow from anode to cathode within the cell (Figure 21.20). The cell reactions are

Anode (oxidation): $Li_xC_6 \longrightarrow xLi^+ + xe^- + C_6(s)$

Cathode (reduction): $Li_{1-x}Mn_2O_4(s) + xLi^+ + xe^- \longrightarrow LiMn_2O_4(s)$

Overall (cell) reaction: $Li_xC_6 + Li_{1-x}Mn_2O_4(s) \longrightarrow LiMn_2O_4(s) + C_6(s)$

$E_{cell} = 3.7\ V$

The cell reaction is reversed during recharging. The lithium-ion battery powers countless laptop computers, tablets, cell phones, and camcorders. Its key drawbacks are cost and flammability of the organic solvent.

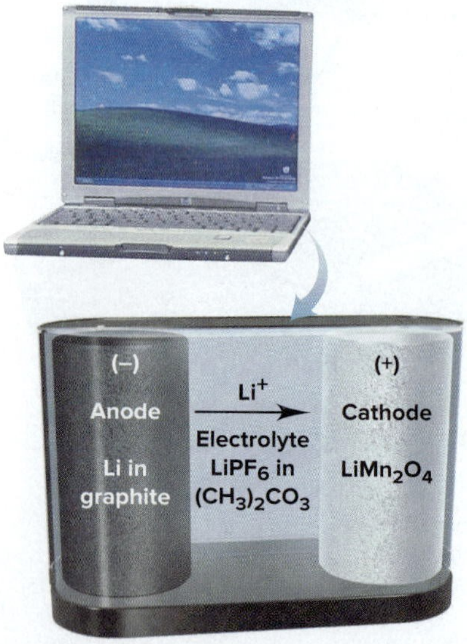

Figure 21.20 Lithium-ion battery.

Source: © AP/Wide World Photos

Fuel Cells

In contrast to primary and secondary batteries, a **fuel cell,** sometimes called a *flow battery,* is not self-contained. The reactants (usually a fuel and oxygen) enter the cell, and the products leave, *generating electricity through controlled combustion (oxidation)* of the fuel. The fuel does not burn because, as in other voltaic cells, the half-reactions are separated, and the electrons move through an external circuit.

The most common fuel cell being developed for use in cars is the *proton exchange membrane (PEM) cell,* which uses H_2 as the fuel and has an operating temperature of around 80°C (Figure 21.21). The cell reactions are

Anode (oxidation): $2H_2(g) \longrightarrow 4H^+(aq) + 4e^-$

Cathode (reduction): $O_2(g) + 4H^+(aq) + 4e^- \longrightarrow 2H_2O(g)$

Overall (cell) reaction: $2H_2(g) + O_2(g) \longrightarrow 2H_2O(g) \qquad E_{cell} = 1.2\ V$

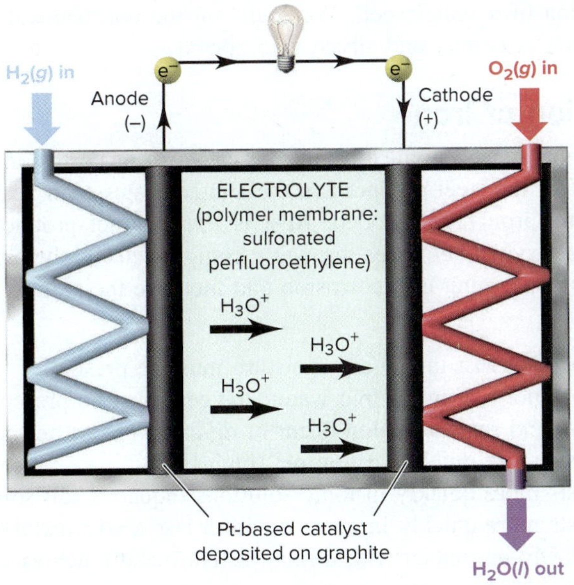

Figure 21.21 Proton exchange membrane cell: a hydrogen fuel cell.

How H_2 Fuel Cells Work Reaction rates are lower in fuel cells than in other batteries, so an *electrocatalyst* is used to decrease the activation energy (Section 16.7). The PEM cell's electrodes are made of a nanocomposite consisting of a Pt-based catalyst deposited on graphite. These are embedded in a polymer electrolyte membrane having a perfluoroethylene backbone ($-\!\![F_2C\!-\!CF_2]_n\!\!-$) with attached sulfonic acid groups (RSO_3^-) that play a key role in ferrying protons from anode to cathode.

- *At the anode,* two H_2 molecules adsorb onto the catalyst and are split and oxidized. From each H_2, two e^- travel through the wire to the cathode, while two H^+ are hydrated and migrate through the electrolyte as H_3O^+.
- *At the cathode,* an O_2 molecule is believed to adsorb onto the catalyst, which provides an e^- to form O_2^-. One H_3O^+ donates its H^+ to the O_2^-, forming HO_2 (that is, HO—O). The O—O bond stretches and breaks as another H_3O^+ gives its H^+ and the catalyst provides another e^-: the first H_2O has formed. In similar fashion, a third H^+ and e^- attach to the freed O atom to form OH, and a fourth H^+ and e^- are transferred to form the second H_2O. Both water molecules desorb and leave the cell.

Applications of Fuel Cells Hydrogen fuel cells have been used for years to provide electricity *and* water during space flights. Similar ones have begun to supply electric power for residential needs, and every major car manufacturer has a fuel-cell model. Fuel cells produce no pollutants and convert about 75% of a fuel's bond energy into power, compared to 40% for a coal-fired power plant and 25% for a gasoline engine.

› Summary of Section 21.5

- Batteries are voltaic cells arranged in series and are classified as primary (e.g., alkaline, mercury, silver, and lithium), secondary (e.g., lead-acid, nickel–metal hydride, and lithium-ion), or fuel cells.
- Supplying electricity to a rechargeable (secondary) battery reverses the redox reaction, re-forming reactant.
- Fuel cells generate a current through the controlled oxidation of a fuel such as H_2.

21.6 CORROSION: AN ENVIRONMENTAL VOLTAIC CELL

If you think all spontaneous electrochemical processes are useful, like those in batteries and fuel cells, consider the problem of **corrosion,** which causes tens of billions of dollars of damage to cars, ships, buildings, and bridges each year. This natural process, which oxidizes metals to their oxides and sulfides, shares many similarities with the operation of a voltaic cell. We focus on the corrosion of iron, but several other metals, such as copper and silver, also corrode.

The Corrosion of Iron

The most common and economically destructive form of corrosion is the rusting of iron. About 25% of the steel produced in the United States each year is for replacing steel in which the iron has corroded. Rust is *not* a direct product of the reaction between iron and oxygen but arises through a complex electrochemical process. Let's look at the facts concerning iron corrosion and then use the features of a voltaic cell to explain them:

Fact 1. Iron does not rust in dry air; moisture must be present.
Fact 2. Iron does not rust in air-free water; oxygen must be present.
Fact 3. Iron loss and rust formation occur at *different* places on the *same* object.
Fact 4. Iron rusts more quickly at low pH (high $[H^+]$).
Fact 5. Iron rusts more quickly in ionic solutions (aqueous salt solutions).
Fact 6. Iron rusts more quickly in contact with a less active metal (such as Cu) and more slowly in contact with a more active metal (such as Zn).

Two separate redox processes occur during corrosion:

1. *The loss of iron.* Picture the surface of a piece of iron (Figure 21.22). A strain, ridge, or dent in contact with water is typically the site of iron loss (fact 1). This site is called an *anodic region* because the following oxidation half-reaction occurs there:

$$Fe(s) \longrightarrow Fe^{2+}(aq) + 2e^- \qquad \text{[anodic region; oxidation]}$$

Once the iron atoms lose electrons, the damage to the object has been done, and a pit forms where the iron is lost.

The freed electrons move through the external circuit—the piece of iron itself—until they reach a region of relatively high $[O_2]$ (fact 2), usually the air near the edge of a water droplet that surrounds the newly formed pit. At this *cathodic region,* the electrons released from the iron atoms reduce O_2:

$$O_2(g) + 4H^+(aq) + 4e^- \longrightarrow 2H_2O(l) \qquad \text{[cathodic region; reduction]}$$

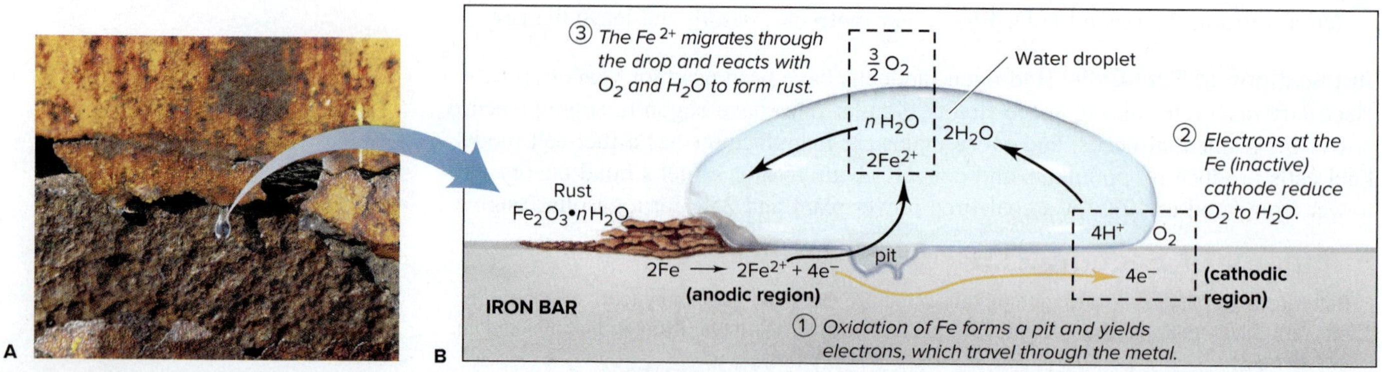

Figure 21.22 **The corrosion of iron. A,** Close-up view of an iron surface. Corrosion usually occurs at a surface irregularity. **B,** A small area of the surface, showing the steps in the corrosion process.

Source: A: © Vincent Roy mastweb/Alamy

This portion of the corrosion process (the sum of these two half-reactions) occurs without any rust forming:

$$2Fe(s) + O_2(g) + 4H^+(aq) \longrightarrow 2Fe^{2+}(aq) + 2H_2O(l) \qquad \text{[overall]}$$

2. *The rusting process.* Rust forms in another redox reaction. The Fe^{2+} ions formed at the anodic region disperse through the water and react with O_2, often away from the pit, to form the Fe^{3+} in rust (fact 3). The overall reaction for this step is

$$2Fe^{2+}(aq) + \tfrac{1}{2}O_2(g) + (2 + n)H_2O(l) \longrightarrow Fe_2O_3{\cdot}nH_2O(s) + 4H^+(aq)$$

[The coefficient n for H_2O appears because rust, $Fe_2O_3{\cdot}nH_2O$, may have a variable number of waters of hydration.] The rust deposit is incidental to the real damage, which is the loss of iron that weakens the strength of the object. Adding the two previous equations gives the overall equation for the loss and rusting of iron:

$$2Fe(s) + \tfrac{3}{2}O_2(g) + nH_2O(l) + \cancel{4H^+(aq)} \longrightarrow Fe_2O_3{\cdot}nH_2O(s) + \cancel{4H^+(aq)}$$

Other species ($2Fe^{2+}$ and $2H_2O$) also cancel, but we showed the canceled H^+ ions to emphasize that they act as a catalyst: they speed the process as they are used up in one step and created in another. For this reason, rusting is faster at low pH (high $[H^+]$) (fact 4). Ionic solutions speed rusting by improving the conductivity of the aqueous medium near the anodic and cathodic regions (fact 5). The effect of ions is especially evident on ocean-going vessels (Figure 21.23) and on the underbodies and around the wheel wells of cars driven in cold climates, where salts are used to melt ice on slippery roads. (We discuss fact 6 in the next subsection.)

Thus, in some key ways, corrosion resembles the operation of a voltaic cell:

- Anodic and cathodic regions are physically separated.
- The regions are connected via an external circuit through which the electrons travel.
- In the anodic region, iron behaves like an active electrode, whereas in the cathodic region, it is inactive.
- The moisture surrounding the pit functions somewhat like an electrolyte and salt bridge, a solution of ions and a means for them to move and keep the solution neutral.

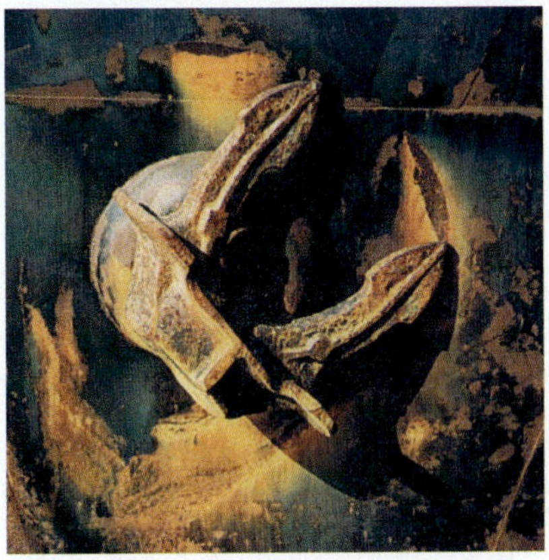

Figure 21.23 Enhanced corrosion at sea. The high ion concentration of seawater enhances the corrosion of iron in hulls and anchors.

Source: © David Weintraub/Science Source

Protecting Against the Corrosion of Iron

Corrosion is prevented by eliminating corrosive factors. Washing road salt off automobile bodies removes ions. Painting an object keeps out O_2 and moisture. Plating chromium on plumbing fixtures is a more permanent method, as is "blueing" of gun barrels and other steel objects, in which a coating of Fe_3O_4 (magnetite) is bonded to the surface.

The final point regarding corrosion (fact 6) concerns the relative activity of other metals in contact with iron. The essential idea is that *iron functions as both anode and cathode in the rusting process, but it is lost only at the anode.* Thus,

1. *Corrosion increases when iron behaves more like the anode.* When iron is in contact with a *less* active metal (weaker reducing agent), such as copper, it loses electrons more readily (its anodic function is enhanced; Figure 21.24A). For example, when iron plumbing is connected directly to copper plumbing, the iron pipe corrodes

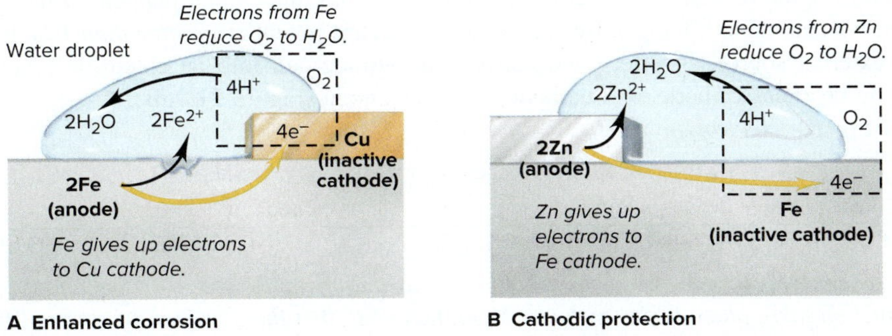

Figure 21.24 The effect of metal-metal contact on the corrosion of iron. A, Fe in contact with Cu corrodes faster. **B,** Fe in contact with Zn does not corrode. This method of preventing corrosion is known as *cathodic protection.*

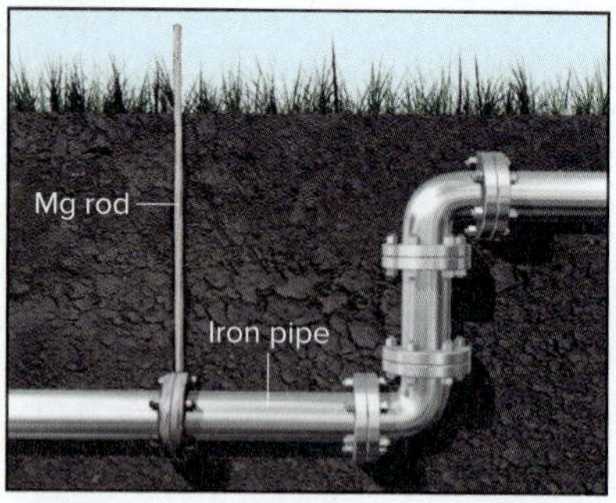

Figure 21.25 The use of a sacrificial anode to prevent iron corrosion. In cathodic protection, an active metal, such as zinc, magnesium, or aluminum, acts as the anode and is sacrificed instead of the iron.

rapidly. Nonconducting rubber or plastic spacers are placed between the metals to avoid this problem.

2. *Corrosion decreases when iron behaves more like the cathode.* In *cathodic protection,* the most effective way to prevent corrosion, iron makes contact with a *more* active metal (stronger reducing agent), such as zinc. The iron becomes the cathode and remains intact, while the zinc acts as the anode and loses electrons (Figure 21.24B). Coating steel with a "sacrificial" layer of zinc is called *galvanizing*. In addition to blocking physical contact with H_2O and O_2, the zinc (or other active metal) is "sacrificed" (oxidized) instead of the iron. Sacrificial anodes are placed underwater and underground to protect iron and steel pipes, tanks, oil rigs, and so on. Magnesium and aluminum are often used because they are much more active than iron and, thus, act as the anode (Figure 21.25). Moreover, they form adherent oxide coatings, which slow their own corrosion.

› Summary of Section 21.6

- Corrosion damages metal structures through a natural electrochemical process.
- Iron corrosion occurs in the presence of oxygen and moisture and is increased by high $[H^+]$, high [ion], or contact with a less active metal, such as Cu.
- Fe is oxidized and O_2 is reduced in one redox reaction, while Fe^{2+} is oxidized and O_2 is reduced to form rust (hydrated form of Fe_2O_3) in another redox reaction that often takes place at a different location.
- Because Fe functions as both anode and cathode in the corrosion process, an iron object can be protected by physically coating it or by joining it to a more active metal (such as Zn, Mg, or Al), which acts as the anode in place of the Fe.

21.7 ELECTROLYTIC CELLS: USING ELECTRICAL ENERGY TO DRIVE NONSPONTANEOUS REACTIONS

In contrast to a voltaic cell, which generates electrical energy from a spontaneous reaction, *an electrolytic cell requires electrical energy from an external source to drive a* ***non****spontaneous redox reaction.* (Industrial electrolysis is discussed in Chapter 22.)

Construction and Operation of an Electrolytic Cell

Let's see how an electrolytic cell operates by constructing one from a voltaic cell. Consider a tin-copper voltaic cell (Figure 21.26A). The Sn anode will gradually be oxidized to Sn^{2+} ions, which enter the electrolyte, and the Cu^{2+} ions will gradually be reduced and plate out on the Cu cathode because the cell reaction is spontaneous in that direction.

For the voltaic cell:

$$\mathrm{Sn}(s) \longrightarrow \mathrm{Sn}^{2+}(aq) + 2e^- \qquad \text{[anode; oxidation]}$$
$$\mathrm{Cu}^{2+}(aq) + 2e^- \longrightarrow \mathrm{Cu}(s) \qquad \text{[cathode; reduction]}$$
$$\mathrm{Sn}(s) + \mathrm{Cu}^{2+}(aq) \longrightarrow \mathrm{Sn}^{2+}(aq) + \mathrm{Cu}(s) \qquad E^\circ_{\text{cell}} = 0.48\text{ V and }\Delta G^\circ = -93\text{ kJ}$$

Therefore, the *reverse* cell reaction is *non*spontaneous and cannot happen on its own. But, we can make it happen by supplying an electric potential *greater than* E°_{cell} from an external source. In effect, we convert the voltaic cell into an electrolytic cell—anode becomes cathode, and cathode becomes anode (Figure 21.26B).

For the electrolytic cell:

$$\mathrm{Cu}(s) \longrightarrow \mathrm{Cu}^{2+}(aq) + 2e^- \qquad \text{[anode; oxidation]}$$
$$\mathrm{Sn}^{2+}(aq) + 2e^- \longrightarrow \mathrm{Sn}(s) \qquad \text{[cathode; reduction]}$$
$$\mathrm{Cu}(s) + \mathrm{Sn}^{2+}(aq) \longrightarrow \mathrm{Cu}^{2+}(aq) + \mathrm{Sn}(s) \qquad E^\circ_{\text{cell}} = -0.48\text{ V and }\Delta G^\circ = 93\text{ kJ}$$

In an electrolytic cell, as in a voltaic cell, *oxidation takes place at the anode and reduction takes place at the cathode.* Note, however, that *the direction of electron flow and the signs of the electrodes are reversed.*

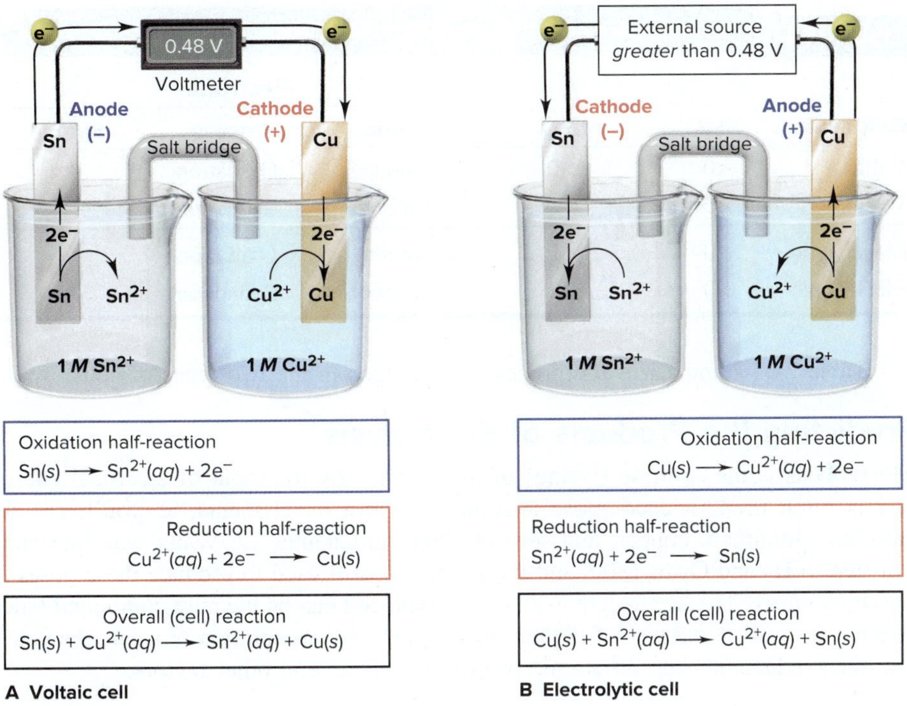

Figure 21.26 **The tin-copper reaction as the basis of a voltaic and an electrolytic cell. A,** The spontaneous reaction between Sn and Cu^{2+} generates 0.48 V in a voltaic cell. **B,** If more than 0.48 V is supplied, the nonspontaneous (reverse) reaction between Cu and Sn^{2+} occurs. Note the changes in electrode charges and direction of electron flow.

To understand these differences, we focus on the *cause* of the electron flow:

- In a voltaic cell, electrons are generated *in* the anode, so it is *negative,* and they are *removed from* the cathode, so it is *positive.*
- In an electrolytic cell, an external power source supplies electrons *to* the cathode, so it is *negative,* and removes them *from* the anode, so it is *positive.*

A rechargeable battery is a voltaic cell when it is discharging and an electrolytic cell when it is recharging. Let's compare these two functions in a lead-acid car battery (Figure 21.27; see also Figure 21.18):

- In the discharge mode (voltaic cell), oxidation occurs at electrode I, thus making that *negative* electrode the anode and the *positive* electrode (electrode II) the cathode.
- In the recharge mode (electrolytic cell), reduction occurs at electrode I, making that *negative* electrode the cathode and the *positive* electrode (electrode II) the anode.

Figure 21.27 **The processes occurring during the discharge *(top)* and recharge *(bottom)* of a lead-acid battery.**
Source: © Jill Braaten

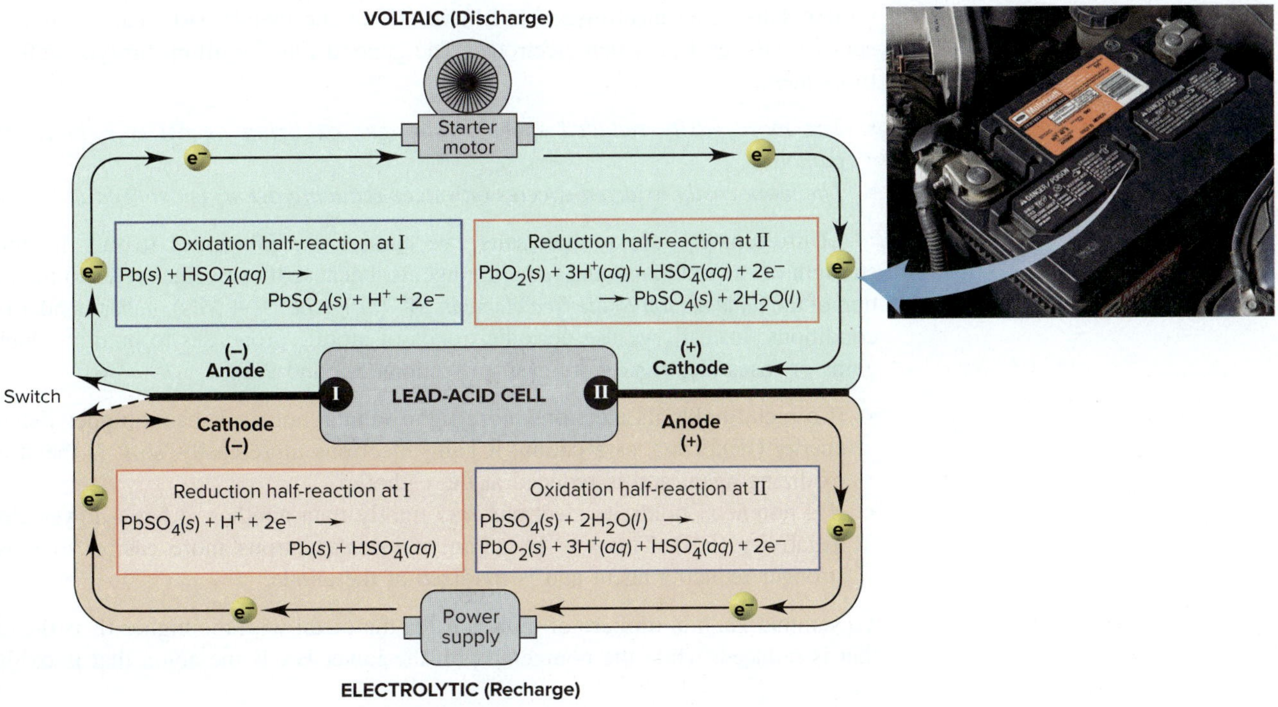

Table 21.4 Comparison of Voltaic and Electrolytic Cells

			Electrode		
Cell Type	**ΔG**	**E_{cell}**	**Name**	**Process**	**Sign**
Voltaic	<0	>0	Anode	Oxidation	–
Voltaic	<0	>0	Cathode	Reduction	+
Electrolytic	>0	<0	Anode	Oxidation	+
Electrolytic	>0	<0	Cathode	Reduction	–

Table 21.4 summarizes the processes and signs in the two types of cells.

Predicting the Products of Electrolysis

Electrolysis is the splitting (lysing) of a substance by the input of electrical energy and is often used to decompose a compound into its elements, as you'll see for chlorine, aluminum, copper, and several others in Chapter 22. Water was first electrolyzed to H_2 and O_2 in 1800, and the process is still used to produce these gases in ultrahigh purity. The electrolyte in an electrolytic cell can be the pure compound (such as a molten salt), a mixture of molten salts, or an aqueous solution of a salt. As you'll see, the products depend on atomic properties and several other factors.

Electrolysis of Pure Molten Salts If the salt is pure, predicting the products of its electrolysis is straightforward: *the cation will be reduced and the anion oxidized.* The electrolyte is the molten salt itself, and the ions are attracted by the oppositely charged electrodes.

Consider the electrolysis of molten (fused) calcium chloride. The two species present are Ca^{2+} and Cl^-, so Ca^{2+} ion is reduced and Cl^- ion is oxidized:

$$2Cl^-(l) \longrightarrow Cl_2(g) + 2e^- \qquad \text{[anode; oxidation]}$$
$$Ca^{2+}(l) + 2e^- \longrightarrow Ca(s) \qquad \text{[cathode; reduction]}$$
$$Ca^{2+}(l) + 2Cl^-(l) \longrightarrow Ca(s) + Cl_2(g) \qquad \text{[overall]}$$

Calcium is prepared industrially this way, as are several other active metals, such as Na and Mg, and the halogens Cl_2 and Br_2.

Electrolysis of Mixed Molten Salts More typically, the electrolyte is a mixture of molten salts being electrolyzed to obtain one of the metals. How can we tell which species will react at which electrode? The general rule for all electrolytic cells identifies them:

- *The more easily reduced species (stronger oxidizing agent) is reduced at the cathode.*
- *The more easily oxidized species (stronger reducing agent) is oxidized at the anode.*

Unfortunately, for molten salts, we *cannot* use $E°$ values to tell the relative strength of the oxidizing and reducing agents, because those values refer to the reduction of the *aqueous ion to the free element,* $M^{n+}(aq) + ne^- \longrightarrow M(s)$, under standard-state conditions. Instead, we use periodic trends of atomic properties to predict which ion gains or loses electrons more easily (Sections 8.3 and 9.5):

- If a metal holds its electrons *more* tightly than another, it has a higher ionization energy (IE). Thus, as a cation, it gains electrons more easily, so it is the stronger oxidizing agent and is reduced at the cathode.
- If a nonmetal holds its electrons *less* tightly than another, it has a lower electronegativity (EN). Thus, as an anion, it loses electrons more easily, so it is the stronger reducing agent and is oxidized at the anode.

To summarize, in a mixture of molten salts, the metal with the higher IE is the cation that is reduced, while the nonmetal with the lower EN is the anion that is oxidized.

SAMPLE PROBLEM 21.9 Predicting the Electrolysis Products of a Molten Salt Mixture

Problem A chemical engineer melts a naturally occurring mixture of NaBr and $MgCl_2$ and decomposes it in an electrolytic cell. Predict the substance formed at each electrode, and write balanced half-reactions and the overall cell reaction.

Plan We have to determine which metal and nonmetal will form more easily at each electrode. We list the ions as oxidizing or reducing agents and then use periodic trends to determine which metal ion is more easily reduced and which nonmetal ion more easily oxidized.

Solution Listing the ions as oxidizing or reducing agents:

The possible oxidizing agents are Na^+ and Mg^{2+}; these cations may undergo reduction.

The possible reducing agents are Br^- and Cl^-; these anions may undergo oxidation.

Determining the cathode product (more easily reduced cation): Mg is to the right of Na in Period 3. IE increases from left to right, so it is harder to remove e^- from Mg. Thus, Mg^{2+} has a greater attraction for e^- and is more easily reduced (stronger oxidizing agent). Mg^{2+} will be reduced preferentially at the cathode:

$$Mg^{2+}(l) + 2e^- \longrightarrow Mg(l) \qquad \text{[cathode; reduction]}$$

Determining the anode product (more easily oxidized anion): Br is below Cl in Group 7A(17). EN decreases down the group, so Br accepts e^- less readily. Thus, Br^- loses its e^- more easily and is more easily oxidized (stronger reducing agent). Br^- will be oxidized preferentially at the anode:

$$2Br^-(l) \longrightarrow Br_2(g) + 2e^- \qquad \text{[anode; oxidation]}$$

Writing the overall cell reaction:

$$Mg^{2+}(l) + 2Br^-(l) \longrightarrow Mg(l) + Br_2(g) \qquad \text{[overall]}$$

Comment The cell temperature must be high enough to keep the salt mixture molten. In this case, the temperature is greater than the melting point of Mg, so it appears as a liquid in the equation, and greater than the boiling point of Br_2, so it appears as a gas.

FOLLOW-UP PROBLEMS

21.9A The most ionic and least ionic of the common alkali halides are, respectively, CsF and LiI. A solid mixture of these compounds is melted and electrolyzed. Determine which metal and nonmetal form at the electrodes, and write the overall cell reaction.

21.9B A sample of $AlBr_3$ contaminated with KF is melted and electrolyzed. Determine the electrode products, and write the overall cell reaction.

SOME SIMILAR PROBLEMS 21.89–21.92

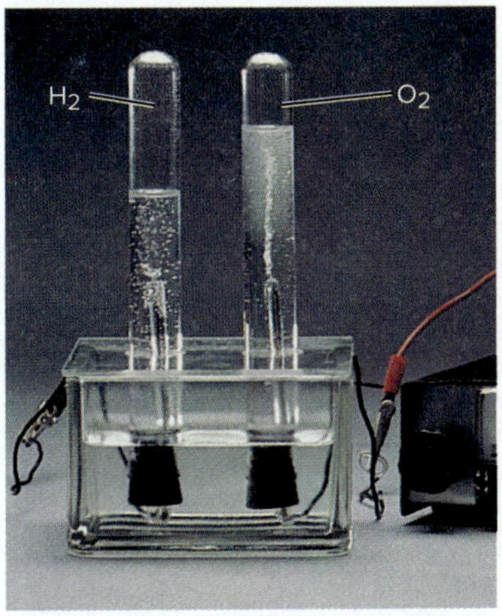

Oxidation half-reaction
$2H_2O(l) \longrightarrow O_2(g) + 4H^+(aq) + 4e^-$

Reduction half-reaction
$2H_2O(l) + 2e^- \longrightarrow H_2(g) + 2OH^-(aq)$

Overall (cell) reaction
$2H_2O(l) \longrightarrow 2H_2(g) + O_2(g)$

Figure 21.28 The electrolysis of water. Twice as much H_2 forms as O_2.

Source: © McGraw-Hill Education/Stephen Frisch, photographer

Electrolysis of Water and Nonstandard Half-Cell Potentials Before we analyze the electrolysis products of aqueous salt solutions, let's examine the electrolysis of water itself. Very pure water is difficult to electrolyze because so few ions are present to conduct a current. However, if we add a small amount of a salt that cannot be electrolyzed (such as Na_2SO_4), electrolysis proceeds rapidly. An electrolytic cell with separate compartments for H_2 and O_2 is used (Figure 21.28). At the anode, water is oxidized; note that the oxidation number (O.N.) of O changes from −2 in H_2O to 0 in O_2:

$$\overset{+1\,-2}{2H_2O(l)} \longrightarrow \overset{0}{O_2(g)} + 4\overset{+1}{H^+}(aq) + 4e^- \qquad E = 0.82\ \text{V} \qquad \text{[anode; oxidation]}$$

At the cathode, water is reduced; note that the O.N. of H changes from +1 in H_2O to 0 in H_2:

$$\overset{+1\,-2}{2H_2O(l)} + 2e^- \longrightarrow \overset{0}{H_2(g)} + \overset{-2\,+1}{2OH^-(aq)} \qquad E = -0.42\ V \qquad \text{[cathode; reduction]}$$

Doubling the cathode half-reaction to make e^- lost equal e^- gained, adding the half-reactions (which involves combining the H^+ and OH^- into H_2O and canceling e^- and excess H_2O), and calculating E_{cell} gives the overall reaction:

$$2H_2O(l) \longrightarrow 2H_2(g) + O_2(g) \qquad E_{cell} = -0.42\ V - 0.82\ V = -1.24\ V \qquad \text{[overall]}$$

Note that, because $[H^+]$ and $[OH^-]$ are 1.0×10^{-7} *M* rather than the standard-state value of 1 *M*, these electrode potentials are *not* standard electrode potentials and are designated with *E*, not *E*°. In aqueous ionic solutions, $[H^+]$ and $[OH^-]$ are also approximately 10^{-7} *M*, so we use these nonstandard $E_{half\text{-}cell}$ values to predict electrode products.

Electrolysis of Aqueous Salt Solutions and the Effect of Overvoltage Aqueous salt solutions are mixtures of ions *and* water, so we have to compare the electrode potentials of the ions and of water to predict the electrode products.

1. *Predicting the electrode products.* When two half-reactions are possible at an electrode:

- *The reduction with the less negative (more positive) electrode potential occurs.*
- *The oxidation with the less positive (more negative) electrode potential occurs.*

What happens, for instance, when a solution of potassium iodide is electrolyzed?

- The possible *oxidizing agents* are K^+ and H_2O; their reduction half-reactions are

$$K^+(aq) + e^- \longrightarrow K(s) \qquad E° = -2.93\ V$$
$$2H_2O(l) + 2e^- \longrightarrow H_2(g) + 2OH^-(aq) \qquad E = -0.42\ V \qquad \text{[reduction]}$$

The less *negative* electrode potential for water means that it is much easier to reduce H_2O than K^+, so H_2 forms at the cathode. No potassium metal forms at the cathode.

- The possible *reducing agents* are I^- and H_2O; their oxidation half-reactions are

$$2I^-(aq) \longrightarrow I_2(s) + 2e^- \qquad E° = 0.53\ V \qquad \text{[oxidation]}$$
$$2H_2O(l) \longrightarrow O_2(g) + 4H^+(aq) + 4e^- \qquad E = 0.82\ V$$

The less *positive* electrode potential for iodide ion means that it is easier to oxidize I^- than H_2O, so I_2 forms at the anode.

2. *The effect of overvoltage.* The products predicted from a comparison of electrode potentials are not always the actual products. For gases to be produced at metal electrodes, additional voltage is required. This increment above the expected voltage is the **overvoltage.** It is 0.4–0.6 V for $H_2(g)$ or $O_2(g)$ and is due to the large activation energy (Section 16.5) required for a gas to form at the electrode.

Overvoltage has major practical significance. A multibillion-dollar example is the industrial production of chlorine from concentrated NaCl solution. Water is easier to reduce than Na^+, so H_2 forms at the cathode, even *with* an overvoltage of 0.6 V:

$$Na^+(aq) + e^- \longrightarrow Na(s) \qquad E° = -2.71\ V$$
$$2H_2O(l) + 2e^- \longrightarrow H_2(g) + 2OH^-(aq) \qquad E = -0.42\ V\ (\approx -1\ V \text{ with overvoltage}) \qquad \text{[reduction]}$$

But Cl_2 *does* form at the anode, even though a comparison of the electrode potentials alone would lead us to predict that O_2 should form:

$$2H_2O(l) \longrightarrow O_2(g) + 4H^+(aq) + 4e^- \qquad E = 0.82\ V\ (\sim 1.4\ V \text{ with overvoltage})$$
$$2Cl^-(aq) \longrightarrow Cl_2(g) + 2e^- \qquad E° = 1.36\ V \qquad \text{[oxidation]}$$

An overvoltage of ~0.6 V for O_2 makes Cl_2 the product that is easier to form. Keeping $[Cl^-]$ high also favors its formation. Thus, Cl_2, one of the 10 most heavily produced chemicals, is formed from plentiful natural sources of aqueous sodium chloride (Chapter 22).

3. *A summary: which product at which electrode.* Experiments have shown the elements that can be prepared electrolytically from aqueous solutions of their salts:

- Cations of less active metals, including gold, silver, copper, chromium, platinum, and cadmium, *are* reduced to the metal.
- Cations of more active metals, including those in Groups 1A(1) and 2A(2) and Al from 3A(13), *are not* reduced. Water is reduced to H_2 and OH^- instead.
- Anions that *are* oxidized, because of overvoltage from O_2 formation, include the halides ($[Cl^-]$ must be high), except for F^-.
- Anions that *are not* oxidized include F^- and common oxoanions, such as SO_4^{2-}, CO_3^{2-}, NO_3^-, and PO_4^{3-} because the central nonmetal in these oxoanions is already in its highest oxidation state. Water is oxidized to O_2 and H^+ instead.

Student Hot Spot

Student data indicate that you may struggle with determining the products of electrolysis. Access the Smartbook to view additional Learning Resources on this topic.

SAMPLE PROBLEM 21.10 Predicting the Electrolysis Products of Aqueous Salt Solutions

Problem Use half-reactions to show which product forms at each electrode during the electrolysis of aqueous solutions of the following salts:
(a) KBr **(b)** $AgNO_3$ **(c)** $MgSO_4$

Plan We identify the reacting ions and compare their electrode potentials with those of water, taking the additional 0.4–0.6 V overvoltage into account. The reduction half-reaction with the less negative electrode potential occurs at the cathode, and the oxidation half-reaction with the less positive electrode potential occurs at the anode.

Solution

(a)

$$K^+(aq) + e^- \longrightarrow K(s) \qquad E° = -2.93\ V$$
$$2H_2O(l) + 2e^- \longrightarrow H_2(g) + 2OH^-(aq) \qquad E = -0.42\ V$$

Despite the overvoltage, which makes E for the reduction of water between –0.8 V and –1.0 V, H_2O is still easier to reduce than K^+, so $H_2(g)$ forms at the cathode.

$$2Br^-(aq) \longrightarrow Br_2(l) + 2e^- \qquad E° = 1.07\ V$$
$$2H_2O(l) \longrightarrow O_2(g) + 4H^+(aq) + 4e^- \qquad E = 0.82\ V$$

Because of the overvoltage, which makes E for the oxidation of water between 1.2 V and 1.4 V, Br^- is easier to oxidize than water, so $Br_2(l)$ forms at the anode *(see photo).*

(b)

$$Ag^+(aq) + e^- \longrightarrow Ag(s) \qquad E° = 0.80\ V$$
$$2H_2O(l) + 2e^- \longrightarrow H_2(g) + 2OH^-(aq) \qquad E = -0.42\ V$$

As the cation of an inactive metal, Ag^+ is a better oxidizing agent than H_2O, so $Ag(s)$ forms at the cathode. NO_3^- cannot be oxidized, because N is already in its highest (+5) oxidation state. Thus, $O_2(g)$ forms at the anode:

$$2H_2O(l) \longrightarrow O_2(g) + 4H^+(aq) + 4e^-$$

(c)

$$Mg^{2+}(aq) + 2e^- \longrightarrow Mg(s) \qquad E° = -2.37\ V$$
$$2H_2O(l) + 2e^- \longrightarrow H_2(g) + 2OH^-(aq) \qquad E = -0.42\ V$$

Like K^+ in part (a), Mg^{2+} cannot be reduced in the presence of water, so $H_2(g)$ forms at the cathode. The SO_4^{2-} ion cannot be oxidized because S is in its highest (+6) oxidation state. Thus, H_2O is oxidized, and $O_2(g)$ forms at the anode:

$$2H_2O(l) \longrightarrow O_2(g) + 4H^+(aq) + 4e^-$$

Electrolysis of aqueous KBr.

Source: © McGraw-Hill Education/Stephen Frisch, photographer

FOLLOW-UP PROBLEMS

21.10A Use half-reactions to show which product forms at each electrode when an aqueous solution of lead(II) nitrate is electrolyzed.

21.10B Use half-reactions to show which product forms at each electrode in the electrolysis of aqueous $AuBr_3$.

SOME SIMILAR PROBLEMS 21.97–21.100

Stoichiometry of Electrolysis: The Relation Between Amounts of Charge and Products

Since charge flowing through an electrolytic cell yields products at the electrodes, it makes sense that more product forms when more charge flows. This relationship was first determined experimentally in the 1830s by Michael Faraday:

- *Faraday's law of electrolysis: the amount of substance produced at each electrode is directly proportional to the quantity of charge flowing through the electrolytic cell.*

Each balanced half-reaction shows the amounts (mol) of reactant, electrons, and product involved in the change, so it contains the information we need to answer such questions as "How much material will form from a given quantity of charge?" or, conversely, "How much charge is needed to produce a given amount of material?"

Measuring Current to Find Charge We cannot measure charge directly, but we *can* measure *current,* the charge flowing per unit time. The SI unit of current is the **ampere (A),** which is defined as a charge of 1 coulomb (C) flowing through a conductor in 1 second (s):

$$\text{Current (A)} = \text{charge (C)/time (s)} \qquad \text{or} \qquad 1\text{ A} = 1\text{ C/s} \tag{21.11}$$

Thus, the current multiplied by the time gives the charge:

$$\text{Current} \times \text{time} = \text{charge} \qquad \text{or} \qquad \text{A} \times \text{s} = \frac{\text{C}}{\text{s}} \times \text{s} = \text{C}$$

Therefore, by measuring the current *and* the time during which the current flows, we find the charge, which relates to the amount of product (Figure 21.29).

Faraday's Law and the Stoichiometry of Electrolysis Problems based on Faraday's law often ask you to calculate current, mass of material, or time. As we said, the electrode half-reaction provides the key to solving these problems because it relates the mass of product to a certain quantity of charge. To apply Faraday's law:

1. Balance the half-reaction to find the amount (mol) of electrons needed per mole of product.
2. Use the Faraday constant ($F = 9.65\times10^4$ C/mol e^-) to find the quantity of charge.
3. Depending on the question, do one of the following:
   a. Use the molar mass to find the charge needed for a given mass of product, or vice versa.
   b. Use the mass of product and the given current to find the time needed, or vice versa.

Here is a typical problem in practical electrolysis: how long does it take to produce 3.0 g of $Cl_2(g)$ by electrolysis of aqueous NaCl using a power supply with a current of 12 A? The problem asks for the time needed to produce a certain mass, so let's first relate mass to amount (mol) of electrons to find the charge needed. Then, we'll relate the charge to the current to find the time.

The half-reaction tells us that 2 mol of electrons are lost to form 1 mol of Cl_2 and we'll use this relationship as a conversion factor:

$$2Cl^-(aq) \longrightarrow Cl_2(g) + 2e^-$$

We convert the given mass of Cl_2 to amount (mol) of Cl_2, use the conversion factor from the half-reaction, and multiply by the Faraday constant to find the total charge:

$$\text{Charge (C)} = 3.0\text{ g } Cl_2 \times \frac{1\text{ mol } Cl_2}{70.90\text{ g } Cl_2} \times \frac{2\text{ mol } e^-}{1\text{ mol } Cl_2} \times \frac{9.65\times10^4\text{ C}}{1\text{ mole } e^-}$$

$$= 8.2\times10^3\text{ C}$$

Figure 21.29 A summary diagram for the stoichiometry of electrolysis.

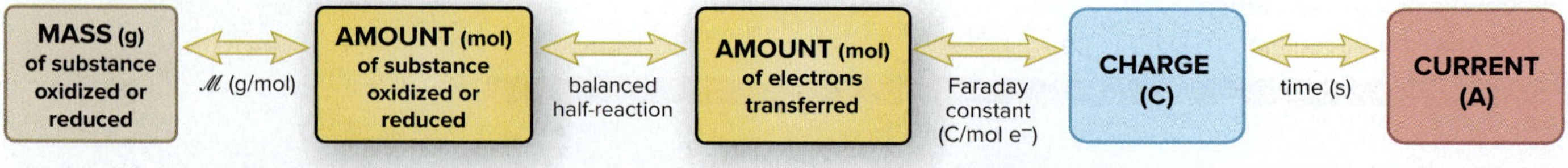

Now we use the relationship between charge and current to find the time needed:

$$\text{Time (s)} = \frac{\text{charge (C)}}{\text{current (A, or C/s)}} = 8.2{\times}10^3\ \text{C} \times \frac{1\ \text{s}}{12\ \text{C}}$$
$$= 6.8{\times}10^2\ \text{s}\ (\sim 11\ \text{min})$$

Note that the entire calculation follows Figure 21.29 until the last step, which asks for the time that the given current must flow:

grams of Cl_2 ⟹ moles of Cl_2 ⟹ moles of e^- ⟹ coulombs ⟹ seconds

Sample Problem 21.11 applies these ideas in an important industrial setting.

SAMPLE PROBLEM 21.11 Applying the Relationship Among Current, Time, and Amount of Substance

Problem A technician plates a faucet with 0.86 g of Cr metal by electrolysis of aqueous $Cr_2(SO_4)_3$. If 12.5 min is allowed for the plating, what current is needed?

Plan To find the current, we divide charge by time, so we need to find the charge. We write the half-reaction for the Cr^{3+} reduction to get the amount (mol) of e^- transferred per mole of Cr. To find the charge, we convert the mass of Cr needed (0.86 g) to amount (mol) of Cr. Then, we use the Faraday constant ($9.65{\times}10^4$ C/mol e^-) to find the charge and divide by the time (12.5 min, converted to seconds) to obtain the current (see the road map).

Road Map

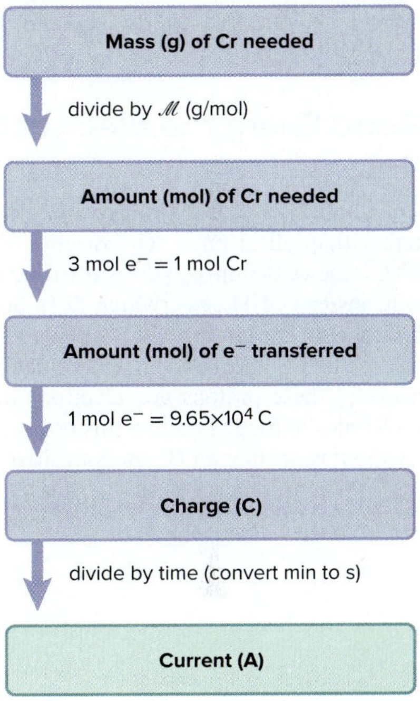

Solution Writing the balanced half-reaction [$Cr_2(SO_4)_3$ contains the Cr^{3+} ion]:

$$Cr^{3+}(aq) + 3e^- \longrightarrow Cr(s)$$

Combining steps to find amount (mol) of e^- transferred for the mass of Cr needed:

$$\text{Amount (mol) of } e^- \text{ transferred} = 0.86\ \text{g Cr} \times \frac{1\ \text{mol Cr}}{52.00\ \text{g Cr}} \times \frac{3\ \text{mol } e^-}{1\ \text{mol Cr}} = 0.050\ \text{mol } e^-$$

Calculating the charge using the Faraday constant as a conversion factor between moles of electrons transferred and charge:

$$\text{Charge (C)} = 0.050\ \text{mol } e^- \times \frac{9.65{\times}10^4\ \text{C}}{1\ \text{mol } e^-} = 4.8{\times}10^3\ \text{C}$$

Calculating the current:

$$\text{Current (A)} = \frac{\text{charge (C)}}{\text{time (s)}} = \frac{4.8{\times}10^3\ \text{C}}{12.5\ \text{min}} \times \frac{1\ \text{min}}{60\ \text{s}} = 6.4\ \text{C/s} = 6.4\ \text{A}$$

Check Rounding gives

$$(\sim 0.9\ \text{g})(1\ \text{mol Cr}/50\ \text{g})(3\ \text{mol } e^-/1\ \text{mol Cr}) = 5{\times}10^{-2}\ \text{mol } e^-$$

Then

$$(5{\times}10^{-2}\ \text{mol } e^-)(\sim 1{\times}10^5\ \text{C/mol } e^-) = 5{\times}10^3\ \text{C}$$

and

$$(5{\times}10^3\ \text{C}/12\ \text{min})(1\ \text{min}/60\ \text{s}) = 7\ \text{A}$$

Comment In order to introduce Faraday's law, we have neglected some details about actual electroplating. In practice, electroplating chromium has an efficiency of only 30–40% and must be run at a specific temperature range for the plate to appear bright. Nearly 10,000 metric tons ($2{\times}10^8$ mol) of chromium are used annually for electroplating.

FOLLOW-UP PROBLEMS

21.11A Using a current of 4.75 A, how many minutes does it take to plate a sculpture with 1.50 g of Cu from a $CuSO_4$ solution?

21.11B To protect it from corrosion, a section of iron pipe is electrogalvanized with zinc from a basic solution of $Na_2Zn(OH)_4$. The half-reaction is

$$Zn(OH)_4^{2-}(aq) + 2e^- \longrightarrow Zn(s) + 4OH^-(aq)$$

If the process takes 8.75 min with a current of 7.03 A, what mass of Zn is deposited on the pipe?

SOME SIMILAR PROBLEMS 21.101–21.106

The following Chemical Connections essay links several themes of this chapter to the generation of energy in a living cell.

Cellular Electrochemistry and the Production of ATP

Biological cells apply the principles of electrochemical cells to generate energy. The complex multistep process can be divided into two parts:

1. Bond energy in food generates an electrochemical potential.
2. That potential creates the bond energy of ATP (see Chemical Connections in Chapter 20.)

These steps are part of the *electron-transport chain* (ETC), which lies on the inner membranes of *mitochondria,* the subcellular particles that produce energy (Figure B21.1).

The ETC is a series of large molecules (mostly proteins), each of which contains a *redox couple* (the oxidized and reduced forms of a species), such as Fe^{3+}/Fe^{2+}, that passes electrons along the chain. At three points in this chain, large potential differences supply enough free energy to convert adenosine diphosphate (ADP) into ATP.

Bond Energy to Electrochemical Potential

Cells utilize the energy in food by releasing it in controlled steps rather than all at once. The reaction that ultimately powers the ETC is the oxidation of hydrogen to form water: $H_2 + \frac{1}{2}O_2 \longrightarrow H_2O$. But, instead of H_2 gas, which does not occur in organisms, the hydrogen takes the form of two H^+ ions and two e^-. The biological oxidizing agent NAD^+ (*n*icotinamide *a*denine *d*inucleotide) acquires these protons and electrons in the process of oxidizing molecules in food. To show this process, we use this half-reaction (without canceling an H^+ on both sides):

$$NAD^+(aq) + 2H^+(aq) + 2e^- \longrightarrow NADH(aq) + H^+(aq)$$

At the mitochondrial inner membrane, the NADH and H^+ transfer the two e^- to the first redox couple of the ETC and release the two H^+. The electrons are transported down the chain of redox couples, where they finally reduce O_2. The overall process, with standard electrode potentials,* is

$$NADH(aq) + H^+(aq) \longrightarrow NAD^+(aq) + 2H^+(aq) + 2e^- \qquad E^{\circ\prime} = -0.315\text{ V}$$

$$\tfrac{1}{2}O_2(aq) + 2e^- + 2H^+(aq) \longrightarrow H_2O(l) \qquad E^{\circ\prime} = 0.815\text{ V}$$

$$NADH(aq) + H^+(aq) + \tfrac{1}{2}O_2(aq) \longrightarrow NAD^+(aq) + H_2O(l)$$

$$E^{\circ\prime}_{\text{overall}} = 0.815\text{ V} - (-0.315\text{ V}) = 1.130\text{ V}$$

Thus, for every 2 mol of e^-, 1 mol of NADH enters the ETC, and the free energy equivalent of 1.130 V is available:

$$\begin{aligned}\Delta G^{\circ\prime} &= -nFE^{\circ\prime}\\ &= -(2\text{ mol e}^-/\text{mol NADH})(96.5\text{ kJ/V·mol e}^-)(1.130\text{ V})\\ &= -218\text{ kJ/mol NADH}\end{aligned}$$

Note that this aspect of the process functions like a voltaic cell: a spontaneous reaction, the reduction of O_2 to H_2O, is used to generate a potential. In contrast to the operation of a voltaic cell, this process occurs in many small steps. Figure B21.2 is a highly simplified diagram showing the three steps in the ETC that generate a high enough potential to form ATP from ADP.

Each step consists of several components, in which electrons are passed from one redox couple to the next. Since most of the ETC components are iron-containing proteins, the redox change

*In biological systems, standard potentials are designated $E^{\circ\prime}$ and the standard states include a pH of 7.0 ($[H^+] = 1\times10^{-7}$ *M*).

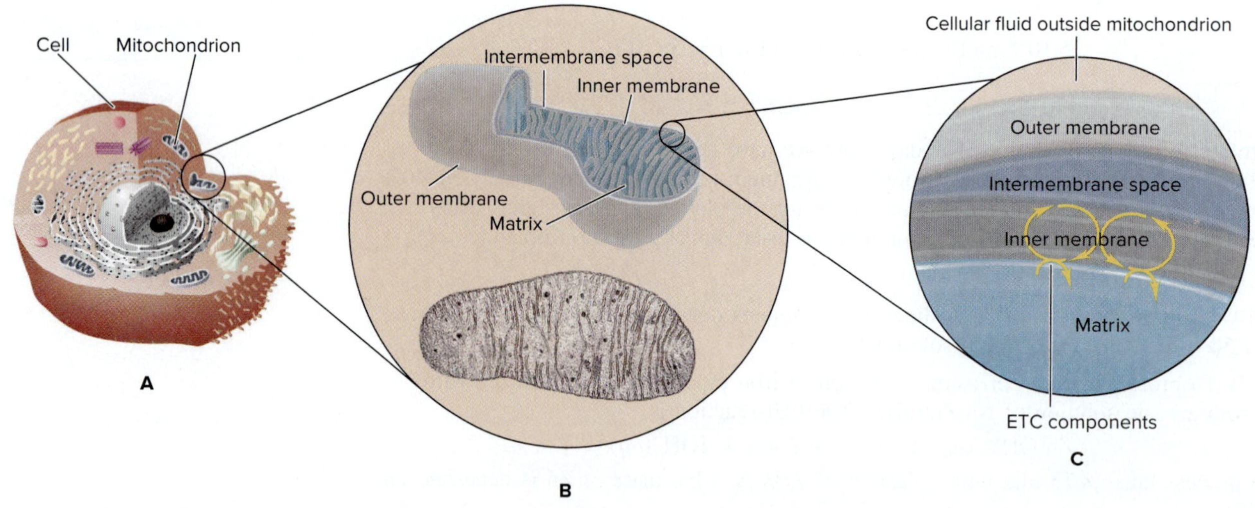

Figure B21.1 **The mitochondrion. A,** Mitochondria are subcellular particles outside the cell nucleus. **B,** They have a smooth outer membrane and a highly folded inner membrane, shown schematically and in an electron micrograph. **C,** The components of the electron-transport chain are attached to the inner membrane, which is in contact with the matrix, a relatively viscous aqueous mixture of small organic molecules, enzymes, and other macromolecules.

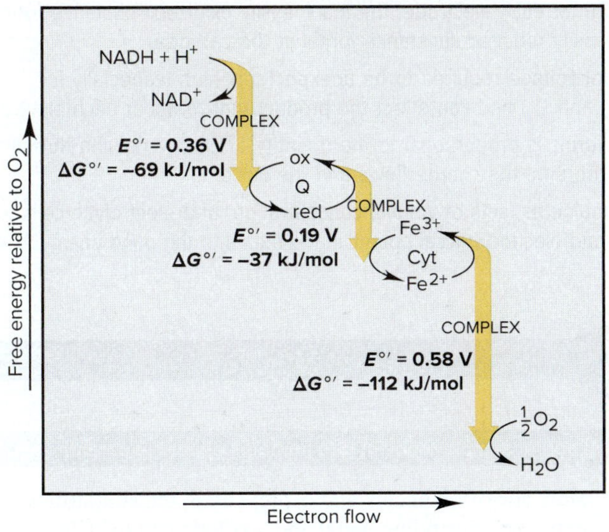

Figure B21.2 The main energy-yielding steps in the electron-transport chain (ETC). At the three points shown, $E°'$ (or $\Delta G°'$) is large enough to form ATP from ADP. (A complex consists of many components; Q is a large organic molecule; Cyt stands for a cytochrome, a protein that contains a metal-ion redox couple.)

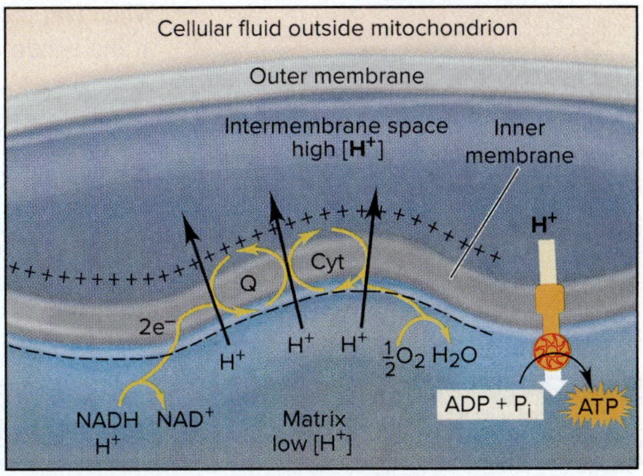

Figure B21.3 Coupling electron transport to proton transport to ATP synthesis. Electrons are transported along the ETC *(curved yellow arrows).* Protons pumped into the intermembrane space create an $[H^+]$ difference that generates a potential across the inner membrane. When that difference is large enough, H^+ flows back into the inner space, and the free energy drives the formation of ATP.

consists of the oxidation of Fe^{2+} to Fe^{3+} in one component accompanied by the reduction of Fe^{3+} to Fe^{2+} in the other:

$$Fe^{2+}\text{ (in A)} + Fe^{3+}\text{ (in B)} \longrightarrow Fe^{3+}\text{ (in A)} + Fe^{2+}\text{ (in B)}$$

In other words, metal ions bound within the ETC proteins are the actual species undergoing the redox reactions.

Electrochemical Potential to Bond Energy

Formation of ATP requires a lot of energy:

$$ADP^{3-}(aq) + HPO_4^{2-}(aq) + H^+(aq) \longrightarrow ATP^{4-}(aq) + H_2O(l)$$
$$\Delta G°' = 30.5 \text{ kJ/mol}$$

Note that, at each of the three points in Figure B21.2, the free energy *released* exceeds the 30.5 kJ needed to form 1 mol of ATP. Thus, as in an electrolytic cell, an electrochemical potential is supplied to drive a nonspontaneous reaction.

We've followed the flow of *electrons* through the ETC. To see how electrochemical potential is converted to bond energy in ATP, we focus on the released *protons*. The free energy released at the three key steps is used to force the H^+ ions into the intermembrane space; as a result, $[H^+]$ of the intermembrane space becomes higher than $[H^+]$ of the matrix, the aqueous mixture that fills the spaces between the folds of the inner membrane (Figure B21.3). This part of the process uses the free energy released by the three steps to *create an H^+ concentration cell across the inner membrane.*

When the $[H^+]$ difference across the inner membrane reaches a "trigger" point of about 2.5-fold, H^+ ions spontaneously flow back through the membrane (in effect, closing the switch and allowing the concentration cell to operate). The free energy released in this spontaneous process drives the nonspontaneous ATP formation via an enzyme-catalyzed mechanism.

Thus, the mitochondrion uses the "electron-motive force" of redox couples on its inner membrane to generate a "proton-motive force" across the membrane, which converts a potential difference into bond energy.

Problems

B21.1 In the final steps of the ETC, iron and copper ions in a large protein complex interact with each other.
(a) Write balanced equations for the one-electron half-reactions of Fe^{3+} and of Cu^+.
(b) Write a balanced overall equation for this redox reaction.

B21.2 As the ETC proceeds, a difference in $[H^+]$ develops across the inner mitochondrial membrane. The "proton-motive force," which is used to form ATP, is based largely on this difference and is equivalent to an electrical potential of 0.224 V. What is the free energy change (in kJ/mol) associated with the "proton-motive force"?

› Summary of Section 21.7

- An electrolytic cell uses electrical energy to drive a nonspontaneous reaction.
- Oxidation occurs at the anode and reduction at the cathode, but the direction of electron flow and the charges of the electrodes are opposite those in voltaic cells.
- When two products could form at each electrode, the more easily oxidized substance forms at the anode and the more easily reduced substance forms at the cathode.
- Overvoltage causes the actual voltage required to be unexpectedly high (especially for producing gases, such as H_2 and O_2) and can affect the product that forms at each electrode.
- The amount of product that forms is proportional to the quantity of charge flowing through the cell, which is related to the time the charge flows and the current.
- Biological redox systems combine aspects of voltaic, concentration, and electrolytic cells to convert bond energy in food into electrochemical potential and then into the bond energy of ATP.

CHAPTER REVIEW GUIDE

Learning Objectives

Relevant section (§) and/or sample problem (SP) numbers appear in parentheses.

Understand These Concepts

1. The meanings of *oxidation* and *reduction;* why an oxidizing agent is reduced and a reducing agent is oxidized (§21.1)
2. How the half-reaction method is used to balance redox reactions in acidic or basic solution (§21.1)
3. The distinction between voltaic and electrolytic cells in terms of the sign of ΔG (§21.1)
4. How voltaic cells use a spontaneous reaction to release electrical energy (§21.2)
5. The physical makeup of a voltaic cell: arrangement and composition of half-cells, relative charges of electrodes, and purpose of a salt bridge (§21.2)
6. How the difference in reducing strength of the electrodes determines the direction of electron flow in a voltaic cell (§21.2)
7. The correspondence between a positive E_{cell} and a spontaneous cell reaction (§21.3)
8. The usefulness and significance of standard electrode potentials ($E^\circ_{\text{half-cell}}$) (§21.3)
9. How $E^\circ_{\text{half-cell}}$ values are combined to give E°_{cell} (§21.3)
10. How the standard reference electrode is used to find an unknown $E^\circ_{\text{half-cell}}$ (§21.3)
11. How an emf series (e.g., Table 21.2 or Appendix D) is used to write spontaneous redox reactions (§21.3)
12. How the relative reactivity of a metal is determined by its reducing power and is related to the negative of its $E^\circ_{\text{half-cell}}$ (§21.3)
13. How E_{cell} (the nonstandard cell potential) is related to ΔG (maximum work) and the charge (moles of electrons times the Faraday constant) flowing through the cell (§21.4)
14. The interrelationship of ΔG°, E°_{cell}, and K (§21.4)
15. How E_{cell} changes as the cell operates (as Q changes) (§21.4)
16. Why a voltaic cell can do work until $Q = K$ (§21.4)
17. How a concentration cell does work until the half-cell concentrations are equal (§21.4)
18. The distinction between primary (nonrechargeable) and secondary (rechargeable) batteries and the nature of fuel cells (§21.5)
19. How corrosion occurs and is prevented; the similarities between a corroding metal and a voltaic cell (§21.6)
20. How electrolytic cells use nonspontaneous redox reactions driven by an external source of electricity (§21.7)
21. How atomic properties (ionization energy and electronegativity) determine the products of the electrolysis of molten salt mixtures (§21.7)
22. How the electrolysis of water influences the products of aqueous electrolysis; the importance of overvoltage (§21.7)
23. The relationship between the quantity of charge flowing through the cell and the amount of product formed (§21.7)

Master These Skills

1. Balancing redox reactions by the half-reaction method (§21.1 and SP 21.1)
2. Describing a voltaic cell with a diagram and notation (§21.2 and SP 21.2)
3. Combining $E^\circ_{\text{half-cell}}$ values to obtain E°_{cell} (§21.3 and SP 21.3)
4. Using E°_{cell} and a known $E^\circ_{\text{half-cell}}$ to find an unknown $E^\circ_{\text{half-cell}}$ (SP 21.4)
5. Manipulating half-reactions to write a spontaneous redox reaction and calculate its E°_{cell} (SP 21.5)
6. Ranking the relative strengths of oxidizing and reducing agents in a redox reaction (SP 21.5)
7. Predicting whether a metal can displace hydrogen or another metal from solution (§21.3)
8. Using the interrelationship of ΔG°, E°_{cell}, and K to calculate two of the three given the third (§21.4 and SP 21.6)
9. Using the Nernst equation to calculate the nonstandard cell potential (E_{cell}) (SP 21.7)
10. Calculating E_{cell} of a concentration cell (SP 21.8)
11. Predicting the products of the electrolysis of a mixture of molten salts (SP 21.9)
12. Predicting the products of the electrolysis of aqueous salt solutions (SP 21.10)
13. Calculating the current (or time) needed to produce a given amount of product by electrolysis (SP 21.11)

Key Terms

Page numbers appear in parentheses.

anode (944)
ampere (A) (980)
battery (968)
cathode (944)
cell potential (E_{cell}) (950)
concentration cell (964)
coulomb (C) (950)
corrosion (972)
electrochemical cell (939)
electrochemistry (939)
electrolytic cell (944)
electrode (944)
electrolyte (944)
electromotive force (emf) (950)
electrolysis (976)
Faraday constant (F) (959)
fuel cell (970)
half-cell (946)
half-reaction method (940)
Nernst equation (962)
overvoltage (978)
salt bridge (947)
standard cell potential (E°_{cell}) (950)
standard electrode (half-cell) potential ($E^\circ_{half\text{-}cell}$) (950)
standard reference half-cell (standard hydrogen electrode) (952)
volt (V) (950)
voltaic (galvanic) cell (944)
voltage (950)

Key Equations and Relationships

Page numbers appear in parentheses.

21.1 Relating the spontaneity of a process to the sign of the cell potential (950):

$$E_{cell} > 0 \text{ for a spontaneous process}$$

21.2 Relating electric potential to energy and charge in SI units (950):

$$\text{Potential} = \text{energy/charge} \quad \text{or} \quad 1\text{ V} = 1\text{ J/C}$$

21.3 Relating standard cell potential to standard electrode potentials in a voltaic cell (951):

$$E^\circ_{cell} = E^\circ_{cathode(reduction)} - E^\circ_{anode(oxidation)}$$

21.4 Defining the Faraday constant (960):

$$F = 9.65\times10^4 \frac{\text{J}}{\text{V·mol e}^-} \quad (3\text{ sf})$$

21.5 Relating the free energy change to the cell potential (960):

$$\Delta G = -nFE_{cell}$$

21.6 Finding the standard free energy change from the standard cell potential (960):

$$\Delta G^\circ = -nFE^\circ_{cell}$$

21.7 Finding the equilibrium constant from the standard cell potential (960):

$$E^\circ_{cell} = \frac{RT}{nF}\ln K$$

21.8 Substituting known values of R, F, and T into Equation 21.7 and converting to a common logarithm (960):

$$E^\circ_{cell} = \frac{0.0592\text{ V}}{n}\log K \quad \text{or} \quad \log K = \frac{nE^\circ_{cell}}{0.0592\text{ V}} \quad (\text{at } 298.15\text{ K})$$

21.9 Calculating the nonstandard cell potential (Nernst equation) (962):

$$E_{cell} = E^\circ_{cell} - \frac{RT}{nF}\ln Q$$

21.10 Substituting known values of R, F, and T into the Nernst equation and converting to a common logarithm (962):

$$E_{cell} = E^\circ_{cell} - \frac{0.0592\text{ V}}{n}\log Q \quad (\text{at } 298.15\text{ K})$$

21.11 Relating current to charge and time (980):

$$\text{Current (A)} = \text{charge (C)/time (s)} \quad \text{or} \quad 1\text{ A} = 1\text{ C/s}$$

BRIEF SOLUTIONS TO FOLLOW-UP PROBLEMS

21.1A $6KMnO_4(aq) + 6KOH(aq) + KI(aq) \longrightarrow 6K_2MnO_4(aq) + KIO_3(aq) + 3H_2O(l)$

21.1B $2Cr(OH)_3(s) + NaIO_3(aq) + 4NaOH(aq) \longrightarrow 2Na_2CrO_4(aq) + NaI(aq) + 5H_2O(l)$

21.2A

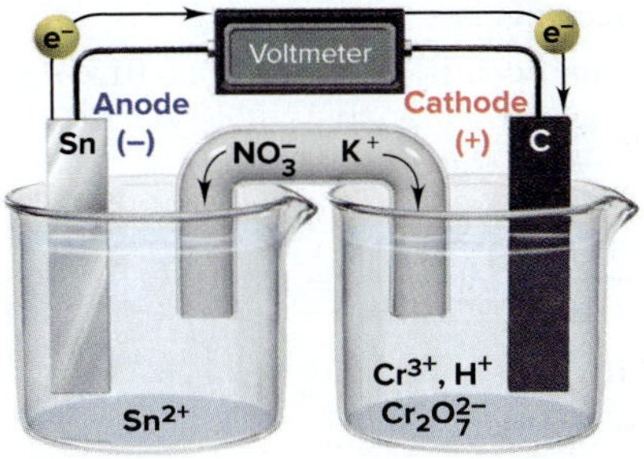

$Sn(s) \longrightarrow Sn^{2+}(aq) + 2e^-$ [anode; oxidation]

$6e^- + 14H^+(aq) + Cr_2O_7^{2-}(aq) \longrightarrow 2Cr^{3+}(aq) + 7H_2O(l)$ [cathode; reduction]

$3Sn(s) + Cr_2O_7^{2-}(aq) + 14H^+(aq) \longrightarrow 3Sn^{2+}(aq) + 2Cr^{3+}(aq) + 7H_2O(l)$ [overall]

Cell notation:

$Sn(s) \mid Sn^{2+}(aq) \parallel H^+(aq), Cr_2O_7^{2-}(aq), Cr^{3+}(aq) \mid$ graphite

21.2B

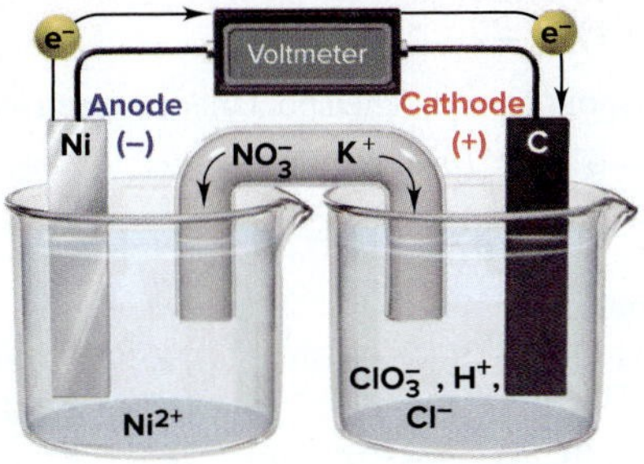

$Ni(s) \longrightarrow Ni^{2+}(aq) + 2e^-$ [anode; oxidation]

$ClO_3^-(aq) + 6H^+(aq) + 6e^- \longrightarrow Cl^-(aq) + 3H_2O(l)$ [cathode; reduction]

$3Ni(s) + ClO_3^-(aq) + 6H^+(aq) \longrightarrow 3Ni^{2+}(aq) + Cl^-(aq) + 3H_2O(l)$ [overall]

Cell notation:

$Ni(s) \mid Ni^{2+}(aq) \parallel H^+(aq), ClO_3^-(aq), Cl^-(aq) \mid$ graphite

21.3A $2Ag(s) + Cu^{2+}(aq) \longrightarrow 2Ag^+(aq) + Cu(s)$

$$E^\circ_{cell} = E^\circ_{cathode(reduction)} - E^\circ_{anode(oxidation)}$$
$$= E^\circ_{copper} - E^\circ_{silver} = 0.34\text{ V} - 0.80\text{ V}$$
$$= -0.46\text{ V; not spontaneous}$$

BRIEF SOLUTIONS TO FOLLOW-UP PROBLEMS (continued)

21.3B $Cl_2(g) + 2Fe^{2+}(aq) \longrightarrow 2Cl^-(aq) + 2Fe^{3+}(aq)$

$$E^\circ_{cell} = E^\circ_{cathode\ (reduction)} - E^\circ_{anode\ (oxidation)}$$
$$= E^\circ_{chlorine} - E^\circ_{iron} = 1.36\ V - 0.77\ V$$
$$= 0.59\ V;\ \text{spontaneous}$$

21.4A $Br_2(aq) + 2e^- \longrightarrow 2Br^-(aq)$ $\quad E^\circ_{bromine} = 1.07\ V$ [cathode]

$2V^{3+}(aq) + 2H_2O(l) \longrightarrow 2VO^{2+}(aq) + 4H^+(aq) + 2e^-$ $\quad E^\circ_{vanadium} = ?$ [anode]

$$E^\circ_{cell} = E^\circ_{cathode} - E^\circ_{anode} = E^\circ_{bromine} - E^\circ_{vanadium}$$
$$E^\circ_{vanadium} = E^\circ_{bromine} - E^\circ_{cell} = 1.07\ V - 0.73\ V = 0.34\ V$$

21.4B $V^{3+}(aq) + H_2O(l) \longrightarrow VO^{2+}(aq) + 2H^+(aq) + e^-$ $\quad E^\circ_{vanadium} = 0.34\ V$ [anode]

$NO_3^-(aq) + 4H^+(aq) + 3e^- \longrightarrow NO(g) + 2H_2O(l)$ $\quad E^\circ_{nitrate} = ?$ [cathode]

$$E^\circ_{cell} = E^\circ_{cathode} - E^\circ_{anode} = E^\circ_{nitrate} - E^\circ_{vanadium} = 0.62\ V$$
$$E^\circ_{nitrate} = E^\circ_{cell} + E^\circ_{vanadium} = 0.62\ V + 0.34\ V = 0.96\ V$$

21.5A (a) Combine (2) and reversed (1) to obtain equation A:

$BrO_3^-(aq) + 6Ag(s) \longrightarrow Br^-(aq) + 3Ag_2O(s)$

$E^\circ_{cell} = 0.61\ V - 0.34\ V = 0.27\ V$

Combine (1) and reversed (3) to obtain equation B:

$Ag_2O(s) + H_2O(l) + Zn(s) \longrightarrow 2Ag(s) + Zn(OH)_2(s)$

$E^\circ_{cell} = 0.34\ V - (-1.25\ V) = 1.59\ V$

Combine (2) and reversed (3) to obtain equation C:

$BrO_3^-(aq) + 3H_2O(l) + 3Zn(s) \longrightarrow Br^-(aq) + 3Zn(OH)_2(s)$

$E^\circ_{cell} = 0.61\ V - (-1.25\ V) = 1.86\ V$

(b) Oxidizing agents: $BrO_3^- > Ag_2O > Zn(OH)_2$

Reducing agents: $Zn > Ag > Br^-$

21.5B
$$Fe^{2+}(aq) + 2e^- \longrightarrow Fe(s) \qquad E^\circ = -0.44\ V$$
$$2[Fe^{2+}(aq) \longrightarrow Fe^{3+}(aq) + e^-] \qquad E^\circ = 0.77\ V$$
$$\overline{3Fe^{2+}(aq) \longrightarrow 2Fe^{3+}(aq) + Fe(s)}$$

$E^\circ_{cell} = -0.44\ V - 0.77\ V = -1.21\ V$

The reaction is nonspontaneous. The spontaneous reaction is

$2Fe^{3+}(aq) + Fe(s) \longrightarrow 3Fe^{2+}(aq)$ $\quad E^\circ_{cell} = 1.21\ V$

Reducing agents: $Fe > Fe^{2+} > Fe^{3+}$

21.6A From Appendix D,

$$2[MnO_4^-(aq) + 2H_2O(l) + 3e^- \longrightarrow MnO_2(s) + 4OH^-(aq)] \qquad E^\circ = 0.59\ V$$
$$3[Cu(s) \longrightarrow Cu^{2+}(aq) + 2e^-] \qquad E^\circ = 0.34\ V$$
$$\overline{2MnO_4^-(aq) + 4H_2O(l) + 3Cu(s) \longrightarrow 2MnO_2(s) + 8OH^-(aq) + 3Cu^{2+}(aq)}$$

$E^\circ_{cell} = 0.59\ V - 0.34\ V = 0.25\ V$

We use Equations 21.6 and 21.8 with $n = 6$ mol e$^-$/mol rxn and $E^\circ_{cell} = 0.25$ V.

$$\Delta G^\circ = -nFE^\circ_{cell} = -\frac{6\ \text{mol e}^-}{\text{mol rxn}} \times \frac{96.5\ \text{kJ}}{\text{V·mol e}^-} \times 0.25\ V$$
$$= -1.4{\times}10^2\ \text{kJ/mol rxn}$$
$$E^\circ_{cell} = 0.25\ V = \frac{0.0592\ V}{n}\log K$$
$$\log K = \frac{0.25\ V \times 6}{0.0592\ V} = 25.34;\ K = 10^{25.34} = 2.2{\times}10^{25}$$

21.6B $Cd(s) + Cu^{2+}(aq) \longrightarrow Cd^{2+}(aq) + Cu(s)$ $\quad n = 2$ mol e$^-$/mol rxn

$$\Delta G^\circ = -RT\ln K$$
$$\ln K = -\frac{\Delta G^\circ}{RT} = -\frac{-1.43{\times}10^5\ \text{J/mol rxn}}{8.314\ \text{J/mol rxn·K} \times 298\ K} = 57.72$$
$$K = e^{57.72} = 1.2{\times}10^{25}$$
$$E^\circ_{cell} = \frac{0.0592\ V}{n}\log K = \frac{0.0592\ V}{2}\log(1.2{\times}10^{25}) = 0.742\ V$$

21.7A From Appendix D,

$$2[Cr(s) \longrightarrow Cr^{3+}(aq) + 3e^-] \qquad E^\circ = -0.74\ V$$
$$3[Sn^{2+}(aq) + 2e^- \longrightarrow Sn(s)] \qquad E^\circ = -0.14\ V$$
$$\overline{2Cr(s) + 3Sn^{2+}(aq) \longrightarrow 2Cr^{3+}(aq) + 3Sn(s)}$$

$E^\circ_{cell} = -0.14\ V - (-0.74\ V) = 0.60\ V;\ n = 6$ mol e$^-$/mol rxn

$$Q = \frac{[Cr^{3+}]^2}{[Sn^{2+}]^3} = \frac{(1.60)^2}{(0.20)^3} = 320$$
$$E_{cell} = E^\circ_{cell} - \frac{0.0592\ V}{n}\log Q = 0.60\ V - \frac{0.0592\ V}{6}\log 320$$
$$= 0.58\ V$$

21.7B From Appendix D,

$$Fe(s) \longrightarrow Fe^{2+}(aq) + 2e^- \qquad E^\circ = -0.44\ V$$
$$Cu^{2+}(aq) + 2e^- \longrightarrow Cu(s) \qquad E^\circ = 0.34\ V$$
$$\overline{Fe(s) + Cu^{2+}(aq) \longrightarrow Fe^{2+}(aq) + Cu(s)}$$

So, $E^\circ_{cell} = 0.34\ V - (-0.44\ V) = 0.78\ V$

Therefore, $E_{cell} = 0.78\ V + 0.25\ V = 1.03\ V;\ n = 2$ mol e$^-$/mol rxn.

$$E_{cell} = E^\circ_{cell} - \frac{0.0592\ V}{n}\log Q$$
$$1.03\ V = 0.78\ V - \frac{0.0592\ V}{2}\log\frac{[Fe^{2+}]}{[0.30]}$$
$$-8.45 = \log\frac{[Fe^{2+}]}{[0.30]}$$
$$10^{-8.45} = 3.6 \times 10^{-9} = \frac{[Fe^{2+}]}{[0.30]}$$
$$[Fe^{2+}] = 3.6{\times}10^{-9} \times 0.30\ M = 1.1{\times}10^{-9}\ M$$

21.8A The cell reaction is

$Ni^{2+}(aq;\ 0.40\ M)$[half-cell B] $\longrightarrow Ni^{2+}(aq;\ 0.015\ M)$[half-cell A]

$$E_{cell} = E^\circ_{cell} - \frac{0.0592\ V}{n}\log\frac{[Ni^{2+}]_{dil}}{[Ni^{2+}]_{conc}}$$
$$= 0\ V - \frac{0.0592\ V}{2}\log\frac{0.015}{0.40} = 0.042\ V$$

21.8B The cell reaction is

$Au^{3+}(aq;\ 2.5{\times}10^{-2}\ M)$[half-cell B] $\longrightarrow Au^{3+}(aq;\ 7.0{\times}10^{-4}\ M)$[half-cell A]

$$E_{cell} = E^\circ_{cell} - \frac{0.0592\ V}{n}\log\frac{[Au^{3+}]_{dil}}{[Au^{3+}]_{conc}}$$
$$= 0\ V - \frac{0.0592\ V}{3}\log\frac{7.0{\times}10^{-4}}{2.5{\times}10^{-2}} = 0.031\ V$$

In a concentration cell, the anode half-cell contains the more dilute solution, which is the solution in half-cell A. Since the anode has the negative charge, the electrode in A is negative.

21.9A Oxidizing agents: Cs^+ and Li^+. Li, at the top of Group 1A(1), has a higher IE, so it is more easily reduced:

$Li^+ + e^- \longrightarrow Li$ [cathode; reduction]

Reducing agents: F^- and I^-. I, near the bottom of Group 7A(17), has a lower EN, so it is more easily oxidized:

$2I^- \longrightarrow I_2 + 2e^-$ [anode; oxidation]

$2Li^+(l) + 2I^-(l) \longrightarrow 2Li(l) + I_2(g)$ [overall]

21.9B Oxidizing agents: K^+ and Al^{3+}. Al is above and to the right of K in the periodic table, so it has a higher IE:

$Al^{3+}(l) + 3e^- \longrightarrow Al(s)$ [cathode; reduction]

Reducing agents: F^- and Br^-. Br is below F in Group 7A(17), so it has a lower EN:

$2Br^-(l) \longrightarrow Br_2(g) + 2e^-$ [anode; oxidation]

$2Al^{3+}(l) + 6Br^-(l) \longrightarrow 2Al(s) + 3Br_2(g)$ [overall]

21.10A The possible reduction (cathode) half-reactions are

$Pb^{2+}(aq) + 2e^- \longrightarrow Pb(s)$ $E° = -0.13$ V

$2H_2O(l) + 2e^- \longrightarrow H_2(g) + 2OH^-$ $E = -0.42$ V

The possible oxidation (anode) half-reactions are

$NO_3^-(aq) \longrightarrow$ no reaction

$2H_2O(l) \longrightarrow O_2(g) + 4H^+(aq) + 4e^-$ $E = 0.82$

Thus, Pb^{2+} is reduced and Pb forms at the cathode, and H_2O is oxidized and O_2 forms at the anode.

21.10B The possible reduction (cathode) half-reaction are

$Au^{3+}(aq) + 3e^- \longrightarrow Au(s)$ $E° = 1.50$ V [cathode; reduction]

$2H_2O(l) + 2e^- \longrightarrow H_2(g) + 2OH^-(aq)$ $E = -0.42$ V

Since Au^{3+} has the more positive reduction potential, Au forms at the cathode.

Because of overvoltage, O_2 will not form at the anode, so Br_2 will form:

$2H_2O(l) \longrightarrow O_2(g) + 4H^+(aq) + 4e^-$
$E = 0.82$ V (~1.4 V with overvoltage)

$2Br^-(aq) \longrightarrow Br_2(l) + 2e^-$ $E° = 1.07$ V [cathode; oxidation]

21.11A $Cu^{2+}(aq) + 2e^- \longrightarrow Cu(s)$; therefore,
2 mol e^-/1 mol Cu = 2 mol e^-/63.55 g Cu

$$\text{Time (min)} = 1.50 \text{ g Cu} \times \frac{2 \text{ mol e}^-}{63.55 \text{ g Cu}} \times \frac{9.65\times10^4 \text{ C}}{1 \text{ mol e}^-} \times \frac{1 \text{ s}}{4.75 \text{ C}} \times \frac{1 \text{ min}}{60 \text{ s}} = 16.0 \text{ min}$$

21.11B The half-reaction shows 2 mol e^-/1 mol Zn. Thus,

$$\text{Charge (C)} = 8.75 \text{ min} \times \frac{60 \text{ s}}{1 \text{ min}} \times \frac{7.03 \text{ C}}{\text{s}} = 3691 \text{ C}$$

$$\text{Mass (g) of Zn} = 3691 \text{ C} \times \frac{1 \text{ mol e}^-}{9.65\times10^4 \text{ C}} \times \frac{1 \text{ mol Zn}}{2 \text{ mol e}^-} \times \frac{65.38 \text{ g Zn}}{1 \text{ mol Zn}} = 1.25 \text{ g Zn}$$

PROBLEMS

Problems with **colored** numbers are answered in Appendix E and worked in detail in the Student Solutions Manual. Problem sections match those in the text and give the numbers of relevant sample problems. Most offer Concept Review Questions, Skill-Building Exercises (grouped in pairs covering the same concept), and Problems in Context. The Comprehensive Problems are based on material from any section or previous chapter.

Note: Unless stated otherwise, all problems refer to systems at 298 K (25°C).

Redox Reactions and Electrochemical Cells

(Sample Problem 21.1)

Concept Review Questions

21.1 Define *oxidation* and *reduction* in terms of electron transfer and change in oxidation number.

21.2 Why must an electrochemical process involve a redox reaction?

21.3 Can one half-reaction in a redox process take place independently of the other? Explain.

21.4 Water is used to balance O atoms in the half-reaction method. Why can't O^{2-} ions be used instead?

21.5 During the redox balancing process, what step is taken to ensure that e^- lost equals e^- gained?

21.6 How are protons removed when balancing a redox reaction in basic solution?

21.7 Are spectator ions used to balance the half-reactions of a redox reaction? At what stage might spectator ions enter the balancing process?

21.8 Which type of electrochemical cell has $\Delta G_{sys} < 0$? Which type shows an increase in free energy?

21.9 Which statements are true? Correct any that are false.
(a) In a voltaic cell, the anode is negative relative to the cathode.
(b) Oxidation occurs at the anode of a voltaic or electrolytic cell.
(c) Electrons flow into the cathode of an electrolytic cell.
(d) In a voltaic cell, the surroundings do work on the system.
(e) A metal that plates out of an electrolytic cell appears on the cathode.
(f) In an electrochemical cell, the electrolyte provides a solution of mobile electrons.

Skill-Building Exercises (grouped in similar pairs)

21.10 Consider the following balanced redox reaction:

$$16H^+(aq) + 2MnO_4^-(aq) + 10Cl^-(aq) \longrightarrow 2Mn^{2+}(aq) + 5Cl_2(g) + 8H_2O(l)$$

(a) Which species is being oxidized?
(b) Which species is being reduced?
(c) Which species is the oxidizing agent?
(d) Which species is the reducing agent?
(e) From which species to which does electron transfer occur?
(f) Write the balanced molecular equation, with K^+ and SO_4^{2-} as the spectator ions.

21.11 Consider the following balanced redox reaction:

$$2CrO_2^-(aq) + 2H_2O(l) + 6ClO^-(aq) \longrightarrow 2CrO_4^{2-}(aq) + 3Cl_2(g) + 4OH^-(aq)$$

(a) Which species is being oxidized?
(b) Which species is being reduced?
(c) Which species is the oxidizing agent?

(d) Which species is the reducing agent?
(e) From which species to which does electron transfer occur?
(f) Write the balanced molecular equation, with Na^+ as the spectator ion.

21.12 Balance the following skeleton reactions and identify the oxidizing and reducing agents:
(a) $ClO_3^-(aq) + I^-(aq) \longrightarrow I_2(s) + Cl^-(aq)$ [acidic]
(b) $MnO_4^-(aq) + SO_3^{2-}(aq) \longrightarrow MnO_2(s) + SO_4^{2-}(aq)$ [basic]
(c) $MnO_4^-(aq) + H_2O_2(aq) \longrightarrow Mn^{2+}(aq) + O_2(g)$ [acidic]

21.13 Balance the following skeleton reactions and identify the oxidizing and reducing agents:
(a) $O_2(g) + NO(g) \longrightarrow NO_3^-(aq)$ [acidic]
(b) $CrO_4^{2-}(aq) + Cu(s) \longrightarrow Cr(OH)_3(s) + Cu(OH)_2(s)$ [basic]
(c) $AsO_4^{3-}(aq) + NO_2^-(aq) \longrightarrow AsO_2^-(aq) + NO_3^-(aq)$ [basic]

21.14 Balance the following skeleton reactions and identify the oxidizing and reducing agents:
(a) $Cr_2O_7^{2-}(aq) + Zn(s) \longrightarrow Zn^{2+}(aq) + Cr^{3+}(aq)$ [acidic]
(b) $Fe(OH)_2(s) + MnO_4^-(aq) \longrightarrow MnO_2(s) + Fe(OH)_3(s)$ [basic]
(c) $Zn(s) + NO_3^-(aq) \longrightarrow Zn^{2+}(aq) + N_2(g)$ [acidic]

21.15 Balance the following skeleton reactions and identify the oxidizing and reducing agents:
(a) $BH_4^-(aq) + ClO_3^-(aq) \longrightarrow H_2BO_3^-(aq) + Cl^-(aq)$ [basic]
(b) $CrO_4^{2-}(aq) + N_2O(g) \longrightarrow Cr^{3+}(aq) + NO(g)$ [acidic]
(c) $Br_2(l) \longrightarrow BrO_3^-(aq) + Br^-(aq)$ [basic]

21.16 Balance the following skeleton reactions and identify the oxidizing and reducing agents:
(a) $Sb(s) + NO_3^-(aq) \longrightarrow Sb_4O_6(s) + NO(g)$ [acidic]
(b) $Mn^{2+}(aq) + BiO_3^-(aq) \longrightarrow MnO_4^-(aq) + Bi^{3+}(aq)$ [acidic]
(c) $Fe(OH)_2(s) + Pb(OH)_3^-(aq) \longrightarrow Fe(OH)_3(s) + Pb(s)$ [basic]

21.17 Balance the following skeleton reactions and identify the oxidizing and reducing agents:
(a) $NO_2(g) \longrightarrow NO_3^-(aq) + NO_2^-(aq)$ [basic]
(b) $Zn(s) + NO_3^-(aq) \longrightarrow Zn(OH)_4^{2-}(aq) + NH_3(g)$ [basic]
(c) $H_2S(g) + NO_3^-(aq) \longrightarrow S_8(s) + NO(g)$ [acidic]

21.18 Balance the following skeleton reactions and identify the oxidizing and reducing agents:
(a) $As_4O_6(s) + MnO_4^-(aq) \longrightarrow AsO_4^{3-}(aq) + Mn^{2+}(aq)$ [acidic]
(b) $P_4(s) \longrightarrow HPO_3^{2-}(aq) + PH_3(g)$ [acidic]
(c) $MnO_4^-(aq) + CN^-(aq) \longrightarrow MnO_2(s) + CNO^-(aq)$ [basic]

21.19 Balance the following skeleton reactions and identify the oxidizing and reducing agents:
(a) $SO_3^{2-}(aq) + Cl_2(g) \longrightarrow SO_4^{2-}(aq) + Cl^-(aq)$ [basic]
(b) $Fe(CN)_6^{3-}(aq) + Re(s) \longrightarrow Fe(CN)_6^{4-}(aq) + ReO_4^-(aq)$ [basic]
(c) $MnO_4^-(aq) + HCOOH(aq) \longrightarrow Mn^{2+}(aq) + CO_2(g)$ [acidic]

Problems in Context

21.20 In many residential water systems, the aqueous Fe^{3+} concentration is high enough to stain sinks and turn drinking water light brown. The iron content is analyzed by first reducing the Fe^{3+} to Fe^{2+} and then titrating with MnO_4^- in acidic solution. Balance the skeleton reaction of the titration step:

$$Fe^{2+}(aq) + MnO_4^-(aq) \longrightarrow Mn^{2+}(aq) + Fe^{3+}(aq)$$

21.21 *Aqua regia*, a mixture of concentrated HNO_3 and HCl, was developed by alchemists as a means to "dissolve" gold. The process is a redox reaction with this simplified skeleton reaction:

$$Au(s) + NO_3^-(aq) + Cl^-(aq) \longrightarrow AuCl_4^-(aq) + NO_2(g)$$

(a) Balance the reaction by the half-reaction method.
(b) What are the oxidizing and reducing agents?
(c) What is the function of HCl in aqua regia?

Voltaic Cells: Using Spontaneous Reactions to Generate Electrical Energy

(Sample Problem 21.2)

Concept Review Questions

21.22 Consider the following general voltaic cell:

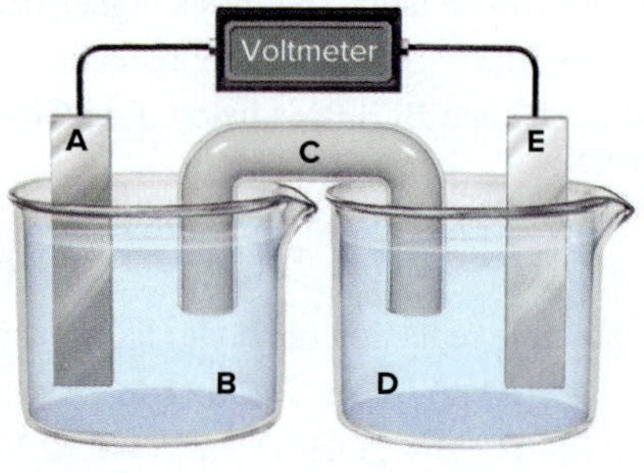

Identify the following:
(a) Anode
(b) Cathode
(c) Salt bridge
(d) Electrode from which e^- leave the cell
(e) Electrode with a positive charge
(f) Electrode that gains mass as the cell operates (assuming that a metal plates out)

21.23 Why does a voltaic cell not operate unless the two compartments are connected through an external circuit?

21.24 What purpose does the salt bridge serve in a voltaic cell, and how does it accomplish this purpose?

21.25 What is the difference between an active and an inactive electrode? Why are inactive electrodes used? Name two substances commonly used for inactive electrodes.

21.26 When a piece of metal A is placed in a solution containing ions of metal B, metal B plates out on the piece of A.
(a) Which metal is being oxidized?
(b) Which metal is being displaced?
(c) Which metal would you use as the anode in a voltaic cell incorporating these two metals?
(d) If bubbles of H_2 form when B is placed in acid, will they form if A is placed in acid? Explain.

Skill-Building Exercises (grouped in similar pairs)

21.27 Consider the following voltaic cell:

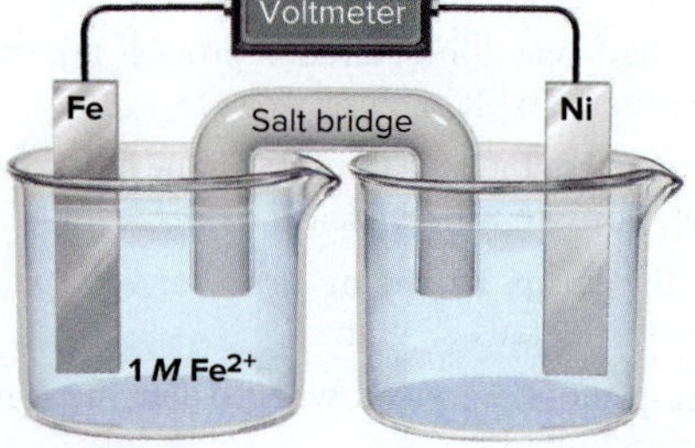

(a) In which direction do electrons flow in the external circuit?
(b) In which half-cell does oxidation occur?
(c) In which half-cell do electrons enter the cell?

(d) At which electrode are electrons consumed?
(e) Which electrode is negatively charged?
(f) Which electrode decreases in mass during cell operation?
(g) Suggest a solution for the electrolyte in the cathode compartment.
(h) Suggest a pair of ions for the salt bridge.
(i) For which electrode could you use an inactive material?
(j) In which direction do anions within the salt bridge move to maintain charge neutrality?
(k) Write balanced half-reactions and the overall cell reaction.

21.28 Consider the following voltaic cell:

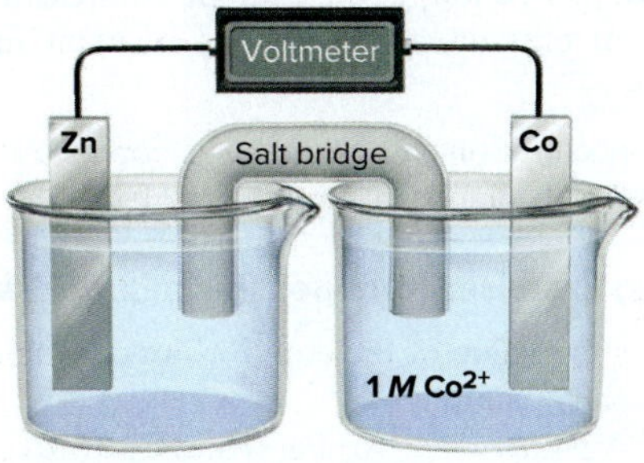

(a) In which direction do electrons flow in the external circuit?
(b) In which half-cell does reduction occur?
(c) In which half-cell do electrons leave the cell?
(d) At which electrode are electrons generated?
(e) Which electrode is positively charged?
(f) Which electrode increases in mass during cell operation?
(g) Suggest a solution for the electrolyte in the anode compartment.
(h) Suggest a pair of ions for the salt bridge.
(i) For which electrode could you use an inactive material?
(j) In which direction do cations within the salt bridge move to maintain charge neutrality?
(k) Write balanced half-reactions and the overall cell reaction.

21.29 A voltaic cell is constructed with an Sn/Sn^{2+} half-cell and a Zn/Zn^{2+} half-cell. The zinc electrode is negative.
(a) Write balanced half-reactions and the overall cell reaction.
(b) Diagram the cell, labeling electrodes with their charges and showing the directions of electron flow in the circuit and of cation and anion flow in the salt bridge.

21.30 A voltaic cell is constructed with an Ag/Ag^+ half-cell and a Pb/Pb^{2+} half-cell. The silver electrode is positive.
(a) Write balanced half-reactions and the overall cell reaction.
(b) Diagram the cell, labeling electrodes with their charges and showing the directions of electron flow in the circuit and of cation and anion flow in the salt bridge.

21.31 A voltaic cell is constructed with an Fe/Fe^{2+} half-cell and an Mn/Mn^{2+} half-cell. The iron electrode is positive.
(a) Write balanced half-reactions and the overall cell reaction.
(b) Diagram the cell, labeling electrodes with their charges and showing the directions of electron flow in the circuit and of cation and anion flow in the salt bridge.

21.32 A voltaic cell is constructed with a Cu/Cu^{2+} half-cell and an Ni/Ni^{2+} half-cell. The nickel electrode is negative.
(a) Write balanced half-reactions and the overall cell reaction.
(b) Diagram the cell, labeling electrodes with their charges and showing the directions of electron flow in the circuit and of cation and anion flow in the salt bridge.

21.33 Write the cell notation for the voltaic cell that incorporates each of the following redox reactions:
(a) $Al(s) + Cr^{3+}(aq) \longrightarrow Al^{3+}(aq) + Cr(s)$
(b) $Cu^{2+}(aq) + SO_2(g) + 2H_2O(l) \longrightarrow Cu(s) + SO_4^{2-}(aq) + 4H^+(aq)$

21.34 Write a balanced equation from each cell notation:
(a) $Mn(s) \mid Mn^{2+}(aq) \| Cd^{2+}(aq) \mid Cd(s)$
(b) $Fe(s) \mid Fe^{2+}(aq) \| NO_3^-(aq) \mid NO(g) \mid Pt(s)$

Cell Potential: Output of a Voltaic Cell

(Sample Problems 21.3 to 21.5)

Concept Review Questions

21.35 How is a standard reference electrode used to determine unknown $E^\circ_{\text{half-cell}}$ values?

21.36 What does a negative E°_{cell} indicate about a redox reaction? What does it indicate about the reverse reaction?

21.37 The standard cell potential is a thermodynamic state function. How are E° values treated similarly to ΔH°, ΔG°, and S° values? How are they treated differently?

Skill-Building Exercises (grouped in similar pairs)

21.38 In basic solution, Se^{2-} and SO_3^{2-} ions react spontaneously:
$$2Se^{2-}(aq) + 2SO_3^{2-}(aq) + 3H_2O(l) \longrightarrow 2Se(s) + 6OH^-(aq) + S_2O_3^{2-}(aq) \qquad E^\circ_{\text{cell}} = 0.35\text{ V}$$
(a) Write balanced half-reactions for the process.
(b) If E°_{sulfite} is −0.57 V, calculate $E^\circ_{\text{selenium}}$.

21.39 In acidic solution, O_3 and Mn^{2+} ions react spontaneously:
$$O_3(g) + Mn^{2+}(aq) + H_2O(l) \longrightarrow O_2(g) + MnO_2(s) + 2H^+(aq) \qquad E^\circ_{\text{cell}} = 0.84\text{ V}$$
(a) Write the balanced half-reactions.
(b) Using Appendix D to find E°_{ozone}, calculate $E^\circ_{\text{manganese}}$.

21.40 Use the emf series (Appendix D) to arrange each set of species.
(a) In order of *decreasing* strength as *oxidizing* agents: Fe^{3+}, Br_2, Cu^{2+}
(b) In order of *increasing* strength as *oxidizing* agents: Ca^{2+}, $Cr_2O_7^{2-}$, Ag^+

21.41 Use the emf series (Appendix D) to arrange each set of species.
(a) In order of *decreasing* strength as *reducing* agents: SO_2, $PbSO_4$, MnO_2
(b) In order of *increasing* strength as *reducing* agents: Hg, Fe, Sn

21.42 Balance each skeleton reaction, use Appendix D to calculate E°_{cell}, and state whether the reaction is spontaneous:
(a) $Co(s) + H^+(aq) \longrightarrow Co^{2+}(aq) + H_2(g)$
(b) $Hg_2^{2+}(aq) \longrightarrow Hg^{2+}(aq) + Hg(l)$

21.43 Balance each skeleton reaction, use Appendix D to calculate E°_{cell}, and state whether the reaction is spontaneous:
(a) $Mn^{2+}(aq) + Co^{3+}(aq) \longrightarrow MnO_2(s) + Co^{2+}(aq)$ [acidic]
(b) $AgCl(s) + NO(g) \longrightarrow Ag(s) + Cl^-(aq) + NO_3^-(aq)$ [acidic]

21.44 Balance each skeleton reaction, use Appendix D to calculate E°_{cell}, and state whether the reaction is spontaneous:
(a) $Cd(s) + Cr_2O_7^{2-}(aq) \longrightarrow Cd^{2+}(aq) + Cr^{3+}(aq)$
(b) $Ni^{2+}(aq) + Pb(s) \longrightarrow Ni(s) + Pb^{2+}(aq)$

21.45 Balance each skeleton reaction, use Appendix D to calculate E°_{cell}, and state whether the reaction is spontaneous:

(a) $Cu^+(aq) + PbO_2(s) + SO_4^{2-}(aq) \longrightarrow PbSO_4(s) + Cu^{2+}(aq)$ [acidic]

(b) $H_2O_2(aq) + Ni^{2+}(aq) \longrightarrow O_2(g) + Ni(s)$ [acidic]

21.46 Use the following half-reactions to write three spontaneous reactions, calculate E°_{cell} for each reaction, and rank the strengths of the oxidizing and reducing agents:

(1) $Al^{3+}(aq) + 3e^- \longrightarrow Al(s)$ $E^\circ = -1.66$ V

(2) $N_2O_4(g) + 2e^- \longrightarrow 2NO_2^-(aq)$ $E^\circ = 0.867$ V

(3) $SO_4^{2-}(aq) + H_2O(l) + 2e^- \longrightarrow SO_3^{2-}(aq) + 2OH^-(aq)$ $E^\circ = 0.93$ V

21.47 Use the following half-reactions to write three spontaneous reactions, calculate E°_{cell} for each reaction, and rank the strengths of the oxidizing and reducing agents:

(1) $Au^+(aq) + e^- \longrightarrow Au(s)$ $E^\circ = 1.69$ V

(2) $N_2O(g) + 2H^+(aq) + 2e^- \longrightarrow N_2(g) + H_2O(l)$ $E^\circ = 1.77$ V

(3) $Cr^{3+}(aq) + 3e^- \longrightarrow Cr(s)$ $E^\circ = -0.74$ V

21.48 Use the following half-reactions to write three spontaneous reactions, calculate E°_{cell} for each reaction, and rank the strengths of the oxidizing and reducing agents:

(1) $2HClO(aq) + 2H^+(aq) + 2e^- \longrightarrow Cl_2(g) + 2H_2O(l)$ $E^\circ = 1.63$ V

(2) $Pt^{2+}(aq) + 2e^- \longrightarrow Pt(s)$ $E^\circ = 1.20$ V

(3) $PbSO_4(s) + 2e^- \longrightarrow Pb(s) + SO_4^{2-}(aq)$ $E^\circ = -0.31$ V

21.49 Use the following half-reactions to write three spontaneous reactions, calculate E°_{cell} for each reaction, and rank the strengths of the oxidizing and reducing agents:

(1) $I_2(s) + 2e^- \longrightarrow 2I^-(aq)$ $E^\circ = 0.53$ V

(2) $S_2O_8^{2-}(aq) + 2e^- \longrightarrow 2SO_4^{2-}(aq)$ $E^\circ = 2.01$ V

(3) $Cr_2O_7^{2-}(aq) + 14H^+(aq) + 6e^- \longrightarrow 2Cr^{3+}(aq) + 7H_2O(l)$ $E^\circ = 1.33$ V

Problems in Context

21.50 When metal A is placed in a solution of a salt of metal B, the surface of metal A changes color. When metal B is placed in acid solution, gas bubbles form on the surface of the metal. When metal A is placed in a solution of a salt of metal C, no change is observed in the solution or on the surface of metal A.

(a) Will metal C cause formation of H_2 when placed in acid solution?

(b) Rank metals A, B, and C in order of *decreasing* reducing strength.

21.51 When a clean iron nail is placed in an aqueous solution of copper(II) sulfate, the nail becomes coated with a brownish-black material.

(a) What is the material coating the iron?

(b) What are the oxidizing and reducing agents?

(c) Can this reaction be made into a voltaic cell?

(d) Write the balanced equation for the reaction.

(e) Calculate E°_{cell} for the process.

Free Energy and Electrical Work

(Sample Problems 21.6 to 21.8)

Concept Review Questions

21.52 (a) How do the relative magnitudes of Q and K relate to the signs of ΔG and E_{cell}? Explain.

(b) Can a cell do work when $Q/K > 1$ or $Q/K < 1$? Explain.

21.53 A voltaic cell consists of A/A^+ and B/B^+ half-cells, where A and B are metals and the A electrode is negative. The initial $[A^+]/[B^+]$ is such that $E_{cell} > E^\circ_{cell}$.

(a) How do $[A^+]$ and $[B^+]$ change as the cell operates?

(b) How does E_{cell} change as the cell operates?

(c) What is $[A^+]/[B^+]$ when $E_{cell} = E^\circ_{cell}$? Explain.

(d) Is it possible for E_{cell} to be less than E°_{cell}? Explain.

21.54 Explain whether E_{cell} of a voltaic cell will increase or decrease with each of the following changes:

(a) Decrease in cell temperature

(b) Increase in [active ion] in the anode compartment

(c) Increase in [active ion] in the cathode compartment

(d) Increase in pressure of a gaseous reactant in the cathode compartment

21.55 In a concentration cell, is the more concentrated electrolyte in the cathode or the anode compartment? Explain.

Skill-Building Exercises (grouped in similar pairs)

21.56 What is the value of the equilibrium constant for the reaction between each pair at 25°C?

(a) $Ni(s)$ and $Ag^+(aq)$ (b) $Fe(s)$ and $Cr^{3+}(aq)$

21.57 What is the value of the equilibrium constant for the reaction between each pair at 25°C?

(a) $Al(s)$ and $Cd^{2+}(aq)$ (b) $I_2(s)$ and $Br^-(aq)$

21.58 What is the value of the equilibrium constant for the reaction between each pair at 25°C?

(a) $Ag(s)$ and $Mn^{2+}(aq)$ (b) $Cl_2(g)$ and $Br^-(aq)$

21.59 What is the value of the equilibrium constant for the reaction between each pair at 25°C?

(a) $Cr(s)$ and $Cu^{2+}(aq)$ (b) $Sn(s)$ and $Pb^{2+}(aq)$

21.60 Calculate ΔG° for each of the reactions in Problem 21.56.

21.61 Calculate ΔG° for each of the reactions in Problem 21.57.

21.62 Calculate ΔG° for each of the reactions in Problem 21.58.

21.63 Calculate ΔG° for each of the reactions in Problem 21.59.

21.64 What are E°_{cell} and ΔG° of a redox reaction at 25°C for which $n = 1$ and $K = 5.0\times10^4$?

21.65 What are E°_{cell} and ΔG° of a redox reaction at 25°C for which $n = 1$ and $K = 5.0\times10^{-6}$?

21.66 What are E°_{cell} and ΔG° of a redox reaction at 25°C for which $n = 2$ and $K = 65$?

21.67 What are E°_{cell} and ΔG° of a redox reaction at 25°C for which $n = 2$ and $K = 0.065$?

21.68 A voltaic cell consists of a standard reference half-cell and a Cu/Cu^{2+} half-cell. Calculate $[Cu^{2+}]$ when E_{cell} is 0.22 V.

21.69 A voltaic cell consists of an Mn/Mn^{2+} half-cell and a Pb/Pb^{2+} half-cell. Calculate $[Pb^{2+}]$ when $[Mn^{2+}]$ is 1.4 M and E_{cell} is 0.44 V.

21.70 A voltaic cell with Ni/Ni^{2+} and Co/Co^{2+} half-cells has the following initial concentrations: $[Ni^{2+}] = 0.80$ M; $[Co^{2+}] = 0.20$ M.

(a) What is the initial E_{cell}?

(b) What is $[Ni^{2+}]$ when E_{cell} reaches 0.03 V?

(c) What are the equilibrium concentrations of the ions?

21.71 A voltaic cell with Mn/Mn^{2+} and Cd/Cd^{2+} half-cells has the following initial concentrations: $[Mn^{2+}] = 0.090$ M; $[Cd^{2+}] = 0.060$ M.

(a) What is the initial E_{cell}?
(b) What is E_{cell} when $[Cd^{2+}]$ reaches 0.050 *M*?
(c) What is $[Mn^{2+}]$ when E_{cell} reaches 0.055 V?
(d) What are the equilibrium concentrations of the ions?

21.72 A voltaic cell consists of two H_2/H^+ half-cells. Half-cell A has H_2 at 0.95 atm bubbling into 0.10 *M* HCl. Half-cell B has H_2 at 0.60 atm bubbling into 2.0 *M* HCl. Which half-cell houses the anode? What is the voltage of the cell?

21.73 A voltaic cell consists of two Sn/Sn^{2+} half-cells, A and B. The electrolyte in A is 0.13 *M* $Sn(NO_3)_2$. The electrolyte in B is 0.87 *M* $Sn(NO_3)_2$. Which half-cell houses the cathode? What is the voltage of the cell?

Electrochemical Processes in Batteries

Concept Review Questions

21.74 What is the direction of electron flow with respect to the anode and the cathode in a battery? Explain.

21.75 In the everyday batteries used in flashlights, toys, and so forth, no salt bridge is evident. What is used in these cells to separate the anode and cathode compartments?

21.76 Both a D-sized and an AAA-sized alkaline battery have an output of 1.5 V. What property of the cell potential allows this to occur? What is different about these two batteries?

Problems in Context

21.77 Many common electrical devices require the use of more than one battery.
(a) How many alkaline batteries must be placed in series to light a flashlight with a 6.0-V bulb?
(b) What is the voltage requirement of a camera that uses six silver batteries?
(c) How many volts can a car battery deliver if two of its anode/cathode cells are shorted?

Corrosion: An Environmental Voltaic Cell

Concept Review Questions

21.78 During reconstruction of the Statue of Liberty, Teflon spacers were placed between the iron skeleton and the copper plates that cover the statue. What purpose do these spacers serve?

21.79 Why do steel bridge-supports rust at the waterline but not above or below it?

21.80 After the 1930s, chromium replaced nickel for corrosion resistance and appearance on car bumpers and trim. How does chromium protect steel from corrosion?

21.81 Which of the following metals are suitable for use as sacrificial anodes to protect against corrosion of underground iron pipes? If any are not suitable, explain why:
(a) Aluminum (b) Magnesium (c) Sodium (d) Lead
(e) Nickel (f) Zinc (g) Chromium

Electrolytic Cells: Using Electrical Energy to Drive Nonspontaneous Reactions
(Sample Problems 21.9 to 21.11)

Concept Review Questions

Note: Unless stated otherwise, assume that the electrolytic cells in the following problems operate at 100% efficiency.

21.82 Consider the following general electrolytic cell:

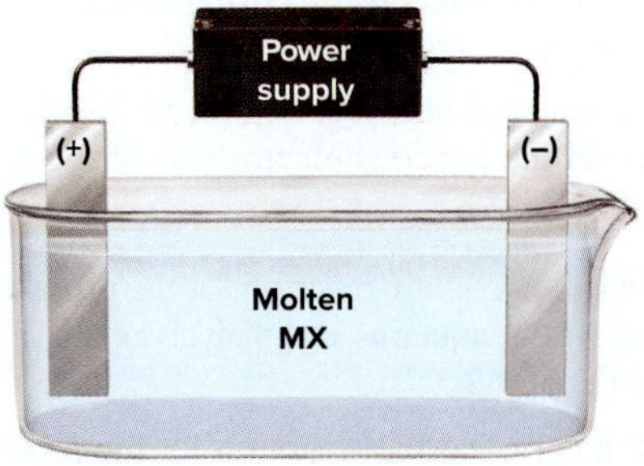

(a) At which electrode does oxidation occur?
(b) At which electrode does elemental M form?
(c) At which electrode are electrons being released by ions?
(d) At which electrode are electrons entering the cell?

21.83 A voltaic cell consists of Cr/Cr^{3+} and Cd/Cd^{2+} half-cells with all components in their standard states. After 10 minutes of operation, a thin coating of cadmium metal has plated out on the cathode. Describe what will happen if you attach the negative terminal of a dry cell (1.5 V) to the cathode and the positive terminal to the anode.

21.84 Why are $E_{half\text{-}cell}$ values for the oxidation and reduction of water different from $E^\circ_{half\text{-}cell}$ values for the same processes?

21.85 In an aqueous electrolytic cell, nitrate ions never react at the anode, but nitrite ions do. Explain.

21.86 How does overvoltage influence the products in the electrolysis of aqueous salts?

Skill-Building Exercises (grouped in similar pairs)

21.87 In the electrolysis of molten NaBr:
(a) What product forms at the anode?
(b) What product forms at the cathode?

21.88 In the electrolysis of molten BaI_2:
(a) What product forms at the negative electrode?
(b) What product forms at the positive electrode?

21.89 In the electrolysis of a molten mixture of KI and MgF_2, identify the product that forms at the anode and at the cathode.

21.90 In the electrolysis of a molten mixture of CsBr and $SrCl_2$, identify the product that forms at the negative electrode and at the positive electrode.

21.91 In the electrolysis of a molten mixture of NaCl and $CaBr_2$, identify the product that forms at the anode and at the cathode.

21.92 In the electrolysis of a molten mixture of RbF and $CaCl_2$, identify the product that forms at the negative electrode and at the positive electrode.

21.93 Which of the following elements can be prepared by electrolysis of their aqueous salts: copper, barium, aluminum, bromine?

21.94 Which of the following elements can be prepared by electrolysis of their aqueous salts: strontium, gold, tin, chlorine?

21.95 Which of the following elements can be prepared by electrolysis of their aqueous salts: lithium, iodine, zinc, silver?

21.96 Which of the following elements can be prepared by electrolysis of their aqueous salts: fluorine, manganese, iron, cadmium?

21.97 Write a balanced half-reaction for the product that forms at each electrode in the aqueous electrolysis of the following salts: (a) LiF; (b) $SnSO_4$.

21.98 Write a balanced half-reaction for the product that forms at each electrode in the aqueous electrolysis of the following salts: (a) $ZnBr_2$; (b) $Cu(HCO_3)_2$.

21.99 Write a balanced half-reaction for the product that forms at each electrode in the aqueous electrolysis of the following salts: (a) $Cr(NO_3)_3$; (b) $MnCl_2$.

21.100 Write a balanced half-reaction for the product that forms at each electrode in the aqueous electrolysis of the following salts: (a) FeI_2; (b) K_3PO_4.

21.101 Electrolysis of molten $MgCl_2$ is the final production step in the isolation of magnesium from seawater by the Dow process (Section 22.4). Assuming that 45.6 g of Mg metal forms:
(a) How many moles of electrons are required?
(b) How many coulombs are required?
(c) How many amps will produce this amount in 3.50 h?

21.102 Electrolysis of molten NaCl in a Downs cell is the major isolation step in the production of sodium metal (Section 22.4). Assuming that 215 g of Na metal forms:
(a) How many moles of electrons are required?
(b) How many coulombs are required?
(c) How many amps will produce this amount in 9.50 h?

21.103 How many grams of radium can form by passing 235 C through an electrolytic cell containing a molten radium salt?

21.104 How many grams of aluminum can form by passing 305 C through an electrolytic cell containing a molten aluminum salt?

21.105 How many seconds does it take to deposit 65.5 g of Zn on a steel gate when 21.0 A is passed through a $ZnSO_4$ solution?

21.106 How many seconds does it take to deposit 1.63 g of Ni on a decorative drawer handle when 13.7 A is passed through a $Ni(NO_3)_2$ solution?

Problems in Context

21.107 A professor adds Na_2SO_4 to water to facilitate its electrolysis in a lecture demonstration. (a) What is the purpose of the Na_2SO_4? (b) Why is the water electrolyzed instead of the salt?

21.108 Subterranean brines in parts of the United States are rich in iodides and bromides and serve as an industrial source of these elements. In one recovery method, the brines are evaporated to dryness and then melted and electrolyzed. Which halogen is more likely to form from this treatment? Why?

21.109 Zinc plating (galvanizing) is an important means of corrosion protection. Although the process is done customarily by dipping the object into molten zinc, the metal can also be electroplated from aqueous solutions. How many grams of zinc can be deposited on a steel tank from a $ZnSO_4$ solution when a 0.855-A current flows for 2.50 days?

Comprehensive Problems

21.110 The MnO_2 used in alkaline batteries can be produced by an electrochemical process of which one half-reaction is

$$Mn^{2+}(aq) + 2H_2O(l) \longrightarrow MnO_2(s) + 4H^+(aq) + 2e^-$$

If a current of 25.0 A is used, how many hours are needed to produce 1.00 kg of MnO_2? At which electrode is the MnO_2 formed?

21.111 Car manufacturers are developing engines that use H_2 as fuel. In Iceland, Sweden, and other parts of Scandinavia, where hydroelectric plants produce inexpensive electric power, the H_2 can be made industrially by the electrolysis of water.
(a) How many coulombs are needed to produce 3.5×10^6 L of H_2 gas at 12.0 atm and 25°C? (Assume that the ideal gas law applies.)
(b) If the coulombs are supplied at 1.44 V, how many joules are produced?
(c) If the combustion of oil yields 4.0×10^4 kJ/kg, what mass of oil must be burned to yield the number of joules in part (b)?

21.112 The overall cell reaction occurring in an alkaline battery is

$$Zn(s) + MnO_2(s) + H_2O(l) \longrightarrow ZnO(s) + Mn(OH)_2(s)$$

(a) How many moles of electrons flow per mole of reaction?
(b) If 4.50 g of zinc is oxidized, how many grams of manganese dioxide and of water are consumed?
(c) What is the total mass of reactants consumed in part (b)?
(d) How many coulombs are produced in part (b)?
(e) In practice, voltaic cells of a given capacity (coulombs) are heavier than the calculation in part (c) indicates. Explain.

21.113 An inexpensive and accurate method of measuring the quantity of electricity flowing through a circuit is to pass the current through a solution of a metal ion and weigh the metal deposited. A silver electrode immersed in an Ag^+ solution weighs 1.7854 g before the current has passed and weighs 1.8016 g after the current has passed. How many coulombs have passed?

21.114 Brass, an alloy of copper and zinc, can be produced by simultaneously electroplating the two metals from a solution containing their 2+ ions. If 65.0% of the total current is used to plate copper, while 35.0% goes to plating zinc, what is the mass percent of copper in the brass?

21.115 A thin circular-disk earring 4.00 cm in diameter is plated with a coating of gold 0.25 mm thick from an Au^{3+} bath.
(a) How many days does it take to deposit the gold on one side of one earring if the current is 0.013 A (d of gold = 19.3 g/cm^3)?
(b) How many days does it take to deposit the gold on both sides of the pair of earrings?
(c) If the price of gold is $1595 per troy ounce (31.10 g), what is the total cost of the gold plating?

21.116 (a) How many minutes does it take to form 10.0 L of O_2 measured at 99.8 kPa and 28°C from water if a current of 1.3 A passes through the electrolytic cell?
(b) What mass of H_2 forms?

21.117 Trains powered by electricity, including subways, use direct current. One conductor is the overhead wire (or "third rail" for subways), and the other is the rails upon which the wheels run. The rails are on supports in contact with the ground. To minimize corrosion, should the overhead wire or the rails be connected to the positive terminal? Explain.

21.118 A silver button battery used in a watch contains 0.75 g of zinc and can run until 80% of the zinc is consumed.
(a) For how many days can the battery produce a current of 0.85 microamp (10^{-6} amp)?
(b) When the battery dies, 95% of the Ag_2O has been consumed. How many grams of Ag were used to make the battery?
(c) If Ag costs $23.00 per troy ounce (31.10 g), what is the cost of the Ag consumed each day the watch runs?

21.119 Like any piece of apparatus, an electrolytic cell operates at less than 100% efficiency. A cell depositing Cu from a Cu^{2+} bath operates for 10 h with an average current of 5.8 A. If 53.4 g of copper is deposited, at what efficiency is the cell operating?

21.120 Commercial electrolysis is performed on both molten NaCl and aqueous NaCl solutions. Identify the anode product, cathode product, species reduced, and species oxidized for the electrolysis of (a) molten NaCl and (b) an aqueous solution of NaCl.

21.121 To examine the effect of ion removal on cell voltage, a chemist constructs two voltaic cells, each with a standard hydrogen electrode in one compartment. One cell also contains a Pb/Pb^{2+} half-cell; the other contains a Cu/Cu^{2+} half-cell.
(a) What is $E°$ of each cell at 298 K?
(b) Which electrode in each cell is negative?
(c) When Na_2S solution is added to the Pb^{2+} electrolyte, solid PbS forms. What happens to the cell voltage?
(d) When sufficient Na_2S is added to the Cu^{2+} electrolyte, CuS forms and $[Cu^{2+}]$ drops to 1×10^{-16} *M*. Find the cell voltage.

21.122 Electrodes used in electrocardiography are disposable, and many of them incorporate silver. The metal is deposited in a thin layer on a small plastic "button," and then some is converted to AgCl:

$$Ag(s) + Cl^-(aq) \rightleftharpoons AgCl(s) + e^-$$

(a) If the surface area of the button is 2.0 cm^2 and the thickness of the silver layer is 7.5×10^{-6} m, calculate the volume (in cm^3) of Ag used in one electrode.
(b) The density of silver metal is 10.5 g/cm^3. How many grams of silver are used per electrode?
(c) If Ag is plated on the button from an Ag^+ solution with a current of 12.0 mA, how many minutes does the plating take?
(d) If bulk silver costs \$28.93 per troy ounce (31.10 g), what is the cost (in cents) of the silver in one disposable electrode?

21.123 Commercial aluminum production is done by electrolysis of a bath containing Al_2O_3 dissolved in molten Na_3AlF_6. Why isn't it done by electrolysis of an aqueous $AlCl_3$ solution?

21.124 Comparing the standard electrode potentials ($E°$) of the Group 1A(1) metals Li, Na, and K with the negative of their first ionization energies reveals a discrepancy:

Ionization process reversed: $M^+(g) + e^- \rightleftharpoons M(g)$ (−IE)

Electrode reaction: $M^+(aq) + e^- \rightleftharpoons M(s)$ ($E°$)

Metal	−IE (kJ/mol)	E°(V)
Li	−520	−3.05
Na	−496	−2.71
K	−419	−2.93

Note that the electrode potentials do not decrease smoothly down the group, while the ionization energies do. You might expect that if it is more difficult to remove an electron from an atom to form a gaseous ion (larger IE), then it would be less difficult to add an electron to an aqueous ion to form an atom (smaller $E°$), yet $Li^+(aq)$ is *more* difficult to reduce than $Na^+(aq)$. Applying Hess's law, use an approach similar to a Born-Haber cycle to break down the process occurring at the electrode into three steps and label the energy involved in each step. How can you account for the discrepancy?

21.125 In Appendix D, standard electrode potentials range from about +3 V to −3 V. Thus, it might seem possible to use a half-cell from each end of this range to construct a cell with a voltage of approximately 6 V. However, most commercial aqueous voltaic cells have $E°$ values of 1.5–2 V. Why are there no aqueous cells with significantly higher potentials?

21.126 Tin is used to coat "tin" cans used for food storage. If the tin is scratched, exposing the iron of the can, will the iron corrode more or less rapidly than if the tin were not present? Inside the can, the tin itself is coated with a clear varnish. Explain.

21.127 Commercial electrolytic cells for producing aluminum operate at 5.0 V and 100,000 A.
(a) How long does it take to produce exactly 1 metric ton (1000 kg) of aluminum?
(b) How much electrical power (in kilowatt-hours, kW·h) is used (1 W = 1 J/s; 1 kW·h = 3.6×10^3 kJ)?
(c) If electricity costs \$0.123 per kW·h and cell efficiency is 90.%, what is the cost of electricity to produce exactly 1 lb of aluminum?

21.128 Magnesium bars are connected electrically to underground iron pipes to serve as sacrificial anodes.
(a) Do electrons flow from the bar to the pipe or the reverse?
(b) A 12-kg Mg bar is attached to an iron pipe, and it takes 8.5 yr for the Mg to be consumed. What is the average current flowing between the Mg and the Fe during this period?

21.129 Bubbles of H_2 form when metal D is placed in hot H_2O. No reaction occurs when D is placed in a solution of a salt of metal E, but D is discolored and coated immediately when placed in a solution of a salt of metal F. What happens if E is placed in a solution of a salt of metal F? Rank metals D, E, and F in order of *increasing* reducing strength.

21.130 In addition to reacting with gold (see Problem 21.21), aqua regia is used to bring other precious metals into solution. Balance the skeleton equation for the reaction with Pt:

$$Pt(s) + NO_3^-(aq) + Cl^-(aq) \longrightarrow PtCl_6^{2-}(aq) + NO(g)$$

21.131 The following reactions are used in batteries:

I. $2H_2(g) + O_2(g) \longrightarrow 2H_2O(l)$ $E_{cell} = 1.23$ V

II. $Pb(s) + PbO_2(s) + 2H_2SO_4(aq) \longrightarrow 2PbSO_4(s) + 2H_2O(l)$ $E_{cell} = 2.04$ V

III. $2Na(l) + FeCl_2(s) \longrightarrow 2NaCl(s) + Fe(s)$ $E_{cell} = 2.35$ V

Reaction I is used in fuel cells, II in the automobile lead-acid battery, and III in an experimental high-temperature battery for powering electric vehicles. The aim is to obtain as much work as possible from a cell, while keeping its weight to a minimum.
(a) In each cell, find the moles of electrons transferred and ΔG.
(b) Calculate the ratio, in kJ/g, of w_{max} to mass of reactants for each of the cells. Which has the highest ratio, which the lowest, and why? (*Note:* For simplicity, ignore the masses of cell components that do not appear in the cell as reactants, including electrode materials, electrolytes, separators, cell casing, wiring, etc.)

21.132 A current is applied to two electrolytic cells in series. In the first, silver is deposited; in the second, a zinc electrode is consumed. How much Ag is plated out if 1.2 g of Zn dissolves?

21.133 You are investigating a particular chemical reaction. State all the types of data available in standard tables that enable you to calculate the equilibrium constant for the reaction at 298 K.

21.134 In an electric power plant, personnel monitor the O_2 content of boiler feed water to prevent corrosion of the boiler tubes. Why does Fe corrode faster in steam and hot water than in cold water?

21.135 A voltaic cell using Cu/Cu^{2+} and Sn/Sn^{2+} half-cells is set up at standard conditions, and each compartment has a volume of 345 mL. The cell delivers 0.17 A for 48.0 h.

(a) How many grams of Cu(*s*) are deposited?
(b) What is the $[Cu^{2+}]$ remaining?

21.136 If the E_{cell} of the following cell is 0.915 V, what is the pH in the anode compartment?

$$Pt(s) \mid H_2\ (1.00\ atm) \mid H^+(aq) \parallel Ag^+\ (0.100\ M) \mid Ag(s)$$

21.137 From the skeleton equations below, create a list of balanced half-reactions in which the strongest oxidizing agent is on top and the weakest is on the bottom:

$$U^{3+}(aq) + Cr^{3+}(aq) \longrightarrow Cr^{2+}(aq) + U^{4+}(aq)$$
$$Fe(s) + Sn^{2+}(aq) \longrightarrow Sn(s) + Fe^{2+}(aq)$$
$$Fe(s) + U^{4+}(aq) \longrightarrow \text{no reaction}$$
$$Cr^{3+}(aq) + Fe(s) \longrightarrow Cr^{2+}(aq) + Fe^{2+}(aq)$$
$$Cr^{2+}(aq) + Sn^{2+}(aq) \longrightarrow Sn(s) + Cr^{3+}(aq)$$

21.138 You are given the following three half-reactions:
(1) $Fe^{3+}(aq) + e^- \rightleftharpoons Fe^{2+}(aq)$
(2) $Fe^{2+}(aq) + 2e^- \rightleftharpoons Fe(s)$
(3) $Fe^{3+}(aq) + 3e^- \rightleftharpoons Fe(s)$
(a) Calculate $\Delta G°$ for (1) and (2) from their $E°_{half\text{-}cell}$ values.
(b) Calculate $\Delta G°$ for (3) from (1) and (2).
(c) Calculate $E°_{half\text{-}cell}$ for (3) from its $\Delta G°$.

21.139 Use the half-reaction method to balance the equation for the conversion of ethanol to acetic acid in acid solution:

$$CH_3CH_2OH + Cr_2O_7^{2-} \longrightarrow CH_3COOH + Cr^{3+}$$

21.140 When zinc is refined by electrolysis, the desired half-reaction at the cathode is

$$Zn^{2+}(aq) + 2e^- \longrightarrow Zn(s)$$

A competing reaction, which lowers the yield, is the formation of hydrogen gas:

$$2H^+(aq) + 2e^- \longrightarrow H_2(g)$$

If 91.50% of the current flowing results in zinc being deposited, while 8.50% produces hydrogen gas, how many liters of H_2, measured at STP, form per kilogram of zinc?

21.141 A chemist designs an ion-specific probe for measuring $[Ag^+]$ in an NaCl solution saturated with AgCl. One half-cell has an Ag wire electrode immersed in the unknown AgCl-saturated NaCl solution. It is connected through a salt bridge to the other half-cell, which has a calomel reference electrode [a platinum wire immersed in a paste of mercury and calomel (Hg_2Cl_2)] in a saturated KCl solution. The measured E_{cell} is 0.060 V.
(a) Given the following standard half-reactions, calculate $[Ag^+]$.

Calomel: $Hg_2Cl_2(s) + 2e^- \longrightarrow 2Hg(l) + 2Cl^-(aq)$ $E° = 0.24$ V

Silver: $Ag^+(aq) + e^- \longrightarrow Ag(s)$ $E° = 0.80$ V

(*Hint:* Assume that $[Cl^-]$ is so high that it is essentially constant.)
(b) A mining engineer wants an ore sample analyzed with the Ag^+-selective probe. After pretreating the ore sample, the chemist measures the cell voltage as 0.53 V. What is $[Ag^+]$?

21.142 Use Appendix D to calculate the K_{sp} of AgCl.

21.143 Black-and-white photographic film is coated with silver halides. Because silver is expensive, the manufacturer monitors the Ag^+ content of the waste stream, $[Ag^+]_{waste}$, from the plant with an Ag^+-selective electrode at 25°C. A stream of known Ag^+ concentration, $[Ag^+]_{standard}$, is passed over the electrode in turn with the waste stream and the data recorded by a computer.
(a) Write the equations relating the nonstandard cell potential to the standard cell potential and $[Ag^+]$ for each solution.
(b) Combine these into a single equation to find $[Ag^+]_{waste}$.
(c) Rewrite the equation from part (b) to find $[Ag^+]_{waste}$ in ng/L.
(d) If E_{waste} is 0.003 V higher than $E_{standard}$, and the standard solution contains 1000. ng/L, what is $[Ag^+]_{waste}$?
(e) Rewrite the equation from part (b) to find $[Ag^+]_{waste}$ for a system in which T changes and T_{waste} and $T_{standard}$ may be different.

21.144 Calculate the K_f of $Ag(NH_3)_2^+$ from

$Ag^+(aq) + e^- \rightleftharpoons Ag(s)$ $E° = 0.80$ V

$Ag(NH_3)_2^+(aq) + e^- \rightleftharpoons Ag(s) + 2NH_3(aq)$ $E° = 0.37$ V

21.145 Even though the toxicity of cadmium has become a concern, nickel-cadmium (nicad) batteries are still used commonly in many devices. The overall cell reaction is

$$Cd(s) + 2NiO(OH)(s) + 2H_2O(l) \longrightarrow 2Ni(OH)(s) + Cd(OH)_2(s)$$

A certain nicad battery weighs 18.3 g and has a capacity of 300. mA·h (that is, the cell can store charge equivalent to a current of 300. mA flowing for 1 h).
(a) What is the capacity of this cell in coulombs?
(b) What mass of reactants is needed to deliver 300. mA·h?
(c) What percentage of the cell mass consists of reactants?

21.146 The zinc-air battery is a less expensive alternative to silver batteries for use in hearing aids. The cell reaction is

$$2Zn(s) + O_2(g) \longrightarrow 2ZnO(s)$$

A new battery weighs 0.275 g. The zinc accounts for exactly $\frac{1}{10}$ of the mass, and the oxygen does not contribute to the mass because it is supplied by the air.
(a) How much electricity (in C) can the battery deliver?
(b) How much free energy (in J) is released if E_{cell} is 1.3 V?

21.147 Use Appendix D to create an activity series of Mn, Fe, Ag, Sn, Cr, Cu, Ba, Al, Na, Hg, Ni, Li, Au, Zn, and Pb. Rank these metals in order of decreasing reducing strength, and divide them into three groups: those that displace H_2 from water, those that displace H_2 from acid, and those that cannot displace H_2.

21.148 Both Ti and V are reactive enough to displace H_2 from water; of the two metals, Ti is the stronger reducing agent. The difference in their $E°_{half\text{-}cell}$ values is 0.43 V. Given

$$V(s) + Cu^{2+}(aq) \longrightarrow V^{2+}(aq) + Cu(s) \quad \Delta G° = -298\ kJ/mol$$

use Appendix D to calculate the $E°_{half\text{-}cell}$ values for V^{2+}/V and Ti^{2+}/Ti half-cells.

21.149 For the reaction

$$S_4O_6^{2-}(aq) + 2I^-(aq) \longrightarrow I_2(s) + S_2O_3^{2-}(aq) \quad \Delta G° = 87.8\ kJ/mol$$

(a) Identify the oxidizing and reducing agents. (b) Calculate $E°_{cell}$. (c) For the reduction half-reaction, write a balanced equation, give the oxidation number of each element, and calculate $E°_{half\text{-}cell}$.

21.150 Two concentration cells are prepared, both with 90.0 mL of 0.0100 *M* $Cu(NO_3)_2$ and a Cu bar in each half-cell.
(a) In the first concentration cell, 10.0 mL of 0.500 *M* NH_3 is added to one half-cell; the complex ion $Cu(NH_3)_4^{2+}$ forms, and E_{cell} is 0.129 V. Calculate K_f for the formation of the complex ion.
(b) Calculate E_{cell} when an additional 10.0 mL of 0.500 *M* NH_3 is added.
(c) In the second concentration cell, 10.0 mL of 0.500 *M* NaOH is added to one half-cell; the precipitate $Cu(OH)_2$ forms ($K_{sp} = 2.2\times10^{-20}$). Calculate $E°_{cell}$.
(d) What would the molarity of NaOH have to be for the addition of 10.0 mL to result in an $E°_{cell}$ of 0.340 V?

21.151 Two voltaic cells are to be joined so that one will run the other as an electrolytic cell. In the first cell, one half-cell has Au foil in 1.00 *M* $Au(NO_3)_3$, and the other half-cell has a Cr bar in 1.00 *M* $Cr(NO_3)_3$. In the second cell, one half-cell has a Co bar in 1.00 *M* $Co(NO_3)_2$, and the other half-cell has a Zn bar in 1.00 *M* $Zn(NO_3)_2$.
(a) Calculate E°_{cell} for each cell.
(b) Calculate the total potential if the two cells are connected as voltaic cells in series.
(c) When the electrode wires are switched in one of the cells, which cell will run as the voltaic cell and which as the electrolytic cell?
(d) Which metal ion is being reduced in each cell?
(e) If 2.00 g of metal plates out in the voltaic cell, how much metal ion plates out in the electrolytic cell?

21.152 A voltaic cell has one half-cell with a Cu bar in a 1.00 *M* Cu^{2+} salt, and the other half-cell with a Cd bar in the same volume of a 1.00 *M* Cd^{2+} salt.
(a) Find E°_{cell}, ΔG°, and *K*.
(b) As the cell operates, $[Cd^{2+}]$ increases; find E_{cell} and ΔG when $[Cd^{2+}]$ is 1.95 *M*.
(c) Find E_{cell}, ΔG, and $[Cu^{2+}]$ at equilibrium.

21.153 Gasoline is a mixture of hydrocarbons, but the heat released when it burns is close to that of octane, $C_8H_{18}(l)$ ($\Delta H^\circ_f = -250.1$ kJ/mol). Research is underway to use H_2 from the electrolysis of water in fuel cells to power cars instead of gasoline.
(a) Calculate ΔH° when 1.00 gal of gasoline ($d = 0.7028$ g/mL) burns to produce carbon dioxide gas and water vapor.
(b) How many liters of H_2 at 25°C and 1.00 atm must burn to produce this quantity of energy?
(c) How long would it take to produce this amount of H_2 by electrolysis with a current of 1.00×10^3 A at 6.00 V?
(d) How much power in kilowatt-hours (kW·h) is required to generate this amount of H_2 (1 W = 1 J/s, 1 J = 1 C·V, and 1 kW·h = 3.6×10^6 J)?
(e) If the cell is 88.0% efficient and electricity costs $0.123 per kW·h, what is the cost of producing the amount of H_2 equivalent to 1.00 gal of gasoline?

22 The Elements in Nature and Industry

22.1 How the Elements Occur in Nature
Earth's Structure and Elements' Abundance
Sources of the Elements

22.2 The Cycling of Elements Through the Environment
Carbon Cycle
Nitrogen Cycle
Phosphorus Cycle

22.3 Metallurgy: Extracting a Metal from Its Ore
Pretreating the Ore
Converting Mineral to Element
Refining and Alloying

22.4 Tapping the Crust: Isolation and Uses of Selected Elements
Sodium and Potassium
Iron, Copper, and Aluminum
Magnesium
Hydrogen

22.5 Chemical Manufacturing: Two Case Studies
Sulfuric Acid
Chlor-Alkali Process

Source: © Oleksiy Maksymenko Photography/Alamy

Concepts and Skills to Review Before You Study This Chapter

- trends in atomic properties (Sections 8.3, 8.4, 9.2, and 9.5)
- catalysts and reaction rate (Section 16.7)
- Le Châtelier's principle (Section 17.6)
- acid-base equilibria (Sections 18.3 and 18.4)
- solubility and complex-ion equilibria (Sections 19.3 and 19.4)
- temperature and reaction spontaneity (Section 20.3)
- free energy and equilibrium (Section 20.4)
- standard electrode potentials (Section 21.3)
- electrolysis of molten salts and aqueous salt solutions (Section 21.7)

To make that sports car you may have dreamed of owning, manufacturers need chemical elements. Iron, in the form of steel, accounts for about 55% of the weight of a typical car and aluminum about 8%. Zinc is used to galvanize some of the metal parts, lead is used in the battery, and copper is needed for the wiring. To make your dream car (or your old, trusty, college-student car) a reality, we need to extract elements from their source, Earth. Chemical principles aid in answering practical questions about this process: how much of the element is present in nature, in what form does it occur, how do organisms affect its distribution, and how do we isolate it for our own use? We'll apply concepts of kinetics, equilibrium, thermodynamics, and electrochemistry from previous chapters to explain some major chemical processes in nature and industry.

IN THIS CHAPTER . . . *We examine the distribution of elements on Earth and the natural cycles of three key elements. Then, we apply concepts of kinetics, equilibrium, thermodynamics, and electrochemistry from previous chapters to explain the methods that have been developed to extract, purify, and utilize several important elements.*

- We identify the abundances and sources of elements in the various regions of Earth.
- We consider how three essential elements—carbon, nitrogen, and phosphorus—cycle through the environment and focus on how humans influence these processes.
- We focus on the isolation and purification of metals, discussing general redox and metallurgical procedures for extracting an element from its ore.
- We examine in detail the isolation and uses of seven key elements: sodium, potassium, iron, copper, aluminum, magnesium, and hydrogen.
- We take a close look at two of the most important processes in chemical manufacturing: the production of sulfuric acid and the isolation of chlorine.

22.1 HOW THE ELEMENTS OCCUR IN NATURE

To begin our examination of how we use the elements, let's take inventory of our elemental stock—the distribution and relative amounts of the elements on Earth, especially that thin outer portion of our planet that we can reach.

Earth's Structure and the Abundance of the Elements

Any attempt to isolate an element must begin with a knowledge of its **abundance,** the amount of the element in a particular region of the natural world. The abundances of the elements result from details of our planet's evolution.

Formation and Layering of Earth About 4.5 billion years ago, vast clouds of cold gases and interstellar debris from exploded older stars gradually coalesced into the Sun and planets. At first, Earth was a cold, solid sphere of uniformly distributed elements and simple compounds. In the next billion years or so, heat from radioactive decay and continuous meteor impacts raised the planet's temperature to around 10^4 K, sufficient to form an enormous molten mass. Any

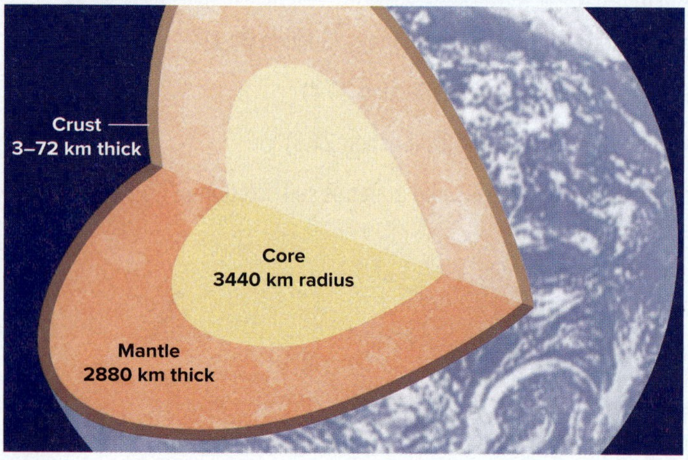

Figure 22.1 The layered internal structure of Earth.

remaining gaseous elements, such as the cosmically abundant and very light hydrogen and helium, were lost to space.

As Earth cooled, chemical and physical processes resulted in its **differentiation,** the formation of regions of different composition and density. Differentiation gave Earth an internal structure consisting of three layers (Figure 22.1):

- The dense **core** (d = 10–15 g/cm^3) is composed of a molten outer core and a solid Moon-sized inner core. Remarkably, the inner core is nearly as hot as the surface of the Sun and spins within the molten outer core slightly faster than Earth itself does!
- The thick, homogeneous **mantle** lying around the core has an overall density of 4–6 g/cm^3.
- A thin, heterogeneous **crust** lies on top of the mantle. The average density of the crust, on which all the comings and goings of life take place, is 2.8 g/cm^3.

Table 22.1 compares the abundance of some key elements in the universe, the whole Earth (actually core plus mantle only, which account for more than 99% of Earth's mass), and Earth's three regions. (Because the deepest terrestrial sampling can penetrate only a few kilometers into the crust, some of these data represent extrapolations from meteor samples and from seismic studies of earthquakes.) Several points stand out:

- Cosmic and whole-Earth abundances are very different, particularly for H and He; these two elements account for the majority of the mass of the universe but are not abundant in elemental form on Earth.
- The elements O, Si, Fe, and Mg are abundant both cosmically and on Earth. Together, they account for more than 90% of Earth's mass.
- The core is particularly rich in the dense Group 8B(8) through 8B(10) metals: Co, Ni, and especially Fe, the most abundant element in the whole Earth.

Table 22.1 Cosmic and Terrestrial Abundances (Mass %) of Selected Elements*

Element	Universe	Earth	Crust	Mantle	Core
O	1.07	29.5	49.5	43.7	—
Si	0.06	15.2	25.7	21.6	—
Al	—	1.1	7.5	1.8	—
Fe	0.19	34.6	4.7	13.3	88.6
Ca	0.007	1.1	3.4	2.1	—
Mg	0.06	12.7	2.8	16.6	—
Na	—	0.6	2.6	0.8	—
K	—	0.07	2.4	0.2	—
H	73.9	—	0.87	—	—
Ti	—	0.05	0.58	0.18	0.01
Cl	—	—	0.19	—	—
P	—	—	0.12	—	—
Mn	—	—	0.09	—	—
C	0.46	—	0.08	—	—
S	0.04	1.9	0.06	>2	—
Ni	0.006	2.4	0.008	0.3	8.5
Co	—	0.13	0.003	0.04	0.6
He	24.0	—	—	—	—

*A missing abundance value indicates either that reliable data are not available or that the value is less than 0.001% by mass.

- Crustal abundances are very different from whole-Earth abundances. The crust makes up only 0.4% of Earth's mass, but has the largest share of nonmetals, metalloids, and light, active metals: Al, Ca, Na, and K. The mantle contains much smaller proportions of these, and the core has none. Oxygen is the most abundant element in the crust and mantle but is absent from the core.

These differences in Earth's major layers arose from the effects of thermal energy. When Earth was molten, gravity and convection caused more dense materials to sink and less dense materials to rise, yielding several compositional *phases:*

- Most of the Fe sank to form the core, or *iron phase.*
- In the light outer phase, oxygen combined with Si, Al, Mg, and some Fe to form silicates, the material of rocks. This *silicate phase* later separated into the mantle and crust.
- The *sulfide phase,* intermediate in density and insoluble in the other two, consisted mostly of iron sulfide and mixed with parts of the silicate phase above and the iron phase below.
- The thin, primitive atmosphere, probably a mixture of water vapor (which gave rise to the oceans), carbon monoxide, and nitrogen (or ammonia), was produced by *outgassing* (the expulsion of trapped gases).

The distribution of the remaining elements (discussed here using only the new periodic table group numbers) was controlled by their chemical affinity for one of the three phases. In general terms, as Figure 22.2 shows:

- Elements with low or high electronegativity—active metals (those in Groups 1 through 5, Cr, and Mn) and nonmetals (O, lighter members of Groups 13 to 15, and all of Group 17)—tended to congregate in the silicate phase as ionic compounds.
- Metals with intermediate electronegativities (many from Groups 6 to 10) dissolved in the iron phase.
- Lower-melting transition metals and many metals and metalloids in Groups 11 to 16 became concentrated in the sulfide phase.

The Impact of Life on Crustal Abundances At present, the crust is the only physically accessible layer of Earth, so only crustal abundances have practical significance. The crust is divided into the **lithosphere** (the solid region of the crust), the **hydrosphere** (the liquid region), and the **atmosphere** (the gaseous region). Over billions of years, weathering and volcanic upsurges have dramatically altered the composition of the crust. The **biosphere,** which consists of the living systems that inhabit and have

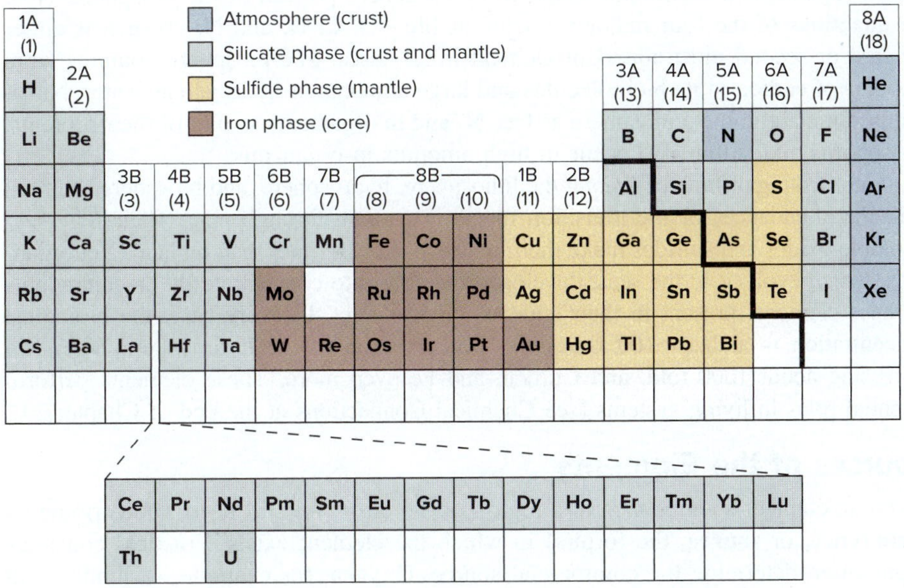

Figure 22.2 Geochemical differentiation of the elements.

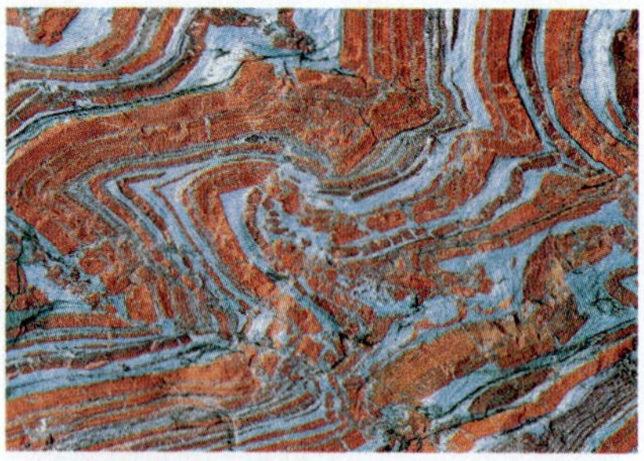

A Banded-iron formations containing Fe_2O_3

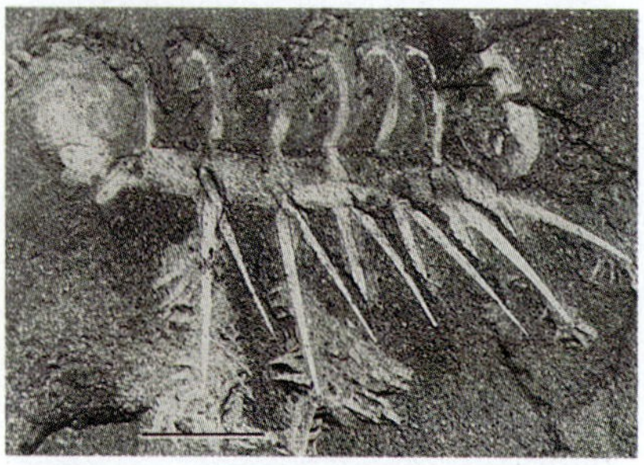

B Fossil of early multicellular organism

Figure 22.3 Ancient effects of an O_2-rich atmosphere.

Source: (A) © Doug Sherman/Geofile RF; (B) © University of Cambridge, Department of Earth Sciences

inhabited the planet, has been another *major influence on crustal element composition and distribution.*

When the earliest rocks were forming and the ocean basins were filling with water, binary inorganic molecules in the atmosphere were reacting to form first simple, and then more complex, organic molecules. The energy for these (mostly) endothermic changes was supplied by lightning, solar radiation, geologic heating, and meteoric impacts. In an amazingly short period of time, probably no more than 500 million years, the first organisms appeared. It took less than another billion years for these to evolve into simple algae that could derive metabolic energy from photosynthesis, converting CO_2 and H_2O into organic molecules and releasing O_2 as a byproduct.

Let's look at some of the essential ways the evolution of organisms affected the chemistry of the crust:

1. *Increase in O_2.* Due to photosynthesis, the atmosphere gradually became richer in O_2 over the 300 million years following the emergence of algae. Thus, *oxidation became the major source of free energy in the crust and biosphere.* As a result of this oxidizing environment, the predominating Fe(II)-containing minerals were oxidized to Fe(III)-containing minerals, such as hematite, Fe_2O_3, shown in Figure 22.3A in ancient banded-iron formations (red beds). The increase in oxygen also led to an explosion of O_2-utilizing life forms, a few of which evolved into the organisms of today. Figure 22.3B shows a fossil of one of the multicellular organisms whose appearance coincided with the increase in O_2.

2. *K^+ vs. Na^+.* The K^+ concentration of the oceans is much lower than the Na^+ concentration. Forming over eons from outgassed water vapor condensing into rain and streaming over the land, the oceans became complex ionic solutions with 30 times as much Na^+ as K^+. One reason for this is that plants require K^+ for growth, so they absorbed dissolved K^+ that would otherwise have washed down to the sea.

3. *Buried organic carbon.* Enormous subterranean deposits of organic carbon formed when ancient plants, buried deeply and decomposing under high pressure and temperature in the absence of free O_2, gradually turned into coal. Animals buried under similar conditions turned into petroleum. Crustal deposits of this organic carbon provide the fuels that move our cars, heat our homes, and electrify our cities.

4. *Fossilized $CaCO_3$.* Carbon and oxygen, together with calcium, the fifth most abundant element in the crust, occur in vast sedimentary deposits all over the world as limestone, dolomite *(see photo),* marble, and chalk, all of which are the fossilized skeletal remains of countless early marine organisms.

Dolomite Mountains, Italy.

Source: © Ruth Melnick

Table 22.2 compares selected elemental abundances in the whole crust, the three crustal regions, and the human body, a representative portion of the biosphere. Note the quantities of the four major elements of life—O, C, H, and N. Oxygen is either the first or second most abundant element in all cases. The biosphere contains large amounts of carbon in its biomolecules and large amounts of hydrogen as water. Nitrogen is abundant in the atmosphere as free N_2 and in organisms as part of their proteins. Phosphorus and sulfur also occur in high amounts in organisms.

One striking difference among the lithosphere, hydrosphere, and biosphere (human) is in the abundances of the transition metals vanadium through zinc. Their relatively insoluble oxides and sulfides make them much scarcer in water than on land. Yet organisms, which evolved in the seas, developed the ability to concentrate the trace amounts of these elements present in their aqueous environment. In every case, the biological concentration is at least 100 times the concentration in the hydrosphere, with Mn increasing about 1000-fold, and Cu, Zn, and Fe even more. These elements perform essential roles in living systems (see Chemical Connections at the end of Chapter 23).

Sources of the Elements

Given an element's abundance in a region of the crust, we still need to determine its **occurrence,** or **source,** the form(s) in which the element exists. Practical considerations often determine the commercial source. Oxygen, for example, is abundant in

Table 22.2 **Abundances of Selected Elements in the Crust, Its Regions, and the Human Body as Representative of the Biosphere (Mass %)**

		Crustal Regions			
Element	**Crust**	**Lithosphere**	**Hydrosphere**	**Atmosphere**	**Human**
O	49.5	45.5	85.8	23.0	65.0
C	0.08	0.018	—	0.01	18.0
H	0.87	0.15	10.7	0.02	10.0
N	0.03	0.002	—	75.5	3.0
P	0.12	0.11	—	—	1.0
Mg	1.9	2.76	0.13	—	0.50
K	2.4	1.84	0.04	—	0.34
Ca	3.4	4.66	0.05	—	2.4
S	0.06	0.034	—	—	0.26
Na	2.6	2.27	1.1	—	0.14
Cl	0.19	0.013	2.1	—	0.15
Fe	4.7	6.2	—	—	0.005
Zn	0.013	0.008	—	—	0.003
Cr	0.02	0.012	—	—	3×10^{-6}
Co	0.003	0.003	—	—	3×10^{-6}
Cu	0.007	0.007	—	—	4×10^{-4}
Mn	0.09	0.11	—	—	1×10^{-4}
Ni	0.008	0.010	—	—	3×10^{-6}
V	0.015	0.014	—	—	3×10^{-6}

all three crustal regions, but the atmosphere is its primary industrial source because it occurs there as the free element. Nitrogen and the noble gases (except helium) are also obtained from the atmosphere. Several other elements occur uncombined, formed in large deposits by prehistoric biological action, such as sulfur in caprock salt domes and nearly pure carbon in coal. The relatively unreactive elements gold and platinum also occur in an uncombined (or *native*) state.

The overwhelming majority of elements, however, occur in **ores,** natural compounds or mixtures of compounds from which an element can be extracted economically *(see photo): the financial costs of mining, isolating, and purifying must be considered when choosing a process to obtain an element.*

Deposit of borate (tufa), the ore of boron.
Source: © Robert Holmes/Corbis

Figure 22.4 on the next page shows the most useful sources of the elements. Note that elements with the same types of ores tend to be grouped together in the periodic table:

- Alkali metal halides are ores for both groups of their component elements, Group 1A(1) and Group 7A(17).
- Group 2A(2) metals occur as carbonates in the marble and limestone of mountain ranges, although magnesium is very abundant in seawater (making that the preferred source, as we discuss later).
- Even though many elements occur as silicates, most of these compounds are very stable thermodynamically. Thus, the cost that would be incurred in processing them prohibits their use—aside from silicon, only lithium and beryllium are obtained from their silicates.
- The ores of most industrially important metals are either oxides, which dominate the left half of the transition series, or sulfides, which dominate the right half of the transition series and a few of the main groups beyond.

The reasons for the prominence of oxide and sulfide ores of metallic elements are complex and include weathering processes, selective precipitation (Section 19.3), and relative solubilities. Certain atomic properties are relevant as well. Metals toward the left side of the periodic table have lower ionization energies and electronegativities,

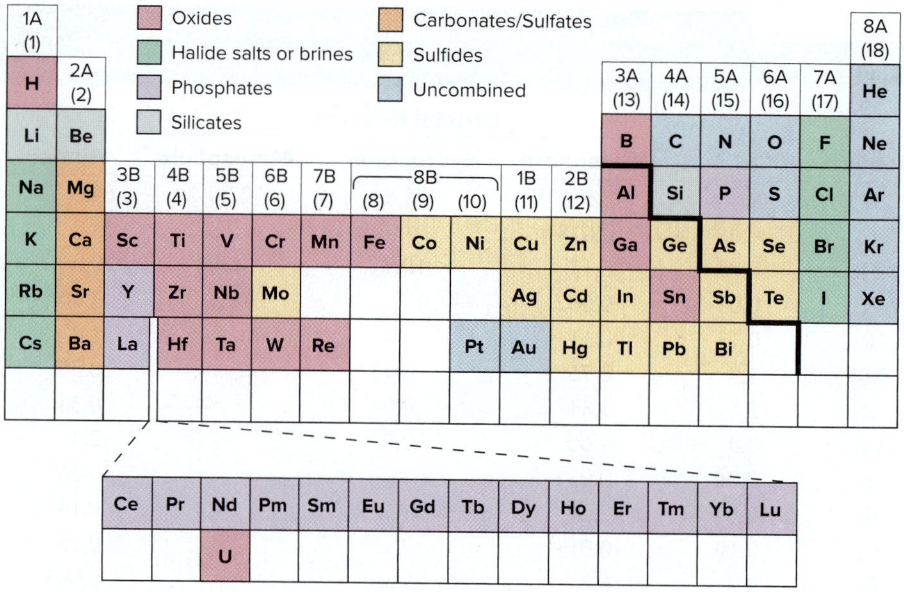

Figure 22.4 Sources of the elements.

so they tend to give up electrons or hold them loosely in bonds. The O^{2-} ion is small enough to approach a metal cation closely, which results in a high lattice energy for the metal oxide. In contrast, metals toward the right side have higher ionization energies and electronegativities. Thus, they tend to form bonds that are more covalent, which suits the larger, more polarizable S^{2-} ion.

› Summary of Section 22.1

- › As the young Earth cooled, the elements became differentiated into a dense metallic core, a silicate-rich mantle, and a low-density crust.
- › High abundances of light metals, metalloids, and nonmetals are concentrated in the crust, which has three regions: lithosphere (solid), hydrosphere (liquid), and atmosphere (gaseous).
- › The biosphere (living systems) profoundly affected crustal chemistry by producing free O_2, and thus an oxidizing environment.
- › Some elements occur in their native state, but most are combined in ores. The ores of most important metallic elements are oxides or sulfides.

22.2 THE CYCLING OF ELEMENTS THROUGH THE ENVIRONMENT

The distributions of many elements change at widely differing rates. The physical, chemical, and biological paths that atoms of an element take through the three regions of the crust—atmosphere, hydrosphere, and lithosphere—constitute the element's **environmental cycle.** In this section, we consider the cycles for carbon, nitrogen, and phosphorus and highlight the effects of humans on them.

The Carbon Cycle

Carbon is one of a handful of elements that appear in all three regions of Earth's crust.

- In the lithosphere, it occurs in elemental form as graphite and diamond, in fully oxidized form in carbonate minerals, in fully reduced form in petroleum hydrocarbons, and in complex mixtures, such as coal and living matter.
- In the hydrosphere, it occurs in living matter, in carbonate minerals formed by the action of coral-reef organisms, and in dissolved CO_2 and related oxoanions.
- In the atmosphere, it occurs principally in gaseous CO_2, a minor (0.04%) but essential component that exists in equilibrium with the aqueous fraction.

Key Interactions Within the Cycle Figure 22.5 depicts the complex interplay of sources and the effect of the biosphere on carbon's environmental cycle. It provides

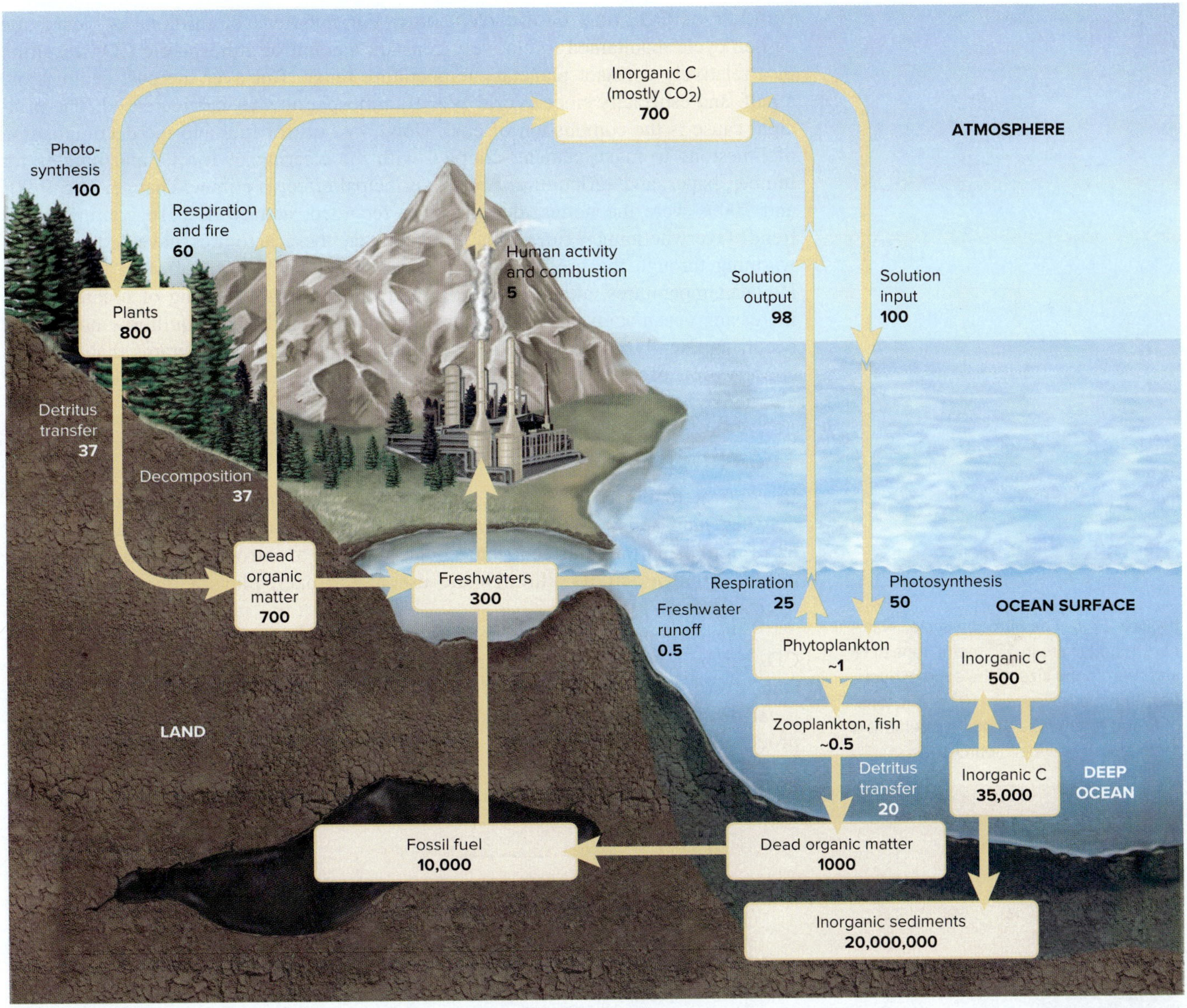

Figure 22.5 **The carbon cycle.** Numbers in boxes refer to the size of the source; numbers along arrows refer to the annual movement of the element from source to source. Both sets of values are in 10^9 t of C (1 t = 1000 kg = 1.1 tons).

an estimate of the size of each source and, where possible, the amount of carbon moving annually between sources. Note these key points:

- The portions of the cycle are linked by the atmosphere—one link between oceans and air, the other between land and air. The 2.6×10^{12} metric tons of CO_2 in air cycles through the oceans and atmosphere about once every 300 years but spends a much longer time in carbonate minerals.
- Atmospheric CO_2 accounts for only 0.04% of crustal carbon but moves the element through the other regions. A CO_2 molecule spends, on average, 3.5 years in the atmosphere.
- The land and oceans are in contact with the largest sources of carbon, immobilized as carbonates, coal, and oil in rocky sediment beneath the soil.
- *Photosynthesis, respiration,* and *decomposition* are major factors in the cycle. **Fixation** is the process of converting a gaseous substance into a condensed form. Via photosynthesis, marine plankton and terrestrial plants use sunlight to fix atmospheric CO_2 into carbohydrates. Plants release CO_2 by respiration at night and, much more slowly, when they decompose. Animals eat the plants and release CO_2 by respiration and decomposition.
- Fires and volcanoes also release CO_2 into the air.

Buildup of CO_2 and Global Warming For hundreds of millions of years, the carbon cycle maintained a relatively constant amount of atmospheric CO_2, resulting in a relatively constant planetary temperature range. But over the past century and a half, and especially since World War II, atmospheric CO_2 has increased. The principal cause is the combustion of coal, wood, and oil for fuel and the decomposition of limestone to make cement, coupled with the clearing of forests and jungles for lumber, paper, and agriculture. And the principal effect is climate change. The 1990s and 2000s were the hottest decades ever recorded, and 2011–2016 continued this trend. Overwhelming evidence is showing that these activities have led to global warming through the greenhouse effect (see Chemical Connections in Chapter 6). Higher temperatures, alterations in dry and rainy seasons, melting of polar ice, and increasing ocean acidity, with associated changes in carbonate equilibria, are clearly occurring. Nearly all science policy experts are pressing for programs that combine conservation of carbon-based fuels, an end to deforestation, and development of alternative energy sources.

The Nitrogen Cycle

In contrast to the carbon cycle, the nitrogen cycle includes a *direct* interaction between land and sea (Figure 22.6). All nitrites and nitrates are soluble, so rain and runoff from the land contribute huge amounts of nitrogen to lakes, rivers, and oceans. Human activity, through the use of fertilizer, plays a major role in this cycle.

Like carbon dioxide, atmospheric nitrogen must be fixed. However, whereas CO_2 can be incorporated by plants in either its gaseous or aqueous form, the great stability of N_2 prevents plants from using it directly. Fixation of N_2 requires a great deal of energy and occurs through atmospheric, industrial, and biological processes.

Figure 22.6 The nitrogen cycle. Numbers in boxes are in 10^9 metric tons of N and refer to the size of the source; numbers along arrows are in 10^6 metric tons of N and refer to the annual movement of the element between sources.

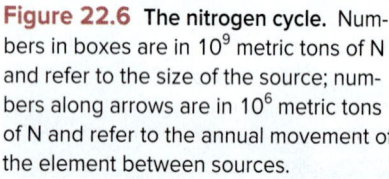

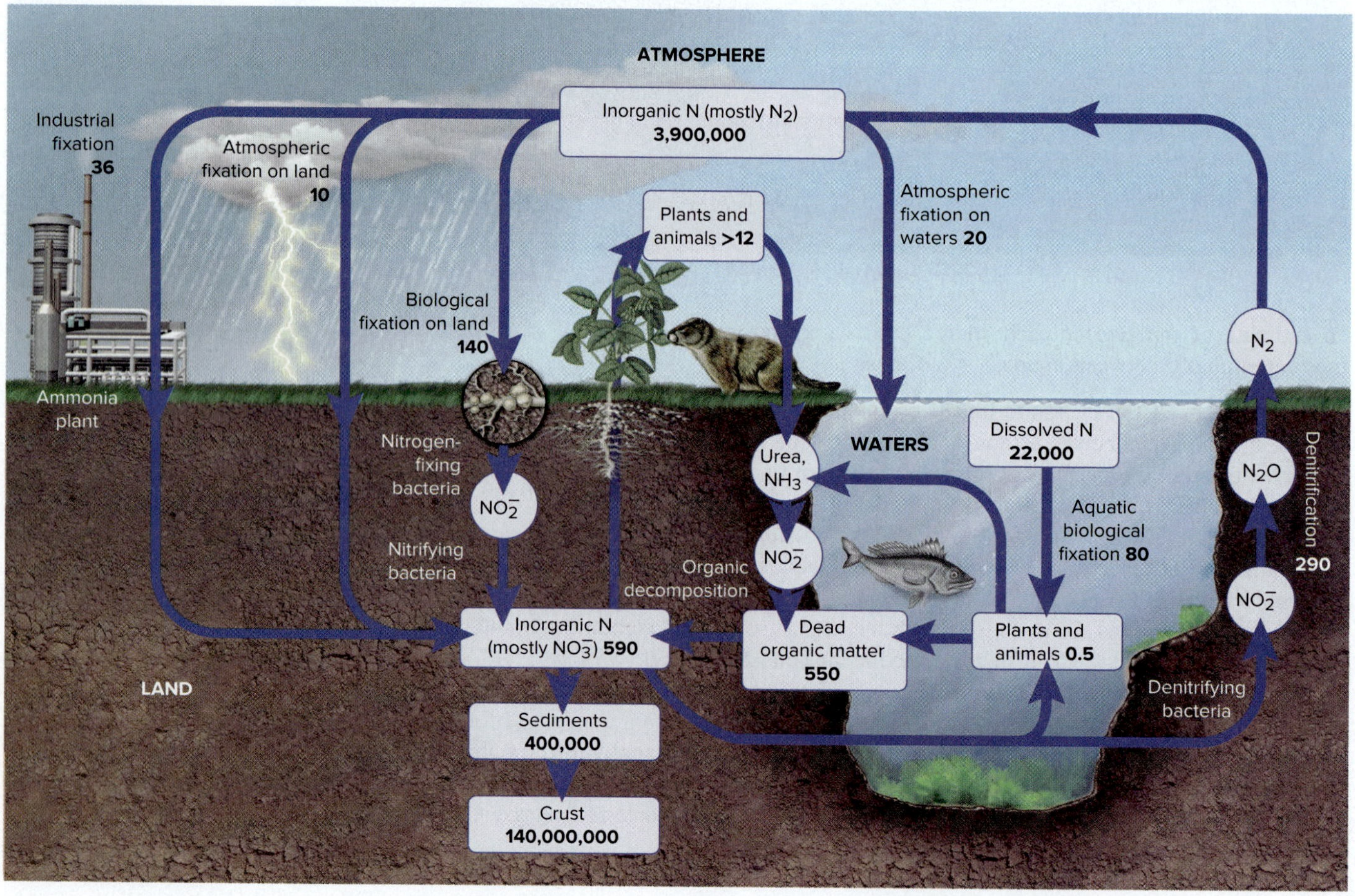

Atmospheric Fixation Lightning causes the high-temperature endothermic reaction of N_2 and O_2 to form NO, which is then oxidized by ozone exothermically to NO_2:

$$N_2(g) + O_2(g) \longrightarrow 2NO(g) \qquad \Delta H^\circ = 180.6\ \text{kJ}$$
$$NO(g) + O_3(g) \longrightarrow NO_2(g) + O_2(g) \qquad \Delta H^\circ = -200.1\ \text{kJ}$$

During the day, NO_2 reacts with hydroxyl radical to form HNO_3:

$$NO_2(g) + HO\cdot(g) \longrightarrow HNO_3(g)$$

At night, a multistep reaction between NO_2 and ozone is involved. In either case, rain dissociates the nitric acid, and $NO_3^-(aq)$ enters both sea and land to be utilized by plants.

Industrial Fixation Most human-caused fixation occurs industrially during ammonia synthesis via the Haber process (Section 17.6):

$$N_2(g) + 3H_2(g) \rightleftharpoons 2NH_3(g)$$

The process takes place on an enormous scale, and NH_3 ranks first, on a mole basis, among compounds produced industrially. Some of this NH_3 is converted to HNO_3 in the Ostwald process (Section 14.7), but most is used as fertilizer, either directly or in the form of urea and ammonium salts (sulfate, phosphate, and nitrate), which enter the biosphere when taken up by plants (as discussed in the next subsection).

In recent decades, high-temperature combustion in electric power plants and in car, truck, and plane engines has become an important contributor to total fixed nitrogen. High engine operating temperatures mimic lightning to form NO from the N_2 and O_2 in the air taken in to burn the hydrocarbon fuel. The NO in exhaust gases reacts to form nitric acid in the atmosphere, which adds to the nitrate load that reaches the ground.

The overuse of fertilizers *and* automobiles presents an increasingly serious water pollution problem in many areas. Leaching of the land by rain causes nitrate from fertilizer use and vehicle operation to enter lakes, rivers, and coastal estuaries and leads to *eutrophication,* the depletion of O_2 and the death of aquatic animal life from excessive algal and plant growth and decomposition. Excess nitrate also spoils nearby drinkable water.

Biological Fixation The biological fixation of atmospheric N_2 occurs in blue-green algae and in nitrogen-fixing bacteria that live on the roots of leguminous plants (such as peas, alfalfa, and clover). These microbial processes dwarf the previous two, fixing more than seven times as much nitrogen as the atmosphere and six times as much as industry. Root bacteria fix N_2 by enzymatically reducing it to NH_3 and NH_4^+. Enzymes in other soil bacteria catalyze the multistep oxidation of NH_4^+ to NO_2^- and finally NO_3^-, which the plants reduce again to make proteins. When the plants die, still other soil bacteria oxidize the proteins to NO_3^-.

Animals eat the plants, break down the plant proteins to make their own proteins, and excrete nitrogenous wastes, such as urea [$(H_2N)_2C{=}O$]. The nitrogen in the proteins is released when the animals die and decompose, and is converted by soil bacteria to NO_2^- and NO_3^- again. This central pool of inorganic nitrate has three main fates: some enters marine and terrestrial plants, some enters the enormous sediment store of mineral nitrates, and some is reduced by denitrifying bacteria to NO_2^- and then reduced further to N_2O and N_2, which re-enter the atmosphere to complete the cycle.

The Phosphorus Cycle

Virtually all the mineral sources of phosphorus contain the phosphate group, PO_4^{3-}. The most commercially important ores are **apatites,** compounds of general formula $Ca_5(PO_4)_3X$, where X is usually F, Cl, or OH. The cycling of phosphorus through the environment involves three interlocking subcycles (Figure 22.7, *next page*). Two rapid biological cycles—a land-based cycle *(yellow arrows in figure)* completed in a matter of years and a water-based cycle *(blue arrows)* completed in weeks to years—are superimposed on an inorganic cycle *(pink arrows)* that takes millions of years to complete. Unlike the carbon and nitrogen cycles, the phosphorus cycle has no gaseous component and thus does *not* involve the atmosphere.

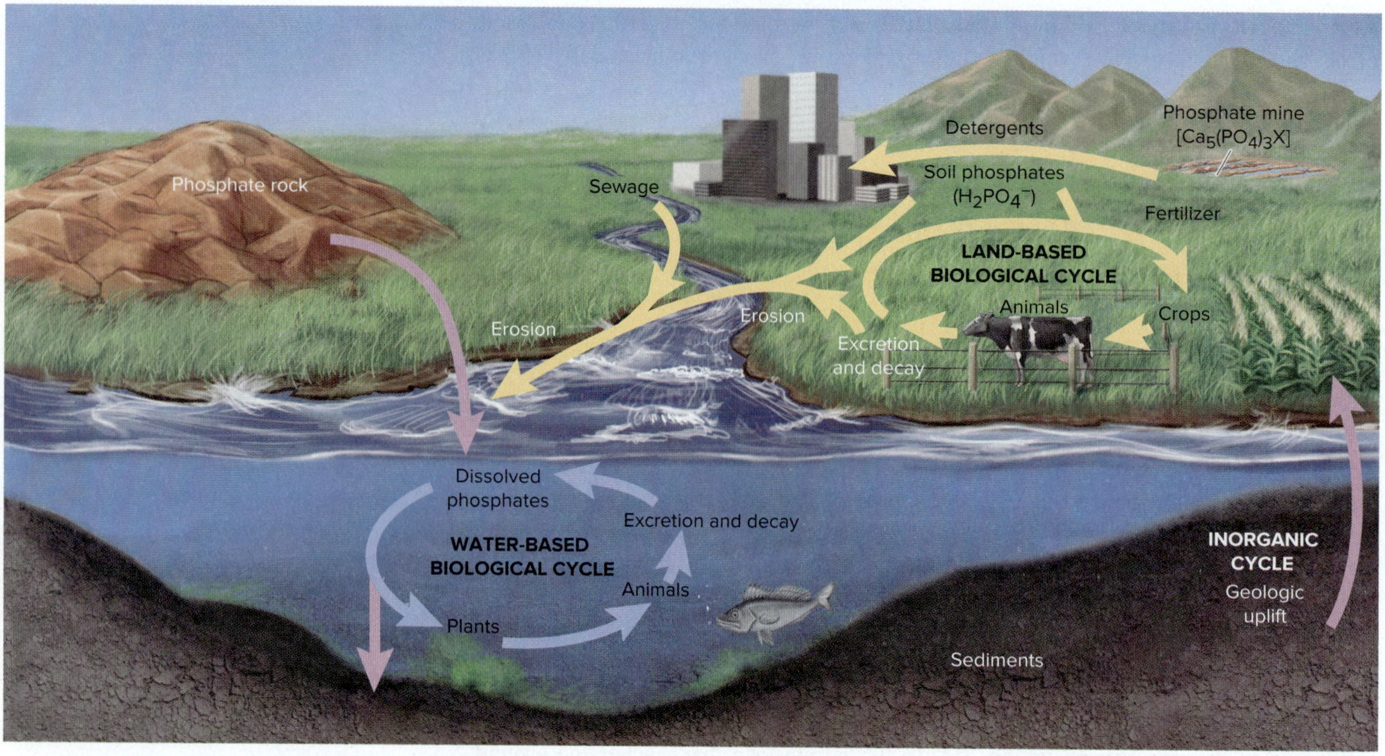

Figure 22.7 The phosphorus cycle.

The Inorganic Cycle Although most phosphorus occurs in phosphate rock formed when Earth's lithosphere solidified, a sizeable amount has come from outer space in meteorites. About 100 metric tons of meteorites enter Earth's atmosphere every day, and with an average P content of 0.1% by mass, they have contributed about 10^{11} metric tons of phosphorus to the crust over Earth's lifetime. The most important components of phosphate rock are nearly insoluble phosphate salts:

$$K_{sp} \text{ of } Ca_3(PO_4)_2 \approx 10^{-29}$$
$$K_{sp} \text{ of } Ca_5(PO_4)_3OH \approx 10^{-51}$$
$$K_{sp} \text{ of } Ca_5(PO_4)_3F \approx 10^{-60}$$

Weathering slowly leaches phosphates from the soil and carries the ions through rivers to the sea. Plants in the land-based biological cycle speed up this process. In the ocean, some phosphate is absorbed by organisms in the water-based biological cycle, but the majority is precipitated again by Ca^{2+} ion and deposited on the continental shelf. Geologic activity lifts the continental shelves, returning the phosphate to the land.

The Land-Based Biological Cycle The biological cycles involve the incorporation of phosphate into organisms (biomolecules, bones, teeth, and so forth) and the release of phosphate through their excretion and decomposition. In the land-based cycle, plants continually remove phosphate from the inorganic cycle. Recall that there are three phosphate oxoanions, which exist in equilibrium in aqueous solution:

$$H_2PO_4^- \rightleftharpoons \underset{+\,H^+}{HPO_4^{2-}} \rightleftharpoons \underset{+\,H^+}{PO_4^{3-}}$$

In topsoil, phosphates occur as insoluble compounds of Ca^{2+}, Fe^{3+}, and Al^{3+}. Because they can absorb only the water-soluble dihydrogen phosphate, plants have evolved the ability to secrete acids near their roots to convert the insoluble salts gradually into the soluble ion:

$$Ca_3(PO_4)_2(s;\text{ soil}) + 4H^+(aq;\text{ secreted by plants}) \longrightarrow 3Ca^{2+}(aq) + 2H_2PO_4^-(aq;\text{ absorbed by plants})$$

Animals that eat the plants excrete soluble phosphate, which is used by newly growing plants. As plants and animals excrete, die, and decompose, some phosphate is washed into rivers and the oceans. Most of the 2 million metric tons of phosphate that washes from land to sea each year comes from this biological source. Thus, the biosphere greatly increases the movement of phosphate from lithosphere to hydrosphere.

The Water-Based Biological Cycle Various phosphate oxoanions continually enter the aquatic environment. Studies that incorporate trace amounts of radioactive phosphorus into $H_2PO_4^-$ and monitor its uptake show that, within 1 minute, 50% of the ion is taken up by photosynthetic algae, which use it to synthesize their biomolecules (denoted by the top curved arrow in the equation, which shows the overall process):

$$106CO_2(g) + 16NO_3^-(aq) + H_2PO_4^-(aq) + 122H_2O(l) + 17H^+(aq) \underset{\text{decomposition}}{\overset{\text{synthesis}}{\rightleftharpoons}} C_{106}H_{263}O_{110}N_{16}P(aq;\ \text{in algal cell fluid}) + 138O_2(g)$$

(The complex formula on the right represents the total composition of algal biomolecules, not some particular compound.)

As on land, animals eat the plants and are eaten by other animals, and they all excrete phosphate, die, and decompose (denoted by the bottom curved arrow in the preceding equation). Some of the released phosphate is used by other aquatic organisms, and some returns to the land when fish-eating animals (mostly humans and birds) excrete phosphate, die, and decompose. In addition, some aquatic phosphate precipitates with Ca^{2+}, sinks to the seabed, and returns to the long-term mineral deposits of the inorganic cycle.

Human Effects on Phosphorus Movement From prehistoric through preindustrial times, these interlocking phosphorus cycles were balanced. Modern human activity, however, alters the movement of phosphorus considerably. Annually, we use more than 100 million tons of phosphate rock. Figure 22.8 shows the major end products, with box height proportional to the mass % of phosphate rock used annually. Removal of phosphate rock has minimal influence on the inorganic cycle, but the end products have unbalanced the two biological cycles.

The imbalance arises from our overuse of *soluble phosphate fertilizers,* such as $Ca(H_2PO_4)_2$ and $NH_4H_2PO_4$, the end products from about 85% of the phosphate rock

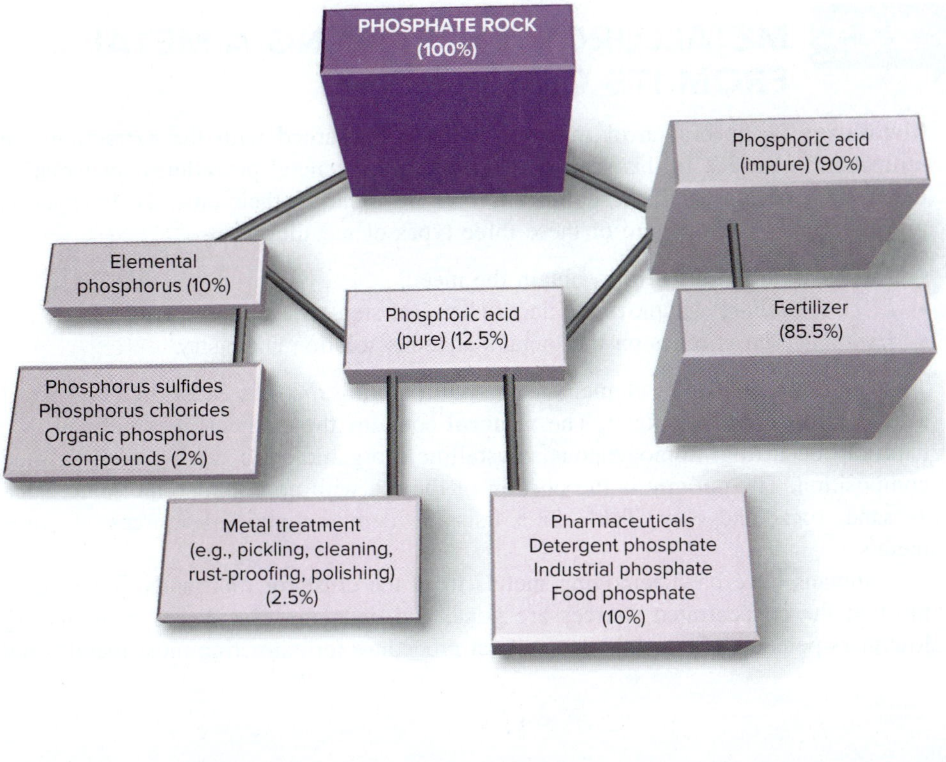

Figure 22.8 Industrial uses of phosphorus.

mined. Some fertilizer finds its way into rivers, lakes, and oceans to enter the water-based cycle. Much larger pollution sources, however, are the crops grown with the fertilizer and the detergents made from phosphate rock. The great majority of this phosphate arrives eventually in cities as crops, as animals that were fed crops, and as consumer products, such as the tripolyphosphates in detergents. Human garbage, excrement, wash water with detergents, and industrial wastewater containing phosphate are carried through sewers to the aquatic system. This human contribution equals the natural contribution—another 2 million metric tons of phosphate per year. The increased concentration in rivers and lakes causes eutrophication, which robs the water of O_2 so that it cannot support life. Such "dead" rivers and lakes are no longer usable for fishing, drinking, or recreation.

A century ago, in 1912, Lake Zurich in Switzerland had been choked with algae and become devoid of fish. Treatment with $FeCl_3$ precipitated enough phosphate to gradually return the lake to its natural state. Soluble aluminum compounds have the same effect:

$$FeCl_3(aq) + PO_4^{3-}(aq) \longrightarrow FePO_4(s) + 3Cl^-(aq)$$

$$KAl(SO_4)_2 \cdot 12H_2O(aq) + PO_4^{3-}(aq) \longrightarrow AlPO_4(s) + K^+(aq) + 2SO_4^{2-}(aq) + 12H_2O(l)$$

› Summary of Section 22.2

- The environmental distribution of many elements changes cyclically with time and is affected in major ways by organisms.
- Carbon occurs in all three regions of Earth's crust, with atmospheric CO_2 linking the other two regions. Photosynthesis and decomposition of organisms alter the amount of carbon in the land and oceans. Human activity has increased atmospheric CO_2, leading to climate change.
- Nitrogen is fixed by lightning, by industry, and primarily by microorganisms. When plants or animals decompose, bacteria convert proteins to nitrites and nitrates and eventually to N_2, which returns to the atmosphere. Through extensive use of fertilizers, humans have added excess nitrogen to freshwaters.
- The phosphorus cycle has no gaseous component. Inorganic phosphates leach slowly into land and water, where biological cycles interact. When plants and animals decompose, they release phosphate to natural waters, where it is then absorbed by other plants and animals. Through overuse of phosphate fertilizers and detergents, humans double the amount of phosphorus entering the aqueous environment.

22.3 METALLURGY: EXTRACTING A METAL FROM ITS ORE

Metallurgy is the branch of materials science concerned with the extraction and utilization of metals. In this section, we discuss the general procedures for isolating metals and, occasionally, the nonmetals that are found in their ores. The extraction process applies one or more of these three types of metallurgy:

- *Pyrometallurgy* uses heat to obtain the metal.
- *Electrometallurgy* employs an electrochemical step.
- *Hydrometallurgy* relies on the metal's aqueous solution chemistry.

The extraction of an element begins with *mining the ore.* Most ores consist of a mineral attached to gangue. The **mineral** contains the element; it is defined as a naturally occurring, homogeneous, crystalline inorganic solid, with a well-defined composition. The **gangue** is the portion of the ore with no commercial value, such as sand, rock, and clay. Table 22.3 lists the common mineral sources of some metals.

Humans have been removing metals from the crust for thousands of years, so most of the concentrated sources are gone, and, in many cases, ores contain very low mass percents of a metal. The general procedure for extracting most metals (and

Table 22.3 Common Mineral Sources of Some Elements

Element	Mineral, Formula
Al	Gibbsite (in bauxite), $Al(OH)_3$
Ba	Barite, $BaSO_4$
Be	Beryl, $Be_3Al_2Si_6O_{18}$
Ca	Limestone, $CaCO_3$
Fe	Hematite, Fe_2O_3
Hg	Cinnabar, HgS
Na	Halite, NaCl
Pb	Galena, PbS
Sn	Cassiterite, SnO_2
Zn	Sphalerite, ZnS

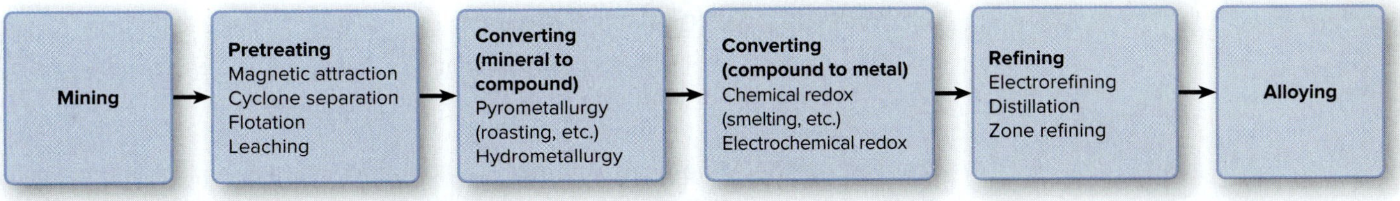

Figure 22.9 Steps in metallurgy.

some nonmetals) involves a few basic steps, described in the upcoming subsections (Figure 22.9).

Pretreating the Ore

Following a crushing, grinding, or pulverizing step, which can be very expensive, pretreatment uses some physical or chemical difference to separate mineral from gangue. The following are common pretreatment techniques:

- *Magnetic attraction.* For magnetic minerals, such as magnetite (Fe_3O_4), a magnet can remove the mineral and leave the gangue behind.
- *Density separation.* When large density differences exist, a *cyclone separator* blows high-pressure air through the pulverized mixture to separate the particles. The lighter silicate-rich gangue is blown away, while the denser mineral-rich particles hit the walls of the separator and fall through the open bottom (Figure 22.10).
- *Flotation.* In **flotation,** an oil-detergent mixture is stirred with the pulverized ore in water to form a slurry (Figure 22.11). The surfaces of mineral and of gangue become wet to different extents with water and detergent. Rapid mixing with air produces an oily, mineral-rich froth that floats, while the silicate particles sink. Skimming, followed by solvent removal, isolates the concentrated mineral fraction.
- *Leaching.* Some metals are extracted by a hydrometallurgical process known as **leaching,** usually accomplished via formation of a complex ion. The modern extraction of gold is a good example of this technique. The crushed ore, which often contains as little as 25 ppm of gold, is contained in a plastic-lined pool, treated with a cyanide ion solution, and aerated. In the presence of CN^-, the O_2 in air oxidizes gold metal to gold(I) ion, Au^+, which forms the soluble complex ion, $Au(CN)_2^-$:

$$4Au(s) + O_2(g) + 8CN^-(aq) + 2H_2O(l) \longrightarrow 4Au(CN)_2^-(aq) + 4OH^-(aq)$$

  (This method is controversial because cyanide enters streams and lakes, where it poisons fish and birds.)

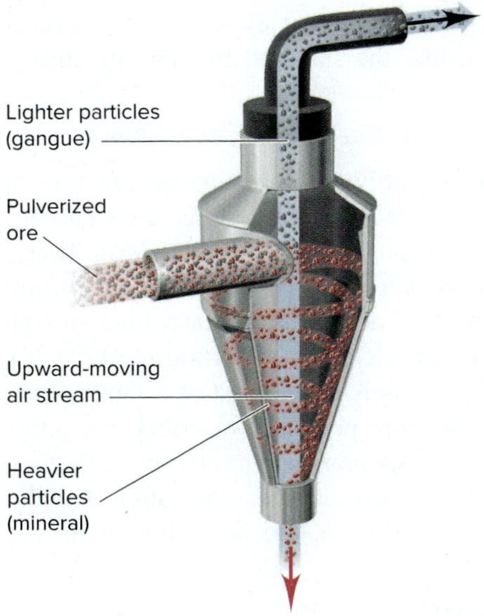

Figure 22.10 The cyclone separator.

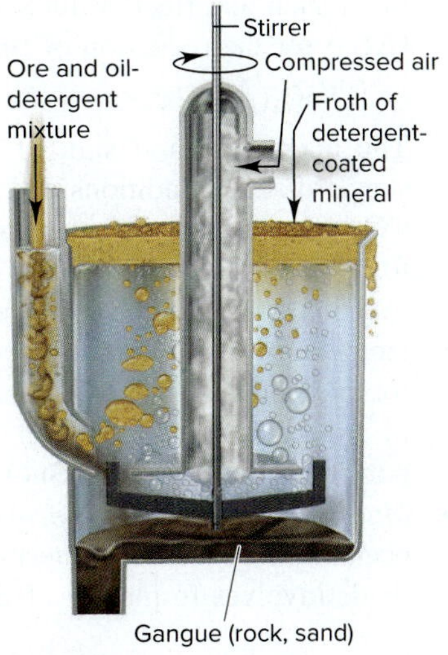

Figure 22.11 The flotation process.

Converting Mineral to Element

After the mineral has been freed of debris and concentrated, it may undergo several chemical steps during its conversion to the element.

Converting the Mineral to Another Compound First, the mineral is often converted to another compound, one that has different solubility properties, is easier to reduce, or is free of an impurity. Conversion to an oxide is common because oxides can be reduced easily. Carbonates are heated to convert them to the oxide:

$$CaCO_3(s) \xrightarrow{\Delta} CaO(s) + CO_2(g)$$

Metal sulfides, such as ZnS, can be converted to oxides by **roasting** in air:

$$2\,ZnS(s) + 3O_2(g) \xrightarrow{\Delta} 2\,ZnO(s) + 2SO_2(g)$$

In most countries, hydrometallurgical methods are now used to avoid the atmospheric release of SO_2 from roasting. One way of processing copper, for example, is by bubbling air through an acidic slurry of insoluble Cu_2S, which completes both the pretreatment and conversion steps:

$$2Cu_2S(s) + 5O_2(g) + 4H^+(aq) \longrightarrow 4Cu^{2+}(aq) + 2SO_4^{2-}(aq) + 2H_2O(l)$$

Converting Compound to Element Through Chemical Redox The next step converts the new mineral form (usually an oxide) to the free element by either chemical or electrochemical redox methods. Let's consider the chemical methods first. In these processes, a reducing agent reacts directly with the compound. The most common reducing agents are carbon, hydrogen, and a more active metal.

1. *Reduction with carbon.* Because of its low cost and ready availability, carbon, in the form of coke (a porous material obtained from incomplete combustion of coal) or charcoal, is a common reducing agent. Heating an oxide with a reducing agent such as coke to obtain the metal is called **smelting.** Many metal oxides, such as zinc oxide and tin(IV) oxide, are smelted with carbon to free the metal, which then may need to be condensed and solidified (note the physical states of the metal products):

$$ZnO(s) + C(s) \longrightarrow Zn(g) + CO(g)$$
$$SnO_2(s) + 2C(s) \longrightarrow Sn(l) + 2CO(g)$$

Several nonmetals that occur with positive oxidation states in minerals can be reduced with carbon as well. Phosphorus, for example, is produced from calcium phosphate:

$$2Ca_3(PO_4)_2(s) + 10C(s) + 6SiO_2(s) \longrightarrow 6CaSiO_3(s) + 10CO(g) + P_4(s)$$

(Metallic calcium is a much stronger reducing agent than carbon, so it is not formed.)

- *Thermodynamic considerations.* Thermodynamic principles explain why carbon is often such an effective reducing agent. Consider the standard free energy change (ΔG°) for the reduction of tin(IV) oxide:

  $$SnO_2(s) + 2C(s) \longrightarrow Sn(s) + 2CO(g) \qquad \Delta G^\circ = 245 \text{ kJ at } 25^\circ\text{C (298 K)}$$

  The magnitude and sign of ΔG° indicate a highly *non*spontaneous reaction at standard-state conditions and 298 K. However, because solid C becomes gaseous CO, the standard molar entropy change is very positive ($\Delta S^\circ \approx 380$ J/K). Therefore, *the $-T\Delta S^\circ$ term of ΔG° becomes more negative with higher temperature* (Section 20.3). As the temperature increases, ΔG° decreases, and at a high enough temperature, the reaction becomes spontaneous ($\Delta G^\circ < 0$). At 1000°C (1273 K), for example, $\Delta G^\circ = -62.8$ kJ. The actual process is carried out at a temperature of around 1250°C (1523 K), so ΔG° is even more negative. In effect, the nonspontaneous reduction of SnO_2 is coupled to the spontaneous oxidation of C.
- *Multistep nature of the process.* As you know, overall reactions often actually occur through several intermediate steps. For instance, the reduction of tin(IV) oxide involves formation of tin(II) oxide:

  $$SnO_2(s) + C(s) \longrightarrow \cancel{SnO(s)} + CO(g)$$
  $$\cancel{SnO(s)} + C(s) \longrightarrow Sn(l) + CO(g)$$

[Molten tin (mp = 232 K) is obtained at this temperature.] The second step is thought to be composed of others, in which CO, not C, is the actual reducing agent:

$$SnO(s) + \cancel{CO(g)} \longrightarrow Sn(l) + \cancel{CO_2(g)}$$
$$\cancel{CO_2(g)} + C(s) \rightleftharpoons \cancel{2}CO(g)$$

Other pyrometallurgical processes are similarly complex. For instance, shortly you'll see that the overall reaction for the smelting of iron is

$$2Fe_2O_3(s) + 3C(s) \longrightarrow 4Fe(l) + 3CO_2(g)$$

However, the actual process is a multistep one with CO as the reducing agent. The advantage of CO over C is the far greater physical contact the gaseous reducing agent can make with the other reactant, which speeds the process.

2. *Reduction with hydrogen.* For oxides of some metals, especially some members of Groups 6B(6) and 7B(7), reduction with carbon forms metal carbides that are difficult to convert further, so hydrogen is used as the reducing agent instead, especially for less active metals, such as tungsten:

$$WO_3(s) + 3H_2(g) \longrightarrow W(s) + 3H_2O(g)$$

One step in the purification of the metalloid germanium also uses hydrogen:

$$GeO_2(s) + 2H_2(g) \longrightarrow Ge(s) + 2H_2O(g)$$

3. *Reduction with a more active metal.* If a metal oxide forms both an undesirable carbide and an undesirable hydride, it is reduced by a more active metal. In the *thermite reaction,* aluminum powder reduces the metal oxide in a spectacular exothermic reaction to give the molten metal (Figure 22.12). The reaction for chromium is

$$Cr_2O_3(s) + 2Al(s) \longrightarrow 2Cr(l) + Al_2O_3(s) \qquad \Delta H^\circ << 0$$

Reduction by an active metal is also used when the mineral of the element is not an oxide. In the extraction of gold, after the pretreatment by leaching, the gold(I) complex ion is reduced with zinc:

$$2Au(CN)_2^-(aq) + Zn(s) \longrightarrow 2Au(s) + Zn(CN)_4^{2-}(aq)$$

In some cases, an active metal, such as calcium, is used to recover an even more active metal, such as rubidium, from its molten salt:

$$Ca(l) + 2RbCl(l) \longrightarrow CaCl_2(l) + 2Rb(g)$$

A similar process (detailed in Section 22.4) is used to recover sodium.

4. *Oxidation to obtain a nonmetal.* Just as *reduction* of a mineral is used to obtain the metal, *oxidation* of a mineral can be used to obtain a nonmetal. A stronger oxidizing agent removes electrons from the nonmetal anion to give the free nonmetal, as in the production of iodine from concentrated brines:

$$2I^-(aq) + Cl_2(g) \longrightarrow 2Cl^-(aq) + I_2(s)$$

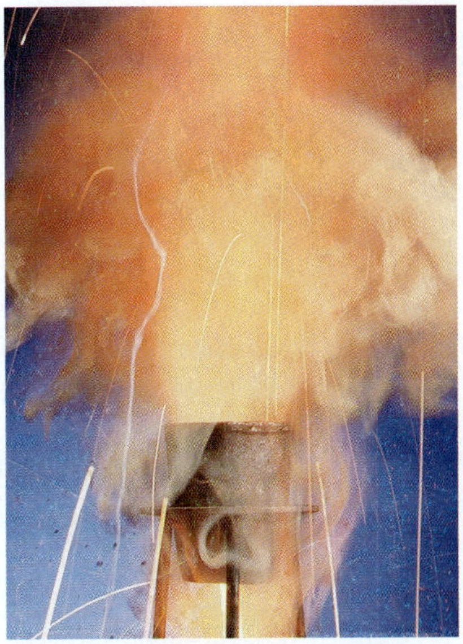

Figure 22.12 The thermite reaction.
Source: © Richard Megna/Fundamental Photographs, NYC

Converting Compound to Element Through Electrochemical Redox In electrochemical redox processes, the minerals are converted to the elements in an electrolytic cell (Section 21.7). Sometimes, the *pure* mineral, in the form of the molten halide or oxide, is used to prevent unwanted side reactions. The cation is reduced to the metal at the cathode, and the anion is oxidized to the nonmetal at the anode:

$$BeCl_2(l) \longrightarrow Be(s) + Cl_2(g)$$

High-purity hydrogen gas is prepared by electrochemical reduction:

$$2H_2O(l) \longrightarrow 2H_2(g) + O_2(g)$$

Specially designed cells separate the products to prevent recombination. An inexpensive source of electricity is used for large-scale methods. The current and voltage requirements depend on the electrochemical potential and any necessary overvoltage (Section 21.7).

Figure 22.13 on the next page shows the most common redox step used in obtaining each free element from its mineral.

Figure 22.13 The redox step in converting a mineral to the element.

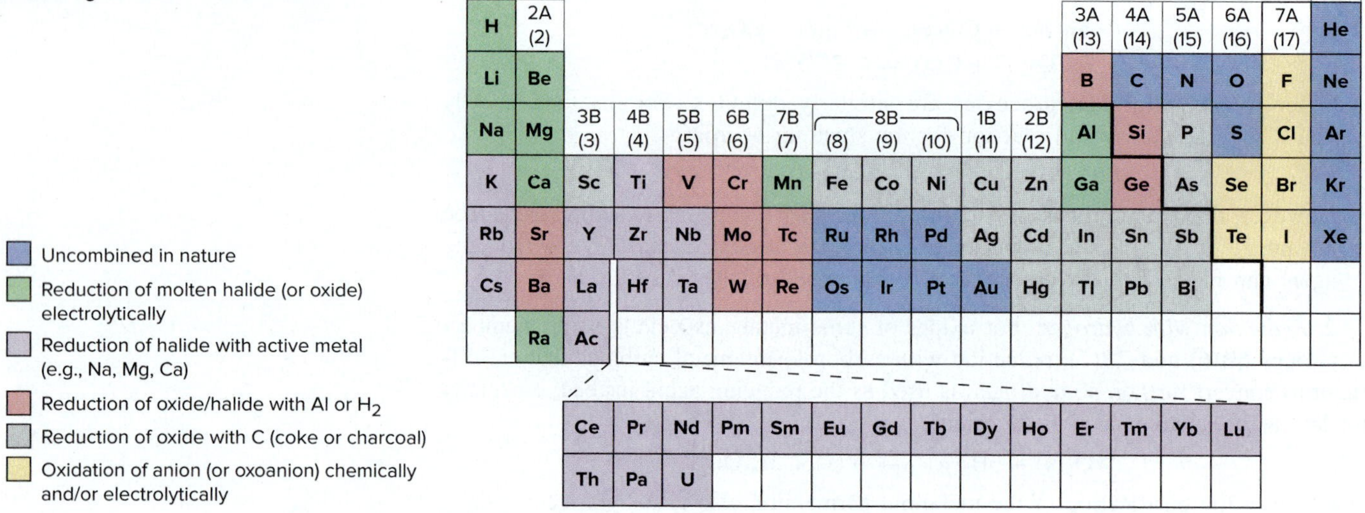

Refining and Alloying the Element

After being isolated, the element typically still contains impurities, so it must be refined to purify it further. Then, metallic elements are often alloyed to improve their properties.

Refining (Purifying) the Element Refining is often carried out by one of three common methods:

1. *Electrorefining.* An electrolytic cell is employed in **electrorefining,** in which the *impure metal acts as the anode and the pure metal as the cathode.* As the reaction proceeds, the anode disintegrates, and the metal ions are reduced to deposit on the cathode. In some cases (as with copper, discussed shortly), impurities that fall beneath the anode are the source of several valuable and less abundant elements.
2. *Distillation.* Metals with relatively low boiling points, such as zinc and mercury, are refined by *distillation.*
3. *Zone refining.* In the process of **zone refining,** a rod of an impure metal or metalloid (such as silicon) is passed slowly through a heating coil in an inert atmosphere. The first narrow zone of the impure solid melts, and as the next zone melts, the dissolved impurites from the first zone lower the freezing point, so the solvent (purer solid of first zone) refreezes. The process continues, zone by zone, for the entire rod. After several passes through the coil, in which each zone's impurities move into the adjacent zone, the refrozen solid at the end of the rod becomes extremely pure (Figure 22.14). Metalloids used in electronic semiconductors, such as silicon and germanium, are zone refined to greater than 99.999999% purity.

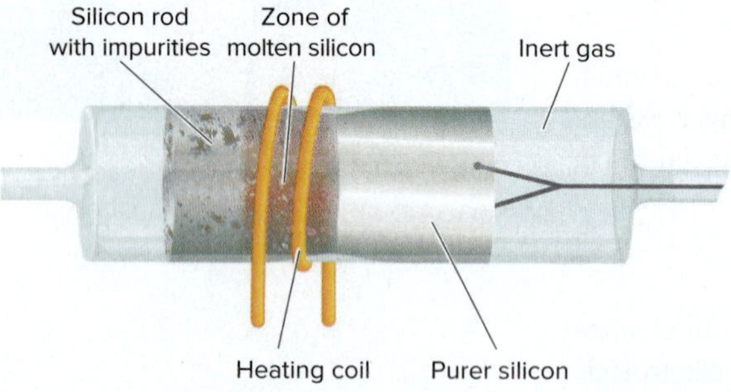

Figure 22.14 Zone refining of silicon.

Alloying the Purified Element An **alloy** is a metal-like mixture consisting of solid phases of two or more pure elements (a solid solution) or, in some cases, distinct intermediate phases (these alloys are sometimes referred to as *intermetallic compounds*) that are so finely divided they can only be distinguished microscopically. Alloying a metal with other metals (and, in some cases, nonmetals) is done to alter the metal's melting point and to enhance properties such as luster, conductivity, malleability, ductility, and strength.

Iron is used only when alloyed. In pure form, it is soft and corrodes easily. However, when alloyed with carbon and other metals, such as Mo for hardness and Cr and Ni for corrosion resistance, it forms various steels. Copper stiffens when zinc is added to it to make brass. Mercury solidifies when alloyed with sodium, and

Table 22.4 Some Important Alloys and Their Composition

Name	Composition (Mass %)	Uses
Mostly Fe		
Stainless steel	73–79 Fe, 14–18 Cr, 7–9 Ni	Cutlery, instruments
Nickel steel	96–98 Fe, 2–4 Ni	Cables, gears
High-speed steels	80–94 Fe, 14–20 W (or 6–12 Mo)	Cutting tools
Mostly Cu		
Bronzes	70–95 Cu, 1–25 Zn, 1–18 Sn	Statues, castings
Brasses	50–80 Cu, 20–50 Zn	Plating, ornamental objects
Mostly precious metals		
Sterling silver	92.5 Ag, 7.5 Cu	Jewelry, tableware
14-carat gold	58 Au, 4–28 Ag, 14–28 Cu	Jewelry
18-carat white gold	75 Au, 12.5 Ag, 12.5 Cu	Jewelry
Dental amalgam	69 Ag, 18 Sn, 12 Cu, 1 Zn (dissolved in Hg)	Dental fillings
Other		
Permalloy	78 Ni, 22 Fe	Ocean cables
Typical tin solder	67 Pb, 33 Sn	Electrical connections

vanadium becomes extremely tough with some carbon added. Table 22.4 shows the composition and uses of some common alloys.

The simplest alloys, called *binary alloys,* contain only two elements, which can combine in two ways:

- *The added element enters interstices,* spaces between the parent metal atoms in its crystal structure (Section 12.6). In vanadium carbide, for instance, carbon atoms occupy holes of a face-centered cubic vanadium unit cell (Figure 22.15A).
- The *added element substitutes for atoms of the parent.* In many types of brass, for example, relatively few zinc atoms substitute randomly for copper atoms in the face-centered cubic copper unit cell (see Figure 13.4A). On the other hand, β-brass is an intermediate phase that crystallizes as the Zn/Cu ratio approaches 1/1 in a body-centered cubic structure with a central Zn atom surrounded by eight Cu atoms (Figure 22.15B). In one alloy of copper and gold, Cu atoms occupy the faces of a face-centered cube, and Au atoms lie at the corners (Figure 22.15C).

Atomic properties, including size, electron configuration, and number of valence electrons, determine which metals form stable alloys. For example, transition metals from the left half of the *d* block [Groups 3B(3) to 5B(5)] often form alloys with metals from the right half [Groups 8B(9) to 1B(11)]: electron-poor Zr, Nb, and Ta form strong alloys with electron-rich Ir, Pt, and Au; an example is the very stable $ZrPt_3$.

Figure 22.15 Three binary alloys.

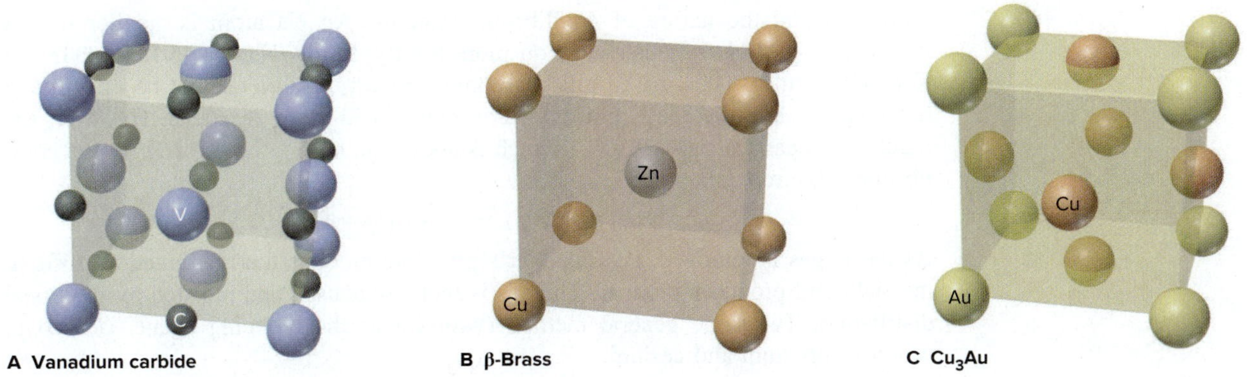

› Summary of Section 22.3

› Metallurgy involves mining an ore, separating it from debris, pretreating it to concentrate the mineral source, converting the mineral to another compound that is easier to process further, reducing this compound to the metal, purifying the metal, and in many cases, alloying it to obtain a more useful material.

22.4 TAPPING THE CRUST: ISOLATION AND USES OF SELECTED ELEMENTS

The isolation process depends on the physical and chemical properties of the source: we isolate an element that occurs uncombined in the air differently from one dissolved in the sea or one found in a rocky ore. In this section, we detail methods for recovering some important elements.

Producing the Alkali Metals: Sodium and Potassium

The alkali metals are among the most reactive elements and thus are always found as ions in nature, either in solid minerals or in aqueous solution. The two most important of these metals are sodium and potassium. Their abundant, water-soluble compounds are used throughout industry and research, and the Na^+ and K^+ ions are essential to organisms.

Industrial Production of Sodium and Potassium Different methods are employed in the industrial production of these two metals:

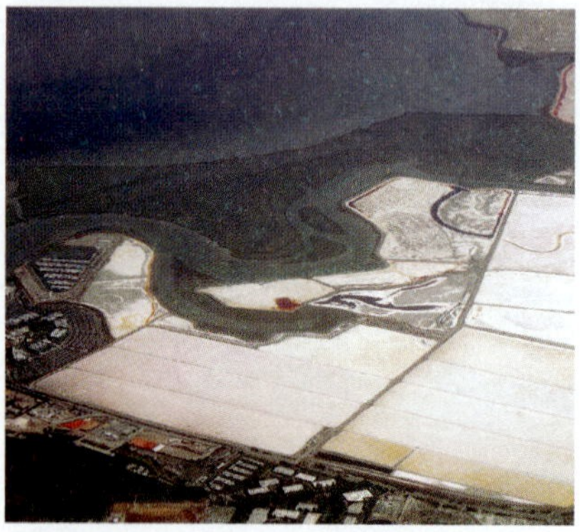

Evaporative salt beds near Redwood City, California.

Source: © Natasha Kramskaya/Shutterstock.com

1. *Production of sodium.* The sodium ore is *halite* (largely NaCl), which is obtained either by evaporation of concentrated salt solutions called *brines (see photo)* or by mining vast salt deposits formed from the evaporation of prehistoric seas. The Cheshire salt field in Britain, for example, is 60 km by 24 km by 400 m thick and contains more than 90% NaCl. Other large deposits occur in New Mexico, Michigan, New York, and Kansas.

The brine is evaporated and the solid crushed and fused (melted) for use in an electrolytic apparatus called the **Downs cell** (Figure 22.16). To reduce heating costs, 1 part NaCl (mp = 801°C) is mixed with $1\frac{1}{2}$ parts $CaCl_2$ to form a mixture that melts at only 580°C. Reduction of the metal ions to Na and Ca takes place at a cylindrical steel cathode, with the molten metals floating on the denser molten salt mixture. As they rise through a short collecting pipe, the liquid Na is siphoned off, while a higher melting Na/Ca alloy solidifies and falls back into the molten electrolyte. Chloride ions are oxidized to Cl_2 gas at a large anode within an inverted cone-shaped chamber that separates the metals from the Cl_2 to prevent their explosive recombination. The Cl_2 is purified and sold as a valuable byproduct.

2. *Production of potassium.* Sylvite (mostly KCl) is the major ore of potassium. The metal is too soluble in molten KCl to be obtained by a method similar to that used for sodium. Instead, chemical reduction of K^+ ions by Na is based on atomic properties and the nature of equilibrium systems. An Na atom is smaller than a K atom, so it holds its outer electron more tightly: IE_1 of Na = 496 kJ/mol; IE_1 of K = 419 kJ/mol. Thus, based on this atomic property, Na would not be effective at reducing K^+ because the equilibrium constant would favor reactants. However, the reduction is carried out at 850°C, which is above the boiling point of K, so the equilibrium mixture contains *gaseous* K:

$$Na(l) + K^+(l) \rightleftharpoons Na^+(l) + K(g)$$

As the K gas is removed, Le Châtelier's principle predicts that the reaction shifts to the right and produces more K. The gas is then condensed and purified by fractional distillation. The same general method (with Ca as the reducing agent) is used to produce rubidium and cesium.

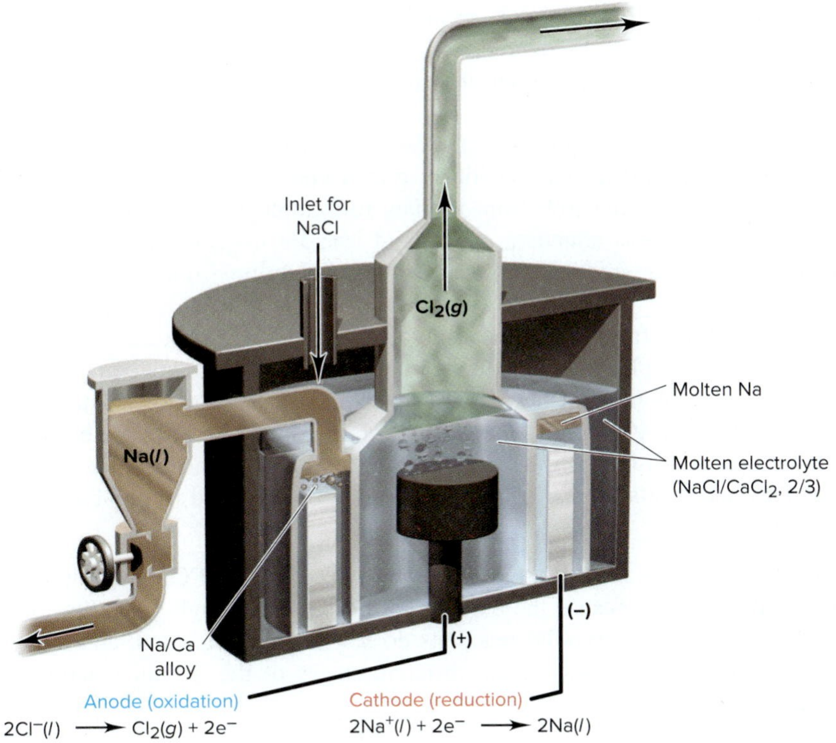

Figure 22.16 The Downs cell for production of sodium.

Uses of Sodium and Potassium The compounds of Na and K (and indeed, of all the alkali metals) have many more uses than the elements themselves. Nevertheless, there are some interesting uses of the metals that take advantage of their strong reducing power:

1. *Uses of sodium.* Large amounts of Na were used as an alloy with lead to make gasoline antiknock additives, such as tetraethyllead:

$$4C_2H_5Cl(g) + 4Na(s) + Pb(s) \longrightarrow (C_2H_5)_4Pb(l) + 4NaCl(s)$$

The toxic effects of environmental lead have limited this use to aviation fuel in the United States. Moreover, although leaded gasoline is still used in several developing countries, the global market for tetraethyllead is declining more than 15% annually.

If certain types of nuclear reactors, called *breeder reactors* (Chapter 24), become a practical way to generate energy in the United States, Na production would increase enormously. Its low melting point, viscosity, and interaction with neutrons, combined with its high thermal conductivity and heat capacity, make it perfect for cooling the reactor and exchanging heat to the steam generator.

2. *Uses of potassium.* The major use of potassium at present is in an alloy with sodium for use as a heat exchanger in chemical and nuclear reactors. Another application relies on production of its superoxide, which it forms by direct contact with O_2:

$$K(s) + O_2(g) \longrightarrow KO_2(s)$$

This material is employed as an emergency source of O_2 in breathing masks for miners, divers, submarine crews, and firefighters *(see photo):*

$$4KO_2(s) + 4CO_2(g) + 2H_2O(g) \longrightarrow 4KHCO_3(s) + 3O_2(g)$$

Firefighter with KO_2 breathing mask.
Source: © Bruce Ando/Photolibrary/Getty Images

The Indispensable Three: Iron, Copper, and Aluminum

Their innumerable applications make iron (in the form of steel), copper, and aluminum indispensable metals in an industrial society.

Metallurgy of Iron: From Ore to Pig Iron Although people have practiced iron smelting for more than 3000 years, it was not even 250 years ago that iron assumed

Table 22.5 Important Minerals of Iron

Mineral Type	Mineral, Formula
Oxide	Hematite, Fe_2O_3 Magnetite, Fe_3O_4 Ilmenite, $FeTiO_3$
Carbonate	Siderite, $FeCO_3$
Sulfide	Pyrite, FeS_2 Pyrrhotite, FeS

its current dominant role. In 1773, an inexpensive process to convert coal to carbon in the form of coke was discovered, and the coke process made iron smelting cheap and efficient and led to large-scale iron production, which ushered in the Industrial Revolution.

Modern society rests, quite literally, on the various alloys of iron known as **steel.** Although steel production has obviously grown enormously since the 18th century, the process of recovering iron from its ores still employs *reduction by carbon in a blast furnace.* The most abundant minerals of iron are listed in Table 22.5. Only the first four are used for steelmaking, because traces of sulfur from the sulfide minerals make steel brittle.

The conversion from ore to metal involves a series of redox and acid-base reactions that appear simple, even though their detailed chemistry is not completely understood even today. A modern **blast furnace** (Figure 22.17) is a tower made of brick, about 14 m wide by 40 m high. The *charge,* which usually consists of hematite, coke, and limestone, is fed through the top. More coke is burned in air at the bottom. The charge falls and meets a *blast* of rapidly rising hot air created by the burning coke:

$$2C(s) + O_2(g) \longrightarrow 2CO(g) + \textbf{\textit{heat}}$$

At the bottom of the furnace, the temperature exceeds 2000°C, while at the top, it reaches only 200°C. Because the blast of hot air passes through the entire furnace in only 10 s, the various gas-solid reactions do *not* reach equilibrium, and many intermediate products form. As a result, different stages of the overall reaction process occur at different heights:

1. In the upper part of the furnace (200°C to 700°C), the charge is preheated, and a *partial reduction* occurs. The hematite (Fe_2O_3) is reduced to magnetite (Fe_3O_4) and then to iron(II) oxide (FeO) by CO, which is the actual reducing agent. Carbon dioxide is formed as well. Limestone ($CaCO_3$) also decomposes to form CO_2 and the basic oxide CaO.
2. Lower down, at 700°C to 1200°C, a *final reduction* step occurs, as some of the coke reduces CO_2 to form more CO, which reduces the FeO to Fe.

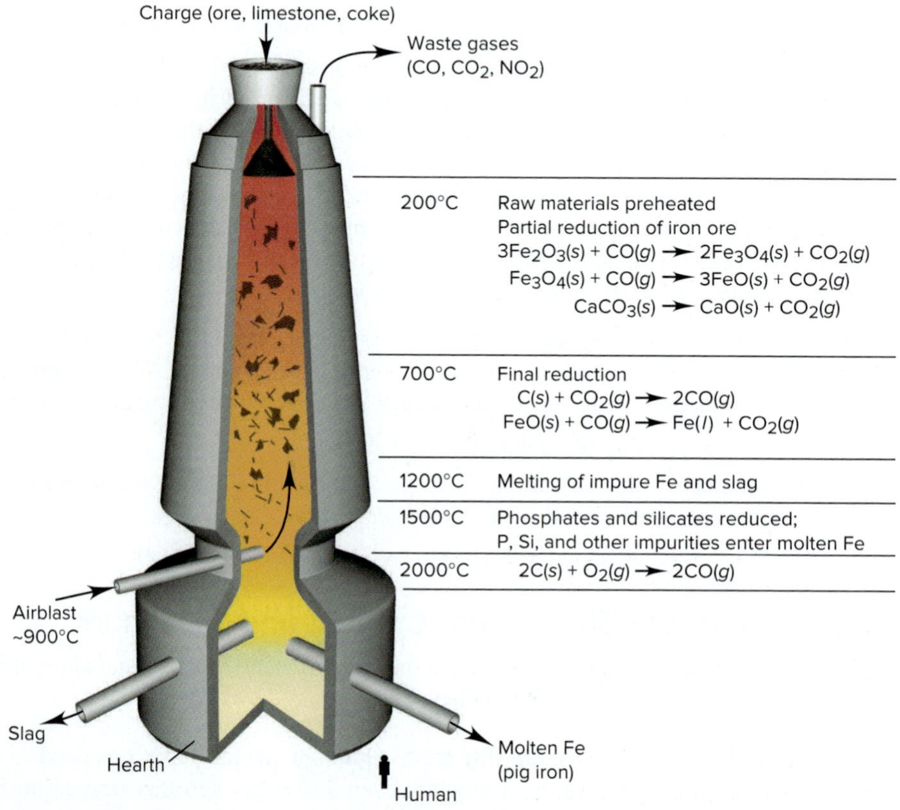

Figure 22.17 The major reactions in a blast furnace.

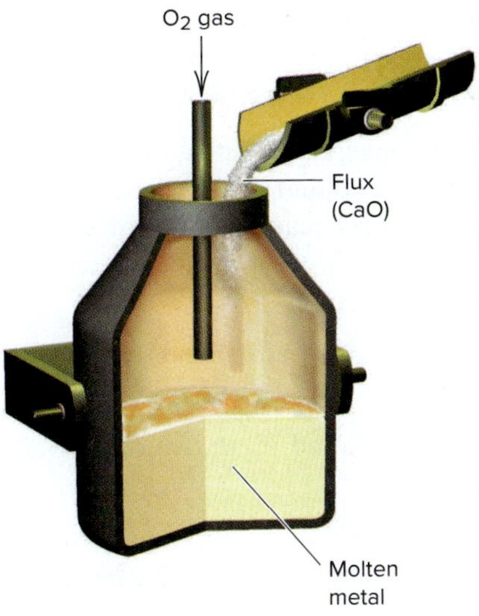

Figure 22.18 The basic-oxygen process for making steel.
Source: © Luis Veiga/Getty Images

3. At temperatures between 1200°C and 1500°C, the iron melts and drips to the bottom of the furnace. Acidic silica particles from the gangue react with the CaO in a Lewis acid-base reaction to form a molten waste product called **slag:**

$$CaO(s) + SiO_2(s) \longrightarrow CaSiO_3(l)$$

The slag drips down and floats on the denser iron.

4. Some unwanted reactions occur between 1500°C and 2000°C. Any remaining phosphates and silicates are reduced to P and Si, and some Mn and traces of S dissolve into the molten iron along with carbon. The resulting impure product is called *pig iron* and contains about 3% to 4% C. A small amount of pig iron is used to make *cast iron,* but most is purified and alloyed to make various kinds of steel.

Metallurgy of Iron: From Pig Iron to Steel Pig iron is converted to steel in a separate furnace by means of the **basic-oxygen process** (Figure 22.18). High-pressure O_2 is blown over and through the molten iron, so that impurities (C, Si, P, Mn, and S) are oxidized rapidly. The highly negative standard enthalpies of formation of the carbon and silicon oxides (ΔH_f° of CO_2 = −394 kJ/mol; ΔH_f° of SiO_2 = −911 kJ/mol) raise the temperature, which speeds the reaction. A lime (CaO) flux is added, which converts the oxides to a molten slag [primarily $CaSiO_3$ and $Ca_3(PO_4)_2$] that is decanted from the molten steel. The product is **carbon steel,** which contains 1% to 1.5% C and other impurities. It is alloyed with metals that prevent corrosion and increase its strength or flexibility.

Metallurgy of Copper: From Ore to Impure Metal After many centuries of being mined to make bronze and brass, copper ores have become much less plentiful and less rich in copper. Despite this, more than 2.5 billion pounds of copper is produced in the United States annually. The most common ore is chalcopyrite, $CuFeS_2$, a mixed sulfide of FeS and CuS *(see photo).* Most remaining deposits contain less than 0.5% Cu by mass. To "win" this small amount of copper requires several metallurgical steps:

Mining chalcopyrite.
Source: © Lee Prince/Shutterstock.com

1. *Flotation.* The low copper content in chalcopyrite must be enriched by removing the iron. The first step is pretreatment by flotation (see Figure 22.11), which concentrates the ore to around 15% Cu by mass.

2. *Roasting.* The next step in many processing plants is a controlled roasting step, which oxidizes the FeS but not the CuS:

$$2CuFeS_2(s) + 3O_2(g) \longrightarrow 2CuS(s) + 2FeO(s) + 2SO_2(g)$$

3. *Heating with sand.* To remove the FeO and convert the CuS to a more convenient form, the mixture is heated to 1100°C with sand and more of the concentrated ore. Several reactions occur in this step. The FeO reacts with sand to form a molten slag:

$$FeO(s) + SiO_2(s) \longrightarrow FeSiO_3(l)$$

The CuS is thermodynamically unstable at the elevated temperature and decomposes to yield Cu_2S, which is drawn off as a liquid.

4. *Smelting.* In the final smelting step, the Cu_2S is roasted in air, which converts some of it to Cu_2O:

$$2Cu_2S(s) + 3O_2(g) \longrightarrow 2Cu_2O(s) + 2SO_2(g)$$

The two copper(I) compounds then react, with sulfide ion acting as the reducing agent:

$$Cu_2S(s) + 2Cu_2O(s) \longrightarrow 6Cu(l) + SO_2(g)$$

The copper obtained at this stage is usable for plumbing, but it must be purified further for electrical wiring applications by removal of the unwanted Fe and Ni as well as the valuable byproducts Ag, Au, and Pt.

Metallurgy of Copper: Electrorefining Achieving the desired 99.99% purity needed for wiring (copper's most important use) is accomplished by *electrorefining,* which involves the oxidation of Cu to form Cu^{2+} ions in solution, followed by their reduction and the plating out of Cu metal (Figure 22.19). The impure copper obtained from smelting is cast into plates to be used as anodes, and cathodes are made from already purified copper. The electrodes are immersed in acidified $CuSO_4$ solution, and a voltage is applied that accomplishes two tasks simultaneously:

- Copper and the more active impurities (Fe, Ni) are oxidized to their cations, while the less active ones (Ag, Au, Pt) are not. As the anode slabs react, these unoxidized metals fall off as valuable "anode mud" and are purified separately. Sale of the precious metals in the anode mud nearly offsets the cost of electricity to operate the cell, making Cu wire inexpensive.
- Because Cu is much less active than the Fe and Ni impurities, Cu^{2+} ions are reduced at the cathode, but Fe^{2+} and Ni^{2+} ions remain in solution:

$$Cu^{2+}(aq) + 2e^- \longrightarrow Cu(s) \qquad E° = 0.34\text{ V}$$
$$Ni^{2+}(aq) + 2e^- \longrightarrow Ni(s) \qquad E° = -0.25\text{ V}$$
$$Fe^{2+}(aq) + 2e^- \longrightarrow Fe(s) \qquad E° = -0.44\text{ V}$$

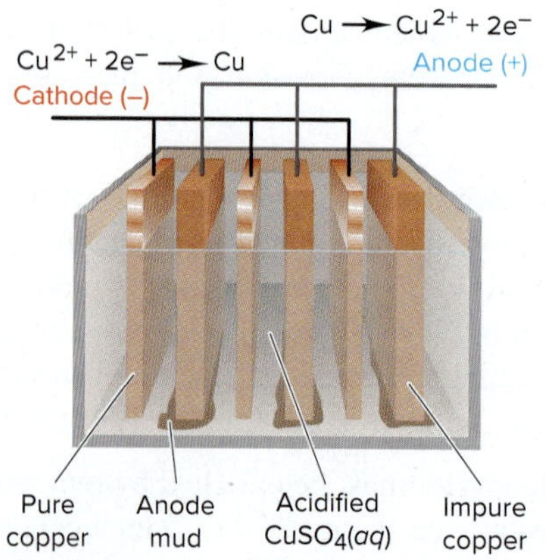

Figure 22.19 The electrorefining of copper.
Source: © Tom Hollyman/Science Source

Metallurgy of Aluminum: From Ore to Metal Aluminum is the most abundant metal in Earth's crust by mass and the third most abundant element (after O and Si). It is found in numerous aluminosilicate minerals, such as feldspars, micas, and clays, and in the rare gems garnet, beryl, spinel, and turquoise. Corundum, pure aluminum oxide (Al_2O_3), is extremely hard; mixed with traces of transition metals, it exists as ruby and sapphire. Impure Al_2O_3 is used in sandpaper and other abrasives.

Mining bauxite.
Source: © Howard Davies/Alamy

Through eons of weathering, certain clays became *bauxite,* the major ore of aluminum. This mixed oxide-hydroxide occurs in enormous surface deposits in Mediterranean and tropical regions *(see photo),* but with world aluminum production near 100 million tons annually, it may someday be scarce. In addition to hydrated Al_2O_3 (about 75%), industrial-grade bauxite also contains Fe_2O_3, SiO_2, and TiO_2, which are removed during the extraction. The overall two-step process combines hydro- and electrometallurgical techniques:

1. *Isolating Al_2O_3 from bauxite.* After mining, bauxite is pretreated by boiling in 30% NaOH in the *Bayer process,* which involves acid-base, solubility, and complex-ion equilibria. The acidic SiO_2 and the amphoteric Al_2O_3 dissolve in the base, but the basic Fe_2O_3 and TiO_2 do not:

$$SiO_2(s) + 2NaOH(aq) + 2H_2O(l) \longrightarrow Na_2Si(OH)_6(aq)$$
$$Al_2O_3(s) + 2NaOH(aq) + 3H_2O(l) \longrightarrow 2NaAl(OH)_4(aq)$$
$$Fe_2O_3(s) + NaOH(aq) \longrightarrow \text{no reaction}$$
$$TiO_2(s) + NaOH(aq) \longrightarrow \text{no reaction}$$

Further heating precipitates the $Na_2Si(OH)_6$ as an aluminosilicate, which is filtered out with the insoluble Fe_2O_3 and TiO_2 ("red mud").

Acidifying the filtrate precipitates Al^{3+} as $Al(OH)_3$. Recall that the aluminate ion, $Al(OH)_4^-$ *(aq),* is actually the complex ion $Al(H_2O)_2(OH)_4^-$ (see Figure 19.19). Weakly acidic CO_2 is added to produce a small amount of H^+ ion, which reacts with this complex ion to form $Al(H_2O)_3(OH)_3$. Cooling supersaturates the solution, and the solid that forms is filtered out:

$$CO_2(g) + H_2O(l) \rightleftharpoons H^+(aq) + HCO_3^-(aq)$$
$$Al(H_2O)_2(OH)_4^-(aq) + H^+(aq) \longrightarrow Al(H_2O)_3(OH)_3(s)$$

[Recall that we usually write $Al(H_2O)_3(OH)_3$ more simply as $Al(OH)_3$.]

Drying at high temperature converts the hydroxide to the oxide:

$$2Al(H_2O)_3(OH)_3(s) \xrightarrow{\Delta} Al_2O_3(s) + 9H_2O(g)$$

2. *Converting Al_2O_3 to the free metal: the Hall-Heroult process.* Aluminum is an active metal, much too strong a reducing agent to be formed at the cathode from aqueous solution (Section 21.7), so the oxide itself is electrolyzed.

The melting point of Al_2O_3 is very high (2030°C), so it is dissolved in molten *cryolite* (Na_3AlF_6) to give a mixture that electrolyzes at ~1000°C. Using cryolite provides a major energy (and thus, cost) savings, but the only sizeable cryolite mines cannot supply enough of the mineral, so production of synthetic cryolite has become a major subsidiary industry in aluminum manufacture.

The electrolytic step, called the *Hall-Heroult process,* takes place in a graphite-lined furnace, with the lining itself acting as the cathode. Anodes of graphite dip into the molten Al_2O_3-Na_3AlF_6 mixture (Figure 22.20, *next page*). The cell typically operates at a moderate voltage of 4.5 V, but with an enormous current flow of 1.0×10^5 to 2.5×10^5 A.

The process is complex and its details are still not entirely known. The specific reactions shown below are among several other possibilities. Molten cryolite contains several ions (including AlF_6^{3-}, AlF_4^-, and F^-), which react with Al_2O_3 to form fluoro-oxy ions (including $AlOF_3^{2-}$, $Al_2OF_6^{2-}$, and $Al_2O_2F_4^{2-}$) that dissolve in the mixture. For example,

$$2Al_2O_3(s) + 2AlF_6^{3-}(l) \longrightarrow 3Al_2O_2F_4^{2-}(l)$$

Al forms at the cathode (reduction), shown here with AlF_6^{3-} as reactant:

$$AlF_6^{3-}(l) + 3e^- \longrightarrow Al(l) + 6F^-(l) \qquad \text{[cathode; reduction]}$$

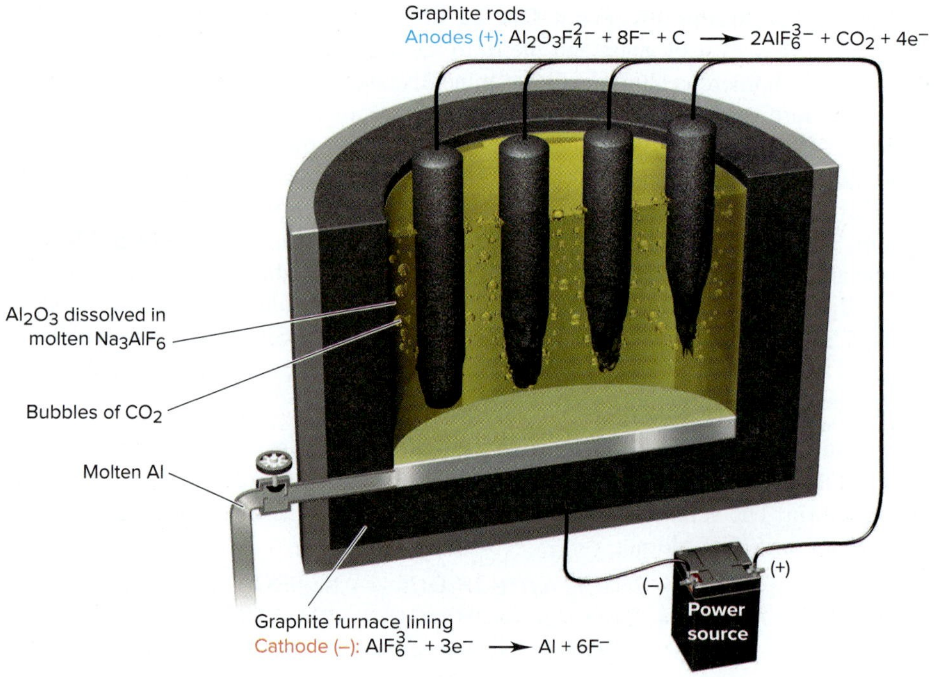

Figure 22.20 The electrolytic cell in the manufacture of aluminum.

The graphite anodes are oxidized and form carbon dioxide gas. Using one of the fluoro-oxy species as an example, the anode reaction is

$$Al_2O_2F_4^{2-}(l) + 8F^-(l) + C(graphite) \longrightarrow 2AlF_6^{3-}(l) + CO_2(g) + 4e^- \quad \text{[anode; oxidation]}$$

Thus, the *graphite anodes themselves are consumed* in this half-reaction and must be replaced frequently.

Combining the three preceding equations and making sure that e^- gained at the cathode equal e^- lost at the anode gives the overall reaction:

$$2Al_2O_3(\text{in } Na_3AlF_6) + 3C(graphite) \longrightarrow 4Al(l) + 3CO_2(g) \quad \text{[overall (cell) reaction]}$$

3. *Energy considerations.* The Hall-Heroult process uses an enormous quantity of energy: aluminum production accounts for more than 5% of total U.S. electrical usage. The most basic reason for such high energy consumption is the electron configuration of Al ([Ne] $3s^23p^1$). Each Al^{3+} ion needs $3e^-$ to form an Al atom, and the atomic mass of Al is so low (~27 g/mol) that 1 mol of e^- (96,500 C) produces only 9 g of Al. An aluminum-air battery can turn this disadvantage around. Once produced, Al represents a concentrated form of electrical energy that can deliver a high output per gram: 1 mol of e^- for every 9 g of Al consumed in the battery.

Metallurgy of Aluminum: Uses and Recycling Aluminum is a superb decorative, functional, and structural metal. It is lightweight, attractive, easy to work, and forms strong alloys. Although Al is very active, it does not corrode readily because of an adherent oxide layer that forms rapidly in air and prevents more O_2 from penetrating. Nevertheless, when it is in contact with less active metals such as Fe, Cu, and Pb, aluminum becomes the anode in a voltaic-like arrangement and deteriorates rapidly (Section 21.6). To prevent this, aluminum objects are often *anodized,* that is, made to act as the anode in an electrolytic cell that coats them with an oxide layer. The object is immersed in a 20% H_2SO_4 bath and connected to a graphite cathode:

$$6H^+(aq) + 6e^- \longrightarrow 3H_2(g) \quad \text{[cathode; reduction]}$$
$$2Al(s) + 3H_2O(l) \longrightarrow Al_2O_3(s) + 6H^+(aq) + 6e^- \quad \text{[anode; oxidation]}$$
$$2Al(s) + 3H_2O(l) \longrightarrow Al_2O_3(s) + 3H_2(g) \quad \text{[overall (cell) reaction]}$$

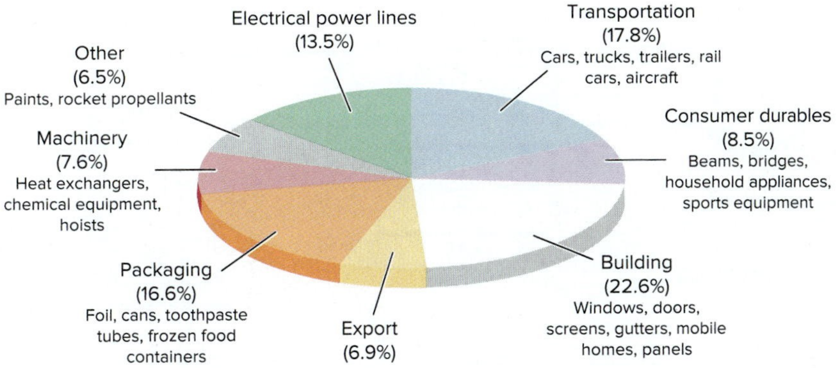

Figure 22.21 **The many familiar and essential uses of aluminum.**

The Al_2O_3 layer deposited is typically from 10 μ to 100 μ thick, depending on the object's intended use. Figure 22.21 presents a breakdown of the many uses of aluminum.

More than 4.0 billion pounds (1.8 million metric tons) of aluminum cans and packaging are discarded each year—a waste of one of the most useful materials in the world *and* the energy used to make it. A quick calculation of the energy needed to prepare 1 mol of Al from *purified* Al_2O_3, compared with the energy needed for recycling, conveys a clear message. For the overall cell reaction in the Hall-Heroult process, $\Delta H° = 2272$ kJ and $\Delta S° = 635.4$ J/K. If we consider *only* the standard free energy change of the reaction at 1000.°C, for 1 mol of Al, we obtain

$$\Delta G° = \Delta H° - T\Delta S° = \frac{2272 \text{ kJ}}{4 \text{ mol Al}} - \left(1273 \text{ K} \times \frac{0.6354 \text{ kJ/K}}{4 \text{ mol Al}}\right) = 365.8 \text{ kJ/mol Al}$$

The molar mass of Al is nearly 2.5 times the mass of a soft-drink or beer can, so the electrolysis step requires nearly 150 kJ of energy for each can!

When aluminum is recycled, the major energy input (which is for melting the cans and foil) has been calculated as ~26 kJ/mol Al. The ratio of these energy inputs is

$$\frac{\text{Energy to recycle 1 mol Al}}{\text{Energy for electrolysis of 1 mol Al}} = \frac{26 \text{ kJ}}{365.8 \text{ kJ}} = 0.071$$

Based on just these portions of the process, recycling uses about 7% as much energy as electrolysis. Recent energy estimates for the entire manufacturing process (including mining, pretreating, maintaining operating conditions, electrolyzing, and so forth) are about 6000 kJ/mol Al, which means recycling requires less than 1% as much energy as manufacturing. The economic advantages, not to mention the environmental ones, are obvious, and recycling of aluminum has become common in the United States *(see photo)*.

Crushed aluminum cans ready for recycling.

Source: © Monty Rakusen/Cultura/Getty Images RF

Mining the Sea for Magnesium

As terrestrial sources of certain elements become scarce or too costly to mine, the oceans will become an important source. In fact, despite the abundant distribution of magnesium on land, it is already obtained from the sea.

Isolation of Magnesium The *Dow process* for isolating magnesium from the sea involves steps similar to those used for rocky ores (Figure 22.22, *next page*):

1. *Mining.* Intake of seawater and straining the debris are the "mining" steps. No pretreatment is needed.

2. *Converting to mineral.* The dissolved Mg^{2+} ion is converted to the mineral $Mg(OH)_2$ with $Ca(OH)_2$, which is generated on-site (at the plant). Seashells ($CaCO_3$) are crushed, decomposed to CaO by heating, and mixed with water to make slaked lime [$Ca(OH)_2$]. This is pumped into the intake tank to precipitate the Mg^{2+} as the hydroxide ($K_{sp} \approx 10^{-9}$):

$$Ca(OH)_2(aq) + Mg^{2+}(aq) \longrightarrow Mg(OH)_2(s) + Ca^{2+}(aq)$$

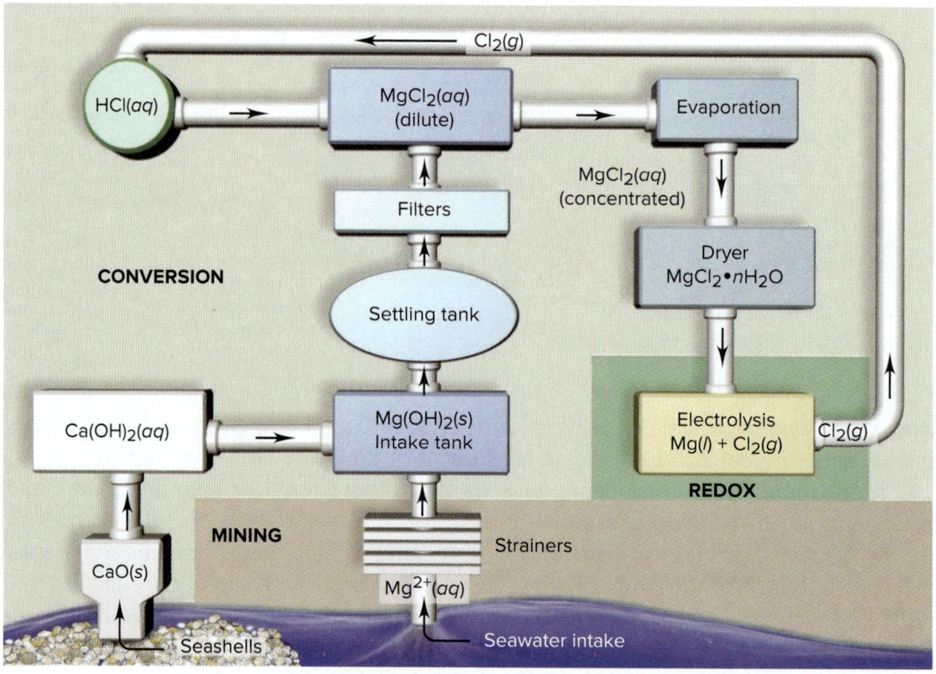

Figure 22.22 The Dow process for isolation of elemental Mg from seawater.

3. *Converting to compound.* The solid $Mg(OH)_2$ is filtered and mixed with excess HCl, which is also made on-site, to form aqueous $MgCl_2$:

$$Mg(OH)_2(s) + 2HCl(aq) \longrightarrow MgCl_2(aq) + 2H_2O(l)$$

The water is evaporated in stages to give solid, hydrated $MgCl_2{\cdot}nH_2O$.

4. *Electrochemical redox.* Heating above 700°C drives off the water of hydration and melts the $MgCl_2$. Electrolysis gives chlorine gas and the molten metal, which floats on the denser molten salt:

$$MgCl_2(l) \longrightarrow Mg(l) + Cl_2(g)$$

The Cl_2 that forms is recycled to make the HCl used in step 3.

Uses of Magnesium Magnesium is the lightest structural metal available (about one-half the density of Al and one-fifth that of steel). Although Mg is quite reactive, it forms an extremely adherent, high-melting oxide layer (MgO) that prevents further reaction with the free metal beneath it; thus, magnesium finds many uses in metal alloys and can be machined into any form. Magnesium alloys occur in everything from aircraft bodies to camera bodies, and from luggage to auto engine blocks. The pure metal is a strong reducing agent, which makes it useful for sacrificial anodes (Section 21.6) and in the metallurgical extraction of other metals, such as Be, Ti, Zr, Hf, and U. For example, titanium, an essential component of jet engines *(see photo),* is made from its major ore, ilmenite, in two steps:

$$(1)\quad 2FeTiO_3(s) + 7Cl_2(g) + 6C(s) \longrightarrow 2TiCl_4(l) + 2FeCl_3(s) + 6CO(g)$$

$$(2)\quad TiCl_4(l) + 2Mg(l) \longrightarrow Ti(s) + 2MgCl_2(l)$$

Jet engine assembly.
Source: © Ken Whitmore/Getty Images

The Sources and Uses of Hydrogen

Although hydrogen accounts for 90% of the atoms in the universe, it makes up only 15% of the atoms in Earth's crust. Moreover, hydrogen exists elsewhere in the universe mostly as H_2 molecules and free H atoms, but virtually all hydrogen in the crust is combined with other elements, either oxygen in natural waters or carbon in biomass and petroleum.

Industrial Production of Hydrogen Hydrogen gas (H_2) is produced on an industrial scale worldwide: 250,000 metric tons ($3{\times}10^{12}$ L at STP; equivalent to ${\sim}1{\times}10^{11}$ mol) are produced annually in the United States alone. All production methods are energy intensive, so the choice is determined by energy costs. In Scandinavia, where hydroelectric power is plentiful, electrolysis is used; on the other hand, where natural gas from oil refineries is plentiful, as in the United States and Great Britain, thermal methods are used.

1. *Electrolysis.* Very pure H_2 is prepared through electrolysis of water with Pt (or Ni) electrodes:

$$2H_2O(l) + 2e^- \longrightarrow H_2(g) + 2OH^-(aq) \qquad E = -0.42\text{ V} \qquad \text{[cathode; reduction]}$$
$$H_2O(l) \longrightarrow \tfrac{1}{2}O_2(g) + 2H^+(aq) + 2e^- \qquad E = 0.82\text{ V} \qquad \text{[anode; oxidation]}$$
$$H_2O(l) \longrightarrow H_2(g) + \tfrac{1}{2}O_2(g) \qquad E_{cell} = -0.42\text{ V} - 0.82\text{ V} = -1.24\text{ V}$$

Overvoltage makes the cell potential about −2 V (Section 21.7). Therefore, under typical operating conditions, it takes about 400 kJ of energy to produce 1 mol of H_2:

$$\Delta G = -nFE = (-2\text{ mol e}^-)\left(\frac{96.5\text{ kJ}}{\text{V}\cdot\text{mol e}^-}\right)(-2\text{ V}) = 4\times10^2\text{ kJ/mol H}_2$$

High-purity O_2 is a valuable byproduct of this process that offsets some of the costs of isolating H_2.

2. *Thermal methods.* The most common thermal methods use water and a simple alkane like methane, which has the highest H/C atom ratio of any hydrocarbon. The reactants are heated to around 1000°C over a nickel-based catalyst in the endothermic *steam-reforming process:*

$$CH_4(g) + H_2O(g) \longrightarrow CO(g) + 3H_2(g) \qquad \Delta H^\circ = 206\text{ kJ}$$

Heat is supplied by burning methane. To generate more H_2, the product mixture (called *water gas*) is heated with steam at 400°C over an iron or cobalt oxide catalyst in the exothermic *water-gas shift reaction,* and the CO reacts (see also Chemical Connections at the end of Section 6.6):

$$H_2O(g) + CO(g) \rightleftharpoons CO_2(g) + H_2(g) \qquad \Delta H^\circ = -41\text{ kJ}$$

The reaction mixture is recycled several times, which decreases the CO to around 0.2% by volume. Passing the mixture through liquid water removes the more soluble CO_2 (solubility = 0.034 mol/L), leaving a higher proportion of H_2 (solubility < 0.001 mol/L). Calcium oxide can also be used to remove CO_2 by formation of $CaCO_3$. By removing CO_2, these steps shift the equilibrium position to the right and produce H_2 that is about 98% pure. To filter out nearly all molecules larger than H_2 and attain greater purity (~99.9%), the gas mixture is passed through a *synthetic zeolite*. This material is an aluminosilicate whose framework of polyhedra contains small cavities of various sizes that trap some molecules and allow others to pass through.

Industrial Uses of Hydrogen Typically, an industrial plant produces H_2 to make some other product. More than 95% of the H_2 produced is consumed in ammonia or petrochemical facilities.

1. *Ammonia synthesis.* In a plant that synthesizes NH_3, the reactant gases, N_2 and H_2, are formed through a series of reactions that involve methane. Here is a typical reaction series in an ammonia plant:

- The steam-reforming reaction is performed with excess CH_4, which depletes the reaction mixture of H_2O:

$$CH_4(g;\text{ excess}) + H_2O(g) \longrightarrow CO(g) + 3H_2(g)$$

- An excess of the product mixture (CH_4, CO, and H_2) is burned in an amount of air ($N_2 + O_2$) that is insufficient for complete combustion, but is enough to consume the O_2, heat the mixture to 1100°C, and form additional H_2O:

$$4CH_4(g;\text{ excess}) + 7O_2(g) \longrightarrow 2CO_2(g) + 2CO(g) + 8H_2O(g)$$
$$2H_2(g;\text{ excess}) + O_2(g) \longrightarrow 2H_2O(g)$$
$$2CO(g;\text{ excess}) + O_2(g) \longrightarrow 2CO_2(g)$$

- Any remaining CH_4 reacts by the steam-reforming reaction, the remaining CO reacts by the water-gas shift reaction ($H_2O + CO \rightleftharpoons CO_2 + H_2$) to form more H_2, and then the CO_2 is removed with CaO ($CaO + CO_2 \longrightarrow CaCO_3$).
- The amounts are carefully adjusted to produce a final mixture that contains a 1/3 mole ratio of N_2 (from the added air) to H_2 (with traces of CH_4, Ar, and CO). This mixture is used directly in the synthesis of ammonia (Section 17.6).

2. *Hydrogenation.* A second major use of H_2 is *hydrogenation* of C═C bonds in liquid oils to form C—C bonds in solid fats and margarine. The process uses H_2 in contact with transition metal catalysts, such as powdered nickel (see Figure 16.25). Solid fats are also used in commercial baked goods. Look at the list of ingredients on most packages of bread, cake, and cookies, and you'll see the "partially hydrogenated vegetable oils" made by this process.

3. *Bulk chemicals and energy production.* Hydrogen is also essential in the manufacture of numerous "bulk" chemicals, those produced in large amounts because they have many further uses. One application that has been gaining great attention is the production of methanol. In this process, carbon monoxide reacts with hydrogen over a copper–zinc oxide catalyst:

$$CO(g) + 2H_2(g) \xrightarrow{\text{Cu-ZnO catalyst}} CH_3OH(l)$$

Methanol is already being used as a gasoline additive. As less expensive sources become available, hydrogen will also be used increasingly in fuel cells on a major scale (Section 21.5).

Production and Uses of Deuterium; Kinetic Isotope Effect In addition to ordinary hydrogen (^1H), or protium, there are two other naturally occurring isotopes of significantly lower abundance. Deuterium (^2H or D) has one neutron, and rare, radioactive tritium (^3H or T) has two. Like H_2, both occur as diatomic molecules: D_2 (or $^2H_2$) and T_2 (or $^3H_2$). Table 22.6 compares some of their molecular and physical properties. As you can see from these data, the heavier the isotope is and, thus, the higher the molar mass of the molecule, the higher the melting point, boiling point, and heats of phase change.

Because hydrogen is so light, the relative difference in mass of its isotopes is enormous compared with that of isotopes of other common, heavier elements. For example, ^2H has 2 times the mass of ^1H, whereas ^{13}C has only 1.08 times the mass of ^{12}C. *The mass difference leads to different bond energies, which affect reactivity.* As a result, an H atom bonded to a given atom vibrates at a higher frequency than a D atom does, so the bond to H is higher in energy. Thus, *it is weaker and easier to break.* Therefore, any reaction that includes breaking a bond to H or D in the rate-determining step occurs *faster with H than with D.* This phenomenon, called a *kinetic isotope effect,* is also used to isolate deuterium.

Deuterium and its compounds are produced from D_2O (heavy water), which is present as a minor component (0.016 mol %) in normal water and is isolated on the

Table 22.6 Some Molecular and Physical Properties of Diatomic Protium, Deuterium, and Tritium

Property	H_2	D_2	T_2
Molar mass (g/mol)	2.016	4.028	6.032
Bond length (pm)	74.14	74.14	74.14
Melting point (K)	13.96	18.73	20.62
Boiling point (K)	20.39	23.67	25.04
ΔH°_{fus} (kJ/mol)	0.117	0.197	0.250
ΔH°_{vap} (kJ/mol)	0.904	1.226	1.393
Bond energy (kJ/mol at 298 K)	432	443	447

multiton scale by *electrolytic enrichment*. Due to the kinetic isotope effect, there is a *higher rate of bond breaking for O—H bonds compared to O—D bonds,* and thus a higher rate of electrolysis of H_2O compared with D_2O.

For example, with Pt electrodes, H_2O is electrolyzed about 14 times faster than D_2O. As some of the liquid decomposes to the elemental gases, the remainder becomes enriched in D_2O. Thus, by the time the volume of water has been reduced to 1/20,000 of its original volume, the remaining water is around 99% D_2O. By combining samples and repeating the electrolysis, water that is more than 99.9% D_2O is obtained.

Deuterium gas is produced by electrolysis of D_2O or by any of the chemical reactions that produce hydrogen gas from water, such as

$$2Na(s) + 2D_2O(l) \longrightarrow 2Na^+(aq) + 2OD^-(aq) + D_2(g)$$

Compounds containing deuterium (or tritium) are produced from reactions that give rise to the corresponding hydrogen-containing compound; for example,

$$SiCl_4(l) + 2D_2O(l) \longrightarrow SiO_2(s) + 4DCl(g)$$

Compounds with acidic protons undergo hydrogen/deuterium exchange:

$$CH_3COOH(l) + D_2O(l;\ \text{excess}) \longrightarrow CH_3COOD(l) + DHO(l;\ \text{small amount})$$

We discuss the natural and synthetic formation of tritium in Chapter 24.

› Summary of Section 22.4

Production highlights of key elements are as follows:

- Na is isolated by electrolysis of molten NaCl in the Downs process; Cl_2 is a byproduct.
- K is produced by reduction with Na in a thermal process.
- Fe is produced through a multistep high-temperature process in a blast furnace. The crude pig iron is converted to carbon steel in the basic-oxygen process and then alloyed with other metals to make different steels.
- Cu is produced by concentration of the ore through flotation, reduction to the metal by smelting, and purification by electrorefining. The metal has extensive electrical and plumbing uses.
- Al is extracted from bauxite by pretreating the ore with concentrated base, followed by electrolysis of the product Al_2O_3 mixed with molten cryolite. Al alloys are widely used in homes and industry. The total energy required to extract Al from its ore is over 100 times that needed for recycling it.
- The Mg^{2+} in seawater is converted to $Mg(OH)_2$ and then to $MgCl_2$, which is electrolyzed to obtain the metal; Mg forms strong, lightweight alloys.
- H_2 is produced by electrolysis of water or in the formation of gaseous fuels from hydrocarbons. It is used in NH_3 production, in hydrogenation of vegetable oils, and in energy production. The isotopes of hydrogen differ significantly in atomic mass and thus in the rate at which their bonds to other atoms break (kinetic isotope effect). This difference is used to obtain D_2O from water.

22.5 CHEMICAL MANUFACTURING: TWO CASE STUDIES

In this final section, we examine the interplay of theory and practice in two of the most important processes in the inorganic chemical industry: the contact process for the production of sulfuric acid and the chlor-alkali process for the production of chlorine.

Sulfuric Acid, the Most Important Chemical

Because it has countless uses, sulfuric acid is produced throughout the world on a gigantic scale—more than 150 million tons a year. The modern **contact process** is based on the *catalyzed oxidation of SO_2*. Here are the key steps.

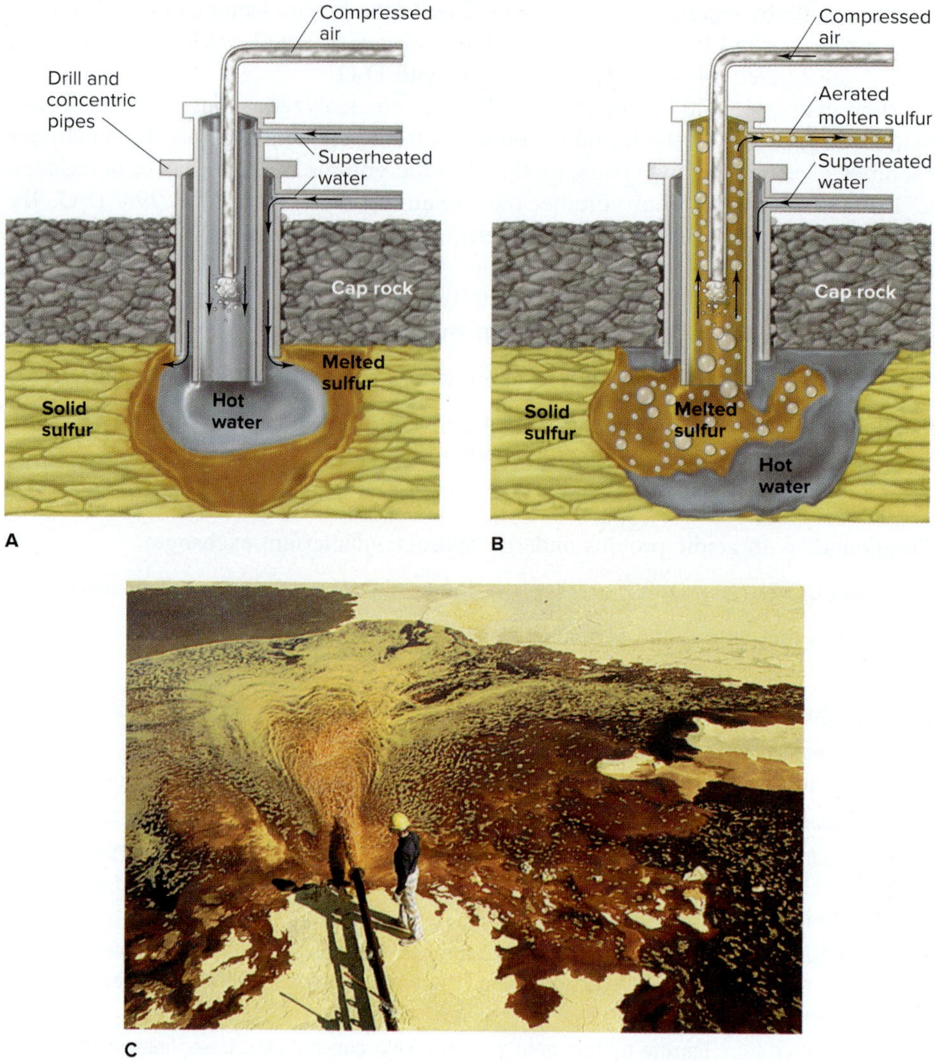

Figure 22.23 The Frasch process for mining elemental sulfur.
Source: (C) © Nathan Benn/Corbis

1. *Obtaining sulfur.* In most countries today, the production of sulfuric acid starts with the production of elemental sulfur, often by the *Claus process,* in which the H_2S in "sour" natural gas is chemically separated and then oxidized:

$$2H_2S(g) + 2O_2(g) \xrightarrow{\text{low temperature}} \tfrac{1}{8}S_8(g) + SO_2(g) + 2H_2O(g)$$

$$2H_2S(g) + SO_2(g) \xrightarrow{\text{Fe}_2\text{O}_3\text{ catalyst}} \tfrac{3}{8}S_8(g) + 2H_2O(g)$$

Where natural deposits of sulfur are found, it is obtained by a nonchemical method called the *Frasch process.* A hole is drilled to the deposit, and superheated water (about 160°C) is pumped down two outer concentric pipes to melt the sulfur (Figure 22.23A). Then, a combination of the hydrostatic pressure in an outermost pipe and the pressure of compressed air sent through a narrow inner pipe forces the sulfur to the surface (Figure 22.23B and C). The costs of drilling, pumping, and supplying water (5×10^6 gallons per day) are balanced somewhat by the purity of the product (~99.7% sulfur). About 90% of processed sulfur is used in making sulfur dioxide for production of sulfuric acid. Indeed, this end product is so important that a nation's level of sulfur production is a reliable indicator of its overall industrial capacity: the United States, Canada, China, Russia, Japan, and Germany are the top six sulfur producers.

2. *From sulfur to sulfur dioxide.* The sulfur is burned in air to form SO_2:

$$\tfrac{1}{8}S_8(s) + O_2(g) \longrightarrow SO_2(g) \qquad \Delta H^\circ = -297\text{ kJ}$$

Some SO_2 is also obtained by roasting metal sulfide ores and from the Claus process.

3. *From sulfur dioxide to trioxide.* The *contact process* oxidizes SO_2 to SO_3:

$$SO_2(g) + \frac{1}{2}O_2(g) \rightleftharpoons SO_3(g) \qquad \Delta H^\circ = -99 \text{ kJ}$$

The reaction is *exothermic* and very *slow* at room temperature. From Le Châtelier's principle (Section 17.6), the yield of SO_3 can be increased by (1) changing the temperature, (2) increasing the pressure (more moles of gas are on the left than on the right), or (3) adjusting the concentrations (adding excess O_2 and removing SO_3). In this case, the pressure effect is small and economically not worth exploiting. Let's examine the other two effects:

- *Effect of changing the temperature.* Adding heat (raising the temperature) increases the frequency and energy of SO_2–O_2 collisions and thus increases the *rate* of SO_3 formation. However, because the formation is exothermic, removing heat (lowering the temperature) shifts the equilibrium position to the right and thus increases the *yield* of SO_3. This is a classic situation that calls for use of a catalyst. By lowering the activation energy, *a catalyst allows equilibrium to be reached more quickly and at a lower temperature;* thus, rate *and* yield are optimized (Section 16.7). The catalyst in the contact process is V_2O_5 on inert silica, which is active between 400°C and 600°C.
- *Effect of changing the concentration.* Providing an excess of O_2 as a 5/1 mixture of air to SO_2, which is about 1/1 O_2 to SO_2, supplies about twice as much O_2 as in the balanced equation. The mixture is passed over catalyst beds in four stages, and the SO_3 is removed to shift the reaction toward more SO_3 formation. The overall yield of SO_3 is 99.5%.

4. *From sulfur trioxide to acid.* Sulfur trioxide is the anhydride of sulfuric acid. However, SO_3 cannot be added to water because, at the operating temperature, it would first meet water vapor, which catalyzes its polymerization to $(SO_3)_x$, and results in a smoke of solid particles that yields little acid (reaction 1). To avoid this, previously formed H_2SO_4 absorbs the SO_3 (reaction 2) and forms pyrosulfuric acid (or disulfuric acid, $H_2S_2O_7$; *see margin*), which is then hydrolyzed with sufficient water (reaction 3):

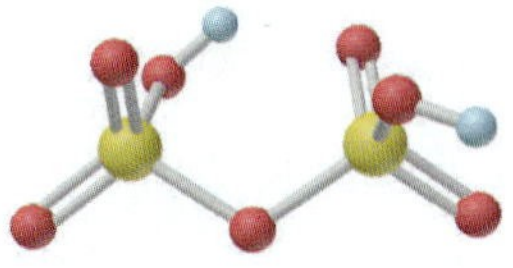

disulfuric acid

$$\begin{aligned}
&(1) \quad SO_3(g) + H_2O(l) \longrightarrow H_2SO_4(l) && \text{[lower yield]}\\
&(2) \quad SO_3(g) + H_2SO_4(l) \longrightarrow H_2S_2O_7(l) &&\\
&(3) \quad H_2S_2O_7(l) + H_2O(l) \longrightarrow 2H_2SO_4(l) && \text{[higher yield]}
\end{aligned}$$

The industrial uses of sulfuric acid are legion, as Figure 22.24 indicates.

Sulfuric acid is remarkably inexpensive (about \$150/ton), largely because each step in the process is exothermic—burning S ($\Delta H^\circ = -297$ kJ/mol), oxidizing SO_2 ($\Delta H^\circ = -99$ kJ/mol), hydrating SO_3 ($\Delta H^\circ = -132$ kJ/mol)—and the heat is a valuable

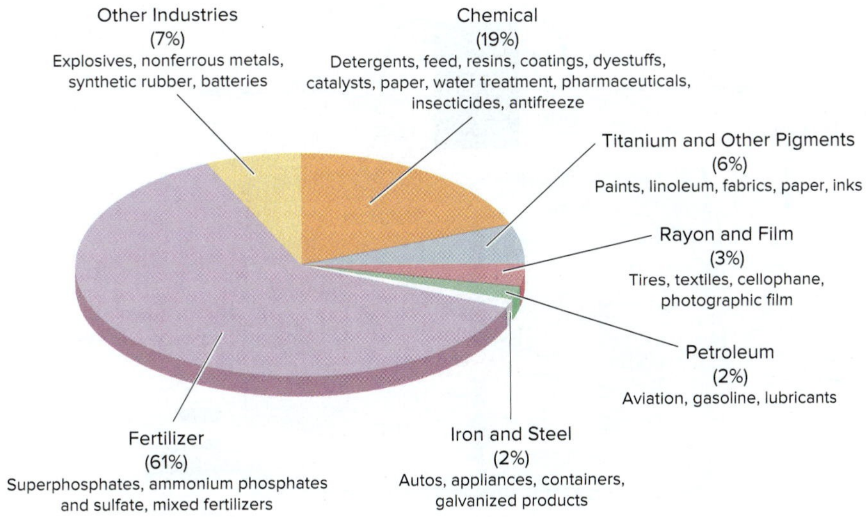

Figure 22.24 The many indispensable applications of sulfuric acid.

byproduct. Three-quarters of the heat is sold as steam, and the rest of it is used to pump gases through the plant. A typical plant making 825 tons of H_2SO_4 per day produces enough steam to generate 7×10^6 watts of electric power.

The Chlor-Alkali Process

Chlorine is produced and used in amounts many times greater than all the other halogens combined, ranking among the top 10 chemicals produced in the United States. The **chlor-alkali process,** which forms the basis of one of the largest inorganic chemical industries, *electrolyzes concentrated aqueous NaCl to produce Cl_2,* along with several other important chemicals. There are three versions of this process.

Diaphragm-Cell Method As you learned in Section 21.7, the electrolysis of aqueous NaCl does not yield both of the component elements. Because of the overvoltage, Cl^- ions rather than H_2O molecules are oxidized at the anode. However, Na^+ ions are not reduced at the cathode because the half-cell potential (−2.71 V) is much more negative than that for reduction of H_2O (−0.42 V), even with the normal overvoltage (around −0.6 V). Therefore, the half-reactions for electrolysis of aqueous NaCl are

$$2Cl^-(aq) \longrightarrow Cl_2(g) + 2e^- \qquad E° = 1.36\text{ V} \quad [\text{anode; oxidation}]$$
$$2H_2O(l) + 2e^- \longrightarrow 2OH^-(aq) + H_2(g) \qquad E \approx -1.0\text{ V} \quad [\text{cathode; reduction}]$$
$$2Cl^-(aq) + 2H_2O(l) \longrightarrow 2OH^-(aq) + H_2(g) + Cl_2(g) \qquad E_{cell} = -1.0\text{ V} - 1.36\text{ V} = -2.4\text{ V}$$

To obtain commercially meaningful amounts of Cl_2, however, a voltage almost twice this value and a current in excess of 3×10^4 A are used.

When we include the spectator ion Na^+, the total ionic equation shows that the process yields another important product, NaOH:

$$2Na^+(aq) + 2Cl^-(aq) + 2H_2O(l) \longrightarrow 2Na^+(aq) + 2OH^-(aq) + H_2(g) + Cl_2(g)$$

As Figure 22.25 shows, the sodium salts in the cathode compartment exist as an aqueous mixture of NaCl and NaOH; the NaCl is removed by fractional crystallization. Thus, in this version of the chlor-alkali process, which uses an *asbestos diaphragm* to separate the anode and cathode compartments, electrolysis of NaCl brines yields Cl_2, H_2, and industrial-grade NaOH, a useful base. Like other reactive products, H_2 and Cl_2 are kept apart to prevent explosive recombination. Note the higher liquid level in the anode *(left)* compartment. This slight hydrostatic pressure difference minimizes backflow of NaOH, which prevents the disproportionation reactions of Cl_2 that occur in the presence of OH^- (Section 14.9), such as

$$Cl_2(g) + 2OH^-(aq) \longrightarrow Cl^-(aq) + ClO^-(aq) + H_2O(l)$$

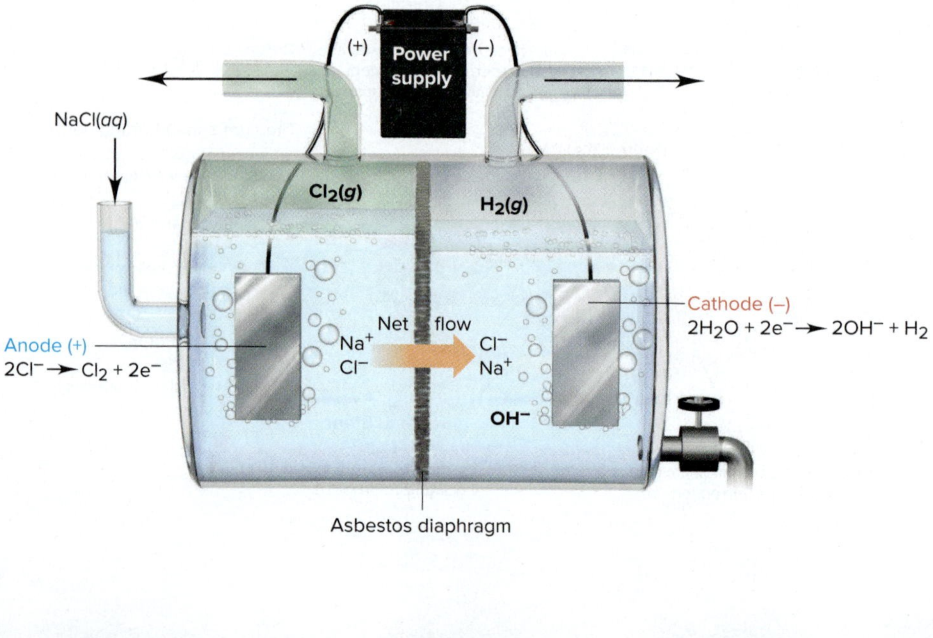

Figure 22.25 A diaphragm cell for the chlor-alkali process.

Mercury-Cell Method If high-purity NaOH is desired, a slightly different version, called the *chlor-alkali mercury-cell process,* is employed. Mercury is used as the cathode, which creates such a large overvoltage for reduction of H_2O to H_2 that the process *does* favor reduction of Na^+. The sodium dissolves in the mercury to form sodium amalgam, Na(Hg). In the mercury-cell version, the half-reactions are

$$2Cl^-(aq) \longrightarrow Cl_2(g) + 2e^- \quad \text{[anode; oxidation]}$$

$$2Na^+(aq) + 2e^- \xrightarrow{Hg} 2Na(Hg) \quad \text{[cathode; reduction]}$$

To obtain sodium hydroxide, the sodium amalgam is pumped out of the system and treated with H_2O, which is reduced by the Na:

$$2Na(Hg) + 2H_2O(l) \xrightarrow{-Hg} 2Na^+(aq) + 2OH^-(aq) + H_2(g)$$

The mercury released in this step is recycled back to the electrolysis bath. Therefore, *the products are the same in both versions* of the chlor-alkali process, but the purity of NaOH from the mercury-cell method is much higher.

The mercury-cell method is being steadily phased out because, as the mercury is recycled, small amounts are lost in the industrial wastewater. On average, 200 g of Hg is lost per ton of Cl_2 produced. In the 1980s, annual U.S. production via the mercury-cell method was 2.75 million tons of Cl_2; thus, during that decade, 550,000 kg of mercury, a toxic heavy metal, was flowing into U.S. waterways each year!

Membrane-Cell Method The *chlor-alkali membrane-cell process* replaces the asbestos diaphragm used in the diaphragm-cell method to separate the cell compartments with a polymeric membrane. The membrane allows only cations to move through it and only from anode to cathode compartments. Thus, as Cl^- ions are removed at the anode through oxidation to Cl_2, Na^+ ions in the anode compartment move through the membrane to the cathode compartment and form an NaOH solution. In addition to forming purer NaOH than the older diaphragm-cell method, the membrane-cell method uses less electricity and eliminates the problem of Hg pollution. As a result, it has been adopted throughout much of the industrialized world.

› Summary of Section 22.5

- Sulfuric acid production starts with the extraction of sulfur, either by the oxidation of H_2S or the mining of sulfur deposits. The sulfur is roasted to SO_2 and oxidized to SO_3 by the catalyzed contact process, which optimizes the yield at lower temperatures. Absorption of the SO_3 into H_2SO_4, followed by hydration, forms sulfuric acid.
- In the diaphragm-cell method for the chlor-alkali process, aqueous NaCl is electrolyzed to form Cl_2, H_2, and low-purity NaOH. The mercury-cell method produces high-purity NaOH but has been almost completely phased out because of mercury pollution. The membrane-cell method requires less electricity and does not use Hg.

CHAPTER REVIEW GUIDE

Learning Objectives

Relevant section (§) numbers appear in parentheses.

Understand These Concepts

1. How gravity, thermal convection, and elemental properties led to the silicate, sulfide, and iron phases and the predominance of certain elements in Earth's crust, mantle, and core (§22.1)
2. How organisms affect crustal abundances of elements, especially oxygen, carbon, calcium, and some transition metals; the onset of oxidation as an energy source (§22.1)
3. How atomic properties influence which elements have oxide ores and which have sulfide ores (§22.1)
4. The central role of CO_2 and the importance of photosynthesis, respiration, and decomposition in the carbon cycle (§22.2)
5. The central role of N_2 and the importance of atmospheric, industrial, and biological fixation in the nitrogen cycle (§22.2)
6. The absence of a gaseous component and the interactions of the inorganic and biological cycles in the phosphorus cycle; the impact of humans on this cycle (§22.2)
7. How pyro-, electro-, and hydrometallurgical processes are employed to extract a metal from its ore; the importance of the reduction step from compound to metal; refining and alloying processes (§22.3)
8. Functioning of the Downs cell for Na production and the application of Le Châtelier's principle to K production (§22.4)

Learning Objectives (continued)

Relevant section (§) numbers appear in parentheses.

9. How iron ore is reduced in a blast furnace and how pig iron is purified by the basic-oxygen process (§22.4)
10. How Fe is removed from copper ore and impure Cu is electrorefined (§22.4)
11. The importance of amphoterism in the Bayer process for isolating Al_2O_3 from bauxite; the significance of cryolite in the electrolytic step; the energy advantage of Al recycling (§22.4)
12. The steps in the Dow process for the extraction of Mg from seawater (§22.4)
13. How H_2 production, whether by chemical or electrolytic means, is tied to NH_3 production (§22.4)
14. How the kinetic isotope effect is applied to produce deuterium (§22.4)
15. How the Frasch process is used to obtain sulfur from natural deposits (§22.5)
16. The importance of equilibrium and kinetic factors in H_2SO_4 production (§22.5)
17. How overvoltage allows electrolysis of aqueous NaCl to form Cl_2 gas in the chlor-alkali process; coproduction of NaOH; comparison of diaphragm-cell, mercury-cell, and membrane-cell methods (§22.5)

Key Terms

Page numbers appear in parentheses.

abundance (997)
alloy (1012)
apatites (1005)
atmosphere (999)
basic-oxygen process (1017)
biosphere (999)
blast furnace (1016)
carbon steel (1017)
chlor-alkali process (1028)
contact process (1025)
core (998)
crust (998)
differentiation (998)
Downs cell (1014)
electrorefining (1012)
environmental cycle (1002)
fixation (1003)
flotation (1009)
gangue (1008)
hydrosphere (999)
leaching (1009)
lithosphere (999)
mantle (998)
metallurgy (1008)
mineral (1008)
occurrence (source) (1000)
ore (1001)
roasting (1010)
slag (1017)
smelting (1010)
steel (1016)
zone refining (1012)

PROBLEMS

Problems with **colored** numbers are answered in Appendix E and worked in detail in the Student Solutions Manual. Problem sections match those in the text and offer Concept Review Questions and Problems in Context. The Comprehensive Problems are based on material from any section or previous chapter.

How the Elements Occur in Nature

Concept Review Questions

22.1 Hydrogen is by far the most abundant element cosmically. In interstellar space, it exists mainly as H_2. In contrast, on Earth, it exists very rarely as H_2 and is ninth in abundance in the crust. Why is hydrogen so abundant in the universe? Why is hydrogen so rare as a diatomic gas in Earth's atmosphere?

22.2 Metallic elements can be recovered from ores that are oxides, carbonates, halides, or sulfides. Give an example of each type.

22.3 The location of elements in the regions of Earth has enormous practical importance. (a) Define *differentiation*, and explain which physical property of a substance is primarily responsible for this process. (b) What are the four most abundant elements in the crust? (c) Which element is abundant in the crust and mantle but not the core?

22.4 How does the position of a metal in the periodic table relate to whether it occurs primarily as an oxide or as a sulfide in nature?

Problems in Context

22.5 What material is the source for commercial production of each of the following elements: (a) aluminum; (b) nitrogen; (c) chlorine; (d) calcium; (e) sodium?

22.6 Aluminum is widely distributed throughout the world in the form of aluminosilicates. What property of these minerals prevents them from being a source of aluminum?

22.7 Describe two ways in which the biosphere has influenced the composition of Earth's crust.

The Cycling of Elements Through the Environment

Concept Review Questions

22.8 Use atomic and molecular properties to explain why life is based on carbon rather than some other element, such as silicon.

22.9 Define *fixation*. Name two elements that undergo environmental fixation. What natural forms of them are fixed?

22.10 Carbon dioxide enters the atmosphere by natural processes and from human activity. Why is the latter a cause of concern?

22.11 Diagrams of environmental cycles are simplified to omit minor contributors. For example, the production of lime from limestone is not shown in the cycle for carbon (Figure 22.5). Which labeled category in the figure includes this process? Name two other processes that contribute to this category.

22.12 Describe three pathways for the fixation of atmospheric nitrogen. Is human activity a significant factor? Explain.

22.13 Why don't the N-containing species in Figure 22.6 include rings or long chains with N—N bonds?

22.14 (a) Which region of Earth's crust is not involved in the phosphorus cycle? (b) Name two roles organisms play in the cycle.

Problems in Context

22.15 Nitrogen fixation requires a great deal of energy because the N_2 bond is strong.
(a) How do the processes of atmospheric and industrial fixation reflect this energy requirement?
(b) How do the thermodynamics of the two processes differ? (*Hint:* Examine the respective heats of formation.)
(c) In view of the mild conditions for biological fixation, what must be the source of the "great deal of energy"? (d) What would be the most obvious environmental result of a low activation energy for N_2 fixation?

22.16 The following steps are *unbalanced* half-reactions involved in the nitrogen cycle. Balance each half-reaction to show the number of electrons lost or gained, and state whether it is an oxidation or a reduction (all occur in acidic conditions):
(a) $N_2(g) \longrightarrow NO(g)$ (b) $N_2O(g) \longrightarrow NO_2(g)$
(c) $NH_3(aq) \longrightarrow NO_2^-(aq)$ (d) $NO_3^-(aq) \longrightarrow NO_2^-(aq)$
(e) $N_2(g) \longrightarrow NO_3^-(aq)$

22.17 The use of silica to form slag in the production of phosphorus from phosphate rock was introduced by Robert Boyle more than 300 years ago. When fluorapatite [$Ca_5(PO_4)_3F$] is used in phosphorus production, most of the fluorine atoms appear in the slag, but some end up in the toxic and corrosive gas SiF_4.
(a) If 15% by mass of the fluorine in 100. kg of $Ca_5(PO_4)_3F$ forms SiF_4, what volume of this gas is collected at 1.00 atm and the industrial furnace temperature of 1450.°C?
(b) In some facilities, the SiF_4 is used to produce sodium hexafluorosilicate (Na_2SiF_6) which is sold for water fluoridation:

$$2SiF_4(g) + Na_2CO_3(s) + H_2O(l) \longrightarrow Na_2SiF_6(aq) + SiO_2(s) + CO_2(g) + 2HF(aq)$$

How many cubic meters of drinking water can be fluoridated to a level of 1.0 ppm of F^- using the SiF_4 produced in part (a)?

22.18 An impurity sometimes found in $Ca_3(PO_4)_2$ is Fe_2O_3, which is removed during the production of phosphorus as *ferrophosphorus* (Fe_2P). (a) Why is this impurity troubling from an economic standpoint? (b) If 50. metric tons (t) of crude $Ca_3(PO_4)_2$ contain 2.0% Fe_2O_3 by mass and the overall yield of phosphorus is 90.%, how many metric tons of P_4 can be isolated?

Metallurgy: Extracting a Metal from Its Ore

Concept Review Questions

22.19 Define: (a) ore; (b) mineral; (c) gangue; (d) brine.

22.20 Define: (a) roasting; (b) smelting; (c) flotation; (d) refining.

22.21 What factors determine which reducing agent is selected for producing a specific metal?

22.22 Use atomic properties to explain the reduction of a less active metal by a more active one: (a) in aqueous solution; (b) in the molten state. Give a specific example of each process.

22.23 What class of element is obtained by oxidation of a mineral? What class of element is obtained by reduction of a mineral?

Problems in Context

22.24 Which set of elements gives each of the following alloys: (a) brass; (b) stainless steel; (c) bronze; (d) sterling silver?
1. Cu, Ag 2. Cu, Sn, Zn 3. Ag, Au
4. Fe, Cr, Ni 5. Fe, V 6. Cu, Zn

Tapping the Crust: Isolation and Uses of Selected Elements

Concept Review Questions

22.25 How are each of the following involved in iron metallurgy: (a) slag; (b) pig iron; (c) steel; (d) basic-oxygen process?

22.26 What are the distinguishing features of each extraction process: pyrometallurgy, electrometallurgy, and hydrometallurgy? Explain briefly how the types of metallurgy are used in the production of (a) Fe; (b) Na; (c) Au; (d) Al.

22.27 What property allows copper to be purified in the presence of iron and nickel impurities? Explain.

22.28 Why is cryolite used in the electrolysis of aluminum oxide?

22.29 (a) What is a kinetic isotope effect?
(b) Do compounds of hydrogen exhibit a relatively large or small kinetic isotope effect? Explain.
(c) Carbon compounds also exhibit a kinetic isotope effect. How do you expect it to compare in magnitude with that for hydrogen compounds? Why?

22.30 How is Le Châtelier's principle involved in the production of elemental potassium?

Problems in Context

22.31 Elemental Li and Na are prepared by electrolysis of a molten salt, whereas K, Rb, and Cs are prepared by chemical reduction.
(a) In general terms, explain why the alkali metals cannot be prepared by electrolysis of their aqueous salt solutions.
(b) Use ionization energies (see the Family Portraits for Group 1A(1) in Section 14.3 and for Group 2A(2) in Section 14.4) to explain why calcium should *not* be able to isolate Rb from molten RbX (X = halide).
(c) Use physical properties to explain why calcium *is* used to isolate Rb from molten RbX.
(d) Can Ca be used to isolate Cs from molten CsX? Explain.

22.32 A Downs cell operating at 77.0 A produces 31.0 kg of Na.
(a) What volume of $Cl_2(g)$ is produced at 1.0 atm and 540.°C?
(b) How many coulombs were passed through the cell?
(c) How long did the cell operate?

22.33 (a) In the industrial production of iron, what is the reducing substance loaded into the blast furnace?
(b) In addition to furnishing the reducing power, what other function does this substance serve?
(c) What is the formula of the active reducing agent in the process?
(d) Write equations for the stepwise reduction of Fe_2O_3 to iron in the furnace.

22.34 One of the substances loaded into a blast furnace is limestone, which produces lime in the furnace.
(a) Give the chemical equation for the reaction forming lime.
(b) Explain the purpose of lime in the furnace. The term *flux* is often used as a label for a substance acting as the lime does. What is the derivation of this word, and how does it relate to the function of the lime?
(c) Write a chemical equation describing the action of lime flux.

22.35 The last step in the Dow process for the production of magnesium metal involves electrolysis of molten $MgCl_2$.
(a) Why isn't the electrolysis carried out with aqueous $MgCl_2$? What are the products of this aqueous electrolysis?
(b) Do the high temperatures required to melt $MgCl_2$ favor products or reactants? (*Hint:* Consider the ΔH_f° of $MgCl_2$.)

22.36 Iodine is the only halogen that occurs in a positive oxidation state, in $NaIO_3$ impurities within Chile saltpeter, $NaNO_3$.
(a) Is this mode of occurrence consistent with iodine's location in the periodic table? Explain.
(b) In the production of I_2, IO_3^- reacts with HSO_3^-:

$$IO_3^-(aq) + HSO_3^-(aq) \longrightarrow HSO_4^-(aq) + SO_4^{2-}(aq) + H_2O(l) + I_2(s) \quad \text{[unbalanced]}$$

Identify the oxidizing and reducing agents.
(c) If 0.78 mol % of an $NaNO_3$ deposit is $NaIO_3$, how much I_2 (in g) can be obtained from 1.000 ton (2000. lb) of the deposit?

22.37 Selenium is prepared by the reaction of H_2SeO_3 with gaseous SO_2.
(a) What redox process does the sulfur dioxide undergo? What is the oxidation state of sulfur in the product?
(b) Given that the reaction occurs in acidic aqueous solution, what is the formula of the sulfur-containing species?
(c) Write the balanced redox equation for the process.

22.38 F_2 and Cl_2 are produced by electrolytic oxidation, whereas Br_2 and I_2 are produced by chemical oxidation of the halide ions in a concentrated aqueous solution (brine) by a more electronegative halogen. Give two reasons why Cl_2 isn't prepared this way.

22.39 Silicon is prepared by the reduction of K_2SiF_6 with Al. Write the equation for this reaction. (*Hint:* Can F^- be oxidized in this reaction? Can K^+ be reduced?)

22.40 What is the mass percent of iron in each of the following iron ores: Fe_2O_3, Fe_3O_4, FeS_2?

22.41 Phosphorus is one of the impurities present in pig iron that is removed in the basic-oxygen process. Assuming that phosphorus is present as P atoms, write equations for its oxidation and subsequent reaction in the basic slag.

22.42 The final step in the smelting of $CuFeS_2$ is

$$Cu_2S(s) + 2Cu_2O(s) \longrightarrow 6Cu(l) + SO_2(g)$$

(a) Give the oxidation states of copper in Cu_2S, Cu_2O, and Cu.
(b) What are the oxidizing and reducing agents in this reaction?

22.43 Use equations to show how acid-base properties are used to separate Fe_2O_3 and TiO_2 from Al_2O_3 in the Bayer process.

22.44 A piece of Al with a surface area of 2.5 m^2 is anodized to produce a film of Al_2O_3 that is 23 μm (23×10^{-6} m) thick. (a) How many coulombs flow through the cell in this process (assume that the density of the Al_2O_3 layer is 3.97 g/cm^3)? (b) If it takes 18 min to produce this film, what current must flow through the cell?

22.45 The production of H_2 gas by the electrolysis of water typically requires about 400 kJ of energy per mole.
(a) Use the relationship between work and cell potential (Section 21.4) to calculate the minimum work needed to form 1.0 mol of H_2 gas at a cell potential of 1.24 V.
(b) What is the energy efficiency of the cell operation?
(c) Find the cost of producing 500. mol of H_2 if electricity is $0.06 per kilowatt-hour (1 watt-second = 1 joule).

22.46 (a) What are the components of the reaction mixture following the water-gas shift reaction? (b) Explain how zeolites are used to purify the H_2 formed.

22.47 Metal sulfides are often first converted to oxides by roasting in air and then reduced with carbon to produce the metal. Why aren't the metal sulfides reduced directly by carbon to yield CS_2? Give a thermodynamic analysis of both processes for ZnS.

Chemical Manufacturing: Two Case Studies

Concept Review Questions

22.48 Explain in detail why a catalyst is used to produce SO_3.

22.49 Among the exothermic steps in the manufacture of sulfuric acid is the process of hydrating SO_3. (a) Write two chemical reactions that show this process. (b) Why is the direct reaction of SO_3 with water not feasible?

22.50 Why is commercial H_2SO_4 so inexpensive?

22.51 (a) What are the three commercial products formed in the chlor-alkali process?
(b) State an advantage and a disadvantage of using the mercury-cell method for this process.

Problems in Context

22.52 Consider the reaction of SO_2 to form SO_3 at standard conditions.
(a) Calculate $\Delta G°$ at 25°C. Is the reaction spontaneous?
(b) Why is the reaction not performed at 25°C?
(c) Is the reaction spontaneous at 500.°C? (Assume that $\Delta H°$ and $\Delta S°$ are constant with changing T.)
(d) Compare K at 500.°C and at 25°C.
(e) What is the highest T at which the reaction is spontaneous?

22.53 If a chlor-alkali cell used a current of 3×10^4 A, how many pounds of Cl_2 would be produced in a typical 8-h operating day?

22.54 The products of the chlor-alkali process, Cl_2 and NaOH, are kept separated.
(a) Why is this necessary when Cl_2 is the desired product?
(b) ClO^- or ClO_3^- may form by disproportionation of Cl_2 in basic solution. What determines which product forms?
(c) What mole ratio of Cl_2 to OH^- will produce ClO^-? ClO_3^-?

Comprehensive Problems

22.55 The key step in the manufacture of sulfuric acid is the oxidation of sulfur dioxide in the presence of a catalyst, such as V_2O_5. At 727°C, 0.010 mol of SO_2 is injected into an empty 2.00-L container ($K_p = 3.18$).
(a) What is the equilibrium pressure of O_2 that is needed to maintain a 1/1 mole ratio of SO_3 to SO_2?
(b) What is the equilibrium pressure of O_2 needed to maintain a 95/5 mole ratio of SO_3 to SO_2?

22.56 Tetraphosphorus decoxide (P_4O_{10}) is made from phosphate rock and used as a drying agent in the laboratory.
(a) Write a balanced equation for its reaction with water.
(b) What is the pH of a solution formed from the addition of 8.5 g of P_4O_{10} to sufficient water to form 0.750 L? (See Table 18.5, for additional information.)

22.57 Heavy water (D_2O) is used to make deuterated chemicals.
(a) What major species, aside from the starting compounds, do you expect to find in a solution of CH_3OH and D_2O?
(b) Write equations to explain how these various species arise. (*Hint:* Consider the autoionization of both components.)

22.58 A blast furnace uses Fe_2O_3 to produce 8400. t of Fe per day.
(a) What mass of CO_2 is produced each day?
(b) Compare this amount of CO_2 with that produced by 1.0 million automobiles, each burning 5.0 gal of gasoline a day. Assume that gasoline has the formula C_8H_{18} and a density of 0.74 g/mL, and that it burns completely. (Note that U.S. gasoline consumption is over 4×10^8 gal/day.)

22.59 A major use of Cl_2 is in the manufacture of vinyl chloride, the monomer of poly(vinyl chloride). The two-step sequence for formation of vinyl chloride is depicted below.

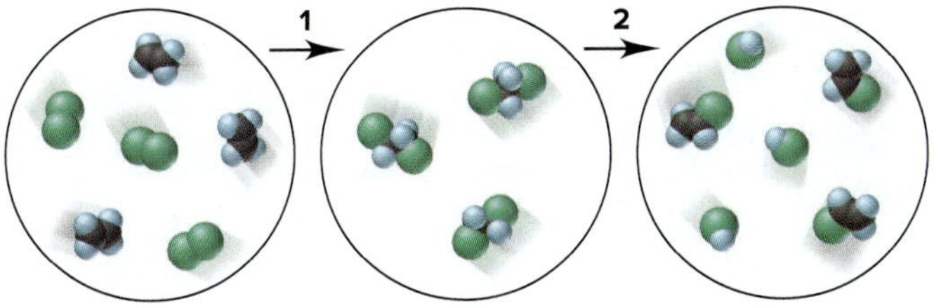

(a) Write a balanced equation for each step.
(b) Write the overall equation.
(c) What type of organic reaction is shown in step 1?
(d) What type of organic reaction is shown in step 2?
(e) If each molecule depicted in the initial reaction mixture represents 0.15 mol of substance, what mass (in g) of vinyl chloride forms?

22.60 In the production of magnesium, $Mg(OH)_2$ is precipitated by using $Ca(OH)_2$, which itself is "insoluble." (a) Use K_{sp} values to show that $Mg(OH)_2$ can be precipitated from seawater in which $[Mg^{2+}]$ is initially 0.052 M. (b) If the seawater is saturated with $Ca(OH)_2$, what fraction of the Mg^{2+} is precipitated?

22.61 Step 1 of the Ostwald process for nitric acid production is

$$4NH_3(g) + 5O_2(g) \xrightarrow{\text{Pt/Rh catalyst}} 4NO(g) + 6H_2O(g)$$

An unwanted side reaction for this step is

$$4NH_3(g) + 3O_2(g) \longrightarrow 2N_2(g) + 6H_2O(g)$$

(a) Calculate K_p for these two NH_3 oxidations at 25°C.
(b) Calculate K_p for these two NH_3 oxidations at 900.°C.
(c) The Pt/Rh catalyst is one of the most efficient in the chemical industry, achieving 96% yield in 1 millisecond of contact with the reactants. However, at normal operating conditions (5 atm and 850°C), about 175 mg of Pt is lost per metric ton (t) of HNO_3 produced. If the annual U.S. production of HNO_3 is 1.01×10^7 t and the market price of Pt is \$1557/troy oz, what is the annual cost of the lost Pt (1 kg = 32.15 troy oz)?
(d) Because of the high price of Pt, a filtering unit composed of ceramic fiber is often installed, which recovers as much as 75% of the lost Pt. What is the value of the Pt captured by a recovery unit with 72% efficiency?

22.62 Several transition metals are prepared by reduction of the metal halide with magnesium. Titanium is prepared by the Kroll method, in which the ore (ilmenite) is converted to the gaseous chloride, which is then reduced to Ti metal by molten Mg (see the discussion on the isolation of magnesium in Section 22.4). Assuming yields of 84% for step 1 and 93% for step 2, and an excess of the other reactants, what mass of Ti metal can be prepared from 21.5 metric tons of ilmenite?

22.63 The production of S_8 from the $H_2S(g)$ found in natural gas deposits occurs through the Claus process (Section 22.5):
(a) Use these two unbalanced steps to write an overall balanced equation for this process:
(1) $H_2S(g) + O_2(g) \longrightarrow S_8(g) + SO_2(g) + H_2O(g)$
(2) $H_2S(g) + SO_2(g) \longrightarrow S_8(g) + H_2O(g)$
(b) Write the overall reaction with Cl_2 as the oxidizing agent instead of O_2. Use thermodynamic data to show whether $Cl_2(g)$ can be used to oxidize $H_2S(g)$.
(c) Why is oxidation by O_2 preferred to oxidation by Cl_2?

22.64 Acid mine drainage (AMD) occurs when geologic deposits containing pyrite (FeS_2) are exposed to oxygen and moisture. AMD is generated in a multistep process catalyzed by acidophilic (acid-loving) bacteria. Balance each step and identify those that increase acidity:
(1) $FeS_2(s) + O_2(g) \longrightarrow Fe^{2+}(aq) + SO_4^{2-}(aq)$
(2) $Fe^{2+}(aq) + O_2(g) \longrightarrow Fe^{3+}(aq) + H_2O(l)$
(3) $Fe^{3+}(aq) + H_2O(l) \longrightarrow Fe(OH)_3(s) + H^+(aq)$
(4) $FeS_2(s) + Fe^{3+}(aq) \longrightarrow Fe^{2+}(aq) + SO_4^{2-}(aq)$

22.65 Below 912°C, pure iron crystallizes in a body-centered cubic structure (ferrite) with d = 7.86 g/cm^3; from 912°C to 1394°C, it adopts a face-centered cubic structure (austenite) with d = 7.40 g/cm^3. Both types of iron form interstitial alloys with carbon. The maximum amount of carbon is 0.0218 mass % in ferrite and 2.08 mass % in austenite. Calculate the density of each alloy.

22.66 Why isn't nitric acid produced by oxidizing N_2 as follows?

$$\begin{array}{lrcl} (1) & N_2(g) + 2O_2(g) & \longrightarrow & 2NO_2(g) \\ (2) & 3NO_2(g) + H_2O(l) & \longrightarrow & 2HNO_3(aq) + NO(g) \\ (3) & 2NO(g) + O_2(g) & \longrightarrow & 2NO_2(g) \\ \hline & 3N_2(g) + 6O_2(g) + 2H_2O(l) & \longrightarrow & 4HNO_3(aq) + 2NO(g) \end{array}$$

(*Hint:* Evaluate the thermodynamics of each step.)

22.67 Before the development of the Downs cell, the Castner cell was used for the industrial production of Na metal. The Castner cell was based on the electrolysis of molten NaOH.
(a) Write balanced cathode and anode half-reactions for this cell.
(b) A major problem with this cell was that the water produced at one electrode diffused to the other and reacted with the Na. If all the water produced reacted with Na, what would be the maximum efficiency of the Castner cell expressed as moles of Na produced per mole of electrons flowing through the cell?

22.68 When gold ores are leached with CN^- solutions, gold forms a complex ion, $Au(CN)_2^-$.
(a) Find E_{cell} for the oxidation in air (P_{O_2} = 0.21) of Au to Au^+ in basic (pH 13.55) solution with $[Au^+]$ = 0.50 M. Is the reaction $Au^+(aq) + e^- \longrightarrow Au(s)$, $E°$ = 1.68 V, spontaneous?
(b) How does formation of the complex ion change $E°$ so that the oxidation can be accomplished?

22.69 Nitric oxide occurs in the tropospheric nitrogen cycle, but it destroys ozone in the stratosphere.
(a) Write equations for its reaction with ozone and for the reverse reaction.
(b) Given that the forward and reverse steps are first order in each component, write general rate laws for them.
(c) Calculate $\Delta G°$ for this reaction at 280. K, the average temperature in the stratosphere. (Assume that the $\Delta H°$ and $S°$ values in Appendix B do not change with temperature.)
(d) What ratio of rate constants is consistent with K at this temperature?

22.70 A key part of the carbon cycle is the fixation of CO_2 by photosynthesis to produce carbohydrates and oxygen gas.
(a) Using the formula $(CH_2O)_n$ to represent a carbohydrate, write a balanced equation for the photosynthetic reaction.
(b) If a tree fixes 48 g of CO_2 per day, what volume of O_2 gas measured at 1.0 atm and 78°F does the tree produce per day?
(c) What volume of air (0.040 mol % CO_2) at the same conditions contains this amount of CO_2?

22.71 Farmers use ammonium sulfate as a fertilizer. In the soil, nitrifying bacteria oxidize NH_4^+ to NO_3^-, a groundwater contaminant that causes methemoglobinemia ("blue baby" syndrome). The World Health Organization standard for maximum $[NO_3^-]$ in groundwater is 45 mg/L. A farmer adds 210. kg of $(NH_4)_2SO_4$ to a field and 37% is oxidized to NO_3^-. What is the groundwater $[NO_3^-]$ (in mg/L) if 1000. m^3 of the water is contaminated?

22.72 The key reaction (unbalanced) in the manufacture of synthetic cryolite for aluminum electrolysis is

$$HF(g) + Al(OH)_3(s) + NaOH(aq) \longrightarrow Na_3AlF_6(aq) + H_2O(l)$$

Assuming a 95.6% yield of dried, crystallized product, what mass (in kg) of cryolite can be obtained from the reaction of 365 kg of $Al(OH)_3$, 1.20 m^3 of 50.0% by mass aqueous NaOH ($d = 1.53$ g/mL), and 265 m^3 of gaseous HF at 305 kPa and 91.5°C? (Assume that the ideal gas law holds.)

22.73 Because of their different molar masses, H_2 and D_2 effuse at different rates (Section 5.5).
(a) If it takes 16.5 min for 0.10 mol of H_2 to effuse, how long does it take for 0.10 mol of D_2 to do so in the same apparatus at the same T and P?
(b) How many effusion steps does it take to separate an equimolar mixture of D_2 and H_2 to 99 mol % purity?

22.74 The disproportionation of CO to graphite and CO_2 is thermodynamically favored but slow.
(a) What does this mean in terms of the magnitudes of the equilibrium constant (K), rate constant (k), and activation energy (E_a)?
(b) Write a balanced equation for the disproportionation of CO.
(c) Calculate K_c at 298 K.
(d) Calculate K_p at 298 K.

22.75 The overall cell reaction for aluminum production is

$$2Al_2O_3(\text{in } Na_3AlF_6) + 3C(\text{graphite}) \longrightarrow 4Al(l) + 3CO_2(g)$$

(a) Assuming 100% efficiency, how many metric tons (t) of Al_2O_3 are consumed per metric ton of Al produced?
(b) Assuming 100% efficiency, how many metric tons of the graphite anode are consumed per metric ton of Al produced?
(c) Actual conditions in an aluminum plant require 1.89 t of Al_2O_3 and 0.45 t of graphite per metric ton of Al. What is the percent yield of Al with respect to Al_2O_3?
(d) What is the percent yield of Al with respect to graphite?
(e) What volume of CO_2 (in m^3) is produced per metric ton of Al at operating conditions of 960.°C and exactly 1 atm?

22.76 World production of chromite ($FeCr_2O_4$), the main ore of chromium, was 1.97×10^7 metric tons in 2006. To isolate chromium, a mixture of chromite and sodium carbonate is heated in air to form sodium chromate, iron(III) oxide, and carbon dioxide. The sodium chromate is dissolved in water, and this solution is acidified with sulfuric acid to produce the less soluble sodium dichromate. The sodium dichromate is filtered out and reduced with carbon to produce chromium(III) oxide, sodium carbonate, and carbon monoxide. The chromium(III) oxide is then reduced to chromium with aluminum metal.
(a) Write balanced equations for each step.
(b) What mass of chromium (in kg) could be prepared from the 2006 world production of chromite?

22.77 Like heavy water (D_2O), "semi-heavy" water (HDO) undergoes H/D exchange. The molecular scenes depict an initial mixture of HDO and H_2 reaching equlibrium.

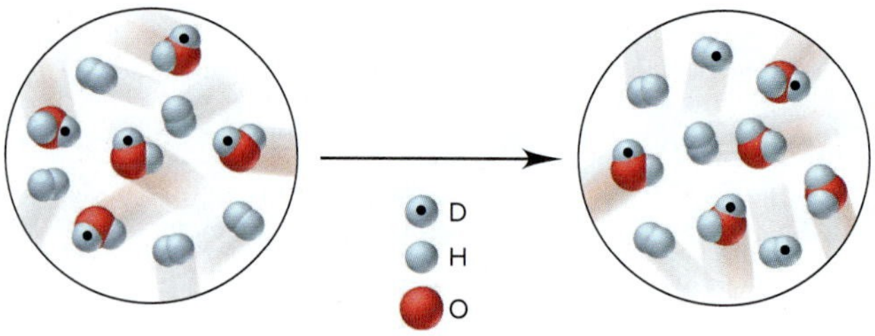

(a) Write the balanced equation for the reaction.
(b) Is the value of K greater or less than 1?
(c) If each molecule depicted represents 0.10 M, calculate K.

22.78 Even though most metal sulfides are sparingly soluble in water, their solubilities differ by several orders of magnitude. This difference is sometimes used to separate the metals by controlling the pH. Use the following data to find the pH at which you can separate 0.10 M Cu^{2+} and 0.10 M Ni^{2+}:
Saturated $H_2S = 0.10$ M
K_{a1} of $H_2S = 9\times10^{-8}$ K_{a2} of $H_2S = 1\times10^{-17}$
K_{sp} of $NiS = 1.1\times10^{-18}$ K_{sp} of $CuS = 8\times10^{-34}$

22.79 Ores with as little as 0.25% by mass of copper are used as sources of the metal.
(a) How many kilograms of such an ore would be needed to construct another Statue of Liberty, which contains 2.0×10^5 lb of copper?
(b) If the mineral in the ore is chalcopyrite ($CuFeS_2$), what is the mass % of chalcopyrite in the ore?

22.80 How does acid rain affect the leaching of phosphate into groundwater from terrestrial phosphate rock? Calculate the solubility of $Ca_3(PO_4)_2$ in each of the following:
(a) Pure water, pH 7.0 (Assume that PO_4^{3-} does not react with water.)
(b) Moderately acidic rainwater, pH 4.5 (*Hint:* Assume that all the phosphate exists in the form that predominates at this pH.)

22.81 The lead(IV) oxide used in car batteries is prepared by coating an electrode plate with PbO and then oxidizing it to lead dioxide (PbO_2). Despite its name, PbO_2 has a nonstoichiometric mole ratio of lead to oxygen of about 1/1.98. In fact, the holes in the PbO_2 crystal structure due to missing O atoms are responsible for the oxide's conductivity.
(a) What is the mole % of O missing from the PbO_2 structure?
(b) What is the molar mass of the nonstoichiometric compound?

22.82 Chemosynthetic bacteria reduce CO_2 by "splitting" $H_2S(g)$ instead of the $H_2O(g)$ used by photosynthetic organisms. Compare the free energy change for splitting H_2S with that for splitting H_2O. Is there an advantage to using H_2S instead of H_2O?

22.83 Silver has a face-centered cubic structure with a unit cell edge length of 408.6 pm. Sterling silver is a substitutional alloy that contains 7.5% copper atoms. Assuming the unit cell remains the same, find the density of silver and of sterling silver.

22.84 Earth's mass is estimated to be 5.98×10^{24} kg, and titanium represents 0.05% by mass of this total.
(a) How many moles of Ti are present?
(b) If half of the Ti is found as ilmenite ($FeTiO_3$), what mass of ilmenite is present?
(c) If the airline and auto industries use 1.00×10^5 tons of Ti per year, how many years will it take to use up all the Ti (1 ton = 2000 lb)?

22.85 In 1790, Nicolas Leblanc found a way to form Na_2CO_3 from NaCl. His process, now obsolete, consisted of three steps:

$$2NaCl(s) + H_2SO_4(aq) \longrightarrow Na_2SO_4(aq) + 2HCl(g)$$
$$Na_2SO_4(s) + 2C(s) \longrightarrow Na_2S(s) + 2CO_2(g)$$
$$Na_2S(s) + CaCO_3(s) \longrightarrow Na_2CO_3(s) + CaS(s)$$

(a) Write a balanced overall equation for the process.
(b) Calculate the ΔH_f° of CaS if ΔH_{rxn}° is 351.8 kJ/mol.
(c) Is the overall process spontaneous at standard-state conditions and 298 K?
(d) How many grams of Na_2CO_3 form from 250. g of NaCl if the process is 73% efficient?

22.86 Limestone ($CaCO_3$) is the second most abundant mineral on Earth after SiO_2. For many uses, it is first decomposed thermally to quicklime (CaO). MgO is prepared similarly from $MgCO_3$.
(a) At what T is each decomposition spontaneous?
(b) Quicklime reacts with SiO_2 to form a slag ($CaSiO_3$), a by-product of steelmaking. In 2010, the total steelmaking capacity of the U.S. steel industry was 2,236,000 tons per week, but only 84% of this capacity was utilized. If 50. kg of slag is produced per ton of steel, what mass (in kg) of limestone was used to make slag in 2010?

23 Transition Elements and Their Coordination Compounds

23.1 Properties of the Transition Elements
Electron Configurations
Atomic and Physical Properties
Chemical Properties

23.2 The Inner Transition Elements
The Lanthanides
The Actinides

23.3 Coordination Compounds
Complex Ions
Formulas and Names
Isomerism

23.4 Theoretical Basis for the Bonding and Properties of Complex Ions
Valence Bond Theory
Crystal Field Theory

Source: © PjrStudio/Alamy

Concepts and Skills to Review Before You Study This Chapter

- properties of light (Section 7.1)
- electron shielding of nuclear charge (Section 8.1)
- electron configuration, ionic size, and magnetic behavior (Sections 8.2 to 8.4)
- valence bond theory (Section 11.1)
- constitutional, geometric, and optical isomerism (Section 15.2)
- Lewis acid-base concepts (Section 18.9)
- complex-ion formation (Section 19.4)
- redox behavior and standard electrode potentials (Section 21.3)

Though almost at the end of the text, we still haven't discussed the majority of the elements *and* some of the most familiar. The **transition elements** *(transition metals)* make up the *d* block (B groups) and *f* block *(inner transition elements)* of the periodic table (Figure 23.1, *next page*) and have crucial uses in industry and biology.

Aside from essential copper and iron (Chapter 22), some other indispensable transition elements are chromium (plumbing fixtures), gold and silver (jewelry and electronics), platinum (automobile catalytic converters), titanium (bicycle and aircraft parts and artificial joints), nickel (coins and catalysts), and zinc (batteries), to mention a few of the better known ones. There are also the lesser known zirconium (nuclear-reactor liners), vanadium (axles and crankshafts), molybdenum (boiler plates), tantalum (organ-replacement parts), palladium (telephone-relay contacts)—the list goes on and on.

Transition metal cations occupy the centers of many complex ions. In addition to providing fundamental insights into bonding and structure, these species play vital roles in organisms and give color to gems—Cr^{3+} ions make emeralds green and rubies red, and a mixture of Fe^{2+}, Fe^{3+}, and Ti^{4+} ions makes sapphires blue *(see photo)*.

IN THIS CHAPTER . . . *We focus on the transition elements and their properties, and especially on the bonding and structure of their coordination compounds.*

- We discuss key atomic, physical, and chemical properties of transition and inner transition elements—with attention to the effects of filling inner sublevels—and contrast them with properties of the main-group elements.
- For most of the chapter, we concentrate on the nomenclature, structure, and isomerism of transition elements' coordination compounds, species that contain complex ions.
- We use valence bond theory to explain the bonding and molecular shape of complex ions.
- We use crystal field theory to explain the colors and magnetic properties of coordination compounds, which relate directly to structure.
- We end with some examples of the roles of transition metal ions in biochemistry.

23.1 PROPERTIES OF THE TRANSITION ELEMENTS

The transition elements differ considerably in physical and chemical behavior from the main-group elements:

- *All transition elements are metals,* whereas the main-group elements in each period change from metal to nonmetal.
- *Many transition metal compounds are colored and paramagnetic,* whereas most main-group ionic compounds are colorless and diamagnetic.

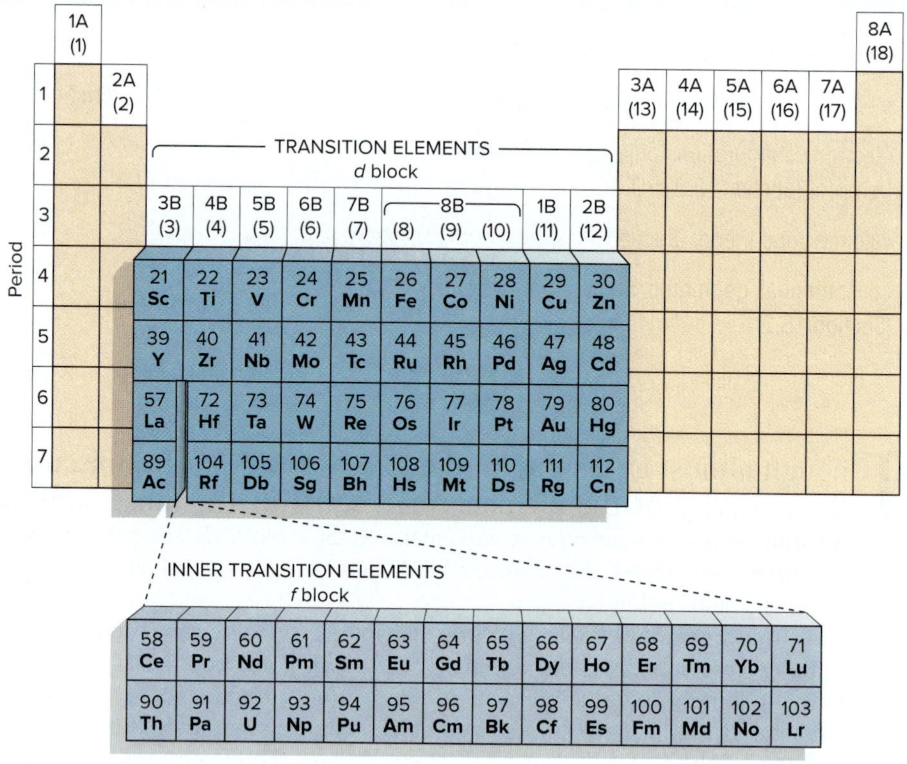

Figure 23.1 The transition elements (*d* block) and inner transition elements (*f* block) in the periodic table.

We first discuss electron configurations of the atoms and ions of transition elements and then examine these elements' key properties to see how they contrast with the same properties of main-group elements.

Electron Configurations of the Transition Metals and Their Ions

As with all the elements, the properties of the transition elements and their compounds arise largely from the electron configurations of their atoms (Section 8.2) and ions (Section 8.4). As Figure 23.1 shows, *d*-block (B-group) elements occur in four series that lie within Periods 4 through 7 between the last *ns*-block element [Group 2A(2)] and the first *np*-block element [Group 3A(13)]. Each of the four series contains ten elements based on the filling of five *d* orbitals, for a total of 40 transition elements. Lying between the first and second members of the series in Periods 6 and 7 are the inner transition elements, in which *f* orbitals are filled. Several points are important to review:

1. *Electron configurations of the atoms*. Despite several exceptions, in general, the *condensed* ground-state electron configuration for the atoms in each *d*-block series is

$$[\text{noble gas}]\ ns^2(n-1)d^x, \text{ with } n = 4 \text{ to } 7 \text{ and } x = 1 \text{ to } 10$$

In Periods 6 and 7, the condensed configuration includes the *f* sublevel:

$$[\text{noble gas}]\ ns^2(n-2)f^{14}(n-1)d^x, \text{ with } n = 6 \text{ or } 7$$

The *partial* (valence-level) electron configuration for the *d*-block elements excludes the noble gas core and the filled inner *f* sublevel:

$$ns^2(n-1)d^x$$

Figure 23.2 The Period 4 transition metals. The ten elements appear in periodic-table order across pages 1038–1039.

Source: © McGraw-Hill Education/Stephen Frisch, photographer

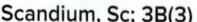

Scandium, Sc; 3B(3) **Titanium, Ti; 4B(4)** **Vanadium, V; 5B(5)** **Chromium, Cr; 6B(6)** **Manganese, Mn; 7B(7)**

Table 23.1 Orbital Occupancy of the Period 4 Transition Metals

Element	Partial Orbital Diagram 4*s*	3*d*	4*p*	Unpaired Electrons
Sc	[↑↓]	[↑][ ][ ][ ][ ]	[ ][ ][ ]	1
Ti	[↑↓]	[↑][↑][ ][ ][ ]	[ ][ ][ ]	2
V	[↑↓]	[↑][↑][↑][ ][ ]	[ ][ ][ ]	3
Cr	[↑]	[↑][↑][↑][↑][↑]	[ ][ ][ ]	6
Mn	[↑↓]	[↑][↑][↑][↑][↑]	[ ][ ][ ]	5
Fe	[↑↓]	[↑↓][↑][↑][↑][↑]	[ ][ ][ ]	4
Co	[↑↓]	[↑↓][↑↓][↑][↑][↑]	[ ][ ][ ]	3
Ni	[↑↓]	[↑↓][↑↓][↑↓][↑][↑]	[ ][ ][ ]	2
Cu	[↑]	[↑↓][↑↓][↑↓][↑↓][↑↓]	[ ][ ][ ]	1
Zn	[↑↓]	[↑↓][↑↓][↑↓][↑↓][↑↓]	[ ][ ][ ]	0

2. *Filling pattern in Period 4.* The first (Period 4) transition series consists of scandium (Sc) through zinc (Zn) (Figure 23.2 and Table 23.1). Scandium has the electron configuration [Ar] $4s^2 3d^1$, and the addition of one electron at a time first half-fills, then fills, the 3*d* orbitals, up to zinc. Recall that chromium and copper are exceptions to this pattern:

- The 4*s* and 3*d* orbitals in Cr are half-filled to give [Ar] $4s^1 3d^5$.
- The 4*s* orbital in Cu is half-filled to give [Ar] $4s^1 3d^{10}$.

These exceptions are due to the relative energies of the 4*s* and 3*d* sublevels as electrons are added and to the unusual stability of half-filled and filled sublevels (see Figure 8.26).

3. *Electron configurations of the ions.* Transition metal ions form through *the loss of ns electrons before (n − 1)d electrons.* Thus, Ti loses the two 4*s* electrons to form the Ti^{2+} ion:

$$\text{Ti: [Ar] } 4s^2 3d^2 \qquad \text{Ti}^{2+}\text{: [Ar] } 3d^2,\ \textit{not}\ \text{[Ar] } 4s^2$$

Ti^{2+} is referred to as a d^2 ion. Ions of different metals with the same configuration often have similar properties. For example, Mn^{2+} and Fe^{3+} are d^5 ions:

$$\text{Mn: [Ar] } 4s^2 3d^5 \qquad \text{Mn}^{2+}\text{: [Ar] } 3d^5 \qquad \text{Fe: [Ar] } 4s^2 3d^6 \qquad \text{Fe}^{3+}\text{: [Ar] } 3d^5$$

Both have pale colors in solution and form complex ions with similar magnetic properties.

Table 23.1 shows partial orbital box diagrams for the Period 4 elements. In general for all periods of the transition elements, the number of *unpaired* electrons (or half-filled orbitals) *increases in the first half* of the series and, when pairing begins, *decreases in the second half.* A key point to note in upcoming discussions is that the electron configuration of the *atom* (O.N. = 0) correlates with the physical properties

Iron, Fe; 8B(8)

Cobalt, Co; 8B(9)

Nickel, Ni; 8B(10)

Copper, Cu; 1B(11)

Zinc, Zn; 2B(12)

of the *element,* whereas the electron configuration of the *ion* (and other species of higher O.N.) correlates with the chemical properties of the *compounds.*

SAMPLE PROBLEM 23.1 **Writing Electron Configurations of Transition Metal Atoms and Ions**

Problem Write *condensed* electron configurations for the following: **(a)** Zr; **(b)** V^{3+}; **(c)** Mo^{3+}. (Assume that transition elements in higher periods behave like those in Period 4.)

Plan We locate the element in the periodic table and count its position in the respective transition series. These elements are in Periods 4 and 5, so the general configuration is [noble gas] $ns^2(n-1)d^x$. For the ions, we recall that *ns* electrons are lost first.

Solution **(a)** Zr is the second element in the 4*d* series: [Kr] $5s^24d^2$.

(b) V is the third element in the 3*d* series: [Ar] $4s^23d^3$. In forming V^{3+}, three electrons are lost (two 4*s* and one 3*d*), so V^{3+} is a d^2 ion: [Ar] $3d^2$.

(c) Mo lies below Cr in Group 6B(6), so we expect the same exception as for Cr. Thus, Mo has the electron configuration [Kr] $5s^14d^5$. To form the ion, Mo loses the one 5*s* and two of the 4*d* electrons, so Mo^{3+} is a d^3 ion: [Kr] $4d^3$.

Check Figure 8.9 shows we're correct for the atoms. Be sure that charge plus number of *d* electrons in the ion equals the sum of outer *s* and *d* electrons in the atom.

FOLLOW-UP PROBLEMS

Brief Solutions for ***all*** *Follow-up Problems appear at the end of the chapter.*

23.1A Write *partial* electron configurations (no noble gas core or filled inner sublevels) for the following: **(a)** Ag^+; **(b)** Cd^{2+}; **(c)** Ir^{3+}.

23.1B Find the charge of each of the following ions, given the *condensed* electron configuration: **(a)** Ta^{x+}: [Xe] $4f^{14}5d^2$; **(b)** Mn^{x+}: [Ar] $3d^3$; **(c)** Os^{x+}: [Xe] $4f^{14}5d^5$.

SOME SIMILAR PROBLEMS 23.10–23.15

Atomic and Physical Properties of the Transition Elements

Properties of the transition elements contrast in several ways with properties of the main-group elements.

Trends Across a Period Consider the variations in atomic size, electronegativity, and ionization energy across Period 4 (Figure 23.3):

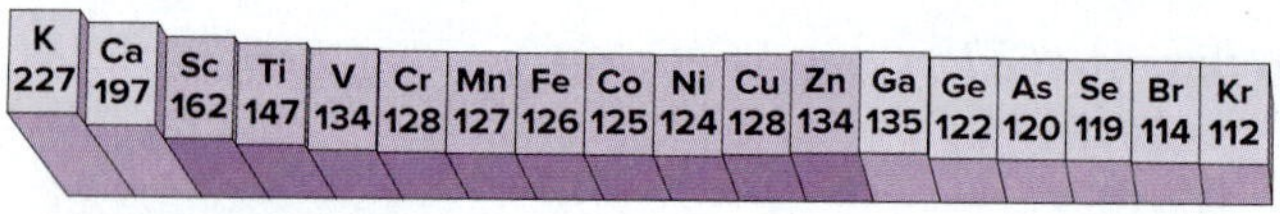

A Atomic radius (pm)

- *Atomic size.* Atomic size decreases overall across the period (Figure 23.3A). There is a smooth, steady decrease across the main groups because the electrons are added to *outer ns* or *np* orbitals, which shield the increasing nuclear charge poorly. This decrease is not steady in the transition series, where *atomic size decreases at first but then remains relatively constant.* The reason is that the *d* electrons fill *inner* orbitals, so they shield outer electrons very efficiently; therefore, the outer 4*s* electrons are *not* pulled closer.

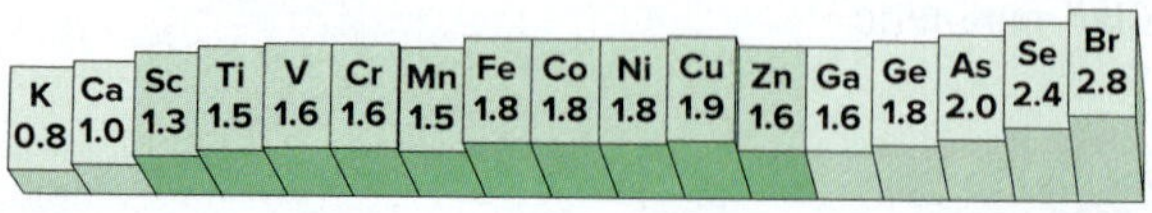

B Electronegativity

- *Electronegativity (EN).* Electronegativity generally increases across the period, but, once again, the main groups show a steady, steep increase between the metal potassium (0.8) and the nonmetal bromine (2.8), whereas the transition elements have *relatively constant EN* (Figure 23.3B), consistent with their relatively constant size. The transition elements all have intermediate electronegativity values, much like the large, metallic members of Groups 3A(13) to 5A(15).

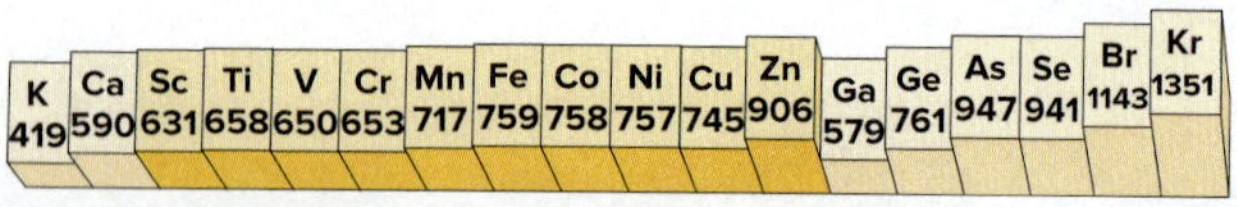

C First ionization energy (kJ/mol)

Figure 23.3 **Trends in key atomic properties of Period 4 elements.** Relative sizes of atomic radius **(A)**, electronegativity **(B)**, and first ionization energy **(C)** for all Period 4 elements are indicated by the heights of the bars, with darker shading on the transition series.

- *Ionization energy (IE_1).* The ionization energies of the Period 4 main-group elements rise steeply from left to right, more than tripling from potassium (419 kJ/mol) to krypton (1351 kJ/mol), as electrons become more difficult to remove from the poorly shielded, increasing nuclear charge. In the transition metals, *IE_1 values increase relatively little* because the inner 3*d* electrons shield more effectively (Figure 23.3C). [Recall from Section 8.3 that the drop at Group 3A(13) occurs because it is relatively easy to remove the first electron from the outer *np* orbital.]

Trends Within a Group Vertical trends for transition elements are also different from main-group trends.

1. *Atomic size.* As expected, atomic size increases from Period 4 to 5, as it does for the main-group elements, but there is virtually *no size increase from Period 5 to 6* (Figure 23.4A). Since the lanthanides, with their buried 4*f* sublevel, appear between the 4*d* (Period 5) and 5*d* (Period 6) series, an element in Period 6 is separated from the one above it by 32 elements (ten 4*d*, six 5*p*, two 6*s*, and fourteen 4*f* electrons) instead of just 18. The fourteen 4*f* electrons shield the outer electrons poorly from the increase in nuclear charge due to fourteen additional protons, resulting in a shrinkage in atomic size called the **lanthanide contraction.** By coincidence, this size *decrease* is about equal to the normal *increase* between periods, so Periods 5 and 6 transition elements have about the same atomic sizes.

2. *Electronegativity.* The vertical trend in electronegativity is opposite the decreasing trend in the main groups. Electronegativity *increases* from Period 4 to Period 5, but there is no further increase in Period 6 (Figure 23.4B). The heavier elements, especially gold (EN = 2.4), become quite electronegative, with values higher than most metalloids and even some nonmetals (e.g., EN of Te and of P = 2.1). (In fact, with the very electropositive cesium, which has EN = 0.7, gold forms the saltlike CsAu.) Although atomic size increases slightly from top to bottom in a group, the nuclear charge increases much more. Therefore, the heavier transition metals form bonds with more covalent character because they attract electrons more strongly than do main-group metals.

3. *Ionization energy.* The small increase in size combined with the large increase in nuclear charge also explains why *IE_1 values generally increase* down a transition group (Figure 23.4C). This trend also runs counter to the main-group trend, where heavier members are so much larger that the outer electron is easier to remove.

4. *Density.* Atomic size, and therefore volume, is inversely related to density. Across a period, densities increase, then level off, and finally dip a bit at the end of

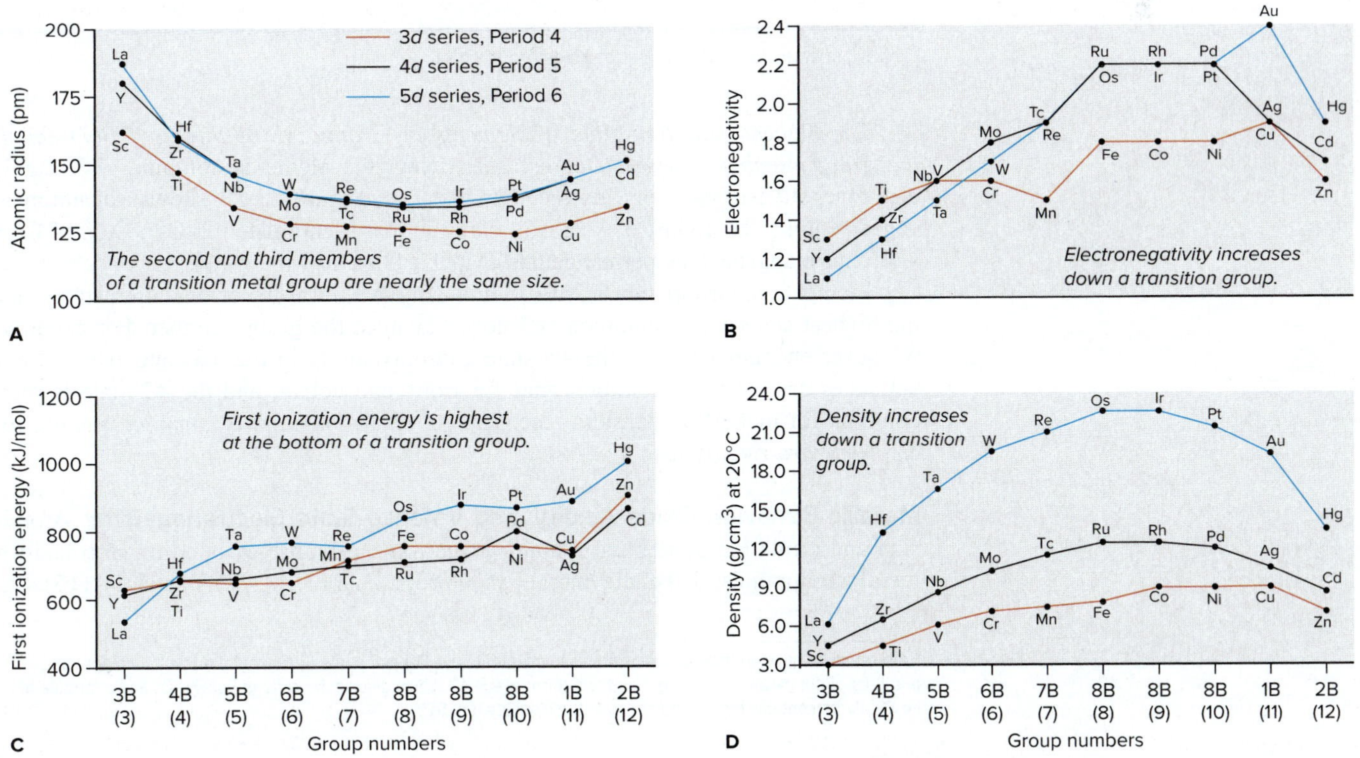

Figure 23.4 **Vertical trends in key properties within the transition elements.** The trends are unlike those for the main-group elements.

Figure 23.5 Aqueous oxoanions of transition elements.

Source: © McGraw-Hill Education/Stephen Frisch, photographer

a series (Figure 23.4D). Down a transition group, densities increase dramatically because atomic volumes change little from Period 5 to 6, but atomic masses increase significantly. As a result, the Period 6 series contains some of the densest elements known: tungsten, rhenium, osmium, iridium, platinum, and gold have densities about 20 times that of water and twice that of lead.

Chemical Properties of the Transition Elements

Like their atomic and physical properties, the chemical properties of transition elements are very different from those of main-group elements. Let's examine key properties in the Period 4 series and then see how behavior changes within a group.

Multiple Oxidation States One of the most characteristic chemical properties of the transition metals is the occurrence of *multiple oxidation states;* main-group metals display one or, at most, two states. For example, chromium and manganese (Figure 23.5A) have three common oxidation states and many others that are less common (Table 23.2). Since the *ns* and $(n - 1)d$ electrons are so close in energy, transition elements can use all or most of these electrons in bonding.

Table 23.2 Oxidation States and *d*-Orbital Occupancy of the Period 4 Transition Metals*

Oxidation State	3B (3) Sc	4B (4) Ti	5B (5) V	6B (6) Cr	7B (7) Mn	8B (8) Fe	8B (9) Co	8B (10) Ni	1B (11) Cu	2B (12) Zn
0	d^1	d^2	d^3	d^5	d^5	d^6	d^7	d^8	d^{10}	d^{10}
+1			d^3	d^5	d^5	d^6	d^7	d^8	d^{10}	
+2		d^2	d^3	d^4	d^5	d^6	d^7	d^8	d^9	d^{10}
+3	d^0	d^1	d^2	d^3	d^4	d^5	d^6	d^7	d^8	
+4		d^0	d^1	d^2	d^3	d^4	d^5	d^6		
+5			d^0	d^1	d^2		d^4			
+6				d^0	d^1	d^4				
+7					d^0					

*The most important orbital occupancies are in color.

The highest oxidation state of elements in Groups 3B(3) through 7B(7) equals the group number. These states are seen when the elements combine with highly electronegative oxygen or fluorine. For instance, Figure 23.5B shows vanadium as vanadate ion (VO_4^{3-}; O.N. of V = +5), chromium as dichromate ($Cr_2O_7^{2-}$; O.N. of Cr = +6), and manganese as permanganate (MnO_4^-; O.N. of Mn = +7).

Elements in Groups 8B(8), 8B(9), and 8B(10) exhibit fewer oxidation states, and the highest state is less common and never equal to the group number. For example, we never encounter iron in the +8 state and only rarely in the +6 state. The +2 and +3 states are the most common ones for iron* and cobalt, and the +2 state is most common for nickel, copper, and zinc. *The +2 oxidation state is common because ns^2 electrons are readily lost.*

Metallic Behavior, Oxide Acidity, and Valence-State Electronegativity Atomic size and oxidation state have a major effect on the nature of bonding in transition metal compounds. Like the metals in Groups 3A(13), 4A(14), and 5A(15), the

*Iron *seems* to have unusual oxidation states in magnetite (Fe_3O_4) and pyrite (FeS_2), but it doesn't. In magnetite, one-third of the metal ions are Fe^{2+} and two-thirds are Fe^{3+}, which gives a 1/1 ratio of FeO/Fe_2O_3 and a formula of Fe_3O_4. Pyrite contains Fe^{2+} combined with the disulfide ion, S_2^{2-}.

transition elements in their *lower* oxidation states behave chemically more like metals. That is, *ionic bonding is more prevalent for the lower oxidation states, and covalent bonding is more prevalent for the higher states.* For example, at room temperature, $TiCl_2$ (O.N. = +2) is an ionic solid, whereas $TiCl_4$ (O.N. = +4) is a molecular liquid (see also Figure 14.11). In the higher oxidation states, the atoms have higher charge densities (higher ratio of charge to volume, Section 13.3), so they polarize the electron clouds of the nonmetal ions more strongly and the bonding becomes more covalent. For the same reason, the oxides become less basic (more acidic) as the oxidation state increases: TiO is weakly basic in water, whereas TiO_2 is amphoteric (reacts with both acid and base).

Why does oxide acidity increase with oxidation state? And, how can a metal like chromium or manganese, in addition to forming oxides, form an oxoanion in which it uses covalent bonds? The answers involve a type of "effective" electronegativity called *valence-state electronegativity,* which also has numerical values. A metal atom with a positive oxidation state has a greater attraction for the bonded electrons, that is, a higher electronegativity, than it does when it has a zero oxidation state. Thus, for example, the electronegativity of elemental chromium is 1.6, close to that of aluminum (1.5), another active metal. For chromium(III), the value increases to 1.7, still characteristic of a metal. But, for chromium(VI), the value is 2.3, close to the values of some nonmetals, such as phosphorus (2.1) and sulfur (2.5). Thus, like P in PO_4^{3-} and S in SO_4^{2-}, Cr in CrO_4^{2-} is covalently bonded at the center of an oxoanion of a relatively strong acid, H_2CrO_4, and manganese(VII) in MnO_4^- behaves similarly in the strong acid $HMnO_4$.

Reducing Strength Table 23.3 shows the standard electrode potentials of the Period 4 transition metals in their +2 oxidation state in acid solution. Note that, in general, reducing strength decreases across the series. All the Period 4 transition metals, except copper, are active enough to reduce H^+ from aqueous acid to form hydrogen gas. In contrast to the rapid reaction at room temperature of the Group 1A(1) and 2A(2) metals with water, however, most transition metals have an oxide coating that allows rapid reaction only with hot water or steam.

Table 23.3 Standard Electrode Potentials of M^{2+} Ions of Period 4 Transition Metals

Half-Reaction	$E°$ (V)
$Ti^{2+}(aq) + 2e^- \rightleftharpoons Ti(s)$	−1.63
$V^{2+}(aq) + 2e^- \rightleftharpoons V(s)$	−1.19
$Cr^{2+}(aq) + 2e^- \rightleftharpoons Cr(s)$	−0.91
$Mn^{2+}(aq) + 2e^- \rightleftharpoons Mn(s)$	−1.18
$Fe^{2+}(aq) + 2e^- \rightleftharpoons Fe(s)$	−0.44
$Co^{2+}(aq) + 2e^- \rightleftharpoons Co(s)$	−0.28
$Ni^{2+}(aq) + 2e^- \rightleftharpoons Ni(s)$	−0.25
$Cu^{2+}(aq) + 2e^- \rightleftharpoons Cu(s)$	0.34
$Zn^{2+}(aq) + 2e^- \rightleftharpoons Zn(s)$	−0.76

Color and Magnetism of Compounds *Most main-group ionic compounds are colorless* because the metal ion has a filled outer level (noble gas electron configuration, ns^2 or ns^2np^6). With only *much* higher energy orbitals available to receive an excited electron, the ion does not absorb visible light. In contrast, electrons in a partially filled *d* sublevel can absorb visible wavelengths and move to *slightly* higher energy orbitals. As a result, *many transition metal compounds have striking colors.* Exceptions are the compounds of scandium, titanium(IV), and zinc, which are colorless because their metal ions have either an empty *d* sublevel (Sc^{3+} or Ti^{4+}: [Ar] $3d^0$) or a filled one (Zn^{2+}: [Ar] $3d^{10}$) (Figure 23.6).

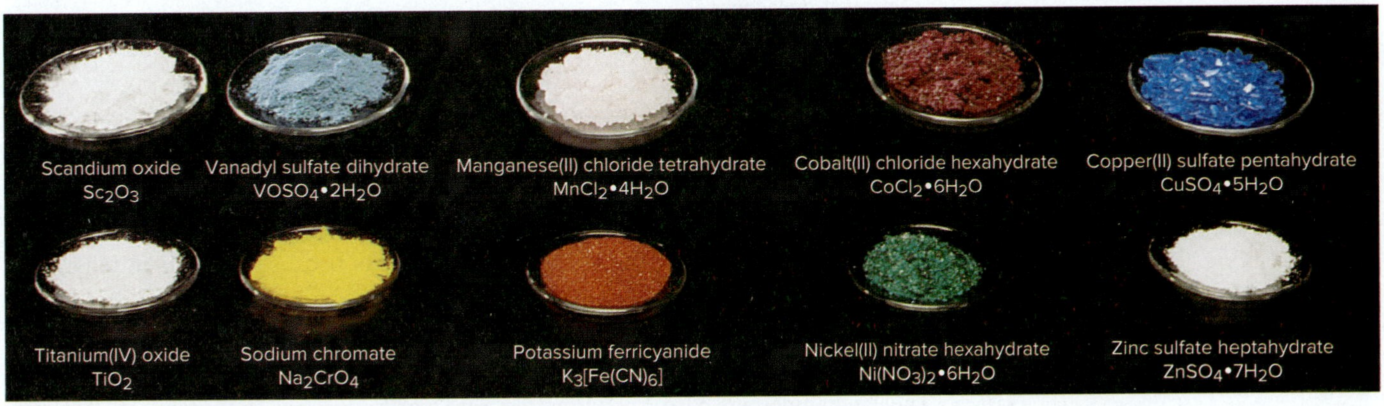

Figure 23.6 Colors of representative compounds of the Period 4 transition metals.
Source: © McGraw-Hill Education/Stephen Frisch, photographer

Table 23.4 Some Properties of Group 6B(6) Elements

Element	Atomic Radius (pm)	IE_1 (kJ/mol)	$E°$ (V) for $M^{3+}(aq) \mid M(s)$
Cr	128	653	−0.74
Mo	139	685	−0.20
W	139	770	−0.11

Magnetic properties are also related to sublevel occupancy (Section 8.4). *Most main-group metal ions are diamagnetic* for the same reason they are colorless: all their electrons are paired. In contrast, *many transition metal compounds are paramagnetic because of their unpaired d electrons.* Transition metal ions with a d^0 or d^{10} configuration are also colorless and diamagnetic.

Chemical Behavior Within a Group The *increase* in reactivity going down a group of main-group metals due to the *decrease* in IE_1 does *not* occur going down a group of transition metals. The chromium (Cr) group [6B(6)] shows a typical pattern (Table 23.4). IE_1 *increases* down the group, which makes the two heavier members *less* reactive than the lightest one. Chromium is also a much stronger reducing agent than molybdenum (Mo) or tungsten (W), as shown by the standard electrode potentials.

The similarity in atomic size of members of Periods 5 and 6 leads to similar chemical behavior, which has some important consequences. Because Mo and W compounds behave similarly, their ores often occur together in nature, which makes these elements very difficult to separate. The same situation occurs with zirconium and hafnium in Group 4B(4) and with niobium and tantalum in Group 5B(5).

› Summary of Section 23.1

- All transition elements are metals.
- Atoms of the *d*-block elements have $(n - 1)d$ orbitals being filled, and their ions have an empty *ns* orbital.
- Unlike the trends for the main-group elements, the atomic size, electronegativity, and first ionization energy change relatively little across a transition series. Because of the lanthanide contraction, atomic size changes little from Period 5 to 6 in a transition metal group; thus, electronegativity, first ionization energy, and density *increase* down such a group.
- Transition metals typically have several oxidation states, with the +2 state most common. The elements exhibit less metallic behavior in their higher states, as they have higher valence-state electronegativity.
- Most Period 4 transition metals are active enough to reduce hydrogen ion from acid solution.
- Many transition metal compounds are colored and paramagnetic because the metal ion has unpaired *d* electrons.
- In contrast to main-group metals, transition metals show decreasing reactivity down a group.

23.2 THE INNER TRANSITION ELEMENTS

The 14 **lanthanides** in Period 6 [cerium (Ce; $Z = 58$) through lutetium (Lu; $Z = 71$)] and the 14 **actinides** in Period 7 [thorium (Th; $Z = 90$) through lawrencium (Lr; $Z = 103$)] are called **inner transition elements** because their seven inner 4*f* or 5*f* orbitals are being filled.

The Lanthanides

The lanthanides are also known as the *rare earth elements,* but many are not rare: cerium (Ce), for instance, ranks 26th in abundance (by mass %) and is five times more abundant than lead. All the lanthanides are silvery, high-melting metals.

As with other transition elements, atomic properties vary little across the period. As a result, chemical properties are so similar that the two ores of lanthanides, ceria and yttria, are mixtures of compounds of all 14 of them. This natural co-occurrence arises because the elements exist as M^{3+} ions of very similar radii. Most lanthanides have the ground-state electron configuration [Xe] $6s^2 4f^x 5d^0$, where x varies across the series. The three exceptions (Ce, Gd, and Lu) have a single electron in one of their $5d$ orbitals: cerium ([Xe] $6s^2 4f^1 5d^1$) forms a stable 4+ ion with an empty (f^0) sublevel, and the Gd^{3+} and Lu^{3+} ions have a stable half-filled (f^7) or filled (f^{14}) sublevel.

Lanthanide compounds are used in tinted glass, electronic devices (magnets, batteries, and lasers), and high-quality camera lenses, but two industrial applications account for over 60% of the total usage. In gasoline refining, catalysts used to "crack" hydrocarbons into smaller molecules are 5% by mass rare earth oxides. In steelmaking, a mixture of lanthanides, called *misch metal,* removes carbon impurities from molten iron and steel. Recently, shortages of some lanthanides used for certain electronic applications have been a concern.

SAMPLE PROBLEM 23.2 **Finding the Number of Unpaired Electrons**

Problem The alloy $SmCo_5$ forms a permanent magnet because both samarium and cobalt have unpaired electrons. How many unpaired electrons are in Sm ($Z = 62$)?

Plan We write the condensed electron configuration of Sm and then, using Hund's rule and the aufbau principle, place the electrons in a partial orbital diagram and count the unpaired electrons.

Solution Samarium is the eighth element after Xe. Two electrons go into the $6s$ sublevel. In general, the $4f$ sublevel fills before the $5d$ (among the lanthanides, only Ce, Gd, and Lu have $5d$ electrons), so the remaining electrons go into the $4f$. Thus, the condensed configuration of Sm is [Xe] $6s^2 4f^6$. There are seven f orbitals, so each of the six f electrons enters a separate orbital:

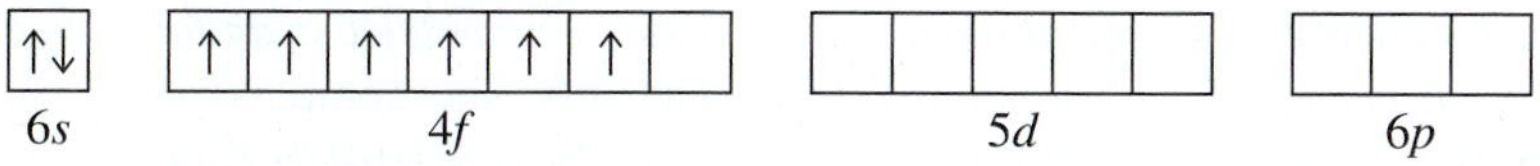

Thus, Sm has six unpaired electrons.

Check Six $4f$ e^- plus two $6s$ e^- plus the 54 e^- in Xe gives 62, the atomic number of Sm.

FOLLOW-UP PROBLEMS

23.2A How many unpaired electrons are in the Er^{3+} ion?

23.2B How many unpaired electrons are in the Dy^{3+} ion?

SOME SIMILAR PROBLEMS 23.30–23.34

The Actinides

All actinides are radioactive. Like the lanthanides, they have very similar physical and chemical properties. Thorium and uranium occur in nature, but the transuranium elements, those with Z greater than 92, have been synthesized in particle accelerators (as we discuss in Section 24.3). The actinides are silvery and chemically reactive and, like the lanthanides, form highly colored compounds. The actinides and lanthanides have similar outer-electron configurations. Although the +3 oxidation state is characteristic of the actinides, as it is for the lanthanides, other states also occur. For example, uranium exhibits +3 through +6 states, with the +6 state the most prevalent; thus, the most common oxide of uranium is UO_3.

› Summary of Section 23.2

› There are two series of inner transition elements. The lanthanides ($4f$ series) have a common +3 oxidation state and exhibit very similar properties.

› The actinides ($5f$ series) are radioactive. All actinides have a +3 oxidation state; several, including uranium, have higher states as well.

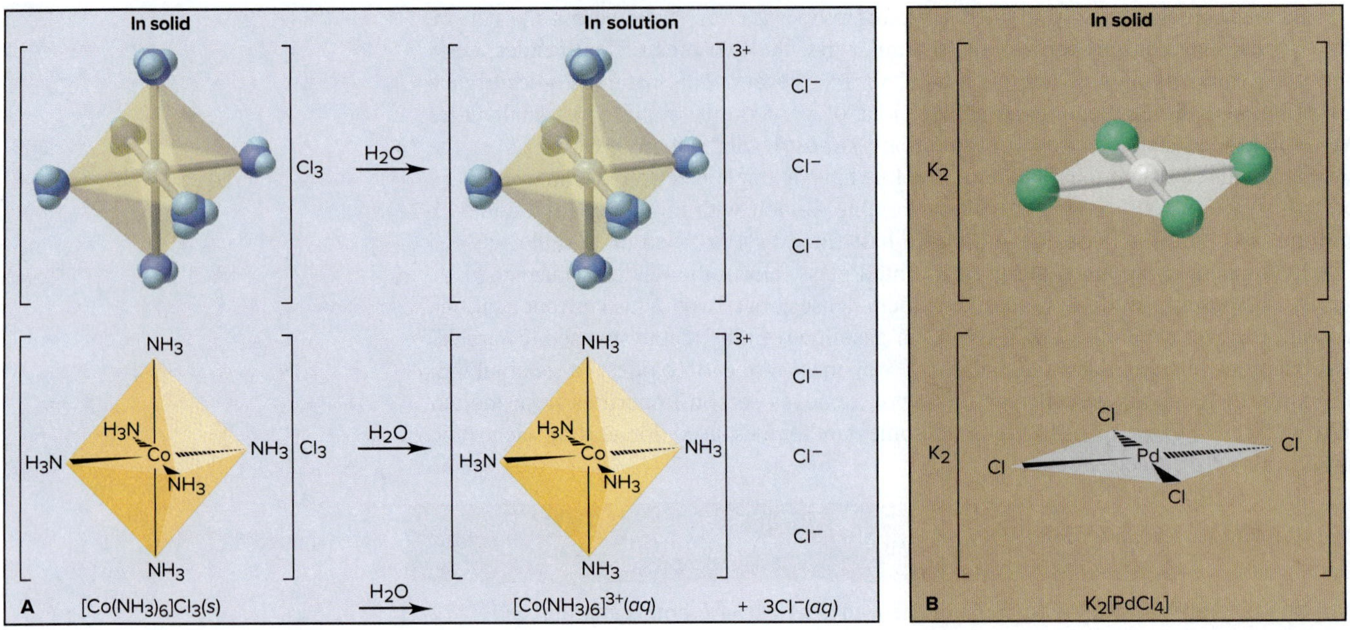

Figure 23.7 Components of a coordination compound. Coordination compounds are shown as models *(top)*, wedge-bond perspective drawings *(middle)*, and formulas *(bottom)*. **A,** When $[Co(NH_3)_6]Cl_3$ dissolves in water, the six ligands remain bound in the complex ion. **B,** $K_2[PdCl_4]$ has four Cl^- ligands and two K^+ counter ions.

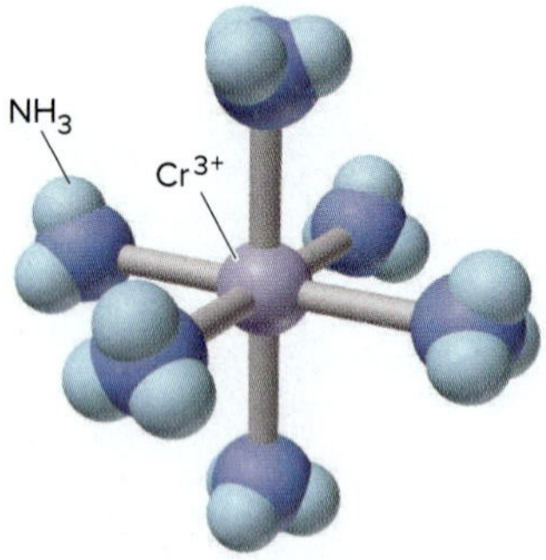

The complex ion $[Cr(NH_3)_6]^{3+}$ has a central Cr^{3+} ion bonded to six NH_3 ligands.

23.3 COORDINATION COMPOUNDS

The most distinctive feature of transition metal chemistry is the common occurrence of **coordination compounds** (or *complexes*). These species contain at least one **complex ion,** which consists of *a central metal cation bonded to molecules and/or anions called* ***ligands*** *(see margin)*. To maintain charge neutrality, the complex ion is associated with **counter ions.** In the coordination compound $[Co(NH_3)_6]Cl_3$ (Figure 23.7A), the complex ion (in square brackets) is $[Co(NH_3)_6]^{3+}$, the six NH_3 molecules bonded to the Co^{3+} are ligands, and the three Cl^- ions are counter ions. The coordination compound $K_2[PdCl_4]$ in Figure 23.7B consists of two K^+ counter ions and the complex ion $[PdCl_4]^{2-}$, in which Pd^{2+} is bonded to four Cl^- ion ligands.

A coordination compound behaves like an electrolyte in water: the complex ion and counter ions separate, but the complex ion behaves like a polyatomic ion because *the ligands and central metal ion remain attached.* Thus, as Figure 23.7 shows, 1 mol of $[Co(NH_3)_6]Cl_3$ yields 1 mol of $[Co(NH_3)_6]^{3+}$ ions and 3 mol of Cl^- counter ions while 1 mol of $K_2[PdCl_4]$ yields 2 mol of K^+ counter ions and 1 mol of $[PdCl_4]^{2-}$ ions. This section covers the structure, naming, and properties of complex ions.

Complex Ions: Coordination Numbers, Geometries, and Ligands

A complex ion is described by the metal ion and the number and types of ligands attached to it. The ion's structure is related to coordination number and geometry.

Coordination Numbers The **coordination number** is the *number of ligand atoms* bonded directly to the central metal ion in a complex ion. It is *specific* for a given metal ion in a particular oxidation state and in a particular complex ion. The coordination number of the Co^{3+} ion in $[Co(NH_3)_6]^{3+}$ is 6 because six ligand atoms (N atoms of NH_3 molecules) are bonded to it. The coordination number of the Pt^{2+} ion in many of its complex ions is 4, whereas that of the Pt^{4+} ion in its complex ions is 6. Copper(II) may have a coordination number of 2, 4, or 6 in different complex ions. In general, *the most common coordination number in complex ions is 6,* but 2 and 4 are often seen, and some higher ones are known.

Student Hot Spot

Student data indicate that you may struggle with the concept of coordination compounds and their features. Access the Smartbook to view additional Learning Resources on this topic.

Table 23.5 Coordination Numbers and Shapes of Some Complex Ions

Coordination Number	Shape		Examples
2	Linear		$[CuCl_2]^-$, $[Ag(NH_3)_2]^+$, $[AuCl_2]^-$
4	Square planar	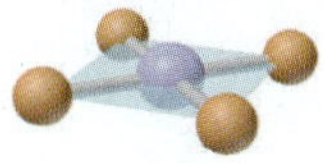	$[Ni(CN)_4]^{2-}$, $[PdCl_4]^{2-}$, $[Pt(NH_3)_4]^{2+}$, $[Cu(NH_3)_4]^{2+}$
4	Tetrahedral	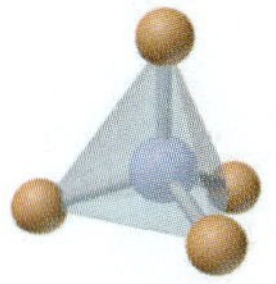	$[Cu(CN)_4]^{3-}$, $[Zn(NH_3)_4]^{2+}$, $[CdCl_4]^{2-}$, $[MnCl_4]^{2-}$
6	Octahedral	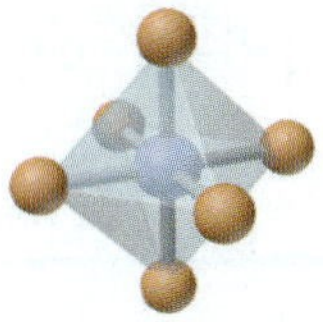	$[Ti(H_2O)_6]^{3+}$, $[V(CN)_6]^{4-}$, $[Cr(NH_3)_4Cl_2]^+$, $[Mn(H_2O)_6]^{2+}$, $[FeCl_6]^{3-}$, $[Co(en)_3]^{3+}$

Geometries The geometry (shape) of a complex ion depends on *the coordination number and the metal ion.* Table 23.5 shows the geometries for coordination numbers 2, 4, and 6. A complex ion whose metal ion has a coordination number of 2, such as $[Ag(NH_3)_2]^+$, is *linear.* The coordination number 4 gives rise to either of two geometries. Most d^8 metal ions form *square planar* complex ions (Figure 23.7B). The d^{10} ions form *tetrahedral* complex ions. A coordination number of 6 results in an *octahedral* geometry, as shown by $[Co(NH_3)_6]^{3+}$ (Figure 23.7A). Note the similarity with some of the molecular shapes in VSEPR theory (Section 10.2).

Ligands The ligands of complex ions are *molecules or anions* with one or more **donor atoms.** Each *donates a lone pair of electrons* to the metal ion, thus forming a covalent bond. Because they must have at least one lone pair, donor atoms often come from Group 5A(15), 6A(16), or 7A(17). (As you learned in Section 18.9, ligands are Lewis bases.)

Ligands are classified in terms of their number of donor atoms, or "teeth":

- *Monodentate* (Latin, "one-toothed") ligands bond through a single donor atom.
- *Bidentate* ligands have two donor atoms, each of which bonds to the metal ion.
- *Polydentate* ligands have more than two donor atoms.

Bidentate and polydentate ligands give rise to *rings* in the complex ion. For instance, ethylenediamine (abbreviated *en*) has a chain of four atoms (:N—C—C—N:), so it forms a five-membered ring, with the two electron-donating N atoms bonding to the metal ion (Figure 23.8A). Such ligands seem to grab the metal ion like claws, so a complex ion that contains them is also called a **chelate** (pronounced "KEY-late"; Greek *chela,* "crab's claw"). With its six donor atoms, the ethylenediaminetetraacetate ($EDTA^{4-}$) ion forms very stable complexes with metal ions (Figure 23.8B). This property makes EDTA useful in treating *heavy-metal poisoning.* Ingested by the patient, the $EDTA^{4-}$ ion removes lead and other heavy-metal ions from the blood and other body fluids.

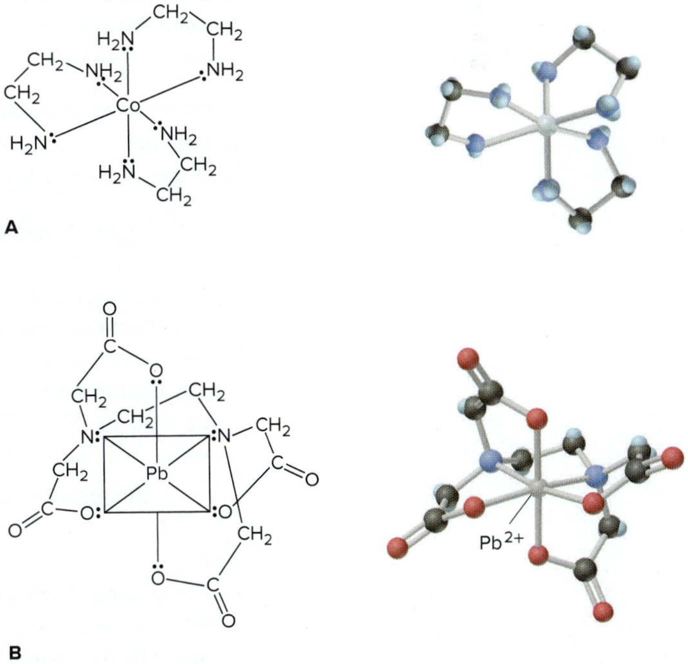

Figure 23.8 Chelates. Structure (*left*) and model (*right*) of (**A**) the bidentate ligand ethylenediamine (en) complexed with a Co^{3+} ion and (**B**) the chelate $[Pb(EDTA)]^{2-}$.

Table 23.6 Some Common Ligands in Coordination Compounds

Ligand Type	Examples			
Monodentate	$H_2\ddot{O}$: water	$:\ddot{F}:^-$ fluoride ion	$[:C\equiv N:]^-$ cyanide ion	$[:\ddot{O}-H]^-$ hydroxide ion
	$:NH_3$ ammonia	$:\ddot{Cl}:^-$ chloride ion	$[:\ddot{S}=C=\ddot{N}:]^-$ (or) thiocyanate ion	$[:\ddot{O}-\ddot{N}=\ddot{O}:]^-$ (or) nitrite ion
Bidentate	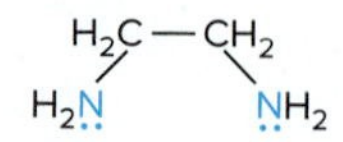 ethylenediamine (en)	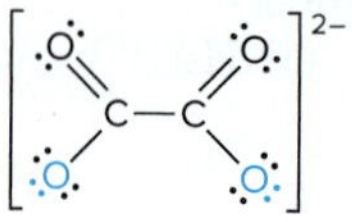 oxalate ion		
Polydentate	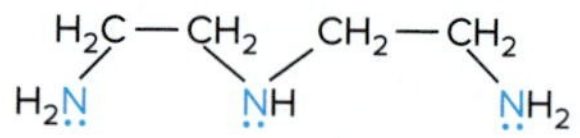 diethylenetriamine	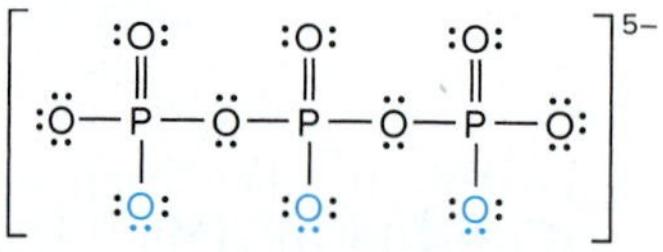 triphosphate ion	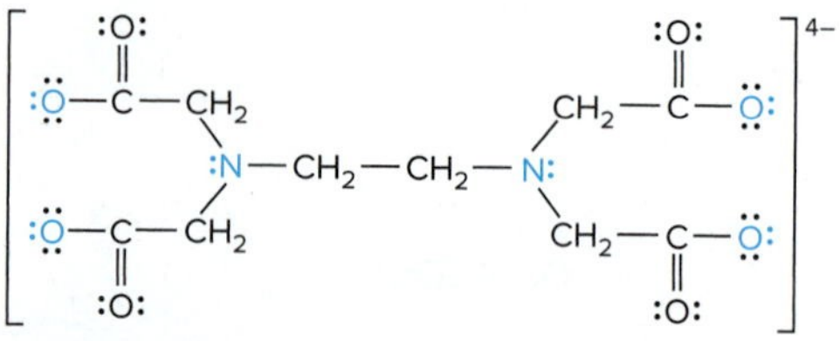 ethylenediaminetetraacetate ion ($EDTA^{4-}$)	

The $EDTA^{4-}$ ion is often added to products such as shampoos as it removes Mg^{2+} and Ca^{2+} ions from hard water, allowing the shampoo to clean more efficiently.

Table 23.6 shows some common ligands in coordination compounds. Note that each ligand has one or more donor atoms (colored type), each with a lone pair of electrons to donate, thereby forming a covalent bond to the metal ion. Since the bidentate ligand ethylenediamine has two atoms that form bonds to the metal ion, the complex ion $[Co(en)_3]^{3+}$ has a coordination number of 6 (two covalent bonds per ligand); the complex ion $[Pb(EDTA)]^{2-}$ has one polydentate ligand with six donor atoms, so it also has a coordination number of 6.

Formulas and Names of Coordination Compounds

The combination of ions in a coordination compound is the key to writing its formula and name. A coordination compound can consist of a complex cation with simple anionic counter ions, a complex anion with simple cationic counter ions, or even a complex cation with complex anion as counter ion.

Determining the Charge of the Metal Ion The charge of the complex ion can be determined from the charge of its counter ion(s) and basic arithmetic can be used to find the charge of the metal ion:

$$\text{Charge of metal ion} = \text{charge of complex ion} - \text{total charge of ligands} \quad \textbf{(23.1)}$$

- For a compound with a *complex anion,* say, $K_2[Co(NH_3)_2Cl_4]$, the two K^+ counter ions (written outside the brackets) have a total charge of $2 \times 1+ = 2+$ and balance the charge of the complex anion, which must have a charge of 2−; the complex ion contains two NH_3 molecules and four Cl^- ions as ligands. The two NH_3 are neutral, the four Cl^- have a total charge of 4–. We can use Equation 23.1 to determine the charge of the central metal ion:

$$\begin{aligned}\text{Charge of metal ion} &= \text{charge of complex ion} - \text{total charge of ligands}\\ &= (2-) - [(2 \times 0) + (4 \times 1-)]\\ &= (2-) - (4-) = 2+;\ \text{that is, } Co^{2+}\end{aligned}$$

- For a compound with a *complex cation,* say $[Co(NH_3)_4Cl_2]Cl$, the Cl^- ion outside the brackets is the counter ion and the complex cation has a charge of 1+ to balance the 1– charge of that counter ion. The complex ion $[Co(NH_3)_4Cl_2]^+$ has four NH_3 ligands that are neutral and two Cl^- ligands with a total charge of 2–, so the central metal ion must be Co^{3+}:

$$\begin{aligned}\text{Charge of metal ion} &= \text{charge of complex ion} - \text{total charge of ligands}\\ &= (1+) - [(4 \times 0) + (2 \times 1-)] = 3+\end{aligned}$$

Rules for Writing Formulas There are four rules for writing formulas of coordination compounds (the first two are the same as those for the formula of any ionic compound):

1. *The cation is written before the anion.*
2. *The charge of the cation(s) is balanced by the charge of the anion(s).*
3. *For the complex ion, neutral ligands are written (in alphabetical order) before anionic ligands (written in alphabetical order), and the formula of the whole ion is placed in brackets.*
4. *Counter ions are written outside the brackets.*

SAMPLE PROBLEM 23.3 Finding the Coordination Number and Charge of the Central Metal Ion in a Coordination Compound

Problem Give the coordination number and the charge of the central metal ion in each coordination compound:
(a) $Na_2[Zn(OH)_4]$ **(b)** $K[Co(H_2O)_2(C_2O_4)_2]$ **(c)** $[Ru(H_2O)_2(NH_3)_2Cl_2]Br$

Plan To find the coordination number, we use Table 23.6 to determine the number of ligand atoms bonded to the central metal ion in the complex ion (in square brackets). Counter ions (those ions outside the brackets) are not included when determining the coordination number. We know that the charge of the complex ion must balance the charge of the counter ions; the charge of the metal ion is equal to the charge of the complex ion minus the total charge of the ligands.

Solution (a) Each of the four monodentate OH^- ligands forms one bond to the metal ion for a coordination number of 4. Since the two Na^+ counter ions have a total charge of 2+, the charge of the complex ion is 2−; the four OH^- ligands have a total charge of 4−. The charge of the zinc ion is

$$\text{Charge of metal ion} = \text{charge of complex ion} - \text{total charge of ligands}$$
$$= (2-) - (4-) = 2+$$

(b) There are two monodentate H_2O ligands, each forming one bond to the metal ion; since the two $C_2O_4^{2-}$ ligands are bidentate, each of them forms two bonds to the metal ion, for a total of four bonds. Thus, the coordination number is 6. Since the K^+ counter ion has a charge of 1+, the charge of the complex ion is 1−. The H_2O ligands are neutral, and the two $C_2O_4^{2-}$ ligands have a total charge of 4−. The charge of the cobalt ion is

$$\text{Charge of metal ion} = (1-) - [0 + (4-)] = 3+$$

(c) The complex has two H_2O, two NH_3, and two Cl^- ligands, each of which forms one bond to the metal ion for a coordination number of 6. Since the Br^- counter ion has a 1− charge, the complex ion has a charge of 1+. The H_2O and NH_3 ligands are neutral, and the two Cl^- ligands have a total charge of 2−. The charge of the ruthenium ion is

$$\text{Charge of metal ion} = (1+) - [0 + 0 + (2-)] = 3+$$

Check Be sure that the coordination number equals the sum of the number of monodentate ligands and twice the number of bidentate ligands. Be sure that the charges of the metal ion, ligands, and counter ions sum to zero.

FOLLOW-UP PROBLEMS

23.3A Give the coordination number and the charge of the central metal ion in each coordination compound:
(a) $[Co(en)_2Br_2]NO_3$ **(b)** $Mg_3[CrCl_5Br]_2$

23.3B Write the formula of the coordination compound composed of the following:
(a) A Mn^{2+} central ion with a coordination number of 6, cyanide ligands, and potassium counter ions
(b) A Cu^{2+} central ion with a coordination number of 4, ethylenediamine ligands, and sulfate as the counter ion

SOME SIMILAR PROBLEMS 23.47, 23.48, 23.51, and 23.52

Table 23.7 Names of Some Neutral and Anionic Ligands

Neutral		Anionic	
Name	**Formula**	**Name**	**Formula**
Aqua	H_2O	Fluoro	F^-
Ammine	NH_3	Chloro	Cl^-
Carbonyl	CO	Bromo	Br^-
Nitrosyl	NO	Iodo	I^-
		Hydroxo	OH^-
		Cyano	CN^-

Table 23.8 Names of Some Metal Ions in Complex Anions

Metal	Name in Anion
Iron	Ferrate
Copper	Cuprate
Lead	Plumbate
Silver	Argentate
Gold	Aurate
Tin	Stannate

Rules for Naming Coordination Compounds Originally named after their discoverer or color, coordination compounds are named systematically according to a set of rules. Let's see how to name $[Co(NH_3)_4Cl_2]Cl$ and $K_2[Co(H_2O)Br_3Cl_2]$. As we go through the naming steps, refer to Tables 23.7 and 23.8.

1. *The cation is named before the anion.* Thus, we name the $[Co(NH_3)_4Cl_2]^+$ ion before the Cl^- counter ion.
2. *Within the complex ion, the ligands are named in alphabetical order* ***before*** *the metal ion.* In the $[Co(NH_3)_4Cl_2]^+$ ion, four NH_3 and two Cl^- are named before Co^{3+}.
3. *Neutral ligands generally have the molecule name,* but there are a few exceptions (Table 23.7). *Anionic ligands drop the -ide and add -o after the root name;* thus, the anion name *fluoride* for an F^- ion becomes the ligand name *fluoro*. The two ligands in $[Co(NH_3)_4Cl_2]^+$ are *ammine* (NH_3) and *chloro* (Cl^-) with *ammine* coming before *chloro* alphabetically.
4. *A numerical prefix indicates the number of ligands of a particular type.* For example, *tetra*ammine denotes *four* NH_3, and *di*chloro denotes *two* Cl^-. Other prefixes are *tri-*, *penta-*, and *hexa-*. But prefixes do *not* affect the alphabetical order: *tetraammine* comes before *dichloro*. For ligand names that include a numerical prefix (such as ethylene*di*amine), we use *bis* (2), *tris* (3), or *tetrakis* (4) to indicate the number of such ligands, followed by the ligand name in parentheses. For example, a complex ion that has two ethylenediamine ligands has *bis(ethylenediamine)* in its name.
5. *The oxidation state of the metal ion has a roman numeral (in parentheses) only if the metal ion can have more than one state.* Since cobalt can have +2 and +3 states, we add a III to name the complex ion. Thus, the compound is

   tetraamminedichlorocobalt(III) chloride

   The only typographical space in the name comes between cation and anion.
6. *If the complex ion is an anion, we drop the ending of the metal name and add -ate.* Thus, the name for $K_2[Co(H_2O)Br_3Cl_2]$ is potassium aquatribromodichlorocobaltate(III). There are two K^+ counter ions, so the complex anion has a charge of 2–. The three Br^- and two Cl^- ligands have a total charge of 5–, so Co must be in the +3 oxidation state. For some metals, we use the Latin root with the *-ate* ending (Table 23.8). For example, the name for $Na_4[FeBr_6]$ is sodium hexabromoferrate(II).

Student Hot Spot

Student data indicate that you may struggle with naming coordination compounds. Access the Smartbook to view additional Learning Resources on this topic.

SAMPLE PROBLEM 23.4 Writing Names and Formulas of Coordination Compounds

Problem **(a)** What is the systematic name of $Na_3[AlF_6]$?
(b) What is the systematic name of $[Co(en)_2Cl_2]NO_3$?
(c) What is the formula of tetraamminebromochloroplatinum(IV) chloride?
(d) What is the formula of hexaamminecobalt(III) tetrachloroferrate(III)?

Plan We use the rules presented above and the information in Tables 23.7 and 23.8.

Solution **(a)** The complex ion is $[AlF_6]^{3-}$. There are six *(hexa-)* F^- ions *(fluoro)* as ligands, so we have *hexafluoro*. The complex ion is an anion, so the ending of the metal ion (aluminum) must be changed to *-ate:* hexafluoroaluminate. Aluminum has only the +3 oxidation state, so we do *not* use a roman numeral. The positive counter ion is named first and separated from the name of the anion by a space: sodium hexafluoroaluminate.

(b) Listed alphabetically, there are two Cl^- *(dichloro)* and two en [*bis(ethylenediamine)*] as ligands. The complex ion is a cation, so the metal name is unchanged, but we specify its oxidation state because cobalt can have several. One NO_3^- balances the 1+ cation charge. With 2− for two Cl^- and 0 for two en, the metal must be *cobalt(III)*. The word *nitrate* follows a space: dichlorobis(ethylenediamine)cobalt(III) nitrate.

(c) The central metal ion is written first, followed by the neutral ligands and then (in alphabetical order) by the negative ligands. *Tetraammine* is four NH_3, *bromo* is one Br^-, *chloro* is one Cl^-, and *platinate(IV)* is Pt^{4+}, so the complex ion is $[Pt(NH_3)_4BrCl]^{2+}$. Its 2+ charge is the sum of 4+ for Pt^{4+}, 0 for four NH_3, 1− for one Br^-, and 1− for one Cl^-. To balance the 2+ charge, we need two Cl^- counter ions: $[Pt(NH_3)_4BrCl]Cl_2$.

(d) This compound consists of two different complex ions. In the cation, *hexaammine* is six NH_3 and *cobalt(III)* is Co^{3+}, so the cation is $[Co(NH_3)_6]^{3+}$. The 3+ charge is the sum of 3+ for Co^{3+} and 0 for six NH_3. In the anion, *tetrachloro* is four Cl^-, and *ferrate(III)* is Fe^{3+}, so the anion is $[FeCl_4]^-$. The 1− charge is the sum of 3+ for Fe^{3+} and 4− for four Cl^-. In the neutral compound, one 3+ cation must be balanced by three 1− anions: $[Co(NH_3)_6][FeCl_4]_3$.

Check Reverse the process to be sure you obtain the name or formula in the problem.

FOLLOW-UP PROBLEMS

23.4A **(a)** What is the name of $[Cr(H_2O)_5Br]Cl_2$? **(b)** What is the formula of barium hexacyanocobaltate(III)?

23.4B **(a)** What is the formula of diaquabis(ethylenediamine)cobalt(III) sulfate? **(b)** What is the name of $Mg[Cr(NH_3)_2Cl_4]_2$?

SOME SIMILAR PROBLEMS 23.45, 23.46, 23.49, 23.50, 23.53, 23.54, 23.57, and 23.58

Isomerism in Coordination Compounds

Isomers are compounds with the same chemical formula but different properties. Recall the discussion of isomerism in organic compounds (Section 15.2); coordination compounds exhibit the same two broad categories—constitutional isomers and stereoisomers.

Constitutional Isomers: Atoms Connected Differently Compounds with the same formula, but with the atoms connected differently, are **constitutional (structural) isomers.** Coordination compounds exhibit two types, *coordination* and *linkage* isomers:

1. **Coordination isomers** occur when the composition of the complex ions, but not of the compounds, is different. This type of isomerism occurs in two ways:

- *Exchange of ligand and counter ion.* The roles of the Cl^- ions and the NO_2^- ions are reversed in the following example:

  In $[Pt(NH_3)_4Cl_2](NO_2)_2$, the Cl^- ions are the ligands, and the NO_2^- ions are counter ions;
  In $[Pt(NH_3)_4(NO_2)_2]Cl_2$, the NO_2^- ions are the ligands, and the Cl^- ions are counter ions.

  A common test to see whether Cl^- is a ligand or a counter ion is to treat a solution of the compound with $AgNO_3$; the Cl^- counter ion will form a white precipitate of AgCl, but the Cl^- ligand won't. Thus, a solution of $[Pt(NH_3)_4Cl_2](NO_2)_2$ does not form AgCl because Cl^- is bound to the metal ion, but a $[Pt(NH_3)_4(NO_2)_2]Cl_2$ solution forms 2 mol of AgCl per mole of compound.
- *Exchange of ligands.* This type of isomerism occurs in compounds consisting of two complex ions. For example, in $[Cr(NH_3)_6][Co(CN)_6]$ and $[Co(NH_3)_6][Cr(CN)_6]$, NH_3 is a ligand of Cr^{3+} in one compound and of Co^{3+} in the other.

Figure 23.9 **A pair of linkage (constitutional) isomers.**

Source: © McGraw-Hill Education/Charles Winters/Timeframe Photography, Inc.

2. **Linkage isomers** occur when the composition of the complex ion is the same but the ligand donor atom is different. Some ligands can bind to the metal ion through *either of two donor atoms.* For example, the nitrite ion can bind through the N atom (*nitro,* O_2N:) or either of the O atoms (*nitrito,* ONO:) to give linkage isomers, such as pentaammine*nitro*cobalt(III) chloride $[Co(NH_3)_5(NO_2)]Cl_2$ and pentaammine*nitrito*cobalt(III) chloride $[Co(NH_3)_5(ONO)]Cl_2$ (Figure 23.9).

Other examples of ligands that have more than one donor atom are the cyanate ion, which can attach via the O atom (*cyanato,* NCO:) or the N atom (*isocyanato,* OCN:), and the thiocyanate ion, which can attach via the S atom (*thiocyanato,* NCS:) or the N atom (*isothiocyanato,* SCN:):

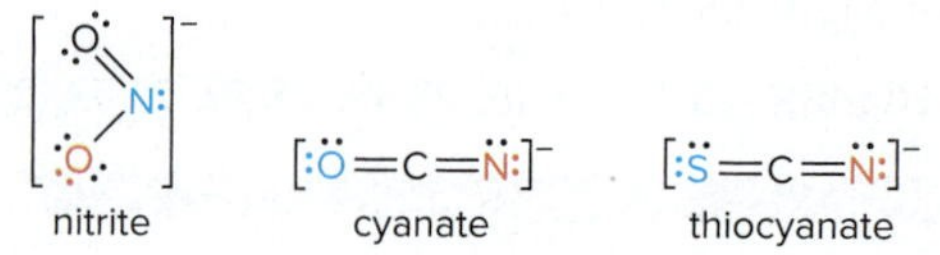

Stereoisomers: Atoms Arranged Differently in Space

Stereoisomers are compounds that have the same atomic connections but different spatial arrangements of the atoms. *Geometric* and *optical* isomers, which we discussed for organic compounds, occur with coordination compounds as well:

1. **Geometric isomers** (also called ***cis-trans*** **isomers,** or *diastereomers*) occur when atoms or groups of atoms are arranged differently in space relative to the central metal ion. For example, the square planar $[Pt(NH_3)_2Cl_2]$ has two arrangements, which give rise to two different compounds (Figure 23.10A). The isomer with identical ligands

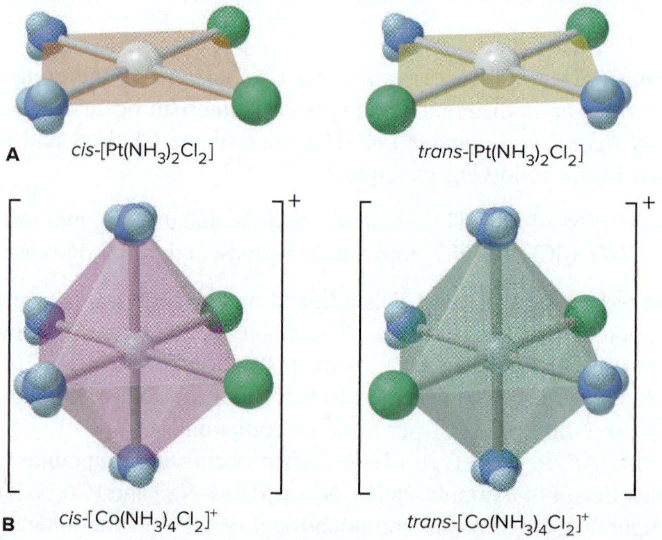

Figure 23.10 **Geometric** ***(cis-trans)*** **isomerism. A,** The *cis* and *trans* isomers of $[Pt(NH_3)_2Cl_2]$. **B,** The *cis* and *trans* isomers of $[Co(NH_3)_4Cl_2]^+$. The colored shapes indicate the actual colors of the species.

next to each other is *cis*-diamminedichloroplatinum(II), and the one with identical ligands *across* from each other is *trans*-diamminedichloroplatinum(II).

These isomers have remarkably different biological behaviors. In the mid-1960s, Barnett Rosenberg and his colleagues found that *cis*-$[Pt(NH_3)_2Cl_2]$ (cisplatin) was an antitumor agent. It and several closely related platinum(II) complexes are still used to treat certain types of cancer. In contrast, *trans*-$[Pt(NH_3)_2Cl_2]$ has no antitumor effect. Cisplatin may work by becoming oriented within a cancer cell so that a donor atom on each DNA strand can replace a Cl^- ligand and bind to the platinum strongly, preventing DNA replication (Section 15.6).

Octahedral complexes also exhibit *cis-trans* isomerism (Figure 23.10B). The *cis* isomer of the $[Co(NH_3)_4Cl_2]^+$ ion has the two Cl^- ligands at any two adjacent positions of the ion's octahedral shape, whereas the *trans* isomer has these ligands across from each other. Tetrahedral complexes do not exhibit this type of isomerism.

2. **Optical isomers** (also called *enantiomers*) occur when a molecule and its mirror image cannot be superimposed (see Figures 15.8 to 15.10). Unlike other types of isomers, which have distinct physical properties, optical isomers are physically identical except for *the direction in which they rotate the plane of polarized light.* Many octahedral complex ions show optical isomerism, which we can determine by rotating one isomer and seeing if it is superimposable on the other isomer (its mirror image). For example, in Figure 23.11A, the two structures (I and II) of *cis*-$[Co(en)_2Cl_2]^+$, the *cis*-dichlorobis(ethylenediamine)cobalt(III) ion, are mirror images of each other. Rotate structure I clockwise 180° around a vertical axis, and you obtain III. The Cl^- ligands of III match those of II, but the en ligands do not: II and III (which is I rotated) are not superimposable: they are optical isomers. One isomer is designated *d*-$[Co(en)_2Cl_2]^+$ and the other is *l*-$[Co(en)_2Cl_2]^+$, indicating whether the isomer rotates polarized light to the right (*d*- for "dextro-") or to the left (*l*- for "levo-"). (The *d*- or *l*- designation can only be determined by experiment.)

In contrast, as shown in Figure 23.11B, the two structures of the *trans*-dichlorobis(ethylenediamine)cobalt(III) ion are *not* optical isomers: rotate I 90°

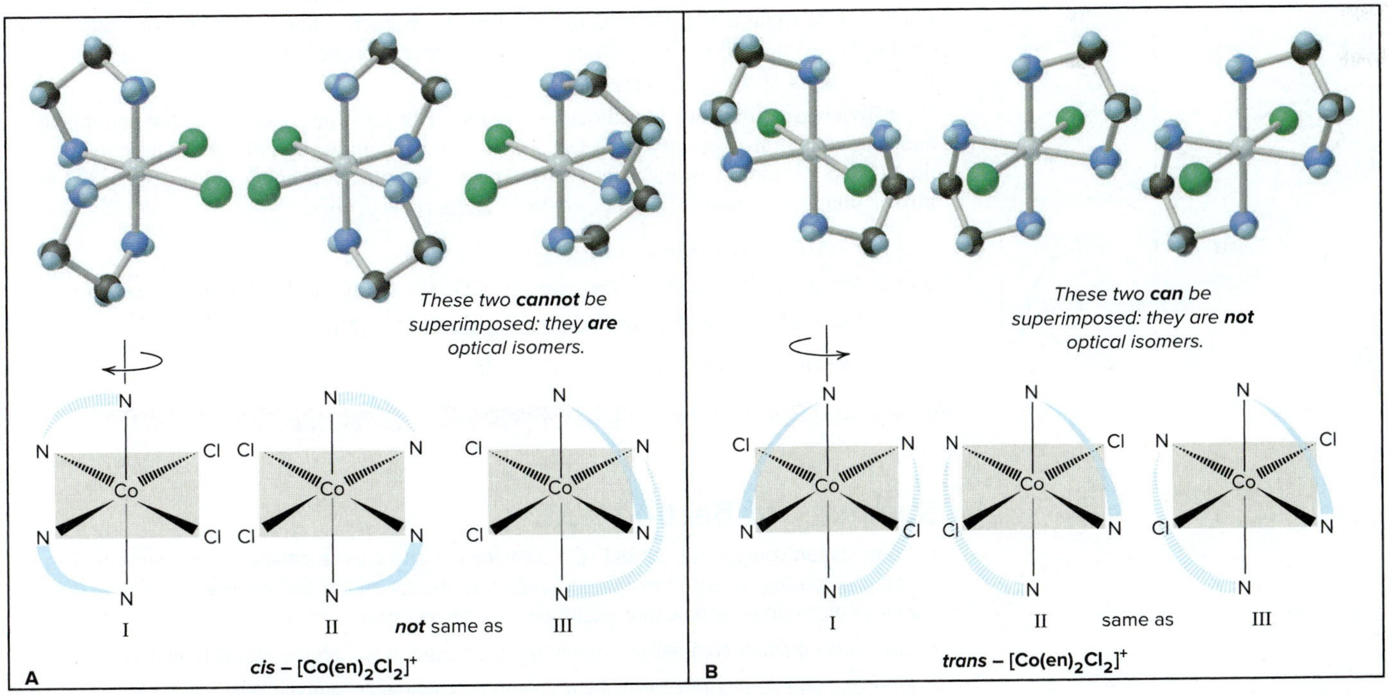

Figure 23.11 **Optical isomerism in an octahedral complex ion. A,** Structure I and its mirror image, structure II, are optical isomers of *cis*-$[Co(en)_2Cl_2]^+$. (The curved wedges represent the bidentate ligand ethylenediamine, $H_2N—CH_2—CH_2—NH_2$.) **B,** The *trans* isomer does *not* have optical isomers.

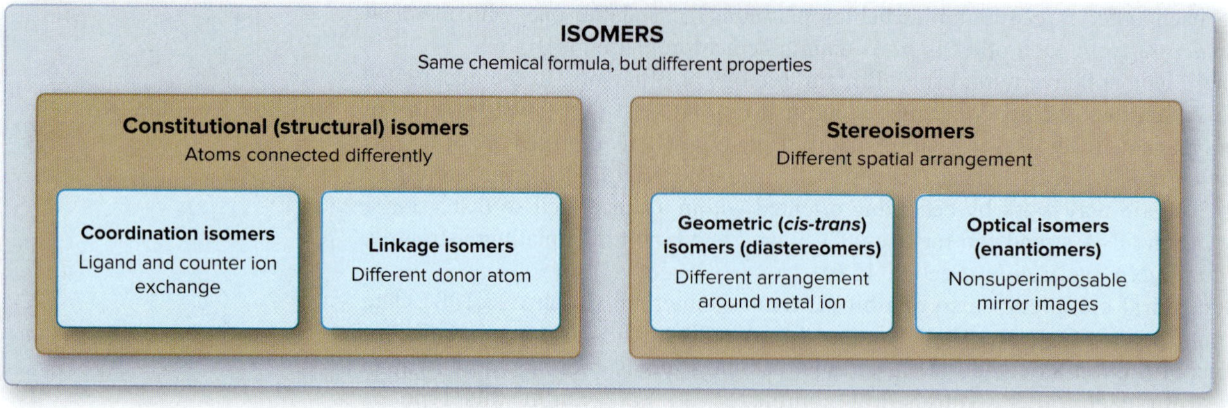

Figure 23.12 Important types of isomerism in coordination compounds.

counterclockwise around a vertical axis and you obtain III, which *is* superimposable on II.

Figure 23.12 is an overview of the most common types of isomerism in coordination compounds.

SAMPLE PROBLEM 23.5 Determining the Type of Stereoisomerism

Problem Draw stereoisomers for each of the following and state the type of isomerism:
(a) $[Pt(NH_3)_2Br_2]$ (square planar) **(b)** $[Cr(en)_3]^{3+}$ (en = $H_2\ddot{N}CH_2CH_2\ddot{N}H_2$)

Plan We determine the geometry around each metal ion and the nature of the ligands. If there are different ligands that can be placed in different positions relative to each other, geometric *(cis-trans)* isomerism occurs. Then, we see whether the mirror image of an isomer is superimposable on the original. If it is *not,* optical isomerism also occurs.

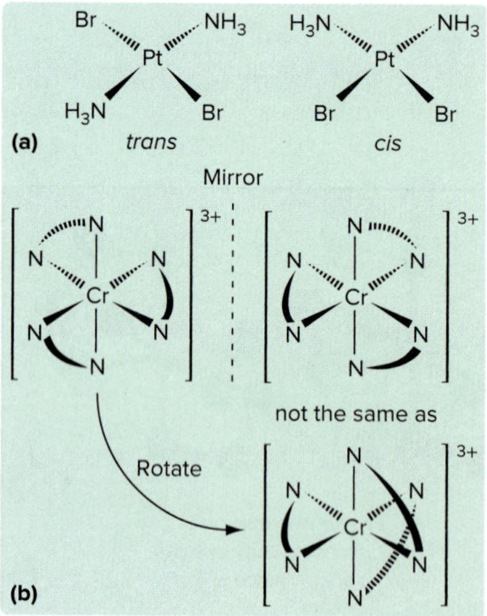

Solution **(a)** The square planar Pt(II) complex has two different monodentate ligands. Each pair of ligands can lie next to or across from each other *(see margin)*. Thus, geometric isomerism occurs. Each isomer *is* superimposable on a mirror image of itself, so there is no optical isomerism.

(b) Ethylenediamine (en) is a bidentate ligand. The Cr^{3+} has a coordination number of 6 and an octahedral geometry, like Co^{3+}. The three bidentate ligands are identical, so there is no geometric isomerism. However, the complex ion has a nonsuperimposable mirror image *(see margin)*. Thus, optical isomerism occurs.

FOLLOW-UP PROBLEMS

23.5A What stereoisomers, if any, are possible for the $[Co(NH_3)_2(en)Cl_2]^+$ ion?

23.5B Draw all of the stereoisomers of the $[Ru(H_2O)_2(NH_3)_2Cl_2]^+$ ion.

SOME SIMILAR PROBLEMS 23.63–23.66

› Summary of Section 23.3

- Coordination compounds consist of a complex ion and charge-balancing counter ions. The complex ion has a central metal ion bonded to neutral and/or anionic ligands, which have one or more donor atoms that each have a lone pair of electrons.
- The most common complex-ion geometry is octahedral (six ligand atoms bonding).
- Formulas and names of coordination compounds follow systematic rules.
- Coordination compounds can exhibit constitutional isomerism (coordination and linkage) and stereoisomerism (geometric and optical).

23.4 THEORETICAL BASIS FOR THE BONDING AND PROPERTIES OF COMPLEX IONS

In this section, we consider valence bond theory and crystal field theory, which address different features of complex ions: how metal-ligand bonds form, why certain geometries are preferred, and why these ions are brightly colored and often paramagnetic.

Applying Valence Bond Theory to Complex Ions

Valence bond (VB) theory, which helps explain bonding and structure in main-group compounds (Section 11.1), can also be used to describe bonding in complex ions. In the formation of a complex ion, the *filled* ligand orbital overlaps an *empty* metal-ion orbital: *the ligand (Lewis base) donates an electron pair, and the metal ion (Lewis acid) accepts it to form a covalent bond in the complex ion (Lewis adduct)* (Section 18.9). A bond in which one atom contributes both electrons is a **coordinate covalent bond;** once formed, it is identical to any covalent single bond.

Recall that the VB concept of hybridization proposes mixing particular combinations of *s*, *p*, and *d* orbitals to obtain sets of hybrid orbitals, which have specific geometries. For coordination compounds, the model proposes that *the type of metal-ion orbital hybridization determines the geometry of the complex ion.* Let's discuss orbital combinations that lead to octahedral, square planar, and tetrahedral geometries.

Octahedral Complexes The hexaamminechromium(III) ion, $[Cr(NH_3)_6]^{3+}$, is an *octahedral complex* (Figure 23.13). The six lowest-energy, *empty* orbitals of the Cr^{3+} ion—two 3*d*, one 4*s*, and three 4*p*—mix and become six equivalent d^2sp^3 hybrid orbitals that point toward the corners of an octahedron.* Six NH_3 molecules donate lone pairs from their N atoms to form six metal-ligand bonds. Three unpaired 3*d* electrons of the central Cr^{3+} ion ([Ar] $3d^3$) remain in unhybridized orbitals and make the complex ion paramagnetic.

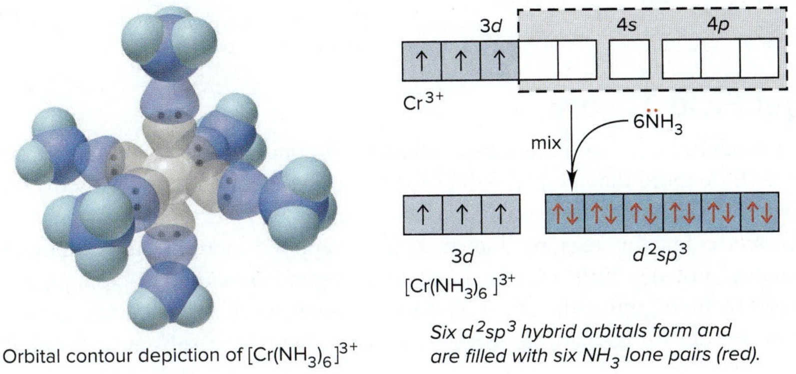

Figure 23.13 Hybrid orbitals and bonding in the octahedral $[Cr(NH_3)_6]^{3+}$ ion.

Square Planar Complexes Metal ions with a d^8 configuration usually form *square planar complexes* (Figure 23.14, *next page*). In the $[Ni(CN)_4]^{2-}$ ion, for example, the model proposes that one 3*d*, one 4*s*, and two 4*p* orbitals of Ni^{2+} mix and form four dsp^2 hybrid orbitals, which point to the corners of a square and accept one electron pair from each of four CN^- ligands.

A look at the ground-state electron configuration of the Ni^{2+} ion, however, raises a key question: how can the Ni^{2+} ion ([Ar] $3d^8$) offer an empty 3*d* orbital for accepting a lone pair, if its eight 3*d* electrons lie in three filled and two half-filled orbitals? Apparently, in the d^8 configuration of Ni^{2+}, electrons in the half-filled orbitals *pair up* and leave one 3*d* orbital empty. This explanation is consistent with the fact that

*Note the difference between hybrid-orbital designations here and in Chapter 11, even though both designations give the orbitals in energy order. In $[Cr(NH_3)_6]^{3+}$, the 3*d* orbitals have a *lower n* value than the 4*s* and 4*p* orbitals, so the hybrid orbitals are d^2sp^3. But, for SF_6, the 3*d* orbitals have the same *n* value as the 3*s* and 3*p* orbitals of S, so the designation is sp^3d^2.

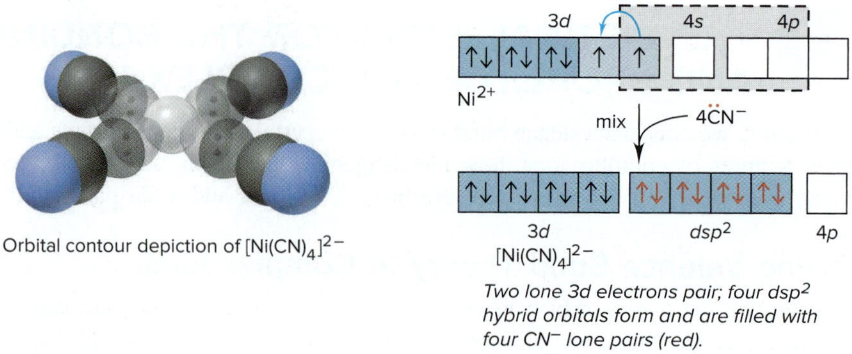

Figure 23.14 Hybrid orbitals and bonding in the square planar $[Ni(CN)_4]^{2-}$ ion.

the complex is diamagnetic (no unpaired electrons). Moreover, it means that the energy *gained* by using a 3*d* orbital for bonding in the hybrid orbital is greater than the energy *required* to overcome repulsions from pairing the 3*d* electrons.

Tetrahedral Complexes Metal ions that have a filled *d* sublevel, such as Zn^{2+} ([Ar] $3d^{10}$), often form diamagnetic *tetrahedral complexes* (Figure 23.15). In the $[Zn(OH)_4]^{2-}$ ion, for example, VB theory proposes that the lowest *empty* Zn^{2+} orbitals—one 4*s* and three 4*p*—mix to become four sp^3 hybrid orbitals that point to the corners of a tetrahedron and are occupied by lone pairs, one from each of four OH^- ligands.

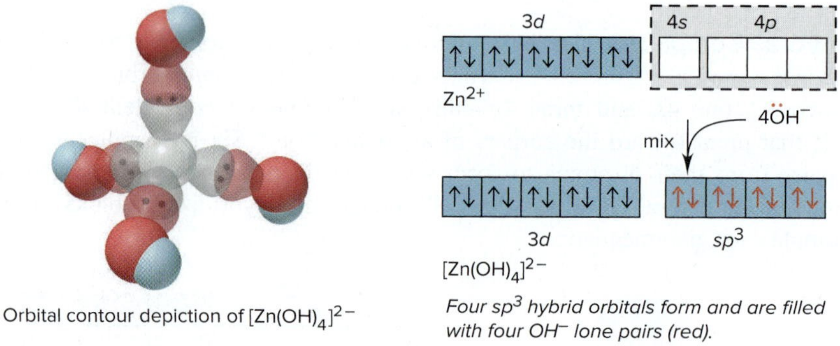

Figure 23.15 Hybrid orbitals and bonding in the tetrahedral $[Zn(OH)_4]^{2-}$ ion.

Crystal Field Theory

The VB model is easy to picture and rationalizes bonding and shape, but it treats the orbitals as little more than empty "slots" for accepting electron pairs. Moreover, it gives no insight into the colors of complex ions and sometimes predicts their magnetic properties incorrectly. For this reason, other models are applied more often. Although **crystal field theory** provides little insight about metal-ligand bonding, it explains color and magnetism by highlighting the *effect on d-orbital energies of the metal ion as the ligands approach.* Before we discuss the theory, let's consider why a substance is colored.

What Is Color? White light consists of all wavelengths (λ) in the visible range (Section 7.1) and can be dispersed into colors of a narrower wavelength range. Objects appear colored in white light because they absorb only certain wavelengths: an opaque object *reflects* the other wavelengths, and a clear one *transmits* them. If an object absorbs all visible wavelengths, it appears black; if it reflects all, it appears white.

Each color has a *complementary* color; for example, green and red are complementary colors. Figure 23.16 shows these relationships on an artist's color wheel in which complementary colors are wedges opposite each other. A mixture of complementary colors absorbs all visible wavelengths and appears black.

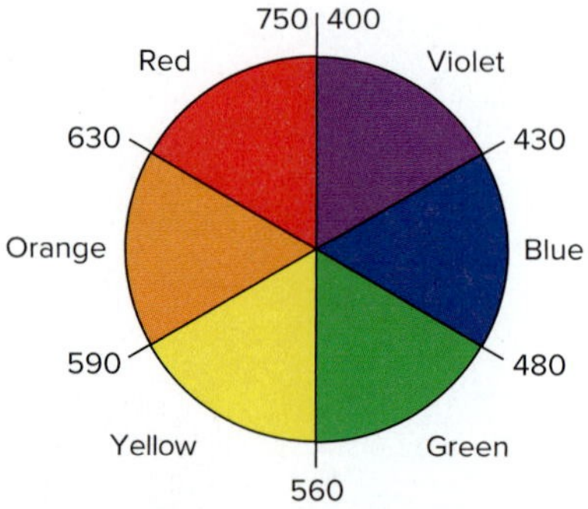

Figure 23.16 An artist's wheel. Colors, with approximate wavelength ranges (in nm), are shown as wedges.

An object has a particular color for one of two reasons:

- It reflects (or transmits) light of *that* color. Thus, if an object absorbs all wavelengths *except* green, the reflected (or transmitted) light is seen as green.
- It absorbs light of the *complementary* color. Thus, if the object absorbs only red, the *complement* of green, the remaining mixture of reflected (or transmitted) wavelengths is also seen as green.

Table 23.9 Relation Between Absorbed and Observed Colors

Absorbed Color	λ (nm)	Observed Color	λ (nm)
Violet	400	Green-yellow	560
Blue	450	Orange	600
Blue-green	490	Red	620
Yellow-green	570	Violet	410
Yellow	580	Dark blue	430
Orange	600	Blue	450
Red	650	Green	520

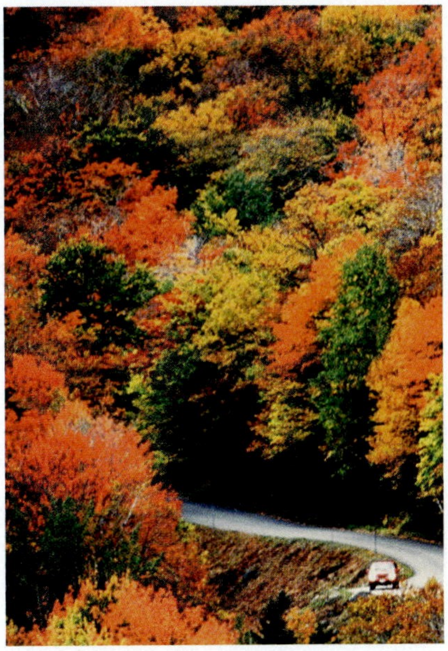

Foliage changing color in autumn.
Source: © Pete Turner/The Image Bank/Getty Images

Table 23.9 lists the color absorbed and the resulting color observed. The relation between absorbed and observed colors is demonstrated by the seasonal colors of deciduous trees. In the spring and summer, leaves contain high concentrations of the photosynthetic pigment *chlorophyll* and lower concentrations of other pigments called *xanthophylls.* Chlorophyll absorbs strongly in the blue and red regions, reflecting mostly green. In the fall, photosynthesis slows, so the leaf no longer makes chlorophyll. The green fades as the chlorophyll decomposes, revealing the xanthophylls that were present but masked by the chlorophyll. Xanthophylls absorb green and blue strongly, reflecting bright yellows and reds *(see photo).*

Splitting of *d* Orbitals in an Octahedral Field of Ligands The crystal field model explains that the properties of complex ions result from the splitting of *d*-orbital energies, which arises from *electrostatic attractions between the metal cation and the negative charge of the ligands.* This negative charge is either partial, as in a polar covalent ligand like NH_3, or full, as in an anionic ligand like Cl^-. Picture six ligands approaching a metal ion along the mutually perpendicular *x*, *y*, and *z* axes, to form an octahedral arrangement (Figure 23.17A). Let's follow the orientation of ligand and orbital and how the approach affects orbital energies.

1. *Orientation of ligand and metal-ion orbitals.* As ligands approach, their electron pairs repel electrons in the five *d* orbitals of the metal ion. In the isolated ion, the *d* orbitals have different orientations but *equal* energies. But, in the negative field of

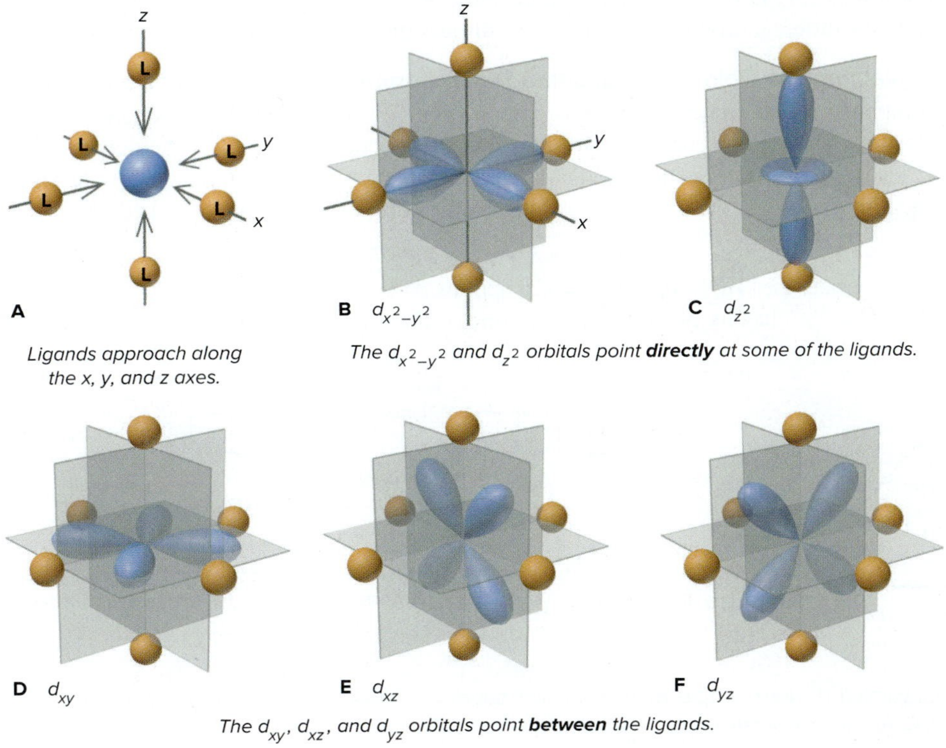
z
L
y
x
A
B $d_{x^2-y^2}$
C d_{z^2}
Ligands approach along the x, y, and z axes.
The $d_{x^2-y^2}$ and d_{z^2} orbitals point **directly** at some of the ligands.
D d_{xy}
E d_{xz}
F d_{yz}
The d_{xy}, d_{xz}, and d_{yz} orbitals point **between** the ligands.

Figure 23.17 The five *d* orbitals in an octahedral field of ligands.

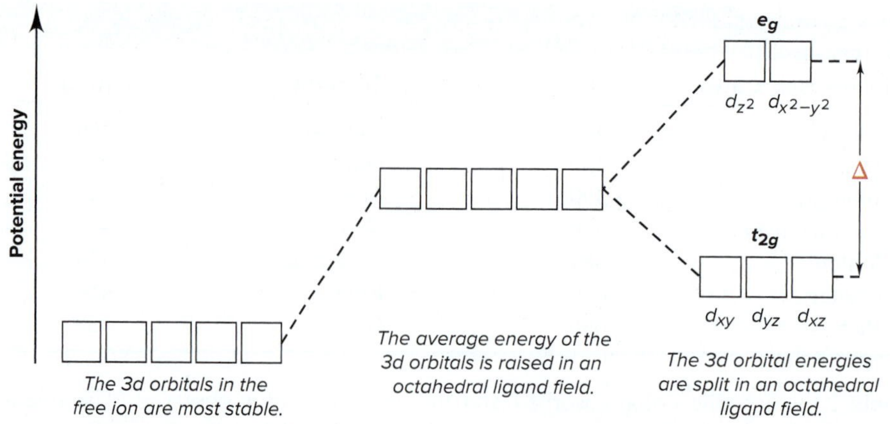

Figure 23.18 Splitting of *d*-orbital energies in an octahedral field of ligands.

ligands, the *d* electrons are *repelled unequally because of their different orbital orientations.* The ligands moving *along* the *x*, *y*, and *z* axes approach

- *Directly toward* the lobes of the $d_{x^2-y^2}$ and d_{z^2} orbitals (Figure 23.17B and C).
- *Between* the lobes of the d_{xy}, d_{xz}, and d_{yz} orbitals (Figure 23.17D to F).

2. *Effect on d-orbital energies.* As a result of these different orientations, electrons in the $d_{x^2-y^2}$ and d_{z^2} orbitals experience *stronger* repulsions than do electrons in the d_{xy}, d_{xz}, and d_{yz} orbitals. An energy diagram shows that the five *d* orbitals have the same energy and are most stable in the free ion and their *average* energy is higher in the ligand field. But *the orbital energies split, with two d orbitals higher in energy and three lower* (Figure 23.18):

- The two higher energy orbitals are **e_g orbitals** and arise from the $d_{x^2-y^2}$ and d_{z^2} orbitals.
- The three lower energy orbitals are **t_{2g} orbitals** and arise from the d_{xy}, d_{xz}, and d_{yz} orbitals.

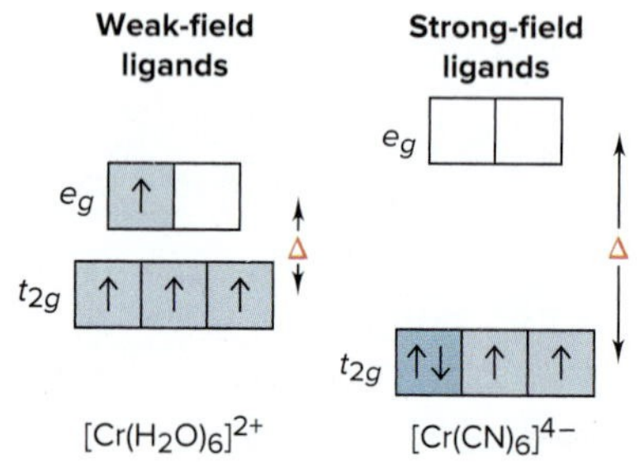

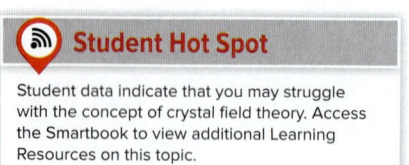

Figure 23.19 The effect of ligands on splitting energy and orbital occupancy.

Student Hot Spot

Student data indicate that you may struggle with the concept of crystal field theory. Access the Smartbook to view additional Learning Resources on this topic.

3. *The crystal field effect.* This splitting of orbital energies is called the *crystal field effect,* and the energy difference between e_g and t_{2g} orbitals is the **crystal field splitting energy (Δ).** Different ligands create crystal fields of different strength:

- **Strong-field ligands** lead to a *larger* splitting energy (larger Δ).
- **Weak-field ligands** lead to a *smaller* splitting energy (smaller Δ).

For instance, H_2O is a weak-field ligand, and CN^- is a strong-field ligand (Figure 23.19). Note the different orbital occupancies; we discuss the reason for these differences shortly.

Explaining the Colors of Transition Metal Complexes The color of a coordination compound is determined by Δ of its complex ion. When the ion absorbs radiant energy, electrons can use that energy to move from the lower energy t_{2g} level to the higher energy e_g level. Recall from Section 7.2 that the *difference* between two atomic energy levels is equal to the energy (and inversely related to the wavelength) of the absorbed photon:

$$\Delta E_{\text{electron}} = E_{\text{photon}} = h\nu = hc/\lambda$$

Consider the $[Ti(H_2O)_6]^{3+}$ ion, which appears purple in aqueous solution (Figure 23.20). Hydrated Ti^{3+} has its one *d* electron in one of the three lower energy t_{2g} orbitals. The

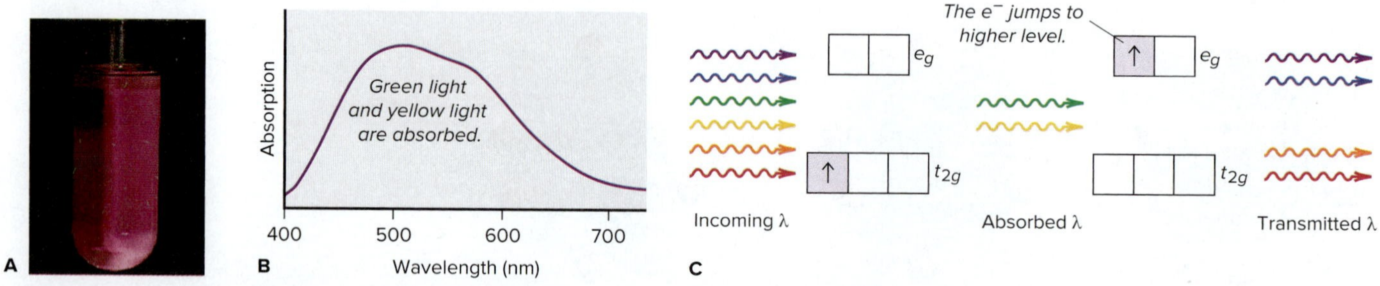

Figure 23.20 The color of $[Ti(H_2O)_6]^{3+}$. **A,** The hydrated Ti^{3+} ion is purple. **B,** An absorption spectrum shows that green light and yellow light are absorbed and other wavelengths are transmitted. **C,** An orbital diagram depicts the colors absorbed when the *d* electron jumps to the higher level.

Source: (A) © McGraw-Hill Education/Pat Watson, photographer energy difference (Δ) between the t_{2g} and e_g orbitals in this ion corresponds to the energy of photons in the green-to-yellow range of visible light. When white light shines on the solution, these colors of light are absorbed, and the electron uses that energy to jump to one of the e_g orbitals. Red, blue, and violet light are transmitted, so the solution appears purple.

Absorption spectra can show the wavelengths absorbed by (1) different metal ions with the same ligand and (2) a metal ion with different ligands. Such data allow us to relate the energy of the absorbed light to Δ and make two key observations:

- *For a given ligand,* color depends on the *oxidation state of the metal ion.* The higher a metal's oxidation state, the larger the crystal field splitting energy and the greater the energy of visible light that must be absorbed to move an electron from the t_{2g} level to the e_g level. For example, $[V(H_2O)_6]^{2+}(aq)$, with vanadium in an oxidation state of +2, absorbs in the yellow range and is violet in color, while $[V(H_2O)_6]^{3+}(aq)$ with vanadium in an oxidation state of +3 absorbs in the higher energy violet range and is yellow in color (Figure 23.21A).
- *For a given metal ion,* color depends on the *ligand.* A single ligand substitution can affect the wavelengths absorbed and, thus, the color. For example, replacing one NH_3 ligand in the yellow $[Cr(NH_3)_6]^{3+}$ ion with a Cl^- ligand results in a purple $[Cr(NH_3)_5Cl]^{2+}$ ion (Figure 23.21B).

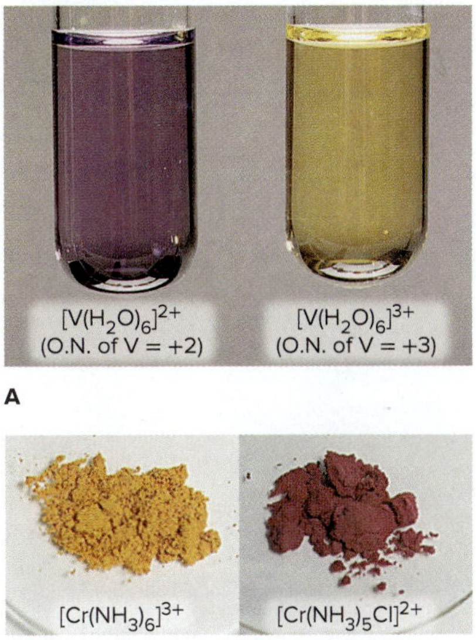

Figure 23.21 Effects of oxidation state and ligand on color. A, Solutions of two hydrated vanadium ions. **B,** A change in one ligand can influence the color.

Source: © McGraw-Hill Education/Stephen Frisch, photographer

The Spectrochemical Series The fact that color depends on the ligand allows us to create a **spectrochemical series,** which ranks the ability of a ligand to split *d*-orbital energies (Figure 23.22). Using this series, we can predict the *relative* magnitude of Δ for a series of octahedral complexes of a *given* metal ion. As Δ increases, higher energies (shorter wavelengths) of light must be absorbed to excite electrons. Although we cannot predict the actual color of a given complex ion, we can determine whether it will absorb longer or shorter wavelengths than other complexes with different ligands.

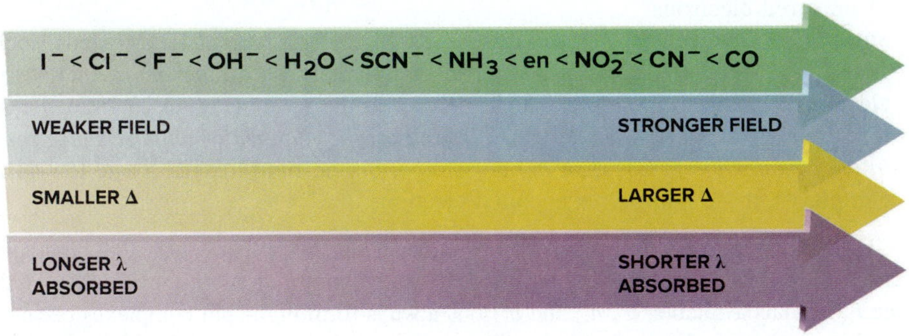

Figure 23.22 The spectrochemical series. For reference, water is a weak-field ligand.

SAMPLE PROBLEM 23.6 Ranking Crystal Field Splitting Energies (Δ) for Complex Ions of a Metal

Problem Rank $[Ti(H_2O)_6]^{3+}$, $[Ti(CN)_6]^{3-}$, and $[Ti(NH_3)_6]^{3+}$ in terms of Δ and of the energy of visible light absorbed.

Plan The formulas show that Ti has the same oxidation state (+3) in the three ions. Using Figure 23.22, we rank the ligands by crystal field strength: the stronger the ligand, the greater the splitting, and the higher the energy of light absorbed.

Solution The ligand field strength is in the order $CN^- > NH_3 > H_2O$, so the relative size of Δ and energy of light absorbed is

$$[Ti(CN)_6]^{3-} > [Ti(NH_3)_6]^{3+} > [Ti(H_2O)_6]^{3+}$$

FOLLOW-UP PROBLEMS

23.6A Which complex ion absorbs visible light of higher energy, $[V(H_2O)_6]^{3+}$ or $[V(NH_3)_6]^{3+}$?

23.6B Some of the complex ions that Co^{3+} forms are $[Co(en)_3]^{3+}$, $[Co(NH_3)_5Cl]^{2-}$, $[Co(NH_3)_4Cl_2]^+$, and $[Co(NH_3)_5(H_2O)]^{3+}$. The observed colors of these ions (listed in arbitrary order) are red, green, purple, and yellow. Match each complex ion with its color.

SOME SIMILAR PROBLEMS 23.92–23.95, 23.97, and 23.98

Explaining the Magnetic Properties of Transition Metal Complexes Splitting of energy levels gives rise to magnetic properties based on the number of *unpaired* electrons in the metal ion's *d* orbitals. Based on Hund's rule, electrons occupy orbitals of equal energy one at a time. When all lower energy orbitals are half-filled, one of two situations occurs:

- The next electron can enter a half-filled orbital and pair up by overcoming a repulsive *pairing energy* ($E_{pairing}$).
- The next electron can enter an empty, higher energy orbital by overcoming Δ.

Thus, *the relative sizes of $E_{pairing}$ and Δ determine the occupancy of d orbitals,* which determines the number of unpaired electrons and, thus, the magnetic behavior of the ion.

As an example, the isolated Mn^{2+} ion ([Ar] $3d^5$) has five unpaired electrons of equal energy (Figure 23.23A). In an octahedral field of ligands, orbital occupancy is affected by the ligand in one of two ways:

- *Weak-field ligands and high-spin complexes.* Weak-field ligands, such as H_2O in $[Mn(H_2O)_6]^{2+}$, cause a *small* splitting energy, so it takes *less* energy for *d* electrons to jump to the e_g set and stay unpaired than to pair up in the t_{2g} set (Figure 23.23B). Thus, for weak-field ligands, $E_{pairing} > \Delta$. Therefore, *the number of unpaired electrons in the complex ion is the **same** as in the free ion:* weak-field ligands create **high-spin complexes,** those with the *maximum* number of unpaired electrons.
- *Strong-field ligands and low-spin complexes.* In contrast, strong-field ligands, such as CN^- in $[Mn(CN)_6]^{4-}$, cause a *large* splitting energy, so it takes *more* energy for electrons to jump to the e_g set than to pair up in the t_{2g} set (Figure 23.23C). Thus, for strong-field ligands, $E_{pairing} < \Delta$. Therefore, *the number of unpaired electrons in the complex ion is **less** than in the free ion.* Strong-field ligands create **low-spin complexes,** those with *fewer* unpaired electrons.

Student Hot Spot

Student data indicate that you may struggle with determining the number of unpaired electrons in high-spin vs. low-spin complex ions. Access the Smartbook to view additional Learning Resources on this topic.

Orbital diagrams for d^1 through d^9 ions in octahedral complexes show that high-spin *and* low-spin options are possible only for d^4, d^5, d^6, and d^7 ions (Figure 23.24). With three t_{2g} orbitals available, d^1, d^2, and d^3 ions always form high-spin complexes because there is no need to pair up. Similarly, d^8 and d^9 ions always form high-spin complexes because the t_{2g} set is filled with six electrons, so the e_g orbitals *must* have either two (d^8) or one (d^9) unpaired electron(s).

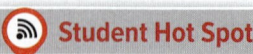

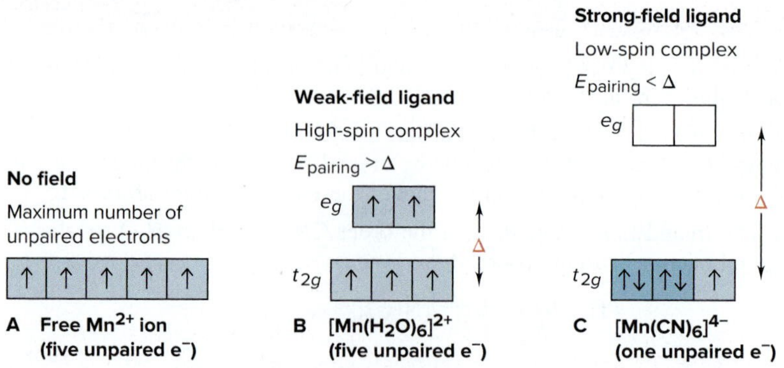

Figure 23.23 High-spin and low-spin octahedral complex ions of Mn^{2+}.

SAMPLE PROBLEM 23.7 **Identifying High-Spin and Low-Spin Complex Ions**

Problem Iron(II) forms a complex in hemoglobin. For each of the two octahedral complex ions $[Fe(H_2O)_6]^{2+}$ and $[Fe(CN)_6]^{4-}$, draw an energy diagram showing orbital splitting, predict the number of unpaired electrons, and identify the ion as low spin or high spin.

Plan The Fe^{2+} electron configuration shows the number of *d* electrons, and the spectrochemical series (Figure 23.22) shows the relative ligand strengths. We draw energy diagrams, in which the t_{2g} and e_g orbital sets are farther apart (larger energy difference) for the strong-field ligand. Then we add electrons, noting that a weak-field ligand gives the *maximum* number of unpaired electrons and a high-spin complex, whereas a strong-field ligand gives the *minimum* number of unpaired electrons and a low-spin complex.

Solution Fe^{2+} has the [Ar] $3d^6$ configuration. H_2O produces smaller splitting than CN^-. The energy diagrams are shown below. The $[Fe(H_2O)_6]^{2+}$ ion has four unpaired electrons (high spin), and the $[Fe(CN)_6]^{4-}$ ion has no unpaired electrons (low spin).

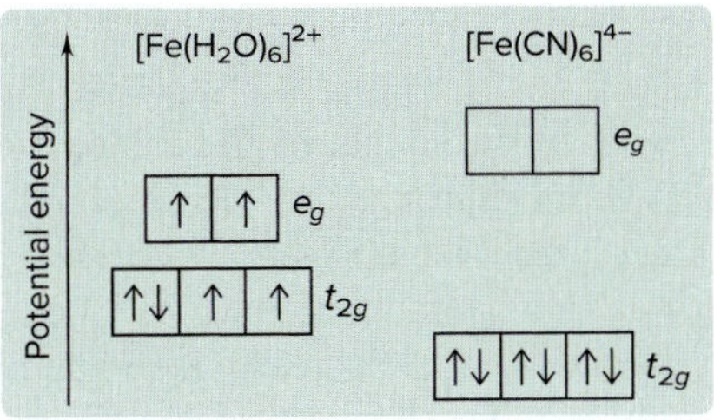

Comment **1.** H_2O is a weak-field ligand, so it forms high-spin complexes. **2.** We cannot confidently predict the spin of a complex without having actual values for Δ and $E_{pairing}$. **3.** Cyanide ions and carbon monoxide are toxic because they bind to the iron complexes in proteins involved in cellular energy (see the Chemical Connections at the end of Chapter 21 and the upcoming Chemical Connections).

FOLLOW-UP PROBLEMS

23.7A How many unpaired electrons do you expect for $[Mn(CN)_6]^{3-}$? Is this a high-spin or low-spin complex ion?

23.7B How many unpaired electrons do you expect for $[CoF_6]^{3-}$? Is this a high-spin or low-spin complex ion?

SOME SIMILAR PROBLEMS 23.88–23.91

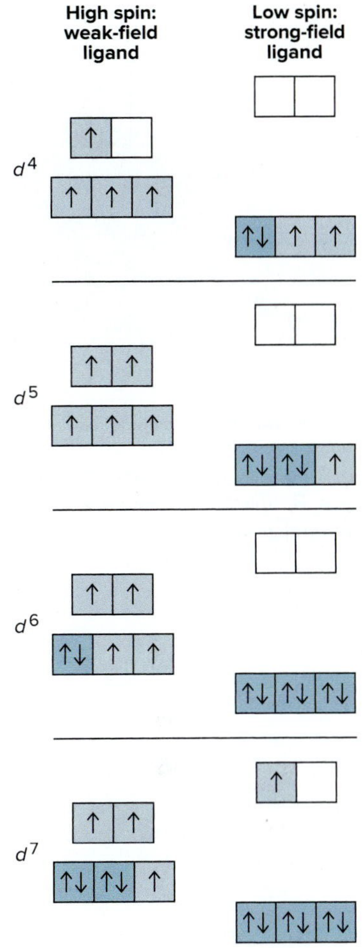

Figure 23.24 Orbital occupancy for high-spin and low-spin octahedral complexes of d^4 through d^7 metal ions.

Crystal Field Splitting in Tetrahedral and Square Planar Complexes Four ligands around a metal ion also cause *d*-orbital splitting, but the magnitude and pattern of the splitting depend on whether the ligands approach from a tetrahedral or a square planar orientation.

1. *Tetrahedral complexes.* When the ligands approach from the corners of a tetrahedron, none of the five *d* orbitals is directly in their paths (Figure 23.25A, *next page*). Thus, the overall attraction of ligand and metal ion is weaker, so the splitting of *d*-orbital energies is *less* in a tetrahedral than in an octahedral complex with the same ligands:

$$\Delta_{tetrahedral} < \Delta_{octahedral}$$

Repulsions are minimized when the ligands approach the d_{xy}, d_{xz}, and d_{yz} orbitals closer than they do the $d_{x^2-y^2}$ and d_{z^2} orbitals. This situation is the *opposite of the octahedral case,* so the relative *d*-orbital energies are reversed: the d_{xy}, d_{xz}, and d_{yz} orbitals become *higher* in energy than the $d_{x^2-y^2}$ and d_{z^2} orbitals. *Only high-spin tetrahedral complexes are known* because the magnitude of Δ is always smaller than $E_{pairing}$, regardless of the ligand.

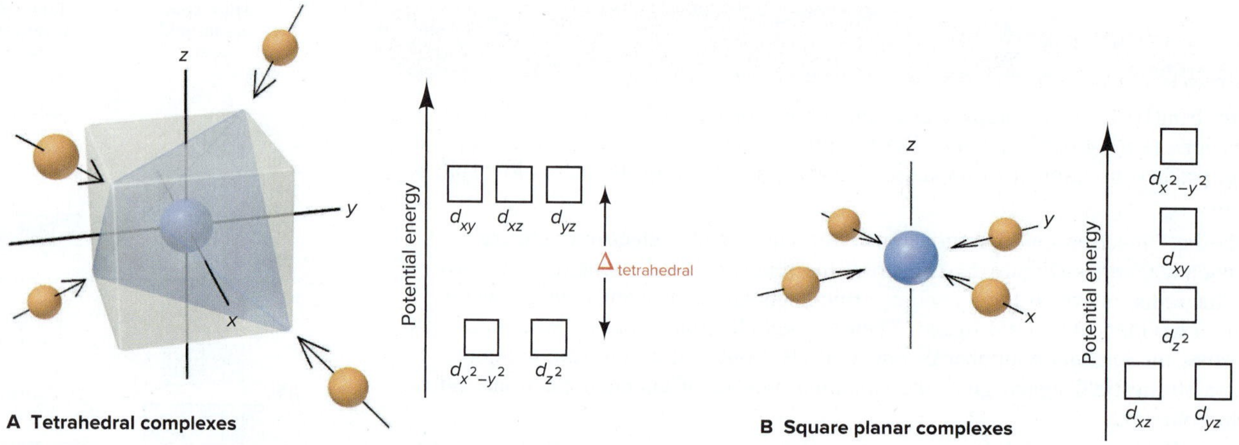

Figure 23.25 Splitting of *d*-orbital energies. A, The pattern of splitting by a tetrahedral field of ligands is the opposite of the octahedral pattern. **B,** Splitting by a square planar field of ligands decreases the energies of d_{xz}, d_{yz}, and especially dz^2 orbitals relative to the octahedral pattern.

2. *Square planar complexes.* The effects of the ligand field in the square planar case are easier to picture if we imagine starting with an octahedral geometry and then removing the two ligands along the *z*-axis (Figure 23.25B). With no *z*-axis interactions present, any *d* orbital with a *z*-axis component has lower energy, with the d_{z^2} orbital decreasing most, and the d_{xz} and d_{yz} also decreasing. In contrast, the two *d* orbitals in the *xy*-plane interact strongly with the ligands, and because the $d_{x^2-y^2}$ orbital has its lobes *on* the axes, its energy is highest. The d^8 metal ions form square planar complexes, such as $[PdCl_4]^{2-}$. They are *low spin* and usually *diamagnetic* because the four pairs of *d* electrons fill the four lowest-energy orbitals.

A Final Word About Bonding Theories As you have seen in several other cases, no one model is satisfactory in every respect. For complex ions, VB theory offers a simple picture of bond formation but does not even attempt to explain spectral and magnetic behavior of complexes. Crystal field theory predicts those two behaviors but offers no insight into the covalent nature of metal-ligand bonding. Thus, chemists now rely on the more comprehensive and quantitative *ligand field theory,* which combines aspects of the previous two models with MO theory (Section 11.3). While beyond the scope of this text, this powerful model predicts bond properties from the overlap of metal ion and ligand orbitals as well as the spectral and magnetic properties resulting from the splitting of the metal ion's *d* orbitals.

In addition to their important chemical applications, complexes of the transition elements play vital roles in living systems, as the following Chemical Connections essay describes.

› Summary of Section 23.4

- According to valence bond theory, complex ions have coordinate covalent bonds between ligands (Lewis bases) and metal ions (Lewis acids).
- Ligand lone pairs occupy hybridized metal-ion orbitals, leading to the characteristic shapes of complex ions.
- According to crystal field theory, the surrounding field of ligands splits a metal ion's *d*-orbital energies. The crystal field splitting energy (Δ) depends on the charge of the metal ion and the crystal field strength of the ligands.
- Δ influences the color (energy of the photons absorbed) and paramagnetism (number of unpaired *d* electrons) of a complex ion. Strong-field ligands create a large Δ and produce low-spin complexes that absorb light of higher energy (shorter wavelength); the reverse is true of weak-field ligands.
- Several transition metals, including iron and zinc, are essential dietary components in trace amounts because they are crucial to the biochemical function of key proteins.

Transition Metals as Essential Dietary Trace Elements

Living things consist primarily of water and complex organic compounds made of four *building-block elements* C, O, H, and N. They also contain seven other elements, known as *macro*nutrients because of their fairly high concentrations. These macronutrients are (listed in order of atomic number): Na, Mg, P, S, Cl, K, and Ca. In addition, organisms contain a large number of elements in low concentrations, and most of these *micro*nutrients, or *trace elements,* are transition metals.

With the exception of scandium and titanium, all Period 4 transition elements are essential to organisms (Table B23.1, *next page*), and plants also require molybdenum (from Period 5). A transition metal ion typically lies next to a protein chain covalently bonded to surrounding amino acid R groups whose N and O atoms act as ligands. Despite the complexity of biomolecules, the principles of bonding and *d*-orbital splitting are the same as in simple inorganic systems. In this discussion, we focus on iron and zinc.

Iron and the Heme Group

Iron plays a crucial role in oxygen transport in all vertebrates. The oxygen-transporting protein hemoglobin (Figure B23.1A) consists of four folded protein chains called *globins,* each cradling the iron-containing complex *heme*. Each heme complex has an Fe^{2+} ion at the center of a tetradentate ring ligand *porphin* and bonded to the ligand's four N lone pairs to form a *square planar* complex. When heme is bound in hemoglobin, the complex becomes *square pyramidal,* with an N atom from a nearby amino acid (histidine) as the fifth ligand. Hemoglobin exists in two forms, depending on the presence of O_2, which acts as a sixth ligand:

- *In the lungs*, where $[O_2]$ is high, heme binds O_2 to form *oxyhemoglobin*, which is transported in the arteries to tissues. With O_2 bound, heme exists as part of an *octahedral* complex (Figure B23.1B). Since O_2 is a strong-field ligand, the d^6 Fe^{2+} ion is part of a low-spin complex. Because of the relatively large *d*-orbital splitting, oxyhemoglobin absorbs blue (high-energy) light, so arterial blood is bright red.
- *At the tissues*, where $[O_2]$ is low, the bound O_2 is released, and *deoxyhemoglobin* forms, which is transported in the veins back to the lungs. Without the O_2, the Fe^{2+} ion is part of a high-spin complex, which has relatively small *d*-orbital splitting. Thus, deoxyhemoglobin absorbs red (low-energy) light, and venous blood is dark, purplish red.

The relative position of Fe^{2+} in the plane of the porphin ring also depends on the presence of the sixth ligand. Bound to O_2, Fe^{2+} moves *into* the porphin plane; when O_2 is released, it moves slightly *out of* the plane. This tiny (60 pm) change due to the binding or release of O_2 alters the shape of the globin chain. And this change in shape alters the shape of the next globin chain, triggering the release or attachment of *its* O_2, and so on to the other two globin chains. This "cooperation" among the four globin chains allows hemoglobin to rapidly pick up O_2 from the lungs and unload it rapidly in the tissues.

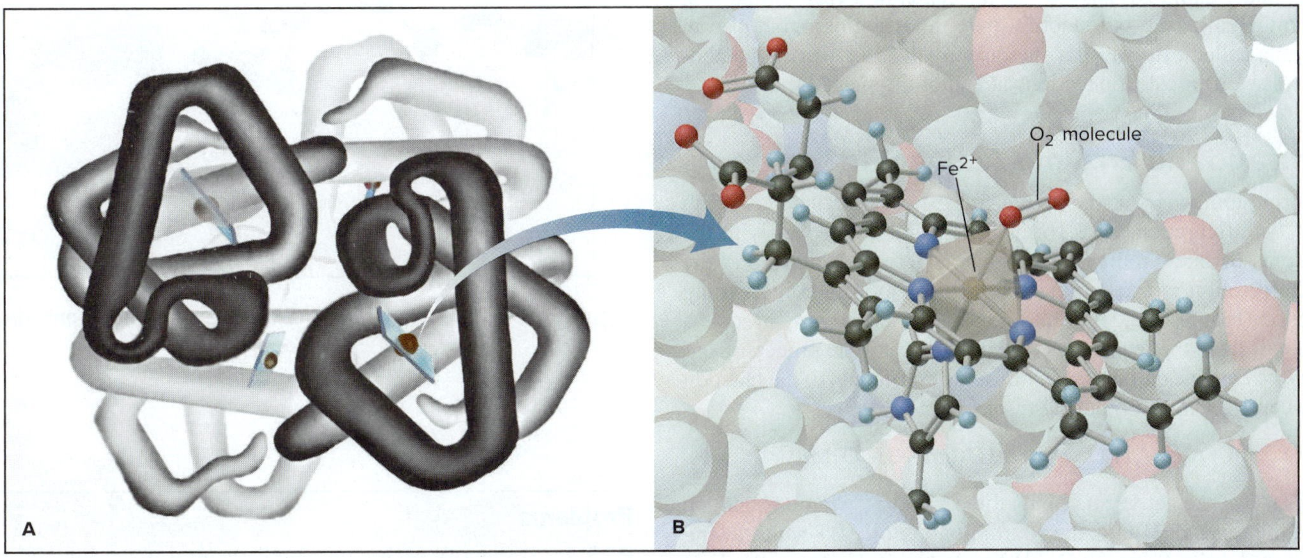

Figure B23.1 Hemoglobin and the octahedral complex in heme. A, Hemoglobin consists of four protein chains, each with a bound heme. **B,** In oxyhemoglobin, the octahedral complex in heme has an O_2 molecule as the sixth ligand of iron(II).

(continued)

Table B23.1 Some Transition Metal Trace Elements in Humans

Element	Biomolecule(s) Containing Element	Function(s) of Biomolecule(s)
Vanadium	Protein (?)	Redox couple in fat metabolism (?)
Chromium	Glucose tolerance factor	Glucose utilization
Manganese	Isocitrate dehydrogenase	Cell respiration
Iron	Hemoglobin and myoglobin	Oxygen transport
	Cytochrome *c*	Cell respiration; ATP formation
	Catalase	Decomposition of H_2O_2
Cobalt	Cobalamin (vitamin B_{12})	Development of red blood cells
Copper	Ceruloplasmin	Hemoglobin synthesis
	Cytochrome oxidase	Cell respiration; ATP formation
Zinc	Carbonic anhydrase	Elimination of CO_2
	Carboxypeptidase A	Protein digestion
	Alcohol dehydrogenase	Metabolism of ethanol

Heme binding of O_2 is an equilibrium process. Carbon monoxide is toxic because it binds to Fe^{2+} about 200 times more tightly than O_2 does; each CO that binds eliminates one heme group from functioning. Like O_2, CO is a strong-field ligand and produces a bright red, "healthy" look in the poisoned individual. Because binding is reversible, breathing high concentrations of O_2 displaces CO from the heme:

$$\text{heme—CO} + O_2 \rightleftharpoons \text{heme—}O_2 + CO$$

Porphin is a very common biological ligand. Chlorophyll has Mg^{2+} and vitamin B_{12} has Co^{3+} at the center of a very similar ring system. Heme itself is found not only in hemoglobin, but also in proteins called *cytochromes* that are involved in energy metabolism (see Chemical Connections at end of Chapter 21).

Zinc and the Enzyme Active Site

The zinc ion occurs at the active sites of several enzymes. With its d^{10} configuration, Zn^{2+} forms tetrahedral complexes with the N atoms of three amino acid groups at three positions, and the fourth position is free to interact with the molecule undergoing reaction (Figure B23.2). In its role as Lewis acid, Zn^{2+} accepts a lone pair from the reactant as a step in the catalytic process. Consider the enzyme carbonic anhydrase, which catalyzes the reaction between H_2O and CO_2 during respiration:

$$CO_2(g) + H_2O(l) \rightleftharpoons H^+(aq) + HCO_3^-(aq)$$

The Zn^{2+} at the active site binds three histidine N atoms, and the O of the reactant H_2O is the fourth donor atom. By withdrawing electron density from the O—H bonds, the Zn^{2+} makes the H_2O acidic enough to lose a proton. In the rate-determining step, the bound OH^- ion is attracted to the partially positive C of CO_2 much more strongly than the lone pair of a free water molecule would be; thus, the reaction rate is higher. One reason Cd^{2+} is toxic is that it competes with Zn^{2+} for the carbonic anhydrase active site.

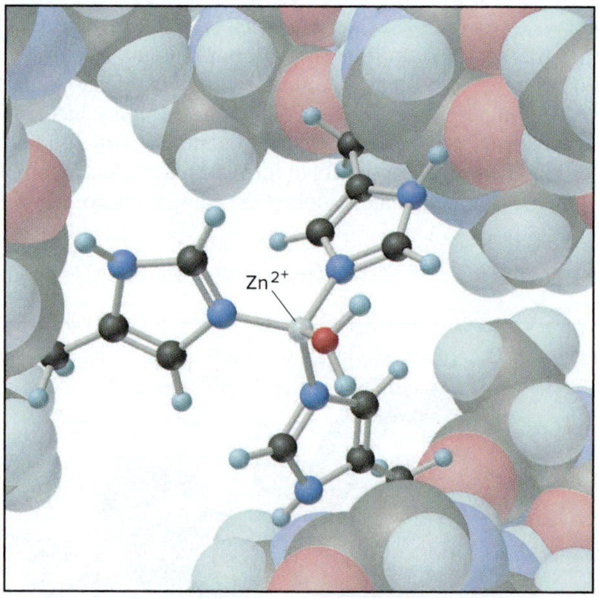

Figure B23.2 The tetrahedral Zn^{2+} complex in carbonic anhydrase.

Problems

B23.1 (a) Identify each compound's central metal ion and give its oxidation state: chlorophyll, heme, and vitamin B_{12}. (b) What similarity in structure do these compounds have?

B23.2 Suggest a structural reason why carbonic anhydrase synthesized with Ni^{2+}, Fe^{2+}, or Mn^{2+} in place of Zn^{2+} has less catalytic efficiency.

CHAPTER REVIEW GUIDE

Learning Objectives

Relevant section (§) and/or sample problem (SP) numbers appear in parentheses.

Understand These Concepts

1. The positions of the *d*- and *f*-block elements and the general forms of their atomic and ionic electron configurations (§23.1)
2. How atomic size, ionization energy, and electronegativity vary across a period and down a group of transition elements and how these trends differ from those of the main-group elements; why the densities of Period 6 transition elements are so high (§23.1)
3. Why the transition elements often have multiple oxidation states and why the +2 state is common (§23.1)
4. Why metallic behavior (prevalence of ionic bonding and basic oxides) of transition elements decreases as oxidation state increases; how valence-state electronegativity explains transition metal atoms in oxoanions (§23.1)
5. Why many transition metal compounds are colored and paramagnetic (§23.1)
6. The common +3 oxidation state of the lanthanides and the similarity of the radii of their M^{3+} ions; the radioactivity of the actinides (§23.2)
7. The coordination numbers, geometries, and ligand structures of complex ions (§23.3)
8. How coordination compounds are named and their formulas written (§23.3)
9. The types of constitutional isomerism (coordination and linkage) and stereoisomerism (geometric and optical) of complex ions (§23.3)
10. How valence bond theory uses hybridization to account for the shapes of octahedral, square planar, and tetrahedral complexes (§23.4)
11. How crystal field theory explains that approaching ligands cause *d*-orbital energies to split (§23.4)
12. How the relative crystal field strength of ligands (spectrochemical series) affects the *d*-orbital splitting energy (Δ) (§23.4)
13. How the magnitude of Δ accounts for the energy of light absorbed and, thus, the color of a complex ion (§23.4)
14. How the relative sizes of pairing energy and Δ determine the occupancy of *d* orbitals and, thus, the magnetic properties of complex ions (§23.4)
15. How *d*-orbital splitting in tetrahedral and square planar complexes differs from that in octahedral complexes (§23.4)

Master These Skills

1. Writing electron configurations of transition metal atoms and ions (SP 23.1)
2. Using a partial orbital diagram to determine the number of unpaired electrons in a transition metal atom or ion (SP 23.2)
3. Recognizing the structural components of complex ions (§23.3)
4. Determining the coordination number and charge of the central metal ion in a coordination compound (SP 23.3)
5. Naming and writing formulas of coordination compounds (SP 23.4)
6. Determining the type of stereoisomerism in complexes (SP 23.5)
7. Correlating a complex ion's shape with the number and type of hybrid orbitals of the central metal ion (§23.4)
8. Using the spectrochemical series to rank complex ions in terms of Δ and the energy of visible light absorbed (SP 23.6)
9. Using the spectrochemical series to determine if a complex ion is high spin or low spin (SP 23.7)

Key Terms

Page numbers appear in parentheses.

actinides (1044)
chelate (1047)
complex ion (1046)
constitutional (structural) isomers (1051)
coordinate covalent bond (1055)
coordination compound (1046)
coordination isomers (1051)
coordination number (1046)
counter ion (1046)
crystal field splitting energy (Δ) (1058)
crystal field theory (1056)
donor atom (1047)
e_g orbital (1058)
geometric *(cis-trans)* isomers (1052)
high-spin complex (1060)
inner transition elements (1044)
isomer (1051)
lanthanide contraction (1041)
lanthanides (1044)
ligand (1046)
linkage isomers (1052)
low-spin complex (1060)
optical isomers (1053)
spectrochemical series (1059)
stereoisomers (1052)
strong-field ligand (1058)
t_{2g} orbital (1058)
transition elements (1037)
weak-field ligand (1058)

Key Equation and Relationship

Page number appears in parentheses.

23.1 Finding the charge of the metal ion in a complex anion or cation (1048):

Charge of metal ion = charge of complex ion − total charge of ligands

BRIEF SOLUTIONS TO FOLLOW-UP PROBLEMS

23.1A (a) Ag: $5s^1 4d^{10}$; Ag^+: $4d^{10}$ (b) Cd: $5s^2 4d^{10}$; Cd^{2+}: $4d^{10}$
(c) Ir: $6s^2 5d^7$; Ir^{3+}: $5d^6$

23.1B (a) Ta: [Xe] $6s^2 4f^{14} 5d^3$; three electrons have been lost: Ta^{3+}
(b) Mn: [Ar] $4s^2 3d^5$; four electrons have been lost: Mn^{4+}
(c) Os: [Xe] $6s^2 4f^{14} 5d^6$; three electrons have been lost: Os^{3+}

23.2A Er: [Xe] $6s^2 4f^{12}$; Er^{3+} is [Xe] $4f^{11}$; three unpaired electrons

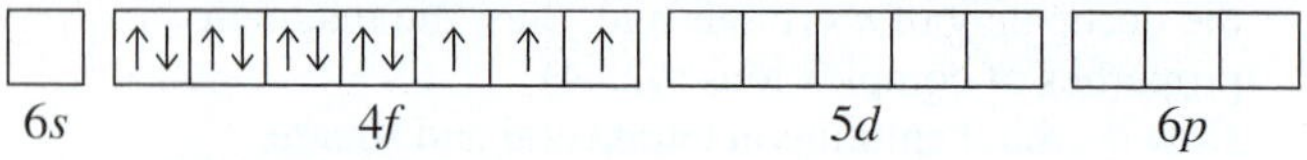

23.2B Dy: [Xe] $6s^2 4f^{10}$; Dy^{3+} is [Xe] $4f^9$; five unpaired electrons

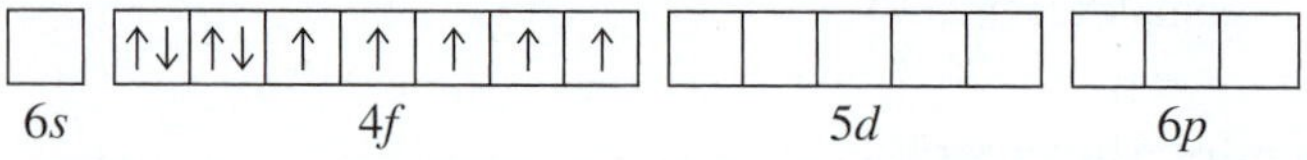

23.3A (a) Each en bidentate ligand forms two bonds to the metal ion, for a total of four bonds; the two monodentate Br^- ligands form a total of two bonds: coordination number = 6. The complex ion has a charge of 1+:

Charge of metal ion = charge of complex ion − total charge of ligands
$= (1+) - [(2 \times 0) + (2 \times 1-)] = 3+$; Co^{3+}

(b) The complex has six monodentate ligands: coordination number = 6; three Mg^{2+} counter ions have a total charge of 6+; each of the two complex ions has a charge of 3−:

Charge of metal ion = charge of complex ion − total charge of ligands
$= (3-) - [(6 \times 1-)] = 3+$; Cr^{3+}

23.3B (a) $K_4[Mn(CN)_6]$ (b) $[Cu(en)_2]SO_4$

23.4A (a) Five (*penta-*) H_2O (*aqua*) ligands, one Br^- (*bromo*) ligand, and the Cr^{3+} ion in the complex ion with Cl^- counter ions: Pentaaquabromochromium(III) chloride
(b) Hexacyano: $(CN)_6$; charge of complex ion = 3+ [from Co(III)] + (6−) [from $(CN)_6$] = 3−; three 2+ barium counter ions are needed to balance the charge of two 3− complex ions: $Ba_3[Co(CN)_6]_2$

23.4B (a) Two (*di-*) water (*aqua*) and two (*bis*) en ligands; charge of complex ion = 3+ [from Co(III)] + 2(0) [from $(H_2O)_2$] and 2(0) [from $(en)_2$] = 3+; three SO_4^{2-} counter ions are needed to balance the charge of two 3+ complex ions: $[Co(H_2O)_2(en)_2]_2(SO_4)_3$
(b) Two (*di-*) NH_3 (*ammine*) and four (*tetra-*) Cl^- (*chloro*) ligands and the Cr^{3+} ion with Mg^{2+} as the counter ion: magnesium diamminetetrachlorochromate(III)

23.5A Two sets of *cis-trans* isomers, and the two *cis* isomers are optical isomers.

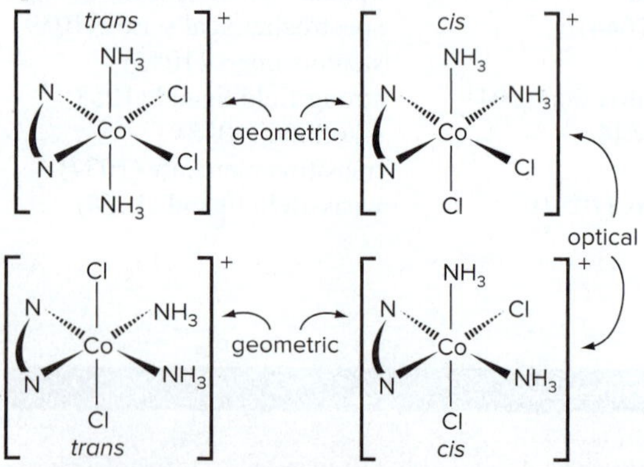

23.5B Five geometric isomers; the *cis* isomer has an optical isomer.

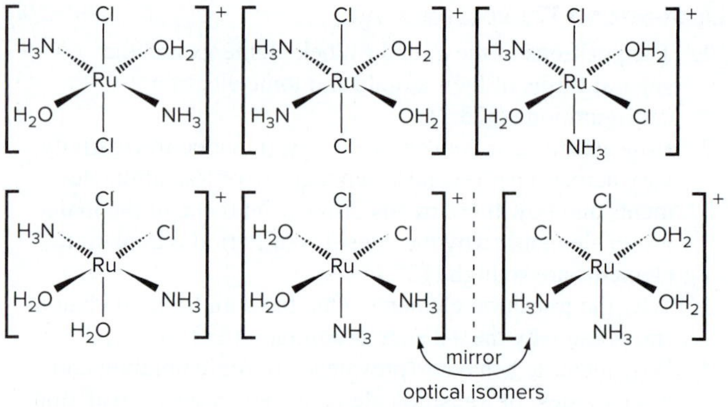

23.6A Both metal ions are V^{3+}; in terms of ligand field energy, $NH_3 > H_2O$, so $[V(NH_3)_6]^{3+}$ absorbs light of higher energy.

23.6B In terms of ligand field energy and energy of light absorbed, en > NH_3 > H_2O > Cl^-; $[Co(en)_3]^{3+}$ is yellow (violet light absorbed), $[Co(NH_3)_5(H_2O)]^{3+}$ is red (green light absorbed), $[Co(NH_3)_5Cl]^{2+}$ is purple (yellow light absorbed), and $[Co(NH_3)_4Cl_2]^+$ is green (red light absorbed). As weaker field ligands replace stronger field ligands, the energy of light absorbed decreases.

23.7A The metal ion is Mn^{3+}: [Ar] $3d^4$; CN^- is a strong-field ligand:

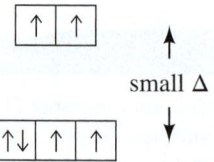

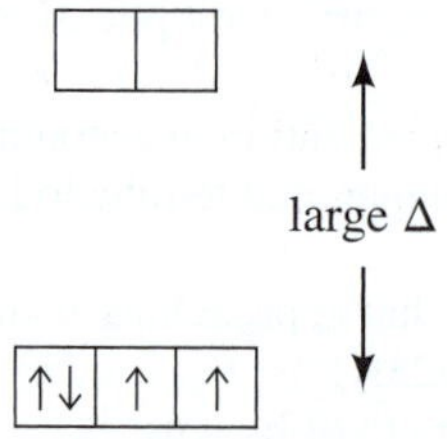

Two unpaired *d* electrons; low-spin complex

23.7B The metal ion is Co^{3+}: [Ar] $3d^6$; F^- is a weak-field ligand:

Four unpaired *d* electrons; high-spin complex

PROBLEMS

Problems with **colored** numbers are answered in Appendix E and worked in detail in the Student Solutions Manual. Problem sections match those in the text and give the numbers of relevant sample problems. Most offer Concept Review Questions, Skill-Building Exercises (grouped in pairs covering the same concept), and Problems in Context. The Comprehensive Problems are based on material from any section or previous chapter.

Note: In these problems, the term *electron configuration* refers to the condensed, ground-state electron configuration.

Properties of the Transition Elements

(Sample Problem 23.1)

Concept Review Questions

23.1 How is the *n* value of the *d* sublevel of a transition element related to the period number of the element?

23.2 Write the general electron configuration of a transition element (a) in Period 5; (b) in Period 6.

23.3 (a) What is the general rule concerning the order in which electrons are removed from a transition metal atom to form an ion? Give an example from Group 5B(5). (b) Name two types of measurements used to study electron configurations of ions.

23.4 What is the maximum number of unpaired *d* electrons that an atom or ion can possess? Give an example of an atom and an ion that have this number.

23.5 How does the variation in atomic size across a transition series contrast with the change across the main-group elements of the same period? Why?

23.6 (a) What is the lanthanide contraction? (b) How does it affect atomic size down a group of transition elements? (c) How does it influence the densities of the Period 6 transition elements?

23.7 (a) What is the range in electronegativity across the first (3*d*) transition series? (b) What is the range across Period 4 of the main-group elements? (c) Explain the difference.

23.8 (a) Explain the major difference between the number of oxidation states of most transition elements and that of most main-group elements. (b) Why is the +2 oxidation state so common among transition elements? (c) What is valence-state electronegativity? Is the electronegativity of Cr different in CrO, Cr_2O_3, and CrO_3? Explain.

23.9 (a) What behavior distinguishes paramagnetic and diamagnetic substances? (b) Why are paramagnetic ions common among transition elements but not main-group elements? (c) Why are colored solutions of metal ions common among transition elements but not main-group elements?

Skill-Building Exercises (grouped in similar pairs)

23.10 Using the periodic table to locate each element, write the electron configuration of (a) V; (b) Y; (c) Hg.

23.11 Using the periodic table to locate each element, write the electron configuration of (a) Ru; (b) Cu; (c) Ni.

23.12 Using the periodic table to locate each element, write the electron configuration of (a) Os; (b) Co; (c) Ag.

23.13 Using the periodic table to locate each element, write the electron configuration of (a) Zn; (b) Mn; (c) Re.

23.14 Give the electron configuration and the number of unpaired electrons for (a) Sc^{3+}; (b) Cu^{2+}; (c) Fe^{3+}; (d) Nb^{3+}.

23.15 Give the electron configuration and the number of unpaired electrons for (a) Cr^{3+}; (b) Ti^{4+}; (c) Co^{3+}; (d) Ta^{2+}.

23.16 What is the highest oxidation state for (a) Ta; (b) Zr; (c) Mn?

23.17 What is the highest oxidation state for (a) Nb; (b) Y; (c) Tc?

23.18 Which transition metals have a maximum O.N. of +6?

23.19 Which transition metals have a maximum O.N. of +4?

23.20 In which compound does Cr exhibit greater metallic behavior, CrF_2 or CrF_6? Explain.

23.21 VF_5 is a liquid that boils at 48°C, whereas VF_3 is a solid that melts above 800°C. Explain this difference in properties.

23.22 Is it more difficult to oxidize Cr or Mo? Explain.

23.23 Is MnO_4^- or ReO_4^- a stronger oxidizing agent? Explain.

23.24 Which oxide, CrO_3 or CrO, is more acidic in water? Why?

23.25 Which oxide, Mn_2O_3 or Mn_2O_7, is more basic in water? Why?

Problems in Context

23.26 The green patina of Cu-alloy roofs results from corrosion in the presence of O_2, H_2O, CO_2, and sulfur compounds. The other members of Group 1B(11), Ag and Au, do not form a patina. Corrosion of Cu and Ag in the presence of sulfur compounds leads to a black tarnish, but Au does not tarnish. This pattern is different from that in Group 1A(1), where ease of oxidation *increases* down the group. Explain these different group patterns.

The Inner Transition Elements

(Sample Problem 23.2)

Concept Review Questions

23.27 What atomic property of the lanthanides leads to their remarkably similar chemical properties?

23.28 (a) What is the maximum number of unpaired electrons in a lanthanide ion? (b) How does this number relate to occupancy of the 4*f* subshell?

23.29 Which of the actinides are radioactive?

Skill-Building Exercises (grouped in similar pairs)

23.30 Give the electron configuration of (a) La; (b) Ce^{3+}; (c) Es; (d) U^{4+}.

23.31 Give the electron configuration of (a) Pm; (b) Lu^{3+}; (c) Th; (d) Fm^{3+}.

23.32 Only a few lanthanides show an oxidation state other than +3. Two of these, europium (Eu) and terbium (Tb), are found near the middle of the series, and their unusual oxidation states can be associated with a half-filled *f* subshell. (a) Write the electron configurations of Eu^{2+}, Eu^{3+}, and Eu^{4+}. Why is Eu^{2+} a common ion, whereas Eu^{4+} is unknown? (b) Write the electron configurations of Tb^{2+}, Tb^{3+}, and Tb^{4+}. Would you expect Tb to show a +2 or a +4 oxidation state? Explain.

23.33 Cerium (Ce) and ytterbium (Yb) exhibit some oxidation states in addition to +3. (a) Write the electron configurations of Ce^{2+}, Ce^{3+}, and Ce^{4+}. (b) Write the electron configurations of Yb^{2+}, Yb^{3+}, and Yb^{4+}. (c) In addition to the 3+ ions, the ions Ce^{4+} and Yb^{2+} are stable. Suggest a reason for this stability.

Problems in Context

23.34 Which lanthanide has the maximum number of unpaired electrons in both its atom and its 3+ ion? Give the number of unpaired electrons in the atom and in the ion.

Coordination Compounds

(Sample Problems 23.3 to 23.5)

Concept Review Questions

23.35 Describe the makeup of a complex ion, including the nature of the ligands and their interaction with the central metal ion. Explain how a complex ion can be positive or negative and how it occurs as part of a neutral coordination compound.

23.36 What electronic feature must a donor atom of a ligand have?

23.37 What is the coordination number of a metal ion in a complex ion? How does it differ from the oxidation number?

23.38 What structural feature is characteristic of a chelate?

23.39 What geometries are associated with the coordination numbers 2, 4, and 6?

23.40 What are the coordination numbers of cobalt(III), platinum(II), and platinum(IV) in complexes?

23.41 How is a complex ion a Lewis adduct?

23.42 What does the ending *-ate* in the name of a complex ion signify?

23.43 In what order are the metal ion and ligands identified in the name of a complex ion?

23.44 Is a linkage isomer a type of constitutional isomer or stereoisomer? Explain.

Skill-Building Exercises (grouped in similar pairs)

23.45 Give systematic names for the following formulas:
(a) $[Ni(H_2O)_6]Cl_2$ (b) $[Cr(en)_3](ClO_4)_3$ (c) $K_4[Mn(CN)_6]$

23.46 Give systematic names for the following formulas:
(a) $[Co(NH_3)_4(NO_2)_2]Cl$ (b) $[Cr(NH_3)_6][Cr(CN)_6]$
(c) $K_2[CuCl_4]$

23.47 What are the charge and coordination number of the central metal ion(s) in each compound in Problem 23.45?

23.48 What are the charge and coordination number of the central metal ion(s) in each compound in Problem 23.46?

23.49 Give systematic names for the following formulas:
(a) $K[Ag(CN)_2]$ (b) $Na_2[CdCl_4]$ (c) $[Co(NH_3)_4(H_2O)Br]Br_2$

23.50 Give systematic names for the following formulas:
(a) $K[Pt(NH_3)Cl_5]$ (b) $[Cu(en)(NH_3)_2][Co(en)Cl_4]$
(c) $[Pt(en)_2Br_2](ClO_4)_2$

23.51 What are the charge and coordination number of the central metal ion(s) in each compound in Problem 23.49?

23.52 What are the charge and coordination number of the central metal ion(s) in each compound in Problem 23.50?

23.53 Give formulas corresponding to the following names:
(a) Tetraamminezinc sulfate
(b) Pentaamminechlorochromium(III) chloride
(c) Sodium bis(thiosulfato)argentate(I)

23.54 Give formulas corresponding to the following names:
(a) Dibromobis(ethylenediamine)cobalt(III) sulfate
(b) Hexaamminechromium(III) tetrachlorocuprate(II)
(c) Potassium hexacyanoferrate(II)

23.55 What is the coordination number of the metal ion and the number of individual ions per formula unit in each of the compounds in Problem 23.53?

23.56 What is the coordination number of the metal ion and the number of individual ions per formula unit in each of the compounds in Problem 23.54?

23.57 Give formulas corresponding to the following names:
(a) Hexaaquachromium(III) sulfate
(b) Barium tetrabromoferrate(III)
(c) Bis(ethylenediamine)platinum(II) carbonate

23.58 Give formulas corresponding to the following names:
(a) Potassium tris(oxalato)chromate(III)
(b) Tris(ethylenediamine)cobalt(III) pentacyanoiodomanganate(II)
(c) Diamminediaquabromochloroaluminum nitrate

23.59 Give the coordination number of the metal ion and the number of ions per formula unit in each compound in Problem 23.57.

23.60 Give the coordination number of the metal ion and the number of ions per formula unit in each compound in Problem 23.58.

23.61 Which of these ligands can participate in linkage isomerism: (a) NO_2^-; (b) SO_2; (c) NO_3^-? Explain with Lewis structures.

23.62 Which of these ligands can participate in linkage isomerism: (a) SCN^-; (b) $S_2O_3^{2-}$ (thiosulfate); (c) HS^-? Explain with Lewis structures.

23.63 For any of the following that can exist as isomers, state the type of isomerism and draw the structures:
(a) $[Pt(CH_3NH_2)_2Br_2]$ (b) $[Pt(NH_3)_2FCl]$ (c) $[Pt(H_2O)(NH_3)FCl]$

23.64 For any of the following that can exist as isomers, state the type of isomerism and draw the structures:
(a) $[Zn(en)F_2]$ (b) $[Zn(H_2O)(NH_3)FCl]$ (c) $[Pd(CN)_2(OH)_2]^{2-}$

23.65 For any of the following that can exist as isomers, state the type of isomerism and draw the structures:
(a) $[PtCl_2Br_2]^{2-}$ (b) $[Cr(NH_3)_5(NO_2)]^{2+}$ (c) $[Pt(NH_3)_4I_2]^{2+}$

23.66 For any of the following that can exist as isomers, state the type of isomerism and draw the structures:
(a) $[Co(NH_3)_5Cl]Br_2$ (b) $[Pt(CH_3NH_2)_3Cl]Br$
(c) $[Fe(H_2O)_4(NH_3)_2]^{2+}$

Problems in Context

23.67 Chromium(III), like cobalt(III), has a coordination number of 6 in many of its complex ions. Before Alfred Werner, in the 1890s, established the idea of a complex ion, coordination compounds had traditional formulas. Compounds are known that have the traditional formula $CrCl_3{\cdot}nNH_3$, where $n = 3$ to 6. Which of these compounds has an electrical conductivity in aqueous solution similar to that of an equimolar NaCl solution?

23.68 When $MCl_4(NH_3)_2$ is dissolved in water and treated with $AgNO_3$, 2 mol of AgCl precipitates immediately for each mole of $MCl_4(NH_3)_2$. Give the coordination number of M in the complex.

23.69 Palladium, like its group neighbor platinum, forms four-coordinate Pd(II) and six-coordinate Pd(IV) complexes. Write formulas for the complexes with these compositions:
(a) $PdK(NH_3)Cl_3$ (b) $PdCl_2(NH_3)_2$ (c) PdK_2Cl_6 (d) $Pd(NH_3)_4Cl_4$

Theoretical Basis for the Bonding and Properties of Complexes

(Sample Problems 23.6 and 23.7)

Concept Review Questions

23.70 (a) What is a coordinate covalent bond?
(b) Is such a bond involved when $FeCl_3$ dissolves in water? Explain.
(c) Is such a bond involved when HCl gas dissolves in water? Explain.

23.71 According to valence bond theory, what set of orbitals is used by a Period 4 metal ion in forming (a) a square planar complex; (b) a tetrahedral complex?

23.72 A metal ion uses d^2sp^3 orbitals when forming a complex. What is its coordination number and the shape of the complex?

23.73 A complex in solution absorbs green light. What is the color of the solution?

23.74 In terms of the theory of color absorption, explain two ways that a solution can be blue.

23.75 (a) What is the crystal field splitting energy (Δ)?
(b) How does it arise for an octahedral field of ligands?
(c) How is it different for a tetrahedral field of ligands?

23.76 What is the distinction between a weak-field ligand and a strong-field ligand? Give an example of each.

23.77 Is a complex with the same number of unpaired electrons as the free gaseous metal ion termed high spin or low spin?

23.78 How do the relative magnitudes of $E_{pairing}$ and Δ affect the paramagnetism of a complex?

23.79 Why are there both high-spin and low-spin octahedral complexes but only high-spin tetrahedral complexes?

Skill-Building Exercises (grouped in similar pairs)

23.80 Give the number of *d* electrons (*n* of d^n) for the central metal ion in (a) $[TiCl_6]^{2-}$; (b) $K[AuCl_4]$; (c) $[RhCl_6]^{3-}$.

23.81 Give the number of *d* electrons (*n* of d^n) for the central metal ion in (a) $[Cr(H_2O)_6](ClO_3)_2$; (b) $[Mn(CN)_6]^{2-}$; (c) $[Ru(NO)(en)_2Cl]Br$.

23.82 How many *d* electrons (*n* of d^n) are in the central metal ion in (a) $Ca[IrF_6]$; (b) $[HgI_4]^{2-}$; (c) $[Co(EDTA)]^{2-}$?

23.83 How many *d* electrons (*n* of d^n) are in the central metal ion in (a) $[Ru(NH_3)_5Cl]SO_4$; (b) $Na_2[Os(CN)_6]$; (c) $[Co(NH_3)_4CO_3I]$?

23.84 Sketch the orientation of the orbitals relative to the ligands in an octahedral complex to explain the splitting and the relative energies of the d_{xy} and the $d_{x^2-y^2}$ orbitals.

23.85 The two e_g orbitals are identical in energy in an octahedral complex but have different energies in a square planar complex, with the d_{z^2} orbital being much lower in energy than the $d_{x^2-y^2}$. Explain with orbital sketches.

23.86 Which of these ions *cannot* form both high- and low-spin octahedral complexes: (a) Ti^{3+}; (b) Co^{2+}; (c) Fe^{2+}; (d) Cu^{2+}?

23.87 Which of these ions *cannot* form both high- and low-spin octahedral complexes: (a) Mn^{3+}; (b) Nb^{3+}; (c) Ru^{3+}; (d) Ni^{2+}?

23.88 Draw orbital-energy splitting diagrams and use the spectrochemical series to show the orbital occupancy for each of the following (assuming that H_2O is a weak-field ligand):
(a) $[Cr(H_2O)_6]^{3+}$ (b) $[Cu(H_2O)_4]^{2+}$ (c) $[FeF_6]^{3-}$

23.89 Draw orbital-energy splitting diagrams and use the spectrochemical series to show the orbital occupancy for each of the following (assuming that H_2O is a weak-field ligand):
(a) $[Cr(CN)_6]^{3-}$ (b) $[Rh(CO)_6]^{3+}$ (c) $[Co(OH)_6]^{4-}$

23.90 Draw orbital-energy splitting diagrams and use the spectrochemical series to show the orbital occupancy for each of the following (assuming that H_2O is a weak-field ligand):
(a) $[MoCl_6]^{3-}$ (b) $[Ni(H_2O)_6]^{2+}$ (c) $[Ni(CN)_4]^{2-}$

23.91 Draw orbital-energy splitting diagrams and use the spectrochemical series to show the orbital occupancy for each of the following (assuming that H_2O is a weak-field ligand):
(a) $[Fe(C_2O_4)_3]^{3-}$ ($C_2O_4^{2-}$ creates a weaker field than H_2O does.)
(b) $[Co(CN)_6]^{4-}$ (c) $[MnCl_6]^{4-}$

23.92 Rank the following in order of *increasing* Δ and energy of light absorbed: $[Cr(NH_3)_6]^{3+}$, $[Cr(H_2O)_6]^{3+}$, $[Cr(NO_2)_6]^{3-}$.

23.93 Rank the following in order of *decreasing* Δ and energy of light absorbed: $[Cr(en)_3]^{3+}$, $[Cr(CN)_6]^{3-}$, $[CrCl_6]^{3-}$.

23.94 A complex ion, $[ML_6]^{2+}$, is violet. The same metal forms a complex with another ligand, Q, that creates a weaker field. What color might $[MQ_6]^{2+}$ be expected to show? Explain.

23.95 The complex ion $[Cr(H_2O)_6]^{2+}$ is violet. Another CrL_6 complex is green. Can ligand L be CN^-? Can it be Cl^-? Explain.

Problems in Context

23.96 Octahedral $[Ni(NH_3)_6]^{2+}$ is paramagnetic, whereas planar $[Pt(NH_3)_4]^{2+}$ is diamagnetic, even though both metal ions are d^8 species. Explain.

23.97 The hexaaqua complex $[Ni(H_2O)_6]^{2+}$ is green, whereas the hexaammonia complex $[Ni(NH_3)_6]^{2+}$ is violet. Explain.

23.98 Three of the complex ions that are formed by Co^{3+} are $[Co(H_2O)_6]^{3+}$, $[Co(NH_3)_6]^{3+}$, and $[CoF_6]^{3-}$. These ions have the observed colors (listed in arbitrary order) yellow-orange, green, and blue. Match each complex with its color. Explain.

Comprehensive Problems

23.99 When neptunium (Np) and plutonium (Pu) were discovered, the periodic table did not include the actinides, so these elements were placed in Groups 7B(7) and 8B(8). When americium (Am) and curium (Cm) were synthesized, they were placed in Groups 8B(9) and 8B(10). However, during chemical isolation procedures, Glenn Seaborg and his colleagues, who had synthesized these elements, could not find their compounds among other compounds of members of the same groups, which led Seaborg to suggest they were part of a new inner transition series.
(a) How do the electron configurations of these elements support Seaborg's suggestion?
(b) The highest fluorides of Np and Pu are hexafluorides, and the highest fluoride of uranium is also the hexafluoride. How does

this chemical evidence support the placement of Np and Pu as *inner* transition elements rather than transition elements?

23.100 How many different formulas are there for octahedral complexes with a metal M and four ligands A, B, C, and D? Give the number of isomers for each formula and describe the isomers.

23.101 At one time, it was common to write the formula for copper(I) chloride as Cu_2Cl_2, instead of CuCl, analogously to Hg_2Cl_2 for mercury(I) chloride. Use electron configurations to explain why Hg_2Cl_2 and CuCl are both correct.

23.102 For the compound $[Co(en)_2Cl_2]Cl$, give:
(a) The coordination number of the metal ion
(b) The oxidation number of the central metal ion
(c) The number of individual ions per formula unit
(d) The moles of AgCl that precipitate when 1 mol of compound is dissolved in water and treated with $AgNO_3$

23.103 Hexafluorocobaltate(III) ion is a high-spin complex. Draw the energy diagram for the splitting of its *d* orbitals.

23.104 A salt of each of the ions in Table 23.3 is dissolved in water. A Pt electrode is immersed in each solution and connected to a 0.38-V battery. All of the electrolytic cells are run for the same amount of time with the same current.
(a) In which cell(s) will a metal plate out? Explain.
(b) Which cell will plate out the least mass of metal? Explain.

23.105 Criticize and correct the following statement: strong-field ligands always give rise to low-spin complexes.

23.106 Some octahedral complexes have distorted shapes. In some, two metal-ligand bonds that are 180° apart are shorter than the other four. In $[Cu(NH_3)_6]^{2+}$, for example, two Cu—N bonds are 207 pm long, and the other four are 262 pm long. (a) Calculate the longest distance between two N atoms in this complex. (b) Calculate the shortest distance between two N atoms.

23.107 In many species, a transition metal has an unusually high or low oxidation state. Write balanced equations for the following and find the oxidation state of the transition metal in the product:
(a) Iron(III) ion reacts with hypochlorite ion in basic solution to form ferrate ion (FeO_4^{2-}), Cl^-, and water.
(b) Potassium hexacyanomanganate(II) reacts with K metal to form $K_6[Mn(CN)_6]$.
(c) Heating sodium superoxide (NaO_2) with Co_3O_4 produces Na_4CoO_4 and O_2 gas.
(d) Vanadium(III) chloride reacts with Na metal under a CO atmosphere to produce $Na[V(CO)_6]$ and NaCl.
(e) Barium peroxide reacts with nickel(II) ions in basic solution to produce $BaNiO_3$.
(f) Bubbling CO through a basic solution of cobalt(II) ion produces $[Co(CO)_4]^-$, CO_3^{2-}, and water.
(g) Heating cesium tetrafluorocuprate(II) with F_2 gas under pressure gives Cs_2CuF_6.
(h) Heating tantalum(V) chloride with Na metal produces NaCl and Ta_6Cl_{15}, in which half of the Ta is in the +2 state.
(i) Potassium tetracyanonickelate(II) reacts with hydrazine (N_2H_4) in basic solution to form $K_4[Ni_2(CN)_6]$ and N_2 gas.

23.108 Draw a Lewis structure with lowest formal charges for MnO_4^-.

23.109 The coordination compound $[Pt(NH_3)_2(SCN)_2]$ displays two types of isomerism. Name the types, and give names and structures for the six possible isomers.

23.110 An octahedral complex with three different ligands (A, B, and C) can have formulas with three different ratios of the ligands:
$[MA_4BC]^{n+}$, such as $[Co(NH_3)_4(H_2O)Cl]^{2+}$
$[MA_3B_2C]^{n+}$, such as $[Cr(H_2O)_3Br_2Cl]$
$[MA_2B_2C_2]^{n+}$, such as $[Cr(NH_3)_2(H_2O)_2Br_2]^+$
For each example, give the name, state the type(s) of isomerism present, and draw all isomers.

23.111 In $[Cr(NH_3)_6]Cl_3$, the $[Cr(NH_3)_6]^{3+}$ ion absorbs visible light in the blue-violet range, and the compound is yellow-orange. In $[Cr(H_2O)_6]Br_3$, the $[Cr(H_2O)_6]^{3+}$ ion absorbs visible light in the red range, and the compound is blue-gray. Explain these differences in light absorbed and compound color.

23.112 The actinides Pa, U, and Np form a series of complex ions, such as the anion in the compound $Na_3[UF_8]$, in which the central metal ion has an unusual geometry and oxidation state. In the crystal structure, the complex ion can be pictured as resulting from interpenetration of simple cubic arrays of uranium and fluoride ions. (a) What is the coordination number of the metal ion in the complex ion? (b) What is the oxidation state of uranium in the compound? (c) Sketch the complex ion.

23.113 Several coordination isomers, with both Co and Cr as 3+ ions, have the molecular formula $CoCrC_6H_{18}N_{12}$. (a) Give the name and formula of the isomer in which the Co complex ion has six NH_3 groups. (b) Give the name and formula of the isomer in which the Co complex ion has one CN and five NH_3 groups.

23.114 Consider the square planar complex shown at right. Which of the structures A–F are geometric isomers of it?

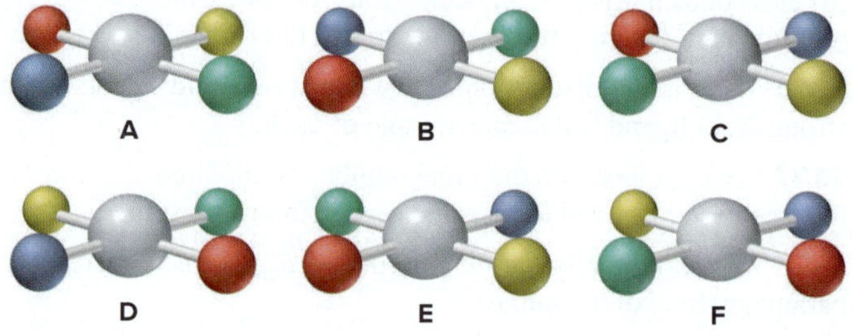

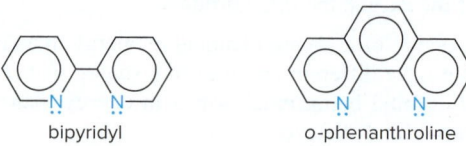

23.115 A shortcut to finding optical isomers is to see if the complex has a *plane of symmetry*—a plane passing through the metal atom such that every atom on one side of the plane is matched by an identical one at the same distance from the plane on the other side. Any planar complex has a plane of symmetry, since all atoms lie in one plane. Use this approach to determine whether these exist as optical isomers: (a) $[Zn(NH_3)_2Cl_2]$ (tetrahedral); (b) $[Pt(en)_2]^{2+}$; (c) *trans*-$[PtBr_4Cl_2]^{2-}$; (d) *trans*-$[Co(en)_2F_2]^+$; (e) *cis*-$[Co(en)_2F_2]^+$.

23.116 Alfred Werner (see Problem 23.67) prepared two compounds by heating a solution of $PtCl_2$ with triethyl phosphine, $P(C_2H_5)_3$, which is a ligand for Pt. Both compounds have, by mass, Pt, 38.8%; Cl, 14.1%; C, 28.7%; P, 12.4%; and H, 6.02%. Write formulas, structures, and systematic names for the two isomers.

23.117 Two bidentate ligands used extensively in analytical chemistry are bipyridyl (bipy) and *ortho*-phenanthroline (*o*-phen):

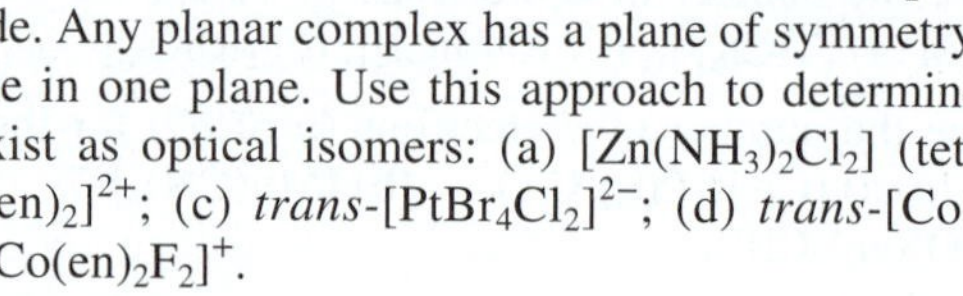

bipyridyl o-phenanthroline

Draw structures and discuss the possibility of isomers for

(a) $[Pt(bipy)Cl_2]$ (b) $[Fe(o\text{-phen})_3]^{3+}$

(c) $[Co(bipy)_2F_2]^+$ (d) $[Co(o\text{-phen})(NH_3)_3Cl]^{2+}$

23.118 The effect of entropy on reactions is demonstrated by the stabilities of certain complexes. (a) Based on the numbers of reactant and product particles, predict which of the following reactions will be favored in terms of ΔS°_{rxn}:

$$[Cu(NH_3)_4]^{2+}(aq) + 4H_2O(l) \longrightarrow [Cu(H_2O)_4]^{2+}(aq) + 4NH_3(aq)$$

$$[Cu(H_2NCH_2CH_2NH_2)_2]^{2+}(aq) + 4H_2O(l) \longrightarrow [Cu(H_2O)_4]^{2+}(aq) + 2en(aq)$$

(b) Given that the Cu—N bond strength is approximately the same in both complexes, which complex will be more stable with respect to ligand exchange in water? Explain.

23.119 You know the following about a coordination compound:

(1) The partial empirical formula is $KM(CrO_4)Cl_2(NH_3)_4$.

(2) It has A (red) and B (blue) crystal forms.

(3) When 1.0 mol of A or B reacts with 1.0 mol of $AgNO_3$, 0.50 mol of a red precipitate forms immediately.

(4) After the reaction in (3), 1.0 mol of A reacts very slowly with 1.0 mol of silver oxalate ($Ag_2C_2O_4$) to form 2.0 mol of a white precipitate. (Oxalate can displace other ligands.)

(5) After the reaction in (3), 1.0 mol of B does not react further with 1.0 mol of $AgNO_3$.

From this information, determine the following:

(a) The coordination number of M

(b) The group(s) bonded to M ionically and covalently

(c) The stereochemistry of the red and blue forms

23.120 The extent of crystal field splitting is often determined from spectra. (a) Given the wavelength (λ) of maximum absorption, find the crystal field splitting energy (Δ), in kJ/mol, for each of the following complex ions:

Ion	λ (nm)	Ion	λ (nm)
$[Cr(H_2O)_6]^{3+}$	562	$[Fe(H_2O)_6]^{2+}$	966
$[Cr(CN)_6]^{3-}$	381	$[Fe(H_2O)_6]^{3+}$	730
$[CrCl_6]^{3-}$	735	$[Co(NH_3)_6]^{3+}$	405
$[Cr(NH_3)_6]^{3+}$	462	$[Rh(NH_3)_6]^{3+}$	295
$[Ir(NH_3)_6]^{3+}$	244		

(b) Write a spectrochemical series for the ligands in the Cr complexes. (c) Use the Fe data to state how oxidation state affects Δ. (d) Use the Co, Rh, and Ir data to state how period number affects Δ.

23.121 Ionic liquids have many applications in engineering and materials science. The dissolution of the metavanadate ion in chloroaluminate ionic liquids has been studied:

$$VO_3^- + AlCl_4^- \longrightarrow VO_2Cl_2^- + AlOCl_2^-$$

(a) What is the oxidation number of V and Al in each ion?

(b) In reactions of V_2O_5 with HCl, acid concentration affects the product. At low acid concentration, $VO_2Cl_2^-$ and VO_3^- form:

$$V_2O_5 + HCl \longrightarrow VO_2Cl_2^- + VO_3^- + H^+$$

At high acid concentration, $VOCl_3$ forms:

$$V_2O_5 + HCl \longrightarrow VOCl_3 + H_2O$$

Balance each equation, and state which, if either, is a redox process.

(c) What mass of $VO_2Cl_2^-$ or $VOCl_3$ can form from 12.5 g of V_2O_5 and the appropriate concentration of acid?

23.122 The orbital occupancies for the *d* orbitals of several complex ions are diagrammed below.

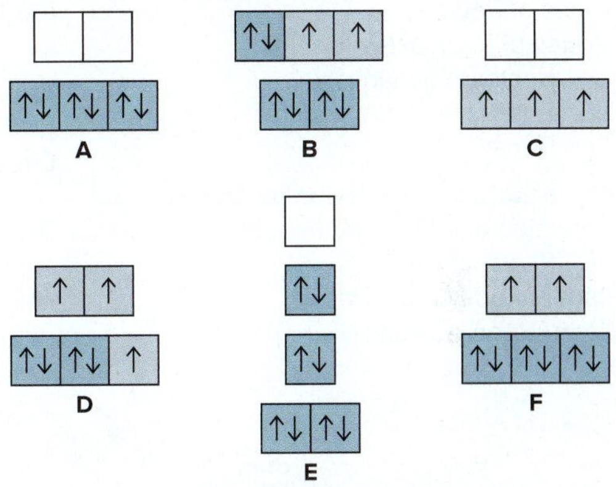

(a) Which diagram corresponds to the orbital occupancy of the cobalt ion in $[Co(CN)_6]^{3-}$?

(b) If diagram D depicts the orbital occupancy of the cobalt ion in $[CoF_6]^n$, what is the value of *n*?

(c) $[NiCl_4]^{2-}$ is paramagnetic and $[Ni(CN)_4]^{2-}$ is diamagnetic. Which diagrams correspond to the orbital occupancies of the nickel ions in these species?

(d) Diagram C shows the orbital occupancy of V^{2+} in the octahedral complex VL_6. Can you determine whether L is a strong- or weak-field ligand? Explain.

24 Nuclear Reactions and Their Applications

24.1 Radioactive Decay and Nuclear Stability
Chemical vs. Nuclear Change
Components of the Nucleus
Types of Radioactive Emissions
Modes of Radioactive Decay; Nuclear Equations
Nuclear Stability and Mode of Decay

24.2 The Kinetics of Radioactive Decay
Detection and Measurement of Radioactivity
Rate of Radioactive Decay
Radioisotopic Dating

24.3 Nuclear Transmutation: Induced Changes in Nuclei
Early Transmutation Experiments; Nuclear Shorthand Notation
Particle Accelerators and Transuranium Elements

24.4 Ionization: Effects of Nuclear Radiation on Matter
Effects of Ionizing Radiation on Living Tissue
Background Sources of Ionizing Radiation
Assessing the Risk from Ionizing Radiation

24.5 Applications of Radioisotopes
Radioactive Tracers
Additional Applications of Ionizing Radiation

24.6 The Interconversion of Mass and Energy
Mass Difference Between a Nucleus and Its Nucleons
Nuclear Binding Energy and Binding Energy per Nucleon

24.7 Applications of Fission and Fusion
Nuclear Fission
Nuclear Fusion

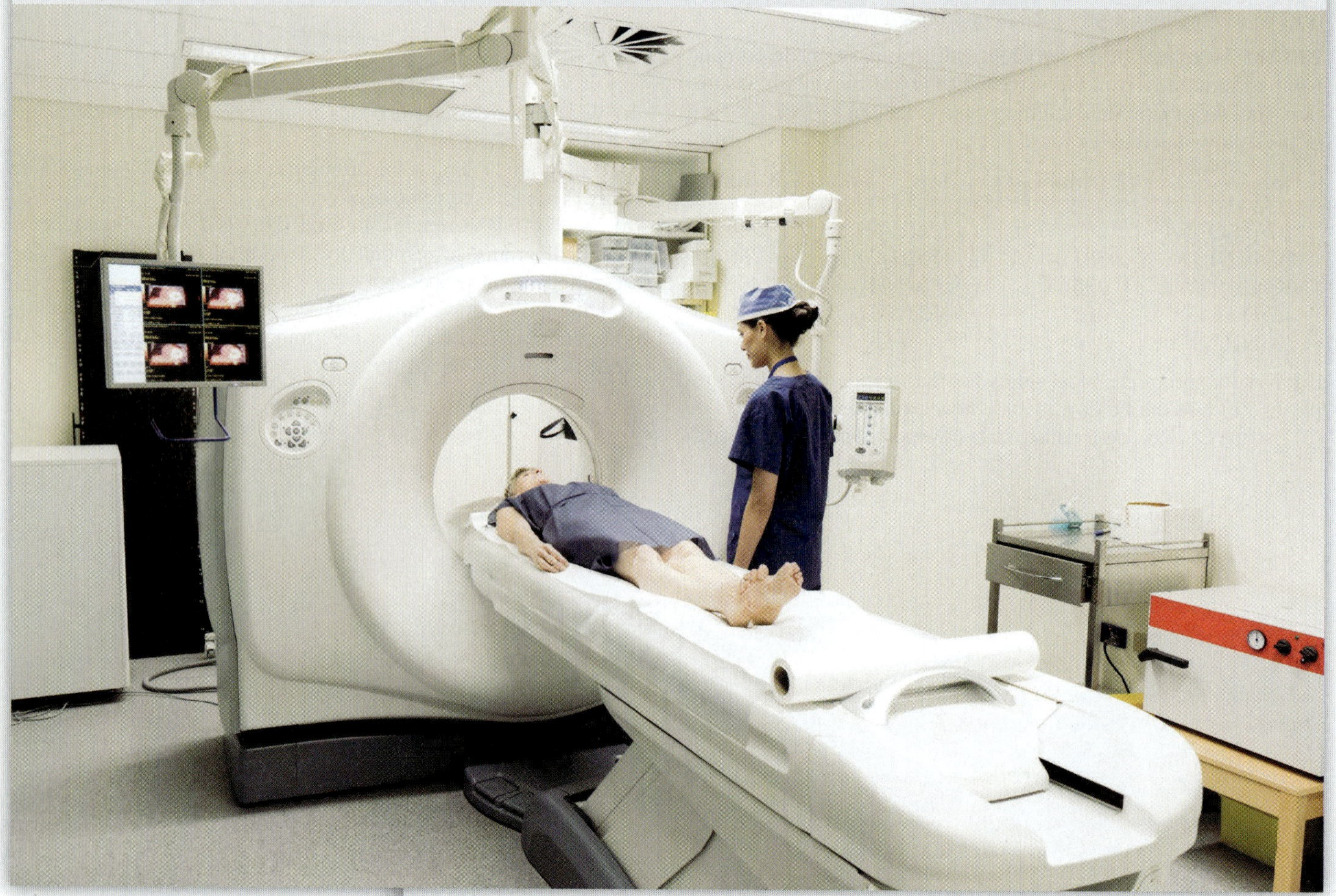

Source: © Johnny Greig/E+/Getty Images

Concepts and Skills to Review Before You Study This Chapter

- discovery of the atomic nucleus (Section 2.4)
- protons, neutrons, mass number, and the $^{A}_{Z}X$ notation (Section 2.5)
- half-life and first-order reaction rate (Section 16.4)

Up to this point in the text, we have focused on the atom's electrons, treating the nucleus as their electrostatic anchor and examining the effect of its positive charge on atomic properties and, ultimately, chemical behavior. In this chapter, for the most part, we ignore the surrounding electrons and focus on the atom's tiny, dense core, which is held together by the strongest force in the universe. Scientists studying the structure and behavior of the nucleus encounter great mystery and wonder but also uncover a great number of applications. To cite just one example, modern medicine relies, to an ever-increasing extent, on diagnoses based on the emission of radiation from unstable nuclei. In fact, more than one in four hospital admissions now involve imaging tests such as stress tests, bone scans, and PET scans *(see photo)* that utilize *radioisotopes.*

But society is ambivalent about some applications of nuclear research. The promise of abundant energy and effective treatments for diseases is offset by the threat of nuclear waste contamination, reactor accidents, and unimaginable destruction from nuclear war or terrorism. Can the power of the nucleus be harnessed for our benefit, or are the risks too great? In this chapter, you'll learn the principles that can help you consider this vital question.

IN THIS CHAPTER . . . *We survey the field of nuclear chemistry, examining radioactive nuclei and their decay processes, especially the effects and applications of radioactivity and the interconversion of matter and energy.*

- We investigate nuclear stability to determine why some nuclei are stable and others undergo radioactive decay.
- We learn how radioactivity is detected and how the kinetics of decay is applied.
- We see how nuclei synthesized in particle accelerators have extended the periodic table.
- We consider the effects of radioactive emissions on matter, especially living tissue, and focus on some major uses of radioisotopes in science, technology, and medicine.
- We discuss the mass difference that arises when a nucleus forms from its subatomic particles and the energy that is equivalent to this difference.
- We explore current and future attempts to harness the energy released when heavy nuclei split or lighter ones fuse.
- We look at the nuclear processes in stars that continually create chemical elements.

24.1 RADIOACTIVE DECAY AND NUCLEAR STABILITY

A stable nucleus remains intact indefinitely, but *the great majority of nuclei are unstable.* An unstable nucleus exhibits **radioactivity,** the spontaneous disintegration of a nucleus by the emission of radiation. In Section 24.2, you'll see that each type of unstable nucleus has its own characteristic *rate* of radioactive decay. In this section, we cover important terms and notation for nuclei, discuss some key events in the discovery of radioactivity, define the types of emission, describe various modes of radioactive decay, and see how to predict which occurs for a given nucleus. Let's begin, however, with a brief comparison of chemical change and nuclear change.

Table 24.1 Comparison of Chemical and Nuclear Reactions

Chemical Reactions	Nuclear Reactions
1. One substance is converted into another, but atoms never change identity.	1. Atoms of one element typically are converted into atoms of another element.
2. Electrons in orbitals are involved as bonds break and form; nuclear particles do not take part.	2. Protons, neutrons, and other nuclear particles are involved; electrons in orbitals take part much less often.
3. Reactions are accompanied by relatively small changes in energy and no measurable changes in mass.	3. Reactions are accompanied by relatively large changes in energy and measurable changes in mass.
4. Reaction rates are influenced by temperature, concentration, catalysts, and the compound in which an element occurs.	4. Reaction rates depend on number of nuclei, but are not affected by temperature, catalysts, or, except on rare occasions, the compound in which an element occurs.

Comparing Chemical and Nuclear Change

The changes that occur in atomic nuclei differ strikingly from chemical changes (Table 24.1). In chemical reactions, electrons are shared or transferred to form *compounds,* while nuclei remain unchanged. In nuclear reactions, the roles are reversed: electrons take part much less often, while nuclei undergo changes that, in nearly every case, form different *elements.* Nuclear reactions are often accompanied by energy changes a million times greater than those for chemical reactions, energy changes so large that the accompanying changes in mass *are* detectable. Moreover, nuclear reaction yields and rates are *not* subject to the effects of pressure, temperature, and catalysis that influence chemical reactions.

The Components of the Nucleus: Terms and Notation

Recall from Chapter 2 that the nucleus contains essentially all the atom's mass but accounts for only about 10^{-5} times its radius and, therefore, 10^{-15} times its volume, making the nucleus incredibly dense: about 10^{14} g/mL. ‹ *Protons* and *neutrons,* the elementary particles that make up the nucleus, are called **nucleons.** A **nuclide** is a nucleus with a particular composition, that is, with specific numbers of the two types of nucleons. Most elements occur in nature as a mixture of **isotopes,** atoms with the characteristic number of protons of the element but different numbers of neutrons. For example, oxygen has three naturally occurring isotopes, and carbon also has three, but tin has ten, the most of any element. *Thus, each isotope of an element has a different nuclide.*

Big Atom with a Tiny, Massive Core

If you could strip the electrons from the atoms in an object and compress the nuclei together, the object would lose only a fraction of a percent of its mass, but it would shrink to 0.0000000000001% (10^{-13}%) of its volume!

The relative mass and charge of a particle—elementary particle or nuclide—are described by the notation ${}^{A}_{Z}X$ where X is the *symbol* for the particle, *A* is the *mass number,* or the total number of nucleons making up the particle, and *Z* is the *charge* of the particle; for nuclei, *A* is the *sum of protons and neutrons* and *Z* is the *number of protons* (atomic number). With this notation, the three subatomic elementary particles are represented as follows:

$$ {}^{0}_{-1}e \text{ (electron)}, \quad {}^{1}_{1}p \text{ (proton), and } {}^{1}_{0}n \text{ (neutron)} $$

(A proton is also sometimes represented as ${}^{1}_{1}H^{+}$.) The number of neutrons (*N*) in a nucleus is the mass number (*A*) minus the atomic number (*Z*): $N = A - Z$. For example, the two naturally occurring stable isotopes of chlorine are

${}^{35}_{17}Cl$ (or ${}^{35}Cl$)	with 17 protons and 35 − 17 = 18 neutrons
${}^{37}_{17}Cl$ (or ${}^{37}Cl$)	with 17 protons and 37 − 17 = 20 neutrons

Nuclides can also be designated with the element name followed by the mass number, for example, chlorine-35 and chlorine-37. In naturally occurring samples of an element or its compounds, *the isotopes of the element are present in specific proportions* that vary only very slightly (see Chapter 2). Thus, in a sample of sodium chloride (or any Cl-containing substance), 75.77% of the Cl atoms are chlorine-35 and the remaining 24.23% are chlorine-37.

To understand this chapter, you need to be comfortable with nuclear notations, so please take a moment to review Sample Problem 2.4 and Problems 2.39 to 2.46.

The Discovery of Radioactivity and the Types of Emissions

In 1896, the French physicist Antoine-Henri Becquerel accidentally discovered that uranium minerals emit a penetrating radiation that can expose a photographic plate. Becquerel also found that the radiation creates an electric discharge in air, thus providing a means for measuring its intensity.

Curie and Rutherford: The Nature of Radioactivity and Element Identity Based on Becquerel's discovery, in 1898, a young doctoral student named Marie Sklodowska Curie began a search for other minerals that behaved like uranium minerals. She found that thorium minerals also emit radiation and, most importantly, that *the intensity of the radiation is directly proportional to the concentration of the element in the mineral, not to the formula of the mineral or compound.* Curie named the emissions *radioactivity* and showed that they are *unaffected by temperature, pressure, or other physical and chemical conditions.*

After months of painstaking chemical work, Curie and her husband, physicist Pierre Curie, showed clear evidence of two new elements in pitchblende, the principal ore of uranium: polonium (Po; $Z = 84$), the most metallic member of Group 6A(16), and radium (Ra; $Z = 88$), the heaviest alkaline earth metal.

Then, focusing on obtaining a measurable amount of radium, Curie started with several tons of pitchblende residues from which the uranium had been extracted and worked arduously for four years to isolate 0.1 g of radium chloride, which she melted and electrolyzed to obtain pure metallic radium. Curie is the only person to have received two Nobel Prizes in science, one in physics for her research into radioactivity and the other in chemistry for her discoveries of polonium and radium.

During the next few years, Becquerel, the Curies, and Paul Villard in France and Ernest Rutherford and his coworkers in England studied the nature of radioactive emissions. A key finding was the observation of Rutherford and his colleague Frederick Soddy that elements other than radium were formed when radium decayed. In 1902, they proposed that radioactive emissions result in the change of one element into another. This explanation seemed like a return to alchemy and was met with disbelief and ridicule. We now know it to be true: *when a nuclide of one element decays, it emits radiation and usually changes into a nuclide of a different element.*

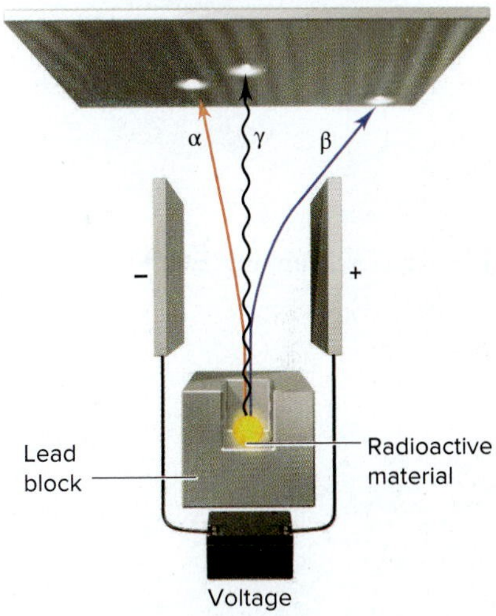

Figure 24.1 How the three types of radioactive emissions behave in an electric field.

Types of Radioactive Emissions There are three natural types of radioactive emissions:

- **Alpha particles** (symbolized α, ${}^{4}_{2}\alpha$, or ${}^{4}_{2}He^{2+}$) are identical to helium-4 nuclei.
- **Beta particles** (symbolized β, β^-, or sometimes ${}^{0}_{-1}\beta$) are high-speed electrons. (The emission of an electron from a nucleus may seem strange, but as you'll see shortly, it results from a nuclear reaction.)
- **Gamma rays** (symbolized γ, or sometimes ${}^{0}_{0}\gamma$) are very high-energy photons.

Figure 24.1 illustrates the behavior of these emissions in an electric field: the positively charged α particles curve to a small extent toward the negative plate, the negatively charged β particles curve to a greater extent toward the positive plate (because they have lower mass), and the uncharged γ rays are not affected by the electric field.

Modes of Radioactive Decay; Balancing Nuclear Equations

When a nuclide decays, it becomes a nuclide of lower energy, and the excess energy is carried off by the emitted radiation and the recoiling nucleus. The decaying, or reactant, nuclide is called the *parent;* the product nuclide is called the *daughter.* Nuclides can decay in several ways. As each of the major modes of decay is introduced (Table 24.2, *next page*), we'll show examples of that mode and apply the key principle used to balance nuclear reactions: *the total Z (charge, number of protons) and the total A (sum of protons and neutrons) of the reactants equal those of the products:*

$$\frac{\text{Total } A}{\text{Total } Z}\text{ Reactants} = \frac{\text{Total } A}{\text{Total } Z}\text{ Products} \quad \textbf{(24.1)}$$

Table 24.2 Modes of Radioactive Decay*

Mode	Emission	Decay Process	Change in A	Change in Z	Change in N
α Decay	α ($^4_2He^{2+}$)	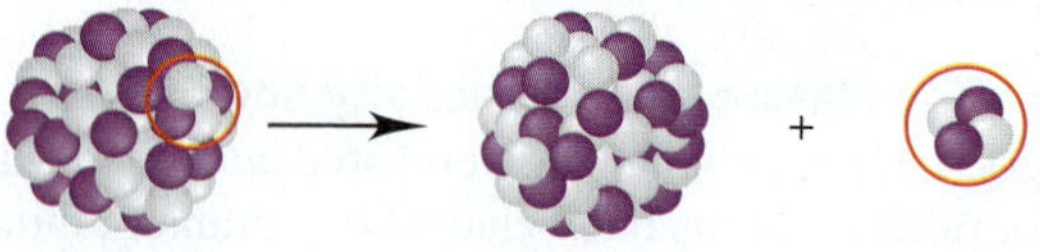 + α expelled	−4	−2	−2
β^- Decay†	β^- ($^{0}_{-1}\beta$)	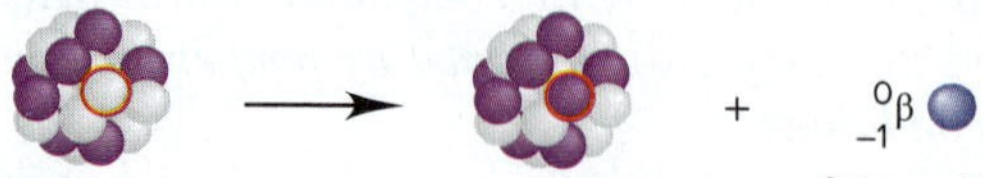 nucleus with xp^+ and yn^0 → nucleus with $(x+1)p^+$ and $(y-1)n^0$ + $^{0}_{-1}\beta$ (β^- expelled)	0	+1	−1
	Net:	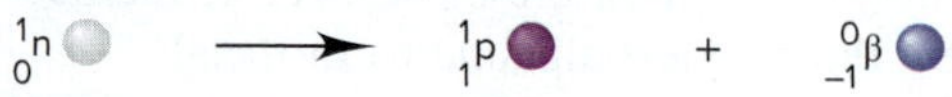 ^1_0n (in nucleus) → ^1_1p (in nucleus) + $^{0}_{-1}\beta$ (β^- expelled)			
Positron (β^+) emission†	β^+ ($^0_1\beta$)	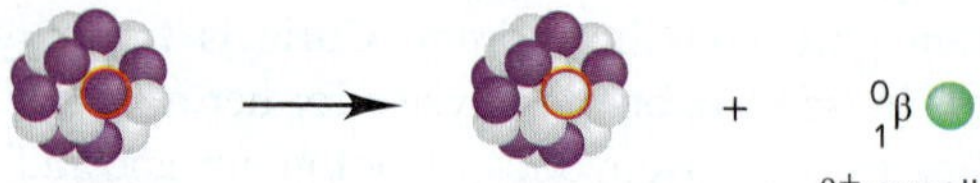 nucleus with xp^+ and yn^0 → nucleus with $(x-1)p^+$ and $(y+1)n^0$ + $^0_1\beta$ (β^+ expelled)	0	−1	+1
	Net:	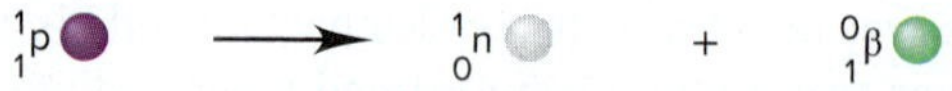 ^1_1p (in nucleus) → ^1_0n (in nucleus) + $^0_1\beta$ (β^+ expelled)			
Electron (e^-) capture (EC)†	x-ray	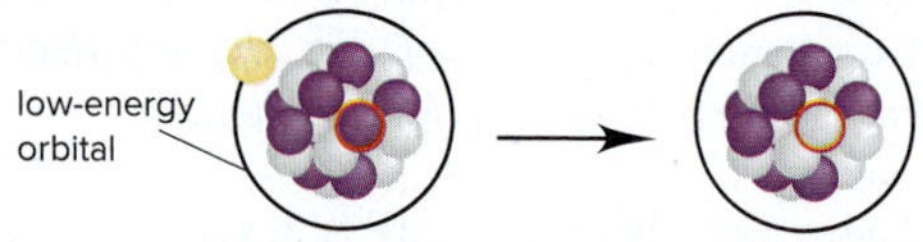 low-energy orbital; nucleus with xp^+ and yn^0 → nucleus with $(x-1)p^+$ and $(y+1)n^0$	0	−1	+1
	Net:	 $^{0}_{-1}e$ (absorbed from low-energy orbital) + ^1_1p (in nucleus) → ^1_0n (in nucleus)			
Gamma (γ) emission	γ	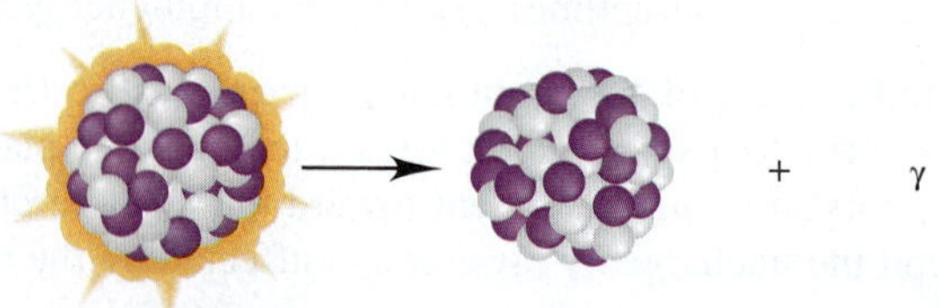 excited nucleus → stable nucleus + γ (γ photon radiated)	0	0	0

*Nuclear chemists consider β^- decay, positron emission, and electron capture to be three decay modes of the more general process known as beta decay (see text).
†Neutrinos (ν) or antineutrinos ($\bar{\nu}$) are also formed during the three modes of beta decay. Although we will not include them in other equations in the chapter, keep in mind that antineutrinos are always expelled during β^- decay, and neutrinos are expelled during β^+ emission and e^- capture. The mass of a neutrino or antineutrino is estimated to be less than 10^{-4} the mass of an electron.

1. **Alpha (α) decay** involves the loss of an α particle from a nucleus. For each α particle emitted by the parent, *A decreases by 4 and Z decreases by 2* in the daughter. Every element beyond bismuth (Bi; $Z = 83$) is radioactive and exhibits α decay, which is *the most common means for a heavy, unstable nucleus to become more stable.* For example, radium undergoes α decay to yield radon (Rn; $Z = 86$):

$$^{226}_{88}\text{Ra} \longrightarrow \, ^{222}_{86}\text{Rn} + \, ^{4}_{2}\alpha$$

Note that the A value for Ra equals the sum of the A values for Rn and α ($226 = 222 + 4$), and that the Z value for Ra equals the sum of the Z values for Rn and α ($88 = 86 + 2$).

2. **Beta (β) decay** is a general class of radioactive decay that encompasses three modes: β^- decay, β^+ emission, and electron capture.

- **β^- decay** (or *negatron emission*) occurs through the ejection of a β^- particle (high-speed electron) from the nucleus. This change does not involve expulsion of a β^- particle that was in the nucleus; rather, *a neutron is converted into a proton, which remains in the nucleus, and a β^- particle, which is expelled immediately:*

$$^{1}_{0}\text{n} \longrightarrow \, ^{1}_{1}\text{p} + \, ^{0}_{-1}\beta$$

As always, the totals of the A and the Z values for reactant and products are equal. Radioactive nickel-63 becomes stable copper-63 through β^- decay:

$$^{63}_{28}\text{Ni} \longrightarrow \, ^{63}_{29}\text{Cu} + \, ^{0}_{-1}\beta$$

Another example is the β^- decay of carbon-14, used in radiocarbon dating:

$$^{14}_{6}\text{C} \longrightarrow \, ^{14}_{7}\text{N} + \, ^{0}_{-1}\beta$$

Note that *β^- decay results in a product nuclide with the same A but with Z one greater (one more proton) than in the reactant nuclide.* In other words, an atom of the element with the next *higher* atomic number is formed. Although equations in this chapter do not include it, a neutral particle called an antineutrino ($\bar{\nu}$) is also emitted during β^- decay. The mass of the antineutrino is thought to be much less than 10^{-4} times that of an electron.

- **Positron (β^+) emission** is the emission of a β^+ particle from the nucleus. A key idea of modern physics is that most fundamental particles have corresponding *antiparticles* with the same mass but opposite charge. The **positron** is the antiparticle of the electron. Positron emission occurs through a process in which *a proton in the nucleus is converted into a neutron and a positron, which is expelled:*

$$^{1}_{1}\text{p} \longrightarrow \, ^{1}_{0}\text{n} + \, ^{0}_{1}\beta$$

Also emitted during this process is a *neutrino (ν)*, the antiparticle of the antineutrino. In terms of the effect on A and Z, *positron emission has the opposite effect of β^- decay: the daughter has the same A but Z is one less (one fewer proton) than for the parent.* Thus, an atom of the element with the next *lower* atomic number forms. Carbon-11, a synthetic radioisotope, decays to a stable boron isotope through β^+ emission:

$$^{11}_{6}\text{C} \longrightarrow \, ^{11}_{5}\text{B} + \, ^{0}_{1}\beta$$

- **Electron (e^-) capture (EC)** occurs when the nucleus draws in an electron from a low atomic energy level. The net effect is that *a proton is transformed into a neutron:*

$$^{1}_{1}\text{p} + \, ^{0}_{-1}\text{e} \longrightarrow \, ^{1}_{0}\text{n}$$

(We use the symbol e^-, or $^{0}_{-1}$e, to distinguish an orbital electron from a beta particle β^-, or $^{0}_{-1}\beta$.) The orbital vacancy is quickly filled by an electron that moves down from a higher energy level, and that process continues through still higher energy levels, with x-ray photons and neutrinos carrying off the energy difference in each step. Radioactive iron becomes stable manganese through electron capture:

$$^{55}_{26}\text{Fe} + \, ^{0}_{-1}\text{e} \longrightarrow \, ^{55}_{25}\text{Mn} + h\nu \text{ (x-rays and neutrinos)}$$

Even though the processes are different, *electron capture has the same net effect as positron emission:* Z reduced by 1, A unchanged.

3. **Gamma (γ) emission** involves the radiation of high-energy γ photons (also called γ rays) from an excited nucleus. Just as an atom in an excited *electronic* state reduces its energy by emitting photons, usually in the UV and visible ranges (see Section 7.2), a nucleus in an excited state lowers its energy by emitting γ photons, which are of much higher energy (much shorter wavelength) than UV photons. Many nuclear processes leave the nucleus in an excited state, so *γ emission accompanies many other (mostly β) types of decay.* Several γ photons of different energies can be emitted from an excited nucleus as it returns to the ground state. Some of Marie Curie's experiments involved the release of γ rays, as in the decay of polonium-215:

$$^{215}_{84}\text{Po} \longrightarrow {}^{211}_{82}\text{Pb} + {}^{4}_{2}\alpha \text{ (several } \gamma \text{ emitted)}$$

Gamma emission often accompanies β^- decay:

$$^{99}_{43}\text{Tc} \longrightarrow {}^{99}_{44}\text{Ru} + {}^{0}_{-1}\beta \text{ (several } \gamma \text{ emitted)}$$

Because γ rays have no mass or charge, *γ emission does not change A or Z.* Two gamma rays are emitted when a particle and an antiparticle annihilate each other. In the medical technology known as *positron-emission tomography* (see Section 24.5), a positron and an electron annihilate each other (with all *A* and *Z* values shown):

$$^{0}_{1}\beta + {}^{0}_{-1}\text{e} \longrightarrow 2\,{}^{0}_{0}\gamma$$

Table 24.3 summarizes the changes in mass number and atomic number that result from the various modes of decay.

Table 24.3 Changes in *A* and *Z* During Radioactive Decay

Mode of Decay	Change in *A*	Change in *Z*
α decay	−4	−2
β^- decay	0	+1
β^+ decay	0	−1
Electron capture	0	−1
Gamma emission	0	0

SAMPLE PROBLEM 24.1 Writing Equations for Nuclear Reactions

Problem Write balanced equations for the following nuclear reactions:
a. Naturally occurring thorium-232 undergoes α decay.
b. Zirconium-86 undergoes electron capture.

Plan We first write a skeleton equation that includes the mass numbers, atomic numbers, and symbols of all the particles on the correct sides of the equation, showing the unknown product particle as $^{A}_{Z}\text{X}$. Then, because the total of mass numbers and the total of atomic numbers must be equal on the left and right sides, we solve for *A* and *Z*, and use *Z* to determine X from the periodic table.

Solution **(a)** Writing the skeleton equation, with the α particle as a product:

$$^{232}_{90}\text{Th} \longrightarrow {}^{A}_{Z}\text{X} + {}^{4}_{2}\alpha$$

Solving for *A* and *Z* and balancing the equation: For *A*, 232 = *A* + 4, so *A* = 228. For *Z*, 90 = *Z* + 2, so *Z* = 88. From the periodic table, we see that the element with *Z* = 88 is radium (Ra). Thus, the balanced equation is

$$^{232}_{90}\text{Th} \longrightarrow {}^{228}_{88}\text{Ra} + {}^{4}_{2}\alpha$$

(b) Writing the skeleton equation, with the captured electron as a reactant:

$$^{86}_{40}\text{Zr} + {}^{0}_{-1}\text{e} \longrightarrow {}^{A}_{Z}\text{X}$$

Solving for *A* and *Z* and balancing the equation: For *A*, 86 + 0 = *A*, so *A* = 86. For *Z*, 40 + (−1) = *Z*, so *Z* = 39. The element with *Z* = 39 is yttrium (Y), so we have

$$^{86}_{40}\text{Zr} + {}^{0}_{-1}\text{e} \longrightarrow {}^{86}_{39}\text{Y}$$

Check Always read across superscripts and then across subscripts, with the yield arrow as an equal sign, to check your arithmetic. In part (a), for example, 232 = 228 + 4, and 90 = 88 + 2.

FOLLOW-UP PROBLEMS

Brief Solutions for ***all*** *Follow-up Problems appear at the end of the chapter.*

24.1A Write a balanced equation for the nuclear reaction in which fluorine-20 undergoes β^- decay.

24.1B Write a balanced equation for the reaction in which a nuclide undergoes β^+ decay and changes to tellurium-124.

SOME SIMILAR PROBLEMS 24.8–24.15

Student Hot Spot

Student data indicate that you may struggle with writing nuclear reactions. Access the Smartbook to view additional Learning Resources on this topic.

Nuclear Stability and the Mode of Decay

Can we predict how, and whether, an unstable nuclide will decay? Our knowledge of the nucleus is much less complete than that of the whole atom, but some patterns emerge by observing naturally occurring nuclides.

The Band of Stability Two key factors determine the stability of a nuclide:

1. The number of neutrons (*N*), the number of protons (*Z*), and their ratio (*N/Z*), which we calculate from (*A* − *Z*)/*Z*. This factor relates primarily to nuclides that undergo one of the three modes of β decay.
2. The total mass of the nuclide, which mostly relates to nuclides that undergo α decay.

A plot of number of neutrons vs. number of protons for all stable nuclides (Figure 24.2A) shows:

- The points form a narrow **band of stability** that gradually curves above the line for $N = Z$ ($N/Z = 1$).
- The only stable nuclides with $N/Z < 1$ are ${}^{1}_{1}\text{H}$ and ${}^{3}_{2}\text{He}$.
- Many lighter nuclides with $N = Z$ are stable, such as ${}^{4}_{2}\text{He}$, ${}^{12}_{6}\text{C}$, ${}^{16}_{8}\text{O}$, and ${}^{20}_{10}\text{Ne}$; the heaviest of these is ${}^{40}_{20}\text{Ca}$. Thus, for lighter nuclides, one neutron for each proton ($N = Z$) is enough to provide stability.
- The *N/Z* ratio of stable nuclides gradually increases as *Z* increases. For example, for ${}^{56}_{26}\text{Fe}$, $N/Z = 1.15$; for ${}^{107}_{47}\text{Ag}$, $N/Z = 1.28$; for ${}^{184}_{74}\text{W}$, $N/Z = 1.49$, and, finally, for ${}^{209}_{83}\text{Bi}$, $N/Z = 1.52$. Thus, for heavier stable nuclides, $N > Z$ ($N/Z > 1$), and *N* increases faster than *Z*. As we discuss shortly, and show in Figure 24.2B, if *N/Z* of a nuclide is either too high (above the band) or not high enough (below the band), the nuclide is unstable and undergoes one of the three modes of beta decay.

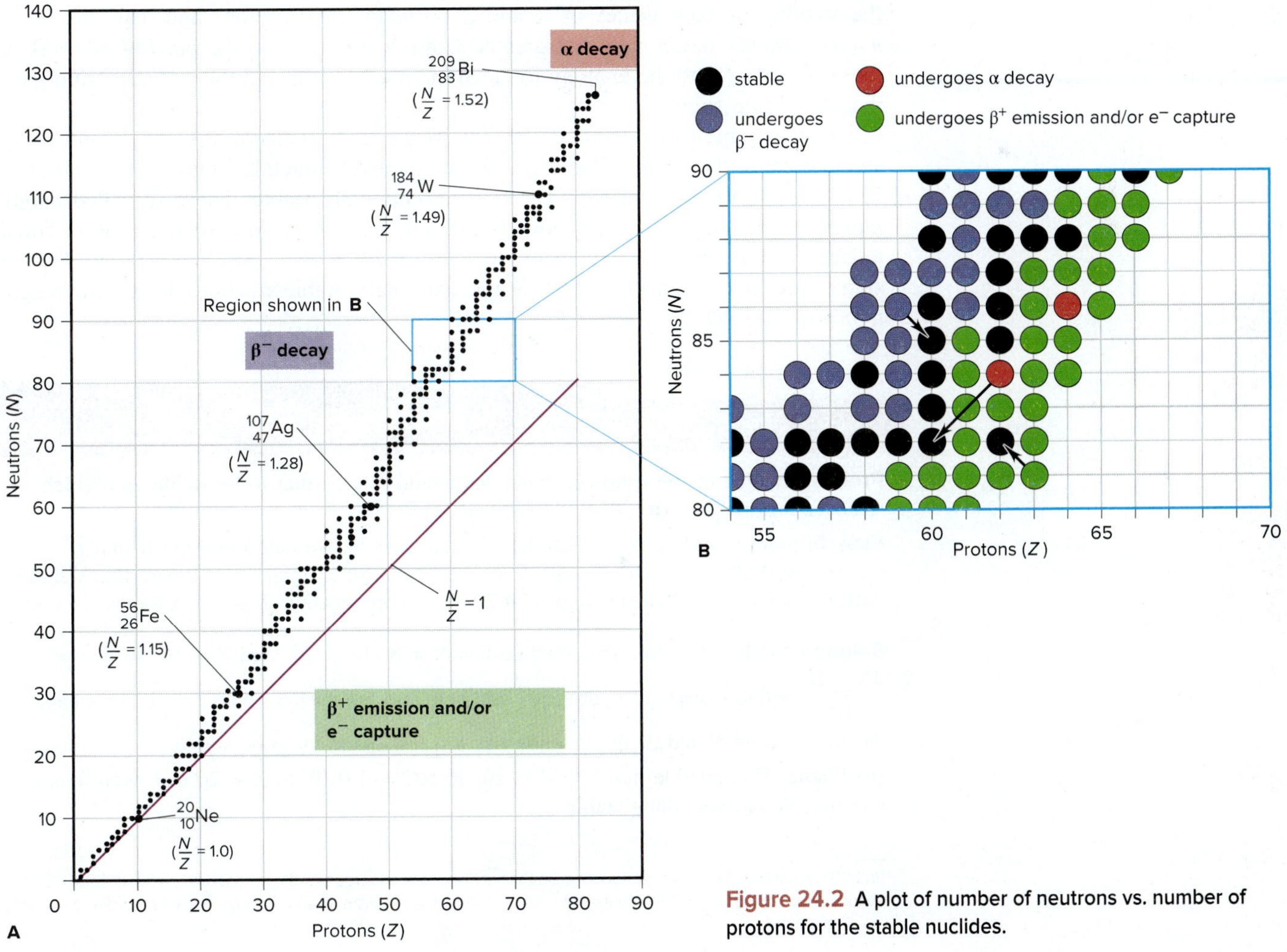

Figure 24.2 A plot of number of neutrons vs. number of protons for the stable nuclides.

Table 24.4 Number of Stable Nuclides for Elements 48 to 54*

Element	Atomic Number (*Z*)	Number of Nuclides
Cd	**48**	**8**
In	49	2
Sn	**50**	**10**
Sb	51	2
Te	**52**	**8**
I	53	1
Xe	**54**	**9**

*Even values of *Z* are shown in **boldface**.

Table 24.5 An Even-Odd Breakdown of the Stable Nuclides

Z	*N*	Number of Nuclides	Example
Even	Even	157	$^{84}_{38}Sr$
Even	Odd	53	$^{29}_{14}Si$
Odd	Even	50	$^{39}_{19}K$
Odd	Odd	4	$^{14}_{7}N$
	Total	264	

- All nuclides with $Z > 83$ are unstable.* Thus, the largest members of Groups 1A(1) through 8A(18), actinium and the actinides ($Z = 89$–103), and the other elements of the fourth (6*d*) transition series ($Z = 104$–112), are radioactive and (as discussed shortly) undergo α decay.

Stability and Nuclear Structure Nuclear structure is related to nuclear stability in several ways:

1. *The effect of the strong force.* Given that protons are positively charged and neutrons are uncharged, the first question that arises is: why does the nucleus stay together? Nuclear scientists answer this question and explain the importance of *N* and *Z* values in terms of two opposing forces: electrostatic repulsive forces between protons would break the nucleus apart if not for an attractive force called the **strong force,** which exists between all nucleons (protons and neutrons). This force is 137 times stronger than the repulsive force but *operates only over the short distances within the nucleus.* Thus, the *attractive* strong force overwhelms the much weaker *repulsive* electrostatic force and keeps the nucleus together.

2. *Odd and even numbers of nucleons.* The oddness or evenness of *N* and *Z* values is related to some important patterns of nuclear stability. Two interesting points become apparent when we classify the known stable nuclides:

- Elements with an even *Z* (number of protons) usually have a larger number of stable nuclides than elements with an odd *Z*. Table 24.4 demonstrates this point for cadmium ($Z = 48$) through xenon ($Z = 54$).
- Well over half the stable nuclides have *both* even *N* and even *Z* (Table 24.5). Only four nuclides with odd *N* and odd *Z* are stable: $^{2}_{1}H$, $^{6}_{3}Li$, $^{10}_{5}B$, and $^{14}_{7}N$.

3. *Nucleon energy levels.* One model of nuclear structure that attempts to explain the stability of even values of *N* and *Z* postulates that protons and neutrons lie in *nucleon energy levels,* and that greater stability results from the *pairing of spins* of like nucleons. (Note the analogy to electron energy levels and the stability from pairing of electron spins.)

Just as noble gases—with 2, 10, 18, 36, 54, and 86 electrons—are exceptionally stable because they have filled *electron* energy levels, nuclides with *N* or *Z* values of 2, 8, 20, 28, 50, 82 (and $N = 126$) are exceptionally stable. These so-called *magic numbers* are thought to correspond to the numbers of protons or neutrons in filled *nucleon* energy levels. A few examples are $^{50}_{22}Ti$ ($N = 28$), $^{88}_{38}Sr$ ($N = 50$), and the ten stable nuclides of tin ($Z = 50$). Some extremely stable nuclides have two magic numbers: $^{4}_{2}He$, $^{16}_{8}O$, $^{40}_{20}Ca$, and $^{208}_{82}Pb$ ($N = 126$).

SAMPLE PROBLEM 24.2 Predicting Nuclear Stability

Problem Which of the following nuclides would you predict to be stable and which radioactive: **(a)** $^{18}_{10}Ne$; **(b)** $^{32}_{16}S$; **(c)** $^{236}_{90}Th$; **(d)** $^{123}_{56}Ba$? Explain.

Plan In order to evaluate the stability of each nuclide, we determine the *N* and *Z* values, the *N*/*Z* ratio from $(A - Z)/Z$, and the value of *Z*; then we compare the *N*/*Z* ratio with the stable values (from Figure 24.2A) and note whether *Z* and *N* are even or odd.

Solution **(a)** Radioactive. This nuclide has $N = 8$ $(18 - 10)$ and $Z = 10$, so $N/Z = \frac{18 - 10}{10} = 0.8$. Except for hydrogen-1 and helium-3, no nuclides with $N < Z$ are stable; despite its even *N* and *Z*, this nuclide has too few neutrons to be stable.

(b) Stable. This nuclide has $N = Z = 16$, so $N/Z = 1.0$. With $Z < 20$ and even *N* and *Z*, this nuclide is most likely stable.

*In 2003, French nuclear physicists discovered that $^{209}_{83}Bi$ actually undergoes α decay, but the half-life is 1.9×10^{19} yr, about a billion times longer than the estimated age of the universe. Bismuth ($Z = 83$) can thus be considered stable.

(c) Radioactive. This nuclide has $Z = 90$, and every nuclide with $Z > 83$ is radioactive.

(d) Radioactive. This nuclide has $N = 67$ and $Z = 56$, so $N/Z = 1.20$. For Z values of 55 to 60, Figure 24.2A shows $N/Z \geq 1.3$ for stable nuclides, so this nuclide has too few neutrons to be stable.

Check By consulting a table of isotopes, such as the one in the *CRC Handbook of Chemistry and Physics,* we find that our predictions are correct.

FOLLOW-UP PROBLEMS

24.2A Predict whether each of the following nuclides is stable or radioactive, and explain your prediction: **(a)** $^{10}_{5}B$; **(b)** $^{58}_{23}V$.

24.2B Why is $^{31}_{15}P$ stable but $^{30}_{15}P$ unstable?

SOME SIMILAR PROBLEMS 24.16–24.19

Predicting the Mode of Decay An unstable nuclide generally decays in a mode that shifts its *N/Z* ratio toward the band of stability. This fact is illustrated in Figure 24.2B, which expands the small region of $Z = 54$ to 70 in Figure 24.2A to show the stable *and* many of the unstable nuclides in that region, as well as their modes of decay. Note the following points:

1. *Neutron-rich nuclides.* Nuclides with too many neutrons for stability (a high *N/Z* ratio) lie above the band of stability. They undergo *β^- decay,* which converts a neutron into a proton, thus reducing the value of *N/Z*. For example, C-14 undergoes β^- decay to produce a nuclide with a lower *N/Z* ratio:

$$^{14}_{6}C \longrightarrow {}^{14}_{7}N + {}^{0}_{-1}\beta$$

$$\frac{n}{p} = \frac{8}{6} = 1.33 \qquad \frac{n}{p} = \frac{7}{7} = 1$$

2. *Proton-rich nuclides.* Nuclides with too many protons for stability (a low *N/Z* ratio) lie below the band of stability. They undergo *β^+ emission* and/or *e^- capture,* both of which convert a proton into a neutron, thus increasing the value of *N/Z*. (The rate of e^- capture increases with *Z*, so β^+ emission is more common among lighter elements and e^- capture more common among heavier elements.) The following nuclear reactions show the decay processes for two nuclides with low *N/Z* ratios:

$$^{23}_{12}Mg \longrightarrow {}^{23}_{11}Na + {}^{0}_{1}\beta$$

$$\frac{n}{p} = \frac{11}{12} = 0.92 \qquad \frac{n}{p} = \frac{12}{11} = 1.09$$

$$^{81}_{36}Kr + {}^{0}_{-1}e \longrightarrow {}^{81}_{35}Br$$

$$\frac{n}{p} = \frac{45}{36} = 1.25 \qquad \frac{n}{p} = \frac{46}{35} = 1.31$$

3. *Heavy nuclides.* Nuclides with $Z > 83$ (shown in Figure 24.2A) are too heavy to be stable and undergo *α decay,* which reduces their *Z* and *N* values by two units per emission.

With the information in Figure 24.2, predicting the mode of decay of an unstable nuclide involves simply comparing its *N/Z* ratio with those in the nearby region of the band of stability. But, even without Figure 24.2, we can often make an educated guess about mode of decay. The atomic mass of an element is the weighted average of its naturally occurring isotopes. Therefore,

- The mass number *A* of a *stable* nuclide will be relatively close to the atomic mass.
- If an *unstable* nuclide of an element (given *Z*) has an *A* value much higher than the atomic mass, *it is neutron rich and will probably decay by β^- emission.*
- If, on the other hand, an unstable nuclide has an *A* value much lower than the atomic mass, *it is proton rich and will probably decay by β^+ emission and/or e^- capture.*

In the next sample problem, we predict the mode of decay of some unstable nuclides.

Student Hot Spot

Student data indicate that you may struggle with predicting a radioactive nuclide's mode of decay. Access the Smartbook to view additional Learning Resources on this topic.

SAMPLE PROBLEM 24.3 Predicting the Mode of Nuclear Decay

Problem Use the atomic mass of the element to predict the mode(s) of decay of the following radioactive nuclides: **(a)** $^{12}_{5}B$; **(b)** $^{234}_{92}U$; **(c)** $^{81}_{33}As$; **(d)** $^{127}_{57}La$.

Plan If the nuclide is too heavy to be stable ($Z > 83$), it undergoes α decay. For other cases, we use the Z value to obtain the atomic mass from the periodic table. If the mass number of the nuclide is higher than the atomic mass, the nuclide has too many neutrons: N too high ⇒ β^- decay. If the mass number is lower than the atomic mass, the nuclide has too many protons: Z too high ⇒ β^+ emission and/or e^- capture.

Solution **(a)** This nuclide has $Z = 5$, which is boron (B), and the atomic mass is 10.81. The nuclide's A value of 12 is significantly higher than its atomic mass, so this nuclide is neutron rich. It will probably undergo β^- decay.

(b) This nuclide has $Z = 92$, so it will undergo α decay and decrease its total mass.

(c) This nuclide has $Z = 33$, which is arsenic (As), and the atomic mass is 74.92. The A value of 81 is much higher, so this nuclide is neutron rich and will probably undergo β^- decay.

(d) This nuclide has $Z = 57$, which is lanthanum (La), and the atomic mass is 138.9. The A value of 127 is much lower, so this nuclide is proton rich and will probably undergo β^+ emission and/or e^- capture.

Check To confirm our predictions in parts (a), (c), and (d), let's compare each nuclide's N/Z ratio to those in the band of stability: (a) This nuclide has $N = 7$ and $Z = 5$, so $N/Z = 1.40$, which is too high for this region of the band, so it will undergo β^- decay. (c) This nuclide has $N = 48$ and $Z = 33$, so $N/Z = 1.45$, which is too high for this region of the band, so it will undergo β^- decay. (d) This nuclide has $N = 70$ and $Z = 57$, so $N/Z = 1.23$, which is too low for this region of the band, so it will undergo β^+ emission and/or e^- capture. Our predictions based on N/Z values were the same as those based on atomic mass.

Comment Both possible modes of decay are observed for the nuclide in part (d).

FOLLOW-UP PROBLEMS

24.3A Use the A value for the nuclide and the atomic mass from the periodic table to predict the mode of decay of **(a)** $^{61}_{26}Fe$; **(b)** $^{241}_{95}Am$.

24.3B Use the A value for the nuclide and the atomic mass from the periodic table to predict the mode of decay of **(a)** $^{40}_{22}Ti$; **(b)** $^{65}_{27}Co$.

SOME SIMILAR PROBLEMS 24.20–24.23

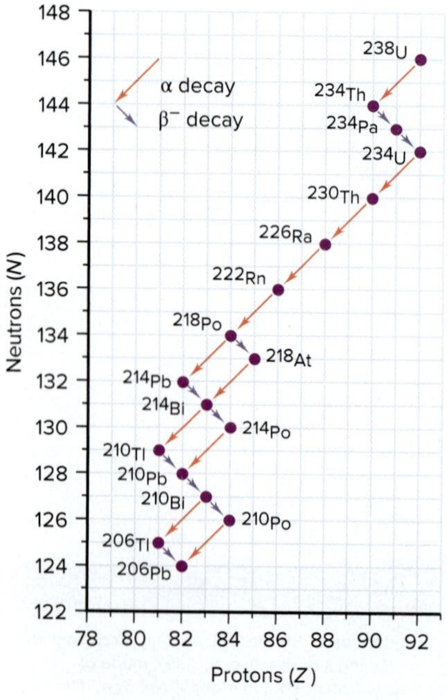

Figure 24.3 The ^{238}U decay series.

Decay Series A parent nuclide may undergo a series of decay steps before a stable daughter nuclide forms. The succession of steps is called a **decay series,** or **disintegration series,** and is typically depicted on a gridlike display. Figure 24.3 shows the decay series from uranium-238 to lead-206. Number of neutrons (N) is plotted against number of protons (Z) to form the grid, which displays a series of α and β^- decays. The typical zigzag pattern arises because $N > Z$, which means that α decay, which reduces both N and Z by two units, decreases Z slightly more than it does N. Therefore, α decays result in neutron-rich daughters, which undergo β^- decay to gain more stability. Note that a given nuclide can undergo both modes of decay. (Gamma emission accompanies many of these steps but does not affect the type of nuclide.)

The series in Figure 24.3 is one of three that occur in nature. All end with isotopes of lead whose nuclides all have one ($Z = 82$) or two ($N = 126$, $Z = 82$) magic numbers. A second series begins with uranium-235 and ends with lead-207, and a third begins with thorium-232 and ends with lead-208. (Neptunium-237 began a fourth series, but its half-life is so much less than the age of Earth that only traces of it remain today.)

› Summary of Section 24.1

- In dramatic contrast to chemical reactions, nuclear reactions involve particles *within* the nucleus, usually result in atoms of different elements, are accompanied by measurable changes in mass, and are not affected by temperature, pressure, catalysis, or the nature of the compound.
- To become more stable, a radioactive nuclide may emit α particles ($^4_2He^{2+}$; helium-4 nuclei), β particles (β^- or $^{\ 0}_{-1}\beta$; high-speed electrons), positrons (β^+ or $^0_1\beta$), or γ rays (high-energy photons) or may capture an orbital electron ($^{\ 0}_{-1}e$).
- A narrow band of neutron-to-proton ratios *(N/Z)* includes those of all the stable nuclides.
- Even values of *N* and *Z* are associated with stable nuclides, as are certain "magic numbers" of neutrons and protons.
- By comparing a nuclide's mass number with the atomic mass and its *N/Z* ratio with those in the band of stability, we can predict that, in general, neutron-rich nuclides undergo β^- decay and proton-rich nuclides undergo β^+ emission and/or e^- capture. Heavy nuclides ($Z > 83$) undergo α decay.
- Three naturally occurring decay series all end in isotopes of lead.

24.2 THE KINETICS OF RADIOACTIVE DECAY

Both chemical and nuclear systems tend toward maximum stability. Just as the concentrations in a chemical system change in a predictable direction to give a stable equilibrium ratio, the type and number of nucleons in an unstable nucleus change in a predictable direction to give a stable *N/Z* ratio. As you know, however, the tendency of a chemical system to become more stable tells nothing about how long that process will take, and the same holds true for nuclear systems. In this section, we first see how radioactivity is detected and measured and then examine the kinetics of nuclear change; later, we'll examine the energetics of nuclear change.

Detection and Measurement of Radioactivity

Radioactive emissions interact with surrounding atoms. To determine the rate of nuclear decay, we measure radioactivity by observing the effects of these interactions over time. Because these effects can be electrically amplified billions of times, it is possible to detect the decay of a single nucleus. Ionization counters and scintillation counters are two devices used to measure radioactive emissions.

Ionization Counters An *ionization counter* detects radioactive emissions as they ionize a gas. Ionization produces free electrons and gaseous cations, which are attracted to electrodes that conduct a current to a recording device. The most common type of ionization counter is a **Geiger-Müller counter** (Figure 24.4, *next page*). It consists of a tube filled with an argon-methane mixture; the tube housing acts as the cathode, and a thin wire in the center of the tube acts as the anode. Emissions from the sample enter the tube through a thin window and strike argon atoms, producing Ar^+ ions that migrate toward the cathode and free electrons that are accelerated toward the anode. These electrons collide with other argon atoms and free more electrons in an *avalanche effect.* The current created is amplified and appears as a meter reading and/or an audible click. An initial release of 1 electron can release 10^{10} electrons in a microsecond, giving the Geiger-Müller counter great sensitivity.

Scintillation Counters In a **scintillation counter,** radioactive emissions are detected by their ability to excite atoms and cause them to emit light. The light-emitting substance in the counter, called a *phosphor,* is coated onto part of a *photomultiplier tube,* a device that increases the original electrical signal. Incoming radioactive particles strike the phosphor, which emits photons. Each photon, in turn, strikes a cathode, releasing an electron through the photoelectric effect (Section 7.1). This electron hits other portions of the tube that release increasing numbers of electrons, and the

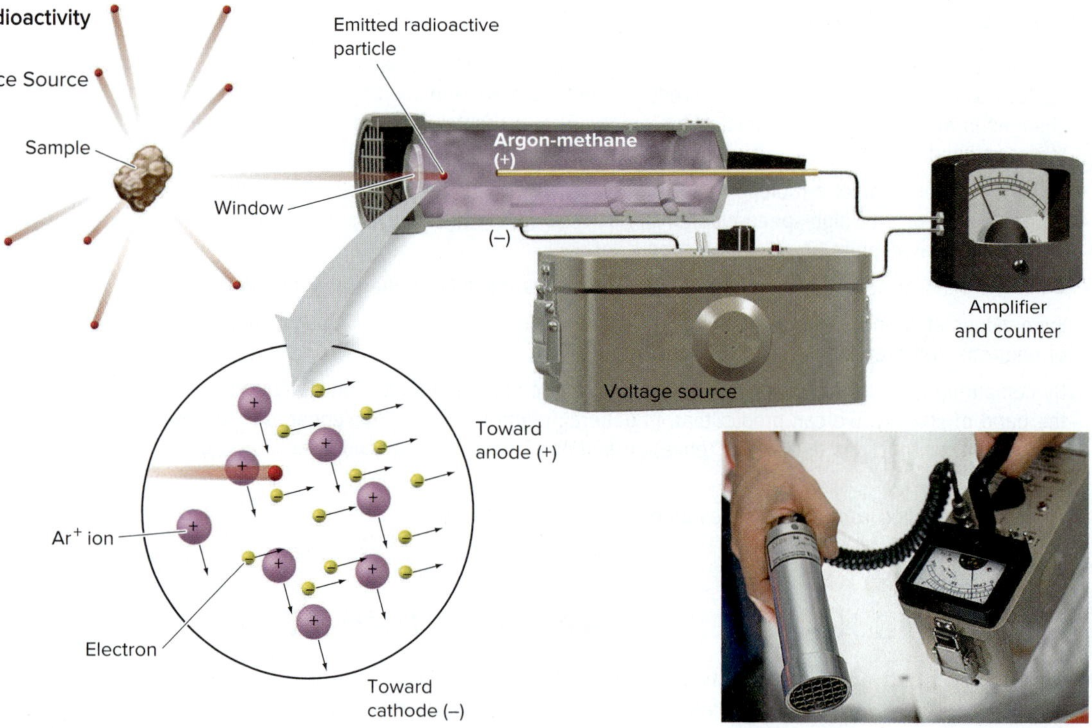

Figure 24.4 **Detection of radioactivity by an ionization counter.**
Source: © Hank Morgan/Science Source

Figure 24.5 **A scintillation "cocktail" in tubes to be placed in the counter.**
Source: © Meridian Biotechnologies Ltd.

resulting current is recorded. Liquid scintillation counters employ an organic mixture that contains a phosphor and a solvent (Figure 24.5). This "cocktail" dissolves the radioactive sample *and* emits pulses of light when excited by the emission. The number of pulses is proportional to the concentration of the radioactive substance. These counters are often used to measure β^- emissions from dissolved biological samples, particularly those containing compounds of ^3H and ^{14}C.

The Rate of Radioactive Decay

Radioactive nuclei decay at a characteristic rate, regardless of the chemical substance in which they occur.

Activity and the Decay Constant The *decay rate,* or **activity ($\mathscr{A}$),** of a radioactive sample is the change in number of nuclei ($\mathscr{N}$) divided by the change in time (t). As with a chemical reaction rate (Section 16.2), because the number of nuclei is *decreasing,* a minus sign precedes the expression to obtain a positive decay rate:

$$\text{Decay rate } (\mathscr{A}) = -\frac{\Delta\mathscr{N}}{\Delta t}$$

There are two common units of radioactivity:

- The SI unit of radioactivity is the **becquerel (Bq),** defined as one disintegration per second (d/s): 1 Bq = 1 d/s.
- A much larger and more common unit of radioactivity is the **curie (Ci),** which was originally defined as the number of disintegrations per second in 1 g of radium-226, but is now a fixed quantity: 1 Ci = 3.70×10^{10} d/s. Because the curie is so large, the millicurie (mCi) and microcurie (μCi) are commonly used.

The radioactivity of a sample is often given as a *specific activity,* the decay rate per gram.

An activity is meaningful only when we consider the large number of nuclei in a macroscopic sample. Suppose there are 1×10^{15} radioactive nuclei of a particular type in a sample and they decay at a rate of 10% per hour. Although any particular nucleus in the sample might decay in a microsecond or in a million hours, the *average* of all decays results in 10% of the entire collection of nuclei disintegrating each hour.

During the first hour, 10% of the *original* number, or 1×10^{14} nuclei, will decay. During the next hour, 10% of the remaining 9×10^{14} nuclei, or 9×10^{13} nuclei, will decay. During the next hour, 10% of those remaining will decay, and so forth. Thus, for a large collection of radioactive nuclei, *the number decaying per unit time is proportional to the number present:*

$$\text{Decay rate } (\mathscr{A}) \propto \mathscr{N} \qquad \text{or} \qquad \mathscr{A} = k\mathscr{N}$$

where k is called the **decay constant** and is characteristic of each type of nuclide. The larger the value of k, the higher the decay rate: larger $k \Longrightarrow$ higher $\mathscr{A}$.

Combining the two expressions for decay rate just given, we obtain

$$\mathscr{A} = -\frac{\Delta\mathscr{N}}{\Delta t} = k\mathscr{N} \qquad \textbf{(24.2)}$$

Note that the activity depends only on $\mathscr{N}$ raised to the first power (and on the constant value of k). Therefore, *radioactive decay is a first-order process* (Section 16.4), but first order with respect to the *number* of nuclei rather than their concentration.

SAMPLE PROBLEM 24.4 Calculating the Specific Activity and the Decay Constant of a Radioactive Nuclide

Problem A sample of 2.6×10^{-12} mol of antimony-122 (^{122}Sb) emits 2.76×10^{8} β^- particles per minute.

a. Calculate the specific activity of the sample (in Ci/g).

b. Find the decay constant for ^{122}Sb.

Plan **(a)** The activity ($\mathscr{A}$) is the number of nuclei ($\mathscr{N}$) emitting a particle in a given time (t), and the specific activity is the activity per gram of sample. We are given $\mathscr{A}$, which is equal to the number of particles emitted per minute (2.76×10^{8} β^- particles), and we know the amount (mol) of sample. Because the sample is an isotope, we use the mass number in g/mol (122 g/mol) to find the mass (g). Converting $\mathscr{A}$ (in d/min) to curies (in d/s) and dividing by the mass gives the specific activity (in Ci/g). **(b)** The decay constant (k) relates the activity to the number of nuclei ($\mathscr{N}$). We know $\mathscr{A}$ (number of nuclei disintegrating/min) and the amount (in mol), so we use Avogadro's number to find $\mathscr{N}$ (number of nuclei) and apply Equation 24.2 to get k.

Solution **a.** Finding the mass of the sample:

$$\text{Mass (g)} = (2.6\times10^{-12}\ \text{mol}\ ^{122}\text{Sb})\left(\frac{122\ \text{g}\ ^{122}\text{Sb}}{1\ \text{mol}\ ^{122}\text{Sb}}\right) = 3.2\times10^{-10}\ \text{g}\ ^{122}\text{Sb}$$

Using the activity and the mass to find the specific activity:

$$\text{Specific activity (Ci/g)} = \left(\frac{2.76\times10^{8}\ \text{d}}{\text{min}}\right)\left(\frac{1\ \text{min}}{60\ \text{s}}\right)\left(\frac{1\ \text{Ci}}{3.70\times10^{10}\ \text{d/s}}\right)\left(\frac{1}{3.2\times10^{-10}\ \text{g}}\right) = 3.9\times10^{5}\ \text{Ci/g}$$

b. Using the activity and the amount to find the decay constant (Equation 24.2):

$$\text{Decay constant } (k) = \frac{\mathscr{A}\text{(nuclei/min)}}{\mathscr{N}\text{(nuclei)}} = \frac{2.76\times10^{8}\ \text{nuclei/min}}{(2.6\times10^{-12}\ \text{mol})(6.022\times10^{23}\ \text{nuclei/mol})} = 1.8\times10^{-4}\ \text{min}^{-1}$$

Check With so many exponents in the values, rounding is a good way to check the math. In part (b), for example,

$$k = \frac{3\times10^{8}\ \text{nuclei/min}}{(3\times10^{-12}\ \text{mol})(6\times10^{23}\ \text{nuclei/mol})} = 2\times10^{-4}\ \text{min}^{-1}$$

FOLLOW-UP PROBLEMS

24.4A Arsenic-76 decays by β^- emission. If 3.4×10^{-8} mol of ^{76}As emits 1.53×10^{11} β^- particles per second, find the specific activity in **(a)** Ci/g; **(b)** Bq/g.

24.4B Chlorine-36 disintegrates by β^- decay. If 6.50×10^{-2} mol of ^{36}Cl emits 9.97×10^{12} β^- particles per hour, find the decay constant.

SOME SIMILAR PROBLEMS 24.33–24.40

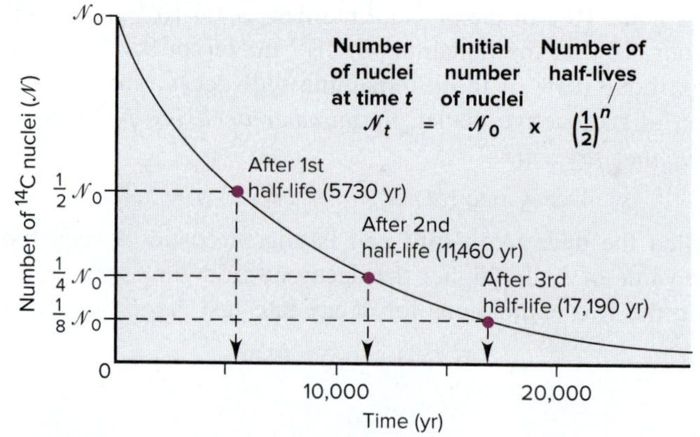

Figure 24.6 Decrease in number of ^{14}C nuclei over time.

Half-Life of Radioactive Decay Decay rates are also commonly expressed in terms of the fraction of nuclei that decay over a given time interval. The **half-life ($t_{1/2}$)** of a nuclide has the same meaning as for a chemical change (Section 16.4) and can be expressed in terms of number of nuclei, mass of radioactive substance, and activity:

- *Number of nuclei.* Half-life is the time it takes for half the nuclei in a sample to decay—*the number of nuclei remaining is halved after each half-life.* Figure 24.6 shows the decay of carbon-14, which has a half-life of 5730 years (yr), in terms of number of ^{14}C nuclei remaining:

$$^{14}_{6}\text{C} \longrightarrow\ ^{14}_{7}\text{N} +\ ^{0}_{-1}\beta$$

- *Mass.* As ^{14}C decays, the mass of ^{14}C decreases while the mass of ^{14}N increases. If we start with 1.0 g of ^{14}C, half that mass of ^{14}C (0.50 g) will be left after 5730 yr, half of that mass (0.25 g) after another 5730 yr, and so on.
- *Activity.* The activity depends on the number of nuclei, so the activity is halved after each succeeding half-life.

We determine the half-life of a nuclear reaction from its rate constant. Rearranging Equation 24.2 and integrating over time gives an expression for finding the number of nuclei remaining, $\mathcal{N}_t$, after a given time t:

$$\ln \frac{\mathcal{N}_t}{\mathcal{N}_0} = -kt \quad \text{or} \quad \mathcal{N}_t = \mathcal{N}_0 e^{-kt} \quad \text{and} \quad \ln \frac{\mathcal{N}_0}{\mathcal{N}_t} = kt \qquad \textbf{(24.3)}$$

where $\mathcal{N}_0$ is the number of nuclei at $t = 0$. (Note the similarity to Equation 16.4.) To calculate the half-life ($t_{1/2}$), we set $\mathcal{N}_t$ equal to $\frac{1}{2}\mathcal{N}_0$ and solve for $t_{1/2}$:

$$\ln \frac{\mathcal{N}_0}{\frac{1}{2}\mathcal{N}_0} = kt_{1/2} \quad \text{so} \quad t_{1/2} = \frac{\ln 2}{k} \qquad \textbf{(24.4)}$$

Exactly analogous to the half-life of a first-order chemical change, *this half-life is **not** dependent on the number of nuclei and is inversely related to the decay constant:*

large $k \Longrightarrow$ short $t_{1/2}$ and small $k \Longrightarrow$ long $t_{1/2}$

The decay constants and half-lives of many radioactive nuclides vary over a very wide range, even for a given element (Table 24.6).

Table 24.6 Decay Constants (k) and Half-Lives ($t_{1/2}$) of Beryllium Isotopes

Nuclide	k	$t_{1/2}$
$^{7}_{4}$Be	1.30×10^{-2}/day	53.3 days
$^{8}_{4}$Be	1.0×10^{16}/s	6.7×10^{-17} s
$^{9}_{4}$Be	Stable	
$^{10}_{4}$Be	4.3×10^{-7}/yr	1.6×10^{6} yr
$^{11}_{4}$Be	5.02×10^{-2}/s	13.8 s

SAMPLE PROBLEM 24.5 Finding the Number of Radioactive Nuclei

Problem Strontium-90 is a radioactive byproduct of nuclear reactors that behaves biologically like calcium, the element above it in Group 2A(2). When ^{90}Sr is ingested by mammals, it is found in their milk and eventually in the bones of those drinking the milk. If a sample of ^{90}Sr has an activity of 1.2×10^{12} d/s, what are the activity and the fraction of nuclei that have decayed after 59 yr ($t_{1/2}$ of ^{90}Sr = 29 yr)?

Plan The fraction of nuclei that have decayed is the change in number of nuclei, expressed as a fraction of the starting number. The activity of the sample ($\mathcal{A}$) is proportional to the number of nuclei ($\mathcal{N}$), so we know that

$$\text{Fraction decayed} = \frac{\mathcal{N}_0 - \mathcal{N}_t}{\mathcal{N}_0} = \frac{\mathcal{A}_0 - \mathcal{A}_t}{\mathcal{A}_0}$$

We are given $\mathcal{A}_0$ (1.2×10^{12} d/s), so we find $\mathcal{A}_t$ from the integrated form of the first-order rate equation (Equation 24.3), in which t is 59 yr. To solve that equation, we first need k, which we can calculate from the given $t_{1/2}$ (29 yr) using Equation 24.4.

Solution Calculating the decay constant k:

$$t_{1/2} = \frac{\ln 2}{k} \quad \text{so} \quad k = \frac{\ln 2}{t_{1/2}} = \frac{0.693}{29\ \text{yr}} = 0.024\ \text{yr}^{-1}$$

Applying Equation 24.3 to calculate $\mathcal{A}_t$, the activity remaining at time t:

$$\ln \frac{\mathcal{N}_0}{\mathcal{N}_t} = \ln \frac{\mathcal{A}_0}{\mathcal{A}_t} = kt \quad \text{or} \quad \ln \mathcal{A}_0 - \ln \mathcal{A}_t = kt$$

So, $\ln \mathcal{A}_t = -kt + \ln \mathcal{A}_0 = -(0.024\ \text{yr}^{-1} \times 59\ \text{yr}) + \ln (1.2\times10^{12}\ \text{d/s})$

$\ln \mathcal{A}_t = -1.4 + 27.81 = 26.4$

$\mathcal{A}_t = 2.9\times10^{11}$ d/s

(All the data contain two significant figures, so we retained two in the answer.)
Calculating the fraction decayed:

$$\text{Fraction decayed} = \frac{\mathcal{A}_0 - \mathcal{A}_t}{\mathcal{A}_0} = \frac{1.2\times10^{12}\ \text{d/s} - 2.9\times10^{11}\ \text{d/s}}{1.2\times10^{12}\ \text{d/s}} = 0.76$$

Check The answer is reasonable: t is about 2 half-lives, so $\mathcal{A}_t$ should be about $\frac{1}{4}\mathcal{A}_0$, or about 0.3×10^{12}; therefore, the activity should have decreased by about $\frac{3}{4}$.

Comment 1. A substitution in Equation 24.3 is useful for finding $\mathcal{A}_t$, the activity at time t: $\mathcal{A}_t = \mathcal{A}_0 e^{-kt}$.

2. Another way to find the fraction of activity (or nuclei) remaining uses the number of half-lives ($t/t_{1/2}$). By combining Equations 24.3 and 24.4 and substituting $(\ln 2)/t_{1/2}$ for k, we obtain

$$\ln \frac{\mathcal{N}_0}{\mathcal{N}_t} = \left(\frac{\ln 2}{t_{1/2}}\right) t = \frac{t}{t_{1/2}} \ln 2 = \ln 2^{\frac{t}{t_{1/2}}}$$

Inverting the ratio gives

$$\ln \frac{\mathcal{N}_t}{\mathcal{N}_0} = \ln \left(\frac{1}{2}\right)^{\frac{t}{t_{1/2}}}$$

Taking the antilog gives

$$\text{Fraction remaining} = \frac{\mathcal{N}_t}{\mathcal{N}_0} = \left(\frac{1}{2}\right)^{\frac{t}{t_{1/2}}} = \left(\frac{1}{2}\right)^{\frac{59}{29}} = 0.24$$

So, $\text{Fraction decayed} = 1.00 - 0.24 = 0.76$

FOLLOW-UP PROBLEMS

24.5A Sodium-24 ($t_{1/2} = 15$ h) is used to study blood circulation. If a patient is injected with an aqueous solution of $^{24}NaCl$ whose activity is 2.5×10^{9} d/s, how much of the activity is present in the patient's body and excreted fluids after 4.0 days?

24.5B Iron-59 ($t_{1/2} = 44.5$ days) is a β^- emitter used to monitor spleen function during disease or after trauma. If an accident victim receives a 5.6-μCi dose, what fraction of the ^{59}Fe nuclei will have decayed after 17 days?

SOME SIMILAR PROBLEMS 24.41 and 24.42

Radioisotopic Dating

The historical record fades rapidly with time and is virtually nonexistent for events of more than a few thousand years ago. Much knowledge of prehistory comes from **radioisotopic dating,** which uses unstable isotopes, or **radioisotopes,** to determine the ages of objects. This technique supplies data in fields such as art history, archeology, geology, and paleontology.

The technique of *radiocarbon dating,* for which the American chemist Willard F. Libby won the Nobel Prize in chemistry in 1960, is based on measuring the amounts of ^{14}C and ^{12}C in materials of biological origin. The accuracy of the method falls off after about six half-lives of ^{14}C ($t_{1/2}$ = 5730 yr), so it is used to date objects up to about 36,000 years old.

Here is how radiocarbon dating works:

1. High-energy cosmic rays, consisting mainly of protons, enter the atmosphere from outer space and initiate a cascade of nuclear reactions, some of which produce neutrons that bombard ordinary ^{14}N atoms to form ^{14}C:
$$^{14}_{7}N + ^{1}_{0}n \longrightarrow ^{14}_{6}C + ^{1}_{1}p$$
Through the competing processes of formation and radioactive decay, the amount of ^{14}C in the atmosphere has remained nearly constant.*
2. The ^{14}C atoms combine with O_2, diffuse throughout the lower atmosphere, and enter the total carbon pool as gaseous $^{14}CO_2$ and aqueous $H^{14}CO_3^-$. They mix with ordinary $^{12}CO_2$ and $H^{12}CO_3^-$, reaching a constant $^{12}C/^{14}C$ ratio of about $10^{12}/1$.
3. The CO_2 is taken up by plants during photosynthesis, and then taken up and excreted by animals that eat the plants. Thus, the $^{12}C/^{14}C$ ratio of a living organism is the same as the ratio in the environment.
4. When an organism dies, it no longer absorbs or releases CO_2, so the $^{12}C/^{14}C$ ratio steadily increases because *the amount of ^{14}C decreases as it decays:*
$$^{14}_{6}C \longrightarrow ^{14}_{7}N + ^{0}_{-1}\beta$$
The difference between the $^{12}C/^{14}C$ ratio in a dead organism and the ratio in living organisms reflects the time elapsed since the organism died.

As you saw in Sample Problem 24.5, the first-order rate equation can be expressed in terms of a ratio of activities:

$$\ln \frac{\mathcal{N}_0}{\mathcal{N}_t} = \ln \frac{\mathcal{A}_0}{\mathcal{A}_t} = kt$$

We use this expression in radiocarbon dating, where $\mathcal{A}_0$ is the activity in a living organism and $\mathcal{A}_t$ is the activity in the object whose age is unknown. Solving for t gives the age of the object:

$$t = \frac{1}{k} \ln \frac{\mathcal{A}_0}{\mathcal{A}_t} \qquad \textbf{(24.5)}$$

A graphical method used in radioisotopic dating shows a plot of the natural logarithm of the specific activity vs. time, which gives a straight line with a slope of $-k$, the negative of the decay constant. Using such a plot and measuring the ^{14}C specific activity of an object, we can determine its age; examples appear in Figure 24.7.

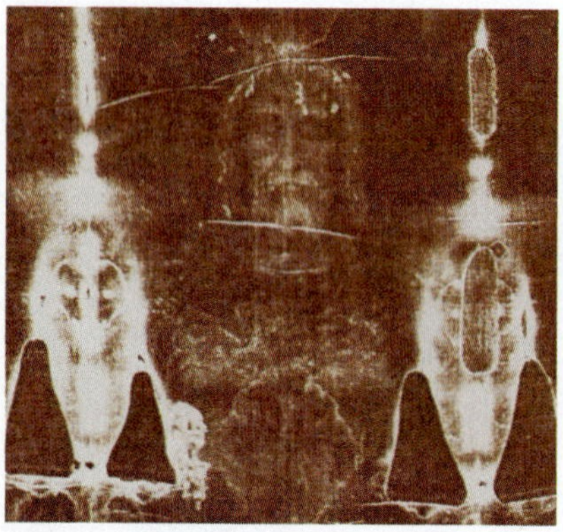

The Shroud of Turin.

One well-known example of the use of radiocarbon dating was in the determination of the age of the Shroud of Turin *(see photo)*. One of the holiest Christian relics, the shroud is a piece of linen that bears a faint image of a man's body and was thought to be the burial cloth used to wrap the body of Jesus Christ. However, three labs in Europe and the United States independently measured the $^{12}C/^{14}C$ ratio of a 50-mg piece of the linen and determined that the flax from which the cloth was made was grown between 1260 A.D. and 1390 A.D.

To determine the ages of more ancient objects or of objects that do not contain carbon, different radioisotopes are measured. For example, comparing the ratio of ^{238}U ($t_{1/2}$ = 4.5×10^9 yr) to its final decay product, ^{206}Pb, in meteorites gives 4.7 billion years for the age of the Solar System, and therefore Earth. From this and other isotope ratios, such as $^{40}K/^{40}Ar$ ($t_{1/2}$ of ^{40}K = 1.3×10^9 yr) and $^{87}Rb/^{87}Sr$ ($t_{1/2}$ of ^{87}Rb = 4.9×10^{10} yr), Moon rocks collected by Apollo astronauts have been dated at 4.2 billion years old.

*Cosmic ray intensity does vary slightly with time, which affects the amount of atmospheric ^{14}C. From ^{14}C activity in ancient trees, we know the amount fell slightly about 3000 years ago to current levels. Recently, nuclear testing and fossil-fuel combustion have also altered the fraction of ^{14}C slightly. Taking these factors into account improves the accuracy of the dating method.

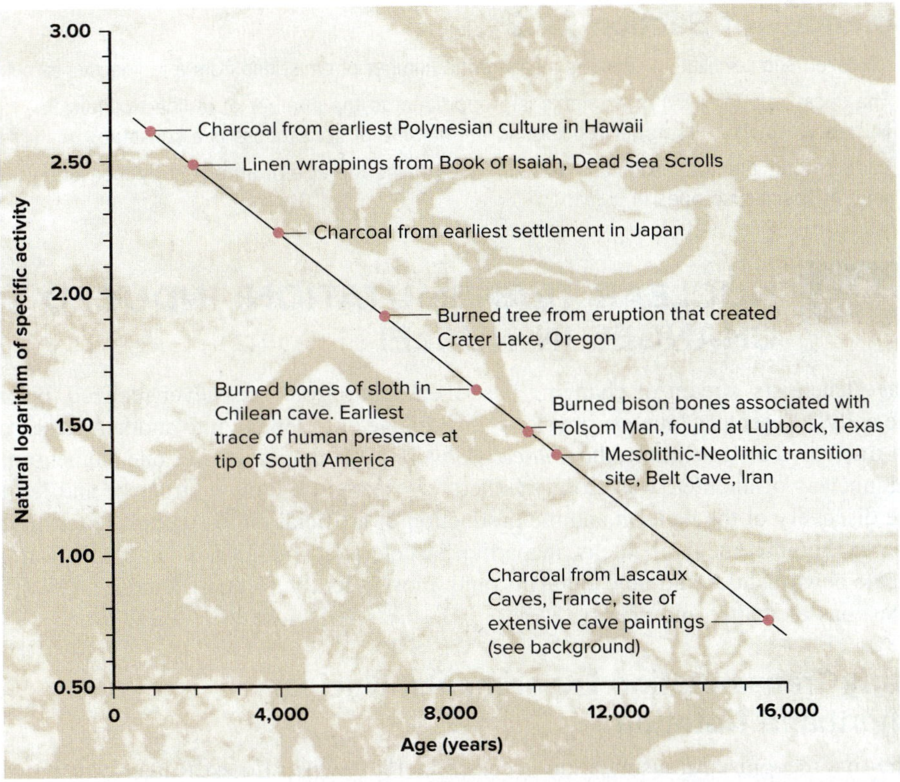

Figure 24.7 Ages of several objects determined by radiocarbon dating.

SAMPLE PROBLEM 24.6 Applying Radiocarbon Dating

Problem The charred bones of a sloth in a cave in Chile represent the earliest evidence of human presence at the southern tip of South America. A sample of the bone has a specific activity of 5.22 disintegrations per minute per gram of carbon (d/min·g). If the $^{12}C/^{14}C$ ratio for living organisms results in a specific activity of 15.3 d/min·g, how old are the bones ($t_{1/2}$ of ^{14}C = 5730 yr)?

Plan We calculate k from the given $t_{1/2}$ (5730 yr). Then we apply Equation 24.5 to find the age (t) of the bones, using the given activities of the bones ($\mathscr{A}_t$ = 5.22 d/min·g) and of a living organism ($\mathscr{A}_0$ = 15.3 d/min·g).

Solution Calculating k for ^{14}C decay:

$$k = \frac{\ln 2}{t_{1/2}} = \frac{0.693}{5730\ \text{yr}}$$
$$= 1.21\times10^{-4}\ \text{yr}^{-1}$$

Calculating the age (t) of the bones:

$$t = \frac{1}{k}\ln\frac{\mathscr{A}_0}{\mathscr{A}_t} = \frac{1}{1.21\times10^{-4}\ \text{yr}^{-1}}\ln\left(\frac{15.3\ \text{d/min·g}}{5.22\ \text{d/min·g}}\right)$$
$$= 8.89\times10^{3}\ \text{yr}$$

The bones are about 8890 years old.

Check The activity of the bones is between $\frac{1}{2}$ and $\frac{1}{4}$ of the activity of a living organism, so the age of the bones should be between one and two half-lives of ^{14}C (from 5730 to 11,460 yr).

FOLLOW-UP PROBLEMS

24.6A A sample of wood from an Egyptian mummy case has a specific activity of 9.41 d/min·g. How old is the case ($t_{1/2}$ of ^{14}C = 5730 yr)?

24.6B A fragment of a woolen tunic found in an ancient cemetery in Africa has a specific activity of 12.87 d/min·g. How old is the tunic ($t_{1/2}$ of ^{14}C = 5730 yr)?

SOME SIMILAR PROBLEMS 24.43 and 24.44

› Summary of Section 24.2

- Ionization and scintillation counters measure the number of emissions from a radioactive sample.
- The decay rate (activity) of a sample is proportional to the number of radioactive nuclei. Nuclear decay is a first-order process, so the half-life of the process is a constant.
- Radioisotopic methods, such as ^{14}C dating, determine the age of an object by measuring the ratio of specific isotopes in it.

24.3 NUCLEAR TRANSMUTATION: INDUCED CHANGES IN NUCLEI

The alchemists' dream of changing base metals into gold was never realized, but in the early 20th century, nuclear physicists *did* change one element into another. Research on **nuclear transmutation,** the *induced* conversion of the nucleus of one element into the nucleus of another, was closely linked to research on atomic structure and led to the discovery of the neutron and the production of artificial radioisotopes. Later, high-energy bombardment of nuclei in particle accelerators began the ongoing effort to create new nuclides and new elements, and, most recently, to understand fundamental questions of matter and energy.

Early Transmutation Experiments; Nuclear Shorthand Notation

The first recognized transmutation occurred in 1919, when Ernest Rutherford showed that α particles emitted from radium bombarded atmospheric nitrogen to form a proton and oxygen-17:

$$^{14}_{7}\text{N} + ^{4}_{2}\alpha \longrightarrow ^{1}_{1}\text{p} + ^{17}_{8}\text{O}$$

By 1926, experimenters had found that α bombardment transmuted most elements with low atomic numbers to the next higher element, with ejection of a proton.

Notation for Transmutation Reactions A shorthand notation used specifically for nuclear transmutation reactions shows the reactant (target) nucleus to the left and the product nucleus to the right of a set of parentheses, within which a comma separates the projectile particle from the ejected particle(s):

$$\text{reactant nucleus (particle in, particle(s) out) product nucleus} \qquad \textbf{(24.6)}$$

Using this notation, we write the previous reaction as $^{14}\text{N}\ (\alpha,\text{p})\ ^{17}\text{O}$.

The Neutron and the First Artificial Radioisotope An unexpected finding in a transmutation experiment led to *the discovery of the neutron.* When lithium, beryllium, and boron were bombarded with α particles, they emitted highly penetrating radiation that could not be deflected by a magnetic or electric field. Unlike γ radiation, these emissions were massive enough to eject protons from the substances they penetrated. In 1932, James Chadwick, a student of Rutherford, proposed that the emissions consisted of neutral particles with a mass similar to that of a proton, which he named *neutrons.* He received the Nobel Prize in physics in 1935 for his discovery.

In 1933, Irene and Frederic Joliot-Curie *(see photo),* daughter and son-in-law of Marie and Pierre Curie, created the first artificial radioisotope, phosphorus-30. When they bombarded aluminum foil with α particles, phosphorus-30 and neutrons were formed:

$$^{27}_{13}\text{Al} + ^{4}_{2}\alpha \longrightarrow ^{1}_{0}\text{n} + ^{30}_{15}\text{P} \quad \text{or} \quad ^{27}\text{Al}\ (\alpha,\text{n})\ ^{30}\text{P}$$

Since then, other techniques for producing artificial radioisotopes have been developed. In fact, the majority of the nearly 1000 known radioactive nuclides have been produced artificially.

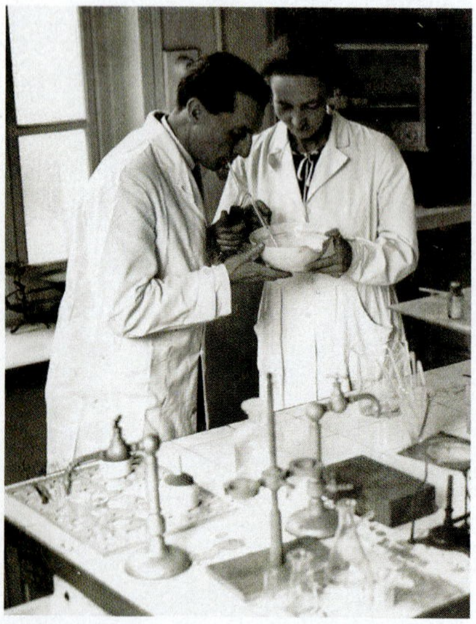

The Joliot-Curies in their laboratory.
Source: © Hulton Archive/Getty Images

Particle Accelerators and the Transuranium Elements

During the 1930s and 1940s, researchers probing the nucleus bombarded elements with neutrons, α particles, protons, and **deuterons** (nuclei of the stable hydrogen isotope deuterium, ^{2}H). Neutrons are especially useful as projectiles because they have no charge and thus are not repelled as they approach a target nucleus. The other particles are all positive, so early researchers found it difficult to give them enough energy to overcome the repulsion of the target nuclei. Beginning in the 1930s, however, **particle accelerators** were able to impart high kinetic energies to particles by placing them in an electric field, usually in combination with a magnetic field. In the simplest and earliest design, protons are introduced at one end of a tube and attracted to the other end by a potential difference.

Linear Accelerator A major advance was the *linear accelerator,* a series of separated tubes of increasing length that, through a source of alternating voltage, change their charge from positive to negative in synchrony with the movement of the particle through them (Figure 24.8).

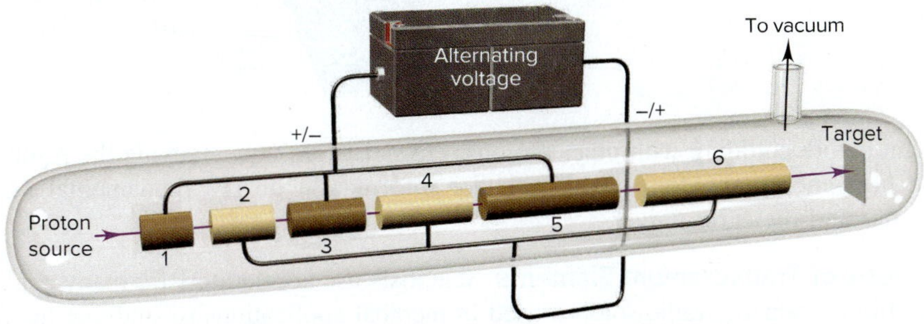

Figure 24.8 Schematic diagram of a linear accelerator.

A proton, for example, exits the first tube just when that tube becomes positive and the next tube negative. Repelled by the first tube and attracted by the second, the proton accelerates across the gap between them. The process is performed in stages to achieve high particle energies without having to apply a single very high voltage. A 40-ft linear accelerator with 46 tubes, built in California after World War II, accelerated protons to speeds several million times faster than earlier accelerators.

Other Early Particle Accelerators Two other early accelerators represented important design advances:

- The *cyclotron* (Figure 24.9, *next page*) was invented by E. O. Lawrence in 1930. Typically 1–2 m in diameter, it applies the principle of the linear accelerator but uses electromagnets to give the particle a spiral path to save space. The magnets lie within an evacuated chamber above and below two "dees," open, D-shaped electrodes that act like the tubes in the linear design. The particle is accelerated as it passes from one dee, which is momentarily positive, to the other, which is momentarily negative. Its speed and path radius increase until it is deflected toward the target nucleus.
- The *synchrotron* uses a synchronously increasing magnetic field to make the particle's path circular rather than spiral.

Powerful Modern Accelerators Some very powerful accelerators that are used to study the physics of high-energy particle collisions include both a linear section and a synchrotron section. The *tevatron* at the Fermi Lab near Chicago has a circumference of 3.9 miles and can accelerate particles to an energy slightly less than 1×10^{12} electron volts (1 TeV) before colliding them.

A small section of the Large Hadron Collider.
Source: © James Brittain/View Pictures/age fotostock

The world's most powerful accelerator is the Large Hadron Collider (LHC) near Geneva, Switzerland *(see photo)*. The first successful collision of protons in the LHC was achieved in early 2010. When operating at peak capacity, the LHC accelerates protons to an energy of up to 13 TeV as they travel around the main ring at a speed of 11,000 revolutions per second, attaining a final speed about 99.99% of the speed

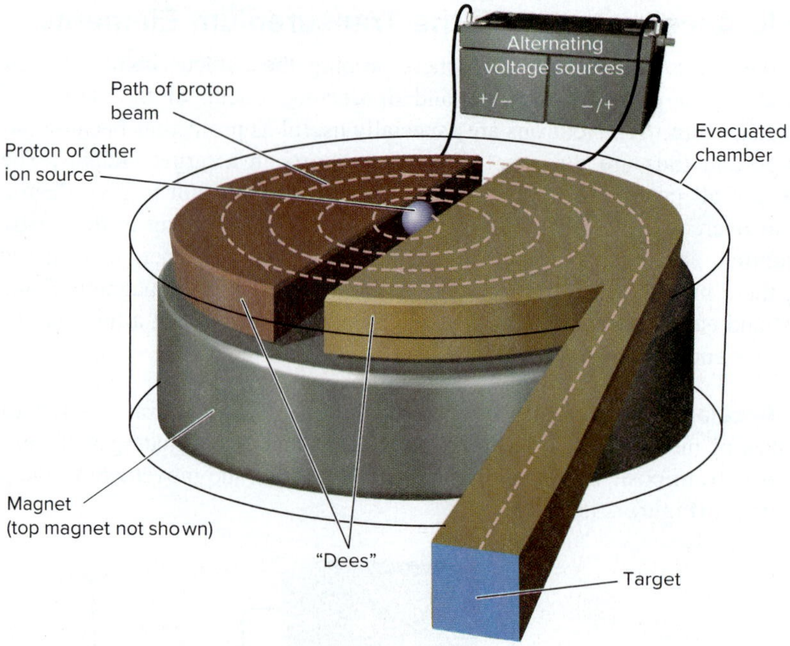

Figure 24.9 Schematic diagram of a cyclotron.

of light before colliding with other protons. Physicists are now studying the results of these high-energy subatomic collisions for information on the fundamental nature of matter and the early universe.

Synthesis of Transuranium Elements Scientists use accelerators for many applications, from producing radioisotopes used in medical applications to studying the fundamental nature of matter. Perhaps the most specific application for chemists is the synthesis of **transuranium elements,** those with atomic numbers higher than uranium, the heaviest naturally occurring element. Some reactions that were used to form several of these elements appear in Table 24.7. The transuranium elements include the remaining actinides (Z = 93 to 103), in which the 5f sublevel is filled; the elements in the fourth transition series (Z = 104 to 112), in which the 6d sublevel is filled; and the remaining elements in the 7p sublevel (Z = 113 to 118).

Table 24.7 Formation of Some Transuranium Nuclides*

Reaction							Half-life of Product
$^{239}_{94}\text{Pu}$	+	2^{1}_{0}n	$\longrightarrow$	$^{241}_{95}\text{Am}$	+	$^{0}_{-1}\beta$	432 yr
$^{239}_{94}\text{Pu}$	+	$^{4}_{2}\alpha$	$\longrightarrow$	$^{242}_{96}\text{Cm}$	+	$^{1}_{0}\text{n}$	163 days
$^{241}_{95}\text{Am}$	+	$^{4}_{2}\alpha$	$\longrightarrow$	$^{243}_{97}\text{Bk}$	+	2^{1}_{0}n	4.5 h
$^{242}_{96}\text{Cm}$	+	$^{4}_{2}\alpha$	$\longrightarrow$	$^{245}_{98}\text{Cf}$	+	$^{1}_{0}\text{n}$	45 min
$^{253}_{99}\text{Es}$	+	$^{4}_{2}\alpha$	$\longrightarrow$	$^{256}_{101}\text{Md}$	+	$^{1}_{0}\text{n}$	76 min
$^{243}_{95}\text{Am}$	+	$^{18}_{8}\text{O}$	$\longrightarrow$	$^{256}_{103}\text{Lr}$	+	5^{1}_{0}n	28 s

*Like chemical reactions, nuclear reactions may occur in several steps. For example, the first reaction here is actually an overall process that occurs in three steps:
(1) $^{239}_{94}\text{Pu} + ^{1}_{0}\text{n} \longrightarrow ^{240}_{94}\text{Pu}$ (2) $^{240}_{94}\text{Pu} + ^{1}_{0}\text{n} \longrightarrow ^{241}_{94}\text{Pu}$ (3) $^{241}_{94}\text{Pu} \longrightarrow ^{241}_{95}\text{Am} + ^{0}_{-1}\beta$

Z	Name (Symbol)
104	rutherfordium (Rf)
105	dubnium (Db)
106	seaborgium (Sg)
107	bohrium (Bh)
108	hassium (Hs)
109	meitnerium (Mt)
110	darmstadtium (Ds)
111	roentgenium (Rg)
112	copernicium (Cn)
113	nihonium (Nh)
114	flerovium (Fl)
115	moscovium (Mc)
116	livermorium (Lv)
117	tennessine (Ts)
118	oganesson (Og)

Conflicting claims of discovery by scientists in the United States and the former Soviet Union led to controversies about names for some of the more recently synthesized elements. To provide interim names until the disputes could be settled, the International Union of Pure and Applied Chemistry (IUPAC) adopted a system that uses the atomic number as the basis for a Latin name. Thus, for example, element 104 was named unnilquadium (un = 1, nil = 0, quad = 4, ium = metal suffix), with the symbol Unq. After much compromise, IUPAC has finalized the names of the new elements *(margin).*

› Summary of Section 24.3

- › One nucleus can be transmuted to another through bombardment with high-energy particles.
- › Accelerators increase the kinetic energy of particles in nuclear bombardment experiments and are used to produce transuranium elements and radioisotopes for medical use.

24.4 IONIZATION: EFFECTS OF NUCLEAR RADIATION ON MATTER

In 1986, an accident at the Chernobyl nuclear facility in the former Soviet Union released radioactivity that, according to the World Health Organization, will eventually cause thousands of cancer deaths. In the same year, isotopes used in medical treatment emitted radioactivity that prevented thousands of cancer deaths. In this section and Section 24.5, we examine radioactivity's harmful and beneficial effects.

The key to both types of effects is that *nuclear changes cause chemical changes in surrounding matter.* In other words, even though the nucleus of an atom may undergo a reaction with little or no involvement of the atom's electrons, the emissions from that reaction *do* affect the electrons of nearby atoms.

Virtually all radioactivity causes **ionization** in surrounding matter, as the emissions collide with atoms and dislodge electrons:

$$\text{Atom} \xrightarrow{\text{ionizing radiation}} \text{ion}^+ + e^-$$

From each ionization event, a cation and a free electron result, and the number of such *cation-electron pairs* produced is directly related to the energy of the incoming **ionizing radiation.**

Many applications of ionizing radiation depend not only on the ionizing event itself, but also on secondary processes. For example, in a Geiger-Müller counter, the free electron of the cation-electron pair often collides with another atom, which ejects a second electron, and in a scintillation counter, the initial ionization eventually results in the emission of light. In nuclear power plants (Section 24.7), the initial process as well as several secondary processes cause the release of heat that makes the steam used to generate electricity.

Effects of Ionizing Radiation on Living Tissue

Ionizing radiation has a destructive effect on living tissue, and if the ionized atom is part of a key biological macromolecule or cell membrane, the results can be devastating to the cell and perhaps the organism.

The danger from a radioactive nuclide depends on three factors:

- the type of radiation emitted,
- the half-life of the radioactive nuclide, and
- most importantly, the biological behavior of the radioactive nuclide.

For example, both ^{235}U and ^{239}Pu emit α particles and have long half-lives, but uranium is rapidly excreted by the body, whereas plutonium behaves like calcium and, thus, is incorporated into bones and teeth. One of the most dangerous radioactive nuclides, even worse than plutonium, is $^{90}Sr^{2+}$ (formed during a nuclear explosion) because it behaves very similarly to Ca^{2+} and is absorbed rapidly by bones. Let's first see how to measure radiation dose and then look in more detail at the damaging power of ionizing radiation.

Units of Radiation Dose and Its Effects To measure the effects of ionizing radiation, we need a unit for radiation dose. Units of radioactive decay, such as the becquerel and curie, measure the number of decay events in a given time but not their energy or absorption by matter. However, the number of cation-electron pairs produced in a given amount of living tissue *is* a measure of the energy absorbed by the tissue. Two pairs of related units are used to quantify this absorbed energy:

- The SI unit for energy absorption is the **gray (Gy);** it is equal to 1 joule of energy absorbed per kilogram of body tissue:

$$1 \text{ Gy} = 1 \text{ J/kg}$$

- A more widely used unit is the **rad (*r*adiation-*a*bsorbed *d*ose),** which is one-hundredth as large as the gray:

$$1 \text{ rad} = 0.01 \text{ J/kg} = 0.01 \text{ Gy}$$

To measure actual tissue damage, we must account for differences in the strength of the radiation, the exposure time, and the type of tissue. To do this, we multiply the absorbed dose by a *r*elative *b*iological *e*ffectiveness (RBE) factor, which depends on the effect a given type of radiation has on a given tissue or body part. The product is expressed in one of two units:

- The **rem (*r*oentgen *e*quivalent for *m*an)** is the unit of radiation dosage equivalent to a given amount of tissue damage in a human:

$$\text{no. of rems} = \text{no. of rads} \times \text{RBE}$$

  Doses are often expressed in millirems (10^{-3} rem).

- The SI unit for dosage equivalent is the **sievert (Sv).** It is defined in the same way as the rem but with absorbed dose in grays; thus, 1 Sv = 100 rem.

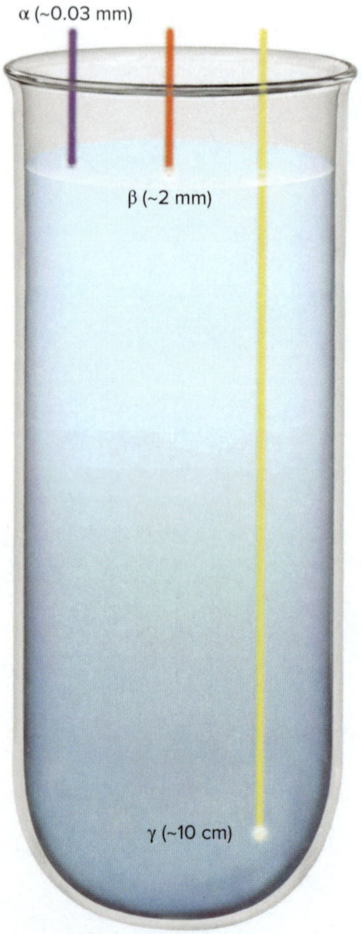

Figure 24.10 Penetrating power of radioactive emissions.

Penetrating Power of Emissions The effect on living tissue of a radiation dose depends on the penetrating power *and* ionizing ability of the radiation. Since water is the main component of living tissue, penetrating power is often measured in terms of the depth of water that stops 50% of incoming radiation. In Figure 24.10, the average values of the penetrating distances are shown in actual size for the three common types of emissions. Note, in general, that *penetrating power is inversely related to the mass, charge, and energy of the emission.* In other words, if a particle interacts strongly with matter, it penetrates only slightly, and vice versa.

1. *Alpha particles.* Alpha particles are massive and highly charged, which means that they interact with matter most strongly of the three common emissions. As a result, they penetrate so little that a piece of paper, light clothing, or the outer layer of skin can stop α radiation from an external source. However, if ingested, an α emitter can cause grave localized internal damage through extensive ionization. For example, in the early 20th century, wristwatch and clock dials were painted by hand with paint containing radium so that the numbers would glow in the dark. To write numbers clearly, the women who applied the paint "tipped" fine brushes repeatedly between their lips. Small amounts of ingested $^{226}Ra^{2+}$ were incorporated into their bones along with normal Ca^{2+}, which led to numerous cases of bone fracture and jaw cancer.

2. *Beta particles and positrons.* Beta particles (β^-) and positrons (β^+) have less charge and much less mass than α particles, so they interact less strongly with matter. Even though a given particle has less chance of causing ionization, a β^- (or β^+) emitter is a more destructive external source because the particles penetrate deeper. Specialized heavy clothing or a thick (0.5 cm) piece of metal is required to stop these particles. The relative danger from α and β^- emissions is shown by two international incidents involving poisoning. In 1957, an espionage agent was poisoned by thallium-204, a β^- emitter, but he survived because β^- particles do relatively little radiation damage. In contrast, in 2006, another agent was poisoned with polonium-210, an α emitter, and died in three weeks because of the much greater radiation damage.

3. *Gamma rays*. Neutral, massless γ rays interact least with matter and, thus, penetrate most. A block of lead several inches thick is needed to stop them. Therefore, an external γ ray source is the most dangerous because the energy can ionize many layers of living tissue.

The extent of interaction with matter is variable for emitted *neutrons,* which are neutral and massive. Outside a nucleus, neutrons decay with a half-life of 10.6 min into a proton and an electron (and an antineutrino). With energies ranging from very low to very high, they may be scattered or captured by interaction with nuclei.

Molecular Interactions How does the damage take place on the molecular level? When ionizing radiation interacts with molecules, it causes the *loss of an electron from a bond or a lone pair.* The resulting charged species go on to form **free radicals,** molecular or atomic species with one or more unpaired electrons, which, as you know, are very reactive (Section 10.1). They form electron pairs by attacking bonds in other molecules, often forming more free radicals. Two types of materials are especially vulnerable to free-radical attack:

1. *Water*. When γ radiation strikes living tissue, for instance, the most likely molecule to absorb it is water, yielding an electron and a water ion-radical:

$$H_2O + \gamma \longrightarrow H_2O\bullet^+ + e^-$$

The products collide with other water molecules to form more free radicals:

$$H_2O\bullet^+ + H_2O \longrightarrow H_3O^+ + \bullet OH \qquad \text{and} \qquad e^- + H_2O \longrightarrow H\bullet + OH^-$$

These free radicals attack more H_2O molecules and surrounding biomolecules, whose bonding and structure are intimately connected with their function (Section 15.6).

2. *Membranes and macromolecules*. The double bonds in membrane lipids are highly susceptible to free-radical attack:

$$H\bullet + RCH{=}CHR' \longrightarrow RCH_2{-}\dot{C}HR'$$

In this reaction, one electron of the π bond forms a C—H bond between one of the double-bonded carbons and the H•, and the other electron resides on the other carbon to form a free radical. Changes to lipid structure cause damage that results in leakage through cell membranes and destruction of the protective fatty tissue around organs. Changes to critical bonds in enzymes lead to their malfunction as catalysts of metabolic reactions. Changes in the nucleic acids and proteins that govern the rate of cell division can cause cancer. Genetic damage and mutations may occur when bonds in the DNA of sperm and egg cells are altered by free radicals.

Background Sources of Ionizing Radiation

We are continuously exposed to ionizing radiation from natural and artificial sources (Table 24.8). Indeed, life evolved in the presence of natural ionizing radiation, called **background radiation.** Ionizing radiation can alter bonds in DNA and cause harmful mutations but also causes the beneficial ones that allow species to evolve.

Background radiation has several natural sources:

1. *Cosmic radiation* increases with altitude because of decreased absorption by the atmosphere. Thus, people in Denver absorb twice as much cosmic radiation as people in Los Angeles; even a single jet flight involves measurable absorption.

2. Thorium and uranium minerals are present in rocks and soil. Radon, the heaviest noble gas [Group 8A(18)], is a radioactive product of uranium and thorium decay. Its concentration in the air we breathe is related to the presence of trace minerals in building materials and to the uranium content of local soil and

Table 24.8 Typical Radiation Doses from Natural and Artificial Sources

Source of Radiation	Average Adult Exposure
Natural	
Cosmic radiation	30–50 mrem/yr
Radiation from the ground	
From clay soil and rocks	~25–170 mrem/yr
In wooden houses	10–20 mrem/yr
In brick houses	60–70 mrem/yr
In concrete (cinder block) houses	60–160 mrem/yr
Radiation from the air (mainly radon)	
Outdoors, average value	20 mrem/yr
In wooden houses	70 mrem/yr
In brick houses	130 mrem/yr
In concrete (cinder block) houses	260 mrem/yr
Internal radiation from minerals in tap water and daily intake of food (^{40}K, ^{14}C, Ra)	~40 mrem/yr
Artificial	
Diagnostic x-ray methods	
Lung (local)	0.04–0.2 rad/film
Kidney (local)	1.5–3 rad/film
Dental (dose to the skin)	≤1 rad/film
Therapeutic radiation treatment	Locally ≤10,000 rad
Other sources	
Jet flight (4 h)	~1 mrem
Nuclear testing	<4 mrem/yr
Nuclear power industry	<1 mrem/yr
Total average value	100–200 mrem/yr

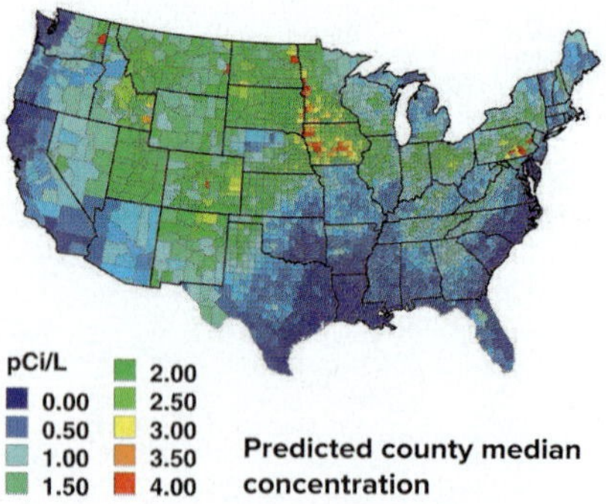

Figure 24.11 Continental U.S. radon distribution.

rocks (Figure 24.11). Once it enters the body, radon poses a serious potential hazard as it decays to radioactive nuclides of Po, Pb, and Bi, through α, β^-, and γ emissions. The emissions damage lung tissue, and the heavy-metal atoms formed aggravate the problem. Recent EPA estimates indicate that radon contributes to 15% of annual lung cancer deaths.

3. About 150 g of K^+ ions is dissolved in the water in the tissues of an average adult, and 0.0118% of these ions are radioactive ^{40}K. The presence of these substances and of atmospheric $^{14}CO_2$ makes *all* food, water, clothing, and building materials slightly radioactive.

The largest artificial source of radiation, and the one that is easiest to control, is from medical diagnostic techniques, especially x-rays. The radiation dosage from nuclear testing and radioactive waste disposal is miniscule for most people, but exposures for those living near test sites or disposal areas may be much higher.

Assessing the Risk from Ionizing Radiation

How much radiation is too much? To answer this question, we must ask several others: How strong is the exposure? How long is the exposure? Which tissue is exposed? Are children being exposed? One reason we lack clear data to answer these questions is that scientific ethical standards forbid the intentional exposure of humans in an experimental setting. However, accidentally exposed radiation workers and Japanese atomic bomb survivors have been studied extensively.

1. *Human studies.* Table 24.9 summarizes the immediate effects on humans of an acute single dose of ionizing radiation to the whole body. The severity of the effects increases with dose; a dose of 500 rem will kill about 50% of the exposed population within a month.

Table 24.9 Acute Effects of a Single Dose of Whole-Body Irradiation

Dose (rem)	Effect	Lethal Dose: Population (%)	Lethal Dose: Number of Days
5–20	Possible late effect; possible chromosomal aberrations	—	—
20–100	Temporary reduction in white blood cells	—	—
50+	Temporary sterility in men (100+ rem = 1-yr duration)	—	—
100–200	"Mild radiation sickness": vomiting, diarrhea, tiredness in a few hours Reduction in infection resistance Possible bone growth retardation in children	—	—
300+	Permanent sterility in women	—	—
500	"Serious radiation sickness": marrow/intestine destruction	50–70	30
400–1000	Acute illness, early deaths	60–95	30
3000+	Acute illness, death in hours to days	100	2

2. *Animal studies*. Most data on radiation risk come from laboratory animals, whose biological systems may differ greatly from ours. Nevertheless, studies with mice and dogs show that lesions and cancers appear after massive whole-body exposure, with rapidly dividing cells affected first. In an adult animal, these are cells of the bone marrow, organ linings, and reproductive organs, but many other tissues are affected in an immature animal or fetus.

3. *Models of risk assessment*. Studies in both animals and humans show an increase in the incidence of cancer from either a high, single exposure or a low, chronic exposure. Two current models of radiation risk are shown in Figure 24.12:

- The *linear response model* proposes that radiation effects, such as cancer risks, accumulate over time regardless of dose and that populations should not be exposed to any radiation above background levels.
- The *S-shaped response model* implies there is a threshold above which the effects are more significant: very low risk at low dose and high risk at high dose.

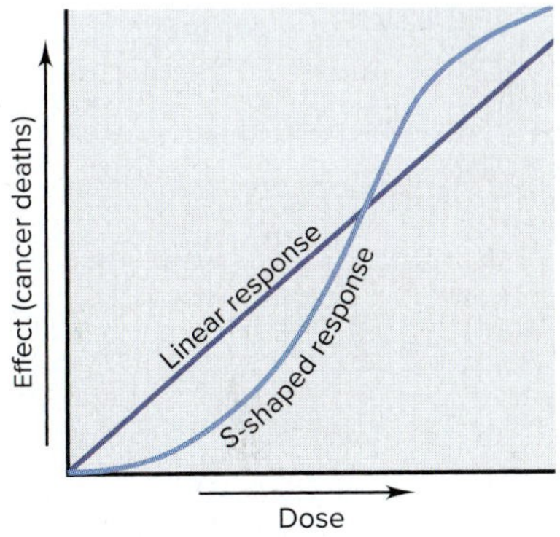

Figure 24.12 Two models of radiation risk.

4. *Hereditary effects*. Reliable data on hereditary (genetic) effects of radiation are few. Studies on fruit flies show a linear increase in genetic defects with both dose and exposure time. However, in the mouse, whose genetic system is obviously much more similar to ours than is the fruit fly's, a total dose given over a long period created one-third as many genetic defects as the same dose given over a short period. Therefore, rate of exposure is a key factor. The children of atomic bomb survivors show higher-than-normal childhood cancer rates, implying that their parents' reproductive systems were affected.

› Summary of Section 24.4

- All radioactive emissions cause ionization.
- The effect of ionizing radiation on living matter depends on the quantity of energy absorbed and the extent of ionization in a given type of tissue. Radiation dose for the human body is measured in rem.
- Ionization forms free radicals, some of which proliferate and destroy biomolecular function.
- All organisms are exposed to varying quantities of natural ionizing radiation.
- Studies show that either a large acute dose or a chronic small dose is harmful.

24.5 APPLICATIONS OF RADIOISOTOPES

Radioisotopes are powerful tools for studying processes in biochemistry, medicine, materials science, environmental studies, and many other scientific and industrial fields. Such uses depend on the fact that *isotopes of an element exhibit* ***very*** *similar chemical and physical behavior.* In other words, except for having a less stable nucleus, a radioisotope has nearly the same properties as a nonradioactive isotope of the same element.* For example, the fact that $^{14}CO_2$ is utilized by a plant in the same way as $^{12}CO_2$ forms the basis of radiocarbon dating.

Radioactive Tracers

Just think how useful it could be to follow a substance through a complex process or from one region of a system to another. A tiny amount of a radioisotope mixed with a large amount of a stable isotope of the same element can act as a **tracer,** a chemical "beacon" emitting radiation that signals the presence of the substance.

Reaction Pathways Tracers are used to study simple and complex reaction pathways.

1. *Inorganic systems: the periodate-iodide reaction.* One well-studied pathway is that of the reaction between periodate and iodide ions:

$$IO_4^-(aq) + 2I^-(aq) + H_2O(l) \longrightarrow I_2(s) + IO_3^-(aq) + 2OH^-(aq)$$

Is IO_3^- the result of IO_4^- reduction or I^- oxidation? When "cold" (nonradioactive) IO_4^- is added to a solution of I^- that contains some "hot" (radioactive, indicated in red) $^{131}I^-$, the I_2 is the radioactive product, not the IO_3^-:

$$IO_4^-(aq) + 2\,^{131}I^-(aq) + H_2O(l) \longrightarrow {}^{131}I_2(s) + IO_3^-(aq) + 2OH^-(aq)$$

These results show that IO_3^- forms through the reduction of IO_4^-, and that I_2 forms through the oxidation of I^-. To confirm this pathway, IO_4^- containing some $^{131}IO_4^-$ is added to a solution of I^-. As expected, in this case, the IO_3^- is radioactive, not the I_2:

$$^{131}IO_4^-(aq) + 2I^-(aq) + H_2O(l) \longrightarrow I_2(s) + {}^{131}IO_3^-(aq) + 2OH^-(aq)$$

Thus, tracers act like "handles" we can "hold" to follow the changing reactants.

2. *Biochemical pathways: photosynthesis.* Far more complex pathways can be followed with tracers as well. The photosynthetic pathway, in which energy from sunlight is used to form the chemical bonds of glucose, is the most essential and widespread metabolic process on Earth. It has an overall reaction that looks quite simple:

$$6CO_2(g) + 6H_2O(l) \xrightarrow[\text{chlorophyll}]{\text{light}} C_6H_{12}O_6(s) + 6O_2(g)$$

However, the actual process is extremely complex: 13 enzyme-catalyzed steps are required to incorporate each C atom from CO_2, so the six CO_2 molecules needed to form a molecule of $C_6H_{12}O_6$ require six repetitions of the pathway. Melvin Calvin and his coworkers took seven years to determine the pathway, using ^{14}C in CO_2 as the tracer and chromatography as the means of separating the products formed after different times of light exposure. Calvin won the Nobel Prize in chemistry in 1961 for this remarkable achievement.

Physiological Studies Tracers are used in many studies of physiological function. Some recent studies examine challenges of living in outer space. In an animal study of red blood cell loss during extended space flight, blood plasma volume was measured with ^{125}I-labeled albumin (a blood protein), and red blood cells labeled with

*Although this statement is generally correct, differences in isotopic mass *can* influence bond strengths and therefore reaction rates. Such behavior is called a *kinetic isotope effect* and is particularly important for isotopes of hydrogen—^1H, ^2H, and ^3H—because their masses differ by such large proportions. Section 22.4 discussed how the kinetic isotope effect is employed in the industrial production of heavy water, D_2O.

^{51}Cr as a trace substitute for Fe were used to assess the survival rate of blood cells. In another study, blood flow in skin under long periods of microgravity was monitored using injected ^{133}Xe.

Material Flow Tracers are used in studies of solid surfaces and the flow of materials. Metal atoms hundreds of layers deep within a solid have been shown to exchange with metal ions from the surrounding solution within a matter of minutes. Chemists and engineers use tracers to study material movement in semiconductor chips, paint, and metal plating, in detergent action, and in the process of corrosion, to mention just a few of many applications.

Hydrologic engineers use tracers to study the volume and flow of large bodies of water. By following radioactive nuclides that formed during atmospheric nuclear bomb tests (^3H in H_2O, $^{90}Sr^{2+}$, and $^{137}Cs^+$), scientists have mapped the flow of water from land to lakes and streams to oceans. They also use tracers to study the surface and deep ocean currents that circulate around the globe, the mechanisms of hurricane formation, and the mixing of the troposphere and stratosphere.

Various industries employ tracers to study material flow during manufacturing processes, such as the flow of ore pellets in smelting kilns, the paths of wood chips and bleach in paper mills, the diffusion of fungicide into lumber, and in a particularly important application, the porosity and leakage of oil and gas wells in geological formations.

Activation Analysis Another use of tracers is in *neutron activation analysis* (NAA). In this method, neutrons bombard a nonradioactive sample, converting a small fraction of its atoms to radioisotopes, which exhibit characteristic decay patterns, such as γ-ray spectra, that reveal the elements present. Unlike chemical analysis, NAA leaves the sample virtually intact, so the method can be used to determine the composition of a valuable object or a very small sample. For example, a painting thought to be a 16th-century Dutch masterpiece was shown through NAA to be a 20th-century forgery, because a microgram-sized sample of its pigment contained much less silver and antimony than the pigments used by the Dutch masters. Forensic chemists use NAA to detect traces of ammunition on a suspect's hand or traces of arsenic in the hair of a victim of poisoning.

Automotive engineers employ NAA and γ-ray detectors to measure friction and wear of moving parts without having to take an engine apart. For example, when a steel surface that has been neutron-activated to form some radioactive ^{59}Fe moves against a second steel surface, the amount of radioactivity on the second surface indicates the amount of material rubbing off. The radioactivity appearing in a lubricant placed between the surfaces can demonstrate the lubricant's ability to reduce wear.

Medical Diagnosis The largest use of radioisotopes is in medical science. Tracers with half-lives of a few minutes to a few days are employed to observe specific organs and body parts.

For example, a healthy thyroid gland incorporates dietary I^- into iodine-containing hormones at a known rate. To assess thyroid function, the patient drinks a solution containing a trace amount of $Na^{131}I$; the thyroid gland absorbs $^{131}I^-$ ions, which undergo β^- decay, and the emissions produce an image of the gland (Figure 24.13A, *next page*). Technetium-99 ($Z = 43$) is also used for imaging the thyroid (Figure 24.13B, *next page*), as well as the heart, lungs, and liver. Technetium does not occur naturally, so the radioisotope (actually a metastable form, ^{99m}Tc) is prepared from radioactive molybdenum just before use:

$$^{99}_{42}Mo \longrightarrow {}^{99m}_{43}Tc + {}^{0}_{-1}\beta$$

Tracers are also used to measure physiological processes, such as blood flow. The rate at which the heart pumps blood, for example, can be observed by injecting ^{59}Fe, which becomes incorporated into the hemoglobin of blood cells. Several radioisotopes used in medical diagnosis are listed in Table 24.10.

Table 24.10 Some Radioisotopes Used as Tracers in Medical Diagnosis

Isotope	Body Part or Process
^{11}C, ^{18}F, ^{13}N, ^{15}O	PET studies of brain, heart
^{60}Co, ^{192}Ir	Cancer therapy
^{64}Cu	Metabolism of copper
^{59}Fe	Blood flow, spleen
^{67}Ga	Tumor imaging
^{123}I, ^{131}I	Thyroid
^{111}In	Brain, colon
^{42}K	Blood flow
^{81m}Kr	Lung
^{99m}Tc	Heart, thyroid, liver, lung, bone
^{201}Tl	Heart muscle
^{90}Y	Cancer, arthritis

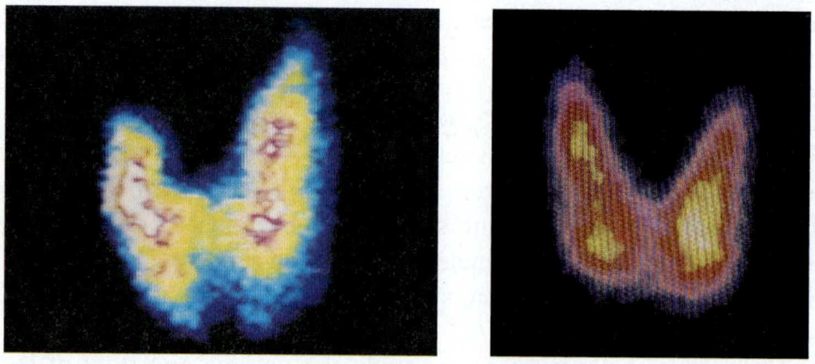

A B

Figure 24.13 The use of radioisotopes to image the thyroid gland. A, This ^{131}I scan shows an asymmetric image that is indicative of disease. **B,** A ^{99}Tc scan of a healthy thyroid.

Source: (A) © Scott Camazine/Science Source; (B) © Chris Priest/Science Source

Positron-emission tomography (PET) is a powerful imaging method for observing brain structure and function. A biological substance is synthesized with one of its atoms replaced by an isotope that emits positrons. The substance is injected into a patient's bloodstream, from which it is taken up into the brain. The isotope emits positrons, each of which annihilates a nearby electron. In this process, two γ photons are emitted simultaneously 180° from each other:

$$ {}^{0}_{1}\beta + {}^{0}_{-1}\text{e} \longrightarrow 2\,{}^{0}_{0}\gamma $$

An array of detectors around the patient's head pinpoints the sites of γ emission, and the image is analyzed by computer. Two of the isotopes used are ^{15}O, injected as $H_2{}^{15}O$ to measure blood flow, and ^{18}F bonded to a glucose analog to measure glucose uptake, which is an indicator of energy metabolism.

Among many fascinating findings revealed by PET are those that show how changes in blood flow and glucose uptake accompany normal or abnormal brain activity (Figure 24.14). Also, substances incorporating ^{11}C and ^{15}O are being investigated using PET to learn how molecules interact with and move along the surface of a catalyst.

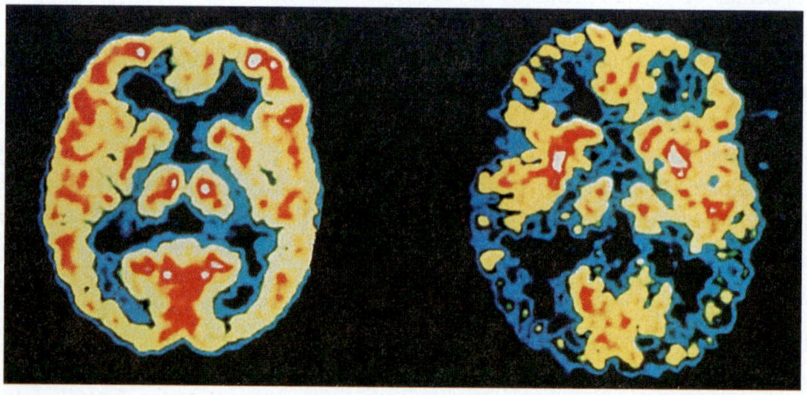

Figure 24.14 PET and brain activity. These PET scans show brain activity in a normal person *(left)* and in a patient with Alzheimer's disease *(right)*. Red and yellow indicate relatively high activity within a region.

Source: © Dr. Robert Friedland/SPL/Science Source

Additional Applications of Ionizing Radiation

Many other uses of radioisotopes involve higher-energy ionizing radiation.

1. *Radiation therapy.* Cancer cells divide more rapidly than normal cells, so radioisotopes that interfere with cell division kill more cancer cells than normal ones. Implants of ^{198}Au or of ^{90}Sr, which decays to the γ-emitting ^{90}Y, have been used to destroy pituitary and breast tumor cells, and γ rays from ^{60}Co have been used to destroy tumors of the brain and other organs.

Figure 24.15 The increased shelf life of irradiated food.

Source: Iowa State University

2. *Destruction of microbes.* Irradiation of food increases shelf life by killing microorganisms that cause rotting or spoilage (Figure 24.15), but the practice is controversial. Advocates point to the benefits of preserving fresh foods, grains, and seeds for

long periods, whereas opponents suggest that irradiation might lower the food's nutritional content or produce harmful byproducts. Irradiation also provides a way to destroy newer, more resistant bacterial strains that survive the increasing use of the more common antibiotics in animal feed. The United Nations has approved irradiation for potatoes, wheat, chicken, and strawberries, and the U.S. Food and Drug Administration has approved it for these and other food items as well.

3. *Insect control.* Ionizing radiation has been used to control harmful insects. Captured males are sterilized by radiation and released to mate, thereby reducing the number of offspring. This method has been used to control the Mediterranean fruit fly in California and certain disease-causing insects, such as the tsetse fly and malarial mosquito, in other parts of the world.

4. *Power for spacecraft instruments.* A nonharmful use of ionizing radiation relies on its secondary processes. Most spacecraft use solar energy to provide power for instruments. In deep-space missions, when solar energy is too weak, however, a radioisotope heater unit (RHU) has been used. It consists of a $^{238}PuO_2$ fuel pellet the size of a pencil eraser that is clad within a multilayered graphite and metal shell a little smaller than a flashlight battery. The whole RHU weighs only 40 g.

› Summary of Section 24.5

- Radioisotopic tracers have been used to study reaction mechanisms, material flow, elemental composition, and medical conditions.
- Ionizing radiation has been used in devices that destroy cancer tissue, kill organisms that spoil food, control insect populations, and power spacecraft instruments.

24.6 THE INTERCONVERSION OF MASS AND ENERGY

Most of the nuclear processes we've considered so far have involved radioactive decay, in which a nucleus emits one or a few small particles or photons to become a more stable and, usually, slightly lighter nucleus. Two other nuclear processes cause much greater mass changes. In nuclear **fission,** *a heavy nucleus splits into two much lighter nuclei*. In nuclear **fusion,** the opposite process occurs: *two lighter nuclei combine to form a heavier one*. Both fission and fusion release enormous quantities of energy. Let's look at the origins of this energy by examining the change in mass that accompanies the breakup of a nucleus into its nucleons and then determine the energy that is equivalent to this mass change.

The Mass Difference Between a Nucleus and Its Nucleons

For most of the 20th century, we knew that mass and energy are interconvertible. The separate mass and energy conservation laws are combined to state that *the total quantity of mass-energy in the universe is constant.* Therefore, when *any* reacting system releases or absorbs energy, it also loses or gains mass.

Mass Difference in Chemical Reactions The interconversion of mass and energy is not important for chemical reactions because the energy changes involved in breaking or forming chemical bonds are so small that the mass changes are negligible. For example, when 1 mol of water breaks up into its atoms, heat is absorbed and we have:

$$H_2O(g) \longrightarrow 2H(g) + O(g) \qquad \Delta H^\circ_{rxn} = 2 \times \text{BE of O—H} = 934 \text{ kJ}$$

We find the mass that is equivalent to this energy from *Einstein's equation:*

$$E = mc^2 \qquad \text{or} \qquad \Delta E = \Delta mc^2 \qquad \text{so} \qquad \Delta m = \frac{\Delta E}{c^2} \tag{24.7}$$

where Δm is the mass difference between reactants and products:

$$\Delta m = m_{\text{products}} - m_{\text{reactants}}$$

Substituting the enthalpy of reaction (in J/mol) for ΔE and the numerical value for c (2.9979×10^8 m/s), we obtain

$$\Delta m = \frac{9.34 \times 10^5\ \text{J/mol}}{(2.9979 \times 10^8\ \text{m/s})^2} = 1.04 \times 10^{-11}\ \text{kg/mol} = 1.04 \times 10^{-8}\ \text{g/mol}$$

(Units of kg/mol are obtained because the joule includes the kilogram: 1 J = 1 kg·m^2/s^2.) The mass of 1 mol of H_2O molecules (reactant) is about 10 ng *less* than the combined masses of 2 mol of H atoms and 1 mol of O atoms (products), a change difficult to measure with even the most advanced balance. Such minute mass changes when bonds break or form allow us to assume that, for all practical purposes, mass is conserved in *chemical* reactions.

Mass Difference in Nuclear Reactions The much larger mass change that accompanies a *nuclear* process is related to the enormous energy required to bind the nucleons together in a nucleus. In an analogy with the calculation above involving the water molecule, consider the change in mass that occurs when one ^{12}C nucleus breaks apart into its nucleons—six protons and six neutrons:

$$^{12}\text{C} \longrightarrow 6\,^1_1\text{p} + 6\,^1_0\text{n}$$

We calculate this mass difference in a special way. By combining the mass of six H *atoms* and six neutrons and then subtracting the mass of one ^{12}C *atom,* the masses of the electrons cancel: six e^- (in six ^1H atoms) cancel six e^- (in one ^{12}C atom). The mass of one ^1H atom is 1.007825 amu, and the mass of one neutron is 1.008665 amu, so

$$\begin{aligned} \text{Mass of six } ^1\text{H atoms} &= 6 \times 1.007825\ \text{amu} = 6.046950\ \text{amu} \\ \text{Mass of six neutrons} &= 6 \times 1.008665\ \text{amu} = 6.051990\ \text{amu} \\ \hline \text{Total mass} &= 12.098940\ \text{amu} \end{aligned}$$

The mass of the reactant, one ^{12}C atom, is 12 amu (exactly). The mass difference (Δm) we obtain is the total mass of the nucleons minus the mass of the nucleus:

$$\begin{aligned} \Delta m &= 12.098940\ \text{amu} - 12.000000\ \text{amu} \\ &= 0.098940\ \text{amu}/^{12}\text{C} = 0.098940\ \text{g/mol}\ ^{12}\text{C} \end{aligned}$$

Two key points emerge from these calculations:

- *The mass of the nucleus is* ***less*** *than the combined masses of its nucleons:* there is *always* a mass decrease when nucleons form a nucleus.
- The mass change of this nuclear process (9.89×10^{-2} g/mol $\approx 10. \times 10^{-2}$, or 0.10 g/mol) is nearly 10 million times what we found earlier for the chemical process of breaking the bonds in water (10.4×10^{-9} g/mol) and could be measured easily on any laboratory balance.

Nuclear Binding Energy and Binding Energy per Nucleon

Einstein's equation for the relation between mass and energy allows us to find the energy equivalent of any mass change. For 1 mol of ^{12}C, after converting grams to kilograms, we have

$$\begin{aligned} \Delta E = \Delta mc^2 &= (9.8940 \times 10^{-5}\ \text{kg/mol})(2.9979 \times 10^8\ \text{m/s})^2 \\ &= 8.8921 \times 10^{12}\ \text{J/mol} = 8.8921 \times 10^9\ \text{kJ/mol} \end{aligned}$$

This quantity of energy is called the **nuclear binding energy** for carbon-12, and the positive value means *energy is absorbed. The nuclear binding energy is the energy required to break 1 mol of nuclei of an element into individual nucleons:*

$$\text{Nucleus} + \text{nuclear binding energy} \longrightarrow \text{nucleons}$$

Thus, the nuclear binding energy is *qualitatively* analogous to the sum of bond energies of a covalent compound or the lattice energy of an ionic compound. But, *quantitatively*, nuclear binding energies are typically several million times greater.

The Electron Volt as the Unit of Binding Energy We use joules to express the binding energy per mole of nuclei, but the joule is much too large a unit to express the binding energy of a single nucleus. Instead, nuclear scientists use the **electron**

volt (eV), the energy an electron acquires when it moves through a potential difference of 1 volt:

$$1\ \text{eV} = 1.602\times10^{-19}\ \text{J}$$

Binding energies are commonly expressed in millions of electron volts, that is, in *mega–electron volts* (MeV):

$$1\ \text{MeV} = 10^6\ \text{eV} = 1.602\times10^{-13}\ \text{J}$$

A particularly useful factor converts the atomic mass unit to its energy equivalent in electron volts:

$$1\ \text{amu} = 931.5\times10^6\ \text{eV} = 931.5\ \text{MeV} \quad \textbf{(24.8)}$$

Nuclear Stability and Binding Energy per Nucleon Earlier we found the mass change when ^{12}C breaks apart into its nucleons to be 0.098940 amu. The binding energy per ^{12}C nucleus, expressed in MeV, is

$$\frac{\text{Binding energy}}{^{12}\text{C nucleus}} = 0.098940\ \text{amu} \times \frac{931.5\ \text{MeV}}{1\ \text{amu}} = 92.16\ \text{MeV}$$

We can compare the stability of nuclides of different elements by determining the *binding energy per nucleon.* For ^{12}C, we have

$$\text{Binding energy per nucleon} = \frac{\text{binding energy}}{\text{no. of nucleons}} = \frac{92.16\ \text{MeV}}{12\ \text{nucleons}} = 7.680\ \text{MeV/nucleon}$$

SAMPLE PROBLEM 24.7 Calculating the Binding Energy per Nucleon

Problem Iron-56 is an extremely stable nuclide. Compute the binding energy per nucleon for ^{56}Fe and compare it with that for ^{12}C (mass of ^{56}Fe atom = 55.934939 amu; mass of ^{1}H atom = 1.007825 amu; mass of neutron = 1.008665 amu).

Plan Iron-56 has 26 protons and 30 neutrons. We calculate the mass difference, Δm, when the nucleus forms by subtracting the given mass of one ^{56}Fe atom from the sum of the masses of 26 ^{1}H atoms and 30 neutrons. To find the binding energy per nucleon, we multiply Δm by the equivalent in MeV (931.5 MeV/amu) and divide by the number of nucleons (56).

Solution Calculating the mass difference, Δm:

$$\begin{aligned}\text{Mass difference} &= [(26 \times \text{mass } ^{1}\text{H atom}) + (30 \times \text{mass neutron})] - \text{mass } ^{56}\text{Fe atom}\\ &= [(26)(1.007825\ \text{amu}) + (30)(1.008665\ \text{amu})] - 55.934939\ \text{amu}\\ &= 0.52846\ \text{amu}\end{aligned}$$

Calculating the binding energy per nucleon:

$$\text{Binding energy per nucleon} = \frac{0.52846\ \text{amu} \times 931.5\ \text{MeV/amu}}{56\ \text{nucleons}} = 8.790\ \text{MeV/nucleon}$$

An ^{56}Fe nucleus would require more energy per nucleon to break up into its nucleons than would a ^{12}C nucleus (7.680 MeV/nucleon), so ^{56}Fe is more stable than ^{12}C.

Check The answer is consistent with the great stability of ^{56}Fe. Given the number of decimal places in the values, rounding to check the math is useful only to find a *major* error. The number of nucleons (56) is exact, so we retain four significant figures.

FOLLOW-UP PROBLEMS

24.7A Enormous amounts of nickel and iron make up Earth's core, and nickel-58 is a very stable nuclide. Find the binding energy per nucleon for ^{58}Ni (mass of ^{58}Ni atom = 57.935346 amu), and compare it with the value for ^{56}Fe (8.790 MeV/nucleon).

24.7B Uranium-235 is an essential component of the fuel in nuclear power plants. Calculate the binding energy per nucleon for ^{235}U (mass of ^{235}U atom = 235.043924 amu). Is this nuclide more or less stable than ^{12}C (whose value is 7.680 MeV/nucleon)?

SOME SIMILAR PROBLEMS 24.79–24.82

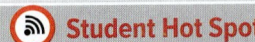

Student Hot Spot

Student data indicate that you may struggle with binding energy calculations. Access the Smartbook to view additional Learning Resources on this topic.

Figure 24.16 The variation in binding energy per nucleon.

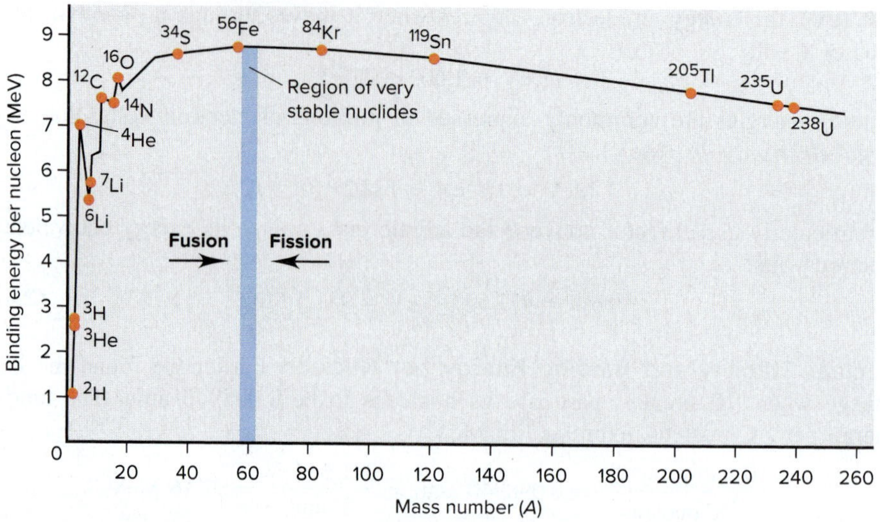

Fission or Fusion: Increasing the Binding Energy per Nucleon Calculations similar to those in Sample Problem 24.7 for other nuclides show that the binding energy per nucleon varies considerably. The essential point is that *the greater the binding energy per nucleon, the more stable the nuclide is.*

Figure 24.16 shows a plot of the binding energy per nucleon vs. mass number. It provides information about nuclide stability and the two possible processes—fission and fusion—that nuclides undergo to form more stable nuclides. Most nuclides with fewer than 10 nucleons have a relatively small binding energy per nucleon. The ^{4}He nucleus is an exception—it is stable enough to be emitted intact as an α particle. Above A = 12, the binding energy per nucleon varies from about 7.6 to 8.8 MeV.

The most important observation is that *the binding energy per nucleon peaks at elements with $A \approx 60$.* In other words, nuclides become more stable with increasing mass number up to around 60 nucleons and then become less stable with higher numbers of nucleons. The existence of a peak of stability suggests that there are two ways nuclides can increase their binding energy per nucleon:

- *Fission.* A heavier nucleus can *split into lighter nuclei (closer to $A \approx 60$)* by undergoing fission. The product nuclei have greater binding energy per nucleon (are more stable) than the reactant nucleus, and the difference in *energy is released.* Nuclear power plants generate energy through fission, as do atomic bombs (Section 24.7).
- *Fusion.* Lighter nuclei, on the other hand, can *combine to form a heavier nucleus (closer to $A \approx 60$)* by undergoing fusion. Once again, the product is more stable than the reactants, and *energy is released.* The Sun and other stars generate energy through fusion, as do thermonuclear (hydrogen) bombs. In these examples and in all current research efforts for developing fusion as a useful energy source, hydrogen nuclei fuse to form the very stable helium-4 nucleus.

In Section 24.7, we examine fission and fusion and their applications.

› Summary of Section 24.6

› The mass of a nucleus is less than the sum of the masses of its nucleons. The energy equivalent to this mass difference is the nuclear binding energy, often expressed in units of MeV.

› The binding energy per nucleon is a measure of nuclide stability and varies with the number of nucleons. Nuclides with $A \approx 60$ are most stable.

› Lighter nuclides join (fusion) or heavier nuclides split (fission) to create more stable products.

24.7 APPLICATIONS OF FISSION AND FUSION

Of the many beneficial applications of nuclear reactions, the greatest is the potential for abundant quantities of energy, which is based on the multimillion-fold increase in energy yield of nuclear reactions over chemical reactions. Our experience with nuclear energy from power plants in the last quarter of the 20th century, however, has shown that we must improve ways to tap this energy source safely and economically and deal with the waste generated. In this section, we discuss how fission and fusion occur and how we are applying them.

The Process of Nuclear Fission

During the mid-1930s, Enrico Fermi and coworkers bombarded uranium ($Z = 92$) with neutrons in an attempt to synthesize transuranium elements. Many of the unstable nuclides produced were tentatively identified as having $Z > 92$, but other scientists were skeptical. Four years later, the German chemist Otto Hahn and his associate F. Strassmann showed that one of these unstable nuclides was an isotope of barium ($Z = 56$). The Austrian physicist Lise Meitner, a coworker of Hahn, and her nephew Otto Frisch proposed that barium resulted from the *splitting* of the uranium nucleus into *smaller* nuclei, a process that they called *fission* as an analogy to cell division in biology. Element 109 was named *meitnerium* in honor of this extraordinary physicist who proposed the correct explanation of nuclear fission and discovered the element protactinium (Pa; $Z = 91$) as well as numerous radioisotopes.

The ^{235}U nucleus can split in many different ways, giving rise to various daughter nuclei, but all routes have the same general features. Figure 24.17 depicts *one* of these fission patterns, the production of isotopes of Kr and Ba:

$$^{235}_{92}\text{U} + {}^{1}_{0}\text{n} \longrightarrow {}^{92}_{36}\text{Kr} + {}^{141}_{56}\text{Ba} + 3\,{}^{1}_{0}\text{n}$$

Neutron bombardment results in a highly excited ^{236}U nucleus, which splits apart in 10^{-14} s. The products are two nuclei of unequal mass, two to four neutrons (average of 2.4), and a large quantity of energy. A single ^{235}U nucleus releases 3.5×10^{-11} J when it splits; 1 mol of ^{235}U (about $\frac{1}{2}$ lb) releases 2.1×10^{13} J—a billion times as much energy as burning $\frac{1}{2}$ lb of coal (2×10^{4} J)!

Chain Reaction and Critical Mass We harness the energy of nuclear fission, much of which eventually appears as heat, by means of a **chain reaction** (Figure 24.18, *next page*): the few neutrons that are released by the fission of one nucleus collide with other fissionable nuclei and cause them to split, releasing more neutrons, and so on, in a self-sustaining process (in Figure 24.18, vertical dashed lines separate the steps). In this manner, the energy released increases rapidly because each fission event in a chain reaction releases about two-and-a-half times as much energy as the preceding one.

Whether a chain reaction occurs depends on the mass (and thus the volume) of the fissionable sample. If the piece of uranium is large enough, the product neutrons strike another fissionable nucleus *before* flying out of the sample, and a chain reaction takes place. The mass required to achieve a chain reaction is called the **critical mass.** If the sample has less than the critical mass (a *subcritical mass*), too many product

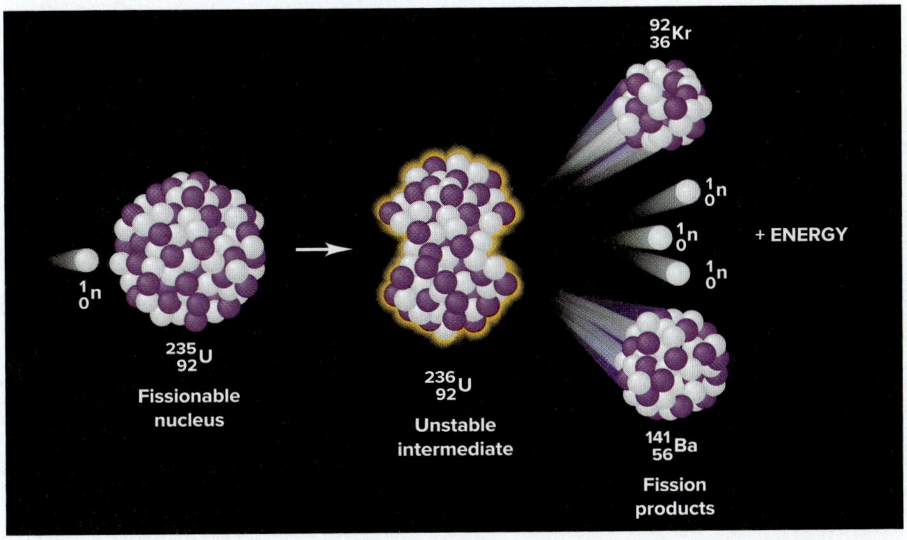

Figure 24.17 Fission of ^{235}U caused by neutron bombardment.

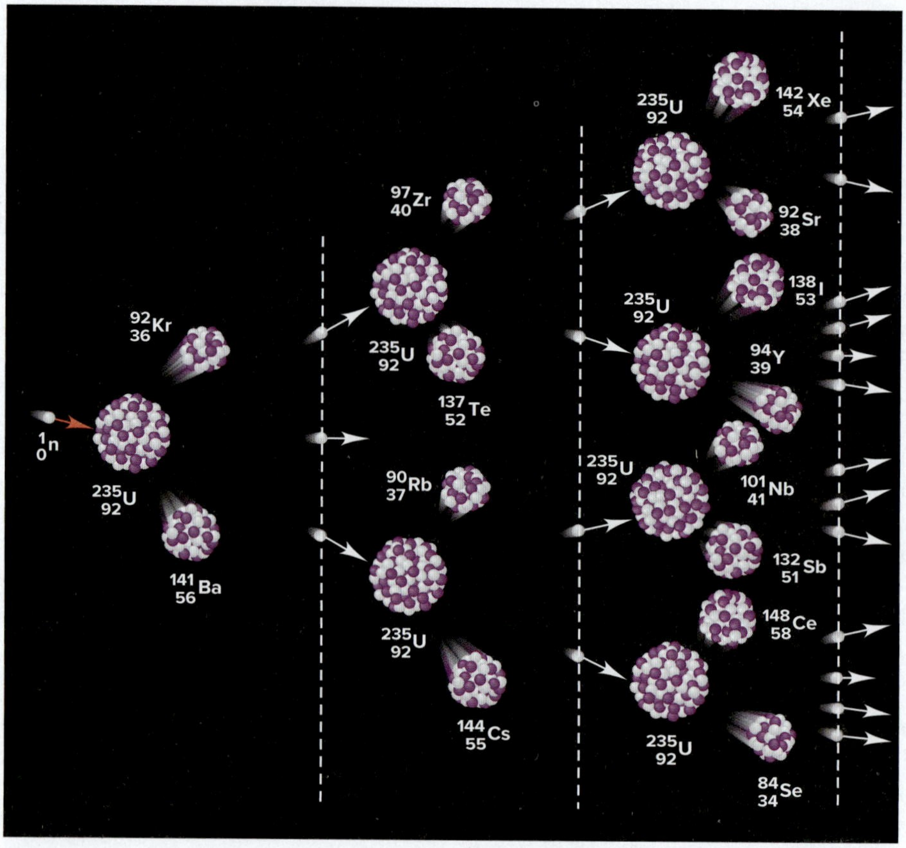

Figure 24.18 A chain reaction involving fission of ^{235}U.

neutrons leave the sample before they collide with, and cause the fission of, another ^{235}U nucleus, and thus a chain reaction does not occur.

Uncontrolled Fission: The Atomic Bomb An uncontrolled chain reaction can be adapted to make an extremely powerful explosive, as several of the world's leading atomic physicists hypothesized just prior to World War II. In August 1939, Albert Einstein wrote the president of the United States, Franklin Delano Roosevelt, to this effect, warning of the danger of allowing the Nazi government of Germany to develop this power first. It was this concern that led to the Manhattan Project, an enormous scientific effort to develop a bomb based on nuclear fission, which was initiated in 1941.* In August 1945, the United States detonated two atomic bombs over Japan, and the horrible destructive power of these bombs was a major factor in the surrender of the Japanese a few days later.

In an atomic bomb, small explosions of trinitrotoluene (TNT) bring subcritical masses of fissionable material together to exceed the critical mass, and the ensuing chain reaction brings about the explosion (Figure 24.19). The proliferation of nuclear power plants, which use fissionable materials to generate energy for electricity, has increased concern that more countries (and unscrupulous individuals) may have access to such material for making bombs. Since the devastating terrorist attacks of September 11, 2001 in the United States, this concern has been heightened.

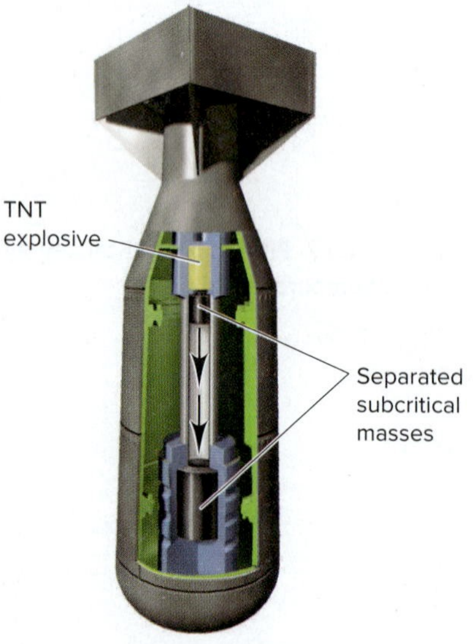

Figure 24.19 An atomic bomb based on ^{235}U.

Controlled Fission: Nuclear Energy Reactors Like a coal-fired power plant, *a nuclear power plant generates heat to produce steam, which turns a turbine attached to an electric generator*. But a nuclear plant has the potential to produce electric power much more cleanly than by the combustion of coal.

1. *Operation of a nuclear power plant.* Heat generation takes place in the **reactor core** of a nuclear plant (Figure 24.20). The core contains the *fuel rods,* which consist

*For an excellent scientific and historical account of the development of the atomic bomb, see R. Rhodes, *The Making of the Atomic Bomb*, New York, Simon and Schuster, 1986.

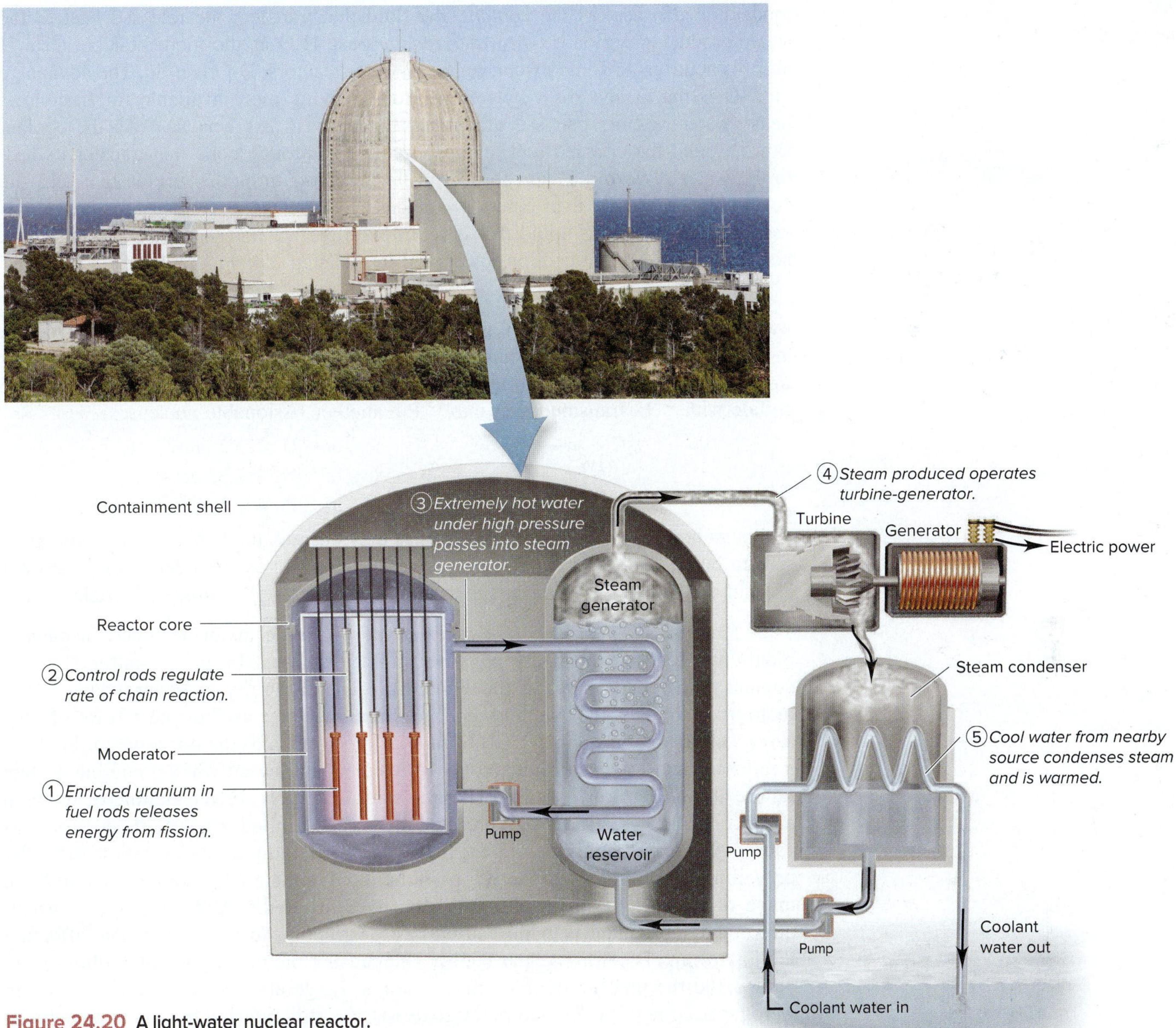

Figure 24.20 A light-water nuclear reactor.
Source: © Philip Lange/Shutterstock.com

of fuel enclosed in tubes of a corrosion-resistant zirconium alloy. The fuel is uranium(IV) oxide (UO_2) that has been *enriched* from 0.7% ^{235}U, the natural abundance of this fissionable isotope, to the 3% to 4% ^{235}U required to sustain a chain reaction in a practical volume. (Enrichment of nuclear fuel is the most important application of Graham's law; see Section 5.5.) Sandwiched between the fuel rods are movable *control rods* made of cadmium or boron (or, in nuclear submarines, hafnium), substances that absorb neutrons very efficiently. For example, a neutron absorption reaction for boron is

$$^{10}_{5}B + ^{1}_{0}n \longrightarrow ^{7}_{3}Li + ^{4}_{2}He$$

When the control rods are lowered between the fuel rods, the chain reaction slows because fewer neutrons are available to bombard uranium atoms; when they are raised, the chain reaction speeds up. Neutrons that leave the fuel-rod assembly collide with a *reflector,* usually made of a beryllium alloy, which absorbs very few neutrons. Reflecting the neutrons back to the fuel rods speeds the chain reaction.

Flowing around the fuel and control rods in the reactor core is the *moderator,* a substance that slows the neutrons, making them much better at causing fission than the fast ones emerging directly from the fission event. In most modern reactors, the moderator also acts as the *coolant,* the fluid that transfers the released heat to the steam-producing region. *Light-water reactors* use H_2O as the moderator because ^1H absorbs neutrons to some extent; in heavy-water reactors, D_2O is used. The advantage of D_2O is that it absorbs very few neutrons, leaving more available for fission, so heavy-water reactors can use uranium that has been *less enriched.* As the coolant flows around the encased fuel, pumps circulate it through coils that transfer its heat to the water reservoir. Steam formed in the reservoir turns the turbine that runs the generator. The steam then enters a condenser, where water from a lake or river is used to cool it back to liquid that is returned to the water reservoir. Often (although not shown here), large cooling towers aid in this step (see Figure 13.20).

2. *Breeder reactors.* Because ^{235}U is not an abundant isotope, the *breeder reactor* was designed to consume one type of nuclear fuel as it produces another. Outside the moderator, fuel rods are surrounded by natural U_3O_8, which contains 99.3% *nonfissionable* ^{238}U atoms. As neutrons formed during ^{235}U fission escape the fuel rod, they collide with ^{238}U, transmuting it into ^{239}Pu, another fissionable nucleus:

$$^{238}_{92}U + ^{1}_{0}n \longrightarrow ^{239}_{92}U \qquad (t_{1/2} \text{ of } ^{239}_{92}U = 23.5 \text{ min})$$
$$^{239}_{92}U \longrightarrow ^{239}_{93}Np + ^{0}_{-1}\beta \qquad (t_{1/2} \text{ of } ^{239}_{93}Np = 2.35 \text{ days})$$
$$^{239}_{93}Np \longrightarrow ^{239}_{94}Pu + ^{0}_{-1}\beta \qquad (t_{1/2} \text{ of } ^{239}_{94}Pu = 2.4\times10^4 \text{ yr})$$

Although breeder reactors can make fuel as they operate, they are difficult and expensive to build, and ^{239}Pu is extremely toxic and long lived. Breeder reactors are not used in the United States, although several are operating in Europe and Asia.

3. *Power plant accidents and other concerns.* Some major accidents at nuclear plants have caused decidedly negative public reactions. In 1979, malfunctions of coolant pumps and valves at the Three-Mile Island facility in Pennsylvania led to melting of some of the fuel and damage to the reactor core, but the release of only a very small amount (about 1 Ci) of radioactive gases into the atmosphere. In 1986, a million times as much radioactivity (1 MCi) was released when a cooling system failure at the Chernobyl plant in Ukraine caused a much greater melting of fuel and an uncontrolled reaction. High-pressure steam and ignited graphite moderator rods caused the reactor building to explode and expel radioactive debris. Carried by prevailing winds, the radioactive particles contaminated vegetables and milk in much of Europe. Severe health problems have resulted.* And, in 2011, an earthquake and subsequent tsunami caused vast destruction in northern Japan, affecting the Fukushima Dai-ichi nuclear facility and causing the breakdown of cooling-water pumps and the melting of three reactor cores. Large amounts of radioactive steam were released into the air, and radioactive water is still entering the sea, with serious, long-term health effects expected.

4. *Current use of nuclear power.* Despite potential safety problems, nuclear power remains an important source of electricity. In the late 1990s, nearly every European country employed nuclear power, and it is the major power source in some countries—Sweden creates 50% of its electricity this way and France almost 80%. Currently, the United States obtains about 20% of its electricity from nuclear power, and Canada slightly less. As our need for energy grows and climate change from fossil-fuel consumption worsens, safer reactors will be designed and built. However, as a result of the disaster in Japan, many countries are reevaluating their use of nuclear energy. For example, Germany has commited to closing all its nuclear power plants by 2022.

5. *Thermal pollution and waste disposal.* Even a smoothly operating nuclear power plant has certain inherent problems. The problem of *thermal pollution* is common to all power plants. Water used to condense the steam is several degrees warmer when returned to its source, which can harm aquatic organisms (Section 13.4). A more serious problem

---

*An even greater release of radioactivity (1.1 MCi), involving several radioisotopes including the extremely dangerous ^{90}Sr, occurred in 1957 in Kyshtym, a town in the South Ural Mountains of Russia. As a result, radioactivity spread globally, but the accident was kept secret by the former Soviet government until 1980.

is *nuclear waste disposal.* Many of the fission products formed in nuclear reactors have long half-lives, and no satisfactory plan for their permanent disposal has yet been devised. Proposals to place the waste in containers and bury them in deep bedrock cannot possibly be field-tested for the thousands of years the material will remain harmful. Leakage of radioactive material into groundwater is a danger, and earthquakes can occur even in geologically stable regions. It remains to be seen whether we can operate fission reactors *and* dispose of the waste safely and economically.

The Promise of Nuclear Fusion

Nuclear fusion in the Sun is the ultimate source of nearly all the energy—and chemical elements—on Earth. In fact, *all the elements heavier than hydrogen were formed in fusion and decay processes within stars,* as the upcoming Chemical Connections essay describes.

Much research is being devoted to making nuclear fusion a practical, direct source of energy on Earth. To understand the advantages of fusion, let's consider one of the most discussed fusion reactions, in which deuterium and tritium react:

$$^{2}_{1}\text{H} + ^{3}_{1}\text{H} \longrightarrow ^{4}_{2}\text{He} + ^{1}_{0}\text{n}$$

This reaction produces 1.7×10^{9} kJ/mol, an enormous quantity of energy with no radioactive byproducts. Moreover, the reactant nuclei are relatively easy to come by. We obtain deuterium from the electrolysis of water (Section 22.4). In nature, tritium forms through the cosmic (neutron) irradiation of ^{14}N:

$$^{14}_{7}\text{N} + ^{1}_{0}\text{n} \longrightarrow ^{3}_{1}\text{H} + ^{12}_{6}\text{C}$$

However, this process results in a natural abundance of only 10^{-7}% ^{3}H. More practically, tritium is produced in nuclear accelerators by bombarding lithium-6 or by surrounding the fusion reactor itself with material containing lithium-6:

$$^{6}_{3}\text{Li} + ^{1}_{0}\text{n} \longrightarrow ^{3}_{1}\text{H} + ^{4}_{2}\text{He}$$

Thus, in principle, fusion *seems* promising and may represent an ideal source of power. However, some extremely difficult technical problems remain. Fusion requires enormous energy in the form of heat to give the positively charged nuclei enough kinetic energy to force themselves together. The fusion of deuterium and tritium, for example, occurs at practical rates at about 10^{8} K, a temperature hotter than the Sun's core! How can such conditions be achieved?

Two current research approaches have promise. In one, atoms are stripped of their electrons at high temperatures, which results in a gaseous *plasma,* a neutral mixture of positive nuclei and electrons. Because of the extreme temperatures needed for fusion, no *material* can contain the plasma. The most successful approach to date has been to enclose the plasma within a magnetic field. The *tokamak* design has a donut-shaped container in which a helical magnetic field confines the plasma and prevents it from contacting the walls (Figure 24.21). Scientists at the Princeton University Plasma Physics facility have achieved some success in generating energy from fusion this way. In another approach, the required high temperature is reached by using many focused lasers to compress and heat the fusion reactants. In any event, one or more major breakthroughs are needed before fusion will be a practical, everyday source of energy.

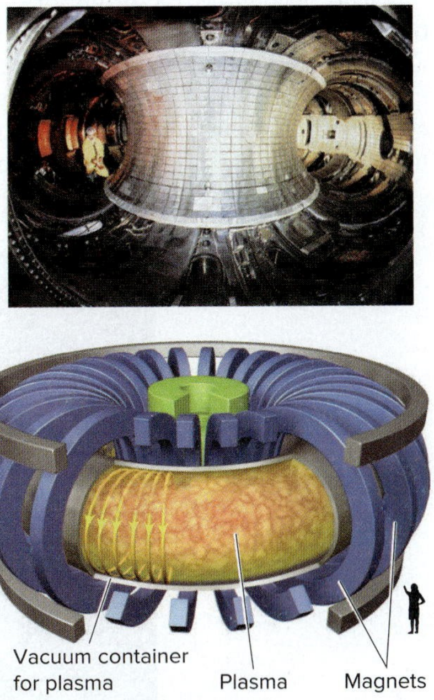

Figure 24.21 **The tokamak design for magnetic containment of a fusion plasma.**
Source: Dietmar Krause/Princeton Plasma Physics Lab

› Summary of Section 24.7

- In nuclear fission, neutron bombardment causes a nucleus to split into two smaller nuclei and release neutrons that split other nuclei, giving rise to a chain reaction.
- A nuclear power plant controls the rate of the chain reaction to produce heat that creates steam, which is used to generate electricity.
- Potential hazards, such as radiation leaks, thermal pollution, and disposal of nuclear waste, remain current concerns.
- Nuclear fusion holds great promise as a source of clean abundant energy, but it requires extremely high temperatures and is not yet practical.
- The elements were formed through a complex series of nuclear reactions in evolving stars.

Origin of the Elements in the Stars

How did the universe begin? Where did matter come from? How were the elements formed? The most accepted model proposes that a sphere of unimaginable properties—diameter of 10^{-28} cm, density of 10^{96} g/mL (density of a nucleus $\approx 10^{14}$ g/mL), and temperature of 10^{32} K—exploded in a "Big Bang," for reasons not yet known, and distributed its contents through the void of space. Cosmologists consider this moment the beginning of time.

One second later, the universe was an expanding mixture of neutrons, protons, and electrons, denser than rock and hotter than an exploding hydrogen bomb (about 10^{10} K). In the next few minutes, it became a gigantic fusion reactor creating the first atomic nuclei other than ^{1}H: ^{2}H, ^{3}He, and ^{4}He. After 10 min, more than 25% of the mass of the universe existed as ^{4}He, and only about 0.025% as ^{2}H. About 100 million years later, or almost 14 billion years ago, gravitational forces pulled this cosmic mixture into primitive, contracting stars.

This account of the origin of the universe is based on the observation of spectra from the Sun, other stars, nearby galaxies, and cosmic (interstellar) dust. Spectral analysis of planets and chemical analysis of Earth and Moon rocks, meteorites, and cosmic-ray particles furnish data about isotope abundance. From these, a model has been developed for **stellar nucleosynthesis,** the creation of the elements in stars. The overall process occurs in several stages during a star's evolution, and the entire sequence of steps occurs only in very massive stars, having 10 to 100 times the mass of the Sun. Each step involves a contraction of the volume of the star that results in higher temperature and yields heavier nuclei. Such events are forming elements in stars today. The key stages in the process are shown in Figure B24.1 and described below:

1. *Hydrogen burning produces He.* The initial contraction of a star heats its core to about 10^7 K, at which point a fusion process called *hydrogen burning* begins. Through three possible reactions (one is shown below), helium nuclei are produced from the abundant protons:

$$4\,^{1}_{1}\text{H} \longrightarrow \,^{4}_{2}\text{He} + 2\,^{0}_{1}\beta + 2\nu + \text{energy}$$

2. *Helium burning produces C, O, Ne, and Mg.* After several billion years of hydrogen burning, about 10% of the ^{1}H is consumed, and the star contracts further. The ^{4}He forms a dense core, hot enough (2×10^{8} K) to fuse ^{4}He nuclei. The energy released during *helium burning* expands the remaining ^{1}H into a vast envelope: the star becomes a *red giant,* more than 100 times its original diameter. Within its core, pairs of ^{4}He nuclei (α particles) fuse into

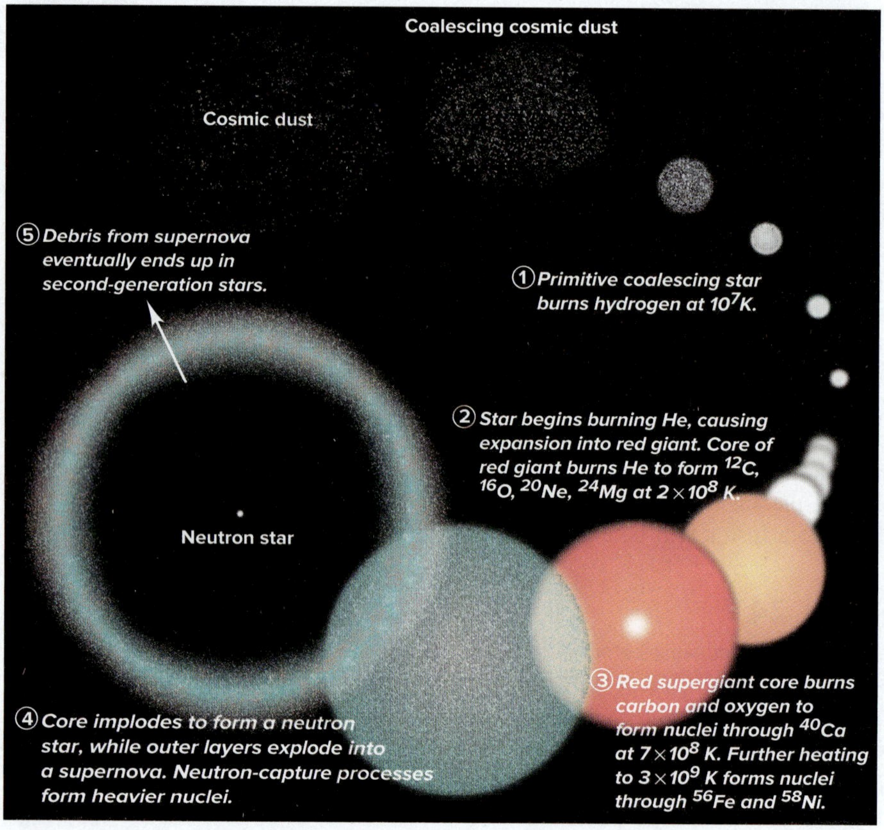

Coalescing cosmic dust
Cosmic dust
⑤ Debris from supernova eventually ends up in second-generation stars.
① Primitive coalescing star burns hydrogen at 10⁷K.
② Star begins burning He, causing expansion into red giant. Core of red giant burns He to form ¹²C, ¹⁶O, ²⁰Ne, ²⁴Mg at 2×10⁸ K.
Neutron star
③ Red supergiant core burns carbon and oxygen to form nuclei through ⁴⁰Ca at 7×10⁸ K. Further heating to 3×10⁹ K forms nuclei through ⁵⁶Fe and ⁵⁸Ni.
④ Core implodes to form a neutron star, while outer layers explode into a supernova. Neutron-capture processes form heavier nuclei.

Figure B24.1 Element synthesis in the life cycle of a star.

unstable ^{8}Be nuclei ($t_{1/2} = 7\times10^{-17}$ s), which play the role of an activated complex in chemical reactions. A ^{8}Be nucleus collides with another alpha particle to form ^{12}C. Then, further fusion with more alpha particles creates nuclei up to ^{24}Mg:

$$^{12}C \xrightarrow{\alpha} {}^{16}O \xrightarrow{\alpha} {}^{20}Ne \xrightarrow{\alpha} {}^{24}Mg$$

3. *Elements through Fe and Ni form.* For another 10 million years, ^{4}He is consumed, and the heavier nuclei created form a core. This core contracts and heats, expanding the star into a *supergiant.* Within the hot core (7×10^{8} K), *carbon and oxygen burning* occur:

$$^{12}C + {}^{12}C \longrightarrow {}^{23}Na + {}^{1}H$$
$$^{12}C + {}^{16}O \longrightarrow {}^{28}Si + \gamma$$

Absorption of α particles forms nuclei up to ^{40}Ca:

$$^{12}C \xrightarrow{\alpha} {}^{16}O \xrightarrow{\alpha} {}^{20}Ne \xrightarrow{\alpha} {}^{24}Mg \xrightarrow{\alpha}$$
$$^{28}Si \xrightarrow{\alpha} {}^{32}S \xrightarrow{\alpha} {}^{36}Ar \xrightarrow{\alpha} {}^{40}Ca$$

Further contraction and heating to a temperature of 3×10^{9} K allow reactions in which nuclei release neutrons, protons, and α particles and then recapture them. As a result, nuclei with lower binding energies supply nucleons to create those with higher binding energies. In stars of moderate mass, less than 10 times the mass of the Sun, this process stops at iron-56 and nickel-58, the nuclei with the highest binding energies.

4. *Heavier elements form.* In very massive stars, the next stage is the most spectacular. With all the fuel consumed, the core collapses within a second. Many Fe and Ni nuclei break down into neutrons and protons. Protons capture electrons to form neutrons, and the entire core forms an incredibly dense *neutron star.* (An Earth-sized star that became a neutron star would fit in the Houston Astrodome!) As the core implodes, the outer layers explode in a *supernova,* which expels material throughout space. A supernova occurs an average of every few hundred years in each galaxy or once every second somewhere in the universe. The heavier elements are formed during supernova events and are found in *second-generation stars,* those that coalesce from interstellar ^{1}H and ^{4}He and the debris of exploded first-generation stars.

Heavier elements form through *neutron-capture* processes. In the *s-process* (*slow* neutron-capture process), a nucleus captures a neutron, at some point over a period of 10 to 1000 years. If the resulting nucleus is unstable, it undergoes β^- decay to form the next element, as in this conversion of ^{68}Zn to ^{70}Ge:

$$^{68}Zn \xrightarrow{n} {}^{69}Zn \xrightarrow{\beta^-} {}^{69}Ga \xrightarrow{n} {}^{70}Ga \xrightarrow{\beta^-} {}^{70}Ge$$

The stable isotopes of most heavy elements up to ^{209}Bi form by the s-process.

Less stable isotopes and those with A greater than 230 cannot form by the s-process because their half-lives are too short. These form by the *r-process* (*rapid* neutron-capture process) during the fury of the supernova. Many neutron captures, followed by many β^- decays, occur in a second, as when ^{56}Fe is converted to ^{79}Br:

$$^{56}_{26}Fe + 23^{1}_{0}n \longrightarrow {}^{79}_{26}Fe \longrightarrow {}^{79}_{35}Br + 9\,{}^{0}_{-1}\beta$$

We know from the heavy elements present in the Sun that it is at least a second-generation star presently undergoing hydrogen burning. Together with its planets, it was formed from the dust of exploded stars about 4.6×10^{9} years ago. This means that many of the atoms on Earth, including some within you, came from exploded stars and are older than the Solar System itself!

Any theory of element formation must be consistent with the element abundances we observe (Section 22.1). Although local compositions, such as those of Earth and the Sun, differ, and mineral grains in meteorites may have abnormal isotopic compositions, large regions of the universe have, on average, similar compositions. Scientists believe element forming reaches a dynamic equilibrium, resulting in *relatively constant amounts of the isotopes.*

Problems

B24.1 Compare the s- and r-processes of stellar nucleosynthesis in terms of rate, number of neutrons absorbed, and types of isotopes formed.

B24.2 The overall reaction taking place during hydrogen burning in a young star is

$$4^{1}_{1}H \longrightarrow {}^{4}_{2}He + 2^{0}_{1}\beta + 2\nu + \text{energy}$$

How much energy (in MeV) is released per He nucleus formed? Per mole of He formed? (Mass of $^{1}_{1}H$ atom = 1.007825 amu; mass of $^{4}_{2}He$ atom = 4.00260 amu; mass of positron = 5.48580×10^{-4} amu.)

B24.3 Cosmologists modeling the origin of the elements postulate nuclides with very short half-lives. One of these short-lived nuclides, ^{8}Be ($t_{1/2} = 7\times10^{-17}$ s), is thought to play a key role in stellar nucleosynthesis by fusing with ^{4}He to form ^{12}C. Another model proposes the simultaneous fusion of three ^{4}He nuclei to form ^{12}C. Comment on the validity of this alternative mechanism.

B24.4 The s-process produces isotopes of elements up to ^{209}Bi. This process is terminated when ^{209}Bi captures a neutron to produce ^{210}Bi, which subsequently undergoes β^- decay to produce nuclide A, which then decays by emitting an α particle to form nuclide B. This nuclide captures three neutrons to produce nuclide C, which undergoes β^- decay to produce nuclide D, starting the cycle again. Write balanced nuclear equations for the formation of nuclides A through D.

CHAPTER REVIEW GUIDE

Learning Objectives

Relevant section (§) and/or sample problem (SP) numbers appear in parentheses.

Understand These Concepts

1. How nuclear changes differ, in general, from chemical changes (§24.1)
2. The meanings of *radioactivity, nucleon, nuclide,* and *isotope* (§24.1)
3. Characteristics of three types of radioactive emissions: α, β, and γ (§24.1)
4. The various modes of radioactive decay and how each changes the values of A and Z (§24.1)
5. How the N/Z ratio, the even-odd nature of N and Z, and magic numbers correlate with nuclear stability (§24.1)
6. How an unstable nuclide's mass number or N/Z ratio correlates with its mode of decay (§24.1)
7. How a decay series combines numerous decay steps and ends with a stable nuclide (§24.1)
8. How ionization and scintillation counters detect and measure radioactivity (§24.2)
9. Why radioactive decay is a first-order process; the meanings of *decay rate* and *specific activity* (§24.2)
10. The meaning of *half-life* in the context of radioactive decay (§24.2)
11. How the specific activity of an isotope in an object is used to determine the object's age (§24.2)
12. How particle accelerators and reactors are used to synthesize new nuclides (§24.3)
13. The units of radiation dose; the effects on living tissue of various dosage levels; the inverse relationship between the mass and charge of an emission and its penetrating power (§24.4)
14. How ionizing radiation creates free radicals that damage tissue; sources and risks of ionizing radiation (§24.4)
15. How radioisotopes are used in research, analysis, and medical diagnosis (§24.5)
16. Why the mass of a nucleus is less than the total mass of its nucleons and how this mass difference is related to the nuclear binding energy (§24.6)
17. How nuclear stability is related to binding energy per nucleon (§24.6)
18. How heavy nuclides undergo fission and lighter ones undergo fusion to increase the binding energy per nucleon (§24.6)
19. The current application of fission and potential application of fusion to produce energy (§24.7)

Master These Skills

1. Expressing the mass and charge of a particle with the ${}^{A}_{Z}X$ notation (§24.1; see also §2.5)
2. Using changes in the values of A and Z to write and balance nuclear equations (SP 24.1)
3. Using the N/Z ratio, the even-odd nature of N and Z, and the presence of magic numbers to predict nuclear stability (SP 24.2)
4. Using the atomic mass of the element to predict the mode of decay of an unstable nuclide and checking the prediction with the N/Z ratio (SP 24.3)
5. Converting units of radioactivity (§24.2)
6. Calculating specific activity, decay constant, half-life, and number of radioactive nuclei (§24.2 and SPs 24.4 and 24.5)
7. Estimating the age of an object from the specific activity and half-life of carbon-14 (SP 24.6)
8. Writing notations for nuclear transmutations (§24.3)
9. Calculating radiation dose and converting units (§24.4)
10. Calculating the mass difference between a nucleus and its nucleons and the energy equivalent (§24.6)
11. Calculating the binding energy per nucleon and using it to compare stabilities of nuclides (SP 24.7)

Key Terms

Page numbers appear in parentheses.

activity ($\mathscr{A}$) (1084)
alpha (α) decay (1077)
alpha (α) particle (1075)
background radiation (1095)
band of stability (1079)
becquerel (Bq) (1084)
beta (β) decay (1077)
β^- decay (1077)
beta (β) particle (1075)
chain reaction (1105)
critical mass (1105)
curie (Ci) (1084)
decay constant (1085)
decay (disintegration) series (1082)
deuteron (1091)
electron (e^-) capture (EC) (1077)
electron volt (eV) (1102)
fission (1101)
free radical (1095)
fusion (1101)
gamma (γ) emission (1078)
gamma ray (γ) (1075)
Geiger-Müller counter (1083)
gray (Gy) (1094)
half-life ($t_{1/2}$) (1086)
ionization (1093)
ionizing radiation (1093)
isotope (1074)
nuclear binding energy (1102)
nuclear transmutation (1090)
nucleon (1074)
nuclide (1074)
particle accelerator (1091)
positron (1077)
positron (β^+) emission (1077)
radioactivity (1073)
radioisotope (1087)
radioisotopic dating (1087)
rad (*r*adiation-*a*bsorbed *d*ose) (1094)
reactor core (1106)
rem (*r*oentgen *e*quivalent for *m*an) (1094)
scintillation counter (1083)
sievert (Sv) (1094)
stellar nucleosynthesis (1110)
strong force (1080)
tracer (1098)
transuranium element (1092)

Key Equations and Relationships

Page numbers appear in parentheses.

24.1 Balancing a nuclear equation (1075):

$$^{\text{Total } A}_{\text{Total } Z}\text{Reactants} = {}^{\text{Total } A}_{\text{Total } Z}\text{Products}$$

24.2 Expressing the decay rate (activity) for radioactive nuclei (1085):

$$\text{Decay rate } (\mathscr{A}) = -\frac{\Delta \mathscr{N}}{\Delta t} = k\mathscr{N}$$

24.3 Finding the number of nuclei remaining after a given time, $\mathcal{N}_t$ (1086):

$$\ln \frac{\mathcal{N}_t}{\mathcal{N}_0} = -kt \quad \text{or} \quad \mathcal{N}_t = \mathcal{N}_0 e^{-kt}$$

24.4 Finding the half-life of a radioactive nuclide (1086):

$$t_{1/2} = \frac{\ln 2}{k}$$

24.5 Calculating the time to reach a given specific activity (age of an object in radioisotopic dating) (1088):

$$t = \frac{1}{k} \ln \frac{\mathcal{A}_0}{\mathcal{A}_t}$$

24.6 Writing a transmutation in shorthand notation (1090):

reactant nucleus (particle in, particle(s) out) product nucleus

24.7 Adapting Einstein's equation to calculate mass difference and/or nuclear binding energy (1101):

$$\Delta m = \frac{\Delta E}{c^2} \quad \text{or} \quad \Delta E = \Delta mc^2$$

24.8 Relating the atomic mass unit to its energy equivalent in MeV (1103):

$$1 \text{ amu} = 931.5\times10^6 \text{ eV} = 931.5 \text{ MeV}$$

BRIEF SOLUTIONS TO FOLLOW-UP PROBLEMS

24.1A $^{20}_{9}\text{F} \longrightarrow {}^{20}_{10}\text{Ne} + {}^{0}_{-1}\beta$

24.1B $^{124}_{53}\text{I} \longrightarrow {}^{124}_{52}\text{Te} + {}^{0}_{1}\beta$

24.2A (a) Stable. Like many light nuclides (those with $Z < 20$) for which $N = Z$, this one ($Z = 5$) is stable (one of four stable nuclides with both an odd Z and an odd N). (b) Radioactive. This nuclide has both odd N (35) and odd Z (23) and $N/Z = 1.52$, which is too high for this region of the band of stability.

24.2B ^{31}P has an even N (16), but ^{30}P has both N and Z odd. ^{31}P also has a slightly higher N/Z ratio that is closer to the band of stability.

24.3A (a) Iron (Fe) has an atomic mass of 55.85 amu. The A value of 61 is higher: β^- decay. (b) $Z > 83$, which is too high for stability: α decay will occur.

24.3B (a) Titanium's atomic mass is 47.87 amu, which is much higher than the A value of 40; this nuclide most likely decays by electron capture or β^+ emission. (b) Cobalt's atomic mass is 58.93 amu, which is much lower than the A value of 65; it will probably undergo β^- decay.

24.4A (a)
$$\text{Mass (g)} = 3.4\times10^{-8} \text{ mol } ^{76}\text{As} \times 76 \text{ g/mol} = 2.6\times10^{-6} \text{ g } ^{76}\text{As}$$

$$\text{Specific activity (Ci/g)} = 1.53\times10^{11} \text{ d/s} \times \frac{1 \text{ Ci}}{3.70\times10^{10} \text{ d/s}} \times \frac{1}{2.6\times10^{-6} \text{ g}} = 1.6\times10^6 \text{ Ci/g}$$

(b)
$$\text{Specific activity (Bq/g)} = 1.53\times10^{11} \text{ d/s} \times \frac{1 \text{ Bq}}{1 \text{ d/s}} \times \frac{1}{2.6\times10^{-6} \text{ g}} = 5.9\times10^{16} \text{ Bq/g}$$

24.4B
$$k = \frac{\mathcal{A} \text{ (nuclei/h)}}{\mathcal{N} \text{(nuclei)}} = \frac{9.97\times10^{12} \text{ nuclei/h}}{(6.50\times10^{-2} \text{ mol})(6.022\times10^{23} \text{ nuclei/mol})} = 2.55\times10^{-10} \text{ h}^{-1}$$

24.5A
$$\ln \mathcal{A}_t = -kt + \ln \mathcal{A}_0 = -\left(\frac{\ln 2}{15 \text{ h}} \times 4.0 \text{ days} \times \frac{24 \text{ h}}{1 \text{ day}}\right) + \ln (2.5\times10^9 \text{ d/s}) = 17.20$$
$$\mathcal{A}_t = 3.0\times10^7 \text{ d/s}$$

24.5B
$$\mathcal{A}_0 = 5.6\times10^{-6} \text{ Ci} \times \frac{3.70\times10^{10} \text{ d/s}}{1 \text{ Ci}} = 2.1\times10^5 \text{ d/s}$$
$$\ln \mathcal{A}_t = -kt + \ln \mathcal{A}_0 = -\left(\frac{\ln 2}{44.5 \text{ days}} \times 17 \text{ days}\right) + \ln (2.1\times10^5 \text{ d/s}) = -0.265 + 12.25 \text{ d/s} = 11.99 \text{ d/s}$$
$$\mathcal{A}_t = 1.6\times10^5 \text{ d/s}$$
$$\text{Fraction decayed} = \frac{2.1\times10^5 \text{ d/s} - 1.6\times10^5 \text{ d/s}}{2.1\times10^5 \text{ d/s}} = 0.23$$

24.6A
$$t = \frac{1}{k} \ln \frac{\mathcal{A}_0}{\mathcal{A}_t} = \frac{5730 \text{ yr}}{\ln 2} \ln \left(\frac{15.3 \text{ d/min·g}}{9.41 \text{ d/min·g}}\right) = 4.02\times10^3 \text{ yr}$$

The mummy case is about 4000 years old.

24.6B
$$t = \frac{1}{k} \ln \frac{\mathcal{A}_0}{\mathcal{A}_t} = \frac{5730 \text{ yr}}{\ln 2} \ln \left(\frac{15.3 \text{ d/min·g}}{12.87 \text{ d/min·g}}\right) = 1430 \text{ yr}$$

24.7A ^{58}Ni has 28 $^{1}_{1}$p and 30 $^{1}_{0}$n.

$$\Delta m = [(28 \times 1.007825 \text{ amu}) + (30 \times 1.008665 \text{ amu})] - 57.935346 \text{ amu} = 0.543704 \text{ amu}$$

$$\frac{\text{Binding energy}}{\text{nucleon}} = \frac{0.543704 \text{ amu} \times \dfrac{931.5 \text{ MeV}}{1 \text{ amu}}}{58 \text{ nucleons}} = 8.732 \text{ MeV/nucleon}$$

Therefore, ^{56}Fe (8.790 MeV/nucleon) is slightly more stable than ^{58}Ni.

24.7B ^{235}U has 92 $^{1}_{1}$p and 143 $^{1}_{0}$n.

$$\Delta m = [(92 \times 1.007825 \text{ amu}) + (143 \times 1.008665 \text{ amu})] - 235.043924 \text{ amu} = 1.9151 \text{ amu}$$

$$\frac{\text{Binding energy}}{\text{nucleon}} = \frac{1.9151 \text{ amu} \times \dfrac{931.5 \text{ MeV}}{1 \text{ amu}}}{235 \text{ nucleons}} = 7.591 \text{ MeV/nucleon}$$

Therefore, ^{235}U is less stable than ^{12}C (7.680 MeV/nucleon).

PROBLEMS

Problems with **colored** numbers are answered in Appendix E and worked in detail in the Student Solutions Manual. Problem sections match those in the text and give the numbers of relevant sample problems. Most offer Concept Review Questions, Skill-Building Exercises (grouped in pairs covering the same concept), and Problems in Context. The Comprehensive Problems are based on material from any section or previous chapter.

Radioactive Decay and Nuclear Stability

(Sample Problems 24.1 to 24.3)

Concept Review Questions

24.1 How do chemical and nuclear reactions differ in
(a) Magnitude of the energy change?
(b) Effect on rate of increasing temperature?
(c) Effect on rate of higher reactant concentration?
(d) Effect on yield of higher reactant concentration?

24.2 Sulfur has four naturally occurring stable isotopes. The one with the lowest mass number is sulfur-32, which is also the most abundant (95.02%).
(a) What percentage of the S atoms in a matchhead are ^{32}S?
(b) The isotopic mass of ^{32}S is 31.972070 amu. Is the atomic mass of S larger, smaller, or equal to this mass? Explain.

24.3 What led Marie Curie to draw the following conclusions? (a) Radioactivity is a property of the element and not the compound in which it is found. (b) A highly radioactive element, aside from uranium, occurs in pitchblende.

24.4 Which of the following processes produce an atom of a *different* element: (a) α decay; (b) β^- decay; (c) γ emission; (d) β^+ emission; (e) e^- capture? Show how *Z* and *N* change, if at all, with each process.

24.5 Why is $^{3}_{2}He$ stable, but $^{2}_{2}He$ has never been detected?

24.6 How do the modes of decay differ for a neutron-rich nuclide and a proton-rich nuclide?

24.7 Why might it be difficult to use only a nuclide's *N*/*Z* ratio to predict whether it will decay by β^+ emission or by e^- capture? What other factor is important?

Skill-Building Exercises (grouped in similar pairs)

24.8 Write balanced nuclear equations for the following:
(a) Alpha decay of $^{234}_{92}U$
(b) Electron capture by neptunium-232
(c) Positron emission by $^{12}_{7}N$

24.9 Write balanced nuclear equations for the following:
(a) β^- decay of sodium-26
(b) β^- decay of francium-223
(c) Alpha decay of $^{212}_{83}Bi$

24.10 Write balanced nuclear equations for the following:
(a) β^- emission by magnesium-27
(b) β^+ emission by $^{23}_{12}Mg$
(c) Electron capture by $^{103}_{46}Pd$

24.11 Write balanced nuclear equations for the following:
(a) β^- decay of silicon-32
(b) Alpha decay of polonium-218
(c) Electron capture by $^{110}_{49}In$

24.12 Write balanced nuclear equations for the following:
(a) Formation of $^{48}_{22}Ti$ through positron emission
(b) Formation of silver-107 through electron capture
(c) Formation of polonium-206 through α decay

24.13 Write balanced nuclear equations for the following:
(a) Formation of $^{241}_{95}Am$ through β^- decay
(b) Formation of $^{228}_{89}Ac$ through β^- decay
(c) Formation of $^{203}_{83}Bi$ through α decay

24.14 Write balanced nuclear equations for the following:
(a) Formation of ^{186}Ir through electron capture
(b) Formation of francium-221 through α decay
(c) Formation of iodine-129 through β^- decay

24.15 Write balanced nuclear equations for the following:
(a) Formation of ^{52}Mn through positron emission
(b) Formation of polonium-215 through α decay
(c) Formation of ^{81}Kr through electron capture

24.16 Which nuclide(s) would you predict to be stable? Why?
(a) $^{20}_{8}O$ (b) $^{59}_{27}Co$ (c) $^{9}_{3}Li$

24.17 Which nuclide(s) would you predict to be stable? Why?
(a) $^{146}_{60}Nd$ (b) $^{114}_{48}Cd$ (c) $^{88}_{42}Mo$

24.18 Which nuclide(s) would you predict to be stable? Why?
(a) ^{127}I (b) tin-106 (c) ^{68}As

24.19 Which nuclide(s) would you predict to be stable? Why?
(a) ^{48}K (b) ^{79}Br (c) argon-32

24.20 What is the most likely mode of decay for each nuclide?
(a) $^{238}_{92}U$ (b) $^{48}_{24}Cr$ (c) $^{50}_{25}Mn$

24.21 What is the most likely mode of decay for each nuclide?
(a) $^{111}_{47}Ag$ (b) $^{41}_{17}Cl$ (c) $^{110}_{44}Ru$

24.22 What is the most likely mode of decay for each nuclide?
(a) ^{15}C (b) ^{120}Xe (c) ^{224}Th

24.23 What is the most likely mode of decay for each nuclide?
(a) ^{106}In (b) ^{141}Eu (c) ^{241}Am

24.24 Why is $^{52}_{24}Cr$ the most stable isotope of chromium?

24.25 Why is $^{40}_{20}Ca$ the most stable isotope of calcium?

Problems in Context

24.26 ^{237}Np is the parent nuclide of a decay series that starts with α emission, followed by β^- emission, and then two more α emissions. Write a balanced nuclear equation for each step.

24.27 Why is helium found in deposits of uranium and thorium ores? What kind of radioactive emission produces it?

24.28 In a natural decay series, how many α and β^- emissions per atom of uranium-235 result in an atom of lead-207?

The Kinetics of Radioactive Decay

(Sample Problems 24.4 to 24.6)

Concept Review Questions

24.29 What electronic process is the basis for detecting radioactivity in (a) a scintillation counter; (b) a Geiger-Müller counter?

24.30 What is the reaction order of radioactive decay? Explain.

24.31 After 1 min, three radioactive nuclei remain from an original sample of six. Is it valid to conclude that $t_{1/2}$ equals 1 min? Is this conclusion valid if the original sample contained 6×10^{12} nuclei and 3×10^{12} remain after 1 min? Explain.

24.32 Radioisotopic dating depends on the constant rate of decay and formation of various nuclides in a sample. How is the proportion of ^{14}C kept relatively constant in living organisms?

Skill-Building Exercises (grouped in similar pairs)

24.33 What is the specific activity (in Ci/g) if 1.65 mg of an isotope emits 1.56×10^6 α particles per second?

24.34 What is the specific activity (in Ci/g) if 2.6 g of an isotope emits 4.13×10^8 β^- particles per hour?

24.35 What is the specific activity (in Bq/g) if 8.58 μg of an isotope emits 7.4×10^4 α particles per minute?

24.36 What is the specific activity (in Bq/g) if 1.07 kg of an isotope emits 3.77×10^7 β^- particles per minute?

24.37 If one-trillionth of the atoms of a radioactive isotope disintegrate each day, what is the decay constant of the process?

24.38 If 2.8×10^{-10}% of the atoms of a radioactive isotope disintegrate in 1.0 yr, what is the decay constant of the process?

24.39 If 1.00×10^{-12} mol of ^{135}Cs emits 1.39×10^5 β^- particles in 1.00 yr, what is the decay constant?

24.40 If 6.40×10^{-9} mol of ^{176}W emits 1.07×10^{15} β^+ particles in 1.00 h, what is the decay constant?

24.41 The isotope $^{212}_{83}$Bi has a half-life of 1.01 yr. What mass (in mg) of a 2.00-mg sample will remain after 3.75×10^3 h?

24.42 The half-life of radium-226 is 1.60×10^3 yr. How many hours will it take for a 2.50-g sample to decay to the point where 0.185 g of the isotope remains?

24.43 A rock contains 270 μmol of ^{238}U ($t_{1/2} = 4.5\times10^9$ yr) and 110 μmol of ^{206}Pb. Assuming that all the ^{206}Pb comes from decay of the ^{238}U, estimate the rock's age.

24.44 A fabric remnant from a burial site has a ^{14}C/^{12}C ratio of 0.735 of the original value. How old is the fabric?

Problems in Context

24.45 Due to decay of ^{40}K, cow's milk has a specific activity of about 6×10^{-11} mCi per milliliter. How many disintegrations of ^{40}K nuclei are there per minute in an 8.0-oz glass of milk?

24.46 Plutonium-239 ($t_{1/2} = 2.41\times10^4$ yr) represents a serious nuclear waste hazard. If seven half-lives are required to reach a tolerable level of radioactivity, how long must ^{239}Pu be stored?

24.47 A rock that contains 3.1×10^{-15} mol of ^{232}Th ($t_{1/2} = 1.4\times10^{10}$ yr) has 9.5×10^4 fission tracks, each track representing the fission of one atom of ^{232}Th. How old is the rock?

24.48 A volcanic eruption melts a large chunk of rock, and all gases are expelled. After cooling, $^{40}_{18}$Ar accumulates from the ongoing decay of $^{40}_{19}$K in the rock ($t_{1/2} = 1.25\times10^9$ yr). When a piece of rock is analyzed, it is found to contain 1.38 mmol of ^{40}K and 1.14 mmol of ^{40}Ar. How long ago did the rock cool?

Nuclear Transmutation: Induced Changes in Nuclei

Concept Review Questions

24.49 Irene and Frederic Joliot-Curie converted $^{27}_{13}$Al to $^{30}_{15}$P in 1933. Why was this transmutation significant?

24.50 Early workers mistakenly thought neutron beams were γ radiation. Why? What evidence led to the correct conclusion?

24.51 Why must the electrical polarity of the tubes in a linear accelerator be reversed at very short time intervals?

24.52 Why does bombardment with protons usually require higher energies than bombardment with neutrons?

Skill-Building Exercises (grouped in similar pairs)

24.53 Determine the missing species in these transmutations, and write a full nuclear equation from the shorthand notation:
(a) ^{10}B (α,n) __
(b) ^{28}Si (d,__) ^{29}P (where d is a deuteron, ^{2}H)
(c) __ (α,2n) ^{244}Cf

24.54 Name the unidentified species, and write each transmutation process in shorthand notation: (a) gamma irradiation of a nuclide yields a proton, a neutron, and ^{29}Si; (b) bombardment of ^{252}Cf with ^{10}B yields five neutrons and a nuclide; (c) bombardment of ^{238}U with a particle yields three neutrons and ^{239}Pu.

Problems in Context

24.55 Elements 104, 105, and 106 have been named rutherfordium (Rf), dubnium (Db), and seaborgium (Sg), respectively. These elements are synthesized from californium-249 by bombarding with carbon-12, nitrogen-15, and oxygen-18 nuclei, respectively. Four neutrons are formed in each reaction as well. (a) Write balanced nuclear equations for the formation of these elements. (b) Write the equations in shorthand notation.

Ionization: Effects of Nuclear Radiation on Matter

Concept Review Questions

24.56 The effects on matter of γ rays and α particles differ. Explain.

24.57 What is a cation-electron pair, and how does it form?

24.58 Why is ionizing radiation more harmful to children than adults?

24.59 Why is •OH more dangerous than OH^- in an organism?

Skill-Building Exercises (grouped in similar pairs)

24.60 A 135-lb person absorbs 3.3×10^{-7} J of energy from radioactive emissions. (a) How many rads does she receive? (b) How many grays (Gy) does she receive?

24.61 A 3.6-kg laboratory animal receives a single dose of 8.92×10^{-4} Gy. (a) How many rads did the animal receive? (b) How many joules did the animal absorb?

24.62 A 70.-kg person exposed to ^{90}Sr absorbs 6.0×10^5 β^- particles, each with an energy of 8.74×10^{-14} J. (a) How many grays does the person receive? (b) If the RBE is 1.0, how many millirems is this? (c) What is the equivalent dose in sieverts (Sv)?

24.63 A laboratory rat weighs 265 g and absorbs 1.77×10^{10} β^- particles, each with an energy of 2.20×10^{-13} J. (a) How many rads does the animal receive? (b) What is this dose in Gy? (c) If the RBE is 0.75, what is the equivalent dose in Sv?

Problems in Context

24.64 If 2.50 pCi [1 pCi (picocurie) = 1×10^{-12} Ci] of radioactivity from ^{239}Pu is emitted in a 95-kg human for 65 h, and each disintegration has an energy of 8.25×10^{-13} J, how many grays does the person receive?

24.65 A small region of a cancer patient's brain is exposed for 24.0 min to 475 Bq of radioactivity from ^{60}Co for treatment of a tumor. If the brain mass exposed is 1.858 g and each β^- particle emitted has an energy of 5.05×10^{-14} J, what is the dose in rads?

Applications of Radioisotopes

Concept Review Questions

24.66 What two ways are radioactive tracers used in organisms?

24.67 Why is neutron activation analysis (NAA) useful to art historians and criminologists?

24.68 Positrons cannot penetrate matter more than a few atomic diameters, but positron emission of radiotracers can be monitored in medical diagnosis. Explain.

24.69 A steel part is treated to form some iron-59. Oil used to lubricate the part emits 298 β^- particles (with the energy characteristic of ^{59}Fe) per minute per milliliter of oil. What other information would you need to calculate the rate of removal of the steel from the part during use?

Problems in Context

24.70 The oxidation of methanol to formaldehyde can be accomplished by reaction with chromic acid:

$$6H^+(aq) + 3CH_3OH(aq) + 2H_2CrO_4(aq) \longrightarrow 3CH_2O(aq) + 2Cr^{3+}(aq) + 8H_2O(l)$$

The reaction can be studied with the stable isotope tracer ^{18}O and mass spectrometry. When a small amount of $CH_3{}^{18}OH$ is present in the alcohol reactant, $CH_2{}^{18}O$ forms. When a small amount of $H_2Cr^{18}O_4$ is present, $H_2{}^{18}O$ forms. Does chromic acid or methanol supply the O atom to the aldehyde? Explain.

The Interconversion of Mass and Energy

(Sample Problem 24.7)

Note: Data for problems in this section: mass of ^1H atom = 1.007825 amu; mass of neutron = 1.008665 amu.

Concept Review Questions

24.71 Many scientists at first reacted skeptically to Einstein's equation, $E = mc^2$. Why?

24.72 How does a change in mass arise when a nuclide forms from nucleons?

24.73 When a nucleus forms from nucleons, is energy absorbed or released? Why?

24.74 What is the binding energy per nucleon? Why is the binding energy per nucleon, rather than per nuclide, used to compare nuclide stability?

Skill-Building Exercises (grouped in similar pairs)

24.75 A ^3H nucleus decays with an energy of 0.01861 MeV. Convert this energy into (a) electron volts; (b) joules.

24.76 Arsenic-84 decays with an energy of 1.57×10^{-15} kJ per nucleus. Convert this energy into (a) eV; (b) MeV.

24.77 How many joules are released when 1.5 mol of ^{239}Pu decays, if each nucleus releases 5.243 MeV?

24.78 How many MeV are released per nucleus when 3.2×10^{-3} mol of chromium-49 releases 8.11×10^5 kJ?

24.79 Oxygen-16 is one of the most stable nuclides. The mass of a ^{16}O atom is 15.994915 amu. Calculate the binding energy (a) per nucleon in MeV; (b) per atom in MeV; (c) per mole in kJ.

24.80 Lead-206 is the end product of ^{238}U decay. One ^{206}Pb atom has a mass of 205.974440 amu. Calculate the binding energy (a) per nucleon in MeV; (b) per atom in MeV; (c) per mole in kJ.

24.81 Cobalt-59 is the only stable isotope of this transition metal. One ^{59}Co atom has a mass of 58.933198 amu. Calculate the binding energy (a) per nucleon in MeV; (b) per atom in MeV; (c) per mole in kJ.

24.82 Iodine-131 is one of the most important isotopes used in the diagnosis of thyroid cancer. One atom has a mass of 130.906114 amu. Calculate the binding energy (a) per nucleon in MeV; (b) per atom in MeV; (c) per mole in kJ.

Problems in Context

24.83 The ^{80}Br nuclide decays either by β^- decay or by electron capture. (a) What is the product of each process? (b) Which process releases more energy? (Masses of atoms: ^{80}Br = 79.918528 amu; ^{80}Kr = 79.916380 amu; ^{80}Se = 79.916520 amu; neglect the mass of electrons involved because these are atomic, not nuclear, masses.)

Applications of Fission and Fusion

Concept Review Questions

24.84 What is the minimum number of neutrons from each fission event that must be absorbed by other nuclei for a chain reaction to be sustained?

24.85 In what main way is fission different from radioactive decay? Are all fission events in a chain reaction identical? Explain.

24.86 What is the purpose of enrichment in the preparation of fuel rods? How is it accomplished?

24.87 Describe the nature and purpose of these components of a nuclear reactor: (a) control rods; (b) moderator; (c) reflector.

24.88 State an advantage and a disadvantage of heavy-water reactors compared to light-water reactors.

24.89 What are the expected advantages of fusion reactors over fission reactors?

24.90 Why is iron the most abundant element in Earth's core?

Problems in Context

24.91 The reaction that will probably power the first commercial fusion reactor is

$$^3_1H + ^2_1H \longrightarrow ^4_2He + ^1_0n$$

How much energy would be produced per mole of reaction? (Mass of ^3_1H = 3.01605 amu; mass of ^2_1H = 2.0140 amu; mass of ^4_2He = 4.00260 amu; mass of ^1_0n = 1.008665 amu.)

Comprehensive Problems

24.92 Some $^{243}_{95}Am$ was present when Earth formed, but it all decayed in the next billion years. The first three steps in this decay series are emissions of an α particle, a β^- particle, and another α particle. What other isotopes were present on the young Earth in a rock that contained some $^{243}_{95}Am$?

24.93 Curium-243 undergoes α decay to plutonium-239:

$$^{243}Cm \longrightarrow ^{239}Pu + \alpha$$

(a) Find the change in mass, Δm (in kg). (Mass of ^{243}Cm = 243.0614 amu; mass of ^{239}Pu = 239.0522 amu; mass of ^4He = 4.0026 amu; 1 amu = 1.661×10^{-24} g.)
(b) Find the energy released in joules.

(c) Find the energy released in kJ/mol of reaction, and comment on the difference between this value and a typical heat of reaction for a chemical change, which is a few hundred kJ/mol.

24.94 Plutonium "triggers" for nuclear weapons were manufactured at the Rocky Flats plant in Colorado. An 85-kg worker inhaled a dust particle containing 1.00 μg of $^{239}_{94}Pu$, which resided in his body for 16 h ($t_{1/2}$ of $^{239}Pu = 2.41\times10^4$ yr; each disintegration released 5.15 MeV). (a) How many rads did he receive? (b) How many grays?

24.95 Archeologists removed some charcoal from a Native American campfire site, burned it in O_2, and bubbled the CO_2 formed into $Ca(OH)_2$ solution (limewater). The $CaCO_3$ that precipitated was filtered and dried. If 4.58 g of the $CaCO_3$ had a radioactivity of 3.2 d/min, how long ago was the campfire burning?

24.96 A 5.4-μg sample of $^{226}RaCl_2$ has a radioactivity of 1.5×10^5 Bq. Calculate $t_{1/2}$ of ^{226}Ra.

24.97 How many rads does a 65-kg human receive each year from the approximately 10^{-8} g of $^{14}_{6}C$ naturally present in her body ($t_{1/2} = 5730$ yr; each disintegration releases 0.156 MeV)?

24.98 A sample of AgCl emits 175 nCi/g. A saturated solution prepared from the solid emits 1.25×10^{-2} Bq/mL due to radioactive Ag^+ ions. What is the molar solubility of AgCl?

24.99 The scene below depicts a neutron bombarding ^{235}U:

(a) Is this an example of fission or of fusion?
(b) Identify the other nuclide formed.
(c) What is the most likely mode of decay of the nuclide with $Z = 55$?

24.100 What fraction of the ^{235}U ($t_{1/2} = 7.0\times10^8$ yr) created when Earth was formed would remain after 2.8×10^9 yr?

24.101 ^{238}U ($t_{1/2} = 4.5\times10^9$ yr) begins a decay series that ultimately forms ^{206}Pb. The scene below depicts the relative number of ^{238}U atoms *(red)* and ^{206}Pb atoms *(green)* in a mineral. If all the Pb comes from ^{238}U, calculate the age of the sample.

24.102 Technetium-99m is a metastable nuclide used in numerous cancer diagnostic and treatment programs. It is prepared just before use because it decays rapidly through γ emission:

$$^{99m}Tc \longrightarrow {}^{99}Tc + \gamma$$

Use the data below to determine (a) the half-life of ^{99m}Tc; (b) the percentage of the isotope that is lost if it takes 2.0 h to prepare and administer the dose.

Time (h)	γ Emission (photons/s)
0	5000.
4	3150.
8	2000.
12	1250.
16	788
20	495

24.103 How many curies are produced by 1.0 mol of ^{40}K ($t_{1/2} = 1.25\times10^9$ yr)? How many becquerels?

24.104 The fraction of a radioactive isotope remaining at time t is $(\frac{1}{2})^{t/t_{1/2}}$, where $t_{1/2}$ is the half-life. If the half-life of carbon-14 is 5730 yr, what fraction of carbon-14 in a piece of charcoal remains after (a) 10.0 yr; (b) 10.0×10^3 yr; (c) 10.0×10^4 yr? (d) Why is radiocarbon dating more reliable for the fraction remaining in part (b) than that in part (a) or in part (c)?

24.105 The isotopic mass of $^{210}_{86}Rn$ is 209.989669 amu. When this nuclide decays by electron capture, it emits 2.368 MeV. What is the isotopic mass of the resulting nuclide?

24.106 Exactly 0.1 of the radioactive nuclei in a sample decay per hour. Thus, after n hours, the fraction of nuclei remaining is $(0.900)^n$. Find the value of n equal to one half-life.

24.107 In neutron activation analysis (NAA), stable isotopes are bombarded with neutrons. Depending on the isotope and the energy of the neutron, various emissions are observed. What are the products when the following neutron-activated species decay? Write an overall equation in shorthand notation for the reaction starting with the stable isotope before neutron activation.
(a) $^{52}_{23}V^* \longrightarrow [\beta^-$ emission$]$ (b) $^{64}_{29}Cu^* \longrightarrow [\beta^+$ emission$]$
(c) $^{28}_{13}Al^* \longrightarrow [\beta^-$ emission$]$

24.108 In the 1950s, radioactive material was spread over the land from aboveground nuclear tests. A woman drinks some contaminated milk and ingests 0.0500 g of ^{90}Sr, which is taken up by bones and teeth and not eliminated. (a) How much ^{90}Sr ($t_{1/2}$ = 29 yr) is present in her body after 10 yr? (b) How long will it take for 99.9% of the ^{90}Sr that she ingested to decay?

24.109 The scene below represents a reaction (with neutrons gray and protons purple) that occurs during the lifetime of a star. (a) Write a balanced nuclear equation for the reaction. (b) If the mass difference is 7.7×10^{-2} amu, find the energy (in kJ) released.

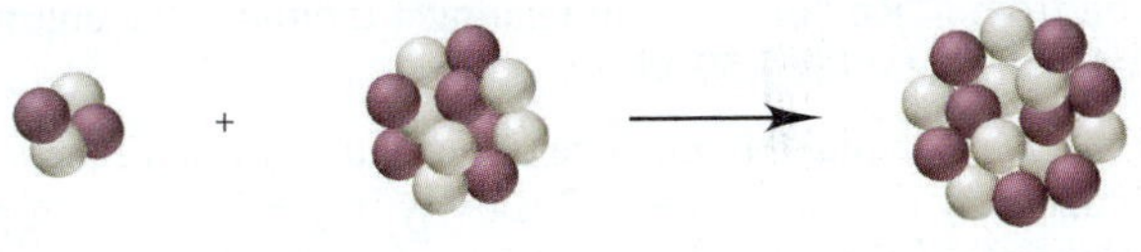

24.110 What volume of radon will be produced per hour at STP from 1.000 g of ^{226}Ra ($t_{1/2}$ = 1599 yr; 1 yr = 8766 h; mass of one ^{226}Ra atom = 226.025402 amu)?

24.111 ^{90}Kr ($t_{1/2}$ = 32 s) is used to study respiration. How soon after being made must a sample be administered to the patient if the activity must be at least 90% of the original activity?

24.112 Which isotope in each pair is more stable? Why?
(a) $^{140}_{55}Cs$ or $^{133}_{55}Cs$ (b) $^{79}_{35}Br$ or $^{78}_{35}Br$
(c) $^{28}_{12}Mg$ or $^{24}_{12}Mg$ (d) $^{14}_{7}N$ or $^{18}_{7}N$

24.113 A bone sample containing strontium-90 ($t_{1/2}$ = 29 yr) emits 7.0×10^4 β^- particles per month. How long will it take for the emission to decrease to 1.0×10^4 particles per month?

24.114 The 23rd-century starship *Enterprise* uses a substance called "dilithium crystals" as its fuel.
(a) Assuming that this material is the result of fusion, what is the product of the fusion of two ^6Li nuclei?
(b) How much energy is released per kilogram of dilithium formed? (Mass of one ^6Li atom is 6.015121 amu.)
(c) When four ^1H atoms fuse to form ^4He, how many positrons are released?

(d) To determine the energy potential of the fusion processes in parts (b) and (c), compare the changes in mass per kilogram of dilithium and of ^{4}He.
(e) Compare the change in mass per kilogram in part (b) to that for the formation of ^{4}He by the method used in current fusion reactors (see Section 24.7). (For masses, see Problem 24.91.)
(f) Using early 21st-century fusion technology, how much tritium can be produced per kilogram of ^{6}Li in the following reaction: $^6_3\text{Li} + ^1_0\text{n} \longrightarrow ^4_2\text{He} + ^3_1\text{H}$? When this amount of tritium is fused with deuterium, what is the change in mass? How does this quantity compare with the use of dilithium in part (b)?

24.115 Uranium and radium are found in many rocky soils throughout the world. Both undergo radioactive decay, and one of the products is radon-222, the heaviest noble gas ($t_{1/2}$ = 3.82 days). Inhalation of this gas contributes to many lung cancers. According to the Environmental Protection Agency, the level of radioactivity from radon in homes should not exceed 4.0 pCi/L of air.
(a) What is the safe level of radon in Bq/L of air?
(b) A home has a radon measurement of 41.5 pCi/L. The owner vents the basement in such a way that no more radon enters the living area. What is the activity of the radon remaining in the room air (in Bq/L) after 9.5 days?
(c) How many more days does it take to reach the EPA recommended level?

24.116 Nuclear disarmament could be accomplished if weapons were not "replenished." The tritium in warheads decays to helium with a half-life of 12.26 yr and must be replaced or the weapon is useless. What fraction of the tritium is lost in 5.50 yr?

24.117 A decay series starts with the synthetic isotope $^{239}_{92}$U. The first four steps are emissions of a β^- particle, another β^-, an α particle, and another α. Write a balanced nuclear equation for each step. Which natural series could start by this sequence?

24.118 How long can a 54-lb child be exposed to 1.0 mCi of radiation from ^{222}Rn before accumulating 1.0 mrad if the energy of each disintegration is 5.59 MeV?

24.119 An earthquake in the area of present-day San Francisco is to be dated by measuring the ^{14}C activity ($t_{1/2}$ = 5730 yr) of parts of a tree uprooted during the event. The tree parts have an activity of 12.9 d/min·g C, and a living tree has an activity of 15.3 d/min·g C. How long ago did the earthquake occur?

24.120 Were organisms a billion years ago exposed to more or less ionizing radiation than similar organisms today? Explain.

24.121 Tritium (^{3}H; $t_{1/2}$ = 12.26 yr) is continually formed in the upper troposphere by interaction of solar particles with nitrogen. As a result, natural waters contain a small amount of tritium. Two samples of wine are analyzed, one known to be made in 1941 and another made earlier. The water in the 1941 wine has 2.23 times as much tritium as the water in the other. When was the other wine produced?

24.122 Even though plutonium-239 ($t_{1/2} = 2.41\times10^4$ yr) is one of the main fission fuels, it is still a radiation hazard present in spent uranium fuel from nuclear power plants. How many years does it take for 99% of the plutonium-239 in spent fuel to decay?

24.123 Carbon from the remains of an extinct Australian marsupial, called *Diprotodon,* has a specific activity of 0.61 pCi/g. Modern carbon has a specific activity of 6.89 pCi/g. How long ago did the *Diprotodon* apparently become extinct?

24.124 The reaction that allows radiocarbon dating to be performed is the continual formation of carbon-14 in the upper atmosphere:

$$^{14}_7\text{N} + ^1_0\text{n} \longrightarrow ^{14}_6\text{C} + ^1_1\text{H}$$

What is the energy change that is associated with this process in eV/reaction and in kJ/mol reaction? (Masses of atoms: $^{14}_7$N = 14.003074 amu; $^{14}_6$C = 14.003241 amu; $^1_1$H = 1.007825 amu; mass of $^1_0$n = 1.008665 amu.)

24.125 What is the nuclear binding energy of a lithium-7 nucleus in units of kJ/mol and eV/nucleus? (Mass of a lithium-7 atom = 7.016003 amu.)

24.126 Gadolinium-146 undergoes electron capture. Identify the product, and use Figure 24.2 to find the modes of decay and the two intermediate nuclides in the series:

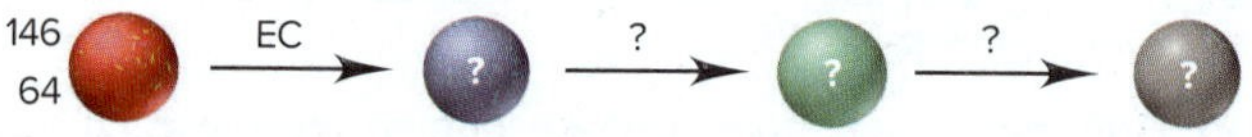

24.127 Using 21st-century technology, hydrogen fusion requires temperatures around 10^8 K. But, lower initial temperatures are used if the hydrogen is compressed. In the late 24th century, the starship *Leinad* uses such methods to fuse hydrogen at 10^6 K.
(a) What is the kinetic energy of an H atom at 1.00×10^6 K?
(b) How many H atoms are heated to 1.00×10^6 K from the energy of one H and one anti-H atom annihilating each other?
(c) If the heated H atoms of part (b) fuse into ^{4}He atoms (with the loss of two positrons per ^{4}He formed), how much energy (in J) is generated?
(d) How much more energy is generated by the fusion in part (c) than by the hydrogen-antihydrogen collision in part (b)?
(e) Should the captain of the *Leinad* change the technology and produce ^{3}He (mass = 3.01603 amu) instead of ^{4}He?

24.128 A metastable (excited) form of ^{50}Sc changes to its stable form by emitting γ radiation with a wavelength of 8.73 pm. What is the change in mass of 1 mol of the isotope when it undergoes this change?

24.129 A sample of cobalt-60 ($t_{1/2}$ = 5.27 yr), a powerful γ emitter used to treat cancer, was purchased by a hospital on March 1, 2012. The sample must be replaced when its activity reaches 70.% of the original value. On what date must it be replaced?

24.130 Uranium-233 decays to thorium-229 by α decay, but the emissions have different energies and products: 83% emit an α particle with an energy of 4.816 MeV and give ^{229}Th in its ground state; 15% emit an α particle of 4.773 MeV and give ^{229}Th in excited state I; and 2% emit a lower energy α particle and give ^{229}Th in the higher excited state II. Excited state II emits a γ ray of 0.060 MeV to reach excited state I.
(a) Find the γ-ray energy and wavelength that would convert excited state I to the ground state. (b) Find the energy of the α particle that would raise ^{233}U to excited state II.

24.131 Uranium-238 undergoes a slow decay step ($t_{1/2} = 4.5\times10^9$ yr) followed by a series of fast steps to form the stable isotope ^{206}Pb. Thus, on a time scale of billions of years, ^{238}U effectively decays "directly" to ^{206}Pb, and the relative amounts of these isotopes are used to find the age of some rocks. Two students derive equations relating number of half-lives (n) since the rock formed to the amounts of the isotopes:

Student 1: $\left(\frac{1}{2}\right)^n = \dfrac{^{238}_{92}\text{U}}{^{206}_{82}\text{Pb}}$

Student 2: $$\left(\tfrac{1}{2}\right)^n = \frac{^{238}_{92}\text{U}}{^{238}_{92}\text{U} + {}^{206}_{82}\text{Pb}}$$

(a) Which equation is correct, and why?
(b) If a rock contains exactly twice as much ^{238}U as ^{206}Pb, what is its age in years?

24.132 In the naturally occurring thorium-232 decay series, the steps emit this sequence of particles: α, β^-, β^-, α, α, α, α, β^-, β^-, and α. Write a balanced equation for each step.

24.133 At death, a nobleman in ancient Egypt was mummified and his body contained 1.4×10^{-3} g of ^{40}K ($t_{1/2} = 1.25\times10^9$ yr), 1.2×10^{-8} g of ^{14}C ($t_{1/2} = 5730$ yr), and 4.8×10^{-14} g of ^{3}H ($t_{1/2} = 12.26$ yr). Which nuclide would give the most accurate estimate of the mummy's age? Explain.

24.134 Assuming that many radioactive nuclides can be considered safe after 20 half-lives, how long will it take for each of the following nuclides to be safe: (a) ^{242}Cm ($t_{1/2} = 163$ days); (b) ^{214}Po ($t_{1/2} = 1.6\times10^{-4}$ s); (c) ^{232}Th ($t_{1/2} = 1.39\times10^{10}$ yr)?

24.135 An ancient sword has a blade from the early Roman Empire, around 100 AD, but the wooden handle, inlaid wooden decorations, leather ribbon, and leather sheath have different styles. Given the following activities, estimate the age of each part. Which part was made near the time of the blade ($t_{1/2}$ of ^{14}C = 5730 yr; $\mathscr{A}_0 = 15.3$ d/min·g)?

Part	$\mathscr{A}_t$ (d/min·g)
Handle	10.1
Inlaid wood	13.8
Ribbon	12.1
Sheath	15.0

24.136 The starship *Voyager,* like many other vessels of the newly designed 24th-century fleet, uses antimatter as fuel.
(a) How much energy is released when 1.00 kg each of antimatter and matter annihilate each other?
(b) When the antimatter is atomic antihydrogen, a small amount of it is mixed with excess atomic hydrogen (gathered from interstellar space during flight). The annihilation releases so much heat that the remaining hydrogen nuclei fuse to form ^{4}He. If each hydrogen-antihydrogen collision releases enough heat to fuse 1.00×10^5 hydrogen atoms, how much energy (in kJ) is released per kilogram of antihydrogen?
(c) Which produces more energy per kilogram of antihydrogen, the procedure in part (a) or that in part (b)?

24.137 Use Einstein's equation, the mass in grams of 1 amu, and the relation between electron volts and joules to find the energy equivalent (in MeV) of a mass difference of 1 amu.

24.138 Determine the age of a rock containing 0.065 g of uranium-238 ($t_{1/2} = 4.5\times10^9$ yr) and 0.023 g of lead-206. (Assume that all the lead-206 came from ^{238}U decay.)

24.139 Plutonium-242 decays to uranium-238 by emission of an α particle with an energy of 4.853 MeV. The ^{238}U that forms is unstable and emits a γ ray ($\lambda = 0.02757$ nm). (a) Write balanced equations for these reactions. (b) What would be the energy of the α particle if ^{242}Pu decayed directly to the more stable ^{238}U?

24.140 Seaborgium-263 (Sg), the first isotope of element 106 synthesized, was produced, along with four neutrons, by bombarding californium-249 with oxygen-18. The ^{263}Sg then underwent a series of decays starting with three α emissions. Write balanced equations for the synthesis and the three α emissions of ^{263}Sg.

24.141 Some nuclear power plants use plutonium-239, which is produced in breeder reactors. The rate-determining step is the second β^- emission. How long does it take to make 1.00 kg of ^{239}Pu if the reaction is complete when the product is 90.% ^{239}Pu?

24.142 A random-number generator can be used to simulate the probability of a given atom decaying over a given time. For example, the formula "=RAND()" in an Excel spreadsheet returns a random number between 0 and 1; thus, for one radioactive atom and a time of one half-life, a number less than 0.5 means the atom decays and a number greater than 0.5 means it doesn't.
(a) Place the "=RAND()" formula in cells A1 to A10 of an Excel spreadsheet. In cell B1, place "=IF(A1<0.5, 0, 1)." This formula returns 0 if A1 is <0.5 (the atom decays) and 1 if A1 is >0.5 (the atom does not decay). Place analogous formulas in cells B2 to B10 (using the "Fill Down" procedure in Excel). To determine the number of atoms remaining after one half-life, sum cells B1 to B10 by placing "=SUM(B1:B10)" in cell B12. To create a new set of random numbers, click on an empty cell (e.g., B13) and hit "Delete." Perform 10 simulations, each time recording the total number of atoms remaining. Do half of the atoms remain after each half-life? If not, why not?
(b) Increase the number of atoms to 100 by placing suitable formulas in cells A1 to A100, B1 to B100, and B102. Perform 10 simulations, and record the number of atoms remaining each time. Are these results more realistic for radioactive decay? Explain.

24.143 In the following Excel-based simulation, the fate of 256 atoms is followed over five half-lives. Set up formulas in columns A and B, as in Problem 24.142, and simulate the fate of the sample of 256 atoms over one half-life. Cells B1 to B256 should contain 1 or 0. In cell C1, enter "=IF(B1=0, 0, RAND())." This returns 0 if the original atom decayed in the previous half-life or a random number between 0 and 1 if it did not. Fill the formula in C1 down to cell C256. Column D should have formulas similar to those in B, but with modified references, as should columns F, H, and J. Columns E, G, and I should have formulas similar to those in C, but with modified references. In cell B258, enter "=SUM(B1:B256)." This records the number of atoms remaining after the first half-life. Put formulas in cells D258, F258, H258, and J258 to record atoms remaining after subsequent half-lives.
(a) Ideally, how many atoms should remain after each half-life?
(b) Make a table of the atoms remaining after each half-life in four separate simulations. Compare these outcomes to the ideal outcome. How would you make the results more realistic?

24.144 Representations of three nuclei (with neutrons *gray* and protons *purple*) are shown below. Nucleus 1 is stable, but 2 and 3 are not. (a) Write the symbol for each isotope. (b) What is (are) the most likely mode(s) of decay for 2 and 3?

1 2 3

Common Mathematical Operations in Chemistry

In addition to basic arithmetic and algebra, four mathematical operations are used frequently in general chemistry: manipulating logarithms, using exponential notation, solving quadratic equations, and graphing data. Each is discussed briefly in this appendix.

MANIPULATING LOGARITHMS

Meaning and Properties of Logarithms

A *logarithm* is an exponent. Specifically, if $x^n = A$, we can say that the logarithm to the base x of the number A is n, and we can denote it as

$$\log_x A = n$$

Because logarithms are exponents, they have the following properties:

$$\log_x 1 = 0$$
$$\log_x (A \times B) = \log_x A + \log_x B$$
$$\log_x \frac{A}{B} = \log_x A - \log_x B$$
$$\log_x A^y = y \log_x A$$

Types of Logarithms

Common and natural logarithms are used in chemistry and the other sciences.

1. For *common* logarithms, the base (x in the examples above) is 10, but they are written without specifying the base; that is, $\log_{10} A$ is written more simply as $\log A$; thus, the notation *log* means base 10. The common logarithm of 1000 is 3; in other words, you must raise 10 to the 3rd power to obtain 1000:

$$\log 1000 = 3 \qquad \text{or} \qquad 10^3 = 1000$$

Similarly, we have

$$\log 10 = 1 \qquad \text{or} \qquad 10^1 = 10$$
$$\log 1{,}000{,}000 = 6 \qquad \text{or} \qquad 10^6 = 1{,}000{,}000$$
$$\log 0.001 = -3 \qquad \text{or} \qquad 10^{-3} = 0.001$$
$$\log 853 = 2.931 \qquad \text{or} \qquad 10^{2.931} = 853$$

The last example illustrates an important point about significant figures with all logarithms: *the number of significant figures in the number equals the number of digits to the right of the decimal point in the logarithm.* That is, the number 853 has three significant figures, and the logarithm 2.931 has three digits to the right of the decimal point.

To find a common logarithm with an electronic calculator, press the LOG button, enter the number, and press ENTER.

2. For *natural* logarithms, the base is the number e, which is 2.71828 . . . , and $\log_e A$ is written $\ln A$; thus, the notation *ln* means base e. The relationship between the common and natural logarithms is easily obtained:

$$\log 10 = 1 \qquad \text{and} \qquad \ln 10 = 2.303$$

Therefore, we have

$$\ln A = 2.303 \log A$$

To find a natural logarithm with an electronic calculator, press the LN button, enter the number, and press ENTER. If your calculator does not have an LN button, press the LOG button, enter the number, press ENTER, and multiply that result by 2.303.

Antilogarithms

The *antilogarithm* is the base raised to the logarithm:

$$\text{antilogarithm (antilog) of } n \text{ is } 10^n$$

Using two of the earlier examples, the antilog of 3 is 1000, and the antilog of 2.931 is 853. To obtain the antilog with a calculator, press the 10^x button, enter the number, and press ENTER. Similarly, to obtain the natural antilogarithm, press the e^x button, enter the number, and press ENTER. [On some calculators, enter the number and first press INV and then the LOG (or LN) button.]

USING EXPONENTIAL (SCIENTIFIC) NOTATION

Many quantities in chemistry are very large or very small. For example, in the conventional way of writing numbers, the number of gold atoms in 1 gram of gold is

59,060,000,000,000,000,000,000 atoms (to four significant figures)

As another example, the mass in grams of one gold atom is

0.0000000000000000000003272 g (to four significant figures)

Exponential (scientific) notation provides a much more practical way of writing such numbers. In exponential notation, we express numbers in the form

$$A \times 10^n$$

where A (the coefficient) is greater than or equal to 1 and less than 10 (that is, $1 \leq A < 10$), and n (the exponent) is an integer.

If the number we want to express in exponential notation is larger than 1, the exponent is positive ($n > 0$); if the number is smaller than 1, the exponent is negative ($n < 0$). The size of n tells the number of places the decimal point (in conventional notation) must be moved to obtain a coefficient A greater than or equal to 1 and less than 10 (in exponential notation). In exponential notation, 1 gram of gold contains 5.906×10^{22} atoms, and each gold atom has a mass of 3.272×10^{-22} g.

Changing Between Conventional and Exponential Notation

In order to use exponential notation, you must be able to convert to it from conventional notation, and vice versa.

1. To change a number from conventional to exponential notation, move the decimal point to the left for numbers equal to or greater than 10 and to the right for numbers between 0 and 1:

   75,000,000 changes to 7.5×10^7 (decimal point moves 7 places to the left)

   0.006042 changes to 6.042×10^{-3} (decimal point moves 3 places to the right)

2. To change a number from exponential to conventional notation, move the decimal point the number of places indicated by the exponent to the right for numbers with positive exponents and to the left for numbers with negative exponents:

   1.38×10^5 changes to 138,000 (decimal point moves 5 places to the right)

   8.41×10^{-6} changes to 0.00000841 (decimal point moves 6 places to the left)

3. An exponential number with a coefficient greater than 10 or less than 1 can be changed to the standard exponential form by converting the coefficient to the standard form and adding the exponents:

$$582.3\times10^{6} \text{ changes to } 5.823 \times 10^{2} \times 10^{6} = 5.823\times10^{(2+6)} = 5.823\times10^{8}$$

$$0.0043\times10^{-4} \text{ changes to } 4.3 \times 10^{-3} \times 10^{-4} = 4.3\times10^{[(-3)+(-4)]} = 4.3\times10^{-7}$$

Using Exponential Notation in Calculations

In calculations, you can treat the coefficient and exponents separately and apply the properties of exponents (see earlier discussion of logarithms).

1. To multiply exponential numbers, multiply the coefficients, add the exponents, and reconstruct the number in standard exponential notation:

$$(5.5\times10^{3})(3.1\times10^{5}) = (5.5 \times 3.1)\times10^{(3+5)} = 17\times10^{8} = 1.7\times10^{9}$$

$$(9.7\times10^{14})(4.3\times10^{-20}) = (9.7 \times 4.3)\times10^{[14+(-20)]} = 42\times10^{-6} = 4.2\times10^{-5}$$

2. To divide exponential numbers, divide the coefficients, subtract the exponents, and reconstruct the number in standard exponential notation:

$$\frac{2.6\times10^{6}}{5.8\times10^{2}} = \frac{2.6}{5.8} \times 10^{(6-2)} = 0.45\times10^{4} = 4.5\times10^{3}$$

$$\frac{1.7\times10^{-5}}{8.2\times10^{-8}} = \frac{1.7}{8.2} \times 10^{[(-5)-(-8)]} = 0.21\times10^{3} = 2.1\times10^{2}$$

3. To add or subtract exponential numbers, change all numbers so that they have the same exponent, then add or subtract the coefficients:

$$(1.45\times10^{4}) + (3.2\times10^{3}) = (1.45\times10^{4}) + (0.32\times10^{4}) = 1.77\times10^{4}$$

$$(3.22\times10^{5}) - (9.02\times10^{4}) = (3.22\times10^{5}) - (0.902\times10^{5}) = 2.32\times10^{5}$$

SOLVING QUADRATIC EQUATIONS

A *quadratic equation* is one in which the highest power of x is 2. The general form of a quadratic equation is

$$ax^2 + bx + c = 0$$

where a, b, and c are numbers. For given values of a, b, and c, the values of x that satisfy the equation are called *solutions* of the equation. We calculate x with the quadratic formula:

$$x = \frac{-b \pm \sqrt{b^2 - 4ac}}{2a}$$

We commonly require the quadratic formula when solving for some concentration in an equilibrium problem. For example, we might have an expression that is rearranged into the quadratic equation

$$4.3x^2 + 0.65x - 8.7 = 0$$

a b c

Applying the quadratic formula, with $a = 4.3$, $b = 0.65$, and $c = -8.7$, gives

$$x = \frac{-0.65 \pm \sqrt{(0.65)^2 - 4(4.3)(-8.7)}}{2(4.3)}$$

The "plus or minus" sign ($\pm$) indicates that there are always two possible values for x. In this case, they are

$$x = 1.3 \qquad \text{and} \qquad x = -1.5$$

In any real physical system, however, only one of the values will have any meaning. For example, if x were $[H_3O^+]$, the negative value would give a negative concentration, which has no physical meaning.

GRAPHING DATA IN THE FORM OF A STRAIGHT LINE

Visualizing changes in variables by means of a graph is used throughout science. In many cases, it is most useful if the data can be graphed in the form of a straight line. Any equation will appear as a straight line if it has, or can be rearranged to have, the following general form:

$$y = mx + b$$

where y is the dependent variable (typically plotted along the vertical axis), x is the independent variable (typically plotted along the horizontal axis), m is the slope of the line, and b is the intercept of the line on the y axis. The intercept is the value of y when $x = 0$:

$$y = m(0) + b = b$$

The slope of the line is the change in y for a given change in x:

$$\text{Slope } (m) = \frac{y_2 - y_1}{x_2 - x_1} = \frac{\Delta y}{\Delta x}$$

The *sign* of the slope tells the *direction* in which the line slants. If y increases as x increases, m is positive, and the line slopes upward with higher values of x; if y decreases as x increases, m is negative, and the line slopes downward with higher values of x. The *magnitude* of the slope indicates the *steepness* of the line. A line with $m = 3$ is three times as steep (y changes three times as much for a given change in x) as a line with $m = 1$.

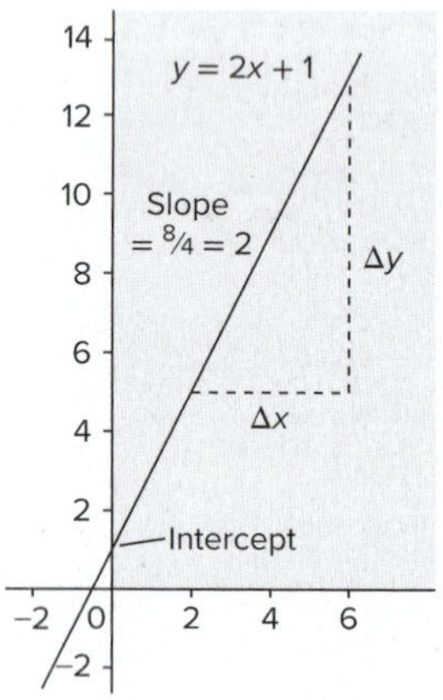

Figure A.1

Consider the linear equation $y = 2x + 1$. A graph of this equation is shown in Figure A.1. In practice, you can find the slope by drawing a right triangle for which the line is the hypotenuse. Then, one leg gives Δy, and the other gives Δx. In the figure, $\Delta y = 8$ and $\Delta x = 4$.

At several places in the text, an equation is rearranged into the form of a straight line in order to determine information from the slope and/or the intercept. For example, in Chapter 16, we obtained the following expression:

$$\ln \frac{[A]_0}{[A]_t} = kt$$

Based on the properties of logarithms, we have

$$\ln [A]_0 - \ln [A]_t = kt$$

Rearranging into the form of an equation for a straight line gives

$$\ln [A]_t = -kt + \ln [A]_0$$
$$y \quad = mx + \quad b$$

Thus, a plot of $\ln [A]_t$ vs. t is a straight line, from which you can see that the slope is $-k$ (the negative of the rate constant) and the intercept is $\ln [A]_0$ (the natural logarithm of the initial concentration of A).

At many other places in the text, linear relationships occur that were not shown in graphical terms. For example, the conversion of temperature scales in Chapter 1 can also be expressed in the form of a straight line:

$$T(\text{in } ^\circ\text{F}) = \tfrac{9}{5}T(\text{in } ^\circ\text{C}) + 32$$
$$y \quad = mx \quad + b$$

Standard Thermodynamic Values for Selected Substances*

Substance or Ion	ΔH_f° (kJ/mol)	ΔG_f° (kJ/mol)	S° (J/mol·K)
$e^-(g)$	0	0	20.87
Aluminum			
$Al(s)$	0	0	28.3
$Al^{3+}(aq)$	−524.7	−481.2	−313
$AlCl_3(s)$	−704.2	−628.9	110.7
$Al_2O_3(s)$	−1676	−1582	50.94
Barium			
$Ba(s)$	0	0	62.5
$Ba(g)$	175.6	144.8	170.28
$Ba^{2+}(g)$	1649.9	—	—
$Ba^{2+}(aq)$	−538.36	−560.7	13
$BaCl_2(s)$	−806.06	−810.9	126
$BaCO_3(s)$	−1219	−1139	112
$BaO(s)$	−548.1	−520.4	72.07
$BaSO_4(s)$	−1465	−1353	132
Boron			
B(β-rhombohedral)	0	0	5.87
$BF_3(g)$	−1137.0	−1120.3	254.0
$BCl_3(g)$	−403.8	−388.7	290.0
$B_2H_6(g)$	35	86.6	232.0
$B_2O_3(s)$	−1272	−1193	53.8
$H_3BO_3(s)$	−1094.3	−969.01	88.83
Bromine			
$Br_2(l)$	0	0	152.23
$Br_2(g)$	30.91	3.13	245.38
$Br(g)$	111.9	82.4	174.90
$Br^-(g)$	−218.9	—	—
$Br^-(aq)$	−120.9	−102.82	80.71
$HBr(g)$	−36.3	−53.5	198.59
Cadmium			
$Cd(s)$	0	0	51.5
$Cd(g)$	112.8	78.2	167.64
$Cd^{2+}(aq)$	−72.38	−77.74	−61.1
$CdS(s)$	−144	−141	71
Calcium			
$Ca(s)$	0	0	41.6
$Ca(g)$	192.6	158.9	154.78
$Ca^{2+}(g)$	1934.1	—	—
$Ca^{2+}(aq)$	−542.96	−553.04	−55.2
$CaF_2(s)$	−1215	−1162	68.87
$CaCl_2(s)$	−795.0	−750.2	114
$CaCO_3(s)$	−1206.9	−1128.8	92.9
$CaO(s)$	−635.1	−603.5	38.2
$Ca(OH)_2(s)$	−986.09	−898.56	83.39
$Ca_3(PO_4)_2(s)$	−4138	−3899	263
$CaSO_4(s)$	−1432.7	−1320.3	107
Carbon			
C(graphite)	0	0	5.686
C(diamond)	1.896	2.866	2.439
$C(g)$	715.0	669.6	158.0
$CO(g)$	−110.5	−137.2	197.5
$CO_2(g)$	−393.5	−394.4	213.7
$CO_2(aq)$	−412.9	−386.2	121
$CO_3^{2-}(aq)$	−676.26	−528.10	−53.1
$HCO_3^-(aq)$	−691.11	587.06	95.0
$H_2CO_3(aq)$	−698.7	−623.42	191
$CH_4(g)$	−74.87	−50.81	186.1
$C_2H_2(g)$	227	209	200.85
$C_2H_4(g)$	52.47	68.36	219.22
$C_2H_6(g)$	−84.667	−32.89	229.5
$C_3H_8(g)$	−105	−24.5	269.9
$C_4H_{10}(g)$	−126	−16.7	310
$C_6H_6(l)$	49.0	124.5	172.8
$CH_3OH(g)$	−201.2	−161.9	238
$CH_3OH(l)$	−238.6	−166.2	127
$HCHO(g)$	−116	−110	219
$HCOO^-(aq)$	−410	−335	91.6
$HCOOH(l)$	−409	−346	129.0
$HCOOH(aq)$	−410	−356	164
$C_2H_5OH(g)$	−235.1	−168.6	282.6
$C_2H_5OH(l)$	−277.63	−174.8	161
$CH_3CHO(g)$	−166	−133.7	266
$CH_3COOH(l)$	−487.0	−392	160
$C_6H_{12}O_6(s)$	−1273.3	−910.56	212.1
$C_{12}H_{22}O_{11}(s)$	−2221.7	−1544.3	360.24
$CN^-(aq)$	151	166	118
$HCN(g)$	135	125	201.7
$HCN(l)$	105	121	112.8
$HCN(aq)$	105	112	129
$CS_2(g)$	117	66.9	237.79
$CS_2(l)$	87.9	63.6	151.0
$CH_3Cl(g)$	−83.7	−60.2	234
$CH_2Cl_2(l)$	−117	−63.2	179
$CHCl_3(l)$	−132	−71.5	203
$CCl_4(g)$	−96.0	−53.7	309.7

*All values at 298 K.

(continued)

Substance or Ion	ΔH_f° (kJ/mol)	ΔG_f° (kJ/mol)	S° (J/mol·K)	Substance or Ion	ΔH_f° (kJ/mol)	ΔG_f° (kJ/mol)	S° (J/mol·K)
$CCl_4(l)$	−139	−68.6	214.4	$FeCl_3(s)$	−399.5	−334.1	142
$COCl_2(g)$	−220	−206	283.74	$FeO(s)$	−272.0	−251.4	60.75
Cesium				$Fe_2O_3(s)$	−825.5	−743.6	87.400
$Cs(s)$	0	0	85.15	$Fe_3O_4(s)$	−1121	−1018	145.3
$Cs(g)$	76.7	49.7	175.5	Lead			
$Cs^+(g)$	458.5	427.1	169.72	$Pb(s)$	0	0	64.785
$Cs^+(aq)$	−248	−282.0	133	$Pb^{2+}(aq)$	1.6	−24.3	21
$CsF(s)$	−554.7	−525.4	88	$PbCl_2(s)$	−359	−314	136
$CsCl(s)$	−442.8	−414	101.18	$PbO(s)$	−218	−198	68.70
$CsBr(s)$	−395	−383	121	$PbO_2(s)$	−276.6	−219.0	76.6
$CsI(s)$	−337	−333	130	$PbS(s)$	−98.3	−96.7	91.3
Chlorine				$PbSO_4(s)$	−918.39	−811.24	147
$Cl_2(g)$	0	0	223.0	Lithium			
$Cl(g)$	121.0	105.0	165.1	$Li(s)$	0	0	29.10
$Cl^-(g)$	−234	−240	153.25	$Li(g)$	161	128	138.67
$Cl^-(aq)$	−167.46	−131.17	55.10	$Li^+(g)$	687.163	649.989	132.91
$HCl(g)$	−92.31	−95.30	186.79	$Li^+(aq)$	−278.46	−293.8	14
$HCl(aq)$	−167.46	−131.17	55.06	$LiF(s)$	−616.9	−588.7	35.66
$ClO_2(g)$	102	120	256.7	$LiCl(s)$	−408	−384	59.30
$Cl_2O(g)$	80.3	97.9	266.1	$LiBr(s)$	−351	−342	74.1
Chromium				$LiI(s)$	−270	−270	85.8
$Cr(s)$	0	0	23.8	Magnesium			
$Cr^{3+}(aq)$	−1971	—	—	$Mg(s)$	0	0	32.69
$CrO_4^{2-}(aq)$	−863.2	−706.3	38	$Mg(g)$	150	115	148.55
$Cr_2O_7^{2-}(aq)$	−1461	−1257	214	$Mg^{2+}(g)$	2351	—	—
Copper				$Mg^{2+}(aq)$	−461.96	−456.01	118
$Cu(s)$	0	0	33.1	$MgCl_2(s)$	−641.6	−592.1	89.630
$Cu(g)$	341.1	301.4	166.29	$MgCO_3(s)$	−1112	−1028	65.86
$Cu^+(aq)$	51.9	50.2	−26	$MgO(s)$	−601.2	−569.0	26.9
$Cu^{2+}(aq)$	64.39	64.98	−98.7	$Mg_3N_2(s)$	−461	−401	88
$Cu_2O(s)$	−168.6	−146.0	93.1	Manganese			
$CuO(s)$	−157.3	−130	42.63	$Mn(s, \propto)$	0	0	31.8
$Cu_2S(s)$	−79.5	−86.2	120.9	$Mn^{2+}(aq)$	−219	−223	−84
$CuS(s)$	−53.1	−53.6	66.5	$MnO_2(s)$	−520.9	−466.1	53.1
Fluorine				$MnO_4^-(aq)$	−518.4	−425.1	190
$F_2(g)$	0	0	202.7	Mercury			
$F(g)$	78.9	61.8	158.64	$Hg(l)$	0	0	76.027
$F^-(g)$	−255.6	−262.5	145.47	$Hg(g)$	61.30	31.8	174.87
$F^-(aq)$	−329.1	−276.5	−9.6	$Hg^{2+}(aq)$	171	164.4	−32
$HF(g)$	−273	−275	173.67	$Hg_2^{2+}(aq)$	172	153.6	84.5
Hydrogen				$HgCl_2(s)$	−230	−184	144
$H_2(g)$	0	0	130.6	$Hg_2Cl_2(s)$	−264.9	−210.66	196
$H(g)$	218.0	203.30	114.60	$HgO(s)$	−90.79	−58.50	70.27
$H^+(aq)$	0	0	0	Nitrogen			
$H^+(g)$	1536.3	1517.1	108.83	$N_2(g)$	0	0	191.5
Iodine				$N(g)$	473	456	153.2
$I_2(s)$	0	0	116.14	$N_2O(g)$	82.05	104.2	219.7
$I_2(g)$	62.442	19.38	260.58	$NO(g)$	90.29	86.60	210.65
$I(g)$	106.8	70.21	180.67	$NO_2(g)$	33.2	51	239.9
$I^-(g)$	−194.7	—	—	$N_2O_3(g)$	86.6	142.4	314.7
$I^-(aq)$	−55.94	−51.67	109.4	$N_2O_4(g)$	9.16	97.7	304.3
$HI(g)$	25.9	1.3	206.33	$N_2O_5(g)$	11	118	346
Iron				$N_2O_5(s)$	−43.1	114	178
$Fe(s)$	0	0	27.3	$NH_3(g)$	−45.9	−16	193
$Fe^{3+}(aq)$	−47.7	−10.5	−293	$NH_3(aq)$	−80.83	26.7	110
$Fe^{2+}(aq)$	−87.9	−84.94	113	$N_2H_4(l)$	50.63	149.2	121.2
$FeCl_2(s)$	−341.8	−302.3	117.9	$NO_3^-(aq)$	−206.57	−110.5	146

(continued)

Substance or Ion	ΔH_f° (kJ/mol)	ΔG_f° (kJ/mol)	S° (J/mol·K)
$HNO_3(l)$	−173.23	−79.914	155.6
$HNO_3(aq)$	−206.57	−110.5	146
$NF_3(g)$	−125	−83.3	260.6
$NOCl(g)$	51.71	66.07	261.6
$NH_4Cl(s)$	−314.4	−203.0	94.6
Oxygen			
$O_2(g)$	0	0	205.0
$O(g)$	249.2	231.7	160.95
$O_3(g)$	143	163	238.82
$OH^-(aq)$	−229.94	−157.30	−10.54
$H_2O(g)$	−241.826	−228.60	188.72
$H_2O(l)$	−285.840	−237.192	69.940
$H_2O_2(l)$	−187.8	−120.4	110
$H_2O_2(aq)$	−191.2	−134.1	144
Phosphorus			
$P_4(s, \text{white})$	0	0	41.1
$P(g)$	314.6	278.3	163.1
$P(s, \text{red})$	−17.6	−12.1	22.8
$P_2(g)$	144	104	218
$P_4(g)$	58.9	24.5	280
$PCl_3(g)$	−287	−268	312
$PCl_3(l)$	−320	−272	217
$PCl_5(g)$	−402	−323	353
$PCl_5(s)$	−443.5	—	—
$P_4O_{10}(s)$	−2984	−2698	229
$PO_4^{3-}(aq)$	−1266	−1013	−218
$HPO_4^{2-}(aq)$	−1281	−1082	−36
$H_2PO_4^-(aq)$	−1285	−1135	89.1
$H_3PO_4(aq)$	−1277	−1019	228
$H_3PO_4(l)$	−1271.7	−1123.6	150.8
Potassium			
$K(s)$	0	0	64.672
$K(g)$	89.2	60.7	160.23
$K^+(g)$	514.197	481.202	154.47
$K^+(aq)$	−251.2	−282.28	103
$KF(s)$	−568.6	−538.9	66.55
$KCl(s)$	−436.7	−409.2	82.59
$KBr(s)$	−394	−380	95.94
$KI(s)$	−328	−323	106.39
$KOH(s)$	−424.8	−379.1	78.87
$KClO_3(s)$	−397.7	−296.3	143.1
$KClO_4(s)$	−432.75	−303.2	151.0
Rubidium			
$Rb(s)$	0	0	69.5
$Rb(g)$	85.81	55.86	169.99
$Rb^+(g)$	495.04	—	—
$Rb^+(aq)$	−246	−282.2	124
$RbF(s)$	−549.28	—	—
$RbCl(s)$	−435.35	−407.8	95.90
$RbBr(s)$	−389.2	−378.1	108.3
$RbI(s)$	−328	−326	118.0
Silicon			
$Si(s)$	0	0	18.0
$SiF_4(g)$	−1614.9	−1572.7	282.4
$SiO_2(s)$	−910.9	−856.5	41.5
Silver			
$Ag(s)$	0	0	42.702
$Ag(g)$	289.2	250.4	172.892
$Ag^+(aq)$	105.9	77.111	73.93

Substance or Ion	ΔH_f° (kJ/mol)	ΔG_f° (kJ/mol)	S° (J/mol·K)
$AgF(s)$	−203	−185	84
$AgCl(s)$	−127.03	−109.72	96.11
$AgBr(s)$	−99.51	−95.939	107.1
$AgI(s)$	−62.38	−66.32	114
$AgNO_3(s)$	−45.06	19.1	128.2
$Ag_2S(s)$	−31.8	−40.3	146
Sodium			
$Na(s)$	0	0	51.446
$Na(g)$	107.76	77.299	153.61
$Na^+(g)$	609.839	574.877	147.85
$Na^+(aq)$	−239.66	−261.87	60.2
$NaF(s)$	−575.4	−545.1	51.21
$NaCl(s)$	−411.1	−384.0	72.12
$NaBr(s)$	−361	−349	86.82
$NaOH(s)$	−425.609	−379.53	64.454
$Na_2CO_3(s)$	−1130.8	−1048.1	139
$NaHCO_3(s)$	−947.7	−851.9	102
$NaI(s)$	−288	−285	98.5
Strontium			
$Sr(s)$	0	0	54.4
$Sr(g)$	164	110	164.54
$Sr^{2+}(g)$	1784	—	—
$Sr^{2+}(aq)$	−545.51	−557.3	−39
$SrCl_2(s)$	−828.4	−781.2	117
$SrCO_3(s)$	−1218	−1138	97.1
$SrO(s)$	−592.0	−562.4	55.5
$SrSO_4(s)$	−1445	−1334	122
Sulfur			
S(rhombic)	0	0	31.9
S(monoclinic)	0.3	0.096	32.6
$S(g)$	279	239	168
$S_2(g)$	129	80.1	228.1
$S_8(g)$	101	49.1	430.211
$S^{2-}(aq)$	41.8	83.7	22
$HS^-(aq)$	−17.7	12.6	61.1
$H_2S(g)$	−20.2	−33	205.6
$H_2S(aq)$	−39	−27.4	122
$SO_2(g)$	−296.8	−300.2	248.1
$SO_3(g)$	−396	−371	256.66
$SO_4^{2-}(aq)$	−907.51	−741.99	17
$HSO_4^-(aq)$	−885.75	−752.87	126.9
$H_2SO_4(l)$	−813.989	−690.059	156.90
$H_2SO_4(aq)$	−907.51	−741.99	17
Tin			
Sn(white)	0	0	51.5
Sn(gray)	3	4.6	44.8
$SnCl_4(l)$	−545.2	−474.0	259
$SnO_2(s)$	−580.7	−519.7	52.3
Titanium			
$Ti(s)$	0	0	30.7
$TiCl_4(l)$	−804.2	−737.2	252.3
$TiO_2(s)$	−944.0	−888.8	50.6
Zinc			
$Zn(s)$	0	0	41.6
$Zn(g)$	130.5	94.93	160.9
$Zn^{2+}(aq)$	−152.4	−147.21	−106.5
$ZnO(s)$	−348.0	−318.2	43.9
$ZnS(s, \text{zinc blende})$	−203	−198	57.7

APPENDIX C

Equilibrium Constants for Selected Substances*

Dissociation (Ionization) Constants (K_a) of Selected Acids

Name and Formula	Lewis Structure†	K_{a1}	K_{a2}	K_{a3}
Acetic acid CH_3COOH	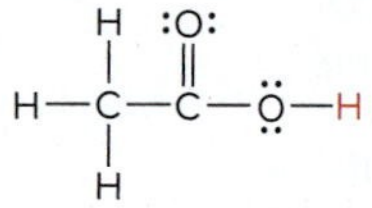	1.8×10^{-5}		
Acetylsalicylic acid $CH_3COOC_6H_4COOH$	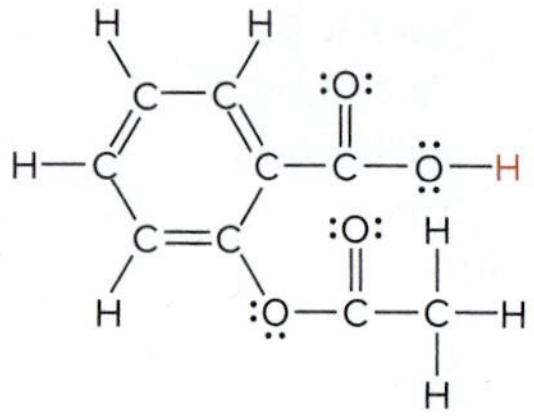	3.6×10^{-4}		
Adipic acid $HOOC(CH_2)_4COOH$	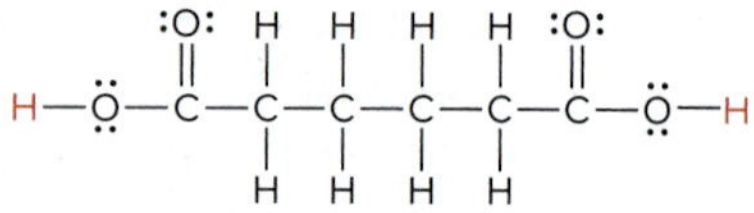	3.8×10^{-5}	3.8×10^{-6}	
Arsenic acid H_3AsO_4	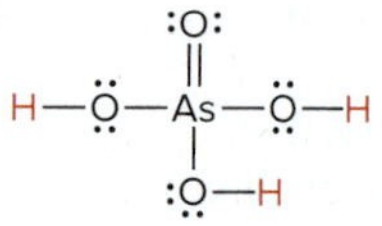	6×10^{-3}	1.1×10^{-7}	3×10^{-12}
Ascorbic acid $H_2C_6H_6O_6$	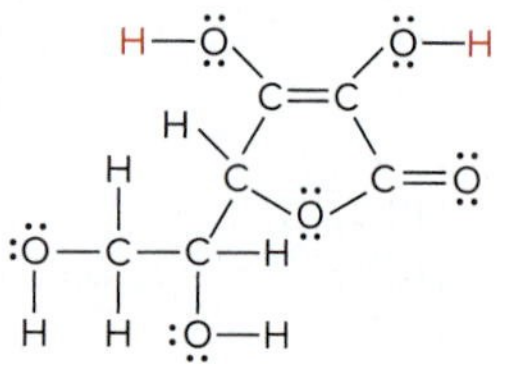	1.0×10^{-5}	5×10^{-12}	
Benzoic acid C_6H_5COOH	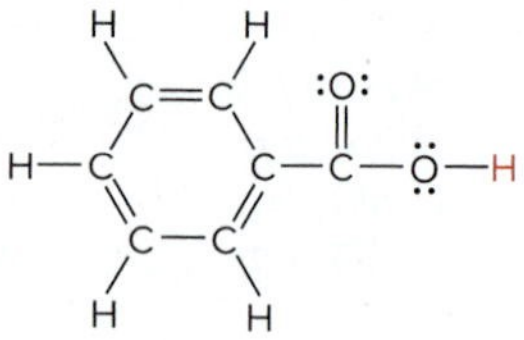	6.3×10^{-5}		
Carbonic acid H_2CO_3	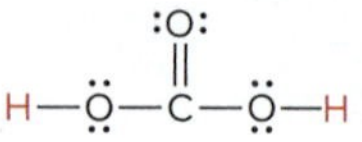	4.5×10^{-7}	4.7×10^{-11}	
Chloroacetic acid $ClCH_2COOH$	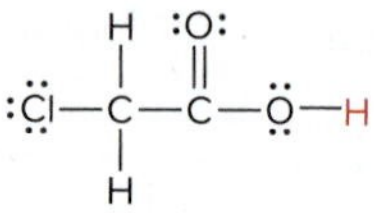	1.4×10^{-3}		
Chlorous acid $HClO_2$	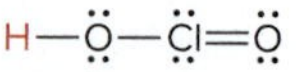	1.1×10^{-2}		

*All values at 298 K, except for acetylsalicylic acid, which is at 37°C (310 K) in 0.15 *M* NaCl.

†Acidic (ionizable) proton(s) shown in red. Structures have lowest formal charges. Benzene rings show one resonance form.

(continued)

Dissociation (Ionization) Constants (K_a) of Selected Acids

Name and Formula	Lewis Structure†	K_{a1}	K_{a2}	K_{a3}
Citric acid $HOOCCH_2C(OH)(COOH)CH_2COOH$	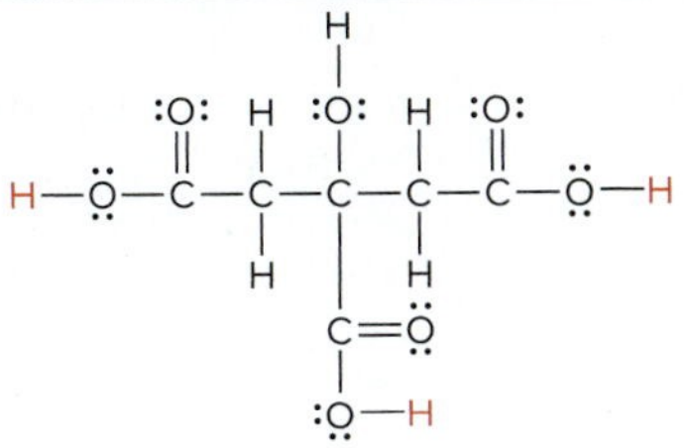	7.4×10^{-4}	1.7×10^{-5}	4.0×10^{-7}
Formic acid $HCOOH$	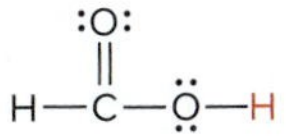	1.8×10^{-4}		
Glyceric acid $HOCH_2CH(OH)COOH$	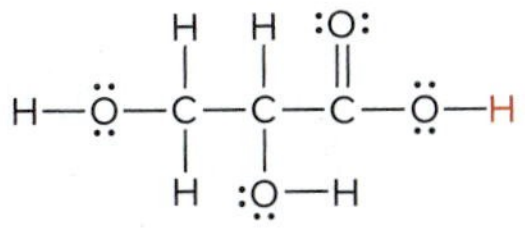	2.9×10^{-4}		
Glycolic acid $HOCH_2COOH$	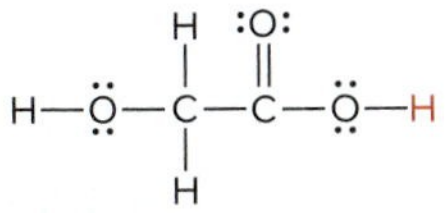	1.5×10^{-4}		
Glyoxylic acid $HC(O)COOH$	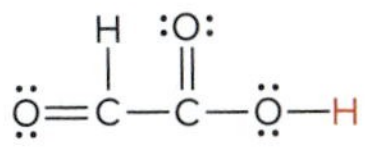	3.5×10^{-4}		
Hydrocyanic acid HCN	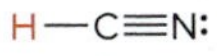	6.2×10^{-10}		
Hydrofluoric acid HF	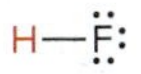	6.8×10^{-4}		
Hydrosulfuric acid H_2S	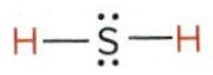	9×10^{-8}	1×10^{-17}	
Hypobromous acid $HBrO$	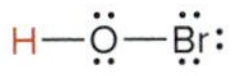	2.3×10^{-9}		
Hypochlorous acid $HClO$	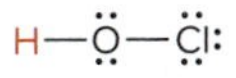	2.9×10^{-8}		
Hypoiodous acid HIO	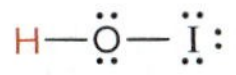	2.3×10^{-11}		
Iodic acid HIO_3	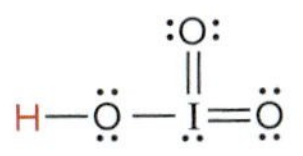	1.6×10^{-1}		
Lactic acid $CH_3CH(OH)COOH$	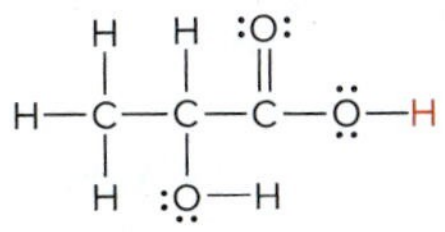	1.4×10^{-4}		
Maleic acid $HOOCCH{=}CHCOOH$	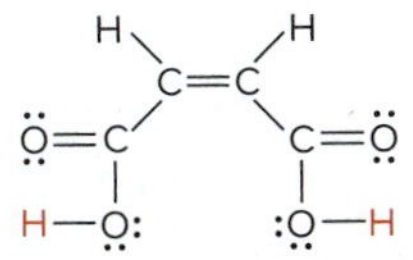	1.2×10^{-2}	4.7×10^{-7}	

(continued)

Dissociation (Ionization) Constants (K_a) of Selected Acids (*continued*)

Name and Formula	Lewis Structure†	K_{a1}	K_{a2}	K_{a3}
Malonic acid $HOOCCH_2COOH$	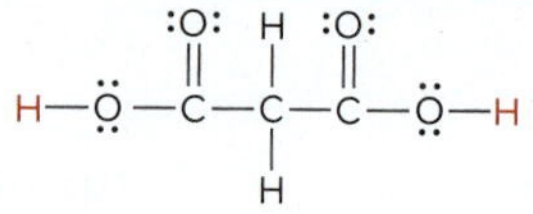	1.4×10^{-3}	2.0×10^{-6}	
Nitrous acid HNO_2	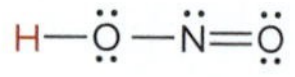	7.1×10^{-4}		
Oxalic acid HOOCCOOH	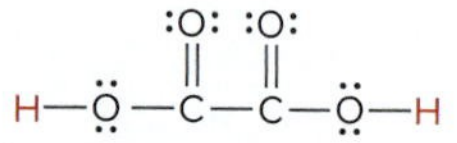	5.6×10^{-2}	5.4×10^{-5}	
Phenol C_6H_5OH	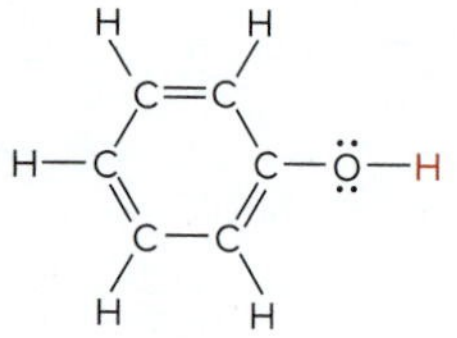	1.0×10^{-10}		
Phenylacetic acid $C_6H_5CH_2COOH$	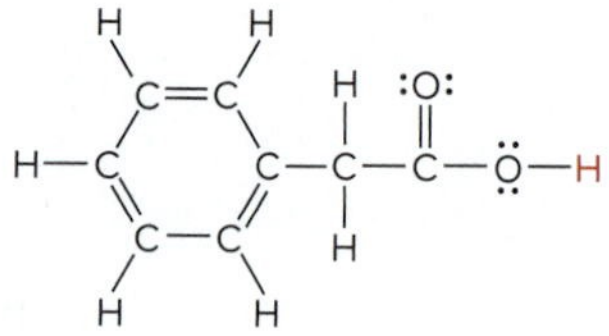	4.9×10^{-5}		
Phosphoric acid H_3PO_4	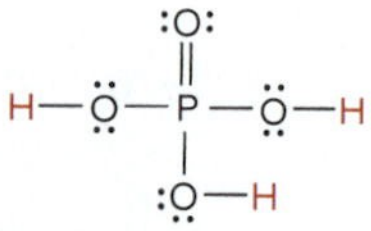	7.2×10^{-3}	6.3×10^{-8}	4.2×10^{-13}
Phosphorous acid $HPO(OH)_2$	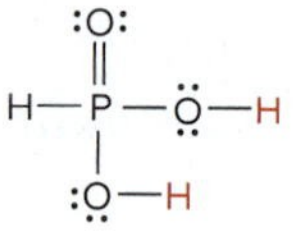	3×10^{-2}	1.7×10^{-7}	
Propanoic acid CH_3CH_2COOH	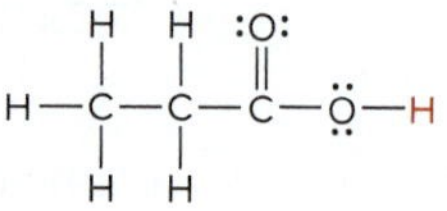	1.3×10^{-5}		
Pyruvic acid $CH_3C(O)COOH$	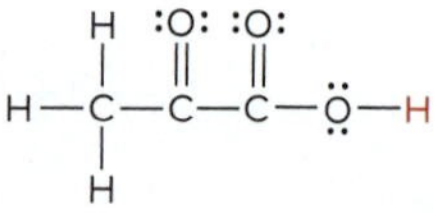	2.8×10^{-3}		
Succinic acid $HOOCCH_2CH_2COOH$	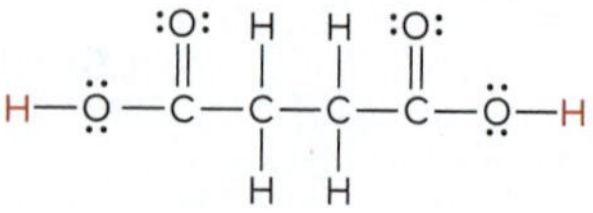	6.2×10^{-5}	2.3×10^{-6}	
Sulfuric acid H_2SO_4	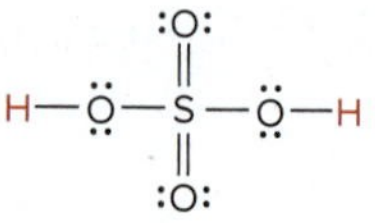	Very large	1.0×10^{-2}	
Sulfurous acid H_2SO_3	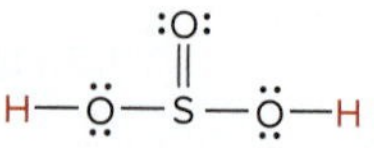	1.4×10^{-2}	6.5×10^{-8}	

(*continued*)

Dissociation (Ionization) Constants (K_b) of Selected Amine Bases

Name and Formula	Lewis Structure[†]	K_{b1}	K_{b2}
Ammonia NH_3	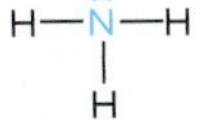	1.76×10^{-5}	
Aniline $C_6H_5NH_2$	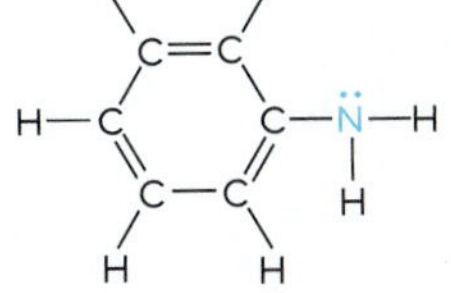	4.0×10^{-10}	
Diethylamine $(CH_3CH_2)_2NH$	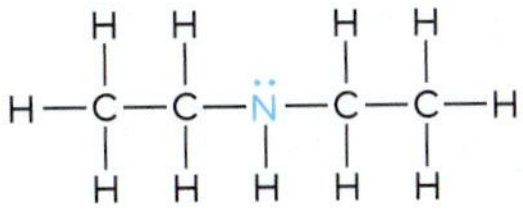	8.6×10^{-4}	
Dimethylamine $(CH_3)_2NH$	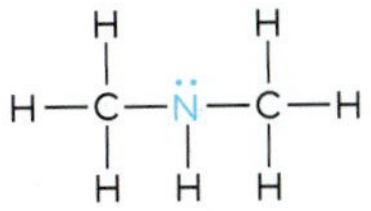	5.9×10^{-4}	
Ethanolamine $HOCH_2CH_2NH_2$	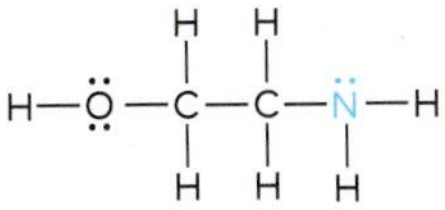	3.2×10^{-5}	
Ethylamine $CH_3CH_2NH_2$	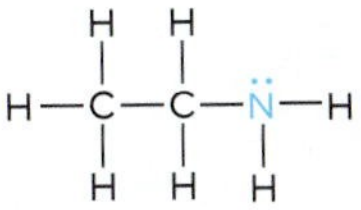	4.3×10^{-4}	
Ethylenediamine $H_2NCH_2CH_2NH_2$	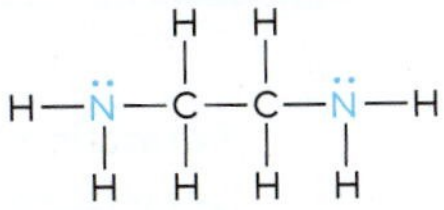	8.5×10^{-5}	7.1×10^{-8}
Methylamine CH_3NH_2	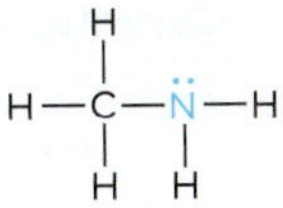	4.4×10^{-4}	
tert-Butylamine $(CH_3)_3CNH_2$	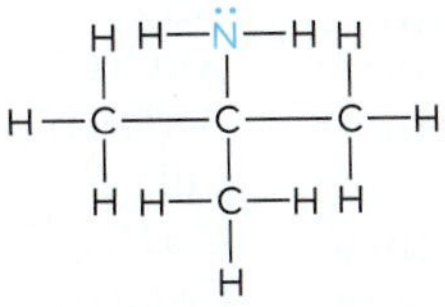	4.8×10^{-4}	
Piperidine $C_5H_{10}NH$	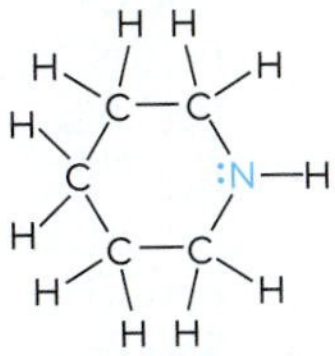	1.3×10^{-3}	
n-Propylamine $CH_3CH_2CH_2NH_2$	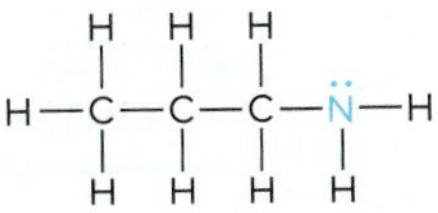	3.5×10^{-4}	

[†]Blue type indicates the basic nitrogen and its lone pair.

(continued)

Dissociation (Ionization) Constants (K_b) of Selected Amine Bases (*continued*)

Name and Formula	Lewis Structure†	K_{b1}	K_{b2}
Isopropylamine $(CH_3)_2CHNH_2$	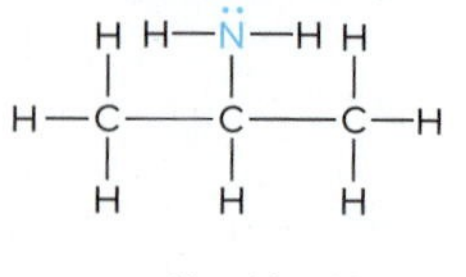	4.7×10^{-4}	
1,3-Propylenediamine $H_2NCH_2CH_2CH_2NH_2$	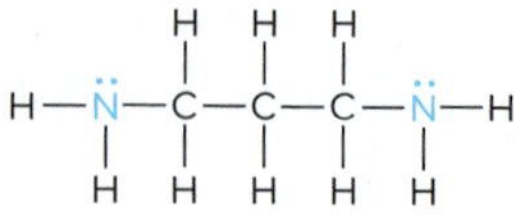	3.1×10^{-4}	3.0×10^{-6}
Pyridine C_5H_5N	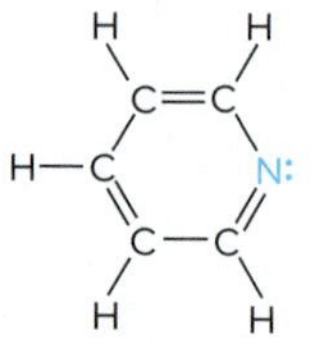	1.7×10^{-9}	
Triethylamine $(CH_3CH_2)_3N$	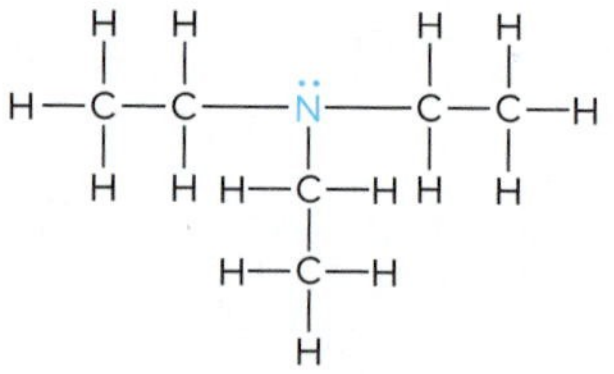	5.2×10^{-4}	
Trimethylamine $(CH_3)_3N$	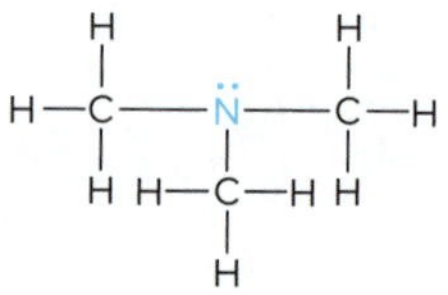	6.3×10^{-5}	

Dissociation (Ionization) Constants (K_a) of Some Hydrated Metal Ions

Free Ion	Hydrated Ion	K_a
Fe^{3+}	$Fe(H_2O)_6^{3+}(aq)$	6×10^{-3}
Sn^{2+}	$Sn(H_2O)_6^{2+}(aq)$	4×10^{-4}
Cr^{3+}	$Cr(H_2O)_6^{3+}(aq)$	1×10^{-4}
Al^{3+}	$Al(H_2O)_6^{3+}(aq)$	1×10^{-5}
Cu^{2+}	$Cu(H_2O)_6^{2+}(aq)$	3×10^{-8}
Pb^{2+}	$Pb(H_2O)_6^{2+}(aq)$	3×10^{-8}
Zn^{2+}	$Zn(H_2O)_6^{2+}(aq)$	1×10^{-9}
Co^{2+}	$Co(H_2O)_6^{2+}(aq)$	2×10^{-10}
Ni^{2+}	$Ni(H_2O)_6^{2+}(aq)$	1×10^{-10}

Formation Constants (K_f) of Some Complex Ions

Complex Ion	K_f
$Ag(CN)_2^-$	3.0×10^{20}
$Ag(NH_3)_2^+$	1.7×10^{7}
$Ag(S_2O_3)_2^{3-}$	4.7×10^{13}
AlF_6^{3-}	$4\ \times10^{19}$
$Al(OH)_4^-$	$3\ \times10^{33}$
$Be(OH)_4^{2-}$	$4\ \times10^{18}$
CdI_4^{2-}	$1\ \times10^{6}$
$Co(OH)_4^{2-}$	$5\ \times10^{9}$
$Cr(OH)_4^-$	8.0×10^{29}
$Cu(NH_3)_4^{2+}$	5.6×10^{11}
$Fe(CN)_6^{4-}$	$3\ \times10^{35}$
$Fe(CN)_6^{3-}$	4.0×10^{43}
$Hg(CN)_4^{2-}$	9.3×10^{38}
$Ni(NH_3)_6^{2+}$	2.0×10^{8}
$Pb(OH)_3^-$	$8\ \times10^{13}$
$Sn(OH)_3^-$	$3\ \times10^{25}$
$Zn(CN)_4^{2-}$	4.2×10^{19}
$Zn(NH_3)_4^{2+}$	7.8×10^{8}
$Zn(OH)_4^{2-}$	$3\ \times10^{15}$

Solubility-Product Constants (K_{sp}) of Slightly Soluble Ionic Compounds

Name, Formula	K_{sp}
Carbonates	
Barium carbonate, $BaCO_3$	2.0×10^{-9}
Cadmium carbonate, $CdCO_3$	1.8×10^{-14}
Calcium carbonate, $CaCO_3$	3.3×10^{-9}
Cobalt(II) carbonate, $CoCO_3$	1.0×10^{-10}
Copper(II) carbonate, $CuCO_3$	3×10^{-12}
Lead(II) carbonate, $PbCO_3$	7.4×10^{-14}
Magnesium carbonate, $MgCO_3$	3.5×10^{-8}
Mercury(I) carbonate, Hg_2CO_3	8.9×10^{-17}
Nickel(II) carbonate, $NiCO_3$	1.3×10^{-7}
Strontium carbonate, $SrCO_3$	5.4×10^{-10}
Zinc carbonate, $ZnCO_3$	1.0×10^{-10}
Chromates	
Barium chromate, $BaCrO_4$	2.1×10^{-10}
Calcium chromate, $CaCrO_4$	1×10^{-8}
Lead(II) chromate, $PbCrO_4$	2.3×10^{-13}
Silver chromate, Ag_2CrO_4	2.6×10^{-12}
Cyanides	
Mercury(I) cyanide, $Hg_2(CN)_2$	5×10^{-40}
Silver cyanide, $AgCN$	2.2×10^{-16}
Halides	
Fluorides	
Barium fluoride, BaF_2	1.5×10^{-6}
Calcium fluoride, CaF_2	3.2×10^{-11}
Lead(II) fluoride, PbF_2	3.6×10^{-8}
Magnesium fluoride, MgF_2	7.4×10^{-9}
Strontium fluoride, SrF_2	2.6×10^{-9}
Chlorides	
Copper(I) chloride, $CuCl$	1.9×10^{-7}
Lead(II) chloride, $PbCl_2$	1.7×10^{-5}
Silver chloride, $AgCl$	1.8×10^{-10}
Bromides	
Copper(I) bromide, $CuBr$	5×10^{-9}
Silver bromide, $AgBr$	5.0×10^{-13}
Iodides	
Copper(I) iodide, CuI	1×10^{-12}
Lead(II) iodide, PbI_2	7.9×10^{-9}
Mercury(I) iodide, Hg_2I_2	4.7×10^{-29}
Silver iodide, AgI	8.3×10^{-17}
Hydroxides	
Aluminum hydroxide, $Al(OH)_3$	3×10^{-34}
Cadmium hydroxide, $Cd(OH)_2$	7.2×10^{-15}
Calcium hydroxide, $Ca(OH)_2$	6.5×10^{-6}
Cobalt(II) hydroxide, $Co(OH)_2$	1.3×10^{-15}
Copper(II) hydroxide, $Cu(OH)_2$	2.2×10^{-20}
Iron(II) hydroxide, $Fe(OH)_2$	4.1×10^{-15}
Iron(III) hydroxide, $Fe(OH)_3$	1.6×10^{-39}
Magnesium hydroxide, $Mg(OH)_2$	6.3×10^{-10}
Manganese(II) hydroxide, $Mn(OH)_2$	1.6×10^{-13}
Nickel(II) hydroxide, $Ni(OH)_2$	6×10^{-16}
Zinc hydroxide, $Zn(OH)_2$	3×10^{-16}
Iodates	
Barium iodate, $Ba(IO_3)_2$	1.5×10^{-9}
Calcium iodate, $Ca(IO_3)_2$	7.1×10^{-7}
Lead(II) iodate, $Pb(IO_3)_2$	2.5×10^{-13}
Silver iodate, $AgIO_3$	3.1×10^{-8}
Strontium iodate, $Sr(IO_3)_2$	3.3×10^{-7}
Zinc iodate, $Zn(IO_3)_2$	3.9×10^{-6}
Oxalates	
Barium oxalate dihydrate, $BaC_2O_4\cdot2H_2O$	1.1×10^{-7}
Calcium oxalate monohydrate, $CaC_2O_4\cdot H_2O$	2.3×10^{-9}
Strontium oxalate monohydrate, $SrC_2O_4\cdot H_2O$	5.6×10^{-8}
Phosphates	
Calcium phosphate, $Ca_3(PO_4)_2$	1.2×10^{-29}
Magnesium phosphate, $Mg_3(PO_4)_2$	5.2×10^{-24}
Silver phosphate, Ag_3PO_4	2.6×10^{-18}
Sulfates	
Barium sulfate, $BaSO_4$	1.1×10^{-10}
Calcium sulfate, $CaSO_4$	2.4×10^{-5}
Lead(II) sulfate, $PbSO_4$	1.6×10^{-8}
Radium sulfate, $RaSO_4$	2×10^{-11}
Silver sulfate, Ag_2SO_4	1.5×10^{-5}
Strontium sulfate, $SrSO_4$	3.2×10^{-7}
Sulfides	
Cadmium sulfide, CdS	1.0×10^{-24}
Copper(II) sulfide, CuS	8×10^{-34}
Iron(II) sulfide, FeS	8×10^{-16}
Lead(II) sulfide, PbS	3×10^{-25}
Manganese(II) sulfide, MnS	3×10^{-11}
Mercury(II) sulfide, HgS	2×10^{-50}
Nickel(II) sulfide, NiS	3×10^{-16}
Silver sulfide, Ag_2S	8×10^{-48}
Tin(II) sulfide, SnS	1.3×10^{-23}
Zinc sulfide, ZnS	2.0×10^{-22}

APPENDIX D

Standard Electrode (Half-Cell) Potentials*

Half-Reaction	$E°$(V)
$F_2(g) + 2e^- \rightleftharpoons 2F^-(aq)$	+2.87
$O_3(g) + 2H^+(aq) + 2e^- \rightleftharpoons O_2(g) + H_2O(l)$	+2.07
$Co^{3+}(aq) + e^- \rightleftharpoons Co^{2+}(aq)$	+1.82
$H_2O_2(aq) + 2H^+(aq) + 2e^- \rightleftharpoons 2H_2O(l)$	+1.77
$PbO_2(s) + 3H^+(aq) + HSO_4^-(aq) + 2e^- \rightleftharpoons PbSO_4(s) + 2H_2O(l)$	+1.70
$Ce^{4+}(aq) + e^- \rightleftharpoons Ce^{3+}(aq)$	+1.61
$MnO_4^-(aq) + 8H^+(aq) + 5e^- \rightleftharpoons Mn^{2+}(aq) + 4H_2O(l)$	+1.51
$Au^{3+}(aq) + 3e^- \rightleftharpoons Au(s)$	+1.50
$Cl_2(g) + 2e^- \rightleftharpoons 2Cl^-(aq)$	+1.36
$Cr_2O_7^{2-}(aq) + 14H^+(aq) + 6e^- \rightleftharpoons 2Cr^{3+}(aq) + 7H_2O(l)$	+1.33
$MnO_2(s) + 4H^+(aq) + 2e^- \rightleftharpoons Mn^{2+}(aq) + 2H_2O(l)$	+1.23
$O_2(g) + 4H^+(aq) + 4e^- \rightleftharpoons 2H_2O(l)$	+1.23
$Br_2(l) + 2e^- \rightleftharpoons 2Br^-(aq)$	+1.07
$NO_3^-(aq) + 4H^+(aq) + 3e^- \rightleftharpoons NO(g) + 2H_2O(l)$	+0.96
$2Hg^{2+}(aq) + 2e^- \rightleftharpoons Hg_2^{2+}(aq)$	+0.92
$Hg^{2+}(aq) + 2e^- \rightleftharpoons Hg(l)$	+0.85
$Ag^+(aq) + e^- \rightleftharpoons Ag(s)$	+0.80
$Hg_2^{2+}(aq) + 2e^- \rightleftharpoons 2Hg(l)$	+0.79
$Fe^{3+}(aq) + e^- \rightleftharpoons Fe^{2+}(aq)$	+0.77
$O_2(g) + 2H^+(aq) + 2e^- \rightleftharpoons H_2O_2(aq)$	+0.68
$MnO_4^-(aq) + 2H_2O(l) + 3e^- \rightleftharpoons MnO_2(s) + 4OH^-(aq)$	+0.59
$I_2(s) + 2e^- \rightleftharpoons 2I^-(aq)$	+0.53
$O_2(g) + 2H_2O(l) + 4e^- \rightleftharpoons 4OH^-(aq)$	+0.40
$Cu^{2+}(aq) + 2e^- \rightleftharpoons Cu(s)$	+0.34
$AgCl(s) + e^- \rightleftharpoons Ag(s) + Cl^-(aq)$	+0.22
$SO_4^{2-}(aq) + 4H^+(aq) + 2e^- \rightleftharpoons SO_2(g) + 2H_2O(l)$	+0.20
$Cu^{2+}(aq) + e^- \rightleftharpoons Cu^+(aq)$	+0.15
$Sn^{4+}(aq) + 2e^- \rightleftharpoons Sn^{2+}(aq)$	+0.13
$2H^+(aq) + 2e^- \rightleftharpoons H_2(g)$	0.00
$Pb^{2+}(aq) + 2e^- \rightleftharpoons Pb(s)$	−0.13
$Sn^{2+}(aq) + 2e^- \rightleftharpoons Sn(s)$	−0.14
$N_2(g) + 5H^+(aq) + 4e^- \rightleftharpoons N_2H_5^+(aq)$	−0.23
$Ni^{2+}(aq) + 2e^- \rightleftharpoons Ni(s)$	−0.25
$Co^{2+}(aq) + 2e^- \rightleftharpoons Co(s)$	−0.28
$PbSO_4(s) + H^+(aq) + 2e^- \rightleftharpoons Pb(s) + HSO_4^-(aq)$	−0.31
$Cd^{2+}(aq) + 2e^- \rightleftharpoons Cd(s)$	−0.40
$Fe^{2+}(aq) + 2e^- \rightleftharpoons Fe(s)$	−0.44
$Cr^{3+}(aq) + 3e^- \rightleftharpoons Cr(s)$	−0.74
$Zn^{2+}(aq) + 2e^- \rightleftharpoons Zn(s)$	−0.76
$2H_2O(l) + 2e^- \rightleftharpoons H_2(g) + 2OH^-(aq)$	−0.83
$Cr^{2+}(aq) + 2e^- \rightleftharpoons Cr(s)$	−0.91
$Mn^{2+}(aq) + 2e^- \rightleftharpoons Mn(s)$	−1.18
$Al^{3+}(aq) + 3e^- \rightleftharpoons Al(s)$	−1.66
$Mg^{2+}(aq) + 2e^- \rightleftharpoons Mg(s)$	−2.37
$Na^+(aq) + e^- \rightleftharpoons Na(s)$	−2.71
$Ca^{2+}(aq) + 2e^- \rightleftharpoons Ca(s)$	−2.87
$Sr^{2+}(aq) + 2e^- \rightleftharpoons Sr(s)$	−2.89
$Ba^{2+}(aq) + 2e^- \rightleftharpoons Ba(s)$	−2.90
$K^+(aq) + e^- \rightleftharpoons K(s)$	−2.93
$Li^+(aq) + e^- \rightleftharpoons Li(s)$	−3.05

*All values at 298 K. Written as reductions; $E°$ value refers to all components in their standard states: 1 *M* for dissolved species; 1 atm pressure for the gas behaving ideally; the pure substance for solids and liquids.

APPENDIX E

Answers to Selected Problems

Chapter 1

1.2 Gas molecules fill the entire container; the volume of a gas is the volume of the container. Solids and liquids have a definite volume. The volume of the container does not affect the volume of a solid or liquid. (a) gas (b) liquid (c) liquid **1.4** Physical property: a characteristic shown by a substance itself, without any interaction with or change into other substances. Chemical property: a characteristic of a substance that appears as it interacts with, or transforms into, other substances. (a) Color (yellow-green and silver to white) and physical state (gas and metal to crystals) are physical properties. The interaction between chlorine gas and sodium metal is a chemical change. (b) Color and magnetism are physical properties. No chemical changes. **1.6**(a) Physical change; there is only a temperature change. (b) Chemical change; the change in appearance indicates an irreversible chemical change. (c) Physical change; there is only a change in size, not composition. (d) Chemical change; the wood (and air) become different substances with different compositions. **1.8**(a) fuel (b) wood **1.13** Lavoisier measured the total mass of the reactants and products, not just the mass of the solids, and this total mass remained constant. His measurements showed that a gas was involved in the reaction. He called this gas *oxygen* (one of his key discoveries). **1.16** A well-designed experiment must have the following: (1) at least two variables that are related; (2) a way to control all the variables, so that only one at a time may be changed; (3) reproducible results. **1.20**(a) increases (b) remains the same (c) decreases (d) increases (e) remains the same **1.23** An extensive property depends on the amount of material present. An intensive property is the same regardless of how much material is present. (a) extensive property (b) intensive property (c) extensive property (d) intensive property **1.24**(a) $(2.54\ \text{cm})^2/(1\ \text{in})^2$ and $(1\ \text{m})^2/(100\ \text{cm})^2$ (b) $(1000\ \text{m})^2/(1\ \text{km})^2$ and $(100\ \text{cm})^2/(1\ \text{m})^2$ (c) $(1.609\times10^3\ \text{m/mi})$ and (1 h/3600 s) (d) (1000 g/2.205 lb) and $(3.531\times10^{-5}\ \text{ft}^3/\text{cm}^3)$ **1.26** 1.43 nm **1.28** 3.94×10^3 in **1.30** (a) $2.07\times10^{-9}\ \text{km}^2$ (b) \$10.43 **1.32** Answers will vary, depending on the person's mass. **1.34**(a) $5.52\times10^3\ \text{kg/m}^3$ (b) 345 lb/ft^3 **1.36**(a) $2.56\times10^{-9}\ \text{mm}^3$ (b) 10^{-10} L **1.38**(a) 9.626 cm^3 (b) 64.92 g **1.40** 2.70 g/cm^3 **1.42**(a) 20.°C; 293 K (b) 109 K; −263°F (c) −273°C; −460.°F **1.45**(a) 2.47×10^{-7} m (b) 67.6 Å **1.52**(a) none (b) none (c) $0.041\underline{0}$ (d) $4.\underline{0}1\underline{00}\times10^4$ **1.54**(a) 0.00036 (b) 35.83 (c) 22.5 **1.56** 6×10^2 **1.58**(a) 1.34 m (b) 21,621 mm^3 (c) 443 cm **1.60**(a) 1.310000×10^5 (b) 4.7×10^{-4} (c) 2.10006×10^5 (d) 2.1605×10^3 **1.62**(a) 5550 (b) 10,070. (c) 0.000000885 (d) 0.003004 **1.64**(a) 8.025×10^4 (b) 1.0098×10^{-3} (c) 7.7×10^{-11} **1.66**(a) 4.06×10^{-19} J (b) 1.61×10^{24} molecules (c) 1.82×10^5 J/mol **1.68**(a) Height measured, not exact. (b) Planets counted, exact. (c) Number of grams in a pound is not a unit definition, not exact. (d) Definition of a millimeter, exact. **1.70** 7.50 ± 0.05 cm

1.72(a) $I_{avg} = 8.72$ g; $II_{avg} = 8.72$ g; $III_{avg} = 8.50$ g; $IV_{avg} = 8.56$ g; sets I and II are most accurate. (b) Set III is the most precise but is the least accurate. (c) Set I has the best combination of high accuracy and high precision. (d) Set IV has both low accuracy and low precision.

1.74(a)

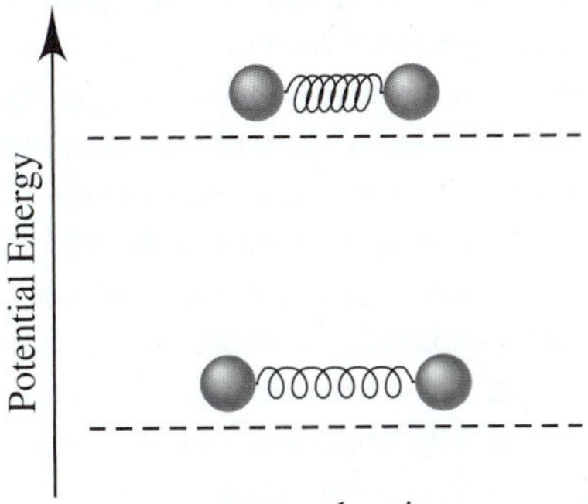

compressed spring
less stable—energy
stored in spring

(b)

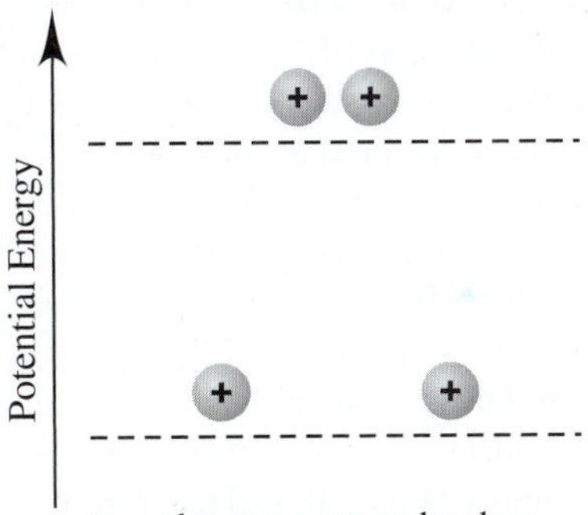

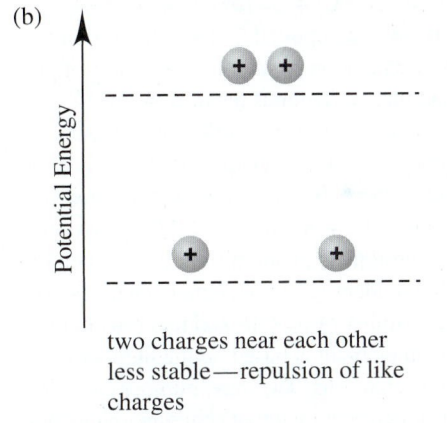

two charges near each other
less stable—repulsion of like
charges

1.76 7.7/1 **1.78**(a) density = 0.21 g/L, will float (b) CO_2 is denser than air, will sink (c) density = 0.30 g/L, will float (d) O_2 is denser than air, will sink (e) density = 1.38 g/L, will sink (f) 0.55 g for empty ball; 0.50 g for ball filled with hydrogen **1.80**(a) 8.0×10^{12} g (b) $4.1\times10^5\ \text{m}^3$ (c) \$$4.1\times10^{14}$ **1.82**(a) −195.79°C (b) −320.42°F (c) 5.05 L **1.83**(a) 2.6 m/s (b) 16 km (c) 12:45 pm **1.85** freezing point = −3.7°X; boiling point = 63.3°X **1.86** 2.3×10^{25} g oxygen; 1.4×10^{25} g silicon; 5×10^{15} g ruthenium and rhodium

Chapter 2

Answers to Boxed Reading Problems: B2.1(a) 5 peaks (b) *m/e* value for heaviest particle = 74; *m/e* value for lightest particle = 35 **B2.3**(a) Since salt dissolves in water and pepper does not, add water to mixture and filter to remove solid pepper. Evaporate water to recover solid salt. (b) Distillation purifies the water and leaves the soot behind. (c) Warm the mixture; filter to remove glass; freeze the filtrate to obtain ice. (d) Heat the mixture; the ethanol will boil off (distill), while the sugar will remain behind. (e) Use column chromatography to separate the pigments.

• **2.1** Compounds contain different types of atoms; there is only one type of atom in an element. **2.4**(a) The presence of more than one element makes pure calcium chloride a compound. (b) There is only one kind of atom, so sulfur is an element. (c) The presence of more than one compound makes baking powder a mixture. (d) The presence of more than one type of atom means cytosine cannot be an element. The specific, not variable, arrangement means it is a compound. **2.12**(a) elements, compounds, and mixtures (b) compounds (c) compounds **2.14**(a) Law of definite composition: the composition is the same regardless of its source. (b) Law of mass conservation: the total quantity of matter does not change. (c) Law of multiple proportions: two elements can combine to form two different compounds that have different proportions of those elements. **2.16**(a) No, the percent by mass of each element in a compound is fixed. (b) Yes, the *mass* of each element in a compound depends on the amount of compound. **2.18** The two experiments demonstrate the law of definite composition. The unknown compound decomposes the same way both times. The experiments also demonstrate the law of conservation of mass since the total mass before reaction equals the total mass after reaction. **2.20**(a) 1.34 g F (b) 0.514 Ca; 0.486 F (c) 51.4 mass % Ca; 48.6 mass % F **2.22**(a) 0.603 (b) 322 g Mg **2.24** 3.498×10^6 g Cu; 1.766×10^6 g S **2.26** compound 1: 0.905 S/Cl; compound 2: 0.451 S/Cl; ratio: 2.00/1.00 **2.29** Coal A **2.31** Dalton postulated that atoms of an element are identical and that compounds result from the chemical combination of specific ratios of different elements. **2.32** If you know the ratio of any two quantities and the value of one of them, the other can always be calculated; in this case, the charge and the mass/charge ratio were known. **2.36** The atomic number is the number of protons in an atom's nucleus. When the atomic number changes, the identity of the element changes. The mass number is the total number of protons and neutrons in the nucleus. The identity of an element is based on the number of protons, not the number of neutrons. The mass number can vary (by a change in number of neutrons) without changing the identity of the element. **2.39** All three isotopes have 18 protons and 18 electrons. Their respective mass numbers are 36, 38, and 40, with the respective numbers of neutrons being 18, 20, and 22. **2.41**(a) These have the same number of protons and electrons, but different numbers of neutrons; same *Z*. (b) These have the same number of neutrons, but different numbers of protons and electrons; same *N*. (c) These have different numbers of protons, neutrons, and electrons; same *A*. **2.43**(a) $^{38}_{18}Ar$ (b) $^{55}_{25}Mn$ (c) $^{109}_{47}Ag$ **2.45**(a) $^{48}_{22}Ti$ (b) $^{79}_{34}Se$ (c) $^{11}_{5}B$

2.47 69.72 amu **2.49** $^{35}Cl = 75.774\%$, $^{37}Cl = 24.226\%$ **2.52**(a) In the modern periodic table, the elements are arranged in order of increasing atomic number. (b) Elements in a group (or family) have similar chemical properties. (c) Elements can be classified as metals, metalloids, or nonmetals. **2.55** The alkali metals [Group 1A(1)] are metals and readily lose one electron to form cations; the halogens [Group 7A(17)] are nonmetals and readily gain one electron to form anions. **2.56**(a) germanium; Ge; 4A(14); metalloid (b) phosphorus; P; 5A(15); nonmetal (c) helium; He; 8A(18); nonmetal (d) lithium; Li; 1A(1); metal (e) molybdenum; Mo; 6B(6); metal **2.58**(a) Ra; 88 (b) Si; 14 (c) Cu; 63.55 amu (d) Br; 79.90 amu **2.60** Atoms of these two kinds of substances will form ionic bonds in which one or more electrons are transferred from the metal atom to the nonmetal atom to form a cation and an anion, respectively. **2.63** Coulomb's law states the energy of attraction in an ionic bond is directly proportional to the product of charges and inversely proportional to the distance between charges. The product of charges in MgO [(2+)(2−)] is greater than that in LiF [(1+)(1−)]. Thus, MgO has stronger ionic bonding. **2.66** The Group 1A(1) elements form cations, and the Group 7A(17) elements form anions. **2.68** Each potassium atom loses one electron to form an ion with a 1+ charge. Each sulfur atom gains two electrons to form an ion with a 2− charge. Two potassium atoms, losing one electron each, are required for each sulfur atom, which gains two electrons. The oppositely charged ions attract each other to form an ionic solid, K_2S. **2.70** K^+; Br^- **2.72**(a) oxygen; 17; 6A(16); 2 (b) fluorine; 19; 7A(17); 2 (c) calcium; 40; 2A(2); 4 **2.74** Lithium forms the Li^+ ion; oxygen forms the O^{2-} ion. Number of O^{2-} ions = 4.2×10^{21} O^{2-} ions. **2.76** NaCl **2.78** The subscripts in a formula give the numbers of ions in a formula unit of the compound. The subscripts indicate that there are two F^- ions for each Mg^{2+} ion. Using this information and the mass of each element, we can calculate percent mass of each element in the compound. **2.80** The two samples are similar in that both contain 20 billion oxygen atoms and 20 billion hydrogen atoms. They differ in that they contain different types of molecules: H_2O_2 molecules in the hydrogen peroxide sample, and H_2 and O_2 molecules in the mixture. In addition, the mixture contains 20 billion molecules (10 billion H_2 and 10 billion O_2), while the hydrogen peroxide sample contains 10 billion molecules. **2.84**(a) Na_3N, sodium nitride (b) SrO, strontium oxide (c) $AlCl_3$, aluminum chloride **2.86**(a) MgF_2, magnesium fluoride (b) ZnS, zinc sulfide (c) $SrCl_2$, strontium chloride **2.88**(a) $SnCl_4$ (b) iron(III) bromide (c) CuBr (d) manganese(III) oxide **2.90**(a) sodium hydrogen phosphate (b) NH_4ClO_4 (c) lead(II) acetate trihydrate (d) $NaNO_2$ **2.92**(a) BaO (b) $Fe(NO_3)_2$ (c) MgS **2.94**(a) H_2CO_3; carbonic acid (b) HIO_4; periodic acid (c) HCN; hydrocyanic acid (d) H_2S; hydrosulfuric acid **2.96**(a) ammonium ion, NH_4^+; ammonia, NH_3 (b) magnesium sulfide, MgS; magnesium sulfite, $MgSO_3$; magnesium sulfate, $MgSO_4$ (c) hydrochloric acid, HCl; chloric acid, $HClO_3$; chlorous acid, $HClO_2$ (d) cuprous bromide, CuBr; cupric bromide, $CuBr_2$ **2.98** Disulfur tetrafluoride, S_2F_4 **2.100**(a) calcium chloride (b) copper(I) oxide (c) stannic fluoride (d) hydrochloric acid **2.102**(a) $(NH_4)_2SO_4$; 132.14 amu (b) NaH_2PO_4; 119.98 amu (c) $KHCO_3$; 100.12 amu **2.104**(a) 108.02 amu (b) 331.2 amu (c) 72.08 amu **2.106**(a) 12 oxygen atoms; 342.14 amu (b) 9 hydrogen atoms; 132.06 amu (c) 8 oxygen atoms; 344.69 amu **2.108**(a) SO_3; sulfur trioxide; 80.06 amu (b) C_3H_8; propane; 44.09 amu **2.112** Separating the components of a mixture requires physical methods only; that is, no chemical changes (no changes in composition) take place, and the components maintain their chemical identities and properties throughout. Separating the components of a compound requires a chemical change (change in composition). **2.115**(a) compound (b) homogeneous mixture (c) heterogeneous mixture (d) homogeneous mixture (e) homogeneous mixture **2.117**(a) filtration (b) chromatography **2.119**(a) fraction of volume = 5.2×10^{-13} (b) mass of nucleus = 6.64466×10^{-24} g; fraction of mass = 0.999726 **2.120** strongest ionic bonding: MgO; weakest ionic bonding: RbI **2.124**(a) I = NO; II = N_2O_3; III = N_2O_5 (b) I has 1.14 g O per 1.00 g N; II, 1.71 g O; III, 2.86 g O **2.128**(a) Cl^-, 1.898 mass %; Na^+, 1.056 mass %; SO_4^{2-}, 0.265 mass %; Mg^{2+}, 0.127 mass %; Ca^{2+}, 0.04 mass %; K^+, 0.038 mass %; HCO_3^-, 0.014 mass % (b) 30.72% (c) Alkaline earth metal ions, total mass % = 0.17%; alkali metal ions, total mass % = 1.094% (d) Anions (2.177 mass %) make up a larger mass fraction than cations (1.26 mass %). **2.131** Molecular formula, $C_4H_6O_4$; molecular mass, 118.09 amu; 40.68% by mass C; 5.122% by mass H; 54.20% by mass O **2.134**(a) Formulas and masses in amu: $^{15}N_2{}^{18}O$, 48; $^{15}N_2{}^{16}O$, 46; $^{14}N_2{}^{18}O$, 46; $^{14}N_2{}^{16}O$, 44; $^{15}N^{14}N^{18}O$, 47; $^{15}N^{14}N^{16}O$, 45 (b) $^{15}N_2{}^{18}O$, least common; $^{14}N_2{}^{16}O$, most common **2.136** 58.091 amu **2.138** nitroglycerin, 39.64 mass % NO; isoamyl nitrate, 22.54 mass % NO **2.139** 0.370 lb C; 0.0222 lb H; 0.423 lb O; 0.185 lb N **2.144** (1) chemical change (2) physical change (3) chemical change (4) chemical change (5) physical change

Chapter 3

3.2(a) 12 mol C atoms (b) 1.445×10^{25} C atoms **3.7**(a) left (b) left (c) left (d) neither **3.8**(a) 121.64 g/mol (b) 76.02 g/mol (c) 106.44 g/mol (d) 152.00 g/mol **3.10**(a) 134.7 g/mol (b) 175.3 g/mol (c) 342.14 g/mol (d) 125.84 g/mol **3.12**(a) 22.6 g Zn (b) 3.16×10^{24} F atoms (c) 4.28×10^{23} Ca atoms **3.14**(a) 1.1×10^2 g $KMnO_4$ (b) 0.188 mol O atoms (c) 1.5×10^{20} O atoms **3.16**(a) 9.72 g $MnSO_4$ (b) 44.6 mol $Fe(ClO_4)_3$ (c) 1.74×10^{21} N atoms **3.18**(a) 1.56×10^3 g Cu_2CO_3 (b) 0.0725 g N_2O_5 (c) 0.644 mol $NaClO_4$; 3.88×10^{23} formula units $NaClO_4$ (d) 3.88×10^{23} Na^+ ions; 3.88×10^{23} ClO_4^- ions; 3.88×10^{23} Cl atoms; 1.55×10^{24} O atoms **3.20**(a) 6.375 mass % H (b) 71.52 mass % O **3.22**(a) 0.1252 mass fraction C (b) 0.3428 mass fraction O **3.25**(a) 0.9507 mol cisplatin (b) 3.5×10^{24} H atoms **3.27**(a) 195 mol $Fe_2O_3{\cdot}4H_2O$ (b) 195 mol Fe_2O_3 (c) 2.18×10^4 g Fe **3.29** $CO(NH_2)_2 > NH_4NO_3 > (NH_4)_2SO_4 > KNO_3$ **3.30**(a) 883 mol PbS (b) 1.88×10^{25} Pb atoms **3.34** (b) From the mass percent, determine the empirical formula. Add up the total number of atoms in the empirical formula, and divide that number into the total number of atoms in the molecule. The result is the multiplier that converts the empirical formula into the molecular formula.

Mass (g) of each element (mass percent expressed in grams)
↓ divide by $\mathcal{M}$ (g/mol)
Amount (mol) of each element
↓ use numbers of moles as subscripts
Preliminary empirical formula
↓ change to integer subscripts
Empirical formula
↓ divide total number of atoms in molecule by number of atoms in empirical formula and multiply empirical formula by that factor
Molecular formula

(c) Find the empirical formula from the mass percents. Compare the number of atoms given for the one element to the number in the empirical formula. Multiply the empirical formula by the factor that is needed to obtain the given number of atoms for that element.

First three steps are the same as in road map for part (b).

Empirical formula
↓ divide the number of atoms given for the one element by the number of atoms of that element in the empirical formula and multiply the empirical formula by that factor
Molecular formula

(e) Count the numbers of the various types of atoms in the structural formula and put these into a molecular formula.

Structural formula
↓ use the numbers of atoms of the elements as subscripts
Molecular formula

3.36(a) CH_2; 14.03 g/mol (b) CH_3O; 31.03 g/mol (c) N_2O_5; 108.02 g/mol (d) $Ba_3(PO_4)_2$; 601.8 g/mol (e) TeI_4; 635.2 g/mol **3.38** Disulfur dichloride; SCl; 135.02 g/mol **3.40**(a) C_3H_6 (b) N_2H_4 (c) N_2O_4 (d) $C_5H_5N_5$ **3.42**(a) Cl_2O_7 (b) $SiCl_4$ (c) CO_2 **3.44**(a) NO_2 (b) N_2O_4 **3.46**(a) 1.20 mol F (b) 24.0 g M (c) calcium **3.48** $C_{10}H_{20}O$ **3.51** $C_{21}H_{30}O_5$ **3.53** A balanced equation provides information on 1) the identities of reactants and products by showing their chemical formulas; 2) the physical states of reactants and products by using the symbols for the various phases of matter; 3) the molar ratios between reactants and products with coefficients. **3.56** b

3.58(a) $16Cu(s) + S_8(s) \longrightarrow 8Cu_2S(s)$
(b) $P_4O_{10}(s) + 6H_2O(l) \longrightarrow 4H_3PO_4(l)$
(c) $B_2O_3(s) + 6NaOH(aq) \longrightarrow 2Na_3BO_3(aq) + 3H_2O(l)$
(d) $4CH_3NH_2(g) + 9O_2(g) \longrightarrow 4CO_2(g) + 10H_2O(g) + 2N_2(g)$

3.60(a) $2SO_2(g) + O_2(g) \longrightarrow 2SO_3(g)$
(b) $Sc_2O_3(s) + 3H_2O(l) \longrightarrow 2Sc(OH)_3(s)$
(c) $H_3PO_4(aq) + 2NaOH(aq) \longrightarrow Na_2HPO_4(aq) + 2H_2O(l)$
(d) $C_6H_{10}O_5(s) + 6O_2(g) \longrightarrow 6CO_2(g) + 5H_2O(g)$

3.62(a) $4Ga(s) + 3O_2(g) \longrightarrow 2Ga_2O_3(s)$
(b) $2C_6H_{14}(l) + 19O_2(g) \longrightarrow 12CO_2(g) + 14H_2O(g)$
(c) $3CaCl_2(aq) + 2Na_3PO_4(aq) \longrightarrow Ca_3(PO_4)_2(s) + 6NaCl(aq)$
3.67 Balance the equation for the reaction: $aA + bB \longrightarrow cC$. Since A is the limiting reactant, A is used to determine the amount of C. Divide the mass of A by its molar mass to obtain the amount (mol) of A. Use the molar ratio from the balanced equation to find the amount (mol) of C. Multiply the amount (mol) of C by its molar mass to obtain the mass of C.

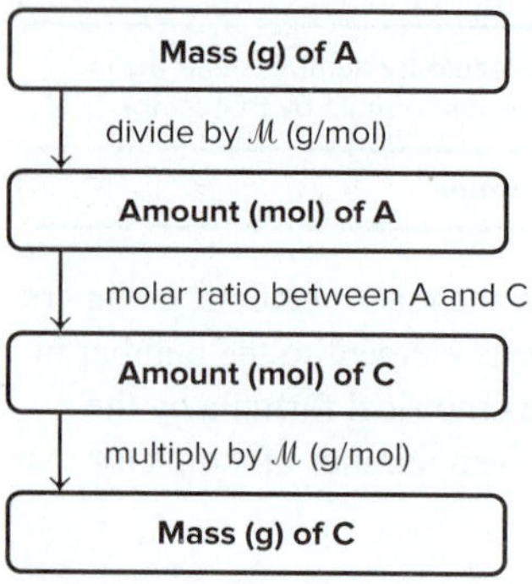

3.69(a) 0.455 mol Cl_2 (b) 32.3 g Cl_2 **3.71**(a) 1.42×10^3 mol KNO_3 (b) 1.43×10^5 g KNO_3 **3.73** 195.8 g H_3BO_3; 19.16 g H_2 **3.75** 2.60×10^3 g Cl_2
3.77(a) $I_2(s) + Cl_2(g) \longrightarrow 2ICl(s)$
$ICl(s) + Cl_2(g) \longrightarrow ICl_3(s)$
(b) $I_2(s) + 3Cl_2(g) \longrightarrow 2ICl_3(s)$
(c) 1.33×10^3 g I_2
3.79(a) B_2 (b) 4 AB_3 molecules **3.81** 3.6 mol **3.83**(a) 0.105 mol CaO (b) 0.175 mol CaO (c) calcium (d) 5.88 g CaO **3.85** 1.36 mol HIO_3, 239 g HIO_3; 44.9 g H_2O in excess **3.87** 4.40 g CO_2; 4.80 g O_2 in excess **3.89** 12.2 g $Al(NO_2)_3$, no NH_4Cl, 48.7 g $AlCl_3$, 30.7 g N_2, 39.5 g H_2O **3.91** 50.% **3.93** 90.5% **3.95** 24.0 g CH_3Cl **3.97** 39.7 g CF_4 **3.98**(a) A (b) $2Cl_2O(g) \longrightarrow 2Cl_2(g) + O_2(g)$ (c) 1.8×10^{23} Cl_2O molecules **3.105** $x = 3$ **3.106** ethane > propane > cetyl palmitate > ethanol > benzene **3.111**(a) $Fe_2O_3(s) + 3CO(g) \longrightarrow 2Fe(s) + 3CO_2(g)$ (b) 3.39×10^7 g CO **3.113** 89.8% **3.115**(a) $2AB_2 + B_2 \longrightarrow 2AB_3$ (b) AB_2 (c) 5.0 mol AB_3 (d) 0.5 mol B_2 **3.116** B, C, and D have the same empirical formula, C_2H_4O; 44.05 g/mol **3.117** 44.3% **3.126**(a) 586 g CO_2 (b) 10.5% CH_4 by mass **3.127** 10./0.66/1.0 **3.132**(a) 192.12 g/mol; $C_6H_8O_7$ (b) 0.549 mol
3.133(a) $N_2(g) + O_2(g) \longrightarrow 2NO(g)$
$2NO(g) + O_2(g) \longrightarrow 2NO_2(g)$
$3NO_2(g) + H_2O(g) \longrightarrow 2HNO_3(aq) + NO(g)$
(b) $2N_2(g) + 5O_2(g) + 2H_2O(g) \longrightarrow 4HNO_3(aq)$
(c) 6.07×10^3 t HNO_3
3.134 A **3.136**(a) 0.039 g heme (b) 6.3×10^{-5} mol heme (c) 3.5×10^{-3} g Fe (d) 4.1×10^{-2} g hemin **3.138**(a) 46.65 mass % N in urea; 31.98 mass % N in arginine; 21.04 mass % N in ornithine (b) 28.45 g N **3.140** 29.54% **3.142**(a) 89.3% (b) 1.47 g ethylene **3.144**(a) 125 g hydrochloride salt (b) 65.6 L H_2O

Chapter 4

4.2 Ionic or polar covalent compounds **4.3** Ions must be present, and they come from electrolytes such as ionic compounds, acids, and bases. **4.6** B **4.13** No, the instructions should read: "Take 100.0 mL of the 10.0 *M* solution and, with stirring, add water until the total volume is 1000. mL." **4.14**(a) Benzene is likely to be insoluble in water because it is nonpolar and water is polar. (b) Sodium hydroxide, an ionic compound, is likely to be very soluble in water. (c) Ethanol (CH_3CH_2OH) is likely to be soluble in water because the alcohol group (—OH) is polar, like the water molecule. (d) Potassium acetate, an ionic compound, is likely to be very soluble in water. **4.16**(a) Yes, CsBr is a soluble salt. (b) Yes, HI is a strong acid. **4.18**(a) 0.64 mol (b) 0.242 mol (c) 1.18×10^{-4} mol **4.20**(a) 3.0 mol (b) 7.57×10^{-5} mol (c) 0.148 mol **4.22**(a) 7.85 g $Ca(C_2H_3O_2)_2$ (b) 0.254 *M* KI (c) 124 mol NaCN **4.24**(a) 4.65 g K_2SO_4 (b) 0.0653 *M* $CaCl_2$ (c) 1.11×10^{20} Mg^{2+} ions **4.26**(a) 0.058 mol Al^{3+}; 3.5×10^{22} Al^{3+} ions; 0.18 mol Cl^-; 1.1×10^{23} Cl^- ions (b) 4.62×10^{-4} mol Li^+; 2.78×10^{20} Li^+ ions; 2.31×10^{-4} mol SO_4^{2-}; 1.39×10^{20} SO_4^{2-} ions (c) 1.50×10^{-2} mol K^+; 9.02×10^{21} K^+ ions; 1.50×10^{-2} mol Br^-; 9.02×10^{21} Br^- ions **4.28**(a) 0.0617 *M* KCl (b) 0.00363 *M* $(NH_4)_2SO_4$ (c) 0.138 *M* Na^+ **4.30**(a) 9.87 g HNO_3/L (b) 15.7 *M* HNO_3 **4.33**(a) Instructions: Be sure to wear goggles to protect your eyes! Pour approximately 2.0 gal of water into the container. Add to the water, slowly, and with mixing, 0.90 gal of concentrated HCl. Dilute to 3.0 gal with more water. (b) 22.6 mL
4.36 Spectator ions do not appear in a net ionic equation because they are not involved in the reaction and serve only to balance charges. **4.40** Assuming that the left beaker contains $AgNO_3$ (because it has gray Ag^+ ions), the right beaker must contain NaCl. Then, NO_3^- is blue, Na^+ is brown, and Cl^- is green.
Molecular equation: $AgNO_3(aq) + NaCl(aq) \longrightarrow AgCl(s) + NaNO_3(aq)$
Total ionic equation: $Ag^+(aq) + NO_3^-(aq) + Na^+(aq) + Cl^-(aq) \longrightarrow AgCl(s) + Na^+(aq) + NO_3^-(aq)$
Net ionic equation: $Ag^+(aq) + Cl^-(aq) \longrightarrow AgCl(s)$
4.41(a) Molecular: $Hg_2(NO_3)_2(aq) + 2KI(aq) \longrightarrow Hg_2I_2(s) + 2KNO_3(aq)$
Total ionic:
$Hg_2^{2+}(aq) + 2NO_3^-(aq) + 2K^+(aq) + 2I^-(aq) \longrightarrow Hg_2I_2(s) + 2K^+(aq) + 2NO_3^-(aq)$
Net ionic: $Hg_2^{2+}(aq) + 2I^-(aq) \longrightarrow Hg_2I_2(s)$
Spectator ions are K^+ and NO_3^-.
(b) Molecular: $FeSO_4(aq) + Sr(OH)_2(aq) \longrightarrow Fe(OH)_2(s) + SrSO_4(s)$
Total ionic: $Fe^{2+}(aq) + SO_4^{2-}(aq) + Sr^{2+}(aq) + 2OH^-(aq) \longrightarrow Fe(OH)_2(s) + SrSO_4(s)$
Net ionic: This is the same as the total ionic equation, because there are no spectator ions.
4.43(a) No precipitate will form. (b) A precipitate will form because silver ions, Ag^+, and bromide ions, Br^-, will combine to form a solid salt, silver bromide, AgBr. The ammonium and nitrate ions do not form a precipitate.
Molecular: $NH_4Br(aq) + AgNO_3(aq) \longrightarrow AgBr(s) + NH_4NO_3(aq)$
Total ionic: $NH_4^+(aq) + Br^-(aq) + Ag^+(aq) + NO_3^-(aq) \longrightarrow AgBr(s) + NH_4^+(aq) + NO_3^-(aq)$
Net ionic: $Ag^+(aq) + Br^-(aq) \longrightarrow AgBr(s)$
4.45(a) No precipitate will form. (b) $BaSO_4$ will precipitate.
Molecular: $(NH_4)_2SO_4(aq) + BaCl_2(aq) \longrightarrow BaSO_4(s) + 2NH_4Cl(aq)$

Total ionic:

$2NH_4^+(aq) + SO_4^{2-}(aq) + Ba^{2+}(aq) + 2Cl^-(aq) \longrightarrow BaSO_4(s) + 2NH_4^+(aq) + 2Cl^-(aq)$

Net ionic: $SO_4^{2-}(aq) + Ba^{2+}(aq) \longrightarrow BaSO_4(s)$

4.47 0.0354 *M* Pb^{2+} **4.49** 0.88 g $BaSO_4$ **4.51**(a) $PbSO_4$ (b) $Pb^{2+}(aq) + SO_4^{2-}(aq) \longrightarrow PbSO_4(s)$ (c) 1.5 g $PbSO_4$

4.53 Potassium carbonate, K_2CO_3 **4.55** 2.206 mass % Cl^-

4.61(a) Formation of a gas, $SO_2(g)$, and of a nonelectrolyte, water, will cause the reaction to go to completion. (b) Formation of a precipitate, $Ba_3(PO_4)_2(s)$, and of a nonelectrolyte, water, will cause the reaction to go to completion. **4.63**(a) 0.035 mol H^+ (b) 6.3×10^{-3} mol H^+ (c) 0.22 mol H^+

4.65(a) Molecular equation:

$KOH(aq) + HBr(aq) \longrightarrow KBr(aq) + H_2O(l)$

Total ionic equation:

$K^+(aq) + OH^-(aq) + H^+(aq) + Br^-(aq) \longrightarrow K^+(aq) + Br^-(aq) + H_2O(l)$

Net ionic equation: $OH^-(aq) + H^+(aq) \longrightarrow H_2O(l)$

The spectator ions are $K^+(aq)$ and $Br^-(aq)$.

(b) Molecular equation: $NH_3(aq) + HCl(aq) \longrightarrow NH_4Cl(aq)$

Total ionic equation:

$NH_3(aq) + H^+(aq) + Cl^-(aq) \longrightarrow NH_4^+(aq) + Cl^-(aq)$

NH_3, a weak base, is written in the molecular (undissociated) form. HCl, a strong acid, is written as dissociated ions. NH_4Cl is a soluble compound because all ammonium compounds are soluble. Net ionic equation: $NH_3(aq) + H^+(aq) \longrightarrow NH_4^+(aq)$

Cl^- is the only spectator ion.

4.67 Total ionic equation: $CaCO_3(s) + 2H^+(aq) + 2Cl^-(aq) \longrightarrow Ca^{2+}(aq) + 2Cl^-(aq) + H_2O(l) + CO_2(g)$

Net ionic equation: $CaCO_3(s) + 2H^+(aq) \longrightarrow Ca^{2+}(aq) + H_2O(l) + CO_2(g)$

4.69 845 mL **4.71** 0.05839 *M* CH_3COOH

4.82(a) S has O.N. = +6 in SO_4^{2-} (i.e., H_2SO_4) and O.N. = +4 in SO_2, so S has been reduced (and I^- oxidized); H_2SO_4 acts as an oxidizing agent. (b) The oxidation numbers remain constant throughout; H_2SO_4 transfers an H^+ to F^- to produce HF, so it acts as an acid. **4.84**(a) +4 (b) +3 (c) +4 (d) −3 **4.86**(a) −1 (b) +2 (c) −3 (d) +3 **4.88**(a) −3 (b) +5 (c) +3 **4.90**(a) +6 (b) +3 (c) +7

4.92(a) MnO_4^- is the oxidizing agent; $H_2C_2O_4$ is the reducing agent. (b) Cu is the reducing agent; NO_3^- is the oxidizing agent.

4.94(a) Oxidizing agent is MnO_4^-; reducing agent is Sn. (b) Oxidizing agent is NO_3^-; reducing agent is Cl^-.

4.96(a) 4.54×10^{-3} mol MnO_4^- (b) 0.0113 mol H_2O_2 (c) 0.386 g H_2O_2 (d) 2.61 mass % H_2O_2 (e) H_2O_2 **4.102** A combination reaction that is also a redox reaction is $2Mg(s) + O_2(g) \longrightarrow 2MgO(s)$. A combination reaction that is not a redox reaction is $CaO(s) + H_2O(l) \longrightarrow Ca(OH)_2(aq)$.

4.103(a) $Ca(s) + 2H_2O(l) \longrightarrow Ca(OH)_2(aq) + H_2(g)$; displacement (b) $2NaNO_3(s) \longrightarrow 2NaNO_2(s) + O_2(g)$; decomposition (c) $C_2H_2(g) + 2H_2(g) \longrightarrow C_2H_6(g)$; combination

4.105(a) $2Sb(s) + 3Cl_2(g) \longrightarrow 2SbCl_3(s)$; combination (b) $2AsH_3(g) \longrightarrow 2As(s) + 3H_2(g)$; decomposition (c) $Zn(s) + Fe(NO_3)_2(aq) \longrightarrow Zn(NO_3)_2(aq) + Fe(s)$; displacement

4.107(a) $Sr(s) + Br_2(l) \longrightarrow SrBr_2(s)$ (b) $2Ag_2O(s) \xrightarrow{\Delta} 4Ag(s) + O_2(g)$ (c) $Mn(s) + Cu(NO_3)_2(aq) \longrightarrow Mn(NO_3)_2(aq) + Cu(s)$

4.109(a) $N_2(g) + 3H_2(g) \longrightarrow 2NH_3(g)$ (b) $2NaClO_3(s) \xrightarrow{\Delta} 2NaCl(s) + 3O_2(g)$ (c) $Ba(s) + 2H_2O(l) \longrightarrow Ba(OH)_2(aq) + H_2(g)$

4.111(a) $2Cs(s) + I_2(s) \longrightarrow 2CsI(s)$ (b) $2Al(s) + 3MnSO_4(aq) \longrightarrow Al_2(SO_4)_3(aq) + 3Mn(s)$ (c) $2SO_2(g) + O_2(g) \longrightarrow 2SO_3(g)$ (d) $2C_4H_{10}(g) + 13O_2(g) \longrightarrow 8CO_2(g) + 10H_2O(g)$ (e) $2Al(s) + 3Mn^{2+}(aq) \longrightarrow 2Al^{3+}(aq) + 3Mn(s)$

4.113 315 g O_2; 3.95 kg Hg **4.115**(a) O_2 is in excess. (b) 0.117 mol Li_2O (c) 0 g Li, 3.49 g Li_2O, and 4.63 g O_2

4.117 67.2 mass % $KClO_3$ **4.119** 2.79 kg Fe **4.120** 99.9 g compound B, which is $FeCl_2$ **4.125** The reaction is $2NO + Br_2 \rightleftharpoons 2NOBr$, which can proceed in either direction. If NO and Br_2 are placed in a container, they will react to form NOBr; if NOBr is placed in the container, it will decompose to form NO and Br_2. In either case, the concentrations of NO, Br_2, and NOBr adjust so that the rates of the forward and reverse reactions become equal, and equilibrium is reached.

4.127(a) $Fe(s) + 2H^+(aq) \longrightarrow Fe^{2+}(aq) + H_2(g)$

O.N.: 0 +1 +2 0

(b) 3.1×10^{21} Fe^{2+} ions

4.129 5.11 g C_2H_5OH; 24.9 L CO_2

4.131(a) $Ca^{2+}(aq) + C_2O_4^{2-}(aq) \longrightarrow CaC_2O_4(s)$ (b) $5H_2C_2O_4(aq) + 2MnO_4^-(aq) + 6H^+(aq) \longrightarrow 10CO_2(g) + 2Mn^{2+}(aq) + 8H_2O(l)$ (c) $KMnO_4$ (d) $H_2C_2O_4$ (e) 55.06 mass % $CaCl_2$

4.134 2.809 mass % $CaMg(CO_3)_2$ **4.136**(a) Step 1: oxidizing agent is O_2; reducing agent is NH_3. Step 2: oxidizing agent is O_2; reducing agent is NO. Step 3: oxidizing agent is NO_2; reducing agent is NO_2. (b) 1.2×10^4 kg NH_3 **4.139** 627 L air

4.141 0.75 kg SiO_2; 0.26 kg Na_2CO_3; 0.18 kg $CaCO_3$

4.144(a) 4 mol IO_3^- (b) 12 mol I_2; IO_3^- is oxidizing agent and I^- is reducing agent. (c) 12.87 mass % thyroxine **4.147**(a) 3.0×10^{-3} mol CO_2 (b) 0.11 L CO_2 **4.149**(a) $C_7H_5O_4Bi$ (b) $C_{21}H_{15}O_{12}Bi_3$ (c) $Bi(OH)_3(s) + 3HC_7H_5O_3(aq) \longrightarrow Bi(C_7H_5O_3)_3(s) + 3H_2O(l)$ (d) 0.490 mg $Bi(OH)_3$ **4.151**(a) Ethanol: $C_2H_5OH(l) + 3O_2(g) \longrightarrow 2CO_2(g) + 3H_2O(l)$; Gasoline: $2C_8H_{18}(l) + 25O_2(g) \longrightarrow 16CO_2(g) + 18H_2O(g)$ (b) 2.50×10^3 g O_2 (c) 1.75×10^3 L O_2 (d) 8.38×10^3 L air **4.153** yes

Chapter 5

Answers to Boxed Reading Problems: B5.1 The density of the atmosphere decreases with increasing altitude. High density causes more drag and frictional heating on the aircraft. At high altitudes, the low density of the atmosphere means that there are relatively few gas particles present to collide with an aircraft. **B5.3** 0.934%, 7.10 torr

• **5.1**(a) The volume of the liquid remains constant, but the volume of the gas increases to the volume of the larger container. (b) The volume of the container holding the gas sample increases when heated, but the volume of the container holding the liquid sample remains essentially constant when heated. (c) The volume of the liquid remains essentially constant, but the volume of the gas is reduced. **5.6** 990 cmH_2O **5.8**(a) 566 mmHg (b) 1.32 bar (c) 3.60 atm (d) 107 kPa **5.10** 0.9408 atm **5.12** 0.966 atm **5.18** At constant temperature and volume, the pressure of a gas is directly proportional to number of moles of the gas. **5.20**(a) Volume decreases to one-third of the original volume. (b) Volume increases by a factor of 3.0. (c) Volume increases by a factor of 4. **5.22**(a) Volume decreases by a factor of 2. (b) Volume increases by a factor of 1.48. (c) Volume decreases by a factor of 3. **5.24** 1.84 L **5.26** −144°C **5.28** 278 atm **5.30** 6.95 L **5.32** 35.3 L **5.34** 0.085 mol Cl_2 **5.36** 1.16 g ClF_3 **5.39** yes **5.41** Beaker is inverted to collect H_2 and upright for CO_2. The molar mass of CO_2 is greater than the molar mass of air, which, in turn, has a greater molar mass than H_2. **5.45** 5.86 g/L **5.47** 1.78×10^{-3} mol AsH_3; 3.48 g/L **5.49** 51.1 g/mol **5.51** 1.33 atm **5.53** 39.2 g P_4 **5.55** 41.2 g PH_3 **5.57** 0.0249 g Al **5.61** C_5H_{12} **5.63**(a) 0.90 mol (b) 6.76 torr **5.64** 286 mL SO_2 **5.66** 0.0997 atm **5.71** At STP, the volume occupied by a mole of any gas is the same. At the same temperature, all gases have the same average kinetic energy, resulting in the same pressure. **5.74**(a) $P_A > P_B > P_C$ (b) $E_A = E_B = E_C$ (c) $\text{rate}_A > \text{rate}_B > \text{rate}_C$ (d) total $E_A >$ total $E_B >$ total E_C (e) $d_A = d_B = d_C$ (f) collision frequency in A > collision frequency in B > collision frequency in C **5.75** 13.21 **5.77**(a) curve 1 (b) curve 1 (c) curve 1; fluorine and argon have about the same molar mass **5.79** 14.9 min **5.81** 4 atoms per molecule **5.84** negative deviations; $N_2 < Kr < CO_2$ **5.86** At 1 atm; at lower pressures, the gas molecules are farther apart and intermolecular forces are less important. **5.89** 6.81×10^4 g/mol **5.92**(a) 22.1 atm (b) 20.9 atm **5.96**(a) N_2, 597 torr; O_2, 159 torr; CO_2, 0.3 torr; H_2O, 3.5 torr (b) 74.9 mol % N_2; 13.7 mol % O_2; 5.3 mol % CO_2; 6.2 mol % H_2O (c) 1.6×10^{21} molecules O_2 **5.98**(a) 4×10^2 mL (b) 0.013 mol N_2 **5.99** 35.7 L NO_2 **5.104** Al_2Cl_6 **5.106** 1.52×10^{-2} mol SO_3 **5.110**(a) 1.95×10^3 g Ni (b) 3.5×10^4 g Ni (c) 63 m^3 CO **5.112**(a) 9 volumes of $O_2(g)$ (b) CH_5N **5.115** The lungs would expand by a factor of 4.86; the diver can safely ascend 52.5 ft to a depth of 73 ft. **5.117** 6.07 g H_2O_2 **5.122** 6.53×10^{-3} g N_2 **5.126**(a) xenon (b) water vapor (c) mercury vapor (d) water vapor **5.130** 17.2 g CO_2; 17.8 g Kr **5.136** Ne, 676 m/s; Ar, 481 m/s; He, 1.52×10^3 m/s **5.138**(a) 0.052 g (b) 1.1 mL **5.145**(a) 16.5 L CO_2 (b) $P_{H_2O} = 48.8$ torr; $P_{O_2} = P_{CO_2} = 3.7\times10^2$ torr **5.150** 332 steps **5.152** 1.4 **5.156** $P_{total} = 245$ torr; $P_{I_2} = 25.2$ torr

Chapter 6

Answers to Boxed Reading Problems: B6.2(a) $2H_2O(g) \longrightarrow CH_4(g) + CO_2(g)$ (b) 12 kJ (c) -3.30×10^4 kJ

• **6.4** Increase: eating food, lying in the sun, taking a hot bath. Decrease: exercising, taking a cold bath, going outside on a cold day. **6.6** The amount of the change in internal energy is the same for both heater and air conditioner. Since both devices consume the same amount of electrical energy, the change in energy of the heater equals that of the air conditioner. **6.9** 0 J **6.11** 1.54×10^3 J **6.13**(a) 6.6×10^7 kJ (b) 1.6×10^7 kcal (c) 6.3×10^7 Btu **6.15** −51 J **6.18** 8.8 h **6.19** Measuring the heat transfer at constant pressure is more convenient than measuring it at constant volume. **6.21**(a) exothermic (b) endothermic (c) exothermic (d) exothermic (e) endothermic (f) endothermic (g) exothermic

6.24

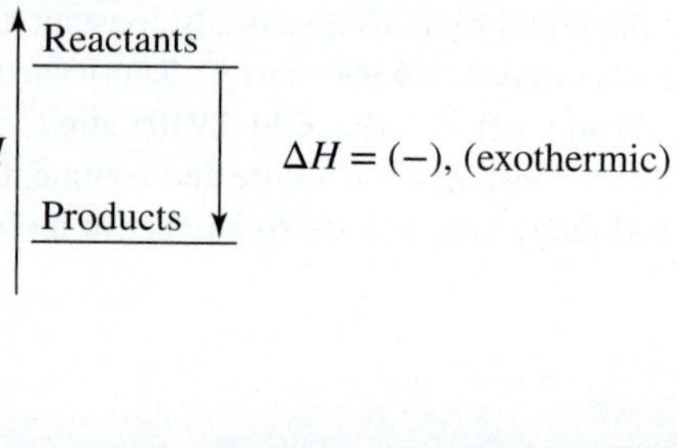

6.26(a) Combustion of ethane:

$$2C_2H_6(g) + 7O_2(g) \longrightarrow 4CO_2(g) + 6H_2O(g) + \text{heat}$$

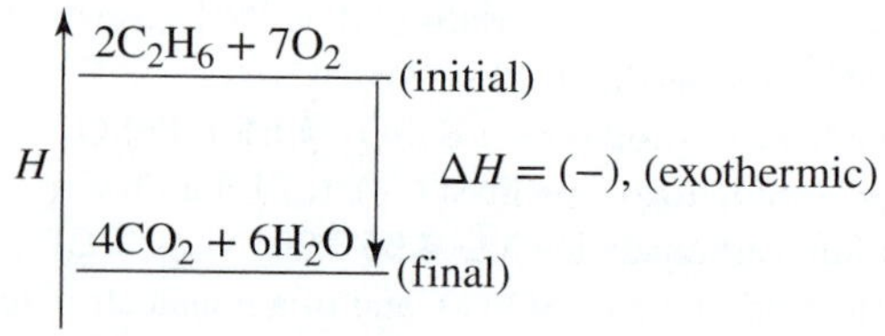

(b) Freezing of water: $H_2O(l) \longrightarrow H_2O(s) + \text{heat}$

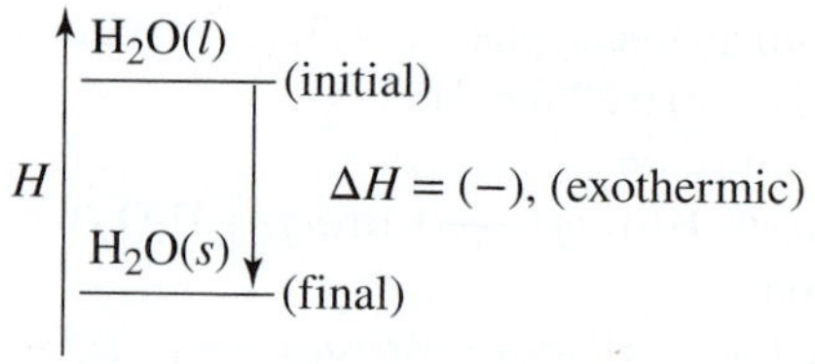

6.28(a) $2CH_3OH(l) + 3O_2(g) \longrightarrow 2CO_2(g) + 4H_2O(g) + \text{heat}$

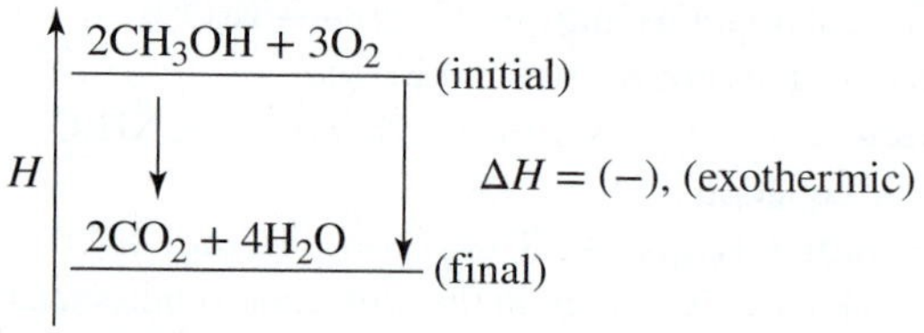

(b) $\frac{1}{2}N_2(g) + O_2(g) + \text{heat} \longrightarrow NO_2(g)$

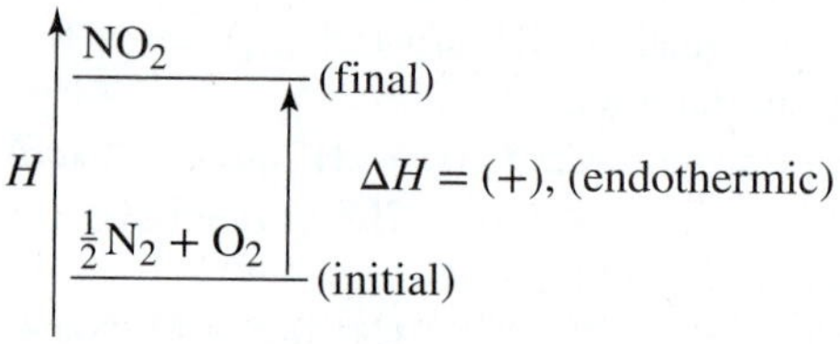

6.30(a) This is a phase change from the solid phase to the gas phase. Heat is absorbed by the system, so q_{sys} is positive. (b) The volume of the system increases, as more moles of gas are present after the phase change than before. So the system has done work of expansion, and w is negative. Since $\Delta E_{sys} = q + w$, q is positive, and w is negative, the sign of ΔE_{sys} cannot be predicted. It will be positive if $q > w$ and negative if $q < w$. (c) $\Delta E_{univ} = 0$. If the system loses energy, the surroundings gain an equal quantity of energy. The sum of the energy of the system and the energy of the surroundings remains constant. **6.33** To determine the specific heat capacity of a substance, you need its mass, the heat added (or lost), and the change in temperature. **6.35** Heat capacity is the quantity of heat required to raise the temperature of an object 1 K. Specific heat capacity is the quantity of heat required to raise 1 g of a substance or material by 1 K. Molar heat capacity is the quantity of heat required to raise the temperature of 1 mol of a substance by 1 K. **6.37** 6.9×10^3 J **6.39** 295°C **6.41** 77.5°C **6.43** 45°C **6.45** 36.6°C **6.47** −55.8 kJ/mol H_2O **6.49** −2805 kJ/mol **6.55** The reaction has a positive ΔH, because it requires an input of energy to break the oxygen-oxygen bond. **6.56** ΔH is negative; it is opposite in sign and half as large as ΔH for the vaporization of 2 mol of H_2O. **6.57**(a) exothermic (b) 20.2 kJ (c) -4.2×10^2 kJ (d) −15.7 kJ **6.59**(a) $\frac{1}{2}N_2(g) + \frac{1}{2}O_2(g) \longrightarrow NO(g)$ $\quad \Delta H = 90.29$ kJ (b) −10.5 kJ **6.61** -1.88×10^6 kJ **6.65**(a) $C_2H_4(g) + 3O_2(g) \longrightarrow 2CO_2(g) + 2H_2O(l)$; $\quad \Delta H_{rxn} = -1411$ kJ (b) 1.39 g C_2H_4

6.69 −110.5 kJ **6.70** −813.4 kJ **6.72** $N_2(g) + 2O_2(g) \longrightarrow 2NO_2(g)$ $\Delta H_{overall} = 66.4$ kJ; equation 1 is A, equation 2 is B, and equation 3 is C. **6.74** 44.0 kJ **6.77** The standard enthalpy of reaction, ΔH°_{rxn}, is the enthalpy change for a reaction when all substances are in their standard states. The standard enthalpy of formation, ΔH°_f, is the enthalpy change that accompanies the formation of 1 mol of a compound in its standard state from elements in their standard states.
6.79(a) $\frac{1}{2}Cl_2(g) + Na(s) \longrightarrow NaCl(s)$
(b) $H_2(g) + \frac{1}{2}O_2(g) \longrightarrow H_2O(g)$ (c) no changes
6.80(a) $Ca(s) + Cl_2(g) \longrightarrow CaCl_2(s)$
(b) $Na(s) + \frac{1}{2}H_2(g) + C(graphite) + \frac{3}{2}O_2(g) \longrightarrow NaHCO_3(s)$
(c) $C(graphite) + 2Cl_2(g) \longrightarrow CCl_4(l)$
(d) $\frac{1}{2}H_2(g) + \frac{1}{2}N_2(g) + \frac{3}{2}O_2(g) \longrightarrow HNO_3(l)$
6.82(a) −1036.9 kJ (b) −433 kJ **6.84** −157.3 kJ/mol
6.86(a) 503.9 kJ (b) $-\Delta H_1 + 2\Delta H_2 = 504$ kJ
6.87(a) $C_{18}H_{36}O_2(s) + 26O_2(g) \longrightarrow 18CO_2(g) + 18H_2O(g)$
(b) −10,488 kJ (c) −36.9 kJ; −8.81 kcal
(d) 8.81 kcal/g × 11.0 g = 96.9 kcal
6.89(a) initial = 23.6 L/mol; final = 24.9 L/mol (b) 187 J (c) -1.2×10^2 J (d) 3.1×10^2 J (e) 310 J (f) $\Delta H = \Delta E + P\Delta V = \Delta E - w = (q + w) - w = q_P$
6.98(a) 1.1×10^3 J (b) 1.1×10^8 J (c) 1.3×10^2 mol CH_4 (d) \$0.0050/mol (e) \$0.85 **6.103**(a) $\Delta H^\circ_{rxn1} = -657.0$ kJ; $\Delta H^\circ_{rxn2} = 32.9$ kJ (b) −106.6 kJ **6.104**(a) -6.81×10^3 J (b) +243 °C **6.105** −22.2 kJ **6.106**(a) 34 kJ/mol (b) −757 kJ
6.108(a) -1.25×10^3 kJ (b) 2.24×10^3 °C

Chapter 7

Answers to Boxed Reading Problems: B7.1(a) slope = $1.3\times10^4/M$; intercept = 0.00

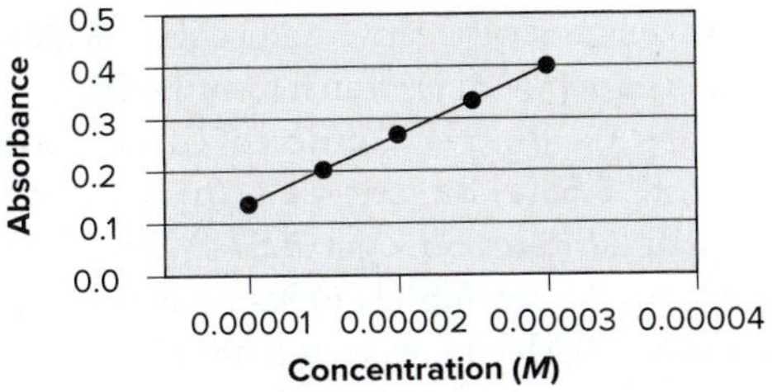

(b) diluted solution = 1.8×10^{-5} M; original solution = 1.4×10^{-4} M

• **7.2**(a) x-ray < ultraviolet < visible < infrared < microwave < radio waves (b) radio < microwave < infrared < visible < ultraviolet < x-ray (c) radio < microwave < infrared < visible < ultraviolet < x-ray **7.7** 316 m; 3.16×10^{11} nm; 3.16×10^{12} Å **7.9** 2.5×10^{-23} J **7.11** red < yellow < blue **7.13** 1.3483×10^7 nm; 1.3483×10^8 Å **7.16**(a) 1.24×10^{15} s^{-1}; 8.21×10^{-19} J (b) 1.4×10^{15} s^{-1}; 9.0×10^{-19} J **7.18** Bohr's key assumption was that the electron in an atom does not radiate energy while in a stationary state, and it can move to a different orbit only by absorbing or emitting a photon whose energy is equal to the difference in energy between two states. These differences in energy correspond to the wavelengths in the known line spectra for the hydrogen atom. A Solar System model would not allow for the movement of electrons between levels. **7.20**(a) absorption (b) emission (c) emission (d) absorption **7.22** Yes, the predicted line spectra are accurate. The energies could be predicted from

$$E_n = \frac{-(Z^2)(2.18\times10^{-18}\text{ J})}{n^2},$$ where Z is the atomic number for the atom or ion. The energy levels for Be^{3+} will be greater by a factor of 16 (Z = 4) than those for the hydrogen atom. This means that the pattern of lines will be similar, but the lines will be at different wavelengths. **7.23** 434.17 nm **7.25** 1875.6 nm **7.27** -2.76×10^5 J/mol **7.29** d < a < c < b **7.31** $n = 4$
7.37 Macroscopic objects do exhibit a wavelike motion, but the wavelength is too small for humans to perceive.
7.39(a) 7.10×10^{-37} m (b) 1×10^{-35} m **7.41** 2.2×10^{-26} m/s
7.43 3.75×10^{-36} kg **7.47** The total probability of finding the 1s electron in any distance r from the H nucleus is greatest when the value of r is 0.529 Å. The probability is greater for the 1s orbital.
7.48(a) principal determinant of the electron's energy or distance from the nucleus (b) determines the shape of the orbital (c) determines the orientation of the orbital in three-dimensional space **7.49**(a) one (b) five (c) three (d) nine
7.51(a) m_l: −2, −1, 0, +1, +2 (b) m_l: 0 (if $n = 1$, then $l = 0$) (c) m_l: −3, −2, −1, 0, +1, +2, +3
7.53(a)

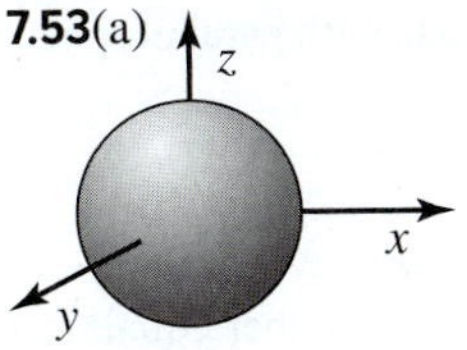

(b)

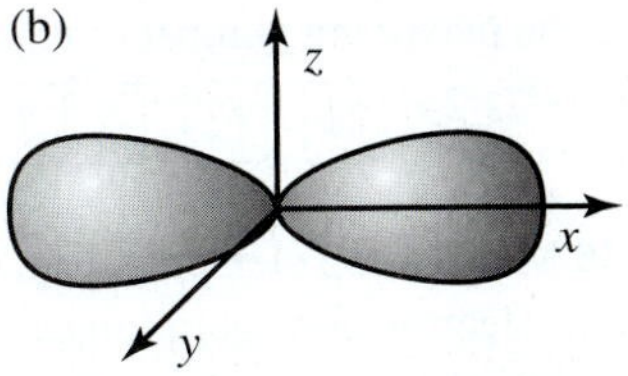

7.55

Sublevel	Allowable m_l values	No. of orbitals
(a) d ($l = 2$)	−2, −1, 0, +1, +2	5
(b) p ($l = 1$)	−1, 0, +1	3
(c) f ($l = 3$)	−3, −2, −1, 0, +1, +2, +3	7

7.57(a) $n = 5$ and $l = 0$; one orbital (b) $n = 3$ and $l = 1$; three orbitals (c) $n = 4$ and $l = 3$; seven orbitals **7.59**(a) no; $n = 2$, $l = 1$, $m_l = -1$; $n = 2$, $l = 0$, $m_l = 0$ (b) allowed (c) allowed (d) no; $n = 5$, $l = 3$, $m_l = +3$; $n = 5$, $l = 2$, $m_l = 0$
7.62(a) $E = -(2.180\times10^{-18}\text{ J})(1/n^2)$. This is identical to the expression from Bohr's theory. (b) 3.028×10^{-19} J (c) 656.1 nm
7.63(a) The attraction of the nucleus for the electrons must be overcome. (b) The electrons in silver are more tightly held by the nucleus. (c) silver (d) Once the electron is freed from the atom, its energy increases in proportion to the frequency of the light.
7.66 Li^{2+} **7.68** 2 ⟶ 1 (b) 5 ⟶ 2 (c) 4 ⟶ 2 (d) 3 ⟶ 2 (e) 6 ⟶ 3 **7.72**(a) $l = 1$ or 2 (b) $l = 1$ or 2 (c) $l = 3$, 4, 5, or 6 (d) $l = 2$ or 3
7.74(a) $\Delta E =$
$$(-2.18\times10^{-18}\text{ J})\left(\frac{1}{\infty^2} - \frac{1}{n^2_{initial}}\right)Z^2\left(\frac{6.022\times10^{23}}{1\text{ mol}}\right)$$
(b) 3.28×10^7 J/mol (c) 205 nm (d) 22.8 nm
7.76(a) 5.293×10^{-11} m (b) 5.293×10^{-9} m **7.78** 6.4×10^{27} photons
7.80(a) no overlap (b) overlap (c) two (d) At longer wavelengths, the hydrogen atom line spectrum begins to become a continuous band. **7.82**(a) 7.56×10^{-18} J; 2.63×10^{-8} m (b) 5.122×10^{-17} J; 3.881×10^{-9} m (c) 1.2×10^{-18} J; 1.66×10^{-7} m **7.84**(a) 1.87×10^{-19} J (b) 3.58×10^{-19} J **7.86**(a) red; green (b) 5.89 kJ (Sr); 5.83 kJ (Ba)
7.88(a) As the wavelength of maximum absorbance, it provides the highest sensitivity. (b) ultraviolet region (c) 1.93×10^{-2} g vitamin A/g oil **7.92** 1.0×10^{18} photons/s **7.95** 3s ⟶ 2p; 3d ⟶ 2p; 4s ⟶ 2p; 3p ⟶ 2s **7.97** 133 nm **7.99**(a) 1.8×10^{-17} J (b) 3.1×10^{-40} J (c) As the mass of the particle in a box increases, the difference in energy between levels decreases. **7.101** Lowest energy: 10^{-4} m; highest energy: 10^{-10} m

Chapter 8

8.1 Elements are listed in the periodic table in a systematic way that correlates with a periodicity of their chemical and physical properties. The theoretical basis for the table in terms of sequential atomic number and electron configuration does not allow for a "new element" between Sn and Sb. **8.3**(a) predicted atomic mass = 54.23 amu (b) predicted melting point = 6.3°C **8.6** The spin quantum number, m_s, relates to the electron only; all the others describe the orbital. **8.9** Shielding occurs when electrons protect, or shield, other electrons from the full nuclear attraction. The effective nuclear charge is the nuclear charge an electron actually experiences. As the number of electrons, especially inner electrons, increases, the effective nuclear charge decreases. **8.11**(a) 6 (b) 10 (c) 2 **8.13**(a) 6 (b) 2 (c) 14 **8.16** Hund's rule states that electrons will occupy empty orbitals in a given sublevel (with parallel spins) before filling half-filled orbitals. The lowest energy arrangement has the maximum number of unpaired electrons with parallel spins.

N: $1s^22s^22p^3$ [↑↓] [↑↓] [↑|↑|↑]
1s 2s 2p

8.18 Main-group elements from the same group have similar outer electron configurations, and the A-group number equals the number of outer electrons. Outer electron configurations vary in a periodic manner within a period, with each succeeding element having an additional electron. **8.20** The maximum number of electrons in any energy level n is $2n^2$, so the $n = 4$ energy level holds a maximum of $2(4^2) = 32$ electrons. **8.21**(a) $n = 5$, $l = 0$, $m_l = 0$, $m_s = +\frac{1}{2}$ (b) $n = 3$, $l = 1$, $m_l = -1$, 0 or +1, and m_s $+\frac{1}{2}$ = or $-\frac{1}{2}$. (c) $n = 5$, $l = 0$, $m_l = 0$, $m_s = +\frac{1}{2}$ (d) $n = 2$, $l = 1$, $m_l = +1$, $m_s = -\frac{1}{2}$
8.23(a) Rb: $1s^22s^22p^63s^23p^64s^23d^{10}4p^65s^1$
(b) Ge: $1s^22s^22p^63s^23p^64s^23d^{10}4p^2$
(c) Ar: $1s^22s^22p^63s^23p^6$
8.25(a) Cl: $1s^22s^22p^63s^23p^5$ (b) Si: $1s^22s^22p^63s^23p^2$
(c) Sr: $1s^22s^22p^63s^23p^64s^23d^{10}4p^65s^2$
8.27(a) Ti: [Ar] $4s^23d^2$

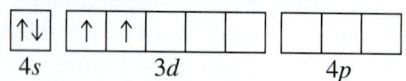

[↑↓] [↑|↑| | |] [| |]
4s 3d 4p

(b) Cl: [Ne] $3s^23p^5$

[↑↓] [↑↓|↑↓|↑]
3s 3p

(c) V: [Ar] $4s^23d^3$

[↑↓] [↑|↑|↑| |] [| |]
4s 3d 4p

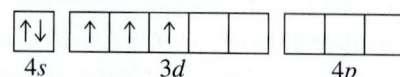

8.29(a) Mn: [Ar] $4s^23d^5$

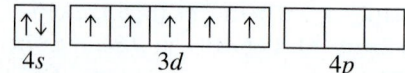

[↑↓] [↑|↑|↑|↑|↑] [| |]
4s 3d 4p

(b) P: [Ne] $3s^23p^3$

[↑↓] [↑|↑|↑]
3s 3p

(c) Fe: [Ar] $4s^23d^6$

[↑↓] [↑↓|↑|↑|↑|↑] [| |]
4s 3d 4p

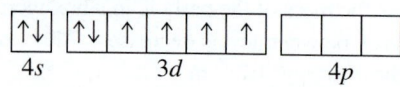

8.31(a) O; Group 6A(16); Period 2

[↑↓] [↑↓|↑|↑]
2s 2p

(b) P; Group 5A(15); Period 3

[↑↓] [↑|↑|↑]
3s 3p

8.33(a) Cl; Group 7A(17); Period 3

[↑↓] [↑↓|↑↓|↑]
3s 3p

(b) As; Group 5A(15); Period 4

[↑↓] [↑|↑|↑]
4s 4p

8.35(a) [Ar] $4s^23d^{10}4p^1$; Group 3A(13) (b) [He] $2s^22p^6$; Group 8A(18)
8.37

	Inner Electrons	Outer Electrons	Valence Electrons
(a) O	2	6	6
(b) Sn	46	4	4
(c) Ca	18	2	2
(d) Fe	18	2	8
(e) Se	28	6	6

8.39(a) B; Al, Ga, In, and Tl (b) S; O, Se, Te, and Po (c) La; Sc, Y, and Ac **8.41**(a) C; Si, Ge, Sn, and Pb (b) V; Nb, Ta, and Db (c) P; N, As, Sb, and Bi
8.43 Na (first excited state): $1s^22s^22p^63p^1$

[↑↓] [↑↓] [↑↓|↑↓|↑↓] [ ] [↑| |]
1s 2s 2p 3s 3p

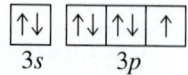

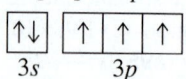

8.50 A high IE_1 and a very negative EA_1 suggest that the elements are halogens, in Group 7A(17), which form 1− ions.
8.53(a) K < Rb < Cs (b) O < C < Be (c) Cl < S < K (d) Mg < Ca < K **8.55**(a) Ba < Sr < Ca (b) B < N < Ne (c) Rb < Se < Br (d) Sn < Sb < As **8.57** $1s^22s^22p^1$ (boron, B) **8.59**(a) Na (b) Na (c) Be **8.61**(1) Metals conduct electricity; nonmetals do not. (2) Metal ions have a positive charge; nonmetal ions have a negative charge. (3) Metal oxides are mostly ionic and act as bases; nonmetal oxides are mostly covalent and act as acids. **8.62** Metallic character increases down a group and decreases to the right across a period. These trends are the same as those for atomic size and opposite those for ionization energy. **8.65** Possible ions are 2+ and 4+. The 2+ ions form by loss of the outermost two p electrons; the 4+ ions form by loss of these and the outermost two s electrons. **8.69**(a) Rb (b) Ra (c) I **8.71**(a) As (b) P (c) Be **8.73** acidic solution; $SO_2(g)$ + $H_2O(l) \longrightarrow H_2SO_3(aq)$ **8.75**(a) Cl^-: $1s^22s^22p^63s^23p^6$ (b) Na^+: $1s^22s^22p^6$ (c) Ca^{2+}: $1s^22s^22p^63s^23p^6$ **8.77**(a) Al^{3+}: $1s^22s^22p^6$ (b) S^{2-}: $1s^22s^22p^63s^23p^6$ (c) Sr^{2+}: $1s^22s^22p^63s^23p^64s^23d^{10}4p^6$ **8.79**(a) 0 (b) 3 (c) 0 (d) 1 **8.81** a, b, and d **8.83**(a) V^{3+}: [Ar] $3d^2$, paramagnetic (b) Cd^{2+}: [Kr] $4d^{10}$, diamagnetic (c) Co^{3+}: [Ar] $3d^6$, paramagnetic (d) Ag^+: [Kr] $4d^{10}$, diamagnetic **8.85** For palladium to be diamagnetic, all of its electrons must be paired. (a) You might first write the condensed electron configuration for Pd as

[Kr] $5s^2 4d^8$. However, the partial orbital diagram is not consistent with diamagnetism.

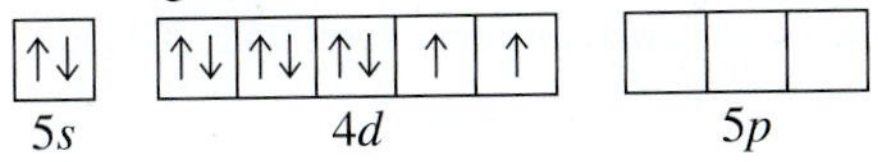

(b) This is the only configuration that supports diamagnetism, [Kr] $4d^{10}$.

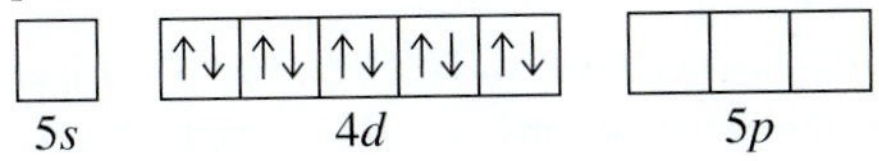

(c) Promoting an *s* electron into the *d* sublevel still leaves two electrons unpaired.

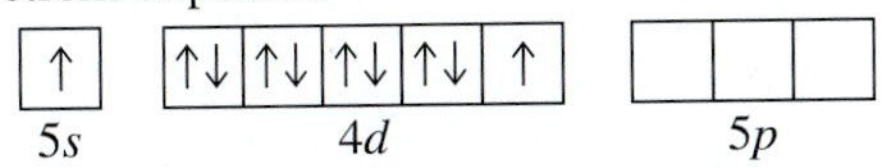

8.87(a) $Li^+ < Na^+ < K^+$ (b) $Rb^+ < Br^- < Se^{2-}$ (c) $F^- < O^{2-} < N^{3-}$ **8.90** Ce: [Xe] $6s^2 4f^1 5d^1$; Ce^{4+}: [Xe]; Eu: [Xe] $6s^2 4f^7$; Eu^{2+}: [Xe] $4f^7$. Ce^{4+} has a noble-gas configuration; Eu^{2+} has a half-filled *f* subshell. **8.91**(a) Cl_2O, dichlorine monoxide (b) Cl_2O_3, dichlorine trioxide (c) Cl_2O_5, dichlorine pentoxide (d) Cl_2O_7, dichlorine heptoxide (e) SO_3, sulfur trioxide (f) SO_2, sulfur dioxide (g) N_2O_5, dinitrogen pentoxide (h) N_2O_3, dinitrogen trioxide (i) CO_2, carbon dioxide (j) P_4O_{10}, tetraphosphorus decoxide **8.94**(a) $SrBr_2$, strontium bromide (b) CaS, calcium sulfide (c) ZnF_2, zinc fluoride (d) LiF, lithium fluoride **8.95**(a) 2009 kJ/mol (b) −549 kJ/mol **8.97** All ions except Fe^{8+} and Fe^{14+} are paramagnetic; Fe^+ and Fe^{3+} would be most strongly attracted.

Chapter 9

Answers to Boxed Reading Problems: B9.1 The C═C bond shows IR absorption at a shorter wavelength (higher energy) because it is a stronger bond than C—C. The C≡C bond would show absorption at a shorter wavelength (higher energy) than the C═C bond, since the triple bond has a higher bond energy than the double bond. **B9.2**(a) 7460 nm (symmetric stretch), 1.50×10^4 nm (bending), 4260 nm (asymmetrical stretch) (b) 2.66×10^{-20} J (symmetric stretch), 1.33×10^{-20} J (bending), 4.67×10^{-20} J (asymmetrical stretch); bending requires the least amount of energy.

• **9.1**(a) Greater ionization energy decreases metallic character. (b) Larger atomic radius increases metallic character. (c) Higher number of outer electrons decreases metallic character. (d) Larger effective nuclear charge decreases metallic character. **9.4**(a) Cs (b) Rb (c) As **9.6**(a) ionic (b) covalent (c) metallic **9.8**(a) covalent (b) ionic (c) covalent **9.10**(a) Rb· (b) ·Ṡi· (c) :Ï· **9.12**(a) ·Sr· (b) :Ṗ· (c) :S̤· **9.14**(a) 6A(16); [noble gas] ns^2np^4 (b) 3(A)13; [noble gas] ns^2np^1 **9.17**(a) Because the lattice energy is the result of electrostatic attractions among the oppositely charged ions, its magnitude depends on several factors, including ionic size, ionic charge, and the arrangement of ions in the solid. For a particular arrangement of ions, the lattice energy increases as the charges on the ions increase and as their radii decrease. (b) A < B < C **9.20**(a) Ba^{2+}, [Xe]; Cl^-, [Ne] $3s^2 3p^6$, $[:\ddot{Cl}:]^-$; $BaCl_2$ (b) Sr^{2+}, [Kr]; O^{2-}, [He] $2s^2 2p^6$, $[:\ddot{O}:]^{2-}$; SrO (c) Al^{3+}, [Ne]; F^-, [He] $2s^2 2p^6$, $[:\ddot{F}:]^-$; AlF_3 (d) Rb^+, [Kr]; O^{2-}, [He] $2s^2 2p^6$, $[:\ddot{O}:]^{2-}$; Rb_2O **9.22**(a) 2A(2) (b) 6A(16) (c) 1A(1) **9.24**(a) 3A(13) (b) 2A(2) (c) 6A(16) **9.26**(a) BaS; the charge on each ion is twice the charge on the ions in CsCl. (b) LiCl; Li^+ is smaller than Cs^+. **9.28**(a) BaS; Ba^{2+} is larger than Ca^{2+}. (b) NaF; the charge on each ion is less than the charge on Mg and O. **9.30** 788 kJ; the lattice energy for NaCl is less than that for LiF, because the Na^+ and Cl^- ions are larger than the Li^+ and F^- ions. **9.33** −336 kJ **9.34** When two chlorine atoms are far apart, there is no interaction between them. As the atoms move closer together, the nucleus of each atom attracts the electrons of the other atom. The closer the atoms, the greater this attraction; however, the repulsions between the two nuclei and between the electrons also increase at the same time. The final internuclear distance is the distance at which maximum attraction is achieved in spite of the repulsions. **9.35** The bond energy is the energy required to break the bond between H atoms and Cl atoms in 1 mol of HCl molecules in the gaseous state. Energy is needed to break bonds, so bond breaking is always endothermic and $\Delta H^\circ_{\text{bond breaking}}$ is positive. The quantity of energy needed to break the bond is released upon its formation, so $\Delta H^\circ_{\text{bond forming}}$ has the same magnitude as $\Delta H^\circ_{\text{bond breaking}}$ but is opposite in sign (always exothermic and negative). **9.39**(a) I—I < Br—Br < Cl—Cl (b) S—Br < S—Cl < S—H (c) C—N < C═N < C≡N **9.41**(a) C—O < C═O; the C═O bond (bond order = 2) is stronger than the C—O bond (bond order = 1). (b) C—H < O—H; O is smaller than C so the O—H bond is shorter and stronger than the C—H bond. **9.43** Less energy is required to break weak bonds. **9.45** Both are one-carbon molecules. Since methane contains no carbon-oxygen bonds, it will have the greater enthalpy of reaction per mole for combustion. **9.47** −168 kJ **9.49** −22 kJ **9.50** −59 kJ **9.51** Electronegativity increases from left to right and increases from bottom to top within a group. Fluorine and oxygen are the two most electronegative elements. Cesium and francium are the two least electronegative elements. **9.53** Ionic bonds occur between two elements of very different electronegativity, generally a metal with low electronegativity and a nonmetal with high electronegativity. Although electron sharing occurs to a very small extent in some ionic bonds, the primary force in ionic bonds is attraction of opposite charges resulting from electron transfer between the atoms. A nonpolar covalent bond occurs between two atoms with identical electronegativity values where the sharing of bonding electrons is equal. A polar covalent bond is between two atoms (generally nonmetals) of different electronegativities so that the bonding electrons are unequally shared. The H–O bond in water is polar covalent. The bond is between two nonmetals so it is covalent and not ionic, but atoms with different electronegativity values are involved. **9.56**(a) Si < S < O (b) Mg < As < P **9.58**(a) N > P > Si (b) As > Ga > Ca

9.60(a) $\overleftarrow{\text{N—B}}$ (b) $\overrightarrow{\text{N—O}}$ (c) $\overset{\text{none}}{\text{C—S}}$

(d) $\overrightarrow{\text{S—O}}$ (e) $\overleftarrow{\text{N—H}}$ (f) $\overrightarrow{\text{Cl—O}}$

9.62 a, d, and e **9.64**(a) nonpolar covalent (b) ionic (c) polar covalent (d) polar covalent (e) nonpolar covalent (f) polar covalent; $SCl_2 < SF_2 < PF_3$

9.66(a) $\overrightarrow{\text{H—I}}$ < $\overrightarrow{\text{H—Br}}$ < $\overrightarrow{\text{H—Cl}}$

(b) $\overrightarrow{\text{H—C}}$ < $\overrightarrow{\text{H—O}}$ < $\overrightarrow{\text{H—F}}$

(c) $\overrightarrow{\text{S—Cl}}$ < $\overrightarrow{\text{P—Cl}}$ < $\overrightarrow{\text{Si—Cl}}$

9.69(a) Shiny, conducts heat, conducts electricity, and is malleable. (b) Metals lose electrons to form positive ions, and metals form basic oxides. **9.73**(a) 800. kJ/mol, which is lower than the value in Table 9.2 (b) -2.417×10^4 kJ (c) 1690. g CO_2 (d) 65.2 L O_2 **9.75**(a) -125 kJ (b) yes, since ΔH_f° is negative (c) -392 kJ (d) No, ΔH_f° for $MgCl_2$ is much more negative than that for MgCl. **9.77**(a) 406 nm (b) 2.93×10^{-19} J (c) 1.87×10^4 m/s **9.80** C—Cl: 3.53×10^{-7} m; bond in O_2: 2.40×10^{-7} m **9.81** XeF_2: 132 kJ/mol; XeF_4: 150. kJ/mol; XeF_6: 146 kJ/mol **9.83**(a) The presence of the very electronegative fluorine atoms bonded to one of the carbons makes the C—C bond polar. This polar bond will tend to undergo heterolytic rather than homolytic cleavage. More energy is required to achieve heterolytic cleavage. (b) 1420 kJ **9.86** 13,286 kJ **9.88** 8.70×10^{14} s^{-1}; 3.45×10^{-7} m; the UV region **9.90**(a) $CH_3OCH_3(g)$: -326 kJ; $CH_3CH_2OH(g)$: -369 kJ (b) The formation of gaseous ethanol is more exothermic. (c) 43 kJ

Chapter 10

Answers to Boxed Reading Problem: B10.1 resonance form on the left: trigonal planar around C, trigonal pyramidal around N; resonance form on the right: trigonal planar around both C and N.

• **10.1** He cannot serve as a central atom because it does not bond. H cannot because it forms only one bond. Fluorine cannot because it needs only one electron to complete its valence level, and it does not have *d* orbitals available to expand its valence level. Thus, it can bond to only one other atom. **10.3** X obeys the octet rule in all the structures except c and g.

10.5(a)
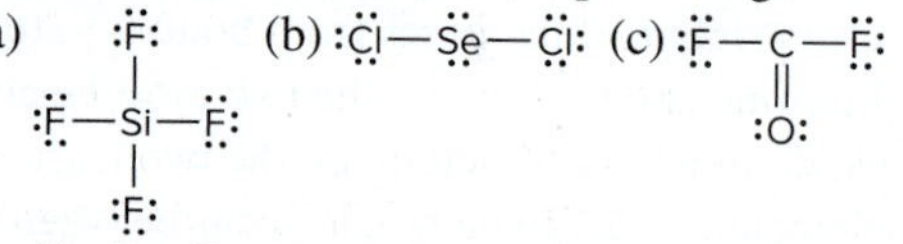

10.7(a)
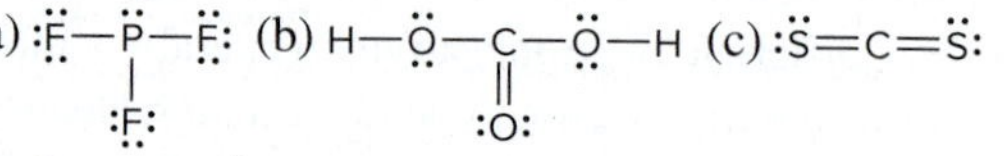

10.9(a)
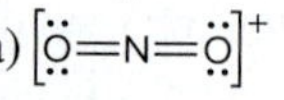

(b)
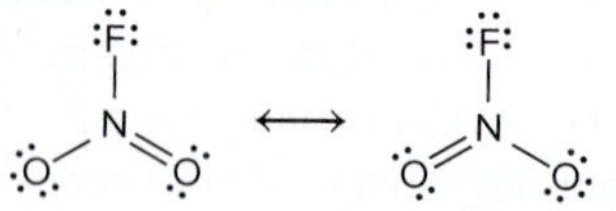

10.11(a)
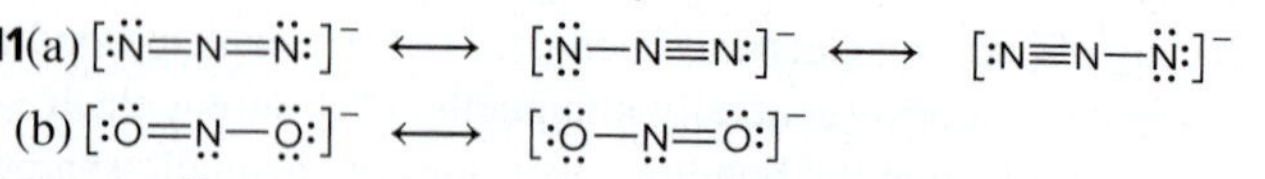

10.13(a)
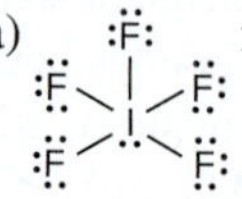
formal charges: I = 0, F = 0

(b)
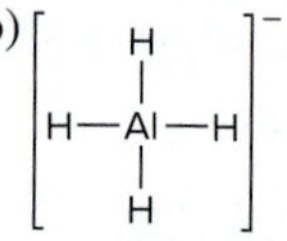
formal charges: H = 0, Al = −1

10.15(a)
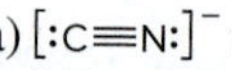
formal charges: C = −1, N = 0

(b)
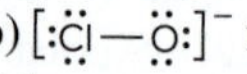
formal charges: Cl = 0, O = −1

10.17(a)
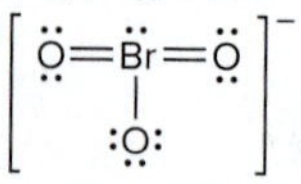
formal charges: Br = 0, doubly bonded O = 0, singly bonded O = −1
O.N.: Br = +5; O = −2

(b)
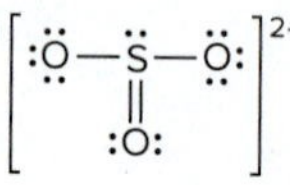
formal charges: S = 0, singly bonded O = −1, doubly bonded O = 0
O.N.: S = +4; O = −2

10.19(a) B has 6 valence electrons in BH_3, so the molecule is electron deficient. (b) As has an expanded valence level with 10 electrons. (c) Se has an expanded valence level with 10 electrons.

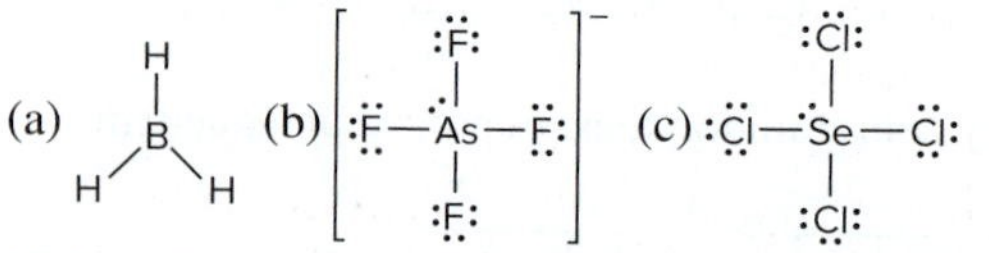

10.21(a) Br expands its valence level to 10 electrons. (b) I has an expanded valence level of 10 electrons. (c) Be has only 4 valence electrons in BeF_2, so the molecule is electron deficient.

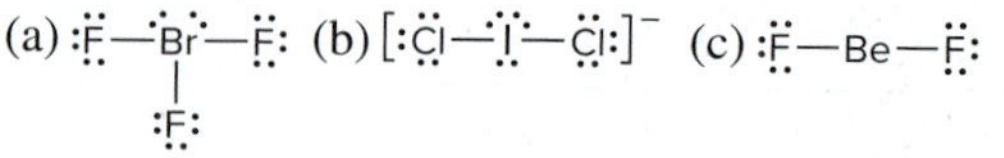

10.23
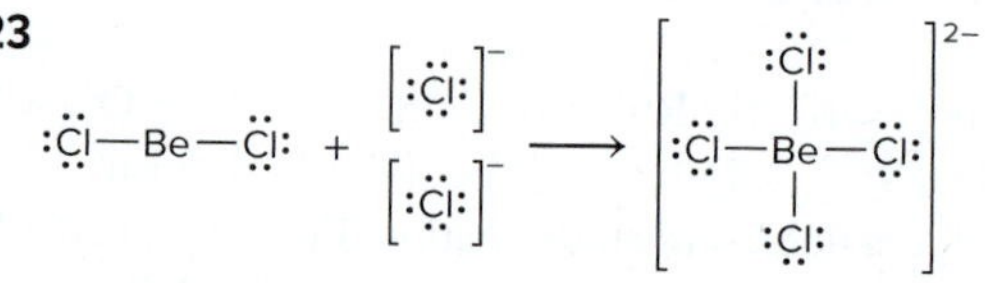

10.26 structure A **10.28** The molecular shape and the electron-group arrangement are the same when no lone pairs are present on the central atom. **10.30** tetrahedral, AX_4; trigonal pyramidal, AX_3E; bent or V shaped, AX_2E_2

10.32
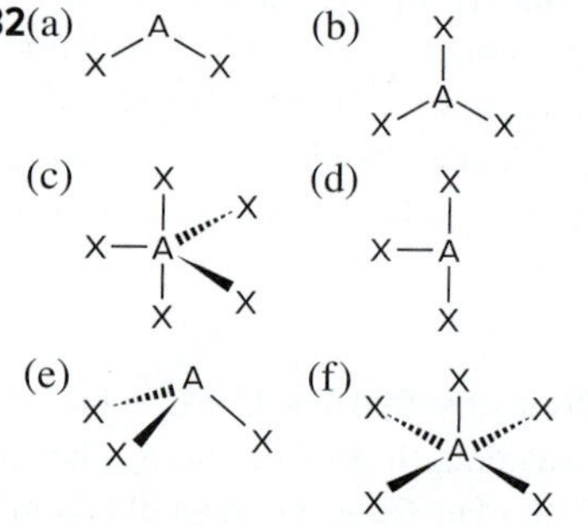

10.34(a) trigonal planar, bent, 120° (b) tetrahedral, trigonal pyramidal, 109.5° (c) tetrahedral, trigonal pyramidal, 109.5° **10.36**(a) trigonal planar, trigonal planar, 120° (b) trigonal planar, bent, 120° (c) tetrahedral, tetrahedral, 109.5° **10.38**(a) trigonal planar, AX_3, 120° (b) trigonal pyramidal, AX_3E, 109.5° (c) trigonal bipyramidal, AX_5, 90° and 120° **10.40**(a) bent, 109.5°, less than 109.5° (b) trigonal bipyramidal, 90° and 120°, angles are ideal (c) seesaw, 90° and 120°, less than ideal (d) linear, 180°, angle is ideal **10.42**(a) C: tetrahedral, 109.5°; O: bent, < 109.5° (b) N: trigonal planar, 120° **10.44**(a) C in CH_3: tetrahedral, 109.5°; C in C=O: trigonal planar, 120°; O with H: bent, < 109.5° (b) O: bent, < 109.5° **10.46** $OF_2 < NF_3 < CF_4 < BF_3 < BeF_2$ **10.48**(a) The C and N each have three electron groups, so the ideal bond angles are 120°; the O has four electron groups, so the ideal bond angle is 109.5°. The N and O have lone pairs, so the bond angles are less than ideal. (b) All central atoms have four electron groups, so the ideal bond angles are 109.5°. The lone pairs on the O reduce this value. (c) The B has three electron groups and an ideal bond angle of 120°. All the O's have four electron groups (ideal bond angles of 109.5°), two of which are lone pairs that reduce the bond angle.

10.51
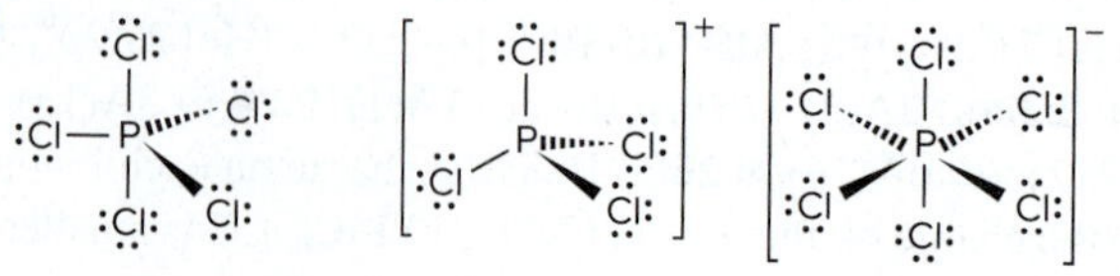

In the gas phase, PCl_5 is AX_5, so the shape is trigonal bipyramidal, and the bond angles are 120° and 90°. The PCl_4^+ ion is AX_4, so the shape is tetrahedral, and the bond angles are 109.5°. The PCl_6^- ion is AX_6, so the shape is octahedral, and the bond angles are 90°. **10.52** Molecules are polar if they have polar bonds that are not arranged to cancel each other. A polar bond is present any time there is a bond between elements with differing electronegativities. **10.55**(a) CF_4 (b) BrCl and SCl_2 **10.57**(a) SO_2, because it is polar and SO_3 is not. (b) IF has a greater electronegativity difference between its atoms. (c) SF_4, because it is polar and SiF_4 is not. (d) H_2O has a greater electronegativity difference between its atoms.

10.59

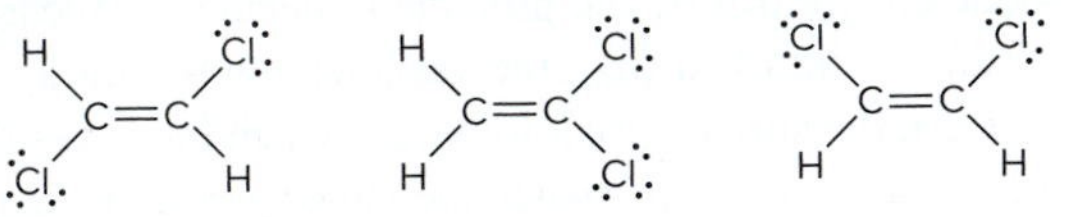

X Y Z

Yes, compound Y has a dipole moment.

10.61(a)

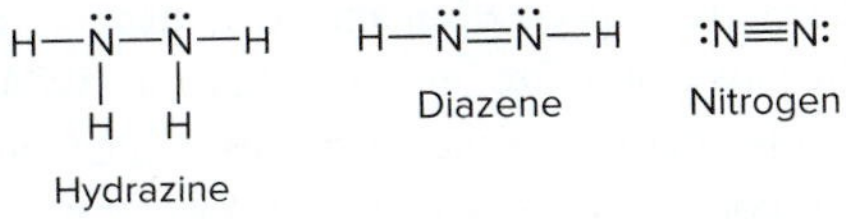

The single N—N bond (bond order = 1) is weaker and longer than the others. The triple bond (bond order = 3) is stronger and shorter than the others. The double bond (bond order = 2) has an intermediate strength and length.

(b) $\Delta H^\circ_{rxn} = -367$ kJ

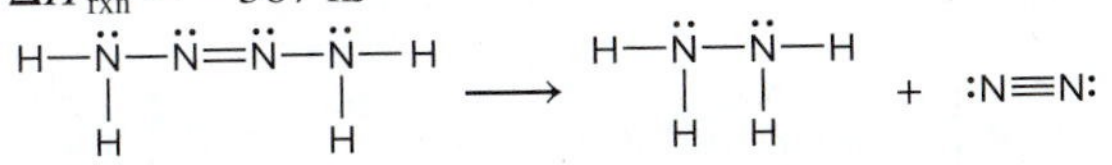

10.65(a) formal charges: Al = −1, end Cl = 0, bridging Cl = +1; I = −1, end Cl = 0, bridging Cl = +1 (b) The iodine atoms are each AX_4E_2, and the shape around each is square planar. These square planar portions are adjacent, giving a planar molecule.

10.70

(a) In propylene oxide, the shape around each C is tetrahedral, with ideal angles of 109.5°. (b) The C that is not part of the three-membered ring should have close to the ideal angle. The atoms in the ring form an equilateral triangle, so the angles around the two C's in the ring are reduced from the ideal 109.5° to nearly 60°. **10.75**(a) −1267 kJ/mol (b) −1226 kJ/mol (c) −1234.8 kJ/mol. The two answers differ by less than 10 kJ/mol. This is very good agreement since average bond energies were used to calculate answers a and b. (d) −37 kJ

10.77

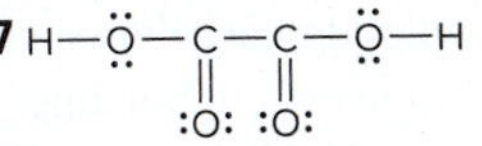

10.80 CH_4: −409 kJ/mol O_2; H_2S: −398 kJ/mol O_2 **10.82**(a) The O in the OH species has only 7 valence electrons, which is less than an octet, and 1 electron is unpaired. (b) 426 kJ (c) 508 kJ **10.84**(a) The F atoms will substitute at the axial positions first. (b) PF_5 and PCl_3F_2

10.86 $H_2C_2O_4$:

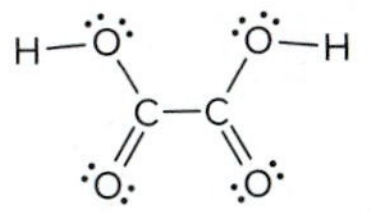

$HC_2O_4^-$:

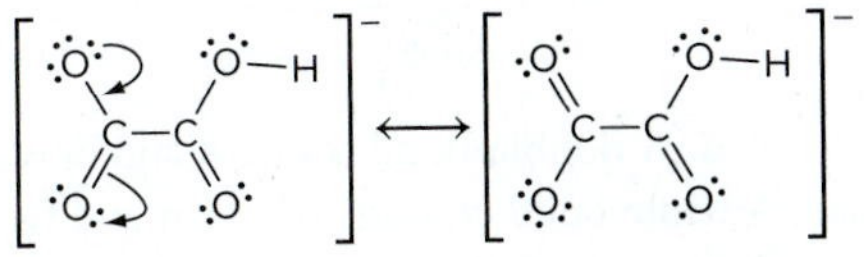

$C_2O_4^{2-}$:

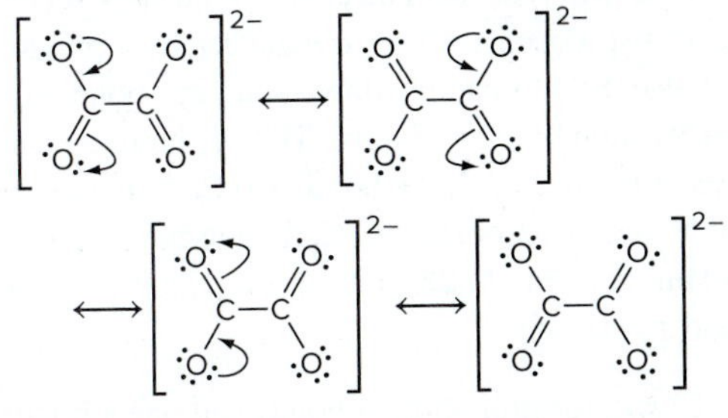

In $H_2C_2O_4$, there are two shorter and stronger C═O bonds and two longer and weaker C—O bonds. In $HC_2O_4^-$, the carbon-oxygen bonds on the side retaining the H remain as one long, weak C—O and one short, strong C═O. The carbon-oxygen bonds on the other side of the molecule have resonance forms with a bond order of 1.5, so they are intermediate in length and strength. In $C_2O_4^{2-}$, all the carbon-oxygen bonds have a bond order of 1.5. **10.90** 22 kJ **10.91** Trigonal planar molecules are nonpolar, so AY_3 cannot have that shape. Trigonal pyramidal molecules and T-shaped molecules are polar, so AY_3 could have either of these shapes. **10.95**(a) 339 pm (b) 316 pm and 223 pm (c) 270 pm

Chapter 11

11.1(a) sp^2 (b) sp^3d^2 (c) sp (d) sp^3 (e) sp^3d **11.3** C has only 2*s* and 2*p* atomic orbitals, allowing for a maximum of four hybrid orbitals. Si has 3*s*, 3*p*, and 3*d* atomic orbitals, allowing it to form more than four hybrid orbitals. **11.5**(a) six, sp^3d^2 (b) four, sp^3 **11.7**(a) sp^2 (b) sp^2 (c) sp^2 **11.9**(a) sp^3 (b) sp^3 (c) sp^3 **11.11**(a) Si: one *s* and three *p* atomic orbitals form four sp^3 hybrid orbitals. (b) C: one *s* and one *p* atomic orbital form two *sp* hybrid orbitals. (c) S: one *s*, three *p*, and one *d* atomic orbital mix to form five sp^3d hybrid orbitals. (d) N: one *s* and three *p* atomic orbitals mix to form four sp^3 hybrid orbitals. **11.13**(a) B ($sp^3 \longrightarrow sp^3$) (b) A ($sp^2 \longrightarrow sp^3$)

11.15

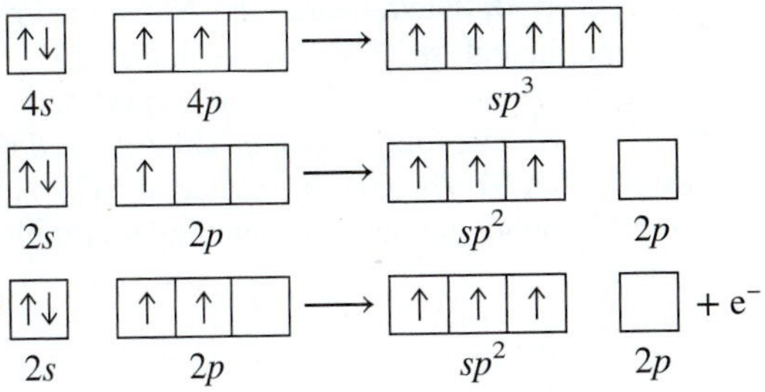

11.17

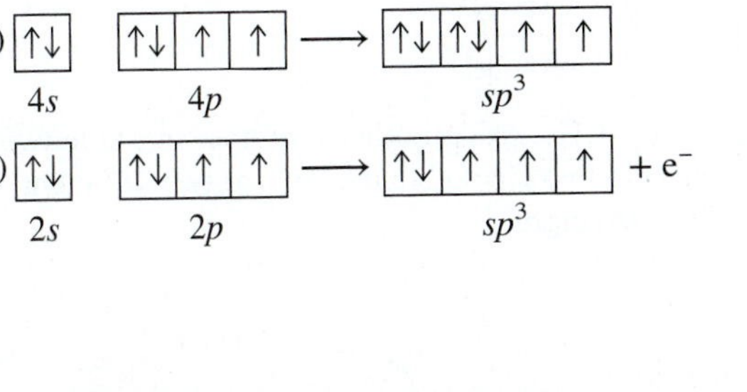

(c) 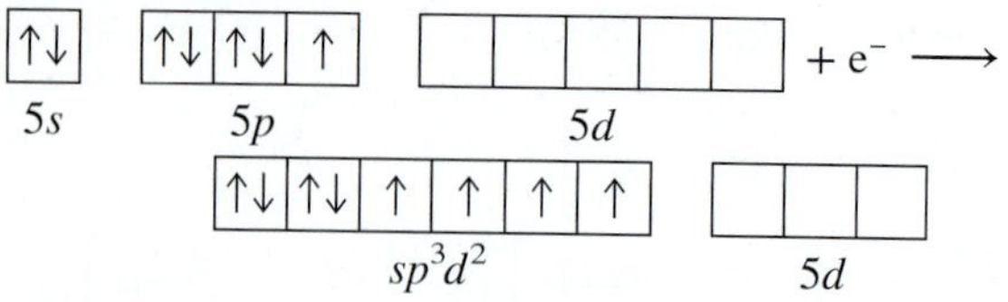

11.20(a) False. A double bond is one σ and one π bond. (b) False. A triple bond consists of one σ and two π bonds. (c) True (d) True (e) False. A π bond consists of one pair of electrons; it occurs after a σ bond has been previously formed. (f) False. End-to-end overlap results in a bond with electron density along the bond axis. **11.21**(a) Nitrogen uses sp^2 to form three σ bonds and one π bond. (b) Carbon uses sp to form two σ bonds and two π bonds. (c) Carbon uses sp^2 to form three σ bonds and one π bond. **11.23**(a) N: sp^2, forming two σ bonds and one π bond :F̤—N̈=Ö:

(b) C: sp^2, forming three σ bonds and one π bond 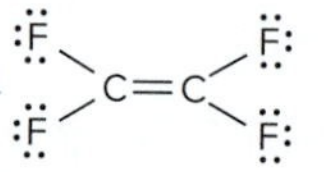

(c) C: sp, forming two σ bonds and two π bonds :N≡C—C≡N:

11.25 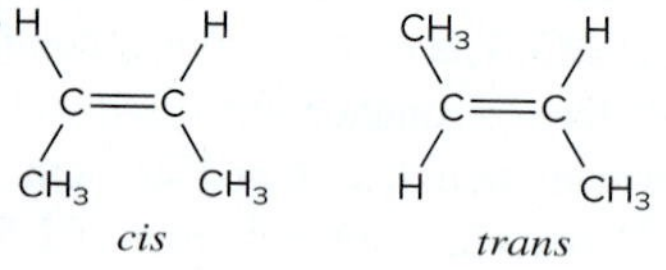

The single bonds are all σ bonds. The double bond is one σ bond and one π bond. **11.26** Four MOs form from the four p atomic orbitals. The total number of MOs must equal the number of atomic orbitals. **11.28**(a) Bonding MOs have lower energy than antibonding MOs. Lower energy means more stable. (b) Bonding MOs do not have a node between the nuclei. (c) Bonding MOs have higher electron density between the nuclei than antibonding MOs do. **11.30**(a) two (b) two (c) four **11.32**(a) A is π^*_{2p}, B is σ_{2p}, C is π_{2p}, and D is σ^*_{2p}. (b) π^*_{2p} in A, σ_{2p} in B, and π_{2p} in C have at least one electron. (c) π^*_{2p} in A has only one electron. **11.34**(a) stable (b) paramagnetic (c) $(\sigma_{2s})^2(\sigma^*_{2s})^1$ **11.36**(a) $C_2^+ < C_2 < C_2^-$ (b) $C_2^- < C_2 < C_2^+$ **11.40**(a) C (ring): sp^2; C (all others): sp^3; O (all): sp^3; N: sp^3 (b) 26 (c) 6 **11.42**(a) 17 (b) All carbons are sp^2; the ring N is sp^2, the other N's are sp^3. **11.44**(a) B changes from sp^2 to sp^3. (b) P changes from sp^3 to sp^3d. (c) C changes from sp to sp^2. Two electron groups surround C in C_2H_2, and three electron groups surround C in C_2H_4. (d) Si changes from sp^3 to sp^3d^2. (e) no change for S **11.46** P: tetrahedral, sp^3; N: trigonal pyramid, sp^3; C_1 and C_2: tetrahedral, sp^3; C_3: trigonal planar, sp^2 **11.51**(a) B and D are present. (b) Yes, sp hybrid orbitals. (c) Two sets of sp orbitals, four sets of sp^2 orbitals, and three sets of sp^3 orbitals. **11.52** Through resonance, the C—N bond gains some double-bond character, which hinders rotation about that bond.

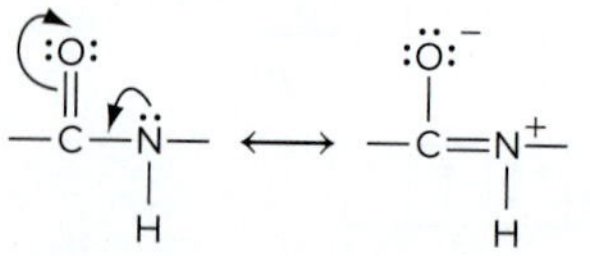

11.56(a) C in —CH_3: sp^3; all other C atoms: sp^2; O in two C—O bonds: sp^3; O in two C=O bonds: sp^2 (b) two (c) eight; one **11.57**(a) four (b) eight

Chapter 12

Answer to Boxed Reading Problem: B12.1 1.76×10^{-10} m

• **12.1** In a solid, the energy of attraction of the particles is greater than their energy of motion; in a gas, it is less. Gases have high compressibility and the ability to flow, while solids have neither. **12.4**(a) Because the intermolecular forces are only partially overcome when fusion occurs but need to be totally overcome in vaporization. (b) Because solids have greater intermolecular forces than liquids do. (c) $\Delta H_{vap} = -\Delta H_{cond}$ **12.5**(a) intermolecular (b) intermolecular (c) intramolecular (d) intramolecular **12.7**(a) condensation (b) fusion (c) vaporization **12.9** The gas molecules slow down as the gas is compressed. Therefore, much of the kinetic energy lost by the propane molecules is released to the surroundings. **12.13** At first, the vaporization of liquid molecules from the surface predominates, which increases the number of gas molecules and hence the vapor pressure. As more molecules enter the gas phase, more gas molecules hit the surface of the liquid and "stick" more frequently, so the condensation rate increases. When the vaporization and condensation rates become equal, the vapor pressure remains constant. **12.14** As the strength of intermolecular forces increases, (a) critical temperature increases, (b) boiling point increases, (c) vapor pressure decreases, and (d) heat of vaporization increases. **12.18** because the condensation of the vapor supplies an additional 40.7 kJ/mol **12.19** 7.67×10^3 J **12.21** 0.777 atm **12.23** 2×10^4 J/mol
12.25

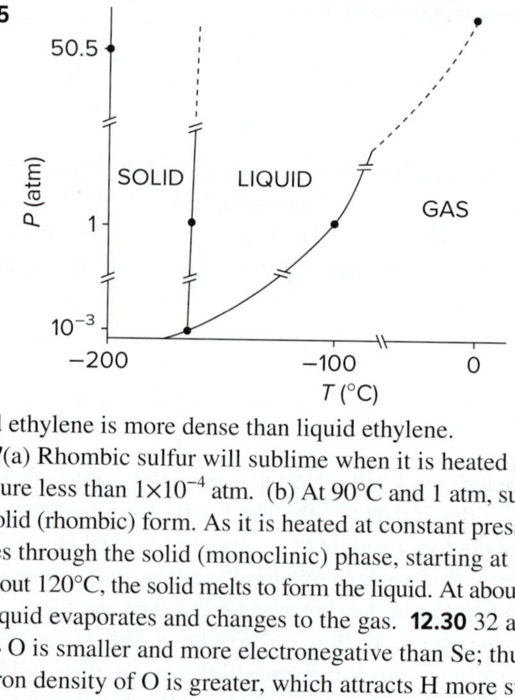

Solid ethylene is more dense than liquid ethylene. **12.27**(a) Rhombic sulfur will sublime when it is heated at a pressure less than 1×10^{-4} atm. (b) At 90°C and 1 atm, sulfur is in the solid (rhombic) form. As it is heated at constant pressure, it passes through the solid (monoclinic) phase, starting at 114°C. At about 120°C, the solid melts to form the liquid. At about 445°C, the liquid evaporates and changes to the gas. **12.30** 32 atm **12.34** O is smaller and more electronegative than Se; thus, the electron density of O is greater, which attracts H more strongly. **12.36** All particles (atoms and molecules) exhibit dispersion forces, but the total force is weak for small molecules. Dipole-dipole forces between small polar molecules dominate dispersion forces between those molecules. **12.39**(a) hydrogen bonding (b) dispersion forces (c) dispersion forces **12.41**(a) dipole-dipole forces (b) dispersion forces (c) H bonding
12.43(a) 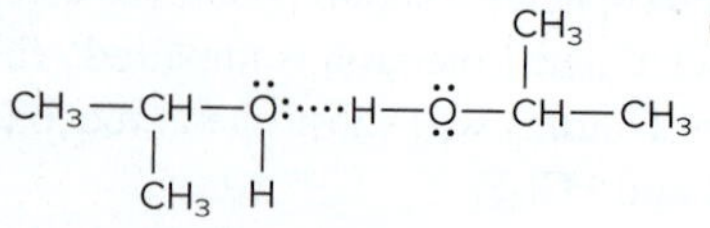 (b) 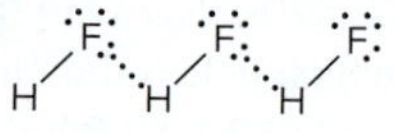

12.45(a) dispersion forces (b) H bonding (c) dispersion forces **12.47**(a) I^- (b) $CH_2{=}CH_2$ (c) H_2Se. In (a) and (c) the larger particle has the higher polarizability. In (b), the less tightly held π electron clouds are more easily distorted. **12.49**(a) C_2H_6; it is a smaller molecule exhibiting weaker dispersion forces than C_4H_{10}. (b) CH_3CH_2F; it has no H—F bonds, so it exhibits only dipole-dipole forces, which are weaker than the H bonds of CH_3CH_2OH. (c) PH_3; it has weaker intermolecular forces (dipole-dipole) than NH_3 (hydrogen bonding). **12.51**(a) HCl; it has dipole-dipole forces, and there is stronger ionic bonding in LiCl. (b) PH_3; it has dipole-dipole forces, and there is stronger H bonding in NH_3. (c) Xe; it exhibits weaker dispersion forces since its smaller size results in lower polarizability than for the larger I_2 molecules. **12.53**(a) C_4H_8 (cyclobutane), because it is more compact than C_4H_{10}. (b) PBr_3; the dipole-dipole forces in PBr_3 are weaker than the ionic bonds in NaBr. (c) HBr; the dipole-dipole forces in HBr are weaker than the H bonds in water. **12.55** As atomic size decreases and electronegativity increases, the electron density of an atom increases. Thus, the attraction to an H atom on another molecule increases while its bonded H atom becomes more positive. Fluorine is the smallest of the three and the most electronegative, so the H bonds in hydrogen fluoride are the strongest. Oxygen is smaller and more electronegative than nitrogen, so H bonds in water are stronger than H bonds in ammonia. **12.59** The cohesive forces in water and mercury are stronger than the adhesive forces to the nonpolar wax on the floor. Weak adhesive forces result in spherical drops. The adhesive forces overcome the even weaker cohesive forces in the oil, and so the oil drop spreads out. **12.61** Surface tension is defined as the energy needed to increase the surface area by a given amount, so units of energy per area are appropriate. **12.63** $CH_3CH_2CH_2OH < HOCH_2CH_2OH < HOCH_2CH(OH)CH_2OH$. More H bonding means greater attraction between molecules, so more energy is needed to increase surface area. **12.65** $HOCH_2CH(OH)CH_2OH > HOCH_2CH_2OH > CH_3CH_2CH_2OH$. More H bonding means greater attraction between molecules, so the liquid flows less easily. **12.70** Water is a good solvent for polar and ionic substances and a poor solvent for nonpolar substances. Water is a polar molecule and dissolves polar substances because their intermolecular forces are of similar strength. **12.71** A single water molecule can form four H bonds. The two hydrogen atoms each form one H bond to oxygen atoms on neighboring water molecules. The two lone pairs on the oxygen atom form H bonds with hydrogen atoms on two neighboring molecules. **12.74** Water exhibits strong capillary action, which allows it to be easily absorbed by the plant's roots and transported upward to the leaves. **12.80** simple cubic **12.83** The energy gap is the energy difference between the highest filled energy level (valence band) and the lowest unfilled energy level (conduction band). In conductors and superconductors, the energy gap is zero because the valence band overlaps the conduction band. In semiconductors, the energy gap is small. In insulators, the gap is large. **12.85** atomic mass and atomic radius **12.86**(a) face-centered cubic (b) body-centered cubic (c) face-centered cubic **12.88** 1.54 g/cm^3 **12.90**(a) The change in unit cell is from a sodium chloride structure in CdO to a zinc blende structure in CdSe. (b) Yes, the coordination number of Cd changes from 6 in CdO to 4 in CdSe. **12.92** 524 pm **12.94**(a) Nickel forms a metallic solid since it is a metal whose atoms are held together by metallic bonds. (b) Fluorine forms a molecular solid since the F_2 molecules are held together by dispersion forces. (c) Methanol forms a molecular solid since the CH_3OH molecules are held together by H bonds. (d) Tin forms a metallic solid since it is a metal whose atoms are held together by metallic bonds. (e) Silicon is under carbon in Group 4A(14); it exhibits similar bonding properties to carbon. Since diamond and graphite are both network covalent solids, it makes sense that Si forms a network covalent solid as well. (f) Xe is an atomic solid since its individual atoms are held together by dispersion forces. **12.96** four **12.98**(a) four Se^{2-} ions, four Zn^{2+} ions (b) 577.48 amu (c) 1.77×10^{-22} cm^3 (d) 5.61×10^{-8} cm **12.100**(a) insulator (b) conductor (c) semiconductor **12.102**(a) Conductivity increases. (b) Conductivity increases. (c) Conductivity decreases. **12.104** 1.68×10^{-8} cm **12.111** A substance whose properties are the same in all directions is isotropic; otherwise, the substance is anisotropic. Liquid crystals have a degree of order only in certain directions, so they are anisotropic. **12.117**(a) n-type semiconductor (b) p-type semiconductor **12.119** $n = 3.4\times10^3$ **12.121** 8.4 pm **12.124**(a) 19.8 torr (b) 0.0485 g **12.130** 259 K (−14 °C) **12.132**(a) 2.23×10^{-2} atm (b) 6.23 L

12.135(a) furfuryl alcohol

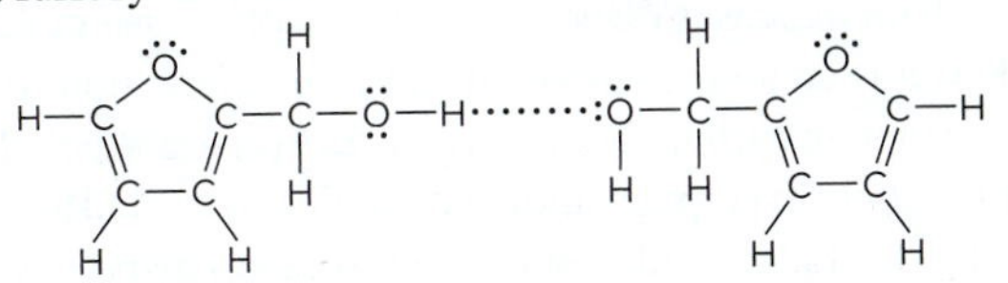

2-furoic acid

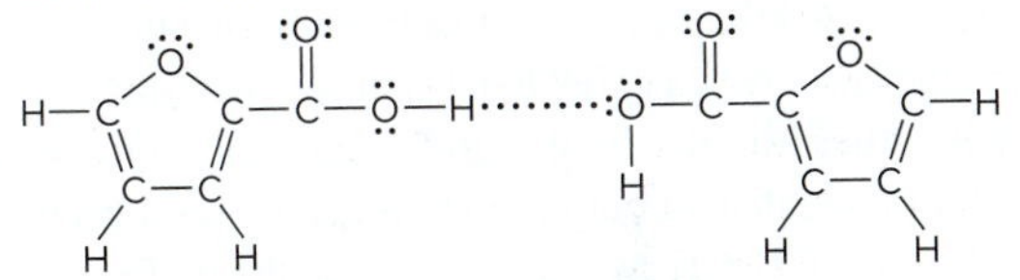

(b) furfuryl alcohol

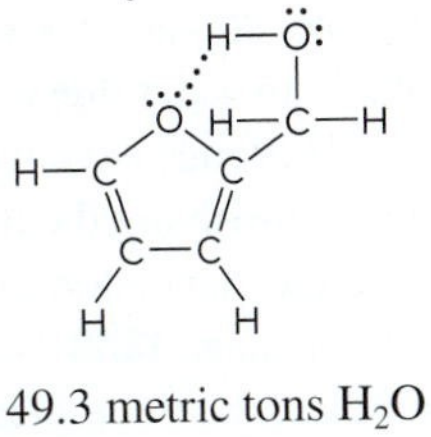

2-furoic acid

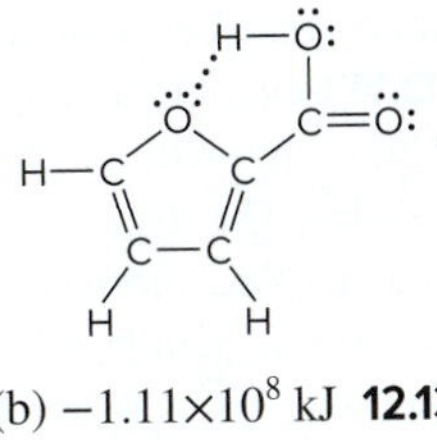

12.136(a) 49.3 metric tons H_2O (b) -1.11×10^8 kJ **12.137** 2.9 g/m^3 **12.142**(a) 1.1 min (b) 10. min

(c)

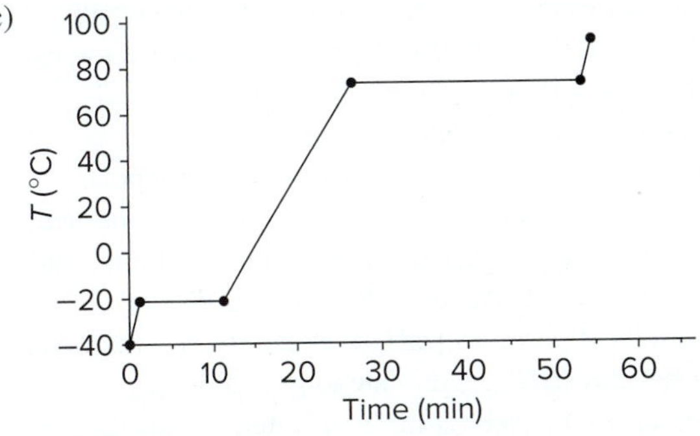

12.143 2.98×10^5 g BN **12.146** 45.98 amu

Chapter 13

Answer to Boxed Reading Problem: B13.1(a) The colloidal particles in water generally have negatively charged surfaces and so repel each other, slowing the settling process. Cake alum, $Al_2(SO_4)_3$, is added to coagulate the colloids. The Al^{3+} ions neutralize the negative surface charges on the particles, allowing them to aggregate and settle. (b) Water that contains large amounts of divalent cations (such as Ca^{2+} and Mg^{2+}) is called hard water. If this water is used for cleaning, these ions combine with the fatty-acid anions in soaps to produce insoluble deposits. (c) In reverse osmosis, a pressure greater than the osmotic pressure is applied to the solution, forcing the water back through the membrane, leaving the ions behind. (d) Chlorine may give the water an unpleasant odor, and it can form carcinogenic chlorinated organic compounds. (e) The high concentration of Na^+ in the saturated solution of NaCl displaces the divalent and polyvalent ions from the ion-exchange resin.

• **13.2** When a salt such as NaCl dissolves, ion-dipole forces cause the ions to separate, and many water molecules cluster around each ion in hydration shells. Ion-dipole forces bind the first shell to an ion. The water molecules in that shell form H bonds to others to create the next shell, and so on. **13.4** Sodium stearate is a more effective soap because the hydrocarbon chain in the stearate ion is longer than that in the acetate ion. The longer chain increases dispersion forces with grease molecules. **13.7** KNO_3 is an ionic compound and is therefore more soluble in water than in carbon tetrachloride. **13.9**(a) ion-dipole forces (b) H bonding (c) dipole–induced dipole forces **13.11**(a) H bonding (b) dipole–induced dipole forces (c) dispersion forces **13.13**(a) $HCl(g)$, because the molecular interactions (dipole-dipole forces) in ether are more like those in HCl than is the ionic bonding in NaCl. (b) $CH_3CHO(l)$, because the molecular interactions with ether (dipole-dipole forces) can replace those between CH_3CHO molecules, but not the H bonds between water molecules. (c) $CH_3CH_2MgBr(s)$, because the molecular interactions (dipole-dipole and dispersion forces) with ether are greater than those between ether molecules and the ions in $MgBr_2$. **13.16** Gluconic acid is soluble in water due to the extensive H bonding involving the —OH groups attached to five of its carbons. The dispersion forces involving the nonpolar tails of caproic acid are more similar to the dispersion forces in hexane; thus, caproic acid is soluble in hexane. **13.18** The nitrogen-containing bases form H bonds to their complementary bases. The flat, N-containing bases stack above each other, which allows extensive dispersion forces. The exterior, negatively charged, sugar-phosphate chains experience ion-dipole forces and form H bonds with water molecules in the aqueous surroundings, which also stabilizes the structure.
13.21 Dispersion forces are present between the nonpolar tails of the lipid molecules within the bilayer. The polar heads interact with the aqueous surroundings through H bonds and ion-dipole forces. **13.25** The enthalpy changes needed to separate the solvent particles ($\Delta H_{solvent}$) and to mix the solvent and solute particles (ΔH_{mix}) combine to give $\Delta H_{solvation}$.
13.29 The compound is very soluble in water, because a decrease in enthalpy and an increase in entropy both favor the formation of a solution.

13.30

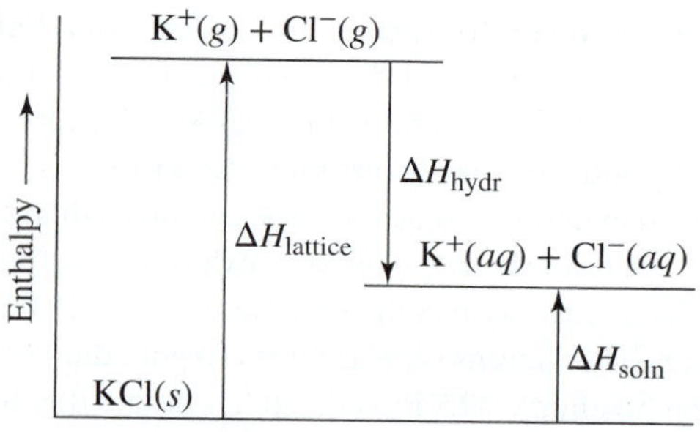

13.32(a) The volume of Na^+ is smaller. (b) Sr^{2+} has a larger ionic charge and a smaller volume. (c) Na^+ is smaller than Cl^-. (d) O^{2-} has a larger ionic charge. (e) OH^- has a smaller volume than SH^-. (f) Mg^{2+} has a smaller volume. (g) Mg^{2+} has both a smaller volume and a larger ionic charge. (h) CO_3^{2-} has a larger ionic charge. **13.34**(a) Na^+ (b) Sr^{2+} (c) Na^+ (d) O^{2-} (e) OH^- (f) Mg^{2+} (g) Mg^{2+} (h) CO_3^{2-} **13.36**(a) −704 kJ/mol (b) The K^+ ion contributes more because it is smaller and, therefore, has a greater charge density. **13.38**(a) increases (b) decreases (c) increases **13.41** Add a pinch of X to each solution. Addition of a "seed" crystal of solute to a supersaturated solution causes the excess solute to crystallize immediately, leaving behind a saturated solution. The solution in which the added X dissolves is the unsaturated solution. The solution in which the added X remains undissolved is the saturated solution. **13.44**(a) increase (b) decrease **13.46**(a) 0.102 g O_2 (b) 0.0214 g O_2 **13.49** 0.20 mol/L **13.52**(a) molarity and % w/v or % v/v (b) parts-by-mass (% w/w) (c) molality **13.54** With just this information, you can convert between molality and molarity, but you need to know the molar mass of the solvent to convert to mole fraction.
13.56(a) 0.944 M $C_{12}H_{22}O_{11}$ (b) 0.167 M $LiNO_3$ **13.58**(a) 0.0749 M NaOH (b) 0.36 M HNO_3 **13.60**(a) Add 4.25 g KH_2PO_4 to enough water to make 365 mL of aqueous solution. (b) Add 125 mL of 1.25 M NaOH to enough water to make 465 mL of solution.
13.62(a) Weigh out 48.0 g KBr, dissolve it in about 1 L distilled water, and then dilute to 1.40 L with distilled water. (b) Measure 82.7 mL of the 0.264 M $LiNO_3$ solution and add distilled water to make a total of 255 mL. **13.64**(a) 0.896 m glycine (b) 1.21 m glycerol **13.66** 4.48 m C_6H_6 **13.68**(a) Add 2.39 g $C_2H_6O_2$ to 308 g H_2O. (b) Add 0.0508 kg of 52.0% HNO_3 by mass to 1.15 kg H_2O to make 1.20 kg of 2.20% HNO_3 by mass. **13.70**(a) 0.29 (b) 58 mass % (c) 23 m C_3H_7OH **13.72** 42.6 g CsBr; mole fraction = 7.16×10^{-3}; 7.84% by mass **13.74** 5.11 m NH_3; 4.53 M NH_3; mole fraction = 0.0843 **13.76** 2.5 ppm Ca^{2+}; 0.56 ppm Mg^{2+} **13.80** Its solution conducts a "strong" current. A strong electrolyte dissociates completely into ions in solution.
13.82 The boiling point is higher and the freezing point is lower for the solution than for the pure solvent. **13.85** A dilute solution of an electrolyte behaves more ideally than a concentrated one. With increasing concentration, the effective concentration deviates from the molar concentration because of ionic attractions. Thus, 0.050 m NaF has a boiling point closer to its predicted value.
13.88(a) strong electrolyte (b) strong electrolyte (c) nonelectrolyte (d) weak electrolyte **13.90**(a) 0.6 mol of solute particles (b) 0.13 mol (c) 2×10^{-4} mol (d) 0.06 mol **13.92**(a) CH_3OH in H_2O (b) H_2O in CH_3OH solution **13.94**(a) $\Pi_{II} < \Pi_I < \Pi_{III}$ (b) $bp_{II} < bp_I < bp_{III}$ (c) $fp_{III} < fp_I < fp_{II}$ (d) $vp_{III} < vp_I < vp_{II}$ **13.96** 23.4 torr
13.98 −0.467°C **13.100** 79.5°C **13.102** 1.18×10^4 g $C_2H_6O_2$

13.104 2.44×10^{-3} atm **13.106** 342 g/mol **13.108** -14.6°C **13.110**(a) NaCl: 0.173 *m* and $i = 1.84$ (b) CH_3COOH: 0.0837 *m* and $i = 1.02$ **13.113** 209 torr for CH_2Cl_2; 48.1 torr for CCl_4 **13.114** The fluid inside a bacterial cell is both a solution and a colloid. It is a solution of ions and small molecules and a colloid of large molecules (proteins and nucleic acids). **13.118** Soap micelles have nonpolar tails pointed inward and anionic heads pointed outward. The like charges on the heads of one micelle repel those on the heads of a neighboring micelle. This repulsion between micelles keeps them from coagulating. Soap is more effective in freshwater than in seawater because the divalent cations in seawater combine with the anionic heads to form a precipitate. **13.122** 3.4×10^9 L **13.126** 0.0°C: 4.53×10^{-4} *M* O_2; 20.0°C: 2.83×10^{-4} *M* O_2; 40.0°C: 2.00×10^{-4} *M* O_2 **13.128**(a) 89.9 g/mol (b) C_2H_5O; $C_4H_{10}O_2$

(c) Forms H bonds: 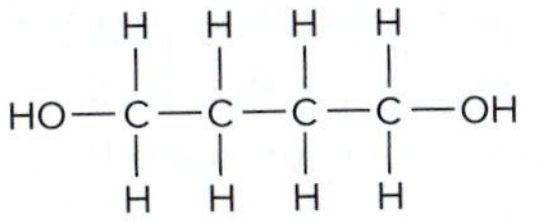

Does not form H bonds: 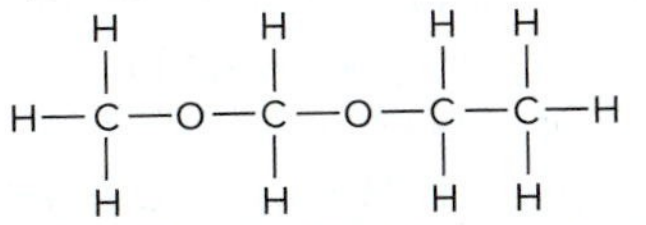

13.132(a) 9.45 g NaF (b) 0.0017 g F^- **13.135**(a) 68 g/mol (b) 2.1×10^2 g/mol (c) The molar mass of CaN_2O_6 is 164.10 g/mol. This value is less than the 2.1×10^2 g/mol calculated when the compound is assumed to be a strong electrolyte and is greater than the 68 g/mol calculated when the compound is assumed to be a nonelectrolyte. Thus, the compound forms a nonideal solution because the ions interact and therefore do not dissociate completely in solution. (d) $i = 2.4$ **13.139**(a) 1.82×10^4 g/mol (b) 3.41×10^{-5} °C **13.140** 8.2×10^5 ng/L **13.144**(a) 0.02 (b) 5×10^{-3} atm·L/mol (c) yes **13.146** Weigh 3.11 g of $NaHCO_3$ and dissolve in 247 g of water. **13.150** $M = m$(kg solvent/L solution) $= m \times d_{\text{solution}}$. Thus, for very dilute solutions, molality × density = molarity. For an aqueous solution, the number of liters of solution is approximately the same as the kg of solvent because the density of water is close to 1 kg/L, so $m = M$. **13.152**(a) 2.5×10^{-3} *M* SO_2 (b) The base reacts with the sulfur dioxide to produce calcium sulfite. This removal of sulfur dioxide allows more sulfur dioxide to dissolve. **13.157**(a) 7.83×10^{-3} mol/L·atm (b) 4×10^{-5} *M* (c) 3×10^{-6} (d) 1 ppm **13.161**(a) 0.0212 mol/L·atm (b) 8.86 ppm (c) k_H: $C_6F_{14} > C_6H_{14}$ > ethanol > water. To dissolve oxygen in a solvent, the solvent molecules must be moved apart to make room for the gas. The stronger the intermolecular forces in the solvent, the more difficult it is to separate solvent particles and the lower the solubility of the gas. Both C_6F_{14} and C_6H_{14} have weak dispersion forces, with C_6F_{14} having the weaker forces due to the electronegative fluorine atoms repelling each other. In both ethanol and water, the molecules are held together by strong H bonds with those bonds being stronger in water, as the boiling point indicates. **13.163**(a) Yes, the phases of water can still coexist at some temperature and can therefore establish equilibrium. (b) The triple point would occur at a lower pressure and lower temperature because the dissolved air would lower the vapor pressure of the solvent. (c) Yes, this is possible because the gas-solid phase boundary exists below the new triple point. (d) No; at both temperatures, the presence of the solute lowers the vapor pressure of the liquid. **13.165**(a) 2.8×10^{-4} g/mL (b) 81 mL

Chapter 14

14.1 The outermost electron is attracted by a smaller effective nuclear charge in Li because of shielding by the inner electrons, and it is farther from the nucleus than the electron of H is. Both of these factors lead to a lower ionization energy for Li.
14.3(a) NH_3 will hydrogen bond.

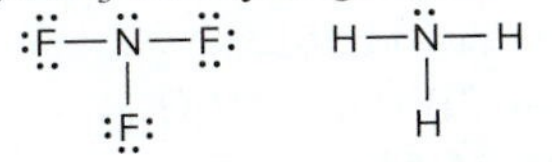

(b) CH_3CH_2OH will hydrogen bond.

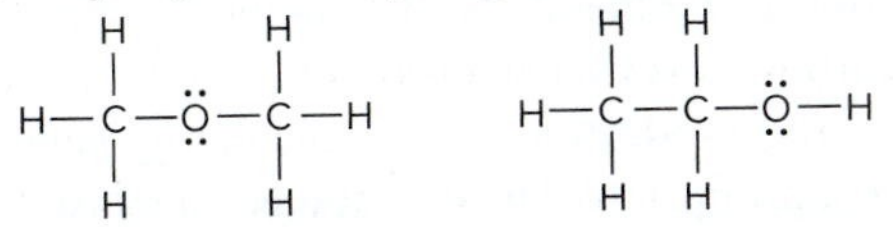

14.5(a) $2Al(s) + 6HCl(aq) \longrightarrow 2AlCl_3(aq) + 3H_2(g)$
(b) $LiH(s) + H_2O(l) \longrightarrow LiOH(aq) + H_2(g)$
14.7(a) $NaBH_4$: +1 for Na, +3 for B, −1 for H; $Al(BH_4)_3$: +3 for Al, +3 for B, −1 for H; $LiAlH_4$: +1 for Li, +3 for Al, −1 for H (b) tetrahedral 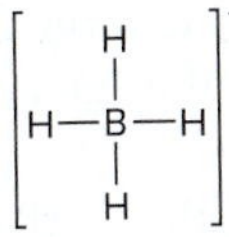

14.10 For Groups 1A(1) to 4A(14), the number of covalent bonds equals the A-group number. For Groups 5A(15) to 7A(17), it equals 8 minus the A-group number. There are exceptions in Period 3 to Period 6 because it is possible for the 3A(13) to 7A(17) elements to use *d* orbitals and form more bonds. **14.13**(a) Group 3A(13) or 3B(3) (b) If E is in Group 3B(3), the oxide will be more basic and the fluoride will be more ionic; if E is in Group 3A(13), the oxide will be more acidic and the fluoride will be more covalent. **14.16**(a) reducing agent (b) Alkali metals have relatively low ionization energies, which means they easily lose the outermost electron.
(c) $2Na(s) + 2H_2O(l) \longrightarrow 2Na^+(aq) + 2OH^-(aq) + H_2(g)$
$2Na(s) + Cl_2(g) \longrightarrow 2NaCl(s)$
14.18 Density and ionic size increase down a group; the other three properties decrease down a group.
14.20 $2Na(s) + O_2(g) \longrightarrow Na_2O_2(s)$
14.22 $K_2CO_3(s) + 2HI(aq) \longrightarrow 2KI(aq) + H_2O(l) + CO_2(g)$
14.26 Group 2A(2) metals have an additional valence electron to increase the strength of metallic bonding, which leads to higher melting points, and also higher boiling points, greater hardness, and greater density.
14.27(a) $CaO(s) + H_2O(l) \longrightarrow Ca(OH)_2(s)$
(b) $2Ca(s) + O_2(g) \longrightarrow 2CaO(s)$
14.30(a) $BeO(s) + H_2O(l) \longrightarrow$ no reaction
(b) $BeCl_2(l) + 2Cl^-(\text{solvated}) \longrightarrow BeCl_4^{2-}(\text{solvated})$
Be behaves like other Group 2A(2) elements in reaction (b).
14.33 The electron removed from Group 2A(2) atoms occupies the outer *s* orbital, whereas in Group 3A(13) atoms, the electron occupies the outer *p* orbital. For example, the electron configuration for Be is $1s^22s^2$ and for B it is $1s^22s^22p^1$. It is easier to remove the *p* electron of B than an *s* electron of Be, because the energy of a *p* orbital is higher than that of the *s* orbital of the same level. Even though atomic size decreases because of increasing Z_{eff}, IE decreases from 2A(2) to 3A(13).
14.34(a) Most atoms form stable compounds when they complete their outer shell (octet). Some compounds of Group 3A(13) elements, like boron, have only six electrons around the central

atom. Having fewer than eight electrons is called *electron deficiency*.
(b) $BF_3(g) + NH_3(g) \longrightarrow F_3B{-}NH_3(g)$
$B(OH)_3(aq) + OH^-(aq) \longrightarrow B(OH)_4^-(aq)$
14.36 $In_2O_3 < Ga_2O_3 < Al_2O_3$ **14.38** Apparent O.N., +3; actual O.N., +1. 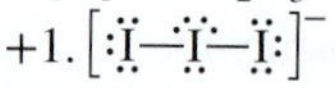 The anion I_3^- has the general formula AX_2E_3 and bond angles of 180°. $(Tl^{3+})(I^-)_3$ does not exist because of the low strength of the Tl—I bond.
14.43(a) $B_2O_3(s) + 2NH_3(g) \longrightarrow 2BN(s) + 3H_2O(g)$
(b) 1.30×10^2 kJ (c) 5.3 kg borax **14.44** Basicity in water is greater for the oxide of a metal. Tin(IV) oxide is more basic in water than carbon dioxide because tin has more metallic character than carbon. **14.46**(a) Ionization energy generally decreases down a group. (b) The deviations (increases) from the expected trend are due to the presence of the first transition series between Si and Ge and of the lanthanides between Sn and Pb. (c) Group 3A(13) **14.49** Atomic size increases down a group. As atomic size increases, ionization energy decreases, and so it is easier to form a positive ion. An atom that is easier to ionize exhibits greater metallic character.
14.51(a) 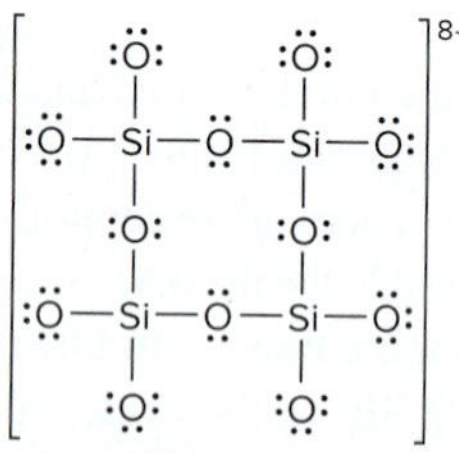 (b) 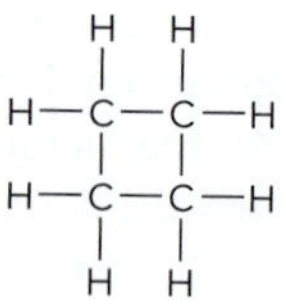
14.54(a) diamond, C (b) calcium carbonate, $CaCO_3$ (c) carbon dioxide, CO_2 (d) carbon monoxide, CO (e) silicon, Si
14.58(a) −3 to +5 (b) For a group of nonmetals, the oxidation states range from the lowest, equal to the A-group number − 8, or 5 − 8 = −3 for Group 5A, to the highest, equal to the A-group number, or +5 for Group 5A.
14.61 $H_3AsO_4 < H_3PO_4 < HNO_3$
14.63(a) $4As(s) + 5O_2(g) \longrightarrow 2As_2O_5(s)$
(b) $2Bi(s) + 3F_2(g) \longrightarrow 2BiF_3(s)$
(c) $Ca_3As_2(s) + 6H_2O(l) \longrightarrow 3Ca(OH)_2(s) + 2AsH_3(g)$
14.65(a) $N_2(g) + 2Al(s) \xrightarrow{\Delta} 2AlN(s)$
(b) $PF_5(g) + 4H_2O(l) \longrightarrow H_3PO_4(aq) + 5HF(g)$
14.67 trigonal bipyramidal, with axial F atoms and equatorial Cl atoms; 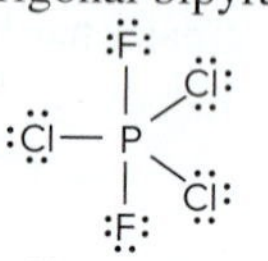
14.71(a) 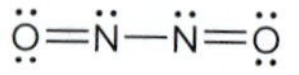 (b) 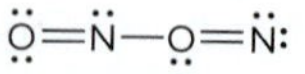
(c) 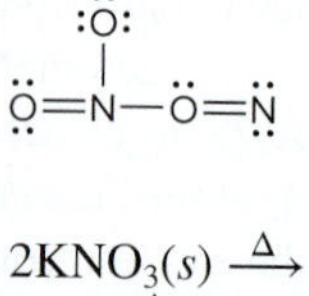 (d) 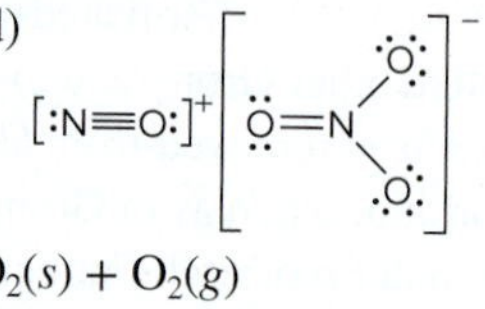
14.73(a) $2KNO_3(s) \xrightarrow{\Delta} 2KNO_2(s) + O_2(g)$
(b) $4KNO_3(s) \xrightarrow{\Delta} 2K_2O(s) + 2N_2(g) + 5O_2(g)$
14.75(a) Boiling point and conductivity vary in similar ways down both groups. (b) Degree of metallic character and types of bonding vary in similar ways down both groups. (c) Both P and S have allotropes, and both bond covalently with almost every other nonmetal. (d) Both N and O are diatomic gases at normal temperatures and pressures. (e) O_2 is a reactive gas, whereas N_2 is not. Nitrogen can have any of six oxidation states, whereas oxygen has two.

14.77(a) $NaHSO_4(aq) + NaOH(aq) \longrightarrow Na_2SO_4(aq) + H_2O(l)$
(b) $S_8(s) + 24F_2(g) \longrightarrow 8SF_6(g)$
(c) $FeS(s) + 2HCl(aq) \longrightarrow H_2S(g) + FeCl_2(aq)$
(d) $Te(s) + 2I_2(s) \longrightarrow TeI_4(s)$
14.79(a) acidic (b) acidic (c) basic (d) amphoteric (e) basic
14.81 $H_2O < H_2S < H_2Te$ **14.84**(a) O_3, ozone (b) SO_3, sulfur trioxide (c) SO_2, sulfur dioxide (d) H_2SO_4, sulfuric acid
(e) $Na_2S_2O_3{\cdot}5H_2O$, sodium thiosulfate pentahydrate
14.86 $S_2F_{10}(g) \longrightarrow SF_4(g) + SF_6(g)$; O.N. of S in S_2F_{10} is +5; O.N. of S in SF_4 is +4; O.N. of S in SF_6 is +6. **14.88**(a) −1, +1, +3, +5, +7 (b) The electron configuration for Cl is [Ne] $3s^23p^5$. By gaining one electron, Cl achieves an octet. When forming covalent bonds, Cl completes or expands its valence level by maintaining electron pairs in bonds or as lone pairs. (c) Fluorine has only the −1 oxidation state because its small size and absence of *d* orbitals prevent it from forming more than one covalent bond. **14.90**(a) Cl—Cl bond is stronger than Br—Br bond. (b) Br—Br bond is stronger than I—I bond. (c) Cl—Cl bond is stronger than F—F bond. The fluorine atoms are so small that electron-electron repulsions between the lone pairs decreases the strength of the bond.
14.92(a) $3Br_2(l) + 6OH^-(aq) \longrightarrow 5Br^-(aq) + BrO_3^-(aq) + 3H_2O(l)$
(b) $ClF_5(l) + 6OH^-(aq) \longrightarrow 5F^-(aq) + ClO_3^-(aq) + 3H_2O(l)$
14.93(a) $2Rb(s) + Br_2(l) \longrightarrow 2RbBr(s)$
(b) $I_2(s) + H_2O(l) \longrightarrow HI(aq) + HIO(aq)$
(c) $Br_2(l) + 2I^-(aq) \longrightarrow I_2(s) + 2Br^-(aq)$
(d) $CaF_2(s) + H_2SO_4(l) \longrightarrow CaSO_4(s) + 2HF(g)$
14.95 HIO < HBrO < HClO < $HClO_2$ **14.99** $I_2 < Br_2 < Cl_2$, because Cl_2 is able to oxidize Re to the +6 oxidation state, Br_2 only to the +5 state, and I_2 only to the +4 state **14.100** helium; argon **14.102** Only dispersion forces hold atoms of the noble gases together. **14.106** $\Delta H_{soln} = -411$ kJ **14.110**(a) Second ionization energies for alkali metals are so high because the electron being removed is from the next lower energy level and these are very tightly held by the nucleus. Also, the alkali metal would lose its noble gas electron configuration.
(b) $2CsF_2(s) \longrightarrow 2CsF(s) + F_2(g)$; $\Delta H_{rxn} = -405$ kJ/mol
14.112(a) hyponitrous acid, $H_2N_2O_2$; nitroxyl, HNO
(b) 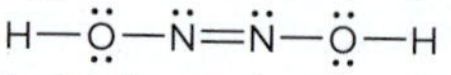 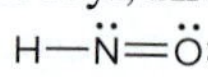
(c) In both species the shape is bent about the N atoms.
(d) 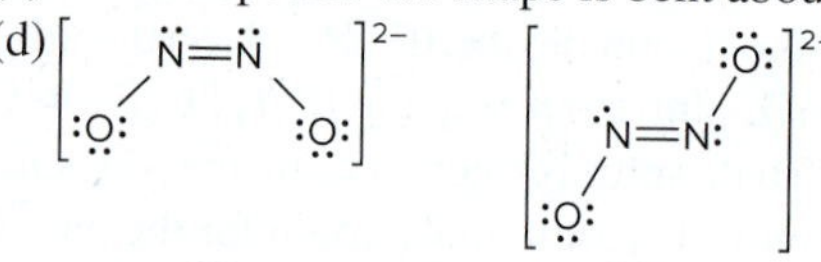
cis *trans*
14.116 13 t **14.118** In a disproportionation reaction, a substance acts as both a reducing agent and an oxidizing agent because atoms of an element within the substance attain both higher and lower oxidation states in the products. The disproportionation reactions are b, c, d, e, and f. **14.120**(a) Group 5A(15)
(b) Group 7A(17) (c) Group 6A(16) (d) Group 1A(1)
(e) Group 3A(13) (f) Group 8A(18) **14.122** 117.2 kJ
14.127 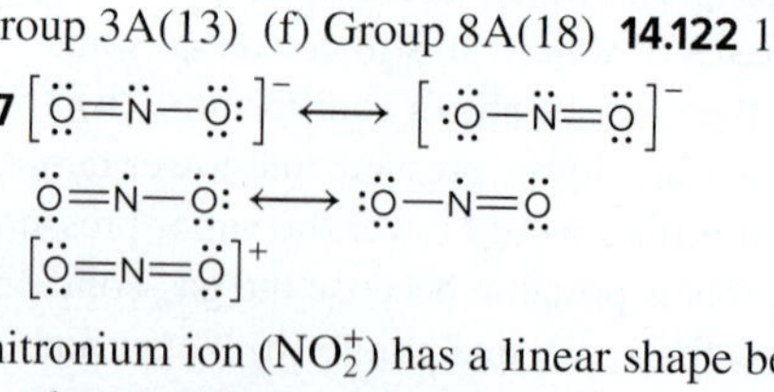
The nitronium ion (NO_2^+) has a linear shape because the central N atom has two surrounding electron groups, which achieve minimum repulsion at 180°. The nitrite ion (NO_2^-) bond angle is

more compressed than the nitrogen dioxide (NO_2) bond angle because the lone pair of electrons takes up more space than the lone electron. **14.129** 2.29×10^4 g UF_6 **14.133** O^+, O^-, and O^{2+} **14.134**(a) 39.96 mass % in $CuHAsO_3$; As, 62.42 mass % in $(CH_3)_3As$ (b) 0.35 g $CuHAsO_3$

Chapter 15

Answers to Boxed Reading Problems: B15.1 butane, 2 peaks; 2-methylpropane, 2 peaks; pentane, 3 peaks; 2-methylbutane, 4 peaks; 2,2-dimethylpropane, 1 peak **B15.2** 0.08 J/mol

• **15.2**(a) Carbon's electronegativity is midway between the most metallic and most nonmetallic elements of Period 2. To attain a filled outer shell, carbon forms covalent bonds to other atoms in molecules, network covalent solids, and polyatomic ions. (b) Since carbon has four valence shell electrons, it forms four covalent bonds to attain an octet. (c) To reach the He electron configuration, a carbon atom must lose four electrons, requiring too much energy to form the C^{4+} cation. To reach the Ne electron configuration, the carbon atom must gain four electrons, also requiring too much energy to form the C^{4-} anion. (d) Carbon is able to bond to itself extensively because its small size allows for close approach and great orbital overlap. The extensive orbital overlap results in a strong, stable bond. (e) The C—C σ bond is short enough to allow sideways overlap of unhybridized *p* orbitals of neighboring C atoms. The sideways overlap of *p* orbitals results in the π bonds that are part of double and triple bonds. **15.3**(a) H, O, N, P, S, and halogens (b) Heteroatoms are atoms of any element other than carbon and hydrogen. (c) More electronegative than C: N, O, F, Cl, and Br; less electronegative than C: H and P. Sulfur and iodine have the same electronegativity as carbon. (d) Since carbon can bond to a wide variety of heteroatoms and to other carbon atoms, it can form many different compounds. **15.6** The C—H and C—C bonds are unreactive because electronegativities are close and the bonds are short. The C—I bond is somewhat reactive because it is long and weak. The C=O bond is reactive because oxygen is more electronegative than carbon and the partially positive C atom attracts electron-rich atoms. The C—Li bond is also reactive because the bond polarity results in an electron-rich region around carbon and an electron-poor region around lithium. **15.7**(a) An alkane and a cycloalkane are organic compounds that consist of carbon and hydrogen and have only single bonds. A cycloalkane has a ring of carbon atoms. An alkene is a hydrocarbon with at least one double bond. An alkyne is a hydrocarbon with at least one triple bond. (b) alkane = C_nH_{2n+2}, cycloalkane = C_nH_{2n}, alkene = C_nH_{2n}, alkyne = C_nH_{2n-2} (c) Alkanes and cycloalkanes are saturated hydrocarbons. **15.10** a, c, and f

15.13(a)

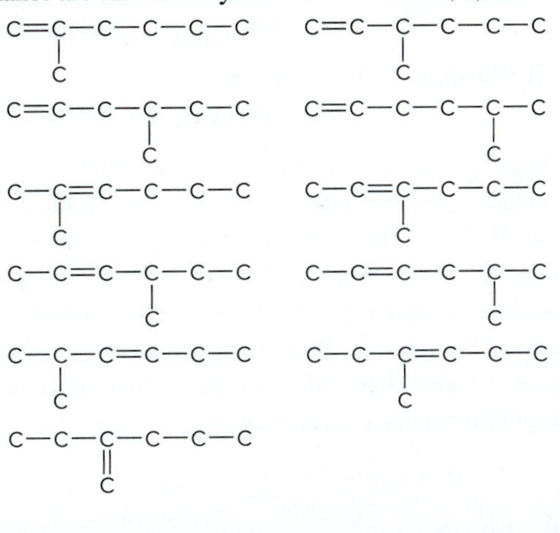

(b)

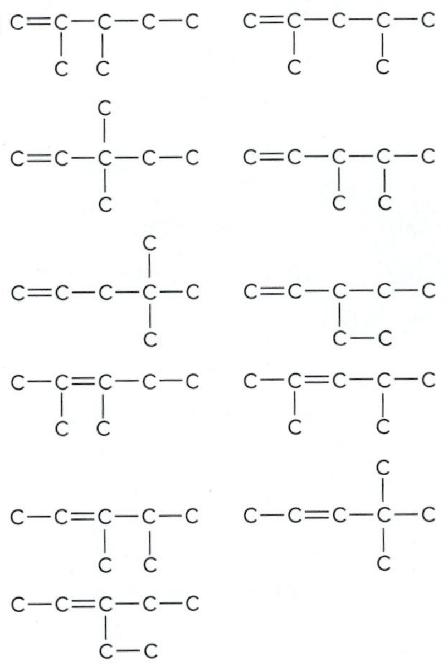

(c)

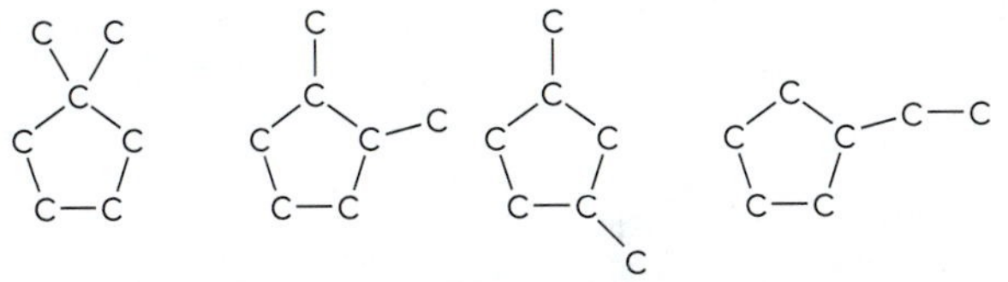

15.15(a)

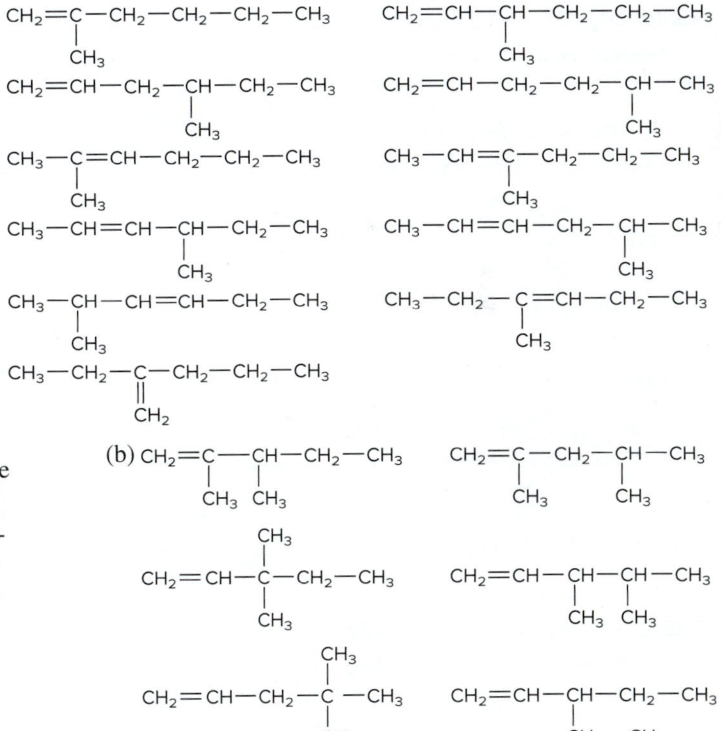

(b)

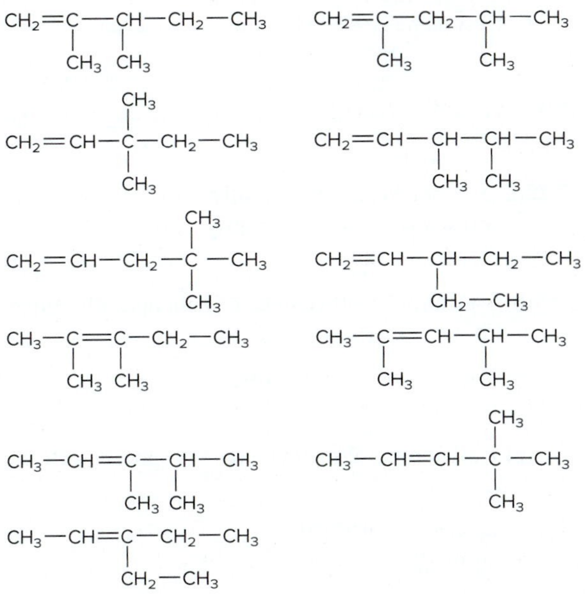

(c)

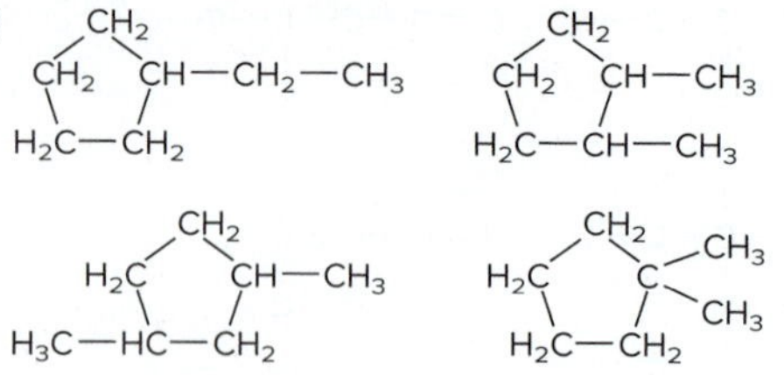

15.17(a)

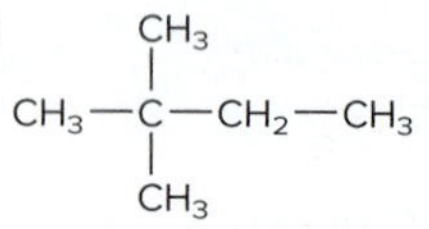

(b) $H_2C{=}CH{-}CH_2{-}CH_3$

(c)

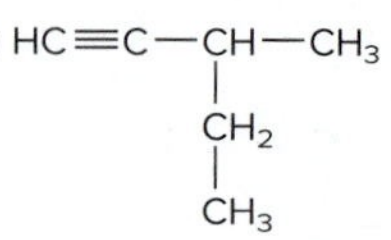

(d) Structure is correct.

15.19(a)

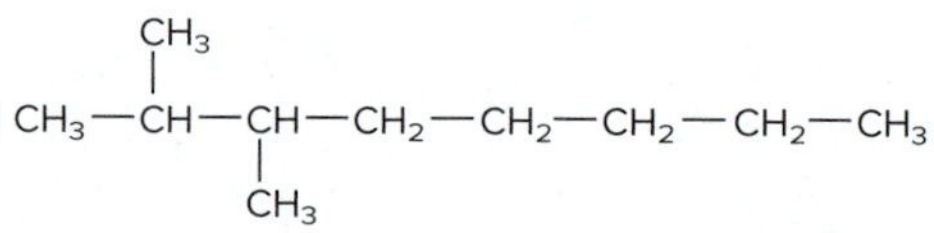

(b)

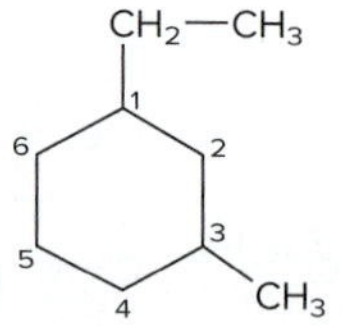

(c) 3,4-dimethylheptane (d) 2,2-dimethylbutane

15.21(a)

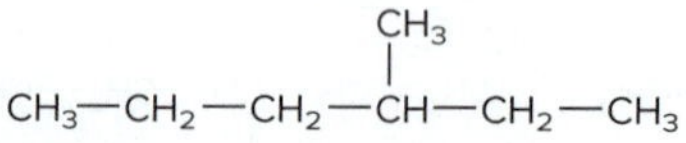

Correct name is 3-methylhexane.

(b)

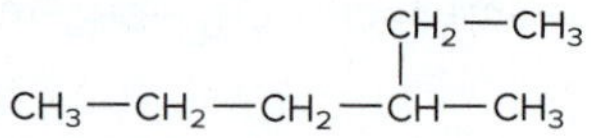

Correct name is 3-methylhexane.

(c)

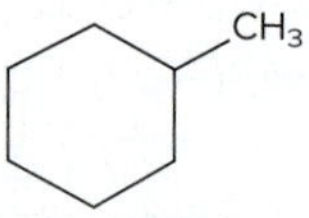

Correct name is methylcyclohexane.

(d)

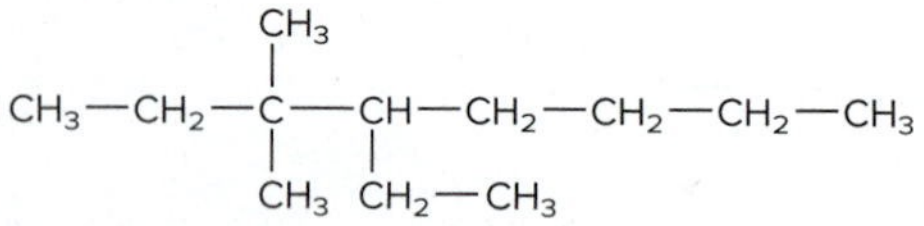

Correct name is 4-ethyl-3,3-dimethyloctane.

15.23(a)

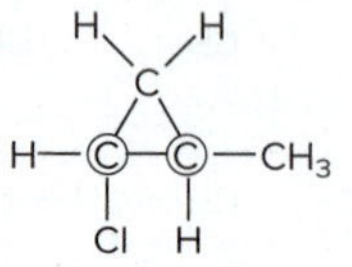

(b)

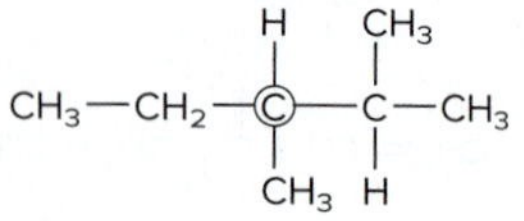

15.25(a) 3-Bromohexane is optically active.

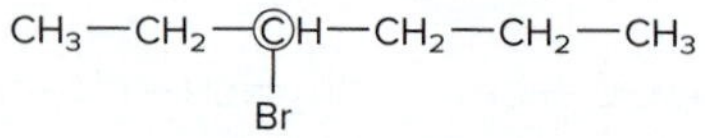

(b) 3-Chloro-3-methylpentane is not optically active.

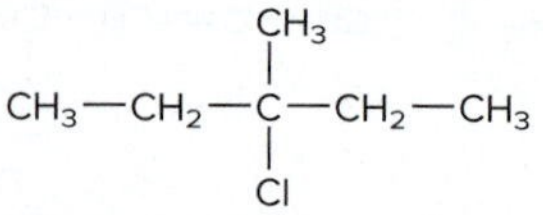

(c) 1,2-Dibromo-2-methylbutane is optically active.

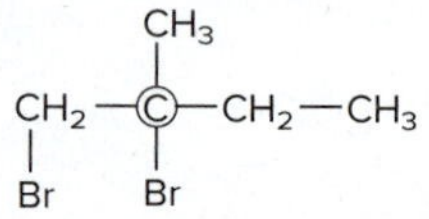

15.27(a)

cis-2-pentene *trans*-2-pentene

(b)

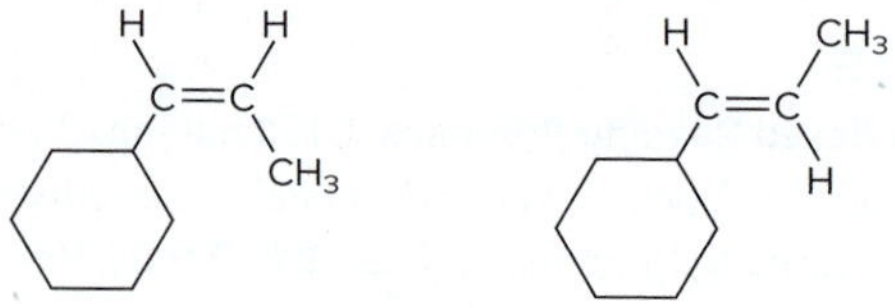

cis-1-cyclohexylpropene *trans*-1-cyclohexylpropene

(c) no geometric isomers

15.29(a) no geometric isomers

(b)

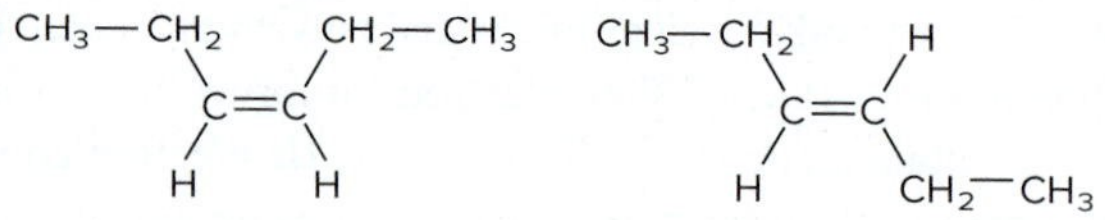

cis-3-hexene *trans*-3-hexene

(c) no geometric isomers

(d)

cis-1,2-dichloroethene *trans*-1,2-dichloroethene

15.31

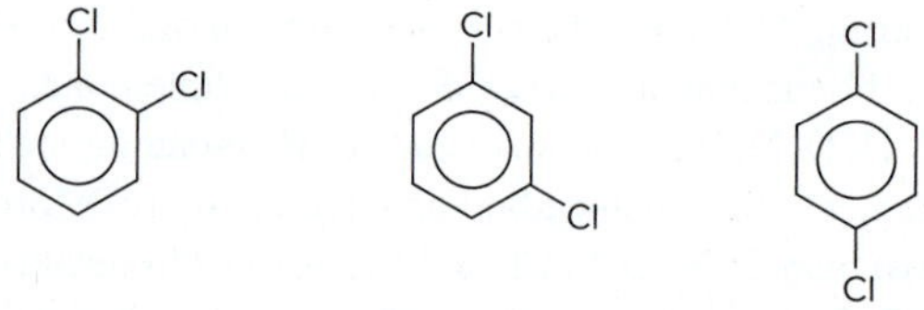

1,2-dichlorobenzene (*o*-dichlorobenzene) 1,3-dichlorobenzene (*m*-dichlorobenzene) 1,4-dichlorobenzene (*p*-dichlorobenzene)

15.33

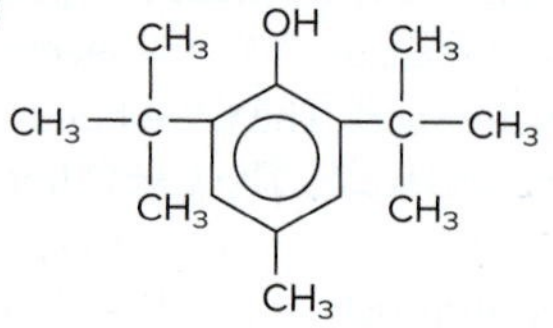

15.34

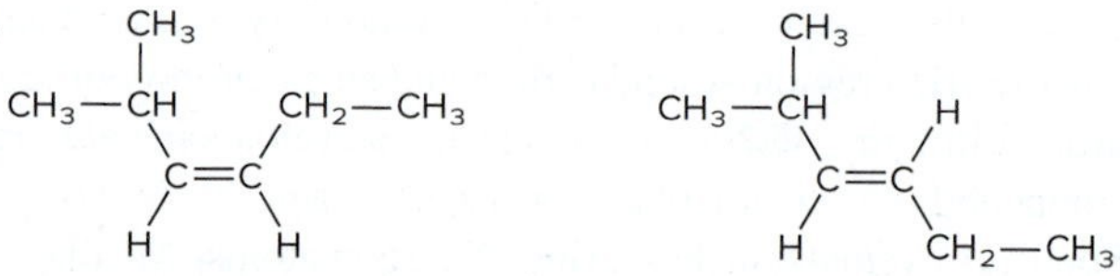

cis-2-methyl-3-hexene *trans*-2-methyl-3-hexene

The compound 2-methyl-2-hexene does not have *cis-trans* isomers. **15.38** π bond **15.40**(a) elimination (b) addition

15.42(a) $CH_3CH_2CH{=}CHCH_2CH_3 + H_2O \xrightarrow{H^+} CH_3CH_2CH_2CH(OH)CH_2CH_3$

(b) $CH_3CHBrCH_3 + CH_3CH_2OK \longrightarrow CH_3CH{=}CH_2 + CH_3CH_2OH + KBr$

(c) $CH_3CH_3 + 2Cl_2 \xrightarrow{hv} CHCl_2CH_3 + 2HCl$

15.44(a) oxidation (b) reduction (c) reduction **15.46**(a) oxidized (b) oxidized **15.49**(a) Methylethylamine is more soluble because it has the ability to form H bonds with water molecules. (b) 1-Butanol has a higher melting point because it can form intermolecular H bonds. (c) Propylamine has a higher boiling point because it contains N—H bonds that allow H bonding, and trimethylamine cannot form H bonds.

15.52 Both groups react by addition to the π bond. The very polar C=O bond attracts the electron-rich O of water to the partially positive C. There is no such polarity in the alkene, so either C atom can be attacked, and two products result.

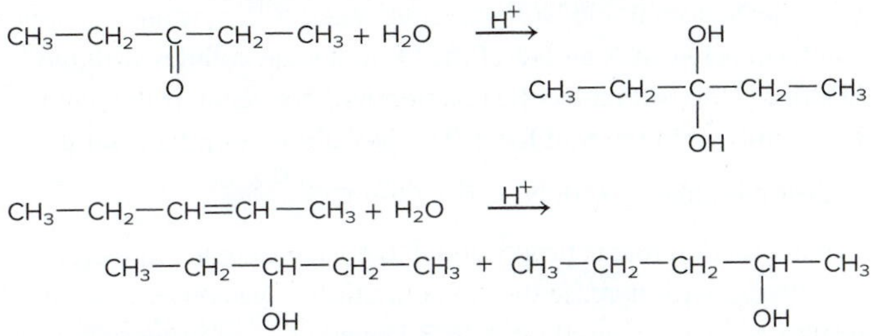

15.55 Esters and acid anhydrides form through dehydration-condensation reactions, and water is the other product.
15.57(a) alkyl halide (b) nitrile (c) carboxylic acid (d) aldehyde

15.59(a)
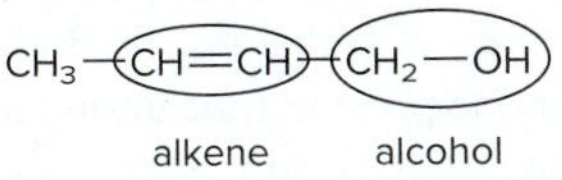

(b)
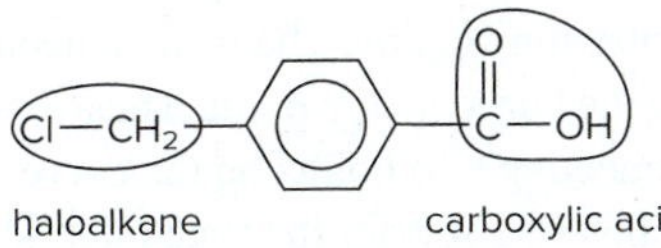

(c)
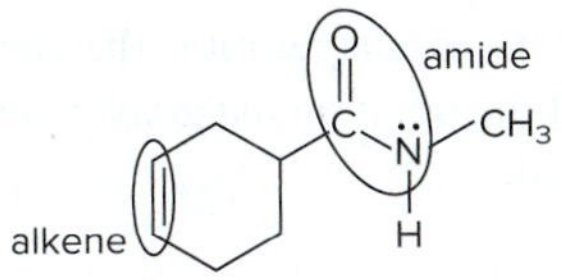

(d)
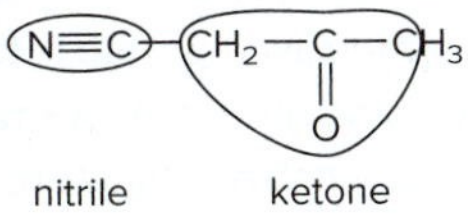

(e)
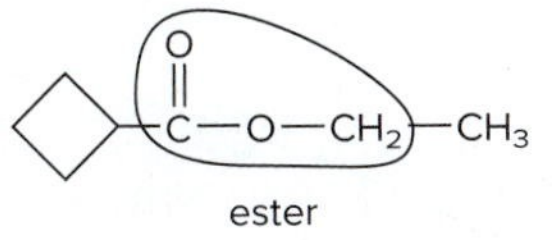

15.61
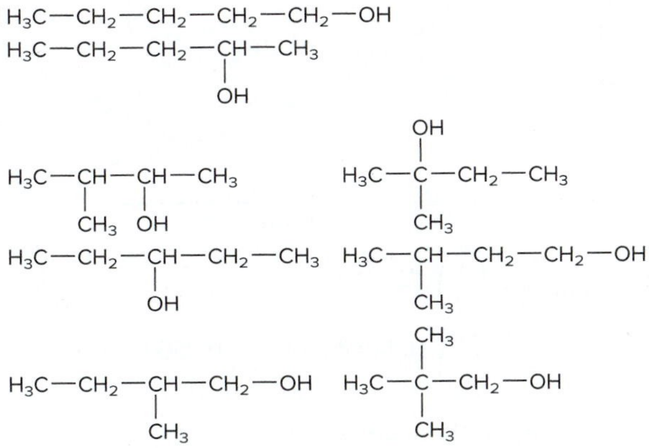

15.63
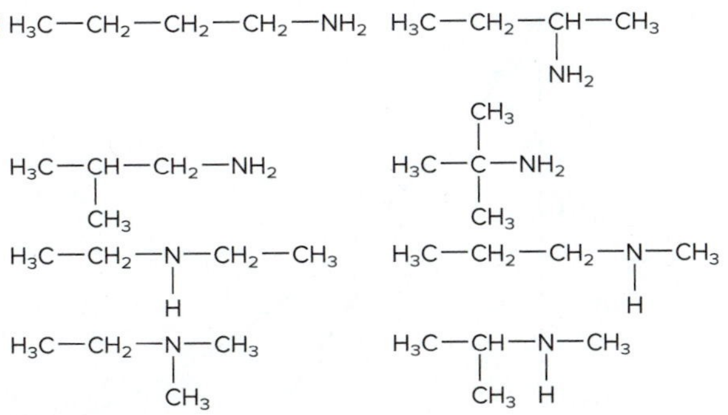

15.65(a)
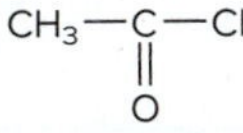
(b)
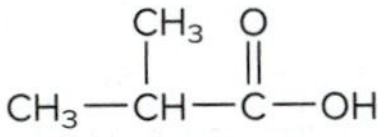
(c)
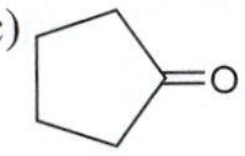

15.67(a)
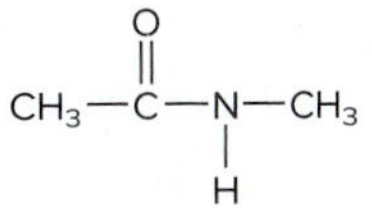
(b)
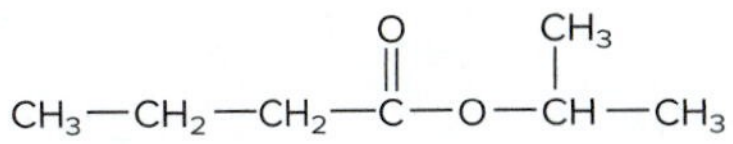
(c)
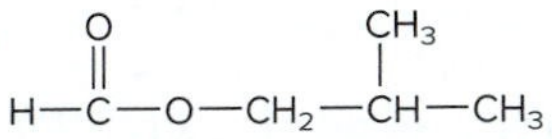

15.69(a)
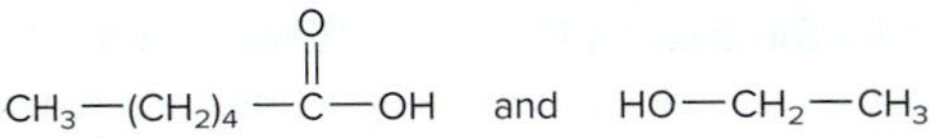
(b)
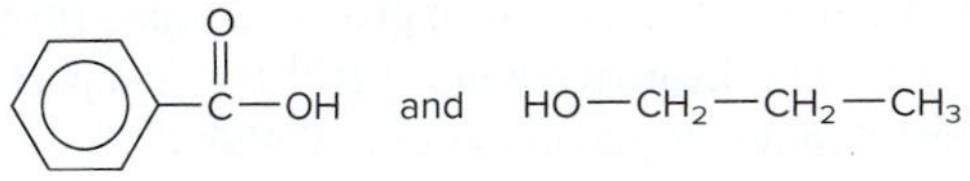
(c)
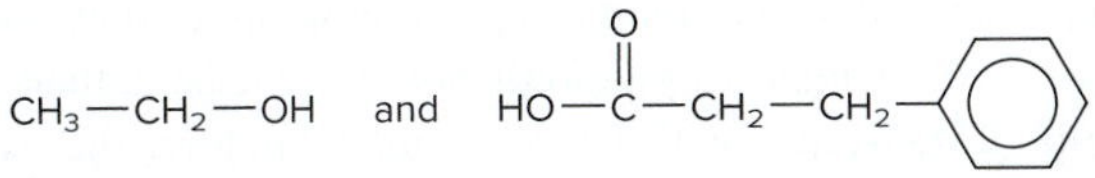

15.71(a) $CH_3—CH_2—OH$
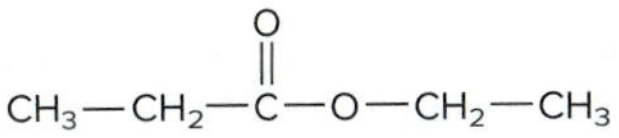
(b)
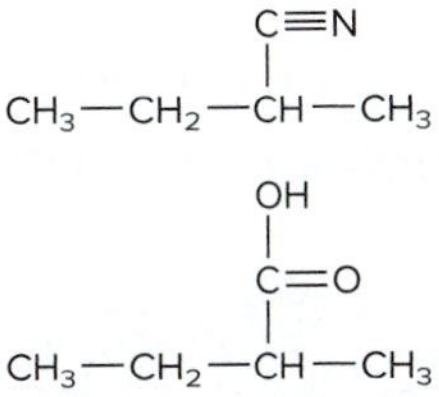

15.73(a) propanoic acid (b) ethylamine **15.77** addition reactions and condensation reactions **15.80** Dispersion forces are responsible for strong attractions between the long, unbranched chains of high-density polyethylene (HDPE). Low-density polyethylene (LDPE) has branching in the chains that prevents packing and weakens these attractions. **15.82** An amine and a carboxylic acid react to form a nylon; a carboxylic acid and an alcohol form a polyester.

15.83(a)
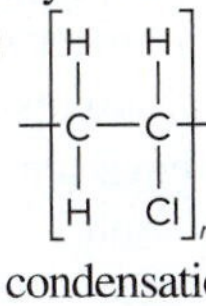
(b)
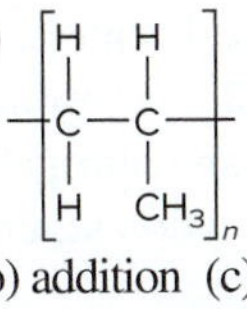

15.87(a) condensation (b) addition (c) condensation (d) condensation **15.89** The amino acid sequence in a protein determines its shape and structure, which determine its function. **15.92** The DNA base sequence determines the RNA base sequence, which determines the protein amino acid sequence.

15.93(a)
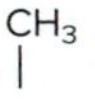
(b)
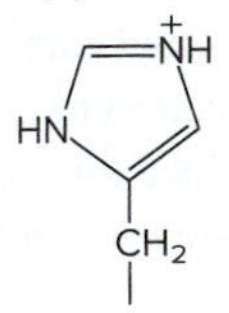
(c)
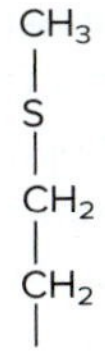

15.95(a)
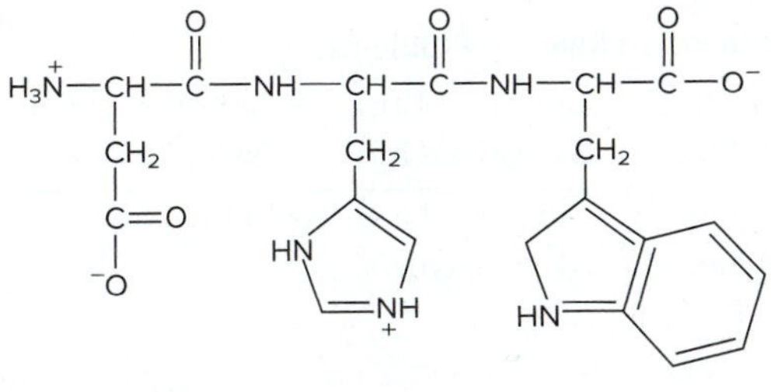

(b)

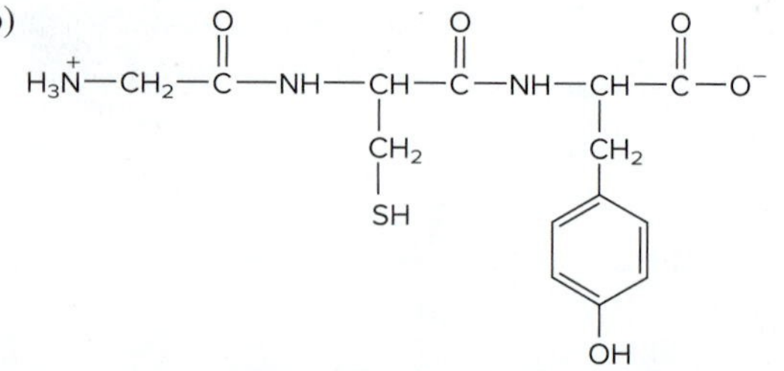

15.97(a) AATCGG (b) TCTGTA **15.99** ACAATGCCT; this sequence codes for three amino acids. **15.101**(a) Both R groups are from cysteine, which can form a disulfide bond (covalent bond). (b) Lysine and aspartic acid give a salt link. (c) Asparagine and serine will hydrogen bond. (d) Valine and phenylalanine interact through dispersion forces. **15.106**(a) Perform an acid-catalyzed dehydration of the alcohol followed by addition of Br_2. (b) Oxidize 1 mol of ethanol to acetic acid, then react 1 mol of acetic acid with 1 mol of ethanol to form the ester.

15.108

$$H_2N{-}CH_2{-}CH_2{-}CH_2{-}CH_2{-}CH_2{-}NH_2$$

cadaverine

$$H_2N{-}CH_2{-}CH_2{-}CH_2{-}CH_2{-}NH_2$$

putrescine

$$Br{-}CH_2CH_2{-}Br + 2CN^- \longrightarrow N{\equiv}C{-}CH_2CH_2{-}C{\equiv}N + 2Br^-$$

$$N{\equiv}C{-}CH_2CH_2{-}C{\equiv}N \xrightarrow{\text{reduction}} H_2N{-}CH_2CH_2CH_2CH_2{-}NH_2$$

15.109 (a)

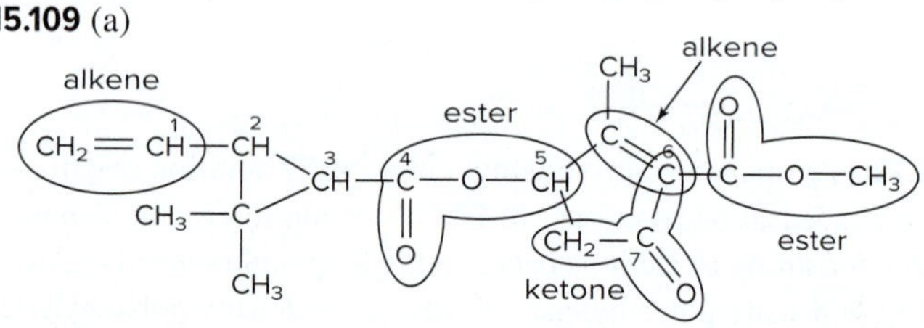

(b) Carbon 1 is sp^2 hybridized. Carbon 2 is sp^3 hybridized. Carbon 3 is sp^3 hybridized. Carbon 4 is sp^2 hybridized. Carbon 5 is sp^3 hybridized. Carbons 6 and 7 are sp^2 hybridized.

(c) Carbons 2, 3, and 5 are chiral centers, as they are each bonded to four different groups. **15.113** The resonance structures show that the bond between carbon and nitrogen will have some double bond character that restricts rotation around the bond.

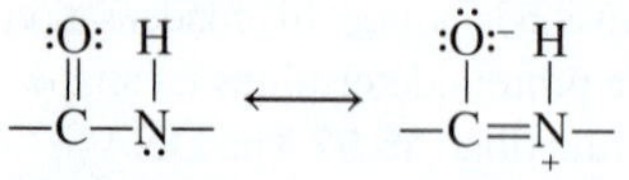

15.116(a) Hydrolysis requires the addition of water to break the peptide bonds. (b) glycine, 4; alanine, 1; valine, 3; proline, 6; serine, 7; arginine, 49 (c) 10,700 g/mol **15.117** When 2-butanone is reduced, equal amounts of both isomers are produced because the reaction does not favor the production of one over the other. The mixture is not optically active since the rotation of light to the right by one isomer cancels the rotation of light to the left by the other isomer.

Chapter 16

Answers to Boxed Reading Problems:

B16.2 (a)

$$\begin{array}{ll} \text{Step 1} & \cancel{NO(g)} + O_3(g) \longrightarrow \cancel{NO_2(g)} + O_2(g) \\ \text{Step 2} & \cancel{NO_2(g)} + O(g) \longrightarrow \cancel{NO(g)} + O_2(g) \\ \hline \text{Overall} & O_3(g) + O(g) \longrightarrow 2O_2(g) \end{array}$$

(b) rate = $3{\times}10^7$ molecule/s

B16.3 (a) Step 1 with Cl will have the higher value for the p factor. Since Cl is a single atom, no matter how it collides with the ozone molecule, the two particles should react, if the collision has enough energy. NO is a molecule. If the O_3 molecule collides with the N atom in the NO molecule, reaction can occur as a bond can form between N and O; if the O_3 molecule collides with the O atom in the NO molecule, reaction will not occur as the bond between N and O cannot form. The probability of a successful collision is smaller with NO. (b) 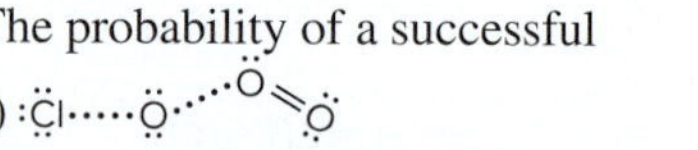

• **16.2** Reaction rate is proportional to concentration. An increase in pressure will increase the concentration of gas-phase reactants, resulting in an increased rate. **16.3** The addition of water will lower the concentrations of all dissolved solutes, and the rate will decrease. **16.5** An increase in temperature increases the rate by increasing the number of collisions between particles, but more importantly, it increases the energy of collisions. **16.8**(a) The instantaneous rate is the rate at one point in time during the reaction. The average rate is the average over a period of time. On a graph of reactant concentration vs. time, the instantaneous rate is the slope of the tangent to the curve at any point. The average rate is the slope of the line connecting two points on the curve. The closer together the two points (the shorter the time interval), the closer the average rate is to the instantaneous rate. (b) The initial rate is the instantaneous rate at the point on the graph where $t = 0$, that is, when reactants are mixed.

16.10

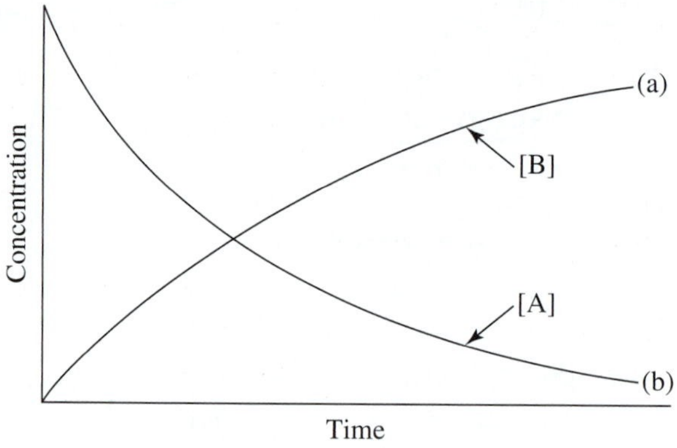

16.12(a)

$$\text{rate} = -\left(\frac{1}{2}\right)\frac{\Delta[AX_2]}{\Delta t}$$

$$= -\left(\frac{1}{2}\right)\frac{(0.0088\ \text{mol/L} - 0.0500\ \text{mol/L})}{(20.0\ \text{s} - 0\ \text{s})}$$

$$= 0.0010\ \text{mol/L·s}$$

(b)

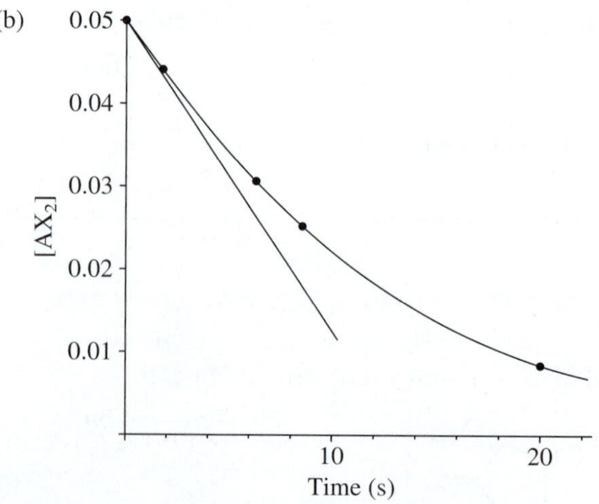

Estimated initial rate = 0.004 mol/L·s. The initial rate is higher than the average rate because the rate decreases as reactant concentration decreases.

16.14 rate $= -\dfrac{1}{2}\dfrac{\Delta[\text{A}]}{\Delta t} = \dfrac{\Delta[\text{B}]}{\Delta t} = \dfrac{\Delta[\text{C}]}{\Delta t}$; 4 mol/L·s

16.16 rate $= -\dfrac{\Delta[\text{A}]}{\Delta t} = -\dfrac{1}{2}\dfrac{\Delta[\text{B}]}{\Delta t} = \dfrac{\Delta[\text{C}]}{\Delta t}$; 0.2 mol/L·s

16.18 $2N_2O_5(g) \longrightarrow 4NO_2(g) + O_2(g)$

16.21 rate $= -\dfrac{\Delta[N_2]}{\Delta t} = -\dfrac{1}{3}\dfrac{\Delta[H_2]}{\Delta t} = \dfrac{1}{2}\dfrac{\Delta[NH_3]}{\Delta t}$

16.22(a) rate $= -\dfrac{1}{3}\dfrac{\Delta[O_2]}{\Delta t} = \dfrac{1}{2}\dfrac{[\Delta O_3]}{\Delta t}$

(b) 1.45×10^{-5} mol/L·s

16.23(a) k is the rate constant, the proportionality constant in the rate law; it is specific to the reaction and that temperature at which the reaction occurs. (b) m represents the order of the reaction with respect to [A], and n represents the order of the reaction with respect to [B]. The order of a reactant does not necessarily equal its stoichiometric coefficient in the balanced equation. (c) L^2/mol^2·min **16.25**(a) Rate doubles. (b) Rate decreases by a factor of 4. (c) Rate increases by a factor of 9. **16.26** first order in BrO_3^-; first order in Br^-; second order in H^+; fourth order overall **16.28**(a) Rate doubles. (b) Rate is halved. (c) The rate increases by a factor of 16. **16.30** second order in NO_2; first order in Cl_2; third order overall **16.32**(a) Rate increases by a factor of 9. (b) Rate increases by a factor of 8. (c) Rate is halved. **16.34**(a) second order in A; first order in B (b) rate = $k[A]^2[B]$ (c) $5.00\times10^3\ L^2/mol^2$·min **16.36**(a) time^{-1} (b) L/mol·time (c) L^2/mol^2·time (d) $L^{3/2}/mol^{3/2}$·time **16.39** The integrated rate law can be used to plot a graph. If the plot of [reactant] versus time is linear, the order is zero. If the plot of ln[reactant] versus time is linear, the order is first. If the plot of inverse concentration (1/[reactant]) versus time is linear, the order is second. (a) first order (b) second order (c) zero order **16.41** 7 s **16.43** 22 minutes **16.45**(a) $k = 0.0660\ \text{min}^{-1}$ (b) 21.0 min

16.47(a)

x-axis (time, s)	$[NH_3]$	*y*-axis ($\ln [NH_3]$)
0	4.000 *M*	1.38629
1.000	3.986 *M*	1.38279
2.000	3.974 *M*	1.37977

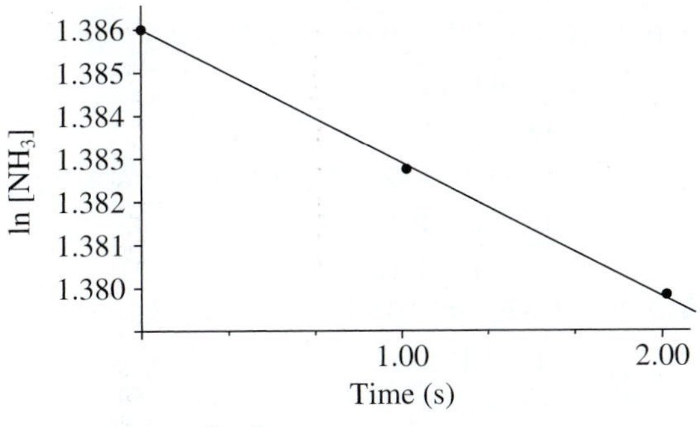

$k = 3\times10^{-3}\ s^{-1}$

(b) $t_{1/2} = 2\times10^2$ s

16.49 No, other factors that affect the rate are the energy and orientation of the collisions. **16.52** Measure the rate constant at a series of temperatures and plot ln k versus $1/T$. The slope of the line equals $-E_a/R$. **16.55** No, reaction is reversible and will eventually reach a state where the forward and reverse rates are equal. When this occurs, there are no concentrations equal to zero. Since some reactants are reformed from EF, the amount of EF will be less than 4×10^{-5} mol. **16.56** At the same temperature, both reaction mixtures have the same average kinetic energy, but the reactant molecules do not have the same average velocity. The trimethylamine molecule has greater mass than the ammonia molecule, so trimethylamine molecules will collide less often with HCl molecules. Moreover, the bulky groups bonded to nitrogen in trimethylamine mean that collisions with HCl having the correct orientation occur less frequently. Therefore, the rate of the reaction between NH_3 and HCl is higher. **16.57** 12 unique collisions **16.59** 2.96×10^{-18} **16.61** 0.033 s^{-1}

16.63(a)

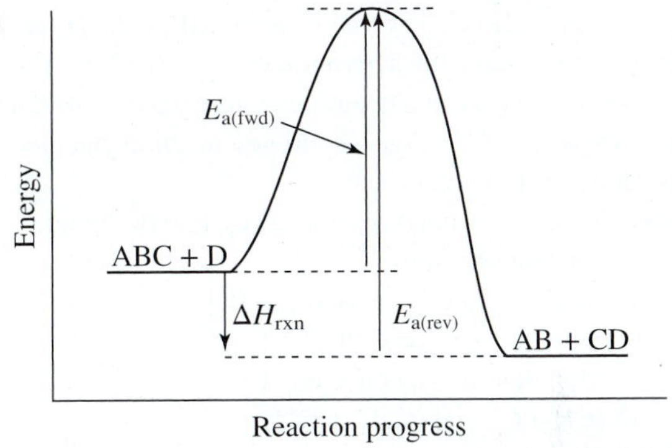

(b) 2.70×10^2 kJ/mol

(c)

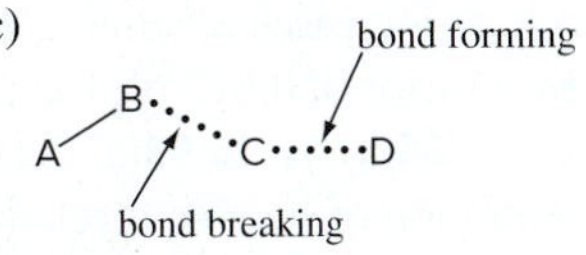

16.66(a) Because the enthalpy change is positive, the reaction is endothermic.

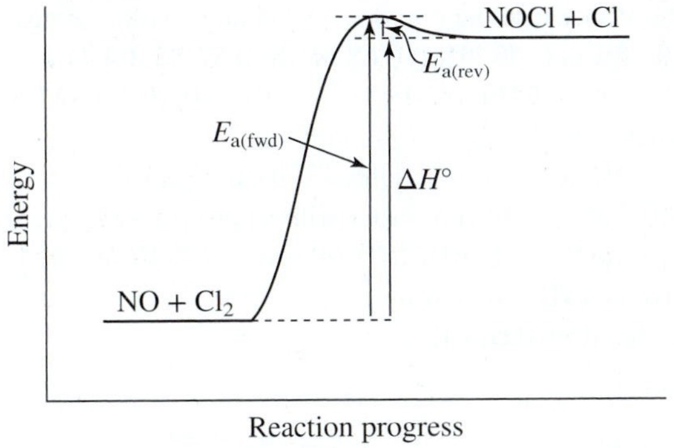

(b) 3 kJ (c)

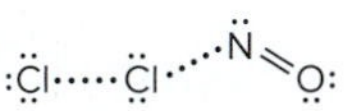

16.67 The rate of an overall reaction depends on the rate of the slowest step. The rate of the overall reaction will be lower than the average of the individual rates because the average includes higher rates as well. **16.71** The probability of three particles colliding with one another with the proper energy and orientation is much less than the probability of two particles doing so. **16.72** No, the overall rate law must contain only reactants (no intermediates), and the overall rate is determined by the slow step. **16.74**(a) $A(g) + B(g) + C(g) \longrightarrow D(g)$

(b) X and Y are intermediates.

(c)

Step	Molecularity	Rate Law
$A(g) + B(g) \rightleftharpoons X(g)$	bimolecular	$rate_1 = k_1[A][B]$
$X(g) + C(g) \longrightarrow Y(g)$	bimolecular	$rate_2 = k_2[X][C]$
$Y(g) \longrightarrow D(g)$	unimolecular	$rate_3 = k_3[Y]$

(d) yes (e) yes **16.76** The proposed mechanism is valid because the individual steps are chemically reasonable, they add to give the overall equation, and the rate law for the mechanism matches the observed rate law. **16.79** No. A catalyst changes the mechanism of a reaction to one with lower activation energy. Lower activation energy means a faster reaction. An increase in temperature does not influence the activation energy, but increases the fraction of collisions with sufficient energy to equal or exceed the activation energy. **16.80**(a) No. The spark provides energy that is absorbed by the H_2 and O_2 molecules to achieve the activation energy. (b) Yes. The powdered metal acts as a heterogeneous catalyst, providing a surface on which the oxygen-hydrogen reaction can proceed with a lower activation energy.

16.85(a) Water does not appear as a reactant in the rate-determining step.

(b) Step 1: $rate_1 = k_1[(CH_3)_3CBr]$

Step 2: $rate_2 = k_2[(CH_3)_3C^+]$

Step 3: $rate_3 = k_3[(CH_3)_3COH_2^+]$

(c) $(CH_3)_3C^+$ and $(CH_3)_3COH_2^+$

(d) The rate-determining step is step 1. The rate law for this step agrees with the rate law observed with $k = k_1$.

16.87 $4.61{\times}10^4$ J/mol **16.91**(a) Rate increases 2.5 times. (b) Rate is halved. (c) Rate decreases by a factor of 0.01. (d) Rate does not change. **16.92** second order **16.95** 57 yr **16.98**(a) 0.21 h^{-1}; 3.3 h (b) 6.6 h (c) If the concentration of sucrose is relatively low, the concentration of water remains nearly constant even with small changes in the amount of water. This gives an apparent zero-order reaction with respect to water. Thus, the reaction is first order overall because the rate does not change with changes in the amount of water. **16.101**(a) 0.68 M (b) 0.57 **16.104** 71 kPa **16.108** $7.3{\times}10^3$ J/mol **16.110**(a) $2.4{\times}10^{-15}$ M (b) $2.4{\times}10^{-11}$ mol/L·s **16.113**(a) 2.8 days (b) 7.4 days (c) 4.5 mol/m^3

16.116(a) $rate_1 = 1.7{\times}10^{-5}\ M\ s^{-1}$; $rate_2 = 3.4{\times}10^{-5}\ M\ s^{-1}$; $rate_3 = 3.4{\times}10^{-5}\ M\ s^{-1}$ (b) zero order with respect to $S_2O_8^{2-}$; first order with respect to I^- (c) $4.3{\times}10^{-4}\ s^{-1}$ (d) rate $= (4.3{\times}10^{-4}\ s^{-1})[KI]$

16.118(a) $7{\times}10^4$ cells/L (b) 20. min

16.121(a) Use the Monod equation.

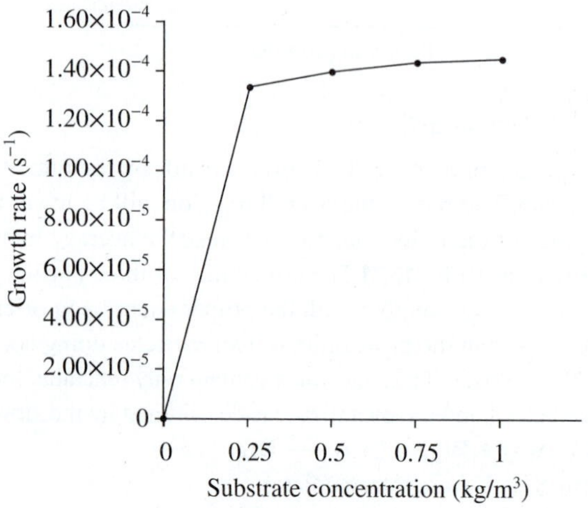

(b) $8.2{\times}10^3$ cells/m^3 (c) $8.4{\times}10^3$ cells/m^3

16.126(a) $8.0{\times}10^1$ s (b) 59 s (c) $2.6{\times}10^2$ s **16.129**(a) $3.5{\times}10^{-5}$ s

16.131(a) $\text{Rate} = k_2[NO_2Cl][Cl] = \dfrac{k_1k_2[NO_2Cl]^2}{k_{-1}[NO_2] + k_2[NO_2Cl]}$

(b) $\text{Rate} = k[NO_2Cl]$ (c) $\text{Rate} = k\dfrac{[NO_2Cl]^2}{[NO_2]}$

16.132(a) 0.1 mol/L·s (b) 0.01 M

Chapter 17

Answers to Boxed Reading Problems: B17.1(a) Enzyme 3 (b) Enzyme 6 (c) If F inhibited enzyme 1, neither branch of the pathway would occur when enough F was made and product I would be lacking. (d) If F inhibited enzyme 6, the second branch would not take place when enough F was made, and more F would build up.

• **17.1** If the change is in concentration, it results temporarily in more products and less reactants. After equilibrium is re-established, K_c remains unchanged because the ratio of products and reactants remains the same. **17.7** The equilibrium constant expression is $K = [O_2]$. If the temperature remains constant, K remains constant. If the initial amount of Li_2O_2 present is sufficient to reach equilibrium, the amount of O_2 obtained will be constant.

17.8(a) $Q = \dfrac{[HI]^2}{[H_2][I_2]}$

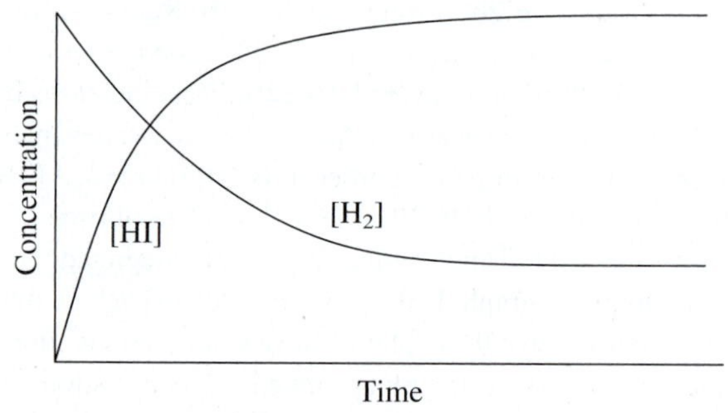

The value of Q increases as a function of time until it reaches the value of K. (b) no **17.11** Yes. If Q_1 is for the formation of 1 mol NH_3 from H_2 and N_2, and Q_2 is for the formation of NH_3 from H_2 and 1 mol of N_2, then $Q_2 = Q_1^2$.

17.12(a) $4NO(g) + O_2(g) \rightleftharpoons 2N_2O_3(g)$; $Q_c = \dfrac{[N_2O_3]^2}{[NO]^4[O_2]}$

(b) $SF_6(g) + 2SO_3(g) \rightleftharpoons 3SO_2F_2(g)$; $Q_c = \dfrac{[SO_2F_2]^3}{[SF_6][SO_3]^2}$

(c) $2SClF_5(g) + H_2(g) \rightleftharpoons S_2F_{10}(g) + 2HCl(g)$; $Q_c = \dfrac{[S_2F_{10}][HCl]^2}{[SClF_5]^2[H_2]}$

17.14(a) $2NO_2Cl(g) \rightleftharpoons 2NO_2(g) + Cl_2(g)$; $Q_c = \dfrac{[NO_2]^2[Cl_2]}{[NO_2Cl]^2}$

(b) $2POCl_3(g) \rightleftharpoons 2PCl_3(g) + O_2(g)$; $Q_c = \dfrac{[PCl_3]^2[O_2]}{[POCl_3]^2}$

(c) $4NH_3(g) + 3O_2(g) \rightleftharpoons 2N_2(g) + 6H_2O(g)$; $Q_c = \dfrac{[N_2]^2[H_2O]^6}{[NH_3]^4[O_2]^3}$

17.16(a) 7.9 (b) $3.2{\times}10^{-5}$

17.18(a) $2Na_2O_2(s) + 2CO_2(g) \rightleftharpoons 2Na_2CO_3(s) + O_2(g)$;

$$Q_c = \frac{[O_2]}{[CO_2]^2}$$

(b) $H_2O(l) \rightleftharpoons H_2O(g)$; $Q_c = [H_2O(g)]$

(c) $NH_4Cl(s) \rightleftharpoons NH_3(g) + HCl(g)$; $Q_c = [NH_3][HCl]$

17.20(a) $2NaHCO_3(s) \rightleftharpoons Na_2CO_3(s) + CO_2(g) + H_2O(g)$;

$Q_c = [CO_2][H_2O]$

(b) $SnO_2(s) + 2H_2(g) \rightleftharpoons Sn(s) + 2H_2O(g)$;

$$Q_c = \frac{[H_2O]^2}{[H_2]^2}$$

(c) $H_2SO_4(l) + SO_3(g) \rightleftharpoons H_2S_2O_7(l)$;

$$Q_c = \frac{1}{[SO_3]}$$

17.23(a)

(1)	$Cl_2(g) + F_2(g) \rightleftharpoons 2ClF(g)$
(2)	$2ClF(g) + 2F_2(g) \rightleftharpoons 2ClF_3(g)$
overall:	$Cl_2(g) + 3F_2(g) \rightleftharpoons 2ClF_3(g)$

(b) $$Q_{overall} = Q_1Q_2 = \frac{\cancel{[ClF]^2}}{[Cl_2][F_2]} \times \frac{[ClF_3]^2}{\cancel{[ClF]^2}[F_2]^2} = \frac{[ClF_3]^2}{[Cl_2][F_2]^3}$$

17.26 K_c and K_p are equal when $\Delta n_{gas} = 0$; K_c and K_p are not equal when Δn_{gas} does not equal 0. **17.27**(a) smaller (b) Assuming that $RT > 1$ ($T > 12.2$ K), $K_p > K_c$ if there are more moles of products than reactants at equilibrium, and $K_p < K_c$ if there are more moles of reactants than products. **17.28**(a) 3 (b) −1 (c) 3 **17.30**(a) 3.2 (b) 28.5 **17.32**(a) 0.15 (b) 3.6×10^{-7} **17.34** The reaction quotient (Q) and equilibrium constant (K) are determined by the ratio [products]/[reactants]. When $Q < K$, the reaction proceeds to the right to form more products. **17.36** no; to the left **17.39** At equilibrium, equal concentrations of $CFCl_3$ and HCl exist, regardless of starting reactant concentrations, because the product coefficients are equal. **17.41**(a) The approximation applies when the change in concentration from initial concentration to equilibrium concentration is so small that it is insignificant; this occurs when K is small and initial concentration is large. (b) This approximation should not be used when the change in concentration is greater than 5%. This can occur when $[\text{reactant}]_{initial}$ is very small or when change in [reactant] is relatively large due to a large K. **17.42** 50.8

17.44

Concentration (M)	$PCl_5(g)$ $\rightleftharpoons$	$PCl_3(g)$ +	$Cl_2(g)$
Initial	0.075	0	0
Change	$-x$	$+x$	$+x$
Equilibrium	$0.075 - x$	x	x

17.46 28 atm **17.48** 0.33 atm **17.50** 3.5×10^{-3} M
17.52 [HI] = 0.0152 M; $[I_2]$ = 0.00328 M
17.54 $[I_2]_{eq} = [Cl_2]_{eq} = 0.0199$ M; $[ICl]_{eq} = 0.0601$ M
17.56 6.01×10^{-6} **17.59** Equilibrium position refers to the specific concentrations or pressures of reactants and products that exist at equilibrium, whereas equilibrium constant is the overall ratio of equilibrium concentrations or pressures. Equilibrium position changes as a result of a change in reactant (or product) concentration. **17.60**(a) B, because the amount of product increases with temperature (b) A, because the lowest temperature will give the least product **17.64**(a) shifts toward products (b) shifts toward products (c) does not shift (d) shifts toward reactants **17.66**(a) more F and less F_2 (b) more C_2H_2 and H_2 and less CH_4 **17.68**(a) no effect (b) less H_2 and O_2 and more H_2O **17.70**(a) no change (b) increase volume **17.72**(a) amount decreases (b) amounts increase (c) amounts increase (d) amount decreases **17.74** 2.0 **17.77**(a) lower temperature, higher pressure (b) Q decreases; no change in K (c) Reaction rates are lower at lower temperatures, so a catalyst is used to speed up the reaction. **17.79**(a) $P_{N_2} = 31$ atm; $P_{H_2} = 93$ atm; $P_{total} = 174$ atm (b) $P_{N_2} = 18$ atm; $P_{H_2} = 111$ atm; $P_{total} = 179$ atm; not a valid suggestion **17.82** 0.204 atm **17.85**(a) 3×10^{-3} atm (b) high pressure; low temperature (c) 2×10^5 (d) No, because water condenses at a higher temperature. **17.87**(a) 0.016 atm (b) $K_c = 5.6\times10^2$; $P_{SO2} = 0.16$ atm **17.89** 12.5 g $CaCO_3$
17.93 Both concentrations increased by a factor of 2.2.
17.95(a) 3.0×10^{-14} atm (b) 0.013 pg CO/L
17.97(a) 98.0% (b) 99.0% (c) 2.60×10^5 J/mol
17.99(a) $2CH_4(g) + O_2(g) + 2H_2O(g) \rightleftharpoons 2CO_2(g) + 6H_2(g)$ (b) 1.76×10^{29} (c) 3.19×10^{23} (d) 48 atm **17.100**(a) 4.0×10^{-21} atm (b) 5.5×10^{-8} atm (c) 29 N atoms/L; 4.0×10^{14} H atoms/L (d) The more reasonable step is $N_2(g) + H(g) \longrightarrow NH(g) + N(g)$. With only 29 N atoms in 1.0 L, the first reaction would produce virtually no $NH(g)$ molecules. There are orders of magnitude more N_2 molecules than N atoms, so the second reaction is the more reasonable step. **17.103**(a) $P_{N_2} = 0.780$ atm; $P_{O_2} = 0.210$ atm; $P_{NO} = 2.67\times10^{-16}$ atm (b) 0.990 atm (c) $K_c = K_p = 4.35\times10^{-31}$ **17.105**(a) 1.26×10^{-3} (b) 794 (c) −51.8 kJ (d) 1.2×10^4 J/mol **17.109**(a) 1.52 (b) 0.9626 atm (c) 0.2000 mol CO (d) 0.01097 M

Chapter 18

18.2 All Arrhenius acids have H in their formula and produce hydronium ion (H_3O^+) in aqueous solution. All Arrhenius bases have OH in their formula and produce hydroxide ion (OH^-) in aqueous solution. Neutralization occurs when each H_3O^+ ion combines with an OH^- ion to form two molecules of H_2O. Chemists found the reaction of any strong base with any strong acid always had $\Delta H = -56$ kJ/mol H_2O produced. **18.4** Strong acids and bases dissociate completely into ions when dissolved in water. Weak acids and bases dissociate only partially. The characteristic property of all weak acids is that a great majority of the acid molecules are undissociated in aqueous solution.
18.5 a, c, and d **18.7** b and d

18.9(a) $$K_a = \frac{[CN^-][H_3O^+]}{[HCN]}$$

(b) $$K_a = \frac{[CO_3^{2-}][H_3O^+]}{[HCO_3^-]}$$

(c) $$K_a = \frac{[HCOO^-][H_3O^+]}{[HCOOH]}$$

18.11(a) $$K_a = \frac{[NO_2^-][H_3O^+]}{[HNO_2]}$$

(b) $$K_a = \frac{[CH_3COO^-][H_3O^+]}{[CH_3COOH]}$$

(c) $$K_a = \frac{[BrO_2^-][H_3O^+]}{[HBrO_2]}$$

18.13 $CH_3COOH < HF < HIO_3 < HI$
18.15(a) weak acid (b) strong base (c) weak acid (d) strong acid

18.17(a) strong base (b) strong acid (c) weak acid (d) weak acid **18.22**(a) The acid with the smaller K_a (4×10^{-5}) has the higher pH, because less dissociation yields fewer hydronium ions. (b) The acid with the larger pK_a (3.5) has the higher pH, because it has a smaller K_a and, thus, lower $[H_3O^+]$. (c) The solution of lower concentration (0.01 *M*) contains fewer hydronium ions. (d) A 0.1 *M* weak acid solution contains fewer hydronium ions. (e) The 0.01 *M* base solution contains more hydroxide ions, so fewer hydronium ions. (f) The solution that has pOH = 6.0 has the higher pH: pH = 14.0 − 6.0 = 8.0. **18.23**(a) 12.05; basic (b) 11.13; acidic **18.25**(a) 2.212; acidic (b) −0.708; basic **18.27**(a) $[H_3O^+] = 1.4\times10^{-10}\ M$, $[OH^-] = 7.1\times10^{-5}\ M$, pOH = 4.15 (b) $[H_3O^+] = 2.7\times10^{-5}\ M$, $[OH^-] = 3.7\times10^{-10}\ M$, pH = 4.57 **18.29**(a) $[H_3O^+] = 1.7\times10^{-5}\ M$, $[OH^-] = 5.9\times10^{-10}\ M$, pOH = 9.23 (b) $[H_3O^+] = 4.5\times10^{-9}\ M$, $[OH^-] = 2.2\times10^{-6}\ M$, pH = 8.36 **18.31** 4.8×10^{-4} mol OH^-/L **18.33** 1.4×10^{-4} mol OH^- **18.36**(a) $2H_2O(l) + \text{heat} \rightleftharpoons H_3O^+(aq) + OH^-(aq)$. As the temperature increases, the reaction shifts to the formation of products; rising temperature increases the value of K_w. (b) $K_w = 2.5\times10^{-14}$; pOH = 6.80; $[OH^-] = 1.6\times10^{-7}\ M$ **18.37** The Brønsted-Lowry model defines acids as proton donors and bases as proton acceptors, while the Arrhenius model defines acids as species containing ionizable hydrogen atoms and bases as species containing hydroxide ions. In both definitions, an acid produces H_3O^+ ions and a base produces OH^- ions when added to water. Ammonia and carbonate ion are two Brønsted-Lowry bases that are not Arrhenius bases because they do not contain OH^- ions. Brønsted-Lowry acids must contain an ionizable hydrogen atom in order to be proton donors, so a Brønsted-Lowry acid is also an Arrhenius acid. **18.40** An amphiprotic species is one that can lose a proton to act as an acid or gain a proton to act as a base. The dihydrogen phosphate ion, $H_2PO_4^-$, is an example.

$$H_2PO_4^-(aq) + OH^-(aq) \longrightarrow H_2O(l) + HPO_4^{2-}(aq)$$

$$H_2PO_4^-(aq) + HCl(aq) \longrightarrow H_3PO_4(aq) + Cl^-(aq)$$

18.41(a) $H_3PO_4(aq) + H_2O(l) \rightleftharpoons H_2PO_4^-(aq) + H_3O^+(aq)$

$$K_a = \frac{[H_3O^+][H_2PO_4^-]}{[H_3PO_4]}$$

(b) $C_6H_5COOH(aq) + H_2O(l) \rightleftharpoons C_6H_5COO^-(aq) + H_3O^+(aq)$

$$K_a = \frac{[H_3O^+][C_6H_5COO^-]}{[C_6H_5COOH]}$$

(c) $HSO_4^-(aq) + H_2O(l) \rightleftharpoons SO_4^{2-}(aq) + H_3O^+(aq)$

$$K_a = \frac{[H_3O^+][SO_4^{2-}]}{[HSO_4^-]}$$

18.43(a) Cl^- (b) HCO_3^- (c) OH^- **18.45**(a) NH_4^+ (b) NH_3 (c) $C_{10}H_{14}N_2H^+$

18.47(a) $HCl + H_2O \rightleftharpoons Cl^- + H_3O^+$
acid base base acid
Conjugate acid-base pairs: HCl/Cl^- and H_3O^+/H_2O
(b) $HClO_4 + H_2SO_4 \rightleftharpoons ClO_4^- + H_3SO_4^+$
acid base base acid
Conjugate acid-base pairs: $HClO_4/ClO_4^-$ and $H_3SO_4^+/H_2SO_4$
(c) $HPO_4^{2-} + H_2SO_4 \rightleftharpoons H_2PO_4^- + HSO_4^-$
base acid acid base
Conjugate acid-base pairs: H_2SO_4/HSO_4^- and $H_2PO_4^-/HPO_4^{2-}$

18.49(a) $NH_3 + H_3PO_4 \rightleftharpoons NH_4^+ + H_2PO_4^-$
base acid acid base
Conjugate acid-base pairs: $H_3PO_4/H_2PO_4^-$ and NH_4^+/NH_3
(b) $CH_3O^- + NH_3 \rightleftharpoons CH_3OH + NH_2^-$
base acid acid base
Conjugate acid-base pairs: NH_3/NH_2^- and CH_3OH/CH_3O^-
(c) $HPO_4^{2-} + HSO_4^- \rightleftharpoons H_2PO_4^- + SO_4^{2-}$
base acid acid base
Conjugate acid-base pairs: HSO_4^-/SO_4^{2-} and $H_2PO_4^-/HPO_4^{2-}$

18.51(a) $OH^-(aq) + H_2PO_4^-(aq) \rightleftharpoons H_2O(l) + HPO_4^{2-}(aq)$
Conjugate acid-base pairs: $H_2PO_4^-/HPO_4^{2-}$ and H_2O/OH^-
(b) $HSO_4^-(aq) + CO_3^{2-}(aq) \rightleftharpoons SO_4^{2-}(aq) + HCO_3^-(aq)$
Conjugate acid-base pairs: HSO_4^-/SO_4^{2-} and HCO_3^-/CO_3^{2-}

18.53 $K_c > 1$: $HS^- + HCl \rightleftharpoons H_2S + Cl^-$
$K_c < 1$: $H_2S + Cl^- \rightleftharpoons HS^- + HCl$

18.55 $K_c > 1$ for both a and b **18.57** $K_c < 1$ for both a and b **18.59**(a) A strong acid is 100% dissociated, so the acid concentration will be very different after dissociation. (b) A weak acid dissociates to a very small extent, so the acid concentration before and after dissociation is nearly the same. (c) The acid concentration is nearly the same, but the extent of dissociation is greater than for a concentrated solution. (d) The acid concentration will be very different after dissociation. **18.60** No. HCl is a strong acid and dissociates to a greater extent than the weak acid CH_3COOH. The K_a of the acid, not the concentration of H_3O^+, determines the strength of the acid. **18.61** scene 1 **18.64** 1.5×10^{-5} **18.66** $[H_3O^+] = [NO_2^-] = 2.1\times10^{-2}\ M$; $[OH^-] = 4.8\times10^{-13}\ M$ **18.68** $[H_3O^+] = [ClCH_2COO^-] = 0.041\ M$; $[ClCH_2COOH] = 1.21\ M$; pH = 1.39 **18.70**(a) $[H_3O^+] = 6.0\times10^{-3}\ M$; pH = 2.22; $[OH^-] = 1.7\times10^{-12}\ M$; pOH = 11.78 (b) 1.9×10^{-4} **18.72** 2.2×10^{-7} **18.74**(a) 2.37 (b) 11.53 **18.76**(a) 2.290 (b) 12.699 **18.78** 1.1% **18.80** $[H_3O^+] = [HS^-] = 9\times10^{-5}\ M$; pH = 4.0; $[OH^-] = 1\times10^{-10}\ M$; pOH = 10.0; $[H_2S] = 0.10\ M$; $[S^{2-}] = 1\times10^{-17}\ M$ **18.83** 1.5% **18.84** As a nonmetal becomes more electronegative, the acidity of its binary hydride increases. The electronegative nonmetal attracts the electrons more strongly in the polar bond, shifting the electron density away from H, thus making H^+ more easily transferred to a water molecule to form H_3O^+. **18.87** Chlorine is more electronegative than iodine, and $HClO_4$ has more oxygen atoms than HIO. **18.88**(a) H_2SeO_4 (b) H_3PO_4 (c) H_2Te **18.90**(a) H_2Se (b) $B(OH)_3$ (c) $HBrO_2$ **18.92**(a) 0.5 *M* $AlBr_3$ (b) 0.3 *M* $SnCl_2$ **18.94**(a) 0.2 *M* $Ni(NO_3)_2$ (b) 0.35 *M* $Al(NO_3)_3$ **18.96** All Brønsted-Lowry bases contain at least one lone pair of electrons, which binds an H^+ and allows the base to act as a proton acceptor.

18.99(a) $C_5H_5N(aq) + H_2O(l) \rightleftharpoons OH^-(aq) + C_5H_5NH^+(aq)$

$$K_b = \frac{[C_5H_5NH^+][OH^-]}{[C_5H_5N]}$$

(b) $CO_3^{2-}(aq) + H_2O(l) \rightleftharpoons OH^-(aq) + HCO_3^-(aq)$

$$K_b = \frac{[HCO_3^-][OH^-]}{[CO_3^{2-}]}$$

18.101(a) $HONH_2(aq) + H_2O(l) \rightleftharpoons OH^-(aq) + HONH_3^+(aq)$

$$K_b = \frac{[HONH_3^+][OH^-]}{[HONH_2]}$$

(b) $HPO_4^{2-}(aq)+H_2O(l) \rightleftharpoons H_2PO_4^-(aq)+OH^-(aq)$

$$K_b = \frac{[H_2PO_4^-][OH^-]}{[HPO_4^{2-}]}$$

18.103 11.79 **18.105** 11.45 **18.107**(a) 5.6×10^{-10} (b) 2.5×10^{-5} **18.109**(a) 12.04 (b) 10.77 **18.111**(a) 11.19 (b) 5.56 **18.113**(a) 8.78 (b) 4.66 **18.115** $[OH^-] = 5.5\times10^{-4}$ *M*; pH = 10.74 **18.118** NaF contains the anion of the weak acid HF, so F^- acts as a base. NaCl contains the anion of the strong acid HCl.

18.120(a) $KBr(s) \xrightarrow{H_2O} K^+(aq) + Br^-(aq)$; neutral

(b) $NH_4I(s) \xrightarrow{H_2O} NH_4^+(aq) + I^-(aq)$

$NH_4^+(aq) + H_2O(l) \rightleftharpoons NH_3(aq)+H_3O^+(aq)$; acidic

(c) $KCN(s) \xrightarrow{H_2O} K^+(aq) + CN^-(aq)$

$CN^-(aq)+H_2O(l) \rightleftharpoons HCN(aq) + OH^-(aq)$; basic

18.122(a) $Na_2CO_3(s) \xrightarrow{H_2O} 2Na^+(aq) + CO_3^{2-}(aq)$

$CO_3^{2-}(aq) + H_2O(l) \rightleftharpoons HCO_3^-(aq) + OH^-(aq)$; basic

(b) $CaCl_2(s) \xrightarrow{H_2O} Ca^{2+}(aq) + 2Cl^-(aq)$; neutral

(c) $Cu(NO_3)_2(s) \xrightarrow{H_2O} Cu^{2+}(aq) + 2NO_3^-(aq)$

$Cu(H_2O)_6^{2+}(aq) + H_2O(l) \rightleftharpoons Cu(H_2O)_5OH^+(aq) + H_3O^+(aq)$; acidic

18.124(a) $SrBr_2(s) \xrightarrow{H_2O} Sr^{2+}(aq) + 2Br^-(aq)$; neutral

(b) $Ba(CH_3COO)_2(s) \xrightarrow{H_2O} Ba^{2+}(aq) + 2CH_3COO^-(aq)$

$CH_3COO^-(aq) + H_2O(l) \rightleftharpoons CH_3COOH(aq) + OH^-(aq)$; basic

(c) $(CH_3)_2NH_2Br(s) \xrightarrow{H_2O} (CH_3)_2NH_2^+(aq) + Br^-(aq)$

$(CH_3)_2NH_2^+(aq) + H_2O(l) \rightleftharpoons (CH_3)_2NH(aq) + H_3O^+(aq)$; acidic

18.126(a) $NH_4^+(aq) + H_2O(l) \rightleftharpoons NH_3(aq) + H_3O^+(aq)$

$PO_4^{3-}(aq) + H_2O(l) \rightleftharpoons HPO_4^{2-}(aq)+OH^-(aq)$

$K_b > K_a$; basic

(b) $SO_4^{2-}(aq)+H_2O(l) \rightleftharpoons HSO_4^-(aq)+OH^-(aq)$

Na^+ gives no reaction; basic

(c) $ClO^-(aq) + H_2O(l) \rightleftharpoons HClO(aq) + OH^-(aq)$

Li^+ gives no reaction; basic

18.128(a) $Fe(NO_3)_2 < KNO_3 < K_2SO_3 < K_2S$

(b) $NaHSO_4 < NH_4NO_3 < NaHCO_3 < Na_2CO_3$

18.130 Since both bases produce OH^- ions in water, both bases appear equally strong. $CH_3O^-(aq) + H_2O(l) \longrightarrow OH^-(aq) + CH_3OH(aq)$ and $NH_2^-(aq) + H_2O(l) \longrightarrow OH^-(aq) + NH_3(aq)$. **18.132** Ammonia, NH_3, is a more basic solvent than H_2O. In a more basic solvent, weak acids such as HF act like strong acids and are 100% dissociated.

18.134 A Lewis acid is an electron-pair acceptor, while a Brønsted-Lowry acid is a proton donor. The proton of a Brønsted-Lowry acid is a Lewis acid because it accepts an electron pair when it bonds with a base. All Lewis acids are not Brønsted-Lowry acids. A Lewis base is an electron-pair donor, and a Brønsted-Lowry base is a proton acceptor. All Brønsted-Lowry bases are Lewis bases, and vice versa. **18.135**(a) No, $Ni(H_2O)_6^{2+}(aq) + 6NH_3(aq) \rightleftharpoons Ni(NH_3)_6^{2+} + 6H_2O(l)$; NH_3 is a weak Brønsted-Lowry base, but a strong Lewis base.

(b) cyanide ion and water (c) cyanide ion **18.138**(a) Lewis acid (b) Lewis base (c) Lewis acid (d) Lewis base **18.140**(a) Lewis acid (b) Lewis base (c) Lewis base (d) Lewis acid **18.142**(a) Lewis acid: Na^+; Lewis base: H_2O (b) Lewis acid: CO_2; Lewis base: H_2O (c) Lewis acid: BF_3; Lewis base: F^- **18.144**(a) Lewis (b) Brønsted-Lowry and Lewis (c) none (d) Lewis **18.147** 3.5×10^{-8} to 4.5×10^{-8} *M* H_3O^+; 5.2×10^{-7} to 6.6×10^{-7} *M* OH^- **18.148**(a) Acids vary in the extent of dissociation depending on the acid-base character of the solvent. (b) Methanol is a weaker base than water since phenol dissociates less in methanol than in water.

(c) $C_6H_5OH(solvated) + CH_3OH(l) \rightleftharpoons CH_3OH_2^+(solvated) + C_6H_5O^-(solvated)$

(d) $CH_3OH(l) + CH_3OH(l) \rightleftharpoons CH_3O^-(solvated) + CH_3OH_2^+(solvated)$; $K = [CH_3O^-][CH_3OH_2^+]$

18.151(a) $SnCl_4$ is the Lewis acid; $(CH_3)_3N$ is the Lewis base. (b) 5*d* **18.152** pH = 5.00, 6.00, 6.79, 6.98, 7.00 **18.158** 3×10^{18} **18.161** 10.48 **18.163** 2.41 **18.166** amylase, 2×10^{-7} *M*; pepsin, $1\times10^{2-}$ *M*; trypsin, 3×10^{-10} *M* **18.170** 1.47×10^{-3} **18.172**(a) Ca^{2+} does not react with water; $CH_3CH_2COO^-(aq) + H_2O(l) \rightleftharpoons CH_3CH_2COOH(aq) + OH^-(aq)$; basic (b) 9.08 **18.178** 4.5×10^{-5} **18.181**(a) The concentration of oxygen is higher in the lungs, so the equilibrium shifts to the right. (b) In an oxygen-deficient environment, the equilibrium shifts to the left to release oxygen. (c) A decrease in $[H_3O^+]$ shifts the equilibrium to the right. More oxygen is absorbed, but it will be more difficult to remove the O_2. (d) An increase in $[H_3O^+]$ shifts the equilibrium to the left. Less oxygen is bound to Hb, but it will be easier to remove it. **18.183**(a) 1.012 *M* (b) 0.004 *M* (c) 0.4% **18.184**(a) 10.0 (b) The pK_b for the N in the 3° amine group is much smaller than that for the N in the aromatic ring; thus, the K_b for the former is significantly larger (yielding a much greater amount of OH^-). (c) 4.7 (d) 5.1

Chapter 19

Answers to Boxed Reading Problems: B19.2(a) 65 mol (b) 6.28 (c) 4.0×10^3 g HCO_3^-

• **19.2** The acid component neutralizes added base, and the base component neutralizes added acid, so the pH of the buffer solution remains relatively constant. The components of a buffer do not neutralize one another because they are a conjugate acid-base pair. **19.6** The buffer range, the pH range over which the buffer acts effectively, is consistent with a buffer-component concentration ratio from 10 to 0.1. This gives a pH range equal to $pK_a \pm 1$. **19.8**(a) Ratio and pH increase; added OH^- reacts with HA. (b) Ratio and pH decrease; added H^+ reacts with A^-. (c) Ratio and pH increase; added A^- increases $[A^-]$. (d) Ratio and pH decrease; added HA increases [HA]. **19.10**(a) buffer 3 (b) All of the buffers have the same pH range. (c) buffer 2 **19.12** $[H_3O^+] = 5.6\times10^{-6}$ *M*; pH = 5.25 **19.14** $[H_3O^+] = 5.2\times10^{-4}$ *M*; pH = 3.28 **19.16** 3.89 **19.18** 10.03 **19.20** 9.47 **19.22**(a) K_{a2} (b) 10.55 **19.24** 3.6 **19.26** 0.20 **19.28** 3.37 **19.30** 8.82 **19.32**(a) 4.81 (b) 0.66 g KOH **19.34**(a) $HOOC(CH_2)_4COOH/HOOC(CH_2)_4COO^-$ or $C_6H_5NH_3^+/C_6H_5NH_2$ (b) $H_2PO_4^-/HPO_4^{2-}$

or $H_2AsO_4^-/HAsO_4^{2-}$ **19.36**(a) $HOCH_2CH(OH)COOH/HOCH_2CH(OH)COO^-$ or $CH_3COOC_6H_4COOH/CH_3COOC_6H_4COO^-$ (b) $C_5H_5NH^+/C_5H_5N$ **19.39** 1.6 **19.41**(a) initial pH: *strong acid*–strong base < *weak acid*–strong base < *weak base*–strong acid (b) pH at equivalence point: *weak base*–strong acid < *strong acid*–strong base < *weak acid*–strong base **19.43** At the midpoint of the buffer region, the concentrations of weak acid and conjugate base are equal, so the pH = pK_a of the acid. **19.45** To see a distinct color in a mixture of two colors, you need one color to have about 10 times the intensity of the other. For this to be the case, the concentration ratio $[HIn]/[In^-]$ has to be greater than 10/1 or less than 1/10. This occurs when pH = $pK_a - 1$ or pH = $pK_a + 1$, respectively, giving a pH range of about 2 units. **19.47** The equivalence point in a titration is the point at which the number of moles of base is stoichiometrically equivalent to the number of moles of acid. The end point is the point at which the indicator changes color. If an appropriate indicator is selected, the end point is close to the equivalence point, but they are not usually the same. The pH at the end point, or at the color change, may precede or follow the pH at the equivalence point, depending on the indicator chosen. **19.49**(a) 1.0000 (b) 1.6368 (c) 2.898 (d) 3.903 (e) 7.00 (f) 10.10 (g) 12.05 **19.51**(a) 2.91 (b) 4.81 (c) 5.29 (d) 6.09 (e) 7.41 (f) 8.76 (g) 10.10 (h) 12.05 **19.53**(a) 8.54 and 59.0 mL (b) 4.39 and 66.0 mL, 9.68 and total 132.1 mL **19.55**(a) 5.17 and 123 mL (b) 5.80 and 194 mL **19.57** pH range from 7.5 to 9.5 **19.59**(a) bromthymol blue (b) thymol blue or phenolphthalein **19.61**(a) methyl red (b) bromthymol blue **19.64** Fluoride ion is the conjugate base of a weak acid and reacts with H_2O: $F^-(aq) + H_2O(l) \rightleftharpoons HF(aq) + OH^-(aq)$. As the pH increases, the equilibrium shifts to the left and $[F^-]$ increases. As the pH decreases, the equilibrium shifts to the right and $[F^-]$ decreases. The changes in $[F^-]$ influence the solubility of BaF_2. Chloride ion is the conjugate base of a strong acid, so it does not react with water and its concentration is not influenced by pH. **19.66** The ionic compound precipitates. **19.67**(a) $K_{sp} = [Ag^+]^2[CO_3^{2-}]$ (b) $K_{sp} = [Ba^{2+}][F^-]^2$ (c) $K_{sp} = [Cu^{2+}][S^{2-}]$ **19.69**(a) $K_{sp} = [Ca^+][CrO_4^{2-}]$ (b) $K_{sp} = [Ag^+][CN^-]$ (c) $K_{sp} = [Ag^+]^3[PO_4^{3-}]$ **19.71** 1.3×10^{-4} **19.73** 2.8×10^{-11} **19.75**(a) 2.3×10^{-5} *M* (b) 4.2×10^{-9} *M* **19.77**(a) 1.7×10^{-3} *M* (b) 2.0×10^{-4} *M* **19.79**(a) $Mg(OH)_2$ (b) PbS (c) Ag_2SO_4 **19.81**(a) $CaSO_4$ (b) $Mg_3(PO_4)_2$ (c) $PbSO_4$ **19.83**(a) $AgCl(s) \rightleftharpoons Ag^+(aq) + Cl^-(aq)$. The chloride ion is the anion of a strong acid, so it does not react with H_3O^+. No change with pH. (b) $SrCO_3(s) \rightleftharpoons Sr^{2+}(aq) + CO_3^{2-}(aq)$ The strontium ion is the cation of a strong base, so pH will not affect its solubility. The carbonate ion acts as a base: $CO_3^{2-}(aq) + H_2O(l) \rightleftharpoons HCO_3^-(aq) + OH^-(aq)$ also $CO_2(g)$ forms and escapes: $CO_3^{2-}(aq) + 2H_3O^+(aq) \longrightarrow CO_2(g) + 3H_2O(l)$. Therefore, the solubility of $SrCO_3$ will increase with addition of H_3O^+ (decreasing pH).
19.85(a) $Fe(OH)_2(s) \rightleftharpoons Fe^{2+}(aq) + 2OH^-(aq)$ The OH^- ion reacts with added H_3O^+: $OH^-(aq) + H_3O^+(aq) \longrightarrow 2H_2O(l)$. The added H_3O^+ consumes the OH^-, driving the equilibrium toward the right to dissolve more $Fe(OH)_2$. Solubility increases with addition of H_3O^+ (decreasing pH).
(b) $CuS(s) \rightleftharpoons Cu^{2+}(aq) + S^{2-}(aq)$ S^{2-} is the anion of a weak acid, so it reacts with added H_3O^+. Solubility increases with addition of H_3O^+ (decreasing pH). **19.87** yes **19.89** yes **19.92**(a) $Fe(OH)_3$ (b) The two metal ions are separated by adding just enough NaOH to precipitate iron(III) hydroxide. (c) 2.0×10^{-7} *M* **19.95** No, because it indicates that a complex ion forms between the zinc ion and hydroxide ions, such as $Zn^{2+}(aq) + 4OH^-(aq) \rightleftharpoons Zn(OH)_4^{2-}(aq)$
19.96 $Hg(H_2O)_4^{2+}(aq) + 4CN^-(aq) \rightleftharpoons Hg(CN)_4^{2-}(aq) + 4H_2O(l)$
19.98 $Ag(H_2O)_2^+(aq) + 2S_2O_3^{2-}(aq) \rightleftharpoons Ag(S_2O_3)_2^{3-}(aq) + 2H_2O(l)$
19.100 8.1×10^{-15} *M* **19.102** 0.05 *M* **19.104** 1.0×10^{-16} *M* Zn^{2+}; 0.025 *M* $Zn(CN)_4^{2-}$; 0.049 *M* CN^- **19.106** 0.035 L of 2.00 *M* NaOH and 0.465 L of 0.200 *M* HCOOH **19.108**(a) 0.99 (b) assuming the volumes are additive: 0.468 L of 1.0 *M* HCOOH and 0.232 L of 1.0 *M* NaOH (c) 0.34 *M* **19.111** 1.3×10^{-4} *M* **19.114**(a) 14 (b) 1 g from the second beaker **19.117**(a) 0.088 (b) 0.14 **19.118** 0.260 *M* TRIS; pH = 8.53 **19.123** 8×10^{-5} **19.125**(a)

V (mL)	pH	$\Delta pH/\Delta V$	V_{avg} (mL)
0.00	1.00		
10.00	1.22	0.022	5.00
20.00	1.48	0.026	15.00
30.00	1.85	0.037	25.00
35.00	2.18	0.066	32.50
39.00	2.89	0.18	37.00
39.50	3.20	0.62	39.25
39.75	3.50	1.2	39.63
39.90	3.90	2.7	39.83
39.95	4.20	6	39.93
39.99	4.90	18	39.97
40.00	7.00	200	40.00
40.01	9.40	200	40.01
40.05	9.80	10	40.03
40.10	10.40	10	40.08
40.25	10.50	0.67	40.18
40.50	10.79	1.2	40.38
41.00	11.09	0.60	40.75
45.00	11.76	0.17	43.00
50.00	12.05	0.058	47.50
60.00	12.30	0.025	55.00
70.00	12.43	0.013	65.00
80.00	12.52	0.009	75.00

(b)

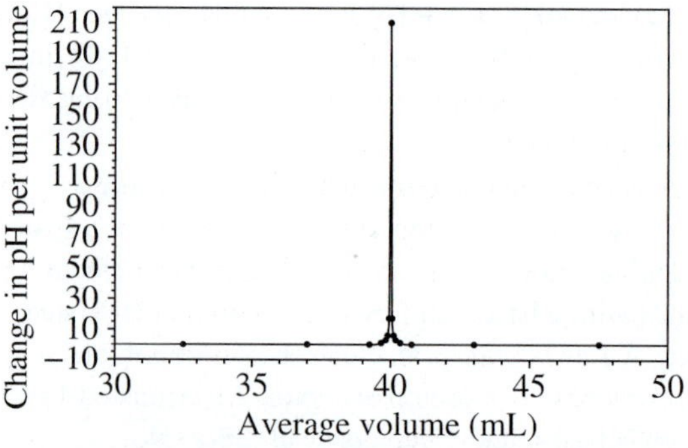

Maximum slope (equivalence point) is at V_{avg} = 40.00 mL. **19.127** 4.05 **19.133** H_2CO_3/HCO_3^- and $H_2PO_4^-/HPO_4^{2-}$;

$[HPO_4^{2-}]/[H_2PO_4^-] = 5.8$ **19.135**(a) 58.2 mL (b) 7.85 mL (c) 6.30 **19.137** 170 mL **19.141** 5.63 **19.143** 3.9×10^{-9} μg Pb^{2+}/100 mL blood **19.145** No NaCl will precipitate. **19.146**(a) A and D (b) $pH_A = 4.35$; $pH_B = 8.67$; $pH_C = 2.67$; $pH_D = 4.57$ (c) C, A, D, B (d) B

Chapter 20

Answer to Problem: B20.2 −12.6 kJ/mol

• **20.2** A spontaneous process occurs by itself, whereas a nonspontaneous process requires a continuous input of energy to make it happen. It is possible to cause a nonspontaneous process to occur, but the process stops once the energy input is removed. A reaction that is nonspontaneous under one set of conditions may be spontaneous under a different set of conditions.
20.5 The transition from liquid to gas involves a greater increase in dispersal of energy and freedom of motion than does the transition from solid to liquid. **20.6** For an exothermic reaction, $\Delta S_{surr} > 0$. For an endothermic reaction, $\Delta S_{surr} < 0$. A chemical cold pack for injuries is an example of an application using a spontaneous endothermic process. **20.8** a, b, and c
20.10 a and b **20.12**(a) positive (b) negative (c) negative
20.14(a) positive (b) positive (c) positive
20.16(a) negative (b) negative (c) positive
20.18(a) positive (b) negative (c) positive
20.20(a) positive (b) negative (c) positive
20.22(a) Butane; the double bond in 2-butene restricts freedom of rotation. (b) Xe(*g*); it has the greater molar mass. (c) $CH_4(g)$; gases have greater entropy than liquids. **20.24**(a) $C_2H_5OH(l)$; it is a more complex molecule. (b) $KClO_3(aq)$; ions in solution have their energy more dispersed than ions in a solid. (c) K(*s*); it has a greater molar mass. **20.26**(a) Diamond < graphite < charcoal. Freedom of motion is least in the network solid, increases for graphite sheets, and is greatest in the amorphous solid. (b) Ice < liquid water < water vapor. Entropy increases as a substance changes from solid to liquid to gas. (c) O atoms $< O_2 < O_3$. Entropy increases with molecular complexity.
20.28(a) $ClO_4^-(aq) > ClO_3^-(aq) > ClO_2^-(aq)$; decreasing molecular complexity (b) $NO_2(g) > NO(g) > N_2(g)$. N_2 has lower standard molar entropy because it consists of two of the same atoms; the other species have two different types of atoms. NO_2 is more complex than NO. (c) $Fe_3O_4(s) > Fe_2O_3(s) > Al_2O_3(s)$. Fe_3O_4 is more complex and more massive. Fe_2O_3 is more massive than Al_2O_3. **20.31** For a system at equilibrium, $\Delta S_{univ} = \Delta S_{sys} + \Delta S_{surr} = 0$. For a system moving to equilibrium, $\Delta S_{univ} > 0$.
20.32 $S^\circ_{Cl_2O(g)} = 2S^\circ_{HClO(g)} - S^\circ_{H_2O(g)} - \Delta S^\circ_{rxn}$
20.33(a) negative; $\Delta S^\circ = -172.4$ J/K (b) positive; $\Delta S^\circ = 141.6$ J/K (c) negative; $\Delta S^\circ = -837$ J/K
20.35 $\Delta S^\circ = 93.1$ J/K; yes, the positive sign of ΔS is expected because there is a net increase in the number of gas molecules.
20.37 $\Delta S^\circ = -311$ J/K; yes, the negative entropy change matches the decrease in moles of gas. **20.39**(a) −75.6 J/K; (b) 490. J/K; because ΔS_{univ} is positive, the reaction is spontaneous at 298 K. **20.41**(a) −242 J/K (b) 559 J/K; because ΔS_{univ} is positive, the reaction is spontaneous at 298 K.
20.44 −97.2 J/K **20.46** A spontaneous process has $\Delta S_{univ} > 0$. Since the absolute temperature is always positive, ΔG_{sys} must be negative ($\Delta G_{sys} < 0$) for a spontaneous process.
20.48(a) ΔH_{sys} is negative. Bonds are being formed, which is an exothermic process. (b) ΔS_{sys} is negative. As molecules form from atoms, the total number of particles decreases and, with it, the entropy of the system. (c) Since ΔH_{sys} is negative, heat is released, so ΔS_{surr} will be positive. (d) Because both ΔH_{sys} and ΔS_{sys} are negative, this reaction will become more spontaneous (ΔG_{sys} will become more negative) as temperature decreases. **20.49** ΔH° is positive and ΔS° is positive. Melting is an example. **20.50** The entropy changes little within a phase. As long as a substance does not change phase, the value of ΔS° is relatively unaffected by temperature. **20.51**(a) −1138.0 kJ (b) −1379.4 kJ (c) −224 kJ **20.53**(a) −1138 kJ (b) −1379 kJ (c) −226 kJ **20.55**(a) Entropy decreases (ΔS° is negative) because the number of moles of gas decreases. The combustion of CO releases energy (ΔH° is negative). (b) −257.2 kJ (using Equation 20.8) or −257.3 kJ (using Equation 20.7).
20.57(a) −0.409 kJ/mol·K (b) −197 kJ/mol **20.59**(a) $\Delta H^\circ_{rxn} = 90.7$ kJ; $\Delta S^\circ_{rxn} = 221$ J/K (b) At 28°C, $\Delta G^\circ = 24.3$ kJ; at 128°C, $\Delta G^\circ = 2.2$ kJ; at 228°C, $\Delta G^\circ = -19.9$ kJ (c) For the substances in their standard states, the reaction is nonspontaneous at 28°C, near equilibrium at 128°C, and spontaneous at 228°C.
(d) 410. K **20.61**(a) $\Delta H^\circ_{rxn} = -241.826$ kJ, $\Delta S^\circ_{rxn} = -44.4$ J/K, $\Delta G^\circ_{rxn} = -228.60$ kJ (b) Yes. The reaction will become nonspontaneous at higher temperatures. (c) The reaction is spontaneous below 5.45×10^3 K. **20.63**(a) ΔG° is a relatively large positive value. (b) $K >> 1$. Q depends on initial conditions, not equilibrium conditions. **20.66** The standard free energy change, ΔG°, applies when all components of the system are in their standard states; $\Delta G^\circ = \Delta G$ when all concentrations equal 1 *M* and all partial pressures equal 1 atm. **20.67**(a) 2.0×10^{-8} (b) 1.27×10^{21} **20.69**(a) 3.46×10^4 (b) 8.2×10^{-7} **20.71** $\Delta H^\circ = 30910$ J, $\Delta S^\circ = 93.15$ J/K, $T = 331.8$ K **20.73** 1.7×10^{-49}
20.75 3.36×10^5 **20.77** 2.7×10^4 J/mol; no **20.79**(a) 2.9×10^4 J/mol (b) The reverse direction, formation of reactants, is spontaneous, so the reaction proceeds to the left. (c) 7.0×10^3 J/mol; the reaction proceeds to the left to reach equilibrium. **20.81**(a) scene 1 (b) scene 3 > scene 1 > scene 2 **20.83**(a) no T (b) 163 kJ (c) 1×10^2 kJ/mol **20.86**(a) spontaneous (b) + (c) + (d) − (e) −, not spontaneous (f) − **20.90**(a) 2.3×10^2 (b) The treatment is to administer oxygen-rich air; increasing the concentration of oxygen shifts the equilibrium to the left, in the direction of $Hb{\cdot}O_2$. **20.92** −370. kJ **20.94**(a) $2N_2O_5(g) + 6F_2(g) \longrightarrow 4NF_3(g) + 5O_2(g)$ (b) $\Delta G^\circ_{rxn} = -569$ kJ (c) $\Delta G_{rxn} = -5.60\times10^2$ kJ/mol
20.96 $\Delta H^\circ_{rxn} = -137.14$ kJ; $\Delta S^\circ_{rxn} = -120.3$ J/K; $\Delta G^\circ_{rxn} = -101.25$ kJ **20.104**(a) 1.67×10^3 J/mol (b) 7.37×10^3 J/mol (c) -4.04×10^3 J/mol (d) 0.19 **20.108**(a) 465 K (b) 6.59×10^{-4} (c) The reaction rate is higher at the higher temperature. The shorter time required (kinetics) overshadows the lower yield (thermodynamics). **20.112** 1.18 J/K **20.114** 7.63 J/K **20.116** 24.1 J/K

Chapter 21

Answers to Boxed Reading Problems: B21.1(a) reduction: $Fe^{3+} + e^- \longrightarrow Fe^{2+}$; oxidation: $Cu^+ \longrightarrow Cu^{2+} + e^-$ (b) Overall: $Fe^{3+} + Cu^+ \longrightarrow Fe^{2+} + Cu^{2+}$

• **21.1** Oxidation is the loss of electrons and results in a higher oxidation number; reduction is the gain of electrons and results

in a lower oxidation number. **21.3** No, one half-reaction cannot take place independently because there is a transfer of electrons from one substance to another. If one substance loses electrons, another substance must gain them. **21.6** To remove H^+ ions from an equation, add an equal number of OH^- ions to both sides to neutralize the H^+ ions and produce water. **21.8** Spontaneous reactions, for which $\Delta G_{sys} < 0$, take place in voltaic cells (also called galvanic cells). Nonspontaneous reactions, for which $\Delta G_{sys} > 0$, take place in electrolytic cells. **21.10**(a) Cl^- (b) MnO_4^- (c) MnO_4^- (d) Cl^- (e) from Cl^- to MnO_4^-
(f) $8H_2SO_4(aq) + 2KMnO_4(aq) + 10KCl(aq) \longrightarrow 2MnSO_4(aq) + 5Cl_2(g) + 8H_2O(l) + 6K_2SO_4(aq)$

21.12(a) $ClO_3^-(aq) + 6H^+(aq) + 6I^-(aq) \longrightarrow Cl^-(aq) + 3H_2O(l) + 3I_2(s)$
Oxidizing agent is ClO_3^- and reducing agent is I^-.
(b) $2MnO_4^-(aq) + H_2O(l) + 3SO_3^{2-}(aq) \longrightarrow 2MnO_2(s) + 3SO_4^{2-}(aq) + 2OH^-(aq)$
Oxidizing agent is MnO_4^- and reducing agent is SO_3^{2-}.
(c) $2MnO_4^-(aq) + 6H^+(aq) + 5H_2O_2(aq) \longrightarrow 2Mn^{2+}(aq) + 8H_2O(l) + 5O_2(g)$
Oxidizing agent is MnO_4^- and reducing agent is H_2O_2.

21.14(a) $Cr_2O_7^{2-}(aq) + 14H^+(aq) + 3Zn(s) \longrightarrow 2Cr^{3+}(aq) + 7H_2O(l) + 3Zn^{2+}(aq)$
Oxidizing agent is $Cr_2O_7^{2-}$ and reducing agent is Zn.
(b) $MnO_4^-(aq) + 3Fe(OH)_2(s) + 2H_2O(l) \longrightarrow MnO_2(s) + 3Fe(OH)_3(s) + OH^-(aq)$
Oxidizing agent is MnO_4^- and reducing agent is $Fe(OH)_2$.
(c) $2NO_3^-(aq) + 12H^+(aq) + 5Zn(s) \longrightarrow N_2(g) + 6H_2O(l) + 5Zn^{2+}(aq)$
Oxidizing agent is NO_3^- and reducing agent is Zn.

21.16(a) $4NO_3^-(aq) + 4H^+(aq) + 4Sb(s) \longrightarrow 4NO(g) + 2H_2O(l) + Sb_4O_6(s)$
Oxidizing agent is NO_3^- and reducing agent is Sb.
(b) $5BiO_3^-(aq) + 14H^+(aq) + 2Mn^{2+}(aq) \longrightarrow 5Bi^{3+}(aq) + 7H_2O(l) + 2MnO_4^-(aq)$
Oxidizing agent is BiO_3^- and reducing agent is Mn^{2+}.
(c) $Pb(OH)_3^-(aq) + 2Fe(OH)_2(s) \longrightarrow Pb(s) + 2Fe(OH)_3(s) + OH^-(aq)$
Oxidizing agent is $Pb(OH)_3^-$ and reducing agent is $Fe(OH)_2$.

21.18(a) $5As_4O_6(s) + 8MnO_4^-(aq) + 18H_2O(l) \longrightarrow 20AsO_4^{3-}(aq) + 8Mn^{2+}(aq) + 36H^+(aq)$
Oxidizing agent is MnO_4^- and reducing agent is As_4O_6.
(b) $P_4(s) + 6H_2O(l) \longrightarrow 2HPO_3^{2-}(aq) + 2PH_3(g) + 4H^+(aq)$
P_4 is both the oxidizing agent and the reducing agent.
(c) $2MnO_4^-(aq) + 3CN^-(aq) + H_2O(l) \longrightarrow 2MnO_2(s) + 3CNO^-(aq) + 2OH^-(aq)$
Oxidizing agent is MnO_4^- and reducing agent is CN^-.

21.21(a) $Au(s) + 3NO_3^-(aq) + 4Cl^-(aq) + 6H^+(aq) \longrightarrow AuCl_4^-(aq) + 3NO_2(g) + 3H_2O(l)$
(b) Oxidizing agent is NO_3^- and reducing agent is Au. (c) HCl provides chloride ions that combine with the gold(III) ion to form the stable $AuCl_4^-$ complex ion. **21.22**(a) A (b) E (c) C (d) A (e) E (f) E **21.25** An active electrode is a reactant or product in the cell reaction. An inactive electrode does not take part in the reaction and is present only to conduct a current. Platinum and graphite are commonly used for inactive electrodes. **21.26**(a) metal A (b) metal B (c) metal A (d) Hydrogen bubbles will form when metal A is placed in acid. Metal A is a better reducing agent than metal B, so if metal B reduces H^+ in acid, then metal A will reduce it also. **21.27**(a) left to right (b) left (c) right (d) Ni (e) Fe (f) Fe (g) 1 *M* $NiSO_4$ (h) K^+ and NO_3^- (i) neither (j) from right to left
(k) Oxidation: $Fe(s) \longrightarrow Fe^{2+}(aq) + 2e^-$
Reduction: $Ni^{2+}(aq) + 2e^- \longrightarrow Ni(s)$
Overall: $Fe(s) + Ni^{2+}(aq) \longrightarrow Fe^{2+}(aq) + Ni(s)$

21.29(a) Oxidation: $Zn(s) \longrightarrow Zn^{2+}(aq) + 2e^-$
Reduction: $Sn^{2+}(aq) + 2e^- \longrightarrow Sn(s)$
Overall: $Zn(s) + Sn^{2+}(aq) \longrightarrow Zn^{2+}(aq) + Sn(s)$
(b)
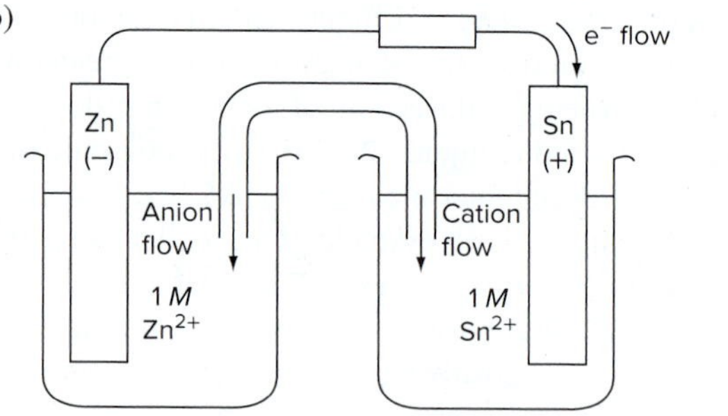

21.31(a) Reduction: $Fe^{2+}(aq) + 2e^- \longrightarrow Fe(s)$
Oxidation: $Mn(s) \longrightarrow Mn^{2+}(aq) + 2e^-$
Overall: $Fe^{2+}(aq) + Mn(s) \longrightarrow Fe(s) + Mn^{2+}(aq)$
(b)
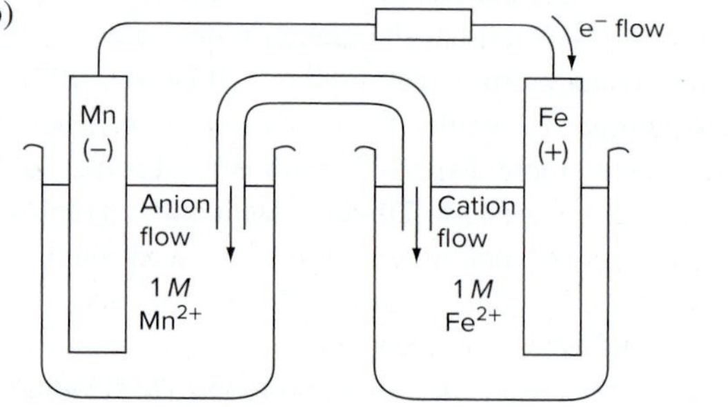

21.33(a) $Al(s) \mid Al^{3+}(aq) \parallel Cr^{3+}(aq) \mid Cr(s)$
(b) $Pt(s) \mid SO_2(g) \mid SO_4^{2-}(aq), H^+(aq) \parallel Cu^{2+}(aq) \mid Cu(s)$
21.36 A negative E°_{cell} indicates that the redox reaction is not spontaneous at the standard state, that is, $\Delta G^\circ > 0$. The reverse reaction is spontaneous with $E^\circ_{cell} > 0$.
21.37 Similar to other state functions, E° changes sign when a reaction is reversed. Unlike ΔG°, ΔH°, and S°, E° (the ratio of energy to charge) is an intensive property. When the coefficients in a reaction are multiplied by a factor, the values of ΔG°, ΔH°, and S° are multiplied by that factor. However, E° does not change because both the energy and charge are multiplied by the factor and thus their ratio remains unchanged.

21.38(a) Oxidation: $Se^{2-}(aq) \longrightarrow Se(s) + 2e^-$
Reduction: $2SO_3^{2-}(aq) + 3H_2O(l) + 4e^- \longrightarrow S_2O_3^{2-}(aq) + 6OH^-(aq)$
(b) $E^\circ_{anode} = E^\circ_{cathode} - E^\circ_{cell} = -0.57 \text{ V} - 0.35 \text{ V} = -0.92 \text{ V}$

21.40(a) $Br_2 > Fe^{3+} > Cu^{2+}$ (b) $Ca^{2+} < Ag^+ < Cr_2O_7^{2-}$

21.42(a) $Co(s) + 2H^+(aq) \longrightarrow Co^{2+}(aq) + H_2(g)$
$E^\circ_{cell} = 0.28$ V; spontaneous
(b) $Hg_2^{2+}(aq) \longrightarrow Hg^{2+}(aq) + Hg(l)$
$E^\circ_{cell} = -0.13$ V; not spontaneous

21.44(a) $Cr_2O_7^{2-}(aq) + 3Cd(s) + 14H^+(aq) \longrightarrow 2Cr^{3+}(aq) + 3Cd^{2+}(aq) + 7H_2O(l)$

$E^\circ_{cell} = 1.73$ V; spontaneous

(b) $Pb(s) + Ni^{2+}(aq) \longrightarrow Pb^{2+}(aq) + Ni(s)$

$E^\circ_{cell} = -0.12$ V; not spontaneous

21.46 $3N_2O_4(g) + 2Al(s) \longrightarrow 6NO_2^-(aq) + 2Al^{3+}(aq)$

$E^\circ_{cell} = 0.867\ V - (-1.66\ V) = 2.53$ V

$2Al(s) + 3SO_4^{2-}(aq) + 3H_2O(l) \longrightarrow 2Al^{3+}(aq) + 3SO_3^{2-}(aq) + 6OH^-(aq)$

$E^\circ_{cell} = 2.59$ V

$SO_4^{2-}(aq) + 2NO_2^-(aq) + H_2O(l) \longrightarrow SO_3^{2-}(aq) + N_2O_4(g) + 2OH^-(aq)$

$E^\circ_{cell} = 0.06$ V

Oxidizing agents: $Al^{3+} < N_2O_4 < SO_4^{2-}$

Reducing agents: $SO_3^{2-} < NO_2^- < Al$

21.48 $2HClO(aq) + Pt(s) + 2H^+(aq) \longrightarrow Cl_2(g) + Pt^{2+}(aq) + 2H_2O(l)$

$E^\circ_{cell} = 0.43$ V

$2HClO(aq) + Pb(s) + SO_4^{2-}(aq) + 2H^+(aq) \longrightarrow Cl_2(g) + PbSO_4(s) + 2H_2O(l)$

$E^\circ_{cell} = 1.94$ V

$Pt^{2+}(aq) + Pb(s) + SO_4^{2-}(aq) \longrightarrow Pt(s) + PbSO_4(s)$

$E^\circ_{cell} = 1.51$ V

Oxidizing agents: $PbSO_4 < Pt^{2+} < HClO$

Reducing agents: $Cl_2 < Pt < Pb$

21.50(a) yes (b) C > A > B **21.53** $A(s) + B^+(aq) \longrightarrow A^+(aq) + B(s)$ with $Q = [A^+]/[B^+]$. (a) $[A^+]$ increases and $[B^+]$ decreases. (b) E_{cell} decreases. (c) $E_{cell} = E^\circ_{cell} - (RT/nF) \ln ([A^+]/[B^+])$; $E_{cell} = E^\circ_{cell}$ when $(RT/nF) \ln ([A^+]/[B^+]) = 0$. This occurs when $\ln ([A^+]/[B^+]) = 0$, that is, $[A^+]$ equals $[B^+]$. (d) Yes, when $[A^+] > [B^+]$. **21.55** In a concentration cell, the overall reaction decreases the concentration of the more concentrated electrolyte because that electrolyte is reduced in the cathode compartment. **21.56**(a) 3×10^{35} (b) 4×10^{-31} **21.58**(a) 1×10^{-67} (b) 6×10^{9} **21.60**(a) -2.03×10^5 J (b) 1.74×10^5 J **21.62**(a) 3.82×10^5 J (b) -5.6×10^4 J **21.64** $E^\circ_{cell} = 0.28$ V; $\Delta G^\circ = -2.7\times10^4$ J **21.66** $E^\circ_{cell} = 0.054$ V; $\Delta G^\circ = -1.0\times10^4$ J **21.68** 8.8×10^{-5} M **21.70**(a) 0.05 V (b) 0.50 M (c) $[Co^{2+}] = 0.91$ M; $[Ni^{2+}] = 0.09$ M **21.72** half-cell A; 0.083 V **21.74** Electrons flow from the anode, where oxidation occurs, to the cathode, where reduction occurs. The electrons always flow from the anode to the cathode no matter what type of battery. **21.76** A D alkaline battery is larger than an AAA one, so it contains greater amounts of the cell components. The cell potential is an intensive property and does not depend on the amounts of the cell components. The total charge, however, does depend on the amount of cell components, so the D battery produces more charge. **21.78** The Teflon spacers keep the two metals separated so that the copper cannot conduct electrons that would promote the corrosion (rusting) of the iron skeleton. **21.81** Sacrificial anodes are made of metals that have E° less than that of iron, -0.44 V, so they are more easily oxidized than iron. Only (b), (f), and (g) will work for iron: (a) will form an oxide coating that prevents further oxidation; (c) will react with groundwater quickly; (d) and (e) are less easily oxidized than iron. **21.83** To reverse the reaction requires 0.34 V with the cell in its standard state. A 1.5-V cell supplies more than enough potential, so the Cd metal is oxidized to Cd^{2+} and Cr metal plates out. **21.85** The oxidation number of N in NO_3^- is +5, the maximum O.N. for N. In the nitrite ion, NO_2^-, the O.N. of N is +3, so nitrogen can be further oxidized. **21.87**(a) Br_2 (b) Na **21.89** I_2 gas forms at the anode; magnesium (liquid) forms at the cathode. **21.91** Bromine gas forms at the anode; calcium metal forms at the cathode.

21.93 copper and bromine **21.95** iodine, zinc, and silver

21.97(a) Anode: $2H_2O(l) \longrightarrow O_2(g) + 4H^+(aq) + 4e^-$

Cathode: $2H_2O(l) + 2e^- \longrightarrow H_2(g) + 2OH^-(aq)$

(b) Anode: $2H_2O(l) \longrightarrow O_2(g) + 4H^+(aq) + 4e^-$

Cathode: $Sn^{2+}(aq) + 2e^- \longrightarrow Sn(s)$

21.99(a) Anode: $2H_2O(l) \longrightarrow O_2(g) + 4H^+(aq) + 4e^-$

Cathode: $NO_3^-(aq) + 4H^+(aq) + 3e^- \longrightarrow NO(g) + 2H_2O(l)$

(b) Anode: $2Cl^-(aq) \longrightarrow Cl_2(g) + 2e^-$

Cathode: $2H_2O(l) + 2e^- \longrightarrow H_2(g) + 2OH^-(aq)$

21.101(a) 3.75 mol e^- (b) 3.62×10^5 C (c) 28.7 A **21.103** 0.275 g Ra **21.105** 9.20×10^3 s **21.107**(a) The sodium and sulfate ions conduct a current, facilitating electrolysis. Pure water, which contains very low (10^{-7} M) concentrations of H^+ and OH^-, conducts electricity very poorly. (b) The reduction of H_2O has a more positive half-potential than does the reduction of Na^+; the oxidation of H_2O is the only reaction possible because SO_4^{2-} cannot be oxidized. Thus, it is easier to reduce H_2O than Na^+ and easier to oxidize H_2O than SO_4^{2-}. **21.109** 62.6 g Zn **21.111**(a) 3.3×10^{11} C (b) 4.8×10^{11} J (c) 1.2×10^4 kg **21.114** 64.3 mass % Cu **21.115**(a) 8 days (b) 32 days (c) \$940 **21.118**(a) 2.4×10^4 days (b) 2.1 g (c) 6.1×10^{-5} dollars **21.121**(a) Pb/Pb^{2+}: $E^\circ_{cell} = 0.13$ V; Cu/Cu^{2+}: $E^\circ_{cell} = 0.34$ V (b) The anode (negative electrode) is Pb. The anode in the other cell is platinum in the standard hydrogen electrode. (c) The precipitation of PbS decreases $[Pb^{2+}]$, which increases the potential. (d) -0.13 V **21.124** The three steps equivalent to the overall reaction

$M^+(aq) + e^- \longrightarrow M(s)$:

(1) $M^+(aq) \longrightarrow M^+(g)$	ΔH is $-\Delta H_{hydration}$
(2) $M^+(g) + e^- \longrightarrow M(g)$	ΔH is $-IE$
(3) $M(g) \longrightarrow M(s)$	ΔH is $-\Delta H_{atomization}$

The energy for step 3 is similar for all three elements, so the difference in energy for the overall reaction depends on the values for $\Delta H_{hydration}$ and IE. The Li^+ ion has a greater hydration energy than Na^+ and K^+ because it is smaller, with a larger charge density that holds the water molecules more tightly. The energy required to remove the waters surrounding Li^+ offsets the lower ionization energy, making the overall energy for the reduction of Li^+ ion larger than expected. **21.125** The very high and very low standard electrode potentials involve extremely reactive substances, such as F_2 (a powerful oxidizer) and Li (a powerful reducer). These substances react directly with water because any aqueous cell with a voltage of more than 1.23 V has the ability to electrolyze water into hydrogen and oxygen. **21.127**(a) 1.073×10^5 s (b) 1.5×10^4 kW·h (c) 0.92¢ **21.129** If metal E and a salt of metal F are mixed, the salt is reduced, producing metal F because E has the greatest reducing strength of the three metals; F < D < E.

21.131(a) Cell I: 4 mol electrons; $\Delta G = -4.75\times10^5$ J

Cell II: 2 mol electrons; $\Delta G = -3.94\times10^5$ J

Cell III: 2 mol electrons; $\Delta G = -4.53\times10^5$ J

(b) Cell I: −13.2 kJ/g
Cell II: −0.613 kJ/g
Cell III: −2.63 kJ/g
Cell I has the highest ratio (most energy released per gram) because the reactants have very low mass, while cell II has the lowest ratio because the reactants have large masses.

21.135(a) 9.7 g Cu (b) 0.56 *M* Cu^{2+} **21.136** 2.94
21.137 $Sn^{2+}(aq) + 2e^- \longrightarrow Sn(s)$
$Cr^{3+}(aq) + e^- \longrightarrow Cr^{2+}(aq)$
$Fe^{2+}(aq) + 2e^- \longrightarrow Fe(s)$
$U^{4+}(aq) + e^- \longrightarrow U^{3+}(aq)$
21.141(a) 3.6×10^{-9} *M* (b) 0.3 *M*
21.143(a) Nonstandard cell:
$E_{waste} = E^\circ_{cell} - (0.0592\text{ V}/1)\log[Ag^+]_{waste}$
Standard cell:
$E_{standard} = E^\circ_{cell} - (0.0592\text{ V}/1)\log[Ag^+]_{standard}$

(b) $[Ag^+]_{waste} = \text{antilog}\left(\frac{E_{standard} - E_{waste}}{0.0592}\right)[Ag^+]_{standard}$

(c) $C_{Ag^+,\,waste} = \text{antilog}\left(\frac{E_{standard} - E_{waste}}{0.0592}\right)C_{Ag^+_,\,standard}$,

where *C* is concentration in ng/L
(d) 900 ng/L
(e) $[Ag^+]_{waste} =$

$$\text{antilog}\left[\frac{(E_{standard} - E_{waste})\dfrac{nF}{2.303R} + T_{standard}\log[Ag^+]_{standard}}{T_{waste}}\right]$$

21.145(a) 1.08×10^3 C (b) 0.629 g Cd, 1.03 g NiO(OH), 0.202 g H_2O; total mass of reactants = 1.86 g (c) 10.1%
21.147 Li > Ba > Na > Al > Mn > Zn > Cr > Fe > Ni > Sn > Pb > Cu > Ag > Hg > Au. Metals with potentials lower than that of water (−0.83 V) can displace H_2 from water: Li, Ba, Na, Al, and Mn. Metals with potentials lower than that of hydrogen (0.00 V) can displace H_2 from acid: Li, Ba, Na, Al, Mn, Zn, Cr, Fe, Ni, Sn, and Pb. Metals with potentials greater than that of hydrogen (0.00 V) cannot displace H_2: Cu, Ag, Hg, and Au.
21.150(a) 5.3×10^{-11} (b) 0.20 V (c) 0.43 V (d) 8.2×10^{-4} *M* NaOH **21.153**(a) -1.18×10^5 kJ (b) 1.20×10^4 L (c) 9.43×10^4 s (d) 157 kW·h (e) 22.0¢

Chapter 22

22.2 Fe from Fe_2O_3; Ca from $CaCO_3$; Na from NaCl; Zn from ZnS **22.3**(a) Differentiation refers to the processes involved in the formation of Earth into regions (core, mantle, and crust) of differing composition. Substances separated according to their densities, with the more dense material in the core and the less dense in the crust. (b) O, Si, Al, and Fe (c) O **22.7** Plants produced O_2, slowly increasing the oxygen concentration in the atmosphere and creating an environment for oxidizing metals. The oxygen-free decay of plant and animal material created large fossil fuel deposits. **22.9** Fixation refers to the process of converting a substance in the atmosphere into a form more readily usable by organisms. Carbon and nitrogen; fixation of carbon dioxide gas by plants and fixation of nitrogen gas by nitrogen-fixing bacteria. **22.12** Atmospheric nitrogen is fixed by three pathways: atmospheric, industrial, and biological. Atmospheric fixation requires high-temperature reactions (e.g., initiated by lightning) to convert N_2 into NO and other oxidized species. Industrial fixation involves mainly the formation of ammonia, NH_3, from N_2 and H_2. Biological fixation occurs in nitrogen-fixing bacteria that live on the roots of legumes. Human activity is an example of industrial fixation. It contributes about 17% of the fixed nitrogen. **22.14**(a) the atmosphere (b) Plants excrete acid from their roots to convert PO_4^{3-} ions into more soluble $H_2PO_4^-$ ions, which the plant can absorb. Through excretion and decay, organisms return soluble phosphate compounds to the cycle. **22.17**(a) 1.1×10^3 L (b) 4.2×10^2 m^3 **22.18**(a) The iron(II) ions form an insoluble salt, $Fe_3(PO_4)_2$, that decreases the yield of phosphorus. (b) 8.8 t **22.20**(a) Roasting consists of heating a mineral sulfide in air at high temperatures to convert it to the oxide. (b) Smelting is the reduction of the metal oxide to the free metal using heat and a reducing agent such as coke. (c) Flotation is a separation process in which the ore is removed from the gangue by exploiting the difference in density in the presence of detergent. The gangue sinks to the bottom and the lighter ore-detergent mix is skimmed off the top. (d) Refining, using electricity or heat, is the final step in the process to yield the pure element. **22.25**(a) Slag is a byproduct of steel-making and contains the impurity SiO_2. (b) Pig iron is the impure product of iron metallurgy (containing 3–4% C and other impurities). (c) Steel refers to iron that has been alloyed with other elements to give the metal desirable properties. (d) The basic-oxygen process is used to purify pig iron and obtain carbon steel. **22.27** Iron and nickel are more easily oxidized and less easily reduced than copper. They are separated from copper in the roasting step and converted to slag. In the electrorefining process, all three metals are in solution, but only Cu^{2+} ions are reduced at the cathode to form Cu(*s*). **22.30** Le Châtelier's principle says that the system shifts toward formation of K as the gaseous metal leaves the cell. **22.31**(a) $E^\circ_{half\text{-}cell} = -3.05$ V, −2.93 V, and −2.71 V for Li^+, K^+, and Na^+, respectively. In all of these cases, it is energetically more favorable to reduce H_2O to H_2 than to reduce M^+ to M. (b) $2RbX + Ca \longrightarrow CaX_2 + 2Rb$ where $\Delta H = IE_1(Ca) + IE_2(Ca) - 2IE_1(Rb) = 929$ kJ/mol. Based on the IEs and positive ΔH for the forward reaction, it seems more reasonable that Rb metal will reduce Ca^{2+} than the reverse. (c) If the reaction is carried out at a temperature greater than the boiling point of Rb, the product mixture will contain gaseous Rb, which can be removed from the reaction vessel; this would cause a shift in equilibrium to form more Rb as product. (d) $2CsX + Ca \longrightarrow CaX_2 + 2Cs$, where $\Delta H = IE_1(Ca) + IE_2(Ca) - 2IE_1(Cs) =$ 983 kJ/mol. This reaction is less favorable than for Rb, but Cs has a lower boiling point. **22.32**(a) 4.5×10^4 L (b) 1.30×10^8 C (c) 1.69×10^6 s **22.35**(a) Mg^{2+} is more difficult to reduce than H_2O, so $H_2(g)$ would be produced instead of Mg metal. $Cl_2(g)$ forms at the anode due to overvoltage. (b) The ΔH°_f of $MgCl_2(s)$ is −641.6 kJ/mol. High temperature favors the reverse (endothermic) reaction, the formation of magnesium metal and chlorine gas. **22.37**(a) Sulfur dioxide is the reducing agent and is oxidized to the +6 state (SO_4^{2-}). (b) $HSO_4^-(aq)$ (c) $H_2SeO_3(aq) + 2SO_2(g) + H_2O(l) \longrightarrow Se(s) + 2HSO_4^-(aq) + 2H^+(aq)$
22.42(a) O.N. for Cu: in Cu_2S, +1; in Cu_2O, +1; in Cu, 0 (b) Cu_2S is the reducing agent, and Cu_2O is the oxidizing agent.

22.47 $2ZnS(s) + C(graphite) \longrightarrow 2Zn(s) + CS_2(g)$; $\Delta G^\circ_{rxn} =$ 463 kJ. Since ΔG°_{rxn} is positive, this reaction is not spontaneous at standard-state conditions. $2ZnO(s) + C(s) \longrightarrow 2Zn(s) + CO_2(g)$; $\Delta G^\circ_{rxn} = 242.0$ kJ. This reaction is also not spontaneous, but is less unfavorable. **22.48** The formation of sulfur trioxide is very slow at ordinary temperatures. Increasing the temperature can speed up the reaction, but because the reaction is exothermic, increasing the temperature decreases the yield. Adding a catalyst increases the rate of the reaction, allowing a lower temperature to be used to enhance the yield. **22.51**(a) Cl_2, H_2, and NaOH (b) The mercury-cell method yields higher purity NaOH, but releases some Hg, which is harmful in the environment. **22.52**(a) $\Delta G^\circ = -142$ kJ; yes (b) The rate of the reaction is very low at 25°C. (c) $\Delta G^\circ_{500} = -53$ kJ, so the reaction is spontaneous. (d) $K_{25} = 7.8\times10^{24} > K_{500} = 3.8\times10^3$ (e) 1.05×10^3 K **22.53** 7×10^2 lb Cl_2 **22.56**(a) $P_4O_{10}(s) + 6H_2O(l) \longrightarrow 4H_3PO_4(l)$ (b) 1.52 **22.58**(a) 9.006×10^9 g CO_2 (b) The 4.3×10^{10} g CO_2 produced by automobiles is much greater than that from the blast furnace. **22.60**(a) If $[OH^-] > 1.1\times10^{-4}$ *M* (i.e., if pH > 10.04), $Mg(OH)_2$ will precipitate. (b) 1 (To the correct number of significant figures, all the magnesium precipitates.) **22.61**(a)K_{25}(step 1) = 1×10^{168}; K_{25}(side rxn) = 7×10^{228} (b)K_{900}(step 1) = 4.5×10^{49}; K_{900}(side rxn) = 1.4×10^{63} (c) 8.8×10^7 dollars (d) 6.4×10^7 dollars **22.64**(1) $2H_2O(l) + 2FeS_2(s) + 7O_2(g) \longrightarrow 2Fe^{2+}(aq) + 4SO_4^{2-}(aq) + 4H^+(aq)$ increases acidity (2) $4H^+(aq) + 4Fe^{2+}(aq) + O_2(g) \longrightarrow 4Fe^{3+}(aq) + 2H_2O(l)$ (3) $Fe^{3+}(aq) + 3H_2O(l) \longrightarrow Fe(OH)_3(s) + 3H^+(aq)$ increases acidity (4) $8H_2O(l) + FeS_2(s) + 14Fe^{3+}(aq) \longrightarrow 15Fe^{2+}(aq) + 2SO_4^{2-}(aq) + 16H^+(aq)$ increases acidity **22.65** density of ferrite: 7.86 g/cm^3; density of austenite: 7.55 g/cm^3 **22.67**(a) Cathode: $Na^+(l) + e^- \longrightarrow Na(l)$ Anode: $4OH^-(l) \longrightarrow O_2(g) + 2H_2O(g) + 4e^-$ (b) 50% **22.70**(a) $nCO_2(g) + nH_2O(l) \longrightarrow (CH_2O)_n(s) + nO_2(g)$ (b) 27 L (c) 6.7×10^4 L **22.71** 73 mg/L **22.72** 891 kg Na_3AlF_6 **22.73**(a) 23.2 min (b) 14 effusion steps **22.75**(a) 1.890 t Al_2O_3 (b) 0.3339 t C (c) 100% (d) 74% (e) 2.813×10^3 m^3 **22.80** Acid rain increases the leaching of phosphate into the groundwater, due to the protonation of PO_4^{3-} to form HPO_4^{2-} and $H_2PO_4^-$. As shown in calculations for parts (a) and (b), solubility of $Ca_3(PO_4)_2$ increases more than 10^4 from the value in pure water to the value in acidic rainwater. (a) 6.4×10^{-7} *M* (b) 1.1×10^{-2} *M* **22.81**(a) 1.00 mol % (b) 238.9 g/mol **22.83** density of silver = 10.51 g/cm^3; density of sterling silver = 10.2 g/cm^3

Chapter 23

Answers to Boxed Reading Problems B23.2 Zinc forms tetrahedral complex ions. The other ions listed form ions of other geometries (Ni^{2+} forms square planar complex ions while Fe^{2+} and Mn^{2+} form octahedral complex ions). If these ions are placed in a tetrahedral environment in place of the zinc, they cannot function as well as zinc since the enzyme catalyst would have a different shape.

• **23.2**(a) $1s^22s^22p^63s^23p^64s^23d^{10}4p^65s^24d^x$ (b) $1s^22s^22p^63s^23p^64s^23d^{10}4p^65s^24d^{10}5p^66s^24f^{14}5d^x$ **23.4** Five; examples are Mn: [Ar] $4s^23d^5$ and Fe^{3+}: [Ar] $3d^5$. **23.6**(a) The elements should increase in size as they increase in mass from Period 5 to Period 6. Because there are 14 inner transition elements in Period 6, the effective nuclear charge increases significantly; so the atomic size decreases, or "contracts." This effect is significant enough that Zr^{4+} and Hf^{4+} are almost the same size but differ greatly in atomic mass. (b) The atomic size increases from Period 4 to Period 5, but stays fairly constant from Period 5 to Period 6. (c) Atomic mass increases significantly from Period 5 to Period 6, but atomic radius (and thus volume) increases slightly, so Period 6 elements are very dense. **23.9**(a) A paramagnetic substance is attracted to a magnetic field, while a diamagnetic substance is slightly repelled by one. (b) Ions of transition elements often have half-filled *d* orbitals whose unpaired electrons make the ions paramagnetic. Ions of main-group elements usually have a noble-gas configuration with no partially filled levels. (c) Some *d* orbitals in the transition element ions are empty, which allows an electron from one *d* orbital to move to a slightly higher energy one. The energy required is small and falls in the visible range, so the ion is colored. All orbitals are filled in ions of main-group elements, so enough energy would have to be added to move an electron to the next principal energy level, not just another orbital within the same energy level. This amount of energy is very large and much greater than the visible range of wavelengths. **23.10**(a) $1s^22s^22p^63s^23p^64s^23d^3$ or [Ar] $4s^23d^3$ (b) $1s^22s^22p^63s^23p^64s^23d^{10}4p^65s^24d^1$ or [Kr] $5s^24d^1$ (c) [Xe] $6s^24f^{14}5d^{10}$ **23.12**(a) [Xe] $6s^24f^{14}5d^6$ (b) [Ar] $4s^23d^7$ (c) [Kr] $5s^14d^{10}$ **23.14**(a) [Ar], no unpaired electrons (b) [Ar] $3d^9$, one unpaired electron (c) [Ar] $3d^5$, five unpaired electrons (d) [Kr] $4d^2$, two unpaired electrons **23.16**(a) +5 (b) +4 (c) +7 **23.18** Cr, Mo, and W **23.20** in CrF_2, because Cr is in a lower oxidation state in that compound **23.22** Atomic size increases slightly down a group of transition elements, but nuclear charge increases much more, so the first ionization energy generally increases. Since the ionization energy of Mo is higher than that of Cr, so it is more difficult to remove electrons from Mo. Also, the reduction potential for Mo is less negative, so it is more difficult to oxidize Mo than Cr. **23.24** CrO_3, with Cr in a higher oxidation state, yields a more acidic aqueous solution. **23.28**(a) seven (b) The number corresponds to a half-filled *f* subshell. **23.30**(a) [Xe] $6s^25d^1$ (b) [Xe] $4f^1$ (c) [Rn] $7s^25f^{11}$ (d) [Rn] $5f^2$ **23.32**(a) Eu^{2+}: [Xe] $4f^7$; Eu^{3+}: [Xe] $4f^6$; Eu^{4+}: [Xe] $4f^5$. The stability of the half-filled *f* subshell makes Eu^{2+} most stable. (b) Tb^{2+}: [Xe] $4f^9$; Tb^{3+}: [Xe] $4f^8$; Tb^{4+}: [Xe] $4f^7$. Tb should show a +4 oxidation state because that corresponds to a half-filled subshell. **23.34** Gd has the electron configuration [Xe] $6s^24f^75d^1$ with eight unpaired electrons. Gd^{3+} has seven unpaired electrons: [Xe] $4f^7$. **23.37** The coordination number indicates the number of ligand atoms bonded to the metal ion. The oxidation number represents the number of electrons lost to form the ion. The coordination number is unrelated to the oxidation number. **23.39** 2, linear; 4, tetrahedral or square planar; 6, octahedral **23.42** The complex ion has a negative charge. **23.45**(a) hexaaquanickel(II) chloride (b) tris(ethylenediamine) chromium(III) perchlorate (c) potassium hexacyanomanganate(II) **23.47**(a) 2+, 6 (b) 3+, 6 (c) 2+, 6 **23.49**(a) potassium dicyanoargentate(I) (b) sodium tetrachlorocadmate(II) (c) tetraammineaquabromocobalt(III) bromide **23.51**(a) 1+, 2 (b) 2+, 4 (c) 3+, 6 **23.53**(a) $[Zn(NH_3)_4]SO_4$ (b) $[Cr(NH_3)_5Cl]Cl_2$ (c) $Na_3[Ag(S_2O_3)_2]$ **23.55**(a) 4, two ions (b) 6, three ions

(c) 2, four ions **23.57**(a) $[Cr(H_2O)_6]_2(SO_4)_3$ (b) $Ba[FeBr_4]_2$ (c) $[Pt(en)_2]CO_3$ **23.59**(a) 6, five ions (b) 4, three ions (c) 4, two ions **23.61**(a) The nitrite ion forms linkage isomers because it can bind to the metal ion through the lone pair on the N atom or any lone pair on either O atom. $[:\ddot{O}-\ddot{N}=\ddot{O}]^-$ (b) Sulfur dioxide molecules form linkage isomers because the lone pair on the S atom or any lone pair on either O atom can bind the central metal ion. $\ddot{O}=\ddot{S}=\ddot{O}$ (c) Nitrate ions have an N atom with no lone pair and three O atoms, all with lone pairs that can bind to the metal ion. But all of the O atoms are equivalent, so these ions do not form linkage isomers.

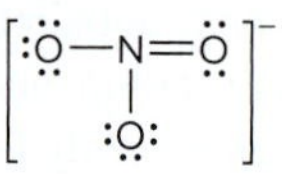

23.63(a) geometric isomerism

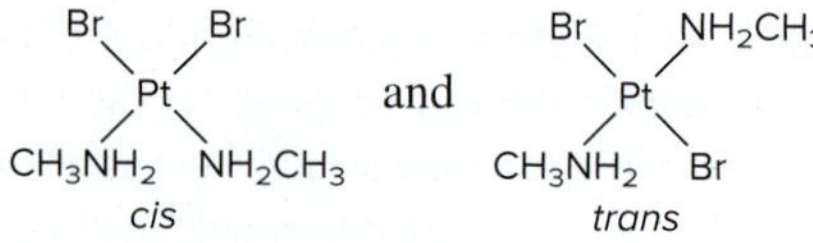

(b) geometric isomerism

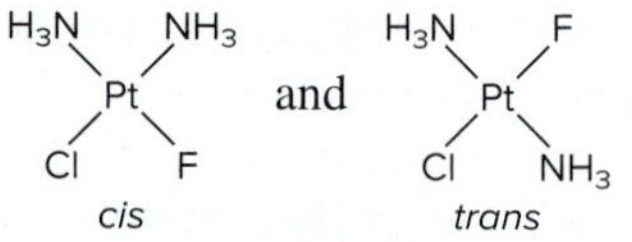

(c) geometric isomerism

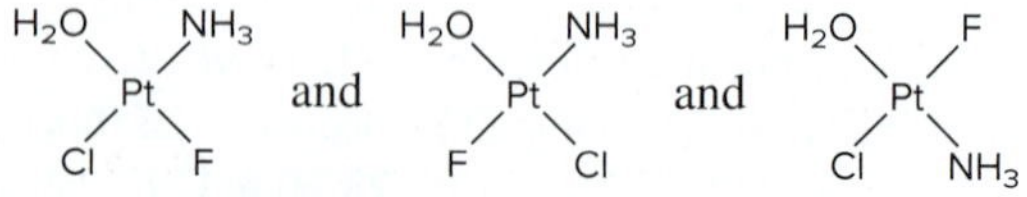

23.65(a) geometric isomerism

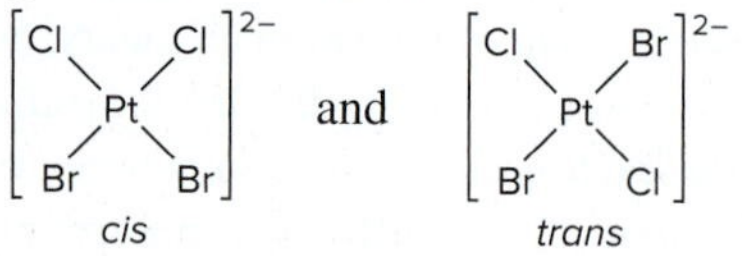

(b) linkage isomerism

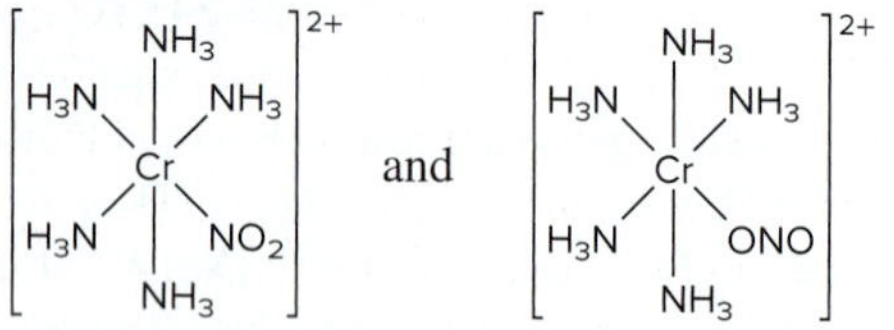

(c) geometric isomerism

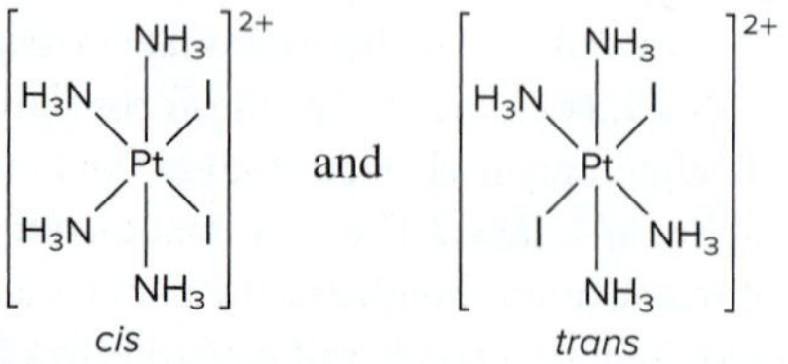

23.67 The compound with the traditional formula is $CrCl_3 \cdot 4NH_3$; the modern formula is $[Cr(NH_3)_4Cl_2]Cl$. **23.69**(a) $K[Pd(NH_3)Cl_3]$ (b) $[PdCl_2(NH_3)_2]$ (c) $K_2[PdCl_6]$ (d) $[Pd(NH_3)_4Cl_2]Cl_2$ **23.71**(a) dsp^2 (b) sp^3 **23.74** Absorption of orange light or absorption of all colors of light except blue will give a solution a blue color. **23.75**(a) The crystal field splitting energy (Δ) is the energy difference between the two sets of *d* orbitals that result from electrostatic effects of ligands on a central transition metal atom. (b) In an octahedral field of ligands, the ligands approach along the *x*, *y*, and *z* axes. The $d_{x^2-y^2}$ and d_{z^2} orbitals are located *along* the *x*, *y*, and *z* axes, so ligand interactions there are higher in energy. The other ligand interactions are lower in energy because the d_{xy}, d_{yz}, and d_{xz} orbitals are located *between* the *x*, *y*, and *z* axes. (c) In a tetrahedral field of *ligands*, the ligands do not approach along the *x*, *y*, and *z* axes. The ligand interaction is greater for the d_{xy}, d_{yz}, and d_{xz} orbitals and lesser for the $d_{x^2-y^2}$ and dz^2 orbitals. Therefore, the crystal field splitting is *reversed*, and the d_{xy}, d_{yz}, and d_{xz} orbitals are higher in energy than the $d_{x^2-y^2}$ and d_{z^2} orbitals. **23.78** If Δ is greater than $E_{pairing}$, electrons will pair their spins in the lower energy set of *d* orbitals before entering the higher energy set of *d* orbitals as unpaired electrons. If Δ is less than $E_{pairing}$, electrons will occupy the higher energy set of *d* orbitals as unpaired electrons before pairing in the lower energy set of *d* orbitals. **23.80**(a) no *d* electrons (b) eight *d* electrons (c) six *d* electrons **23.82**(a) five (b) ten (c) seven

23.84

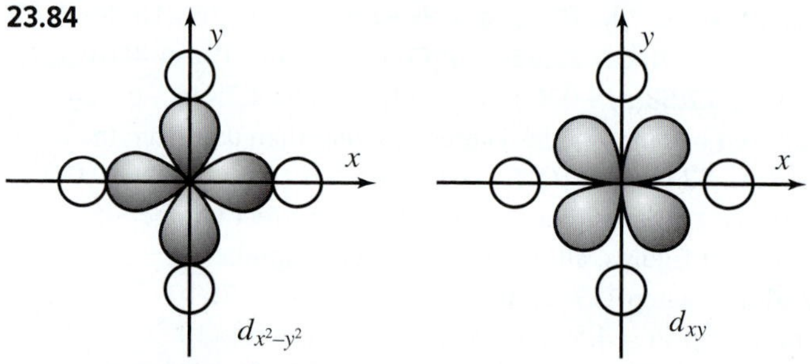

In an octahedral field of ligands, the ligands approach along the *x*, *y*, and *z* axes. The lobes of the $d_{x^2-y^2}$ orbital lie *along* the *x* and *y* axes, so ligand interaction is greater. The lobes of the d_{xy} orbital lie *between* the *x* and *y* axes, so ligand interaction is less. The greater interaction of the $d_{x^2-y^2}$ orbital results in its higher energy. **23.86** a and d

23.88(a)

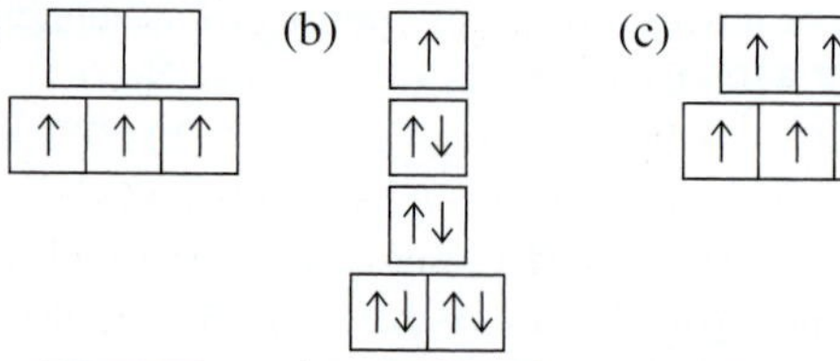

23.90(a)

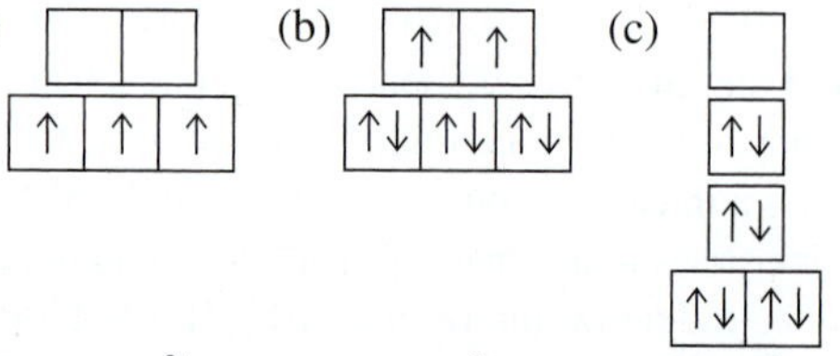

23.92 $[Cr(H_2O)_6]^{3+} < [Cr(NH_3)_6]^{3+} < [Cr(NO_2)_6]^{3-}$ **23.94** A violet complex absorbs yellow-green light. The light absorbed by a complex with a weaker field ligand would have a lower energy and a higher wavelength. Light of lower energy than yellow-green light is yellow, orange, or red. The color observed would be blue or green. **23.97** H_2O is a weak-field ligand, somewhat weaker than NH_3. A weak-field ligand results in a lower splitting energy, and the complex absorbs visible light of lower energy. The hexaaqua complex appears green because it absorbs red light. The hexaammine complex appears violet because it absorbs yellow light, which has higher energy (shorter λ) than red light. **23.101** Hg^+ is [Xe] $6s^1 4f^{14} 5d^{10}$ and Cu^+ is [Ar] $3d^{10}$. The mercury(I) ion has one electron in the 6*s* orbital that can form a covalent bond with the electron in the 6*s* orbital of another mercury(I) ion. In the copper(I) ion there are no electrons in the *s* orbital, so these ions cannot bond with one another. **23.102**(a) 6 (b) +3 (c) two (d) 1 mol

23.109 geometric (*cis-trans*) and linkage isomerism

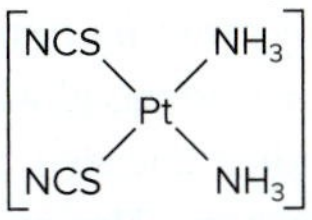

cis-diamminedithiocyanatoplatinum(II)

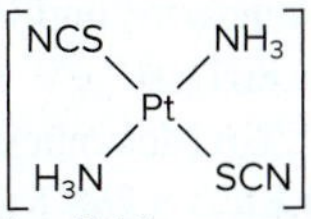

trans-diamminedithiocyanatoplatinum(II)

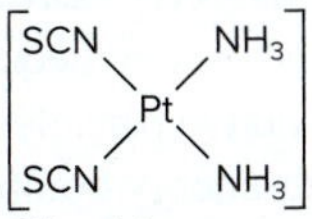

cis-diamminediisothiocyanatoplatinum(II)

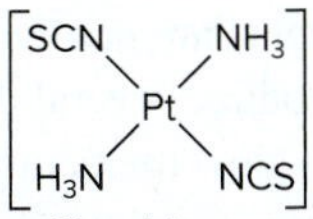

trans-diamminediisothiocyanatoplatinum(II)

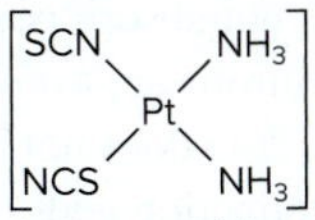

cis-diammineisothiocyanatothiocyanatoplatinum(II)

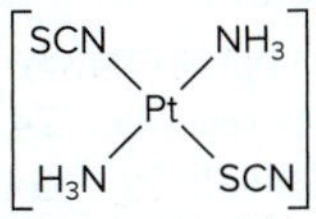

trans-diammineisothiocyanatothiocyanatoplatinum(II)

23.110 $[Co(NH_3)_4(H_2O)Cl]^{2+}$
tetraammineaquachlorocobalt(III) ion
2 geometric isomers

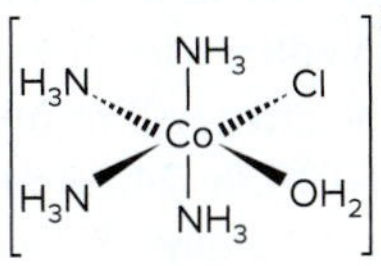

cis

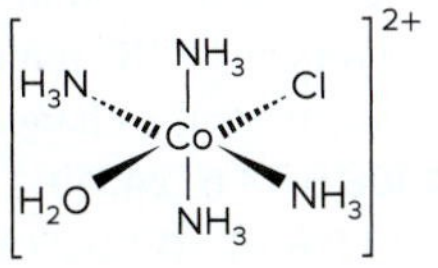

trans

$[Cr(H_2O)_3Br_2Cl]$
triaquadibromochlorochromium(III)
3 geometric isomers

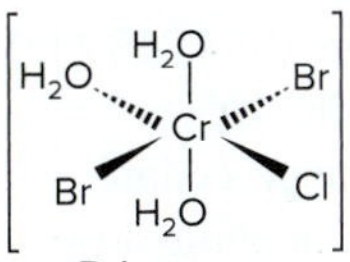

Br's are *trans*

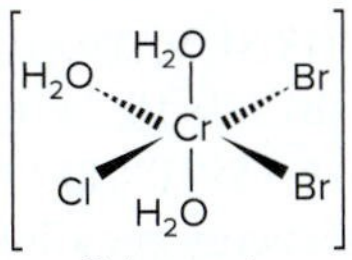

Br's are *cis*
H_2O's are *cis* and *trans*

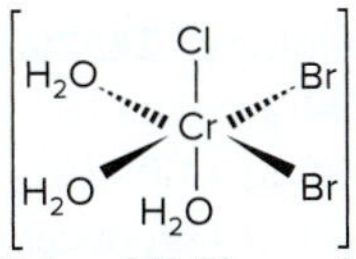

Br's and H_2O's are *cis*

$[Cr(NH_3)_2(H_2O)_2Br_2]^+$
diamminediaquadibromochromium(III) ion
6 isomers (5 geometric)

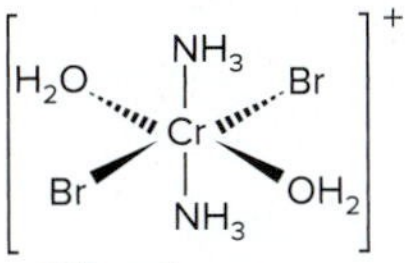

all ligands are *trans*

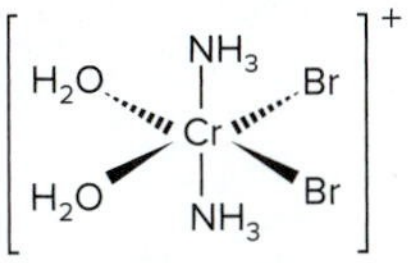

only NH_3 is *trans*

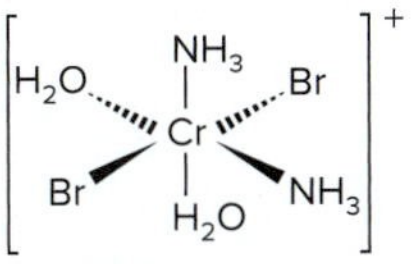

only Br is *trans*

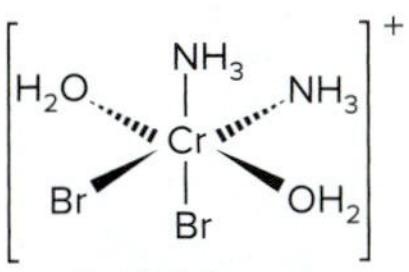

only H_2O is *trans*

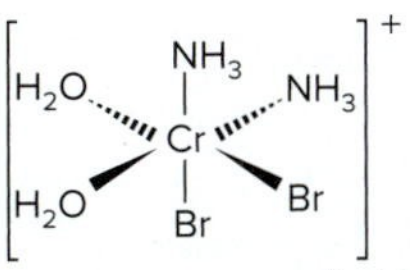

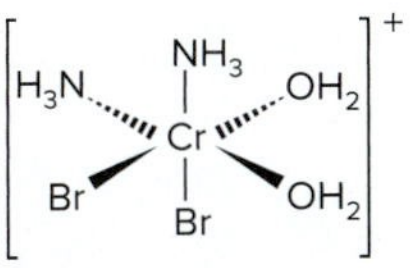

optical isomers
all three ligands are *cis*

23.115(a) no optical isomers (b) no optical isomers (c) no optical isomers (d) no optical isomers (e) optical isomers
23.116 $Pt[P(C_2H_5)_3]_2Cl_2$

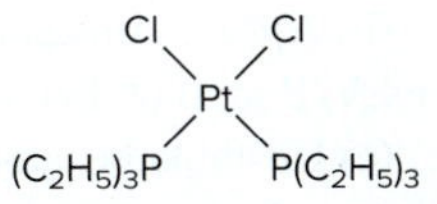

cis-dichlorobis(triethylphosphine)platinum(II)

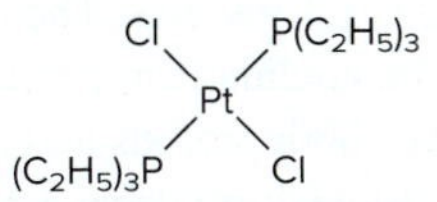

trans-dichlorobis(triethylphosphine)platinum(II)

23.118(a) The first reaction shows no change in the number of particles. In the second reaction, the number of reactant particles is greater than the number of product particles. A decrease in the number of particles means a decrease in entropy. Based on entropy change only, the first reaction is favored. (b) The ethylenediamine complex will be more stable with respect to ligand exchange in water because the entropy change for that exchange is unfavorable (negative).

Chapter 24

Answers to Boxed Reading Problems: B24.1 In the *s*-process, a nucleus captures a neutron at some point over a long period of time. Then the nucleus emits a β^- particle to form another element. The stable isotopes of most heavy elements up to ${}^{209}Bi$ form by the *s*-process. The *r*-process very quickly forms less stable isotopes and those with *A* greater than 230 by multiple neutron captures, followed by multiple β^- decays. **B24.3** The simultaneous fusion of three nuclei is a termolecular process. Termolecular processes have a very low probability of occurring. The bimolecular fusion of ${}^{8}Be$ with ${}^{4}He$ is more likely.
B24.4 ${}^{210}_{83}Bi \longrightarrow {}^{0}_{-1}\beta + {}^{210}_{84}Po$ (nuclide A); ${}^{210}_{84}Po \longrightarrow {}^{4}_{2}\alpha + {}^{206}_{82}Pb$ (nuclide B); ${}^{206}_{82}Pb + 3{}^{1}_{0}n \longrightarrow {}^{209}_{82}Pb$ (nuclide C); ${}^{209}_{82}Pb \longrightarrow {}^{0}_{-1}\beta + {}^{209}_{83}Bi$ (nuclide D)

• **24.1**(a) Chemical reactions are accompanied by relatively small changes in energy; nuclear reactions are accompanied by relatively large changes in energy. (b) Increasing temperature increases the rate of a chemical reaction but has no effect on a nuclear reaction. (c) Chemical reaction rates increase with higher reactant concentrations, while nuclear reaction rates increase with increasing number of nuclei. (d) If the reactant is limiting in a chemical reaction, then more reactant produces more product and the yield increases. The presence of more radioactive reactant results in more decay product, so a higher reactant concentration increases the yield. **24.2**(a) 95.02% (b) The atomic mass is larger than the isotopic mass of ${}^{32}S$. Sulfur-32 is the lightest isotope. **24.4**(a) *Z* down by 2, *N* down by 2 (b) *Z* up by 1, *N* down by 1 (c) no change in *Z* or *N* (d) *Z* down by 1, *N* up by 1 (e) *Z* down by 1, *N* up by 1. A different element is produced in all cases except (c). **24.6** A neutron-rich nuclide decays by β^- decay. A neutron-poor nuclide undergoes positron decay or electron capture. **24.8**(a) ${}^{234}_{92}U \longrightarrow {}^{4}_{2}\alpha + {}^{230}_{90}Th$ (b) ${}^{232}_{93}Np + {}^{0}_{-1}e \longrightarrow {}^{232}_{92}U$ (c) ${}^{12}_{7}N \longrightarrow {}^{0}_{1}\beta + {}^{12}_{6}C$
24.10(a) ${}^{27}_{12}Mg \longrightarrow {}^{0}_{-1}\beta + {}^{27}_{13}Al$ (b) ${}^{23}_{12}Mg \longrightarrow {}^{0}_{1}\beta + {}^{23}_{11}Na$ (c) ${}^{103}_{46}Pd + {}^{0}_{-1}e \longrightarrow {}^{103}_{45}Rh$ **24.12**(a) ${}^{48}_{23}V \longrightarrow {}^{48}_{22}Ti + {}^{0}_{1}\beta$ (b) ${}^{107}_{48}Cd + {}^{0}_{-1}e \longrightarrow {}^{107}_{47}Ag$ (c) ${}^{210}_{86}Rn \longrightarrow {}^{206}_{84}Po + {}^{4}_{2}\alpha$

24.14(a) $^{186}_{78}Pt + ^{0}_{-1}e \longrightarrow ^{186}_{77}Ir$ (b) $^{225}_{89}Ac \longrightarrow ^{221}_{87}Fr + ^{4}_{2}\alpha$ (c) $^{129}_{52}Te \longrightarrow ^{129}_{53}I + ^{0}_{-1}\beta$ **24.16**(a) Appears stable because its N and Z values are both magic numbers, but its N/Z ratio (1.50) is too high; it is unstable. (b) Appears unstable because its Z value is an odd number, but its N/Z ratio (1.19) is in the band of stability, so it is stable. (c) Unstable because its N/Z ratio is too high. **24.18**(a) The N/Z ratio for ^{127}I is 1.4; it is stable. (b) The N/Z ratio for ^{106}Sn is 1.1; it is unstable because this ratio is too low. (c) The N/Z ratio is 1.1 for ^{68}As. The ratio is within the range of stability, but the nuclide is most likely unstable because there are odd numbers of both protons and neutrons.
24.20(a) alpha decay (b) positron decay or electron capture (c) positron decay or electron capture **24.22**(a) β^- decay (b) positron decay or electron capture (c) alpha decay
24.24 Stability results from a favorable N/Z ratio, even N and/or Z, and the occurrence of magic numbers. The N/Z ratio of ^{52}Cr is 1.17, which is within the band of stability. The fact that Z is even does not account for the variation in stability because all isotopes of chromium have the same Z. However, ^{52}Cr has 28 neutrons, so N is both an even number and a magic number for this isotope only. **24.28** seven alpha emissions and four β^- emissions
24.31 No, it is not valid to conclude that $t_{1/2}$ equals 1 min because the number of nuclei is so small. Decay rate is an average rate and is only meaningful when the sample is macroscopic and contains a large number of nuclei. For the sample containing 6×10^{12} nuclei, the conclusion is valid. **24.33** 2.56×10^{-2} Ci/g **24.35** 1.4×10^{8} Bq/g **24.37** 1×10^{-12} day^{-1} **24.39** 2.31×10^{-7} yr^{-1} **24.41** 1.49 mg **24.43** 2.2×10^{9} yr **24.45** 30 d/min **24.47** 1.0×10^{6} yr
24.50 Neither γ radiation nor neutron beams have charge, so neither is deflected by a magnetic or electric field. Neutron beams differ from γ radiation in that a neutron has a mass approximately equal to that of a proton. It was observed that a neutron beam induces the emission of protons from a substance; γ radiation does not cause such emission. **24.52** Protons are repelled from the target nuclei due to interaction with like (positive) charges. Higher energy is required to overcome the repulsion. **24.53**(a) $^{10}_{5}B + ^{4}_{2}\alpha \longrightarrow ^{1}_{0}n + ^{13}_{7}N$ (b) $^{28}_{14}Si + ^{2}_{1}H \longrightarrow ^{1}_{0}n + ^{29}_{15}P$ (c) $^{242}_{96}Cm + ^{4}_{2}\alpha \longrightarrow 2^{1}_{0}n + ^{244}_{98}Cf$ **24.58** Ionizing radiation is more dangerous to children because they have rapidly dividing cells, so there is more chance for radiation to cause cell damage or mutation. **24.60**(a) 5.4×10^{-7} rad (b) 5.4×10^{-9} Gy
24.62(a) 7.5×10^{-10} Gy (b) 7.5×10^{-5} mrem (c) 7.5×10^{-10} Sv
24.65 1.86×10^{-3} rad **24.67** NAA does not destroy the sample, while chemical analyses do. Neutrons bombard a nonradioactive sample, inducing some atoms within the sample to be radioactive. The radioisotopes decay by emitting radiation characteristic of each isotope. **24.73** Energy is released when a nucleus forms from nucleons. The nuclear binding energy is the quantity of energy holding 1 mol of nuclei together. This energy must be absorbed to break up the nucleus into nucleons and is released when nucleons come together. **24.75**(a) 1.861×10^{4} eV (b) 2.981×10^{-15} J
24.77 7.6×10^{11} J **24.79**(a) 7.976 MeV/nucleon (b) 127.6 MeV/atom (c) 1.231×10^{10} kJ/mol **24.81**(a) 8.768 MeV/nucleon (b) 517.3 MeV/atom (c) 4.99×10^{10} kJ/mol
24.85 Radioactive decay is a spontaneous process in which unstable nuclei emit radioactive particles and energy. Fission occurs as the result of high-energy bombardment of nuclei that cause them to break into smaller nuclei, radioactive particles, and energy. All fission events are not the same. The nuclei split in a number of ways to produce several different products.
24.88 The water serves to slow the neutrons so that they are better able to cause a fission reaction. Heavy water is a better moderator because it does not absorb neutrons as well as light water does; thus, more neutrons are available to initiate the fission process. However, D_2O does not occur naturally in great abundance, so its production adds to the cost of a heavy-water reactor. **24.93**(a) 1.1×10^{-29} kg (b) 9.9×10^{-13} J (c) 5.9×10^{8} kJ/mol; this is approximately 1 million times larger than a typical heat of reaction. **24.95** 8.0×10^{3} yr
24.98 1.35×10^{-5} M **24.100** 6.2×10^{-2} **24.102**(a) 5.99 h (b) 21% **24.104**(a) 0.999 (b) 0.298 (c) 5.58×10^{-6} (d) Radiocarbon dating is more reliable for the fraction in part (b) because a significant amount of ^{14}C has decayed and a significant amount remains. Therefore, a change in the amount of ^{14}C will be noticeable. For the fraction in part (a), very little ^{14}C has decayed, and for (c) very little ^{14}C remains. In either case, it will be more difficult to measure the change, so the error will be relatively large. **24.106** 6.58 h **24.110** 4.904×10^{-9} L/h **24.113** 81 yr
24.115(a) 0.15 Bq/L (b) 0.27 Bq/L (c) 3.3 days; for a total of 12.8 days **24.118** 7.4 s **24.121** 1926 **24.124** 6.27×10^{5} eV/reaction, 6.05×10^{7} kJ/mol reaction **24.127**(a) 2.07×10^{-17} J (b) 1.45×10^{7} H atoms (c) 1.4960×10^{-5} J (d) 1.4959×10^{-5} J (e) No, the captain should continue using the current technology. **24.130**(a) 0.043 MeV, 2.9×10^{-11} m (b) 4.713 MeV
24.134(a) 3.26×10^{3} days (b) 3.2×10^{-3} s (c) 2.78×10^{11} yr
24.136(a) 1.80×10^{17} J (b) 6.15×10^{16} kJ (c) The procedure in part (b) produces more energy per kilogram of antihydrogen.
24.137 9.316×10^{2} MeV **24.141** 7.81 days

A

absolute scale (See *Kelvin scale.*)

absorption spectrum The spectrum produced when atoms absorb specific wavelengths of incoming light and become excited from lower to higher energy levels.

abundance The amount of an element in a particular region of the natural world.

accuracy The closeness of a measurement to the actual value.

acid In common laboratory terms, any species that produces H^+ ions when dissolved in water. (See also *Arrhenius, Brønsted-Lowry,* and *Lewis acid-base* definitions.)

acid anhydride A compound, usually formed by a dehydration-condensation reaction of an oxoacid, that yields two molecules of the acid when it reacts with water.

acid-base buffer (also *buffer*) A solution that resists changes in pH when a small amount of either strong acid or strong base is added.

acid-base indicator An organic molecule whose color is different in acid and in base; the color is used to monitor the equivalence point of a titration or the pH of a solution.

acid-base reaction Any reaction between an acid and a base. (See also *neutralization reaction.*)

acid-base titration curve A plot of the pH of a solution of acid (or base) versus the volume of base (or acid) added to the solution.

acid-dissociation (acid-ionization) constant (K_a) An equilibrium constant for the dissociation of an acid (HA) in H_2O to yield the conjugate base (A^-) and H_3O^+:

$$K_a = \frac{[H_3O^+][A^-]}{[HA]}$$

actinides The Period 7 elements that constitute the second inner transition series (5*f* block), which includes thorium (Th; $Z = 90$) through lawrencium (Lr; $Z = 103$).

activated complex (See *transition state.*)

activation energy (E_a) The minimum energy with which molecules must collide to react.

active site The region of an enzyme formed by specific amino acid side chains at which catalysis occurs.

activity ($\mathscr{A}$) (also *decay rate*) The change in number of nuclei ($\mathscr{N}$) of a radioactive sample divided by the change in time (t).

activity series of the metals A listing of metals arranged in order of decreasing strength of the metal as a reducing agent in aqueous reactions.

actual yield The amount of product actually obtained in a chemical reaction.

addition polymer (also *chain-reaction,* or *chain-growth, polymer*) A polymer formed when monomers (usually containing C═C) combine through an addition reaction.

addition reaction A type of organic reaction in which atoms linked by a multiple bond become bonded to more atoms.

adduct The product of a Lewis acid-base reaction, a species that contains a new covalent bond.

adenosine triphosphate (ATP) A high-energy molecule that serves most commonly as a store and source of energy in organisms.

alcohol An organic compound (ending, *-ol*) that contains a —C—Ö—H functional group.

aldehyde An organic compound (ending, *-al*) that contains the carbonyl functional group (C═Ö) in which the carbonyl C is also bonded to H.

alkane A hydrocarbon that contains only single bonds (general formula, C_nH_{2n+2}).

alkene A hydrocarbon that contains at least one C═C bond (general formula, CnH_2n).

alkyl group A saturated hydrocarbon chain with one bond available.

alkyl halide (See *haloalkane.*)

alkyne A hydrocarbon that contains at least one C≡C bond (general formula, C_nH_{2n-2}).

allotrope One of two or more crystalline or molecular forms of an element. In general, one allotrope is more stable than another at a particular pressure and temperature.

alloy A mixture with metallic properties that consists of solid phases of two or more pure elements, a solid-solid solution, or distinct intermediate phases.

alpha (α) decay A radioactive process in which an alpha particle is emitted from a nucleus.

alpha particle (α, $^4_2\alpha$, or $^4_2He^{2+}$) A positively charged particle, identical to a helium-4 nucleus, that is one of the common types of radioactive emissions.

amide An organic compound that contains the —C(═Ö:)—N̈— functional group.

amine An organic compound (general formula, —C—N̈—) derived structurally by replacing one or more H atoms of ammonia with organic groups; a weak organic base.

amino acid An organic compound [general formula, H_2N—CH(R)—COOH] with at least one carboxyl and one amine group on the same molecule; the monomer unit of a protein.

amorphous solid A solid that has a poorly defined shape because its particles do not have an orderly arrangement throughout a sample.

ampere (A) The SI unit of electric current; 1 ampere of current results when 1 coulomb of charge flows through a conductor in 1 second.

amphoteric Able to act as either an acid or a base.

amplitude The height of the crest (or depth of the trough) of a wave; related to the intensity of the energy (brightness of the light).

angular momentum quantum number (l) An integer from 0 to $n - 1$ that is related to the shape of an atomic orbital.

anion A negatively charged ion.

anode The electrode at which oxidation occurs in an electrochemical cell. Electrons are given up by the reducing agent and leave the cell at the anode.

antibonding MO A molecular orbital formed when wave functions are subtracted from each other, which decreases electron density between the nuclei and leaves a node. Electrons occupying such an orbital destabilize the molecule.

apatite A compound of general formula $Ca_5(PO_4)_3X$, where X is generally F, Cl, or OH; a source of phosphorus.

aqueous solution A solution in which water is the solvent.
aromatic hydrocarbon A compound of C and H with one or more rings of C atoms (often drawn with alternating C—C and C=C bonds), in which there is extensive delocalization of π electrons.
Arrhenius acid-base definition A model of acid-base behavior in which an acid is a substance that has H in its formula and dissociates in water to yield H_3O^+, and a base is a substance that has OH in its formula and produces OH^- in water.
Arrhenius equation An equation that expresses the exponential relationship between temperature and the rate constant: $k = Ae^{-E_a/RT}$.
atmosphere The mixture of gases that extends from a planet's surface and eventually merges with outer space; the gaseous region of Earth's crust. (For the unit, see *standard atmosphere.*)
atom The smallest unit of an element that retains the chemical nature of the element. A neutral, spherical entity composed of a positively charged central nucleus surrounded by one or more negatively charged electrons.
atomic mass (also *atomic weight*) The average of the masses of the naturally occurring isotopes of an element weighted according to their abundances.
atomic mass unit (amu) (also *dalton, Da*) A mass exactly equal to $\frac{1}{12}$ the mass of a carbon-12 atom.
atomic number (*Z*) The unique number of protons in the nucleus of each atom of an element (equal to the number of electrons in the neutral atom). An integer that expresses the positive charge of a nucleus or subatomic particle in multiples of the electronic charge.
atomic orbital A wave function for an electron that describes the region of space in which there is a high probability of finding the electron.
atomic size A measure of how closely one atom lies next to another, atomic size is determined from the distances between nuclei of adjacent atoms. (See also *metallic radius* and *covalent radius.*)
atomic solid A solid consisting of individual atoms held together by dispersion forces; the frozen noble gases are the only examples.
atomic symbol (also *element symbol*) A one- or two-letter abbreviation for the English, Latin, or Greek name of an element.
aufbau principle The conceptual approach for building up atoms by adding one proton at a time to the nucleus and one electron to the lowest energy sublevel that is available, to obtain the ground-state electron configurations of the elements.
autoionization (also *self-ionization*) A reaction in which two molecules of a substance react to give ions. The most important example is for water: $2H_2O(l) \rightleftharpoons H_3O^+(aq) + OH^-(aq)$
average rate The change in concentration of reactants (or products) divided by a finite time period.
Avogadro's law The gas law stating that, at fixed temperature and pressure, equal volumes of any ideal gas contain equal numbers of particles, and, therefore, the volume of a gas is directly proportional to its amount (mol): $V \propto n$.
Avogadro's number A number (6.022×10^{23} to four significant figures) equal to the number of atoms in exactly 12 g of carbon-12; the number of atoms, molecules, or formula units in one mole of an element or compound.
axial group An atom (or group) that lies above or below the trigonal plane of a trigonal bipyramidal molecule, or a similar structural feature in a molecule.

B

background radiation Natural ionizing radiation, the most important form of which is cosmic radiation.
balancing coefficient (also *stoichiometric coefficient*) A numerical multiplier of all the atoms in the formula immediately following it in a balanced chemical equation.
band of stability The band of stable nuclides that appears on a plot of number of neutrons vs. number of protons for all nuclides.
band theory An extension of molecular orbital (MO) theory that explains many properties of metals and other solids, in particular, the differences in conductivity of metals, metalloids, and nonmetals.
barometer A device used to measure atmospheric pressure. Most commonly, a tube open at one end, which is filled with mercury and inverted into a dish of mercury.
base In common laboratory terms, any species that produces OH^- ions when dissolved in water. (See also *Arrhenius, Brønsted-Lowry,* and *Lewis acid-base* definitions.)
base pair Two complementary mononucleotide bases that are H bonded to each other; guanine (G) always pairs with cytosine (C), and adenine (A) always pairs with thymine (T) (or uracil, U, in RNA).
base unit (also *fundamental unit*) A unit that defines the standard for one of the seven physical quantities in the International System of Units (SI).
base-dissociation (base-ionization) constant (K_b) An equilibrium constant for the reaction of a base (B) with H_2O to yield the conjugate acid (BH^+) and OH^-:

$$K_b = \frac{[BH^+][OH^-]}{[B]}$$

basic-oxygen process The method used to convert pig iron to steel, in which O_2 is blown over and through molten pig iron to oxidize impurities and decrease the content of carbon.
battery A group of voltaic cells arranged in series; primary and secondary types are self-contained, but flow batteries are not.
becquerel (Bq) The SI unit of radioactivity; 1 Bq = 1 d/s (disintegration per second).
bent shape (also *V shape*) A molecular shape that arises when a central atom is bonded to two other atoms and has one or two lone pairs; occurs as the AX_2E shape class (bond angle < 120°) in the trigonal planar arrangement and as the AX_2E_2 shape class (bond angle < 109.5°) in the tetrahedral arrangement.
beta (β) decay A radioactive change that encompasses any of three specific processes: β^- decay, β^+ emission, or e^- capture.
beta particle (β, β^-, or ${}^{0}_{-1}\beta$) A negatively charged particle, identified as a high-speed electron, that is one of the common types of radioactive emissions.
bimolecular reaction An elementary reaction involving the collision of two reactant species.
binary covalent compound A compound that consists of atoms of two elements, typically nonmetals, in which bonding occurs primarily through electron sharing.
binary ionic compound A compound that consists of the oppositely charged ions of two elements.
biomass conversion The process of applying chemical and biological methods to convert plant or animal matter into fuels.
biosphere The living systems that inhabit Earth.
blast furnace A tower-shaped furnace made of brick material in which intense heat and blasts of air are used to convert iron ore and coke to iron metal and carbon dioxide.

body-centered cubic unit cell A unit cell in which a particle lies at each corner and in the center of a cube.

boiling point (bp or T_b) The temperature at which the vapor pressure inside bubbles forming in a liquid equals the external (atmospheric) pressure.

boiling point elevation (ΔT_b) The increase in the boiling point of a solvent caused by the presence of dissolved solute.

bond angle The angle formed by the bonds joining the nuclei of two surrounding atoms to the nucleus of the central atom, which is at the vertex.

bond energy (BE) (also *bond enthalpy* or *bond strength*) The standard enthalpy change (always > 0) accompanying the breakage of a given bond in 1 mol of gaseous molecules.

bond length The distance between the nuclei of two bonded atoms.

bond order The number of electron pairs shared by two bonded atoms.

bonding MO A molecular orbital formed when wave functions are added to each other, which increases electron density between the nuclei. Electrons occupying such an orbital stabilize the molecule.

bonding pair (also *shared pair*) An electron pair shared by two nuclei; the mutual attraction between the nuclei and the electron pair forms a covalent bond.

Born-Haber cycle A series of hypothetical steps and their enthalpy changes that converts elements to an ionic compound; it is used to calculate the lattice energy.

Boyle's law The gas law stating that, at constant temperature and amount of gas, the volume occupied by a gas is inversely proportional to the applied (external) pressure: $V \propto 1/P$.

branch A side chain appended to a polymer backbone or to the longest sequence of atoms in an organic compound.

bridge bond (also *three-center, two-electron bond*) A covalent bond in which three atoms are held together by two electrons.

Brønsted-Lowry acid-base definition A model of acid-base behavior based on proton transfer, in which an acid and a base are defined, respectively, as a species that donates a proton and one that accepts a proton.

buffer (See *acid-base buffer.*)

buffer capacity A measure of the ability of a buffer to resist a change in pH; related to the total concentrations and relative proportions of buffer components.

buffer range The pH range over which a buffer acts effectively.

C

calibration The process of correcting for systematic error of a measuring device by comparing it to a known standard.

calorie (cal) A unit of energy defined as exactly 4.184 joules; originally defined as the heat needed to raise the temperature of 1 g of water 1°C (from 14.5°C to 15.5°C).

calorimeter A device used to measure the heat released or absorbed by a physical or chemical process taking place within it.

capillarity (or *capillary action*) The rising of a liquid through a narrow space against the pull of gravity.

carbon steel The steel that is produced by the basic-oxygen process, contains about 1% to 1.5% C and other impurities, and is alloyed with metals that prevent corrosion and increase strength or flexibility.

carbonyl group The C=O grouping of atoms.

carboxylic acid An organic compound (ending, *-oic acid*) that contains the 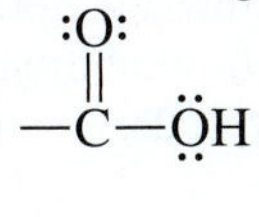 group.

catalyst A substance or mixture that increases the rate of a reaction without being used up in the process.

catenation The process by which atoms of an element bond to each other in chains; most common with carbon in organic compounds but also occurs with boron, silicon, sulfur, and several other elements.

cathode The electrode at which reduction occurs in an electrochemical cell. Electrons enter the cell and are acquired by the oxidizing agent at the cathode.

cathode ray The ray of light emitted by the cathode (negative electrode) in a gas discharge tube; travels in a straight line, unless deflected by magnetic or electric fields.

cation A positively charged ion.

cell potential (E_{cell}) (also *electromotive force,* or *emf; cell voltage*) The difference in electrical potential between the two electrodes of an electrochemical cell.

Celsius scale (formerly *centigrade scale*) A temperature scale in which the freezing and boiling points of water are defined as 0°C and 100°C, respectively.

ceramic A nonmetallic, nonpolymeric solid that is hardened by heating it to high temperatures and, in most cases, consists of silicate microcrystals suspended in a glassy cementing medium.

chain reaction In nuclear fission, a self-sustaining process in which neutrons released by splitting of one nucleus cause other nuclei to split, which releases more neutrons, and so on.

change in enthalpy (ΔH) The change in the system's internal energy plus the product of the constant pressure and the change in volume: $\Delta H = \Delta E + P\Delta V$; alternatively, the heat lost or gained at constant pressure: $\Delta H = q_P$.

charge density The ratio of the charge of an ion to its volume.

Charles's law The gas law stating that at constant pressure, the volume occupied by a fixed amount of gas is directly proportional to its absolute temperature: $V \propto T$.

chelate A complex ion in which the metal ion is bonded to a bidentate or polydentate ligand.

chemical bond The force that holds two atoms together in a molecule (or formula unit).

chemical change (also *chemical reaction*) A change in which one or more substances are converted into one or more substances with different composition and properties.

chemical equation A statement that uses chemical formulas to express the identities and quantities of the substances involved in a chemical or physical change.

chemical formula A notation of atomic symbols and numerical subscripts that shows the type and number of each atom in a molecule or formula unit of a substance.

chemical kinetics The study of the rates of reactions and the factors that affect them.

chemical property A characteristic of a substance that appears as it interacts with, or transforms into, other substances.

chemical reaction (See *chemical change.*)

chemistry The scientific study of matter and its properties, the changes it undergoes, and the energy associated with those changes.

chiral molecule One that is not superimposable on its mirror image; an optically active molecule. In organic compounds, a chiral molecule typically contains a C atom bonded to four different groups (an asymmetric carbon).

chlor-alkali process An industrial method that electrolyzes concentrated aqueous NaCl and produces Cl_2, H_2, and NaOH.

chromatography A separation technique in which a mixture is dissolved in a fluid (gas or liquid), and the components are separated through differences in adsorption to (or solubility in) a solid surface (or viscous liquid).

***cis-trans* isomers** (See *geometric isomers.*)
Clausius-Clapeyron equation An equation that expresses the linear relationship between vapor pressure P of a liquid and temperature T; in two-point form, it is

$$\ln \frac{P_2}{P_1} = \frac{-\Delta H_{vap}}{R}\left(\frac{1}{T_2} - \frac{1}{T_1}\right)$$

coal gasification An industrial process for altering the large molecules in coal to sulfur-free gaseous fuels.
colligative property A property of a solution that depends on the number, not the identity, of solute particles. (See also *boiling point elevation, freezing point depression, osmotic pressure,* and *vapor pressure lowering.*)
collision frequency The average number of collisions per second that a particle undergoes.
collision theory A model that explains reaction rate as based on the number, energy, and orientation of colliding particles.
colloid A heterogeneous mixture in which a dispersed (solute-like) substance is distributed throughout a dispersive (solvent-like) substance.
combustion The process of burning in air, often with release of heat and light.
combustion analysis A method for determining the formula of a compound from the amounts of its combustion products; used commonly for organic compounds.
common-ion effect The shift in the position of an ionic equilibrium away from an ion involved in the process that is caused by the addition or presence of that ion.
complex (See *coordination compound.*)
complex ion An ion consisting of a central metal ion covalently bonded to two or more anions or molecules, called ligands.
composition The types and amounts of simpler substances that make up a sample of matter.
compound A substance composed of two or more elements that are chemically combined in fixed proportions.
concentration A measure of the quantity of solute dissolved in a given quantity of solution (or of solvent).
concentration cell A voltaic cell in which both compartments contain the same components but at different concentrations.
condensation The process of a gas changing into a liquid.
condensation polymer A polymer formed from monomers with two functional groups that are linked together in a dehydration-condensation reaction.
conduction band In band theory, the empty, higher energy portion of the band of molecular orbitals into which electrons move when conducting heat and electricity.
conductor A substance (usually a metal) that conducts an electric current well.
conjugate acid-base pair Two species related to each other through the gain or loss of a proton; the acid has one more proton than its conjugate base.
constitutional isomers (also *structural isomers*) Compounds with the same molecular formula but different arrangements of atoms.
contact process An industrial process for the manufacture of sulfuric acid based on the catalyzed oxidation of SO_2.
controlled experiment An experiment that measures the effect of one variable at a time by keeping other variables constant.
conversion factor A ratio of equivalent quantities that is equal to 1 and used to express a quantity in different units.
coordinate covalent bond A covalent bond formed when one atom donates both electrons to provide the shared pair; once formed, this bond is identical to any covalent single bond.
coordination compound (also *complex*) A substance containing at least one complex ion and counter ion(s).
coordination isomers Two or more coordination compounds with the same composition in which the complex ions have different ligand arrangements.
coordination number In a crystal, the number of nearest neighbors surrounding a particle. In a complex ion, the number of ligand atoms bonded to the central metal ion.
copolymer A polymer that consists of two or more types of monomer.
core The dense, innermost region of Earth.
core electrons (See *inner electrons.*)
corrosion The natural redox process that results in unwanted oxidation of a metal.
coulomb (C) The SI unit of electric charge. One coulomb is the charge of 6.242×10^{18} electrons; one electron possesses a charge of 1.602×10^{-19} C.
Coulomb's law A law stating that the electrostatic energy between particles A and B is directly proportional to the product of their charges and inversely proportional to the distance between them:

$$\text{electrostatic energy} \propto \frac{\text{charge A} \times \text{charge B}}{\text{distance}}$$

counter ion A simple ion associated with a complex ion in a coordination compound.
coupling of reactions The pairing of reactions of which one releases enough free energy for the other to occur.
covalent bond A type of bond in which atoms are bonded through the sharing of electrons; the mutual attraction of the nuclei and an electron pair that holds atoms together in a molecule.
covalent bonding The idealized bonding type that is based on localized electron-pair sharing between two atoms with little difference in their tendencies to lose or gain electrons (most commonly nonmetals).
covalent compound A compound that consists of atoms bonded together by shared electron pairs.
covalent radius One-half the shortest distance between nuclei of identical covalently bonded atoms.
critical mass The minimum mass of a fissionable substance needed to achieve a chain reaction.
critical point The point on a phase diagram above which the vapor cannot be condensed to a liquid; the end of the liquid-gas curve.
crosslink A branch that covalently joins one polymer chain to another.
crust The thin, light, heterogeneous outer layer of Earth, which consists of gasous, liquid, and solid regions.
crystal defect Any of a variety of disruptions in the regularity of a crystal structure.
crystal field splitting energy (Δ) The difference in energy between two sets of metal-ion *d* orbitals that results from electrostatic interactions with the surrounding ligands.
crystal field theory A model that explains the color and magnetism of complex ions and their coordination compounds based on the effects of ligands on metal-ion *d*-orbital energies.
crystalline solid Solid with a well-defined shape because of the orderly arrangement of the atoms, molecules, or ions.
crystallization A technique used to separate and purify the components of a mixture through differences in solubility, resulting in a component coming out of solution as crystals.

cubic closest packing A crystal structure based on the face-centered cubic unit cell in which the layers have an *abcabc. . .* pattern.
cubic meter (m^3) The derived SI unit of volume.
curie (Ci) The most common unit of radioactivity, originally defined as the number of nuclei disintegrating each second in 1 g of radium-226; now a fixed quantity, 1 Ci = 3.70×10^{10} d/s (disintegrations per second).
cyclic hydrocarbon A hydrocarbon with one or more rings in its structure.

D

***d* orbital** An atomic orbital with $l = 2$.
dalton (Da) A unit of mass identical to *atomic mass unit (amu).*
Dalton's law of partial pressures A gas law stating that, in a mixture of unreacting gases, the total pressure is the sum of the partial pressures of the individual gases: $P_{total} = P_1 + P_2 + P_3 + \cdots$.
data Pieces of quantitative information obtained by observation.
de Broglie wavelength The wavelength of a moving particle obtained from the de Broglie equation: $\lambda = h/mu$.
decay constant The rate constant k for radioactive decay.
decay series (also *disintegration series*) The succession of steps a parent nuclide undergoes as it decays into a stable daughter nuclide.
degree of polymerization (*n*) The number of repeat units in a polymer chain.
dehydration-condensation reaction A reaction in which an H_2O molecule is lost for every pair of groups that join.
delocalization (See *electron-pair delocalization.*)
density (*d*) An intensive physical property of a substance at a given temperature and pressure, defined as the ratio of the mass to the volume: $d = m/V$.
deposition The process of changing directly from a gas to a solid.
derived unit Any of various combinations of the seven SI base units.
desalination A process used to remove large amounts of ions from seawater, usually by reverse osmosis.
deuterons Nuclei of the stable hydrogen isotope deuterium, ^2H.
diagonal relationship Physical and chemical similarities between a Period 2 element and one located diagonally down and to the right in Period 3.
diamagnetism The tendency of a species not to be attracted (or to be slightly repelled) by a magnetic field as a result of its electrons being paired.
diastereomers (See *geometric isomers.*)
differentiation The geochemical process by which regions in Earth were formed based on differences in composition and density.
diffraction The phenomenon in which a wave striking the edge of an object bends around it. A wave passing through a slit as wide as its wavelength forms a circular wave.
diffusion The movement of one fluid through another.
dimensional analysis (also *factor-label method*) A calculation method in which arithmetic steps are accompanied by canceling units that represent physical dimensions.
dipole moment (μ) A measure of molecular polarity; the magnitude of the partial charges on the ends of a molecule (in coulombs) times the distance between them (in meters).
dipole-dipole force The intermolecular attraction between oppositely charged poles of nearby polar molecules.
dipole–induced dipole force The intermolecular attraction between a polar molecule and the oppositely charged pole it induces in a nearby molecule.
disaccharide An organic compound formed by a dehydration-condensation reaction between two simple sugars (monosaccharides).
disintegration series (See *decay series.*)
dispersion force (also *London force*) The intermolecular attraction between all particles as a result of instantaneous polarizations of their electron clouds; the intermolecular force primarily responsible for the condensed states of nonpolar substances.
disproportionation reaction A redox reaction in which one substance acts as both the oxidizing and reducing agents.
distillation A separation technique in which a more volatile component of a mixture vaporizes and condenses separately from the less volatile components.
donor atom An atom that donates a lone pair of electrons to form a covalent bond, usually from ligand to metal ion in a complex ion.
doping Adding small amounts of other elements into the crystal structure of a semiconductor to enhance a specific property, usually conductivity.
double bond A covalent bond that consists of two bonding pairs; two atoms sharing four electrons in the form of one σ and one π bond.
double helix The two intertwined polynucleotide strands held together by H bonds that form the structure of DNA (deoxyribonucleic acid).
Downs cell An industrial apparatus that electrolyzes molten NaCl to produce sodium and chlorine.
dynamic equilibrium The condition at which the forward and reverse reactions are taking place at the same rate, so there is no net change in the amounts of reactants or products.

E

effective collision A collision in which the particles meet with sufficient energy and an orientation that allows them to react.
effective nuclear charge (Z_{eff}) The nuclear charge an electron actually experiences as a result of shielding effects due to the presence of other electrons.
effusion The process by which a gas escapes from its container through a tiny hole into an evacuated space.
e_g orbitals The set of orbitals (composed of $d_{x^2-y^2}$ and d_{z^2}) that results when the energies of the metal-ion d orbitals are split by a ligand field. This set is higher in energy than the other (t_{2g}) set in an octahedral field of ligands and lower in energy in a tetrahedral field.
elastomer A polymeric material that can be stretched and springs back to its original shape when released.
electrochemical cell A system that incorporates a redox reaction to produce or use electrical energy.
electrochemistry The study of the relationship between chemical change and electrical work.
electrode The part of an electrochemical cell that conducts the electricity between the cell and the surroundings.
electrolysis The nonspontaneous lysing (splitting) of a substance, often to its component elements, by the input of electrical energy.
electrolyte A substance that conducts a current when it dissolves in water. A mixture of ions, in which the electrodes of an electrochemical cell are immersed, that conducts a current.
electrolytic cell An electrochemical system that uses electrical energy to drive a nonspontaneous chemical reaction ($\Delta G > 0$).
electromagnetic radiation (also *electromagnetic energy* or *radiant energy*) Oscillating, perpendicular electric and magnetic fields moving simultaneously through space as waves and manifested as visible light, x-rays, microwaves, radio waves, and so on.
electromagnetic spectrum The continuum of radiant energy arranged in order of increasing wavelength.

electromotive force (emf) (See *cell potential.*)
electron (e^-) capture (EC) A type of radioactive decay in which a nucleus draws in an orbital electron, usually one from the lowest energy level, and releases energy.
electron (e^-) A subatomic particle that possesses a unit negative charge (-1.60218×10^{-19} C) and occupies the space around the atomic nucleus.
electron affinity (EA) The energy change (in kJ) accompanying the addition of 1 mol of electrons to 1 mol of gaseous atoms or ions.
electron cloud depiction An imaginary representation of an electron's rapidly changing position around the nucleus over time.
electron configuration The distribution of electrons within the levels and sublevels of the atoms of an element; also the notation for such a distribution.
electron deficient Referring to a bonded atom, such as Be or B, that has fewer than eight valence electrons.
electron density diagram (also *electron probability density diagram*) The pictorial representation for a given energy sublevel of the quantity ψ^2 (the probability density of the electron lying within a particular tiny volume) as a function of r (distance from the nucleus).
electron volt (eV) The energy (in joules, J) that an electron acquires when it moves through a potential difference of 1 volt; 1 eV = 1.602×10^{-19} J.
electron-pair delocalization (also *delocalization*) The process by which electron density is spread over several atoms rather than remaining between two.
electron-sea model A qualitative description of metallic bonding proposing that metal atoms pool their valence electrons in a delocalized "sea" of electrons in which the metal ions (nuclei and core electrons) are submerged in an orderly array.
electronegativity (EN) The relative ability of a bonded atom to attract shared electrons.
electronegativity difference (ΔEN) The difference in electronegativities between two bonded atoms.
electrorefining An industrial, electrolytic purification process in which a sample of impure metal acts as the anode and a sample of the pure metal acts as the cathode.
element The simplest type of substance with unique physical and chemical properties. An element consists of only one kind of atom, so it cannot be broken down into simpler substances.
element symbol (See *atomic symbol.*)
elementary reaction (also *elementary step*) A simple reaction that describes a single molecular event in a proposed reaction mechanism.
elimination reaction A type of organic reaction in which C atoms are bonded to fewer atoms in the product than in the reactant, which creates a multiple bond between the C atoms.
emission spectrum The line spectrum produced when excited atoms return to lower energy levels and emit photons characteristic of the element.
empirical formula A chemical formula that shows the lowest relative number of atoms of each element in a compound.
enantiomers (See *optical isomers.*)
end point The point in a titration at which the indicator changes color permanently.
endothermic process A process that occurs with an absorption of heat from the surroundings and therefore an increase in the enthalpy of the system ($\Delta H > 0$).
energy The ability to do work, that is, to move matter. [See also *kinetic energy* (E_k) and *potential energy* (E_p).]
enthalpy (*H*) A thermodynamic quantity that is equal to the system's internal energy plus the product of the pressure and volume.
enthalpy diagram A graphic depiction of the enthalpy change of a system.
Enthalpy of formation (ΔH_f) (also *heat of formation*) The enthalpy change occurring when 1 mol of a compound forms from its elements. When all components are in their standard states, this is called the standard enthalpy of formation (ΔH°_f).
Enthalpy of fusion (See *heat of fusion.*)
Enthalpy of hydration (See *heat of hydration.*)
Enthalpy of reaction (ΔH_{rxn}) (also *heat of reaction*) The enthalpy change that occurs during a reaction. When all components are in their standard states, this is called the *standard enthalpy of reaction* (ΔH°_{rxn}).
Enthalpy of solution (See *heat of solution.*)
Enthalpy of sublimation (See *heat of sublimation.*)
Enthalpy of vaporization (See *heat of vaporization.*)
entropy (*S*) A thermodynamic quantity related to the number of ways the energy of a system can be dispersed through the motions of its particles.
environmental cycle The physical, chemical, and biological paths through which the atoms of an element move within Earth's crust.
enzyme A biological macromolecule (usually a protein) that acts as a catalyst.
enzyme-substrate complex (ES) The intermediate in an enzyme-catalyzed reaction, which consists of enzyme and substrate(s) and whose concentration determines the rate of product formation.
equatorial group An atom (or group) that lies in the trigonal plane of a trigonal bipyramidal molecule, or a similar structural feature in a molecule.
equilibrium constant (*K*) The value obtained when equilibrium concentrations are substituted into the reaction quotient.
equilibrium vapor pressure (See *vapor pressure.*)
equivalence point The point in a titration when the number of moles of the added species is stoichiometrically equivalent to the original number of moles of the other species.
ester An organic compound that contains the 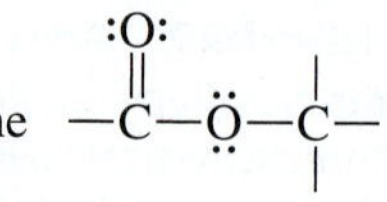 group.
exact number A quantity, usually obtained by counting or based on a unit definition, that has no uncertainty associated with it and, therefore, contains as many significant figures as a calculation requires.
exchange reaction (See *metathesis reaction.*)
excited state Any electron configuration of an atom (or ion or molecule) other than the lowest energy (ground) state.
exclusion principle A principle developed by Wolfgang Pauli stating that no two electrons in an atom can have the same set of four quantum numbers. The principle arises from the fact that an orbital has a maximum occupancy of two electrons and their spins are paired.
exothermic process A process that occurs with a release of heat to the surroundings and therefore a decrease in the enthalpy of the system ($\Delta H < 0$).
expanded valence shell A valence level that can accommodate more than eight electrons by using available *d* orbitals; occurs only with central nonmetal atoms from Period 3 or higher.
experiment A set of procedural steps that tests a hypothesis.
extensive property A property, such as mass, that depends on the quantity of substance present.

F

***f* orbital** An atomic orbital with $l = 3$.
face-centered cubic unit cell A unit cell in which a particle occurs at each corner and in the center of each face of a cube.

Faraday constant (*F*) The physical constant representing the charge of 1 mol of electrons: $F = 96{,}485$ C/mol e^-.
fatty acid A carboxylic acid that has a long hydrocarbon chain and is derived from a natural source.
filtration A method of separating the components of a mixture on the basis of differences in particle size.
first law of thermodynamics (See *law of conservation of energy.*)
fission The process by which a heavier nucleus splits into two much lighter nuclei, with the release of a large quantity of energy.
fixation A chemical or biochemical process that converts a gaseous substance in the environment into a condensed form that can be used by organisms.
flame test A procedure for identifying the presence of metal ions in which a granule of a compound or a drop of its solution is placed in a flame to observe a characteristic color.
flotation A metallurgical process in which an oil-detergent mixture is stirred with pulverized ore in water to create a slurry, followed by rapid mixing of the slurry with air to produce an oily, mineral-rich froth that floats, separating the mineral from the gangue.
formal charge The hypothetical charge on an atom in a molecule or ion, equal to the number of valence electrons minus the sum of all the unshared and half the shared valence electrons.
formation constant (K_f) An equilibrium constant for the formation of a complex ion from the hydrated metal ion and ligands.
formation equation An equation in which 1 mole of a compound forms from its elements.
formula mass The sum (in amu) of the atomic masses of a formula unit of a (usually ionic) compound.
formula unit The chemical unit of a compound that contains the relative numbers of the types of atoms or ions expressed in the chemical formula.
fossil fuel Any fuel, including coal, petroleum, and natural gas, derived from the products of the decay of dead organisms.
fraction by mass (also *mass fraction*) The portion of a compound's mass contributed by an element; the mass of an element in a compound divided by the mass of the compound.
fractional distillation A physical process involving numerous vaporization-condensation steps used to separate two or more volatile components.
free energy (*G*) A thermodynamic quantity that is the difference between the system's enthalpy and the product of the absolute temperature and the system's entropy: $G = H - TS$.
free energy change (ΔG) The change in free energy that occurs during a reaction.
free radical A molecular or atomic species with one or more unpaired electrons, which typically make it very reactive.
freezing The process of cooling a liquid until it solidifies.
freezing point depression (ΔT_f) The lowering of the freezing point of a solvent caused by the presence of dissolved solute particles.
frequency (ν) The number of complete waves, or cycles, that pass a given point per second, expressed in units of 1/second, or s^{-1} [also called *hertz* (Hz)]; related inversely to wavelength.
frequency factor (*A*) The product of the collision frequency *Z* and an orientation probability factor *p* that is specific for a reaction.
fuel cell (also *flow battery*) A battery that is not self-contained and in which electricity is generated by the controlled oxidation of a fuel.
functional group A specific combination of bonded atoms, typically containing a carbon-carbon multiple bond and/or a carbon-heteroatom bond, that reacts in a characteristic way no matter what molecule it occurs in.
fundamental unit (See *base unit.*)
fusion (See *melting.*)
fusion (nuclear) The process by which light nuclei combine to form a heavier nucleus with the release of energy.

G

galvanic cell (See *voltaic cell.*)
gamma emission The type of radioactive decay in which gamma rays are emitted from an excited nucleus.
gamma ray (γ) A very high-energy photon.
gangue In an ore, the debris, such as sand, rock, and clay, attached to the mineral.
gas One of the three states of matter. A gas fills its container regardless of the shape because its particles are far apart.
Geiger-Müller counter An ionization counter that detects radioactive emissions through their ionization of gas atoms within the instrument.
genetic code The set of three-base sequences that is translated into specific amino acids during the process of protein synthesis.
geometric isomers (also *cis-trans isomers* or *diastereomers*) Stereoisomers in which the molecules have the same connections between atoms but differ in the spatial arrangements of the atoms. The *cis* isomer has identical groups on the same side of a double bond (or of a central metal atom); the *trans* isomer has them on opposite sides.
Graham's law of effusion A gas law stating that the rate of effusion of a gas is inversely proportional to the square root of its density (or molar mass):

$$\text{rate} \propto \frac{1}{\sqrt{\mathcal{M}}}$$

gray (Gy) The SI unit of absorbed radiation dose; 1 Gy = 1 J/kg tissue.
green chemistry Field that is focused on developing methods to synthesize compounds efficiently and reduce or prevent the release of harmful products into the environment.
ground state The electron configuration of an atom (or ion or molecule) that is lowest in energy.
group A vertical column in the periodic table; elements in a group usually have the same outer electron configuration and, thus, similar chemical behavior.

H

H bond (See *hydrogen bond.*)
Haber process An industrial process used to form ammonia from its elements.
half-cell A portion of an electrochemical cell in which a half-reaction takes place.
half-life ($t_{1/2}$) In chemical processes, the time required for the reactant concentration to reach half of its initial value. In nuclear processes, the time required for half the initial number of nuclei in a sample to decay.
half-reaction method A method of balancing redox reactions by treating the oxidation and reduction half-reactions separately.
haloalkane (also *alkyl halide*) A hydrocarbon with one or more halogen atoms (X) in place of H; contains a $-\overset{|}{\underset{|}{\mathrm{C}}}-\ddot{\underset{..}{\mathrm{X}}}\!:$ group.
hard water Water that contains large amounts of divalent cations, especially Ca^{2+} and Mg^{2+}.
heat (*q*) (also *thermal energy*) The energy transferred between objects because of a difference in their temperatures only.
heat capacity (*C*) The quantity of heat required to change the temperature of an object by 1 K.

heat of formation (See *enthalpy of formation.*)
heat of fusion (ΔH°_{fus}) (also *enthalpy of fusion*) The enthalpy change occurring when 1 mol of a solid substance melts; designated (ΔH°_{fus}) at the standard state.
heat of hydration (ΔH_{hydr}) (also *enthalpy of hydration*) The enthalpy change occurring when 1 mol of a gaseous species (often an ion) is hydrated. The sum of the enthalpies from separating water molecules and mixing the gaseous species with them; designated ΔH°_{hydr} at the standard state.
heat of reaction (ΔH_{rxn}) (See *enthalpy of reaction.*)
heat of solution (ΔH_{soln}) (also *enthalpy of hydration*) The enthalpy change occurring when a solution forms from solute and solvent. The sum of the enthalpies from separating solute and solvent substances and mixing them; designated (ΔH°_{soln}) at the standard state.
heat of sublimation (ΔH°_{subl}) (also *enthalpy of sublimation*) The enthalpy change occurring when 1 mol of a solid substance changes directly to a gas. The sum of the heats of fusion and vaporization; designated at the standard state.
heat of vaporization (ΔH°_{vap}) (also *enthalpy of vaporization*) The enthalpy change occurring when 1 mol of a liquid substance vaporizes; designated (ΔH°_{vap}) at the standard state.
heating-cooling curve A plot of temperature vs. time for a sample when heat is absorbed or released at a constant rate.
Henderson-Hasselbalch equation An equation for calculating the pH of a buffer system:

$$\text{pH} = \text{p}K_a + \log\left(\frac{[\text{base}]}{[\text{acid}]}\right)$$

Henry's law A law stating that the solubility of a gas in a liquid is directly proportional to the partial pressure of the gas above the liquid: $S_{gas} = k_H \times P_{gas}$.
Hess's law A law stating that the enthalpy change of an overall process is the sum of the enthalpy changes of the individual steps.
heteroatom Any atom in an organic compound other than C or H.
heterogeneous catalyst A catalyst that occurs in a different phase from the reactants, usually a solid interacting with gaseous or liquid reactants.
heterogeneous mixture A mixture that has one or more visible boundaries among its components.
hexagonal closest packing A crystal structure based on the hexagonal unit cell in which the layers have an *abab*. . . pattern.
high-spin complex Complex ion that has the same number of unpaired electrons as in the isolated metal ion; contains weak-field ligands.
homogeneous catalyst A catalyst (gas, liquid, or soluble solid) that exists in the same phase as the reactants.
homogeneous mixture (also *solution*) A mixture that has no visible boundaries among its components.
homologous series A series of organic compounds in which each member differs from the next by a —CH_2— (methylene) group.
homonuclear diatomic molecule A molecule composed of two identical atoms.
Hund's rule A principle stating that when orbitals of equal energy are available, the electron configuration of lowest energy has the maximum number of unpaired electrons with parallel spins.
hybrid orbital An atomic orbital postulated to form during bonding by the mathematical mixing of specific combinations of nonequivalent orbitals in a given atom.
hybridization A postulated process of orbital mixing to form hybrid orbitals.
hydrate A compound in which a specific number of water molecules are associated with each formula unit.
hydration Solvation in water.
hydration shell The oriented cluster of water molecules that surrounds an ion in aqueous solution.
hydrocarbon An organic compound that contains only H and C atoms.
hydrogen bond (H bond) A type of dipole-dipole force that arises between molecules that have an H atom bonded to a small, highly electronegative atom with lone pairs, usually N, O, or F.
hydrogenation The addition of hydrogen to a carbon-carbon multiple bond to form a carbon-carbon single bond.
hydrolysis Cleaving a molecule by reaction with water; one part of the molecule bonds to water's —OH and the other to water's other H.
hydronium ion (H_3O^+) A proton covalently bonded to a water molecule.
hydrosphere The liquid region of Earth's crust.
hypothesis A testable proposal made to explain an observation. If inconsistent with experimental results, a hypothesis is revised or discarded.

I

ideal gas A hypothetical gas that exhibits linear relationships among volume, pressure, temperature, and amount (mol) at all conditions; approximated by simple gases at ordinary conditions.
ideal gas law (also *ideal gas equation*) An equation that expresses the relationships among volume, pressure, temperature, and amount (mol) of an ideal gas: $PV = nRT$.
ideal solution A solution that follows Raoult's law at any concentration.
indicator (See *acid-base indicator.*)
induced-fit model A model of enzyme action that pictures the binding of the substrate as inducing the active site to change its shape and become catalytically active.
infrared (IR) The region of the electromagnetic spectrum between the microwave and visible regions.
infrared (IR) spectroscopy An instrumental technique for determining the types of bonds in a covalent molecule by measuring the absorption of IR radiation.
initial rate The instantaneous rate at the moment the reactants are mixed, that is, at $t = 0$.
inner electrons (also *core electrons*) Electrons that fill all the energy levels of an atom except the valence level; electrons also present in atoms of the previous noble gas and any completed transition series.
inner transition elements The elements of the periodic table in which the seven inner *f* orbitals are being filled; the lanthanides and the actinides.
instantaneous rate The reaction rate at a particular time, given by the slope of a tangent to a plot of reactant concentration vs. time.
insulator A substance (usually a nonmetal) that does not conduct an electric current.
integrated rate law A mathematical expression for reactant concentration as a function of time.
intensive property A property, such as density, that does not depend on the quantity of substance present.
interhalogen compound A compound consisting entirely of halogens.
intermolecular forces (also *interparticle forces*) The attractive and repulsive forces among the particles—molecules, atoms, or ions—in a sample of matter.

internal energy (*E*) The sum of the kinetic and potential energies of all the particles in a system.
ion A charged particle that forms from an atom (or covalently bonded group of atoms) when it gains or loses one or more electrons.
ion exchange A process of softening water by exchanging one type of ion (usually Ca^{2+}) for another (usually Na^+) by binding the ions on a specially designed resin.
ion pair A gaseous ionic molecule, formed when an ionic compound vaporizes.
ion-dipole force The intermolecular attractive force between an ion and a polar molecule (dipole).
ion-product constant for water (K_w) The equilibrium constant for the autoionization of water; equal to 1.0×10^{-14} at 298 K. $K_w = [H_3O^+][OH^-]$
ionic atmosphere A cluster of ions of net opposite charge surrounding a given ion in solution.
ionic bonding The idealized type of bonding based on the attraction of oppositely charged ions that arise through electron transfer between atoms with large differences in their tendencies to lose or gain electrons (typically metals and nonmetals).
ionic compound A compound that consists of oppositely charged ions.
ionic radius The size of an ion as measured by the distance between the nuclei of adjacent ions in a crystalline ionic compound.
ionic solid A solid whose unit cell contains cations and anions.
ionization In nuclear chemistry, the process by which an atom absorbs energy from a high-energy radioactive particle and loses an electron to become ionized.
ionization energy (IE) The energy (in kJ) required for complete removal of 1 mol of electrons from 1 mol of gaseous atoms or ions.
ionizing radiation The high-energy radiation from natural and artificial sources that forms ions in a substance by causing electron loss.
ion–induced dipole force The attractive force between an ion and the dipole it induces in the electron cloud of a nearby nonpolar molecule.
isoelectronic Having the same number and configuration of electrons as another species.
isomer One of two or more compounds with the same molecular formula but different properties, often as a result of different arrangements of atoms.
isothermal process A process that takes place at constant temperature.
isotopes Atoms of a given atomic number (that is, of a specific element) that have different numbers of neutrons and therefore different mass numbers.
isotopic mass The mass (in amu) of an isotope relative to the mass of carbon-12.

J

joule (J) The SI unit of energy; $1\ J = 1\ kg\cdot m^2/s^2$.

K

kelvin (K) The SI base unit of temperature. The kelvin is the same size as the Celsius degree.
Kelvin scale (also *absolute scale*) The preferred temperature scale in scientific work, which has absolute zero (0 K, or −273.15°C) as the lowest temperature.
ketone An organic compound (ending, *-one*) that contains a carbonyl group bonded to two other C atoms, 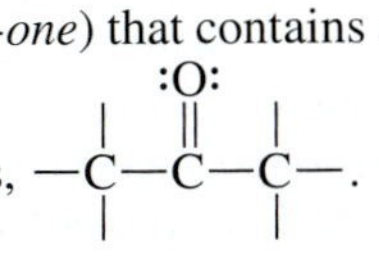.
kilogram (kg) The SI base unit of mass.
kinetic energy (E_k) The energy an object has because of its motion.
kinetic-molecular theory The model that explains macroscopic gas behavior in terms of particles in random motion whose volumes and interactions are negligible.

L

lanthanide contraction The additional decrease in atomic and ionic size, beyond the expected trend, caused by the poor shielding of the increasing nuclear charge by *f* electrons in the elements following the lanthanides.
lanthanides (also *rare earth elements*) The Period 6 (4*f*) series of inner transition elements, which includes cerium (Ce; $Z = 58$) through lutetium (Lu; $Z = 71$).
lattice The three-dimensional arrangement of points created by choosing each point to be at the same location within each particle of a crystal; thus, the lattice consists of all points with identical surroundings.
lattice energy ($\Delta H^\circ_{lattice}$) The enthalpy change (always positive) that accompanies the separation of 1 mol of a solid ionic compound into gaseous ions.
law (See *natural law.*)
law of chemical equilibrium (also *law of mass action*) The law stating that when a system reaches equilibrium at a given temperature, the ratio of quantities that make up the reaction quotient has a constant numerical value.
law of conservation of energy (also *first law of thermodynamics*) A basic observation that the total energy of the universe is constant; thus, $\Delta E_{universe} = \Delta E_{system} + \Delta E_{surroundings} = 0$.
law of definite (or constant) composition A mass law stating that, no matter what its source, a particular compound is composed of the same elements in the same parts (fractions) by mass.
law of mass action (See *law of chemical equilibrium.*)
law of mass conservation A mass law stating that the total mass of substances does not change during a chemical reaction.
law of multiple proportions A mass law stating that if elements A and B react to form two or more compounds, the different masses of B that combine with a fixed mass of A can be expressed as a ratio of small whole numbers.
Le Châtelier's principle A principle stating that if a system in a state of equilibrium is disturbed, it will undergo a change that shifts its equilibrium position in a direction that reduces the effect of the disturbance.
leaching A hydrometallurgical process that extracts a metal selectively, usually through formation of a complex ion.
level (also *shell*) A specific energy state of an atom given by the principal quantum number *n*.
leveling effect The inability of a solvent to distinguish the strength of an acid (or base) that is stronger than the conjugate acid (or conjugate base) of the solvent.
Lewis acid-base definition A model of acid-base behavior in which acids and bases are defined, respectively, as species that accept and donate an electron pair.
Lewis electron-dot symbol A notation in which the element symbol represents the nucleus and inner electrons and surrounding dots represent the valence electrons.
Lewis structure (also *Lewis formula*) A structural formula consisting of electron-dot symbols, with lines as bonding pairs and dots as lone pairs.
ligand A molecule or anion bonded to a central metal ion in a complex ion.

like-dissolves-like rule An empirical observation stating that substances having similar kinds of intermolecular forces dissolve in each other.

like-dissolves-like rule An empirical observation stating that substances having similar kinds of intermolecular forces dissolve in each other.

limiting reactant (also *limiting reagent*) The reactant that is consumed when a reaction occurs and, therefore, the one that determines the maximum amount of product that can form.

line spectrum A series of separated lines of different colors representing photons whose wavelengths are characteristic of an element. (See also *emission spectrum.*)

linear arrangement The geometric arrangement obtained when two electron groups maximize their separation around a central atom.

linear shape A molecular shape formed by three atoms lying in a straight line, with a bond angle of 180° (shape class AX_2 or AX_2E_3).

linkage isomers Coordination compounds in which the complex ions have the same composition but with different ligand donor atoms linked to the central metal ion.

lipid Any of a class of biomolecules, including fats, that are soluble in nonpolar solvents and not soluble in water.

lipid bilayer An extended sheetlike double layer of phospholipid molecules that forms in water and has the charged heads of the molecules on the surfaces of the bilayer and the nonpolar tails within the interior.

liquid One of the three states of matter. A liquid fills a container to the extent of its own volume and thus forms a surface.

liquid crystal A substance that flows like a liquid but packs like a crystalline solid at the molecular level.

liter (L) A non-SI unit of volume equivalent to 1 cubic decimeter ($0.001\ m^3$).

lithosphere The solid region of Earth's crust.

lock-and-key model A model of enzyme function that pictures the enzyme active site and the substrate as rigid shapes that fit together as a lock and key, respectively.

London force (See *dispersion force.*)

lone pair (also *unshared pair*) An electron pair that is part of an atom's valence level but not involved in covalent bonding.

low-spin complex Complex ion that has fewer unpaired electrons than in the free metal ion because of the presence of strong-field ligands.

M

macromolecule (See *polymer.*)

magnetic quantum number (m_l) An integer from $-l$ through 0 to $+l$ that specifies the orientation of an atomic orbital in the three-dimensional space about the nucleus.

manometer A device used to measure the pressure of a gas in a laboratory experiment.

mantle A thick homogeneous layer of Earth's internal structure that lies between the core and the crust.

mass The quantity of matter an object contains. Balances are designed to measure mass.

mass-action expression (See *reaction quotient.*)

mass fraction (See *fraction by mass.*)

mass number (*A*) The total number of protons and neutrons in the nucleus of an atom.

mass percent (also *mass %* or *percent by mass*) The fraction by mass expressed as a percentage. (45) A concentration term [% (w/w)] expressed as the mass of solute dissolved in 100. parts by mass of solution.

mass spectrometry An instrumental method for measuring the relative masses of particles in a sample by creating charged particles and separating them according to their mass/charge ratios.

matter Anything that possesses mass and occupies volume.

mean free path The average distance a particle travels between collisions at a given temperature and pressure.

melting (also *fusion*) The change of a substance from a solid to a liquid.

melting point (mp or T_f) The temperature at which the solid and liquid forms of a substance are at equilibrium.

metabolic pathway A biochemical reaction sequence that flows in one direction, with each reaction catalyzed by an enzyme.

metal A substance or mixture that is relatively shiny and malleable and is a good conductor of heat and electricity. In reactions, metals tend to transfer electrons to nonmetals and form ionic compounds.

metallic bonding An idealized type of bonding based on the attraction between metal ions and their delocalized valence electrons. (See also *electron-sea model.*)

metallic radius One-half the shortest distance between the nuclei of adjacent individual atoms in a crystal of an element.

metallic solid A solid whose individual atoms are held together by metallic bonding.

metalloid (also *semimetal*) An element with properties between those of metals and nonmetals.

metallurgy The branch of materials science concerned with the extraction and utilization of metals.

metathesis reaction (also *double-displacement reaction*) A reaction in which atoms or ions of two compounds exchange bonding partners. Precipitation and acid-base reactions are examples.

meter (m) The SI base unit of length. The distance light travels in a vacuum in 1/299,792,458 second.

methanogenesis The process of producing methane by anaerobic biodegradation of plant and animal waste.

Michaelis constant (K_M) A combination of rate constants in enzyme kinetics.

Michaelis-Menten equation An equation that describes the kinetics of an enzyme-catalyzed reaction in terms of the initial concentrations of enzyme and substrate.

$$\text{Rate} = \frac{k_2[\text{E}]_0[\text{S}]}{K_\text{M} + [\text{S}]}$$

microstate An instantaneous, quantized state of a system of particles throughout which the total energy of the system is dispersed.

milliliter (mL) A volume (0.001 L) equivalent to 1 cm^3.

millimeter of mercury (mmHg) A unit of pressure based on the difference in the heights of mercury in a barometer or manometer. Renamed the *torr* in honor of Torricelli.

mineral The portion of an ore that contains the element of interest; a naturally occurring, homogeneous, crystalline inorganic solid, with a well-defined composition.

miscible Soluble in any proportion.

mixture Two or more elements and/or compounds that are physically intermingled but not chemically combined.

MO bond order One-half the difference between the number of electrons in bonding MOs and the number in antibonding MOs.

model (also *theory*) A simplified conceptual picture based on experiment that explains how a natural phenomenon occurs.

molality (*m*) A concentration term expressed as number of moles of solute dissolved in 1000 g (1 kg) of solvent.

molar heat capacity (C_m) The quantity of heat required to change the temperature of 1 mol of a substance by 1 K.
molar mass ($\mathcal{M}$) The mass of 1 mol of entities (atoms, molecules, or formula units) of a substance, in units of g/mol.
molarity (*M*) A unit of concentration expressed as the moles of solute dissolved in 1 L of solution.
mole (mol) The SI base unit for amount of a substance. The amount that contains a number of entities equal to the number of atoms in exactly 12 g of carbon-12 (which is 6.022×10^{23}).
mole fraction (*X*) A concentration term expressed as the ratio of number of moles of solute to the total number of moles (solute plus solvent).
molecular equation A chemical equation showing a reaction in solution in which reactants and products appear as intact, undissociated compounds.
molecular formula A formula that shows the actual number of atoms of each element in a molecule of a compound.
molecular mass (also *molecular weight*) The sum (in amu) of the atomic masses of the elements in a molecule (or formula unit) of a compound.
molecular orbital (MO) An orbital of given energy and shape that extends over a molecule and can be occupied by no more than two paired electrons.
molecular orbital (MO) diagram A depiction of the relative energy and number of electrons in each MO, as well as the atomic orbitals from which the MOs form.
molecular orbital (MO) theory A model that describes a molecule as a collection of nuclei and electrons in which the electrons occupy orbitals that extend over the entire molecule.
molecular polarity The overall distribution of electronic charge in a molecule, determined by its shape and bond polarities.
molecular shape The three-dimensional arrangement of the atomic nuclei in a molecule.
molecular solid A solid held together by intermolecular forces between individual molecules.
molecular weight (See *molecular mass.*)
molecularity The number of reactant particles involved in an elementary step.
molecule A structure consisting of two or more atoms that are bound chemically and behave as an independent unit.
monatomic ion An ion derived from a single atom.
monomer A small molecule, linked covalently to others of the same or similar type to form a polymer; the repeat unit of the polymer.
mononucleotide A monomer unit of a nucleic acid, consisting of an N-containing base, a sugar, and a phosphate group.
monosaccharide A simple sugar; a polyhydroxy ketone or aldehyde with three to nine C atoms.

N

nanotechnology The science and engineering of nanoscale systems (size range of 1–100 nm).
natural law (also *law*) A summary, often in mathematical form, of a universal observation.
Nernst equation An equation stating that the voltage of an electrochemical cell under any conditions depends on the standard cell voltage and the concentrations of the cell components:

$$E_{\text{cell}} = E^\circ_{\text{cell}} - \frac{RT}{nF} \ln Q$$

net ionic equation A chemical equation of a reaction in solution in which spectator ions have been eliminated to show the actual chemical change.
network covalent solid A solid in which all the atoms are bonded covalently so that individual molecules are not present.
neutralization Process that occurs when an H^+ ion from an acid combines with an OH^- ion from a base to form H_2O.
neutralization reaction An acid-base reaction that yields water and a solution of a salt; when the H^+ ions of a strong acid react with an equivalent amount of the OH^- ions of a strong base, the solution is neutral.
neutron (n^0) An uncharged subatomic particle found in the nucleus, with a mass slightly greater than that of a proton.
nitrile An organic compound containing the $-C{\equiv}N{:}$ group.
node A point at which a wave function passes through zero (has zero amplitude). The probability of finding the electron at a node is zero.
nonbonding MO A molecular orbital that is not involved in bonding.
nonelectrolyte A substance whose aqueous solution does not conduct an electric current.
nonmetal An element that lacks metallic properties. In reactions, nonmetals tend to share electrons with each other to form covalent compounds or accept electrons from metals to form ionic compounds.
nonpolar covalent bond A covalent bond between identical atoms in which the bonding pair is shared equally.
nuclear binding energy The energy required to break 1 mol of nuclei of an element into individual nucleons.
nuclear magnetic resonance (NMR) spectroscopy An instrumental technique used to determine the molecular environment of a given type of nucleus, most often ^1H, from its absorption of radio waves in a magnetic field.
nuclear transmutation The induced conversion of one nucleus into another by bombardment with a particle.
nucleic acid An unbranched polymer of mononucleotides that occurs as two types, DNA and RNA (deoxyribonucleic and ribonucleic acids), which differ chemically in the nature of the sugar portion of their mononucleotides.
nucleon An elementary particle found in the nucleus of an atom; a proton or neutron.
nucleus The tiny central region of the atom that contains all the positive charge and essentially all the mass.
nuclide A nuclear species with specific numbers of protons and neutrons.

O

observation A fact obtained with the senses, often with the aid of instruments. Quantitative observations provide data that can be compared.
occurrence (also *source*) The form(s) in which an element exists in nature.
octahedral arrangement The geometric arrangement obtained when six electron groups maximize their space around a central atom; when all six groups are bonding groups, the molecular shape is octahedral (AX_6; ideal bond angle = 90°).
octet rule The observation that when atoms bond, they often lose, gain, or share electrons to attain a filled outer level of eight electrons (or two for H and Li).
optical isomers (also *enantiomers*) A pair of stereoisomers consisting of a molecule and its mirror image that cannot be superimposed on each other.
optically active Able to rotate the plane of polarized light.
orbital diagram A depiction of orbital occupancy in terms of electron number and spin shown by means of arrows in a series of small boxes or on a series of short lines.

ore A naturally occurring compound or mixture of compounds from which an element can be profitably extracted.
organic compound A compound in which carbon is nearly always bonded to one or more other carbons and to hydrogen, and often to other elements.
organometallic compound An organic compound in which a carbon (in an R group) has a polar covalent bond to a metal atom.
osmosis The process by which solvent flows through a semipermeable membrane from a dilute to a concentrated solution.
osmotic pressure (Π) The pressure that results from the ability of solvent, but not solute, particles to cross a semipermeable membrane. The pressure required to prevent the net movement of solvent across the membrane.
outer electrons Electrons that occupy the highest energy level (highest *n* value) and are, on average, farthest from the nucleus.
overall (net) equation A chemical equation that is the sum of two or more balanced sequential equations in which a product of one becomes a reactant for the next.
overvoltage The additional voltage, usually associated with gaseous products forming at an electrode, that is required above the standard cell voltage to accomplish electrolysis.
oxidation The loss of electrons by a species, accompanied by an increase in oxidation number.
oxidation number (O.N.) (also *oxidation state*) A number equal to the magnitude of the charge an atom would have if its shared electrons were transferred to the atom that attracts them more strongly.
oxidation-reduction reaction (also *redox reaction*) A process in which there is a net movement of electrons from one reactant (reducing agent) to another (oxidizing agent).
oxidation state (See *oxidation number.*)
oxidizing agent The substance that accepts electrons in a reaction and undergoes a decrease in oxidation number.
oxoanion An anion in which an element, usually a nonmetal, is bonded to one or more oxygen atoms.

P

***p* orbital** An atomic orbital with $l = 1$.
packing efficiency The percentage of the total volume occupied by atoms, ions, or molecules in a unit cell.
paramagnetism The tendency of a species with unpaired electrons to be attracted by an external magnetic field.
partial ionic character An estimate of the actual charge separation in a bond (caused by the electronegativity difference of the bonded atoms) relative to complete separation.
partial pressure The portion of the total pressure contributed by a gas in a mixture of gases.
particle accelerator A device used to impart high kinetic energies to nuclear particles.
pascal (Pa) The SI unit of pressure; 1 Pa = 1 N/m^2.
penetration The phenomenon in which an outer electron moves through the region occupied by the core electrons to spend part of its time closer to the nucleus; penetration increases the average effective nuclear charge for that electron.
percent by mass (mass %) (also *mass percent*) The fraction by mass expressed as a percentage.
percent yield (% yield) The actual yield of a reaction expressed as a percentage of the theoretical yield.
period A horizontal row of the periodic table.
periodic law A law stating that when the elements are arranged by atomic mass, they exhibit a periodic recurrence of properties.
periodic table of the elements A table in which the elements are arranged by atomic number into columns (groups) and rows (periods).
pH The negative of the common logarithm of $[H_3O^+]$.
phase A physically distinct and homogeneous part of a system.
phase change A physical change from one phase to another, usually referring to a change in physical state.
phase diagram A diagram used to describe the stable phases and phase changes of a substance as a function of temperature and pressure.
photoelectric effect The observation that when monochromatic light of sufficient frequency shines on a metal, an electric current is produced.
photon A quantum of electromagnetic radiation.
photovoltaic cell A device capable of converting light directly into electricity.
physical change A change in which the physical form (or state) of a substance, but not its composition, is altered.
physical property A characteristic shown by a substance itself, without interacting with or changing into other substances.
pi (π) bond A covalent bond formed by sideways overlap of two atomic orbitals that has two regions of electron density, one above and one below the internuclear axis.
pi (π) MO A molecular orbital formed by combination of two atomic (usually *p*) orbitals whose orientations are perpendicular to the internuclear axis.
Planck's constant (*h*) A proportionality constant relating the energy and frequency of a photon, equal to 6.626×10^{-34} J·s.
plastic A material that, when deformed, retains its new shape.
polar covalent bond A covalent bond in which the electron pair is shared unequally, so the bond has partially negative and partially positive poles.
polar molecule A molecule with an unequal distribution of charge as a result of its polar bonds and shape.
polarimeter A device used to measure the rotation of plane-polarized light by an optically active compound.
polarizability The ease with which a particle's electron cloud can be distorted.
polyatomic ion An ion in which two or more atoms are bonded covalently.
polymer (also *macromolecule*) An extremely large molecule that results from the covalent linking of many simpler molecular units (monomers).
polyprotic acid An acid with more than one ionizable proton.
polysaccharide A macromolecule composed of many monosaccharides linked covalently.
positron (β^+) The antiparticle of an electron.
positron emission A type of radioactive decay in which a positron is emitted from a nucleus.
potential energy (E_p) The energy an object has as a result of its position relative to other objects or because of its composition.
precipitate The insoluble product of a precipitation reaction.
precipitation reaction A reaction in which two soluble ionic compounds form an insoluble product, a precipitate.
precision (also *reproducibility*) The closeness of a measurement to other measurements of the same phenomenon in a series of experiments.
pressure (*P*) The force exerted per unit of surface area.
pressure-volume work (*PV* work) A type of mechanical work done when a volume change occurs against an external pressure.
principal quantum number (*n*) A positive integer that specifies the energy and relative size of an atomic orbital; a number that specifies an energy level in an atom.
probability contour A shape that defines the volume around an atomic nucleus within which an electron spends a given percentage of its time.

product A substance formed in a chemical reaction.
property A characteristic that gives a substance its unique identity.
protein A natural, unbranched polymer composed of any of about 20 types of amino acid monomers linked together by peptide bonds.
proton (p^+) A subatomic particle found in the nucleus that has a unit positive charge (1.60218×10^{-19} C).
proton acceptor A species that accepts an H^+ ion; a Brønsted-Lowry base.
proton donor A species that donates an H^+ ion; a Brønsted-Lowry acid.
pseudo-first-order A reaction with a rate law that appears to be first-order because all reactants except one are present in such large concentrations.
pseudo–noble gas configuration The $(n-1)d^{10}$ configuration of a p-block metal ion that has an empty outer energy level.

Q

quantum A packet of energy equal to $h\nu$. The smallest quantity of energy that can be emitted or absorbed.
quantum mechanics The branch of physics that examines the wave nature of objects on the atomic scale.
quantum number A number that specifies a property of an orbital or an electron.

R

rad (*r*adiation-*a*bsorbed *d*ose) The quantity of radiation that results in 0.01 J of energy being absorbed per kilogram of tissue; 1 rad = 0.01 J/kg tissue = 10^{-2} Gy.
radial probability distribution plot The graphic depiction of the total probability distribution (sum of ψ^2) of an electron in the region near the nucleus.
radioactivity The emissions resulting from the spontaneous disintegration of an unstable nucleus.
radioisotope An isotope with an unstable nucleus that decays through radioactive emissions.
radioisotopic dating A method for determining the age of an object based on the rate of decay of a particular radioactive nuclide relative to a stable nuclide.
radius of gyration (R_g) A measure of the size of a coiled polymer chain, expressed as the average distance from the center of mass of the chain to its outside edge.
random coil The shape adopted by most polymer chains and caused by random rotation about the bonds joining the repeat units.
random error Error that occurs in all measurements (with its size depending on the measurer's skill and the instrument's precision) and results in values *both* higher and lower than the actual value.
Raoult's law A law stating that the vapor pressure of solvent above a solution equals the mole fraction of solvent times the vapor pressure of pure solvent: $P_{\text{solvent}} = X_{\text{solvent}} \times P^{\circ}_{\text{solvent}}$.
rare earths (See *lanthanides.*)
rate constant (k) The proportionality constant that relates reaction rate to reactant (and product) concentrations.
rate law (also *rate equation*) An equation that expresses the rate of a reaction as a function of reactant (and product) concentrations and temperature.
rate-determining step (also *rate-limiting step*) The slowest step in a reaction mechanism and therefore the step that limits the overall rate.
reactant A starting substance in a chemical reaction.
reaction energy diagram A graph that shows the potential energy of a reacting system as it progresses from reactants to products.
reaction intermediate A substance that is formed and used up during the overall reaction and therefore does not appear in the overall equation.
reaction mechanism A series of elementary steps that sum to the overall reaction and is consistent with the rate law.
reaction order The exponent of a reactant concentration in a rate law that shows how the rate is affected by changes in that concentration.
reaction quotient (Q) (also *mass-action expression*) A ratio of terms for a given reaction consisting of product concentrations multiplied together and divided by reactant concentrations multiplied together, with each concentration raised to the power of its balancing coefficient. The value of Q changes until the system reaches equilibrium, at which point it equals K.
reaction rate The change in the concentrations of reactants (or products) with time.
reactor core The part of a nuclear reactor that contains the fuel rods and generates heat from fission.
redox reaction (See *oxidation-reduction reaction.*)
reducing agent The substance that donates electrons in a redox reaction and undergoes an increase in oxidation number.
reduction The gain of electrons by a species, accompanied by a decrease in oxidation number.
refraction A phenomenon in which a wave changes its speed and therefore its direction as it passes through a phase boundary into a different medium.
rem (*r*oentgen *e*quivalent for *m*an) The unit of radiation dosage for a human based on the product of the number of rads and a factor related to the biological tissue; 1 rem = 10^{-2} Sv.
reproducibility (See *precision.*)
resonance hybrid The weighted average of the resonance structures for a species.
resonance structure (also *resonance form*) One of two or more Lewis structures for a species that cannot be adequately depicted by a single structure. Resonance structures differ only in the position of bonding and lone electron pairs.
reverse osmosis A process for preparing drinkable water that uses an applied pressure greater than the osmotic pressure to remove ions from an aqueous solution, typically seawater.
reversible process A process that occurs in such tiny increments that the system remains at equilibrium and the direction of the change can be reversed by an infinitesimal reversal of conditions.
rms (root-mean-square) speed (u_{rms}) The speed of a molecule having the average kinetic energy; very close to the most probable speed.
roasting A pyrometallurgical process in which metal sulfides are converted to oxides.
round off The process of removing digits based on a series of rules to obtain an answer with the proper number of significant figures (or decimal places).

S

***s* orbital** An atomic orbital with $l = 0$.
salt An ionic compound that results from an acid-base reaction after solvent is removed.
salt bridge An inverted U tube containing a solution of nonreacting ions that connects the compartments of a voltaic cell and maintains neutrality by allowing ions to flow between compartments.
saturated hydrocarbon A hydrocarbon in which each C is bonded to four other atoms.
saturated solution A solution that contains the maximum amount of dissolved solute at a given temperature (prepared with undissolved solute present).
scanning tunneling microscopy An instrumental technique that uses electrons moving across a minute gap to observe the topography of a surface on the atomic scale.

Schrödinger equation An equation that describes how the electron matter-wave changes in space around the nucleus. Solutions of the equation provide energy states associated with the atomic orbitals.
scientific method A process of creative proposals and testing aimed at objective, verifiable discoveries of the causes of natural events.
scintillation counter A device used to detect radioactive emissions through their excitation of atoms, which subsequently emit light.
screening (See *shielding.*)
second (s) The SI base unit of time.
second law of thermodynamics A law stating that a process occurs spontaneously in the direction that increases the entropy of the universe.
seesaw shape A molecular shape caused by the presence of one equatorial lone pair in a trigonal bipyramidal arrangement (AX_4E).
selective precipitation The process of separating ions through differences in the solubility of their compounds with a given precipitating ion.
self-ionization (See *autoionization.*)
semiconductor A substance whose electrical conductivity is poor at room temperature but increases significantly with rising temperature.
semimetal (See *metalloid.*)
semipermeable membrane A membrane that allows solvent, but not solute, to pass through.
shared pair (See *bonding pair.*)
shell (See *level.*)
shielding (also *screening*) The ability of other electrons, especially those occupying inner orbitals, to lessen the nuclear attraction for an outer electron.
SI unit A unit composed of one or more of the base units of the Système International d'Unités, a revised form of the metric system.
side reaction An undesired chemical reaction that consumes some of the reactant and reduces the overall yield of the desired product.
sievert (Sv) The SI unit of human radiation dosage; 1 Sv = 100 rem.
sigma (σ) bond A type of covalent bond that arises through end-to-end orbital overlap and has most of its electron density along an imaginary line joining the nuclei.
sigma (σ) MO A molecular orbital that is cylindrically symmetrical about an imaginary line that runs through the nuclei of the component atoms.
significant figures The digits obtained in a measurement. The greater the number of significant figures, the greater the certainty of the measurement.
silicate A type of compound found throughout rocks and soil and consisting of repeating —Si—O— groupings and, in most cases, metal cations.
silicone A type of synthetic polymer containing —Si—O repeat units, with organic groups and crosslinks.
simple cubic unit cell A unit cell in which a particle occupies each corner of a cube.
single bond A bond that consists of one electron pair.
slag A molten waste product formed in a blast furnace by the reaction of acidic silica with a basic metal oxide.
smelting Heating an oxide form of a mineral with a reducing agent, such as coke, to obtain a metal.
soap The salt formed in a reaction between a fatty acid and a strong base, usually a Group 1A(1) or 2A(2) hydroxide.
solid One of the three states of matter. A solid has a fixed shape that does not conform to the container shape.
solubility (*S*) The maximum amount of solute that dissolves in a fixed quantity of a particular solvent at a specified temperature.
solubility-product constant (K_{sp}) An equilibrium constant for a slightly soluble ionic compound dissolving in water.
solute The substance that dissolves in the solvent.
solution (See *homogeneous mixture.*)
solvated Surrounded closely by solvent molecules.
solvation The process of surrounding a solute particle with solvent particles.
solvent The substance in which one or more solutes dissolve.
source (See *occurrence.*)
***sp* hybrid orbital** An orbital formed by the mixing of one *s* and one *p* orbital of a central atom.
sp^2 hybrid orbital An orbital formed by the mixing of one *s* and two *p* orbitals of a central atom.
sp^3 hybrid orbital An orbital formed by the mixing of one *s* and three *p* orbitals of a central atom.
sp^3d hybrid orbital An orbital formed by the mixing of one *s*, three *p*, and one *d* orbital of a central atom.
sp^3d^2 hybrid orbital An orbital formed by the mixing of one *s*, three *p*, and two *d* orbitals of a central atom.
specific heat capacity (*c*) The quantity of heat required to change the temperature of 1 gram of a substance or material by 1 K.
spectator ion An ion that is present as part of a reactant but is not involved in the chemical change.
spectrochemical series A ranking of ligands in terms of their ability to split *d*-orbital energies.
spectrometry Any instrumental technique that uses a portion of the electromagnetic spectrum to measure the atomic and molecular energy levels of a substance.
speed of light (*c*) A fundamental constant giving the speed at which electromagnetic radiation travels in a vacuum: $c = 2.99792458\times10^8$ m/s.
spin quantum number (m_s) A number, either $+\frac{1}{2}$ or $-\frac{1}{2}$, that indicates the direction of electron spin.
spontaneous change A change that occurs under specified conditions without an ongoing input of external energy.
square planar shape A molecular shape (AX_4E_2) caused by the presence of two lone pairs at opposite vertices in an octahedral electron-group arrangement.
square pyramidal shape A molecular shape (AX_5E) caused by the presence of one lone pair in an octahedral electron-group arrangement.
standard atmosphere (atm) The average atmospheric pressure measured at sea level and 0°C, defined as 1.01325×10^5 Pa.
standard cell potential (E°_{cell}) The potential of a cell measured with all components in their standard states and no current flowing.
standard electrode potential ($E^\circ_{half\text{-}cell}$) (also *standard half-cell potential*) The standard potential of a half-cell, with the half-reaction written as a reduction.
standard enthalpy of formation (ΔH°_f) (also *standard heat of formation*) The enthalpy change occurring when 1 mol of a compound forms from its elements with all components in their standard states.
standard enthalpy of reaction (ΔH°_{rxn}) (also *standard heat of reaction*) The enthalpy change that occurs during a reaction when all components are in their standard states.
standard enthalpy of reaction (ΔS°_{rxn}) The entropy change that occurs when all components are in their standard states.

standard free energy change ($\Delta G°$) The free energy change that occurs when all components of a system are in their standard states.
standard free energy of formation ($\Delta G°_f$) The standard free energy change that occurs when 1 mol of a compound is made from its elements with all components in their standard states.
standard half-cell potential (See *standard electrode potential.*)
standard heat of formation (See *standard enthalpy of formation.*)
standard heat of reaction (See *standard enthalpy of reaction.*)
standard hydrogen electrode (See *standard reference half-cell.*)
standard molar entropy ($S°$) The entropy of 1 mol of a substance in its standard state.
standard molar volume The volume of 1 mol of an ideal gas at standard temperature and pressure: 22.4141 L.
standard reference half-cell (also *standard hydrogen electrode*) A specially prepared platinum electrode immersed in 1 M $H^+(aq)$ through which H_2 gas at 1 atm is bubbled. $E°_{half\text{-}cell}$ is defined as 0 V.
standard state A set of specific conditions used to compare thermodynamic data: 1 atm for gases behaving ideally, 1 M for dissolved species, or the pure substance for liquids and solids.
standard temperature and pressure (STP) The reference conditions for a gas: 0°C (273.15 K) and 1 atm (760 torr)
state function A property of a system determined only by the system's current state, regardless of how it arrived at that state.
state of matter One of the three physical forms of matter: solid, liquid, or gas.
stationary state In the Bohr model, one of the allowable energy levels of the atom in which it does not release or absorb energy.
steady-state approximation The approximation that the concentrations of reaction intermediates remain constant during reaction because the net rate of formation = 0.
steel An alloy of iron with small amounts of carbon and usually other metals.
stellar nucleosynthesis The process by which elements are formed in the stars through nuclear fusion.
stereoisomers Molecules with the same arrangement of atoms but different orientations of groups in space. (See also *geometric isomers* and *optical isomers.*)
stoichiometric coefficient (See *balancing coefficient.*)
stoichiometry The study of the mass-mole-number relationships of chemical formulas and reactions.
strong force An attractive force that exists between all nucleons and is many times stronger than the electrostatic repulsive force.
strong-field ligand A ligand that causes larger crystal field splitting energy and therefore is part of a low-spin complex.
structural formula A formula that shows the actual number of atoms, their relative placement, and the bonds between them.
structural isomers (See *constitutional isomers.*)
sublevel (also *subshell*) An energy substate of an atom within a level. Given by the n and l values, the sublevel designates the size and shape of the atomic orbitals.
sublimation The process by which a solid changes directly into a gas.
substance A type of matter, either an element or a compound, that has a fixed composition.
substitution reaction An organic reaction that occurs when an atom (or group) from one reactant substitutes for one attached to a carbon in another reactant.
substrate A reactant that binds to the active site in an enzyme-catalyzed reaction.
superconductivity The ability to conduct a current with no loss of energy to resistive heating.
supersaturated solution An unstable solution in which more solute is dissolved than in a saturated solution.
surface tension The energy required to increase the surface area of a liquid by a given amount.
surroundings All parts of the universe other than the system being considered.
suspension A heterogeneous mixture containing particles that are visibly distinct from the surrounding fluid.
synthetic natural gas (SNG) A gaseous fuel mixture, mostly methane, formed from coal.
system The defined part of the universe under study.
systematic error A type of error producing values that are all either higher or lower than the actual value, often caused by faulty equipment or a consistent flaw in technique.

T

T shape A molecular shape caused by the presence of two equatorial lone pairs in a trigonal bipyramidal arrangement (AX_3E_2).
t_{2g} orbitals The set of orbitals (composed of d_{xy}, d_{yz}, and d_{xz}) that results when the energies of the metal-ion d orbitals are split by a ligand field. This set is lower in energy than the other (e_g) set in an octahedral field and higher in energy in a tetrahedral field.
temperature (T) A measure of how hot or cold a substance is relative to another substance. A measure of the average kinetic energy of the particles in a sample.
tetrahedral arrangement The geometric arrangement formed when four electron groups maximize their separation around a central atom; when all four groups are bonding groups, the molecular shape is tetrahedral (AX_4; ideal bond angle 109.5°).
theoretical yield The amount of product predicted by the stoichiometrically equivalent molar ratio in the balanced equation.
theory (See *model.*)
thermochemical equation A balanced chemical equation that includes the enthalpy change for the reaction.
thermochemistry The branch of thermodynamics that focuses on the heat involved in chemical and physical change.
thermodynamics The study of energy and its transformations.
thermometer A device for measuring temperature that contains a fluid that expands or contracts within a graduated tube.
third law of thermodynamics A law stating that the entropy of a perfect crystal is zero at 0 K.
titration A method of determining the concentration of a solution by monitoring the amount of a solution of known concentration needed to react with it.
torr A unit of pressure identical to 1 mmHg.
total ionic equation An equation for an aqueous reaction that shows all the soluble ionic substances dissociated into ions.
tracer A radioisotope that signals the presence of the species of interest via radioactive emissions.
transcription The process of producing messenger RNA from DNA.
transition element An element that occupies the d block or the f block (inner transition element) of the periodic table.
transition state (also *activated complex*) An unstable species formed in an effective collision of reactants that exists momentarily when the system is highest in energy and that can either form products or re-form reactants.
transition state theory A model that explains how the energy of reactant collisions is used to form a high-energy transitional species that can change to reactant or product.
translation The process by which messenger RNA specifies the sequence of amino acids to synthesize a protein.
transuranium element An element with atomic number higher than that of uranium ($Z = 92$).

trigonal bipyramidal arrangement The geometric arrangement formed when five electron groups maximize their separation around a central atom. When all five groups are bonding groups, the molecular shape is trigonal bipyramidal (AX_5; ideal bond angles, axial-center-equatorial 90° and equatorial-center-equatorial 120°).
trigonal planar arrangement The geometric arrangement formed when three electron groups maximize their separation around a central atom. When all three groups are bonding groups, the molecular shape is trigonal planar (AX_3; ideal bond angle 120°).
trigonal pyramidal shape A molecular shape (AX_3E) caused by the presence of one lone pair in a tetrahedral arrangement.
triple bond A covalent bond that consists of three bonding pairs, two atoms sharing six electrons; one π and two bonds.
triple point The pressure and temperature at which three phases of a substance are in equilibrium. In a phase diagram, the point at which three phase-transition curves meet.
Trouton's Rule The observation that most liquids have an entropy of vaporization value near 88 J/mol·K
Tyndall effect The scattering of light by a colloid.

U

ultraviolet (UV) Radiation in the region of the electromagnetic spectrum between the visible and the x-ray regions.
uncertainty A characteristic of every measurement that results from the inexactness of the measuring device and the need to estimate when taking a reading.
uncertainty principle The principle stated by Heisenberg that it is impossible to know simultaneously the exact position and velocity of a particle; the principle becomes important only for particles of very small mass.
unimolecular reaction An elementary reaction that involves the decomposition or rearrangement of a single particle.
unit cell The smallest portion of a crystal that, if repeated in all three directions, yields the crystal.
universal gas constant (*R*) A proportionality constant that relates the energy, amount of substance, and temperature of a system; $R = 0.0820578$ atm·L/mol·K $= 8.31447$ J/mol·K.
unsaturated hydrocarbon A hydrocarbon with at least one carbon-carbon multiple bond; one in which at least two C atoms are bonded to fewer than four atoms.
unsaturated solution A solution in which more solute can be dissolved at a given temperature.
unshared pair (See *lone pair.*)

V

V shape (See *bent shape.*)
valence band In band theory, the lower energy portion of the band of molecular orbitals, which is filled with valence electrons.
valence bond (VB) theory A model that attempts to reconcile the shapes of molecules with those of atomic orbitals through the concepts of orbital overlap and hybridization.
valence electrons The electrons involved in compound formation; in main-group elements, the electrons in the valence (outer) level.
valence-shell electron-pair repulsion (VSEPR) theory A model explaining that the shapes of molecules and ions result from minimizing electron-pair repulsions around a central atom.
van der Waals constants Experimentally determined positive numbers used in the van der Waals equation to account for the interparticle attractions and particle volume of real gases.
van der Waals equation An equation that accounts for the behavior of real gases.
van der Waals radius One-half of the closest distance between the nuclei of identical nonbonded atoms.
vapor pressure (also *equilibrium vapor pressure*) The pressure exerted by a vapor at equilibrium with its liquid in a closed system.
vapor pressure lowering (Δ*P*) The lowering of the vapor pressure of a solvent caused by the presence of dissolved solute particles.
vaporization The process of changing from a liquid to a gas.
variable A quantity that can have more than a single value. (See also *controlled experiment.*)
viscosity A measure of the resistance of a fluid to flow.
volatility The tendency of a substance to become a gas.
volt (V) The SI unit of electric potential: 1 V = 1 J/C.
voltage (See *cell potential.*)
voltaic cell (also *galvanic cell*) An electrochemical cell that uses a spontaneous redox reaction to generate electric energy.
volume (*V*) The amount of space occupied by a sample of matter.
volume percent [% (v/v)] A concentration term defined as the volume of solute in 100. volumes of solution.

W

wastewater Used water, usually containing industrial and/or residential waste, that is treated before being returned to the environment.
water softening The process of replacing the hard-water ions Ca^{2+} and Mg^{2+} with Na^+ ions.
wave function A mathematical expression that describes the motion of the electron's matter-wave in terms of time and position in the region of the nucleus. The term is used qualitatively to mean the region of space in which there is a high probability of finding the electron.
wave function A mathematical description of the electron's matter-wave in three dimensions.
wave-particle duality The principle stating that both matter and energy have wavelike and particle-like properties.
wavelength (λ) The distance between any point on a wave and the corresponding point on the next wave, that is, the distance a wave travels during one cycle.
weak-field ligand A ligand that causes smaller crystal field splitting energy and therefore is part of a high-spin complex.
weight The force that is exerted by a gravitational field on an object and is directly proportional to the object's mass.
work (*w*) The energy transferred when an object is moved by a force.

X

x-ray diffraction analysis An instrumental technique used to determine dimensions of a crystal structure by measuring the diffraction patterns caused by x-rays impinging on the crystal.

Z

zone refining A process used to purify metals and metalloids in which impurities are removed from a bar of the element by concentrating them in a thin molten zone.
β^- decay (also *negatron emission*) A radioactive process in which a high-speed electron is emitted from a nucleus.

INDEX

Page numbers followed by *f* indicate figures; *mn,* marginal notes; *n,* footnotes; and *t,* tables. Page numbers preceded by "A-" indicate pages in the *Appendix.* Page numbers preceded by "G-" indicate pages in the *Glossary.*

A

AB block copolymers, 518
Absolute temperature scale. *See* Kelvin temperature scale
Absolute zero, 26, 213
Absorption spectrum, 308–309, G-1
Abundance of elements, 997–1000, 998*t,* 1001*t,* G-1
Accuracy, 32–33, 33*f,* G-1
Acetaldehyde. *See* Ethanal
Acetamide, 655*t*
Acetaminophen, 664*f*
Acetate ions, 72*t,* 844–845, 844*t,* 845*f*
Acetic acid
  acid-dissociation constant for, 796–797*t,* 801*t,* A-8
  as base, 827
  boiling point elevation constant, 560*t*
  buffer action and, 844–845, 844*t,* 845*f*
  formula and model, 110*t*
  freezing point depression constant, 560*t*
  functions of, 110*t*
  molecular structure of, 655*t,* 662
  reaction with bases, 170, 170*t,* 171*f*
  reversible dissociation of, 189
  uses of, 793*t*
Acetone
  bonds and orbitals of, 454
  molecular shape of, 427
  molecular structure of, 655*t,* 660*f*
  nuclear magnetic resonance spectroscopy of, 649, 649*f*
Acetonitrile, 484*f,* 655*t*
Acetylene
  bonds in, 451, 453, 453*f*
  electron density and bond order in, 454*f*
  molecular shape, 451
  molecular structure of, 655*t*
  production of, 228
Acetylsalicylic acid, A-8
Acid anhydrides, 663, 666, 666*f,* G-1
Acid-base behavior of oxides, 354–355, 354*f*
Acid-base buffers, 843–852
  actions of, 843–844*f,* 843–846
  buffer capacity, 849, 849*f*
  buffer range, 849–850
  common-ion effect and, 844
  concentration of components, 845–846, 845*f,* 848, 848*f*
  defined, 843, G-1
  Henderson-Hasselbalch equation and, 848–849
  molecular scene representation of, 850–851
  preparing, 851–852
Acid-base equilibria, 792–830
  acid-base indicators and, 803, 803*f,* 860–862
  acid-dissociation constant and, 796–797*t,* 796–798, A-8–A-10
  acidic salt solutions and, 823–824, 825*t*
  amphiprotic anions and, 825
  autoionization of water and, 798–803, 799*f,* 827
  base-dissociation constant and, 818–822, 819*t,* A-11–A-12
  basic salt solutions and, 824, 825*t*
  Brønsted-Lowry acid-base definition and, 803–808, 804*f,* 827, 830
  concentration and, 809–813
  conjugate acid-base pairs and, 804–807, 805*t,* 806*f,* 844, 844*f,* 851
  hydrated metal ions and, 817, A-12
  hydronium ions and, 794, 799–803, 799*f*
  ion-product constant for water, 799–803, 799*f*
  Lewis acid-base definition and, 827–830, 877
  neutral salt solutions and, 823, 825*t*
  pH and, 800–803, 800*f,* 802*f*
  polyprotic acids and, 813–815, 814*t*
  relationship between weak bases and weak acids, 818–822
  salt solutions and, 823–826
  variation in acid strength, 795–796*f,* 795–798
  weak-acid equilibria problem solving, 808–815
  of weakly acidic cations and weakly basic anions, 824–825
Acid-base indicators
  acid-base titration and, 173–174, 173*f*
  acid-base titration curves and, 860–862
  color changes of, 860–861, 861*f*
  defined, 803, 860, G-1
  monitoring pH with, 860–862
  pH paper, 803, 803*f*
Acid-base reactions, 165–174
  acid-base indicators and, 173–174, 173*f,* 860–862
  defined, 165, G-1, G-11
  end point in, 173, 173*f*
  equivalence point in, 173
  gas-forming reactions, 169–170, 171*f,* 189–190
  heat change during, 270–271
  key events in, 167
  of oxides, 354–355, 354*f*
  proton transfer in, 168–171, 169*f*
  reaction direction in, 805–808
  salt formation in, 167
  stoichiometry of, 172–174
  titration curves and, 853–863
  titrations, 172–174, 173*f*
  water formation in, 167
  writing ionic equations for, 167–168
Acid-base titration, 172–174, 173*f*
Acid-base titration curves, 853–863
  acid-base indicators and, 860–862
  defined, 853, G-1
  end point and, 861
  equivalence point and, 853*f,* 854–857, 855*f,* 859*f,* 863*f*
  for polyprotic acids, 862–863, 863*f*
  for strong acid–strong base, 853–855, 853*f*
  for weak acid–strong base, 855–859, 855*f*
  for weak base–strong acid, 859–860, 859*f*
Acid-dissociation constant
  defined, 796, G-1
  of hydrated metal ions, A-12
  range of values for, 796, 796–797*t*
  relation with base-dissociation constant in conjugate pairs, 821–822
  relation with pK_a, 801*t*
  tables of, A-8–A-10
  weak acid equilibria problem solving and, 808–815
Acidic solutions, balancing redox reactions in, 941–942
Acid rain
  carbonates as protection from effects of, 603, 876
  components of, 281
  effects of, 875–876
  formation of, 875, 876*f*
  Hess's law as applied to, 275–276
  pH of, 875
  preventing, 876
  sulfuric acid and, 616, 616*mn,* 875, 876
Acids. *See also* Acid-base reactions
  Arrhenius, 794, 830
  balancing redox reactions in, 941–942
  Brønsted-Lowry acids, 169, 803–808, 804*f,* 827, 830
  buffers, 843–852
  concentration, finding from titration, 173–174
  conjugate acid-base pairs, 804–807, 805*t,* 806*f,* 844, 844*f,* 851
  defined, 165, 169, 794, 828, G-1
  dissociation of, 165–166, 189, 747, 795–798, 795*f,* 811–813, 844*t*
  as electrolytes, 165–166, 166*f*
  equilibria (*See* Acid-base equilibria)
  indicators (*See* Acid-base indicators)
  leveling effect on, 827
  Lewis acids, 827–830, 877
  monoprotic, 796, 797*t*
  names, 74
  polyprotic, 813–815, 814*t,* 862–863, 863*f*
  sour taste from, 793
  strong (*See* Strong acids)
  structural features of, 166
  titration curves, 853–863
  titrations, 173–174
  uses of, 793*t*
  in water, 747, 794–798
  weak (*See* Weak acids)
Acid strength. *See also* Strong acids; Weak acids
  acid-dissociation constant and, 796–797*t,* 796–798
  classifying, 797–798
  direction of reaction and, 805–808
  of hydrated metal ions, 817, 817*f*
  molecular properties and, 816–818
  of nonmetal hydrides, 816, 816*f*
  of oxoacids, 816–817, 817*f*
Acrylonitrile, 385*f,* 602*f*
Actin, arrangement of filaments, 511*f*
Actinides
  defined, 1044, G-1
  electron configuration, 341*f,* 342
  properties of, 1045
Actinium, 342
Activated complex, 719
Activated state, 717, 717*f*
Activation energy
  calculating, 718, 718*f,* 718*t*
  catalysis and, 730, 730*f,* 731, 735
  defined, 716, G-1
  importance of, 716–717
  transition state theory and, 719–722
Active metals
  activity series of metals, 184–186, 958–959, G-1
  displacing hydrogen from water, 184, 185*f*
  reactivity of, 958
  reduction with, 1011, 1011*f*
Active site, 732–733, 733*f,* 781, 781*f,* G-1
Activity in radioactive decay, 1084, G-1
Activity series of halogens, 186, 186*f*
Activity series of metals, 184–186, 958–959, G-1
Actual yield, 128, G-1
Addition, significant figures and, 30–31
Addition polymers, 668, 668*f,* 669*t,* G-1
Addition reactions
  with benzene, 660
  defined, 651, G-1
  organic, 651, 660, 668
  polymers in, 668
Adduct, 828–829, G-1
Adenine, 657*f,* 675
Adenosine triphosphate (ATP)
  defined, G-1
  production of, 611, 982–983
  universal role of, 925–926
Adhesive forces, 489–490, 490*f*
Adhesives, 518
Adipic acid, A-8
Advanced materials, 507–520
  ceramic materials, 512–513, 512*t,* 513*f*
  electronic materials, 507–509, 508*f*
  liquid crystals, 509–511
  nanotechnology and, 519–520
  polymeric materials, 514–518, 514*t,* 515–517*f,* 518*t*
Aerosol, 568*t*
Air bags, 205
Air composition, 205, 207, 239, 239*t*
Airplane de-icers, 566
Air pollution
  acid rain (*See* Acid rain)
  automobile exhaust and, 276–277
  chlorofluorocarbons and, 602, 602*f,* 735
  smog and, 224, 414, 609, 609*mn*
  thermal pollution, 550, 551*f,* 1108–1109
Alanine, 643, 643*f,* 644, 672*f*
Alchemy, 10, 10*f,* 50
Alcohols
  defined, 654, G-1
  as functional group, 654, 654*f,* 655*t,* 656
  grain alcohol density, 24*t*
  solubility of, 535–536, 536*f,* 536*t*

Aldehydes, 655*t*, 660–661, 660*f*, G-1
Alizarin, 861*f*
Alizarin yellow R, 861*f*
Alkali metals. *See also* Halides
anomalous behavior of, 587, 590
atomic radius of, 348*f*
comparison with alkaline earth metals, 593
electron configuration, 343, 355, 591, 592
industrial production and uses of, 1014–1015, 1015*f*
ionization energy of, 349*f*
melting point of, 396, 396*f*
oxidation number for, 176*t*
in periodic table, 63
properties of, 590–592
reactivity of, 230, 343*f*, 591, 592, 592*mn*
sources of elements, 1001, 1002*f*
Alkaline batteries, 968, 968*f*
Alkaline earth metals
anomalous behavior of, 590, 590*f*
comparison with alkali metals, 593
diagonal relationships and, 595, 595*f*
electron configuration, 355, 593, 594
melting point of, 396, 396*f*
oxidation number for, 176*t*
in periodic table, 63
properties of, 592–595
reactivity of, 594
Alkanes
chiral molecules and, 643–644
constitutional isomerism and, 641–642, 642*t*
cycloalkanes, 640, 641*f*
defined, G-1
depicting with formulas and models, 639–641*f*
functions of, 76
general formula for, 639
as homologous series, 639
naming, 76, 76*t*, 639, 639–640*t*, 646–647
optical isomers of, 642–644
properties of, 641, 641*f*
as saturated hydrocarbons, 639
straight-chain, 76, 76*t*
Alkenes
defined, 644, G-1
as functional group, 655*t*, 659–660
general formula for, 644
geometric isomerism and, 645–646, 645*t*, 646*f*
naming, 644, 646–647
reactivity of, 659–660
Alkyl group, 651, G-1
Alkyl halides, 656–657, 659, G-1
Alkynes, 646, 655*t*, 666, G-1
Allotropes
of carbon, 600–601, 600*f*
defined, 600, G-1
entropy and, 909
of oxygen, 612
of phosphorus, 606, 606*f*
of selenium, 612, 614
of sulfur, 612, 650
of tin, 601
Alloying, 507
Alloys
binary, 1013, 1013*f*
composition and uses of, 1013*t*
crystal defects in, 507
defined, 396, 1012, G-1
interstitial, 538, 538*f*
magnesium, 1022
metallurgy, 1012–1013, 1013*f*, 1013*t*
samarium-cobalt, 1045
as solid-solid solutions, 538, 538*f*
substitutional, 538, 538*f*
Alpha decay, 1076*t*, 1077, G-1
Alpha helix, 673*f*, 674
Alpha particles
behavior in electric field, 1075, 1075*f*
de Broglie wavelength of, 311*t*
defined, G-1
nuclear reactions and, 1090
penetrating power of, 1094, 1094*f*
radioactive decay, 1076*t*, 1077
Rutherford's α-scattering experiment, 54–55, 54*f*
transmutation by, 1090
-al (suffix), 655*t*, 660
Aluminate, 598
Aluminosilicates, 876
Aluminum
acid strength of hydrated ions and, 817, 817*f*
from bauxite, 274, 1019
bond type, 598*t*
density of, 24*t*
diagonal relationship with beryllium, 598
diffraction pattern, 311, 311*f*
electron configuration, 338*t*, 339
hydrogen displacement by, 185
isolation of, 1019
melting point of, 598*t*
metallic radius, 345*f*
metallurgy of, 1019–1021, 1020*f*
mineral sources of, 1008*t*
properties of, 354, 355, 595–598
recycling, 1021, 1021*mn*
specific heat capacity of, 268*t*
uses of, 1021, 1021*f*
Aluminum-air batteries, 1020
Aluminum chloride
dimeric structure of, 595, 595*f*
dissolution of, 828, 908
electron density distribution, 394*f*
ionic properties, 394
Aluminum hydroxide, 866*t*, 881, 882*f*
Aluminum ions, 69*t*
Aluminum oxide, 181, 355, 1019–1020
Aluminum salts, 793*t*
Aluminum sulfate, 570
Amides, 655*t*, 662, 664–665, 664*f*, G-1
-amide (suffix), 655*t*, 664
Amines
base-dissociation constant of, 818–819, 819*t*
defined, 657, 818, G-1
as functional group, 655*t*, 657–658, 657*f*
as weak bases, 797
-amine (suffix), 655*t*, 657
Amino acids. *See also* Proteins
defined, 539, 672, G-1
optical isomers of, 643, 643*f*, 644
physiological form of, 539*f*
polarity of side chains, 539–540, 539*f*
as polyprotic acids, 863
structure of, 539–540, 672–673, 672*f*
Ammonia
base-dissociation constant of, 818, 819*t*, A-11
in complex ion formation, 877–878
as covalent hydride, 586
dissociation of, 167, 189
fixed mass ratio of, 44
formation of, 181, 607, 751, 911, 913–914
Haber process for synthesis of, 779–780, 779*t*, 780*f*, 1005
hybrid orbitals and, 447, 447*f*
hydrogen sulfide and, 804
Lewis structure of, 819*t*
molecular shape of, 429
mole fraction of air at sea level, 239*t*
nitric acid from, 280
polarity of, 429
properties of, 607
synthesis of, 1023–1024
titration of, 859–860, 859*f*
uses of, 779, 793*t*
as weak base, 797
Ammonium carbonate, 101–102
Ammonium cyanide, 824–825
Ammonium ions, 72*t*, 421, 804
Ammonium nitrate
dissolving, 546, 546*f*
mass of elements in, 102–104
solubility of, 550
spontaneous change in, 897, 916
in water, 549
Ammonium perchlorate, 621–622
Amontons's law, 213
Amorphous solids, 493, 504, 504*f*, G-1
Amount-mass-number conversions
of chemical entities, 97–102
in solutions, 151–152, 151*f*, 162*f*
Amount-mass-number relationships
of chemical entities, 97–102
in chemical equations, 118*f*
for compounds, 100*f*
for elements, 98*f*
overview, 127*f*
Amount (mol) of gases
Avogadro's law and, 214–215, 214*f*, 223, 231, 234, 234*f*
pressure and, 218–219
volume and, 214–215, 214*f*, 218–219, 223
Amount of substance, 14*t*
Ampere (A), 14*t*, 980, G-1
Amphiprotic anions, 825
Amphoteric, defined, 354, G-1
Amphoteric hydroxides, 881–882, 882*f*
Amphoteric nature of water, 804
Ampicillin, 111, 111*f*
Amplitude, 296, 296*f*, G-1
Analysis. *See* Chemical analysis
-ane (suffix), 76, 639, 640*t*
Angstroms (Å), 296
Angular momentum quantum number, 320, 320*t*, 332*t*, G-1
Anhydrides, acid, 663, 666, 666*f*
Aniline, 189, 819*t*, A-11
Animal radiation studies, 1097
Anionic ligands, 1050, 1050*t*
Anions
amphiprotic, 825
complex, 1048, 1050, 1050*t*
defined, 64, G-1
electron configuration and, 355
ionic radius of, 359, 359*f*
monatomic, 69*t*
names, 74
oxoanions, 72, 72*f*
polyatomic, 72*t*
salts of weakly acidic cations and weakly basic anions, 824–825
of weak acids as weak bases, 820–821
Anisotropic, 509
Anodes
defined, 944, G-1
half-reactions at, 944, 944*mn*
sacrificial, 974, 974*f*
Antacids, 172
Antarctic ozone hole, 735, 735*f*
Antibiotics, dual polarity of, 542, 543*f*
Antibonding MOs, 456, 456*f*, 457, G-1
Anticancer drugs, 1053
Antifluorite structure, 503
Antifreeze, 561, 566, 567
Antilogarithms, A-2
Antimony
bond type, 598*t*
melting point of, 598*t*
properties of, 353, 355, 604–607
Antineutrino, 1077, 1095
Antioxidants, 414
Apatites, 1005, G-1
Aqueous equilibria, 842–882
of acid-base buffers, 843–852
acid-base titration curves and, 853–863
complex ions and, 877–882
of slightly soluble ionic compounds, 864–875
Aqueous heats of solution, 545–547, 546*f*
Aqueous ionic reactions, 155–157, 156*f*
Aqueous solutions
defined, 81, G-2
electrolysis of, 978–979
molecular scenes to depict ionic compounds in, 148–149
salts in, 823–824, 825*t*
standard state for, 277
Arginine, 672*f*
Argon
cubic closest packing in, 502, 502*f*
electron configuration, 338*t*, 339
heats of vaporization and fusion, 471*f*
in ionization counters, 1083, 1084*f*
mole fraction in air at sea level, 239*t*
properties of, 622–624
Aristotle, 43
Arithmetic operations, 18, 30–31
Aromatic hydrocarbons, 647–648, 648*f*, 659–660, 660*f*, G-2
Array, 4

Arrhenius, Svante, 716, 794–795
Arrhenius acid-base definition, 794–795, 830, G-2
Arrhenius acids, 794, 830
Arrhenius bases, 794, 830
Arrhenius equation, 716, 718, 719, G-2
Arsenic
bond type, 598*t*
electron configuration, 340*t*
melting point of, 598*t*
properties of, 353, 355, 604–607
Arsenic acid, 814*t*, A-8
Aryl halides, 657
Asbestos, 604
Asbestos diaphragm, 1028, 1028*f*
Ascorbic acid, 793*t*, 814–815, A-8
Asparagine, 672*f*
Aspartic acid, 672*f*
Aspirin, 80*t*
Astatine, 617–622
-ate (suffix), 72, 72*f*, 74, 1050
Atmosphere (atm), 209, 209*t*
Atmosphere of Earth
abundancy of elements in, 1001*t*
carbon cycle and, 1002–1004, 1003*f*
composition of, 205, 239, 239*t*
convection in, 240
defined, 239, 999, G-2
impact of life on, 999–1000
mixing of gases in, 470
nitrogen cycle and, 1004–1005, 1004*f*
ozone (*See* Ozone)
pressure variation with altitude, 239, 239*f*
regions of, 239–240, 239*f*
temperature of, 239–240, 239*f*, 282–283
Atmospheric pressure
altitude and, 208
boiling point and, 479
effects of, 207, 207*f*
measurement of, 208, 208–209*f*
units of measurement for, 209–210, 209*t*
variation in, 239, 239*f*
Atom economy, 129–130
Atomic bomb, 1106, 1106*f*, 1106*n*
Atomic clock, 18
Atomic mass, 57–59, 61, 96, G-2
Atomic-mass interval, 59, 61
Atomic mass unit (amu), 56*n*, 57, 96, G-2
Atomic nucleus. *See* Nucleus
Atomic number
defined, 56, G-2
identification of elements from, 63–64
periodic table organization by, 61, 331
Atomic orbitals, 314–326
defined, 318, G-2
d orbitals (*See d* orbitals)
energy levels and, 320–321
filling order, 341–342, 341*f*
f orbitals, 325, 325*f*, 326*t*
g orbitals, 325
hybrid orbitals, 444–451
particle-in-a-box model and, 315–317, 315*f*
Pauli exclusion principle and, 333
p orbitals, 324, 324*f*, 326*t*, 458–459*f*, 458–460
quantum numbers and, 319–321, 320*t*
Schrödinger equation and, 314–315, 317–318
shape of, 323–325, 323–325*f*
s orbitals, 323–324, 323*f*, 326*t*, 444
wave function and, 314, 314*f*, 317, 317*f*
Atomic properties
of alkali metals, 591
of alkaline earth metals, 594
of boron family elements, 595–597
of carbon family elements, 599
as central theme in chemistry, 8
chemical reactivity and, 353–360
of halogens, 618
of nitrogen family elements, 605
of noble gases, 623
of oxygen family elements, 613
of Period 2 elements, 587–590
periodic trends in, 345–352, 352*f*
of transition elements, 1040–1041*f*, 1040–1042
Atomic radius. *See also* Atomic size
of alkali metals, 591
of alkaline earth metals, 594
of boron family elements, 596
calculation from bond length, 381
of carbon family elements, 599
covalent, 345, 345*f*
crystal structure, determining from, 498
of halogens, 618
ionic radius comparison, 359–360, 360*f*
main-group elements, 346–347, 346*f*
metallic, 345, 345*f*
of nitrogen family elements, 605
of noble gases, 623
of oxygen family elements, 613
of Period 2 elements, 589*t*
periodic trends in, 346–347, 346*f*
transition elements, 346*f*, 347, 1040–1041*f*
unit cell, determining from, 499
van der Waals radius comparison, 482, 482*f*
Atomic size. *See also* Atomic radius
bond length relationship with, 381, 381*f*
comparison with ionic size, 359–360, 360*f*
covalent radius and, 482, 482*f*
defined, 345, 345*f*, G-2
electronegativity and, 390, 391*f*
entropy and, 909
of main-group elements, 346–347, 346*f*
periodic trends in, 345–346*f*, 345–347, 348*f*, 352*f*
ranking elements by, 347
of transition elements, 346*f*, 347, 1040, 1040–1041*f*, 1041
Atomic solids, 501*t*, 502, 502*f*, G-2
Atomic spectra, 302–307
Bohr model of the hydrogen atom and, 303–305, 305*f*
energy levels of hydrogen atom and, 305–307, 305*f*
excited state and, 304
ground state and, 304
line spectra, 302–305
quantum staircase analogy and, 304, 304*f*
Rydberg equation and, 302–303
stationary states and, 304
Atomic structure
Bohr model of, 303–305, 305*f*
general features of, 55–56, 56*f*, 56*t*
quantum-mechanical model and, 335–345
quantum-mechanical model of, 314–326
Atomic symbol, 56, 57*f*, G-2
Atomic theory
Dalton's theory, 50–51
electron discovery and, 52–54
law of definite (constant) composition and, 47–48, 48*f*, 50
law of mass conservation and, 46–47, 47*f*, 50
law of multiple proportions and, 49–51
Millikan's oil drop experiment and, 53–54, 53*f*
modern, 55–59, 61
nuclear atom model, 52–55
nucleus discovery and, 54–55, 54*f*
Rutherford's α-scattering experiment and, 54–55, 54*f*
Atomic wave functions, 455–456, 456*f*
Atomic weight. *See* Atomic mass
Atomization, 374
Atoms
billiard-ball model of, 52
chemical bonding and properties of, 369–372, 370*f*
defined, 50, 55, G-2
of elements, 44, 44*f*
energy of, 301
general features of, 55–56, 56*f*, 56*t*
history of concept, 43
mass of (*See* Atomic mass)
plum-pudding model of, 54
structure of (*See* Atomic structure)
ATP. *See* Adenosine triphosphate
Aufbau principle, 335, G-2
Aurora borealis, 53, 53*mn*
Autoionization of water, 798–803, 799*f*, 827, G-2
Automobiles
batteries in, 969–970, 970*f*, 975, 975*f*
catalytic converter in, 731–732, 876
exhaust from, 276–277
motor oil viscosity in, 491
tire pressure, 213
Autumn foliage, 1057, 1057*mn*
Avalanche effect, 1083
Average reaction rate, 695, G-2
Avogadro, Amedeo, 96
Avogadro's law
breathing and, 215, 215*f*
defined, 214, G-2
Earth's atmosphere and, 239
ideal gas law and, 216*f*
kinetic-molecular theory and, 231, 234, 234*f*
molecular view of, 234, 234*f*
volume-amount relationship and, 214–215, 214*f*, 223
Avogadro's number
Boltzmann constant and, 898
as conversion factor, 98–102, 98*f*, 100*f*
defined, 96, G-2
symbol for, 235
Axel, Richard, 431
Axial groups, in VSEPR theory, 421, G-2

B

Background radiation, 1095–1096, G-2
Bacterium, approximate composition of, 533, 533*t*
Baking soda, 171
Balanced equations
gas laws to determine, 221–222
information contained in, 117, 117*t*
from molecular scenes, 115–116
nuclear equations, 1078
reaction quotient and, 751, 752–753
redox reactions, 940–943
sequence to obtain overall equation, 120–121
states of matter and, 113–114
steps for balancing, 111–115
stoichiometrically equivalent molar ratios from, 116–120
Balances, 17
Balancing (stoichiometric) coefficient, 112–113, G-2
Ball-and-stick models, 79, 80*t*, 639, 640*f*
Band of stability, 1079, G-2
Band theory, 504–506, G-2
Barite, 493*f*
Barium
atomic radius of, 498
ionic compounds of, 353
isotope of, 1105
mineral sources of, 1008*t*
properties of, 353, 593–594
Barium hydroxide, 167, 897, 916
Barium ions, 69*t*
Barium nitrate, 76
Barium sulfate, 869, 869*mn*
Barometers, 208, 208*f*, G-2
Bartlett, Neil, 624
Bar (unit of measurement), 209, 209*t*
Baseball quantum numbers, 333*mn*
Base-dissociation constant
of ammonia and amines, 818–820, 819*t*
anions of weak acids as weak bases, 820–821
defined, 818, G-2
pH determination from, 818–820
relation to acid-dissociation constant in conjugate pairs, 821–822
tables of, A-11–A-12
Base pairs, 675, G-2
Bases. *See also* Acid-base reactions
Arrhenius, 794, 830
bitter taste from, 793
Brønsted-Lowry acids, 169, 803–808, 804*f*, 827, 830
buffers, 843–852
conjugate acid-base pairs, 804–807, 805*t*, 806*f*, 844, 844*f*, 851
defined, 165, 169, 794, 828, G-2
dissociation of, 165–166
as electrolytes, 165–166, 166*f*
equilibria (*See* Acid-base equilibria)
indicators (*See* Acid-base indicators)
leveling effect on, 827
Lewis bases, 827–830, 877
strong (*See* Strong bases)
structural features of, 167
titration curves, 853–863
uses of, 793*t*
in water, 794–798
weak (*See* Weak bases)
Base units of SI system, 13, 14*t*, G-2
Basic-oxygen process, 1017, 1017*f*, G-2

Catalysts
  biological (enzymes), 732–734, 733f, 734f
  defined, 729, G-3
  heterogeneous, 731–732, 732f, 735
  homogeneous, 730–731, 731f, 735
  lack of effect on equilibrium, 777, 777t
  Ziegler-Natta, 668
Catalytic converter, 731–732, 876
Catalytic hydrogenation, 731, 732f
Catenation, 602, 634, 648, 650, G-3
Cathode rays, 52–53, 52f, G-3
Cathodes, 944, 944mn, G-3
Cathodic protection, 973–974f, 974
Cation-electron pairs, 1093
Cations
  complex, 1048
  defined, 64, G-3
  electron configuration and, 355
  ionic radius of, 359, 359f
  metal cations as Lewis acids, 829
  monatomic, 69t
  polyatomic, 72t
  salts of weakly acidic cations and weakly basic anions, 824–825
Caves, limestone, 870–871, 870f
CCS (carbon capture and storage), 283
Cell membrane structure, 542, 542f
Cell potential, 950–967
  concentration and, 961–967
  defined, 950, G-3
  standard cell potential, 950–953, 950f
  of voltaic cells, 950–959
Cell shape, osmotic pressure and, 567, 567f
Cellular electrochemistry, 982–983
Cellular metabolism, 781, 781f
Cellulose, 671
Celsius temperature scale, 25–27, 26f, 27t, G-3
Cement, specific heat capacity of, 268t
Centi- (prefix), 14t
Ceramic materials
  defined, 512, G-3
  preparing, 512–513
  structure of, 513, 513f
  uses of, 512t, 513
Cerium, 1044, 1045
Cesium
  atomic clock and, 18
  electronegativity of, 392
  properties of, 590–592
Cesium ions, 69t
Cetyl palmitate, 663f
CFCs (chlorofluorocarbons), 103, 602, 602f, 657, 735
CFLs (compact fluorescent lamps), 285
Chadwick, James, 55, 1090
Chain-reaction (chain-growth) polymers, 668, 668f, 669t
Chain reactions, 1105–1106, 1106f, G-3
Chair conformation, of cyclohexane, 640
Chalcocite, 118, 120, 121
Chalcopyrite, 1017, 1017mn
Change in enthalpy. *See* Enthalpy change
Charcoal, as amorphous solid, 504
Charge
  of electrons, 53–54, 53f, 55, 56f, 56t
  in ionic bonding, 65
  of protons, 55, 56f, 56t
Charge density, 545, G-3
Charge distribution, unequal, 146, 146f
Charge-induced dipole forces, 486, 534
Charles, J. A. C., 212, 224
Charles's law
  breathing and, 215, 215f
  combined gas law and, 214
  defined, 213, G-3
  Earth's atmosphere and, 240
  ideal gas law and, 216f
  kinetic-molecular theory and, 231, 233, 233f
  molecular view of, 233, 233f
  pressure-temperature relationship in, 213–214
  volume-temperature relationship in, 212–213, 212f, 223
Chelate, 1047, 1047f, G-3
Chemical analysis
  nuclear magnetic resonance spectroscopy, 649–650
  scanning tunneling microscopy, 500–501, 501f
  spectrometry in, 308–309
  x-ray diffraction analysis, 500, 500f
Chemical bonds, 64, G-3. *See also* Bonding
Chemical changes. *See also* Chemical reactions
  bond energy and, 386–390
  comparison with physical changes, 5, 5f, 6, 7–8
  defined, 5, G-3
Chemical equations
  amount-mass-number relationships in, 118f
  aqueous ionic reactions, 155–157, 156f
  calculating amounts of reactant and product, 116–130
  defined, 111, G-3
  formation, 277–278
  molecular (*See* Molecular equations)
  molecular scenes and, 115–116
  net ionic (*See* Net ionic equations)
  thermochemical, 273–274
  total ionic (*See* Total ionic equations)
  writing and balancing, 111–116
Chemical equilibrium. *See* Equilibrium
Chemical formulas
  alkanes, 639–641f
  for binary covalent compounds, 74–76
  combustion analysis, determination from, 108–109, 108f
  for coordination compounds, 1049, 1050–1051
  defined, 68, G-3
  empirical formulas, 104–108, 110, 370
  for ionic compounds, 68–73
  isomers and, 110–111, 110t
  mass percent and, 102–103
  molecular (*See* Molecular formulas)
  molecular masses from, 76–78
  molecular structure and, 110–111, 110t
  representation of molecules with, 78–79, 80t
  for straight-chain alkanes, 76t
  structural, 79, 105, 110–111
  of unknown compounds, 104–111
Chemical kinetics
  defined, 691, G-3
  reactions rate (*See* Reaction rate)
  theories of, 716–722
Chemical names. *See* Nomenclature
Chemical problem solving. *See* Problem solving
Chemical properties. *See also specific elements*
  defined, 5, G-3
  of Period 2 elements, 587–590
  of transition metals, 1042–1043f, 1042–1044
Chemical reactions. *See also* Chemical equations; *specific reactions*
  acid-base reactions, 165–174
  addition reactions, 651, 660, 668
  alkyl group, 651
  aqueous ionic reactions, 155–157, 156f
  bimolecular, 723, 723t
  combination reactions, 181–182, 182f
  comparison with nuclear reactions, 1074, 1074t
  coupling of reactions to drive nonspontaneous change, 924, 925, 925f
  decomposition reactions, 182–183, 183f
  defined, 5, G-3
  dehydration-condensation (*See* Dehydration-condensation reactions)
  displacement reactions, 159, 167, 184–185f, 184–186
  disproportionation, 609, 614, 618, 619, 621
  elementary, 722–724, 723t
  elimination reactions, 651–652
  enthalpy of reaction, 386–390, 388f
  equilibrium and, 188–190, 189f
  ideal gas law and stoichiometry, 228–230, 229f
  information sequence to determine kinetic parameters of, 704, 705f
  limiting reactants and, 122–127, 123f
  mass conservation during, 46–47, 47f, 50
  mass difference in, 1101–1102
  metathesis, 159, 167, 184
  neutralization, 793, 794
  organic (*See* Organic reactions)
  overall (net) equation and, 120–121, 121f
  precipitation reactions, 157–164, 190
  in radioactive tracer use, 1098
  redox (*See* Redox reactions)
  reversibility of, 188–190, 747
  in sequences, 120–121
  side reactions, 128, 128f
  spontaneous direction of, 896
  standard entropy of reaction, 910–912
  substitution reactions, 652, 660
  termolecular, 723, 723t
  unimolecular, 723, 723t
  yield in, 128–129, 129f
Chemical reactivity, atomic structure and, 353–360
Chemical shift, 649
Chemical system, 258, 258f
Chemistry
  central theme in, 8
  defined, 4, G-3
  electrochemistry (*See* Electrochemistry)
  in everyday life, 3
  inorganic, 63
  matematical operations in, A-1–A-4
  modern materials and applications, 12, 12f
  organic, 63, 633
  prechemical traditions, 10–11, 10f
  quantitative theories of, 12
  thermochemistry (*See* Thermochemistry)
Chernobyl disaster, 1093, 1108
Chiral molecules, 643–644, G-3
Chitin, 671, 671mn
Chlor-alkali process, 1028–1029, 1028f, G-3
Chlorate ions, 72t
Chloride ions, 69t
Chlorides, 394, 394–395f
Chlorine
  abundance of, 1001t
  chemistry of, 619–622
  covalent radius, 345, 345f
  electron configuration, 338t, 339
  industrial production of, 1028–1029, 1028f
  oxoacids, 621–622, 622t
  ozone depletion and, 735, 735f
  properties of, 44, 45t, 354, 617–622, 617f
  reactivity, 343f
  standard enthalpies of formation, 278t
  standard state of, 278
  in water treatment, 570
Chlorine dioxide, 621, 621f
Chlorine gas, 230, 230mn
Chlorine oxides, 621–622, 621f
Chlorine trifluoride, 123–124, 415
Chlorite ions, 72t
Chloroacetic acid, A-8
Chloroethane, 658
Chlorofluorocarbons (CFCs), 103, 602, 602f, 657, 735
Chloroform
  boiling point elevation constant, 560t
  formation of, 389
  freezing point depression constant, 560t
  intermolecular forces, 534f
  polarity of, 428–429
Chloromethane, 655t
Chlorophyll, 308–309, 309f, 829, 829f, 1057
Chlorous acid, 796–797t, A-8
Cholesteric phase, 510, 510f
Cholesterol, molecular structure of, 654f
Chromate ions, 72t, 868, 868f
Chromates, A-13
Chromatography, 83, 83f, G-3
Chromium
  abundance of, 1001t
  appearance of, 1038f
  chromium(II) and chromium(III), 71t
  as dietary trace element, 1064t
  electron configuration, 339, 340t, 343, 1039, 1039t
  oxidation state, 1042, 1042t
  properties of, 1044, 1044t
  purification of, 1011
  in voltaic cells, 948–949
Chromium(III) ions, 941–942, 941mn
Chromosomes, 675
CIGS (copper indium gallium selenide) cells, 284
Cisplatin, 1053
cis-trans isomers, 430, 645, 645t, 646f, 1052, 1052f
Citric acid, 793t, A-9
Clausius, Rudolf, 900
Clausius-Clapeyron equation, 478–479, G-4
Claus process, 281, 1026
Clay ceramics, 512
Climate change, 282–283
Clock, atomic, 18
Closed-end manometers, 208, 209f
Coal, 281, 876, 1000, 1001, 1003
Coal gasification, 281, G-4

Cobalt
abundance of, 1001*t*
appearance of, 1039*f*
cobalt(II) and cobalt(III), 71*t*
as dietary trace element, 1064*t*
electron configuration, 340*t*, 1039*t*
oxidation state, 1042, 1042*t*
Cobalt(II) carbonate, 866*t*
Cobalt(II) chloride hexahydrate, 1043*t*
Cocaine, molecular structure of, 657*f*
Coefficients, balancing in thermochemical equations, 274
Coffee-cup calorimeter, 269–271, 269*f*
Cohesive forces, 489–490, 490*f*
Coke, 1016
Cola, carbon dioxide in can of, 551
Cold packs, 546
Collagen, 674, 674*f*
Colligative properties of solutions, 557–567
applications of, 566–567, 567*f*
boiling point elevation, 559–560, 560*f*, 560*t*
defined, 557, G-4
electrolytes, 557, 557*f*, 564–566, 565*f*
fractional distillation and, 564, 567, 567*f*
freezing point depression, 560–561, 560*t*, 566–567
nonelectrolytes, 557, 557*f*, 558–566
osmotic pressure, 562, 562*f*, 567, 567*f*, 571
solute molar mass and, 563
underlying theme of, 562–563
vapor pressure lowering, 558–559, 558*f*
Collision frequency
defined, G-4
in Earth's atmosphere, 240
kinetic-molecular theory and, 238, 238*mn*
reaction rate and, 692, 693, 717
Collisions
effective, 719, 719*f*
elasticity of, 231
energy of, 693, 693*f*, 717–718, 717*f*, 718*t*
gas laws and, 231–234
Collision theory, 716–719, 718*t*, G-4
Colloids
Brownian motion in, 569
defined, 533, G-4
stability of, 569
structure and properties of, 568–569
Tyndall effect and, 569, 569*f*
types of, 568, 568*t*
in water purification, 570–571
Color, 1056–1057, 1056*f*, 1057*t*, 1058–1059
Column chromatography, 83, 83*f*
Combination reactions, 181–182, 182*f*
Combined gas law, 214
Combustion. *See also* Fuels
biodiesel from, 281–282
calculating heat of reaction, 272–273
of coal, 281
defined, 11, 186, G-4
of gasoline, 114–115, 262, 919–920
of hydrogen, 281–282, 697–698
Lavoisier's theory of, 11, 13
mass conservation and, 46
of methane, 387–388, 388*f*
of octane, 114–115, 263, 263*f*, 919–920
phlogiston theory and, 11
of propane, 117, 117*t*, 911–912
respiration and, 187, 914
spontaneous change and, 897
Combustion analysis, 108–109, 108*f*, G-4
Combustion reactions, 186–187
Common-ion effect, 844, 868–869, 868*f*, G-4
Compact fluorescent lamps (CFLs), 285
Complementary color, 1056
Complex ions
of amphoteric hydroxides, 881–882, 882*f*
calculating concentration of, 878–879
in coordination compounds, 1046–1054, 1046*f*, 1047*t*
defined, 877, 1046, G-4
formation of, 877–878, 878*f*, 878*t*
geometry of, 1047, 1047*t*
isomerism and, 1051–1054
names and formulas for, 1048–1051, 1050*t*
octahedral complexes and, 1055, 1055*f*
solubility of precipitates and, 879–881
square planar complexes and, 1055, 1056*f*
structure of, 877, 877*f*
table of formation constants, A-12
tetrahedral complexes and, 1056, 1056*f*
valence bond theory for, 1055–1056
Composition of matter, 4, G-4
Compounds
amount-mass-number conversions of, 100–102, 100*f*
atomic-scale view of, 44, 44*f*
bonding, 64–67
calculating molecular mass of, 76–78
comparison with mixtures, 45, 81, 81*f*
covalent (*See* Covalent compounds)
defined, 44, G-4
determining formula of unknown compounds, 104–111
elemental mass in, 48–49
fixed ratio, 44
formulas, names, and masses, 68–79
inorganic, 602–603, 659, 662
ionic (*See* Ionic compounds)
law of definite composition and, 47–48, 48*f*, 50
law of multiple proportions and, 49–51
mass fraction and, 47–48
mass of elements in, 104
mass percent and, 47–48, 48*f*, 102–103
molar mass of, 97
molecular mass and, 76–78
molecules of, 44, 44*f*
organic (*See* Organic compounds)
properties of, 44
specific heat capacity of, 268*t*
standard state for, 277
Compton, Arthur, 312
Computer-aided magnetic resonance imaging, 649–650, 650*f*
Concentration
acid dissociation and, 811–813
of buffer components, 845–846, 845*f*, 848, 848*f*
cell potential and, 961–967
defined, 552, 553*t*, G-4
dilution, 152–154, 153*f*
effect of change on equilibrium, 770–773, 772*t*, 777*t*
equilibrium constant and, 756–757
expressing as molarity, 150
as intensive property, 150
interconverting terms, 556–557
mass percent and, 554
molality and, 553–554, 553*t*
molarity and, 552–553, 553*t*
mole fraction and, 553*t*, 554–555
parts by mass, 553*t*, 554, 555
parts by volume, 553*t*, 554, 555
reaction rate and, 692, 696–698, 696*f*
redox titration for finding, 180
visualizing changes in, 154–155
volume percent and, 554
weak-acid equilibria problem solving and, 809–813
Concentration cells
action of, 965, 965*f*
applications of, 966–967, 967*f*
calculating potential of, 964–966
defined, 964, G-4
Condensation. *See also* Dehydration-condensation reactions
defined, 471, G-4
as exothermic and spontaneous process, 896
liquid-gas equilibria and, 476–477, 476*f*
process of, 474
Condensation polymers, 669–670, 670*f*, G-4
Condensed electron configuration, 338*t*, 339, 340*t*
Condensed formula, 639, 640*f*
Condensed states, 469
Conduction bands, 505, 505–506*f*, 506, G-4
Conductivity. *See* Electrical conductivity
Conductometric methods for reaction rate measurement, 699
Conductors, 397, 506, 506*f*, G-4
Conjugate acid-base pairs
acid-base buffers and, 844, 844*f*, 851
Brønsted-Lowry acid-base definition and, 804, 805*t*
defined, 804, G-4
identifying, 805
ranking, 806–807
relation between acid- and base-dissociation constants, 821–822
strength of, 805–808, 806*f*
Conservation of energy, 9, 261, 284–285, 896, 896*n*
Constant-pressure calorimetry, 269–271, 269*f*
Constant-pressure molar heat capacity, 904
Constant-volume calorimetry, 271–273, 272*f*
Constant-volume molar heat capacity, 904
Constitutional (structural) isomers
in alkanes, 641–642, 642*t*
in coordination compounds, 1051–1052, 1052*f*, 1054*f*
defined, 1051, G-4
properties of, 110–111, 110*t*
Constructive interference, 443
Contact process, 1025–1026, G-4
Controlled experiments, 12, G-4
Controlled orientation, in polymers, 519
Control rods, 1107
Convective mixing, 240
Conversion factors
Avogadro's number as, 98–102, 98*f*, 100*f*
choosing, 18–19
constructing, 18
converting between unit systems, 19
defined, 18, G-4
interconverting wavelength and frequency, 297–298, 301–302
for units raised to a power, 23
Coordinate covalent bond, 1055, G-4
Coordination compounds
bidentate ligands and, 1047, 1048*t*
chelate and, 1047, 1047*f*
complex ions in, 1046–1054, 1046*f*, 1047*t*
components of, 1046, 1046*f*
constitutional isomers in, 1051–1052, 1052*f*, 1054*f*
coordination isomers in, 1051
coordination numbers and, 1046, 1047*t*, 1049
counter ions in, 1046, 1049, 1050
crystal field theory and, 160–1062*f*, 1056–1059*f*, 1056–1062, 1057*t*
defined, 1046, G-4
donor atoms and, 1047
formulas and names of, 1048–1051, 1050*t*
geometric isomers and, 1052–1053, 1052*f*
isomerism in, 1051–1054
ligands and, 1046, 1047–1048, 1048*t*
linkage isomers in, 1052, 1052*f*
magnetic properties of, 1060–1061
monodentate ligands and, 1047, 1048*t*
optical isomers and, 1053–1054, 1053*f*
polydentate ligands and, 1047, 1048*t*
stereoisomers and, 1052–1054*f*
valence bond theory and, 1055–1056
Coordination isomers, 1051, G-4
Coordination number
in coordination compounds, 1046, 1047*t*, 1049
in crystalline solids, 494, 495–496, 495*f*
defined, G-4
Copernicium, 1092
Copolymers, 518, 669, G-4
Copper
abundance of, 1000, 1001*t*
alloys, 1013, 1013*f*, 1013*t*
appearance of, 1039*f*
in concentration cells, 965, 965*f*
conductivity of, 397, 538
copper(I) and copper(II), 71*t*, 72, 72*f*
cubic crystal structure of, 499, 503*f*
as dietary trace element, 1064*t*
in displacement reactions, 185, 185*f*
in electrolytic cells, 974, 975*f*
electron configuration, 339–340, 340*t*, 358, 1039, 1039*t*
electrorefining, 1018, 1018*f*
flame tests with, 308*f*
isolation of, 1010, 1017
metallurgy of, 1017–1018, 1018*f*
oxidation state, 1042, 1042*t*
properties of, 6, 7*t*
roasting, 118–120
sources of, 118
specific heat capacity of, 268*t*
in voltaic cells, 945–946*f*, 945–949, 953, 963–964, 964*f*
Copper(II) chloride, 874
Copper(II) hydroxide, 874
Copper(II) sulfate pentahydrate, 1043*f*
Copper indium gallium selenide (CIGS) cells, 284
Copper(I) oxide, 118, 120, 121, 923–924
Copper(I) sulfide, 118, 119, 1010, 1017–1018
Core electrons, 342, 350
Core of Earth, 998, 998–999*f*, 998*t*, G-4
Corrosion, 972–974, G-4

Corundum, 1019
CO-shift reactions, 281
Cosmic radiation, 1088, 1095
Cosmology, element synthesis and, 1110–1111, 1110*f*
Cottrell precipitator, 569
Coulomb (C), 54, 56*n*, 950, G-4
Coulomb's law
  charge density and, 545
  defined, 65, G-4
  on intermolecular forces, 469–470, 482, 535
  lattice energy and, 375–376
  nuclear charge and, 333
Counter ions, 1046, 1049, 1050, G-4
Counters, for detection of radioactive emissions, 1083–1084, 1084*f*
Coupling of reactions, 924, 925, 925*f*, G-4
Covalent bonds
  bonding pairs, 380
  bond order in, 380, 382*t*, 411
  coordinate, 1055
  defined, 66, 370, G-4
  double (*See* Double bonds)
  electron sharing and, 370, 370*f*, 379
  formation of, 66–67, 66*f*, 175, 175*f*, 379, 379*f*
  in intermolecular force comparison, 483*t*
  lone pairs, 380
  model of, 379–383, 386
  nonpolar, 392, 392*f*
  pi bond, 452–454, 453*f*, 455, 455*f*, 644
  polar, 390, 390*f*, 392, 393*f*
  in polyatomic ions, 67, 67*f*
  sigma bond, 452, 452–453*f*, 453, 455
  single (*See* Single bonds)
  triple (*See* Triple bonds)
Covalent bond theories
  molecular orbital theory, 455–463
  valence bond theory, 443–451
Covalent compounds
  binary, 73*t*, 74–76
  comparison with ionic compounds, 67
  defined, 64, G-4
  formation of, 66–67, 66*f*
  names and formulas, 74–76
  network covalent solids, 383, 383*f*, 386
  nonelectrolytes, 150
  oxidation number and, 176
  properties of, 383, 386
  redox reactions and, 175, 175*f*, 181–182, 182*f*
  in water, 150
Covalent hydrides, 586
Covalent radius, 345, 345*f*, 381*f*, 482, 482*f*, G-4
Cracking, of ionic compounds, 377, 377*f*
Cristobalite, 504*f*
Critical mass, 1106, G-4
Critical point, 480, 480*f*, G-4
Critical pressure, 480
Critical temperature, 480
Crosslinks, in polymers, 518, G-4
Crude oil refining, 567, 567*f*
Crust of Earth, 998–999*f*, 998–1000, 998*t*, 1001*t*, G-4
Crutzen, Paul J., 735
Cryolite, 619, 1019
Crystal defects, in electronic materials, 507, G-4
Crystal field splitting energy, 1058, 1059–1060, G-4
Crystal field theory, 1056–1062
  chlorophyll in, 1057
  color and, 1056–1057, 1056*f*, 1057*t*, 1058–1059
  complementary color and, 1056
  defined, 1056, G-4
  high-spin complexes and, 1060–1061
  low-spin complexes and, 1060–1061
  magnetic properties and, 1060–1061
  octahedral complexes in, 1057–1058
  spectrochemical series and, 1059, 1059*f*
  splitting *d* orbitals in octahedral field of ligands, 1057–1058
  square planar complexes in, 1062, 1062*f*
  strong-field ligands in, 1058, 1058*f*, 1060, 1060–1061*f*
  tetrahedral complexes in, 1061, 1062*f*
  weak-field ligands in, 1058, 1058*f*, 1060, 1060–1061*f*
Crystal lattice, 493–494, 493*f*
Crystalline solids
  atomic, 501*t*, 502, 502*f*
  atomic radius determination, 498
  coordination number in, 494, 495–496, 495*f*
  crystal lattice of, 493–494, 493*f*
  defined, G-4
  dissolution and, 908, 908*f*
  entropy and, 908, 908*f*
  ionic, 501*t*, 502–503
  metallic, 501*t*, 503, 503*f*
  molecular, 501*t*, 502, 502*f*
  network covalent, 501*t*, 503–504, 504*t*
  packing efficiency in, 496–498, 497*f*
  structural features of, 493, 493*f*
  types and properties of, 501–504, 501*t*
  unit cell of, 493–499, 497*f*, 499*f*
  x-ray diffraction analysis of, 500, 500*f*
Crystallinity of polymers, 516, 516*f*
Crystallization, 83, G-4
Crystals and crystal structures. *See* Crystalline solids
Crystal violet, 861*f*
Cubic centimeter (cm^3), 15, 16*f*
Cubic closest packing, 497*f*, 498, G-5
Cubic decimeter (dm^3), 15, 16*f*
Cubic meter (m^3), 15, G-5
Cubic unit cell, 494–499, 495*f*, 497*f*, 499*f*
Curie, Marie Sklodowska, 1075, 1078, 1090
Curie, Pierre, 1075, 1090
Curie (Ci), 1084, G-5
Current. *See also* Electrical conductivity
  electrical, 14*t*, 52
  measuring to find charge, 980
  in photoelectric effect, 301, 301*f*
  in strong electrolytes, 557, 557*f*
Cyanate ions, 412, 824
Cyanide ions
  formula for, 72*t*
  in gold leaching, 1009, 1011
  isoelectronic nature of, 603
  triple bonds in, 666
Cyanides, A-13
Cyclic hydrocarbons, 640, 641*f*, G-5
Cycloalkanes, 640, 641*f*
Cyclobutane, 641*f*, 708–709
Cyclobutanol, 661
Cyclobutanone, 661
Cyclohexane, 640, 641*f*, 660*f*
Cyclohexanol, 656
Cyclohexene, 656, 660, 660*f*
Cyclone separator, 1009, 1009*f*
Cyclopentane, 641*f*
Cyclopropane, 641*f*, 714, 911
Cyclo-S_8, 612, 612*f*
Cyclotron accelerator, 1091, 1092*f*
Cysteine, 540, 672*f*
Cytochromes, 1064
Cytosine, 675

D

Dacron, 670
DAF (dissolved air flotation), 570
Dalton, John, 50–51, 225
Dalton (Da), 57, G-5
Dalton's law of partial pressures
  application of, 226–227
  defined, 225–226, G-5
  discovery of, 225
  Earth's atmosphere and, 239
  kinetic-molecular theory and, 231, 233, 233*f*
  molecular view of, 233, 233*f*
Darmstadtium, 343, 1092
Data, 12, G-5
Davisson, C., 311
d block, 595, 1037, 1038, 1038*f*. *See also* Transition elements
De Broglie, Louis, 310
De Broglie wavelength, 311, 311*t*, G-5
Debye, Peter, 428*n*
Debye (D), 428
Decane, 76*t*, 641*f*
Deca- (prefix), 72, 73*t*
Decay constant, 1085, 1086*t*, G-5
Decay rate, 1084–1087, 1086*f*, 1086*t*
Decay series, 1082, 1082*f*, G-5
Decimal places, significant figures and, 29
Decimal prefixes, 14, 14*n*, 14*t*
Deci- (prefix), 14*t*
Dec- (numerical root), 639*t*
Decomposition reactions, 182–183, 183*f*, 1003, 1005, 1006
Decompression sickness, 551, 551*mn*
Definite composition, law of, 47–48, 48*f*, 50
Degree of polymerization, 514, 514*t*, G-5
Dehydration-condensation reactions
  acid anhydrides and, 663
  defined, 611, G-5
  disacchardie formation from, 671
  esters and, 664
  halogens and, 621
  in monomers of condensation polymers, 669
  nucleoside triphosphates and, 675
  peptide bond formation from, 673
  polyphosphates and, 611
Dehydration of alcohols, 656
De-icers, 566, 567
Deka- (prefix), 14*t*
Delocalized electrons, 371, 397, 410, 647, 659
Democritus, 43, 50
Dendrimers, 518
Denitrification, 571, 1005
Density
  of alkali metals, 590, 591
  of alkaline earth metals, 593, 594
  of boron family elements, 596
  calculating from mass and volume, 24–25
  of carbon family elements, 599
  of common substances, 24*t*
  defined, 23, G-5
  electron density, 379, 380*f*, 394, 635
  gases and, 24, 24*t*, 206, 222–224
  of halogens, 618
  as intensive property, 27, 27*f*
  of nitrogen family elements, 605
  of noble gases, 623
  of oxygen family elements, 613
  SI units, 24
  of transition elements, 1041–1042, 1041*f*
  of water, 24*t*, 492
Dental amalgam, 1013*t*
Dental fillings, 959, 959*f*
Deoxyhemoglobin, 1063
Deoxyribonucleic acid. *See* DNA
Deoxyribose, 675
Deposition, 471, 471*f*, G-5
Derived units of SI system, 13–14, G-5
Desalination, 571, G-5
Desulfurization, 281
Detergents, 569, 1008
Deuterium, 1024–1025, 1024*t*, 1091, 1109
Deuterons, 1091, G-5
Dextrorotatory isomers, 643
Diagonal relationships in periodic table, 595, 595*f*, 598, 604, G-5
Diamagnetism, 357, 358*f*, 460–461, 460*f*, 1037, 1044, G-5
Diamond
  allotropism, 600
  covalent bonding in, 383, 383*f*
  light refraction and dispersion in, 298
  properties of, 504, 504*t*
  standard molar entropy and, 909
  standard state of, 277
Diaphragm-cell method, 1028–1029, 1028*f*
Diastereomers. *See* Geometric isomers
Diatomic molecules
  beryllium, 458, 458*f*
  bond polarity of, 428
  defined, 44
  of halogens, 343
  heteronuclear, 462–463, 462*f*
  homonuclear, 458–462
  lithium, 458, 458*f*
  orbital overlap in, 443*f*, 444
  oxygen, 44, 97
  vibration of, 384, 384*f*
Diberyllium, 458, 458*f*
Diborane, 597, 598*f*
Dichlorine heptoxide, 621, 621*f*
Dichlorine monoxide, 621, 621*f*
Dichloroethylene, 430, 455, 455*f*
Dichromate ions, 72*t*, 941–942, 941*mn*
Dietary trace elements, 1063–1064, 1064*t*
Diethylamine, 657, 658, 819*t*, A-11

Diethyl ether
boiling point elevation constant, 560*t*
freezing point depression constant, 560*t*
heats of vaporization and of fusion, 471*f*
infrared spectrum of, 385*f*
surface tension of, 489*t*
vapor pressure of, 478, 478*f*
viscosity of, 489*t*
Differentiation, geochemical, 998–999, 999*f*, G-5
Diffraction
defined, 299, G-5
electron, 311, 311*f*
of waves, 299, 299*f*
of x-rays, 311, 311*f*, 500, 500*f*
Diffraction pattern, 299, 299*f*, 311, 311*f*
Diffusion, 237–238, 238*f*, G-5
Dihydrogen phosphate ions, 72*t*
Dilithium, 458, 458*f*
Dilution, 152–154, 153*f*
Dimensional analysis, 19, G-5
Dimers, 595, 595*f*
Dimethoxymethane, 649, 649*f*
Dimethylamine, 657, 819–820, 819*t*, A-11
Dimethylbenzene, 385*f*, 648
Dimethylbutane, 487–488, 640
Dimethyl ether, 110*t*, 484*f*
Dimethylformamide, 664*f*
Dimethylmethanamide, 664*f*
Dinitrogen monoxide, 239*t*, 608, 608*t*
Dinitrogen pentoxide, 354, 608*t*, 711, 711–712*f*, 752–753
Dinitrogen tetroxide
decomposition of, 899
equilibrium with nitrogen dioxide, 748–751, 748*f*, 750*t*, 751*f*, 759–760
Lewis structure of, 414
properties of, 608*t*, 609
in rocket fuel, 125–127
Dinitrogen trioxide, 608*t*
Dinitrophenol, 861*f*
Diphosphate ions, 611, 611*f*
Diphosphoric acid, 666
Dipole-dipole forces, 483–484, 483*t*, 534, 534*f*, G-5
Dipole-induced dipole forces, 483*t*, 486, 534*f*, 535, G-5
Dipole moment
boiling point and, 483–484, 484*f*
defined, 428, G-5
molecular polarity and, 428–429
Di- (prefix), 73*t*
Direction of reaction. *See* Reaction direction
Disaccharides, 671, 671*f*, G-5
Discoveries, chronology of, 295*mn*
Disintegration series, 1082, 1082*f*
Disodium hydrogen phosphate, 825
Dispersion, 298
Dispersion (London) forces
alkanes and, 641
boiling points and, 641
defined, G-5
intermolecular, 483*t*, 486–487
in proteins, 540, 540*f*
in soaps, 541
in solutions, 534*f*, 535, 548
Displacement reactions
double, 159, 167, 184
in redox reactions, 184–185*f*, 184–186
single, 184
Disproportionation reaction, 609, 614, 618, 619, 621, G-5
Dissociation. *See also* Acid-dissociation constant; Base-dissociation constant
of acids, 165–166, 189, 747, 795–798, 795*f*, 811–813, 844*t*
of bases, 165–166, 170, 189
of ionic compounds, 146–147, 146*f*
Dissolution
of covalent compounds, 150
enthalpy change and, 544, 545–546*f*, 548–549, 548*f*
entropy change and, 547–549, 897, 908, 908*f*
of ionic compounds, 146–147, 146*f*
solution process and, 544–549
solvent properties of water and, 491
Dissolved air flotation (DAF), 570
Dissolved polymers, 516–517
Distillation
defined, G-5
fractional, 83, 564, 567, 567*f*
in metallurgy, 1012
as separation technique, 83, 83*f*
Disulfide bridge, 540*f*, 541
Disulfuric acid, 666, 1027, 1027*mn*
Division, significant figures and, 30
DNA (deoxyribonucleic acid)
molecular depiction, 80*t*
replication of, 678, 678*f*
structure of, 542–544, 543*f*, 675, 676*f*
transcription functions of, 677, 677*f*
DNA fingerprinting, 680, 680*f*
DNA polymerase, 678
DNA sequencing, 679–680, 679*f*
Dolomite, 1000, 1000*mn*
Donor atoms, 1047, G-5
Doped semiconductors, 507–509, 508*f*
Doping, defined, 508, G-5
d orbitals
defined, G-5
hybridization of, 451
orbital overlap and, 444
quantum mechanics and, 324–325, 325*f*, 326*t*
splitting in octahedral field of ligands, 1057–1058
Double bonds
in alkenes, 644
bond angle and, 420
bond energy of, 382, 382*t*
bond order and, 380
cis and *trans* structures of, 455, 455*f*
color test for, 651, 651*f*
defined, G-5
in fatty acids, 663
functional groups with, 659–666
inorganic compounds with, 662
Lewis acids with, 829
Lewis structures for, 408–409
pi bond and, 452, 453*f*
restricted rotation by, 637, 644
sigma bond and, 452, 453*f*
strength of, 453–454
Double-displacement reactions, 159, 167, 184
Double helix in DNA, 543–544, 543*f*, 675, 676*f*, G-5
Downs cell, 1014, 1015*f*, G-5
Dow process, 1021–1022, 1022*f*
Dry cell batteries, 968
Dry ice, 471, 479, 899
Dual polarity
of alcohols, 535–536
of antibiotics, 542, 543*f*
effects on solubility, 535–537
of membranes, 542, 542*f*
of soaps, 541, 541*f*
Dubnium, 1092
Ductility, 396–397
Dynamic equilibrium, 188–190, 477, G-5
Dynamite, 215

E

EA. *See* Electron affinity
Earth
abundance of elements on, 997–1000, 998*t*, 1001*t*
atmosphere of (*See* Atmosphere of Earth)
biosphere, 999–1000, 1001*t*, 1007
core of, 998, 998–999*f*, 998*t*
crust of, 998–999*f*, 998–1000, 998*t*, 1001*t*
de Broglie wavelength of, 311*t*
differentiation of, 998–999, 999*f*
element cycling on, 1002–1008
formation and layering of, 997–999, 998*f*, 998*t*
hydrosphere, 999, 1001*t*
impact of life on, 999–1000
lithosphere, 999, 1001*t*
mantle of, 998, 998–999*f*, 998*t*
oceans, Earth without, 268, 268*mn*
water and temperature of, 268
EDTA (ethylenediaminetetraacetate) ion, 1047–1048
Effective collisions, 719, 719*f*, G-5
Effective nuclear charge, 333–334, 334*f*, 346–347, 352, 359, G-5
Effusion, 236–237, 236*f*, G-5
e_g orbitals, 1058, G-5
Einstein, Albert
on atomic bomb, 1106
on mass-energy relationship, 47, 1101
photon theory and, 301, 310
theory of relativity, 310
Elastomers, 518, 518*t*, G-5
Electrical charges, effects of, 52, 52*mn*
Electrical conductivity
band theory and, 505, 506, 506*f*
colligative of solutions, 557, 557*f*
of covalent substances, 386
of ionic substances, 147–148, 147*f*, 378, 378*f*
ion mobility and, 378*f*
of metals, 397
of Period 3 chlorides, 395*f*
Electrical currents, 14*t*, 52
Electrical potential, 949, 949*mn*
Electrocatalysts, 971
Electrochemical cells, 939, 944, 944*f*, G-5
Electrochemical potential, 982–983
Electrochemical redox, 1011, 1022
Electrochemistry, 938–984
in batteries, 968–971
biological energetics and, 982–983
cellular, 982–983
corrosion and, 972–974
defined, 939, G-5
electrochemical cells, 939, 944, 944*f*
electrolysis, 976–981, 977*f*, 979–980*f*
free energy and electrical work, 959–967
half reactions in, 940–943, 944*mn*, 965, 965*f*
microanalysis and, 967*f*
Electrodes
active vs. inactive, 947, 947*f*
charges of, 947
defined, 944, G-5
half-reactions occurring at, 944*mn*
ion-specific, 967, 967*f*
microelectrode for microanalysis, 967*f*
pH, 803, 967, 967*f*
Electrolysis
in aluminum production, 1019–1020, 1020*f*
of aqueous ionic solutions, 978–979
in chlorine production, 1028–1029, 1028*f*
defined, 976, G-5
of mixed molten salts, 976–977
overvoltage in, 978
of pure molten salts, 976
stoichiometry of, 980–981, 980*f*
of water, 183, 283, 977–978, 977*f*, 1023
Electrolytes
acids and bases as, 165–166, 166*f*
defined, 148, 944, G-5
electrical conductivity of, 557, 557*f*
ionic compounds as, 148
strong, 557, 557*f*, 564–566, 565*f*
weak, 557, 557*f*
Electrolytic cells
in aluminum manufacture, 1020*f*
aqueous ionic solutions and, 978–979
comparison with voltaic cells, 944, 944*f*, 974–975, 976*t*
construction and operation of, 974–975
defined, 944, G-5
in lead-acid batteries, 975, 975*f*
mixed molten salts, 976–977
nonstandard half-cell potentials, 977–978
overvoltage and, 978
pure molten salts, 976
stoichiometry of electrolysis and, 980–981, 980*f*
water and, 977–978, 977*f*
Electrolytic decomposition, 183
Electrolytic enrichment, 1025
Electromagnetic radiation, 295–297, G-5
Electromagnetic spectrum, 296–297, 297*f*, G-5
Electrometallurgy, 1008
Electromotive force, 950
Electron affinity (EA)
comparison with electronegativity, 390*n*
defined, 351, G-6
magnesium oxide and, 377
as positive value, 352*n*
trends in, 351–352, 352*f*
Electron capture, 1076*t*, 1077, G-6
Electron cloud depictions, 318, 323, 323–325*f*, G–6

Electron configuration
alkali metals, 343, 355, 591, 592
alkaline earth metals, 355, 593, 594
atomic properties and, 345–352
aufbau principle and, 335
boron family elements, 596
carbon family elements, 599, 634*f*
chemical reactivity and, 343*f*
condensed, 338*t*, 339, 340*t*
defined, 331, G-6
determining, 344–345
filling order and, 335*f*, 339–342, 341*f*, 342*mn*, 343, 456–457
within groups, 343
halogens, 618
Hund's rule and, 336–337, 457, 587
magnetic properties and, 357–358, 358*f*
of main-group ions, 355–356, 355*f*
in many-electron atoms, 332–335
of molecules, 456–457
nitrogen family elements, 605
noble gases, 623
notation for, 335, 338
orbital diagrams and, 335–337, 336*f*, 338*f*, 338*t*, 340*t*
oxygen family elements, 613
Period 1 elements, 335–336, 338*f*, 341*f*
Period 2 elements, 336, 336*f*, 338*f*, 341*f*
Period 3 elements, 338, 338*t*, 341*t*
Period 4 elements, 339–340, 340*t*, 341*f*
periodic table and, 335–345, 335*f*, 341*f*, 356–357
principles of, 340–342
transition elements, 339–340, 340*t*, 341*f*, 342–343, 1038–1040, 1039*t*
of transition metal ions, 356–357
Electron deficiency
boron and, 597
defined, G-6
Lewis acids and, 828–829
Lewis structures and, 413–414
Electron density, 379, 380*f*, 394, 635
Electron density contour maps, 379, 380*f*
Electron density diagram, 318, 319*f*, G-6
Electron density relief maps, 323, 323*f*, 379, 380*f*, 392, 392*f*, 394, 394*f*
Electron-dot formulas, 79
Electron-dot symbols. *See* Lewis electron-dot symbols
Electronegativity difference, 392–393, 393*f*, G-6
Electronegativity (EN)
of alkali metals, 591
of alkaline earth metals, 594
atomic size and, 390, 391*f*
bond polarity and, 392–393
of boron family elements, 596
of carbon family elements, 599, 634
comparison with electron affinity, 390*n*
defined, 390, 390*n*, G-6
of halogens, 617, 618, 622
of nitrogen family elements, 605
of noble gases, 623
oxidation number and, 392, 412
of oxygen family elements, 613
Pauling scale and, 390, 391*f*
of Period 2 elements, 589*t*
of transition elements, 1040, 1040–1041*f*, 1041
trends in, 390, 392
valence-state, 1042–1043
Electron-group arrangements, 418, 418*f*
Electronic (analytical) balances, 17
Electronic materials
crystal defects in, 507
doped semiconductors, 507–509, 508*f*
doping, 508
p-n junction manufacture, 508–509, 508*f*
solar cells, 509
Electron microscope, 311–312, 312*f*
Electron-pair delocalization, 410, G-6
Electron pair donation, 827–830
Electron pooling, 370*f*, 371
Electrons
categories of, 342
cathode rays and, 52–53, 52*f*
charge of, 53–54, 53*f*, 55, 56*f*, 56*t*, 176
as current, 301, 301*f*
de Broglie wavelength of, 311, 311*t*
defined, G-6
delocalized, 371, 397, 410, 647, 659
diffraction pattern, 311, 311*f*
discovery of, 52–54
energy of, 304
filling order, 335*f*, 339–342, 341*f*, 342*mn*, 343, 456–457
inner (core), 342, 350
mass of, 53–54, 56*t*
in molecular orbitals, 455–457
movement in oxidation-reduction reactions, 174–175, 175*f*
outer, 342, 350
predicting number lost or gained, 65
probable location of, 318–319, 319*f*
properties of, 52–54, 56*t*
quantum numbers and, 319–320, 320*t*
quantum numbers of, 332, 332*t*
repulsions, 333–334
sharing, 64
transferring, 64, 65
valence, 342, 350, 396, 405, 406
wave nature of, 310–312, 310*f*
Electron-sea model, 395–396, G-6
Electron sharing, covalent bonding and, 370, 370*f*, 379
Electron spin, 443, 443*f*
Electron-spin quantum number, 332–333, 332*f*
Electron transfer, ionic bonding and, 175, 175*f*, 370, 370*f*, 372–373, 373*f*
Electron transport chain (ETC), 982–983, 983*f*
Electron volt (eV), 1102–1103, G-6
Electrophoresis, 680
Electrorefining, 1012, 1018, 1018*f*, G-6
Electrostatic interactions
in bonding, 65
Coulomb's law and, 375–376
electronegativity and, 390
energy-level splitting and, 333–335
lattice energy and, 375
nuclear stability and, 1080
potential and kinetic energy in, 9–10
Elementary reactions, 722–724, 723*t*, G-6
Elements. *See also specific elements*
abundance of, 997–1000, 998*t*, 1001*t*
alloying, 1012–1013, 1013*f*, 1013*t*
amount-mass-number conversions involving, 98–100, 98*f*
atomic masses of, 57–59, 61, 96
atomic number of, 56, 61, 63–64
atomic-scale view of, 44, 44*f*
atomic theory and, 50–51
atoms of, 44, 44*f*
in carbon cycle, 1002–1004, 1003*f*
classification of, 62–63, 63*f*
converting minerals to, 1010–1011, 1012*f*
defined, 43, 44, G-6
emission and absorption spectra of, 308, 308*f*
in environmental cycles, 1002–1008
formation in nuclear reaction, 1074
geochemical differentiation of, 998–999, 999*f*
history of, 43
ionization energy of, 348–350, 349*f*, 351*t*
isolation and uses of, 1014–1025
isotopes of, 57
law of multiple proportions and, 49–51
line spectra of, 302, 303*f*
mass determination from mass fraction, 103–104
mass-mole-number relationships for, 98*f*
mass of element in compound, 48–49
mass percent and, 47–48, 48*f*
mass percent in compounds, 102–103
metallurgy and, 1008–1013
minerals as sources of, 1008, 1008*t*
molar mass of, 97
as molecules, 44, 44*f*, 66, 67*f*, 97
in nitrogen cycle, 1004–1005, 1004*f*
number of, 61
oxidation number and, 176
oxides of, 354–355, 354*f*
periodic table of (*See* Periodic table of the elements)
in phosphorus cycle, 1005–1008, 1006*f*
properties of, 44
ranking by atomic size, 347
in redox reactions, 181–188
refining, 1012, 1012*f*
sources of, 1000–1001, 1002*f*, 1008, 1008*t*
specific heat capacity of, 268*t*
standard state for, 277
in stars, 1110–1111, 1110*f*
symbols for, 56, 57*f*
trace, 1063–1064, 1064*t*
Element symbol, 56, 57*f*
Elimination reactions, 651–652, G-6
Emission of photons, 304, 304*f*
Emission spectrum, 308, 308*f*, G-6
Empirical formulas
defined, 104, G-6
determination of, 105–106
different compounds with same formula, 110, 110*t*
of ionic compounds, 370
molecular formulas obtained from, 106–108
Emulsions, 568*t*, 569
EN. *See* Electronegativity
Enantiomers. *See* Optical isomers
Endothermic processes
bond energy, 381, 387
defined, 266, G-6
enthalpy and, 266–267, 266*f*
entropy change in surroundings, 912–913
phase changes, 471
solution process, 544, 545*f*
spontaneous reactions, 896–897, 897*f*, 915–916, 915*f*
End point, 173, 173*f*, 179, 179*f*, 861, G-6
End-product feedback inhibition, 781, 781*f*
End-to-end overlap, 452
Energy
activation (*See* Activation energy)
of atoms, 301, 305–307, 305*f*
from blackbody radiation, 299–300, 300*f*, 300*n*
of collisions, 693, 693*f*, 717–718, 717*f*, 718*t*
common quantities of, 262*f*
conservation of, 9, 261, 284–285, 896, 896*n*
conversion of mass into, 47
conversion of potential energy to kinetic energy, 9–10, 9*f*
crystal-field splitting, 1058, 1059–1060
defined, 8, G-6
dispersal of, 897–898, 912–913
distinction from matter, 298–299
electromagnetic, 295–297
as extensive property, 28
flow to and from a system, 258–260
forms and interconversion, 257–264
free (*See* Free energy)
future use of, 281–285
heat, 258, 259, 260*f*
importance in study of matter, 8–10
interconversion of mass and, 1101–1104
internal, 258–260, 262, 896
ionization (*See* Ionization energy (IE))
kinetic (*See* Kinetic energy)
lattice (*See* Lattice energy)
levels of, 320–321
nuclear, 284 (*See also* Nuclear fission; Nuclear fusion)
nuclear binding energy, 1102–1104, 1104*f*
potential (*See* Potential energy)
quantization of, 898
quantum theory of, 299–300
radiant, 295–297
renewable, 281
solar, 284
as state functions, 262–263
of systems, 258–260, 306*mn*
thermal, 258
units of, 261–262
wave-particle duality of, 310–313, 310*f*
work, 258, 259–260, 260*f*
Energy efficiency, 920
Energy-level diagrams, 717*f*
Energy-level splitting, 333–335
Energy threshold, 716
Energy transfer
as heat, 258, 259, 260*f*
path independence of, 262–263, 263*f*
sign conventions, 261*t*
as work, 259–260, 260*f*
-ene (suffix), 644, 655*t*
English unit system, 14, 15*t*, 19
Enthalpy
defined, 265, G-6
endothermic processes and, 266–267, 266*f*
exothermic processes and, 266, 266*f*
Hess's law of heat summation and, 275–277

lattice energy and, 373–376
as state function, 266
Enthalpy change
comparison with energy change, 265–266
defined, 265, G-3
free energy change and, 920–924, 921*t*, 924*f*
Hess's law of heat summation and, 275–277
phase changes and, 470–472
in solution process, 544, 545–546*f*, 548–549, 548*f*
spontaneous, 896–897
standard enthalpy of formation and, 277–280, 278*t*, 279*f*
standard enthalpy of reaction and, 277–280
in thermochemical equations, 273–274
thermodynamic variables in, 900–905
using to find amount of substances, 274
Enthalpy diagrams, 266, 266*f*, 267, 546, 546*f*, G-6
Enthalpy of formation, 277–280, 278*t*, 279*f*
Enthalpy of hydration. *See* Heat of hydration
Enthalpy of reaction, 386–390, 388*f*
Enthalpy of solution. *See* Heat of solution
Enthalpy of sublimation. *See* Heat of sublimation
Enthalpy of vaporization. *See* Heat of vaporization
Entropy
defined, 898, G-6
as extensive property, 906
number of microstates and, 898–900
second law of thermodynamics and, 905
standard molar entropy, 905–910
as state function, 898–899
Entropy change
with atomic size or molecular complexity, 909, 909*f*
calculating in reactions, 910–916
equilibrium state and, 914–915
with gases dissolved in liquids, 908, 908*f*
gas expansion and, 899–901
during isothermal process, 900–902
phase changes and, 547, 898–899, 902–904, 903*t*, 906–907, 907*f*
predicting relative values, 909–910
quantitative meaning of, 899–905
signs of, 911, 920–924, 921*t*, 924*f*
with solids and liquids dissolved in water, 908, 908*f*
in solution process, 547–549
spontaneous, 913–916, 915*f*
standard entropy of reaction, 910–912
in surroundings, 912–914
temperature and, 904–906, 907*f*
Environmental cycles, 1002–1008
carbon, 1002–1004, 1003*f*
defined, 1002, G-6
nitrogen, 1004–1005, 1004*f*
phosphorus, 1005–1008, 1006*f*
Environmental issues
acid rain (*See* Acid rain)
chlorofluorocarbons, 602, 602*f*, 735
fuels and, 281–285
global warming, 282–283, 384, 1004
greenhouse effect, 282–283, 1004
ozone depletion, 735, 735*f*
phosphates and pollution, 1008
pollution (*See* Pollution)
polychlorinated biphenyls and, 602
smog, 224, 414, 609, 609*mn*
Enzymes
as catalysts, 732–734, 733*f*, 734*f*
defined, 732, G-6
hydrogen bonding in, 485
mechanisms of action, 734
in metabolic pathways, 781, 781*f*
molecular shape of, 431, 432*f*
optical isomers and, 644, 644*f*
Enzyme-substrate complex (ES), 733–734, 733*f*, G-6
Epinephrine, molecular structure of, 657*f*
Epsom salt, 72, 593
Equations. *See* Chemical equations
Equatorial groups, in VSEPR theory, 421, G-6
Equilibrium, 746–781
comparison with kinetics, 747
concentration changes and, 770–773, 772*t*, 777*t*
dynamic, 188–190, 477
entropy change and, 914–915
expressing with pressure terms, 756–757
free energy change and, 926–932
heterogeneous, 752, 752*f*
homogeneous, 752
lack of catalysts and, 777, 777*t*
law of chemical equilibrium, 750
Le Châtelier's principle and, 770–780
liquid-gas equilibria, 476–478, 476*f*
macroscopic view of, 748, 748*f*
in metabolic pathways, 781
molecular view of, 748, 748*f*
phase changes and, 476–479
pressure changes and, 773–775, 774*f*, 777*t*
problem solving involving, 760–769, 770*f*
quantitative view of, 748–749
reaction direction and, 753, 757–760, 758*f*, 768–769
reaction quotient and, 750–755
reversible reactions and, 188–190
simultaneous, 874–875
solid-gas equilibria, 479
solid-liquid equilibria, 479
solubility and, 549–552
state of, 188–190, 189*f*
temperature changes and, 775–777, 777*t*
Equilibrium constant
concentration and, 756–757
defined, 749, G-6
free energy change and, 926–932, 928*t*, 959–961, 960*f*
as measure of reaction extent, 749
pressure and, 756–757
problem solving involving, 760–769, 770*f*
range of, 749, 749*f*
reaction direction and, 753, 757–760, 758*f*, 768–769
reaction quotient and, 750–755, 926–927, 926*t*
standard cell potential and, 959–961, 960*f*
tables of, A-8–A-13
as unitless number, 751
in van't Hoff equation, 776–777
Equilibrium vapor pressure. *See* Vapor pressure
Equivalence point, 173, 179, 853*f*, 854–857, 855*f*, 859*f*, 863*f*, G-6
Error, 12, 32–33, 33*f*
Erythrose, 110*t*
ES (enzyme-substrate complex), 733–734, 733*f*, G-6
Esterification, 664
Esters, 655*t*, 662, 663–664, 663*f*, 664*mn*, G-6
ETC (electron transport chain), 982–983, 983*f*
Ethanal
boiling point of, 484, 484*f*
dipole moment of, 484, 484*f*
intermolecular forces, 534*f*
molecular structure of, 655*t*, 660, 660*f*
Ethanamide, 655*t*
Ethanamine, 655*t*, 657, 664
Ethane
boiling point of, 641*f*
bonds in, 451, 452, 452*f*
electron density and bond order in, 454*f*
formula and model, 76*t*
molecular shape of, 426, 427*f*, 451
oxidation of, 653
Ethanenitrile, 655*t*
Ethanoic acid. *See* Acetic acid
Ethanol
boiling point elevation constant, 560*t*
constitutional isomers of, 110*t*
dissolution in water, 908, 908*f*
entropy and, 909
freezing point depression constant, 560*t*
gasohol and, 281
heats of vaporization and of fusion, 471*f*
molecular shape of, 426–427, 427*f*
solubility of, 536*t*
specific heat capacity of, 268*t*
surface tension of, 489*t*
vapor pressure of, 478, 478*f*
viscosity of, 489*t*, 490
Ethanolamine, 281, A-11
Ethene. *See* Ethylene
Eth- (numerical root), 639*t*
Ethyl alcohol. *See* Ethanol
Ethylamine, 655*t*, 658, 664, A-11
Ethylammonium chloride, 658
Ethylene
bonds in, 451, 452, 453*f*
from cyclobutane decomposition, 708–709
double and single bonds of, 380
electron density and bond order in, 454*f*
free-radical polymerization of, 668, 668*f*
hydrogenation of, 731, 732*f*
Lewis structure for, 409
molar mass of, 514
molecular shape, 451
molecular structure of, 655*t*
ozone and, 694–696
reaction with HCl, 651
Ethylenediamine, A-11
Ethylenediaminetetraacetate (EDTA) ion, 1047–1048
Ethylene glycol
antifreeze, 561, 566, 567
de-icers, 566, 567
molecular structure of, 654*f*
solubility of, 150
specific heat capacity of, 268*t*
Ethyne. *See* Acetylene
Eutrophication, 1005
Evaporative salt beds, 1014, 1014*mn*
Exact numbers, 31, G-6
Excited state, 304, 319, G-6
Exclusion principle, 333, 443, G-6
Exosphere, 239*f*, 240
Exothermic processes
bond energy, 381, 387
collision theory and, 717, 717*f*
defined, 266, G-6
enthalpy and, 266, 266*f*
entropy change in surroundings, 912–913
phase changes, 471
in sodium bromide formation, 373*f*
solution process, 544, 545*f*
spontaneous reactions, 896–897, 915–916, 915*f*
temperature effects on, 912–913
Expanded formula, 639, 640*f*
Expanded valence shells, 414–416, G-6
Expansion of gases, 206
Experiments, 12, 13*f*, G-6
Exponential (scientific) notation, 14, 14*t*, A-2–A-3
Extensive properties, 27–28, 27*f*, G-6

F

Face-centered cubic unit cell, 494, 495*f*, 497–498, G-6
Fahrenheit temperature scale, 25–27, 26*f*, 27*t*
Faraday, Michael, 959, 980
Faraday constant, 959, 980, G-6
Fats, 390, 390*t*, 663, 663*f*
Fatty acids, 541, 542, 663, G-6
f block, 595, 1037, 1038*f*. *See also* Inner transition elements
Feedback inhibition, 781, 781*f*
Feedstocks, hydrocarbons as, 636
Feldspar, 604
Femto- (prefix), 14*t*
Fenn, John B., 60
Fermi, Enrico, 1105
Ferrofluid, 519–520, 520*f*
Fertilizer, 586, 609, 611, 779, 1005, 1007–1008
Fibrous proteins, 674, 674*f*
Filtration, 83, 570, G-6
Fire extinguishers, 224
Fireworks, 308, 308*f*, 621, 917–918, 917*mn*
First ionization energy
of alkali metals, 591
of alkaline earth metals, 594
of boron family elements, 596
of carbon family elements, 599
defined, 348
of halogens, 618
in ionic bonding, 373
of nitrogen family elements, 605
of noble gases, 623
of oxygen family elements, 613
of Period 2 elements, 589*t*
periodicity of, 348–350
of transition elements, 1040, 1040–1041*f*, 1041
First law of thermodynamics, 261, 896, 914, G-6
First-order reactions
comparison with zero- and second-order reactions, 715*t*
defined, 699
half-life of, 712–714
integrated rate law and, 708, 710, 711*f*, 715*t*
radioactive decay, 1085
rate constant for, 704*t*
rate law and, 699, 700, 700*f*
rate vs. reactant concentration in, 700*f*
reactant concentration vs. time in, 700*f*

- Fischer-Tropsch process, 281
- Fission. *See* Nuclear fission
- Fixation
  - of carbon, 1003
  - defined, 1003, G-6
  - of nitrogen, 605, 779, 1004*f*, 1005
- Fixed mass ratio, 44
- Flame tests, 308, 308*f*, G-7
- Flasks, volumetric, 15, 16*f*
- Flerovium, 598–601, 1092
- Floc, 570
- Flotation process, 1009, 1009*f*, 1017, G-7
- Flow, environmental, 470*mn*
- Flow batteries, 970
- Fluoride ions, 69*t*
- Fluorinating agents, 620
- Fluorine
  - chemistry of, 619–622
  - diatomic molecules of, 443*f*, 444, 460*f*, 461
  - electron configuration, 337, 338*f*, 355
  - electron density of, 392, 392*f*
  - ionic bonding and, 372–376, 372*f*, 374*f*, 376*f*
  - ionization energy, 351*t*
  - nitrogen dioxide and, 725, 728, 729*f*
  - orbital overlap, 443*f*, 444
  - oxidation number for, 176*t*
  - as oxidizing agent, 619
  - oxoacids, 622*t*
  - properties of, 589*t*, 590, 617–622, 617*f*
  - reactivity, 343*f*
- Fluorite, 503, 503*f*, 867*mn*
- Fluoro-oxy ions, 1020
- Foams, 568*t*, 569
- Food
  - bond energy and, 390, 390*t*
  - cooking at elevation, 479
  - freeze-dried, 471
  - irradiation of, 1100–1101, 1100*f*
  - preservation of, 567
- *f* orbitals, 325, 325*f*, 326*t*, G-6
- Formal charge, 411–416, G-7
- Formaldehyde
  - formula and model, 110*t*
  - functions of, 110*t*
  - molecular shape of, 420
  - molecular structure of, 660*f*
- Formate ions, 851
- Formation constant, 877–878, 878*t*, A-12, G-7
- Formation equations, 277–278, G-7
- Formic acid, 662, 662*f*, 797*t*, 851, A-9
- Formula mass, 76, G-7
- Formulas. *See* Chemical formulas; Molecular formulas; Structural formulas
- Formula unit, 68, 78–79, G-7
- Fossil fuels, 281–283, G-7
- Fossils, 1000, 1000*f*
- Fractional bond orders, 411
- Fractional distillation, 83, 564, 567, 567*f*, G-7
- Fractional reaction order, 701
- Fraction by mass (mass fraction), 47–48, G-7
- Francium, 590–592, 590*n*
- Frasch process, 1026, 1026*f*
- Free energy
  - in biological systems, 925–926
  - defined, 916, G-7
  - electrical work and, 959–967
  - in electrochemistry, 959–967
  - Gibbs free energy, 916
- Free energy change
  - calculating, 917–919, 930–931
  - defined, 916
  - equilibrium constant and, 926–932, 928*t*
  - extent of reaction and, 931–932, 932*f*
  - reaction direction and, 931–932, 932*f*
  - reaction spontaneity and, 916–917, 920–924, 921*t*, 924*f*
  - standard cell potential and, 959–961, 960*f*
  - standard free energy change, 917–919, 927–928, 928*t*
  - standard free energy of formation, 918–919
  - temperature and, 920–924, 921*t*, 924*f*
  - work and, 919–920
- Free-radical polymerization, 668, 668*f*
- Free radicals
  - defined, 414, 668, 1095, G-7
  - ionizing radiation and, 1095
  - Lewis structures and, 414
  - ozone depletion and, 735
- Freezing
  - defined, 471, G-7
  - as exothermic and spontaneous process, 896
  - process of, 474
- Freezing point depression, 560–561, 560*t*, 566–567, G-7
- Freezing point of water, 26, 26*f*, 27*t*
- Freons, 602, 602*f*
- Frequency
  - defined, G-7
  - of electromagnetic radiation, 296, 296–297*f*, 297–298
  - interconverting wavelength and, 297–298, 301–302
  - threshold, 301
- Frequency factor, 719, G-7
- Frisch, Otto, 1105
- Fructose, 671, 671*f*
- Fuel cells, 970–971, 971*f*, G-7
- Fuel rods, 1106–1107
- Fuels. *See also* Combustion
  - biomass conversion to, 281–282
  - bond energy and, 389, 389*f*
  - coal, 281
  - fossil, 281–283
  - gasoline, 114–115, 262
  - greenhouse effect and global warming due to, 282–283
  - hydrogen, 283–284, 697–698
  - methanol, 408, 449–450
  - nuclear, 237
  - rocket, 125–127
  - syngas, 281, 603
  - wood, 281
- Fuller, R. Buckminster, 600
- Fullerene ends, 600
- Functional groups
  - alcohols, 654, 654*f*, 655*t*, 656
  - aldehydes, 655*t*, 660–661, 660*f*
  - alkenes, 655*t*, 659–660
  - alkynes, 646, 655*t*, 666
  - amides, 655*t*, 662, 664–665, 664*f*
  - amines, 655*t*, 657–658, 657*f*
  - carbonyl, 660–661
  - carboxylic acids, 655*t*, 662–663, 662*f*
  - defined, 635, G-7
  - with double bonds, 659–666
  - esters, 655*t*, 662, 663–664, 663*f*, 664*mn*
  - haloalkanes, 655*t*, 656–657, 657*f*
  - ketones, 655*t*, 660–661, 660*f*
  - nitriles, 655*t*, 666
  - overview, 655*t*
  - recognizing, 667
  - with single bonds, 654, 656–659, 662–666
  - with triple bonds, 666
- Fundamental units of SI system, 13, 14*t*
- Fusion. *See* Melting; Nuclear fusion

G

- GAC (granular activated carbon), 570
- Galileo Galilei, 208
- Gallium
  - bond type, 598*t*
  - conversion between number of entities and amount, 99
  - as doped semiconductor, 508, 508*f*
  - electron configuration, 340*t*
  - melting point of, 598*t*
  - properties of, 396, 595–598
  - quantum dots, 519
- Gallium arsenide cells, 284
- Galvanic cells. *See* Voltaic cells
- Galvanizing process, 974
- Gamma emission, 1076*t*, 1078, G-7
- Gamma rays
  - behavior in electric field, 1075, 1075*f*
  - defined, G-7
  - penetrating power of, 1094*f*, 1095
  - radioactive decay, 1076*t*, 1078
  - wavelength of, 296, 297*f*
- Gangue, 1008, 1009, 1009*f*, 1017, G-7
- Gases. *See also* Gas laws
  - Amontons's law and, 213
  - Avogadro's law and, 214–215, 214*f*
  - behavior at standard conditions, 215
  - Boyle's law and, 211–212, 211*f*
  - Charles's law and, 212–213, 212*f*
  - collecting over water, 227–228, 227*f*
  - comparison with physical states of solids and liquids, 205–207, 206*f*
  - Dalton's law of partial pressures and, 225–227, 231, 233, 233*f*, 239
  - defined, 4, G-7
  - density and, 24, 24*t*, 206, 222–224
  - deviation from ideal behavior, 241–244
  - diffusion, 237–238, 238*f*
  - dissolution and entropy change, 908, 908*f*
  - effusion, 236–237, 236*f*
  - entropy of, 547
  - expansion of, 206, 899–901
  - extreme conditions and, 241–243
  - free flowing nature of, 206
  - ideal, 211, 215, 215*f*, 901–902
  - kinetic-molecular theory and, 231–238, 470
  - molar mass and, 223, 224–225
  - molar volume of common gases, 241*t*
  - noble (*See* Noble gases)
  - particle arrangement in, 4, 4*f*
  - pressure and, 206, 207–210, 218, 239, 241–243, 242*f*
  - properties of, 470, 470*t*
  - solubility and, 550, 551–552
  - solutions, ability to form, 207
  - solving for unknown variable at fixed conditions, 220–221
  - spontaneous expansion of, 899–900
  - standard state for, 277, 277*n*
  - temperature and, 206, 212–213, 212*f*, 218, 239–240, 239*f*
  - universal gas constant, 216
  - van der Waals equation and constant, 243–244, 243*t*
  - volume-amount relationship and, 214–215, 214*f*, 218–219
  - volume-pressure relationship, 206, 211–212, 211*f*, 217–218
  - volume-temperature relationship, 206, 212–213, 212*f*, 218
- Gas-forming reactions, 169–170, 171*f*, 189–190
- Gas-gas solutions, 537–538
- Gasification, 281
- Gas laws, 210–238
  - Amontons's, 213
  - Avogadro's (*See* Avogadro's law)
  - Boyle's (*See* Boyle's law)
  - Charles's (*See* Charles's law)
  - combined, 214
  - Dalton's law of partial pressures, 225–227, 231, 233, 233*f*, 239
  - Henry's, 551–552
  - ideal, 211, 216, 216*f*, 222–230
  - kinetic-molecular theory, 231–238
  - molecular view of, 232–234
  - overview, 210–211
  - solving problems involving, 217–222
- Gas-liquid chromatography (GLC), 83, 83*f*
- Gas-liquid equilibria, 476–478, 476*f*
- Gas-liquid solutions, 537
- Gasohol, 281
- Gasoline, combustion of, 114–115, 262, 919–920
- Gas pressure. *See* Pressure
- Gas-solid equilibria, 479
- Gas-solid solutions, 538
- Gay-Lussac, J. L., 212, 224
- Geiger-Müller counter, 1083, 1084*f*, G-7
- Genes, 675
- Genetic code, 677, G-7
- Geochemical differentiation, 998–999, 999*f*
- Geometric isomers, 645–646, 645*t*, 646*f*, 1052–1053, 1052*f*, G-7
- Germanium
  - bond type, 598*t*
  - electron configuration, 340*t*
  - melting point of, 598*t*
  - properties of, 598–601
  - purification of, 1011
  - zone refining of, 1012
- Germer, L., 311
- Gibbs, Josiah Willard, 916
- Gibbs equation, 916–917
- Gibbs free energy, 916
- Giga- (prefix), 14*t*
- Glass
  - as amorphous solid, 504, 504*f*
  - borosilicate, 597, 597*mn*
  - polymer, 517
  - specific heat capacity of, 268*t*
- Glass electrodes, 967, 967*f*
- Glass transition temperature, 517
- Glassware, volumetric, 15, 16*f*
- GLC (gas-liquid chromatography), 83, 83*f*
- Global warming, 282–283, 384, 1004

Globular proteins, 674
Glucose
combustion and, 187
formula and model, 110*t*
functions of, 110*t*
information contained in chemical formula of, 97, 97*t*
metabolism of, 46, 121, 925, 925*f*
optical isomers of, 644
oxidation of, 915
structure of, 671, 671*f*
Glutamic acid, 540, 672*f*, 863
Glutamine, 672*f*
Glyceric acid, A-9
Glycerol, 559, 567, 663, 664
Glycine, 150, 672*f*, 863
Glycogen, 671
Glycolic acid, A-9
Glyoxylic acid, A-9
Gold
alchemy and, 10
alloys, 1013, 1013*f*, 1013*t*
density of, 24*t*, 1042
leaching, 1009, 1011
malleability of, 396–397
molar mass of, 97
sources of, 1001
specific heat capacity of, 268*t*
Gold leaf, 397*mn*
g orbitals, 325
Graham, Thomas, 236
Graham's law of effusion, 236–237, 1107, G-7
Grain alcohol, density of, 24*t*
Gramicidin A, 542, 543*f*
Gram-molecular weight. *See* Molar mass
Grams (g), 96
Granite, specific heat capacity of, 268*t*
Granular activated carbon (GAC), 570
Graphene, 600*f*, 601
Graphing data, A-4
Graphite
allotropism, 600
carbon monoxide formation from, 761
in electrolytic cell for aluminum manufacture, 1020, 1020*f*
as inactive electrode, 947
properties of, 504, 504*t*
specific heat capacity of, 268*t*
standard molar entropy, 909
standard state of, 277
in voltaic cells, 947
Gray (Gy), 1094, G-7
Grease, 541
Green chemistry
applications of, 223–224
atom economy and, 129–130
defined, 129, G-7
examples of, 121
Greenhouse effect, 282–283, 1004
Ground state, 304, 319, G-7
Group 1A(1) elements. *See* Alkali metals
Group 2A(2) elements. *See* Alkaline earth metals
Group 3A(13) elements. *See* Boron family elements
Group 4A(14) elements. *See* Carbon family elements
Group 5A(15) elements. *See* Nitrogen family elements
Group 6A(16) elements. *See* Oxygen family elements
Group 7A(17) elements. *See* Halogens
Group 8A(18) elements. *See* Noble gases
Group 1B(11) elements, mechanical properties of, 396–397
Groups. *See also specific groups*
defined, G-7
electron affinities of, 352, 352*f*
electron configurations within, 342, 343
ionization energy trends in, 348–349, 349*f*
metallic behavior, 353–354*f*, 353–355
of periodic table, 61, 63
reactivities within, 338
Guanine, 675
Guitar string vibrations, 310, 310*f*
Guldberg, Cato, 750
Gypsum, 281, 593
Gyration, radius of, 515–516, 515*n*

H

Haber, Fritz, 779
Haber process, 605, 779–780, 779*t*, 780*f*, 1005, G-7
Hafnium, 1044
Hahn, Otto, 1105
Halates, 621, 622*t*
Half-cell potentials, 950–956, 952*f*, 954*t*, 977–978, A-14
Half-cells, 944, 946, G-7
Half-life
defined, 1086, G-7
of first-order reactions, 712–714
of radioactive decay, 1086, 1086*f*, 1086*t*
of reactions, 712–715, 712*f*
of second-order reactions, 715
of zero-order reactions, 715
Half-reaction method for balancing redox reactions, 940–943, G-7
Half reactions, 940–943, 944*mn*, 965, 965*f*
Halic acid, 622, 622*t*
Halides
alkyl, 656–657, 659
formation of, 182
halogen, 620
hydrogen, 619–620, 762
ionic, 591
lattice energy of alkali-metal halides, 376
nitrogen family elements, 605, 607
oxygen family elements, 613, 615
reactions involving, 659
solubility-product constant for, A-13
sources of elements, 1001, 1002*f*
Halites, 621, 622*t*, 1014
Hall-Heroult process, 1019–1020
Haloalkanes, 655*t*, 656–657, 657*f*, G-7
Halogens
activity series of, 186, 186*f*
chemistry of, 619–622
combination reactions and, 182
diatomic molecules of, 343
electron configuration, 343
in hypohalous acids, 816
oxidation number for, 176*t*
in periodic table, 63
properties of, 617–622, 617*f*
reactivity of, 230, 343*f*, 617–619
standard state of, 278
Halo- (prefix), 655*t*
Halous acid, 622*t*
Hamiltonian operator, 315
Hard water, 570–571, G-7
Hassium, 1092
H-bonds. *See* Hydrogen bonds
HDPE (high-density polyethylene), 518
Heat
calorimetry and, 268–273
defined, 25, 258, G-7
endothermic processes and, 266–267, 266*f*
energy transfer as, 258, 259, 260*f*
exothermic processes and, 266, 266*f*
as extensive property, 27–28
finding quantity from temperature change, 269
in phase changes, 473–476
release with gasoline combustion, 262
sign conventions, 261*t*
in surroundings, 912–914, 912*mn*
Heat capacity
defined, 268, G-7
molar, 268
specific heat capacity, 268, 268*t*, 270, 491
of water, 491
Heating-cooling curve, 473–474, 473*f*, G-8
Heating systems, residential, 284
Heat of formation, standard, 277–280, 278*t*, 279*f*
Heat of fusion
in boron, carbon, and nitrogen family elements, 598*t*, 600
defined, G-7
phase changes and, 471, 471–472*f*, 474
Heat of hydration, 545–546, 545*t*, 592, G-7
Heat of reaction
standard, 277–280
stoichiometry of thermochemical equations and, 273–274
Heat of solution, 544, 545–546*f*, 545–547, G-7
Heat of sublimation, 472, 472*f*, G-8
Heat of vaporization, 471, 471–472*f*, 491–492, G-8
Heat summation, Hess's law of, 275–277
Heavy-metal ions, 571
Heavy-metal poisoning, 1047
Hecto- (prefix), 14*t*
Heisenberg, Werner, 313
Heisenberg's uncertainty principle, 313
Helium
boiling point of, 537*t*
burning of, 1110–1111
effusion rate of, 236–237
electron configuration, 333, 336, 338*f*
liquid, 506
molecular orbitals of, 457, 457*f*
mole fraction in air at sea level, 239*t*
properties of, 622–624
solubility in water, 537*t*
Hematite, 1000, 1000*f*, 1016, 1016*t*
Heme group, 1063–1064, 1063*f*
Hemimorphite, 603, 603*f*
Hemoglobin, 563, 863, 1061, 1063–1064, 1063*f*
Henderson-Hasselbalch equation, 848–849, G-8
Henry's law, 551–552, G-8
Heptane, 76*t*, 641*f*
Hepta- (prefix), 72, 73*t*
Hept- (numerical root), 639*t*
Hereditary effects of radiation, 1097
Hertz (Hz), 296
Hess's law
Born-Haber cycle and, 374
defined, G-8
heat of fusion and, 474
heat of solution and, 544
heat of sublimation and, 472
standard entropy of reaction and, 911
standard free energy of formation and, 918
in thermochemistry, 275–277
Heteroatoms, 635, 635*f*, G-8
Heterogeneous catalysis, 731–732, 732*f*, 735, G-8
Heterogeneous equilibrium, 752, 752*f*
Heterogeneous mixtures, 81, 533, 568, G-8
Heteronuclear diatomic molecules, 462–463, 462*f*
Hexaamminechromium(III) ion, 1055, 1055*f*
Hexagonal closest packing, 497*f*, 498, G-8
Hexagonal cubic unit cell, 497–498, 497*f*
Hexane
boiling point of, 641*f*
formula and model, 76*t*
intermolecular forces, 487–488, 534*f*
as solvent, 535, 536, 536*t*, 548, 548*f*
Hexanol, 536*t*
Hexa- (prefix), 73*t*
Hex- (numerical root), 639*t*
High-density polyethylene (HDPE), 518
High-performance liquid chromatography (HPLC), 83
High-resolution polyacrylamide-gel electrophoresis, 680
High-spin complexes, 1060–1061, G-8
High-surface-area materials, 520
Histidine, 672*f*
Homogeneous catalysis, 730–731, 731*f*, 735, G-8
Homogeneous equilibrium, 752
Homogeneous mixtures, 81, 533, 568, G-8
Homologous series, 639, G-8
Homonuclear diatomic molecules, 458–462
atomic *p* orbital combinations, 458–459*f*, 458–460
bonding in *p*-block, 460–461, 460*f*
defined, G-8
elements of, 458
s-block bonding of, 458, 458*f*
Homopolymers, 518
Hot packs, 546
HPLC (high-performance liquid chromatography), 83
Human radiation studies, 1096
Hund's rule, 336–337, 457, 587, G-8
Hybridization, 444–452, G-8
Hybrid orbitals, 444–451
complex ions and, 1055–1056
composition and orientation of, 449*t*
defined, 444, G-8
molecular formulas and, 449, 449*f*
sp hybridization, 444–446, 445*f*, 449*t*
sp^2 hybridization, 446, 446*f*, 449*t*
sp^3 hybridization, 447, 447*f*, 449*t*
sp^3d hybridization, 447–448, 448*f*, 449*t*
sp^3d^2 hybridization, 448, 449*f*, 449*t*
Hydrated ions, 592
Hydrated metal ions
acid strength of, 817, 817*f*
dissociation constants table for, A-12
formation of, 88*f*, 877–878
Hydrates, 72, 73*t*, G-8
Hydration, defined, 545, G-8
Hydration, heat of, 545–546, 545*t*, 592

Hydration shells, 534, 535*f*, G-8
Hydrazine, 125–127, 607
Hydride ions, 69*t*, 586
Hydrides
boranes, 590, 597–598, 598*f*, 604
bridge bonds, 598, 598*f*
catenated inorganic, 648, 650
covalent, 586
ionic, 586, 591, 592
metal, 586–587, 586*f*
nitrogen family elements, 605, 607
oxygen family elements, 614
trends in acid strength, 816, 816*f*
VSEPR theory and, 451
Hydriodic acid, 171
Hydrocarbons, 636–650
alkanes (*See* Alkanes)
alkenes, 644–646, 645*t*, 646*f*, 655*t*, 659–660
alkynes, 646, 655*t*, 666
aromatic, 647–648, 648*f*, 659–660, 660*f*
constitutional isomerism and, 641–642, 642*t*
cyclic, 640, 641*f*
defined, 636, G-8
double bonds in, 644, 651, 651*f*
drawing, 637–638, 637*f*
entropy and, 909
geometric isomerism and, 645–646, 645*t*, 646*f*
naming, 639, 639–640*t*, 646–648
optical isomers of, 642–644
saturated, 639
single bonds in, 639–640
structure and classes of, 636–637*f*, 636–638
triple bonds in, 646, 666
unsaturated, 644
Hydrochloric acid. *See* Hydrogen chloride
Hydrocyanic acid, 796, 796–797*t*, A-9
Hydrofluoric acid, 797*t*, 806, A-9
Hydrogen
abundance of, 1001*t*
burning of, 1110
chemistry of, 586–587
combustion of, 281–282, 697–698
covalent bonding and, 66–67, 66*f*
covalent bonding of, 379–380
density of, 24*t*
deuterium, 1024–1025, 1024*t*
displacement by metals, 184, 185*f*
electron density of, 379, 380*f*, 392, 392*f*
electron spin and, 443, 443*f*
as fuel, 283–284, 697–698
in fuel cells, 971, 971*f*
iodine and, 696*f*, 697, 911
isotopes of, 1024–1025, 1024*t*
molecular orbitals of, 456, 456*f*
mole fraction in air at sea level, 239*t*
nitric oxide and, 700
orbital overlap, 443, 443*f*
oxidation number for, 176*t*
periodic table placement, 585–586, 585*f*
production of, 283–284, 1022–1023
reduction with, 1011
standard enthalpies of formation, 278*t*
transportation and storage of, 284
tritium, 1024, 1024*t*
uses of, 1023–1024
Hydrogenation
of benzene, 660
catalytic, 731, 732*f*
defined, 731, G-8
of liquid oils, 1024
reactions involving, 274
Hydrogen atoms
Bohr model of, 303–305, 305*f*
electron configuration, 336, 338*f*
energy levels of, 305–307, 305*f*, 325, 325*f*
ionization energy of, 348
line spectra of, 302–303, 303*f*, 304–305, 305*f*
quantum number, 332–333, 336
Schrödinger equation for, 317–318
stationary states of, 304
Hydrogen bisulfate ions, 616
Hydrogen bonds
in biological systems, 485
boiling point and, 484–485, 484*f*
defined, 484, G-8
in DNA, 543–544, 543*f*
intermolecular forces, 483*t*, 484–485, 484*f*
in proteins, 540, 540*f*
significance of, 484–485
in solutions, 534, 534*f*
surface tension and, 489, 489*t*
in water, 491, 491*f*
Hydrogen bromide, 165, 699
Hydrogen carbonate ions, 72*t*
Hydrogen chloride
in acid-base reactions, 167–170
formation of, 175, 175*f*, 619
reaction with ethene, 651
in stomach, 171, 172
titration of, 853*f*, 854, 859–860, 859*f*
uses of, 620
in water, 804, 804*f*
Hydrogen fluoride
bonding pair in, 380
bond polarity of, 392, 392*f*
covalent bonding in, 66–67
electronegativity of, 390
formation of, 111, 112*f*, 386, 387, 388*f*
molecular orbitals of, 462–463, 462*f*
orbital overlap, 443*f*, 444
orientation in electric field, 428*f*
polarity of, 428, 428*f*
properties of, 619
Hydrogen halides, 619–620, 762
Hydrogen iodide, 696*f*, 697, 760–762
Hydrogen ions
in acids, 165
concentration measurement of, 966–967, 967*f*
formula for, 69*t*
number in solution, 166
release in neutralization, 794–795
Hydrogen peroxide
concentration calculations, 556
decomposition of, 699, 731, 731*f*
disproportionation of, 614
empirical formula of, 104–105
molecular formula for, 105, 106
structural formula for, 105
uses of, 614
Hydrogen phosphate ions, 72*t*
Hydrogen "skin" on carbon skeletons, 636, 637, 637*f*
Hydrogen sulfate ions, 72*t*, 616, 801*t*
Hydrogen sulfide
acid rain and, 876
ammonia and, 804
formation of, 614
on Io (Jupiter's moon), 7
mole fraction in air at sea level, 239*t*
removal from coal and natural gas, 772–773
simultaneous equilibria and, 874–875
VSEPR theory and, 451
Hydrogen sulfide ions, 804, 874–875
Hydrohalic acids, 619–620, 797
Hydrolases, 734
Hydrolysis, 664, 732–733, G-8
Hydrolysis constant, 820*n*
Hydrometallurgy, 1008
Hydronium ions
alculating concentration in aqueous solution, 800
balancing redox reactions using, 942
defined, 794, G-8
formation of, 165, 616
formula for, 72*t*
in ion-product constant for water, 799, 799*f*
pH scale and, 800–803, 802*f*
water and, 794
Hydro- (prefix), 74
Hydrosphere, 999, 1001*t*, G-8
Hydrosulfuric acid, 814*t*, A-9
Hydroxide ions
in bases, 165
calculating concentration in aqueous solution, 800
formula for, 72*t*
in ion-product constant for water, 799, 799*f*
number in solution, 166
pOH, 801–802, 802*f*
release in neutralization, 794–795
Hydroxides, 881–882, 882*f*, A-13
Hydroxyapatite, 519
Hypertonic solutions, 567, 567*f*
"Hypo," 880, 880*mn*
Hypobromous acid, 801*t*, A-9
Hypochlorite ions, 72*t*
Hypochlorous acid, 797*t*, 812–813, A-9
Hypofluorous acid, 619
Hypohalites, 621, 622*t*
Hypohalous acids, 622*t*, 816
Hypoiodous acid, A-9
Hypo- (prefix), 72, 72*f*, 74
Hypothesis, 12, 13*f*, G-8
Hypotonic solutions, 567, 567*f*

I

Ice, 473*f*, 492, 492*f*
-ic (suffix), 70, 74
Ideal gases
defined, 211, G-8
isothermal volume change of, 901–902
root-mean-square speed for, 235*n*
standard molar volume and, 215, 215*f*
Ideal gas law
defined, 211, 216, G-8
equation for, 216, 216*f*
reaction stoichiometry and, 228–230, 229*f*
rearrangements of, 222–230
relationship to individual gas laws, 216*f*
van der Waals equation and, 243–244
Ideal solutions, 558, G-8
-ide (suffix), 68, 74
IE. *See* Ionization energy
IGCC (integrated gasification combined cycle), 281
Ilmenite, 1016*t*
India, Union Carbide disaster in, 224
Indicators, 173–174, 173*f*, 179
Indium
bond type, 598*t*
electron configuration, 356
melting point of, 598*t*
properties of, 595–598
Induced dipole forces, 485–486
Induced-fit model of enzyme action, 733, 733*f*, G-8
Inert pair configuration, 356
Inert-pair effect, 597
Infrared (IR) radiation, 296, 297*f*, G-8
Infrared spectroscopy, 384–385, 385*f*, G-8
Initial reaction rate, 696, 699, G-8
Inner (core) electrons, 342, 350, G-8
Inner transition elements
actinides, 341*f*, 342, 1044, 1045
defined, 1044, G-8
electron configuration, 341*f*, 342
lanthanides, 341*f*, 342, 1044–1045
in periodic table, 61, 62*f*, 1038, 1038*f*
Inorganic chemistry, 63
Inorganic compounds, 602–603, 659, 662, 666
Inorganic polymers, 670
Insects, irradiation of, 1101
Insoluble compounds, 147, 159*t*
Instantaneous reaction rate, 695–696, G-8
Instrument calibration, 33, 33*f*
Insulators, 506, 506*f*, G-8
Integrated gasification combined cycle (IGCC), 281
Integrated rate law, 708–715
defined, 707, G-8
first-order reactions and, 708, 710, 711*f*, 715*t*
half-life and, 712–715, 712*f*
reaction order determination from, 710–711, 711*f*
second-order reactions and, 708–711, 711*f*, 715*t*
zero-order reactions and, 709, 711, 711*f*, 715*t*
Intensive properties, 27–28, 27*f*, 150, G-8
Interference, 299, 311
Interhalogen compounds, 620, 620*f*, G-8
Intermetallic compounds, 1012
Intermolecular forces, 468–520
advanced materials and, 507–520
biological macromolecules and, 539–544
comparison with bonding forces, 482, 483*t*
in covalent compounds, 383, 383*f*
defined, 469, G-8
dipole-dipole forces, 483–484, 483*t*
dipole-induced dipole forces, 483*t*, 486
dispersion (London) forces, 483*t*, 486–487
enzyme action and, 732
hydrogen bonds, 483*t*, 484–485, 484*f*
identifying types of, 487–488

ion-dipole forces, 483, 483*t*, 491
ion-induced dipole forces, 483*t*, 485–486
liquids, 488–491
membrane structure and, 542, 542*f*
phase changes and, 469–472
polarizability and, 486
solids, 493–499, 501–506
in solution, 534–535
strength of, 482
summary diagram for analyzing samples, 487*f*
types of, 482–488
van der Waals radius, 482, 482*f*
vapor pressure and, 477*f*, 478
viscosity and, 489*t*, 490
Internal energy, 258–260, 262, 896, G-8
International System of Units. *See* SI units
International Union of Pure and Applied Chemistry (IUPAC), 59, 1092
Interparticle attractions, 241–243, 242*f*
Interstitial alloys, 538, 538*f*
Interstitial hydrides, 586–587, 586*f*
Intramolecular forces, 470, 482. *See also* Bonding
Iodates, A-13
Iodic acid, A-9
Iodide ions, 69*t*, 941–942, 941*mn*
Iodine
electron configuration, 356
hydrogen and, 696*f*, 697, 911
oxoacids, 622*t*
properties of, 617–622, 617*f*
radioactive tracers and, 1099, 1100*f*
reactivity, 343*f*
sublimation of, 479, 479*f*
Iodine pentafluoride, 423
Io (Jupiter's moon), 7
Ion-dipole forces
attractions in, 483, 483*t*
defined, 483, G-8
in proteins, 540, 540*f*
in soaps, 541
in solutions, 534, 534–535*f*
solvent properties of water and, 491
Ion exchange, 570–571, 571*f*, G-8
Ionic atmosphere, 565, 565*f*, G-8
Ionic bonds
Born-Haber cycle and, 374–375, 374*f*
charge in, 65
Coulomb's law and, 375–376
Coulomb's law on, 65
defined, 370, G-9
electron transfer and, 175, 175*f*, 370, 370*f*, 372–373, 373*f*
in intermolecular force comparison, 483*t*
lattice energy and, 373–377, 376*f*
model of, 372–378
strength of, 65, 66*f*
Ionic charge, 176, 376
Ionic compounds
binary, 64–65
bonding (*See* Ionic bonds)
comparison with covalent compounds, 67
crystal structure of, 501*t*, 502–503
defined, 64, G-9
electrical conductivity of, 147–148, 147*f*
entropy change with dissolution, 908, 908*f*
equilibria of slightly soluble ionic compounds, 864–875
formation of, 64–66, 175, 175*f*
formula mass of, 76–78
heats of solution, 545–546, 546*f*
hydrated, 72, 73*t*
molar mass of, 97
names and formulas of, 68–73
properties of, 377, 377–378*f*, 378*t*
redox reactions and, 175, 175*f*, 181, 182*f*
solubility-product constant for, A-13
solubility rules, 159*t*
spontaneous change in, 897
temperature and solubility, 550, 550*f*
in water, 146–149, 146*f*, 159*t*
Ionic equations. *See also* Net ionic equations; Total ionic equations
for acid-base reactions, 167–171, 170*t*, 171*f*
for aqueous ionic reactions, 155–157, 156*f*
Ionic equilibria, 842–882
of acid-base buffers, 843–852
acid-base titration curves and, 853–863
complex ions and, 877–882
of slightly soluble ionic compounds, 864–875
Ionic halides, 591
Ionic hydrides, 586, 591, 592
Ionic radius
of alkali metals, 591, 592
of alkaline earth metals, 594
of anions, 359, 359*f*
of boron family elements, 596
of carbon family elements, 599
of cations, 359, 359*f*
comparison with atomic size, 359–360, 360*f*
defined, 359, G-9
lattice energy and, 376*f*
of nitrogen family elements, 605
of oxygen family elements, 613
Ionic size, 375–376
Ionic solids, 501*t*, 502–503, G-9. *See also* Ionic compounds
Ionic solutions
corrosion of iron and, 973, 973*f*
electrical conductivity of, 147–148, 147*f*
Ion-induced dipole forces, 483*t*, 485–486, 534, 534*f*, G-9
Ionization, defined, G-9
Ionization counters, 1083, 1084*f*
Ionization energy (IE)
of alkali metals, 591, 592
of alkaline earth metals, 593, 594
of boron family elements, 595, 596
of carbon family elements, 599
defined, 348, G-9
first (*See* First ionization energy)
of halogens, 618
of hydrogen atoms, 306
of nitrogen family elements, 605
of noble gases, 623
of oxygen family elements, 613
periodic trends in, 348–350, 349*f*, 351*t*, 352*f*
successive, 350–351, 350*f*, 351*t*
of transition elements, 1040, 1040–1041*f*, 1041
Ionizing radiation
applications of, 1100–1101
background sources of, 1095–1096, 1096*f*, 1096*t*
defined, 1093, G-9
effects on matter, 1093–1095
molecular interactions, 1095
penetrating power of emissions, 1094–1095, 1094*f*
risk from, 1096–1097, 1097*f*, 1097*t*
units of radiation dose, 1093–1094
Ion pairs, 378, 378*f*, 541, G-8
Ion-product constant for water, 799–803, 799*f*, G-8
Ion-product expression, 864–865
Ions. *See also* Anions; Cations
calculating amount in solution, 149
charge density of, 545
complex (*See* Complex ions)
counter, 1046, 1049, 1050
defined, 64, G-8
depicting formation of, 372–373, 372*f*
formation of, 357
heats of hydration, 545, 545*t*
monatomic (*See* Monatomic ions)
names and formulas for, 68–69, 69*f*, 69*t*
noble gases and formation of, 65–66, 66*f*
polyatomic (*See* Polyatomic ions)
predicting formation from elements, 66
selective precipitation and, 874–875
solvated, 147, 157, 165, 165*f*
spectator, 156, 159, 160, 167
Ion-selective (ion-specific) electrodes, 967, 967*t*
Iridium, 1042
IR (infrared) radiation, 296, 297*f*, G-8
Iron
abundance of, 1000, 1001*t*
alloy, 1012, 1013*t*
appearance of, 1039*f*
changes in state, 7
corrosion of, 972–974
as dietary trace element, 1064*t*
electron configuration, 340*t*, 358, 1039, 1039*t*
hematite, 1000, 1000*f*
in heme, 1063–1064, 1063*f*
iron(II) and iron(III), 70, 71*t*
mass-number conversion of, 99
metallurgy, 1015–1017, 1016*f*, 1016*t*
minerals, 1008*t*
minerals of, 1016*t*
oxidation reactions involving, 897
oxidation state, 1042, 1042*n*, 1042*t*
radioactive, 1103
smelting, 1011
specific heat capacity of, 268*t*
Iron(II) hydroxide, 865, 866*t*
Iron(III) chloride, 163–164, 570
Iron(III) hydroxide, 163–164
Iron(II) sulfide, 879
Iron phase, 999, 999*f*
Irradiation of food, 1100–1101, 1100*f*
Isobutane, 642, 642*t*, 735
Isoelectronic ions, 355, 358, G-9
Isoleucine, 672*f*, 781, 781*f*
Isomers
chirality of, 643–644
cis-trans isomerism, 430, 645, 645*t*, 646*f*, 1052, 1052*f*
constitutional, 110–111, 110*t*, 641–642, 642*t*, 1051–1052, 1052*f*, 1054*f*
coordination, 1051
in coordination compounds, 1051–1054
defined, 110, 641, G-9
geometric, 645–646, 645*t*, 646*f*, 1052–1053, 1052*f*
linkage, 1052, 1052*f*
molecular polarity and, 430
optical, 642–644, 1053–1054, 1053*f*
stereoisomers, 642–643, 1052–1054*f*
Isopentane, 642, 642*t*
Isopropylamine, A-12
Isothermal process, 900–902, G-9
Isotonic solutions, 567, 567*f*
Isotope enrichment, 237
Isotopes. *See also* Radioisotopes
calculating atomic mass from, 58–59
defined, 57, 1074, G-9
hydrogen, 1024–1025, 1024*t*
mass spectrometry and, 60, 60*f*
uranium, 237
Isotopic mass, 58, G-9
Isotropic, 509
-ite (suffix), 72, 72*f*, 74
IUPAC (International Union of Pure and Applied Chemistry), 59, 1092

J

Jet engines, 1022, 1022*mn*
Joliot-Curie, Frederic, 1090, 1090*mn*
Joliot-Curie, Irene, 1090, 1090*mn*
Joule (J), 261, G-9
Jupiter (planet), 7

K

Kaolinite, 512
Kelvin (absolute) temperature scale, 25–27, 26*f*, 27*t*, 213, G-9
Kelvin (K), 14*t*, 25, G-9
Ketobutyrate, 781
Ketones, 655*t*, 660–661, 660*f*, G-9
Kevlar, 510
Kilocalorie (kcal), 262
Kilogram (kg), 14*t*, 17, G-9
Kilojoule (kJ), 262
Kilo- (prefix), 14, 14*t*
Kinetic energy
in chemical reactions, 386
defined, 8, G-9
dispersal of, 470, 897–898
kinetic-molecular theory and, 231, 234–235, 234*f*
potential energy converted to, 9–10, 9*f*, 257
temperature and, 234–235, 240, 470
Kinetic isotope effect, 1024, 1098*n*
Kinetic-molecular theory, 231–238
collision frequency and, 238, 238*mn*
defined, 231, G-9
diffusion and, 237–238, 238*f*
effusion and, 236–237, 236*f*
mean free path, 238
molecular speed and, 232, 232*f*, 234, 234*f*
molecular view of gas laws, 232–234
on origin of pressure, 231, 232
phase changes and, 470, 470*t*, 473–474
postulates of, 231–232, 232*f*
root-mean-square speed, 235–236, 235*n*
Kinetics, 732, 747. *See also* Chemical kinetics

Krypton
electron configuration, 340, 340*t*
mole fraction in air at sea level, 239*t*
properties of, 622–624

L

Laboratory glassware, 15, 16*f*
Lactic acid, 107, 110*t*, A-9
Lanthanide contraction, 1041, G-9
Lanthanides
defined, G-9
electron configuration, 341*f*, 342
properties of, 1044–1045
uses of, 1045
Lanthanum, 342
Large Hadron Collider, 1091–1092, 1092*f*
Lattice, 493–494, 493*f*, G-9
Lattice energy
of alkali metal chlorides, 592, 592*f*
of alkaline earth metals, 592*f*, 593
defined, 374, G-9
ionic bonding and, 373–377, 376*f*
periodic trends in, 375–376, 375*f*
in solution process, 545–546, 546*f*
Lavoisier, Antoine
classification of elements by, 61
combustion theory developed by, 11, 13, 46, 914
on law of mass conservation, 46, 50
metric system and, 13
on scientific method, 12, 13
thermal decomposition and, 183
Law of chemical equilibrium, 750, G-9
Law of conservation of energy, 261, 896, G-9
Law of definite (constant) composition, 47–48, 48*f*, 50, G-9
Law of heat summation, Hess's, 275–277
Law of mass action, 750, G-9
Law of mass conservation, 46–47, 47*f*, 50, G-9
Law of multiple proportions, 49–51, G-9
Lawrence, E. O., 1091
Laws of thermodynamics
first, 261, 896, 914
living systems and, 914
second, 905, 914
third, 905–906
LCDs (liquid crystal displays), 511, 511*f*
LDPE (low-density polyethylene), 518
Leaching, 1005, 1006, 1009, 1011, G-9
Lead
bond type, 598*t*
density of, 24*t*
electron configuration, 344
lead(II) and lead(IV), 71*t*
melting point of, 598*t*
mineral sources of, 1008*t*
properties of, 598–601, 601*f*
silver displacement by, 961
Lead-acid batteries, 969–970, 970*f*, 975, 975*f*
Lead(II) chloride, 865, 873
Lead(II) chromate, 868, 868*f*
Lead(II) fluoride, 866–867, 866*t*
Lead(II) iodide, 158–159, 159*f*
Lead(II) nitrate, 159, 159*f*
Lead(II) sulfate, 866, 866*t*
Le Châtelier, Henri, 916
Le Châtelier's principle, 770–780
in ammonia synthesis, 779–780, 779*t*, 780*f*
applications of, 770, 770*mn*, 1027
common-ion effect and, 868
concentration changes and, 770–773, 772*t*, 777*t*
defined, 770, G-9
lack of catalysts and, 777, 777*t*
pressure changes and, 773–775, 774*f*, 777*t*
reaction direction and, 1014
solubility and, 869
temperature changes and, 775–777, 777*t*
van't Hoff equation and, 776–777
Lecithin, 542*f*, 663, 663*f*
LEDs (light-emitting diodes), 285
Length
converting units of, 20–21
English equivalents, 15*t*
examples of, 17*f*
SI units, 14–15*t*, 15
Leucine, 540, 672*f*
Leveling effect, 827, G-9
Levels of energy, 320–321, 334–335, G-9
Levorotatory isomers, 643
Lewis, Gilbert Newton, 371, 372, 827
Lewis acid-base definition, 827–830, 877, G-9
Lewis acids, 827–830, 877
Lewis bases, 827–830, 877
Lewis electron-dot symbols, 371–372, 371*f*, G-9
Lewis structures, 405–417
defined, 405, G-9
electron-deficient atoms and, 413–414
electron-pair delocalization and, 410
exceptions to octet rule for, 413–417
expanded valence shells and, 414–416
formal charge and, 411–416
free radicals and, 414
limitations of, 416
for multiple bond molecules, 408–409
octet rule for, 405–409
odd-electron atoms and, 414
resonance and, 409–413
for single bond molecules, 406–408
terminology considerations, 405*n*
valence-shell electron-pair repulsion theory and, 417, 421, 421*f*
Libby, Willard F., 1087
Ligand field theory, 1062
Ligands
anionic, 1050, 1050*t*
bidentate, 1047, 1048*t*
coordination compounds and, 1046, 1047–1048, 1048*t*
defined, 877, 1046, G-9
donor atoms and, 1047
monodentate, 1047, 1048*t*
polydentate, 1047, 1048*t*
spectrochemical series and, 1059, 1059*f*
splitting *d* orbitals in octahedral field of, 1057–1058
strong-field, 1058, 1058*f*, 1060, 1060–1061*f*
weak-field, 1058, 1058*f*, 1060, 1060–1061*f*
Light
as electromagnetic radiation, 295–297
particle nature of, 299–301, 300*f*
photon theory of, 301
rainbows and, 298, 298*mn*
refraction and dispersion of, 298, 298*f*
speed of, 296
wave nature of, 296–297*f*, 296–298
Light bulbs, efficiency of, 285
Light-emitting diodes (LEDs), 285
Lightning, 1005
Light-water reactors, 1107*f*, 1108
Like-dissolves-like rule, 534, 535, 536*f*, 544, G-9
Lime, 593, 752, 1021
Limestone
acid rain and, 876
caves, 870–871, 870*f*
decomposition of, 752
in desulfurization, 281
gases released from, 210
Liming, 876
Limiting-reactant problems
precipitation reactions and, 163–164
reactant and product amount calculations, 125–127
reaction table use in, 122–123
solving with molecular depictions, 123–124
Limiting reactants
chemical reactions involving, 122–127, 123*f*
defined, 122, G-9
determining, 122
in everyday life situations, 122*mn*
Linear accelerators, 1091, 1091*f*
Linear arrangement
of complex ions, 1047, 1047*t*
defined, 419, G-9
of hybrid orbitals, 444, 446, 449*t*
VSEPR theory and, 418–419*f*, 419, 422, 422*f*, 424*f*
Linear response model of radiation risk, 1097
Linear shape, G-10
Line spectra, 302–305, G-9
Linkage isomers, 1052, 1052*f*, G-10
Lipid bilayers, 542, 542*f*, G-10
Lipids, 663, 663*f*, G-10
Liquid crystal displays (LCDs), 511, 511*f*
Liquid crystals
applications of, 511, 511*f*
defined, 509, G-10
molecular characteristics of, 509–510, 509*f*
ordering of particles in, 509–511
phases, 510–511
Liquid-gas equilibria, 476–478, 476*f*
Liquids
capillarity, 489–490, 489*f*
comparison with physical states of solids and gases, 205–207, 206*f*
defined, 4, G-10
density of, 24, 24*t*
diffusion of gas through, 237–238
entropy change with dissolution, 908, 908*f*
entropy of, 547
intermolecular forces, 488–491
kinetic-molecular view of, 470
particle arrangement in, 4, 4*f*
properties of, 470, 470*t*, 488–491
solutions and molecular polarity, 535–537
surface tension of, 489, 489*f*, 489*mn*, 489*t*
viscosity, 490–491, 490*t*
volatile, 224–225
Liquid-solid equilibria, 479
Liter (L), 15, G-10
Lithium
anomalous behavior of, 587, 590
in batteries, 969, 969*f*
density of, 24–25
diagonal relationship with magnesium, 595, 595*f*
diatomic, 458, 458*f*
electron configuration, 333, 334, 336, 336*f*, 338*f*
hydrogen displacement from water, 184*f*
ionic bonding and, 372–376, 372*f*, 374*f*, 376*f*
ionization energy, 351*t*
molecular orbitals in, 505, 505*f*
nuclear fusion and, 1109
in organometallic compounds, 661
properties of, 588*t*, 590–592, 590*mn*
reactivity, 343*f*
soaps, 541
Lithium batteries, 969, 969*f*
Lithium fluoride
Born-Haber cycle and, 374–375, 374*f*
ionic bonding and, 372–376, 372*f*, 374*f*, 376*f*
lattice energy and, 373–376, 376*f*
percent ionic character of, 393*f*
Lithium hydroxide, 229
Lithium-ion batteries, 970, 970*f*
Lithium ions, 69*t*
Lithosphere, 999, 1001*t*, G-10
Livermorium, 612–615, 1092
Lock-and-key model of enzyme action, 733, 733*f*, G-10
Logarithms, A-1–A-2
London, Fritz, 486
London forces. *See* Dispersion forces
Lone pairs
defined, 380, G-10
Lewis structures and, 408
orbital hybridization and, 446–448
in VSEPR theory, 419–423
Low-density polyethylene (LDPE), 518
Lowry, Thomas M., 169, 803
Low-spin complexes, 1060–1061, G-10
LSD (lysergic acid diethylamide), 664*f*
Luminous intensity, 14*t*
Luster, 505
Lyotropic phases, 510
Lysergic acid diethylamide (LSD), 664*f*
Lysine, 540, 602*f*, 657*f*, 672*f*

M

Macromolecules. *See also* Biological macromolecules; Polymers
defined, 79, 514
free radical attack on, 1095
synthetic, 668–670
Macroscopic properties, 8, 44
Magnesium
abundance of, 1001, 1001*t*
diagonal relationship with lithium, 595, 595*f*
electron configuration, 338*t*, 339
hexagonal closest packing in, 503*f*
industrial extraction of, 162–163
isolation of, 1021–1022, 1022*f*
magnesium oxide formed from, 112–114, 113*f*
metallurgy of, 1021–1022, 1022*f*
properties of, 353, 593–595
uses of, 1022
Magnesium carbonate, 865

- Magnesium chloride
  - depicting solutions of, 565–566
  - electrolysis and, 183
  - electron density distribution, 394*f*
  - intermolecular forces, 487–488
  - ionic properties of, 394
  - in precipitation reactions, 162–163
  - selective precipitation and, 874
- Magnesium hydroxide, 172, 874
- Magnesium ions, 69*t*, 570–571, 829, 829*f*
- Magnesium oxide, 112–114, 113*f*, 175, 175*f*, 377
- Magnetic properties
  - of coordination compounds, 1060–1061
  - crystal field theory and, 1060–1061
  - diamagnetism, 357, 358*f*, 460–461, 460*f*, 1037, 1044
  - paramagnetism, 357, 358–359, 358*f*, 460–461, 460*f*, 1037, 1044
  - of transition elements, 1037, 1044
  - of transition metal ions, 357–358, 358*f*
- Magnetic quantum number, 320, 320*t*, 332*t*, G-10
- Magnetic resonance imaging (MRI), 649–650, 650*f*
- Magnetite, 520, 520*f*, 973, 1009, 1016*t*, 1017
- Main-group elements. *See also specific elements*
  - acid-base behavior of oxides, 354–355, 354*f*
  - atomic size of, 346–347, 346*f*
  - electron affinity of, 352*f*
  - electron configuration, 341*f*
  - ionization energy of, 349*f*
  - Lewis electron-dot symbols for, 371–372, 371*f*
  - metallic behavior of, 353, 354, 354*f*
  - monatomic ion names and formulas, 68–69, 69*f*, 69*t*
  - oxidation number of, 354, 354*f*
  - periodic patterns in, 584–624
  - in periodic table, 61, 62*f*
  - redox behavior of, 354
- Main-group ions, electron configuration of, 355–356, 355*f*
- Malathion, 611
- Maleic acid, A-9
- Malleability, 396–397, 505
- Malonic acid, A-10
- Manganese
  - abundance of, 1000, 1001*t*
  - appearance of, 1038*f*
  - in batteries, 968, 968*f*
  - complex ions of, 1060, 1060*f*
  - as dietary trace element, 1064*t*
  - electron configuration, 339, 340*t*, 1039, 1039*t*
  - oxidation state, 1042, 1042*t*
- Manganese(II) chloride tetrahydrate, 1043*f*
- Manhattan Project, 1106
- Manometers, 208, 208*mn*, 209*f*, 699, G-10
- Mantle of Earth, 998, 998–999*f*, 998*t*, G-10
- Many-electron atoms, 332–335, 348
- Mass
  - amount-mass-number conversions, 97–98, 151–152, 151*f*, 162*f*
  - atomic, 57–59, 61
  - conservation of, 46–47, 47*f*, 50
  - conversion to energy, 47
  - converting units of, 22
  - defined, 17, G-10
  - English equivalents, 15*t*
  - examples of, 17*f*
  - as extensive property, 27, 27*f*
  - interconversion of energy and, 1101–1104
  - isotopic, 58
  - law of definite composition and, 47–48, 48*f*, 50
  - molality and, 553
  - molar (*See* Molar mass)
  - molecular, 76–78
  - SI units, 14–15*t*, 17, 17*f*, 17*n*
  - weight vs., 17
- Mass action, law of, 750
- Mass-action expression. *See* Reaction quotient
- Mass/charge ratio (m/e), 53, 60, 60*f*
- Mass difference
  - in chemical reactions, 1101–1102
  - explaination by atomic theory, 50–51
  - mass conservation and, 46–47, 47*f*, 50
  - multiple proportions and, 49–51
  - in nuclear reactions, 1102
  - visualization of, 51
- Mass-energy equivalence, 896*n*
- Mass fraction, 47–48
- Mass number
  - atomic notation for, 57*f*
  - defined, 56, G-10
  - nuclide, 1074, 1081
- Mass percent (mass %), 47–48, 48*f*, 102–103, 554, G-10
- Mass ratio, fixed, 44
- Mass spectrometry, 58, 60, 60*f*, G-10
- Matches, 611, 621
- Material flow tracers, 1099
- Materials, advanced. *See* Advanced materials
- Mathematical operations, A-1–A-4
- Matter
  - ancient Greek views of, 43
  - classification of, 81, 82*f*
  - composition of, 4
  - conservation of, 46
  - de Broglie wavelength of, 311, 311*t*
  - defined, 4, G-10
  - distinction from, 298–299
  - effects of nuclear radiation on, 1093–1095
  - energy, importance in study of, 8–10
  - entropy and states of, 547
  - extensive and intensive properties of, 27–28, 27*f*
  - properties of, 5–7, 7*t*
  - states of (*See* States of matter)
  - types of, 44–46
  - wave-particle duality of, 310–313, 310*f*
- Maxwell, James Clerk, 231, 916
- Mean free path, 238, G-10
- Measurement. *See also* SI units
  - accuracy and, 32–33, 33*f*
  - density, 23–25, 24*t*
  - historical standards of, 13
  - length, 14–15*t*, 15, 17*f*
  - mass, 14–15*t*, 17, 17*f*
  - precision and, 32–33, 33*f*
  - significant figures in, 28–32, 29*f*
  - temperature, 25–27, 26*f*, 27*t*
  - time, 14*t*, 18
  - uncertainty in, 28–33
  - volume, 15, 15*t*, 16–17*f*
- Mechanical balances, 17
- Medical diagnosis, radioactive tracer use in, 1099–1100, 1099*t*, 1100*f*
- Medical tradition, history of, 11
- Mega- (prefix), 14*t*
- Meitner, Lise, 1105
- Meitnerium, 1092, 1105
- Melting
  - defined, 471, G-10
  - as endothermic and spontaneous process, 897
  - solid-liquid equilibria and, 479
  - sublimation and, 472, 472*f*
- Melting point
  - of alkali metals, 396, 396*f*, 591
  - alkaline earth metals, 593, 594
  - of alkaline earth metals, 396, 396*f*
  - of boron family elements, 596, 598*t*
  - of carbon family elements, 598*t*, 599
  - defined, 479, G-10
  - of halogens, 618
  - of ionic compounds, 378, 378*t*
  - of metals, 396, 396*f*, 396*t*
  - of nitrogen family elements, 598*t*, 605
  - of noble gases, 623
  - of oxygen family elements, 613
  - of Period 3 chlorides, 395*f*
- Membrane-cell process, of chlorine production, 1029
- Membranes
  - dual polarity of, 542, 542*f*
  - free radical attack on, 1095
  - proton exchange, 971, 971*f*
  - semipermeable, 562, 571, G-14
  - structure of, 542, 542*f*
- Mendeleev, Dmitri, 61, 331
- Meniscus, 16*f*, 489–490, 490*f*
- Menten, Maud, 733
- Mercury
  - in barometers, 208, 208*f*
  - in batteries, 968–969
  - Boyle's experiments with, 211–212, 211*f*
  - calx, 11
  - entropy change resulting from change in temperature, 904–905
  - heats of vaporization and of fusion, 471*f*
  - Lavoisier's experiments with, 11
  - line spectra of, 303*f*
  - in manometers, 208, 209*f*
  - meniscus shape in glass, 490, 490*f*
  - mercury(I) and mercury(II), 71*t*
  - mineral sources of, 1008*t*
  - as neurotoxin, 281
  - standard enthalpies of formation, 278*t*
  - standard state of, 278
  - surface tension of, 489*t*
  - viscosity of, 489*t*
- Mercury-cell process, 1029
- Mercury(I) iodide, 866*t*
- Mercury(II) oxide, 183, 183*f*
- Mesosphere, 239*f*, 240
- Messenger RNA (mRNA), 677–678, 677*f*
- Metabolic pathways, 121, 781, 781*f*, G-10
- Metallic bonds
  - alkali metals and, 591
  - alkaline earth metals and, 594
  - band theory and, 506
  - defined, 371, G-10
  - electron pooling and, 370*f*, 371
  - electron-sea model and, 395–396
  - in intermolecular force comparison, 483*t*
  - in mercury, 490
- Metallic (interstitial) hydrides, 586–587, 586*f*
- Metallic radius, 345, 345*f*, G-10
- Metallic solids, 501*t*, 503, 503*f*, G-10
- Metalloids
  - defined, G-10
  - examples of, 63*f*
  - oxides of, 355
  - in periodic table, 62, 62*f*
  - properties of, 62, 353
  - as semiconductors, 506
  - zone refining of, 1012
- Metallurgy, 1008–1022
  - alloying, 1012–1013, 1013*f*, 1013*t*
  - aluminum, 1019–1021, 1020*f*
  - converting minerals to elements, 1010–1011, 1012*f*
  - copper, 1017–1018, 1018*f*
  - defined, 1008, G-10
  - iron, 1015–1017, 1016*f*, 1016*t*
  - magnesium, 1021–1022, 1022*f*
  - ore pretreatment, 1009
  - potassium, 1014–1015
  - process overview, 1009*f*
  - redox reactions in, 1010–1011, 1012*f*
  - refining (purifying) elements, 1012, 1012*f*
  - sodium, 1014–1015, 1015*f*
- Metal oxide (calx), 11
- Metals
  - acid-base behavior of oxides, 354–355, 354*f*
  - acid strength of hydrated metal ions, 817, 817*f*
  - activity series, 184–186, 958–959
  - alloys (*See* Alloys)
  - boiling point of, 396, 396*t*
  - bonding (*See* Metallic bonds)
  - calx, 11
  - combination reactions and, 181, 182*f*
  - comparison with nonmetals, 370–371
  - conductivity of, 397, 506
  - crystal structure of, 503, 503*f*
  - defined, G-10
  - displacement reactions and, 184–186
  - electron affinity of, 352
  - electron configuration of, 356–359
  - electronegativity of, 390, 394
  - electron-sea model and, 395–396
  - examples of, 63*f*
  - forming more than one ion, 70, 71*t*
  - ionic bonding and, 181, 370
  - ionization energy of, 352
  - Lewis electron-dot symbols for, 371
  - magnetic properties of, 357–358, 358*f*
  - mechanical properties of, 396–397, 397*f*
  - melting point of, 396, 396*f*, 396*t*
  - metal cations as Lewis acids, 829
  - metallic radius for, 345, 345*f*
  - in periodic table, 62, 62*f*
  - properties of, 62, 353
  - tendency to lose or gain electrons, 353–354, 353*f*
  - trends in metallic behavior, 353–354*f*, 353–355
- Metal sulfides, 616, 1010
- Metathesis reactions, 159, 167, 184, G-10

Meteorites, 1006, 1088
Meter (m), 14–15t, 15, 296, G-10
Methanal. *See* Formaldehyde
Methane
  boiling point of, 641f
  carbon dioxide and water formed from, 8mn
  chlorination of, 389
  combustion of, 387–388, 388f
  conversion to hydrogen fuel, 763
  cubic closest packing in, 502, 502f
  effusion rate of, 236–237
  entropy and, 909
  formation equation for, 277
  formula and model, 76t
  heats of vaporization and of fusion, 471f
  hybrid orbitals and, 444, 447, 447f
  for hydrogen production, 1023
  molecular shape of, 420
  mole fraction in air at sea level, 239t
  in residential heating systems, 284
  syngas and, 281
Methanogenesis, 281, G-10
Methanoic acid, 662, 662f
Methanol
  hybrid orbitals and, 449–450
  intermolecular forces, 534f
  Lewis structure for, 408
  molecular structure of, 654f, 655t, 664
  production of, 1024
  solubility of, 536, 536f, 536t
Methionine, 672f
Meth- (numerical root), 639t
Methyl acetate, 655t, 664
Methylamine, 189, 818, 819f, 819t, A-11
Methylbenzene, 647–648
Methyl bromide, 719–720, 720f
Methyl chloride, 483–484, 484f, 655t
Methyl cyanide, 655t
Methyl ethanoate, 655t, 664
Methyl ethyl ketone, 660f
Methylhexane, 643, 643f
Methylisocyanate, 224
Methylpropane, 110t, 642, 642t
Methylpropene, 656
Methyl red, 861–862
Metric system, 13
Mica, 604
Micelles, 569
Michaelis, Leonor, 733
Michaelis constant, 734, 734f, G-10
Michaelis-Menten equation, 734, G-10
Microanalysis, 967f
Micro- (prefix), 14t
Microscope, electron, 311–312, 312f
Microstates, 898–900, 912–913, G-10
Millikan, Robert, 53–54
Millikan's oil-drop experiment, 53–54, 53f
Milliliter (mL), 15, G-10
Millimeter of mercury (mmHg), 208, 209, 209t, G-10
Milli- (prefix), 14, 14t
Minerals
  converting to elements, 1010–1011, 1012f
  defined, 1008, G-10
  as element source, 1008, 1008t
  of iron, 1016t
Misch metal, 1045
Miscibility, 534, G-10
Mitochondrion, 982, 982f
Mixing, convective, 240
Mixtures, 532–571. *See also* Colloids; Solutions
  atomic-scale view of, 44f, 45
  classification of, 81, 82f
  comparison with compounds, 45, 81, 81f
  defined, 45, 533, G-10
  heterogeneous, 81, 533, 568
  homogeneous, 81, 533, 568
  partial pressure of gas in mixture of gas, 225–227
  particle size and, 568
  racemic, 643
  reaction direction and, 768–769
  separation techniques for, 83, 83f
m- (meta), 648
mmHg (millimeter of mercury), 208, 209, 209t, G-10
Mobile phase of chromatography, 83, 83f
MO bond order, 457, G-10
Model (theory), in the scientific method, 13, 13f, G-10
Molal boiling point elevation constant, 560, 560t
Molal freezing point depression constant, 560t, 561
Molality, 553–554, 553t, G-10
Molar heat capacity, 268, G-10
Molarity, 150, 552–553, 553t, G-10
Molar mass
  boiling point and, 486, 486f
  of compounds, 97
  defined, 96, G-10
  dispersion (London) forces and, 486, 486f
  of elements, 97
  entropy and, 909
  gases and, 223, 224–225
  Graham's law of effusion and, 236–237
  in molecular formula determination, 107–108
  molecular speed and, 234, 234f
  of polymers, 514, 514t
  solute molar mass from colligative properties, 563
Molar ratio, stoichiometrically equivalent, 116–120
Molar solubility, 866–867, 868t
Molar volume, of common gases, 241t
Molecular complexity, entropy and, 909, 909f
Molecular depictions
  ball-and-stick models, 79, 80t
  determination of formula, name, and mass using, 77–78
  of DNA, 80t
  in limiting-reactant problems, 123–124
  in precipitation reactions, 160–161
  space-filling models, 79, 80t
  of straight-chain alkanes, 76t
Molecular equations
  for acid-base reactions, 167–171, 170t, 171f
  for aqueous ionic reactions, 156, 156f, 158f, 159
  defined, 156, G-10
Molecular formulas
  combustion analysis, determination from, 108–109, 108f
  of covalent compounds, 370
  defined, 78–79, 105, G-10
  different compounds with same formula, 110–111, 110t
  elemental analysis and molar mass, determination from, 107–108
  hybrid orbitals and, 449, 449f
  Lewis structures, converting to, 406–408, 406f
  molecular shape, converting to, 423, 424f, 425–426
Molecular hydrides, 586
Molecularity, 722–724, 723t, G-11
Molecular mass, 76–78, G-11
Molecular orbital band theory, 504–506
Molecular orbital diagrams, 457–458, 457f, G-11
Molecular orbitals (MOs)
  antibonding, 456, 456f, 457
  bonding, 456, 456f
  bond order, 457
  defined, 455, G-11
  electrons in, 455–457
  energy and shape of, 456, 456f
  formation of, 455–456
  of hydrogen, 456, 456f
  of lithium, 505, 505f
  nonbonding, 462
Molecular orbital theory, 455–463
  antibonding MOs in, 456, 456f, 457
  atomic 2*p* orbital combinations in, 458–459f, 458–460
  bonding MOs in, 456, 456f
  central themes of, 455–457
  defined, 455, G-11
  formation of molecular orbitals in, 455–456
  heteronuclear diatomic molecules in, 462–463, 462f
  homonuclear diatomic molecules in, 458–462
  MO bond order, 457
  nonbonding MOs in, 462
  pi MOs in, 458–459f, 458–460
  polyatomic molecules in, 463, 463f
  sigma MOs, 456, 456f
Molecular polarity
  bond angle and, 428–429
  bond polarity and, 428–429
  defined, 428, G-11
  dipole moment and, 428–429
  effect on behavior, 430, 430f
  liquid solutions and, 535–537
  molecular shape and, 146, 428–430
  pi bonds and, 455
Molecular scene problems
  balancing equations from, 115–116
  buffers and, 850–851
  enthalpy change and free energy change from, 921–922
  equilibrium parameter determination from, 778–779
  finding quantities at various times with, 713–714
  free energy change calculation from, 928–930
  heat of phase change and, 475–476
  ionic compounds in aqueous solutions, 148–149
  precipitate formation and, 872–873
  predicting net direction of acid-base reactions using, 807–808
  reaction direction determination and, 758–759
  reaction order determination from, 706–707
Molecular shape, 404–432
  bent, 146, 146f, 419, 419–420f, 421, 424f, 447
  biological significance of, 431, 431–432f
  boiling point and, 487, 487f
  converting molecular formula to, 423, 424f, 425–426
  defined, 417, G-11
  of electron-deficient atoms, 413–414
  electron-group arrangements and, 418, 418f
  expanded valence shells and, 414–416
  formal charge and, 411–416
  Lewis structures and (*See* Lewis structures)
  linear arrangement, 418–419f, 419, 422, 422f, 424f
  molecular polarity and, 146, 428–430
  of molecules with multiple central atoms, 426–427, 427f
  of multiple bond molecules, 408–409
  octahedral arrangement, 418f, 422–424f
  of odd-electron atoms, 414
  of organic molecules, 634
  resonance structures and, 409–413
  seesaw, 422, 422f, 424f
  of single bond molecules, 406–408
  square planar, 423, 423–424f, 448
  square pyramidal, 423, 423–424f, 448
  tetrahedral arrangement, 418f, 420–421, 420f, 424f
  trigonal bipyramidal arrangement, 418f, 421–422, 422f, 424f
  trigonal planar arrangement, 418–419f, 419–420, 424f
  trigonal pyramidal shape, 420–421, 420f, 424f, 447
  T shape, 422, 422f, 424f
  using VSEPR to determine, 423, 425–426
  viscosity and, 490
  VSEPR theory and (*See* Valence-shell electron-pair repulsion theory)
Molecular solids, 501t, 502, 502f, G-11
Molecular speed
  molar mass and, 234, 234f
  temperature and, 232, 232f, 477–478, 477f
Molecular wave functions, 456, 456f
Molecular weight. *See* Molecular mass
Molecules
  bonding (*See* Bonding)
  chemical formula and structure of, 110–111, 110t
  of compounds, 44, 44f
  defined, 44, G-11
  depictions of (*See* Molecular depictions)
  diatomic, 44
  electron configuration of, 456–457
  elements as, 44, 44f, 66, 67f, 97
  as Lewis acids, 828–829
  Lewis structures of (*See* Lewis structures)
  orbitals of (*See* Molecular orbitals (MOs))
  organic (*See* Organic molecules)
  polarity of (*See* Molecular polarity)
  polyatomic, 463, 463f
  reaction rate and structure of, 719, 719f
  shape of (*See* Molecular shape)
Mole fraction, 226, 553t, 554–555, G-10
Mole (mol)
  in amount-mass-number relationships, 97–102, 98f, 100f, 118f
  Avogadro's number and, 96, 98–102, 98f, 100f
  of common substances, 96, 96f, 96mn
  defined, 95–96, G-10
  ions in aqueous ionic solutions, 149
  molar mass, 96–97, 97n
  relationship with energy transferred as heat during reactions, 274, 274f

as SI base unit, 14*t*
standard molar volume, 215, 215*f*
Molina, Mario J., 735
Molybdenum, 343, 1044, 1044*t*
Monatomic elements, molar mass of, 97
Monatomic ions
defined, 65, G-11
metals that form more than one monatomic ion, 70, 71*t*
names and formulas, 68–69, 69*f*, 69*t*
properties of, 355–360
Monochromatic light, 296
Monodentate ligands, 1047, 1048*t*
Monomers, 514, G-11
Mononucleotides, 543, 675, G-11
Mono- (prefix), 73*t*
Monoprotic acids, 796, 797*t*
Monosaccharides, 671, 671*f*, G-11
Moon rocks, radioisotopic dating of, 1088
MOs. *See* Molecular orbitals
Moscovium, 1092
Moseley, Henry G. J., 331
Motor oil, viscosity of, 491
Motor vehicles. *See* Automobiles
MRI (magnetic resonance imaging), 649–650, 650*f*
mRNA (messenger RNA), 677–678, 677*f*
Mulliken, Robert S., 392*n*
Multiple proportions, law of, 49–51
Multiplication, significant figures and, 30
Mylar, 670
Myosin, arrangement of filaments, 511*f*

N

NAA (neutron activation analysis), 1099
NAD (nicotinamide adenine dinucleotide), 982
Naming. *See* Nomenclature
Nanocomposites, 519
Nanomachines, 520, 520*f*
Nanometers (nm), 296
Nano- (prefix), 14*t*
Nanostructuring, 519–520
Nanotechnology, 519–520, G-11
Nanotubes, 520, 600–601, 600*f*
Naproxen, 644, 644*mn*
Natta, Giulio, 668
Natural gas, 281
Natural law, 12, G-11
Negative reaction order, 701
Negatron emission, 1077
Nematic phase, 510, 510*f*
Neon
boiling point of, 537*t*
diatomic molecules of, 460*f*, 461
electron configuration, 337, 338*f*
ionization energy, 351*t*
line spectra of, 302, 303*f*
mass/charge ratio of, 60, 60*f*
molar mass of, 97
mole fraction in air at sea level, 239*t*
properties of, 589*t*, 622–624
solubility in water, 537*t*
Neon signs, 53
Neopentane
constitutional isomers of, 642, 642*t*
formula and model, 642*t*
molecular shape of, 487, 487*f*
Neoprene, 518*t*
Nernst, Walther Hermann, 962
Nernst equation, 962–963, G-11
Nerve cells, concentration cells in, 967
Net ionic equations
for acid-base reactions, 167–171, 170*t*, 171*f*
for aqueous ionic reactions, 156*f*, 157, 158–159, 158*f*
defined, 157, G-11
Network covalent solids, 383, 383*f*, 386, 501*t*, 503–504, 504*t*, G-11
Neutralization, 793, 794, 852, G-11
Neutralization reactions. *See* Acid-base reactions
Neutrino, 1076*n*, 1077
Neutron activation analysis (NAA), 1099
Neutron-capture processes, 1111
Neutrons
defined, G-11
discovery of, 55, 1090
mass of, 56*t*
nuclide and, 1074
properties of, 56*t*
Neutron star, 1110*f*, 1111
Neutron-to-proton (N/Z) ratio, 1079, 1079*f*
Nickel
abundance of, 1001*t*
appearance of, 1039*f*
displacement of hydrogen from acid by, 185, 185*f*
electron configuration, 339, 340*t*, 343, 1039*t*
oxidation state, 1042, 1042*t*
Nickel-cadmium (nicad) batteries, 970
Nickel(II) hydroxide, 873
Nickel(II) nitrate hexahydrate, 152, 1043*t*
Nickel–metal hydride (Ni-MH) batteries, 970, 970*f*
Nicotinamide adenine dinucleotide (NAD), 982
Nihonium, 1092
Nile River, 569, 569*f*
Niobium, 1044
Nitrate ions
formula for, 72*t*
molecular shape of, 419
properties of, 609, 609*f*
resonance structure of, 412–413
Nitrates, 609–610
Nitric acid
acid rain and, 875
fixation and, 1005
number of ions in solution, 166
production from ammonia, 280
properties of, 609–610, 609*f*
Nitric oxide
boiling point of, 537*t*
hydrogen and, 700
molecular orbitals of, 462*f*, 463
mole fraction in air at sea level, 239*t*
oxygen and, 703–704, 703*t*, 725–728, 729*f*
ozone and, 692, 693, 693*f*, 699, 700
properties of, 608–609, 608*t*
redox reactions involving, 182
solubility in water, 537*t*
Nitride ions, 69*t*
Nitrides, 611
Nitrification, 571
Nitriles, 655*t*, 666, G-11
-nitrile (suffix), 655*t*
Nitrite ions, 72*t*, 609*f*, 610
Nitrogen
abundance of, 1000, 1001*t*
bond type, 598*t*
chemistry of, 607–610
critical temperature of, 480
diatomic molecules of, 460*f*, 461
electron configuration, 337, 338*f*
fixation of, 605, 779, 1004*f*, 1005
ionization energy, 351*t*
irradiation of, 1109
Lewis structure for, 409
Lewis symbol for, 371
liquid, 506
mean free path of, 238
melting point of, 598*t*
mole fraction in air at sea level, 239*t*
pollutants, 876
properties of, 353, 589*t*, 604–607
sources of, 1001
standard enthalpies of formation, 278*t*
triple bond in, 380
Nitrogen cycle, 1004–1005, 1004*f*
Nitrogen dioxide
atmospheric concentration of, 609
carbon monoxide and, 705–706, 724–725
equilibrium with dinitrogen tetroxide, 748–751, 748*f*, 750*t*, 751*f*, 759–760
fluorine and, 725, 728, 729*f*
formation of, 115–116, 609
formula and model, 80*t*
Lewis structure of, 414
mass-number conversion of, 100–101
mass of oxygen in, 104
mole fraction in air at sea level, 239*t*
structures and properties of, 608*t*, 609
Nitrogen family elements
acid-base behavior of, 354–355, 354*f*
comparison with oxygen family, 612, 614–615
halides, 605, 607
heat of fusion of, 598*t*
oxides, 606–607
oxoacids, 605
properties of, 604–611
reactivity of, 605, 606–607
tendency to lose or gain electrons, 353–354, 353*f*
Nitrogen fixation, 605, 779, 1004*f*, 1005
Nitrogen gas, 537*t*, 551
Nitrogen monoxide. *See* Nitric oxide
Nitrogen oxides, 608–609, 608*t*, 875, 908*f*, 909
Nitrogen trifluoride, 406–407
Nitrous acid, 609*f*, 610, 797*t*, 801*t*, 823, A-10
Nitrous oxide, 276–277, 608, 608*t*
NMR (nuclear magnetic resonance) spectroscopy, 649–650, G-11
Noble gases
atomic radius of, 348*f*
as atomic solids, 502
electron affinity of, 352
electron configuration, 343
electron configuration of, 355–356, 355*f*
ion formation and, 65–66, 66*f*
ionization energy of, 349*f*, 352
in periodic table, 63
properties of, 622–624
reactivity of, 624
sources of, 1001
Nodes, 314, 324, G-11
Nomenclature
acids, 74
alkanes, 76, 76*t*, 639, 639–640*t*, 646–647
alkenes, 644, 646–647
aromatic hydrocarbons, 647–648
binary covalent compounds, 74–76
binary ionic compounds, 68–71
coordination compounds, 1050–1051, 1050*t*
hydrates, 72, 73*t*
ionic compounds, 68–73
metals forming more than one ion, 70, 71*t*
numerical roots, 639, 639–640*t*
organic compounds, 76, 76*t*, 639, 639–640*t*
oxoanions, 72, 72*f*
Roman numerals for metal ions, 70, 71*t*
transuranium elements, 1092, 1092*t*
Nonane, 76*t*, 641*f*
Nona- (prefix), 73*t*
Nonbonding MOs, 462, G-11
Nonelectrolytes
colligative properties of, 557, 557*f*, 558–566
defined, 150, G-11
non-conductivity of, 557, 557*f*
nonvolatile solutions, 558–563
volatile solutions, 564
Nonmetal hydrides, acid strength of, 816, 816*f*
Nonmetals
comparison with metals, 370–371
covalent bonding in, 181, 370, 370*f*
covalent radius for, 345, 345*f*
defined, G-11
electron affinity of, 352
electronegativity of, 390, 394
examples of, 63*f*
as insulators, 506
ionic bonding in, 181, 370, 370*f*
ionization energy of, 352
Lewis electron-dot symbols for, 371
oxides of, 355
oxoacids, esters, and amides of, 666, 666*f*
in periodic table, 62, 62*f*
properties of, 62, 353
Non- (numerical root), 639*t*
Nonpolar covalent bonds, 392, 392*f*, G-11
Nonrechargeable batteries, 968–969
Nonstandard half-cell potentials, 977–978
Nonvolatile nonelectrolyte solutions, 558–563
Northern lights, 53, 53*mn*
n-type semiconductors, 508–509, 508*f*
Nuclear atom model, 52–55
Nuclear binding energy, 1102–1104, 1104*f*, G-11
Nuclear charge, effect on sublevel energy, 333, 333*f*
Nuclear energy, 284. *See also* Nuclear fission; Nuclear fusion
Nuclear energy reactors, 1106–1109, 1107*f*
Nuclear equations, 1078

- Photons
  - defined, 301, G-12
  - emission of, 304, 304*f*
  - energy of, 304
  - particle nature of, 312
  - ping-pong analogy for, 301*mn*
- Photon theory of light, 301
- Photosynthesis, 261, 1000, 1003, 1098
- Photovoltaic cells, 284, G-12
- Photovoltaic thermal (PVT) hybrid cells, 284
- pH paper, 803, 803*f*
- pH scale, 800, 800*f*
- Physical changes
  - comparison with chemical changes, 5, 5*f*, 6, 7–8
  - defined, 5, G-12
  - on Io (Jupiter's moon), 7
  - reversal of, 7
  - temperature and, 6–7
- Physical properties
  - of alkali metals, 591
  - alkaline earth metals, 593, 594
  - bonding, impact on, 598, 600–601
  - of carbon family elements, 598–601
  - defined, 5, G-12
  - of halogens, 617, 617*f*, 618
  - of nitrogen family elements, 605–606
  - of noble gases, 622–624
  - of oxygen family elements, 612–614
  - of Period 2 elements, 587–590
  - of transition elements, 1040–1041*f*, 1040–1042
- Physical states
  - characteristics of, 4, 4*f*, 205–207, 206*f*
  - kinetic-molecular view of, 470, 470*t*
  - reaction rate and, 693
- Pi bond, 452–454, 453*f*, 455, 455*f*, 644, G-12
- Picometers (pm), 296
- Pico- (prefix), 14*t*
- Pig iron, 1017
- Pi MOs, 458–460, G-12
- Piperidine, A-11
- Pipets, 15, 16*f*
- Piston-cylinder assembly, 214–215, 221–222, 774
- Pitchblende, 48–49, 1075
- Planck, Max, 300
- Planck's constant, 300, G-12
- Plasma, fusion of, 1109, 1109*f*
- Plastics, 517, 619, G-12
- Platinum
  - density of, 1042
  - as electrode, 947
  - electron configuration, 343
  - sources of, 1001
- Pleated sheet structure, 673*f*, 674
- Plum-pudding model of atoms, 54
- Plutonium, 1093
- p-n junction, 508–509, 508*f*
- pOH, 801–802, 802*f*
- Polar arrow, 146, 392
- Polar bonds, 146
- Polar covalent bonds, 390, 390*f*, 392, 393*f*, G-12
- Polarimeter, 643, 643*f*, G-12
- Polarity, 539–540. *See also* Molecular polarity
- Polarizability, 486, G-12
- Polar molecules
  - defined, G-12
  - dipole-dipole forces and, 483, 483*f*
  - water as, 146
- Pollution
  - air (*See* Air pollution)
  - phosphates and, 1008
  - thermal, 550, 551*f*, 1108–1109
- Polonium
  - decay of, 1078
  - discovery of, 1075
  - properties of, 612–615
  - radioactive poison from, 1094
- Polyamides, 669–670, 670*f*
- Polyatomic ions
  - compounds containing, 71–73
  - defined, 67, G-12
  - hydrates, 72, 73*t*
  - Lewis structures for, 406, 407
  - oxoanions, 72, 72*f*
- Polyatomic molecules, 463, 463*f*
- Polybutadiene, 518*t*
- Polychlorinated biphenyls (PCBs), 602, 602*f*, 657, 657*f*
- Polychloroprene, 518*t*
- Polychromatic light, 296
- Polydentate ligands, 1047, 1048*t*
- Poly(dimethyl siloxane), 518*t*, 604
- Polyesters, 670
- Polyethylene, 514–516, 514*t*, 515*f*, 518, 668, 669*t*
- Polyisoprene, 518*t*
- Polymer glass, 517
- Polymeric viscosity improvers, 491
- Polymers
  - addition polymers, 668, 668*f*, 668*t*
  - biopolymers, 514
  - branches of, 517–518
  - chain dimensions, 514–516
  - condensation polymers, 669–670, 670*f*
  - copolymers, 518, 669
  - crosslinks, 518
  - crystallinity of, 516, 516*f*
  - defined, 514, G-12
  - degree of polymerization, 514, 514*t*
  - dendrimers, 518
  - dissolved, 516–517
  - elastomers, 518, 518*t*
  - homopolymers, 518
  - inorganic, 670
  - liquid crystal and, 511
  - molar mass of, 514, 514*t*
  - monomers, 514
  - naming, 668
  - nucleic acids, 542–543, 674–680
  - plasticity of, 517
  - proteins (*See* Proteins)
  - pure, 517
  - radius of gyration, 515–516, 515*n*
  - random coil, 515, 515*f*
  - silicone, 604
  - synthetic, 668–670
  - thermoplastic, 518
  - thermoset, 518, 518*mn*
  - viscosity of, 516–517, 517*f*
- Polyphosphates, 611, 611*f*
- Polyphosphazene, 611, 670
- Poly- (prefix), 668
- Polypropylenes, 668, 669*t*
- Polyprotic acids
  - acid-base titration curves for, 862–863, 863*f*
  - acid-dissociation constant for, 813, 814*t*
  - amino acids as, 863
  - calculating equilibrium concentrations for, 814–815
  - defined, 813, G-12
- Polysaccharides, 671–672, 671*mn*, G-12
- Polystyrene, 223–224, 668, 669*t*
- Polysulfanes, 650
- Poly(vinylchloride) (PVC), 619, 668, 669*t*
- *p* orbitals
  - defined, G-12
  - molecular orbitals from, 458–459*f*, 458–460
  - orbital overlap and, 443*f*, 444
  - quantum mechanics and, 324, 324*f*, 326*t*
- Porphin, 1063
- Positron emission, 1076*t*, 1077, 1094, G-12
- Positron-emission tomography (PET), 1078, 1100, 1100*f*
- Positrons, 1077, 1094, G-12
- Potassium
  - abundance of, 1000, 1001*t*
  - distribution of, 1000
  - effective nuclear charge of, 334, 334*t*
  - electron configuration, 339, 340*t*, 344, 356
  - industrial production of, 1014
  - properties of, 590–592, 592*mn*
  - radioactive, 1096
  - reaction with chlorine gas, 230, 230*mn*
  - reactivity, 343*f*
  - soaps, 541
  - uses of, 1015
- Potassium bromide, 148, 159, 979, 979*mn*
- Potassium carbonate, 169
- Potassium chlorate, 183, 917–918
- Potassium chloride, 181, 182*f*
- Potassium dichromate, 653, 656
- Potassium ethoxide, 656
- Potassium ferricyanide, 1043*t*
- Potassium hydrogen sulfite, 826
- Potassium hydroxide, 171
- Potassium ions, 69*t*, 967
- Potassium nitrate, 158
- Potassium sulfate, 148–149
- Potassium sulfide, 97
- Potassium superoxide, 1015, 1015*mn*
- Potential energy
  - bonding and, 369, 379, 381
  - in chemical reactions, 386
  - conversion to kinetic energy, 9–10, 9*f*, 257
  - defined, 8, G-12
  - intermolecular forces as, 469
- Pounds per square inch (lb/in^2 or psi), 209*t*
- p- (para), 648
- Prechemical traditions, 10–11, 10*f*
- Precipitates
  - complex ions and solubility of, 879–881
  - defined, 157, G-12
  - predicting formation of, 871–873, 871*mn*
- Precipitation, selective, 874–875
- Precipitation reactions, 157–164
  - in aqueous ionic reactions, 157–161, 158*f*
  - defined, 157, G-12
  - key events in, 157
  - metathesis reactions, 159
  - molecular depictions in, 160–161
  - predicting, 159–160, 159*t*
  - reversibility of, 190
  - stoichiometry of, 162–164
- Precipitator, Cottrell, 569
- Precision, 32–33, 33*f*, G-12
- Prefixes
  - acids, 74
  - alkanes, 76
  - coordination compounds, 1050
  - decimal, 14, 14*n*, 14*t*
  - hydrates, 72, 73*t*
  - numerical, 72, 73*t*
  - organic compound, 639, 639–640*t*
  - oxoanions, 72, 72*f*
  - for units of mass, 17, 17*n*
- Pressure
  - Amontons's law and, 213
  - amount and, 218–219
  - atmospheric, 207–208*f*, 207–209, 209*t*, 239, 239*f*
  - Boyle's law and, 211–212, 211*f*, 217–218, 223, 902
  - critical, 480
  - defined, 207, G-12
  - effect of change on equilibrium, 773–775, 774*f*, 777*t*
  - equilibrium constant based on, 756–757
  - extreme, 241–243, 242*f*
  - gases and, 206, 207–210, 218, 239, 241–243, 242*f*
  - origin of, 231, 232
  - pressure-volume work, 263–264, 264*f*
  - solubility and, 551–552, 551*f*
  - temperature and, 213–214, 218
  - units of, 209–210, 209*t*
  - volume and, 206, 211–212, 211*f*, 217–218
- Pressure-volume work (PV work), 263–264, 264*f*, G-12
- Primary batteries, 968–969
- Primary structure of proteins, 673, 673*f*
- Principal energy level, 335
- Principal quantum number, 319, 320*t*, 332*t*, G-12
- Prisms, 298, 302
- Probability contour, 319, 319*f*, 323–325*f*, G-12
- Probability density, 314–315, 314*f*, 317, 317*f*, 323, 323*f*
- Problem solving
  - conversion factors in, 18–23
  - dilution problems, 153–154
  - equilibrium and, 760–769, 770*f*
  - gas laws and, 217–222
  - systematic approach to, 19–20
  - weak-acid equilibria and, 808–815
- Product-favored reactions, 931, 932*f*
- Products
  - calculating amount of, 116–130
  - calculating amounts in limiting-reactant problems, 125–127
  - defined, 112, G-12
  - determining heat of reaction from heat of formation for, 279–280, 279*f*
  - expressing reaction rate in terms of concentration of, 696–698, 696*f*
  - gaseous, 227–228, 227*f*
  - gas variables to find amounts of, 229–230
  - in precipitation reactions, 162–163

Proline, 672*f*
Propane
boiling point of, 484*f*, 641*f*
combustion of, 117, 117*t*, 911–912
dipole moment of, 484*f*
formula and model, 76*t*, 640
Propanoic acid
acid-dissociation constant of, 797*t*, A-10
determining concentration and pH, 810–811
titration of, 855–859, 855*f*
Propanol, 536*t*, 653
Propanone, 653, 655*t*, 660*f*, 661
Properties
chemical (*See* Chemical properties)
defined, 5, G-12
of matter, 5–7, 7*t*
physical (*See* Physical properties)
Prop- (numerical root), 639*t*
Propylamine, A-11
Propylenediamine, A-12
Proteins. *See also* Amino acids
defined, 539, 672, G-12
fibrous, 674, 674*f*
globular, 674
intermolecular forces and shape of, 540–541, 540*f*
mass spectrometry studies of, 60
polarity of amino acid side chains, 539–540
structure of, 539–540*f*, 539–541, 672–673*f*, 672–674
synthesis of, 677–678, 677*f*
Protein synthesis, 677–678, 677*f*
Protium, 1024, 1024*t*
Proton acceptors, 803, G-12
Proton donors, 803, G-12
Proton exchange membrane (PEM), 971, 971*f*
Protons
atomic number and, 56
charge of, 55, 56*f*, 56*t*
defined, 55, G-12
discovery of, 55
in Large Hadron Collider, 1091–1092
in linear accelerator, 1091, 1091*f*
mass of, 56*t*
nuclide and, 1074
properties of, 56*t*
solvated, 165
transfer in acid-base reactions, 168–171, 169*f*
Proton transfer, in Brønsted-Lowry acid-base definition, 803–808
p-scale, 800, 801
Pseudo–first-order reactions, 709, G-12
Pseudo-noble gas configuration, 355–356, G-12
p-type semiconductors, 508–509, 508*f*
Purines, 675
PVC, 619, 668, 669*t*
PVT (photovoltaic thermal) hybrid cells, 284
PV work (pressure-volume work), 263–264, 264*f*, G-12
Pyridine, 819*t*, 820, A-12
Pyrimidines, 675
Pyrite, 1016*t*
Pyrometallurgy, 1008
Pyrrhotite, 1016*t*
Pyruvic acid, A-10

Q

Quadratic formula, 765, A-3
Quantitative theories of chemistry, 12
Quantized energy of particles, 316, 318
Quantum, 300, G-13
Quantum dots, 284, 519, 519*f*
Quantum mechanics
atomic orbital, 314–326
defined, 314, G-13
electron cloud depictions and, 318, 323, 323–325*f*
electron density diagram and, 318, 319*f*
energy levels in hydrogen atoms and, 325, 325*f*
level (shell), 321
molecular orbital model and, 455
nodes and, 324
orbital overlap and, 443
particle-in-a-box model and, 315–317, 315*f*
periodic table and, 335–345
probability contour and, 319, 319*f*, 323–325*f*
radial probability distribution plot and, 319, 319*f*, 319*mn*
Schrödinger equation and, 314–315, 317–318
sublevel (subshell), 321–322
wave function and, 314, 314*f*, 317, 317*f*
Quantum numbers
angular momentum, 320, 320*t*, 332*t*
of atomic orbital, 319–321, 320*t*
baseball, 333*mn*
correlation with orbital diagrams, 337
defined, G-13
electron orbit and, 304
of electrons, 332, 332*t*
identifying incorrect numbers, 322
magnetic, 320, 320*t*, 332*t*
in particle-in-a-box model, 316
principal, 319, 320*t*, 332*t*
in quantum theory, 300
spin, 332–333, 332*f*, 332*t*
Quantum staircase, 304, 304*f*
Quantum theory, 299–300, 312*f*
Quartz
covalent bonding in, 383, 383*f*
crystal structure of, 493*f*, 504, 504*f*, 604, 604*f*
Quaternary structure of proteins, 673*f*, 674

R

Racemic mixtures, 643
Radial probability distribution plot
defined, G-13
d orbital and, 325*f*
order of filling and, 339
p orbital and, 324*f*
quantum mechanics and, 319, 319*f*, 319*mn*
s orbital and, 323*f*, 324
sublevel energy and, 334, 334*f*
Radiant energy, 295–297
Radiation
applications of, 1100–1101
background, 1095–1096
background radiation, 1096
blackbody, 299–300, 300*f*, 300*n*
cosmic, 1088, 1095
counters for detection of, 1083–1084, 1084*f*
effects on matter, 1093–1095
electromagnetic, 295–297
free radical formation and, 1095
infrared, 296, 297*f*
intensity of, 296
ionizing (*See* Ionizing radiation)
penetrating power of, 1094–1095, 1094*f*
risk from, 1096–1097, 1097*f*, 1097*t*
sources of, 1095–1096, 1096*t*
ultraviolet (*See* Ultraviolet radiation)
units of radiation dose, 1093–1094
Radiation-absorbed dose (rad), 1094, G-13
Radiation therapy, 1100
Radioactive decay
decay series, 1082, 1082*f*
as first-order process, 712
half-life of, 1086, 1086*f*, 1086*t*
kinetics of, 1083–1090
nuclear equations for, 1078
nuclear stability and, 1079–1082, 1079*f*, 1080*t*
predicting mode of, 1081–1082
radioisotopic dating, 1087–1089, 1089*f*
rate of, 1084–1087, 1086*f*, 1086*t*
types of, 1075–1078, 1076*t*
Radioactive emissions
alpha particles (*See* Alpha particles)
behavior in electric field, 1075, 1075*f*
beta particles (*See* Beta particles)
detection of, 1083–1084, 1084*f*
gamma rays (*See* Gamma rays)
ionization of, 1093–1097
penetrating power of, 1094–1095, 1094*f*
positron emission, 1076*n*, 1076*t*, 1077, 1094
Radioactive tracers, 1098–1100, 1099*t*, 1100*f*
Radioactivity. *See also* Nuclear reactions
of actinides, 1045
defined, 1073, G-13
detection and measurement of, 1083–1084, 1084*f*
discovery of, 1075
Radiocarbon dating, 1087–1089, 1088*n*, 1089*f*
Radioisotope heater unit (RHU), 1101
Radioisotopes
applications of, 1098–1101
artificial, 1090
dating, 1087–1089, 1089*f*
defined, 1087, G-13
ionizing radiation and, 1100–1101
tracers, 1098–1100, 1099*t*, 1100*f*
Radioisotopic dating, 1087–1089, 1089*f*, G-13
Radium
discovery of, 1075
in paint, 1094
properties of, 593–594
radioactive decay, 1077
Rutherford's experiment and, 54
Radius
atomic (*See* Atomic radius)
covalent, 345, 345*f*, 381*f*, 482, 482*f*
of gyration, 515–516, 515*n*, G-13
ionic (*See* Ionic radius)
van der Waals, 482, 482*f*
Radon, 622–624, 1095–1096, 1096*f*, 1096*t*
Rad (radiation-absorbed dose), 1094, G-13
Rainbows, 298, 298*mn*
Random coil, 515, 515*f*, G-13
Random error, 32–33, 33*f*, G-13
Raoult's law, 558–559, 564, G-13
Rare earth elements. *See* Lanthanides
Rate constant
defined, 698, G-13
determining, 704–706
temperature and, 716, 716*f*
units, 704, 704*t*
Rate-determining step, 724–725, 724*mn*, G-13
Rate law (rate equation)
collision theory and, 716–719, 718*t*
defined, 698, G-13
for elementary reactions, 723*t*
integrated, 708–715, 711*f*
rate constant and, 698, 704–706
reaction mechanisms and, 725–728, 729*f*
reaction order and, 698–707
steady-state approximation and, 727–728, 733
Rate-limiting step, 724–725
Rate of reaction. *See* Reaction rate
Reactant-favored reactions, 931–932, 932*f*
Reactants
calculating amount of, 116–130
calculating amounts in limiting-reactant problems, 125–127
concentration determination after a given time, 708–709
defined, 111–112, G-13
determining heat of reaction from heat of formation for, 279–280, 279*f*
expressing reaction rate in terms of concentration of, 696–698, 696*f*
gas variables to find amounts of, 229–230
limiting, 122–127, 123*f*
in precipitation reactions, 162–163
Reaction direction
in acid-base reactions, 805–808
equilibrium state and, 753, 757–760, 758*f*, 768–769
Reaction energy diagrams, 720–722, 721*f*, 728, 729–730*f*, G-13
Reaction intermediate, 724, G-13
Reaction mechanisms
defined, 722, G-13
elementary reactions and, 722–724, 723*t*
molecularity and, 722–724, 723*t*
rate-determining step of, 724–725
rate law and, 725–728, 729*f*
Reaction order
changing reactant concentrations for determining, 702–703*t*, 702–704
defined, 699, G-13
first-order reactions, 699, 700, 700*f*, 704*t*, 708, 710, 711*f*, 715*t*
fractional, 701
integrated rate law in determination of, 710–711, 711*f*
molecular scene problems for determining, 706–707
negative, 701
rate constant and, 704, 704*t*
rate data for determining, 705–706
rate law and, 698–707
rate laws for determining, 701–702
second-order reactions, 699, 700, 700*f*, 704*t*, 708–711, 711*f*, 715*t*
terminology, 699–700
third-order reactions, 704, 704*t*
zero-order reactions, 700, 700*f*, 704*t*, 709, 711, 711*f*, 715*t*

Solutions. *See also* Solubility; Solutes; Solution process; Solvents
acidic solutions, balancing redox reactions in, 941–942
amount-mass-number conversions for, 151–152, 151*f*
aqueous (*See* Aqueous solutions)
basic solutions, balancing redox reactions in, 942
boiling point elevation and, 559–560, 560*f*, 560*t*
colligative properties of, 557–567
concentration of, 552–557
covalent, 150
defined, 533
diluting, 152–154, 153*f*
electrolyte, 557, 557*f*, 564–566, 565*f*
formation from gases, 207
freezing point depression and, 560–561, 560*t*, 566–567
gas-gas, 537–538
gas-liquid, 537
gas-solid, 538
heat of, 544, 545–546*f*, 545–547
ideal, 558
intermolecular forces in, 534–535
ionic, 146–149
mixtures (*See* Mixtures)
molarity of, 150
molecular polarity and, 535–537
nonelectrolyte, 557, 557*f*, 558–566
osmotic pressure and, 562, 562*f*, 567, 567*f*, 571
particle size and, 568
preparing, 152–154, 152*f*
salt solutions, acid-base properties of, 823–824, 825*t*
saturated, 549, 549*f*
solid-solid, 538, 538*f*
solute molar mass from colligative properties, 563
solution process, 544–549
supersaturated, 549, 549*f*
tonicity of, 567
unsaturated, 549
vapor pressure lowering and, 558–559, 558*f*
water purification and, 570–571
Solvation
defined, 545, G-14
of ions, 147, 157, 165, 165*f*
of protons, 165
Solvents
defined, 145, 534, G-14
hexane as, 535, 536, 536*t*, 548, 548*f*
water as, 145–155, 491, 535–536, 536*f*, 536*t*, 908, 908*f*
s orbitals, 323–324, 323*f*, 326*t*, 444, G-13
Sources of elements, 1000–1001, 1002*f*, 1008, 1008*t*
Sour taste from acids, 793
Space-filling models, 79, 80*t*, 639, 640*f*
Specific heat capacity, 268, 268*t*, 270, 491, G-14
Specific rotation of optical isomers, 643
Spectator ions, 156, 159, 160, 167, G-14
Spectra
absorption, 308–309
atomic, 302–307
emission, 308, 308*f*
line, 302–305
Spectrochemical series, 1059, 1059*f*, G-14
Spectrometers, 308, 308*f*
Spectrometry, 308–309, 699, G-14
Spectroscopy, infrared, 384–385, 385*f*
Speed of light, 296, G-14
sp hybridization, 444–446, 445*f*, 449*t*
sp^2 hybridization, 446, 446*f*, 449*t*
sp^3 hybridization, 447, 447*f*, 449*t*
sp^3d hybridization, 447–448, 448*f*, 449*t*
sp^3d^2 hybridization, 448, 449*f*, 449*t*
sp hybrid orbitals, 444–446, 445*f*, G-14
sp^2 hybrid orbitals, 446, 446*f*, 449*t*, G-14
sp^3 hybrid orbitals, 447, 447*f*, 449*t*, G-14
sp^3d hybrid orbitals, 447–448, 448*f*, 449*t*, G-14
sp^3d^2 hybrid orbitals, 448, 449*f*, 449*t*, G-14
Spin quantum number, 332–333, 332*f*, 332*t*, G-14
Spontaneous change
defined, 895, 931–932, G-14
determining, 913–914
enthalpy change and, 896–897
exothermic and endothermic reactions, 896–897, 897*f*, 915–916, 915*f*
free energy change and, 916–917
predicting, 895–910
redox reactions and, 945–949, 945*f*, 954–957
second law of thermodynamics and, 905
temperature and, 920–924, 921*t*, 924*f*
in voltaic cells, 945–949, 945*f*, 954–957
s-process, 1111
Square brackets [ ], 694, 751
Square planar complexes
in crystal field splitting, 1062, 1062*f*
hemoglobin, 1063
in valence bond theory, 1055, 1056*f*
Square planar shape, 423, 423–424*f*, 448, 1047, 1047*t*, G-14
Square pyramidal shape, 423, 423–424*f*, 448, G-14
S-shaped response model of radiation risk, 1097
Stalactites, 871
Stalagmites, 871
Standard atmosphere (atm), 209, 209*t*, G-14
Standard cell potential, 950–953, 950*f*, 959–966, G-14
Standard electrode potential, 950–956, 952*f*, 954*t*, A-14, G-14
Standard enthalpy of formation, 277–280, 278*t*, 279*f*, G-14
Standard enthalpy of reaction, 277–280, G-14
Standard entropy of reaction, 910–912, G-14
Standard free energy change, 917–919, 927–928, 928*t*, G-14
Standard free energy of formation, 918–919, G-14
Standard heat of reaction. *See* Standard enthalpy of reaction
Standard hydrogen electrode, 951–952, 952*f*, G-14
Standard molar entropy, 905–910, G-14
Standard molar volume, 215, 215*f*, G-14
Standard reference half-cell, 951–952, 952*f*, G-14
Standard states, 277, 277*n*, G-15
Standard temperature and pressure (STP), 215, G-15
Starch, 671–672
Stars, element synthesis in, 1110–1111, 1110*f*
State functions, 262–263, 263*mn*, 266, G-15
States of matter
balanced equations and, 113–114
changes in, 5–7, 5*f*, 7*mn*
defined, 4, G-15
entropy and, 547
kinetic-molecular view of, 470, 470*t*
physical, 4, 4*f*, 205–207, 206*f*
Stationary phase of chromatography, 83, 83*f*
Stationary states, 304, G-15
Steady-state approximation, 727–728, 733, G-15
Steam-carbon reactions, 281
Steam-reforming process of hydrogen production, 283, 1023
Stearic acid, 662*f*
Steel
alloy composition, 1012, 1013*t*
defined, 1016, G-15
metallurgy, 1016–1017
specific heat capacity of, 268*t*
Stellar nucleosynthesis, 1110, G-15
Stereoisomers, 642–643, 1052–1054*f*, G-15
Stereoselective catalysts, 668
Sterling silver, 1013*t*
Stoichiometry
of acid-base reactions, 172–174
balancing (stoichiometric) coefficient, 112–113
defined, 95, G-15
of electrolysis, 980–981, 980*f*
equivalent molar ratios from balanced equations, 116–120
ideal gas law and, 228–230, 229*f*
of precipitation reactions, 162–164
of redox reactions, 179–180
of thermochemical equations, 273–274
STP (standard temperature and pressure), 215, G-15
Straight-chain alkanes, 76, 76*t*
Strassmann, F., 1105
Stratosphere, 239*f*, 240
Strong acids. *See also* Acid strength
calculating pH for, 801–803
categorization of, 165, 165*t*, 797
concentration of buffer components in, 845, 845*f*, 848, 848*f*
direction of reactions involving, 806V
dissociation extent, 795, 795*f*
leveling effect on, 827
reactions involving, 170–171, 170*t*
strong acid–strong base titration curves, 853–855, 853*f*
structural features of, 166–167
in water, 794–798
weak base–strong acid titration curves, 859–860, 859*f*
Strong bases
calculating pH for, 801–803
categorization of, 165, 165*t*, 797
concentration of buffer components in, 845–846, 845*f*, 848, 848*f*
leveling effect on, 827
reactions involving, 170–171, 170*t*
strong acid–strong base titration curves, 853–855, 853*f*
structural features of, 166–167
in water, 794–798
weak acid–strong base titration curves, 855–859, 855*f*
Strong electrolyte solutions, 557, 557*f*, 564–566, 565*f*
Strong-field ligands, 1058, 1058*f*, 1060, 1060–1061*f*, G-15
Strong force, 1080, G-15
Strontium
flame tests with, 308*f*
line spectra of, 303*f*
properties of, 593–594
radioisotopes of, 1086–1087
Strontium ions, 69*t*
STR (short tandem repeat) analysis, 680, 680*f*
Structural formulas
defined, 79, 105, G-15
isomers and, 110–111
Structural isomers. *See* Constitutional isomers
Styrofoam, 569
Subcritical mass, 1106
Sublevel energy
aufbau principle and, 335
crossover of, 357, 357*f*
defined, G-15
electron repulsion and, 333–334
nuclear charge and, 333, 333*f*
orbital shape and, 334
penetration and, 334, 334*t*, 339
shielding and, 333–334, 334*f*, 339
Sublevels
defined, 321
order of filling, 335*f*, 339–342, 341*f*, 342*mn*, 456–457
quantum numbers and energy level in, 321–322
splitting levels into, 334–335
Sublimation
defined, 471, G-15
heat of, 472, 472*f*
of iodine, 479, 479*f*
Submicroscopic properties, 44
Substances
defined, 4, 44, G-15
mixtures distinguished from, 45
standard state of, 277
Substituents, 647
Substitutional alloys, 538, 538*f*
Substitution reactions, 652, 660, G-15
Substrates, 732–734, 733*f*, 734*f*, G-15
Subtraction, significant figures and, 30–31
Succinic acid, A-10
Sucrose, 671, 671*f*
Suction pumps, 208
Suffixes
acids, 74
alkanes, 76
binary ionic compounds, 68
metal ions, 70
organic compound, 639, 639–640*t*
oxoanions, 72, 72*f*
Sugars. *See also* Carbohydrates
dehydration by sulfuric acid, 616, 616*f*
properties of, 670–672, 671*f*
Sulfanilamide, 666*f*
Sulfate ions, 72*t*, 415, 616
Sulfates, A-13
Sulfide ions, 69*t*
Sulfide phase, 999, 999*f*
Sulfides, 611, 616, 1001, 1002*f*, 1010, A-13
Sulfite ions, 72*t*, 616
Sulfites, reaction with acids, 169–170
Sulfonamides, 666, 666*f*
Sulfur
abundance of, 1000, 1001*t*
acid-base behavior of, 355
allotropes of, 612, 650
catenation of, 650
chemistry of, 615–616
in coal, 281
electron configuration, 338*t*, 339

on Io (Jupiter's moon), 7
oxidation to sulfur trioxide, 275–276
pollutants, 876
production of, 1025–1026, 1026*f*
properties of, 354, 612–615
sources of, 1001
standard enthalpies of formation, 278*t*
standard state of, 278

Sulfur dioxide
acid rain and, 875, 876
from coal burning, 281
formation of, 615
Lewis structure of, 416
molar mass of, 97
mole fraction in air at sea level, 239*t*
oxidation of, 753, 922–923, 930–931
release during roasting, 1010
roasting copper and, 118, 119

Sulfur hexafluoride
hybrid orbitals and, 448, 449*f*
Lewis structure of, 415
molecular shape of, 423

Sulfuric acid
acid-dissociation constant for, A-10
in acid rain, 616, 616*mn*, 875, 876
chemical manufacturing, 1025–1028
formation of, 616
Lewis structure of, 415
manufacture of, 730–731
production of, 922–923, 930–931
properties of, 616
uses of, 616, 1027–1028, 1027*f*

Sulfurous acid
acid-dissociation constant for, 814*t*, A-10
acid rain and, 875
Lewis acid and, 829
properties of, 615–616
titration of, 862, 863*f*

Sulfur tetrafluoride, 415, 422, 449–450
Sulfur trioxide, 275–276, 416, 615, 753
Superconducting oxides, 506, 512, 513
Superconductivity, 506, G-15
Supercooled liquids, 504
Supercritical fluid (SCF), 480
Supernova, 1110*f*, 1111
Supersaturated solutions, 549, 549*f*, G-15
Surface area, reaction rate and, 693, 693*f*
Surface tension, 489, 489*f*, 489*mn*, 489*t*, 492, G-15
Surfactants, 489
Surroundings
defined, 258, G-15
entropy change in, 912–914
Suspensions, 568, G-15
Sweating, 471, 471*f*
Sylvite, 1014
Synchrotron, 1091
Synthesis gas (syngas), 281, 603
Synthetic macromolecules, 668–670
Synthetic natural gas (SNG), 281, G-15
System
defined, 258, G-15
determining energy change in, 262, 896, 910–912
energy flow to and from, 258–260
Systematic error, 32–33, 33*f*, G-15
System energy, 258–260, 306*mn*

T

Table salt, density of, 24*t*
Talc, 604
Tanaka, Koichi, 60
Tantalum, 1044
TCDD (tetrachlorodibenzodioxin), 554
Technetium, 344, 1099, 1100*f*
Technological tradition, history of, 11
Tellurium, 612–615
Temperature. *See also* Endothermic processes; Exothermic processes
absolute scale, 213
Amontons's law and, 213
atmospheric, 239–240, 239*f*, 282–283
Celsius scale, 25–27, 26*f*, 27*t*
changes in matter and, 6–7
Charles's law and, 212–213, 212*f*
collision energy and, 717–718, 717*f*, 718*t*
converting units of, 26–27
critical, 480
defined, 25, G-15
effect of change on equilibrium, 775–777, 777*t*
entropy change due to change in, 904–906, 907*f*
entropy of surroundings and, 912–913
examples of, 25*f*
Fahrenheit scale, 25–27, 26*f*, 27*t*
gases and, 206, 212–213, 212*f*, 218, 239–240, 239*f*
glass transition, 517
heat vs., 25
as intensive property, 27–28
Kelvin scale, 25–27, 26*f*, 27*t*
kinetic energy and, 234–235, 240, 470
molecular speed and, 232, 232*f*, 477–478, 477*f*
phase changes and, 470–471, 474
pressure and, 213–214, 218
reaction rate and, 693, 716, 716*f*, 718*t*
reaction spontaneity, effect on, 920–924, 921*t*, 924*f*
SI units, 14*t*, 25–27, 26*f*, 27*t*
solubility and, 549–550, 550*f*
standard molar entropy and, 906, 907*f*
vapor pressure and, 477–478
vapor pressure of water and, 227, 227*t*
viscosity and, 490, 490*t*
volume and, 212–213, 212*f*, 218, 219–220
Temperature inversion, 240
TEM (transmission electron microscope), 311
Tennessine, 1092
Tera- (prefix), 14*t*
Termolecular reactions, 723, 723*t*
Tertiary structure of proteins, 673*f*, 674
Tetrachlorodibenzodioxin (TCDD), 554
Tetrafluoroborate ions, 406–407
Tetrahedral arrangement
of complex ions, 1047, 1047*t*
defined, 420, G-15
hybrid orbitals and, 447, 449*t*
VESPR theory and, 418*f*, 420–421, 420*f*, 424*f*
Tetrahedral complexes
carbonic anhydrase, 1064, 1064*t*
in crystal field splitting, 1061, 1062*f*
in valence bond theory, 1056, 1056*f*
Tetramethylsilane, 649
Tetraphosphorus decoxide, 354, 354*f*, 610
Tetraphosphorus hexoxide, 610
Tetra- (prefix), 73*t*
Tevatron, 1091
t_{2g} orbitals, 1058, G-15
Thalidomide, 644
Thallium
bond type, 598*t*
electron configuration, 356
melting point of, 598*t*
properties of, 595–598
radioactive poison from, 1094
Theoretical yield, 128, G-15
Theory, in the scientific method, 13, 13*f*
Thermal conductivity, 397, 505
Thermal decomposition, 183
Thermal energy, 258
Thermal pollution, 550, 551*f*, 1108–1109
Thermals, 240
Thermite, 119, 120
Thermite reaction, 1011, 1011*f*
Thermochemical equation, 273–274, G-15
Thermochemical equivalence, 274
Thermochemical solution cycle, 544
Thermochemistry, 256–285
calorimetry, 268–273
defined, 257, G-15
energy forms and interconversion, 257–264
enthalpy, 265–267
future use of energy and, 281–285
Hess's law of heat summation in, 275–277
standard heats of reaction in, 277–280
stoichiometry of thermochemical equations, 273–274
Thermodynamics, 894–932
of carbon reduction, 1010–1011
defined, 257, G-15
first law of, 261, 896, 914
kitchen appliances and, 259*mn*
living things and, 914
predicting spontaneous change and, 895–910
second law of, 905, 914
standard values for common substances, A-5–A-7
third law of, 905–906
Thermometers, 25, 29*f*, 511, G-15
Thermoplastic elastomers, 518
Thermoplastic polymers, 518
Thermoset polymers, 518, 518*mn*
Thermosphere, 239*f*, 240
Thermotropic phases, 510
Thiosulfate ion, 613
Third law of thermodynamics, 905–906, G-15
Third-order reactions, 704, 704*t*
Thomson, J. J., 53, 54
Thomson, William (Lord Kelvin), 213
Thorium, 1045, 1078, 1095
Three-Mile Island nuclear facility, 1108
Threonine, 672*f*, 781, 781*f*
Threshold frequency, 301
Thymine, 675
Thymol blue, 861*f*
Tie-line, in redox equations, 177
Time, 14*t*, 18
Tin
allotropes of, 601
bond type, 598*t*
in electrolytic cells, 974, 975*f*
electron configuration, 356
melting point of, 598*t*
mineral sources of, 1008*t*
properties of, 598–601, 601*f*
solder, 1013*t*
tin(II) and tin(IV), 71*t*
Tin(II) chloride, 419
Tin(IV) oxide, 1010
Tire pressure, 213
Titanium
appearance of, 1038*f*
electron configuration, 340*t*, 356–358, 1039, 1039*t*
oxidation state, 1042*t*
Titanium(IV) oxide, 1043*f*
Titration. *See also* Acid-base titration curves
acid-base, 172–174, 173*f*
defined, 172, G-15
end point and, 861
equivalence point in, 853*f*, 854–857, 855*f*, 859*f*, 863*f*
redox, 179–180, 179*f*
TNT (trinitrotoluene), 648, 1106
Tobacco mosaic virus, 511*f*
Tokamak design, 1109, 1109*f*
Toluene, 564, 648, 829
Tonicity, 567
Torr, 209, 209*t*, G-15
Torricelli, Evangelista, 208
Total ionic equations
for acid-base reactions, 167–171, 170*t*, 171*f*
for aqueous ionic reactions, 156, 156*f*, 158*f*, 159
defined, 156, G-15
Trace elements, 1063–1064, 1064*t*
Tracers, radioactive, 1098–1100, 1099*t*, 1100*f*, G-15
Traffic flow, rate-determining step for, 724*mn*
Transcription, 677, 677*f*, G-15
Transfer RNA (tRNA), 677*f*, 678
trans isomers. *See cis-trans* isomers
Transistors, 508
Transition elements
appearance of, 1038–1039*f*
atomic size of, 346*f*, 347
behavior within groups, 1044, 1044*t*
chemical properties of, 1042–1043*f*, 1042–1044
color of, 1043, 1043*f*
complexes (*See* Coordination compounds)
coordination compounds, 1046–1054
defined, 339, 1037, G-15
as dietary trace elements, 1063–1064, 1064*t*
electron configuration, 339–340, 340*t*, 341*f*, 342–343, 1038–1040, 1039*t*
influence on Boron family properties, 595
inner (*See* Inner transition elements)
lanthanide contraction and, 1041
metallic behavior of, 1042–1043
metallic hydrides and, 586
oxidation state of, 1042, 1042–1043*t*
in periodic table, 339–340, 340*t*, 341*f*, 342–343, 1038, 1038*f*
physical and atomic properties, 1040–1041*f*, 1040–1042
properties of, 1037–1044
reducing strength, 1043, 1043*t*

Transition metal ions, 356–358, 358*f*
Transition states, 719–722, G-15
Transition state theory, 719–722, G-15
Translation, 677–678, 677*f*, G-15
Transmission electron microscope (TEM), 311
Transuranium elements, 1092, 1092*t*, 1105, G-15
Trees, absorption of water by, 567
Trends, 12. *See also* Periodic trends
Trial and error, 12
Triatomic molecules, vibration of, 384, 384*f*
Triethylamine, A-12
Triglycerides, 663, 664
Trigonal bipyramidal arrangement
  defined, 421, G-15
  hybrid orbitals and, 447, 449*t*
  VSEPR theory and, 418*f*, 421–422, 422*f*, 424*f*
Trigonal planar arrangement
  of alkenes, 644
  defined, 419, G-15
  of hybrid orbitals, 446, 449*t*
  VSEPR theory and, 418–419*f*, 419–420, 424*f*
Trigonal pyramidal shape, 420–421, 420*f*, 424*f*, 447, 657, G-15
Triiodide ions, 422
Trimethylamine, 657, A-12
Trinitrotoluene (TNT), 648, 1106
Triple bonds
  in alkynes, 646
  bond energy of, 382, 382*t*
  bond order and, 380
  defined, G-15
  functional groups with, 666
  molecular shape of, 453
  pi bond and, 453, 453*f*
  rotation and, 455
  sigma bond and, 453, 453*f*
  strength of, 453–454
Triple point, 480, 480*f*, 481, G-15
Tri- (prefix), 73*t*
Tristearin, 663*f*
Tritium, 1024, 1024*t*, 1109
tRNA (transfer RNA), 677*f*, 678
Troposphere, 239, 239*f*
Trouton's Rule, 902–903, G-16
Tryptophan, 672*f*
T shape, in VSEPR theory, 422, 422*f*, 424*f*, G-15
Tungsten
  density of, 1042
  electron configuration, 343
  properties of, 1044, 1044*t*
  purification of, 1011
Twisted nematic phase, 510, 510*f*
Tyndall effect, 569, 569*f*, G-16
Tyrosine, 672*f*

U

Ultraviolet (UV) radiation
  absorption by stratospheric ozone, 240
  defined, G-16
  on electromagnetic spectrum, 296, 297*f*
  ozone as absorber of, 409, 735
  in water purification, 570
Uncertainty in measurement, 28–33, G-16
Uncertainty principle, 313, 314, G-16
Unimolecular reactions, 723, 723*t*, G-16
Union Carbide plant, release of poisonous gas from, 224
Unit cells, 493–499
  body-centered cubic, 494, 495*f*, 496–497, 497*f*
  defined, 494, G-16
  edge length, 499, 499*f*
  face-centered cubic, 494, 495*f*, 497–498
  hexagonal, 497–498, 497*f*
  packing efficiency, 496–498, 497*f*
  simple cubic, 494, 495*f*, 496, 497*f*
Units, in conversions and calculations, 18–19, 23
Universal gas constant, 216, G-16
Universe, origins of, 1110
Unsaturated hydrocarbons, 644, G-16
Unsaturated solutions, 549, G-16
Unshared electron pairs, 380, G-16
Uracil, 675
Uranium
  in atomic bomb, 1106*f*
  calculating mass in pitchblende, 48–49
  decay series, 1082, 1082*f*
  enrichment, 237
  fission of, 1105, 1105–1106*f*
  isotopes, 237
  in nuclear energy reactors, 1107–1108, 1107*f*
  oxidation state, 1045
  pitchblende, 1075
  radioactivity of, 712, 1075, 1093
  sources of, 1095
Urea, 1005
Urease, 732–733
UV radiation. *See* Ultraviolet radiation

V

Vacancies and crystal defects, 507
Valence bands, 505, 505–506*f*, 506, G-16
Valence bond (VB) theory, 443–451
  central themes of, 443–444
  complex ions and, 1055–1056
  defined, 443, G-16
  hybrid orbitals and, 444–451, 449*t*
  molecular rotation and, 455, 455*f*
  orbital overlap and, 443–444, 443*f*, 451–455
  single and multiple bonds in, 451–454
Valence electrons
  defined, 342, G-16
  ionization energies and, 350
  in metallic bonding, 396
  octet rule and, 405, 406
Valence-shell electron-pair repulsion (VSEPR) theory, 417–427
  bent shape in, 419, 419–420*f*, 421, 424*f*, 447
  defined, 417, G-16
  electron-group arrangements, 418, 418*f*
  hydrides and, 451
  limitations of, 443
  linear arrangement, 418–419*f*, 419, 422, 422*f*, 424*f*
  molecules with multiple central atoms, 426–427, 427*f*
  octahedral arrangement, 418*f*, 422–424*f*
  seesaw shape in, 422, 422*f*, 424*f*
  square planar shape, 423, 423–424*f*
  square pyramidal shape, 423, 423–424*f*
  tetrahedral arrangement, 418*f*, 420–421, 420*f*, 424*f*
  trigonal bipyramidal arrangement, 418*f*, 421–422, 422*f*, 424*f*
  trigonal planar arrangement, 418–419*f*, 419–420, 424*f*
  trigonal pyramidal shape, 420–421, 420*f*, 424*f*
  T shape in, 422, 422*f*, 424*f*
  using to determine molecular shape, 423, 425–426
Valence shell expansion, 414–416
Valence-state electronegativity, 1042–1043
Valine, 672*f*, 863
Vanadium
  abundance of, 1000, 1001*t*
  appearance of, 1038*f*
  as dietary trace element, 1064*t*
  electron configuration, 339, 340*t*, 1039*t*
  oxidation state, 1042, 1042*t*
Vanadium carbide, 1013, 1013*f*
Vanadyl sulfate dihydrate, 1043*f*
van der Waals, Johannes, 243
van der Waals constants, 243, 243*t*, G-16
van der Waals equation, 243–244, G-16
van der Waals forces. *See* Intermolecular forces
van der Waals radius, 482, 482*f*, G-16
van't Hoff, Jacobus, 564
van't Hoff equation, 776–777
van't Hoff factor, 564, 565*f*
Vaporization
  defined, 471, G-16
  as endothermic and spontaneous process, 897
  heat of, 471, 471–472*f*, 491–492
  liquid-gas equilibria and, 476–478, 476*f*
  molar enthalpy/entropy of, 902, 903*t*
  sweating and, 471, 471*f*
Vapor pressure
  boiling point and, 479
  defined, 477, G-16
  equilibrium, 476*f*, 477
  solute effect on vapor pressure of solution, 558–559, 558*f*
  temperature and intermolecular forces impacting, 477–478
  of water, 227, 227*t*
Vapor pressure lowering, 558–559, 558*f*, G-16
Variables, 12, G-16
VB theory. *See* Valence bond theory
Vehicles. *See* Automobiles
Vibration, infrared spectroscopy and, 384, 384–385*f*
Vibrational motion, entropy and, 909, 909*f*
Villard, Paul, 1075
Vinegar, 171, 793
Vinyl chloride, 619
Viscosity
  defined, 490, G-16
  factors affecting, 490, 490*t*
  intermolecular forces and, 489*t*, 490
  of liquid crystal phase, 509–510
  of polymers, 516–517, 517*f*
Vision, chemistry of, 645–646, 646*f*
Vitalism, 633
Vitamin B_{12}, 829
Vitamin C, 108–109, 814–815
Vitamin E, 414
Volatile liquids, finding molar mass of, 224–225
Volatile nonelectrolyte solutions, 564
Volatility, defined, 83, G-16
Voltage, 950, 950*t*. *See also* Cell potential
Voltaic cells. *See also* Batteries
  action of, 949, 949*mn*
  activity series of metals and, 958–959
  cell potential of, 950–959
  comparison with electrolytic cells, 944, 944*f*, 974–975, 976*t*
  construction and operation of, 946–947
  corrosion of iron and, 972–974
  defined, 944, G-16
  dental fillings and, 959, 959*f*
  diagramming, 948–949
  electrodes and, 944, 944*mn*
  notation for, 948–949
  relative strengths of oxidizing and reducing agents and, 953–957
  spontaneous redox reactions and, 945–949, 945*f*, 954–957
  standard cell potential of, 950–953, 950*f*, 959–966
  standard electrode potential, 950–956, 952*f*, 954*t*
  standard hydrogen electrode and, 951–952, 952*f*
Voltmeters, 950*n*
Volt (V), 950, G-16
Volume
  amount and, 214–215, 214*f*, 218–219, 223
  Avogadro's law and, 214–215, 214*f*, 223
  Boyle's law and, 211–212, 211*f*, 217–218, 223, 902
  Charles's law and, 212–213, 212*f*
  converting units of, 21–22
  defined, 15, G-16
  effect of change on equilibrium, 773–775, 774*f*, 777*t*
  English equivalents, 15*t*
  examples of, 17*f*
  as extensive property, 27, 27*f*
  gases and, 206, 211–212, 211*f*, 217–218
  laboratory glassware for working with, 15, 16*f*
  pressure and, 206, 211–212, 211*f*, 217–218
  SI units, 15, 15*t*, 16*f*
  standard molar, 215, 215*f*
  temperature and, 212–213, 212*f*, 218, 219–220
Volume percent, 554, G-16
Volumetric glassware, 15, 16*f*
VSEPR theory. *See* Valence-shell electron-pair repulsion theory
V shape, in VSEPR theory, 419, 419–420*f*, 421, 424*f*, 447

W

Waage, Peter, 750
Wastewater, 571, G-16
Water
  acid-base reactions and formation of, 167
  acids and bases in, 747, 794–798
  alkali metal reactions with, 592, 592*mn*
  amphiprotic nature of, 804
  aqueous ionic reactions, 155–157, 156*f*
  autoionization of, 798–803, 799*f*, 827
  behavior of salts in, 823–824, 825*t*
  bent shape of, 146, 146*f*, 447, 447*f*
  boiling point elevation constant, 560*t*
  boiling point of, 26, 26*f*, 27*t*
  capillarity of, 489–490, 490*f*, 492
  chemical changes in, 5, 5*f*
  collecting gas over, 227–228, 227*f*
  cooling curve for, 473*f*
  covalent bonds in, 67

covalent compounds in, 150
density of, 24*t*, 492
dipole moment, 428
dissociation of, 167
Earth's temperature and, 268
electrolysis of, 183, 283, 977–978, 977*f*, 1023
formation from methane and oxygen, 8*mn*
freezing point depression constant, 560*t*
freezing point of, 26, 26*f*, 27*t*
gas solubility in, 550
hard, 570–571
heat of fusion, 471, 471*f*
heat of vaporization, 471, 471*f*
heat of vaporization of, 491–492
in hydrates, 72
hydrogen bonds in, 491, 491*f*
hydrogen displacement from, 184, 184*f*
intermolecular forces, 534*f*
ionic compounds in, 146–149, 146*f*, 159*t*
ion-product constant for, 799–803, 799*f*
ion-radical, 1095
meniscus shape in glass, 489–490, 490*f*
molecular mass of, 76
molecular shape of, 421, 428
phase changes of, 471, 473–474, 473*f*
phase diagram for, 480–481, 480*f*
physical changes in, 5, 5*f*, 6–7
polar nature of, 146, 146*f*, 428
solvent properties of, 145–155, 491, 535–536, 536*f*, 536*t*, 908, 908*f*
specific heat capacity of, 268, 268*t*, 491
supercritical, 480
surface tension of, 489, 489*t*, 492
thermal properties of, 491–492
triple point of, 481
uniqueness of, 491–492, 493*f*
vapor pressure of, 227, 227*t*
viscosity of, 489–490*t*
Water gas, 281, 1023
Water-gas shift reactions, 281, 1023
Water purification, 570–571
Water softening, 570–571, 571*f*, G-16
Water striders, 489*mn*
Water treatment plants, 570, 570*f*
Water vapor, 6, 472
Wave behavior, comparison of particle behavior to, 299, 299*f*
Wave function, 314, 314*f*, 317, 317*f*, G-16
Wavelength
de Broglie, 311, 311*t*
defined, G-16
in electromagnetic radiation, 296, 296–297*f*, 297–298
of electron transition, 307
interconverting frequency and, 297–298, 301–302
line spectra and, 302–303, 306
Wave motion in restricted systems, 310, 310*f*
Wave nature of electrons, 310–312, 310*f*
Wave nature of light, 296–298
Wave-particle duality, 312, 312*f*, G-16
Waxes, 538
Weak-acid equilibria problem solving, 808–815
concentration and, 809–813
general approach for, 808
notation system for, 809
polyprotic acids in, 813–815, 814*t*
Weak acids. *See also* Acid strength
acid-dissociation constant values for, 796, 796*t*
anions as weak bases, 820–821
categorization of, 165, 165*t*, 797
direction of reactions involving, 806
dissociation of, 189, 795–796, 795*f*
reactions involving, 170–171, 170*t*
relation to weak bases, 818–822
structural features of, 166–167
weak-acid equilibria, 808–815
weak acid–strong base titration curves, 855–859, 855*f*
Weak bases
anions of weak acids as, 820–821
categorization of, 165, 165*t*, 797
dissociation of, 189
reactions involving, 170–171, 170*t*
relation to weak acids, 818–822
structural features of, 166–167
weak base–strong acid titration curves, 859–860, 859*f*
Weak-field ligands, 1058, 1058*f*, 1060, 1060–1061*f*, G-16
Weathering, 1006
Weight, 17, G-16
Welding, 507
Whiskers of silicon carbide, 513
White phosphorus, 606, 606*f*
Wöhler, Friedrich, 633
Wood
cellulose in, 671
as fuel, 281
specific heat capacity of, 268*t*
Work
defined, 259, G-16
done on (or by) a system, 259–260, 260*f*
energy transfer as, 259–260, 260*f*
free energy and electrical work, 959–967
free energy change and, 919–920
pressure-volume work, 263–264, 264*f*
sign conventions, 261*t*
units of, 261
Wulfenite, 493*f*

X

Xanthophylls, 1057
Xenon
intermolecular forces, 534*f*
mole fraction of air at sea level, 239*t*
properties of, 622–624
Xenon tetrafluoride, 423, 624, 624*f*
X-ray diffraction analysis, 500, 500*f*, G-16
X-ray photons, 1077
X-rays
diffraction, 311, 311*f*, 500, 500*f*
opacity of barium sulfate suspension, 869, 869*mn*
radiation dose from, 1096, 1096*t*
wavelength of, 296, 297*f*
Xylenes, 648

Y

Yield
actual, 128
green chemistry perspective on, 129–130
in multistep syntheses, 129, 129*f*
percent, 128–129
side reactions and, 128, 128*f*
theoretical, 128
-yl (suffix), 639, 640*t*
-yne (suffix), 646, 655*t*

Z

Zeolites, 1023
Zero, absolute, 26
Zero-order reactions
comparison with first- and second-order reactions, 715*t*
defined, 700
half-life of, 715
integrated rate law and, 709, 711, 711*f*, 715*t*
rate constant for, 704*t*
rate law and, 700, 700*f*
rate vs. reactant concentration in, 700*f*
reactant concentration vs. time in, 700*f*
Zewail, Ahmed H., 719
Ziegler, Karl, 668
Ziegler-Natta catalysts, 668
Zinc
abundance of, 1000, 1001*t*
alloys, 1013, 1013*f*, 1013*t*
appearance of, 1039*f*
in batteries, 968, 968*f*
as dietary trace element, 1064*t*
electron configuration, 340, 340*t*, 358, 1039, 1039*t*
enzyme function and, 1064
in galvanizing process, 974
mineral sources of, 1008*t*
oxidation state, 1042, 1042*t*
reaction with a strong and a weak acid, 796, 796*f*
redox reactions involving, 940*f*
in voltaic cells, 945–946*f*, 945–949, 952–953, 952*f*, 962–964, 964*f*
Zinc blende, 503, 503*f*
Zinc chloride, 151–152
Zinc formate, 826
Zinc hydroxide, 882
Zinc iodate, 866*t*
Zinc ions, 69*t*
Zinc oxide, 512, 1010
Zinc sulfate heptahydrate, 1043*f*
Zircon, 603, 603*f*
Zirconia, 513
Zirconium, 1044, 1078
Zone refining, 599, 1012, 1012*f*, G-16
Zwitterion, 863

Useful Data and Information

Atomic and Molecular Properties

Atomic radii	Figure 8.13, p. 346
Bond energies	Table 9.2, p. 381
Bond lengths	Table 9.2, p. 381
Ground-state electron configurations	Figure 8.9, p. 341
Electronegativity values	Figure 9.21, p. 391
Ionic radii	Figure 8.29, p. 360
First ionization energies	Figure 8.16, p. 349
Molecular shapes	Figure 10.10, p. 424

Equilibrium Constants and Thermodynamic Data

K_a of hydrated metal ions	Appendix C, p. A-12
K_a of selected acids	Appendix C, pp. A-8 to A-10
Strengths of conjugate acid-base pairs	Figure 18.8, p. 806
K_b of amine bases	Appendix C, p. A-11
K_f of complex ions	Table 19.4, p. 878, and Appendix C, p. A-12
K_{sp} of slightly soluble ionic compounds	Appendix C, p. A-13
Standard electrode potentials, $E^\circ_{half\text{-}cell}$	Appendix D, p. A-14
Standard free energies of formation, ΔG°_f	Appendix B, pp. A-5 to A-7
Standard heats of formation, ΔH°_f	Appendix B, pp. A-5 to A-7
Standard molar entropies, S°	Appendix B, pp. A-5 to A-7

Names and Formulas

Ligands	Table 23.7, p. 1050
Metal ions in complex anions	Table 23.8, p. 1050
Metals with more than one monatomic ion	Table 2.4, p. 71
Monatomic ions	Table 2.3, p. 69
Organic functional groups	Table 15.5, p. 655
Polyatomic ions	Table 2.5, p. 72

Properties of the Elements

Group 1A(1): Alkali metals	p. 591
Group 2A(2): Alkaline earth metals	p. 594
Group 3A(13): Boron family	p. 596
Group 4A(14): Carbon family	p. 599
Group 5A(15): Nitrogen family	p. 605
Group 6A(16): Oxygen family	p. 613
Group 7A(17): Halogens	p. 618
Group 8A(18): Noble gases	p. 623
Period 4 transition metals, atomic properties	Figure 23.3, p. 1040
Period 4 transition metals, oxidation states	Table 23.2, p. 1042

Miscellaneous

Rules for assigning an oxidation number	Table 4.4, p. 176
SI-English equivalent quantities	Table 1.4, p. 15
Solubility rules for ionic compounds in water	Table 4.1, p. 159
Vapor pressure of water	Table 5.2, p. 227

The Elements

Name	Symbol	Atomic Number	Atomic Mass*
Actinium	Ac	89	(227)
Aluminum	Al	13	26.98
Americium	Am	95	(243)
Antimony	Sb	51	121.8
Argon	Ar	18	39.95
Arsenic	As	33	74.92
Astatine	At	85	(210)
Barium	Ba	56	137.3
Berkelium	Bk	97	(247)
Beryllium	Be	4	9.012
Bismuth	Bi	83	209.0
Bohrium	Bh	107	(267)
Boron	B	5	10.81
Bromine	Br	35	79.90
Cadmium	Cd	48	112.4
Calcium	Ca	20	40.08
Californium	Cf	98	(249)
Carbon	C	6	12.01
Cerium	Ce	58	140.1
Cesium	Cs	55	132.9
Chlorine	Cl	17	35.45
Chromium	Cr	24	52.00
Cobalt	Co	27	58.93
Copernicium	Cn	112	(285)
Copper	Cu	29	63.55
Curium	Cm	96	(247)
Darmstadtium	Ds	110	(281)
Dubnium	Db	105	(262)
Dysprosium	Dy	66	162.5
Einsteinium	Es	99	(254)
Erbium	Er	68	167.3
Europium	Eu	63	152.0
Fermium	Fm	100	(253)
Flevorium	Fl	114	(289)
Fluorine	F	9	19.00
Francium	Fr	87	(223)
Gadolinium	Gd	64	157.3
Gallium	Ga	31	69.72
Germanium	Ge	32	72.61
Gold	Au	79	197.0
Hafnium	Hf	72	178.5
Hassium	Hs	108	(277)
Helium	He	2	4.003
Holmium	Ho	67	164.9
Hydrogen	H	1	1.008
Indium	In	49	114.8
Iodine	I	53	126.9
Iridium	Ir	77	192.2
Iron	Fe	26	55.85
Krypton	Kr	36	83.80
Lanthanum	La	57	138.9
Lawrencium	Lr	103	(257)
Lead	Pb	82	207.2
Lithium	Li	3	6.941
Livermorium	Lv	116	(293)
Lutetium	Lu	71	175.0
Magnesium	Mg	12	24.31
Manganese	Mn	25	54.94
Meitnerium	Mt	109	(268)

Name	Symbol	Atomic Number	Atomic Mass*
Mendelevium	Md	101	(256)
Mercury	Hg	80	200.6
Molybdenum	Mo	42	95.94
Moscovium	Mc	115	(288)
Neodymium	Nd	60	144.2
Neon	Ne	10	20.18
Neptunium	Np	93	(244)
Nickel	Ni	28	58.70
Nihonium	Nh	113	(284)
Niobium	Nb	41	92.91
Nitrogen	N	7	14.01
Nobelium	No	102	(253)
Oganesson	Og	118	(294)
Osmium	Os	76	190.2
Oxygen	O	8	16.00
Palladium	Pd	46	106.4
Phosphorus	P	15	30.97
Platinum	Pt	78	195.1
Plutonium	Pu	94	(242)
Polonium	Po	84	(209)
Potassium	K	19	39.10
Praseodymium	Pr	59	140.9
Promethium	Pm	61	(145)
Protactinium	Pa	91	(231)
Radium	Ra	88	(226)
Radon	Rn	86	(222)
Rhenium	Re	75	186.2
Rhodium	Rh	45	102.9
Roentgenium	Rg	111	(272)
Rubidium	Rb	37	85.47
Ruthenium	Ru	44	101.1
Rutherfordium	Rf	104	(263)
Samarium	Sm	62	150.4
Scandium	Sc	21	44.96
Seaborgium	Sg	106	(266)
Selenium	Se	34	78.96
Silicon	Si	14	28.09
Silver	Ag	47	107.9
Sodium	Na	11	22.99
Strontium	Sr	38	87.62
Sulfur	S	16	32.07
Tantalum	Ta	73	180.9
Technetium	Tc	43	(98)
Tellurium	Te	52	127.6
Tennessine	Ts	117	(294)
Terbium	Tb	65	158.9
Thallium	Tl	81	204.4
Thorium	Th	90	232.0
Thulium	Tm	69	168.9
Tin	Sn	50	118.7
Titanium	Ti	22	47.88
Tungsten	W	74	183.9
Uranium	U	92	238.0
Vanadium	V	23	50.94
Xenon	Xe	54	131.3
Ytterbium	Yb	70	173.0
Yttrium	Y	39	88.91
Zinc	Zn	30	65.41
Zirconium	Zr	40	91.22

*All atomic masses are given to four significant figures. Values in parentheses represent the mass number of the most stable isotope.

Data Tables

Fundamental Physical Constants (six significant figures)

Avogadro's number	N_A	$= 6.02214\times10^{23}$/mol
atomic mass unit	amu	$= 1.66054\times10^{-27}$ kg
charge of the electron (or proton)	e	$= 1.60218\times10^{-19}$ C
Faraday constant	F	$= 9.64853\times10^{4}$ C/mol
mass of the electron	m_e	$= 9.10939\times10^{-31}$ kg
mass of the neutron	m_n	$= 1.67493\times10^{-27}$ kg
mass of the proton	m_p	$= 1.67262\times10^{-27}$ kg
Planck's constant	h	$= 6.62607\times10^{-34}$ J·s
speed of light in a vacuum	c	$= 2.99792\times10^{8}$ m/s
standard acceleration of gravity	g	$= 9.80665$ m/s^2
universal gas constant	R	$= 8.31447$ J/(mol·K)
		$= 8.20578\times10^{-2}$ (atm·L)/(mol·K)

SI Unit Prefixes

p	n	μ	m	c	d	k	M	G
pico-	nano-	micro-	milli-	centi-	deci-	kilo-	mega-	giga-
10^{-12}	10^{-9}	10^{-6}	10^{-3}	10^{-2}	10^{-1}	10^{3}	10^{6}	10^{9}

Conversions and Relationships

Length

SI unit: meter, m

1 km	= 1000 m
	= 0.62 mile (mi)
1 inch (in)	= 2.54 cm
1 m	= 1.094 yards (yd)
1 pm	= 10^{-12} m = 0.01 Å

Volume

SI unit: cubic meter, m^3

1 dm^3	= 10^{-3} m^3
	= 1 liter (L)
	= 1.057 quarts (qt)
1 cm^3	= 1 mL
1 m^3	= 35.3 ft^3

Pressure

SI unit: pascal, Pa

1 Pa	= 1 N/m^2
	= 1 kg/m·s^2
1 atm	= 1.01325×10^{5} Pa
	= 760 torr
1 bar	= 1×10^{5} Pa

Mass

SI unit: kilogram, kg

1 kg	= 10^3 g
	= 2.205 lb
1 metric ton (t)	= 10^3 kg

Energy

SI unit: joule, J

1 J	= 1 kg·m^2/s^2
	= 1 coulomb·volt (1 C·V)
1 cal	= 4.184 J
1 eV	= 1.602×10^{-19} J

Math relationships

$\pi = 3.1416$

volume of sphere $= \frac{4}{3}\pi r^3$

volume of cylinder $= \pi r^2 h$

Temperature

SI unit: kelvin, K

0 K	= −273.15°C
mp of H_2O	= 0°C (273.15 K)
bp of H_2O	= 100°C (373.15 K)
T (K)	= T (°C) + 273.15
T (°C)	= [T (°F) − 32]$\frac{5}{9}$
T (°F)	= $\frac{9}{5}T$ (°C) + 32